BURGER'S
MEDICINAL CHEMISTRY
AND
DRUG DISCOVERY

BURGER'S MEDICINAL CHEMISTRY AND DRUG DISCOVERY

BURGER'S MEDICINAL CHEMISTRY AND DRUG DISCOVERY

Sixth Edition

Volume 3: Cardiovascular Agents and Endocrines

Edited by

Donald J. Abraham

Department of Medicinal Chemistry
School of Pharmacy
Virginia Commonwealth University
Richmond, Virginia

Burger's Medicinal Chemistry and Drug Discovery
is available Online in full color at
www.mrw.interscience.wiley.com/bmcdd.

WILEY-
INTERSCIENCE

A John Wiley and Sons, Inc., Publication

Cover Description   The molecule on the cover is hemoglobin with the allosteric effector RSR 13 attached. Three groups initiated structure-based drug design in the middle to late 1970s. Two of the groups, Peter Goodford's at the Burroughs Wellcome Laboratories in London and the editors' at the School of Pharmacy at the University of Pittsburgh, worked on hemoglobin while David Matthews at Agouron Pharmaceuticals worked on dihydrofolate reductase. Max Perutz and his coworkers' solution of the phase problem produced the first three-dimensional structure of a protein whose coordinates were of interest for drug design. The Editor worked with Max Perutz from 1980 to 1988, attempting to design antisickling agents. One of the active antisickling molecules, clofibric acid, would not lead to a sickle cell drug but to an allosteric effector RSR 13 designed and synthesized at Virginia Commonwealth University, which has been studied clinically for treatment of metastatic brain cancer. Max Perutz, whose work would provide the underpinnings for structure-based drug design, passed away in February 2002. His spirit and love for science that transferred to his students, postdoctoral fellows, visiting scientists, and colleagues worldwide, like Professor Burger's, is the leaven and inspiration for new discoveries.

Published by John Wiley & Sons, Inc., Hoboken, New Jersey.
Published simultaneously in Canada.

For general information on our other products and services please contact our Customer Care Department within the U.S. at (877) 762-2974, outside the U.S. at (317) 572-3993 or fax (317) 572-4002.

Wiley also publishes its books in a variety of electronic formats. Some content that appears in print, however, may not be available in electronic format.

For ordering and customer service, call 1-800-CALL-WILEY.

Library of Congress Cataloging-in-Publication Data:

Burger's medicinal chemistry and drug discovery.—6th ed., Volume 3: cardiovascular agents and endocrines/ Donald J. Abraham, editor

ISBN 0-471-37029-0 (v. 3: acid-free paper)

Printed in the United States of America.

10 9 8 7 6 5 4 3 2 1

BURGER MEMORIAL EDITION

The Sixth Edition of Burger's Medicinal Chemistry and Drug Discovery is being designated as a Memorial Edition. Professor Alfred Burger was born in Vienna, Austria on September 6, 1905 and died on December 30, 2000. Dr. Burger received his Ph.D. from the University of Vienna in 1928 and joined the Drug Addiction Laboratory in the Department of Chemistry at the University of Virginia in 1929. During his early years at UVA, he synthesized fragments of the morphine molecule in an attempt to find the analgesic pharmacophore. He joined the UVA chemistry faculty in 1938 and served the department until his retirement in 1970. The chemistry department at UVA became the major academic training ground for medicinal chemists because of Professor Burger.

Dr. Burger's research focused on analgesics, antidepressants, and chemotherapeutic agents. He is one of the few academicians to have a drug, designed and synthesized in his laboratories, brought to market [Parnate, which is the brand name for tranylcypromine, a monoamine oxidase (MAO) inhibitor]. Dr. Burger was a visiting Professor at the University of Hawaii and lectured throughout the world. He founded the *Journal of Medicinal Chemistry, Medicinal Chemistry Research,* and published the first major reference work *"Medicinal Chemistry"* in two volumes in 1951. His last published work, a book, was written at age 90 (*Understanding Medications: What the Label Doesn't Tell You,* June 1995). Dr. Burger received the Louis Pasteur Medal of the Pasteur Institute and the American Chemical Society Smissman Award. Dr. Burger played the violin and loved classical music. He was married for 65 years to Frances Page Burger, a genteel Virginia lady who always had a smile and an open house for the Professor's graduate students and postdoctoral fellows.

PREFACE

The Editors, Editorial Board Members, and John Wiley and Sons have worked for three and a half years to update the fifth edition of Burger's Medicinal Chemistry and Drug Discovery. The sixth edition has several new and unique features. For the first time, there will be an online version of this major reference work. The online version will permit updating and easy access. For the first time, all volumes are structured entirely according to content and published simultaneously. Our intention was to provide a spectrum of fields that would provide new or experienced medicinal chemists, biologists, pharmacologists and molecular biologists entry to their subjects of interest as well as provide a current and global perspective of drug design, and drug development.

Our hope was to make this edition of Burger the most comprehensive and useful published to date. To accomplish this goal, we expanded the content from 69 chapters (5 volumes) by approximately 50% (to over 100 chapters in 6 volumes). We are greatly in debt to the authors and editorial board members participating in this revision of the major reference work in our field. Several new subject areas have emerged since the fifth edition appeared. Proteomics, genomics, bioinformatics, combinatorial chemistry, high-throughput screening, blood substitutes, allosteric effectors as potential drugs, COX inhibitors, the statins, and high-throughput pharmacology are only a few. In addition to the new areas, we have filled in gaps in the fifth edition by including topics that were not covered. In the sixth edition, we devote an entire subsection of Volume 4 to cancer research; we have also reviewed the major published Medicinal Chemistry and Pharmacology texts to ensure that we did not omit any major therapeutic classes of drugs. An editorial board was constituted for the first time to also review and suggest topics for inclusion. Their help was greatly appreciated. The newest innovation in this series will be the publication of an academic, "textbook-like" version titled, "Burger's Fundamentals of Medicinal Chemistry." The academic text is to be published about a year after this reference work appears. It will also appear with soft cover. Appropriate and key information will be extracted from the major reference.

There are numerous colleagues, friends, and associates to thank for their assistance. First and foremost is Assistant Editor Dr. John Andrako, Professor emeritus, Virginia Commonwealth University, School of Pharmacy. John and I met almost every Tuesday for over three years to map out and execute the game plan for the sixth edition. His contribution to the sixth edition cannot be understated. Ms. Susanne Steitz, Editorial Program Coordinator at Wiley, tirelessly and meticulously kept us on schedule. Her contribution was also key in helping encourage authors to return manuscripts and revisions so we could publish the entire set at once. I would also like to especially thank colleagues who attended the QSAR Gordon Conference in 1999 for very helpful suggestions, especially Roy Vaz, John Mason, Yvonne Martin, John Block, and Hugo

Kubinyi. The editors are greatly indebted to Professor Peter Ruenitz for preparing a template chapter as a guide for all authors. My secretary, Michelle Craighead, deserves special thanks for helping contact authors and reading the several thousand e-mails generated during the project. I also thank the computer center at Virginia Commonwealth University for suspending rules on storage and e-mail so that we might safely store all the versions of the author's manuscripts where they could be backed up daily. Last and not least, I want to thank each and every author, some of whom tackled two chapters. Their contributions have provided our field with a sound foundation of information to build for the future. We thank the many reviewers of manuscripts whose critiques have greatly enhanced the presentation and content for the sixth edition. Special thanks to Professors Richard Glennon, William Soine, Richard Westkaemper, Umesh Desai, Glen Kellogg, Brad Windle, Lemont Kier, Malgorzata

Dukat, Martin Safo, Jason Rife, Kevin Reynolds, and John Andrako in our Department of Medicinal Chemistry, School of Pharmacy, Virginia Commonwealth University for suggestions and special assistance in reviewing manuscripts and text. Graduate student Derek Cashman took able charge of our web site, *http://www.burgersmedchem.com*, another first for this reference work. I would especially like to thank my dean, Victor Yanchick, and Virginia Commonwealth University for their support and encouragement. Finally, I thank my wife Nancy who understood the magnitude of this project and provided insight on how to set up our home office as well as provide John Andrako and me lunchtime menus where we often dreamed of getting chapters completed in all areas we selected. To everyone involved, many, many thanks.

DONALD J. ABRAHAM
Midlothian, Virginia

Dr. Alfred Burger

Photograph of Professor Burger followed by his comments to the American Chemical Society 26th Medicinal Chemistry Symposium on June 14, 1998. This was his last public appearance at a meeting of medicinal chemists. As general chair of the 1998 ACS Medicinal Chemistry Symposium, the editor invited Professor Burger to open the meeting. He was concerned that the young chemists would not know who he was and he might have an attack due to his battle with Parkinson's disease. These fears never were realized and his comments to the more than five hundred attendees drew a sustained standing ovation. The Professor was 93, and it was Mrs. Burger's 91st birthday.

Opening Remarks

ACS 26th Medicinal Chemistry Symposium

June 14, 1998
Alfred Burger
University of Virginia

It has been 46 years since the third Medicinal Chemistry Symposium met at the University of Virginia in Charlottesville in 1952. Today, the Virginia Commonwealth University welcomes you and joins all of you in looking forward to an exciting program.

So many aspects of medicinal chemistry have changed in that half century that most of the new data to be presented this week would have been unexpected and unbelievable had they been mentioned in 1952. The upsurge in biochemical understandings of drug transport and drug action has made rational drug design a reality in many therapeutic areas and has made medicinal chemistry an independent science. We have our own journal, the best in the world, whose articles comprise all the innovations of medicinal researches. And if you look at the announcements of job opportunities in the pharmaceutical industry as they appear in *Chemical & Engineering News,* you will find in every issue more openings in medicinal chemistry than in other fields of chemistry. Thus, we can feel the excitement of being part of this medicinal tidal wave, which has also been fed by the expansion of the needed research training provided by increasing numbers of universities.

The ultimate beneficiary of scientific advances in discovering new and better therapeutic agents and understanding their modes of action is the patient. Physicians now can safely look forward to new methods of treatment of hitherto untreatable conditions. To the medicinal scientist all this has increased the pride of belonging to a profession which can offer predictable intellectual rewards. Our symposium will be an integral part of these developments.

CONTENTS

xiii

BURGER'S
MEDICINAL CHEMISTRY
AND
DRUG DISCOVERY

CHAPTER ONE

Cardiac Drugs: Antianginal, Vasodilators, and Antiarrhythmics

GAJANAN S. JOSHI
Allos Therapeutics, Inc.
Westminster, Colorado

JAMES C. BURNETT
Virginia Commonwealth University
Richmond, Virginia

DONALD J. ABRAHAM
Institute for Structural Biology and Drug Discovery
School of Pharmacy and Department of Medicinal Chemistry
Virginia Commonwealth University
Richmond, Virginia

Contents

Burger's Medicinal Chemistry and Drug Discovery
Sixth Edition, Volume 3: Cardiovascular Agents and Endocrines
Edited by Donald J. Abraham
ISBN 0-471-37029-0   © 2003 John Wiley & Sons, Inc.

1 INTRODUCTION

It is an exciting time for drug discovery, because we are in the midst of a rapid evolution towards one of medicine's ultimate goals—moving from treating the symptoms of diseases to the absolute prevention of diseases. One of the major milestones that will aid in realizing this goal is the first draft of the human genome map, which was recently completed. The announcement of this milestone marked what will be seen in the future as a turning point in the search for new medicines that will address the cause, versus the symptoms, of many human ailments.

Over the last several decades, tremendous advances in basic and clinical research on cardiovascular disease have greatly improved the prevention and treatment of this, the nation's number one killer of men and women of all races. It is estimated that approximately 40% of Americans (approximately 60 million between the ages of 40–70 years) suffer from some degree of this disease (1–3). During the second half of the 20th century, the problem of treating heart disease has been at the forefront of the international medical communities' consciousness. This is reflected in the World Health Organizations 1967 classification of cardiovascular disease as the world's most serious epidemic.

The development of unique, novel, and tissue-specific cardiac drugs to replace or supplement existing therapies for various cardiac disorders continues to generate significant and growing attention, and has evolved hand-in-hand with research that has facilitated a better understanding of the underlying causes of cardiac disease states.

The subject of cardiovascular disorders and their treatment is vast and diverse. This chapter focuses on areas relevant to the antianginal, vasodilating, and antiarrhythmic drugs. The cardiac physiology, pathophysiology, and causes of these common diseases are reviewed before considering the drugs used in their treatment. For additional information, the reader is referred to other chapters in this series that cover advances and updates on therapeutics and treatments of other cardiovascular ailments such as myocardial infarction, antithromobotics, antihyperlipidemic agents, oxygen delivery, nitric oxide, angiogenesis, and adrenergics and adrenergic blocking agents. This chapter makes no attempt to provide comprehensive reviews of literature related to these fields.

2 CARDIAC PHYSIOLOGY

The human heart and physiological processes that are altered during cardiovascular disease are reviewed as background to the mechanisms of action of therapeutics used to treat angina and arrhythmia. However, for in-depth details about heart anatomy and physiology, the reader is referred to textbooks and reviews (4).

2.1 Heart Anatomy

The human heart consists of four chambers: the right and left atria and the right and left ventricles. Blood returning from the body collects in the right atrium, passes into the right ventricle, and is pumped to the lungs. Blood returning from the lungs enters the left atrium, passes into the left ventricle, and is pumped into the aorta. Valves in the heart prevent the backflow of blood from the aorta to the ventricle, the atrium, and the veins.

Heart muscle (the myocardium) is composed of three types of fibers or cells. The first type of muscle cells, found in the sinus and atrioventricular node, are weakly contractile, autorhythmic, and exhibit slow intercellular conduction. The second type, located in the ventricles, are the largest myocardial cells, and are specialized for fast impulse conduction. These cells constitute the system for propagating excitation over the heart. The remaining myocardial cells (the third type) are strongly contractile and make up the bulk of the heart.

Muscle cells in the heart abut very tightly from end to end and form fused junctions known as intercalated discs. This serves two functions. First, when one muscle cell contracts, it pulls on cells attached to its ends. Second, when cardiac cells depolarize, the wave of depolarization travels along the cell membrane until it reaches the intercalated disc, where it moves on to the next cell. Thus, heart muscle contracts in a unified and coordinated fashion. Large channels, referred to as gap junctions, pass through the intercalated discs and connect adjacent cells. These connections play an important role in transmitting the action potential from one cell to another.

Myocardial cells receive nutrients from coronary arteries that branch from the base of the aorta and spread over the surface of the organ. Blockage of sections of these coronary arteries occurs during coronary artery disease (CAD). This leads to myocardial ischemia, which is the cause of myocardial infarction (heart attack) and angina pectoris.

2.2 Electrophysiology

With the exception of differences in calcium ion uptake and release, the mechanisms of contraction of human skeletal and cardiac muscle are generally the same. However, unlike skeletal muscle, which requires neuronal stimulation, heart muscle contracts automatically. A heartbeat is composed of a rhythmic contraction and relaxation of the heart muscle mass, and is associated with an action potential in each cell. The constant pumping action of the heart depends on the precise integration of electrical impulse generation, transmission, and myocardial tissue response.

A heartbeat involves three principle electrical events. First, an electrical signal to contract is initiated. This is followed by the propagation of the impulse signal from its point of origin over the rest of the heart. Finally, the signal abates, or dies away. Cardiac arrhythmias develop when any of these three events are disrupted or impaired.

Figure 1.1 displays the principle components of the heart involved in cardiac impulse generation and conduction. In a normal healthy heart, the electrical impulse signal to contract is initiated in the sinoatrial (SA) node, which is located at the top of the right atrium (Fig. 1.1). Following depolarization of the SA node, the impulse spreads out into the atria through membrane junctions in an orderly fashion from cell to cell. The atria contract first. Following, as the impulse for contraction spreads over this part of the heart toward the ventricles, it is focused through specialized automatic fibers in the atria known as the atrioventricular (AV) node (Fig. 1.1). At this node, the impulse is slowed so that the atria finish contracting before the impulse is propagated to myocardial tissue of the ventricles. This allows for the rhythmic pumping action that allows blood to pass from the atria to the ventricles.

After the electrical impulse emerges from the AV node, it is propagated by tissue known as the bundle of His, which passes the signal on to fast-conducting myocytes known as Purkinje fibers. These fibers conduct the impulse

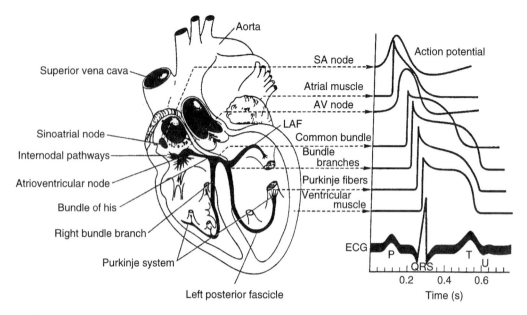

Figure 1.1. Action potentials and the conducting system of the heart. Shown are typical transmembrane action potentials of the SA and AV nodes, specialized conducting myocardial cells, and non-specialized myocardial cells. Also shown is the ECG plotted on the same time scale. Courtesy of Appleton & Lange.

to surrounding, nonspecialized myocardial cells. The transmission of the impulse results in a characteristic electrocardiographic pattern that can be equated to predictable myocardial cell membrane potentials and Na⁺ and K⁺ fluxes in and out of cells.

Following contraction, heart muscle fibers enter a refractory period during which they will not contract, nor will they accept a signal to contract. Without this resting period, the initial contraction impulse that originated in the SA node would not abate, but would continue to propagate over the heart, leading to disorganized contraction (known as fibrillation).

2.3 Excitation and Contraction Coupling

Myocardial pacemaker cells, usually in the SA node, initiate an action potential that travels from cell to cell through the intercalated discs. This opens calcium channels and leads to a small influx of extracellular calcium ions, which triggers events leading to muscle contraction.

The biophysical property that connects excitation impulse and muscle contraction is

based on the electrical potential differences that exist across cell membranes. These potentials arise because of several factors: (1) intracellular fluid is rich in potassium (K⁺) and poor in sodium (Na⁺) (the reverse is true of extracellular fluid); (2) the cell membrane is more permeable to K⁺ than it is to Na⁺; (3) anions in the intracellular fluid are mostly organic and fixed, and do not diffuse out through the membrane; and (4) cells use active transport to maintain gradients of Na⁺ and K⁺. In most cardiac cells the transmembrane potential difference is approximately −90 mV.

Stimulation, either electrical or chemical, can depolarize the cell membranes by causing conformational changes that open selective membrane ion channels. This allows Na⁺ to flow into the cell and reduce the negative intracellular charge. The transmembrane potential is reduced to a threshold value, which produces an action potential that is transmitted in an all-or-none fashion along the cellular membrane. As the action potential travels along the cell membrane, it induces a rise in the levels of free, or activator, calcium (Ca²⁺) within the cell. This, in turn, initiates the in-

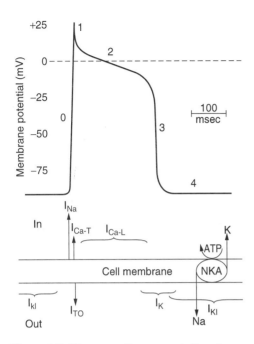

Figure 1.2. Diagrammatic representation of an action potential of a nonautomatic ventricular cell, showing the principle ion fluxes involved in membrane depolarization and repolarization. The membrane potential in millivolts is given on the vertical axis. This denotes the electrical potential of the inner face of the membrane relative to the outer face. Phases of the potential are numbered 0, 1, 2, 3, and 4 and are described in detail in the text.

teraction between actin and myosin, which leads to muscle contraction.

The action potential of a non-automatic ventricular myocyte is shown in Fig. 1.2. It is divided into five phases (0–4). The rapid membrane depolarization, phase 0 (also referred to as the upstroke), results from the opening of fast sodium channels, and is augmented by Ca^{2+} entering through calcium channels. Following depolarization there is a brief initial repolarization (phase 1; termed early repolarization), caused by the closing of the sodium channels, and a brief outward movement of K^+ ions. This is followed by a plateau period (phase 2), during which the slow influx of Ca^{2+} through an L-type calcium channel occurs (Fig. 1.2). This phase is most notable because it creates a prolonged refractory period during which the muscle cannot be re-excited. Phase 3 is the repolarization period and is caused

primarily by the opening of an outward-rectifying K^+ channel and the closure of the calcium channels. The repolarization that occurs during this phase involves the interplay of several different types of potassium channels. Following phase 3, the transmembrane potential is restored to its resting value (phase 4; Fig. 1.2).

Cells of the nodal tissue and specialized conducting myocytes, such as Purkinje fibers, can spontaneously depolarize and generate action potentials that propagate over myocardial tissue. This is referred to as automaticity, and all of these cells have pacemaker potential. In automatic cells, the outward leak of K^+ slows after repolarization, whereas Na^+ continues to leach into the cell. This results in a steady-state increase in intracellular cations and leads to depolarization. The action potential phase 4 of such cells is not flat, as observed in Fig. 1.2, but becomes less negative until it reaches a threshold that triggers the opening of an L-type calcium channel in nodal tissue, or the sodium channel in conducting tissue. Thus, phase 0 in nodal tissue is caused by the influx of Ca^{2+} and not Na^+. Figure 1.1 displays the action potentials for a selection of cardiac cells having spontaneous and non-spontaneous depolarizability. The electrical activity of myocardial cells produce an electrical current that can be measured and recorded as an electrocardiogram (ECG).

The time taken by an automatic myocyte to depolarize spontaneously is dependent on the maximum negative value of the resting membrane potential and the slope of phase 4. Under normal circumstances, cells of the SA node depolarize before other potential pacemaker cells, because the maximum value of the transmembrane potential is approximately −60 mV and the upward slope of phase 4 is steep. Thus, the SA node is normally the pacemaker for the rest of the heart. However, if the impulse from the SA node is slowed or blocked, or if the process of depolarization is accelerated in other automatic cells, non-SA cells may initiate a wave of depolarization that either replaces the SA node impulse or interferes with it. Heartbeats that originate from non-SA pacemaker activity are referred to as ectopic beats.

The spontaneous impulse rate of automatic cells depends on the slope of action potential phase 4, the magnitude of the maximum diastolic potential, and the threshold potential. Changes in any of these values can occur during disease states, or from the effects of small "drug" molecules. β1-Adrenergic receptor agonists increase heart rate by increasing phase 4 of the pacemaker cell action potential. Cholinergic drugs that are agonists of muscarinic receptors slow the heart by decreasing the phase 4 slope. Thus, compounds that block muscarinic receptors (atropine-like) increase heart rate, while compounds that block beta receptors slow the heart.

3 ION CHANNELS

The ion channel transmembrane protein consists of subunits designated as α, β, γ, and δ. The α subunit is the major component of Na^+, K^+, and Ca^{2+} channels that spans the membrane and is tetrameric in nature. Each unit of this tetramer is designated as a domain, and each domain is made up of six segments designated as S1, S2, S3, S4, S5, and S6. The S5 and S6 segments are linked to each other in a specific arrangement so as to form the lining of the ion channels. The S4 segment of each domain contains many lysine and arginine residues that act in response to changes in the membrane potential, and are thus involved in the opening (voltage gating) of the channel. It is believed that the S4 segment constitutes the "m" gate (5–11), whereas a polypeptide chain that links the S6 segment of domain III to the S1 segment of domain IV constitutes the "h" gate (12, 13). Other transmembrane subunits such as β, γ, and δ are believed to play a regulatory role and mainly contribute to the positioning and conformation of the α subunit in the membrane.

Many of the drugs used to treat angina and cardiac arrhythmias exert their therapeutic effects by blocking Na^+, K^+, and Ca^{2+} ion channels. In the case of sodium channel blockers, this results in a decrease in the slope of phase 0 of the action potential, and thereby decreases the V_{max}, or rate of conduction of the impulse. Sodium channel blockade can also prolong the refractory period by increas-

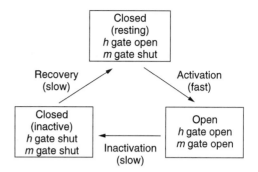

Figure 1.3. Simplified representation of the gating mechanism in voltage-activated sodium, potassium, and calcium channels. The model hypothesizes three states, closed resting, open active, and closed inactive, and two gates, h and m. The figure depicts what are generally considered to be the essential features of gating, which include a closed "resting" state that is capable of rapidly opening in response to changes in membrane potential followed by a refractory period in which the channel slowly returns to the resting state.

ing the time that the channel is in the inactivated state, before returning to the resting state. Potassium channel blockers increase the duration of the action potential, because potassium currents are responsible for repolarizing the membrane during the action potential. Calcium channel blockers slow impulse conduction through the SA and AV nodes.

3.1 Channel Gates

The term gating refers to the process during which external stimuli cause conformational changes in membrane proteins, leading to the opening and closing of ion channels. It has been theorized that ion channels have at least two gates, referred to as m and h, and that both gates must be open for ions to pass through the channel (14). According to this model, channel gates cycle through three states: (*1*) closed resting (R); (*2*) open active (A); and (*3*) closed inactive (I).

The gating model shown in Fig. 1.3 is a basic outline of the channel gating mechanism and is useful for describing drug action (15, 16). In the closed resting state, the h gate is open and the m gate is shut. During depolarization the m gate switches to the open position and the channel is activated, allowing the fast passage of ions through the channel. Depolarization also initiates the channel inactivation so that the channel begins to move from the open to the closed inactive state. In the closed inactive state both the m and h gates are shut, and the channel does not respond to further repolarization until it has moved back to the closed resting state, during which the h gate is again open and the m gate is shut.

3.2 Sodium Channels

There is strong evidence indicating that the amino acid sequence of sodium channels has been conserved over a long period. As indicated earlier, the inward voltage dependent Na^+ channel consists of four protein subunits designated as α, β, γ, and δ. The α subunit, the major component of Na^+ channel made up of 200 amino acids, is subdivided into four covalently bound domains and contains binding sites for a number of antiarrhythmic compounds and other drugs. Na^+ channels are found in neurons, vertebrate skeletal muscle, and cardiac muscle. Electrophysiological studies, indicating that Na^+ channels favor the passage of Na^+ over K^+, point to the fact that Na^+ channels must be narrow, and ion conductance depends on the ionic size (ionic radius of Na^+ is 0.95 Å compared with that of K^+ being 1.33 Å) and possibly steric factors (7). There is also some evidence indicating that voltage-dependent cardiac Na^+ channels exist in two isoforms: fast and slow (17). Activation of the fast Na^+ channels in cardiac cells produces the rapid influx of Na^+ and depolarizes the membrane in all cardiac myocytes, except nodal tissue, where Na^+ channels are either absent or relatively few in number (18). Na^+ channels are almost all voltage-gated, and gates open in response to changes in membrane potential.

3.3 Potassium Channels

Potassium channels are outward voltage-dependent channels. Similar to Na^+ channels, the major K^+ channel subunit consists of four domains, but unlike the α subunit of Na^+ channels, the domains of the K^+ channels are not covalently linked. At least 10 genes code for the K^+ channel domains, which means that hundreds of combinations of four-domain channels can be constructed. Recently, it has been reported that some of the K^+ channels contain only two transmembrane segments (9, 10). Many K^+ channels are classified as rectifying, which means that they are unidirectional (or transport ions in one direction only), and that their ability to pass current varies with membrane potential.

The inward-rectifying K^+ current (usually designated I_{KI}) allows K^+ to move out of the cell during phase 4 of the action potential, but is closed by depolarization; the outward-rectifying K^+ current (usually designated I_K) is opened by depolarization. Hence, the I_{KI} makes a substantial contribution to the value of the resting (phase 4) membrane potential; the I_K is the major outward current contributing to repolarization (19). ATP-sensitive K^+ channels are activated if the ATP level in the heart decreases as observed in myocardial ischemia (20). This leads to the inward flow of Ca^{2+} ions, thereby reducing myocardial contractility and conserving energy for basic cell survival processes.

The K^+ channels are highly selective and are 100-fold more permeable to K^+ than to Na^+ (21). Certain types of molecules bind extracellularly and block the voltage-gated K^+ channels (16). These include several peptide toxins, as well as small charged organic molecules such as tetraethylammonium, 4-aminopyridine, and quinine. Other molecules have been found to be K^+ channel openers. These compounds act on the ATP-sensitive K^+ current and provide cardioprotection during ischemia. K^+ channel opener molecules also relax smooth muscle cells and may increase the coronary blood flow during angina (22–28).

3.4 Calcium Channels

Calcium ions are essential for the chain of events that lead to myocardial contraction. The role of calcium in the cardiac cycle has been studied extensively for years. Four types of voltage-dependent calcium channels with specific function and location have been identified. These include (1) the L-type (found mainly in skeletal, cardiac, and smooth muscle cells); (2) the T-type (located in pacemaker cells); (3) the N-type (found in neuronal cells); and (4) the P-type (located at neuromuscular junctions) (29–31). L-type Ca^{2+} channels are

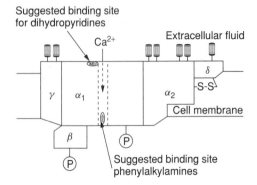

Figure 1.4. Suggested structure of an L-type calcium channel from skeletal muscle, showing the five protein subunits that comprise the channel. Phosphorylation sites are indicated by P. Binding sites for phenylalkylamine and dihydropyridine calcium channel blockers are also shown. Courtesy of *Trend. Pharmacol. Sci.*

formed by a complex arrangement of five protein subunits designated as the $\alpha1$, $\alpha2$, β, γ, and δ subunits, which are comprised of polypeptide chains of different lengths. The arrangement of these subunits is shown in Fig. 1.4. The tetrameric $\alpha1$ subunit is the most important functional component forming the Ca^{2+} channel and is responsible for producing the pharmacological effects of calcium channel blockers. Ca^{2+} channels play an important role in cellular excitability by allowing the rapid influx of Ca^{2+}, which depolarizes the cell. The resulting increase in intracellular Ca^{2+} is essential for the regulation of Ca^{2+}-dependent processes including excitation–contraction coupling, excitation–secretion coupling, and gene regulation. It has been suggested that dihydropyridine calcium channel blockers exert their effects by binding at the top of the channel near the cytoplasmic entrance, whereas arylalkylamines bind at the bottom of the channel between domains III and IV of the $\alpha1$ subunit. The other hydrophobic $\alpha2$, β, γ, and δ subunits may play a role in positioning the $\alpha1$ subunit in the membrane. L-type channels are activated slowly by partial depolarization of the cell membrane and inactivated by full depolarization and by increasing Ca^{2+} concentration. In nodal tissues, where fast sodium channels are absent or

sparse, L- and T-type Ca^{2+} channels are responsible for depolarization.

4 ANTIANGINAL AGENTS AND VASODILATORS

Angina pectoris is the principle symptom of ischemic heart disease and is caused by an imbalance between myocardial oxygen demand and oxygen supply by coronary vessels. Such an imbalance can result from increased myocardial oxygen demand caused by exercise, decreased myocardial oxygen delivery, or both. Angina pectoris is always associated with sudden, severe chest pain and discomfort, although some individuals do not experience pain with ischemia. Thus, angina occurs because the blood supply to the myocardium through coronary vessels is insufficient to meet the metabolic needs of the heart muscle for oxygen (32).

4.1 Factors Affecting Myocardial Oxygen Supply

Blood Oxygenation and Oxygen Extraction Involving Tissue Ischemia. A normal and healthy heart extracts about 75% of blood oxygen at rest; however, increased coronary blood flow and extraction results in an increase in oxygen supply. Ischemic heart disease develops when there is a deficiency in the supply of blood and oxygen to the heart and is typically caused by a narrowing of the coronary arteries, a condition known as coronary artery disease (CAD) or coronary heart disease (CHD). CAD is a consequence of the complicated pathological process involving the development of atherosclerotic lesions in the coronary arteries, in which cholesterol, triglycerides, and other substances in the blood deposit in the walls of arteries, narrowing them. The narrowing limits the extraction and flow of oxygen rich blood to the heart.

Pulmonary Conditions. Sometimes acute and chronic bronchopulmonary disorders such as pneumonia, bronchitis, emphysema, tracheobronchitis, chronic asthmatic bronchitis, tuberculosis, and primary amyloidosis of the lung affect oxygen extraction and its supply to the heart, causing severe ischemia. Also, if the heart does not work as efficiently as it should, it reduces the cardiac output. This causes the congestion of fluid in the tissues and leads to swelling (edema). Occasionally, the fluid collects in the lungs and interferes with breathing, causing shortness of breath at rest or during exertion. Edema is also exacerbated by a reduced ability of the kidneys to dispose of sodium and water. The retained water further increases the edema (swelling).

Coronary Vascular Conditions. Various conditions such as coronary collateral blood flow, coronary arterial resistance affected by the nervous system, accumulation of local metabolites, tissue death, endothelial function, diastolic blood pressure, and endocardial–epicardial blood flow contribute significantly to the pathogenesis of angina.

4.2 Factors That Govern Myocardial Oxygen Demand

Heart rate. A significant change in the regular beat (fast or slow) or rhythm of the heart (arrhythmias) may affect the myocardial oxygen demand. Excessive slowing of heartbeat is called bradycardia and is sometimes associated with fatigue, dizziness, and lightheadedness or fainting. The various symptoms of bradycardia have been categorized as sinus bradycardia, junctional rhythm, and heart block. These symptoms can easily be corrected with an electrical pacemaker, which is implanted under the skin and takes over the functioning of the natural pacemaker. Conversely, a rapid heartbeat is referred to as tachycardia. Tachycardias are classified into two types: supraventricular and ventricular. Different types of abnormal rapid heart beats have been categorized as sinus tachycardia (normal response to exercise), atrial tachycardia, atrial fibrillation, atrial flutter, AV nodal re-entry, AV reciprocating tachycardia, premature atrial contractions, ventricular tachycardia, and premature ventricular contractions. Electrocardiographic monitoring is needed for the correct diagnosis of arrhythmias.

Because exercise stimulates the heart to beat faster and more forcefully, more blood, and hence more oxygen, is needed by the myocardium to meet this increased workload. Normally this is accomplished by dilation of coronary blood vessels; however, sometimes

atherosclerosis may inhibit the flow of oxygen-rich blood, resulting in ischemia.

Cardiac Contractility (Inotropic State). Reduction in the cardiac output causes a reflex activation of the sympathetic nervous system to stimulate heart rate and contractility, further leading to greater oxygen demand. If the coronary arteries are occluded and incapable of delivering the needed oxygen, an ischemia will occur.

Preload-Venous Pressure and Its Impact on Diastolic Ventricular Wall Tension and Ventricular Volumes. It has been suggested that an important strategy in the treatment of cardiac function is reduction of the work load of the heart, by reducing the number of heart beats per minute and the work required per heart beat defined by preload and afterload. Preload is defined as the volume of blood that fills the heart before contraction. Contraction of the great veins increases preload, whereas dilation of veins reduces preload.

Afterload-Systolic Pressure Required to Pump Blood Out. Afterload is defined as the force that the heart must generate to eject blood from the ventricles. It largely depends on the resistance of arterial vessels. Contraction of these vessels increases afterload, whereas dilation reduces afterload.

4.3 Types of Angina

Stable Angina. Stable angina is also called chronic angina, exertional angina, typical or classic angina, angina of effort, or atherosclerotic angina. The main underlying pathophysiology of this, the most common type of angina, is usually atherosclerosis, i.e., plaques that occlude the vessels or coronary thrombi that block the arteries. This type of angina usually develops by "exertion", exercise, emotional stress, discomfort, or cold exposure and can be diagnosed using EKG. Therapeutic approaches to treat this type of angina include increasing the myocardial blood flow and decreasing the cardiac preload and afterload.

Vasospastic Angina. It is also called variant angina or Prinzmetal's angina. It is usually caused by a transient vasospasm of coronary blood vessels or atheromas at the site of plaque. This can easily be seen by EKG changes in ST elevation that tend to occur at rest. Sometimes chest pain develops even at rest. A therapeutic approach to treat this type of angina is to decrease vasospasm of coronary arteries, normally provoked by α-adrenergic activation in coronary vasculature. However, α-adrenergic activation is not the only cause of vasospasm.

Unstable Angina. It is also called preinfarction angina, crescendo angina, or angina at rest. It is usually characterized by recurrent episodes of prolonged attacks at rest and results from small platelet clots (platelet aggregation) at an atherosclerotic plaque site that may also induce local vasospasm. This type of angina requires immediate medical intervention such as cardiac bypass surgery or angioplasty, because it could ultimately lead to myocardial infarction (MI). Treatment regimens include inhibition of platelet aggregation and thrombus formation, vasodilation of coronary arteries (angioplasty), or decrease in cardiac load.

4.4 Etiology and Causes of Angina

The risk factors for the development of CHD and angina pectoris are genetic predisposition, age, male sex, and a series of reversible risk factors. The most important factors include high-fat and cholesterol-rich diets (32, 33, 34), lack of exercise, inability to retain normal cardiac function under increased exercise tolerance (35, 36), tobacco and smoking (because nicotine is a vasoconstrictor) (37), excessive alcohol drinking, carbohydrate and fat metabolic disorders, diabetes, hypertension (38, 39), obesity (40, 41), and the use of drugs that produce vasoconstriction or enhanced oxygen demand. The increased cholesterol levels caused by the consumption of a diet rich in saturated fat stimulates the liver to produce cholesterol, a lipid needed by all cells for the synthesis of cell membranes and in some cells for the synthesis of other steroids, is the principal reversible determinant of risk of heart disease. Low density lipoproteins (LDLs, also referred to as "bad" cholesterol) transport cholesterol from liver to other tissues, whereas high density lipoproteins (HDLs, also referred as "good" cholesterol) transport cholesterol from tissues back to the liver to be metabolized. Triglycerides are transported from the liver to the tissues mainly as very low density lipoproteins (VLDLs). VLDLs are the

precursors of the LDLs. The LDLs are characterized by high levels of cholesterol, mainly in the form of highly insoluble cholesteryl esters. However, there is a strong relationship between high LDL levels and coronary heart disease and a negative correlation between HDL and heart disease. Total blood cholesterol is the most common measurement of blood cholesterol, and various total blood cholesterol levels and risk factors accepted by most physicians and the American Heart Association are discussed next. In general, for people who have total cholesterol levels lower than 200 mg/dL, heart attack risk is relatively low (unless a person has other risk factors). If the total cholesterol level is 240 mg/dL, the person has twice the risk of heart attack as someone who has a cholesterol level of 200 mg/dL. Cholesterol levels of 240 mg/dL are considered high, and the risk of heart attack and, indirectly, of stroke is greater. About 20% of the U.S. population has high blood cholesterol levels. The LDL cholesterol level also greatly affects the risk of heart attack, and indirectly, of stroke. Lower LDL cholesterol levels correlate with a lower risk. Sometimes the ratio of total cholesterol to HDL cholesterol is used as another measure. In this case, the goal is to keep the ratio below 5:1; the optimum ratio is 3.5:1. It is assumed that people with high triglycerides (more than 200 mg/dL) have underlying diseases or genetic disorders. In such cases, the main treatment is to change the lifestyle by controlling weight and limiting carbohydrate intake, because carbohydrates raise triglyceride levels and lower HDL cholesterol levels.

During the last few years, there has been reliable evidence that coronary artery disease (CAD) is a complex genetic disease. In fact, a number of genes associated with lipoprotein abnormalities and genes influencing hypertension, diabetes, obesity, immune, and clotting systems play important roles in atherosclerotic cardiac disorders. Researchers have identified genes regulating LDL cholesterol, HDL cholesterol, and triglyceride levels based on common apo E genetic variation (42–46). Many genes linked to CAD are involved in determining how the body removes low density lipoprotein (LDL) cholesterol from the bloodstream. If LDL is not properly removed, it accumulates in the arteries and can lead to CAD. The protein that removes LDL from the bloodstream is called the LDL receptor (LDLR). In 1985, Michael Brown and Joseph Goldstein were awarded a Nobel prize for determining that a mutation in this gene was responsible for familial hypercholesterolemia (FH). People with FH have abnormally high blood levels of LDL (47).

As with LDLR, mutations in the apo E gene affect blood levels of LDL. Although, more than 30 mutant forms of apo E have been identified, people carrying the E4 version of the gene tend to have higher cholesterol levels than the general population, whereas cholesterol levels in people with the E2 version are significantly lower. The apo E gene has also been implicated in Alzheimer's disease (48).

4.5   Treatment

In general, the action of various therapeutic drugs occurs through (1) alteration of myocardial contractility or heart rate, (2) modification of conduction of the cardiac action potential, or (3) vasodilatation of coronary and peripheral vessels. Therefore, the contents of this section focus on therapeutics that apply to the treatment of angina and/or act as vasodilators.

4.5.1 Treatment of Angina. The various treatment modalities of different kinds of angina include the following: (1) prevention of precipitating factors; (2) use of nitrates as vasodilators to treat acute symptoms; (3) use of prophylactic treatment using a choice of drugs among antianginal agents, calcium channel blockers, and β-blockers; (4) surgeries such as angioplasty, coronary stenting, and coronary artery bypass surgery; and (5) anticoagulants and the use of antithromobolytic agents.

4.5.2 Prevention. Even though cardiovascular disease remains the leading cause of death in the United States, most risk reduction strategies have traditionally focused on detection and treatment of the disease. However, some of the risk factors of cardiac diseases are reversible and changes in lifestyle could significantly contribute towards decreasing mortality from CHD. One can reduce the risk of hypercholesterolemia by reducing the total amount of fat in the diet, being physically active (because exercise can help to increase HDL), avoiding cigarette smoking and exposure to secondhand smoke, and also by reducing sodium intake (49). In individuals whose cholesterol level does not respond to dietary intervention and in those having genetic predisposition to high cholesterol levels, drug therapy may be necessary. There are now several very effective medications available which have been proven to be effective for treating elevated cholesterol and preventing heart attacks and death. These include statins such as atorvastatin, cerevastatin, fluvastatin, lovastatin, pravastatin, and simvastatin (which lower LDL cholesterol by 30–50% and increase HDL) and fibrates such as bezafibrate, fenofibrate, and gemfibrozil (which lower elevated levels of blood triglycerides and increase HDL).

Several bile acid sequestrant antilipemic agents such as questran, colestid, and welchol are also used as an adjunct therapy to decrease elevated serum and LDL cholesterol levels in the management of type IIa and IIb hyperlipoproteinemia. These drugs are known to reduce the risks of coronary heart disease (CHD) and myocardial infarction. The bile acid sequestrant antilipemic drugs are known to have reduced or no GI absorption and are normally regarded as safe in pregnant patients.

4.6   Vasodilators

A number of the simple organic nitrates and nitrites find application in both the short- and long-term prophylactic treatment of angina pectoris, myocardial infarction, and hypertension. Most of these nitrates and nitrites are formulated by mixing with suitable inert excipients such as lactose, dextrose, mannitol, alcohol, and propylene glycol for safe handling, because some of these compounds are heat sensitive, very flammable, and powerful explosives. The onset, duration of action, and potency of organic nitrates could be attributed to structural differences. However, there is no relationship between the number of nitrate groups and the activity.

4.6.1 Mechanism of Action. The nitrates and nitrites are simple organic compounds that metabolize to a free radical nitric oxide

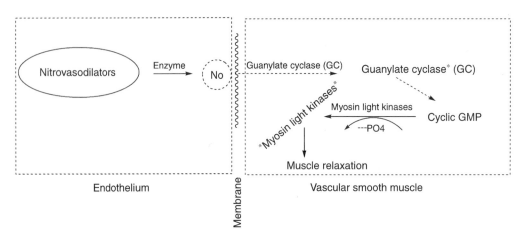

Figure 1.5. Suggested mechanism of action of nitrate and nitrites used as vasodilators to generate NO, the most potent (endogenous) vasodilator that induces a cascade of reactions resulting in smooth muscle relaxation and vasodilation.

(NO) at or near the plasma membrane of vascular smooth muscle cells. In 1980, Furchgott and Zawadzski first discovered that NO is the most potent endogenous vasodilator (50). NO is a highly reactive species with a very short half-life of few seconds. It is an endothelium derived relaxing factor (EDRF) that influences vascular tone. Nitric oxide induces vasodilation by stimulating soluble guanylate cyclase to produce cyclic GMP (cGMP) as shown in Fig. 1.5. The latter eventually leads to dephosphorylation of the light chains of myosin (51). The resulting hemodynamic effect produces dilation of epicardial coronary arteries, systemic resistance vessels, and veins (52, 53). It is this dilation that causes a reduction in the coronary vascular resistance and is responsible for the efficacy of these compounds (54–56). Thus, the main action of the nitrates and nitrites involves peripheral vasodilation, either venous (low doses), or both venous and arterial (higher doses). A result of pooling of blood in the veins is a reduction in the venous return and the ventricular volume (preload). This reduction in reduction in the volume decreases oxygen demand on the heart and the pain of angina is relieved quickly. Furthermore, it has been found that nitrates exert their effect only on the large coronary vessels. This is because minor vessels lack the ability to convert nitrate to NO. (Fig. 1.5).

The use of nitrates leads to reflex activation of the sympathetic nervous system, which increases the heart rate and the myocardial contractility. Reduction in the ventricular wall tension decreases the myocardial oxygen consumption. At the same time, nitrates improve myocardial oxygen supply by increasing the coronary blood flow to the endocardium. Thus, nitrates alter the imbalance of myocardial oxygen consumption and supply, which is the basis of angina pectoris. The main pharmacological effect of organic nitrates is relaxation of vascular smooth muscle, which results in vasodilation. Organic nitrates provide an exogenous source of NO that augments the actions of EDRF, which is impaired during coronary artery diseases (55). It has been suggested that nitrates may be useful as antiplatelet and antithrombic agents in the management of intracoronary thrombi (56). Although the exact mechanism of action of nitrates on antiplatelet aggregation is unknown, it is postulated that activation of cGMP inhibits the calcium influx, resulting in fibrinogen binding to glycoprotein IIb/IIIa receptors.

All vasodilators can be divided into three types depending on their pharmacological site of action. These include cerebral, coronary, and peripheral vasodilators. In this chapter, only coronary and peripheral vasodilators, represented in Fig. 1.6, are reviewed.

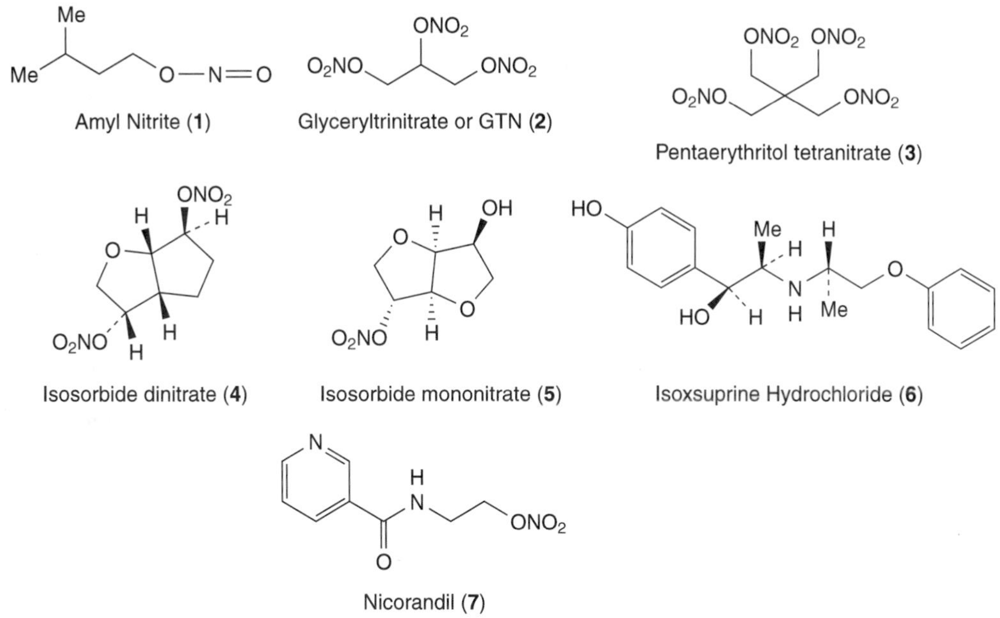

Figure 1.6. Chemical structures of various currently used vasodilators for the treatment of angina.

4.6.2 Vasodilating Agents. Various compounds that are currently used as vasodilators are described below. Structures of these compounds are depicted in Fig. 1.6, and some of their properties such as bioavailability, half-life, and some possible side effects are illustrated in Table 1.1.

Amyl Nitrite (**1**) (Fig. 1.6, Table 1.1): This drug is an aliphatic compound with an unpleasant odor. It is a volatile and inflammable liquid and is immiscible in water. Amyl nitrate can be administered to patients with coronary artery disease by nasal inhalation for acute relief of angina pectoris. It has also been used to treat heart murmurs resulting from stenosis and aortic or mitral valve irregularities. Amyl nitrate acts within 30 s after administration, and its duration of action is about 3–5 min. However, this drug has a number of adverse side effects such as tachycardia and headache.

Glyceryl trinitrate or GTN (**2**) (Fig. 1.6, Table 1.1): Glyceryl trinitrate is a short-acting trinitrate ester of glycerol, with a duration of

Table 1.1 Currently Used Vasodilators

Name	Uses	Side Effects
Amyl nitrite (1)	Angina pectoris, cyanide poisoning, heart murmurs	Tachycardia, CNS[a]
Glyceryl trinitrate (2)	Angina pectoris, hypertension, acute MI, refractory heart failure	CNS[a]
Pentaerythritol tetranitrate (3)	Prophylactic anginal attacks	CNS[a]
Isosorbide dinitrate (4)	Angina pectoris, congestive heart failure, dysphasia	Reflex tachycardia, CNS[a]
Isosorbide mononitrate (5)	Angina pectoris, congestive heart failure	CNS,[a] GI intolerance
Isoxsuprine HCl (6)	Peripheral vascular diseases (Burger's disease, Raynaud's disease), arteriosclerosis obliterans	CNS,[a] Tachycardia
Nicorandil (7)	Antianginal (not available in USA)	Hypotension

[a]CNS adverse effects include headache, dizziness, nausea, vomiting, diarrhea, flushing, weakness, rash, and syncope.

action of approximately 30 min. GTN is easily absorbed through the skin and has a strong vasodilating effect. In fact, glyceryl trinitrate is the only vasodilator known to enhance coronary collateral circulation and is capable of preventing myocardial infarction induced by coronary occlusion. As a result, it is widely used in preventing attacks of angina versus stopping these attacks once started. GTN can be administered by sublingual, transdermal, or intravenous route. However, about 40–80% of the dose is normally lost during IV administration because of the absorption by plastic material used to administer the dose.

Pentaerythritol Tetanitrate or PETN (3) (Fig. 1.6, Table 1.1): PETN is a nitric acid ester of the tetrahydric alcohol pentaerythritol. Because PETN is a powerful explosive, it is normally mixed and diluted with other inert materials for safe handling purposes and to prevent accidental explosions. PETN is mainly used in the prophylactic management of angina to reduce the severity and frequency of attacks. It has a similar mechanism of action as GTN and effects vascular smooth muscle cells to induce vasodilation as other nitrates. PETN's duration of action can be prolonged by using a sustained release formulation.

Isosorbide Dinitrate (4) (Fig. 1.6, Table 1.1): This compound forms white crystalline rosettes that are soluble in water. Isosorbide dinitrate can be administrated by oral, sublingual, or intrabuccal routes, and the approximate onset and duration of action depends on the administration route and various dosage forms. The approximate onset and duration of action of various dosage forms of isosorbide dinitrate are as follows:

Dosage Forms	Onset	Duration of Action
Oral	1 h	5–6 h
Extended release	30 min	6–8 h
Chewable	within 3 min	0.5–2 h
Sublingual	within 3 min	2 h

Isosorbide dinitrate is metabolized to the corresponding mononitrates (2 and 5 mononitrate) within several minutes to hours, depending on the route of administration. This drug is routinely used for the relief of acute angina pectoris, as well as in the short-and long-term prophylactic management of angina. It can also be used in combination with cardiac glycosides or diuretics for the possible treatment of congestive heart failure (57–59).

Isosorbide mononitrate (5) (Fig. 1.6, Table 1.1): Isosorbide mononitrate is the major active metabolite of isosorbide dinitrate and occurs as a white, crystalline, odorless powder. Similar to dinitrate, mononitrate is freely soluble in water and alcohol. The mononitrate is available commercially as conventional tablets, as extended release formulation capsules, or as controlled release coated pellets. The extended release formulations and tablets should be stored in tight, light resistance containers at room temperature.

Isosorbide mononitrate is readily absorbed from the GI tract and is principally metabolized in the liver. But unlike isosorbide dinitrate, it does not undergo first pass hepatic metabolism, and therefore the bioavailability of isosorbide mononitrate in conventional or extended release tablets is very high (100% and 80%, respectively). About 50% of a dose of isosorbide mononitrate undergoes denitration to form isosorbide, followed by partial dehydration to form sorbitol. Mononitrate also undergoes glucuronidation to form 5-mononitrate glucuronide. None of the indicated metabolites show pharmacological activity.

Similar to isosorbide dinitrate, the mononitrate is used for the acute relief of angina pectoris, for prophylactic management in situations likely to provoke angina attacks, and also for the long-term management of angina pectoris (60–61).

Isoxsuprine Hydrochloride (6) (Fig. 1.6, Table 1.1): This vasodilator is structurally related to nylidrin and occurs as a white crystalline powder that is sparingly soluble in water (62). Isoxsuprine causes vasodilation by direct relaxation of vascular smooth muscle cells, which decreases the peripheral resistance. It also stimulates β-adrenergic receptors, and at high doses can reduce blood pressure. This drug is used as an adjunct therapy in the management of peripheral vascular diseases such as Burger's disease, Raynaud's disease, arteriosclerosis obliterans, and for the relief of cerebrovascular insufficiency (63–66).

Nicorandil (**7**) (Fig. 1.6, Table 1.1): Nicorandil is a nicotinamide analog possessing a nitrate moiety. It exhibits a dual mechanism of action, acting as both a nitrovasodilator and a potassium channel activator (67, 68). Nicorandil offers cardioprotection and has been shown to improve the myocardial blood flow. This results in decreased systemic vascular resistance and blood pressure, pulmonary capillary wedge, and left ventricular end-diastolic pressure (69, 70). It is relatively well tolerated when used orally or intravenously in patients with stable angina.

However, Falase et al. reported that the use of nicorandil in patients undergoing cardiopulmonary bypass surgery needs further evaluation as severe vasodilation and hypotension requiring significant vasoconstrictor support has been observed (71).

4.6.3 Pharmacokinetics and Tolerance of Organic Nitrates. All organic nitrates exhibit similar pharmacological effects. The foremost factor contributing to the pharmacokinetics of glycerol trinitrates (GTN), and other longer acting organic nitrates, is the existence of high capacity hepatic nitrate reductase in the liver. This enzyme eliminates the nitrate groups in a stepwise process. But in serum, nitrates are metabolized independent of glutathione (54, 72, 73). In general, the organic nitrates are well absorbed from the oral mucosa following administration lingually, sublingually, intrabuccally, or as chewable tablets. The organic nitrates are also well absorbed from the GI tract and then undergo first pass metabolism in the liver. Nitroglycerin is well absorbed through the skin if applied topically as an ointment or transdermal system. Orally administered nitrates and topical nitroglycerin are relatively long acting. However, the rapid development of tolerance to the hemodynamic and antianginal effects of various dosage forms is known to occur with continuous therapy. Therefore, an approximately 8 h/day nitrate-free period is needed to prevent tolerance. Slow release transdermal patches of GTN are the most favored dosage form for achieving prolonged nitrate levels. Highly lipophilic nitrates, following IV infusion, are widely distributed in to vascular and peripheral tissues, whereas less lipophilic nitrates

are not as widely distributed. At plasma concentrations of 50–500 ng/mL, approximately 30–60% are bound to plasma proteins.

4.6.4 Side Effects. The principle side effects of nitrates include dilation of cranial vessels. This causes headaches, and it can be a limiting factor in the doseage used. More serious side effects are tachycardia and hypotension, which result in a corresponding increase in myocardial oxygen demand and decreased coronary perfusion—both of which have an adverse effect on the myocardial oxygen balance. Another well-documented problem is the development of tolerance to nitrates. Blood vessels become hypo- or non-reactive to the drugs, particularly if large doses, frequent dosing regimens, or long-acting formulations are used. To avoid this, nitrates are best used intermittently, allowing a few hours without treatment during a 24-h period.

4.7 Calcium Channel Blockers

Verapamil was the first calcium-channel blocker (CCB). It was first used in Europe (1962) and then in Japan for its antiarrhythmic and coronary vasodilator effects. The CCBs have become prominent cardiovascular drugs during the last 40 years. Many experimental and clinical studies have defined their mechanism of action, the effects of new drugs in this therapeutic class, and their indications and interactions with other drugs.

Calcium plays a significant role in the excitation–contraction coupling processes of the heart and vascular smooth muscle cells, as well as in the conduction of the heart cells. The membranes of these cells contain a network of numerous inward channels that are selective for calcium. The activation of these channels leads to the plateau phase of the action potential of cardiac muscle cells. Please refer to Section 3.4 for a detailed discussion on calcium channels, their mechanism of action, and their role in cardiovascular diseases.

4.7.1 Applications. Calcium channel blocking agents are the first drugs of choice for the management of Prinzmetal angina. Because of fewer adverse side effects on glucose homeostasis, lipid, and renal function, it has also been suggested that extended release or intermediate-long acting calcium channel blocking agents may be useful in the management of hypertension in patients with diabetes mellitus. However, data from limited clinical studies indicate that patients with impaired glucose metabolism receiving calcium channel blockers are at higher risks of nonfatal MI and other adverse cardiovascular events than those receiving ACE inhibitor or β-adrenergic agents (74).

A number of recent reviews describing the use of Ca^{2+} channel blockers in the treatment of hypertension are available (75). New Ca^{2+} channel blockers have greater selectivity and can be used to treat hypertension in the presence of concomitant diseases, such as angina pectoris, hyperlipidemia, diabetes mellitus, or congestive heart failure. Reflex tachycardia and vasodilator-induced headache are the major side effects that limit the use of these agents as antihypertensives (75).

Based on their pharmacophore and chemical structure, the calcium channel blockers can be divided into three different classes of compounds. These include (1) arylalkylamines, (2) benzothiazepines, and (3) 1–4 dihydropyridines. These drugs have broad applications in cardiovascular therapy because of their effects such as (1) arterial vasodilation resulting in reduced afterload, (2) slowing of impulse generation and conductance in nodal tissue, and (3) reduction in cardiac work and sometimes myocardial contractility, i.e., negative inotropic effect to improve myocardial oxygen balance. Each of the above classes of compounds and their pharmacological action will be discussed in detail.

4.7.2 Arylalkylamines and Benzothiazepines.
These drugs vary in their relative cardiovascular effects and clinical doses but have the most pronounced direct cardiac effects (e.g., verapamil, Fig. 1.7).

Bepridil Hydrochloride (8) (Fig. 1.7, Table 1.2): Bepridil is a nondihydropyridine calcium channel blocking agent with antianginal and antiarrhythmic properties. This compound inhibits calcium ion influx across L-type (slow, low voltage) calcium channels (76). However, unlike other agents, it also inhibits calcium ion influx across receptor operated channels and inhibits intracellular calmodulin-dependent processes by hindering the release of calcium and sodium influx across fast sodium channels. Thus, bepridil exhibits both calcium and sodium channel blocking activity and also possesses electrophysiological properties similar to those of class I antiarrhythmic agents, which prolong QT and QTc intervals (77). Although the precise mechanism of action remains to be fully determined, this drug reduces (in a dose-dependent manner) heart rate and arterial pressure by dilating peripheral arterioles and reducing total peripheral resistance. This leads to a modest decrease (less than 5 mm Hg) in systolic and diastolic blood pressure. When administered IV, it also reduces left ventricular contractility and increases filling pressure.

Although bepridil hydrochloride is usually administered orally for the treatment of chronic stable angina, it is not the first drug of choice because of its arrhythmogenic potential and associated agranulocytosis. Consequently, it is administered only in patients that have failed to respond to other antianginal agents (78, 79).

When used alone or in combination with other antianginal agents, it is as effective as β-adrenergic blocking agents or other dihydropyridine calcium channel blockers. However, bepridil can aggravate existing arrhythmias or induce new arrhythmias to the extent of potentially severe and fatal ventricular tachyarrhythmias, related to an increase in QT and QTc interval (80). Bepridil is rapidly and completely absorbed after oral administration and is 99% bound to plasma proteins.

Diltiazem Hydrochloride (9) (Fig. 1.7, Table 1.2): Like bepridil, diltiazem is also a nondihydropyridine calcium channel blocker, but it belongs to a benzothiazepine family of compounds (81, 82) Diltiazem is a light sensitive crystalline powder that is soluble in water and formulated as either a hydrochloride or malate salt. Diltiazem has a pharmacologic profile that is similar to other calcium channel blockers, i.e., it acts by inhibiting the transmembrane influx of extracellular calcium ions across the myocardial cell membrane and vascular smooth muscle cells (83, 84). However, unlike dihydropyridine calcium channel blockers, diltiazem exhibits inhibitory effects on the cardiac conduction system—mainly at

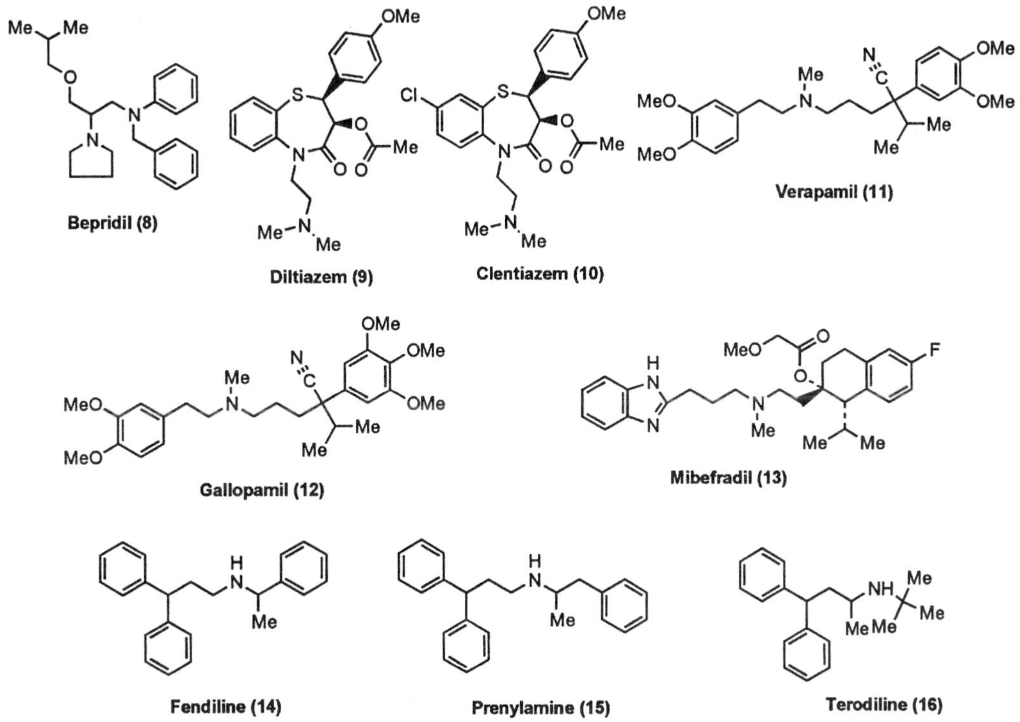

Bepridil (8)

Diltiazem (9)

Clentiazem (10)

Verapamil (11)

Gallopamil (12)

Mibefradil (13)

Fendiline (14)

Prenylamine (15)

Terodiline (16)

Drugs 10, 12, 14, 15 and 16 are not available in USA and drug # 13 has been discontinued in USA in 1998.

Figure 1.7. Chemical structures of various currently used arylalkylamines and benzothiazepines used as antianginal agents and vasodilators.

the atrioventricular (AV) node and minor sinus (SA) node. The frequency-dependent effect of diltiazem on AV nodal conduction selectively decreases the heart's ventricular rate during tachyarrhythmias involving the AV node. However, in patients with SA node dysfunction, it decreases the heart rate and prolongs sinus cycle length, resulting in sinus arrest. Diltiazem has little to no effect on the QT interval.

Diltiazem is administered orally as hydrochloride salt tablets or extended release capsules for the treatment of printzmetal angina, chronic stable angina, and hypertension. IV infusion is the preferred formulation for the treatment of supraventricular tachyarrhythmias. A controlled study also indicated that the simultaneous use of diltiazem and a β-adrenergic blocking agent in patients with chronic stable angina reduced the frequency of attacks and increased exercise tolerance (85).

Clentiazem (10) (Fig. 1.7, Table 1.2): Clentiazem is a chlorinated derivative of diltiazem and is currently undergoing clinical evaluation for the treatment of angina pectoris and hypertension in Europe. The primary mechanism of clentiazem responsible for the antihypertensive effects seems to be reduction in the peripheral arterial resistance caused by calcium channel blockade (86, 87).

Verapamil Hydrochloride (11) (Fig. 1.7, Table 1.2): Like diltiazem, verapamil is also a non-dihydropyridine calcium channel blocker. It is available as a racemic mixture and occurs as a crystalline powder that is soluble in water. The L-isomer of verapamil, which is 2–3 times more active than the corresponding D-isomer for its pharmacodynamic response on AV conduction, has been shown to inhibit the ATP dependent calcium transport mechanism of the sarcolemma (88). This drug has a pharmacological mechanism of action that is similar to other calcium channel blocking agents—it

Table 1.2 Properties of Arylalkylamines and Benzothiazepines

Name	Oral Bioavailability (%)	Half-Life (h)	Uses	Side Effects
Bepridil HCl (8)	Rapid and good	26–64	Chronic stable angina	CNS,[b] Ventricular arrhythmias
Dilitazem HCl (9)	80	2–11	Prinzmetal angina, MI, hypertension Supraventricular arrhythmias	Hypotension, GI, CNS,[a] bradycardia
Clentiazem (10)	ND[b]	ND[b]	Hypertension (not available in USA)	ND[b]
Verapamil HCl (11)	90	2–8	Supraventricular tachyarrhythmias Angina, MI, hypertension, hypertrophic cardiomyopathy	Bradycardia, AV block, edema, CNS,[a] hepatic
Gallopamil (12)	ND[b]	ND[b]	Angina pectoris, hypertension, supraventricular tachycardia, ischemia (not available in USA)	ND[b]
Mibefradil (13)	ND[b]	ND[b]	Discontinued use in USA in 1998 for safety reasons (drug–drug interaction)	
Fendiline (14)	ND[b]	ND[b]	Angina pectoris (not available in USA)	ND[b]
Prenylamine (15)	ND[b]	ND[b]	Prophylactic angina pectoris (not in USA)	Hypotension
Terodiline (16)	ND[b]	ND[b]	Suspended caused by cholinergic activity in addition to Ca^{2+} channel antagonist activity (not available in USA)	

[a]CNS adverse effects include headache, dizziness, nausea, vomiting, diarrhea, flushing, weakness, rash, and syncope.
[b]ND: No data obtained due to limited literature information.

reduces afterload and myocardial contractility. However, verapamil also exerts negative dromotropic effects on the AV nodal conduction and is also classified as a class IV antiarrhythmic agent (89). The effects of verapamil on nodal impulse generation and conduction are useful in treating certain types of arrhythmias. However, its effects on myocardial contractility may cause complications in patients with heart failure. Therefore, verapamil is used in the treatment and prevention of supraventricular tachyarrhythmia and in hypentensive patients not affected by cardiodepressent effects (90).

Verapamil is also administered orally in the treatment of prinzmetal angina and chronic stable angina and is as effective as any other β-adrenergic blocking agent or calcium channel blocker. IV verapamil is the drug of choice for the management of supraventricular tachyarrhythmias including rapid conversion to sinus rhythm of paroxysmal supraventricular tachycardias (PSVT) (those associated with Wolff-Parkinson-White or Lown-Ganong-Levine syndrome) and temporary relief of atrial fibrillation. It is also used as a monotherapy or in combination with other antihypertensive agents for the treatment of hypertension.

Gallopamil (12) (Fig. 1.7, Table 1.2): Gallopamil is a more potent methoxy analog of verapamil and has demonstrated efficacy in both effort and rest angina, hypertension, and supraventricular tachycardia (91–94). Furthermore, intracoronary administration of gallopamil may be useful in treating myocardial ischemia during percutaneous transluminal coronary angioplasty (95).

Intrarenal gallopamil has shortened the course of acute renal failure. It has been suggested that the role of inhaled gallopamil in asthma remains to be defined, and well-controlled potential comparisons with verapamil are needed to define the place in therapy of gallopamil for all indications.

Mibefradil (13) (Fig. 1.7, Table 1.2): Mibefradil is a T- and L-type calcium channel blocker (CCB) that was FDA approved in for the management of hypertension and chronic stable angina (96–99). However, postmarketing surveillance discovered potential severe life-threatening drug–drug interactions be-

tween mibefradil and β-blockers, digoxin, verapamil, and diltiazem, especially in elderly patients, resulting in one death and three cases of cardiogenic shock with intensive support of heart rate and blood pressure. Therefore, the manufacturer voluntarily withdrew mibefradil from U.S. market in 1998 (100).

Fendiline (14) (Fig. 1.7, Table 1.2): Fendiline is used in the long-term treatment of coronary heart disease. This agent is a coronary vasodilator and clinical studies have established that it is as therapeutically effective as both isosorbide and diltiazem in the treatment of angina pectoris (101–105). Recently, the action of fendiline on cardiac electrical activity has also been investigated in guinea pig papillary muscle. Results from these studies suggest that a frequency- and concentration-dependent block of Na^+ and L-type Ca^{2+} channels occurs in the presence of fendiline, leading to inhibition of fast and slow conduction and inactivation of Ca^{2+} channels (106).

Further studies have shown that fendiline also induces an increase in Ca^{2+} concentration in Chang liver cells by releasing stored Ca^{2+} in an inositol 1,4,5-triphosphate independent manner and by causing extracellular Ca^{2+} influx (107).

Prenylamine (15) (Fig. 1.7, Table 1.2): Prenylamine is a homolog of fendiline and is used in the treatment of chronic coronary insufficiency and prophylaxis of anginal paroxysms. The latter is recognized by a disturbance in brain blood circulation and sometimes hypertension, but prenylamine is not sufficiently effective in very acute anginal paroxysms (108). Because it is a coronary vasodilator, it acts as a calcium antagonist, but without any substantial effect on the contractility of the myocardium. However, it improves the vascular blood circulation and thereby oxygen supply of the myocardium. It also decreases the amount of norepinephrine and serotonin in the myocardium and brain and therefore possesses a slight blocking effect on β-adrenergic receptors. Because this agent enhances the antihypertensive effect of β blockers, its dosage must be closely monitored. If given in high doses during tachycardia, it can lead to deceleration of cardiac activity (109).

Terodiline (16) (Fig. 1.7, Table 1.2): Terodiline is an alkyl analog of fendiline and is used as a calcium channel antagonist. This agent also possesses anticholinergic and vasodilator activity (110). When administered twice daily, terodiline is effective in the treatment of urinary urge incontinence (110). Comparative studies with other agents used in urge incontinence are required to determine if the dual mechanism of action and superior absorption of terodiline offer clinical advantages. Safety concerns about ventricular arrhythmias have suspended general clinical investigations.

4.7.3 1–4 Dihydropyridine Derivatives.

This is an important class of drugs that are broadly used as vasodilators (Figs. 1.8 and 1.9). In general, 1,4-dihydropyridines demonstrate slight selectivity towards vascular versus myocardial cells and therefore have greater vasodilatory effects than other calcium channel blockers. 1,4-dyhydropyrdines are also known to possess insignificant electrophysiological and negative inotropic effects compared with verapamil or diltiazem. The dihydropyridines have no significant direct effects on the heart, although they may cause reflex tachycardia. Some of the properties of first and second generation dihydropyridines are given in Table 1.3. Representative drugs from this class are shown in Figs. 1.8 and 1.9. Most of the newer drugs have longer elimination half-lives, but also show higher rates of hepatic clearance and hence low bioavailabilities. The only exception is amlodipine, which has a much higher bioavailability (60%) and a long elimination half-life. Several metabolic pathways of DHP-type calcium channel blockers have been identified in humans. The most important metabolic pathway seems to be the oxidation of the 1,4-dihydropyridine ring into pyridine catalyzed by the cytochrome P450 (CYP) 3A4 isoform and the oxidative cleavage of carboxylic acid (111). Calcium antagonists are known to block calcium influx through the voltage-operated calcium channels into smooth muscle cells. Several of the compounds in the 1,4-DHP category such as nifedipine, nisoldipine, or isradipine have been shown to be useful in the management of coronary artery diseases. However, these calcium antagonists have some major disadvantages: they are photosensitive and decompose rapidly; they are not soluble in water; and because of their depressive effects on myocardium, they have negative inotropic effects. CCBs account for almost $4 billion in sales, and dihydropyridines like lercanidipine are the fastest growing class of CCB. There are 13 derivatives of DHP calcium channel blockers currently licensed for the treatment of hypertension. Examples of the most prescribed drugs include amlodipine, felodipine, isradipine, lacidipine, lercanidipine, nicardipine, nifedipine, and nisoldipine. Currently, thiazide diuretics or β blockers are recommended as first-line therapeutics for hypertension. Calcium channel blockers, ACE inhibitors, or α blockers may be considered when first-line therapy is not tolerated, contraindicated, or ineffective.

Amlodipine Besylate (**17**) (Fig. 1.8, Table 1.3): Amlodipine belongs to a 1,4-dihydropyridine family of compounds possessing structural resemblance to nifedipine, felodipine, nimodipine, and others. This drug is a calcium channel blocking agent with a long duration of action. It is mainly used orally either alone or in combination with other antihypertensive agents to treat hypertension and prinzmetal and chronic stable angina (112, 113).

Aranidipine (**18**) (Fig. 1.8, Table 1.3): Aranidipine is a 1,4-dihydropyridine calcium channel blocker with vasodilating and antihypertensive activity, and therefore is used for the treatment of hypertension (114–116). This compound is used either alone or in combination with a diuretic or β blocker, for the once-daily treatment of mild-to-moderate essential hypertension. Aranidipine is under investigation for the treatment of angina pectoris, but available data are limited to preclinical animal studies. It decreases T-type and L-type calcium currents in a concentration-dependent manner. The duration of aranidipine's antihypertensive effect is longer than that of nifedipine and nicardipine. Aranidipine does not significantly affect heart rate, cardiac output, or stroke volume index at rest or after exercising in patients with mild-to-moderate hypertension. However, it significantly increases left ventricular fractional shortening (FS) and left ventricular ejection fraction (EF) at rest. It does not adversely affect the hemodynamics of lipoprotein or carbohydrate me-

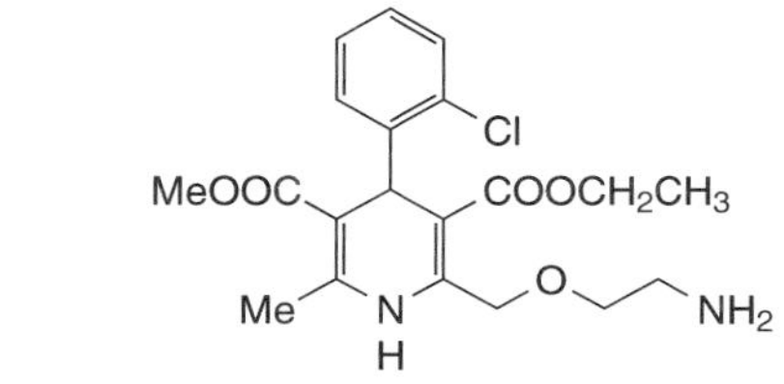

Amlodipine (**17**)

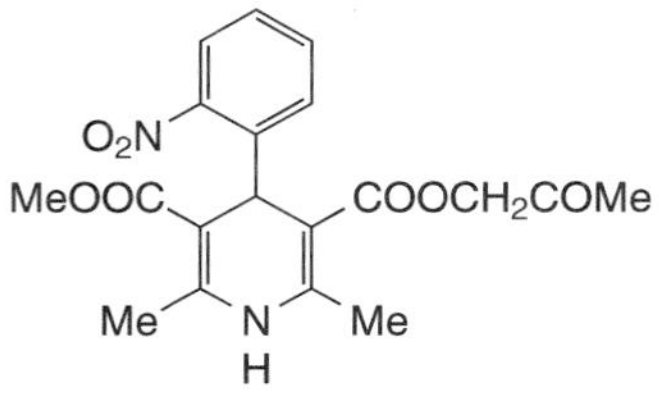

Aranidipine (**18**)

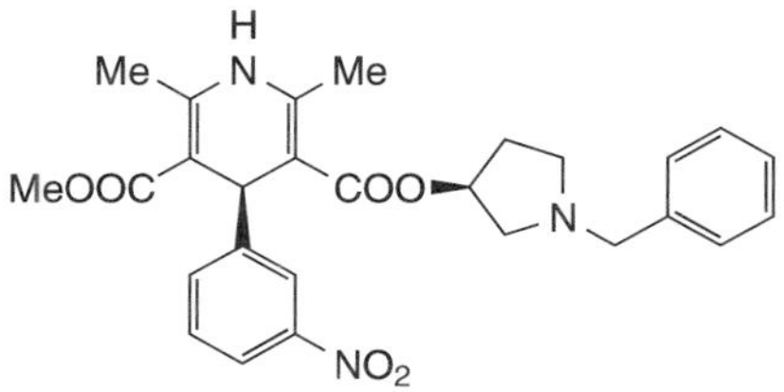

Barnidipine (**19**)

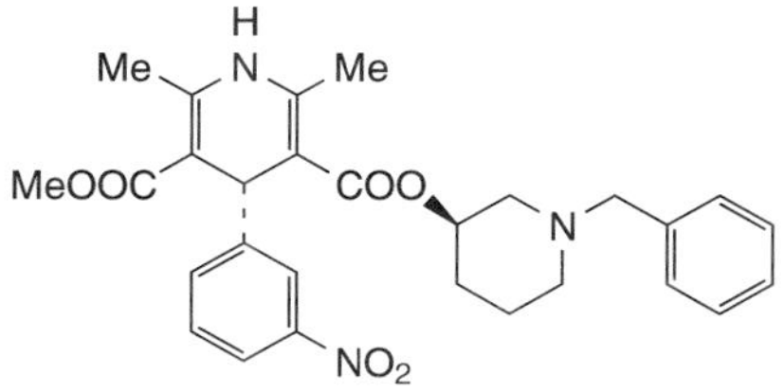

Benidipine (**20**)

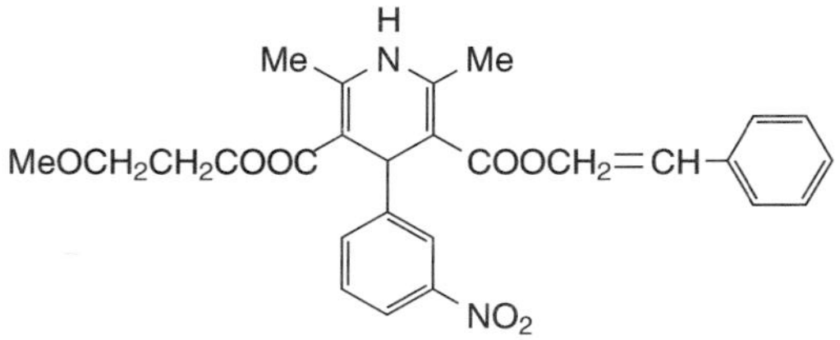

Cilnidipine (**21**)

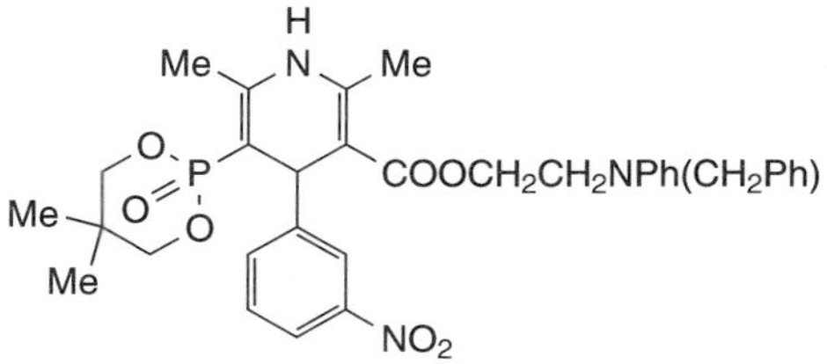

Efonidipine (**22**)

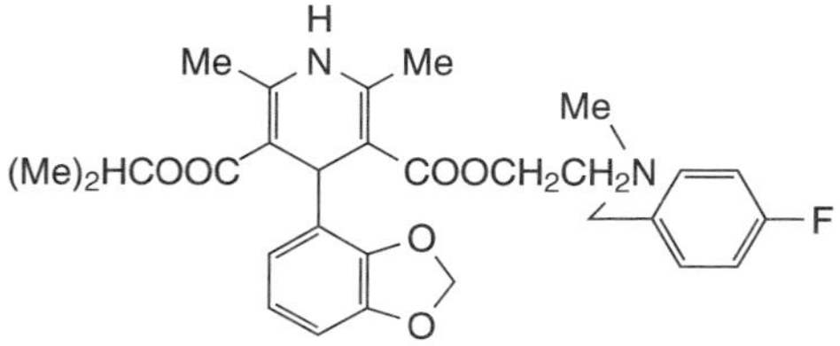

Elgodipine (**23**)

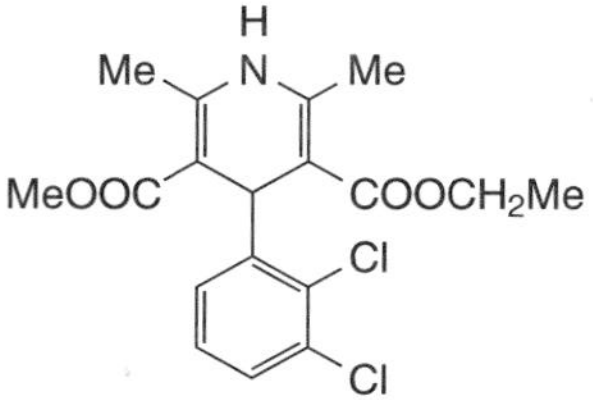

Felodipine (**24**)

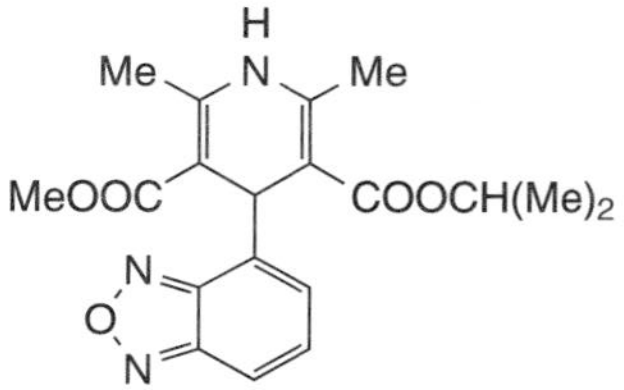

Isradipine (**25**)

Drugs 18, 19, 20, 21, 22 and 23 are not available in USA.

Figure 1.8. Chemical structures of various 1,4-dihydropyridines, currently used as calcium channel blockers that are used as antianginal and antihypertensive agents, which cause vasodilation.

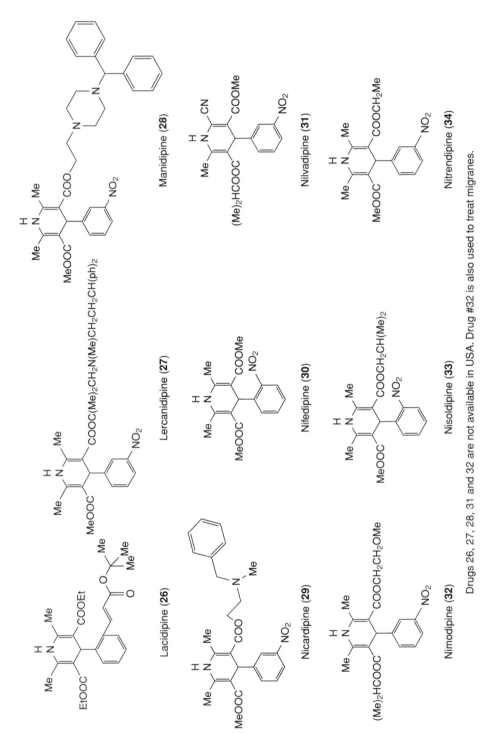

Lacidipine (26)

Lercanidipine (27)

Manidipine (28)

Nicardipine (29)

Nifedipine (30)

Nilvadipine (31)

Nimodipine (32)

Nisoldipine (33)

Nitrendipine (34)

Drugs 26, 27, 28, 31 and 32 are not available in USA. Drug #32 is also used to treat migranes.

Figure 1.9. Chemical structures of various 1,4-dihydropyridines, currently used as calcium channel blockers that are used as antianginal and antihypertensive agents, which cause vasodilation.

Table 1.3 Properties of Commonly Used 1,4-Dihydropyridine (Calcium Channel Blockers)

Name	Oral Bioavailability	Half-Life (h)	Uses	Possible Side Effects
Amlodipine besylate (17)	64	.934–58	Hypertension, prinzmetal angina	Hemodynamic, renal
Aranidipine (18)	ND[a]	ND[a]	Hypertension, angina pectoris	LVFS,[b] LVEF[c] at rest
Barnidipine (19)	ND[a]	ND[a]	Angina pectoris, hypertension	CNS[d]
Benidipine (20)	ND[a]	ND[a]	Hypertension, angina pectoris	CNS[d]
Cilnidipine (21)	ND[a]	ND[a]	Hypertension	CNS[d]
Efonidipine (22)	ND[a]	Long	Angina pectoris	CNS[d]
Elgodipine (23)	ND[a]	ND[a]	Angina pectoris, vasodilator during PTCA[e]	CNS[d]
Felodipine (24)	16	10–18	Angina, mild hypertension	Reflex tachycardia, angina
Isradipine (25)	19	8	Stable angina	Antiatherogenic effects
Lacidipine (26)	ND[a]	ND[a]	Hypertension	CNS[d]
Lercanidipine (27)	ND[a]	Short	Hypertension	CNS,[d] peripheral edema
Manidipine (28)	ND[a]	ND[a]	Hypertension	CNS[d]
Nicardipine (29)	15–40	11–12	Hypertension	
Nifedipine (30)	45	3–4	Prinzmetal and chronic angina, hypertension	Slight (−)inotropic effect
Nilvadipine (31)	ND[a]	ND[a]	Hypertension, angina pectoris	CNS[d]
Nimodipine (32)[f]	12	1	Subarachnoid hemorrhage	CNS[d]
Nisoldipine (33)	4	15–16	Hypertension, angina pectoris	CNS[d]
Nitredipine (34)	16	8–10	Hypertension	CNS[d]

[a]ND: No data obtained because of limited literature information. Compounds 18, 19, 20, 21, 22, and 23 are not available in USA.
[b]LVFS: Left Ventricular Fractional Shortening.
[c]LVEF: Left Ventricular Ejection Fraction.
[d]CNS adverse effects include headache, dizziness, nausea, vomiting, diarrhea, flushing, weakness, rash, and syncope.
[e]PTCA: Percutaneous Transluminal Coronary Angioplasty.
[f]Nimodipine's selectivity for cerebral arterioles makes it useful to treat subarachnoid hemorrhage and not hypertension as with other dihydropyridines. It is also used to treat migraines.

tabolism, and the pharmacokinetics of arani-dipine are not altered in the elderly or in patients with renal failure (114–116).

Barnidipine (**19**) (Fig. 1.8, Table 1.3): Barnidipine is a long-acting calcium antagonist that was launched in Japan in September 1992 under the brand name Hypoca. Barnidipine is used in Europe under the brand name Vasexten. As a long-acting calcium antagonist, this drug requires administration only once a day for the treatment of angina and hypertension, and it is available as a modified release formulation with a gradual and long duration of action (117–119). It is a selective calcium channel antagonist that reduces peripheral vascular resistance secondary to its vasodilatory action (120). Recently it was suggested that Barnidipine administration for a week decreased the blood pressure and made the sodium balance negative by increasing urinary sodium excretion in patients with essential hypertension. The natriuretic effect of this drug could contribute at least in part to its antihypertensive effect (121). Also, the possible use of benidipine for protection against cerebrovascular lesions in salt-loaded stroke-prone spontaneously hypertensive rats was evaluated by magnetic resonance imaging (MRI) (122).

Benidipine (**20**) (Fig. 1.8, Table 1.3): Benidipine is a new dihydropyridine with potent, long-lasting calcium antagonism (123). The administration of benidipine once daily effectively decreases blood pressure and attenuates blood pressure response to mental stress. Reflex tachycardia, deterioration of diurnal blood pressure change, and excessive lowering of nighttime blood pressure have not been observed after benidipine administration. Therefore, it has been suggested that benidipine may be useful for the treatment of elderly hypertensive patients with cardiovascular disease and as an antianginal medication. However no clinical data is currently available (124).

Cilnidipine (**21**) (Fig. 1.8, Table 1.3): Cilnidipine is a unique calcium antagonist that has both L-type and N-type voltage-dependent calcium channel blocking activity (125–129). Cilnidipine is under investigation for the treatment of hypertension in Europe.

Recently, cilnidipine, its analogs, and other dihydropyridine derivatives were evaluated for their state dependent inhibition of L-type Ca^{2+} channels, and revealed that structurally related DHPs act in distinct ways to inhibit the L-type channel in the resting, open, and inactivated states. Cilnidipine and related DHPs seem to exert their blocking action on the open channel by binding to a receptor distinct from the known DHP-binding site (130). Furthermore, the effect of cilnidipine on left ventricular (LV) diastolic function in hypertensive patients, as assessed by pulsed doppler echocardiography and pulsed tissue doppler imaging, has been examined. These studies suggest that changes in LV diastolic performance in patients with essential hypertension, following cilnidipine treatment, were biphasic, displaying an initial increase in early diastolic transmitral flow velocity and a later increase in early diastolic LV wall motion velocity. The initial and later changes can be related to an acute change in afterload and improvement in LV relaxation (131).

Efonidipine (**22**) (Fig. 1.8, Table 1.3): Efonidipine is a new, long-acting dihydropyridine calcium channel blocker derivative used in the treatment of hypertension (132–135). When the effect of efonidipine on endothelin-1 (ET-1) in open-chest anesthetized dogs was studied, it was concluded that efonidipine attenuates ET-1-induced coronary vasoconstriction, and therefore would be useful for some patients with variant angina, in which ET-1 is involved in the genesis of coronary vasoconstriction (136).

Recently, to gain insight into the renoprotective mechanism of efonidipine hydrochloride, the acute effects of efonidipine on proteinuria, glomerular hemodynamics, and the tubuloglomerular feedback (TGF) mechanism in anesthetized 24- to 25-week-old spontaneously hypertensive rats (SHR) with glomerular injury were evaluated. The results indicate that efonidipine attenuates the TGF response in SHR by dilating the afferent arteriole, thus maintaining the level of renal plasma flow (RPF) and glomerular filtration rate (GFR) despite reduced renal perfusion pressure (137).

Elgodipine (**23**) (Fig. 1.8, Table 1.3): This compound is a novel type of DHP that is very selective and a potent coronary vasodilator calcium channel blocker (138). Elgodipine is very stable to light (2% degradation after 1 year of exposure to room light and temperature) compared with other currently available compounds, which are water soluble and decompose within 24 h. It is very selective for vascular smooth muscle, in particular, coronary vessels. Because elgodipine is more than 100-fold more selective for coronary vessels versus cardiac fibers, it has few negative inotropic effects. Elgodipine seems to be potentially useful as a coronary vasodilator during PTCA (percutaneous transluminal coronary angioplasty). Its stability and solubility allows for intracoronary administration in patients with stable angina. Furthermore, because of a lack of negative inotropic effects, it is also administered in patients with moderate heart failure (139–141).

Some of the preliminary electrophysiological data in volunteers have shown that elgodipine differs from other calcium channel blockers in its effects on atria-ventricular conduction.

The chemical stability of elgodipine allows for its incorporation into suitable polymeric matrixes for transdermal administration. Preliminary data *in vitro* and *in vivo* in volunteers have shown that elgodipine penetrates into the skin. Studies are in progress to determine the daily effective dose, and therefore, the feasibility of transdermal patches (142).

Felodipine (**24**) (Fig. 1.8, Table 1.3): Felodipine is a member of the DHP calcium channel blocker family. This compound is insoluble in water but is freely soluble in dichloromethane and ethanol. Felodipine exists as a racemic mixture and is used to treat high blood pressure, Raynaud's syndrome, and congestive heart failure (143, 144). It reversibly competes with nitrendipine and/or other calcium channel blockers for dihydropyridine binding sites and blocks voltage-dependent Ca^{2+} currents in vascular smooth muscle.

Following oral administration, felodipine is almost completely absorbed and undergoes extensive first-pass metabolism. However, following intravenous administration, the plasma concentration of felodipine declines triexponentially, with mean disposition half-lives of 4.8 min, 1.5 h, and 9.1 h. Following oral administration of the immediate-release for-

mulation, the plasma level of felodipine also declines polyexponentially, with a mean terminal half-life of 11–16 h.

The bioavailability of felodipine is influenced by the presence of high fat or carbohydrates and increases approximately twofold when taken with grapefruit juice. A similar finding has been seen with other dihydropyridine calcium antagonists, but to a lesser extent than that seen with felodipine (145–147).

Felodipine produces dose-related decreases in systolic and diastolic blood pressure, which correlates with the plasma concentration of felodipine. Felodipine can lead to increased excretion of potassium, magnesium, and calcium (148). It has been recommended that to prevent side effects, individuals who are taking felodipine should avoid grapefruit and its juice (149). This is because grapefruit (juice) is an inhibitor of cytochrome P450 isoforms 3A4 and 1A2, which are needed for the normal metabolism of felodipine.

Isradipine (**25**) (Fig. 1.8, Table 1.3): Isradipine is a calcium antagonist that is available for oral administration and is used in the management of hypertension, either alone or concurrently with thiazide-type diuretics (150–153). Isradipine binds to calcium channels with high affinity and specificity and inhibits calcium flux into cardiac and smooth muscle. In patients with normal ventricular function, isradipine's afterload reducing properties lead to some increase in cardiac output. Effects in patients with impaired ventricular function have not been fully studied.

In humans, peripheral vasodilation produced by isradipine results from decreased systemic vascular resistance and increased cardiac output. In general, no detrimental effects on the cardiac conduction system were seen with the use of isradipine.

Isradipine is 90–95% absorbed and is subject to extensive first-pass metabolism, resulting in a bioavailability of about 15–24%. Isradipine is completely metabolized before excretion, and no unchanged drug is detected in the urine. Six metabolites have been characterized in blood and urine, with the mono acids of the pyridine derivative and a cyclic lactone product accounting for >75% of the material identified. The reaction mechanism ultimately leading to metabolite transformation to the cyclic lactone is complex.

Lacidipine (**26**) (Fig. 1.9, Table 1.3): Lacidipine also belongs to the DHP class of calcium channel blockers (154). Recently it was shown that lacidipine can slow the progression of atherosclerosis more effectively than atenolol, according to the results of the European lacidipine study on atherosclerosis (155). The improvement of focal cerebral ischemia by lacidipine may be partly caused by long-lasting improvement of collateral blood supply to the ischemic area (156).

When comparative effects of both lacidipine and nifedipine were measured, both drugs reduced blood pressure significantly during a 24-h period with one dosage daily; only lacidipine reduced left ventricular mass significantly after 12 weeks of treatment (157).

Lercanidipine (**27**) (Fig. 1.9, Table 1.3): Lercanidipine is a member of the dihydropyridine calcium channel blocker class of drugs. Recently, a New Drug Application with the U.S. FDA to market lercanidipine for the treatment of hypertension has been submitted. This drug has been available in European countries for more than 4 years, with an established record of anti-hypertensive effect and safety in millions of patients (158, 159). In fact, lercanidipine has grown to be the third most prescribed CCB in Italy. Lercanidipine prevents calcium from entering the muscle cells of the heart and blood vessels, which enables the blood vessels to relax, thereby lowering blood pressure. It has a short plasma half-life, but its high lipophilicity allows accumulation in cell membranes, resulting in long duration of action. It has been suggested that lercanidipine causes fewer vasodilatory adverse side effects than other CCBs and is therefore being promoted for the treatment of isolated systolic hypertension (ISH) in elderly patients (160).

Manidipine (**28**) (Fig. 1.9, Table 1.3): Manidipine is effective in the treatment of essential hypertension (161, 162). When the effect of manidipine hydrochloride on isoproterenol-induced LV hypertrophy and the expression of the atrial natriuretic peptide (ANP) transforming growth factor was evaluated, it was found that manidipine prevented cardiac hypertrophy, and changed the expression of genes for ANP and interstitial components of extracellular matrix induced by isoproterenol (163).

Nicardipine (**29**) (Fig. 1.9, Table 1.3): Nicardipine belongs to the 1,4-dihydropyridine calcium channel blocking family of compounds. It is usually administered either orally or by slow continuous IV infusion (when oral administration is not viable), for the treatment of chronic stable angina and the short-term management of hypertension. It is used as a monotherapy or in combination with other antianginal or antihypertensive drugs.

Nifedipine (**30**) (Fig. 1.9, Table 1.3): The principle physiological action of nifedipine is similar to other 1,4-dihydropyridine derivatives. This drug functions by inhibiting the transmembrane influx of extracellular calcium ions across the myocardial membrane and vascular smooth muscle cells, without affecting plasma calcium concentrations. Although the exact mechanism of action of nifedipine is unknown, it is believed to deform the slow calcium channel and hinder the ion-control gating mechanism of the calcium channel by interfering with the release of calcium ions from the sarcoplasmic reticulum. The inhibition of calcium influx dilates the main coronary and systemic arteries because of the impediment of the contractile actions of cardiac and smooth muscle. This reduced myocardial contractility results in increased myocardial oxygen delivery, while decreasing the total peripheral resistance associated by a modest lowering of systemic blood pressure, small increase in heart rate, and reduction in the afterload, ultimately leading to reduced myocardial oxygen consumption.

Unlike verapamil and diltiazem, nifedipine does not exert any effect on SA or AV nodal conduction at therapeutic dosage levels. Nifedipine is administered orally through extended release tablets in various dosage forms. It is mainly used in the treatment of Prinzmetal angina and chronic stable angina. In the latter case, it is as effective as β-adrenergic agents or oral nitrates, but is used only when the patient has low tolerability for adequate doses of these drugs.

Nilvadipine (**31**) (Fig. 1.9, Table 1.3): Nilvadipine is marketed as a racemic mixture for the treatment of hypertension and angina (164–166). Nilvadipine also provides protection against cerebral ischemia in rats having chronic hypertension. These effects are dependent on the duration of treatment (167). Results from a clinical study in the United States, during which a combination of imidapril and a diuretic, β-adrenoceptor antagonist, or a calcium channel blocker (such as nilvadipine) were administered, indicated a reasonable and safe treatment option when striving for additive pharmacodynamic effects not accompanied by relevant pharmacokinetic interactions (168).

Nimodipine (32) (Fig. 1.9, Table 1.3): Nimodipine is a structural analog of nifedipine, and the S-($-$)-enantiomer is primarily responsible for the calcium channel blocking activity. Substitution of a nitro substituent on the aryl ring and planarity of the 1,4-dihydropyridine moiety contribute greatly to the pharmacological effect of nimodipine. Nimodipine is a light sensitive yellowish crystalline powder. The mechanism of action of nimodipine is similar to other calcium channel blockers; however, the preferential binding affinity of nimodipine towards the cerebral tissue is yet to be fully understood. Nimodipine functions by binding to the stereoselective high affinity receptor sites on the cell membrane, in or near the calcium channel, and inhibits the influx of calcium ions. The vasodilatory effect of nimodipine also seems to arise partly from the inhibition of the activities of sodium-potassium activated ATPase, an enzyme required for the active transport of sodium across the myocardial cell membranes.

Nisoldipine (33) (Fig. 1.9, Table 1.3): Nisoldipine is a dihydropyridine, similar to nifedipine, but is 5–10 times more potent as a vasodilator and has little effect on myocardial contractility. Nisoldipine is available as a long-acting extended release preparation and seems effective in treating mild-to-moderate hypertension and angina with once-daily oral administration (169–171). Nisoldipine selectively relaxes the muscles of small arteries causing them to dilate, but has little or no effect on muscles or the veins of the heart.

In vitro studies show that the effects of nisoldipine on contractile processes are selective, with greater potency on vascular smooth muscle than on cardiac muscle. The effect of nisoldipine on blood pressure is principally a consequence of a dose-related decrease of peripheral vascular resistance. Whereas nisoldipine, like other dihydropyridines, exhibits a mild diuretic effect, most of the antihypertensive activity is attributed to its effect on peripheral vascular resistance.

Nisoldipine is metabolized into five major metabolites that are excreted in the urine. The major biotransformation pathway seems to be the hydroxylation of the isobutyl ester. A hydroxylated derivative of the side-chain, present in plasma at concentrations approximately equal to the parent compound, seems to be the only active metabolite and has about 10% of the activity of the parent compound. Cytochrome P_{450} enzymes play a key role in the metabolism of nisoldipine. The particular isoenzyme system responsible for its metabolism has not been identified, but other dihydropyridines are metabolized by cytochrome P_{450} 3A4. Nisoldipine should not be administered with grapefruit juice because it interferes with nisoldipine metabolism. Because there is very little information available about this drug's use in patients with severe congestive heart failure, it should be administered with caution to these patients.

Recently, antianginal and anti-ischemic effects of nislodipine and ramipril in patients with Syndrome X (typical angina pectoris, positive treadmill exercise test but negative intravenous ergonovine test and angiographically normal coronary arteries) suggested that they have similar anti-ischemic and antianginal effects in patients with Syndrome X (172).

Nitrendipine (34) (Fig. 1.9, Table 1.3): This drug is used to treat mild to moderate hypertension (173, 174).

In summary, calcium antagonists inhibit the influx of extracellular calcium ions into cells. This results in decreased vascular smooth muscle tone and vasodilation leads to a reduction in blood pressure. The 1,4-dihydropyridine derivatives (aranidipine, cilnidipine, amlodipine, nisoldipine, nifedipine, felodipine, nitrendipine, and nimodipine) differ from the benzothiazepine (e.g., diltiazem) and phenylalkylamine (e.g., verapamil) classes of calcium antagonists with regard to potency, tissue selectivity, and antiarrhythmic effects.

In general, dihydropyridine agents are the most potent arteriolar vasodilators, producing the least negative inotropic and electrophysiological effects; in contrast, verapamil and diltiazem slow AV conduction and exhibit negative inotropic activity while also maintaining some degree of arteriolar vasodilatation.

Calcium channel blockers are commonly used to treat high blood pressure, angina, and some forms of arrhythmia. In the treatment of hypertension and chronic heart failure, a combination therapy enhances therapeutic efficacy. Pharmacodynamically, combinations of ACE inhibitor plus a diuretic, β-adrenoreceptor antagonist, or calcium channel blocker are the most promising.

4.7.4 Other Therapeutics. For the past two decades, the cardiovascular drug market has lead drug discovery efforts and sales in the pharmaceutical industry. A constant flow of new and effective drugs has kept this sector in its number one position and will continue to do so in the future. In addition to the above indicated classes of drugs, a number of highly effective drugs have been introduced and are routinely used either alone or in combination therapy. Some of these categories are discussed below:

4.7.5 Cardiac Glycosides. Glycosides are a distinct class of compounds that are either found in nature or can be synthetically prepared. The natural glycosides are isolated from various plant species namely digitalis purpurea Linne, digitalis lanata Ehrhart, strophanthus gratus, or acokanthea schimperi. Therefore, these compounds are also named as digitalis, digoxin, and digitoxin. Currently, digoxin is the only cardiac glycoside commercially available in United States. Glycosides have a characteristic steroid (aglycone) structure complexed with a sugar moiety at C-3 position of the steroid through the β-hydroxyl group.

Glycosides are mainly used in the prophylactic management and treatment of congestive heart failure and atrial fibrillation. They are known to relieve the symptoms of systemic venous congestion (right-sided heart failure or peripheral edema) and pulmonary congestion (left-sided heart failure). However, glycosides also find applications to treat and prevent sinus and supraventricular tachycardia and symptoms of angina pectoris and myocardial infarction, but only in combination with β-adrenergic blocking agents and in patients with congestive heart failure.

The exact mechanism of pharmacological action of glycosides has not been fully elucidated. However, glycosides exhibit a positive inotropic effect accompanied by reduction in peripheral resistance and enhancement of myocardial contractility resulting in increased myocardial oxygen consumption. They also inhibit the activities of sodium-potassium activated ATPase, an enzyme required for the active transport of sodium across the myocardial cell membranes.

Glycosides are normally administered either orally or by IV injection and possess a half-life of 36 h to 5–7 days in normal patients, depending on the choice of drugs.

4.7.6 Angiotension-Converting Enzyme (ACE) Inhibitors and β blockers. ACE inhibitors prevent the conversion of angiotension I to angiotension II, a potent vasoconstrictor. This consequentially reduces plasma concentrations of angiotension II and hence vasodilation, and results in attenuation of blood pressure. ACE inhibitors also affect the release of renin from the kidneys and increase plasma renin activity (PRA). It has been suggested that the hypotensive effect of ACE inhibitors may decrease vascular tone because of angiotension-induced vasoconstriction and increased sympathetic activity. The reduced production of angiotension II lowers the plasma aldosterone concentration (caused by less secretion of aldosterone from the adrenal cortex). Aldosterone is known to decrease the sodium extraction concentration and water retention, resulting in a desired hypotensive effect. β Blockers reduce the oxygen demand of the heart by slowing the heart rate and lowering arterial pressure. Such drugs include propranolol (Inderal), α and β blockers labetalol (Normodyne, Trandate), acebutolol (Sectral), atenolol (Tenormin), metoprolol (Toprol), and bisoprolol (Zebeta). These drugs are equally effective as calcium channel blockers and have fewer adverse effects.

4.7.7 Glycoprotein IIb/IIIa Receptor Antagonists.
These compounds are also called blood thinners because they block platelet activity. These drugs are very beneficial for many patients with angina and do not seem to pose an increased risk for stroke, including strokes caused by bleeding. Some of the most widely used drugs include Abciximab (ReoPro, Centocor), eptifibatide (Integrelin), lamifiban, and tirofiban (Aggrastat). Glycoprotein IIb/IIIa receptor antagonists are used to reduce the risk for heart attack or death in many patients with unstable angina and non-Q-wave myocardial infarctions when used in combination with heparin or aspirin. Patients with unstable angina showing elevated levels of troponin T factor are good candidates for these drugs.

4.7.8 Anti-Clotting Agents.
Anti-clotting agents, either anticoagulants or anti-platelet drugs, are being used to treat unstable angina, to protect against heart attacks, and to prevent blood clots during heart surgeries. They can be used alone or in combinations, depending on the severity of the condition. Clopidogrel (Plavix), a platelet inhibitor, has been shown to be 20% more effective than aspirin for reducing the incidence of a heart attack. Other promising anti-clotting drugs comprise argatroban (Novastan), danaparoid (Organ), and forms of hirudin (bivalirudin lepidrudin or desirudi), a substance derived from the saliva of leeches. One study suggested that the hirudin agents may be superior to heparin in preventing angina and heart attack, although bleeding is a greater risk with hirudin.

5 ANTIARRHYTHMIC AGENTS

5.1 Mechanisms of Cardiac Arrhythmias

The pumping action of the heart involves three principle electrical events: the generation of a signal; the conduction or propagation of the signal; and the fading away of the signal. When one or more of these events is disrupted, cardiac arrhythmias may arise.

5.1.1 Disorders in the Generation of Electrical Signals.
In normal heart, cells located in the right atrium, referred to as the SA node or pacemaker cells, initiate a cardiac impulse. The spontaneous electrical depolarization of the SA pacemaker cells is independent of the nervous system; however, these cells are innervated by both sympathetic and parasympathetic fibers, which can cause increases or decreases in heart rate as a result of nervous system stimulation. Other special cells in the heart also possess the ability to generate an impulse, and may influence cardiac rhythm, but are normally surpassed by the dominant signal generation of SA pacemaker cells. When normal pacemaker function is suppressed—caused by pathological changes occurring from infarction, digitalis toxicity, or excessive vagal tone—or when excessive release of catecholamines from sympathomimetic nerve fibers occurs, these other automatic cells (including special atrial cells, certain AV node cells, the bundle of His, and Purkinje fibers) have the potential to become ectopic pacemakers, which can dominant cardiac rhythm and consequently lead to arrhythmias.

5.1.2 Disorders in the Conduction of the Electrical Signal.
Disorders in the transmission of the electrical impulse can lead to conduction block and reentry phenomenon. Conduction block may be complete (no impulses pass through the block), partial (some impulses pass through the block), and bi-directional or unidirectional. During bi-directional block, an impulse is blocked regardless of the direction of entry; a unidirectional block occurs when an impulse from one direction is completely blocked, while impulses from the opposite direction are propagated (although usually at a slower than normal rate).

5.1.3 Heart Block.
Heart block occurs when the impulse signal from the SA node is not transmitted through either the AV node or lower electrical pathways properly. Heart block is classified by degree of severity: (1) first degree heart block, all impulses moving through the AV node are conducted, but at a slower than normal rate; (2) second degree heart block, some impulses fully transit the AV node, whereas others are blocked (as a result, the ventricles fail to beat at the proper moment); (3) third degree heart block, no im-

pulses reach the ventricles (automatic cells in the ventricles initiate impulses, but at a slower rate, and as a result the atria and ventricles beat at somewhat independent rates).

5.1.4 Reentry Phenomenon. The most important cause of life-threatening cardiac arrhythmias results from a condition known as reentry, which occurs when an impulse wave circles back, reenters previously excited tissue, and reactivates these cells. Under normal conditions, reentry does not occur, as cells become refractory (unable to accept a signal) for a period of time that is sufficient for the original signal to die away. Hence the cells will not contract again until a new impulse emerges from the SA node. However, there are certain conditions during which this does not happen, and the impulse continues to circulate. The essential condition for reentry to occur involves the development of a cellular refractory period that is shorter than the conduction velocity. Consequently, any circumstance that shortens the refractory period or lengthens the conduction time can lead to reentry.

Nearly all tachycardias, including fibrillation, are caused by reentry. The length of the refractory period depends mainly on the rate of activation of the potassium current; the rate of conduction depends on the rate of activation of the calcium current in nodal tissue, and the sodium current in other myocytes. The channels controlling these currents are the targets for suppressing reentry.

While many conditions can lead to reentry, the most common is shown in Fig. 1.10. The conditions needed for this type of reentry are as follows. First, the existence of an obstacle, around which the impulse wave front can propagate, is needed. The obstacle may be infracted or scarred tissue that cannot conduct the impulse. The second condition needed is the existence of a pathway that allows conduction at the normal rate around one side of the obstacle, whereas the other side of the pathway is impaired. The impairment may be such that it allows conduction in only one direction (unidirectional block) or it may allow conduction to proceed at a greatly reduced rate, such that when the impulse emerges from the impaired tissue, the normal tissue is no longer refractory. These pathways are usually local-

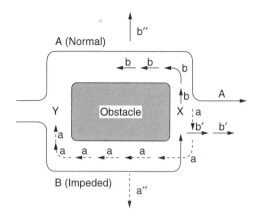

Figure 1.10. Model for reentrant activity. A depolarization impulse approaches an obstacle (nonconducting region of the myocardium) and splits into two pathways (A and B) to circumvent the obstacle. If pathway B has impeded ability to conduct the action potential the following may occur. (1) If impulse B is slowed and arrives at cross junction X after the absolute refractory period of cells depolarized by A, the impulse may continue around the obstacle as shown by path b and/or follow A along path b'. In both cases, the impulse is said to be reflected. (2) If pathway B shows unidirectional block, impulse A may continue around the obstacle as shown by path a. If the obstacle is large enough, so that cells in cross-region Y are repolarized before the return of a or b, then a circus movement may be established. Both (1) and (2) may propagate daughter impulses (a'' and b'') to other parts of the myocardium. These effects can give rise to coupled beats and fibrillation.

ized, for example, within the AV node or the end branches of part of the Purkinje system. Alternatively, they can be more extensive and may give rise to daughter impulses capable of spreading to the rest of the myocardium.

Unidirectional block occurs in tissue that has been impaired, such that its ability to conduct an impulse is completely blocked in one direction but only slowed in the other. As a result of unidirectional block, the impulse cannot proceed forward along path B (Fig. 1.10), and the cells on this path remain in a polarized state. However, when the impulse traveling along path A reaches a suitable cross-junction (point X in Fig. 1.10), the impulse proceeds back along path a, although at a slower rate than normal (indicated by the dotted path line, Fig. 1.10). When this impulse reaches another cross junction (Y), it is picked up by path A and conducted around the circle. If the obstacle is large enough, the cells in path A will have repolarized and the path will again be followed, giving rise to a continuous circular movement. When reentry occurs randomly in the myocardium, it results in random impulses that lead to cardiac fibrillation.

5.2 Types of Cardiac Arrhythmias

Arrhythmias can be divided into two categories—ventricular and supraventricular arrhythmias. Within these two categories, arrhythmias are further defined by the speed of the heartbeats. Bradycardia indicates a very slow heart rate of less than 60 beats/min; tachycardia refers to a very fast heart rate of more than 100 beats/min. Fibrillation refers to fast, uncoordinated heartbeats.

Listed below are common forms of arrhythmias grouped according to their origin in the heart. Supraventricular arrhythmias include the following: (1) sinus arrhythmia (cyclic changes in heart rate during breathing); (2) sinus tachycardia (the SA node emits impulses faster than normal); (3) sick sinus syndrome (the SA node fires improperly, resulting in either slowed or increased heart rate); (4) premature supraventricular contractions (a premature impulse initiation in the atria causes the heart to beat prior to the time of the next normal heartbeat); (5) supraventricular tachycardia (early impulse generation in the atria speed up the heart rate); (6) atrial flutter (rapid firing of signals in the atria cause atrial myocardial cells to contract quickly, leading to a fast and steady heartbeat); (7) atrial fibrillation (electrical impulses in the atria are fired in a fast and uncontrolled manner, and arrive in the ventricles in an irregular fashion); and (8) Wolff-Parkinson-White Syndrome (abnormal conduction paths between the atria and ventricles causes electrical signals to arrive in the ventricles too early, and subsequently reenter the atria).

Arrhythmias originating in the ventricles include the following: (1) premature ventricular complexes (electrical signals from the ventricles cause an early heartbeat, after which the heart seems to pause before the next normal contraction of the ventricles occurs); (2) ventricular tachycardia (increased heart rate caused by ectopic signals from the ventricles); and (3) ventricular fibrillation (electrical impulses in the ventricles are fired in a fast and uncontrolled manner, causing the heart to quiver).

5.3 Classification of Antiarrhythmic Drugs

The classification of antiarrhythmic agents is important for clinical application; however, there is no single classification system that has gained universal endorsement. At this time, the method proposed by Singh and Vaughan Williams (175) continues to be the most enduring classification scheme. Since its initial conception, this classification method has undergone several modifications—calcium channel blockers have been added as a fourth class of compounds (176) and class I agents have been subdivided into three groups to account for their sodium channel blocking kinetics (177).

In general, the Singh and Vaughan Williams classification system is based on results obtained from microelectrode studies conducted on individual heart cells *in vitro*. Under this system, class I antiarrhythmic agents include drugs that block sodium channels (these compounds have local anesthetic properties) (178). Compounds in this class are further subdivided into three groups: IA, IB, and IC (179–181). Class IA drugs are moderately potent sodium channel blockers and usually prolong repolarization; class IB drugs have the lowest potency as sodium channel blockers, produce little to no change in action potential duration, and usually shorten repolarization; and class IC drugs, which are the most potent of the sodium channel blockers, have little to no effect on repolarization (182).

Class II drugs act indirectly on the electrophysiology of the heart by blocking β-adrenergic receptors. Class III compounds are agents that prolong the duration of the action potential (increase refractoriness). The mechanism of action of these drugs often involves inhibition of both sodium and potassium channels. Class IV antiarrhythmic agents are calcium channel blockers.

The Singh and Vaughan Williams classification scheme has received broad application and is used in most textbooks (183–187). In fact, based on a Medline search with the key

Table 1.4   Singh and Vaughan Williams Classification of Antiarrhythmic Drugs

Class	Drugs	Mechanism of Action
IA	quinidine, procainamide, disopyramide	Sodium channel blockade, lengthen refractory period
IB	lidocaine, phenytoin, tocainide, mexiletine	Sodium channel blockade, shorten duration of action potential
IC	encainide, flecainide, lorcainide, moricizine, propafenone	Sodium channel blockade, conduction slowed
II	propranolol, esmolol, acebutolol, l-sotalol	Blockade of β-adrenergic receptors, AV conduction time slowed, automaticity suppressed
III	amiodarone, bretylium, sotalol (d, l), dofetilide, ibutilide	Potassium channel blockade, prolonged refractoriness
IV	verapamil, diltiazem, bepridil	Blockade of slow inward Ca^{2+}
Miscellaneous	adenosine, digitoxin, digoxin	Miscellaneous

words "antiarrhythmic drug," it was found that of 50 consecutive articles (published in 1998) 56% used the Singh and Vaughan Williams classification in the title and/or the abstract (188). Table 1.4 lists examples of drugs in each of these classes and summarizes the mechanisms of action of each class. Note that miscellaneous drugs (189) have been added to Table 1.4 to account for compounds with mechanisms of action that do not fit within the four standard classes.

The Singh and Vaughan Williams method of classification has received strong criticism from the Task Force of the Working Group on Arrhythmias of the European Society of Cardiology. In reports published simultaneously in *Circulation* (190) and the *European Heart Journal* (191), criticisms were listed in detail. Key points of contention with regard to the Singh and Vaughn Williams classification are listed below:

1. The classification is incomplete; there is no class for miscellaneous drugs such as digoxin and adenosine, which are both clinically used to treat arrhythmia.
2. The effects of a drug in a particular class can result from more than one mechanism (see Table 1.5), and it is difficult to determine which of the multiple actions are responsible for the antiarrhythmic activity.
3. Antiarrhythmic drugs that are channel blockers are listed, but channel activators are not covered.

4. The metabolites of many of the antiarrhythmic drugs also contribute to their potency (for example, lorcainide and its metabolite, norlorcainide, are both active antiarrhythmic agents).
5. The classification is based on studies of normal myocardial cells and may not represent cell behavior during disease states.
6. The classification may lead to inappropriate administration, because it implies that drugs in the same class have similar favorable and unfavorable effects.

In response to these limitations, a new classification system, known as the Sicilian Gambit, was devised by the Task Force of the Working Group on Arrhythmias (190, 191). This system draws its name from an opening chess move termed the "Queen's Gambit," which was designed to provide several aggressive options to the player using it, and as the Task Force was meeting in Sicily at the time, this point of origin was also included in the title of the classification system. In general, the Sicilian Gambit is more flexible than the Singh and Vaughan Williams method and takes into account that individual antiarrhythmic agents may have more than one mechanism of action. In Table 1.5, antiarrhythmic agents have been grouped according to the traditional Singh and Vaughan Williams method of classification, but also displayed are the multiple mechanisms of action

Table 1.5 Multiple Inhibitory Mechanisms of Antiarrhythmic Drugs

Drug Name and Class*	Channels: Na Fast	Channels: Na Med	Channels: Na Slow	Channels: Ca	Channels: K	Channels: I_f	Receptors: α	Receptors: β	Receptors: M_2	Receptors: P	Pumps: Na/K ATPase
Class IA											
Disopyramide		High			Medium						
Procainamide		High			Medium						
Quinidine		High			Medium		Low		Low		
Class IB											
Lidocaine	Low										
Mexiletine	Low										
Tocainide	Low										
Class IC											
Encainide			High								
Flecainide			High		Low						
Moricizine	High										
Propafenone		High						Medium			
Class II											
Propranolol	Low							High			
Class III											
Amiodarone	Low			Low	High		Medium	Medium			
Bretylium					High		Agonist/Antagonist	Agonist/Antagonist			
Sotalol (*d*, *l*)					High			High			
Class IV											
Bepridil	Low			High	Medium						
Diltiazem				Medium							
Verapamil	Low			High			Medium				
Miscellaneous											
Adenosine										Agonist	
Digoxin									Agonist		High

Blocking Potency

Low:
Medium:
High:
Agonist:
Agonist/Antagonist:

that would clearly lead to substantially different clinical effects for compounds within each class, as pointed out by proponents of the Sicilian Gambit (190).

Criticisms and rebuttals with regard to developing an optimal method for classifying antiarrhythmic agents are ongoing (192, 193) and will continue as novel research discoveries shed new light on the causes of cardiac arrhythmias and the mechanisms of action of antiarrhythmic drugs. For the scope of this chapter, the Singh and Vaughan Williams method of categorizing antiarrhythmic drugs will serve as a benchmark for organizing and describing these agents, but with the caveat that this system has limitations.

5.4 Perspective: Treatment of Arrhythmias

In recent years there have been many changes in the way that arrhythmia is treated; new technologies, including radiofrequency ablation and implantable devices for atrial and ventricular arrhythmias have proven to be remarkably successful mechanical treatments. In addition, Cardiac Suppression Trials (CAST) and numerous other studies have provided evidence indicating that drugs which act mainly by blocking sodium ion channels—class I agents under the Singh and Vaughn Williams system of classification—may have the potential to increase mortality in patients with structural heart disease (194, 195). Since the CAST results were released, the use of class I drugs has decreased, and attention has shifted to developing new class III agents, which prolong the action potential and refractoriness by acting on potassium channels. Of the class III antiarrhythmic agents, amiodarone has been studied extensively and has proven to be a highly effective drug for treating life-threatening arrhythmias.

In addition, new studies have indicated that combination therapies, for example administration of amiodarone and class II β-blockers, or concomitant treatment with implantable mechanical devices and drug therapies are effective avenues for treating arrhythmias.

These topics are discussed in greater detail in Section 6. The remainder of Section 5 covers individual antiarrhythmic agents as they are categorized under the Singh and Vaughn Williams classification.

5.5 Class I: Membrane-Depressant Agents

Antiarrhythmic agents in this class bind to sodium channels and inhibit or block sodium conductance. This inhibition interferes with charge transfer across the cell membrane. Investigations into the effects of class I antiarrhythmics on sodium channel activity have resulted in the division of this class into three separate subgroups—referred to as IA, IB, and IC (196).

The basis for dividing the class I drugs into subclasses resulted from measured differences in the quantitative rates of drug binding to, and dissociation from, sodium ion channels (196). Class IB drugs, which include lidocaine, tocainide, and mexiletine, rapidly dissociate from sodium channels and consequently have the lowest potencies of the class I drugs—these molecules produce little to no change in the action potential duration and shorten repolarization. Class IC drugs, which include encainide and lorcainide, are the most potent of the class I antiarrhythmics; drugs in this class display a characteristically slow dissociation rate from sodium ion channels, causing a reduction in impulse conduction time. Agents in

this class have been observed to have modest effects on repolarization. Drugs in class IA—quinidine, procainamide, and disopyramide—have sodium ion channel dissociation rates that are intermediate between class IB and IC compounds.

The affinity of the class I antiarrhythmic agents for sodium channels vary with the state of the channel or with the membrane potential (197). As indicated in Section 3.1, sodium channels exist in at least three states: R = closed resting, or closed near the resting potential, but able to be opened by stimulation and depolarization; A = open activated, allowing Na^+ ions to pass selectively through the membrane; and I = closed inactivated, and unable to be opened (196). Under normal resting conditions, the sodium channels are predominantly in the resting or R state. When the membrane is depolarized, the sodium channels are active and conduct sodium ions. Following, the inward sodium current rapidly decays as the channels move to the inactivated (I) state. The return of the I state to the R state is referred to as channel reactivation and is voltage-and-time-dependent. Class I antiarrhythmic drugs have a low affinity for R channels, and a relatively high affinity for both the A and I channels (198).

An overview of the uses and side effects of class I antiarrhythmics are displayed in Table 1.6. In addition to blocking sodium channels, the class I compounds have also been observed to effect other ion channels and receptors (Table 1.5).

5.5.1 Class IA Antiarrhythmics.

Quinidine (**35**) (Fig. 1.11, Table 1.6): Quinidine is the prototype of the class IA antiarrhythmic agents. It is obtained from species of the genus *Cinchona*, and is the D-isomer of quinine. This molecule contains two basic nitrogens: one in the quinoline ring and one in the quinuclidine moiety. The nitrogen in the quinuclidine moiety is more basic. Three salt formulations are available: quinidine gluconate, quinidine polygalacturonate (199), and quinidine sulfate (200). Of the three, the gluconate formulation is the most soluble in water.

Quinidine binds to open sodium ion channels, decreasing the entry of sodium into myocardial cells. This depresses phase 4 diastolic

Table 1.6   Class I Antiarrhythmic Agents: Uses and Side Effects

Drug Name	Use	Side Effects
Class IA		
Quinidine (35)	Atrial and ventricular tachycardia	Hematological, GI, liver
Procainamide (36)	Ventricular arrhythmias	Hematological, GI, CNS,[a] sensitivity
Disopyramide (37)	Ventricular arrhythmias	Anticholinergic, hematological
Class IB		
Lidocaine (38)	Ventricular arrhythmias	CNS
Tocainide (39)	Ventricular arrhythmias	CNS, GI, Hypotension
Mexiletine (40)	Ventricular arrhythmias	CNS, GI
Phenytoin (41)	Improved atrioventricular conduction	CNS, gingival hyperpsia, blood dyscrasias
Class IC		
Encainide (42)	Ventricular arrhythmias	CNS, Ocular
Flecainide (43)	Ventricular and supraventricular arrhythmias	CNS, Ocular
Lorcainide (44)	Ventricular arrhythmia and tachycardia, Wolff-Parkinson-white Syndrome	CNS, GI
Propafenone (45)	Supraventricular arrhythmias	CNS, GI
Moricizine (46)	Ventricular arrhythmias	CNS, GI, Ocular

[a]CNS adverse effects include headache, dizziness, nausea, vomiting, diarrhea, flushing, weakness, rash, and syncope.

depolarization (shifting the intracellular threshold potential toward zero), decreases transmembrane permeability to the passive influx of sodium (slowing the process of phase 0 depolarization, which decreases impulse velocity), and increases action potential duration (201). Physiologically, this results in a reduction in SA node impulse initiation and depression of the automaticity of ectopic cells. Quinidine is also thought to act, at least in part, by binding to potassium channels (Table 1.5).

Quinidine is used to treat supraventricular and ventricular arrhythmias including atrial flutter and fibrillation, and atrial and ventricular premature beats and tachycardias. It is primarily metabolized by the liver; a hydroxylated metabolite, 2-hydroxyquinidine, is equal in potency to the parent compound (202).

Procainamide (**36**) (Fig. 1.11, Table 1.6): Procainamide is an amide derivative of procaine. Replacement of the ether oxygen in procaine with an amide nitrogen (in procainamide) decreases CNS side effects, rapid hydrolysis, and instability in aqueous solution that results from the ester moiety in procaine. Procainamide is formulated as a hydrochloride salt of its tertiary amine. The metabolite of procainamide is N-acetylprocainamide (NAPA), which possesses 25% of the parent drugs activity (203, 204).

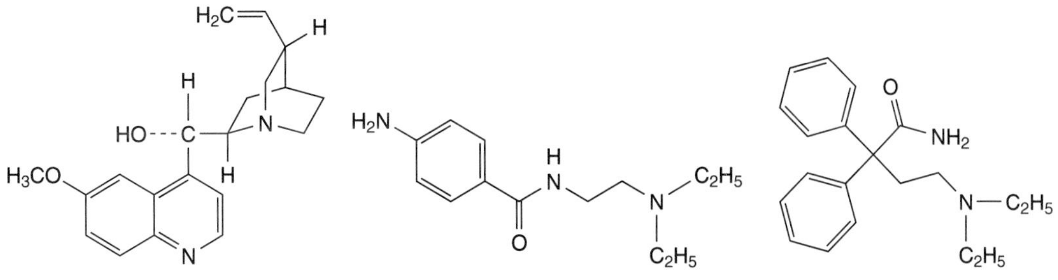

Quinidine (**35**)   Procainamide (**36**)   Disopyramide (**37**)

Figure 1.11. Chemical structures of class IA antiarrhythmic agents: sodium channel blockers.

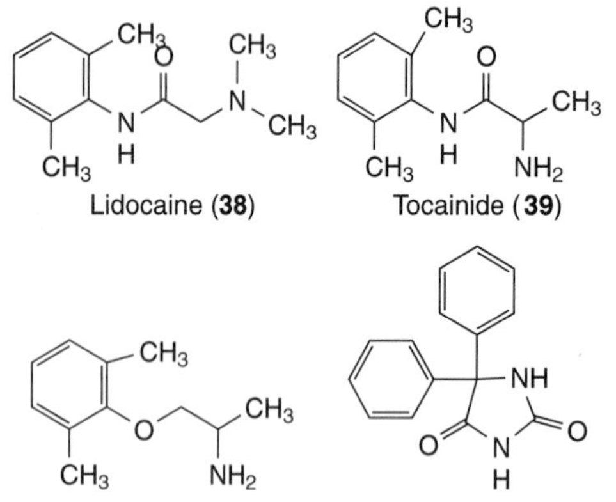

Figure 1.12. Chemical structures of class IB antiarrhythmic agents: sodium channel blockers.

Mechanistically, procainamide has the same cardiac electrophysiological effects as quinidine. It decreases automaticity and impulse conduction velocity, and increases the duration of the action potential (205). This compound may be used to treat all of the arrhythmias indicated for treatment with quinidine, including atrial flutter and fibrillation, and atrial and ventricular premature beats and tachycardias.

Disopyramide (**37**) (Fig. 1.11, Table 1.6): The electrophysiological effects of this drug are similar to those of quinidine and procainamide—decreased phase 4 depolarization and decreased conduction velocity (206). This molecule contains the ionizable tertiary amine that is characteristic of compounds in this class, is formulated as a phosphate salt, and is administered both orally and intravenously (207).

Because of its structural similarity to anticholinergic drugs, disopyramide produces side effects that are characteristic of these compounds, including dry mouth, urinary hesitancy, and constipation. Clinically it is used to treat life-threatening ventricular tachyarrhythmias.

5.5.2 Class IB Antiarrhythmics. Lidocaine (**38**) (Fig. 1.12, Table 1.6): Lidocaine is formulated as a hydrochloride salt that is soluble in both water and alcohol. It binds to inactive

sodium ion channels, decreasing diastolic depolarization and prolonging the resting period (208, 209).

Lidocaine is administered intravenously for suppression of ventricular cardiac arrhythmias and is the prototype compound for class IB. Its first-pass metabolite, monoethylglycinexylidide, results from deethylation of the tertiary amine and is an active antiarrhythmic agent (210, 211).

Tocainide (**39**) (Fig. 1.12, Table 1.6): Tocainide is an analogue of lidocaine, but structurally differs in that it possesses a primary, versus a tertiary, terminal side-chain amine. In addition, a methyl substituent on the carbon atom that is adjacent to the side-chain amide carbonyl may partially protect this moiety against hydrolysis.

Tocainide has a similar mechanism of action to that of lidocaine (212, 213). It is orally active, and the presence of a primary amine allows for formulation as a hydrochloride salt. Therapeutically it is used to prevent or treat ventricular tachycardias.

Mexiletine (**40**) (Fig. 1.12, Table 1.6): Structurally, mexiletine resembles lidocaine and tocainide in that it contains a xylyl moiety. However, it differs in that it possesses a side-chain ether versus an amide moiety (as found in lidocaine and tocainide). As a result, mexiletine is not vulnerable to hydrolysis and has a longer half-life than lidocaine (214).

Mexiletine possesses a primary amine and is formulated as a hydrochloride salt that is orally active. Its effects on cardiac electrophysiology are similar to that of lidocaine (215). It is used in the treatment of ventricular arrhythmias, including ventricular tachycardias that are life-threatening. However, because of the proarrhythmic effects of this compound, it is generally not used with lesser arrhythmias (216).

Phenytoin (**41**) (Fig. 1.12, Table 1.6): Phenytoin is a hydantoin derivative of the anticonvulsants that does not possess the sedative properties of the central nervous system depressants. It is structurally dissimilar to class I antiarrhythmic compounds and is the only non-basic member of this family of compounds. However, its effects on cardiac cells are similar to those of lidocaine. Mechanistically, it depresses ventricular automaticity

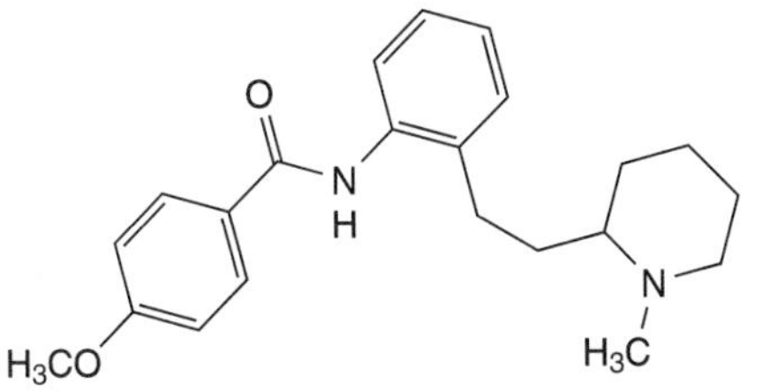

Encainide (**42**)

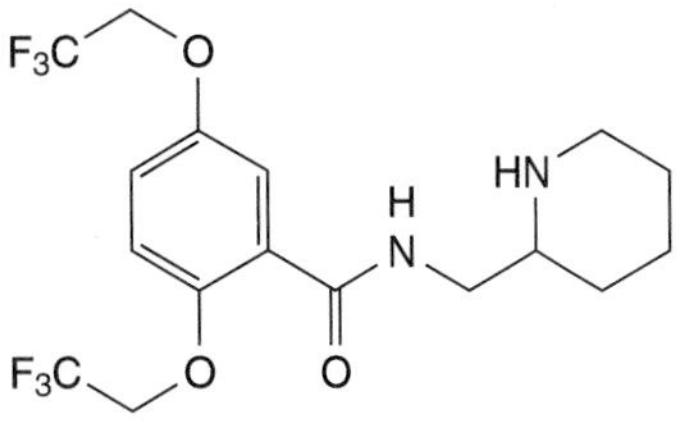

Flecainide (**43**)

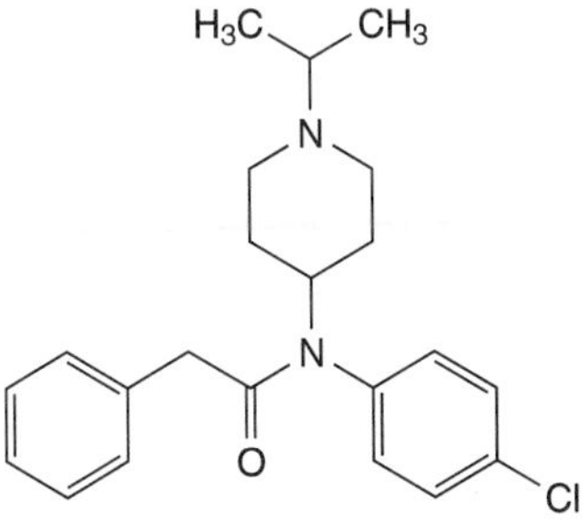

Lorcainide (**44**)

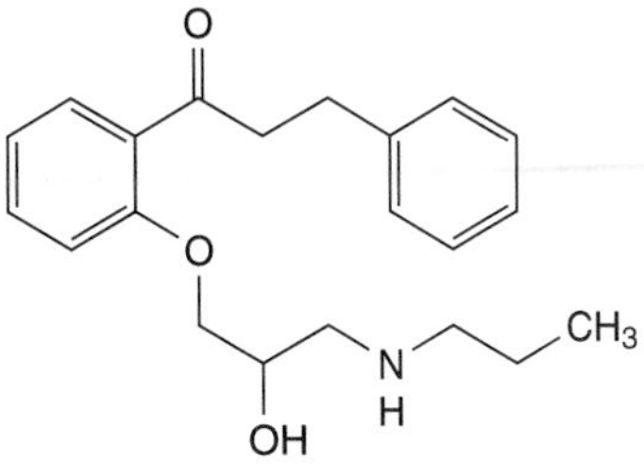

Propafenone (**45**)

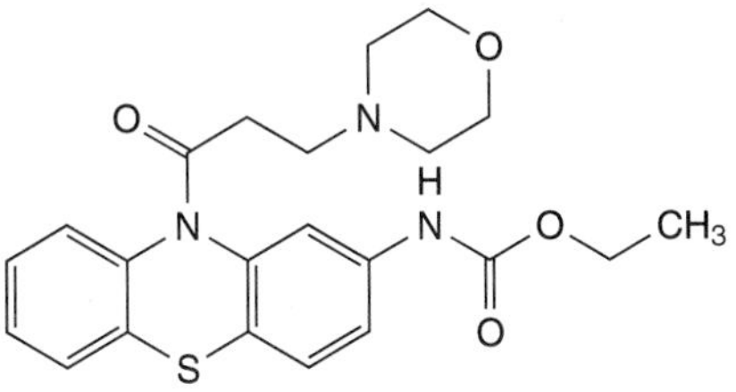

Moricizine (**46**)

Figure 1.13. Chemical structures of class IC antiarrhythmic agents: sodium channel blockers.

and prolongs the effective refractory period relative to the action potential duration. It decreases the force of contraction, depresses pacemaker action, and improves atrioventricular conduction, especially when administered in conjunction with digitalis (217).

5.5.3 Class IC Antiarrhythmics. Encainide (**42**) (Fig. 1.13, Table 1.6): This compound is a benzanilide derivative containing a piperidine ring. Like other class I compounds, it blocks sodium channels, depressing the upstroke velocity of phase 0 of the action potential and increasing the recovery period after repolarization (218). Encainide has also been shown to block the delayed potassium rectifier current (219). Like other class I antiarrhythmics, encainide contains a terminal tertiary amine and is formulated as a chloride salt. It is used to treat life-threatening ventricular arrhythmias.

Metabolites of encainide also display activity. The metabolite ODE, resulting from demethylation of the methoxy moiety, is more potent than encainide (220). A second metabolite, NDE, results from *N*-demethylation.

Flecainide (**43**) (Fig. 1.13, Table 1.6): Flecainide is a benzamide/piperidine derivative. However, it is structurally dissimilar from encainide in that it contains one less benzyl group, possesses two lipophilic trifluoroethoxy substituents at the 1 and 4 positions on the benzamide ring (versus a single methoxy substituent at the 4 position of the benzamide in encainide), and lacks a methyl substituent on the piperidine nitrogen. It is formulated as an acetate salt, and like encainide, its metabolites are active. It possesses cardiac electrophysiological effects that are similar to those of encainide, i.e., it slows cardiac impulse conduction, and it is used to treat life-threatening ventricular arrhythmias and supraventricular tachyarrhythmias (221, 222).

Lorcainide (**44**) (Fig. 1.13, Table 1.6): Lorcainide is another benzamide/piperidine derivative in this class. Its mechanism of action is similar to that of encainide—it slows conduction in myocardial tissue and reduces the speed of depolarization of myocardial fibers, suppressing impulse conduction in the heart (223, 224).

Lorcainide is formulated as a hydrochloride salt and is orally active. Metabolism of this drug results in *N*-dealkylation of the piperidyl nitrogen to yield norlorcainide (225). This metabolite is as potent as its parent compound, but its half-life is approximately three times longer. Lorcainide is used to treat ventricular arrhythmia, ventricular tachycardia, and Wolff-Parkinson-White Syndrome.

Propafenone (**45**) (Fig. 1.13, Table 1.6): Structurally, propafenone is unlike other compounds in this class—encainide, flecainide, and lorcainide. Instead, it is an ortho substituted aryloxy propanolamine similar in structure to the major class of β blockers. The racemic mixture possesses good Na^+ channel blocking action, whereas the S-(−)-isomer is the potent β blocker. Mechanistically, this compound has a direct stabilizing effect on myocardial membranes, which manifests in a reduction in upstroke velocity (phase 0) of the action potential (226). In Purkinje fibers, and to a lesser extent myocardial fibers, propafenone decreases the fast inward current carried by sodium ions, prolongs the refractory period, reduces spontaneous automaticity, and depresses triggered activity (227, 228).

Propafenone is indicated in the treatment of paroxysmal atrial fibrillation/flutter and paroxysmal supraventricular tachycardia. It is also used to treat ventricular arrhythmias, such as sustained ventricular tachycardias that are life-threatening.

Moricizine (**46**) (Fig. 1.13, Table 1.6): Moricizine is a phenothiazine derivative and is a structurally unique member of the class IC antiarrhythmic agents. Like other agents in this subclass, it decreases the speed of cardiac conduction by lengthening the refractory period and shortening the length of the action period of cardiac tissue (229). Moricizine is formulated as a hydrochloride salt, and is used to treat life-threatening ventricular arrhythmias.

5.6 Class II: β-Adrenergic Blocking Agents

The competitive inhibitors in this class are all β-adrenergic antagonists that have been found to produce membrane-stabilizing or depressant effects on myocardial tissue at concentrations above normal therapeutic doses. It is hypothesized that their antiarrhythmic properties are mainly caused by inhibition of adrenergic stimulation of the heart (230, 231) by the endogenous catecholamines epinephrine and norepinephrine, which increase the slow, inward movement of Ca^{2+} during phase 2 of the action potential. The principle electrophysiological effects of the β blocking agents manifest as a reduction in the phase 4 slope potential of sinus or ectopic pacemaker cells, which decreases heart rate and slows ectopic

Table 1.7   Class II Antiarrhythmic Agents: Uses and Side Effects

Drug Name	Use	Side Effects
Propranolol (47)	Cardiac arrhythmias	Cardiovascular, CNS[a]
Nadolol (48)	Atrial fibrillation	Cardiovascular, CNS
l-Sotalol (49)	Ventricular arrhythmias	Arrhythmogenic, cardiovascular, CNS
Atenolol (50)	Atrial fibrillation	CNS, GI
Acebutolol (51)	Ventricular arrhythmias	CNS, GI
Esmolol (52)	Supraventricular tachyarrhythmias	CNS, GI
Metoprolol (53)	Atrial tachyarrhythmias	Cardiovascular, CNS

[a]CNS adverse effects include headache, dizziness, nausea, vomiting, diarrhea, flushing, weakness, rash, and syncope.

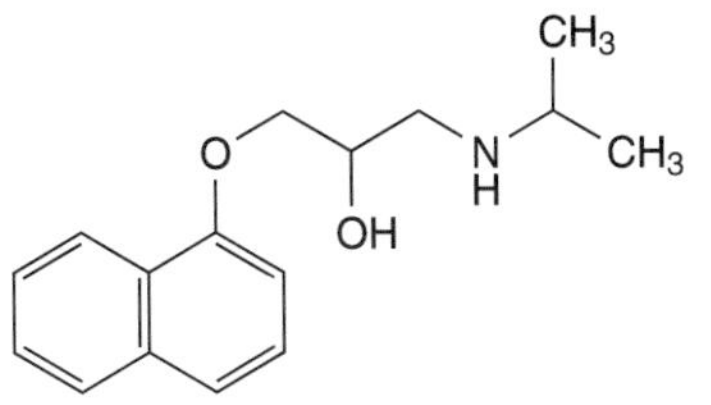

Propranolol (**47**)

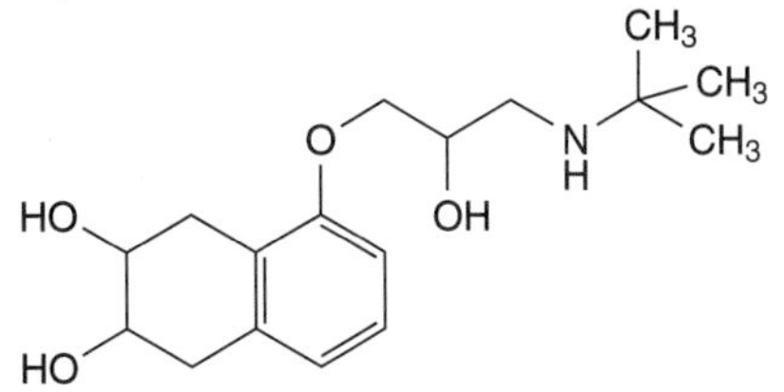

Nadolol (**48**)

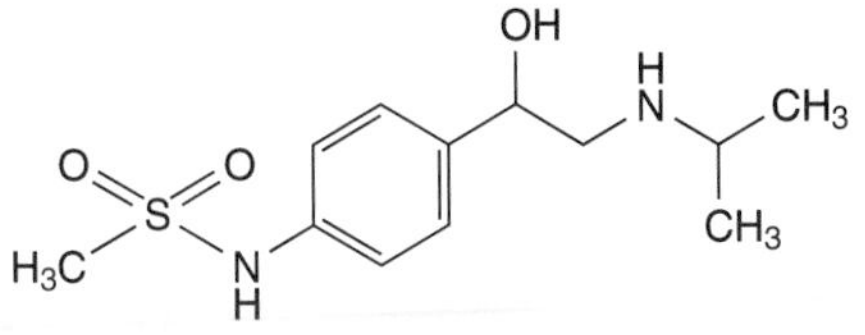

Sotalol (**49**)

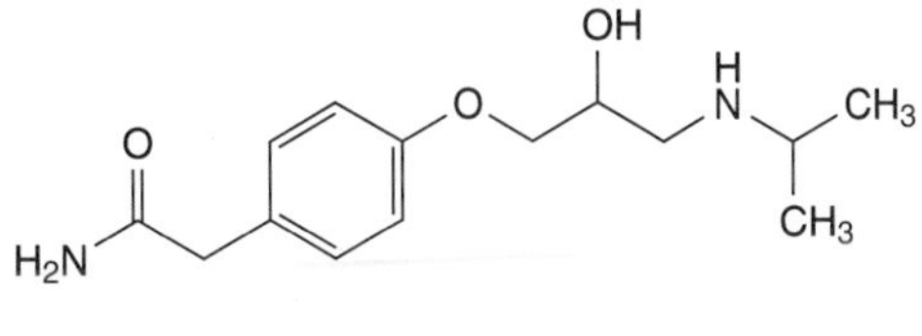

Atenolol (**50**)

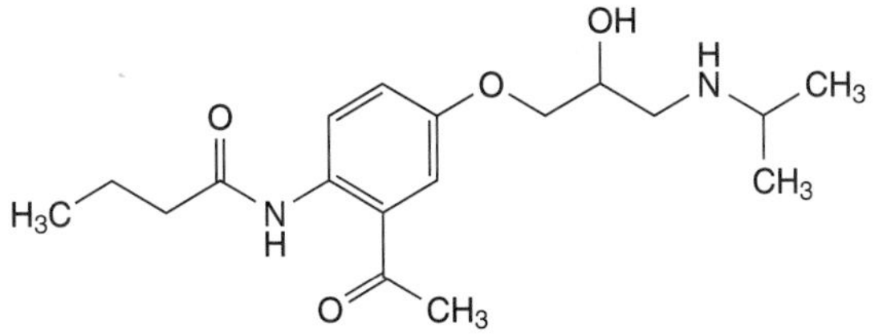

Acetobutolol (**51**)

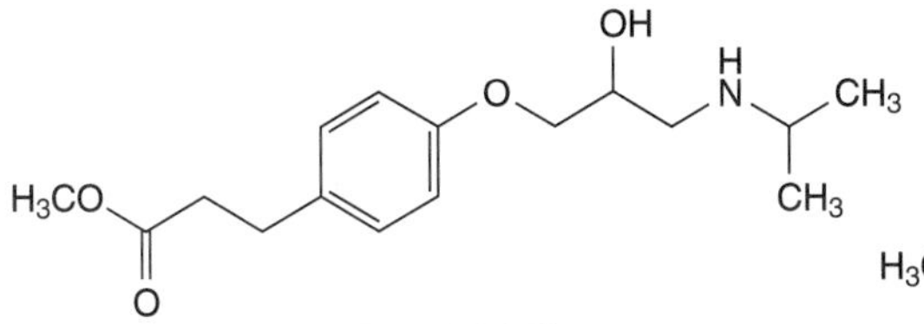

Esmolol (**52**)

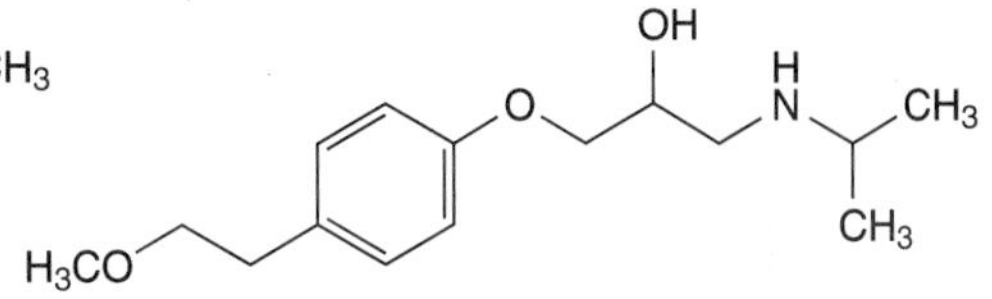

Metoprolol (**53**)

Figure 1.14. Chemical structures of class II antiarrhythmic agents: β-adrenergic blocking agents.

tachycardias. A list of compounds in the class, along with uses and side effects are shown in Table 1.7.

With the exception of sotalol (232), the compounds in this class are all structurally similar agents known as aryloxypropanolamines (Fig. 1.14). This name originates from the presence of an —OCH_2— group located between a substituted benzene ring and an ethylamino side-chain. The aromatic ring and its substituents are the primary determinants of β-antagonist selectivity. Substitution of the *para* position of the benzene ring, in tandem with the absence of *meta* position substitution, seems to confer selectivity for β1 cardiac receptors. Sotalol differs from other members of this class in that it lacks the —OCH_2— group. This results in a shortening of the characteristic ethylamino side-chain (Fig. 1.14).

Propranolol (**47**) is the prototype agent for this class of compounds. Because of the substitution pattern on its aromatic ring, it is not a selective β-adrenergic blocking agent (Fig. 1.14). During propranolol-mediated β-receptor block, the chronotropic, ionotropic, and vasodilator responses to β-adrenergic stimulation are decreased. Propranolol exerts its antiarrhythmic effects in concentrations associated with β-adrenergic blockade, and this seems to be its principal antiarrhythmic

Table 1.8 Class III Antiarrhythmic Agents: Uses and Side Effects

Drug Name	Use	Side Effects
Sotalol *(d, l)* (49)	Ventricular arrhythmias	Arrhythmogenic, cardiovascular, CNS[a]
Amiodarone (54)	Ventricular arrhythmias	Pulmonary toxicity
Bretylium (56)	Ventricular arrhythmias and ventricular fibrillation	Cardiovascular, arrhythmogenic, CNS, GI
Ibutilide (57)	Atrial fibrillation, ventricular tachycardia	Arrhythmogenic, slowed heart rate, heart failure
Dofetilide (59)	Ventricular arrhythmias, atrial fibrillation and flutter	Arrhythmogenic, tores de pointes, CNS, GI
Azimilide (60)	Supraventricular tachycardia, atrial flutter	

[a]CNS adverse effects include headache, dizziness, nausea, vomiting, diarrhea, flushing, weakness, rash, and syncope.

mechanism of action (233). It has also been shown to possess membrane-stabilizing activity that is similar to quinidine and other anesthetic-like drugs. However, the significance of this membrane action in the treatment of arrhythmias is uncertain, as the concentrations required to produce this effect are greater than required for the observance of its β-blocking effects.

Nadolol (**48**) (234) and L-Sotalol (**49**) (235) are both nonspecific β blockers (Fig. 1.14), whereas *para* substitutions on the aromatic rings of atenolol (**50**) (236), acetobutolol (**51**) (237), esmolol (**52**) (238), and metoprolol (**53**) (239) all confer β_1 antagonist selectivity (Fig. 1.14). Each of these agents exerts electrophysiological effects that result in slowed heart rate, decreased AV nodal conduction, and increased AV nodal refractoriness.

5.7 Class III: Repolarization Prolongators

The drugs in this class—amiodarone, bretylium, dofetilide, ibutilide, and (D,L) or racemic sotalol—all produce electrophysical changes in myocardial tissue by blocking ion channels; however, some are selective, whereas others are multi-channel blockers (this is not surprising, because there is a high degree of sequence homology between the different ion channels). Importantly, all class III drugs have one common effect—that of prolonging the action potential, which increases the effective refractory period without altering the depolarization or the resting membrane potential (240). A list of compounds in this class, along with

uses and some side effects are shown in Table 1.8.

Racemic sotalol, dofetilide, and ibutilide are potassium channel blockers. Potassium channels, particularly the channel giving rise to the "delayed rectifier current," are activated during repolarization phase 3 of the action potential. Sotalol also possesses β-adrenergic blocking properties (see above), whereas ibutilide is also a sodium channel blocker. The mechanisms of action of amiodarone and bretylium, which also prolong the action potential, remain unclear. However, both have sodium channel-blocking properties.

Of the compounds listed in this class, sotalol, dofetilide, and ibutilide are structurally similar (Fig. 1.15). All three drugs contain a central aromatic ring with a sulfonamide moiety and a *para*-substituted alkylamine sidechain. Dofetilide, unlike sotalol and ibutilide, is nearly symmetrical, with two methanesulfonamides at either end of the molecule.

Amiodarone (**54**) (Fig. 1.15, Table 1.8): Amiodarone is structurally unique in this class, and because it has received a great deal of attention over the past several years for its ability to treat arrhythmias, it is considered to be the prototype compound for this class. Amiodarone is currently the most used drug for patients with life-threatening arrhythmias—approximately one-half of the patients currently receiving antiarrhythmic drug therapy are treated with amiodarone (241).

It is a benzofuranyl derivative with a central diiodobenzoyl substituent and an alkyl

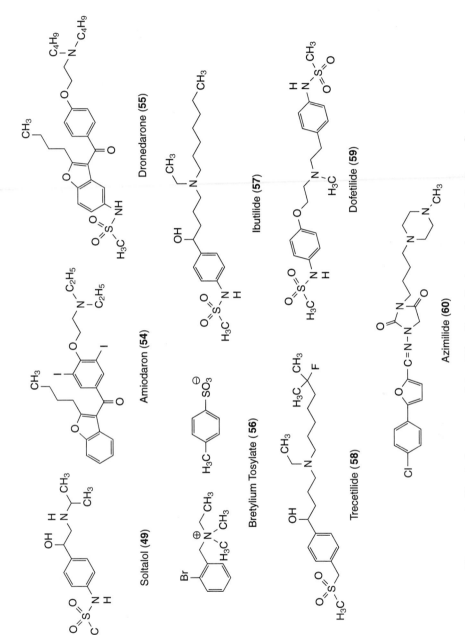

Figure 1.15. Chemical structures of class III antiarrhythmic agents: repolarization prolongators.

Soltalol (**49**)

Amiodaron (**54**)

Dronedarone (**55**)

Bretylium Tosylate (**56**)

Ibutilide (**57**)

Trecetilide (**58**)

Dofetilide (**59**)

Azimilide (**60**)

41

amine side-chain. Mechanistically, this molecule prolongs the duration of the action potential and effective refractory period, with minimal effect on resting membrane potential (242–244). Amiodarone exhibits mechanisms of activity from each of the four Singh and Vaughan Williams classes. In addition, it also displays non-competitive α- and β-adrenergic inhibitory properties. It is effective in the treatment of life-threatening recurrent ventricular arrhythmias and atrial fibrillation (245) and is orally available as a chloride salt.

Amiodarone contains two iodine substituents, and consequently does effect thyroid hormones (246). However, its most serious side effects involve both the exacerbation of arrhythmias and pulmonary toxicity.

A new, experimental noniodinated benzofuranyl derivative of amiodarone, dronedarone (**55**) (247, 248), has emerged as a potential new member of the class III antiarrhythmics. It has been found to have similar electrophysiological effects as amiodarone, but with fewer side effects (249).

(D,L)-Sotalol (**49**) (Fig. 1.15, Table 1.8): Sotalol has been classified as both a class II and a class III antiarrhythmic agent. The L isomer is classified as a β blocker and is 50 times more active than the D isomer in this capacity; the racemic mixture of this drug is considered to be a class III agent, because it inhibits the rapidly activating component of the potassium channel involved in the rectifier potassium current.

Sotalol is used to treat and prevent life-threatening ventricular arrhythmias (250). Additionally, because of its class II and III activity, it is also effective against supraventricular arrhythmias (251). The mode of action, pharmacokinetics, and therapeutic uses of sotalol have been reviewed extensively, and in 1993, a report on this drug was published (252).

Sotalol is formulated as a hydrochloride salt and can be administered orally. In terms of efficacy, clinical trials have indicated that this compound is at least as effective or more effective in the management of life-threatening ventricular arrhythmia than other available drugs (253).

Bretylium Tosylate (**56**) (Fig. 1.15, Table 1.8): Bretylium tosylate is a bromobenzyl qua-

ternary ammonium salt. It is formulated as a tosylate salt and is soluble in water and alcohol. It is administered by intravenous or intramuscular injection, and is used to treat ventricular fibrillation and ventricular arrhythmias that are resistant to other therapy.

The mechanism of antiarrhythmic action of this drug has not been determined. It does not suppress phase 4 depolarization, but it does prolong the effective refractory period (254). This compound also selectively accumulates in neurons and inhibits norepinephrine release, and it has been suggested that its adrenergic neuronal-blocking properties are responsible for its antiarrhythmic activity (255).

Ibutilide (**57**) (Fig. 1.15, Table 1.8): Ibutilide is formulated as a fumarate salt and is administered by intravenous injection (256). This agent prolongs repolarization of cardiac tissue by increasing the duration of the action potential and the effective refractory period in cardiac cells. It blocks both sodium and potassium channels (257–259), but unlike sotalol, does not possess β-adrenergic blocking activity. Ibutilide is used in the treatment of supraventricular tachyarrhythmias, such as atrial flutter and atrial fibrillation (260, 261). However, it may cause ventricular arrhythmias and is not recommended for the treatment of this condition.

Trecetilide (**58**), a congener of ibutilide, is currently being investigated for intravenous and oral treatment of atrial flutter and atrial fibrillation. In addition to blocking potassium channel receptors, this compound also seems to prolong repolarization through other mechanisms as well.

Dofetilide (**59**) (Fig. 1.15, Table 1.8): Dofetilide is one of the newest members of the class III antiarrhythmic agents and has recently received U.S. FDA approval. It prolongs repolarization and refractoriness without affecting cardiac conduction velocity, and is a relatively selective blocking agent of the delayed rectifier potassium current (262, 263). Unlike ibutilide and sotalol, this agent does not inhibit sodium channels or β-adrenergic receptors.

It is used to treat supraventricular tachyarrhythmias and to restore normal sinus rhythm during atrial fibrillation and atrial

Table 1.9 Class IV Antiarrhythmic Agents: Uses and Side Effects

Drug Name	Use	Side Effects
Bepridil (8)	Supraventricular arrhythmias	GI, CNS,[a] cardiovascular
Diltiazem (9)	Supraventricular arrhythmias	GI, CNS, cardiovascular
Verapamil (11)	Supraventricular arrhythmias	GI, CNS, hepatic

[a]CNS adverse effects include headache, dizziness, nausea, vomiting, diarrhea, flushing, weakness, rash, and syncope.

flutter (264, 265). Dofetilide is formulated as a hydrochloride salt and is administered orally.

Azimilide (**60**) (Fig. 1.15, Table 1.8): This agent is a novel class III antiarrhythmic agent that has been shown to block both the slow activating and rapidly activating components of the delayed rectifier potassium current (266). Structurally it is unlike other molecules in this class, containing both imidazolidione and piperazine moieties. Azimilide has recently received much attention and is being evaluated for its ability to treat atrial flutter, atrial fibrillation, and paroxysmal supraventricular tachycardia (267–269).

5.8 Class IV: Calcium Channel Blockers

All of the calcium channel blockers in this class of agents—verapamil, diltiazem, and bepridil—also possess antianginal activity and are also covered in detail under antianginal agents and vasodilators in this chapter. With respect to cardiac arrhythmias, these agents affect calcium ion flux, which is integral for the propagation of an electrical impulse through the AV node (270). By decreasing this influx, the calcium channel blockers slow conduction. This, in turn, slows the ventricular rate. Furthermore, many of the calcium channel blockers also have the ability to block sodium channels, which is not surprising because of the close homologies in the amino acid sequences of calcium channels and sodium channels.

Drugs in this class also decrease SA node automaticity, depress myocardial contractility, and reduce peripheral vascular resistance. The prototype drug in this class is verapamil. A list of uses and side effects for compounds in this class are shown in Table 1.9.

Verapamil (**11**) (Fig. 1.16, Table 1.9): Verapamil blocks the influx of calcium ions across cell membranes (271). Structurally it is not related to other antiarrhythmic drugs. It is formulated as a hydrochloride salt and is readily soluble in water. Verapamil is used to treat supraventricular arrhythmias including

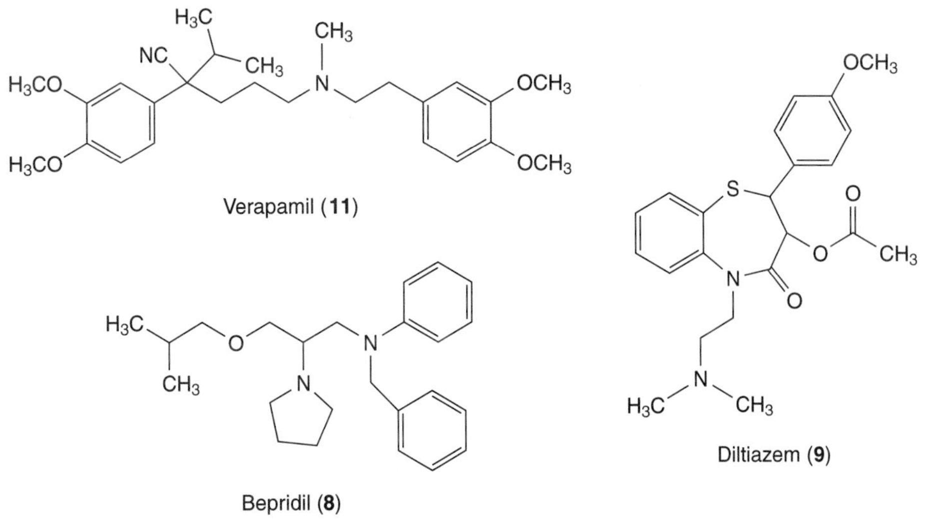

Verapamil (**11**)

Bepridil (**8**)

Diltiazem (**9**)

Figure 1.16. Chemical structures of class IV antiarrhythmic agents: calcium channel blockers.

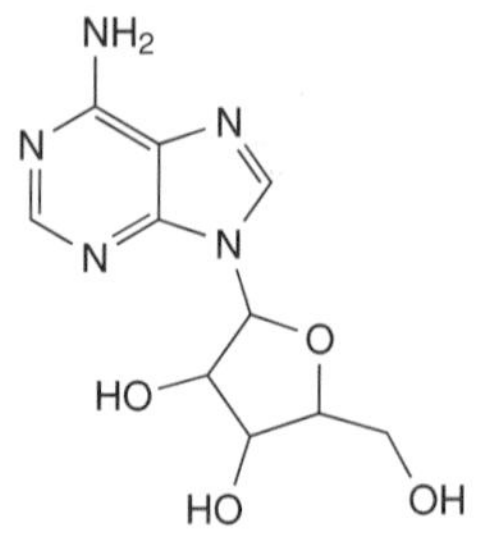

Adenosine (**61**) 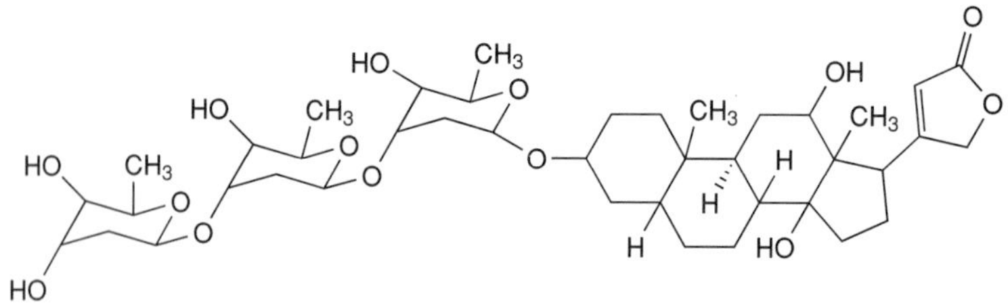   Digoxin (**62**)

Figure 1.17. Chemical structures of miscellaneous antiarrhythmic agents.

atrial tachycardias and fibrillations (272) and is administered both orally and through intravenous injection.

The mechanism of action of this compound arises from the blockade of calcium ion channels, which inhibits the influx of extracellular Ca^{2+} across the cell membranes of myocardial cells and vascular smooth muscle cells. Its activity is voltage dependent, with receptor affinity increasing as the cardiac cell membrane potential is reduced; and frequency dependent, with an increase in affinity resulting from an increase in the frequency of depolarizing stimulus.

Verapamil is rapidly metabolized to at least 12 dealkylated metabolites. Norverapamil, a major and active metabolite, has 20% of the cardiovascular activity of verapamil, and reaches plasma concentrations that are almost equal to those of verapamil within 4–6 h after administration.

Diltiazem (**9**) (Fig. 1.16, Table 1.9): Diltiazem is a benzothiazepine derivative and is formulated as a hydrochloride salt. It may be administered orally or through injection.

Similar to verapamil, diltiazem inhibits the influx of Ca^{2+} during the depolarization of cardiac smooth muscle. This decreases atrioventricular conduction and prolongs the refractory period. Therapeutically, this drug prolongs AV nodal refractoriness and exhibits effects on AV nodal conduction such that the heart rate during tachycardias is reduced. Diltiazem also slows the ventricular rate during atrial fibrillation or atrial flutter (273–274).

Bepridil (**8**) (Fig. 1.16, Table 1.9): Bepridil inhibits the transmembrane influx of Ca^{2+} into cardiac and vascular smooth muscle. Like

diltiazem, it slows the heart rate by prolonging both the effective refractory periods of the atria and ventricles, and the refractory period of the AV node (275).

It is formulated as a hydrochloride salt and is orally available. It is used in treating AV reentrant tachyarrhythmias and in the management of high ventricular rates secondary to atrial flutter or fibrillation (276, 277).

5.9 Miscellaneous Antiarrhythmic Agents

Two antiarrhythmic agents that do not fall within the Singh and Vaughan Williams classification are shown in Fig. 1.16 and briefly described in the text.

Adenosine (**61**) (Fig. 1.17): Adenosine is chemically unrelated to other antiarrhythmic drugs. It is soluble in water but practically insoluble in alcohol. For the treatment of arrhythmias, it is administered through intravenous injection.

Adenosine reduces SA node automaticity, slows conduction time through the AV node, and can interrupt reentry pathways. It is used to restore normal sinus rhythm in patients with paroxysmal supraventricular tachycardia, including Wolff-Parkinson-White Syndrome (278–281).

Digoxin (**62**) (Fig. 1.17): Digoxin belongs to the family of compounds known as the cardiac glycosides. The natural glycosides are isolated from various plant species: *Digitalis purpurea Linne, Digitalis lanata Ehrhart, Strophanthus gratu,* or *Acokanthea schimperi.*

Digoxin inhibits sodium-potassium ATPase, which is responsible for regulating the quantity of sodium and potassium inside cells. Inhibition of this enzyme results in an increase in the intracellular concentration of sodium and calcium and a decrease in intracellular potassium. Because of decreased intracellular potassium concentrations, phase 4 of the action potential becomes more positive, which in turn reduces the height of phase 0. As a result, the conduction rate in cardiac cells is slowed—SA node discharge and AV node conduction are slowed. It is available both orally and through intravenous injection and is used to treat and prevent sinus and supraventricular fibrillation, flutter, and tachycardia (282).

6 FUTURE TRENDS AND DIRECTIONS

6.1 Antiarrhythmics: Current and Future Trends

Following the discovery that lidocaine was useful for treating cardiac arrhythmias, early drug discovery and development of antiarrhythmic agents focused on compounds that were structurally similar to lidocaine and possessed similar mechanisms of action—that of blocking sodium channels. This led to the initial identification of lidocaine congeners such as tocainide and mexiletine, and later to encainide and flecainide. The long standing hypotheses for treating arrhythmias with sodium ion channel blockers was based on the belief that these molecules could effectively prevent or suppress the onset of arrhythmias and/or terminate this condition when it became persistent (283). However, the Cardiac Arrhythmia Suppression Trial (CAST) (284), which evaluated the effects of well-established sodium channel blockers on mortality in post-myocardial patients (with frequent premature ventricular arrhythmias), dispelled this hypothesis. In fact, these studies found that both encainide and flecainide increased mortality. Since the CAST studies, other trials with mexiletine, propafenone, and moricizine (285) (CAST II) (286) have also shown similar results, and a correlation between increased mortality and the use of sodium ion channel blocking agents in post-myocardial infarction patients has been established.

Based on the findings of CAST and related studies, the treatment of antiarrhythmias has shifted away from class I sodium channel blockers and now focuses on class III drugs (287), which act by prolonging the action potential duration and the refractory period. Class III agents lack many of the negative side effects observed in other classes of antiarrhythmics, affect both atrial and ventricular tissue, and can be administered orally or intravenously. Members of this class, such as amiodarone (which has proven to be a clinically efficient therapeutic for the treatment of a wide variety of arrhythmias) and racemic sotalol, have been the center of much attention in recent years and have led to the search for new class III drugs with improved safety profiles (288). New and investigational class III agents that are more selective for potassium channel subtypes include azimilide (289), dofetilide, dronedarone, ersentilide, ibutilide, tedisamil, and trecetilide (290).

There have also been numerous reports on the synthesis and evaluation of new antiarrhythmic compounds; several of these are briefly described: Matyus et al. (291) have reported the synthesis and biological evaluation of novel phenoxyalkyl amines that exhibit both class IB and class III type electrophysiological properties; Tripathi et al. (292) have performed synthesis and SAR studies on 1-substituted-N-(4-alkoxycarbonylpiperidin-1-yl) alkanes that showed potent antiarrhythmic activity comparable with quinidine; Bodor et al. (293) reported a novel tryptamine analog that was found to selectively bind to the heart (and within the heart to have tissue specificity) and to possess effects on vital signs of the cardiovascular system that indicated antiarrhythmic activity; Morey et al. (294) designed a series of amiodarone homologs that resulted in an SAR that will have implications for the future development of amiodarone-like antiarrhythmic agents; Himmel et al. (295) synthesized and evaluated the activities of thiadiazinone derivatives that are potent and selective for potassium ion channels and show class III antiarrhythmic activity; Thomas et al. (296) have developed a novel antiarrhythmic agent—BRL-32872—that inhibits both potassium and calcium channels; and Levy et al. (297) have described novel dibenzoazepine and 11-oxo-dibenzodiazepine derivatives that are effective ventricular defibrillating drug candidates.

Along with advances in the understanding and development of new therapeutic agents, the development of technological devices to treat arrhythmias has also evolved. One of the most important achievements has been the implantable cardioverter defibrillator (ICD) (241). This device has been an option for treating arrhythmias since the early 1980s, and in the treatment of ventricular tachycardia and fibrillation, no other therapy has been as effective in prolonging patient survival (283). However, an important point regarding ICD treatment is that it is often used in combination with antiarrhythmic drug therapy (241). For frequent symptomatic episodes of ventricular tachycardia, administration of an adjuvant drug therapy is often required to provide maximum prevention and treatment of life-threatening arrhythmias. In particular, combination therapy with ICD and both β blockers and amiodarone have received the most attention (241).

Finally, new evidence suggests that combinations of therapeutics may be more effective at treating and controlling arrhythmias than using any single agent alone (288). In particular, clinical sources have indicated that the pharmacological properties of amiodarone and β blockers may be additive or even synergistic for treating arrhythmias (288). Details of the analysis of amiodarone interaction with β blockers in the European Myocardial Infarct Amiodarone Trial (EMIAT) and in the Canadian Amiodarone Myocardial Infarction Trial (CAMIAT) have recently been reported (298). Data from randomized patients in these trials were analyzed by multivariate proportional hazard models and indicated that combination therapy consisting of amiodarone and β-blockers led to a significantly better survival rate. Hence, the possibility of administering combination therapies will be an important aspect in the future development of therapeutic techniques for treating arrhythmias.

6.2 Antianginal Agents and Vasodilators: Future Directions

There is increasing evidence of a relationship between apoptosis and pathophysiology in both ischemic and nonischemic cardiomyopathies, and a large number of papers published since 1997 suggest a link between some of the major genetic and biochemical regulators of apoptosis in the heart. There has been a quest for a therapeutic agent that would delay the onset of apoptosis in the ischemic heart. In the future, several therapeutic interventions can be developed to prolong the survival of smooth muscle and endothelial cells and to enhance the vascular contractibility, tone, and eventually delay the process of atherosclerosis (299, 300).

Elucidation of the phenomenon of myocardial preconditioning may hold the key to the development of a drug to the treatment of ischemic heart disease (301).

NO is a unique moiety implicated in the regulation of various physiological processes, including smooth muscle contractibility and platelet reactivity. Consequently, it has been suggested that NO may have a significant cardioprotection role in hypercholesterolemia, atherosclerosis, hypertension, or inhibition of platelet aggregation. As a result, the development of selective NO synthase inhibitors will address potential beneficial therapeutic outcomes of NO modulation to the pathophysiology of these disorders (302, 303).

Over the past few years, a number of potent asymmetric aza analogs of dihydropyrimidine (DHPM), possessing a similar pharmacological profile to classical dihydropyridine calcium channel blockers, have been studied extensively to evaluate their molecular interactions at the receptor level. Some of the lead compounds (SQ 32926 or SQ 32547) are superior in potency and duration of antihypertensive activity to classical DHP analogs and compare favorably with second generation drugs such as nicardipine and amlodipine. This class of compounds (DHPM) might be the next generation of calcium channel blockers for the treatment of cardiovascular diseases (304).

Clinical administration of drugs with negative inotropic activity is not desirable because of their cardiosuppressive effects, especially in patients with a tendency toward heart failure. Therefore, there has been a search for cardioprotective agents acting through entirely different mechanisms. It has been suggested that re-evaluation of dihydropyridine calcium channel blockers might lead to the discovery of therapeutic agents that also have effects on other membrane channels. Efonidipine, possessing inhibitory effects on both L-type and T-type Ca^{2+} channels, shows potent bradycardic effects through a characteristic prolongation of the phase 4 depolarization, leading to minimum reflex tacycardia or to bradycardia. AHC-52 and AHC-93 seem to be interesting prototypes of cardioprotective drugs that act through modification of anion homeostatis (305).

Also, there are number of novel potential drug candidates undergoing various clinical studies. One of the most promising candidates is ranolazine. This compound represents the first in a new class of drugs, called pFOX inhibitors, with the potential for treating angina. pFox inhibitors possess a unique mechanism of action, and therefore patients may be able to find relief from the painful attacks of angina without many of the unwanted effects of current anti-anginal drugs.

REFERENCES

1. A. M. Minino and B. L. Smith, *Natl. Vital Stat. Rep.*, **49**, 1–40 (2001).

2. S. L. Huston, E. J. Lengerich, E. Conlisk, and K. Passaro, *Morb. Mortal. Wkly. Rep.*, **47**, 945–949 (1998).

3. M. A. Marano, *Vital Health Stat. 10*, **199**, 1–428 (1998).

4. R. E. Thomas in M. E. Wolf, Ed., *Burger's Medicinal Chemistry and Drug Discovery*, 5th ed., vol. **2**, Wiley, New York, 1996, pp. 153–261.

5. A. M. Katz, *N. Engl. J. Med.*, **328**, 1244–1251 (1993).

6. W. A. Catterall, *Science*, **242**, 50–61 (1988).

7. B. Hille, *Ionic Channels in Excitable Membranes*, 2nd ed., Sinauer Associates, Sunderland, MA, 1992, pp. 1–607.

8. M. Strong, K. G. Chandy, and G. A. Gutman, *Mol. Biol. Evol.*, **10**, 31–38 (1993).

9. K. Ho, C. G. Nichols, W. J. Lederer, J. Lutton, P. M. Vaassiler, M. V. Kanazirska, and S. C. Herbert, *Nature*, **362**, 31–38 (1993).

10. Y. Kubo, T. J. Baldwin, Y. N. Jan, and L. Y. Jan, *Nature*, **362**, 127–133 (1993).

11. M. Noda, T. Ikeda, T. Kayono, H. Suzuki, H. Takeshima, M. Kurasaki, H. Takahashi, and S. Numa, *Nature*, **320**, 188–189 (1986).

12. W. Stuhmer, et al., *Nature*, **339**, 597–603 (1989).

13. R. W. Aldrich, *Nature*, **339**, 578–579 (1989).

14. A. L. Hokin and A. F. Huxley, *J. Physiol. (Lond.)*, **117**, 500–544 (1952).

15. B. Bean, *Nature*, **348**, 192–193 (1990).

16. O. Pongs, *Trends Pharmacol. Sci.*, **13**, 359–365 (1992).

17. Task Force for the Working Group on Arrhythmias of the European Society of Cardiology, *Eur. Heart J.*, **12**, 1112–1131 (1991).

18. H. F. Brown, *Physiol. Rev.*, **62**, 505–530 (1992).

19. W. R. Giles and Y. Imaizumi. *J. Physiol. (Lond.)*, **405**, 123–145 (1988).

20. E. Carmeliet, L. Storms, and J. Vereecke. in D. P. Zipes and H. Jaife, Eds., *Cardiac Electrophysiology from Cell to Bedside*, W. B. Saunders, Philadelphia, 1990, pp. 103–108.

21. T. Hoshi, W. N. Zogotta, and R. Aldrich, *Science*, **250**, 533–538 (1990).

22. D. Escande and L. Cavero. *Trends Pharmacol. Sci.*, **13**, 269–272 (1992).

23. W. C. Cole, C. D. Mcpherson, and D. Sontag, *Circ. Res.*, **69**, 571–581 (1991).

24. G. J. Gross, et al., *Am. J. Cardiol.*, **63**, 11J–17J (1989).

25. G. J. Grover, J. Newburger, P. G. Sleph, S. Dzwonczlyk, S. C. Taylor, S. Z. Ahmed, and K. S. Atwal, *J. Pharmacol. Exp. Ther.*, **257**, 156–162 (1991).

26. J. A. Auchampach, M. Maruyama, L. Cavero, and G. J. Gross, *J. Pharmacol. Exp. Ther.*, **259**, 961–967 (1991).

27. D. Thuringer and D. Escande, *Mol. Pharmacol.*, **36**, 897–902 (1989).

28. V. Mitrovic, E. Oehm, J. Thormann, H. Pitschner, and C. Hamm, *Herz*, **25**, 130–142 (2000).

29. D. Pelzer, S. Pelzer, and T. F. MacDonald, *Rev. Physiol. Biochem. Pharmacol.*, **114**, 107–207 (1990).

30. D. F. Slish, D. Schulz, and A. Schwartz, *Hypertension*, **19**, 19–24 (1992).

31. O. Krizanova, R. Diebold, P. Lory, and A. Schwartz, *Circulation*, **87**, VII44–VII48 (1993).

32. C. T. Sempos, J. I. Cleeman, M. K. Carroll, et al., *J. Am. Med. Assoc.*, **269**, 3009–3014 (1993).

33. R. F. Gillum, *Am. Heart J.*, **126**, 1042–1047 (1993).

34. F. M. Sacks, M. A. Pfeffer, L. A. Moye, et al., *N. Engl. J. Med.*, **335**, 1001–1009 (1996).

35. A. L. Dunn, B. H. Marcus, J. B. Kampert, M. E. Garcia, H. W. Kohl, and S. N. Blair, *J. Am. Med. Assoc.*, **281**, 327–334 (1999).

36. A. J. Manson, F. Hu, J. W. Rich-Edwards, G. Colditz, M. J. Stampfer, W. H. Willett, F. Speizer, and C. Hennekens, *N. Engl. J. Med.*, **341**, 650–658 (1999).

37. CDC, Morb. Mortal. Wkly. Rep., 48(45) (1999).

38. B. M. Pasty, N. L. Smith, D. S. Siscovick, et al., *J. Am. Med. Assoc.*, **277**, 739–745 (1997).

39. V. L. Burt, J. A. Culter, M. Higgins, et al., *Hypertension*, **26**, 60–69 (1995).

40. D. S. Freedman, W. H. Dietz, S. R. Srinivasan, and G. S. Berenson, *Pediatrics*, **103**, 1175–1182 (1999).

41. K. Karason, I. Wallentin, B. Larson, and L. Sjöstrom, *Obes. Res.*, **6**, 422–429 (1998).

42. J. L. Breslow, *Annu. Rev. Genet.*, **34**, 233–254 (2000).

43. M. A. Austin, *Proc. Nutr. Soc.*, **56**, 667–670 (1997).

44. L. W. Castelini, A. Weinreb, J. Bodnar, A. M. Goto, and M. Doolittle, *Nat. Genet.*, **18**, 374–377 (1998).

45. G. M. Dallinga-Thie, X. D. Bu, M. V. L. S. Trip, J. I. Rotter, A. J. Lusis, and T. W. de Bruin, *J. Clin. Invest.*, **99**, 953–961 (1997).

46. J. Davingnon, R. E. Gregg, and C. F. Sing., *Arteriosclerosis*, **8**, 1–21 (1988).

47. R. M. Fisher, S. E. Humphries, and P. J. Talmund, *Atherosclerosis*, **135**, 145–159 (1997).

48. H. Knoblauch, B. Muller-Myhsok, A. Busjahn, L. B. Avi, and S. Bahring, *Am. J. Hum. Genet.*, **66**, 157–166 (2000).

49. C. Glaser, *J. Clin. Hypertens.*, **2**, 204–209 (2000).

50. R. F. Furchgott and J. V. Zawadzaki, *Nature*, **288**, 373–376 (1980).

51. S. Moncada, R. M. J. Palmer, and E. A. Higgs, *Pharmacol. Rev.*, **43**, 109–142 (1991).

52. J. P. Cooke and V. J. Dzau, *Annu. Rev. Med.*, **48**, 489–509 (1997).

53. J. O. Parker, *Am. J. Cardiol.*, **72**, 3C–6C (1993).

54. J. Ahlner, R. G. Andersson, K. Torfgard, and K. L. Axelsson, *Pharmacol. Rev.*, **43**, 351–423 (1991).

55. H. L. Fung, *Br. J. Clin. Pharmacol.*, **34**, 5S–9S (1992).

56. J. P. Cooke and V. J. Dzau, *Annu. Rev. Med.*, **48**, 489–509 (1997).

57. Goldeberg, et al., *Acta. Physiol. Scand.*, **15**, 173 (1948).

58. H. Laufen, M. Aumann, and M. Leitold, *Arzneimittel-Forsch.*, **33**, 980 (1983).

59. L. A. Silviveri and N. J. DeAngelis, *Anal. Profiles Drug Subs.*, **4**, 225–244 (1975).

60. *Chem. Abstr.*, **42**, 5564 (1948).

61. Kochergin and Titkova, *Chem. Abstr.*, **54**, 8647h (1960).

62. M. V. Dijk, *Rec. Trav. Chim.*, **75**, 1215 (1956).

63. Moed, inventor, N. Am. Philips, assignee, US patent 3,056,836, 1962.

64. E. I. Goldenthal, *Toxicol. Appl. Pharmacol.*, **18**, 185 (1971).

65. H. Wesseling, G. Hovinga, A. Verslois, J. Broring, K. van Aken, and F. Moolenaar, *Eur. J. Clin. Pharmacol.*, **27**, 615–618 (1984).

66. S. H. Skotnicki, G. van Gall, and P. F. Wijn, *Angiology*, **35**, 685 (1984).

67. H. Nagano, T. Mori, S. Takaku, I. Matsunaga, T. Kugira, T. Osasawara, S. Sugano, M. Shindo, inventors, Chugai Seiyaku Kabushik Kaisha, assignee, US patent 4200640, April 29, 1980.

68. N. Tairar and M. Endoh, *Naunyn Schmiedebergs Arch. Pharmacol.*, **322**, 319–321 (1983).

69. F. Yoneyama, et al., *Cardiovasc. Drugs Ther.*, **4**, 1119 (1990).

70. H. Purcell and K. Fox, *Br. J. Clin. Pract.*, **47**, 150–154 (1993).

71. B. A. Falase, B. S. Bajaj, T. J. Wall, V. Argano, and A. Y. Youhana, *Ann. Thorac. Surg.*, **67**, 1158–1159 (1999).

72. H. L. Fung, *Am. J. Cardiol.*, **72**, 9C–15C (1993).

73. S. Tsuchida, T. Maki, and T. Sata, *J. Biol. Chem.*, **265**, 7150–7157 (1990).

74. V. Burt, P. Whelton, and E. J. Roccella, *Hypertension*, **25**, 305–313 (1995).

75. K. Kato, *Eur. Heart J.*, **14**, 40–47 (1993).

76. N. Busch, et al., inventors, US patent reissued 30577, 1981.

77. M. T. Michelin, et al., *Therapie*, **32**, 485 (1977).

78. C. Labrid, et al., *J. Pharmacol. Exp. Ther.*, **211**, 546 (1979).

79. M. K. Sharmaa, et al., *Am. J. Cardiol.*, **611**, 1210 (1988).

80. S. Vogel, et al., *J. Pharmacol. Exp. Ther.*, **210**, 378 (1979).

81. K. Igarashi, and T. Honma, inventors, Shionogi and Co., Ltd., assignee, US patent 4552695, November 12, 1985.

82. M. Sato, et al., *Arzneimittel-Forsch.*, **21**, 1338 (1971).

83. T. Nagano, et al., *Jpn. J. Pharmacol.*, **22**, 467 (1972).

84. M. Chaffmann and R. N. Brogden, *Drugs*, **29**, 387–454 (1985).

85. R. S. Gibson, et al., *N. Engl. J. Med.*, **315**, 423 (1986).

86. S. Kawakita, et al., *Clin. Cardiol.*, **14**, 53 (1991).

87. H. Narita, et al., *Arzneimittel-Forsch.*, **38**, 515 (1988).

88. L. J. Theodore and W. L. Nelson, *J. Org. Chem.*, **52**, 1309 (1987).

89. D. D. Waters, et al., *Am. J. Cardiol.*, **47**, 179 (1981).

90. D. J. Triggle and V. C. Swamy, *Circ. Res.*, **52**, 117–128 (1983).

91. L. J. Theodore and W. L. Nelson, *J. Org. Chem.*, **52**, 1309 (1987).

92. H. Haas and E. Busch, *Arzneimittel-Forsch.*, **17**, 257 (1967); H. Haas and E. Busch, *Arzneimittel-Forsch.*, **18**, 401 (1968).

93. A. Fleckenstein, *Arzneimittel-Forsch.*, **20**, 1317 (1970).

94. N. S. Khurmi, et al., *Am. J. Cardiol.*, **53**, 684 (1984).

95. R. N. Brogden and P. Benfield, *Drugs*, **47**, 93–115 (1994).

96. F. Hefti, et al., *Arzneimittel-Forsch.*, **40**, 417 (1990).

97. G. Mehrke, et al., *J. Pharmacol. Exp. Ther.*, **271**, 1483 (1994).

98. M. C. M. Portegies, et al., *J. Cardiovasc. Pharmacol.*, **18**, 746 (1991).

99. J. P. Clozel, et al., *Cardiovasc. Drug Rev.*, **9**, 4–17 (1991).

100. M. E. Mullins, Z. Horowitz, D. H. J. Linden, G. W. Smith, R. L. Norton, and J. Stump., *JAMA*, **280**, 157–158 (1998).

101. R. Weyhenmyer, et al., *Arzneimittel-Forsch.*, **37**, 58 (1987).

102. W. R. Kukovetz, et al., *Arzneimittel-Forsch.*, **26**, 1321 (1976).

103. A. Fleckenstein, *Arzneimittel-Forsch.*, **27**, 562 (1977).

104. Z. Antaloczy and I. Preda, *Ther. Hung.*, **27**, 71 (1979).

105. M. Spedding, *Arch. Pharmacol.*, **318**, 234 (1982).

106. M. Gautam, A. Tewari, S. Singh, C. Dixit, K. G. Raghu, P. Prakash, and O. Tripathi, *Jpn. J. Pharmacol.*, **83**, 175–181 (2000).

107. J. S. Cheng, K. J. Chou, J. L. Wang, K. C. Lee, L. L. Tseng, K. Y. Tang, J. K. Huang, H. T. Chang, W. Su, Y. P. Law, and C. R. Jane, *Clin. Exp. Pharmacol. Physiol.*, **28**, 729–733 (2001).

108. G. Ehrhart, et al., inventors, DE patent 1100031, CA patent 56,3413h, 1962, US patent 3152173, 1964.

109. J. E. Murphy, *J. Int. Med. Res.*, **1**, 204–209 (1973).

110. B. Karlen, et al., *Eur. J. Clin. Pharmacol.*, **23**, 267 (1982).

111. U. Ulmsten, et al., *Am. J. Obstet. Gynecol.*, **153**, 619 (1985).

112. S. Ohno, K. Mizukoshi, O. Komatsu, K. Ichihara, T. Morishima, inventors, Maruko Seiyaku Co., Ltd., assignee, US patent 4446325, May 1, 1984.

113. K. Miyoshi, et al., *Eur. J. Pharmacol.*, **238**, 139 (1993).

114. A. Kanda, et al., *J. Cardiovasc. Pharmacol.*, **22**, 167 (1993)

115. S. Suzuki, et al., *Arzneimittel-Forsch.*, **43**, 1152 (1993)

116. S. Nakano, et al., *Yakuri to Chiryo.*, **21**, S931 (1993).

117. T. Kojima and T. Takenaka, inventors, Yamanouchi Pharmaceutical Co., Ltd., assignee, US patent 4220649, September 2, 1980.

118. K. Tamazawa, et al., *J. Med. Chem.*, **29**, 2504 (1986).

119. H. Satoh, *Cardiovasc. Drug Rev.*, **9**, 340–356 (1991).

120. H. S. Malhotra and G. L. Plosker, *Drugs*, **61**, 989–996 (2001).

121. Y. Ohya, I. Abe, Y. Ohta, U. Onaka, K. Fujii, S. Kagiyama, Y. Fujishima-Nakao, and M. Fujishima, *Int. J. Clin. Pharmacol. Ther.*, **38**, 304–308 (2000).

122. H. Ueno, T. Hara, A. Ishi, and K. Shuto, *Jpn. J. Pharmacol.*, **84**, 56–62 (2000).

123. S. Munetal, K. Kohara, and K. Hiwada, *Int. J. Clin. Pharmacol. Ther.*, **37**, 141–147 (1999).

124. O. Nakajima, H. Akioka, and M. Miyazaki, *Arzneimittel-Forsch.*, **50**, 620–625 (2000).

125. T. Kutsuma, H. Ikawa, and Y. Sato, inventors, Fujirebio Kabushiki Kaisha, assignee, US patent 4672068, June 9, 1987.

126. K. Ikeda, et al., *Oyo. Yakuri.*, **44**, 433 (1992).

127. M. Hosona, et al., *J. Pharmacobio-Dyn.*, **15**, 547 (1992).

128. M. Ishi, *Jpn. Pharmacol. Ther.*, **21**, 59 (1993).

129. S. Wada, *Chem. Abstr.*, **118**, 32711 (1992).

130. R. Uchida, J. Yamazaki, S. Ozeki, and K. Kitamura, *Jpn. J. Pharmacol.*, **85**, 260–270 (2001).

131. Y. Onose, T. Oki, H. Yamada, K. Manabe, Y. Kageji, M. Matsuoka, T. Yamamoto, T. Tabata, T. Wakatsuki, and S. Ito, *Jpn. Circ. J.*, **65**, 305–309 (2001).

132. K. Seto, S. Tanaka, and R. Sakoda, inventors, Nissan Chemical Industries, Ltd., assignee, US patent 4885284, December 5, 1989.

133. C. Shudo, et al., *Jpn. Pharm. Pharmacol.*, **45**, 525 (1993).

134. T. Yamashita, et al., *Jpn. J. Pharmacol.*, **57**, 337 (1991).

135. T. Saito, et al., *Curr. Ther. Res.*, **52**, 113 (1992).

136. T. Yokoyama, K. Ichihara, and Y. Abiko, *Jpn. J. Pharmacol.*, **72**, 291–297 (1996).

137. M. Kawabata, T. Ogawa, W. H. Han, and T. Takabatake, *Clin. Exp. Pharmacol. Physiol.*, **26**, 674–679 (2001).

138. C. F. Torija and J. A. Galiano-Ramos, inventors, Instituto de Investigacion Y Desarrollo Quimicobiologico S. A., assignee, US patent 4952592, August 28, 1990.

139. J. Tamargo, et al., *Arzneimittel-Forsch.*, **41**, 895 (1991).

140. H. Suryapranata, et al., *Am. J. Cardiol.*, **69**, 1171 (1992).

141. D. U. Acharya, et al., *Eur. Heart J.*, **15**, 665 (1994).

142. S. Motte, X. Alberich, and F. Harrison, *Int. J. Clin. Pharmacol. Ther.*, **37**, 20–27 (2001).

143. P. Decoster, et al., *Eur. J. Clin. Invest.*, **12**, 43 (1982).

144. B. Edgar, et al., *Biopharm. Drug Dispos.*, **8**, 235 (1987).

145. A. Miniscalco, J. Lundahl, and C. G. Regardh, *J. Pharmacol. Exp. Ther.*, **261**, 1195–1199 (1992).

146. D. G. Bailey, J. Malcolm, O. Arnold, and J. D. Spence, *Br J. Clin. Pharmacol.*, **46**, 101–110 (1998).

147. D. G. Bailey, J. D. Spence, and B. Edgar, *Clin. Invest. Med.*, **12**, 357–362 (1989).

148. U. L. Hulthen and P. L. Katzman, *J. Hypertens.*, **6**, 231–237 (1988).

149. M. R. Werbach, *Foundations of Nutritional Medicine*, Third Line Press, Inc., Tarzana, CA, 1997, p. 208.

150. F. L. S. Tee and J. M. Jaffe, *Eur. J. Clin. Pharmacol.*, **32**, 361 (1987).

151. C. E. Handler and E. Sowton, *Eur. J. Clin. Pharmacol.*, **27**, 415 (1984).

152. E. B. Nelson, et al., *Clin. Pharmacol. Ther.*, **40**, 694 (1986).

153. R. P. Hof, et al., *J. Cardiovasc. Pharmacol.*, **8**, 221 (1986).

154. M. Safar, et al., *Clin. Pharmacol. Ther.*, **46**, 94 (1989).

155. A. Zanchetti, *Pharm. J.*, **266**, 842–845 (2001).

156. H. Funato, H. Kawano, Y. Akada, Y. Katsuki, M. Sato, and A. Uemura, *Jpn. J. Pharmacol.*, **75**, 415–423 (1997).

157. R. H. Hernández, D. M. Castillo, M. J. A. Hernández, M. C. A. Padilla, and J. G. Pajuelo, *Am. J. Hypertens.*, **10**, 108A (1997).

158. D. Nardi, A. Leonadi, G. Graziani, and G. Bianchi, inventor, Recordati S. A. Chemical and Pharmaceutical Co., assignee, US patent 4705797, November 10, 1987.

159. D. Policicchio, R. Magliocca, and A. Malliani, *J. Cardiovasc. Pharmacol.*, **29**, S31–S35 (1997).

160. J. A. Staessen, et al., *Lancet*, **350**, 757–764 (1997).

161. K. K. Maguro, et al., *Chem. Pharm. Bull. (Tokyo)*, **33**, 3787 (1985).

162. K. Mizuno, et al., *Curr. Ther. Res.*, **52**, 248 (1992).

163. M. Yoshiyama, K. Takeuchi, S. Kim, A. Hanatani, T. Omura, I. Toda, K. Akioka, M. Teragaki, H. Iwao, and J. Yoshikawa, *Jpn. Circ. J.*, **62**, 47–52 (1998).

164. P. A. Molyvdas and N. Sperelakis, *J. Cardiovasc. Pharmacol.*, **8**, 449 (1986).

165. G. J. Gross, et al., *Gen. Pharmacol.*, **14**, 677 (1983).

166. K. Mizuno, et al., *Res. Commun. Chem. Pathol. Pharmacol.*, **52**, 3 (1986).

167. S. O. Kawamura, Y. Li, M. Shirasawa, N. Yasui, and H. Fukasawa, *Tohoku J. Exp. Med.* **185**, 239–246 (1998).

168. K. B. Grögler, W. Ungethüm, B. M. Witt, and G. G. Belz, *Eur. J. Clin. Pharmacol.*, **57**, 275–284 (2001).

169. S. Kazda, et al., *Arzneimittel-Forsch.*, **30**, 2144 (1980).

170. H. A. Friedel and E. M. Sorkin, *Drugs*, **36**, 682–731 (1988).

171. J. Mitchell, et al., *J. Clin. Pharmacol.*, **33**, 46–52 (1993).

172. F. Özçelik, A. Altun, and G. Özbay, *Clin. Cardiol.*, **22**, 361–365 (1999).

173. H. Meyer, *Arzneimittel-Forsch.*, **31**, 407 (1981).

174. U. Brugmann, et al., *Herz*, **10**, 53 (1985).

175. B. N. Singh and E. M. Vaughan Williams, *Br. J. Pharmacol.*, **39**, 675–687 (1970).

176. B. N. Singh and E. M. Vaughan Williams, *Cardiovasc. Res.*, **6**, 109–119 (1972).

177. B. N. Singh and O. Hauswirth, *Am. Heart J.*, **87**, 367–382 (1974).

178. B. N. Singh, *J. Cardiovasc. Electrophysiol.*, **10**, 283–301 (1999).

179. B. N. Singh and E. M. Vaughan Williams, *Circ. Res.*, **29**, 286–295 (1971).

180. J. D. Allen, F. J. Brennan, and A. L. Witt, *Circ. Res.*, **43**, 470–481 (1978).

181. D. C. Harrison, R. A. Winkle, M. Sami, J. W. Mason, Eds. *Cardiac Arrythmias: A Decade of Progress*. G. K. Hall, Boston, 1981, pp. 315–330.

182. T. J. Campbell, *Cardiovasc. Res.*, **17**, 344–352 (1983).

183. D. P. Zipes in E. Braunwald, Ed., *A Textbook of Cardiovascular Medicine*, W. B. Saunders, Philadelphia, 1997, pp. 593–639.

184. R. L. Woosley in R. W. Alexander, R. C. Schlant, and V. Fuster, Eds., *Hurst's The Heart Arteries and Veins*, McGraw-Hill, New York, 1998, pp. 969–994.

185. M. E. Josephson, et al., Eds., *Harrison's Principles of Internal Medicine*, 14th ed., McGraw-Hill, New York, 1998, pp. 1261–1277.

186. J. J. McNeil in T. M. Sleight, Ed., *Avery's Drug Treatment. Principles and Practice of Clinical Pharmacology and Therapeutics*, ADIS Press, Auckland, New Zealand, 1987, pp. 591–675.

187. L. M. Hondeghem in B. G. Katzung, Ed., *Basic and Clinical Pharmacology*, 7th ed. Appelton and Lange, Stanford, USA, 1998, pp. 216–241.

188. S. Nattel and B. N. Singh *Am. J. Cardiol.*, **84**, 11R–19R (1999).

189. S. Nattel, *Drugs*, **41**, 672–701 (1991).

190. Task Force of the Working Group on Arrhythmias of the European Society of Cardiology, *Circulation*, **84**, 1831–1851 (1991).

191. Task Force of the Working Group on Arrhythmias of the European Society of Cardiology, *Eur. Heart J.*, **12**, 112–113 (1991).

192. E. M. Vaughn Williams, *J. Clin. Pharmacol.*, **32**, 964–977 (1992).

193. S. Nattel and A. Arenae, *Drugs*, **45**, 9–14 (1993).

194. A. J. Camm and Y. G. Yap, *J. Cardiovasc. Electrophysiol.*, **10**, 307–317 (1999).

195. W. Law, D. Newman, and P. Dorian, *Drugs*, **60**, 1315–1328 (2000).

196. E. M. Vaughn Williams, *J. Clin. Pharmacol.*, **24**, 129–147 (1984).

197. T. J. Campbell, *Cardiovasc. Res.*, **17**, 251–258 (1983).

198. T. J. Campbell, *Cardiovasc. Res.*, **17**, 344–352 (1983).

199. A. Sjoerdsma, et al., *Circulation*, **28**, 492 (1963).

200. M. A. Loutfy, et al., *Anal. Profiles Drug Subs.*, **12**, 483–546 (1983).

201. J. W. Mason and L. M. Hondeghem, *Ann. NY Acad. Sci.*, **432**, 162–176 (1984).

202. D. C. Harrison in J. Morganroth and E. N. Moore, Eds., *Cardiac Arrythmias*, Martinus Nijhoff, Boston, 1985, p. 36.

203. J. Koch-Wester, *Ann. NY Acad. Sci.*, **179**, 139 (1971).

204. E. V. Giardinia, et al., *Clin. Pharmacol. Ther.*, **19**, 339 (1976).

205. R. B. Poet and H. Kadin, *Anal. Profiles Drug Subs.*, **4**, 333–383 (1975).

206. B. Befeler, et al., *Am. J. Cardiol.*, **35**, 282 (1975).

207. L. A. Vismara and D. T. Mason, *Clin. Pharmacol. Ther.*, **16**, 330 (1974).

208. J. T. Bigger and C. C. Jaffe, *Am. J. Cardiol.*, **27**, 82 (1971).

209. M. F. Powell, *Anal. Profiles Drug Subs.*, **15**, 761–779 (1986).

210. G. Hollunger, *Acta Pharmacol. Toxicol.*, **17**, 356–373 (1960).

211. G. Hollunger, *Acta Pharmacol. Toxicol.*, **17**, 374 (1960).

212. J. L. Anderson, *Circulation*, **57**, 685 (1978).

213. D. M. Roden and R. L. Woolsey, *N. Engl. J. Med.*, **315**, 41–45 (1986).

214. A. H. Beckett and E. C. Chiodomere, *Postgrad. Med. J.*, **64**, 60, (1977).

215. C. Y. C. Chew, et al., *Drugs*, **17**, 161–181 (1979).

216. M. A. Abounassif, et al., *Anal. Profiles Drug Subs.*, **20**, 433–474 (1991).

217. R. H. Helfant, et al., *Am. Heart J.*, **77**, 315 (1969).

218. M. Sami, et al., *Am. J. Cardiol.*, **44**, 526 (1979).

219. E. Carmeliet, *Cardiovasc. Drugs Ther.*, **7**, 599–604 (1993).

220. R. E. Kates, et al., *Am. J. Cardiol.*, **53**, 248 (1983).

221. P. Somani, *Clin. Pharmacol. Ther.*, **27**, 464 (1980).

222. J. L. Anderson, et al., *N. Engl. J. Med.*, **305**, 473 (1981).

223. U. Klotz, et al., *Int. J. Clin. Pharmacol. Biopharmacol.*, **17**, 152 (1979).

224. C. E. Erickson and R. N. Brogden, *Drugs*, **27**, 279–300 (1984).

225. R. Woestenborghs, *J. Chromatogr.*, **164**, 169 (1979).

226. H. J. Hapke and E. Prigge, *Arzneunuttek-Forsh.*, **26**, 1849 (1976).

227. J. Mergenthaler, et al., *Naunyn Schmiedebergs Arch. Pharmacol.*, **363**, 472–480 (2001).

228. F. Bellandi, et al., *Am. J. Cardiol.*, **88**, 640–645 (2001).

229. T. Yamane, et al., *Br. J. Pharmacol.*, **108**, 812 (1993).

230. F. H. Leenen, *Can. J. Cardiol.*, **15**, 2A–12A (1999).

231. K. O. Ogunyankin and B. N. Singh, *Am. J. Cardiol.*, **84**, 76R–82R (1999).

232. D. J. MacNeil, *Am. J. Cardiol.*, **80**, 90G–98G (1997).

233. J. D. Fitzgerald in A. Scriabine, Ed., *Pharmacology of Antihypertensive Drugs*, Raven Press, New York, 1980, pp. 195–208.

234. L. Slusarek and K. Florey, *Anal. Profiles Drug Subs.*, **9**, 455–485 (1980).

235. B. N. Singh, et al., *Drugs*, **34**, 311–349 (1987).

236. W. K. Sriwatanak and S. R. Nahorski, *Eur. J. Pharmacol.*, **66**, 169–178 (1980).

237. B. N. Singh, et al., *Drugs*, **29**, 531–569 (1985).

238. R. J. Gorczynski, et al., *J. Cardiovasc. Pharmacol.*, **5**, 668 (1983).

239. P. Benfield, et al., *Drugs*, **31**, 376–429 (1986).

240. B. N. Singh and J. S. Sarma, *Curr. Cardiol. Rep.*, **3**, 314–323 (2001).

241. F. E. Marchlinski, et al., *Am. J. Cardiol.*, **84**, 69R–75R (1999).

242. T. A. Plomp, *Anal. Profiles Drug Subs.*, **20**, 1–120 (1991).

243. M. Chow, *Ann. Pharmacother.*, **30**, 637–643 (1996).

244. B. N. Singh, *Clin. Cardiol.*, **20**, 608–618 (1997).

245. D. Roy, et al., *N. Engl. J. Med.*, **342**, 913 (2000).

246. H. C. van Beeren, O. Bakker, and W. M. Wiersinga, *Endocrinology*, **137**, 2807–2814 (1996).

247. A. S. Manning, et al., *J. Cardiovasc. Pharmacol.*, **26**, 453–461 (1995).

248. A. S. Manning, et al., *J. Cardiovasc. Pharmacol.*, **25**, 252–261 (1995).

249. B. N. Singh and K. Nademanee, *Am. Heart J.*, **109**, 421–430 (1985).

250. J. P. Saul, et al., *Clin. Pharmacol. Ther.*, **69**, 145–147 (2001).

251. C. P. Lau, et al., *Am. J. Cardiol.*, **88**, 371–375 (2001).

252. A. Fitton and E. M. Sorkin, *Drugs*, **46**, 678–719 (1993).

253. J. Morganroth, *Am. J. Cardiol.*, **72**, 3A–7A (1993).

254. R. H. Heissenbuttel and J. T. Bigger, *Ann. Intern. Med.*, **91**, 229–238 (1979).

255. M. R. Rosen and A. L. Wit, *Am. Heart J.*, **106**, 829–839 (1983).

256. K. A. Ellenbogen, et al., *J. Am. Coll. Cardiol.*, **28**, 130–136 (1996).

257. J. B. Hester, et al., *J. Med. Chem.*, **34**, 308–315 (1991).

258. T. Yang, D. J. Snyders, and D. M. Roden, *Circulation*, **91**, 1799–1806 (1995).

259. JJ. Lynch Jr., *J. Cardiovasc. Pharmacol.*, **25**, 336–340 (1995).

260. L. V. Buchanan, et al., *J. Cardiovasc. Pharmacol.*, **19**, 256–263 (1992).

261. G. S. Friedrichs, et al., *J. Pharmacol. Exp. Ther.*, **266**, 1348–1354 (1993).

262. E. Carmeliet, *Cardiovasc. Drugs Ther.*, **7**, 599–604 (1993).

263. J. Kiehn, A. E. Lacerda, B. Wible, and A. M. Brown, *Circulation*, **94**, 2572–2579 (1996).

264. R. H. Falk, A. Pollak, S. N. Singh, and T. Friedrich, *J. Am. Coll. Cardiol.*, **29**, 385–390 (1997).

265. D. S. Echt, et al., *J. Cardiovasc. Electrophysiol.*, **6**, 687–699 (1995).

266. R. Karam, et al., *Am. J. Cardiol.*, **81**, 40D–46D (1998).

267. E. L. C. Pritchett, et al., *Eur. Heart J.*, **20**, 352 (1999).

268. S. Connelly, et al., *Eur. Heart J.*, **20**, 351 (1999).

269. E. L. C. Pritchett, et al., *Circulation*, **98**, 633 (1999).

270. K. C. Yedinak, *Am. Pharm.*, **33**, 49–64 (1993).

271. B. N. Singh, et al., *Drugs*, **25**, 125–153 (1983).

272. M. J. Niebauer and M. K. Chung, *Cardiol. Rev.*, **9**, 253–258 (2001).

273. A. Gabrielli, et al., *Crit. Care Med.*, **29**, 1874–1897 (2001).

274. H. F. Tse, et al., *Am. J. Cardiol.*, **88**, 568–570 (2001).

275. H. E. Wang, et al., *Ann. Emerg. Med.*, **37**, 38–45 (2001).

276. H. Ozaki, et al., *J. Cardiovasc. Pharmacol.*, **33**, 492–499 (1999).

277. M. J. Apostolakos and M. E. Varon, *New Horiz.*, **4**, 45–57 (1996).

278. S. Viskin, et al., *J. Am. Coll. Cardiol.*, **38**, 173–177 (2001).

279. H. L. Tan, et al., *Pacing Clin. Electrophysiol.*, **24**, 450–455 (2001).

280. U. Stark, M. Brodmann, A. Lueger, and G. Stark, *J. Crit. Care*, **16**, 8–16 (2001).

281. S. Luber, et al., *Am. J. Emerg. Med.*, **19**, 40–42 (2001).

282. E. O. Robles de Medina and A. Algra, *Lancet*, **354**, 882–883 (1999).

283. B. N. Singh, *Am. J. Cardiol.*, **84**, 3R–10R (1999).

284. The Cardiac Arrhythmia Suppression Trial (CAST) Investigators, *N. Engl. J. Med.*, **321**, 406–412 (1989).

285. IMPACT Research Group, *J. Am. Coll. Cardiol.*, **4**, 1148–1163 (1984).

286. The Cardiac Arrhythmia Suppression Trial II Investigators, *N. Engl. J. Med.*, **327**, 227–233 (1992).

287. I. Kodama, et al., *Am. J. Cardiol.*, **84**, 20R–28R (1999).

288. K. O. Ogunyankin and B. N. Singh, *Am. J. Cardiol.*, **84**, 76R–79R (1999).

289. S. J. Connolly, et al., *Am. J. Cardiol.*, **88**, 974–979 (2001).

290. A. K. Gupta, *Indian Heart J.*, **53**, 354–360 (2001).

291. P. Matyus, et al., *Med. Res. Rev.*, **20**, 294–303 (2000).

292. R. C. Tripathi, et al., *Bioorg. Med. Chem. Lett.*, **9**, 2693–2698 (1999).

293. N. Bodor, H. H. Farag, and P. Polgar, *J. Pharm. Pharmacol.*, 889–894 (2001).

294. T. E. Morey, et al., *J. Pharmacol. Exp. Ther.*, **297**, 260–266 (2001).

295. H. M. Himmel, et al., *J. Cardiovasc. Pharmacol.*, **38**, 438–449 (2001).

296. D. Thomas, et al., *J. Pharmacol. Exp. Ther.*, **297**, 753–761 (2001).

297. O. Levy, M. Erez, D. Varon, and E. Keinan, *Bioorg. Med. Chem. Lett.*, **11**, 2921–2926 (2001).

298. F. Boutitie, et al., *Circulation*, **99**, 2268–2275 (1999).

299. G. Duque, *Am. J. Geriatr. Cardiol.*, **9**, 263–270 (2000).

300. S. Williams, *N. Engl. J. Med.*, **341**, 709–717 (1999).

301. M. V. Cohen and J. M. Downey, *Annu. Rev. Med.*, **47**, 21–29 (1996).

302. A. J. Hobbs, A. Higgs, and S. Moncada, *Annu. Rev. Pharmacol. Toxicol.*, **39**, 191–220 (1999).

303. J. P. Cooke and V. J. Dzau, *Annu. Rev. Med.*, **48**, 489–509 (1997).

304. C. O. Kappe, *Molecules*, **3**, 1–9 (1998).

305. H. Tanaka and K. Shigenobu, *Cardiovasc. Drug Rev.*, **18**, 93–102 (2000).

CHAPTER TWO

Diuretic and Uricosuric Agents

Cynthia A. Fink
Jeffrey M. McKenna
Lincoln H. Werner
Novartis Biomedical Research Institute
Metabolic and Cardiovascular Diseases Research
Summit, New Jersey

Contents

Burger's Medicinal Chemistry and Drug Discovery
Sixth Edition, Volume 3: Cardiovascular Agents and Endocrines
Edited by Donald J. Abraham
ISBN 0-471-37029-0   © 2003 John Wiley & Sons, Inc.

1   INTRODUCTION

Diuretics are among the most frequently pre-scribed therapeutic agents for the treatment of edema, hypertension, and congestive heart fail-ure and act primarily by inhibiting the reabsorp-tion of sodium ions from the renal tubules in the kidney. This diverse class of therapeutic agents includes organomercurials, polyols, sugars, thia-zides, phenoxyacetic acids, aminomethylphe-nols, xanthines, aromatic sulfonamides, pteri-dines, pyrazines, and steroids. These agents have been classified in a variety of ways includ-ing; chemical structure, mechanism of action, tubular site of action, magnitude of natriuretic effect, and by their effect on electrolyte deple-tion. Today no single classification system is commonly used; however, diuretics are often grouped as loop, potassium-sparing, thiazide, and osmotic diuretics.

Uricosuric agents increase the excretion of uric acid, one of the principal products of purine metabolism. These compounds are used in the treatment of gout, a condition in which plasma levels of uric acid are elevated and, as a result, deposits of crystalline sodium urate form in con-

nective tissues. Hyperuricemia is an adverse ef-fect sometimes observed with diuretic treat-ment and arises from decreased extracellular volume and increased urate reabsorption. To-day, a heterogeneous array of diuretic com-pounds possessing different structures and sites of action are available for safe and effective treatment of edema and cardiovascular diseases including compounds that have combined di-uretic and uricosuric properties. This modern era of diuretic therapy began in 1949, when sul-fanilamide was discovered to possess diuretic and natriuretic properties (1). Since the 1950s, significant advances have been made in the dis-covery of new diuretic agents and their precise cellular mechanism of action.

The kidneys are the principal organs of ex-cretions within the body and perform three major functions in maintaining homeostasis:

1. Remove water, electrolytes, products of metabolic waste, drugs, and other materi-als from the blood.

2. Possess endocrine functions; that is, secrete erythropoietin, renin, and 1,2,5-hydroxy-cholecalciferol.

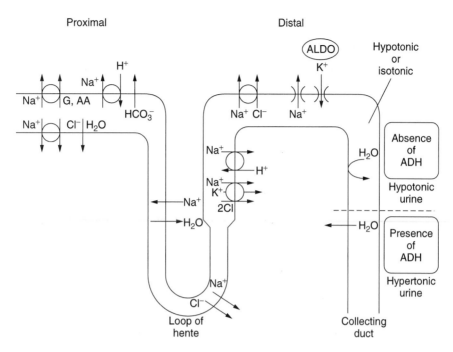

Figure 2.1. Major transport mechanisms in the apical membrane of the tubule cells along the nephron. G, glucose; AA, amino acids; ADH, antidiuretic hormone; ALDO, aldosterone (5).

3. Selectively reabsorb water, electrolytes, and needed nutrients from the urine.

Together the kidneys weigh about 300 g, which is about 0.4% of the total body weight. The kidneys can be divided into three major regions: the pelvis, the cortex, and the medulla. The working unit of the kidney is the nephron, with each kidney containing about 1.2 million such structural units (2). There are two types of nephrons: the cortical nephrons and the juxtamedullary nephrons, with about 80% of the nephrons found in the human kidney being cortical nephrons (3). The fundamental components that make up the nephron are the glomerulus and the tubules. The glomerulus is composed of a convoluted capillary network that is joined with connective tissue. The diameter of the afferent arterioles is larger than the efferent arterioles and as a result the glomerular filtration pressure is estimated to be about 50 mm of mercury. This facilitates the rapid clearance of water and a variety of low to medium molecular weight solutes from the blood.

The Bowman's capsule surrounds the capillary network of the glomerulus and its function is to collect the filtrate. An estimated 180 L of glomerular filtrate forms daily, which is about 60 times the total plasma content (4). Fortunately, the reabsorption process begins immediately. Approximately 99% of the water and electrolytes are reabsorbed in the renal tubules. The glomerular filtrate is composed of water, electrolytes (NH_4^+, Na^+, K^+, Ca^{2+}, Mg^{2+}, Cl^-, and HPO_4^{2-}), glucose, amino acids, and nitrogenous wastes of metabolism. It actually has a profile similar to that of blood plasma, except it contains no blood cells and little or no plasma proteins. Reabsorption of the water and solutes occurs through the walls of the proximal and distal convoluted tubules, in the loop of Henle, and in the collecting tubules by active and passive transport systems (Fig. 2.1) (5). From the use of micropuncture and isolated tubule techniques, much is known about the cellular and molecular mechanism of tubular reabsorption. Each particular segment of the nephron possesses its own characteristic ion-transport systems. In gen-

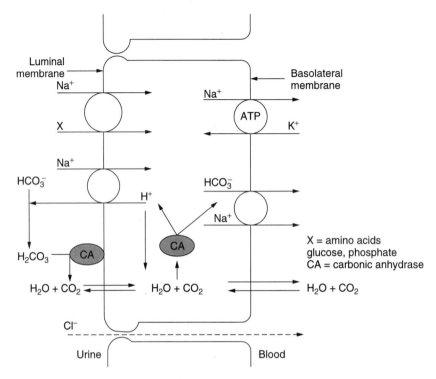

Figure 2.2. Cell model of the proximal tubule.

eral, these cells all contain a rate-limiting sodium entry system on the luminal membrane, which is coupled to a Na^+/K^+-ATPase on the basolateral membrane for sodium removal.

The reabsorption process begins in the proximal tubules. Approximately 50–55% of the filtered sodium and water along with about 90% of the filtered amino acids, bicarbonate, glucose, and phosphate are reclaimed here (Fig. 2.2) (5a). Glucose, phosphate, and the amino acids enter proximal tubule cells through electrogenic cotransport with sodium. The major route of sodium reentry into this tubule cell is the Na^+/H^+ exchanger. This transport system is also responsible for most of the proximal tubular reabsorption of bicarbonate and creates a favorable gradient that allows for both the active and passive transport of about 50% of the filtered chloride ion (6). The Na^+/H^+ exchanger has recently been cloned by a gene-transfer approach (7). A Na^+/K^+-ATPase pump located on the basolateral side of the proximal cell removes the sodium to maintain a low intracellular sodium concentration (approximately one-tenth that of the luminal fluid). Carbonic anhydrase in the cytoplasm indirectly catalyzes the intracellular formation of protons, which keeps the Na^+/H^+ exchanger active. These excreted protons also neutralize the bicarbonate in the tubule to form carbonic acid. Carbonic anhydrase located in the luminal brush border dehydrates the carbonic acid to form carbon dioxide and water.

As mentioned, chloride ion is removed from the proximal tubule by passive and active transport systems. As solutes are removed from the tubule, the osmotic gradient facilitates the reabsorption of water. This effectively increases the concentration of chloride ion above that found in the lateral intercellular space. This space is permeable to chloride, and it is passively absorbed across the junction (8). Chloride can also enter the cell through a chloride-formate exchanger (Fig. 2.3) (9, 10). The basolateral membrane is also believed to contain a Na^+/HCO_3^- cotransporter (11).

About 35–40% of the filtered sodium is reabsorbed in the loop of Henle. The major luminal transporter of sodium in this region of

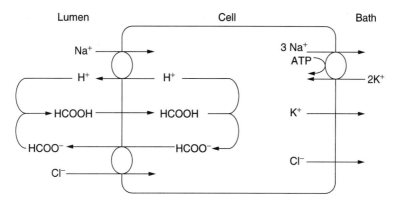

Figure 2.3. Cell model of the chloride-formate exchanger in the proximal tubule. [After G. Giebisch, *J. Clin. Invest.*, **79**, 32 (1987).]

the renal tubule is the $Na^+/K^+/2Cl^-$ electroneutral cotransporter located in the thick ascending region of the loop. The energy that drives the cotransporter arises from a concentration gradient generated from the Na^+/K^+-ATPase pump located on the basolateral membrane (Fig. 2.4). Chloride exits the basolateral side through a chloride channel and/or electroneutral KCl cotransporter (12). The potassium that enters the cell through the $Na^+/K^+/2Cl^-$ cotransporter can be recycled back through the lumen, through a potassium channel, to keep the tubule concentration of this ion high enough for the cotransporter to continue to function. The result of potassium leaving the luminal membrane and chloride at the basolateral side by conductive pathways generates a lumen-positive potential. This positive potential drives the flow of sodium ions out through a paracellular pathway. There is also a Na^+/H^+ exchanger on the apical membrane that plays a minor role in the reabsorption of sodium.

The distal convoluted tubule reabsorbs approximately 5–8% of the sodium contained in the glomerular filtrate. The major luminal transporter of sodium in this region is the neutral sodium chloride cotransporter (Fig. 2.5). A Na^+/K^+-ATPase pump is located on the basolateral membrane to remove sodium from the cell. Potassium can reenter the tubule through a barium-sensitive potassium channel. This region of the renal tubule is the site at which calcium excretion and reabsorption are regulated. Parathyroid hormone and cal-

citriol are the main mediators of calcium reabsorption. They both increase distal calcium reabsorption, although the exact mechanism is not well understood (13). Calcium is believed to exit the cell through a Ca^{2+}/ATPase or a Na^+/Ca^{2+} exchanger on the basolateral membrane (14, 15).

The collecting tubule is the last section of the renal tubule in which filtrate modification occurs. This region is responsible for 2–3% of sodium reabsorption. There are two major cell types in this region of the nephron: the principal cells and the intercalated cells. The principal cells are the predominant cell type, and they are responsible for sodium reabsorption and potassium secretion. Sodium enters the cell by way of a sodium channel and exits through the basolateral Na^+/K^+-ATPase pump (Fig. 2.6). Potassium can exit this cell on the luminal and basolateral sides through conductance channels. The primary site of action of aldosterone is also in the principal cells. Aldosterone increases sodium reabsorption by opening sodium channels. The intercalated cells control hydrogen ion secretion and potassium reabsorption (Fig. 2.7). Protons are generated in the cell by the catalytic actions of carbonic anhydrase and are transported into the lumen through H^+/translocating ATPase (16). An ATP-dependent K^+/H^+ exchanger may be responsible for potassium reabsorption in times of potassium depletion (17). Two recent reviews have appeared on the molecular mechanisms of the actions of diuretics (18, 19).

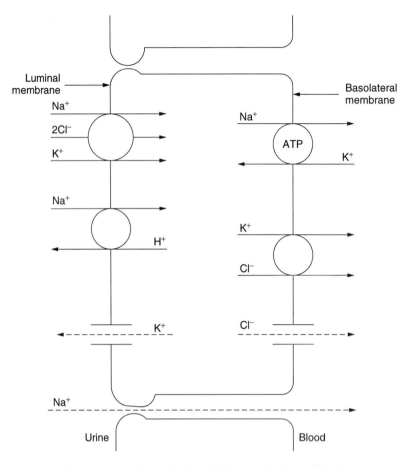

Figure 2.4. Cell model of the thick ascending loop of Henle.

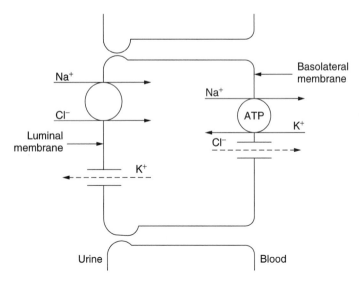

Figure 2.5. Cell model of the distal convoluted tubule.

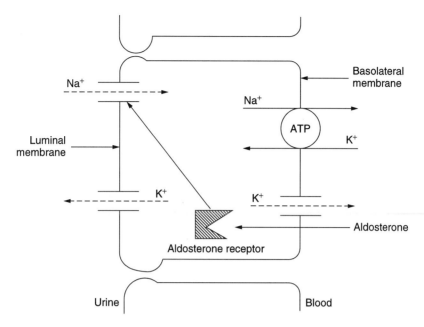

Figure 2.6. Cell model of the principal cells in the collecting tubule.

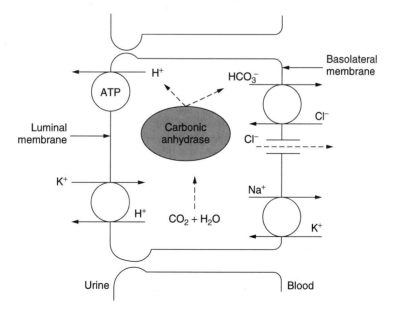

Figure 2.7. Cell model of the intercalated cell of the collecting tubule.

1.1 Pharmacological Evaluation of Diuretics

The following three statements or generalizations are direct quotes from a study by K. H. Beyer and refer to the pharmacological evaluation of drugs in general and to diuretics in particular (20).

> The more closely one can approximate under controlled laboratory conditions the physiological correlates of clinically defined disease, the more likely one will be able to modulate it effectively.
>
> *In vitro* experiments are apt to be inadequate and hence misleading when employed solely to anticipate the physiological correlates of complex clinical situations.
>
> Often aberrations in function we call toxicity, relate more to changes a compound induced physiologically than to a direct, inherent, destructive effect of the agent, per se, on tissues.

Diuretics are generally evaluated in two species: the rat and the dog. The rat is used, as a rule, in initial screening for convenience and economy. Results obtained with diuretic agents in dogs, however, are generally more predictive of the response in humans than those obtained in the rat. However, many compounds exhibit diuretic activity in the rat (e.g., antihistamines, such as tripelennamine) but are inactive in the dog and in humans. On the other hand, mercurial diuretics and ethacrynic acid are inactive in the rat; furosemide exhibits diuretic activity in the rat only at doses several times higher than effective doses in the dog. Various modifications of the experimental procedures of Lipshitz et al. (21, 22) are frequently used to measure urine volume and Na^+, K^+, and Cl^- excretion (e.g., male rats, fasted for 18 h, given 5 mL of 0.2% NaCl solution/100 g body weight by stomach tube). The diuretic drugs are given by stomach tube at the time of fluid loading. The rats are placed in metabolism cages and urine volumes are measured at 30-min intervals over a 3- to 5-h period. Total amounts of Na^+ and K^+ excreted over the time period are determined by flame photometry; chloride can be determined in the Technicon Autoanalyzer using Skegg's modification of Zall's method (23). A number of standard diuretics have also been tested in the mouse (24). In this animal, ethacrynic acid is a potent diuretic in contrast to the very low diuretic activity seen with this compound in the rat.

The chimpanzee (25) has also been used to evaluate certain diuretics. In this animal, the effect of the compound on uric acid excretion can also be studied, because apes, like humans, are devoid of hepatic uricase and therefore maintain a relatively high level of circulating serum urate (26). Chimpanzees have also been used in the study of uricosuric agents (25), but tests with such large animals obviously present numerous problems. Attempts have also been made to block the enzyme uricase in rats by administering potassium oxonate and thus to obtain higher serum uric acid levels (27).

Considerable evidence is available that demonstrates the tendency of certain benzothiadiazine diuretics to elevate blood glucose values in seemingly normal, as well as diabetic or prediabetic, individuals (28). A method using high doses of the compounds injected intraperitoneally into rats and determining blood glucose levels, as compared to control values, has been used to estimate a possible hyperglycemic effect of diuretics (29). Today the molecular and cellular mechanism of actions of diuretics can also be investigated. Micropuncture and single nephron studies can provide insight into the exact site of action in the renal tubule and yield information regarding the specific transport mechanisms that are blocked by a drug.

1.2 Clinical Aspects of Diuretics

In a healthy human subject, changes in dietary intake or variations in the extrarenal loss of fluid and electrolytes are followed relatively rapidly by adjustments in the rate of renal excretion, thus maintaining the normal volume and composition of extracellular fluid in the body. Edema is an increase in extracellular fluid volume. In almost every case of edema encountered in clinical medicine, the underlying abnormality involves a decreased rate of renal excretion. One of the factors influencing the normal relationship between the volume of interstitial fluid and the circulating plasma is the pressure within the small blood vessels. In diseases of hepatic origin (e.g., cirrhosis), the pressure relationships are disturbed primarily within the portal circulation and ascites results. In congestive heart failure, pressure-flow relationships may be disturbed more in the pulmonary or systemic circulation and edema may be localized accordingly.

There is overwhelming evidence to indicate that the primary disturbance of the kidney is in its ability to regulate sodium excretion, which underlies the pathogenesis of edema. Three approaches are available when edema fluid accumulates because of excessive reabsorption of sodium and other electrolytes by the renal tubules. First, one can attempt to correct the primary disease if possible; second, one can reduce renal absorption of electrolytes by the use of drugs; and third, one can restrict sodium intake to a level that corresponds to the diminished renal capacity for sodium excretion. Cardiac decompensation is one of the most common causes of edema. Treatment consists of full digitalization, which should be considered the primary therapeutic agent. Diuretic drugs have a secondary though very important role because it has been shown that blocking excessive electrolyte reabsorption in the renal tubule alleviates the symptoms of cardiac failure and also improves cardiac function. Diuretics are also used in the treatment of hypertension.

Diuretic therapy may lead to a number of metabolic and electrolyte disorders. In general, these disturbances are mild but can be life-threatening in certain cases. Some of the common adverse effects observed with diuretic treatment are hypokalemia, hyperuricemia, and glucose intolerance. Diuretics that possess a site of action proximal to the collecting tubules, such as the loop and thiazide diuretics, induce potassium loss; an average loss of 0.5–0.7 meq/L generally results with long-term therapy (30). In most young hypertensive patients, this reduction does not present any problem; however, in older patients or patients with preexisting heart disease, it may lead to the occurrence of ventricular arrhythmias. Combinations with potassium-sparing diuretics are frequently used to minimize the effect. Dietary potassium supplements may also be prescribed. In patients receiving long-term diuretic therapy, serum uric acid concentrations increase on average 1.3 mg/L as the result of a decrease in extracellular volume

and increased urate reabsorption from the filtrate. Some patients may experience an attack of gout, or those with preexisting gout or excessive uric acid production may experience more frequent attacks. A number of studies have shown that some patients who have undergone long-term diuretic therapy have elevated blood levels of glucose and that their tolerance to glucose decreases (31). The mechanism of the diuretic-induced glucose intolerance is unknown. Some other side effects that are observed with diuretic treatment are increases in cholesterol levels in men and postmenopausal women (32), ototoxicity, and hypomagnesemia.

2 CLINICAL APPLICATIONS

2.1 Current Drugs

2.1.1 Osmotic Diuretics. Osmotic diuretics all have several key features in common:

1. They are passively filtered at the glomerulus.
2. They undergo limited reabsorption in the renal tubules.
3. They are usually metabolically and pharmacologically inert.
4. They have a high degree of water solubility.

These agents all function as diuretic agents by preventing the reabsorption of water and sodium from the renal tubules. The addition of a nonreabsorbable solute prevents water from being passively reabsorbed from the tubule, which in turn prevents a sodium gradient from forming and thereby limiting sodium reabsorption. These actions hinder salt and water reabsorption from the proximal tubules; however, it has also been proposed that these agents have multiple sites and mechanisms of action (33, 34).

Most of the osmotic diuretics are sugars and polyols (Table 2.1). Mannitol (Table 2.1, **5**) is the prototype of the osmotic diuretics and has been studied extensively. The compound is poorly absorbed after oral administration and is therefore administered by intravenous (i.v.) infusion. It is freely filtered at the glo-

Table 2.1 Osmotic Diuretics

No.	Generic Name	Trade Name	Structure
(1)	Ammonium chloride		NH_4Cl
(2)	Glycerine	Osmoglyn	
(3)	Glucose		
(4)	Isosorbide	Ismotic	
(5)	Mannitol	Osmitrol	
(6)	Sorbitol		
(7)	Sucrose		
(8)	Urea	Ureaphil	H_2NCONH_2

merulus and reabsorption is quite limited. The usual diuretic dose is 50–100 g given as a 25% solution.

These compounds are not prescribed as primary diuretic agents to edematous patients. One of the most important indications for the use of mannitol is the prophylaxis of acute renal failure. After cardiovascular operations or severe traumatic injury, for instance, a precipitous fall in urine flow may be anticipated. Administration of mannitol, under such conditions, exerts an osmotic effect within the tubular fluid, inhibiting water reabsorption. A reasonable flow of urine can thus be maintained, and the kidneys can be protected from damage. Osmotic diuretics have also been prescribed in the relief of cerebral edema after neurosurgery, to lower intraocular pressure in ophthalmologic procedures, and after a drug overdose to maintain urine flow.

2.1.2 Mercurial Diuretics. For approximately 30 years, mercurial diuretics were the most important diuretic agents. Since the introduction of orally active, potent, less toxic, nonmercuiral diuretics, beginning in 1950 with acetazolamide, their use has greatly declined. Today they represent only a small fraction of the injectable diuretics used, and injectable diuretics in turn are only a small portion of the total diuretic market. Organomercurials are generally given intramuscularly. The usual dose is a 1-mL solution containing 40 mg Hg. In responsive, edematous patients, an increase in urine flow is evident in 1–2 h and reaches a maximum in 6–9 h. The effect is usually over within 24 h. A loss of about 2.5% of body weight represents an average response (35). Mercury is eliminated from the body in the urine, as a complex with cysteine. These diuretics therefore should not be prescribed to patients with renal insufficiency marked by adequate excretion of the mercury-cysteine complex.

2.1.2.1 Pharmacology. Before the development of the loop diuretics, the organomercurials were the most potent diuretics available. Further studies have found that the major effect of organomercurials appears to be in the ascending limb of Henle (36–38). Organomercurials inhibit active chloride reabsorption in the thick ascending limb of Henle. During diuresis, the urine contains a high concentration of chloride ion matched by almost equivalent amounts of sodium ion (35). The effect of mercurial diuretics on potassium excretion is complex. They depress the tubular secretion of potassium and, for this reason, the diuresis is accompanied by significantly less potassium loss than occurs with other diuretics that do not inhibit the secretory mechanism. However, mercurials can have a paradoxical effect of increasing potassium excretion when initial excretory rates are low. Inhibition of organic acid secretion is seen in humans but not in the dog. In the chimpanzee and to a lesser extent in humans, mercurials (e.g., mersalyl, **9**) have an intense uricosuric action (39).

Side effects, such as diarrhea, gingivitis, proteinuria, and stomatitis, can occur with organomercurial treatment. As with other diuretics, electrolyte imbalances are common after long-term use (hypochloremic alkalosis, hypokalemia, hyponatremia). In some cases after intravenous administration, severe hypersensitivity reactions may develop.

2.1.2.2 History. The medicinal use of mercurial diuretics dates back to 400 B.C., when Hippocrates administered metallic mercury to increase urine excretion. Calomel (mercurial

$$CO_2 + H_2O \rightleftharpoons H_2CO_3 \rightleftharpoons H^+ + HCO_3^-$$

(9)

chloride) was used by Paracelsus (1493–1541, A.D.) as a diuretic. This information was lost until the nineteenth century, when Jendrassik rediscovered the use of calomel as a diuretic agent (40). Calomel was an ingredient of the famous Guy's Hospital pill (calomel, squill, and digitalis). Calomel exerted a cathartic effect, and its absorption from the intestine was unpredictable. In 1919, Vogl (41) discovered the diuretic effect of merbaphen (**10**) after parenteral administration of this

(10)

antisyphilitic agent. General use of the drug as a diuretic was short-lived because of its toxicity. However, it did lead to the synthesis of a large number of organomercurials between 1920 and 1950.

2.1.2.3 Structure-Activity Relationship. It is believed that the mechanism of action for these diuretic agents involves the *in vivo* release of mercuric ion in the renal tubules (42–44). This ion is thought to bind a sulfhydryl enzyme in the tubule membrane that is involved in sodium reabsorption. The mercuric ion reacts with a sulfhydryl group on the enzyme and another nucleophilic group in close proximity, to form a bidentate complex that inactivates the enzyme and, as a result, the

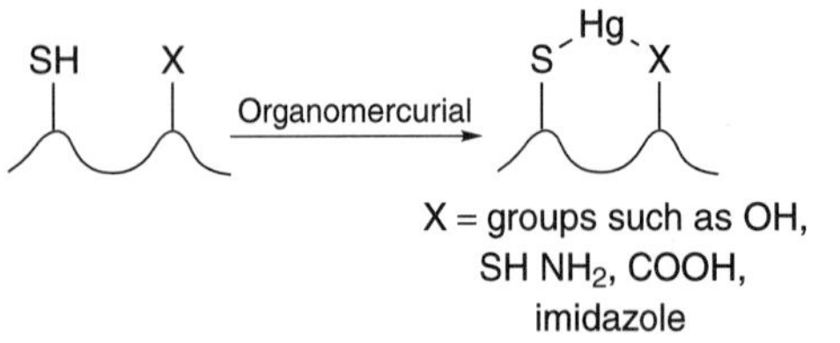

X = groups such as OH,
SH NH$_2$, COOH,
imidazole

Figure 2.8. Mercury-enzyme complex.

sodium reabsorption process (Fig. 2.8). Most mercurial diuretics have the general structure (**11**), in which Y is usually CH$_3$ and R is a

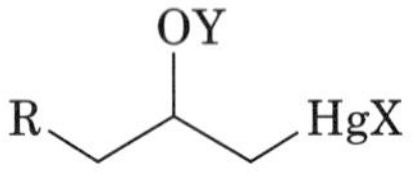

(**11**)

complex organic moiety, usually incorporating an amide function or a urea group. All organic mercurial diuretics thus far examined are acid-labile *in vitro*. It is interesting to note that compounds of the related structure (**12**) with an unsubstituted β-carbon atom are acid-stable and do not exhibit diuretic activity. Formula (**13**) shows the structural characteristics

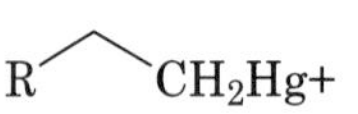

(**12**)

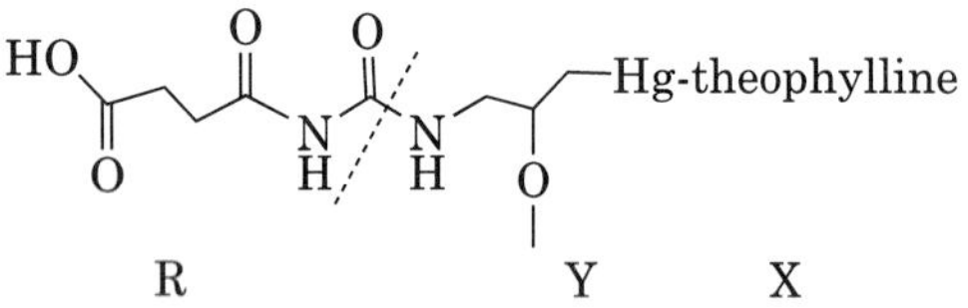

(**13**)

of mercurial diuretics and the most important mercurial diuretics (**14–17**) are shown in Table 2.2. The R substituent largely determines the distribution and rate of excretion of the compound. The Y substituent, determined by the solvent in which the mercuration is carried out, generally has little effect on the properties of the compound (45, 46). Among others, Y substituents such as H, CH$_3$, CH$_2$CH$_2$OH, and CH$_2$CH$_2$OCH$_3$ have been studied.

The nature of the X substituent affects the toxicity of the compound, irritation at the site of injection, and rate of absorption (45, 46). Theophylline has been used commonly as an X substituent (47) or is commonly added by itself with the organomercurials. Because of its peripheral vasodilating effects, it can increase absorption of the mercurial diuretics at the site of injection. Theophylline is also weakly diuretic. When X is a thiol, such as mercaptoacetic acid or thiosorbitol, cardiac toxicity and local irritation are reduced (48, 49). Diglucomethoxane (**18**, Mersoben) (50) should be

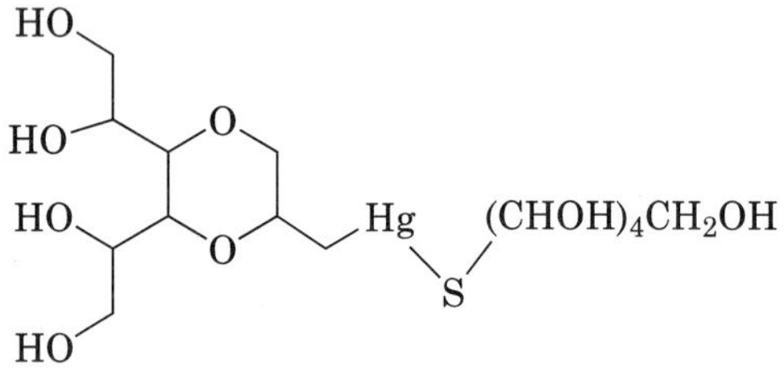

(**18**)

cited as a well-tolerated, potent mercurial diuretic that does not conform to the general structure (**11**). Generally, mercurial diuretics are administered parenterally; chloromerodrin (Table 2.2, **14**), which lacks a carboxylic acid group, is orally effective but gastric irritation precludes its widespread use (51).

2.1.3 Carbonic Anhydrase Inhibitors. Acetazolamide (**19**) is the prototypical carbonic anhydrase inhibitor. It is rapidly absorbed from the stomach, reaches a peak plasma level within 2 h, and is eliminated unchanged in the urine within 8–12 h. The efficacious dose is 250 mg to 1 g daily. During continuous administration of acetazolamide, the excretion of HCO$_3^-$ leads to the development of metabolic acidosis. Under such acidic conditions, the diuretic effect of carbonic anhydrase inhibition is much reduced or completely absent, and therefore the effect of the drug is self-limiting (52). This is attributed to the fall in the level of plasma and filtered bicarbonate, given that the latter is lost in the urine. A state of equilibrium is reached when the small amount of hydrogen ion that is secreted in spite of carbonic anhydrase blockade is sufficient to reab-

Table 2.2 Mercurial Diuretics

No.	Generic Name	Trade Name	Structure
(14)	Chloromerodrin	Neohydrin	$H_2NOCHN-CH_2CH(OCH_3)CH_2HgCl$
(15)	Meralluride USP	Mercuhydrin	

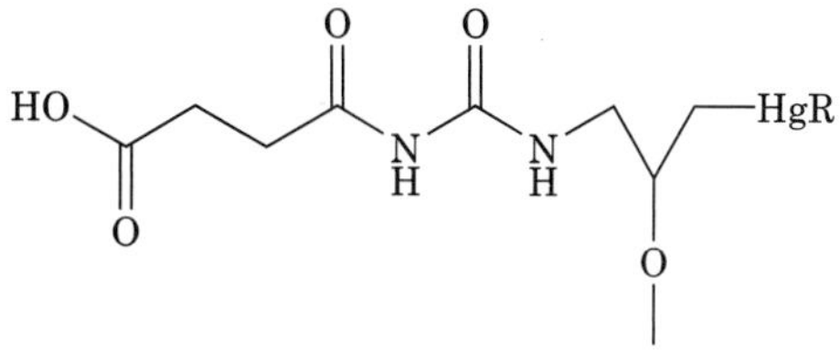

R = theophylline

| (16) | Sodium mercaptomerin | Thiomerin | |

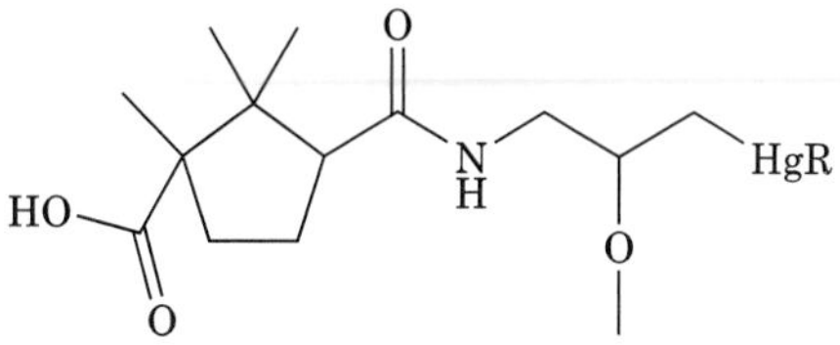

R = SCH_2CO_2Na

| (17) | Mercurophylline NF XII | Mercupurin, Novurit | |

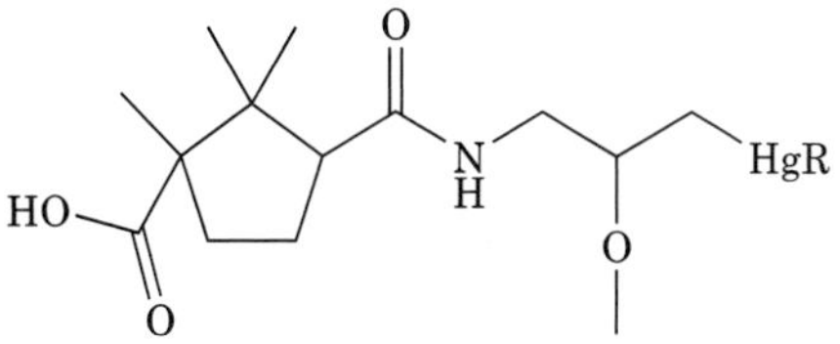

R = theophylline

(20)

sorb the reduced amount of filtered bicarbonate ion. Because of the development of kaliuresis, metabolic acidosis, and their self-limiting nature, these inhibitors are not generally prescribed for diuretic therapy. Their most common use today is to lower intraocular pressure in the treatment of glaucoma. There is a great deal of current research in this area (53–55). Two new agents are available for the treatment of glaucoma dorzolamide (20, Trusopt) and brinzolamide (21, Azopt) (56, 57). Carbonic anhydrase inhibitors also have some therapeutic utility in epilepsy, congestive heart failure, mountain sickness, gastric and duodenal ulcers, and more recently in the area of osteroporosis, antitumor agents, and as diagnostic tools (58, 59).

2.1.3.1 History. Building on earlier work by Strauss and Southworth in 1937 (60),

Mann and Keilin (61), Davenport and Wilhelmi (62), and Pitts and Alexander in 1945 (63) proposed that the normal acidification of the urine results from the secretion of hydrogen ions by tubular cells. They confirmed that in dogs sulfanilamide (22) renders the urine alkaline, perhaps because of the reduction of the availability of H^+ for secretion brought about by the inhibition of the enzyme carbonic

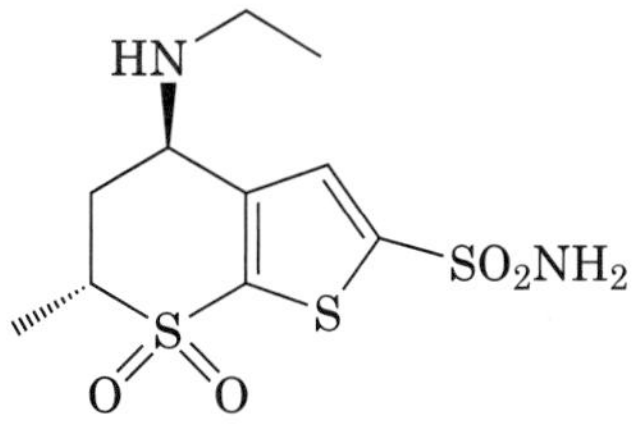

(21)

(22)

anhydrase. The resulting increase in Na^+ and HCO_3^- excretion suggested to Schwartz (64) the diuretic potential of sulfanilamide. However, this was of no practical significance because of the very high doses required to achieve diuresis.

2.1.3.2 Pharmacology. Carbonic anhydrase is a zinc-containing enzyme that was first discovered in erythrocytes by Roughton in the early 1930s. This enzyme was subsequently found in many tissues, including the renal cortex, gastric mucosa, pancreas, eye, and central nervous system. Carbonic anhydrase catalyzes the reversible hydration of carbon dioxide and the dehydration of carbonic acid. These reactions can occur in the absence of the enzyme, although the rates are too slow for normal physiological function to occur. Normally the enzyme is present in the tissue in high excess. Because of the levels of this enzyme in the kidney, approximately 99% of the enzyme's activity must be inhibited for physiological activity to be observed. To date, 14 different isozymes or carbonic anhydrase-related proteins have been identified in humans and higher vertebrates.

Hydrogen ion secretion takes place in the proximal tubule, the distal tubule, and the collection duct. The driving force for H^+ secretion in the distal portions is the *trans*-tubular negative potential. In the proximal tubule, protons are actively excreted by the H^+/Na^+ exchanger. The source of cellular hydrogen ion is the hydration of carbon dioxide within the proximal tubular cells catalyzed by the ac-

tion of cytosolic carbonic anhydrase, to produce cellular H^+ and HCO_3^-. The hydrogen ion is secreted into the tubular lumen through the Na^+/H^+ exchanger, and then the Na^+ is reabsorbed and enters the peritubular fluid as $NaHCO_3$. In the proximal tubule, the secreted H^+ combines with HCO_3^- to form H_2CO_3, which is then dehydrated to CO_2 and H_2O. This reaction is also catalyzed by carbonic anhydrase located on the luminal border of the proximal tubular cells. The carbon dioxide diffuses back into the cell, where it is again hydrated and used as a source of hydrogen ion to drive the H^+/Na^+ exchanger (see Fig. 2.2). Carbonic anhydrase inhibitors are among the most well understood class of diuretics. Their major function is to inhibit the enzyme carbonic anhydrase, although they are also believed to decrease cell membrane permeability for carbon dioxide and also to inhibit glucose 6-phosphate dehydrogenase (65). The administration of an inhibitor of carbonic anhydrase promptly leads to an increase in urine volume. The urinary concentrations of HCO_3^-, Na^+, and K^+ increase, whereas the normally acidic urine becomes alkaline and the concentration of chloride ion drops. In addition, there is a fall in titratable acid and ammonium ion excretion. The net effect of the inhibitors in the proximal tubule is to prevent the reabsorption of bicarbonate. This can effect volume reabsorption in a number of ways:

1. Fewer protons are available for the H^+/Na^+ exchanger on the luminal membrane; thus sodium reabsorption is slowed.

2. The passive reabsorption process in the late proximal tubule is indirectly inhibited by the decreased chloride gradient.

3. Bicarbonate effectively becomes a nonreabsorbable anion and osmotically adds to the diuretic effect.

However, more than half of the bicarbonate that passes through the proximal tubule is reabsorbed in later segments of the renal tubule, thus attenuating the effectiveness of this class of diuretic agents. Carbonic anhydrase inhibitors also cause a significant kaliuresis, which can be attributed to the inhibition of distal proton secretion and high aldosterone levels, which result from volume depletion.

2.1.3.3 Structure-Activity Relationship.

There are two main classes of compounds that inhibit carbonic anhydrase: unsubstituted sulfonamides and metal-complexing inorganic anions (e.g. azide, cyanate, cyanide, hydrogen sulfide, etc.). The inorganic anions have served as tools for better understanding of the catalytic and inhibitory mechanism of carbonic anhydrase inhibition. The sulfonamide class has led to the development of several useful therapeutic agents. After the discovery of the carbonic anhydrase inhibitory activity of sulfanilamide, a variety of aromatic sulfonamides were found to exhibit the same type of activity (66). Aliphatic sulfonamides were much less active; substitution of the sulfonamide nitrogen in aromatic sulfonamides eliminated the activity. Roblin and coworkers (67, 68), following the work of Schwarts (64), investigated a series of heterocyclic sulfonamides. Compounds up to 800 times more active *in vitro* than sulfanilamide as carbonic anhydrase inhibitors were found. An attempt to correlate pK_a values and *in vitro* carbonic anhydrase inhibitory activity in a series of closely related, 1,3,4-thiadiazole-2-sulfonamides (23) was not successful (69). The rela-

(23) R' = lower alkyl, phenyl
R'' = H, CH$_3$CO

tionship between *in vitro* enzyme inhibition and *in vivo* diuretic potency was not very predictable because of variation in drug distribution, binding, and metabolism (70), especially when different types of aromatic or heterocyclic sulfonamides were compared (Table 2.3).

Certain derivatives of 1,3,4-thiadiazole sulfonamides were among the most active *in vitro* inhibitors of carbonic anhydrase, with potencies several hundred times that of sulfanilamide (67, 68). One of these, 5-acetylamine-1,3,4-thiadiazole-2-sulfonamide (acetazolamide, **19**, Tables 2.3 and 2.4) was studied in detail by Maren et al. (71) and became the first clinically effective diuretic of the carbonic an-

hydrase inhibitor class. A number of structural modifications of acetazolamide have been studied. An increase in the number of carbons in the acyl group is accompanied by retention of *in vitro* enzyme inhibitor activity and diuretic activity, but the side effects become more pronounced. Removal of the acyl groups leads to a markedly lower activity *in vitro* (72, 73). Substitution on the sulfonamide nitrogen abolishes the enzyme-inhibitory activity *in vitro*, but diuretic activity in animals is still present if the substituent is removable by metabolism (74). Two isomeric products, (**33**) and (**31**) (methazolamide), are obtained

(31)

(33)

on methylation of acetazolamide (**19**) (69). Both these compounds are somewhat more active *in vitro* than acetazolamide but offer no advantages as diuretics over the parent compound.

A related sulfamoylthiadizolesulfonamide, benzolamide (**32**, Table 2.4), is about five times more active than acetazolamide. Clinical studies showed that 3 mg/kg p.o. of (**32**) produces a full bicarbonate diuresis, with increased excretion of sodium and potassium (75, 76). Ethoxzolamide, a benzothiazole derivative (Table 2.4, **30**) is a clinically effective diuretic carbonic anhydrase inhibitor (77). The compound lacking the ethoxy group is inactive as a diuretic when given orally to dogs, although it is a potent carbonic anhydrase inhibitor *in vitro* (78). Dichlorophenamide (**26**, Tables 2.3 and 2.4), a benzenedisulfonamide derivative, is as active as acetazolamide *in vitro* as a carbonic anhydrase inhibitor and equally active as a diuretic (70). A large num-

Table 2.3 Carbonic Anhydrase Diuretics

No.	Generic Name	Trade Name	Structure	Ref.
(19)	Acetazolamide USP	Neohydrin		71
(26)	Dichlorophenamide USP	Daranide, Oratrol		70
(30)	Ethoxzolamide USP	Cardrase, Ethamide		77
(31)	Methazolamide USP	Neptazane		77
(32)	Benzolamide			75

ber of benzenedisulfonamide derivatives have been prepared and studied as diuretics. Some of these are very active as diuretics, although they are weak carbonic anhydrase inhibitors. In contrast to the compound just discussed, Na^+ and HCO_3^- excretion is not increased; instead, an approximately equal amount of chloride ion accompanies the sodium. These are described in the section on aromatic sulfonamides.

2.1.4 Aromatic Sulfonamides. Beyer and Baer (79), in a study published in 1975, discussed their early findings on the natriuretic and chloruretic activity of *p*-carboxybenzenesulfonamide (**24**, Table 2.4). Although the compound is considerably less active as a carbonic anhydrase inhibitor than acetazolamide, the way the kidney handled the carboxybenzenesulfonamide was considered to be more important and served to define the sal-

uretic properties sought and ultimately found in chlorothiazide (**28**, Table 2.4). A key discovery by Sprague (80) was that the introduction of a second sulfamoyl group *meta* to the first can markedly increase not only the natriuretic effect but also the chloruretic action of the compound. This is evident when the data for compounds (**22**) or (**24**) are compared with compound (**25**) (Table 2.4). Interestingly, the introduction of a second chlorine substituent as in compound (**26**) (dichlorphenamide, Table 2.4) produces a compound that is considerably less chloruretic, with an excretion pattern that is typical of a carbonic anhydrase inhibitor.

The activity of 6-chlorobenzene-1,3-disulfonamide (**25**, Table 2.4) is further enhanced by the introduction of an amino group *ortho* to the second sulfonamide group as in (**27**). Thus, 4-amino-6-chloro-1,3-benzenedisulfonamide (**27**) is an effective diuretic agent, with a more

Table 2.4 Dissociation of Carbonic Anhydrase Inhibitory Activity *In Vitro* and Renal Electrolyte Effects in the Dog[a]

No.	Structure	Concentration Causing 50% Inh. of Carbonic Anhydrase (M)	Dose (i.v.)[b] (mg/kg)	Urine Excretion Rate (μeq/min)			
				Na^+	K^+	Cl^-	pH
(19)	CH_3CONH-thiophene-SO_2NH_2	7.2×10^{-8}	C	43	22	21	6.3
			2.5	186	128	53	8.0
(22)	NH_2SO_2–C$_6$H$_4$–NH_2	1.3×10^{-5}	C	14	14	3	6.8
			25	51	25	13	6.9
(24)	NH_2SO_2–C$_6$H$_4$–COOH	4.4×10^{-6}	C	26	53	7	6.6
			25	102	136	32	7.7
(25)	NH_2SO_2–C$_6$H$_3$(Cl)(SO_2NH_2)	1.4×10^{-7}	C	42	24	15	5.9
			2.5	468	101	303	7.4
(26)	NH_2SO_2–C$_6$H$_2$(Cl)(Cl)(SO_2NH_2)	7.5×10^{-6}	C	23	37	21	5.9
			2.5	188	110	64	7.7
(27)	NH_2SO_2–C$_6$H$_2$(Cl)(NH_2)(SO_2NH_2)	3.8×10^{-6}	C	5	20	19	5.5
			0.05	14	46	63	5.4
			0.25	54	57	96	5.5

favorable electrolyte excretion pattern than that of (25). Chloride is the major anion excreted, and HCO_3^- excretion is low because the urinary pH does not increase. The carbonic anhydrase inhibitory activity of (27) is only about three times that of sulfanilamide (22). The chlorine in the 6-position of (27) can be replaced by a bromine, trifluoromethyl, or nitro group without much change in activity; whereas a fluoro, amino, methyl, or methoxy group is less effective (81).

Substitution on the nitrogen atoms of 4-amino-6-chloro-1,3-benzenedisulfonamide gives compounds (34). Methyl or allyl substitution of the aromatic amino group yields

(34)

R1 = R3 = H, R4 = CH_3 or CH_2CHCH_2
R1 = R3 = H, R4 = $CO(CH_2)_{2-4}CH_3$
R1 = R3 = R4 = CH_3CO
R1 = R3 = CH_3 or Ethyl, R4 = H

compounds with reduced oral activity. Acylation of the anilino group leads to an increase in activity when R_4 is a simple aliphatic acyl rad-

Table 2.4 *(Continued)*

No.	Structure	Concentration Causing 50% Inh. of Carbonic Anhydrase (M)	Dose (i.v.)[b] (mg/kg)	Urine Excretion Rate (μeq/min)			
				Na^+	K^+	Cl^-	pH
(28)	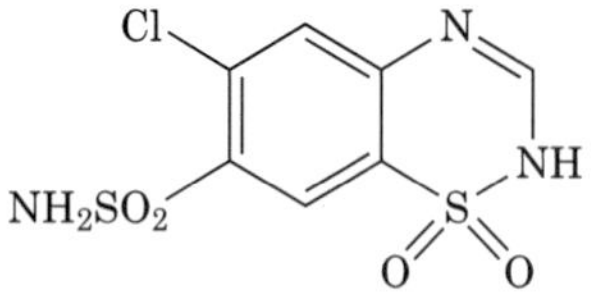	1.7×10^{-6}	C	11	11	7	6.1
			0.05	20	24	7	6.6
			0.25	115	38	80	6.7
			1.25	308	65	236	6.9
(29)	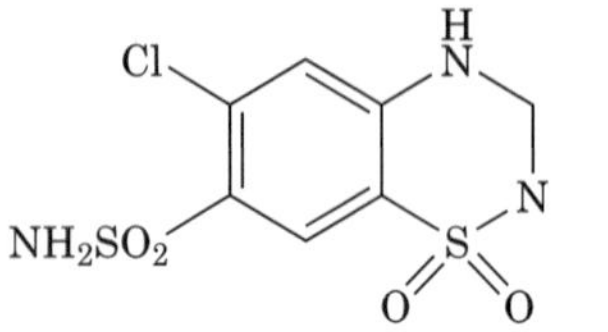	2.3×10^{-5}	C	62	34	43	6.5
			0.01	126	32	122	5.9
			0.05	265	33	291	5.5
			0.25	414	39	427	5.9

[a]From Ref. 70.
[b]C, control phase (no drug), average of two or three 10-min clearances.

ical and reaches a maximum at four to six carbon atoms. Aromatic acyl derivatives are less active. Acylation with formic acid results in a cyclized product, 6-chloro-1,2,4-benzothiadiazine-7-sulfonamide-1,1-dioxide, chlorothiazide (**28**, Table 2.4) (82). The compound was the starting point for the development of the thiazide diuretics, one of the most important group of diuretics, which are discussed later. Complete acetylation (R_1 = R_3 = R_4 = CH_3CO) lowers the activity (82). Methylation of both sulfonamide groups (R_1 = R_3 = CH_3 or CH_3CH_2; R_4 = H) gives a compound that has

diuretic activity in the rat, but the observed activity is attributable to *in vivo* dealkylation of the sulfonamide function (83).

2.1.4.1 Mefruside. Horstmann and coworkers (84) at a later date realized that 6-chloro-1,3-benzene-disulfonamide (**25**, Table 2.4) had shown an excretion pattern combining Na^+, HCO_3^-, and a substantial amount of chloride ions, even though it was an active carbonic anhydrase inhibitor. Investigation of derivatives substituted on the sulfonamide nitrogen *para* to the chloro substituent led to compounds with high diuretic activity. Intro-

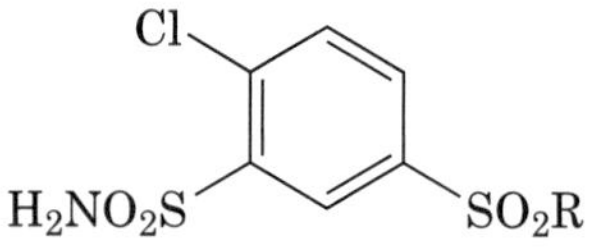

(**25**) R = NH_2
(**35**) R = $NHCH_3$
(**36**) R = $N(CH_3)_2$

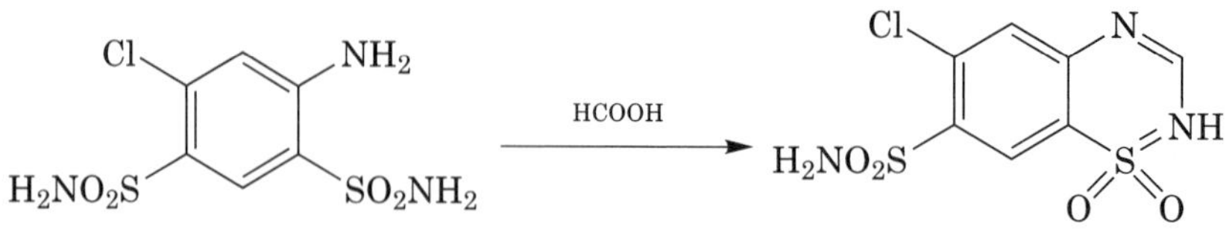

(**27**)

(**28**)

duction of a single methyl group (**35**) does not reduce the HCO_3^- excretion as compared to (**25**), but disubstitution as in (**36**) leads to a substantially lower excretion of HCO_3^- and improved Na^+/Cl^- ratio because of less carbonic anhydrase activity. A large number of *N*-substituted benzenesulfonamide derivatives were prepared; the *N*-disubstituted compounds were of more interest because they exhibited primarily a saluretic effect. A selected number of these compounds are shown in Table 2.5. The threshold dose level and the increase in sodium excretion over control values in the rat are also given. These compounds are relatively weak carbonic anhydrase inhibitors, with little effect on HCO_3^- excretion, particularly those that are disubstituted on the sulfonamide nitrogen. A particularly favorable type of substituent is the tetrahydrofurfuryl group (**41–44**, Table 2.5). Activity is enhanced when the tetrahydrofuran ring bears a methyl substituent in the 2-position. A diuretic effect in rats is obtained with this compound (**43**, mefruside) at the low dose of 0.04 mg/kg. This compound has an asymmetric carbon atom; however, the difference in diuretic activity between the more active form and the racemate is not of practical significance. The action of mefruside is characterized by a prolonged increase in the rate of excretion of NaCl and water (85). The corresponding pyrrolidine derivatives are also active diuretics (**45** and **46**, Table 2.5).

Studies with ^{14}C-labeled mefruside have shown that the compound is almost completely metabolized *in vivo* to the lactone (**44**, Table 2.5) (86). This lactone has about the same diuretic activity as that of the parent compound and may be responsible for much of the observed activity of mefruside (84). Mefruside has been compared with chlorothiazide in the dog, and the findings suggest that mefruside has a mechanism of action similar to that of the thiazide diuretics without any unique features (87). Mefruside has also been studied in humans at doses of 25 and 100 mg. In volunteers undergoing water diuresis, the drug caused natriuresis and chloruresis extending for 20 h. Bicarbonate excretion was also increased, whereas the acute excretion of potassium was slightly increased. *In vivo* carbonic anhydrase studies revealed 50% inhibition at

7.3×10^{-7} *M* concentration (chlorothiazide; 1.7×10^{-6} *M*). The potency of mefruside and its effect on the renal concentrating/diluting mechanisms suggest that its action is similar to that of the thiazide diuretics (88, 89). Studies in hypertensive patients also showed a close correlation with the thiazide diuretics in terms of both desirable and undesirable effects (90, 91).

2.1.5 Thiazides. Thiazides are a major class of diuretic agents that have been used for over 30 years. It was this group of therapeutic agents that first challenged and then replaced the mercurial diuretics that were used in the first half of the twentieth century. Treatment of 4-amino-6-chlorobenzene-1,3-disulfonamide with formic acid resulted in the formation of the cyclic benzothiadiadiazine derivative, chlorothiazide (**28**, Table 2.4) (82). This compound was the first orally active, potent diuretic that could be used to the full extent of its functional capacity as a natriuretic agent without upsetting the normal acid-base balance. Chlorothiazide is saluretic with minimal side effects and fulfilled a clinical need. In addition to diuretic activity, chlorothiazide and its congeners exert a mild blood pressure-lowering effect in hypertensive patients.

2.1.5.1 Pharmacology. This is discussed in relation to the hydrothiazides in Section 2.1.6.1.

2.1.5.2 History. Thiazide diuretics arose as an outgrowth of the carbonic anhydrase inhibitor area, in particular from the work of Novello and Sprague. Research in this area was based on the conviction of Beyer (92) that it should be possible to find a sulfonamide derivative that was saluretic and that increased Na^+ and Cl^- excretion in approximately equal quantities; in other words, a compound that did not act as a classical carbonic anhydrase inhibitor, which increased water, Na^+, and HCO_3^- elimination. A saluretic diuretic should permit a substantial reduction of edema without affecting the normal acid-base balance.

2.1.5.3 Structure-Activity Relationship. An SAR has been developed for the thiazides. The effect of varying the 3 and 6 substituents is shown in Table 2.6. Interestingly, compound (**48**) ($R_6 = H$) has very little diuretic activity,

Table 2.5   Diuretic Activity of Some 4-Chloro-3-sulfamoyl-*N*-substituted Benzenesulfonamides[a]

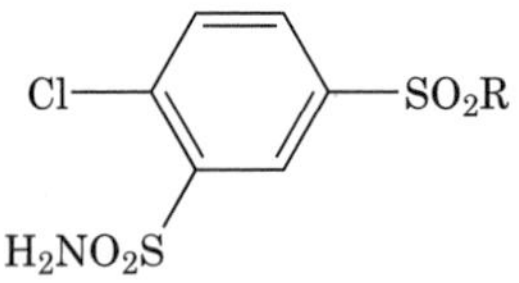

Cl—⟨benzene ring⟩—SO$_2$R

H$_2$NO$_2$S

No.	R	Threshold Dose, p.o. (mg/kg)	Increase in Na$^+$ Excretion at 80 mg/kg p.o. [μeq/kg/6 h (Rat)]
(37)	NHCH$_2$CH(CH$_3$)$_2$	2.4	4615
(38)	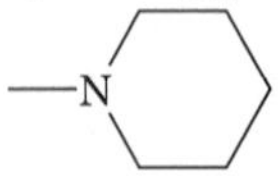 —N⟨piperidine⟩	2.1	3000
(39)	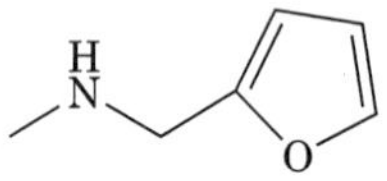 H N-CH$_2$-(furyl)	6	3910
(40)	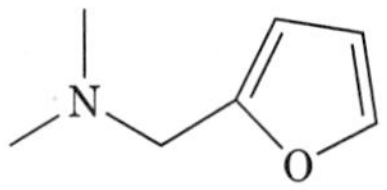 CH$_3$-N-CH$_2$-(furyl)	18	2900
(41)	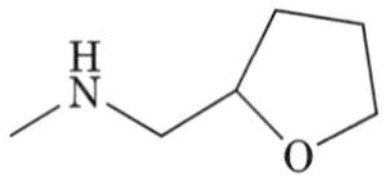 H N-CH$_2$-(tetrahydrofuryl)	0.9	3150
(42)	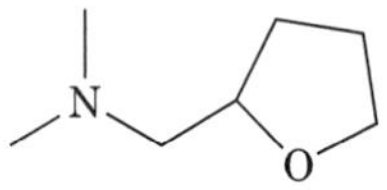 CH$_3$-N-CH$_2$-(tetrahydrofuryl)	1.0	3000
(43)	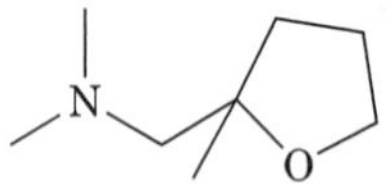 CH$_3$-N-CH$_2$-(2-methyltetrahydrofuryl)	0.04	4100
(44)	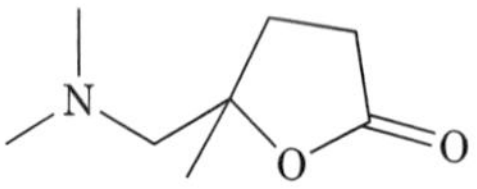 CH$_3$-N-CH$_2$-(methyl-γ-butyrolactone)	0.04 i.v. 0.15 p.o.	3700 i.v. 3000 p.o.
(45)	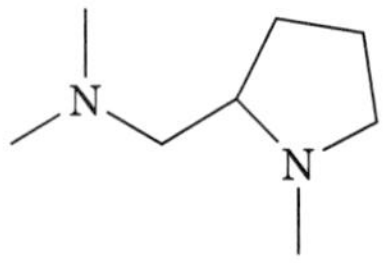 CH$_3$-N-CH$_2$-(N-methylpyrrolidine)	0.15	3300
(46)	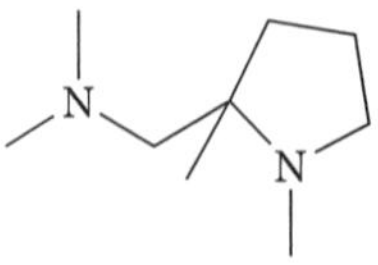 CH$_3$-N-CH$_2$-(methyl-N-methylpyrrolidine) · HCl	0.3	2515
(47)	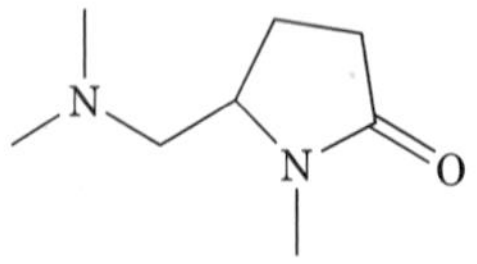 CH$_3$-N-CH$_2$-(N-methylpyrrolidinone)	0.04	3000

[a]From Ref. 84.

Table 2.6 Comparative Effects of 3- and 6-Substituted Thiazides on Electrolyte Excretion in the Dog upon Oral Administration[a]

No.	R_6	R_3	Na$^+$	K$^+$	Cl$^-$	Reference
(28)	Cl	H	++++	+	++++	chlorothiazide
(48)	H	H	+/−		+/−	
(49)	Br	H	++++	+	++++	
(50)	Me	H	+	+/−	+	
(51)	OMe	H	++	+/−	++	
(52)	NO$_2$	H	+++	+/−	+++	
(53)	NH$_2$	H	+/−		+/−	
(54)	Cl	Me	+++	+/−	+++	
(55)	Cl	n-Pr	+++	+/−	+++	
(56)	Cl	n-C$_5$H$_{11}$	+++	+/−	+++	
(57)	Cl	C$_6$H$_5$	+/−	+/−	+/− +/−	
(58)	CF$_3$	H	++++		++++	93
(59)	Cl	CH$_2$SBn	++++	+	++++	96, 97
(60)	Cl	CHCl$_2$	++++	+	++++	98
(61)	Cl	CH$_2$(C$_5$H$_9$)	++++	+	++++	98

[a]From Ref. 80.

whereas compounds where R_6 = Cl, Br, or CF$_3$ are highly active; and alkyl groups in the 3-position decrease the activity slightly. The 3-oxo derivative of chlorothiazide also has weak diuretic activity (80). Interchanging the chlorine and sulfamoyl groups at positions 6 and 7 in chlorothiazide lowers the activity. Replacement of the 7-sulfamoyl group by CH$_3$SO$_2$ or

H gives compounds with little activity (81). The degree of activity observed with compounds bearing an acyl or alkyl group on the 7-sulfamoyl group is in accord with the hypothesis that metabolic cleavage of the N-substituent occurs to yield the free sulfamoyl function (93, 94). In the case of N$_7$-caproyl-chlorothiazide, urinary bioassay indicated

Table 2.7 Pyrido-1,2,4-thiadiazines 1,1-Dioxides[a]

No.	R	No.	R	No.
(62)	H	(64)	H	(69)
(63)	Me	(65)	Me	
		(66)	NH$_2$	
		(67)	OH	
		(68)	Cl	

[a]From Ref. 99.

Table 2.8 Benzothiadiazine Diuretics

No.	Generic Name	Trade Name	Clinical Dose p.o. (mg/kg)	Structure	Reference
(28)	Chlorothiazide NF	Diuril	500–2000		81
(29)	Hydrochlorothiazide	Esidrix, HydroDiurit	25–100		101, 115
(59)	Benzthiazide NF	Aquatag, Exna	25–50		96, 97
(71)	Bendroflume thiazide NF	Bristuron, Naturetin	2–5		125
(72)	Buthiazide	Saltucin, Eunephran		115, 117
(73)	Cyclopenthizide	Navidrix	0.5–1.5		123

76

(74)	Cyclothiazide NF	Anhydron	1–6	123
(75)	Hydroflumethiazide	Saluron	25–50	125
(76)	Methyclothiazide	Enduron	5–10	122
(77)	Polythiazide	Renese	4–8	16, 164
(78)	Trichlormethiazide	Methahydrin, Naqua	4–8	165, 166

that 50% of the excreted drug was present as chlorothiazide (70), whereas the N_7-acetyl derivative showed only weak saluretic activity and no detectable cleavage of the acetyl group (95). Substitution of the ring nitrogen atoms at position 2 or 4 with a methyl group reduced the activity and makes the heterocyclic ring more vulnerable to hydrolytic cleavage (81).

The introduction of a more complex substituent in the 3 position [e.g., $R_3 = CH_2SCH_2C_6H_5$ (59, Table 2.6)] led to a compound that was 8–10 times more potent on a weight basis than chlorothiazide (96, 97). Similarly, the dichloromethyl and cyclopentylmethyl analogs (60 and 61, Table 2.6) were 10–20 times more potent, respectively, than chlorothiazide on a weight basis when tested in experimental animals (98). A number of "aza" analogs of chlorothiazide, derived from 2-aminopyridine-3,5-disulfonamide and 4-aminopyridine-3,5-disulfonamide, have been prepared (62–69, Table 2.7). In general, the activity of each compound was comparable with, although somewhat less potent than, that of its 1,2,4-benzothiadiadiazine analog (99). Initially, the antihypertensive effect was thought to be a consequence of the diuretic action. It was subsequently found, however, that removal of the 7-sulfonamide group from compounds of the chlorothiazide class eliminated the diuretic effect but not the antihypertensive action (100, 101). A compound of this type, diazoxide (70), is

(70)

a much more effective antihypertensive agent. Surprisingly, salt and water retention has been observed with this compound (101).

2.1.6 Hydrothiazides. The thiazide diuretics have their greatest usefulness in the management of edema of chronic cardiac decompensation. In hypertensive disease, with or without overt edema, the thiazides have a mild antihypertensive effect. They are used with caution in patients with significantly impaired renal function. In some patients with nephro-

sis they have been effective, but their therapeutic usefulness in such cases has been unpredictable. The side effects observed with thiazide treatment can be divided into two types: hypersensitivity reactions and metabolic complications. Some of the common metabolic complications of these diuretics are hypokalemia, magnesium depletion, hypercalcemia, hyperuricemia, and hyperlipidemia. Thiazides may also induce hyperglycemia and can aggravate a preexisting diabetic state. With respect to hypersensitivity, dermatitis, purpura, and necrotizing vasculitis have been observed. The thiazide diuretics are available as tablets; the wide range of dosages of the individual preparations are shown in Table 2.8. To minimize the possibility of potassium depletion, fixed combinations with potassium-sparing diuretics have been made available (e.g., hydrochlorothiazide-triamterene and hydrochlorothiazide-amiloride).

2.1.6.1 Pharmacology. Chlorothiazide and hydrochlorothiazide are the prototypes of a group of related heterocyclic sulfonamides that differ among themselves mainly in regard to the dosage required for natriuretic activity. Examples of these compounds are shown in Tables 2.8 and 2.9. The unique property of these drugs is their ability to produce a much larger chloruresis associated with a greater natriuretic potency than that of the carbonic anhydrase inhibitor acetazolamide and its congeners.

After oral administration to normal subjects, hydrochlorothiazide is rapidly absorbed. Peak plasma levels are reached after 2.6 ± 0.8 h, and the drug is still detectable after 9 h. Approximately 70% of a 65 mg dose was accounted for in urine after 48 h (102). Hydrochlorothiazide and other benzothiazine diuretics are excreted by the kidneys both through glomerular filtration and tubular secretion. The latter is shared with other organic acids and is specifically inhibited by probenecid. Concurrent administration of hydrochlorothiazide and probenecid did not modify the effects of hydrochlorothiazide on the urinary excretion of calcium, magnesium, and citrate. This combined therapy also prevented or abolished the increased serum uric acid levels associated with the use of thiazide diuretics (103). Thiazides are absorbed in the intestine

Table 2.9 Hydrochlorothiazide Derivatives Structure-Activity Relationships of Canine Studies

No.	Structure	Dose, p.o. (μg/kg)	Urine Excretion (mL over 6 h)	Na$^+$ Excretion (meq over 6 h)	K$^+$ Excretion (meq over 6 h)	Natriuretic Activity (Approx.)	Partition Coefficient[a] (Ether/Water)
Cont. avg.		0	42.5–53.2	7.1–10.0	4.75		
(28)		1250	102	18.5	6.5	1	0.08
(29)		20	69	12.1	4.35	10	0.37
		310	100	20.3	6.0		
		1250	125	26.1	8.6		
(72)	R = CH$_2$CH(Me)$_2$	1.3	59	13.8	4.3		
		20	100	20.7	6.8		
(73)	CH$_2$(C$_5$H$_9$)	1.3	95	22.7	4.9	1000	10.2
(79)	CH$_2$Cl	20	83	17.2	4.0		
		310	131	30.6	6.1		
(80)	CHCl$_2$	1.3	65	11.6	5.1	100	1.53
(81)	CH$_2$C$_6$H$_5$	20	113	22.5	6.5		
			74	18.6	4.0		

[a]From ref. 79.

in varying degrees after oral administration. Chlorothiazide is absorbed to the extent of only about 10%, although other members of this family have a much higher bioavailability. Bile acid-binding resins, such as cholestyramine, have been reported to bind to these drugs and therefore prevent their absorption (104). Generally these diuretics are highly plasma protein bound, primarily to albumin. Thiazides gain access to the renal tubule principally through proximal tubular secretion and to a small extent by glomerular filtration. Thiazide diuretics work by inhibiting the electroneutral Na^+/Cl^- cotransporter in the distal convoluted renal tubules. It is believed that these agents compete for the Cl^- binding site on the cotransporter (105).

As a class, the thiazides have an important action on potassium excretion. In most patients, a satisfactory chloruretic and natriuretic response is accompanied by significant kaliuresis; this is also seen in dog diuretic studies (see Table 2.9) (106, 107). At low doses, with some selected thiazides, a separation of natriuretic and kaliuretic effects have been observed, although at higher doses and repeated administration these differences disappear. The kaliuresis, although enhanced by the carbonic anhydrase activity of many of these compounds, is probably a consequence of increased delivery of sodium and fluid to the distal segment of the nephron. Elevated serum uric acid levels, which may be associated with gout resulting from decreased uric acid excretion during chronic thiazide administration, have been well documented (107). Calcium excretion is decreased, and the excretion of magnesium is enhanced by the administration of thiazide diuretics in normal subjects and in patients (103). Thiazide treatment has also been observed to increase plasma levels of cholesterol and triglycerides (108).

The metabolic fate of thiazides varies significantly. Chlorothiazide and hydrochlorothiazide undergo very little metabolism, whereas the more lipid-soluble drug, indapamide (**92**, Table 2.10), undergoes extensive degradation. Given that 90% of the sodium is reabsorbed before it reaches the distal convoluted tubule, the effectiveness of this class of diuretics is limited.

2.1.6.2 History. The new phase in the development of the thiazide diuretics was opened by the findings of de Stevens et al. (109), that condensation of 4-amino-6-chloro-1,3-benzenedisulfonamide (**27**) with 1 mole of formaldehyde gives 6-chloro-3,4-dihydro-2*H*-1,2,4-benzothiadiadiazin-7-sulfonamide-1,1-dioxide (**29**), a stable crystalline compound. This compound has been given the generic name hydrochlorothiazide. It was surprising that saturation of the 3,4-double bond in chlorothiazide leads to a compound that is 10 times more active in dogs (109, 110) and humans (111–113) as a diuretic. Hydrochlorothiazide (**29**) has less than one-tenth the carbonic anhydrase inhibiting activity of chlorothiazide (**28**, Table 2.4). Like chlorothiazide, it also exerts a mild antihypertensive effect in hypertensive subjects (114).

2.1.6.3 Structure-Activity Relationship. Structure-activity relationships have been extensively investigated by various groups (81, 115–129). Substitution in the 6-position of hydrochlorothiazide follows the same rules as found for chlorothiazide; that is, compounds of approximately equal activity result when the substituent in the 6-position is Cl, Br, or CF_3. Compounds where R_6 = H or NH_2 are only weakly active. Substitution in the 3-position of hydrochlorothiazide has a pronounced effect on the diuretic potency, and compounds that are more than 100 times as active as hydrochlorothiazide on a weight basis have been

(**27**)   (**29**)

obtained. It should be noted, however, that the maximal diuretic and saluretic effect that can be achieved with any of the thiazide diuretics is of the same magnitude, although the dose required may vary considerably.

Substituents in the 3-position of hydrochlorothiazide having the most favorable effect on activity were alkyl, cycloalkyl, haloalkyl, and arylalkyl, all of which may be classified as hydrophobic in character. This is illustrated in Table 2.9, where the structures of derivatives of hydrochlorothiazide and the respective diuretic responses in the dog are shown. Table 2.8 lists the commercially available thiazide diuretics and the respective optimally effective doses per day in humans. Beyer and Baer (79) have studied four thiazides covering a 1000-fold increase in saluretic activity on a log-stepwise basis from chlorothiazide to hydrochlorothiazide, to trichlormethiazide, to cyclopenthiazide. This increase in activity appears to correlate with their lipid solubility (in terms of their phosphate buffer partition coefficient), rather than their carbonic anhydrase inhibitory effect (Table 2.9).

2.1.7 Other Sulfonamide Diuretics. This group of diuretics includes compounds that produce a pharmacological response similar to that seen with the thiazide diuretics (i.e., they are saluretics), and the maximally attainable level of urinary sodium excretion is in the same range as that of hydrochlorothiazide. The compounds in this group differ in chemical structure; however, most of them are derivatives of *m*-sulfamoylbenzoic acid.

2.1.7.1 Chlorothalidone. An interesting class of compounds was developed from certain substituted benzophenones (130). Optimum diuretic properties were found in 3-(4-chloro-3-sulfamoylphenyl)-3-hydroxy-1-oxoisoindoline (**82**, chlorthalidone, Table 2.10), which is the isomeric form of an *ortho*-substituted benzophenone. The related compounds, (**83**) and (**84**), are more potent carbonic anhydrase inhibitors than chlorthalidone but are less active as diuretics and have a shorter duration of action.

Chlorthalidone has shown good diuretic activity in dogs (131) and is characterized by an unusually long duration of action. It is about 70 times as active as hydrochlorothiazide as a carbonic anhydrase inhibitor *in vitro* and, al-

(83)

(84)

though it induces primarily a saluresis, there is a increased output of K^+ and HCO_3^- at higher doses (70). Clinical studies have substantiated the pharmacological properties (132–135). As with the thiazide diuretics, a mild antihypertensive effect was seen (136). The recommended clinical dosage is 50–200 mg daily or every other day.

2.1.7.2 Hydrazides of m-Sulfamoylbenzoic Acids. The diuretic properties of a large series of compounds derived from 2-chlorobenzenesulfonamide, with a wide variety of functional groups in the 5-position (**85**), have been stud-

(85)

ied in the rat and in the dog (137). The R group includes such functional groups as substituted amines, hydrazines, pyrazoles, ketones, ester groups, substituted carboxamides, and hydrazides. The hydrazides are the most active group of compounds. One of these, *cis-N*-(2,6-dimethyl-1-piperidyl)-3-sulfamoyl-4-chlorobenzamide (clopamide, **86**, Table 2.10), has been studied in greater detail (137–139).

A dose-related diuretic response was seen in rats at doses of 0.01–1 mg/kg. In unanesthetized dogs, a diuretic response was seen with oral doses as low as 2 μg/kg. In anesthetized

Table 2.10 Other Sulfonamide Diuretics

No.	Generic Name	Trade Name	Clinical Dose, p.o. (mg)	Structure	Reference
(43)	Mefruside	Baycaron	25–100		84
(82)	Chlorthalidone	Hygroton	50–200		130
(86)	Clopamide	Aquex, Brinaldix	10–40		137
(87)	Alipamide		20–80		140

82

No.	Name	Trade name	Dose	Structure	Ref.
(88)	Clorexolone	Flonatril, Nefrolan	25–100	H_2NSO_2; Cl (cyclohexyl)	148
(89)	Diapamide	Vectren	500	Cl; CH_3HNSO_2	142
(90)	Metolazone	Zaroxolyn	2.5–20	H_2NSO_2; Cl	153
(91)	Quinethazone	Hydromox	50–200	H_2NSO_2; Cl	152b
(92)	Indapamide	Ipamix, Natrilix	2.5–5	Cl; SO_2NH_2	144

83

dogs, the natriuretic response observed was dose related between 0.01 and 1 mg/kg administered i.v. The drug produced a prompt increase in urine flow and increased the excretion of sodium, potassium, and chloride. A small increase in bicarbonate excretion, which did not significantly alter plasma or urinary pH, was also noted. During maximal diuresis produced by hydrochlorothiazide, administration of clopamide had no effect on sodium excretion. Conversely, after a maximally effective dose of clopamide, hydrochlorothiazide was without effect, although in both cases an additional response to furosemide and spironolactone was observed. This suggests that, although clopamide is not a thiazide diuretic, its natriuretic action closely resembles that of the thiazides. The recommended clinical dose is 10–40 mg/day.

A related hydrazide, alipamide (**87**, Table 2.10), is an effective diuretic agent in rats, dogs, monkeys, and humans (140). The suitable therapeutic dosage in humans is 20–80 mg/day. Alipamide exhibits primarily a saluretic action; carbonic anhydrase inhibition becomes an important factor only at high dose levels. A structure-activity study, in rats, showed that a hydroxamic acid moiety could replace a hydrazide group without loss of activity (141).

Similarly, diapamide (**89**) is an effective saluretic agent in rats, dogs, and monkeys (142). In humans, the compound is comparable to the thiazides in terms of urine volume and electrolyte excretion (143). Elevated plasma urate and glucose levels accompany chronic administration. The clinically effective dose is 500 mg/day.

Indapamide (**92**, Table 2.10), a related sulphamoylbenzamide, possesses both saluretic

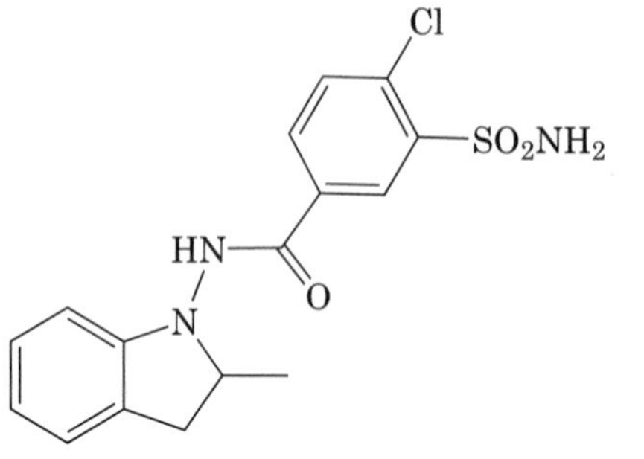

and antihypertensive activity in rats, dogs, and humans after oral administration. In humans, 5 mg of indapamide administered daily produces a greater and more consistent lowering of blood pressure than 500 mg of chlorothiazide given daily (144). Indapamide at a daily dose of 2.5 mg decreases serum potassium levels 0.5 meq/L and increases uric acid levels to about 1.0 mg/100 mL. Replacement of the indoline moiety of indapamide has also yielded compounds that possess potent diuretic activity. The 1-methylisoindoline analog (**93**) had a diuretic activity similar

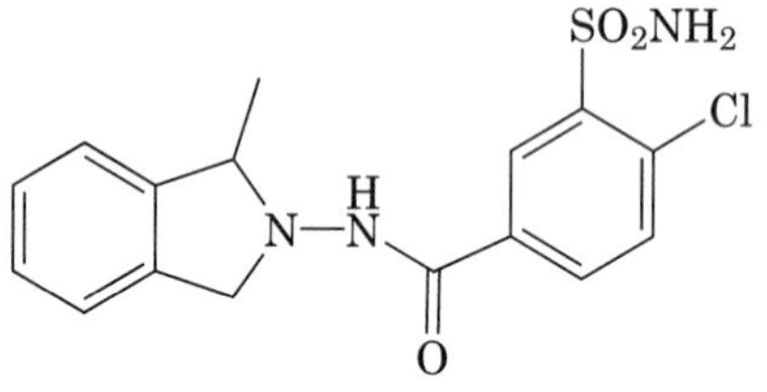

(93)

to that of indapamide; however, it possessed an improved Na^+/K^+ excretion ratio (145). This compound was later found to cause blue pigmentation of the fur and some internal organs in rats and mice at 500 mg/kg p.o (146). The *trans*-pyrrolidine derivative (**94**) has been found to be a

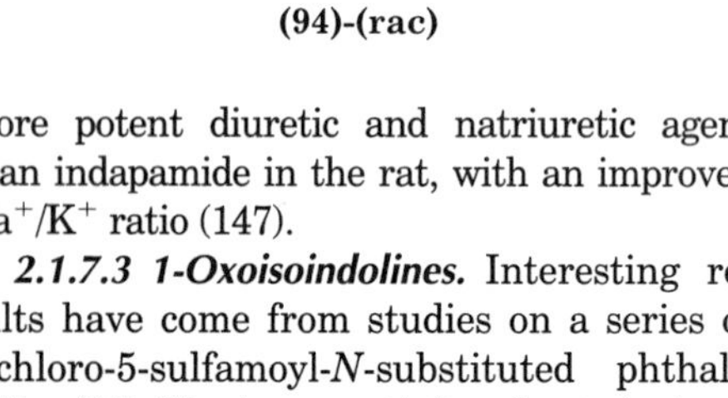

(94)-(rac)

more potent diuretic and natriuretic agent than indapamide in the rat, with an improved Na^+/K^+ ratio (147).

2.1.7.3 1-Oxoisoindolines. Interesting results have come from studies on a series of 4-chloro-5-sulfamoyl-*N*-substituted phthalimides (**95**). Maximum activity, about six times that of chlorothiazide on a weight basis, was seen when the *N*-substituent was a saturated ring containing six to eight carbons. Compounds in which R represented a smaller or larger ring were less active. When R was lower

(92)

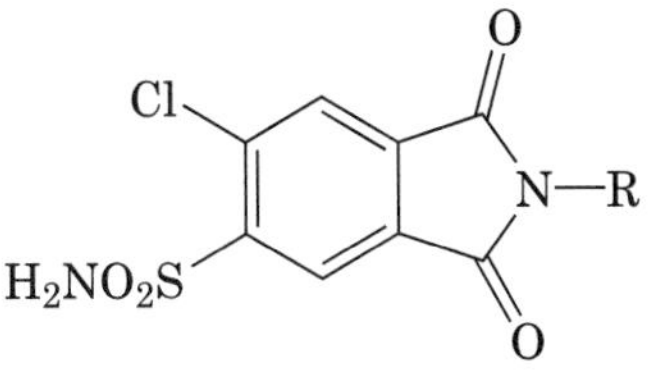

(95)

alkyl, decreased activity was also found. Reduction of one of the carbonyl groups to yield the corresponding 3-hydroxy-1-oxoisoindoline (96, R = cyclohexyl) resulted in a 10-fold in-

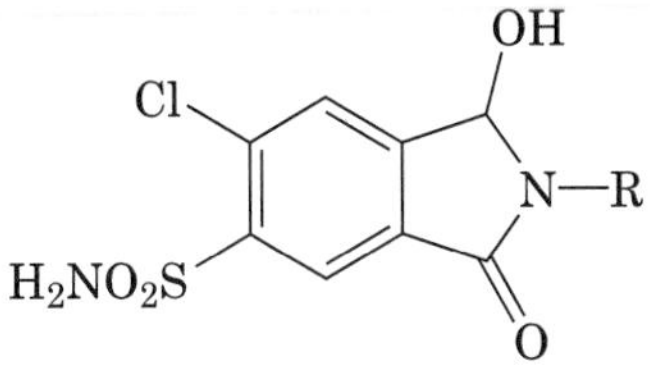

(96)

crease in potency. Complete reduction of the carbonyl function to a methylene yielded clorexolone (88, Table 2.10), which was 300 times more active than chlorothiazide on a weight basis when tested in the rat (148). Interestingly, reduction of the other oxo group to yield the isomeric 6-chloro-5-sulfamoyl-1-oxoioindoline (97) resulted in complete loss of activ-

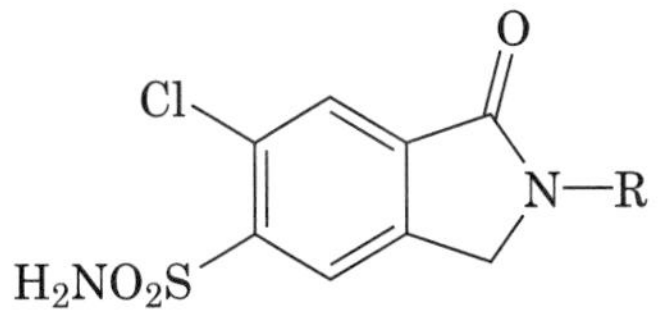

(97)

ity. Structure-activity relationships for a number of 2-substituted 1-oxoisoindolines are shown in Table 2.11. Methylation or acetylation of the sulfamoyl group of clorexolone decreases the activity by at least a factor of 10. In humans, clorexolone (88) is a potent diuretic. The clinical dose is 25–100 mg/day; the pattern of water and electrolyte excretion is similar to that caused by the benzothiadiazine diuretics. Urinary pH and HCO_3^- excretion remain unchanged after administration of clorexolone, indicating that there is no significant

Table 2.11   Diuretic Activity in Rats of 2-Substituted 1-Oxoisoindolines[a]

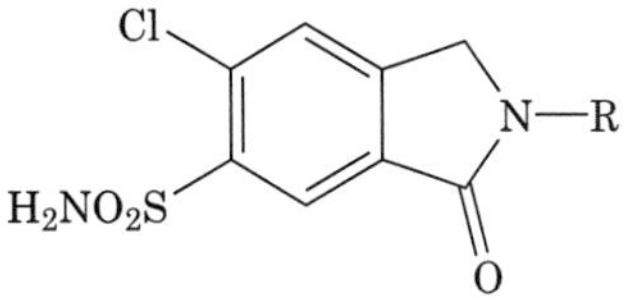

No.	R	Diuretic Activity (Chlorothiazide = 1)
(88)	Cyclohexyl	300
(98)	Isobutyl	75–100
(99)	Cyclopentyl	100
(100)	4-Methylcyclohexyl	100
(101)	3-Methylcyclohexyl	100
(102)	3,4-Dimethylcyclohexyl	200
(103)	Cycloheptyl	50–100
(104)	Cyclooctyl	100
(105)	Norborn-2-yl	100
(106)	Cyclohexylmethyl	200

[a]From ref. 148.

in vivo involvement of renal carbonic anhydrase inhibition. As is the case with the thiazide diuretics, elevated serum uric acid levels are seen, but there may well be less propensity for hyperglycemia (149–151). In contrast to the thiazide diuretics, insignificant amounts of the drug are excreted unchanged in humans and in dogs. The metabolites are compounds monohydroxylated on the cyclohexane ring. Neither the compound nor its metabolites are stored in the body tissues (152a).

2.1.7.4 Quinazolinone Sulfonamides. Replacement of the ring sulfone group in the thiazides by a carbonyl yields quinazolinones and dihydroquinazolinones (107) and (108),

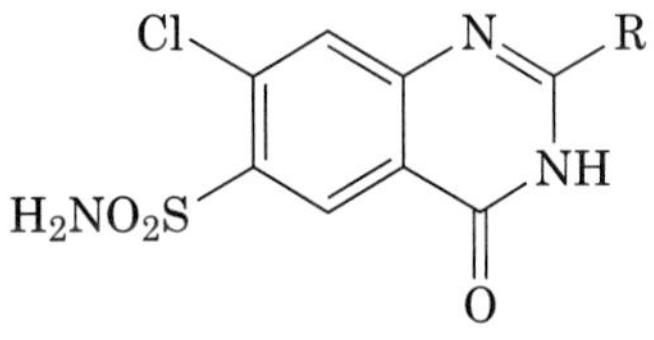

(107)

respectively. These compounds produce nearly the same diuretic response as that of the parent thiazide derivatives; the dihydro-derivatives again are more active on a dose/kg basis.

(108)

Substitution at R_3 by alkyl is disadvantageous, in that it reverses the favorable Cl^- and K^+ excretion patterns seen in the few examples studied (152b). The preferred member of the series was quinethazone (91, Table 2.10; 108, R_2 = Et, R_3 = H), which in humans has the same order of potency as that of hydrochlorothiazide with a high Na^+/K^+ excretion ratio. The duration of activity appears to be about 24 h (154). The recommended clinical dose is 50–200 mg/day.

A series of dihydroquinazolinones substituted in the 3-position (108, R_3 = aryl and arylalkyl) were studied and it was shown that some of these compounds are highly active diuretics. The more active derivatives have at least one hydrogen in the 2-position, a primary SO_2NH_2 group in the 6-position, and an *ortho* or *para* lower alkyl or CF_3-substituted aromatic ring in the 3-position of the quinazoline nucleus. The most interesting member of the series is metolazone (90, Table 2.10; 108, R_2 = Me, R_3 = o-Me-C_6H_4) (153). Studies in normal volunteers led to the conclusion that metolazone exerts its effect in the proximal tubule and in the cortical segment of the ascending limb of Henle of the early distal convoluted tubule. The absence of significant bicarbonatriuria is evidence against carbonic anhydrase inhibition. Metolazone did not impair the ability to acidify the urine normally in response to an oral load of NH_4Cl, and it was concluded that metolazone has no effect on the distal H^+ secretory mechanism (155–157). In dogs, metolazone was found to be excreted by glomerular filtration and renal tubular secretion. The secretory mechanism was antagonized by probenecid; however, this did not affect the diuretic action of metolazone (158). A 10–15 mg dose of metolazone was approximately equivalent to 50 mg hydrochlorothiazide, and the time course of diuretic action was similar to that of hydrochlorothiazide. No acute eleva-

tion of urate or glucose or signs of toxicity were seen in a short-term study (159). The recommended clinical dose is 5–20 mg/day. In hypertensive patients a double-blind study compared a dose of 50 mg of hydrochlorothiazide with 2.5 and 5.0 mg of metolazone, and similar effects on blood pressure were observed. The effects on other parameters (e.g., body weight, electrolytes, serum uric acid, and blood sugar levels) were also comparable (160). In a study in patients with nonedematous, stable chronic renal failure, a high dosage of metolazone (20–150 mg) increased urine flow significantly. Its activity was greater than that of the thiazides, which are ineffective at glomerular filtration rates of less than 15–20 mL/min.

Fenquizone (109), which is structurally related to metolazone, has a thiazide-like diuretic

(109)

profile, but it has less of an effect than that of the thiazides on carbohydrate and lipid metabolism (161). In patients with mild essential hypertension, a 12-month study with fenquizone (10 mg/day) showed that the compound significantly lowered systolic and diastolic blood pressure. Serum levels of glucose, triglycerides, and cholesterol remained unchanged; however, uric acid levels increased slightly.

2.1.7.5 Mixed Sulfamide Diuretic-Antihypertensive Agents. Recent attempts have been made to combine an o-chlorobenzenesulfonamic diuretic moiety with another molecule with known antihypertensive activity. Mencel and coworkers synthesized compounds in which they covalently joined enalaprilat, a known angiotensin-converting enzyme (ACE) inhibitor, to several known thiazide diuretics and arylsulfonamides (162). Compound (110) was found to be a potent ACE inhibitor *in vitro*. In a sodium-depleted spontaneously hypertensive rat (100 mg/kg i.p.), (110) reduced

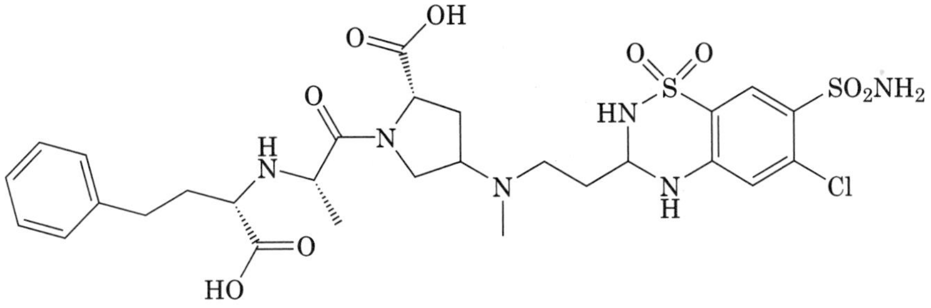

(110)

blood pressure by 41–42%. Compound (110) did elevate potassium and sodium excretion in the rat but it was found to be less potent than chlorothiazide in this model.

Fravolini and coworkers have covalently linked an o-chlorobenzesulfonamic diuretic to a propanolamine β-blocking pharmacophore (163). Compounds (111) and (112) possess both β-adrenergic antagonist and diuretic activity in the rat. At an equimolar dose, (111) produced a saluretic effect similar to that of hydrochlorothiazide, but as a β-blocker it was weaker than cartelol and propranolol.

BMY-15037–1 (113) is a chlorosulfamoyl-isoindolone derivative that has both diuretic and α-adrenoceptor antagonistic effects. In

spontaneously hypertensive rats, oral administration of 0.330 mg/kg of this compound decreased mean arterial pressure and induced saliuresis. The duration of action of BMY-15037–1 was similar to that of prazosine.

2.1.7.6 Tizolemide. A structurally novel type of sulfonamide diuretic was developed by Lang and coworkers at Hoechst (164). Tizolemide (114) was selected from a series of compounds for further investigation. Optimal activity was associated with an unsubstituted sulfamoyl group. In dogs the diuretic activity was similar to that of hydrochlorothiazide. Interestingly, tizolemide lowered serum uric acid levels in the cebus monkey, indicating a possible uricosuric effect.

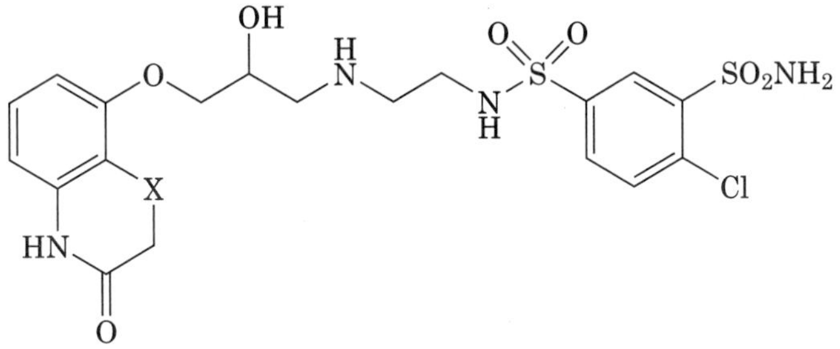

(111) X = S
(112) X = CH$_2$

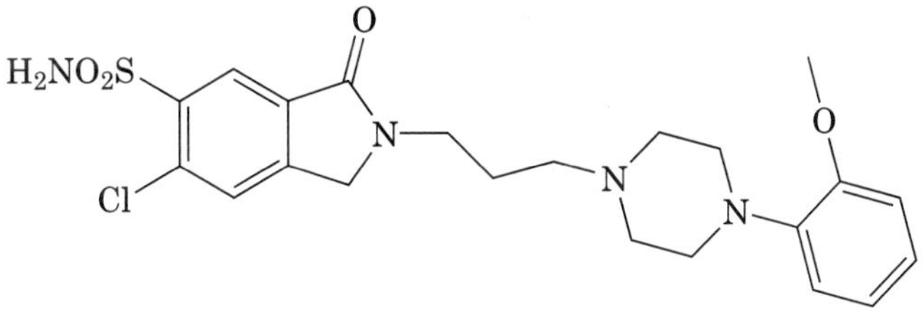

(113)

Table 2.12 High Ceiling Diuretics

No.	Generic Name	Trade Name	Clinical Dose, p.o. (mg)	Structure
(116)	Ethacrynic Acid	Edecrin	50–200	
(117)	Bumetanide	Burinex, Lunetron	1–5	
(118)	Furosemide	Lasix	20–80	
(119)	Piretanide			
(120)	Xipamide	Aquaphor	40–80	
(121)	Azosemide	Luret		

Table 2.12 (*Continued*)

No.	Generic Name	Trade Name	Clinical Dose, p.o. (mg)	Structure
(122)	Torasemide	Unat, Toradiur	2.5–20	

2.1.7.7 Bemitradine. Workers at Searle extended work on a series of previously described tetrazolopyrimidines that were shown to be antihypertensive agents in rats and in humans (165, 166). A series of triazolopyrimidines were prepared and SC-33643 was found to be the most potent. Bemitradine (SC-33643, **115**) has a thiazide-like profile of diuresis but

is not a sulfonamide. Bemitradine (**115**) was 5.5 times more potent than hydrochlorothiazide after oral administration in the unanesthetized dog (167) and significantly increased renal blood flow and glomerular filtration rates. Bemitradine is well absorbed after oral administration; however, the bioavailability is low because of hepatic first-pass metabolism.

2.1.8 High Ceiling Diuretics. The term *high ceiling diuretics* has been used to denote a group of diuretics that have a distinctive action on renal tubular function. As suggested by their name, these drugs produced a peak diuresis far greater than that observed with other diuretics. These agents act primarily by inhibiting the reabsorption of sodium in the thick ascending loop of Henle and, thus, they are also commonly referred to as *loop diuretics*. This class of diuretic agents holds few structural features in common. Because they are most alike in potency and with respect to their renal site of action, they represent more a pharmacological rather than a chemical class of agents. The high ceiling or loop diuretics now in use or currently being studied are shown in Table 2.12.

2.1.8.1 Ethacrynic Acid. The clinical dose of ethacrynic acid (**116**) lies between 50 and 200 mg/day. In long-term studies, the antihypertensive effects of 100 mg of ethacrynic acid were similar to 50 mg hydrochlorothiazide in patients with mild hypertension (168). Ethacrynic acid continues to be an effective diuretic even at very low glomerular filtration rates, and therefore is useful in the treatment of patients with chronic renal failure (169). Ototoxicity has been reported that manifests itself as transient deafness (169). Permanent deafness has also been observed after treatment with high doses of ethacrynic acid in re-

Table 2.13 Structure–Activity Relationships of Ethacrynic Acid Analogs

No.	Structure	Diuretic Activity (Dog i.v.)	$t_{1/2}$ (min)^a	Reference
(116)		+6	<1	174
(125)		+/−	90	174
(126)		+2	11	174
(127)		+1	27	174
(128)		+3	1	174
(129)		+5	210	174

nal failure (170). Ethacrynic acid possesses excellent oral absorption and rapid onset of action is seen when it is administered orally or intravenously (171). Ethacrynic acid is extensively metabolized to its cysteine adduct after oral administration. It is believed that it is this metabolite that represents the pharmacologically active form of the drug because it has a much greater activity than that of the parent compound (172). About two-thirds of the compound is excreted by the kidney in the cysteine-adduct form along with parent compound and some unidentified metabolites. The remaining one-third undergoes biliary elimination, again as its cysteine adduct.

It was soon after the introduction of organomercurials as diuretics that the idea arose that their biological activity resulted from the blockade of essential sulfhydryl groups. From these early considerations, several series of

Table 2.13 (*Continued*)

No.	Structure	Diuretic Activity (Dog i.v.)	$t_{1/2}$ (min)[a]	Reference
(130)	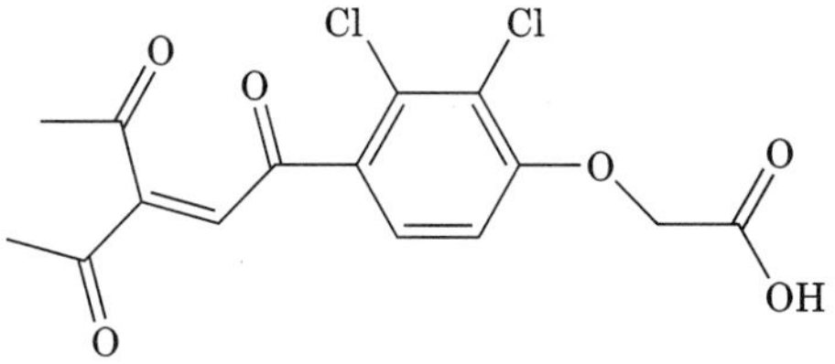	(+7)	2	174,175
(131)	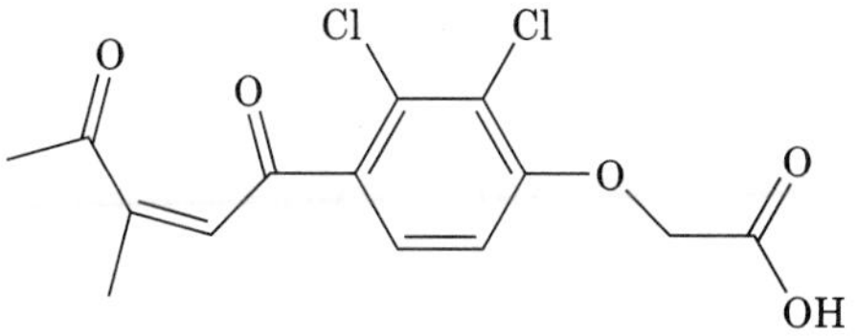			176
(132)	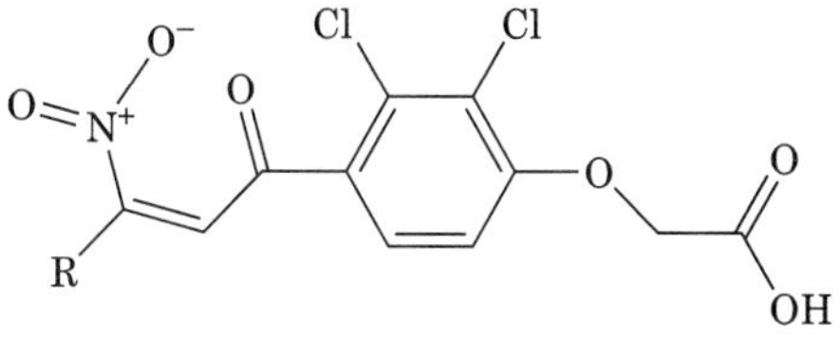 R = CH₃ or Et	+6	2	

R = CH$_3$ or Et

[a] $t_{1/2}$ = time in minutes required for one-half of a standard amount of test compound to react with excess mercaptoacetic acid at pH 7.4 and 25°C in DMF-phosphate buffer, analogous to the procedure of Duggan and Noll (176).

highly active diuretics were developed that were thought to react selectively with functionally important sulfhydryl groups, or possibly other nucleophilic groups, that were essential for sodium transport in the nephron. These compounds generally contain an activated double bond attached to a moiety containing a carboxylic acid group of a type expected to assist transport into, or excretion by, the kidney. The general structure of these compounds is exemplified by formulas (123) and (124). They are highly active in the dog when administered orally or parenterally, but are inactive in the rat.

A marked increase in diuretic activity was observed when chlorine was introduced *ortho* to the carbonyl group of the aryl side chain. Not only was the diuretic activity better, but the rate at which chemical addition of sulfhydryl compounds across the double bond in an *in vitro* system increased. The presence of two chlorines in positions 2 and 3 of the phenoxy-

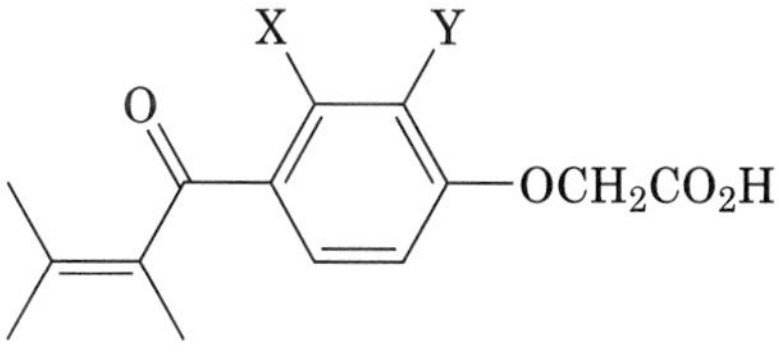

(123) 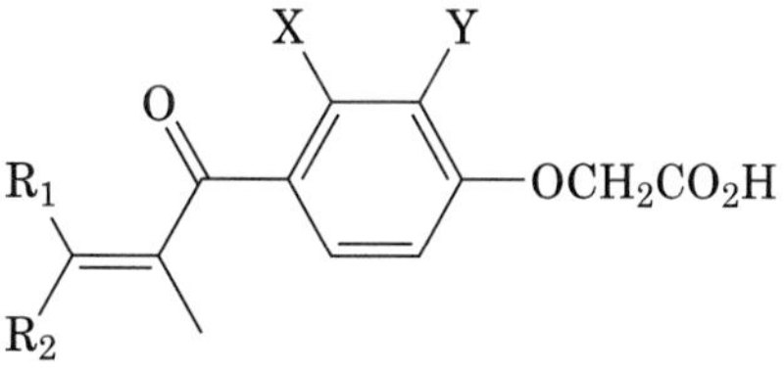   (124)

X, Y = H, Cl, R$_1$, R$_2$ = CH$_3$CO, NO$_2$, CN, alkyl

acetic acid further increased the activity. A number of compounds corresponding to structures (**123**) and (**124**) and their diuretic activity in dogs are shown in Table 2.13 (173, 174). Ethacrynic acid (**116**) is the most interesting compound of this series and has been studied extensively. A report on (diacylvinylaryloxy)-acetic acids has shown that compound (**130**) (Table 2.13) is approximately three times as active as ethacrynic acid (175). The corresponding (acylvinylaryloxy)acetic acids (e.g., **131**) are less active (176). The 4-(2-nitropropenyl)phenoxyacetic acid derivatives (e.g., **132**) are also three to five times as active as ethacrynic acid (177).

Ethacrynic acid does not inhibit carbonic anhydrase *in vitro*. It has a steeper dose-response curve than that of hydrochlorothiazide, and the magnitude of its maximum saluretic effect is several times that of hydrochlorothiazide (171). The renal corticomedulary electrolyte gradient, after administration of ethacrynic acid and other high ceiling diuretics, is virtually eliminated as a result of nearly total inhibition of Na^+ transport in the ascending limb of Henle (172).

2.1.8.2 Indacrinone. Indacrinone or MK 196 (**133**) is an indanyl derivative of ethacrynic

(133)

acid. In clinical studies in healthy subjects consuming a standard diet, 10 mg of MK 196 produced a slightly smaller diuresis than 40 mg of furosemide, and MK 196 did not influence uric acid excretion or 24-h urate clearance. A single dose of 40 mg of furosemide caused uric acid retention with a significant decrease of 24-h urate clearance; prolonged administration caused a statistically significant increase in plasma uric acid levels. Prolonged administration of MK 196 did not increase plasma uric acid levels, and the ratio of urate-creatinine clearance was indistinguish-

able from the values found in the placebo group; MK 196 thus appears to be a diuretic without uric acid-retaining properties (178). In a comparison of the diuretic effects of MK 196 and furosemide in normal volunteers receiving the drug every day for 14 days, an oral dose of 10 mg of MK 196 caused a gradual diuretic and saluretic response, resulting in a maximal plateau during the period 4–7 h after drug administration followed by a slow return to baseline during the next 16–18 h. Although at the doses studied (10 and 20 mg, MK 196), the maximal response to furosemide (40 mg) was always higher than the maximal response to MK 196, the total 24-h saluresis after 10 mg of MK 196 was equivalent to that produced by furosemide. After 20 mg of MK 196 the 24-h response was greater than that with furosemide (179). A double-blind pilot study was conducted to compare the antihypertensive efficacy of two doses of MK 196 (10 and 15 mg) with 50 mg hydrochlorothiazide in patients with mild to moderate hypertension. Both doses of MK 196 lowered blood pressure as much as or more than 50 mg hydrochlorothiazide during the 24-h period after drug administration (180).

Clinical results indicated that MK 196 is a highly active diuretic with a gradual onset of action that reaches a plateau that persists for 4–7 h, then gradually returns to baseline values over the next 16–18 h. This is probably attributable to the long half-life of this compound as observed in animals. The antihypertensive effects are comparable to those observed with hydrochlorothiazide.

Workers at Merck discovered that annulation of the unsaturated ketone side chain on the aromatic ring yielded compounds that retained their diuretic activity but that also possessed uricosuric properties. Both enantiomers of indacrinone possess uricosuric activity but the (−)-enantiomer is the more potent diuretic (181). Clearance studies in the rat indicated that MK 196 is a potent diuretic acting in the ascending limb of Henle, which results in significant increases in urinary excretion of sodium, calcium, magnesium, water (182), and uric acid (183).

The physiological disposition of ^{14}C-labeled MK 196 was studied in the rat, dog, monkey, and chimpanzee. The drug was well absorbed and showed minimal metabolism in the rat, dog, and monkey. Triphasic rates of elimination of drug and radioactivity were observed in these three species. In dogs, the terminal half-life was estimated to be about 68 h; in monkeys, there was a longer terminal half-life of approximately 105 h. The long terminal half-life of this compound may result in part from binding to plasma proteins. The major route of radioactivity elimination is through the feces for the rat (~80%). In contrast, the monkey and the chimpanzee eliminate the majority of the dose through the urine. Minimal metabolism of MK 196 was observed in the rat, dog, and monkey; however, in humans and in chimpanzees, there was extensive biotransformation. The major metabolite resulted from *para* hydroxylation of the phenyl group to yield [6,7-dichloro-2-(4-hydroxyphenyl)-2-methyl-1-oxo-5-indanyloxy]-acetic acid (**134**).

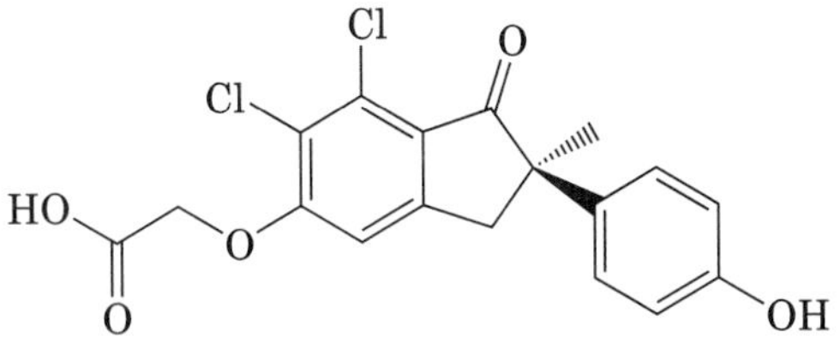

rat, dog, rhesus monkey, and baboon (187). The half-life of the compound in the rat and dog is about 2 h. In the monkey and baboon it was found to have a much longer half-life, 18 and 40 h, respectively. In all species less than 5% of MK-473 is excreted intact in the urine. In humans, this compound is well absorbed, but it undergoes extensive metabolism. Related compounds from this series have been studied for their therapeutic utility in the area of brain injury (188).

2.1.8.3 Other Aryloxy Acetic Acid High Ceiling Diuretics. Since the discovery of ethacrynic acid and indacrinone many new members of the phenoxyacetic acid family of diuretics have been reported.

A series of [(3-aryl-1,2-benzisoxazol)-6-yloxy] acetic acids were described by Shutske and coworkers at Hoechst (189). Of this group, HP 522 (**136**) was found to be a potent diuretic

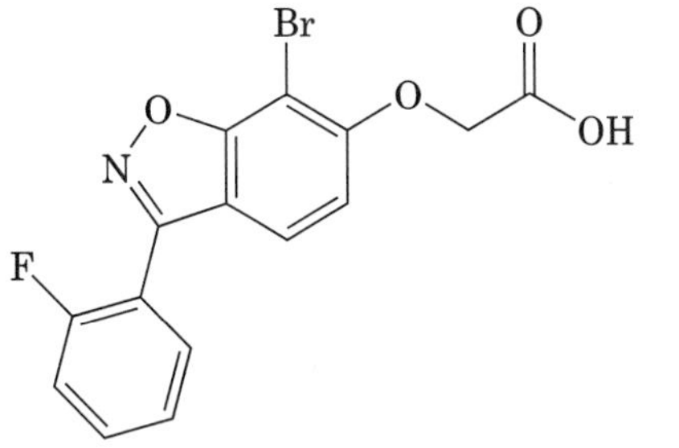

(134)

(136)

This metabolite accounted for more than 40% of the 0- to 48-h urinary radioactivity; about 20% of the radioactivity was accounted for as unchanged drug (184, 185).

The cyclopentyl analog MK-473 (**135**) has diuretic and uricosuric activity similar to that

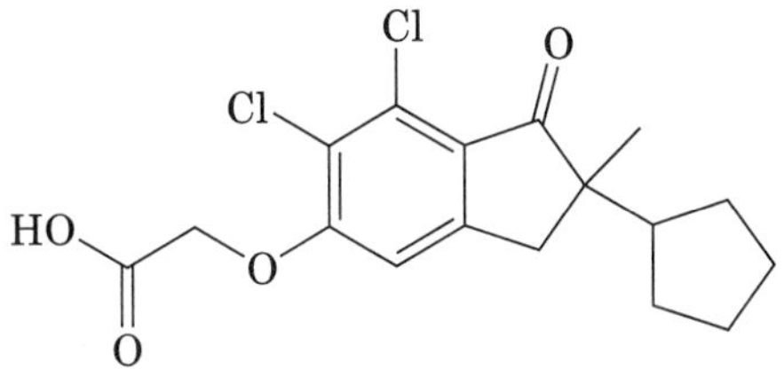

in mice and dogs. It was found to lower plasma uric acid levels in chimpanzees after chronic administration (10 mg kg^{-1} day^{-1}, p.o.) (190).

Workers at Merck reported on a series of 2,3-dihydro-5-acylbenzofurancarboxylic acids (191). The 5(2-thienylcarbonyl)-2-benzofurancarboxylic acid derivative (**137**) was found to have a higher natriuretic ceiling than that of hydrochlorothiazide and furosemide in the

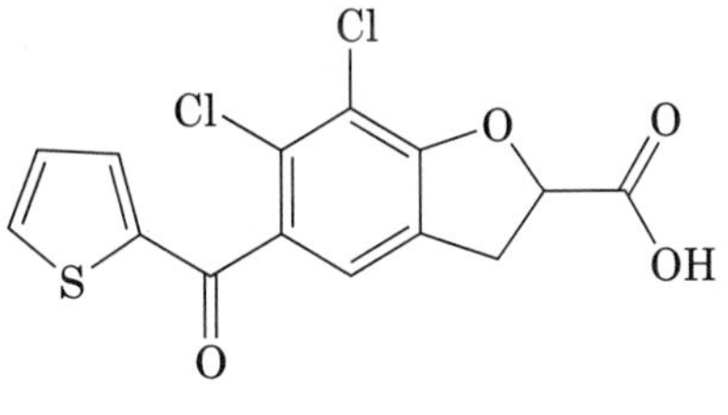

(135)

of indacrinone (**133**), but it also possesses substantial antihypertensive activity (186). The physiological disposition of this compound has been studied in several species including the

(137)

rat. Resolution of the enantiomers and testing in the chimpanzee revealed that the S-enantiomer is responsible for the compounds' diuretic and saluretic activity.

Plattner and coworkers at Abbott have disclosed a series of 5,6-dihydrofuro[3,2-*f*]-1,2-benzisoxazole-6-carboxylic acid high ceiling diuretics (192). Abbott 53385 (138) had di-

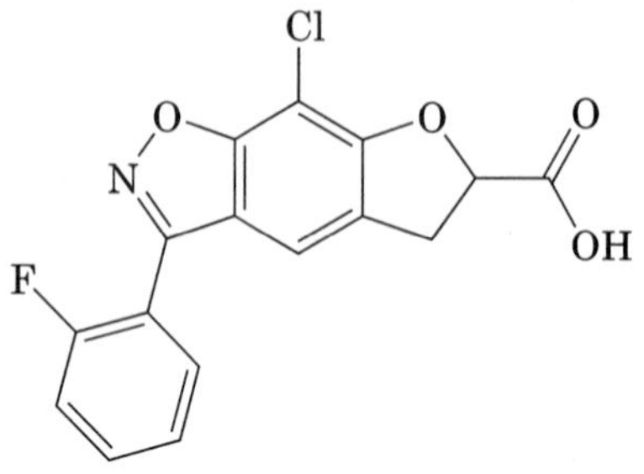

(138)

uretic effects similar to those of furosemide in the saline-loaded mouse. In the conscious dog, it was about six times more potent than furosemide. Resolution of the enantiomers and pharmacological evaluation showed that only the S-isomer displays the diuretic and saluretic activity. In humans the drug was well tolerated and no adverse effects were noted (192b). At 20, 40, 60, and 80 mg a significant dose-related increase in urine volume, sodium, and chloride excretion were produced. Hepatic clearance is the main route of excretion in humans. Very little of the drug is eliminated in the urine.

2.1.8.4 Furosemide.

At the time work on ethacrynic was proceeding at Merck, Sharp & Dohme, furosemide was being developed in the Hoechst laboratories in Germany. Investigation of a series of 5-sulfamoylanthranilic acids (139), substituted on the aromatic amino

group, showed that these compounds were effective diuretics. The isomeric series (140) did not show saluretic properties (176, 193).

More than 100 variously substituted derivatives were studied pharmacologically, but only those that corresponded to the general structure (139) exhibited outstanding saluretic activity. The most active was furosemide (118, R = furfuryl; Table 2.12). In contrast to the dihydrobenzothiadiazine diuretics, where the substitutent in the 3-position of the heterocyclic ring can be varied to a considerable degree, the requirements for high activity in the 5-sulfamoylanthranilic acid series are much more stringent. On parenteral and oral administration to different species and to humans, the degree of diuretic effect elicited, as measured by urine flow and Na^+ and Cl^- excretion, was several times that obtainable with the thiazide diuretics (194, 195).

In a study undertaken to explore the effect of furosemide on water excretion during hydration and hydropenia in dogs, it was found that as much as 38% of filtered sodium was excreted during furosemide diuresis and both free water clearance (C_{H_2O}) and solute-free water reabsorption (TC_{H_2O}) were inhibited, indicating a marked effect in the ascending loop of Henle. Furosemide is largely excreted unchanged in the urine, but a metabolite, 4-chloro-5-sulfamoylanthranilic acid, has been identified (196, 197). Studies by Hook et al. (198) and Ludens et al. (199) indicated that furosemide reduces renal vascular resistance and thus enhances total renal blood flow in dogs. Clinical studies in normal subjects and in patients with edema of various etiologies have clearly shown that furosemide is an extremely potent saluretic drug (200, 201).

Furosemide is rapidly absorbed after oral administration in humans. It is highly bound

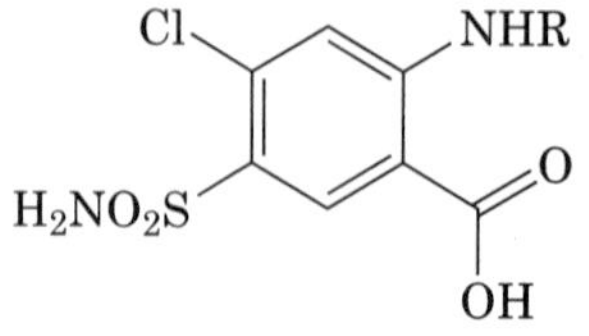

(139)

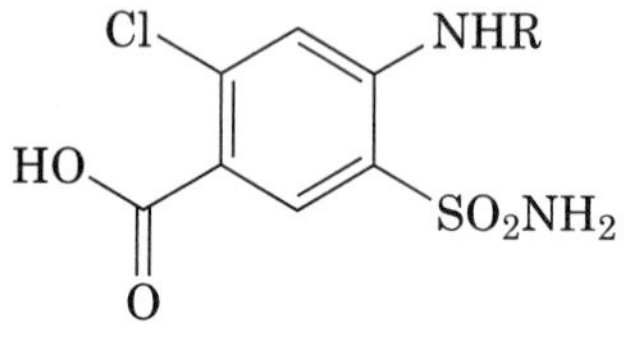

(140)

R = CH₂Ph, ——CH₂(2-furyl), ——CH₂(2-thienyl)

to plasma protein, primarily to albumin, with bound fractions averaging about 97%. The bioavailability of furosemide is 50–70% in normal subjects; it is eliminated by renal, biliary, and intestinal pathways; it is excreted as its glucuronide; and it is unchanged by the renal route.

The antihypertensive effects of furosemide were shown to be qualitatively and quantitatively similar to those of chlorothiazide in nonedematous patients with essential hypertension (202). Furosemide has been reported to produce moderate diuretic response in patients with renal disease and resistant edematous states when other diuretics (e.g., thiazides, mercurials, triamterene, and spironolactone) have failed. Doses up to 1.4 g/day may be required (203, 204). Ototoxicity has also been reported after large doses of furosemide (205).

2.1.8.5 Bumetanide. In a series of studies starting in 1970, Feit and coworkers in Denmark investigated the diuretic activity of derivatives of 3-amino-5-sulfamoylbenzoic acid (141). In the initial series, R_2 was chlorine and

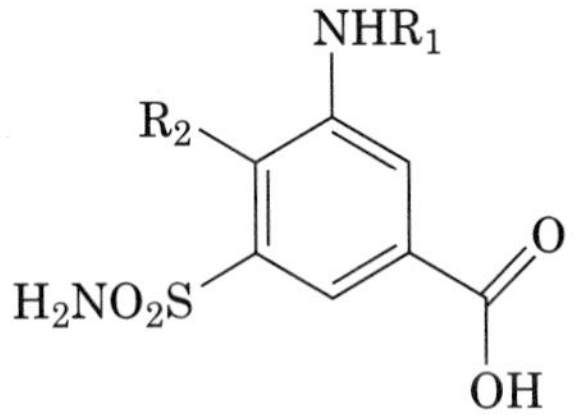

(142)

(141)

R_1 was varied widely from alkyl to substituted benzyl. One of the most interesting compounds in this series was (142) (3-butylamino-4-chloro-5-sulfamoylbenzoic acid), which approached the activity of furosemide when given i.v. (10 mg/kg in a NaOH solution) to dogs. Interestingly, whereas in the anthranilic acid-furosemide series the N-furfuryl substituent afforded outstanding activity, this was not the case in this series (206).

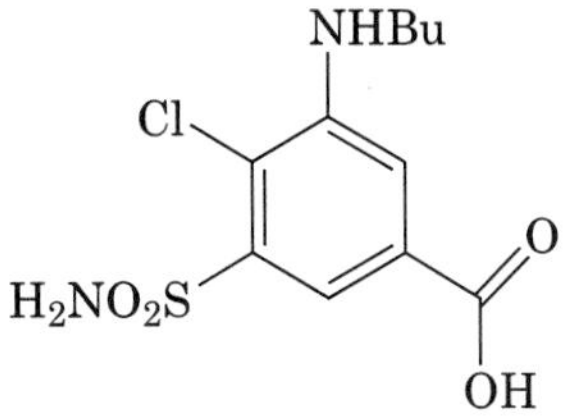

Further investigation showed that compounds, in which $R_2 = -OC_6H_5$, $-NHC_6H_5$, or $-SC_6H_5$ in the generic structure (141), were very active diuretics; the structures and saluretic activity in dogs of the more active derivatives are shown in Table 2.14.

Compound (148) (bumetanide, Table 2.14) was the most interesting drug and has been studied extensively (207). Further structure-activity studies uncovered related compounds with equally high diuretic potency; compounds (150–166) (Table 2.15) are representative of the series studied. It was found that a phenoxy group in the 4-position enhances activity in the anthranilic acid series as well as in the 3-amino-5-sulfamoylbenzoic acid series (e.g., compound 150) (208). The phenoxy group could be replaced by C_6H_5CO, $C_6H_5CH_2$ (209), and even a directly bonded C_6H_5 group (210) (e.g., compounds 151, 152, and 158) (Table 2.15). Interestingly, an equilibrium appears to exist between (151) and the corresponding benzoyl derivative (167).

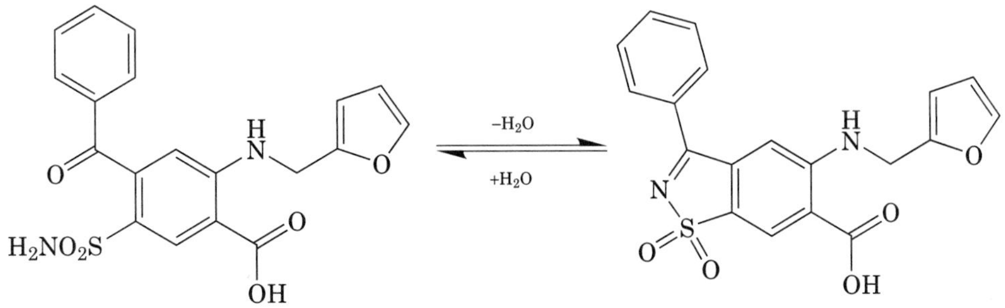

(167)

(151)

Table 2.14 Compounds Related to Bumetanide[a]

No.	Structure	Dose, p.o. (mg/kg)	Volume (mL/kg Urine)	Urinary Excretion[b]		
				Na$^+$	K$^+$	Cl$^-$
(118)		0.5	8	1.5	0.2	1.8
(143)		0.1	39	3.7	0.6	4.8
(144)		0.1	31	3.2	0.9	4.5
(145)		0.1	38	3.1	0.9	5.0
(146)		0.25	40	4.0	1.1	5.7

It is postulated, however, that at physiological pH the compound is present as its ring-opened benzoyl derivative (167). Compounds (161) and (162) (Table 2.15), which do not have an amino substituent attached to the benzene ring of the sulfamoylbenzoic acid, were more stable and only cyclodehydrated on heating (210).

In the 3-amino-5-sulfamoylbenzoic acid series, the 3-amino substituent can be replaced

Table 2.14 *(Continued)*

No.	Structure	Dose, p.o. (mg/kg)	Volume (mL/kg Urine)	Urinary Excretion[b]		
				Na+	K+	Cl−
(147)		0.1	36	4.0	2.1	5.9
(148)		0.25	51	4.8	1.0	7.0
		0.1	31	3.2	0.49	4.5
		0.01	13	1.2	0.3	1.4
(149)		0.05	27	3.2	0.6	3.9

[a]From Ref. 204.
[b]Per 6 h in dog (meq/kg).

by an -OR or -SR group (**155, 159, 162**, Table 2.15) (210, 211); however, oxidation of the -SR group to -SO$_2$R (**156**, Table 2.15) eliminates the diuretic activity (212). Compound (**162**) (Table 2.15) is one of the most potent benzoic acid diuretics ever reported. It shows significant diuretic activity in dogs at 1 μg/kg, which represents a potency approximately five times as high as that of bumetanide (210). In the anthranilic acid series, the structural requirements are more exacting and the thiosalicyclic acid analog (**157**, Table 2.15) is only weakly active.

A series of compounds in which the sulfamoyl group was replaced by a methylsulfonyl group was also investigated (213). Many of the 5-methylsulfonylbenzoic acid derivatives showed considerable diuretic activity (e.g., **163** and **165**, Table 2.15). The diuretic pat-terns of these compounds resemble those of previously discussed sulfamoylbenzoic acids. However, substitution of the sulfamoyl group by the spatially and sterically similar methyl-sulfonyl group generally led to decreased potency. Substitution of methylthio or methyl-sulfinyl for the methylsulfonyl group reduced the potency considerably (e.g., **164**, Table 2.15). The anthranilic acid analog (**166**, Table 2.15) of the highly active 3,4-substituted methylsulfonylbenzoic acid derivative (**163**) (Table 2.15) was inactive at the dose tested, again confirming that the structural requirements in the anthranilic acid series are more demanding.

Replacement of the chloro group in hydro-chlorothiazide, by a C$_6$H$_5$S group (**168**) elimi-nated the diuretic activity (214). Similar results were found in the case of quinethazone

Table 2.15 Compounds Related to Bumetanide

No.	Structure	Dose, i.v. (mg/kg)	Volume (mL/kg Urine)	Na^{+a}	K^{+a}	Cl^{-a}	Reference
(150)		1.0 0.1	43 25	5.0 2.9	0.8 0.45	6.4 3.8	205
(151)		1.0 0.1	44 26	4.8 2.7	0.7 0.8	5.9 3.3	206
(152)		0.25	19	2.1	0.4	2.7	206

(153)		209	2.9	0.4	2.4	18	0.25
(154)		209	4.2	0.7	3.1	28	0.25
			4.1	0.6	3.3	29	0.25[b]
(155)		208	3.5	0.57	2.5	25.7	0.1
			3.8	0.69	3.3	29.8	0.1[b]
(156)		208	—[c]	—[c]	—[c]		1.0

(153) NHBu; 4-phenyl; CO_2H; benzisothiazole S, O, O, N

(154) NHBu; $PhCH_2$; H_2NSO_2; CO_2H

(155) SBu; phenoxy; H_2NSO_2; CO_2H

(156) SO_2Bu; phenoxy; H_2NSO_2; CO_2H

Table 2.15 (*Continued*)

No.	Structure	Dose, i.v. (mg/kg)	Volume (mL/kg Urine)	Na^{+a}	K^{+a}	Cl^{-a}	Reference
(157)		1.0	8	0.8	0.3	0.8	208
(158)		1.0	23	3.3	0.81	3.4	207
(159)		1.0	28	3.2	0.64	4.1	207

207	4.1	0.59	3.0	26	0.1	
207	4.4	0.89	3.6	33	0.25	
	2.0	0.42	1.0	12	0.1	
207	4.9	0.97	4.4	38	0.1	
	3.5	0.53	2.5	23	0.01	

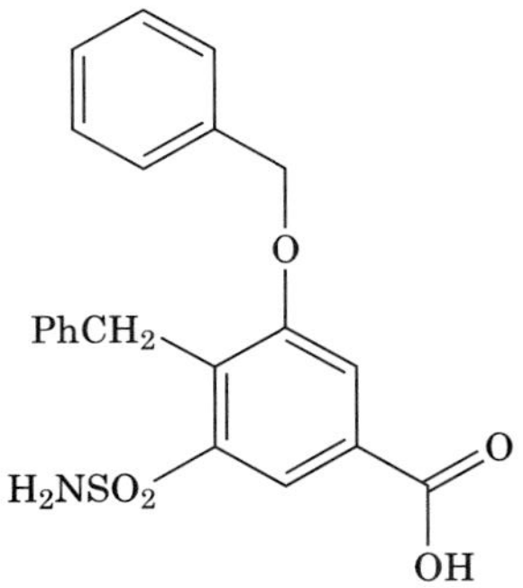

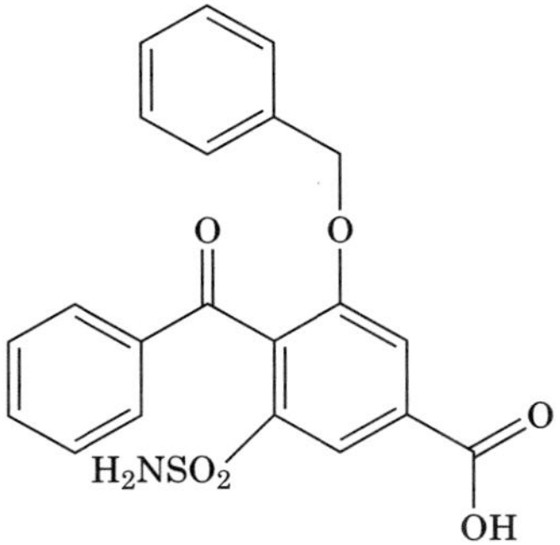

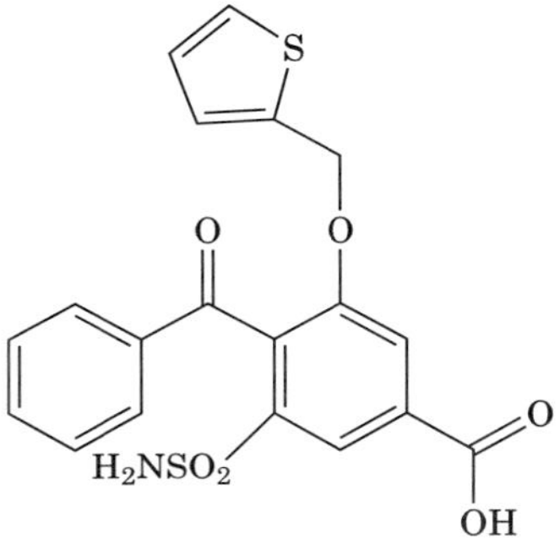

(160) PhCH$_2$ — H$_2$NSO$_2$

(161) H$_2$NSO$_2$

(162) H$_2$NSO$_2$

Table 2.15 (*Continued*)

No.	Structure	Dose, i.v. (mg/kg)	Volume (mL/kg Urine)	Na^{+a}	K^{+a}	Cl^{-a}	Reference
(163)		1.0	31	4.8	0.98	5.1	210
(164)		1.0	7	0.9	0.17	1.1	210
(165)		1.0^b	33	3.9	0.64	4.7	210
(166)		10		$—^c$	$—^c$	$—^c$	210

[a] Urinary excretion per 3 h meq/kg.
[b] p.o.
[c] Same as control.

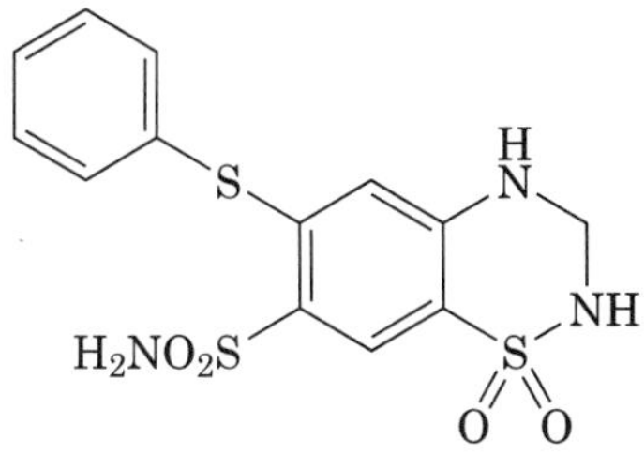

(168)

and clopamide (**91**, **86**, Table 2.10). Structural modifications of bumetanide (**148**, Table 2.14) were further explored by Nielsen and Feit (215). It was found that the carboxyl group in the 1-position of bumetanide could be replaced by a sulfinic or sulfonic acid group or converted to an aminobenzyl group (216) (**169** to

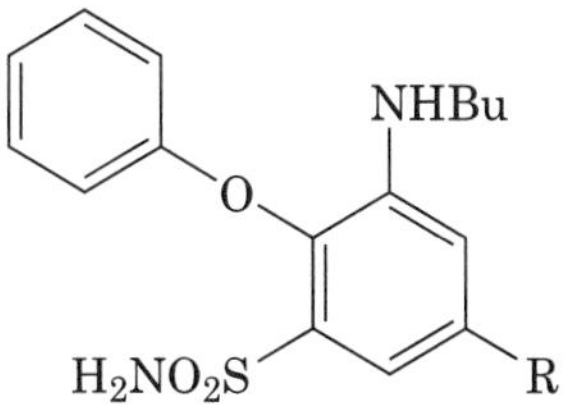

(**169**) R = SO_2H or SO_3H
(**170**) R = $NHCH_2Ph$

170) with retention of diuretic activity. In a further modification, the sulfamoyl group in the 5-position was replaced by a formamido group (**171**) (217); this compound has approx-

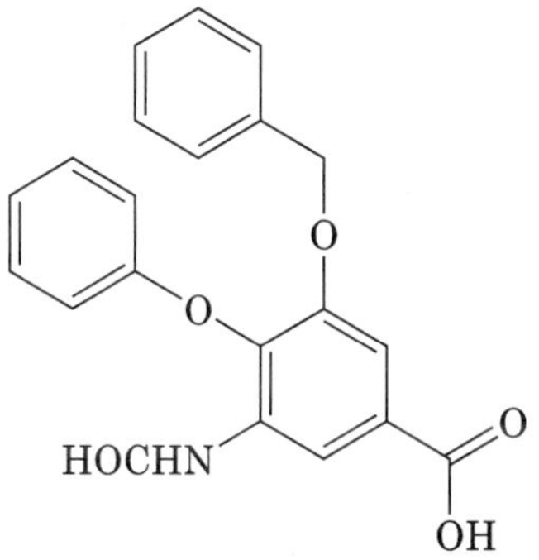

(171)

imately one-tenth the activity of bumetanide. The electrolyte excretion pattern is still similar to that of bumetanide.

Bumetanide is a potent diuretic in dogs after both oral and intravenous administration. It is comparable to furosemide in its type of action and its maximum effect, but when given orally bumetanide is approximately 100 times more active. In dogs the drug is excreted rapidly by glomerular filtration and tubular secretion. No metabolites were detected in dogs (218). A parallel between bumetanide excretion and saluretic action in humans over the total period of response has been shown (219). The drug is a highly potent diuretic in patients with congestive heart failure (220, 221) and in subjects with liver cirrhosis (222). Studies in humans indicate a major site of action in the ascending limb of Henle; a significant phosphaturia induced during the period of maximum diuresis also suggests additional action in the proximal tubule (223, 224). Bumetanide produces a rapid diuretic response, with a pattern of salt and water excretion resembling that of furosemide. At the time of maximal diuresis, 13–23% of the filtered load of sodium is excreted; urinary calcium and magnesium also increase. As with other sulfonamide diuretics, hyperuricemia occurs after prolonged therapy. The bioavailability of bumetanide after oral administration is 72–96% in normal subjects and diuresis usually begins within 30–60 min. Bumetanide is highly bound to plasma protein (94–97%) and thus probably gains access to its site of action by secretion into the proximal tubule.

Probenecid, however, fails to alter bumetanide-induced diuresis in cats and in humans (225, 226). Bumetanide is quickly eliminated by metabolism and urinary excretion from the body and has a plasma half-life of 1.5 h. Several metabolites have been identified after oral administration of ^{14}C-labeled bumetanide in humans (227). All of the metabolites isolated, with exception of the conjugates, involved oxidation of the *n*-butyl side chain. About 80% of bumetanide is excreted from the body through the renal route, with about 50% of the drug excreted unchanged by this route. The major metabolite detected is the 3′-alcohol (**172**). The remaining 20% of the drug is excreted by intestinal elimination. The 2′-alcohol (**173**) is the major metabolite found in the bile and feces.

2.1.8.6 Piretanide. Workers at Hoechst synthesized a number of 3,4-disubstituted

(172)

(173)

5-sulfamoyl benzoic acids similar to bumetanide but using a new synthetic method to incorporate the amine group in the presence of the other functional groups (228). Piretanide (**119**, HOE 118) is the most interesting compound in the series. Micropuncture and clearance studies indicate that the primary site of action is in the thick ascending loop of Henle (229). Tritiated piretanide was prepared and was found to bind to two sites in purified membranes of dog kidney medulla (230). Studies have shown that a 6 mg dose of piretanide provides equipotent diuresis as 40 mg of furosemide or 1 mg of bumetanide. However, compared to furosemide and bumetanide, piretanide causes a lower level of potassium excretion and, like bumetanide and furosemide, piretanide causes an increase in serum uric acid levels. Piretanide is well absorbed in the gastrointestinal tract after oral administration and its bioavailability is greater than 95% in normal subjects and in patients with renal failure (231). The kinetics of absorption were studied in normal volunteers that were immobilized (232). The drug was found to be absorbed directly from the stomach when administered through a gastroscope.

In patients with hypertension, long-term treatment with piretanide was shown to significantly lower blood pressure (233). From animal studies it appears that piretanide has a direct effect on the vascular tissue, but the mechanism is unknown (234). In the spontaneously hypertensive rat, chronic treatment of drug (30 mg/kg) attenuated the pressor response to angiotensin II and phenylephrine (235). Piretanide also lowered blood pressure in the anesthetized dogs without significant cardiac effects (236). At high doses negative chronotropic and inotropic effects were observed. These effects were not observed for furosemide given at 1–3 mg/kg.

2.1.8.7 Azosemide. Azosemide (**121**) is a sulfamoyl diuretic that was developed at Boehringer Mannheim (243). It is about five times more potent than furosemide after i.v. dosing, although it is equipotent after oral administration. Azosemide is poorly absorbed in the gastrointestinal tract and has a bioavailability of about 10% (231). Clearance studies have shown that the main site of action of azosemide is in the loop of Henle and to a lesser extent in the proximal tubule (238). In normal subjects, azosemide has a slower onset of action than that of furosemide but at 4, 8, and 12 h, volume and sodium excretion levels were similar. The compound is extensively metabolized and only about 2% of the drug is excreted unchanged after oral dosing (239). Glucuronide conjugates and 5-(2-amino-4-chloro-5-sulphamoylphenyl)tetrazole have been the identified metabolites (240). Azosemide is stable in solutions of various pH values, ranging from 2 to 13, up to 48 h (241). The drug shows some instability at pH 1. The recommended dose to treat fluid retention is 80–160 mg orally or 15 mg i.v. The compound's poor bioavailability and slower onset of action may limit its use when rapid diuresis is required.

2.1.8.8 Xipamide. Xipamide (**120**, Table 2.12) is a derivative of 5-sulfamoyl salicyclic acid (242); 4-chloro-5-sulfamoylsalicyclic acid itself had previously been reported by Feit and coworkers (211) to be a high ceiling diuretic. Investigation of esters; aliphatic, cycloaliphatic, aromatic, and heterocyclic amides; ureides; and hydrazides of 4-chloro-5-sulfamoylsalicyclic acid showed that 4-chloro-2',6'-dimethyl-5-sulfamoylsalicylanilide is the most active derivative.

Table 2.16   Analogs of Xipamide[a]

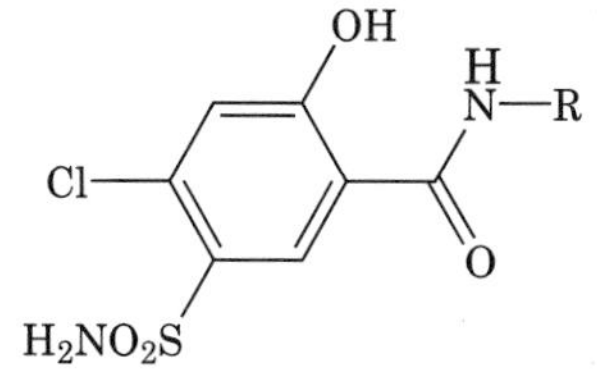

OH

Cl —

H₂NO₂S

$\overset{H}{N}$—R

O

(structure of xipamide core)

No.	R Group	Urine Volume in Rats (mL/kg at 1 mg/kg)	Urine Volume in Rats (mL/kg at 100 mg/kg)
Control		4–5	4–5
(120)	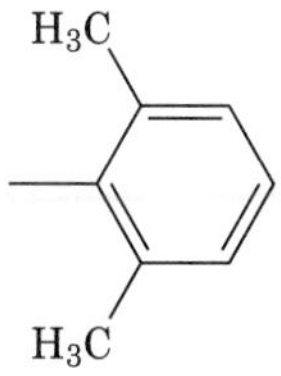 H₃C ... H₃C	22.0	41.6
(175)	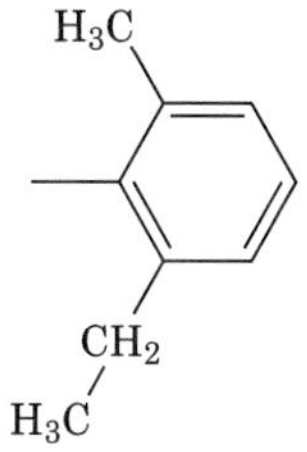 H₃C ... CH₂ H₃C	8.1	16.4
(176)	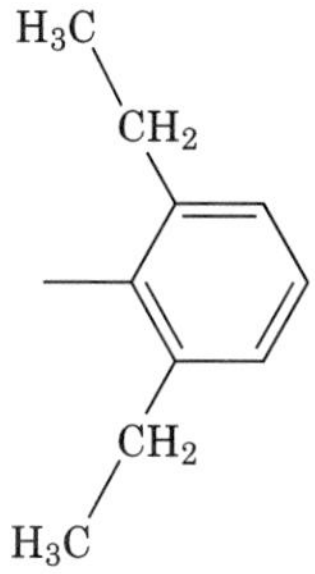 H₃C CH₂ ... CH₂ H₃C	3.2	11.1
(177)	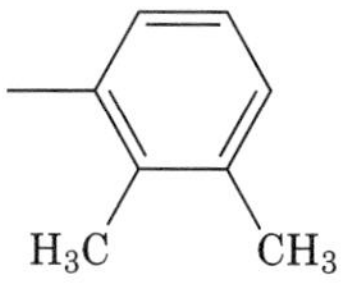 H₃C CH₃	>10[b]	11.4
(178)	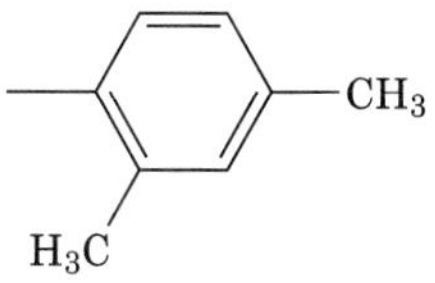 CH₃ H₃C	0.56[b]	13.8

Replacement of the Cl group by Br, F, or CF₃ led to compounds with lower activity. The effects of modification of the anilide group are shown in Table 2.16 (242). Compounds methylated on oxygen and/or nitrogen are less active. Interestingly, the 2,6-dimethylpiperidino- derivative (183, Table 2.16) related to clopamide (86) is inactive (243).

Xipamide is active in rats and dogs after oral or intravenous administration. Applica-

Table 2.16 (*Continued*)

No.	R Group	Urine Volume in Rats (mL/kg at 1 mg/kg)	Urine Volume in Rats (mL/kg at 100 mg/kg)
(179)		0.60^b	31.2
(180)		0.74^b	17.8
(181)		$\sim5^b$	19.6
(182)		1.1	11.5
(183)			0

^{a}From Ref. 242.
^{b}Dose (in mg/kg) that increases 5-h urine volume by 50%.

tion of 0.2 mg/kg i.v. in the dog, accompanied by a continuous infusion of 5% mannitol solution, led to a diuretic effect starting 40 min postinjection. After 2 h, a diuretic effect could still be detected. An antihypertensive effect in spontaneous hypertensive rats was observed after 1 mg/kg p.o (244). Xipamide is a weak carbonic anhydrase inhibitor ($ED_{50} = 1.1 \times 10^{-5}\,M$) and is comparable to sulfanilamide ($ED_{50} = 1.3 \times 10^{-5}\,M$) or hydrochlorothiazide ($ED_{50} = 2.3 \times 10^{-5}\,M$ (Table 2.4) (245).

Xipamide is absorbed quickly from the gastrointestinal tract and has a bioavailability of about 73%. At its therapeutic plasma level, it is 99% bound to plasma proteins (246). Twenty-four hours after oral administration, about 50% of the drug is excreted unchanged and about 25% as its glucuronide conjugate. The plasma radioactivity half-life of ^{35}S-xipamide was 5 h after intravenous administration and between 5.8 and 8.2 h after oral administration (247).

The diuretic profile of xipamide is mixed; it behaves both as a loop diuretic and as a thiazide diuretic. In normal volunteers, doses of 0.5 mg/kg p.o. of xipamide are more effective than 0.5 mg/kg of chlorothalidone. In patients with mild to moderate hypertension, xipamide at 20 to 40 mg/day p.o. is as effective as 1 mg bumetanide or 50 mg of hydrochlorothiazide. Xipamide produces a maximal natriuresis and kaliuresis similar to that of furosemide. The diuretic effects of xipamide last for about 24 h, with the maximum effect occurring during the first 12 h (248). In an evaluation in patients with edema of cardiac origin, xipamide was found to be an effective diuretic at 40 mg/day p.o. Serum potassium levels were slightly lowered (249). In rats, serum potassium depletion could be avoided, and an equilibrated potassium balance was achieved in a 13-day study by combining xipamide with triamterene (250).

Xipamide is generally well tolerated. Mild upper gastrointestinal symptoms are the most frequently reported side effects (2–4%) (251) and fatigue (3–8%) (252). Although ototoxicity has been seen with salicyclic acid and with furosemide, studies with xipamide in guinea pigs did not reveal any ototoxicological properties (253). Xipamide does cause small increases in serum urate and blood urate concentrations and occasional increased glucose and plasma lipid levels in diabetic individuals (247).

2.1.8.9 Triflocin. The discovery of triflocin (184) resulted from a study of derivatives of flufenamic acid for possible anti-inflammatory activity. Compounds incorporating a nicotinic acid moiety unexpectedly exhibited diuretic activity. Triflocin is structurally a novel and highly efficacious diuretic agent, capable of promoting the excretion of as much as 30% of the sodium chloride filtered at the glomerulus. It was effective in the rat, rabbit, guinea pig, dog, and monkey. Triflocin is characterized by

(184)

excellent oral absorption, rapid onset of action, and short duration of effect. The magnitude of diuresis produced by the compound is similar to that seen with furosemide and ethacrynic acid. The renal sites of action are interpreted to be the proximal tubule and the ascending limb of Henle (254, 255). Interestingly, a study of rats and dogs indicated that triflocin has no propensity for evoking hyperglycemia (256). The drug was studied in normal volunteers and found to be a markedly potent natriuretic agent; free water clearance (C_{H_2O}) was inhibited during water diuresis and solute-free water reabsorption (TC_{H_2O}) reduced hydropenia, indicating a major site of action in the ascending limb of Henle. In addition a fall in the glomerular filtration rate of 10–15% was found at doses of 1 g given orally (257). Long-term toxicity studies revealed adverse effects; clinical studies were therefore discontinued (258). In rats, doses of 100 and 200 mg kg^{-1} day^{-1} caused no adverse effects (259). Doses of 2000 and 5000 mg kg^{-1} day^{-1} generally caused death within 1 week.

2.1.8.10 Torasemide. Torasemide (122) is a pyridylsulfonylurea, high ceiling diuretic that is structurally similar to triflocine (184). Its site of action is the Na$^+$/2Cl$^-$/K$^+$ cotransporter located in the loop of Henle. It also blocks chloride channels located on the basolateral side of the thick ascending limb (260). Torasemide is about 8–10 times more potent in dogs and in humans, has a longer duration of action, and causes less potassium excretion than furosemide.

Torasemide is quickly absorbed from the gastrointestinal tract after oral administration and has a bioavailability of about 90%. It is highly bound to plasma protein (>95%) and has a half-life of about 2 h after intravenous dosing and 3 h after oral dosing in humans (261). The compound undergoes extensive metabolism in several species. In rats, less than 1% of the drug is excreted unchanged: most is excreted as a variety of hydroxylated metabolites (262). In humans, only 20% of the unchanged drug is excreted in the urine. Its volume of distribution in humans was determined to be 0.2 L/kg (263). In normal volunteers torasemide was administered orally in doses ranging from 10 to 100 mg (264). At the highest doses (80 and 100 mg) some volunteers complained of knee, calf, and foot cramps. These were generally short in duration. In hypertensive patients, a 2.5–20 mg dose of torasemide is effective. Torasemide was first introduced into the market in 1993 in Germany and Italy by Boehringer Mannheim. It is available in 2.5, 5, 10, and 20 mg tablets and 10 mg/2 mL ampules for injection. Recently, some additional analogs of torasemide were reported where the sulfonyl urea was replaced with other isosteric groups including sulfonyl-thiourea, cyanoguanidine, and 1,1-diaminonitroethylene (265). These analogs had diminished diuretic activity in respect to torasemide.

2.1.8.11 Muzolimine. Workers at Bayer synthesized a series of 1-substituted pyrazol-5-ones. Some of the compounds prepared in this series were disclosed to be highly active diuretics. Muzolimine (Bay g2821, **185**) was selected for further study (266).

(185)

The structure of this compound differs considerably from that of other high ceiling di-

uretics because it contains neither a sulfonamide nor a carboxyl group. It has a pK_a value of 9.2 and is very lipophilic. Clearance studies in dogs indicate that muzolimine does not increase the glomerular filtration rate, but has a saluretic effect similar to that of furosemide, induced by inhibition of tubular reabsorption in the ascending limb of the loop of Henle (267).

Micropuncture studies in rat kidneys showed that muzolimine was effective only when given as a peritubular perfusion and not when administered intraluminally, in contrast to furosemide and bumetanide, which were effective when applied either peritubularly or intraluminally (267). Renal Na^+/K^+-ATPase activity *in vitro* is inhibited only at high concentrations: Mg^{2+}-ATPase activity was not affected (268).

Muzolimine is rapidly absorbed after oral administration and is estimated to have a bioavailability of greater than 90%. The plasma protein binding is 65%, which is lower than many of the other high ceiling diuretics, and this may be the reason that the drug is effective in patients with advanced renal failure (269). Muzolimine also has a half-life of 10 to 20 h. It undergoes extensive metabolism in the liver; its major route of excretion is through the bile (270), with only about 10% of the drug being excreted unchanged.

Preliminary studies with muzolimine in patients showed that the drug is a high ceiling diuretic with an onset of action and a peak diuresis similar to that of furosemide. The duration of action was 6 to 8 h as compared to 3 to 5 h after furosemide; 40 mg of muzolimine was more potent than 40 mg furosemide in all parameters investigated (271). In normal volunteers, the threshold dose was 10 mg and the dose response curve for sodium was practically linear for doses up to 80 mg (272). Acute water diuresis and hydropenic studies carried out in seven normal volunteers suggested that muzolimine acts in the proximal tubule and in the medullary portion of the ascending limb of the loop of Henle (273). In July 1987, muzolimine was withdrawn from the market for toxicological reasons (polyneuropathy).

2.1.8.12 MK 447. Through screening procedures, workers at Merck found that 2-aminomethyl-3,4,6-trichlorophenol (**186**) (274)

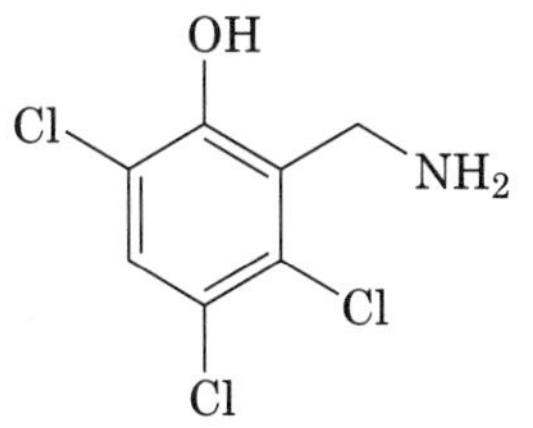

(186)

displayed significant saluretic-diuretic properties. Exploration of structure-activity relationships showed that alkyl substitution, preferably α-branched, in position-4 and halo substitution in position-6 resulted in greatly enhanced activity. Substitution of the nitrogen and oxygen with groups resistant to hydrolysis greatly reduced the saluretic effects of these compounds (275). Also, reorientation of the 2-(aminomethyl) group from the position *ortho* to the phenolic hydroxyl group to the *meta* and *para* positions results in loss of activity (276). Optimal activity was displayed by 2-amino-methyl-4-(1,1-dimethylethyl)-6-iodophenol (MK 447, **187** (277). The 5-aza analog

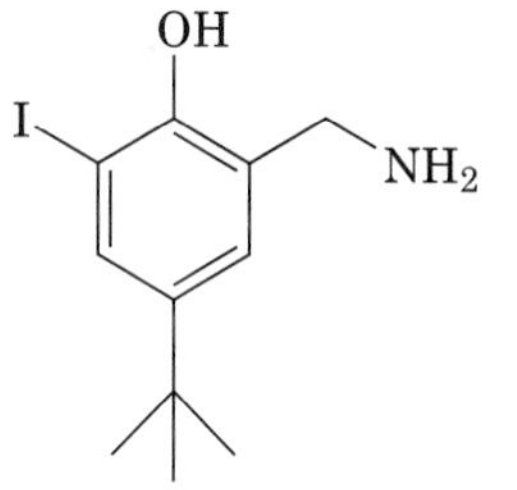

After oral administration, MK 447 is rapidly absorbed from the gastrointestinal tract. It has a plasma half-life in rats and dogs of 1 and 7.5 h, respectively. In humans, the half-life is 4–8 h (281). The compound undergoes extensive metabolism in rats, dogs, and humans. The major metabolite identified in rat and dog urine is the *O*-sulfate conjugate (282). In humans, 17% of the activity of radiolabeled MK 447 is ascribed to the *O*-sulfate conjugate. The major metabolite in human urine was tentatively identified as the *N*-glucuronide. It is believed that the *O*-sulfo derivative is an active metabolite of MK 447 and may be responsible for the salidiuretic activity observed (283, 284). MK 447 also possesses anti-inflammatory activity. It is believed that the drug's anti-inflammatory activity is attributed to its ability to inhibit the endoperoxide PGG_2 (285).

2.1.8.13 Etozolin. During the investigation of a series of 4-thiazolidones, some of which have choleretic properties (286), a number of compounds with high ceiling diuretic activity were found (287). Compound (**188**)

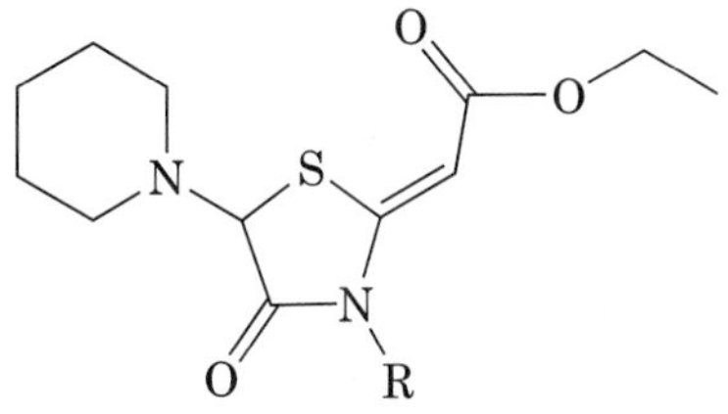

(187)

(188) R = ethyl
(189) R = methyl

displayed similar activity to MK 447 in the rat, but was less active in the dog (278).

The saluretic effects of MK 447 in rats and dogs are generally superior, both qualitatively and quantitatively, to those of earlier high ceiling loop diuretics. MK 447 was more effective than furosemide at 0.1–10 mg/kg p.o. in rats and dogs (279). A study in normal volunteers confirmed the high potency of the compound. Despite copious diuresis and natriuresis, no significant change in the elimination rate of potassium was observed (280). In humans, a 100 mg dose is equipotent to 80 mg of furosemide, although its duration of action is longer.

(piprozoline) is a choleretic compound without diuretic activity. Compound (**189**) (etozolin) is a highly active diuretic with weak choleretic properties. Minor deviations from structure (**189**) lead to a loss of diuretic activity. The different pharmacodynamic properties of (**188**) and (**189**) could not be traced to thermodynamic factors but rather must be related to closely defined receptor interactions (286).

Long-term toxicity studies in rats and dogs have shown that etozolin is well tolerated, and that it has a wide margin of safety (288). Studies in rats and dogs indicate that the compound is a potent saluretic agent, with a relatively slow onset of action and prolonged activity. The maximal diuretic effect of etozolin lies between that of the thiazides and furosemide. Antihypertensive effects occur in the spontaneous hypertensive rat, DOCA, and Goldblatt rats. Etozolin does not appear to influence glucose tolerance in rats and dogs; these results are of particular interest because the tests were carried out in animals that had been treated with high doses of drug for 18 and 12 months, respectively (289).

Clearance and micropuncture studies have shown that an initial dose of 50 mg/kg i.v. followed by 50 mg kg^{-1} h^{-1} i.v. results in a markedly increased urinary flow and sodium excretion, combined with a decreased glomerular filtration rate. Reabsorption in the proximal tubule is not affected significantly; however, fluid and electrolyte reabsorption in the loop of Henle is definitely decreased. Although etozolin differs chemically from furosemide and ethacrynic acid, it appears to share the same site of action in the nephron (290).

Absorption and metabolic studies with ^{14}C etozolin in rats, dogs, and humans indicate that at least 90% is absorbed after oral administration in humans. Etozolin is approximately 45% bound to plasma proteins. In the rat, blood levels could be described with a two-compartment body model, the absorption half-life was 0.6 h, and the elimination half-life was determined to be approximately 6 h. In humans the elimination half-life was 8.5 h; the blood levels followed with high probability a one-compartment body model (291).

The main metabolite of etozolin is the free acid ozolinone (**190**) and its glucuronide.

(190)

Other metabolites have also been detected (292, 293). In rats, etozolin is a more potent diuretic than hydrochlorothiazide, although

in dogs it is less potent. In subjects with normal renal function, a 800 mg dose of the drug increased the excretion of water, chlorine, magnesium, potassium, and sodium without altering creatine clearance (294). In normal volunteers, a dose of 400 mg of etozolin was equipotent to 75 mg of a thiazide diuretic; 1200 mg was 2.8 times more effective than the 75 mg of the thiazide. Diuresis starts within 1–2 h after dosing, reaches a peak after 2–4 h, and then gradually decreases over the next 6 h (295). Because of the long-lasting effect of etozolin, the compound would seem indicated for the treatment of cardiac and renal edema as well as for the treatment of hypertension (294). In hypertensive patients after a period of 2 weeks, treatment with 400 mg of etozolin daily significantly reduces systolic and diastolic blood pressure. The drug was introduced in 1977 as Elkpain.

2.1.8.14 Ozolinone. Ozolinone (**190**), as stated above, is the major metabolite of etozoline, which is formed by enzymatic cleavage of the ethyl ester. It is reported to be more potent and less toxic than etozoline (296). Resolution of the enantiomers and examination of their biological activity revealed that the (−)-enantiomer possesses the diuretic activity (298). The (+)-enantiomer possesses little diuretic activity and actually antagonized furosemide activity (295). Ozolinone has a half-life of 6–10 h in humans (299) and has a plasma protein binding of 35%.

2.1.8.15 Pharmacology of High Ceiling Diuretics. It is currently believed that the renal site of action for these diuretics is the Na$^+$/K$^+$/2Cl$^-$ cotransporter located in the thick ascending loop of Henle. By binding to the chloride site of the cotransporter, these drugs inhibit the reabsorption of sodium, thus promoting their diuretic action (300). The carboxyl group common to many of these diuretics is essential for the binding activity, and it has only been successfully replaced by a sulfonamide. Most of these drugs are readily absorbed in the gastrointestinal tract. For example, furosemide and bumetanide have bioavailabilities of 65 and 100%, respectively. Generally these compounds are secreted from the blood to the urine through the organic acid transport system in the proximal tubule and travel through the renal tubule to their more distal site of action. Increased potassium excretion and the elevation of plasma uric acid levels, as was observed with the thiazides, are also seen with the loop diuretics. These diuretics also increase calcium and magnesium excretion. The calciuric action of these agents has led to their use in symptomatic hypercalcemia (301). Many of the hypersensitivity and metabolic disorders seen with the thiazides are also seen with loop diuretics. The development of transient or permanent deafness is a serious but rare complication observed with this class of agents. It is believed to arise from the changes in electrolyte composition of the endolymph (302). It usually occurs when blood levels of these drugs are very high.

2.1.9 Steroidal Aldosterone Antagonist. Spironolactone (**191**) is the most extensively

(191)

studied aldosterone antagonist. It promotes diuresis by competing with aldosterone at the receptor sites responsible for sodium ion reabsorption and best clinical results have been obtained from those patients suffering from cirrhosis and nephrotic syndrome, in whom the aldosterone secretion rate is very high. After administration for several weeks to hypertensive patients, the compound exhibits a modest antihypertensive effect; however, in normotensive subjects no reduction in blood pressure is seen.

Most recently, spironolactone (**191**) has been the subject of clinical investigations in heart failure (HF) (303). Indeed, the Randomized Aldactone Evaluation Study (RALES)

trial has now vastly heightened interest within this class of compound (304). The study looked at the effect of adding aldactone (spironolactone, **191**) as a potential therapy for reducing death in patients with heart failure. It also defied current thinking, in that spironolactone should not be administered in conjunction with an ACE inhibitor because of the possibility of hyperkalemia and the misconception that, by using an ACE inhibitor, aldosterone will be as effectively blocked as angiotensin II. Within the trial, investigators compared a standard treatment regimen of an ACE inhibitor and a diuretic, with or without digoxin added to this regimen, plus (**191**) or placebo in patients suffering with severe heart failure. The RALES study, conducted in 15 countries, was a randomized, double-blind, placebo-controlled trial in which 1663 patients with systolic left ventricular dysfunction were enrolled; these patients were classified as either Class III or Class IV, a classification for the most severe of cases as defined by the New York Heart Association (NYHA). Although originally scheduled to conclude in December 1999, the trial was halted 18 months early, given that the results were statistically and clinically significant and failure to terminate the trial would have been unethical.

Among the 1663 patients there were 386 deaths (46%, $n = 841$) in the placebo group and 284 deaths (35%, $n = 822$) in the spironolactone (**191**) group; this figure represented a 30% decrease in mortality. Also observed during the trial were 336 placebo-treated and 260 spironolactone-treated patients who had at least one nonfatal hospitalization, representing 753 hospitalizations for the placebo group and 515 for the spironolactone group. This represents a 30% decrease in nonfatal cardiac hospitalizations. Also of significance within the study were two observations of differences between the placebo and spironolactone-treated groups. It should be noted that within both groups the average blood pressure levels during the course of the study were normal and unchanged and that the incidence of hyperkalemia was no different, as defined by potassium ion concentration $[K^+] \gg 6$ meq/L. What did differ was the mean plasma $[K^+]$, which was 0.2–0.3 meq/L higher in the (**191**)-treated group as compared to the placebo and there was a 10% incidence of gynecomastia within the spironolactone-treated group compared to 1.5% in the placebo group. Thus, given the widespread use of potassium supplementation and a difficult explanation by the antiandrogen effect, it is difficult to understand the exact benefit that spironolactone had in this trial. However, it is clear that these findings suggest that the gold-standard treatment for severe heart failure should now include an aldosterone receptor antagonist. The RALES trial also helped to confirm that aldosterone plays in important role in the pathophysiology of heart failure. It also directs research and development strategies toward a more effective treatment option based on aldosterone blockade.

Consequently, within this context and especially with respect to the side-effect profile of spironolactone (191), the equally efficacious and far more selective compounds (192–194) can now be considered as potential cardioprotectant and antihypertensive therapies. Cur-

(194)

rently, eplerenone (194) appears to be the desired specific antimineralocorticoid that is required (305). It is currently in Phase III, having been licensed from Ciba-Geigy (now Novartis) by G. D. Searle. The efficacy and tolerability of eplerenone (194) has been evaluated in a clinical setting involving 417 patients suffering mild to moderate hypertension, in which it was found to have a profile similar to that of spironolactone (191) (306). In a related dose study involving 321 patients with heart failure (NYHA, II-IV), (194) was compared again with (191) in conjunction with standard therapy (ACE inhibitor, diuretic, and/or dixogin) (307). It was found that blood natriuretic peptide decreased significantly with either active treatment after 12 weeks and the urinary aldosterone and renin levels increased compared to placebo in patients receiving doses of 50 mg/day or higher of (194). Interestingly, the incidence of hyperkalemia was significantly higher with 100 mg/day eplerenone when compared to spironolactone (12.0 versus 8.7%); however, the testosterone in male patients increased more with spironolactone than with eplerenone. The latter effect was probably attributable to a feedback mechanism in response to blockade of the androgen receptors by (191).

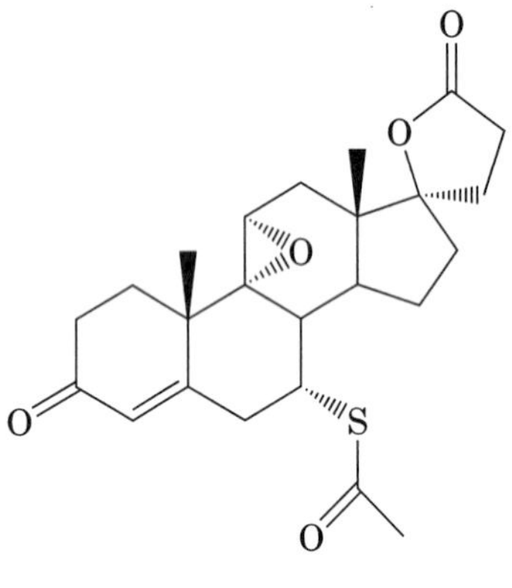

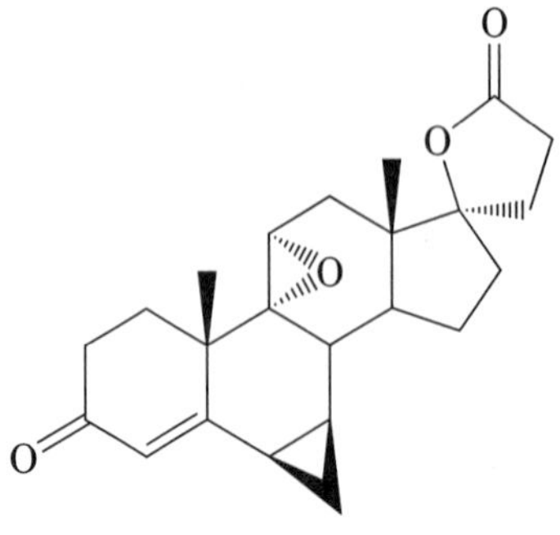

Thus, although the use of antialdosterone agents would now appear to be part of good clinical practice for heart failure, it may well transpire that its most widespread use will be as a cardioprotectant for patients who have hypertension. Expectations are that eplerenone (194) will be marketed in 2002.

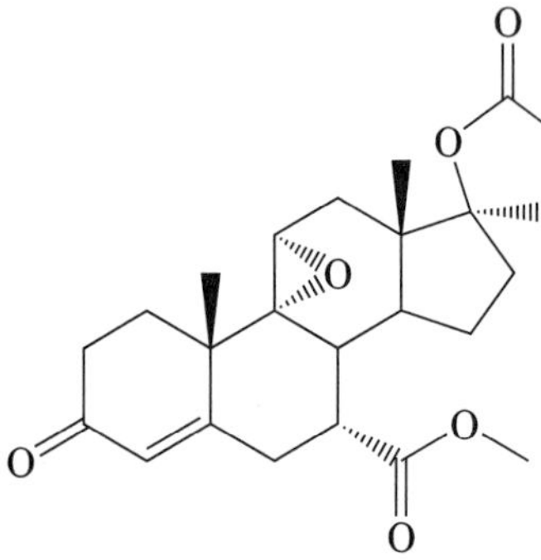

2.1.9.1 Pharmacology. The adrenal cortex is responsible for the biosynthesis of a number

(192)

(193)

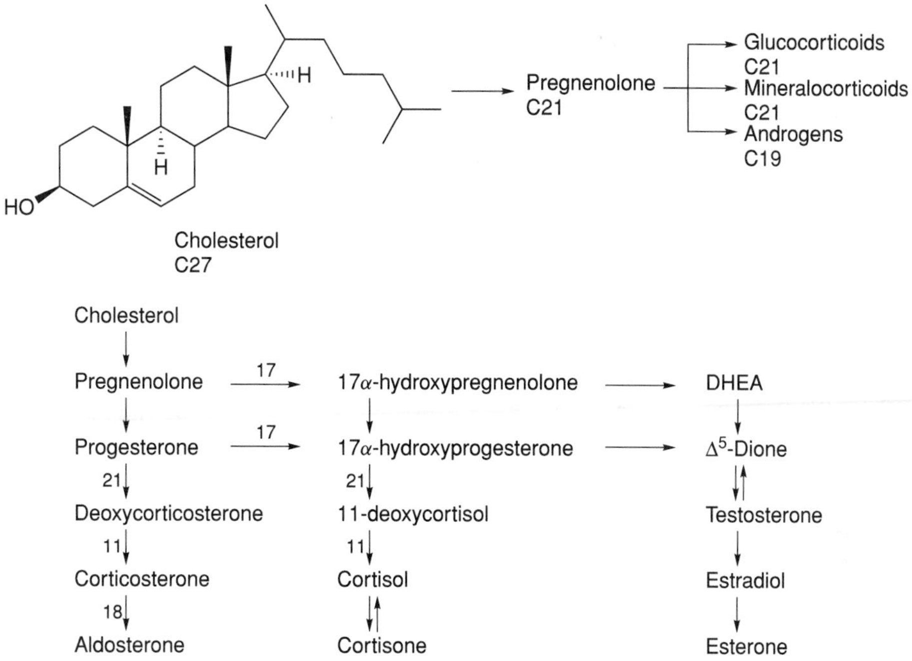

Figure 2.9. Pathway of adrenal steroidogenesis. $11 = 11\beta$-hydroxylase; $17 = 17\alpha$-hydroxylase; $18 = 18$-hydroxylase; $21 = 21\alpha$-hydroxylase.

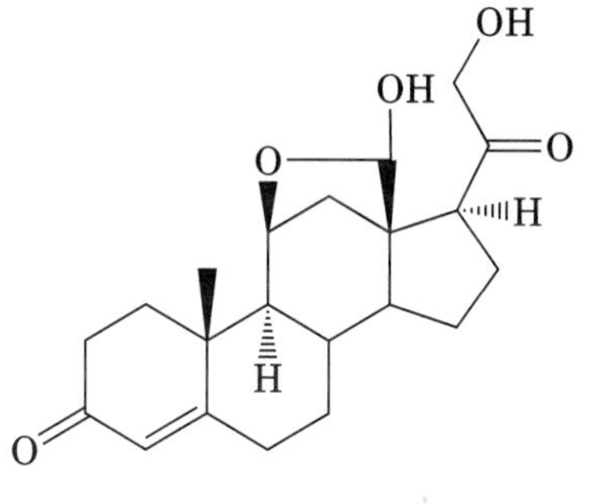

(195)

of important steroids, using cholesterol as its feedstock and cytochrome P450 hydroxylases as catalysts (Fig. 2.9). The principal adrenocortical products are aldosterone (**195**), dehydroepiandrosterone sulfate (DHEA), and cortisol, each arising specifically from cells in the zona glomerulosa, zona fasciculata, and zona reticularis, respectively. The overproduction of aldosterone, deoxycorticosterone (DOC; a precursor to aldosterone in its biosynthetic pathway), or cortisol all have the potential to cause hypertension. Aldosterone and DOC are mineralocorticoids and produce their effects

through sodium and water retention, whereas cortisol (a glucocorticoid) is thought to impart its effects through its incomplete metabolism within target tissues. The aldosterone content in the adrenal gland is 1–2 μg from which it is secreted at a rate of 70–250 μg/day, yielding plasma levels between 5 and 100 pg/mL, indicating that the adrenal does not store the hormone but is capable of rapidly synthesizing it when needed. Greater than 85% of this hormone is metabolized by first pass through the liver; thus, its degradation is intrinsically linked to hepatic blood flow. Reduction in hepatic blood flow is very possible in heart failure, which would then generate a vicious circle of elevated levels of aldosterone, which increases the extracellular volume through sodium retention and potassium excretion, which itself then leads to a decrease in cardiac output. Recent evidence has been amassed that indicates a far greater role for aldosterone in heart failure than previously indicated (308–310). It would now appear that a direct link has been established between increased plasma levels of aldosterone and mortality.

These data also indicate that the detrimental effects of aldosterone in HF can be attributed not only to the increased load on the heart by way of sodium retention but also: (1) hypokalemia and hypomagnesemia with concomitant promotion of arrythmias; and (2) increased sympathetic tone arising from baroreceptor desensitization, which blocks neuronal norepinephrine reuptake and increases myocardial toxicity from catecholamines and myocardial fibrosis. Thus, aldosterone blockade appears to be logical in the treatment of heart failure and potentially hypertension, thus serving as a cardioprotectant strategy.

Aldosterone binds to a cytoplasmic receptor from the basolateral side located in the principal cells of the collecting tubule. Translocation of this hormone-receptor complex to the nucleus leads to the generation of specific transport proteins. These proteins then directly or indirectly increase reabsorption of sodium and the excretion of potassium (311).

The renin-angiotensin-aldosterone system (RAAS) is an important regulatory system for the modulation of arterial blood pressure together with fluid and electrolyte homeostasis. A reduction in renal blood flow stimulates renin production and excretion from the juxtaglomerular cells of the kidney and into the systemic circulation. This enzyme converts angiotensinogen to angiotensin I; thereafter angiotensin-converting enzyme (ACE) converts this into angiotensin II. Angiotensin II is a potent vasoconstictor that also stimulates the production of aldosterone. Thus, with the use of ACE inhibitors it was thought that aldosterone production would be modulated; however, it would now appear that there is an escape phenomenon possibly leading to increased plasma levels of the hormone rather than the anticipated decrease (312). Hence, blockade of aldosterone and its action through antagonism at the mineralocorticoid receptor (MR) or through the inhibition of its biosynthesis (see Section 2.1.10) is necessary to reduce elevated levels of the hormone.

2.1.9.2 History and Structure-Activity Relationship. During the late 1950s Cella et al. reported the synthesis and structure-activity relationships in a series of steroidal-spirofused lactones that possessed aldosterone antagonist activity (313–317). This activity was as-

cribed to the compounds, given that they had no effect in adrenalectomized animals unless aldosterone or another mineralocorticoid was administered before the spirolactone (314), and also because the spirolactone produced the same effect as impaired aldosterone synthesis (318). The first compound of interest was 3-(3-oxo-17β-hydroxy-4-androsten-17-α-yl)-propanoic acid lactone (**196**), which when

(196)

administered subcutaneously showed the desired aldosterone antagonistic effect. Subsequent studies established the importance for both the five-membered spirolactone and the 3-keto-Δ^4 α,β unsaturated enone system within the A-ring. Interestingly, the diastereoisomer possessing the opposite configuration of the spirolactone at C17 was devoid of all activity (318), and the 19-normethyl derivative (**197**) was more active than (**196**) in rats.

(197)

During experiments conducted with compound (**197**), when it had been administered to rats maintained on a low sodium diet, it was found that compensation for the renal loss of sodium was mediated through upregulation of aldosterone secretion (319). Likewise, sodium diuresis in the clinical environment was accompanied by increased aldosterone excretion in the urine (320). The 19-normethyl derivative (197) has been successfully employed in primary aldosteronism (321), nephrotic edema (322, 323), and hepatic cirrhosis (323, 324). Patients with cardiac failure did not respond well. The effect of (197) is determined by the degree to which sodium reabsorption is controlled by aldosterone; consequently, it is not well tolerated in cases of untreated Addison's disease (325) and normal subjects on a low sodium diet (326). Because the compound has a direct effect on the kidney by way of inhibition of aldosterone, sodium and chloride ion excretions increase and potassium, hydrogen, and ammonium ion excretions decrease, representing an electroyte excretion pattern different from that of most other diuretics that increase potassium ion loss and hence lead to hypokalemia. Both (196) and (197), when administered parenterally, showed better activity when compared to an oral regimen, indicating an absorption problem within this class.

Although success in increasing bioavailability was encountered through incorporation of unsaturation at positions 1, 6, or both simultaneously (198–200) (316), an enhance-

(199)

(200)

(201)

(198)

ment was also seen when a thioacetyl group was incorporated into the 1α orientation (201). This compound itself was then superseded through incorporation of the same thioacetyl moiety, however, into the 7 position of the steroid nucleus, interestingly again with the α-orientation. This latter compound, 3-(3-oxo-7α-acetylthio-17β-hydroxy-4-androsten-17α-yl)-propanoic acid lactone (191; spirono-

lactone; aldactone), has undergone extensive clinical trials. Inversion of the 7-thioacetyl group into the β-configuration reduced both oral and parenteral activity by 90% (327).

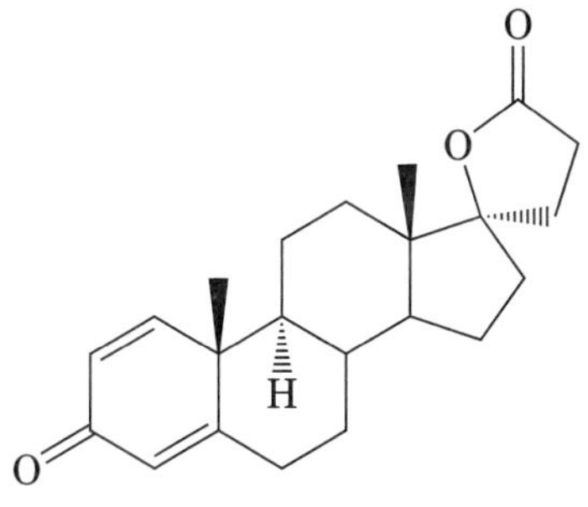

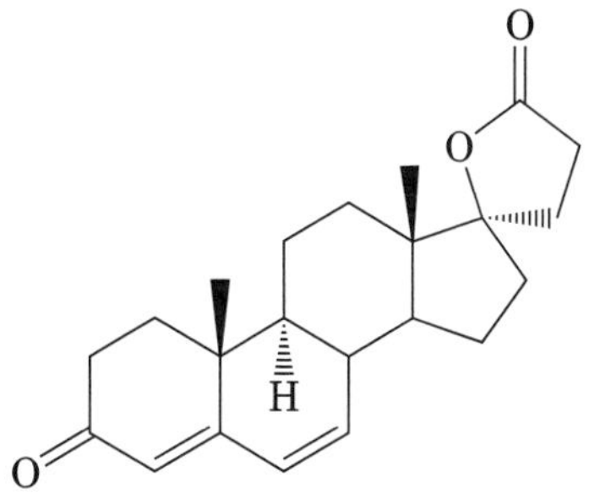

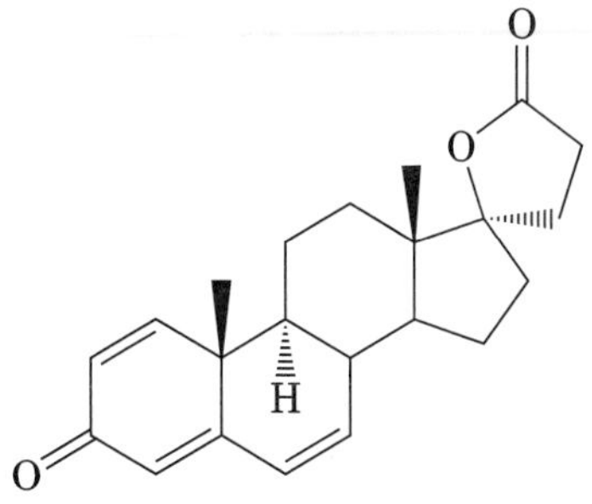

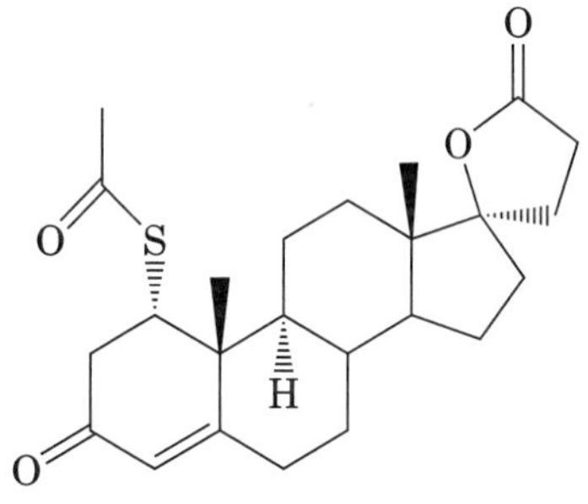

Introductions of methyl groups at the 2, 4, 6, 7, and 16 positions as well as a keto- or hydroxyl incorporation at 11 position or a fluoro at the 9 position did not improve the compound profile (327, 328). The spirolactams (202–203) have also been synthesized (329, 330) but have little aldosterone antagonist activity (331).

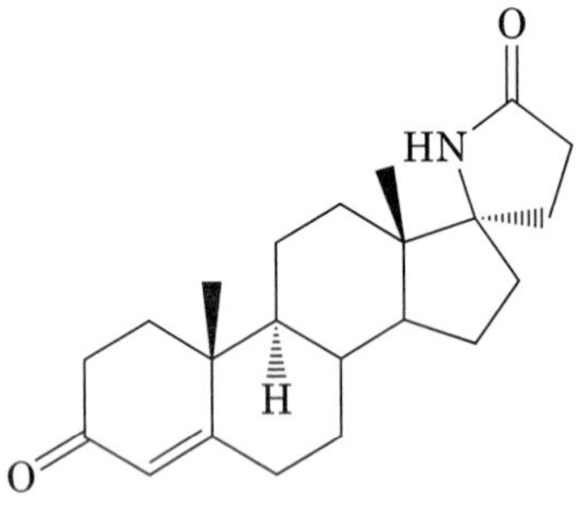

(202)

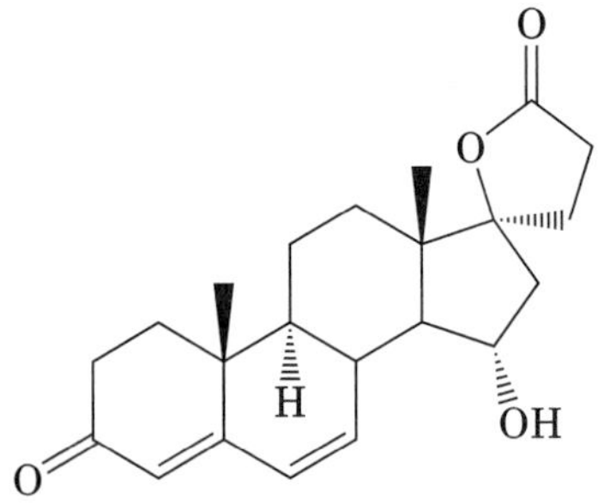

(205)

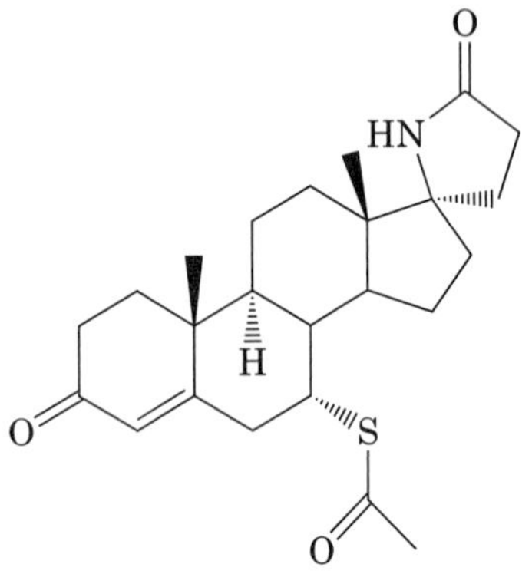

(203)

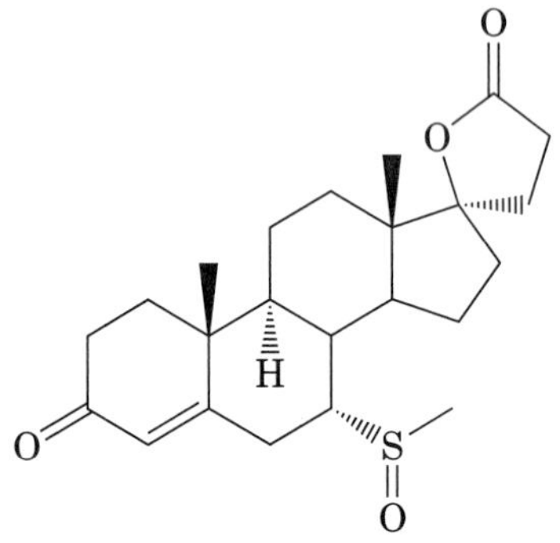

(206)

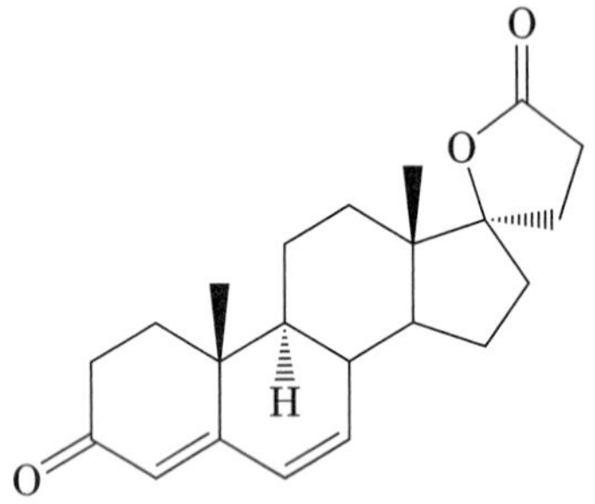

(204)

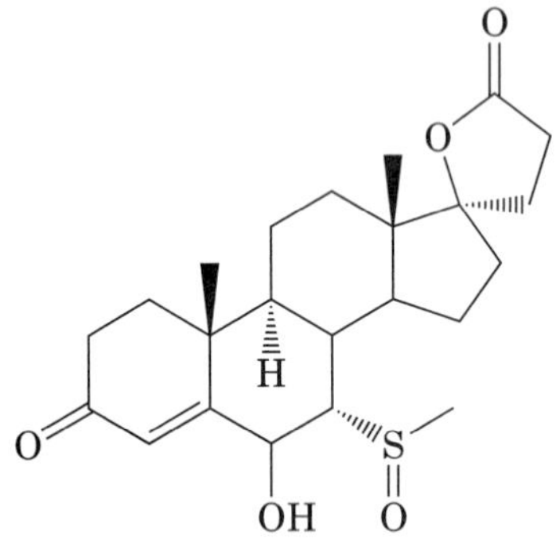

(207)

The metabolism of spironolactone has been studied in detail (332–334); several metabolites have been isolated from the urine of normal subjects, indicating that the compound is subject to elimination, oxidation, and rearrangements (204–208). After oral administration, about 70% of the drug is absorbed (335); however, extensive first-pass metabolism and enterohepatic circulation greatly reduces circulating levels. No unmetabolized compound is detected in the urine and it is also known to induce CYP450s within the liver, thus possibly altering the metabolic profile of coadministered drugs. The compound is also greatly plasma protein bound (95%).

In further efforts (336) to gain water solubility the γ-hydroxy potassium carboxylate was prepared (easily generated upon saponification of 199) and it was found to be approxi-

Table 2.17 Steroidal Aldosterone Antagonists

No.	Generic Name	Trade Name	Structure	Reference
(191)	Spironolactone	Aldactone		316
(194)	Epoxymexrenone	Eplerenone		305
(209)	Potassium canrenoate	Soldactone		336
(210)	Potassium prorenoate			337

mately equipotent with spironolactone (191). It was equally efficacious when dosed orally or parenterally; however, it was found to be ineffective in the absence of mineralocorticoids (MC). Thus, potassium canrenoate (209; soldactone; Table 2.17) is a specific antagonist of mineralocorticoids with pharmacodynamic properties similar to those of spironolactone.

Table 2.17   (*Continued*)

No.	Generic Name	Trade Name	Structure	Reference
(**211**)	Potassium mexrenoate			339

(**208**)

Cyclopropanation of the Δ^6 double bond of (**209**) generated potassium prorenoate (**210**), a water-soluble steroid with the ability to antagonize the sodium-retaining and, when apparent, potassium-dissipating effects of the mineralocorticoids (337). Within the aldosterone-treated dog model, the latter compound was three times more potent than (**191**); however, it was similarly relatively inactive at the renal level in the adrenalectomized rat without MC replacement. Further investigations showed that the compound possessed no more than 2% of the natiuretic activity of hydrochlorothiazide in the intact animal. Clearance studies within dogs showed a direct renal tubular interaction between (**210**) and aldosterone (337). The relative potency of (**210**) and (**191**) was compared in a double-blind, balance, cross-over study in normal subjects (338). The compound (**210**), as related to elevation of urinary log[sodium/potassium ion] ratio and as related to potassium retention

was significantly higher than that of spironolactone. Prorenoate (**210**) also produced greater natriuresis but this effect was not significant. Interestingly, the *in vivo* experiment converts the hydroxy carboxylic acid salt to the spironolactone (338).

As previously mentioned, the 7α-thioacetyl derivatives were found to possess increased bioavailability, and within these latter open lactone potassium salt derivatives, enhanced activity was now seen through the inclusion of a 7α-carbalkoxy group. The result was potassium mexrenoate (**211**) (339), a water-soluble compound whose oral activity was better than that of its spirolactone variant (Table 2.17).

Two further cyclopropyl-containing derivatives within these series have been highlighted as potent analogs that possess lower relative binding affinities for the progesterone and androgen receptors (PR and AR) compared to that of spironolactone. This desired progression toward more selective compounds is the result of the side-effect profile of spironolactone, which is a factor that also limits its utility. These two derivatives both contain the β-configured cyclopropane of the Δ^{15} double bond. Mespirenone (**212**) is an analog of (**200**) and possesses three times more potency within the adrenolactomized rat treated with glucocorticoid or *d*-aldosterone (340) and six times more potency within normal volunteers (341), and ZK 91587 (**213**) is twice as potent as spironolactone in the former model (342).

Modification of the steroid nucleus has also been attempted, and the resulting compounds examined for mineralocorticoid-blocking ac-

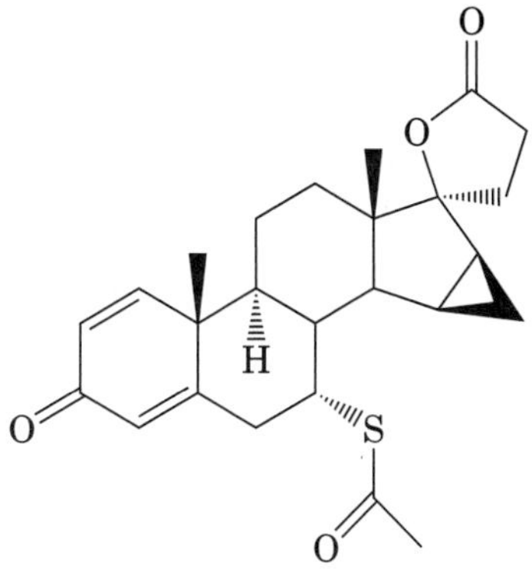

(212)

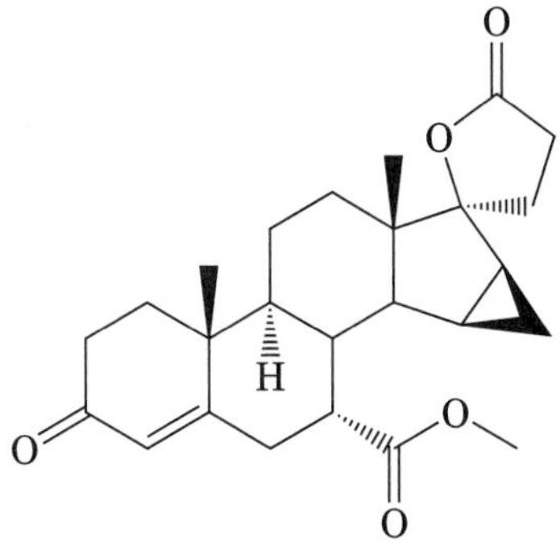

(213)

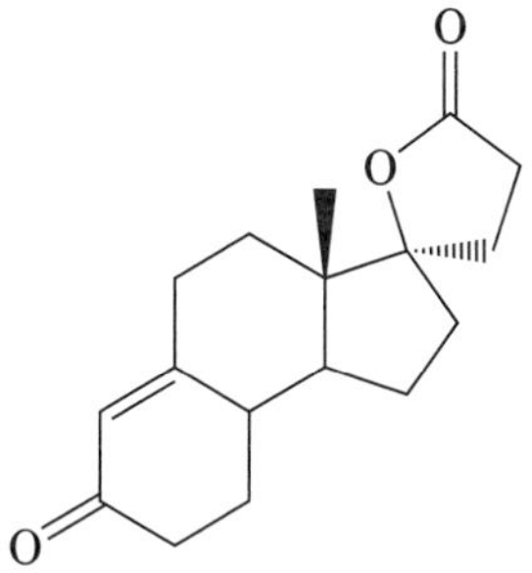

(214)

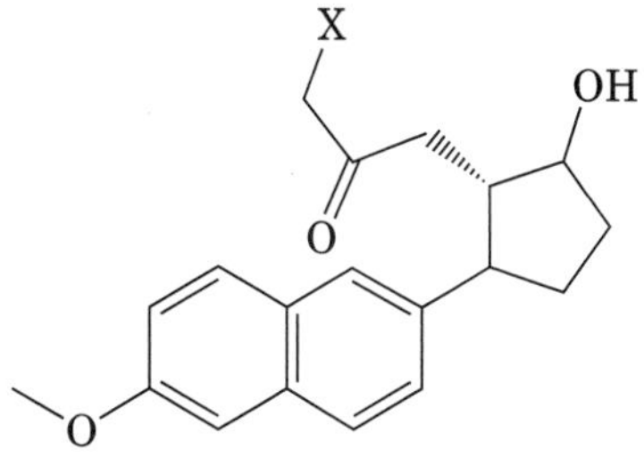

(215) X: OH
      Cl
      F

Although cyclopropanation has thus far been seen to be advantageous, within a series of progesterone derivatives it was seen to abolish the desired antialdosterone activity (344). Progestereone blocks aldosterone at high doses, and this effect was further enhanced through the introduction of oxygenation within the steroid D-ring, specifically at C15; indeed, insertion of unsaturation at Δ^1 or Δ^6 also enhanced activity, as exemplified by (216). However, cyclopropanation of the C6-C7 double bond yielded (217), a compound of greatly reduced antialdosterone activity. Interestingly, and not analogously to the progesterone series, oxygenation of spironolactone

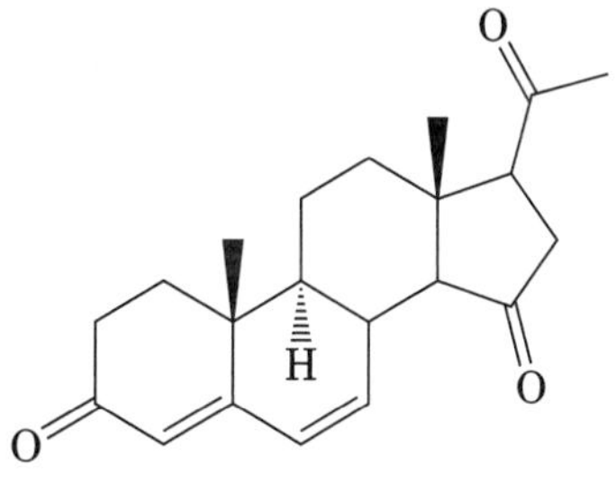

(216)

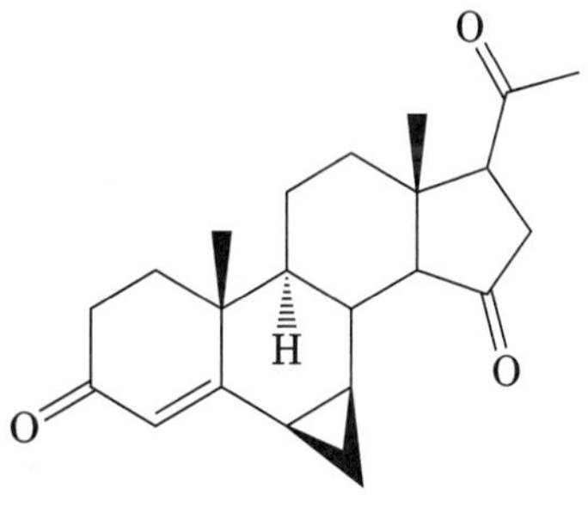

(217)

tivities. An analog of (196) in which the A-ring has been removed (214) retained the sodium-excretive properties to a degree (343). During more extensive studies a series of naphthylcyclopentanol ketones were prepared. The α-hydroxy-ketone (215) showed the best properties, whereas the chloro- and fluoro-derivatives were less effective.

within the D-ring reduced activity to approximately 14% of that of the parent (345); however, this was at C16 and not C15 as before (**218**).

(**218**)

Workers at Ciba-Geigy also introduced oxygenation within the steroid nucleus, although in an alternative fashion, through epoxide formation (346). They prepared a series of 9,11-α-epoxy steroids, which generated as a result of this modification derivatives of spironolactone (**192**), prorenone (**193**), and mexrenone (**194**). In general the epoxide functionality had only a small effect on the binding of these compounds to the MR, and *in vivo* at a dose of 3 mg/kg all these derivatives were twice as potent as (**191**). However, most interestingly, these compounds were shown to have decreased affinity for the PRs and ARs, now exhibiting selectivities between 10- and 500-fold (347). Indeed, it is the binding to these latter receptors that accounts for the side effects of spironolactone (**191**), evidenced by menstrual irregularities in women and gynecomastia in men (348).

Equally interestingly, when comparing eplerenone, epoxymexrenone (**194**) to spironolactone (**191**), and their respective *in vitro* binding affinities to their *in vivo* potencies, there is a marked difference; (**194**) has 5% the affinity for the MR and 100% potency *in vivo* when compared to (**191**) (349). A contributing factor to this is undoubtedly the fact that (**194**) is minimally plasma-protein bound as compared to (**191**), which is 95% plasma-protein bound. However, potentially compounding this is the metabolic fate of eplerenone: it has not been well studied in either rat or human and consequently these data could aid the explanation of these differences seen *in vivo*.

2.1.10 Aldosterone Biosynthesis Inhibitors. The alternative approach to an antialdosterone diuretic rather than through blockade of aldosterone and its action through antagonism at the mineralocorticoid receptor (MR) is to inhibit the biosynthesis of aldosterone within the adrenal zona glomerulosa. The previously described spironolactones in addition to their effects on the aldosterone receptor have also been shown to have an effect directly on aldosterone biosynthesis (350). This observation was made in the adrenal tissues of sodium-depleted rats at 10^{-4}–10^{-5} M concentrations. It is believed that spironolactone, carenone, and potassium carenoate inhibit the mitochondrial 11β- and 18-hydroxylase activity and hence prevent aldosterone synthesis (351). Spironolactone may even inhibit 21-hydroxylase (352); however, it is believed that the major site of action of these steroids is the aldosterone receptor because the systemic concentrations needed to affect biosynthesis are greatly elevated when compared to those required for the receptor antagonist activity.

There are few small molecules that exhibit the inhibition of aldosterone through inhibition of its biosynthetic pathway, one such compound of which is (**219**). Metyrapone (**219**) has

(**219**)

undergone extensive biological profiling and clinical trials (353). In moderate doses it blocks 11β-hydroxylation within the steroid nucleus, thus inhibiting the biosynthesis and secretion of cortisol, corticosterone, and aldosterone. However, the consequential increased secretion of 11-deoxycorticosterone, a potent salt-retaining hormone, negates the effects of reduced aldosterone secretion. Another synthetic small molecule that can inhibit the biosynthesis of aldosterone is CGS16949A (**220**) (354). This interesting property was shown during profiling studies carried out with the aromatase inhibitor. The

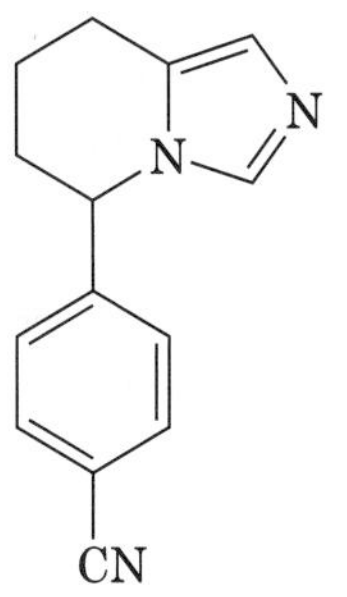

(220)

inhibitory effect is specific within the aldosterone biosynthetic pathway, in that the compound inhibits the second hydroxylation step (CMO II) of C18, the angular methyl group of corticosterone, and hence the subsequent production of aldosterone. Thus when compared to (219) and its inhibitory effects attributed to 11β-hydroxylase inhibition, this latter approach would appear to be more specific and selective. Closely related to this compound and implicitly its generic structure is a series of compounds (exemplified by 221) covered in

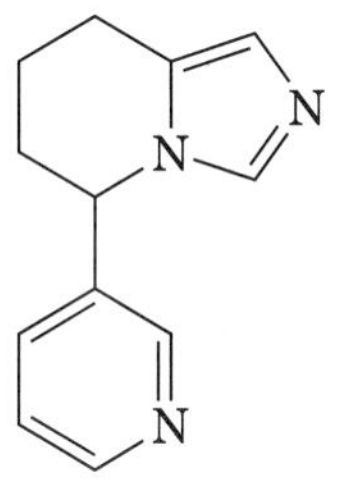

(221)

a patent filed by Yamanouchi Pharmaceuticals (355), these bicyclic imidazoles are also claimed as aldosterone biosynthesis inhibitors affecting the identical cytochrome P450 enzyme, as previously detailed.

2.1.11 Cyclic Polynitrogen Compounds
2.1.11.1 Xanthines. The diuretic actions of alkylxanthines, such as caffeine (222) and theophyline (223), have been known for more than a century. They have been of limited clinical utility because of their low potency, development of tolerance after repeated administration, and side effects such as psychomotor effects and cardiac stimulation. It has been

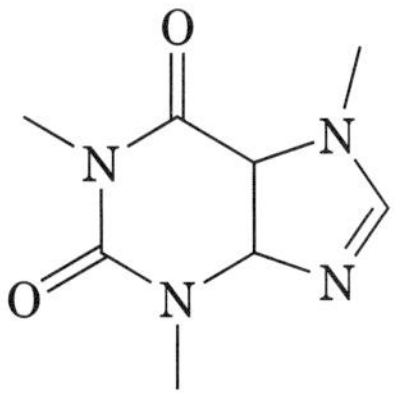

(222)

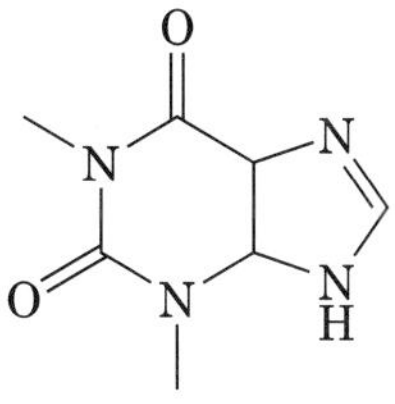

(223)

proposed that the pharmacological basis for their mechanism of action is adenosine receptor antagonism (356). Adenosine produces antidiuretic and antinatriuretic responses in several species. These effects can be competitively antagonized by theophylline (357). There are four adenosine receptor subtypes, A_1, A_{2A}, A_{2B}, and A_3 (358). It is believed that the renal actions of adenosine are the result of its stimulation of the adenosine A_1 receptor (359). It has been shown that compounds that antagonize the A_1 receptor exhibit diuretic effects (360, 361). Compounds that are adenosine A_2 antagonists do not show any diuretic or natriuretic properties (362).

A series of 8-substituted 1,3-dipropylxanthines were reported by Suzuki and coworkers (363). Many of these compounds were potent adenosine A_1 receptor ligands with diuretic activity. One of the most potent analogs was 8-(dicyclopropylmethyl)-1,3-dipropylxanthine (224; K_i, A_1 = 6.4 nM). This compound also

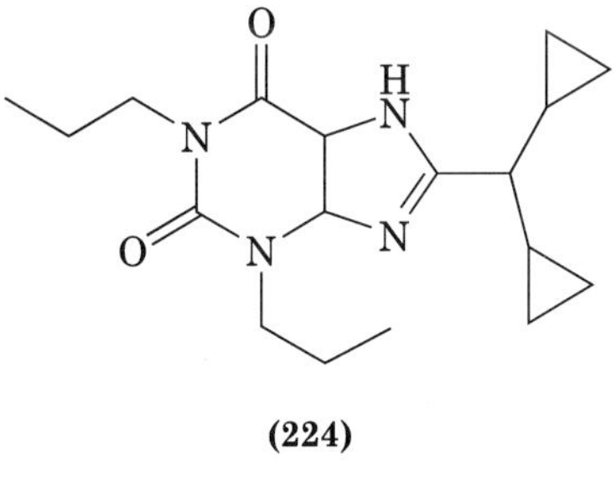

(224)

increased urine volume and sodium excretion in the rat after oral administration.

8-Cyclopentyl-1,3-dipropylxanthine (**225**) was also reported to be a potent selective adenosine A_1 receptor antagonist with diuretic action (363). At 0.1 mg/kg, i.v., (**225**) signifi-

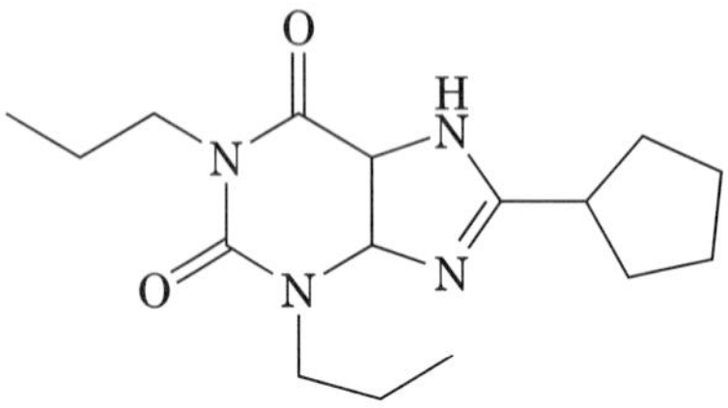

(**225**)

cantly increased urine volume and sodium excretion in the rat with no significant change in potassium excretion (364). The tubular site of action is thought to be in the proximal tubule.

8-(Noradamantan-3-yl)-1,3-dipropylxanthine (KW-3902, **226**) has been described as

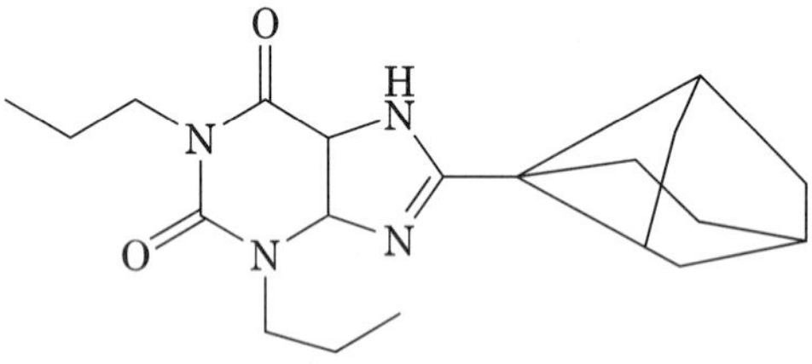

(**226**)

being a potent adenosine A_1 receptor antagonist (K_i, A_1 = 1.1 nM) (365). In saline-loaded normal rats, a 0.001–1 mg/kg oral dose of KW-3902 caused a significant increase in urine volume and sodium excretion, with little effect on potassium excretion (366). In the saline-loaded conscious dog, KW-3902 exhibited a longer-lasting natriuresis than that of either furosemide or trichlormethiazide (367). It also induced less hypokalemia and hyperuricemia in rats when compared to furosemide or trichlormethiazide (368). No attenuation in its pharmacological action was observed after repeated oral dosing (0.1 mg kg^{-1} day^{-1}) for 24 days (369). KW-3902 possessed renal protective effects against glycerol-, cisplatin-, and cephaloride-induced acute renal failure (366,

370, 371). From stop-flow methods and lithium clearance studies, it appears that the site of action of this drug is in the proximal tubule (372). KW-3902 had little effect on renal hemodynamics. More recently, CVT-124, 1,3-dipropyl-8-[2-(5,6-epoxy)norbornyl]xanthine (**227**) has been reported to be a very potent

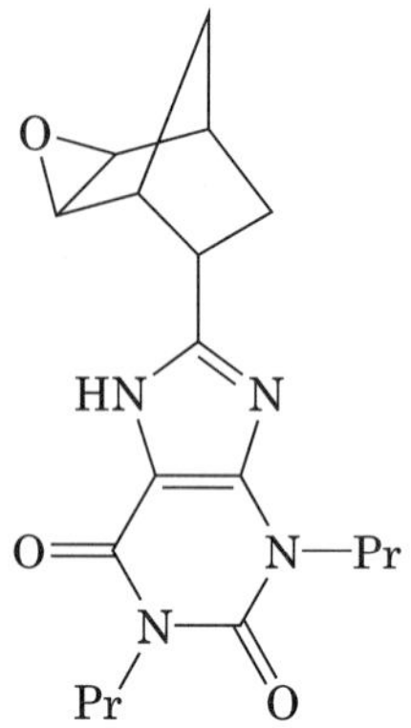

(**227**)

and selective A_1 antagonist [K_i, A_1 = 0.45 nM; K_i, A_{2a} = 1100 nM; human (373)]. In anesthetized rats CVT-124 (0.1–1 mg/kg) resulted in a dose-dependent increase in urine flow and sodium excretion (374). No changes in heart rate or blood pressure were observed. In conscious chronically instrumented rats, the administration of CVT-124 led to a significant increase in urine and sodium excretion without affecting renal hemodynamics or potassium excretion (375). The natriuretic effects of this compound have been demonstrated in normal volunteers and in humans with congestive heart failure (376).

2.1.11.2 Aminouracils. During an extensive study of compounds related to the xanthines, it was discovered that certain of the intermediate substituted 6-aminouracils were orally active diuretics in animals (377). The 1,3-disubstituted derivatives of 6-aminouracil (**228**) are diuretics, whereas the monosubstituted compounds are not. The 1-*n*-propyl-3-ethyl derivative, which was the most potent diuretic in the series, was unsuitable for clinical use because of gastrointestinal side effects. Compounds that were investigated clinically were 1-allyl-3-ethyl-6-aminouracil, aminometradine (**229**, Table 2.18), and the

(228)

1-methallyl-3-methyl analog (**230**, Table 2.18). Clinical studies in edematous subjects indicated that mercurial diuretics usually have a greater and more reliable effect (378).

At the time of their development, they represented some advance over the xanthines, but they were displaced by more effective oral diuretics within a relatively short time.

2.1.11.3 Triazines. Recognition of the triazines as a class of diuretic agents stemmed from the work of Lipschitz and Hadidian (379), who tested a group of compounds of this type in rats [e.g., melamine (**231**) and formo-

(**231**) R1 = NH$_2$, R2 = R3 = H
(**232**) R1 = R2 = R3 = H
(**233**) R1 = H, R2 = R3 = Ac

guanamine (**232**)]. Formoguanamine was effective orally as a diuretic in humans (380), but subsequent clinical studies revealed side effects such as crystalluria (379) and poor Na$^+$ excretion (381), which precluded its further use. A structural variant, prepared in an attempt to overcome the side effects, was diacetylformoguanamine (**233**), which was active as an oral diuretic in dogs (382) but still caused crystalluria and inadequate Na$^+$ excretion. The extraordinary potency of the triazines in rats does not carry over into the dog, and the compounds are only moderately active in humans.

Among other derivatives of formoguanamine, those with other substituted amino groups had a particularly favorable diuretic effect in rats (383). The most potent compounds in the series were 2-anilino-*s*-triazine (**234**, amanozine) and 2-amino-4-(*p*-chloroanilino)-*s*-triazine (**235**, chlorazanil). The diuretic activity of the last two compounds was confirmed in dogs (384, 385) and in humans (386, 387). The *m*-chloro isomer of chlorazanil, 2-amino-4-(*m*-chloroanilino)-*s*-triazine, was a potent, orally effective diuretic in rats, dogs, and humans (388, 389) and may be significantly more active than chlorazanil, with an enhanced saluretic effect. 2-Amino-4-(*p*-fluoroanilino)-*s*-traizine was twice as active as chlorazanil (386). Replacement of the halogen in chlorazanil with acetyl, carbethoxy, or sulfamoyl groups reduced activity (390), but replacement with alkylmercapto groups led to a twofold increase in activity (391). Both the incidence and degree of crystalluria in dogs were greater with alkylmercapto compounds than with chlorazanil, but the oral toxicity in mice was reduced (391).

The only triazine to achieve any degree of clinical use is chlorazanil. It has a more pronounced effect on water excretion than on Na$^+$ and Cl$^-$ (387) and has little effect on K$^+$ excretion, which is probably linked to a lack of marked enhancement of Na$^+$ excretion. Because diuresis is not accompanied by changes in glomerular filtration rate (392), the drug probably exerts its action through inhibition of tubular reabsorption. The effects of deoxycorticosterone and chlorazanil on Na$^+$ and K$^+$ excretion are mutually antagonistic, which may mean that the natriuretic and diuretic properties of the drug are attributed to inhibition of Na$^+$ reabsorption in the distal segment (393).

(**234**) R = H
(**235**) R = Cl

Table 2.18 Cyclic Polynitrogen Compounds

No.	Generic Name	Trade Name	Structure	Reference
(223)	Theophylline (aminophylline)		In combination with diaminoethane 2:1	—
(229)	Aminometradine	Mictine		376
(230)	Amisometradine	Rolicton		376
(235)	Chlorazanil	Daquin, Diurazine, Orpisin		382
(240)	Triameterene	Dyrenium		396
(249)	Amiloride	Collectril		—
(255)	Clazolimine			433

The triazines, in particular chlorazanil, have been used clinically mainly in Europe. Interest in this type of compound declined with the advent of the more effective thiazide diuretics.

2.1.12 Potassium-Sparing Diuretics. There are three structurally different classes of potassium-sparing diuretics: steroids, pyrazines, and pteridines. The steroidal aldosterone antagonists and inhibitors have been discussed in Sections 2.1.10 and 2.1.11. These diuretic agents inhibit sodium reabsorption mainly in the principal cells of the collecting tubules. The aldosterone antagonists interact with the aldosterone cytoplasmic receptors, which interaction causes a series of events leading to sodium excretion and potassium reabsorption. The pyrazines and pteridines, on the other hand, inhibit the actions of the sodium channels located on the luminal side of the principal cells. These compounds are thought either to block this channel or to switch them from an open to a closed state (394). As a result, the apical cell membrane becomes hyperpolarized, and the transepithelial potential is decreased. This reduces the driving force for potassium to exit the luminal side by way of potassium channels, thus decreasing renal potassium loss. On average, the human body contains about 140 g of potassium. A calculation has shown that 98% of potassium is intracellular and only about 2% is in the extracellular compartment. Therefore, removal or addition of a small amount of potassium to this extracellular pool is very evident (395).

2.1.12.1 Triamterene. The triamterene ring system is found in many naturally occurring compounds, such as folic acid and riboflavin. These compounds are important in the regulation of metabolism in humans. The observation that xanthopterin (**236**) was capable of affecting renal tissue led Wiebelhaus, Weinstock, and associates to test a series of pteri-

dines in a simple rat diuretic screening procedure (396). One compound, 2,4-diamino-6,7-dimethyl-pteridine (**237**), showed sufficient

(237)

diuretic activity to encourage further investigation of the diuretic potential of the pteridines. A number of related 2,4-diaminopteridines were studied, but only (**237**) showed good activity in both the saline-loaded and saline-deficient rat. Changes in the 2,4-diamino part of the molecule resulted in a marked decrease in diuretic activity (397).

A class of related pteridines, of which 4,7-diamino-2-phenyl-6-pteridine-carboxamide (**238**) is the prototype, has been investigated.

(238)

This derivative is active in both the saline-loaded and sodium-deficient rat, but in contrast to (**237**), it causes substantial potassium loss in the sodium-deficient rat. In structure-activity studies, particular attention was directed toward modification of the carboxamide function (397–399).

One of the more interesting compounds was 4,7-diamino-*N*-(2-morpholinoethyl)-2-phenyl-6-pteridinecarboxamide (**239**). In pharmacological investigations (400), this compound was an orally active diuretic agent, generating about the same maximum degree of response in dogs as did hydrochlorothiazide. The urinary excretion of Na^+ and Cl^- was markedly enhanced, with minimal augmentation of K^+ excretion and

(236)

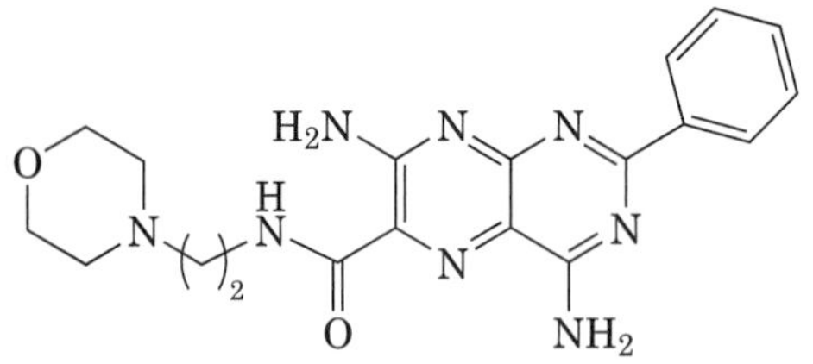

(239)

(240)

little effect on urine pH. Onset of action was rapid, with the greatest saluretic effect occurring within 2 h of oral administration to saline-loaded dogs. The compound showed diuretic activity in both normal and adrenalectomized rats, which, with the absence of K^+ retention, indicated that aldosterone antagonism is not a major component of its saluretic activity.

A consideration of the structural features of 2,4-diamino-6,7-dimethylpteridine (**237**) and 4,7-diamino-2-phenyl-6-pteridinecarboxamide (**238**) led to the investigation of 2,4,7-triamino-6-phenylpteridine (triamterene, **240**) as a potential diuretic agent (397).

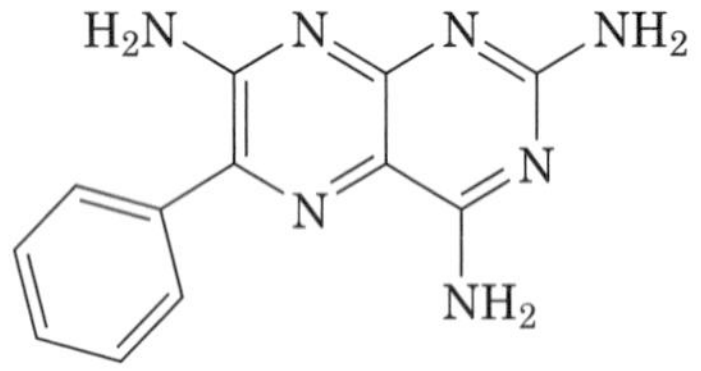

The compound was very potent in the saline-loaded rat, and in the sodium-deficient rat it not only caused a marked excretion of sodium but simultaneously decreased K^+ excretion. In structure-activity studies of compounds related to triamterene, replacement of one of the primary amino groups by lower alkylamino groups led to compounds that retained triamterene-like diuretic activity. More extensive changes generally led to substantially less active compounds. Table 2.19 lists the activities of some 2,4,7-triamino-6-substituted pteridines. The activity of triamterene is very sensitive to substitution of the phenyl group with only small changes possible if diuretic activity is to be retained. The p-tolyl compound, for example, is only about half as

Table 2.19 2,4,7-Triamino-6-substituted Pteridines[a]

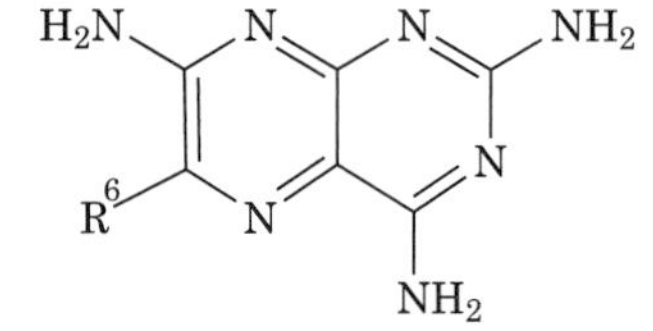

R6	Diuretic Activity in Saline-Loaded Rat[b]	Diuretic Activity in Sodium-Deficient Rat[c]
Phenyl (triamterene)	3	3
2-Me-C_6H_4	2	1
3-Me-C_6H_4	1	1
4-Me-C_6H_4	2	1
2-F-C_6H_4	2	
4-F-C_6H_4	1	
4-MeO-C_6H_4	1	
4-Ph-C_6H_4	0	
2-Furyl	0	1
3-Furyl	1	2
2-Thienyl	2	2
3-Thienyl	1	2
4-Thiazolyl	0	
2-Pyridyl	0	
3-Pyridyl	0	
4-Pyridyl	0	
H	2	3
Methyl	3	2
$CH(CH_3)_2$	1	
Butyl	2	2
Δ^3-Cyclohexenyl	1	
Benzyl	2	2

[a]From Ref. 396.
[b]Rating scheme for saline-loaded rat assay: maximum response at any dose, in volume percent of urine, compared to volume of 0.9% saline load, less that of untreated control. <22% = 0; 22–45% = 1; 46–69% = 2, and >69% = 3.
[c]Rating scheme for sodium-deficient rat assay: maximum response of any dose, in milligrams of sodium excreted in the urine per rat. <3 mg = 0; 3–6 mg = 2; and >9 mg = 3.

active as triamterene. In general, *ortho* isomers seem to be more active than the other isomers. The p-hydroxyphenyl analog of triamterene, a metabolite of the latter, is essentially inactive. The 4-deuterophenyl analog of triamterene was prepared in an attempt to increase activity by reducing the rate of metabolism, but its activity is very similar to that of the parent compound. When the phenyl group of triamterene is replaced by a heterocyclic nucleus, the size of the group appears to be important and high activity is seen only in the case of small, nonbasic groups. The low activity of compounds containing basic centers in this position, such as thiazole and pyridine, may be rationalized by assuming that the basic centers are highly solvated and are, in effect, large substituents. The 6-alkyl analogs are active diuretics; however, size is important. Although good activity is seen in the 6-*n*-butyl homolog the isopropyl and cyclohexenyl derivatives have only modest activity. Isomers of triamterene were also studied; the 7-phenyl isomer was one of the most potent K^+ blockers found in the pteridines, even though it is only a weak natriuretic agent. The 2-phenyl isomer is very similar to triamterene in its biological properties. Among pyrimidopyrimidines related to triamterene, 2,4,7-triamino-5-phenylpyrimido[4,5-*d*]pyrimidine (**241**) was in-

(241)

vestigated in some detail. It resembles (**238**), in that it does not block potassium excretion in the sodium-deficient rat.

The structure-activity relationships of pteridine diuretics may be rationalized by assuming that the pteridines bind to some active site at two points (397). The more important site involves a basic center of the drug, which in triamterene may be N-1, N-8, or both. Groups that decrease the base strength of the pteridine nucleus reduce activity. The other site probably involves the phenyl substituent of triamterene and may be hydrophobic in nature. There appear to be critical size limitations at the site, as shown by the change in activity in relation to methyl substitution. Because compounds such as 2,4,7-triaminopteridine are active, the phenyl group is not a primary requirement for activity and apparently acts in a reinforcing capacity, such as increas-

ing the degree of binding and establishing the correct orientation of the molecule at the receptor site.

Triamterene (**240**) is a potent, orally effective diuretic in both the saline-loaded and sodium-deficient rat, and is accompanied by no increase in potassium excretion. Also, the effect of aldosterone on the excretion of electrolytes in the adrenalectomized rat are completely antagonized by triamterene. Similar results were obtained in dogs, and it appeared that the compound might be functioning as an aldosterone antagonist (401). Initial clinical studies (402, 403) established the natriuretic properties of triamterene in humans in cases when aldosterone excretion might be at an elevated level, and evidence was obtained for inhibition of the nephrotropic effect of aldosterone. However, triamterene possessed natriuretic activity in adrenalectomized dogs and rats (404–406) and in adrenalectomized patients (407); this was inconsistent with an aldosterone antagonism mechanism. Thus, although triamterene reverses the end results of aldosterone, its activity does not depend on the displacement of aldosterone. The compound acts directly on the renal transport of sodium. Stop-flow studies in dogs pointed to an effect on the distal site of Na^+/K^+ exchange, and there was no evidence for a proximal renal tubular effect (408). Triamterene acts at the apical cell membrane in the collecting tubule where it blocks sodium channels and thus leads to reduced potassium excretion (409). It is also believed to act at the peritubular side as well (410).

The overall effect of triamterene on electrolytes is to increase moderately the excretion of Na^+ and, to a lesser extent, of Cl^- and HCO_3^-, and to reduce K^+ and NH_4^+ excretion (403, 411, 412). Triamterene is a more active natriuretic agent than spironolactone and is well absorbed after administration of a single oral dose of 50–300 mg/day (402, 403). It is about 50–60% bound to plasma proteins. Triamterene is extensively metabolized and only about 3–5% of the drug is excreted unchanged in the urine (413). The compound undergoes hepatic hydroxylation and sulfate ester formation. The sulfate ester is also biologically active and is excreted in the urine. After intravenous administration of triamterene in humans, the concentration of the sulfate ester was 10 times that of the parent compound, and it has been concluded that the pharmacologically active form of the drug in humans is the ester (414). Triamterene is also metabolized to its glucuronide adduct, which undergoes biliary elimination. The duration of action in humans is about 16 h (415).

An increased diuresis ensued when triamterene was administered to patients who were receiving the aldosterone antagonist spironolactone (416, 417), thus further emphasizing the fundamental differences in the mechanism of action of these two drugs. Triamterene potentiates the natriuretic action of the thiazides while reducing the kaliuretic effect (416), and other clinical studies with this combination showed that normal serum potassium levels could be maintained without potassium supplements (417). The natriuretic potency of triamterene does not approach that of the thiazide diuretics, and the main value of the drug would appear to be its use in combination with thiazides in clinical situations in which K$^+$ loss is a problem. Triamterene should not be prescribed to patients with renal insufficiency or hyperkalemia, nor should it be administered concurrently with potassium supplements. Triamterene increases serum uric acid levels and cases of hyperuricemia have been reported. Some other side effects that are observed are skin rashes, gastointestinal disturbances, hyperkalemia, weakness, and dry mouth.

2.1.12.2 Other Bicyclic Polyaza Diuretics. A group of workers at Takeda synthesized and studied the diuretic activity of a large series of polynitrogen heterocycles (418). The ring systems prepared are shown in Table 2.20, documenting the extensive effort made in this investigation. Of the 219 compounds studied in this series, two compounds, DS 210 (**242**) and DS 511 (**243**), were selected for more extensive evaluation.

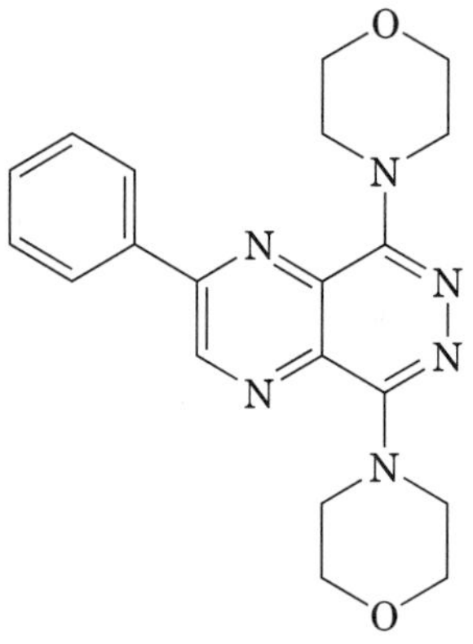

The compounds were initially screened in rats, and hydrochlorothiazide was used as the reference compound. DS 210 (**242**) produces a maximal natriuretic effect similar to that of hydrochlorothiazide in rats without affecting potassium excretion. It shows additive activity with hydrochlorothiazide, acetazolamide, amiloride, and furosemide. Potassium excre-

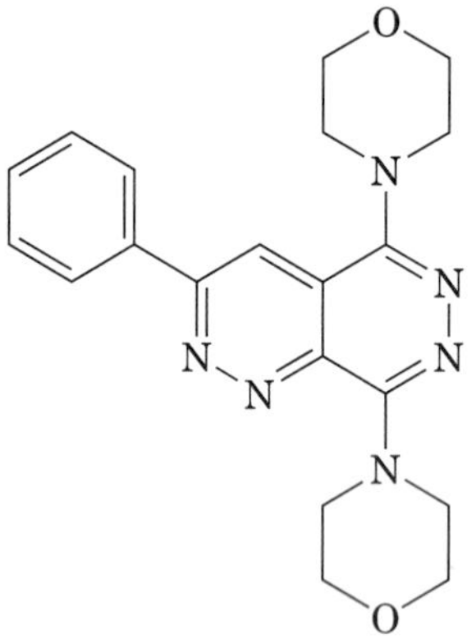

(242)

(243)

tion induced by other diuretics is not modified by DS 210. The diuretic effect is lost in adrenalectomized rats and restored by cortisol treatment (419). DS 511 (**243**) has shown diuretic activity comparable to that of hydrochlorothiazide in rats, dogs, and humans, and seems to have a unique mode and site of action in the nephron (420).

Hawes and coworkers at the University of Saskatchewan prepared a series of 2,3-disubstituted 1,8-naphthyridines containing a phenyl group at the 3-, 4-, or 7-positions (421). In this study, compounds containing a phenyl group at the 7-position had no diuretic activity. Compounds (**244**) and (**245**), which contained a phenyl group at the 4-position, have similar diuretic and natriuretic properties when compared to those of triamterene. These compounds also lack kaliuretic properties.

Table 2.20   Polyaza Heterocyclic Systems Studied as Diuretics[a]

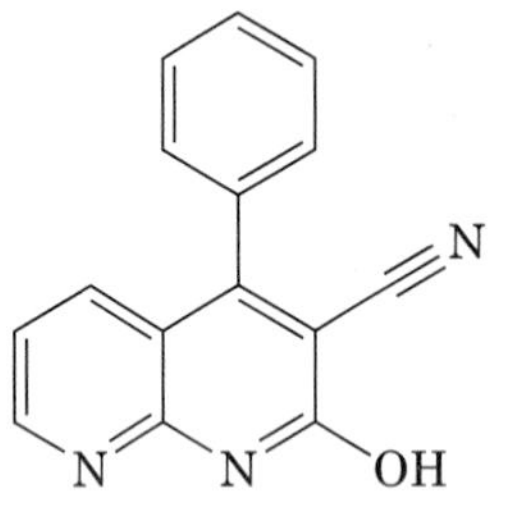

(244)

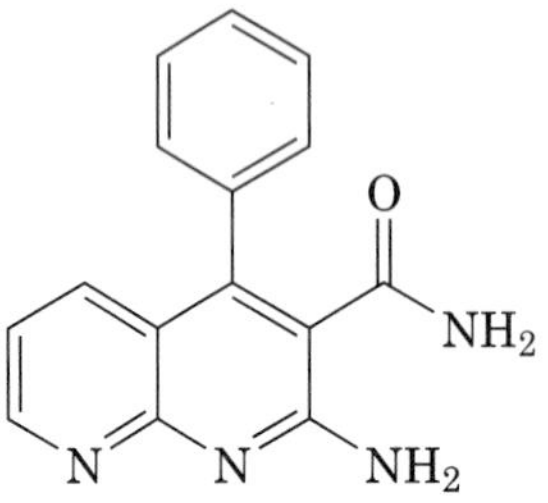

(245)

Parish and coworkers synthesized a number of 2-pyrido[2,3-*d*]-pyrimidin-4-ones (422). 1,2-Dihydro-2-(3-pyridyl)-3*H*-pyrido>[2,3-*d*]-pyrimidin-4-one (**246**) was a potent diuretic

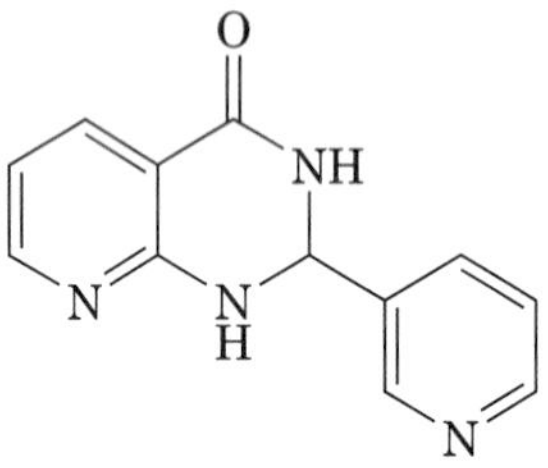

(246)

agent in the rat, although it was not as potent as hydrochlorothiazide. An oral dose of 81 mg/kg in the rat resulted in potassium excretion levels that were the same as those of the saline controls; the sodium and chloride ion excretion levels were the same as those of the saline controls; and the sodium and chloride ion excretion levels had doubled. Monge and coworkers at the University of Navarra further studied the structure-activity relationships in this series (423). Placement of nitro or amino groups at the 6-position resulted in compounds that had marginal or no diuretic

activity. Methylation of both the amide and amine nitrogens gave a compound with similar diuretic activity but with a poorer Na^+/K^+ ratio.

Hester and coworkers at Upjohn prepared a series of 1-(2-amino-1-phenylethyl)-6-phenyl-4*H*-[1,2,4]triazolo[4,3-*a*][1,4]-benzodiazepines and evaluated their diuretic activity (424). Several of these compounds possessed diuretic and natriuretic activity, after oral administration to rats, with no kaliuretic activity. The most potent benzodiazepine in this series is (**247**); however, it is considerably less

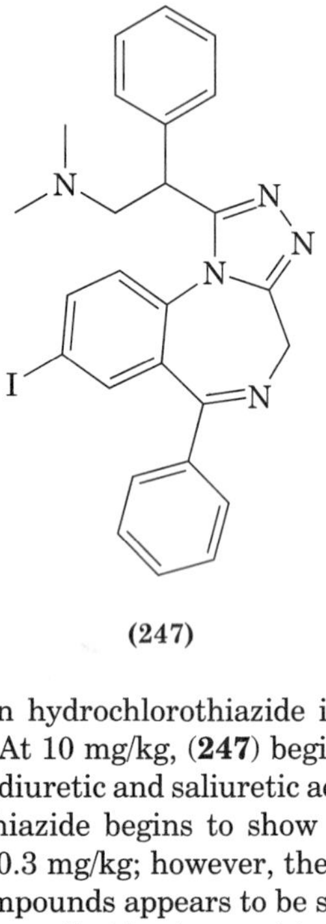

(247)

potent than hydrochlorothiazide in the conscious rat. At 10 mg/kg, (**247**) begins to show significant diuretic and saliuretic activity. Hydrochlorothiazide begins to show significant activity at 0.3 mg/kg; however, the efficacy of the two compounds appears to be similar.

2.1.12.3 Amiloride. An empirical approach was taken by a group at Merck, Sharp & Dohme seeking compounds with no or minimal kaliuretic effects. Screening procedures indicated that *N*-amido-3-amino-6-bromopyrazinecarboxamide (**248**), a compound available through previous work in the folic acid series, was of interest. The introduction of an amino group in the 5-position markedly increased the sodium and chloride excretion: *N*-amidino-3,5-diamino-6-chloropyrazine-carboxamide (amiloride, **249**) was among the most promising in animals and in humans (174). The *N*-amidinopyrazinecarboximides produce

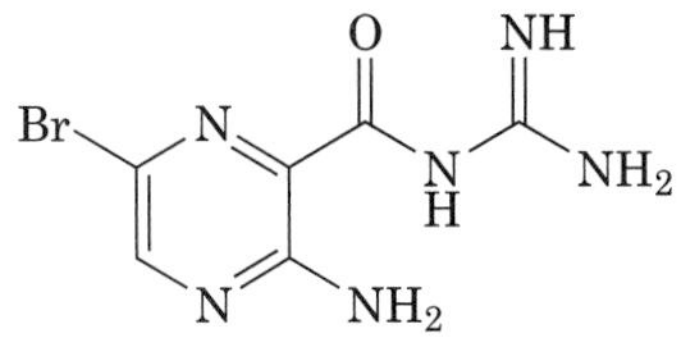

(248)

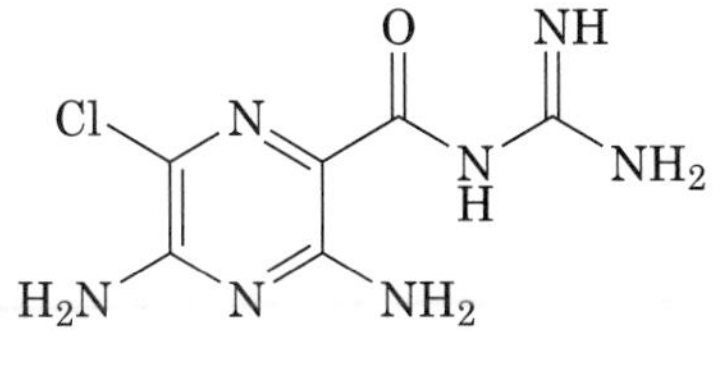

(249)

(251)

a pronounced diuresis in normal rats and leave potassium excretion unaffected or repressed. In the adrenalectomized rat they antagonize the renal actions of exogenous aldosterone, DOCA, and hydrocortisone. In dogs, the compounds are less potent, but the relative activity in the series is the same as those in rats.

Structure-activity relationships in this series have been investigated in considerable detail and some representative compounds from these studies are listed in Table 2.21. The activities of the compounds were determined on the basis of their DOCA-inhibitory activity, which closely paralleled the diuretic activity in intact rats and dogs (425–427).

Workers at Ciba investigated the replacement of the acylguanidine moiety with a 1,2,4-oxadiazol-3-amino- group (428). This compound, CGS 4270 (250), has a similar profile to that of amiloride in rats and dogs.

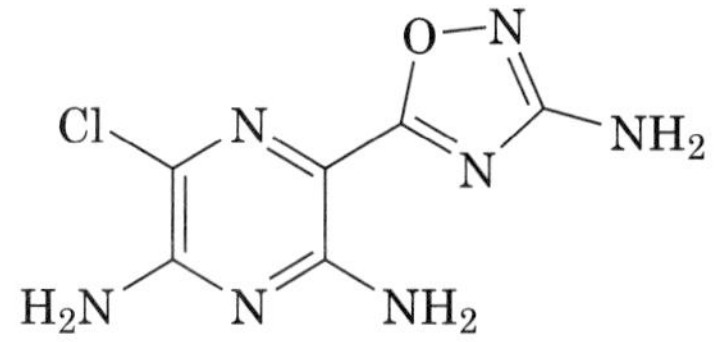

(250)

More recently, Ried and coworkers prepared a number of amiloride analogs that were modified at the 2- and 6-positions (429). In general, replacement of the chlorine at the 6-position was detrimental to the antikaliuretic, pharmacodynamic, and pharmacokinetic properties of this class. A number of the N-substituted 3,5-diamino-6-chloropyrazine-2-carboxamides were found to be as potent as triamterene. The N-(dimethylaminoethyl) amide analog (251) was more potent than triamterene as a diuretic, natriuretic, and antikaliuretic agent after oral administration to the rat. The compound was well absorbed and was excreted without significant metabolism.

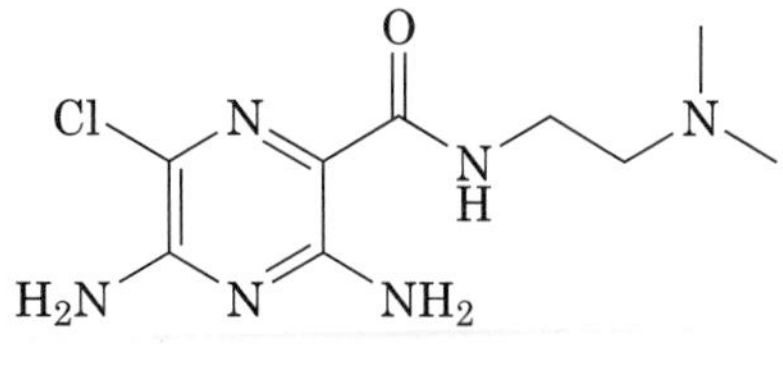

Workers at ICI have synthesized amiloride analogs that have diuretic activity combined with calcium channel-blocking activity or β-adrenoceptor blocking activity. ICI 147798 (252) is a single-molecule derivative of amiloride that was discovered to possess both diuretic and β-adrenoreceptor blocking properties (430). At doses of 1–20 mg/kg p.o., the natriuretic activity of ICI 147798 was similar to that of hydrochlorothiazide in dogs; at 1 mg/kg it produced significantly less kaliuresis than did hydrochlorothiazide. ICI 147798 blocked adrenergic receptors in vitro and in vivo, and it also inhibited isoproternol-induced tachycardia in rats, guinea pigs, cats, and dogs. ICI 206970 (253), another analog of amiloride, possessed diuretic and calcium channel-blocking activity (431). In the dog, after oral administration, ICI 206970 was less potent than hydrochlorothiazide with respect to its diuretic and saliuretic effects. In contrast to hydrochlorothiazide, no significant changes were observed in plasma potassium levels after 14 days of dosing.

The structural similarity of amiloride and triamterene (249 and 240, Table 2.18) and their similar biological actions have raised the question of whether the pteridines are, in fact, closed-ring versions of the N-amidinopyrazine-carboxamides. The open-chain analogs of triamterene and the bicyclic analogs of amiloride

Table 2.21. DOCA Inhibitory Activity in Adrenalectomized Rats of Some N-Amidino-3-aminopyrazinescarboxamides[a]

R$_1$	R$_2$	R$_3$	R$_4$	R$_5$	R$_6$	DOCA inhibition score[b]
H	H	H	H	H	Cl	+++
H	H	H	H	H	Br	++
H	H	H	H	H	I	+
CH$_3$	CH$_3$	H	H	H	Cl	+++
CH$_3$	H	H	H	H	Cl	+
H	H	H	H	H	H	0
H	H	H	H	H	CF$_3$	+/−
H	H	H	H	H	CH$_3$	++
H	H	H	H	CH$_3$	H	+
H	H	H	H	H	Ph	+
H	H	H	H	NH$_2$	Cl	++++
H	H	H	H	MeNH	Cl	+++
H	H	H	H	(Me)$_2$CHCH$_2$NH	Cl	+
H	H	H	H	(Me)$_3$CNH	Cl	+/−
H	H	H	H	(Me)$_2$N	Cl	+++
H	H	H	H	PhNH	Cl	++
CH$_3$	H	H	H	NH$_2$	Cl	++++
CH$_3$	CH$_3$	H	H	NH$_2$	Cl	++++
3,4-Cl$_2$C$_6$H$_5$CH$_2$	H	H	H	NH$_2$	Cl	+++
H	H	H	H	NH$_2$	Br	+++
H	H	H	H	Cl	Cl	0
H	H	H	H	OH	Cl	+/−
H	H	H	H	OMe	Cl	+/−
H	H	H	H	SH	Cl	+/−
H	H	H	H	SMe	Cl	+
CH$_3$CO	H	H	CH$_3$CO	H	Cl	+/−
H	H	H	H	H	SMe	+/−

[a]From Refs. 426–428.

[b]This score is related to the dose of each compound that produces a 50% reversal of electrolyte effect from the administration of 12 μg of DOCA to adrenalectomized rats and is scored as follows: 10 μg (+++), 10–50 μg (+++), 51–100 μg (++), 101–800 μg (+), >800 μg (+/−).

that have been studied are generally less active than the drugs themselves. Triamterene is a weaker base (pK_a = 6.2) than amiloride (pK_a = 8.67). Amiloride as its hydrochloride is readily water soluble, whereas triamterene is slightly soluble. After oral administration, approximately 50% of amiloride is absorbed in humans (432). It is approximately 23% bound to plasma proteins and is not metabolized in humans. It is excreted in the urine mainly unchanged.

In the usual dosage, amiloride has no important pharmacological actions except those related to the renal tubular transport of electrolytes. Clinically, it is used extensively in combination with hydrochorothiazide.

2.1.12.4 Azolimine and Clazolimine. A series of imidazolones was studied by a group at Lederle Laboratories in their search for a nonsteroidal antagonist of the renal effects of mineralocorticoids. Azolimine (**254**) and clazolimine (**255**) were the most interesting in this

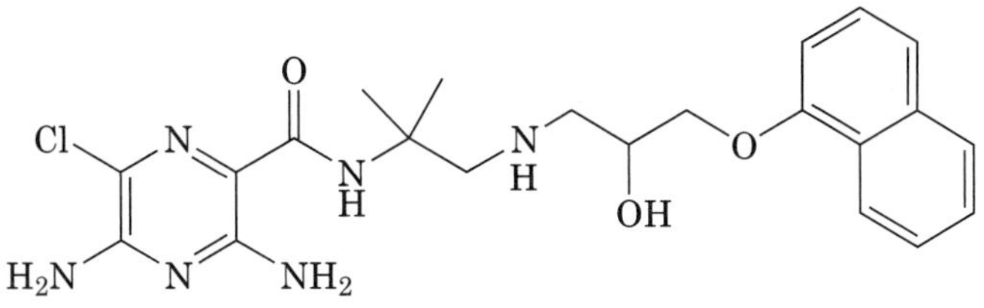

(252)

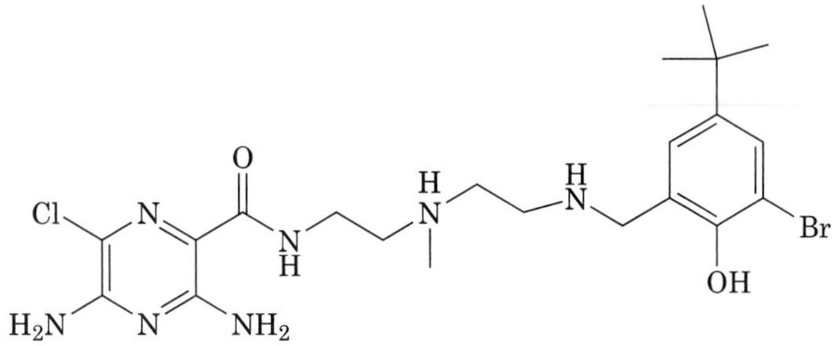

(253)

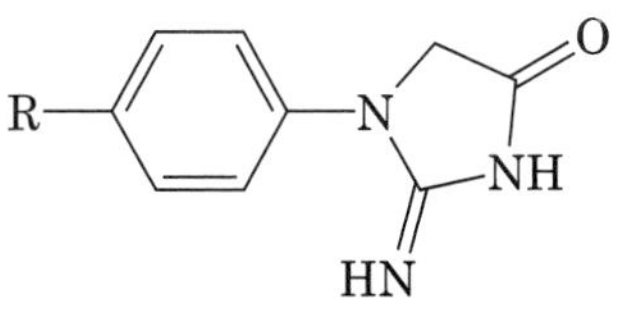

(254) R = H
(255) R = Cl

series. Azolimine antagonized the effects of mineralocorticoids on renal electrolyte excretion in several animal models. Large doses of azolimine produced natriuresis in adrenalectomized rats in the absence of exogeneous mineralocorticoid, but its effectiveness was greater in the presence of a steroid agonist. In conscious dogs, azolimine was effective only when deoxycorticosterone was administered. Azolimine significantly improved the urinary Na^+/K^+ ratio when used in combination with thiazides and other classical diuretics in both

adrenalectomized, deoxycorticosterone-treated rats and sodium-deficient rats (433). Similar effects were found for clazolimine (434). The compound may be useful in combination with the classical diuretics as an aldosterone antagonist diuretic in humans.

2.1.13 Atrial Natriuretic Peptide. Atrial natriuretic peptide (ANP, **256**) is a 28-amino acid peptide that is released into circulation from the heart after atrial distension and increases in heart rate. It is synthesized and stored in specific atrial secretory granules as a 126-amino acid precursor molecule. ANP exerts natriuretic, diuretic, and vasorelaxant properties upon administration and suppresses renin and aldosterone levels (435, 436). Through interaction with its receptor it promotes the generation of cGMP by guanylate cyclase activation. The pharmacological properties produced by this peptide suggest

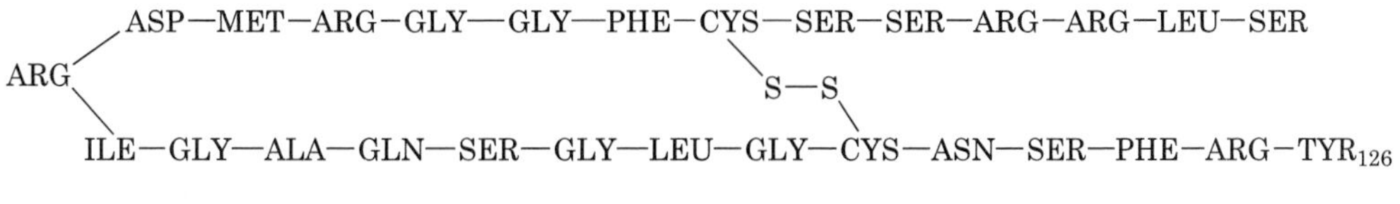

(256)

that it may be beneficial in the treatment of several cardiovascular disorders. The therapeutic potential of ANP, however, is limited by its poor oral absorption and extremely short biological half-life of less than 60 s in the rat and only a few minutes in humans (437, 438).

The mechanism of action for the natriuretic activity of ANP has been the subject of much research over the past few years. It is believed that ANP-induced natriuresis results from its effect on renal hemodynamics. ANP is able to increase the glomerular filtration rate significantly (439, 440). This effect seems to be brought about by vasodilation of the afferent arterioles with vasoconstriction of the efferent vessels. This effect increases glomerular capillary pressure and, therefore, the glomerular filtration rate. Further studies have shown that ANP may also inhibit sodium reabsorption in the collecting tubules and duct (441).

ANP has also been demonstrated to inhibit both basal and angiotensin-stimulated secretions of aldosterone in $vitro$ in adrenal preparations and in $vivo$ after infusion in animals and humans (442–444). Because ANP-induced natriuresis occurs very rapidly, the reduction in aldosterone secretion contributes to a longer-term modulation of sodium excretion.

ANP is eliminated from the circulation by way of two major pathways. Studies have shown that ANP is eliminated from the circulation by enzymatic degradation. The enzyme most responsible for its degradation is neutral endopeptidase (NEP, EC 3.4.24.11) (445, 446). NEP is a zinc metallopeptidase that cleaves the α-amino bond of hydrophobic amino acids. Other enzymes in the renin-angiotensin and kallikrein-kinin systems have also been shown to degrade ANP.

ANP is also removed from circulation through a receptor-mediated clearance pathway (448). ANP clearance receptor (c-receptor) can be found in several tissues, including kidney cortex, vascular, and smooth muscle cells (448, 449). ANP binds to this receptor, and then this receptor-ANP complex is internalized. ANP is transported to the lysosome, where it undergoes extensive hydrolysis. The clearance receptor is recycled to the cell's surface, where it can repeat this process.

Because infusion of ANP was shown to produce several potentially therapeutic benefits,

much of the current research in this area has been focused on methods of potentiating the activity of ANP in $vivo$ by preventing its degradation (436).

2.1.13.1 ANP Clearance Receptor Blockers. Several groups have prepared ligands for the ANP c-receptor that have prolonged the $t_{1/2}$ of ANP. SC 46542 {des-[Phe^{106}, Gly^{107}, Ala^{115}, Gln^{116}]ANP (103–126)} is a biologically inactive analog of ANP that has similar affinity for the c-receptor as ANP (450). In the normal, conscious rat and spontaneously hypertensive rat, however, SC46542 did not significantly increase immunoreactive ANP concentrations in plasma.

Two linear peptides, Ala$_7$-rat-ANP$_{8-17}$-NH$_2$ and naptoxyacetyl isonipecotyl-Arg-Ile-Asp-Arg-Ile-NH$_2$, were shown to increase plasma immunoreactive ANP concentrations in anesthetized rats (451). In response to the infusion of these compounds, a significant increase in glomerular filtration rate and sodium excretion was observed.

Infusion of C-ANP$_{4-23}$ was also shown to increase plasma immunoreactive ANP concentrations in anesthetized and conscious rats (452). C-ANP$_{4-23}$ increased the urinary excretion of water and sodium in the conscious DOCA/salt-hypertensive rats when administered i.v. (453).

More recently, workers at AstraZeneca have reported on new series of nonbasic ANP c-receptor antagonists (454). They have made modifications to C-ANP$_{4-23}$ and AP-811 (**257**), which have retained good affinity for the c-receptor and have improved physical properties. Either of the arginines of AP-811 could be replaced with alanine.

2.1.13.2 Neutral Endopeptidase Inhibitors. Originally, NEP inhibitors were designed and studied for their analgesic properties, given that this enzyme was known to degrade enkephalins. When it was discovered that NEP also degraded ANP, many of the known NEP inhibitor compounds, such as thiorphan (**258**) and phosphoramidon (**259**), were evaluated for their potential diuretic and cardiovascular activities. Both thiorphan and phosphoramidon increased the half-life of exogenous ANP in the rat (455). Phosphoramidon, when infused into rats with reduced renal mass, significantly increased diuresis, natriuresis, and

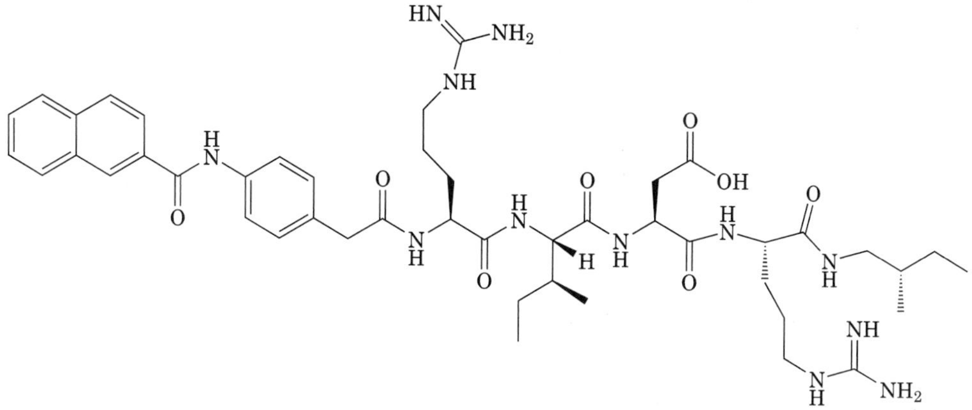

(257)

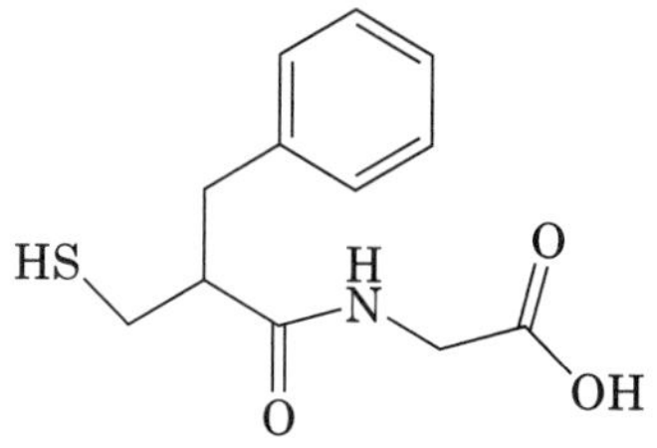

(258)

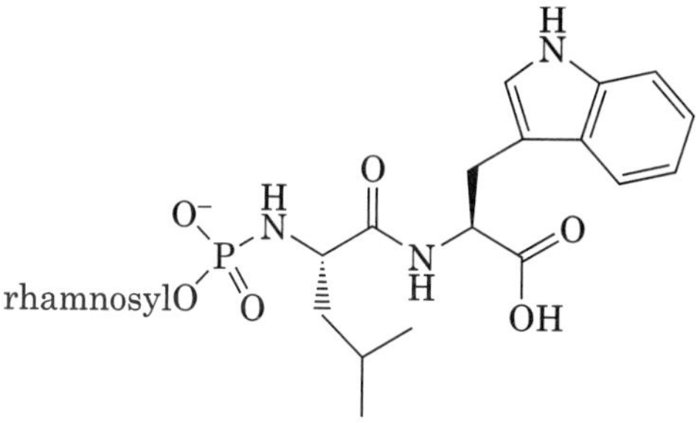

(259)

dition, there is also an increase of plasma endothelin-1 levels (463, 464) that may also explain the limited effectiveness of this class.

In recent years, there have been many reports on new potent inhibitors of NEP. Many of these compounds are di- or tripeptides that contain a group that binds to the zinc atom in the active site of the enzyme. There are four different classes of NEP inhibitors: thiols, carboxylates, phosphoryl-containing, and hydroxamates.

Thiorphan was the first reported thiol inhibitor of NEP. Both the R- and S-enantiomers of thiorphan have the same enzyme inhibitory potency, $IC_{50} = 4$ nM. Extensive structure-activity relationship studies have shown that it is possible to replace the glycine residue with an O-benzyl serine and still retain potency (465). This compound, ES 37 (**260**), has an IC_{50} for NEP of 4 nM and is a

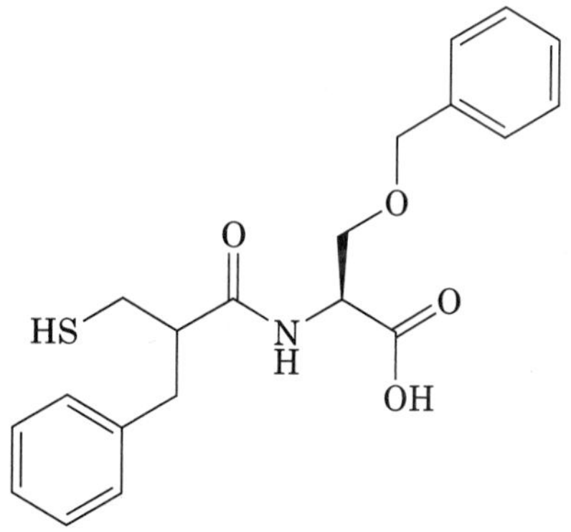

glomerular filtration rate (456). Thiorphan was also shown to increase sodium excretion in anesthetized and conscious normal rats (457).

Several selective inhibitors of NEP have been evaluated in clinical trials and have been found to have little or no efficacy in lowering blood pressure (458–460). It is believed that NEP inhibition may slow the metabolism or clearance of angiotensin II (460–462). In ad-

(260)

potent inhibitor of angiotensin-converting enzyme (ACE), IC_{50} = 12 nM. Reduction of the phenyl ring also affords a potent NEP inhibitor, IC_{50} = 32 nM.

Investigators at Squibb replaced the glycine in thiorphan with an aminoheptanoic acid (466). This compound, SQ 29072 (**261**), is

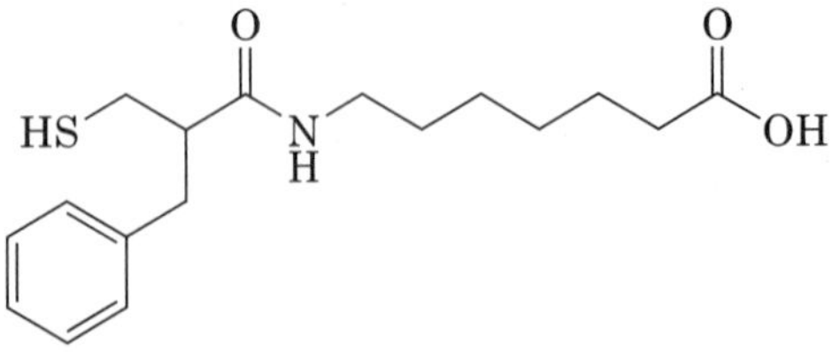

(**261**)

a potent NEP inhibitor with an IC_{50} value of 26 nM. When administered intravenously (300 µg/kg) to a conscious SHR, SQ 29072 produced a modest diuretic and natriuretic response (467). In the DOCA/salt-hypertensive rat, when equidepressor doses of SQ 29072 and ANF(99–126) were administered, there was a prolonged urinary excretion of sodium. Another structurally related analog, SQ 28603 (**262**), was also reported to be a potent and

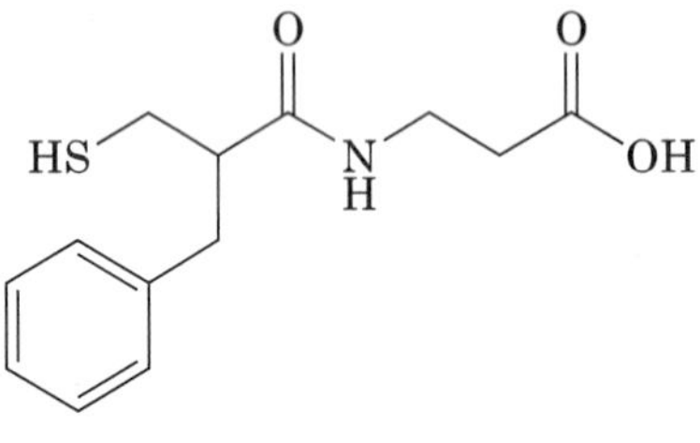

(**262**)

selective inhibitor of NEP. When infused in conscious, DOCA/salt-hypertensive rats, SQ 28603 caused an increase in plasma ANP concentration and in sodium excretion and significantly lowered mean arterial pressure (468).

Workers at Schering-Plough also prepared thiorphan-type analogs. SCH 42495 (**263**) elevated plasma ANF concentrations in animal models (469, 470). In a study with eight patients with essential hypertension, plasma ANF levels increased (+123%, P < 0.01) and later remained elevated (+34%, P < 0.01)

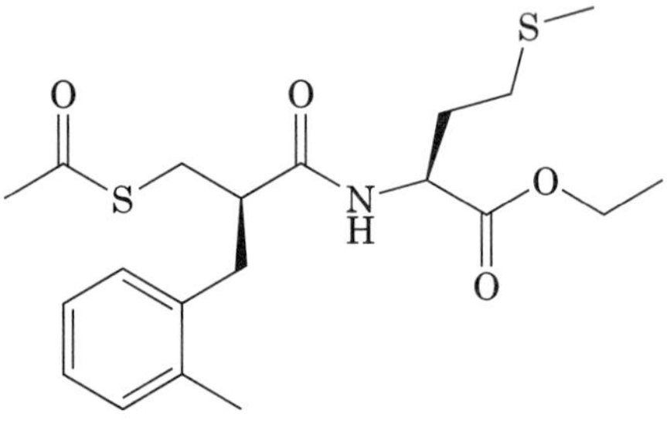

(**263**)

(471). The compounds also led to a significant natriuresis in the initial 24 h of treatment. This effect attenuated over time.

A number of carboxyl-containing NEP inhibitors were prepared by workers at Schering-Plough and Pfizer. Candoxatrilat (UK 69578, **264**) was a potent NEP inhibitor (K_i =

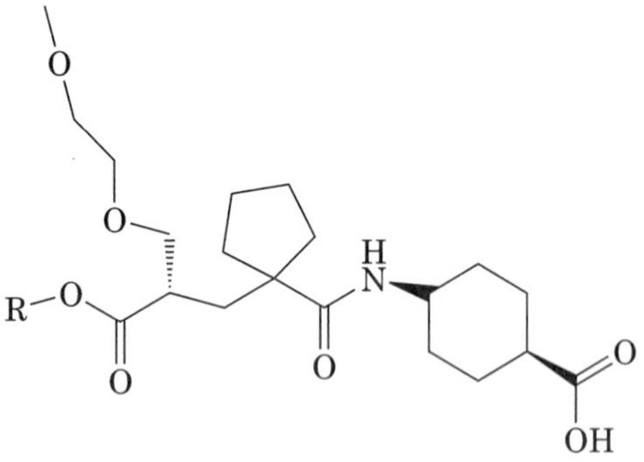

(**264**) R = H, candoxatrilat
(**265**) R =

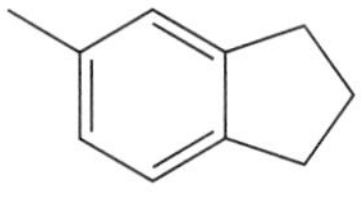

candoxatril

2.8×10^{-8} M) designed by workers at Pfizer (472). When given i.v. to mice, it increased endogenous levels of ANP and produced diuretic and natriuretic responses. When the prodrug candoxatril (**265**) was administered to human subjects, doses of 10–200 mg caused a rise in basal ANP levels. Natriuresis was observed only at the highest dose (473).

Workers at Schering-Plough also prepared a number of carboxyl-containing NEP inhibitors. SCH 39370 (**266**) was discovered to be a potent NEP inhibitor (IC_{50} = 11 nM) and, when administered to rats with congestive heart failure, it caused an increase in urinary

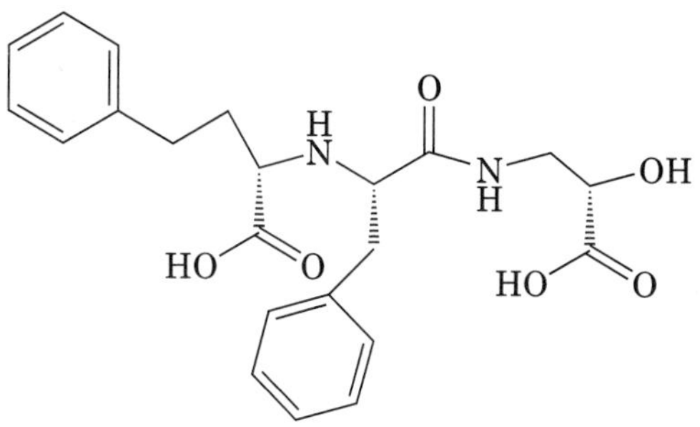

(266)

(268)

volume and plasma ANP levels (474, 475). In an ovine heart failure model, SCH 39370, when given as a bolus injection, caused significant natriuresis and diuresis (476). A structurally related compound, SCH34826 (**267**),

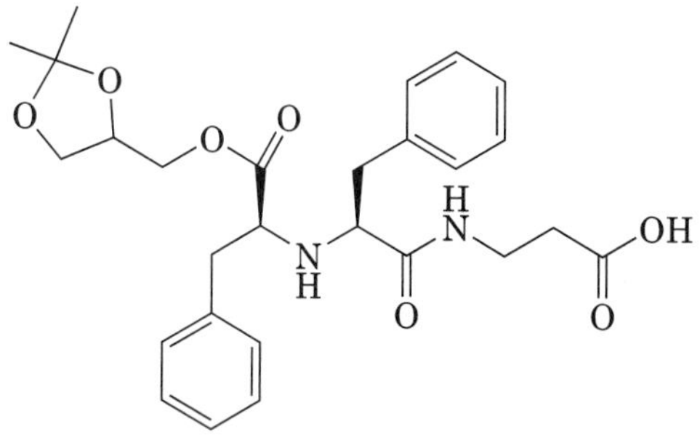

potent NEP inhibitor ($IC_{50} = 2$ nM), is a natural competitive inhibitor produced by *Streptomyces tanashiensis*. Vogel and coworkers reported that phosphoryl-Leu-Phe (**269**) is a

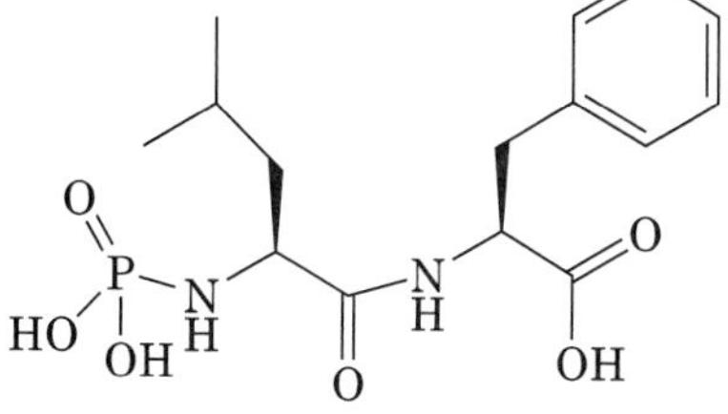

(267)

(269)

produced a significant natriuretic effect in DOCA/salt-hypertensive rats (477). In normal volunteers maintained on a high sodium diet for 5 days, SCH 34826 promoted a significant increase in sodium, calcium, and phosphate excretion (478).

Hydroxamates form strong bidentate ligands to zinc. Compounds containing this functional group are potent NEP inhibitors. The prototype of this class is *RS*-kelatorphan. This compound strongly inhibits NEP ($IC_{50} = 1.8$ nM) and aminopeptidase N(ANP) ($IC_{50} = 380$ nM) (479). The *SS*-isomer of kelatorphan, RB 45 (**268**), is a more selective NEP inhibitor [$IC_{50} = 1.8$ nM, $IC_{50} = 29,000$ nM (APN)]. In rats, when given at 10 mg/kg i.v., RB 45 increased the half-life of ANP (480).

Phosphoryl-containing inhibitors also interact strongly with zinc and are potent NEP inhibitors. Phosphoramidon (**259**), which is a

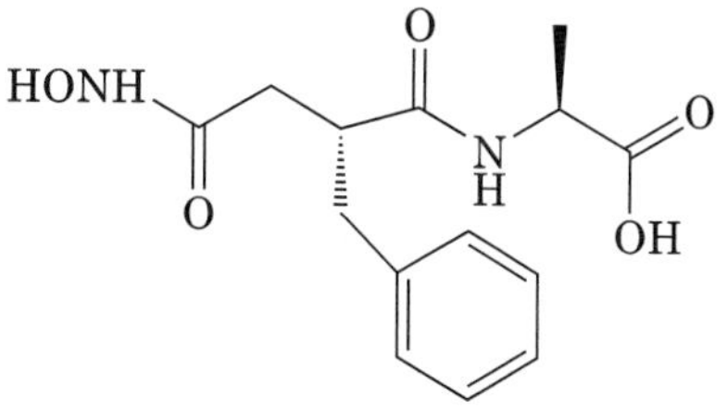

potent NEP inhibitor ($IC_{50} = 0.3$ nM) (481). This is about an order of magnitude more potent than thiorphan.

Workers at Ciba reported on a series of potent phosphorous-containing inhibitors of NEP (482). CGS 24592 (**270**) had an IC_{50}

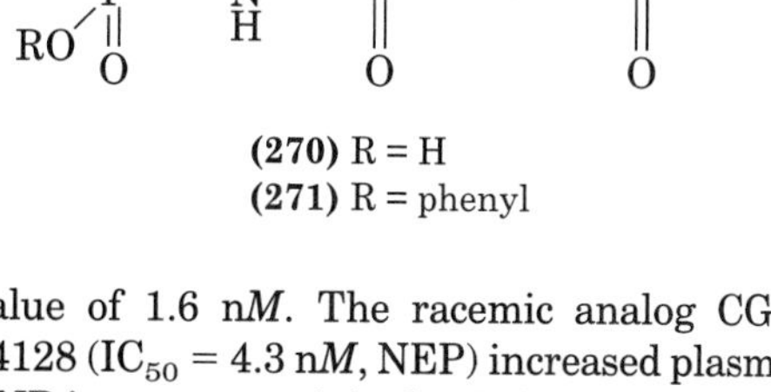

(270) R = H
(271) R = phenyl

value of 1.6 nM. The racemic analog CGS 24128 ($IC_{50} = 4.3$ nM, NEP) increased plasma ANP immunoreactivity levels by 191% in rats administered exogenous ANP(99–126) (483). CGS 24128 also potentiated the natriuretic ac-

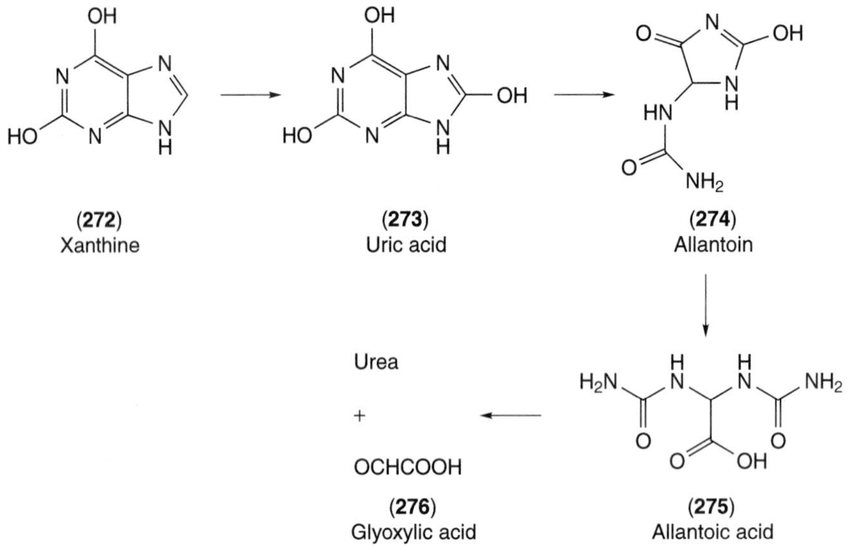

Figure 2.10. Purine metabolism.

tivity of exogenous administered ANP(99–126). Because of the poor bioavailability of CGS 24592, a series of prodrugs were investigated. CGS 25462 (**271**) provided significant and sustained antihypertensive effect in the DOCA/salt-hypertensive rat after oral administration.

2.1.14 Uricosuric Agents. In humans, one of the principal products of purine metabolism (i.e., uric acid) is implicated in several human diseases such as gout. Guanine and adenine are both converted to xanthine (**272**); oxidation, catalyzed by xanthine oxidase, yields uric acid (**273**). In humans, uric acid is the excretory product and most of it is excreted by the kidney. In most mammals, uric acid is further hydrolyzed by uricase to allantoin (**274**), a more soluble excretory product. Allantoin, in turn, is further degraded to allantoic acid (**275**) by allantoinase, and then to urea and glyoxylic acid (**276**) by allantoinase (Fig. 2.10). Uric acid is not the major pathway of nitrogen excretion in humans. Instead, the ammonia nitrogen of most amino acids, the major nitrogen source, is shunted into the urea cycle. Uric acid is mostly insoluble in acidic solutions, although alkalinity increases its solubility. At the pH of blood (pH 7.44), uric acid is present

as the monosodium salt, which is also very highly soluble and tends to form supersaturated solutions.

Uric acid forms from purines, which are liberated as a result of enzymatic degradation of tissue and dietary nucleoproteins and nucleotides, but it is also formed by purine synthesis (484). When the level of monosodium urate in the serum exceeds the point of maximum solubility, urate crystals may form, particularly in the joints and connective tissues. These deposits are responsible for the manifestations of gout. Serum urate levels can be lowered by decreasing the rate of production of uric acid or by increasing the rate of elimination of uric acid. The most common method of reducing uric acid levels is to administer uricosuric drugs, which increase the rate of elimination of uric acids by the kidneys.

2.1.14.1 Sodium Salicylate. The uricosuric properties of sodium salicylate (**277**, Table 2.22) were noted before 1890, and its use continued through 1950. As late as 1955, sodium salicylate was used for the long-term treatment of gout (485). For adequate uricosuric activity, however, salicylate must be administered in doses greater than 5 g/day, often resulting in serious side effects, so that its usage has gradually declined.

Table 2.22 Uricosuric Agents

No.	Generic Name	Trade Name	Structure	Reference
(277)	Salicylic acid			485
(278)	Probenecid	Benemid		487
(283)	Sulfinpyrazone	Anturane		490
(286)	Allopurinol	Zyloprim		497
(288)	Benzbromarone	Desuric, Minuric, Narcaricin		502

2.1.14.2 Probenecid. Probenecid (278, Table 2.22) was developed as a result of a search for a compound that would depress the renal tubular secretion of penicillin (486) at a time when the supply of penicillin was limited. Recognition of the uricosuric properties of probenecid resulted from prior experience with the uricosuric effects of the related compound carinamide (279) in normal subjects and in gouty subjects (487). Carinamide had been introduced as an agent for increasing penicillin blood levels by blocking its rapid excretion through the kidney. Its biological half-life was relatively short, and the search for compounds with a longer half-life that would not have to be administered so frequently led to probenecid.

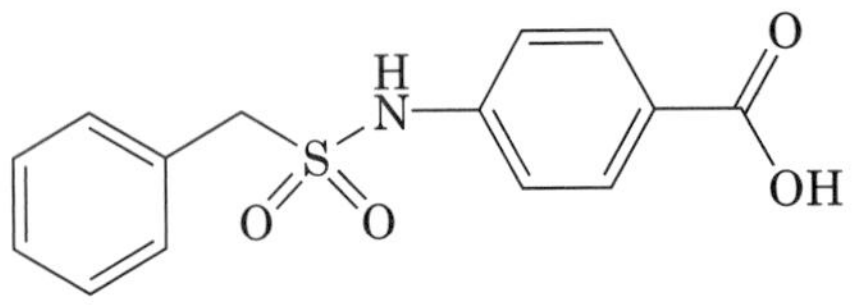

(279)

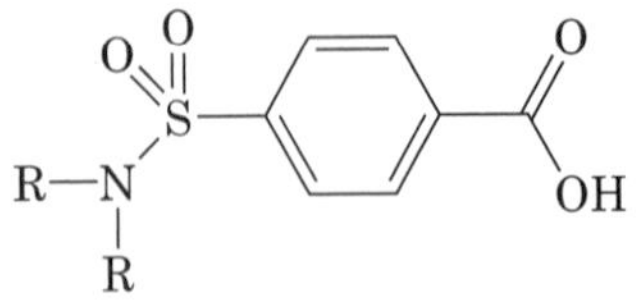

(280) R = H, Methyl, Ethyl, Propyl

In a study of a series of *N*-dialkylsulfamoyl-benzoates (280), Beyer (488) found that as the length of the *N*-alkyl groups increased, the renal clearance of the compounds decreased. This most likely results from the enhanced lipid solubility imparted by the longer alkyl groups, which would account for their complete back diffusion in acidic urine. Optimal activity was found in probenecid, the *N*-dipropyl derivative. The structure-activity relationship of probenecid congeners and that of other uricosuric agents has been reviewed in detail by Gutman (487).

Normally, a high percentage of the uric acid filtered by the glomerulus is reabsorbed by an active transport process in the proximal tubule. It is now clear that the human proximal tubule also secretes uric acid, as does the proximal tubule of many lower animals. Small doses of probenecid depresses the excretion of uric acid by blocking tubular secretion, whereas high doses lead to greatly enhanced excretion of uric acid by depressing proximal reabsorbtion of uric acid (489).

Probenecid is completely absorbed after oral administration; peak plasma levels are reached in 2–4 h. The half-life of the drug in plasma for most patients is 6–12 h. The drug is 85–95% bound to plasma proteins. The small unbound portion is filtered at the glomerulus; a much larger portion is actively secreted by the proximal tubule. The high lipid solubility of the undissociated form results in virtually complete reabsortion by back diffusion unless the urine is markedly alkaline.

Probenecid is insoluble in water, but the sodium salt is freely soluble. In the treatment of chronic gout, a single daily dose of 250 mg is given for 1 week, followed by 500 mg administered twice daily. A daily dose of up to 2 g may be required.

2.1.14.3 Sulfinpyrazone. Despite the therapeutic efficacy of phenylbutazone (281) as an

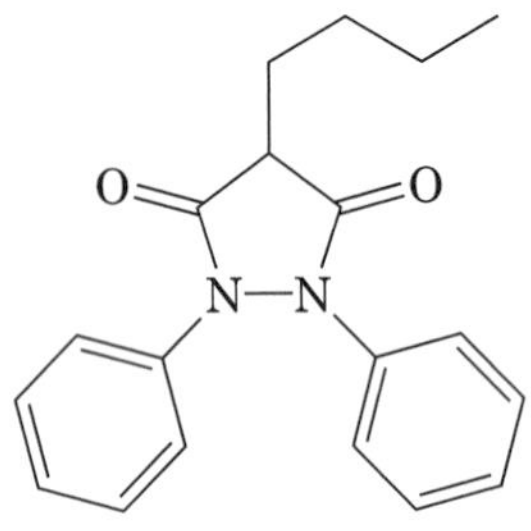

(281)

anti-inflammatory and uricosuric agent, its side effects were severe enough to preclude its continuous use in the treatment of chronic gout. Evaluation of several chemical congeners indicated that the phenylthioethyl analog of phenylbutazone (282) had promising anti-

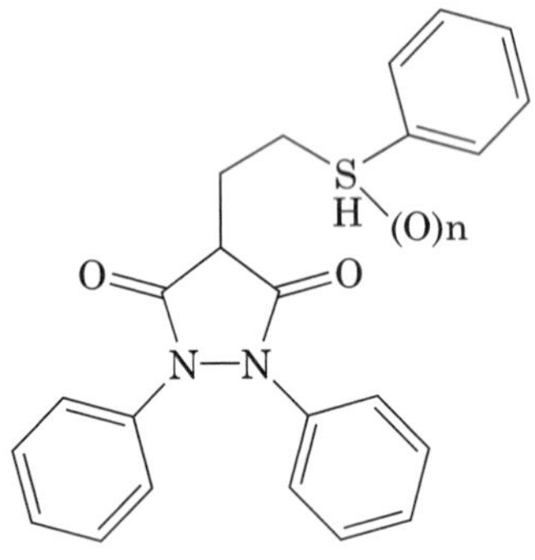

(282) n = 0
(283) n = 1
(284) n = 2

inflammatory and uricosuric activity (490). A metabolite, the sulfoxide pyrazone (283), exhibited enhanced uricosuric activity (491, 492). Interestingly, the corresponding sulfone (284) does not appear to be a metabolite (490). Sulfinpyrazone lacks the clinically striking anti-inflammatory and analgesic properties of phenylbutazone.

Sulfinpyrazone is a strong acid ($pK_a = 2.8$) and readily forms soluble salts. Evaluation of a number of congeners indicated that a low pK_a and polar side chain substituents favor uricosuric activity (493) and increase the rate of renal excretion (494). The inverse relationship between uricosuric potency and pK_a has also been confirmed in a number of 2-substituted analogs of probenecid (285) (probenecid R =

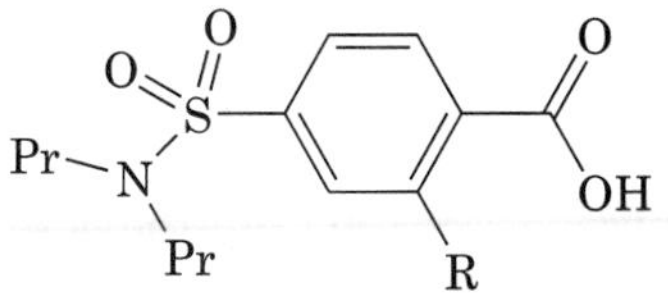

(285) R = OH, Cl, NO_2

H; 278). All three compounds were considerably stronger acids than probenecid. Evaluation in the *Cebus albifrons* monkey indicated that these compounds were about 10 times as potent as probenecid when compared on the basis of concentration of drug in plasma (495).

In small doses, as seen with other uricosuric agents, sulfinpyrazone may reduce the excretion of uric acid, presumably by inhibiting secretion but not tubular reabsorbtion. Its uricosuric action is additive to that of probenecid and phenylbutazone but antagonizes that of the salicylates. Sulfinpyrazone can displace to an unusual degree other organic anions that are bound extensively to plasma protein (e.g., sulfonamides and salicylates), thus altering their tissue distribution and renal excretion (489, 496). Depending on concomitant medication, this may be a clinical asset or liability.

For the treatment of chronic gout, the initial dosage is 100–200 mg/day. After the first week the dose may be increased up to 400 mg/day until a satisfactory lowering of plasma uric acid is achieved.

2.1.14.4 Allopurinol. Allopurinol (286) does not reduce serum uric acid levels by increasing renal uric acid excretion; instead it lowers plasma urate levels by inhibiting the final steps in uric acid biosynthesis.

Uric acid in humans is formed primarily by xanthine oxidase-catalyzed oxidation of hypoxanthine and xanthine (272) to uric acid

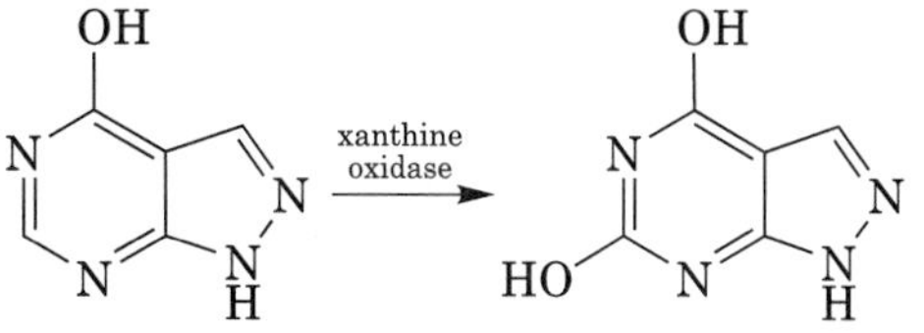

(286)   (287)

(273). Allopurinol (286) and its primary metabolite, alloxanthine (287), are inhibitors of xanthine oxidase.

Inhibition of the last two steps in uric acid biosynthesis by blocking xanthine oxidase reduces the plasma concentration and urinary excretion of uric acid and increases the plasma levels and renal excretion of the more soluble oxypurine precursors. Normally, in humans the urinary purine content is almost solely uric acid; treatment with allopurinol results in the urinary excretion of hypoxanthine, xanthine, and uric acid, each with its independent solubility. Lowering the uric acid concentration in plasma below its limit of solubility facilitates the dissolution of uric acid deposits. The effectiveness of allopurinol in the treatment of gout and hyperuricemia that results from hematogical disorders and antineoplastic therapy has been demonstrated (497–499).

For the control of hyperuricemia in gout, an initial daily dose of 100 mg is increased weekly at intervals by 100 mg. The usual daily maintenance dose for adults is 300 mg.

2.1.14.5 Benzbromarone. Benzbromarone (288) is a benzofuran derivative that has been

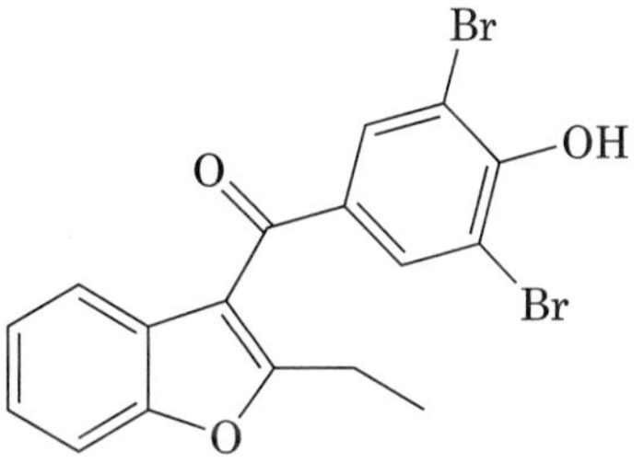

(288)

reported to lower serum urate levels in animals and human studies. In normal and hyperuricaemic subjects, benzbromarone reduced serum uric acid levels by one-third to one-half (500, 501). In comparison with other urate-lowering drugs, 80 mg of micronized or 100 mg of nonmicronized benzbromarone had equal urate-lowering activity to 1–1.5 g of probenecid or 400–800 mg of sulfinpyrazone (500, 502).

The mechanism of the urate-lowering activity of benzbromarone appears to be attributable to its uricosuric activity. In rats, benzobromarone inhibited urate reabsorption in the proximal tubules when given at 10 mg/kg i.v. (503). In isolated rat liver preparation, benzbromarone inhibits xanthine oxidase *in vitro* but not *in vivo* (504). In humans, this compound only weakly inhibits xanthine oxidase and no increase in urinary excretion of xanthine or hypoxanthine was observed (505). After oral administration, about 50% of benzbromarone is absorbed. The drug undergoes extensive dehalogenation in the liver and is excreted mainly in the bile and feces. For control of gout the usual therapeutic dose is 100–200 mg daily. Benzbromarone has few side effects and is usually well tolerated.

3 CONCLUSION

The development and therapeutic use of diuretic agents constitutes one of the most significant advances in medicine made during the twentieth century. Continuous progress has been made during this time on the development of safer and more effective diuretic agents. Between 1920 and 1950, a large number of organic mercurials were prepared and evaluated as diuretics. Because of the lack of oral activity and toxicity of these compounds, research efforts were focused on the development of orally effective nonmercurial diuretics. The carbonic anhydrase inhibitors, developed in 1950 and later, were orally active but upset the acid-base balance and could be given only intermittently. The thiazide diuretics, developed in the late 1950s, represented a true advance in the treatment of edema. They were remarkably nontoxic and effective in most cases. It very soon became apparent that not only were they effective diuretics, they were also useful in the treatment of hypertension by themselves or in combination with other antihypertensive drugs.

Four side effects were noticed after the widespread and prolonged use of the thiazide diuretics:

1. potassium depletion
2. uric acid retention
3. hyperglycemia
4. increased plasma lipids

Potassium depletion has been encountered most frequently. The kaliuretic effect of the thiazides can be compensated for by supplementary dietary potassium; nevertheless, research was directed toward the development of potassium-sparing diuretics. Amiloride (1965), spironolactone (1959), and triamterene (1965) were discovered as a result of this effort; these compounds are weak diuretics, however, and are generally used in combination with other diuretics (e.g., hydrochlorothiazide).

The next step was the discovery of the high ceiling or loop diuretics [e.g., ethacrynic acid (1962), furosemide (1963), and bumetanide (1971)], which are shorter acting and more potent than the thiazide diuretics. They too have the same potential side effects as the thiazides. One advantage of the loop diuretics is their efficacy in chronic renal insufficiency, particularly in cases with low glomerular filtration rates.

A large volume of highly technical information has been published over the past 15 years regarding this therapeutic area. More sensitive analytical techniques have been developed, so that data regarding bioavailability and pharmokinetics are now available for diuretics that are currently prescribed and that are in development. Advances in renal and ion-transport research have led to a more precise understanding of the cellular mechanisms of actions of the various classes of diuretic agents. This has aided in the design of newer, more effective agents.

Diuretics introduced into more recent clinical studies include (*1*) newer, more potent loop diuretics such as torasemide and azosemide, (*2*) development of uricosuric diuretics, (*3*) newer-generation sulfamoyl diuretics, and (*4*) development of neutral endopeptidase inhibitors.

Since the 1960s, diuretics have been used to treat millions of patients with hypertension. With the number of adverse effects seen with long-term diuretic treatment, such as hypokalemia, hypercholesterolemia, and hyperglycemia, and because diuretic-based antihypertensive drug trials have failed to show a reduction in the incidence of myocardial infarction, this practice has become controversial. Today many other therapies exist for the treatment of hypertension, such as calcium channel blockers, angiotensin-converting enzyme, and angiotensin receptor blocker inhibitors. However, because diuretics are convenient, inexpensive, and generally are well tolerated, they will probably continue to play a role in the treatment of hypertension.

REFERENCES

1. W. B. Schwartz, N. Engl. J. Med., 240, 173 (1949).

2. C. Rouiller, The Kidney, Morphology, Biochemistry, Physiology, Academic Press, New York, 1969.

3. B. Brenner and F. Rector, The Kidney, Saunders, Philadelphia, 1986, p. 9.

4. W. Foye, Principles of Medicinal Chemistry, Lea and Febiger, Philadelphia, 1989, p. 447.

5. G. Giebisch, Eur. J. Clin. Pharmacol., 44 (Suppl. 1), S3 (1993); (a) B. O. Rose, Kidney Int., 337 (1991).

6. P. A. Preisig and F. C. Rector, Am. J. Physiol. Renal Fluid Electrolyte Physiol., 255, F461 (1988).

7. C. Sardet, A. Franchi, and J. Pouyssegur, Cell, 56, 271 (1989).

8. R. Green and G. Giebisch, Am. J. Physiol. Renal Fluid Electrolyte Physiol., 257, F669 (1989).

9. C. A. Berry and F. C. Rector, Kidney Int., 36, 403 (1989).

10. L. Schild, G. Giebisch, and L. P. Karniski, J. Clin. Invest., 79, 32 (1987).

11. W. F. Boron and E. L. Boulpaep, Kidney Int., 36, 392 (1989).

12. R. Greger, Pfluegers Arch., 390, 38 (1981).

13. F. Bronner, Am. J. Physiol. Renal Fluid Electrolyte Physiol., 257, F707 (1989).

14. J. L. Borke, A. Minami, A. Verma, J. T. Penniston, and R. Kumas, Kidney Int., 34, 262 (1988).

15. T. Shimuzu, K. Yoshitouri, M. Nakamura, and M. Imai, Am. J. Physiol. Renal Fluid Electrolyte Physiol., 259, F408 (1990).

16. L. Palmer and I. Edelman, Ann. N. Y. Acad. Sci., 372, 1 (1981).

17. M. Martinez-Maldorado and H. Cordovc, Kidney Int., 38, 632 (1990).

18. S. C. Hebert, Semin. Nephrol., 19, 504 (1999).

19. R. Greger, Nephrol. Dial. Transplant., 14, 536 (1999).

20. K. H. Beyer, Perspect. Biol. Med., 19, 500 (1976).

21. N. L. Lipshitz, Z. Hadidian, and A. Kerpcsar, J. Pharmacol. Exp. Ther., 79, 97 (1943).

22. A. A. Renzi, J. J. Chart, and R. Gaunt, Toxicol. Appl. Pharmacol., 1, 406 (1959).

23. D. M. Zall, D. Fisher, and M. Q. Garner, Anal. Chem., 28, 1665 (1956).

24. T. W. K. Hill and P. J. Randall, J. Pharm. Pharmacol., 28, 552 (1976).

25. G. M. Fanelli Jr., D. I. Bohn, A. Scriabine, and K. H. Beyer Jr., J. Pharmacol. Exp. Ther., 200, 402 (1977).

26. G. M. Fanelli, D. L. Bohn, and H. F. Russo, Comp. Biochem. Physiol., 33, 459 (1970).

27. B. Stavric, W. J. Johnson, and H. C. Grice, Proc. Soc. Exp. Biol. Med., 130, 512 (1969); B. Stravric, E. A. Nera, W. J. Johnson, and F. A. Salem, Invest. Urol., 11, 3 (1973).

28. F. W. Wolff, W. W. Parmley, K. White, and R. Okun, J. Am. Med. Assoc., 185, 568 (1963).

29. I. I. A. Tabachnick, A. Gulbenkian, and A. Yannell, Life Sci., 4, 1931 (1965).

30. D. Morgan and C. Davison, Br. Med. J., 280, 295 (1980).

31. A. Amery and C. Bulpitt, Lancet, 1, 681 (1978).

32. S. MacMahon and G. MacDonald, Am. J. Med., 80, 40 (1986).

33. J. F. Seely and J. H. Dirks, J. Clin. Invest., 48, 2330 (1969).

34. R. C. Blantz, J. Clin. Invest., 54, 1135 (1974).

35. G. H. Mudge in L. S. Goodman and A. Gilman, Eds., Pharmacological Basis of Therapeutics, 5th ed., Section VII, Macmillan, New York, 1975, p. 809.

36. R. W. Berliner, J. H. Dirks, and W. J. Cirksena, Ann. N. Y. Acad. Sci., 139, 424 (1966).

37. J. R. Clapp and R. R. Robinson, Am. J. Physiol., 215, 228 (1968).

38. R. L. Evanson, E. A. Lockhart, and J. H. Dirks, Am. J. Physiol., 222, 282 (1972).

39. G. M. Fanelli Jr., D. L. Bohn, S. S. Reily, and I. M. Weiner, *Am. J. Physiol.*, **224**, 985 (1973).

40. E. Jendrassik, *Arch. Klin. Med.*, **38**, 499 (1886).

41. A. Vogl, *Am. Heart J.*, **39**, 881 (1950).

42. I. M. Weiner, R. I. Levy, and G. H. Mudge, *J. Pharmacol. Exp. Ther.*, **138**, 96 (1962).

43. T. W. Clarkson and J. J. Vostal, *Modern Diuretic Therapy in the Treatment of Cardiovascular and Renal Disease*, Excerpta Medica, Amsterdam, 1973, pp. 229–240.

44. E. J. Cafruny, K. C. Cho, V. Nigrovic, and A. Small in ref. 38, pp. 124–134.

45. G. de Stevens, *Diuretics: Chemistry and Pharmacology*, Academic Press, New York, 1963, p. 38.

46. R. H. Kessler, R. Lozano, and R. F. Pitts, *J. Clin. Invest.*, **36**, 656 (1957).

47. R. C. Batterman, D. Unterman, and A. C. DeGraff, *J. Am. Med. Assoc.*, **140**, 1268 (1949).

48. W. Modell, *Am. J. Med. Sci.*, **231**, 564 (1956).

49. L. H. Werner and C. R. Scholz, *J. Am. Chem. Soc.*, **76**, 2453 (1954).

50. R. H. Chaney and R. F. Maronde, *Am. J. Med. Sci.*, **231**, 26 (1956).

51. J. Moyer, S. Kinard, and R. Herschberger, *Antibiot. Med. Clin. Ther.*, **3**, 179 (1956).

52. T. H. Maren, *Bull. Johns Hopkins Hosp.*, **98**, 159 (1956); T. H. Maren in H. Herkin, Ed., *Handbook of Experimental Pharmacology: Diuretics*, Vol. **24**, Springer-Verlag, Berlin/Heidelberg/New York, 1969, p. 195.

53. T. H. Scholz, J. M. Sondey, W. C. Randall, H. Schwam, W. J. Thompson, P. J. Mallorga, M. F. Sugrue, and S. L. Graham, *J. Med. Chem.*, **36**, 2134 (1993).

54. K. L. Shepard, S. L. Graham, R. J. Hudcosky, S. R. Michelson, T. H. Scholz, H. Schwam, A. M. Smith, J. M. Sondey, K. M. Strohmaier, R. L. Smith, and M. F. Sugrue, *J. Med. Chem.*, **34**, 3098 (1991).

55. J. J. Baldwin, G. S. Ponticello, and M. F. Sugrue, *Drugs Future*, **15**, 350 (1990).

56. M. F. Sugrue, *Prog. Ret. Eye Res.*, **19**, 87 (2000).

57. L. H. Silver, *Surv. Ophthalmol.*, **44** (Suppl. 2), 147 (2000).

58. C. T. Supuran and A. Scozzafava, *Expert Opin. Ther. Patents*, **10**, 575 (2000).

59. E. Larson, R. Roach, R. Schoene, and T. Hornbein, *J. Am. Med. Assoc.*, **248**, 328 (1982).

60. M. B. Strauss and H. Southworth, *Bull. Johns Hopkins Hosp.*, **63**, 41 (1938).

61. T. Mann and D. Keilin, *Nature*, **146**, 164 (1940).

62. H. W. Davenport and A. E. Wilhelmi, *Proc. Soc. Exp. Biol. Med.*, **48**, 53 (1941).

63. R. F. Pitts and R. S. Alexander, *Am. J. Physiol.*, **144**, 239 (1945).

64. W. B. Schwarts, *N. Engl. J. Med.*, **240**, 173 (1949).

65. M. Laski, *Semin. Nephrol.*, **6**, 210 (1986).

66. H. A. Krebs, *Biochemistry*, **43**, 525 (1948).

67. R. O. Roblin Jr. and J. W. Clapp, *J. Am. Chem. Soc.*, **72**, 4890 (1950).

68. W. H. Miller, A. M. Desser, and R. O. Roblin Jr., *J. Am. Chem. Soc.*, **72**, 4893 (1950).

69. R. W. Young, K. H. Wood, J. A. Eichler, J. R. Vaughan Jr., and G. W. Anderson, *J. Am. Chem. Soc.*, **78**, 4649 (1956).

70. K. H. Beyer and J. E. Baer, *Pharmacol. Rev.*, **13**, 517 (1961).

71. T. H. Maren, E. Mayer, and B. C. Wadsworth, *Bull. Johns Hopkins Hosp.*, **95**, 199 (1954).

72. R. V. Ford, C. L. Spurr, and J. H. Moyer, *Circulation*, **16**, 394 (1957).

73. J. R. Vaughan Jr., J. A. Eichler, and G. W. Anderson, *J. Org. Chem.*, **21**, 700 (1956).

74. T. H. Maren, *J. Pharmacol. Exp. Ther.*, **117**, 385 (1956).

75. D. M. Travis, *J. Pharmacol. Exp. Ther.*, **167**, 253 (1969).

76. R. T. Kunan Jr., *J. Clin. Invest.*, **51**, 294 (1972).

77. A. Posner, *Am. J. Ophthalmol.*, **45**, 225 (1958).

78. T. W. K. Hill and P. J. Randall, *J. Pharm. Pharmacol.*, **28**, 552 (1976).

79. K. H. Beyer Jr. and J. E. Baer, *Med. Clin. North Am.*, **59**, 735 (1975).

80. J. M. Sprague, *Ann N. Y. Acad Sci.*, **71**, 328 (1958).

81. F. C. Novello, S. C. Bell, L. A. Abrams, C. Ziegler, and J. M. Sprague, *J. Org. Chem.*, **25**, 965 (1960).

82. F. C. Novello and J. M. Sprague, *J. Am. Chem. Soc.*, **79**, 2028 (1957).

83. F. J. Lund and W. Kobinger, *Acta Pharmacol. Toxicol.*, **16**, 297 (1960).

84. H. Horstmann, H. Wollweber, and K. Meng, *Arzneim.-Forsch.*, **17**, 659 (1967).

85. K. Meng and G. Kroneberg, *Arzneim.-Forsch.*, **17**, 659 (1967).

86. B. Duhm, W. Maul, H. Medenwald, P. Patzchke, and L. A. Wengner, *Arzneim.-Forsch.*, **17**, 672 (1972).

87. C. B. Wilson and W. M. Kirkendall, *J. Pharmacol. Exp. Ther.*, **173**, 422 (1970).

88. R. J. Santos, V. Paz-Martinez, J. K. Lee, and J. H. Nodine, *Int. J. Clin. Pharmacol.*, **3**, 14 (1970).

89. C. B. Wilson and W. M. Kirkendall, *J. Pharmacol. Exp. Ther.*, **171**, 288 (1970).

90. W. H. R. Auld and W. R. Murdoch, *Br. Med. J.*, **4**, 786 (1971).

91. S. J. Jachuck, *Br. Med. J.*, **3**, 590 (1972).

92. K. H. Beyer Jr., *Perspect. Biol. Med.*, **19**, 500 (1976).

93. F. J. Lund and W. Kobinger, *Acta Pharmacol. Toxicol.*, **16**, 297 (1960).

94. E. H. Wiseman, E. C. Schreiber, and R. Pinson Jr., *Biochem. Pharmacol.*, **11**, 881 (1962).

95. T. H. Maren and C. E. Wiley, *J. Pharmacol. Exp. Ther.*, **143**, 230 (1964).

96. R. L. Hauman and J. M. Weller, *Clin. Pharmacol. Ther.*, **1**, 175 (1960).

97. S. Y. P'an, A. Scriabine, D. E. McKersie, and W. M. McLamore, *J. Pharmacol. Exp. Ther.*, **128**, 122 (1960).

98. G. deStevens, *Diuretics: Chemistry and Pharmacology*, Academic Press, New York, 1963, p. 100.

99. E. J. Cragoe Jr., J. A. Nicholson, and J. M. Sprague, *J. Med. Pharm. Chem.*, **4**, 369 (1961).

100. J. G. Topliss, M. H. Sherlock, H. Reimann, L. M. Konzelman, E. P. Shapiro, B. W. Pettersen, H. Schneider, and N. Sperber, *J. Med. Chem.*, **6**, 122 (1963).

101. A. A. Rubin, F. E. Roth, R. M. Taylor, and H. Rosenkilde, *J. Pharmacol. Exp. Ther.*, **136**, 344 (1962).

102. B. Beerman, M. Groschinski-Grind, and B. Lindstrom, *Eur. J. Clin. Pharmacol.*, **11**, 203 (1977).

103. D. A. Garcia and E. R. Yendt, *Can. Med. Assoc. J.*, **103**, 473 (1970).

104. J. M. Tran, M. A. Farrel, and P. P. Fanestil, *Am. J. Physiol. Renal Fluid Electrolyte Physiol.*, **258**, F908 (1990).

105. D. B. Hunninghake, S. King, and K. LaCroix, *Int. J. Clin. Pharmacol.*, **20**, 151 (1982).

106. M. Hohenegger, *Adv. Clin. Pharmacol.*, **9**, 1 (1975).

107. M. Goldberg in J. Orloff and R. W. Berlinger, Eds., *Handbook of Physiology, Section 8*, American Physiology Society, Washington, DC, 1973, pp. 1003–1031.

108. E. Perez-Stable and P. V. Caralis, *Am. Heart J.*, **106**, 245 (1983).

109. G. de Stevens, L. H. Werner, A. Halamandaris, and S. Ricca Jr., *Experientia*, **14**, 463 (1958).

110. J. E. Baer, H. F. Russo, and K. H. Beyer, *Proc. Soc. Exp. Biol. Med.*, **100**, 442 (1959).

111. A. F. Esch, I. M. Wilson, and E. D. Freis, *Med. Ann. Dist. Columbia*, **28**, 9 (1959).

112. C. W. H. Havard and J. C. B. Fenton, *Br. Med. J.*, **1**, 1560 (1959).

113. H. Losse, H. Wehmeyer, W. Strobel, and H. Wesselkock, *Muench Med. Wochenschr.*, **101**, 677 (1959).

114. W. Hollander, A. V. Chobanian, and R. W. Wilkins in J. H. Moyer, Ed., *Hypertension*, Saunders, Philadelphia, 1959, p. 570.

115. L. H. Werner, A. Halamandaris, S. Ricca Jr., L. Dorfman, and G. deStevens, *J. Am. Chem. Soc.*, **82**, 1161 (1960).

116. E. J. Cragoe Jr., O. W. Woltersdorf Jr., J. E. Baer, and J. M. Sprague, *J. Med. Chem.*, **5**, 896 (1962).

117. J. G. Topliss, M. H. Sherlock, F. H. Clarke, M. C. Daly, B. W. Pettersen, J. Lipski, and N. Sperber, *J. Org. Chem.*, **26**, 3842 (1961).

118. J. Klosa and H. Voigt, *J. Prakt. Chem.*, **16**, 264 (1962).

119. J. Klosa, *J. Prakt. Chem.*, **18**, 225 (1962).

120. J. Klosa, *J. Prakt. Chem.*, **33**, 298 (1966).

121. J. Klosa, *J. Prakt. Chem.*, **21**, 176 (1963).

122. W. J. Close, L. R. Swett, L. E. Brady, J. H. Short, and M. Vernsten, *J. Am. Chem. Soc.*, **82**, 1132 (1960).

123. J. H. Short and U. Biermacher, *J. Am. Chem. Soc.*, **82**, 1135 (1960).

124. J. H. Short and L. R. Swett, *J. Org. Chem.*, **26**, 3428 (1961).

125. C. T. Holdrege, R. B. Babel, and L. C. Cheney, *J. Am. Chem. Soc.*, **81**, 4807 (1959).

126. W. M. McLamore and G. D. Laubach, U.S. Pat. 3,111,517 (1963).

127. J. M. McManus, U.S. Pat. 3,009,911 (1961).

128. C. W. Whitehead, J. J. Traverso, H. R. Sullivan, and F. J. Marshall, *J. Org. Chem.*, **26**, 2814 (1961).

129. C. W. Whitehead and J. J. Traverso, *J. Org. Chem.*, **27**, 951 (1962).

130. W. Graf, E. Girod, E. Schmid, and W. G. Stoll, *Helv. Chim. Acta*, **42**, 1085 (1959).

131. E. G. Stenger, H. Witz, and R. Pulver, *Schweiz. Med. Wochenschr.*, **89**, 1130 (1959).

132. R. Veyrat, E. F. Arnold, and A. Duckert, *Schweiz. Med. Wochenschr.*, **89**, 1133 (1959).

133. F. Reutter and Schaub, *Schweiz. Med. Wochenschr.*, **89**, 1158 (1959).

134. W. Leppla, H. Buch, and G. A. Jutzler, *Ger. Med. Monthly*, **5**, 402 (1960).

135. M. Fuchs, B. E. Newman, S. Irie, R. Maranoff, E. Lippman, and J. H. Moyer, *Curr. Ther. Res.*, **2**, 11 (1960).

136. W. E. Bowlus and H. G. Langford, *Clin. Pharmacol. Ther.*, **5**, 708 (1964).

137. E. Jucker, A. Lindenmann, E. Schenker, E. Fluckiger, and M. Taeschler, *Arzneim.-Forsch.*, **13**, 269 (1963).

138. B. Terry and J. B. Hook, *J. Pharmacol. Exp. Ther.*, **160**, 367 (1968).

139. V. Parsons and R. Kemball Price, *Practitioner*, **195**, 648 (1965).

140. D. H. Kaump, R. L. Fransway, L. T. Blouin, and D. Williams, *J. New Drugs*, **4**, 21 (1964).

141. M. L. Hoefle, L. T. Blouin, H. A. DeWald, A. Holmes, and D. Williams, *J. Med. Chem.*, **11**, 970 (1968).

142. L. T. Blouin, D. H. Kaump, R. L. Fransway, and D. Williams, *J. New Drugs*, **3**, 302 (1963).

143. E. V. Mackay and S. K. Khoo, *Med. J. Aust.*, **1**, 607 (1969).

144. P. Milliez and P. Tcherdakoff, *Curr. Med. Res. Opin.*, **3**, 9 (1975).

145. V. Anania, M. S. Desole, and E. Miele, *Riv. It. Biol. Med.*, **2**, 135 (1982).

146. G. Cignarella, P. Sanna, E. Miele, A. Anania, and M. Desole, *J. Med. Chem.*, **24**, 1003 (1981).

147. G. Cignarella, D. Barlocco, D. Landriania, G. Pinna, G. Andrivoli, and G. Dona, *Farmaco*, **46**, 527 (1991).

148. E. J. Cornish, G. E. Lee, and W. R. Wragg, *J. Pharm. Pharmacol.*, **18**, 65 (1966).

149. A. F. Lant, W. I. Baba, and G. M. Wilson, *Clin. Pharmacol. Ther.*, **7**, 196 (1966).

150. W. I. Baba, A. F. Lant, and G. M. Wilson, *Clin. Pharmacol. Ther.*, **7**, 212 (1966).

151. J. L. Verov, D. S. Tunstall-Pedoe, and T. J. C. Cooke, *Br. J. Clin. Pract.*, **20**, 351 (1966).

152. (a) K. Corbett, S. A. Edwards, G. E. Lee, and T. L. Threlfall, *Nature*, 208, 286 (1965); (b) E. Cohen, B. Klarberg, and J. R. Vaughan Jr., *J. Am. Chem. Soc.*, **82**, 2731 (1960).

153. B. V. Shetty, L. A. Campanella, T. L. Thomas, M. Fedorchuk, T. A. Davidson, L. Michelson, H. Volz, and S. E. Zimmerman, *J. Med. Chem.*, **13**, 886 (1970).

154. R. H. Sellers, M. Fuchs, G. Onesti, C. Swartz, A. N. Brest, and J. H. Moyer, *Clin. Pharmacol. Ther.*, **3**, 180 (1962).

155. W. N. Suki, F. Dawoud, G. Eknoyan, and M. Martinez-Maldonado, *J. Pharmacol. Exp. Ther.*, **180**, 6 (1972).

156. J. W. Smiley, G. Onesti, and C. Swatz, *Clin. Pharmacol. Ther.*, **13**, 336 (1972).

157. M. F. Michelis, F. DeRubertis, N. P. Beck, R. H. McDonald Jr., and B. B. Davis, *Clin. Pharmacol. Ther.*, **11**, 821 (1970).

158. E. J. Belair, A. I. Cohen, and J. Yelnoski, *Br. J. Pharm.*, **45**, 476 (1972).

159. B. J. Materson, J. L. Hotchkiss, J. S. Barkin, B. H. Rietberg, K. Bailey, and E. C. Perez-Stable, *Curr. Ther. Res.*, **14**, 545 (1972).

160. R. M. Pilewski, E. T. Scheib, J. R. Misage, E. Kessler, E. Krifcher, and A. P. Shapiro, *Clin. Pharmacol. Ther.*, **12**, 843 (1971).

161. F. Costa, R. Caldari, E. Ambrosion, and B. Magnani, *Curr. Ther. Res.*, **32**, 359 (1982).

162. J. J. Mencel, J. R. Regan, J. Barton, P. R. Menard, J. G. Bruno, R. R. Calvo, B. E. Kornberg, A. Schwab, E. S. Neiss, and J. T. Suh, *J. Med. Chem.*, **33**, 1606 (1990).

163. V. Cecchetti, A. Fravolini, F. Schiaffella, O. Tabarrini, G. Bruni, and G. Segre, *J. Med. Chem.*, **36**, 157 (1993).

164. H. J. Lang, B. Knabe, R. Muschaweck, M. Hropot, and E. Linder in E. J. Cragoe, Ed., *Diuretic Agents, ACS Symposium Series* **83**, American Chemical Society, Washington, DC, 1978, p. 24.

165. L. Hofman, *Arch. Int. Pharmacodyn. Ther.*, **169**, 189 (1967).

166. B. Johnson, *Clin. Pharmacol. Ther.*, **11**, 77 (1970).

167. *Drugs Future*, **10**, 298 (1985).

168. C. T. Dollery, E. H. O. Parry, and D. S. Young, *Lancet*, **1**, 947 (1964).

169. J. F. Maher and G. E. Schreiner, *Ann. Intern. Med.*, **62**, 15 (1965).

170. V. K. G. Pillay, F. D. Schwarts, K. Aimi, and R. M. Kark, *Lancet* (1969).

171. K. H. Beyer, J. E. Baer, J. K. Michaelson, and H. F. Russo, *J. Pharmacol. Exp. Ther.*, **147**, 1 (1965).

172. M. Goldberg, *Ann. N. Y. Acad. Sci.*, **139**, 443 (1966).

173. E. M. Schultz, E. J. Cragoe Jr., J. B. Bicking, W. A. Bolhofer, and J. M. Sprague, *J. Med. Chem.*, **5**, 660 (1962).

174. J. M. Sprague, *Ann. Rep. Med. Chem.*, **5**, X1 (1970).

175. J. B. Bicking, W. J. Holtz, L. S. Watson, and E. J. Cragoe Jr., *J. Med. Chem.*, **19**, 530 (1976).

176. D. E. Duggan and R. M. Noll, *Arch. Biochem. Biophys.*, **109**, 388 (1965).

177. E. M. Schultz, J. B. Bicking, A. A. Deana, N. P. Gould, T. P. Strobaugh, L. S. Watson, and E. J. Cragoe Jr., *J. Med. Chem.*, **19**, 783 (1976).

178. Z. E. Dziewanowsak, K. F. Tempero, F. Perret, G. Hitzenberger, and G. H. Besselaar, *Clin. Res.*, **24**, 253A (1976).

179. K. F. Tempero, G. Hitzenberger, Z. E. Dziewanowska, and H. Halkin, *Clin. Pharmacol.*, **19**, 116 (1976).

180. K. F. Tempero, J. A. Vedin, C. E. Wilhelmsson, P. Lund-Johansen, C. Vorburger, C. Bolongnese, and Z. E. Dziewanowska, *Clin. Pharmacol. Ther.*, **21**, 97 (1977).

181. E. Blain, G. Fanelli, and J. Irvin, *Clin. Exp. Hypertens.*, **4**, 161 (1982).

182. R. McKenzie, T. Knight, and E. J. Weinman, *Proc. Soc. Exp. Biol. Med.*, **153**, 202 (1976).

183. E. J. Weinman, T. Knight, R. M. McKenzie, and G. Eknoyan, *Clin. Res.*, **24**, 416A (1976).

184. A. G. Zacchei, T. I. Wishousky, B. H. Arison, and G. M. Fanelli Jr., *Drug Metab. Dispos.*, **4**, 479 (1976).

185. A. G. Zacchei and T. I. Wishousky, *Drug Metab. Dispos.*, **4**, 490 (1976).

186. O. Woltersdorf, S. deSolm, E. Schults, and E. Cragoe, *J. Med. Chem.*, **20**, 1400 (1977).

187. A. G. Zacchei, T. I. Wishousky, and L. S. Watson, *Drug Metab. Dispos.*, **6**, 313 (1978).

188. E. J. Cragoe, N. P. Gould, O. W. Woltersdorf, C. Ziegler, R. S. Bourke, L. R. Nelson, H. K. Kimelberg, J. B. Waldman, A. J. Popp, and N. Sedransk, *J. Med. Chem.*, **25**, 567 (1982).

189. G. Shutske, L. Setescak, R. Allen, L. Davis, R. Effland, K. Ranborn, J. Kitzen, J. Wilken, and W. Novick, *J. Med. Chem.*, **25**, 36 (1982).

190. J. M. Kitzen, M. A. Schwenkler, P. R. Bixby, S. J. Wilson, G. Shutske, L. Setescak, R. Allen, and I. Rosenblum, *Life Sci.*, **27**, 2547 (1980).

191. W. Hoffman, O. Woltersdorf, F. Novello, E. Cragoe, J. Springer, L. Watson, and G. Fanelli, *J. Med. Chem.*, **24**, 865 (1981).

192. (a) J. Plattner, A. Fung, J. Parks, R. Pariza, S. Crowley, A. Pernet, P. Runnel, and P. Dodge, *J. Med. Chem.*, **27**, 1016 (1984); (b) R. R. Luther, G. L. Ringham, E. W. Thomas, K. J. Patterson, and K. G. Tolman, *J. Clin. Pharmacol.*, **28**, 795 (1988).

193. W. Siedel, K. Strum, and W. Scheurich, *Chem. Ber.*, **99**, 345 (1966).

194. R. J. Timmerman, F. R. Springman, and R. K. Thoms, *Curr. Ther. Res.*, **6**, 88 (1964).

195. R. Muschaweck and P. Hajdu, *Arzneim.-Forsch.*, **14**, 46 (1964).

196. A. Haussler and P. Hajdu, *Arzneim.-Forsch.*, **14**, 710 (1964).

197. A. Haussler and H. Wicha, *Arzneim.-Forsch.*, **15**, 81 (1965).

198. J. B. Hook, A. H. Blatt, M. J. Brody, and H. E. Williamson, *J. Pharmacol. Exp. Ther.*, **154**, 667 (1966).

199. J. H. Ludens, J. B. Hook, M. J. Brody, and H. E. Williamson, *J. Pharmacol. Exp. Ther.*, **163**, 456 (1968).

200. W. Stokes and L. C. A. Nunn, *Br. Med. J.*, **2**, 910 (1964).

201. W. M. Kirkendall and J. H. Stein, *Am. J. Cardiol.*, **22**, 162 (1968).

202. C. R. Bariso, I. B. Hanenson, and T. E. Gaffney, *Curr. Ther. Res.*, **12**, 333 (1970).

203. R. G. Muth, *J. Am. Med. Assoc.*, **195**, 1066 (1966).

204. D. S. Silverberg, R. A. Ulan, M. A. Baltzan, and R. B. Baltzan, *Can. Med. Assoc. J.*, **103**, 129 (1970).

205. O. H. Morelli, L. I. Moledo, E. Alanis, O. L. Gaston, and O. Terzaghi, *Postgrad. Med. J.*, **47** (April Suppl.), 29 (1972).

206. P. W. Feit, H. Bruun, and C. K. Nielsen, *J. Med. Chem.*, **13**, 1071 (1970).

207. P. W. Feit, *J. Med. Chem.*, **14**, 432 (1971).

208. P. W. Feit and O. B. Tvaermose Nielsen, *J. Med. Chem.*, **15**, 79 (1972).

209. P. W. Feit, O. B. Tvaermose Nielsen, and N. Rastrup-Andersen, *J. Med. Chem.*, **16**, 127 (1973).

210. O. B. Tvaermose Nielsen, H. Bruun, C. Bretting, and P. W. Feit, *J. Med. Chem.*, **18**, 41 (1975).

211. P. W. Feit, O. B. Tvaermose Nielsen, and H. Bruun, *J. Med. Chem.*, **17**, 572 (1974).

212. O. B. Tvaermose Nielsen, C. K. Nielsen, and P. W. Feit, *J. Med. Chem.*, **16**, 1170 (1973).

213. P. W. Feit and O. B. Tvaermose Nielsen, *J. Med. Chem.*, **19**, 402 (1976).

214. P. W. Feit, O. B. Tvaermose Nielsen, and H. Bruun, *J. Med. Chem.*, **15**, 437 (1972).

215. O. B. Tvaermose Nielsen and P. W. Feit in E. J. Cragoe, Ed., *Diuretic Agents, ACS Symposium Series 83*, American Chemical Society, Washington, DC, 1978, p. 12.

216. P. W. Feit, O. B. Tvaermose Nielsen, C. Bretting, and H. Bruun, U.S. Pat. 4,082,851 (1978).

217. P. W. Feit and O. B. Tvaermose Nielsen, *J. Med. Chem.*, **20**, 1687 (1977).

218. E. H. Ostergaard, M. P. Magnussen, C. Kaergaard Nielsen, E. Eilertsen, and H. H. Frey, *Arzneim.-Forsch.*, **22**, 66 (1972).

219. P. W. Feit, K. Roholt, and H. Sorensen, *J. Pharm. Sci.*, **62**, 375 (1973).

220. M. J. Asbury, P. B. B. Gatenby, S. O'Sullivan, and E. Bourke, *Br. Med. J.*, **1**, 211 (1972).

221. K. H. Olesen, B. Sigurd, E. Steiness, and A. Leth, *Acta Med. Scand.*, **193**, 119 (1973).

222. K. H. Olessen, B. Sigurd, E. Steiness, and A. Leth in A. F. Lant and G. M. Wilson, Eds., *Modern Diuretic Therapy in the Treatment of Cardiovascular and Renal Disease*, Excerpta Medica, Amsterdam, 1973, p. 155.

223. E. Bourke, M. J. A. Asbury, S. O'Sullivan, and P. B. B. Gatenby, *Eur. J. Pharmacol.*, **23**, 283 (1973).

224. S. Carriere and R. Dandavino, *Clin. Pharmacol. Ther.*, **20**, 428 (1976).

225. P. Friedman, Roch-Ramel, *J. Pharmacol. Exp. Ther.*, **203**, 82 (1977).

226. C. Brater and P. Chennavasi, *J. Clin. Pharmacol.*, **21**, 311 (1981).

227. S. Halladay, G. Sipes, and D. Carter, *Clin. Pharmacol. Ther.*, **22**, 179 (1977).

228. W. Merkel, D. Bormann, D. Mania, R. Muschaweck, and M. Hropot, *Eur. J. Med. Chem.*, **11**, 399 (1976).

229. W. McNabb, F. Nourmahamed, B. Brooks, and A. Lant, *Clin. Pharm. Ther.*, **35**, 328 (1984).

230. E. M. Giesen-Crouse, P. Fandeleur, and J. L. Imbs, *J. Pharmacol.*, **17**, 146 (1986).

231. B. Beerman and M. Grind, *Clin. Pharmacokinet.*, **13**, 254 (1987).

232. D. Brockmeier, H. G. Grigoleit, H. Heptner, and B. H. Meyer, *Meth. Find. Exp. Clin. Pharmacol.*, **8**, 731 (1986).

233. S. Clissold and Brogden, *Drugs*, **29**, 489 (1985).

234. E. Klaus, H. Alpermann, G. Caspritz, W. Linz, and B. Scholken, *Arzneim.-Forsch.*, **33**, 1273 (1983).

235. K. Kawashima, T. Hayakawa, H. Oohata, K. Fujimoto, and T. Suzuki, *Gen. Pharmacol.*, **20**, 213 (1989).

236. S. Chiba, Y. Furukawa, K. Saegusa, and Y. Ogiwara, *Jpn. Heart J.*, **28**, 783 (1987).

237. W. Merkel, D. Bormann, D. Mania, R. Muschaweck, and M. Hropot, *Eur. J. Med. Chem.*, **11**, 399 (1976).

238. D. Brater, *Clin. Pharmacol. Ther.*, **25**, 428 (1979).

239. D. Brater, B. Day, S. Anderson, and R. Serwell, *Clin. Pharmacol. Ther.*, **34**, 454 (1983).

240. S. H. Lee and M. G. Lee, *J. Chromatogr. B.*, **656**, 367 (1994).

241. S. Lee and M. G. Lee, *Biopharm. Drug Dispos.*, **16**, 547 (1995).

242. W. Liebenow and F. Leuschner, *Arzneim.-Forsch.*, **25**, 240 (1975).

243. F. Krueck, W. Bablok, E. Bensenfelder, G. Betzien, and B. Kaufmann, *Eur. J. Clin. Pharm.*, **14**, 153 (1978).

244. F. Leuschner, W. Neumann, and H. Barhmann, *Arzneim.-Forsch.*, **25**, 245 (1975).

245. F. W. Hempelmann, *Arzneim.-Forsch.*, **25**, 259 (1975).

246. F. W. Hempelmann, *Arzneim.-Forsch.*, **25**, 258 (1975).

247. B. N. C. Prichard and R. N. Brogden, *Drugs*, **30**, 313 (1985).

248. F. W. Hempelmann, F. Leuschner, and W. Liebenow, *Arzneim.-Forsch.*, **25**, 252 (1975).

249. G. Voltz, *Arzneim.-Forsch.*, **25**, 256 (1975).

250. M. Hohenegger and F. Holzer, *Int. J. Clin. Pharmacol.*, **13**, 298 (1975).

251. A. Brochez, M. Castro, A. Odegaard, and J. Thomis, *Int. J. Clin. Pharm. Ther. Tox.*, **21**, 394 (1983).

252. O. Hammer and U. Dembowski, *Med. Klin.*, **41**, 1862 (1969).

253. P. Federspil and H. Mausen, *Int. J. Clin. Pharmacol.*, **9**, 326 (1974).

254. R. Z. Gussin, J. R. Cummings, E. H. Stokey, and M. A. Ronsberg, *J. Pharmacol. Exp. Ther.*, **167**, 194 (1969).

255. E. A. Lockhart, J. H. Dirks, and S. Carriere, *Am. J. Physiol.*, **223**, 89 (1972).

256. R. Z. Gussin and M. A. Ronsberg, *Proc. Soc. Exp. Biol. Med.*, **131**, 1258 (1969).

257. Z. S. Agus and M. Goldberg, *J. Lab. Clin. Med.*, **76**, 280 (1970).

258. *FDC Rep.*, **36** (39), A6 (Sept. 30, 1974).

259. S. Saito, Y. Tokunaga, Y. Takagi, M. Torizuka, and K. Fukushima, *Nippon Univ. J. Med.*, **12**, 27 (1970).

260. M. Wittner, A. DiStefano, E. Schlatter, J. Delarge, and R. Greger, *Pfluegers Arch.*, **407**, 611 (1986).

261. M. Lesne, F. Clerck-Braun, F. Duhoux, and C. vanYpersele, *Arch. Int. Pharmacodyn.*, **249**, 322 (1981).

262. A. Ghys, J. Denef, J. deSuray, M. Gerin, J. Delarge, and J. Willem, *Arzneim.-Forsch.*, **35**, 1520 (1985).

263. H. Knauf and E. Mutschler, *Clin. Pharmacokinet.*, **34**, 1 (1998).

264. R. Lambe, O. Kennedy, M. Kenny, and A. Darragh, *Eur. J. Clin. Pharmacol.*, **31** (Suppl.), 9 (1986).

265. J. Wouters, C. Michaux, F. Durant, J. Dogne, J. Delarge, and B. Masereel, *Eur. J. Med. Chem.*, **35**, 923 (2000).

266. E. Moller, H. Horstmann, K. Meng, and D. Loew, *Experientia*, **33**, 382 (1977).

267. D. Loew and K. Meng, *Pharmatherapeutica*, **1**, 333 (1977).

268. H. J. Kramer, *Pharmatherapeutica*, **1**, 353 (1977).

269. A. Canton, D. Russo, and R. Gallo, *Br. Med. J.*, **282**, 595 (1981).

270. W. Ritter, *Clin. Nephrol.*, **19**, 26 (1983).

271. K. J. Berg, S. Jorstad, and A. Tromsdal, *Pharmatherapeutica*, **1**, 319 (1977).

272. D. Loew, *Curr. Med. Res. Opin.*, **4**, 455 (1977).

273. M. Mussche and N. Lamerie, *Curr. Med. Res. Opin.*, **4**, 462 (1977).

274. G. E. Stokker, A. A. Deana, S. J. deSolms, E. M. Schultz, R. L. Smith, E. J. Cragoe Jr., J. E. Baer, C. T. Ludden, H. F. Russo, A. Scriabine, C. S. Sweet, and L. S. Watson, *J. Med. Chem.*, **23**, 1414 (1980).

275. G. Stokker, A. Deana, S. deSolms, E. Schultz, R. Smith, E. Cragoe, J. Baer, H. Russo, and L. Watson, *J. Med. Chem.*, **25**, 735 (1982).

276. G. Stokker, A. Deana, S. deSolms, E. Schultz, R. Smith, and E. Cragoe, *J. Med. Chem.*, **24**, 1063 (1981).

277. R. L. Smith, G. E. Stokker, and E. Cragoe Jr., *J. Med. Chem.*, **000**, 000 (0000).

278. G. Stokker, R. Smith, E. Cragoe, C. Ludden, H. Russo, C. Sweet, and L. Watson, *J. Med. Chem.*, **24**, 115 (1981).

279. D. Tocco, G. Stokker, R. Smith, R. Walker, B. Arison, and W. Vandenheuvel, *Pharmacologist*, **20**, 214 (1978).

280. M. B. Affrime, D. T. Lowenthal, G. Onesti, P. Busby, C. Swartz, and B. Lei, *Clin. Pharmacol. Ther.*, **21**, 97 (1977).

281. D. Lowenthal, G. Onesti, A. Pfrimem, J. Schrogie, K. Kim, D. Busby, and R. Swartz, *J. Clin. Pharmacol.*, **18**, 414 (1978).

282. D. J. Tocco, F. DeLuna, A. E. W. Duncan, R. W. Walker, B. H. Arison, and W. J. A. Vandenheuvel, *Drug Metab. Dispos.*, **7**, 330 (1979).

283. R. P. Garay, C. Nazaret, and E. J. Cragoe, *Eur. J. Pharmacol.*, **200**, 141 (1991).

284. E. Schlatter, R. Greger, and C. Weidtke, *Pfluegers Arch.*, **396**, 210 (1983).

285. *Drugs Future*, **2**, 317 (1977).

286. G. Satzinger, *Arzneim.-Forsch.*, **27**, 466 (1977).

287. G. Satzinger, *Arzneim.-Forsch.*, **27**, 1742 (1977).

288. M. Herrmann, J. Wiegleb, and F. Leuschner, *Arzneim.-Forsch.*, **27**, 1758 (1977).

289. M. Herrmann, H. Bahrmann, E. Berkenmayer, V. Ganser, W. Heldt, and W. Steinbrecher, *Arzneim.-Forsch.*, **27**, 1745 (1977).

290. J. Greven and O. Heidenreich, *Arzneim.-Forsch.*, **27**, 1755 (1977).

291. V. Gladigau and K. O. Vollmer, *Arzneim.-Forsch.*, **27**, 1786 (1977).

292. K. O. Vollmer, A. V. Hodenberg, A. Poission, A. Gladigau, and H. Hengy, *Arzneim.-Forsch.*, **27**, 1767 (1977).

293. A. V. Hoenberg, K. O. Vollmer, W. Klemisch, and B. Liedtke, *Arzneim.-Forsch.*, **00**, 0000 (0000).

294. E. Scheitza, *Arzneim.-Forsch.*, **27**, 1804 (1977).

295. G. Biamino, *Arzneim.-Forsch.*, **27**, 1786 (1977).

296. G. Satzinger, M. Herrman, and K. Vollmer, Ger. Pat. 2,414,345 (1979).

297. J. Greven, W. Pefrain, N. Glaser, K. Maywald, and O. Heidenreich, *Pfluegers Arch.*, **384**, 57 (1980).

298. J. Greven and O. Heidenreich, *Med. Welt*, **30**, 1014 (1979).

299. V. Gladigau and K. Vollmer, *Arzneim.-Forsch.*, **27**, 1785 (1977).

300. S. M. O'Grady, H. C. Palfrey, and M. Field, *J. Membrane Biol.*, **96**, 11 (1987).

301. W. N. Suki, J. J. Yium, M. VonMinden, C. Saller-Hebert, G. Eknoyan, and M. Martinez-Maldonado, *N. Engl. J. Med.*, **283**, 836 (1970).

302. L. P. Rybak, *J. Otolaryngol.*, **11**, 127 (1982).

303. B. Pitt, F. Zannad, W. J. Remme, R. Cody, A. Castaigne, A. Perez, J. Palensky, and J. Wittes, *N. Engl. J. Med.*, **341**, 709 (1999).

304. B. Pitt, Ed., *Eur. Heart J.*, **2** (Suppl. A), A1 (2000).

305. X. Rabasseda, J. Silvestre, and J. Castaner, *Drugs Future*, **24**, 488 (1999).

306. M. Epstein, J. Menard, J. C. Alexander, and B. Roniker, *Circulation*, **98**, Abstr. 498 (1998).

307. B. Pitt and B. Roniker, *J. Am. Coll. Cardiol.*, **33** (Suppl. A), 188A (1999).

308. K. Swedberg, P. Eneroth, J. Kjekshus, and L. Wilhelmsen, *Circulation*, **82**, 1730 (1990).

309. D. Vaughan, G. Lamas, and M. Pfeffer, *Am. J. Cardiol.*, **66**, 529 (1990).

310. C. R. Benedict, D. E. Johnstone, D. H. Wiener, M. G. Bourassa, V. Bittner, R. Kay, P. Kirlin, B. Greenberg, R. M. Kohn, and J. M. Nicklas, *J. Am. Coll. Cardiol.*, **23**, 1410 (1994).

311. P. Corvol, M. Claire, M. Oblin, K. Geering, and B. Rossier, *Kidney Int.*, **20**, 1 (1981).

312. B. Pitt, *Cardiovasc. Drugs Ther.*, **9**, 145 (1995).

313. J. A. Cella and C. M. Kagawa, *J. Am. Chem. Soc.*, **79**, 4808 (1957).

314. C. M. Kagawa, J. A. Cella, and C. G. Van Arman, *Science*, **126**, 1015 (1957).

315. J. A. Cella, E. A. Brown, and R. R. Butner, *J. Org. Chem.*, **24**, 743 (1959).

316. J. A. Cella and R. C. Tweit, *J. Org. Chem.*, **24**, 1109 (1959).

317. E. A. Brown, R. D. Muir, and J. A. Cella, *J. Org. Chem.*, **25**, 96 (1960).

318. G. W. Liddle in F. C. Barter, Ed., *The Clinical Use of Aldosterone Antagonist*, Thomas, Springfield, IL, 1960, p. 14.

319. B. Spinger, *Endocrinology*, **65**, 512 (1959).

320. E. Bolte, M. Verdy, J. Marc-Aurele, J. Broullier, P. Beauregard, and J. Genest, *Can. Med. Assoc. J.*, **79**, 881 (1958).

321. R. M. Salassa, V. R. Mattox, and M. H. Power, *J. Clin. Endocrinol. Metab.*, **18**, 787 (1958).

322. G. W. Liddle, *Arch. Intern. Med.*, **102**, 998 (1958).

323. J. D. H. Slater, A. Moxham, R. Hunter, and J. D. N. Nabarro, *Lancet*, **11**, 931 (1958).

324. D. N. S. Kerr, A. E. Read, R. M. Haslam, and S. Sherlock, *Lancet*, **11**, 1084 (1959).

325. G. W. Little, *Science*, **126**, 1016 (1957).

326. E. J. Ross and J. E. Bethune, *Lancet*, **1**, 127 (1959).

327. R. C. Tweit, F. B. Colton, N. L. McNiven, and W. Klyne, *J. Org. Chem.*, **27**, 3325 (1962).

328. J. A. Cella in J. H. Moyer and M. Fuchs, Eds., *Edema*, Saunders, Philadelphia, 1960, p. 303.

329. L. N. Nysted and R. R. Butner, *J. Org. Chem.*, **27**, 3175 (1962).

330. A. A. Patchett, F. Hoffman, F. F. Giarrusso, H. Schwam, and G. E. Arth, *J. Org. Chem.*, **27**, 3822 (1962).

331. G. de Stevens, *Diuretics: Chemistry and Pharmacology*, Academic Press, New York, 1963, p. 130.

332. A. Karim and E. A. Brown, *Steroids*, **20**, 41 (1972).

333. L. J. Chinn, E. A. Brown, S. S. Mizuba, and A. Karim, *J. Med. Chem.*, **20**, 352 (1977).

334. U. Abshagen, H. Rennekamp, K. Koch, M. Senn, and W. Steingross, *Steroids*, **28**, 467 (1976).

335. I. Weiner in L. S. Goodman and A. Gilman, Eds., *The Pharmacological Basis of Therapeutics*, **8th ed.**, Pergamon, New York, 1990, pp. 713–731.

336. C. M. Kagawa, D. J. Bouska, M. L. Anderson, and W. F. Krol, *Arch. Int. Pharmacodyn.*, **149**, 8 (1964).

337. L. M. Hofmann, L. J. Chinn, H. A. Padrera, M. I. Krupnick, and O. D. Suleymanov, *J. Pharmacol. Exp. Ther.*, **194**, 450 (1975).

338. L. Ramsay, I. Harrison, J. Shelton, and M. Tidd, *Clin. Pharmacol. Ther.*, **18**, 391 (1975).

339. R. M. Weirer and L. M. Hofmann, *J. Med. Chem.*, **18**, 817 (1975).

340. M. Haberey, P. Buse, W. Losert, and Y. Nishino, *Naunyn Schmiedebergs Arch. Pharmacol.*, **334** (Suppl.), Abstr. 109 (1986).

341. M. Hildebrand and W. Siefert, Proceedings of the 3rd World Conference on Clinical Pharmacological Therapy, July 27-August 1, 1986, Stockholm, 1986, Abstr. 138.

342. W. Losert, P. Buse, J. Casais-Stenzel, M. Haberey, H. Laurent, K. Nickish, E. Schillinger, and R. Wiechert, *Arzneim.-Forsch.*, **36**, 1583 (1986).

343. L. J. Chinn, *J. Org. Chem.*, **27**, 1741 (1962).

344. L. J. Chinn and B. N. Desai, *J. Med. Chem.*, **18**, 268 (1975).

345. L. J. Chinn and L. M. Hofmann, *J. Med. Chem.*, **16**, 839 (1973).

346. L. J. Chinn, H. L. Dryden Jr., and R. R. Burtner, *J. Org. Chem.*, **26**, 3910 (1961).

347. M. De Gasparo, U. Joss, H. P. Ramjoue, S. E. Whitebread, H. Haenni, L. Schenkel, C. Kraehenduehl, M. Biollaz, and J. Grob, *J. Pharmacol. Exp. Ther.*, **240**, 650 (1987).

348. D. Loriaux, R. Menard, A. Taylor, J. Pita, and R. Santin, *Ann. Int. Med.*, **85**, 630 (1976).

349. J. W. Funder, K. Myles, J. Delyani, M. Ward, P. Kanellakis, and A. Bobik, Proceedings of the 25th International Aldosterone Conference, 1999, p. 23.

350. H. Erbler, *Naunyn Schmiedebergs Arch. Pharmacol.*, **273**, 366 (1972).

351. B. Aupetit, J. Duchier, and J. Legrand, *Ann. Endocrinol.*, **39**, 355 (1978).

352. J. Greiner, R. Kramer, J. Jarrel, and H. Colby, *J. Pharmacol. Exp. Ther.*, **198**, 709 (1976).

353. J. C. Frolich, T. W. Wilson, B. J. Sweetman, M. Smigel, A. S. Nies, K. Carr, J. T. Watson, and J. A. Oates, *J. Clin. Invest.*, **55**, 763 (1975).

354. L. M. Demers, J. C. Melbry, T. E. Wilson, A. Lipton, H. A. Harvey, and R. J. Santen, *J. Clin. Endocrinol. Metab.*, **70**, 1162 (1990).

355. Jpn. Pat. 09071586.

356. C. Persson, I. Erjefalt, L. Edholm, J. Karlsson, and C. Lamm, *Life Sci.*, **31**, 2673 (1982).

357. H. Osswald, *Naunyn Schmiedebergs Arch. Pharmacol.*, **288**, 79 (1975).

358. S. M. Kaiser and R. J. Quinn, *Drug Discovery Today*, **4**, 542 (1999).

359. W. Spielman and L. Arend, *Hypertension*, **17**, 117 (1991).

360. M. Collis, G. Baxter, and J. Keddie, *J. Pharm. Pharmacol.*, **38**, 850 (1986).

361. M. Collis, G. Shaw, and J. Keddie, *J. Pharm. Pharmacol.*, **43**, 138 (1991).

362. J. Shimada, F. Suzuki, H. Nonaka, and A. Ishii, *J. Med. Chem.*, **35**, 924 (1992).

363. J. Shimada, F. Suzuki, H. Nonaka, A. Karasawa, H. Mizuumoto, T. Ohno, K. Kubo, and A. Ishii, *J. Med. Chem.*, **34**, 469 (1991).

364. R. Knight, C. Bowmer, and M. Yates, *Br. J. Pharmacol.*, **109**, 272 (1993).

365. F. Suzuki, J. Shimada, H. Mizumoto, A. Karasawa, K. Kubo, H. Nonaka, A. Ishii, and T. Kawakita, *J. Med. Chem.*, **35**, 3066 (1992).

366. H. Mizumoto, A. Karasawa, and K. Kubo, *J. Pharmacol. Exp. Ther.*, **266**, 200 (1993).

367. T. Kobayashi, H. Mizumoto, and A. Karasawa, *Biol. Pharm. Bull.*, **16**, 1231 (1993).

368. T. Kobayashi, H. Mizumoto, A. Karasawa, and K. Kubo, *Jpn. J. Pharm.*, **58** (Suppl. 1), 195 (1992).

369. H. Kusaka and A. Karasawa, *Jpn. J. Pharm.*, **63**, 513 (1993).

370. H. Mizumoto, T. Kobayashi, A. Karasawa, H. Nonaka, A. Ishii, K. Kubo, J. Shimada, and F. Suzuki, *Jpn. J. Pharm.*, **58** (Suppl. 1), 194 (1992).

371. K. Nagashima, H. Kusaka, K. Sato, and A. Karasawa, *Jpn. J. Pharm.*, **64**, 9 (1994).

372. H. Mizumoto and A. Karasawa, *Jpn. J. Pharm.*, **61**, 251 (1993).

373. J. R. Pfister, L. Belardinelli, G. Lee, R. T. Lum, P. Milner, W. C. Stanley, J. Linden, S. P. Baker, and G. Schreiner, *J. Med. Chem.*, **40**, 1773 (1997).

374. C. S. Wilcox, W. J. Welch, G. F. Schreiner, and L. Belardinelli, *J. Am. Soc. Nephrol.*, **10**, 714 (1999).

375. M. Gellai, G. F. Schreiner, R. R. Ruffolo, T. Fletcher, R. DeWolf, and D. P. Brooks, *J. Pharmacol. Exp. Ther.*, **286**, 1191 (1998).

376. A. A. Wolff, S. L. Skettino, E. Beckman, and L. Belardinelli, *Drug Dev. Res.*, **45**, 166 (1998).

377. V. Papesch and E. F. Schroeder, *J. Org. Chem.*, **16**, 1879 (1951).

378. A. Kattus, T. M. Arrington, and E. V. Newman, *Am. J. Med.*, **12**, 319 (1952).

379. W. L. Lipschitz and Z. Hadidian, *J. Pharmacol. Exp. Ther.*, **81**, 84 (1944).

380. V. Papesch and E. F. Schroeder in F. F. Blicke and R. H. Cox, Eds., *Medicinal Chemistry*, Vol. **111**, John Wiley & Sons, New York, 1956, p. 175.

381. T. Turchetti, *Riforma Med.*, **64**, 405 (1950).

382. E. V. Newman, J. Franklin, and J. Genest, *Bull. Johns Hopkins Hosp.*, **82**, 409 (1948).

383. O. Clauder and G. Bulcsu, *Magy. Kem. Foly.*, 57, 68 (1951); *Chem. Abstr.*, **46**, 4023 (1952).

384. G. Szabo, O. Clauder, and Z. Magyar, *Magy. Belorv. Arch.*, **6**, 156 (1953).

385. C. M. Kagawa and C. G. Van Arman, *J. Pharmacol. Exp. Ther.*, **124**, 318 (1958).

386. D. V. Miller and R. V. Ford, *Am. J. Med. Sci.*, **236**, 32 (1958).

387. R. V. Ford, J. B. Rochelle, A. C. Bullock, C. L. Spurr, C. Handley, and J. H. Moyer, *Am. J. Cardiol.*, **3**, 148 (1959).

388. M. H. Sha, M. Y. Mhasalker, and C. V. Deliwala, *J. Sci. Ind. Res. (India)*, **19c**, 282 (1960); D. J. Mehta, U. K. Sheth, and C. V. Deliwala, *Nature*, **187**, 1034 (1960).

389. K. N. Modi, C. V. Deliwala, and U. K. Sheth, *Arch. Int. Pharmacodyn.*, **151**, 13 (1964).

390. L. Szabo, L. Szporny, and O. Clauder, *Acta Pharm. Hung.*, 31, 163 (1961); *Chem. Abstr.*, **55**, 24780I (1961).

391. W. B. McKeon Jr., *Arch. Int. Pharmacodyn.*, **151**, 225 (1964).

392. D. A. LeSher and F. E. Shideman, *J. Pharmacol. Exp. Ther.*, **116**, 38 (1956).

393. H. E. Williamson, F. E. Shideman, and D. A. LeSher, *J. Pharmacol. Exp. Ther.*, **126**, 82 (1959).

394. M. Burg and S. Sariban-Sohraby in J. B. Puschett, Ed., *Diuretics: Chemistry, Pharmacology and Clinical Application*, Elsevier, New York, 1984, pp. 329–334.

395. J. E. Baer in A. F. Lant and G. M. Wilson, Eds., *Modern Diuretic Therapy in the Treatment of Cardiovascular and Renal Disease*, Excerpta Medica, Amsterdam, 1973, p. 148.

396. V. D. Wiebelhaus, J. Winstock, A. R. Maass, F. T. Brennan, G. Sosnowski, and T. Larsen, *J. Pharmacol. Exp. Ther.*, **149**, 397 (1965).

397. J. Weinstock, J. W. Wilson, V. D. Wiebelhaus, A. R. Maass, F. T. Brennan, and G. Sosnowski, *J. Med. Chem.*, **11**, 573 (1968).

398. T. S. Osdene, A. A. Santilli, L. E. McCardle, and M. E. Rosenthale, *J. Med. Chem.*, **9**, 697 (1966).

399. T. S. Osdene, A. A. Santilli, L. E. McCardle, and M. E. Rosenthale, *J. Med. Chem.*, **10**, 165 (1967).

400. M. E. Rosenthale and C. G. Van Arman, *J. Pharmacol. Exp. Ther.*, **142**, 111 (1963).

401. V. D. Wiebelhaus, J. Weinstock, F. T. Brennan, G. Sosnowski, and T. J. Larsen, *Fed. Proc.*, **20**, 409 (1961).

402. A. P. Crosley Jr., L. Ronquillo, and F. Alexander, *Fed. Proc.*, **20**, 410 (1961).

403. J. H. Laragh, E. B. Reilly, T. B. Stites, and M. Angers, *Fed. Proc.*, **20**, 410 (1961).

404. V. D. Wiebelhaus, J. Weinstock, F. T. Brennan, G. Sosnowski, T. Larsen, and K. Gahagan, *Pharmacologist*, **3**, 59 (1961).

405. W. Schaumann, *Klin. Wochenschr.*, **40**, 756 (1962).

406. W. I. Baba, G. R. Tudhope, and G. M. Wilson, *Br. Med. J.*, **2**, 756 (1962).

407. G. W. Liddle, *Metab. Clin. Exp.*, **10**, 1021 (1961).

408. G. M. Ball and J. A. Greene Jr., *Proc. Soc. Exp. Biol. Med.*, **113**, 326 (1963).

409. J. Crabbe, *Arch. Int. Pharmacodyn. Ther.*, **173**, 474 (1968).

410. J. Gatry, *J. Pharmacol. Exp. Ther.*, **176**, 586 (1971).

411. A. P. Crosley Jr., L. Ronquillo, W. S. Strickland, and F. Alexander, *Ann. Intern. Med.*, **56**, 241 (1962).

412. D. J. Ginsberg, A. Saad, and G. J. Gabuzda, *N. Engl. J. Med.*, **271**, 1229 (1964).

413. E. Mutschler, H. Gilfrich, and H. Knauf, *Clin. Exp. Hypertens.*, **4**, 249 (1983).

414. H. Gilfrich, G. Kremer, and W. Mohrke, *Eur. J. Clin. Pharmacol.*, **25**, 237 (1983).

415. P. Baume, F. J. Radcliffe, and C. R. Corry, *Am. J. Med. Sci.*, **245**, 668 (1963).

416. W. R. Cattell and C. W. H. Havard, *Br. Med. J.*, **2**, 1362 (1962).

417. R. A. Thompson and M. F. Crowley, *Postgrad. Med. (Oxford)*, **41**, 706 (1965).

418. K. Nishikawa, H. Shimakawa, Y. Inada, Y. Shibouta, S. Kikuchi, S. Yurugi, and Y. Oka, *Chem. Pharm. Bull.*, **24**, 2057 (1976).

419. K. Nishikawa and S. Kikuchi, *Jpn. J. Pharm.*, 22 (Suppl.), 103 (1972); Y. Shirakawa and T. Fujita, *Jpn. J. Pharm.*, **22** (Suppl.), 102 (1972).

420. H. Kawaki, R. Tsukuda, K. Nishikawa, S. Kikuchi, and T. Hirano, *J. Takeda Res. Labs.*, 32, 299 (1973); Y. Inada, K. Nishikawa, A. Nagaoka, and S. Kikuchi, *Arzneim.-Forsch.*, **27**, 1663 (1977).

421. Davis, R. Gedir, E. Hawes, and G. Wibberley, *Eur. J. Med. Chem.*, **20**, 381 (1985).

422. H. Parish, R. Gilliom, W. Purcell, R. Browne, R. Spirk, and H. White, *J. Med. Chem.*, **25**, 98 (1982).

423. A. Monge, V. Martinez-Merion, M. Simon, and C. Sanmartin, *Arzneim.-Forsch.*, **43**, 1322 (1993).

424. J. Hester, S. Luden, D. Emmert, and B. West, *J. Med. Chem.*, **32**, 1157 (1989).

425. J. B. Bicking, J. W. Mason, O. W. Woltersdorf Jr., J. H. Jones, S. F. Kwong, C. M. Robb, and E. J. Cragoe Jr., *J. Med. Chem.*, **8**, 638 (1965).

426. J. B. Bicking, C. M. Robb, S. F. Kwong, and E. J. Cragoe Jr., *J. Med. Chem.*, **10**, 589 (1967).

427. J. H. Jones, J. B. Bicking, and E. J. Cragoe Jr., *J. Med. Chem.*, **10**, 899 (1967).

428. J. Watthey, M. Desai, R. Rutledge, and R. Dotson, *J. Med. Chem.*, **23**, 690 (1980).

429. T. Russ, W. Ried, F. Ullrich, and E. Mutschler, *Arch. Pharm.*, **325**, 761 (1992).

430. S. Kau, B. Howe, B. Li, L. Smith, J. Keddie, J. Barlow, R. Giles, and M. Goldberg, *J. Pharmacol. Exp. Ther.*, **242**, 818 (1987).

431. S. Kau, P. Johnson, J. Li, J. Zuzack, K. Lescznskak, C. Yochim, J. Schwartz, and R. Giles, *Methods Find. Exp. Clin. Pharmacol.*, **15**, 357 (1993).

432. B. Beerman and M. Groschinsky-Grind, *Clin. Pharmacokinet.*, **5**, 221 (1980).

433. R. Z. Gussin, M. A. Ronsberg, E. H. Stokey, and J. R. Cummings, *J. Pharmacol. Exp. Ther.*, **195**, 8 (1975).

434. M. A. Ronsberg, A. Z. Gussin, E. H. Stokey, and P. S. Chan, *Pharmacologist*, **18**, 150 (1976).

435. V. Ackermann, *Clin. Chem.*, **32**, 241 (1986).

436. A. Raine, J. Firth, and J. Ledingham, *Clin. Sci.*, **76**, 1 (1989).

437. T. Yandle, A. Richards, M. Nicholls, R. Cuneo, E. Espiner, and J. Livesey, *Life Sci.*, **38**, 827 (1986).

438. F. Luft, R. Lang, and G. Arnoff, *J. Pharmacol. Exp. Ther.*, **236**, 416 (1986).

439. M. Cogan, *Am. J. Physiol. Renal Fluid Electrolyte Physiol.*, **250**, F710 (1986).

440. M. Camargo, S. Atlas, and T. Maack, *Life Sci.*, **38**, 2397 (1986).

441. A. Kenny and S. Stephenson, *FEBS Lett.*, **232**, 1 (1988).

442. W. Oelkers, S. Kleiner, and V. Bahr, *Hypertension*, **12**, 462 (1988).

443. R. Cuneo, E. Espiner, M. Nichols, T. Yandle, and J. Livesey, *J. Clin. Endocrinol. Metab.*, **65**, 765 (1986).

444. K. Atarashi, P. Mulrow, and R. Franco-Saenz, *J. Clin. Invest.*, **76**, 1807 (1985).

445. J. Almenoff and M. Orlowski, *Biochemistry*, **22**, 590 (1983).

446. S. Stephenson and A. Kenny, *J. Biochem.*, **241**, 237 (1987).

447. T. Maack, F. Almeida, M. Suzuki, and D. Nussenzveig, *Contrib. Nephrol.*, **68**, 58 (1988).

448. P. Nussenzveig, J. Lewicki, and T. Maack, *J. Biol. Chem.*, **265**, 20952 (1990).

449. P. Leitman, J. Resen, T. Kuno, Y. Kamisaki, J. Chang, and F. Munad, *J. Biol. Chem.*, **261**, 11650 (1986).

450. J. Koepke, L. Tyler, A. Trapani, P. Bovy, K. Spear, G. Olins, and E. Blaine, *J. Pharmacol. Exp. Ther.*, **249**, 172 (1989).

451. J. Okolicany, G. McEnroe, L. Gregory, J. Lewicki, and T. Maack, *Can. J. Physiol.*, **69**, 1561 (1991).

452. T. Maack, M. Suzuki, F. Almeida, P. Nussenzveig, R. M. Scarborough, G. McEnroe, and J. Lewicki, *Science*, **238**, 675 (1987).

453. S. Vemulapalli, P. Chiv, A. BrownGrisctik, and E. Sybertz, *Life Sci.*, **49**, 383 (1991).

454. C. A. Veale, V. C. Alford, D. Aharony, D. L. Banville, R. A. Bialecki, F. J. Brown, J. R. Damewood, C. L. Dantzman, P. D. Edwards, R. T. Jacobs, R. C. Mauger, M. M. Murphy, W. E. Palmer, K. K. Pine, W. L. Rumsey, L. E. Garcia-Davenport, A. Shaw, G. B. Steelman, J. M. Surian, and E. P. Vacek, *Bioorg. Med. Chem. Lett.*, **10**, 1949 (2000).

455. R. Webb, G. Yasay, C. McMartin, R. McNeal, and M. Zimmerman, *J. Cardiovasc. Pharmacol.*, **14**, 285 (1989).

456. H. Lafferty, M. Gunning, P. Silva, M. Zimmerman, B. Brenner, and S. Anderson, *Circ. Res.*, **65**, 640 (1989).

457. A. Trapani, G. Smits, D. McGraw, K. Spear, S. Koepke, G. Olins, and E. Blane, *J. Cardiovasc. Pharmacol.*, **14**, 419 (1989).

458. M. Richards, E. Espiner, C. Frampton, H. Ikram, T. Yandle, M. Sopwith, and N. Cussans, *Hypertension*, **16**, 269 (1990).

459. E. G. Bevan, J. M. C. Connell, J. Doyle, H. A. Carmichael, D. L. Davies, and A. B. Lorimer, *J. Hypertension*, **10**, 607 (1992).

460. A. M. Richards, G. A. Wittert, I. G. Crozier, E. A. Espiner, T. G. Yandle, H. Ikram, and C. Frampton, *J. Hypertension*, **11**, 407 (1993).

461. A. M. Richards, G. A. Wittert, E. A. Espiner, T. G. Yandle, H. Ikram, and C. Frampton, *Circ. Res.*, **71**, 1501 (1992).

462. J. E. O'Connell, A. G. Jardine, D. L. Davies, J. McQueen, and J. M. C. O'Connell, *Clin. Sci.*, **85**, 19 (1993).

463. S. I. Ando, M. A. Rahman, G. C. Butler, B. L. Senn, and J. S. Floras, *Hypertension*, **26**, 1160 (1995).

464. G. McDowell, W. Coutie, C. Shaw, K. D. Buchanan, A. D. Struthers, and D. P. Nicholls, *Br. J. Clin. Pharmacol.*, **43**, 329 (1997).

465. M. Fournie-Zaluski, E. Lucas, G. Waksman, and B. Roques, *Eur. J. Biochem.*, **139**, 267 (1984).

466. A. Seymore, S. Fennell, and J. Swerdel, *Hypertension*, **14**, 87 (1989).

467. A. Seymore, J. Norman, M. Asaad, S. Fennel, J. Swerdel, D. Little, and C. Dorso, *J. Cardiovasc. Pharmacol.*, **16**, 163 (1990).

468. A. Seymore, J. Norman, M. Asaad, S. Fennel, D. Little, V. Kratunis, and W. Rogers, *J. Cardiovasc. Pharmacol.*, **17**, 296 (1991).

469. S. Vemulapalli, P. J. S. Chiu, R. W. Watkins, C. Foster, and E. J. Sybertz, *Am. J. Hypertens.*, **4**, 15A–16A (1991).

470. R. W. Watkins, P. J. S. Chiu, S. Vemulapalli, C. Foster, M. Chatterjee, E. M. Smith, B. Neustadt, M. Hastlanger, and E. Sybertz, *Am. J. Hypertens.*, **4**, 32A (1991).

471. A. M. Richards, I. Crozier, T. Kosoglou, M. Rallings, E. Espiner, M. G. Nicholls, T. Vandle, H. Ikram, and C. Frampton, *Hypertension*, **22**, 119 (1993).

472. J. Danilewicz, P. Barclay, I. Barish, P. Brown, S. Campbell, K. James, G. Samuels, N. Terrett, and M. Wythes, *Biochem. Biophys. Res. Commun.*, **164**, 58 (1989).

473. J. O'Connell, A. Jardine, and G. Davidson, *J. Hypertension*, **10**, 271 (1992).

474. E. Sybertz, *J. Pharmacol. Exp. Ther.*, **250**, 624 (1989).

475. K. Helin, I. Tikkanen, O. Saijonmaa, E. Sybertz, S. Vemulapall, H. Sariolatt, and F. Fyhrquist, *Eur. J. Pharmacol.*, **198**, 23 (1991).

476. M. Fitzpatrick, M. Rademaker, C. Charles, T. Vandle, E. Espiner, H. Ikram, and E. Sybertz, *J. Cardiovasc. Pharmacol.*, **19**, 635 (1992).

477. S. Vemulapalli, P. Chiu, A. Brown, K. Griscti, and E. Sybertz, *Life Sci.*, **49**, 383 (1991).

478. M. Burnier, M. Ganslmayer, F. Perret, M. Porchet, T. Kosoglou, A. Gould, J. Nussberger, J. Waeber, and H. Brunner, *Clin. Pharmacol. Ther.*, **50**, 181 (1991).

479. M. Fournie-Zaluski, A. Coulaud, R. Bouboton, P. Chailler, J. Devin, G. Waksman, J. Costentin, and B. Roques, *J. Med. Chem.*, **28**, 1158 (1985).

480. G. Olins, P. Krieter, A. Trapani, K. Spear, and P. Bovy, *Mol. Cell. Endocrinol.*, **61**, 201 (1989).

481. M. Altstein, S. Blumberg, and Z. Vogel, *Eur. J. Pharmacol.*, **76**, 299 (1982).

482. S. Delombaert, M. Erion, J. Tan, L. Blanchard, L. El-Chehabi, R. Ghai, C. Berry, and A. Trapani, *J. Med. Chem.*, **37**, 498 (1994).

483. A. J. Trapani, M. E. Beil, D. T. Cote, S. DeLombaert, M. D. Erion, T. E. Gerlock, R. D. Ghai, M. F. Hopkins, J. V. Peppard, R. L. Webb, R. W. Lappe, and M. Worcel, *J. Cardiovasc. Pharmacol.*, **23**, 358 (1994).

484. R. Walter and P. L. Hoffman in J. R. Brobeck, Ed., *Best and Taylor's Physiological Basis of Medical Practice*, **9th ed.**, Section 1(6), Williams & Wilkins, Baltimore, 1973.

485. F. G. W. Marson, *Lancet*, **11**, 360 (1955).

486. K. H. Beyer, H. F. Russo, E. K. Tillson, A. K. Miller, W. F. Verwey, and S. R. Gass, *Am. J. Physiol.*, **166**, 625 (1951).

487. A. B. Gutman, *Adv. Pharmacol.*, **4**, 91 (1966).

488. K. H. Beyer, *Arch. Int. Pharmacodyn.*, **98**, 97 (1954).

489. P. Brazeau in L. S. Goodman and A. Gilman, Eds., *The Pharmacological Basis of Therapeutics*, **5th ed.**, Section VIII, Macmillan, New York, 1975, p. 860.

490. R. Pfister and F. Hafliger, *Helv. Chim. Acta*, **44**, 232 (1961).

491. J. J. Burns, T. F. Yu, A. Ritterband, J. M. Perel, A. B. Gutman, and B. B. Brodie, *J. Pharmacol. Exp. Ther.*, **119**, 418 (1957).

492. T. F. Yu, J. J. Burns, and A. B. Gutman, *Arth. Rheum.*, **1**, 352 (1958).

493. J. J. Burns, T. F. Yu, P. Dayton, L. Berger, A. B. Gutman, and B. B. Brodie, *Nature*, **182**, 1162 (1958).

494. A. B. Gutman, P. G. Dayton, T. F. Yu, L. Berger, W. Chen, L. E. Sicam, and J. J. Burns, *Am. J. Med.*, **29**, 1017 (1960).

495. K. C. Blanchard, D. Maroske, D. G. May, and I. M. Weiner, *J. Pharmacol. Exp. Ther.*, **180**, 397 (1972).

496. A. H. Anton, *J. Pharmacol. Exp. Ther.*, **134**, 291 (1961).

497. T. F. Yu and A. B. Gutman, *Am. J. Med.*, **37**, 885 (1964).

498. R. W. Rundles, E. N. Metz, and H. R. Silberman, *Ann. Intern. Med.*, **64**, 229 (1966).

499. D. M. Woodbury and E. Fingl in L. S. Goodman and A. Gilman, Eds., *The Pharmacological Basis of Therapeutics*, **5th ed.**, Section 11, Macmillan, New York, 1975, p. 352.

500. T. Yu, *J. Rheumatology*, **3**, 305 (1976).

501. F. Matzkies, F. Berg, and R. Minzlaff, *Fortsch. Med.*, **95**, 1748 (1977).

502. N. Zoller, W. Dofel, and W. Grobner, *Klin. Wochensch.*, **48**, 426 (1970).

503. R. Kramp, *Eur. J. Clin. Invest.*, **3**, 245 (1973).

504. R. Kramer and M. Muller, *Experientia*, **29**, 391 (1973).

505. J. Broehuysen, M. Pacco, R. Sion, L. Demeulenaere, and M. vanHee, *Eur. J. Clin. Pharm.*, **4**, 125 (1972).

CHAPTER THREE

Myocardial Infarction Agents

George E. Billman
Ruth A. Altschuld
The Ohio State University
Columbus, Ohio

Contents

Burger's Medicinal Chemistry and Drug Discovery
Sixth Edition, Volume 3: Cardiovascular Agents and
Endocrines
Edited by Donald J. Abraham
ISBN 0-471-37029-0   © 2003 John Wiley & Sons, Inc.

1 INTRODUCTION

Acute myocardial infarction was called the quintessential disease of the 20th century (1). Before the introduction of coronary care units, short-term in-hospital mortality was approximately 30%. Coronary care units halved mortality in the early 1960s, primarily because of the use of β-adrenergic antagonists, continuous electrocardiography (ECG) monitoring, and direct current defibrillators. The advent in the 1980s of thrombolytic therapy for dissolving occlusive blood clots again halved mortality, but in the past few years, incremental improvements in reperfusion therapy have produced only small further reductions in mortality. Acute myocardial infarction remains the most important cause of death in the United States (1), and post-infarction remodeling in survivors is contributing to the growing congestive heart failure epidemic of the late 20th and early 21st centuries.

There have been four classical goals in the pharmacologic treatment of an acute myocardial infarction: (1) pain relief, (2) reperfusion and maintenance of vessel patency, (3) prevention and treatment of arrhythmias, and (4) prevention of post-infarction ventricular remodeling, a leading cause of congestive heart failure (1). A fifth goal has begun to emerge, i.e., the prevention of reperfusion damage following successful thrombolysis, percutaneous intervention (e.g., angioplasty) to open an occlusion, or coronary artery bypass surgery.

2 PATHOPHYSIOLOGY OF MYOCARDIAL INFARCTION

2.1 Coronary Occlusion

The typical myocardial infarction begins with the rupture of an atherosclerotic plaque (2). A thrombus or blood clot forms at the site and over time fills the lumen of the coronary artery, interfering with or abolishing blood flow. Thromboemboli and vasospasm may also pre-cipitate coronary artery occlusion. Regardless of the initiating event, tissue downstream from an occlusion is deprived of arterial blood with its life sustaining oxygen and nutrients, and metabolic wastes accumulate. The lack of oxygen inhibits mitochondrial oxidative phosphorylation, the major source of the adenosine triphosphate (ATP) used to power excitation-contraction coupling and maintain intracellular homeostasis. The affected muscle cells are briefly able to regenerate ATP from the high energy phosphate storage pool, phosphocreatine, but with no-flow ischemia, this high energy phosphate store is depleted within minutes, and ATP begins to decline. This is accompanied by the accumulation of the ATP breakdown products, adenosine diphosphate, adenosine monophosphate, and inorganic phosphate, which activate glycogenolysis and anaerobic glycolysis (3). An increase in circulating catecholamines also accelerates glycogen breakdown and glycolytic flux (4).

Anaerobic glycolysis can generate some ATP, but the conversion of glycogen to lactic acid yields only a small percentage of the energy that could otherwise be obtained from the complete oxidation of glycogen's glucose moieties to carbon dioxide and water. As a result, the tissue becomes energy starved and contractile function declines. This down-regulation of contractility may help preserve the limited energy reserves of the ischemic myocardium, but there continues to be a mismatch between energy production and consumption. The lack of blood flow also allows for the buildup of metabolic wastes, particularly lactic acid and amphiphilic fatty acid metabolites, and the tissue becomes acidotic. The increased intracellular H^+ concentration favors intracellular Na^+ accumulation through the sarcolemmal Na^+/H^+ exchanger and this, in turn, favors excess Ca^{2+} accumulation by the reverse mode of the electrogenic sarcolemmal Na^+/Ca^{2+} exchanger (5, 6). Intracellular free Ca^{2+} concentrations gradually increase, and cytosolic Ca^{2+} overload activates proteases

(7–9) and lipases (10–12), which, in turn, degrade important cellular components.

2.1.1 Apoptosis versus Necrosis. If the tissue downstream from a coronary occlusion is not reperfused, affected cells will eventually die and be replaced by scar tissue. This impairs overall cardiac pump function because cardiac myocytes are terminally differentiated and have very limited ability to replicate. The period of time from the onset of thrombosis until myocyte death varies depending on the degree of myocardial ischemia and on the contractile state of the myocardium. Some occlusions "stutter" and there can be intermittent blood flow past the blockage (13). There can also be considerable individual variation in the extent of native coronary collateral blood vessels. Thus, portions of the myocardial tissue distal to an occlusion can remain viable and potentially salvageable for up to 12 h in some cases (2).

Cells in tissue that is not reperfused eventually undergo necrosis, which initiates an inflammatory response and scar formation. Recent studies indicate that some cells in the infarct border zone and some that are successfully reperfused before necrotic cell death may subsequently die through programmed cell death or apoptosis (14–16). Although many of the apoptotic cells that have been detected in and adjacent to an infarct area may be non-muscle cells, which are more abundant but much smaller in size than the contractile cardiomyocytes (15), strategies to prevent cardiomyocyte apoptosis should reduce infarct size.

2.1.2 Preconditioning. In experimental settings, several brief periods of ischemia and reperfusion before a 30–90 min coronary occlusion reduce the size of the subsequent infarct (17). This "preconditioning" phenomenon has been the subject of intense investigation and can be replicated by a variety of pharmacologic agents that activate protein kinase C (18–20) and/or the mitochondrial ATP-sensitive potassium channel (21–23). In reperfused rat hearts, ischemic preconditioning reduced apoptosis by inhibiting neutrophil accumulation and down-regulating the expression of the pro-apoptotic protein, Bax (24).

The need to treat the myocardium with a preconditioning agent before a sustained ischemic period has limited the clinical usefulness of this process to elective ischemia such as that associated with cardiac surgery. If a preconditioning-like effect could be achieved pharmacologically at reperfusion, it undoubtedly would have a beneficial effect.

2.2 Malignant Arrhythmias

As noted above, myocardial ischemia provokes abnormalities in the biochemical homeostasis of individual cardiac cells. These intracellular changes culminate in the disruption of cellular electrophysiologic properties, and life-threatening alterations in cardiac rhythm, such as ventricular fibrillation, frequently occur. Various chemical substances have been proposed as possible causative factors in the genesis of ventricular fibrillation during myocardial ischemia, including catecholamines, amphiphilic products of lipid metabolism, various peptides, cytosolic calcium accumulation, and increases in extracellular potassium (25–30). The following sections shall focus on the role that changes in cellular calcium and potassium play in the induction of cardiac arrhythmias during myocardial ischemia.

2.2.1 Role of Cytosolic Free Calcium. Under normal conditions ventricular muscle cells maintain resting levels of cytosolic calcium approximately 5000 times lower than the extracellular calcium concentration (31). Several important regulatory mechanisms are responsible for maintaining the low cytosolic calcium levels vital for normal cardiac function. In brief, calcium influx is restricted by voltage-sensitive calcium channels that are activated by the cardiac action potential and regulated by intracellular messengers (e.g., phosphorylation) (31–33). Calcium is also extruded from the cell by an electrogenic Na^+/Ca^{2+} exchanger (forward mode, 3 Na^+ in, 1 Ca^{2+} out) and sarcolemmal Ca^{2+} adenosine triphosphatase (ATPase). Inside the cell, a second Ca^{2+} ATPase pumps calcium into the lumen of the sarcoplasmic reticulum. These systems rapidly decrease the elevations in cytosolic free calcium concentration brought about by excitation and induce relaxation during diastole. In addition, mitochondria can take up calcium, and a number of calcium-binding proteins also serve to buffer intracellular calcium levels (31). Under steady-state conditions, calcium influx across the cell membrane (primarily through L-type calcium channels) during systole is matched by an equal calcium efflux (mediated by the Na^+/Ca^{2+} exchanger and to a lesser extent by the sarcolemmal Ca^{2+} ATPase) during diastole. As a result, there is no net increase or decrease in intracellular free calcium concentration. Disturbances in this intracellular calcium homeostasis can profoundly alter a variety of cellular functions, including the myocyte electrical stability.

Intracellular calcium rises dramatically with the induction of ischemia, exceeding peak systolic calcium levels within 5–10 min after ischemia onset (34–37). Myocardial ischemia may provoke large increases in cytosolic calcium both directly (by alteration of the cellular calcium homeostatic mechanisms) and indirectly (by activation of the autonomic nervous system). Myocardial ischemia profoundly affects the autonomic regulation of the heart (38). Coronary artery occlusion elicits reflex increases in cardiac sympathetic activity, accompanied by reductions in parasympathetic tone (38–40). In fact, Billman and co-workers (39, 40) demonstrated that acute myocardial ischemia provoked larger increases in sympathetic activity, coupled with greater reductions in cardiac vagal tone, in animals subsequently shown to be susceptible to ventricular fibrillation.

Alterations in autonomic regulation trigger a cascade of intracellular events that ultimately increase cytosolic calcium levels. Release of catecholamines from sympathetic nerve terminals activates α-, β_1-, and β_2-adrenergic receptors on cardiac myocytes. Stimulation of the β_1-adrenergic receptor activates adenylyl cyclase, which in turn, increases cellular levels of cyclic adenosine monophosphate (cAMP) (41). This cyclic nucleotide activates a cAMP-dependent protein kinase (PKA) that phosphorylates a variety of proteins, including the voltage-dependent calcium channel (33) and the calcium release channel of the sarcoplasmic reticulum (42–45). It also phosphorylates the sarcoplasmic reticulum Ca^{2+}-ATPase inhibitor, phospholamban, relieving its inhibition of calcium

sequestration (46). These reactions culminate in increased calcium entry into cardiac cells and increased uptake and release from intracellular stores. The activation of myocardial β_2-adrenergic receptors can also contribute to cytosolic calcium increase induced by sympathetic neural activation. Until recently, myocardial β-adrenergic receptors were thought to be primarily of the β_1-adrenergic receptor subtype (47, 48). However, it is now apparent that ventricular myocytes also contain functional β_2-adrenergic receptors, which may become particularly important in cardiac disease (47–51). For example, β_1-adrenergic receptor density decreases as a consequence of heart failure, whereas the number of β_2-adrenergic receptors remains relatively constant (51). As such, the failing heart becomes more dependent on β_2-adrenergic receptors for inotropic support. The activation of β_2-adrenergic receptors (using the selective agonist, zinterol) has, in fact, been shown to provoke significantly greater increases in calcium transient amplitude in myocytes isolated from animals susceptible to ventricular fibrillation than in myocytes obtained from animals resistant to malignant arrhythmias (50). This activation of the β_2-adrenergic receptors produced large increases in the calcium current with little or no increase in whole cell cAMP or phospholamban phosphorylation (52). Thus, β_2-adrenergic receptor activation may elicit a localized cAMP-independent increase confined to the sarcolemma.

α-Adrenergic receptor stimulation of the heart results in activation of a phospholipase that hydrolyzes phosphatidyl inositol into two second messengers, diacylglycerol and inositol trisphosphate (53). Inositol trisphosphate facilitates calcium release from the sarcoplasmic reticulum, whereas diacylglycerol activates the important regulatory protein, protein kinase C (PKC). Thus, α- and β-adrenergic stimulations act synergistically to increase cytosolic calcium during ischemia.

Conversely, parasympathetic nerve activation, which decreases during ischemia, opposes the action of sympathetic nerve stimulation, reduces cAMP levels, and increases levels of cyclic guanosine monophosphate (cGMP) (54). cGMP, in turn, decreases the open time of calcium channels independently of changes in cAMP levels (55) and activates a sarcolemmal calcium pump (56). Parasympathetic stimulation, therefore, lowers intracellular calcium and halts the response to sympathetic stimulation. Thus, the alterations in autonomic function elicited by myocardial ischemia would tend to favor the accumulation of cytosolic calcium.

Myocardial ischemia also directly alters several of the important calcium regulatory pathways noted above. As ischemia progresses, cellular ATP levels decline. As a consequence of ATP depletion, several energy-dependent functions are impaired. Sodium (Na^+)/potassium (K^+) ATPase (sodium pump) can no longer function properly, and cellular Na^+ levels increase (31). Increased Na^+ reverses the normal direction of the Na^+/Ca^{2+} exchanger so that sodium is extruded and calcium is taken up by the cell (31). In addition, the Ca^{2+} ATPases (calcium pumps) of the sarcolemma and sarcoplasmic reticulum are impaired so that less calcium is pumped out of the cell or into the sarcoplasmic reticulum during diastole (relaxation is delayed). The net result of this impairment of cellular calcium homeostatic mechanisms and enhanced sympathetic outflow to the heart is a significant rise in cytosolic calcium levels (35–37, 57). These increases in cytosolic calcium could, in turn, provoke alterations in ion fluxes across the sarcolemma that ultimately culminate in malignant ventricular arrhythmias (see below). Indeed, Billman et al. (58) indirectly demonstrated that cytosolic calcium may be elevated in animals particularly susceptible to ventricular fibrillation. They found that calcium-dependent kinase activity was significantly greater in tissue obtained from animals that developed life-threatening arrhythmias during myocardial ischemia than in tissue obtained from animals resistant to arrhythmia formation. Specifically, calcium-calmodulin–dependent phosphorylation was two- to threefold higher in ventricular tissue obtained from animals that had ventricular fibrillation compared with animals that did not develop arrhythmias during ischemia.

2.2.2 Calcium and Arrhythmia Formation.
Disturbances in cardiac rhythm may result from perturbations in impulse generation, impulse conduction, or a combination of both (59). Elevations in cellular calcium induced by myocardial ischemia, as described above, can produce abnormalities in these cardiac electrical properties and thereby trigger malignant arrhythmias.

It is well established that during coronary artery occlusion the resting potential of ischemic cardiac tissue becomes progressively less negative than the resting potential of surrounding non-ischemic tissue (59). The spread of this injury current tends to depolarize the surrounding tissue. Under normal conditions ventricular cells do not display a spontaneous rhythm; when such cells become partially depolarized, however, they may display an automatic rhythm (59–61). This ischemia-induced ectopic rhythm is critically dependent on calcium entry and can be abolished by lowering extracellular calcium (61) or exposing the cardiac cells to a calcium channel antagonist (60). Therefore, some forms of ventricular ectopic automaticity seem to depend on a slow inward calcium current.

As noted above, myocardial ischemia also results in elevations of cytosolic calcium, which in turn, have been shown to provoke oscillations in membrane potential (62, 63). These oscillations or fluctuations in membrane voltage are known as afterdepolarizations because their generation critically depends on the presence of a preceding action potential (59). There are two types of afterdepolarizations: delayed afterdepolarizations (DADs) that occur after repolarization of the preceding action potential and early afterdepolarizations (EADs) that occur either at the plateau (phase 2) or later during repolarization (phase 3) of the cardiac action potential (59, 64). When the amplitude of the afterpotential is large enough to reach threshold, repetitively sustained action potentials are generated. This form of ectopic automaticity is known as triggered activity, because it does not occur unless preceded by at least one action potential. These abnormal afterdepolarizations, particularly EADs, can also enhance the electrical heterogeneity between neighboring regions of the myocardium (64). The resulting differences in repolarization can lead to the formation of new action potentials through electrotonic (passive electrical) interactions between areas that have repolarized (i.e., recovered excitability) and those regions that have not (i.e., remain depolarized). This latter mechanism represents a form of re-entrant excitation (see below). Thus, under appropriate conditions, afterdepolarizations could provide both the trigger (premature ectopic beats) and the substrate (electrical heterogeneity, non-uniform repolarization) for the initiation and propagation of the lethal arrhythmias.

The membrane currents responsible for these oscillations remain to be fully elucidated. However, it is generally agreed that DADs result from spontaneous calcium release from the sarcoplasmic reticulum and a calcium-activated inward depolarizing current (65). At least three candidates have been proposed to carry this inward current: Na^+/Ca^{2+} exchanger current, a Ca^{2+}-activated Cl^- current, and a Ca^{2+}-activated non-selective cation current (66–70). Accumulating evidence favors the Na^+/Ca^{2+} exchanger current as the most important current for DAD formation. Schlotthauer and Bers (66), in an elegant series of studies, showed that caffeine-induced DADs resulted almost entirely from the Na^+/Ca^{2+} exchanger current, and also that only small (40 μM) changes in cytosolic calcium were necessary to provoke the afterdepolarizations. Thus, one would predict that drugs that selectively inhibit this exchanger should also prevent arrhythmias induced during ischemia (see below). In a similar manner, EADs are known to result when repolarization has been prolonged as the result of either decreasing the outward potassium current, increasing the inward current (either sodium or calcium), or some combination of changes in these currents (64). However, it has proven to be difficult to ascertain which of the individual currents is responsible for these membrane oscillations during the prolonged repolarization. It is now clear that reactivation of both the sodium and L-type calcium channel contribute significantly to the upstroke of the oscillation, whereas the inward mode of the Na^+/Ca^{2+} exchanger current plays an important role in the initial delay in repolarization (64). As was noted for DADs, EADs are critically dependent on elevations in cytosolic calcium (64). Both early and delayed afterdepo-

larizations have been recorded in isolated cardiac cells or tissue in response to interventions that favor calcium loading (hypoxia, cocaine, catecholamines, digitalis, calcium channel agonist BAY K 8644), and each can be suppressed by calcium channel antagonists and the intracellular calcium chelator, BAPTA-AM (51, 58, 62, 63, 71, 72).

The initiation of ventricular fibrillation may depend on inward movement of calcium (73). Ryanodine, a plant alkaloid that renders the sarcoplasmic reticulum leaky and unable to retain normal amounts of calcium (74, 75), suppresses cytosolic calcium oscillations but fails to prevent electrically induced ventricular fibrillation in isolated rabbit hearts (73). In contrast, verapamil and nifedipine, L-type calcium channel blockers, terminate ventricular fibrillation (73). In related studies, Billman (76–78) demonstrated that several organic (verapamil, flunarizine, nifedipine, diltiazem, mibefradil) and inorganic (magnesium [Mg^{2+}]) calcium channel antagonists prevent malignant ventricular arrhythmias induced by ischemia. Conversely, the L-type calcium channel agonist, BAY K 8644, induced ventricular fibrillation in animals resistant to the development of arrhythmias (76). Ryanodine failed to prevent malignant arrhythmias despite large reductions in peak cytosolic calcium, indicated by corresponding reductions in contractile force development (79). These data strongly suggest that calcium influx across the sarcolemma, rather than calcium release from the sarcoplasmic reticulum, may be critical for the induction of ventricular fibrillation.

Calcium also may contribute to changes in impulse conduction. Conduction abnormalities may result from simple conduction block or more complex forms of re-entry (59). In the normal heart, action potentials generated in the sinus node terminate after the sequential activation of the atria and the ventricles, because the surrounding tissue has become refractory or non-excitable after depolarization. If, however, the impulse conduction is slowed in one region of the heart and the surrounding tissue has repolarized, it may be possible to re-excite the surrounding tissue before the next impulse is conducted from the sinus region. This phenomenon, known as re-entrant excitation, is responsible for the generation of extrasystoles (re-entrant arrhythmias). Calcium channel antagonists exert their most obvious effects on the conduction of action potentials through the atrioventricular (A-V) node. Because A-V nodal tissue generates slow-response (i.e., calcium-dependent) action potentials, calcium antagonists prolong A-V conduction time and refractory period (80, 81). These actions attenuate the ventricular response to rapid atrial arrhythmias (atrial flutter or fibrillation) and terminate supraventricular tachycardias in which the A-V node forms part of the re-entrant circuit (80).

The effects of calcium on conduction abnormalities in the ventricles are, however, equivocal. Ordered or simple re-entrant arrhythmias in which the impulse is conducted in a finite and well-circumscribed loop may occur in an ischemic heart (59). Re-entrant arrhythmias require decremental conduction and unidirectional block as preconditions for arrhythmia formation (59). Conduction velocity in cardiac tissue depends on the rate of depolarization (dV/dt_{max} or V_{max}) and action potential amplitude, factors primarily mediated by the fast sodium channels (33). As noted above, myocardial ischemia results in depolarization of the resting membrane potential, which may lead to inactivation of sodium channels (59, 82). Consequently, conduction velocity decreases, and unidirectional block may occur. In acute ischemia, many conduction disturbances that produce re-entrant arrhythmias are not mediated by slow-response action potentials but rather by reduced sodium entry through fast channels (83). It is therefore not surprising that calcium channel antagonists are not effective against ordered re-entrant arrhythmias (81, 84). However, conduction velocity also depends on a low electrical resistance between cells (59, 82). As ischemia progresses, intracellular Ca^{2+} and hydrogen (H^+) increase (31, 35–37, 57, 85). High concentrations of these ions reduce conductance across the gap junctions, which form the low electrical resistance pathway that facilitates cell-to-cell coupling (86). Thus, during later stages of ischemia or in chronically ischemic hearts, conduction disturbances may result from the uncoupling of cardiac cells due to the cellular accumulation of calcium. Calcium channel antagonists can diminish calcium accumulation and thereby improve conduction in ischemic hearts. Verapamil reduces, whereas BAY K 8644 exacerbates, the slowing of ventricular conduction induced by global ischemia in the isolated rabbit heart (87).

In contrast to ordered re-entrant circuits, calcium may contribute significantly to random or irregular re-entrant circuits. Random re-entry is characterized by multiple irregular pathways that change continuously, producing an unpredictable, chaotic conduction pattern. Ventricular fibrillation is the epitome of random re-entry. A major factor contributing to ventricular fibrillation, particularly during myocardial ischemia, is a spatial dispersion or nonuniformity of the refractory period (59); this allows impulse conduction to become fragmented during ensuing heartbeats and thus sets the stage for random re-entry. Dispersion of refractory periods results, at least in part, from disturbances in action potential duration, which can be recorded as alterations in the S-T segment (electrical alternans) (59, 88–91). For example, Lee et al. (36) showed that alterations in the amplitude of calcium transients accompanied corresponding changes in action potential duration. The pattern of alternans was stable at a given recording site but varied from site to site in a given preparation. They concluded that "the alternans behavior of the calcium transients in a particular region is independent of the behavior of other regions, which results in spatial heterogeneity of the calcium transients during ischemia" (36). Calcium channel antagonists have been shown to reduce calcium transient and electrical alternans (92) and the spatial dispersion of refractory period from the endocardium to epicardium during ischemia (93). These data indicate that nonhomogeneity of refractory periods may result from a calcium-mediated oscillation of action potential duration, and in turn, form a substrate for irregular re-entry.

In summary, abnormalities in cellular Ca^{2+} may contribute significantly to the development of malignant ventricular arrhythmias by inducing various forms of ectopic automaticity, by changing conduction, or by a combination of both automaticity and conduction disturbances. If, for example, an extrasystole occurs in a region of nonuniform refractory period, irregular re-entrant pathways and ventricular fibrillation may result.

2.2.3 Extracellular Potassium Accumulation During Myocardial Ischemia.

In addition to changes in cytosolic calcium as described above, myocardial ischemia will elicit profound changes in extracellular potassium. The resulting depolarization of the surrounding tissue, decreases in action potential duration, and nonuniformities of repolarization (as well as refractory period) could all contribute to the induction of the life-threatening arrhythmias associated with myocardial ischemia. It is now generally accepted that disruptions in coronary blood flow elicit both rapid increases in extracellular potassium and reductions in action potential duration. Harris and co-workers (94, 95) were the first to show that extracellular potassium rises dramatically after coronary artery ligation, correlating with the onset of ventricular arrhythmias. They further demonstrated that intracoronary injections of KCl provoked electrocardiographic changes and triggered ventricular arrhythmias similar to those induced by myocardial ischemia (94, 95). They proposed that changes in extracellular potassium represented a major factor in the development of malignant arrhythmias during ischemia. In recent years, a number of studies using ion selective electrodes to measure potassium activity directly have largely confirmed these earlier observations (96–98). Extracellular potassium has been found to increase within the first 15 s and reach a plateau within 5–10 min after the interruption of coronary perfusion (96, 97, 99, 100). Furthermore, regional differences or inhomogeneities of potassium accumulation were recorded, accompanied by corresponding differences in ventricular electrical activity (98–100). The increase in extracellular potassium results primarily from increases in potassium efflux rather than from decreased potassium influx due to inhibition of the Na^+/K^+-ATPase (101–103). Several mechanisms have been proposed to explain the enhanced potassium efflux, including an increased potassium outward conductance due to the direct activation of one or more potassium channels (104–106) or a passive potassium efflux

coupled with anion (lactate or inorganic phosphate) conductance to balance transmembrane charge (97). The latter hypothesis stipulates that potassium efflux results secondarily to the movement of intracellularly generated anions during ischemia to balance charge movement as these negatively charged ions diffuse across the sarcolemma. Thus, potassium efflux would result from a passive redistribution of potassium ions in response to the net inward current resulting from anion efflux, rather than from an active ion-anion linked process. Weiss et al. (107) have tested this hypothesis. In particular, the contribution of inorganic phosphate and lactate ion to potassium efflux during ischemia and hypoxia was investigated. They found that under a variety of conditions, a major component of cellular potassium loss was not related to the efflux of these anions. They concluded that this "non–anion-coupled" potassium efflux during metabolic inhibition was most likely to result from an increase in membrane potassium conductance.

A growing body of evidence suggests that ischemically induced potassium accumulation and the corresponding reductions in action potential duration result primarily from the opening of ATP-sensitive potassium channels. Using the patch clamp technique, Trube and Hescheler (108) were the first to record single ATP-sensitive potassium channel activity. Noma (104) and Hescheler et al. (109) further demonstrated that reductions in cellular ATP induced by cyanide exposure evoked an outward potassium current. They, therefore, proposed that the activation of an ATP-sensitive potassium channel might be responsible for the reductions in action potential duration induced by hypoxia. Several studies have since further implicated the activation of this current in the changes in cardiac action potential and extracellular potassium accumulation during myocardial ischemia (110–120). The ATP-sensitive potassium channel inhibitor, glibenclamide, for example, has been shown either to attenuate or abolish reductions in action potential duration in hypoxic myocytes (115, 117), isolated cardiac tissue (110, 112, 113, 116, 120), and regionally or globally ischemic hearts (118, 121, 122). This sulphonylurea drug has also been shown to reduce extracellular potassium accumulation induced by ischemia (110–112, 114). Conversely, ATP-sensitive potassium channel agonists (pinacidil, cromakalim) exacerbated ischemically induced reductions in action potential duration, as well as promoted extracellular potassium accumulation (110, 115–120, 122–125). However, the ATP-sensitive potassium channel is activated only at low ATP concentrations with half-maximum suppression of channel opening at 20–100 μM (104, 126, 127), yet intracellular concentrations are normally much higher (5–10 mM). Furthermore, cytosolic ATP levels remain in the millimolar range for the first 10 min of hypoxia, well after potassium accumulation begins (103, 128). The role that this channel plays in the response to myocardial ischemia has therefore been questioned. Recently, a number of investigators have shown that, because of the high density of cardiac ATP-sensitive potassium channels, only a small increase in the open-state probability (<1% of maximum) was sufficient to shorten action potential duration during ischemia (129–131). For example, Faivre and Findlay (130) found during patch clamp studies of guinea pig myocytes that the opening of only 30 channels (less than 1% of the population) provoked a 50% reduction in action potential duration. They concluded that "physiologically relevant activity of the K_{ATP} channel in cardiac membrane is confined to a very small percentage of the possible cell K_{ATP} current, and thus, intracellular ATP would not have to fall very far before the opening of K_{ATP} channels would influence cardiac excitability." Deutsch et al. (132) investigated the potassium current induced by hypoxia in an intact rabbit papillary muscle. They showed that during hypoxia, a significant shortening of action potential duration (blocked by glibenclamide) occurred when tissue ATP levels fell by approximately 25%. They concluded that only modest changes in cellular ATP were required to induce major changes in cardiac electrical properties.

The activation of the ATP-sensitive potassium channel may also contribute significantly to the S-T segment changes associated with myocardial ischemia. In anesthetized open chest dogs (133), the intracoronary injection of the ATP-sensitive potassium channel opener, pinacidil, elicited elevations in the S-T segment

very similar to those induced by myocardial ischemia. Conversely, glibenclamide attenuated the S-T segment elevations induced by the occlusion of the left anterior descending coronary artery (134). Similar results were also obtained in conscious dogs. Billman et al. (135) demonstrated that either glibenclamide or the cardioselective ATP-sensitive potassium channel antagonist HMR 1098 attenuated ischemically induced S-T segment changes. These drugs also prevented ischemically induced increases in the descending portion of the T wave (an index of the transmural dispersion of repolarization) (136). Finally, myocardial ischemia failed to alter the ECG of mice in which the Kir 6.2 gene (the gene responsible for the pore forming subunit of the cardiac ATP-sensitive potassium channel) (137) had been removed (138). Furthermore, the large changes in the ECG that were induced by ligation of the left coronary artery in the wild-type control mice could be suppressed by prior treatment with the cardioselective ATP-sensitive potassium channel antagonist HMR 1098 (138). It therefore seems likely that the activation of the ATP-sensitive potassium channel contributes significantly to alterations in cardiac electrical stability induced by myocardial ischemia, leading to the formation of malignant arrhythmias.

2.2.4 Extracellular Potassium and Cardiac Arrhythmias. As noted above, cardiac arrhythmias can result from either abnormalities in impulse generation or impulse conduction. The extracellular accumulation of potassium induced by myocardial ischemia can provoke these perturbations in cardiac electrical activity. The accumulation of extracellular potassium promotes the depolarization of the tissue surrounding the ischemic regions, as noted above. The flow of this injury current (electrotonic current flow between ischemic and normal cells) has been implicated as a potential cause for the initiation of premature ventricular beats. Under normal conditions, ventricular cells do not display a spontaneous rhythm. However, when the cells are partially depolarized an automatic rhythm can be produced (59). Coronel et al. (99) demonstrated an increased excitability of normal tissue near the border of the ischemia, a region of the heart in which extracellular potassium concentrations were also increased.

Changes in action potential duration induced by alterations in potassium efflux can also provoke abnormalities of impulse conduction. As noted above, increased potassium efflux from ischemic tissue triggers a reduction in action potential duration. A major factor contributing to ventricular fibrillation, particularly during myocardial ischemia, is a dispersion or inhomogeneity of the refractory period (59). This allows for the fragmentation of impulse conduction during ensuing beats. Recent evidence suggests that a major factor contributing to the dispersion of refractory period is regional differences in action potential duration (59). As previously noted, the activation of the ATP-sensitive potassium channel produced large reductions in action potential duration (110–113, 115–120, 122), which are inhibited by glibenclamide (115–120, 122) but exacerbated by ATP-sensitive potassium channel agonists (110, 124, 139–141). A differential ATP sensitivity of the ATP-dependent potassium channel has also been reported between endocardial and epicardial cells during ischemia, such that smaller reductions in ATP were necessary to activate potassium channels located in epicardial tissue (142). As a result, an inhomogeneity in extracellular potassium and refractory period, as well as a gradient in action potential duration, was recorded between the epicardial and endocardial tissue (142–144). Nonuniformities in refractory period could set the stage for the formation of irregular reentrant pathways and ventricular arrhythmias. In a similar manner, the ATP-sensitive potassium channel antagonist glibenclamide was shown to attenuate ischemically induced reductions in the refractory period (145, 146). In contrast, activation of the ATP-sensitive potassium current with pinacidil elicited a marked dispersion of repolarization and refractory period between the epicardium and endocardium, leading to the development of extrasystoles (147). These effects could be abolished by glibenclamide (147). Thus, the activation of the ATP-sensitive potassium channel could contribute significantly to the induction of malignant arrhythmias by changing impulse generation (depolarization induced changes in automaticity), conduction (refractory period dispersion), or a combination of both.

2.3 Ventricular Remodeling

In addition to sudden death from malignant arrhythmias, an important negative consequence of myocardial infarction is the development of chronic congestive heart failure. A clinical history of myocardial infarction increases the age-adjusted risk of developing heart failure roughly 10-fold, and the 5-year survival rate for patients diagnosed as suffering from congestive heart failure is only about 30% (148). When viable myocardium is lost to an infarct, the added contractile demands on the remaining myocardial tissue elicit a series of responses that produce hypertrophy, interstitial fibrosis, and remodeling of the residual muscle mass. These processes can at first normalize the depressed cardiac output, but with a sufficient long-standing strain on the heart, there is decompensation, dilatation of the ventricles, and eventual heart failure (148).

Ventricular remodeling begins within the early hours of an acute myocardial infarction (149). First there is infarct expansion because of changes in the extracellular matrix and myocyte apoptosis. Dilatation of the noninfarcted tissue occurs later, but much of this occurs during the first few days post-infarct. At 1 week post-infarction, left ventricular end systolic volume may be 80% above normal and is dependent on infarct size and patency of the infarct-related artery (149). The increase in myocardial wall stress as a result of ventricular dilatation activates both the plasma and intracardiac renin-angiotensin system (148). Angiotensin has been directly implicated in ventricular remodeling and may also increase norepinephrine release, further increasing the metabolic demands on injured but still viable myocardium (149). An important goal of the management of myocardial infarction is to minimize this maladaptive response.

3 TREATMENT FOR MYOCARDIAL INFARCTION

3.1 Pain Relief

Relief of pain is important for the patient with an acute myocardial infarction. Pain and anxiety heighten adrenergic responses, and these in turn, are arrhythmogenic and place additional metabolic demands on the heart. Morphine sulfate is the drug of choice in the United States (1). It has the added benefit of increasing parasympathetic outflow to the heart, which as noted above, could indirectly alter cardiac metabolic demand and the resulting ionic changes associated with myocardial ischemia. Morphine is used judiciously, however, because the resolution of pain is often the hallmark of successful thrombolysis, whereas persistent pain signifies the need for additional intervention (1). Intravenous nitroglycerin is also used for the treatment of persistent chest pain in some patients with acute myocardial infarction. Nitroglycerin is a prodrug that provides nitric oxide (NO), an activator of smooth muscle guanylyl cyclase, the enzyme that produces cGMP (150). Activation of a cGMP-dependent protein kinase initiates a cascade of reactions that ultimately result in smooth muscle relaxation. In particular, venodilation reduces preload and thereby lessens metabolic demand. NO may also inhibit platelet aggregation, but the therapeutic significance of this effect in acute myocardial infarction is uncertain (150).

3.2 Thrombolysis

Thrombolysis, or the dissolving of an occlusive blood clot, has become standard treatment for myocardial infarctions associated with S-T segment elevations. Restoration of blood flow is evaluated angiographically in terms of achieving thrombolysis in myocardial infarction (TIMI) flow grades (151). TIMI grade 0 indicates no antegrade flow; grade 1 indicates partial contrast penetration and incomplete distal filling; grade 2 indicates a patent infarct related artery with opacifications of the entire distal artery with delayed contrast filling or washout; and grade 3 indicates normal flow. Significant reductions in morbidity and mortality are associated with restoration of TIMI grade 3. In the GUSTO I study, the mortality rate in patients with TIMI grade 3 flow 90 min after the initiation of thrombolysis was less than one-half that in patients with TIMI grades 0 or 1 flow (152).

Thrombolysis has been found to benefit those patients treated within 12 h of the onset of chest pain, but improvements in survival are greatest when treatment is begun within the first few hours (2, 153). In the GISSI trial, thrombolytic therapy (streptokinase; see below) begun within the first hour decreased in-hospital mortality by 51%, therapy begun within 3 h reduced mortality by 26%, and that begun after 3–6 h reduced mortality by 20% (2, 153). In the LATE trial, tissue plasminogen activator (t-PA) administered 6–12 h after the onset of symptoms reduced 35-day all-cause mortality by 27% (2). In the EMERAS study, streptokinase administered after 6–12 h gave an 11% nonsignificant reduction of in-hospital mortality (153). It should be noted that comparisons among the various large clinical trials are complicated by differences in adjunctive therapies, treatment endpoints, and patient selection criteria.

3.2.1 Streptokinase. Streptokinase and anisoylated plasminogen streptokinase-activated complex (APSAC) are effective thrombolytic agents (154). Streptokinase is a 47,000-dalton protein produced by hemolytic streptococci. It forms a 1:1 noncovalent complex with circulating plasminogen, producing a conformational change that facilitates the cleavage of the inactive 790 amino acid plasminogen at arginine 560 to form free plasmin. Plasmin is a relatively nonspecific protease that digests fibrin clots and other plasma proteins, including several coagulation factors. APSAC is a streptokinase-plasminogen complex prepared *in vitro* from purified acylated human lys-plasminogen and purified streptokinase. The acyl group must be hydrolyzed *in vivo* before activation; this allows time for the plasminogen to bind to fibrin before activation by streptokinase.

3.2.2 Plasminogen Activators. These thrombolytic drugs have exploited or been patterned after the plasmin-based fibrinolytic system that normally dissolves small intravascular clots *in vivo* (154). This fibrinolytic system is highly regulated such that unwanted fibrin thrombi are removed while fibrin in wounds is unaffected. Fibrinolysis is normally initiated when t-PA is released from endothelial cells in response to such signals as the stasis produced by a vascular occlusion. This endothelial cell–derived t-PA is rapidly cleared from blood or inhibited by a circulating inhibitor, plasminogen activator inhibitor-1, and thus exerts little effect on circulating plasminogen. By contrast, the t-PA that binds to fibrin converts fibrin-bound plasminogen to plasmin. Plasminogen and plasmin both bind to fibrin at sites near their lysine-rich amino termini, sites that are also required for binding of plasmin to α_2-antiplasmin, a 452 amino acid glycoprotein that instantaneously inactivates plasmin. Therefore, fibrin-bound plasmin is highly effective because it is shielded from inhibition, whereas plasmin that escapes into the circulation is rapidly inhibited (154).

t-PA (alteplase) is a first generation tissue plasminogen activator produced commercially by recombinant DNA technology. It is synthesized using human cDNA for natural human tissue type plasminogen activator and is expressed in and harvested from cultured Chinese hamster ovary cells. t-PA is a 527 amino acid serine protease with minimal activity in the absence of fibrin, but when bound to the fibrin in a thrombus it converts occluded plasminogen to plasmin, initiating local thrombolysis (154). Accelerated t-PA administration achieves TIMI grade 3 more rapidly than streptokinase and has been associated with more favorable clinical outcomes (155).

Alteplase must be continuously infused because of its short half-life. Variants of t-PA with deletion of specific domains or specific amino acid substitutions have shown reduced plasma clearance and are suitable for bolus injections (155). Reteplase (rPA) is a single peptide chain molecule consisting of 355 amino acids, starting with serine1 and ending with proline527 of the original t-PA sequence. It also lacks amino acids valine4 through glutamate175. Tenecteplase (TNK-tPA) contains a threonine to asparagine substitution at position 103 of t-PA, an asparagine to glycine substitution at position 117, and the replacement of lysine296, histidine297, arginine298, and arginine299 by alanines. Reteplase and tenecteplase have reduced plasma clearance and can be administered as a single bolus (TNK-tPA) or as a double bolus 30 min apart (rPA), offering the possibility for pre-hospital initiation of thrombolysis (155).

When exogenous thrombolytic agents are applied at their recommended dosages, excessive bleeding and hemorrhagic stroke can ensue. Yet in large clinical trials, the benefits of timely thrombolysis have vastly outweighed the drawbacks (2, 153).

3.2.3 Anticoagulants. The benefits of thrombolytic therapy are improved significantly with the concomitant use of the antiplatelet agent, aspirin, which is an irreversible inactivator of platelet cyclooxygenase, and the classical anticoagulant, heparin, which is a thrombin inhibitor (156). Antiplatelet agents and thrombin inhibitors help maintain vessel patency by preventing reocclusion. These anticoagulants have proved beneficial even in the absence of thrombolysis, presumably because of the inhibition of new thrombotic events (153, 156). It is currently recommended that all aspirin-tolerant individuals suffering from an apparent acute myocardial infarction be immediately given chewable aspirin, and daily low dose aspirin is also recommended for most postmyocardial infarction patients (157). The thienopyridines, clopidogrel or ticlopidine, which inhibit platelet function by inhibiting the binding of fibrinogen to activated platelets, should be used in patients who are allergic to aspirin or who suffer from gastrointestinal bleeding (156). In conjunction with thrombolysis for an acute myocardial infarction, adequate heparinization is particularly important in patients receiving t-PA because this drug has far less of a systemic anticoagulant effect than streptokinase or APSAC (154, 156, 158). t-PA plus heparin has been found to be slightly but significantly more efficacious at improving survival of patients with acute myocardial infarction, despite a slightly increased risk of stroke (153, 156).

3.2.4 Glycoprotein IIb/IIIa Receptor Blockers. When platelets are activated, the glycoprotein IIb/IIIa (GP IIb/IIIa) receptor undergoes a change in configuration that increases its affinity for binding to fibrinogen and other ligands (159). Fibrinogen binding to receptors on different platelets results in platelet aggregation. GP IIb/IIIa receptor antagonists prevent fibrinogen binding and therefore prevent platelet aggregation (159). The available GP

IIb/IIIa antagonists have different pharmaco-kinetic and pharmacodynamic properties. Abciximab is a Fab fragment of a humanized murine antibody that has a short plasma half-life but strong affinity for the receptor. Some receptor occupancy can persist for weeks, and platelet aggregation gradually returns to normal 24–48 h after the drug is discontinued.

Eptifibatide is a cyclic heptapeptide that contains a lysine-glycine-aspartate sequence. Tirofiban is a nonpeptide mimetic of the arginine-glycine-aspartate sequence of fibrinogen. These two synthetic antagonists bind less tightly than abciximab, with receptor occupancy in equilibrium with plasma levels. They have a half-life of 2–3 h and are more specific for the GP IIb/IIIa receptor than abciximab, which can also bind to the vitronectin receptor on endothelial cells and the MAC-1 receptor on leukocytes (159).

GP IIb/IIIa receptor antagonists have proved especially beneficial in patients with unstable angina or non-S-T segment elevation myocardial infarctions. These patients are not treated with thrombolytics but may undergo percutaneous interventions. The GP IIb/IIIa receptor antagonists reduce the incidence of restinosis following percutaneous interventions (159).

The GUSTO V trial compared the effects of standard dose reteplase (rPA) with half-dose reteplase plus full-dose abciximab (160). Several pilot studies had suggested that the combination of a low dose plasminogen activator with a IIb/IIIa receptor antagonist improved speed, durability, and completeness of myocardial reperfusion. Preliminary data from the GUSTO V trial showed that at 30 days there were fewer nonfatal reinfarctions in the combination group and there was less need for urgent revascularization (rescue angioplasty). In the reteplase group, 5.9% of patients had died at 30 days compared with 5.6% in the half-dose reteplase plus abciximab group (p = 0.43) (160). Bleeding complications were more prevalent in the combination therapy group, but there was no increase in intracranial hemorrhage (160). Data for 1-year mortality are widely anticipated; these will establish whether more rapid and complete reperfusion has lasting benefits.

3.3 Treatment of Arrhythmias Induced by Myocardial Ischemia

Interventions that alter ion flux across the sarcolemma should also alter the potential for arrhythmias induced by myocardial ischemia. In particular, one would predict that either ATP-sensitive potassium channel antagonists or calcium channel antagonists would protect against the formation of malignant arrhythmias under these conditions. The following section will briefly review the experimental and clinical experience with these ion channel antagonists. We shall first begin with a brief description of the major types of anti-arrhythmic drugs.

3.3.1 Classification of Anti-Arrhythmic Drugs.
Anti-arrhythmic drugs are widely classified by their effects on cardiac electrical properties using the system devised by Vaughn Williams (161, 162). In this schema, drugs are assigned to one of four classes. The class I drugs block sodium channels and are subclassified based on their effects on action potential conduction. The class IA drugs produce a moderate prolongation of conduction and repolarization. Representative examples of class IA drugs include quinidine, procainamide, and disopyramide. The class IB drugs, which include lidocaine and tocainide, exert little or no effect on conduction and repolarization. In contrast, the class IC drugs (flecainide, encainide) exhibit a marked prolongation of conduction and repolarization. The class IC drugs have been shown to increase cardiac mortality in patients recovering from myocardial infarction most likely as a result of an increased incidence of sudden death (a proarrhythmic effect fibrillation), and as such, should no longer be used to suppress arrhythmias in this patient population (163). Class II anti-arrhythmic drugs block β-adrenergic receptors and are the only agents widely accepted to reduce the incidence of sudden cardiac death in post-infarction patients (see below). Class III drugs prolong repolarization and effective refractory period, most likely through modulation of potassium channels. Examples include amiodarone, dofetilide, ibutilide, bretylium, and d-sotalol. Pro-arrhythmic properties have also been reported for some (d-sotalol, dofedilide), but not all,

(amiodarone) class III anti-arrhythmic drugs (164, 165). These drugs have been shown to prolong QT interval, promote life-threatening tachyarrhythmia, torsades de pointes, and increase cardiac mortality in some patient populations. Amiodarone is, in fact, the only class III agent that has been shown to reduce cardiac mortality in post-infarction patients (166, 167). Because this drug blocks a number of ion channels (e.g., L-type calcium channels and a number of potassium channels including the ATP-sensitive potassium channel) (168), it has been difficult to ascertain the mechanism responsible for this cardioprotection (169). Finally, class IV drugs inhibit calcium channels (see below).

As described above, myocardial ischemia elicits profound changes in the regulation of intracellular calcium and potassium. In the following section, we shall focus on agents that may be particularly useful in the modulation of calcium and the ATP-sensitive potassium currents during myocardial ischemia and could thereby protect against malignant arrhythmias. The section will close with a brief discussion of the well-established effects of β-adrenergic receptor antagonists on cardiac mortality after myocardial infarction. It should be emphasized that the list of agents described in the following sections represents only a few of the many agents used to treat cardiac arrhythmias. For more detailed presentation of specific agents, particularly class I and class III agents, please refer to refs. 170 and 171.

3.3.2 Calcium Channel Antagonists.

Calcium antagonists represent a large, structurally diverse group of chemicals that share the ability to inhibit calcium entry into muscle cells through actions on the calcium-selective ion channels. Cardiac tissue contains two types of calcium channels: channels with a long activation (L-type) and channels that activate transiently (T-type). The L-type calcium channels are located in both atrial and ventricular tissue, whereas the T-type channels are more predominate in atrial (i.e., pacemaker) cells (172, 173). However, pathological conditions including ventricular hypertrophy (174), dilated cardiomyopathy (175), or myocardial infarction (176) have been associated with an increased expression of functional T-type calcium channels. Vascular smooth muscle contains both L- and T-type calcium channels that each contribute to the regulation of smooth muscle calcium, and thereby, vascular tone (172, 173). Given the widespread distribution of calcium channels within the cardiovascular system, it is not surprising that calcium channel antagonists have received considerable therapeutic interest. Indeed, since their discovery in the 1960s (177–179), calcium antagonists have become increasingly more important in the treatment of various cardiovascular diseases, most notably hypertension and angina pectoris (180). The use of calcium channel antagonists has been largely restricted to the treatment of supraventricular arrhythmias. In contrast, the therapeutic potential in the management of life-threatening ventricular arrhythmias, particularly during myocardial ischemia, has received less attention. The following sections of this chapter will summarize experimental and clinical studies that illustrate the anti-arrhythmic potential of individual calcium channel antagonists.

3.3.3 Verapamil.

Verapamil, a benzeneacetonitrite, is structurally similar to papaverine and was first synthesized in 1962 (181). Verapamil blocks calcium entry equally in both cardiac and smooth muscle through actions on both L- and T-type calcium channels (30, 182). Verapamil, like most calcium channel antagonists, is not completely selective for calcium channels and has been shown to block a variety of potassium channels (30), including the ATP-sensitive potassium channel, transient outward current, and the rapidly activating component of the delayed-rectifier current (IKr). Of particular note, verapamil has been shown to block human ether-a-go-go-related gene (HERG)-encoded channels (IKr) transfected into various cell lines at clinically relevant concentrations (183). However, whether or not the verapamil-mediated inhibition of these potassium currents contributes to the clinical effects of this drug has not yet been determined. The blockade of calcium-entry in smooth muscle elicits vasodilation with corresponding reductions in arterial pressure and increases in coronary blood flow. In cardiac muscle, verapamil tends to depress both contractile force and impulse conduction. The second action is particularly obvious at the A-V node.

In 1968, Kaumann and Aramendia (184) were the first to demonstrate that verapamil can protect against ventricular fibrillation induced by ligation of a coronary artery. These findings have been confirmed by many other animal studies (76, 184–187). In intact animals, verapamil prevents ventricular arrhythmias induced by myocardial ischemia (76, 187) and attenuates the reduction in ventricular fibrillation threshold that accompanies coronary artery occlusion (185, 188, 189). Billman further demonstrated that verapamil completely suppresses ventricular fibrillation induced by either cocaine (190) or the combination of exercise and acute myocardial ischemia (76). In a similar manner, verapamil, in combination with the angiotensin-converting enzyme inhibitor trandolapril, prevented ventricular tachyarrhythmias induced by myocardial ischemia followed by reperfusion in anesthetized pigs (191). In contrast, verapamil failed to prevent ordered re-entrant arrhythmias induced by programmed electrical stimulation (80, 84, 191, 192).

Clinical experience with verapamil to date has been somewhat inconsistent. For example, verapamil is generally ineffective in the treatment of either stable re-entrant arrhythmias or arrhythmias induced by programmed electrical stimulation (80, 84, 191, 192), which also probably result from re-entry. In contrast, some reports indicate that arrhythmias associated with acute myocardial ischemia or exercise-induced ventricular tachycardia respond favorably to verapamil (191, 193). Verapamil also significantly decreased the frequency and severity of ventricular arrhythmias in patients with left ventricular hypertrophy (194). Verapamil has been shown to be effective in the treatment of idiopathic left ventricular tachycardia (a relatively rare but distinct entity in young, mostly male, patients) and to a lesser extent in the treatment of right ventricular outflow ventricular tachycardia (which is more common in female subjects) (195–197). Finally, in a large Danish multicenter study (198) of patients recovering from myocardial infarction, verapamil significantly reduced the frequency of major cardiac events (cardiac death or second myocardial infarction). Sudden death was reduced by 20–26% in patients treated with verapamil—results comparable with those obtained with β-adrenergic antagonists (199, 200). This protection, however, was noted only in patients without evidence of heart failure. Verapamil had no beneficial effects on major cardiac events in patients with heart failure, and in fact, may have increased cardiac mortality (198). The reduction in sudden death may reflect inhibition or reduction of calcium loading in ventricular cells, but no direct data yet support this suggestion.

3.3.4 Diltiazem.
Diltiazem, a benzodiazepine derivative, is structurally quite distinct from verapamil and was first identified as a calcium antagonist in 1971 (201). Diltiazem has pronounced effects on both vascular smooth and cardiac muscle. This drug elicits relaxation of the vascular smooth muscle, increasing coronary blood flow and reducing arterial pressure. Diltiazem depresses A-V conduction, but has a lesser negative inotropic effect than verapamil. Diltiazem can also inhibit the delayed rectifier potassium current but not at clinically useful concentrations (30, 183). Less consistent antiarrhythmic actions have been noted for diltiazem than for verapamil, perhaps because of its less potent cardiac actions. Diltiazem reduced regional differences in impulse conduction in ischemic tissue (60, 202), prevented slow-response action potentials (33), and abolished calcium transient or electrical alternans (36, 92). In a similar manner, diltiazem decreased the amplitude of T wave alternans and the incidence of ventricular tachyarrhythmias in chloralose-anesthetized dogs (203). Diltiazem also protected against ventricular fibrillation in anesthetized animals (92, 204). In unanesthetized canine preparations, diltiazem delayed the time to onset of malignant arrhythmias, but it failed to prevent ventricular fibrillation induced by irreversible coronary artery occlusion (205). Billman further demonstrated that diltiazem prevented ventricular fibrillation induced by cocaine (190) or acute myocardial ischemia (76). In contrast, diltiazem failed to prevent ischemic changes in the ventricular fibrillation threshold (206), as well as arrhythmias associated with reperfusion (203, 207).

Similar mixed results have been reported in clinical studies of the antiarrhythmic potential of diltiazem. Diltiazem was found to reduce the reinfarction rate by approximately 51% in patients with myocardial infarction, but in contrast to verapamil, did not affect overall mortality rates (208). Furthermore, diltiazem failed to reduce stable ventricular arrhythmias during a 12-h recording period (209). However, in a study involving over 2400 patients with myocardial infarction, diltiazem produced a significant decrease in mortality in a subgroup of patients without radiographic evidence of pulmonary congestion (i.e., heart failure) (210). It is unclear whether the reduction in mortality arises from blockade of cardiac calcium channels and the resultant prevention of cellular calcium overload. The subgroup analysis further demonstrated that diltiazem provoked a dramatic increase in mortality in patients with myocardial infarction complicated by pulmonary congestion (210). Calcium antagonists, therefore, may have deleterious effects in patients with compromised cardiac function (211).

3.3.5 Nifedipine. Nifedipine, a dihydropyridine, was first synthesized in 1971 (212). In contrast to either diltiazem or verapamil, at therapeutic concentrations, nifedipine acts primarily on vascular smooth muscle. Nifedipine acts as a more potent vasodilator than either diltiazem or verapamil; cardiac actions are noted only at much higher concentrations. Therefore, it is not surprising that nifedipine exhibits few antiarrhythmic properties in intact animals or patients. Nifedipine, with a few exceptions (see below), is generally ineffective in the treatment of experimentally induced ischemic arrhythmias (76, 213, 214). Nifedipine failed to alter the reduction in ventricular fibrillation threshold induced by coronary artery occlusion (214) and to prevent reperfusion arrhythmias (207). However, it has been reported that nifedipine prevents ventricular fibrillation in the ischemic rat heart. This protection is attributed to improved coronary perfusion (i.e., less myocardial ischemia) rather than to direct cardiac actions of the drug (215). Nifedipine protected against malignant arrhythmias induced by exercise plus myocardial ischemia, but only at

very high concentrations (76). Nifedipine also failed to reduce mortality in patients recovering from myocardial infarction (216). A reflex tachycardia induced by nifedipine's reduction of arterial blood pressure may, in fact, worsen the condition of such patients by placing a higher metabolic demand on the damaged heart. Indeed, much of the recent controversy (217) surrounding its use may be related to this reflex action. The increased metabolic demand placed on the heart may counteract the beneficial actions of reduced arterial blood pressure (211).

3.3.6 Flunarizine. Flunarizine is a difluoronated piperazine derivative that, under physiologic conditions, interacts weakly with both sodium and calcium (L- and T-type) channels (218, 219). However, during hypoxia, flunarizine inhibits increases in cytosolic calcium and prevents tissue damage, particularly in neural tissue (218). It seems reasonable that flunarizine may act as a calcium overload antagonist and thereby prevent malignant arrhythmias during myocardial ischemia. Flunarizine protected against isoproterenol-induced cardiac lesions and ischemic injury of the rat heart (218). It reduced ventricular fibrillation by 100% and ectopic beats during occlusion and reperfusion of the left anterior descending coronary artery (218); however, a much higher dose was required for protection equivalent to that noted with verapamil (218). Flunarizine prevented ventricular tachycardia due to delayed afterdepolarizations induced by ouabain toxicity but not arrhythmias due to activation of re-entrant circuits (220, 221). Based on these findings, Vos et al. (220) proposed that flunarizine may be used to differentiate between arrhythmias arising from calcium overload (triggered activity) and arrhythmias due to re-entry. Flunarizine has also been shown to prevent torsades de pointes (a severe polymorphic ventricular tachycardia associated with early afterdepolarizations) induced by the selective IKr blocker almokalant in dogs with chronic A-V block (222) or anesthetized rabbits (almokalant combined with β-adrenergic receptor stimulation with methoxamine) (223). In the latter study, neither the intracellular calcium chelator BAPTA-AM nor the sarcoplasmic reticulum calcium release disruptor ryanodine could prevent the arrhythmias. The authors, therefore, concluded that calcium entry, particularly through the T-type calcium channels, played crucial role in the induction of torsades de pointes (223). Finally, Billman (77, 190) showed that flunarizine completely suppresses ventricular fibrillation induced by either cocaine or exercise plus myocardial ischemia; however, large reductions in the inotropic state were noted. The authors are unaware of any clinical reports of the use of flunarizine in the management of ventricular arrhythmias. However, given the potential for severe cardiac mechanical depression, it is likely that flunarizine could produce deleterious effects, particularly in patients with compromised cardiac function. The long-term administration of flunarizine should also be avoided, as this drug can produce a high incidence of neuromotor symptoms (224).

3.3.7 Magnesium. It is well established that magnesium is an important regulator of cytosolic calcium and has, in fact, been called "nature's physiologic calcium channel blocker" (225). This inorganic calcium channel antagonist has been reported to correct a variety of rhythm disorders caused by calcium overload (226–229). For example, magnesium terminated digitalis-induced arrhythmias (227), prevented the early afterdepolarizations elicited by cesium chloride (228), and suppressed reperfusion arrhythmias in the isolated rat heart (229). Billman (76) further showed that the intravenous injection of magnesium sulfate prevented the ventricular fibrillation induced by acute myocardial ischemia in conscious dogs with healed myocardial infarctions. In contrast, Euler (230) reported that magnesium infusion failed to prevent either ventricular tachycardia or ventricular fibrillation induced during ischemia or reperfusion in anesthetized dogs.

Similar conflicting results have been reported in humans (231). Oral magnesium supplements have been advocated as a nonpharmacological means of controlling frequent ventricular arrhythmias for a number of years (231, 232) but have yielded only modest results. For example, Zehender et al. (232) found that daily magnesium supplements reduced arrhythmia frequency by only 17% in patients

with frequent ventricular arrhythmias. In contrast, intravenous magnesium significantly reduced the incidence of arrhythmias during acute myocardial infarction (233–235). The most consistent results from magnesium therapy have been reported for the treatment of torsades de pointes. Tzivonis and co-workers (236, 237) reported that intravenous magnesium sulfate completely abolished torsades de pointes within 1–5 min after either the first or second bolus injection of 2 g (8 mM) magnesium sulfate. However, the same regimen failed to prevent recurrent polymorphic ventricular tachycardia in patients with ischemic heart disease (236). Given these often conflicting results, the therapeutic value of magnesium in the management of ventricular arrhythmias has been questioned (231).

3.3.8 Mibefradil. Experimental and clinical evidence described above suggests that calcium antagonists have the potential to protect against malignant arrhythmias induced by myocardial ischemia. Calcium antagonists that exert demonstrable direct cardiac actions in therapeutic concentrations (verapamil, flunarizine) are, in general, more effective antiarrhythmic agents in myocardial ischemia than antagonists that are more selective for vascular smooth muscle (nifedipine). However, the more effective calcium channel antagonists also depress cardiac mechanical function and often produce deleterious effects in patients with compromised cardiac function (198, 210)—the very patients at greatest risk for malignant arrhythmias (238). The challenge, therefore, is to develop calcium antagonists that modulate myocardial calcium, particularly during ischemia, without compromising cardiac function.

Mibefradil (Ro 40-5967), a benzimidazolyl-substituted tetraline derivative, represents a new class of calcium channel antagonists with these clinically important characteristics (239, 240). Mibefradil is relatively selective for the T-type calcium channel but also blocks the L-type channel (241) and the delayed rectifier potassium current (242). Mibefradil has also been reported to inhibit sarcolemmal ATP-potassium channels in adrenal zona fasciulata cells (243) but to activate mitochondrial ATP-sensitive potassium channels, thereby facilitating ischemic pre-conditioning and limiting myocardial infarction size in isolated rat hearts (244, 245). With regards to calcium channels, mibefradil was found to have a 10- to 15-fold greater affinity *in vitro* for the T-type calcium channel than for the L-type calcium channel (246, 247). At therapeutic concentrations *in vivo*, the cardiovascular actions of mibefradil seem to be mediated primarily through actions on the T-type calcium channel ($\sim$80%) with lesser actions on the L-type calcium channel ($\sim$20%). Mibefradil was also found to be highly selective for vascular, as opposed to cardiac, tissue (173, 239, 240, 248). Sarsero et al. (248) investigated the vascular-to-cardiac selectivity of a variety of calcium channels antagonists using human small arteries (aortic vaso vasorum) and right atrial trabeculae. They reported the following vascular-to-cardiac selectivity ratios: mibefradil, 41; felodipine, 12; nifedipine, 7; amlodipine, 5; and verapamil, 0.2. It is therefore not surprising that mibefradil has been reported to increase coronary blood flow and reduce arterial blood pressure with little or no negative inotropic effects (240, 249, 250). Indeed, mibefradil binds at or near the same membrane sites as verapamil but exerts few negative inotropic effects on the isolated heart (240, 251–254). Mibefradil has the same affinity for the [^{3}H]desmethoxyverapamil binding site as verapamil, but it elicits an approximately 10-fold less reduction in myocardial force (239). In addition, mibefradil improves myocardial function during experimental myocardial ischemia, whereas in the same study, verapamil not only failed to protect against ischemia but also produced severe cardiac depression (240). In marked contrast to either diltiazem or verapamil, mibefradil does not seem to alter contractile function in animal models of heart failure or myocardial infarction (78, 240, 250–256). Initial studies indicated that mibefradil was well tolerated by patients (257), reducing arterial pressure without adverse effects on cardiac contractile function, even in patients with heart failure (258). A number of clinical studies demonstrated that this drug was effective in treating patients with hypertension (259–262) and angina pectoris (263–268). Indeed, mibefradil was also found to be particularly effective in suppressing both exercise-induced and "silent" ischemic episodes in patients with ischemic heart disease (267, 268). Given the high bioavailability ($\sim$90%) and long plasma half-life (17–25 h) (173, 240), this drug was a strong candidate for once-a-day dosing. However, mibefradil did not alter cardiovascular mortality in patients with congestive heart failure (MACH I study) (269). In fact, patients co-medicated with antiarrhythmic drugs including amiodarone exhibited a significantly increased rate of death (269). It was also discovered that mibefradil inhibits cytochrome P450, the major enzyme involved in the metabolism of many cardiovascular drugs (270). As a result, a number of adverse interactions (including pronounced bradycardia) between mibefradil and other medications were reported, which prompted the withdrawal of this drug from the market in 1998 (271–273).

These adverse reactions not withstanding, it should be noted that mibefradil has also been reported to have favorable actions on cardiac rhythm. For example, mibefradil reduced regional differences in action potential duration induced by myocardial ischemia (274) and significantly reduced the incidence of ventricular fibrillation brought about by the combination of exercise and acute ischemia (78). In contrast to either verapamil or diltiazem, this protection was afforded without adversely affecting A-V conduction or inotropic state (78). Mibefradil prevented ischemically induced reductions in ventricular fibrillation threshold without depressing the maximal rate of left ventricular pressure development. However, only higher doses of mibefradil, which depressed ventricular contractile function, could prevent arrhythmias induced by reperfusion in the anesthetized pig (189). These authors, therefore, concluded that both the L- and T-type calcium currents contribute to arrhythmia formation during ischemia and reperfusion. Mibefradil was also found to be less effective than verapamil in the prevention of ventricular fibrillation induced by reperfusion in the isolated rat heart (275). In contrast to ventricular arrhythmias, mibefradil was found to be particularly effective in the treatment of atrial fibrillation (276, 277). The T-type calcium channel number does not change as a consequence of chronic atrial fibrillation, whereas the L-type calcium channel density decreases (278), and as such, the T-type calcium current may contribute to maintenance of this arrhythmia. Indeed, mibefradil reduced regional differences in atrial refractory period (i.e., decreased heterogeneity in repolarization), terminated atrial fibrillation, and prevented the ion channel remodeling associated with chronic atrial fibrillation (276, 277). T-type selective calcium channel antagonists may, therefore, represent a novel approach for the management of this clinically important and often intractable atrial arrhythmia.

It is still unclear how mibefradil, in contrast to verapamil or diltiazem, can protect against ischemic arrhythmias with minimal adverse actions on contractile function. It is possible that a unique electropharmacologic property of mibefradil may be responsible for this cardioprotection. At physiologic cell membrane resting potentials, verapamil is 20 times more potent than mibefradil in the inhibition of calcium entry into cardiac myocytes (279). However, in depolarized cells, verapamil and mibefradil are equipotent in the inhibition of calcium entry (279). More recent studies have confirmed these findings (247, 280–283). Bezprozvanny and Tsien (282) demonstrated that mibefradil is a more effective inhibitor of the L-type calcium current at reduced (depolarized) membrane potentials. They concluded that this voltage dependency resulted from a preferential binding to the open and inactivated state of the L-type calcium channel. As noted above, significant numbers of ventricular cells depolarize during myocardial ischemia (59). Therefore, mibefradil may be particularly selective for ischemic tissue—the tissue most vulnerable to calcium overload and arrhythmia formation. As such, mibefradil would exert little negative effect on normal tissue, but during ischemia, would become an effective myocardial calcium channel antagonist. Perhaps because of its membrane potential sensitivity, mibefradil may act as an ischemia-selective calcium channel antagonist. Indeed, mibefradil completely suppresses ventricular tachycardia induced by programmed electrical stimulation during myocardial ischemia but fails to prevent electrically induced arrhythmias under nonischemic conditions (284). In fact, mibefradil, as well as other calcium channel antagonists, exacer-

bated the nonischemic arrhythmias induced by programmed electrical stimulation (284). For example, mibefradil (as well as verapamil and diltiazem) increased both duration and rate of ventricular tachycardia in approximately 50% of the animals tested (284). Finally as noted above, mibefradil inhibits preferentially calcium entry through T-type calcium channels (283, 285). The expression of the T-type calcium channels has been shown to increase as a consequence of a variety of cardiac diseases including myocardial infarction (174–176). Therefore, it is possible that an increased activation of T-type calcium channels after myocardial infarction also may contribute significantly to the induction of ventricular fibrillation, particularly during myocardial ischemia.

3.3.9 Sodium/Calcium Exchanger Antagonists. As previously noted, the electrogenic sodium/calcium exchanger plays a major role in the regulation of intracellular calcium in both normal and diseased hearts (286, 287). Because this protein can move ions (and charge) in either direction, mode-selective antagonists may have important therapeutic applications. For example, during myocardial ischemia, the sodium pump can no longer function properly (because of reductions in ATP levels), and as a result, cellular sodium levels increase. This elevation in intracellular sodium reverses the direction of the sodium/calcium exchanger so that sodium is extruded and calcium is taken up by the myocytes. As we have seen, this increased calcium entry exacerbates calcium overload and may lead to afterdepolarizations that trigger arrhythmias. The up-regulation of this sodium/calcium exchanger that occurs as a consequence of heart failure has been linked to systolic dysfunction and arrhythmias in these patients (288, 289). Therefore, one might speculate that drugs that suppress sodium/calcium exchanger activity may be beneficial, reducing calcium overload and thereby improving the electrical stability of the heart.

Recently, KB-R7943, an amphiphilic molecule that contains a positively charged isothiourea group, has been reported to be a potent and selective sodium/calcium exchanger antagonist. KB-R7943 dose-dependently inhibited the whole cell Na^+/Ca^{2+} exchanger current recorded in rat (290), guinea pig ventricular myocytes (291, 292), smooth muscle cells (290), and Na^+/Ca^{2+} exchanger-transfected fibroblasts (290) in low (IC_{50} = 0.3–2.4 μM) concentrations. Importantly, KB-R7943 exerted a more potent inhibition of the reverse mode than on the forward mode of the Na^+/Ca^{2+} exchanger (290, 292–294). For example, in guinea pig myocytes, lower concentrations of KB-R7943 were required to block the outward Na^+/Ca^{2+} exchanger current (IC_{50} approximately 0.32 μM) than that required for the inward current (IC_{50} = 17 μM). Similar results were obtained in cultured neonatal rat cardiomyocytes (290) and with the canine Na^+/Ca^{2+} exchanger expressed in *Xenopus* oocytes (293).

KB-R7943 has also been reported to be a particularly effective Na^+/Ca^{2+} exchanger antagonist under conditions that would favor myocardial calcium overload. This drug blocked calcium entry into rat ventricular myocytes induced by either removing extracellular sodium or by the cardiac glycoside strophanthidin, thereby reducing diastolic calcium and abolishing spontaneous calcium oscillation (291). Interestingly, this reduction in calcium overload did not alter the positive inotropic response to the glycoside (291). KB-R7943 suppressed ouabain-induced arrhythmias in isolated guinea pig atria and significantly increased the dose of ouabain required to provoke arrhythmias in anesthetized guinea pigs (295). Inhibition of the Na^+/Ca^{2+} exchanger with KB-R7943 reduced calcium overload and cell death associated with hypoxia/reoxygenation of isolated cardiac cells or ischemia/reperfusion in a variety of different preparations including rat cardiomyocytes (290, 296–298), guinea pig papillary muscle (299, 300), and isolated perfused rat heart (296–298, 301) or rabbit hearts (302). KB-R7943 also reduced the infarction size in isolated perfused rabbit hearts (30 min of global ischemia followed by 2 h of reperfusion) (302) or in intact pigs subjected to 48 min of ischemia followed by 10 min of reperfusion compared with placebo-treated animals (a 34% reduction in infarct size) (296). KB-R7943 also protected against the calcium overload induced by lysophatidyl choline in isolated rat

myocytes (303). Furthermore, KB-R7943 prevented the cell-to-cell propagation of calcium overload mediated hypercontraction in 16 of 20 rat myocyte cell pairs (304). KB-R7943 also reduced both the incidence and the duration of arrhythmias provoked by hypoxia/reoxygenation in guinea pig papillary muscle preparations (299). Finally, KB-R7943 (5 mg/kg iv bolus) significantly prolonged atrial effective refractory period in a time-dependent fashion but did not alter either ventricular refractory period or A-V nodal conduction (AH and HV interval) nor did it affect sinus cycle length, QT interval, or mean arterial pressure in anesthetized dogs (305). This dose of KB-R7943 also prevented tachycardia-induced shortening of atrial refractory period in the anesthetized dogs (306). Because both calcium overload and reductions of atrial refractory period have been implicated in the maintenance of atrial fibrillation (276), the authors proposed that KB-R7943 might be potentially useful in the treatment of this atrial arrhythmia.

It should be noted, however, that KB-R7943 could also inhibit a number of other ion channels in cardiac tissue obtained from a variety of species. At concentrations up to 30 μM, KB-R7943 had little effect on the Na^+/H^+ exchanger, sarcolemmal Ca^{2+} ATPase, or Na^+/K^+ ATPase (290). However, at 30 μM but not 10 μM, KB-R7943 could block voltage-dependent Na^+ and Ca^+ channels (290). In guinea pig ventricular myocytes, KB-R7943 inhibited the Na^+ current, Ca^{2+} current, and the inward rectifier K^+ current with an IC_{50} of approximately 14, 8, and 7 μM, respectively (292). KB-R7943 may also inhibit the calcium-activated chloride current (307). In isolated blood perfused canine right atrium and left ventricle, KB-R7943 (0.03–3 μM) elicited a negative inotropic response as well as a negative followed by positive chronotropic effect (308). These responses may have resulted from the nonspecific inhibition of ion channels in addition to actions on the Na^+/Ca^{2+} exchanger current. In contrast, KB-R7943 (5 μM) did not alter rat ventricular myocyte Ca^{2+}, sarcoplasmic reticulum calcium levels, or steady-state twitch amplitude (291). Furthermore, the resting potential was not altered by KB-R7943 treatment but the action potential plateau duration increased, an observation that is consistent with inhibition of K^+ currents (291). KB-R7943 has been reported to inhibit nicotinic and glutamate [N-methyl-D-aspartate (NMDA)] receptor-mediated calcium entry in neural tissue (309, 310). Finally, the reverse mode selectivity of KB-R7943 has recently been questioned. The preferential inhibition of the reverse mode of the Na^+/Ca^{2+} exchanger by KB-R7943 has been shown to disappear under certain conditions (301, 311). For example, KB-R7943 equipotently blocked both direction of the exchanger current under ionic conditions that favored bidirectional conduction (301). However, the selectivity for the outward current was maintained for the canine sodium/calcium exchanger expressed in frog oocytes even during conditions that favored bidirectional transport (293).

In summary, the preliminary studies described above suggest that KB-R7943 may act selectively to block the reverse mode of the sodium calcium exchanger. As such, KB-R7943 could be particularly effective against alterations in cardiac function induced by ischemic calcium overload. Indeed, this drug was shown to reduce arrhythmias, protect against cell death, and reduce the myocardial infarction size induced by ischemia/reperfusion. It should be noted that side effects of this drug caused by nonspecific actions on a number of other ion channels and neurotransmitter receptors could limit its application in the clinic. However, it seems reasonable that selective sodium/calcium exchanger (particularly the reverse mode) antagonists offer an exciting and novel approach for the management of malignant arrhythmias.

3.3.10 Calcium Channel Agonists. As previously noted, elevations in intracellular calcium can uncouple cardiac cells delaying electrical conduction (see above), thereby contributing to the formation of re-entrant circuits. It has recently been proposed that an enhancement of the L-type calcium current could lead to further decreases in the gap junction conduction, increasing the electrical resistance between cells and thereby eliminating re-entrant pathways (312). In other words, calcium channel agonists could convert a unidirectional conduction block into a bidirec-

tional conduction block and thereby remove the substrate for re-entry in much the same manner as sodium channel (class I anitarrhythmic drugs) antagonists. Thus, calcium channel agonists could be antiarrhythmic in certain settings. Cabo et al. (312), in fact, reported that the calcium channel agonist Bay Y 5959 prevented the initiation of electrically induced ventricular tachycardia that resulted from re-entry in nearly one-half of the animals tested (anesthetized dogs with a healing, 4-day-old infarction). The effects of this drug on ischemically induced arrhythmias were not investigated, nor was this drug completely successful even under conditions that favored re-entry. In contrast, the calcium channel agonist Bay K 8644 has been shown to induce afterdepolarizations (313), and ventricular fibrillation during myocardial ischemia in animals previously shown to be resistant to malignant arrhythmias (76). Furthermore, elevations in intracellular calcium would both directly (increased heart rate and inotropic state) and indirectly (increased arterial pressure, afterload) place a greater metabolic demand on the heart, and could thereby exacerbate ischemia in diseased hearts. Although it is possible that calcium agonists may terminate re-entrant arrhythmias in certain settings (electrically induced arrhythmias), this limited protection is achieved at considerable risk. The resulting calcium overload would not only increase the risk for afterdepolarizations and triggered automaticity (see above) but would also place a severe metabolic burden on what may be an already compromised heart. As such, these agents would be clearly counterindicated in the vast majority of patients with existing cardiac disease—the very patients at the greatest risk for lethal ventricular arrhythmias (314). It is, therefore, reasonable to conclude that the risks associated with calcium channel agonists greatly exceed the putative benefits for this therapy. Unless calcium channel agonists can be developed that selectively target regions of the heart where conduction is compromised without affecting normal cardiac or vascular muscle, it is unlikely that these agents will gain widespread clinical acceptance.

3.3.11 ATP-Sensitive Potassium Channel Antagonists. It is now generally accepted that the activation of the ATP-sensitive potassium channel during myocardial ischemia provokes a potassium efflux and reductions in action potential duration that lead to dispersion of repolarization. Because heterogeneity of repolarization plays a crucial role in the induction of ventricular fibrillation, drugs that prevent ATP-sensitive potassium channel activation should be particularly effective in the suppression of malignant arrhythmias induced by ischemia.

The sulfonylurea drug glibenclamide has been shown to prevent hypoxia induced reductions in action potential duration in single cells, isolated hearts, and intact animals (110, 112, 113, 115–120, 122, 315–320). This drug can attenuate the ischemically induced changes in the S-T segment in intact anesthetized or unanesthetized animals (134, 135). Glibenclamide also prevented arrhythmias induced by ischemia in isolated hearts (118, 321–324). Furthermore, both glibenclamide and glimepride reduced blood glucose and decreased the incidence of irreversible ventricular fibrillation induced by reperfusion (after a 6-min coronary occlusion) in anesthetized rats (325). Glibenclamide has also been reported to reduce the incidence of life-threatening arrhythmias induced by coronary artery ligation in the conscious rat (326) and the anesthetized rabbit (327), as well as improve survival in the anesthetized rat during ischemia and reperfusion (328). Billman et al. (146, 329) further demonstrated that glibenclamide prevented ventricular fibrillation induced by the combination of acute ischemia during submaximal exercise in animals previously shown to be susceptible to sudden death. It should be noted, however, that in these animals glibenclamide significantly reduced both the exercise- and reactive hyperemia–induced increases in coronary blood flow as well as depressed ventricular function (large reductions in left ventricular dP/dt maximum) (146, 329). In the isolated working rabbit heart, glibenclamide provoked an immediate decrease in coronary blood flow reducing forward flow to zero (140). Thus, glibenclamide has potent vasoconstrictor effects as a result of

the inhibition of ATP-sensitive potassium channels located on the coronary vascular smooth muscle.

There are a few clinical studies that illustrate the anti-arrhythmic potential of ATP-sensitive potassium channels (330–332, 397). Cacciapuoti et al. (397) found that glibenclamide significantly reduced the frequency and severity of ventricular arrhythmias recorded during transient ischemia in non–insulin-dependent diabetic patients with coronary artery disease. This drug, however, did not affect non-ischemic arrhythmias nor did it change the number or length of the ischemic episodes.

Glibenclamide significantly reduced the incidence of ventricular fibrillation in non–insulin-dependent patients with acute myocardial infarction (332). Glibenclamide also reduced the S-T segment elevation and prevented ventricular fibrillation induced by chest impact (a baseball delivered at 30 mph to the chest of anesthetized swine) (333). The authors concluded that "selective ATP-sensitive potassium channel activation may be a pivotal mechanism in sudden death resulting from low energy chest wall trauma" (i.e., Commotio Cordis) "in young people during sporting activities" (333).

Glibenclamide is not selective for cardiac tissue. As noted above, this drug can profoundly reduce coronary blood flow through actions on vascular smooth muscle and can also promote insulin release, thereby provoking hypoglycemia (146). These non-cardiac actions would limit the anti-arrhythmic potential of glibenclamide in the clinic. Cardioselective compounds should have fewer side effects and would therefore provide a better therapeutic option than the nonselective ATP-sensitive potassium channel antagonist glibenclamide.

Several different ATP-sensitive potassium channel subtypes have been identified. The ATP-sensitive potassium channel consists of a pore-forming subunit coupled to a sulfonylurea receptor (334–336). The functional channel forms as a hetero-octomer composed of a tetramer of the pore and four sulfonyl receptor subunits. At present, two different pore-forming subunits have been identified, both of which produce an inward rectifier potassium current (Kir 6.1 and Kir 6.2) (138, 337). Three different sulfonylurea receptor subtypes have been isolated: SUR1 (on pancreatic islet cells), SUR2A (on cardiac tissue), and SUR2B (on vascular smooth muscle) (335, 336, 338, 339). Thus, six different potassium channel pore and sulfonylurea receptor combinations are possible. Suzuki et al. (137) recently showed that Kir 6.2 and Kir 6.1 were required for cardiac and vascular smooth muscle ATP-sensitive potassium channel activity, respectively. Thus, Kir 6.2/SUR2A most likely forms the cardiac cell membrane ATP-sensitive potassium channel, whereas Kir 6.1/SUR2B is located on vascular smooth muscle. It should therefore be possible to develop compounds that selectively inhibit (or activate) a particular ATP-sensitive potassium channel subtype. A drug that selectively blocks the Kir 6.2/SUR2A subtype should prevent ischemically induced changes in cardiac electrical properties (e.g., reductions in action potential duration) and thereby prevent arrhythmias without the side effects noted for the nonselective ATP-sensitive channel antagonist glibenclamide.

The sulfonylthiourea drug HMR 1883 and its sodium salt HMR 1098 were recently developed to block the cardiac ATP-sensitive potassium channel (340). HMR 1883 inhibited the sarcolemmal cardiac ATP-sensitive potassium channel activated by the channel opener rilmakalin at a much lower concentration (guinea pig myocytes $IC_{50} = 0.6–2.2\ \mu M$) than was required to promote insulin release (9- to 50-fold higher concentration was required to block pancreatic RIN m5F cells) (341). Furthermore, and in contrast to glibenclamide, HMR 1883 did not alter hypoxia-induced increases in coronary blood flow in Langendorff-perfused guinea pig hearts (341). Importantly, HMR 1883 did not alter flavoprotein autoflorescence, an index of mitochondrial redox state (342). In contrast, 5-hydroxydecanoic acid, a selective blocker of mitochondrial channels, completely inhibited the flavoprotein florescence (22). In agreement with these in vitro findings, HMR 1883 did not prevent the cardioprotective effects induced by ischemic preconditioning in either rat (343) or rabbit (344). Accumulating evidence suggests that the activation of mitochondrial ATP-sensitive potassium channels plays a crucial role in the mechanical protection that results from ischemic preconditioning (21, 345). Recently, Liu et al. (337) showed that the mitochondrial ATP-sensitive potassium channel most closely resembles the Kir6.1/SUR1 subtype. Thus, HMR 1883 could inhibit cardiac membrane ATP-sensitive potassium channels with minimal effects on mitochondrial channels.

HMR 1883 also prevented ischemically induced changes in the S-T segment in anesthetized mice (138), anesthetized swine (346), and conscious dogs (135). In the conscious dogs with healed myocardial infarctions, both HMR 1883 and glibenclamide prevented ischemically induced reductions in effective refractory period (146). HMR 1883 significantly reduced monophasic action potential shortening induced by coronary artery occlusion in anesthetized pigs (347). HMR 1883 also reduced cardiac mortality in anesthetized pigs (346, 348) and prevented ventricular fibrillation induced by myocardial ischemia and reperfusion in rats (349). Finally, both glibenclamide and HMR 1883 significantly reduced the incidence of ventricular fibrillation induced by myocardial ischemia in conscious dogs with healed anterior wall myocardial infarctions (146). In contrast to glibenclamide, HMR 1883 did not alter plasma insulin or blood glucose levels in these animals (146). Furthermore, glibenclamide, but not HMR 1883, significantly reduced exercise-induced increase in mean coronary blood flow and provoked large reductions in left ventricular dP/dt maximum (both at rest and during exercise) (146). Thus, HMR 1883 seems to act selectively on the cardiac cell membrane ATP-sensitive potassium channel and thereby prevents ischemically induced arrhythmias with minimal effects on either pancreatic or smooth muscle channels.

In contrast to ATP-sensitive potassium channel antagonists, channel agonists have been reported to induce arrhythmias during ischemia in both isolated hearts and intact animals (322, 350, 351). The channel agonist pinacidil facilitated ventricular fibrillation during myocardial ischemia in isolated rat (322) or rabbit (350) hearts with reduced potassium levels. Di Diego and Antzelevitch (147) found that the activation of the ATP-sensitive potassium channel with pinacidil caused marked inhomogeneities of refractory period, which provoked extrasystoles. These extrasystoles were blocked by glibenclamide in strips of isolated canine myocardium. They concluded that this dispersion of repolarization and refractory period formed a substrate for reentry. Indeed, the ATP-sensitive potassium channel opener cromakalim reduced effective refractory period and increased vulnerability for re-entry (139). Furthermore, this drug was also shown to increase interventricular dispersion of refractory period and induced ventricular fibrillation in 5 of 12 isolated rabbit hearts under normoxic conditions (140). Chi et al. (351) further demonstrated that pinacidil increased the frequency of ventricular fibrillation during myocardial ischemia in unanesthetized dogs. This drug had no effect on arrhythmias induced by electrical stimulation in normally perfused tissue. Cromakalim also failed to reduce arrhythmias induced by ischemia in either isolated rabbit hearts (140) or intact dog (352). It should be noted that the pro-arrhythmic effects of these agents have been reported only at high doses, doses that often produce hypotension and corresponding reflex tachycardia (351). More recently, pinacidil failed to alter the dispersion of refractory period or promote arrhythmia formation in animals with healed myocardial infarctions (353). Clinical trials with ATP-sensitive potassium channel agonists as antihypertensive agents have been reported not to induce arrhythmias (354, 355). However, Goldberg (121) reported that 30% of patients taking pinacidil as an antihypertensive medication showed abnormal ECG (T wave) changes suggestive of alterations of repolarization. It is possible that in the presence of ischemia these cardiac actions of channel agonists become more pronounced, increasing the propensity for life-threatening arrhythmias.

It must be emphasized, however, that under certain conditions, potassium channel activation may also have anti-arrhythmic properties, particularly against arrhythmias resulting from abnormalities in repolarization (356–358). ATP-sensitive potassium channel agonists have been reported to show anti-arrhythmic properties in some *nonischemic* and a few ischemic experimental models (326, 328, 359–363). Arrhythmias that result from early afterdepolarizations (oscillations of membrane potential that occur during phase 3 of the cardiac action potential) are dependent on delayed repolarizations (59) and may be prevented by ATP-sensitive channel activators (359–363). For example, the ATP-sensitive potassium channel agonists pinacidil and nicorandil can abolish early afterdepolarizations induced by cesium chloride (363) or BAY K 8644 (362). These agents have also been reported to suppress delayed afterdepolarizations (oscillations that occur after repolarization of the preceding action potential) produced by cardiac glycosides (361) in isolated tissue. It is also important to note that some ATP-sensitive potassium channel agonists have been reported to promote ischemic preconditioning and thereby reduce the mechanical dysfunction induced by ischemia. Indeed, the ATP-sensitive potassium channel opener diazoxide may act somewhat selectively on mitochondrial ATP-sensitive potassium channels (22, 23).

In summary, activation of cardiac cell membrane ATP-sensitive potassium channels during myocardial ischemia promotes potassium efflux, reduction in action potential duration, and inhomogenieties in repolarization, thereby creating a substrate for re-entry. Drugs that block this channel should be particularly effective anti-arrhythmic agents. Indeed, it is interesting to note that many currently available anti-arrhythmic drugs including verapamil, mibefradil, quinidine, lidocaine, and amiodarone have also been reported to inhibit ATP-sensitive potassium channels at therapeutic concentrations (168, 356, 364, 365). Therefore, the inhibition of the ATP-sensitive potassium channel may be required for anti-arrhythmic actions during myocardial ischemia. HMR 1883 (or its sodium salt HMR 1098) selectively blocks cardiac sarcolemmal ATP-sensitive potassium channels. As such, this drug attenuates ischemically induced changes in cardiac electrical properties, thereby preventing malignant arrhythmias without the untoward effects of other drugs. Because, as noted above, the ATP-sensitive potassium channel only becomes active as ATP levels fall, HMR 1883 has the added advantage that this drug would only exert actions on ischemic tissue with little or no effect noted on normal tissue. Thus, selective antagonists of the cardiac cell surface ATP-sensitive potassium channel may represent the first truly ischemia selective anti-arrhythmic medications and should be free of the pro-arrhythmic effects that have plagued many of the currently available anti-arrhythmic drugs. One would predict that HMR 1883 should reduce cardiac mortality even in high risk patients with advanced ischemic heart disease.

3.3.12 β-adrenergic Receptor Antagonists.

As noted above, myocardial ischemia and infarction can provoke pronounced changes in the autonomic regulation of the heart that, in turn, have been implicated in the formation of ventricular fibrillation. As a rule, cardiac electrical stability is reduced by activation of the sympathetic nervous system, whereas increased parasympathetic activity may protect against the development of malignant arrhythmias (see Ref. 54). Both clinical (4) and experimental (39) studies have shown that acute myocardial ischemia elicits increases in sympathetic tone coupled with reductions in parasympathetic activity. Furthermore, the individuals with the greatest changes in autonomic balance also exhibited the greatest propensity for sudden cardiac death (39). Myocardial infarction has also been reported to elicit similar changes in the autonomic control of the heart (366–368). In fact, the relative mortality has been calculated to be more than fivefold higher in patients with increased sympathetic/reduced parasympathetic tone (measured by R-R interval variability), compared with those patients with more normal autonomic balance (309, 366). Therefore, it is not surprising that interventions that reduce cardiac effects of sympathetic activation, particularly β-adrenergic receptor antagonists, have been the subject of considerable investigation.

Since the early 1970s, at least 25 randomized trials (enrolling over 23,000 patients) of β-adrenergic blockade on the secondary prevention of cardiac events have been completed (see Refs. 200, 369, 370). The pooled data reveal that this treatment resulted in a 23% reduction in mortality with a 32% reduction in sudden deaths. Recent studies show that despite the obvious benefits of long-term β-adrenergic therapy, a large number of post-in-

farction patients are not placed on these drugs (371–373). For example, Viskin et al. (373) found that only 58% of infarct survivors with *no* contraindication to β-adrenergic blockers actually received these drugs at the time of hospital discharge. More importantly, only 11% of the patients received doses clinically proven to reduce cardiac mortality; the majority of patients received substantially lower doses! It has been conservatively estimated that if cardiologists would adhere to the guidelines for the use of β-adrenergic blockers as suggested by the American College of Cardiology, nearly 1900 deaths would be averted annually nationwide (373).

In contrast to the long-term benefits of β-adrenergic receptor antagonists, the influence of these agents during acute phase myocardial infarction is less definitive. There have been at least 32 trials involving approximately 29,000 patients in which β-adrenergic receptor blockers have been initiated within the first few hours of either confirmed or suspected myocardial infarction (200, 369, 370). The pooled data from these studies suggest that early treatment with β-adrenergic receptor blockers results in a 20–30% reduction of infarction size, and a 15% decrease in cardiac mortality, probably as the result of a corresponding reduction in the incidence of ventricular fibrillation (369). However, a careful analysis reveals that results, particularly with regards to ventricular fibrillation, vary depending on the agent used. β-Adrenergic receptor antagonists with intrinsic sympathomimetic activity failed to alter the incidence of ventricular fibrillation (200, 374, 375). The majority of the studies using β-adrenergic receptor antagonists failed to report significant reductions in the incidence of ventricular fibrillation during acute myocardial infarction (376–380). For example, metoprolol did not reduce the number of patients that experienced ventricular fibrillation (377, 378). Although atenolol reduced overall mortality 15%, this drug (380) did not affect the number of patients that died as the result of malignant arrhythmias. In contrast, propranolol almost completely eliminated death due to ventricular fibrillation during acute myocardial infarction (381, 382). These data suggest a better anti-arrhythmic protection can be achieved with complete (i.e., β_1- and β_2-) rather than selective (i.e., β_1-) β-adrenergic receptor blockade. These data further indicate that the activation of myocardial β-adrenergic receptors may play a particularly important role in the induction of malignant arrhythmias and acute myocardial infarction.

There is now a growing body of evidence demonstrating that in addition to β_1-adrenergic receptors, ventricular myocytes also contain functional β_2-adrenergic receptors (48). Furthermore, these receptors may become particularly important in the regulation of contractile function in certain disease states (28, 48, 49). For example, as a consequence of chronic heart failure, there is a substantial loss of β_1-adrenergic receptors with little or no loss in β_2-adrenergic receptors (51). The failing heart therefore may become more dependent upon the activation of β_2-adrenergic receptors for inotropic support during sympathetic stimulation. Indeed, recent studies demonstrated that the β_2-adrenergic agonist zinterol elicited significantly larger calcium transients (increases in cytosolic Ca^{2+} in response to electrical field stimulation) in ventricular myocytes prepared from failing hearts as compared to normal tissue (49). As noted above, cytosolic calcium plays a critical role in the induction of ventricular fibrillation. β_2-Adrenergic receptor–mediated changes in Ca^{2+} influx could lead to the formation of arrhythmias and sudden death in myocardial infarction patients. It is interesting to note that heart failure patients and patients with poor left ventricular contractile function are at a substantially greater risk of sudden death compared with patients with a more normal pump function (314). One might speculate that changes in cytosolic Ca^{2+} mediated by a greater activation of myocardial β_2-adrenergic receptors may provoke lethal cardiac arrhythmias in these patients. Recently, the β_2-adrenergic receptor antagonist ICI 118,552 was shown to prevent canine ventricular fibrillation induced by myocardial ischemia during the last minute of submaximal exercise (50). Furthermore, the β_2-adrenergic receptor agonist zinterol elicited much larger Ca^{2+} transients in ventricular myocytes prepared from the hearts of animals susceptible to ventricular fibrillation compared with cells from

hearts of animals resistant to these malignant arrhythmias (50). Thus, activation of β_2-adrenergic receptors during myocardial ischemia could culminate in ventricular fibrillation. In this regard, it is interesting to note that there are isolated clinical reports in which β_2-adrenergic agonists have precipitated sudden death as a consequence of cardiac actions in asthmatic patients (370, 383).

In summary, the data described above clearly demonstrate that β-adrenergic receptor antagonists reduce cardiac mortality in post-infarction patients. This reduction in mortality largely results from a reduction in the incidence of arrhythmogenic deaths. Furthermore, the data indicate that nonselective β-adrenergic receptor antagonists offer a more complete protection than the so-called cardioselective β- (β_1-) adrenergic receptor antagonists. It would seem that the activation of ventricular β_2-adrenergic receptors could provoke life-threatening arrhythmias, particularly during myocardial ischemia or in patients with compromised cardiac function.

3.4 Prevention of Remodeling

3.4.1 Ace Inhibitors. Large multicenter trials have shown conclusively that angiotensin-converting enzyme (ACE) inhibitors significantly improve post-infarction survival (13, 148, 158, 383–386). Unlike the case with the calcium channel antagonists, this effect seems to be generic and unrelated to the use of any particular drug. It is currently recommended that ACE inhibitor therapy be initiated 1-2 days post-infarct (13, 148, 158, 383–386). ACE inhibitors are especially important in patients with a reduced ejection fraction (a measure of the efficiency of cardiac pumping) who are most prone to develop symptomatic congestive heart failure after the infarct has healed (13, 148, 158, 383–386). In patients with more normal cardiac function, it is currently believed that ACE inhibitors can be discontinued after 6 months, at which time injury-related remodeling of the heart should have ceased (383). However, it has been argued that the anti-thrombotic effects of ACE inhibitors may contribute significantly to their salutary effects on post-infarction mortality, in which case more prolonged treatment may prove beneficial (383). Studies where ACE inhibitors were given very early in the treatment of acute myocardial infarction (before stabilization) failed to show any additional benefit, and there may have been a slight increase in mortality most likely as a result of hypotension in some patients (149). There is debate as to whether early ACE inhibitor therapy might reduce infarct expansion and reduce long-term mortality if restricted to patients without borderline low blood pressure (149).

The mechanisms responsible for the protective effects of ACE inhibitors in the post-infarct myocardium are not fully understood. ACE, as the name implies, converts inactive angiotensin I to active angiotensin II; it also inactivates bradykinin. Thus, ACE inhibitors prevent the synthesis of angiotensin II and prevent the degradation of bradykinin. Decreases in angiotensin II decrease aldosterone production and reduce blood pressure. This is clearly beneficial in the setting of hypertension, and unloading of the heart may account for some of the effects of ACE inhibition post-infarction. There are also direct cardiac effects of the ACE inhibitors that cannot be replicated with equivalent reductions in blood pressure using other means (148). Angiotensin II binds to specific receptors on cardiac myocytes and fibroblasts (387, 388) and induces myocyte hypertrophy (heart muscle cell growth), fibroblast hyperplasia (387), and remodeling of the extracellular connective tissue matrix (148). Myocyte hypertrophy, an increase in the size of individual heart muscle cells, can be an appropriate response when the myocardium must replace the muscle mass lost to an infarct in order to perform its normal contractile work. But for reasons that are not yet adequately understood, hypertrophy can lead to eventual heart failure. Extracellular remodeling realigns individual myocytes, converting a normal heart into a more spherical, dilated organ (eccentric hypertrophy) (148). This process is poorly understood, but eccentric hypertrophy is a very poor prognostic sign. By blocking myocyte hypertrophy and extracellular remodeling, ACE inhibitors seem to be beneficial for the treatment of post-infarct patients, especially those with relatively large infarcts who are at the greatest

risk for developing congestive heart failure (148, 158, 383, 385, 389, 390).

3.4.2 Glucose/Insulin/Potassium. Because ischemic myocardium is profoundly deenergized, there has long been the view that metabolic support should help to preserve viable tissue. In experimental animal models of ischemia and reperfusion (391) the infusion of clinically relevant concentrations of glucose plus insulin plus potassium led to greater recovery of mechanical function and reduced cells necrosis as monitored by enzyme release (391). A pilot trial of glucose/insulin/potassium demonstrated reduced mortality in patients with acute myocardial infarctions undergoing reperfusion therapy (392) and Apstein and Taegtmeyer have written an editorial urging a large prospective clinical trial of this therapy (393). However, the costs of such a trials would be prohibitive, especially because thrombolytic therapy in GUSTO V reduced 30-day mortality to less than 6% (160). Nevertheless, salvage of viable myocardium may have longer-term effects to prevent cardiac dilatation and failure. In this context, it is noteworthy that insulin administered at reperfusion inhibited apoptosis in cultured heart cells (394). Insulin activates the serine-threonine kinase Akt, and Akt activation has been shown experimentally to prevent ischemic (395) and reperfusion-induced (396) cardiomyocyte injury.

4 SUMMARY AND CONCLUSIONS

Great advances have been made in the treatment of acute myocardial infarction. Defibrillation, β-adrenergic antagonists, and thrombolysis have significantly reduced in-hospital mortality. Long-term treatment with nonspecific β-blockers, ACE inhibitors, and aspirin benefit large numbers of post-infarct patients by preventing arrhythmias, ventricular remodeling, and reocclusion of the infarct-related vessel. Nevertheless, sudden cardiac death continues to be a leading cause of mortality. The possibility that β_2-adrenergic receptors are involved in ventricular fibrillation suggests that β_1-selective antagonists may not be the drugs of choice for patients at risk for lethal arrhythmias. Additional work is needed to develop appropriate therapies for the prevention of sudden cardiac death in patients with a healed myocardial infarction.

Myocardial infarction continues to be a leading cause of congestive heart failure. Therapies aimed at reducing infarct size (rapid thrombolysis, GP IIb/IIIa receptor antagonists, β-blockers) and ventricular remodeling (ACE inhibitors, glucose/insulin/potassium) can retard or prevent the late-phase deterioration of pump function. Reocclusion of the infarct-related vessel remains a problem, however. Improved strategies for preventing thrombosis and reducing or eliminating atherosclerotic plaques would greatly benefit the post-infarct patient. Such strategies could also prevent the initial acute myocardial infarction in at risk patients with coronary artery disease. Finally, therapeutics designed to prevent coronary artery disease would dramatically reduce the incidence of acute myocardial infarction, sudden cardiac death, and congestive heart failure.

REFERENCES

1. E. M. Antman in D. Julian and E. Braunwald, Eds., *Management of Acute Myocardial Infarction*, W. B. Saunders Co., London, 1994, pp. 29–70.

2. G. V. Martin and J. W. Kennedy in D. Julian and E. Braunwald, Eds., *Management of Acute Myocardial Infarction*, W. B. Saunders Co., London, 1994, pp. 71–105.

3. R. B. Jennings and K. A. Reimer, *Annu. Rev. Med.*, **42**, 225–246 (1991).

4. A. Schomig, *Circulation*, **82**, II13–II22 (1990).

5. M. Karmazyn and M. P. Moffat, *Cardiovasc. Res.*, **27**, 915–924 (1993).

6. N. Khandoudi, J. Ho, and M. Karmazyn, *Circ. Res.*, **75**, 369–378 (1994).

7. F. Di Lisa, R. De Tullio, F. Salamino, R. Barbato, E. Melloni, N. Siliprandi, S. Schiaffino, and S. Pontremoli, *Biochem. J.*, **308**, 57–61 (1995).

8. D. E. Atsma, E. M. L. Bastiaanse, A. Jerzewski, L. J. M. Van der Valk, and A. Van der Laarse, *Circ. Res.*, **76**, 1071–1078 (1995).

9. K. Suzuki, H. Sorimachi, T. Yoshizawa, K. Kinbara, and S. Ishiura, *Biol. Chem. Hoppe Seyler*, **376**, 523–529 (1995).

10. S. L. Hazen, D. A. Ford, and R. W. Gross, *J. Biol. Chem.*, **266**, 5629–5633 (1991).

11. S. C. Armstrong and C. E. Ganote, *Am. J. Pathol.*, **138**, 545–555 (1991).

12. K. Y. Hostetler and E. J. Jellison, *Mol. Cell. Biochem.*, **88**, 77–82 (1989).

13. P. A. Poole-Wilson, *Am. J. Cardiol.*, **75**, 4E–9E (1995).

14. L. Hofstra, I. H. Liem, E. A. Dumont, H. H. Boersma, W. L. van Heerde, P. A. Doevendans, E. De Muinck, H. J. Wellens, G. J. Kemerink, C. P. Reutelingsperger, and G. A. Heidendal, *Lancet*, **356**, 209–212 (2000).

15. H. Yaoita, K. Ogawa, K. Maehara, and Y. Maruyama, *Cardiovasc. Res.*, **45**, 630–641 (2000).

16. Z. Q. Zhao, M. Nakamura, N. P. Wang, J. N. Wilcox, S. Shearer, R. S. Ronson, R. A. Guyton, and J. Vinten-Johansen, *Cardiovasc. Res.*, **45**, 651–660 (2000).

17. K. A. Reimer and R. B. Jennings, *Basic Res. Cardiol.*, **91**, 1–4 (1996).

18. P. P. Ping, J. Zhang, Y. M. Qiu, X. L. Tang, S. Manchikalapudi, X. N. Cao, and R. Bolli, *Circ. Res.*, **81**, 404–414 (1997).

19. P. P. Ping, H. Takano, J. Zhang, X. L. Tang, Y. M. Qiu, R. X. Li, S. Banerjee, B. Dawn, Z. Balafonova, and R. Bolli, *Circ. Res.*, **84**, 587–604 (1999).

20. S. C. Armstrong, D. B. Hoover, M. H. Delacey, and C. E. Ganote, *J. Mol. Cell Cardiol.*, **28**, 1479–1492 (1996).

21. G. J. Gross and R. M. Fryer, *Circ. Res.*, **84**, 973–979 (1999).

22. Y. Liu, T. Sato, B. O'Rourke, and E. Marban, *Circulation*, **97**, 2463–2469 (1998).

23. K. D. Garlid, P. Paucek, V. Yarov-Yarovoy, H. N. Murray, R. B. Darbenzio, A. J. D'Alonzo, N. J. Lodge, M. A. Smith, and G. J. Grover, *Circ. Res.*, **81**, 1072–1082 (1997).

24. M. Nakamura, N. P. Wang, Z. Q. Zhao, J. N. Wilcox, V. Thourani, R. A. Guyton, and J. Vinten-Johansen, *Cardiovasc. Res.*, **45**, 661–670 (2000).

25. L. H. Opie, D. Nathan, and W. F. Lubbe, *Am. J. Cardiol.*, **43**, 131–148 (1979).

26. M. J. Curtis, M. K. Pugsley, and M. J. Walker, *Cardiovasc. Res.*, **27**, 703–719 (1993).

27. G. E. Billman, *J. Cardiovasc. Pharmacol.*, **18**, S107–S117 (1991).

28. L. D. Horwitz, K. M. VanBenthuysen, F. M. Sheridan, E. J. Lesnefsky, I. M. Dauber, and I. F. McMurtry, *Free Rad. Biol. Med.*, **8**, 381–386 (1990).

29. E. Carmeliet and K. Mubagwa, *Prog. Biophys. Mol. Biol.*, **70**, 1–72 (1998).

30. A. A. Grace and A. J. Camm, *Cardiovasc. Res.*, **45**, 43–51 (2000).

31. W. H. Barry, *Trends Cardiovasc. Med.*, **1**, 162–166 (1991).

32. H. Reuter, *Nature*, **301**, 569–574 (1983).

33. N. Sperelakis, G. M. Wahler, and G. Bkaily, *Curr. Topics Membranes Transport*, **25**, 43–76 (1985).

34. L. A. Ransnas and P. A. Insel, *Anal. Biochem.*, **176**, 185–190 (1989).

35. Y. Koretsune and E. Marban, *Circulation*, **80**, 369–379 (1989).

36. H. C. Lee, R. Mohabir, N. Smith, M. R. Franz, and W. T. Clusin, *Circulation.*, **78**, 1047–1059 (1988).

37. C. Steenbergen, E. Murphy, L. Levy, and R. E. London, *Circ. Res.*, **60**, 700–707 (1987).

38. P. B. Corr, K. Yamada, and F. X. Witkowski in H. A. Fozzard, E. Haber, R. B. Jennings, A. M. Katz, and H. E. Morgan, Eds., *The Heart and Cardiovascular System: Scientific Foundations*, Raven Press, New York, 1986, pp. 1343–1404.

39. M. N. Collins and G. E. Billman, *Am. J. Physiol.*, **257**, H1886–H1894 (1989).

40. M. S. Houle and G. E. Billman, *Am. J. Physiol. Heart Circ. Physiol.*, **276**, H215–H223 (1999).

41. D. B. Evans, *J. Cardiovasc. Pharmacol.*, **8**, S22–S29 (1986).

42. A. Yoshida, M. Takahashi, T. Imagawa, M. Shigekawa, H. Takisawa, and T. Nakamura, *J. Biochem. (Tokyo)*, **111**, 186–190 (1992).

43. M. A. Strand, C. F. Louis, and J. R. Mickelson, *Biochim. Biophys. Acta Mol. Cell Res.*, **1175**, 319–326 (1993).

44. C. Hawkins, A. Xu, and N. Narayanan, *J. Biol. Chem.*, **269**, 31198–31206 (1994).

45. J. Hain, H. Onoue, M. Mayrleitner, S. Fleischer, and H. Schindler, *J. Biol. Chem.*, **270**, 2074–2081 (1995).

46. E. G. Kranias, R. C. Gupta, G. Jakab, H. W. Kim, N. A. Steenaart, and S. T. Rapundalo, *Mol. Cell Biochem.*, **82**, 37–44 (1988).

47. R. A. Altschuld and G. E. Billman, *Pharmacol. Ther.*, **88**, 1–14 (2000).

48. O. E. Brodde, *Pharmacol. Rev.*, **43**, 203–242 (1991).

49. R. A. Altschuld, R. C. Starling, R. L. Hamlin, G. E. Billman, J. Hensley, L. Castillo, R. H. Fertel, C. M. Hohl, P. M. L. Robitaille, L. R. Jones, R. P. Xiao, and E. G. Lakatta, *Circulation*, **92**, 1612–1618 (1995).

50. G. E. Billman, L. C. Castillo, J. Hensley, C. M. Hohl, and R. A. Altschuld, *Circulation*, **96**, 1914–1922 (1997).

51. M. R. Bristow, R. Ginsburg, V. Umans, M. Fowler, W. Minobe, R. Rasmussen, P. Zera, R. Menlove, P. Shah, S. Jamieson, and E. B. Stinson, *Circ. Res.*, **59**, 297–309 (1986).

52. A. M. Brown, A. Yatani, Y. Imoto, J. Codina, R. Mattera, and L. Birnbaumer, *Ann. N.Y. Acad. Sci.*, **560**, 373–386 (1989).

53. M. J. Berridge, *Annu. Rev. Biochem.*, **56**, 159–193 (1987).

54. A. M. Watanabe in W. C. Randall, Ed., *The Nervous Control of Cardiovascular Function*, Oxford University Press, New York, 1984, pp. 130–164.

55. G. M. Wahler, N. J. Rusch, and N. Sperelakis, *Can. J. Physiol. Pharmacol.*, **68**, 531–534 (1990).

56. S. S. Rashatwar, T. L. Cornwell, and T. M. Lincoln, *Proc. Natl. Acad. Sci. U.S.A.*, **84**, 5685–5689 (1987).

57. Y. Kihara and J. P. Morgan, *Circ. Res.*, **68**, 1378–1389 (1991).

58. G. E. Billman, B. McIlroy, and J. D. Johnson, *FASEB J.*, **5**, 2586–2592 (1991).

59. A. L. Wit and M. J. Janse, *The Ventricular Arrhythmias of Ischemia and Infarction*, Futura, Mount Kisco, NY, 1993.

60. W. T. Clusin, M. Buchbinder, A. K. Ellis, R. S. Kernoff, J. C. Giacomini, and D. C. Harrison, *Circ. Res.*, **54**, 10–20 (1984).

61. B. G. Katzung, *Circ. Res.*, **37**, 118–127 (1975).

62. R. S. Kass and R. W. Tsien, *Biophys. J.*, **38**, 259–269 (1982).

63. G. R. Ferrier, *Prog. Cardiovasc. Dis.*, **19**, 459–474 (1977).

64. P. G. Volders, M. A. Vos, B. Szabo, K. R. Sipido, S. H. de Groot, A. P. Gorgels, H. J. Wellens, and R. Lazzara, *Cardiovasc. Res.*, **46**, 376–392 (2000).

65. R. S. Kass, W. J. Lederer, R. W. Tsien, and R. Weingart, *J. Physiol.*, **281**, 187–208 (1978).

66. K. Schlotthauer and D. M. Bers, *Circ. Res.*, **87**, 774–780 (2000).

67. R. M. Egdell and K. T. MacLeod, *J. Mol. Cell Cardiol.*, **32**, 85–93 (2000).

68. A. C. Zygmunt, R. J. Goodrow, and C. M. Weigel, *Am. J. Physiol. Heart Circ. Physiol.*, **275**, H1979–H1992 (1998).

69. G. Szigeti, Z. Rusznak, L. Kovacs, and Z. Papp, *Exp. Physiol.*, **83**, 137–153 (1998).

70. X. Han and G. R. Ferrier, *J. Mol. Cell Cardiol.*, **22**, 871–882 (1990).

71. M. M. Adamantidis, J. F. Caron, and B. A. Dupuis, *J. Mol. Cell Cardiol.*, **18**, 1287–1299 (1986).

72. W. A. Coetzee, S. C. Dennis, L. H. Opie, and C. A. Muller, *J. Mol. Cell Cardiol.*, **19**, 77–97 (1987).

73. J. C. Merillat, E. G. Lakatta, O. Hano, and T. Guarnieri, *Circ. Res.*, **67**, 1115–1123 (1990).

74. R. G. Hansford and E. G. Lakatta, *J. Physiol. (Lond.)*, **390**, 453–467 (1987).

75. B. Lewartowski, R. G. Hansford, G. A. Langer, and E. G. Lakatta, *Am. J. Physiol. Heart Circ. Physiol.*, **259**, H1222–H1229 (1990).

76. G. E. Billman, *J. Pharmacol. Exp. Ther.*, **248**, 1334–1342 (1989).

77. G. E. Billman, *Eur. J. Pharmacol.*, **212**, 231–235 (1992).

78. G. E. Billman, *Eur. J. Pharmacol.*, **229**, 179–187 (1992).

79. M. D. Lappi and G. E. Billman, *Cardiovasc. Res.*, **27**, 2152–2159 (1993).

80. M. Akhtar, P. Tchou, and M. Jazayeri, *Circulation*, **80**, IV31–IV39 (1989).

81. B. N. Singh and K. Nademanee, *Am. J. Cardiol.*, **59**, 153B–162B (1987).

82. V. Elharrar and D. P. Zipes in D. P. Zipes, J. C. Bailey, and V. Elharrar, Eds., *The Slow Inward Current and Cardiac Arrhythmias*, Martinus Nijhoff, The Hague, 1980, pp. 357–373.

83. A. G. Kleber, *Experientia*, **43**, 1056–1061 (1987).

84. R. J. Sung, W. A. Shapiro, E. N. Shen, F. Morady, and J. Davis, *J. Clin. Invest.*, **72**, 350–360 (1983).

85. L. S. Gettes, in H. A. Fozzard, E. Haber, R. B. Jennings, A. M. Katz, and H. E. Morgan, Eds., *The Heart and Cardiovascular System: Scientific Foundations*, Raven Press, New York, 1986, pp. 1239–1257.

86. D. C. Spray and M. V. Bennett, *Annu. Rev. Physiol.*, **47**, 281–303 (1985).

87. G. Kabell, *Circulation*, **77**, 1385–1394 (1988).

88. S. G. Dilly and M. J. Lab, *J. Physiol. (Lond.)*, **402**, 315–333 (1988).

89. H. K. Hellerstern and I. M. Liebow, *J. Physiol. (Lond.)*, **160**, 366–374 (1950).

90. D. C. Russell, J. H. Smith, and M. F. Oliver, *Br. Heart J.*, **42**, 88–96 (1979).

91. A. G. Kleber, M. J. Janse, F. J. van Capelle, and D. Durrer, *Circ. Res.*, **42**, 603–613 (1978).

92. H. Hashimoto, K. Suzuki, S. Miyake, and M. Nakashima, *Circulation*, **68**, 667–672 (1983).

93. S. Kimura, A. L. Bassett, T. Kohya, P. L. Kozlovskis, and R. J. Myerburg, *Circulation*, **76**, 1146–1154 (1987).

94. A. S. Harris, *Am. Heart J.*, **71**, 797–802 (1966).

95. A. S. Harris, A. Bisteni, and R. A. Russell, *Science*, **119**, 200–203 (1954).

96. J. L. Hill and L. S. Gettes, *Circulation*, **61**, 768–778 (1980).

97. A. G. Kleber, *J. Mol. Cell Cardiol.*, **16**, 389–394 (1984).

98. R. Coronel, J. W. Fiolet, F. J. Wilms-Schopman, A. F. Schaapherder, T. A. Johnson, L. S. Gettes, and M. J. Janse, *Circulation*, **77**, 1125–1138 (1988).

99. R. Coronel, F. J. Wilms-Schopman, T. Opthof, F. J. van Capelle, and M. J. Janse, *Circ. Res.*, **68**, 1241–1249 (1991).

100. T. A. Johnson, C. L. Engle, L. M. Boyd, G. G. Koch, M. Gwinn, and L. S. Gettes, *Circulation*, **83**, 622–634 (1991).

101. A. G. Kleber, *Circ. Res.*, **52**, 442–450 (1983).

102. E. E. Rau, K. I. Shine, and G. A. Langer, *Am. J. Physiol.*, **232**, H85–H94 (1977).

103. K. I. Shine, A. M. Douglas, and N. Ricchiuti, *Am. J. Physiol.*, **232**, H564–H570 (1977).

104. A. Noma, *Nature*, **305**, 147–148 (1983).

105. M. Kameyama, M. Kakei, R. Sato, T. Shibasaki, H. Matsuda, and H. Irisawa, *Nature*, **309**, 354–356 (1984).

106. D. Kim and R. A. Duff, *Circ. Res.*, **67**, 1040–1046 (1990).

107. J. N. Weiss, S. T. Lamp, and K. I. Shine, *Am. J. Physiol.*, **256**, H1165–H1175 (1989).

108. G. Trube and J. Hescheler, *Naunyn-Schmiedberg's Arch. Pharmacol.*, **322**, R64 (1983).

109. J. Hescheler, D. Pelzer, G. Trube, and W. Trautwein, *Pflugers Arch.*, **393**, 287–291 (1982).

110. B. Vanheel and A. De Hemptinne, *Cardiovasc. Res.*, **26**, 1030–1039 (1992).

111. M. N. Hicks and S. M. Cobbe, *Cardiovasc. Res.*, **25**, 407–413 (1991).

112. J. N. Weiss, N. Venkatesh, and S. T. Lamp, *J. Physiol. (Lond.)*, **447**, 649–673 (1992).

113. A. A. Wilde, D. Escande, C. A. Schumacher, D. Thuringer, M. Mestre, J. W. Fiolet, and M. J. Janse, *Circ. Res.*, **67**, 835–843 (1990).

114. N. Venkatesh, S. T. Lamp, and J. N. Weiss, *Circ. Res.*, **69**, 623–637 (1991).

115. M. Fosset, J. R. De Weille, R. D. Green, H. Schmid-Antomarchi, and M. Lazdunski, *J. Biol. Chem.*, **263**, 7933–7936 (1988).

116. E. Rui Zpetrich, N. Leblanc, F. deLorenzi, Y. Allard, and O. F. Schanne, *Br. J. Pharmacol.*, **106**, 924–930 (1992).

117. K. Benndorf, M. Friedrich, and H. Hirche, *Pflugers Arch.*, **419**, 108–110 (1991).

118. P. F. Kantor, W. A. Coetzee, E. E. Carmeliet, S. C. Dennis, and L. H. Opie, *Circ. Res.*, **66**, 478–485 (1990).

119. S. S. Bekheit, M. Restivo, M. Boutjdir, R. Henkin, K. Gooyandeh, M. Assadi, S. Khatib, W. B. Gough, and N. El-Sherif, *Am. Heart J.*, **119**, 1025–1033 (1990).

120. H. Nakaya, Y. Takeda, N. Tohse, and M. Kanno, *Br. J. Pharmacol.*, **103**, 1019–1026 (1991).

121. M. R. Goldberg, *J. Cardiovasc. Pharmacol.*, **12** (Suppl 2), S41–S47 (1988).

122. J. K. Smallwood, P. J. Ertel, and M. I. Steinberg, *Naunyn Schmiedebergs Arch. Pharmacol.*, **342**, 214–220 (1990).

123. N. Venkatesh, J. S. Stuart, S. T. Lamp, L. D. Alexander, and J. N. Weiss, *Circ. Res.*, **71**, 1324–1333 (1992).

124. G. Edwards and A. H. Weston, *Annu. Rev. Pharmacol. Toxicol.*, **33**, 597–637 (1993).

125. A. Mitani, K. Kinoshita, K. Fukamachi, M. Sakamoto, K. Kurisu, Y. Tsuruhara, F. Fukumura, A. Nakashima, and K. Tokunaga, *Am. J. Physiol. Heart Circ. Physiol.*, **261**, H1864–H1871 (1991).

126. I. Findlay, *Pflugers Arch.*, **412**, 37–41 (1988).

127. C. G. Nichols and W. J. Lederer, *J. Physiol. (Lond.)*, **423**, 91–110 (1990).

128. M. J. Rovetto, J. T. Whitmer, and J. R. Neely, *Circ. Res.*, **32**, 699–711 (1973).

129. C. G. Nichols, C. Ripoll, and W. J. Lederer, *Circ. Res.*, **68**, 280–287 (1991).

130. J. F. Faivre and I. Findlay, *Biochim. Biophys. Acta.*, **1029**, 167–172 (1990).

131. C. G. Nichols and W. J. Lederer, *Am. J. Physiol.*, **261**, H1675–H1686 (1991).

132. N. Deutsch, T. S. Klitzner, S. T. Lamp, and J. N. Weiss, *Am. J. Physiol. Heart Circ. Physiol.*, **261**, H671–H676 (1991).

133. I. Kubota, M. Yamaki, T. Shibata, E. Ikeno, Y. Hosoya, and H. Tomoike, *Circulation.*, **88**, 1845–1851 (1993).

134. T. Kondo, I. Kubota, H. Tachibana, M. Yamaki, and H. Tomoike, *Cardiovasc. Res.*, **31**, 683–687 (1996).

135. G. E. Billman, M. S. Houle, H. C. Englert, and H. Goegelein. *Circulation*, **100** (Suppl I), 1–52 (1999).

136. G. X. Yan and C. Antzelevitch, *Circulation*, **98**, 1928–1936 (1998).

137. M. Suzuki, R. A. Li, T. Miki, H. Uemura, N. Sakamoto, Y. Ohmoto-Sekine, M. Tamagawa, T. Ogura, S. Seino, E. Marban, and H. Nakaya, *Circ Res.*, **88**, 570–577 (2001).

138. R. A. Li, M. Leppo, T. Miki, S. Seino, and E. Marban, *Circ. Res.*, **87**, 837–839 (2000).

139. T. Uchida, M. Yashima, M. Gotoh, Z. Qu, A. Garfinkel, J. N. Weiss, M. C. Fishbein, W. J. Mandel, P. S. Chen, and H. S. Karagueuzian, *Circulation*, **99**, 704–712 (1999).

140. R. Wolk, S. M. Cobbe, K. A. Kane, and M. N. Hicks, *J. Cardiovasc. Pharmacol.*, **33**, 323–334 (1999).

141. E. Krause, H. Englert, and H. Gogelein, *Pflugers Arch.*, **429**, 625–635 (1995).

142. T. Furukawa, S. Kimura, N. Furukawa, A. L. Bassett, and R. J. Myerburg, *Circ. Res.*, **70**, 91–103 (1992).

143. R. F. Gilmour Jr. and D. P. Zipes, *Circ. Res.*, **46**, 814–825 (1980).

144. S. Kimura, A. L. Bassett, T. Kohya, P. L. Kozlovskis, and R. J. Myerburg, *Circulation*, **74**, 401–409 (1986).

145. D. Tweedie, C. Henderson, and K. Kane, *Eur. J. Pharmacol.*, **240**, 251–257 (1993).

146. G. E. Billman, H. C. Englert, and B. A. Scholkens, *J. Pharmacol. Exp. Ther.*, **286**, 1465–1473 (1998).

147. J. M. Di Diego and C. Antzelevitch, *Circulation*, **88**, 1177–1189 (1993).

148. J. M. Pfeffer, T. A. Fischer, and M. A. Pfeffer, *Annu. Rev. Physiol.*, **57**, 805–826 (1995).

149. B. Pitt, in D. Julian and E. Braunwald, Eds., *Management of Acute Myocardial Infarction*, W. B. Saunders Co, London, 1994, pp. 253–266.

150. J. Abrams, *Arch. Intern. Med.*, **155**, 357–364 (1995).

151. R. J. Simes, E. J. Topol, D. R. Holmes Jr., H. D. White, W. R. Rutsch, A. Vahanian, M. L. Simoons, D. Morris, A. Betriu, and R. M. Califf, *Circulation*, **91**, 1923–1928 (1995).

152. S. A. Spinler and S. M. Inverso, *Pharmacotherapy*, **21**, 691–716 (2001).

153. G. B. Habib, *Chest*, **107**, 225–232 (1995).

154. D. B. Wilson, T. E. Bross, S. L. Hofmann, and P. W. Majerus, *J. Biol. Chem.*, **259**, 11718–11724 (1984).

155. J. Llevadot, R. P. Giugliano, and E. M. Antman, *JAMA*, **286**, 442–449 (2001).

156. B. J. Meyer and J. H. Chesebro in D. Julian and E. Braunwald, Eds., *Management of Acute Myocardial Infarction*, W. B. Saunders Co., London, 1994, pp. 163–192.

157. G. B. Habib, *Chest*, **107**, 809–816 (1995).

158. G. S. Reeder, *Mayo Clin. Proc.*, **70**, 464–468 (1995).

159. E. Braunwald, E. M. Antman, J. W. Beasley, R. M. Califf, M. D. Cheitlin, J. S. Hochman, R. H. Jones, D. Kereiakes, J. Kupersmith, T. N. Levin, C. J. Pepine, J. W. Schaeffer, E. E. Smith III, D. E. Steward, P. Theroux, R. J. Gibbons, J. S. Alpert, K. A. Eagle, D. P. Faxon, V. Fuster, T. J. Gardner, G. Gregoratos, R. O. Russell, and S. C. Smith Jr., *Circulation*, **102**, 1193–1209 (2000).

160. E. J. Topol, *Lancet*, **357**, 1905–1914 (2001).

161. A. Bindoli, E. Barzon, and M. P. Rigobello, *Cardiovasc. Res.*, **30**, 821–824 (1995).

162. E. M. Vaughan Williams, *Eur. Heart J.*, **5**, 96–98 (1984).

163. D. S. Echt, P. R. Liebson, L. B. Mitchell, R. W. Peters, D. Obias-Manno, A. H. Barker, D. Arensberg, A. Baker, L. Friedman, and H. L. Greene, *N. Engl. J. Med.*, **324**, 781–788 (1991).

164. P. T. Sager, *Curr. Opin. Cardiol.*, **14**, 15–23 (1999).

165. A. L. Waldo, A. J. Camm, H. deRuyter, P. L. Friedman, D. J. MacNeil, J. F. Pauls, B. Pitt, C. M. Pratt, P. J. Schwartz, and E. P. Veltri, *Lancet*, **348**, 7–12 (1996).

166. Amiodarone Trials Meta-Analysis Investigators, *Lancet*, **350**, 1417–1424 (1997).

167. I. Sim, K. M. McDonald, P. W. Lavori, C. M. Norbutas, and M. A. Hlatky, *Circulation*, **96**, 2823–2829 (1997).

168. D. S. Holmes, Z. Q. Sun, L. M. Porter, N. E. Bernstein, L. A. Chinitz, M. Artman, and W. A. Coetzee, *J. Cardiovasc. Electrophysiol.*, **11**, 1152–1158 (2000).

169. I. Kodama, K. Kamiya, and J. Toyama, *Cardiovasc. Res.*, **35**, 13–29 (1997).

170. *Cardiovascular Drug Therapy*, W. B. Saunders Co., Philadelphia, 1990.

171. R. D. Bremel and A. Weber, *Nature*, **238**, 97–101 (1972).

172. A. M. Katz, *Eur. Heart J.*, **1** (Suppl H), H18–H23 (1999).

173. S. I. Ertel, E. A. Ertel, and J. P. Clozel, *Cardiovasc. Drugs Ther.*, **11**, 723–739 (1997).

174. H. B. Nuss and S. R. Houser, *Circ. Res.*, **73**, 777–782 (1993).

175. L. Sen and T. W. Smith, *Circ. Res.*, **75**, 149–155 (1994).

176. B. Huang, D. Qin, L. Deng, M. Boutjdir, and N. Sherif, *Cardiovasc. Res.*, **46**, 442–449 (2000).

177. A. Fleckenstein, *Circ. Res.*, **52**, I3–16 (1983).

178. A. Fleckenstein, H. Kammermeier, H. J. Doring, and H. J. Freund, *Z. Kreislaufforsch.*, **56**, 839–858 (1967).

179. M. Kohlhardt, B. Bauer, H. Krause, and A. Fleckenstein, *Pflugers Arch.*, **335**, 309–322 (1972).

180. H. A. Struyker-Boudier, J. F. Smits, and J. G. De Mey, *J. Cardiovasc. Pharmacol.*, **15** (Suppl 4), S1–S10 (1990).

181. H. Haas and G. Hartefelder, *Arzneimittelforschung*, **12**, 549–558 (1962).

182. T. J. Kamp, Z. Zhou, and S. Zhang in D. P. Zipes and J. Jalife, Eds., *Cardiac Electrophysiology: From Cell to Bedside*, W. B. Saunders, Philadelphia, 2000, pp. 141–155.

183. S. Zhang, Z. Zhou, Q. Gong, J. C. Makielski, and C. T. January, *Circ. Res.*, **84**, 989–998 (1999).

184. A. J. Kaumann and P. Aramendia, *J. Pharmacol. Exp. Ther.*, **164**, 326–332 (1968).

185. J. D. Fondacaro, J. Han, and M. S. Yoon, *Am. Heart J.*, **96**, 81–86 (1978).

186. R. Mohabir and G. R. Ferrier, *J. Cardiovasc. Pharmacol.*, **17**, 74–82 (1991).

187. P. N. Temesy-Armos, M. Legenza, S. R. Southworth, and B. F. Hoffman, *J. Am. Coll. Cardiol.*, **6**, 674–681 (1985).

188. W. W. Brooks, R. L. Verrier, and B. Lown, *Cardiovasc. Res.*, **14**, 295–302 (1980).

189. C. A. Muller, L. H. Opie, J. McCarthy, D. Hofmann, C. A. Pineda, and M. Peisach, *J. Am. Coll. Cardiol.*, **32**, 268–274 (1998).

190. G. E. Billman, *J. Pharmacol. Exp. Ther.*, **266**, 407–416 (1993).

191. C. A. Muller, L. H. Opie, C. A. Pineda, J. McCarthy, and V. Kraljevic, *Cardiovasc. Drugs Ther.*, **12**, 449–455 (1998).

192. A. E. Buxton, H. L. Waxman, F. E. Marchlinski, and M. E. Josephson, *Am. J. Cardiol.*, **53**, 738–744 (1984).

193. L. Schamroth, D. M. Krikler, and C. Garrett, *Br. Med. J.*, **1**, 660–662 (1972).

194. F. H. Messerli, B. D. Nunez, H. O. Ventura, and D. W. Snyder, *Arch. Intern. Med.*, **147**, 1725–1728 (1987).

195. T. Tsuchiya, K. Okumura, T. Honda, A. Iwasa, and K. Ashikaga, *J. Am. Coll. Cardiol.*, **37**, 1415–1421 (2001).

196. B. Belhassen, H. H. Rotmensch, and S. Laniado, *Br. Heart J.*, **46**, 679–682 (1981).

197. S. H. Lee, S. A. Chen, C. T. Tai, C. E. Chiang, T. J. Wu, C. C. Cheng, C. W. Chiou, K. C. Ueng, S. P. Wang, B. N. Chiang, and M. S. Chang, *Cardiology*, **87**, 33–41 (1996).

198. J. Fischer Hansen and Danish Group for Verapamil Myocardial Infarction, *J. Cardiovasc. Pharmacol.*, **18**, S20–S25 (1991).

199. S. Yusuf, J. Wittes, and L. Friedman, *JAMA*, **260**, 2088–2093 (1988).

200. P. H. Held and S. Yusuf, *Eur. Heart J.*, **14** (Suppl F), 18–25 (1993).

201. M. Sato, T. Nagao, I. Yamaguchi, H. Nakajima, and A. Kiyomoto, *Arzneimittelforschung*, **21**, 1338–1343 (1971).

202. T. Peter, T. Fujimoto, H. Hamamoto, A. McCullen, and W. J. Mandel, *Am. J. Cardiol.*, **49**, 602–605 (1982).

203. B. D. Nearing, J. J. Hutter, and R. L. Verrier, *J. Cardiovasc. Pharmacol.*, **27**, 777–787 (1996).

204. M. Anastasiou-Nana, J. Nanas, R. L. Menlove, and J. L. Anderson, *J. Cardiovasc. Pharmacol.*, **6**, 780–787 (1984).

205. E. Patterson, B. T. Eller, and B. R. Lucchesi, *J. Pharmacol. Exp. Ther.*, **225**, 224–233 (1983).

206. J. J. Lynch, D. G. Montgomery, and B. R. Lucchesi, *J. Pharmacol. Exp. Ther.*, **239**, 340–345 (1986).

207. F. H. Sheehan and S. E. Epstein, *Am. Heart J.*, **103**, 973–977 (1982).

208. D. S. Echt, P. R. Liebson, and L. B. Mitchell, *N. Engl. J. Med.*, **319**, 385–392 (1988).

209. J. T. Bigger Jr., J. Coromilas, L. M. Rolnitzky, J. L. Fleiss, and R. E. Kleiger, *Am. J. Cardiol.*, **65**, 539–546 (1990).

210. W. E. Boden, R. J. Krone, R. E. Kleiger, D. Oakes, H. Greenberg, E. J. Dwyer Jr., J. P. Miller, J. Abrams, J. Coromilas, and R. Goldstein, *Am. J. Cardiol.*, **67**, 335–342 (1991).

211. M. Epstein in M. Epstein, Ed., *Calcium Antagonists in Clinical Medicine*, Hanley & Belfus, Inc., Philadelphia, 1998, pp. 553–571.

212. F. Bossert and W. Vater, *Naturwissenschaften*, **58**, 578 (1971).

213. J. L. Bergey, R. L. Wendt, K. Nocella, and J. D. McCallum, *Eur. J. Pharmacol.*, **97**, 95–103 (1984).

214. M. Naito, E. L. Michelson, J. J. Kmetzo, E. Kaplinsky, and L. S. Dreifus, *Circulation*, **63**, 70–79 (1981).

215. W. F. Lubbe, J. A. McLean, and T. Nguyen, *Am. Heart J.*, **105**, 331–333 (1983).

216. Israeli SPRINT Study Group, Secondary Prevention Reinfarction Israeli Nifedipine Trial (SPRINT), *Eur. Heart J.*, **5**, 516–528 (1984).

217. C. D. Furberg, B. M. Psaty, and J. V. Meyer, *Circulation*, **92**, 1326–1331 (1995).

218. B. Holmes, R. N. Brogden, R. C. Heel, T. M. Speight, and G. S. Avery, *Drugs*, **27**, 6–44 (1984).

219. J. Tytgat, J. Vereecke, and E. Carmeliet, *Eur. J. Pharmacol.*, **296**, 189–197 (1996).

220. M. A. Vos, A. P. Gorgels, J. D. Leunissen, and H. J. Wellens, *Circulation*, **81**, 343–349 (1990).

221. M. A. Vos, A. P. Gorgels, J. D. Leunissen, N. T. van der, F. J. Halbertsma, and H. J. Wellens, *J. Cardiovasc. Pharmacol.*, **19**, 682–690 (1992).

222. S. C. Verduyn, M. A. Vos, A. P. Gorgels, J. van der Zande, J. D. Leunissen, and H. J. Wellens, *J. Cardiovasc. Electrophysiol.*, **6**, 189–200 (1995).

223. L. Carlsson, L. Drews, and G. Duker, *J. Pharmacol. Exp. Ther.*, **279**, 231–239 (1996).

224. J. L. Montastruc, M. E. Llau, O. Rascol, and J. M. Senard, *Fundam. Clin. Pharmacol.*, **8**, 293–306 (1994).

225. L. T. Iseri and J. H. French, *Am. Heart J.*, **108**, 188–193 (1984).

226. A. Keren and D. Tzivoni, *Pacing Clin. Electrophysiol.*, **13**, 937–945 (1990).

227. E. Laban and G. A. Charbon, *J. Am. Coll. Nutr.*, **5**, 521–532 (1986).

228. D. S. Bailie, H. Inoue, S. Kaseda, J. Ben David, and D. P. Zipes, *Circulation*, **77**, 1395–1402 (1988).

229. A. Bril and L. Rochette, *Can. J. Physiol. Pharmacol.*, **68**, 694–699 (1990).

230. D. E. Euler, *Cardiovasc. Drugs Ther.*, **9**, 565–571 (1995).

231. T. J. Campbell, *Electrophysiol. Rev.*, **4**, 251–254 (2000).

232. M. Zehender, T. Meinertz, T. Faber, A. Caspary, A. Jeron, K. Bremm, and H. Just, *J. Am. Coll. Cardiol.*, **29**, 1028–1034 (1997).

233. H. S. Rasmussen, P. McNair, P. Norregard, V. Backer, O. Lindeneg, and S. Balslev, *Lancet*, **1**, 234–236 (1986).

234. M. Shechter, H. Hod, N. Marks, S. Behar, E. Kaplinsky, and B. Rabinowitz, *Am. J. Cardiol.*, **66**, 271–274 (1990).

235. L. F. Smith, A. M. Heagerty, R. F. Bing, and D. B. Barnett, *Int. J. Cardiol.*, **12**, 175–183 (1986).

236. D. Tzivoni, A. Keren, A. M. Cohen, H. Loebel, I. Zahavi, A. Chenzbraun, and S. Stern, *Am. J. Cardiol.*, **53**, 528–530 (1984).

237. D. Tzivoni, S. Banai, C. Schuger, J. Benhorin, A. Keren, S. Gottlieb, and S. Stern, *Circulation*, **77**, 392–397 (1988).

238. G. W. De Keulenaer, P. Fransen, D. L. Brutsaert, and S. U. Sys, *Cardiovasc. Res.*, **30**, 646–647 (1995).

239. W. Osterrieder and M. Holck, *J. Cardiovasc. Pharmacol.*, **13**, 754–759 (1989).

240. J.-P. Clozel, L. Banken, and W. Osterrieder, *J. Cardiovasc. Pharmacol.*, **14**, 713–721 (1989).

241. J. M. B. Pinto, E. A. Sosunov, R. Z. Gainullin, M. R. Rosen, and P. A. Boyden, *J. Cardiovasc. Electrophysiol.*, **10**, 1224–1235 (1999).

242. L. Perchenet and O. Clement-Chomienne, *J. Pharmacol. Exp. Ther.*, **295**, 771–778 (2000).

243. J. C. Gomora, J. A. Enyeart, and J. J. Enyeart, *Mol. Pharmacol.*, **56**, 1192–1197 (1999).

244. M. M. Mocanu, S. Gadgil, D. M. Yellon, and G. F. Baxter, *Cardiovasc. Drugs Ther.*, **13**, 115–122 (1999).

245. R. Schulz, H. Post, A. Jalowy, U. Backenkohler, H. Dorge, C. Vahlhaus, and G. Heusch, *Circulation*, **99**, 305–311 (1999).

246. R. L. Martin, J. H. Lee, L. L. Cribbs, E. Perez-Reyes, and D. A. Hanck, *J. Pharmacol. Exp. Ther.*, **295**, 302–308 (2000).

247. A. Rutledge and D. J. Triggle, *Eur. J. Pharmacol.*, **280**, 155–158 (1995).

248. D. Sarsero, T. Fujiwara, P. Molenaar, and J. A. Angus, *Br. J. Pharmacol.*, **125**, 109–119 (1998).

249. B. Cremers, M. Flesch, M. Suedkamp, and M. Boehm, *J. Cardiovasc. Pharmacol.*, **29**, 692–696 (1997).

250. S. Sandmann, J. Y. Min, A. Meissner, and T. Unger, *Cardiovasc. Res.*, **44**, 67–80 (1999).

251. A. Ezzaher, N. el Houda Bouanani, J. B. Su, L. Hittinger, and B. Crozatier, *J. Pharmacol. Exp. Ther.*, **257**, 466–471 (1991).

252. J. Su, N. Renaud, A. Carayon, B. Crozatier, L. Hittinger, and M. Laplace, *Br. J. Pharmacol.*, **113**, 395–402 (1994).

253. J. P. Clozel, M. Veniant, and W. Osterrieder, *Cardiovasc. Drugs Ther.*, **4**, 731–736 (1990).

254. E. I. Ratner, V. N. Bochkov, and V. A. Tkachuk, *Arzneimittelforschung*, **46**, 953–955 (1996).

255. M. Veniant, J. P. Clozel, P. Hess, and R. Wolfgang, *J. Cardiovasc. Pharmacol.*, **17**, 277–284 (1991).

256. M. Veniant, J. P. Clozel, P. Hess, and R. Wolfgang, *J. Cardiovasc. Pharmacol.*, **18** (Suppl 10), S55–S58 (1991).

257. R. Schmitt, C. H. Kleinbloesem, G. G. Belz, V. Schroeter, U. Feifel, H. Pozenel, W. Kirch, A. Halabi, A. J. Woittiez, and H. A. Welker, *Clin. Pharmacol. Ther.*, **52**, 314–323 (1992).

258. M. F. Rousseau, W. Hayashida, C. van Eyll, O. M. Hess, C. R. Benedict, S. Ahn, F. Chapelle, I. Kobrin, and H. Pouleur, *J. Am. Coll. Cardiol.*, **28**, 972–979 (1996).

259. P. J. Bernink, G. Prager, A. Schelling, and I. Kobrin, *Hypertension*, **27**, 426–432 (1996).

260. B. M. Massie, Y. Lacourciere, R. Viskoper, A. Woittiez, and I. Kobrin, *Am. J. Cardiol.*, **80**, 27C–33C (1997).

261. S. Oparil, P. Bernink, M. Bursztyn, S. Carney, and I. Kobrin, *Am. J. Cardiol.*, **80**, 12C–19C (1997).

262. R. J. Viskoper, P. J. Bernink, A. Schelling, A. B. Ribeiro, I. M. Kantola, M. R. Wilkins, and I. Kobrin, *J. Hum. Hypertens.*, **11**, 387–393 (1997).

263. A. L. Bakx, E. E. Van der Wall, S. Braun, H. Emanuelsson, A. V. Bruschke, and I. Kobrin, *Am. Heart J.*, **130**, 748–757 (1995).

264. S. Braun, E. E. Van der Wall, H. Emanuelsson, and I. Kobrin, *J. Am. Coll. Cardiol.*, **27**, 317–322 (1996).

265. W. H. Frishman, N. Bittar, S. Glasser, G. Habib, W. Smith, and R. Pordy, *Clin. Cardiol.*, **21**, 483–490 (1998).

266. G. J. Davies, I. Kobrin, A. Caspi, L. H. Reisin, D. C. de Albuquerque, D. Armagnijan, O. R. Coelho, and A. Schneeweiss, *Am. Heart J.*, **134**, 220–228 (1997).

267. D. Tzivoni, Z. Gilula, M. W. Klutstein, L. Reisin, S. Botvin, and I. Kobrin, *Cardiovasc. Drugs Ther.*, **14**, 503–509 (2000).

268. D. Tzivoni, H. Kadr, S. Braat, W. Rutsch, J. A. Ramires, and I. Kobrin, *Circulation*, **96**, 2557–2564 (1997).

269. T. B. Levine, P. J. Bernink, A. Caspi, U. Elkayam, E. M. Geltman, B. Greenberg, W. J. McKenna, J. K. Ghali, T. D. Giles, A. Marmor, L. H. Reisin, S. Ammon, and E. Lindberg, *Circulation*, **101**, 758–764 (2000).

270. M. Spoendlin, J. Peters, H. Welker, A. Bock, and G. Thiel, *Nephrol. Dial. Transplant.*, **13**, 1787–1791 (1998).

271. K. A. Reimer and R. M. Califf, *Circulation*, **99**, 198–200 (1999).

272. R. SoRelle, *Circulation*, **98**, 831–832 (1998).

273. A. L. Po and W. Y. Zhang, *Lancet*, **351**, 1829–1830 (1998).

274. A. Benardeau, J. Weissenburger, L. Hondeghem, and E. A. Ertel, *J. Pharmacol. Exp. Ther.*, **292**, 561–575 (2000).

275. A. Farkas, A. Qureshi, and M. J. Curtis, *Br. J. Pharmacol.*, **128**, 41–50 (1999).

276. S. Fareh, A. Benardeau, B. Thibault, and S. Nattel, *Circulation*, **100**, 2191–2197 (1999).

277. S. Fareh, A. Benardeau, and S. Nattel, *Cardiovasc. Res.*, **49**, 762–770 (2001).

278. L. Yue, J. Feng, R. Gaspo, G. R. Li, Z. Wang, and S. Nattel, *Circ. Res.*, **81**, 512–525 (1997).

279. L. M. Fang and W. Osterrieder, *Eur. J. Pharmacol.*, **196**, 205–207 (1991).

280. J. Hensley, G. E. Billman, J. D. Johnson, C. M. Hohl, and R. A. Altschuld, *J. Mol. Cell Cardiol.*, **29**, 1037–1043 (1997).

281. K. Bian and K. Hermsmeyer, *Naunyn Schmiedebergs Arch. Pharmacol.*, **348**, 191–196 (1993).

282. I. Bezprozvanny and R. W. Tsien, *Mol. Pharmacol.*, **48**, 540–549 (1995).

283. G. Mehrke, X. G. Zong, V. Flockerzi, and F. Hofmann, *J. Pharmacol. Exp. Ther.*, **271**, 1483–1488 (1994).

284. G. E. Billman and R. L. Hamlin, *J. Pharmacol. Exp. Ther.*, **277**, 1517–1526 (1996).

285. S. K. Mishra and K. Hermsmeyer, *J. Cardiovasc. Pharmacol.*, **24**, 1–7 (1994).

286. M. Shigekawa and T. Iwamoto, *Circ. Res.*, **88**, 864–876 (2001).

287. M. P. Blaustein and W. J. Lederer, *Physiol. Rev.*, **79**, 763–854 (1999).

288. S. M. Pogwizd, M. Qi, W. L. Yuan, A. M. Samarel, and D. M. Bers, *Circ. Res.*, **85**, 1009–1019 (1999).

289. S. M. Pogwizd, K. Schlotthauer, L. Li, W. Yuan, and D. M. Bers, *Circ. Res.*, **88**, 1159–1167 (2001).

290. W. Trautwein, A. Cavalie, V. Flockerzi, F. Hofmann, and D. Pelzer, *Circ. Res.*, **61**, I17–I23 (1987).

291. H. Satoh, K. S. Ginsburg, K. Qing, H. Terada, H. Hayashi, and D. M. Bers, *Circulation*, **101**, 1441–1446 (2000).

292. T. Watano, J. Kimura, T. Morita, and H. Nakanishi, *Br. J. Pharmacol.*, **119**, 555–563 (1996).

293. C. L. Elias, A. Omelchenko, G. J. Gross, M. Hnatowich, and L. V. Hryshko, *Biophys. J.*, **78**, 54A (2000).

294. J. Kimura, T. Watano, M. Kawahara, E. Sakai, and J. Yatabe, *Br. J. Pharmacol.*, **128**, 969–974 (1999).

295. T. Watano, Y. Harada, K. Harada, and N. Nishimura, *Br. J. Pharmacol.*, **127**, 1846–1850 (1999).

296. J. Inserte, M. Ruiz-Meana, D. Garcia-Dorado, F. Padilla, J. A. Barrades, Y. Puigfel, and J. Soler-Soler, *Circulation*, **102** (Suppl II), 137 (2000).

297. Y. Ladilov, S. Haffner, and C. Balser, *Circulation*, **98** (Suppl), 821 (1998).

298. Y. Ladilov, S. Haffner, C. Balser-Schaefer, H. Maxeiner, and H. M. Piper, *Am. J. Physiol. Heart Circ. Physiol.*, **276**, H1868–H1876 (1999).

299. M. Mukau, H. Terada, S. Sugiyam, H. Satoh, and H. Hayashi. *Eur. Heart J.*, **20** (Suppl), 554 (1999).

300. M. Mukai, H. Terada, S. Sugiyama, H. Satoh, and H. Hayashi, *J. Cardiovasc. Pharmacol.*, **35**, 121–128 (2000).

301. S. Seki, H. Takeda, T. Onodera, K. Horikoshi, M. Nagai, M. Taniguchi, and S. Mochizuki, *Eur. Heart J.*, **21** (Suppl), 476 (2000).

302. T. Matsumoto, T. Miura, A. Tsuchida, Y. Nishino, S. Genda, A. Kuna, and K. Shimamoto, *Circulation*, **102** (Suppl II), 137 (2000).

303. H. Ma, H. Hashizume, A. Hara, K. Yazawa, F. Ushikubi, and Y. Abiko, *Jpn. J. Pharmacol.*, **79** (Suppl I), 124 (1999).

304. M. Ruiz-Meana, D. Garcia-Dorado, B. Hofstaetter, H. M. Piper, and J. Soler-Soler, *Circ. Res.*, **85**, 280–287 (1999).

305. A. Miyata, D. P. Zipes, and M. Rubart, *Circulation*, **102** (Suppl II), 92 (2000).

306. A. Miyata, D. P. Zipes, and M. Rubart, *Circulation*, **102** (Suppl II), 154 (2000).

307. A. Kuruma, M. Hiraoka, and S. Kawano, *Pflugers Arch.*, **436**, 976–983 (1998).

308. F. Kurogouchi, Y. Furukawa, D. Zhao, M. Hirose, K. Nakajima, M. Tsuboi, and S. Chiba, *Jpn. J. Pharmacol.*, **82**, 155–163 (2000).

309. M. T. La Rovere, J. T. Bigger Jr., F. I. Marcus, A. Mortara, and P. J. Schwartz, *Lancet*, **351**, 478–484 (1998).

310. A. I. Sobolevsky and B. I. Khodorov, *Neuropharmacology*, **38**, 1235–1242 (1999).

311. S. H. Woo and M. Morad, *Proc. Natl. Acad. Sci. U.S.A.*, **98**, 2023–2028 (2001).

312. C. Cabo, H. Schmitt, and A. L. Wit, *Circulation*, **102**, 2417–2425 (2000).

313. E. Marban, S. W. Robinson, and W. G. Wier, *J. Clin. Invest.*, **78**, 1185–1192 (1986).

314. J. T. Bigger Jr., J. L. Fleiss, R. Kleiger, J. P. Miller, and L. M. Rolnitzky, *Circulation*, **69**, 250–258 (1984).

315. I. MacKenzie, V. L. Saville, and J. F. Waterfall, *Br. J. Pharmacol.*, **110**, 531–538 (1993).

316. J. Bellemin-Baurreau, A. Poizot, P. E. Hicks, L. Rochette, and J. M. Armstrong, *Eur. J. Pharmacol.*, **256**, 115–124 (1994).

317. S. Dhein, P. Pejman, and K. Krusemann, *Eur. J. Pharmacol.*, **398**, 273–284 (2000).

318. K. Hamada, J. Yamazaki, and T. Nagao, *Jpn. J. Pharmacol.*, **76**, 149–154 (1998).

319. H. L. Zhang, Y. S. Li, S. X. Fu, and X. P. Yang, *Zhongguo Yao Li Xue. Bao.*, **12**, 398–402 (1991).

320. S. I. Koumi, R. L. Martin, and R. Sato, *Am. J. Physiol.*, **272**, H1656–H1665 (1997).

321. G. Pogatsa, M. Z. Koltai, I. Balkanyi, I. Devai, V. Kiss, and A. Koszeghy, *Acta Physiol. Hung.*, **71**, 243–250 (1988).

322. C. D. Wolleben, M. C. Sanguinetti, and P. K. Siegl, *J. Mol. Cell Cardiol.*, **21**, 783–788 (1989).

323. M. Gwilt, C. G. Henderson, J. Orme, and J. D. Rourke, *Eur. J. Pharmacol.*, **220**, 231–236 (1992).

324. A. Tosaki, D. T. Engelman, R. M. Engelman, and D. K. Das, *J. Pharmacol. Exp. Ther.*, **275**, 1115–1123 (1995).

325. N. E. El Reyani, O. Bozdogan, I. Baczko, I. Lepran, and J. G. Papp, *Eur. J. Pharmacol.*, **365**, 187–192 (1999).

326. I. Lepran, I. Baczko, A. Varro, and J. G. Papp, *J. Pharmacol. Exp. Ther.*, **277**, 1215–1220 (1996).

327. T. D. Barrett and M. J. Walker, *J. Mol. Cell Cardiol.* **30**, 999–1008 (1998).

328. I. Baczko, I. Lepran, and J. G. Papp, *Eur. J. Pharmacol.*, **324**, 77–83 (1997).

329. G. E. Billman, C. E. Avendano, J. R. Halliwill, and J. M. Burroughs, *J. Cardiovasc. Pharmacol.*, **21**, 197–204 (1993).

330. T. M. Davis, R. W. Parsons, R. J. Broadhurst, M. S. Hobbs, and K. Jamrozik, *Diabetes Care.*, **21**, 637–640 (1998).

331. A. Lomuscio and C. Fiorentini, *Diabetes Res Clin. Pract.*, **31** (Suppl), S21–S26 (1996).

332. A. Lomuscio, D. Vergani, L. Marano, M. Castagnone, and C. Fiorentini, *Coron. Artery Dis.*, **5**, 767–771 (1994).

333. M. S. Link, P. J. Wang, B. A. VanderBrink, E. Avelar, N. G. Pandian, B. J. Maron, and N. A. Estes III, *Circulation*, **100**, 413–418 (1999).

334. N. Inagaki, T. Gonoi, J. P. Clement, N. Namba, J. Inazawa, G. Gonzalez, L. Aguilar-Bryan, S. Seino, and J. Bryan, *Science*, **270**, 1166–1170 (1995).

335. H. Gogelein, *Curr. Opin. Invest. Drugs*, **2**, 71–80 (2001).

336. H. Gogelein, J. Hartung, and H. C. Englert, *Cell Physiol. Biochem.*, **9**, 227–241 (1999).

337. Y. Liu, G. Ren, B. O'Rourke, E. Marban, and J. Seharaseyon, *Mol. Pharmacol.*, **59**, 225–230 (2001).

338. F. Reimann, P. Proks, and F. M. Ashcroft, *Br. J. Pharmacol.*, **132**, 1542–1548 (2001).

339. F. M. Gribble, S. J. Tucker, S. Seino, and F. M. Ashcroft, *Diabetes*, **47**, 1412–1418 (1998).

340. H. C. Englert, U. Gerlach, H. Goegelein, J. Hartung, H. Heitsch, D. Mania, and S. Scheidler, *J. Med. Chem.*, **44**, 1085–1098 (2001).

341. H. Gogelein, J. Hartung, H. C. Englert, and B. A. Scholkens, *J. Pharmacol. Exp. Ther.*, **286**, 1453–1464 (1998).

342. T. Sato, *Circ. Res.*, **85**, 1113–1114 (1999).

343. R. M. Fryer, J. T. Eells, A. K. Hsu, M. M. Henry, and G. J. Gross, *Am. J. Physiol. Heart Circ. Physiol.*, **278**, H305–H312 (2000).

344. O. Jung, H. C. Englert, W. Jung, H. Gogelein, B. A. Scholkens, A. E. Busch, and W. Linz, *Naunyn Schmiedebergs Arch. Pharmacol.*, **361**, 445–451 (2000).

345. B. O'Rourke, *Circ. Res.*, **87**, 845–855 (2000).

346. K. J. Wirth, B. Rosenstein, J. Uhde, H. C. Englert, A. E. Busch, and B. A. Scholkens, *J. Pharmacol. Exp. Ther.*, **291**, 474–481 (1999).

347. K. J. Wirth, J. Uhde, B. Rosenstein, H. C. Englert, H. Gogelein, B. A. Scholkens, and A. E. Busch, *Naunyn Schmiedebergs Arch. Pharmacol.*, **361**, 155–160 (2000).

348. H. Bohn, H. C. Englert, and B. A. Schoelkens. *Br. J. Pharmacol.*, **124**, 23P (1998).

349. K. J. Wirth, E. Klaus, H. G. Englert, B. A. Scholkens, and W. Linz, *Naunyn Schmiedebergs Arch. Pharmacol.*, **360**, 295–300 (1999).

350. L. Chi, S. C. Black, P. I. Kuo, S. O. Fagbemi, and B. R. Lucchesi, *J. Cardiovasc. Pharmacol.*, **21**, 179–190 (1993).

351. L. Chi, A. C. Uprichard, and B. R. Lucchesi, *J. Cardiovasc. Pharmacol.*, **15**, 452–464 (1990).

352. A. J. D'Alonzo, J. C. Sewter, R. B. Darbenzio, and T. A. Hess, *Naunyn Schmiedebergs Arch. Pharmacol.*, **352**, 222–228 (1995).

353. A. J. D'Alonzo, R. B. Darbenzio, T. A. Hess, J. C. Sewter, P. G. Sleph, and G. J. Grover, *Cardiovasc. Res.*, **28**, 881–887 (1994).

354. H. A. Friedel and R. N. Brogden, *Drugs*, **39**, 929–967 (1990).

355. M. Krumenacker and E. Roland, *J. Cardiovasc. Pharmacol.*, **20** (Suppl 3), S93–S102 (1992).

356. T. J. Colatsky, C. H. Follmer, and C. F. Starmer, *Circulation*, **82**, 2235–2242 (1990).

357. J. J. Lynch Jr., M. C. Sanguinetti, S. Kimura, and A. L. Bassett, *FASEB J.*, **6**, 2952–2960 (1992).

358. A. A. M. Wilde, *Cardiovasc. Drugs Ther.*, **7**, 521–526 (1993).

359. B. Lui, F. Gaylon, J. R. McCullough, and M. Bassali, *Drug Dev. Res.*, **14**, 123–139 (1988).

360. F. A. Fish, C. Prakash, and D. M. Roden, *Circulation*, **82**, 1362–1369 (1990).

361. D. A. Lathrop, P. P. Nanasi, and A. Varro, *Br. J. Pharmacol.*, **99**, 119–123 (1990).

362. S. Sorota, M. S. Siegal, and B. F. Hoffman, *J. Mol. Cell Cardiol.*, **23**, 1191–1198 (1991).

363. N. Takahashi, M. Ito, T. Saikawa, and M. Arita, *Cardiovasc. Res.*, **25**, 445–452 (1991).

364. R. A. Haworth, A. B. Goknur, and H. A. Berkoff, *Circ. Res.*, **65**, 1157–1160 (1989).

365. A. Olschewski, M. E. Brau, H. Olschewski, G. Hempelmann, and W. Vogel, *Circulation*, **93**, 656–659 (1996).

366. R. E. Kleiger, J. P. Miller, J. T. Bigger Jr., and A. J. Moss, *Am. J. Cardiol.*, **59**, 256–262 (1987).

367. J. T. Bigger Jr., R. E. Kleiger, J. L. Fleiss, L. M. Rolnitzky, R. C. Steinman, and J. P. Miller, *Am. J. Cardiol.*, **61**, 208–215 (1988).

368. G. E. Billman, P. J. Schwartz, and H. L. Stone, *Circulation*, **66**, 874–880 (1982).

369. P. Held and S. Yusuf, *Cardiology*, **76**, 132–143 (1989).

370. J. Wikstrand and M. Kendall, *Eur. Heart J.*, **13** (Suppl D), 111–120 (1992).

371. D. A. Brand, L. N. Newcomer, A. Freiburger, and H. Tian, *J. Am. Coll. Cardiol.*, **26**, 1432–1436 (1995).

372. R. A. O'Rourke, *J. Am. Coll. Cardiol.*, **26**, 1437–1439 (1995).

373. S. Viskin, I. Kitzis, and E. Lev, *J. Am. Coll. Cardiol.*, **25**, 1327–1332 (1995).

374. J. M. Barber, D. M. Boyle, N. C. Chaturvedi, N. Singh, and M. J. Walsh, *Acta Med. Scand.-Suppl.*, **587**, 213–219 (1976).

375. K. L. Evemy and B. L. Pentecost, *Eur. J. Cardiol.*, **7**, 391–398 (1978).

376. A. Hjalmarson, P. Armitage, and D. Chamberlain, *Eur. Heart J.*, **6**, 199–226 (1985).

377. K. S. Salathia, J. M. Barber, E. L. McIlmoyle, J. Nicholas, A. E. Evans, J. H. Elwood, G. Cran, R. G. Shanks, and D. M. Boyle, *Eur. Heart J.*, **6**, 190–198 (1985).

378. A. Hjalmarson, D. Elmfeldt, J. Herlitz, S. Holmberg, I. Malek, G. Nyberg, L. Ryden, K. Swedberg, A. Vedin, F. Waagstein, A. Waldenstrom, J. Waldenstrom, H. Wedel, L. Wilhelmsen, and C. Wilhelmsson, *Lancet*, **2**, 823–827 (1981).

379. A. Hjalmarson, J. Herlitz, S. Holmberg, L. Ryden, K. Swedberg, A. Vedin, F. Waagstein, A. Waldenstrom, J. Waldenstrom, H. Wedel, L. Wilhelmsen, and C. Wilhelmsson, *Circulation*, **67**, I26–I32 (1983).

380. ISIS1 (First International Study of Survival Collaborative Group), *Lancet*, **1**, 921–923 (1988).

381. R. M. Norris, P. F. Barnaby, M. A. Brown, G. G. Geary, E. D. Clarke, R. L. Logan, and D. N. Sharpe, *Lancet*, **2**, 883–886 (1984).

382. K. Shiukumar, L. Schultz, S. Goldstein, and M. Gheorghiade, *Am. Heart J.*, **135**, 261–267 (1998).

383. E. D. Robin and R. McCauley, *Chest*, **101**, 1699–1702 (1992).

384. R. Gorlin, in D. Julian and E. Braunwald, Eds., *Management of Myocardial Infarction*, W. B. Saunders, London, 1994, pp. 343–359.

385. J. B. Young, *Cardiovasc. Drugs Ther.*, **9**, 89–102 (1995).

386. J. G. Cleland, *Eur. Heart J.*, **16**, 153–159 (1995).

387. J. B. Young, *Coron. Artery Dis.*, **6**, 272–280 (1995).

388. J. Sadoshima and S. Izumo, *Circ. Res.*, **73**, 413–423 (1993).

389. J. Sadoshima, Z. Qiu, J. P. Morgan, and S. Izumo, *Circ. Res.*, **76**, 1–15 (1995).

390. G. S. Reeder, *Mayo Clin. Proc.*, **70**, 1185–1190 (1995).

391. F. R. Eberli, E. O. Weinberg, W. N. Grice, G. L. Horowitz, and C. S. Apstein, *Circ. Res.*, **68**, 466–481 (1991).

392. R. Diaz, E. A. Paolasso, L. S. Piegas, C. D. Tajer, M. G. Moreno, R. Corvalan, J. E. Isea, and G. Romero, *Circulation*, **98**, 2227–2234 (1998).

393. C. S. Apstein and H. Taegtmeyer, *Circulation*, **96**, 1074–1077 (1997).

394. A. K. Jonassen, B. K. Brar, O. D. Mjos, M. N. Sack, D. S. Latchman, and D. M. Yellon, *J. Mol. Cell Cardiol.*, **32**, 757–764 (2000).

395. T. Matsui, J. Tao, F. Del Monte, K. H. Lee, L. Li, M. Picard, T. L. Force, T. F. Franke, R. J. Hajjar, and A. Rosenzweig, *Circulation*, **104**, 330–335 (2001).

396. Y. Fujio, T. Nguyen, D. Wencker, R. N. Kitsis, and K. Walsh, *Circulation*, **101**, 660–667 (2000).

397. F. Cacciapuoti, R. Spiezia, M. Bianchi, D. Lama, M. D'Auino, and M. Varrichio, *Am. J. Cardiol.*, **67**, 843–847 (1991).

CHAPTER FOUR

Endogenous Vasoactive Peptides

James L. Stanton
Randy L. Webb
Metabolic and Cardiovascular Diseases
Novartis Institute for Biomedical Research
Summit, New Jersey

Contents

Burger's Medicinal Chemistry and Drug Discovery
Sixth Edition, Volume 3: Cardiovascular Agents and
Endocrines
Edited by Donald J. Abraham
ISBN 0-471-37029-0   © 2003 John Wiley & Sons, Inc.

1   INTRODUCTION

The control of vasomotor tone regulates the overall cardiovascular system and its responses to physiological and pathophysiological stress. It provides differential blood flow to regional beds at times of normal stress, such as exercise, and during conditions of pathological stress, such as congestive heart failure. Changes in vascular tone permit rapid and sometimes large shifts of blood volume to meet the needs of the body, for example, during hemorrhage. Although the vascular system regulates many of its functions at the site of the blood vessel itself, other factors that may control vascular tone include the autonomic nervous system, circulating hormones functioning in an endocrine role, and reflex metabolic control. For many years, studies on neurohormonal control of the vasculature have been dominated by the role of catecholamines and acetylcholine. Over the past two decades, new and improved techniques in immunohistochemistry, molecular biology, chromatography, electrophysiology, and pharmacology have led to a wealth of discoveries that have dramatically reshaped our understanding of the autonomic nervous system. With the isolation and characterization of neuropeptides such as substance P and vasoactive intestinal peptide in the 1970s and 1980s, there came revitalization in attempts to identify nonadrenergic, noncholinergic transmitters. Indeed, many newly discovered peptides were found to be widespread in autonomic and sensory neurons innervating the vasculature, and they had potent vasoactive effects on many blood vessels (1–3). As more neuropeptides were isolated throughout the 1980s, the number of peptides found to occur in peripheral autonomic or sensory neurons continued to increase. The existence of more than one transmitter substance in some nerves, cotransmission, is now also widely recognized (4, 5).

The major role of neuropeptides in the vasculature appears to be mainly modulatory, such that they may produce short- or long-term influences on the release or action of other neurotransmitters, either directly or indirectly. Furthermore, there is increasing evidence that endothelial cells, forming the innermost layer of blood vessels, play a crucial role in the vasodilatory response of the vessel to acetylcholine and to a growing number of other substances (6–9). Considerable research into vascular control mechanisms has led to the concept of dual regulation of blood vessel tone, whereby both nerves and the endothelium are involved. There is also considerable evidence supporting the involvement of endogenous vasoactive peptides in a variety of other physiological activities, including pain transmission, behavioral actions, and metabolic processes. Vasoactive peptides have been shown to play an important role in many pathological situations, for example, asthma, arthritis, vascular headaches, hypertension, and other cardiovascular diseases. In this chapter we outline the evidence for the different actions of various vasoactive peptides, their interaction with other systems, the possible role they have in diseases, and the present and future therapeutic use of vasoactive peptide agonists and antagonists. Attention will be given largely to those peptides most prominent in their vascular effects as vasoconstrictors, including angiotensin, arginine vasopressin, endothelins, neuropeptide Y, and urotensin II, or as vasodilators, including natriuretic peptides, calcitonin gene-related peptide, tachykinins, bradykinin, and vasoactive intestinal peptide. Other endogenous peptides, which possess vascular activity, are also discussed. These include somatostatin, gastrin-releasing peptide, neurotensin, and relaxin.

2 ANGIOTENSIN II

The octapeptide angiotensin II (Ang II) is the principal effector of the renin–angiotensin system (RAS) and plays an important role in the control of blood pressure and in the pathogenesis of a variety of cardiovascular diseases, including hypertension. Aspects of the RAS were discovered more than 100 years ago by Tigerstedt and Bergmann, who described a pressor system originating in the kidney (10).

2.1 Biosynthesis and Metabolism

The synthesis and metabolism of Ang II are outlined in Fig. 4.1 (11–13). The process is initiated when the enzyme renin acts on circulating angiotensinogen to release the decapeptide angiotensin I (Ang I). Ang I, which has no intrinsic activity, is rapidly converted to the octapeptide Ang II through the action of angiotensin-converting enzyme (ACE). Ang II is subsequently metabolized by aminopeptidases to yield [angiotensin$_{2-8}$] (Ang III) and [angiotensin$_{3-8}$] (Ang IV). Ang I can also be cleaved to produce [angiotensin$_{1-7}$]. Ang III, Ang IV, and [angiotensin$_{1-7}$] show some biological activity, but their relative importance is still under investigation (13–18).

Various stimuli, such as hyponatremia, hypovolemia, and hypotension, cause renin to be released within the kidney and into the circulation. Renin release is mediated by means of intrarenal vascular and tubular mechanisms, as well as the sympathetic nervous system. Ang II itself exerts a negative feedback inhibition on renin release and also synthesis (19). Extrarenal tissues, such as the brain, heart, adrenals, and blood vessels, have been shown to contain renin or renin-like enzyme or mRNA for renin, but at much lower levels than those in the kidney (20–22). The kidney is the major source of plasma renin. Renin specifically hydrolyzes the amide bond between the amino acids in positions 10 and 11 of the N-terminus of angiotensinogen to form Ang I. Angiotensinogen, the only known physiological substrate for renin, is an α-2-globulin that is synthesized in the liver and released into the circulation. Ang I is converted into the biologically active octapeptide, Ang II, by angiotensin-converting enzyme (ACE, also called kininase II), a zinc-containing enzyme found in many tissues. ACE, unlike renin, has a broad substrate specificity and cleaves several other peptides (e.g., bradykinin, substance P).

2.2 Structural Features

Many analogs of Ang II have been synthesized, and considerable information on the struc-

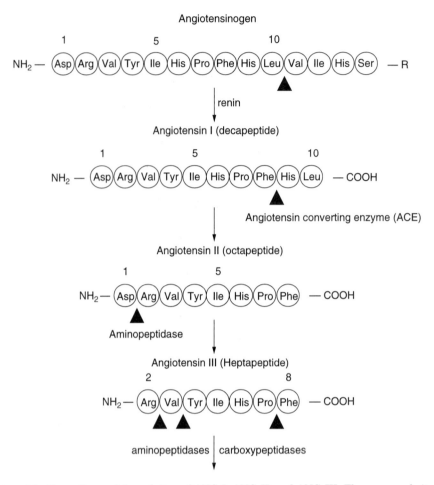

Figure 4.1. Formation and degradation of ANG I, ANG II, and ANG III. The presumed sites of action of the specific peptidase enzymes are indicated by arrowheads.

ture–activity relationship (SAR) is available (23, 24). As is evident from the activity of Ang III, most of the essential information resides in the C-terminal heptapeptide. Phenylalanine in position 8 is critical. Its removal from any of the angiotensins abolishes agonist activity. The other aromatic residues in positions 4 and 6, the guanido group in position 2, and the C-terminal carboxyl are involved mainly in binding to the receptor site. Position 1 is not critical, but replacement of aspartic acid in position 1 with sarcosine (*N*-methylglycine) enhances binding and slows hydrolysis by aminopeptidases. Substitution at either position 4 or position 8 within the Ang II sequence has produced analogs with antagonistic properties.

2.3 Receptors

Ang II exerts its wide variety of physiological actions by binding to specific receptors located in the plasma membranes of various tissues (25). In humans, two receptor subtypes have been identified, designated AT_1 and AT_2.

2.3.1 AT_1 Receptors. The AT_1 receptor is a member of the seven transmembrane G-protein–coupled receptor family. Binding of Ang II to the AT_1 receptor initiates a series of events, including stimulation of phospholipase C and subsequent activation of protein kinase C and release of intracellular calcium (26, 27). Specific AT_1 receptors have been identified in multiple tissues, including vascular smooth muscle, adrenal cortex and medulla, kidney, uterus, brain, liver, and heart (25, 26). The AT_1 receptor has been cloned (28, 29). Mice lacking the AT_1 gene (30) show reduced blood pressure, increased water uptake, and increased urine excretion. Mice overexpressing the AT_1 gene (31, 32) show myocyte hyperplasia and lethal heart block. Most of the physiological responses of Ang II are mediated through the AT_1 receptor (25, 33, 34). In rodents, but not humans, two forms of AT_1 (AT_{1A} and AT_{1B}) have been identified and are pharmacologically indistinguishable (35).

2.3.2 AT_2 Receptors. The AT_2 receptor subtype was identified with radioligand binding studies (36, 37). The AT_2 receptor is also a member of the seven transmembrane G protein-coupled receptor family (38). It is coupled to various tyrosine and serine/threonine phosphatases and mediates protein dephosphorylation (39). The AT_1 and AT_2 receptors have only 34% sequence homology (38). The AT_2 receptor is expressed during fetal development but not during normal adult life (40, 41), and may play a role in fetal growth and differentiation. The receptor has been identified at low density in adults in adrenal glands, brain, uterus, kidney, and heart (42, 43). Under pathological conditions, such as vascular injury, heart failure, and myocardial infarction, AT_2 receptors are reexpressed in adults, and may oppose the pathologic effects induced by the AT_1 receptors. AT_2-deleted mice have been generated and show increased blood pressure and salt retention (44, 45).

2.4 Biological Actions

Ang II influences blood pressure by acting at various sites, including vascular smooth muscle, kidneys, adrenals, and the central and peripheral nervous systems (11, 12). It is a very potent vasoconstrictor, particularly of the renal vasculature, and, therefore, can directly increase vascular resistance. The effects of Ang II on the kidney include modulation of renal blood flow, glomerular filtration rate, tubular sodium and water reabsorption, and inhibition of renin release (46, 47). In the adrenal cortex Ang II acts directly on the zona glomerulosa to stimulate aldosterone, a hormone with potent sodium-retaining effects on renal tubules. The effects of Ang II on the peripheral autonomic nervous system include sympathetic stimulation, facilitation of adrenergic transmission, and adrenal medullary release of noradrenaline. These multiple neuroendocrine actions of Ang II contribute to arterial vasoconstriction and blood pressure regulation. In the central nervous system Ang II mediates a direct pressor response by increasing efferent sympathetic activity and stimulating the release of vasopressin (48, 49). Ang II may also influence blood pressure by contributing to both vascular and cardiac hypertrophy (50). In addition, Ang II has a direct positive inotropic action on the heart. It also has been linked to cardiac fibrosis (51, 52), oxidative stress (53), and atherosclerosis (54, 55). All of these effects are mediated through the AT_1 receptor. The cardiovascular functions produced by the AT_2 receptor include suppression of cell proliferation (56), apoptosis (57, 58), and stimulation of renal NO production (59).

2.5 Inhibitors and Antagonists of the Renin-Angiotensin System

Angiotensin-converting enzyme (ACE) inhibitors were the first commercially available blockers of the RAS. More recently, AT_1 receptor antagonists have been marketed and renin inhibitors have been studied clinically. These agents have been used primarily for the treatment of hypertension and congestive heart failure.

2.5.1 Renin Inhibitors. The enzyme renin, an aspartyl endoprotease, is highly specific for the hydrolysis of the Leu^{10}–Val^{11} bond of angiotensinogen, so inhibition of renin produces selective blockade of the RAS (60). A number of orally active, nonpeptidic renin inhibitors have been described (61) and several have been tested in humans. Ro42-5892 (**1**) and CGP 38560A (**2**) (Fig. 4.2) have been shown to lower blood pressure in primates (62, 63) and humans (63–65). To date, no renin inhibitors have reached the market, primarily because of rapid elimination *in vivo* and cost of synthesis. However, recent reports indicate that these issues may be surmountable (66). See Fig. 4.2, structures (**3**) and (**4**).

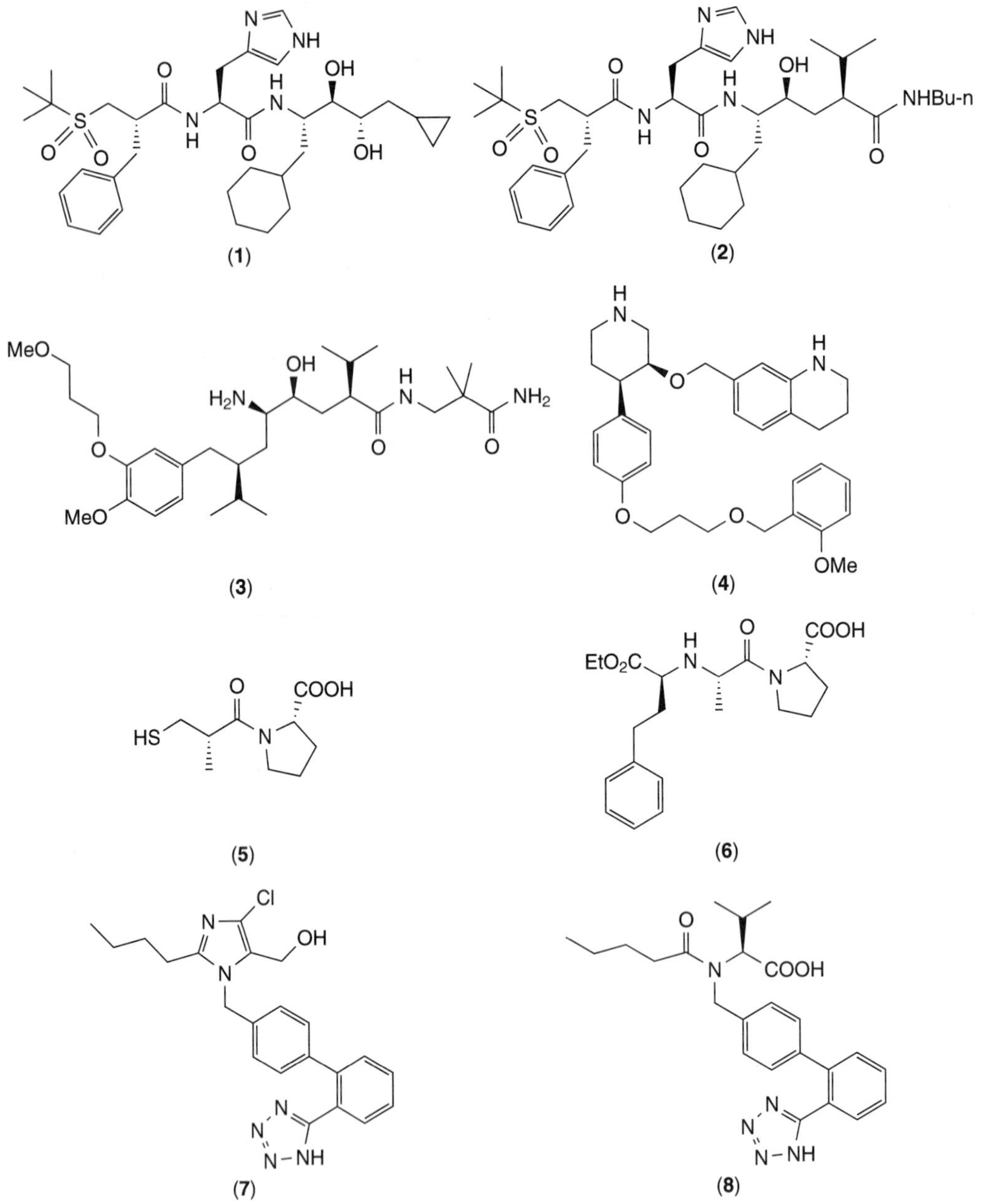

Figure 4.2. Chemical structures of Ro42-5892 (**1**), CGP 38560A (**2**), compound 10 (**3**), compound rac-30 (**4**), captopril (**5**), enalapril (**6**), losartan (**7**), and valsartan (**8**).

2.5.2 ACE Inhibitors. The enzyme ACE is a zinc-containing metalloprotease. ACE inhibitors block the conversion of Ang I to Ang II. The first commercially available ACE inhibitors were captopril (67) and enalapril (68)

(Fig. 4.2). Since then, many additional ACE inhibitors have become available (69).

2.5.3 AT_1 Receptor Antagonists. Saralasin, a peptide analog of Ang II, was the first re-

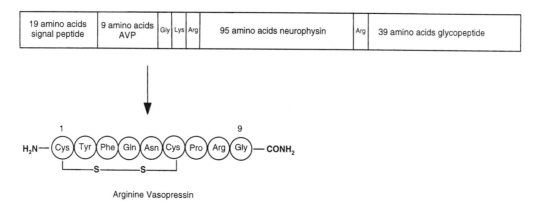

Figure 4.3. Schematic illustration of bovine prepro-vasopressin structure and the primary structure of arginine vasopressin.

ported angiotensin receptor antagonist (70). However, because of poor bioavailability, short duration of action, and partial agonist effects, saralasin is not used clinically. In 1982 Furukawa and coworkers discovered some benzyl-substituted imidazoles that were specific Ang II antagonists (71). These compounds served as leads to develop losartan (72), the first commercially available AT_1 antagonist, and valsartan (73) (Fig. 4.2). Since then, several other AT_1 antagonists have appeared (73–75). These agents are as effective as ACE inhibitors, but avoid some of their mechanism-related side effects, such as cough, rash, and angioedema.

3   ARGININE VASOPRESSIN

Arginine vasopressin (AVP) was originally identified more than 100 years ago as a pituitary gland extract that was shown to be a potent vasoconstrictor in the anesthetized dog (76). Vasopressin is a nonapeptide (Fig. 4.3), was first isolated and characterized 40 years ago (77), and possesses potent antidiuretic and vasopressor activity (78). It was the first peptide hormone to be chemically synthesized (79). Since then, significant progress has been made in our understanding of the actions of vasopressin as a result of the development of specific receptor agonists and antagonists (80).

3.1   Biosynthesis, Metabolism, and Structural Features

AVP is synthesized as a larger prohormone in the hypothalamus and is carried by axonal transport to the posterior pituitary, where it is stored for later release. The vasopressin gene is divided into three exons by two short intervening sequences (81). Gene transcription leads to a long pre-mRNA, which is then spliced across the two introns to yield the final translatable mRNA of about 750–800 bases. The mRNA is then translated to give the initial precursor or "preprohormone." This preprohormone consists of the following peptide sequences beginning at the N-terminus: a signal peptide required for further processing; vasopressin; a tripeptide-bridging structure; neurophysin; and a single arginine residue joining the neurophysin sequence to a 39 amino acid residue that will become a glycopeptide (Fig. 4.3) (81). This precursor is packaged into secretory granules, is subjected to specific cleavage and modification, and is transported from the cell body to the nerve terminal (82). Cleavage occurs at the paired basic amino acids -Lys-Arg- separating Gly^{10}-vasopressin from neurophysin. The C-terminal glycine then acts as a nitrogen donor to furnish the amidation of the physiologically active nonapeptide vasopressin. Vasopressin deficiency is the hallmark feature of human familial diabetes insipidus. This disease, which was originally thought to result from an inability to synthesize vasopressin, more specifically occurs because of gene mutations of the signal peptide and in neurophysin, thus indicating the importance of these peptide fragments in the appropriate expression and secretion of a physiologically active vasopressin (83). Recently, it has become apparent that vasopressin also may act in a paracrine or autocrine manner. Vasopressin and its mRNA have been identified in various peripheral tissues, including the adrenal gland, testis, ovary, thyroid gland, spleen, pancreas, heart, and intestine (84, 85).

Once released, AVP is subject to proteolysis catalyzed by peptidases. In the periphery, AVP is converted into inactive metabolites by peptidases of kidney and liver attacking its C-terminal portion or by an N-terminal cleaving enzyme in plasma (86, 87). In the brain N- and C-terminal cleaving activities appear to occur in different cellular compartments. C-terminal cleavage of AVP (Arg-Gly-NH$_2$) was seen in soluble fractions of brain tissue (88), and N-terminal cleavage (release of Tyr^2) predominated in membrane preparations (89). Similar to AVP's proteolytic susceptibility in the periphery, N-terminal cleavage in the brain is mediated by aminopeptidase activity. Trypsin and carboxamido-peptidase are candidates for cleavage of the C-terminal Arg-Gly-NH$_2$ bond (90, 91); the postproline cleaving enzyme can split the Pro^7–Arg^8 bond (92). *In vivo* studies have shown that the putative vasopressin fragments are relatively most abundant in extrahypothalamic brain regions and [vasopressin$_{4-9}$] and [vasopressin$_{5-9}$] amino acid fragments of vasopressin can account for up to 30% of the AVP content (93). The products of AVP processing have potent and selective central activities (94, 95).

The structure of AVP can be modified in a variety of ways to give analogs exhibiting increased antidiuretic potency, different antidiuretic/pressor selectivity, and increased duration of action. Examples of analogs exhibiting these different characteristics are reviewed elsewhere (96). One such modified peptide, desmopressin (dDAVP), has become the cornerstone for the treatment of bleeding disorders such as hemophilia A and von Willebrand disease (97). Other structural modifications

yield terlipressin, glypressin, and ornipressin, novel therapeutics to treat acute bleeding episodes, variceal bleeding associated with portal hypertension, and to act as a local vasoconstrictor during surgery (98–100). Recently, the first known vasopressin analogs to possess selective vasodepressor/hypotensive activity have been described (101). These compounds may prove useful in treating hypertension.

3.2 Receptors

Three distinct vasopressin receptors have been cloned, sequenced, and expressed and are designated V_{1a}, V_{1b}, and V_2 (102–104). Vasopressin receptors belong to the seven transmembrane G protein-coupled receptor family (105). Binding of vasopressin to its receptor results in receptor internalization and subsequent desensitization. Receptor recycling occurs in the case of the V_1 receptor but not with the V_2 receptor (105). The vasopressin V_2 receptor is expressed mainly in the medullary renal tubules, coupled to a cholera toxin-sensitive G protein Gs and activates cAMP (106). Activation of the V_2 receptor induces insertion of water channels (aquaporins) into the luminal surface of renal collecting tubules with a resultant increase in water reabsorption and is responsible for the well-known antidiuretic effects. The V_{1a} receptor, known as the vascular receptor, is expressed on vascular smooth muscle, spleen, adipocytes, brain, testis, liver, reproductive organs, and adrenal cortex (106). Vasopressin binding to the V_{1a} receptor activates phospholipases A$_2$, C, and D, generates inositol phosphates and diacylglycerol, stimulates protein kinase C, and mobilizes intracellular calcium and Na$^+$–H$^+$ exchanger (106). The classical vasopressor actions of vasopressin are mediated by the V_{1a} receptor. The pituitary V_{1b} receptor, also known as the V_3 receptor, potentiates the release of ACTH through activation of phospholipase C and protein kinase C (106). Finally, evidence has recently been put forward describing a novel dual angiotensin II/vasopressin receptor possessing equal affinity for vasopressin and angiotensin II and is expressed on medullary tubules (107).

3.3 Biological Actions

It is generally recognized that AVP plays a role in normal blood pressure regulation as well as in the pathogenesis of certain forms of experimental hypertension, especially low renin models such as the deoxycorticosterone acetate–salt model (108). On a molar basis AVP is a more powerful vasoconstrictor, through its action on endothelial vascular smooth muscle cells, than angiotensin II or catecholamines (109, 110). Pharmacological doses are required to achieve a significant increase in arterial blood pressure. It appears that the rise in blood pressure in response to physiological concentrations of AVP is blunted by an increase in total peripheral resistance and a fall in cardiac output (111–113). Baroreflex mechanisms appear to contribute to lowering of cardiac output in conscious dogs (112). AVP could also have negative inotropic actions that contribute to a fall in cardiac output (114). Activation of the V_2 receptor effectively opposes V_{1a}-mediated vasoconstriction and could account, in part, for the lack of sensitivity of blood pressure in response to increasing doses of AVP. Specifically, the integrated control of the renal medullary circulation appears to be intimately involved in the overall systemic hypertensive response to AVP (115). There are regional differences in blood flow responses to exogenous AVP. AVP infusion in dogs produced a decrease in blood flow to the skin, skeletal muscle, pancreas, thyroid gland, and fat, and to a lesser extent to myocardium, gastrointestinal tract, and brain. Some organs did not show any fall in blood flow at all, particularly kidney and liver (116). Long-term exposure to vasopressin has growth-promoting effects on the vasculature as well (117).

In addition to its peripheral vasoconstrictor and renal effects, AVP has been shown to produce cardiovascular effects that are mediated by the central nervous system (118). Vasopressin-containing cells and AVP receptors are located in areas known to be involved in cardiovascular regulation (119). Experimental evidence thus supports a central role for AVP in hypertension through its ability to increase sympathetic outflow and suggests an involvement in the circadian rhythm of blood pressure (119). Furthermore, a significant interaction between vasopressin and angiotensin II has been demonstrated in the central control of blood pressure (120).

The role of AVP in long-term regulation of blood pressure and in the pathogenesis of hypertension is still controversial. Although plasma levels of AVP have been reported to be increased in many models of hypertension, it has not been possible to demonstrate that chronic administration of AVP, at levels similar to those seen in hypertension, can sustain an elevated arterial pressure (121). Furthermore, it has not been possible to consistently demonstrate that administration of pressor antagonists of AVP causes more than a transient fall in blood pressure in any model of experimental hypertension (121). AVP acting on the V_2 receptor and/or alterations in medullary blood flow could account for an escape from chronic V_{1a} blockade (115, 122).

3.4 Antagonists

There have been substantial improvements in the design of specific antagonists for AVP acting through V_1 and V_2 receptors. Early success with the design of subtype-specific antagonists was achieved through the synthesis of peptidic analogs of AVP (101, 123). The design and synthesis of nonpeptidic AVP antagonists, either nonselective or subtype selective, have been only a recent innovation and initially were hampered by the high species differences observed in ligand selectivity (124). The first reported orally active, nonpeptidic antagonist was the V_{1a} antagonist, OPC-21268 (125). Subsequent to this disclosure, additional nonpeptidic antagonists have been reported with V_{1a}, V_2, and dual V_{1a}/V_2 antagonistic activity (Fig. 4.4) (126, 127). Notably, SR 121463A, a high affinity antagonist for the V_2 receptor (K_i = 0.26 ± 0.04 nM), is highly selective and orally active (127). To date, none of these nonpeptidic antagonists has reached the market, although numerous agents are in advanced stages of clinical testing. These antagonists already have been shown to be effective in animal models of heart failure and hyponatremic states, and are powerful aquaretic agents (128–131).

4 ENDOTHELIN PEPTIDE FAMILY

4.1 Structural Features

Endothelin (ET), identified several years ago as a potent vasoconstrictor, constitutes a fam-

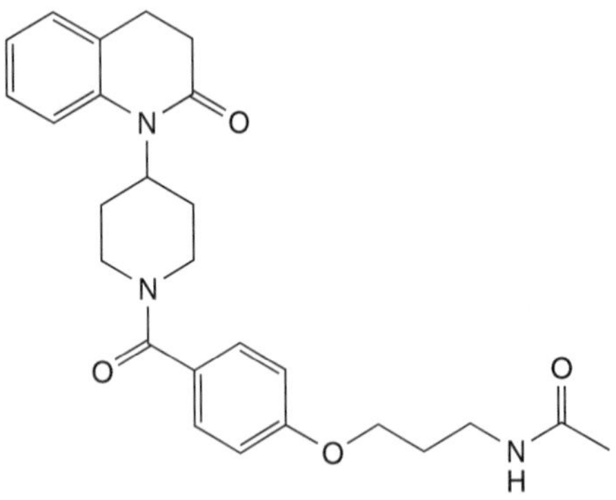

OPC-21268

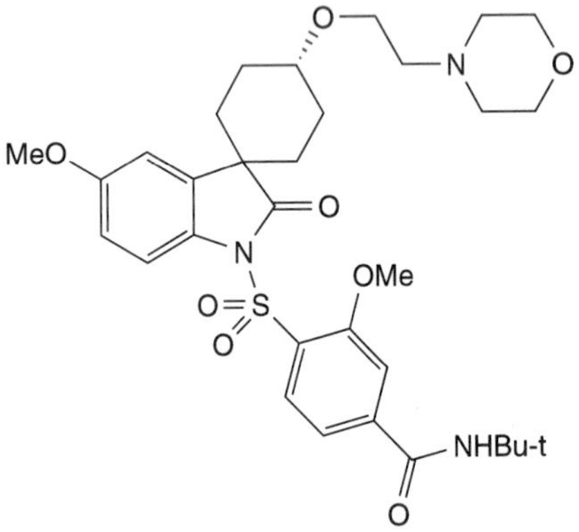

SR 121463

Figure 4.4. Chemical structures of OPC-21268 and SR 121463.

ily of structurally very closely related peptides (132–133), originally isolated from rodent sources (ET-3) as well as human and porcine sources (ET-1). To date, four distinct members of the family, endothelin-1 (ET-1), endothelin-2 (ET-2), endothelin-3 (ET-3), and vasoactive intestinal constrictor (VIC), have been identified in mammalian systems (Fig. 4.5) (134). Endothelins have a common structure of 21 amino acids with two cysteine–cysteine disulfide bonds between positions 1–15 and 3–11 and a hydrophobic C-terminus (residues 16–21) (Fig. 4.5). These peptides share sequence homology with sarafotoxins (Fig. 4.5) isolated from the venom of the burrowing asp, *Atractaspis engaddensis* (135, 136), which also shares most of their biological characteristics. The main differences between the endothelin isopeptides is in their amino *N*-terminal segments. The biochemistry and biological activity of the ETs can be found in several recent reviews (137–140).

4.2 Biosynthesis and Metabolism

Endothelin is encoded by a well-characterized gene located on human chromosome 6 (141). In endothelial cells, endothelin is synthesized as a 203 amino acid precursor protein called prepro-ET-1, which is cleaved by a dibasic pair-specific endopeptidase to form first pro-ET-1 or "big ET-1" a 38 (human) or 39 (por-

Endothelin-1 Cys-Ser-Cys-Ser-Ser-Leu-Met-Asp-Lys-Glu-Cys-Val-Tyr-Phe-Cys-His-Leu-Asp-Ile-Ile-Trp (porcine/human/rat/canine)

Endothelin-2 Cys-Ser-Cys-Ser-Ser-**Trp**-**Leu**-Asp-Lys-Glu-Cys-Val-Tyr-Phe-Cys-His-Leu-Asp-Ile-Ile-Trp (human)

Endothelin-3 Cys-**Thr**-Cys-**Phe**-**Thr**-**Tyr**-**Lys**-Asp-Lys-Glu-Cys-Val-Tyr-**Tyr**-Cys-His-Leu-Asp-Ile-Ile-Trp (human/rat)

Vasoactive Intestinal Constrictor (VIC)
(mouse)

Cys-Ser-Cys-**Asp**-Ser-**Trp**-**Leu**-Asp-Lys-Glu-Cys-Val-Tyr-Phe-Cys-His-Leu-Asp-Ile-Ile-Trp

Sarafotoxin-6c (representative of the sarafotoxin family of peptides; S6-a,b,c,d)

Cys-**Thr**-Cys-**Lys**-**Asp**-**Met**-**Thr**-Asp-Lys-Glu-Cys-**Leu**-Tyr-Phe-Cys-His-**Gln**-Asp-Ile-Ile-Trp

Figure 4.5. Amino acid sequences of endothelin-1, endothelin-2, endothelin-3, vasoactive intestinal constrictor, and sarafotoxin-6c, one of the sarafotoxin family of peptides. The endothelin peptides all consist of 21 amino acids, with considerable homology between other members of this family of peptides, in which each peptide contains two essential intramolecular disulfide bonds. Amino acids differing from endothelin-1 are designated in bold.

cine) amino acid peptide (132) (Fig. 4.6). Big ET-1 is then converted to active ET-1 (21 amino acid residues) by a neutral metalloprotease termed endothelin-converting enzyme (ECE-1) (142). ECE has been cloned and several ECE-1 isoforms have been identified, designated ECE-1a, ECE-1b, and ECE-1c, and apparently a single gene encodes the various isoforms (135–138). Whereas ECE-1a is primarily located on the plasma membrane, ECE-1b is within the intracellular compartment and ECE-1c is expressed in the Golgi (146). ECE-1 is a highly glycosylated protein, is contained within the zinc metalloendopeptidase family that includes NEP 24.11, and may also be involved in the degradation of bradykinin (147).

The lack of secretory storage granules in endothelial cells suggests that ET-1 is made upon demand (148). In most cases, induction of ET-1 secretion above basal levels requires 2–5 h and most results from enhanced preproET-1 gene expression (149, 150). Endothelin mRNA in cultured endothelial cells is induced by several vasoactive substances including thrombin, angiotensin II, vasopressin, and epinephrine. The interaction between the endothelin and renin–angiotensin systems appears to be particularly crucial in the regulation of cardio-renal function (151). In addition, ET-1 release is induced by hemodynamic stimuli such as shear stress, as well as by ET-1 itself (for reviews, see Refs. 140, 152–154). In human coronary arteries, a dual-processing pathway has been postulated. In addition to the well-characterized constitutive pathway, a regulated secretory pathway may exist where endothelin is contained within Weibel–Palade bodies (155). Various factors activate phospholipase C, which results in the activation of protein kinase C through the formation of 1,2-diacylglycerol, and in mobilization of calcium ions through the production of inositol-1,4,5-trisphosphate. The production of ET-1 is also regulated by a negative pathway involving a guanosine 3′,5′-cyclic monophosphate (cGMP) mechanism. It was found that agents such as atrial natriuretic peptide, nitric oxide, and nitrovasodilators, which increase cGMP levels, reduce the secretion of ET-1 (156–158). Following intravenous administration to animals, radiolabeled ET-1

and ET-3 were found to be rapidly removed from the circulation with a half-life of less than 2 min. No ET catabolites were found in the blood, and the radioactivity was localized primarily in lungs, kidneys, and liver, suggesting that these organs are the principal sites of metabolism of ET peptides (159, 160). *In vitro* experiments have demonstrated that ET peptides can be degraded by a purified neutral endopeptidase 24.11. Circulating ET-1 can also be cleared by receptor-mediated uptake (161).

4.3 Receptors

Endothelin's actions are mediated by at least two subtypes of ET receptor. Two distinct ET_A and ET_B receptors have been cloned and exhibit 60% homology (162, 163). The predicted structure of both receptors is similar to seven transmembrane G-protein–coupled receptors. For the ET_A receptor, the rank order of potency is ET-1 > ET-2 ≫ ET-3 and the affinity of binding to ET-1 is about 100 times greater than that of ET-3 and subserves vasoconstriction (164, 165). ET_B receptors show a similar affinity for all three ETs and for the sarafotoxins and subserve vasodilatation as well as vasoconstriction (164, 166, 167). The distinction between ET_A and ET_B receptors has been confirmed by the recent development of selective agonists and antagonists. Radioligand binding studies and *in situ* hybridization studies with cDNA receptor probes have demonstrated that ET receptors are widely distributed, in keeping with the multiple actions of these peptides (132, 168, 169). Recently, the ET_B receptor has been shown to act as a clearance receptor for removing endothelin from the circulation (170). Subsequent to endothlein receptor stimulation, a complex cascade of second-messenger pathways becomes activated and contributes to the diversity of endothelin actions (171).

4.4 Biological Actions

4.4.1 Cardiovascular. After the discovery of ET-1 it became apparent that the endothelins were the most potent vasoconstrictor substances yet identified in vascular preparations from humans and experimental animals (132).

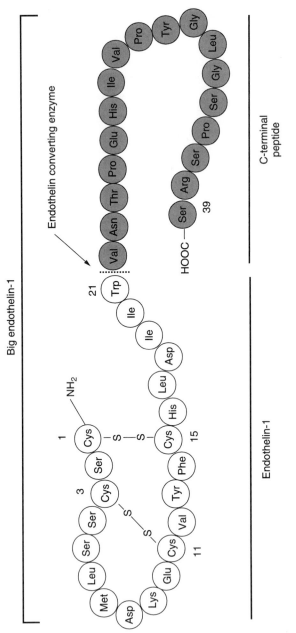

Figure 4.6. Structures of porcine big endothelin-1 and endothelin-1.

The vascular effects of endothelin are mediated by ET_A and ET_B receptors located primarily on vascular smooth muscle and endothelial cells, respectively (172–174). After intravenous administration, ET-1 elicits a characteristic initial, short-lived depressor response followed by a secondary and prolonged pressor response (132, 174). Prostacyclin and nitric oxide are predominantly responsible for the vasodilation, whereas the vasocontriction following ET_A activation is mediated by increases in intracellular calcium flux (175, 176). In addition to the well-characterized effects on vascular tone, additional vascular effects of endothelin include thrombogenesis, neutrophil adhesion, mitogenic effects, and alterations of vascular permeability (177, 178).

4.4.2 Endocrine Actions.

Endothelins appear in low concentrations in the blood circulation and act in a paracrine manner. Locally synthesized endothelins are secreted abluminally into the interstitial space, where they may act on nearby cells. For example, the vasoconstrictor actions of ET-1 *in vivo* are opposed by the concomitant release of circulating atrial natriuretic peptide (ANP) (179, 180), and locally produced prostacyclin and endothelium-derived relaxing factor (181), all of which are potent vasodilators and may play a role in modulating the pressor response to ETs. Endothelin gene expression is regulated by several vasoactive agents such as thrombin and adrenaline, which are known to stimulate phosphoinositide hydrolysis in endothelial cells (132). Angiotensin, vasopressin, calcium ionophore A23187, endotoxin, and hypoxia are also known to stimulate gene expression and release of endothelin (181). ET, in turn, can increase plasma levels of various vasoactive hormones including aldosterone and catecholamines (179, 180). Several studies have reported the inhibitory effect of ET on the release of renin *in vitro* (182, 183). *In vivo* studies, however, have shown a significant increase in plasma renin activity after intravenous infusions of pharmacological doses of ET (179, 180). There appears to be a complex interaction between the renin–angiotensin and endothelin systems under normal and pathological conditions (184).

4.4.3 Renal Actions.

Endothelins can cause a profound increase in renal vascular resistance, decrease in renal blood flow, and decrease in glomerular filtration in association with a decrease in sodium excretion and an increase in plasma renin activity (179, 180), effects that resemble those seen in acute renal failure. Infusion of antiendothelin antibodies has been found to exert a renal-protective effect in ischemia (185, 186). The renal vasculature is several times more sensitive to the constrictor effects of ET than are other vascular regions (187). Interestingly, in addition to ET_A-mediated vasoconstriction, activation of the ET_B receptor also has been shown to elicit renal vasoconstriction (188). A more detailed review of the renal actions of endothelin has been published (189).

4.5 Pathological Roles of Endothelin

The powerful vasoconstrictive and mitogenic properties of endothelins suggest that they may have an important role in the development of numerous cardiovascular diseases. Plasma endothelin concentrations are elevated in many pathological conditions including cardiac hypertrophy and failure, myocardial infarction, and in various acute and chronic renal pathologies (171, 190–192). A growing body of evidence supports a contributory role for endothelin in types I and II diabetes mellitus and its associated vascular complications (193, 194). The endothelin system is activated in septic and cardiogenic shock (195). The plasma ET-1 level associated with essential hypertension remains controversial. Higher plasma ET-1 levels have been measured in patients with essential hypertension than in normal individuals (196). However, endothelin acts primarily in an autocrine and paracrine manner where the levels determined in plasma may represent only spillover from tissues and, consequently, may serve as an unreliable diagnostic marker. Tissue levels would appear to provide a more accurate assessment. Increased endothelin production has been shown in endothelium of small arteries from hypertensive patients (197). Circulating and tissue levels of endothelin are elevated in patients with atherosclerosis and correlate with the severity of disease (198). Furthermore, endothelial cells, macrophages, and

Table 4.1   Endothelin Receptor Antagonists

Antagonist	Reference	Receptor	Rank Order of Affinity for ETs	References
BQ-123	165	ET_A	$ET\text{-}1 \geq ET\text{-}2 \gg ET\text{-}3$	(165, 166)
BQ-153	165	ET_A		
BQ-485	212	ET_A		
BQ-610	213	ET_A		
A-127722	214	ET_A		
A-216546	215	ET_A		
BE-18257A	165, 216	ET_A		
BE-18257B	165, 216	ET_A		
BMS 182874	217	ET_A		
EMD 122946	218	ET_A		
FR 139317	219	ET_A		
50-235	220	ET_A		
TAK-044	221	ET_A		
TTA-386	222	ET_A		
LU135252 Darusentan	223	ET_A		
ZD 1611	224	ET_A		
TBC-11251 Sitaxsentan	225	ET_A		
RPR 111844	226	ET_A		
IRL 1038	227	ET_B	$ET\text{-}1 = ET\text{-}2 = ET\text{-}3$	(165, 167)
BQ-788	162	ET_B		
RES 701-1	228	ET_B		
A192621	229	ET_B		
A308165	230	ET_B		
IRL 3630	231	Nonspecific ET_A/ET_B		
L-749,329	232	Nonspecific ET_A/ET_B		
PD 145065	233	Nonspecific ET_A/ET_B		
PD 142893	234	Nonspecific ET_A/ET_B		
Ro-46-2005	235	Nonspecific ET_A/ET_B		
Ro-47-0203 Bosentan	236	Nonspecific ET_A/ET_B		
Ro-61-0612 Tezosentan	237	Nonspecific ET_A/ET_B		
SB 209670	238	Nonspecific ET_A/ET_B		
SB 217242 Enrasentan	239	Nonspecific ET_A/ET_B		

smooth muscle cells, all of which are cellular components of atherosclerotic lesions, can express endothelin (199, 200). A primary role has been suggested for endothelin in the pathogenesis of pulmonary hypertension. Plasma levels of endothelin are elevated in primary pulmonary hypertension (201) and ET-1 expression is increased in small muscular pulmonary arteries from patients with pulmonary hypertension (190, 202). The observation that endothelin levels are increased in various forms of cancer and the demonstration of a potent mitogenic effect of endothelin has led to speculation of ET-1 involvement in angiogenesis (203). In addition to direct effects on cell growth, endothelin may alter mitogenesis

indirectly through stimulation of vascular endothelial growth factor (204).

4.6   Antagonists and Inhibitors

Considering the potential importance of ETs in various physiologic and pathophysiologic conditions, a major effort has been made to develop specific receptor antagonists to modify the actions of ET. More recently, the characterization of the enzymes responsible for generating ET has led to the design of specific enzyme inhibitors.

4.6.1 Antagonists. Various compounds have been reported in recent years as ET_A- or ET_B-selective and as mixed $ET_{A/B}$ receptor an-

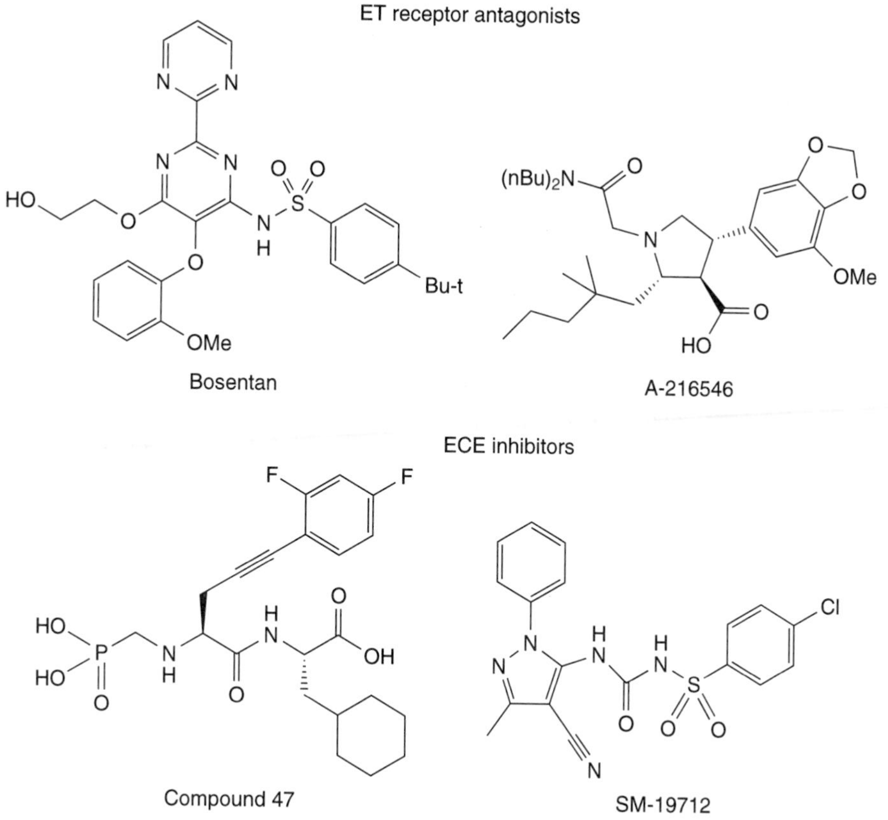

Figure 4.7. Structures of key ET receptor antagonists and ECE inhibitors.

tagonists (Table 4.1 and Fig. 4.7). Several comprehensive reviews that define the structural requirements necessary for receptor subtype specificity and their overall activity profiles have been published (205–207). The most clinically advanced compound, bosentan, is a mixed antagonist and is being developed for heart failure and pulmonary hypertension (208). Bosentan competitively antagonizes the binding of radiolabeled endothelin to ET_A and ET_B receptors with K_i values of 4.7 and 95 nM, respectively. Bosentan is well absorbed (50%) and peak plasma levels are achieved in humans between 2 and 3 h after oral administration (208). Bosentan reduced systemic vascular resistance in hypertensive and heart failure patients and had a significant benefit in the pulmonary vascular bed. Bosentan has recently been approved for the treatment of pulmonary hypertension. Bosentan is also being

evaluated clinically for the treatment of hypertension, congestive heart failure, and subarachnoid hemorrhage (208). Evidence for a role of endothelin in various cardio-renal diseases, including hypertension, heart failure, diabetes, and renal failure, has been strengthened by results from extensive preclinical and clinical testing of endothelin antagonists (209–211).

There also are other classes of drugs that interfere with endothelin release or modify its action, such as angiotensin-converting enzyme inhibitors, calcium-channel blockers, and angiotensin II receptor antagonists (140). Endothelin antagonists have great potential for use in the treatment of patients with cardiovascular pathology, including systemic hypertension, myocardial ischemia, restenosis after angioplasty, and congestive heart failure. Results of various ongoing clinical trials are expected to further clarify our understanding of the role of endothelin in disease processes.

4.6.2 Endothelin-Converting Enzyme Inhibitors. Although a specific enzyme processing step was hypothesized at the time of the initial description of endothelin (132), the isolation and characterization of ECE occurred some years later. ECE was described as a membrane-associated, glycosylated metalloprotease of greater than 100 kDa (240, 241). ECE has been purified from bovine endothelial cells, denoted as ECE-1, and is expressed intra- and extracellularly (242). ECE is highly related to another metalloprotease, neutral endopeptidase 24.11 (NEP) (243). Consequently, the development of selective endothelin-converting enzyme (ECE) inhibitors has proven to be a significant challenge. The first enzyme inhibitors to be produced were peptidic and were isolated from fermentation broths, FR901533 (244), B-90063 (245), and TMC-66 (246). Initial efforts with nonpeptidic ECE inhibitors demonstrated nanomolar potency; however, significant inhibitory activity against NEP 24.11 was retained (247). More recently, a nonpeptidic ECE inhibitor, SM-19712, has been described (248) with high potency against ECE (IC$_{50}$ = 42 nM) and high selectivity (inactive against ACE and NEP at 100 μM). See Figure 4.7. Early results from *in vitro* and *in vivo* test systems confirm the expected biological efficacy of ECE inhibition (249–251).

5 NEUROPEPTIDE Y

5.1 Structural Features

Neuropeptide Y (NPY), a 36 amino acid peptide isolated in 1982 (252), is a potent vasoconstrictor as well as having significant effects on feeding, memory, intestinal secretions, and adrenocortical function. It is a member of the pancreatic polypeptide family (PP-family). It derives its name from having amino- and carboxy-terminal tyrosine residues (single letter code Y) (Fig. 4.8). Neuropeptide Y is widely distributed in the central and peripheral nervous system of numerous mammalian species including humans (253) and the peptide is co-

stored with different transmitters such as noradrenaline (254). Two structurally related peptides in mammals, peptide YY (PYY) and pancreatic polypeptide (PP) (Fig. 4.8), are also members of the PP-family. PYY was originally purified from pig intestine (255). PYY is produced by endocrine cells of the intestine and the pancreas of all vertebrates investigated (see Ref. 256 and references therein). Pancreatic polypeptide (PP) was originally discovered in chicken pancreas (257) and has subsequently been isolated or cloned in several tetrapods. PP appears to be expressed exclusively in endocrine cells of the pancreas (258). Each member of the PP-family consists of two antiparallel helices: a polyproline-like helix (residues 1–8) that through hydrophobic residues is connected with an amphiphilic α-helix (residues 15–30), creating a hairpinlike loop. The three-dimensional structures are supported by X-ray crystallographic data of avian PP (259) as well as NMR, circular dichroism, and molecular modeling (260, 261).

NPY and PYY are only about 50% homologous with PP in the primary structure; nevertheless, all three peptides retain the key amino acids necessary to hold the distinct tertiary structure (260). The structure of the PP-family is stable in aqueous solution (260, 262). The folded structure results in close association of the carboxy- and amino-terminal parts of the molecule, which is important for receptor recognition (263). The amino acid composition of the individual peptides is well conserved, both during evolution and among species (264, 265). In the case of NPY, the only identified substitution between species is that the methionine at position 17 in human, rats, guinea pigs, and rabbits is replaced by leucine in pigs (265). Shortly after their discovery, it was observed that the entire NPY and PYY molecules were necessary to evoke vasoconstriction, whereas N-terminally truncated forms of NPY and PYY were quite effective in suppressing sympathetic nerve activity.

5.2 Biosynthesis and Metabolism

The DNA that codes for NPY is located on chromosome 7 in humans (266). The corresponding mRNA contains a single open reading frame for a prepro-NPY sequence consisting of 97 amino acids (267). Cleavage of the

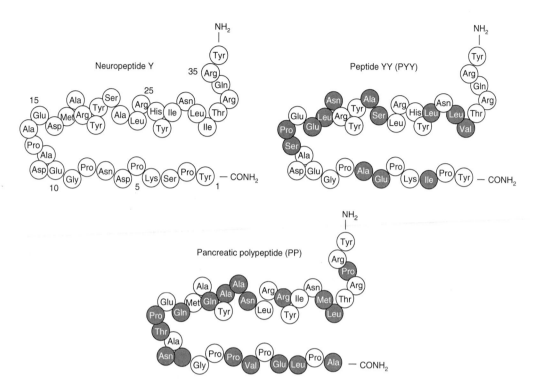

Figure 4.8. Amino acid sequences of human neuropeptide Y, peptide YY, and pancreatic polypeptide. Amino acids differing from neuropeptide Y are indicated by filled circles.

signal peptide yields pro-NPY consisting of a 69 amino acid sequence. This prohormone is further processed by a convertase, which splits off the 30 amino acid sequence containing the CPON (C-flanking peptide of NPY). Two additional enzymatic modifications lead to the mature NPY, which is C-terminally amidated (Fig. 4.9).

Like other peptide hormones, NPY is synthesized in the endoplasmic reticulum of the neuronal Golgi apparatus. NPY migrates by way of specific vesicles along the axon and is released into the synaptic cleft as a result of nerve impulses. Peripherally, it is released from predominantly postganglionic, sympathetic neurons. NPY is localized in larger vesicles; high frequency stimulation leads to the migration of these vesicles within the presynaptic membranes and the release of NPY (268). This release is regulated by presynaptic NPY autoreceptors, which can also inhibit the release of noradrenaline, and by presynaptic α_2-adrenergic receptors. Released NPY and PYY may also be exposed to further degrada-

tion to a N-terminal truncated NPY_{3-36} by dipeptidyl peptidase IV. This cleavage of NPY and PYY results in inactivation of these peptides with respect to biological effects mediated by the Y_1 receptor subtype, but not by Y_2 receptor subtype (269).

5.3   Receptors

NPY elicits its physiological effects by activating at least six receptor subtypes, Y_1–Y_5 and y_6 (270). These receptors belong to the family of seven transmembrane G-protein–coupled receptors and their activation leads to inhibition of adenylate cyclase (271). Receptors Y_1, Y_2, Y_4, Y_5, and y_6 have been cloned (270).

5.3.1 Y_1 Receptors. The NPY Y_1 receptor is located in distinct regions of the brain and in vascular smooth muscle cells of most blood vessels. [Pro^{34}] NPY has high affinity for Y_1, whereas PP and N-terminally truncated NPY derivatives bind with low affinity (270–274). Y_1 elicits pressor and contractile effects on

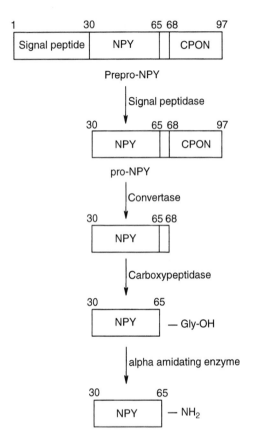

Figure 4.9. Biosynthesis of NPY. CPON, C-flanking peptide of NPY.

diates presynaptic actions involving the suppression of neurotransmitter release (287).

5.3.3 Y_3 Receptor. The Y_3 receptor recognizes NPY but not PYY (288). Central administration of NPY and related peptides into the nucleus of the tractus solitarius of rat brain evoked changes in cardiovascular responses associated with activation of Y_3 receptors (289, 290). Y_3 receptors have been identified in the bovine adrenal medulla (291), in rat cardiac membranes (292), heart, adrenal gland, brain stem, and hippocampus (269, 271, 272).

5.3.4 Y_4 Receptor. The Y_4 receptor has a high affinity for PP and low affinity for NPY, and is probably not an NPY receptor (271, 272).

5.3.5 Y_5 Receptor. The Y_5 receptor has a high affinity for NPY, PP, and N-terminally truncated forms of NPY (269, 270, 293). The Y_5 receptor is localized in the brain as well as in testis and pancreas (293, 294). With Y_1, the Y_5 receptor appears to mediate NPY effects on food intake (277, 293, 295).

5.3.6 y_6 Receptor. This receptor is written in lowercase to indicate that it is a nonfunctional pseudogene in rats and primates because of a single base deletion in the sixth transmembrane region leading to a truncated receptor (296, 297).

5.4 Biological Actions

5.4.1 Cardiovascular. NPY can regulate cardiovascular function through the central nervous system by a specific action on various nuclei (298). In the periphery NPY can modulate cardiovascular function by acting on the vasculature, heart, and kidney. In general, NPY can act at a pre- or postsynaptic level. Presynaptic NPY receptors exist in the vasculature, heart, and kidney, where they mediate inhibition of noradrenaline release as well as their own release from sympathetic nerve endings (299, 300). Presynaptic NPY receptors may also exist on parasympathetic nerve endings (e.g., in the heart), where they inhibit the release of acetylcholine (301, 302). The postsynaptic effects of NPY at the neurovas-

vascular smooth muscle (275, 276), as well as mediating (with the Y_5 receptor) NPY effects on feeding (277).

5.3.2 Y_2 Receptor. The Y_2 receptor is predominantly localized on the presynaptic membranes of postganglionic, sympathetically innervated neurons of the peripheral nervous system (278, 279). Y_2 receptors have also been identified in the proximal tubule of kidney (280), in parasympathetic neurons (281), and in red blood cells (282). Centrally, Y_2 receptors seem to be the dominant NPY receptor, which are particularly numerous in the hippocampus (283, 284). The Y_2 receptor has high affinity for PYY and N-terminally truncated forms of NPY (285). As for the Y_1 receptor, C-terminal amidation of NPY is essential for activity (286). Among its actions, the Y_2 receptor me-

cular junction can be further divided into direct and indirect (i.e., potentiating effects) (303). Thus, NPY can cause direct vasoconstriction of some vessels, most notably the coronary, cerebral, mesenteric, and renal vascular beds, but can also potentiate the vasoconstricting effects of agents such as noradrenaline, angiotensin, or serotonin in some vessels, including those in which NPY does not cause direct vasoconstriction. Direct vasoconstricting effects of NPY have been documented in isolated vessel preparations, isolated tissues as well as *in vivo* in conscious, anesthetized, and pithed animals, and *in situ* in the human forearm (289, 304–306). Direct vasoconstricting effects of NPY have been found in arteries as well as veins, although NPY may be a more effective vasoconstrictor of arteries than of veins (307). The direct vasoconstricting effects of NPY are predominantly mediated by Y_1 receptors and frequently involve the influx of extracellular calcium through voltage-operated channels. In view of its powerful vasopressor effect *in vivo*, NPY is a surprisingly poor vasoconstrictor *in vitro*. Large blood vessels usually respond poorly to NPY (308), but it is possible that the responsiveness to NPY may increase with reduced vessel diameter (309). Alternatively, a vasoconstrictor tone may be required (e.g., by activation of α-adrenoceptors) to enable isolated blood vessels to respond to NPY (289, 310); this is compatible with the observation of reciprocal facilitation of NPY and noradrenaline (310). The effectiveness of the two agents may be enhanced at the sympathetic neuroeffector junction. NPY is also capable of inhibiting the vasorelaxing effects of agents such as acetylcholine (289, 311). Thus, NPY may be involved in the regulation of vascular tone at several autonomic neuroeffector junctions where the peptide seems to act in concert with other transmitters.

The cardiac effects of NPY are complex (304). NPY can affect cardiac function indirectly through central effects, by alterations in afterload secondary to its direct and potentiating vasoconstricting effects, and by alterations in preload, which may be enhanced because of venous vasoconstriction or reduced secondary to histamine release. At least three possible sites of action exist within the heart.

First, NPY can cause powerful coronary vasoconstriction. Second, NPY may act presynaptically to inhibit the release of cardiac autonomic neurotransmitters. Finally, NPY may act directly on cardiac muscle to alter contraction, chronotropy, and electrical conduction acutely, and to affect protein synthesis and hypertrophy chronically (312). It is well documented that systemic administration of NPY reduces cardiac output, and all of the above cardiac and extracardiac effects of NPY may potentially participate in this reduction (304). NPY binding at the Y_1 receptor has been implicated in blood pressure control (313), and appears to exert its hypertensive effects under conditions of high sympathetic nervous system activity, particularly in response to stress (314). Stress, which is generally thought to contribute to the development of primary hypertension, raises the plasma levels of NPY and noradrenaline in humans (315). Sympathetic nerve activity is often increased in hypertensive patients and plasma levels of NPY and noradrenaline are elevated compared with those of normotensive controls (316). These observations, suggesting differences in central and peripheral NPY system between normotensive and hypertensive individuals, indicate that NPY may play an important role in primary hypertension.

5.4.2 Renal. Modulation of renal function by NPY can also occur at multiple levels, which include indirect effects through systemic hemodynamics, the central nervous system, or modulation of release of hormones affecting renal function, and direct effects on renal blood flow and tubular function (317). NPY can reduce renin release and plasma renin activity (318–322). Systemic NPY administration has been reported to elevate plasma levels of atrial natriuretic peptide (323). NPY reduces renal blood flow in a variety of species and experimental settings (321, 324, 325), an effect that, at least in rats, occurs by way of a Y_1-like receptor (322, 326, 327). In contrast to the antidiuretic effects of other renal vasoconstricting neurotransmitters and hormones, NPY enhances diuresis and natriuresis *in vivo* in anesthetized rats (325, 328).

Together, these data demonstrate that NPY can modulate cardiovascular function in multiple ways by acting on specific brain areas as well as on peripheral tissues. The net effect of these actions depends on the spatial and temporal release pattern of NPY, and on the absence or presence of other neurotransmitters and hormones that may be potentiated in their cardiovascular effects or the release of which may be inhibited by NPY. Moreover, NPY not only affects acute cardiovascular events, but might also be involved in chronic processes such as the development of hypertrophy, hyperplasia, or both.

5.4.3 Other Effects. NPY elicits other significant biological effects, showing antiepileptic (329), anxiolytic (330), and angiogenic (331) activity. It regulates a variety of pituitary hormones (332) and adrenocorticoid function (333). In particular, NPY stimulates feeding and appears to play a key role in the regulation of eating behavior (334, 335).

5.5 Antagonists

Primarily because of their potential antiobesity properties, industrial and academic laboratories have exerted considerable efforts to discover nonpeptidic NPY antagonists, with selectivity for the Y_1 or Y_5 receptor. The first reported Y_1-selective agent, BIBP3226, blocked NPY-induced vasoconstriction and pressor response (336) (see Fig. 4.10). CGP 71683A, a selective Y_5 antagonist, blocked NPY-induced feeding by 50% in rats (337). Since then, many more Y_1 and Y_5 antagonists have been reported (334, 338), and these should generate valuable data on the role of NPY in cardiovascular and metabolic diseases.

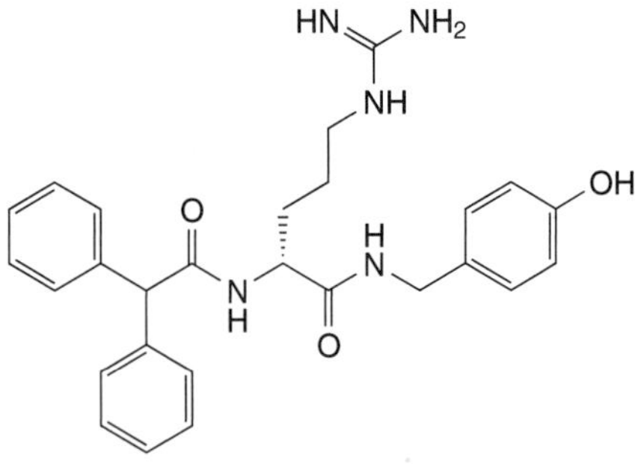

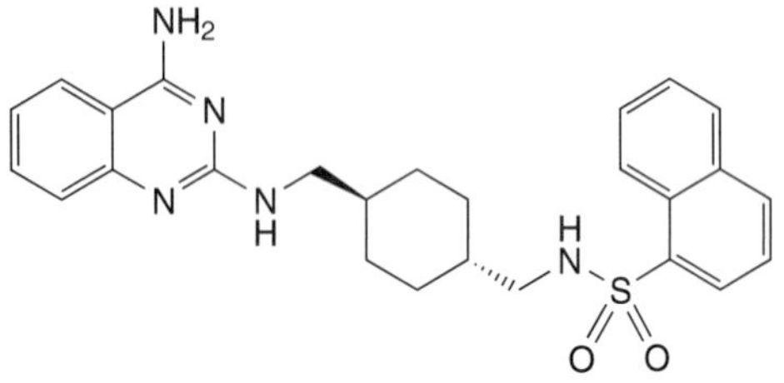

6 UROTENSIN-II

Urotensin-II (U-II), one of the most potent vasoconstrictors yet identified, was first characterized over 30 years ago in certain species of fish, where it was isolated from spinal cord tissue (339). At first U-II was thought to reside only in fish, but a human isoform was identified in 1998 (340).

6.1 Biosynthesis, Metabolism, and Structural Features

Human U-II is synthesized from prepro-U-II, a 124 amino acid peptide (340, 341), by a series

Figure 4.10. NPY Y_1 and Y_5 antagonists.

of posttranslational enzymatic cleavages at basic processing sites (Fig. 4.11) (342). U-II is located at the C-terminal portion of prepro-U-II. Expression of prepro-U-II mRNA has been detected in human brain (medulla oblongata), spinal cord, adrenal glands, and kidney (343–345). U-II, but not its mRNA, has been found in vascular and cardiac tissue (344). Rat, mouse, porcine, and human U-II have been cloned (346). No reports describing the metabolism and inactivation of U-II have appeared. However, pretreatment of fish U-II with carboxypeptidase A destroys bioactivity in vitro, presumably by cleavage of the C-terminal amino acid (Val) (347).

U-II consists of 12 amino acids in various fish species and 11 amino acids in the human isoform (Fig. 4.12). All isoforms of U-II contain a hexapeptide loop formed by a disulfide bridge spanning two cysteines and the amino acids within the loop (positions 5–10 in human U-II) are conserved in all species (348). In contrast, the N-terminal region of U-II has considerable variation; for example, rat and human U-II have 48% sequence identity (Fig. 4.12). U-II has some structural similarity to somatostatin-14, but these two peptides are probably not derived from a common ancestral

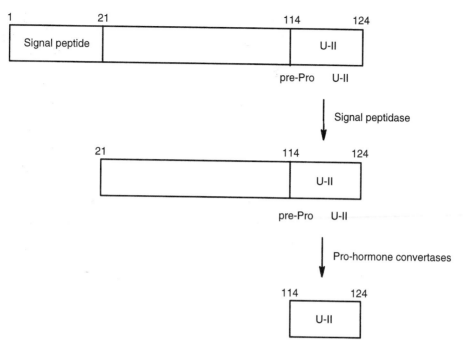

Figure 4.11. Biosynthesis of U-II.

gene (349). The solution conformation of U-II has been studied by NMR (350) and a computer model of the three-dimensional structure has been developed (351). These studies suggest that the hexapeptide loop region has a well-defined structure, whereas the N-terminal portion is disordered.

Experiments have shown that the hexapeptide C-terminal region of U-II is required for biological activity (348). This conclusion is supported by NMR studies suggesting a well-defined loop structure as well as by the complete conservation of this region across all species. SAR studies have indicated that the intact hexapeptide ring plus its flanking amino acids, each containing an acidic group, is the minimum core structure that exhibits contractile potency similar to that of U-II (352, 353). Removal of the C-terminal amino acid results in complete loss of activity (347).

6.2 Receptor

U-II was identified as the endogenous ligand for the orphan receptor GPR14, a seven transmembrane G-protein–coupled receptor identified in the rat (354). The human receptor has

75% homology with GPR14 (344). U-II binds selectively to its receptor, and although somatostatin-14 has some structural similarities to U-II, it does not bind (344). The human U-II receptor has been detected in heart and peripheral vascular tissue as well as brain, skeletal muscle, and spinal cord (344, 355, 356). On binding to its receptor, U-II stimulates phospholipase C, which produces an increase in inositol phosphate (357), and results in an influx of extracellular calcium and an intracellular release of calcium (344, 348, 358).

6.3 Biological Actions

Human U-II is the most potent vasoconstrictor yet identified, and is more than 10 times more potent than endothelin-I (356). Its vasoconstrictor properties are seen in a variety of mammalian vascular tissues, including airway and pulmonary tissues (359–361). Administration of U-II to anesthetized monkeys led to myocardial depression and circulatory collapse (344). Thus, U-II may influence cardiovascular homeostasis and be involved in the pathology of ischemic heart disease and congestive heart failure. U-II and CGRP are the only two neuropeptides found

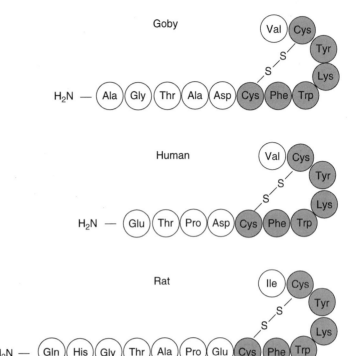

Figure 4.12. Structures of urotensin-II from goby (fish), human, and rat. Shaded amino acids indicate the hexapeptide sequence that is conserved in all species.

in motoneurons of the spinal cord, suggesting that U-II might also have a modulating role at neuromuscular junctions (362).

7 NATRIURETIC PEPTIDE FAMILY

The natriuretic peptide family currently consists of three structurally similar but genetically distinct endogenous peptides: atrial natriuretic peptide (ANP), brain natriuretic peptide (BNP), and C-type natriuretic peptide (CNP). They are involved in the regulation of cardiovascular function causing natriuresis, diuresis, and vasorelaxation.

7.1 Biosynthesis, Metabolism, and Structural Features

The three natriuretic peptides ANP, BNP, and CNP each have a 17 amino acid ring structure, in which some of the amino acid residues are conserved (Fig. 4.13).

7.1.1 ANP. The circulating, biologically active form of human ANP is composed of a 28 amino acid peptide with a 17 amino acid ring

closed by a disulfide bond between two cysteine residues (363). This disulfide crosslink between cysteines 7 and 23 is essential for all known biological activity of ANP (364). ANP is synthesized predominantly in the cardiac atria. ANP is encoded by a single gene, which in humans is located on chromosome 1, band p36, and is composed of three exons and two introns (365). Human ANP preprohormone is initially translated from mRNA within the atrial muscle and in a variety of other tissues (366, 367) as a 151 amino acid peptide (Fig. 4.14). This preprohormone is processed within the endoplasmic reticulum by a serine protease to form a prohormone (pro-ANP, also known as [ANP$_{1-126}$] or γ-ANP) consisting of 126 amino acids after removal of the 25 amino acid signal peptide from its N-terminal. This peptide, pro-ANP, is the major form of ANP stored in the atrial granules of rats and other species (368, 369). Upon stimulation, these granules move to the cell surface, releasing the stored pro-ANP. In most other regulated hormone systems, the mature hormone is processed from the precursor before storage in granules.

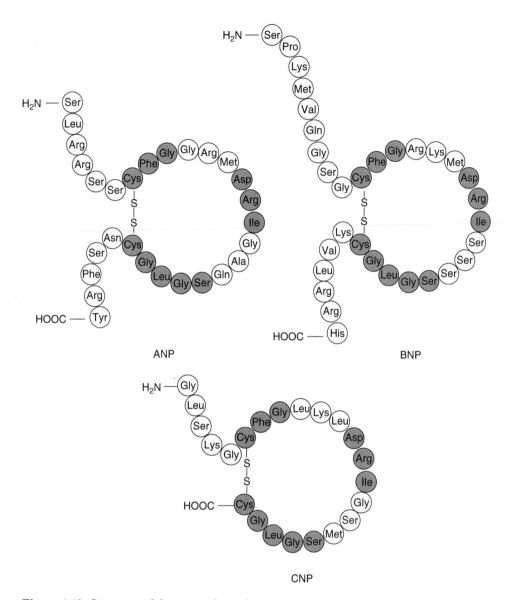

Figure 4.13. Structures of the mature form of the human natriuretic peptide family. The shaded circles indicate identical amino acids in the ring of the three peptides.

ANP is therefore distinct from most other hormones in being packaged in the prohormone form, and therefore requires a processing mechanism that acts during or after secretion. Pro-ANP is cleaved into the biologically active hormone ([ANP$_{99-126}$], also known as α-ANP) which is released together with the N-terminal [ANP$_{1-98}$] (370–372). Humans differ from other species in having β-ANP, an antiparallel dimer of ANP, present in atrial tissue, especially in subjects with heart failure (373). The release of ANP from the heart is enhanced by an increase in central volume, which stretches the atrium of the heart (370, 372), and by an increase in heart rate (374). Accordingly, increased circulating levels of ANP are seen in congestive heart failure (375), chronic renal failure (376), and in severe hypertension (377). Circulating ANP is cleared rapidly, by both an enzymatic hydrolysis and a

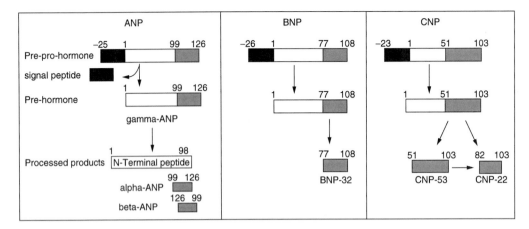

Figure 4.14. Processing pathways and major molecular forms of the three natriuretic peptides in humans.

receptor mechanism (clearance receptor). Plasma half-life is approximately 3 min (378). Endopeptidase 24.11 localized to the brush border of the proximal tubule of the kidney, lung, and vascular endothelium (379–381) cleaves ANP between residues of Cys (105) and Phe (106) and between Ser (123) and Phe (124) disrupting the 17-membered ring, yielding biologically inactive peptides (382, 383). The cleaved product has been identified in human coronary sinus plasma (384), indicating that endopeptidase 24.11 most likely acts on ANP *in vivo*. ANP is also removed from the circulation by binding to the natriuretic peptide receptor-C (NPR-C) in vascular endothelium (385).

7.1.2 BNP. Brain natriuretic peptide was originally isolated in 1988 from porcine brain (386). Although it was called brain natriuretic peptide, subsequent studies in pigs (387, 388) and in humans (389–391) showed that BNP levels in the heart are much higher than levels in the brain. The human form consists of a 32 amino acid peptide and, like ANP, has a 17 residue ring structure closed by a disulfide bridge (Fig. 4.13). BNP shows significant variation in structure across species. Separate genes encode BNP. This peptide is also synthesized as a prohormone, in which the biologically active species is contained at the C-terminus (392). In porcine cardiac atria, there appear to be two major storage forms of BNP, consisting of a 131 amino acid residue (134

amino acids in human) precursor called pre-pro-BNP and a 106 amino acid residue peptide called pro-BNP (in humans it has 108 amino acids) (Fig. 4.14). Both prepro-BNP and pro-BNP have BNP at their C-terminal ends. Pro-BNP is probably generated from prepro-BNP by proteolytic cleavage of a 26 amino acid residue signal peptide from its *N*-terminus. Pro-BNP, like ANP, is the major tissue form in porcine cardiocytes. Thereafter, pro-BNP is cleaved to produce biologically active BNP. In humans, BNP is produced primarily in ventricular myocytes in response to increases in ventricular pressure and volume and is released constitutively into the blood (393). Also BNP is secreted in response to physical exercise (394).

In contrast to ANP, human BNP is the major storage form in the heart, which indicates that posttranslational processing of BNP precursor occurs in the heart and not during secretion as in the case of ANP (395). BNP has a plasma half-life of 22 min (396–398) and is removed from circulation by cleavage with endopeptidase 24.11 and by uptake by the receptor NPR-C.

7.1.3 CNP. Structurally, CNP is the most conserved member of the natriuretic peptide family. It is considered to be the more primitive natriuretic peptide, from which ANP and BNP evolved through gene duplication (399). The homology between rat and pig precursors is 97% and between human and pig prepro-

CNP is 96% (400–402). A biologically active peptide has two mature forms, one containing 22 amino acids and the other 53 amino acids, both of which have been identified in porcine brain (403). The porcine CNP precursor molecule consists of 126 amino acid residues and is termed prepro-CNP (Fig. 4.14). Cleavage of prepro-CNP yields a 103-residue peptide pro-CNP. Thereafter pro-CNP is cleaved to produce active CNP-22. Processing after this residue generates another endogenous CNP (CNP-53). The C-terminus of CNP-53 is identical to CNP-22, and CNP-53 is often referred to as the *N*-terminal extended form of CNP (401, 403, 404). CNP differs from other family members ANP and BNP in a number of respects, most notably that the C-terminus of CNP ends at the second cysteine residue and lacks any further C-terminal extension from the ring structure (Fig. 4.13). The stimuli for synthesis and release of CNP appear to be cytokines, growth factors, and mechanical stretch (405, 406). CNP is released mainly from vascular endothelial cells, but has also been found in the kidney, brain, and intestine, and does not circulate in plasma. The degradation of CNP, like ANP and BNP, occurs by way of endopeptidase 24.11 and the NPR-C receptor (407).

7.2 Receptors

Autoradiographic studies have shown a widespread distribution of ANP-binding sites in many tissues, including endothelium, vascular smooth muscle, adrenal glands, lung, and brain (408). Three natriuretic peptide receptors have been identified. Two of these, natriuretic peptide receptor A and B (NPR-A and NPR-B), have guanylate cyclase activity and mediate the biological activity of the natriuretic peptides (409, 410). A natural product polysaccharide, HS-142-1, was shown to bind competively at the ANP receptor, blocking ANP binding (411). NPR-A and NPR-B each share about 30% sequence homology with tyrosine kinases, but possess no kinase activity (412). Extensive studies have shown that ANP and BNP act through the NPR-A receptor and CNP acts through the NPR-B receptor (385, 413, 414). The third receptor, NPR-C, does not contain the guanylyl cyclase and protein kinase domains and does not mediate cGMP ac-

tions (415). All three peptides are bound by the NPR-C receptor (385, 416, 417). NPR-C internalizes the bound natriuretic peptides and delivers them to lysosomes for degradation (409, 410, 418). Thus, this receptor has been characterized as a "clearance receptor."

7.3 Biological Actions

7.3.1 ANP. ANP has diverse actions affecting cardiovascular, renal and, endocrine homeostasis. The main effects occurring during infusion of ANP to healthy humans are a decrease in blood pressure, by direct arteriolar vasodilation (419, 420) and from a decrease in cardiac output (421, 422); an increase in natriuresis and diuresis attributed to changes in renal hemodynamics, especially an increase in glomerular filtration rate (423); a reduction in intravascular volume both by inducing diuresis and by shifting intravascular fluid to the interstitial space (424); inhibition of plasma renin activity and aldosterone secretion from the adrenal cortex (425–427); and antagonism of antidiuretic hormone (428). Genetic mouse models presenting alterations in expression of genes for ANP or its receptors indicate that genetic defects in ANP or natriuretic receptor activity may play a role in salt-sensitive hypertension as well as other sodium-retaining diseases such as congestive heart failure (CHF) and cirrhosis (429). ANP infusion also influences the sympathetic nervous system, probably by inhibiting the baroreceptor-mediated activation of renal sympathetic nerves as well as increasing noradrenaline release (421, 430, 431). ANP causes antimitogenic and antiproliferative effects in glomerular mesangial cells, in vascular smooth muscle cells, and in the heart (432) and arterial wall (433). ANP inhibits the synthesis of the potent vasoconstrictor endothelin I (434) and induces apoptosis in neonatal rat cardiomyocytes (436). ANP also appears to be a local mediator of intestinal function (435).

7.3.2 BNP. The natriuretic, endocrine, and hemodynamic effects of BNP are comparable to those of ANP (437–439). Infusion of BNP into healthy humans at physiological doses produces effects that are similar to those seen with ANP. BNP had no effect on heart rate and blood pressure, but potently inhibited the renin–angiotensin axis and had a large natriuretic effect. Transgenic mice overexpressing BNP have lower blood pressure (440). Antimitogenic effects of BNP appear to be similar to those of ANP (441). Like ANP, BNP inhibits vascular smooth muscle cell proliferation by blocking the proliferative actions of angiotensin II and endothelin I in the heart and other tissues (442–444). Thus overall ANP and BNP appear to function as an integrated cardiac natriuretic peptide system and may play a significant role in vascular remodeling and resetting vascular compliance.

7.3.3 CNP. Unlike ANP and BNP, CNP has little effect on natriuresis and diuresis (445) but acts in a local paracrine manner on vascular endothelial cells as a vasorelaxant and inhibitor of cell proliferation (446–448). CNP may also act as a local neuroendocrine regulator because it is found in high concentrations in the hypothalamus (449, 450) and in the anterior pituitary (402), as well as other endocrine glands (451). In addition CNP interacts with other endothelial modulators, the nitric oxide system (452), and endothelin (453, 454). When these observations are taken together, CNP appears to serve as a local paracrine and autocrine factor in the regulation of vascular wall biology.

7.4 Therapeutic Potential

Because they influence renal, hemodynamic, and endocrine function, are vasorelaxant, and inhibit cardiovascular cell growth, natriuretic peptides have been suggested as playing an important role in hypertension (455, 456), congestive heart failure (457–467), renal failure (462–464), restenosis (444, 465), and atherosclerosis (466, 467). Given their peptidic nature and rapid degradation, the natriuretic peptides are not themselves suitable as therapeutic agents. ANP gene therapy has been explored in a rat model (468). Inhibition of neutral endopeptidase 24.11, which should prolong the duration of action of the natriuretic peptides, has been studied both preclinically and clinically for hypertension and congestive heart failure. In clinical studies, although they did not lower blood pressure in most hypertensive patients (469, 470), neutral

endopeptidase 24.11 inhibitors showed efficacy in treating mild to moderate heart failure. The inhibitor candoxatril is marketed for this indication (471). Dual inhibitors of neutral endopeptidase 24.11 and angiotensin-converting enzyme could be more efficacious for hypertension and CHF, and clinical studies have been carried out (472, 473). In addition, measurement of plasma levels of natriuretic peptides or the N-terminal ANP prohormone, [ANP_{31-67}], has been suggested as a diagnostic and prognostic tool to determine the severity of CHF (459, 474) and acute myocardial infarction (475, 476).

8   CALCITONIN GENE–RELATED PEPTIDE FAMILY

The calcitonin gene–related peptide family consists of calcitonin (CT), calcitonin gene–related peptide (CGRP), adrenomedullin (ADM), and amylin. Human CGRP shows 19% homology with calcitonin, 35% homology with adrenomedullin, and 51% homology with amylin. CT maintains skeletal muscle mass during calcium stress and plays a central role in modulating calcium homeostasis. Amylin stimulates the breakdown of glycogen in skeletal muscles and thus helps to regulate blood glucose levels. Amylin does not produce significant vasoactive effects and will not be discussed further. CGRP and ADM are potent vasodilators, and are described below.

8.1 Structure, Biosynthesis, and Metabolism

8.1.1 CGRP. The existence of this peptide was predicted on the basis of the alternative processing of the calcitonin gene (477, 478). Calcitonin gene–related peptide, a 37 amino acid neuropeptide (Fig. 4.15), first identified in 1982 (478, 479) as an extremely potent vasodilator, is approximately 1000 times more potent than acetylcholine or substance P (480, 481). This peptide is distributed throughout the central and peripheral nervous systems (482–484) and is located in areas involved in cardiovascular function (485–487).

Calcitonin and both of the very similar forms of CGRP (alpha and beta) are encoded on chromosome 11 (488). Alternate splicing of the primary RNA transcript of the calcitonin

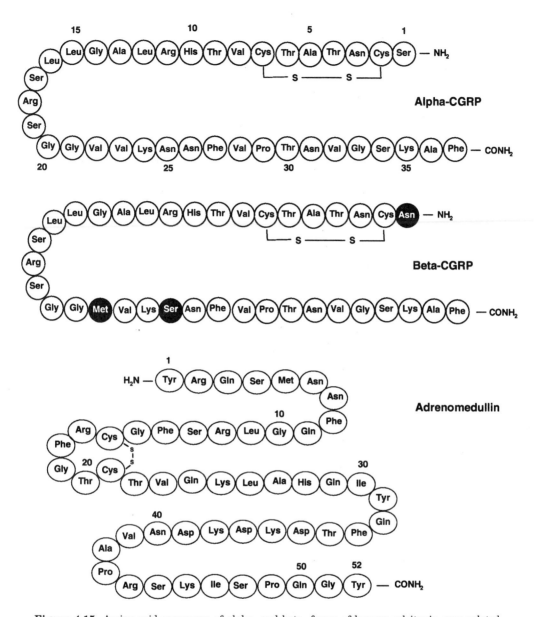

Figure 4.15. Amino acid sequences of alpha- and beta- forms of human calcitonin gene-related peptide (hCGRP) and adrenomedullin. Amino acids differing from alpha-CGRP are indicated by filled circles in beta-CGRP.

gene results in the production of two biologically active peptides, calcitonin and a 37 amino acid peptide, now designated αCGRP (Fig. 4.16). A second gene on chromosome 11, unrelated to calcitonin, produces βCGRP (Fig. 4.15) (489). Human αCGRP and βCGRP differ by only three amino acids. The structure of CGRP is well conserved among species, with 75% identity between salmon and human forms. Whereas calcitonin is largely confined to thyroid cells, CGRP is widely distributed throughout the central and peripheral nervous systems (481, 482). Only αCGRP mRNA is found in the heart (490) and the cardiovascular effects of αCGRP are generally more potent than those of the β-form (491). The CGRP

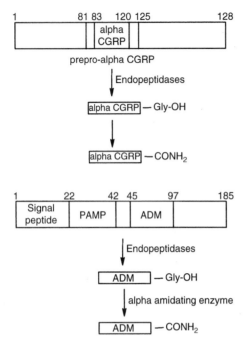

Figure 4.16. Biosynthesis of alpha-CGRP and ADM.

structure contains a Cys^2–Cys^7 disulfide bridge and C-terminal carboxamide, both essential for activity. A three-dimensional computer model of αCGRP has been constructed (492). Cleavage or derivatization of the disulfide bond destroys agonist activity (493). The region of the peptide between Val-8 and Arg-18 has the potential to form an amphipathic α-helix, and this is evident by NMR studies (494, 495). The fragment [CGRP$_{8-37}$] acts as an antagonist (496). The N-terminal amino acids are essential for agonist activity (493). The removal of the C-terminal amino acid results in a significant loss of activity (497). CGRP mRNA produces prepro-CGRP consisting of 128 amino acids, which is subsequently cleaved at paired dibasic amino acids (498) and finally converted enzymatically to the active 37 amino acid carboxamide (499).

CGRP is abundant in sensory neurons and present at lower levels in motor neurons, colocalized with acetylcholine (500). CGRP is also found in many endocrine cells and organs, where it is stored in secretory vesicles (501). CGRP is often colocalized with other neuro-

transmitters in sensory neurons, particularly substance P (502). In the heart, the amount of CGRP is higher in the atria than in the ventricles (503). Release of CGRP is stimulated by electrical stimulation as well as by prostaglandins, potassium, bradykinin, capsaicin, lactic acid, low pH, and ischemia (504). The degradation of CGRP has not been fully elucidated, but a tryptase enzyme and chymotrypsin may be involved (505, 506).

8.1.2 ADM. Adrenomedullin, a 52 amino acid peptide discovered in 1993 (507), is expressed in a wide range of tissues and has a wide range of action, including vasodilation, natriuresis, and regulation of cell growth. ADM has 24% homology with CGRP. It contains a disulfide bridge between cysteines 16 and 21 and a C-terminal amidated tyrosine, both features of which are necessary for biological activity. The gene for ADM is located on chromosome 11 (508) and produces a 185 amino acid precursor, prepro-adrenomedullin, which is cleaved by proteases and amidated, like CGRP, by the enzyme peptidylglycine C-amidating monoxygenase to generate active ADM (Fig. 4.16) (509, 510). ADM is expressed in the adrenal medulla, heart, lung, kidney, and pituitary gland (511, 512), as well as in plasma and endothelial cells (513). The plasma half-life of ADM in humans is about 22 min (514) and ADM appears to be degraded by neutral endopeptidase 24.11 and amino peptidases (515, 516).

8.2 Receptors

CGRP receptors have been studied in humans and in the rat and are widely distributed in the nervous system (517, 519), including discrete brain areas, and in the cardiovascular system (520). CGRP receptors in general match peptide distribution (518). Two cell surface receptors, CGRP$_1$ and CGRP$_2$, have been postulated (489, 521, 522). CGRP receptors belong to the family of seven transmembrane G-protein–coupled receptors, and binding leads to activation of adenylate cyclase, which increases cyclic adenosine monophosphate production (523–526). ADM shows potent binding at CGRP receptors (527) but also has a distinct receptor (528–530). It is also a member of the

G-protein–coupled receptor family and, like CGRP, stimulates adenylate cyclase.

8.3 Biological Actions

8.3.1 CGRP. Intravenous administration of CGRP produces strong, dose-dependent hypotensive and tachycardia responses both in the rat (531–533) and in human subjects (534). Immunochemical studies have shown that CGRP is distributed broadly within the cardiovascular system as a dense peripheral sensory network that innervates the arteries, veins, heart, and viscera (531, 535–538). CGRP has significant and selective regional hemodynamic effects and has been shown to increase blood flow and/or decrease vascular resistance in the coronary, common carotid, renal, mesenteric, and hindquarter vascular beds (481, 539), in addition to causing an increase in the rate and force of contraction in the heart (540). The coronary vasculature has been demonstrated to be a particularly sensitive target (480, 481). Systemic administration of CGRP decreases blood pressure in a dose-dependent manner in both normotensive animals and humans and in spontaneously hypertensive rats (482, 539), whereas in patients with chronic heart failure, the decreases in right atrial and pulmonary wedge pressure and increases in cardiac output and contractility were more pronounced (541). The primary cause for this blood pressure reduction is peripheral arterial dilation (481, 539). On the basis of these potent vasodilator effects and the perivascular localization of CGRP, it has been postulated that this peptide plays a role in the regulation of blood pressure and regional organ blood flows both under normal physiological conditions and in the pathophysiology of hypertension (482, 483).

8.3.2 ADM. ADM has a biological profile similar to that of CGRP, showing potent vasodilating (542) and natriuretic activity (543) and regulates local and systemic vascular tone (544). It has potent hypotensive effects without tachycardia (545) as well as renal activity (481, 546). In many of its actions ADM appears to act like a physiological antagonist to endothelin-1.

8.4 Therapeutic Potential and Antagonists

8.4.1 CGRP. CGRP elicited various beneficial effects in congestive heart failure patients (547, 548) and showed improvements in renal disease (549, 550) as well as in Raynaud's syndrome (551, 552). It was active in a rat model of pregnancy-induced hypertension (553). CGRP might also contribute to inflammation (554), migraine (555), and gastrointestinal disorders (556); for these indications a CGRP antagonist would be needed. The fragment [$CGRP_{8-37}$] has antagonistic properties (492), and several nonpeptidic antagonists have been reported (Fig. 4.17) (557–559).

8.4.2 ADM. ADM shows potential in pulmonary hypertension (560, 561), congestive heart failure (562), and renal failure (563).

9 TACHYKININS

Tachykinins consist of three related peptides: substance P, neurokinin A (substance K; neuromedin L; NKA), and neurokinin B (Table 4.2) (564, 565). Substance P is the most potent of the tachykinins in inducing vasodilatation and microvascular permeability, two characteristic features of substance P's action.

9.1 Biosynthesis, Structure, and Metabolism

Substance P is a peptide containing 11 amino acids (Table 4.2) derived from two distinct genes encoding for prepro-tachykinin A and B (Fig. 4.18) (566, 567). Prepro-tachykinin A, depending on alternative RNA splicing, will yield either prepro-tachykinin α, β, or γ (568, 569). The α- and γ-prepro-tachykinin precursors possess a single substance P product. The β-prepro-tachykinin isoform additionally possesses a 10 amino acid neurokinin A peptide sequence at positions 98–107 (568) (Fig. 4.18). The tachykinins are degraded by several proteolytic enzymes that may vary regionally. The main degradative enzymes are angiotensin-converting enzyme (ACE), neutral endopeptidase 24.11 (NEP), and dipeptidyl (amino) peptidase (DPPIV) (554). Although numerous enzymes contribute to the metabolism of the tachykinins, both ACE and NEP are mem-

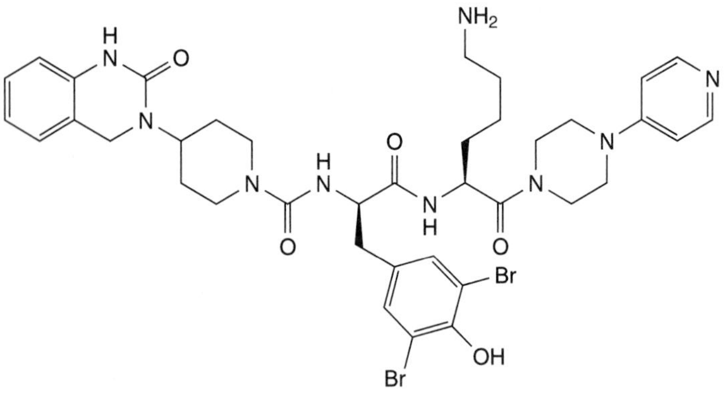

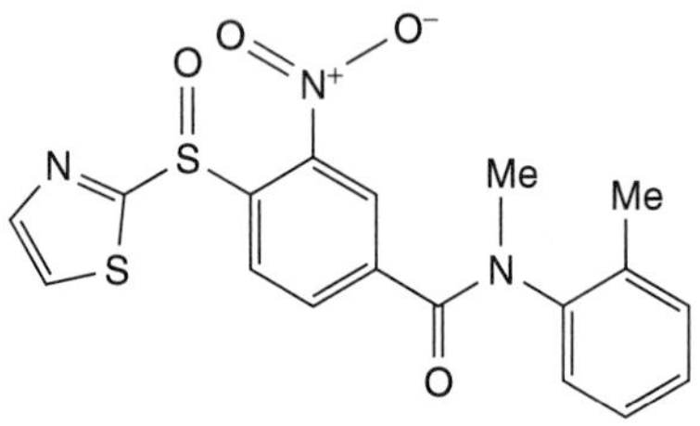

Figure 4.17. Chemical structures of BIBN4096BS (**1**) and SB-273779 (**2**).

brane-bound peptidases optimally located to effectively subserve a metabolic role.

9.2 Receptors

The three tachykinin receptor types are designated NK_1 (substance P receptor), NK_2 (neurokinin A receptor), and NK_3 (neurokinin B receptor) and belong to the seven transmembrane G-protein–coupled receptor family (569, 570–572). Individual tachykinin peptides preferentially activate different tachykinin receptor subtypes. For example, substance P is more potent at NK_1 receptors than at NK_2 or NK_3 sites, and displays the following rank order of potency: substance P > neurokinin A > neurokinin B. Alternatively, neurokinin A is more potent at activating NK_2 receptors (NK_2 potency: neurokinin A > neurokinin B >

substance P), whereas neurokinin B preferentially stimulates NK_3 receptors (NK_3 potency: neurokinin B > neurokinin A > substance P) (571, 572). NK_1 receptors are located both in the central nervous system and in peripheral tissues, whereas NK_2 receptors are predominantly found in the periphery and NK_3 receptors are mainly restricted to the brain (564, 573).

9.3 Biological Actions

Substance P is thought to act principally on endothelial cells (574). By stimulating NK_1 receptors, substance P affects G-proteins and phospholipase C (564) to induce several downstream signaling events that include mobilization of intracellular calcium and nitric oxide (NO) (575, 576). Activation of NK_1 receptors

Table 4.2 Tachykinin Receptors and Peptides

Receptor	Endogenous Peptide	Structure
NK_1	Substance P	Arg-Pro-Lys-Pro-Gln-Gln-Phe-Phe-Gly-Leu-Met-NH_2
NK_2	Neurokinin A	His-Lys-Thr-Asp-Ser-Phe-Val-Gly-Leu-Met-NH_2
NK_3	Neurokinin B	Asp-Met-His-Asp-Phe-Phe-Val-Gly-Leu-Met-NH_2

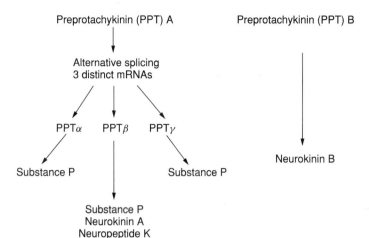

Preprotachykinin (PPT) A

Preprotachykinin (PPT) B

Alternative splicing
3 distinct mRNAs

PPTα   PPTβ   PPTγ

Substance P   Substance P

Neurokinin B

Substance P
Neurokinin A
Neuropeptide K

Figure 4.18. Schematic illustration of substance P and related tachykinin formation.

in the periphery may also be brought about indirectly by release of substance P from varicosities of afferent neurons (577).

Importantly, tachykinins contribute to the localized tissue response to injury. Injury may be induced by physical, chemical, or thermal stimuli to evoke the release of substance P from sensory neurons (578). This phenomenon, referred to as neurogenic inflammation, although occurring in many vascular beds, predominates in cutaneous and splanchnic vessels and is a consequence of NK_1 receptor activation (579). Plasma protein extravasation is a hallmark feature of neurogenic inflammation and occurs primarily in cutaneous and splanchnic vessels (580, 581). Substance P has been shown to be a potent vasodilator (582). However, its effects, which are contingent on neuronal release, appear to be highly species dependent and inconsistent (579). Thus, evidence to support a role for substance P in neurogenic vasodilatation is less compelling than that for demonstrating its effects to increase vascular permeability (579). The edema caused by substance P is attenuated by selective NK_1 antagonists, such as SR140333 (583), but is minimally blocked by selective NK_2 antagonists and not blocked by selective NK_3 antagonists, further suggesting a role for the NK_1 receptor type (583–585).

In addition to vascular leakiness and local edema, substance P also appears to modulate blood pressure through both central and peripheral neural mechanisms (586). Importantly, substance P also regulates vascular permeability within the respiratory system and thereby is important in respiratory function. In the airways, as in other regions, selective NK_1 agonists increase the leakage of vascular proteins across the endothelium and mucus hypersecretion (585). This results in edema within the interstitial tissues, as well as extravasation of exudate into the airway lumen. Moreover, substance P's effects are not mimicked by NK_2 or NK_3 agonists, suggesting that NK_2 and NK_3 receptor types do not affect vascular permeability in the respiratory system as in other systems (587).

Substance P also indirectly regulates cardiovascular function through its actions on other physiological systems. For example, substance P is coreleased with calcitonin gene–related peptide and together, they cause vasodilatation in guinea pig submucosal arterioles and coronary vasculature of numerous species (588, 589), whereas vasoactive intestinal peptide may potentiate the effects of substance P on the microvasculature in some systems (590). Importantly, substance P opposes the actions of the potent vasoactive peptide endothelin-1 (591). At least some of substance P's actions result from its ability to suppress the synthesis and release of catecholamines from the adrenal medulla (592), as well as alterations in ACTH-corticosterone production (593). It has recently been shown that the final α-amidation step necessary for full activity of substance P can occur directly within endothelial cells and can be released by shear stress (594, 595). This endothelial-derived substance

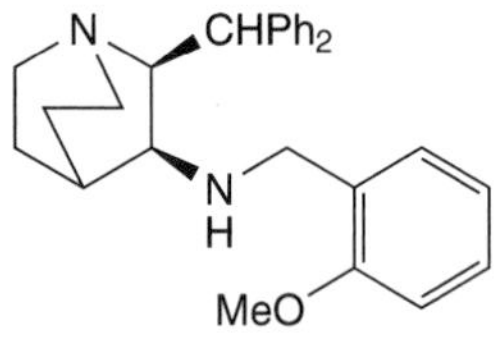

Figure 4.19. Chemical structures of CP-96,345 (**1**), L-733,060 (**2**), and SR 142801 (**3**).

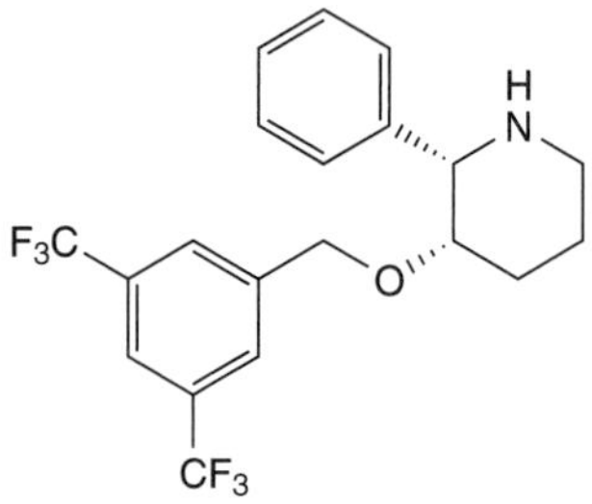

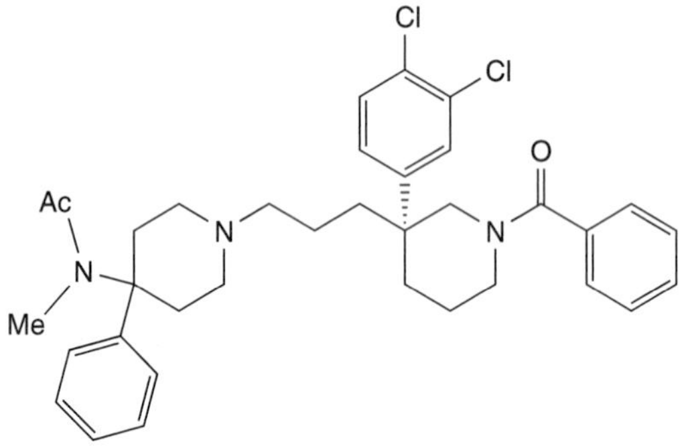

P can then evoke vasorelaxation through a nitric oxide dependent mechanism (594). Interestingly, the immediate precursor before the α-amidation, substance P-Gly, also possesses a vasorelaxant potency similar to that of the mature peptide (596).

9.4 Antagonists

As with the other vasoactive peptide systems, initial synthetic attempts to design receptor antagonists focused on modifications of the endogenous peptide. Spantide, [D-Arg^1, D-Trp^{7,9}, Leu^{11}]-substance P, is a competitive NK$_1$ peptide antagonist created by a general strategy of replacing the N- or C-terminal amino acids of substance P (597). Spantide, although a potent substance P antagonist, induces histamine release and is neurotoxic at high doses (597). Additional peptide analogs that act as antagonists at NK$_1$, NK$_2$, and NK$_3$ receptors are described elsewhere (564, 597, 598).

The first nonpeptidic NK$_1$ receptor antagonist reported was CP-96,345 (599, 600). Subsequently, nonpeptidic antagonists of NK$_2$ and NK$_3$ receptors were reported (Fig. 4.19) (571, 600). The NK$_1$ antagonist, L-733,060,

was shown to inhibit plasma extravasation without affecting blood pressure and heart rate in rats (601). The cardiovascular (rise in blood pressure and heart rate) and behavioral reactions that occurred in response to a noxious stimulus were attenuated by central administration of NK$_1$ antagonists (602). SR 142801, a novel nonpeptidic NK$_3$ antagonist, devoid of agonist properties, has proved useful in defining the functional role of tachykinin receptors in the periphery. The pressor effects evoked by systemic administration of NK$_3$ agonists and neurokinin B were inhibited by SR 142801 (603). These findings underscore the important role subserved by subtype-specific antagonists in defining the constellation of tachykinin-induced effects.

10 BRADYKININ

Bradykinin was first identified in 1949 as an extract from ox blood (604). It was more than 10 years later that the complete peptide sequence was unequivacally described (605, 606). Since that time, major strides have been made in our understanding of this peptide

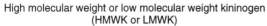

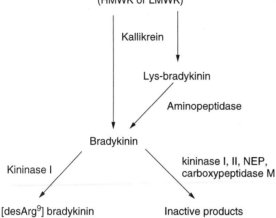

High molecular weight or low molecular weight kininogen
(HMWK or LMWK)

Kallikrein

Lys-bradykinin

Aminopeptidase

Bradykinin

Kininase I

kininase I, II, NEP,
carboxypeptidase M

[desArg⁹] bradykinin

Inactive products

Figure 4.20. Formation and degradation of bradykinin.

through the use of molecular techniques and the discovery of selective receptor antagonists.

10.1 Biosynthesis, Structure, and Metabolism

Bradykinin (BK) is a nonapeptide with the following structure: Arg-Pro-Pro-Gly-Phe-Ser-Pro-Phe-Arg [BK$_{1-9}$]. Bradykinin is formed in the plasma as a component of the inflammatory response at sites of tissue damage (607) (Fig. 4.20). The initiating event is the activation of Factor XII in the blood, which occurs at sites of tissue injury. Activated Factor XII (XIIa) converts prekallikrein to the active protease kallikrein. Kallikrein can then enzymatically digest high molecular weight kininogen to yield bradykinin. Alternative splicing of the primary transcript of the kininogen gene results in the synthesis of low molecular weight kininogen. In humans, this protein is a substrate for tissue kallikrein and yields lysyl-BK ([Lys⁰]-BK), also known as kallidin (607, 608). Tissue kallikrein and plasma kallikrein are structurally unrelated enzymes. Once formed, the kinins are short-lived with a half-life of less than 30 s (609). Although numerous enzymes are capable of degrading the kinins, ACE and NEP 24.11 are the predominant proteases responsible for BK's short half-life in the circulation and in tissues (610, 611). ACE preferentially cleaves BK between the Pro⁷–Phe⁸ and Phe⁵–Ser⁶ bonds, whereas NEP digests BK between Gly⁴–Phe⁵ and Pro⁷–Phe⁸ bonds (611). The relative distribution and proportion of these enzymes within tissue com-

partments can alter significantly the overall degradation pattern of BK.

10.2 Receptors

The biological actions of BK in mammals are brought about by the activation of two distinct receptors. Originally, the two-receptor classification of B$_1$ and B$_2$ was based on an opposing pharmacological pattern of responses (612). The rank order of potency of a series of ligands for the B$_1$ receptor is as follows: [desArg⁹]-BK > [Tyr(Me)⁸]-BK > BK, whereas at the B$_2$ receptor: [Tyr(Me)⁸]-BK > BK > [desArg⁹]-BK. This two-receptor distinction later was confirmed with the cloning and expression of the B$_1$ and B$_2$ genes (613–615). The kinin B$_1$ and B$_2$ receptors belong to the seven transmembrane G-protein–coupled receptor family (616). The B$_2$ receptor is responsible for the most notable physiologic effects of BK in mammals and this receptor is considered to be constitutively expressed (617). In contrast, the B$_1$ receptor is inducible and its expression is upregulated at the site and time of tissue injury (617, 618).

10.3 Biological Actions

The B$_1$ and the B$_2$ receptors are members of the seven transmembrane G-protein–coupled receptor family. The B$_2$ receptor is coupled to Gαi and Gαq, whereas the B$_1$ receptor couples to Gαq/11 and Gαi$_{1,2}$ (619, 620). Kinin receptor activation, through the actions of the G proteins, can stimulate various intracellular pathways including phospholipases A_2, C, and D, leading to protein phosphorylation. Selective protein phosphorylation then results in the generation of intracellular calcium and prostaglandin and nitric oxide release (621). BK increases vascular permeability at the site of tissue injury and also possesses potent vasodilator activity, two critical components of the local inflammatory response (622). In addition to nitric oxide–mediated BK vasodilatation, an endothelium-derived hyperpolarizing factor, resistant to inhibitors of the nitric oxide system, has been shown to contribute to BK-induced vascular relaxation (623). A role for the mitogen-activated protein kinase family has also recently been demonstrated as a mediator of BK activity (624).

Bradykinin has been implicated as an important factor in mediating numerous physiological and pathophysiological processes, especially those within the cardiovascular and renal systems (625). Genetic manipulations of the BK receptor system (knockouts, transgenic animals) suggest that BK may be important in the development of the blood pressure phenotype (626). A role for kinins in blood pressure regulation, cardiac ischemia, myocardial infarction and remodeling, and renal disease has been shown (625–627). The clinical effects of the ACE inhibitors in cardiovascular disease, in part, are attributed to BK (628, 629).

10.4 Antagonists

Efforts to identify selective and specific antagonists for BK receptors have been pursued for more than 20 years. Numerous chemical and amino acid substitutions have been made in an attempt to increase potency, selectivity, and duration of action (610, 630, 631). The carboxy terminal arginine appears to be particularly important, in that its presence or absence exerts a dramatic impact on agonist/antagonist receptor selectivity (630). One notable B_2 antagonist is HOE-140, a peptidic antagonist with high potency and long duration of action [$_D$Arg-Arg-Pro-Hyp-Gly-Thi-Ser-$_D$Tic-Oic-Arg] (631). Through the use of a series of antagonists for B_1 and B_2 receptors, evidence for species differences and for novel non-B_1/B_2 receptors has been put forth (632).

Recently, nonpeptidic antagonists have been described (Fig. 4.21) (633). These novel antagonists will no doubt further delineate BK receptor subtypes and aid in clarifying the role of BK in various pathophysiological conditions.

11 VASOACTIVE INTESTINAL PEPTIDE AND RELATED PEPTIDES

Vasoactive intestinal peptide (VIP) was first isolated from porcine intestine (634). The peptide derives its name from the profound and long-lasting vasodilatory action seen upon systemic administration (635). VIP is a highly basic, single-chain linear polypeptide, containing 28 amino acid residues in its sequence with a C-terminal asparagine amide (636). The primary sequence of VIP is identical in most mammals, with the guinea pig being the one notable exception (637).

VIP is derived from a 170 amino acid precursor, prepro-VIP (638). The prepro-VIP peptide contains another biologically active peptide, referred to as PHI (peptide with N-terminal histidine and a C-terminal isoleucine amide). The human equivalent of PHI has a C-terminal methionine and is referred to as PHM. PHI/PHM is structurally related to VIP, and shares many of its biological actions, although it is generally less potent than VIP (639, 640). VIP appears to be coreleased with PHI/PHM (641); VIP also can be released with acetylcholine and together they act synergistically on peripheral vascular targets (642).

VIP is a member of a family of regulatory peptides that also includes pituitary adenylate cyclase-activating peptide (PACAP). PACAP is a basic 38 amino acid α-amidated peptide structurally related to VIP (643). The receptors for these peptides are members of the seven transmembrane G-protein–coupled superfamily that also includes glucagon, glucagon-like peptide, secretin, and growth hormone-releasing factor (644, 645). PACAP binds with high affinity to three distinct receptors, whereas VIP interacts specifically with only two of these receptors (646). The receptors for these peptides have been designated by different names based on the binding characteristics of various ligands; however, the recommended nomenclature is PAC_1, $VPAC_1$,

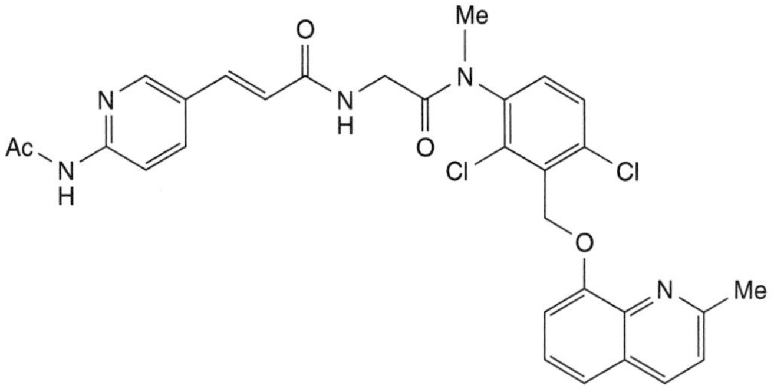

FR 173,657

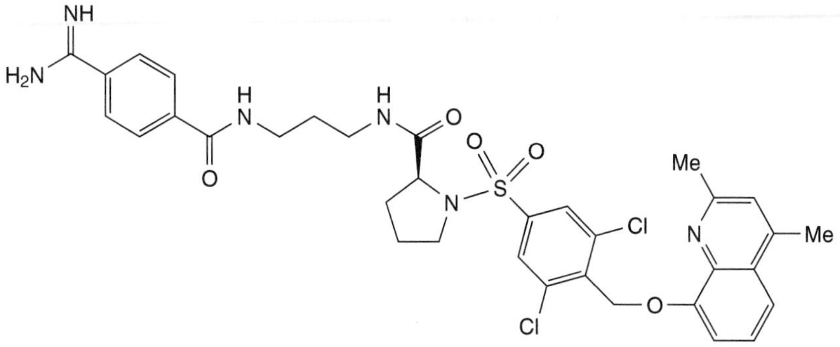

LF16-0687

Figure 4.21. Chemical structures of FR 173657 and LF16-0687.

and VPAC$_2$ (647). The VPAC$_1$ and VPAC$_2$ receptors bind VIP and PACAP with similar affinity, whereas the PAC$_1$ receptor binds PACAP preferentially (646).

At least three distinct receptors for PACAP and VIP have been cloned and expressed (647). The VPAC$_1$ receptor was first isolated from rat lung (648), the VPAC$_2$ receptor initially was cloned from the rat olfactory lobe (649), whereas the PAC$_1$ receptor was cloned originally from a rat carcinoma cell line (650). The intracellular events that occur subsequent to ligand binding by these receptors predominantly involve stimulation of cAMP through stimulation of G$_s$ protein (646). Additional second-messenger systems such as production of inositol triphosphate and calcium mobilization by stimulation of phospholipase C are activated by PAC$_1$ receptors (651).

Among CNS regions involved in cardiovascular function, VIP is present within the nucleus of the tractus solitarius and in the intermediolateral spinal cord, especially within the lumbosacral spinal cord (652). VIP receptors have been localized within cerebral microvessels (653). In the peripheral nervous system, VIP is present in pre- and postganglionic fibers and in nerve terminals of the autonomic nervous system in humans. Autonomic VIP nerve fibers tend to be widely but somewhat nonuniformly distributed among blood vessels (639, 654, 655). Whereas both VIP and PACAP can be found in the hypothalamus, only VIP is synthesized in the pituitary gland (637). The overall localization and distribution of PACAP and VIP within the central nervous system are quite different (645). In the periphery, VIP and PACAP are often colocalized to the same cells (645). VIP and PACAP preferentially are associated with cerebral blood vessels (656–658), vagal projections to the heart (659), and with nerve fibers innervating the smaller diameter blood vessels (660, 661).

VIP is a potent vasodilator (634, 662). VIP and PACAP elicit vasodilatation in cerebral blood vessels (663). Electrical stimulation of the

Table 4.3 VIP Receptors and Ligands

Receptor	Agonist[a]	Antagonist
PAC$_1$	Maxadilan (667)	PACAP (6–27) (671)
VPAC$_1$	[Lys^{15}, Arg^{16}, Leu^{27}]VIP$_{1-7}$ GRF$_{8-27-}$ NH$_2$ (668)	[Ac-His^1, D-Phe^2, Lys^{15}, Arg^{16}]VIP$_{3-7}$ GRF$_{8-27-}$ NH$_2$ (672)
VPAC$_2$	Ro 25-1553 (669) Ro 25-1392 (670)	

[a]GRF, growth hormone releasing factor.

cerebral cortex or mesencephalic reticular activating system, which innervates the cortex, causes local release of VIP and vasodilatation of arterioles and venules at the cortical surface (664). Importantly, VIP directly causes artery vasodilatation in the absence of endothelial cells, suggesting that VIP acts directly on the smooth muscle (640). Moreover, in the cat hindlimb, VIP- and PACAP-induced vasodilatation does not require nitric oxide, prostaglandins, or K$^+$$_{ATP}$ channels (665). Although VIP evokes a depressor response in the cat, the hemodynamic pattern of responses to PACAP administration is more complex, initially manifesting as a depressor response that is subsequently followed by a more prolonged rise in arterial pressure (665). VIP circulates in the plasma and VIP is released into coronary vessels during vagal stimulation to cause coronary artery dilation (666). VIP concentrations in the coronary sinus have been shown to be elevated during coronary artery occlusion and reperfusion (666).

Several peptides have been proposed as VIP receptor agonists and antagonists (Table 4.3). However, the development of selective, high affinity agonists and antagonists for VIP and PACAP receptors is still in an early stage as none of the available peptidic fragments possesses sufficient absolute specificity and selectivity. There are no currently available antagonists for the VPAC$_2$ receptor. Further progress in our understanding of the VIP/PACAP receptor family awaits the development of nonpeptidic antagonists with both high selectivity and specificity.

12 OTHER PEPTIDES

12.1 Somatostatin

Somatostatin is a 14 amino acid peptide that has two cysteines linked by a disulfide bridge

(Table 4.4). It was identified in 1973 and originally characterized for its actions as a hypothalamic inhibitor of pituitary growth hormone release (673). Subsequently, somatostatin has been found to be a regulatory hormone that inhibits the release of a variety of peptide hormones, including glucagon, growth hormone, insulin, and gastrin (674), and inhibits cell proliferation (675). Somatostatin is a potent vasoconstrictor and has a negative inotropic action on noradrenaline-mediated atrial muscle contractions in humans (676, 677). Its antiarrhythmic action is thought to be caused, in part, by a reduction in calcium influx across the sarcoplasmic reticulum of atrial myocytes (678). However, somatostatin's actions may also result from a reduction in calcium influx in atrioventricular node cells (679). Because of its action as a potent vasoconstrictor, somatostatin may have a useful role in stopping uncontrolled bleeding with esophageal varices (680–682). Within the CNS, somatostatin-containing cell bodies and/or afferent fibers are present in the rostral portion of the ventrolateral medulla (VLM) and nucleus of the tractus solitarius (NTS) and project to the intermediolateral column of the spinal cord (683, 684). Direct stimulation of the VLM affects blood pressure. Peripherally, somatostatin-containing nerve fibers are generally of limited distribution. Somatostatin is present in subpopulations of pre- and postganglionic autonomic fibers, and is evident in autonomic ganglia including mesenteric and superior cervical ganglia (685). Somatostatin immunoreactivity is localized within the fibers innervating the heart as well as the intrinsic parasympathetic neurons in the heart (677, 686). Five distinct seven transmembrane G-protein–coupled receptor subtypes for somatostatin have been cloned and characterized

Table 4.4   Structures of Somatostatin, Gastrin-Releasing Peptide, Neurotensin, and Relaxin

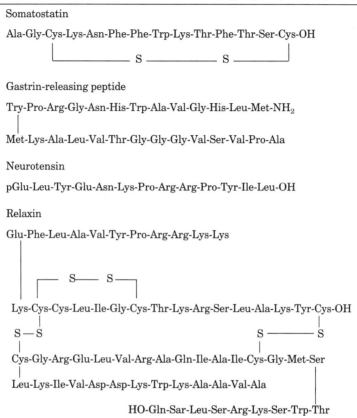

Somatostatin

Ala-Gly-Cys-Lys-Asn-Phe-Phe-Trp-Lys-Thr-Phe-Thr-Ser-Cys-OH

Gastrin-releasing peptide

Try-Pro-Arg-Gly-Asn-His-Trp-Ala-Val-Gly-His-Leu-Met-NH₂

Met-Lys-Ala-Leu-Val-Thr-Gly-Gly-Gly-Val-Ser-Val-Pro-Ala

Neurotensin

pGlu-Leu-Tyr-Glu-Asn-Lys-Pro-Arg-Arg-Pro-Tyr-Ile-Leu-OH

Relaxin

Glu-Phe-Leu-Ala-Val-Tyr-Pro-Arg-Arg-Lys-Lys

Lys-Cys-Cys-Leu-Ile-Gly-Cys-Thr-Lys-Arg-Ser-Leu-Ala-Lys-Tyr-Cys-OH

Cys-Gly-Arg-Glu-Leu-Val-Arg-Ala-Gln-Ile-Ala-Ile-Cys-Gly-Met-Ser

Leu-Lys-Ile-Val-Asp-Asp-Lys-Trp-Lys-Ala-Ala-Val-Ala

HO-Gln-Sar-Leu-Ser-Arg-Lys-Ser-Trp-Thr

(687, 688). Several potent peptidic somatostatin analogs have been identified, including octreotide (689), used clinically to relieve symptoms associated with gastro-entero pancreatic endocrine tumors and to stop bleeding from gastro-esophageal varices in patients with cirrhosis (690, 691). In addition, selective nonpeptidic agonists have been identified for the five receptor subtypes (692, 693).

12.2   Gastrin-Releasing Peptide

Human gastrin-releasing peptide contains 27 amino acids (694) and belongs to a family of peptides, including bombesin and neuromedin-B, that share homology in their C-terminal sequences (Table 4.4). Bombesin, a tetradecapeptide, causes vascular relaxation in the gut (699). Gastrin-releasing peptide is present in some nerve fibers innervating the respiratory system (695) and in the gastrointestinal tract (696). Although traditionally the role of gastrin-releasing peptide in vascular function has been uncertain (697), this peptide is related to bombesin, which is a potent vasoactive substance in amphibians (694, 698). Bombesin, neuromedin B, or gastrin-releasing peptide causes increases in phosphoinositol turnover and elevations in intracellular calcium in isolated rat endothelial cells (700). These actions are mediated through seven transmembrane G-protein–coupled receptors, which are described as bombesin-like peptide receptor subtypes 1, 2, and 3 (701–705). A variety of selective peptide ligands for these receptors have been developed (706). In addition, several selective nonpeptidic antagonists have been identified (707, 708). Infusion of a peptidic bombesin antagonist into a man with pulmonary hypertension led to acute hemodynamic effects, including a decrease in systolic pressure (709).

12.3 Neurotensin

Neurotensin (NT), a 13 amino acid peptide originally isolated from bovine hypothalamic extracts in 1973 (710), is found in brain, gastrointestinal, and cardiovascular tissues (Table 4.4) (711–718). NT acts as a neurotransmitter and neuromodulator through specific receptors. Three NT receptor subtypes have thus far been identified and cloned (719), two belonging to the family of seven transmembrane G-protein–coupled receptors. A variety of peptidic analogs of NT have been prepared, and these indicate that only the last six amino acids [NT_{8-13}] are needed for biological activity (720). A nonpeptidic NT antagonist has been identified and used as a tool to study the central effects of NT receptor blockade (721). In the CNS, NT acts as a neuromodulator associated with dopamine (721) and can modulate the activity of various cholinergic neurons and corticotropin-releasing factor cells (722–724). In the periphery, NT is involved in the control of gastrointestinal and cardiovascular systems (718). NT has been shown to cause vasoconstriction in a number of vessels (725). Intravenous or intraperitoneal injections of NT in guinea pigs elicit dose-dependent increases in blood pressure and heart rate (726, 727), resulting in part from activation of the sympathetic nervous system innervating resistance blood vessels and the heart (728). However, administration of an NT analog to normal rats led to hypotension (729).

12.4 Relaxin

Relaxin was first identified as a substance contained in the serum of pregnant guinea pigs that evoked relaxation of the interpubic ligament of the female guinea pig after acute administration (730). Further progress in understanding the physiologic actions of relaxin was impeded by the lack of reliable bioassays and methods for purification and isolation. Significant amounts of purified relaxin became available only after 1974, when methods for its purification were identified (731). Relaxin is a 52 amino acid peptide, belonging to the insulin family of proteins with a molecular weight of approximately 6000 Da (Table 4.4). Relaxin bears structural resemblance to insulin and is similarly composed of two peptide chains,

termed A and B, and is linked through three disulfide bonds (732, 733). Two molecular forms of relaxin have been identified, H1 and H2, and are encoded by two distinct genes (734, 735). Relaxin is derived from a precursor protein, preprorelaxin, after proteolytic digestion of a signal and connecting peptide (736). The highest concentrations of relaxin are found in the female reproductive system (737), although relaxin is also produced in males, primarily in the prostate gland (738). Interestingly, relaxin also can be synthesized by atrial cardiocytes (739). To date, specific receptors that bind relaxin have not been identified and characterized. Knockout mice have confirmed the well-known pregnancy-related effects of relaxin, such as preparation of the birth canal for delivery and mammary gland development (737–740). Relaxin also modulates collagen deposition (741, 742) and has antifibrotic effects in a number of different animal models (743, 744). With regard to the cardiovascular system, relaxin is a potent vasodilator that acts through a nitric oxide–dependent/cGMP pathway (735, 745, 746). Chronic administration of relaxin produces renal hyperfiltration and vasodilation mediated, in part, by activation of the endothelin ET_B receptor (747). Relaxin also has been shown to inhibit platelet aggregation (748). Thus, agents that modulate or simulate the physiologic effects of relaxin may prove useful as novel therapies for vascular (749) and renal diseases (744).

REFERENCES

1. S. Said, Ed., *Vasoactive Intestinal Peptide*, Advances in Peptide Hormone Research Series, Raven Press, New York, 1982.

2. R. Porter and M. O'Connor, Eds., *Substance P in the Nervous System*, Ciba Foundation Symposium **91**, 1982.

3. I. L. Gibbins in S. Holmgren, Ed., *The Comparative Physiology of Regulatory Peptides*, Chapman & Hall, London, 1989, pp. 308–343.

4. M. P. Nusbaum, D. M. Blitz, A. M. Swensen, D. Wood, and E. Marder, *Trends Neurosci.*, **24**, 146 (2001).

5. I. L. Gibbins and J. L. Morris, *Regul. Pept.*, **93**, 93 (2000).

6. V. J. Dzau, *Am. J. Med.*, **77**, 31 (1984).

7. R. F. Furchgott and J. V. Zawadzki, *Nature*, **299**, 373 (1980).

8. G. M. Rubanyi and P. M. Vanhoutte, *J. Physiol. (Lond.)*, **364**, 45 (1985).

9. J. V. Mombouli and P. M. Vanhoutte, *J. Mol. Cell. Cardiol.*, **31**, 61 (1999).

10. R. Tigerstedt and P. G. Bergman, *Skand. Arch. Physiol.*, **8**, 223 (1898).

11. M. J. Peach, *Physiol. Rev.*, **57**, 313 (1997).

12. J. E. Sealey and J. H. Laragh in J. H. Laragh and B. M. Brenner, Eds., *Hypertension: Pathophysiology, Diagnosis, and Management*, Raven Press, New York, 1990, p. 1287.

13. R. A. Santos, M. J. Campagnole–Santos, and S. P. Andrade, *Regul. Pept.*, **91**, 45 (2000).

14. L. Hunyady, K. J. Catt, A. J. Clark, and Z. Gaborik, *Regul. Pept.*, **91**, 29 (2000).

15. B. K. Brosnihan, *Am. J. Cardiol.*, **82**, 17S, 1998.

16. C. M. Ferrario and N. S. Iyer, *Regul. Pept.*, **78**, 13 (1998).

17. I. Moeller, A. M. Allen, S. Y. Chai, J. Zhuo, and F. A. Mendelsohn, *J. Hum. Hypertens.*, **12**, 289 (1998).

18. R. Ardaillou and D. Chansel, *Kidney Int.*, **52**, 1458 (1997).

19. D. W. Jons, M. J. Peach, R. A. Gomez, T. Inagami, and R. M. Carey, *Am. J. Physiol. Renal Fluid Electrolyte Physiol.*, **259**, F882 (1990).

20. D. J. Campbell, *J. Clin. Invest.*, **79**, 1 (1987).

21. N. J. Samani, W. J. Brammer, and J. D. Swales, *Clin. Sci.*, **80**, 339 (1991).

22. M. I. Phillips, E. A. Speakman, and B. Kimura, *Regul. Pept.*, **43**, 1 (1993).

23. D. Regoli, W. K. Park, and F. Rioux, *Pharmacol. Rev.*, **26**, 69 (1974).

24. F. M. Bumpus, *Fed. Proc.*, **36**, 2128 (1977).

25. M. J. Peach and D. E. Dostal, *J. Cardiovasc. Pharmacol.*, **16** (Suppl. 4), S25 (1990).

26. P. B. Timmermans, P. C. Wong, A. T. Chiu, W. F. Herblin, P. Benefield, D. J. Carini, R. J. Lee, R. R. Wexler, J. A. Saye, and R. D. Smith, *Pharmacol. Rev.*, **45**, 208 (1993).

27. K. Helin, M. Stoll, S. Meffert, U. Stroth, and T. Unger, *Am. Med. (Helsinki)*, **29**, 23 (1997).

28. K. Sasaki, Y. Yamano, S. Bardham, N. Iwai, J. J. Murray, M. Hasegawa, Y. Matsuda, and T. Inagami, *Nature*, **351**, 230 (1991).

29. T. J. Murphy, R. W. Alexander, K. K. Griendling, M. S. Runge, and K. E. Bernstein, *Nature*, **351**, 233 (1991).

30. M. Oliverio, H. S. Kim, M. Ito, T. Le, L. Audoly, C. F. Best, S. Hiller, K. Kluckman, N. Maeda, O. Smithies, and T. M. Coffman, *Proc. Natl. Acad. Sci. USA*, **95**, 15496 (1998).

31. L. Hein, M. E. Stevens, G. S. Barsh, R. W. Pratt, B. K. Kobilka, and V. J. Dzau, *Proc. Natl. Acad. Sci. USA*, **94**, 6391 (1997).

32. L. P. Audoly, M. I. Oliverio, and T. M. Coffman, *Trends Endocrinol. Metab.*, **11**, 263 (2000).

33. L. Criscione, H. Thomann, S. Whitebread, M. de Gasparo, P. Buehlmayer, P. Herold, F. Ostermayer, and B. Kamber, *J. Cardiovasc. Pharmacol.*, **16** (Suppl. 4), S56 (1990).

34. A. L. Scott, R. S. L. Chang, V. J. Lotti, and P. K. S. Siegl, *J. Pharmacol. Exp. Ther.*, **261**, 931 (1992).

35. R. C. Speth, S. M. Thompson, and S. J. Johns, *Adv. Exp. Med. Biol.*, **377**, 169 (1995).

36. S. Whitebread, M. Mele, B. Kamber, and M. de Gasparo, *Biochem. Biophys. Res. Commun.*, **163**, 284 (1989).

37. A. T. Chiu, W. F. Herblin, D. E. McCall, R. J. Ardecky, D. J. Carini, J. V. Duncia, L. J. Pease, P. C. Wong, R. R. Wexler, A. L. Johnson, and P. B. Timmermans, *Biochem. Biophys. Res. Commun.*, **165**, 196 (1989).

38. M. de Gasparo and H. M. Siragy, *Regul. Pept.*, **81**, 11 (1999).

39. M. de Gasparo and N. Levens, *Pharmacol. Toxicol.*, **82**, 257 (1998).

40. M. Horiuchi, M. Akishita, and V. J. Dzau, *Hypertension*, **33**, 613 (1999).

41. E. F. Grady, L. A. Sechi, C. A. Griffin, M. Schambelan, and M. Kalinyak, *J. Clin. Invest.*, **88**, 921 (1991).

42. R. Ozono, Z. Q. Wang, A. F. Moore, T. Iragani, H. M. Siragy, and R. M. Carey, *Hypertension*, **30**, 1238 (1997).

43. Z. Q. Wang, A. F. Moore, R. Ozono, H. M. Siragy, and R. M. Carey, *Hypertension*, **32**, 78 (1998).

44. W. F. Herblin, S. M. Diamond, and P. B. Timmermans, *Peptides*, **12**, 581 (1991).

45. R. E. Gibson, H. H. Thorpe, M. E. Cartwright, J. D. Frank, T. W. Schorn, P. B. Bunting, and P. K. Siegl, *Am. J. Physiol. Renal Fluid Electrolyte Physiol.*, **261**, F512 (1991).

46. N. R. Levens, A. E. Freedlender, M. J. Peach, and R. M. Carey, *Endocrinology*, **112**, 43 (1983).

47. W. J. Arendshorst, C. Chatziantoniou, and F. H. Daniels, *Kidney Int.*, **38**(Suppl. 30), S92 (1990).

48. B. G. Zimmerman, *Clin. Sci.*, **60**, 343 (1981).

49. M. I. Phillips, *Ann. Rev. Physiol.*, **49**, 413 (1987).

50. C. L. Jackson and S. M. Schwartz, *Hypertension*, **20**, 713 (1992).

51. J. M. Schnee and W. A. Hsueh, *Cardiovasc. Res.*, **46**, 264 (2000).

52. T. Matsusaka, H. Katori, T. Homma, and I. Ichikawa, *Trends Cardiovasc. Med.*, **9**, 180 (1999).

53. J. C. Romero and J. F. Reckelhoff, *Braz. J. Med. Biol. Res.*, **33**, 653 (2000).

54. B. C. Berk, J. Haendeler, and J. Sottile, *J. Clin. Invest.*, **105**, 1525 (2000).

55. S. Keider, *Life Sci.*, **63**, 1 (1998).

56. M. Stoll, U. M. Steckelings, M. Paul, S. P. Bottari, R. Metzgei, and T. Unger, *J. Clin. Invest.*, **95**, 651 (1995).

57. T. Yamada, M. Horiuchi, and V. J. Dzau, *Proc. Natl. Acad. Sci. USA*, **93**, 156 (1996).

58. H. Matsubara, *Circ. Res.*, **83**, 1182 (1998).

59. H. M. Siragy and R. M. Carey, *J. Clin. Invest.*, **100**, 264 (1997).

60. E. Haber, *Clin. Sci.*, **59**(Suppl. 6), 7S (1980).

61. S. H. Rosenberg and S. A. Boyd in P. A. Van Zwieten and W. J. Greenlee, Eds., *Antihypertensive Drugs*, Harwood, Amsterdam, 1997, p. 77.

62. W. Fischli, J.-P. Clozel, K. El Amrani, W. Wostl, W. Neidhart, H. Stadler, and Q. Branca, *Hypertension*, **18**, 22 (1991).

63. J. M. Wood, L. Criscione, M. de Gasparo, P. Buehlmayer, H. Rueeger, J. L. Stanton, R. A. Jupp, and J. Kay, *J. Cardiovasc. Pharmacol.*, **14**, 221 (1989).

64. A. H. van den Meiracker, P. J. J. Admiraal, A. J. Man in 't Veld, F. H. M. Derkx, H. J. Ritsema van Eck, P. Mulder, P. van Brummelen, and M. A. D. H. Schalekamp, *Br. Med. J.*, **301**, 205 (1990).

65. X. Jeunemaitre, J. Menard, J. Nussberger, T. T. Guyene, H. R. Brunner, and P. Corvol, *Am. J. Hypertens.*, **2**, 819 (1989).

66. (a) J. Rahuel, V. Rasetti, J. Maibaum, H. Rueger, R. Goschke, N. C. Cohen, S. Stutz, F. Cumin, W. Fuhrer, J. M. Wood, and M. G. Grutter, *Chem. Biol.*, **7**, 493 (2000); (b) R. Gueller, A. Binggeli, V. Breu, D. Bur, G. Hirth, C. Jenny, M. Kansy, F. Montavon, M. Mueller, C. Oefner, H. Sradler, E. Vieira, M. Wilhelm, W. Wostl, and H. P. Maerki. *Bioorg. Med. Chem. Lett.*, **9**, 1403 (1999).

67. M. A. Ondetti, B. Rubin, and D. W. Cushman, *Science*, **196**, 441 (1977).

68. A. A. Patchett, E. Harris, E. W. Tristram, M. J. Wyvratt, M. T. Wu, D. Taub, E. R. Peterson, T. J. Ikeler, J. ten Broeke, L. G. Payne, D. L. Ondeyka, E. D. Thorsett, W. J. Greenlee, N. S. Lohr, R. D. Hoffsommer, H. Joshua, W. V. Ruyle, J. W. Rothrock, S. D. Aster, A. L. Maycock, F. M. Robinson, R. Hirschman, C. S. Sweet, E. H. Ulm, D. M. Gross, T. C. Vassil, and C. A. Stone, *Nature*, **288**, 280 (1980).

69. L. M. Burrell and C. I. Johnston, *Victor Chang. Mol. Cardiol. Ser.*, **1**, 191 (2000).

70. G. R. Marshall, W. Vine, and P. Needleman, *Proc. Natl. Acad. Sci. USA*, **67**, 1624 (1970).

71. Y. Furukawa, S. Kishimoto, and S. Nishikawa, U.S. Pat. 4,340,598 (1982).

72. D. J. Carini, J. V. Duncia, and M. E. Pierce, *Pharm. Biotechnol.*, **11**, 29 (1998).

73. P. A. Thurmann, *Expert Opin. Pharmacother.*, **1**, 337 (2000).

74. R. Dina and M. Jafari, *Am. J. Health Syst. Pharm.*, **57**, 1231 (2000).

75. M. Burnier and H. R. Brunner, *Lancet*, **355**, 637 (2000).

76. G. Oliver and E. A. Schafer, *J. Physiol.*, **18**, 277 (1895).

77. V. Du Vigneaud, H. C. Lawler, and E. A. Popenoe, *J. Am. Chem. Soc.*, **75**, 4880 (1953).

78. M. Manning, L. Balaspiri, M. Acosta, and W. H. Sawyer, *J. Med. Chem.*, **16**, 975 (1973).

79. V. Du Vigneaud, D. T. Gish, and P. G. Katsoyannis, *J. Amer. Chem. Soc.*, **76**, 4751 (1954).

80. M. Manning and W. H. Sawyer, *J. Recept. Res.*, **13**, 195 (1993).

81. W. G. North in T. Hökfelt, K. Fuxe, and B. Pernow, Eds., *Coexistence of Neuronal Messengers: A New Principle in Chemical Transmission* (Progress in Brain Research series), Elsevier, Amsterdam, 1986, p. 175.

82. D. Murphy, S. Waller, K. Fairhall, D. A. Carter, and I. C. A. F. Robinson, *Prog. Brain Res.*, **119**, 137 (1998).

83. F. D. Grant, *Exp. Physiol.*, **85S**, 203S (2000).

84. G. Guillon, E. Grazzini, M. Andrez, C. Breton, M. Trueba, C. Serradeil-LeGal, G. Boccara, S. Derick, L. Chouinard, and N. Gallo-Payet. *Endo. Res.*, **24**, 703 (1998).

85. L. M. Burrell, J. Risvanis, C. I. Johnston, M. Naitoh, and L. C. Balding, *Exp. Physiol.*, **85S**, 259S (2000).

86. H. D. Lauson in R. O. Greep, E. B. Astwood, E. Kobil, and W. H. Sawyer, Eds., Handbook of Physiology, Vol. **4**, American Physiological Society, Washington, DC, 1974, p. 287.

87. R. Walter and W. H. Simmons in A. M. Moses and L. Share, Eds., *Neurohypophysis*, Karger, Basel, 1977, p. 167.

88. N. Marks, L. Abrash, and R. Walter, *Proc. Soc. Exp. Biol. Med.*, **142**, 455 (1973).

89. J. P. H. Burbach and J. L. M. Lebouille, *J. Biol. Chem.*, **258**, 1487 (1983).

90. W. H. Simmons and R. Walter, *Biochemistry*, **19**, 39 (1980).

91. W. H. Simmons and R. Walter in D. H. Schlesinger, Ed., *Neurohypophyseal Peptide Hormones and Hormones and other Biological Active Peptides*, Elsevier/North-Holland, Amsterdam, 1981, p. 151.

92. M. Koida and R. Walter, *J. Biol. Chem.*, **251**, 7593 (1976).

93. J. P. H. Burbach, X.-C. Wang, J. A. Ten Haaf, and D. De Wied, *Brain Res.*, **306**, 384 (1984).

94. J. P. H. Burbach, G. L. Kovacs, X.-C. Wang and D. De Wied in G. Koch and K. Richter, Eds., *Biochemical and Clinical Aspects of Neuropeptides: Biosynthesis, Processing and Gene Structure*, Academic Press, New York, 1983, p. 211.

95. D. De Wied, O. Gaffori, J. P. H. Burbach, G. L. Kovacs, and J. M. Van Ree, *J. Pharmacol. Exp. Ther.*, **241**, 268 (1987).

96. M. Manning, K. Bankowski, and W. H. Sawyer in D. M. Gash and G. J. Boer, Eds., *Vasopressin, Principles and Properties*, Plenum Press, New York, 1987, p. 335.

97. P. M. Mannucci, *Blood*, **90**, 2515 (1997).

98. A. Nader, and N. D. Grace, *Gastrointest. Endosc. Clin. North Am.*, **9**, 287 (1999).

99. B. S. Anand, *Natl. Med. J. India*, **11**, 173 (1998).

100. P. C. A. Kam and T. M. Tay, *Eur. J. Anaesthesiol.*, **15**, 133 (1998).

101. W. Y. Chan, N. C. Wo, S. Stoev, L. L. Cheng, and M. Manning, *Exp. Physiol.*, **85S**, 7S (2000).

102. A. Morel, A. O'Carroll, M. J. Brownstein, and S. J. Lolait, *Nature*, **356**, 523 (1992).

103. Y. De Keyzer, C. Auzan, F. Lenne, C. Beldjord, M. Thibonnier, X. Bertagna, and E. Clauser, *FEBS Lett.*, **356**, 215 (1994).

104. S. J. Lolait, A. M. O'Carroll, O. W. McBride, M. Konig, A. Morel, and M. J. Brownstein, *Nature*, **357**, 336 (1992).

105. M. Birnbaumer, *Trends Endocrinol. Metab.*, **11**, 406 (2000).

106. M. Thibonnier, L. N. Berti-Mattera, N. Dulin, D. M. Conarty, and R. Mattera, *Prog. Brain Res.*, **119**, 147 (1998).

107. N. Ruiz-Opazo, *Néphrologie*, **19**, 417 (1998).

108. J. Liang, K. Toba, Y. Ouchi, K. Nagano, M. Akishita, K. Kozaki, M. Ishikawa, M. Eto, and H. Orimo, *J. Auton. Nerv. Syst.*, **62**, 133 (1997).

109. B. M. Altura, *Fed. Proc.*, **36**, 1840 (1977).

110. E. Monos, R. H. Cox, and C. H. Peterson, *Am. J. Physiol. Heart Circ. Physiol.*, **243**, H167 (1978).

111. J. W. Osborn Jr., M. M. Skelton, and A. W. Cowley Jr., *Am. J. Physiol. Heart Circ. Physiol.*, **252**, H628 (1987).

112. J.-P. Montani, J.-F. Liard, J. Schoun, and J. Möhr, *Circ. Res.*, **47**, 346 (1980).

113. T. J. Ebert, A. W. Cowley, and M. Skelton, *J. Clin. Invest.*, **77**, 1136 (1986).

114. G. R. Heyndrickx, D. H. Boettcher, and S. F. Vatner, *Am. J. Physiol.*, **231**, 1579 (1976).

115. A. W. Cowley Jr., *Exp. Physiol.*, **85S**, 223S (2000).

116. J. F. Liard, O. Deriaz, P. Schelling, and M. Thibonnier, *Am. J. Physiol. Heart Circ. Physiol.*, **243**, H663 (1982).

117. R. A. Nemenoff, *Front. Biosci.*, **3**, 194 (1998).

118. O. Schoots, F. Hernando, N. V. Knoers, and J. P. H. Burbach, *Results Probl. Cell. Differ.*, **26**, 107 (1999).

119. K. Toba, M. Ohta, T. Kimura, K. Nagano, S. Ito, and Y. Ouchi, *Prog. Brain Res.*, **119**, 337 (1998).

120. E. Szcepanska-Sadowska, *Regul. Pept.*, **66**, 65 (1996).

121. A. W. Cowley Jr. and J. F. Liard, *Hypertension*, **11**, I25 (1988).

122. S. R. Goldsmith, *J. Card. Fail.*, **5**, 347 (1999).

123. M. Manning, S. Stoev, W. Y. Chan, and W. H. Sawyer, *Ann. N. Y. Acad. Sci.*, **689**, 219 (1993).

124. S. Jard, *Vasopressin and Oxytocin*, Plenum Press, New York, 1998, p. 1.

125. Y. Yamamura, H. Ogawa, T. Chihara, K. Kondo, T. Onogawa, S. Nakamura, T. Mori, M. Tominaga, and Y. Yabuuchi, *Science*, **252**, 572 (1991).

126. R. M. Freidinger and D. J. Pettibone, *Med. Res. Rev.*, **17**, 1 (1997).

127. C. Serradeil-LeGal, *Vasopressin and Oxytocin*, Plenum Press, New York, 1998, p. 427.

128. T. Yatsu, Y. Tomura, A. Tahara, K. Wada, T. Kusayama, J. Tsukada, T. Tokioka, W. Uchida, O. Inagaki, Y. Iizumi, A. Tanaka, and K. Honda, *Eur. J. Pharmacol.*, **376**, 239 (1999).

129. P. Gross and C. Palm, *Exp. Physiol.*, **85S**, 253S (2000).

130. B. Mayinger and J. Hensen, *Exp. Clin. Endocrinol. Diabetes*, **107**, 157 (1999).

131. P. S. Chan, J. Coupet, H. C. Park, F. Lai, D. Hartupee, P. Cervoni, J. P. Dusza, J. D. Albright, X. Ru, H. Mazandarani, T. Tanikella, C. Shepherd, L. Ochalski, T. Bailey, T. Y. W. Lock, X. Ning, J. R. Taylor, and W. Spinelli, *Vasopressin and Oxytocin*, Plenum Press, New York, 1998.

132. M. Yanagisawa, H. Kurihara, S. Kimura, Y. Tomobe, M. Kobayashi, Y. Mitsui, Y. Yazaki, K. Goto, and T. Masaki, *Nature*, **332**, 411 (1988).

133. A. Inoue, M. Yanagisawa, S. Kimura, Y. Kasuya, T. Miyauchi, K. Goto, and T. Masaki, *Proc. Natl. Acad. Sci. USA*, **86**, 2863 (1989).

134. K. Saida, Y. Mitsui, and N. Ishida, *J. Biol. Chem.*, **264**, 14613 (1989).

135. Y. Kloog, I. Ambar, M. Sokolovsky, E. Kochva, Z. Wollberg, and A. Bdolah, *Science*, **242**, 268 (1988).

136. A. Bdolah, Z. Wollberg, G. Fleminger, and E. Kochva, *FEBS Lett.*, **256**, 1 (1989).

137. K. Goto, H. Hama, and Y. Kasuya, *Jpn. J. Pharmacol.*, **72**, 261 (1996).

138. A. Ortegao Mateo and A. A. de Artiñano, *Pharmacol. Res.*, **36**, 339 (1997).

139. B. Hocher, C. Thöne-Reineke, C. Bauer, M. Raschack, and H. H. Neumayer, *Eur. J. Clin. Chem. Biochem.*, **35**, 175 (1997).

140. G. M. Rubanyi and M. A. Polokoff, *Pharmacol. Rev.*, **46**, 325 (1994).

141. T. Arinami, M. Ishikawa, A. Inoue, M. Yanagisawa, T. Masaki, M. C. Yoshida, and H. Hamaguchi, *Am. J. Hum. Genet.*, **48**, 990 (1991).

142. D. Xu, N. Emoto, A. Giaid, C. Slaughter, S. Kaw, D. de Wit, and M. Yanagisawa, *Cell*, **78**, 473 (1994).

143. K. Shimada, M. Takahashi, M. Ikeda, and K. Tanzawa, *FEBS Lett.*, **371**, 140 (1995).

144. O. Valdenaire, E. Rohrbacher, and M-G. Mattei, *J. Biol. Chem.*, **270**, 29794 (1995).

145. A. J. Turner, K. Barnes, A. Schweizer, and O. Valdenaire, *Trends Pharmacol. Sci.*, **19**, 483 (1998).

146. A. Schweizer, O. Valdenaire, P. Nelböck, U. Deuschle, M. Dumas, J. P. Edwards, J. G. Stompf, and B. M. Löffler, *Biochem. J.*, **328**, 871 (1997).

147. B. M. Löffler, *J. Cardiovasc. Pharmacol.*, **35**, S79 (2000).

148. S. Nakamura, M. Naruse, K. Naruse, H. Demura, and H. Uemura, *Histochemistry*, **94**, 475 (1990).

149. T. J. Opgenorth, J. R. Wu-Wong, and K. Shiosaki, *FASEB J.*, **6**, 2653 (1992).

150. M. S. Simonson, *Physiol. Rev.*, **73**, 375 (1993).

151. G. P. Rossi, A. Sacchetto, M. Cesari, and A. C. Pessina, *Cardiovasc. Res.*, **43**, 300 (1999).

152. T. Masaki, *Endocrinol. Rev.*, **14**, 256 (1993).

153. T. F. Lüscher, B. S. Oemar, C. M. Boulanger, and A. W. A. Hahn, *J. Hypertens.*, **11**, 7 (1993).

154. T. F. Lüscher, B. S. Oemar, C. M. Boulanger, and A. W. A. Hahn, *J. Hypertens.*, **11**, 121 (1993).

155. F. D. Russell and A. P. Davenport, *Br. J. Pharmacol.*, **126**, 391 (1999).

156. C. M. Boulanger and T. F. Lüscher, *J. Clin. Invest.*, **85**, 587 (1990).

157. O. Saijonmaa, A. Ristimaki, and F. Fyhrquist, *Biochem. Biophys. Res. Commun.*, **173**, 514 (1990).

158. D. M. Pollock, *Clin. Exp. Pharmacol. Physiol.*, **26**, 258 (1999).

159. R. Shiba, M. Yanagisawa, T. Miyauchi, Y. Ishii, S. Kimura, Y. Uchiyama, T. Masaki, and K. Goto, *J. Cardiovasc. Pharmacol.*, **13**(Suppl. 5), S98 (1989).

160. J. Pernow, A. Hemsen, and J. M. Lundberg, *Biochem. Biophys. Res. Commun.*, **161**, 647 (1989).

161. T. Fukuroda, T. Fujikawa, S. Ozaki, K. Ishikawa, M. Yano, and M. Nishikibe, *Biochem. Biophys. Res. Commun.*, **199**, 1461 (1994).

162. H. Arai, S. Hori, I. Aramori, H. Ohkubo, and S. Nakanishi, *Nature*, **348**, 730 (1990).

163. T. Sakurai, M. Yanagisawa, Y. Takuwa, H. Miyazaki, S. Kimura, K. Goto, and T. Masaki, *Nature*, **348**, 732 (1990).

164. M. J. Summer, T. R. Cannon, J. W. Mundin, D. G. White, and I. S. Watts, *Br. J. Pharmacol.*, **107**, 858 (1992).

165. M. Ihara, K. Noguchi, T. Saeki, T. Fukuroda, S. Tsuchida, S. Kimura, T. Fukami, K. Ishikawa, M. Nishikibe, and M. Yano, *Life Sci.*, **50**, 247 (1992).

166. R. Takayanagi, K. Kitazumi, C. Takasaki, et al., *Fed. Eur. Biochem. Soc. Lett.*, **282**, 103 (1991).

167. W. K. Samson, K. D. Skala, B. D. Alexander, and F. S. Huang, *Biochem. Biophys. Res. Commun.*, **169**, 737 (1990).

168. C. Koseki, M. Imai, Y. Hirata, M. Yanagisawa, and T. Masaki, *Am. J. Physiol. Regul. Integr. Comp. Physiol.*, **256**, R858 (1989).

169. T. Masaki, H. Ninomiya, A. Sakamoto, and Y. Okamoto, *Mol. Cell. Biochem.*, **190**, 153 (1999).

170. F. Brunner and A. M. Doherty, *FEBS Lett.*, **396**, 238 (1996).

171. E. L. Schiffrin and R. M. Touyz, *J. Cardiovasc. Pharmacol.*, **32**(Suppl. 3), S2 (1998).

172. M. Kirchengast and K. Münter, *Proc. Soc. Exp. Biol. Med.*, **221**, 313 (1999).

173. T. Miyauchi and T. Masaki, *Annu. Rev. Physiol.*, **61**, 391 (1999).

174. S. Schmitz-Spanke and J. D. Schipke, *Basic Res. Cardiol.*, **95**, 290 (2000).

175. E. Bassenge, *Basic Res. Cardiol.*, **90**, 125 (1995).

176. R. Marsault, P. Vigne, J. P. Breittmayer, and C. Frelin, *Am. J. Physiol. Cell Physiol.*, **261**, C986 (1991).

177. G. A. Gray and D. J. Webb, *Pharmacol. Ther.*, **72**, 109 (1996).

178. J. G. Filep, A. Fournier, and E. Földes-Filep, *Br. J. Pharmacol.*, **112**, 963 (1994).

179. K. L. Goetz, B. C. Wang, J. B. Madwed, J. L. Zhu, and R. J. Leadley Jr., *Am. J. Physiol. Regul. Integr. Comp. Physiol.*, **255**, R1064 (1988).

180. W. L. Miller, M. M. Redfield, and J. C. Bennett, *J. Clin. Invest.*, **83**, 317 (1989).

181. G. M. Rubanyi and L. H. Parker Botelho, *FASEB J.*, **5**, 2713 (1991).

182. Y. Matsumura, K. Nakase, R. Ikegawa, K. Hayashi, T. Ohyama, and S. Morimoto, *Life Sci.*, **44**, 149 (1989).

183. M. Takagi, H. Matsuoka, K. Atarashi, and S. Yagi, *Biochem. Biophys. Res. Commun.*, **157**, 1164 (1988).

184. G. P. Rossi, A. Sacchetto, M. Cesari, and A. C. Pessina, *Cardiovasc. Res.*, **44**, 449 (1999).

185. Y. Shibouta, N. Suzuki, A. Shino, H. Matsumoto, Z-I. Terashita, K. Kondo, and K. Nishikawa, *Life Sci.*, **46**, 1611 (1990).

186. K. Yamada and S. Yoshida, *Am. J. Physiol. Renal Fluid Electrolyte Physiol.*, **260**, F34 (1991).

187. J. Pernow, F-J. Bloutier, A. Franco-Cereceda, J. S. Lacroix, R. Matran, and J. M. Lundberg, *Acta Physiol. Scand.*, **134**, 573 (1988).

188. D. A. Pollock and T. J. Opgenorth, *Am. J. Physiol. Regul. Integr. Comp. Physiol.*, **264**, R222 (1993).

189. D. E. Kohan, *Am. J. Kidney Dis.*, **29**, 2 (1997).

190. A. Giaid, M. Yanagisawa, D. Langleben, R. P. Michel, R. Levy, H. Shennib, S. Kimura, T. Masaki, W. P. Duguid, and D. J. Stewart, *N. Engl. J. Med.*, **328**, 1732 (1993).

191. B. Battistini, P. D'Orléans-Juste, and P. Sirois, *Lab. Invest.*, **68**, 600 (1993).

192. E. L. Schiffrin, H. D. Intengan, G. Thibault, and R. M. Touyz, *Curr. Opin. Cardiol.*, **12**, 354 (1997).

193. J.-C. Dussaule, J.-J. Boffa, P.-L. Tharaux, F. Fakhouri, R. Ardaillou, and C. Chatziantoniou, *Adv. Nephrol.*, **30**, 281 (2000).

194. R. L. Hopfner and V. Gopalakrishnan, *Diabetologia*, **42**, 1383 (1999).

195. M. Wanecek, E. Weitzberg, A. Rudehill, and A. Oldner, *Eur. J. Pharmacol.*, **407**, 1 (2000).

196. Y. Saito, K. Nakao, M. Mukoyama, and H. Imura, *N. Engl. J. Med.*, **322**, 205 (1990).

197. E. L. Schiffrin, L. Y. Deng, P. Sventek, and R. Day, *J. Hypertens.*, **15**, 57 (1997).

198. A. Lerman, B. S. Edwards, J. W. Hallett, D. M. Heublein, S. M. Sandberg, and J. C. Burnett Jr., *N. Engl. J. Med.*, **325**, 997 (1991).

199. D. Hasdai and A. Lerman, *Coron. Artery Dis.*, **6**, 901 (1995).

200. C. Ihling, H. R. Gobel, A. Lippoldt, S. Wessels, M. Paul, H. E. Schaefer, and A. M. Zeiher, *J. Pathol.*, **179**, 303 (1996).

201. W. Druml, H. Steltzer, W. Waldhausl, K. Lenz, A. Hammerle, H. Vierhapper, S. Gasic, and O. F. Wagner, *Am. Rev. Respir. Dis.*, **148**, 1169 (1993).

202. M. R. MacLean, *Pulm. Pharmacol. Ther.*, **11**, 125 (1998).

203. K. Dawas, M. Loizidou, A. Shankar, H. Ali, and I. Taylor, *Ann. R. Coll. Surg. Engl.*, **81**, 306 (1999).

204. A. Pedram, M. Razandi, R. M. Hu, and E. R. Levin, *J. Biol. Chem.*, **272**, 17097 (1997).

205. B. Battistini and P. Dussault, *Pulm. Pharmacol. Ther.*, **11**, 97 (1998).

206. C. F. van der Walle and D. J. Barlow, *Curr. Med. Chem.*, **5**, 321 (1998).

207. M. L. Webb and T. D. Meek, *Med. Chem. Rev.*, **17**, 17 (1997).

208. S. Roux, V. Breu, S. I. Ertel, and M. Clozel, *J. Mol. Med.*, **77**, 364 (1999).

209. H. H. Dao and P. Moreau, *Expert Opin. Invest. Drugs*, **8**, 1807 (1999).

210. A. Benigni and G. Remuzzi, *Lancet*, **353**, 133 (1999).

211. E. E. Ohlstein, J. D. Elliott, G. Z. Feuerstein, and R. R. Ruffolo Jr., *Med. Chem. Rev.*, **16**, 365 (1996).

212. S. Itoh, T. Sasaki, K. Ide, K. Ishikawa, M. Nishikibe, and M. Yano, *Biochem. Biophys. Res. Commun.*, **195**, 969 (1993).

213. M. Zuccarello, A. I. Lewis, and R. M. Rapoport, *Eur. J. Pharmacol.*, **259**, R1 (1994).

214. T. J. Openorth, A. L. Adler, S. V. Calzadilla, W. J. Chiou, B. D. Dayton, D. B. Dixon, L. J. Gehrke, L. Hernandez, S. R. Magnuson, K. C. Marsh, E. I. Novosad, T. W. von Geldern, J. L. Wessale, M. Winn, and J. R. Wu-Wong, *J. Pharmacol. Exp. Ther.*, **276**, 473 (1996).

215. G. Liu, K. J. Henry Jr., B. G. Szczepankiewicz, M. Winn, N. S. Kozmina, S. A. Boyd, J. Wasicak, T. W. von Geldern, J. R. Wu-Wong, W. J. Chiou, et al., *J. Med. Chem.*, **41**, 3261 (1998).

216. M. Hihara, T. Fukuroda, T. Saeki, M. Nishikibe, K. Kojiri, H. Suda, and M. Yano, *Biochem. Biophys. Res. Commun.*, **178**, 132 (1991).

217. P. D. Stein, J. T. Hunt, D. M. Floyd, S. Moreland, K. E. J. Dickinson, C. Mitchell, E. C. K. Liu, M. L. Webb, N. Murugesan, J. Dickey, D. McMullen, R. Zhang, V. G. Lee, R. Serafino, C. Denaley, T. R. Schaeffer, and M. J. Dozlowski, *J. Med. Chem.*, **37**, 329 (1994).

218. W. W. K. R. Mederski, D. Dorsch, M. Osswald, S. Anzali, M. Christadler, C.-J. Schmitges, P. Schelling, C. Wilm, and M. Fluck, *Bioorg. Med. Chem. Lett.*, **8**, 1771 (1998).

219. H. Nirei, K. Hamada, M. Shoubo, K. Sogabe, Y. Notsu, and T. Ono, *Life Sci.*, **52**, 1869 (1993).

220. S.-I. Mihara and M. Fujimoto, *Eur. J. Pharmacol.*, **246**, 33 (1993).

221. Y. Masuda, T. Sugo, T. Kikuchi, A. Kawata, M. Satoh, Y. Fujisawa, Y. Itoh, M. Wakimasu, and T. Ohtaki, *J. Pharmacol. Exp. Ther.*, **279**, 675 (1996).

222. M. Takeda, M. D. Breyer, T. D. Noland, T. Homma, R. L. Hoover, T. Inagami, and V. Kon, *Kidney Int.*, **42**, 1713 (1992).

223. S. Prié, T. K. Leung, P. Cernacek, J. W. Ryan, and J. Dupuis, *J. Pharmacol. Exp. Ther.*, **282**, 1312 (1997).

224. C. Wilson, S. J. Hunt, E. Tang, N. Wright, E. Kelly, S. Palmer, C. Heys, S. Mellor, R. James, and R. Bialecki, *J. Pharmacol. Exp. Ther.*, **290**, 1085 (1999).

225. C. Wu, M. F. Chan, F. Stavros, B. Raju, I. Okun, S. Mong, K. M. Keller, T. Brock, T. P. Kogan, and R. A. Dixon, *J. Med. Chem.*, **40**, 1690 (1997).

226. P. C. Astles, C. Brealey, T. J. Brown, V. Facchini, C. Handscombe, N. V. Harris, C. McCarthy, I. M. McLay, B. Porter, A. G. Roach, C.

Sargent, C. Smith, and R. J. Walsh, *J. Med. Chem.*, **41**, 2732 (1998).

227. Y. Urade, Y. Fujitani, K. Oda, T. Watakabe, I. Umemura, M. Takai, T. Okada, K. Sakata, and H. Karaki, *FEBS Lett.*, **311**, 12 (1992).

228. T. Tanaka, E. Tsukuda, M. Nozawa, H. Nonaka, T. Ohno, H. Kase, K. Yamada, and Y. Matsuda, *Mol. Pharmacol.*, **45**, 724 (1994).

229. J. P. Cullen, D. Bell, E. J. Kelso, and B. J. McDermott, *Eur. J. Pharmacol.*, **417**, 157 (2001).

230. G. Liu, N. S. Kozmina, M. Winn, T. W. von Geldern, W. J. Chiou, D. B. Dixon, B. Nguyen, K. C. Marsh, and T. J. Opgenorth, *J. Med. Chem.*, **42**, 3679 (1999).

231. J. Sakaki, T. Murata, Y. Yuumoto, I. Nakamura, and K. Hayakawa, *Bioorg. Med. Chem. Lett.*, **8**, 2247 (1998).

232. T. F. Walsh, K. Fitch, D. L. Williams, K. L. Murphy, N. A. Nolan, D. J. Pettibone, R. S. L. Chang, S. S. O'Malley, B. V. Clineschmidt, D. F. Veber, and W. J. Greenlee, *Bioorg. Med. Chem. Lett.*, **5**, 1155 (1995).

233. W. L. Cody, A. M. Doherty, J. X. He, P. L. DePue, L. A. Waite, J. G. Topliss, S. J. Haleen, D. Ladouceur, M. A. Flynn, K. E. Hill, and E. E. Reynolds, *Med. Chem. Res.*, **3**, 154 (1993).

234. W. L. Cody, A. M. Doherty, J. X. He, P. L. DePue, S. T. Rapundalo, G. A. Hingorani, T. C. Major, R. L. Panek, D. T. Dudley, S. J. Haleen, D. LaDouceur, K. E. Hill, M. A. Flynn, and E. E. Reynolds, *J. Med. Chem.*, **35**, 3301 (1992).

235. M. Clozel, V. Breu, K. Burri, J.-M. Cassal, W. Fischli, G. A. Gray, G. Hirth, B.-M. Löffler, M. Müller, W. Neidhart, and H. Ramuz, *Nature*, **365**, 759 (1993).

236. M. Clozel, V. Breu, G. A. Gray, B. Kalina, B. M. Löffler, K. Burri, J. M. Cassal, G. Hirth, M. Müller, N. Neidhart, and H. Ramuz, *J. Pharmacol. Exp. Ther.*, **279**, 228 (1994).

237. M. Clozel, H. Ramuz, J. P. Clozel, V. Breu, P. Hess, B. M. Löffler, P. Coassolo, and S. Roux, *J. Pharmacol. Exp. Ther.*, **290**, 840 (1999).

238. E. H. Ohlstein, P. Nambi, S. A. Douglas, R. M. Edwards, M. Gellai, A. Lago, J. D. Leber, R. D. Cousins, A. Gao, J. S. Frazee, et al., *Proc. Natl. Acad. Sci. USA*, **91**, 8052 (1994).

239. E. H. Ohlstein, P. Nambi, A. Logo, D. W. Hay, G. Beck, K. L. Fong, E. P. Eddy, P. Smith, H. Ellens, and J. D. Elliott, *J. Pharmacol. Exp. Ther.*, **276**, 609 (1996).

240. K. Ohnaka, R. Takayanagi, M. Nishikawa, M. Haji, and H. Nawata, *J. Biol. Chem.*, **268**, 26579 (1993).

241. M. Takahashi, Y. Matsushita, Y. Iijima, and K. Tanzawa, *J. Biol. Chem.*, **268**, 21394 (1993).

242. D. Xu, N. Emoto, A. Giaid, C. Slaughter, S. Kaw, D. deWit, and M. Yanagisawa, *Cell*, **78**, 473 (1994).

243. A. J. Turner, L. J. Murphy, M. S. Medeiros, and K. Barnes, *Adv. Exp. Med. Biol.*, **389**, 141 (1996).

244. Y. Tsurumi, K. Fujie, M. Nishikawa, S. Kiyoto, and M. Okuhara, *J. Antibiot. (Tokyo)*, **48**, 169 (1995).

245. S. Takaishi, N. Tuchiya, A. Sato, T. Negishi, Y. Takamatsu, Y. Matsushita, T. Watanabe, Y. Iijima, H. Haruyama, T. Kinoshita, M. Tanaka, and K. Kodama, *J. Antibiot. (Tokyo)*, **51**, 805 (1998).

246. Y. Asai, N. Nonaka, S. Suzuki, M. Nishio, K. Takahashi, H. Shima, K. Ohmori, T. Ohnuki, and S. Komatsubara, *J. Antibiot. (Tokyo)*, **52**, 607 (1999).

247. E. M. Wallace, J. A. Moliterni, M. A. Moskal, A. D. Neubert, N. Marcopulos, L. B. Stamford, A. J. Trapani, P. Savage, M. Chou, and A. Y. Jeng., *J. Med. Chem.*, **41**, 1513 (1998).

248. K. Umekawa, H. Hasegawa, Y. Tsutsumi, K. Sato, Y. Matsumura, and N. Ohnashi, *Jpn. J. Pharmacol.*, **84**, 7 (2000).

249. S. De Lombaert, L. Blanchard, L. B. Stamford, J. Tan, E. M. Wallace, Y. Satoh, J. Fitt, D. Hoyer, D. Simonsbergen, J. Moliterni, N. Marcopoulos, P. Savage, M. Chou, A. J. Trapani, and A. Y. Jeng, *J. Med. Chem.*, **43**, 488 (2000).

250. C. A. Fink, M. Moskal, F. Firooznia, D. Hoyer, D. Symonsbergen, D. Wei, Y. Qiao, P. Savage, M. E. Beil, A. J. Trapani, and A. Y. Jeng, *Bioorg. Med. Chem. Lett.*, **10**, 2037 (2000).

251. P. Martin, A. Tzanidis, A. Stein-Oakley, and H. Krum, *J. Cardiovasc. Pharmacol.*, **36**(5 Suppl. 1), S367 (2000).

252. K. Tatemoto, M. Carlquist, and V. Mutt, *Nature*, **296**, 659 (1982).

253. J. K. McDonald, *CRC Crit. Rev. Neurobiol.*, **4**, 97 (1988).

254. T. Hökfelt, J. M. Lundberg, H. Lagercrantz, K. Tatemoto, V. Mutt, J. Lindberg, L. Terenius, B. J. Everitt, K. Fuxe, L. Agnati, and M. Goldstein, *Neuroscience Lett.*, **36**, 217 (1983).

255. K. Tatemoto, *Proc. Natl. Acad. Sci. USA*, **79**, 5485 (1982).

256. G. Böttcher, J. Sjöberg, R. Ekman, R. Håkanson, and F. Sundler, *Regul. Pept.*, **43**, 115 (1993).

257. J. R. Kimmel, L. J. Hayden, and H. G. Pollock, *J. Biol. Chem.*, **250**, 9369 (1975).

258. D. Larhammar, C. Soderberg, and A. G. Blomqvist in C. Wahlestedt and W. F. Colmers, Eds., *The Neurobiology of Neuropeptide Y and Related Peptides*, Humana Press, Clifton, NJ, 1993, p. 1.

259. T. L. Blundell, J. E. Pitts, I. J. Tickle, S. P. Wood, and C. W. Wu, *Proc. Natl. Acad. Sci. USA*, **78**, 4175 (1981).

260. I. D. Glover, D. J. Barlow, J. E. Pitts, S. P. Wood, I. J. Tickle, T. L. Blundell, K. Tatemoto, J. R. Kimmel, A. Wollmer, W. Strassburger, and Y. S. Zhang, *Eur. J. Biochem.*, **142**, 379 (1985).

261. Y. Boulanger, Y. Chen, F. Commodari, L. Senecal, A. M. Laberge, A. Fournier, and S. St-Pierre, *Int. J. Peptide Protein Res.*, **45**, 86 (1995).

262. J. L. Krstenansky and S. H. Buck, *Neuropeptides*, **10**, 77 (1987).

263. J. Fuhlendorff, N. L. Johansen, S. G. Melberg, H. Thogersen, and T. W. Schwartz, *J. Biol. Chem.*, **265**, 11706 (1990).

264. M. M. T. O'Hare, S. Tenmoku, L. Aakerlund, L. Hilsted, A. Johnsen, and T. W. Schwartz, *Regul. Pept.*, **20**, 293 (1988).

265. T. W. Schwartz, J. Fuhlendorff, N. Langeland, H. Thogersen, J. C. Jorgensen, and S. P. Sheikh in V. Mutt, K. Fuxe, T. Hökfelt, and J. M. Lundberg, Eds., *Neuropeptide Y*, Raven Press, New York, 1989, p. 143.

266. T. Takeuchi, D. L. Gumucio, T. Yamada, M. H. Meisler, C. D. Minth, J. E. Dixon, R. E. Eddy, and T. B. Shows, *J. Clin. Invest.*, **77**, 1038 (1986).

267. C. D. Minth, S. R. Bloom, J. M. Polak, and J. E. Dixon, *Proc. Natl. Acad. Sci. USA*, **81**, 4577 (1984).

268. M. Martire and G. Pistritto, *Pharmacol. Res.*, **25**, 203–215 (1992).

269. R. Mentlein, P. Dahms, D. Grandt, and R. Krüger, *Regul. Pept.*, **49**, 133 (1993).

270. M. C. Michel, A. Beck-Sickinger, H. Cox, H. N. Doods, H. Herzog, D. Larhammar, R. Quirion, T. Schwartz, and T. Westfall, *Pharmacol. Rev.*, **50**, 143 (1998).

271. A. Balasubramaniam, *Peptides*, **18**, 445 (1997).

272. D. Larhammer, *Regul. Pept.*, **65**, 165 (1996).

273. D. R. Gehlert, *Proc. Soc. Exp. Biol. Med.*, **218**, 7 (1998).

274. J. Fuhlendorff, U. Gether, L. Aakerlund, N. Langeland-Johansen, H. Thogersen, S. G. Melberg, U. B. Olsen, O. Thastrup, and, T. W. Schwartz, *Proc. Natl. Acad. Sci. USA*, **87**, 182 (1990).

275. T. Peetrazzini, J. Seydoux, P. Kunstner, J. F. Aubert, E. Grouzmann, F. Beermann, and H. R. Brunner, *Nat. Med.*, **4**, 722 (1998).

276. H. N. Doods, H. A. Wieland, W. Engel, W. Eberlein, K. D. Willim, M. Entzeroth, W. Wienen, and K. Rudolf, *Regul. Pept.*, **65**, 71 (1996).

277. A. Inui, *Trends Pharmacol. Sci.*, **20**, 43 (1999).

278. S. P. Sheikh, E. Roach, J. Fuhlendorff and J. A. Williams, *Am. J. Physiol. Gastrointest. Liver Physiol.*, **260**, G250 (1991).

279. C. Wahlestedt, N. Yanaihara, and R. Håkanson, *Regul. Pept.*, **13**, 307 (1986).

280. S. Nielsen, S. P. Sheikh, M. I. Sheikh, and E. I. Christensen, *Am. J. Physiol. Renal Fluid Electrolyte Physiol.*, **260**, F359 (1991).

281. M. Stjernquist and C. Owman, *Acta Physiol. Scand.*, **138**, 95 (1990).

282. A. K. Myers, M. Y. Farhat, G. H. Shen, W. Debinski, C. Wahlestedt, and Z. Zukowska-Grojec, *Ann. N. Y. Acad. Sci.*, **611**, 408 (1990).

283. Y. Dumont, A. Fournier, S. St-Pierre, T. W. Schwartz, and R. Quirion, *Eur. J. Pharmacol.*, **191**, 501 (1990).

284. A. Inui, M. Okita, T. Inoue, N. Sakatani, M. Oya, H. Morioka, K. Shii, K. Yokono, N. Mizuno, and S. Baba, *Endocrinology*, **124**, 402 (1989).

285. W. F. Colmers, G. J. Klapstein, A. Fournier, S. St. Pierre, and K A. Treherne, *Br. J. Pharmacol.*, **102**, 41 (1991).

286. Y. Dumont, J. C. Martel, A. Fournier, S. St. Pierre, and R. Quirion, *Prog. Neurobiol.*, **38**, 125 (1992).

287. C. H. Broberger, M. Landry, H. Wong, J. N. Walsh, and T. Hoekfelt, *Neuroendocrinology*, **66**, 393 (1997).

288. A. G. Blomqvist and H. Herzog, *Trends Neurosci.*, **20**, 294 (1997).

289. L. Grundemar and R. Håkanson, *Gen. Pharmacol.*, **24**, 785 (1993).

290. L. Grundemar, C. Wahlestedt, and D. J. Reis, *Neurosci. Lett.*, **122**, 135 (1991).

291. A. Wahlestedt, S. Regunathan, and D. J. Reis, *Life Sci.*, **50**, PL-7 (1992).

292. A. Balasubramaniam, S. Sheriff, D. F. Rigel, and J. E. Fischer, *Peptides*, **11**, 545 (1990).

293. C. Gerald, M. W. Walker, L. Criscione, E. L. Gustafson, C. Batzl-Hartmann, K. E. Smith, P. Vaysse, M. M. Durkin, T. M. Laz, D. L. Linemeyer, A. O. Schaffhauser, S. Whitebread, K. G. Hofbauer, R. I. Taber, T. A. Brauchek, and R. L. Weinshank, *Nature*, **382**, 168 (1996).

294. M. A. Statnick, D. A. Schober, S. Gackenheimer, D. Johnson, L. Beavers, N. G. Mayne, J. P. Burnett, R. Gadski, and D. R. Gehlert, *Brain Res.*, **810**, 16 (1998).

295. A. Bischoff and M. C. Michel, *Trends Pharmacol. Sci.*, **20**, 104 (1999).

296. M. Matsumoto, T. Nomura, K. Momose, Y. Ikeda, Y. Kondou, H. Akiho, J. Togami, Y. Kimura, M. Okada, and T. Yamaguchi, *J. Biol. Chem.*, **271**, 27217 (1996).

297. P. Gregor, Y. Feng, L. B. Decarr, J. Cornfield, and M. L. McCaleb, *J. Biol. Chem.*, **271**, 27776 (1996).

298. M. A. McAuley, X. Chen, and T. C. Westfall in W. F. Colmers and C. Wahlestedt, Eds., *The Biology of Neuropeptide Y and Related Peptides*, Humana Press, Clifton, NJ, 1993, pp. 389.

299. M. Haass, B. Cheng, G. Richardt, R. E. Lang, and A. Schömig, *Naunyn-Schmiedebergs Arch. Pharmacol.*, **339**, 71 (1989).

300. J. Pernow and J. M. Lundberg, *Naunyn-Schmiedebergs Arch. Pharmacol.*, **340**, 379 (1989).

301. L.-M. Ren, Y. Furukawa, Y. Karasawa, M. Murakami, M. Takei, M. Narita, and S. Chiba, *J. Pharmacol. Exp. Ther.*, **259**, 38 (1991).

302. M. Moriarty, I. L. Gibbins, E. K. Potter, and D. I. McCloskey, *Neurosci. Lett.*, **139**, 275 (1992).

303. L. Edvinsson, R. Håkanson, C. Wahlestedt, and R. Uddman, *Trends Pharmacol. Sci.*, **8**, 231 (1987).

304. B. J. McDermott, B. C. Millar, and H. M. Piper, *Cardiovas. Res.*, **27**, 893–905 (1993).

305. Z. Zukowska-Grojec and C. Wahlestedt in W. F. Colmers and C. Wahlestedt, Eds., *The Biology of Neuropeptide Y and Related Peptides*, Humana Press, Clifton, NJ, 1993, p. 315.

306. P. Walker, E. Grouzmann, M. Burnier, and B. Waeber, *Trends Pharmacol. Sci.*, **12**, 111 (1991).

307. I. G. Joshua, *Peptides*, **12**, 37 (1991).

308. L. Edvinsson, E. Ekblad, R. Håkanson, and C. Wahlestedt, *Br. J. Pharmacol.*, **83**, 519 (1984).

309. M. P. Owen, *J. Pharmacol. Exp. Ther.*, **265**, 887 (1993).

310. C. Wahlestedt, R. Håkanson, C. A. Vaz, and Z. Zukowska-Grojec, *Am. J. Physiol. Regul. Integr. Comp. Physiol.*, **258**, R736 (1990).

311. C. Han and P. W. Abel, *J. Cardiovasc. Pharmacol.*, **9**, 675 (1987).

312. B. C. Millar, K.-O. Schlüter, X.-J. Zhou, B. J. McDermott, and H. M. Piper, *Am. J. Physiol. Cell Physiol.*, **266**, C1271 (1994).

313. L. Quadri, M. Gobbini, and L. Monti, *Curr. Pharm. Des.*, **4**, 489 (1998).

314. A. Franco-Cereda and J. Liska, *Eur. J. Pharmacol.*, **349**, 1 (1998).

315. J. Pernow, J. M. Lundberg, L. Kaijser, P. Hjemdahl, E. Theodorsson-Norheim, A. Martinsson, and B. Pernow, *Clin. Physiol.*, **6**, 561 (1986).

316. M. R. Brown, et al. in V. Mutt, K. Fuxe, T. Hökfelt, and J. M. Lundberg, Eds., *Neuropeptide Y*, New York, Raven Press, 1989, pp. 201–214.

317. A. Bischoff and M. C. Michel, *Pfluegers Arch.-Eur. J. Physiol.*, **435**, 443 (1998).

318. E. Hackenthal, K. Aktories, K. H. Jakobs, and R. E. Lang, *Am. J. Physiol. Renal Fluid Electrolyte Physiol.*, **252**, F543 (1987).

319. J.-F. Aubert, P. Walker, E. Grouzmann, J. Nussberger, H. R. Brunner, and B. Waeber, *Clin. Exp. Pharmacol. Physiol.*, **19**, 223 (1992).

320. R. Corder, M. B. Vallotton, P. J. Lowry, and A. G. Ramage, *Neuropeptides*, **14**, 111 (1989).

321. S. F. Echtenkamp and P. F. Dandridge, *Am. J. Physiol. Renal Fluid Electrolyte Physiol.*, **256**, F524 (1989).

322. A. Bischoff, K. Munter, and M. C. Michel, *Naunyn-Schmiedebergs Arch. Pharmacol.*, **351** (Suppl.), R150 (1995).

323. B. Baranowska, J. Gutkowska, A. Lemire, M. Cantin, and J. Genest, *Biochem. Biophys. Res. Commun.*, **145**, 680 (1987).

324. P. B. Persson, H. Ehmke, B. Nafz, R. Lang, E. Hackenthal, R. Nobiling, M. S. Dietrich, and H. R. Kirchheim, *J. Physiol. (Lond.)*, **444**, 289 (1991).

325. A. Bischoff, W. Erdbrugger, J. Smits, and M. C. Michel, *Naunyn-Schmiedebergs Arch. Pharmacol.*, **349**(Suppl.), R39 (1994).

326. M. M. M. El-Din and K. U. Malik, *J. Pharmacol. Exp. Ther.*, **246**, 479 (1988).

327. W. F. Oellerich and K. U. Malik, *J. Pharmacol. Exp. Ther.*, **266**, 1321 (1993).

328. D. D. Smyth, D. E. Blandford, and S. L. Thom, *Eur. J. Pharmacol.*, **152**, 157 (1988).

329. J. C. Erickson, K. E. Clegg, and R. D. Palmiter, *Nature*, **381**, 415 (1996).

330. B. Wahlestedt, E. M. Pich, G. F. Koob, F. Yee, and M. Heilig, *Science*, **359**, 528 (1993).

331. Z. Zukowska-Grojec, *Drug News Perspect.*, **10**, 587 (1997).

332. C. J. Small, D. G. Morgan, K. Meeran, M. M. Heath, I. Gunn, C. M. Edwards, J. Gardiner, G. M. Taylor, J. D. Hurley, M. Rossi, A. P. Goldstone, D. O'Shea, D. M. Smith, M. A. Ghatei, and S. R. Bloom, *Proc. Natl. Acad. Sci. USA*, **94**, 11686 (1997).

333. G. G. Nussdorfer and G. Gottardo, *Horm. Metab. Res.*, **30**, 368 (1998).

334. H. A. Wieland, B. S. Hamilton, B. Krist, and H. N. Doods, *Expert Opin. Invest. Drugs*, **9**, 1327 (2000).

335. F. Kokot and R. Ficek, *Miner. Electrolyte Metab.*, **25**, 303 (1999).

336. K. Rudolf, W. Eberlein, W. Engel, H. A. Wieland, K. D. Willim, M. Entzeroth, W. Wienen, A. G. Beck-Sickinger, and H. N. Doods, *Eur. J. Pharmacol.*, **271**, R11 (1994).

337. L. Criscione, P. Rigollier, C. Batzl-Hartmann, H. Rueger, A. Stricker-Krongrad, P. Wyss, L. Brunner, S. Whitebread, Y. Yamaguchi, C. Gerald, R. O. Heurich, M. W. Walker, M. Chiesi, W. Schilling, K. G. Hofbauer, and N. Levens, *J. Clin. Invest.*, **102**, 2136 (1998).

338. A. W. Stamford and E. M. Parker, *Annu. Rep. Med. Chem.*, **34**, 31 (1999).

339. G. Moore, A. Letter, M. Tesanovic, and K. Lederis, *Can. J. Biochem.*, **53**, 242 (1975).

340. Y. Coulourn, I. Lihrmann, S. Jegou, Y. Anour, H. Tostivint, J. C. Beauvillain, J. M. Conlon, H. A. Bern, and H. Vaudry, *Proc. Natl. Acad. Sci. USA*, **95**, 15803 (1998).

341. J. S. Culp, D. E. McNulty, C. E. Ellis, S. A. Douglas, R. N. Willette, N. V. Aiyar, A. R. Arnold, N. Khandoudi, B. Gout, and K. Al-Barazanji, PCT World Pat. WO9935266 (1999).

342. J. M. Conlon, D. Arnold-Reed, and R. J. Balment, *FEBS Lett.*, **266**, 37 (1990).

343. H. P. Nothacker, Z. Wang, A. M. McNeill, Y. Saito, S. Merten, B. O'Dowd, S. P. Duckles, and O. Civelli, *Nat. Cell Biol.*, **1**, 383 (1999).

344. R. S. Ames, H. M. Sarau, J. K. Chambers, R. N. Willette, N. V. Aiyar, A. M. Romanic, C. S. Louden, J. J. Foley, C. F. Sauermelch, R. W. Coatney, Z. Ao, J. Disa, S. D. Holmes, J. M. Stadel, J. D. Martin, W. S. Liu, G. I. Glover, S. Wilson, D. E. McNulty, C. E. Ellis, N. A. Eishourbagy, U. Stabon, J. J. Trill, D. W. Hay, E. H. Ohlstein, D. J. Bergsma, and S. A. Douglas, *Nature*, **401**, 282 (1999).

345. M. Mori, T. Sugo, M. Abe, Y. Shimomura, M. Kurihara, C. Kitada, K. Kikuchi, Y. Shintani, T. Kurokawa, H. Onda, O. Nishimura, and M. Fujino, *Biochem. Biophys. Res. Commun.*, **265**, 123 (1999).

346. Y. Coulouarn, S. Jegou, H. Tostivint, H. Vaudry, and I. Lihrmann, *FEBS Lett.*, **457**, 28 (1999).

347. D. K. Chan, R. Gunther, and H. A. Bern, *Gen. Comp. Endocrinol.*, **34**, 347 (1978).

348. H. Itoh, Y. Itoh, J. Rivier, and K. Lederis, *Am. J. Physiol. Regul. Integr. Comp. Physiol.*, **252**, R361 (1987).

349. J. M. Conlon, H. Tostivint, and H. Vaudry, *Regul. Pept.*, **69**, 95 (1997).

350. R. Bhaskaran, A. I. Arunkumar, and C. Yu, *Biochem. Biophys. Acta*, **1199**, 115 (1994).

351. T. D. Perkins, S. Bansal, and D. J. Barlow, *Biochem. Soc. Trans.*, **18**, 918 (1990).

352. B. McMaster, Y. Kobayashi, J. Rivier, and K. Lederis, *Proc. West. Pharmacol. Soc.*, **29**, 205 (1986).

353. H. Itoh, D. McMaster, and K. Lederis, *Eur. J. Pharmacol.*, **149**, 61 (1988).

354. A. Marchese, M. Heiber, T. Nguyen, H. H. Heng, V. R. Saldivia, R. Cheng, P. M. Murphy, L. C. Tsui, X. Shi, P. Gregor, S. R. George, B. F. O'Dowd, and J. M. Docherty, *Genomics*, **29**, 335 (1995).

355. Q. Liu, S. S. Pong, Z. Zeng, Q. Zhang, A. D. Howard, D. L. Williams, M. Davidoff, R. Wang, C. P. Austin, T. P. McDonald, C. Bai, S. R. George, J. F. Evans, and C. T. Caskey, *Biochem. Biophys. Res. Commun.*, **266**, 174 (1999).

356. J. J. Maguire, R. E. Kuc, and A. P. Davenport, *Br. J. Pharmacol.*, **131**, 441 (2000).

357. O. S. Opgaard, H. P. Notracker, F. J. Ehlert, and D. N. Krause, *Eur. J. Pharmacol.*, **406**, 265 (2000).

358. A. Gibson, S. Conyers, and H. A. Bern, *J. Pharm. Pharmacol.*, **40**, 893 (1988).

359. S. A. Douglas, A. C. Sulpizio, V. Piercy, H. M. Sarau, R. S. Ames, N. V. Aiyar, E. H. Ohlstein, and R. A. Willette, *Br. J. Pharmacol.*, **131**, 1262 (2000).

360. D. W. Hay, M. A. Luttmann, and S. A. Douglas, *Br. J. Pharmacol.*, **131**, 10 (2000).

361. M. R. MacLean, D. Alexander, A. Stirrat, M. Gallagher, S. A. Douglas, E. H. Ohlstein, I. Morecroft, and K. Polland, *Br. J. Pharmacol.*, **130**, 201 (2000).

362. J. P. Changeux, A. Duclert, and S. Sekine, *Ann. N. Y. Acad. Sci.*, **657**, 361 (1992).

363. T. G. Flynn, M. L. de Bold, and A. J. de Bold, *Biochem. Biophys. Res. Commun.*, **117**, 859 (1983).

364. S. A. Atlas, H. D. Kleiner, M. J. Camargo, A. Januszewicz, J. E. Sealey, J. H. Laragh, J. W. Schilling, J. A. Lewicki, L. K. Johnson, and T. Maack, *Nature*, **309**, 717 (1984).

365. T. L. Yang-Feng, G. Floyd-Smith, M. Nemer, J. Droun, and U. Francke, *Am. J. Hum. Genet.*, **37**, 1117 (1985).

366. D. G. Gardner, C. F. Deschepper, W. F. Ganong, S. Hane, J. Fiddes, J. D. Baxter, and J. Lewicki, *Proc. Natl. Acad. Sci. USA*, **83**, 6697 (1986).

367. D. L. Vesely, P. A. Palmer, and A. T. Giordano, *Peptides*, **13**, 165 (1992).

368. G. Thibault, R. Garcia, J. Gutkowska, J. Bilodeau, C. Lazure, N. G. Seidah, M. Chretien, J. Genest, and M. Cantin, *Biochem. J.*, **241**, 265 (1987).

369. M. Miyata, K. Kangawa, T. Toshimori, T. Hatoh, and H. Matsuo, *Biochem. Biophys. Res. Commun.*, **129**, 248 (1985).

370. D. L. Vesely, P. Norsk, C. J. Winters, D. M. Rico, A. L. Sallman, and M. Epstein, *Proc. Soc. Exp. Biol. Med.*, **192**, 230 (1989).

371. A. L. Gerbes and A. M. Vollmar, *Biochem. Biophys. Res. Commun.*, **156**, 228 (1988).

372. J. R. Dietz, S. J. Nazian, and D. L. Vesely, *Am. J. Physiol. Heart Circ. Physiol.*, **260**, H1774 (1991).

373. A. Sugawara, K. Nakao, N. Morii, T. Yamada, H. Itoh, S. Shiono, Y. Saito, M. Mukoyama, H. Arai, K. Nishimura, K. Obata, H. Yasue, T. Ban, and H. Imura, *J. Clin. Invest.*, **81**, 1962 (1988).

374. L. Ngo, R. P. Wyeth, J. K. Bissett, W. L. Hester, M. T. Newton, A. L. Sallman, C. J. Winters, and D. Vesely, *Am. Heart J.*, **117**, 385 (1989).

375. A. M. Richards, J. G. F. Cleland, G. Tonolo, G. D. McIntyre, B. J. Leckie, H. J. Dargie, S. G. Ball, and J. I. S. Robertson, *Br. Med. J.*, **293**, 409 (1986).

376. A. M. Richards, G. Tonolo, G. D. McIntyre, B. J. Leckie, and J. I. S. Robertson, *J. Hypertens.*, **5**, 227 (1987).

377. P. Montorsi, G. Tonolo, J. Polonia, D. Hepburn, and A. M. Richards, *Hypertension*, **10**, 570 (1987).

378. M. P. Gnädinger, R. E. Lang, L. Hasler, D. E. Uehlinger, S. Shaw, and P. Weidmann, *Miner. Electrolyte Metab.*, **12**, 371 (1986).

379. J. A. Koehn, J. A. Norman, B. N. Jones, L. LeSueur, Y. Sakane, and R. D. Ghai, *J. Biol. Chem.*, **262**, 11623 (1987).

380. A. S. Hollister, R. J. Rodeheffer, F. J. White, J. R. Potts, T. Imada, and T. Inagami, *J. Clin. Invest.*, **83**, 623 (1989).

381. Y. Hashimoto, K. Nakao, N. Hama, H. Imura, S. Mori, M. Yamaguchi, M. Yasuhara, and R. Hori, *Pharm. Res.*, **11**, 60 (1994).

382. A. A. Seymour, J. N. Swerdel, S. A. Fennell, and N. G. Delaney, *Life Sci.*, **43**, 2265 (1988).

383. G. M. Olins, K. L. Spear, N. R. Siegel, and H. A. Zurcher-Neely, *Biochem. Biophys. Acta*, **901**, 97 (1987).

384. T. Yandle, I. Crozier, G. Nicholls, E. Espiner, A. Carne, and S. Brennan, *Biochem. Biophys. Res. Commun.*, **146**, 832 (1987).

385. S.-I. Suga, K. Nakao, K. Hosoda, M. Mukoyama, Y. Ogawa, G. Shirakami, H. Arai, Y. Saito, Y. Kambayashi, K. Inouye, and H. Imura, *Endocrinology*, **130**, 229 (1992).

386. T. Sudoh, K. Kangawa, N. Minamino, and H. Matsuo, *Nature*, **332**, 78 (1988).

387. S. Ueda, N. Minamino, T. Sudoh, K. Kangawa, and H. Matsuo, *Biochem. Biophys. Res. Commun.*, **155**, 733 (1988).

388. N. Minamino, M. Aburaya, S. Ueda, K. Kangawa, and H. Matsuo, *Biochem. Biophys. Res. Commun.*, **155**, 740 (1988).

389. H. Imura, K. Nakao, and H. Itoh, *Front. Neuroendocrinol.*, **13**, 217 (1992).

390. H. Tateyama, J. Hino, N. Minamino, K. Kangawa, T. Ogihara, and H. Matsuo, *Biochem. Biophys. Res. Commun.*, **166**, 1080 (1990).

391. K. Takahashi, K. Totsune, M. Sone, M. Ohneda, O. Murakami, K. Itoi, and T. Mouri, *Peptides*, **13**, 121 (1992).

392. T. Sudoh, K. Maekawa, M. Kojima, N. Minamino, K. Kangawa, and H. Matsuo, *Biochem. Biophys. Res. Commun.*, **159**, 1427 (1989).

393. K. Hosoda, K. Nakao, M. Mukoyama, Y. Saito, M. Jougasaki, G. Shirakami, S. Suga, Y. Ogawa, H. Yasue, and H. Imura, *Hypertension*, **17**, 1152 (1991).

394. G. Wambach and J. Koch, *Clin. Exp. Hypertens.*, **17**, 619 (1995).

395. Y. Kambayashi, K. Nakao, M. Mukoyama, Y. Saito, Y. Ogawa, S. Shiono, K. Inouye, N. Yoshida, and H. Imura, *FEBS Lett.*, **259**, 341 (1990).

396. S. J. Holmes, E. A. Espiner, A. M. Richards, T. G. Yandle, and C. Frampton, *J. Clin. Endocrinol. Metab.*, **76**, 91 (1993).

397. M. Mukoyama, K. Nakao, K. Hosoda, S. Suga, Y. Saito, Y. Ogawa, G. Shirakami, M. Jougasaki, K. Obata, H. Yasue, Y. Kambayashi, K. Inouye, and H. Imura, *J. Clin. Invest.*, **87**, 1402 (1991).

398. A. J. Kenny, A. Bourne, and J. Ingram, *Biochem. J.*, **291**, 83 (1993).

399. Y. Ogawa, H. Itoh, Y. Yoshitake, M. Inoue, T. Yoshimasa, T. Serikawa, and K. Nakao, *Genomics*, **24**, 383 (1994).

400. T. Sudoh, N. Minamino, K. Kangawa, and H. Matsuo, *Biochem. Biophys. Res. Commun.*, **168**, 863 (1990).

401. Y. Tawaragi, K. Fuchimura, S. Tanaka, N. Minamino, K. Kangawa, and H. Matsuo, *Biochem. Biophys. Res. Commun.*, **175**, 645 (1991).

402. Y. Komatsu, K. Nakao, S. Suga, Y. Ogawa, M. Mukoyama, H. Arai, G. Shirakami, K. Hosoda, O. Nakagawa, N. Hama, I. Kishimoto, and H. Imura, *Endocrinology*, **129**, 1104 (1991).

403. N. Minamino, K. Kangawa, and H. Matsuo, *Biochem. Biophys. Res. Commun.*, **170**, 973 (1990).

404. M. Kojima, N. Minamino, K. Kangawa, and H. Matsuo, *FEBS Lett.*, **276**, 209 (1990).

405. S. Suga, H. Itoh, Y. Komatsu, Y. Ogawa, N. Hama, Y. Yoshimasa, and K. Nakao, *Endocrinology*, **133**, 3038 (1993).

406. S. Suga, K. Nakao, H. Itoh, Y. Komatsu, Y. Ogawa, N. Hama, and H. Imura, *J. Clin. Invest.*, **90**, 1145 (1992).

407. J. C. Dussaule, A. Stefanski, M. L. Bea, P. Ronco, and R. Ardailou, *Am. J. Physiol. Renal Fluid Electrolyte Physiol.*, **264**, F45 (1993).

408. R. E. Stewart, S. E. Swithers, L. M. Plunkett, and R. McCarty, *Neurosci. Biobehav. Rev.*, **12**, 151 (1988).

409. S. Chang, D. G. Lowe, M. Lewis, R. Hellmiss, E. Chen, and D. V. Goeddel, *Nature*, **341**, 68 (1989).

410. S. Schulz, S. Singh, R. A. Bellet, G. Singh, D. J. Tubb, H. Chin, and D. L. Garbers, *Cell*, **58**, 1155 (1989).

411. Y. Morishita, T. Sano, K. Ando, Y. Saitoh, H. Kase, K. Yamada, and Y. Matsuda, *Biochem. Biophys. Res. Commun.*, **176**, 949 (1991).

412. S. K. Hanks, A. M. Quinn, and T. Hunter, *Science*, **241**, 42 (1988).

413. K. J. Koller, D. G. Lowe, G. L. Bennett, N. Minamino, K. Kangawa, H. Matsuo, and D. V. Goeddel, *Science*, **252**, 120 (1991).

414. B. D. Bennett, G. L. Bennett, R. V. Vitangcol, J. R. S. Jewett, J. Burnier, W. Henzel, and D. G. Lowe, *J. Biol. Chem.*, **266**, 23060 (1991).

415. F. Fuller, J. G. Porter, A. E. Arfsten, J. Muller, J. W. Schilling, R. M. Scarborough, J. A. Lewicki, and D. B. Schenk, *J. Biol. Chem.*, **263**, 9395 (1988).

416. T. Maack, *Ann. Rev. Physiol.*, **54**, (1992).

417. M. B. Anand-Srivastava and G. J. Trachte, *Pharmacol. Rev.*, **45**, 455 (1993).

418. D. R. Nusseinzveig, J. Lewicki, and T. Maack, *J. Biol. Chem.*, **265**, 20952 (1990).

419. A. Hughes, S. Thom, P. Goldberg, G. Martin, and P. Sever, *Clin. Sci.*, **74**, 207 (1988).

420. R. J. Winquist, *Fed. Proc.*, **45**, 2371 (1986).

421. T. J. Ebert, M. M. Skelton, and A. W. Cowley, *Hypertension*, **11**, 537 (1988).

422. A. Morice, J. Pepke-Zaba, E. Loysen, R. Lapworth, M. Ashby, T. Higenbottam, and M. Brown, *Clin. Sci.*, **74**, 359 (1988).

423. D. De Zeeuw, W. M. T. Janssen, and P. E. de Jong, *Kidney Int.*, **41**, 115 (1992).

424. G. Christensen, *Scand. J. Clin. Lab. Invest.*, **53**, 203 (1993).

425. B. A. Clark, D. Elahi, R. P. Shannon, J. Y. Wei, and F. H. Epstein, *Am. J. Hypertens.*, **4**, 500 (1991).

426. D. J. Grandis, B. F. Uretsky, S. M. Ray, L. Vassilaros, J. G. Verbalis, and J. B. Puschett, *Am. J. Hypertens.*, **4**, 219 (1991).

427. R. L. Solomon, J. C. Atherton, H. Bobinski, V. Hillier, and R. Green, *Clin. Sci.*, **75**, 403 (1988).

428. B. M. Brenner, B. J. Ballermann, M. E. Gunning, and M. L. Zeidel, *Physiol. Rev.*, **70**, 665 (1990).

429. L. G. Melo, M. E. Steinhelper, S. C. Pang, Y. Tse, and U. Ackermann, *Physiol. Genomics*, **3**, 45 (2000).

430. C. C. Lang and A. D. Struthers, *Clin. Auton. Res.*, **1**, 329 (1991).

431. P. Thorén, A. L. Mark, D. A. Morgan, T. P. O'Neill, P. Needleman, and M. J. Brody, *Am. J. Physiol. Heart Circ. Physiol.*, **251**, H1252 (1986).

432. H. Fujisaki, H. Itoh, Y. Hirata, M. Tanaka, M. Hata, M. Lin, S. Adachi, H. Akimoto, F. Marumo, and M. Hiroe, *J. Clin. Invest.*, **96**, 1059 (1995).

433. R. R. Brandt, D. M. Heublein, M. T. Mattingly, M. R. Pittelkow, and J. C. Burnett, *Am. J. Physiol. Heart Circ. Physiol.*, **268**, H921 (1995).

434. M. Kohno, K. Yokokawa, T. Horio, K. Yasunari, K. I. Murakawa, and T. Takeda, *Circ. Res.*, **70**, 241 (1992).

435. L. V. Gonzalez, M. P. Majowicz, and N. A. Vidal, *Peptides*, **21**, 875 (2000).

436. C. Wu, N. H. Bishopric, and R. E. Pratt, *J. Biol. Chem.*, **272**, 14860 (1997).

437. A. McGregor, M. Richards, E. A. Espiner, T. Yandle, and H. Ikram, *J. Clin. Endocrinol. Metab.*, **70**, 1103 (1990).

438. S. J. Holmes, E. A. Espiner, A. M. Richards, T. G. Yandle, and C. Frampton, *J. Clin. Endocrinol. Metab.*, **76**, 91 (1993).

439. G. La Villa, L. Stefani, C. Lazzeri, C. Zurli, C. Tosti Guerra, G. Barletta, R. Bandinelli, G. Strazzulla, and F. Franchi, *Hypertension*, **26**, 628 (1995).

440. Y. Ogawa, H. Itoh, N. Tamura, S. Suga, T. Yoshimasa, M. Uehira, S. Matsuda, S. Shiono, H. Nishimoto, and K. Nakao, *J. Clin. Invest.*, **93**, 1911 (1994).

441. L. Cao and D. G. Gardner, *Hypertension*, **25**, 227 (1995).

442. E. A. Espiner in W. K. Samson and E. R. Levin, Eds., *Contemporary Endocrinology: Natriuretic Peptide in Health and Disease*, Humana Press, Totowa, NJ, 1997, p. 123.

443. K. T. Weber, *Semin. Nephrol.*, **17**, 467 (1997).

444. J. Yamashita, H. Itoh, Y. Ogawa, N. Tamura, K. Takaya, T. Igaki, T. Doi, T. Chun, M. Inoue, K. Masatsugu, and K. Nakao, *Hypertension*, **29** (Part 2), 381 (1997).

445. P. J. Hunt and T. G. Yandle, *J. Clin. Endocrinol. Metab.*, **78**, 1428 (1994).

446. A. J. Stingo, A. L. Clavell, D. M. Heublein, C. M. Wei, M. R. Pittelkow, and J. C. Burnett, *Am. J. Physiol. Heart Circ. Physiol.*, **263**, H1318 (1992).

447. M. Furuya, M. Yoshida, Y. Hayashi, N. Ohnuma, N. Minamino, K. Kangawa, and H. Matsuo, *Biochem. Biophys. Res. Commun.*, **177**, 927 (1991).

448. H. H. Chen and J. C. Burnett, *J. Cardiovasc. Pharm.*, **32**(Suppl. 3), S22 (1998).

449. N. Minamino, M. Aburaya, M. Kojima, K. Miyamoto, K. Kangawa, and H. Matsuo, *Biochem. Biophys. Res. Commun.*, **197**, 326 (1993).

450. J. P. Herman, M. C. Langub, and R. E. Watson, *Endocrinology*, **133**, 1903 (1993).

451. R. C. Fowkes and C. A. McArdle, *Trends Endocrinol. Metab.*, **11**, 333 (2000).

452. T. Marumo, T. Nakaki, K. Hishikawa, J. Hirahashi, H. Suzuki, R. Kato, and T. Surata, *Endocrinology*, **136**, 2135 (1995).

453. M. Kohno, T. Horio, K. Yakokawa, N. Kurihara, and T. Takeda, *Hypertension*, **19**, 320 (1992).

454. P. Vigne, L. Lund, and C. Frelin, *J. Neurochem.*, **62**, 2269 (1994).

455. A. M. Richards, *J. Intern. Med.*, **235**, 543 (1994).

456. L. G. Melo, S. C. Pang, and U. Ackermann, *News Physiol. Sci.*, **15**, 143 (2000).

457. M. G. Nichols, *J. Intern. Med.*, **235**, 515 (1994).

458. D. F. Davila, J. H. Donis, G. Bellabarba, A. Torres, J. Casado, and C. Mazzei de Davila, *Med. Hypotheses*, **54**, 242 (2000).

459. D. L. Vesely, *Congestive Heart Fail.*, **5**, 171 (1999).

460. J. A. Grantham and J. C. Burnett, Contemporary Endocrinology: Natriuretic Peptides in Health and Disease, Vol. **5**, Humana Press, Totowa, NJ, 1997, p. 309.

461. R. O. Bonow, *Circulation*, **93**, 1946 (1996).

462. J. G. McDougall, R. DeMatteo, C. N. May, and N. A. Yates, *Adv. Organ Biol.*, **9** (Renal Circulation), 157 (2000).

463. M. K. Dishart and J. A. Kellum, *Drugs*, **59**, 79 (2000).

464. R. L. Allgren, T. C. Marbury, S. N. Rahman, L. S. Weisburg, A. Z. Fenves, R. A. Lafayette, R. M. Sweet, F. C. Genter, B. R. Kurnik, J. D. Conger, and M. H. Sayegh, *N. Engl. J. Med.*, **336**, 828 (1997).

465. M. Furuya, K. Aisaka, T. Miyazaki, N. Honbou, K. Kawashima, and T. Ohno, *Biochem. Biophys. Res. Commun.*, **193**, 248 (1993).

466. K. Kugiyama, S. Sugijama, T. Matsumura, and O. Yasutaka, *Circulation*, **90**, 1 (1994).

467. K. Kugiyama, S. Sugijama, T. Matsumura, and T. Yasue, *Circulation*, **88**, 1 (1993).

468. K. Lin, J. Chao, and L. Chao, *Hypertension*, **26**, 847 (1995).

469. J. A. Lewicki and A. A. Protter in J. H. Zaragh and B. M. Brenner, Eds., *Hypertension: Pathophysiology, Diagnosis, and Management*, Raven Press, New York, 1995, p. 1029.

470. B. Favrat, M. Burnier, J. Nussberger, J. M. Lecomte, R. Brouard, B. Waeber, and H. R. Brunner, *J. Hypertens.*, **13**, 797 (1995).

471. B. G. Firth, R. Perna, J. F. Bellomo, and R. D. Toto, *Am. J. Med. Sci.*, **297**, 203 (1989).

472. J. A. Robb and D. E. Ryono, *Exp. Opin. Ther. Pat.*, **9**, 1665 (1999).

473. J. Bralet and J. C. Schwartz, *Trends Pharmacol. Sci.*, **22**, 106 (2001).

474. H. H. Chen and J. C. Burnett, *Proc. Assoc. Am. Physicians*, **111**, 406 (1999).

475. T. Omland, V. V. S. Bonarjee, R. T. Lie, and K. Caidahl, *Am. J. Cardiol.*, **76**, 230 (1995).

476. N. Arakawa, M. Nakamura, H. Aoki, and K. Hiramori, *J. Am. Coll. Cardiol.*, **27**, 1656 (1996).

477. M. G. Rosenfeld, J.-J. Mermod, S. G. Amara, L. W. Swanson, P. E. Sawchenko, J. Rivier, W. W. Vale, and R. M. Evans, *Nature*, **304**, 129 (1983).

478. L. H. Breimer, I. MacIntyre, and M. Zaidi, *Biochem. J.*, **255**, 377 (1988).

479. M. G. Rosenfeld, C. R. Lin, S. G. Amara, L. S. Stolarsky, B. A. Roos, E. S. Ong, and R. M. Evans, *Proc. Natl. Acad. Sci. USA*, **79**, 1717 (1982).

480. S. D. Brain, T. J. Williams, J. R. Tippins, H. R. Morris, and I. MacIntyre, *Nature*, **313**, 54 (1985).

481. G. K. Asimakis, D. J. DiPette, V. R. Conti, O. B. Holland, and J. B. Zwischenberger, *Life Sci.*, **41**, 597 (1987).

482. Y. Lee, K. Tokami, Y. Kawai, S. Girgis, C. J. Hillyard, I. MacIntyre, P. C. Emson, and M. Tohyama, *Neuroscience*, **15**, 1227 (1985).

483. G. Skofitsch and D. M. Jacobowitz, *Peptides*, **6**, 721 (1985).

484. L. Kruger, W. Mantyh, C. Sternini, N. C. Brecha, and C. R. Mantyh, *Brain Res.*, **463**, 223 (1988).

485. D. J. DiPette and S. J. Wimalawansa in J. Cross III and L. V. Aveoli, Eds., *Cardiovascular Actions of Calcitropic Hormones*, CRC Press, Baltimore, 1995, p. 239.

486. J. McEvan, S. Legon, S. J. Wimalawansa, M. Zaidi, C. T. Dollery, and I. MacIntyre in J. H. Laragh, B. M. Brenner, and N. M. Kaplan, Eds., *Endocrine Mechanisms in Hypertension*, Raven Press, New York, 1989, p. 287.

487. S. J. Winalawansa and I. MacIntyre, *Int. J. Cardiol.*, **20**, 29 (1988).

488. D. Przepiorka, S. B. Baylin, O. W. McBride, J. R. Testa, A. de Bustros, and B. D. Nelkin, *Biochem. Biophys. Res. Commun.*, **120**, 493 (1984).

489. S. G. Amara, J. L. Arriza, S. E. Leff, L. W. Swanson, R. M. Evans, and M. G. Rosenfeld, *Science*, **229**, 1094 (1985).

490. S. Gulbenkian, A. Merighi, J. Wharton, I. M. Varndell, and J. M. Polak, *J. Neurocytol.*, **15**, 535 (1986).

491. C. Beglinger, W. Born, R. Münch, A. Kurtz, J.-P. Gutzwiller, K. Jäger, and J. A. Fischer, *Peptides*, **12**, 1347 (1991).

492. C. Saldanha and D. Mahadevan, *Protein Eng.*, **4**, 539 (1991).

493. T. Dennis, A. Fournier, S. St. Pierre, and R. Quirion, *J. Pharmacol. Exp. Ther.*, **251**, 718 (1989).

494. B. Lynch and E. T. Kaiser, *Biochemistry*, **27**, 7600 (1998).

495. Y. Boulanger, A. Khaiat, Y. Chen, S. St.-Pierre, and A. Fournier, *Peptide Res.*, **8**, 206 (1995).

496. T. Chiba, Y. Yamaguchi, T. Yamatani, A. Nakamura, T. Morishita, T. Inui, M. Fukase, T. Noda, and T. Fujita, *Am. J. Physiol. Endocrinol. Metab.*, **256**, E331 (1989).

497. J. P. O'Connell, S. M. Kelly, D. P. Raleigh, J. A. Hubbard, N. C. Price, C. M. Dobson, and B. J. Smith, *Biochem. J.*, **291**, 205 (1993).

498. V. Jonas, C. R. Lin, E. Kawashima, D. Semon, L. W. Swanson, J. J. Mermod, R. M. Evans, and M. G. Rosenfeld, *Proc. Natl. Acad. Sci. USA*, **82**, 1994 (1985).

499. J. W. Hoppener, P. H. Steenberg, R. J. Slebos, A. Visser, C. J. Lips, H. S. Janz, J. M. Bechet, M. Lenoirg, W. Born, and S. Haller-Berm, *J. Clin. Endocrinol. Metab.*, **64**, 809 (1987).

500. M. G. Rosenfeld, J. J. Mermod, S. G. Amara, L. W. Swanson, P. E. Sawchenko, J. Rivier, W. W. Vale, and R. M. Evans, *Nature*, **304**, 129 (1983).

501. G. Hofle, R. Weiler, R. Fischer-Colbrie, C. Humpel, A. Laslop, T. Wohlfarter, R. Hogue-Angeletti, A. Saria, P. J. Fleming, and H. Winkler, *Regul. Pept.*, **32**, 321 (1991).

502. S. Gulbenkian, A. Merighi, and J. Wharton, *J. Neurocytol.*, **15**, 535 (1986).

503. A. Franco-Cereceda, *Br. J. Pharmacol*, **102**, 506 (1991).

504. A. Franco-Cereceda, *Acta Physiol. Scand.*, **133**(Suppl. 569), 1 (1988).

505. P. L. Greves, F. Nyberg, T. Hokfelt, and L. Terenius, *Regul. Pept.*, **25**, 277 (1989).

506. S. D. Brain and T. J. Williams, *Br. J. Pharmacol.*, **86**, 855 (1985).

507. K. Kitamura, K. Kangawa, M. Kawamoto, Y. Ichiki, S. Nakamura, H. Matsuo, and T. Eto, *Biochem. Biophys. Res. Commun.*, **192**, 553 (1993).

508. T. Ishimitsu, M. Kojima, K. Kangawa, J. Hino, H. Matsuoka, K. Kitamura, T. Eto, and H. Matsuo, *Biochem. Biophys. Res. Commun.*, **203**, 631 (1994).

509. K. Kitamura, J. Kato, M. Kawamoto, M. Tanaka, N. Chino, K. Kangawa, and T. Eto, *Biochem. Biophys. Res. Commun.*, **244**, 551 (1998).

510. K. Kitamura, K. Kangawa, H. Matsuo, and T. Eto, *Drugs*, **49**, 485 (1995).

511. F. Satoh, K. Takahashi, O. Murakami, K. Totsune, M. Sone, M. Ohneda, K. Abe, Y. Miura, Y. Hayashi, H. Sassano, and T. Mouri, *J. Clin. Endocrinol. Metab.*, **80**, 1750 (1995).

512. Y. Ichiki, K. Kitamura, K. Kangawa, M. Kawamoto, H. Matsuo, and T. Eto, *FEBS Lett.*, **338**, 6 (1994).

513. S. Sugo, N. Minamino, K. Kangawa, K. Miyamoto, K. Kitamura, J. Sakata, T. Eto, and H. Matsuo, *Biochem. Biophys. Res. Commun.*, **201**, 1160 (1994).

514. K. Meeran, K. O'Shea, P. D. Upton, C. J. Small, M. A. Ghatei, P. H. Byfield, and S. R. Bloom, *J. Clin. Endocrinol. Metab.*, **82**, 95 (1997).

515. L. K. Lewis, M. W. Smith. S. O. Brennan, T. G. Yandle, A. M. Richards, and M. G. Nicholls, *Peptides*, **18**, 733 (1997).

516. O. Lisy, M. Jougasaki, J. A. Schirger, H. H. Chen, P. T. Barclay, and J. C. Burnett, *Am. J. Physiol. Renal Fluid Electrolyte Physiol.*, **44**, F410 (1998).

517. S. Inagaki, S. Kito, Y. Kubota, S. Girgis, C. J. Hillyard, and I. MacIntyre, *Brain Res.*, **374**, 287 (1986).

518. F. A. Tschopp, H. Henke, J. B. Petermann, P. H. Tobler, R. Janzer, T. Hoekfelt, J. M. Lundberg, C. Cuello, and J. A. Fischer, *Proc. Natl. Acad. Sci. USA*, **82**, 248 (1995).

519. S. J. Wimalawansa, P. C. Emson, and I. MacIntyre, *Neuroendocrinology*, **46**, 131 (1987).

520. M. Zaidi, L. H. Breimer, and I. MacIntyre, *J. Exp. Physiol.*, **72**, 371 (1987).

521. R. Quirion, D. Van Rossum, Y. Dumont, S. St.-Pierre, and A. Fournier, *Ann. N. Y. Acad. Sci.*, **657**, 88 (1992).

522. R. Quirion and Y. Dumont, *Mol. Biol. Intell.*, **10**, 1 (2000).

523. S. Sigrist, A. Franco-Cereceda, R. Muff, H. Henke, J. M. Lundberg, and J. A. Fischer, *Endocrinology*, **119**, 381 (1986).

524. T. Ishikawa, N. Okamura, A. Saito, T. Masaki, and K. Goto, *Circ. Res.*, **63**, 726 (1988).

525. A. Hughes, S. Thom, G. Martin, and P. Sever, *Clin. Sci.*, **13**(Suppl.), 88P (1986).

526. L. Edvinsson, B. B. Fredholm, E. Hamel, I. Jansen, and C. Verrecchia, *Neurosci. Lett.*, **58**, 213 (1985).

527. U. Zimmerman, J. A. Fischer, and R. Muff, *Peptides*, **16**, 421 (1995).

528. D. Van Rossum, D. P. Menard, J. K. Chang, and R. Quirion, *Can. J. Physiol. Pharmacol.*, **73**, 1084 (1995).

529. S. Kapas, K. J. Catt, and A. J. Clark, *J. Biol. Chem.*, **270**, 25344 (1995).

530. U. Zimmerman, J. A. Fischer, K. Frei, A. H. Fischer, R. K. Reinscheid, and R. Muff, *Brain Res.*, **724**, 238 (1996).

531. J. M. Lundberg, A. Franco-Cereceda, X. Hua, T. Hökfelt, and J. A. Fischer, *Eur. J. Pharmacol.*, **108**, 315 (1985).

532. K. Ando, B. L. Pegram, and E. D. Frohlich, *Am. J. Physiol. Regul. Integr. Comp. Physiol.*, **258**, R425 (1990).

533. S. M. Gardiner, A. M. Compton, P. A. Kemp, T. Bennett, R. Foulkes, and B. Hughes, *Br. J. Pharmacol.*, **103**, 1509 (1991).

534. A. Franco-Cereceda, C. Gennari, R. Nami, D. Agnusdei, J. Pernow, J. M. Lundberg, and J. A. Fischer, *Circ. Res.*, **60**, 393 (1987).

535. R. Uddman, L. Edvinsson, E. Ekblad, R. Håkanson, and F. Sundler, *Regul. Pept.*, **15**, 1 (1986).

536. M. O. Coupe, J. C. W. Mak, M. Yacoub, P. J. Oldershaw, and P. J. Barnes, *Circulation*, **81**, 741 (1990).

537. A. Franco-Cereceda, A. Saria, and J. M. Lundberg, *Acta Physiol. Scand.*, **135**, 173 (1989).

538. K. Ono, M. Delay, T. Nakajima, H. Irisawa, and W. Giles, *Nature*, **340**, 721 (1989).

539. D. J. DiPette, K. Schwarzenberger, N. Kerr, and O. B. Holland, *Am. J. Med. Sci.*, **297**, 65 (1989).

540. X. Wang and R. R. Fiscus, *Am. J. Physiol. Regul. Integr. Comp. Physiol.*, **256**, R421 (1989).

541. Y. C. Shekar, I. S. Anand, R. Sarma, et al., *Am. J. Cardiol.*, **67**, 732 (1991).

542. D. A. Barber, Y. S. Park, J. C. Burnett, and V. M. Miller, *J. Cardiovasc. Pharmacol.*, **30**, 695 (1997).

543. J. C. Lainchbury, G. J. Cooper, D. H. Coy, N. Y. Jiang, L. K. Lewis, and T. G. Yandle, *Clin. Sci.*, **92**, 467 (1997).

544. J. J. Lah and W. H. Frishman, *Heart Dis.*, **2**, 259 (2000).

545. Y. Ishiyama, K. Kitamura, Y. Ichiki, S. Nakamura, O. Kida, K. Kangawa, and T. Eto, *Eur. J. Pharmacol.*, **241**, 271 (1993).

546. B. L. Jensen, B. K. Kramer, and A. Kurtz, *Hypertension*, **29**, 1148 (1997).

547. Y. C. Shekhan, I. S. Anand, R. Sarma, R. Ferrari, P. L. Wahi, and P. A. Poole-Wilson, *Am. J. Cardiol.*, **67**, 732 (1991).

548. N. G. Uren, D. Seydoux, and G. J. Davies, *Cardiovasc. Res.*, **27**, 1477 (1993).

549. M. P. Gnadinger, D. E. Uchlinger, P. Weidmann, S. G. Sha, R. Muff, W. Born, W. Rascher, and J. A. Fischer, *Am. J. Physiol. Endocrinol. Metab.*, **257**, E848 (1989).

550. A. Kurtz, R. Muff, W. Bonn, J. M. Lundberg, B. I. Millberg, M. P. Graedinger, D. E. Vehlinger, P. Weidmann, T. Hoekfelt, and J. A. Fischer, *J. Clin. Invest.*, **82**, 538 (1988).

551. S. Shawket, C. Dickerson, B. Hayleman, and M. J. Brown, *Br. J. Clin. Pharmacol.*, **32**, 209 (1991).

552. C. B. Bunker, G. Terenghi, D. R. Springall, J. M. Polak, and P. M. Dowd, *Lancet*, **336**, 1530 (1990).

553. C. Yallampalli and S. J. Wimalawansa, *Trends Endocrinol. Metab.*, **9**, 113 (1998).

554. S. D. Brain and T. J. William, *Nature*, **335**, 73 (1985).

555. P. J. Goodsby, L. Edvinson, and R. Ekman, *Ann. Neurol.*, **28**, 183 (1990).

556. M. Chovet, *Curr. Opin. Chem. Biol.*, **4**, 428 (2000).

557. H. Doods, G. Hallermayer, D. Wu, M. Entzeroth, K. Rudolf, W. Engel, and W. Eberlein, *Br. J. Pharmacol.*, **129**, 420 (2000).

558. R. A. Daines, K. K. Sham, J. J. Taggert, W. D. Kingsbury, J. Chan, A. Breen, J. Disa, and N. Aiyar, *Biorg. Med. Chem. Lett.*, **7**, 2673 (1997).

559. R. G. Hill, A. A. Patchett, and L. Young, World Pat. Appl. PCT 18764 (2000).

560. J. Kato, K. Kitamura, K. Kuwasako, M. Tanaka, Y. Ishiyama, T. Shimokubo, Y. Ichiki, S. Nakamura, K. Kangawa, and T. Eto, *Am. J. Hypertens.*, **8** (10 Part 1), 997 (1995).

561. F. Yoshihara, T. Nishikimi, T. Horio, C. Yutani, S. Takishita, H. Matsuo, T. Ohe, and K. Kangawa, *Eur. J. Pharmacol.*, **335**, 33 (1998).

562. M. Jougasaki, R. Rodeheffer, M. Redfield, K. Yamamoto, C. M. Wei, L. J. McKinley, and J. C. Burnett, *J. Clin. Invest.*, **97**, 2370 (1996).

563. N. Nagaya, T. Satoh, T. Nishikini, M. Uematsu, S. Furuicki, F. Sakamaki, H. Oya, Kyotani, N. Nakanishi, Y. Goto, Y. Masuda, K. Miyatake, and K. Kangawa, *Circulation*, **101**, 498 (2000).

564. A. A. M. Khawaja and D. F. Rogers, *Int. J. Cell. Biol.*, **28**, 721 (1996).

565. T. Hökfelt, B. Pernow, and J. Warren, *J. Intern. Med.*, **249**, 27 (2001).

566. M. M. Chang, S. E. Leeman, and H. D. Niall, *Nature (New Biol.)*, **232**, 86 (1971).

567. N. D. Boyd, S. G. MacDonald, R. Kage, J. Luber-Narod, and S. E. Leeman, *Ann. N. Y. Acad. Sci.*, **632**, 79 (1991).

568. H. Nawa, H. Kotani, and S. Nakanishi, *Nature*, **312**, 729 (1984).

569. J. V. Broeck, H. Torfs, J. Poels, W. Van Poyer, E. Swinnen, K. Ferket, and A. De Loof, *Ann. N. Y. Acad. Sci.*, **897**, 374 (1999).

570. D. Regoli, A. Boudon, and J.-L. Fauchére, *Pharmacol. Rev.*, **46**, 551 (1994).

571. C. J. Swain, *Prog. Med. Chem.*, **35**, 57 (1998).

572. S. McLean, *Med. Res. Rev.*, **16**, 297 (1996).

573. G. G. Nussdorfer and L. K. Malendowicz, *Peptides*, **19**, 949 (1998).

574. K. Egashira, S. Suzuki, Y. Hirooka, H. Kai, M. Sugimachi, T. Imaizumi, and A. Takeshita, *Hypertension*, **25**, 201 (1995).

575. C. E. Hill and D. J. Gould, *J. Pharmacol. Exp. Ther.*, **273**, 918 (1995).

576. N. R. Sharma and M. J. Davis, *Am. J. Physiol. Heart Circ. Physiol.*, **268**, H962 (1995).

577. C. A. Maggi, *Prog. Neurobiol.*, **45**, 1 (1995).

578. D. M. White, *J. Peripher. Nerv. Syst.*, **2**, 191 (1997).

579. P. Holzer, *Gen. Pharmacol.*, **30**, 5 (1998).

580. P. Holzer, *Rev. Physiol. Biochem. Pharmacol.*, **121**, 49 (1992).

581. S. D. Brain in P. Geppetti and P. Holzer, Eds., *Neurogenic Inflammation*, CRC Press, Boca Raton, FL, 1996, p. 229.

582. R. F. Furchgott, *Annu. Rev. Pharmacol. Toxicol.*, **24**, 175 (1984).

583. S. C. Tang, F. Fend, L. Müller, H. Braunsteiner, and C. J. Wiedermann, *Lab. Invest.*, **69**, 86 (1993).

584. L. Matsson, L.-I. Norevall, and S. Forsgren, *Eur. J. Oral Sci.*, **103**, 70 (1995).

585. P. R. Germonpré, G. F. Joos, E. Everaert, J. C. Kips, and R. A. Pauwels, *Am. J. Respir. Crit. Care Med.*, **152**, 1796 (1995).

586. J. M. Lundberg, T. Hökfelt, A. Änggard, L. Lundblad, A. Saria, J. Fahrenkrug, and L. Terenius in P. M. Vanhoutte and S. F. Vatner, Eds., *Vasodilator Mechanisms*, Karger, Basel, 1984, p. 60.

587. L. Abelli, F. Nappi, C. A. Maggi, P. Rovero, M. Astolfi, D. Regoli, G. Drapeau, and A. Giachetti, *Ann. N. Y. Acad. Sci.*, **632**, 358 (1991).

588. S. Vanner, *Am. J. Physiol. Gatsrointest. Liver Physiol.*, **267**, G650 (1994).

589. D. B. Hoover, Y. Chang, J. C. Hancock, and L. Zhang, *Jpn. J. Pharmacol.*, **84**, 367 (2000).

590. X.-P. Gao, H. A. Jaffe, C. O. Olopade, and I. Rubinstein., *J. Appl. Physiol.*, **79**, 968 (1995).

591. N. Kaito, H. Onoue, and T. Abe, *Peptides*, **16**, 1127 (1995).

592. L. W. Role, S. E. Leeman, and R. L. Perlman, *Neuroscience*, **6**, 1813 (1981).

593. J. Donnerer, R. Amann, G. Skofitsch, and F. Lembeck, *Ann. N. Y. Acad. Sci.*, **632**, 296 (1991).

594. G. A. Abou-Mohamed, J. Huang, C. D. Oldham, T. A. Taylor, L. Jin, R. B. Caldwell, S. W. May, and R. W. Caldwell, *J. Cardiovasc. Pharmacol.*, **35**, 871 (2000).

595. G. Burnstock, *J. Anat.*, **194**, 335 (1999).

596. C. D. Oldham, C. Li, J. Feng, R. O. Scott, W. Z. Wang, A. B. Moore, P. R. Girard, J. Huang, R. B. Caldwell, R. W. Caldwell, and S. W. May, *Am. J. Physiol. Cell Physiol.*, **273**, C1908 (1997).

597. L. Quartara and C. A. Maggi, *Neuropeptides*, **31**, 537 (1997).

598. L. Quartara, P. Rovero, and C. A. Maggi, *Med. Res. Rev.*, **15**, 139 (1995).

599. R. M. Snider, J. W. Constantine, J. A. Lowe 3rd, K. P. Longo, W. S. Lebel, H. A. Woody, S. E. Drozda, M. C. Desai, F. J. Vinick, R. W. Spencer, and H. J. Hess, *Science*, **251**, 435 (1991).

600. V. Leroy, P. Mauser, Z. Gao, and N. P. Peet, *Exp. Opin. Invest. Drugs*, **9**, 735 (2000).

601. G. R. Seabrook, S. L. Shepheard, D. J. Williamson, P. Tyrer, M. Rigby, M. A. Cascieri, T. Harrison, R. J. Hargreaves, and R. G. Hill, *Eur. J. Pharmacol.*, **317**, 129 (1996).

602. J. Culman, S. Klee, C. Ohlendorf, and T. Unger, *J. Pharmacol. Exp. Ther.*, **280**, 238 (1997).

603. A. Roccon, D. Marchionni, and D. Nisato, *Br. J. Pharmacol.*, **118**, 1095 (1996).

604. M. Rocha e Silva, W. T. Beraldo, and G. Rosenfeld, *Am. J. Physiol.*, **156**, 261 (1949).

605. D. A. Elliott, G. P. Lewis, and E. W. Horton, *Biochem. Biophys. Res. Commun.*, **3**, 87 (1960).

606. R. A. Boissonas, S. Guttmann, P. A. Jaquenoud, H. Konzett, and E. Struermer, *Experientia*, **16**, 326 (1960).

607. A. P. Kaplan, K. Joseph, Y. Shibayama, Y. Nakazawa, B. Ghebrehiwet, S. Reddigari, and M. Silverberg, *Clin. Rev. Allergy Immunol.*, **16**, 403 (1998).

608. J. M. Conlon, *Ann. N. Y. Acad. Sci.*, **839**, 1 (1998).

609. F. Marceau and D. R. Bachvarov, *Clin. Rev. Allergy Immunol.*, **16**, 385 (1998).

610. F. Marceau, J. F. Hess, and D. R. Bachvarov, *Pharmacol. Rev.*, **50**, 357 (1998).

611. D. J. Campbell, *Braz. J. Med. Biol. Res.*, **33**, 665 (2000).

612. D. Regoli and J. Barabé, *Pharmacol. Rev.*, **32**, 1 (1980).

613. J. G. Menke, J. A. Borkowski, K K. Bierilo, T. MacNeil, A. W. Derrick, K. A. Schneck, R. W.

Ransom, C. D. Strader, D. L. Linemeyer, and J. F. Hess, *J. Biol. Chem.*, **2698**, 21583 (1994).

614. D. Eggerickx, E. Raspe, D. Bertrand, G. Vassart, and M. Parmentier, *Biochem. Biophys. Res. Commun.*, **187**, 1306 (1992).

615. J. F. Hess, J. A. Borkowski, G. S. Young, C. D. Strader, and R. W. Ransom, *Biochem. Biophys. Res. Commun.*, **184**, 260 (1992).

616. D. Regoli, S. N. Allogho, A. Rizzi, and F. J. Gobeil, *Eur. J. Pharmacol.*, **348**, 1 (1998).

617. M. Altamura, S. Meini, L. Quartara, and C. A. Maggi, *Regul. Pept.*, **80**, 13 (1999).

618. J. M. Hall, *Gen. Pharmacol.*, **28**, 1 (1997).

619. J. K. Liao and C. J. Homcy, *J. Clin. Invest.*, **92**, 2168 (1993).

620. C. E. Austin, A. Faussner, H. E. Robinson, S. Chakravarty, D. J. Kyle, J. M. Bathon, and D. Proud, *J. Biol. Chem.*, **272**, 11420 (1997).

621. J. P. Schanstra, C. Alric, M. E. Marin-Castano, J.-P. Girolami, and J.-L. Bascands, *Int. J. Mol. Med.*, **3**, 185 (1999).

622. M. Wahl, E. T. Whalley, A. Unterberg, L. Schilling, A. A. Parsons, A. Baethmann, and A. R. Young, *Immunopharmacology*, **33**, 257 (1996).

623. J.-V. Mombouli, I. Bissiriou, V. Agboton, and P. M. Vanhoutte, *Immunopharmacology*, **33**, 46 (1996).

624. A. A. Jaffa, B. S. Miller, S. A. Rosenzweig, P. S. Naidu, V. Velarde, and R. K. Mayfield, *Am. J. Physiol. Renal Fluid Electrolyte Physiol.*, **273**, F916 (1997).

625. B. A. Schölkens, *Immunopharmacology*, **33**, 209 (1996).

626. P. Madeddu, C. Emanueli, L. Gaspa, B. Salis, A. F. Milia, L. Chao, and J. Chao, *Immunopharmacology*, **44**, 9 (1999).

627. P. Gohlke, C. Tschöpe, and T. Unger in A. Zanchetti, Ed., *Hypertension and the Heart*, Plenum Press, New York, 1997, p. 159.

628. K. Ito, Y.-Z. Zhu, Y.-C. Zhu, P. Gohlke, and T. Unger, *Jpn. J. Pharmacol.*, **75**, 311 (1997).

629. G. Bönner, *Drugs*, **54**, 24 (1997).

630. J. M. Stewart, L. Gera, D. C. Chan, E. T. Whalley, W. L. Hanson, and J. S. Zuzack, *Can. J. Physiol. Pharmacol.*, **75**, 719 (1997).

631. J. M. Stewart, L. Gera, E. J. York, D. C. Chan, and P. Bunn, *Immunopharmacology*, **43**, 155 (1999).

632. D. Regoli, G. Calo, A. Rizzi, G. Bogoni, F. Gobeil, C. Campobasso, G. Mollica, and L. Beani, *Regul. Pept.*, **65**, 83 (1996).

633. (a) D. Pruneau, J. A. Paquet, J. M. Luccarini, E. Defrene, C. Fouchet, R. M. Franck, B. Loillier, C. Robert, P. Belichard, H. Duclos, B. Cremers, and P. Dodey, *Immunopharmacology*, **43**, 187 (1999); (b) T. Griesbacher and F. J. Legat, *Inflamm. Res.*, **49**, 535 (2000).

634. S. I. Said and V. Mutt, *Science*, **169**, 1217 (1970).

635. S. I. Said and V. Mutt, *Nature*, **225**, 863 (1970).

636. V. Mutt and S. I. Said, *Eur. J. Biochem.*, **42**, 581 (1974).

637. G. G. Nussdorfer and L. K. Malendowicz, *Peptides*, **19**, 1443 (1998).

638. N. Itoh, K.-I. Obata, N. Yanaihara, and H. Okamoto, *Nature*, **304**, 547 (1983).

639. J. M. Lundberg, J. Fahrenkrug, T. Hökfelt, C.-R. Martling, O. Larsson, K. Tatemoto, and A. Änggård, *Peptides*, **5**, 593 (1984).

640. B. Greenberg, K. Rhoden, and P. J. Barnes, *Blood Vessels*, **24**, 45 (1987).

641. J. Pearson, L. Brandeis, and A. C. Cuello, *Nature*, **295**, 61 (1982).

642. T. Hökfelt, O. Johansson, Å. Ljungdahl, J. M. Lundberg, and M. Schultzberg, *Nature*, **284**, 515 (1980).

643. A. Miyata, A. Arimura, R. R. Dahl, N. Minamino, A. Uehara, L. Jiang, M. D. Culler, and D. H. Coy, *Biochem. Biophys. Res. Commun.*, **164**, 567 (1989).

644. M. Laburthe, A. Couvineau, P. Gaudin, J.-J. Maoret, C. Rouyer-Fessard, and P. Nicole, *Ann. N. Y. Acad. Sci.*, **805**, 94 (1996).

645. D. Vaudry, B. J. Gonzalez, M. Basille, L. Yon, A. Fournier, and H. Vaudry, *Pharmacol. Rev.*, **52**, 269 (2000).

646. M. Laburthe and T. Voisin in G. H. Greeley, Ed., *Gastrointestinal Endocrinology*, Humana Press, Totowa, NJ, 1998, p. 125.

647. I. Gozes, M. Fridkin, J. M. Hill, and D. E. Brenneman, *Curr. Med. Chem.*, **6**, 1019 (1999).

648. T. Ishihara, R. Shigemoto, K. Mori, K. Takahashi, and S. Nagata, *Neuron*, **8**, 811 (1992).

649. E. M. Lutz, W. J. Sheward, K. M. West, J. A. Morrow, C. Fink, and A. J. Harmar, *FEBS Lett.*, **334**, 3 (1993).

650. J. R. Pisegna and S. A. Wank, *Proc. Natl. Acad. Sci. USA*, **90**, 6345 (1993).

651. J. Christophe, *Biochim. Biophys. Acta*, **1241**, 45 (1995).

652. A. J. Harmar, A. Arimura, I. Gozes, L. Journot, M. Laburthe, J. R. Pisegna, S. R. Rawlings, P. Robberecht, S. I. Said, S. P. Sreedharan, S. A. Wank, and J. A. Waschek, *Pharmacol. Rev.*, **50**, 265 (1998).

653. J.-L. Martin, D. L. Feinstein, N. Yu, O. Sorg, C. Rossier, and P. J. Magistretti, *Brain Res.*, **587**, 1 (1992).

654. J. M. Lundberg, A. Änggård, J. Fahrenkrug, T. Hökfelt, and V. Mutt, *Proc. Natl. Acad. Sci. USA*, **77**, 1651 (1980).

655. K. Törnebrandt, A. Nobin, and C. Owman in P. M. Vanhoutte, Ed., *Vasodilatation: Vascular Smooth Muscle, Peptides, Autonomic Nerves and Endothelium*, Raven Press, New York, 1988, p. 65.

656. L. Edvinsson, R. Juul, and I. Jansen, *Acta Neurol. Scand.*, **90**, 324 (1994).

657. I. Jansen-Olesen, P. J. Goadsby, R. Uddman, and L. Edvinsson, *J. Auton. Nerv. Syst.*, **49**, S97 (1994).

658. R. Uddman, P. J. Goadsby, I. Jansen, and L. Edvinsson, *J. Cereb. Blood Flow Metab.*, **13**, 291 (1993).

659. H. Takagi, Y. Kubota, and Y. Morishima in K. Nakamura, Ed., *Brain and Blood Pressure Control*, Exerpta Medica, Amsterdam, 1985, p. 79.

660. R. Uddman, J. Alumets, L. Edvinsson, R. Håkanson, and F. Sundler, *Acta Physiol. Scand.*, **112**, 65 (1981).

661. N. G. Della, R. E. Papka, J. B. Furness, and M. Costa, *Neuroscience*, **9**, 605 (1983).

662. N. Minamino, K. Kangawa, A. Fukuda, and H. Matsuo, *Neuropeptides*, **4**, 157 (1984).

663. M. Anzai, Y. Suzuki, M. Takayasu, Y. Kajita, Y. Mori, Y. Seki, K. Saito, and M. Shibuya, *Eur. J. Pharmacol.*, **285**, 173 (1995).

664. T. Yaksh and V. L. W. Go in P. M. Vanhoutte, Ed., *Vasodilatation: Vascular Smooth Muscle, Peptides, Autonomic Nerves, and Endothelium*, Raven Press, New York, 1988, p. 65.

665. H. C. Champion, J. A. Santiago, E. A. Garrison, D. Y. Cheng, D. H. Coy, W. A. Murphy, R. J. Ascuitto, N. T. Ross-Ascuitto, D. B. McNamara, and P. J. Kadowitz, *Ann. N. Y. Acad. Sci.*, **805**, 429 (1996).

666. R. J. Henning and D. R. Sawmiller, *Cardiovasc. Res.*, **49**, 27 (2001).

667. I. Tatsuno, D. Uchida, T. tanaka, N. Saeki, A. Hirai, Y. Saito, O. Moro, and M. Tajima, *Brain Res.*, **889**, 138 (2001).

668. P. Gourlet, A. Vandermeers, P. Vertongen, J. Rathé, P. De Neef, J. Cnudde, M. Waelbroeck, and P. Robberecht, *Peptides*, **18**, 1539 (1997).

669. P. Gourlet, P. Vertongen, A. Vandermeers, M. C. Vandermeers-Piret, J. Rathé, P. De Neef, and P. Robberecht, *Peptides*, **18**, 403 (1997).

670. M. Xia, S. P. Sreedharan, D. R. Bolin, G. O. Gaufo, and E. J. Goetzl, *J. Pharmacol. Exp. Ther.*, **281**, 629 (1997).

671. S. Lamouche and N. Yamaguchi, *Am. J. Physiol. Regul. Integr. Comp. Physiol*, **280**, R510 (2001).

672. P. Gourlet, P. De Neef, J. Cnudde, M. Waelboeck, and P. Robberecht, *Peptides*, **18**, 1555 (1997).

673. P. Brazeau, W. Vale, R. Burgus, N. Ling, M. Butcher, J. Rivier, and R. Guillemin, *Science*, **179**, 77 (1973).

674. D. J. Koerker, L. A. Harker, and C. J. Goodner, *N. Engl. J. Med.*, **96**, 749 (1975).

675. J. Epelbaum, P. Dournaud, M. Fodor, and C. Viollet, *Crit. Rev. Neurobiol.*, **8**, 25 (1994).

676. A. Franco-Cereceda, L. Bengtsson, and J. M. Lundberg, *Eur. J. Pharmacol.*, **134**, 69 (1987).

677. A. Franco-Cereceda, J. M. Lundberg, and T. Hökfelt, *Eur. J. Pharmacol.*, **132**, 101 (1986).

678. S. C. Webb, D. M. Krikler, W. G. Hendry, T. E. Adrian, and S. R. Bloom, *Br. Heart J.*, **56**, 236 (1986).

679. A. V. Greco, A. Ghirlanda, C. Barone, A. Bertoli, S. Caputo, L. Uccioli, and R. Manna, *Br. Med. J.*, **288**, 28 (1984).

680. K. F. Binmoeller and N. Soehendra, *Am. J. Gastroenterol.*, **90**, 1923 (1995).

681. J. P. Cello, *Int. Surg.*, **80**, 82 (1995).

682. J. N. Baxter and S. A. Jenkins, *Scand. J. Gastroenterol.*, **29**(Suppl. 207), 17 (1994).

683. M. Kalia, K. Fuxe, T. Hökfelt, O. Johansson, R. Lang, D. Ganten, C. Cuello, and L. Terenius, *J. Comp. Neurol.*, **222**, 409 (1984).

684. C. Bouras, P. J. Magistretti, J. H. Morrison, and J. Constantinidis, *Neuroscience*, **22**, 781 (1987).

685. A. G. Karczmar, K. Koketsu, and S. Nishi, *Autonomic and Enteric Ganglia: Transmission and Its Pharmacology*, Plenum Press, New York, 1986.

686. S. M. Day, J. Gu, J. M. Polak, and S. R. Bloom, *Br. Heart J.*, **53**, 153 (1985).

687. Y. C. Patel and C. B. Srikant, *Trends Endocrinol. Metab.*, **8**, 398 (1997).

688. Y. C. Patel, *Front. Neuroendocrinol.*, **20**, 157 (1999).

689. W. Bauer, U. Briner, W. Dopfner, R. Haller, R. Huguenin, P. Marbach, T. Petcher, and J. Pless, *Life Sci.*, **31**, 1133 (1982).

690. E. T. Janson and K. Oberg, *Curr. Pharm. Des.*, **5**, 693 (1999).

691. A. Hadengue, *Digestion*, **60**(Suppl. 2), 31 (1999).

692. S. P. Rohrer, E. T. Birgin, R. T. Mosley, S. C. Berk, S. M. Hutchins, D. M. Shen, Y. Xiong, E. C. Hayes, R. M. Parmar, F. Foor, S. W. Mitra, S. J. Degrado, M. Shu, J. M. Klopp, S. J. Cai, A. Blake, W. W. Chan, A. Pasternak, L. Yang, A. A. Patchett, R. G. Smith, K. T. Chapman, and J. M. Schaeffer, *Science*, **282**, 737 (1998).

693. S. P. Rohrer and S. C. Berk, *Curr. Opin. Drug Discov. Dev.*, **2**, 293 (1999).

694. E. R. Spindel, W. W. Chin, J. Price, L. H. Rees, G. M. Besser, and J. F. Habener, *Proc. Natl. Acad. Sci. USA*, **81**, 5699 (1984).

695. R. Uddman, E. Moghimzadeh, and F. Sundler, *Arch. Otolaryngol. Head Neck Surg.*, **239**, 145 (1984).

696. G. Burnstock, *Cephalalgia*, **5** (Suppl. 2), 25 (1985).

697. P. J. Barnes in J. M. Polak, Ed., *Regulatory Peptides*, Burkhäuser, Basel, Switzerland, 1989, pp. 317–333.

698. C. Heym, R. Webber, and D. Adler, *Arch. Oral Biol.*, **39**, 213 (1994).

699. T. N. Luu, A. H. Chester, G. S. O'Neil, S. Tadjkarimi, J. R. Pepper, and M. H. Yacoub, *Am. J. Physiol. Heart Circ. Physiol.*, **264**, H583 (1993).

700. P. Vigne, E. Feolde, C. Van Renterghem, J. P. Breittmayer, and C. Frelin, *Eur. J. Biochem.*, **233**, 414 (1995).

701. J. Battey and E. Wade, *Trends Neurosci.*, **14**, 524 (1991).

702. E. R. Spindel, E. Giladi, P. Brehm, P. H. Goodman, and T. P. Segerson, *Mol. Endocrinol.*, **4**, 1956 (1990).

703. E. Wada, J. Way, H. Shapira, K. Kusano, A. M. Lebacq-Verheyden, D. Coy, R. Jensen, and J. Battey, *Neuron*, **6**, 421 (1991).

704. Z. Fathi, M. H. Corjay, H. Shapira, E. Wada, R. Benya, R. Jensen, J. Viallet, E. A. Sausville, and J. F. Battey, *J. Biol. Chem.*, **268**, 5979 (1993).

705. V. Garbulev, A. Akhundovo, H. Buechmer, and F. Fahrenholy, *Eur. J. Biochem.*, **208**, 405 (1992).

706. S. A. Mantey, H. C. Weber, E. Sainz, M. Akeson, R. R. Ryan, T. K. Pradhan, R. P. Searles, E. R. Spendel, J. F. Battey, D. H. Coy, and R. T. Jensen, *J. Biol. Chem.*, **272**, 26062 (1997).

707. D. C. Horwell, M. C. Pritchard, and J. Raphy, *Adv. Amino Acid Mimet. Peptomimet.*, **2**, 165 (1999).

708. T. W. Moody and R. T. Jensen, *Drugs Future*, **23**, 1305 (1998).

709. S. J. Main, A. Gasgoigne, A. Batchelor, and P. C. Adams, *Lancet*, **348**, 1243 (1996).

710. R. Carraway and S. E. Leeman, *J. Biol. Chem.*, **248**, 6854 (1973).

711. G. R. Uhl and S. H. Snyder, *Life Sci.*, **19**, 1827 (1976).

712. R. M. Kobayashi, M. Brown, and W. Vale, *Brain Res.*, **126**, 584 (1977).

713. L. Jennes, W. E. Stumpf, and P. W. Kalivas, *J. Comp. Neurol.*, **210**, 211 (1982).

714. P. E. Cooper, M. H. Fernstrom, O. P. Rorstad, S. E. Leeman, and J. B. Martin, *Brain Res.*, **218**, 219 (1981).

715. P. J. Manberg, W. W. Youngblood, C. B. Nemeroff, M. N. Rossor, L. L. Iversen, A. J. Prange Jr., and J. S. Kizer, *J. Neurochem.*, **38**, 1777 (1982).

716. J. K. Mai, J. Triepel, and J. Metz, *Neuroscience*, **22**, 499 (1987).

717. P. C. Emson, M. Goedert, and P. W. Mantyh in A. Bjorklund and T. Hökfelt, Eds., Handbook of Chemical Neuroanatomy, Vol. **4**, Elsevier, Amsterdam, 1985, p. 355.

718. P. Kitabgi, F. Checler, J. Mazella, and J. P. Vincent, *Rev. Clin. Basic Pharmacol.*, **5**, 397 (1985).

719. J. P. Vincent, J. Mazella, and P. Kitabgi, *Trends Pharmacol. Sci.*, **20**, 302 (1999).

720. B. M. Tyler-McMahon, M. Boules, and E. Richelson, *Regul. Pept.*, **93**, 125 (2000).

721. W. Rostene, M. Azzi, H. Boudin, I. Lepee, F. Souaze, M. Mendez-Ubach, C. Betancur, and D. Gully, *Ann. N. Y. Acad. Sci.*, **814**, 125 (1997).

722. R. Quirion, *Peptides*, **4**, 609 (1983).

723. C. B. Nemeroff and S. T. Cain, *Trends Pharmacol.*, **56**, 201 (1985).

724. C. B. Nemeroff, *Psychoneuroendocrinology*, **11**, 15 (1986).

725. R. Quirion, F. Rioux, S. St-Pierre, and D. Regoli, *Life Sci.*, **25**, 1969 (1979).

726. F. Rioux, M. Lemieux, and M. Lebel, *Peptides*, **11**, 921 (1990).

727. F. Rioux and D. Paré, *Peptides*, **14**, 227 (1993).

728. F. Rioux and M. Lemieux, *Br. J. Pharmacol.*, **106**, 187 (1992).

729. E. D. Di Paola and E. Richelson, *Eur. J. Pharmacol.*, **175**, 279 (1990).

730. F. L. Hisaw, *Proc. Soc. Exp. Biol. Med.*, **23**, 661 (1926).

731. O. D. Sherwood and E. M. O'Byrne, *Arch. Biochem. Biophys.*, **160**, 185 (1974).

732. C. Schwabe, J. K. McDonald, and B. G. Steinetz, *Biochem. Biophys. Res. Commun.*, **70**, 397 (1976).

733. C. Schwabe, J. K. McDonald, and B. G. Steinetz, *Biochem. Biophys. Res. Commun.*, **75**, 503 (1977).

734. P. Hudson, J. Haley, M. John, M. Cronk, R. Crawford, J. Haralambidis, G. Tregear, J. Shine, and H. Niall, *Nature*, **301**, 628 (1983).

735. D. Bani, *Gen. Pharmacol.*, **38**, 13 (1997).

736. B. E. Kemp and H. D. Niall, *Vitam. Horm.*, **41**, 79 (1984).

737. O. D. Sherwood in E. Knobil and J. Neill, Eds., *The Physiology of Reproduction*, 2nd ed., Raven Press, New York, 1994, p. 861.

738. R. Ivell, N. Hunt, F. Khan-Dawood, and M. Y. Dawood, *Mol. Cell Endocrinol.*, **66**, 251 (1989).

739. M. J. Taylor and C. L. Clark, *J. Endocrinol.*, **143**, R5 (1994).

740. L. Zhao, P. J. Roche, J. M. Gunnersen, V. E. Hammond, G. W. Tregear, E. M. Wintour, and F. Berk, *Endocrinology*, **140**, 445 (1999).

741. E. N. Unemori and E. P. Amento, *J. Biol. Chem.*, **265**, 10681 (1990).

742. E. N. Unemori, L. S. Beck, W. P. Lee, Y. Xu, M. Siegel, G. Keller, H. D. Liggitt, E. A. Bauer, and E. P. Amento, *J. Invest. Dermatol.*, **101**, 280 (1993).

743. E. N. Unemori, L. B. Pickford, A. L. Salles, C. E. Piercy, B. H. Grove, M. E. Erikson, and E. P. Amento, *J. Clin. Invest.*, **98**, 2739 (1996).

744. S. L. Garber, Y. Mirochnik, C. S. Brecklin, E. N. Unemori, A. K. Singh, L. Slobodskoy, B. H. Grove, J. A. Arruda, and G. Dunea, *Kidney Int.*, **59**, 876 (2001).

745. T. Bani-Sacchi, M. Bigazzi, D. Bani, P. F. Nannaioni, and E. Masini, *Br. J. Pharmacol.*, **116**, 1589 (1995).

746. J. Novak, L. A. Danielson, L. J. Kerchner, O. D. Sherwood, R. J. Ramirez, P. A. Moalli, and K. P. Conrad, *J. Clin. Invest.*, **107**, 1469 (2001).

747. L. A. Danielson, L. J. Kercher, and K. P. Conrad, *Am. J. Physiol. Regul. Integr. Comp. Physiol.*, **279**, R1298 (2000).

748. D. Bani, M. Bigazzi, E. Masini, G. Bani, and T. B. Sacchi, *Lab. Invest.*, **73**, 709 (1995).

749. M. Ho and J. J. F. Belch, *Scand. J. Rheumatol.*, **27**, 319 (1998).

CHAPTER FIVE

Hematopoietic Agents

MAUREEN HARRINGTON
Indiana University
Walther Oncology Center
Indianapolis, Indiana

Contents

Burger's Medicinal Chemistry and Drug Discovery
Sixth Edition, Volume 3: Cardiovascular Agents and Endocrines
Edited by Donald J. Abraham
ISBN 0-471-37029-0   © 2003 John Wiley & Sons, Inc.

1 INTRODUCTION

Hematopoiesis is a life-long process that involves the continuous formation and turnover of blood cells. Maintaining adequate blood cell production, as well as being able to meet increased demand (sickle cell anemia, infection), is in part under the control of a group of hormone-like glycoproteins referred to collectively as cytokines that includes the hematopoietic growth factors and the interleukins. The hematopoietic growth factors are as follows: erythroprotein (EPO), thrombopoietin (TPO), stem cell factor (SCF; also known as steel factor, kit ligand, and mast cell growth factor), and the colony-stimulating factors (CSF): granulocyte/macrophage-CSF, granulocyte-CSF, and macrophage-CSF (also known as CSF-1). Originally the term interleukin (IL) was operationally defined and implied that production of and the response to the molecule was restricted to leukocytes. As the cell and molecular biology of the interleukins has expanded, it is clear the original definition is not always applicable, e.g., many nonleukocytes produce and/or respond to the interleukins. However the term interleukin has been retained by the International Congress of Immunology. Consequently as new hematopoietic growth factors are identified, an interleukin number is assigned once the amino acid sequence has been determined; currently there are 23 interleukins.

Under steady-state conditions, 2×10^{11} blood cells are produced and destroyed per day. Key to maintaining a high rate of blood cell production is the hematopoietic stem cell, which gives rise to all mature circulating cells: erythrocytes, platelets, lymphocytes, monocytes/macrophages, and neutrophilic, eosinophilic, and basophilic granulocytes (Fig. 5.1). Between the pluripotential hematopoietic stem cells that gives rise to either myeloid or lymphoid cells and the end-stage mature circulating cells are a hierarchy of progenitor cells that differ in degree of lineage restriction. Hematopoietic stem cells self-renew (give rise to identical daughter cells) and/or divide and give rise to progenitor cells that are increasingly restricted in their developmental pathway. As the maturational state between the hematopoietic stem cell and the progenitor cell progresses, the capacity for self-renewal declines, and subsequent divisions will eventually yield an end-stage, fully differentiated cell type that has lost the capacity to proliferate. The balance between self-renewal and maturational divisions of the hematopoietic stem cells and progenitor cells allows one hematopoietic stem cell to yield approximately 1000 mature cells (1).

Key experiments by Jacobsen et al. (2, 3) in the early 1950s, demonstrating restoration of hematopoiesis with spleen and/or bone marrow-derived cells in irradiated animals, combined with Till and McCulloch's (4) work demonstrating that a single bone marrow-derived cell could form a macroscopic nodule in the spleen composed of rapidly proliferating hematopoietic cells, showed the *in vivo* existence of a hematopoietic stem cell. The pivotal development of an *in vitro* assay system by Bradley and Metcalf (5), Ichikawa et al. (6), and Pluznik and Sachs (7) with refinements by Dexter et al. (8) and Whitlock et al. (9) for culturing hematopoietic cells, allowed many of the developmental pathways and the regulatory molecules involved in hematopoietic homeostasis to be identified. In these assay systems, hematopoietic stem and progenitor cells are cultured in a semi-solid matrix, and in the presence of cytokines, colonies of cells initiated from a single cell develop. The Dexter and Whitlock-Witte modifications involve co-culturing hematopoietic stem and progenitor cells with bone marrow—derived stromal cell monolayers. Based on morphological and/or histochemical criteria, the composition of cells within the colony is determined, and the phenotype of the stem/progenitor cell that generated the colony (the colony forming unit or CFU) and therefore the target of cytokine(s) action can be identified. Colony assays have proven to be powerful screening tools for identifying myeloid specific cytokines and somewhat less fruitful in identifying cytokines re-

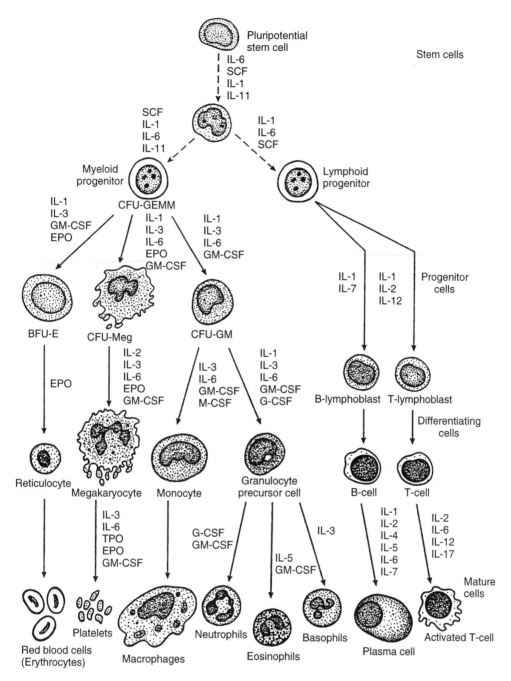

Figure 5.1. Schematic of blood cell development. Blood cell development is a hierarchical process with self-renewal and maturational divisions occurring as a continuum. A pluripotential stem cell will divide, and the daughter cells will either be identical in functional capacity (self-renewal) or the daughter cell will be slightly more mature (maturational division). The number of cell divisions, hence, the number of different cell types between the hematopoietic stem cell and the genration of myeloid or lymphoid progenitors, is not known, nor is the point at which phenotypically distinct lymphoid and myeloid progenitors are generated. The dashed lines in the figure are meant to reflect this. As progenitors, cells can undergo self-renewal and maturation divisions; however, as the cells progress toward the end-stage mature cell, the capacity for self-renewal is lost, and primarily, maturational divisions occur. The colony-forming unit (CFU) and burst-forming unit (BFU) are morphological distinctions. GEMM, granulocyte, erythroid, monocyte, megakaryocyte; E, erythroid; Meg, megakaryocyte; GM, granulocyte, monocyte. These refer to the cell types present in the colonies.

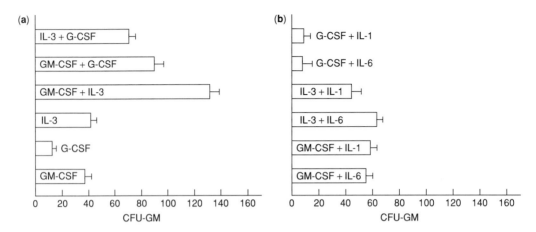

Figure 5.2. Synergistic effects between human growth factors on CFU-GM formation of myeloid progenitors. Bone marrow–derived myeloid progenitor cells were incubated with the indicated cytokines under standard cell culture conditions. Seven days later, the number of CFU-GM present in the cell cultures were quantitated. Results are presented as the mean of duplicate determinations ± SD and are representative of four separate experiments. This figure was reproduced with permission from Jacobsen et al. (11).

quired for the development and maturation of the earliest lymphoid-restricted progenitor cells and end stage B- and T-lymphocytes. In general, identification of lymphoid-specific cytokines has relied more heavily on examining the ability of the cytokine(s) to promote the proliferation of developmentally restricted T- and B-cell lines. The molecular mechanism(s) in which cytokines effect hematopoietic cell lineage restriction, if they do, is not clear and is an area of intense research. However, results obtained with *in vitro* assay systems have elucidated several key features of cytokine action (see Ref. 10 for an in-depth discussion).

Several properties of cytokines, including multiple cell targets, synergistic responses, and overlapping activities, have important biological implications. First, as schematically depicted in Fig. 5.1, many cytokines share the ability to affect the activity of multiple cell types, and dependent on the maturational state of the cell type, may elicit different responses. Targets for GM-CSF include the multipotential myeloid progenitor cell (CFU) that gives rise to the granulocyte, erythroid, monocyte/macrophage, and megakaryocyte cell lineages (CFU-GEMM), and the more restricted CFU-GM progenitor that generates the granulocyte and monocyte/macrophage cell lineages. The CFU-GEMM and CFU-GM progen-

itors proliferate in response to GM-CSF. In contrast, GM-CSF treatment of neutrophils and macrophages, mature end-stage cells that have lost the capacity to proliferate, enhances their functional activity. In both instances, GM-CSF treatment leads to target cell activation, and the difference in biological effects elicited by GM-CSF reflects the functional capacity of the target cell at that point in differentiation pathway. A second key activity some cytokines share is their ability to synergize. Synergistic responses observed between cytokines range from greater that additive responses when used in combination versus individually to a single cytokine with no apparent effect increasing the activity of a functional cytokine. Experimental data representative of these two types of synergistic interactions is presented in Fig. 5.2. The target cell population is highly purified mouse bone marrow-derived progenitor cells (Lin BM cells) and the ability of several cytokines alone and in combination to promote granulocyte/macrophage progenitor cell growth (CFU-GM) is being quantitated. As presented in A, IL-3, GM-CSF, or G-CSF alone support CFU-GM growth; however, greater than additive effects are observed when the cytokines are added in combination. Neither IL-1 or IL-6 alone support CFU-GM growth (Fig. 5.2); however if

GM-CSF or IL-3 plus IL-1 or IL-6 is assayed, progenitor cell growth is greater compared with growth with GM-CSF or IL-3 alone (compare A and B). In contrast, IL-1 or IL-6 has little effect on G-GSF–stimulated progenitor cell growth. Molecular mechanisms that give rise to the synergistic interactions detected with the various cytokine responses are being actively explored in numerous laboratories. Identified mechanisms include one cytokine increasing the expression of another cytokine receptor (11) and one cytokine inducing transcription of genes whose transcripts are stabilized by another cytokine (12). The third key feature of cytokines, also depicted in Fig. 5.1, is the apparent overlapping activities seen with cytokines. In some instances, there are quantitative differences in the responses elicited. IL-3, GM-CSF, or SCF in the presence of EPO will support CFU-GEMM growth; however, colonies grown in the presence of EPO plus SCF contain significantly more cells. Alternatively, a common activity can be the result of one cytokine inducing the expression of the cytokine that mediates the response. IL-3 and M-CSF independently support CFU-GM growth; however, IL-3 can induce M-CSF gene expression in CFU-GM (13). Consequently, in both cases M-CSF may mediate the response. Whether the overlapping activities detected in *in vitro* colony assays implies functional redundancy *in vivo* is less clear. Studies in animals have revealed overlapping as well as distinct activities for the cytokines (see Ref. 14 for a comprehensive discussion). Physiological parameters such as access to the cytokine, local cytokine concentration, and presence of other cytokines will also influence effects seen *in vivo*. In the following sections, four cytokines approved by the Food and Drug Administration (FDA) for use in humans are described; this is followed by a limited discussion of one cytokine, stem cell factor, that currently has orphan drug status, and a brief description of two cytokines currently undergoing clinical evaluation. The field of cytokine research is ever expanding as new interleukins are identified each year. As our understanding of the cell and molecular biology of cytokine action on hematopoiesis increases, the clinical use of these agents will become more refined.

2 ERYTHROPOIETIN

EPO, a key regulator of erythropoiesis, was the first hematopoietic growth factor activity identified. In 1906, Carnot and Deflandre (15) reported that serum derived from an anemic animal, when introduced into a normal animal, led to increased numbers of circulating red blood cells and proposed that an activity present in the serum, hemopoietin, mediated the effect. Subsequent studies demonstrated that anemia itself lead to an increased blood level of hemopoietin and that hemopoietin was produced by the kidney (16, 17). Biologically active EPO was first purified from urine (18), and oligonucleotide probes, based on the EPO amino acid sequence, were used to clone the human gene (19, 20). In 1985, the first patient received recombinant EPO (21).

2.1 Physical Properties

The protein-coding region of the EPO gene is composed of five exons and four introns, and the human gene is located at chromosome 7pter-q22 (22). The 1.6-kilobase (kb) transcript encodes a 193 amino acid protein of which the first 27 amino acids (leader-sequence) and a carboxy terminal arginine are removed before secretion. The mature protein contains two internal disulfide bonds and has a calculated molecular mass of 18.4 kilodaltons (kDa). However, EPO is also glycosylated (adding three N-linked and 1 O-linked carbohydrate chain) before secretion; thus, circulating forms of EPO are larger (34–39 kDa) (23, 24). During fetal/neonatal growth, EPO is produced in the liver, but near birth, production shifts to the kidney. Peritubular cells (fibroblast-like type I interstitial cells) that are present in the kidney cortex and outer medulla are the primary sites of EPO production; within the liver, a subset of hepatocytes, in addition to the fibroblastoid fat storing Ito cells, retain the capacity to produce EPO (25). Under normal physiological conditions, circulating EPO levels are between 10–20 mU/mL in plasma (approximately 0.1 nM). In response to tissue hypoxia or anemia, circulating EPO levels can increase 100- to 1000-fold. The cellular and molecular mechanism(s) that senses and signals for increased EPO gene expression is unclear. There are data that suggest that the putative oxygen sensor in EPO-producing cells is a ferroprotein (26). The ferroprotein may use a hemoglobin-like mechanism in which a heme containing iron would reversibly bind oxygen (27) or alternatively use a nonmitochondrial electron transport mechanism that involves an iron moiety (28, 29). Independent of the mechanism through which the cell senses changes in oxygen status, production of hypoxia inducible factor-1α (HIF1α) is increased. HIF1α heterodimerizes with HIF1β (also known as ARNT), which is constitutively expressed. HIF1α/HIF1β heterodimers bind to hypoxia inducible elements (HREs) present in the promoter region of the EPO gene, leading to an increase in EPO gene transcription. Once there has been an appropriate increase in red blood cell mass, EPO production declines.

2.2 Bioactivity

EPO induces erythrocyte production by stimulating the proliferation and differentiation of erythroid progenitors termed burst forming units-erythroid (BFU-E) (24) and the proliferation and differentiated activity of more mature erythroid precursors (proerythroblast, erythroblast) (30). EPO can stimulate megakaryocyte colony formation *in vitro* (31, 32) and platelet production in mice (33). However, in humans, EPO administration has had inconsistent effects on platelet levels. Consequently the physiological relevance of the *in vitro* and animal studies indicating that EPO induced changes in platelet formation is unclear. EPO effects are mediated through a cell surface receptor composed of a single membrane-spanning domain (34, 35). The EPO receptor is a member of the cytokine receptor superfamily (CKR-SF) that includes receptors for interleukins 2–7, G-CSF, GM-CSF, TPO, and the receptors for two nonhematopoietic ligands, prolactin and growth hormone (36). The CKR-SF is distinguished by common domains present in the extracellular portion of the receptors: a cytokine receptor homologous (CRH) region, and in some but not all receptors, an immunoglobulin-like (Ig) domains and/or a fibronectin type III-like (FNIII) domains. The CRH region contains two conserved sequence motifs: four conserved cysteine residues in the amino terminal half and a

carboxy-terminal Trp-Ser-x-Trp-Ser motif (x = nonconserved amino acid). The EPO receptor contains a CRH region, but no Ig or FNIII domains. In response to EPO binding, EPO receptors homodimerize and may also heterodimerize with other cytokine receptors, such as the receptor for stem cell factor (37). Receptor homo- or heterodimerization leads to receptor activation and activation of intracellular signaling pathways (38, 39). EPO-mediated activation of the EPO receptor leads to phosphorylation of tyrosine residues on the EPO receptor (40) and activation of phosphatidylinositol 3-kinase (41, 42) as well as protein kinase Cε (43). EPO-mediated activation of the tyrosine kinase Jak2, leading to activation of the STAT5 transcription factor, has also been observed (44, 45). EPO-mediated changes in gene expression include increased expression of the c-myc gene (46, 47) and increased expression of erythroid specific genes (48).

2.3 Therapeutic Indications

EPO is used in the treatment of disease-related anemias that are defined by a reduced red blood cell volume for which a blood transfusion is anticipated or needed. EPO would not be used for correctable anemias, e.g., iron, folate, or vitamin B_{12} deficiencies. Current FDA-approved uses for EPO are for anemia associated with chronic renal failure, anemia in acquired immune deficiency syndrome (AIDS) patients receiving azidothymidine (AZT), anemia in cancer patients receiving chemotherapy (49), and for allogenic blood transfusions in patients undergoing elective, noncardiac, nonvascular surgery (50–52). Anemia in patients with chronic renal failure is chiefly the consequence of nonfunctioning kidneys failing to produce enough EPO (53, 54). Other contributing factors include the following: hemolysis, blood loss, aluminum toxicity, hyperparathyroidism, and folate deficiency. In poorly dialyzed patients with end-stage renal disease, accumulation of uremic toxins (polyamines) in the blood can produce a bone marrow suppression (55). The etiology of the AIDS-related anemia is unknown; contributing factors include diminished production of EPO as well as an AZT mediated down-regulation of EPO receptors on bone marrow progenitor cells (56). One of the dose-limiting toxicities associated with AZT therapy in AIDS patients is anemia and nearly one-half of all AZT treated patients require red blood cell transfusions. Anemia in patients receiving chemotherapy is associated with increased circulating levels of EPO, and the underlying etiology may include the following: impaired responsiveness to EPO (57) or elevated circulating levels of cytokines that negatively regulate hematopoiesis (e.g., tumor necrosis factor) (58). Under clinical evaluation is the use of EPO in anemias associated with chronic disease (e.g., rheumatoid arthritis); anemia of prematurity, sickle cell anemia, myelodysplastic syndromes, and in response to surgical blood loss (49).

EPO stimulatory effects on erythroid progenitor proliferation and differentiation generates increased numbers of erythroblasts and reticulocytes, leading to increased numbers of circulating erythrocytes. In general, erythrocyte properties (size and volume) are unaffected. The increased hematocrit is paralleled by an increased hemoglobin level that can lead to a decline in plasma iron and ferritin concentrations. The primary target of EPO action is the erythroid cell lineage, and meaningful changes in leukocyte counts have not been observed (53); however, variable effects on platelet production have been noted. The changes in platelet production may be a consequence of EPO therapy and the result of a reactive thrombocytosis elicited in response to the iron deficiency generated from increased hemoglobin use (59). Some patients respond poorly to EPO therapy, and potential contributing factors include the following: iron deficiency, occult blood loss, and infection or inflammatory conditions (59). The most significant therapeutic benefit of EPO therapy is the reduced need for, and in some cases elimination of, blood transfusions. The reversal of the anemia also leads to a significant improvement in quality of life parameters (e.g., senses of well being) (53).

2.4 Side Effects

Side effects associated with EPO therapy seem to be a consequence of reversing the anemia as opposed to a response to EPO itself. The most common adverse effect seen in patients receiving EPO for the treatment of anemia associated with chronic renal failure is the development or reoccurrence of hypertension. Hypertensive episodes, usually in the first 3 months of EPO therapy, may be related to an increase in total peripheral resistance that occurs in response to reversal of the anemia-induced vasodilatation and the increase in viscosity of whole blood (60). Other side effects observed include clotting at the sit of vascular access in renal dialysis patients, development of iron deficiency that is related to increased use of stores, and seizures possibly linked to increases in blood pressure (see Ref. 61 for an extensive list).

2.5 Pharmacokinetics

Native EPO is a mixture of α and β forms that have identical protein content and effects *in vivo* but differ in carbohydrate content; the α form contains more *N*-acetylneuraminic acid. Glycosylation is required for biological activity *in vitro* and *in vivo* and prolongs the EPO half-life *in vivo*. Increased sugar chain-branching (tetra- versus bi-antennary) reduces the rate of clearance; a decrease in half-life is detected following removal of sialic acid residues because of rapid clearance by the liver (62, 63). Commercially available EPO (recombinant human; rHuEPO) preparations are obtained from Chinese hamster ovary cells engineered to express the human gene. Because a mammalian cell line is used for production, rHuEPO is glycosylated; commercial rHuEPO preparations may differ in the degree of glycosylation and sugar chain-branching. Glycosylation may explain the nonimmunogenicity of rHuEPO preparation; local skin irritations that are occasionally seen may relate to the use of human albumin in the preparations.

rHuEPO is administered parenterally (e.g., intravenous infusion, subcutaneous injection, or intraperitoneal in patients undergoing peritoneal dialysis). There is no clinical advantage to the intravenous versus subcutaneous route of administration unless a venous access device is in place. RHuEPO distributes into a single compartment, and the volume of distribution approximates or slightly exceeds plasma volume. There is limited information on elimination kinetics, but rHuEPO seems to decline in a first-order fashion. The mean plasma elimination half-life ranges from 4 to 16 h in healthy individuals and individuals with chronic renal failure (64). Desialylated EPO is generated and cleared by the liver; approximately 10% of the administered dose appears unchanged in the urine. RHuEPO (intravenous or subcutaneous) is normally administered two to three times weekly. The onset of rHuEPO action is within 1–2 weeks, with the desired effects seen in 8–12 weeks.

2.6 Preparations

There are two commercially available preparations of rHuEPO currently available: epoetin alfa (EPOGEN, Amgen) and PROCRIT (Ortho Biotech). Both preparations are derived from the Chinese hamster ovary cells engineered to express the human EPO cDNA, are nearly identical preparations, and are prepared for parenteral administration for IV or subcutaneous use in sterile preservative-free solutions.

3 GRANULOCYTE-MACROPHAGE COLONY-STIMULATING FACTOR

A biological activity that supported the *in vitro* growth of neutrophil and mononuclear cells was initially detected in culture media conditioned by phytohemagglutin-P stimulated peripheral blood lymphocytes (65). Based on this study and others (66–69), human granulocyte-macrophage colony-stimulating factor (GM-CSF) was purified from culture supernates obtained from a T-lymphoblast cell line infected with human T-cell leukemia virus-II (70). At that time, the novel approach of expression cloning was used to isolate the human GM-CSF cDNA from a cDNA library constructed with mRNA isolated from lectin stimulated T-lymphoblasts and expressed in COS-1 cells (71). The first phase I and II clinical trials with GM-CSF were conducted in 1987 (72, 73).

3.1 Physical Properties

The human GM-CSF gene has been mapped to chromosome 5q21-q32 within 500 kb of several other cytokine genes including IL-3, IL-4, and IL-5 (74, 75). The human GM-CSF gene is approximately 2.5 kb in length, is composed of four exons and three introns, and is expressed as a 1-kb transcript in T-lymphocytes, macrophages, mast cells, endothelial cells, osteoblasts, and fibroblasts (71, 76, 77). Under basal conditions, there is little, if any, GM-CSF gene expression. In response to immune or inflammatory stimulation/mediators such as tumor necrosis factor or IL-1, steady-state levels of GM-CSF transcripts increase dramatically. Changes in the steady-state level of GM-CSF are mediated at the transcriptional as well as the post-transcriptional level. Transcriptional changes in GM-CSF gene expression are mediated through a combination of *trans* acting factors including NF-κB, Elk-1, and AP-1 binding to cognate *cis* acting elements located in the region of the GM-CSF gene that flanks the 5′ end of the coding region (78–81). Post-transcriptional changes in GM-CSF in mRNA levels are mediated through AU-rich elements located in the 3′ untranslated region of the GM-CSF mRNA. The AU-rich sequences serve as binding sites for an RNA activity that may be inactivated in response to IL-1 or tumor necrosis factor-α (for reviews see refs. 77, 82). The human GM-CSF protein is a monomer that contains two internal disulfide bonds and is composed of 144 amino acids; the first 17 amino acids compose the leader sequence. The crystal structure for GM-CSF solved to 2.4 Å revealed a two stranded-antiparallel β sheet with an open bundle of four α helices (83). The predicted molecular mass for the GM-CSF is also a glycoprotein; consequently, the secreted forms of the GM-CSF protein range in size from 18 to 22 kDa. In contrast to EPO, where complete removal of the carbohydrate portion eliminates biological activity *in vitro* because of loss of structure, removal of the carbohydrate on GM-CSF is associated with an increase in specific activity (84). The carbohydrate portion of GM-CSF has been speculated to interfere with GM-CSF receptor binding and/or receptor activation. In normal healthy adults, circulating levels of GM-CSF are near or below the limits of detection (0.1 ng/mL by enzyme-linked immunosorbent assay). Increased circulating levels of GM-CSF have been seen in some, but not all burn patients (85), in transplant recipients preceding an infection (86), and in cancer patients with malignant disease and high platelet levels (87). In general, increased circulating levels of GM-CSF are not seen, which is consistent with the concept that GM-CSF most likely works at the local level in a paracrine or autocrine fashion. Consistent with this model, elevated levels of GM-CSF have been detected in the synovial fluid of patients with chronic inflammatory disease (88). Interestingly, in eosinophils, a GM-CSF target, obtained from asthmatics, the GM-CSF protein is found in granules (89). In these patients, the local release of GM-CSF could enhance in the inflammatory response during an asthmatic attack.

3.2 Bioactivity

In vitro, recombinant GM-CSF supports the proliferation and differentiation of granulocyte/macrophage progenitors (CFU-GM) and the functional activity of mature macrophages, neutrophils, and eosinophils (90). Enhanced functional activities of mature cells include the following: tumoricidal activity, antibody-dependent cell-mediated cytoxicity, superoxide production, phagocytic activity, secretion of other cytokines (e.g., GM-CSF stimulates CSF-1 production by macrophages). In combination with IL-3, IL-6, and SCF, GM-CSF supports the proliferation and differentiation of more primitive myeloid progenitors (CFU-GEMM); in combination with EPO, GM-CSF will support the proliferation and differentiation of erythroid progenitors and the proliferation and differentiation of cells in the megakaryocytic lineage (see Fig. 5.1). GM-CSF effects are initiated at the plasma membrane in response to GM-CSF–mediated receptor activation. The GM-CSF receptor is a member of a subfamily of the cytokine receptor superfamily (CKR-SF) that also includes receptors for IL-3 and IL-5 (91, 92). The receptors for this subfamily are heterodimers, and IL-3, IL-5, or GM-CSF binds to an IL-3–, IL-5–, or GM-CSF-specific ligand-binding α-chain. Like other members of the CKR-SF, the GM-CSF receptor ligand-specific α-chain contains a CRH region and a FNIII domain (see Section 2.2 for a description of the defining features of the CKR-SF family members). In general, the low affinity trans-membrane α-chain does not transduce a signal; however,

GM-CSF binding to its α-chain can signal for an increase in glucose uptake (93). The high affinity signaling receptors are composed of a ligand specific α-chain plus a transmembrane β_c-chain, which are shared by Il-3, IL-5, and GM-CSF. The β_c-chain contains a longer cytoplasmic tail compared with the α-chain. Current models predict that IL-3, IL-5, or GM-CSF binds to high affinity $\alpha\beta_c$ heterodimers; this results in a ligand-dependent interaction between the α- and β_c-chain cytoplasmic regions that initiate the signaling process. Putative post-receptor signaling pathways activated in response to GM-CSF binding include changes in ion fluxes, inositol phosphate mobilization, activation of protein kinase C, and the mitogen-activated protein kinases (MAPK; for review see Ref. 94). GM-CSF–induced changes in gene expression have been linked to activation of the tyrosine kinases Jak1 and Jak2, leading to activation of the STAT5 transcription factor (95, 96), and activation of the MAPK cascade, leading to activation of the Elk and Egr-1 transcription factors (97). Consistent with a role in promoting cellular proliferation, one response to GM-CSF is increased expression of growth-related genes such as c-fos, c-jun, and c-myc.

3.3 Therapeutic Indications

FDA-approved uses for recombinant GM-CSF are to accelerate myeloid recovery in patients with non-Hodgkin's lymphoma, acute lymphoblastic leukemia, or Hodgkin's disease undergoing autologous bone marrow transplantation; to prolong the survival of adults who have undergone allogenic or autologous bone marrow transplantation and in whom engraftment is delayed or has failed; to accelerate neutrophil recovery in patients with acute myeloid leukemia that have received chemotherapy; and to mobilize hematopoietic progenitor cells into circulation for collection by leukapheresis. In general, patients who have received high dose chemotherapy with or without the subsequent replacement of hematopoietic stem and progenitor cells (bone marrow, peripheral blood, or umbilical cord blood transplant) have an initial period of neutropenia that is associated with an increased risk of developing a life-threatening infections. In clinical trials, patients receiving GM-CSF therapy had a more rapid neutrophil recovery, fewer days of antibiotic treatment, fewer infectious episodes, and spent fewer days in the hospital (98–101). Under investigation is the use of GM-CSF to increase neutrophil counts in patients with AIDS; congenital, cyclic, or acquired neutropenias; myelodysplastic syndrome; or severe aplastic anemia. Also under evaluation is the use of GM-CSF to recruit quiescent malignant hematopoietic stem/progenitor cells into cycle, thus rendering them responsive to subsequent chemotherapy. GM-CSF mobilizes hematopoietic stem/progenitor cells from the bone marrow into circulation, and this effect of GM-CSF has been used when peripheral blood is harvested for use in autologous peripheral blood stem cell transplants (102–104).

The initial response to GM-CSF administration is a transient drop in circulating neutrophils and monocytes caused by margination and sequestration of neutrophil and monocytes in the lung (105); within 2–6 h, neutrophil and monocyte counts return. GM-CSF responses are dose-dependent and biphasic. Increased neutrophil counts are seen with low GM-CSF doses; at higher doses, increased numbers of monocyte/macrophages and eosinophils are seen. After initial transient changes, there is a steady increase in numbers of circulating neutrophils, followed by a plateau 3–7 days after initiation of GM-CSF therapy. A second increase in circulating neutrophils is seen 4–5 days after the initial plateau. Redistribution of mature cells from the marrow, a shortening maturation time, and retention of neutrophils in the vasculature (decreased extra vascular migration and increased demargination) accounts for the first phase of neutrophil production. The second phase is caused by GM-CSF recruiting into cycle and decreasing the cell cycle time of the stem/progenitor cell compartment (106); this also leads to increased number of circulating stem/progenitor cells (107, 108). GM-CSF has a primarily myeloid effect; in general, lymphocyte, reticulocyte, and erythrocyte counts remain unchanged with variable effects on platelet counts. Neutrophil counts return to pretreatment levels 2–10 days after GM-CSF therapy is discontinued.

3.4 Side Effects

Sargramostim (Leukine, Immunex), recombinant human GM-CSF (rHuGM-CSF), is produced in yeast and is usually well tolerated at clinically useful doses. rHuGM-CSF differs from native GM-CSF by the substitution of a leucine for a proline at position 23, and potentially, a difference in the nature and amount of carbohydrate present. The most common side effects observed with sargramostim in placebo-controlled studies were diarrhea, asthenia, rash, and malaise. With intravenous opposed to subcutaneous administration, edema, capillary leak syndrome, pleural, and/or pericardial effusion have been seen (109). The fluid retention and capillary leak syndrome may be the result of GM-CSF–induced production of tumor necrosis factor by neutrophils. A first dose effect has been seen with patients receiving sargramostim; more severe first dose effects are seen with molgramostim (rHuGM-CSF expressed in bacteria) and regramostim (rHuGM-CSF expressed in Chinese hamster ovary cells). First dose effects can include transient flushing, tachycardia, hypotension, musculoskeletal pain, nausea, vomiting, dyspnea, and a fall in arterial oxygen saturation. The latter two responses are thought to be caused by sequestration of neutrophils in the pulmonary circulation. The potential for drug—drug interactions may exist if rHuGM-CSF is given simultaneously with drugs known to have myeloproliferative effects, e.g., lithium and corticosteroids.

3.5 Pharmacokinetics

Sargramostim and regramostim are produced as glycoproteins. Both GM-CSF preparations differ from each other and from endogenous GM-CSF in the degree of glycosylation. Analysis of regramostim and molgramostim (nonglycosylated rHuGM-CSF) in humans after subcutaneous administration revealed that nonglycosylated GM-CSF was absorbed more rapidly and to a greater extent, but was also cleared more rapidly than glycosylated rHuGM-CSF. Whether the carbohydrate moiety accounts for the differences observed or whether there are additional differences between the two forms of rHuGM-CSF *in vivo* is not clear. In one study, 4 of 16 patients receiving Sargramostim (rHuGM-CSF produced in yeast) developed antibodies that recognize the rHuGM-CSF. Mapping studies revealed that the antibodies were recognizing epitopes that are glycosylated in native GM-CSF (110). rHuGM-CSF is administered parenterally (intravenous infusion, subcutaneously). When administered subcutaneously, rHuGM-CSF is absorbed rapidly with peak serum concentrations occurring within 2–4 h. The information available on sargramostim suggests it distributes into two compartments with first-order kinetics. It is unclear into which tissues Sargramostim distributes or how it is metabolized and/or eliminated. Serum concentrations decline in a biphasic manner (109). Sargramostim is most commonly administered as an intravenous infusion; it can also be administered subcutaneously, and myelopoietic effects obtained following either route of administration are comparable.

3.6 Preparations

Sargramostim is the only FDA-approved form of rHuGM-CSF available in the United States. It is prepared from a powder, which also contains mannitol, sucrose, and tromethamine, in sterile or bacteriostatic water for parenteral administration. Molgramostim (Gramal, Probiomed; Grow-Gen-Gm, Gautier; Leucomax, Novartis, Sandoz, Schering-Plough, UpJohn, Rhone-Poulenc Rorer) is available outside the United States.

4 GRANULOCYTE-COLONY STIMULATING FACTOR

A biological activity that promoted neutrophilic differentiation of a mouse myelomonocytic cell line was first identified in the serum of endotoxin-treated mice (111). Subsequently, a similar activity present in media conditioned by lungs obtained from endotoxin-treated mice formation was identified and was named granulocyte-colony stimulating activity (112). Initially human granulocyte-colony stimulating factor (G-CSF) activity was detected and purified from media conditioned by a bladder carcinoma and a squamous carcinoma cell line (113, 114). Based on the human G-CSF protein sequence,

degenerate oligonucleotides probes were used to identify and clone the human G-CSF cDNA (115–117). The first clinical trails with recombinant human G-CSF were performed in cancer patients who had received chemotherapy (118–120).

4.1 Physical Properties

The human G-CSF gene locus spans 2.3 kb, contains five exons and four introns, and has been mapped to chromosome 17q21–22 (117, 121, 122). G-CSF is expressed as a 2-kb transcript (122) in monocytes, macrophages, neutrophils, fibroblasts, and endothelial cells. In general, basal G-CSF gene expression is low; in response to inflammatory/immune mediators (e.g., tumor necrosis factor, IL-1, endotoxin, M-CSF, and GM-CSF), steady-state levels of G-CSF mRNA increase because of enhanced G-CSF gene transcription and increased stability of the G-CSF mRNA (123). In parallel to GM-CSF, changes in G-CSF gene transcription are mediated by $trans$ acting factors, including NF-IL6/CEBP and NF-κB, binding to their cognate cis acting elements, located 5′ to the G-CSF protein coding region (124, 125). The G-CSF mRNA also contains AU-rich elements that contribute to the regulation of G-CSF mRNA stability. Structurally, G-CSF, like GM-CSF, has a four α-helical bundle topology (126, 127). The human G-CSF protein contains two internal disulfide bonds and is composed of 204 (or 207, see below) amino acids; before secretion, a 30—amino acid leader sequence is removed. Native G-CSF is a glycoprotein with a molecular mass of approximately 20 kDa; without the carbohydrate, the predicted molecular mass is closer to 18.6 kDa. There are two alternatively spliced variants of G-CSF; one encodes a 177 amino acid mature peptide and the other encodes a 174 amino acid mature peptide. Two splice donor sites are present at the 5′ end of intron 2 with one acceptor site at the 3′ end of intron 2 (117). The splice donor sites are present in tandem, nine base pairs apart. Consequently, the larger protein contains three additional amino acids (val-ser-glu) between Leu[35] and Val[36]. The 174 amino acid form of G-CSF predominates in $vivo$ and is approximately 20-fold more active than the 177 amino acid form. In normal healthy adults with a normal neutrophil count, circulating levels of G-CSF are <78 pg/mL as measured in enzyme-linked immunoassays (128). Increased circulating levels of G-CSF have been detected in some patients with leukemia (129), in patients with cytoxic chemotherapy-induced neutropenia (129), and during the acute phase of an infection (130).

4.2 Bioactivity

Insight into the role of endogenous G-CSF can be gained from the phenotype seen in mice in which the endogenous G-CSF gene has been inactivated ("knock-outs") (131). G-CSF knock-out mice have chronic neutropenia that is reversed after administration of G-CSF, which suggests one function of endogenous G-CSF is to maintain normal neutrophil levels. When challenged with infectious agents, e.g., *Listeria monocytogenes*, the G-CSF knock-out mice developed more severe infections, had a higher mortality rate, and did not respond with neutrophilia. The latter observation indicates endogenous G-CSF may mediate the characteristic neutrophilia seen in response to infections. *In vitro*, recombinant G-CSF is a lineage restricted growth factor that affects the proliferation, differentiation, and activation of progenitors cells committed to the granulocytic/monocytic lineage (CFU-GM; see Fig. 5.1). In combination with other cytokines, e.g., IL-3 or GM-CSF, G-CSF can support the proliferation of more primitive progenitors (132–135). G-CSF stimulates the functional activities of developing and mature neutrophils including chemotaxis, phagocytosis, antibody-dependent cellular cytoxity, and enhanced production of superoxide anion and hydrogen peroxide (see Refs. 123, 136 for reviews). G-CSF–mediated effects are initiated at the plasma membrane in response to G-CSF mediated receptor activation. The G-CSF receptor is a member of the cytokine receptor superfamily (CKR-SF) that includes receptors for interleukins 2–7, GM-CSF, TPO, EPO, and the receptors for two nonhematopoietic ligands, prolactin and growth hormone (36). As discussed previously (see Section 2.2), the CKR-SF is distinguished by common domains present in extracellular portion of the receptor. The G-CSF receptor contains in its extracellular domain, in a carboxy terminal to

plasma membrane spanning domain orientation, a carboxy terminal Ig followed by a CRH region and three FNIII domains. Mutagenesis studies have revealed that the carboxy terminal Ig domain and the CRH region are required for G-CSF binding and subsequent receptor oligomerization (137, 138). Immediate down-stream effectors of ligand-activated G-CSF receptor include the tyrosine kinases, Jak1 and Jak2 (139, 140), that in turn phosphorylate and activate the transcription factor STAT3, thus leading to changes in gene expression ((141); for in depth review see ref. (95)). Increased cAMP levels and release of membrane-associated arachidonic acid are also detected after G-CSF receptor activation (142, 143).

4.3 Therapeutic Indications

Current FDA-approved indications for G-CSF therapy are to decrease the incidence of infection in patients with nonmyeloid malignancies receiving myelosuppressive chemotherapy regiments that are associated with a significant incidence of severe neutropenia with fever and to reduce the duration of neutropenia and neutropenia related clinical sequelae, e.g., febrile neutropenia in patients with nonmyeloid malignancies undergoing myeloablative chemotherapy followed by bone-marrow transplantation. The myelosuppression associated with certain chemotherapy regimens and the myeloablative chemotherapy before bone marrow transplant places patients at risk for developing life-threatening infections and can limit the length, frequency, and/or dose of chemotherapy. Under these conditions, G-CSF therapy can, in general, decrease the duration of severe neutropenia, infectious episodes, and infective therapy and hospitalization. Ongoing studies are aimed at determining whether prophylactic G-CSF therapy is more effective than waiting for the neutropenia and/or neutropenic sequelae to occur. G-CSF therapy is approved for patients with severe chronic neutropenias (congenital neutropenia, cyclic neutropenia, or idiopathic neutropenia). These patients have decreased numbers of circulating neutrophils and are more susceptible to bacterial infections. G-CSF therapy reduces the incidence and duration of neutropenia related sequelae, e.g., fever, infection, and oropharyngeal ulcers, and also leads to improved quality of life. Whether long-term G-CSF therapy should be used in these patients is under evaluation. In comparison with GM-CSF therapy, G-CSF therapy may be more clinically effective, because G-CSF does not also cause an increase in the number of circulating eosinophils. G-CSF is also used to mobilize hematopoietic stem and progenitor cells into the peripheral blood for collection by leukapheresis. The use of G-CSF therapy to increase neutrophil counts in patients with myelodysplastic syndrome and to treat the neutropenia in adults with acute leukemia undergoing myelosuppressive chemotherapy is also under clinical evaluation. These particular uses of G-CSF therapy are controversial because in both instances there is the potential for G-CSF to serve as a mitogen and thus stimulate expansion of leukemia blast cells. G-CSF is under evaluation in patients with AIDS or AIDS-related complex to offset the granulocytopenia associated with zidovudine therapy.

Within the first hour after intravenous or subcutaneous administration of G-CSF, there is a drop in the number of circulating neutrophils and monocytes because of margination of neutrophils/monocytes to blood vessel walls. Within 24 h, the numbers of circulating neutrophil begin to increase because of demargination of cells and release of maturing neutrophils from the bone marrow (144). The response to G-CSF is dose-dependent, and neutrophil numbers generally plateau within a week. G-CSF treatment leads to a significant increase in the rate of neutrophil production as a consequence of recruiting more myeloid progenitors into cycle, stimulating progenitor cell proliferation, accelerating the maturation rate, and enhancing bone marrow release of maturing neutrophils (145). Transient changes in the circulating levels of other hematopoietic cells have been seen in during G-CSF therapy including increased numbers of circulating myeloid progenitors, monocytes, and lymphocytes, and decreased numbers of platelets (118, 146, 147). After discontinuation of G-CSF therapy, circulating neutrophils numbers return to pretreatment levels within 4–7 days.

4.4 Side Effects

In general, recombinant human G-CSF is well tolerated and is not associated with the capillary leak syndrome seen after administration of other cytokines, e.g., rHuGM-CSF. Mild-to-moderate bone pain, specifically in bones rich in marrow, e.g., sternum, spine, pelvis, and long bones, is a common side effect (≤20% patients) and is most likely related to increased blood cell production. Reversible elevations in serum lactate dehydrogenase, alkaline phosphatase, uric acid, and leukocyte alkaline phosphatase are seen and have been related to increased neutrophil turnover (148). Some patients with myelodysplastic syndrome have developed leukemia while receiving G-CSF (149, 150). Whether the leukemia was the result of G-CSF–mediated acceleration of the disease process or expansion of leukemic cells already present is not known. Exacerbation of pre-existing inflammatory conditions (e.g., eczema, cutaneous vasculitis) has been seen in some patients (150). More notable in patients treated for extended periods of time with G-CSF is splenic enlargement, which may be caused by extramedullary hematopoiesis (151).

4.5 Pharmacokinetics

The currently available commercial preparation of recombinant human G-CSF (Filgrastim, Neupogen; Amgen Inc.) is obtained from *Escherichia coli* engineered to express the human G-CSF cDNA. For expression in *E. coli*, a methionine group was added to the amino terminus of the human G-CSF protein consequently the recombinant G-CSF is referred to as recombinant methionyl human G-CSF (r-metHuG-CSF). Bacterially expressed r-metHuG-CSF is not glycosylated, whereas endogenous G-CSF has an *O*-linked carbohydrate moiety at threonine 133. r-metHuG-CSF is functional *in vitro* and *in vivo*, which suggests that glycosylation per se is not required for biological activity. *In vitro* and *in vivo* studies with recombinant human G-CSF expressed in a mammalian cell line (Chinese hamster ovary cells; Lenograstim, Granocyte; Chugai Pharmaceutical Company), suggest that glycosylation may increase the stability and potency of G-CSF. R-metHuG-CSF can be administered subcutaneously or intravenously; subcutaneously administered r-metHuG-CSF is rapidly absorbed with peak serum concentrations occurring within 4–5 h. Absorption and clearance rates are not dose-dependent and follow first-order pharmacokinetic modeling (152). r-metHuG-CSF is rapidly distributed; the highest concentrations appear in the bone marrow, adrenal glands, kidney, and liver. The volume of distribution after subcutaneous or intravenous administration averages 150 mL/kg. It is not known how r-metHuG-CSF is metabolized or eliminated from the body. The serum half-life of r-metHuG-CSF is approximately 3–4 h and varies between patients; contributing factors include differences in disease states and actual numbers of neutrophils that may contribute to r-metHu-G-CSF metabolism.

4.6 Preparations

Filgrastim (Neupogen; Amgen Inc.) is currently the only FDA-approved form of r-metHuG-CSF available in the United States. r-MetHuG-CSF is packaged in liquid form in ready-to use vials for either subcutaneous or intravenous administration. Granocyte (Rhone-Poulenc), Granulokine (Roche), Gran (Kirin), and Grimatin (Sankyo) are available outside the United States. Currently SD/01, a sustained duration Filgrastim, is undergoing clinical evaluation (153). To generate SD/01, a polyethylene glycol (PEG) molecule was added to the amino terminus of filgrastim. In comparison with Filgrastim, the plasma clearance of SD/01 was decreased, and the plasma half-life of SD/01 was increased. Early clinical evaluations did not uncover serious toxicities. A significant advantage of SD/01 over Filgrastim therapy is that the daily Filgrastim injections could be replaced with weekly or biweekly SD/01 injections.

5 INTERLEUKIN-11

Interleukin-11 (IL-11) was originally discovered in an *in vitro* screen for activities produced by bone marrow stromal cells that would support the proliferation of an IL-6-dependent mouse plasmacytoma cell line (154). Kawashima et al. (155) cloned IL-11 based on its ability to inhibit adipogenesis. Further characterization of IL-11 uncovered its ability to synergize with other cytokines to promote hematopoietic stem and progenitor cell growth. IL-11 is a member of the gp130 family of cytokines that includes IL-6, leukemia inhibitory factor, oncostatin M, ciliary neurotrophic factor, and cardiotropin-1 (for in depth review of the gp130 family see Ref. 156). IL-11, like other gp130 family members, seems to be a mediator of the acute phase response (157).

5.1 Physical Properties

The human IL-11 gene is located at chromosome 19q13.3–13.4 and is composed of five exons and four introns (158). The IL-11 gene is expressed by a wide variety of cell types including fibroblasts, trophoblasts, chondrocytes, and endothelial cells. IL-11 transcripts are found in a wide variety of organs in the adult including thymus, spleen, bone marrow, lung, large and small intestine, kidney, brain, testis, and ovary. IL-11 gene transcripts are either 1.5 or 2.5 kb in size, and the differences in transcript sizes are the result of difference in polyadenylation site selection. However, both IL-11 transcripts encode the same sized proteins. IL-11 gene expression is under transcriptional control with *cis* acting elements for the transcription factors AP1 and C/EBP located in the IL-11 gene promoter. Cytokines like tumor necrosis factor-α and IL-1 that initiate the acute phase response enhance IL-11 gene transcription (159, 160), as well as mitogens such as epidermal growth factor. The presence of AU rich regions in the 3' flanking region of the IL-11 mRNA suggest that changes in IL-11 expression may also be mediated at the post-transcriptional level (158). Unlike most cytokines, IL-11 does not contain any cysteine residues or potential *O*- or *N*-linked glycosylation sites; consequently, after removal of a 21-amino acid signal sequence, mature IL-11, with an approximate molecular mass of 23 kDa, is secreted from cells. Whereas IL-11 is widely expressed, the IL-11 protein is rarely detected in blood, which is consistent with the concept that IL-11, like GM-CSF, most likely works at the local level in a paracrine or autocrine fashion.

5.2 Bioactivity

Analysis of mice in which the IL-11 receptor α-subunit has been genetically deleted revealed a requirement for IL-11 in embryonic implantation in which the blastocysts implant, but the decidualization response necessary for development of the chorioallantoic placentas fails; thus, the embryos fail to develop (161–163). Hematopoietic indices, including hematocrit, platelet count, and circulating white blood cell numbers, appear normal in these animals. Thus, IL-11 does not seem to be essential for blood cell development, proliferation, or maturation. Like IL-6, IL-11 alone has little if any effect on hematopoietic cell growth in *in vitro* assays. However, IL-11 will synergize with other cytokines, like SCF, to support hematopoietic stem and progenitor cell growth. In combination with IL-3 and IL-6, IL-11 will also support BFU-E, CFU-Meg, and megakaryocyte maturation (164–170). IL-11 has effects outside the hematopoietic system, including induction of acute phase protein expression in hepatocytes (171), inhibition of adipogenesis in cultured cell lines (172), and stimulation of embryonic hippocampal-derived neuronal differentiation (173). Of clinical significance are the effects of IL-11 on intestinal epithelium where the specific effect of IL-11 depends on the growth state of the epithelium. In animal models in which the epithelium is damaged as a consequence of the chemotherapy and/or radiation, IL-11 seems to promote the proliferation of the damaged epithelium and simultaneously preventing the proliferation of non-damaged epithelium (174).

IL-11 effects are initiated at the plasma membrane in response to IL-11–mediated receptor activation. In parallel to other members of the gp130 family of cytokines, IL-11 binds to a ligand-specific α-subunit, and the IL-11 ligand–IL-11 receptor α-subunit complex interacts with gp130, the signaling subunit shared in common by the different gp130 family members (92, 175). IL-11 has three distinct receptor interacting sites, one site recognizes specifically the IL-11 receptor α-subunit and the other two sites recognize gp130 (176). The IL-11 receptor α-subunit is a single membrane spanning protein, containing two potential *N*-linked glycosylation sites in its extracellular domain. Two isoforms of the human IL-11 receptor α-subunit have been identified that differ with one lacking a cytoplasmic domain. Transcripts that would encode a soluble form of the IL-11 receptor α-subunit have also been described. Post-receptor signaling pathways activated in response to IL-11 include the Jak-STAT pathway, MAP kinases, src family of tyrosine kinases, and phospholipase D (177).

5.3 Therapeutic Indications

The only currently FDA-approved use for rhIL-11 is for the prevention of chemotherapy-induced thrombocytopenia in patients with nonmyeloid malignancies (solid tumors, lymphomas). In general, the greatest benefit from rhIL-11 therapy is the acceleration of platelet recovery, thus reducing the requirement for platelet transfusions. Thus, in adults, administration of rhIL-11 is initiated within 24 h of the last cycle of chemotherapy with daily administration usually spanning 10–21 days until platelet counts of $>50,000/\mu L$ are reached. rhIL-11 is not as effective once patients become severely thrombocytopenic. In animal models, IL-11 exerts a protective effect on intestinal and oral mucosal integrity that is normally lost after intense irradiation and chemotherapy or is compromised in gastrointestinal-inflammatory disease states (178–180). In animal models, IL-11 treatment is associated with increased crypt cell (181) and enterocyte proliferation (182, 183). Thus, rHuIL-11 is currently in clinical trials for the treatment of cytotoxic-associated mucositis (184) and Crohn's disease (185).

In preclinical studies in rodents, larger animals, and nonhuman primates, subcutaneous or intravenous rHuIL-11 administration led to dose-dependent increases in megakaryocyte size and maturation and increased platelet production. IL-11 had little, if any, effect on the circulating levels of mature red or white blood cells; however, increased numbers of circulating myeloid progenitors were detected (186). In phase I clinical trials, rHu-IL-11 administration before chemotherapy led to a dose-dependent increase in the number of circulating platelets with no changes detected in the numbers of circulating mature white blood cells or myeloid progenitors. In mice, rHuIL-11 has been shown to inhibit the release in an NF-κB–dependent manner of pro-inflammatory cytokines (tumor necrosis factor-α, IL-1, IL-12) from activated macrophages. A reduction in the circulating levels of pro-inflammatory cytokines combined with the ability of IL-11 to promote the growth of certain intestinal epithelial cell types may facilitate the repair of damaged mucosal epithelium (180).

5.4 Side Effects

In general, rHuIL-11 is well tolerated, and the majority of side effects detected (extremity edema, dyspnea, and pleural effusions) are thought to be the result of fluid retention and increased plasma volume caused by sodium retention (187). Other side effects include increased serum levels of acute phase proteins, fatigue, myalgias, and arthralgias (188, 189).

5.5 Pharmacokinetics

Oprelvekin is the active ingredient in the currently available commercial preparation of recombinant human IL-11 (Neumega; Genetics Institute). Recombinant human IL-11 (rHuIL-11) is obtained from E. coli engineered to express the human IL-11 cDNA and is nonglycosylated. rHuIL-11 differs from the naturally occurring molecule in that it lacks an amino terminal proline residue, which to date has had no effect on bioactivity either *in vitro* or *in vivo*. rHuIL-11 is administered parenterally (intravenously, subcutaneously). The serum half-life of rHuIL-11 is approximately 7 h. In comparing intravenous versus subcutaneous administration, subcutaneous administration limits bioavailability to approximately 65%. The reason for the decrease in bioavailability is unknown and could be the result of degradation at the site of injection and/or as a consequence of entry into the lymphatics, which bind to target cells in lymph nodes (188). Interestingly, rHuIL-11 seems to be eliminated by metabolism. With daily dosing, an increase in platelet counts can be detected 5–9 days after initiation of treatment and will continue to increase for an additional week after withdrawal of treatment.

5.6 Preparations

Neumega is packaged in 5-mg amounts in a single-use vial as a sterile, lyophilized powder for subcutaneous administration.

6 STEM CELL FACTOR

Stem cell factor (SCF) is also known as c-kit ligand, mast cell growth factor, or steel factor. Instrumental in the identification of SCF were mice with mutations at the steel locus. The steel locus in mice was known to encode a gene essential for development of neural crest-derived melanocytes, primordial germ cells, and hematopoietic stem cells (190); mutations in the human locus have not been identified. In 1990, several groups reported, using a wide variety of experimental approaches, the identification and cloning of the product of the steel locus, which encoded a cytokine eventually termed SCF (191–198).

6.1 Physical Properties

The human SCF gene is located at chromosome 12q22-q24, spans 50 kb, and is composed of eight exons and seven introns (195). The SCF gene is expressed in a variety of cells including fibroblasts, bone marrow stromal cells, vascular endothelial cells, keratinocytes, intestinal epithelial cells, Sertoli's cells, granulosa cells, and various embryonic tissues (199–201). Differential splicing generates two different SCF mRNAs, one with and one without exon 6, which encodes a proteolytic cleavage site. In the absence of the proteolytic cleavage site, SCF is retained as a transmembrane protein at the cell surface that consists of an extracellular domain (157 amino acids), a transmembrane domain (27 amino acids), and a cytoplasmic domain (36 amino acids). SCF containing the proteolytic cleavage site is synthesized as a 248 amino acid protein that is cleaved to generate a 164 amino acid secreted protein (202). Both transcripts are present in cells that express SCF and both forms, secreted and membrane-bound, are biologically active. Membrane-bound SCF seems to maintain its receptor c-kit in a prolonged activation state, presumably because of less rapid downregulation of c-kit protein (203, 204). The regulation of SCF mRNA splicing and SCF protein proteolysis is not well understood. Mature soluble SCF contains two internal disulphide bonds, five N-linked sites of glycosylation, and a molecular mass of 36 kDa. Under normal physiological conditions, circulating SCF levels range between 2 and 5 ng/mL (205). The cellular and molecular mechanisms that mediate changes in SCF gene expression are not well understood and may vary by cell type (see ref. 206 for an in-depth discussion). For example, treatment of bone marrow–derived stromal-cell pro-inflammatory cytokines, such as tumor necrosis factor-1 and IL-1, leads to a decrease in the steady-state level of SCF mRNA (207), whereas IL-1 treatment of human umbilical cord endothelial cells results in an increase SCF mRNA levels (208).

6.2 Bioactivity

Analysis of mice containing mutations in either the steel locus or the dominant white spotting (W) locus, which encodes the SCF receptor, revealed that SCF is important for hematopoietic cell development and survival, melanocyte survival, and proliferation and the survival and proliferation of primordial germ cells (191–198). An intriguing feature of SCF biology is the proposed role for transmembrane SCF during development where it is postulated to be involved in guiding primordial germ cells to the gonad and melanocytes to developing skin (206). While addition of SCF to mature melanocytes will induce a proliferative response (209) and is capable of modulating the growth of cells within the ovary (210), to date the majority of studies on SCF bioactivity have focused on the role of SCF in hematopoiesis. In vitro, SCF supports the proliferation and differentiation of the progenitors that give rise to CFU-GEMM (see Fig. 5.1). Whether the actual target cell is the hematopoietic stem cell is not clear. In the absence of other cytokines, SCF will not support progenitor cell growth in vitro. However SCF will synergize with IL-3, GM-CSF, and EPO to support the growth of CFU-GEMM, BFU-E, CFU-Meg, and CFU-GM (211). A significant feature of SCF bioactivity is stimulation of the proliferation and differentiation of mast cell precursors and functional activity of mature mast cells (212). The latter effect of SCF includes release of mast cell mediators such as histamine that impact the clinical use of SCF.

SCF effects are initiated at the plasma membrane in response to SCF-mediated receptor activation. The SCF receptor is a member of the split tyrosine kinase family of growth factor receptors and is composed of five immunoglobulin-like repeats in its extracellular domain, a hydrophobic membrane spanning domain, and an intracellular tyrosine kinase domain. The first three immunoglobulin repeats constitute the SCF binding domain and the fourth immunoglobulin repeat is required for receptor homodimerization. The role of the fifth immunoglobulin repeat is not clear. Point mutations in the SCF receptor kinase domain, ligand-binding domain, and the domain immediately proximal to the membrane spanning domain have been identified in hematopoietic cells from patients with mastocytosis, idiopathic myelofibrosis, acute leukemia, and gastrointestinal stromal tumors (213).

The SCF receptor, like the receptors for platelet-derived growth factor and colony stimulating factor-1, is a member of the split tyrosine kinase family (also known as the type III receptor tyrosine kinases; for an in-depth review see ref. 214). Characteristic of the growth factor receptor family is an insert region within the cytoplasmic tyrosine kinase. In response to SCF binding, c-kit receptors homodimerize and the tyrosine kinase intrinsic to the receptor is activated, leading to auto- or cross-phosphorylation of the receptor itself, with several of the phosphorylated tyrosine residues clustered in the insert region. Receptor auto/transphosphorylation generates binding sites for proteins containing Src homology 2 (Sh2) domains. Among the Sh2 binding proteins known to bind c-kit are phospholipase Cγ-1 and a variety of adaptor proteins including growth factor receptor-bound 2 and SOS, upstream regulators of ras, which in turn regulate raf-1-MAPK cascade activity. Additionally several kinases, including Src kinases, the p83α subunit of phosphatidylinositol 3′-kinase, and the tyrosine kinase Jak2 are found in association with activated c-kit receptors. The precise role that the individual signaling pathways play in directing cellular responses to SCF is under investigation in many laboratories (for an in-depth review see ref. 206). SCF-dependent Jak activation has been linked to STAT1α- and STAT5-dependent changes in gene expression (215, 216). PI-3 kinase has been shown to be important for mast cell adhesion to fibronectin (217), whereas PLCγ-1–dependent changes have been associated with changes in cell adhesion and chemotaxis.

6.3 Therapeutic Indications

Based on its ability *in vitro* to support the proliferation and survival of hematopoietic cells, SCF has been evaluated clinically for *in vivo* and *ex vivo* expansion of hematopoietic stem cells, to mobilize hematopoietic stem cells, and in the survival of hematopoietic stem/progenitors in aplastic anemia. r-metHuSCF currently has orphan drug status for use in combination with G-CSF in patients who have undergone myelosuppressive or myeloablative therapy. Under these conditions, the combination of G-CSF plus SCF seems to enhance the number of hematopoietic stem/progenitor cells recovered in peripheral blood and to accelerate engraftment times for neutrophils and platelets (213).

SCF alone exhibits little if any effect on the growth of hematopoietic cells *ex vivo*. However the combination of SCF and other cytokines such as IL-3, IL-6, and EPO significantly enhances *ex vivo* expansion of hematopoietic stem/progenitor cells. Administration of SCF will lead to a significant increase in the numbers of circulating hematopoietic stem/progenitor cells *in vivo* (218, 219), an effect that presumably reflects the presence of circulating cytokines. In patients who received r-metHuSCF before chemotherapy, dose-related increases in neutrophils, and at higher r-metHuSCF doses, increases in platelets, reticulocytes, and circulating progenitors (BFU-E, CFU-GM) were detected (220).

6.4 Side Effects

The most frequently observed side effects associated with SCF therapy are consistent with the knowledge that mast cells have receptors for SCF. Adverse responses occur locally at the site-of-injection and include wheal formation with edema, erythema, pruritus, skin hyper-

pigmentation, and uricaria. The responses occur within 90–120 min after injection, can last up to 48 h, and are reversible. Multisystem systemic anaphylactoid-type reactions including skin reactions and respiratory distress have also been observed (221). To counter these side effects, antihistamines, an H_2-receptor antagonist, and β-adrenoreceptor stimulants are administered prophylactically (213).

6.5 Pharmacokinetics

SCF is administered parenterally by subcutaneous injection. Limited information is available regarding the pharmacokinetics of SCF. After daily subcutaneous administration, increased white cell counts were detected within 7–15 days (221). In breast cancer patients, a 2-week course of daily SCF injections resulted in increased numbers of neutrophils, monocytes, lymphocytes, and reticulocytes (222).

6.6 Preparations

Recombinant human SCF is obtained from *E. coli* engineered to express the human SCF gene. For expression in *E. coli*, a methionine group was added to the amino terminus of the human SCF protein; consequently, the recombinant SCF is referred to as recombinant methionyl human SCF (r-metHuSCF) (223). Recombinant human SCF is manufactured by Amgen and is distributed under the name Ancestim.

7 INVESTIGATIONAL AGENTS

The clinical experience gained and successes achieved with EPO, GM-CSF, G-CSF, IL-11, and SCF have enhanced the pursuit for clinically effective cytokines. As new cytokines are identified and their associated biological activities are characterized, many will enter clinical trials. In the following section, the biological properties and potential clinical applications of cytokines currently undergoing clinical evaluation are described.

7.1 Interleukin-6

Interleukin-6 (IL-6) is a member of a larger cytokine family that includes leukemia inhibitory factor, IL-11, oncostatin M, ciliary neutrophic factor, and cardiotropin-1. This cytokine family, collectively termed the gp130 family of cytokines, is a key mediator of the acute phase response, which is a systemic reaction to inflammation and tissue damage (157). The human IL-6 gene has been mapped to chromosome 6p21-p14 and is composed of five exons and four introns (224, 225). The IL-6 gene is expressed by a wide variety of tissues including fibroblasts, keratinocytes, smooth muscle cells, astrocytes, endothelial cells, T- and B-lymphocytes, pancreatic islet cells, and nearly all tissue-associated macrophage. IL-6 gene expression is primarily under transcriptional control and can be activated by a wide variety of pro-inflammatory mediators including IL-1, tumor necrosis factor-α, interferons, viruses, and bacterial products (226). After removal of a signal sequence, mature IL-6 is composed of 183 amino acids, has two sites of N-linked glycosylation, and has a molecular mass of 20–26 kDa. IL-6 alone has no effect on hematopoietic progenitor cell growth in *in vitro* assays. IL-6 synergizes with other cytokines to promote hematopoietic stem and progenitor cell growth, T-cell proliferation and B-cell activity, and of significant clinical interest, CFU-Meg proliferation and megakaryocytic maturation. In addition to affecting hematopoiesis, IL-6 will inhibit the growth of various human tumor cell lines grown *in vitro* and in animal model systems (227). IL-6 effects are initiated at the plasma membrane in response to IL-6–mediated receptor activation. As described previously for IL-11, another gp130 family member (Section 5.2), receptors for the gp130 family of cytokines are composed of two subunits: a ligand-specific subunit and a common signaling subunit, termed gp130 (91, 95). The extracellular portion of the IL-6 ligand-specific subunit contains an Ig-like domain and two FNIII domains. In response to IL-6 binding to its ligand-specific subunit, the complex interacts with either a gp130 monomer that dimerizes with another gp130 monomer or with gp130 homodimers directly. The end result is a covalent linkage between two gp130 molecules, which in turn initiates the cellular signaling pathways(s) (228–230). Post-receptor signaling pathways activated in response to IL-6 in-

clude the tyrosine kinases Jak1, Jak2, and Tyk2 that in turn lead to activate the transcription factors STAT1 and STAT3 (92, 95, 156). Coordinate with the IL-6–dependent activation of the Jaks, activation of the MAPK cascade occurs, resulting in activation of the NF-IL6 transcription factor. Together NF-IL6 and STAT3 enhance the expression of genes whose products mediate the acute phase response.

Recombinant human IL-6 (rHuIL-6) is produced in bacteria; in contrast to r-metHuG-CSF and r-metSCF, the bacterial expression system employed does not result in the addition of an amino terminal methionine (231). In animal models, administration of rHuIL-6 leads to increased numbers of circulating platelets. In individuals with normal marrow, rHuIL-6 administration leads to a dose-dependent increase in numbers of circulating platelets that peak 10–14 days after initiation of therapy with no changes in circulating white cell numbers (232, 233). Dose-independent side effects detected in phase 1 trials included fever, chills, and mild fatigue; at higher doses, acute phase proteins were detected in the sera and dose-limiting hepatic, neurologic, and cardiac effects (232, 234). At the onset of rHuIL-6 therapy, an anemia can develop that resolves after discontinuation of therapy. The etiology of the anemia is unknown, but it may be the result of red blood cell sequestration in the spleen or changes in plasma volume. rHuIL-6 alone and in combination with other cytokines is being evaluated as a thrombopoietic agent in patients. The ability of IL-6 to inhibit the growth of certain human tumor cell lines in addition to its ability to synergize with IL-2 to activate cytotoxic T-lymphocytes lead to the evaluation of rHuIL-6 as an antitumor agent (232). However, preliminary results were not promising (233). rHuIL-6 is currently manufactured for distribution in the United States by Novartis. rHuIL-6 is administered parenterally (subcutaneously, intravenously). After subcutaneous administration, serum levels of rHuIL-6 peak between 2 and 7 h, with an elimination half-life of 4 h after subcutaneous administration and 20 min after intravenous administration.

7.2 Thrombopoietin

Thrombopoietin (TPO) is also referred to as megakaryocyte growth and development factor (MGDF). The human TPO gene spans 6 kb, is located at chromosome 3q26, and contains five exons and four introns (235, 236). The TPO coding region is interrupted by introns present in the same location as those present in the EPO coding region; the amino terminal half of TPO shares 23% identity and 50% homology with the EPO protein. However, the TPO protein contains an additional 181 amino acids, enriched for sequences that direct N- and O-linked glycosylation, which is not present in the EPO protein. The similarities in gene structure and sequence homology suggest that the TPO and EPO genes may be derived from a common ancestral gene (237). The TPO gene is expressed as a 1.8-kb transcript in a variety of tissues, with highest levels present in the liver and smaller amounts in kidney, smooth muscle, and spleen (238–240). The precise cell within these tissues that expresses TPO is not known, but TPO transcripts can be detected in fibroblasts and endothelial cells, cells commonly found in many tissues (237). After removal of a 21 amino acid signal sequence, mature TPO, composed of 332 amino acids, is glycosylated before secretion. TPO may undergo a proteolytic cleavage that would lead to removal of the carbohydrate-rich carboxy terminus before secretion (241). In vitro, TPO alone stimulates CFU-Meg proliferation and maturation and has no effect on the growth of other myeloid progenitors. TPO will synergize with cytokines that normally have no effect on megakaryopoieses (e.g., IL-11, SCF, or EPO) to promote CFU-Meg growth (242). TPO effects are initiated at the plasma membrane in response to TPO-mediated receptor activation. The TPO receptor is the product of the c-mpl cellular proto-oncogene, which is expressed in early hematopoietic progenitors, megakaryocytes, platelets, and endothelial cells (243). Consistent with the biological activity of TPO, mice genetically engineered to not express the c-mpl gene (c-mpl knockouts) have a dramatic decrease in the numbers of circulating platelets with no changes in the circulating levels of red or white blood cells (244). Structurally, c-mpl (TPO-R) is similar to the receptors for EPO and G-CSF; the extracellular domain contains four conserved cysteine residues in the amino-terminal half and carboxy terminal Trp-Ser-x-Trp-Ser motif (x = nonconserved amino acid). Two different c-mpl transcripts that encode proteins with similar extracellular but different intracellular sequences have been identified (245). The functional significance of the two TPO receptor isoforms is not clear. Putative downstream effectors of ligand-activated TPO receptors are Jak2 tyrosine kinase, ras-raf-1-MAPK cascade, and phosphoinositol 3-kinase (246, 247).

Preclinical evaluation of recombinant mouse TPO revealed a dose-dependent increase in the numbers of splenic and marrow CFU-Meg, increased megakaryocyte maturation, and increased numbers of circulating platelets (240, 246), with no change in the numbers of circulating red and white blood cells. Currently recombinant human thrombopoietin (rHuTPO) has orphan drug status for use in accelerating platelet recovery in patients undergoing hematopoietic stem cell transplantation. Potential applications for TPO include HIV-related thrombocytopenia, thrombocytolytic thrombocytopenias, and improvement of ability of peripheral blood progenitors to reestablish bone marrow function (248–250). Evaluation of rHuTPO in clinical trials after parenteral administration (subcutaneous or intravenous) revealed a stimulation of thrombopoiesis within 4 days (peak response at 12 days) of intravenous administration and within 6 days of subcutaneous administration (peak response at 12–18 days) (251, 252). Adverse reactions include a slight 3–5% increase in the incidence of asymptomatic thrombocytosis, deep-vein thrombosis, pulmonary embolism, and thrombophlebitis. Two forms of thrombopoietin are currently under investigation for clinical use: recombinant human TPO (Genetech), which is full-length TPO and a truncated form of human megakaryocyte growth and development factor (lacks the carboxyl terminal 169 amino acids residues) that is conjugated with polyethylene glycol (PEG-rHuMDGF; Amgen). The clinical use of the PEG-rHuMGDF is currently unclear; it has been shown to induce the formation of antibodies that neutralize native thrombopoietin, resulting in the development of thrombocytopenia. Antibodies to rHuTPO have not been detected.

7.3   Interleukin-3

The human interleukin-3 (IL-3) gene is located at chromosome 5q23–31, 9 kb upstream from the GM-CSF gene, and is composed of five exons and four introns (251, 253). Expression of the IL-3 gene is under transcriptional and post-transcriptional control and is expressed as a 1-kb transcript in activated T-lymphocytes, mast cells, and eosinophils (253). After removal of a 19 amino acid signal sequence, mature IL-3, composed of 133 amino acids, is glycosylated before secretion. In normal healthy adults, circulating levels of IL-3 are near or below the limits of detection (<25 pg/mL serum) (254). The estimated molecular mass of IL-3 is 14–30 kDa. *In vitro* IL-3 synergizes with other cytokines (e.g., GM-CSF, EPO) to support the proliferation and differentiation of multilineage progenitors (CFU-GEMM) and committed progenitors (CFU-GM, CFU-Meg, BFU-E) (255, 256). IL-3 alone will stimulate the functional activity of more mature monocytes (including production of other cytokines) and eosinophils (157–260). IL-3 effects are initiated at the plasma membrane in response to IL-3–mediated receptor activation. The IL-3 receptor is a member of a cytokine receptor subfamily that includes the receptors for GM-CSF and IL-5. Like GM-CSF, IL-3 binds to an IL-3–specific α-chain that does not transduce a signal. The high affinity-signaling receptors are composed of a ligand-specific α-chain plus a transmembrane β_c-chain, shared in common by IL-3, IL-5, and GM-CSF. Current models predict that IL-3, IL-5, or GM-CSF binds to a high affinity $\alpha\beta_c$ heterodimer that results in a ligand-dependent interaction between the cytoplasmic regions that initiates the signaling process (91, 156). Post-receptor signaling pathways activated in response to IL-3 include the Jak tyrosine kinases, which in turn, lead to activation of the transcription factor STAT5 (95, 96).

Recombinant human IL-3 (rHuIL-3) is produced in bacteria (nonglycosylated) and in mammalian cells (glycosylated). In individuals with normal bone marrow function, rHuIL-3

administration leads to a dose-dependent increase in neutrophils, basophils, eosinophils, and platelet numbers (261–263); during the onset of rHuIL-3 therapy, increased numbers of early myeloid progenitors (CFU-GEMM, CFU-GM) appear in the peripheral blood (264). rHuIL-3 has been evaluated in patients that have undergone cancer chemotherapy, after bone marrow transplantation, in bone marrow failure and for hematopoietic progenitor cell mobilization from the bone marrow (265). The clinical use of rHuIL-3 as a single agent is limited, because dose-limiting side effects (severe headache, fever, malaise, fatigue, arthralgias) appear at doses lower than those necessary to achieve a clinical advantage. With the goal of minimizing the side effects associated with IL-3 therapy, a synthetic cytokine, Synthokine, was developed. Synthokine is in essence a greatly modified version of native IL-3 with N- and C-terminal deletions and numerous amino acid substitutions in the remaining portion of the protein (266). Synthokine binds with high affinity to the IL-3 receptor and is more potent than IL-3 in promoting the hematopoietic cell proliferation and in colony formation assays (267). Phase I/II studies are underway, and preliminary results indicate that Synthokine may be useful for enhancing platelet and neutrophils recovery in myeloablated patients (268).

The potent synergistic effect of IL-3 and GM-CSF on hematopoietic progenitor cell growth in a variety of *in vitro* and *in vivo* assays led to the development of a genetically engineered GM-CSF-IL-3 fusion protein (PIXY321). PIXY321 contains full-length GM-CSF and IL-3 proteins separated from each other by an amino acid linker $(GGGGS)_3$ that was included to allow sufficient space for protein folding. PIXY321 binds to either the GM-CSF or the IL-3 receptor with equal or greater affinity as the native cytokine and will support BFU-E, CFU-GM, CFU-GEMM, and CFU-Meg growth (269, 270). PIXY321 is more potent that equimolar mixture of IL-3 and GM-CSF in *in vitro* colony assays. In contrast to IL-3 therapy, where side effects limit administration of sufficient rHuIL-3, fewer side effects were observed for PIXY321. However, results of phase III clinical trials revealed that PIXY321 is no more effective than rHuGM-

CSF; consequently, development of PIXY321 as a therapeutic was halted. However, building on the possibility of generating biologically active chimeric proteins, a family of chimeric IL-3-G-CSF proteins, collectively termed myelopoietins, have been generated. Based on promising *in vivo* studies in animals (271), myelopoietins are being evaluated for their ability to stimulate neutrophils and platelet recovery, as well as to mobilize hematopoietic cell progenitors (272).

8 SUMMARY AND CONCLUSION

The development of cytokines as therapeutic agents has had a major impact on the treatment of renal disease and certain cancers. EPO therapy has led to a significant decline in the number of transfusions that patients with end-stage renal disease need to receive. Consequently, the number of transfusion-related complications, e.g., iron overload, transfer of infectious agents like hepatitis B and C, and human immunodeficiency virus has declined. Additionally, assessment of "quality of life" measures revealed significantly less fatigue and severity of physical symptoms in patients receiving EPO therapy (273). The development of cytokines as therapeutic agents has had a major impact on the management of patients receiving chemotherapy. In the past, amount and/or duration of certain chemotherapeutic regimens was limited by severe myelosuppression. Life-threatening infections and/or bleeding episodes were treated with antibiotics and transfusions, placing patients at risk of developing antibiotic resistance/sensitization and/or acquiring blood-born viral-related infections. Despite these ameliorative measures, the severity of the myelosuppression often limited the dose and/or the number of cycles of chemotherapy patients received, thus diminishing its overall effectiveness. rHuG-CSF therapy, as well as rHuGM-CSF therapy, minimizes the neutropenic periods and has led to decreased antibiotic usage and shorter periods of hospitalization. RHuEPO has diminished the requirement for blood cell transfusions, and current indications are that TPO will decrease the requirement for platelet transfusion. Moreover, the development of cytokine

therapy has allowed higher doses of chemotherapeutic agents (approaching and including myeloablative amounts) to be used and/or more cycles of standard chemotherapy, thus increasing the overall effectiveness of chemotherapeutic regimens (274). Mobilization of stem/progenitor cells from the bone marrow in response to cytokine therapy (e.g., GM-CSF, IL-3, SCF, IL-11) has several therapeutic implications. First, cytokine therapy followed by harvesting of peripheral blood progenitors before administration of myeloablative doses of chemotherapy allows for the use of autologous versus allogenic stem/progenitor cell transplants, thus diminishing the potential complications of allogenic transplants (e.g., identifying HLA matched donor, graft-versus-host disease). Recent evidence suggests a period of myelosuppression follows exogenous cytokine therapy (275, 276). Therapeutically, the myelosuppression results in a myeloprotective effect during the ensuing period of chemotherapy. The next phase of cytokine therapies are likely to involve mobilization of stem/progenitors followed by *ex vivo* expansion stem/progenitors before readministration. A clear long-term goal will be to use the expanded stem/progenitor cells as vehicles for gene therapy.

An unforeseen benefit of clinical trails with the cytokines has been the effects observed on other organ systems. Clearly the protective effect on intestinal and oral mucosal integrity seen with IL-11 is of significant clinical interest. In addition to the myelosuppressive/myeloablative effects of irradiation and chemotherapy, the intestinal and oral mucosa are often severely damaged. Colony stimulating factor-1 (CSF-1) is a cytokine that promotes the proliferation and differentiation of monocyte/macrophage progenitors and the activity of mature macrophages, and can synergize with other cytokines to support the proliferation of more primitive progenitors. CSF-1 has been evaluated in several clinical settings, e.g., treatment of fungal infections, antitumor activity, and osteopetrosis, with mixed results (277). However, CSF-1 has also been evaluated in animal models of arteriosclerosis. In response to rHuCSF-1, cholesterol levels in normal and hypercholesterolimic animals declined, and atherosclerotic lesion progression was slowed and the regression enhanced

(278). The mechanism responsible for the enhanced cholesterol clearance has not been defined; however, hypocholesterolemic effects were also seen in humans. *In vitro* studies suggest that G-CSF, as well as GM-CSF, directly promotes endothelial cell angiogenic activity (279). As our understanding of the biological effects of cytokines in a variety of clinical settings and in basic science research paradigms increase, other therapeutically effective uses for cytokines will undoubtedly be uncovered. In parallel to this however, more effective modes of administration need to be developed.

Currently the route of administration for all cytokines is parenteral. The intravenous route of administration is slow because bolus injections cannot be used. Additionally it affords an additional mechanism for introducing infections. Site administration effects are seen with either subcutaneous, e.g., rash seen in response to rHuG-CSF, or intravenous routes, e.g., the clotting seen at site of vascular access seen in patients receiving rHuEPO. The development of longer-acting forms of cytokines and interleukins by polyethylene glycol addition is an important first step. The development of cytokine-mimetics with oral availability is of prime importance (39). To accomplish this goal, we need to have an increased understanding of the nature of ligand—receptor interactions such as the amino acid residues involved, the functional significance of ligand/receptor glycosylation, and the role ligand/receptor homo-/heterodimerizaiton plays in activation of downstream signaling pathways. We need to understand the three-dimensional structure of ligands and receptors individually and bound complexed. As alluded to in the introduction, the identification of new cytokines and interleukins is growing at a rapid pace. Redundancy within the network of cytokines and interleukins is evident; however, the biological significance of the redundancy is less clear (280, 281). Increased understanding of the biological activities of a class of cytokines, termed classes of white blood cells to a target tissue, suggests that there may be levels of specificity not yet discovered (282). Concomitant with the identification of cytokines/interleukins that promote blood cell development has been the identification of myelosuppressive cytokines (see ref. 283 for a comprehensive review). Hopefully, as our understanding of the physical structure and the biological activities of peptide factors that have myeloproliferative, myelosuppressive, and the more cell type–specific effects grows, the identification of more effective therapeutic agents will follow.

REFERENCES

1. L. G. Lajtha, L. V. Pozzi, R. Schofield, and M. Fox, *Cell Tissue Kinet.*, **2**, 39 (1969).
2. L. O. Jacobsen, E. K. Marks, E. O. Gaston, M. Robinson, and R. E. Zirkle, *Proc. Soc. Exp. Biol. Med.*, **70**, 740 (1949).
3. C. E. Ford, J. L. Hamerton, D. W. H. Barnes, and J. T. Loutit, *Nature*, **177**, 452 (1956).
4. J. E. Till and E. A. McCulloch, *Radiat. Res.*, **14**, 213 (1961).
5. T. R. Bradley and D. Metcalf, *Aust. J. Exp. Biol. Med. Sci.*, **4**, 287 (1966).
6. Y. Ichikawa, D. H. Pluznik, and L. Sachs, *Proc. Natl. Acad. Sci. USA*, **56**, 488 (1966).
7. D. H. Pluznik and L. Sachs, *Exp. Cell Res.*, **43**, 553 (1966).
8. T. M. Dexter, T. D. Allen, and L. G. Lajtha, *J. Cell. Physiol.*, **91**, 335 (1976).
9. C. A. Whitlock, D. Robertson, and O. N. Witte, *J. Immunol. Methods*, **67**, 353 (1984).
10. G. C. Bagby Jr. and G. M. Segal in R. Hoffman, E. J. Benz, S. J. Shattil, B. Furie, H. J. Cohen, and L. E. Silberstein, Eds., *Hematology, Basic Principles and Practice*, 2nd ed., Churchill Livingstone, New York, 1995, p. 207.
11. S. E. W. Jacobsen, F. W. Ruscetti, C. M. Dubois, J. Wine, and J. R. Keller, *Blood*, **80**, 678 (1992).
12. M. T. Yip-Schneider, M. Horie, and H. E. Broxmeyer, *Blood*, **85**, 3494 (1995).
13. E. Vellenga, A. Rambaldi, T. J. Ernst, D. Ostapovicz, and J. D. Griffin, *Blood*, **71**, 1529 (1988).
14. H. E. Broxmeyer and D. E. Williams, *Crit. Rev. Oncol. Hematol.*, **8**, 173 (1988).
15. P. Carnot and C. Deflandre, *C. R. Acad. Sci.*, **143**, 384 (1906).
16. A. J. Erslev, *Blood*, **8**, 349 (1953).
17. L. O. Jacobson, E. Goldwasser, W. Freed, and L. Plzak, *Nature*, **179**, 633 (1957).
18. T. Miyake, C. K. Kung, and E. Goldwasser, *J. Biol. Chem.*, **252**, 5558 (1977).

19. K. Jacobs, C. Shoemaker, R. Rudersdorf, S. D. Neill, R. J. Kaufman, A. Mufson, J. Seehra, S. S. Jones, R. Hewick, and E. F. Fritsch, *Nature*, **313**, 806 (1985).

20. F. K. Lin, S. Suggs, C. H. Lin, J. K. Browne, R. Smalling, J. C. Egrie, K. K. Chen, G. M. Fox, F. Martin, and Z. Stabinsky, *Proc. Natl. Acad. Sci. USA*, **82**, 7580 (1985).

21. C. Winearls, *Nephrol. Dial. Transplant.*, **13**(Suppl 2), 3 (1998).

22. J. S. Powell, K. L. Berkner, R. V. Levo, and J. W. Adamson, *Proc. Natl. Acad. Sci. USA*, **83**, 6465 (1986).

23. T. R. J. Lappin, P. C. Winter, G. E. Elder, C. M. McHale, V. H. Hodges, and J. M. Bridges, *Ann. NY Acad. Sci.*, **718**, 191 (1994).

24. J. K. Browne, A. M. Cohen, J. C. Egrie, P. H. Lai, F. K. Lin, T. Strickland, E. Watson, and N. Stebbing, *Cold Spring Harb. Symp. Quant. Biol.*, **51**, 693 (1986).

25. P. H. Maxwell, M. K. Osmond, C. W. Pugh, A. Heryet, L. G. Nicholls, C. C. Tan, B. G. Doe, D. J. Ferguson, M. H. Johnson, and P. J. Ratcliffe, *Kidney Int.*, **44**, 1149 (1993).

26. P. J. Ratcliffe, B. L. Ebert, D. J. P. Ferguson, J. D. Firth, J. M. Gleadle, P. H. Maxwell, and C. W. Pugh, *Nephrol. Dial. Transplant.*, **10**(Suppl 2), 18 (1995).

27. M. A. Goldberg, S. P. Dunning, and H. F. Bunn, *Science*, **242**, 1412 (1988).

28. J. M. Gleadle, B. L. Ebert, J. D. Firth, and P. J. Ratcliffe, *Am. J. Physiol.*, **268**, C1362 (1995).

29. G. L. Semenza, *Ann. Rev. Cell Dev. Biol.*, **15**, 551 (1999).

30. F. Stohlman Jr., S. Ebbe, B. Morse, D. Howard, and J. Donovan, *Ann. NY Acad. Sci.*, **149**, 156 (1968).

31. T. Ishibashi, J. A. Koziol, and S. A. Burstein, *J. Clin. Invest.*, **79**, 286 (1987).

32. E. N. Dessypris, J. H. Gleaton, and O. L. Armstrong, *Br. J. Haematol.*, **65**, 265 (1987).

33. T. P. McDonald, M. B. Cottrell, R. E. Clift, W. C. Cullen, and F. K. Lin, *Exp. Hematol.*, **15**, 719 (1987).

34. A. D. D. Andrea, H. F. M. Lodish, and G. G. Wong, *Cell*, **57**, 277 (1989).

35. J. C. Winkelmann, L. A. Penny, L. L. Deaven, B. G. Forget, and R. B. Jenkins, *Blood*, **76**, 24 (1990).

36. J. F. Bazan, *Proc. Natl. Acad. Sci. USA*, **87**, 6934 (1990).

37. H. Wu, U. Klingmuller, P. Besmer, and H. F. Lodish, *Nature*, **377**, 242 (1995).

38. S. S. Watowich, A. Yoshimura, G. D. Longmore, D. J. Hilton, Y. Yoshimura, and H. F. Lodish, *Proc. Natl. Acad. Sci. USA*, **89**, 2140 (1992).

39. L. K. Jolliffe, S. A. Middleton, F. P. Barbone, D. L. Johnson, F. J. McMahon, W. H. Lee, R. H. Gruninger, and L. S. Mulcahy, *Nephrol. Dial. Tansplant.*, **10**(Suppl 2), 28 (1995).

40. S. Gobert, F. Porteu, S. Pallu, O. Muller, M. Sabbah, L. Dusanter-Fourt, G. Courtois, C. Lacombe, S. Gisselbrecht, and P. Mayeaux, *Blood*, **86**, 598 (1995).

41. J. E. Damen, R. L. Culter, H. Jiao, T. Yi, and G. Krystal, *J. Biol. Chem.*, **270**, 23402 (1995).

42. B. Witthuhn, F. W. Quelle, O. Silvennoinen, T. Yi, B. Tang, O. Miura, and J. N. Ihle, *Cell*, **74**, 227 (1993).

43. Y. Li, K. L. Davis, and A. J. Sytkowski, *J. Biol. Chem.*, **271**, 27025 (1996).

44. O. Miura, N. Nakamura, J. N. Ihle, and N. Aoki, *J. Biol. Chem.*, **269**, 614 (1994).

45. H. Wakao, N. Harada, T. Kitamura, A. L. F. Mui, and A. Miyajima, *EMBO J.*, **14**, 2527 (1995).

46. R. Spangler and A. S. Sytkowski, *Blood*, **79**, 52 (1992).

47. C. Chen and A. J. Sytkowski, *J. Biol. Chem.*, **276**, 38518 (2001).

48. K. S. Prasad, J. E. Jordan, J. J. Koury, M. C. Bondurant, and S. J. Brandt, *J. Biol. Chem.*, **270**, 11603 (1995).

49. L. T. Goodnough, K. C. Anderson, S. Kurtz, T. A. Lane, P. T. Pisciotto, M. H. Sayers, and L. E. Silverstein, *Transfusion*, **33**, 944 (1993).

50. J. R. deAndrade, M. Jove, G. Landon, D. Frei, M. Guilfoyle, and D. C. Young, *Am. J. Orthop.*, **25**, 533 (1996).

51. M. A. Goldberg, J. W. McCutchen, M. Jove, P. DiCesare, R. J. Friedman, R. Poss, M. Guilfoyle, D. Frei, and D. Young, *Am. J. Orthop.*, **25**, 544 (1996).

52. P. M. Faris, M. A. Ritter, and R. I. Abels, *Am. J. Bone Joint Surg.*, **78A**, 62 (1996).

53. J. W. Eschbach, J. C. Egrie, M. R. Downing, J. K. Browne, and J. W. Adamson, *N. Engl. J. Med.*, **316**, 73 (1987).

54. C. G. Winearls, D. O. Oliver, M. J. Pippard, C. Reid, M. R. Downing, and P. M. Cotes, *Lancet*, **2**, 1175 (1986).

55. S. Moncada, M. J. Palmer, and E. A. Higgs, *Pharmacol. Rev.*, **43**, 109 (1991).

56. S. R. Gogu, B. S. Beckman, R. B. Wilson, and K. C. Agrawal, *Biochem. Pharmacol.*, **50**, 413 (1995).

57. C. Miller, R. Jones, S. Piantotadosi, M. D. Abeloff, and J. L. Spivak, *N. Engl. J. Med.*, **322**, 1689 (1990).

58. R. T. Means and S. B. Krantz, *Blood*, **80**, 1639 (1992).

59. I. C. Macdougall, *Nephrol. Dial. Transplant.*, **10**(Suppl 2), 85 (1995).

60. J. W. Eschbach, M. H. Abdulhadi, J. K. Browne, B. G. Delano, M. R. Downing, J. C. Egrie, R. W. Evans, E. A. Friedman, S. E. Graber, and N. R. Haley, *Ann. Intern. Med.*, **111**, 992 (1989).

61. G. K. McEvoy, Ed., *AHFS Drug Information*, American Society of Health-System Pharmacists, Bethesda, MD, 1996, p. 1043.

62. J. L. Spivak and B. B. Hogans, *Blood*, **73**, 90 (1989).

63. M. Takeuchi, N. Inoue, T. W. Strickland, M. Kubota, M. Wada, R. Shimizu, S. Hoshi, H. Kozutsumi, S. Takasaki, and A. Kobata, *Proc. Natl. Acad. Sci. USA*, **86**, 7819 (1989).

64. I. C. Macdougall, D. E. Roberts, G. A. Coles, and J. D. Williams, *Clin. Pharmacokinet.*, **20**, 99 (1991).

65. M. J. Cline and D. W. Golde, *Nature*, **248**, 703 (1974).

66. A. W. Burgess, J. Camakaris, and D. Metcalf, *J. Biol. Chem.*, **252**, 1998 (1977).

67. M. J. Cline and D. W. Golde, *Nature*, **277**, 177 (1979).

68. A. J. Lusis, D. H. Quon, and D. W. Golde, *Blood*, **57**, 13 (1981).

69. A. J. Lusis, D. W. Golde, D. H. Quon, and L. A. Lasky, *Nature*, **298**, 75 (1982).

70. J. C. Gasson, R. H. Weisbart, S. E. Kaufman, S. C. Clark, R. M. Hewick, G. G. Wong, and D. W. Golde, *Science*, **226**, 1339 (1984).

71. G. G. Wong, J. A. Witek, P. A. Temple, K. M. Wilken, A. C. Leary, D. P. Luxengberg, S. S. Jones, E. L. Brown, R. M. Kay, E. C. Orr, C. Shoemaker, D. W. Golde, R. J. Kaufman, R. M. Hewick, E. A. Wang, and S. C. Clark, *Science*, **228**, 810 (1985).

72. J. E. Groopman, R. T. Mitsuyasu, M. J. DeLeo, D. H. Oette, and D. W. Golde, *N. Engl. J. Med.*, **317**, 593 (1987).

73. S. Vadhan-Raj, M. Keating, A. LeMaistre, W. N. Hittelman, K. McCredie, J. M. Trujillo, H. E. Broxmeyer, C. Henney, and J. U. Gutterman, *N. Engl. J. Med.*, **317**, 1545 (1987).

74. K. Huebner, M. Isobe, C. M. Croce, D. W. Golde, S. E. Kaufman, and J. C. Gasson, *Science*, **230**, 1282 (1985).

75. B. H. Van Leeuwen, M. E. Martinson, G. C. Webb, and I. G. Young, *Blood*, **73**, 1142 (1989).

76. K. Kaushansky, P. J. O'Hara, K. Berkner, G. M. Segal, F. S. Hagen, and J. W. Adamson, *Proc. Natl. Acad. Sci. USA*, **83**, 3101 (1986).

77. J. C. Gasson, *Blood*, **77**, 1131 (1991).

78. R. Schreck and P. A. Baeuerle, *Mol. Cell. Biol.*, **10**, 1281 (1990).

79. A. Tsuboi, M. Muramatsu, A. Tsutsumi, K. Arai, and N. Arai, *Biochem. Biophys. Res. Commun.*, **199**, 1064 (1994).

80. C. Y. Wang, A. G. Bassuk, L. H. Boise, C. B. Thompson, R. Bravo, and J. M. Leiden, *Mol. Cell. Biol.*, **14**, 1153 (1994).

81. S. R. Himes, L. S. Coles, R. Katsikeros, R. K. Lang, and M. F. Shannon, *Oncogene*, **8**, 3189 (1993).

82. S. D. Nimer and H. Uchida, *Stem Cells*, **13**, 324 (1995).

83. K. Diederichs, T. Boone, and P. A. Karplus, *Science*, **254**, 1779 (1991).

84. P. Moonen, J. J. Mermod, J. F. Ernst, M. Hirschi, and J. F. DeLamarter, *Proc. Natl. Acad. Sci. USA*, **84**, 4428 (1987).

85. J. Struzyna, Z. Pojda, B. Braun, M. Chomicka, E. Sobicczewska, and J. Wremble, *Burns*, **21**, 437 (1995).

86. V. Daniel, S. Pasker, M. Wiesel, S. Carl, S. Pomer, G. Staehler, R. Schnobel, R. Weimer, and G. Opelz, *Transplant. Proc.*, **27**, 884 (1995).

87. Z. Estrov, M. Talpaz, G. Mavligit, R. Pazdur, D. Harris, S. M. Greenberg, and R. Kurzrock, *Am. J. Med.*, **98**, 551 (1995).

88. W. D. Xu, G. S. Firestein, R. Taetle, K. Kaushansky, and M. J. Zvaifler, *J. Clin. Invest.*, **83**, 876 (1989).

89. F. Levi-Schaffer, P. Lacy, N. J. Severs, T. M. Newman, J. North, B. Gomperts, A. B. Kay, and R. Moqbel, *Blood*, **85**, 2597 (1995).

90. G. D. Demetri and K. H. S. Antman, *Semin. Oncol.*, **19**, 362 (1992).

91. T. Kishimoto, T. Taga, and S. Akira, *Cell*, **76**, 253 (1994).

92. T. Kishimoto, T. Tanaka, K. Yoshida, S. Akira, and T. Taga, *Ann. NY Acad. Sci.*, **766**, 224 (1995).

93. D-X. Ding, C. I. Rivas, M. L. Heaney, M. A. Raines, J. C. Vera, and D. W. Golde, *Proc. Natl. Acad. Sci. USA*, **91**, 2537 (1994).

94. A. Miyajima, A. L. F. Mui, T. Ogorochi, and K. Sakamaki, *Blood*, **82**, 1960 (1993).

95. J. N. Ihle, *Adv. Cancer Res.*, **68**, 23 (1996).

96. A. L. F. Mui, H. Wakao, N. Harada, A. M. Ofarrell, and A. Miyajima, *J. Leukoc. Biol.*, **57**, 799 (1995).

97. J. A. McCubrey, W. S. May, V. Duronio, and A. Mufson, *Leukemia*, **14**, 9 (2000).

98. S. J. Brandt, W. P. Peters, S. K. Atwater, J. Kurtzbert, M. J. Borowitz, R. B. Jones, E. J. Shpall, R. C. Bast, C. J. Gilbert, and D. H. Oette, *N. Engl. J. Med.*, **318**, 869 (1988).

99. B. R. Blazar, J. H. Kersey, P. B. McGlave, D. A. Vallera, L. C. Lasky, R. J. Haake, B. Bostrom, D. R. Weisdorf, C. Epstein, and N. K. Ramsay, *Blood*, **73**, 849 (1989).

100. J. Nemunaitis, S. N. Rabinowe, J. W. Singer, P. J. Bierman, J. M. Vose, A. S. Freedman, N. Onetto, S. Gillis, D. Oette, M. Gold, D. Buckner, J. A. Hansen, J. Ritz, F. R. Appelbaum, J. O. Armitage, and L. N. Nadler, *N. Engl. J. Med.*, **324**, 1773 (1991).

101. H. H. Gerhartz, M. Engelhard, P. Meusers, G. Brittinger, W. Wilmanns, G. Schlimok, P. Mueller.D. Huhn, R. Musch, W. Seigert, D. Gerhartz, J. H. Hartlapp, E. Theil, C. Huber, C. Peschl, W. Spann, B. Emmerich, C. Schadek, M. Westerhausen, H. W. Pecs, H. Radtke, R. Brandmaier, A. C. Stern, T. C. Jones, H. J. Ehrilich, H. Stein, M. Parwaresch, M. Tiemann, and K. Lennert, *Blood*, **82**, 2329 (1993).

102. A. D. Ho, R. Haas, M. Korblin, M. Dietz, and W. Hunstein, *Bone Marrow Transplant.*, **7**(Suppl 1), 13 (1991).

103. M. A. Socinski, S. A. Cannistra, A. Elias, K. H. Antman, L Schnipper, and J. D. Griffin, *Lancet*, **1**, 1194 (1988).

104. J. L. Villeval, U. Duhrsen, G. Morstyn, and D. Metcalf, *Br. J. Haematol.*, **74**, 36 (1990).

105. S. Devereux, H. Bull, D. Campos-Casta, R. Saib, and D. Linch, *Br. J. Haematol.*, **71**, 323 (1989).

106. M. Aglietta, W. Piacibello, F. Sanavio, A. Stachini, F. Apra, M. Schena, C. Mossetti, F. Carnino, F. Caligaris-Cappio, and F. Gavosto, *J. Clin. Invest.*, **83**, 551 (1989).

107. D. Metcalf, *Blood*, **67**, 257 (1986).

108. M. R. Bishop, J. R. Anderson, J. D. Jackson, P. J. Bierman, E. C. Reed, J. M. Vose, J. O. Armitage, P. I. Warkentin, and A. Kessinger, *Blood*, **83**, 610 (1994).

109. S. M. Grant and R. C. Heel, *Drugs*, **43**, 516 (1992).

110. J. G. Gribben, S. Devereux, N. S. B. Thomas, M. Keim, H. M. Jones, A. H. Goldstone, and D. C. Linch, *Lancet*, **335**, 434 (1990).

111. A. W. Burgess and D. Metcalf, *Int. J. Cancer*, **26**, 647 (1980).

112. N. A. Nicola, D. Metcalf, M. Matsumoto, and G. R. Johnson, *J. Biol. Chem.*, **258**, 9017 (1983).

113. K. Welte, E. Platzer, L. Lu, J. Gabrilove, E. Levi, R. Mertelsmann, and M. Moore, *Proc. Natl. Acad. Sci. USA*, **82**, 1526 (1985).

114. H. Nomura, I. Imazeki, M. Oheda, N. Kubota, M. Tamura, M. Ono, Y. Ueyama, and S. Asano, *EMBO J.*, **5**, 871 (1986).

115. L. M. Souza, T. L. Boone, J. Gabrilove, Ph. H. Lai, K. M. Zsebo, D. C. Murdock, V. R. Chazin, J. Bruszewski, H. Lu, K. K. Chen, J. Barendt, E. Platzer, M. A. S. Moore, R. Mertelsmann, and K. Welte, *Science*, **232**, 61 (1986).

116. S. Nagata, M. Tsuchiya, S. Asano, O. Yamaoto, Y. Hirata, N. Kubota, M. Oheda, H. Nomura, and T. Yamazaki, *EMBO J.*, **5**, 575 (1986).

117. S. NagataM. Tsuchiya, S. Asano, O. Yamaoto, Y. Hirata, N. Kubota, M. Oheda, H. Nomura, and T. Yamazaki, *EMBO J.*, **5**, 575 (1986).

118. J. L. Gabrilove, A. Jakubowski, K. Fain, J Grous, A. Scher, C. Sternber, A. Yagoda, B. Clarkson, M. A. Bonilla, H. F. Oettgen, K. Alton, T. Boone, B. Altrock, K. Welte, and L. Souza, *J. Clin. Invest.*, **82**, 1454 (1988).

119. G. Morstyn, L. Campbell, L. M. Souza, N. K. Alton, J. Keech, M. Green, W. Sheridan, D. Metcalf, and R. Fox, *Lancet*, **1**, 667 (1988).

120. M. Bronchud, J. H. Scarffe, N. Thatcher, D. Crowther, L. M. Souza, N. D. Altton, N. G. Testa, and T. M. Dexter, *Br. J. Cancer*, **56**, 809 (1987).

121. M. M. Le Beau, R. Lemons, J. Carrino, M. Pettenati, L. Souza, M. Diaz, and J. Rowley, *Leukemia*, **1**, 795 (1987).

122. R. N. Simmers, L. M. Webber, M. F. Shannon, O. M. Garson, G. Wong, M. A. Vadas, and G. R. Sutherland, *Blood*, **70**, 330 (1987).

123. G. D. Demetri and J. D. Griffin, *Blood*, **78**, 2791 (1991).

124. G. Adler, *J. Exp. Med.*, **186**, 171 (1997).

125. S. M. Dunn, L. S. Coles, R. K. Lang, S. Gerondakis, M. A. Vadas, and M. F. Shannon, *Blood*, **83**, 2469 (1994).

126. K. Diederichs, T. Boone, and P. A. Karplus, *Science*, **254**, 1774 (1991).

127. C. P. Hill.T. D. Osslund, and D. Eisenberg, *Proc. Natl. Acad. Sci. USA*, **90**, 5167 (1993).

128. M. de Haas, J. M. Kerst, C. E. Van der Schoot, J. Calafat, C. E. Hack, H. H. Nuijens, D. Roos, R. H. J. van Oers, and A. E. G. Kr. von dem Borne, *Blood*, **84**, 3885 (1994).

129. K. Watari, S. Asano, N. Shirafuji, H. Kodo, K. Ozawa, F. Takaku, and S. Kamachi, *Blood*, **73**, 117 (1989).

130. J. Cebon, J. Layton, D. Maher, and G. Morstyn, *Br. J. Haematol.*, **86**, 265 (1994).

131. G. J. Lieschke, D. Grail, G. Hodgson, D. Metcalf, E. Stanley, C. Cheers, K. J. Fowler, S. Basu, Y. F. Zhan, and A. R. Dunnm, *Blood*, **84**, 1737 (1994).

132. D. Metcalf and N. A. Nicola, *J. Cell. Physiol.*, **116**, 198 (1983).

133. I. McNiece, R. Anddrews, M. Stewart, S. Clark, T. Boone, and P. Quesenberry, *Blood*, **74**, 609 (1989).

134. K. Ikebuchi, S. C. Clark, J. N. Ihle, L. M. Sourza, and M. Ogawa, *Proc. Natl Acad Sci. USA*, **85**, 3445 (1988).

135. K. Ikebuchi, J. N. Ihle, Y. Hirari, G. G. Wong, S. C. Clark, and M. Ogawa, *Blood*, **72**, 2007 (1988).

136. D. C. Dale, W. C. Liles, W. R. Summer, and S. Nelson, *J. Infect. Dis.*, **172**, 1061 (1995).

137. O. Hiraoka, H. Anaguchi, A. Asakura, and Y. Ota, *J. Biol. Chem.*, **270**, 25928 (1995).

138. R. Fukunaga, E. Ishizaka-Ikeda, C. X. Pan, Y. Seto, and S. Nagata, *EMBO J.*, **10**, 2855 (1991).

139. S. E. Nicholson, A. C. Oates, A. G. Harpur, A. Ziemiecki, A. F. Wilks, and J. E. Layton, *Proc. Natl. Acad. Sci. USA*, **91**, 2985 (1994).

140. K. Shimoda, H. Iwasaki, S. Okamura, Y. Ohno, A. Kubota, F. Arima, T. Otsuka, and Y. Niho, *Biochem. Biophys. Res. Commun.*, **203**, 922 (1994).

141. D. C. Koay and A. C. Sartorelli, *Blood*, **93**, 3774 (1999).

142. R. Sullivan, J. D. Griffin, E. R. Simons, A. I. Schafter, T. Meshulam, J. P. Fredette, A. K. Maas, A. S. Gadenne, J. L. Leavitt, and D. A. Melnick, *J. Immunol.*, **139**, 3422 (1987).

143. S. Matsuda, N. Shirafuji, and S. Asano, *Blood*, **74**, 2343 (1989).

144. G. S. Chatta, T. H. Price, R. C. Allen, and D. C. Dale, *Blood*, **84**, 2923 (1994).

145. D. P. Kerrigan, A. Castillo, K. Foucar, K. Townsend, and J. Neidhart, *Am. J. Clin. Pathol.*, **92**, 280 (1989).

146. U. Duhrsen, I. L. Villeval, J. Boyd, G. Kannourakis, G. Morstyn, and D. Metcalf, *Blood*, **72**, 2074 (1988).

147. S. A. Miles, R. T. Mitsuyasu, K. Lee, J. Moreno, K. Alton, J. C. Egrie, L. Souza, and J. A. Glaspy, *Blood*, **75**, 2137 (1990).

148. J. Crawford, H. Ozer, R. Stoller, D. Johnson, G. Lyman, I. Tabbara, M. Kris, J. Grous, V. Picozzi, G. Rausch, R. Smith, W. Gradishar, A. Ahanda, M. Vincent, M. Stewart, and J. Glaspy, *N. Engl. J. Med.*, **325**, 164 (1991).

149. R. S. Negrin, R. Stein, J. Vardiman, K. Doherty, J. Cornwell, S. Kranz, and P. Greenberg, *Blood*, **82**, 737 (1993).

150. R. S. Negrin, D. H. Haeuber, A. Nagler, Y. Kobayashi, J. Sklar, T. Donlon, M. Vincent, and P. L. Greenberg, *Blood*, **76**, 36 (1990).

151. J. A. Glaspy and D. W. Golde, *Semin. Oncol.*, **19**, 386 (1992).

152. W. P. Petros, *Pharmacotherapy*, **12**, 32S (1992).

153. G. Morstyn, M. Foote, T. Walker, and G. Molineux, *Acta Haematol.*, **105**, 151 (2001).

154. S. R. Paul, F. Bennett, J. A. Calvetti, K. Kelleher, C. R. Wood, R. M. O'Hara Jr., A. C. Leary, B. Sibley, S. C. Clark, D. A. Williams, and Y-C Yang, *Proc. Natl. Acad. Sci. USA*, **87**, 7512 (1990).

155. I. Kawashima, J. Ohsumi, K. Mita-Honjo, K. Shimoda-Takano, H. Ishikawa, S. Sakakibara, K. Miyadai, and Y. Takiguchi, *FEBS Lett.*, **288**, 13 (1991).

156. T. Taga and T. Kishimoto, *Annu. Rev. Immunol.*, **15**, 797 (1997).

157. A. Koj and A. Guzdek, *Ann. NY Acad. Sci.*, **762**, 108 (1995).

158. D. McKinley, Q. Wu, T. Yang-Feng, and Y. C. Yang, *Genomics*, **13**, 814 (1992).

159. Y. C. Yang, *Stem Cells*, **11**, 474 (1993).

160. X. X. Du and D. A. Williams, *Blood*, **83**, 2023 (1994).

161. H. H. Nandurkar, L. Robb, D. Tarlinton, L. Barnett, F. Kontgen, and C. G. Begley, *Blood*, **90**, 2148 (1997).

162. P. Bilinski, D. Roopenian, and A. Gossler, *Genes Dev.*, **12**, 2234 (1998).

163. L. Robb, R. Li, L. Hartley, H. H. Nandurkar, F. Koentgen, and C. G. Begley, *Nat. Med.*, **4**, 303 (1998).

164. M. Musashi, Y. C. Yang, S. R. Paul, S. C. Clark, T. Sudo, and M. Ogawa, *Proc. Natl. Acad. Sci. USA*, **88**, 765 (1991).

165. M. Musashi, S. C. Clark, T. Sudo, D. L. Urdal, and M. Ogawa, *Blood*, **78**, 1448 (1991).

166. F. Hirayama, J. Shih, A. Awgulewitsch, G. W. Warr, S. C. Clark, and M. Ogawa, *Proc. Natl. Acad. Sci. USA*, **89**, 5907 (1992).

167. M. Teramura, S. Kobayashi, S. Hoshino, K. Oshimi, and H. Mizoguchi, *Blood*, **79**, 327 (1992).

168. S. A. Burstein, R. Mei, J. Henthorn, P. Friese, and K. Turner, *J. Cell. Physiol.*, **153**, 305 (1992).

169. E. Bruno, R. A. Briddell, R. J. Cooper, J. E. Brandt, and R. Hoffman, *Exp. Hematol.*, **19**, 378 (1991).

170. V. F. Quesniaux, S. C. Clark, K. Turner, and B. Fagg, *Blood*, **80**, 1218 (1992).

171. H. Baumann and P. Schendel, *J. Biol. Chem.*, **266**, 20424 (1991).

172. I. Kawashima, J. Ohsumi, K. Mita-Honjo, K. Shimoda-Takano, H. Ishikawa, S. Sakakibara, K. Miyadai, and Y. Takiguchi, *FEBS Lett.*, **283**, 199 (1991).

173. M. F. Mehler, R. Rozental, M. Doughterty, D. C. Spray, and J. A. Kessler, *Nature*, **362**, 62 (1993).

174. A. Orazi, Z. Du, Z. Yang, M. Kashai, and D. A. Williams, *Lab. Invest.*, **75**, 33 (1996).

175. D. J. Hilton, A. A. Hilton, A. Raicevic, S. Rakar, M. Harrison-Smith, N. M. Gough, C. G. Begley, D. Metcalf, N. A. Nicola, and T. A. Wilson, *EMBO J.*, **13**, 4765 (1994).

176. V. A. Barton, M. A. Hall, K. R. Hudson, and J. K. Heath, *J. Biol. Chem.*, **275**, 36197 (2000).

177. Y. C. Yang and T. Yin, *Ann. NY Acad. Sci.*, **762**, 31 (1995).

178. X. X. Du, C. M. Doerschuk, A. Orazi, and D. A. Williams, *Blood*, **83**, 33 (1994).

179. S. Sonis, A. Muska, J. O'Brien, A. Van Vugt, P. Langer-Safer, and J. Keith, *Eur. J. Cancer B Oral Oncol.*, **31B**, 261 (1995).

180. T. Berl and U. Schwertschlag, *Oncology*, **14**(9 Suppl 8), 12 (2000).

181. X. X. Du, C. M. Doerschuk, A. Orazi, and D. A. Williams, *Blood*, **83**, 33 (1994).

182. N. F. Fiore, G. Ledniczky, Q. Liu, A. Orazi, X. Du, D. A. Williams, and J. L. Grosfeld, *J. Pediatr. Surg.*, **33**, 24 (1998).

183. Q. Liu, X. X. Du, D. T. Schindel, Z. X. Yang, F. J. Rescorla, D. A. Williams, J. L. Grosfeld, S. Smith, W. H. Hendren, P. C. Guzzetta, and S. W. Bruch, *J. Pediatr. Surg.*, **31**, 1047 (1996).

184. J. Schweroske, L. Schwartzberg, and C. H. Weaver, *Proc. Am. Soc. Clin. Oncol.*, **18**, 584a (1999).

185. B. E. Sands, S. Bank, C. A. Sninsky, M. Robinson, S. Katz, J. W. Singleton, P. B. Miner, M. A. Safdi, S. Galandiuk, S. B. Hanauer, G. W. Varilek, A. L. Buchman, V. D. Rodgers, B. Salzberg, B. Cai, J. Loewy, M. F. DeBruin, H. Rogge, M. Shapiro, and U. S. Schwertschlag, *Gastroenterology*, **117**, 58 (1999).

186. S. J. Goldman, *Stem Cells*, **13**, 462 (1995).

187. J. W. Smith II, *Oncology*, **14**, 41 (2000)

188. K. Aoyama, T. Uchida, F. Takanuki, T. Usui, T. Watanabe, S. Higuchi, T. Toyoki, and H. Mizoguchi, *Br. J. Clin. Pharmacol.*, **43**, 571 (1997).

189. M. S. Gordon, W. J. McCaskill-Stevens, L. A. Battiato, J. Loewy, D. Loesch, E. Breeden, R. Hoffman, K. J. Beach, B. Kuca, J. Kaye, and G. W. Sledge Jr., *Blood*, **87**, 3615 (1996).

190. D. Bennett, *J. Morphol.*, **98**, 199 (1956).

191. D. E. Williams, J. Eisenman, A. Baird, C. Rauch, K. VanNess, C. J. March, L. S. Park, U. Martin, D. Y. Mochizuki, H. S. Boswell, G. S. Burgess, D. Cosman, and S. D. Lyman, *Cell*, **63**, 167 (1990).

192. N. G. Copeland, D. J. Gilbert, B. C. Cho, P. J. Donovan, N. A. Jenkins, D. Cosman, D. Anderson, S. D. Lyman, and D. E. Williams, *Cell*, **63**, 175 (1990).

193. J. G. Flanagan and P. Leder, *Cell*, **63**, 185 (1990).

194. K. M. Zsebo, J. Wypych, I. K. McNiece, H. S. Lu, K. A. Smith, S. B. Karkare, R. K. Sachdev, V. N. Yuschenkoff, N. C. Birkett, L. R. Williams, V. N. Satyagal, W. Tung, R. A. Bosselman, E. A. Mendiaz, and K. E. Langley, *Cell*, **63**, 195 (1990).

195. F. H. Martin, S. V. Suggs, K. E. Langley, H. S. Lu, J. Ting, K. H. Okino, C. F. Morris, I. K. McNiece, F. W. Jacobsen, E. A. Mendiaz, N. C. Birkett, K. A. Smith, M. J. Johnson, V. P. Parker, J. C. Flores, A. C. Patel, E. F. Fisher, H. O. Erjavec, C. J. Herrera, J. Wypych, R. K. Sachdev, J. A. Pope, I. Leslie, D. Wen, C. H. Lin, R. L. Cupples, and K. M. Zsebo, *Cell*, **63**, 203 (1990).

196. K. M. Zsebo, D. A. Williams, E. N. Geissler, V. C. Broudy, F. H. Martin, H. L. Atkins, R.-Y. Hsu, N. C. Birkett, K. H. Okino, D. C. Murdock, F. W. Jacobsen, K. E. Langley, K. A. Smith, T. Takeishi, B. M. Cattanach, S. J. Galli, and S. V. Suggs, *Cell*, **63**, 213 (1990).

197. E. Huang, K. Nocka, D. R. Beier, T.-Y. Chu, J. Buck, H-W. Lahm, D. Wellner, P. Leder, and P. Besmer, *Cell*, **63**, 225 (1990).

198. D. M. Anderson, S. D. Lyman, A. Baird, J. M. Wignall, J. Eisenman, C. Rauch., C. J. March, H. S. Boswell, S. D. Gimpel, D. Cosman, and D. E. Williams, *Cell*, **63**, 235 (1990).

199. M. C. Heinrich, D. C. Dooley, A. C. Freed, L. Band, M. E. Hoatlin, W. W. Keeble, S. T. Peters, K. V. Silvey, F. S. Ey, D. Kabat, R. T. Maziarz, and G. C. Bagby, *Blood*, **82**, 771 (1993).

200. P. Rossi, C. Albanesi, P. Grimaldi, and R. Geremia, *Biochem. Biophys. Res. Commun.*, **176**, 910 (1991).

201. E. Keshet, S. D. Lyman, D. E. Williams, D. M. Anderson, N. A. Jenkins, N. G. Copeland, and L. F. Parada, *EMBO J.*, **10**, 2425 (1991).

202. D. A. Williams and M. K. Majumdar, *Stem Cells*, **12**(Suppl 1), 67 (1994).

203. K. Kurosawa, K. Miyazawa, A. Gotoh, T. Katagiri, J. Nishimaki, L. K. Ashman, and K. Toyama, *Blood*, **87**, 2235 (1996).

204. K. Miyazawa, D. A. Williams, A. Gotoh, J. Nishimaki, H. E. Broxmeyer, and K. Yoyama, *Blood*, **85**, 641 (1995).

205. K. E. Langley, L. G. Bennett, J. Wypych, S. A. Yancik, X-D. Liu, K. R. Westcott, D. G. Chang, K. A. Smith, and K. M. Zsebo, *Blood*, **81**, 656 (1993).

206. S. J. Galli, K. M. Zsebo, and E. N. Geissler, *Adv. Immunol.*, **55**, 1 (1994).

207. D. F. Andrews, J. A. Bianco, D. A. Moran, and J. W. Singer, *Blood*, **78**, 303a, (1991).

208. M. T. Aye, S. Hasemi, B. Leclair, A. Zeibdawi, E. Trudal, M. Halpenny, V. Fuller, and G. Cheng, *Exp. Hematol.*, **20**, 523 (1992).

209. F. Mouriaux, F. Chahud, C. A. Maurage, F. Malecaze, and P. Labalette, *Exp. Eye Res.*, **73**, 151 (2001).

210. M. Ito, T. Harada, M. Tanikawa, A. Fujii, G. Shiota, and N. Terakawa, *Fertil. Steril.*, **75**, 973 (2001).

211. H. E. Broxmeyer, S. Cooper, L. Lu, G. Hangoc, D. Anderson, D. Cosman, S. D. Lyman, and D. E. Williams, *Blood*, **77**, 2142 (1991).

212. S. J. Galli, M. Tsai, and B. K. Wershil, *Am. J. Pathol.*, **142**, 965 (1993).

213. M. A. Smith, C. J. Pallister, and J. G. Smith, *Acta Haematol.*, **105**, 143 (2001).

214. O. Rosnet and D. Birnbaum, *Crit. Rev. Oncog.*, **4**, 595 (1993).

215. C. Deberry, S. Mou, and D. Linnekin, *Biochem. J.*, **327**, 73 (1997).

216. J. J. Ryan, H. Huang, L. J. McReynolds, C. Shelburne, J. Hu-Li, T. F. Huff, and W. E. Paul, *Exp. Hematol.*, **25**, 357 (1997).

217. H. Serve, N. S. Yee, G. Stella, L. Sepp-Lorenzino, J. C. Tan, and P. Besmer, *EMBO J.*, **14**, 473 (1995).

218. R. G. Andrews, S. H. Bartelmez, G. H. Knitter, D. Myerson, I. D. Bernstein, F. R. Appelbaum, and K. M. Zsebo, *Blood*, **80**, 920 (1992).

219. J. Tong, M. S. Gordon, E. F. Srour, R. J. Cooper, A. Orazi, I. McNiece, and R. Hoffman, *Blood*, **82**, 784 (1993).

220. M. S. Gordon and L. M. Schucter in R. Hoffman, E. J. Benz Jr., S. J. Stattil, B. Furie, H. J. Cohen, and L. E. Silberstein, Eds., *Hematology, Basic Principles and Practice*, 2nd ed., Churchill Livingstone, New York, 1995, p. 961.

221. I. K. McNiece and R. A. Briddell, *J. Leukoc. Biol.*, **57**, 14 (1995).

222. A. Orazi, M. S. Gordon, K. John, G. Sledge, R. S. Neiman, and R. Hoffman, *Am. J. Clin. Pathol.*, **103**, 177 (1995).

223. R. G. Andrews, G. H. Knitter, S. H. Bartelmez, K. E. Langley, D. Farrar, R. W. Hendren, F. R. Appelbaum, I. D. Berstein, and K. M. Zsebo, *Blood*, **78**, 1975 (1991).

224. K. Yasukawa, T. Hirano, Y. Watanabe, K. Muratani, T. Matsuda, S. Nakai, and T. Kishimoto, *EMBO J.*, **6**, 2939 (1987).

225. O. Tanabe, S. Akira, T. Kamiya, G. G. Wong, T. Hirano, and T. Kishimoto, *J. Immunol.*, **141**, 3875 (1988).

226. L. T. May, M. I. Ndubuisi, K. Patel, and D. Garcia, *Ann. NY Acad. Sci.*, **762**, 120 (1995).

227. L. Chen, A. Zilberstein, and M. Revel, *Proc. Natl. Acad. Sci. USA*, **85**, 8037 (1988).

228. M. Murakami, M. Hibi, N. Nakagawa, T. Nakagawa, K. Yasukawa, K, Yamanishi, T. Taga, and T. Kishimoto, *Science*, **260**, 1808 (1993).

229. T. Taga, M. Narazaki, K. Yasukawa, T. Saito, D. Miki, M. Hamaguchi, S. David, M. Shoyab, G. D. Yancopoulos, and T. Kishimoto, *Proc. Natl. Acad. Sci. USA*, **89**, 10998 (1992).

230. S. Davis, T. H. Aldrich, N. Stahl, L. Pan, T. Taga, T. Kishimoto, N. Y. Ip, and G. D. Yancopoulos, *Science*, **260**, 1805 (1993).

231. N. Tonouchi, N. Oouchi, N. Kashima, M. Kawai, K. Nagase, A. Okano, H. Matsui, K. Yamada, T. Hirano, and T. O. Kishimoto, *J. Biochem.*, **104**, 30 (1988).

232. J. Weber, J. C. Yang, S. L. Topalian, D. R. Parkinson, D. S. Schwartzentruber, S. E. Ettinghausen, H. Gunn, A. Mixon, H. Kim, D. Cole, R. Levin, and S. A. Rosenberg, *J. Clin. Oncol.*, **11**, 499 (1993).

233. M. Gordon and L. M. Schucter in R. Hoffman, E. J. Benz, S. J. Shattil, B. Furie, H. J. Cohen, and L. E. Silberstein, Eds., *Hematology, Basic Principles and Practice*, 2nd ed., Churchill Livingston, New York, 1995, p. 967.

234. J. S. Weber, *Ann. NY Acad. Sci.*, **762**, 357 (1995).

235. D. C. Foster, C. A. Sprecher, F. J. Grant, J. M. Kramer, J. L. Kuijper, R. D. Holly, T. E. Whitmore, M. D. Heipel, L. A. Bell, A. F. T. Ching,

V. McGrane, C. Hart, P. J. O'Hara, and S. Lok, *Proc. Natl. Acad. Sci. USA*, **91**, 13023 (1994).

236. D. L. Eaton, A. Gurney, B. Malloy, W. J. Kuang, P. E. Hass, M. H. Xie, M. Nagel, and F. J. deSauvage, *Blood*, **84**, 241a (1994).

237. V. C. Broudy and K. Kaushansky, *J. Leukoc. Biol.*, **57**, 719 (1995).

238. F. J. deSauvage, P. E. Hass, S. D. Spencer, B. E. Malloy, A. L. Gurney, S. A. Spencer, W. C. Darbonne, W. J. Henzel, S. C. Wong, W. J. Kuang, K. J. Oles, B. Hultgren, L. A. Solber, D. V. Goeddel, and D. L. Eaton, *Nature*, **369**, 533 (1994).

239. T. D. Bartley, J. Bogenberger, P. Hunt, Y. S. Li, H. S. Lu, F. Martin, M. S. Chang, B. Samal, J. L. Nichol, S. Swift, M. J. Johnson, R. Y. Hsu, V. P. Parker, S. Suggs, J. D. Skrine, L. A. Merewether, C. Clogston, E. Hsu, M. M. Hokom, A. Hornkohl, E. Choi, M. Pangelinan, Y. Sun, V. Mar, J. McNinch, L. Simonet, F. Jacobsen, C. Xie, J. Shutter, H. Chute, R. Basu, L. Selander, D. Trollinger, L. Sieu, D. Padilla, G. Trail, G. Elliott, R. Izumi, T. Covey, J. Crouse, A. Garcia, W. Xu, J. Del Castillo, J. Biron, S. Cole, M. C.-T. Hu, R. Pacifici, I. Ponting, C. Saris, D. Wen, Y. P. Yung, H. Lin, and R. A. Bosselman, *Cell*, **77**, 1117 (1994).

240. S. Lok, K. Kaushansky, R. D. Holly, J. L. Kuijper, C. E. Lofton-Day, P. J. Oort, F. J. Grant, M. D. Heipel, H. Blumberg, R. Johnson, D. Prunkard, A. F. T. Ching, S. L. Mathewes, M. C. Bailey, J. W. Forstrom, M. M. Buddle, S. G. Osborn, S. J. Evans, P. O. Sheppard, S. R. Presnell, P. J. O'Hara, F. S. Hagen, G. J. Roth, and D. C. Forster, *Nature*, **369**, 565 (1994).

241. S. Lok and D. C. Foster, *Stem Cells*, **12**, 586 (1994).

242. K. Kaushansky, *Stem Cells*, **12**(Suppl 1), 91 (1994).

243. N. Methia, F. Louache, W. Vainchenker, and F. Wendling, *Blood*, **82**, 1395 (1993).

244. A. L. Gurney, K. Carver-Moore, F. J. deSauvage, and M. W. Moore, *Science*, **265**, 1445 (1994).

245. I. Vigon, J. P. Mornon, L. Cocault, M. T. Mitjavila, P. Tambourin, S. Gisselbrecht, and M. Souyri, *Proc. Natl. Acad. Sci. USA*, **89**, 5640 (1992).

246. K. Kaushansky, S. Lok, R. D. Holly, V. C. Broudy, N. Lin, M. C. Bailey, J. W. Forstrom, M. M. Buddle, P. J. Oort, F. S. Hage, G. J. Roth, T. Papayannopoulou, and D. C. Foster, *Nature*, **369**, 568 (1994).

247. A. E. Geddis, N. E. Fox, and K. Kaushansky, *J. Biol. Chem.*, **276**, 34473 (2001).

248. R. Carlson, *Inpharma*, **1083**, 9 (1997).

249. J. Levin, *N. Engl. J. Med.*, **336**, 434 (1997).

250. J. E. Ogden, *TIBTECH*, **12**, 389 (1994).

251. M. M. Le Beau, N. D. Estein, S. J. O'brien, A. W. Nienhuis, Y. C. Yang, S. C. Clark, and J. D. Rowley, *Proc. Natl. Acad. Sci. USA*, **84**, 5913 (1987).

252. G. Frendl, *Int. J. Immunopharmacol.*, **14**, 421 (1992).

253. S. Miyatake, T. Yokota, F. Lee, and K. Arai, *Proc. Natl. Acad. Sci. USA*, **82**, 316 (1985).

254. P. Valent, K. C. H. Sillaber, R. Scherrer, K. Geissler, P. Kier, P. Kalhs, M. Kundi, R. Papoian, P. Bettelheim, W. Hinterberger, and K. Lechner, *Bone Marrow Transplant.*, **9**, 331 (1992).

255. A. F. Lopez, L. B. To, Y. C. Yang, J. R. Gamble, M. F. Shannon, G. F. Burns, P. G. Dyuson, C. A. Vadas, and M. A. Vadas, *Proc. Natl. Acad. Sci. USA*, **84**, 2761 (1987).

256. F. J. Bot, L. van Eijk, P Schipper, and B. Lowenberg, *Blood*, **73**, 1157 (1989).

257. M. E. Rothenberg, W. F. Owen, D. S. Silberstein, J. Woods, R. J. Soberman, K. F. Austen, and R. L. Steevens, *J. Clin. Invest.*, **81**, 1986 (1988).

258. M. Carlson, C. Peterson, and P. Venge, *Allergy*, **48**, 437 (1993).

259. S. A. Cannistra, E. Vellenga, P. Grosheck, A. Rambalke, and J. D. Griffin, *Blood*, **71**, 672 (1988).

260. V. Jendrossek, S. Buth, C. Stetter, and M. Gahr, *Agents Actions*, **37**, 127 (1992).

261. V. D. Hondt, P. Weynants, Y. Humblet, T. Guillaume, J. L. Canon, M. Beauduin, P. Duprez, J. Longueville, R. Mull, C. Chatelain, and M. Symann, *J. Clin. Oncol.*, **11**, 2063 (1993).

262. A. Lindemann, A. Ganser, F. Herrmann, J. Frisch, G. Seipelt, G. L. Schulz, D. Hoelzer, and R. Mertelsmann, *J. Clin. Oncol.*, **9**, 2120 (1991).

263. A. Bhatia, T. Olencki, S. Murthy, G. T. Budd, J. Finke, D. Sicca, M. J. Thomassen, S. K. Sandstrom, R. Tubbs, L. Bauer, M. Edinger, D. Young, D. Resta, and R. M. Bukowski, *Blood*, **80**, 1632a, (1992).

264. O. G. Ottmann, A. Ganser, G. Seipelt, M. Eder, G. Schulz, and D. Hoelzer, *Blood*, **76**, 1494 (1990).

265. A. Gianella-Borradori, *Stem Cells*, **12**(Suppl 1), 241 (1994).

266. Y. Feng, B. K. Klein, and C. A. McWherter, *J. Mol. Biol.*, **259**, 524 (1996).

267. A. M. Farese, F. Herodin, J. P. McKearn, C. Baum, E. Burton, and T. J. MacVittie, *Blood*, **87**, 581 (1996).

268. W. Dempke, A. Von Poblozki, A. Grothey, and H-J. Schmoll, *Anticancer Res.*, **20**, 5155 (2000).

269. B. M. Curtis, E. E. Williams, H. E. Broxmeyer, J. Dunn, T. Farrah, E. Jeffery, W. Clevenger, P. L. DeRoos, U. Martin, D. Friend, V. Craig, R. Gayle, V. Price, D. Cosman, C. J. March, and L. S. Park, *Proc. Natl. Acad. Sci. USA*, **88**, 5809 (1991).

270. E. Bruno, R. A. Briddell, R. J. Cooper, J. E. Brandt, and R. Hoffman, *Exp. Hematol.*, **20**, 494 (1992).

271. T. J. MacVittie, A. M. Farese, T. A. Davis, L. B. Lind, and J. P. McKearn, *Exp. Hematol.*, **27**, 1557 (1999).

272. S. Vadhan-Raj, S. Patel, J. Gano, M. Donato, C. Ramos, N. Papadopolous, A. Burgess, C. Plager, C. Baum, J. Sheman, R. Champlin, H. E. Broxmeyer, and R. S. Benjamin, *Blood*, **90**, 312a (1998).

273. Canadian Erythropoietin Study Group, *Br. Med. J.*, **300**, 573 (1990).

274. M. J. George, *Stem Cells*, **12**(Suppl 1), 249 (1994).

275. D. E. Williams, *J. Clin. Immunol.*, **14**, 215 (1994).

276. H. E. Broxmeyer and S. Vadhan-Raj, *Manual for GM-CSF*, Blacwell Science, Oxford, UK, 1996, p. 197.

277. M. Gordon and L. M. Schuchter in R. Hoffman, E. J. Benz, S. J. Shattil, B. Furie, H. J. Cohen, and L. E. Silberstein, Eds., *Hematology, Basic Principles and Practice*, 2nd ed., Churchill Livingstone, New York, 1995, p. 971.

278. R. G. Schaub, M. P. Bree, L. L. Hayes, M. A. Rudd, L. Rabbani, J. Loscalzo, and S. K. Clinton, *Arterioscler. Thromb.*, **14**, 70 (1994).

279. F. Bussolino, E. Bocchietto, F. Silvagno, R. Sokdi, M. Arese, and A. Mantovani, *Pathol. Res. Pract.*, **190**, 834 (1994).

280. H. E. Broxmeyer, *J. Exp. Med.*, **183**, 2411 (1996).

281. D. Metcalf, *Stem Cells*, **12**(Suppl 1), 259, (1994).

282. T. Gura, *Science*, **272**, 954 (1996).

283. H. E. Broxmeyer in T. Wheeton and T. Gordon, Eds., Blood Cell Biochemistry, vol. **7**, *Hemopoietic Growth Factors*, Plenum, London, 1996, p. 121.

CHAPTER SIX

Anticoagulants, Antithrombotics, and Hemostatics

Gregory S. Bisacchi
Bristol-Myers Squibb
Princeton, New Jersey

Contents

Burger's Medicinal Chemistry and Drug Discovery
Sixth Edition, Volume 3: Cardiovascular Agents and
Endocrines
Edited by Donald J. Abraham
ISBN 0-471-37029-0 © 2003 John Wiley & Sons, Inc.

1 INTRODUCTION

Maintenance of proper blood flow is a complex and highly regulated physiological process, with multiple complementary and opposing mechanisms of control. Normally in the vasculature, a precise balance is achieved, permitting free flow while also allowing the trigger of nearly instantaneous clot formation at sites of vascular injury to prevent hemorrhage. Indeed, blood coagulation evolved in humans and animals as a protective mechanism against bleeding. As protective blood clots (hemostatic plugs) form and grow after injury, mechanisms exist to maintain or dissolve them as needed while allowing normal flow in the remainder of the vasculature. Imbalances in the complex regulatory network of coagulation and anticoagulation, however, can lead to a variety of pathological consequences, such as hemorrhage or obstructive clot (thrombus) formation in veins or arteries, leading to stroke, pulmonary embolism, heart attack, and other serious conditions. These imbalances may result from genetic or acquired conditions (1).

This multifaceted regulatory framework of coagulation and anticoagulation inherently presents numerous opportunities for intervention with drugs to modulate one or more pathways, to achieve a therapeutic effect when needed. However, one of the overriding challenges with current or future drug strategies for antithrombotic, thrombolytic (clot-dissolving), and hemostatic therapy is the precise correction of the pathologic imbalance without overcompensating and thus creating safety issues. For example, excessive bleeding, occasionally life-threatening, can be one of the most prevalent side effects of several currently used antithrombotic drugs.

Both thrombi and hemostatic plugs are composed primarily of two structural elements: activated platelets that aggregate within the blood vessel, and a protein fibrin network that stabilizes the platelet mass (Fig. 6.1). Thrombin, a serine protease, is one of the central modulators of clot formation. It is the most potent activator of platelets and acts as the processing enzyme for the production of monomeric protein fibrin units from the precursor protein fibrinogen. Once formed, fibrin monomers self-assemble into long fibrin strands that can be crosslinked by the transglutaminase Factor XIIIa (FXIIIa), to further stabilize the platelet-fibrin mass. FXIIIa is generated by thrombin through activation of the zymogen form of the enzyme FXIII. Thrombin is itself produced by way of a cascade involving many other enzymes and cofactors, and platelets can be activated by many agents other than thrombin. In addition, there are multiple feedback loops and endogenous inhibitors that can either accelerate or dampen these processes. Plasmin, another serine protease, acts to proteolytically degrade fibrin, thus promoting dissolution of the clot. This abbreviated description of coagulation and thrombolysis is expanded on later in this chapter.

2 CLINICAL USE OF AGENTS

2.1 Current Drugs on the Market

Drugs currently marketed as antithrombotics include anticoagulant agents that directly or indirectly interfere with the activity of throm-

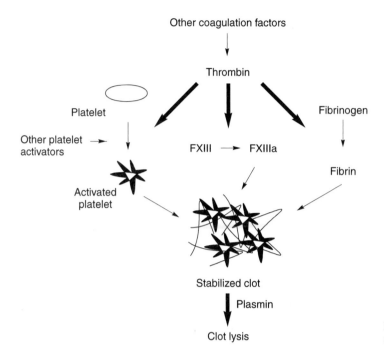

Figure 6.1. Simplified scheme of clot formation and lysis.

bin and/or its precursors, and also include antiplatelet agents that interfere with platelet activation and aggregation (Table 6.1). Heparin and the low molecular weight heparins (LMWH) are heterogeneous mixtures of polyanionic polysaccharides. A certain fraction of these mixtures incorporate a unique pentasaccharide sequence (**1a**, Fig. 6.2), which binds to the endogenous protein antithrombin. This complex can then inactivate thrombin, factor Xa (FXa), the serine protease that activates thrombin, and a few other coagulation enzymes. Fondaparinux (**1b**) consists of just this pentasaccharide sequence. Warfarin (**2**) indirectly limits the activity of thrombin and several other coagulation proteins by inhibiting a required vitamin K-mediated posttranslational modification to these proteins. Warfarin is the only oral anticoagulant currently marketed. Lepirudin (**3a**; Fig. 6.3) is a recombinant 65 amino acid polypeptide that can inactivate thrombin directly by binding simultaneously to its active site and fibrinogen binding site. Desirudin (**3b**; Revasc), which is marketed in Europe, is identical to lepirudin with the exception of two amino acids at the N-terminus. Structurally, lepirudin and desirudin are close analogs of the natural pep-

tide anticoagulant hirudin, originally isolated from the salivary glands of a species of leech. Bivalirudin (**4**), a 20 amino acid peptide, is a highly truncated analog of lepirudin/desirudin which has certain other engineered modifications in the peptide sequence. Like lepirudin and desirudin, bivalirudin binds simultaneously to the active and fibrinogen binding sites of thrombin, thus inactivating it. Argatroban (**5**) reversibly inactivates thrombin by binding to just the active site region. The platelet antagonists, abciximab (a monoclonal antibody), eptifibatide, and tirofiban disrupt aggregation by binding to the platelet glycoprotein (GP) IIIb/IIa receptor in place of the receptor's endogenous ligand, fibrinogen. Tirofiban (**6**) and the cyclic heptapeptide eptifibatide (**7**) both contain a similar pharmacophore: an optimally spaced basic nitrogen and carboxylate group (Fig. 6.4), mimicking the Arg-Gly-Asp (RGD) binding motif of fibrinogen. The structures of the other marketed antiplatelet agents (ticlopidine, clopidogrel, aspirin, and dipyridamole) shown in Fig. 6.4 operate through other mechanisms, discussed later.

Marketed thrombolytic drugs are enzymes or nonenzyme proteins that generate plasmin

Table 6.1 Marketed Antithrombotic Drugs

Generic Name (structure)	Trade Name	Marketed by	Chemical Class	Form. Wt.[a]	Mechanism of Action	Route of Administration
Anticoagulants						
Heparin (**1a**)	Heparin	Eli Lilly; Wyeth-Ayerst	Anionic polysaccharide	15,000 (ave.)	Antithrombin cofactor	IV, SC
Danaparoid	Orgaran	Organon	Anionic polysaccharide (heparinoid)	5500 (ave.)	Antithrombin cofactor	SC
Dalteparin	Fragmin	Pharmacia & Upjohn	Anionic polysaccharide (LMWH[b])	5000 (ave.)	Antithrombin cofactor	SC
Tinzaparin	Innohep	Bristol-Myers Squibb	Anionic polysaccharide (LMWH)	5500–7500 (ave.)	Antithrombin cofactor	SC
Enoxaparin	Lovenox	Aventis	Anionic polysaccharide (LMWH)	4500 (ave.)	Antithrombin cofactor	SC
Fondaparinux (**1b**)	Arixtra	Sanofi-Synthelabo/Organon	Anionic pentasaccharide	1728.1	Antithrombin cofactor	SC
Warfarin (**2**)	Coumadin	Bristol-Myers Squibb	Coumarin	330.3	Vitamin K antagonist	Oral, IV
Lepirudin (**3a**)	Refludan	Aventis	Polypeptide (recombinant hirudin analog)	6979.5	Direct "bivalent" thrombin inhibitor	IV
Bivalirudin (**4**)	Angiomax	The Medicines Company	Polypeptide (recombinant hirudin analog)	2180.3	Direct "bivalent" thrombin inhibitor	IV
Argatroban (**5**)	Novastan	GlaxoSmith-Kline; Texas Biotechnology	Small-molecule Arg-based substrate analog	526.7	Direct active-site thrombin inhibitor	IV

286

Antiplatelet Agents

Drug	Trade name	Company	Description	Formula weight[a]	Class	Route
Abciximab	ReoPro	Eli Lilly	Monoclonal antibody (recombinant Fab)	47,615	GP IIb/IIIa receptor antag.	IV
Eptifibatide (6)	Integrilin	Millennium	Cyclic heptapeptide RGD mimetic	832.0	GP IIb/IIIa receptor antag.	IV
Tirofiban (7)	Aggrastat	Merck	Nonpeptide RGD mimetic	495.1	GP IIb/IIIa receptor antag.	IV
Ticlopidine (8)	Ticlid	Roche	Thienopyridine	300.3	$P2Y_{12}$ receptor antagonist	Oral
Clopidogrel (9)	Plavix	Bristol-Myers Squibb; Sanofi-Synthelabo	Thienopyridine	419.9	$P2Y_{12}$ receptor antagonist	Oral
Aspirin (10)	Ecotrin; Regimen Bayer	GlaxoSmith-Kline; Bayer	Acetyl salicylic acid (ASA)	180.2	COX inhibitor	Oral
Dipyridamole (11)	Persantine	Boehringer Ingelheim	Pyrimidinopyrimidine	504.6	PDE inhibitor	Oral
Aspirin + dipyridamole	Aggrenox	Boehringer Ingelheim	—	—	Combination	Oral

[a]Formula weight, includes salt form and hydrates, where appropriate.
[b]LMWH, low molecular weight heparin.

287

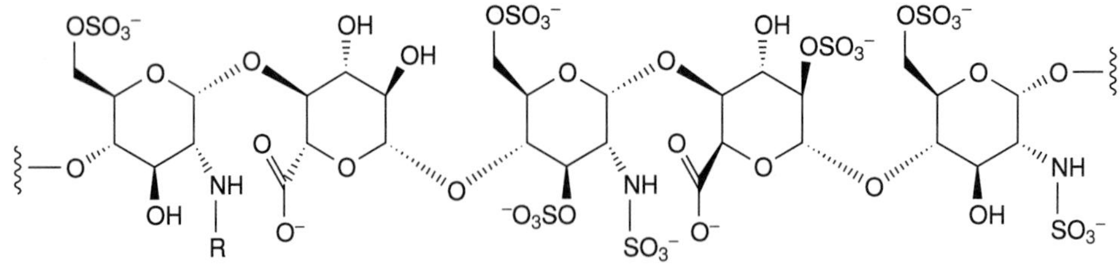

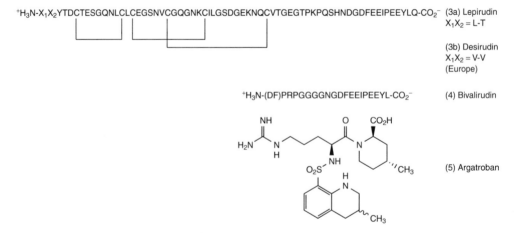

Figure 6.2. Heparin and LMWH (partial structures), fondaparinux, and warfarin.

^+H_3N-X_1X_2YTDCTESGQNLCLCEGSNVCGQGNKCILGSDGEKNQCVTGEGTPKPQSHNDGDFEEIPEEYLQ-CO$_2^-$ (3a) Lepirudin
X_1X_2 = L-T

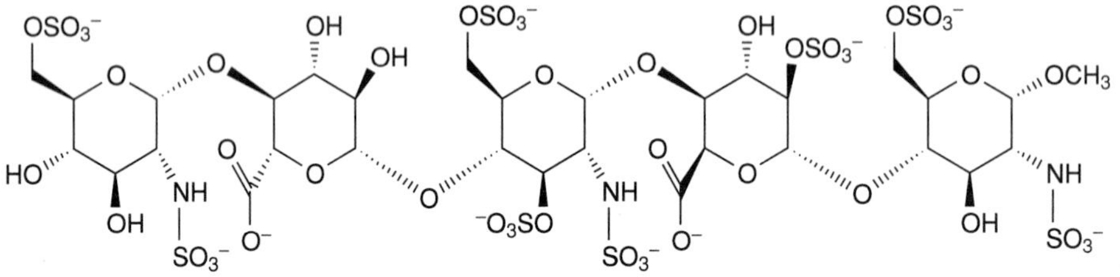

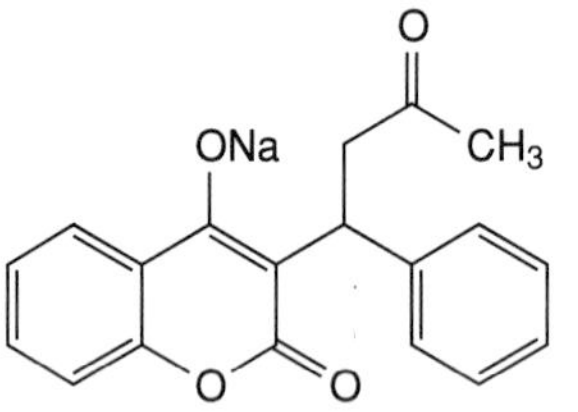

(3b) Desirudin
X_1X_2 = V-V
(Europe)

^+H_3N-(DF)PRPGGGGNGDFEEIPEEYL-CO$_2^-$ (4) Bivalirudin

(5) Argatroban

Figure 6.3. Structures of lepirudin, desirudin, and bivalirudin (single-letter amino acid nomenclature; DF, D-Phe), and argatroban.

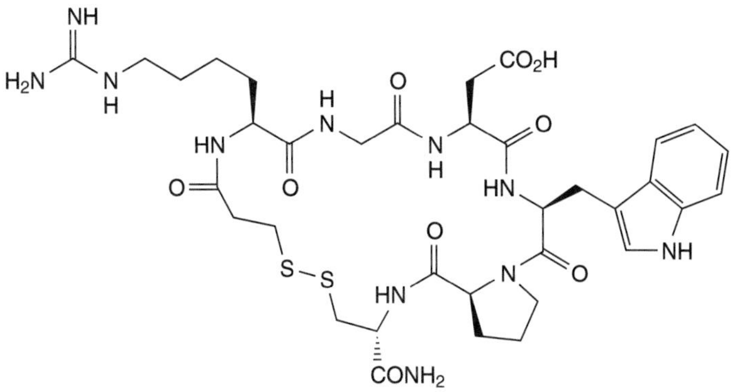

(6) Eptifibatide

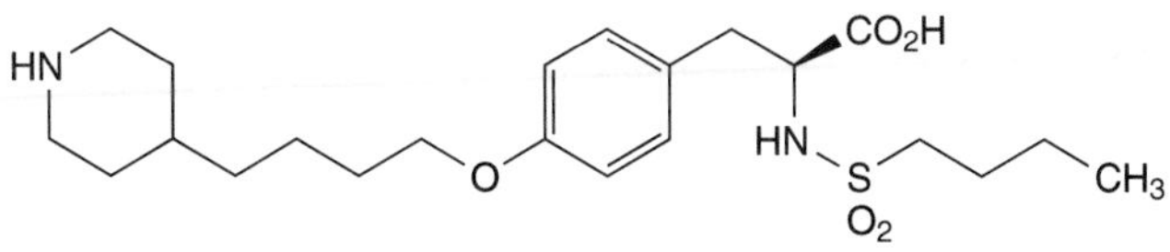

(7) Tirofiban

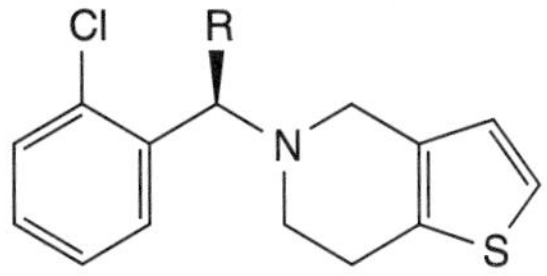

(8) Ticlopidine (R = H)
(9) Clopidogrel (R = CO_2CH_3)

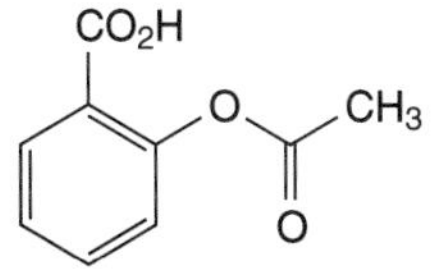

(10) Aspirin

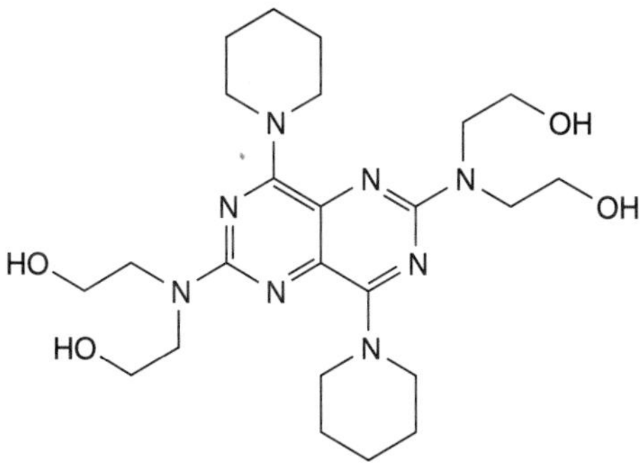

(11) Dipyridamole

Figure 6.4. Structures of marketed antiplatelet agents (abciximab not shown).

from plasminogen, its zymogen precursor (Table 6.2). Because all of these agents promote dissolution of fibrin, they are also referred to as fibrinolytics.

A thrombotic occlusion of a coronary or cerebral artery results in potentially life-threatening acute myocardial infarction or ischemic stroke, respectively, and fibrinolytic therapy can be used to rapidly dissolve the clot and restore flow. Fibrinolytic therapy is also occasionally used to treat dangerous thrombi that form in the venous system, especially the legs. This type of venous thrombosis [deep vein thrombosis (DVT)] often carries the risk that the clot, or pieces thereof, travel to the lung, resulting in life-threatening pulmonary embolism (PE). Fibrinolytic therapy is usually accompanied by and followed up with adminis-

Table 6.2   Marketed Thrombolytic (Fibrinolytic) Agents[a]

Generic Name	Trade Name	Marketed by	Description	Clot Selectivity
Alteplase	Activase	Genentech	Recombinant natural tPA[b]	Selective
Tenecteplase	TNKase	Genentech	Recombinant modified tPA	Selective
Reteplase	Retavase	Centocor	Recombinant modified tPA	Nonselective
Streptokinase	Streptase	AstraZeneca	Purified bacterial protein	Nonselective

[a]All are administered IV.
[b]tPA, tissue plasminogen activator.

tration of anticoagulant or antiplatelet agents, depending on the pathology.

Anticoagulant and antiplatelet agents, sometimes in combination, are also used in patient populations with existing thrombi or who are predisposed to thrombus formation, for example, individuals with atrial fibrillation or stable or unstable angina, and in patients with artificial heart valves. Also, anticoagulant and antiplatelet agents are used as prophylaxis in many surgeries, especially hip and knee replacement surgery and coronary artery bypass grafts, and as prophylaxis in other procedures such as coronary angioplasty. Anticoagulant therapy is also used for treatment of DVT and PE.

Although thrombi can form in either veins or arteries, the structural nature of the clots in each of these settings differs and therefore the optimal agents for drug treatment differ as well. Flow conditions largely dictate the ratio of fibrin to platelets in the growing thrombus. Arterial thrombi, formed under high flow conditions, are platelet-rich aggregates, whereas venous thrombi, formed under more static conditions, are composed largely of fibrin with trapped red blood cells and relatively few platelets. However, as an arterial thrombus occludes the vessel, a condition of stasis results, promoting the original platelet-rich clot to grow a more fibrin-rich component. In this regard, clinical studies have shown that both anticoagulant and antiplatelet agents are efficacious for arterial thrombosis, whereas anticoagulant drugs are superior to antiplatelet agents for venous thrombosis. Arterial thrombi are typically the result of the rupture of atherosclerotic plaques, pathological pockets of lipid and macrophage foam cells that form beneath the vessel endothelium surface (Fig. 6.5). Rupture exposes tissue factor, a

clot-triggering protein, and other thrombogenic material such as collagen and von Willebrand factor (see Section 3.1). The condition "unstable angina" is associated with the pathology of rupture-prone atherosclerotic plaques, whereas stable angina is associated with the presence of plaques, which are not in immediate danger of rupturing.

The currently marketed anticoagulant, antiplatelet, and fibrinolytic agents are, in general, efficacious for the indications in which they are used. On the other hand, many suffer from limitations. As mentioned, unwanted bleeding can be an issue for many of these agents and therefore intensive patient monitoring may be necessary. Several of the current anticoagulant and antiplatelet agents are limited by the requirement for intravenous (IV) administration, which necessitates a hospital stay. All of the fibrinolytics are administered by IV. Many of these agents are not broad-use drugs suitable for a variety of thrombotic indications. The limitations of some of these agents might eventually be over-

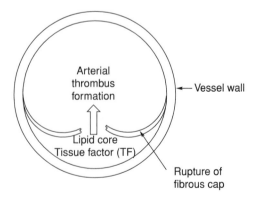

Figure 6.5. Atherosclerotic plaque rupture.

come as ongoing clinical study continues to explore treatment variables (e.g., use of alternative dosing regimens, use of different combinations of existing agents, or examination of their use in different thrombotic indications). At the same time, a great deal of recent research effort is directed toward the discovery and development of next-generation anticoagulants and antiplatelet agents. This research seeks new agents that operate by either established mechanisms of action or by novel mechanisms that current drugs do not target. Whether any of these potential new agents ultimately displaces the drugs in Tables 6.1 and 6.2, only time and clinical experience will tell. In any case, the Holy Grail of antithrombotic drug research is the discovery of safe agents requiring no patient monitoring and having single-agent efficacy over a broad range of thrombotic indications, and, especially for chronic treatment, having good oral bioavailability. As of 2002, we do not yet have such a drug on the market.

Many human bleeding disorders result from inherited defects of certain protein factors involved in coagulation or platelet function and individuals with these diseases may not be able to achieve adequate hemostasis without therapy. Hemophilia A and B are characterized by genetic defects in coagulation factors VIII and IX, respectively, whereas von Willebrand disease is characterized by functionally defective von Willebrand factor, a glycoprotein required for platelet adhesion to the endothelium. Effective therapy for correction of hemostasis in at-risk individuals during acute bleeding episodes or in anticipation of surgery can be achieved by replacement or augmentation of one or more of the natural factors. For this purpose, marketed hemostatics include plasma fractions that are concentrated in multiple factors, as well as natural or recombinant forms of individual factors (Table 6.3).

The remainder of this chapter examines in more detail the current agents on the market that affect thrombosis and hemostasis, and then describes newer antithrombotic agents under development that operate by either known or novel mechanisms.

2.2 Adverse Effects

2.2.1 Bleeding Risks for Antithrombotic and Fibrinolytic Agents. Bleeding is the principal risk associated with antithrombotic and fibrinolytic therapy (2). Bleeding episodes can be major and life-threatening as in cases of intracraneal bleeding, or less severe (minor or nuisance bleeding) as in bleeding gum or nose bleeds. Because thrombotic conditions being treated can range from acute and life-threatening to relatively stable situations, wherein only low level preventive antithrombotic therapy is recommended, the bleeding risks need to be weighed against the benefit of therapy. The bleeding risk is highly dependent on a number of variables, including the drug being used; the thrombotic condition being treated (e.g., stroke vs. myocardial infarction); level and duration of therapy, whether other antithrombotic agents are used in combination; age of patient; or underlying predispositions to bleeding (e.g., past episodes of gastrointestinal bleeding or recent surgery or trauma) (2). Typically the dose of therapeutic agent is adjusted to the level necessary to achieve the desired therapeutic effect without causing unacceptable bleeding. In the case of anticoagulant therapy, the level or "intensity" is often carefully monitored using various *in vitro* tests, which measure the length of time for blood clotting. For example, common clotting tests measure prothrombin time (PT, often used during warfarin therapy) and activated partial thromboplastin time (aPTT, often used during heparin therapy). The international normalized ratio (INR) is calculated from the PT and is meant to be a better comparator among individuals/sites when slightly different *in vitro* clotting reagents are used.

The following describes a few examples of treatments and settings where major bleeding was quantified during anticoagulant therapy. As part of the International Stroke Trial, patients with acute ischemic stroke associated with atrial fibrillation were treated with subcutaneously (SC) administered heparin (two different doses), aspirin, heparin + aspirin, or neither. Heparin at the higher dose caused bleeding complications that offset the antithrombotic benefit. The overall frequency of hemorrhagic stroke in the heparin-treated

Table 6.3 Marketed Hemostatic Preparations[a]

Trade Name	Marketed by	Description
	Factor VIII Preparations	
Monoclate	Aventis Behring	FVIII, purified from human plasma
Hemofil	Baxter Healthcare	FVIII, purified from human plasma
Koate	Bayer Biological	FVIII, purified from human plasma
Kogenate/Helixate	Bayer Biological; Aventis Behring	FVIII, recombinant
Recombinate	Baxter Healthcare	FVIII, recombinant
ReFacto	Genetics Institute	FVIII, recombinant
	Factor IX Preparations	
Mononine	Aventis Behring	FIX, purified from human plasma
Benefix	Genetics Institute	FIX, recombinant
	Other	
NovoSeven[b]	NovoNordisk	FVIIa, recombinant
	Mixtures of Coagulation Factors	
Humate	Aventis Behring	FVIII + von Willibrand Factor; purified from human plasma
Bebulin	Baxter Healthcare	FIX + FII[c] + FX; purified from human plasma
Konyne	Bayer Biological	FIX + FII + FX; purified from human plasma
Proplex	Baxter Healthcare	FIX + FII + FX + FVII; purified from human plasma
Feiba[b]	Baxter Healthcare	FIX + FII + FX + FVIIa; purified from human plasma
Autoplex[b]	Nabi	FXa + FVII + FVIIa + other factors; purified from human plasma

[a]All are administered IV.
[b]For patients with inhibitors (antibodies) to FVIII.
[c]FII (Factor II) is prothrombin, the zymogenic form of thrombin (FIIa).

groups after 14 days was 2.1% compared to 0.4% for the patients not treated with heparin (3). Use of a 7-day IV course of danaparoid for acute ischemic stroke increased the incidence of major bleeding (5.2% vs. 1.8% for placebo) (4). On the other hand, in several trials involving patients with ischemic coronary artery disease, treatment with heparin or LMWHs did not seem to increase the risk of major bleeding (2). In a combined analysis of a number of studies of patients treated for venous thromboembolism (VTE), use of IV heparin was associated with major bleeding rates of 0–7% (fatal bleeding 0–2%) and use of SC LMWHs was associated with major bleeding rates of 0–3% (fatal bleeding 0–0.8%) (2). Use of the pentasaccharide fondaparinux for prophylaxis of DVT in patients undergoing hip or knee surgery resulted in major bleeding rates of 2.2 and 2.1%, respectively, whereas use of

the LMWH enoxaparin in these surgeries led to corresponding bleeding rates of 2.3 and 0.2%, respectively. Although enoxaparin showed a lower bleeding rate in the knee surgery setting, fondaparinux was more efficacious for both types of surgeries (5a,b).

In a stroke prevention trial in patients who had experienced a transient ischemic attack (TIA) or a minor ischemic stroke, warfarin therapy resulted in an 8.1% rate of major bleeding complications (including fatal bleeding) compared to 0.9% for aspirin (6). On the other hand, major bleeding rates associated with warfarin therapy in patients with atrial fibrillation or in patients after a myocardial infarction were relatively lower (7). In a number of studies comparing oral warfarin with SC heparin or SC LMWHs for the treatment of venous thromboembolism, there was a tendency toward higher rates of bleeding with warfarin, although when lower intensity anticoagulation with warfarin was compared with lower doses of LMWH, the frequency of major bleeding did not differ (2). Warfarin plus aspirin results in higher bleeding rates than with warfarin alone (8).

Compared to heparin, lepirudin increased the risk of non-life-threatening major bleeding in patients being treated for unstable angina (UA) or myocardial infarction (MI), but it did not increase bleeding risk in the setting of thromboprophylaxis for elective hip replacement. In both of these settings, however, lepirudin appeared to be more efficacious than either heparin or LMWH (9). Use of bivalirudin in patients with coronary angioplasty resulted in hemorrhagic complications, but at a reduced rate compared to that of heparin (10). A trial with the small-molecule direct thombin inhibitor argatroban in patients undergoing percutaneous coronary intervention (PCI; e.g., coronary angioplasty) showed a slightly lower rate of major bleeds compared to historical heparin rates (11).

Bleeding complications are also associated with the use of antiplatelet agents. Although aspirin is regarded as relatively safe for most patients in terms of lack of major bleeding risk (e.g., hemorrhagic stroke), it can cause gastrointestinal bleeding even at low doses. This effect is believed to be attributed to combined inhibition of platelet cyclooxygenase-1 (COX-1) activity, resulting in an antiplatelet (and therefore antihemostatic) effect, and inhibition of cellular COX-1 (12). Functional cellular COX-1 in the gastric mucosa is important for maintaining mucosal integrity (13). Low dose aspirin (50–100 mg/day) is generally better tolerated and has been shown to be as efficacious as higher doses for antiplatelet therapy. By contrast, the anti-inflammatory activity of aspirin resulting from its inhibition of cellular COX-2 requires larger doses of aspirin. In general, the risk associated with aspirin therapy is acceptable in individuals who are at greater risk of major thrombotic events (12).

The antiplatelet agent dipyridamole, when used alone, appears to have minimal bleeding risk. Thus, in a study of prior stroke patients treated with either dipyridamole, aspirin (25 mg/day), or both, the use of aspirin alone or in combination resulted in increased bleeding, whereas dipyridamole alone was indistinguishable from placebo (12, 14).

In one study with the thienopyridine antiplatelet agent clopidogrel, clinical bleeding event rates were slightly lower compared to those of medium dose aspirin (15), whereas in two other studies that compared aspirin alone to clopidogrel plus aspirin, the clopidogrel-treated patients experienced either the same or slightly higher major bleeding rates compared to thsoe of aspirin (16). The related thienopyridine antiplatelet drug ticlopidine has been associated with increased bleeding, including spontaneous posttraumatic and perioperative bleeding (17).

Abciximab, a monoclonal antibody to the platelet GPIIb/IIIa receptor, was studied in coronary angioplasty patients in combination with standard doses of heparin. Significant increases in major bleeding were attributed to abciximab, though a later study using reduced doses of heparin showed a reduction of the bleeding risk (18). A trial of the heptapeptide GPIIb/IIIa antagonist eptifibatide in patients undergoing PCI showed a modest increase in hemorrhagic complications (19). Moreover, in the setting of PCI, the nonpeptide small molecule GPIIb/IIIa antagonist tirofiban, in combination with heparin or aspirin, was associated with an increase in bleeding compared to that of heparin or aspirin alone (20).

The use of thrombolytic (fibrinolytic) agents in the settings of acute myocardial infarction, ischemic stroke, and VTE carries the risk of major bleeding complications, with the most severe being intracraneal hemorrhage (hemorrhagic stroke). Three trials employing streptokinase for acute ischemic stroke were halted prematurely by safety committees because of unfavorable rates of early mortality and intracranial bleeding (21). Trials using tissue plasminogen activator (tPA) for acute stroke reported intracranial bleeding rates of 3–7% (21). Thrombolytic therapy for MI resulted in noncranial major bleeding rates of 1.1% compared to 0.4% for placebo, and resulted in an excess of 3.7 hemorrhagic strokes per 1000 patients compared to that of placebo (22). Patients who receive thrombolytic therapy for VTE have a 1–2% risk of intracranial bleeding (23).

Combining fibrinolytic therapy with anti-coagulation carries significant risk for bleeding (21). For example, recruitment of patients with acute coronary syndromes (unstable angina or MI) in a trial combining streptokinase or tPA with heparin or hirudin (desirudin) was stopped early because of hemorrhagic stroke, with rates varying from 0.9% (tPA + heparin) to 3.2% (streptokinase + hirudin). It was found that aPTT values were predictive of risk of hemorrhagic stroke in patients receiving these combination therapies (24).

It is hoped that some of the newer agents (see below) will show improvements in therapeutic window and monitoring burden.

2.2.2 Other Adverse Effects for Antithrombotic Drugs.

Thrombocytopenia (low platelet count) can occur during treatment with several anticoagulant and antiplatelet drugs. Heparin-induced thrombocytopenia (HIT) is a consequence of an antibody response to a complex between heparin and endogenous platelet activating factor 4 (PF4). The resulting antibody-heparin-PF4 complex is thought to interact with Fc receptors on platelets, leading to platelet activation and thrombin generation (25, 26). As a result, a general platelet aggregation response ensues with platelet depletion, and an independent thrombotic pathology, called HITTS (heparin-induced thrombocytopenia with thrombosis), often develops. HIT usually develops between day 5 and day 15 of heparin therapy and affects up to 3% of patients (27). In the absence of overt thrombosis, cessation of heparin therapy is often sufficient to relieve HIT. However, to manage the thrombotic condition for which the heparin was originally indicated as well as any thrombotic complications resulting from HIT itself, a switch to a different anticoagulant (e.g., lepirudin) is often necessary (25). Given that a minimum of 12–14 saccharide units is required to form the antibody complex with PF4, LMWHs tend to cause HIT less frequently than does heparin (27). However, as a result of the observation of *in vitro* cross-reactivity to heparin antibodies, LMWHs are contraindicated in patients with known heparin-associated antibodies. On the other hand, the "heparinoid" danaparoid seems not to induce HIT and therefore is used to treat individuals with this condition. Additionally, studies in patients with confirmed HIT demonstrated that the pentasaccharide fondaparinux is not associated with *in vitro* cross-reactivity to heparin antibodies, and therefore this agent may present low risk for HIT (5c).

For all three marketed platelet GPIIb/IIIa antagonists, there have been reports of thrombocytopenia. In a trial with abciximab, approximately 6% of patients developed antibodies to the variable regions of abciximab (28), a human chimeric antibody, and about 1–2% of patients developed low platelet counts as a result (12). Withdrawal of the drug is usually sufficient for recovery (12). Eptifibatide treatment, although not associated with an increased frequency of thrombocytopenia overall, might be associated with a small increase in cases of profound thrombocytopenia (12, 19). Thrombocytopenia arising from treatment with tirofiban has been reported in a small percentage of patients and has been postulated to be the result of an immunologic response to the tirofiban-GPIIb/IIIa complex (29, 30). It has been recommended that all patients receiving parenteral GPIIb/IIIa antagonists be monitored for development of thrombocytopenia within 24 h of the initiation of therapy (30).

Thrombotic thrombocytopenic purpura (TTP) is a relatively rare condition that has been reported in patients treated with ticlopidine and, to a much lesser extent, with clopidogrel (31, 32). TTP is characterized by thrombocytopenia, microangiopathic hemolytic anemia (fragmented red blood cells), and other symptoms. For ticlopicine, the estimated incidence may be as high as one case in every 2000–4000 patients exposed (17). For clopidogrel, TPP has been reported at a rate of about 1 case per 250,000 patients exposed (33).

2.2.3 Adverse Effects for Hemostatic Preparations.

Although the infusion of mixtures of endogenous coagulation factors, especially mixtures containing coagulation factor IX (FIX), into patients with bleeding disorders has historically carried the risk of thrombotic complications (34), such complications are currently not regarded as clinically important. Additionally, the historical risk of infection resulting from viral (e.g., HIV or hepatitis) contamination of plasma products or purified coagulation factors is now reduced as a result of rigorous screening and heat-treating of purified plasma products as well as to the use of recombinant coagulation factors. Nevertheless, it is generally recommended that individuals with hemophilia be vaccinated for hepatitis A and B. Additionally, parvovirus B19, a non-lipid-enveloped DNA virus, cannot currently be removed from plasma products and, although infected adults are usually asymptomatic, this virus can affect pregnant women or immune-compromised individuals (35). The presence of low levels of potentially immunogenic blood group isoagglutins in the case of plasma-purified products, and murine or other nonhuman proteins in the case of recombinant products, is generally regarded as not a clinically significant risk. However, a well-recognized complication of treatment of hemophilia A using replacement FVIII therapy is the development by the patient of neutralizing antibodies to the infused FVIII protein (36). Strategies for overcoming this complication include treatment with higher levels of FVIII when the antibody response is not overwhelming, or augmentation with other coagulation factors, such as FVIIa or products that contain mixtures of coagulation factors (see Table 6.3).

2.3 Absorption, Distribution, Metabolism, Elimination

Not surprisingly, because of its size and charge, heparin is not effective by oral administration, and is given by intermittent IV injection, IV infusion, or SC injection. When given by the IV route, the full pharmacological effect is seen almost immediately. Given SC, heparin peak plasma levels are achieved only after 2–4 h, and the bioavailability is lower than that of the IV route. Subcutaneous administration is either twice or three times daily. Recently, however, heparin formulations intended to facilitate gastrointestinal absorption of heparin ("oral" heparin) are starting to be investigated in humans (37). Heparin efficiently binds plasma proteins and this accounts for some of the variability of anticoagulant response among patients. Patient monitoring, usually using aPTT, is standard with

heparin. Heparin is cleared by two distinct mechanisms (25, 38). In the first mechanism, which is saturable, heparin binds to macrophages and endothelial cells where it is degraded. The second, slower, and nonsaturable mechanism is renal clearance. Heparin pharmacokinetics is therefore not linear, another reason for patient monitoring. In the case of the LMWHs, however, protein and cell binding for these shorter polysaccharides is substantially less compared to that of heparin, and pharmacokinetics are therefore less variable and more predictable. Elimination is almost exclusively by the renal route, with elimination half-lives of about 3–5 h, except for danaparoid, which has a half-life of about 24 h. LMWHs are given by the SC route usually once per day, and patient monitoring, with the exception of individuals with renal impairment, is not routine. This superior pharmacokinetic profile for LMWHs is a well-recognized advantage of LMWHs over heparin and one of the principal reasons for the increasing use of LMWHs instead of heparin (25). Like the LMWHs, the pentasaccharide fondaparinux displays a linear, predictable dose-dependent pharmacokinetic profile. It is 100% bioavailable after SC injection, with a half-life of 14–16 h, and is eliminated primarily by the kidneys (39).

The term "oral anticoagulation" currently refers to warfarin therapy. Warfarin is a racemic mixture and no advantage has been established in administering a single enantiomer. Although R-warfarin is several times more active as an anticoagulant than the S-form, the R-enantiomer has about double the clearance rate compared to that of the S-form. Warfarin is essentially completely absorbed after oral administration, reaching peak levels within 4 h, and has a mean half-life of about 40 h. It is almost entirely protein bound (99%). Because of its mechanism of action, the full pharmacodynamic response to a single dose of warfarin takes approximately 2–5 days (40). Warfarin interferes with the critical vitamin K-mediated γ-carboxylation of glutamic acid residues near the N-terminus of several procoagulant enzymes (prothrombin, FX, FIX, and FVII) as well as two anticoagulant proteins (protein S and protein C). Lacking this critical modification, these enzymes and proteins are essen-

Table 6.4 Plasma Concentrations and Half-Lives of Coagulation Factors

Coagulation Factor	Plasma Concentration (μM)	Plasma Half-Life (h)
	Serine Proteases	
FII (prothrombin)	1.4	100
FX	0.14	65
FIX	0.09	20
FXI	0.03	65
FVII	0.01	5
Protein C	0.06	6
	Cofactors	
FV	0.03	25
FVIII	0.0003	10
Protein S	0.14	60

tially functionally inactive. Because each of these proteins has a different half-life in plasma (Table 6.4), the simultaneous inhibition of their combined biosynthesis leads to a gradual increase in the anticoagulant effect with time. As an example, prothrombin has a half-life of 50 h, and the anticoagulant effect stemming from the loss of this clotting enzyme after initiation of warfarin therapy is not fully realized until the new lower steady state for this protein is achieved.

The cytochrome P450 CYP2C9 is the major enzyme responsible for converting S-warfarin to its inactive oxidative metabolites. Warfarin is also metabolized by reductases to products with minimal anticoagulant activity. Almost the entire dose of warfarin is excreted as metabolites in the urine. One source of patient variability to warfarin is the presence of two variant CYP2C9 allelles in 10–20% of the Caucasian population, but present in less than 5% of African-Americans or Asians (41). Furthermore, there is a long and expanding list of drugs that interfere with warfarin activity (40, 42), either by competing for its major metabolic pathways (e.g., CYP2C9) or displacing it from protein binding sites. Either of these interfering mechanisms can potentially lead to hemorrhagic complications. Other factors, such as hepatic dysfunction or relative abundance or deficiency of vitamin K in the diet, can also influence the effect of warfarin. Regular patient monitoring through measurement of PT (commonly expressed as the INR value) is required during warfarin therapy.

When dosed as an IV bolus, the pharmacokinetics of the polypeptide thrombin inhibitor lepirudin is described by a two-compartment model with an initial half-life of 10 min followed by a terminal elimination half-life of about 1.3 h (43). By SC administration, however, the PK follows a one-compartment model, with a maximum plasma concentration achieved after about 2 h. It is eliminated primarily by the kidneys as a mixture of intact drug and metabolites. Renal impairment can lengthen the elimination half-life up to 150 h (44). Because lepirudin binds nearly irreversibly to thrombin, and there is no antidote, patients are typically monitored (aPTT).

The related peptide bivalirudin has a half-life of 36 min after IV infusion and, unlike lepirudin, is proteolytically cleaved by thrombin (between Arg-Pro) (45) and possibly by the liver, which accounts for a significant fraction of its clearance (46). SC administration results in a more sustained pharmacokinetic profile.

Argatroban has an elimination half-life of 39–51 min after an IV bolus (47) and is extensively metabolized in the liver to four metabolites by hydroxylation and aromatization of the tetrahydroquinoline ring. The major metabolite has been shown to be weaker than argatroban as an anticoagulant. Even though it was demonstrated *in vitro* that CYP3A4/5 converted argatroban to these four metabolites, the lack of an *in vivo* effect by erythromycin (a potent CYP3A4/5 inhibitor) on argatroban pharmacokinetics suggests metabolism by additional routes (48).

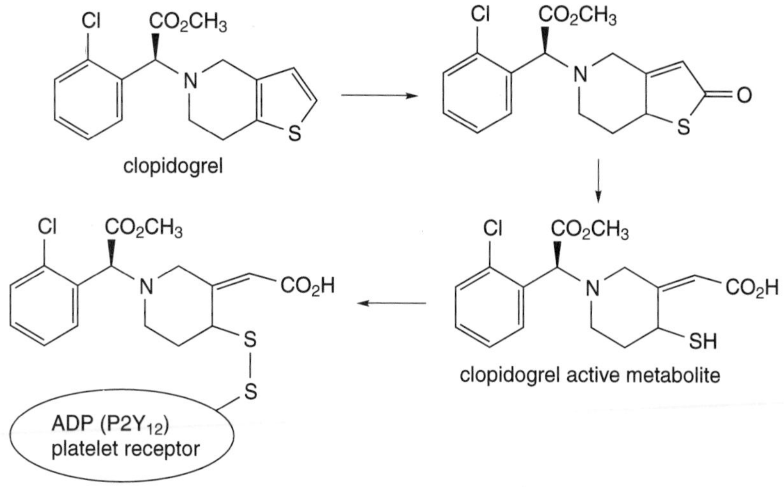

Figure 6.6. Metabolism and proposed mechanism of action of clopidogrel.

After an IV bolus of abciximab (the monoclonal antibody to the platelet GPIIb/IIIa receptor) the free plasma concentrations decrease rapidly over 30 min, probably reflecting its distribution to the platelet receptors. Peak effects on receptor blockade were observed after 2 h (first sampling) with platelet function returning to >50% of baseline after about 24–48 h (49). The pharmacokinetic profile of the cyclic heptapeptide GPIIb/IIIa blocker eptifibatide after IV bolus dosing is linear and dose proportional. Plasma protein binding is low (25%). Renal clearance predominates and the terminal half-life is about 2.5 h (50). Effects on platelet function return to normal within 6–12 h. The nonpeptide GPIIb/IIIa blocker tirofiban has a half-life of 2 h and is cleared mainly by the kidneys, and metabolism appears to be minimal (51). After cessation of tirofiban therapy, platelet function returned to normal within about 4 h (52).

The two ADP receptor antagonists clopidogrel and ticlopidine present an interesting case, in that the parent compounds are inactive as platelet aggregation inhibitors and must undergo hepatic conversion to active metabolites. In the case of clopidogrel, the structure of the active metabolite has been determined to be a sulfhydryl compound, which is postulated to react by disulfide bond formation to a cystein residue within the platelet ADP receptor (Fig. 6.6) (53). Ticlopidine is

likely converted *in vivo* to an analogous active metabolite. After oral dosing of 400 mg of clopidogrel, the inhibition of platelet aggregation was detectable after 2 h and remained relatively stable up to 48 h. On repeated daily dosing of 50–100 mg clopidogrel, platelet aggregation was inhibited from the second day of treatment and reached steady state after 4–7 days. After an oral dose of ^{14}C-labeled clopidogrel, in humans, approximately 50% of the radioactivity was excreted in the urine and approximately 46% in the feces in the 5 days after dosing (33, 54).

Aspirin is rapidly absorbed in the stomach and upper intestine (40–50% oral bioavailability) and peak plasma levels occur 30–40 min after ingestion. Inhibition of platelet function is evident by 1 h. Because aspirin irreversibly inactivates platelet COX-1 by acylation of an active site serine, and because platelets do not synthesize new proteins, this effect lasts the lifetime of the platelet (7–10 days). Because platelet COX-1 is acetylated in the presystemic circulation, the antiplatelet effect of aspirin is largely independent of systemic bioavailability, and the rapid plasma clearance of aspirin (15–20 min) does not impact its therapeutic duration (12). Complete inactivation of platelet COX-1 is achieved when 160 mg of aspirin is taken daily. This dose is lower than that required for the anti-inflammatory effects of aspirin (i.e., levels required to therapeutically affect COX-2).

2.4 Typical Treatment Regimens for Common Thrombotic Conditions

Employing the currently marketed anticoagulant, antiplatelet, and fibrinolytic agents, the following are representative treatment regimens for some of the major classes of thrombotic conditions (55). Of course, treatment modalities will vary based on the individual circumstances of the patient and the biases of the treating physician.

2.4.1 Venous Thromboses. For the prevention of VTE in the setting of major orthopedic surgeries such as hip and knee replacement, either LMWH or warfarin is often employed (heparin being an alternative) with dosing initiated before the surgery. Existing venous thromboembolic disease (e.g., DVT or PE) can result from stasis of blood or abnormalties of the vessel wall. Treatment with LMWH is recommended (unfractionated heparin being an alternative) for the first 5 days, with overlapping oral anticoagulation (warfarin), which is then continued for at least 3 months. Patients with very serious embolism or thromboses and who are at low risk for bleeding may undergo treatment with thrombolytic agents.

Atrial fibrillation is the most common sustained cardiac arrhythmia and can lead to the generation of clots resulting from atrial blood stasis. These clots can be carried to the brain and precipitate ischemic stroke. For long-term prevention of this thrombotic condition in low risk patients, aspirin (325 mg/day) is recommended, whereas in moderate risk groups, either aspirin or warfarin is recommended, and in high risk groups, warfarin is recommended.

2.4.2 Arterial Thromboses. Coronary artery disease (CAD) encompasses stable and unstable angina and acute myocardial infarction. For primary prevention of CAD, the recommended treatments for low, medium, and high risk patients are aspirin, low dose warfarin, and aspirin plus low dose warfarin, respectively. Aspirin is recommended for the management of stable angina. For unstable angina, long-term aspirin therapy (75–162.5 mg) is recommended and for patients hospital-

ized because of UA, heparin or LMWH treatment (at least 48 h) is added. Lepirudin can be substituted for heparin and is indicated instead of heparin for patients with a history of HIT. Patients experiencing acute myocardial infarction (AMI) are typically immediately put on aspirin and treated IV with a fibrinolytic. Fibrinolytics for MI can be contraindicated in certain cases, however, as in patients with a history of intracranial hemorrhage or current active bleeding. Depending on individual circumstances, either streptokinase, alteplase, or TNK-ase can typically be administered. Adjunctive therapy with IV heparin (or IV lepirudin in individuals with a history of HIT) is recommended. For acute MI patients not receiving fibrinolytic therapy, aspirin plus IV heparin followed by a course of warfarin is typically recommended.

Fibrinolytic therapy with tPA is recommended in eligible patients suffering from acute ischemic stroke. Treatment with tPA must be initiated within 3 h of clearly defined symptom onset. Patients with a history of intracranial bleeding, recent surgery, or several other conditions are not eligible for thrombolytic therapy because of the high risk of fatal hemorrhage. For these excluded patients, IV or SC heparin or SC LMWH or danaparoid can be administered, although clinical trials have been inconclusive regarding the benefit of these agents in this setting. For stroke prevention in individuals who have experienced a previous stroke or TIA, aspirin, aspirin plus dipyridamole (which is more effective than aspirin alone), or clopidogrel are typically recommended.

The term PCI refers to various revascularization procedures, such as balloon angioplasty or stent placement, which are usually performed as part of the treatment for CAD. Antithrombotic therapy is administered during these procedures to prevent thrombotic complications (e.g., abrupt closure or late restenosis). Pretreatment is with aspirin, clopidogrel, or aspirin plus clopidogrel (which is more effective than either agent separately). An IV GPIIb/IIIa agent (abciximab, eptifibatide, or tirofiban) along with IV heparin is often used in patients undergoing PCIs, espe-

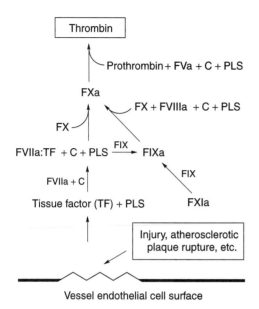

Figure 6.7. Coagulation cascade. C, Ca^{+2}; PLS, phospholipid surface.

cially those at high risk. Bivalirudin is indicated in place of heparin for patients with a history of HIT.

3 PHYSIOLOGY, BIOCHEMISTRY, AND PHARMACOLOGY

3.1 Molecular Mechanisms of Thrombosis and Fibrinolysis

To more fully understand the mechanisms of action of anticoagulant, antiplatelet, and fibrinolytic agents, a more detailed picture of clot formation and lysis is now presented. Clots consist of both fibrin strands and platelets in a ratio dependent on the local vascular environment (e.g., blood flow), as previously discussed. At the site of vascular injury platelets bind to the injured surface, become activated, and aggregate. In parallel, the coagulation cascade (Fig. 6.7) is triggered, leading to the production of thrombin, which both catalyzes the formation of insoluble fibrin strands and further activates platelets. In turn, activated platelets create local environments required for additional thrombin generation. Endogenous regulatory mechanisms directed against thrombin generation and platelet acti-

vation are simultaneously triggered to prevent excessive clot growth, and the processes involved in clot dissolution are also activated. Thus clot growth and dissolution is a highly regulated and delicately balanced set of processes that under normal conditions quickly generates a clot of proper size and maintains it for only the time it is needed.

The coagulation cascade (Fig. 6.7) involves two converging sequences of enzymatic steps involving successive activations of serine proteases in an amplifying manner, ultimately generating a burst of active thrombin. Thus, after vascular injury or atherosclerotic plaque rupture, membrane-bound tissue factor (TF) is exposed to the vascular lumin and noncovalently associates with the circulating serine protease Factor VIIa (FVIIa), which in the presence of calcium on the phospholipid surfaces of subendothelial cells forms the "extrinsic tenase complex" (Fig. 6.8a). The proteolytic activity of FVIIa in this complex with TF is significantly enhanced compared to the activity of free FVIIa. In this complex, FVIIa cleaves a peptide sequence on the inactive zymogen FX, forming the proteolytically active enzyme FXa, which in turn cleaves a peptide sequence on the inactive precursor prothrombin to generate active thrombin. This thrombin-activating step occurs most efficiently when FXa combines with the protein cofactor FVa on platelet phospholipid surfaces in the presence of calcium, forming the "prothrombinase complex" (Fig. 6.8b). An alternative pathway leading to generation of FXa and thrombin involves the proteolytic activation of FIX to FIXa by the serine protease FXIa (Fig. 6.7). The resultant FIXa associates with the protein cofactor FVIIIa in the presence of calcium on platelet phospholipid surfaces, forming the "intrinsic tenase complex" (Fig. 6.8c). As part of this complex, FIXa efficiently activates FX to FXa by the same proteolytic cleavage as was effected by FVIIa. Finally, in a crossover mechanism linking the two pathways, the FVIIa-dependent extrinsic tenase complex can also activate FIX to FIXa. The historical terms "intrinsic," referring to the FXI-mediated pathway, and "extrinsic," referring to the TF-mediated pathway, are still used, although mainly for convenience.

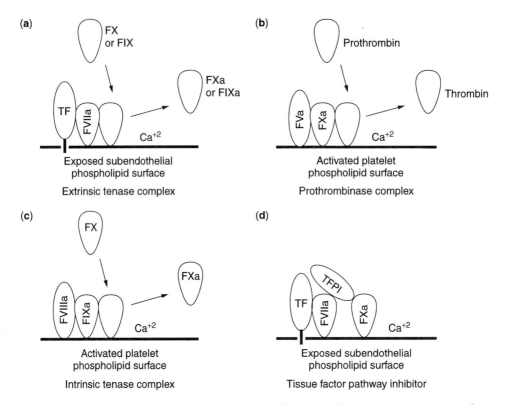

Figure 6.8. (a) Extrinsic tenase complex; (b) prothrombinase complex; (c) intrinsic tenase complex; (d) tissue factor pathway inhibitor complex.

The successive enzymatic catalytic steps in the coagulation cascade result in a great amplification of the overall thrombin "signal." Consequently, only small amounts of upstream zymogens are required compared to those further downstream. For example, the concentration of FVII in plasma is only 0.01 μM, whereas FX is 0.14 μM and prothrombin is 1.4 μM (Table 6.4). FVIIa, FXIa, FIXa, FXa, and thrombin are all trypsinlike serine proteases, and the cofactors TF, FVIIIa, and FVa are nonenzymatic proteins, which serve to allosterically enhance the activities of their serine protease partners (FVIIa, FIXa, and FXa, respectively). The phospholipid surfaces for assembly of the intrinsic tenase and prothrombinase complexes are expressed on activated platelets and serve to "concentrate" and preorganize the factors and cofactors for efficient proteolytic processing. Calcium is required in these complexes for proper binding and orientation of FVII/FVIIa, FIX/FIXa, FX/

FXa, and prothrombin. For these calcium-dependent complexes, binding of the proteins to the anionic phospholipid surfaces is mediated by a cluster of 10–12 γ-carboxyglutamic acid (Gla) residues near the N-terminus of each protein. These Gla domains anchor these proteins to the phospholipid surface, although the precise mechanism of binding is still being investigated. It is believed, however, that calcium ions form salt bridges within the Gla domain, thereby inducing an ordered structure that results in an outward display of certain lipophilic residues. According to this model, binding of the Gla domain with the phospholipid surface results from a combination of lipophilic and Gla-calcium-mediated electrostatic interactions (57). The Gla residues on these proteins are biosynthesized posttranslationally in a vitamin K-dependent manner by γ-carboxylation of glutamic acid residues and therefore these enzymes and their zymogens are collectively referred to as the vitamin K-

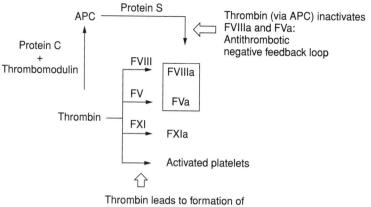

Figure 6.9. Thrombin positive and negative feedback loops.

dependent factors. Proteins C and S are also Gla domain-containing proteins.

The coagulation cascade is regulated, both positively and negatively, at a number of points in the two pathways. Tissue factor pathway inhibitor (TFPI) is an endogeneous protein that is a reversible inhibitor of the extrinsic tenase complex and uses the active site of FXa to bridge to the active site of FVIIa, rendering both proteases inactive (Fig. 6.8d). In a positive feedback loop, FXa can cleave zymogen FVII, thus providing additional FVIIa. Most important, thrombin itself is involved in a number of positive and negative feedback loops, thus regulating its own production (Fig. 6.9). In a positive feedback loop, thrombin cleaves FXI to FXIa, thus activating the intrinsic coagulation pathway toward more thrombin generation. Thrombin also cleaves the proteins FV and FVIII to afford the active procoagulant cofactors FVa and FVIIIa, respectively. Additionally, through cleavage of a peptide fragment on the platelet PARs (protease-activated receptors), thrombin activates platelets that then express prothrombogenic phospholipid surfaces used to assemble the intrinsic tenase and prothrombinase complexes.

In a negative feedback loop, thrombin first complexes with the protein cofactor thrombomodulin (TM) in a manner that effectively alters thrombin's substrate specificity (Fig. 6.10) (58). In particular, TM-complexed thrombin no longer efficiently activates FV or FVIII or PARs on platelets, but does efficiently cleave

protein C, affording activated protein C (APC) (59). APC in combination with its nonenzyme cofactor, protein S, proteolytically inactivates the essential procoagulant factors FVa and FVIIIa. By so doing, APC downregulates thrombin production, effectively acting as an *anticoagulant*. Thrombomodulin-bound thrombin also activates a carboxypeptidase, TAFI (thrombin-activatable fibrinolysis inhibitor, also called plasma carboxypeptidase B or U), which serves an opposing, antifibrinolytic function by slowing the rate of plasmin-mediated clot lysis (see below) (60).

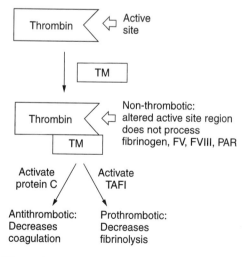

Figure 6.10. Thrombomodulin (TM) alters the activities of thrombin.

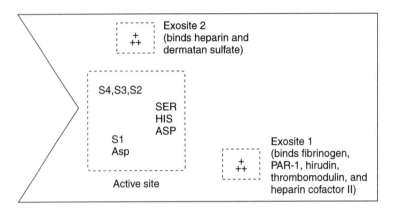

Figure 6.11. Thrombin active site and exosites.

A schematic picture of thrombin is shown in Fig. 6.11. The active site contains the typical trypsinlike serine protease catalytic triad Ser-His-Asp with an adjacent S1 pocket containing an Asp at the bottom, having a specificity for binding Arg and Lys residues by salt bridges. The other substrate-specific binding sites (S2, S3, S4) typically bind hydrophobic residues. Apart from the active site region, thrombin has two distinct anion-binding regions or "exosites" defined by patches of positive charge resulting from grouped Lys/Arg residues. Exosite 1 interacts with negatively charged domains on fibrinogen and platelet PAR-1, serving to orient these substrates for processing at the thrombin active site. Exosite 1 also binds thrombomodulin and heparin cofactor II, and additionally binds the anionic C-terminus of the peptide thrombin inhibitor hirudin as well as other hirudin-like peptides. Exosite 2 typically binds polyanionic polysaccharides such as heparin and dermatan sulfate. This thrombin-heparin interaction serves as the basis for another mechanism by which thrombin activity can be regulated endogenously, namely by its inhibition with antithrombin.

The protein antithrombin belongs to a large class of endogenous enzyme inhibitors called serpins (*serine protease inhibitors*), which includes α1-antitrypsin, α2-antiplasmin, and plasminogen activator inhibitor 1 (PAI-1) among others (61). The mechanism for thrombin inhibition involves initial binding of a specific pentasaccharide sequence of heparin (Fig. 6.12) to antithrombin, which in-

duces a conformational change within antithrombin, priming it for efficient protease inactivation (25). Through its exosite 2 domain, thrombin then binds to part of a 13-saccharide stretch of the anionic heparin molecule adjacent to the pentasaccharide. In this manner, thrombin's active site domain is brought into contact with the inactivating domain of antithrombin, forming an inactive covalent complex. Heparin then releases the complex and can repeat the cycle of thrombin inactivation. The antithrombin-heparin complex also efficiently inactivates FXa, leading to downregulation of thrombin generation. Unlike the thrombin-antithrombin-heparin complex, a separate heparin-FXa interaction is not required for inactivation of FXa by antithrombin-heparin and thus the pentasaccharide interaction with antithrombin is all that is needed for efficient inactivation of FXa. By a similar heparin-dependent mechanism, thrombin is also inactivated by heparin cofactor II (HCII), another serpin. Unlike antithrombin, which can inactivate both thrombin and FXa, HCII inactivates thrombin only and binds to thrombin by a heparin bridge to exosite 2 and by its N-terminus to exosite 1 (62). Dermatin sulfate, another endogenous acidic polysaccharide with anticoagulant activity, catalyzes only HCII, not antithrombin. Physiologically, antithrombin is the more relevant serpin for regulation of intravascular thrombin, whereas HCII is thought to be more important in the extravascular inhibiton of thrombin and in the regulation of thrombin activity during pregnancy (63).

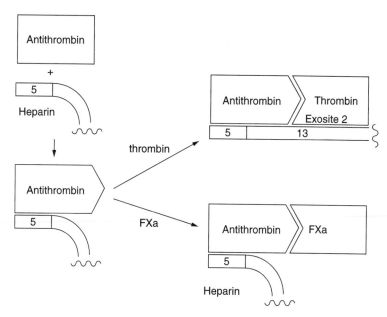

Figure 6.12. The interaction of thrombin and FXa with antithrombin and heparin. "5" represents the heparin pentasaccharide sequence that binds to antithrombin.

Thrombin initiates synthesis of the fibrin network in clots, as shown in Fig. 6.13. Fibrinogen monomer, the precursor to fibrin, can be characterized structually by two D-domains connected to a central E-domain, which contains a set of activation peptides, fibrinopeptides A and B (FPA and FPB). Thrombin binds to FPA and FPB through its exosite 1 and cleaves off these peptides, thus exposing new terminal sequences that promote self-aggregation of the resultant fibrin monomers, leading to a hydrogen-bonded polymeric soluble fibrin strand. Thrombin also generates the active transglutaminase FXIIIa from the inactive zymogen FXIII. FXIIIa stabilizes the fibrin strand by crosslinking glutamine and lysine side-chains on adjacent D domains, forming insoluble fibrin polymer. Additionally, FXIIIa covalently links the serpin α2-antiplasmin to the fibrin stands, thus providing the fibrin strand some protection from the protolytic action of plasmin, the principal fibrin-degrading enzyme. Furthermore, thrombin binds to fibrin within the clot and continues to activate factors V, VIII, XI, and XIII as well as platelets, leading to localized propagation of the thrombus at the site of injury. The crosslinked fibrin network meshes with the aggregated platelet mass to form a highly stabilized clot.

The role of platelets and their activators in clot formation is summarized in Fig. 6.14. After vascular injury, including plaque rupture, proteins such as collagen and von Willebrand factor (VWF) are exposed from the vascular subendothelium. The platelet collagen receptor (Ia) and VWF receptor (Ib) can anchor platelets to these proteins and hence to the vascular surface. Collagen can also serve to activate platelets. Platelet activation triggers conformational changes in the heterodimeric GPIIb/IIIa receptor on the platelet surface, which allows it to bind fibrinogen and VWF. Fibrinogen is a bidentate ligand with the capacity to bridge GPIIb/IIIa receptors across two platelets, thus promoting general platelet aggregation above the endothelium surface. As previously mentioned, thrombin is a powerful activator of platelets. Adenosine diphosphate (ADP), and thromboxane A_2 (TxA$_2$) also activate platelets and, along with thrombin, mediate their activation by stimulation of specific G-protein-coupled receptors, which span the platelet membrane (GPC receptors de-

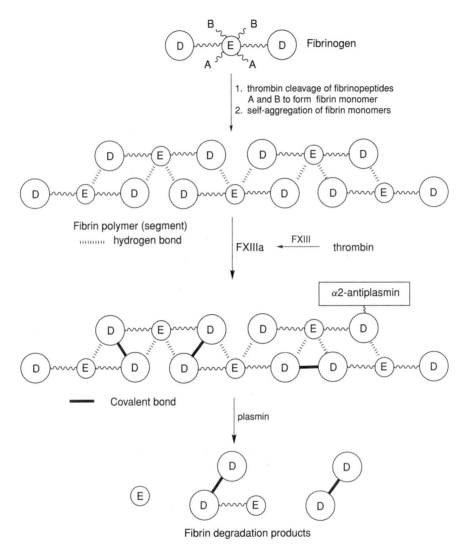

Figure 6.13. Assembly of crosslinked fibrin, mediated by thrombin and FXIIIa, and its degradation, mediated by plasmin.

noted by black squares in Fig. 6.14). Several other, albeit weaker, agonists of platelet activation are recognized [e.g., epinephrine, serotonin, and platelet-activating factor (PAF)] that also employ GPC receptors. The two GPCRs that thrombin employs for platelet activation, platelet-activating receptors 1 and 4 (PAR-1 and -4), are noteworthy because of their unusual mechanism for signal transduction (64). In PAR-1, for example, the N-terminal extracellular domain incorporates an acidic peptide sequence that binds thrombin

by way of exosite 1 (Fig. 6.15). The thrombin active site then cleaves the sequence at Arg-Ser, exposing a new N-terminal domain sequence (SFLLRN) that folds back on the GPCR extracellular loop 2, acting as a tethered agonist, thereby activating its receptor. PAR-4 functions in a similar manner using a different unmasked N-terminal sequence (GYPGQV) and appears to be less sensitive to thrombin, possibly because of the absence of a thrombin exosite-1 binding sequence on PAR-4. Recent studies are consistent with the

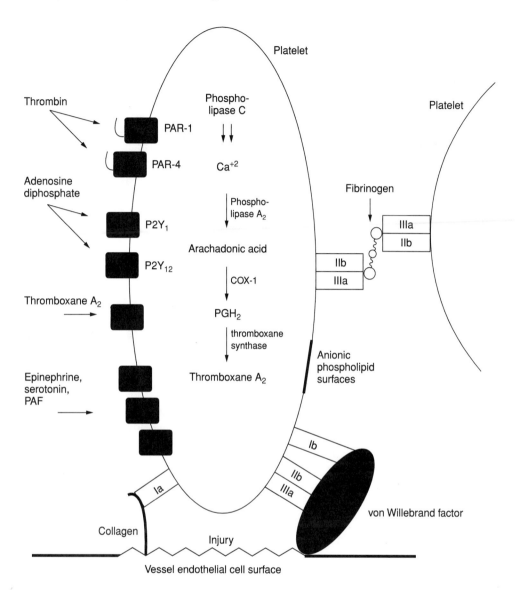

Figure 6.14. Platelets and their activators and receptors.

interpretation that low levels of thrombin activate PAR-1 to provide an immediate state of platelet activation, whereas higher levels of thombin, typically generated during the later stages of clot formation, activate PAR-4 and act to sustain the late phase of the platelet-aggregation process (65).

Depending on the mode of agonism, various chemical pathways leading to platelet activation can be triggered. In a major pathway, phospholipase C within platelets is activated, which then leads to an increase in intracellular calcium concentrations. Increased intracellular calcium then stimulates phospholipase A_2, liberating arachadonic acid from membrane phospholipids. Cyclooxygenase-1 (COX-1) converts arachadonic acid to PGH_2, which is converted by thromboxane synthase to thromboxane A_2, a potent platelet agonist, which is secreted and leads to further activation through its own receptor. Two ADP receptors on the platelet surface modulate plate-

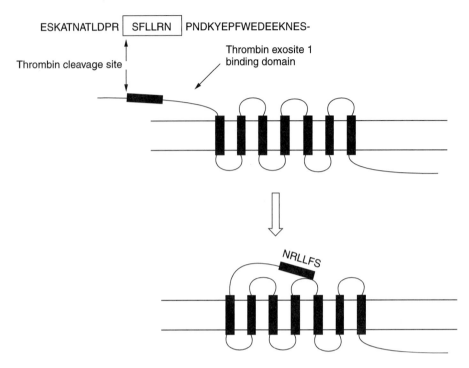

Figure 6.15. The activation of PAR-1 by thrombin.

let activation: $P2Y_1$ and $P2Y_{12}$ ($P2Y_{12}$ was formerly called $P2Y_{AC}$, $P2T_{AC}$, or $P2_{cyc}$). Stimulation of both is required for ADP-dependent platelet activation (66). Stimulation of $P2Y_1$ activates phospholipase C, whereas stimulation of $P2Y_{12}$ downregulates adenylate cyclase activity, leading to lower levels of cAMP, an inhibitor of platelet activation (see below). At the cellular level, stimulation of $P2Y_1$ generates an initial transient aggregation response, whereas agonism of $P2Y_{12}$ generates sustained aggregation (67).

Activation of platelets typically leads to a shape change wherein the platelets lose their usual discoid shape and take on an irregular appearance. Activated platelets express anionic phospholipid surfaces and also receptors for FVa on their outer membrane, which accommodate formation of the intrinsic tenase and prothrombinase complexes, leading to additional thrombin generation. Granules within activated platelets release their contents, among which are serotonin, ADP, fibrinogen, FV, and VWF, which serve to amplify platelet stimulation and clot formation. It has been estimated that the activation of

platelets accelerates thrombin generation by 5–6 orders of magnitude, greatly facilitating fibrin formation. The significant role of platelets in thrombin generation is suggested by the observations that fibrin deposition at sites of vascular injury occurs after platelet adhesion and aggregation, and that fibrin forms in close proximity to the deposited platelets (68).

Platelet activation is inhibited by intracellular cAMP, which initiates a sequence of enzymatic steps leading to a reduction in platelet calcium concentration. Adenylate cyclase, which converts ATP to cAMP, is stimulated by prostaglandin D_2, another product enzymatically generated from arachadonic acid within the platelet.

cGMP inhibits a major cAMP phosphodiesterase that hydrolyzes cAMP, and thus acts indirectly as an inhibitor of platelet activation. Endothelial cells can participate in deactivating platelets by releasing PGI_2 (prostaglandin I_2; prostacyclin), which stimulates platelet adenylate cyclase to generate cAMP, and nitrous oxide (NO), which stimulates platelet guanylate cyclase to generate antiaggregatory cGMP.

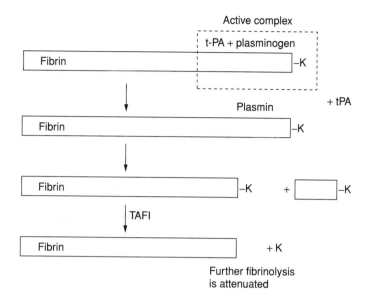

Figure 6.16. TAFI attenuates the acceleration of fibrinolysis mediated by C-terminal lysine residues.

The process of clot dissolution involves plasmin, a trypsinlike serine protease, which proteolytically degrades fibrin, generating smaller fragments (FDPs, fibrin degradation products) such as D-dimer (Fig. 6.13). Plasmin is protolytically activated by tPA, another serine protease, from its inactive zymogen, plasminogen. This activation step can occur in solution but is accelerated more than 100-fold when both plasminogen and tPA bind in a co-ordinated manner on the surface of fibrin (Fig. 6.16). Thus, fibrin can be viewed as a cofactor to plasminogen activation (60). Whereas solution-phase plasmin is inactivated almost instantaneously by α_2-antiplasmin, fibrin-bound plasmin is partially protected from in-activation. Plasminogen and plasmin contain sites within their structures that have affinity for lysine residues, and it is these lysine bind-ing sites (LBS) that facilitate their binding to the surface of fibrin. Also, the LBS mediates initial α2-antiplasmin binding to plasmin. Thus both fibrin (substrate) and α2-antiplas-min (inactivator) compete for the same LBS in plasmin. Simple lysine analogs, such as 6-ami-nocaproic acid are known to inhibit plasmino-gen/plasmin binding to fibrin, and even have been used therapeutically to treat individuals with hemostatic disorders, such as hemophili-acs, although with variable outcomes (69). tPA contains a domain that has affinity for lysine residues, which may mediate its binding to fibrin, and also has a domain that has affin-ity for plasminogen. Details regarding the ex-act nature of the binding and interaction of tPA and plasminogen on the fibrin surface are still being studied.

During the proteolytic degradation of fibrin by plasmin, C-terminal lysine residues are ex-posed on fibrin that induce an additional 2.5-fold rate acceleration of tPA-mediated activa-tion of plasminogen compared to that of uncleaved fibrin (Fig. 6.16). TAFIa (the active form of TAFI) in a negative regulatory man-ner attenuates this rate acceleration by cleav-ing off those C-terminal lysine residues. Con-sequently, TAFIa effectively functions as an antifibrinolytic agent by this unique mecha-nism and promotes persistence of formed fi-brin clots (60, 70). Another negative regula-tory mechanism controlling fibrinolysis is the inactivation of tPA by the serpin PAI-1 (plas-minogen activator inhibitor 1). Recent studies with PAI-1 and α1-antitrypsin have revealed the detailed mechanism by which these, and presumably all, serpins inactivate their target proteases (Fig. 6.17). In this mechanistic de-scription, the tPA active site serine cleaves the PAI-1 reactive center loop between the P1 Arg and P1′ methionine, forming a covalent ester bond with the PAI-1 Arg carbonyl. Normally, within the context of standard substrate turn-

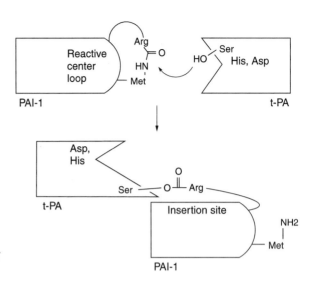

Figure 6.17. Mechanism of inactivation of tPA by the serpin PAI-1.

over by serine proteases, such acylated serine hydroxy groups are readily hydrolyzed in a process mediated by the adjacent His-Asp residues of the catalytic triad. However, acylation by the serpin triggers a dramatic translocation of the protease to the opposite side of the serpin, with the N-terminal segment of the reactive center loop fitting as a strand into an insertion site on the serpin. As a consequence of this serpin-protease alignment, the protease active site is severely deformed, resulting in complete disconnection of the His-Asp away from from the active site (acylated) serine. Thus, normal turnover cannot take place because the catalytic triad is disrupted and the serpin-protease complex is therefore stabilized (71).

Urokinase (uPA) like tPa has the ability to convert plasminogen to plasmin, although unlike tPA, uPA has no affinity for fibrin. Physiologically, uPA mainly plays a role in extracellular matrix degradation (72).

3.2 Mechanisms and Sites of Action of the Classes of Marketed Antithrombotic and Antiplatelet Agents

3.2.1 Heparin and Other Anionic Polysaccharides. As described in Section 3.1, endogenous heparin acts as an antithrombotic by activating the serpin antithrombin (AT), which covalently inactivates both thrombin and FXa by binding to their active sites. To a lesser extent, heparin can also facilitate antithrombin-mediated inhibition of FIXa and FXIa. Pharmaceutical heparin [also called unfractionated heparin (UFH)] is isolated from porcine intestinal mucosa. In addition to its limitations arising from pharmacokinetic variability, potential complications from thrombocytopenia, and the requirement that it be administered IV, heparin is not effective at inhibiting fibrin-bound thrombin, and recent investigations have provided a likely explanation (73). Thrombin binds to fibrin at its exosite 1, and heparin can form a tight bridge from the fibrin surface to exosite 2 of thrombin (Fig. 6.18). This tight ternary complex does not allow the AT-heparin complex access to exosite 2 of thrombin, and therefore inhibition of thrombin cannot proceed. UFH also cannot inactivate FXa in the prothrombinase complex, and for these reasons UHF is not viewed as the ideal agent to prevent clot expansion and propagation, which may explain why hirudin (lepirudin) shows superior efficacy to that of heparin in patients with unstable angina or non-Q wave myocardial infarction (9b). Additionally, it has been noted that upon cessation of heparin treatment, a clustering of thrombotic events occurs in patients (9b, 25). This "heparin rebound" may be partially explained by the ability of heparin to displace and mobilize endogenous TFPI from the vessel wall. It is believed that about one-third of the anti-

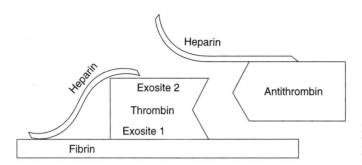

Figure 6.18. Heparin-antithrombin does not inhibit fibrin-bound thrombin.

thrombotic effect of heparin may be attributed to mobilized TFPI. At therapeutic doses, UFH progressively depletes TFPI and when heparin treatment is abruptly halted, the relative lack of available TFPI can result in a transient prothrombotic state (74).

The low molecular weight heparins (LMWH) are prepared by chemical or enzymatic modification of natural heparin and are about one-third the length of heparin. Each of the marketed LMWHs is slightly different structurally, depending on the manner in which it was prepared, and consequently the clinical profiles of each may differ (75). Because of their shorter average length, the LMWHs contain fewer chains of the length required to inhibit thrombin (>17 saccharides, see Fig. 6.12) than does heparin. Consequently, whereas heparin inhibits thrombin and FXa in about a 1:1 ratio by the antithrombin-dependent mechanism, the LMWHs inhibit FXa in ratios of about 2:1 to 4:1 compared to that of thrombin. Additionally, it has been reported that at least one LMWH, enoxaparin, can effectively inhibit platelet prothrombinase activity and the resulting thrombin generation. However, others report that LMWHs and even fondaparinux, the FXa-specific pentasaccharide, are ineffective at inhibiting FXa activity in the prothrombinase complex (76). Also,

LMWHs have been shown to be less efficient than heparin at mobilizing endothelial TFPI from the endothelium (77). As a consequence of their shorter average length, LMWHs appear to offer advantages over heparin in reduced incidences of heparin-induced thrombocytopenia (78).

Despite all of these differences, the principal clinical differentiation between UFH and the LMWHs remains the improved pharmacokinetic profile displayed by LMWHs (25). Danaparoid is a mixture mainly of the anionic polysaccharides heparan sulfate (~84%) and dermatin sulfate (~12%). Danaparoid has a higher ratio of anti-FXa to antithrombin activity (>22-fold), likely attributable to the selectivity of heparan sulfate toward AT-mediated FXa inactivation (79). The dermatin sulfate component can inactivate thrombin through the serpin HCII. Research is continuing in the field of polysulfated polysaccharide antithrombotics to discover new agents with benefits in safety (less bleeding and thrombocytopenia), pharmacokinetics, and utility (80). Some of this research is directed toward further incremental modifications of LMWHs. However, in a somewhat different approach, it was recognized that for selective inhibition of FXa, the heparin AT-binding pentasaccharide (Fig. 6.1a) is the minimally required struc-

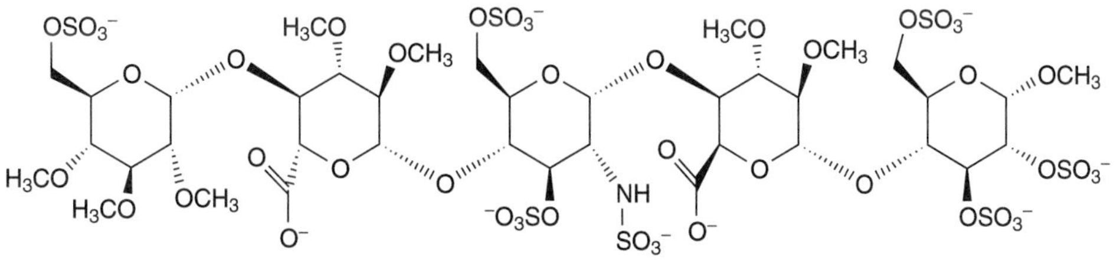

Figure 6.19. Idraparinux.

tural subunit. Such FXa-selective pentasaccharides were synthesized and a few are under clinical investigation. For example, fondaparinux (Arixtra, Fig. 6.1b) and idraparinux (SanOrg 34006, Fig. 6.19), both based on the specific heparin pentasaccharide, are indeed highly selective FXa inhibitors and have shown antithrombotic efficacy in clinical studies (81a). Arixtra was launched in the United States in 2002 for the prophylaxis of DVT in patients undergoing hip and knee surgery. Whereas the pharmacokinetics of fondaparinux allows once-a-day dosing, idraparinux can be administered once a week. Idraparinux binds to AT with a 10-fold greater affinity compared to that of fondaparinux, and it has been suggested that this greater affinity accounts for its longer plasma elimination half-life and resultant longer pharmacodynamic effect (81b,c).

3.2.2 Warfarin and Other Vitamin K-dependent Inhibitors. Warfarin's antithrombotic effect is a consequence of its inhibition of the vitamin K-dependent posttranslational γ-carboxylation of glutamic acid residues in the procoagulant zymogens FVII, FIX, FX, and prothrombin (82). As discussed previously, for these proteins a specific domain containing 10–13 γ-carboxyglutamic acid residues is an essential requirement for permitting their binding to phospholipid surfaces and hence proper assembly into the active tenase and prothrombinase complexes. Lacking Gla domains, FVIIa, FIXa, FXa, and thrombin are physiologically extremely poor procoagulants.

Warfarin and its analogs (e.g., phenprocoumon, acenocoumarol, and dicumarol) inhibit vitamin K epoxide reductase, an enzyme essential for the reductive recycling of vitamin K epoxide to vitamin K hydroquinone (Fig. 6.20). Vitamin K hydroquinone is the cofactor to γ-glutamylcarboxylase, which affects the actual carboxylation of Glu residues. The activity of the anticoagulant Gla-domain-containing proteins, protein C and S, is also inhibited by warfarin, affording a procoagulant activity. Still, the major effect achieved is anticoagulation, primarily because of inhibition of thrombin. The pharmacokinetic and safety liabilities of warfarin have already been discussed (Sections 2.2.1 and 2.3). Dicumerol,

phenprocoumon, and acenocoumarol have also been investigated as anticoagulants in humans, although no significant advantages were found. Some recent attempts have been made to prepare warfarin analogs with improved physical properties (e.g., reduced protein binding), which may have a safety benefit (83). However, little current research is directed toward the discovery of new analogs of warfarin or other indirect or direct inhibitors of Gla biosynthesis.

3.2.3 Direct Thrombin Inhibitors: A Historical Perspective, From Concept to Drug. The direct and selective active-site inhibition of thrombin, especially by small molecules that have potential for oral bioavailablity, has long been seen as an attractive opportunity for the development of therapeutically useful anticoagulant agents. Theoretically, many of the disadvantages of heparin, the LMWHs, and warfarin would be avoided. For example, direct thrombin inhibitors could inactivate fibrin-bound thrombin and could also avoid HIT, both disadvantages of heparin, and also could avoid the unfavorable mechanism-based pharmacodynamics of warfarin. On the other hand, there has been some recent concern that selective, reversible inhibitors of thrombin could have the potential for a so-called thrombin rebound effect after cessation of dosing, especially if the inhibitors are cleared quickly from the plasma. According to this view, after cessation of treatment with a reversible thrombin inhibitor, the inhibitor-thrombin complex dissociates, the drug is eliminated from the vasculature, and the resultant pool of reactivated thrombin could, in theory, trigger clinical thrombotic events. Moreover, reversible inhibition of thrombin during treatment would not be expected to impede the residual *generation* of active thrombin resulting from the ongoing activation of the coagulation pathway. This newly generated thrombin could add to the pool of reactivated thrombin once the inhibitor is cleared from the blood. Clinically, there has in fact been documentation of increases in thrombotic event rates after cessation of treatment with argatroban and inogatran, although these events could not be decisively correlated to a thrombin rebound phenomenon (85). Also, a trial that re-

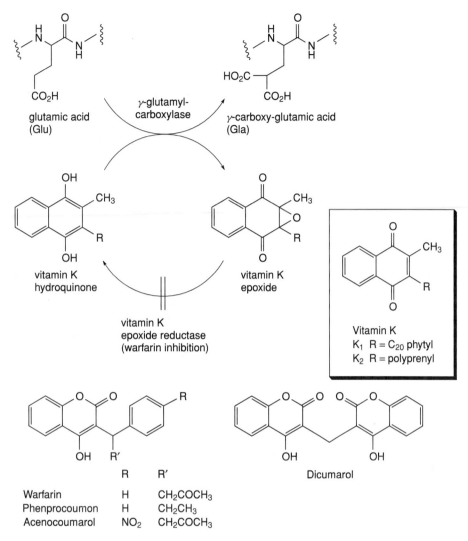

Figure 6.20. Vitamin K–mediated γ-carboxylation of Glu residues in coagulation serine proteases, and several inhibitors of this process.

corded an increased level of thrombin generation during infusion of napsagatran compared to infusion of heparin was not followed by any rebound in thrombin activity after cessation of infusion (86). At present, there does not seem to be a general problem with a rebound phenomenon for selective reversible thrombin inhibitors, although the clinical use of such agents is still relatively limited.

Over the last 20 years, a vast amount of effort has been directed toward the discovery and development of direct thrombin inhibitors, both small-molecule and hirudin-like. X-

ray crystal structures of thrombin complexed with active-site inhibitors have had a major impact in the design of a diversity of highly potent and selective active site-directed agents.

3.2.3.1 Small-Molecule Direct Thrombin Inhibitors. Early insights leading to the discovery of the first synthetic thrombin active-site inhibitors resulted from consideration of the structures of the endogenous substrates of thrombin. Nearly all of the protein substrates for thrombin contain an Arg as the P1 residue, with the P2 residue often being a Pro (Fig. 6.21a). In 1978 Bajusz reported a series of tri-

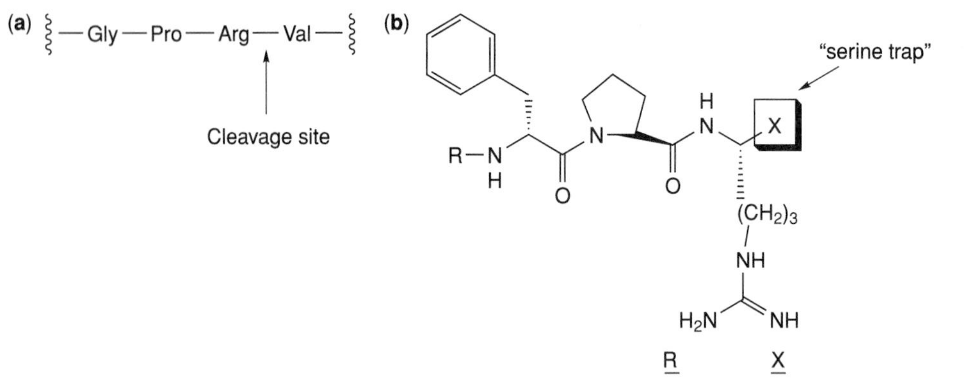

	R	X
Bajusz inhibitor	H	–CHO
PPACK	H	–COCH$_2$Cl
Efegatran	–CH$_3$	–CHO
DuP-714	–COCH$_3$	–B(OH)$_2$

Figure 6.21. (a) A thrombin cleavage site (fibrinogen A); (b) several covalent thrombin inhibitors.

peptide aldehydes modeled after the fibrinogen A peptide cleavage site and one of the more potent compounds, based on clotting times, was the tripeptide aldehyde D-Phe-Pro-Arg-CHO (Fig. 6.21b) (87). At that time, and for many years thereafter, the paradigm of joining a tripeptide-like structure to an electrophilic moiety (an aldehyde, in Bajusz's prototype) led to the preparation of a great diversity of extremely potent thrombin active-site inhibitors, such as PPACK (D-PheProArg-chloromethylketone), efegatran, and DuP-714, to name just a few (Fig. 6.21b). K_i values for many of the inhibitors in this class are in the picomolar range (a). The electrophilic moiety essentially acts as a "serine trap" and accounts for much of the potency of these inhibitors. Other examples of such traps include trifluoromethylketones, α-keto-esters, α-keto-amides, α-keto-heterocycles, and boronic acids, among others.

For endogenous substrates, the catalytic triad serine of thrombin cleaves the arginine amide bond in a manner typical of other serine proteases (88). During this cleavage, an anionic tetrahedral transition state is formed by attack of Ser-195 at the Arg carbonyl, forming a transient covalent bond. The remainder of the substrate is stabilized by a number of other interactions: hydrophobic contacts with the enzyme S2, S3, and S4 pockets; a salt-bridge interaction between the substrate arginine and Arg 189 in the S1 pocket; antiparallel

beta-sheet interactions variously with Ser-214, Trp 215, and Gly 216; and stabilization of the anionic charge (former Arg carbonyl) with H-bond interactions within the so-called oxyanion hole, defined by the Gly-193 and Ser-195 NHs (Fig. 6.22a). The processed protein is then released after normal catalytic deacylation of the arginine-serine ester bond. In 1989 the published X-ray crystal structure of thrombin inhibited by PPACK revealed several critical molecular interactions (Fig. 6.22b) (89). Both Ser-195 and His-57 were covalently attached to the former chloroketone trap, the Pro and D-Phe residues occupied the S2 and S4 pockets, respectively, and several antiparallel beta-sheet hydrogen bond interactions could be observed. In the case of electrophilic carbonyls such as aldehydes, and activated ketones, the analogous hemiacetal or hemiketal bond is formed with Ser 189, whereas in the case of boronic acid-based inhibitors, a boronate ester is formed. As mentioned, many inhibitors in this class display high potency, largely attributable to the effect of forming a covalent bond at the active site of thombin. It came to be appreciated, however, that the serine trap concept had a number of potential drawbacks. For example, these inhibitors typically exhibit slow binding kinetics that may not be sufficiently rapid to achieve desired efficacy. After activation of the coagulation pathway, thrombin is generated in a

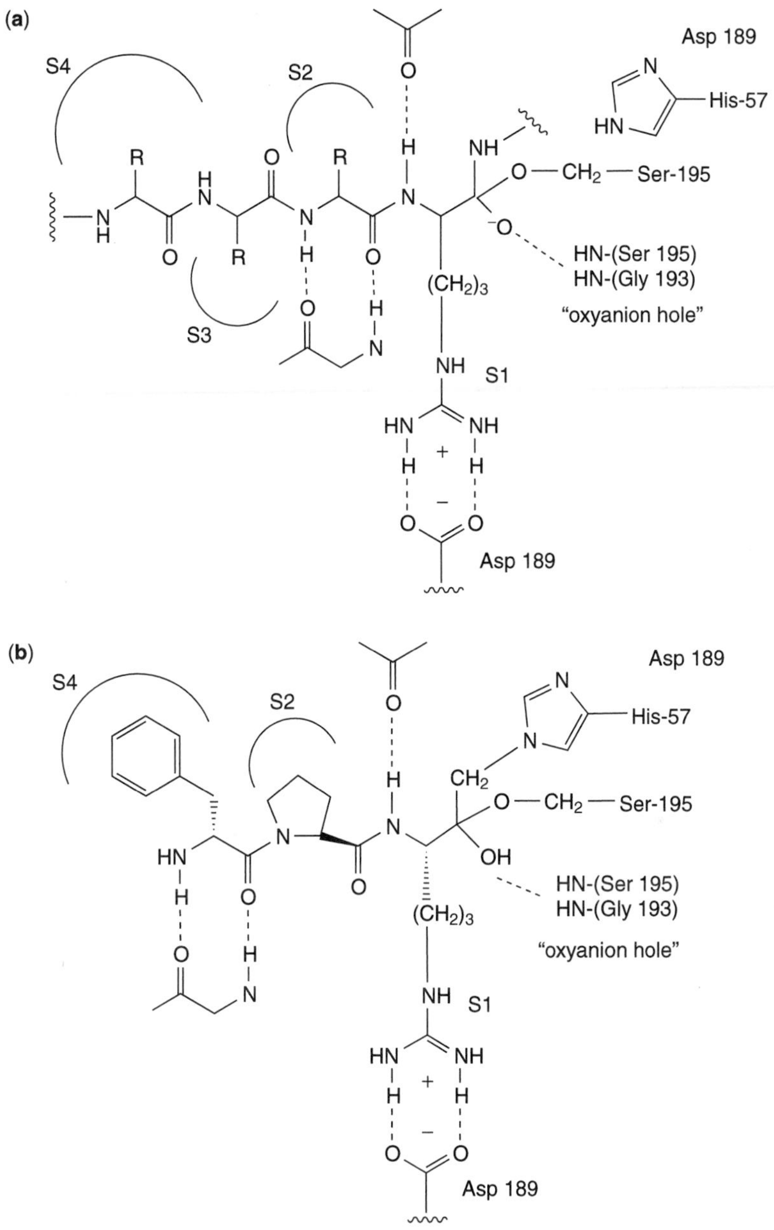

Figure 6.22. (a) Typical interactions between a coagulation serine protease and its substrate during cleavage; (b) covalent inhibition of thrombin by PPACK.

rapid "burst." Studies both *in vitro* and *in vivo* have shown that slow binding inhibitors are less efficacious than fast binding inhibitors of comparable potency. Further, the reactive functionality in this class is viewed as a metabolic liability and the potential for nonspecific

covalent bond formation *in vivo* might lead to immunological reactions (through formation of a hapten) or other undesired side effects (84a, 90).

In parallel to the serine trap approach, other groups were synthesizing potent throm-

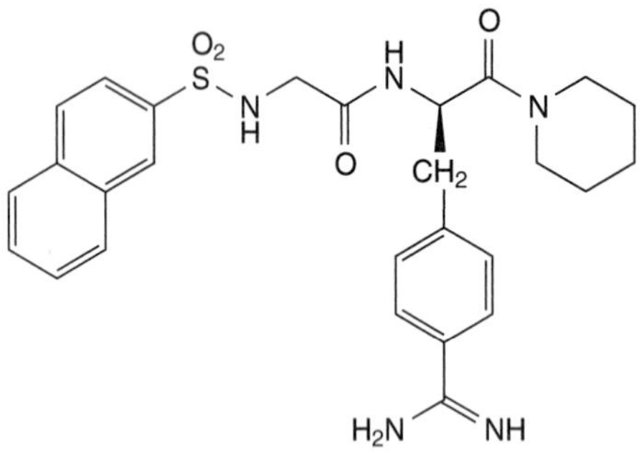

(**12**) NAPAP

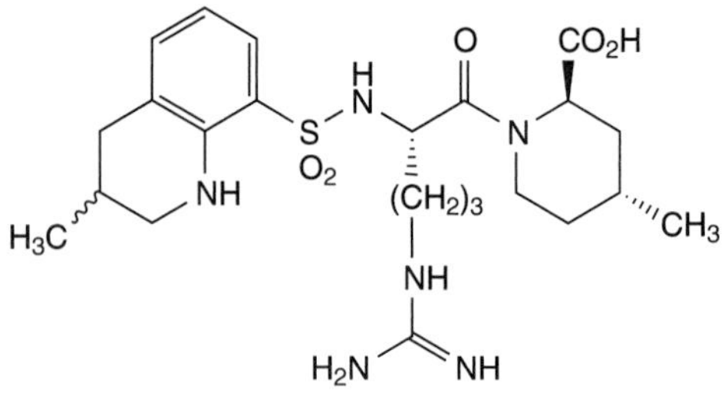

(**5**) argatroban

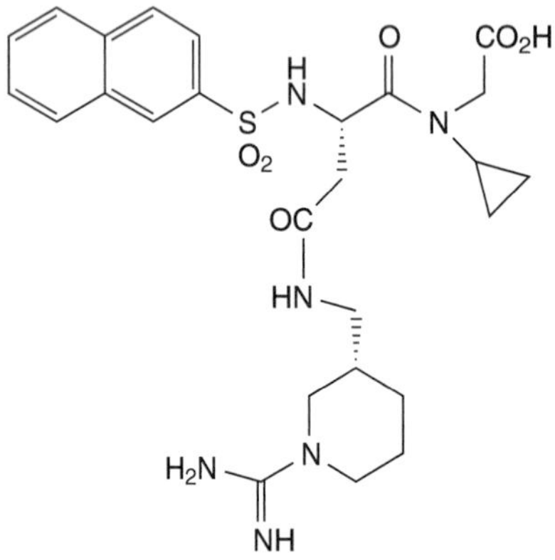

(**13**) napsagatran

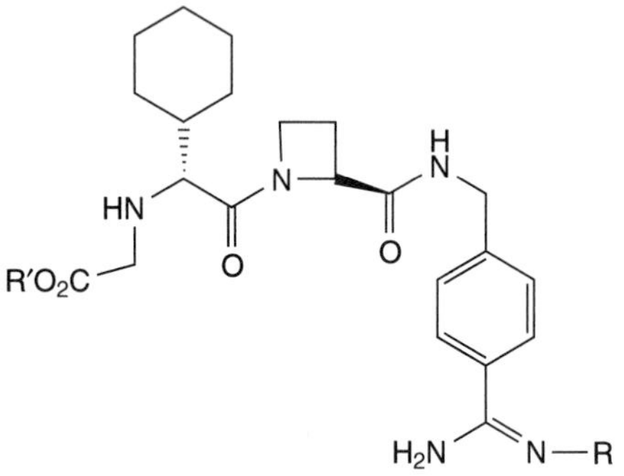

	R	R′
(**14**) melagatran	H	H
(**15**) ximelagatran	-OH	$-CH_2CH_3$

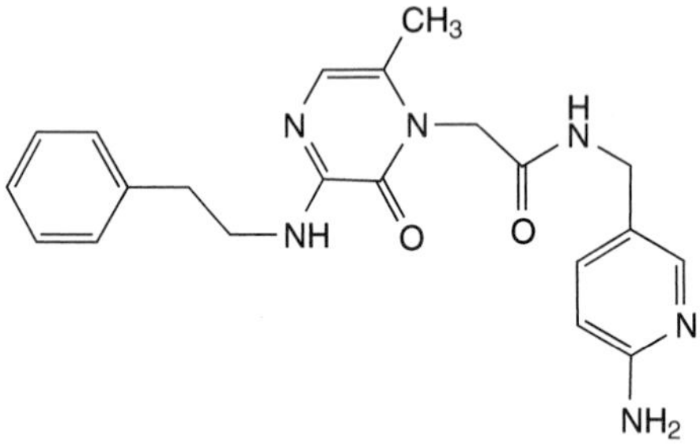

(**16**) L-375,378

Figure 6.23. Examples of reversible active site thrombin inhibitors.

bin inhibitors that did not rely on forming a covalent bond at the thrombin active site. Two early examples, first reported in the early 1980s were NAPAP (naphthylsulfonyl-glycyl-4-AmidinoPhenylAlaninePiperidide, K_i = 6 n*M*; Fig. 6.23, structure **12**) and MD-805, later named argatroban (K_i = 8 n*M*) (91). The design of both of these inhibitors evolved from an earlier prototype, *N*-tosyl-arginine methyl ester (TAME). In the case of argatroban, the methyl ester of TAME was replaced by an amide, the sulfonyl group was optimized, and a carboxylic acid group was appended to solve toxicity problems. X-ray structures for NAPAP and argatroban bound to the active site of thrombin were published in 1991 and 1992 by two different groups (92). The crystal structure of argatroban showed that the arginyl side-chain entered the S1 pocket at an angle different from that of the arginyl chain of the serine trap inhibitor PPACK and consequently only one of the NHs of the guanidine formed an ionic bond to Asp-189. The tetrahydroquinoline inserted into the S4 pocket, with densities for both methyl isomers making acceptable lipophilic interactions. A section of the piperidine ring along with the appended 4-R methyl group inserts tightly into the S2 pocket, and the carboxylate points toward the oxyanion hole, forming a hydrogen bond with Ser-195. Before the X-ray, it was assumed that the piperidine was occupying the S1' site. The stereochemistries of both piperidine groups, carboxylate ($2R$) and methyl ($4R$), were seen from the X-ray structure to be critical and the other possible stereochemical combinations were predicted not to be as well tolerated, which was in agreement with experimental results. The crystal structure of NAPAP shows a generally similar binding motif compared to that of argatroban, in terms of placement of the major residues in the enzyme S pockets, in spite of the fact that the benzamidine and alkylguanidine for the two inhibitors are attached with different stereochemistries.

With a K_i value of 8 nM, argatroban not only is a potent thrombin inhibitor, but it is much less potent against a panel of other coagulation and fibrinolytic enzymes (93). Selectivity against the fibrinolytic enzymes was seen as especially key, in that potent inhibition of plasmin or tPA would essentially lead to a prothrombotic condition. Argatroban is also fairly selective against trypsin (1000-fold), although high selectivity against this digestive enzyme may not necessarily be required for an intravenously dosed agent. (Because of its low oral bioavailability, argatroban is dosed by IV infusion.) Argatroban binds rapidly and reversibly to both fibrin-bound (clot-bound) and soluble thrombin (94). Moreover, it does not induce thrombocytopenia nor does it interact with the antibody that causes HIT (95). Argatroban, originally discovered and developed by Mitsubishi, was approved in Japan in 1990 for treatment of arterial thrombosis and, in 1996, for treatment of acute cerebral thrombosis. In the United States it was approved in 2001 for the treatment of patients with HIT and HIT with

thrombosis (HITTS) and is comarketed by GlaxoSmithKline and Texas Biotechnology. Reviews on argatroban were recently published (96).

Over the years, many other potent reversible active-site thrombin inhibitors were prepared, inspired by the D-Phe-Pro-Arg-like structures of argatroban and NAPAP. Napsagatran (13) was reported in 1994 and is a very potent inhibitor of thrombin (K_i = 0.3 nM) with good selectivity against the fibrinolytic enzymes (97); its development was discontinued, however. The D-Phe-Pro-Arg mimetic, melegatran (K_i = 2 nM; Fig. 6.23, structure 14), is currently in clinical development by AstraZeneca for patients with DVT and for the prevention of stroke in patients with atrial fibrillation. Its poor oral bioavailablity (6%) and potency against trypsin (K_i = 4 nM) necessitates that it be dosed parenterally (98). However, for oral administration, a double prodrug form (ximelagatran, Exanta; 15) is also being developed, wherein the benzamidine moiety is modified by hydroxylation and the carboxylate is the ethyl ester. The bioavailabilty of ximelagatran in humans is moderate (18–24%), although it is rapidly absorbed and metabolized to melagatran (99).

Achieving good oral bioavailability and/or good plasma half-life within the early classes of active-site thrombin inhibitors was frustrated by the peptidic-like nature of the structures and also by the presence of the highly charged guanidine or benzamidine moieties. Further, the requirement for good selectivity against trypsin was also a frequent problem for the development of an oral inhibitor. Over the last decade much research has been directed toward solving these two problems. Today, a number of less polar surrogates for the Arg-like side-chain have been identified and successfully incorporated into nonpeptidic templates that afford very potent active-site thrombin inibitors. Furthermore, by exploiting observed structure-activity relationship (SAR) trends for activities of these thrombin inhibitors against other enzymes such as trypsin and the fibrinolytic enzymes, inhibitors having very high selectivity for thrombin could be identified. X-ray crystallography also aided in the design of more selective inhibitors by revealing differences in the conformations and interactions of inhibitors bound to thrombin compared to other enzymes (especially trypsin). Today, there are numerous examples of nonamidine orally bioavailable thrombin inhibitors having excellent selectivity against trypsin and other serine proteases. As just one example, investigators at Merck have optimized a series of nonamidine pyridinone template-based thrombin inhibitors to provide the pyrazinone L-375,378 (**16**), having a K_i value of 0.8 nM (100). The X-ray crystal structure for this compound shows the aminopyridine occupying S1, with the amino group interacting, through an ordered water molecule, with Asp-189 and also interacting with the carbonyl of Gly-216. The pyridine 6-methyl group makes a productive lipophilic interaction with Val-213 within S1. The pyrazinone methyl occupies S2, whereas the phenyl occupies S4. L-375,378 is selective for thrombin compared to trypsin (2000-fold selectivity) and other serine proteases, including tPA and plasmin (>100,000-fold) and is 90% orally bioavailable in dogs with a half-life of 231 min; in rhesus monkeys it is 60% orally bioavailable.

Thus, it appears that the long-sought goal of identifying orally bioavailable and selective active-site thrombin inhibitors has been achieved. The subject of small-molecule active-site thrombin inhibitors has been extensively reviewed (90, 101).

3.2.3.2 Hirudin and Hirudin-Like Thrombin Inhibitors.

More than a century ago, the anti-coagulating substance hirudin was extracted from the leech *Hirudo medicinalis* (102). Hirudin is a family of more than 20 related 65–66 amino acid peptides containing three disulfide bridges and an *O*-sulfated Tyr near the carboxylate terminus (103). In 1986 the first reports on the preparation of recombinant desulfated hirudins appeared, which allowed the study of single hirudin variants, especially useful for crystallography purposes. Hirudins lacking the sulfate on the C-terminal Tyr have about 10 times reduced activity, but still potently and specifically inhibit thrombin in the subpicomolar range. The origin of this high potency can be explained in the bivalent manner in which the hirudins bind to thrombin, which was first revealed by X-ray crystallography of two recombinant hirudin variants reported in the early 1990s (104). The binding

of recombinant hirudin variant 1 (rHV-1, desirudin, **3b**) to thrombin is representative of the class and is shown in Figure 6.24a. The polyanionic C-terminus tightly binds to thrombin exosite 1 [fibrinogen binding domain (FBD)], whereas the *N*-terminus simultaneously occupies the thrombin active-site region. At the *N*-terminus, the Val-1 and Tyr-3 side-chains occupy roughly the thrombin S2 and S3 sites, making numerous hydrophobic contacts, whereas the terminal amino group makes hydrogen bonds to His-57 and Ser-214. Additionally, because of the manner of insertion of the *N*-terminal hirudin peptide along the active site groove, it forms a short *parallel* set of hydrogen bond contacts to the thrombin backbone (Gly 216 to Gly 218), which is opposite to that seen in the antiparallel interactions of substrates and most inhibitors. The S1 pocket is not occupied by hirudin. Much SAR data have been generated for hirudin involving single and multiple amino acid substitutions as well as other modifications (105). Desirudin (**3b**, Fig. 6.3) is marketed in Europe for the prevention of DVT in hip and knee replacement surgery (106). Lepirudin (**3a**, Refludan; U.S. launch 1998) is structurally similar to desirudin but has Leu-Thr at the first two *N*-terminal positions instead of Val-Val (107). Lepirudin is used as a replacement for heparin in HIT patients.

Even before the crystallographic details of hirudin's binding to thrombin were known, major structural modifications to hirudin were carried out, resulting in the "hirulogs" and the "hirugens" (108). The hirugens are peptide fragments of hirudin containing only the C-terminal FBD and that inhibit thrombin in the low to submicromolar range. The hirulogs are peptide analogs of hirudin, wherein most of the nonbinding peptide core sequence is excised and an active site binding sequence is appended by way of a poly-Gly linker to the FBD, thereby creating essentially a hirugen with an active site binding sequence. Bivalirudin (**4**, hirulog-1; Angiomax) is one such example and contains the familiar D-Phe-Pro-Arg active-site sequence characteristic of the early small-molecule thrombin inhibitors (45, 46, 109). The binding mode of bivalirudin (and other hirulogs) is different from that of hirudin, in that the bivalirudin peptide sequence

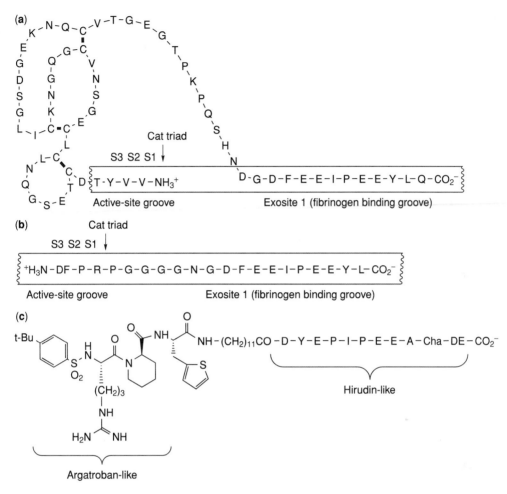

Figure 6.24. (a) Hirudin variant 1 (desirudin) bound to the active site and exosite 1 of thrombin; (b) binding mode of bivalirudin to thrombin; (c) thrombin inhibitor combining structural elements from argatroban and the hirudin C-terminus (DF, D-Phe; DE, D-Glu; Cha, cyclohexyl-Ala).

binds continuously along the FBD and active site grooves, with the active site sequence now making the usual antiparallel contacts with the enzyme (108). One consequence of this inhibitor structure and its binding mode is that the Arg peptide bond to the P1′ Pro is slowly cleaved by the enzyme, affording a less potent inhibitor. Bivalirudin itself has a K_i value of 1.9 nM and upon IV infusion has shown efficacy similar to that of heparin in preventing ischemic complication in patients with unstable angina who underwent angioplasty. Even though both lepirudin and bivalirudin require binding to the thrombin exosite 1 as part of

their mechanism of action, each of these agents has been shown to be active against fibrin-bound thrombin.

Analogs of bivalirudin incorporating different active-site binding domains have been synthesized, with the goal being to stabilize the scissile bond and to increase binding potency at the N-terminus (101a, 105a). One such analog (Fig. 6.24c) contains an argatroban-like active-site binding structure and inhibits thrombin selectively with a K_i value of 0.17 pM (110). However, unlike the small-molecule direct thrombin inhibitors, none of these hirudin-like inhibitors is likely to have substantial oral bioavailability, and currently none of these newer inhibitors is being evaluated in the clinic.

3.2.4 Platelet GPIIb/IIIa Antagonists.

After activation, platelets aggregate by means of their GPIIb/IIIa receptors, binding to bidentate fibrinogen, thus allowing formation of a three-dimensional platelet thrombus. Is has been recognized that, whereas platelet activation is initiated by a number of stimuli (ADP, thrombin, etc.), fibrinogen binding to the GPIIb/IIIa receptor represents the final common step to platelet aggregation. Therefore, by targeting the blockade of this interaction, platelet aggregation should be inhibited, regardless of the source of platelet activation (111). GPIIb/IIIa is a member of a larger family of integrin receptors and it is also referred to as $\alpha_{\text{IIb}}\beta_3$ (integrin nomenclature). GPIIb/IIIa receptors recognize and bind to the tripeptide sequence RGD (Arg-Gly-Asp) of fibrinogen. Fibrinogen has this RGD sequence located at each terminus of its α-chain, thus allowing the bidentate interaction that results in crosslinked platelets (Fig. 6.14). Additionally, evidence has accumulated that fibrinogen can bind to GPIIb/IIIa independently of its RGD sequence through an unrelated dodecapeptide sequence (HHLGGAKQAGDV) located at each terminus of its γ-chain. Although it has long been known that RGD peptides (and RGD mimetics; see below) bind to GPIIb/IIIa and effectively inhibit platelet aggregation, the isolated fibrinogen dodecapeptide can independently bind to the receptor at a location distinct from the RGD binding site and can also inhibit platelet aggregation (112).

Abciximab (c7E3, Reopro) is the Fab fragment of a mouse human chimeric antibody to the GPIIb/IIIa receptor and binds tightly and essentially irreversibly, resulting in potent inhibition of aggregation of activated platelets (111). Interestingly, in spite of this tight binding, it is thought that abciximab continually redistributes from one platetet to another and therefore its effect can persist longer than the 8-day lifetime of an individual platelet. Abciximab also binds to the platelet vitronectin ($\alpha_v\beta_3$) receptor, although the clinical significance of this lack of selectivity has not yet been established. However, it has been shown that blocking both receptors provides an additive effect in the inhibition of platelet-mediated thrombin generation and abciximab achieves a dose-dependent reduction in thrombin generation to a maximum of 45–50% inhibition. Presumably, the decrease in thrombin generation is a consequence, at least in part, from the resultant absence of a concentrated platelet mass and the attendant dilution of soluble activating stimuli. This reduction in platelet-mediated thrombin generation is believed to contribute to the clinical efficacy of abciximab (68, 113).

The structures of eptifibatide (6) and tirofiban (7) mimic the RGD motif and bind tightly to the platelet GPIIb/IIIa receptor. Like abciximab, they have been shown *in vitro* to reduce thrombin generation, although to a lesser extent. The design for eptifibatide was inspired by a KGD-containing snake venom disintegrin protein, which was known to bind to the GPIIb/IIIa receptor both potently and selectively (114). In the SAR leading to the discovery of this drug, it was found that, whereas small cyclic peptides incorporating the KGD sequence were selective, they lacked the potency of their relatively unselective cyclic RGD counterparts. Guanylation of the lysine residue on the KGD analogs, resulting in a homo-Arg residue, fortuitously provided compounds that were both potent and selective. Nonpeptide antagonists such as tirofiban and many other recent analogs essentially follow the design paradigm: (Arg mimetic)-(constrained spacer)-(Asp mimetic), such that the overall length from the basic nitrogen to the acid group is about 16 Å (115).

All three marketed GPIIb/IIIa drugs are poorly absorbed by the oral route and are dosed by continuous IV infusion. Abciximab is approved as an adjunctive therapy with aspirin and heparin for percutaneous coronary interventions (PCI), such as angioplasty, and is being considered as an adjunctive therapy in other settings of arterial (platelet-rich) thrombosis [e.g., acute myocardial infarction (MI) and ischemic stroke]. Eptifibatide is approved as an adjunct in PCI and unstable ischemic syndromes, whereas tirofiban is approved for unstable ischemic syndromes only. Both of

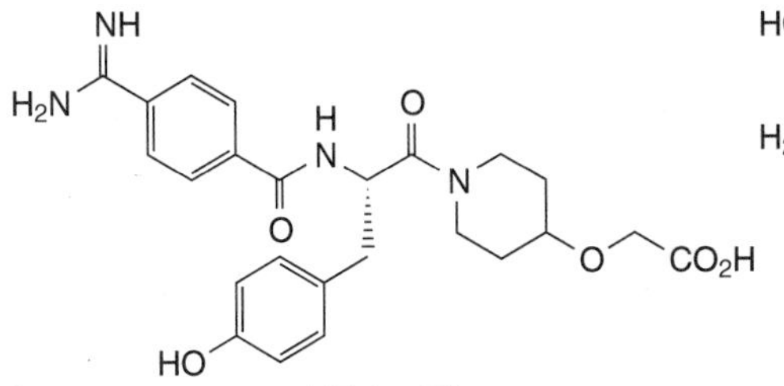

(**17**) lamifiban

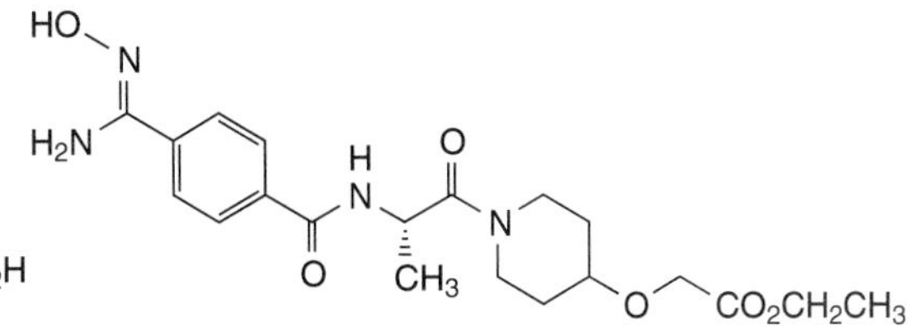

(**18**) sibrafiban

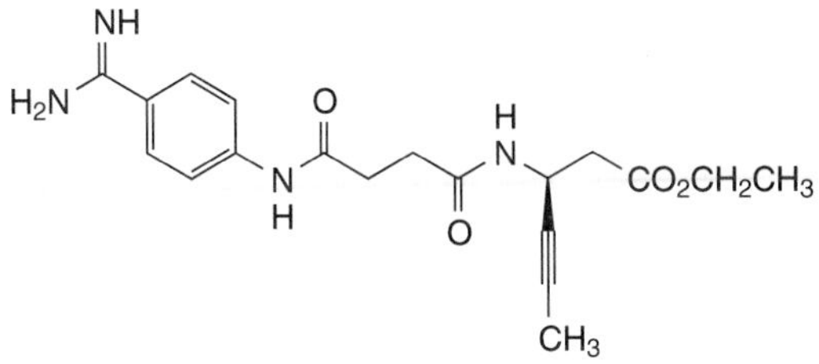

(**19**) xemilofiban

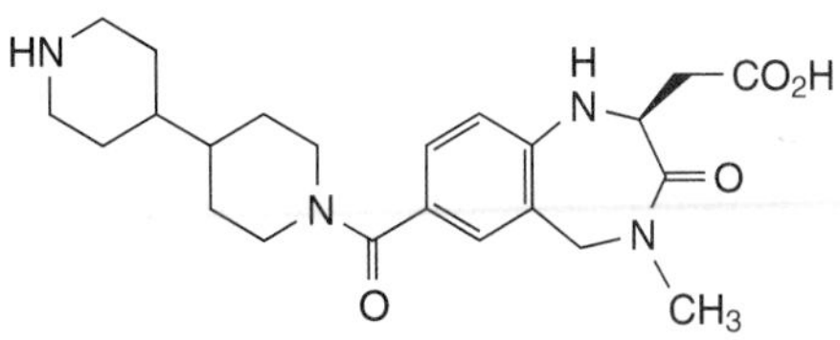

(**20**) lotrafiban

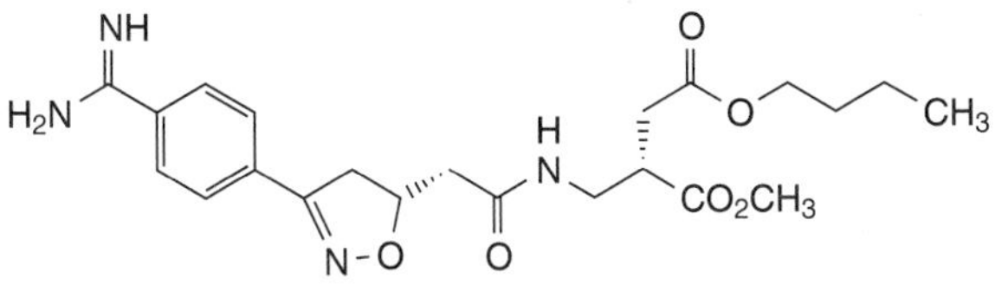

(**21**) roxifiban

Figure 6.25. Several platelet GPIIb/IIIa inhibitors investigated in clinical trials by IV (**17**) and by oral administration (**18–21**).

these agents are also undergoing clinical study for additional indications in arterial thrombosis (111).

Two other parenteral GPIIb/IIIa agents have recently been investigated clinically. YM337, a humanized Fab fragment directed against GPIIb/IIIa, but having less affinity than that of abciximab for the vitronectin receptor, was shown in phase I trials to effectively inhibit GPIIb/IIIa-mediated platelet aggregation (116). The nonpeptide antagonist, lamifiban (**17**, Fig. 6.25), showed a small benefit in reducing acute coronary syndrome events (117).

Over the past decade, a number of pharmaceutical companies have made major efforts to discover and develop orally bioavailable small nonpeptide GPIIb/IIIa antagonists, and most of the effort was directed toward the investigation of RGD mimetics similar in structure to tirofiban and lamifiban (115, 118). Because many of these agents contained highly basic benzamidines, most were prodrugged in various ways to increase oral bioavailability. For example, sibrafiban (**18**) contains a hydroxylated amidine to reduce basicity and an esterified carboxylate, echoing the technique used to prodrug the thrombin inhibitor melagatran. Xemilofiban (**19**), another benzamidine, is prodrugged as the ethyl ester, whereas the piperidine-based RGD mimetic lotrafiban (**20**) was not prodrugged. The human oral bioavailabilities of many of these orally active agents have not been reported, although xemilofiban was reported to have a bioavailability of 13% on oral dosing (119). In contrast to the IV-administered GPIIb/IIIa antagonists, these oral agents have not demonstrated efficacy in patients with acute coronary syndromes, and many of them have been associated with an increase in mortality (111a, 120). The worse clinical outcome with use of these oral agents might be explained by observations that GPIIb/IIIa antagonists can induce the receptor to adopt a ligand-binding conformation that transiently persists after dissociation of drug, allowing fibrinogen to bind and, paradoxically, platelet aggregation to commence (111a, 121). This proaggregation effect may be general for all GPIIb/IIIa antagonists, but the response may be exaggerated for the oral agents, given the periods of trough drug levels that allow receptor occupancy to fall off.

By contrast, the IV agents are maintained at a continuously high plasma concentration with uninterrupted and high receptor occupancy. Further, the IV agents have been typically administered against the background of anticoagulant therapy, which can enhance the clinical response to a GPIIb/IIIa antagonist. Clinical development of most of these oral agents has been terminated. Roxifiban (21, Fig. 6.25), a more recent oral GPIIb/IIIa antagonist, appears to differentiate itself from the earlier oral agents, in that it is bound more tightly to the platelet receptors and thus might be able to maintain sufficient receptor occupancy upon oral dosing to achieve the desired efficacy (118d). It still remains to be established whether clinical outcomes might improve with oral GPIIb/IIIa antagonists having more favorable (tighter) receptor binding properties and/or having pharmacokinetics allowing higher continuous plasma drug levels with less of a trough.

3.2.5 Platelet ADP Receptor Antagonists.
As discussed in section 2.3, the thienotetrahydropyridine (usually referred to simply as thienopyridine) analogs clopidogrel and ticlopidine are inactive per se, requiring hepatic conversion to a ring-opened thiol-active species that irreversibly inhibits the platelet receptor $P2Y_{12}$, presumably by formation of a disulfide linkage to a receptor cystein. The $P2Y_1$ receptor is insensitive to thienopyridines (67). Clopidogrel and ticlopidine are efficacious antiplatelet agents in humans (12, 54b, 122) and, in particular, clopidogrel has shown superiority over aspirin, with comparable safety, in the prevention of myocardial infarction and stroke, and when used in combination with aspirin has shown a reduction in ischemic events compared to that of aspirin alone (123). Although one clinical study comparing clopidogrel to ticlopidine demonstrated

similar efficacy for the two agents at preventing coronary stent thrombosis, the adverse event rate was higher for ticlopidine (124). Also, the historical risk of thrombotic thrombocytopenic purpura is lower with use of clopidogrel compared to that of ticlopidine.

Only the S-isomer of clopidogrel is active as an antiplatelet agent (125); ticlopidine is achiral. A third thienopyridine antiplatelet agent, CS-747 (22, Fig. 6.26), a racemate, is currently undergoing phase I trials (126). As with clopidogrel and ticlopidine, antiplatelet effects for CS-747 require hepatic conversion to an active metabolite, the structure of which has been confirmed and is analogous to the clopidogrel/ticlopidine active metabolites.

Adenosine triphosphate is a competitive antagonist of the action of ADP at the $P2Y_{12}$ receptor, although it is unacceptable as a therapeutic agent as a result of its weak potency and its metabolism to ADP (127). A class of stabilized ATP analog antagonists of $P2Y_{12}$ exhibit selectivity for this receptor and act directly, with no metabolic modification required as for the thienopyridines. Representative of this class of ATP analogs is cangrelor (AR-C69931; 23), which has an IC_{50} value of 0.4 nM against ADP-induced platelet aggregation and >1000-fold selectivity for the $P2Y_{12}$ receptor compared to the other P2-type receptors (127). Structurally, the terminal dichlorophosphonate group, a phosphate mimic, is stabilized toward hydrolysis. Additionally, modifications to the purine 2 and 6 positions provide cangrelor with increased potency over that of ATP.

As an IV agent, cangrelor demonstrated therapeutic efficacy in phase II clinical studies in patients with acute coronary syndromes. Its rapid onset of action and rapid reversal upon cessation of infusion contrasts with the slow onset and reversal of activity of the thienopyridines (128). However, further development of cangrelor has been terminated. Nonphosphate adenosine analog antagonists of $P2Y_{12}$ have been reported, but these are currently less well characterized in terms of their potential as antiplatelet drugs (126c, 129).

The first example of a nonnucleoside reversible selective $P2Y_{12}$ antagonist has been reported, CT50547 (24). This compound displays moderate inhibitory potency in a $P2Y_{12}$

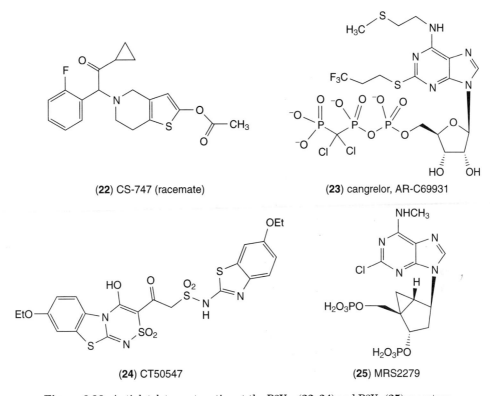

Figure 6.26. Antiplatelet agents active at the $P2Y_{12}$ (**22–24**) and $P2Y_1$ (**25**) receptors.

radioligand binding assay (IC_{50} = 170 nM) and a similar level of potency in an ADP-induced platelet aggregation assay. It is 1000-fold selective for $P2Y_{12}$ compared to the $P2Y_1$ receptor (130).

The activation of $P2Y_1$ receptors in platelets contributes to platelet aggregation and antagonists of this receptor may have potential as antithrombotic agents (67, 131). Naturally occurring adenosine bisphosphates (e.g., adenosine 3′,5′-bisphosphate) act as weak competitive antagonists at the $P2Y_1$ receptor and structural modification at the ribose ring and at the purine 2 and 6 positions have afforded competitive inhibitors with enhanced potency and selectivity (132). For example, MRS2279 (**25**) has an IC_{50} value of 52 nM in a $P2Y_1$ antagonism assay that measured inhibition of phospholipase C induction elicited by 2-thiomethyl-ADP. This compound also potently inhibits platelet aggregation and does not interact with the $P2Y_{12}$ receptor. Non-

phosphate adenosine analog antagonists of the $P2Y_1$ receptor have recently been reported (133).

A question that remains to be answered is whether antagonism of the $P2Y_1$ receptor alone or dual antagonism of the $P2Y_1$ and $P2Y_{12}$ receptors might achieve clinical benefits equivalent to or superior to the antagonism of $P2Y_{12}$ alone.

3.2.6 Aspirin and Dipyridamole. Aspirin is the most common antiplatelet drug in use today (12, 134). In the platelet, aspirin irreversibly inactivates cyclooxygenase-1 (COX-1) by acetylating the hydroxy group of Ser-529 near the active site, thereby blocking the binding of its substrate arachadonic acid. COX-1 in the platelet normally converts arachadonic acid to PGH_2, a precursor of the potent platelet activator thromboxane A_2 (TxA_2). Because platelets lack a nucleus and do not support protein synthesis, they cannot replenish the acetylated COX-1 for the duration of their normal lifetime (about 7–10 days). Moreover, because only about 10% of the platelet pool is replenished each day, once-a-day dosing of aspirin is able to maintain virtually complete inhibition of platelet TxA_2 production. To a lesser extent, aspirin inactivates COX-1 activity in endothelial cells, leading to a decrease in the synthesis of the antiplatelet modulator PGI_2, although this effect can be partially overcome by *de novo* protein biosynthesis. Mucosal COX-1 activity is also inhibited by aspirin, which contributes to gastric bleeding (Section 2.2.1). Aspirin also exhibits anti-inflammatory activity by inhibition of cellular COX-2, although at doses higher than that needed to achieve its COX-1 mediated antiplatelet effects. Other COX-1 inhibitors have been investigated, differing in their antiplatelet and therapeutic profiles compared to those of aspirin, and a few are marketed in other countries (12, 135).

Low dose aspirin is well established at improving outcomes in patients who have had thrombotic events or who may be prone to them. In patients with acute MI, prior MI, unstable angina, or stroke, aspirin reduced the long-term risk of recurrences by 25% and in individuals with stable angina, aspirin reduced the risk of MI by 44% (134).

Dipyridamole is thought to exert antiplatelet effects in part by inhibiting phosphodiesterase-mediated hydrolysis of the platelet-deactivating nucleotides cAMP and cGMP, although the scope of dipyridamole's mechanism or mechansims of action is still not entirely clear. It appears to synergize with aspirin. In patients with a history of transient ischemic attack or ischemic stroke, aspirin and sustained-release dipyridamole decreased risk for stroke by 18 and 16%, respectively, whereas aspirin added to sustained-release dipyridamole decreased risk by 37% (136).

3.3 Thrombolytic Agents: Mechanisms and Improvements

Over the last decade, thrombolytic therapy has had a significant impact on how acute myocardial infarction, and more recently, on how acute ischemic stroke is treated (137). Most thrombolytics, either currently marketed or in trials, are natural or modified forms of tPA, uPA, or bacterial proteins.

These enzyme or protein preparations act on plasminogen either to generate plasmin or to create an activated form of plasminogen having plasminlike activity. Plasmin activity acts to dissolve the fibrin component of clots. There are shortcomings with many of these thrombolytic agents, which include: (*1*) short plasma half-life, which may partly reflect the rate of inactivation by the serpin PAI-1, necessitating either continuous infusion or multiple bolus IV doses; (*2*) induction of a "paradoxical" prothrombotic condition, which may lead to a greater tendency to reocclude, or a systemic lytic condition, or both; and (*3*) immunogenicity (137a, 138). For example, the first-generation thrombolytic, streptokinase, has a reasonably acceptable half-life (18–23 min) but is immunogenic and prone to induce prothrombotic/lytic conditions. Alteplase (natural recombinant human tPA), a second-generation agent, is nonimmunogenic and induces less of a prothrombotic/lytic condition than that of streptokinase, but has a short half-life (4–6 min).

The paradoxical prothrombic and/or lytic condition (see below) induced by several of these agents has been associated with the inability of these agents to exhibit "clot selectivity" (138c,d). The origin of clot selectivity has its basis in the ability of some plasminogen activators (e.g., tenectaplase) to bind efficiently to a unique conformation of plasminogen when the plasminogen molecule is itself bound at the C-terminal lysine sites of partially degraded fibrin. Clot-selective agents do not bind as readily to the solution conformation of plasminogen, which is different from its fibrin-bound conformation (139). This clot selectivity achieves two important physiological results: (*1*) the PA is localized to the site (fibrin + plasminogen), where it will have the most therapeutic effect; and (*2*) localization of the PA to the clot restricts the activator from diffusing into the general vasculature, which can trigger the prothrombotic/lytic states referred to above. In particular, evidence has accumulated that high concentrations of plasminogen activators freely circulating throughout the vasculature can trigger activation of the kallikrein/FXII pathway that leads to generation of FXIa and a resultant prothrombotic state. This may explain, at least in part, why in some settings of thrombolytic therapy, partial or total reocclusion is observed after initial dissolution of the thrombus. Subsequently, after depletion of procoagulant factors and the α2-antiplasmin inhibitor, the circulating plasminogen activator continues to freely generate plasmin within the vasculature and may induce a systemic lytic state, with possible hemorrhagic consequences (138c,d, 140).

The third-generation agent tenectaplase (TNKase) is a recombinant analog of tPA, engineered to render it more clot selective, prolong its half-life, and make it more resistant to PAI-1. The letters TNK in its name indicate some of the amino acid replacements that were made as part of this process. For example, lysine (K), histidine, and two arginines were replaced by four alanines in the catalytic portion of this enzyme to enhance resistance to PAI-1 inhibition. In accord with the interpretation of clot selectivity, it was observed that clinical markers of thrombin generation tracked inversely with the extent of clot selectivity for three PAs, with the level of markers being in the order: streptokinase > alteplase > tenectaplase. Tenectaplase has a longer half-life than that of alteplase (14–18 versus 4–6 min, respectively), allowing less frequent dosing. Therefore, because of its longer half-life and high degree of clot selectivity, tenectaplase seems to represent a significant advance as a fibrinolytic agent over the older agents.

Staphlokinase, another third-generation clot-selective PA undergoing trials, is a recombinant nonenzyme bacterial protein with a mode of action different from that of the serine protease PAs. Staphlokinase binds as a 1:1 complex to the unique C-terminal fibrin-binding conformation of plasminogen, which allows small amounts of circulating plasmin to activate the complexed plasminogen. The resultant staphlokinase-plasmin complex is protected from α1-antiplasmin–mediated inactivation while on the surface of fibrin, but is readily inactivated if it dissociates into the circulation, further accounting for its high clot selectivity (139). The plasma half-life of staphlokinase is short, however (6 min); and it is immunogenic, inducing neutralizing antibodies after 10 days in a majority of patients, and therefore might be restricted to single use.

It has been found that covalently linking polyethylene glycol to staphlokinase enhances its plasma half-life (140).

The mechanism of the non-clot-selective fibrinolytic streptokinase, another nonenzyme bacterial protein, is somewhat different from that of staphlokinase, in that it binds to plasminogen in a way that conformationally opens up and activates the catalytic site, without a proteolytic cleaveage to form plasmin.

Other modifications to tPA or bacterial prothrombolytic proteins continue to be studied, which may result in potential therapeutic advantages (141).

4 ANTITHROMBOTIC AGENTS HAVING ALTERNATIVE MECHANISMS OF ACTION

Treatment of thrombotic conditions using currently marketed agents, although largely effective, have disadvantages. Heparin and warfarin anticoagulant activity must be monitored carefully because of safety concerns. Moreover, warfarin has a delayed onset of action. Most antithombotic and antiplatelet agents must be administered either IV or SC, which is less desirable than oral administration. Aspirin, although ubiquitously used, is not highly efficacious when used alone in many settings. To overcome these drawbacks, a great deal of research is currently directed toward the discovery of new treatment options, many of which exploit alternative mechanisms.

4.1 Inhibitors of Coagulation Factors

The direct inhibition of thrombin represents a logical strategy for achieving an efficacious and reasonably safe therapeutic anticoagulant effect. On the other hand, the inhibition of the serine protease coagulation factors that precede thrombin in the coagulation pathway, FXa, FIXa, or FVIIa, represents an equally viable strategy, and one that may have advantages over the inhibition of thrombin alone. By attenuating the generation of thrombin, rather than inhibiting the catalytic activity of thrombin itself, physiological functions mediated by low levels of thrombin might be spared. Normal hemostasis mediated by thrombin's action at the PAR-1 receptor, for

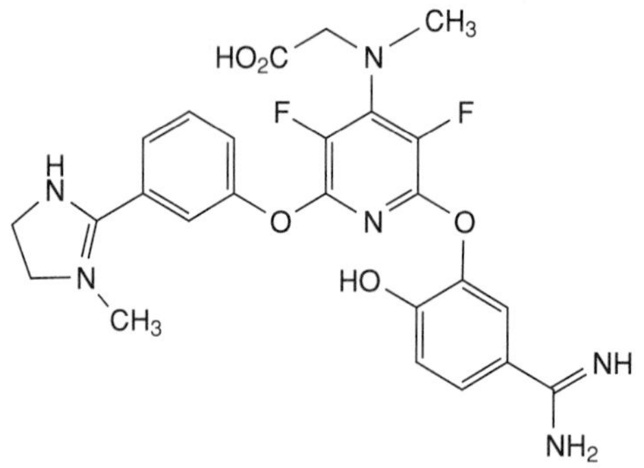

(26) CI-1031, ZK-807834

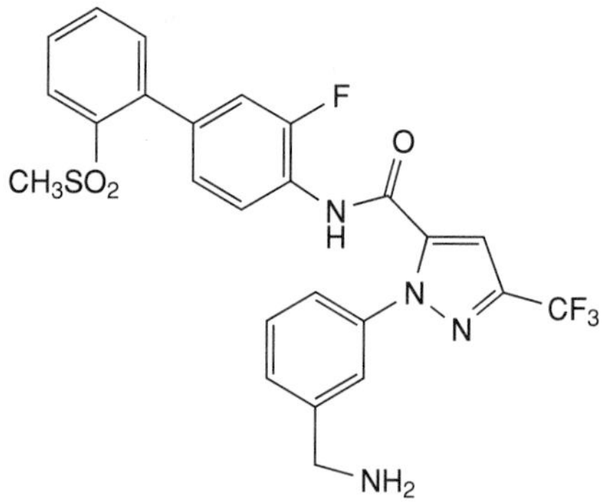

(27) DPC423

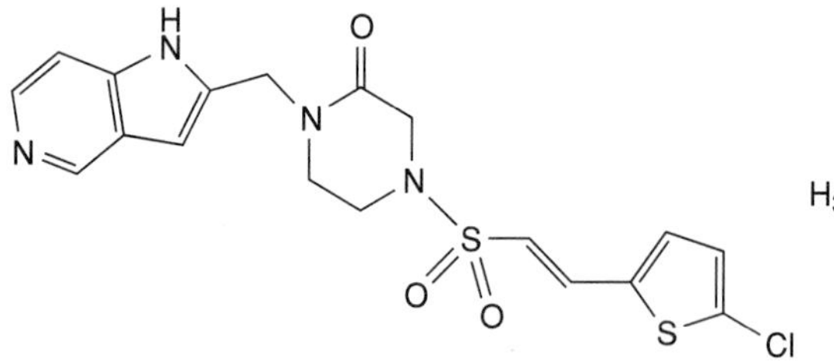

(28) RPR209685

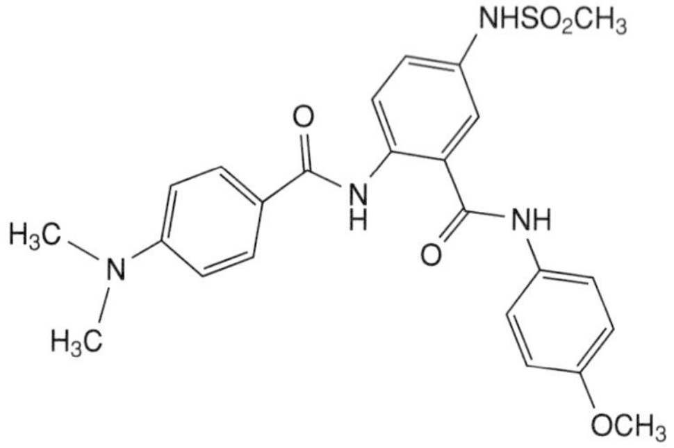

(29)

Figure 6.27. Potent reversible inhibitors of FXa.

example, could remain intact and therefore inhibitors of the earlier coagualtion factors may not only be effective antithrombic agents but may also carry less bleeding risk. Early support for this concept was reported using macromolecular inhibitors of FXa and TF/FVIIa in animal models of thrombosis. These studies concluded that, whereas direct thrombin inhibitors such as hirudin impaired platelet hemostatic function in parallel with their antithrombotic effects, selective inhibition of FXa using TAP (tick anticoagulant peptide) or inhibition of TF/FVIIa using active-site-inhibited FVIIa (FVIIai) could achieve a dose-dependent antithrombotic effect with comparatively less impairment of hemostatic function and hemorrhagic risk (142). Moreover, by inhibiting the earlier coagulation factors that are responsible for the generation of thrombin, the "thrombin rebound" effect (section 3.2.3) might be avoided. Starting in the mid-1990s significant research effort was directed toward the discovery and development of se-

lective inhibitors of alternate coagulation factors, especially Factor Xa. Similar to the search for direct thrombin inhibitors, an emphasis was placed on the development of agents having good oral bioavailability to allow convenient chronic dosing.

4.1.1 Direct Active Site Inhibitors of FXa. Building on the experience gained from the successful development of reversible small-molecule thrombin inhibitors, a number of potent and selective inhibitors of FXa have been discovered that are efficacious in various animal models of thrombosis (143). Some have also shown good bioavailability and plasma half-lives upon oral dosing in animals. Figure 6.27 shows the structures of four representative optimized FXa inhibitors. The most potent of these is the benzamidine CI-1031 (**26**), which has a K_i value for FXa of 0.11 nM and which is >1000-fold less potent for thrombin and trypsin. Surprisingly, in spite of its multiply charged structure, high plasma levels of CI-1031 persisting up to 6 h are achieved after oral dosing in primates (144).

CI-1031 showed efficacy with favorable safety (bleeding) profiles in several animal models of thrombosis, including a tPA fibrinolysis model in dogs (145). An X-ray crystal structure of CI-1031 shows the benzamidine forming a salt bridge with Asp-189 in the S1 pocket, whereas the distal phenyl along with its appended methyl dihydroimidazolidine group occupies the lipophilic S4 pocket defined by Trp-215, Tyr-99, and Phe-174 (146).

The structure of DPC423 (**27**) evolved from related benzamidine-containing analogs and has a less basic benzylamine group that presumably binds in the FXa S1 pocket. DPC423 is potent (K_i = 0.15 nM), selective against thrombin and trypsin, and shows excellent bioavailabilty (57%) and half-life (7.5 h) when dosed orally in dogs. It is efficacious in a canine model of thrombosis, with a good safety profile, and has entered clinical trials (147).

RPR209685 (**28**) and anthranilic amide inhibitor (**29**) are examples of potent noncharged FXa inhibitors. In the case of (**29**) (K_i = 11 nM), both computer modeling and comparisons with the X-ray crystal structures of related amidine-containing analogs suggest that the methoxyphenyl group occupies the S1 pocket of FXa and the dimethylaminophenyl occupies S4 (148). RPR209685 (K_i = 1.1 nM) is >1600-fold selective against trypsin, tPA, plasmin, thrombin, and other serine proteases. It is orally bioavailable in the rat and dog (34% and 97%, respectively) and shows efficacy in a dog model of DVT (149).

TAP, a 60 amino acid peptide isolated from the tick species *Ornithodoros moubata*, is a highly selective inhibitor of FXa (150). An X-ray crystal structure of a recombinant form of this peptide (rTAP) with FXa has revealed a binding motif involving insertion of several N-terminal amino acid residues of the inhibitor into the FXa active site and the C-terminus binding to a separate domain (151), reminiscent of the hirudin-thrombin motif. A number of preclinical *in vivo* studies have validated its antithrombotic efficacy and relative safety. For example, rTAP was more effective than hirudin when given as an adjunctive treatment to dogs undergoing alteplase-induced

thrombolysis. In a baboon model of arterial thrombosis, a 2-h infusion of rTAP resulted in a long-lasting (55 h) antithrombotic effect (152).

4.1.2 Inhibitors of FIXa. In the intrinsic tenase complex, Factor IXa activates FX to FXa (Fig. 6.8). Although hereditary deficiencies of FIX result in hemophilia B, studies in animals suggest that agents that block the activity of FIXa may possess therapeutic utility with an acceptable safety margin. Factor IXa, which is covalently inhibited at its active site with a tripeptide chloromethylketone (DEGR-inactivated FXIa; FXIai), competes with endogenous FIXa for incorporation into the intrinsic tenase complex, resulting in diminished conversion of FX to FXa. FIXai infused into dogs was as efficacious as heparin in a model of coronary artery thrombosis and, moreover, an abnormal bleeding response from a surgical wound was present only with heparin, not FIXai. FIXai was also effective in a guinea pig thrombosis model, with minimal effects on normal hemostasis (153). Monoclonal antibodies against FIX/FIXa that inhibit both the activation of FIX and the activity of FIXa have shown efficacy in models of venous and arterial thrombosis in a number of animal species, with no serious bleeding consequences compared to heparin. Humanized monoclonal antibodies to FIX have been dosed to healthy volunteers (154). No selective small-molecule inhibitors of FIXa have been reported to date, however.

4.1.3 Inhibitors of FVIIa and TF/FVIIa. Four macromolecular inhibitors of TF/FVIIa or FVII/FVIIa are currently being investigated clinically: recombinant TFPI (tissue factor pathway inhibitor), rNAPc2 (recombinant nematode anticoagulant protein c2), a monoclonal antibody to FVII/FVIIa, and active-site-inhibited FVIIa (FVIIai). In their mechanisms of action, TFPI and rNAPc2 both employ FXa as an initial "scaffold," to bridge to and to inactivate the TF/FVIIa complex. Recombinant TFPI (tifocogin) is an effective antithrombotic in several animal models and has been shown in healthy humans to dose-dependently attenuate endotoxin-induced coagulation activation (155). rNAPc2, was shown in patients undergoing elective total knee replacement to be 50% more effective than heparin, with similar bleeding profiles. Trials in patients with unstable angina undergoing percutaneous transluminal coronary angioplasty (PCTA) are scheduled (156). The monoclonal antibody to FVII/FVIIa (Corsevin M) is being studied in clinical trials for arterial thrombosis in the setting of PTCA (157). FVIIa inactivated with Phe-Phe-Pro chloromethyl ketone (FFP-FVIIa) competes with endogenous FVIIa for binding to TF, forming an enzymatically inactive complex. FFP-FVIIa has shown antithrombotic efficacy in several animal models with a good safety profile. An early clinical trial employed FFP-FVIIa with heparin in patients undergoing PCTA and indicated a trend for efficacy, although with some increase in minor bleeds (158). Reports of potent, reversible small-molecule active-site inhibitors of FVIIa (K_i values of 3–12 nM) have begun to emerge (159). All of these potent inhibitors contain a benzamidine group, which presumably binds to Asp-189 in the S1 pocket at the active site of FVIIa. Based on the experience gained with the development of neutral small-molecule inhibitors of thrombin and FXa, selective FVIIa inhibitors having the potential for oral bioavailability should, in theory, also be feasible. A recent study compared the efficacy and safety of small-molecule inhibitors of thrombin, FXa and FVIIa, in a guinea pig model and concluded that at equivalent levels of antithrombotic effect, the FVIIa inhibitor had the smallest bleeding risk (160).

4.2 Antiplatelet Agents with Alternative Modes of Action

4.2.1 Agents That Interfere with the Thromboxane Receptor and Thromboxane Synthase.
Thromboxane A$_2$ (TxA$_2$, Fig. 6.28) activates platelets upon binding to the platelet TxA$_2$ GPCR (Section 3.1). Additionally, TxA$_2$ causes constriction in vascular tissue. Aspirin indirectly inhibits TxA$_2$ biosynthesis by blocking conversion of arachadonic acid to PGH$_2$, the precursors to TxA$_2$. However, aspirin also blocks biosynthesis of the antiplatelet and vasodilating prostaglandin PGI$_2$ in vascular endothelium, a potential disadvantage. In theory, directly inhibiting the generation of TxA$_2$

by blocking the TxA$_2$ synthase-mediated conversion of PGH$_2$ to TxA$_2$ might achieve an advantageous antiplatelet and vasorelaxant effect. Alternatively, or additionally, blocking the TxA$_2$ receptor might also prove therapeutically useful. In reality, however, it was found that inhibition of TxA$_2$ synthase results in build up of the precursor PGH$_2$, which itself activates platelets upon binding to the TxA$_2$ receptor (135, 161). Nevertheless, it was found that dual agents that combine both TxA$_2$ synthase inhibition and TxA$_2$ receptor-blocking activities are potent inhibitors of platelet function and interest continues in the design and testing of such dual agents (135, 161). For example, terbogrel (**30**), a potent antagonist of both the thromboxane receptor (IC$_{50}$ = 11 nM) and thromboxane synthase (IC$_{50}$ = 4 nM) efficiently inhibits collagen-induced platelet aggregation (IC$_{50}$ = 310 nM). Terbogrel is 30% orally bioavailable in rats (162) and is currently in Phase II clinical trials for thrombotic indications. The nonacidic compound BM573 (**31**) prevents platelet aggregation induced by arachidonic acid (IC$_{100}$ = 125 nM) and induced by the receptor agonist U466619 (IC$_{50}$ = 240 nM) (163). Pure TxA$_2$ receptor antagonists having no TxA$_2$ synthase activity [e.g., ifetroban (**32**) and S-18886 (**33**)] are also potent inhibitors of platelet aggregation (164). In particular, S-18886 inhibited U466619-induced platelet aggregation with an IC$_{50}$ value of 230 nM and shows antiplatelet and antithrombotic effects when dosed orally in several animal species (164c–e). Further clinical trials of dual agents and/or pure TxA$_2$ receptor antagonists will be needed to more fully assess whether agents from this class will achieve significant therapeutic usefulness in humans as antithrombotics.

4.2.2 Platelet PAR-1 and -4 Receptor Antagonists.
Thrombin acts as a powerful platelet activator through proteolytic-mediated agonism of the platelet PAR-1 and PAR-4 receptors (Section 3.1). As discussed, PAR-1 agonism provides an immediate activation response at low thrombin levels, whereas PAR-4 agonism provides a response at higher thrombin levels typical of the later stages of clot formation. Peptidic and nonpeptidic PAR-1 antagonists structurally derived from the un-

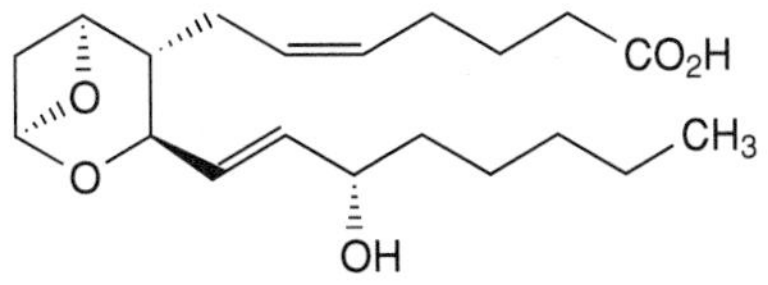

thromboxane A_2 (TxA$_2$)

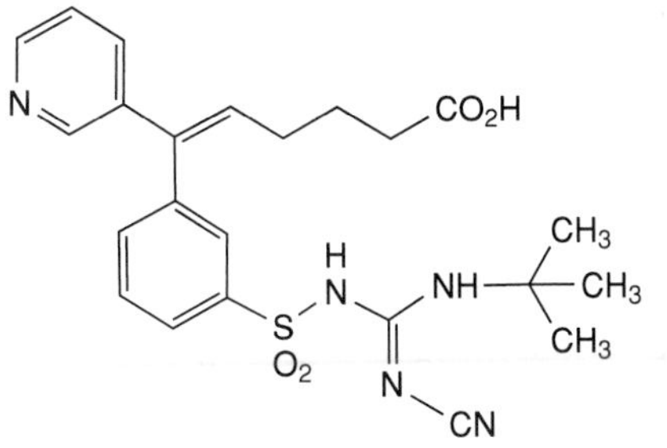

(30) terbogrel

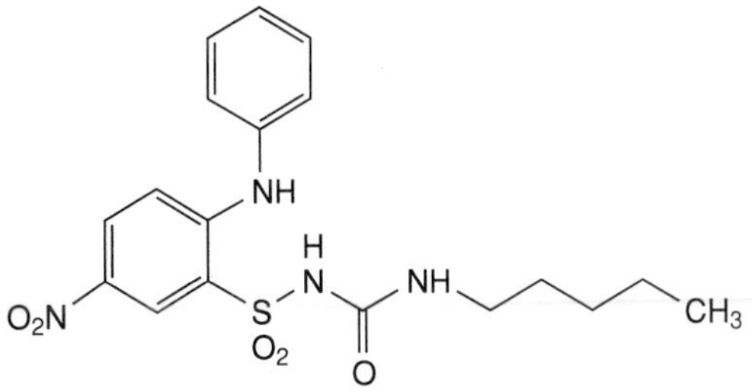

(31) BM573

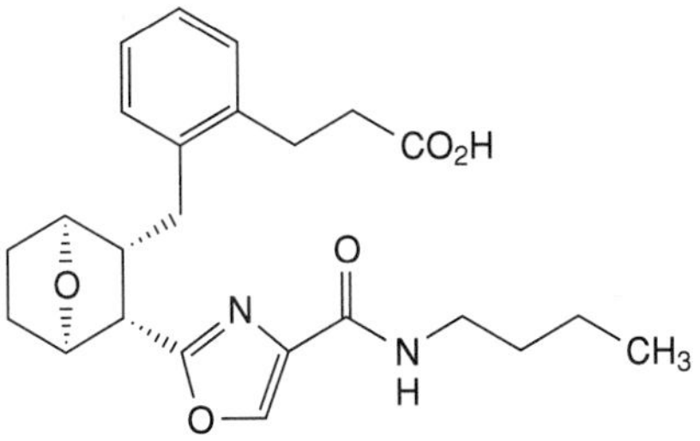

(32) ifetroban

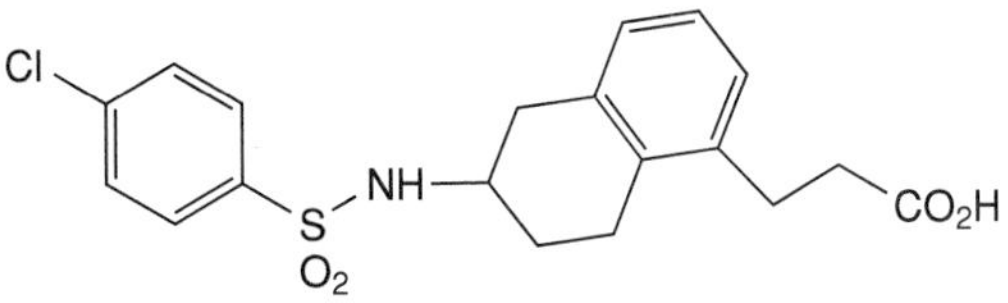

(33) S18886

Figure 6.28. Thromboxane A_2 and agents that block both the TxA$_2$ receptor and TxA$_2$ synthase (**30** and **31**) or the receptor only (**32** and **33**).

masked tethered receptor ligand (SFLLR) or derived from high throughput screening leads have been described (64a, 115, 165). However, to compete with the favorable energetics of a tethered agonist ligand, a soluble small-molecule antagonist is at a theoretical disadvantage. Typically, many of the PAR-1 antagonists reported to date have shown potent blockade of platelet aggregation induced by small peptide agonists (e.g., SFLLR-NH$_2$; a "TRAP" = thrombin receptor-activating peptide), but are less effective at blocking thrombin-mediated platelet activation. Recently, however, druglike antagonists that efficiently block thrombin-induced aggregation have been reported. For example, FR171113 (**34**, Fig. 6.29) is efficacious against both SFLLRN-NH$_2$ and thrombin-mediated platelet aggregation (IC$_{50}$ values of 0.15 and 0.29 μM, respectively) (166a). Also, the PAR-1 selective antagonist RWJ-58259 (**35**) blocked both

TRAP and thrombin-stimulated platelet aggregation (IC$_{50}$ values of 0.11 and 0.37 μM, respectively) *in vitro* and furthermore was shown to be efficacious in a primate model of arterial thrombosis when dosed by IV infusion (166b,c). Primate platelets, similar to human platelets, have both PAR-1 and PAR-4. This result provides some encouragement that a PAR-1-specific antagonist may have the potential to achieve a therapeutically useful antithrombotic effect in humans.

4.2.3 Other Platelet Targets: Serotonin, PGI$_2$, and PAF Receptors. Serotonin (5-hydroxytryptamine, 5-HT) binding to platelet 5-HT$_{2A}$ receptors elicits a weak aggregation response that is enhanced in the presence of collagen at the site of vasculature injury. 5-HT-induced vasoconstriction, mediated by binding to endothelial 5-HT$_{2A}$ and 5-HT$_{1B}$ receptor subtypes, contributes to its thrombotic

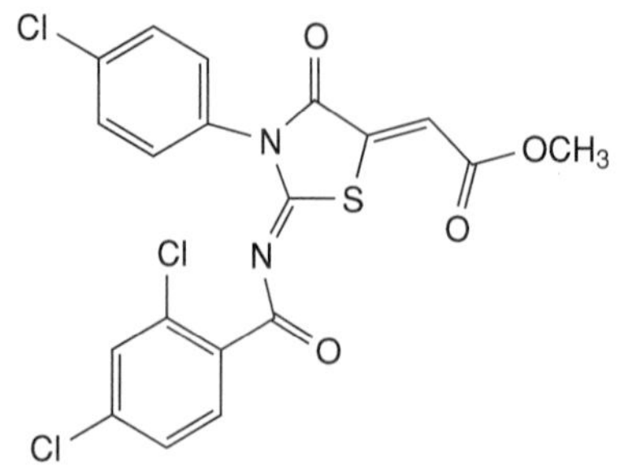

Figure 6.29. Platelet PAR-1 antagonists.

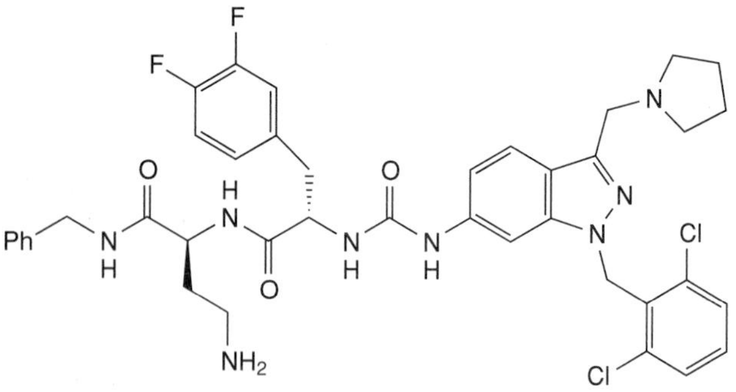

effect *in vivo*. Ketanserin (**36**, Fig. 6.30) and sarpogrelate (**37**) are two well-studied older orally active 5-HT antagonists that show antithrombotic effects *in vivo* (167). Sarpogrelate is marketed in Japan for treatment of peripheral arterial disease. A number of research groups are continuing to investigate peripherally acting serotonin antagonists as potential antithrombotics (168). Whereas ketanserin has high affinity for the 5-HT_{2A} receptor and inhibits collagen-induced platelet aggregation, it has low affinity for the 5-HT_{1B} receptor subtype. A recent analog, SL65.0472 (**38**), was designed to maintain the 5-HT_{2A} blocking activity of ketanserin while also potently inhibiting the endothial 5-HT_{1B} receptor, offering a potential advantage (168a). The acetamide group in SL65.0472 was introduced to limit CNS penetration of this compound. SL65.0472 demonstrated equivalence to ketanserin in human platelet aggregation assays, is efficacious in the Folts model of coronary artery thrombosis at an IV dose of 10–30 µg/kg, and has entered clinical trials. Compound (**39**), an analog of sarpogrelate, is a more potent inhibitor of platelet aggregation *in vitro* than either ketanserin or sarpogrelate, but produced gastric irritation in rats (168c). However, the lauryl ester prodrug (**40**, R-102444) did not produce gastric irritation and, when dosed orally, was more efficacious than sarpogrelate in a rat thrombosis model.

PGI_2 (prostacyclin), produced by endothelial cells, deactivates platelets (see Section 3.1) and also acts as a potent vasodilator. Chemically stable PGI_2 mimetics have been prepared and evaluated for their potential as anti-

thrombotics (169). Stable, orally bioavailable prostanoids such as iloprost and beraprost (**41**) have demonstrated antiplatelet effects in clinical trials, although their half-lives are very short (<1 h). Therefore, some current effort in the field is directed toward improving the pharmacokinetic profile of PGI_2 agonists. For example, FR181877 (**42**), which inhibits ADP-induced platelet aggregation *in vitro* with an IC_{50} value of 81 n*M* (IC_{50} = 4 n*M* for beraprost), has a half-life in male rats of 4.3 h compared to 0.43 h for beraprost (169a).

Platelet-activating factor (PAF), a phospholipid, was discovered in 1971 as a consequence of its platelet-aggregating properties. In spite of its name, PAF has numerous other activities (e.g., proinflammatory and vasodilatory) (170). Many small-molecule antagonists of PAF having *in vitro* antiplatelet activity have been identified, and a few were shown to be efficacious in animal models of thrombosis (171). To date, however, PAF antagonists have not performed as effectively *in vivo* as have other classes of antithrombotics.

To overcome the deficiencies of antiplatelet agents that target single mechanisms, some research has sought to combine two different antiplatelet activities in one molecule. Agents combining thromboxane receptor antagonism with thromboxane synthase inhibition have been mentioned. Additionally, single agents with combined PAF antagonism and thromboxane synthase inhibition have been reported, as have agents combining antagonism for both the thromboxane and 5-HT_{2A} receptors (172). It is still too early to assess whether

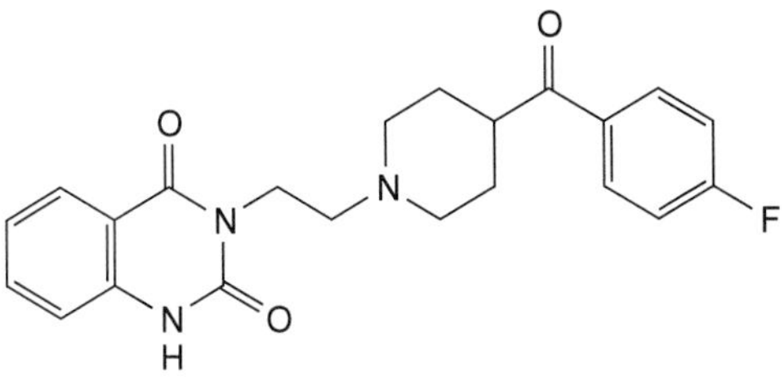

(**36**) ketanserin

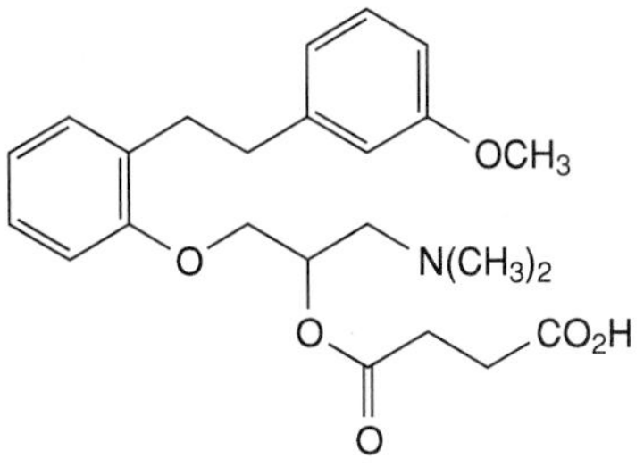

(**37**) sarpogrelate (racemate)

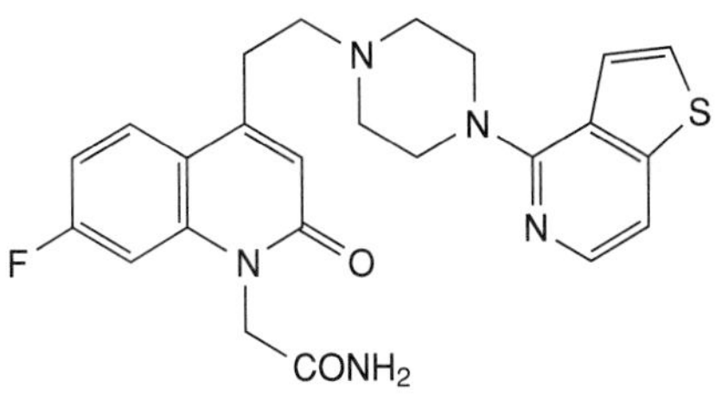

(**38**) SL 65.0472

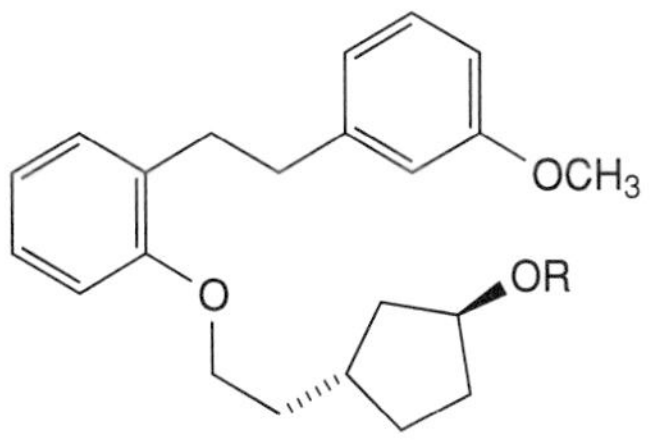

(**39**) R = H
(**40**) R = $CO(CH_2)_{10}CH_3$ R-102444

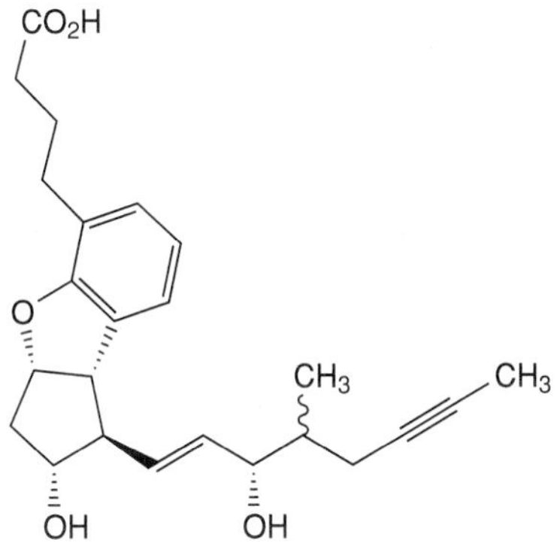

(**41**) beraprost

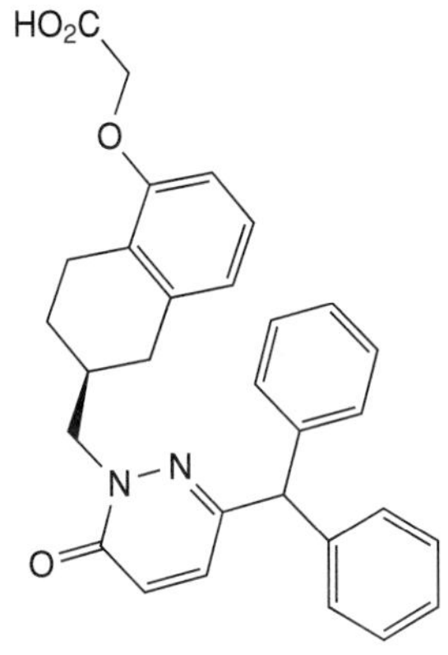

(**42**) FR181877

Figure 6.30. Antagonists of the platelet 5HT receptor (**36–40**); agonists of platelet PGI_2 receptor (**41, 42**).

these newer dual-acting compounds may have potential as efficacious antithrombotics.

4.3 Potential Profibrinolytics: Inhibition of Factor XIIIa, TAFIa, PAI-1, and α2-Antiplasmin

Agents that operate through mechanisms that weaken the structure of fibrin or that enhance the efficacy of endogenous plasmin or its activator tPA should behave as profibrinolytics, and may potentially find therapeutic utility in thrombotic settings. Such agents could be useful for enhancing endogenous fibrinolysis or might be used in conjunction with thrombolytic therapy, allowing the use of lower and perhaps safer doses of recombinant tPA or other fibrinolytics.

The transglutaminase Factor XIIIa strengthens fibrin toward plasmin degradation by covalently crosslinking the fibrin strands and it also protects fibrin from the action of plasmin by covalently attaching the plasmin inhibitor α2-antiplasmin to the surface of fibrin (Section 3.1). Therefore, inhibi-

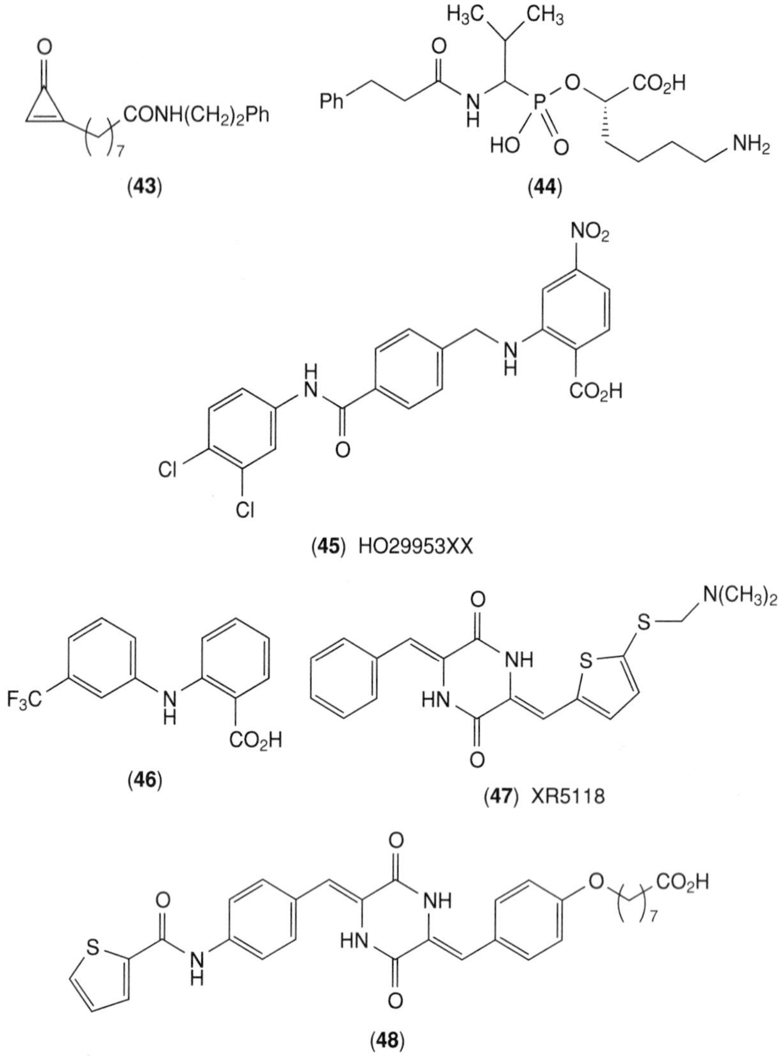

Figure 6.31. Prototype inhibitors of FXIIIa (**43**), TAFIa (**44**), and PAI-1 (**45–48**).

tors of FXIIIa have the potential to increase the susceptibility of clots toward lysis. Tridegin, a peptide isolated from the salivary glands of the giant Amazon leech is a specific inhibitor of Factor XIIIa, and clots formed in the presence of this peptide were shown to be more rapidly lysed *in vitro* (173). A potent (IC$_{50}$ = 26 n*M*) and selective synthetic nonpeptide FXIIIa inhibitor, derived from optimization of a class of FXIIIa inhibitors isolated from a fungus species, contains an unusual cyclopropeneone structure (**43**, Fig. 6.31). Enzyme kinetics indicate that it is a time-depen-

dent (irreversible) inhibitor of FXIIIa and, based on a binding mode analysis using the known crystal structure of FXIIIa, a mechanism was proposed involving attack of the active site cysteine-314 on the cyclopropene double bond (174). A few other peptide and nonpeptide inhibitors of FXIIIa have also been described, although to date, reversible druglike small-molecule inhibitors have not been reported (175).

The plasma carboxypeptidase TAFIa retards the action of plasmin by cleaving C-terminal lysine residues from partially degraded fibrin (Section 3.1). A 39 amino acid peptide inhibitor of TAFIa has been isolated from potatoes. In a rabbit jugular vein model of thrombosis, coinfusion of this peptide with tPA enhanced thrombolysis or markedly reduced the amount of tPA needed to achieve the same amount of lysis (60, 176). A few prototype small-molecule inhibitors of TAFIa have recently been reported [e.g., the phosphonate-based compound **44**]. Because TAFI is activated by thrombin, which is found at the site of a thrombus, inhibition of TAFIa might provide enhanced fibrinolysis with the added advantage of clot selectivity (see Section 3.3).

PAI-1 is the predominant serpin inactivator of tPA and therefore inhibitors of PAI-1 are expected to enhance endogenous fibrinolysis. Several *in vitro* and animal studies support this view. Lysis of human platelet-rich clots is accelerated by antibodies against PAI-1 *in vitro* and Fab fragments against rat PAI-1 inhibit venous and arterial thrombosis *in vivo* (178). A number of small-molecule lead compounds have been reported that allow tPA activity to be preserved in the presence of PAI-1 (179).

The exact mechanism(s) by which these small molecules are able to prevent the inactivation of tPA by PAI-1 is still being investigated. Antibody studies have revealed epitopes on PAI-1 distinct from the reactive center loop that may accommodate the binding of small molecules (179a, 180). In principle, the binding of small molecules to these or other sites might interfere sterically with the interaction of PAI-1 and tPA, or they might convert the PAI-1 inhibitory conformation to alternative conformations that may be either less reactive ("latent" conformation) or that may allow PAI-1 to serve as a normal substrate for tPA with fast turnover kinetics (179a, 181).

Two recent examples of PAI-1 inhibitors, both of which contain carboxylic acid groups (**45** and **48**, Fig. 6.31), are thought to interfere directly with the interaction of PAI-1 and tPA. The modestly potent anthranilic acid inhibitor (**45**) (H029953XX, IC_{50} = 12 μM), derived from flufenamic acid (**46**), is postulated to bind to a region of PAI-1, previously identified as a neutralizing antibody epitope and that contains a hydrophobic cleft flanked by a basic region containing three arginines (179a). Another research group had earlier reported the nonacidic diketopiperizine PAI-1 inhibitor (**47**) (XR5118; IC_{50} = 3.5 μM) and demonstrated its antithrombotic effect in a rabbit thrombolysis model (179b). Systematic modification of structure (**47**) resulted in the carboxylate analog (**48**), one of the more potent small-molecule PAI-1 inhibitors reported to date (IC_{50} = 0.2 μM in a plasmin generation assay, with a similar potency in a functional fibrinolysis assay) (179c). It is hypothesized that (**48**) binds at the same PAI-1 site as inhibitor (**45**) and that the carboxylic acid groups of both (**48**) and (**48**) are interacting with an Arg residue within this binding site.

The serpin for plasmin, α2-antiplasmin (α2AP), has been inhibited by monoclonal antibodies, resulting in enhancements to fibrinolysis. Antibodies to α2AP were shown to amplify the lysis of human plasma clots in the presence of plasminogen activators (182). Separately, it was also shown in a rabbit jugular vein thrombolytic model that specific inhibition of fibrin-linked α2AP by anti-α2AP antibodies increased the rate of spontaneous fibrinolysis (183). α2AP antibodies potentially suitable for therapeutic use in humans have been described (184), although clinical trials have not been carried out to date. Tanninlike molecules have been claimed to have inhibitory activity against α2AP in early reports (185), but small-molecule druglike inhibitors of α2AP have not been reported to date.

5 THE FUTURE OF ANTITHROMBOTIC THERAPY

Unfractionated heparin is being replaced in many of its applications by currently available alternative parenteral agents such as the LMWHs, fondaparinux, lepirudin, bivalirudin, and argatroban (186). These efficacious alternative agents offer advantages in terms of more predictable pharmacokinetics and improved safety, reducing or eliminating the need for patient monitoring, and reducing or eliminating the risk of heparin-induced thrombocytopenia. Orally bioavailable direct thrombin and/or FXa inhibitors now under development seem poised to become available over the next several years and will represent viable alternatives to both the parenteral anticoagulants and to warfarin, the only currently available oral anticoagulant. These agents promise the convenience of an oral drug without the need of intensive patient monitoring, as warfarin requires.

New antiplatelet agents that act by already validated mechanisms (e.g., $P2Y_{12}$ receptor antagonism) or that act by mechanisms not targeted by current agents will also likely emerge in the not too distant future. Additionally, new fibrinolytics having advantages over current agents in terms of greater safety and longer plasma half-life also seem within reach.

The future of antithrombotic therapy appears promising indeed.

6 ABBREVIATIONS

ADP	adenosine diphosphate
AMI	acute myocardial infarction
AP	antiplasmin
AT	antithrombin
APC	activated protein C
aPTT	activated partial thromboplastin time
CAD	coronary artery disease
COX	cyclooxygenase
DVT	deep vein thrombosis
FBD	fibrinogen binding domain
Gla	δ-carboxyglutamic acid
HCII	heparin cofactor II
HIT	heparin-induced thrombocytopenia
HITTS	heparin-induced thrombocytopenia with thrombosis
LBS	lysine binding site
LMWH	low molecular weight heparin
MI	myocardial infarction
PA	plasminogen activator
PAF	platelet-activating factor
PAI	plasminogen activator inhibitor
PAR	protease-activated receptor
PCI	percutanious coronary intervention
PE	pulmonary embolism
PT	prothrombin time
SC	subcutaneous
TAFI	thrombin-activatable fibrinolysis inhibitor
TF	tissue factor
TFPI	tissue factor pathway inhibitor
TIA	transcient ischemic attack
TM	thrombomodulin
UA	unstable angina
UFH	unfractionated heparin
VTE	venous thromboembolism
VWF	von Willebrand factor

REFERENCES

1. R. W. Colman, J. Hirsh, V. J. Marder, A. W. Clowes, J. N. George, Eds., *Hemostasis and Thrombosis*, 4th ed., Lippincott/Williams & Wilkins, Baltimore, 2001. This is a definitive texbook on the biology and biochemistry of blood coagulation, platelet function, human diseases of thomboses, and related topics.

2. M. N. Levine, G. Raskob, S. Landfeld, et al., *Chest*, **119**, 108S (2001).

3. International Stroke Trial Collaborative Group, *Lancet*, **349**, 1569 (1997).

4. Publications Committee for the Trial of ORG10172 in Acute Stroke Treatment (TOAST) Investigators, *JAMA*, **279**, 1265 (1998).

5. (a) K. A. Bauer, B. I. Eriksson, M. R. Lassen, et al., *N. Engl. J. Med.*, **345**1298 (2001); (b) K. A. Bauer, B. I. Eriksson, M. R. Lassen, et al., *N. Engl. J. Med.*, **345**1305 (2001); (c) S. Ahmad, W. P. Jeske, J. M. Walenga, et al., *Clin. Appl. Thromb. Hemostasis*, **5**, 259 (1999).

6. The Stroke Prevention in Reversible Ischemia Trial (SPRIT) Study Group, *Ann. Neurol.*, **42**, 858 (1997).

7. (a) J. B. Segal, R. L. McNamara, M. R. Miller, et al., *J Gen. Intern. Med.*, **15**, 56 (2000); (b) P. Smith, H. Arnesen, and I. Holme, *N. Engl. J. Med.*, **323**, 147 (1990).

8. Coumidin Aspirin Reinfarction Study (CARS) Investigators, *Lancet*, **350**, 389 (1997).

9. (a) B. I. Ericksson, P. Wille-Jorgensen, P. Kalebo, et al., *N. Engl. J. Med.*, **337**, 1329 (1996); (b) The OASIS-2 Investigators, *Lancet*, **353**, 429 (1999); (c) OASIS Investigators, *Circulation*, **96**, 769 (1997); (d) B. I Ericksson, S. Ekman, P. Kalebo, et al., *Lancet*, **347**, 635 (1996).

10. J. A. Bittl and F. Feit, *Am. J. Cardiol.*, **82**, 43 (1998).

11. B. E. Lewis, M. Cohen, F. Leya, et al., *Chest*, **118**, 261S (2000).

12. C. Patrono, B. Coller, J. E. Dalen, et al., *Chest*, **119**, 39S (2001).

13. G. Dannhardt and W. Kiefer, *Eur. J. Med. Chem.*, **36**, 109 (2001)

14. H. C. Diener, L. Cunha, C. Forbes, et al., *J. Neurol. Sci.*, **143**, 1 (1996).

15. CAPRIE Steering Committee, *Lancet*, **348**, 1329 (1996).

16. (a) S. R. Mehta, S. Yusuf, R. J. G. Peters, et al., *Lancet*, **358**, 527 (2001); (b) S. Yusuf, F. Hhao, S. R. Mehta, et al., *N. Engl J. Med.*, **345**, 494 (2001).

17. Product information supplied by Roche Laboratories, Nutley, NJ, June 1999.

18. The EPILOG Investigators, *N. Engl. J. Med.*, **336**, 1689 (1997).

19. The PURSUIT Trial Investigators, *N. Engl. J. Med.*, **339**, 436 (1998).

20. Product information supplied by Merck, West Point, PA, September 2000.

21. G. W. Albers, P. Amarenco, J. D. Easton, et al., *Chest*, **119**, 300S (2001).

22. E. M. Ohman, R. A. Harrington, C. P. Cannon, et al., *Chest*, **119**, 253S (2001).

23. T. M. Hyers, G. Agnelli, R. D. Hull, et al., *Chest*, **119**, 176S (2001).

24. GUSTO IIa Investigators, *Circulation*, **90**, 1631 (1994).

25. J. Hirsh, S. S. Anand, J. L. Halperin, et al., *Circulation*, **103**, 2994 (2001).

26. (a) T. E. Warkentin and A. Greinacher, Eds., Heparin-Induced Thrombocytopenia, Marcel Dekker, New York, 2000; (b) G. T. Gerotziafas, I. Elalamy, C. Lecrubier, et al., *Blood Coagul. Fibrinolysis*, **12**, 511 (2001).

27. T. E. Warkentin, M. N. Levine, J. Hirsh, et al., *N. Engl. J. Med.*, **332**, 1330 (1995).

28. The EPIC Investigators, *N. Engl. J. Med.*, **336**, 1689 (1996).

29. B. Bednar, J. J. Cook, M. A. Holahan, et al., *Blood*, **94**, 587 (1998).

30. M. Madan and S. D. Berkowitz, *Am. Heart J.*, **138**, 317 (1999).

31. C. L. Bennett, C. J. Davidson, D. W. Raisch, et al., *Arch. Intern. Med.*, **159**, 2524 (1999).

32. C. L. Bennett, J. M. Connors, J. M. Carwile, et al., *N. Engl J. Med.*, **342**, 1773 (2000).

33. Product information supplied by Bristol-Myers Squibb (New York, NY) and Sanofi-Synthelabo, New York, NY, April 2000.

34. J. M. Lusher, *Semin. Hematol.*, **28** (Suppl. 6), 3 (1991).

35. (a) T. Guillaume, *Lancet*, **343**, 1101 (1994), (b) Information supplied by Bayer Biological, Elkhart, IN, May 1999.

36. See Ref. 1, Chapter 50.

37. (a) A. Leone-Bay, et al., Blood [Proc. 43rd Annual Meeting Am. Soc. Hematol. (Dec. 7–11, 2001, Orlando, FL) 2001] **98**, Abstr. 177 (2001); (b) M. D. Gonze, K. Salartash, W. C. Sternbergh, et al., *Circulation*, **101**, 2658 (2000); (c) D. Brayden, E. Creed, A. O'Connell, et al., *Pharm. Res.*, **14**, 1772 (1997).

38. (a) C. A. de Swart, B. Nijmeyer, and J. M. Roelogs, *Blood*, **60**, 1251 (1982); (b) P. Olsson, Lagergren, and S. Ek, *Acta Med. Scand.*, **173**, 619 (1963); (c) T. D. Bjornsson, K. M. Worlfram, and B. B. Kitchell, *Clin. Pharmacol. Ther.*, **31**, 104 (1982).

39. (a) B. Boneu, J. Necciari, R. Cariou, et al., *Thromb. Haemost.*, **74**, 1468 (1995); (b) K. A. Bauer, *Am. J. Health-Syst. Pharm.*, **58** (Suppl. 2), S14 (2001).

40. (a) P. W. Majerus and D. M. Tollefsen in J. G. Hardman and L. E. Limbird, Eds., *Goodman & Gilman's The Pharmacological Basis of Therapeutics*, 10th ed., McGraw Hill, New York, 2001, Chapter 55; (b) A. M. Holbrook, P. S. Wells, and N. R. Crowther in L. Poller and J. Hirsh, Eds., *Oral Anticoagulants*, Oxford University Press, New York, 1996. The chapters in these books contain well-written summaries of the pharmacokinetics and pharmacodynamics of warfarin therapy, as well as safety issues.

41. G. P. Aithal, C. P. Day, P. J. L. Kestaven, et al., *Lancet*, **353**, 717 (1999).

42. Product information supplied by supplied by DuPont Pharma (Wilmington, DE), November 1999.

43. (a) A. Greinacher and N. Lubenow, *Circulation*, **103**, 1479 (2001); (b) J. C. Adkins and M. I. Wilde, *BioDrugs*, **10**, 227 (1998); (c) Product information supplied by Aventis, Frankfurt am Main, Germany, October 1998.

44. R. Vanholder and A. Dhondt, *BioDrugs*, **11**, 417 (1999).

45. J. I. Witting, P. Bourdon, and D. V. Brezniak, *Biochem. J.*, **283**, 737 (1992).

46. (a) I. Fox, A. Dawson, P. Loynds, et al., *Thromb. Haemost.*, **69**, 157 (1993); (b) J. J. Nawarskas and J. R. Anderson, *Heart Dis.*, **3**, 131 (2001); (c) R. Scatena, *Exp. Opin. Invest. Drugs*, **9**, 1119 (2000).

47. S. K. Swan and M. J. Hursting, *Pharmacotherapy*, **20**, 318 (2001).

48. J. Q. Tran, R. A. DiCicco, S. B. Sheth, *J. Clin. Pharmacol.*, **39**, 513 (1999).

49. (a) J. E. Tcheng, S. G. Ellis, B. S. George, et al., *Circulation*, **90**, 1757 (1994); (b) M. L. Simoons, M. J. de Boer, M. J. B. M. van der Brand, et al., *Circulation*, **89**, 596 (1994).

50. (a) J. E. Tcheng, J. D. Talley, J. C. O'Shea, et al., *Am. J. Cardiol.*, **88**, 1097 (2001); (b) D. R. Philips and R. M. Scarborough, *Am. J. Cardiol.*, **80**, 11B (1997); (c) Product information supplied by COR Therapeutics, South San Francisco, CA, December 1999.

51. (a) S. Vickers, A. D. Theoharides, B. Arison, et al., *Drug Metab. Dispos.*, **27**, 1360 (1999); (b) Product information supplied by Merck (West Point, PA), September 2000.

52. D. J. Kereiakes, N. S. Kleiman, J. Amborose, et al., *J. Am. Coll. Cardiol.*, **27**, 436 (1996).

53. P. Savi, J. M. Aereillo, M. F. Uzabiaga, et al., *Thromb. Haemost.*, **84**, 891 (2000).

54. (a) J. M. Herbert, D. Frehel, and E. Vallee, *Cardiovasc. Drug Rev.*, **11**, 180 (1993); (b) G. Escolar and M. Heras, *Drugs Today*, **36**, 187 (2000); (c) R. Lins, J. Broekhuysen, J. Necciari, et al., *Semin. Thromb. Hemostasis.*, **25** (Suppl. 2), 29 (1999); (d) H. Caplain, F. Donat, C. Gaud, et al., *Semin. Thromb. Hemost.*, **25** (Suppl. 2), 25 (1999).

55. For authoritative and in-depth overviews of clinical trials using antithrombotic and thrombolytic therapies, and current recommendations for therapy of thrombotic conditions, see the entire issue of *Chest*, **119** (Suppl. 1), (2001).

56. For additional details of the biochemistry and cellular biology of thrombosis and thrombolysis see (a) ref. 1; (b) W. Nieuwenhuizen, M. W. Mosesson, and M. P. M. de Maat, Eds., *Ann. N. Y. Acad. Sci.*, **936** (Fibrinogen) (2001); (c) B. Dahlback, *Lancet*, **355**, 1627 (2000).

57. (a) H. Mizuno, Z. Fujimoto, H. Atoda, et al., *Proc. Natl. Acad. Sci. USA*, **98**, 7230 (2001); (b) L. A. Falls, B. C. Furie, M. Jacobs, et al., *J. Biol. Chem.*, **276**, 23895 (2001).

58. P. Fuentes-Prior, Y. Iwanaga, R. Huber, et al., *Nature*, **404**, 518 (2000).

59. K. Wu and N. Matijevic-Aleksic, *Ann. Med.*, **32** (Suppl. 1), 73 (2000).

60. L. Bajzar, *Arterioscler. Thromb. Vasc. Biol.*, **20**, 2511 (2000).

61. (a) P. I. Bird, R. W. Carrell, F. C. Church, et al., *J. Biol. Chem.*, **276**, 33293 (2001); (b) J. A. Huntington and R. W. Carrell, *Sci. Prog.*, **84**, 125 (2001).

62. P. C. Y. Liaw, D. L. Becker, A. R. Stafford, et al., *J. Biol. Chem.*, **276**, 20959 (2001).

63. L. Liu, L. Dewar, Y. Song, et al., *Thromb. Haemost.*, **73**, 405– 412 (1995).

64. (a) S. R. MacFarlane, M. J. Seatter, T. Kanke, et al., *Pharmacol. Rev.*, **53**, 245 (2001); (b)

M. L. Kahn, M. Nakanishi-Matsui, M. J. Shapiro, et al., *J. Clin. Invest.*, **103**, 879 (1999).

65. S. Coughlin, *Thromb. Haemost.*, **86**, 298 (2001).

66. (a) J. Jin and S. P. Kunapuli, *Proc. Natl. Acad. Sci. USA*, **95**, 8070 (1998); (b) R. A. Nicholas, *Mol. Pharmacol.*, **60**, 416 (2001); (c) G. Hollopeter, H.-M. Jantzen, D. Vincent, et al., *Nature*, **409**, 202 (2001); (d) P. Savi, C. Labouret, N. Delesque, et al., *Biochem. Biophys. Res. Commun.*, **283**, 379 (2001).

67. C. Gachet, *Thromb. Haemost.*, **86**, 222 (2001).

68. J. C. Reverter, S. Beguin, H. Kessels, et al., *J. Clin. Invest.*, **98**, 863 (1996).

69. (a) B. N. Violand, R. Byrne, F. J. Castellina, *Proc. Natl. Acad. Sci. USA*, **84**, 4031 (1978); (b) see ref. 1, Chapter 49.

70. W. Wang, M. B. Boffa, L. Bajzar, et al., *J. Biol. Chem.*, **273**, 27176 (1998).

71. (a) A. Zhou, R. W. Carrell, and J. A Huntington, *J. Biol. Chem.*, **276**, 27541 (2001); (b) J. A. Huntington, R. J. Read, and R. W. Carrell, *Nature*, **407**, 923 (2000).

72. V. Tkachuk, V. Stepanova, and P. J. Little, *Clin. Exp. Pharmacol. Physiol.*, **23**, 759 (1996).

73. D. L. Becker, J. C. Fredenburgh, A. R. Stafford, et al., *J. Biol. Chem.*, **274**, 6226 (1999).

74. (a) R. C. Becker, F. A. Spencer, Y. Li, et al., *J. Am. Coll. Cardiol.*, **34**, 1020 (1999); (b) P. M. Sandset, B. Jjorn, and J.-B. Hansen, *Haemostasis*, **30** (Suppl 2), 48 (2001).

75. O. M. Aguilar and N. S. Kleiman, *Expert Opin. Pharmacother.*, **1**, 1091 (2000).

76. (a) F. A. Spencer, S. P. Ball, Q. Zhang, et al., *J. Thromb. Thrombolysis*, **9**, 223 (2000); (b) A. Rezaie, *Blood*, **97**, 2308 (2001).

77. S. Alban, *Semin. Thromb. Hemost.*, **27**, 503 (2001).

78. T. E. Warkentin, M. N. Levine, J. Hirsh, et al., *N. Engl. J. Med.*, **332**, 1330 (1995).

79. J. M. Acostamadiedo, U. G. Iyer, and J. Owen, *Expert Opin. Pharmacother.*, **1**, 803 (2000).

80. J. A. M. Anderson, J. C. Fredenbrugh, A. R. Stafford, et al., *J. Biol. Chem.*, **276**, 9755 (2001).

81. (a) K. A. Bauer, et al., *Chest*, **120**, 299S (2001); (b) B. M. T. Vogel, M. J. Smit, A. W. M. Princen, et al., Thromb. Haemost. [Proc. 18th Cong. Int. Soc. Thromb. Haemostas. (July 6–12, 2001, Paris) 2001] Suppl., Abstr. P1410 (2001); (c) C. Bal dit Sollier, N. Kang, N. Berge, et al., Thromb. Haemost. [Proc. 18th Cong. Int. Soc. Thromb. Haemostas. (July 6–12,

2001, Paris) 2001] Suppl., Abstr. OC1765 (2001); (d) A. G. G Turpie, *Am. Heart J.*, **142**, S9 (2001); (e) The PINK-SHEET, p. 3, December 17, 2001; (f) J. M. Herert, J. P. Herault, A. Bernat, et al., *Blood*, **91**, 4197 (1998).

82. (a) C. Vermeet and L. J. Schurgers, *Hematol./Oncol. Clin. North Am.*, **14**, 339 (2000); (b) C. Keller, A. C. Matzdorff, and B. Kemkes-Matthes, *Semin. Thromb. Hemost.*, **25**, 13 (1999); (c) L. A. Harker in A. C. G. Uprichard and K. P. Gallagher, Eds., *Handbook of Experimental Pharmacology*, Springer, Berlin, Vol. **132**, 1999, p. 353.

83. J. S. Kerr, H.-Y. Li, R. S. Wexler, et al., *Thromb. Res.*, **88**, 127 (1997).

84. (a) S. D. Kimball, *Curr. Pharm. Des.*, **1**, 441 (1995); (b) M. A. Lauer and A. M. Lincoff in A. C. G. Uprichard and K. P. Gallagher, Eds., *Handbook of Experimental Pharmacology*, Springer, Berlin, Vol. **132**, 1999, p. 331.

85. (a) H. K. Gold, F. W. Torres, H. D. Garabedian, et al., *J. Am. Coll. Cardiol.*, **21**, 1039 (1993); (b) J. T. Willerson and W. Casscells, *J. Am. Coll. Cardiol.*, **21**, 1048 (1993); (c) TRIM Study Group, *Eur. Heart J.*, **18**, 1416 (1997).

86. H. Bounameaux, H. Ehringer, A. Gast, et al., *Thromb. Haemost.*, **81**, 498 (1999).

87. S. Bajusz, E. Barabas, P. Tolnay, et al., *Int. J. Pept. Protein Res.*, **12**, 217 (1978).

88. R. A. Engh, R. Huber, W. Bode, et al., *Tibtechnology*, **13**, 503 (1995).

89. W. Bode, I. Mayr, U. Baumann, et al., *EMBO J.*, **8**, 467 (1989).

90. P. E. J. Sanderson, *Med. Res. Rev.*, **19**, 179 (1999).

91. (a) J. Sturzebecher, F. Markwardt, B. Voigt, et al., *Thromb. Res.*, **29**, 635 (1983); (b) S. Okamoto, A. Hijikata, R. Kikumoto, et al., *Biochem. Biophys. Res. Commun.*, **101**, 440 (1981).

92. (a) D. W. Banner and P. Hadvary, *J. Biol. Chem.*, **266**, 20085 (1991); (b) H. Brandstetter, D. Turk, H. Wolfgang Hoeffken, et al., *J. Mol. Biol.*, **226**, 1985 (1992).

93. R. Kikumoto, Y. Tamao, T. Tezuka, et al., *Biochemistry*, **23**, 85 (1984).

94. C. N. Berry, C. Girardot, C. Lecoffre, et al., *Thromb. Haemost.*, **72**, 381 (1994).

95. (a) W. H. Matthai, *Semin. Thromb. Hemost.*, **25** (Suppl. 1), 57 (1999); (b) S. Suzuki, S. Sakamoto, K. Susumu, et al., *Thromb. Res.*, **88**, 499 (1998).

96. (a) K. McKeage and G. L. Plosker, *Drugs*, **61**, 515 (2001); (b) J. Chen, *Heart Dis.*, **3**, 189 (2001); (c) W. Jeske, J. M. Walenga, B. E. Lewis, et al., *Expert Opin. Invest. Drugs*, **8**, 625 (1999).

97. T. B. Tschopp, J. Ackermann, A. Gast, et al., *Drugs Future*, **20**, 476 (1995).

98. (a) U. Bredbert, U. Ericksson, K. Taure, L. Johansson, et al., *Blood*, **94** (Suppl. 1), Abstr. 10 (1999); (b) F. L. Li-Saw-Hee and G. Y. H. Lip, *Curr. Opin. Cardiovasc. Pulm. Renal Invest. Drugs*, **1**, 88 (1999); (c) D. Gustafsson, T. Antonsson, and R. Bylund, *Thromb. Haemost.*, **79**, 110 (1998).

99. D. Gustafsson, J.-E. Nystrom, S. Carlsson, et al., *Thromb. Res.*, **101**, 171 (2001).

100. (a) P. E. J. Sanderson, T. A. Lyle, K. J. Cutrona, et al., *J. Med Chem.*, **41**, 4466 (1998); (b) P. E. J. Sanderson, K. J. Cutrona, B. D. Dorsey, et al., *Bioorg. Med. Chem. Lett.*, **8**, 817 (1998); (c) P. E. J. Sanderson, D. L. Dyer, A. M. Naylor-Olsen, et al., *Bioorg. Med. Chem. Lett.*, **7**, 1497 (1997).

101. (a) T. Steinmetzer, J. Hauptmann, and J. Sturzebecher, *Expert Opin. Invest. Drugs*, **10**, 845 (2001); (b) C. A. Coburn, *Expert Opin. Ther. Pat.*, **11**, 721 (2001); (c) P. E. J. Sanderson in A. N. Doherty, Ed., *Annual Reports in Medicinal Chemistry*, Academic Press, San Diego, CA, Vol. **36**, 2001, p. 79;(d) J. P. Vacca, *Curr. Opin. Chem. Biol.*, **4**, 394 (2000).

102. J. B. Hycraft, *Naunyn Schmiedebergs Arch. Exp. Pathol. Pharmacol.*, **18**, 209 (1884).

103. M. Salzet, *FEBS Lett.*, **492**, 187 (2001).

104. (a) T. J. Rydel, A. Tulinsky, W. Bode, et al., *J. Mol. Biol.*, **221**, 583 (1991); (b) M. G. Grutter, J. P. Priestle, J. Rahuel, et al., *EMBO J.*, **9**, 2361 (1990); (c) T. J. Rydel, K. G. Ravichandran, A. Tulinsky, et al., *Science*, **249**, 277 (1990).

105. (a) J. Dodt, *Angew. Chem., Int. Ed. Engl.*, **34**, 867 (1995); (b) A. Wallace, S. Dennis, J. Hofsteenge, et al., *Biochemistry*, **28**, 10079 (1989); (c) A. Betz, J. Hofsteenge, and S. R. Stone, *Biochemistry*, **31**, 4557 (1992).

106. A. J. Matheson and K. L. Goa, *Drugs*, **60**, 679 (2000).

107. (a) A. Greinacher and N. Lubenow, *Circulation*, **103**, 1479 (2001); (b) J. C. Adkins and M. I Wilde, *BioDrugs*, **10**, 227 (1998).

108. E. Skrzypczak-Jankun, V. E. Carperos, K. G. Ravichandran, et al., *J. Mol. Biol.*, **221**, 1379 (1991).

109. R. Scatena, *Curr. Opin. Cardiovasc. Pulm. Renal Invest. Drugs*, **2**, 189 (2000).

110. J. J. Slon-Usakiewwicz, J. Sivaraman, Y. Li, et al., *Biochemistry*, **39**, 2384 (2000).

111. (a) B. S. Coller, *Thromb. Haemost.*, **86**, 427 (2001); (b) S. S. Berkowitz, *Haeomostasis*, **30** (Suppl. 3), 27 (2001).

112. (a) M. L. Rand, M. A. Packham, D. Taylor, et al., *Thromb Haemost.*, **82**, 1680 (1999); (b) J. S. Bennett, *Ann. N. Y. Acad. Sci.*, **936**, 340 (2001).

113. Y. Li, F. A. Spencer, S. Ball, et al., *J. Thromb. Thrombolysis*, **10**, 69 (2000).

114. R. M. Scarborough, M. A. Naughton, W. Teng, et al., *J. Biol. Chem.*, **268**, 1066 (1993).

115. J. M. Smallheer, R. E. Olson, and R. R. Wexler in A. M. Doherty, Ed., *Annual Reports in Medicinal Chemistry*, Academic Press, San Diego, CA, Vol. **35**, 2001, p. 103.

116. S. Harder, C. M. Kirechmaier, H. J. Krzywanek, et al., *Circulation*, **100**, 1175 (1999).

117. D. J. Moliterno, *Am. Heart J.*, **139**, 563 (2000).

118. (a) S. A. Mousa, *Curr. Opin. Cardiovasc. Pulm. Renal Invest. Drugs*, **2**, 107 (2000); (b) A. E. P. Adang and J. B. M. Rewinkel, *Drugs Future*, **25**, 369 (2000); (c) K. Konstantopoulos and S. A. Mousa, *Curr. Opin. Invest. Drugs*, **2**, 1086 (2001); (d) S. A. Mousa, J. M. Bozarth, U. P. Naik, et al., *Br. J. Pharmacol.*, **133**, 331 (2001).

119. F. Liu, R. M. Craft, S. A. Morris, et al., *Expert Opin. Invest. Drugs*, **9**, 2673 (2000).

120. S. A. Dogrell, *Drugs Today*, **37**, 509 (2001).

121. C. P. Cannon, *Curr. Opin. Cardiovasc. Pulm. Renal Invest. Drugs*, **2**, 114 (2000).

122. M. Taniuchi, H. I. Kurz, and J. M. Lasala, *Circulation*, **104**, 539 (2001).

123. S. Yusuf, F. Zhao, S. R. Mehta, et al., *N. Engl. J. Med.*, **345**, 494 (2001).

124. (a) P. Urban, A. H. Gershlick, H.-J. Rupprecht, et al., *Circulation*, **100** (Suppl 1), 379 (1999); (b) M. E. Bertrand, H.-J. Rupprecht, and P. Urban, *Circulation*, **100** (Suppl 1), 620 (1999).

125. R. Marianne, R.-D. V. Marieke, J.-P. Montseny, et al., *Drug Metab. Dispos.*, **28**, 1405 (2000).

126. (a) A. Sugidachi, F. Asai, T. Ogawa, et al., *Br. J. Pharmacol.*, **129**, 1439 2000; (b) A. Sugidachi, F. Asai, K. Yoned, et al., *Br. J. Pharmacol.*, **132**, 47 (2001); (c) S. A. Doggrell, N. E. Mealy, and J. Castaner, *Drugs Future*, **26**, 835 (2001).

127. A. H. Ingall, J. Dixon, and A. Bailey, *J. Med. Chem.*, **42**, 23 (1999).

128. (a) R. F. Storey, *Drug Dev. Res.*, **52**, 202 (2001); (b) R. F. Storey, *Platelets*, **12**, 197 (2001); (c) S. C. Chattaraj, *Curr. Opin. Invest. Drugs*, **2**, 250 (2001).

129. (a) AstraZeneca, WO 0136438 (2001); (b) S. D. Guile, *Drug Dev. Res.*, **14** (Special Issue Purines 2000), Abstr. S06-03 (2000); (c) P. A. Willis, R. V. Bonnert, D. Cox, et al., Abstracts of Papers (Proceedings of the 220th National Meeting of the American Chemical Society, 2000), Abstr. MEDI-0189.

130. R. M. Scarborough, A. M. Laibelman, L. A. Clizbe, et al., *Bioorg. Med. Chem. Lett.*, **11**, 1805 (2001).

131. J.-M. Boeynaems, B. Robaye, R. Janssens, et al., *Drug Dev. Res.*, **52**, 187 (2001).

132. (a) N. Erathodiyil, S.-Y. Jang, S. Moro, et al., *J. Med. Chem.*, **43**, 829 (2000); (b) H. S. Kim, D. Barak, T. K. Harden, et al., *J. Med. Chem.*, **44**, 3092 (2001); (c) J. L. Boyer, M. Adams, S. Y. Jang, et al., *Drug Dev. Res.*, **50** (Suppl.), Abstr. S14-02 (2000).

133. K. Sak, A. Uri, E. Enkvist, et al., *Biochem. Biophys. Res. Commun.*, **272**, 327 (2000).

134. J. S. Bennett in C. T. Caskey, C. Austin, and J. Hoxie, Eds., *Annual Review of Medicine*, Vol. **52**, Palo Alto, CA, 2000, p. 161.

135. J. Geiger, *Expert Opin. Invest. Drugs.*, **10**, 865 (2001).

136. H. C. Diener, L. Cunha, C. Forbes, et al., *J. Neurol. Sci.*, **143**, 1 (1996).

137. (a) P. Sinnaeve and F. V. de Werf, *Thromb. Res.*, **103**, S71 (2001); (b) G. Albers, *Neurology*, **57**, S77 (2000).

138. (a) F. J. V de Werf, *Eur. Heart J.*, **20**, 1452 (1999); (b) C. F. Toombs, *Curr. Opin. Pharmacol.*, **1**, 164 (2001); (c) B. E. Sobel, *Coron. Artery Dis.*, **12**, 323 (2001); (d) H. M. Hoffmeister, S. Szabo, U. Helber, et al., *Thromb. Res.*, **103**, S51 (2001).

139. D. C. Rijken and D. V. Sakharov, *Thromb. Res.*, **103**, S41 (2001).

140. S. Vanwetswinkel, S. Plaisance, Z. Xhi-Yong, et al., *Blood*, **95**, 936 (2000).

141. L. A. Robbie, B. Bennett, B. A. Keyt, et al., *Br. J. Haematol.*, **111**, 517 (2000).

142. (a) T. Yokoyama, A. B. Kelley, and U. M. Marzec, *Circulation*, **92**, 485 (1995); (b) L. A. Harker, S. R. Hanson, J. N. Wilcox, et al., *Haemostasis*, **26** (Suppl. 1), 76 (1996).

143. The literature on direct inhibitors of FXa has been extensively reviewed: (a) See refs. 101c and d; (b) R. Rai, P. A. Sprengeler, K. C. Elrod, et al., *Curr. Med. Chem.*, **8**, 101 (2001); (c) A. Betz, *Expert Opin. Ther. Pat.*, **11**, 1007 (2001); (d) M. L. Quan and R. R. Wexler, *Curr. Top. Med. Chem.*, **1**, 137 (2001); (e) B.-Y. Zhu, W. Huang, and T. Su, *Curr. Top. Med. Chem.*, **1**, 101 (2001); (f) H. W. Pauls and W. R. Ewing, *Curr. Top. Med. Chem.*, **1**, 83 (2001).

144. G. B. Phillips, B. O. Buckman, D. D. Davey, et al., *J. Med. Chem.*, **41**, 3557 (1998).

145. (a) D. R. Abenschein, P. K. Baum, P. Verhallen, et al., *J. Pharmacol. Exp. Ther.*, **296**, 567 (2001); (b) T. B. McClanahan, G. W. Hicks, and A. L. Morrison, *Eur. J. Pharmacol.*, **432**, 187 (2001); (c) G. Gibson, R. Totoczak, Y. Peng, et al., Throm. Haemost. [Proc. of the 18th Cong. Int. Soc. Thromb. Haemostas. (July 6–12, 2001, Paris) 2001], Suppl., Abstr. P1604 (2001).

146. M. Adler, D. D. Davey, G. B. Phililips, et al., *Biochemistry*, **39**, 12534 (2000).

147. (a) D. J. P. Pinto, M. J. Orwat, S. Wang, et al., *J. Med. Chem.*, **44**, 566 (2001); (b) A. W. Leamy, J. M. Luettgen, P. C. Wong, et al., Thromb. Haemost. [Proc. of the 18th Cong. Int. Soc. Thromb. Haemostas. (July 6–12, 2001, Paris) 2001], Suppl., Abstr. P1608 (2001); (c) J. S. Barrett, et al., J. Clin. Pharmacol. [Proc. of the 30th Annual Meeting of the American College of Clinical Pharmacologists (September 23–25, 2001, Vienna, VA) 2001], **49**, Abstr. 42 (2001).

148. (a) Y. K. Yee, A. L. Tebbe, J. H. Linebarger, et al., *J. Med. Chem.*, **43**, 873 (2000); (b) D. K. Herron, T. Goodson, M. R. Wiley, et al., *J. Med. Chem.*, **43**, 859 (2000); (c) M. R. Wiley, L. C. Weir, S. Briggs, et al., *J. Med. Chem.*, **43**, 883 (2000).

149. G. B. Poli, Y. M. Choi-Sledeski, R. M. Kearney, et al., Abstracts of Papers, Proceedings of the 221st National Meeting of the American Chemical Society, Abstr. MEDI-068 (2001).

150. G. P. Vlasuk, *Thromb. Haemost.*, **70**, 212 (1993).

151. A. Wei, R. S. Alexander, J. Duke, et al., *J. Mol. Biol.*, **283**, 147 (1998).

152. (a) F. A. Nicolini, P. Lee, J. L. Malycky, et al., *Blood Coagul. Fibrinolysis*, **7**, 39 (1996); (b) H. F. Kotze, S. Lamprecht, P. N. Badenhorst, et al., *Thromb. Haemost.*, **77**, 1137 (1997).

153. (a) J. Himber, C. J. Refino, L. Burcklen, et al., *Thromb. Haemost.*, **85**, 475 (2001); (b) C. R. Benedict, J. Ryan, B. Wolitzky, et al., *J. Clin. Invest.*, **88**, 1760 (1991).

154. (a) S. B. Sheth, D. Wilson, H. M. Davis, et al., *Blood*, **92**, 3629 (Abstr. 1491) (1998); (b) J. R. Toomey, M. N. Blackburn, and G. Z. Feuerstein, *Curr. Opin. Cardiovasc. Pulm. Renal Invest. Drugs*, **2**, 124 (2000).

155. E. De Jong, P. E. P. Dekkers, and A. A. Creasey, *Blood*, **95**, 1124 (2000).

156. (a) G. Vlasuk, A. Bradbury, P. Bergun, et al., *J. Am. Coll. Cardiol.*, **33**, Abstr. 255a–256a (1999); (b) A. Lee, G. Agnelli, H. Buller, J. Ginsberg, et al., *Blood*, **95**, Abstr. 491a (2000).

157. T. J. Girard and N. S. Nicholson, *Curr. Opin. Pharmacol.*, **1**, 159 (2001).

158. A. M. Lincoff, *J. Am. Coll. Cardiol.*, **36**, 312 (2000).

159. (a) W. B. Young, A. Kolesnikov, R. Rai, et al., *Bioorg. Med. Chem. Lett.*, **11**, 2253 (2001); (b) Ono Pharmaceutical, WO 9941231 (1999); Eur. Pat. 1078917 (2001); (c) Genentech, Inc., WO 0041531 (2000).

160. (a) J. Himber, L. Burcklen, T. Tschopp, et al., Thromb. Haemost. [Proc. of the 18th Cong. Int. Soc. Thromb. Haemostas. (July 6–12, 2001, Paris) 2001], Suppl., Abstr. P1387 (2001); (b) J. Fingerle, J. Himber, T. Tschopp, et al., Thromb. Haemost. [Proc. of the 18th Cong. Int. Soc. Thromb. Haemostas. (July 6–12, 2001, Paris) 2001], Suppl., Abstr. OC1766 (2001).

161. J.-M. Dogne, X. Leval, J. Delarge, and B. Masereel, *Expert Opin. Ther. Pat.*, **11**, 1663 (2001).

162. R. Soyka, B. D. Guth, H. M. Weisenberger, et al., *J. Med. Chem.*, **42**, 1235 (1999).

163. S. Rolin, J. M. Dogne, C. Michaux, et al., *Prostaglandins Leukot. Essent. Fatty Acids.*, **65**, 67 (2001).

164. (a) L. Rosenfeld, G. J. Grover, and C. T. Stier Jr., *Cardiovasc. Drug Rev.*, **19**, 97 (2001); (b) R. N. Misra, B. R. Brown, P. M. Sher, et al., *J. Med. Chem.*, **14**, 1401 (1993); (c) B. Cimetiere, T. Dubuffet, O. Muller, et al., *Bioorg. Med. Chem. Lett.*, **8**, 1375 (1998); (d) S. Simonet, J.-J. Descombes, and M.-O. Valleez, *Adv. Exp. Med. Biol.*, **433**, 173 (1997); (e) J. Osende, D. Shimbo, V. Fuster, et al., Thromb. Haemost. [Proc. of the 18th Cong. Int. Soc. Thromb. Haemostas. (July 6–12, 2001, Paris) 2001], Suppl., Abstr. P1629 (2001).

165. (a) S. Chackalamannil, D. Doller, K. Eagen, et al., *Bioorg. Med. Chem. Lett.*, **11**, 2851 (2001); (b) J. C. Barrow, P. G. Nantermet, H. G. Selnick, et al., *Bioorg. Med. Chem. Lett.*, **11**, 2691 (2001); (c) Y. Kato, Y. Kita, M. Nishio, *Eur. J. Pharmacol.*, **384**, 197 (1999); (d) S. Chackalamannil, *Curr. Opin. Drug Discov. Dev.*, **4**, 417 (2001).

166. (a) Y. Kato, Y. Kita, M. Nishio, et al., *Eur. J. Pharmacol.*, **384**, 197 (1999); (b) H.-C. Zhang, C. K. Derian, P Andrade-Gordon, et al., *J. Med. Chem.*, **44**, 1021 (2001); (c) P. Andrade-Gordon, et al., Blood [Proc. of the 43rd Annual Meeting of the American Society of Hematologists (Dec 7–11, 2001, Orlando, FL) 2001], **98**, Abstr. 175 (2001).

167. (a) W. H. Frishman and P. Grewall, *Ann. Med.*, **32**, 195 (2000); (b) F. De Clerck, *J. Cardiovasc. Pharmacol.*, **17** (Suppl. 5), S1 (1991); (c) X. Rabasseda and N. Mealy, *Med. Actual.*, **30**, 43 (1994).

168. (a) G. McCort, C. Hoornaert, M. Alertru, et al., *Bioorg. Med. Chem.*, **9**, 2129 (2001); (b) M. Fujio, T. Kuroita, Y. Sakai, et al., *Bioorg. Med. Chem. Lett.*, **10**, 2457 (2000); (c) N. Tanaka, R. Goto, R. Ito, et al., *Chem. Pharm. Bull.*, **48**, 1729 (2000); (d) H. Kihara, H. Koganei, K. Hirose, et al., *Eur. J. Pharmacol.*, **433**, 157 (2001); (e) D. Pawlak, K. Pawlak, E. Chabielska, et al., *Thromb. Res.*, **90**, 259 (1998).

169. (a) K. Taubaki, K. Taniguchi, S. Tabuchi, et al., *Bioorg. Med. Chem. Lett.*, **10**, 2787 (2000); (b) S. M. Seiler, C. L. Brassard, M. E. Federici, et al., *Prostaglandins*, **53**, 21 (1997); (c) N. Hamanaka, K. Takahashi, Y. Nagao, et al., *Bioorg. Med. Chem. Lett.*, **5**, 1083 (1995).

170. (a) F. von Bruchhausen in F. von Bruchhausen and Ulrich Walter, Eds., *Platelets and Their Factors*, Springer-Verlag, Berlin/New York, 1997, Chapter 28; (b) G. Montrucchio, G. Alloatti, and G. Camussi, *Physiol. Rev.*, **80**, 1669 (2000).

171. (a) D. Handley in F. von Bruchhausen and Ulrich Walter, Eds., *Platelets and Their Factors*, Springer-Verlag, Berlin/New York, 1997, Chapter 24; (b) P. Seth, R. Kumari, M. Dikshit, et al., *Thromb. Res.*, **76**, 503 (1994); (c) S. K. Chung, S. H. Ban, S. H. Kim, et al., *Bioorg. Med. Chem. Lett.*, **10**, 1097 (1995); (d) G. Biagi, I. Giorgi, O. Livi, et al., *Farmaco*, **51**, 761 (1996).

172. (a) M. Fujita, T. Seki, H. Inada, et al., *Bioorg. Med. Chem. Lett.*, **12**, 341 (2002); (b) S. Simonet, J.-J. Descombes, C. Courchay, et al., Thromb. Haemost. [Proc. of the 18th Cong. Int. Soc. Thromb. Haemostas. (July 6–12, 2001, Paris) 2001], Suppl., Abst P1631 (2001).

173. R. B. Wallis, S. Finney, R. T. Sawyer, et al., *Blood Coagul. Fibrinolysis*, **8**, 291 (1997).

174. (a) Y. Iwata, K. Tago, T. Kiho, et al., *J. Mol. Graph. Modell.*, **18**, 591 (2000); (b) H. Kogen, T. Kiho, K. Tago, et al., *J. Am. Chem. Soc.*, **122**, 1842 (2000).

175. (a) K. E. Achyuthan, T. F. Slaughter, M. A. Santiago, et al., *J. Biol. Chem.*, **268**, 21284 (1993); (b) K. F. Freund, K. P. Doshi, S. L. Gaul, et al., *Biochemistry*, **33**, 10109 (1994); (c) A. A. Tymiak, J. G. Tuttle, S. D. Kimball, et al., *J. Antibiot.*, **46**, 204 (1993).

176. (a) M. Nagashima, M. Werner, L. Zhao, et al., *Thromb. Res.*, **98**, 333 (2000); (b) C. J. Refino, D. Schmitt, C. Pater, et al., *Fibrinolysis Proteolysis*, **12**, 29 (1998).

177. (a) Meiji Seika Kaisha, Ltd., WO 0119836 (2000); (b) AstraZeneca, WO 0066550 (2000) and WO 066557 (2000).

178. (a) C. N. Berry, C. Lunven, I Lechaire, et al., *Br. J. Pharmacol.*, **125**, 29 (1998); (b) J. J. van Giezen, V. Nerme, and T. Abrahamsson, *Blood Coagul. Fibrinolysis*, **9**, 8 (1998).

179. (a) P. Bjorquist, J. Ehnebon, T. Inghardt, et al., *Biochemistry*, **37**, 1227 (1998); (b) P. W. Friederich, M. Levi, B. J. Biemond, et al., *Circulation*, **96**, 916 (1997); (c) A. Folkes, M. B. Roe, S. Sohal, et al., *Bioorg. Med. Chem. Lett.*, **11**, 2589 (2001); (d) 3-Dimensional Pharmaceuticals, WO 0174793 (2001); (e) Adir et Compagnie, U.S. Pat. 6,048,875 (2001) and U.S. Pat. 6,302,837 (2001); (f) Corvas International, WO 0136351 (2001).

180. (a) A.-P. Bijnens, A. Gils, J. M. Stassen, et al., *J. Biol. Chem.*, **276**, 44912 (2001); (b) T. Wind, M. A. Jensen, and P. A Andreasen, *Eur. J. Biochem.*, **268**, 1095 (2001).

181. D. T. Eitzman, W. P. Fay, D. A. Lawrence, et al., *J. Clin. Invest.*, **95**, 2416 (1995).

182. G. L. Reed, *Hybridoma*, **16**, 281 (1997).

183. G. L. Reed, G. R. Matsueda, and E. Haber, *Circulation*, **82**, 164 (1990).

184. G. L. Reed, L. Harris, J. Bajorath, et al., WO 9812334 (1999).

185. (a) Y. Fukuyama, M. Kodama, I. Miura, et al., *Chem. Pharm. Bull.*, **38**, 133 (1990); (b) Y. Fukuyama, M. Kodama, I Miura, et al., *Chem. Pharm. Bull.*, **37**, 2438 (1989).

186. (a) C. D. Nicholson and A. W. A. Lensing, *Expert Opin. Invest. Drugs*, **10**, 785 (2001); (b) Editorial, *N. Engl. J. Med.*, **345**, 1340 (2001); (c) The Direct Thrombin Inhibitor Trialists' Collaborative Group, *Lancet*, **359**, 294 (2002).

CHAPTER SEVEN

Antihyperlipidemic Agents

MICHAEL L. SIERRA
Centre de Recherches
Laboratoire GlaxoSmithKline
Les Ulis, France

Contents

Burger's Medicinal Chemistry and Drug Discovery
Sixth Edition, Volume 3: Cardiovascular Agents and
Endocrines
Edited by Donald J. Abraham
ISBN 0-471-37029-0   © 2003 John Wiley & Sons, Inc.

1 INTRODUCTION

Atherosclerosis and the resulting complications including coronary artery disease, peripheral vascular disease, and cerebrovascular disease are responsible for 41% of all deaths in the United States (1a) and remain the leading cause of mortality and morbidity in industrialized nations. Despite the identification of many of the risk factors for the development of atherosclerosis and initiation of behavioral and pharmaceutical intervention to reduce risk, cardiovascular (CV) disease remains a significant health-care burden, estimated by the American Heart Association in 2000 to cost some $326 million (1a). There are over 650,000 new myocardial infarctions and 450,000 recurrent myocardial infarctions each year and about one third of all the people affected die (1a). Although the age-adjusted death rate from CV disease is falling, the overall death rate is little changed because of aging of the population. Of the deaths from CV disease, a large proportion is attributed to coro-

nary heart disease and other direct manifestations of atherosclerosis.

There are a number of nonmodifiable risk factors for CV disease such as age, sex, and family history of disease, but also a number of modifiable risk factors including dyslipidemia, hypertension, obesity, diabetes mellitus, and tobacco smoking. It is estimated, by the use of the current National Cholesterol Education Program (NCEP) guidelines (1b), that more than 250 million people worldwide can be classified as having some form of dyslipidemia.

Dyslipidemia is described in terms of elevation of lipid (cholesterol and triglyceride) in atherogenic lipoprotein particles LDL (low density lipoprotein), IDL (intermediate density lipoprotein), and VLDL (very low density lipoprotein), or reduction in cholesterol carried in protective HDL (high density lipoprotein).

Lipids are carried in the blood in lipoprotein particles with triglyceride (TG) and cholesteryl ester carried in the interior and phospholipid, free cholesterol, and apolipoproteins at the surface. These particles can be classified according to size (density):

1. Chylomicrons are postprandial particles containing mainly TG with apolipoprotein B48, have a density $(d) < 0.94$ g/mL, and are poorly atherogenic.
2. VLDL particles are of hepatic origin, carry mainly TG with some cholesteryl ester, have a density range $d = 0.94-1.006$ g/mL, and are weakly atherogenic.
3. IDL particles are formed from the catabolism of VLDL, are enriched in cholesteryl esters, with a density range $d = 1.006-1.019$ g/mL, and are atherogenic.
4. LDL particles are rich in cholesteryl ester, poor in TG, with a density $d = 1.019-1.063$ g/mL, and are highly atherogenic, especially small dense LDL (sdLDL). VLDL, IDL, and LDL carry apolipoprotein B100.
5. HDL particles are formed from hepatic or intestinal apolipoprotein AI through a process of lipid transfer from other lipoprotein particles and peripheral tissues; they are very rich in cholesterol, with a density $d > 0.063$ g/mL. There is a strong, independent, and inverse association between low concentrations of HDL-cholesterol (HDLc) and coronary heart disease (CHD) risk (2).

Although there is debate about the strength of association between some forms of dyslipidemia and the risk of developing cardiovascular disease, the evidence exists that the reduction in cholesterol and triglyceride in low density lipoprotein classes leads to a reduction in CHD mortality and morbidity in primary and secondary prevention studies and in pa-tients with both severe and borderline dyslipidemia (3–18). The strongest link with atherosclerosis exists for elevated concentrations of LDL-cholesterol (LDLc) and clinical trial data support the benefits of aggressive LDLc lowering in appropriate populations (19, 20). However, recent epidemiological data have reaffirmed that elevated plasma TG and low HDL-cholesterol levels are also important risk factors for atherosclerotic vascular disease. A major function of HDLc is the transport of cholesterol from peripheral tissues to the liver, a process known as reverse cholesterol transport. Low HDLc concentrations may lead to a failure to export lipid from the vessel wall, leading to atherosclerotic plaque development and hence increased CV risk.

Several assessments of the importance of elevated TG as an independent risk factor for CHD have also been completed. Most important, multivariate analyses have demonstrated that elevated TG concentrations: (1) increase CHD risk independent of HDLc (21, 22); (2) increase significantly the number of cardiovascular events when combined with an elevated LDLc/HDLc ratio (23, 24); (3) are an independent risk factor in women aged 50–69 years (19); and (4) are an independent risk factor in type 2 diabetes mellitus (DM) patients (25).

Drug therapy can be considered for patients who, in spite of adequate dietary therapy, regular physical activity, and weight loss, need further treatment for elevated blood cholesterol levels. Current NCEP guidelines (1b) are shown in Table 7.1.

There is also increasing awareness that current lipid-lowering therapies lack suffi-

Table 7.1 Current NCEP Guidelines for Treatment

Patient	LDLc (mg/dL)	
	Start	Goal
With CHD or CHD risk equivalents[a]	≥130	<100
Without CHD and ≥2 risk factors[b]	≥130 (10-year risk: 10–20%)	<130
	≥160 (10-year risk: <10%)	
Without CAD and <2 risk factors[b]	≥190	<160

[a]CHD risk equivalents are: (1) CAD and PVD; (2) diabetes mellitus; (3) multiple risk factors that confer a 10-year risk factor for CHD > 20%.

[b]Risk factors are: (1) male ≥ 45 years/female ≥ 55 years; (2) smoker; (3) hypertension; (4) family history of early CAD; (5) HDLc < 40 mg/dL.

cient efficacy or safety in the treatment of the full spectrum of lipid abnormalities.

2 CLINICAL APPLICATIONS

2.1 The Role of Lipids in the Atherogenic Process

Specific sites in the vasculature associated with decreased shear stress or increased turbulence, such as bifurcations and branches, are favored sites for the development of arterial lesions (26). The triggers for the development of atherosclerosis are not known, but a key initial phase in the development of atherosclerosis is the retention of cholesterol-rich lipoproteins and remnants in the subendothelial space (27, 28).

Lipid deposition is followed by an increase in the number of inflammatory cells, including macrophages in the intimal space. These macrophages become engorged with lipid, forming foam cells, visible as nonobstructive fatty streaks on the endothelial surface of the aorta and coronary arteries. The streaks contain large accumulations of lipid-filled macrophages and smooth muscle cells and fibrous tissue. The main lipid component is intracellular cholesteryl ester. The progression of these lesions from simple fatty streaks to advanced lesions (29) is a process dependent on hemodynamic forces, plasma levels of atherogenic lipoproteins (30), and other risk factors.

Maturing plaques (Starey class IV/V) are areas of intimal thickening, with a fibrous cap covering a central core made up of intracellular and extracellular lipid and necrotic cell debris. These plaques may be angiographically silent, encased within a thickened arterial wall or may project into the lumen and obstruct the arterial flow, and are detectable by angiography.

Historically, it was believed that the plaques gradually encroached into the lumen until flow was either severely restricted or totally occluded. When this occurred in the coronary artery, the patient would present clinically with either angina or an acute myocardial infarction (MI). Recent research has shown, however, the importance of angiographically silent, small plaques (30). These smaller plaques are thought to perforate or rupture through a combination of inflammatory cell destruction of the fibrous cap and physical disruption attributed to blood flow. Rupture is followed by formation of a thrombus, which obstructs the artery and precipitates acute coronary syndromes.

The reasons for the "instability" of these plaques are not clear, but they are known to have an increased lipid content and decreased fibrous cover (31). The most common plaque fissures are the junctions of the plaque and the normal intima (32, 33), a site often containing lipid-laden macrophages, which are thought to play a role in weakening the plaque (34). Recent work suggests that progressive accumulation of lipids may destabilize plaques, thus leading to thinning, weakening, and ultimately destruction of the fibrous cap, which is followed by subsequent rupture of the plaque.

Cholesterol in plasma clearly plays a role both in the development of "stable" and "unstable" plaques. Lipid-lowering therapy in humans is thought to play a role, not only in preventing the formation and growth of plaques but also in reducing the cholesterol content of plaques already formed.

2.2 Current Drugs on the Market

The four major pharmacological classes of drugs routinely used in the treatment of simple hypercholesterolemia (SH), mixed dyslipidemia (MD), and hypertriglyceridemia (HTG)/low HDL are: (1) HMG-CoA reductase inhibitors; (2) fibric acid derivatives; (3) bile acid sequestrants/cholesterol absorption inhibitors; and (4) nicotinic acid derivatives (see Table 7.2). The drug classes and summary effects on lipoprotein classes are summarized in Tables 7.3a and 7.3b.

2.2.1 HMG-CoA Reductase Inhibitors (Statins).

3-Hydroxy-3-methylglutarylcoenzyme A (HMG-CoA) reductase inhibitors were first approved by the U.S. Food and Drug Administration (FDA) in 1987. They have since become the most widely prescribed drug class for the treatment of hypercholesterolemia. Mevastatin (1), the initial drug of this class, was discovered in extracts from *Penicillium citrinum* by Sankyo (Japan) in 1976 (36). Mevastatin (1) was subsequently shown to inhibit

Table 7.2 Marketed Antihyperlipidemic Agents

Generic Name (structure)	Trade Name	Originator	Dose (mg/day)[a]
HMG-CoA reductase inhibitors (statins)			
Lovastatin (**2**)	Mevacor	Merck & Co. Inc	20–80
Fluvastatin (**5**)	Lescol	Novartis AG	20–80
Pravastatin (**3**)	Pravachol	Sankyo/Bristol-Myers Squibb Co.	20–40
Simvastatin (**4**)	Zocor	Merck & Co. Inc.	20–80
Atorvastatin (**6**)	Lipitor	Warner-Lambert (now Pfizer Inc.)	10–80
Cerivastatin[b] (**8**)	Baycol	Bayer AG	0.2–0.8
Pitavastatin (**9**) (previously itavastatin)	TBD[d]	Kowa/Nissan Chemical Industries Ltd/Novartis/ Sankyo	1–4 (?)[c]
Rosuvastatin (**7**)	Crestor	Shionogi/AstraZeneca	10–80
Fibric acid derivatives (fibrates)			
Clofibrate (**10**)	Widely generic	Widely generic	2 × 1000
Gemfibrozil (**11**)	Lopid	Pfizer Inc.	2 × 600
Bezafibrate (**13**)	Bezalip	Roche	400–600
Micronized fenofibrate (**14**)	Tricor/Lipantil	Fournier/Abbott	200
Ciprofibrate (**16**)	Ciprol	Sterling Drug Ltd. (now Sanofi-Synthelabo)	100
Bile acid sequestrants (BAS)/cholesterol absorption inhibitors			
Cholestyramine (**17**)	Questran	Bristol-Myers Squibb Co.	4000–16,000
Colestipol (**18**)	Widely generic	Widely generic	5000–20,000
Colesevelam (**19**)	Welchol	GelTex	2600–3800
Ezetimibe (**20**)	Zetia	Schering Plough Corp.	1–40
Ezetimibe + simvastatin	TBD[d]	Schering Plough Corp./ Merck & Co. Inc.	10 mg each
Nicotinic acid derivatives			
Niacin (**21**)	Niaspan	Kos Pharmaceuticals Inc.	1000–2000
Niacin + lovastatin	Advicor	Kos	500/10–2000/40 (niacin/lovastatin)
Miscellaneous			
Probucol (**22**)	Sinlestal	Hoechst Marion Roussel (now Aventis Pharma)	2 × 500
Tamoxifen (**23**)	Nolvadex	Zeneca (now AstraZeneca)	20
Toremifene (**24**)	Fareston	Orion Corp.	60
Raloxifene (**25**)	Evista	Eli Lilly & Co.	60

[a]Administered orally unless otherwise noted.

[b]Bayer voluntarily withdrew cerivastatin from all markets in August 2001 after increasing reports of adverse effects (35; see Section 2.3.1).

[c]Sanyo announced in November 2001 that its U.S. subsidiary would undertake a second Phase II trial of pitavastatin, using a lower dose, because of cases of muscle pain associated with the elevation of CPK (creatine phosphokinase), a marker of myopathy, at the higher doses (199; see Section 2.3.1.2).

[d]TBD = to be determined.

HMG-CoA reductase (37), the rate-limiting enzyme in cholesterol biosynthesis (38, 39), and to lower serum cholesterol in dogs (40) and cynomolgus monkeys (41). A structurally similar natural product, lovastatin (**2**), was shown to reduce total cholesterol levels in healthy volunteers (42). Both pravastatin (**3**) and simvastatin (**4**) are chemical modifica- tions of lovastatin (**2**). Lovastatin (**2**) and sim- vastatin (**4**) are administered as the inactive lactones, which must be metabolized *in vivo* to their corresponding open hydroxy-acid forms, whereas pravastatin (**3**) is taken as the active open hydroxy-acid form.

The first-generation synthetic statin was fluvastatin (**5**), which is also administered as

Table 7.3a   Current Pharmacological Effects of Therapies in the Treatment of Patients with Mixed Dyslipidemia (characterized by high LDLc and high TGs)

Drug Class	Effects on Lipoproteins			
	Total Cholesterol	LDLc	HDLc	TG
HMG-CoA reductase inhibitors	↓ 15–45%	↓ 20–60%	↑ 2–12%	↓ 10–37%
Fibric acid derivatives	↓ 10–25%	↓ 15–30%	↑ 10–30%	↓ 20–50%
Bile acid sequestrants/cholesterol absorption inhibitors	↓ 8–30%	↓ 15–20%	↑ 0–2%	↑ 5–10%

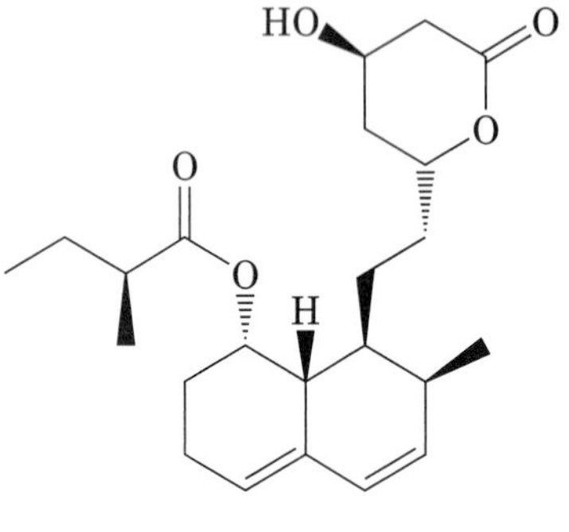

(1)

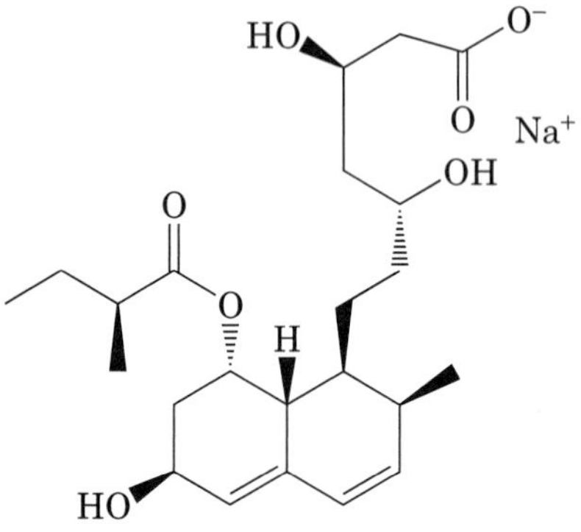

(3)

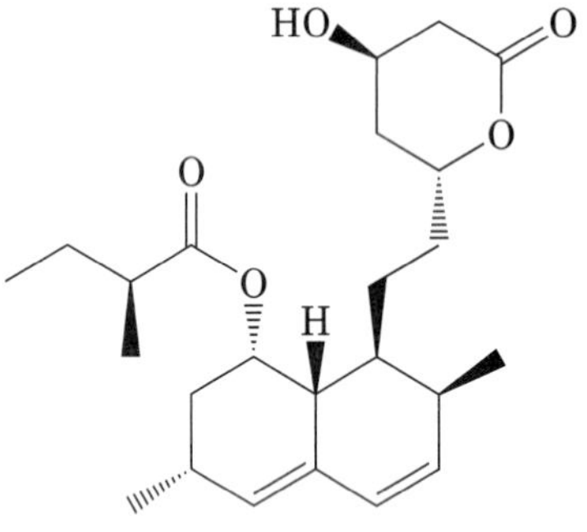

(2)

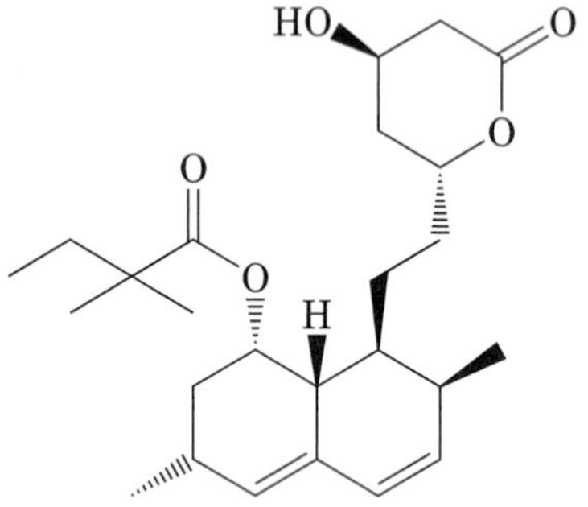

(4)

Table 7.3b   Current Pharmacological Effects of Therapies in the Treatment of Hypertriglyceridemia (characterized by high TGs and low HDL)

Drug Class	Effects on Lipoproteins			
	Total Cholesterol	LDLc	HDLc	TG
HMG-CoA reductase inhibitors	↓ 15–45%	↓ 20–50%	↑ 2–12%	↓ 10–37%
Fibric acid derivatives	NC[a]	↑ 0–15%	↑ 10–35%	↓ 20–60%
Nicotinic acid derivatives	↓ 10–15%	↓ 15–20%	↑ 15–30%	↓ 20–50%

[a]NC, no substantial change.

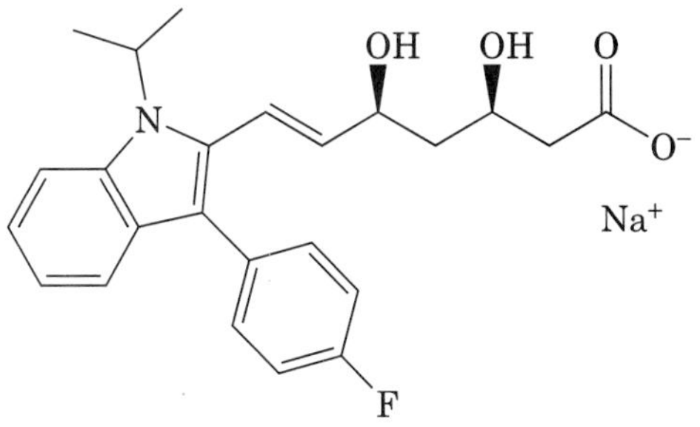

(5)

the active drug substance. The clinical success of the statins has led to newer synthetic molecules, which are more potent [atorvastatin (**6**) and rosuvastatin (**7**)] or equipotent [cerivastatin (**8**) and pitavastatin (**9**)] on a milligram per milligram basis compared with fluvastatin (**5**).

Statins are the drugs of choice for reducing LDL cholesterol (LDLc) and the second-generation statins are moderately effective in reducing TGs at higher doses and when baseline TG concentrations are >150 mg/dL (43). At the

currently approved FDA maximal doses, atorvastatin (**6**) 80 mg is the most efficacious statin, with LDL reductions of 50 to 60% (44), followed by simvastatin (**4**) 80 mg (47%) (44), cerivastatin (**8**) 0.8 mg (42%) (45), lovastatin (**2**) 80 mg (40%) (44), pravastatin (**3**) 40 mg (35%) (44), and fluvastatin (**5**) 40 mg (30%) (44). Recently published Phase III data for rosuvastatin (**7**) suggest that it is better than atorvastatin (**6**) at lowering LDL and in the percentage of patients treated that reach the NCEP goal of less than 100 mg/dL for LDLc (46–48). Pitavastatin (**9**) does not appear to offer any advantage of greater LDL lowering over the existing statins on the basis of two Phase III studies in Japanese patients (49, 50), so only randomized controlled clinical trials to assess the long-term effects on CAD would allow a differentiation. In addition to statin monotherapy, there is increasing interest in combination therapy to obtain greater reductions in LDLc as well as benefits on other risks factors. These combinations are discussed in later sections.

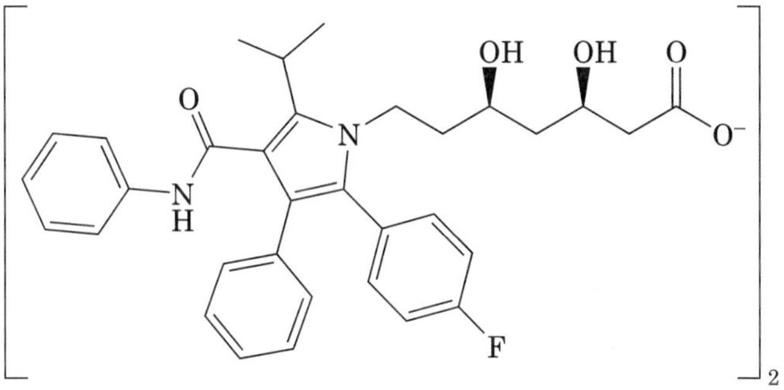

(6)

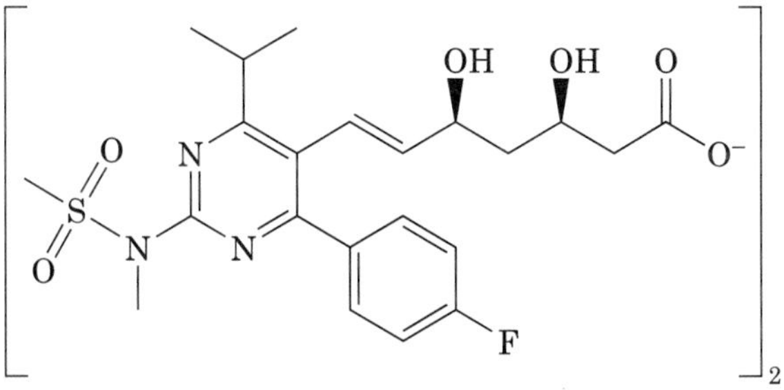

(7)

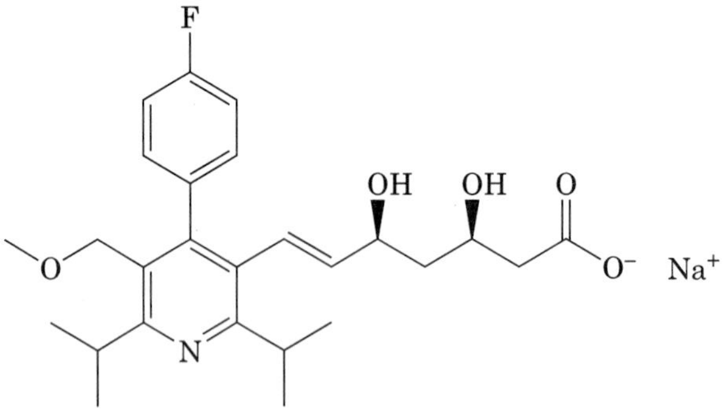

(8)

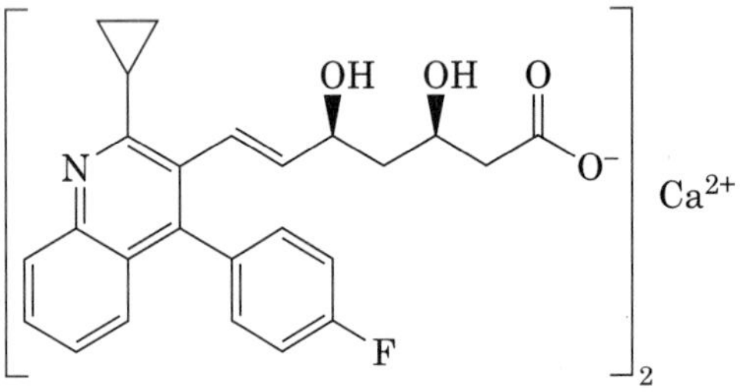

(9)

(11)

2.2.2 Fibric Acid Derivatives (Fibrates). Fibrates have been shown to affect the expression of genes implicated in the regulation of
HDL and TG-rich lipoproteins as well as fatty
acid metabolism through the activation of the
peroxisome proliferator-activated receptor α
(PPARα) (51–53). The current fibrates are all
weak PPARα agonists (i.e., require high micromolar concentrations for receptor activation), which may explain why high doses are
required for their clinical activity (54). They
were developed as hypolipidemic agents
through optimization of their lipid-lowering
activity in rodents before the discovery of the
PPARs. Clofibrate (**10**) and gemfibrozil (**11**)
are two of the older fibrates that have been

shown to moderately lower LDLc and increase
HDLc levels (55–57).

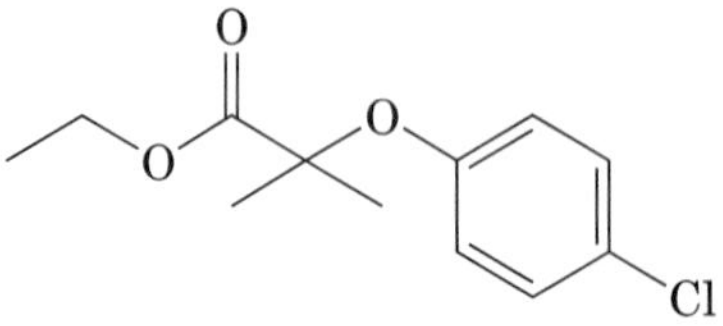

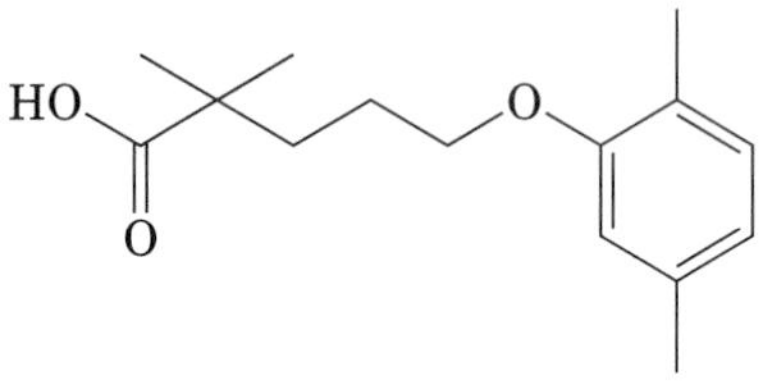

Clofibrate (**10**) is administered as its inactive ester, which is hydrolyzed *in vivo* to the
active drug, clofibric acid (**12**). Although clofi-

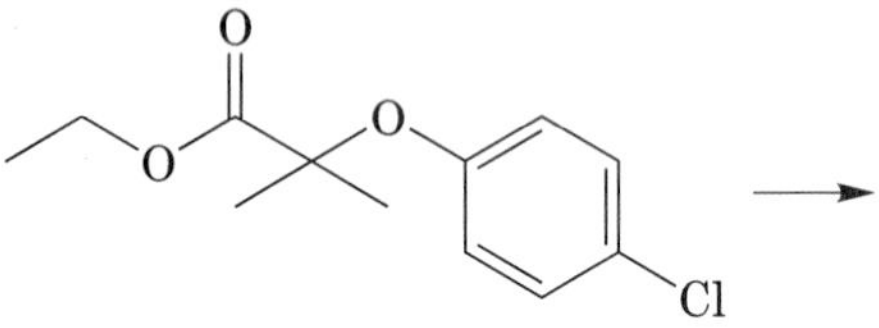

(10)

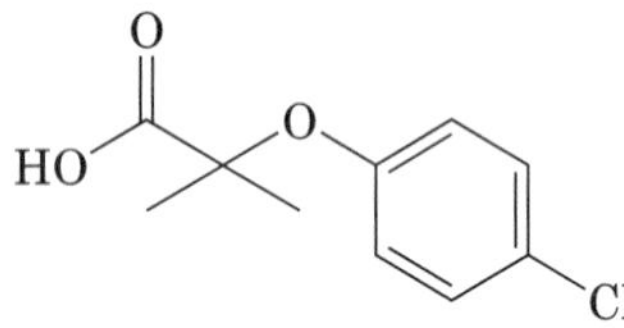

(12)

brate was shown to improve the lipoprotein
profile and cardiac events in several clinical
trials (58–60), there was a greater all-cause
mortality (59, 60). Clofibrate (**10**) is no longer
recommended as a lipid-lowering agent be-

(10)

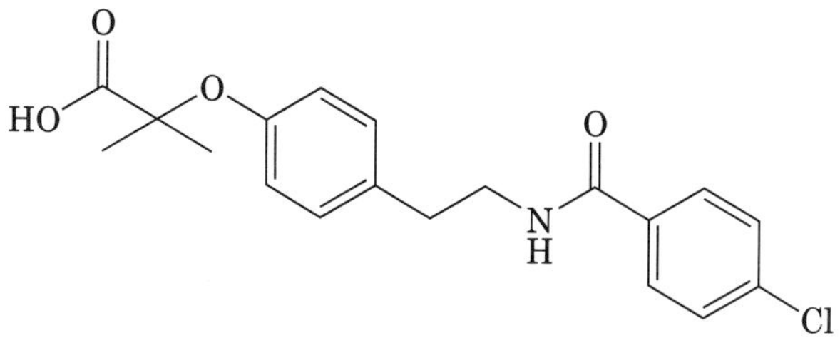

(13)

cause of the increase in overall mortality and the adverse events that are associated with its use.

Gemfibrozil (11), on the other hand, has demonstrated CV benefit in three studies (57, 61, 62). In particular, the VA-HIT trial (57) demonstrated a 22% benefit in CV morbidity and mortality with a 6% increase in HDLc, a 31% decrease in TGs, and no effect on LDLc. The results of this study stress the importance of low HDLc as a major risk factor for CHD. The newer fibrates, bezafibrate (13) and fenofibrate (14), also reduce TGs and increase

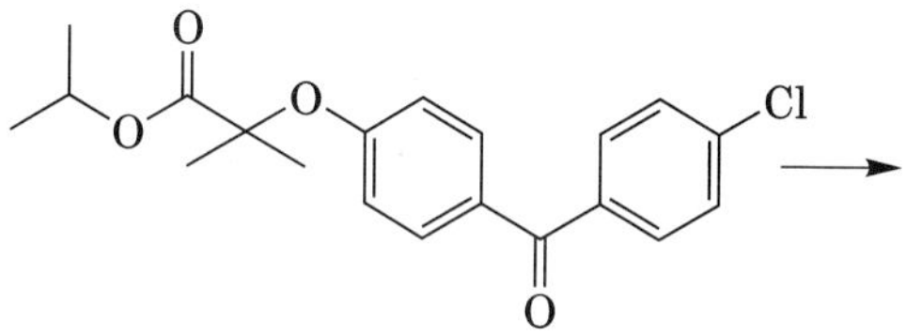

(14)

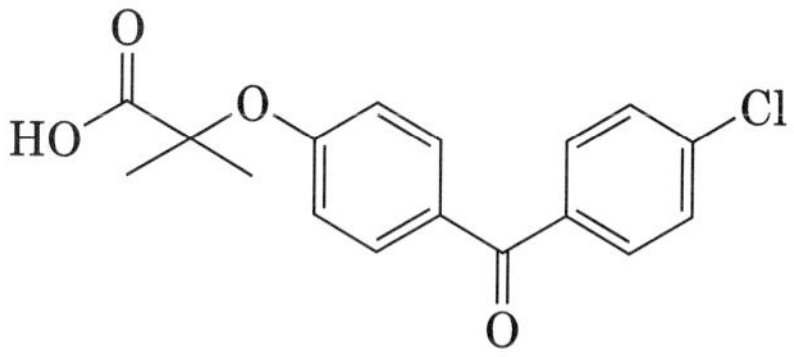

(15)

HDLc but to a greater extent than clofibrate (10) or gemfibrozil (11) because of increased receptor affinity (53). This may also explain the greater extent of LDLc lowering observed with bezafibrate (13) and fenofibrate (14). Like clofibrate, fenofibrate is administered as the inactive ester, which is hydrolyzed *in vivo* to the active form, fenofibric acid (15).

Ciprofibrate (16) was first launched in Europe in 1985 but does not show any advantage over current fibrate treatment (63) at doses

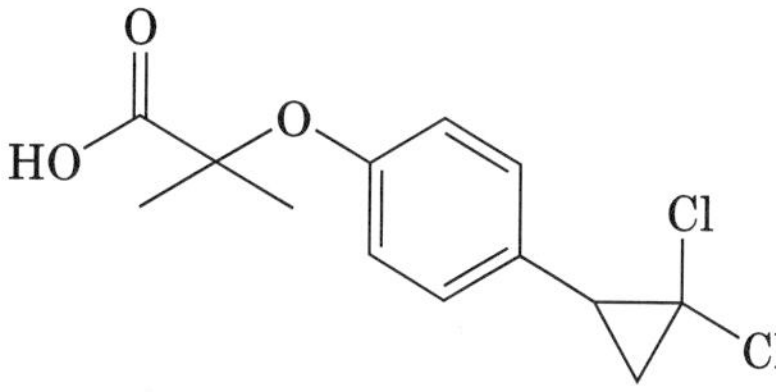

(16)

$\leq$100 mg/day. Because of reports that doses of $\geq$200 mg/day have been linked to rhabdomyolysis (64) the phase III trials in the United States were suspended (65).

All of the current fibrates require multiple daily doses except a micronized formulation of fenofibrate (14), which demonstrates increased absorption and more predictable plasma levels, allowing dose reductions and once-daily administration (66). There are currently three ongoing clinical trials in diabetic patients with fenofibrate (14), to assess the benefit on clinical outcomes (67–69), and an increasing interest in combination therapy with statins (70–78). The results of these studies will be used to forge the future market of these drugs in mono- and combination therapy.

2.2.3 Bile Acid Sequestrants (BAS)/Cholesterol Absorption Inhibitors. The bile acid sequestrants (anion-exchange resins) are nonsystemic drugs, which act by binding bile acids within the intestinal lumen, thus interfering with their reabsorption and enhancing their fecal excretion (79, 80). This leads to the increased hepatic conversion of cholesterol to

$$
\left[\begin{array}{c}
\text{---}CH_2\text{---}CH_2\text{---}CH\text{---}\cdots\cdots \\
\\
\cdots\cdots CH_2\text{---}HC\cdots\cdots \qquad\qquad CH_2\overset{+}{N}(CH_3)_3Cl^-
\end{array} \right]_n
$$

(17)

bile acid through upregulation of cholesterol 7α-hydroxylase activity (81). The liver's increased requirement for cholesterol is partially met through the hepatic removal of circulating LDLc through upregulation of hepatic LDL receptors (79, 80). Bile acid sequestrants have a very slight effect on HDLc and can lead to TG elevations (79–83).

Cholestyramine (**17**) has been in use for 30+ years and has been tested extensively. In both the Lipid Research Clinics Primary Prevention Trial and in the National Heart Lung (84, 85) and Blood Institute Type II Coronary Intervention Study (86, 87), cholestyramine-induced reductions of LDLc were associated with significant reductions in the incidence and progression of CHD, respectively. Cholestyramine (**17**) is a copolymer of 98% polystyrene and 2% divinylbenzene containing about 4 meq of fixed quaternary ammonium groups/gram dry resin. The resin is administered as the chloride salt but exchanges for other anions of higher affinity in the intestinal tract (88).

Colestipol (**18**) is the hydrochloride salt of a copolymer of diethylenetriamine and 1-chloro-2,3-epoxypropane. The functional groups on

colestipol (**18**) are secondary and tertiary amines and its functional anion exchange capacity varies according to the pH in the intestinal tract (89). Both cholestyramine (**17**) and colestipol (**18**) are effective cholesterol-lowering drugs in monotherapy as well as combination with statins (90–92), fibrates (93–97), niacin (92), or probucol (91, 98, 99); however, BAS use is limited because of the need of large doses for efficacy as well as their side-effect profile and interactions with other drugs. Recently, colesevelam (**19**), a third-generation bile acid sequestrant with increased *in vitro* potency (100, 101), has shown similar LDL-lowering efficacy at much lower doses, without the side effects associated with the other bile acid sequestrants (102, 103). The combination of colesevelam (**19**) and an HMG-CoA reductase inhibitor has been shown to be more effective in further lowering serum total cholesterol and LDLc levels beyond that achieved by either agent alone (104–106).

Ezetimibe (**20**) is a cholesterol absorption inhibitor that has just completed phase III trials and is in preregistration. It prevents the absorption of cholesterol by inhibiting the transfer of dietary and biliary cholesterol

$$
\left[\begin{array}{llll}
\cdots\cdots NCH_2\text{---}CH_2\text{---}NCH_2\text{---}CH_2\text{---}NCH_2\text{---}CH_2\text{---}NCH_2\cdots\cdots \\
\quad| \qquad\qquad\qquad | \qquad\qquad\qquad | \qquad\qquad\qquad | \\
\;CH_2 \qquad\qquad\quad CH_2 \qquad\qquad\quad CH_2 \qquad\qquad\quad CH_2 \\
\quad| \qquad\qquad\qquad | \qquad\qquad\qquad | \qquad\qquad\qquad | \\
\;HCOH \qquad\qquad HCOH \qquad\qquad HCOH \qquad\qquad HCOH \\
\quad| \qquad\qquad\qquad | \qquad\qquad\qquad | \qquad\qquad\qquad | \\
\;CH_2 \qquad\qquad\quad CH_2 \qquad\qquad\quad CH_2 \qquad\qquad\quad CH_2 \\
\quad| \qquad\qquad\qquad | \qquad\qquad\qquad | \qquad\qquad\qquad | \\
\cdots\cdots NCH_2\text{-------}CH_2NH \qquad\quad HNCH_2\text{---}CH_2\text{---}NCH_2\cdots\cdots
\end{array} \right]_n
$$

(18)

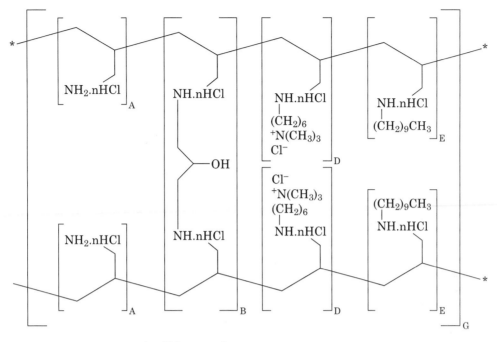

A = Primary amines
B = Cross-linked amines
D = Quaternary ammonium alkylated amines
E = Decyalkylated amines
n = Fraction of protonated amines
G = Extended polymeric network

(19)

(20)

Merck & Co. to develop the fixed combination tablet of ezetimibe (**20**) and simvastatin (**4**). This combination has been shown to reduce LDLc by 52% compared to 35% with simvastatin (**4**) alone (110). Schering-Plough is also investigating combinations with the other statins and the fibrates (111).

2.2.4 Nicotinic Acid Derivatives. The lipid-lowering effects of nicotinic acid (niacin, **21**) have been known for some time (112). Although the exact mechanism of action of nia-

across the intestinal wall. The molecular mechanism by which ezetimibe (**20**) inhibits cholesterol absorption in the intestine remains to be elucidated (107–109). Clinical trials have demonstrated reductions of LDLc by 18% in monotherapy (110, 111). Schering-Plough recently formed a partnership with

(21)

cin (**21**) is unknown, it has been shown to inhibit the mobilization of free fatty acids (FFAs) from adipose tissue, resulting in reduced plasma FFA levels and, thus, a decreased hepatic uptake (113–115). Consequently, hepatic TG synthesis is decreased, leading to a reduction in VLDL secretion and an increase in the intracellular degradation of ApoB (113, 116). As a result of the decreased VLDL production, the plasma LDL level is also reduced because this is the major product of VLDL catabolism. The increase in HDL levels is the result of a decrease in the fractional catabolic rate of ApoAI, the major constituent of the HDL particle (117, 118).

Niacin (**21**) has been studied in six major clinical trials with cardiovascular endpoints (119). The CV endpoints are reduced in monotherapy (120) or in combination (121–126), and over longer time periods all-cause mortality is also decreased (119). Niacin (**21**) is now available in a slow-release formulation (Niaspan; 127–129), designed to decrease the side effects seen on use of this agent that limit its utility (130–133; see Section 2.3.4).

The use of niacin (**21**) and statins in combination has often been avoided because of concerns of myopathy and liver toxicity on the basis of case reports recorded shortly after lovastatin (**2**) was introduced to clinical practice (134, 135). Since then, there have been a number of studies conducted to determine the safety and efficacy of niacin-statin combination therapy (136), including the Arterial Disease Multiple Intervention Trial (ADMIT), which demonstrated that it is both feasible and safe to modify multiple atherosclerotic disease risk factors effectively with intensive combination therapy in patients with peripheral arterial disease (137). More recently, extended-release niacin (**21**) has also been evaluated in combination with various statins (136, 138), including the HDL-Atherosclerosis Treatment Study (HATS) evaluating lipid altering for patients with coronary disease and low HDLc (139). The findings showed extended-release niacin (**21**) with simvastatin (**4**) could reduce cardiac events by 48% in diabetic patients and 65% in nondiabetics (140, 141). Nicostatin, a fixed combination of niacin (**21**) and lovastatin (**2**), has demonstrated efficacy and safety and has just received FDA approval in the United States (142, 143).

2.2.5 Miscellaneous. This section discusses the drugs and classes of drugs used in the treatment of dyslipidemia that fall outside the four major pharmacological classes discussed earlier. Although these compounds exhibit beneficial lipid-lowering effects, they are regarded as second-line drugs for the treatment of hypercholesterolemia and they are often prescribed for other indications.

2.2.5.1 Probucol. Probucol (**22**) was discovered as a lipid-lowering agent in 1964 from a screening program of phenolic antioxidants.

(22)

Its exact mode of action is unclear but it has been shown to reduce both LDLc (8–15%) and HDLc (by as much as 40%) (144–146). Studies have shown an increased fecal loss of bile acids and increased catabolic rate of LDLc (147) as well as a reduction in LDL synthesis (148). Because of the decrease in HDLc and other side effects (see Section 2.3.5.1), probucol (**22**) is rarely used in the treatment of hyperlipidemia and is not marketed in the United States. Probucol's benefit in restenosis is believed to be attributed to its strong antioxidant effects, which may prevent endothelial damage and LDL oxidation secondary to angioplasty (149). However, the Probucol Quantitative Regression Swedish Trial (PQRST) failed to show a clinical benefit of probucol (**22**) in combination with cholestyramine (**17**) compared to cholestyramine (**17**) alone or placebo (150).

2.2.5.2 Hormone Replacement Therapy (HRT). The use of HRT, estrogen and progesterone, in postmenopausal women has increased dramatically over the last 10 years. Not only has it shown benefit for the perimenopausal symptoms but also in several studies, a reduction in cardiovascular events (151). HRT directly stimulates LDL receptor activity, leading to reductions in total cholesterol and LDLc levels, moderate increases in HDLc levels, and a decrease in HDL and LDL oxidation (152). HRT in combination with a statin has also shown to be very effective in lowering LDLc levels (153). The Heart and Estrogen/Progestin Replacement Study (HERS), investigating primary or secondary prevention of HRT, concluded that estrogen plus medroxyprogesterone showed no significant benefit in the prevention of coronary artery disease (CAD) (154). Another recent study from Duke University found similar results to those of the HERS trial (155). There are currently contradictory results from several studies, which is why the FDA has not approved HRT treatment for the indications of: (1) regulation of lipids or (2) reduction of CHD. Recently, the American Heart Association issued a caution on the use of HRT for cardiovascular disease (156).

2.2.5.3 Estrogen Modulators. Several mechanisms are responsible for the cardioprotective effects of estrogen, including beneficial effects on the lipoprotein profile and direct effects on the vascular wall (160). As mentioned in the preceding section, the lipid effects include moderate decreases in LDLc, increases in HDLc, and a decrease in LDL and HDL oxidation. The effects are modulated through the binding of estrogen to its nuclear estrogen receptor (ER) and the regulation of target gene transcription through the ligand-receptor complex (161). The discovery of a second estrogen receptor subtype (ERβ), which may mediate some of estrogen's cardiovascular benefits (161–166), has led to increased ER research to identify selective agonists.

Other estrogen modulators that are used in postmenopausal women are tamoxifen (**23**) and toremifene (**24**) in the treatment of breast cancer and raloxifene (**25**) in the prevention of osteoporosis. All the drugs have been demonstrated to reduce total cholesterol and LDLc.

Furthermore, tamoxifen (**23**) has been associated with lower rates of MI, although higher rates of thromboembolic diseases have been reported (167). Toremifene (**24**) has been associated with increases of ApoAI levels by

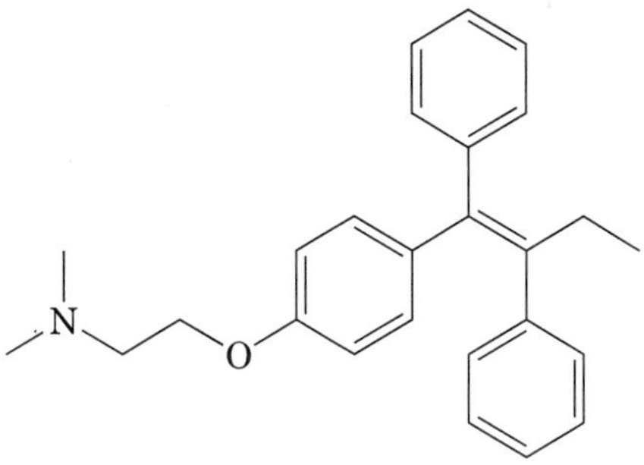

(23)

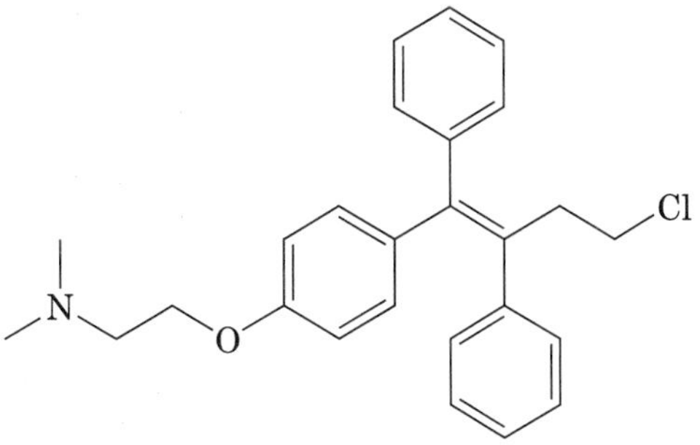

(24)

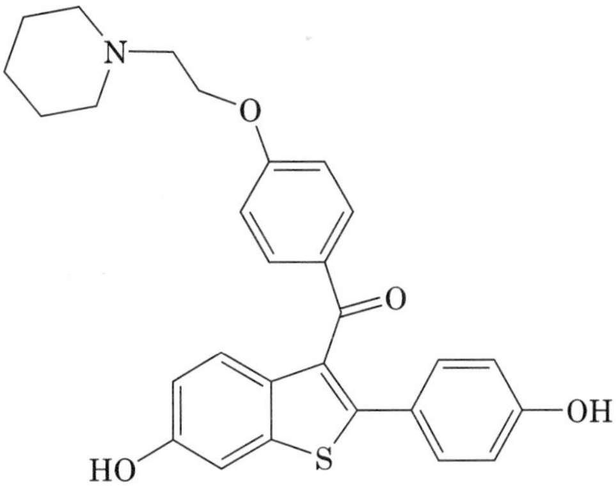

(25)

13% (168). Raloxifene (**25**) has been shown to lower fibrinogen, in addition to increasing HDL levels, without raising TGs (169).

2.2.5.4 Plant Sterols. Plant sterols and stanols inhibit the intestinal absorption of cholesterol and as a result lower plasma LDLc concentrations. They occur, in varying degrees, naturally in almost all vegetables. The most abundant of the phytosterols is β-sitosterol (**26**) and the fully saturated derivative of β-sitosterol is sitostanol (**27**). Plant sterols are absorbed to a small extent, whereas plant stanols are virtually nonabsorbable. Thus, in-

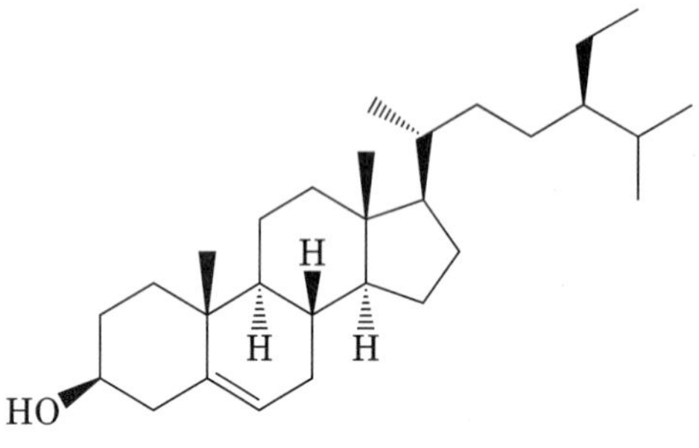

(26)

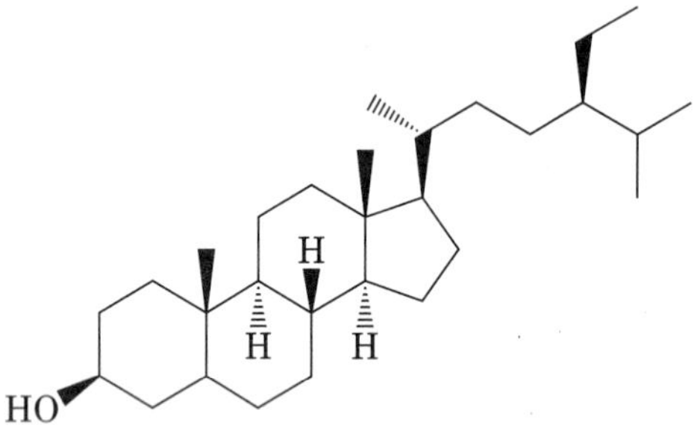

(27)

testinal levels of stanols will be prolonged compared to that of sterols, which may explain why plant stanols appear to be more effective in decreasing cholesterol absorption (170) and reducing serum LDL levels (171). Plant stanols have also been shown to reduce serum cholesterol levels in patients on statin therapy (172). Sterol and stanol esters can be used as food additives, to allow adequate amounts to be consumed without affecting food quality or dietary habits. Low fat stanol or sterol ester-containing margarines in combination with a low fat diet have been shown to reduce LDLc levels in hypercholesterolemic subjects (173, 174).

2.3 Side Effects, Adverse Effects, Drug Interactions, and Contraindications

2.3.1 HMG-CoA Reductase Inhibitors (Statins).
As a class, HMG-CoA reductase inhibitors have been shown to have a low risk of severe side effects after chronic exposure in humans. Most of the known adverse effects are directly related to their biochemical mechanism of action and are the result of potent and reversible inhibition of an enzyme involved in cellular homeostasis (175). The incidence of adverse effects increases with increasing plasma levels of the active drug; however, it should be noted that the increase in risk is not associated with a proportional increase in cholesterol-lowering efficacy (176–180). Toxicology studies have shown that the liver, kidney, muscle, the nonglandular stomach, and lymphatic tissue are potential target organs and tissues (see Refs. 181–188 and Table 7.4). In rodents acanthosis and hyperkeratosis of the nonglandular stomach was observed. These changes were observed only in rodents and appear to be linked to the bio-

Table 7.4 Target Organs in Multidose Toxicology Studies with HMG-CoA Reductase Inhibitors

Target Organ	Lovastatin	Cerivastatin	Fluvastatin	Atorvastatin
Liver	√	√	√	√
Gall bladder	√	—	√	√
GI tract	√	√	√	√
Kidney	√	√	—	—
Muscle	√	√	√	√
Nonglandular stomach	√	√	√	√
Lymphatic system	√	√	√	—
CNS	√	√	—	√
Eyes	√	√	√	—
Testes	√	√	√	—
Thyroid	—	—	√	—

chemical mechanism of this class. The acanthosis and hyperkeratosis side effects require the direct contact of the fore-stomach squamous epithelium, with high concentrations of inhibitor for long periods. It should be noted that dogs appear more sensitive than other species to statin toxicity probably because of the fact that the degree of metabolism is less, resulting in higher systemic exposure to the parent drug.

In addition, the inhibition of HMG-CoA reductase prevents the formation of ubiquinone (**28**) and dolichol (**29**), which are involved in electron transport and glycoprotein synthesis.

(**28**)

(**29**) $n = 15\text{--}20$

Although it is not surprising to find that a potent inhibitor of a key biosynthetic pathway produces toxicity at high doses in animals, it was thought that none of the observed side effects would pose a significant risk to humans at clinical doses. Recently, the voluntary withdrawal of cerivastatin (**8**) from the market because of muscle damage linked to 31 U.S. deaths and at least nine more fatalities abroad, primarily in combination with gemfibrozil (**11**), has focused attention on the safety of statins as a class (35).

2.3.1.1 Hepatotoxicity. Elevations in serum transaminases occur in 2–5% of the patients taking statins, and a level three times the upper reference limit is the point at which treatment should be discontinued (178, 189, 190). These changes are dose dependent, often transient, and return to normal after the drug is discontinued. Small increases in transami-

nases have also been reported with most lipid-lowering drugs and may be a response to changes in lipid metabolism rather than a direct effect of lipid-lowering drugs on the liver (190). Hepatitis is a rare complication of statin therapy (0.01–0.02%) and seems to be an idiosyncratic or cytochrome (CYP) P450-dependent effect (191, 192).

2.3.1.2 Myopathy. Myalgia in patients on statin therapy occurs with an incidence of 2–5% and is not usually associated with an increase in creatine kinase (CK) levels. The symptoms disappear upon discontinuing statin treatment, or some patients may benefit from a clinical supplementation with coenzyme Q_{10} [ubiquinone (**28**)], a mitochondrial electron-shuttle transporter whose levels are depleted by statin therapy (193). Myopathy and myositis are rarer (0.5–1%) and characterized by muscle pain, weakness, or cramps and CK values of at least 10-fold the upper limit of normal (176, 177, 179, 181). There is little correlation between the degree of CK elevation and the severity of the symptoms (194), although this side effect is dose dependent and disappears upon discontinuing treatment. The incidence of myopathy is exacerbated by concomitant therapy with cyclosporin, gemfibrozil, cholestyramine, fenofibrate, niacin, itraconazole, and erythromycin or in the presence of renal insufficiency (176, 177, 181, 194–198). These agents interfere with drug metabolism through the cytochrome P450 system, leading to increased plasma concentrations of unchanged drug over longer periods of time. Rhabdomyolysis with renal dysfunction is a very rare complication of statin treatment and is usually idiosyncratic and dose independent (181). It is characterized by the actual breakdown of the muscle membrane, leading to leakage of the muscle protein myoglobin into the bloodstream (myoglobinemia). The myoglobin travels to the kidneys (myoglobinuria), where it causes the kidney tubules to stop working, thus leading to kidney failure. In addition, the increased potassium levels, released from muscle cell breakdown, can lead to cardiac arrhythmias and death. Recently, Sankyo announced that the launch of pitavastatin (**9**) would be postponed because of the necessity to undertake a second phase II trial, by use of a lower dose. Cases of muscle pain associated with the elevation of CPK (creatine phosphokinase), a marker of myopathy, at the higher doses had been observed (199). Astra-Zeneca also reported that there were incidences of rhabdomyolysis in patients treated with rosuvastatin (7), but only at the high dose.

The exact mechanism through which the statins induce skeletal muscle abnormalities is currently unknown (179, 195, 196). It has been suggested that statins cause intracellular ubiquinone deficiency, as mentioned earlier, which interferes with normal cellular respiration in muscle and results in electron leakage into the tissue, thus causing oxidative stress and ultimate tissue destruction. These changes are reversed by concurrent administration of mevalonic acid in the animal models. However, animal studies have failed to show a relation between tissue ubiquinone (28) levels and the degree of muscle destruction.

Although the incidence of myopathy and rhabdomyolysis has raised concerns over the statins as a class, the benefits of using statins to manage patients' cholesterol far outweigh the risks of serious side effects from their use.

2.3.1.3 Other Effects.
There were concerns over statin-induced cataracts on the basis of toxicology studies in animals but these have now largely disappeared, given that data from clinical studies involving statins have not reported lenticular opacities in patients receiving long-term treatment (177, 178, 200, 201). The most common adverse effects are gastrointestinal, with the occurrence of nausea, bloating, diarrhea, or constipation (189, 194, 202). These are usually transient and resolve spontaneously after 2–3 weeks.

Although the statins as a class have been shown to exhibit favorable hematorheological effects, atorvastatin (6) has been shown to increase fibrinogen, a known CV risk factor, in some patient populations (203–206). Furthermore, the higher doses of atorvastatin (6) have also demonstrated a lower HDL-raising effect and even HDLc reductions compared to those of the other statins (207–210). The underlying mechanisms and eventual clinical relevance and consequences of these effects still need to be elucidated.

2.3.2 Fibric Acid Derivatives (Fibrates).
The side-effect profile is similar for all of the fibrates. The most common side effects are nausea, diarrhea, and indigestion. Other side effects, such as headache, loss of libido, skin rash, and drowsiness, occur less frequently. Toxicological studies have shown that the liver, muscle, and kidney are potential target organs and tissues (211–213). In general, the fibrates potentiate the effects of oral anticoagulants by displacing these drugs from their binding sites on plasma proteins, necessitating a reduction of the dosage of anticoagulant (211, 213–215). The fibrates are also contraindicated in pregnant or lactating women, or patients with severe liver or renal impairment or existing gallbladder disease.

2.3.2.1 Hepatotoxicity.
As with the statins, elevations in serum transaminases occur in 2–13% of the patients (189, 190, 212, 228) taking fibrates, and a level three times the upper reference limit is the point at which treatment should be discontinued. These changes are often transient and return to normal after the drug is discontinued. In rodents, the activation of PPARα by fibrates leads to the induction of hepatic peroxisome proliferation, which is characterized by an increase in peroxisome number and/or peroxisome volume, and hepatomegaly (51). A greater than threefold increase in peroxisome proliferation (PP) is associated with the development of hepatocellar carcinomas in rodents (216–218), whereas the risk of inducing a hepatocarcinogenicity associated with a weak PP response (i.e., two- to threefold increase) is unknown. Although all of the fibrates induce this phenomenon in rodents, it was believed to be species specific (i.e., rodent), given that gemfibrozil (11) (219–221) and fenofibrate (14) (222–225) have been shown not to induce peroxisome proliferative or carcinogenic effects in primate and human livers. On the other hand, clofibrate (10) (226) has been shown to induce PP in human livers, whereas ciprofibrate (16) (227) has been shown to induce PP in primates. One potential explanation for this difference is the very high exposures of both clofibrate (10) and ciprofibrate (16) compared to the other fibrates.

2.3.2.2 Myopathy.
The risk of myopathy with fibrate treatment is increased in patients

with renal impairment (228–231) because of the extended systemic exposure of the drug. This side effect is dose dependent and disappears upon discontinuing treatment.

2.3.3 Bile Acid Sequestrants (BAS)/Cholesterol Absorption Inhibitors.

The use of the bile acid sequestrants is limited by their unpalatability, attributed essentially to the large doses needed for efficacy (3–30 g/day). Gastrointestinal side effects are also associated with the low compliance in a large number of patients. The newer formulations such as tablets, caplets, and flavored granules and the development of more potent sequestrants have been associated with fewer gastrointestinal adverse effects.

The cholesterol absorption inhibitor ezetimibe (**20**) appears to have a very good adverse effect profile, with only limited gastrointestinal side effects. Further large-scale trials will be needed to better define the adverse effect profile.

2.3.4 Nicotinic Acid Derivatives.

Nicotinic acid (**21**) is not very well tolerated (132, 133). Nearly all patients suffer from itching, flushing, and gastrointestinal intolerance, which usually diminishes with prolonged use. Aspirin will prevent the flushing, indicating that this side effect is prostaglandin mediated (232). Current guidelines do not recommend the use of niacin in patients with diabetes because it can exacerbate gout and worsen glycemic control (233–236). Recently, results from the Arterial Disease Multiple Intervention Trial (ADMIT) suggest that lipid-modifying doses of niacin can be safely used in patients with diabetes (237).

2.3.5 Miscellaneous

2.3.5.1 Probucol. Probucol (**22**) is generally well tolerated, with only about 3% of the patients discontinuing treatment because of side effects. The most frequently reported side effect is diarrhea, which may occur in up to one-third of the treated patients (238). Other less frequently reported gastrointestinal side effects include flatulence, abdominal pain, nausea, and vomiting (239). Probucol (**22**) induced ventricular fibrillation and sudden

death in dogs (240), whereas it increased the QT interval in monkeys. It was withdrawn from the U.S. market because of its potential to induce serious ventricular arrhythmias.

2.3.5.2 Hormone Replacement Therapy (HRT). HRT is associated with an increased relative risk of breast and endometrial cancer with each year of treatment, as well as a risk of venous and pulmonary thromboembolism (241–244). Although there is not a strong association of HRT to ovarian cancer, there is a debatable positive correlation (241). Most of the adverse effects of HRT are restricted to current or recent use, and long-term HRT use should be carefully considered on an individual basis, taking into account the patient's existent risk factors for breast and endometrial cancer and for venous/pulmonary thromboembolism vs. the potential benefits of treatment on CV disease and osteoporosis.

2.3.5.3 Estrogen Modulators. There is an increased incidence of hot flashes and leg cramps with all the estrogen modulators, although this side effect does not affect drug compliance. Unlike HRT, tamoxifen (**23**), toremifene (**24**), and raloxifene (**25**) do not increase the risk of breast and endometrial cancer and, in fact, several studies have demonstrated that tamoxifen (**23**) and raloxifene (**25**) are useful in the prevention of breast cancer. Currently, a Study of Tamoxifen Against Raloxifene (STAR) is ongoing to compare the two drugs in the prevention of breast cancer (245, 246).

2.3.5.4 Plant Sterols. Whereas plant sterols are absorbed to a small extent, plant stanols are virtually nonabsorbable. Unless consumed at extraordinarily high levels, practically no side effects have been observed.

2.4 Absorption, Distribution, Metabolism, and Elimination

2.4.1 HMG-CoA Reductase Inhibitors (Statins).

Two thirds of the total cholesterol found in the body is of endogenous origin, with the major site of cholesterol biosynthesis being the liver. Therefore, to minimize the risk of adverse effects associated with high systemic exposures, the statins need to show tissue (liver)-selective inhibition of HMG-CoA reductase, essential for achieving LDLc lowering.

Table 7.5 Pharmacokinetic Properties of the HMG-CoA Reductase Inhibitors

Statin	Plasma Half-Life	Protein Binding	Metabolism	Active Metabolites	Elimination Pathway
Lovastatin (**2**)	1–2 h	>95%	Hydrolysis of inactive lactone and CYP3A4	Hydroxy-acid	83% biliary 10% renal
Fluvastatin (**5**)	~2 h	~99%	Hydroxylation	No active metabolites	93% biliary 6% renal
Pravastatin (**3**)	2–3 h	50%	Hydroxylation	75% as parent compound	70% biliary 20% renal
Simvastatin (**4**)	~2 h	>95%	Hydrolysis of inactive lactone and CYP3A4	β-Hydroxy-acid metabolites	60% biliary
Atorvastatin (**6**)	14 h	>98%	CYP3A4	*ortho*- and *para*-hydroxylated metabolites	Predominantly biliary
Cerivastatin (**8**)	~3 h	>99%	CYP3A4 and CYP2C8	Demethylation of the ether moiety and hydroxylation of the isopropyl group	70% biliary 30% urinary
Pitavastatin (**9**)	11 h	96%	CYP2C9 (major) CYP2C8 (minor)	No active metabolites	>98% biliary >2% urinary
Rosuvastatin (**7**)	20 h	Not reported	CYP2C9 (minor)	No active metabolites reported	Not reported, most likely biliary

Orally administered drugs, once absorbed, are filtered through the liver through the portal vein. The drugs are extracted from the portal venous blood and concentrated in the hepatocyte to an extent that is related to their lipophilicity.

The more lipophilic statins [lovastatin (**2**) > cerivastatin (**8**) ≅ simvastatin (**4**) > fluvastatin (**5**) > atorvastatin (**6**)] exhibit facilitated passive diffusion through hepatocyte cell membranes, leading to selective accumulation in the liver (247, 248). Interestingly, the lactones [lovastatin (**2**) and simvastatin (**4**)] selectively accumulate in the liver in a number of various species studied compared to their respective acid forms (249–251). The lactones are then efficiently metabolized into the active open hydroxy-acid form in the liver. On the other hand, the hepatoselectivity of the hydrophilic statins [pravastatin (**3**) and rosuvastatin (**7**)] can be attributed to their high affinity for a liver-specific transport protein (248, 252, 253).

All of the statins except pravastatin (**3**) are highly protein bound (see Table 7.5), which may limit their use with oral anticoagulant therapy. Interestingly, the more potent statins [i.e., atorvastatin (**6**) and rosuvastatin (**7**)] are associated with long plasma half-lives, indicating the necessity of sustained HMG-CoA reductase inhibition to obtain the desired lipid-lowering efficacy.

Because cytochrome (CYP) P450 metabolism is not important in the elimination of fluvastatin (**5**), pravastatin (**3**), and rosuvastatin (**7**), there is a smaller risk of adverse effects resulting from drug interactions. Although fluvastatin (**5**) is eliminated as hydroxylated inactive metabolites, pravastatin (**3**) and rosuvastatin (**7**) are eliminated as unchanged drug. On the other hand, atorvastatin (**6**) and cerivastatin (**8**) are extensively metabolized by CYP3A4 to active metabolites, which account for 70% and 25%, respectively, of the total HMG-CoA reductase inhibitory activity observed (254, 255).

Table 7.6 Pharmacokinetic Properties of the Fibrates

Fibrate	Plasma Half-Life	Protein Binding	Metabolism
Clofibrate (**10**)	12 h (acid)	>97%	Hydrolysis of the ester and minor metabolite formation conjugated with glucuronide
Gemfibrozil (**11**)	~8 h	>95%	40% unchanged and 60% numerous metabolites (conjugated/benzoic acid/phenol/etc.)
Bezafibrate (**13**)	2 h	>94%	50% unchanged, 22% glucuronide conjugates and 22% other polar metabolites (hydroxy/others)
Micronized fenofibrate (**14**)	20 h (acid)	>95%	Hydrolysis of the ester, 50% conjugated and 50% polar metabolites (phenol/benzhydrol)
Ciprofibrate (**16**)	42 h	>90%	70% glucuronide conjugates

2.4.2 Fibric Acid Derivatives (Fibrates). The fibrates as a class are all well absorbed from the gastointestinal tract (>90%) and display a high degree of protein binding (see Table 7.6). However, the pharmacological response (lipid-lowering action) is more accurately predicted by the dose administered than the plasma concentrations. All the fibrates are metabolized by the hepatic CYP3A4 and primarily excreted through the kidneys (256).

The shorter plasma half-life associated with bezafibrate (**13**) is attributed to the increased renal clearance compared to that of the other fibrates (211). In patients with impaired renal function, the clearance of the fibrates is reduced, leading to prolonged half-lives. Dose adjustment is necessary in patients with mild to moderate renal impairment, whereas treatment in patients with severe renal impairment is precluded.

2.4.3 Bile Acid Sequestrants (BAS)/Cholesterol Absorption Inhibitors. The bile acid sequestrants cholestyramine (**17**), colestipol (**18**), and colesevelam (**19**) are not absorbed from the intestinal tract, with the extent of LDLc lowering being mediated by their ability to bind bile acids (see Section 2.2.3). The resulting solid complex is 100% excreted in feces (257) and, because the bile acid sequestrants are not absorbed, there are neither measurable plasma levels nor metabolism.

The cholesterol absorption inhibitor ezetimibe (**20**) *in vitro* is a CYP3A4 inhibitor, whereas *in vivo*, it has been shown to be primarily metabolized through glucuronidation in the intestine and eliminated through biliary secretion, with no evidence of significant

CYP450-mediated metabolism (258). Interestingly, glucuronidation is generally considered one of the major detoxification processes in xenobiotic metabolism, facilitating biliary and/or urinary elimination. It is also regarded as a process that inactivates the drug. In the case of ezetimibe (**20**), neither of these generalizations holds true. The glucuronidation of ezetimibe (**30**) appears to improve its activity

(30)

in at least two ways: (*1*) the drug is repeatedly delivered back to the site of action (the intestine) through enterohepatic circulation and (*2*) glucuronidation appears to increase the residence time in the gut (107). In addition, once ezetimibe is glucuronidated, >95% is either in the intestinal lumen or wall, indicating that systemic exposure of the glucuronide (**30**) will be very low.

2.4.4 Nicotinic Acid Derivatives. Niacin (**21**) is well absorbed, with >88% of the dose recovered in the urine. At the doses used for the treatment of dyslipidemia, niacin (**21**) is not metabolized to any great extent and the elimination is exclusively renal, largely as unchanged drug. Niacin (21) is only slightly protein bound (<20%) with a short half-life of 0.75–1 h. The extended-release formulation, Niaspan, allows for sustained blood levels of nicotinic acid, thus permitting a reduction in the active dose.

2.4.5 Miscellaneous. Although HRT and estrogen modulators exhibit beneficial lipid-lowering effects, these drugs are not prescribed for the treatment of hypercholesterolemia; thus they are not be treated in this section.

2.4.5.1 Probucol. Probucol (22) is only about 2–8% bioavailable, with a highly variable plasma half-life of 12–500 h. Once absorbed, probucol (22) is 95% incorporated into lipoproteins in the blood and can accumulate in adipose tissue. After stopping treatment, it can take up to 6 months to be removed from all the tissue because of its lipophilicity. Little is known about its metabolism, with only 2% being eliminated in the urine and the remainder in the feces through the bile.

2.4.5.2 Plant Sterols. As mentioned earlier (Section 2.3.5.4), plant sterols and stanols are not absorbed to any great extent from the intestinal tract, with their effects being mediated by their ability to inhibit the intestinal absorption of cholesterol (259). Because the plant sterols and stanols are not absorbed, there are neither measurable plasma levels nor metabolism.

3 PHYSIOLOGY AND PHARMACOLOGY

3.1 Lipid Transport and Lipoprotein Metabolism

To understand the development of the different classes of lipid-lowering drugs discussed in this review, it is necessary to put the different therapeutic targets into their physiological context. The goal of this section is to familarize the reader with the pathways of lipid transport, lipoprotein metabolism, and cholesterol biosynthesis as well as the different tissues involved. It is not meant to be a comprehensive review of the area but only a simplified overview, to give the reader a brief survey of the different drug mechanisms of action.

Cholesterol is essential in the synthesis of cell membranes, bile acids, and hormones, whereas triglycerides are important to peripheral tissue as a source of energy production. Although the liver is the major site of cholesterol biosynthesis, cholesterol and TG from the diet can also be absorbed from the intestine and transported in the form of chylomicrons (Fig. 7.1). The chylomicrons transport cholesterol and TG from the intestine to the adipose tissue for storage and the liver for packaging and resecretion as VLDL or LDL particles. After extensive hydrolysis of TGs, the remaining particles, chylomicron remnants, are taken up by the liver (260). Prolonged uptake of these TG particles (VLDL or chylomicron remnants) by the liver can lead to reduced hepatic production of LDL receptors (LDLr) and to increases in plasma cholesterol levels.

Hydrolysis of TG-rich particles in the liver leads to release of fatty acids. Fatty acids not used for energy generation by the liver are converted to TGs for hepatic storage or packaged into VLDL particles along with cholesteryl esters (CE) to be transported to the peripheral tissue (Fig. 7.2). The VLDL particles are hydrolyzed through the lipoprotein lipase (LPL) to form IDL particles. The liver takes up about 60% of the IDL through the LDLr and the remainder is hydrolyzed by the hepatic lipase (HL) to produce LDL particles.

The major role of LDL is to transport cholesterol to the tissues. When intracellular cholesterol is required, cells may synthesize cholesterol, or acquire exogenous cholesterol through upregulation of LDLr, resulting in the increased uptake of LDL. The LDLr is responsible for removing about 60–80% of the LDL particles (261). Increased intracellular cholesterol inhibits HMG-CoA reductase, the rate-limiting enzyme in cholesterol biosynthesis (Fig. 7.3), and decreases the synthesis of LDLr to limit the further uptake of cholesterol into the cell (260). LDL can be modified by oxidation and glycation, which leads to decreased recognition by the LDLr, increased residence time in the plasma compartment, and increased uptake of modified LDL by scavenger receptors on macrophages. This leads to the accumulation of cholesterol, and lipid in tissue macrophages, which in the arterial sys-

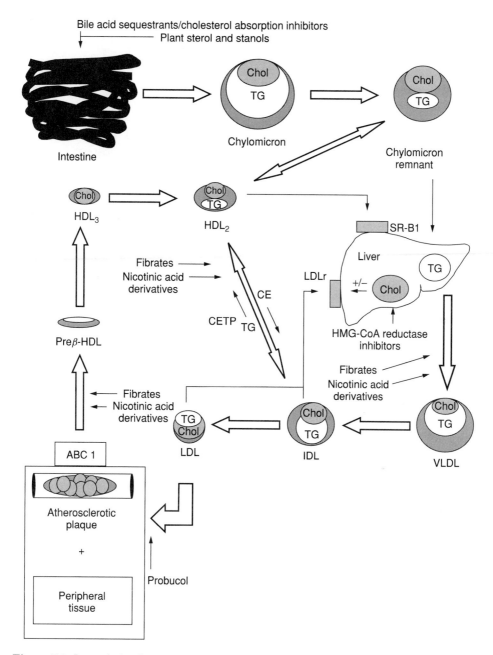

Figure 7.1. Interrelationships of the plasma lipoproteins and the sites of action of the lipid-lowering drugs. HMG-CoA reductase inhibitors upregulate hepatic LDL receptors (LDLr) through their inhibition of cholesterol biosynthesis. Fibrates reduce the hepatic secretion of VLDL and increase lipoprotein lipase (LPL) and hepatic lipase (HL) responsible for hydrolyzing TG-rich lipoproteins. Fibrates also increase reverse cholesterol transport (RCT) by increasing circulating HDL levels through reducing their catabolism by the transfer of cholesteryl esters (CE) from HDL to VLDL by way of the cholesteryl ester transfer protein (CETP) as well as increasing the number of HDL particles. The ATP-binding cassette transporter 1 (ABC 1) mediates the efflux of free cholesterol from cells to plasma, where it is incorporated into HDL. The hepatic uptake of cholesterol from HDL is then mediated by the scavenger receptor class B Type 1 (SR-B1). Bile acid sequestrants, cholesterol absorption inhibitors, and plant sterols and stanols reduce the re-uptake of bile acids and/or cholesterol through the intestine, decreasing the formation of chylomicrons and the transport of TGs and CE to the liver. This leads to the upregulation of hepatic LDLr expression. Nicotinic acid derivatives inhibit the hepatic production of VLDL and increase circulating HDL levels similar to fibrates, although the mechanism is not fully understood. Probucol is an antioxidant in atherosclerotic lesions and possibly in LDL particles.

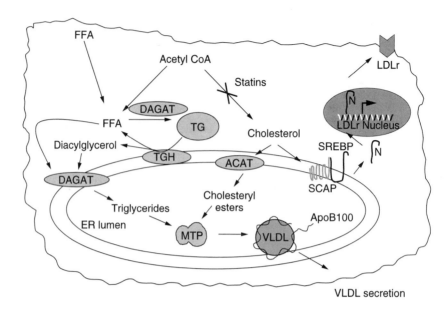

Figure 7.2. Lipid metabolism in the hepatocyte. Free fatty acids (FFAs) that enter the hepatocyte are esterified and stored as triglycerides (TGs). The TGs are hydrolyzed as needed by the triglycerol hydolase (TGH) to FFA and diacylglycerol. The diacylglycerol is further hydrolyzed to produce the TGs that are used in lipid assembly. Acetyl CoA is used by the hepatocyte either for cholesterol biosynthesis or TG synthesis and storage, depending on the hepatocytes' needs. Cholesterol is ester-ified by the acyl-CoA cholesterol acyltransferase (ACAT) to produce cholesterol ester (CE) for use in lipid assembly or for intracellular storage. The microsomal triglyceride transfer protein (MTP) is responsible for the lipidation of the nacent apoB100 with TG and CE to synthesize the VLDL particle, which is then secreted. If the hepatocyte cholesterol is depleted, the sterol regulatory element-binding protein (SREBP)/cleavage-activating protein (SCAP) is activated, sending a signal to the nucleus to upregulate the LDL receptor (LDLr) along with other genes implicated in cholesterol synthesis. Upregulation of the LDLr leads to increased cholesterol uptake from LDL and IDL parti-cles in the plasma.

tem leads eventually to the establishment of atherosclerotic plaques (262).

HDL is regarded as essential for reverse cholesterol transport (RCT), the removal of cholesterol from peripheral tissue and its transport back to the liver. Nascent HDL (pre-β-HDL) is synthesized in the liver and small intestine and enters the plasma compartment. When pre-β-HDL particles come in contact with cells rich in cholesterol, there is a trans-fer of cholesterol to the particle by cell surface proteins [such as the ATP-binding cassette transporter 1 (ABC 1)], which are responsible for the efflux of cholesterol from cells into the plasma (263). Once transferred to HDL, the free cholesterol is esterified by the lecithin-cholesterol acyltransferase (LCAT) and the re-sulting cholesteryl esters are incorporated

into the lipid core of the HDL particle, thus allowing it to increase in size and mature into HDL$_3$. Further addition of CE results in the maturation to HDL$_2$. HDL$_2$ may: (1) deliver cholesterol to the liver through interac-tions with hepatic HDL receptors and may be converted back to HDL$_3$; (2) exchange lipid with other lipoprotein classes through cho-lesteryl ester transfer protein (CETP) medi-ated transfer; or (3) be taken up whole by the liver (264).

The major class of lipid-lowering drugs available are inhibitors of an enzyme (i.e., HMG-CoA reductase inhibitors) implicated in the synthesis of cholesterol or inhibitors of the process of cholesterol absorption (i.e., bile acid sequestrants, cholesterol absorption inhibi-tors, plant sterols, and stanols), except for the

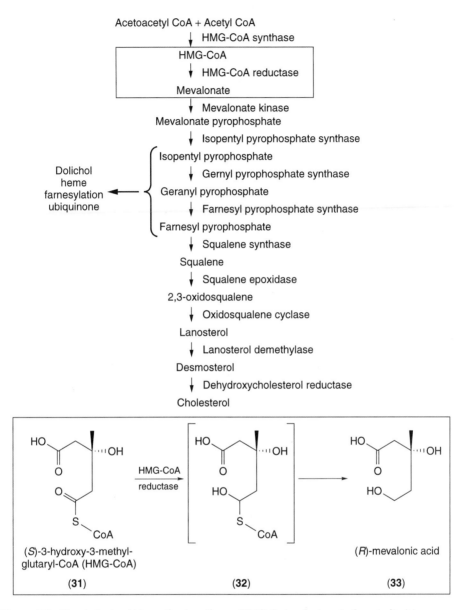

Figure 7.3. The cholesterol biosynthesis pathway. HMG-CoA reductase is the rate-limiting enzyme in the pathway and is involved in the transformation of (S)-3-hydroxy-3-methylglutaryl-CoA (**31**) into (R)-mevalonate (**33**) (see boxed reaction).

fibrates. The fibrates bind into the PPARα ligand-binding domain (LBD), inducing a conformational change of the protein. This enables the complex to bind to the DNA in the nucleus through a PPAR response element (PPRE), thus permitting the regulation of target genes (Fig. 7.4). Upregulation of the apoAI gene leads to increased apoAI-containing HDL particles, whereas the downregulation of apoCIII leads to increased LPL activity, thus reducing the plasma levels of TG-rich lipoproteins (51–53).

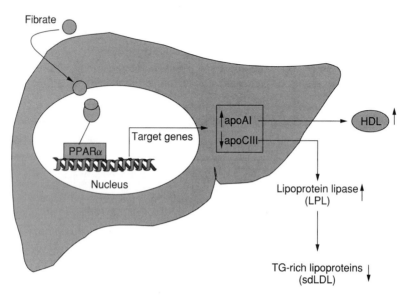

Figure 7.4. The mechanism of action of the fibrates. In the hepatocyte, the drug binds to the ligand-binding domain (LBD) of the PPARα receptor. Once bound, the complex binds to the DNA in the nucleus and transcriptionally regulates target genes in a concerted fashion. ApoAI is upregulated, leading to an increase in HDL, whereas apoCIII is downregulated. The downregulation of apoCIII, the natural inhibitor of lipoprotein lipase (LPL), leads to an increased hydrolysis of TG-rich lipoproteins, including small dense LDL (sdLDL). Genes implicated in fatty acid β-oxidation are also upregulated.

4 HISTORY

4.1 Identification of the Statin Class of Lipid-Lowering Drugs

During the 1970s, epidemiological studies established a relation between elevated serum cholesterol and CHD. This led to increased research in the field of cholesterol biosynthesis inhibitors to identify new mechanisms useful in the treatment of elevated serum cholesterol levels. As already stated (Section 2.2.1), screening of natural product extracts for cholesterol biosynthesis inhibitors led to the identification of mevastatin (**1**) from the extracts of *Penicillium citrinum* by Sankyo (Japan) in 1976 (36). Mevastatin (**1**) was subsequently shown to inhibit HMG-CoA reductase (37), the rate-limiting enzyme in cholesterol biosynthesis (38, 39). This one discovery has led to the development of a $30 billon lipid-lowering market, which is expected to continue to grow in the future. One molecule that helped speed up this growth is atorvastatin (**6**).

4.2 Discovery of Atorvastatin (Lipitor)

An amusing and sensationalized account of the development of atorvastatin appeared January 24, 2000 in the *Dow Jones Business News*, entitled "Birth of a Blockbuster" (265), which is summarized below. Since Lipitor's launch in 1997, analysts have predicted that it will become the biggest-selling prescription medicine in the world. The recent changes to the NCEP guidelines (discussed in Section 1) have also doubled the number of Americans that can be considered for statin treatment, which makes many analysts believe that Lipitor may be the first drug to earn $10 billion annually.

In the early 1980s, Parke-Davis scientists were several months into the development of an HMG-CoA reductase inhibitor, only to learn that Sandoz AG (the former Swiss drug company that is now part of Novartis) had filed a patent that contained their lead molecule. Work then switched to a backup molecule, CI-981, which would later become known as atorvastatin (**6**).

Although 8 years of research had already been invested in developing the compound, animal studies had not demonstrated cholesterol-reducing superiority over the competitors. At the time, there was already one statin on the market and three others in late-stage human studies, so discussions focused on whether atorvastatin should progress into human clinical trials. Discontinuing the development of atorvastatin would have wasted an intense, 2-year effort to come up with a process to manufacture the drug in commercial quantities. The initial molecule was racemic, with only one diastereomer, resulting in the potent *in vitro* inhibition of HMG-CoA reductase. This presented an issue for Parke-Davis scientists: 50% of an inactive compound could cause unwanted toxicity and/or side effects and potentially even jeopardize FDA approval.

The initial large-scale syntheses of the optically pure diastereomer proved difficult because of racemization. Scientists then observed that running the large-scale reactions at $-78°C$ prevented racemization during the reaction. The final manufacturing process took 3 weeks from raw materials to end product and afforded a compound with 100% optical purity.

The first tests of the drug on 24 employee-volunteers at 10 mg decreased LDLc by 38%. That was as good as or better than competing compounds at their recommended maximum doses. At 80 mg, LDLc decreased by 58%, which was more than any other statin at any dose. On the basis of those results, clinical trials were devised that would test 10 mg as a starting dose and 80 mg for patients with especially high cholesterol. Other statins are prescribed in 20 and 40 mg tablets and Parke-Davis's strategy would allow them to demonstrate LDLc-lowering superiority at even the lowest dose of atorvastatin.

Although Parke-Davis officials were confident that atorvastatin would demonstrate superiority, there was no assurance at the time that the market would want a potent low dose pill. Because of several clinical studies in the 1980s linking low cholesterol with an increased risk of death from non–heart-related illnesses, many doctors in the early 1990s were wary of aggressive treatment. The 80 mg strategy was risky too because, if it turned out to have unacceptable side effects, it could taint the drug even at low doses and leave the company with only a 10 mg tablet to take to the market.

The LDLc-lowering efficacy demonstrated by Lipitor in the clinical trials was enough for the FDA to put it on priority review. The company enrolled the last patient in its major trial in October 1995 and filed for approval in June 1996. Six months later, Lipitor was cleared to begin sales.

In the months leading up to approval, the sales of Merck & Co.'s and Bristol-Myers Squibb's statins were beginning to soar. Both had recently completed large-scale, long-term studies showing use of their drugs reduced deaths and heart attacks among high risk patients. Warner-Lambert had no data to support such claims but, thanks to the outcomes from the large statin studies, the market and doctors began to react favorably to the statins as a class.

Lipitor, with its low-dose power and competitive price, hit the market just as doctors and NCEP guidelines began stressing that lower cholesterol levels were healthier.

Merck & Co. and Bristol-Myers Squibb had spent 10 years educating doctors about statins and Pfizer/Warner-Lambert was able to capitalize on Lipitor's superior efficacy. The scientists who developed atorvastatin credit its increased efficacy to the fact that the drug has a much longer half-life than that of its competitors.

5 STRUCTURE-ACTIVITY RELATIONSHIPS

5.1 HMG-CoA Reductase Inhibitors (Statins)

All of the statins mimic the substrate HMG-CoA (**31**) for the HMG-CoA reductase enzyme. The general structure of the inhibitors contains a central hydrophobic ring system, with a hydroxy-acid (**34**) or lactone (**35**; hydroxy-acid prodrug, which is hydrolyzed to the active open acid form *in vivo*) side-chain necessary for anchoring into the active site of the enzyme.

(**31**) (*S*)-3-hydroxy-3-methyl-glutaryl-CoA
(HMG-CoA)

OR

(**34**)   (**35**)

The tight binding of the inhibitors (Table 7.7) into the enzyme blocks access of the substrate, HMG-CoA (**31**), to the active site, although the exact binding mode was unknown for some 25 years. Furthermore, the elucidation of the structure of the catalytic portion of the human HMG-CoA reductase enzyme, cocrystallized either with HMG-CoA, with HMG and CoA, or with HMG, CoA, and NADP$^+$ (nicotinamide adenine dinucleotide phosphate), as well as with mevalonate-derived products (266, 267), permitted little insight into the binding mode of the inhibitors. Recently, the cocrystallization of several statins bound in the catalytic portion of the human enzyme has permitted the elucidation of the binding mode of the inhibitors (268). The statins exploit the conformational flexibility of the enzyme, to create a hydrophobic binding pocket near the active site for the hydrophobic ring system of the inhibitor, thus occupying

both the HMG-binding pocket and a part of the binding surface for the CoA.

Interestingly, all of the synthetic statins contain a 4-fluorophenyl moiety that distinguishes them from the isolated natural product analogs of mevastatin (**1**), which contains the substituted decalin-ring system. There is a stacking between the guanidinium group of Arg[590] and the 4-fluorophenyl moiety and a polar interaction between the arginine ε-nitrogen atom and the fluorine atom (268).

Only atorvastatin (**6**) and rosuvastatin (**7**) contain a polar side chain that makes a hydrogen bond to Ser[565] through either the carbonyl oxygen atom (atorvastatin) or the sulfonamide oxygen atom (rosuvastatin). Furthermore, only rosuvastatin (**7**) forms a polar interaction with Arg[568] because of the presence of the sulfonamide moiety. Now that the binding modes of the different statins can be visualized in the active site, there is the potential to use this information to design third-generation HMG-CoA reductase inhibitors.

5.2 Fibric Acid Derivatives (Fibrates)

The primary structural factor of the fibrates necessary for receptor recognition is the "fibrate head group" (**36**), which is found in all of the synthetic ligands. As mentioned earlier, the fibrates were developed as hypolipidemic agents through optimization of their lipid-lowering activity in rodents before the discovery of the PPARs. Table 7.8 shows the potency of the fibrate drugs on the murine and human PPARα receptors. There is 85% identity at the

Table 7.7 Enzyme Activity of Selected HMG-CoA Reductase Inhibitors

Statin	Potency on Enzyme IC$_{50}$ (nM)^a
Pravastatin (**3**)	44
Simvastatin (**4**)	11
Fluvastatin (**5**)	28
Atorvastatin (**6**)	8
Rosuvastatin (**7**)	5
Cerivastatin (**8**)	10
Pitavastatin (**9**)	7

^{a}IC$_{50}$ is the molar concentration that produces 50% of its maximum possible inhibition (see Refs. 248, 268, 269).

Table 7.8 Activity of PPARα Agonists in Cell-Based Transactivation Assays

Fibrate	Murine PPARα Receptor EC_{50} $(\mu M)^a$	Human PPARα Receptor EC_{50} $(\mu M)^a$
Clofibrate^b (**10**)	50	55
Gemfibrozil (**11**)	>100	>100
Bezafibrate (**13**)	90	50
Fenofibrate^b (**14**)	18	30
Ciprofibrate (**16**)	15	9

^{a}All the data were generated by use of the PPAR-GAL4 transactivation assay with an SPAP reporter, as described in Ref. 270.
^{b}Data are for the active metabolite (i.e., acid).

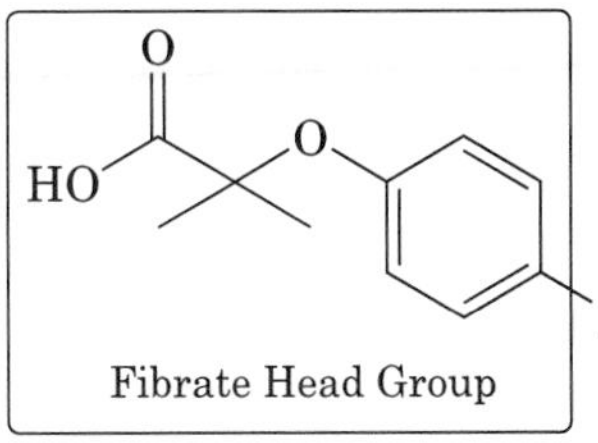

(**36**)

Fibrate Head Group

nucleotide level and 91% identity at the amino acid level between the murine and human PPARα ligand-binding domains (LBDs) (53). These differences may reflect evolutionary adaptation to different dietary ligands and explain the variations in potencies between the species. It should be noted that all of the fibrates currently on the market are weak PPARα agonists and that there are currently numerous pharmaceutical companies working to identify more potent compounds with potentially greater lipid effects in humans.

Fatty acids have been implicated as natural ligands for PPARα by several groups working in the area, although all the fatty acids [i.e., palmitic acid (**37**) and linoleic acid (**38**)] iden-

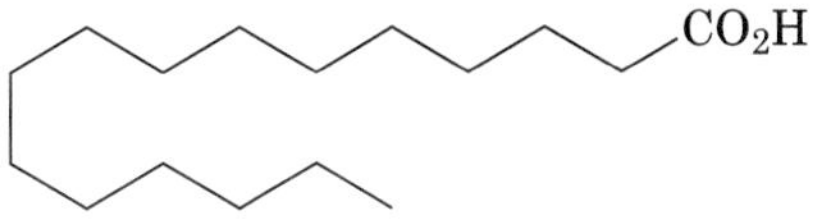

(**37**) Palmitic acid

tified to date bind to PPARα with affinities in the micromolar range. Although total fatty acids levels in serum reach these concentrations (271), it is not known whether the free concentrations of fatty acids in cells are high enough

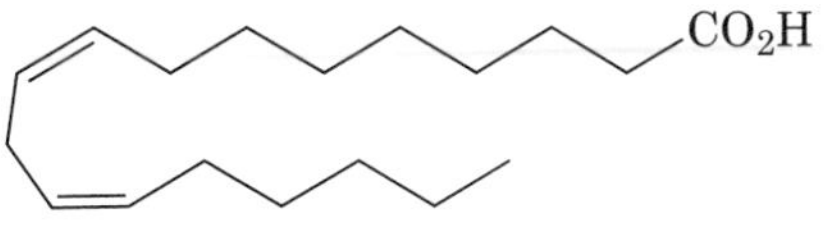

(**38**) linoleic acid

to activate the receptor. This has prompted several groups to search for high affinity natural ligands among the known eicosanoid metabolites of polyunsaturated fatty acids. The lipoxygenase metabolite 8(S)-HETE (**39**) was

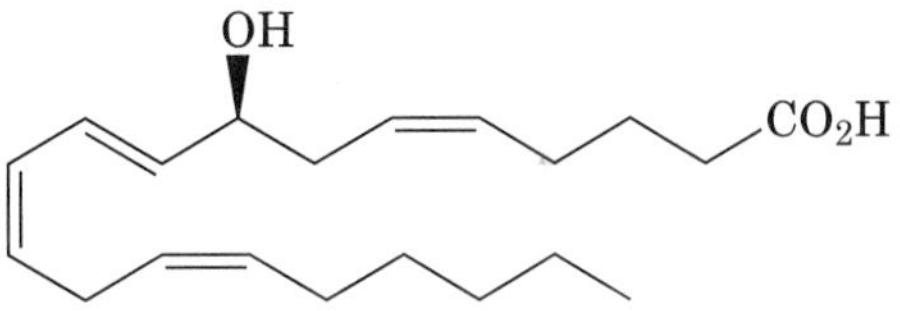

(**39**) 8(S)-HETE

identified as a higher affinity PPARα ligand (272–275), although it is not found in sufficiently high concentrations in the appropriate tissues to be characterized as a natural ligand. Because no single high affinity natural ligand has been identified, it has been proposed that one physiological role of PPARα may be to sense the total flux of fatty acids in metabolically active tissue (272–274, 276, 277).

5.3 Bile Acid Sequestrants (BAS)/Cholesterol Absorption Inhibitors

As mentioned earlier in Section 2.2.3, bile acid sequestrants are cationic resins. The more effective the resin is at selectively binding bile acids at low doses, the lower the chance of observing the gastrointestinal side effects associated with this class of lipid-lowering drugs.

The focus of current work in this area is to increase the loading, that is, the number of quaternary ammonium groups per gram of dry resin that can bind bile acids, thus increasing the efficacy of the resin.

Research scientists at Schering-Plough had identified the first-generation 2-azetidinone (**40**, SCH 48461) as a potent cholesterol ab-

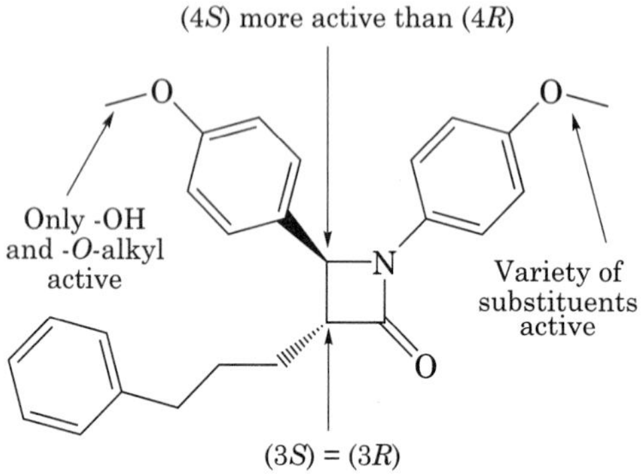

(4S) more active than (4R)

Only -OH and -O-alkyl active

Variety of substituents active

(3S) = (3R)

(40)

sorption inhibitor, starting from a chemistry program to discover conformationally restricted ACAT inhibitors (278, 279). The key structural elements identified for cholesterol absorption inhibition were: (1) the N_1-aryl-substituted 2-azetidinone backbone, (2) a (4S)-alkoxyaryl substituent, and (3) a C_3-aryl-alkyl substituent (278). It was also shown in the same paper that a wide variety of *para*-substituents on the N_1-phenyl were permitted but that the C_4-phenyl required a hydroxyl or alkoxyl residue.

Studies with radiolabeled [³H]-(**40**) in the bile duct-cannulated rat model indicated rapid appearance of a mixture of metabolites in the bile (107). This metabolic mixture was also shown to be a more potent inhibitor than (**40**) of [¹⁴C]-cholesterol absorption, which led the researchers to analyze the putative metabolite structure-activity relationship (SAR) (108). The putative sites of metabolism were identified as (1) demethylation of the methoxy groups, (2) C_3-side-chain benzylic oxidation, and (3) C_3-pendant phenyl oxidation. Finally, the chemistry program identified ezetimibe (**20**), which was designed to block the sites of metabolism and exploit the SAR of the active metabolites.

Interestingly, as mentioned in section 2.4.3, ezetimibe (**20**) is metabolized through glucuronidation of the C_4-phenyl hydroxyl moiety, which improves the cholesterol absorption inhibition activity. This explains the SAR requirement for the substitution of the C_4-phenyl group.

5.4 Nicotinic Acid Derivatives

Nicotinic acid (**21**) is a vitamin of the B family, but its lipid-lowering action is unrelated to its role as a vitamin. It is believed that the lipid effects result from a decrease in fatty acid release from adipocytes, thereby leading to decreased VLDL production. The molecular target of nicotinic acid (**21**) is unknown; however, its identification would facilitate the development of selective nicotinic acid analogs with potentially fewer side effects.

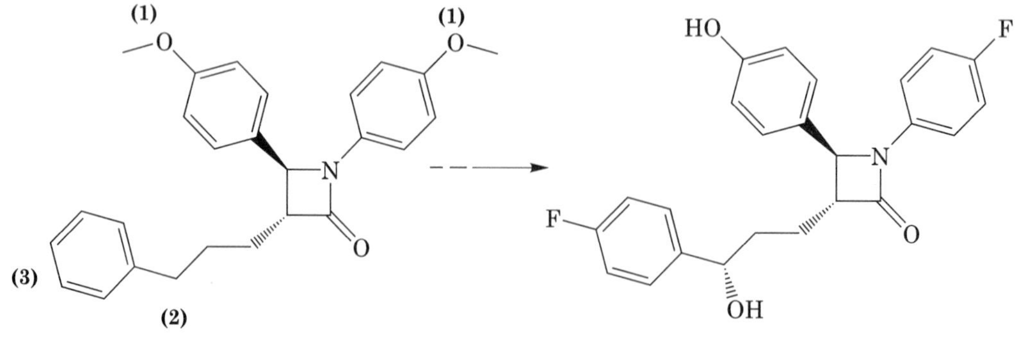

(1)   (1)

(3)

(2)

(40)

(20)

(a)

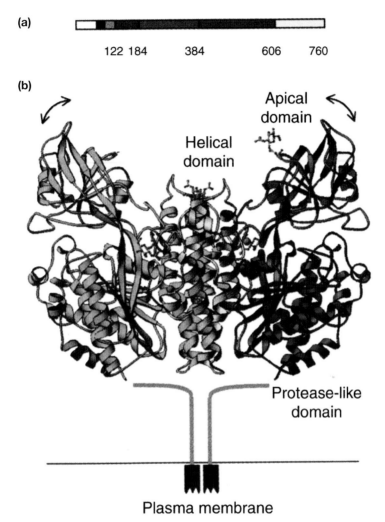

122 184   384   606   760

(b)

Apical
domain

Helical
domain

Protease-like
domain

Plasma membrane

Figure 10.13. Structure of the ectodomain of the transferrin receptor. (a) Domain organization of the transferrin receptor polypeptide chain. The cytoplasmic domain is white; the transmembrane segment is black; the stalk is gray; and the proteaselike, apical and helical domains are red, green and yellow, respectively. Numbers indicate domain residues at domain boundaries. (b) Ribbon diagram of the transferrin receptor dimer depicted in its likely orientation with respect to the plasma membrane. One monomer is colored according to domain (standard coloring as described above), and the other is blue. The stalk region is shown in gray connected to the putative membrane-spanning helices. Pink spheres indicate the location of Sm^{3+} ions in the crystal structure. Arrows show directions of (small) displacements of the apical domain in noncrystallographically related molecules. [Reprinted with permission from C. M. Lawrence, S. Ray, M. Babyonyshev, R. Galluser, D. W. Borhani, and S. C. Harrison, *Science,* **286,** 779–782 (1999). Copyright 1999 American Association for the Advancement of Science.]

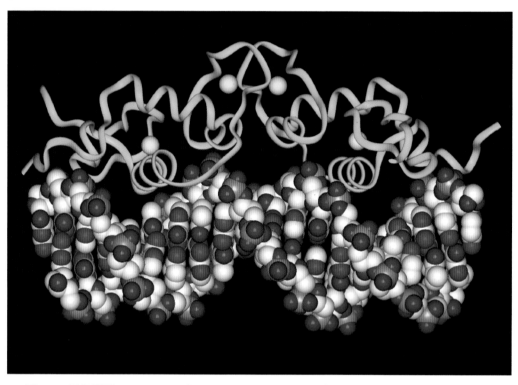

Figure 15.7. DNA-receptor complex structure. Zn^{+2} ions can be seen as spheres within the GR DNA binding domain (ribbon).

5.5 Miscellaneous

Although hormone replacement therapy (HRT) and estrogen modulators exhibit beneficial lipid-lowering effects, these drugs were not optimized for this activity. Therefore, they are not treated in this section.

5.5.1 Probucol. The exact mechanism of action by which probucol (**22**) reduces both LDLc and HDLc is unclear, making a discussion of the SAR impossible. On the other hand, its antioxidant effects are attributed to its free radical scavenger action. The antioxidant effects of the 2,6-di-*tert*-butylphenolic moiety are well known and are not further discussed here.

5.5.2 Plant Sterols. These compounds are nonabsorbable cholesterol analogs that occur naturally in plants and vegetables. Because of their structural similarity with cholesterol (**41**), the plant sterols are able to inhibit the intestinal absorption of cholesterol. Interestingly, the main structural difference between cholesterol, β-sitosterol (**26**), and sitostanol

(**27**) is the ethyl side-chain in position C_{24}. This suggests that there is a very selective mechanism of intestinal cholesterol absorption based on the structural recognition of "cholesterol (**41**)" and that small structural modifications interfere with the normal absorption process. This is understandable, given that β-sitosterol (**26**) and sitostanol (**27**) cannot be further processed by the liver and peripheral tissues for the synthesis of cell membranes and hormones.

6 NEW AND FUTURE TREATMENTS

The new therapeutic options available to clinicians for treating dyslipidemia in the last decade have enabled effective treatment for many patients. Although LDLc is still the major target for therapy, it is likely that over the next several years other lipid and nonlipid parameters will become more generally accepted targets for specific therapeutic interventions. Major pharmaceutical companies are already evaluating new therapeutic agents in human clinical trials.

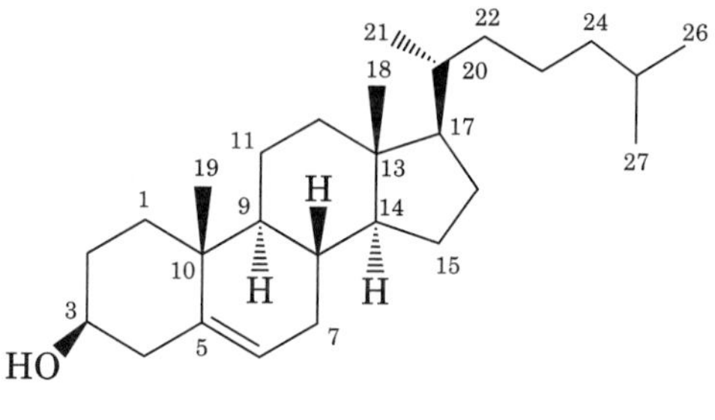

(**41**)

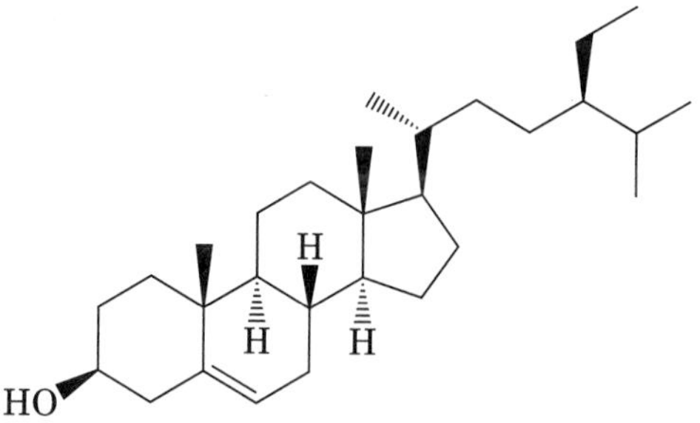

(**26**)

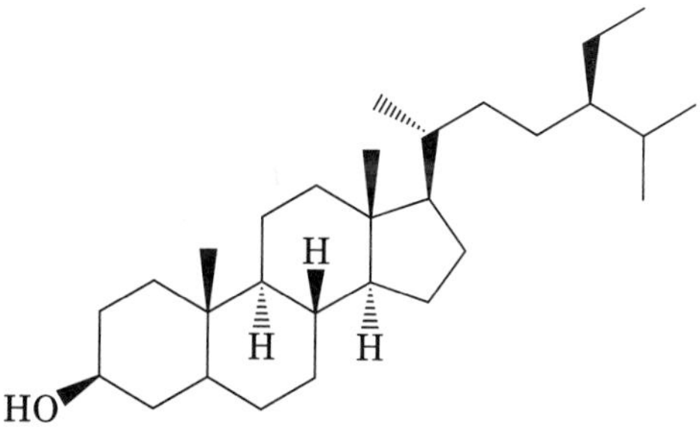

(**27**)

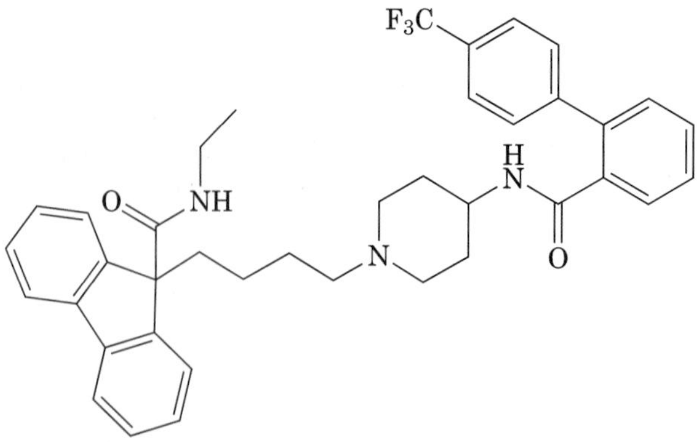

(42) BMS-201038

6.1 New Treatments for Lowering LDLc ± TG Lowering

6.1.1 Novel HMG-CoA Reductase Inhibitors
(Statins). Two new competitors in this area, mentioned earlier in Section 2.2.1, are currently in late-stage development. Crestor [rosuvastatin (**7**), AstraZeneca] is expected to be even more effective than Lipitor [atorvastatin (**6**)] and become a multibillion dollar product (248), and Advicor (nicostatin, Kos), a once-daily combination of Niaspan (**21**) (extended-release niacin) and lovastatin (**2**), lowers LDL and TGs to a greater extent than lovastatin alone, and can raise HDL by as much as 40%. If this product can overcome the safety and tolerability issues associated with niacin (**2**) (see Section 2.3.4) and concerns over myopathy/rhabdomyolysis (see Section 2.3.1.2), it may become a commercial success.

Beyond these, the only other statin of note is pitavastatin (**9**). Novartis has recently licensed this compound in Europe (where it is in phase III) and is in negotiation for U.S. rights.

Recently, it has been reported that both rosuvastatin (**7**) and pitivastatin (**9**) have been associated with rhabdomyolysis at the higher doses evaluated, which will potentially delay their development and complicate regulatory approval.

6.1.2 Microsomal Triglyceride Transfer Protein (MTP) Inhibitors.
MTP, which is found in the liver and intestine, plays a pivotal role in the assembly and secretion of TG-rich lipoproteins (VLDL and chylomicrons), and also catalyzes the transport of TGs, cholesterol esters,

and phospholipids. MTP inhibitors have been shown to significantly reduce (>60%) serum levels of VLDLc, LDLc, and TGs in animal models. The major issue in the prolonged inhibition of hepatic MTP is the potential of an accumulation of TGs in the liver (fatty liver). In addition, BMS discontinued the development of BMS-201038 (**42**), claiming mechanism of action-based adverse events, in the form of liver function. However, Bayer is currently in phase III trials with BAY-139952 (**43**,

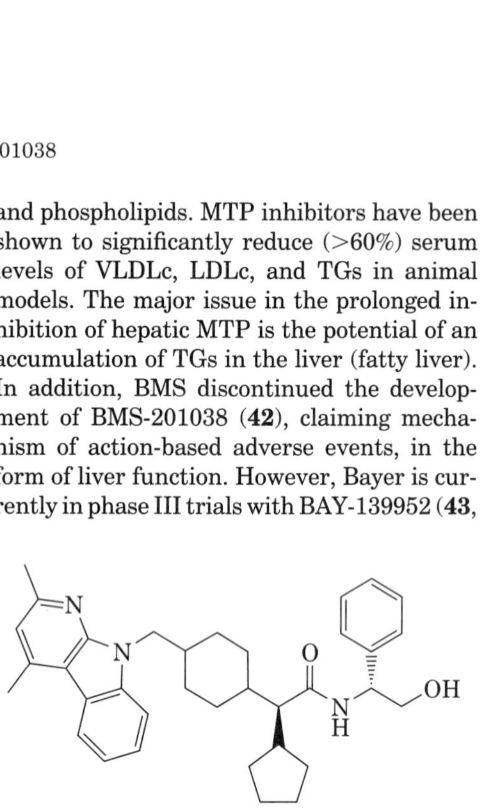

(43) BAY-139952

implitapide), whereas Pfizer is in phase II trials with CP-346086 (structure not published) and Janssen is in phase I with R-103757 (**44**). Interestingly, animal data and results from the completed clinical trials also suggest that the intestinal inhibition of MTP results in decreased fat absorption and weight loss associated with an antiobesity effect. The future of this class of compounds will depend on the ability of these drugs to resolve the potential liver toxicity issues associated with MTP inhibition.

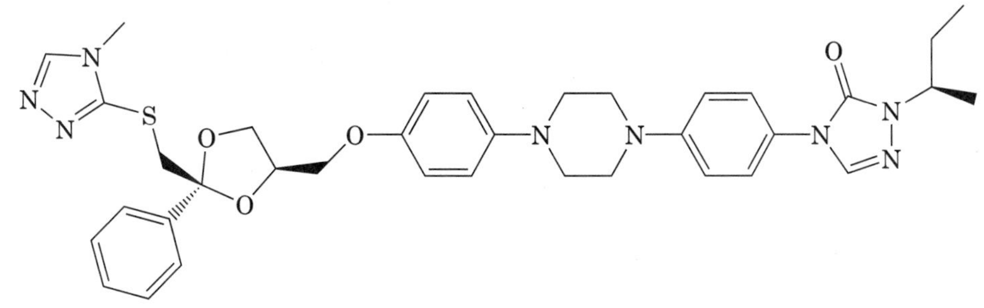

(44) R-103757

6.1.3 LDL-Receptor (LDLr) Upregulators.

The up-regulation of LDL receptors has potential as a novel means of lowering serum LDL. These compounds could be significant competition in the dyslipidemia segment of the market arising from the large unmet need in this area. Pfizer, Tularik, Lilly, and Aventis are all active in this field. Recently, scientists at Glaxo-SmithKline described a new class of compounds that reduce blood levels of cholesterol in an animal model by upregulating the LDLr through a mechanism different from that of the statins (280). Two series of molecules, the steroidlike analogs represented by GW707 (**45**) and the nonsteroidal molecules represented by GW532 (**46**), were identified with nanomolar activities.

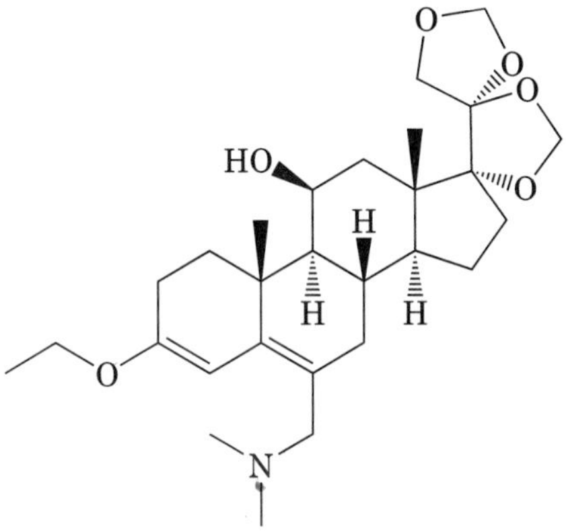

6.1.4 Bile Acid Reabsorption Inhibitors/Bile Acid Sequestrants (BAS)/Cholesterol Absorption Inhibitors.

Although these three classes of molecules can be used as monotherapy, their greatest potential resides in combination therapy. Aventis is currently in phase I trials with the bile acid reabsorption inhibitor HMR-1453 (structure not published) for the treatment of atherosclerosis. BMS is currently in Phase II with the bile acid sequestrant DMP-504 (structure not published), which is reported to be more potent than cholestyramine in reducing serum cholesterol levels but appears to

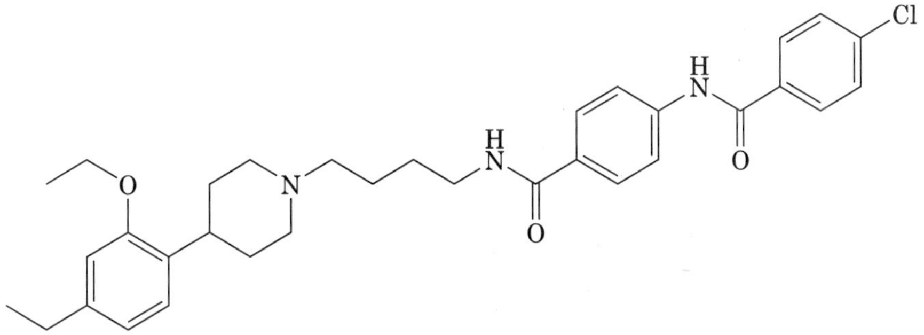

(45) GW707

(46) GW532

have the same side effect profile, which is a disadvantage compared to GelTex's second-generation bile acid sequestrant GT-102279 (structure not published), also in Phase II trials. Ezetimibe (**20**) is the only cholesterol absorption inhibitor currently in clinical trials (see section 2.2.3) and has been shown to reduce LDLc between 10 and 19% in monotherapy. Interestingly, the reduction of LDLc in combination of ezetimibe (**20**) with a statin [i.e., simvastatin (**4**) or atorvastatin (**6**)] is additive.

6.1.5 Acyl-CoA Cholesterol AcylTransferase (ACAT) Inhibitors.

ACAT is a ubiquitous enzyme responsible for esterifying excess intracellular cholesterol. The cholesterol ester is then transferred to lipoprotein particles to be stored in their core and, subsequently, deposited into forming atherosclerotic lesions. The activity of ACAT is enhanced by the presence of intracellular cholesterol; however, whether inhibition of ACAT will prevent atherosclerosis is not yet clear. Furthermore, inhibition of hepatic ACAT decreases the secretion of apoB-containing lipoproteins (VLDL) by the liver. The combination of an ACAT inhibitor and another lipid-lowering agent, particularly a statin, could have added benefit on CV mortality and morbidity. Pfizer is leading the field with Avasimibe (CI-1011, **47**), currently in

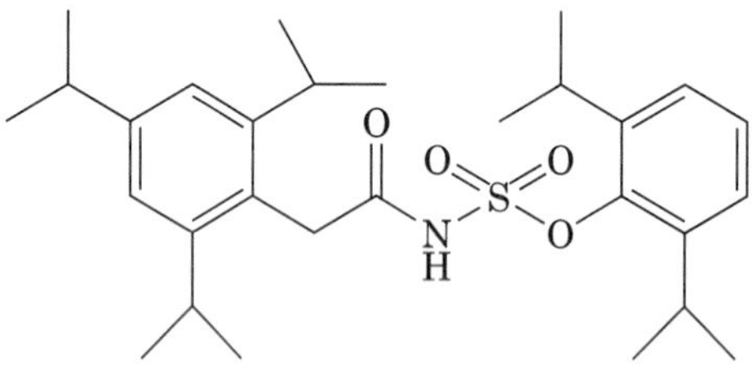

phase III trials, whereas Sankyo (CS 505, structure not published) and bioMerieux-Pierre Fabre (F12511, **48**, or eflucimibe) are both reported in the phase I stage. There are currently many other pharmaceutical companies reported to be working in this field.

6.2 New Treatments for Raising HDLc ± TG Lowering

6.2.1 Peroxisome Proliferator-Activated Receptor (PPAR) Agonists.

Competitors are generally PPARγ or mixed PPARα/γ agonists, focused primarily on diabetes [Dr. Reddy/NovoNordisk (DRF-2725, **49**; phase III), Astra-

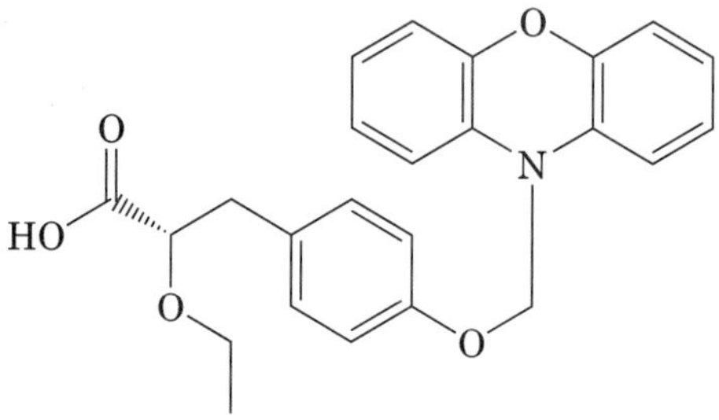

(**49**) DRF-2725

Zeneca (AZ-242, **50**; phase II), BMS (BMS-298585, **51**; phase II), Merck (KRP-297, **52**; phase II), and Ligand/Lilly (LY519818, structure not published; phase I)]. There is also a series of PPARα agonists from Kyorin (**53**), the first of which is in preclinical development for atherosclerosis, whereas GlaxoSmithKline has also reported two PPAR agonists in phase I trials for dyslipidemia [GW 590735 (stucture not published) and GW 501516 (**54**)] as well as Dr. Reddy's DRF-4832 (structure not published), which is to start phase I trials for dyslipidemia later this year.

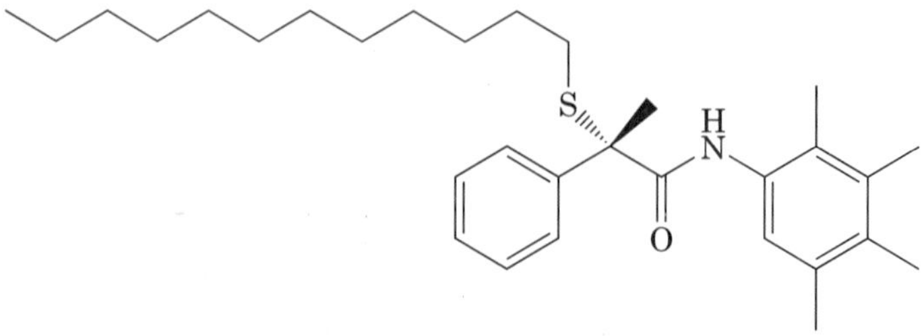

(**47**) CI-1011

(**48**) F12511

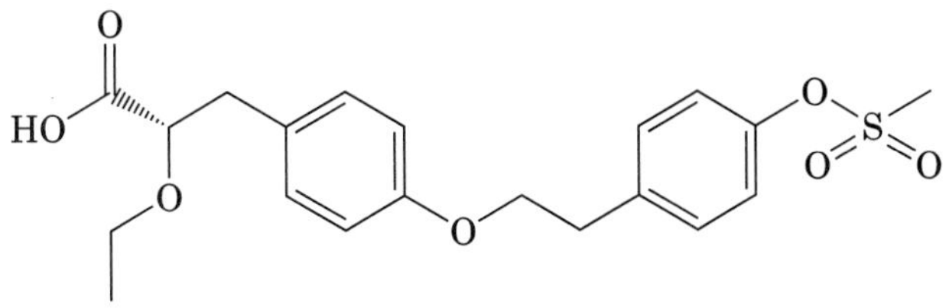

(50) AZ-242

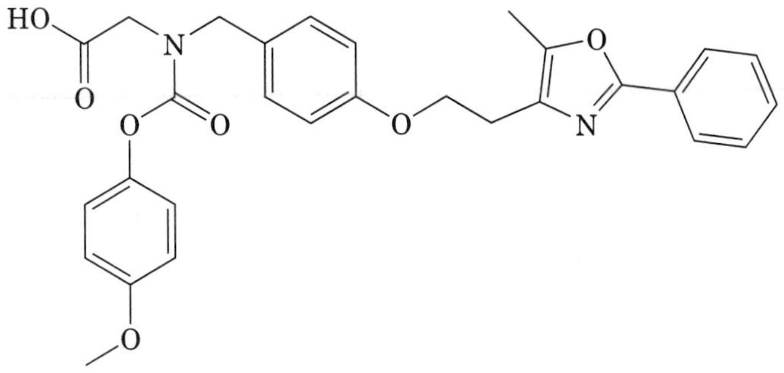

(51) BMS-298585

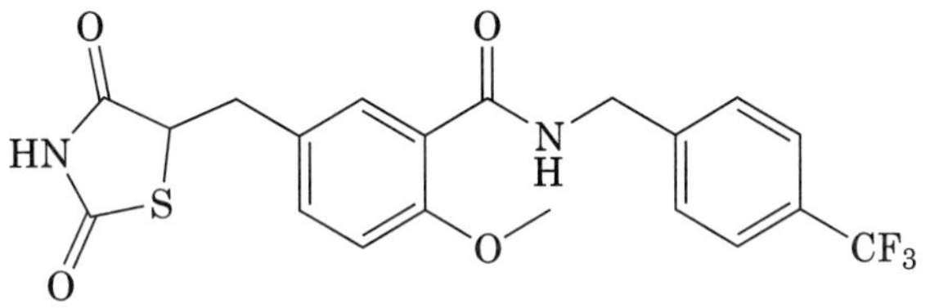

(52) KRP-297

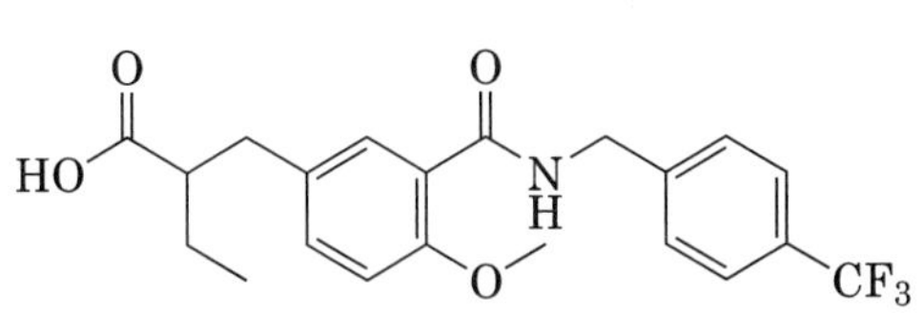

(53) Kyorin

6.2.2 Cholesteryl Ester Transfer Protein (CETP) Inhibitors. CETP is a plasma glycoprotein that mediates the transfer of cholesteryl ester from HDL to VLDL, IDL, and LDL. Compounds that inhibit CETP are expected to increase plasma HDL cholesterol levels and improve the HDL/LDL cholesterol ratio. They could be used as monotherapy, or more likely in combination with statins. Pfizer recently reported excellent phase II results of

CP-529414 (structure not published), a 70% increase in HDLc, indicating these compounds may potentially have a large impact on the dyslipidemia market. Avant Immunotherapeutics is currently in phase II trials with CETi-1 (structure not published), a peptide vaccine against CETP to reduce risk factors for atherosclerosis, and in November 2001 at the American Heart Association 2001 Scientific Sessions, Japan Tobacco presented the phase II data, where oral administration of 900 mg of JTT-705 (**55**) once daily for 4 weeks led to a 34% increase in HDLc levels and a 7% decrease in LDLc.

6.2.3 Liver X-Receptor (LXR) Agonists. LXRα is a nuclear receptor implicated in lipid homeostasis. LXRα can modify the expression of

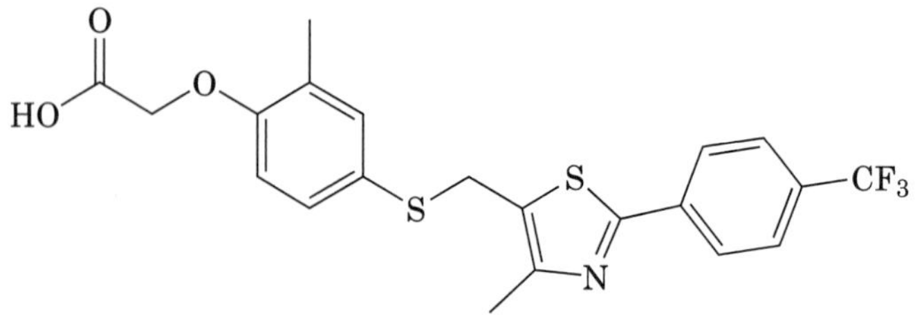

(54) GW 501516

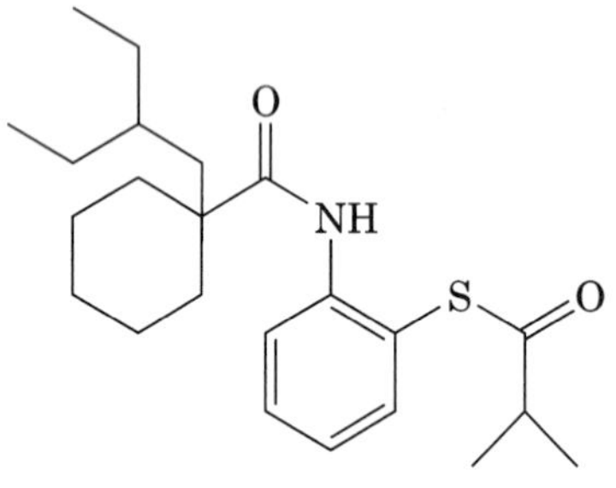

(55) JTT-705

genes for lipogenic enzymes through regulation of the sterol regulatory-element binding Protein 1c (SREBP-1c) expression (281a). On the basis of current animal model data, LXRα agonists will afford large increases in HDL levels, leading to increased reverse cholesterol transport (RCT) (281b). Several pharmaceutical companies are working in this area but none of the molecules is beyond the preclinical stage.

6.3 New Treatments for Atherosclerosis

In the past, the treatment of CV disease was addressed by modifying the major risk factors (i.e., LDLc, HDLc, TG, etc.) associated with the progression of atherosclerosis. Recently, pharmaceutical companies have been investi-

gating new approaches to treat directly the atherosclerotic plaque resulting in plaque stabilization and/or regression.

6.3.1 Chemokines and Cytokines. Chemokines and cytokines may act directly at the atherosclerotic plaque, interfering with the inflammatory process thought to destabilize the plaque. All the research activities seem to be in the preclinical stage, although many large companies seem to be involved in this area, including Merck, Novartis, Aventis, AstraZeneca, and GlaxoSmithKline.

6.3.2 Antioxidants. Oxidative modification of LDL has been accepted as an important event in the development of atherosclerosis. Therefore, antioxidants have been expected to have potential as antiatherogenic agents. However, clinical trials of antioxidants have given rise to controversy regarding their real clinical benefit. Although vitamin E has been reported to reduce the risk of coronary heart disease, two large-scale trials failed to demonstrate the effect of α-tocopherol (56), the most active species of vitamin E, on cardiovascular disease or cerebrovascular mortality (282). This failure of α-tocopherol (56) is probably attributable to the fact that it cannot reach the core of LDL particles. On the other hand,

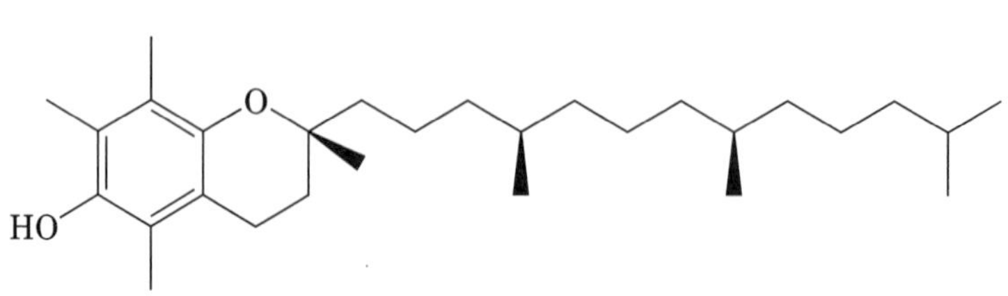

(56) α-Tocopherol

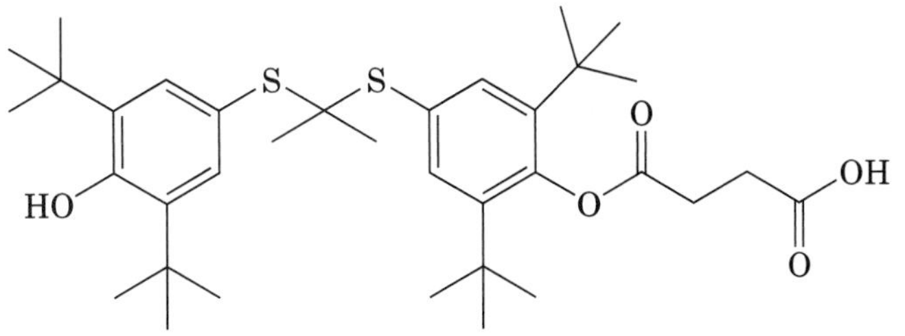

(57) AGI-1067

probucol (22) showed antiatherogenic effects in animal models but had the untoward effect of lowering HDL levels.

Several of these are beyond the preclinical stage, including AGI-1067 (57), from Athero-Genetics (licensed to Schering-Plough, in phase II), which is a structural analog of probucol (22); a compound [BO-653 (58)] from

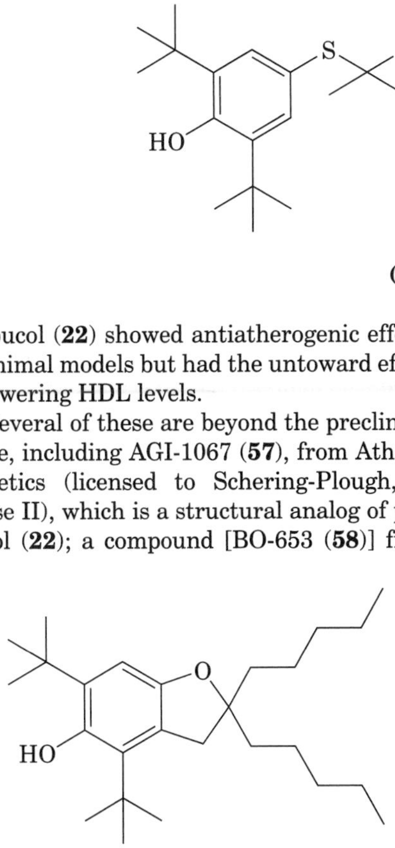

(58) BO-653

Chugai, also in phase II; and a compound (AC-3056) from Aventis that has just completed phase I.

AGI-1067 (57) is a VCAM-1 (vascular cell adhesion molecule 1) gene expression inhibitor under development by AtheroGenics for the potential prevention of atherosclerosis (hypercholesterolemia) and restenosis. VCAM-1 is the surface protein to which various types of leukocyte attach themselves, forming the starting point of new plaques. By inhibiting the expression of VCAM-1, AGI-1067 (57) has the potential to prevent atherosclerosis at the very earliest stage. In November 2001 further data, presented at the American Heart Association 2001 Scientific Sessions, showed that AGI-1067 (57) met its primary endpoint in preventing restenosis in the phase II studies and showed a direct antiatherosclerotic effect on coronary blood vessels, consistent with re-

versing the progression of coronary artery disease. Phase III studies are expected to begin in 2003.

Experimental data have shown that BO-653 (58), currently in phase II studies, is a superior antioxidant to either α-tocopherol (57) or probucol (22) (282). It can penetrate into the core of LDL particles, does not lower HDL levels, and shows antiatherogenic and antirestenosis effects in animal models. However, studies are still needed to determine whether BO-653 (57) has therapeutic utility in humans.

The nonpeptidic compound AC-3056 (structure not published), in-licensed by Amylin Pharmaceuticals, is being developed for the prevention of atherosclerosis and restenosis after angioplasty procedures and metabolic disorders relating to cardiovascular disease. AC-3056 has been characterized in vitro and in animal models as having three different modes of action-targeting steps in the atherosclerosis cascade: (1) lowering of serum LDL cholesterol but not HDL cholesterol; (2) inhibition of lipoprotein oxidation; and (3) inhibition of cytokine-induced expression of cell adhesion molecules in vascular cells (283).

6.3.3 Lipoprotein-Associated Phospholipase A_2 (Lp-PLA$_2$) Inhibitors.

Lipoprotein-associated phospholipase A_2 (Lp-PLA$_2$), an enzyme associated with low density lipoprotein, would appear to be a novel target for therapy to prevent heart attacks on the basis of a study published by scientists from GlaxoSmithKline and Glasgow University (284). In the study, in addition to being a potential drug target, the enzyme could be a new risk factor for cardiovascular disease and as such could serve as a marker, independent of LDL, to predict the occurrence of heart attacks. Lp-PLA$_2$ is found in the bloodstream, bound to LDL. During LDL oxidation, Lp-PLA$_2$ breaks down the fats in LDL, producing substances that attract inflammatory cells, which in turn engulf LDL, eventually contributing to the formation of atherosclerotic plaques. SB-480848 (structure not published), which targets Lp-PLA$_2$, is currently in phase I clinical trials (285) and targets a different rationale from cholesterol reduction in the prevention of heart attack. This would therefore benefit people at risk of a heart attack, but who do not have increased cholesterol levels.

6.3.4 New Miscellaneous Treatments.

Esperion Therapeutics and the University of Milan are developing ETC-216, apolipoprotein AI Milano (also known as apoAI Milano or AIM), a recombinant variant of normal apolipoprotein AI, the major protein component of HDL, which is thought to protect against cardiovascular disease by efficiently removing cholesterol and other lipids from tissues including the arterial wall and transporting them to the liver for elimination. A multiple-dose, multicenter phase II clinical trial has initiated with ETC-216 in patients with acute coronary syndromes (ACS). The trial will assess the efficacy of ETC-216 in regressing coronary atherosclerosis by measuring changes in plaque size of one targeted coronary artery, measured by atheroma volume through the use of intravascular ultrasound. The double-blind, randomized, placebo-controlled study will enroll 50 patients with ACS who are scheduled to undergo coronary angiography and/or angioplasty. ETC-216 offers an attractive mechanism for the treatment of atherosclerosis because it aims to reverse the lipid accumulation already present in atherosclerosis, as well as preventing further accumulation. There are lipid-lowering agents currently available that decrease serum cholesterol levels and stop the progression of atherosclerosis, but no therapies currently exist that selectively remove lipid from atherosclerotic lesions leading to plaque regression in a manner similar to that of apo AI Milano. Because there are currently no human studies available on this compound, the effect that this drug will

have on overall cardiovascular morbidity and mortality is not known. Nonetheless, the ability to reverse lipid accumulation in atherosclerosis with this therapy is claimed to provide substantial benefits compared to those of existing therapies.

Preliminary findings from Esperion Therapeutics' phase IIa clinical study of ETC-588, or LUV (large unilamellar vesicles), for the treatment of ACS indicated that the product met the primary endpoint of demonstrating safety and tolerability in patients with known vascular disease. The study was a double-blind, randomized, placebo-controlled, multiple-dose trial designed to determine the optimal dosing schedule and effect of ETC-588 in 34 patients with stable atherosclerosis and HDL of 45 mg/dL or less. Patients received one of three dose strengths (50, 100, or 200 mg/kg) or placebo every 4 or 7 days. Patients receiving the 100 and 200 mg/kg doses each received seven doses for either 4 or 6 weeks, whereas those on 50 mg/kg received 14 doses for either 8 or 13 weeks. All dose levels and regimens were found to be safe and well tolerated, and an optimal dosing schedule of once every 7 days was defined for future use. Evidence of dose-related cholesterol mobilization was noted. Evaluation is still under way of brachial artery ultrasound measurements and changes in inflammatory markers. ETC-588 is made of naturally occurring lipids that circulate through the arteries and is claimed to remove accumulated cholesterol and other lipids from cells, including those in the arterial wall. It is designed to augment HDL function for the acute or subacute treatment of ischemia. ETC-588 has demonstrated a high capacity to transport cholesterol from peripheral tissues to the liver, improve endothelial function, and regress atherosclerosis in preclinical models.

7 RETRIEVAL OF RELATED INFORMATION

Related information can be retrieved through library online services, especially Current Contents, Medline, and/or SciFinder. Key references can be found by precision searches, by use of combinations of key words or phrases, plus specification of a single year to narrow down the number of matches to be reviewed at one time. The competitor awareness databases [i.e., Competitor Knowledge Base (CKB; 287) and Investigational Drugs database (IDdb3; 288)] have been used as sources for the update of new treatments, whereas general cardiovascular infomation can be found at the following Internet sites:

- American College of Cardiology: www.acc.org
- American Diabetes Association: www.diabetes.org
- American Heart Association: www.americanheart.org
- Doctor's Guide: www.docguide.com
- Healthy Heart Program: www.healthyheart.org
- Heart Information Network: www.heartinfo.org
- MedScape Today: www.medscape.com
- National Heart, Lung and Blood Institute: www.nhlbi.nih.gov
- National Stroke Association: www.stroke.org

A list of general reviews for further reading can also be found at the end of the reference section (Refs. 288–296).

REFERENCES

1. (a) American Heart Association, *2001 Heart and Stroke Statistical Update*, Dallas, TX, American Heart Association, 2000; (b) Expert Panel on Detection, Evaluation, and Treatment of High Blood Cholesterol in Adults, *JAMA*, **285**, 2486–2497 (2001).

2. H. A. Eder and L. I. Gidez, *Med. Clin. North Am.*, **66**, 431–440 (1982).

3. Anonymous, *Lancet*, **344**, 1383–1389 (1994).

4. J. Shepherd, S. M. Cobbe, I. Ford, C. G. Isles, A. R. Lorimer, P. W. MacFarlane, J. H. McKillop, and C. J. Packard, *N. Engl. J. Med.*, **333**, 1301–1307 (1995).

5. F. M. Sacks, M. A. Pfeffer, L. A. Moye, J. L. Rouleau, T. G. Cole, L. Brown, J. W. Warnica, J. M. Arnold, C. C. Wun, B. R. Davis, and E. Braunwald, *N. Engl. J. Med.*, **335**, 1001–1009 (1996).

6. J. R. Downs, M. Clearfield, S. Weis, E. Whitney, D. R. Shapiro, P. A. Beere, A. Langendorfer, E. A. Stein, W. Kruyer, and A. M. Gotto Jr., *JAMA*, **279**, 1615–1622 (1998).

7. Anonymous, *N. Engl. J. Med.*, **339**, 1349–1357 (1998).

8. G. Brown, J. J. Albers, L. D. Fisher, S. M. Schaefer, J. T. Lin, C. Kaplan, X. Q. Zhao, B. D. Bisson, V. F. Fitzpatrick, and H. T. Dodge, *N. Engl. J. Med.*, **323**, 1289–1298 (1990).

9. J. P. Kane, M. J. Malloy, T. A. Ports, N. R. Phillips, J. C. Diehl, and R. J. Havel, *JAMA*, **264**, 3007–3012 (1990).

10. H. Buchwald, R. L. Varco, J. P. Matts, J. M. Long, L. L. Fitch, G. S. Campbell, M. B. Pearce, A. E. Yellin, W. A. Edmiston, and R. D. Smink Jr., *N. Engl. J. Med.*, **323**, 946–955 (1990).

11. D. Ornish, S. E. Brown, L. W. Scherwitz, J. H. Billings, W. T. Armstrong, T. A. Ports, S. M. McLanahan, R. L. Kirkeeide, R. J. Brand, and K. L. Gould, *Lancet*, **336**, 129–133 (1990).

12. G. F. Watts, B. Lewis, J. N. Brunt, E. S. Lewis, D. J. Coltart, L. D. Smith, J. I. Mann, and A. V. Swan, *Lancet*, **339**, 563–569 (1992).

13. L. Cashin-Hemphill, W. J. Mack, J. M. Pogoda, M. E. Sanmarco, S. P. Azen, and D. H. Blankenhorn, *JAMA*, **264**, 3013–3017 (1990).

14. W. L. Haskell, E. L. Alderman, J. M. Fair, D. J. Maron, S. F. Mackey, H. R. Superko, P. T. Williams, I. M. Johnstone, M. A. Champagne, and R. M. Krauss, *Circulation*, **89**, 975–990 (1994).

15. Anonymous, *Lancet*, **344**, 1383–1389 (1994).

16. F. M. Sacks, M. A. Pfeffer, L. A. Moye, J. L. Rouleau, J. D. Rutherford, T. G. Cole, L. Brown, J. W. Warnica, J. Malcolm, et al., *N. Engl. J. Med.*, **335**, 1001–1009 (1996).

17. J. Shepherd, S. M. Cobbe, I. Ford, C. G. Isles, A. R. Lorimer, P. W. Macfarlane, J. H. McKillop, and C. J. Packard, *N. Engl. J. Med.*, **333**, 1301–1307 (1995).

18. J. R. Downs, M. Clearfield, S. Weis, E. Whitney, D. R. Shapiro, P. A. Beere, A. Langendorfer, E. A. Stein, W. Kruyer, and A. M. Gotto Jr., *JAMA*, **279**, 1615–1622 (1998).

19. W. B. Kannel, W. P. Castelli, and T. Gordon, *Ann. Intern. Med.*, **90**, 85–91 (1979).

20. J. P. Desager, Y. Horsmans, C. Vandenplas, and C. Harvengt, *Atherosclerosis*, **124** (Suppl. 1), S65–S73 (1996).

21. T. Gordon, W. P. Castelli, M. C. Hjortland, W. B. Kannel, and T. R. Dawber, *Am. J. Med.*, **62**, 707–714 (1977).

22. N. E. Miller, D. S. Thelle, O. H. Ford, and O. D. Mjos, *Lancet*, **1**, 965–968 (1977).

23. D. J. Gordon and B. M. Rifkind, *N. Engl. J. Med.*, **321**, 1311–1316 (1989).

24. G. Assman and H. Schulte, *Am. J. Cardiol.*, **70**, 733–737 (1992).

25. M. J. Stampfer, F. M. Sacks, S. Salvini, W. C. Willett, and C. H. Hennekens, *N. Engl. J. Med.*, **325**, 373–381 (1991).

26. A. I. Gotlieb and B. L. Langille in V. Fuster, R. Ross, and E. J. Topol, Eds., *Atherosclerosis and Coronary Artery Disease*, Vol. **1**, Lippincott-Raven, Philadelphia, 1996, pp. 595–606.

27. K. J. Williams and I. Tabas, *Atherioscler. Thromb. Vasc. Biol.*, **15**, 551–561 (1995).

28. K. J. Williams and I. Tabas, *Curr. Opin. Lipidol.*, **9**, 471–474 (1998).

29. H. C. Stary, A. B. Chandler, R. E. Dinsmore, V. Fuster, S. Glagov, W. Install Jr., M. E. Rosenfield, C. J. Schwartz, W. D. Wagner, and R. W. Wissler, *Atherioscler. Thromb. Vasc. Biol.*, **15**, 1512–1531 (1995).

30. C. K. Glass and J. L. Witztum, *Cell*, **104**, 503–516 (2001).

31. P. D. Richardson, M. J. Davies, and G. V. Born, *Lancet*, **2**, 941–944 (1989).

32. R. L. Ridolfi and G. M. Hutchins, *Am. Heart J.*, **93**, 468–486 (1977).

33. M. Friedman and G. J. Van den Bovenkamp, *Am. J. Pathol.*, **48**, 19–44 (1966).

34. E. Falk, *Circulation*, **86** (Suppl. 6), III30–III42 (1992).

35. "Bayer Pulls Cholesterol-Lowering Drug, Warns Full-Year Profit Will Miss Views,"*Dow Jones Business News*, Dow Jones & Company, Inc., August 8, 2001.

36. A. Endo, M. Kuroda, and Y. Tsujita, *J. Antibiot.*, **29**, 1346–1348 (1976).

37. A. Endo, M. Kuroda, and K. Tanzawa, *FEBS Lett.*, **72**, 323–326 (1976).

38. H. Rudney and S. R. Panini, *Curr. Opin. Lipidol.*, **4**, 230–237 (1993).

39. D. W. Russell, *Cardiovasc. Drugs Ther.*, **6**, 103–110 (1992).

40. Y. Tsujita, M. Kuroda, K. Tanzawa, N. Kitano, and A. Endo, *Atherosclerosis*, **32**, 307–313 (1979).

41. M. Kuroda, Y. Tsujita, K. Tanzawa, and A. Endo, *Lipids*, **14**, 585–589 (1979).

42. T. A. Tobert, G. D. Bell, J. Birtwell, I. James, W. R. Kukovetz, J. S. Pryor, A. Buntinx, I. B. Holmes, Y. S. Chao, and J. A. Bolognese, *J. Clin. Invest.*, **69**, 913–919 (1982).

43. M. A. Austin, J. E. Hokanson, and K. L. Edwards, *Am. J. Cardiol.*, **81**, 7B–12B (1998).

44. D. J. Maron, S. Fazio, and M. F. Linton, *Circulation*, **101**, 207–213 (2000).

45. W. Insull, J. Isaacsohn, P. Kwiterovich, P. Ra, R. Brazg, C. Dujovne, M. Shan, E. Shurue-Crowley, S. Ripa, and R. Tota, *J. Int. Med. Res.*, **28**, 47–68 (2000).

46. E. A. Stein, G. Lamkin, P. M. Laskarzewski, and M. D. Cressman, *Program and Abstracts of the American College of Cardiology 50th Annual Scientific Session*, March 18–21, 2001, Orlando, FL, Session 1261–173.

47. E. A. Stein, K. L. Strutt, E. Miller, and H. Southworth, *Program and Abstracts of the American College of Cardiology 50th Annual Scientific Session*, March 18–21, 2001, Orlando, FL, Session 1261–176.

48. M. H. Davidson, P. T. S. Ma, E. A. Stein, and H. G. Hutchinson, *Program and Abstracts of the American College of Cardiology 50th Annual Scientific Session*, March 18–21, 2001, Orlando, FL, Session 1261–175.

49. H. Mabuchi and the Hokuriku FH Study Group, *Atherosclerosis*, **151**, 53 (2000); Proceedings of the XIIth International Symposium on Atherosclerosis, June 25–29, 2000, Stockholm, Sweden, Abstr. MoP81:W6.

50. T. Teramoto, Y. Saito, N. Nakaya, and the Japan Itavastatin Clinical Study Group, *Atherosclerosis*, **151**, 53 (2000); Proceedings of the XIIth International Symposium on Atherosclerosis, June 25–29, 2000, Stockholm, Sweden, Abstr. MoP82:W6.

51. I. Issemann and S. Green, *Nature*, **347**, 645–650 (1990).

52. B. P. Neve, J.-C. Fruchart, and Bart Staels, *Biochem. Pharmacol.*, **60**, 1245–1250 (2000).

53. T. M. Willson, P. J. Brown, D. D. Sternbach, and B. R. Henke, *J. Med. Chem.*, **43**, 527–550 (2000).

54. A. Gaw, C. J. Packard, and J. Shepherd, *Handb. Exp. Pharmacol.*, **109**, 325–348 (1994).

55. R. J. Havel and J. P. Kane, *Annu. Rev. Pharmacol.*, **13**, 287–308 (1973).

56. C. R. Sirtori, A. Catapano, and R. Paoletti, *Atheroscler. Rev.*, **2**, 113–153 (1977).

57. S. J. Robins, D. Collins, J. T. Wittes, V. Papademetriou, P. C. Deedwania, J. R. McNamara,

M. L. Kashyap, J. M. Hershman, L. F. Wexler, and H. Bloomfield-Rubins, *JAMA*, **285**, 1585–1591 (2001).

58. Anonymous, *JAMA*, **231**, 360–381 (1975).

59. Anonymous, *Br. Heart J.*, **40**, 1069–1118 (1978).

60. Anonymous, *Lancet*, **2**, 379–385 (1980).

61. M. H. Frick, O. Elo, K. Haapa, O. P. Heinsalma, P. Helo, J. K. Huttunen, P. Kaitaniemi, P. Koskinen, V. Manninen, et al., *N. Engl. J. Med.*, **317**, 1237–1245 (1987).

62. M. H. Frick, M. Syvanne, M. S. Nieminen, H. Kauma, S. Majahalme, V. Virtanen, Y. A. Kesaniemi, A. Pasternack, and M. R. Taskinen, *Circulation*, **96**, 2137–2143 (1997).

63. C. A. Aguilar-Salinas, G. Fanghanel-Salmon, E. Meza, J. Montes, A. Gulias-Herrero, and L. Sanchez, *Metabolism*, **50**, 729–733 (2001).

64. Anonymous, *Pharma J.*, **225**, 115 (1980).

65. Ciprofibrate in *IDdb Drug Report*, November 29, 2000, Investigational Drugs Database, Current Drugs Ltd. May be accessed at www.current-drugs.com

66. J. C. Adkins and D. Faulds, *Drugs*, **54**, 615–633 (1997).

67. G. Steiner, *Diabetes Care*, **23** (Suppl. 2), B49–B53 (2000).

68. G. F. Watts and S. B. Dimmitt, *Curr. Opin. Lipidol.*, **10**, 561–574 (1999).

69. G. Steiner, *Lancet*, **357**, 905–910 (2001).

70. A. C. Hutchesson, A. Moran, and A. F. Jones, *J. Clin. Pharm. Ther.*, **19**, 387–389 (1994).

71. J. W. Smit, G. H. Jansen, T. W. Bruin, and D. W. Erkelens, *Am. J. Cardiol.*, **76**, 126–128 (1995).

72. J. D. Spence, C. E. Munoz, L. Hendricks, L. Latchinian, and H. E. Khouri, *Am. J. Cardiol.*, **76**, 80–83 (1995).

73. M. J. Tikkanen, *Curr. Opin. Lipidol.*, **7**, 385–388 (1996).

74. E. S. Ganotakis, I. A. Jagroop, G. Hamilton, A. F. Winder, and D. P. Mikhailidis, *J. Drug Dev. Clin. Pract.*, **8**, 171–175 (1996).

75. J. A. Papadakis, E. S. Ganotakis, I. A. Jagroop, A. F. Winder, and D. P. Mikhailidis, *Int. J. Cardiol.*, **69**, 237–244 (1999).

76. J.-C. Fruchart and P. Duriez, *Eur. Heart J. Suppl.*, **2** (Suppl. D), D54–D56 (2000).

77. D. Gavish, E. Leibovitz, I. Shapira, and A. Rubinstein, *J. Intern. Med.*, **247**, 563–569 (2000).

78. D. N. Kiortisis, H. Millionis, E. Bairaktari, and M. S. Elisaf, *Eur. J. Clin. Pharmacol.*, **56**, 631–635 (2000).

79. H. E. Bays, C. A. Dujovne, and A. M. Lansing, *Heart Dis. Stroke*, **1**, 357–365 (1992).

80. M. S. Brown and J. L. Goldstein in A. G. Gilman, L. S. Goodman, T. W. Rall, and F. Murad, Eds., *Goodman and Gilman's The Pharmacological Basis of Therapeutics*, 7th ed., MacMillan, New York, 1985, pp. 827–845.

81. H. Danielsson and J. Sjovall, *Annu. Rev. Biochem.*, **44**, 233–253 (1975).

82. Anonymous, *JAMA*, **260**, 359–366 (1988).

83. B. Angelin, K. Einarsson, K. Hellstrom, and B. Leijd, *J. Lipid Res.*, **19**, 1017–1024 (1978).

84. Anonymous, *JAMA*, **251**, 365–374 (1984).

85. Anonymous, *JAMA*, **251**, 351–364 (1984).

86. R. I. Levy, J. F. Brensike, S. E. Epstein, S. F. Kelsey, E. R. Passamani, J. M. Richardson, I. K. Loh, N. J. Stone, R. F. Aldrich, and J. W. Battaglini, *Circulation*, **69**, 325–337 (1984).

87. J. F. Brensike, R. I. Levy, S. F. Kelsey, E. R. Passamani, J. M. Richardson, I. K. Loh, N. J. Stone, R. F. Aldrich, J. W. Battaglini, and D. J. Moriarty, *Circulation*, **69**, 313–324 (1984).

88. J. Blanchard and J. G. Nairn, *J. Phys. Chem.*, **72**, 1204–1208 (1968).

89. N. E. Miller, P. Cliffton-Bligh, P. J. Nestel, and H. M. Whyte, *Med. J. Aust.*, **1**, 1223–1227 (1973).

90. G. L. Vega and S. M. Grundy, *JAMA*, **257**, 33–38 (1987).

91. J. L. Witztum, D. Simmons, D. Steinberg, W. F. Beltz, R. Weinreb, S. G. Young, P. Lester, N. Kelley, and J. Juliano, *Circulation*, **79**, 16–28 (1989).

92. H.-R. Arntz, R. Agrawal, W. Wunderlich, L. Schnitzer, R. Stern, F. Fischer, and H.-P. Schultheiss, *Am. J. Cardiol.*, **86**, 1293–1298 (2000).

93. R. Fellin, G. Baggio, G. Briani, M. R. Baiocchi, E. Manzato, G. Baldo, and G. Crepaldi, *Atherosclerosis*, **29**, 241–249 (1978).

94. A. H. Seplowitz, F. R. Smith, L. Berns, H. A. Eder, and D. S. Goodman, *Atherosclerosis*, **39**, 35–43 (1981).

95. F. R. Heller, J. P. Desager, and C. Harvengt, *Metabolism*, **30**, 67–71 (1981).

96. F. R. Heller, J. P. Desager, and C. Harvengt, *Acta Cardiol.*Suppl., **27**, 103–106 (1981).

97. D. T. Nash, *Postgrad. Med.*, **73**, 75–82 (1983).

98. C. A. Dujovne, P. Krehbiel, and S. B. Chernoff, *Am. J. Cardiol.*, **57**, 36–42 (1986).

99. P. T. Kuo, A. C. Wilson, J. B. Kostis, and A. E. Moreyra, *Am. J. Cardiol.*, **57**, 43–48 (1986).

100. M. H. Davidson, M. A. Dillon, B. Gordon, P. Jones, J. Samuels, S. Weiss, J. Isaacsohn, P. Toth, and S. K. Burke, *Arch. Intern. Med.*, **159**, 1893–1900 (1999).

101. J. M. Donovan, D. Stypinski, M. R. Stiles, T. A. Olson, and S. K. Burke, *Cardiovasc. Drugs Ther.*, **14**, 681–690 (2000).

102. W. H. Mandeville, W. Braunlin, P. Dhal, A. Guo, C. Huval, K. Miller, J. Petersen, S. Polomoscanik, D. Rosenbaum, R. Sacchiero, J. Ward, and S. R. Holmes-Farley, *Mater. Res. Soc. Symp. Proc.*, **550**, 3–15 (1999).

103. W. H. Braunlin, S. R. Holmes-Farley, D. Smisek, A. Guo, W. Appruzese, Q. Xu, P. Hook, E. Zhorov, and H. Mandeville, Proceedings of the 219th National Meeting of the American Chemical Society, San Francisco, CA, March 26–30, 2000, Book of Abstracts, POLY-360.

104. H. H. Knapp, B. Chin, J. M. Gaziano, J. M. Donovan, S. K. Burke, and M. H. Davidson, *Am. J. Med.*, **110**, 352–360 (2001).

105. N. N. Wong, *Heart Dis.*, **3**, 63–70 (2001).

106. Geltex' Welcol Combination with Lipitor Lowers LDL Cholesterol 48% in *Pink-Sheet*, FDC Publications Inc., June 5, 2000, p. 14.

107. M. van Heek, C. F. France, D. S. Compton, R. L. McLeod, N. P. Yumibe, K. B. Alton, E. J. Sybertz, and H. R. Davis Jr., *J. Pharmacol Exp. Ther.*, **283**, 157–163 (1997).

108. S. B. Rosenblum, T. Huynh, A. Afonso, H. R. Davis Jr., N. Yumibe, J. W. Clader, and D. A. Burnett, *J. Med. Chem.*, **41**, 973–980 (1998).

109. M. van Heek, C. Farley, D. S. Compton, L. Hoos, K. B. Alton, E. J. Sybertz, and H. R. Davis Jr., *Br. J. Pharmacol.*, **129**, 1748–1754 (2000).

110. C. Q. Meng, *Curr. Opin. Invest. Drugs*, **2**, 389–392 (2001).

111. "Ezetimibe nears the market," in *Scrip*, PJB Publications Ltd., June 6, 2001.

112. S. M. Grundy, H. Y. Mok, L. Zech, and M. Berman, *J. Lipid Res.*, **22**, 24–36 (1981).

113. R. W. Piepho, *Am. J. Cardiol.*, **86** (Suppl.), 35L–40L (2000).

114. R. W. Butcher, C. E. Baird, and E. W. Sutherland, *J. Biol. Chem.*, **243**, 1705–1712 (1968).

115. L. A. Carlson and L. Oro, *Acta Med. Scand.*, **172**, 641–645 (1962).

116. L. A. Carlson, L. Oro, and J. Ostman, *J. Atherosclerosis Res.*, **8**, 667–677 (1968).

117. F.-Y. Jin, V. S. Kamanna, and M. L. Kashyap, *Arterioscler. Thromb. Vasc. Biol.*, **17**, 2020–2028(1997).

118. J. Shepherd, C. J. Packard, J. R. Patsch, A. M. Gotto Jr., and O. D. Taunton, *J. Clin. Invest.*, **63**, 858–867 (1979).

119. J. R. Guyton, *Am. J. Cardiol.*, **82**, 18U–23U (1998).

120. Anonymous, *JAMA*, **231**, 360–381 (1975).

121. L. A. Carlson and G. Rosenhamer, *Acta Med. Scand.*, **223**, 405–418 (1988).

122. D. H. Blankenhorn, S. A. Nessim, R. L. Johnson, M. E. Sanmarco, S. P. Azen, and L. Cashin-Hemphill, *JAMA*, **257**, 3233–3240 (1987).

123. L. Cashin-Hemphill, W. J. Mack, J. M. Pogoda, M. E. Sanmarco, S. P. Azen, and D. H. Blankenhorn, *JAMA*, **264**, 3013–3017 (1990).

124. G. Brown, J. J. Albers, L. D. Fisher, S. M. Schaefer, J. T. Lin, C. Kaplan, X. Q. Zhao, B. D. Bisson, V. F. Fitzpatrick, and H. T. Dodge, *N. Engl. J. Med.*, **323**, 1289–1298 (1990).

125. J. P. KaneM. J. Malloy, T. A. Ports, N. R. Phillips, J. C. Diehl, and R. J. Havel, *JAMA*, **264**, 3007–3012 (1990).

126. F. M. Sacks, R. C. Pasternak, C. M. Gibson, B. Rosner, and P. H. Stone, *Lancet*, **344**, 1182–1186 (1994).

127. R. H. Knopp, *Am. J. Cardiol.*, **86**, 51L–56L (2000).

128. R. H. Knopp, *Am. J. Cardiol.*, **82**, 24U–28U (1998).

129. R. N. Knobb, P. Alagona, M. Davidson, A. C. Goldberg, S. D. Kafonek, M. Kahyap, D. Sprecher, H. R. Superko, S. Jenkins, and S. Marcovina, *Metabolism*, **47**, 1097–1104 (1998).

130. A. C. Goldberg, P. Alagona Jr., D. M. Capuzzi, J. Guyton, J. M. Morgan, J. Rodgers, R. Sachson, and P. Samuel, *Am. J. Cardiol.*, **85**, 1100–1105 (2000).

131. D. M. Capuzzi, J. Guyton, J. M. Morgan, A. C. Goldberg, R. A. Kriesberg, O. A. Brusco, and J. Brody, *Am. J. Cardiol.*, **82**, 74U–81U (1998).

132. A. C. Goldberg, *Am. J. Cardiol.*, **82**, 35U–38U (1998).

133. J. M. Morgan, D. M. Capuzzi, and J. R. Guyton, *Am. J. Cardiol.*, **82**, 29U–34U (1998).

134. P. Reaven and J. L. Witztum, *Ann. Intern. Med.*, **109**, 597–598 (1988).

135. D. J. Norman, D. R. Illingworth, J. Munson, and J. Hosenpud, *N. Engl. J. Med.*, **318**, 467–477 (1988).

136. J. R. Guyton and D. M. Capuzzi, *Am. J. Cardiol.*, **82**, 82U–84U (1998).

137. R. Garg, M. B. Elam, J. R. Crouse 3rd, K. B. Davis, J. W. Kennedy, D. Egan, J. A. Herd,

D. B. Hunninghake, W. C. Johnson, J. B. Kostis, D. S. Sheps, and W. B. Applegate, *Am. Heart J.*, **140**, 792–803 (2000).

138. J. R. Guyton, A. C. Goldberg, R. A. Kriesberg, D. L. Sprecher, H. R. Superko, and C. M. O'Connor, *Am. J. Cardiol.*, **82**, 737–743 (1998).

139. B. G. Brown, X.-Q. Zhao, A. Chait, J. Frohlich, M. Cheung, N. Heise, A. Dowdy, D. Deangelis, L. D. Fisher, and J. Albers, *Can. J. Cardiol.*, **14** (Suppl. A), 6A–13A (1998).

140. X.-Q. Zhao, et al., *Program and Abstracts of the American College of Cardiology 50th Annual Scientific Session*, March 18–21, 2001, Orlando, FL, Abstr. 842.

141. J. S. Morse, et al., *Program and Abstracts of the American College of Cardiology 50th Annual Scientific Session*, March 18–21, 2001, Orlando, FL, Abstr. 1007.

142. T. A. Pearson, *Am. J. Cardiol.*, **86**, 57L–61L (2000).

143. Niacin/Lovastatin in *IDdb Drug Report*, April 20, 2001, Investigational Drugs Database, Current Drugs Ltd. May be accessed at www.current-drugs.com

144. M. M. T. Buckley, K. L. Goa, A. H. Price, and R. N. Brogden, *Drugs*, **37**, 761–800 (1989).

145. P. N. Durrington and J. P. Miller, *Atherosclerosis*, **55**, 187–194 (1985).

146. J.-C. Tardif, G. Cote, J. Lesperance, M. Bourassa, J. Lambert, S. Doucet, L. Bilodeau, S. Nattel, and P. de Guise, *N. Engl. J. Med.*, **337**, 365–372 (1997).

147. P. J. Nestel and T. Billington, *Atherosclerosis*, **38**, 203–209 (1981).

148. P. J. Nestel, *Artery*, **10**, 95–98 (1982).

149. G. Walldius, U. Erikson, A. G. Olsson, L. Bergstrand, K. Hadell, J. Johansson, L. Kaijser, C. Lassvik, J. Molgaard, and S. Nilsson, *Am. J. Cardiol.*, **74**, 875–883 (1994).

150. M. Setsuda, M. Inden, N. Hiraoka, S. Okamoto, H. Tanaka, T. Okinaka, Y. Nishimura, H. Okano, T. Kouji, and T. Konishi, *Clin. Ther.*, **15**, 374–382 (1993).

151. G. A. Colditz, W. C. Willett, M. J. Stampfer, B. Rosner, F. E. Speizer, and C. H. Hennekens, *N. Engl. J. Med.*, **316**, 1105–1110 (1987).

152. P. T. Ma, T. Yamamoto, J. L. Goldstein, and M. S. Brown, *Proc. Natl. Acad. Sci. USA*, **83**, 792–796 (1986).

153. M. Averbuch, D. Ayalon, N. Eckstein, I. Dotan, I. Shapira, Y. Levo, and A. Pines, *J. Med.*, **29**, 343–350 (1998).

154. S. Hulley, D. Grady, T. Bush, C. Furberg, D. Herrington, B. Riggs, and E. Vittinghoff, *JAMA*, **280**, 605–613 (1998).

155. D. Josefson, *Br. Med. J.*, **318**, 735 (1999).

156. American Heart Association, *NR01–1313 (Circ/Mosca/HRT)*, July 24, 2001, American Heart Association, Dallas, TX, 2001.

157. Anonymous, *Control Clin. Trials*, **19**, 61–109 (1998).

158. N. Crabtree, J. Wright, A. Walgrove, J. Rea, L. Hanratty, M. Lunt, I. Fogelman, R. Palmer, M. Vickers, J. E. Compston, and J. Reev, *Osteoporos. Int.*, **11**, 537–543 (2000).

159. A. H. MacLennan, B. J. Paine, and J. E. Marley, *Aust. Fam. Physician*, **29**, 797–801 (2000).

160. Expert Panel on Detection, Evaluation, and Treatment of High Blood Cholesterol in Adults, *Circulation*, **89**, 1333–1445 (1994).

161. G. G. J. M. Kuiper and J. A. Gustafsson, *FEBS Lett.*, **410**, 87–90 (1997).

162. K. Pettersson, K. Grandien, G. G. J. M. Kuiper, and J. A. Gustafsson, *Mol. Endocrinol.*, **11**, 1486–1496 (1997).

163. T. Watanabe, S. Inoue, S. Ogawa, Y. Ishii, H. Hiroi, K. Ikeda, A. Orimo, and M. Muramatsu, *Biochem. Biophys. Res. Commun.*, **236**, 140–145 (1997).

164. K. Paech, P. Webb, G. G. J. M. Kuiper, S. Nilsson, J. A. Gustafsson, P. J. Kushner, and T. S. Scanlan, *Science*, **277**, 1508–1510 (1997).

165. M. D. Iafrati, R. H. Karas, M. Aronovitz, S. Kim, T. R. Sullivan Jr., D. B. Lubahn, T. F. O'Donnell Jr., K. S. Korach, and M. E. Mendelsohn, *Nat. Med.*, **3**, 545–548 (1997).

166. P. Pace, J. Taylor, S. Suntharalingam, R. C. Coombes, and S. Ali, *J. Biol. Chem.*, **272**, 25832–25838 (1997).

167. C. C. McDonald, F. E. Alexander, B. W. Whyte, A. P. Forrest, and H. J. Stewart, *Br. Med. J.*, **311**, 977–980 (1995).

168. T. Saarto, C. Blomqvist, C. Ehnholm, M.-R. Taskinen, and I. Elomaa, *J. Clin. Oncol.*, **14**, 429–433 (1996).

169. B. W. Walsh, L. H. Kuller, R. A. Wild, S. Paul, M. Farmer, J. B. Lawrence, A. S. Shah, and P. W. Anderson, *JAMA*, **279**, 1445–1451 (1998).

170. T. Heinemann, G. Axtmann, and K. von Bergmann, *Eur. J. Clin. Invest.*, **23**, 827–831 (1993).

171. M. Becker, D. Staab, and K. von Bergmann, *J. Pediatr.*, **122**, 292–296 (1993).

172. S. N. Blair, D. M. Capuzzi, S. O. Gottleib, T. Nguyen, J. M. Morgan, and N. B. Cater, *Am. J. Cardiol.*, **86**, 46–52 (2000).

173. M. Hallikainen and M. Uusitupa, *Am. J. Clin. Nutr.*, **69**, 403–410 (1999).

174. H. A. Neil, G. W. Meijer, and L. Roe, *Atherosclerosis*, **156**, 329–337 (2001).

175. R. J. Gerson, J. S. MacDonald, A. W. Alberts, D. J. Kornbrust, J. A. Majka, R. J. Stubbs, and D. L. Bokelman, *Am. J. Med.*, **87**, 28S–38S (1989).

176. P. A. Todd and K. L. Goa, *Drugs*, **40**, 583–607 (1990).

177. J. M. Henwood and R. C. Heel, *Drugs*, **36**, 429–454 (1988).

178. G. L. Plosker and A. J. Wagstaff, *Drugs*, **51**, 433–459 (1996).

179. (a) D. McTavish and E. M. Sorkin, *Drugs*, **42**, 65–89 (1991); (b) Correction. *ibid.*, **42**, 944 (1991).

180. J. S. MacDonald, R. J. Gerson, D. J. Kornbrust, M. W. Kloss, S. Prahalada, P. H. Berry, A. W. Alberts, and D. L. Bokelman, *Am. J. Cardiol.*, **62**, 16J–27J (1988).

181. D. M. Black, R. G. Bakker-Arkema, and J. W. Nawrocki, *Arch. Intern. Med.*, **158**, 577–584 (1998).

182. (a) C. J. Vaughan, M. B. Murphy, and B. M. Buckley, *Lancet*, **348**, 1079–1082 (1996); (b) Correction. *ibid.*, **349**, 214 (1997).

183. P. H. Chong and J. D. Seeger, *Pharmacotherapy*, **17**, 1157–1177 (1997).

184. R. J. Gerson, J. S. MacDonald, A. W. Alberts, D. J. Kornbrust, J. A. Majka, R. J. Stubbs, and D. L. Bokelman, *Am. J. Med.*, **87**, 28S–38S (1989).

185. S. M. Grundy, *N. Engl. J. Med.*, **319**, 24–33 (1988).

186. H. A. Hartman, L. A. Myers, M. Evans, R. L. Robison, R. G. Engstrom, and F. L. S. Tse, *Fundam. Appl. Toxicol.*, **29**, 48–62 (1996).

187. P. W. Jungnickel, K. A. Cantral, and P. A. Maloley, *Clin. Pharm.*, **11**, 677–689 (1992).

188. E. von Keutz and G. Schluter, *Am. J. Cardiol.*, **82**, 11J–17J (1998).

189. L. A. Jokubaitis, *Am. J. Cardiol.*, **73**, 18D–24D (1994).

190. Anonymous, *JAMA*, **260**, 359–366 (1988).

191. A. Nakad, L. Bataille, V. Hamoir, C. Sempoux, and Y. Horsmans, *Lancet*, **353**, 1763–1764 (1999).

192. M. Ballare, M. Campanini, G. Airoldi, G. Zaccala, M. C. Bertoncelli, G. Cornaglia, M. Porzio, and A. Monteverde, *Minerva Gastroenterol. Dietol.*, **38**, 41–44 (1992).

193. G. F. Watts, C. Castelluccio, C. Rice-Evans, N. A. Taub, H. Baum, and P. J. Quinn, *J. Clin. Pathol.*, **46**, 1055–1057 (1993).

194. W. H. Frishman, P. Zimetbaum, and J. Nadelmann, *J. Clin. Pathol.*, **29**, 975–982 (1989).

195. J. Z. Ayanian, C. S. Fuchs, and R. M. Stone, *Ann. Intern. Med.*, **109**, 682–683 (1988).

196. J. A. Tobert, *Am. J. Med.*, **96**, 300–301 (1994).

197. R. S. Lees and A. M. Lees, *N. Engl. J. Med.*, **333**, 664–665 (1995).

198. C. Chrysanthopoulos and N. Kounis, *Br. Med. J.*, **304**, 1225 (1992).

199. (a) "More US clinical trials with pitavastain," in *Scrip*, PJB Publications Ltd., October 30, 2001; (b) "US setback for pitavastain?," in *Scrip*, PJB Publications Ltd., October 31, 2001; (c) "Pitavastain Japan launch delay expected," in *Scrip*, PJB Publications Ltd., November 6, 2001.

200. R. J. Gerson, J. S. MacDonald, A. W. Alberts, J. Chen, J. B. Yudkovitz, M. D. Greenspan, L. F. Rubin, and D. L. Bokelman, *Exp. Eye Res.*, **50**, 65–78 (1990).

201. J. Schmidt, C. Schmitt, O. Hockwin, U. Paulus, and K. von Bergmann, *Ophthalmic Res.*, **26**, 352–360 (1994).

202. (a) C. B. Blum, *Am. J. Cardiol.*, **73**, 3D–11D (1994); (b) Correction. *ibid.*, **74**, 639 (1994).

203. A. S. Wierzbicki, P. J. Lumb, Y. K. Semra, and M. A. Crook, *Lancet*, **351**, 569–570 (1998).

204. D. R. Nair, J. A. Papadakis, I. A. Jagroop, D. P. Mikhailidis, and A. F. Winder, *Lancet*, **351**, 1430 (1998).

205. A. D. Marais, J. C. Firth, M. E. Bateman, P. Byrnes, C. Martens, and J. Mountney, *Arterioscler. Thromb. Vasc. Biol.*, **17**, 1527–1531 (1997).

206. H. Sinzinger and M. Rodrigues, *Atherosclerosis*, **145**, 415–417 (1999).

207. A. S. Wierzbicki, P. J. Lumb, G. Chik, and M. A. Crook, *Int. J. Clin. Pract.*, **53**, 609–611 (1999).

208. A. S. Wierzbicki, P. J. Lumb, G. Chik, E. R. Christ, and M. A. Crook, *Q. J. Med.*, **92**, 387–394 (1999).

209. D. P. Mikhailidis and A. S. Wierzbicki, *Curr. Med. Res. Opin.*, **16**, 139–146 (2000).

210. (a) J. J. Kastelein, J. L. Isaacsohn, L. Ose, D. B. Hunninghake, J. Frohlich, M. H. Davidson, R. Habib, C. A. Dujovne, J. R. Crouse, M. Liu, M. R. Melino, L. O'Grady, M. Mercuri, and Y. B. Mitchel, *Am. J. Cardiol.*, **86**, 221–223 (2000); (b) Correction. *Am. J. Cardiol.*, **86**, 812 (2000).

211. J. P. Monk and P. A. Todd, *Drugs*, **33**, 539–576 (1987).

212. W. C. Roberts, *Cardiology*, **76**, 169–179 (1989).

213. J. C. Adkins and D. Faulds, *Drugs*, **54**, 615–633 (1997).

214. H. Vinazzer and J. C. Farine, *Atherosclerosis*, **49**, 109–118 (1983).

215. R. Zimmermann, W. Ehlers, E. Walter, A. Hoffrichter, P. D. Lang, K. Andrassy, and G. Schlierf, *Atherosclerosis*, **29**, 477–485 (1978).

216. J. K. Reddy, M. S. Rao, and D. E. Moody, *Cancer Res.*, **36**, 1211–1217 (1976).

217. J. K. Reddy, M. S. Rao, D. L. Azarnoff, and S. Sell, *Cancer Res.*, **39**, 152–161 (1979).

218. J. K. Reddy, D. L. Azarnoff, and C. E. Hignite, *Nature*, **283**, 397–398 (1980).

219. F. A. De La Iglesia, J. E. Lewis, R. A. Buchanan, E. L. Marcus, and G. McMahon, *Atherosclerosis*, **43**, 19–37 (1982).

220. F. A. De La Iglesia in G. Ricci, R. Paoletti, and F. Pocchiari, Eds., *Therapeutic Selection of Risk/Benefit Assessment of Hypolipidemic Drugs*, Raven, New York, 1982, pp. 53–55.

221. R. H. Gray and F. A. De La Iglesia, *Hepatology*, **4**, 520–530 (1984).

222. S. Blumcke, W. Schwartzkopff, H. Lobeck, N. A. Edmondson, D. E. Prentice, and G. F. Blane, *Atherosclerosis*, **46**, 105–116 (1983).

223. P. Gariot, E. Barrat, L. Mejean, J. P. Pointel, P. Drouin, and G. Debry, *Arch. Toxicol.*, **53**, 151–163 (1983).

224. P. Gariot, E. Barrat, P. Drouin, P. Genton, J. P. Pointel, B. Foliquet, M. Kolopp, and G. Debry, *Metabolism*, **36**, 203–210 (1987).

225. G. F. Blane and F. Pinaroli, *Nouv. Presse Méd.*, **9**, 3737–3746 (1980).

226. M. Hanefeld, C. Kemmer, and E. Kadner, *Atherosclerosis*, **46**, 239–246 (1983).

227. J. K. Reddy, N. D. Lalwani, S. A. Qureshi, M. K. Reddy, and C. M. Moehle, *Am. J. Pathol.*, **114**, 171–183 (1984).

228. C. R. Sirtori, L. Calabresi, J. P. Werba, and G. Franceschini, *Pharmacol. Res.*, **26**, 243–260 (1992).

229. T. Langer and R. I. Levy, *N. Engl. J. Med.*, **279**, 856–858 (1968).

230. B. Tomlinson, P. Chan, and W. Lan, *Drugs Aging*, **18**, 665–683 (2001).

231. S. Franc, E. Bruckert, P. Giral, and G. Turpin, *Presse Med.*, **26**, 1855–1858 (1997).

232. J. P. Kane, M. J. Malloy, P. Tun, N. R. Phillips, D. D. Freedman, M. L. Williams, J. S. Rowe, and R. J. Havel, *N. Engl. J. Med.*, **304**, 251–258 (1981).

233. A. Garg, *Diabetes*, **41** (Suppl. 2), 111–115 (1992).

234. J. C. Oki, *Pharmacotherapy*, **15**, 317–337 (1995).

235. Anonymous, *Diabetes Care*, **21**, 179–182 (1998).

236. S. M. Haffner, *Diabetes Care*, **21**, 160–178 (1998).

237. M. B. Elam, D. B. Hunninghake, K. B. Davis, R. Garg, C. Johnson, D. Egan, J. B. Kostis, D. S. Sheps, and E. A. Brinton, *JAMA*, **284**, 1263–1270 (2000).

238. R. C. Heel, R. N. Brogden, T. M. Speight, and G. S. Avery, *Drugs*, **15**, 409–428 (1978).

239. D. McCaughan, *Artery*, **10**, 56–70 (1982).

240. F. N. Marshall and J. E. Lewis, *Toxicol. Appl. Pharmacol.*, **24**, 594–602 (1973).

241. A. Tavani and C. La Vecchia, *Drugs Aging*, **14**, 347–357 (1999).

242. K. Bjarnason, A. Cerin, R. Lindgren, and T. Weber, *Maturitas*, **32**, 161–170 (1999).

243. J. M. Sullivan, *Circulation*, **94**, 2699–2702 (1996).

244. J. E. Rossouw, *Circulation*, **94**, 2982–2985 (1996).

245. V. C. Jordan, *Recent Results Cancer Res.*, **151**, 96–109 (1999).

246. M. N. Prout, *Medscape Womens Health*, **5**, E4 (2000).

247. B. A. Hamelin and J. Turgeon, *Trends Pharmacol. Sci.*, **19**, 26–37 (1998).

248. M. H. Davidson, *Exp. Opin. Invest. Drugs*, **11**, 125–141 (2002).

249. D. E. Duggan, I. W. Chen, W. F. Bayne, R. A. Halpin, C. A. Duncan, M. S. Schwartz, R. J. Stubbs, and S. Vickers, *Drug Metab. Dispos.*, **17**, 166–173 (1989).

250. S. Vickers, C. A. Duncan, I. W. Chen, A. Rosegay, and D. E. Duggan, *Drug Metab. Dispos.*, **18**, 138–145 (1990).

251. T. Koga, K. Fukuda, Y. Shimada, M. Fukami, H. Koike, and Y. Tsujita, *Eur. J. Biochem.*, **209**, 315–319 (1992).

252. A. K. van Vliet, G. C. van Thiel, R. H. Huisman, H. Moshage, S. H. Yap, and L. H. Cohen, *Biochim. Biophys. Acta*, **1254**, 105–111 (1995).

253. K. Ziegler and S. Hummelsiep, *Biochim. Biophys. Acta*, **1253**, 23–33 (1993).

254. H. S. Malhotra and K. L. Goa, *Drugs*, **61**, 1835–1881 (2001).

255. G. L. Plosker, C. J. Dunn, and D. P. Figgitt, *Drugs*, **60**, 1179–1206 (2001).

256. D. B. Miller and J. D. Spence, *Clin. Pharmaco-kinet.*, **34**, 155–162 (1998).

257. (a) K. Einarsson, K. Hellstrom, and M. Kallner, *Eur. J. Clin. Invest.*, **4**, 405–410 (1974); (b) D. P. Rosenbaum, J. S. Petersen, S. Ducharme, P. Markham, and D. I. Goldberg, *J. Pharm. Sci.*, **86**, 591–595 (1997).

258. J. E. Patrick, T. Kosoglou, K. L. Stauber, K. B. Alton, S. E. Maxwell, Y. Zhu, P. Statkevich, R. Iannucci, S. Chowdhury, M. Affrime, and M. N. Cayen, *Drug Metab. Dispos.*, **30**, 430–437 (2002).

259. T. Heinemann, G. A. Kullak-Ublick, B. Pietruck, and K. von Bergmann, *Eur. J. Clin. Pharmacol.*, **40** (Suppl. 1), S59–S63 (1991).

260. J. Sheperd, *Drugs*, **47** (Suppl. 2), 1–10 (1994).

261. M. S. Brown and J. L. Goldstein, *Science*, **232**, 34–47 (1986).

262. R. Ross, *N. Engl. J. Med.*, **340**, 115–126 (1999).

263. S. G. Young and C. J. Fielding, *Nat. Genet.*, **22**, 316–318 (1999).

264. M. Krieger, *Proc. Natl. Acad. Sci. USA*, **95**, 4077–4080 (1998).

265. R. Winslow, "Birth of a Blockbuster: Lipitor's Unlikely Route out of the Lab," in *Dow Jones Business News*, Dow Jones & Company, Inc., January 24, 2000.

266. E. S. Istvan, M. Palnitkar, S. K. Buchanan, and J. Deisenhofer, *EMBO J.*, **19**, 819–830 (2000).

267. E. S. Istvan and J. Deisenhofer, *Biochim. Biophys. Acta*, **1529**, 9–18 (2000).

268. E. S. Istvan and J. Deisenhofer, *Science*, **292**, 1160–1164 (2001).

269. K. Kajinami, H. Mabuchi, and Y. Saito, *Expert Opin. Invest. Drugs*, **9**, 2653–2661 (2000).

270. B. R. Henke, S. G. Blanchard, M. F. Brackeen, K. K. Brown, J. E. Cobb, J. L. Collins, W. W. Harrington Jr., M. A. Hashim, E. A. Hull-Ryde, I. Kaldor, S. A. Kliewer, D. H. Lake, L. M. Leesnitzer, J. M. Lehmann, J. M. Lenhard, L. A. Orband-Miller, J. F. Miller, R. A. Mook Jr., S. A. Noble, W. Oliver Jr., D. J. Parks, K. D. Plunket, J. R. Szewczyk, and T. M. Willson, *J. Med. Chem.*, **41**, 5020–5036 (1998).

271. E. Jungling and H. Kammermeier, *Anal. Biochem.*, **171**, 150–157 (1988).

272. G. Krey, O. Braissant, F. L'Horset, E. Kalkhoven, M. Perroud, M. G. Parker, and W. Wahli, *Mol. Endocrinol.*, **11**, 779–791 (1997).

273. S. A. Kliewer, S. S. Sundseth, S. A. Jones, P. J. Brown, G. B. Wisely, C. S. Koble, P. Devchand, W. Wahli, T. M. Willson, J. M. Lenhard, and J. M. Lehmann, *Proc. Natl. Acad. Sci. USA*, **94**, 4318–4323 (1997).

274. B. M. Forman, J. Chen, and R. M. Evans, *Proc. Natl. Acad. Sci. USA*, **94**, 4312–4317 (1997).

275. K. Yu, W. Bayona, C. B. Kallen, H. P. Harding, C. P. Ravera, G. McMahon, M. Brown, and M. A. Lazar, *J. Biol. Chem.*, **270**, 23975–23983 (1995).

276. H. E. Xu, M. H. Lambert, V. G. Montana, D. J. Parks, S. G. Blanchard, P. J. Brown, D. D. Sternbach, J. M. Lehmann, G. B. Wisely, T. M. Willson, S. A. Kliewer, and M. V. Milburn, *Mol. Cell.*, **3**, 397–403 (1999).

277. M. Gottlicher, E. Widmark, Q. Li, and J. A. Gustafsson, *Proc. Natl. Acad. Sci. USA*, **89**, 4653–4657 (1992).

278. J. W. Clader, D. A. Burnett, M. A. Caplen, M. S. Domalski, S. Dugar, W. Vaccaro, R. Sher, M. E. Browne, H. Zhao, R. E. Burrier, B. Salisbury, and H. R. Davis Jr., *J. Med. Chem.*, **39**, 3684–3693 (1996).

279. D. A. Burnett, M. A. Caplen, H. R. Davis Jr., R. E. Burrier, and J. W. Clader, *J. Med. Chem.*, **37**, 1733–1736 (1994).

280. T. Grand-Perret, A. Bouillot, A. Perrot, S. Commans, M. Walker, and M. Issandou, *Nat. Med.*, **7**, 1332–1338 (2001).

281. (a) S. B. Joseph, B. A. Laffitte, P. H. Patel, M. A. Watson, K. E. Matsukuma, R. Walczak, J. L. Colins, T. F. Osborne, and P. Tontonoz, *J. Biol. Chem.*, **277**, 11019–11025 (2002); (b) J. J. Repa, K. E. Knut, C. Pomajzl, J. A. Richardson, H. Hobbs, and D. J. Mangelsdorf, *J. Biol. Chem.*, **277**, 18793–18800 (2002).

282. E. M. Lonn and S. Yusuf, *Can. J. Cardiol.*, **13**, 957–965 (1997).

283. AC-3056 in *IDdb Drug Report*, May 15, 2002, Investigational Drugs Database, Current Drugs Ltd. May be accessed at www.current-drugs. com

284. O. Cynshi, Y. Kawabe, T. Suzuki, Y. Takashima, H. Kaise, M. Nakamura, Y. Ohba, Y. Kato, K. Tamura, A. Hayasaka, A. Higashida, H. Sakaguchi, M. Takeya, K. Takahashi, K. Inoue, N. Noguchi, E. Niki, and T. Kodama, *Proc. Natl. Acad. Sci. USA*, **95**, 10123–10128 (1998).

285. SB-480848 in *IDdb Drug Report*, April 15, 2002, Investigational Drugs Database, Current Drugs Ltd. May be accessed at www. current-drugs.com

286. C. J. Packard, D. S. J. O'Reilly, M. J. Caslake, A. D. McMahon, I. Ford, J. Cooney, C. H. Macphee, K. E. Suckling, M. Krishna, F. E. Wilkinson, A. Rumley, G. D. O. Lowe, G. Docherty, and J. D. Burczak, *N. Engl. J. Med.*, **343**, 1148–1155 (2000).

287. Competitor Knowledge Base (CKB) is a Glaxo-SmithKline product, available only to GSK employees.

288. Investigational Drugs database (IDdb3) is a product of Current Drugs Ltd. Further information may be accessed at www.current-drugs.com.

289. J. P. F. Chin-Dusting and J. A. Shaw, *Expert Opin. Pharmacother.*, **2**, 419–430 (2001).

290. K. Kajinami, H. Mabuchi, and Y. Saito, *Expert Opin. Invest. Drugs*, **9**, 2653–2661 (2000).

291. G. R. Thompson and R. P. Naoumova, *Expert Opin. Invest. Drugs*, **9**, 2619–2628 (2000).

292. P. H. Chong and B. S. Bachenheimer, *Drugs*, **60**, 55–93 (2000).

293. P. O. Kwiterovich, *Am. J. Cardiol.*, **82**, 3U–17U (1998).

294. H. J. Harwood Jr. and E. S. Hamanaka, *Emerging Drugs*, **3**, 147–172 (1998).

295. D. Bhatnagar, *Pharmacol. Ther.*, **79**, 205–230 (1998).

296. M. Mellies and M. McGovern, *Expert Opin. Invest. Drugs*, **6**, 31–50 (1997).

297. D. R. Feller, L. M. Hagerman, H. A. I. Newman, and D. T. Witiak in W. O. Foye, Ed., *Principles of Medicinal Chemistry*, 3rd ed., Lea & Febiger, Philadelphia, 1984, pp. 481–501.

CHAPTER EIGHT

Oxygen Delivery by Allosteric Effectors of Hemoglobin, Blood Substitutes, and Plasma Expanders

Barbara Campanini
Stefano Bruno
Samanta Raboni
Andrea Mozzarelli
Department of Biochemistry and Molecular Biology
National Institute for the Physics of Matter
University of Parma
Parma, Italy

Contents

Burger's Medicinal Chemistry and Drug Discovery
Sixth Edition, Volume 3: Cardiovascular Agents and Endocrines
Edited by Donald J. Abraham
ISBN 0-471-37029-0 © 2003 John Wiley & Sons, Inc.

1   INTRODUCTION

Oxygen supply is vital for human life. In the lung, oxygen binds to hemoglobin, a protein contained in red blood cells and, in the circulation, it is delivered to the peripheral tissues, where hemoglobin loads carbon dioxide (1). The oxygen content of the air at sea level is 20.95 %, which corresponds to a partial pressure of oxygen of about 160 Torr. In the lung, the oxygen pressure is about 100 Torr and in the mixed venous circulation is about 40 Torr. The corresponding oxygen fractional saturation (i.e., the concentration of oxygenated hemoglobin over the total hemoglobin concentration) is 0.97 and 0.75. Thus, only a small fraction of the oxygen bound to hemoglobin is delivered to the tissues. The affinity of hemoglobin for oxygen is expressed as p50, the oxygen pressure at which 50% of hemes are saturated (Fig. 8.1) (2, 3). Under physiological conditions, pH 7.4 and 37°C, the p50 of human blood is 26 Torr. Molecules that bind to hemoglobin and shift the oxygen binding curve either to the left or to the right are called allosteric effectors. Left shifters increase oxygen affinity and right shifters decrease oxygen affinity. Physiological right shifters are protons, chloride ions, carbonate, and the organic phosphate 2,3-diphosphoglycerate. An increased concentration of these compounds favors the unloading of oxygen to the tissues. In particu-

lar, increased levels of 2,3-diphosphoglycerate are responsible for the adaptation to low oxygen pressures at high altitudes and to low hemoglobin contents in anemia. The sigmoidal shape of the binding curve indicates that oxygen binding increases oxygen affinity and a decrease of the oxygen saturation favors the unloading of more oxygen. This behavior is an indication of hemoglobin positive cooperativity.

Several models have been proposed to explain the allosteric properties of hemoglobin: the Monod-Wyman-Changeux model (MWC) (4), the Koshland-Nemethy-Filmer model (5), the Perutz's stereochemical mechanism (6), and the Ackers's Symmetry rule (7). A modified version of the MWC model, which includes Perutz's stereochemical mechanism, appears to explain most of the functional properties of hemoglobin (8). The fundamental crystallographic study, carried out over more than 50 years by the Nobel laureate Max Perutz, shed light on hemoglobin structure and function (6, 9). Hemoglobin is a tetrameric molecule composed of two α- and β-subunits, arranged in a tetrahedral geometry (Fig. 8.2). Each subunit contains a heme, a tetrapyrrole ring coordinating an iron ion in the center. The iron is also coordinated to a histidine in the heme-binding pocket of hemoglobin. Oxygen, carbon monoxide, and nitric oxide competitively bind to the ferrous iron, whereas carbon dioxide

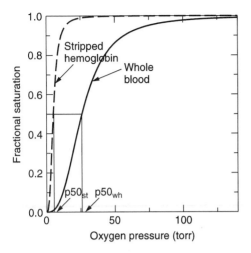

Figure 8.1. Oxygen binding curves of hemoglobin. Whole blood (wh) under physiological conditions exhibits a p50 of 26 Torr (1). "Stripped" hemoglobin (st) (i.e., hemoglobin in the absence of allosteric effectors) exhibits a p50 of 5 Torr at 37°C, pH 7.2 (19).

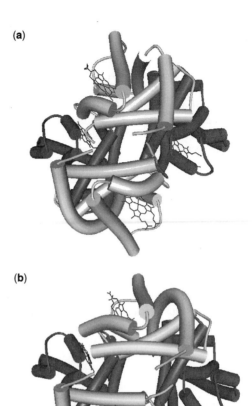

Figure 8.2. Structure of hemoglobin in the R state (a) and in the T state (b). In the T to R transition, one $\alpha\beta$ dimer rotates by about 15°, with respect to the other, and the central cavity narrows.

binds to the amine termini. 2,3-Diphosphoglycerate binds to positively charged residues of the β-chains in the central cavity, whereas chloride ions bind both in the central cavity and at α-chain residues. The deoxyhemoglobin structure is called T (tense) because several salt bridges constrain the molecule, decreasing by 20- to 300-fold the oxygen affinity with respect to the oxyhemoglobin structure, called R (relaxed) (3). The binding of oxygen triggers a series of tertiary and quaternary conformational changes, leading to the breakage of salt bridges and other bonds and to the release of protons and both organic and inorganic anions (6, 9). This description of hemoglobin structure, dynamics, and function is necessarily simplified and mainly aimed at providing key elements of hemoglobin complexity. More detailed descriptions are reported in the above-quoted books (1–3) and papers (4–9) and references therein.

A still controversial function of hemoglobin is nitric oxide (NO) transport. NO is involved in the control of vasoactivity, blood vessels dilation, platelet aggregation, and brain-regulated respiration (10). NO enters the red blood cells through mechanisms still under investigation (11, 12) and predominantly binds to hemoglobin as S-nitrosothiol at β-Cys93 in the R

state and as iron nitrosyl complex in the T state (13, 14). The release of NO from hemoglobin and the reaction with glutathione to form S-nitrosoglutathione have been suggested to be relevant steps in complex mechanisms that regulate NO activities. It is known that, under physiological conditions, Hb concentration is about 1000-fold higher than that of NO (11). Hemoglobin is also involved in NO oxidation, considered the major pathway for NO catabolism (15) and oxidase and peroxidase activities (16).

Within the above-outlined frame of hemoglobin structure and function, three research projects have been developed in the last 30 years. The first investigation is aimed at designing new allosteric effectors of hemoglobin. Compounds that increase the oxygen affinity can be used for the treatment of diseases as sickle-cell anemia, and those that decrease the oxygen affinity can be used for the treatment of ischemia, hypoxia, and to improve the efficiency of radio- and chemotherapy. The second project, strongly stimulated in the 1980s by the risk of HIV contamination in transfused blood, is focused on blood substitutes either by designing new types of oxygen carriers or by constructing genetically and/or chemically modified hemoglobins able to operate in the plasma outside the red blood cells. A third line of research is aimed at formulating solutions able to replace the blood, maintaining the volume and the oncotic pressure of blood fluids in severe hemorrhagic events. Up to now, no allosteric effectors or blood substitutes have been approved for human use in the United States, a clear indication of the difficulties in mimicking the multifaceted roles of hemoglobin.

2 ALLOSTERIC EFFECTORS OF HEMOGLOBIN

2.1 History

In 1967 (17, 18) the endogenous compound 2,3-diphosphoglycerate (2,3-DPG) was discovered to be a powerful allosteric effector of hemoglobin. 2,3-DPG exhibits a physiological concentration in human erythrocytes of 4.6 mM, high enough to form a 1:1 complex with hemoglobin. 2,3-DPG is responsible for the relatively low oxygen affinity of whole blood compared to that of purified hemoglobin. By removing 2,3-DPG, the p50 drops from 26 Torr to approximately 12 Torr. Other natural and synthetic allosteric effectors, structurally related to 2,3-DPG and known as organic phosphates, were discovered over the years. Among them, inositol hexaphosphate (IHP) and its structural analog inositol hexasulfate (IHS) have been extensively studied. It was shown that they all share a common mechanism of action and a common binding site with 2,3-DPG (1). Unfortunately, this class of compounds revealed to be unsuitable as drugs because they do not efficiently cross

the erythrocyte membrane (19). Attempts to engineer red blood cells to make them permeable to IHP were carried out (20).

In the early 1980s two antilipidemic agents, clofibrate and bezafibrate, were shown to lower the affinity of hemoglobin for oxygen as the organic phosphates (19, 21). These compounds were reported to reversibly cross the erythrocyte membrane, thus opening the way to their use as therapeutic agents. However, the more promising of the two molecules, bezafibrate, exhibited a dissociation constant to hemoglobin that was still too low to make it suitable as a drug. Moreover, it was shown that bezafibrate interacts strongly with serum albumin, reducing its *in vivo* activity. Nevertheless, a class of structurally related compounds was developed with the aim of increasing their affinity for hemoglobin and reducing the competition with albumin (see Ref. 22 for a review). From these studies, in 1992, a compound initially called RSR13 emerged as a promising drug candidate. Its approved nonproprietary name is efaproxiral.

In the same period, a completely new class of right shifters of the oxygen-binding curve was discovered. It was known that aromatic monoaldehydes, as vanillin and 12C79 (23), were able to shift the binding curve to the left, thus increasing the overall affinity of hemoglobin for oxygen. This effect might be of clinical interest in the therapy of sickle-cell anemia. The polymerization of the hemoglobin mutant associated with this pathology occurs only when a critical concentration of deoxyhemoglobin is reached upon unloading of oxygen in the peripheral tissues. A shift of the binding curve to the left increases the concentration of oxyhemoglobin, thus reducing the tissue damages caused by the polymerization. In the course of investigating this class of compounds in search of new antisickling agents, surprisingly, some aromatic aldehydes were discovered to lower the oxygen affinity (24). Although they have not yet been developed as drugs, they could be used for the same applications as efaproxiral.

2.2 Clinical Use of Right Shifters

The modulation of oxygen delivery to peripheral tissues through the direct and reversible interaction of drugs with hemoglobin has long been recognized as a possible treatment for several pathological states. Besides antisickling agents, which will be treated elsewhere, drugs interacting with hemoglobin of potential clinical use are essentially those that induce a rightward shift of the oxygen dissociation curve. This approach is beneficial in all conditions that require a transient increase in oxygen delivery in the tissues, either to overcome an insufficient amount to healthy organs (ischemia) or to sensitize solid tumors to radiotherapy and chemotherapy. Up to now, no drug of this class is on the market, mainly because of the poor pharmacokinetic properties of the molecules tested so far. However, efaproxiral is currently being tested in advanced clinical trials.

2.2.1 Organic Phosphates. The interaction of the endogenous allosteric effector 2,3-DPG (**1a**) with hemoglobin is well known. 2,3-DPG binds noncovalently in the cleft between the two β-subunits of deoxyhemoglobin, forming several ionic bonds with positively charged residues and counterbalancing the excess of positive charges in the central cavity (Fig. 8.3a) (25). By preferentially binding to T-state hemoglobin, 2,3-DPG stabilizes the deoxy state with respect to the R state. It was also shown that 2,3-DPG affects the intrinsic affinity of T-state hemoglobin, even in the absence of a switch to the R state (3). Inositol hexaphosphate (**1b**), also known as phytic acid, binds to the same site (26).

2.2.2 Bezafibrate Derivatives. The binding site of the bezafibrate derivatives is different from that of the organic phosphates. Bezafibrate (**2a**) and its analogs bind to twofold symmetry-related sites in the central water cavity of deoxyhemoglobin, approximately 20 Å from the binding site of 2,3-DPG (21, 27) (Fig. 8.3b). Each molecule is engaged in several interactions, mostly hydrophobic contacts and water-mediated hydrogen bonds with both β-subunits and one α-subunit. Efaproxiral and its structural analogs hinder the transition from T to R by preventing the narrowing of the central water cavity. Because the binding sites of efaproxiral and 2,3-DPG are different, they act synergically and not in competition.

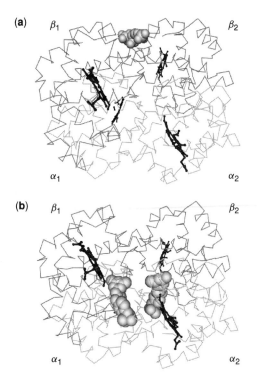

Figure 8.3. Binding sites of 2,3-DPG (a) and RSR13 (b) to hemoglobin. Ligands are shown in space-filling mode and hemes in ball-and-stick mode.

(1)

The bezafibrate derivatives developed so far have the general structure (**2b**) (28).

a

b

(2)

As a rule, the shortening of the four-atom bridge of bezafibrate to a three-atom bridge increases the potency of these compounds as allosteric effectors of hemoglobin. Depending on the X-Y-Z link, the bezafibrate derivatives tested as allosteric effectors can be grouped into five structural classes (28). The substituents of the more interesting molecules are reported in Table 8.1. The R_n groups are hydrophobic moieties and the X-Y-Z system is the bridge between the two substituted phenyl rings.

The shift of p50 induced by these compounds results from two contributions (29). The so-called allosteric factor arises from the perturbation of the equilibrium between the T- and R-state hemoglobin. This perturbation is reflected in a low value of the Hill coefficient, which approaches unity for potent effectors when hemoglobin remains in the low affinity T state even at high oxygen partial pressures. The so-called affinity factor originates from a variation of the intrinsic affinity of both T and R states. If an effector exhibits a pure affinity contribution, it does not affect the T to R equilibrium, and, therefore, the Hill coefficient is the same as the control curve. It was shown that all the bezafibrate derivatives induce, to a different extent, a change both in the affinity and the allosteric equilibrium.

In some cases, an approximated linear correlation was found between the affinity of the allosteric effectors for hemoglobin and the induced variation of p50 (28). However, some of the strongest allosteric effectors show a remarkable variability in effectiveness (change of p50), in spite of similar binding constants to hemoglobin. Abraham and coworkers (28) introduced an intrinsic affinity coefficient and proposed a molecular mechanism. The intrinsic affinity might be influenced by the capacity of the effectors to interact with key residues and, particularly, with αLys99.

Of the five structural classes listed in Table 8.1, only two, A and C, both characterized by an amido group, exhibit high potency. The first bezafibrate derivatives were reported by Lalezari and coworkers (30–32) and they all share a phenylureido-substituted phenoxyisobutylic structure. They differ in the number and position of the chlorine substituents on the phenyl ring (**2b**). The most potent molecule of the series, L345, lowers the oxygen affinity of human erythrocytes 30-fold when it is present at a concentration of 1 mM. It also strongly decreases the cooperativity. These compounds show a partial competition with chloride ions but act synergically with 2,3-DPG. X-ray diffraction studies have revealed that they do not bind just to the bezafibrate site, but also, with lower affinity, to two symmetrically related sites near Arg104β. Unfortunately, all the members of this class lose their activity in the presence of physiological

Table 8.1   Allosteric Effectors of Hemoglobin That Decrease the Oxygen Affinity

Series	Link	Relevant Compounds
Series A ("RSR series")	$-NH-CO-CH_2-$	RSR4 ($R_3,R_5 = Cl$)
		RSR13 ($R_3,R_5 = Me$)
Series B ("MM series")	$-CO-NH-CH_2-$	MM25 ($R_4 = Cl$)
Series C ("L series")	$-NH-CO-NH-$	L35 ($R_3,R_5 = Cl$)
		L345 ($R_3,R_4,R_5 = Cl$)
Series D	$-CH_2-NH-CO-$	
Series E	$-CH_2-CO-NH-$	

concentrations of albumin, even if palmitic acid or tripalmitin is added.

A more successful group of bezafibrate derivatives, named RSR series, was obtained by replacing one amide nitrogen of the urea with a methylene group (27, 33). The reversal of the amide bond results in the much weaker MM series. Although the RSR series is equally or even less potent *in vitro* than the urea derivatives, it is the least affected by the presence of albumin, probably because of its less polar nature. The bezafibrate derivative (2-[4-[[3,5-dimethylanilino]methyl]phenoxy]-2-methylpropionic acid) (**3**), efaproxiral, proved to be the least affected and, therefore, the most promising drug candidate.

(3)

The substitution of the *gem*-dimethyl groups of efaproxiral with different alkyl groups, as well as the substitution of the ether oxygen atom with a methylene group, resulted in a decreased affinity for hemoglobin (34). Unlike the urea analogs, this class seems not to bind to a secondary site.

2.2.2.1 Efaproxiral. Efaproxiral is the only drug belonging to the compounds that right shift the oxygen-binding curve, which is currently being investigated in clinical trials. It is produced by Allos Therapeutics (Denver, CO, USA). It is usually administered as sodium salt.

2.2.2.1.1 Physiology and Pharmacology. The primary pharmacological effect of efaproxiral is a decreased affinity of hemoglobin for oxygen, which implies a steeper gradient of hemoglobin oxygenation between the lung and tissues (Fig. 8.4). Therefore, a higher fraction of oxygen is delivered to the cells, thus increasing both its free concentration and the fraction bound to myoglobin. Early experiments on healthy mice (35) showed that the administration of efaproxiral leads to a significant and reproducible increase in both the p50 of

whole blood *in vivo* and the oxygen pressure in the tissues, measured directly by inserting a microelectrode in the muscle. Experiments on healthy rats (36) showed that the administration of efaproxiral also results in a decreased cardiac output and in an increased vascular resistance. These changes in hemodynamics are unlikely to be caused by a direct interaction of efaproxiral with the vascular tissue. Rather, they are caused by a regulatory adjustment in response to the increased delivery of oxygen to the tissues, probably mediated by oxygen-sensing systems, which are not yet well characterized. This hypothesis is confirmed by the observation that the structurally unrelated compound inositol hexaphosphate produces similar changes in hemodynamics. Because efaproxiral is also used as an enhancer of the effects of radiotherapy, its pharmacological activity was proposed to arise from an additional, unknown mechanism. The hypothesis originates from experiments in which normally oxygenated and hypoxic EMT-6 murine mammary carcinoma cultured cells were exposed to radiation in the absence and presence of efaproxiral and some of its less effective structural analogs (37). Efaproxiral proved to greatly enhance the cytotoxicity of radiation on cultured cells. The effect is greater in cells kept under hypoxic conditions, but it is also present in normally oxygenated cells. Such an effect is obviously hemoglobin independent and demands alternative explanations. The fibric acid class, to which efaproxiral belongs, is known to perform its antilipidemic activity by inhibiting the biosynthesis of cholesterol at a not well-defined level in the steps preceding the synthesis of mevalonate, and altering the synthesis of compounds, such as geranyl-PP, farnesyl-PP, and geranylgeranyl-PP. Because these compounds are involved in signal transduction pathways, they might hinder DNA repair mechanisms, thus making the neoplastic cells more sensitive to DNA-damaging therapies.

2.2.2.1.2 Potential Clinical Use of Efaproxiral

2.2.2.1.2.1 Efaproxiral as Radiation Enhancer. The most promising application of efaproxiral is as an enhancer of the effectiveness of radiation therapy in the treatment of solid neoplasia. The oxygen-dependent sensi-

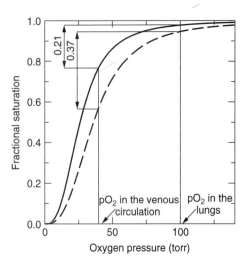

Figure 8.4. Oxygen binding curve of hemoglobin under physiological conditions (solid line) and in the presence of a right shifting agent (dashed line). A 10-Torr right shift of the p50 results in about a two-fold increase in the unloaded oxygen in the peripheral tissues.

tivity of tumors to radiotherapy is attributed to an indirect mechanism of cytotoxicity. The ionizing radiation produces damages in the DNA chain by forming free radicals, mostly derived from oxygen. The efficacy of radiation is therefore strongly reduced by intratumoral hypoxia, which is one of the causes for the failure of therapy, particularly in advanced tumors. To achieve the same toxic effect in hypoxic conditions as in conditions of normal oxygenation, 2.5- to 3-fold the standard dose of radiation should be applied (38). Hypoxia was shown to play an important role in multistep carcinogenesis because it provides a selective pressure for the proliferation of neoplastic cells with a low apoptotic potential (39). Therefore, drugs capable of inducing an increased intratumoral oxygen pressure are potentially useful not only in enhancing the radiation therapy but also in the general treatment of tumors, slowing down the progression toward malignancy.

Intratumoral hypoxia may be caused either by an increased consumption of oxygen due to actively proliferating neoplastic cells or by a poor or inefficient vascolarization of the tumor. The effect of a poor perfusion on radiotherapy can be mimicked *in vitro* by treating

cultured mammal cells with ionizing radiation at different oxygen pressures (38). A low partial pressure was shown to increase the resistance of the cells. *In vivo*, by directly measuring the oxygen pressure in solid tumor, it was possible to correlate low values of oxygen with poor response to radiotherapy, at least for squamous cell carcinoma metastases (40), carcinoma of head and neck (41), and cancer of the uterine cervix (42, 43).

Hyperbaric oxygen is not always effective as enhancer of radiotherapy, possibly because hemoglobin is almost completely loaded at air oxygen pressure; thus higher pressures do increase dissolved oxygen only. An alternative approach to reduce intratumoral hypoxia is based on the liberation of part of the oxygen that remains bound to hemoglobin at peripheral oxygen pressures. Efaproxiral was tested for safety in a phase I clinical trial as radioenhancement agent in 19 patients with newly diagnosed glioblastoma multiform brain cancer treated with standard cranial radiation therapy (44). Phase II and III evaluations are currently underway.

A general advantage of this type of chemo- and radioenhancers is that they do not have to reach the neoplastic cells for efficacy. Therefore, their activity does not depend on the metabolism of the specific involved tissue.

2.2.2.1.2.2 Efaproxiral in the Treatment of Ischemia. Another possible application of efaproxiral currently under investigation is as oxygen-delivery agent in the treatment of ischemia. Ischemia is an imbalance between oxygen supply and demand in peripheral tissues, which can undergo permanent damage if the metabolic needs are not met for a sufficiently long time. It may be caused by trauma, neoplasia, or cardiovascular events. Ischemia is particularly dangerous when either the cardiac tissue or the central nervous system is affected. In the former case, the whole cardiovascular function can be impaired, resulting in possible secondary tissue damages. In the latter case, the clinical consequences can be very different, depending on the involved area of the central nervous system.

Efaproxiral proved effective in reducing the effects of prolonged ischemia in a number of animal models. It was shown to reverse the hypoxia-induced cerebral vasodilatation in rats (45). Preischemic administration of efaproxiral to cats subjected to 5 h of permanent middle cerebral artery occlusion resulted in a significant reduction in the size of the infarcted region (46). However, the administration of efaproxiral proved ineffective in more severe ischemia (47). Preischemic administration of efaproxiral was effective in the treatment of transient focal cerebral ischemia in rats only if it was combined with the administration of the N-methyl-D-aspartate receptor antagonist dizocilpine (48). Efaproxiral proved effective in improving the recovery of the contractile function of stunned myocardium in dogs, a model of ischemic heart disease (49). It was also effective in improving the myocardial mechanical and metabolic recovery after cardiopulmonary bypass, using a dog model of surgically induced myocardial ischemia (50). This application may be of significant clinical importance, given that the risk of ischemia during cardiac surgery is diminished but not eliminated by using hypothermic cardioplegia. Moreover, patients with chronic medical conditions, such as coronary artery disease, diabetes, and hypertension, have a significant risk of experiencing complications associated with hypoxia, even in noncardiac interventions.

The effect of the administration of efaproxiral on metabolism during ischemic events was investigated by monitoring with ^{31}P nuclear magnetic resonance the levels of the so-called high-energy phosphates, particularly creatine phosphate and adenosine triphosphate. During ischemia, because of the impairment of the oxidative metabolism, the level of the high-energy phosphates, phosphocreatine and adenosine triphosphate, decreases and the concentration of inorganic phosphates simultaneously increases. The resulting overall metabolic impairment may cause permanent damages in the tissues. The levels of high-energy phosphates were determined before, during, and after causing myocardial ischemia in dogs (51). It was shown that efaproxiral reduces the decline of high-energy phosphates if administered before ischemia and accelerates the return to normal values if administered after its onset.

Clinical data are also available for efaproxiral. It was administered to patients undergoing general anesthesia and surgery (52). In this study, it was shown that the rightward shift of the oxygen binding curve *in vivo* is dose dependent. A dose between 75 and 100 mg/kg was found to increase the whole blood p50 by approximately 10 Torr. As for animal models, no significant hemodynamic effects were noticed. A small number of patients showed a transient increase in serum creatinine. Clinical trials are currently assessing the benefits of efaproxiral in patients with unstable angina and are planned for the treatment of myocardial infarction and stroke.

2.2.2.1.2.3 Efaproxiral as Performance Enhancer. The observation that efaproxiral increases the aerobic capacity of skeletal muscles in animal models raises the concern of an illicit use as a performance-enhancing substance. This application is motivated by the acceleration of the oxidative muscular metabolism that results from a higher availability of oxygen in the tissues. In view of this use, a gas chromatography/mass spectrometry method for detection of efaproxiral in urine samples has been developed (53).

2.2.2.1.3 Adverse Effects. Because clinical trials are still ongoing, no definitive indication of side effects is known. However, some consequences of the decreased oxygen affinity of hemoglobin can be foreseen. One predictable adverse effect that may result from the lowering of oxygen affinity is the reduced loading of hemoglobin in the lungs, which may cause hypoxia in the tissues. For this reason, the amount of efaproxiral that can be safely administered should not overly reduce the fractional saturation of hemoglobin at atmospheric oxygen pressure. A dose of 100 mg/kg was reported to cause a significant decrease of the arterial oxygen saturation in a limited number of patients (52). In mice (54) a dose of 300 mg/kg resulted in a desensitization of Fsall fibrosarcoma to radiation therapy, likely because of the poor oxygenation of hemoglobin in the lungs. This side effect could be overcome by administering supplemental oxygen. In rat models it was shown that the increase in oxygen pressure in breathed air can compensate for the reduced oxygen affinity (55).

The high oxygen concentration in the tissues may increase the formation of oxygen-derived free-radical species, which are involved in the pathogenesis of the ischemia-reperfusion injury (56). Oxygen reacts with hydroxyl radicals and is also used by several oxide synthases to produce nitric oxide, forming peroxynitrite in the presence of superoxide radicals. These species may further damage the cells. The rat acute subdural hematoma, a model of human head injury, was used to assess the effect of efaproxiral on the production of free radicals, the levels of which were already known to increase during ischemia. It was demonstrated that the administration of efaproxiral does not increase the formation of free radicals (57).

Given that acute renal failure is believed to arise from hypoxic conditions in the outer medulla, a rat model was used to assess the supposedly beneficial effect of efaproxiral (58). However, the opposite effect was observed, associated with an increase of the levels of serum creatinine and a worsening of the overall conditions. Pretreatment with furosemide, which reduces sodium transport and, consequently, diminishes the rate of oxygen consumption, resulted in a less severe dysfunction. This behavior, consequent to efaproxiral administration, if confirmed for humans affected by renal dysfunction, may represent an important side effect of the drug. Other adverse effects, such as nausea, headache, and neurologic symptoms, were reported in clinical trials.

2.2.2.1.4 Pharmacokinetics. Preliminary pharmacokinetic data were obtained from phase I clinical trials on patients undergoing radiation therapy (44, 52, 59). The administration of a dose of 100 mg/kg i.v. resulted in a variation of approximately 10 Torr of the p50 at peak plasma concentration of efaproxiral. This dose is considered to be safe and effective, even when administered daily for several weeks. Higher doses, although not toxic, may not be significantly more effective in raising the tumor oxygenation, as shown for rat models of fibrosarcoma (54).

The plasmatic half-life of efaproxiral ranges between 3.5 and 5 h and the whole blood p50 is linearly related to the plasmatic concentration of the drug. In patients treated up to 6 weeks, no drug accumulation was detected. Efaproxiral is mostly eliminated by glomerular filtration with a mechanism that saturates within the therapeutic dose range. A partial hepatic glucoronidation before renal clearance might take place.

2.2.3 Aromatic Aldehydes. Monoaldehyde allosteric effectors affect the oxygen affinity of hemoglobin in opposite ways. Some of them (23) induce a left shift of the oxygen binding curve and, therefore, are potentially useful for the treatment of sickle-cell anemia. The most promising compounds are 5-(2-formyl-3-hydroxyphenoxy) pentanoic acid, referred to in the literature as 12C79 (**4a**), and vanillin (**4b**).

The aldehydic compounds 5-formylsalicylic acid (**5a**), 2-(benzyloxy)-5-formylbenzoic acid (**5b**), and 2-(phenylethyloxy)-5-formylbenzoic

(4)

acid (**5c**) were designed to increase the affinity for oxygen, but, unexpectedly, showed the opposite effect (24).

(5)

To identify the structural differences that are responsible for this effect, X-ray diffraction studies and molecular modeling techniques were used. By solving the structure of hemoglobin complexed with different aromatic aldehydes of both functional classes, it was discovered that they all form a Schiff base with the terminal amino groups of the two symmetry-related αVal1. The N-termini of the β-subunits are not equally affected, although partial occupancy was observed for some of the complexes. The only exception is vanillin, which binds to two sites, one near αCys104, αHis103, and βGln103 in the central cavity and the other on the surface near βHis116 and βHis117 (60). A possible explanation for the surprising opposite effects brought about by structurally related compounds that bind to the same residue was proposed by Abraham and coworkers (24). In deoxyhemoglobin, α_1Val1 and α_2Arg141 (and α_2Val1 and α_1Arg141) interact through a water molecule. The carboxylate group that characterizes the right-shifting aromatic aldehydes may replace this water-mediated bond by forming an ionic bridge with the positively charged guanidinium group of Arg141 of the other α-subunit. Because this bond is stronger than the water-mediated one, the result is an increase in the stability of the T state. In the R state, Val1 and Arg141 of the opposite α-subunits are too far apart to form any direct interaction, and do not allow the formation of a bridge between them. This is the reason that, although the aldehydes bind to αVal1 both in the T and R states, they stabilize only the T state. Secondary interactions with other residues observed for specific compounds may justify the differences in the strength of these allosteric effectors. They bind parallel to the twofold axis and the substituents in the *para* position point toward Lys99 of the other α-subunit.

The left-shifting aldehydes either do not possess an acidic group or the group is present but not correctly oriented to form an interaction with αArg141. The compound 12C79, for instance, binds hemoglobin parallel to the twofold axis, as the right-shifting aldehydes, but the side chain points in the opposite direction (61). By binding to the α_1Val1, these compounds disrupt the water-mediated ionic bond between the αVal1 and βArg141, but they do not replace it with a stronger bridging interaction.

The binding of both left-shifting and right-shifting aldehydes has been shown to reduce the chloride effect, possibly by narrowing the access to the central water cavity, where chloride ions probably bind in a nonspecific way. Moreover, the effectors interfere with the binding of the endogenous allosteric effector 2,3-DPG, the effect of which is abolished or diminished to a different extent for most of the tested compounds. Therefore, the overall effect on the oxygen affinity of hemoglobin under physiological conditions may depend on different and very small contributions. In the case of the left-shifting compounds, the higher affinity is attributed both to the rupture of the ionic bond between αVal1 and αArg141 and to the inhibition of the binding of 2,3-DPG and chloride ions, the endogenous allosteric effectors. In the case of the right-shifting compounds, the formation of a strong intersubunit interaction prevails and leads to a stabilization of the T state.

Because the side-chain of the monoaldehydes was observed to point toward Lys99 of the opposite α-subunit, it was foreseen that the addition of a terminal group capable of forming a bond with it might result in a further stabilization of the T state. Two groups were tested (62), an aldehyde moiety, which could form a Schiff base with Lys99, and a carboxylic moiety, which could form an ionic bond with the same residue in the protonated form. For each class of compounds, the distance between the two reacting centers was modulated by varying the length of the link that connects the phenyl rings (6). Depending on the linker between the phenyl rings, these compounds are classified as bis(2-carboxy-4-formylphenoxy)-alkanes (**6a**), (2-carboxy-4-formylphenoxy)-(4-carboxyphenoxy)-alkanes (**6b**), and bis(2-carboxy-4-formylphenoxy)-xylenes (**6c**). In (**6c**), the R group can be a *meta* $-CH_2(C_6H_4)CH_2-$, an *ortho* $-CH_2(C_6H_4)-CH_2-$, or a *para* $-CH_2(C_6H_4)CH_2-$, $-CH_2-CH=CHCH_2-$.

These compounds were shown to bind hemoglobin as predicted and the addition of a new bond resulted in a much higher potency with respect to the monoaldehydes tested by Abraham and coworkers (24). Among the compounds capable of binding αLys99, the bisaldehyde allosteric effectors are stronger than the monoaldehyde bisacids. This is expected, given that the former binds covalently to the residue. The increase of p50 depends strongly on the length of the molecule. The shorter the bridging chain, the higher the shift of p50. This is also expected, given that a less flexible chain is likely to produce a greater constraint on the T state.

(6)

3 BLOOD SUBSTITUTES: MODIFIED HEMOGLOBINS AND PERFLUOROCHEMICALS

The transfusion with whole blood is still the most used therapy in emergencies, surgery, and pathologies involving blood loss and insufficient oxygen delivery to organs and tissues. The use of whole blood in therapeutics is always associated with practical and sanitary problems, such as the need for careful cross-matching and blood typing, limitations in the availability from healthy donors, the short shelf life of whole blood, and the concern about contamination by infectious agents, such as hepatitis virus and HIV. The above-mentioned problems have pushed research toward molecules with biochemical and physical properties as close as possible to those of hemoglobin in red blood cells. Up to now two categories of blood substitutes have been developed: modified hemoglobins and perfluorochemicals.

Modified hemoglobins are prepared from human, bovine, or recombinant hemoglobin, by chemical or genetic methods. Modifications of hemoglobin are needed to stabilize the tetrameric form and to increase its p50 to improve oxygen delivery to tissues. In fact, the tetramer-dimer equilibrium of hemoglobin is shifted in red blood cells toward the tetrameric form because of the high hemoglobin concentration (about 5 mM). The tetramer binds 2,3-DPG, present at equivalent concentrations, to form a stoichiometric complex. As a result, the oxygen affinity is low (p50 = 26 Torr) and modulated by allosteric effectors. On the contrary, the concentration of dimeric hemoglobin molecules increases in dilute hemoglobin solutions. The dimer binds 2,3-DPG very weakly. As a result, free hemoglobin exhibits an increase in oxygen affinity and loses cooperative oxygen binding.

Perfluorochemicals are inert, synthetic, linear, or cyclic fluorocarbon compounds, which act as high-solubility solvent for oxygen and do not display any cooperative properties. The amount of dissolved gas increases linearly with increasing oxygen pressure and the total amount of gas carried in the circulation is proportional to the concentration of fluorocarbons in solution.

3.1 History of Blood Substitutes

3.1.1 Development of Modified Hemoglobins. The problem of finding an "artificial" substitute to whole blood to be used in transfusions is not a new aspect in pharmacological and biochemical research. Since the end of the 19th century, some researchers thought that free hemoglobin in solution could be administered as blood substitute to treat severe anemia (63). In 1916 Sellards was the first to administer hemoglobin solutions to humans and to study their adverse effects (64). This study, focused on the renal clearance of hemoglobin solutions, was the first to report a serious renal toxicity induced by hemoglobin administration. It is only in 1937 that Amberson noticed a hypertensive effect as a consequence of intravenous administration of hemoglobin (65). At that time, hemoglobin was quite far from being pure. Potential sources of toxicity were red cell membrane debris contaminants that induce nephrotoxicity and hemolysis. In 1970 Rabiner developed a protocol for the purification of hemoglobin from cell membranes (stroma-free hemoglobin, SFHb) (66). In the 1970s Moss and De Venuto tested SFHb on animals (67, 68). These preliminary tests did not reveal any severe toxicity and phase I clinical trials were allowed. Clinical trials indicated that even stroma-free hemoglobin was highly toxic in humans, mainly because of the effect on kidney and cardiovascular system (69). This study points out that the toxicity of hemoglobin solutions is not exclusively a consequence of cell membrane impurities present in the preparation but also of the biochemical properties of hemoglobin free in solution. In fact, hemoglobin out of red blood cells is mainly dimeric and is, thus, filtered by kidney, causing nephrotoxicity. Furthermore, acellular hemoglobin shows an increased NO scavenging activity, which is the main mechanism accounting for the hypertension, frequently observed upon free hemoglobin administration.

A project aimed at polymerizing hemoglobin to lower its colloid oncotic pressure and to increase its molecular weight was first proposed (70) and further developed by other research groups (see Section 3.3.4). In 1964 a method for crosslinking hemoglobin molecules to be used as an artificial membrane for the preparation of "artificial red blood cells" was developed (70). In this reaction, sebacyl chloride forms an amidic bond with amino groups on the surface of the protein, as shown in Reaction 1. In an attempt to decrease the

$$(8.1)$$

size of the artificial cells, Chang obtained polymerized hemoglobin, containing both intra- and intermolecular linkages. The hemoglobin obtained with this procedure is a stable tetramer with an increased molecular weight compared to that of unmodified hemoglobin. In 1968 Bunn was the first to develop an intramolecular crosslinking procedure aimed at stabilizing the tetrameric form of the protein without polymerizing it (71). A bifunctional agent, bis(N-maleimidomethyl)ether, that crosslinked the two β-chains of each dimer, was used. Afterward, another crosslinking agent, acetylsalicylic acid, was used to stabilize the hemoglobin tetramer (72, 73). A derivative of aspirin, bis(dibromosalicyl)fumarate, which is more reactive in the acylation reac-

tion, was used to crosslink hemoglobin between β-chains (74) and, then, between α-chains (75). These studies led to the development of DBBF-Hb, or $\alpha\alpha$-Hb, by the U.S. Army and DCLHb or HemAssist by Baxter. Baxter and the Army collaborated on the project since 1985, but the negative results of clinical trials led the U.S. Army, at the beginning of the 1990s, and Baxter, in 1998, to drop the project. In 1969 the effect of pyridoxal phosphate (PLP) binding to hemoglobin was reported (76). PLP binds at the 2,3-DPG site, thus increasing the p50 to a value closer to that of hemoglobin in red blood cells. In 1971 Chang, for the first time, used glutaraldehyde to crosslink hemoglobin and catalase (77). Two companies later exploited this procedure, Northfield Inc. with Polyheme, and Biopure Corporation with Hemopure. Both products are glutaraldehyde polymerized hemoglobins. The Northfield product is human pyridoxalated hemoglobin, whereas the Biopure product is bovine hemoglobin. Because of the higher p50 of bovine hemoglobin with respect to the human protein, the oxygen affinity of Polyheme is higher than that of Hemopure. The good rheological properties of polymerized hemoglobin, its high molecular weight, and its modest hypertensive effect explain the positive results in clinical trials and the approval of Hemopure for human use in South Africa.

Since the 1970s many studies on hemoglobin-based blood substitutes also focused on the preparation of conjugated hemoglobin aimed at reducing its potential antigenicity and prolonging its vascular retention. Dextran (78), polyethylene glycol (79), and polyoxyethylene (80) are the most used polymers for the modification of hemoglobin surface.

The problem of hemoglobin dimerization in solution was later approached by exploiting recombinant DNA techniques. In 1984 human β globin was expressed in *Escherichia coli* as a fusion protein with the coding region of bacteriophage lambda repressor protein (81). This expression system was found to be unsuitable for the production of both the α- and β-chains. Further improvement of the expression system in bacteria was attained using a fully synthetic gene with codons optimized for *E. coli* that led to the overexpression of myoglobin (82). In 1990 at Somatogen, the same approach was used to express fully functional human hemoglobin in *E. coli* (83).

Even though the above-mentioned modified hemoglobins have been improved, leading to better performances and to the marketing of some of them as blood substitutes for veterinary use, they are far from exhibiting the same physiological, biochemical, and pharmacological behavior of hemoglobin within red blood cells. Microencapsulation of hemoglobin (70) and polymerization with catalase and superoxide dismutase (77, 84) are advanced approaches to the issue of blood substitutes. Their development might provide a blood substitute with improved half-life, less toxic effects resulting from reperfusion injury (see section 3.3.1.4), and low level of methemoglobin formation and NO scavenging.

3.1.2 Perfluorochemicals as an Alternative to Hemoglobin. The first demonstration that perfluorochemicals (PFC) can sustain life was reported in 1966 when Clark and Gollan (85) showed that mice fully immersed in oxygenated PFC could breathe in the liquid and that the amount of oxygen dissolved was sufficient to support the respiratory function. In 1967 Sloviter and Kamimoto (86) observed that the activity of a rat brain could be maintained for several hours when perfused with emulsified perfluorocarbon. Successively, experiments carried out by Geyer demonstrated that, despite the replacement of their blood with PFCs emulsion to a hematocrit less than 1%, "bloodless" rats survived and grew without apparent abnormalities (87). Fluorocarbons used in these experiments exhibited an organ retention time too long to be feasible in human administration. Other perfluorochemicals were tested to obtain compounds with more favorable excretion rates to be used as blood substitutes. In the early 1980s studies on perfluorodecalin led to Fluosol. Although its ability to deliver oxygen was demonstrated in clinical studies on anemic patients, Fluosol did not receive FDA regulatory approval for treatment of large-volume blood loss because of its short intravascular persistence. In 1989 FDA approved Fluosol for use in conjunction with percutaneous transluminal coronary angioplasty because its efficacy in reducing myocardial ischemia and angina and in maintaining ventricular function was demonstrated (88). The long body retention time of one of its components, the adverse effects attributed to the surfactant, associated with a low fluorocarbon content that conferred only low oxygen-carrying capacity, limited the potential range of its applications. Moreover, a limited stability at room temperature, requiring the product to be shipped and stored in the frozen state and to be formulated as two annex solutions that had to be mixed before use, probably accelerated the decline of Fluosol. Manufacturing and marketing of the product were discontinued in 1994 when improvements in angioplasty technology made it unnecessary for its approved indication (89). Nevertheless, the approval of Fluosol proved that artificial oxygen carriers could be used as alternatives to blood transfusion and it represented the reference point for successive development of second-generation PFCs.

3.2 Clinical Use of Modified Hemoglobins and Perfluorochemicals

The clinical applications of the two categories of blood substitutes are discussed together in this section, whereas other topics are presented separately, because of deep differences in their physical and pharmacological characteristics.

Blood substitutes can be grouped into two main categories on the basis of their potential clinical use:

1. As alternatives to whole blood, oxygen delivery to tissues being their main application. Moreover, blood substitutes could be used for volume maintenance in surgery and trauma with blood loss, treatment of ischemia (stroke, sickle-cell crisis) and refractory anemia. Blood substitutes can also be used in organ preservation before transplantation.

2. As a product with different applications with respect to blood substitute. For example, Optro [recombinant di-alpha human hemoglobin (90)] and HBOC-201 [glutaraldehyde polymerized hemoglobin (91)] can stimulate erythropoiesis and be potentially useful in the treatment of severe anemia.

Another application for blood substitutes is envisaged in the treatment of cancer, in association with chemotherapy and radiotherapy. As already discussed, solid tumors are more susceptible to radiotherapy and chemotherapy if the oxygen delivery to the sick tissues is improved (38). Both modified hemoglobins (92) and perfluorochemicals (93) can be potentially useful for this application. Clinical trials using hemoglobin linked to PEG (174) were carried out, but research halted while on phase I clinical trials. One of the main disadvantages of hemoglobin solutions as blood substitutes is the hypertensive effect, which is discussed in Section 3.3.1.3. On the other hand, in case of hypovolemia or systemic inflammatory response syndrome, this effect could be beneficial and accelerate recovery (94). For this reason, pyridoxalated hemoglobin entered phase II clinical trials in the treatment of septic shock (95). HemAssist, a crosslinked tetrameric hemoglobin (96), was administered during clinical trials to treat hypotension induced by hemodialysis and was found to stabilize the pressure without significant adverse effects. Furthermore, modified hemoglobin solutions could be more convenient than whole blood in emergency cases in which blood typing would be an excessively time-consuming procedure. Perfluorochemicals may be used for the treatment of acute respiratory failure by liquid ventilation because they are thought to help the reopening of collapsed alveoli (97–99). Liquivent from Alliance Pharmaceutical Co. contains perfluorocityl bromide and is currently in phase II/III clinical trials (100). The low molecular weight fluorocarbons, which are gaseous at body temperature, such as perfluoropentane, Echogen from Sonus Pharmaceuticals, and perfluorohexane, Imavist, formerly known as Imagent, from Alliance Pharmaceutical Co., can be used as contrast agents for the assessment of heart function and detection of perfusion deficits by ultrasound imaging (101). Liquid and emulsified perfluorochemicals are currently being evaluated as oxygen-carrying culture media supplements for eukaryotic and prokaryotic cells (102). In addition, perfluorochemicals and, perhaps, recombinant hemoglobin might be envisaged as the only blood supplies administrable to patients who cannot receive donor blood because of religious beliefs.

3.3 Hemoglobin-Based Blood Substitutes on Clinical Trial

In April 2001 Hemopure, a glutaraldehyde-polymerized hemoglobin, received the South Africa Medicines Control Council's approval for its use as a blood substitute to treat acute anemia in adult surgery patients. The same product is under marketing approval in the United States and in Europe. In 1998 Oxyglobin [glutaraldehyde polymerized hemoglobin (HBOC-301)] was the first blood substitute to be approved for the market by the U.S. Food and Drug Administration and the European Commission, in the treatment of anemia in dogs. The product is now commercially available in the United States. Other modified hemoglobin-based blood substitutes are on clinical trials, some of them entering phase III. Clinical trials on three hemoglobin-based blood substitutes have been halted: (1) HemAssist by Baxter on May 1998, because of increased mortality in the test groups with respect to control groups in the therapy of trauma (103) and stroke (104); (2) Optro by Somatogen/Baxter in 1998, after Baxter's acquisition of Somatogen, which was developing the product; and (3) PEG-Hb by Enzon, during phase Ib clinical trials. In Table 8.2 the complete list of hemoglobin-based blood substitutes is reported.

The route of administration is intravenous infusion. Given that the majority of products are not yet available on the market, the administered dose and the infusion rate depend on the settings during clinical trials. The "max administered dose" refers to the highest dose administered to patients during clinical trials without causing severe adverse effects. Whenever in literature the dose was expressed as mg/kg the "max administered dose" was calculated for an average 70-kg subject.

3.3.1 General Side Effects. Side or adverse effects that have been observed upon adminis-

Table 8.2 List of Hemoglobin-Based Blood Substitutes, Including Only the Products That Were in Clinical Trials[a]

Proprietary Name	Nonproprietary Name	Hemoglobin Type	Chemical Class	Originator	Clinical Trial Phase	Max Administered Dose (g)	Reference
Hemolink™	o-Raffinose polymerized crosslinked hemoglobin	Human	Crosslinked polymerized hemoglobin	Hemosol	I II	43 100	193, 194
Hemopure™	HBOC-201 (glutaraldehyde polymerized hemoglobin)	Bovine	Polymerized hemoglobin	Biopure	I II III	45 105 250	179, 181, 177
PolyHeme™	Poly-SHF-P (glutaraldehyde polymerized pyridoxalated hemoglobin)	Human	Polymerized hemoglobin	Northfield	I II III	50 300–500 NA	189, 192
None	PHP (pyridoxalated hemoglobin–polyoxyethylene conjugate)	Human	Conjugated hemoglobin	Apex Biosciences	I I/II II	7 7 ~ 180	95

Hemoglobin-based blood substitutes for which clinical trials were halted or suspended

Optro™	rHb1.1	Recombinant	Crosslinked hemoglobin	Baxter/Somatogen	I II	22.4 50–100	158, 162
None	PEG-Hb (PEG modified hemoglobin)	Bovine	Conjugated hemoglobin	Enzon	Ia Ib	38	202
HemAssist™	DCLHb (diaspirin crosslinked hemoglobin)	Human	Crosslinked hemoglobin	Baxter	I II III	1.75–7 3.5–75 75	145, 146, 148

[a]Other products that may have clinical or biological interest but have never been tested on human subjects are reported in Section 3.3.6.

tration of hemoglobin-based blood substitutes vary in intensity and clinical importance, depending on the type of product, its molecular weight, the chemical modification of the molecule, the viscosity of the solution, and many other factors. In this section, the mechanisms responsible for some of the most important or frequently observed adverse effects are described mainly from a biochemical and physiologic point of view. Specific adverse effects, their relative importance, and their influence on the safety profile of the blood substitute are reported in separate sections dedicated to individual products.

3.3.1.1 Hypertension. The administration of stripped hemoglobin induces an increase in the mean arterial pressure (MAP), accompanied by a decrease in heart rate, as a consequence of systemic vasoconstriction in several vascular beds (105–107 and references therein). In the majority of physiological and pathological scenarios this is considered an adverse effect because it reduces tissue perfusion, a deleterious event in subjects suffering from blood loss or hemodilution. The cause of hemoglobin-induced vasoconstriction is still an open question in blood substitute research and remains one of the most serious drawbacks to the use of hemoglobin solutions as blood substitutes. Currently, the best-characterized and widely accepted mechanism of vasoconstriction induced by acellular hemoglobin is nitric oxide (NO) scavenging. NO is a naturally occurring molecule, released by endothelial cells both in the interstitial space and in the lumen. In the interstitial space NO activates guanylate cyclase on the smooth muscle cells, causing vasorelaxation (Fig. 8.5A). Given that the affinity of hemoglobin for NO is about 200,000-fold higher than that for oxygen, acellular hemoglobin can efficiently scavenge plasmatic NO (see also section 3.3.3.1.4). Furthermore, both dimeric and tetrameric hemoglobin can extravasate by passing through the interendothelial gap junctions (Fig. 8.5B) (108), thus binding NO directly on its site of action, causing vasoconstriction and hypertension (Fig. 8.5C). Administration of cyanomethemoglobin, which does not bind NO, does not induce vasoconstriction. L-Arginine and nitroglyc-

erin, both sources of NO, are effective in reducing the hypertension induced by diaspirin-crosslinked hemoglobin (106).

It was shown that it is possible to reduce the rate of reaction of NO with hemoglobin by introducing mutations in the heme pocket, thereby obtaining hemoglobins with a decreased vasoactivity (109). However, hemoglobins with mutations in the heme pocket are likely to exhibit altered oxygen affinity and their use as blood substitutes is still an open question. If the formation of S-nitrosohemoglobin is confirmed to play a role in the hemodynamic properties of hemoglobin, a new possibility for the design of recombinant molecules with a selective modulation of NO scavenging activity will open (110). In the past much effort was put in the development of polymeric hemoglobins that could not extravasate and that could hopefully give milder hypertensive effects. Some of these products demonstrated to be effective in reducing hemoglobin-induced hypertension. However, no definitive conclusions can be drawn about the relationship existing among extravasation, molecular weight, and hypertensive effect (see Ref. 111 and references therein for an exhaustive critical review on blood substitutes issue).

Recently, many investigators pointed out that hemoglobin molecular weight might have a marginal role in determining hypertensive effects, at least for two reasons. First, hemoglobin free in solution is not subjected to the hydrodynamic separation exerted by blood flow, which accounts for the formation of a "red blood cells free zone" in blood vessels under flow conditions both *in vivo* and *in vitro* (11). As a result hemoglobin free in solution can have a scavenging potential several times higher than that of hemoglobin inside red blood cells. Furthermore, a study carried out on rHb1.1 (recombinant di-alpha human hemoglobin) and its glutaraldehyde-modified forms has questioned the understanding of the exact mechanism of hemoglobin extravasation (112). In fact, polymerization with glutaraldehyde is likely to prevent extravasation not because of the increase in the molecular weight of hemoglobin, but as a consequence of a decrease in the endocytotic transport of the protein attributed to glutaraldehyde decoration. This finding opens new insight in the de-

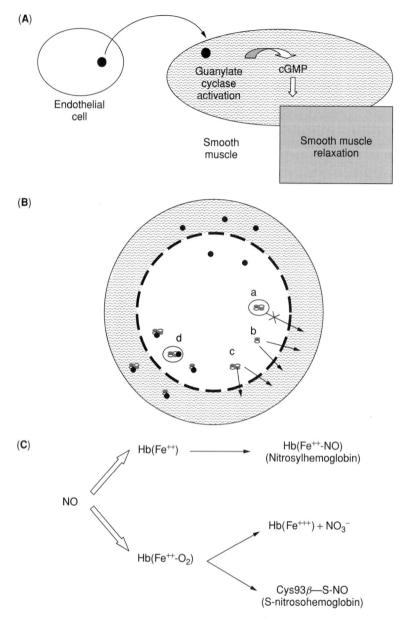

Figure 8.5. Action of NO on blood vessels and NO scavenging by hemoglobin. (A) Mechanism of NO-dependent smooth muscle cells relaxation. NO (●) is produced by endothelial cells and released both into the interstitial space and the lumen of the vessel. In the interstitial space of smooth muscle cells NO activates soluble guanylate cyclase and causes vasorelaxation. (B) Schematic representation of a blood vessel. Hemoglobin inside the red blood cells cannot extravasate (a). Dimeric and tetrameric acellular hemoglobin (b, c) extravasates, passing either through the endothelial intercellular junctions or intracellularly by endocytotic mechanisms. (d) Both cellular and acellular hemoglobin bind NO. Scavenging of NO in the interstitial space results in vasoconstriction. (C) Reaction of deoxygenated and oxygenated hemoglobin with NO. NO binds to deoxyHb, preferentially forming nitrosylHb, and to oxyHb, preferentially forming *S*-nitrosoHb. In addition, NO oxidizes oxyHb to produce metHb and nitrate.

sign of hemoglobin-based blood substitutes and refocuses the role of the polymerization in the prevention of vasoconstriction. Furthermore, some immunohistochemical studies on abdominal aorta of rats show that Dex-BTC-Hb (dextran 10-benzene-tetracarboxylate hemoglobin) is found in the endothelial cells within 30 min after the administration. Dex-BTC-Hb is a high molecular weight polymer formed by stable tetrameric hemoglobin molecules that cannot extravasate (113). Probably, the vasopressor effect of hemoglobin is mediated by other mechanisms, including an increase in endothelin-I production (106) and a direct effect of hemoglobin on smooth muscle cells. In fact, it was reported that hemoglobin can enhance the responses of smooth muscle cells to noradrenaline (107) and yohimbin can reduce the hypertension following diaspirin crosslinked hemoglobin administration (114). In rats the effect mediated by adrenoreceptors is not under central nervous system control, nor it is mediated by the release of catecholamines from the adrenal medulla (107).

More recently, studies aimed at unveiling the relationship between blood substitutes administration and hypertension have focused on two previously underestimated factors: oxygen pressure in the precapillary arterioles and solution viscosity (for a review see Ref. 115). In the past, research on blood substitutes led to the production of high p50-modified hemoglobins with the purpose of enhancing oxygen delivery to the tissues and thereby accelerating the recovery. Works at the beginning of the 1980s (116) demonstrated that vasoconstriction and vasorelaxation of precapillary arterioles depend on oxygen concentration. Vasorelaxation occurs at low local oxygen concentrations, whereas high local oxygen concentrations lead to vasoconstriction. The control of blood flow by oxygen demand represents a strategy to protect tissues from oxidative damage and to ensure, at the same time, a physiological oxygen supply. In this light, low affinity hemoglobin administration is not always correlated with enhanced perfusion. Conversely, lower p50 and lower hemoglobin concentrations might lead to better blood substitutes (117).

Usually, blood substitutes were formulated as low viscosity solutions, based on the assumption that high viscosity can reduce blood flow through the capillary beds and lead to a decrease in cardiac output. Blood viscosity is responsible for the shear forces that are exerted on capillary walls. The lower the viscosity, the lower the shear force exerted. The effect of shear forces is likely to be transduced into chemical mediators of the vascular tone, as, for example, NO and prostacyclins (118). When shear forces are too weak, the production of NO decreases with induction of vasoconstriction. This finding is supported by the demonstration that the administration of low viscosity hemoglobin solutions is correlated with low tissue oxygenation (119).

3.3.1.2 Nephrotoxicity.
Renal toxicity was one of the first adverse effects observed after infusion of unmodified hemoglobin solutions (64). Even when a higher degree of purification of hemoglobin was attained, the administration of unmodified hemoglobin was followed by nephrotoxicity. In 1991 among various recommendations for toxicology tests of blood substitutes, the importance of careful monitoring of nephrotoxicity through both serum creatinine levels tests and evaluation of renal function was stressed (120).

Following hemolysis or destruction of aged red blood cells, hemoglobin released in the plasma is picked up by haptoglobin, a carrier protein that can bind as much as 150 mg/dL of hemoglobin. If free hemoglobin in solution exceeds this binding capacity, it dissociates into $\alpha\beta$ dimers, and is filtered in the glomerules and partially reabsorbed by proximal tubules. Following renal filtration, two possible mechanisms of toxicity can be envisaged: precipitation of hemoglobin in the kidneys and decrease in renal function attributed to local vasoconstriction induced by NO scavenging. Recently, it has been reported that persistent renal damage occurred in rats treated with high (>0.96 g) doses of stroma-free hemoglobin (121). The cause of renal failure was mainly necrosis accompanied by a proliferation of smooth endoplasmatic reticulum in the proximal tubules.

It is worth mentioning that a recent study on the interactions of ultrapure hemoglobin with renal epithelial cells (122) demonstrated that the severe renal toxicity observed in early reports on hemoglobin solutions administration was mainly attributable to the presence of red cell membranes debris and endotoxin contaminants, and that administration of highly purified hemoglobin solutions causes toxicity only at high doses. Progress in blood substitutes research has led to the production of stable tetrameric hemoglobins and highly polymerized hemoglobins with only small renal toxic effects.

3.3.1.3 Gastrointestinal Effects. Hemoglobin-based blood substitutes administration is often accompanied by nausea, vomiting, or other gastrointestinal (GI) side effects related to GI dysmotility.

Gastrointestinal motility depends on blood perfusion and is regulated by a complex interplay of transmitters and mediators acting on nerve fibers. NO is a nonadrenergic, noncholinergic inhibitory transmitter (123) that controls the peristaltic activity and the sphincters' tonus. Depletion of NO can reduce the rate of gastric emptying (124) and attenuate sphincter relaxation (125). On the other hand, a reduced perfusion can alter the electric and contractile activities of gastrointestinal muscle cells (126, 127). As stated in Section 3.3.1.1, many hemoglobin-based blood substitutes act as NO scavengers, leading to smooth muscle cell contraction. Hemoglobin administration can lead to reduced gastrointestinal motility in two ways: through NO scavenging on the smooth muscle cells of the GI tract and, indirectly, through vasoconstriction of the vessels perfusing the GI apparatus. Although it is well established that hemoglobin-induced gastrointestinal dysmotility is mainly caused by NO scavenging (128, 129), it is not clear whether some unknown effects of hemoglobin on different regulatory mechanisms of the gastric motility are also involved. Furthermore, it was found that both NO donors and NO synthesis inhibitors can inhibit gastric motility, likely because of the action of NO on different modulators of GI activity (130). Some hemoglobins are reported to have negligible effects on GI motility, as, for example, PEG hemoglobin (128) and rHb3011, a new generation recombinant hemoglobin (129). Both hemoglobins are known to have a lower affinity for NO compared with that of other hemoglobin-based

blood substitutes, although it is not well established whether a relationship exists between low NO binding and reduced gastrointestinal effects.

3.3.1.4 Hemoglobin Oxidation: Oxidative Toxicity and Reperfusion Injury. Hemoglobin is naturally susceptible to autooxidation, a reaction that leads to the formation of methemoglobin and superoxide radicals, as shown in Reaction 2. Methemoglobin does not bind ox-

$$Hb(Fe^{++})-O_2 \longrightarrow Hb(Fe^{+++}) + O_2^{\cdot-} \quad (8.2)$$

ygen. Furthermore, oxidized hemes have a lower affinity for hemoglobin and, after being released into circulation, they can trigger cytotoxic reactions (16) and contribute to the formation of free radicals (131). Hemoglobin is also endowed with pseudoenzymatic activities, such as peroxidase activity. The reaction of methemoglobin with hydrogen peroxide leads to the formation of ferryl-hemoglobin, a higher oxidation state of hemoglobin (Reaction 3). This species, because of its high redox

$$Hb(Fe^{+++}) + H_2O_2$$
$$\longrightarrow Hb^{(\cdot+)}(Fe^{+++})-O^{2-} + H_2O \quad (8.3)$$

potential, can in turn react with different substrates, causing free-radical formation and oxidative damage (16), particularly lipid peroxidation, protein crosslinking, and degradation of carbohydrates (see Ref. 132 and references therein). In the red blood cells, enzymes such as methemoglobin reductase, catalase, and superoxide dismutase reduce methemoglobin back to its active form, thus eliminating autooxidation products, such as superoxide radicals and hydrogen peroxide. The absence of these enzymatic mechanisms in the plasma compromises both the efficacy of hemoglobin as oxygen carrier and its safety. This situation is even worse when hypoxic tissues are reperfused. In fact, hypoxia activates xanthine oxidase and leads to the accumulation of hypoxanthine. In the presence of oxygen, hypoxanthine is oxidized to xanthine with the release of superoxide (Reaction 4). Superoxide and oxygen radicals generated in the reaction with hemoglobin are potentially toxic for cells and

Hypoxanthine

$2O_2$

Xanthine oxidase

$2O_2^{\cdot-}$ (8.4)

Xanthine

$2O_2$

Xanthine oxidase

Uric acid

$2O_2^{\cdot-}$

tissues (133). Damages to tissues and organs following reperfusion are referred to as reperfusion injury.

Only few studies (134) have been carried out on the oxidative processes taking place upon the administration of blood substitutes and on their consequences on the safety profile of hemoglobin-based products. It has been reported that crosslinked hemoglobins show an increased autooxidation rate compared with that of noncrosslinked hemoglobin and an unchanged or increased peroxidative activity (16). The increased concentration of pancreatic and liver enzymes, frequently observed upon administration of hemoglobin-based blood substitutes, might signal possible oxidative damage of tissues and organs (135). Further investigations are required to assess the biological and pathological effects of oxidation products and metabolites of hemoglobin-based blood substitutes (136). The redox potential of hemoglobin-based blood substitutes is an important parameter in the evaluation of their safety profile. In fact, reducing agents, such as reduced glutathione, are present in the plasma, and they can be effective in reducing the oxidized states of hemoglobin only if the redox potential of modified hemoglobin is sufficiently high to allow electron transfer. The redox potentials of modified hemoglobins vary greatly, depending on the type and degree of modification. Hemoglobin crosslinked with superoxide dismutase and catalase has been proposed as a feasible way to reduce radicals and superoxide production and to decrease the autooxidation rate of hemoglobin (see Ref. 137 and references therein). A new hemoglobin, crosslinked with o-ATP and o-adenosine and conjugated with glutathione, seems to exhibit a significantly reduced prooxidant potential, and anti-inflammatory properties (138).

SynZyme Technologies LLC (La Jolla, CA, USA) is developing a new hemoglobin-based blood substitute, HemoZyme, endowed with superoxide dismutase-like activity that might be useful in preventing and treating reperfusion injury. HemoZyme is a polynitroxylated hemoglobin that also possesses anti-inflammatory and vasodilator activities. The proposed clinical use of HemoZyme is in ischemia and in hemorrhage to prevent reperfusion injury and postreperfusion multiple organ failure (139, 140).

3.3.1.5 Antigenicity. Because hemoglobin is considered only a weak antigen (133), it is unlikely that a single administration of hemoglobin can elicit an immune response. However, the administration of heterologous hemoglobin (e.g., bovine hemoglobin in humans) or repeated administrations, as would be the case in transfusion medicine, might potentially activate an undesired immune response. It was proved that the infusion in rats of autologous hemoglobin, both stroma-free and polymerized, never results in high levels of antibody production or in anaphylactic reactions, whereas infusion of polymerized heterologous hemoglobin in immunized rats produces anaphylactic reactions (133). The problem was further addressed by studying the immune response to repeated administrations of polymerized hemoglobin in hyperimmunized rabbits (141). These results show that the immune system is activated upon repeated administrations of hemoglobin, but the immune response might not be stimulated following a few transfusions. Repeated infusions of human polymerized hemoglobin in rhesus monkeys further confirmed the absence of severe immune response or hypersensitivity (142).

Polymerization seems to enhance hemoglobin immunogenicity, whereas modifications with polyethylene glycol (PEG) or encapsulation are likely to mask the protein from the immune system (143).

3.3.2 Crosslinked Hemoglobins

3.3.2.1 HemAssist (DCLHb). HemAssist is the commercial name for a crosslinked derivative of human hemoglobin developed by Baxter between 1985 and 1998, and in collaboration with the Letterman Army Institute of Research for a few years. DCLHb (Diaspirin Crosslinked Hemoglobin) is the generic name of the product. It is formulated in a balanced physiological electrolyte solution at a concentration of 10 g/dL (144). The mean oncotic pressure of the formulation is 43 mmHg. The p50 of DCLHb ranges from 32 to 33 Torr (145). HemAssist is stable for 1 year at $-20°C$ and for 1 month at 4°C.

The crosslinking agent, bis(3,5-dibromosalicyl)fumarate, is a derivative of aspirin. This molecule (7) was proved to be one of the most effective acylating agents of hemoglobin (74).

(7)

The product obtained from the reaction of bis(3,5-dibromosalicyl)fumarate with hemoglobin depends on the oxygenation state of the protein: oxyhemoglobin is crosslinked between the β-subunits, whereas deoxyhemoglobin in the presence of 2,3-DPG, IHP, or other 2,3-DPG analogs is crosslinked between the α-subunits (75).

HemAssist is produced from outdated human blood. It is deoxygenated and crosslinked in the presence of a 2,3-DPG analog, tripolyphosphate (144). The crosslinking reaction in the presence of tripolyphosphate leads to the stabilization of the T form of hemoglobin, thus increasing its p50 to a value of about 30 Torr. The result of the reaction is a stabilized hemoglobin tetramer, crosslinked between Lys99 residues of the α-chains. This modified hemoglobin can be heat sterilized, through a patented treatment (144), to inactivate viruses and to separate crosslinked molecules from noncrosslinked dimers. The latter are less stable and denature upon heat treatment. DCLHb is one of the most widely studied hemoglobin-based blood substitutes. Clinical trials on this product stopped in 1998 (103, 104) because of the higher mortality observed in the treated groups compared to that of the control groups.

3.3.2.1.1 Clinical Use of DCLHb. HemAssist was tested with a great variety of clinical settings during clinical trials, spanning from stroke to surgery, hypovolemic and hemorrhagic shock, and hemodyalisis (144).

3.3.2.1.1.1 Adverse Effects. The most frequently observed adverse effects following DCLHb administration were mild abdominal pain and increased values in MAP (145, 146). These are reported to be around 15-20 mmHg and mainly affect the diastolic pressure (145). The increase in MAP is usually dose dependent (96), frequently accompanied by a decrease in heart rate or cardiac output, and is transient in nature (147). In some studies, the increase in MAP was reported only after the first infusion and no further increases in systemic and pulmonary arterial pressures were noticed upon subsequent administrations (148). In rats the increase in MAP is not dose dependent at doses higher than 62.5 mg/kg (149), thus suggesting a saturation mechanism responsible for the hypertension. Studies on abdominal aortic aneurysm repair and cardiac surgeries reported a higher systemic vascular resistance (SVR) in treated subjects than that in controls (147, 148). In a phase II study of hemodyalisis patients an improved hemodynamic stability without evidence of renal impairment was associated with HemAssist administration (96).

Levels of lactate dehydrogenase, creatine kinase and its myocardial isoenzyme, and liver-functionality markers vary depending on the experimental setting, but, even in the cases in which an increase was reported, it was not considered a safety concern.

In phase II clinical trials of orthopedic and abdominal surgery significantly higher levels of methemoglobin (metHb) in the treated group were observed, without affecting patients' outcome. The whole blood methemoglobin levels were found to be higher than those of plasma methemoglobin, suggesting a DCLHb-induced conversion of patients' own hemoglobin into metHb (150). Other reported adverse effects were the increase in amylase levels without any signs of pancreatitis, hemoglobinuria, hematuria, hyperbilirubinemia, and abnormal hepatic function.

3.3.2.1.1.2 Pharmacokinetics. Data from phase I clinical trials (145) show a dose-related plasma concentration of HemAssist with a peak at about 30 min after the administration. Studies on rats (151) demonstrated that DCLHb is not metabolized in the circulation and is excreted through urine and feces as low molecular weight compounds.

The $t_{1/2}$ for elimination is dose dependent in human subjects (145) and ranged from 2.5 h for a 25–50 mg/kg dose to 3.3 h for the highest administered dose of 100 mg/kg. According to preclinical studies, the elimination route of DCLHb is species specific and can be described by a two-compartment model. Preclinical studies showed that DCLHb has a widespread tissue distribution, a consequence of extravasation of the protein from the bloodstream (110).

Pharmacokinetics studied on patients undergoing elective orthopedic and abdominal surgery (152), treated with an average dose of 936 mg/kg, showed that the elimination function is monoexponential. The clearance was 4.7 mL h^{-1} kg^{-1}, the harmonic mean half-life was 10.5 h, and the volume of distribution was 66 mL/kg. These findings seem to be quite different from those on healthy volunteers, likely attributable to blood loss and dilution effects.

3.3.2.1.2 Pharmacology. HemAssist is reported to exhibit good oxygen-delivery properties and to lead to a better perfusion of some organs and tissues as a consequence of blood flow redistribution (144). Generally, the hypertension generated by the administration of hemoglobin-based blood substitutes is considered an adverse effect if it is not associated

with an increased oxygen delivery and perfusion of ischemic tissues. Preclinical studies on rats showed that the increase in MAP is followed by an enhanced perfusion of vital organs (153, 154).

Administration of DCLHb is followed both by a reduction of the cardiac output to the skeletal muscles and an increase of cardiac output to the gastrointestinal system. This effect is likely attributable to a redistribution of the blood flow following vasoconstriction (153, 154). The oxygen delivery in the presence of HemAssist is improved with respect to that of the whole blood because of both the increased oxygen diffusion from the cell-free hemoglobin and the lower viscosity of the solution (144).

The interpretation of preclinical and clinical studies seems to be quite controversial on both the perfusion and oxygen-delivery properties of HemAssist or DCLHb. Some authors suggested that the low viscosity of the formulation could result in peripheral vasoconstriction and reduced perfusion to hypoxic tissues (110).

3.3.2.1.3 Structure-Function Relationships. Bis(3,5-dibromosalicyl)fumarate reacts with Lys residues in the hemoglobin 2,3-DPG binding pocket to form a stabilized tetramer. Hemoglobin can be crosslinked through either its β- or α-subunits, depending on crosslinking conditions. Oxyhemoglobin is preferentially crosslinked between β_1Lys82 and β_2Lys82 (74), whereas deoxyhemoglobin in the presence of organic phosphates is preferentially crosslinked between α_1Lys99 and α_2Lys99 (75) (Fig. 8.6). In oxyhemoglobin, the two Lys99 are located in a cluster of polar residues that is inaccessible to the crosslinker. As a consequence, the crosslinking takes place primarily at Lys82 residues in the β-chains. Upon deoxygenation, the site becomes accessible and the reaction between Lys99 residues in the α-chains takes place. The formation of β-crosslinked hemoglobin represents the major side reaction of deoxyhemoglobin crosslinking (75), which is diminished in the presence of 2,3-DPG or IHP.

The specificity of the crosslinking reactions is very high (74) and this is surprising, given the presence of several Lys residues on the surface of hemoglobin, which, for example, re-

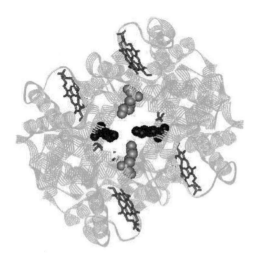

Figure 8.6. Three-dimensional structure of deoxyhemoglobin (PDB code 2HHB, human deoxyhemoglobin). αLys99 residues are shown in black space-filling mode. βLys82 residues are shown in gray space-filling mode.

act with other molecules (e.g., glutaraldehyde) to give polymeric hemoglobins. It is known that diesters are more effective acylating reagents of hemoglobin than their monoester counterparts and that the bromine substituents further increase their efficacy. This is attributed to the strengthening of apolar interactions by benzene rings and bromine atoms. Furthermore, the two negative charges of carboxylic groups may increase the affinity of bis(3,5-dibromosalicyl)fumarate for the positively charged 2,3-DPG binding pocket (74).

Crosslinking between β-subunits has a positive effect in preventing the formation of hemoglobin S fibers. In fact, the crosslinking causes a displacement of helix F and EF corner with consequent burial of two hydrophobic residues, Phe85 and Leu88, in the β-subunits that are involved in the interaction with Val6 in hemoglobin S. As a result, the two residues are less susceptible to interactions with Val6 (155). The quaternary structure is not perturbed by the crosslinker and hemoglobin cooperativity is preserved. The oxygen affinity is only marginally increased as a consequence of the stabilization of the R form of the protein.

Crosslinking between the α-subunits does not prevent the T to R transition (the distance between Lys99 residues increases by only 1 Å

upon oxygenation). The cooperativity is conserved ($n = 2.6$ compared with a value of 2.9 in HbA) and the p50 increases from 6.6 to 13.9 Torr, attributed to a decrease in K_R and K_T and an increase in L (75). All these changes are modest, demonstrating that the perturbation produced by the crosslinking agent on the structure of hemoglobin is small. The main property that is affected by the crosslinker is the alkaline Bohr effect, which significantly decreases by about 50%. Investigators suggested that this is likely caused by a glutamic residue, βGlu101, that becomes an acidic Bohr group (75).

3.3.3 Recombinant Hemoglobins

3.3.3.1 Optro (rHb1.1). Optro is the common name that designates α-fused recombinant Presbyterian hemoglobin (rHb1.1). Optro is the only recombinant hemoglobin that entered clinical trials, although experimentation ended in 1998 when Baxter acquired Somatogen. At the same time, a new research line opened, focusing on the production of recombinant hemoglobin with a reduced rate of reaction with NO. Because clinical trials on this new product have not yet begun, this section deals only with Optro, whereas information about the potential applications of low NO binding recombinant hemoglobins is presented in Section 3.3.3.1.4.

rHb1.1 was cloned at Somatogen in collaboration with the Laboratory of Molecular Biology in Cambridge (156). The molecule has two main features: a low oxygen affinity in the absence of 2,3-DPG and a stable tetrameric structure. The reduced oxygen affinity is mainly attributed to the βAsn108 to Lys mutation. The hemoglobin carrying this mutation is commonly indicated as Presbyterian and it is a naturally occurring form, first reported in 1978 (157). The stable tetrameric structure is obtained by the genetic fusion of the two α-chains with a glycine residue. More details about cloning and structure-function relationships are provided in Section 3.3.3.1.3.

Optro is formulated at a concentration of 5 g/dL. The resulting solution has an oncotic pressure of 12 mmHg and exhibits a p50 of 33 Torr (158).

3.3.3.1.1 Clinical Use of rHb1.1. The proposed applications of Optro are elective surgery and stimulation of erythropoiesis. It was demonstrated that bone marrow depression following administration of AZT to normal and to *murine model of AIDS* mice is alleviated by the administration of rHb1.1 alone or in combination with erythropoietin (159). Furthermore, it was proposed that rHb1.1-induced renal vasoconstriction leads to the production of erythropoietin (90), thereby stimulating erythropoiesis, and that the administration of rHb1.1 together with erythropoietin can increase the production of red blood cells (133).

3.3.3.1.1.1 Adverse Effects. Gastrointestinal discomfort is an adverse effect frequently found upon administration of many hemoglobin-based blood substitutes and was reported for the first time during phase I clinical trials with Optro. For this reason, a phase I clinical study was designed to assess the role of aberrant regulation of the motor function of the esophagus in the GI symptoms reported for rHb1.1 (160). rHb1.1 was reported to alter the esophageal motor function in humans, interfering with swallow-induced relaxation of the low esophageal sphincter, determining an abnormal retrograde peristalsis and increasing the velocity of the normal esophageal peristalsis. These effects were not accompanied by any signs of vasoconstriction or hypertension. The authors suggested that the adverse GI effects induced by rHb1.1 administration were a consequence of an alteration in the fine-tuning of excitatory and inhibitory neurotransmitters of the enteric nervous system. It was proposed that rHb1.1, like many other modified hemoglobins, is an NO scavenger. Its specific action on the GI system without any changes in hemodynamics might be explained by the preferential access of rHb1.1 to the GI neuromuscular apparatus through the vascular system perfusing the myenteric ganglia.

Gastrointestinal side effects were also reported in another phase I study on 48 healthy volunteers. In this case, an increase in MAP, lasting 2 h after rHb1.1 infusion, was observed. The increase in MAP was typically associated with a decrease in heart rate (158). In preclinical studies in rats it was observed that the increase in MAP and systemic vascular resistance was associated with a redistribution of blood flow with enhanced perfusion of heart

and decreased perfusion of liver, spleen, and kidney (161). Other reported adverse effects, both during phase I and phase II clinical trials, are increased amylase and lipase levels without signs of pancreatitis (158, 162), and fever (158). The former effect is likely another consequence of NO scavenging with inhibition of the relaxation of the sphincter of Oddi. In fact, it was reported that, in opossum, rHb1.1 can alter sphincter of Oddi motor activity by binding NO (163). Fever is likely the consequence of a contamination of the hemoglobin solution with bacterial endotoxin. In fact, hemoglobin purified from bacterial cells lysate is never 100% pure and, sometimes, is contaminated by endotoxin, often below the detection limit of the commonly used limulus amebocyte assay. Because high doses of rHb1.1 may be necessary for this blood substitute to be effective, endotoxin contamination has to be carefully evaluated.

3.3.3.1.1.2 Pharmacokinetics. Pharmacokinetic parameters of rHb1.1 were measured during a phase I clinical trial designed to assess the renal toxicity of the product (158). rHb1.1 was administered intravenously over 0.8–1.9 h and the doses administered varied from about 1 to 25 g. Because the largest dose administered in this setting is far lower than that administered in a surgical or emergency setting, pharmacokinetics is expected to vary to some extent with dose scaling up. For a 25-g dose of rHb1.1 the plasma concentration is about 6 mg/mL and the typical volume of distribution is 53 mL/kg. In rats rHb1.1 metabolism takes place in the liver, as for HbA. rHb1.1 elimination is reported to be a saturable process and the half-life is dependent on plasma concentration ranging from 2.4 h for a 0.5 mg/mL dose to 12 h for a 5 mg/mL dose. Clearance is well described by a one-compartment model. The largest amount of rHb1.1 found in the kidney is about 0.04% of the total administered dose, given that α-α fusion prevents dimerization and renal filtration.

3.3.3.1.2 Physiology and Pharmacology. Blood substitutes are generally administered to replace blood lost during surgery or traumatic events. Hemorrhage is often accompanied by vasoconstriction and reduced blood flow through capillaries. One of the main goals of a good blood substitute is to increase perfu-

Table 8.3   Functional Data of HbA$_0$, Recombinant Hb1.1, and the Natural Mutant Hb Presbyterian[a]

Hemoglobin Form	P$_{50}$ (Torr)	Bohr Coefficient	Hill Coefficient (n_{max})
HbA$_0$	4.5	−0.47	2.9
rHb1.1	17.2	−0.32	2.35
Hb Presbyterian[b]	19.8	−0.35	2.53

[a]Data were obtained in a solution containing 50 mM Hepes buffer, 100 mM NaCl, pH 7.4, at 25°C (156).
[b]Data refer to the mutant Val1 to Met in both α and β chains.

sion and oxygen delivery to vital organs. Oxygenation of tissues is a delicate balance between cardiac output, plasma volume, peripheral vascular tone, hemoglobin concentration, oxygen unload, and plasma viscosity. ^{31}P-NMR studies on rats demonstrate directly *in vivo* that rHb1.1 is capable of sustaining metabolic function in the tissues following total blood replacement (164).

3.3.3.1.3 Structure-Function Relationships. rHb1.1 was expressed in *E. coli* using a construct under the control of a single TAC promoter. The construct contained a single operon encoding for two α-chains and one β-chain. The two α-chains are linked by a single codon for a glycine residue, whereas the β-chains contained the mutation Asn108 to Lys that is naturally found in Presbyterian hemoglobin (156). Furthermore, both α- and β-chains contained the substitution Val1 to Met to improve the expression of the protein in the bacterial host. The crystal structure confirmed the presence of the Gly bridge in the deoxy form of the molecule (156). A more recent work on deoxy rHb1.1 at higher resolution showed that crosslinking does not perturb either the quaternary or tertiary structures of rHb1.1 (165). A preliminary characterization of the functional properties of rHb1.1 was in agreement with the structural findings. Even though the Gly bridge slightly constrains both the C- and the N-terminal of the α-chains, the cooperativity of hemoglobin is almost fully retained (Table 8.3), whereas oxygen affinity is drastically reduced (Table 8.3), likely attributable to the presence of Asn108 to Lys mutation. This residue is at the $\alpha_1\beta_1$ interface and, hence, only marginally involved in the conformational changes associated with the T to R transition. The substitution of Asn108 with Lys might lead to the formation of a hydrogen bond between Lys108 and Tyr35. The low oxygen affinity of βAsn108Gly mutant in solution and the unmodified affinity of the T state of this mutant in the crystal suggest a destabilization of the oxygenated state relative to the deoxy state as the origin of the reduced oxygen affinity (166).

The role of the glycine bridge in the equilibrium between unbound and ligated forms is not fully understood. The glycine bridge can be quite easily accommodated in the deoxy form of the protein, but, upon oxygenation, the distance between the two α-chains [≈ 3.3 Å (167)] and the torsion angles of glycine are incompatible with the transition from T to R. In fact, the presence of the glycine forces the protein to adopt a new ligated quaternary structure, the B form (167). This form has a tertiary structure that is almost superimposable with tertiary structures of other ligated hemoglobins (i.e., R, Y, and R2), whereas the quaternary structure has a different organization. The net effect of the genetic crosslinking is still not fully understood. In fact, the functional data are compatible with an increase in the oxygen affinity, whereas the structural data suggest that the glycine bridge stabilizes the T state or eventually the ligand-bound B state (167). Finally, the presence of N-terminal methionines and the glycine bridge, which stabilize the protonated deoxy form of the protein, is responsible for the decrease in the Bohr effect (167).

3.3.3.1.4 Recent Developments. After the suspension of clinical trials on rHb1.1 in 1998, Baxter focused on recombinant hemoglobins with a reduced reaction rate with NO. Usually, the rate of reaction of NO with hemoglobin is reported for the deoxy state. The reaction rate constant is about $10^7 \ M^{-1} s^{-1}$ (95), supporting the idea that hemoglobin could act as an excellent NO scavenger *in vivo*. In precapillary arterioles, hemoglobin is mainly in

Table 8.4　Pharmacological and Biochemical Data of Recombinant Hemoglobins with Low NO-Induced Oxidation Rate Constants[a]

Hemoglobin	Pressure Increase (%)[b]	Percentage Emptying of the Stomach[c]	P_{50} (mmHg)[d]	k' (NO-Induced Oxidation of Hb)[e] ($\mu M^{-1}/s^{-1}$)
rHb1.1[f]	/	30	32.0	58–60
rHb2[f]	75	NA	4.0	24
rHb3[f]	50	NA	3.4	15
rHb4[f]	30	NA	5.2	2
rHb3011[g]	NA	54	46.0	2

[a]Data for rHb1.1 are reported for comparison.

[b]Percentage of the mean arterial pressure increase following administration of rHb1.1 that determines a mean increase of 25.8 mmHg and has the highest NO-induced oxidation rate.

[c]Data refer to a 750 mg/kg dose of hemoglobin. The negative control was an increasing dose of human serum albumin (HSA); 750 mg/kg HSA administration was associated with a gastric emptying of about 60%.

[d]Data were obtained in a solution containing 50 mM Hepes buffer, 100 mM NaCl, pH 7.4 at 37°C.

[e]Data were obtained in a solution containing 100 mM sodium phosphate, pH 7.4 at 20°C.

[f]Ref. 109.

[g]Ref. 129.

the R state and, therefore, the reaction of NO with oxyhemoglobin might have a greater biological relevance than that of the reaction with the deoxy form. NO reacts with oxyhemoglobin, forming both s-nitrosol hemoglobin and methemoglobin and nitrate, through a bimolecular process that is slightly faster than the reaction of NO with deoxyhemoglobin (168). Preliminary studies with sperm whale myoglobin and human hemoglobin showed that substitution of some residues in the distal heme pocket, Leu(B10) and Val(E11), can reduce the reaction rate of NO with oxyhemoglobin resulting from steric hindrance (168). In a subsequent work (109), a series of mutant recombinant di-α-hemoglobins was produced by changing residues Leu(B10), Val(E11), and Leu(G8) in the distal heme pocket to larger hydrophobic residues (Phe, Leu, Trp, and Ile). As a result, changes in both oxygen affinity and rate of NO oxidation were obtained. A good correlation was found between rate of NO scavenging and the pressor effect exerted on conscious rats, whereas no correlation existed between oxygen affinity or autooxidation rate and pressor effect (Table 8.4). This study demonstrated for the first time in a quantitative way that NO scavenging is directly correlated with NO-dependent oxidation rate of oxyhemoglobin and that the rate of NO scavenging, and not the affinity of NO for hemoglobin, correlates with the pressure responses observed upon administration of recombinant hemoglobin.

NO scavenging is responsible not only for the pressor effect of acellular hemoglobin administration but also for many other adverse effects of hemoglobin-based blood substitutes as, for example, gastrointestinal dysmotility and increased plasma levels of pancreatic enzymes. In a preclinical study on rats (129), a recombinant hemoglobin (rHb3011) with reduced rate of NO scavenging and decreased oxygen affinity (Table 8.4) showed a reduced gastrointestinal dysmotility, measured as percentage emptying of the stomach, compared with that of rHb1.1. Presently, Baxter is committed to the development of the above-mentioned hemoglobins with reduced NO scavenging rate.

3.3.3.1.5 Things to Come. The production of recombinant hemoglobins can be achieved in different expression systems, ranging from *E. coli* to yeast, insect cells, plants, and transgenic mammals. Large-scale production and purification of recombinant hemoglobin to be used as a blood substitute have to cope with problems such as high expression yields, easy and cheap purification, and successful elimination of contaminant endotoxins. In spite of Somatogen's efforts in the scaling up of the process, leading to excellent results in terms of expression yield and purity of the final product, it is likely that a bacterial expression system will not be the ideal solution for a large-scale production of recombinant hemoglobin (169). In fact, the low yield of pure proteins expressed in bacteria, even using highly optimized expression vectors, cannot meet the increasing needs of transfusion therapy. Furthermore, given that the administered dose of a blood substitute easily reaches tens of grams, the product has to be extremely pure, the risk associated with endotoxin-induced side effects being proportional to the amount of recombinant product administered.

Two alternatives for the production of pure and endotoxin-free hemoglobin are envisaged in transgenic animals and *in vitro* differentiation of embryonic cells to hematopoietic stem cells. For example, human hemoglobin can be expressed in transgenic mice (170) or in transgenic swine (171, 172), the latter being the more feasible way because of the higher yield of hemoglobin per animal. Recombinant hemoglobin purified from transgenic swine shows some heterogeneity attributed to the formation of hybrid molecules, as a consequence of contamination with swine hemoglobin. Human and swine hemoglobins exhibit high sequence, structure, and function homologies, making feasible the complete substitution of swine hemoglobin with the human protein to improve the homogeneity of the final product (173). Recently, it has been reported that human embryonic stem cells can differentiate, under certain growth conditions, to hematopoietic lineages that can, among others, generate erythroid cells that are identical to those naturally produced in human adult subjects (174). An attractive alternative strategy for Hb production is the preparation of transgenic tobacco containing human α and β genes (175).

3.3.4 Polymerized Hemoglobins.

Polymerization of hemoglobin was initially aimed at increasing the molecular weight of the molecule to prevent renal filtration. When scavenging of NO by hemoglobin was found to be one of the major causes of the frequently observed hypertensive effect associated with hemoglobin-based blood substitutes administration, hemoglobin polymerization was exploited for preventing extravasation of the protein and for reducing its vasoconstrictive effect.

Solutions of polymerized hemoglobin exhibit distinct physical properties and physiological and pharmacological effects. In fact, the oncotic pressure exerted by a solute through an impermeable membrane is proportional to the number of macromolecules in the solution, the higher being the number, the higher the exerted oncotic pressure. Capillary walls are impermeable to plasma proteins. The concentration of plasma proteins is about 7 g/dL and a solution with a concentration higher than 7 g/dL is referred to as hyperoncotic. The infusion of a hyperoncotic solution results in a withdrawal of fluids inside the capillaries followed by an increase in the arterial pressure. Even though cellular hemoglobin has a concentration of about 14 g/dL, it exerts no oncotic pressure because the protein is confined inside the erythrocytes. Hemoglobin polymerization was attempted to optimize the balance between an efficient oxygen transport and a physiological oncotic pressure. Therefore, it is possible to administer hemoglobin solutions with a concentration of hemoglobin up to 13 g/dL, which exert an oncotic pressure of about 25 mmHg, similar to that normally found in human plasma. Furthermore, polymerization increases the half-life of hemoglobin up to about 24 h.

Three types of polymerized hemoglobins have been tested on human subjects so far: Hemopure by Biopure Corp. (Cambridge, MA, USA), PolyHeme by Northfield Laboratories Inc. (Evanston, IL, USA), and Hemolink by Hemosol Inc. (Toronto, Canada). Both Hemopure and PolyHeme are glutaraldehyde-polymerized hemoglobins, whereas Hemolink is an o-raffinose-polymerized hemoglobin.

Glutaraldehyde crosslinks hemoglobin, both inter- and intramolecularly, predominantly through a Schiff base bond with ϵ amino groups of lysines. The Schiff base can be further reduced to obtain an aminic bond (Reaction 5). The polymerization conditions (re-

$$+ \text{Hb} - \text{NH}_2 \longrightarrow$$

$$(8.5)$$

$$\xrightarrow{\text{reduction}} \text{Hb} \underset{\text{H}}{\overset{}{\text{N}}} \diagdown \diagup \underset{\text{H}}{\overset{}{\text{N}}} \text{Hb}$$

action time, glutaraldehyde/protein ratio) can be controlled to obtain the desired molecular weight distribution.

According to recent studies (112, 176), glutaraldehyde reacts with many of the lysine residues located on the surface of hemoglobin and with the N-termini of both α- and β-chains. Polymerization results in a decreased oxygen affinity consequent to the stabilization of the deoxy form of hemoglobin. The constraints introduced by polymerization also lower the cooperativity of the transition from T to R. Furthermore, faster oxidation and autooxidation rates are observed with glutaraldehyde-polymerized hemoglobin. This finding is likely a consequence of the higher exposure of the heme pocket to the solvent, induced by polymerization.

3.3.4.1 Hemopure (HBOC-201). Hemopure is polymerized bovine hemoglobin formulated in modified Ringer's lactate at a concentration ranging from 12 to 14 g/dL. A Hemopure solution exhibits an oncotic pressure of 18 mmHg, an osmolarity of 300 mOsm/kg, a viscosity of 1.3 cP, and a p50 of 38 Torr (177). Hemopure is stable at room temperature for more than 1 year (178).

The purification procedure leads to a controlled, reproducible molecular weight distribution, composed of a main fraction with a molecular weight ranging between 128 and 500 kD and a minor fraction of octamers of about 128 kD, with no traces of dimeric or tetrameric contaminants (179).

Recently, the concern about a potential contamination of bovine blood products and hemoglobin with prions (see Ref. 180 for a review on this topic) has led to the optimization of the sterilization procedure that, according to manufacturer's reports, lowers the content of prions by 5 log units (177).

Biopure is shipping about 2000 units of the product at no cost to some hospitals in South Africa and the product is expected to be freely available on the South African market by the first half of 2002 (see Section 5).

3.3.4.1.1 Clinical Use of Hemopure. Hemopure has been approved in South Africa for the treatment of acute anemia and in surgery as an alternative to allogeneic transfusion. Other potential applications for Hemopure are as an enhancer of chemotherapy for solid hypoxic tumors, in sickle-cell anemia, in autologous blood donation settings, in orthopedic, cardiac, hepatic, vascular, and gastrointestinal surgery, and in trauma and ischemic events.

3.3.4.1.1.1 Adverse Effects. During phase I and II clinical trials no significant adverse effects were noticed. The coagulation profile did not change upon administration of 45 g of HBOC-201 (179, 181). In some cases, leukocytes and reticulocytes counts were higher in the treated groups than those in control groups (182). An increase of IgG antibodies to HBOC-201 was observed during phase II clinical trials at day 7 after administration. The increase in antibodies titer lasted about 1 week. According to the authors, this finding puts into question the feasibility and safety of repeated administrations of HBOC-201 (182). In some clinical trials mild gastrointestinal effects were reported but they did not require any treatment (178). Markers of liver, pancreatic, and renal function did not show any increase above physiological values in any clinical trial.

Hemodynamic effects of HBOC-201 administration vary in different clinical trial settings. A study on 13 abdominal aortic surgery patients demonstrated that oxygen delivery was decreased in the treatment group as a consequence of reduced cardiac output (183). This finding is in contrast with previously reported studies on animals (184). Oxygen delivery to tissues was calculated from cardiac output in human study, whereas in animal studies the oxygen tension in the skeletal muscle was measured directly by a microelectrode histographic technique. Decreased cardiac output can be a direct and positive consequence of an increased tissue oxygenation. Furthermore, other studies demonstrated that HBOC-201 has no negative effects on systemic blood pressure (177) or heart rate, both during anesthesia and in postoperation recovery. These effects, when observed, were only transient and related to only mild hemodynamic perturbations (177, 178).

3.3.4.1.1.2 Pharmacokinetics. Pharmacokinetic data were collected during phase I clinical trials on healthy subjects aimed at evaluating the physiology and pharmacokinetics of HBOC-201 (178) and at monitoring the hematological effects of HBOC-201 both on male and female subjects (179).

It was found that HBOC-201 metabolism is well described by a one-compartment model and does not depend on gender. The elimination is a first-order process, involving the reticuloendothelial system with no evidence of renal filtration. The volume of distribution is dose independent and roughly corresponds to that of plasma, ranging from 3 to 4 L. In contrast, both the half-life and clearance are dose dependent. The half-life for a dose of 45 g is about 20 h and the clearance is about 0.11 L/h. Upon 8 h from administration of HBOC-201, a peak in serum iron was observed, followed, upon 16 h, by an increase up to sixfold over baseline of erythropoietin and, upon 40 h, by a peak in ferritin levels.

3.3.4.1.2 Physiology and Pharmacology. Preclinical studies showed that HBOC-201 is capable of restoring normal levels of oxygen delivery to tissues after complete isovolemic hemodilution (185). The improved oxygenation potential of HBOC-201 with respect to RBC hemoglobin was assessed through direct measurements of skeletal muscle tissue oxygenation (184). In dogs, it was found that the dose of blood substitute required to restore oxygenation at baseline levels was about one-third that of RBC hemoglobin. The increase in muscle oxygenation was a consequence of a high extraction ratio that resulted in a steeper gradient between arterial and venous oxygen pressure. Enhanced oxygen delivery to tissues was also observed in clinical trials on healthy volunteers. HBOC-201 administration in a normovolemic hemodilution setting results in reduced lactate levels and reduced heart rate following exercise stress with respect to controls (186). Furthermore, the higher efficacy of HBOC-201 in restoring the exercise capacity in comparison with autologous transfusion was confirmed: 1 g of HBOC-201 has the same physiological effects as 3 g of RBC hemoglobin. HBOC-201 was effective in increasing the oxygen transport in sickle-cell anemia patients, thereby showing its potential application in sickle-cell anemia crisis (187).

HBOC-201 is also effective in stimulating erythropoiesis, both in men and women (179). In animal studies, HBOC-201 was proved to enhance tumor oxygenation and increase the efficacy of alkylating agents used in chemotherapy (188).

In 2000 Biopure announced preliminary results of a phase III clinical trial carried out on 693 elective surgery patients in Europe, the United States, Canada, and South Africa (see Section 5). During the clinical trial, Hemopure, administered at doses up to 240 g, proved to be effective in reducing the number of patients who received allogeneic red blood cells and confirmed the good safety profile shown during phase I and II clinical trials. During the study more than 35% of patients receiving Hemopure did not need any allogeneic blood donation during the whole follow-up period of 45 days.

3.3.4.2 PolyHeme (Poly SFH-P). PolyHeme is a glutaraldehyde-polymerized pyridoxalated human hemoglobin formulated at a 10 g/dL concentration in physiological solution. The solution is isooncotic with plasma and exhibits a p50 of about 30 Torr (189). The solution is stable for more than 1 year. The steps involved in the production of PolyHeme are the purification of human hemoglobin, pyridoxalation, and subsequent polymerization with glutaraldehyde. After purification of the polymerized protein, the content of tetramers is drastically reduced to less than 1% (189). Differently from Hemopure, PolyHeme is produced from human hemoglobin that has a lower p50 value than that of bovine hemoglobin, but, according to the manufacturer, it is preferable to avoid possible immune responses and transmission of viruses. The pyridoxalation is intended to improve the oxygen-delivery capacity of the product to a physiologically suitable value.

3.3.4.2.1 Pyridoxal Phosphate as a 2,3-DPG Analog. In 1969 pyridoxal phosphate (PLP) (**8**) was proved to bind in the 2,3-DPG binding pocket of hemoglobin and to lower oxygen affinity of stripped hemoglobin to the same extent as that of its natural effector (76). PLP binds preferentially to deoxyhemoglobin through a Schiff base linkage with the N-terminal amino groups of the β-chains in a ratio of one molecule of PLP for one β-chain (190). The Schiff base linkage can be reduced by $NaBH_4$ to obtain a stable pyridoxalated hemoglobin. The preferential binding to the deoxy form of hemoglobin and the covalent nature of the aminic bond between PLP and the protein result in a permanent low oxygen affinity, consequent to the stabilization of the T state (190).

(8)

3.3.4.2.2 Clinical Use of Poly SFH-P. PolyHeme has been tested during phase II clinical trials in acute blood loss and during phase III clinical trials in elective surgery to avoid or decrease the amount of transfused allogeneic blood. Recently, a study of PolyHeme in hemorrhagic shock patients indicated that the product could be effective in lowering the risk of multiple organ failure (191). This is one of the major causes of late postinjury death and seems to be related to the presence of biologic mediators, such as phospholipids and cytokines in stored blood. Administration of PolyHeme appeared to be associated with a decrease of deaths caused by multiple organ failure. However, because of the small number of patients studied, no general and definitive conclusion can be drawn on the efficacy of PolyHeme in lowering the risk of multiple organ failure.

3.3.4.2.2.1 Adverse Effects. No adverse effects were reported during phase I and II clinical trials (189, 192). Heart rate and MAP did not increase significantly after infusion and they remained in physiological ranges up to 3 days after administration. There was no evidence of organ toxicity, as demonstrated by physiological values of creatinine and liver enzymes. PolyHeme administration did not affect temperature. A peak in bilirubin concentration was reported upon 3 days after administration. This finding was interpreted as a consequence of hemoglobin metabolism and not as a sign of liver damage (192).

3.3.4.2.2.2 Pharmacokinetics. The half-life, reported for PolyHeme clearance, is about 24 h (189).

3.3.4.2.3 Pharmacology. A higher hemoglobin concentration can increase the oxygen-carrying capacity of plasma, thereby resulting in a better tissue oxygenation. Unfortunately, this result is achieved only at the expense of an increase in oncotic pressure and thus in plasma volume and MAP. PolyHeme shows the favorable effects of high hemoglobin concentrations up to 10 g/dL, and a physiological oncotic pressure resulting from a controlled polymerization procedure. Clinical trials in acute blood-loss settings demonstrated that PolyHeme is at least as effective as red blood cell hemoglobin in oxygen delivery, and that oxygen extraction ratio from PolyHeme is as high as that from red blood cells (192), because of the increased p50, following the polymerization and pyridoxalation. Furthermore, PolyHeme seems to be effective in reducing the number of allogeneic blood units transfused during emergency following traumatic or acute blood losses (189, 192).

3.3.4.3 Hemolink. Hemolink is the manufacturer's name for *o*-raffinose-crosslinked human hemoglobin. Presently, phase III clinical trials in coronary artery bypass grafting are ongoing for 299 patients (193). Available data refer to phase I and II clinical trials. The Marketing Authorization Application for the product was submitted and was awaiting the review by the UK Health Authority, whereas Health Canada's Therapeutic Products Program was reviewing the product at the end of the year 2000.

Hemolink is produced by treatment of human hemoglobin from outdated blood with *o*-raffinose (194) (Fig. 8.7), followed by elimination of nonpolymerized, dimeric hemoglobin. Hemoglobin is carefully purified by self-displacement chromatography to eliminate contaminants that can have toxic effects (membrane debris, endotoxins). After polymerization, the unpolymerized, dimeric molecules are separated from the final product through ultrafiltration. The distribution of molecular weights (193) is as follows: <5% dimeric hemoglobin; 34–43% tetrameric hemoglobin; 54–62% polymerized hemoglobin up to 500 kD; <3% 500 kD polymerized hemoglobin. The product is formulated in Ringer's lactate solution at a concentration of 10 g/dL.

The formulation exhibits an oncotic pressure of 24 mmHg, a viscosity of 1 cP, and a p50 of 34 Torr (195).

3.3.4.3.1 Clinical Use of Hemolink. Hemolink has potential clinical applications in surgery as an alternative to allogeneic blood donation and to replace patient's blood after harvesting in an autologous blood donation program.

3.3.4.3.1.1 Adverse Effects. Phase I clinical trials were intended to assess the safety of the product and involved 42 subjects to whom up to 600 mg/kg of Hemolink was administered in a top-loading experimental model without causing serious adverse effects (194–196). MAP was not significantly affected by the administration of Hemolink, although a 10% increase was observed in phase I clinical trials. In phase II clinical trials, 13 out of 30 patients (193) exhibited elevated pressure, but, according to investigators' opinion, this was not significantly different from that of the control group. In phase I studies this effect was accompanied by heart rate decrease, a symptom reported for many other hemoglobin-based blood substitutes (194).

Both phase I and phase II studies showed neither renal nor respiratory adverse effects, nor negative interactions with coagulation or other hematological parameters. Increased levels of some liver enzymes, particularly aspartate aminotransferase and amylase, did not appear to be linked to liver or pancreas damage and were not associated with symptoms of acute pancreatitis (193).

The most severe effects were associated with the action of Hemolink on the smooth muscles cells in the GI tract (194) (see Section 3.3.1.3 for details). Some cases of jaundice were reported for patients in phase II clinical trials. The first effect is of mild intensity and probably the result of NO scavenging by hemoglobin in the GI tract, in particular from nonadrenergic, noncholinergic inhibitory neurons (194, 197). Jaundice and the peak in the bilirubin concentration observed in phase II clinical trials are a consequence of hemoglobin metabolism. A peak in the lactate concentration that did not coincide with metabolic acidosis is considered a normal consequence of drug administration and not a symptom of tissue hypoxia (193).

The conclusion of clinical trials I and II is that Hemolink is a safe and efficacious product useful during autologous blood donation programs in decreasing the amount of donors' blood administered in cardiac surgery (193).

3.3.4.3.1.2 Pharmacokinetics. Pharmacokinetic data were reported for phase I clinical trials (194, 195, 198). Hemolink is primarily metabolized by the spleen and liver (194), as assessed by preclinical studies on rats. The decay of Hemolink in the plasma of human subjects is monoexponential, dose dependent, and associated with the molecular weight distribution of hemoglobin polymers in the drug. In humans $t_{1/2}$ is about 1.6 h for a 25 mg/kg dose and increases to a mean value of about 14.2 h for a 500 mg/kg dose (194, 195). This mean value is the result of two contributions: a $t_{1/2}$ of about 7.6 h for tetrameric hemoglobin and of about 18.7 h for polymeric hemoglobin. In the kidney, both low molecular weight products and small amounts of intact hemoglobin are present (194). The low molecular weight molecules are degradation products, whereas traces of intact hemoglobin are attributed to filtration of a small fraction of dissociable, noncrosslinked hemoglobin that is unlikely to cause severe renal damages.

3.3.4.3.2 Physiology and Pharmacology. In contrast with other blood substitutes tested so far, the administration of up to 4 g of Hemolink in rats did not affect the cardiac output and had only mild effects on systemic vascular resistance (199). It is not known to what extent the absence of effects on cardiac output is also true in humans, but it could be important in determining the final oxygen delivery to the tissues. Phase II studies demonstrated that Hemolink is capable of maintaining tissues oxygenation during coronary artery bypass grafting (193).

3.3.4.3.3 Structure-Function Relationships. *o*-Raffinose is a hexaldehyde obtained by oxidation of raffinose with sodium periodate. It forms a Schiff base with the exposed amino groups of the protein (200) (Fig. 8.7). Subsequent reduction of the Schiff base linkage further stabilizes the bond. The polymerization reaction is carried out on the deoxy form of hemoglobin to stabilize the T state and decrease the oxygen affinity of the final product (194). The crosslinking reaction takes place both in the 2,3-DPG pocket, forming a stable tetrameric hemoglobin with reduced oxygen affinity, and on the surface of the protein, forming a wide range of polymerized molecules from about 100 to about 500 kD with traces of higher molecular weight polymers.

Recent studies (199) demonstrated that the peculiar characteristics of Hemolink on systemic and renal hemodynamics are only marginally influenced by the elimination of low molecular weight molecules, but are likely a direct consequence of the o-raffinose decoration of the molecule that reduces NO access to the protein.

3.3.4.4 Recent Developments. Hemosol is presently involved in the development of some new hemoglobin-based blood substitutes with improved pharmacological properties with respect to first-generation modified hemoglobins. o-Raffinose-polymerized hemoglobin conjugated with Troxolon is among these new products. Troxolon (**9**) is a derivative of vitamin E and has strong antioxidant properties. The proposed clinical use is in hemorrhage to prevent reperfusion injury (201).

(9)

3.3.5 Conjugated Hemoglobins. Conjugated or surface-modified hemoglobins are tetrameric hemoglobins decorated with different macromolecules, such as dextran (78), PEG (79), and polyoxyethylene (POE) (80). Conjugation results in an increase in the molecular weight of the protein and in the stabilization of the tetramer, leading to longer intravascular retention times and lower immunogenic potential.

3.3.5.1 PEG-Hb. PEG is a polymer of general formula $H(OCH_2CH_2)_nOH$, where n is greater than or equal to 4. The greater the value of n, the higher the molecular weight of the polymer. PEG-Hb produced by Enzon is obtained by the reaction of succinimidyl PEG

with bovine hemoglobin (202). The product has a molecular weight of 120 kD and it is a stabilized tetramer without evidence of intra- or intermolecular crosslinking (128). PEG-Hb is formulated in 150 mM NaCl, 5 mM sodium phosphate, and 5 mM sodium bicarbonate at a concentration of 6 g/dL (202). The solution has an oncotic pressure of 118 mmHg and a viscosity of about 3 cP. The p50 of PEG-Hb ranges between 9 and 16 Torr (203). Phase Ib clinical trials have been halted and data reported in the following sections mainly refer to preclinical studies and phase Ia clinical trials.

3.3.5.1.1 Clinical Use of PEG-Hb. After phase Ia clinical trials on healthy volunteers, the product was administered to cancer patients during phase Ib clinical trials. In fact, solid hypoxic tumors respond better to radio- and chemotherapy if the malignant cells are well oxygenated (38).

3.3.5.1.1.1 Adverse Effects. PEG-Hb did not show any significant adverse effects both during preclinical trials on animals (202) and during phase I clinical trials (203). In animals, a vacuolization in liver, kidney, and bone marrow cells was reported. It is likely correlated to phagocytosis and subsequent degradation of PEG-Hb in the reticuloendothelial system (204) and seems to be transient without affecting the safety profile of the product (205). The only reported side effect of PEG-Hb administration in phase I clinical trials is gastrointestinal discomfort associated with the highest doses administered. Some investigators proposed that gastrointestinal pain might be a consequence of PEG-Hb extravasation, followed by a transient inflammatory response of the intestine (206).

3.3.5.1.1.2 Pharmacokinetics. Little is known about the pharmacokinetics of PEG-Hb. Its degradation might take place in the reticuloendothelial system (204). As a consequence of polyethylene glycol decoration, PEG-Hb has one of the highest vascular retention times among hemoglobin-based blood substitutes. In humans, the highest administered dose of 38 g exhibits a half-life of about 48 h (203). The slow elimination of PEG-Hb was pointed out to be compatible with a weekly administration in cancer patients to improve hypoxic tumor oxygenation (133).

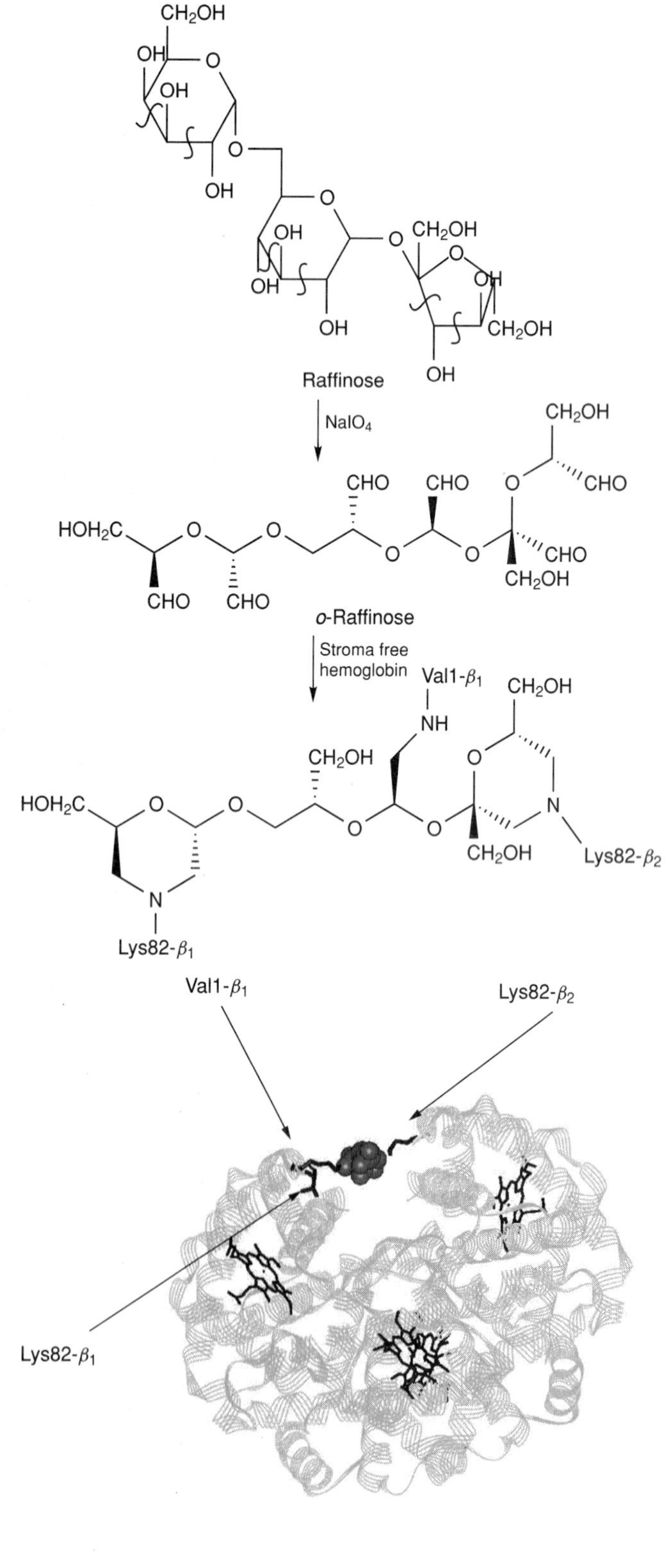

Figure 8.7. Preparation of o-raffinose hemoglobin. Aldehydic groups on o-raffinose react with amino groups of hemoglobin both in the 2,3-DPG pocket and on the protein surface (not shown). The three-dimensional structure of human hemoglobin in the presence of 2,3-DPG (PDB code 1B86) shows the relative positions of β_1Val1, β_1Lys82, and β_2Lys82, shown in ball-and-stick mode, with respect to 2,3-DPG, shown in space-filling mode.

3.3.5.1.2 Physiology and Pharmacology.
PEG decoration, which leads to an increased molecular weight, an increased hydrodynamic volume, and an efficient masking of the protein, contributes to the good pharmacological profile of this product. In fact, differently from other modified hemoglobins, PEG-Hb has only mild vasoconstrictive effects and, thus, only mild consequences on MAP (207, 208), likely resulting from the lower tendency to extravasate or the physical hindrance to NO binding (128). PEG-Hb has proved to sustain physiological functions during exchange-transfusion experiments in rats (209). Furthermore, the high oncotic pressure and viscosity of PEG-Hb solutions make this blood substitute a good volume expander for the treatment of severe blood losses (210). Phase Ib clinical trials evaluated the efficacy of the product as an enhancer in radiotherapy of solid hypoxic tumors in humans (133).

3.3.5.1.3 Things to Come

3.3.5.1.3.1 Hemospan. Sangart Inc. is developing Hemospan, a blood substitute prepared from outdated human blood hemoglobin and maleimide polyethylene glycol. Hemospan has a higher oxygen affinity than that of other blood substitutes presently on clinical trials. It has a higher oncotic pressure, a higher viscosity, and better flow properties. These features are likely to improve the hemodynamic response to the blood substitute administration, decreasing the usually reported vasoconstrictive effect and enhancing tissue perfusion and oxygen extraction (211, 212). The product has been tested on animals, giving encouraging results.

3.3.5.1.3.2 SNO-PEG-Hb. The reduction of the vascular response to the administration of hemoglobin-based blood substitutes has recently been addressed with the preparation of a S-nitrosylated PEG-Hb (SNO-PEG-Hb). The NO group is introduced through reaction of PEG-Hb with nitrosoglutathione, leading to the formation of hemoglobin nitrosylated mainly at Cys93 in the β-chains, with the methemoglobin content kept at less than 10%. Experiments on rats demonstrated that SNO-PEG-Hb is effective in reducing the hypertension normally found upon administration of modified hemoglobins, both attributed to the release of NO and to PEG decoration of the protein. Even though the presence of reduced glutathione in the plasma might accelerate the release of NO, the efficacy of SNO-PEG-Hb proved to be sustained with time (208).

3.3.5.2 PHP (Pyridoxalated Hemoglobin-Polyoxyethylene Conjugate). Iwashita's group was the first to achieve the conjugation of human pyridoxalated hemoglobin with polyoxyethylene to obtain an oxygen carrier in the 1980s (213). Currently, PHP is produced by Apex and is being developed as an NO scavenger in the treatment of shock associated with systemic inflammatory response syndrome (95) and it is going to enter phase III clinical trials (132). The product is only briefly presented in this section, its main pharmaceutical application being shock treatment rather than oxygen delivery.

Differently from PEG-Hb, the polymer used to modify the surface of the protein is a bifunctional reagent (alpha-carboxymethyl, omega-carboxymethoxyl polyoxyethylene) and thus both conjugation and crosslinking reactions take place. The modified hemoglobin is mainly in the conjugated form, but a small amount of crosslinked species is present (214), accounting for about 10% of the total protein, as reported in early works (213). The conjugation reaction takes place on marginally purified hemoglobin solutions, which still contain superoxide dismutase and catalase, thus leading to a final product in which more than 90% of the catalase and superoxide dismutase activities found in the red blood cells is crosslinked to hemoglobin (132, 214). The above-mentioned features are likely responsible for the low toxicity and for the potential applications of PHP. Pyridoxal phosphate acts as a 2,3-DPG analog, lowering oxygen affinity (see Section 3.3.4.2.1).

3.3.5.2.1 Clinical Use of PHP. PHP is being administered in clinical trials to patients suffering from volume refractory shock associated with NO overproduction, but its application in the treatment of reperfusion injury and hemorrhagic shock is envisaged because of its reduced prooxidant properties (132). Phase I clinical trials showed a good safety profile with no evidence of organ toxicity or changes in hemodynamics (95). The safety of the product was also confirmed during phase

II clinical trials up to a dose of 180 g. The plasma concentration of PHP is dose dependent.

3.3.5.2.2 Physiology and Pharmacology. Phase I and II clinical trials on patients with shock secondary to sepsis demonstrated the efficacy of PHP in increasing or maintaining blood pressure with a concomitant decrease in heart rate. Furthermore, PHP is less susceptible to oxidation by hydrogen peroxide and to formation of ferrylhemoglobin (see Section 3.3.1.4), thereby having potential applications in ischemia and subsequent reperfusion injury (132). Notably, PHP has no hypertensive effects on healthy subjects. Differently from other commonly used drugs in septic shock, as, for example, NO synthase (NOS) inhibitors, PHP exerts its hypertensive effects only on subjects suffering from hypotension, having no effects on healthy subjects. This selectivity makes PHP a better choice than NOS inhibitors in the treatment of shock (95). In addition to its NO scavenging activity, PHP is an NO donor because of the formation of *S*-nitrosohemoglobin in a way very similar to that of hemoglobin A (132).

3.3.6 Recent Developments of Modified Hemoglobins.
Modified hemoglobin solutions can be considered oxygen carriers rather than real blood substitutes. In fact, their short half-life limits their use as a short-term resuscitative fluid in emergency and surgery. They cannot replace blood in many pathological situations, such as chronic anemia. Furthermore, hemoglobin solutions still show some adverse effects, hypertension, gastrointestinal discomfort, and potential oxidative damage, which would prevent their application in the clinical routine. Second-generation blood substitutes are being developed with the aim of reducing the pressor effect (Sections 3.3.3.1.4 and 3.3.5.1.3.2) and the oxidative potential of free hemoglobin (Section 3.3.1.4), and optimizing oxygen delivery to tissues through the modulation of the flow properties of the formulation (Section 3.3.5.1.3.1). Even though second-generation blood substitutes have better physiological and pharmacological properties, they still do not possess many of the complex functions of red blood cells and, for this reason, their therapeutic applications will be limited.

3.3.6.1 Encapsulated Hemoglobins.
Encapsulated hemoglobins are generally referred to as third-generation blood substitutes and are expected to have physical and biochemical properties very similar to those of hemoglobin inside red blood cells. Encapsulation of hemoglobin in artificial vesicles, liposome, or nanocapsule is a strategy to mimic red blood cells. These new delivery vehicles were developed to circumvent adverse effects related to administration of solutions of free hemoglobin. In addition, inclusion in artificial vesicles protects hemoglobin from oxidation and degradation, resulting in longer circulation persistence and milder untoward effects. Membranes are semipermeable, allowing free diffusion of gases and small molecules and preventing diffusion of large or charged molecules inside or outside. Hemoglobin is thus maintained at a high concentration inside the vesicles and the equilibrium between tetramers and dimers is shifted toward the former. Moreover, components essential for hemoglobin function, such as allosteric effectors and enzymes, can be included in the formulation (215).

The first study on encapsulated hemoglobin was carried out in 1957 by Chang (216). Artificial red blood cells were prepared using synthetic membranes and encapsulated hemoglobin was reported to have an oxygen dissociation curve similar to that of hemoglobin in red blood cells. Enzymes, such as catalase and carbonic anhydrase, were included in the artificial red blood cells and were shown to retain their activity. Different polymers, polysaccharides, lipids, and proteins have been used to modify the surface characteristics of the membrane and to reduce the removal from the bloodstream. Removal of sialic acid, one of the components of red blood cell membranes, was shown to reduce persistence in the circulation. However, because of their mean diameter that ranges between 1 and 5 μm, microcapsules are rapidly cleared from the bloodstream by the reticuloendothelial system and have too short a circulating time for clinical application (217).

Small lipid-membrane artificial red blood cells have been prepared since 1980 using lipid-membrane liposomes. In the first liposome formulation, vesicles had a mean diameter of about 0.2 μm and a slower clearance rate, but the uptake by RES was still rapid (218). Since then, different formulations of liposome-encapsulated hemoglobin have been developed (217). Phospholipids and sterols are the main components of phospholipids-based liposomes. Cholesterol is added to increase liposome stability and flexibility and to reduce the tendency to fusion of vesicles. Different types of phospholipids were tested to modify liposome surface characteristics and to reduce recognition and removal by the RES. As components of natural membrane, phospholipids are well tolerated but they are rapidly metabolized by lipase and phospholipase. In addition, they are easily subjected to oxidation and are unstable during storage (215). Thus, manufacture of phospholipid-based liposomes is expensive and requires a high degree of mastery. Liposome-encapsulated hemoglobin was tested in animals with minimal adverse effects after administration of small or moderate doses (219).

A different approach uses nonphospholipid amphiphiles to form liposomes. These molecules are produced from a large variety of inexpensive raw materials and a broader spectrum of chemical structures is available. These amphiphiles may be composed of ionic and nonionic polar headgroups, whereas the hydrophobic tails can be formed by many different structural features, such as long-chain fatty acids, acyl derivatives of fatty acids, long-chain alcohols, and amides. Many of these compounds proved to be toxic, but amphiphiles belonging to the alkyl polyoxyethylene ether nonionic surfactant series were found to be suitable for pharmaceutical application (215). Polyoxyethylene-2 cetyl ether was used in a new formulation of nonphospholipid liposome-encapsulated hemoglobin (220), but liposomes were found to aggregate or react with erythrocytes (215).

Hemoglobin-containing nanocapsules with a mean diameter between 80 and 200 nm have been recently produced using different biodegradable polymers (221, 222). Because polymers are stronger and more porous, smaller amounts of membrane material are necessary.

Hemoglobin content is generally similar to that present in red blood cells. The reported values of p50 and Bohr and Hill coefficients for encapsulated bovine hemoglobin were similar to those of free bovine hemoglobin in solution. Enzymes involved in oxygen radical scavenging can be included in nanocapsules. Because nanocapsules are permeable to small hydrophilic molecules, they allow free diffusion of glucose, required by the metHb-reductase system, or reducing agents. *In vitro* studies and preliminary tests in animal models showed promising results (223).

3.3.6.2 Albumin-Heme. Recently, it has been reported that albumin complexed with heme (224) or with a heme derivative (225, 226) is capable of reversibly binding oxygen. The protein has potential applications as a blood substitute, given its oxygen-carrying properties, and as a volume expander, given the high oncotic pressure exerted by its solutions.

3.4 Perfluorochemicals

3.4.1 Current Drugs. Presently, none of the perfluorocarbon-based oxygen carriers is available for clinical use in the United States. Perfluorochemical-based blood substitutes that have entered clinical trials are listed in Table 8.5.

3.4.1.1 First-Generation Fluorocarbon-Based Blood Substitutes. The first injectable perfluorochemical (PFC) emulsion to be produced commercially was Fluosol (Green Cross Corp., Osaka, Japan), which consisted of 14% (w/v) perfluorodecalin (**10**) and 6% (w/v) perfluoro-tri-n-propylamine (**11**), emulsified with Pluronic F-68 (poloxamer 188, a copolymer consisting of two polyoxyethylene chains flanking a polyoxypropylene segment), together with small amounts of egg yolk phospholipids and potassium oleate.

(10)

Table 8.5   List of Perfluorocarbon-Based Blood Substitutes

Trade Name	Generic Name (Structure)	Originator	Dose
Fluosol	Perfluorodecalin (**10**)	Green Cross Co.	5 g/kg[a]
Perftoran	Perfluorodecalin (**10**)	Perftoran Co.	5–30 mL/kg[b]
Oxyfluor	Perfluoro-1,8-dichlorooctane (**16**)	HemaGen/PFC	1.7 g/kg[c]
Oxygent	Perfluorooctyl bromide (**13**)	Alliance Pharmaceuticals Co.	0.9–2.7 g/kg[c]

[a]Maximum dose administered during clinical trials (233).
[b]According to data provided by the company (see Section 5).
[c]The reported dose refers to that used in clinical trials [Oxyfluor (88); Oxygent (239; 246)].

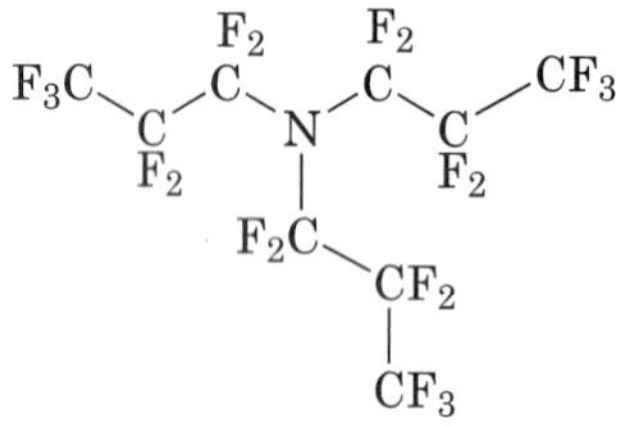

(11)

Other first-generation PFC emulsions are Emulsion No. II developed in China and Perftoran (previously named Ftorosan) from Russia. Emulsion No. II (Institute for Organic Chemistry, Shanghai, China) is a 20% (w/v) emulsion based on the same PFCs used in Fluosol, in which the surfactant is a poloxamer-type compound. It was clinically tested in surgical patients and war casualties. Little is known about the current clinical and commercial status. Perftoran (Perftoran Co., Pushchino, Russia) contains 14% (w/v) perfluorodecalin (**10**) and 6% (w/v) perfluoro-N-(4-methylcyclohexyl)piperidine (**12**) emulsified with the synthetic poloxamer Proxanol 268.

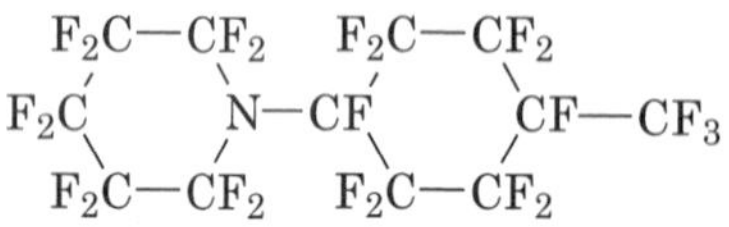

(12)

The Russian health regulatory authorities approved Perftoran in 1995–1996 as a temporary intravascular oxygen carrier in the treatment of traumatic hemorrhagic shock, microcirculatory disturbances, cardioplegia, cardiopulmonary bypass, and perfusion of isolated human organs (89). It is the only blood substitute currently available on the market but no other country, except Russia, has yet approved its use.

3.4.1.2 Second-Generation Perfluorochemicals. Oxygent (Alliance Pharmaceutical Corp., San Diego, CA, USA) is a 60% (w/v) emulsion [58% perfluorooctyl bromide (**13**) and 2% perfluorodecyl bromide (**14**)] stabilized with 3.6% (w/v) egg yolk phospholipids.

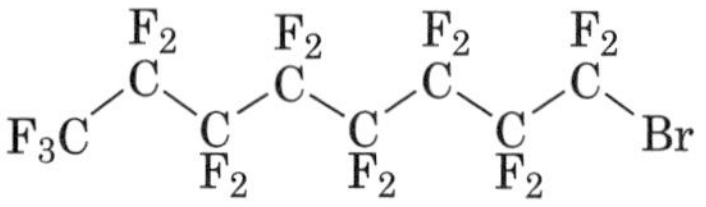

(13)

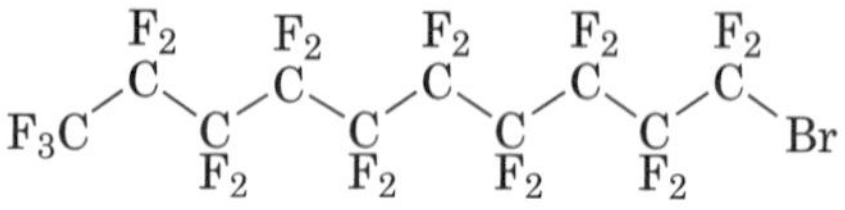

(14)

The average particle size of the emulsion is about 0.16–0.18 μm. It was evaluated in 18 clinical human studies involving more than 1200 patients. Phase II studies showed that Oxygent was more effective than a unit of blood in reversing transfusion triggers and in delaying the need for donor blood usage (89). An international, multicenter European phase III transfusion avoidance study in general surgery patients (492 enrolled subjects) was completed in May 2000 and demonstrated the efficacy of Oxygent in reducing or avoiding the need for allogeneic transfusion when used in conjunction with acute normovolemic hemodilution (227). A second multicenter phase III study in cardiac surgery patients is being conducted primarily in the United States.

Nonclinical studies are ongoing to evaluate the potential use of Oxygent in the treatment of severe decompression sickness (228), sickle-cell anemia (229, 230), and to reduce inflammatory response during extra corporeal circulation (231).

A European research team has recently developed a series of perfluorodecalin (10)-based emulsions, under the provisional commercial name of Fluxon (232). These emulsions use lecithin as surfactant and in some of them 1% (w/v) of perfluoro-1,3-dimorpholinopropane (15) was added.

$$
\begin{array}{ccccc}
F_2C{-}CF_2 & & F_2 & F_2C{-}CF_2 \\
& N{-}C & \overset{C}{\underset{F_2}{}} & \overset{C}{\underset{F_2}{}}{-}N & \\
F_2C{-}CF_2 & & & F_2C{-}CF_2 \\
\end{array}
$$

(15)

At present, no clinical trials have been carried out on these formulations and their potential use as oxygen carriers during perfusion of animal organs is under evaluation.

Oxyfluor (HemaGen, St. Louis, MO, USA) consists of 76% (w/v) of perfluoro-1,8-dichlorooctane (16). Egg yolk phospholipids are added as stabilizer and safflower oil is used as surfactant.

$$
Cl{-}\overset{F_2}{\underset{F_2}{C}}{-}\overset{F_2}{\underset{F_2}{C}}{-}\overset{F_2}{\underset{F_2}{C}}{-}\overset{F_2}{\underset{F_2}{C}}{-}C{-}Cl
$$

(16)

Animal studies assessed its efficacy as a transient oxygen carrier. Phase I/II safety study in low-blood-loss surgical patients were completed. Oxyfluor was under study as an agent to remove microemboli during cardiopulmonary bypass, but the current status of clinical trials has not been reported (88).

3.4.2 General Side Effects. No toxicity or either carcinogenic or teratogenic effects have been established so far for any fluorocarbon (233). No changes in liver and pulmonary function, no immunologic or allergic reaction, and no vasoconstriction or hemodynamic variations were detected (89). In early formula-

tions, toxicity was ascribed to high perfluorochemical vapor pressure, large emulsion particle size, and inadequate osmolarity. The presence of by-products or impurities from the synthesis reaction was also suspected (234). Progress in selective synthesis, purification, and methods of emulsification has led to higher quality and purity product with a more favorable tolerability profile.

Because they are immiscible in water, fluorocarbons need to be in an emulsified form to be injected into the bloodstream. Surfactants or emulsifiers must be added to facilitate the formation of the emulsion. Surfactants and emulsifiers play a key role in determining the characteristics and bioacceptability of the emulsion (235). Many side effects displayed by Fluosol were attributed to the surfactant. Pluronic F-68 was reported to elicit anaphylactic reactions resulting from complement activation and to alter human leucocyte function, causing a depression of host resistance to infection (234). Poor purification, a low cloud point that limited heat sterilization, and the polydisperse nature of Pluronic F-68 might account for these adverse effects. In fact, attenuation or absence of side effects shown by Fluosol were reported when purer poloxamers or other surfactants or emulsifiers were introduced. Egg yolk phospholipids, the emulsifier used in second-generation PFC-based products, do not cause complement activation and are well accepted in medical practice (235).

Extensive toxicology studies demonstrated that PFCs emulsions are well tolerated with no serious adverse events at clinically relevant doses (233). Transient side effects were reported in clinical studies following infusion of Oxygent and Oxyfluor (88). They consisted of an immediate response, characterized by facial flushing, headache, and lower back pain during or immediately after the infusion, and of a delayed increase in temperature, nausea, and chills. At higher doses, a dose-related thrombocytopenia often occurred 2–3 days after administration without any effect on platelet function and bleeding time. Increased platelet clearance was related to alteration of platelet surface caused by PFC emulsion (236). All patients showed spontaneous regression of the symptoms within hours or days.

Temporary splenomegaly and hepatomegaly without any evidence of inflammation and transient alterations in serum levels of liver enzymes were described in patients treated with PFCs emulsion (233). The duration and magnitude of these effects appear to be species specific and dose dependent. No permanent tissue alteration was observed after autopsy in animal studies. All changes were found to be fully reversible and thought to be a consequence of the physiological response to the presence of emulsion particles in the bloodstream.

Phagocytosis of particulate materials results in stimulation of macrophages in the spleen and Kupffer cells in the liver and activation of the arachidonic acid cascade with release of prostaglandins and pyrogenically active cytokines. These effects can be prevented by prophylaxis with long-acting cyclooxygenase inhibitors or corticosteroids (89, 233).

The external appearance, mainly determined by the surfactant, together with the particle size and the particle size distribution of the emulsion droplets influence the recognition of PFC droplets by the reticuloendothelial system (RES) cells and eventually the incidence and magnitude of adverse effects (89). New formulations with smaller particle size have proved to produce milder or no side effects, being less detectable by the cells of the RES (237).

In some studies in animals, pulmonary toxicity was associated with the administration of PFCs emulsions (89, 233). In such cases, lungs were reported not to deflate normally. This phenomenon, referred to as increased pulmonary relative volume (IPRV), was ascribed to retention of air in the alveoli. Interactions between lung surfactant and perfluorochemicals are thought to favor air trapping and be responsible for IPRV. Lung toxicity was found to be species dependent and related to physiologic and morphologic characteristics. Insensitive species generally exhibit large airways dimensions, high transpulmonary pressure, and effective initial RES clearance mechanisms that reduce perfluorochemicals delivery to the lungs. Because human lungs possess all these determinants, IPRV does not seem to be operative in human. No evidence of serious pulmonary toxicity related to IPRV has ever been reported during PFC infusion in humans.

3.4.3 Pharmacokinetics. Once injected into the bloodstream, PFC emulsion particles undergo opsonization (89, 233). They are recognized by phagocytic cells from the RES and removed from circulation. No fecal or urinary elimination was reported, whereas immediate vaporization and exhalation through the lungs contribute in minor extent to initial clearance of PFCs from the bloodstream. In the RES the surfactant is degraded, whereas the PFCs are temporarily stored. No evidence of metabolism, enzymatic cleavage, oxidation, or free-radicals reactions has ever been reported for PFCs. Depending on their lipophilic character, these molecules slowly diffuse from RES cells back in the bloodstream, carried by plasma lipids, and finally reach the lungs where they are excreted as vapor in the expired air.

A four-compartment model representing emulsion of PFCs in blood, in RES tissues, in non-RES tissues, and PFCs dissolved in lipids was proposed to describe the pharmacokinetics of perfluorooctyl bromide emulsions (238). Being slightly soluble in lipids, redistribution from primary sites of storage to adipose tissues occurs during the first week postdosing (233). The removal from circulation operated by cells in the RES is responsible for the limited intravascular residence of these compounds. As previously described, emulsion characteristics were recognized to greatly influence the clearance process. Experiments carried out on rats (237) showed that a decrease in particle size and a narrow particle size distribution reduce the scavenging of PFCs droplets through phagocytosis and prolongs their presence in the bloodstream. Half-life in the bloodstream is further dependent on infused dose. Blood half-life of Fluosol 20% was reported to change from 7.5 to 22 h when the dose increased from 4 to 6 g/kg (111), whereas the circulating half-time of Oxygent was shown to vary from 6 h, when the administered dose is 1.2 g/kg, to 9 h for a 1.8 g/kg dose (239).

A progressive saturation and a consequent loss of efficiency of RES-mediated clearance capacity was suggested to be responsible for

Table 8.6 Perfluorochemical Compounds Used as Blood Substitutes

Chemical Name (Structure)	Oxygen Solubility (mL O_2/100 mL)	Molecular Weight (Da)	Chemical Formula	Organ Retention Time (days)	Reference
Perfluorodecalin (**10**)	40	462	$C_{10}F_{18}$	7	234
Perfluorodecyl bromide (**14**)	NA	599	$C_{10}F_{21}Br$	23	232, 111
Perfluoro-1,8-dichlorooctane (**16**)	43	470	$C_8F_{16}Cl_2$	8	89
Perfluoro-1,3-dimorpholinopropane (**15**)	43	610	$C_{11}F_{22}N_2O_2$	55	232
Perfluoro-N-(4-methylcyclohexyl)piperidine (**12**)	NA	595	$C_{12}F_{23}N$	60	232
Perfluorooctyl bromide (**13**)	50	499	$C_8F_{17}Br$	4	235
Perfluorotri-n-propylamine (**11**)	45	521	$C_9F_{21}N$	65	234

the increase in intravascular half-life and accumulation in non-RES tissues when increasing the loading (233).

The organ persistence increases with increasing molecular weight within a homologous series (240). On the basis of the dependency between excretion rate and molecular weight, a range of molecular weights suitable for PFC administration was established. The lower limit was set at 460 to ensure a vapor pressure high enough to prevent risk of lung emphysema, whereas the value of 520 was chosen as the upper limit to ensure acceptable organ retention times.

No influence of structural properties of PFCs on excretion rate was established, other than that associated with the change they cause in the molecular weight (240) (Table 8.6). Linear and cyclic compounds with similar molecular weight have comparable organ retention time. Addition of heteroatoms to the existing carbon chain increases body persistence because they increase the mass units. However, heteroatoms decrease organ retention when they replace CF_2 or CF groups and cause a loss of mass units. Excretion rates faster than expected on the sole basis of molecular weight effect are observed for compounds that have short hydrogenated fragments at the end of the chain or atoms such as bromine or chlorine on the terminal carbon. In fact, a lipophilic terminus is thought to favor the uptake by plasma lipids and to accelerate excretion (89).

3.4.4 Physiology and Pharmacology. Fluorocarbons are synthetic aromatic or aliphatic compounds in which hydrogen atoms are substituted by fluorine atoms (89). The fluorine-carbon bond is the strongest in organic chemistry. Moreover, fluorine, which is more electronegative than hydrogen, forms a protective electron coating that reduces reactivity of potential functional sites and prevents fluorocarbon molecules from chemical attack. As a result, PFCs are chemically and biologically inert under physiological conditions. Liquid perfluorochemicals are uniquely characterized by weak intermolecular van der Waals forces that facilitate the insertion of gas molecules and account for high dissolving capacity. Thus, the gas is physically dissolved in the PFCs. This property is not specific for oxygen but applies to any low polarity gas, with the solubility decreasing with the decrease in molecular volume of the solute (241).

Among clinically useful PFC compounds, carbon dioxide solubility is reported to vary between 140 and 230 mL per 100 mL, whereas oxygen solubility ranges from 30 to 50 mL per 100 mL, when exposed to a partial pressure of 760 mmHg at 25°C (241). Oxygen solubility values for some of the formulations previously described are reported: Perftoran dissolves 6–7 vol % of oxygen (data provided from the company; see section 5); Fluosol 20% (w/v) dissolves 5.7 vol % (235); Oxygent with 60% PFCs (w/v) dissolves 17 vol % (235). Oxyfluor con-

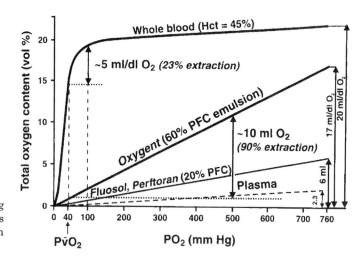

Figure 8.8. Oxygen binding curves of whole blood and various PFC emulsions (with permission from Ref. 89).

taining 76% (w/v) of PFCs dissolves 17.2 vol % of oxygen (242).

The oxygen-carrying capacity of PFCs is linearly related to oxygen partial pressure according to Henry's law (88, 89) (Fig. 8.8). For a given oxygen pressure, solubility increases with increasing concentration of perfluorochemicals. Therefore, the oxygen content of PFC emulsion can be increased by augmenting either the PFC content in the emulsion or the oxygen concentration in the inspired air. A higher partial pressure of oxygen in the inspired air results in higher arterial oxygen pressure and in a larger oxygen tension gradient between plasma and tissues, providing an effective driving force to oxygen diffusion. A higher PFC content increases emulsion viscosity and affects fluidity. Highly concentrated formulations with adequate viscosity for injection were obtained by improved mastery of emulsion technology. Under atmospheric oxygen pressure, even highly concentrated PFC emulsions display lower oxygen capacity than that of hemoglobin on a gram per gram basis, although the absence of chemical bonds between fluorocarbon molecules and oxygen and a large surface area for the gas exchange accelerate the loading and unloading processes. Thus, extraction ratios are higher for PFCs (90%) than for Hb (25–30%) and oxygen carried by PFCs is more readily available. As a consequence, when both PFCs and hemoglobin are present, oxygen is preferentially released from PFCs and oxygen bound to Hb is preserved.

The oxygen solubility of PFCs is not influenced by changes in temperature (88). Hence, at the low temperatures experienced during preservation of isolated organs and hypothermic cardiopulmonary bypass the oxygen-delivery capacity of PFCs does not change, whereas Hb displays an increase in oxygen affinity and lower oxygen release. Oxygen delivery by fluorocarbons is unaffected by changes of pH (234). Administration of PFCs can thus be useful in case of hemorrhagic shock to ensure tissue oxygenation without loss of efficiency resulting from acidosis, and to remove carbon dioxide, thus restoring physiological pH values.

With a mean diameter of about 0.1–0.3 μm, PFC emulsion particles are much smaller than red blood cells (7–8 μm). In larger arteries and arterioles red blood cells circulate in the central portion of the vessel and PFC droplets are assumed to flow in the plasma layer close to vessel wall (near-wall-excess phenomenon) (88, 89). As the vessels get narrower, the average hematocrit decreases and the space among erythrocytes increases, with droplets filling the plasma gaps. Hence, in the microcirculation PFC particles may perfuse even those narrow capillaries that exclude red blood cells, thus enhancing tissue oxygenation and preventing local ischemia resulting from inadequate perfusion during major blood loss or severe anemia. Moreover, treatment with PFCs during cardiopulmonary bypass may have a protective effect against neurological deficits caused by brain hypoxia, because PFCs act by both favoring gaseous microemboli resorption and improving oxygenation of ischemic areas.

In addition to oxygen transport, PFCs are thought to contribute to oxygen transfer between red blood cells and tissues. In fact, gas exchange between blood and tissues requires oxygen to diffuse through the so-called carrier-free-region (plasma, endothelium, and interstitium) (243). Under normal circumstances no specific oxygen carrier is present in this region and the diffusion rate is limited by oxygen solubility in plasma. Because fluorocarbon molecules distribute in the acellular fraction, they increase oxygen solubility in plasma and favor oxygen diffusion from red blood cells to tissues, acting as "stepping stones" (89).

3.4.4.1 Hemodynamics. Hemorrhage, surgical bleeding, anemia, or hemodilution reduce red blood cell mass. Consequently, blood viscosity and arterial oxygen content decrease. Thus, the stroke volume increases, leading to higher cardiac output and the total oxygen consumption is maintained. With more severe levels of anemia there is a critical point at which the compensatory mechanisms cease to be effective and oxygen consumption becomes supply dependent (88).

Infusion of PFCs was proposed to be able to provide adequate tissue oxygenation, normally achieved through blood transfusion. Administration of PFC-based oxygen carriers was seen to increase plasma dissolved oxygen content, tissue oxygenation, and mixed venous oxygen partial pressure. No negative effect on hemodynamics was observed. No change in microcirculation perfusion and vascular resistance was reported. PFCs do not interfere with compensatory cardiac output increase in response to hemodilution or anemia. Oxygen delivery benefits by the increase in cardiac output, given that blood circulates faster and reoxygenation of PFCs in the lungs occurs more frequently (88).

A new strategy to preserve patients' blood and to minimize the need for donor blood transfusion consists of infusion of PFCs in conjunction with acute normovolemic hemodilution. In Alliance's patented procedure, termed "Augmented-Acute Normovolemic Hemodilution A-ANH" (244, 245), blood is withdrawn to reach relatively low hemoglobin levels before surgery and it is successively reinfused during surgery or in the postoperative period. During the operation, solutions of volume expanders are infused to replace collected blood. When surgical bleeding further decreases hemoglobin concentration, patients are treated with a perfluorocarbon emulsion that allows preservation of tissue oxygenation. Preclinical experiments showed that PFC-treated animals can be subjected to more profound hemodilution. Acute normovolemic hemodilution supplemented by administration of a perfluorocarbon-based oxygen carrier and 100% oxygen ventilation was tested in Phase 3 clinical trials for Oxygent and was reported to be effective in reducing the need for blood transfusion (246).

3.4.5 Structure–Activity Relationship. The gas-dissolving capacity of PFCs was attributed to the existence of cavities within the solvent where the solute molecules can be hosted (240, 241). Both molecular weight and chemical structure of fluorocarbons influence oxygen-dissolving capacity (Table 8.6). Oxygen solubility decreases with increasing molecular weight within a homologous series. Linear compounds have higher oxygen-dissolving capacity than that of cyclic and aromatic compounds with similar molecular weight. This reduction in the amount of dissolved oxygen appears to correlate with the decrease in cavity size. In fact, linear structures are thought to create larger channels, whereas inclusion of solute molecules seems to be less favorable in planar structures that form closely interlocked layers. Similarly, the presence of a double bond in the central position confers higher dissolving capacity over that of saturated analogs or unsaturated compounds with double bond at the end of the chain because it seems to create a "notch" in the PFC skeleton and to facilitate cavity formation.

3.4.6 Emulsion Stability. Emulsions do not form spontaneously and may break down into separate phases, when subjected to increasing temperature, alteration in salt composition, interfacial film structure, or decrease in viscosity of the continuous phase that may occur during heat sterilization, handling, and storage. Coalescence is one of the phenomena responsible for emulsion separation in two phases (234, 240). The other major cause of coarsening of emulsion particles is molecular diffusion or Ostwald ripening. Because of the Kelvin effect, individual molecules of PFCs tend to leave the smaller droplets, diffuse through the external phase, and join the larger ones, with an increase in the average particle sizes. The droplet growth essentially depends on the particle size and on the type of PFC. Solubility and diffusibility of PFCs in the continuous phase increase with decreasing molecular weight. In addition, other factors such as a highly dispersed particle size distribution and an increasing temperature are thought to favor Ostwald ripening and contribute to destabilize the emulsion.

Given that emulsion particle size was recognized to play a major role in determining the persistence of PFCs in the bloodstream, causing adverse effects and affecting oxygen delivery efficiency, improvement of emulsion stability by hindering molecular diffusion has become a central issue in the development of the formulation of PFC-based oxygen carriers. Different strategies were adopted (111). Small amounts of high molecular weight PFCs, such as perfluorotri-n-propylamine and perfluoro-N-(4-methylcyclohexyl)piperidine, were added in Fluosol and Perftoran, respectively, to act as stabilizer and to reduce solubility in the aqueous phase. Unfortunately, while enhancing emulsion stability, high molecular weight prolongs PFCs body persistence. In Oxygent, this problem was circumvented by the use of perfluorodecyl bromide as additive PFC. Despite its higher molecular weight, it is rapidly excreted, given the lipophilic character conferred by the bromine atom. Improvement of stability may also be achieved using fluorinated surfactants. These surfactants have one or more F-alkyl hydrophobic chains and are expected to be better emulsifiers for PFC emulsions because they are more surface active than their hydrogenated analogs. Thus, they reduce the interfacial tension between water and PFCs to lower values. Several molecules have been synthesized and proved to be effective, although their pharmacology and toxicity still have to be clearly evaluated (111, 247). Alternatively, fluorocarbon-hydrocarbon

diblock compounds of the $C_nF_{2n+1}C_mH_{2m+1}$ type can be added to phospholipids. The presence of a fluorinated portion and of a hydrogenated segment reinforce the adhesion between phospholipids and PFCs droplets (89, 111, 247) because they act as a "dowel" at the interface.

3.4.7 Recent Developments. Great effort has been made to solve the stability issue of classical fluorocarbon-based emulsions. As previously described, different strategies have been adopted to reduce emulsion degradation. Microemulsions were considered as alternatives to classical formulations because they are thermodynamically stable systems (111, 234, 235). Unlike macroemulsions, they are transparent and form spontaneously when a suitable surfactant is added to a mixture of water and fluorocarbons. Different strategies have been pursued in the attempt to develop microemulsions to be used as blood substitutes. In one case, fluorocarbons were the basis for microemulsions (248). Fluorinated surfactants were required because fluorinated and hydrogenated chains tend to segregate. Many new surfactants were synthesized and examined in pursuit of biocompatible compounds. Different combinations of fluorinated surfactants and fluorocarbons have been tested to elucidate physicochemical and biological properties of these ternary mixtures. Microemulsions of water, fluorocarbon, and perfloroalkylpoly(oxyethylene) surfactants have been reported to dissolve oxygen but their toxicity profile has not been yet clearly established (234).

Recently, a different approach has been followed to circumvent the difficulty of finding a nontoxic fluorinated surfactant. Microemulsions were produced by mixing partially fluorinated compounds with hydrogenated nonionic surfactants that were known to be biocompatible (248). The fluorinated portion conferred to them an oxygen-dissolving capacity, whereas the hydrogenated part enabled them to be microemulsified. Microemulsions containing water, Montanox 80, a widely used, low-toxicity surfactant of the polysorbate family, and $C_8F_{17}CH_2CH{=}CHCHC_4H_9$ were found to dissolve oxygen. The measured amount of dissolved oxygen was higher than the theoretical one, suggesting that micellar organization might enhance oxygen-carrying capacity. The biocompatibility of these disperse systems has still to be completely assessed but preliminary toxicity tests are encouraging and show promise for the development of microemulsions as blood substitutes.

4 PLASMA VOLUME EXPANDERS

4.1 Clinical Use

Plasma volume expanders are infused to replace blood losses and maintain intravascular volume and tissue perfusion. Plasma volume is regulated very finely through several complex mechanisms. Reduction in normal circulating blood volume may occur as a consequence of hemorrhagic trauma, surgical procedures, or blood-salvage strategies, such as acute normovolemic hemodilution. In these situations, changes in blood volume cannot be immediately compensated for by the regulatory mechanisms. Therefore, volume therapy is aimed at temporarily increasing plasma volume until those mechanisms can correct the hypovolemia. Different products are available for the therapy based either on crystalloids or colloids. The choice of the appropriate agent in the treatment of hypovolemia has not yet been univocally settled. Plasma expanders have been evaluated in conjunction with perfluorochemicals and hemoglobin-based oxygen carriers to reduce the need for blood transfusion and ensure tissue oxygenation.

4.1.1 Current Drugs on the Market. Two families of products have been used for this purpose: crystalloids and colloids. Crystalloids consist of electrolytes (e.g., sodium, chloride, and potassium) in water solutions. The most frequently used crystalloids are normal saline and Ringer's solutions. Different commercial formulations are available, varying in electrolyte concentration, tonicity, and osmolarity. The class of colloids includes albumin, starches, dextrans, and gelatins (Table 8.7). Albumin is available as 5% and 25% solutions in isotonic saline. A purified protein fraction is also commercially available and consists of at least 83% albumin, with the remaining frac-

tion being other globulins. All protein solutions are extracted from human donor plasma and heat-treated to eliminate the risk associated with viral or bacterial contamination.

Hydroxyethyl starch (HES) is a synthetic polymer produced by hydroxyethyl substitution at the C2, C3, and C6 positions on the glucose residues of amylopectin polymers. Different degrees of substitution can be obtained by varying the reaction time with ethylene oxide, whereas the duration of acid hydrolysis affects the molecular weight distribution. The solution results in a heterogeneous mixture of particles with different molecular weights and degrees of hydroxylation. Its weight-average molecular weight is about 450,000 Da. It is clinically available as a 6% solution in normal saline. Many different formulations of HES with lower molecular weights and degrees of substitution are under development. Among these, pentastarch (weight-average molecular weight, 264,000 Da) has been used as plasma expander in Canada and some European countries. In the United States, this product is approved only for use in leukapheresis. It is available as a 10% solution.

Dextrans are glucose polymers produced by the bacterium *Leuconostoc mesenteroides* when grown on a sucrose media. The most common preparations for clinical use are dextran 40 (weight-average molecular weight, 40,000 Da) and dextran 70 (weight-average molecular weight, 70,000 Da). They are produced by acid hydrolysis of the parent molecules and are polydisperse. Dextran 40 is available as a 10% solution and dextran 70 is formulated as a 6% solution.

Gelatins used as plasma expanders are chemically modified derivatives of bovine collagen. The thermally degraded and chemically hydrolyzed collagen can be crosslinked with hexamethyl di-isocyanate to form urea-bridged gelatin or can be treated with succinic anhydride to form succinylated gelatins. The former is reported to have a weight-average molecular weight of 35,000 Da and it is available as a 3.5% solution. The weight-average molecular weight of the latter is 30,000 Da and it is formulated as a 4% solution. They are commonly available in Europe and Australia, but not at present in the United States.

Table 8.7   List of Plasma Volume Expanders

Generic Name	Chemical Class	Trade Name	Originator	Dose[a]
Albumin	Protein	Buminate	Baxter Healthcare Co.	NA
		Albuminar	Aventis Behring	
		Albutein	Alpha Therapeutics Co.	
Dextran 40	Modified polysaccharides	Gentran 40	Baxter Healthcare Co.	2 g/kg/day
		LMD	Abbott	
Dextran 70	Modified polysaccharides	Gentran 70	Baxter Healthcare Co.	20 mL/kg/day
Hydroxyethylstarch	Modified polysaccharides	Hespan	B. Braun/McGaw	1500 mL/day
		Hetastarch	Gensia Sicor/ Baxter	
		Hextend	BioTime Inc./ Abbott	
Urea-bridged gelatin	Modified bovine collagen	Haemaccel	Aventis	NA
Succinylated gelatin	Modified bovine collagen	Gelafundin	B. Braun	NA

[a]The reported dose is the maximum recommended dose (251).

4.1.2 Side Effects. Overload is a general adverse effect of administration of plasma expanders and is independent of the infused fluid, given that it is generally the result of inadequate monitoring of the circulatory system. Hypervolemia caused by colloids is more difficult to treat than is crystalloid overload (249). Crystalloids are considered to be safe and devoid of toxic or allergic reactions (250). The principal complications are progression of shock and renal failure when inadequate volumes are infused, and tissue and pulmonary edema consequent to excess volume infusion. When large volumes of crystalloids are infused, electrolyte abnormalities have to be considered. A mild alkalosis may occur with Ringer's lactate solution because of lactate being metabolized to bicarbonate. Administration of sodium chloride solution may cause hyperchloremia or hypernatremia. Adverse effects may occur in specific patients in relation to specific pathological status, such as hyperkalemia in patients with renal failure or lactate accumulation in patients with liver failure because of impairment in metabolism

of lactate to bicarbonate, both following massive infusion of Ringer's lactate solution.

Generally, hypertonic solutions do not cause edema but their higher serum osmolarity and sodium and chloride levels may cause metabolic acidosis and hypokalemia. Colloids are contraindicated for volume replacement in clinical situations where endothelial permeability is altered because colloids may leak into the interstitial space. They exert an osmotic gradient and withdraw water from the vascular space to the interstitium, causing tissue and pulmonary edema (251).

Allergic reactions were reported for the whole class of colloids, with different compounds varying in frequency of occurrence and severity (252). Incidence of allergic reactions for dextran 40 and dextran 70 is reported to be 0.007 and 0.069%, respectively (252), although according to other authors it ranges from 0.03 to 5% (250). Allergy to dextrans is thought to involve immune complex reactions attributed to preexisting antibodies against dextrans stimulated by exposure to these polysaccharides produced by bacteria or present in food products. Monovalent hapten-dextran was used to bind the antibody and reduce risk of allergy. Other factors are probably required for severe reactions to occur, the nature of which remains unclear.

Urticaria, fever, chills, or hypotension have been observed after infusion of albumin. Overall incidence of allergic reaction is 0.011% (252). Because the purified protein fraction contains greater quantities of kinins or prekallikrein activator, it carries a greater risk of hypotension than does human serum albumin (253). Allergic reactions to HES were reported to occur rarely, with an incidence of 0.085% (252). No evidence of any histamine-mediated process was found and the presence of anti-HES antibody was not correlated with allergy (249).

Gelatin was associated with allergic reactions with incidence estimated at 0.146% for urea-bridged gelatins and at 0.066% for modified fluid gelatine (252). Histamine release was associated with rapid infusion, probably because of a direct effect of gelatins on mast cells and basophils (249, 250). True anaphylactic reactions may also result from mediator activation induced by immune complex between gelatin and antigelatin antibody (254).

After infusion of albumin solution in trauma and burn-injured patients, different studies reported a decrease in glomerular filtration rates (GFR) and renal function and a lower urine output, despite a 40% increase of plasma volume (251). The mechanism of albumin-induced reduction in GFR is unknown, although it may be attributable to an increase in the oncotic pressure within the peritubular vessels that may cause a decrease in sodium and water excretion.

The effect of plasma expanders on renal function is still a debated issue. Development of acute renal failure was described after infusion of dextrans 40 (255), which might be because of an accumulation of low molecular weight fractions in the tubule. Consequently, highly viscous urine is produced and dextrans may precipitate, causing renal damage. Renal dysfunctions or clinical conditions that augment urine viscosity might worsen the situation (250). It was suggested that an accumulation in plasma of any unfilterable, osmotically active substance might raise the colloid os-

motic pressure to a level sufficient to offset the hydraulic filtration pressure and suppress filtration (256). The precise mechanism of renal toxicity is not known and it has not been clearly established whether all plasma expanders cause renal dysfunction.

All colloids have been reported to alter coagulation profiles primarily by hemodilution (250, 251). Blood coagulation is mostly impaired by dextran and HES. The effects of the latter was shown to depend on its molecular weight and rate of elimination. High molecular weight and highly substituted HES seem to have detrimental effects on coagulation, given that larger molecules are slowly degraded and persist longer in the plasma. It seem less likely that rapidly degradable HES induces coagulation disorders (257). Thrombocytopenia, prolongation in partial thromboplastin time, bleeding time, and prothrombin time were observed in patients treated with hetastarch. HES is reported to impair the coagulation system through a significant drop in factor VIII/von Willebrand factor complex, to reduce platelet adhesiveness, and to produce less stable thrombus because it accelerates the conversion of fibrin to fibrinogen. However, when the infused dose is lower than 20 mL kg^{-1} day^{-1}, coagulation parameters seem not to be significantly affected (251).

Both dextrans 40 and 70 affect coagulation (250). The high molecular weight dextrans exert more detrimental effects. They are reported to prolong bleeding time and to reduce platelet adhesiveness and aggregation resulting from a decrease of factor VIII. Thus, hemostasis abnormalities are similar to those seen in type I von Willebrand syndrome. In addition, they are reported to coat blood vessel wall and cellular elements and to impair polymerization of fibrin, elasticity, and tensile strength of fibrin clots. Clinically significant disturbance of hemostasis is unlikely to occur when the infused dose is less than 1.5 g kg^{-1} day^{-1} or 20 mL/day. No effect on hemostasis other than hemodilution was observed for gelatins. Hemostatic effects of albumin other than those related to hemodilution are still controversial.

Trace amounts of aluminum may contaminate albumin solutions and aluminum toxicity may represent a risk for premature infants and patients subjected to long-term infusion of this colloid (258). Furthermore, because albumin binds calcium and reduces the fraction of ionized calcium, hypocalcemia and myocardial depression may arise when albumin therapy is used for fluid resuscitation (259).

4.1.3 Pharmacokinetics. The pharmacokinetics profile of exogenous albumin follows that of the endogenous protein (250, 251). Its serum half-life has been reported to be 16 h. Albumin infused in the intravascular space gradually leaves the bloodstream and redistributes to the interstitial sites. About a third of albumin is in the intravascular space, whereas extravascular interstitial sites contain 60% of total albumin. The normal rate of transcapillary escape is 5%/h, but can be higher in many pathological states. Some albumin is tissue bound, whereas the free protein may reenter the intravascular space through lymphatic circulation to maintain serum levels and compensate for transcapillary leak and catabolism. The reticuloendothelial system plays a major role in the removal of albumin from the body and metabolism. Approximately 10% of albumin is degraded daily to amino acids and returns to the liver.

Pharmacokinetics of HES is complex because of its heterogeneity in size and substitution (250, 251, 260). Clearance rate and serum half-life are both affected by hydroxyethylation and molecular weight. The main route of elimination of hydroxyethyl starches is urinary. Smaller molecules are rapidly excreted in urine. Larger particles remain in the circulation longer, where they can be either partially hydrolyzed to small- or medium-sized molecules by serum α-amylase or slowly removed by the reticuloendothelial system, where they accumulate. The extent of enzymatic degradation depends on the degree of hydroxylation because the added alcohol groups protect from enzymatic cleavage. Metabolism of hydroxyethyl starches was reported to occur without formation of glucose or hydroxyethyl glucose (261). Because the dispersion of the molecular weight of the molecules changes over time, several phases of elimination with different corresponding half-lifes have been reported (260). Accumulation of HES within the tissues was reported (262).

Because polydisperse formulations of dextrans are composed of different molecular weight fractions, each of them exhibits its own pharmacokinetic profile (260). Smaller molecules undergo renal excretion or they may enter the interstitial space and return to the bloodstream through lymphatic drainage. Larger polymers persist longer in the bloodstream and are successively stored in hepatic, renal tubular, or reticuloendothelial system cells (250). Dextrans are slowly metabolized to carbon dioxide and water by dextranase localized in spleen, liver, lung, and kidney (260). The variety of routes of elimination and the influence of molecular weight on excretion profiles account for the complexity of pharmacokinetics of dextrans. Following intravenous administration of dextran 40, approximately 60% of the dose given is eliminated within 6 h, whereas administration of dextran 70 results in 35% of the dose cleared within 12 h (251). Similarly to other heterogeneous mixtures, low molecular weight molecules of gelatins are rapidly cleared by glomerular filtration and exhibit short half-lives (251). Small amounts move from the intravascular compartment into the interstitial space, returning to the bloodstream through the lymphatic circulation. Gelatins may also be cleaved by proteases into small peptides and amino acids. No accumulation in the body has been reported (260).

4.2 Physiology and Pharmacology

Body water is divided into compartments: intracellular water accounts for 70% of total body water, whereas the extracellular compartment contains 30% of total body water and includes the interstitial space (75% of the extracellular fluid) and the intravascular space (25%). Intracellular and extracellular fluids have the same osmolarity. Thus, when extracellular osmolarity changes, movement of water between these two sectors occurs. Isotonic solutions have the same osmolarity as that of body fluids and distribute between the intravascular and the interstitial spaces. Generally, the volume of isotonic solution required for fluid replacement has to be four to five times greater than the volume of blood loss (250). The effect is relatively short lasting, so that significant amounts of additional solution are required over time. Hence, the use of isotonic solutions results in the increase of the volume of interstitial fluid and causes a certain degree of edema formation, attributed to accumulation of fluids in the interstitial space that is not completely drained by the lymphatic circulation.

Crystalloids have been tested in different clinical studies and were reported to be effective in resuscitation, even if larger volumes were required and resuscitation times were longer compared to those of colloids. Hypertonic crystalloids increase the osmolarity of extracellular fluids, resulting in the movement of water from the intracellular space to the interstitial and vascular compartments. Thus, they expand the extracellular fluid volume by a greater amount than the volume infused and less fluid has to be administered than that using isotonic crystalloid solutions (250).

The concentration of proteins is much higher in plasma than that in the interstitial space because membranes between the two compartments allow for only limited passage. According to Starling's law, the factor that determines flows of fluid between the intravascular and interstitial spaces and regulates compartment volume is the difference between the oncotic pressure gradient generated by the asymmetric distribution of colloids on each side of the endothelium and the opposing hydrostatic pressure gradient. The primary protein in human plasma, albumin, comprises 75–80% of the normal colloid oncotic pressure, and about 50–60% of the protein content. Albumin is produced in the liver and its synthesis is regulated to keep colloid osmotic pressure constant. At physiological concentrations, albumin ensures an oncotic pressure of about 18 to 22 mmHg. Administration of colloids in the vascular compartment increases the oncotic pressure gradient and draws water from the interstitial space. Because osmotic pressure depends on the number of particles in solution, solutions containing larger amounts of colloids or particles with lower molecular weight exert a higher colloid osmotic pressure and result in a more effective volume restoring (251). Solutions isooncotic with plasma exert an initial volume expansion approximately equal to the volume infused, whereas hyperoncotic ones produce higher volume expansion. Thus, a 25% albumin solution has an oncotic pressure of 100 mmHg and determines an increase of 300–500 mL in intravascular volume for 100 mL of infused solution. A 5% solution has a colloid osmotic pressure of 19 mmHg and increases the intravascular volume by the same amount of volume infused. Reported oncotic pressures for a 6% hetastarch, 10% pentastarch, and 10% dextrans 40 are about 28, 40, and 30 mmHg, respectively (251).

In a polydisperse mixture, the low molecular weight fraction exerts a greater initial oncotic effect that tends to be short lasting because of the rapid renal elimination. Larger molecules generate a smaller colloid osmotic pressure that is sustained longer, given that they persist longer in the circulation (250, 251).

In addition, dextrans have been reported to alter the viscosity of whole blood. The effect depends on molecular weight of the polymers. Aggregation of red blood cells occurs when dextrans with molecular weight higher than 60,000 Da are administered. Below this value, dextrans tend to disaggregate them, thus improving peripheral flow properties especially under abnormal circulatory conditions (251).

5 WEB SITE ADDRESSES

- Alliance Pharmaceutical Corp. (San Diego, CA, USA): http://www.allp.com
- Allos Therapeutics (Denver, CO, USA): http://www.allos.com
- Apex Bioscience Inc. (Research Triangle Park, NC, USA): http://www.apexbioscience.com
- Baxter Healthcare Corporation (Deerfield, IL, USA): http://www.baxter.com
- Biopure® Corporation (Cambridge, MA, USA): http://www.biopure.com
- Enzon Inc. (Piscataway, NJ, USA): http://www.enzon.com
- Hemosol Inc. (Toronto, Canada): http://www.hemosol.com/home_page.html
- Northfield Laboratories Inc. (Evanston, IL, USA): http://www.northfieldlabs.com
- Perftoran (Pushchino, Russia): http://perftoran.ru/

- Sangart (San Diego, CA, USA): http://www.sangart.com/
- SynZyme Technologies LLC (Irvine, CA, USA): http://www.synzyme.com

REFERENCES

1. H. F. Bunn and B. G. Forget, *Hemoglobin: Molecular, Genetic and Clinical Aspects*, Saunders, Philadelphia, 1986.

2. E. Antonini and M. Brunori, *Hemoglobin and Myoglobin in Their Reaction with Ligands*, Amsterdam, 1971.

3. K. Imai, *Allosteric Effects in Hemoglobin*, North-Holland Publishing Co., Cambridge, UK, 1982.

4. J. Monod, J. Wyman, and J. P. Changeaux, *J. Mol. Biol.*, **6**, 306 (1965).

5. D. E. Koshland Jr., G. Nemethy, and D. Filmer, *Biochemistry*, **5**, 365–385 (1966).

6. M. F. Perutz, *Nature*, **228**, 726–739 (1970).

7. G. K. Ackers, M. L. Doyle, D. Myers, and M. A. Daugherty, *Science*, **255**, 54–63 (1992).

8. W. A. Eaton, E. R. Henry, J. Hofrichter, and A. Mozzarelli, *Nat. Struct. Biol.*, **6**, 351–358 (1999).

9. M. F. Perutz, A. J. Wilkinson, M. Paoli, and G. G. Dodson, *Annu. Rev. Biophys. Biomol. Struct.*, **27**, 1–34 (1998).

10. A. J. Lipton, M. A. Johnson, T. Macdonald, M. W. Lieberman, D. Gozal, and B. Gaston, *Nature*, **413**, 171–174 (2001).

11. J. C. Liao, T. W. Hein, M. W. Vaughn, K. T. Huang, and L. Kuo, *Proc. Natl. Acad. Sci. USA*, **96**, 8757–8761 (1999).

12. K. T. Huang, T. H. Han, D. R. Hyduke, M. W. Vaughn, H. Van Herle, T. W. Hein, C. Zhang, L. Kuo, and J. C. Liao, *Proc. Natl. Acad. Sci. USA*, **98**, 11771–11776 (2001).

13. L. Jia, C. Bonaventura, J. Bonaventura, and J. S. Stamler, *Nature*, **380**, 221–226 (1996).

14. J. R. Pawloski, D. T. Hess, and J. S. Stamler, *Nature*, **409**, 622–626 (2001).

15. A. J. Gow, B. P. Luchsinger, J. R. Pawloski, D. J. Singel, and J. S. Stamler, *Proc. Natl. Acad. Sci. USA*, **96**, 9027–9032 (1999).

16. J. Everse and N. Hsia, *Free Radic. Biol. Med.*, **22**, 1075–1099 (1997).

17. R. Benesch and R. E. Benesch, *Biochem. Biophys. Res. Commun.*, **26**, 162–167 (1967).

18. A. Chanutin and R. R. Curnish, *Arch. Biochem. Biophys.*, **121**, 96–102 (1967).

19. M. F. Perutz and C. Poyart, *Lancet*, **2**, 881–882 (1983).

20. C. Ropars, M. Chassaigne, and G. Avenard, *Med. Biol. Eng. Comput.*, **36**, 508–512 (1998).

21. D. J. Abraham, M. F. Perutz, and S. E. Phillips, *Proc. Natl. Acad. Sci. USA*, **80**, 324–328 (1983).

22. C. Poyart, M. C. Marden, and J. Kister, *Methods Enzymol.*, **232**, 496–513 (1994).

23. C. R. Beddell, P. J. Goodford, G. Kneen, R. D. White, S. Wilkinson, and R. Wootton, *Br. J. Pharmacol.*, **82**, 397–407 (1984).

24. D. J. Abraham, M. K. Safo, T. Boyiri, R. E. Danso-Danquah, J. Kister, and C. Poyart, *Biochemistry*, **34**, 15006–15020 (1995).

25. A. Arnone, *Nature*, **237**, 146–149 (1972).

26. A. Arnone and M. F. Perutz, *Nature*, **249**, 34–36 (1974).

27. M. K. Safo, C. M. Moure, J. C. Burnett, G. S. Joshi, and D. J. Abraham, *Protein Sci.*, **10**, 951–957 (2001).

28. D. J. Abraham, J. Kister, G. S. Joshi, M. C. Marden, and C. Poyart, *J. Mol. Biol.*, **248**, 845–855 (1995).

29. M. C. Marden, B. Bohn, J. Kister, and C. Poyart, *Biophys. J.*, **57**, 397–403 (1990).

30. I. Lalezari, S. Rahbar, P. Lalezari, G. Fermi, and M. F. Perutz, *Proc. Natl. Acad. Sci. USA*, **85**, 6117–6121 (1988).

31. I. Lalezari and P. Lalezari, *J. Med. Chem.*, **32**, 2352–2357 (1989).

32. I. Lalezari, P. Lalezari, C. Poyart, M. Marden, J. Kister, B. Bohn, G. Fermi, and M. F. Perutz, *Biochemistry*, **29**, 1515–1523 (1990).

33. R. S. Randad, M. A. Mahran, A. S. Mehanna, and D. J. Abraham, *J. Med. Chem.*, **34**, 752–757 (1991).

34. M. Phelps Grella, R. Danso-Danquah, M. K. Safo, G. S. Joshi, J. Kister, M. Marden, S. J. Hoffman, and D. J. Abraham, *J. Med. Chem.*, **43**, 4726–4737 (2000).

35. S. R. Khandelwal, R. S. Randad, P. S. Lin, H. Meng, R. N. Pittman, H. A. Kontos, S. C. Choi, D. J. Abraham, and R. Schmidt-Ullrich, *Am. J. Physiol. Heart Circ. Physiol.*, **265**, H1450–H1453 (1993).

36. M. P. Kunert, J. F. Liard, and D. J. Abraham, *Am. J. Physiol. Heart Circ. Physiol.*, **271**, H602–H613 (1996).

37. B. A. Teicher, J. S. Wong, H. Takeuchi, L. M. Gravelin, G. Ara, and D. Buxton, *Cancer Chemother. Pharmacol.*, **42**, 24–30 (1998).

38. E. J. Hall, *The Oxygen Effect and Reoxygenation: Radiobiology for the Radiologist,* 4th ed., Lippincott, Philadelphia, 1994.

39. T. G. Graeber, C. Osmanian, T. Jacks, D. E. Housman, C. J. Koch, S. W. Lowe, and A. J. Giaccia, *Nature*, **379**, 88–91 (1996).

40. R. A. Gatenby, H. B. Kessler, J. S. Rosenblum, L. R. Coia, P. J. Moldofsky, W. H. Hartz, and G. J. Broder, *Int. J. Radiat. Oncol. Biol. Phys.*, **14**, 831–838 (1988).

41. M. Nordsmark, M. Overgaard, and J. Overgaard, *Radiother. Oncol.*, **41**, 31–39 (1996).

42. M. Hockel, C. Knoop, K. Schlenger, B. Vorndran, E. Baussmann, M. Mitze, P. G. Knapstein, and P. Vaupel, *Radiother. Oncol.*, **26**, 45–50 (1993).

43. M. Hockel, C. Knoop, K. Schlenger, B. Vorndran, P. G. Knapstein, and P. Vaupel, *Adv. Exp. Med. Biol.*, **345**, 445–450 (1994).

44. L. Kleinberg, S. A. Grossman, S. Piantadosi, J. Pearlman, H. Engelhard, G. Lesser, J. Ruffer, and M. Gerber, *J. Clin. Oncol.*, **17**, 2593–2603 (1999).

45. E. P. Wei, R. S. Randad, J. E. Levasseur, D. J. Abraham, and H. A. Kontos, *Am. J. Physiol. Heart Circ. Physiol.*, **265**, H1439–H1443 (1993).

46. J. C. Watson, E. M. Doppenberg, M. R. Bullock, A. Zauner, M. R. Rice, D. Abraham, and H. F. Young, *Stroke*, **28**, 1624–1630 (1997).

47. H. P. Grocott, R. D. Bart, H. Sheng, Y. Miura, R. Steffen, R. D. Pearlstein, and D. S. Warner, *Stroke*, **29**, 1650–1655 (1998).

48. S. Sarraf-Yazdi, H. Sheng, H. P. Grocott, R. D. Bart, R. D. Pearlstein, R. P. Steffen, and D. S. Warner, *Brain Res.*, **826**, 172–180 (1999).

49. P. S. Pagel, D. A. Hettrick, M. W. Montgomery, J. R. Kersten, R. P. Steffen, and D. C. Warltier, *J. Pharmacol. Exp. Ther.*, **285**, 1–8 (1998).

50. K. S. Kilgore, C. F. Shwartz, M. A. Gallagher, R. P. Steffen, R. S. Mosca, and S. F. Bolling, *Circulation*, **100**, II351–II356 (1999).

51. R. G. Weiss, M. A. Mejia, D. A. Kass, A. F. DiPaula, L. C. Becker, G. Gerstenblith, and V. P. Chacko, *J. Clin. Invest.*, **103**, 739–746 (1999).

52. J. A. Wahr, M. Gerber, J. Venitz, and N. Baliga, *Anesth. Analg.*, **92**, 615–620 (2001).

53. A. Breidbach and D. H. Catlin, *Rapid Commun. Mass Spectrom.*, **15**, 2379–2382 (2001).

54. S. R. Khandelwal, B. D. Kavanagh, P. S. Lin, Q. T. Truong, J. Lu, D. J. Abraham, and R. K. Schmidt-Ullrich, *Br. J. Cancer*, **79**, 814–820 (1999).

55. O. Eichelbronner, A. Sielenkamper, M. D'Almeida, C. G. Ellis, W. J. Sibbald, and I. H. Chin-Yee, *Am. J. Physiol. Heart Circ. Physiol.*, **277**, H290–H298 (1999).

56. H. B. Demopoulos, E. S. Flamm, D. D. Pietronigro, and M. L. Seligman, *Acta Physiol. Scand. Suppl.*, **492**, 91–119 (1980).

57. E. M. Doppenberg, M. R. Rice, B. Alessandri, Y. Qian, X. Di, and R. Bullock, *J. Neurotrauma*, **16**, 123–133 (1999).

58. T. J. Burke, D. Malhotra, and J. I. Shapiro, *Kidney Int.*, **60**, 1407–1414 (2001).

59. B. D. Kavanagh, S. R. Khandelwal, R. K. Schmidt-Ullrich, J. D. Roberts, E. G. Shaw, A. D. Pearlman, J. Venitz, K. E. Dusenbery, D. J. Abraham, and M. J. Gerber, *Int. J. Radiat. Oncol. Biol. Phys.*, **49**, 1133–1139 (2001).

60. D. J. Abraham, A. S. Mehanna, F. C. Wireko, J. Whitney, R. P. Thomas, and E. P. Orringer, *Blood*, **77**, 1334–1341 (1991).

61. F. C. Wireko, G. E. Kellogg, and D. J. Abraham, *J. Med. Chem.*, **34**, 758–767 (1991).

62. T. Boyiri, M. K. Safo, R. E. Danso-Danquah, J. Kister, C. Poyart, and D. J. Abraham, *Biochemistry*, **34**, 15021–15036 (1995).

63. V. Stark, *Dtsch. Med. Wochenschr.*, **24**, 805–808 (1898).

64. A. W. Sellards and G. R. Minot, *J. Med. Res.*, **34**, 469–494 (1916).

65. W. R. Amberson, *Blood Substit. Biol. Rev.*, **12**, 48–52 (1937).

66. S. F. Rabiner, K. O'Brien, G. W. Peskin, and L. H. Friedman, *Ann. Surg.*, **171**, 615–622 (1970).

67. G. S. Moss, R. DeWoskin, A. L. Rosen, H. Levine, and C. K. Palani, *Surg. Gynecol. Obstet.*, **142**, 357–362 (1976).

68. F. De Venuto, T. F. Zuck, A. I. Zegna, and W. Y. Moores, *J. Lab. Clin. Med.*, **89**, 509–516 (1977).

69. J. P. Savitsky, J. Doczi, J. Black, and J. D. Arnold, *Clin. Pharmacol. Ther.*, **23**, 73–80 (1978).

70. T. M. S. Chang, *Science*, **146**, 524 (1964).

71. H. F. Bunn and J. H. Jandl, *Trans. Assoc. Am. Physicians*, **81**, 147–152 (1968).

72. I. M. Klotz and J. W. Tam, *Proc. Natl. Acad. Sci. USA*, **70**, 1313–1315 (1973).

73. K. R. Bridges, G. J. Schmidt, M. Jensen, A. Cerami, and H. F. Bunn, *J. Clin. Invest.*, **56**, 201–207 (1975).

74. J. A. Walder, R. H. Zaugg, R. Y. Walder, J. M. Steele, and I. M. Klotz, *Biochemistry*, **18**, 4265–4270 (1979).

75. R. Chatterjee, E. V. Welty, R. Y. Walder, S. L. Pruitt, P. H. Rogers, A. Arnone, and J. A. Walder, *J. Biol. Chem.*, **261**, 9929–9937 (1986).

76. R. E. Benesch, R. Benesch, and C. I. Yu, *Fed. Proc.*, **28**, 604 (1969).

77. T. M. S. Chang, *Biochem. Biophys. Res. Commun.*, **44**, 1531–1536 (1971).

78. S. C. Tam, J. Blumenstein, and J. T. Wong, *Proc. Natl. Acad. Sci. USA*, **73**, 2128–2131 (1976).

79. K. Iwasaki and Y. Iwashita, *Artif. Organs*, **10**, 411–416 (1986).

80. M. Matsushita, A. Yabuki, P. S. Malchesky, H. Harasaki, and Y. Nose, *Biomater. Artif. Cells Artif. Organs*, **16**, 247–260 (1988).

81. K. Nagai and H. C. Thogersen, *Nature*, **309**, 810–812 (1984).

82. B. A. Springer and S. G. Sligar, *Proc. Natl. Acad. Sci. USA*, **84**, 8961–8965 (1987).

83. S. J. Hoffman, D. L. Looker, J. M. Roehrich, P. E. Cozart, S. L. Durfee, J. L. Tedesco, and G. L. Stetler, *Proc. Natl. Acad. Sci. USA*, **87**, 8521–8525 (1990).

84. F. D'Agnillo and T. M. S. Chang, *Biomater. Artif. Cells Immobil. Biotechnol.*, **21**, 609–621 (1993).

85. L. C. Clark and F. Gollan, *Science*, **152**, 1755–1756 (1966).

86. H. A. Sloviter and T. Kamimoto, *Nature*, **216**, 458–460 (1967).

87. R. P. Geyer, *N. Engl. J. Med.*, **289**, 1077–1082 (1973).

88. P. E. Keipert, in T. M. S. Chang, Ed., *Blood Substitutes: Principles, Methods, Products and Clinical Trials*, Vol. **II**, Karger Landes Systems, Basel, Switzerland, 1998, pp. 127–156.

89. J. G. Riess in T. M. S. Chang, Ed., *Blood Substitutes: Principles, Methods, Products and Clinical Trials.*, Vol. **II**, Karger Landes Systems, Basel, Switzerland, 1998, pp. 101–126.

90. S. Hager, A. Gonzales, M. Gonzales, and R. M. Winslow, *Blood*, **88**, 184A (1996).

91. G. S. Hughes Jr., E. E. Jacobs Jr., S. F. Francom, and The Hemopure Surgical Study Group, *Crit. Care Med.*, **24**, A36 (Abstr. 2) (1996).

92. M. Nozue, I. Lee, J. M. Manning, L. R. Manning, and R. K. Jain, *J. Surg. Oncol.*, **62**, 109–114 (1996).

93. S. F. Flaim in R. M. Winslow, K. D. Vandegriff, and M. Intaglietta, Eds., *Advances in Blood Substitutes: Industrial Opportunities and Medical Challenges*, Birkhauser, Boston, 1997, p. 336.

94. G. Reah, A. R. Bodenham, A. Mallick, E. K. Daily, and R. J. Przybelski, *Crit. Care Med.*, **25**, 1480–1488 (1997).

95. J. De Angelo, *Expert Opin. Pharmacother.*, **1**, 19–29 (1999).

96. S. K. Swan, C. E. Halstenson, A. J. Collins, W. A. Colburn, J. Blue, and R. J. Przybelski, *Am. J. Kidney Dis.*, **26**, 918–923 (1995).

97. S. E. Day and R. G. Gedeit, *Clin. Perinatol.*, **25**, 711–722 (1998).

98. A. Anzueto and J. Melo, *Respir. Care Clin. North Am.*, **4**, 679–694 (1998).

99. J. G. Riess and M. P. Krafft, *Artif. Cells Blood Substit. Immobil. Biotechnol.*, **25**, 43–52 (1997).

100. J. S. Greenspan, M. R. Wolfson, and T. H. Shaffer, *Biomed. Instrum. Technol.*, **33**, 253–259 (1999).

101. R. F. Mattrey, *Artif. Cells Blood Substit. Immobil. Biotechnol.*, **22**, 295–313 (1994).

102. K. C. Lowe, M. R. Davey, and J. B. Power, *Trends Biotechnol.*, **16**, 272–277 (1998).

103. E. P. Sloan, M. Koenigsberg, D. Gens, M. Cipolle, J. Runge, M. N. Mallory, and G. Rodman Jr., *JAMA*, **282**, 1857–1864 (1999).

104. R. Saxena, A. D. Wijnhoud, H. Carton, W. Hacke, M. Kaste, R. J. Przybelski, K. N. Stern, and P. J. Koudstaal, *Stroke*, **30**, 993–996 (1999).

105. J. R. Hess, V. W. MacDonald, and W. W. Brinkley, *J. Appl. Physiol.*, **74**, 1769–1778 (1993).

106. S. C. Schultz, B. Grady, F. Cole, I. Hamilton, K. Burhop, and D. S. Malcolm, *J. Lab. Clin. Med.*, **122**, 301–308 (1993).

107. A. Gulati and S. Rebello, *J. Lab. Clin. Med.*, **124**, 125–133 (1994).

108. R. Motterlini in R. M. Winslow, K. D. Vandegriff, and M. Intaglietta, Eds., *Blood Substitutes: New Challenges*, Birkhaeuser, Boston, 1996, pp. 74–98.

109. D. H. Doherty, M. P. Doyle, S. R. Curry, R. J. Vali, T. J. Fattor, J. S. Olson, and D. D. Lemon, *Nat. Biotechnol.*, **16**, 672–676 (1998).

110. R. M. Winslow, *Vox Sang.*, **79**, 1–20 (2000).

111. J. G. Riess, *Chem. Rev.*, **101**, 2797–2920 (2001).

112. M. P. Doyle, I. Apostol, and B. A. Kerwin, *J. Biol. Chem.*, **274**, 2583–2591 (1999).

113. B. Faivre-Fiorina, A. Caron, C. Fassot, I. Fries, P. Menu, P. Labrude, and C. Vigneron, *Am. J. Physiol. Heart Circ. Physiol.*, **276**, H766–H770 (1999).

114. A. C. Sharma and A. Gulati, *Crit. Care Med.*, **23**, 874–884 (1995).

115. R. M. Winslow, *Br. J. Haematol.*, **111**, 387–396 (2000).

116. L. Lindbom, R. F. Tuma, and K. E. Arfors, *Microvasc. Res.*, **19**, 197–208 (1980).

117. R. M. Winslow and K. D. Vandegriff in R. M. Winslow, K. Vandegriff, and M. Intaglietta, Eds., *Advances in Blood Substitutes: Industrial Opportunities and Medical Challenges*, Birkhaeuser, Boston, 1997, pp. 167–188.

118. M. J. Kuchan and J. A. Frangos, *Am. J. Physiol. Heart Circ. Physiol.*, **264**, H150–H156 (1993).

119. M. Intaglietta, P. C. Johnson and R. M. Winslow, *Cardiovasc. Res.*, **32**, 632–643 (1996).

120. CBER, *Transfusion*, **31**, 369–371 (1991).

121. W. L. Chan, N. L. Tang, C. C. Yim, F. M. Lai, and M. S. Tam, *Toxicol. Pathol.*, **28**, 635–642 (2000).

122. M. Somers, A. I. Piqueras, K. Strange, M. L. Zeidel, W. Pfaller, M. Gawryl, and H. W. Harris, *Am. J. Physiol. Renal Fluid Electrolyte Physiol.*, **273**, F38–F52 (1997).

123. K. M. Sanders and S. M. Ward, *Am. J. Physiol. Gastrointest. Liver Physiol.*, **262**, G379–G392 (1992).

124. V. Plourde, E. Quintero, G. Suto, C. Coimbra, and Y. Tache, *Eur. J. Pharmacol.*, **256**, 125–129 (1994).

125. T. O'Kelly, A. Brading, and N. Mortensen, *Gut*, **34**, 689–693 (1993).

126. R. M. Cabot and S. Kohatsu, *Am. J. Surg.*, **136**, 242–246 (1978).

127. A. Hebra, M. F. Brown, K. McGeehin, D. Broussard, and A. J. Ross 3rd, *J. Pediatr. Surg.*, **28**, 362–365 (1993).

128. C. D. Conover, L. Lejeune, K. Shum, and R. G. Shorr, *Life Sci.*, **59**, 1861–1869 (1996).

129. J. C. Hartman, G. Argoudelis, D. Doherty, D. Lemon, and R. Gorczynski, *Eur. J. Pharmacol.*, **363**, 175–178 (1998).

130. S. K. Sarna, *Curr. Opin. Gastroenterol.*, **12**, 512–516 (1996).

131. G. P. Biro, C. Ou, C. Ryan-MacFarlane, and P. J. Anderson, *Artif. Cells Blood Substit. Immobil. Biotechnol.*, **23**, 631–645 (1995).

132. C. Privalle, T. Talarico, T. Keng, and J. DeAngelo, *Free Radic. Biol. Med.*, **28**, 1507–1517 (2000).

133. T. M. S. Chang, *Blood Substitutes: Principles, Methods, Products and Clinical Trials*, Karger Landes System, Basel, Switzerland, 1997.

134. A. I. Alayash, *Nat. Biotechnol.*, **17**, 545–549 (1999).

135. A. I. Alayash, *Free Radic. Res.*, **33**, 341–348 (2000).

136. C. P. Stowell, J. Levin, B. D. Spiess, and R. M. Winslow, *Transfusion*, **41**, 287–299 (2001).

137. T. M. Chang, F. D'Agnillo, W. P. Yu, and S. Razack, *Adv. Drug Deliv. Rev.*, **40**, 213–218 (2000).

138. J. Simoni, G. Simoni, D. E. Wesson, J. A. Griswold, and M. Feola, *ASAIO J.*, **46**, 679–692 (2000).

139. C. J. C. Hsia, *Artif. Cells Blood Substit. Immobil. Biotechnol.*, **29**, 104 II–7 (2001).

140. R. K. Saetzler, K. E. Arfors, R. F. Tuma, U. Vasthare, L. Ma, C. J. Hsia, and H. A. Lehr, *Free Radic. Biol. Med.*, **27**, 1–6 (1999).

141. W. K. Bleeker, L. M. Zappeij, P. J. den Boer, J. A. Agterberg, G. M. Rigter, and J. C. Bakker, *Artif. Cells Blood Substit. Immobil. Biotechnol.*, **23**, 461–468 (1995).

142. T. N. Estep, J. Gonder, I. Bornstein, and F. Aono, *Biomater. Artif. Cells Immobil. Biotechnol.*, **20**, 603–609 (1992).

143. T. M. S. Chang, C. Lister, T. Nishiya, and R. Varma, *Biomater. Artif. Cells Immobil. Biotechnol.*, **20**, 611–618 (1992).

144. D. J. Nelson in T. M. S. Chang, Ed., *Blood Substitutes: Principles, Methods, Products and Clinical Trials*, Vol. **I**, Karger Landes System, Basel Switzerland, 1998, pp. 39–61.

145. R. J. Przybelski, E. K. Daily, J. C. Kisicki, C. Mattia-Goldberg, M. J. Bounds, and W. A. Colburn, *Crit. Care Med.*, **24**, 1993–2000 (1996).

146. A. Schubert, N. M. Bedocs, J. F. Ohara, J. E. Tetzlaff, K. E. Marks, and A. C. Novick, *Anesthesiology*, **87**, A220 (Abstr.) (1997).

147. E. L. Bloomefield, M. Y. Rady, M. Popovich, S. Estandiari, and N. M. Bedocs, *Anesthesiology*, **85**, A220 (Abstr.) (1996).

148. M. L. Lamy, E. K. Daily, J. F. Brichant, R. P. Larbuisson, R. H. Demeyere, E. A. Vandermeersch, J. J. Lehot, M. R. Parsloe, J. C. Berridge, C. J. Sinclair, J. F. Baron, and R. J. Przybelski, *Anesthesiology*, **92**, 646–656 (2000).

149. D. Malcolm, D. Kissinger, and M. Garrioch, *Biomater. Artif. Cells Immobil. Biotechnol.*, **20**, 495–497 (1992).

150. J. F. O'Hara, J. E. Tetzlaff, D. F. Connors, E. H. Jones, N. M. Bedocs, and A. Schubert, *Anesthesiology*, **87**, A205 (Abstr.) (1997).

151. T. N. Estep, J. Gonder, I. Bornstein, S. Young, and R. C. Johnson, *Biomater. Artif. Cells Immobil. Biotechnol.*, **19**, 378 (Abstr.) (1991).

152. J. F. O'Hara, J. E. Tetzlaff, H. W. Popp, E. H. Jones, N. M. Bedocs, and A. Schubert, *Anesthesiology*, **87**, A344 (Abstr.) (1997).

153. A. C. Sharma, and A. Gulati, *J. Lab. Clin. Med.*, **123**, 299–308 (1994).

154. R. J. Przybelski, G. J. Kant, M. J. Bounds, M. V. Slayter, and R. M. Winslow, *J. Lab. Clin. Med.*, **115**, 579–588 (1990).

155. J. A. Walder, R. Y. Walder, and A. Arnone, *J. Mol. Biol.*, **141**, 195–216 (1980).

156. D. Looker, D. Abbott-Brown, P. Cozart, S. Durfee, S. Hoffman, A. J. Mathews, J. Miller-Roehrich, S. Shoemaker, S. Trimble, G. Fermi, et al., *Nature*, **356**, 258–260 (1992).

157. W. F. Moo-Penn, J. A. Wolff, G. Simon, M. Vacek, D. L. Jue, and M. H. Johnson, *FEBS Lett.*, **92**, 53–56 (1978).

158. M. K. Viele, R. B. Weiskopf, and D. Fisher, *Anesthesiology*, **86**, 848–858 (1997).

159. S. Moqattash, J. D. Lutton, G. Rosenthal, M. F. Abu-Hijleh, and N. G. Abraham, *Acta Haematol.*, **98**, 76–82 (1997).

160. J. A. Murray, A. Ledlow, J. Launspach, D. Evans, M. Loveday, and J. L. Conklin, *Gastroenterology*, **109**, 1241–1248 (1995).

161. A. Loeb, L. J. McIntosh, N. R. Raj, and D. Longnecker, *J. Cardiovasc. Pharmacol.*, **30**, 703–710 (1997).

162. R. Lessen, M. Williams, J. Seltzer, J. Lessin, A. Marr, D. Guvakov, M. Imbing, D. Solanki, K. Grugan, D. Smith, D. Kittel, N. Lawson, R. Caspari, and M. Loveday, *Artif. Cells Blood Substit. Immobil. Biotechnol.*, **24**, A380 (Abstr.) (1996).

163. J. J. Cullen, J. L. Conklin, J. Murray, A. Ledlow, and G. Rosenthal, *Dig. Dis. Sci.*, **41**, 289–294 (1996).

164. L. O. Sillerud, A. Caprihan, N. Berton, and G. J. Rosenthal, *J. Appl. Physiol.*, **86**, 887–894 (1999).

165. E. A. Brucker, *Acta Crystallogr. D Biol. Crystallogr.*, **56 (Pt. 7)**, 812–816 2000.

166. R. W. Noble, H. L. Hui, L. D. Kwiatkowski, P. Paily, A. DeYoung, A. Wierzba, J. E. Colby, S. Bruno, and A. Mozzarelli, *Biochemistry*, **40**, 12357–12368 (2001).

167. K. S. Kroeger and C. E. Kundrot, *Structure*, **5**, 227–237 (1997).

168. R. F. Eich, T. Li, D. D. Lemon, D. H. Doherty, S. R. Curry, J. F. Aitken, A. J. Mathews, K. A. Johnson, R. D. Smith, G. N. Phillips Jr., and J. S. Olson, *Biochemistry*, **35**, 6976–6983 (1996).

169. R. Kumar, *Proc. Soc. Exp. Biol. Med.*, **208**, 150–158 (1995).

170. R. R. Behringer, T. M. Ryan, M. P. Reilly, T. Asakura, R. D. Palmiter, R. L. Brinster, and T. M. Townes, *Science*, **245**, 971–973 (1989).

171. M. E. Swanson, M. J. Martin, J. K. O'Donnell, K. Hoover, W. Lago, V. Huntress, C. T. Parsons, C. A. Pinkert, S. Pilder, and J. S. Logan, *Biotechnology*, **10**, 557–559 (1992).

172. A. C. Sharma, M. J. Martin, J. F. Okabe, R. A. Truglio, N. K. Dhanjal, J. S. Logan, and R. Kumar, *Biotechnology*, **12**, 55–59 (1994).

173. D. S. Katz, S. P. White, W. Huang, R. Kumar, and D. W. Christianson, *J. Mol. Biol.*, **244**, 541–553 (1994).

174. D. S. Kaufman, E. T. Hanson, R. L. Lewis, R. Auerbach, and J. A. Thomson, *Proc. Natl. Acad. Sci. USA*, **98**, 10716–10721 (2001).

175. W. Dieryck, J. Pagnier, C. Poyart, M. C. Marden, V. Gruber, P. Bournat, S. Baudino, and B. Merot, *Nature*, **386**, 29–30 (1997).

176. A. I. Alayash, A. G. Summers, F. Wood, and Y. Jia, *Arch. Biochem. Biophys.*, **391**, 225–234 (2001).

177. L. B. Pearce and M. S. Gawryl in T. M. S. Chang, Ed., *Blood Substitutes: Principles, Methods, Products and Clinical Trials*, Vol. **II**, Karger Landes System, Basel, Switzerland, 1998, pp. 39–61.

178. G. S. Hughes Jr., E. J. Antal, P. K. Locker, S. F. Francom, W. J. Adams, and E. E. Jacobs Jr., *Crit. Care Med.*, **24**, 756–764 (1996).

179. G. S. Hughes Jr., S. F. Francome, E. J. Antal, W. J. Adams, P. K. Locker, E. P. Yancey, and E. E. Jacobs Jr., *J. Lab. Clin. Med.*, **126**, 444–451 (1995).

180. P. R. Foster, *Ann. Med.*, **32**, 501–513 (2000).

181. T. Monk, L. Goodnough, G. Hughes, and E. Jacobs, *Anesthesiology*, **83**, A285 (1995).

182. T. Standl, S. Wilhelm, E. P. Horn, M. Burmeister, and J. S. A. Esch, *Anesthesiology*, **87**, A65 (1997).

183. S. M. Kasper, M. Walter, F. Grune, A. Bischoff, H. Erasmi, and W. Buzello, *Anesth. Analg.*, **83**, 921–927 (1996).

184. T. Standl, P. Horn, S. Wilhelm, C. Greim, M. Freitag, U. Freitag, A. Sputtek, E. Jacobs, and J. Schulte am Esch, *Can. J. Anaesth.*, **43**, 714–723 (1996).

185. R. J. Bosman, J. Minten, H. R. Lu, H. Van Aken, and W. Flameng, *Anesth. Analg.*, **75**, 811–817 (1992).

186. G. S. Hughes Jr., E. P. Yancey, R. Albrecht, P. K. Locker, S. F. Francom, E. P. Orringer,

E. J. Antal, and E. E. Jacobs Jr., *Clin. Pharmacol. Ther.*, **58**, 434–443 (1995).

187. E. P. Orringer, P. M. Gonzalez, and A. C. Hackney in Y. Beuzard, B. Lubin, and J. Rosa, Eds., *Sickle Cell Disease and Thalassaemias: New Trends in Therapy*, Vol. **234**, INSERM/John Libbey Eurotext, Montrouge, France, 1995, pp. 301–302.

188. B. A. Teicher, S. A. Holden, K. Menon, R. E. Hopkins, and M. S. Gawryl, *Cancer Chemother. Pharmacol.*, **33**, 57–62 (1993).

189. S. A. Gould, E. E. Moore, F. A. Moore, J. B. Haenel, J. M. Burch, H. Sehgal, L. Sehgal, R. DeWoskin, and G. S. Moss, *J. Trauma*, **43**, 325–331 (1997).

190. R. E. Benesch, R. Benesch, R. D. Renthal, and N. Maeda, *Biochemistry*, **11**, 3576–3582 (1972).

191. J. L. Johnson, E. E. Moore, P. J. Offner, D. A. Partrick, D. Y. Tamura, G. Zallen, and C. C. Silliman, *J. Trauma*, **50**, 449–455 (2001).

192. S. A. Gould, E. E. Moore, D. B. Hoyt, J. M. Burch, J. B. Haenel, J. Garcia, R. DeWoskin, and G. S. Moss, *J. Am. Coll. Surg.*, **187**, 113–120 (1998).

193. D. C. Cheng, *Can. J. Anaesth.*, **48**, S41–S48 (2001).

194. J. G. Adamson and C. Moore in T. M. S. Chang, Ed., *Blood Substitutes: Principles, Methods, Products and Clinical Trials*, Vol. **II**, Karger Landes System, Basel, Switzerland, 1998, pp. 62–81.

195. N. E. Cutcliffe, F. J. L. Carmichael, and A. G. Greenburg in E. Mathiowitz, Ed., *Encyclopedia of Controlled Drug Delivery*, John Wiley & Sons, New York, 1999.

196. D. Wicks, P. Nakao, P. Champagne, and C. Mihas, *Artif. Cells Blood Substit. Immobil. Biotechnol.* **24**, 460 (Abstr.) (1996).

197. S. Rattan, G. J. Rosenthal, and S. Chakder, *J. Pharmacol. Exp. Ther.*, **272**, 1211–1216 (1995).

198. J. C. Hsia, D. L. Song, S. S. Er, L. T. Wong, P. E. Keipert, C. L. Gomez, A. Gonzales, V. W. Macdonald, J. R. Hess, and R. M. Winslow, *Biomater. Artif. Cells Immobil. Biotechnol.*, **20**, 587–595 (1992).

199. W. Lieberthal, R. Fuhro, J. E. Freedman, G. Toolan, J. Loscalzo, and C. R. Valeri, *J. Pharmacol. Exp. Ther.*, **288**, 1278–1287 (1999).

200. J. C. Hsia, *Biomater. Artif. Cells Immobil. Biotechnol.*, **19**, 402 (1991).

201. J. G. Adamson, G. A. McIntosh, J. B. Stewart, and L. T. Wong, *Artif. Cells Blood Substit. Immobil. Biotechnol.*, **29**, 133 PO-1 (2001).

202. M. L. Nucci, R. G. L. Shorr, and A. Abuchowski, *Drugs of the Future*, **21**, 29–32 (1996).

203. P. Bassett, *Drug Market Dev.*, **7**, 164–168 (1996).

204. K. L. Shum, A. Leon, A. T. Viau, D. Pilon, M. Nucci, and R. G. Shorr, *Artif. Cells Blood Substit. Immobil. Biotechnol.*, **24**, 655–683 (1996).

205. C. Conover, L. Lejeune, R. Linberg, K. Shum, and R. G. Shorr, *Artif. Cells Blood Substit. Immobil. Biotechnol.*, **24**, 599–611 (1996).

206. A. L. Baldwin, L. M. Wilson, and J. E. Valeski, *Am. J. Physiol. Heart Circ. Physiol.*, **275**, H615–H625 (1998).

207. R. B. S. Palaparthy, H. Wang, and Anil Gulati, *Adv. Drug Deliv. Rev.*, **40**, 185–198 (2000).

208. K. Nakai, H. Togashi, T. Yasukohchi, I. Sakuma, S. Fujii, M. Yoshioka, H. Satoh, and A. Kitabatake, *Int. J. Artif. Organs*, **24**, 322–328 (2001).

209. C. Conover, R. Linberg, L. Lejeune, C. Gilbert, K. Shum, and R. G. Shorr, *Artif. Cells Blood Substit. Immobil. Biotechnol.*, **26**, 199–212 (1998).

210. R. M. Winslow, *Annu. Rev. Med.*, **50**, 337–353 (1999).

211. K. D. Vandegriff, J. M. Lohman, A. G. Tsai, M. Intaglietta, D. Drobin, B. T. Kjellstromo, and R. M. Winslow, *Artif. Cells Blood Substit. Immobil. Biotechnol.*, **29**, 103 II-6 (2001).

212. B. T. Kjellstromo, D. Drobin, M. Intaglietta, J. M. Lohman, K. D. Vandegriff, and R. M. Winslow, *Artif. Cells Blood Substit. Immobil. Biotechnol.*, **29**, 97 I-5 (2001).

213. Y. Iwashita, A. Yabuki, K. Yamaji, K. Iwasaki, T. Okami, C. Hirata, and K. Kosaka, *Biomater. Artif. Cells Artif. Organs*, **16**, 271–280 (1988).

214. T. L. Talarico, K. J. Guise, and C. J. Stacey, *Biochim. Biophys. Acta*, **1476**, 53–65 (2000).

215. R. J. Rohlfs and K. D. Vandegriff in R. M. Winslow, K. D. Vandegriff, and M. Intaglietta, Eds., *Blood Substitutes: New Challenges*, Birkhaeuser, Boston, 1996, pp. 163–184.

216. T. M. S. Chang, *Biomater. Artif. Cells Artif. Organs*, **16**, 1–9 (1988).

217. T. M. S. Chang in T. M. S. Chang, Ed., *Blood Substitutes: Principles, Methods, Products and Clinical Trials.*, Vol. **I**, Karger Landes Systems, Basel, Switzerland, 1997, pp. 89–110.

218. L. Djordjevich and I. F. Miller, *Exp. Hematol.*, **8**, 584–592 (1980).

219. A. S. Rudolph and W. T. Phillips in T. M. S. Chang, Ed., *Blood Substitutes: Principles,*

Methods, Products and Clinical Trials, Vol. **II**, Karger Landes Systems, Basel, Switzerland, 1998, pp. 197–215.

220. K. D. Vandegriff, D. F. Wallach, and R. M. Winslow, *Artif. Cells Blood Substit. Immobil. Biotechnol.*, **22**, 849–854 (1994).

221. W. P. Yu and T. M. S. Chang, *Artif. Cells Blood Substit. Immobil. Biotechnol.*, **24**, 169–184 (1996).

222. T. M. S. Chang and W. P. Yu (to McGill University), U.S. Pat. 5,670,173 (September 23, 1997).

223. T. M. S. Chang and W. P. Yu in T. M. S. Chang, Ed., *Blood Substitutes: Principles, Methods, Products and Clinical Trials*, Vol. **II**, Karger Landes Systems, Basel, Switzerland, 1998, pp. 216–231.

224. D. C. Carter, J. X. Ho, and F. Ruker, U.S. Pat. 5,948,609 (September 7, 1999).

225. E. Tsuchida, K. Ando, H. Maejima, N. Kawai, T. Komatsu, S. Takeoka, and H. Nishide, *Bioconjug. Chem.*, **8**, 46–50 (1997).

226. E. Tsuchida, T. Komatsu, K. Hamamatsu, Y. Matsukawa, A. Tajima, A. Yoshizu, Y. Izumi, and K. Kobayashi, *Bioconjug. Chem.*, **11**, 46–50 (2000).

227. P. E. Keipert, *Artif. Cells Blood Substit. Immobil. Biotechnol.*, **29**, 99 II-2 (2001).

228. D. M. Dromsky, A. Fahlman, and B. D. Spiess, *Artif. Cells Blood Substit. Immobil. Biotechnol.*, **29**, 124 VI-1 (2001).

229. M. E. Fabry, M. D. Does, S. M. Suzuka, and R. L. Nagel, *Artif. Cells Blood Substit. Immobil. Biotechnol.*, **29**, 94 I-2 (2001).

230. D. K. Kaul, X. Liu, and R. L. Nagel, *Blood*, **98**, 3128–3131 (2001).

231. P. F. McDonagh, K. Cerney, J. Y. Hokama, G. Lai, R. F. Gonzales, G. Davis-Gorman, and J. G. Copeland, *Artif. Cells Blood Substit. Immobil. Biotechnol.*, **29**, 105 II-8 (2001).

232. K. C. Lowe, *J. Fluor. Chem.*, **109**, 59–65 (2001).

233. S. F. Flaim, *Artif. Cells Blood Substit. Immobil. Biotechnol.*, **22**, 1043–1054 (1994).

234. J. G. Riess and M. Le Blanc in K. C. Lowe, Ed., *Blood Substitutes. Preparation, Physiology and Medical Applications*, Ellis Horwood, Chichester, UK, 1988, pp. 94–129.

235. J. G. Riess, *Vox Sang.*, **61**, 225–239 (1991).

236. M. G. Scott, D. F. Kucik, L. T. Goodnough, and T. G. Monk, *Clin. Chem.*, **43**, 1724–1731 (1997).

237. P. E. Keipert, S. Otto, S. F. Flaim, J. G. Weer, E. A. Schutt, T. J. Pelura, D. H. Klein, and T. L. Yaksh, *Artif. Cells Blood Substit. Immobil. Biotechnol.*, **22**, 1169–1174 (1994).

238. Y. Ni, D. H. Klein, and D. Song, *Artif. Cells Blood Substit. Immobil. Biotechnol.*, **24**, 81–90 (1996).

239. J. G. Riess and P. E. Keipert in E. Tsuchida, Ed., *Blood Substitutes: Present and Future Perspectives*, Elsevier Science SA, Lausanne, Switzerland, 1998, pp. 91–102.

240. J. G. Riess, *Artif. Organs*, **81**, 44–56 (1984).

241. J. G. Riess and M. Le Blanc, *Pure Appl. Chem.*, **54**, 2383–2406 (1982).

242. R. M. Winslow, *Adv. Drug Deliv. Rev.*, **40**, 131–142 (2000).

243. S. F. Flaim in A. S. Rudolph, R. Rabinovici, and G. Z. Feuerstein, Eds., *Red Blood Cells Substitutes: Basic Principles and Clinical Applications*, Marcel Dekker, New York, 1998, pp. 79–117.

244. D. J. Roth, P. E. Keipert, N. S. Faithfull, T. F. Zuck, and J. G. Riess (to Alliance Pharm. Corp.), U.S. Pat. 5,451,205 (September 19, 1995).

245. D. R. Spahn, *Crit. Care*, **3**, R93–R97 (1999).

246. D. R. Spahn, R. van Bremt, G. Theilmeier, J. P. Reibold, M. Welte, H. Heinzerling, K. M. Birck, P. E. Keipert, K. Messmer, and the European Perflubron Emulsion Study Group, *Anesthesiology*, **91**, 1195–1208 (1999).

247. J. G. Riess and M. P. Krafft, *Biomaterials*, **19**, 1529–1539 (1998).

248. A. Lattes and I. Rico-Lattes, *Artif. Cells Blood Substit. Immobil. Biotechnol.*, **22**, 1007–1018 (1994).

249. M. M. Fisher and P. W. Brady, *Drug Safety*, **5**, 86–93 (1990).

250. M. I. Griffel and B. S. Kaufman, *Crit. Care Clin.*, **8**, 235–253 (1992).

251. J. S. Roberts and S. L. Bratton, *Drugs*, **55**, 621–630 (1998).

252. J. Ring and K. Messmer, *Lancet*, **1**, 466–469 (1977).

253. R. W. Colman, *N. Engl. J. Med.*, **299**, 97–98 (1978).

254. J. P. Isbister and M. M. Fisher, *Anaesth. Intens. Care*, **8**, 145–151 (1980).

255. L. Mailloux, C. D. Swartz, R. Capizzi, K. E. Kim, G. Onesti, O. Ramirez, and A. N. Brest, *N. Engl. J. Med.*, **277**, 1113–1118 (1967).

256. M. Moran and C. Kapsner, *N. Engl. J. Med.*, **317**, 150–153 (1987).

257. J. Treib and J. F. Baron, *Intens. Care Med.*, **25**, 258–268 (1999).

258. G. L. Klein, *Nutr. Rev.*, **49**, 74–79 (1991).

259. S. G. Kovalik, A. M. Ledgerwood, C. E. Lucas, and R. F. Higgins, *J. Trauma*, **21**, 275–279 (1981).

260. U. Klotz and H. Kroemer, *Clin. Pharmacokinet.*, **12**, 123–135 (1987).

261. J. D. Hulse and A. Yacobi, *Drug Intell. Clin. Pharm.*, **17**, 334–341 (1983).

262. C. Sirtl, H. Laubenthal, V. Zumtobel, D. Kraft, and W. Jurecka, *Br. J. Anaesth.*, **82**, 510–515 (1999).

RECOMMENDED READINGS

Chang, T. M. S., *Blood Substitutes: Principles, Methods, Products and Clinical Trials*, Vols. **I–II**, Karger Landes System, Basel, Switzerland, 1997, 1998.

Riess, J. G., *Chem. Rev.* **101**, 2797–2920 (2001).

Winslow, R. M., Vandegriff, K. D., and Intaglietta, M., Eds., *Blood Substitutes: New Challenges*, Birkhaeuser, Boston, 1996.

CHAPTER NINE

Inhibition of Sickle Hemoglobin Polymerization as a Basis for Therapeutic Approaches to Sickle-Cell Anemia

Constance Tom Noguchi
Alan N. Schechter
National Institute of Diabetes and Digestive and Kidney Diseases,
    National Institutes of Health
Laboratory of Chemical Biology
Bethesda, Maryland

John D. Haley
OSI Pharmaceuticals Inc.
Farmingdale, New York

Donald J. Abraham
Virginia Commonwealth University
Department of Medicinal Chemistry
Richmond, Virginia

Burger's Medicinal Chemistry and Drug Discovery
Sixth Edition, Volume 3: Cardiovascular Agents and
Endocrines
Edited by Donald J. Abraham
ISBN 0-471-37029-0   © 2003 John Wiley & Sons, Inc.

Contents

1   INTRODUCTION

Understanding the molecular defect in sickle-cell anemia has led to the most important current strategies for disease therapy, based on inhibiting intracellular polymerization of sickle hemoglobin leading to pathophysiology. There have been more than 50 years of investigation of the biochemical and biophysical processes associated with sickle-cell anemia. Under current investigation are new therapeutic strategies based on molecular interactions of the abnormal hemoglobin, the pathophysiologic behavior of the sickle cells in the circulation, and the genetic program of the abnormal erythroid cells.

2   PATHOPHYSIOLOGY OF SICKLE-CELL ANEMIA

Almost a century ago, James B. Herrick noted "peculiar, elongated and sickle shaped" red blood cells in a West Indian student with severe anemia (Fig. 9.1a) (1). Before the understanding of the molecular basis of the disease, this characteristic morphology observed by light microscopy was a useful diagnostic tool in efforts to understand the chronic anemia,

acute painful crisis, organ damage, and genetic inheritance of the disease (2–5). Red blood cells from individuals with sickle-cell anemia undergo morphologic deformation from the normal biconcave-shaped disk when deprived of oxygen, many adopting the characteristic sickle shape (Fig. 9.1b) (6, 7). Although much of the early focus was on the distortion of the red cell, it is now known that the important molecular lesion occurs intracellularly, as shown by electron microscopy (Fig. 9.1c) (6).

2.1   Abnormal Hemoglobin

Hemoglobin makes up 98% of the intracellular protein of the red blood cell. In 1949 Linus Pauling and colleagues determined that the primary defect in sickle-cell anemia, giving rise to the characteristic change in cell shape, was not the result of an inherent membrane defect, but was a consequence of a molecular change on the hemoglobin molecule itself (8). This molecular defect resulted in a change in the overall charge of the hemoglobin molecule that was discernable by a change in electrophoretic mobility (Fig. 9.2a). This change difference between sickle hemoglobin (HbS) and normal hemoglobin A (HbA) has become the

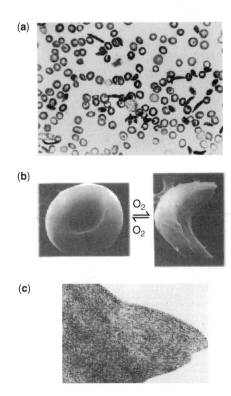

Figure 9.1. Sickle erythrocytes. (a) James Herrick was the first to report elongated and other malformed sickle erythrocytes that were observed in blood taken from a dental student from Grenada. [From J. B. Herrick, *Arch. Intern. Med.*, **6**, 517 (1910).] (b) Scanning electron micrographs of oxygenated (left) and deoxygenated (right) sickle erythrocytes from an individual with homozygous sickle-cell anemia (SS) show the characteristic morphologic difference between the normal discoid shape of the oxygenated erythrocyte and the classic sickle shape of the deoxygenated SS erythrocyte. (c) Transmission electron micrograph of a deoxygenated SS erythrocyte shows long fibers of intracellular hemoglobin polymer.

basis of the current primary diagnostic tool for sickle-cell anemia (9, 10). Vernon Ingram identified the genetic alteration giving rise to HbS as an amino acid substitution on one of two polypeptides making up hemoglobin (11). Hemoglobin is a protein consisting of four polypeptide chains (a pair of α-globin/β-globin dimers). Each globin chain surrounds an iron porphyrin molecule, or heme group, capable of reversible binding to molecular oxygen. The α-globin chain is a polypeptide of 141 amino

acids and is encoded in a gene cluster located on chromosome 16pter-p13.3 (12–14). The β-globin chain is 144 amino acids long and is encoded in a gene cluster on chromosome 11p15.5 (15–17). The substitution of a valine residue for glutamic acid in the sixth amino acid of each of the β-globin chains is the mutation in HbS and accounts for the change in electric charge between HbS and HbA (Fig. 9.2b). However, the mutation also causes a change—a marked decrease in solubility upon deoxygenation (18, 19)—that accounts specifically for the pathophysiology of the disease.

2.2   Abnormal β-Globin Gene

In the adult, red cells are produced in the bone marrow. Hematopoietic stem cells are activated by a variety of cytokines. Stimulation of differentiation along the erythroid lineage requires erythropoietin, a hematopoietic cytokine produced principally by the kidney (20–22). Erythropoietin binding to the surface of erythroid progenitor cells promotes their survival, and proliferation and differentiation leading to the production of red blood cells (Fig. 9.3a). During terminal differentiation, the erythroid precursor cells cease to divide and lose their nuclei, leading to reticulocyte formation, whereas protein synthesis continues with translation of the globin genes and hemoglobin production before formation of the mature red blood cell. With the identification of the messenger RNA (mRNA) for β-globin, a single nucleotide substitution in the sixth codon of the β-globin gene of GTG for GAG was determined as the genetic lesion in sickle-cell anemia (23). Sickle-cell disease corresponds to the homozygous (SS) condition for this β-globin A to T nucleotide mutation; sickle-cell trait (AS) corresponds to the heterozygous or carrier state for this mutation (5).

The HbS mutation is located on the surface of the hemoglobin molecule and HbS exhibits similar binding to oxygen and 2,3-DPG in dilute solution. During hemoglobin synthesis, the mutant β^S chain has a slightly lower affinity for the α-globin chain in formation of α-globin/β-globin dimers. This leads to selective destruction of the β^S chain and a 40:60 ratio for HbS:HbA in individuals with sickle trait (AS) (24).

(a)

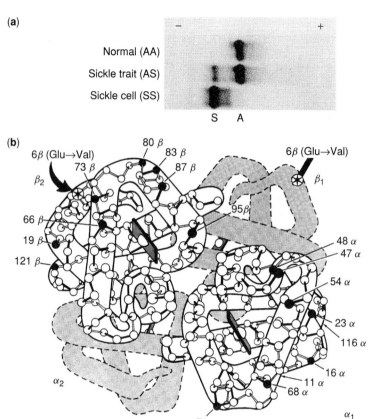

Figure 9.2. Sickle hemoglobin. (a) Electrophoresis pattern of hemolysates from normal (AA), sickle trait (AS), and sickle-cell anemia (SS) red blood cells shows the separation of normal hemoglobin (HbA) from sickle hemoglobin (HbS) based on charge arising from the substitute of a charged glutamic acid residue in HbA by a neutral valine residue in HbS in the β^6 position. AS cells have both types of hemoglobin with more HbA than HbS. (b) Representation of a tetrameric hemoglobin molecule based on the crystal structure shows the position of β^6 and other residues implicated in intermolecular contacts of the deoxygenated hemoglobin S polymer. (From the estate of Irving Geis, with permission.)

(b)

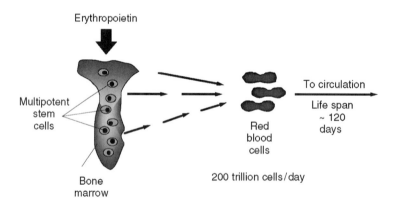

Figure 9.3. Red cell and hemoglobin production. Erythropoietin stimulates the hematopoietic stem cells in the bone marrow to proliferate and differentiate along the erythroid lineage. The multipotential stem cells can self-"renew" as well as differentiate into erythroid progenitor cells. Erythropoietin is required for the viability, proliferation, and differentiation of erythroid progenitor cells into erythroid precursor cells. During differentiation to become red cells, transcription ceases and the erythroid precursor cells lose their nuclei and become reticulocytes. The cells continue to translate protein from globin mRNA. With terminal differentiation, translation ceases and the mature erythrocytes are formed, with hemoglobin making up 98% of the intracellular protein.

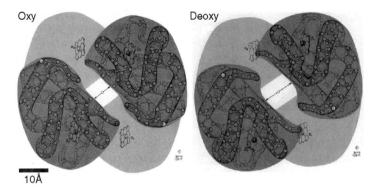

Figure 9.4. Diagram for oxygenated (Oxy) and deoxygenated (Deoxy) hemoglobins based on the crystal structure viewed down the dyad axis of symmetry. The α-carbon backbone of the β-chains is shown. The β^6 position is indicated by open circles. The major conformation change for the deoxygenated T-state from the oxygenated R-state is the shift and rotation of the $\alpha_1\beta_1$ dimer relative to the $\alpha_2\beta_2$ dimer. The β-chains move further apart and create a binding site for 2,3-DPG. (From the estate of Irving Geis, with permission.)

2.3   Hemoglobin and Oxygen Binding

The primary function of the red blood cell is to carry oxygen from the lungs to the tissues (25). Molecular oxygen taken up by the red blood cell is associated with the iron on the heme group. Each hemoglobin molecule made up of four globin chains is capable of cooperative binding of up to four molecules of oxygen. X-ray crystallography provides evidence suggesting that the oxygenated forms of HbS and normal hemoglobin or hemoglobin A (HbA) are isomorphous and similar in overall structure (18, 26, 27). The crystallographic structure of hemoglobin suggests that the molecule itself "breathes" and undergoes a conformation change between its deoxygenated form (T-state) and its fully liganded (or oxygenated) form (R-state) (Fig. 9.4) (28). The allosteric structural changes between the T-state and the R-state are a shift in relative orientation of the α-/β-globin chain dimers. On deoxygenation, the β-globin chains move further apart, providing a binding site for 2,3-DPG in the central cavity of the hemoglobin molecule, resulting in stabilization of the deoxygenated hemoglobin conformation. This structural change accounts for the cooperativity of oxygen binding, leading to the characteristic sigmoidal shape of the oxygen binding isotherm. The differential binding of 2,3-DPG (and pro-

tons) accounts for the allosteric effects of these modulators of oxygenation.

2.4   Hemoglobin Molecular Size and Nonideality

The relatively large size of hemoglobin (64 Å in diameter) and high hemoglobin concentration in the red blood cell of 32–34 g/dL results in extensive crowding. The molecular separation between hemoglobin molecules in an intact red cell is less than one molecular diameter (Fig. 9.5a). At this protein concentration there is a major departure from ideal conditions, and a high degree of nonideal behavior. Based on the fact that the activity coefficient increases exponentially with protein concentration (Fig. 9.5b), hemoglobin acts 50 times more concentrated inside the red blood cell because of its physical size than if it were a point particle. Nevertheless, HbA remains soluble at these high protein concentrations under all physiological conditions.

2.5   Sickle Hemoglobin Polymerization

The allosteric shift in conformation between oxygenated and deoxygenated hemoglobin when there is a substitution of a valine residue for glutamic acid on the β-chain of HbS markedly affects the ability of deoxygenated HbS to

(a)

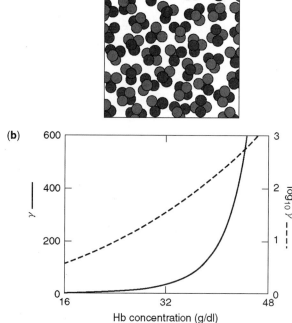

Figure 9.5. Hemoglobin and nonideality. (a) Artistic representation shows the crowding between hemoglobin molecules separated by less than one molecular diameter apart within the red blood cell. (b) The activity coefficient (γ) (solid line: linear scale; dashed line: logarithmic scale) describes the departure from ideal solution behavior and is the ratio of the hemoglobin activity to its measured concentration. At low concentrations, the activity coefficient is close to 1. At physiologic hemoglobin concentrations found inside the red blood cell (32 to 34 g/dL), the activity coefficient indicates that hemoglobin behaves as though it is 50 times more concentrated because of its physical size than if it were a point particle. [From C. T. Noguchi, *Biophys. J.*, **45**, 1153 (1984).]

(b)

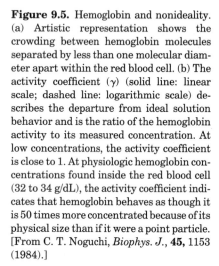

remain in solution (29, 30). Under physiologic conditions, the solubility of deoxygenated HbS is about half the hemoglobin concentration in the red blood cell, or around 16 g/dL (31–33). As a result, when HbS is deoxygenated, the shift in conformation to the deoxygenated conformation favors the aggregated or polymer state and HbS gels. (This process forms semicrystalline structures that are called liquid crystals or tactoids.) Gelation of HbS leads to alteration of the physical properties of HbS, resulting in marked increases in the intracellular red cell viscosity compared to that of cells with oxygenated HbS or deoxygenated or oxygenated HbA (34–36). Growth and alignment of HbS polymer domains also give rise to the marked changes in cell morphology characteristic of sickle-cell anemia (Fig. 9.1b) (6, 7). These intracellular changes also lead to relative rigidity of the cell, making it difficult to traverse small blood vessels (37–39). This ultimately leads to irreversible biochemical and physiological changes to the sickle red blood cell, even though the polymerization process is completely reversible (39).

2.6 Hemoglobin S Polymerization and Disease Manifestations

At physiologic conditions the dramatic decrease in intracellular HbS solubility when deoxygenated and its subsequent intracellular aggregation results in significant changes in the cell's viscoelastic properties (31, 40, 41). Intracellular HbS aggregation decreases red cell flexibility and increases abnormal red cell rheology during transit in the circulation, causing injury to tissues by microvascular occlusion, hemolysis, and decreased oxygen delivery (Fig. 9.6a). The premature red cell destruction results in moderate to severe hemolytic anemia and related clinical complications, with hemoglobin values from 6 to 10 g/dL compared with a normal level of 12 g/dL (9, 42). Clinical symptoms can include chronic as well as acute painful episodes characterized by extreme pain in the extremities, back, chest, or abdomen (Fig. 9.6b) (43). Spleen infarction and resorption decreases immune function and increases infection. Growth and development are delayed. Cardiovascular, pul-

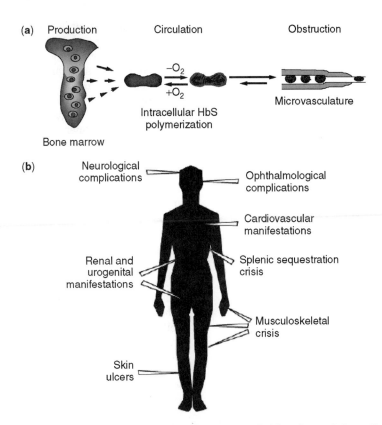

Figure 9.6. Cellular pathophysiology of sickle-cell anemia. (a) Sickle-cell anemia is attributed to the hemoglobin composition in mature red blood cells of virtually total mutant hemoglobin S (HbS) that has low solubility in its deoxygenated form. As HbS gives up oxygen, HbS polymerizes. The increase in intracellular viscosity increases red cell fragility and decreases cell deformability, particularly at low oxygen tensions in the microcirculation, causing occlusion in the arterioles. Occlusion of the red cells in the circulation, which is triggered by mechanisms still poorly understood, but which is primarily dependent on oxygen concentration, results in insufficient oxygen delivery and tissue damage. (b) In sickle-cell anemia almost any organ in the body can be affected by vaso-occlusion and lack of oxygen delivery to the tissues, as shown here.

monary, and renal insufficiency and other organ involvement can lead to end-organ failure and death. The acute and chronic pathophysiology of sickle-cell disease can be explained principally by the polymerization tendency of HbS and impaired transit of SS red cells. However, many other processes related to the red cell and its interaction with other cells may also contribute to these pathophysiological processes, especially to the induction of pathological events.

2.7   Hemoglobin S Polymer Structure

Electron microscopy analysis of aggregated deoxygenated HbS reveals a long fibrous network of aligned hemoglobin molecules (Fig. 9.7a) (44). The aggregated HbS molecules appear as bundles of fibers. The HbS fibers consist of assembled strands of hemoglobin molecules bundled as a 14-strand polymer, with a slight helical twist (350-nm repeat distance) (Fig. 9.7b). X-ray crystallography of deoxygenated HbS in a physiologic buffer provides insight into the intermolecular contact regions between HbS molecules in the fiber structure (Fig. 9.7d) (45–47). The 14-strand polymer is grouped as seven pairs of half-staggered strands of deoxygenated HbS molecules (Fig. 9.7c). In the crystal structure, deoxygenated HbS molecules are oriented as half-staggered

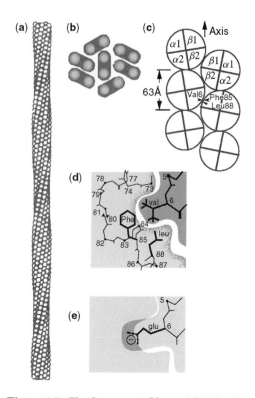

Figure 9.7. The deoxygenated hemoglobin S polymer. (a) A 14-strand model for the HbS polymer was determined from electron micrographs (44, 276). A single gel fiber is about 210 Å in diameter, with a slight helical twist with a repeat of about 350 nm along the fiber axis. (b) A cross section showing possible pairing of hemoglobin molecules in the 14-strand model based on the double strands seen in the crystal of deoxygenated HbS (45, 277). (c) Arrangement of hemoglobin tetramers illustrating the double-strand arrangement observed in the crystal of deoxygenated HbS. Only one β^{S6Val} of each hemoglobin tetramer is involved in an intermolecular contact. (d) Artistic rendition of the β^{S6Val} contact region showing possible arrangement of the hydrophobic pocket involving β^{85Phe} and β^{88Leu} on an adjacent molecule. (e) The normal β^{6Glu} is charged and too bulky to allow the same geometry. (d and e from the estate of Irving Geis, with permission.)

pairs, with only one β^{S6Val} mutation on each HbS molecule fitting in a hydrophobic pocket of an adjacent β-globin chain in the vicinity of β^{85Phe}. A number of other close contact sites are consistent with other studies of naturally occurring hemoglobin mutants that affect HbS aggregation (Fig. 9.7d). Interestingly, de-

oxygenated HbA with glutamic acid in the sixth position of β-globin cannot participate in the hydrophobic interaction with β^{85Phe} because of its size and charge (Fig. 9.7e). However, the availability of β^{85Phe} on HbA is able to provide the hydrophobic pocket for docking of the mutant β^{S6Val} that stabilizes the deoxygenated HbS polymer. Although the HbA tetramer cannot fit into the polymer structure, the hybrid HbA/HbS molecule or hemoglobin tetramer, consisting of mixed dimers of $\alpha/\beta^{\mathrm{A}}$ and $\alpha/\beta^{\mathrm{S}}$, can be accommodated with the normal β^{A}-globin chain providing the β^{85Phe}. Deoxygenated HbA molecules therefore have 1/2 the probability of HbS for participating in the deoxygenated polymer phase (48–50).

2.8 Hemoglobin S Polymerization Kinetics

The polymerization kinetics of deoxygenated HbS solutions exceeding the solubility is preceded by a delay period in which no change in physical parameters such as viscosity, turbidity, heat absorption, or nuclear magnetic resonance (water-proton relaxation) are observed (31, 51, 52). In the absence of preexisting polymer, the delay time for polymerization is inversely dependent on the solubility to a high power (≥ 10) (Fig. 9.8a). Detailed analysis of the exponential progress curves for polymerization kinetics suggests a double-nucleation model (Fig. 9.8b) (53, 54). The formation of a critical nucleus proceeds by homogeneous nucleation. With existing polymer fibers, the polymerization process can proceed by heterogeneous nucleation. Fibers grow by addition of hemoglobin at the ends. As hemoglobin concentrations increase, homogeneous nucleation dominates (55).

2.9 Deoxygenated Hemoglobin S Solubility and Hemoglobin Polymerization

The primary determinants of HbS aggregation or polymerization in the intact red cell are hemoglobin composition, hemoglobin concentration, and oxygen saturation (56). Maximal HbS polymerization is achieved at complete deoxygenation. At 0% oxygen, HbS at physiologic concentration partitions into a polymer-phase fraction of 0.7 and a solution-phase fraction of 0.3 (Fig. 9.9) (33). HbS polymerization can be treated as a phase transition (57, 58).

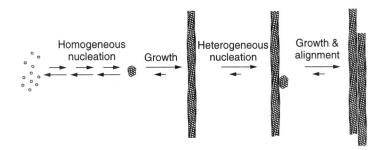

Figure 9.8. Kinetics of hemoglobin S polymerization. The double nucleation model for HbS polymerization suggests two pathways for nucleation (53). The initiation of HbS (circle) polymerization proceeds through the thermodynamically unfavorable aggregates until the formation of a critical nucleus. Subsequent additions of hemoglobin molecules are thermodynamically favorable. The first polymer is formed by homogeneous nucleation. Nucleation of additional polymers can proceed either by homogeneous nucleation or by addition of polymer to preexisting polymer, termed heterogeneous nucleation.

Oxygen binding to HbS in the solution phase proceeds cooperatively as it does for binding to HbA, but the HbS polymer phase does not bind oxygen cooperatively (59, 60). Increasing oxygen saturation decreases the concentration of deoxygenated HbS and the extent of HbS polymerization decreases (29, 33). Although the solubility of deoxygenated HbS is around 16 g/dL, or half of the corpuscular hemoglobin concentration, significant amounts of HbS polymer can be detected at oxygen levels well above 50% oxygen saturation. Because of the broad distribution of corpuscular hemoglobin concentration of red blood cells from an individual homozygous for sickle-cell anemia (SS), HbS polymerization can be detected in some SS red cells, even at oxygen saturation above 90% (61). Although increasing oxygen saturation increases HbS solubility, the increase in solubility is more gradual than would be predicted for an ideal solution. This is a direct consequence of HbS crowding or the high activity (nonideality) of HbS at physiologic concentrations. The low oxygen affinity of the HbS polymer decreases the overall oxygen affinity of SS red blood cells and shifts the oxygen-binding curve to the left (62, 63). This shift can be used as an index of the extent of polymer formation (30, 64).

The low deoxygenated HbS solubility is one of the limiting factors in determining HbS polymerization. Increasing HbS concentration further increases the polymerization tendency (41, 61). Studies of the structure of polymer-

ized HbS led to the appreciation of the relationship between the polymer structure and the morphological abnormality characteristic of sickle-cell anemia (6, 65). Growth and alignment of the HbS polymer fiber can distort the cell, although polymerization can proceed without changes in cell morphology. High hemoglobin concentrations and low deoxygenated HbS solubility favor homogeneous nucleation and small polymer domains (55). Studies of ion transport in SS red blood cells indicated that cycles of polymerization and depolymerization altered membrane transport, leading to changes in ionic and water composition (66, 67).

2.10 Dense SS Erythrocytes and Irreversibly Sickled Cells

In the heterogeneous distribution of corpuscular hemoglobin concentrations of SS red blood cells, the most dense cell fraction, with the greatest hemoglobin concentration (MCHC > 37 g/dL), is the subpopulation with the greatest HbS polymerization tendency (Fig. 9.10) (41, 61). This dense cell fraction contributes disproportionately to the abnormal rheology, deformability, or filterability (Fig. 9.11a) (41, 68). Even when exposed to air, these dense cells contain significant amounts of polymer to decrease filterability *in vitro* that can be further "melted" upon the addition of carbon monoxide (Fig. 9.11b) (39, 68). Irreversible membrane changes lead to membrane rigidity and the formation of irreversibly sickled cells

(a)

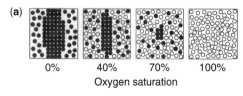

0%   40%   70%   100%

Oxygen saturation

(b)

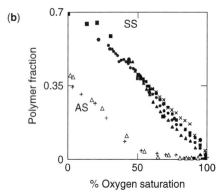

% Oxygen saturation

Figure 9.9. Oxygen content is a primary determinant of hemoglobin S polymerization. (a) Artistic representation of the HbS gel illustrates the relationship between oxygenated hemoglobin tetramers (open circles) and deoxygenated tetramers (filled circles) which may be in the polymer or aligned phase, or in the solution phase in equilibrium with it. (b) ^{13}C-NMR measurements of the fraction of polymerized HbS at equilibrium versus oxygen saturation for SS erythrocytes. Polymer is detected even at very high oxygen saturation in SS erythrocytes, and increases with decreasing oxygen to a fraction of 0.7 at complete deoxygenation. For AS erythrocytes, polymer is detected as oxygen saturation falls below 65% and is maximal at about 0.4 at 0% saturation. For SS and AS erythrocytes, polymer fraction increases with decreasing oxygen and is maximal at complete deoxygenation. AS samples are represented by "Δ" and "+"; all other symbols represent SS. [From C. T. Noguchi et al., *Proc. Natl. Acad. Sci. USA*, **77,** 487 (1980) and C. T. Noguchi, *Biophys. J.,* **45,** 1153 (1984).]

that retain a fixed morphology and are no longer deformable, regardless of the oxygen saturation, even in the absence of HbS polymerization (69–72). In contrast, dense cells are not observed in red blood cell populations of normal individuals, individuals with sickle trait (AS), or individuals with the mild sickle syndrome of S-β^+-thalassemia (73). In these cell populations, loss of filterability correlates

directly with the extent of HbS polymerization for the bulk population (Fig. 9.12).

3   MODIFIERS OF SICKLE-CELL ANEMIA AND THE SICKLE-CELL DISEASE SYNDROMES

3.1   Origins of Sickle-Cell Anemia

Sickle-cell anemia is one of the most prevalent genetic diseases and has the highest frequency in individuals from equatorial Africa. Based on genetic polymorphisms in the β-globin gene cluster associated with the β^S globin gene, the genetic mutation leading to sickle-cell anemia appears to have originated in three independent African regions: Senegal, Benin, and Bantu (Fig. 9.13)(74–76). A group of restriction enzyme sites characteristic of one or more of these genetic polymorphisms is used to define the Senegal, Benin, and Bantu genetic haplotypes, as well as that associated with sickle-cell anemia in Saudi Arabia/India (Fig. 9.13) (77). In the β-globin cluster, the prevalence of the sickle-cell anemia genetic mutation (with some AS frequencies of 25% or higher) overlaps with the geographic distribution of malaria. In spite of the severe clinical manifestations of sickle-cell anemia, the high frequency of the HbS gene correlates with increased resistance to *Plasmodium falciparum* malaria in the AS subpopulation (78–80). Indeed, sickle-cell anemia has low prevalence in nearby arid regions.

3.2   Fetal Hemoglobin

Clinical manifestations of sickle-cell anemia vary markedly and range from mild, with minimal symptoms, to severe, characterized by multiple painful vaso-occlusive crises per year and multiple organ damage (43). Patients with symptomatic disease had the highest frequency of early mortality, and low fetal hemoglobin (HbF) levels correlated with a more severe course in early childhood with increased risk of stroke (81, 82). Fetal hemoglobin is expressed at high levels before birth, and is downregulated after birth as adult hemoglobin (HbA) expression increases, becoming 2% or less in normal adult individuals. Many other modifying factors also contribute to the

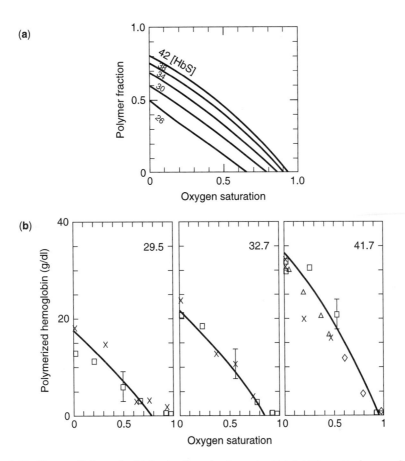

Figure 9.10. Dense cells have the highest polymerization potential. (a) Theoretical curves for illustrating polymer fraction as a function of oxygen saturation for various intracellular HbS concentrations in g/dL as indicated based on a model two-phase model of HbS polymerization. [From C. T. Noguchi, *Biophys. J.*, **45,** 1153 (1984).] (b) ^{13}C-NMR measurements of HbS polymerization of intact SS erythrocytes separated by density gradient to give subpopulations with a narrow density range at the average intracellular hemoglobin concentration (indicated in g/dL) validate the theoretical predictions based on hemoglobin composition, intracellular hemoglobin concentration, and oxygen saturation (solid lines). [From C. T. Noguchi, et al., *J. Clin. Invest.*, **72,** 846 (1983).]

clinical variability of the disease, including physiologic, psychosocial, and environmental conditions (83). In addition to levels of HbF, genetic and other factors, many yet to be identified, contribute to the heterogeneous presentation in clinical severity of sickle-cell disease. These include epistatic genetic modifiers unlinked to the beta- or alpha-globin clusters.

3.3   Genetic Modifiers of Sickle-Cell Anemia

Sickle-cell anemia is not really a monogenetic disease, given that other genes can have very large effects on the severity of the disease and

greatly change its manifestation. The most readily identified genetic effectors are those that modify globin gene expression, that is, the thalassemic syndromes including hereditary persistence of fetal hemoglobin (42). Information on many of these are compiled in a hemoglobinopathy database (http://globin. cse.psu.edu) (84). Variation in hemoglobin composition or hemoglobin concentration within the intact red cell caused by coincident thalassemia or other mutations gives rise to a full range of sickle syndromes (Table 9.1). The sickle syndromes exhibit varying severity,

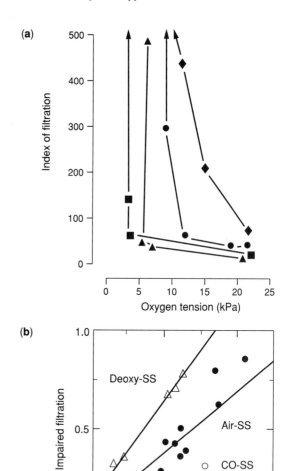

Figure 9.11. Dense SS red cells contribute disproportionately to abnormal rheology. (a) Filtration measurements of subpopulations of density-separated SS erythrocytes shows impairment of filtration detected at higher oxygen saturation for cells with greater corpuscular hemoglobin concentration [from least to most dense, fractions 1 (square), 2 (triangle), 3 (circle), and 4 (diamond)]. [From M. A. Green et al., *J. Clin. Invest.*, **81,** 1669 (1988).] (b) Hemoglobin S polymerization in the dense cell fraction contributes disproportionately to impaired rheology. Dense SS cells have a high polymerization tendency and polymer can still be detected at high oxygen saturation. Dilute SS cell suspensions exposed to room air still exhibit impairment to filtration through 5-micron filters proportional to the percentage of dense cells. In marked contrast, impairment to filtration is significantly reduced to that of near-normal AA cells when cells are exposed to CO that completely melts out the polymerized HbS. [From H. Hiruma et al., *Am. J. Physiol. Heart Circ. Physiol.*, **268,** H2003 (1995).]

ranging from the benign sickle trait (AS) to the most severe African form of homozygous SS sickle-cell anemia. HbS polymerization tendency as calculated from these parameters correlates strongly with the general degree of anemia and relative severity representative of these different sickle syndromes (42).

3.4   Sickle-Cell Anemia and α-Thalassemia

In the α-globin gene cluster, the adult α-globin gene is duplicated (αα). Deletion of one of these α-globin genes (−α) decreases α-globin mRNA expression, a form of α-thalassemia (85). Deletions of two or more α-genes can also

occur and lead to more severe α-thalassemia syndromes. A decrease in α-globin chain production decreases the overall intracellular hemoglobin concentration. This gene deletion is found at a frequency of up to 30% associated with sickle-cell anemia in African populations. The genetic combination of α-thalassemia and homozygous SS results in a reduction of corpuscular hemoglobin concentration, leading to a reduction in the HbS polymerization tendency (86, 87). Increased deformability of the SS erythrocyte with the two α-gene deletion (−α/−α) genotype increases red cell survival and total blood hemoglobin level. The in-

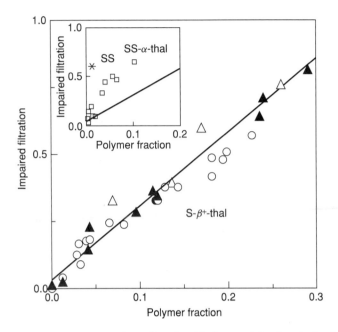

Figure 9.12. HbS polymerization correlates directly with abnormal rheology in the absence of dense cells. Populations of SS cells [with (open squares) or without (star) coexisting α-thalassemia] with significant dense cell fraction show a marked impairment to filtration that does not correlate with polymer fraction determined for the bulk population based on the mean corpuscular hemoglobin concentration (MCHC). When the dense cell fraction is removed, as in sickle trait (AS) erythrocytes (open and filled triangles) or erythrocytes from individuals with S-β^+-thalassemia (open circles), measured impairment of filtration correlates directly with polymerization tendency [predicted polymer fraction based on mean values for hemoglobin composition, mean corpuscular hemoglobin concentration (MCHC), and oxygen saturation]. [From H. Hiruma et al., *Am. J. Hematol.*, **48,** 19 (1995).]

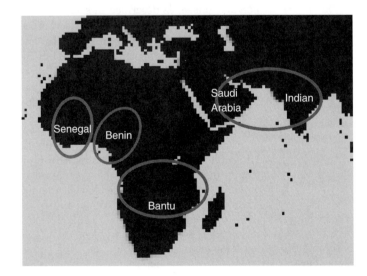

Figure 9.13. Geographic origins of the sickle-cell mutation. Population genetic analyses indicate four major haplotypes associated with sickle-cell anemia that largely localize to four distinct geographic areas (74, 77). These are the Senegal, Benin, Bantu, and Saudi Arabia-Indian haplotypes.

Table 9.1 Some Sickling Disorders

Severity	Condition	Hb (g/dL)	MCHC (g/dL)	HbF (%)	HbS (%)
Asymptomatic	1. HbAS (African)	14.3	33.9	0.8	40.5
	2. HbS-$^{G}\gamma^{A}\gamma\beta^{0}$-HPFH (African)	14.0	33.3	35.9	62.0
Mild	3. HbS-$^{G}\gamma\delta^{\omega}\beta^{0}$-HPFH (African)	11.4	32.7	28.8	69.0
	4. HbS-$^{G}\gamma$-β^{0}-HPFH (African)	11.1	33.0	26.1	71.8
	5. HbSS (Saudi Arabia)	10.9	33.4	28.5	69.4
	6. HbSS (Indian)	10.5	32.0	20.0	77.0
	7. HbS-β^{+}-thalassemia (African)	10.3	31.2	5.0	67.0
	8. HbS-β^{0}-thalassemia (Saudi Arabia)	9.3	32.2	17.3	79.2
Severe	9. HbSS-α-thalassemia (African) ($\alpha-/\alpha-$ genotype)	8.8	32.8	3.8	92.3
	10. HbS-β^{0}-thalassemia (African)	8.6	31.1	6.5	88.5
	11. HbSS-α-thalassemia (African) ($\alpha-/\alpha\alpha$ genotype)	8.1	34.3	4.8	92.1
	12. HbSS (African)	7.8	34.8	5.3	91.9

creased hematocrit and blood viscosity appear to offset the potential benefit of improved red cell parameters (88), and a clear clinical benefit of α-thalassemia in homozygous SS disease has been difficult to demonstrate.

3.5 Sickle-Cell Anemia and β-Thalassemia

Mutations or deletions in the β-globin gene cluster can decrease β-globin gene expression and give rise to β-thalassemia (89, 90). Decreased β-globin gene expression results in decreased intracellular hemoglobin production and intracellular hemoglobin. Two γ-globin genes are also encoded in the β-like globin gene cluster, G-γ- and A-γ-globin. The γ-globin chains in combination with α-globin make up fetal hemoglobin (HbF). In sickle-cell disease, different forms of β-thalassemia can alter the percentage of HbS (%HbS), with resultant increased HbF or HbA$_2$. Because of the decreased polymerization tendency with increased HbF or HbA$_2$ (48, 50), populations with these sickle syndromes present overall a more mild form of sickle-cell disease compared with the African homozygous SS disease (Table 9.1)(42).

3.6 The "Sparing" Effect of Fetal Hemoglobin

Individuals with sickle-cell anemia in Saudi Arabia and India exhibit significantly more mild disease manifestation compared with SS individuals in Africa (76, 91–94). This appears to be a direct consequence of the high level of HbF associated with SS disease in Saudi Arabia and India. The "sparing" effect of HbF has been known since the early 1950s (19). Replacing HbS with HbF reduces deoxygenated hemoglobin polymerization to a greater extent than replacing HbS with HbA (48, 50). Quantitatively, this is related to the formation of mixed hybrids or the combination of HbS/HbF heterodimers to form a hemoglobin tetramer ($\alpha_2\beta^{S}\gamma$). The inability of the HbS/HbF hybrid to participate in the HbS polymer structure further increases the hemoglobin solubility compared with comparable mixtures of HbS and HbA, in which the hybrid can enter the polymer phase, as described in section . This effect can be seen in direct comparison of hemoglobin solubility in mixtures of HbS and non-HbS as the proportion of non-HbS increases (Fig. 9.14). An increase in %HbF to 30% or 40% would increase deoxygenated hemoglobin solubility comparable to a 40:60 mixture of HbS:HbA found in sickle trait (95) and give almost total amelioration of disease manifestation. Note, however, that unlike HbS and HbA, HbF is not necessarily uniformly distributed among the population of erythrocytes (96, 97). In sickle-cell anemia, whereas HbF in a single individual may be up to 8% or greater, HbF can be readily detected in some but not all of the red blood cells. This heterogeneous distribution of HbF compounded with the heterogeneous distribution

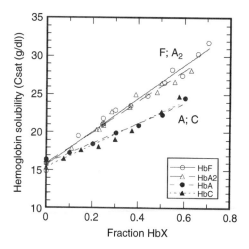

Figure 9.14. Hemoglobin composition modifies the extent of hemoglobin S polymerization. Deoxygenated hemoglobin solubility increases with increasing non-S hemoglobins. The sparing effect of HbF ($\alpha_2\gamma_2$) is greater than HbA ($\alpha_2\beta_2$) because of the differences between the γ-chain and the β-chain. The minor adult hemoglobin, hemoglobin HbA$_2$ ($\alpha_2\delta_2$), has a similar sparing effect on increasing HbS solubility as HbF. Hemoglobin C (HbC; $\alpha_2\beta^C_2$), another mutant hemoglobin with a $\beta^{6Glu>Lys}$ substitution behaves similarly to HbA in mixtures with HbS. Individuals with sickle trait (AS) and SC disease have one β^S-globin gene and one normal β-globin gene or the mutant β^C-globin gene. In sickle trait, the HbS:HbA ratio is about 40:60 because of the higher affinity of the α-chain to the normal β-chain compared with the mutant β^S-chain. In SC disease, the HbS:HbC ratio is about 50:50, and the charge difference of HbC causes red cell dehydration, with an increase in intracellular hemoglobin concentration. These changes result in increased polymerization tendency and disease manifestation in SC disease not observed in sickle trait (AS). [From W. N. Poillon et al., *Proc. Natl. Acad. Sci. USA*, **90**, 5039 (1993).]

of intracellular hemoglobin concentration adds another level of complexity in assessing determinants of disease severity. It should also be noted that HbA$_2$ ($\alpha_2\delta_2$) does not enter the polymer phase and has the same sparing effect as that of HbF (48, 50).

3.7 Sickle-Cell Anemia and Hereditary Persistence of Fetal Hemoglobin

Hereditary persistence of HbF (HPFH) arising from mutations/deletions within the β-glo-

bin gene cluster, giving high levels of γ-globin mRNA, results in high %HbF (98–101). In sickle-cell disease, the resultant reduced HbS polymerization tendency is also correlated with a relatively mild clinical course (42). Some specific genetic polymorphisms in the β-globin gene cluster have been associated with elevated HbF levels. A significant one is the C to T polymorphism at position −158 5′ upstream of the G-γ-globin gene that increases G-γ-globin gene expression (102, 103). This polymorphism is known as the Xmn1 G-γ-polymorphism and can be identified by the ability of the Xmn1 restriction enzyme to cut genomic DNA at this site. Sickle-cell anemias with the Saudi Arabian/Indian and the Senegal genetic background are associated with high levels of HbF and have this Xmn1 G-γ-polymorphism. Other genetic loci not linked to the globin clusters have also been associated with determination of HbF production (104). These include a region on chromosome Xp22.2 and another on chromosome 6q23 (105–107).

3.8 Sickle Trait

As with SS red blood cells, the fraction of polymerized hemoglobin in AS erythrocytes is maximal at complete deoxygenation (49). The hemoglobin solubility in AS erythrocytes decreases to 24 g/dL at complete deoxygenation. The polymerization potential decreases with increasing oxygen saturation with little or no polymer detected at physiologic oxygen levels above 50% oxygen saturation (Fig. 9.9). Although sickle trait or carriers for sickle-cell anemia do not show clinical symptoms, there is a urine-concentrating defect (108). The high osmolality and low oxygen saturation of the renal medulla are conditions that favor HbS polymerization. The urine-concentrating ability correlates inversely with the polymerization potential of AS erythrocytes. The genetic combination of AS with α-thalassemia reduces the %HbS and polymerization potential as well as the urine-concentrating defect (Fig. 9.15). Nevertheless, the absence of pathophysiology associated with AS suggests that reduction of HbS polymerization tendency to that of AS erythrocytes would be an appropriate therapeutic goal.

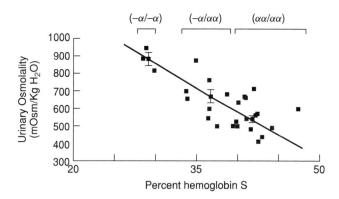

Figure 9.15. Sickle trait (AS) individuals exhibit a urine-concentrating defect correlating with percentage of HbS. The high osmolality and low oxygen saturation of the renal medulla are conditions that favor polymerization. In AS individuals, %HbS (and HbS polymerization tendency) correlates inversely with urine-concentrating ability. Coinheritance of AS with α-thalassemia reduces %HbS as well as the polymerization potential and the urine-concentrating defect. [From A. K. Gupta et al., *J. Clin. Invest.*, **88**, 1963 (1991).]

3.9   Physiologic Modifying Factors

Impaired flow of SS red blood cells arises from the acute and chronic effects of HbS polymerization as cells transit the circulation from the lung, through the tissue microvascular beds, and return. Variation in intracellular hemoglobin composition and intracellular hemoglobin concentration are only two of the modifying factors that give rise to the heterogeneous nature of SS red blood cells and the broad distribution of clinical severity. Microvascular occlusion is attributed primarily to HbS polymerization, which alters SS erythrocyte deformability and causes blockage in the microvasculature. However, the extent of HbS polymerization alone cannot predict the behavior of cells in the macro- and microvascular beds and underscores the importance of direct rheological measurements of SS erythrocytes. Other factors such as plasma proteins and interactions of blood cells (SS red cells, platelets, and leukocytes) with endothelium affect blood flow. The influence of vasomotion and intermittent flow, neural and humoral controls, and adherence of blood cells to the vasculature act to modify passage of SS erythrocytes through the circulation.

3.10   Interactions with Endothelium

Variation in local oxygen saturation as well as tissue demand for oxygen and vascular tone contribute to the variable tissue response of the disease. *In vitro* measurements of SS erythrocyte adherence to endothelium correlate with clinical disease severity (109) and vascular occlusion has been proposed to be initiated by adhesion of the least-dense SS reticulocyte with relatively high endothelium adhesion (110). Several receptors and other molecules on the surface of red blood cells and endothelium have the potential to contribute directly to erythrocyte-endothelial adhesion. These include on the red cell the integrin $\alpha 4\beta 1$ (111), thrombospondin receptor CD36 (112), the very late antigen activator 4 (VLA-4) (113), aggregated band 3 (114), sulfated glycolipids (115), and increased phosphotidylserine (PS) on the surface of SS red blood cells (116). On endothelial cells, association with red cell-endothelium adhesion has been made with the vitronectin receptor integrin $\alpha V\beta 3$ (117), vascular adhesion molecule 1 (VCAM-1) (113), and possibly CD36 and glycoprotein (GP) IB (118). VCAM-1 is not constitutively expressed on endothelium but its expression can be activated by exposure to several agonists such as cytokines and hypoxia (113). Other plasma factors (and their corresponding receptors on erythrocytes and endothelial cells), such as thrombospondin (119, 120), von Willebrand factor (121), and laminin (122), also influence blood cell adherence to endothelium. The role of these interactions in sickle-cell anemia pathophysiology *in vivo* remain uncertain, that is, whether they contribute to chronic events in the microcirculation and organ pathology or whether they contribute more to initiating episodic painful crises. Recently, it has been proposed that it is sickle cell/white cell interactions and not sickle cell/endothelium interactions that are the triggering events in the pathophysiological mechanism (123).

3.11   Animal Model Systems

In addition to *in vitro* systems, animal models and the recent development of mice expressing exclusively human hemoglobins, particularly HbS, have been used to assess SS red cell rheology, pathophysiology, and treatment (124, 125). Intravital microscopy with video image analysis has been used to document the rheological behavior of SS red blood cells and their interaction with vascular endothelium (123, 126, 127). *Ex vivo* preparations of the rat mesocecum vasculature infused with a bolus of red blood cells have been used to determine flow of SS erythrocytes in the microvasculature (110, 115). These studies indicated that vaso-occlusion of oxygenated SS erythrocytes resulted from dense SS cells causing precapillary obstruction and the less-dense reticulocytes and young discocytes preferentially adhering to the immediate postcapillary venules, causing blockage of dense, irreversibly sickled cells (128). Magnetic resonance imaging of the rat leg infused with technetium-99m-labeled erythrocytes provided further evidence of the importance of the densest SS red blood cell in producing vaso-occlusion (129).

4   RATIONAL APPROACHES TO SICKLE-CELL THERAPY

Description of the biophysics of HbS polymerization provides the basis for understanding the pathophysiology of sickle-cell disease. The production of red cells containing a mutant gene product leads to the potential for intraerythrocytic hemoglobin polymerization that can cause obstruction in the microvasculature. The single-nucleotide mutation in the gene for β-globin gives rise to the valine substitution for glutamic acid, resulting in a marked decrease in HbS solubility at physiologic conditions as oxygen is removed from the red blood cell (Fig. 9.6). The resultant transition from the soluble to polymer phase of HbS (Fig. 9.7) alters the viscoelastic properties of HbS inside the SS erythrocyte (Fig. 9.11). Changes in red cell deformability lead to abnormal rheology or blood flow, increased red cell destruction, compromised oxygen delivery, microvascular obstruction, and subsequent downstream clinical manifestations. In addition to symptomatic treatment, therapeutic strategies have targeted hemoglobin polymerization, red cell circulation and the microvasculature, and red cell/hemoglobin production (Table 9.2).

4.1   Inhibition of Hemoglobin S Polymerization

Early efforts at therapeutic strategies focused on increasing deoxygenated HbS solubility (130–132). Rather than the ambitious goal of increasing solubility to match the corpuscular hemoglobin concentration, a reasonable endpoint would be increasing the solubility to mimic that associated with the generally symptom-free AS phenotype (49). Recognizing the hydrophobic substitution of valine for the charged glutamic acid, early efforts focused on disrupting hydrophobic interactions. Agents such as urea, known to perturb solvent interactions, were proposed but their effects were too small to be useful at levels necessary for therapeutic intervention (133–138). Stereospecific competitors such as peptides and modified amino acids were also able to increase solubility (139–142). The hydrophobic aromatic amino acids proved to be the most effective. Chemical modification to increase solubility of the amino acid itself or other oligopeptides increased their potency. X-ray diffraction and solution techniques have been useful in identifying their binding sites (143). However, specific uptake of these compounds by red cells at sufficient concentrations necessary to therapeutically increase HbS solubility remain problematic. The nonideal behavior of even relatively small molecules such as peptides reduced their effectiveness, and changes in solubility were low (142). Furthermore,

Table 9.2   Rational Therapeutic Design for Sickle-Cell Anemia

Mechanism	Target	Examples
Inhibit HbS polymer interactions	Hemoglobin	Noncovalent modifiers (urea, peptides, vanillin) Covalent modifiers to increase solubility and/or oxygen affinity (acetylating agents, cyanate, ethacrynic acid)
Hydration	Erythrocyte	Hyponatremia (DDAVP) Membrane/ion transport modifiers Inhibit Gardos Channel (cetiedil, clotrimazole) Inhibit (KCl) cotransport (Mg^{2+}) Inhibit Cl conductance (NS3623)
Decrease abnormal RBC	Red cell replacement	Exchange transfusion
Decrease microvascular entrapment	Vascular tone Endothelial adhesion	Vasodilators (nifedipine, arginine, nitric oxide) Decrease red cell stasis (Flocor) Decrease endothelial cell adhesion (anti-$\alpha V\beta 3$, anti-P-selectin)
Increase endogenous HbF production	Endogenous γ-globin gene expression	DNA hypomethylation (5-azacytidine) Cell cycle inhibitors (hydroxyurea) Differentiating agents (butyrate, short chain fatty acids, phenylacetate)
Replace HbS production	Hematopoietic stem-cell replacement	Allogeneic bone marrow transplantation
Reduce HbS production	Endogenous hematopoietic stem cells	Increase non-β^S-globin gene expression by retrovirus-mediated gene transfer (γ-globin) Increase non-β^S-globin gene expression by lentivirus-mediated gene transfer (β-mutant-globin) Reduce β^S-globin by ribozymes (β^S-ribozyme, β^A-transribozyme)

several oligopeptides mimicking the region surrounding the β^{S6Val} mutation actually decreased deoxygenated HbS solubility, presumably because of the effects of nonideality and excluded volume (141).

4.2   Cyanate and Sickle-Cell Anemia

In addition to corpuscular hemoglobin concentration and the low deoxygenated HbS solubility, oxygen saturation is one of the major determinants of polymerization tendency. To increase oxygen saturation, strategies aimed at increasing oxygen affinity were developed. Cyanate, a by-product of urea in solution, was found to increase oxygen affinity and reduce sickling of partially deoxygenated SS erythrocytes (144, 145). Cyanate covalently modifies hemoglobin by carbamoylation of the α-amino groups of the globin chains, increasing oxygen affinity (146). Carbamoylation of the β-globin chain specifically also has a small effect on de-

oxygenated HbS solubility. Although *in vitro* assays showed promise in these assays, during clinical treatment of sickle-cell patients with oral administration of potassium cyanate, adverse side effects developed. Cataract formation and peripheral neuropathy as a consequence of oral potassium cyanate administration resulted in discontinued use. Extracorporeal treatment was attempted to overcome these complications, but did not show clear benefit on painful crisis frequency (147–150). More generally, however, it is not clear that increasing oxygen affinity would have overall clinical benefit. Red cells must deliver a constant amount of oxygen to individual tissues and either nonmodified hemoglobin molecules will deliver oxygen (and then polymerize) or oxygen tension will fall sufficiently that even the modified hemoglobin molecules will transfer their oxygen and polymerize.

4.3 Chemical Modifiers of Hemoglobin S

Other chemical modifiers such as nitrogen mustards (151), alkylating agents (152), aldehydes (153), and bis[3,5-dibromo salicyl]fumerate that binds in the 2,3-DPG pocket and cross links the β-82 lysines (154) also increased deoxyhemoglobin S solubility and/or increased oxygen affinity *in vitro*. However, despite their effects, specificity and potency on intact red blood cells are markedly diminished. Covalent modification of hemoglobin in intact erythrocytes is particularly challenging because of potential toxicity and undesirable side reactions. Even with extracorporeal administration, additional problems such as the potential for immunogenic adducts on blood cells arise. Investigation of phenoxy and benzyloxy agents that increased deoxygenated HbS solubility led to studies of the antilipidemic drug clofibrate (155) and the diuretic agent ethacrynic acid (156). The difficulty in identifying potential therapeutic inhibitors of HbS polymerization was exemplified by studies of ethacrynic acid that could inhibit HbS polymerization in cell-free systems. However, as a renal diuretic, treatment of SS cells resulted in ion and water loss. Given that cell shrinkage promotes, rather than inhibits, HbS polymerization the loss of cell water would adversely influence red cell rheology. A mass spectrometry screening method has been proposed as a high throughput methodology to identify new covalent modifiers of hemoglobin (157).

4.4 Vanillin and Hemoglobin S Polymerization

In search for therapeutic agents that could be tolerated at high levels, the food additive vanillin was explored (158). Vanillin was found to inhibit deoxygenated-induced SS cell sickling and possibly increase deoxygenated HbS solubility. Vanillin binds specifically to hemoglobin in the central water cavity and to a region implicated as a contact site in the HbS polymer. Its mode of action is suggested to be a dual mechanism of allosteric modulation to a high oxygen affinity HbS molecule and by stereospecific inhibition of the T-state required

for HbS polymerization. Vanillin would be suitable for further testing in animal models of sickle-cell disease.

4.5 Increasing Sickle Erythrocyte Hydration and Membrane Active Agents

Increasing red cell hydration and decreasing intracellular hemoglobin concentration/modification of intracellular HbS polymerization by decreasing intracellular hemoglobin concentration has been attempted by regimens designed to cause cell swelling. The use of fluid restriction and desmopressin acetate (DDAVP) to induce hyponatremia in a small number of sickle-cell patients under close observation in a metabolic ward resulted in a decrease in MCHC (159). In this limited study there was an apparent decrease in painful crisis frequency. Although impractical for general application because of the severity of this treatment, the feasibility of the approach was demonstrated. Pharmacological agents that affect ion and proton pumps to cause red cell swelling have also been proposed such as cetiedil and, more recently, clotrimazole, an imidazole blocker of the red cell Ca^+-activated K^+ (Gardos) channel (160). Clotrimazole is used to treat mycotic infections through inhibition of cytochrome P450 activity. Preliminary studies in a few sickle-cell anemia patients showed a reduction in MCHC and erythrocyte density, with mostly mild side effects at its lowest dosage (161). Its inhibition of cytochrome P450 may account for the toxicity observed at higher doses. Oral Mg^{2+} supplementation in preliminary studies of SS patients showed a decrease in K-Cl cotransport activity, the principal mediator of red cell dehydration in sickle-cell disease, and a decrease in dense SS red blood cells. However, success and side effects varied with different Mg^{2+} supplements (162, 163). Recent studies of NS3623, an inhibitor of erythrocyte Cl^- conductance, showed *in vivo* hydration of erythrocytes in an animal model (SAD) for sickle-cell disease, with an increase in intracellular Na^+ and K^+ (164). Hematocrit increased and there was a selective loss of the densest erythrocyte population. The highest dose used gave rise to echinocytosis.

5 THERAPEUTIC DECREASE OF MICROVASCULAR ENTRAPMENT

5.1 Antiendothelial Receptor Antibodies

Plasma factors, von Willebrand factor and thrombospondin, contribute to the interaction of SS erythrocytes with vascular endothelium. These factors can bind to receptors on SS erythrocytes and on the surface of endothelial cells. Antibodies to $\alpha V \beta 3$ integrin on the surface of endothelial cells were effective in inhibiting platelet-activating factor-induced SS red blood cell adhesion to endothelium (117). Infusion studies in certain preparations of the rat *ex vivo* mesocecum vasculature demonstrated a reduction in adhesion of SS erythrocytes to the venules and postcapillary occlusion upon pretreatment with monoclonal antibodies (MoAb) 7E3 and LM609. MoAb 7E3 also recognizes $\alpha IIb \beta 3$ (GPII/IIIa), but antibodies specific to $\alpha IIb \beta 3$ had no effect. The resultant decrease in SS red blood cell-endothelium interactions by blockage of $\alpha V \beta 3$ interactions suggests that integrin receptors may be useful targets for therapeutic intervention of blood cell-endothelium interactions. Anti-integrin receptor therapeutics have already been shown to be therapeutically useful in acute coronary syndromes (165).

5.2 P- and E-Selectin and Sickle-Cell Animal Models

In transgenic mouse models expressing human sickle hemoglobins, *in vivo* microcirculatory studies used the cremaster muscle preparation to visualize adhesion of red blood cells to postcapillary venules (166). Transgenic sickle-cell mice demonstrated an inflammatory response to hypoxia/reoxygenation with increased leukocyte adherence, suggesting a model for reperfusion injury associated with human sickle-cell disease (167). This inflammatory response was inhibited by infusion of antibody to P-selectin, but not to anti-E-selectin antibody. Mice deficient in P- and E-selectins exhibit defective leukocyte-endothelium adhesion. When these mice were modeled to express exclusively human HbS, they were protected from vaso-occlusion. These data provide a possible new approach for the pathogenesis of microvascular occlusion and raise

the possibility that drugs targeting leukocyte interactions with endothelium or SS red blood cells may be useful in preventing or treating sickle-cell vaso-occlusion.

5.3 Improved Intravascular Blood Flow and Oxygen Delivery

Oxygenated perflubron-based fluorocarbon emulsion (PFE) has been proposed as a strategy to improve oxygen delivery of partially occluded vessels (127). PFE reduced microvascular obstruction of deoxygenated SS erythrocytes in an *ex vivo* preparation of the rat mesocecum vasculature. In contrast, deoxygenated PFE was not effective in reducing widespread adhesion and postcapillary blockage. PFE has a 10-fold greater capacity to dissolve oxygen compared with that of plasma and appeared to be effective in unsickling SS erythrocytes in partially occluded vessels rather than alter blood cell adhesion or vascular tone.

5.4 Flocor and Sickle-Cell Anemia

Flocor was developed to improve intravascular blood flow by lowering blood viscosity and "friction" between blood cells and the vessel wall. Developed by the company CytRx, Flocor is a polymer shown to reduce slightly the duration and severity of vaso-occlusive crisis in sickle-cell disease patients. Previously examined for treatment of acute myocardial infarction, this rheologic/antithrombotic agent apparently exerts its effects primarily by enhancing blood flow in oxygen-starved tissues. Clinical trials with Flocor suggested a small increase in resolution of vaso-occlusive crisis.

5.5 Nitric Oxide and Sickle-Cell Anemia

The common gas nitric oxide (NO) plays many roles in the body, including relaxation of blood vessels. Researchers studying the effects of nitric oxide inhalation have shown that it can effectively treat several life-threatening lung conditions (168). The gas is successful in expanding constricted blood vessels in the lung without affecting the rest of the body's circulatory system. The effect is limited to the lungs because the gas binds with hemoglobin upon entering the bloodstream, neutralizing

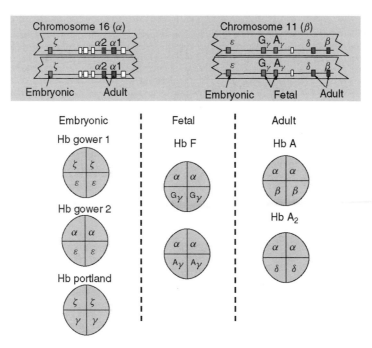

Figure 9.16. Hemoglobin development. Hemoglobin is a tetrameric protein consisting of two α-like and two β-like globin chains. The genetic information for the α-like and β-like globin gene are localized into gene clusters on chromosomes 16 and 11, respectively. The α-globin cluster contains the embryonic ζ- and two adult α-globin genes. The β-globin cluster contains the embryonic ε-, two fetal γ-, and a minor adult δ- and adult β-globin genes. During development, hemoglobin expression exhibits a switch from embryonic to fetal to adult globin genes. Production of the respective globin chains leads to various combinations of hemoglobin tetramers giving rise to the embryonic (Hb Gower and Hb Portland), fetal (HbF), and adult (HbA and HbA$_2$) hemoglobins.

its vessel-expanding properties (169–171). Researchers in Boston reported that it had no effect on normal hemoglobin, but increased the oxygen affinity of sickle hemoglobin, leading to an apparent reduction of sickling (172). Initial studies showed in eight of nine sickle-cell disease patients that breathing nitric oxide caused their red cells to give up oxygen less readily than before, whereas the cells from normal patients showed no change. However, subsequent studies did not confirm these observations and showed no effect of NO treatment on oxygen affinity, other than that attributed to the deleterious formation of methemoglobin (173). Effects of NO inhalation on pulmonary vasculature include a reduction in pulmonary pressures and increased oxygenation, suggesting that NO may be therapeutic for acute chest syndrome and secondary pulmonary hypertension in sickle-cell ane-

mia (174). The observation of low L-arginine and nitric oxide metabolite (NOx) levels during or after vaso-occlusive crisis and acute chest syndrome increased interest in the therapeutic potential of L-arginine as well as NO for sickle-cell anemia (175, 176).

6 THERAPEUTIC INDUCTION OF HEMOGLOBIN F

The hemoglobin tetramer requires two α-like and two β-like globin chains (Fig. 9.3). The α- and β-globin gene clusters encode other like subunits of hemoglobin that are differentially expressed during development (Fig. 9.16). The α-globin gene cluster contains an embryonic form ζ-globin and the two adult α-globin genes. The β-globin gene cluster contains the embryonic ε-globin, the fetal G-γ- and A-γ-globin genes, and the minor adult δ-globin and predominant adult β-globin genes. Sequential expression of these globin chains during development results in production of the embryonic Gower ($\zeta_2\varepsilon_2$, $\alpha_2\varepsilon_2$) and Portland ($\zeta_2\gamma_2$) hemoglobins, fetal hemoglobins ($\alpha_2\gamma_2$), and the adult hemoglobin A ($\alpha_2\beta_2$) and minor adult hemoglobin A_2 ($\alpha_2\delta_2$). Hemoglobin A_2 is generally $\leq 2\%$ and is uniformly distributed among adult erythrocytes. Because these hemoglobins bind oxygen cooperatively, embryonic, fetal, and adult hemoglobins function similarly with possible variation in oxygen affinity and 2,3-DPG binding, and can substitute for adult hemoglobin A in adults. A high throughput screen based on increasing γ-globin promoter activity has been designed to identify new potential inducers of fetal hemoglobin (HbF) (177).

6.1 5-Azacytidine

In sickle-cell anemia, only the β-globin gene is mutated. Substitution of HbF for HbS is not only functional in the adult, but also provides an additional "sparing" effect for HbS polymerization (as discussed above). Activation of fetal globin gene expression has become an important therapeutic strategy in treatment of sickle-cell anemia (178–180). 5-Azacytidine, a DNA methylation inhibitor, was the first such agent to show significant increases in HbF expression in patients with sickle-cell anemia (181, 182). Interestingly, although there was an increase in HbF production, there were no marked increases in MCHC. Rather, there was a normalization of the red cell density distribution and a significant decrease in the dense cell population. The high teratogenic and/or tumorigenic risk associated with 5-azacytidine therapy led to the development of other inducers of HbF (183). Treatment with 5-azacytidine and related compounds remain of interest, particularly in patients who exhibit little or no response to alternative agents such as hydroxyurea (184).

6.2 Hydroxyurea

Available as the anticancer therapeutic for over 30 years, hydroxyurea was found to induce a significant increase in HbF synthesis about 15 years ago (185). In culture, hydroxyurea is known to inhibit progression of the cell cycle into S-phase and inhibit ribonucleotide reductase. Although the exact mechanism of increasing HbF is not known, it is thought to alter the proliferation of early RBC precursors capable of increased HbF synthesis. Hydroxyurea may also increase cellular hydration and MCV. In the late 1980s, several small-scale studies determined the optimal protocols for administering hydroxyurea to sickle-cell patients to elevate HbF (186–190). These studies led to the design and implementation of the Multicenter Study of Hydroxyurea (MSH), which was ended in early 1995 (191). As a result of the MSH study, hydroxyurea has recently been approved by the U.S. Food and Drug Administration (FDA) for use in adults with sickle-cell disease who have had at least three crisis episodes in the preceding year. In the MSH clinical trials of 299 patients, hydroxyurea treatment decreased the number of painful sickle-cell episodes and the frequency of acute chest syndrome by 50%, the number of patients transfused by 30%, and the number of transfusions by 37%. Hydroxyurea's major mechanism of action is to increase HbF (Fig. 9.17). There have been suggestions that other effects, such as decreasing white blood cell levels, may also contribute to clinical benefit, but these have not been proved. In addition, the elevation in VCAM associated with SS disease and suggested to contribute to red cell/endothelial interactions appears to decrease with hydroxyurea. Hydroxyurea also produces nitric oxide both *in vitro* and *in vivo* (192–194). Myelotoxicity and neutropenia were observed with hydroxyurea, and its carcinogenic potential is unknown, particularly for long-term administration in pediatric patients (195, 196).

Preliminary evidence in sickle-cell patients suggested that combination therapy of hydroxyurea and erythropoietin under select conditions could further increase HbF (197), but results appeared to be dependent on treatment regimen (197, 198). Other ribonucleotide reductase inhibitors inducing HbF include Didox, which increased HbF in baboons and in a transgenic mouse model (199, 200), and Resveratrol, which increased HbF in erythroid progenitor cell cultures (201).

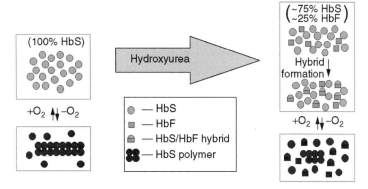

Figure 9.17. Hydroxyurea can induce the production of hemoglobin F in sickle-cell anemia patients that respond to drug treatment. Illustrated is the expected decrease in polymerization potential for HbS at physiologic total intracellular hemoglobin concentration of 34 g/dL if 25% of HbS is replaced with HbF. Because the sparing effect of HbF is greater than that of HbA, induction of 25% HbF with 75% HbS is comparable to the polymerization potential of 60% HbA with 40% HbS expected for sickle trait.

6.3 Butyrate

Butyrate and related short-chain fatty acids stimulate fetal hemoglobin gene expression in erythroid cells (202–205). The butyrates are perhaps the first class of drugs designed to transcriptionally activate fetal globin genes that are developmentally silenced. Although the molecular basis for butyrate action is still being defined, studies have shown that binding of putative regulatory proteins to a specific region of the γ-globin promoter is altered *in vivo* in patients receiving butyrate therapy (206). In initial clinical studies, intravenous infusion of large doses (gram amounts) of butyrate were not consistently efficacious (207). Intermittent or pulse therapy improved the increase in levels of HbF. Recently, greater clinical efficacy has been observed with pulsed intravenous dosing schedules for butyrate (208), although this methodology is likely to be insufficient for general clinical acceptance. An oral form would alleviate problems associated with hospital visits for infusion, and butyrate may offer significant advantage when used in combination with other therapies such as hydroxyurea (178, 179, 209). Although these compounds are relatively safe and without generalized cytotoxicity in patients, drug tolerance develops in some patients after prolonged therapy. An oral form of phenylbutyrate was also found to increase HbF levels (210).

7 BONE MARROW TRANSPLANTATION, GLOBIN GENE EXPRESSION, AND GENE TRANSFER

7.1 Allogeneic Bone Marrow Transplantation

Modern therapy with transfusions and iron chelation, as well as hydroxyurea in most cases of sickle-cell disease, has greatly improved both the quality and length of life for patients with the hemoglobin disorders (43). Nevertheless, progressive iron overload in organs, hepatitis, and other randomly acquired infections increase the risk of mortality with age, especially in β-thalassemia (211). Allogeneic bone marrow transplantation of normal or unaffected (including AS) hematopoietic stem cells from an HLA-matched donor represents the only form of possible cure for these diseases (Fig. 9.18) (212, 213). The potential to be cured decreases with advancing age and risk of death attributed to the procedure increases. Bone marrow transplantation has been investigated most thoroughly for use in β-thalassemia, and more recently for sickle-cell disease, in Europe (214). The degree of morbidity, mortality, and the considerable cost are significant negative factors for widespread application of this approach. Bone marrow transplantation provides a potential cure for only about 5% or fewer of sickle-cell disease patients (215). Patients must have a sickle-cell

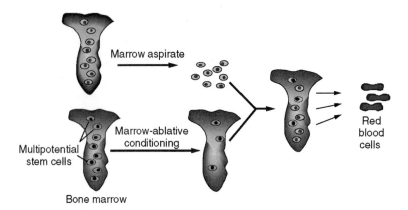

Figure 9.18. Bone marrow transplantation introduces hematopoietic stem cells from a disease-free donor with an AA or AS genotype, as shown in the schematic drawing. Hematopoietic stem cells are harvested from a normal donor. The affected individual is treated with a marrow-conditioning regimen to reduce the pool of affected hematopoietic stem cells. The donor hematopoietic stem cells are then infused into the affected individual. Successful transplantation would result in replacement of the affected hematopoietic stem cells with donor hematopoietic stem cells and the production of normal or unaffected red blood cells.

disease-free sibling who is an exact immunological match. Because of the risk of the procedure (8–10% risk of death attributed to graft vs. host disease and complications), up until now it has been reserved for patients who have failed other treatments. The initial transplant for sickle-cell anemia was carried out as a treatment for coexisting leukemia (216). To date, more than 100 patients with sickle-cell anemia have been treated with bone marrow transplantation.

Transplantation of allogeneic bone marrow through the use of a less-intense marrow-ablative conditioning regimen offers the advantage of reduced toxicity, but increases the potential for hematopoietic mixed chimerism (217). The longer red cell survival time of normal erythrocytes is expected to provide a preferential survival to normal versus SS erythrocytes, and reduce the contribution of remaining SS hematopoietic stem cells to the red cell population (218). The report of a small number (3) of SS patients with stable mixed chimerism (donor myeloid chimerism of 20–75%) after bone marrow transplantation of HbA donor marrow provided evidence for the selective survival of normal erythrocytes (%HbS from 0% to 7%) and a significant ameliorative effect (213). These results offer promise for improvement in the morbidity and mortality associated with bone marrow transplantation from a human leukocyte antigen (HLA)-matched donor for sickle-cell anemia by use of milder marrow-ablative conditioning.

The therapeutic potential of stable mixed chimerism has important implications for gene-transfer approaches in the treatment of sickle-cell anemia and β-thalassemia. Complete modification of the pool of hematopoietic stem cells, a formidable task, no longer appears to be the absolute requirement for therapeutic efficacy. The difficulty in availability of HLA-matched donors and associated toxicity limit treatment by allogeneic bone marrow transplantation. An alternative approach is genetic manipulation of the diseased hematopoietic stem cell by gene replacement/correction or gene addition to restore the normal phenotype. For success, such changes would have to correct a significant proportion of the hematopoietic stem cells with a high level of expression of normal β-like globin genes, resulting in a marked decrease of HbS expression.

7.2 Regulation of Globin Gene Expression

Hemoglobin production proceeds by activation of globin gene transcription (219–222). The coding region for the α-like (encoding 141 amino acids) and β-like (encoding 146 amino acids) globin genes are interrupted by two intervening sequences (introns). Globin transcripts are processed or spliced to remove the two introns, and to add a poly-A tail. Mature globin mRNA transcripts are then transported out of the nucleus and act as templates for globin polypeptide chain synthesis. Appropriate protein folding and incorporation of the iron-containing heme group allows for α-/β-globin dimer formation and association into the hemoglobin tetramer. Globin gene transcription is regulated principally by a proximal promoter located 5' upstream of the coding region. Globin promoters are characterized by an upstream "TATA" box that is located about 30 base pairs (bp) upstream of the start site for transcription, where the transcription initiation complex can assemble. Interactions with other protein complexes assembling upstream on other promoter elements such as CCAAT and CACCC motifs, binding sites for the largely erythroid GATA-1 transcription factor [(A/T)GATA(G/A)] (223), or other enhancers, repressors, or regulatory motifs in distal DNase hypersensitive sites (224) and beyond, contribute to the frequency of transcription initiation of specific globin genes. Nuclear factor-erythroid 2 (NF-E2) (225, 226) and stem cell leukemia (SCL)/Tal-1 transcription factors (227, 228) further contribute to globin gene regulation. Gene-specific transcription factors include erythroid krupple-like factor (EKLF) that binds to the 5' region of the β-globin gene and is required for high level β-globin gene expression in adult erythroid progenitor cells (229), and possibly FKLF and FKLF-2 that are reported to exhibit preferential activation of γ- and ε-globin genes (230, 231). Alteration of transcription factor levels provides the potential for direct manipulation of globin gene transcription.

Naturally occurring mutations and large deletions of the β-globin cluster (Table 9.1) provided initial information on important regulatory elements located *in cis* to the globin genes within the β- or α-globin gene clusters (219). These include point mutations in the 5' flanking region of the γ-globin genes that give rise to the HPFH phenotype by modifying transcription factor binding. Studies of globin gene regulation in transgenic mice provided additional information on DNA regulatory elements that could confer a high level of erythroid-specific gene expression. In transgenic mice, the human β-globin gene that includes the proximal promoter is expressed in a tissue-specific manner, but at low levels (232, 233). DNase hypersensitive sites, particularly hypersensitive site 2 (HS2), located 50 kb 5' of the β-globin gene within the β-globin cluster (234), significantly increase expression of the β-globin or β-like globin transgene (235–238). The five hypersensitive sites spanning 20 kb or more, and referred to as the locus control region (LCR), when combined with the β-globin gene provide a high level of transgene expression in an erythroid-specific manner comparable to the endogenous mouse β-globin gene. The LCR is able to upregulate other *cis*-like erythroid genes in a copy-dependent manner independent of the chromosomal location of the transgene, suggesting that the LCR is critical for a high level of erythroid-specific transcription activity and contributes to determining the chromatin structure of the β-like globin cluster. Construction of vectors for a high level of erythroid-specific expression is likely to incorporate components of the LCR.

The LCR spans 20 kb or more, and its large size limits its utility in vector constructs. To readily use the LCR in expression vectors, core elements have been determined for the DNase hypersensitive sites 2, 3, and 4 (HS2, HS3, HS4) that are able to enhance β-globin gene expression in erythroid cells many fold (239–241). However, in stable cell transfection studies, expression varied from clone to clone, and in some cases, silenced after some time in culture. Although the small truncated LCR was able to provide appropriate gene regulation in transgenic mice, position-effect variation remained problematic in long-term cell culture. "Promoter suppression" can be blocked by use of insulator elements such as the HS4 insulator from the chicken β-globin cluster (242). Inclusion of this insulator in a retrovirus construct significantly improved transduction efficiency in hematopoietic stem cell cultures and in mouse transplantation studies (243, 244). Other strategies for erythroid gene expression have used other enhancer regulatory elements as well as other erythroid-specific promoters (245, 246). The DNase hypersensitive site in the α-globin cluster, HS-40 (247), is another erythroid enhancer that has been incorporated into expression vectors (246, 248).

7.3 Modification of Endogenous Globin Gene Expression

Strategies to reduce globin-specific transcripts include antisense oligonucleotides to block specific globin chain synthesis (249), ribozymes (250), and multiribozymes (251) to reduce the pool of mature transcripts for a specific globin, and transribozyme technology to replace the mutant β^S-globin transcript with a γ-globin transcript (252). In culture studies of globin-producing cells, expression of human β-globin antisense transcripts was able to reduce β-chain biosynthesis and increase γ-chain synthesis fivefold (249). Through the use of hammerhead ribozyme, RNA that is capable of sequence-specific cleavage of other RNA molecules, transgenic mice expressing human β^S-globin and a β^S-globin directed ribozyme reduced the β^S-globin chains by 10% (250). In culture studies, a multiribozyme incorporating five specific globin cleavage sites was able to reduce α-globin mRNA by 50% or more, suggesting a method for further increasing ribozyme activity (251). Other uses of ribozyme technology are based on RNA repair and make use of a *trans*-splicing group I ribozyme to convert β^S-globin transcripts to γ-globin transcripts (252). Initial *in vitro* studies claimed an 8% conversion of β^S-globin transcripts, demonstrating the potential of this approach. Peptide nucleic acids or oligonucleotides capable of forming complexes with the 5' region of the γ-globin gene have been proposed as inducers of HbF (253, 254).

Homologous recombination has been useful in generating genomic mutations or modifications at specific genetic loci. Such modifications in murine embryonic stem cells provide important models of gene-specific diseases. Targeted deletion of the mouse α- and β-globin genes are the basis of mouse models expressing exclusively human hemoglobins, including HbS. However, the current low frequency of homologous recombination diminishes its value as a therapeutic strategy for sickle-cell anemia (255). RNA-DNA oligonucleotides have been proposed as a strategy for site-directed correction of the β^S mutation (256). However, success has been elusive, given the initial reports of a 5–11% conversion rate of GAG in the normal β-globin gene to the GTG β^S-globin mutation in an enriched CD34+ hematopoietic cell population.

7.4 Gene Transfer of Recombinant Globin Gene Vectors

Although large constructs including artificial chromosomes can be used in the construction of transgenic mice, the amount of DNA that can be incorporated into viral vectors is limited (248). Early efforts for genetic-based treatment for sickle-cell anemia and other hemoglobinopathies used replication-defective retroviruses (257–259). Vectors designed to increase expression of normal β-globin were used in mouse studies to target hematopoietic stem cells *in vitro* that were transplanted back into conditioned recipients (Fig. 9.19). The low gene-transfer efficiency and low gene expression in these studies were discouraging. Although sickle-cell anemia was the first genetic disease described more than five decades ago, the difficulty in obtaining a high degree of gene transfer diminished the prospects of treatment of sickle-cell anemia by this modality, and sickle-cell anemia was no longer considered a primary candidate for initial gene therapy trials in humans.

Adenoviral vectors (Ad) were found to be useful in targeting CD34+CD38− hematopoietic cells (260), but their transient nature limits their utility for globin gene expression in late erythroid precursor cells. Recombinant adeno-associated virus (AAV) could provide expression of human γ-globin gene in erythroid cells, particularly in the presence of positive selection, and showed the potential of AAV to transfer globin gene expression (261, 262). Recently, a hybrid adenovirus (Ad)/adeno-associated virus (AAV) vector with a chimeric Ad capsid for efficient gene transfer into human hematopoietic cells and the AAV inverted-terminal repeats for integration of vector genome into the host genome was created to include an 11.6-kb γ-globin expression cassette containing HS2 and HS3 of the LCR (263). This vector was devoid of all viral genes and provided stable transgene expression in hematopoietic cells. Although not yet tested in

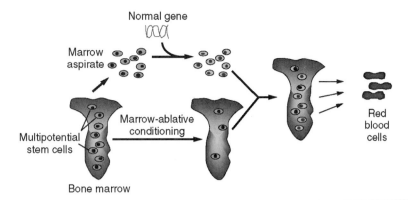

Figure 9.19. A gene-transfer approach to therapy introduces a normal or corrected hemoglobin gene into the hematopoietic stem cell of the affected SS individual, as shown in the schematic drawing. Hematopoietic stem cells are harvested and treated *ex vivo*. The individual is treated with a marrow-conditioning regimen to reduce the pool of affected hematopoietic stem cells. The treated stem cells containing the new genetic material are then infused back into the affected individual. Successful transplantation would result in the replacement of affected hematopoietic stem cells with treated hematopoietic stem cells and the production of normal or unaffected red blood cells.

human hematopoietic stem cells, this vector shows the potential of combining elements from multiple viruses to optimize vector design.

To increase the amount of DNA that can be incorporated into gene-transfer vectors, the human immunodeficiency virus 1 (HIV-1) was modified and used as a basis for vector construction because of its ability to include large DNA fragments and its RNA-splicing potential (246, 264, 265). A large genomic fragment containing the LCR core elements, β-globin gene, and 3′ β-globin enhancer were incorporated into an attenuated HIV-1-derived vector and used successfully in gene-transfer experiments to treat β-thalassemia in a mouse model, by providing therapeutic levels of human normal β-globin (264). However, expression of the transferred normal β-globin gene was heterogeneous and low. As a strategy for treatment of sickle-cell anemia, a modified HIV-1-based lentiviral vector was optimized for expression of a β-globin variant, $\beta^{87\mathrm{Thr}\rightarrow\mathrm{Gln}}$, that prevented HbS polymerization (265). In gene-targeting experiments in sickle-cell anemia mouse models, up to 52% of the hemoglobin was modified and distributed among 99% of circulating erythrocytes, providing normalization of hematological parameters, urine-concentrating ability, and spleen size. These studies demonstrate the ability of lentiviral

vectors to provide long-term expression of globin genes in affected erythroid progenitor cells and reducing the physiologic symptoms in these mouse models of hemoglobinopathies.

7.5 Modification of Hematopoietic Stem Cell Response

The potential for erythropoietin administration to augment the increase in HbF in sickle-cell anemia patients undergoing hydroxyurea therapy suggests an additional strategy for gene-transfer techniques. In animal studies, direct muscle injection of an AAV vector expressing erythropoietin or a DNA plasmid expression vector encoding erythropoietin accompanied by electric pulses to stimulate cell uptake provided long-term expression of erythropoietin *in vivo* (266, 267). Other gene-transfer approaches are based on modifying the surface receptors on hematopoietic stem cells and progenitor cells to enhance drug response (268). Incorporation of drug-resistance genes into the gene-transfer vector provides a potential for competitive selection with drug treatment (269). This strategy biases against untreated endogenous cells or transduced cells not expressing the transferred gene. Inclusion of other genes directed at increasing the pool of transduced hematopoietic stem cells include HOXB4 (270) and possibly the anti-apoptotic Bcl-2 (271). Expression of a truncated erythropoietin receptor provided a selective advantage to transplanted hematopoietic stem cells in the presence of erythropoietin (272). Hybrid receptors have been designed to incorporate some of the hematopoietic cytokine receptors such as receptors for thrombopoietin, erythropoietin, or G-CSF. Through use of a drug-binding extracellular domain with the cytoplasmic domains for thrombopoietin or erythropoietin, the hybrid receptors could be activated by drug binding to mimic cytokine activation (268). In another strategy, cells expressing another hybrid receptor created by fusing the G-CSF cytoplasmic domain with the estrogen receptor extracellular domain became hormone responsive (273). These strategies provide a means of selective stimulation of the transduced hematopoietic stem cells or progenitor cell population. However, host immune response to the expression of new or foreign genes may limit the application of these strategies dependent on the type of ablation/conditioning used.

7.6   Improvement in Gene Transfer Technology

Within the context of gene transfer, efforts have focused on understanding manipulation of hematopoietic stem cells *ex vivo*, on improving vector design for gene delivery, and on optimum design of a globin gene cassette that would provide a high level of globin gene expression in an erythroid-specific manner (274). Cell-marking studies of hematopoietic stem cells in various animal models including nonhuman primates have increased success of gene-transfer technology with the potential of targeting up to 10% or higher. With improved technology and understanding, gene-transfer strategies using a retroviral-derived vector and *ex vivo* infection of hematopoietic stem cells have been useful in providing full correction of the X-linked severe combined immunodeficiency-X1 (SCID-X1) (275). Continued increase in targeting frequency and gene expression have increased the potential of gene-transfer-based therapy for sickle-cell anemia (264, 265). However, a very recent report of leukemia in one of the SCID-X1 children treated has increased caution regarding clinical trials with viral vectors. Studies in nonhuman primates are necessary to judge their ultimate potential for treating human beings with sickle-cell anemia and related hemoglobinopathies.

REFERENCES

1. J. B. Herrick, *Arch. Intern. Med.*, **6**, 517–521 (1910).

2. J. G. Huck, *Bull. Johns Hopkins Hosp.*, **23**, 335–344 (1923).

3. E. V. Hahn and E. B. Gillespie, *Arch. Intern. Med.*, **39**, 233–254 (1927).

4. T. H. Ham and W. B. Castle, *Trans. Assoc. Am. Physicians*, **55**, 127–132 (1940).

5. J. V. Neel, *Science*, **110**, 64–66 (1949).

6. J. F. Bertles and J. Dobler, *Blood*, **33**, 884–898 (1969).

7. W. N. Jensen and L. S. Lessin, *Semin. Hematol.*, **7**, 409–426 (1970).

8. L. Pauling, H. A. Itano, S. J. Singer, and I. C. Wells, *Science*, **110**, 543–548 (1949).

9. T. H. Huisman, *Am. J. Hematol.*, **6**, 173–184 (1979).

10. A. Dubart, M. Goossens, Y. Beuzard, N. Monplaisir, U. Testa, P. Basset, and J. Rosa, *Blood*, **56**, 1092–1099 (1980).

11. V. M. Ingram, *Nature*, **178**, 792–794 (1956).

12. A. Deisseroth, A. Nienhuis, P. Turner, R. Velez, W. F. Anderson, F. Ruddle, J. Lawrence, R. Creagan, and R. Kucherlapati, *Cell*, **12**, 205–218 (1977).

13. D. R. Higgs, J. M. Old, L. Pressley, J. B. Clegg, and D. J. Weatherall, *Nature*, **284**, 632–635 (1980).

14. H. P. Koeffler, R. S. Sparkes, H. Stang, and T. Mohandas, *Proc. Natl. Acad. Sci. USA*, **78**, 7015–7018 (1981).

15. R. V. Lebo, A. V. Carrano, K. Burkhart-Schultz, A. M. Dozy, L. C. Yu, and Y. W. Kan, *Proc. Natl. Acad. Sci. USA*, **76**, 5804–5808 (1979).

16. A. J. Jeffreys, I. W. Craig, and U. Francke, *Nature*, **281**, 606–608 (1979).

17. J. Gusella, A. Varsanyi-Breiner, F. T. Kao, C. Jones, T. T. Puck, C. Keys, S. Orkin, and D. Housman, *Proc. Natl. Acad. Sci. USA*, **76**, 5239–5242 (1979).

18. M. F. Perutz and J. M. Mitchison, *Nature*, **66**, 677–679 (1950).

19. K. Singer and I. Singer, *Blood*, **8**, 1008–1023 (1952).

20. S. B. Krantz, *Blood*, **77**, 419–434 (1991).

21. H. Youssoufian, G. Longmore, D. Neumann, A. Yoshimura, and H. F. Lodish, *Blood*, **81**, 2223–2236 (1993).

22. L. T. Goodnough, B. Skikne, and C. Brugnara, *Blood*, **96**, 823–833 (2000).

23. C. A. Marotta, J. T. Wilson, B. G. Forget, and S. M. Weissman, *J. Biol. Chem.*, **252**, 5040–5053 (1977).

24. H. F. Bunn and M. McDonough, *Biochemistry*, **13**, 988–993 (1974).

25. M. F. Perutz, *Annu. Rev. Biochem.*, **48**, 327–386 (1979).

26. M. F. Perutz, A. M. Liquori, and F. Enrich, *Nature*, **67**, 929–931 (1951).

27. H. F. Bunn and B. Forget, *Hemoglobin: Molecular, Genetic and Clinical Aspects*, Saunders, Philadelphia, 1986.

28. M. F. Perutz, A. J. Wilkinson, M. Paoli, and G. G. Dodson, *Annu. Rev. Biophys. Biomol. Struct.*, **27**, 1–34 (1998).

29. C. T. Noguchi, D. A. Torchia, and A. N. Schechter, *Proc. Natl. Acad. Sci. USA*, **77**, 5487–5491 (1980).

30. M. E. Fabry, L. Desrosiers, and S. M. Suzuka, *Blood*, **98**, 883–884 (2001).

31. J. W. Harris and H. B. Bensusan, *Proc. Soc. Exp. Biol. Med.*, **149**, 826–829 (1975).

32. B. Magdoff-Fairchild, W. N. Poillon, T. Li, and J. F. Bertles, *Proc. Natl. Acad. Sci. USA*, **73**, 990–994 (1976).

33. C. T. Noguchi, D. A. Torchia, and A. N. Schechter, *Proc. Natl. Acad. Sci. USA*, **76**, 4936–4940 (1979).

34. R. W. Briehl, *J. Mol. Biol.*, **123**, 521–538 (1978).

35. E. H. Danish and J. W. Harris, *J. Lab. Clin. Med.*, **101**, 515–526 (1983).

36. J. C. Wang, M. S. Turner, G. Agarwal, S. Kwong, R. Josephs, F. A. Ferrone, and R. W. Briehl, *J. Mol. Biol.*, **315**, 601–612 (2002).

37. S. Usami, S. Chien, and J. F. Bertles, *J. Lab. Clin. Med.*, **86**, 274–279 (1975).

38. N. Mohandas, W. M. Phillips, and M. Bessis, *Semin. Hematol.*, **16**, 95–114 (1979).

39. M. R. Clark, N. Mohandas, and S. B. Shohet, *J. Clin. Invest.*, **65**, 189–196 (1980).

40. R. W. Briehl, *Nature*, **288**, 622–624 (1980).

41. M. A. Green, C. T. Noguchi, A. J. Keidan, S. S. Marwah, and J. Stuart, *J. Clin. Invest.*, **81**, 1669–1674 (1988).

42. G. M. Brittenham, A. N. Schechter, and C. T. Noguchi, *Blood*, **65**, 183–189 (1985).

43. G. R. Serjeant and B. E. Serjeant, *Sickle Cell Disease*, 3rd ed., Oxford University Press, Oxford, UK, 2001.

44. G. W. Dykes, R. H. Crepeau, and S. J. Edelstein, *J. Mol. Biol.*, **130**, 451–472 (1979).

45. B. C. Wishner, K. B. Ward, E. E. Lattman, and W. E. Love, *J. Mol. Biol.*, **98**, 179–194 (1975).

46. E. A. Padlan and W. E. Love, *J. Biol. Chem.*, **260**, 8280–8291 (1985).

47. D. J. Harrington, K. Adachi, and W. E. Royer Jr., *J. Mol. Biol.*, **272**, 398–407 (1997).

48. H. R. Sunshine, J. Hofrichter, and W. A. Eaton, *J. Mol. Biol.*, **133**, 435–467 (1979).

49. C. T. Noguchi, D. A. Torchia, and A. N. Schechter, *J. Biol. Chem.*, **256**, 4168–4171 (1981).

50. W. N. Poillon, B. C. Kim, G. P. Rodgers, C. T. Noguchi, and A. N. Schechter, *Proc. Natl. Acad. Sci. USA*, **90**, 5039–5043 (1993).

51. J. Hofrichter, P. D. Ross, and W. A. Eaton, *Proc. Natl. Acad. Sci. USA*, **71**, 4864–4868 (1974).

52. G. L. Cottam, M. R. Waterman, and B. C. Thompson, *Arch. Biochem. Biophys.*, **181**, 61–65 (1977).

53. F. A. Ferrone, J. Hofrichter, and W. A. Eaton, *J. Mol. Biol.*, **183**, 611–631 (1985).

54. F. A. Ferrone, M. Ivanova, and R. Jasuja, *Biophys. J.*, **82**, 399–406 (2002).

55. R. W. Briehl, *J. Mol. Biol.*, **245**, 710–723 (1995).

56. C. T. Noguchi, *Biophys. J.*, **45**, 1153–1158 (1984).

57. A. P. Minton, *J. Mol. Biol.*, **82**, 483–498 (1974).

58. A. P. Minton, *J. Mol. Biol.*, **110**, 89–103 (1977).

59. R. N. Haire, W. A. Tisel, G. Niazi, A. Rosenberg, S. J. Gill, and B. Richey, *Biochem. Biophys. Res. Commun.*, **101**, 177–182 (1981).

60. H. R. Sunshine, J. Hofrichter, F. A. Ferrone, and W. A. Eaton, *J. Mol. Biol.*, **158**, 251–273 (1982).

61. C. T. Noguchi, D. A. Torchia, and A. N. Schechter, *J. Clin. Invest.*, **72**, 846–852 (1983).

62. A. May and E. R. Huehns, *Br. J. Haematol.*, **30**, 317–335 (1975).

63. R. M. Bookchin, T. Balazs, and L. C. Landau, *J. Lab. Clin. Med.*, **87**, 597–616 (1976).

64. R. E. Benesch, R. Edalji, S. Kwong, and R. Benesch, *Anal. Biochem.*, **89**, 162–173 (1978).

65. C. Acquaye, E. J. Blanchette-Mackie, C. Reindorf, S. Edelstein, and A. N. Schechter, *Blood Cells*, **13**, 359–376 (1988).

66. C. Brugnara, H. F. Bunn, and D. C. Tosteson, *Science*, **232**, 388–390 (1986).

67. R. M. Bookchin, T. Balazs, and V. L. Lew, *J. Mol. Biol.*, **244**, 100–109 (1994).

68. H. Hiruma, C. T. Noguchi, N. Uyesaka, A. N. Schechter, and G. P. Rodgers, *Am. J. Physiol.*, **268**, H2003–H2008 (1995).

69. J. F. Bertles and P. F. Milner, *J. Clin. Invest.*, **47**, 1731–1741 (1968).

70. G. R. Serjeant, B. E. Serjeant, and P. F. Milner, *Br. J. Haematol.*, **17**, 527–533 (1969).

71. S. E. Lux, K. M. John, and M. J. Karnovsky, *J. Clin. Invest.*, **58**, 955–963 (1976).

72. M. R. Clark, R. C. Unger, and S. B. Shohet, *Blood*, **51**, 1169–1178 (1978).

73. H. Hiruma, C. T. Noguchi, N. Uyesaka, S. Hasegawa, E. J. Blanchette-Mackie, A. N. Schechter, and G. P. Rodgers, *Am. J. Hematol.*, **48**, 19–28 (1995).

74. J. Pagnier, J. G. Mears, O. Dunda-Belkhodja, K. E. Schaefer-Rego, C. Beldjord, R. L. Nagel, and D. Labie, *Proc. Natl. Acad. Sci. USA*, **81**, 1771–1773 (1984).

75. Y. Chebloune, J. Pagnier, G. Trabuchet, C. Faure, G. Verdier, D. Labie, and V. Nigon, *Proc. Natl. Acad. Sci. USA*, **85**, 4431–4435 (1988).

76. G. Trabuchet, J. Elion, G. Baudot, J. Pagnier, R. Bouhass, V. M. Nigon, D. Labie, and R. Krishnamoorthy, *Hum. Biol.*, **63**, 241–252 (1991).

77. A. E. Kulozik, B. C. Kar, R. K. Satapathy, B. E. Serjeant, G. R. Serjeant, and D. J. Weatherall, *Blood*, **69**, 1742–1746 (1987).

78. A. C. Allison, *Exp. Parasitol.*, **6**, 418–447 (1957).

79. M. J. Friedman, *Proc. Natl. Acad. Sci. USA*, **75**, 1994–1997 (1978).

80. J. Carlson, G. B. Nash, V. Gabutti, F. al-Yaman, and M. Wahlgren, *Blood*, **84**, 3909–3914 (1994).

81. D. R. Powars, W. A. Schroeder, J. N. Weiss, L. S. Chan, and S. P. Azen, *J. Clin. Invest.*, **65**, 732–740 (1980).

82. M. C. Stevens, R. J. Hayes, S. Vaidya, and G. R. Serjeant, *J. Pediatr.*, **98**, 37–41 (1981).

83. G. R. Serjeant, *Curr. Opin. Hematol.*, **2**, 103–108 (1995).

84. R. C. Hardison, D. H. Chui, C. R. Riemer, W. Miller, M. F. Carver, T. P. Molchanova, G. D. Efremov, and T. H. Huisman, *Hemoglobin*, **22**, 113–127 (1998).

85. J. A. Phillips 3rd, T. A. Vik, A. F. Scott, K. E. Young, H. H. Kazazian Jr., K. D. Smith, V. F. Fairbanks, and H. M. Koenig, *Blood*, **55**, 1066–1069 (1980).

86. K. de Ceulaer, D. R. Higgs, D. J. Weatherall, R. J. Hayes, B. E. Serjeant, and G. R. Serjeant, *N. Engl. J. Med.*, **309**, 189–190 (1983).

87. C. T. Noguchi, G. J. Dover, G. P. Rodgers, G. R. Serjeant, S. E. Antonarakis, N. P. Anagnou, D. R. Higgs, D. J. Weatherall, and A. N. Schechter, *J. Clin. Invest.*, **75**, 1632–1637 (1985).

88. B. E. Serjeant, K. P. Mason, M. W. Kenny, J. Stuart, D. R. Higgs, D. J. Weatherall, R. J. Hayes, and G. R. Serjeant, *Br. J. Haematol.*, **55**, 479–486 (1983).

89. J. A. Kantor, P. H. Turner, and A. W. Nienhuis, *Cell.*, **21**, 149–57 (1980).

90. D. J. Weatherall, *Baillieres Clin. Haematol.*, **11**, 127–146 (1998).

91. G. Brittenham, B. Lozoff, J. W. Harris, V. S. Sharma, and S. Narasimhan, *Am. J. Hematol.*, **2**, 25–32 (1977).

92. R. P. Perrine, M. E. Pembrey, P. John, S. Perrine, and F. Shoup, *Ann. Intern. Med.*, **88**, 1–6 (1978).

93. G. Brittenham, B. Lozoff, J. W. Harris, S. M. Mayson, A. Miller, and T. H. Huisman, *Am. J. Hematol.*, **6**, 107–123 (1979).

94. W. G. Wood, M. E. Pembrey, G. R. Serjeant, R. P. Perrine, and D. J. Weatherall, *Br. J. Haematol.*, **45**, 431–445 (1980).

95. C. T. Noguchi, G. P. Rodgers, G. Serjeant, and A. N. Schechter, *N. Engl. J. Med.*, **318**, 96–99 (1988).

96. G. J. Dover, S. H. Boyer, and M. E. Pembrey, *Science*, **211**, 1441–1444 (1981).

97. G. J. Dover, T. Chan, and F. Sieber, *Blood*, **61**, 1242–1246 (1983).

98. G. Stamatoyannopoulos, W. G. Wood, T. Papayannopoulou, and P. E. Nute, *Blood*, **46**, 683–692 (1975).

99. S. Friedman and E. Schwartz, *Nature*, **259**, 138–140 (1976).

100. S. Ottolenghi, P. Comi, B. Giglioni, P. Tolstoshev, W. G. Lanyon, G. J. Mitchell, R. Williamson, G. Russo, S. Musumeci, et al., *Cell*, **9**, 71–80 (1976).

101. S. Ottolenghi, B. Giglioni, P. Comi, A. M. Gianni, E. Polli, C. T. Acquaye, J. H. Oldham, and G. Masera, *Nature*, **278**, 654–657 (1979).

102. L. E. Lie-Injo, M. L. Lim, Z. Randhawa, T. Vijayasilan, and K. Hassan, *Hemoglobin*, **11**, 231–239 (1987).

103. S. K. Ballas, C. A. Talacki, K. Adachi, E. Schwartz, S. Surrey, and E. Rappaport, *Hemoglobin*, **15**, 393–405 (1991).

104. S. H. Boyer, G. J. Dover, G. R. Serjeant, K. D. Smith, S. E. Antonarakis, S. H. Embury, L. Margolet, A. N. Noyes, M. L. Boyer, et al., *Blood*, **64**, 1053–1058 (1984).

105. G. J. Dover, K. D. Smith, Y. C. Chang, S. Purvis, A. Mays, D. A. Meyers, C. Sheils, and G. Serjeant, *Blood*, **80**, 816–824 (1992).

106. J. E. Craig, J. Rochette, C. A. Fisher, D. J. Weatherall, S. Marc, G. M. Lathrop, F. Demenais, and S. Thein, *Nat. Genet.*, **12**, 58–64 (1996).

107. Y. P. Chang, M. Maier-Redelsperger, K. D. Smith, L. Contu, R. Ducroco, M. de Montalembert, M. Belloy, J. Elion, G. J. Dover, et al., *Br. J. Haematol.*, **96**, 806–814 (1997).

108. A. K. Gupta, K. A. Kirchner, R. Nicholson, J. G. Adams 3rd, A. N. Schechter, C. T. Noguchi, and M. H. Steinberg, *J. Clin. Invest.*, **88**, 1963–1968 (1991).

109. R. P. Hebbel, M. A. Boogaerts, J. W. Eaton, and M. H. Steinberg, *N. Engl. J. Med.*, **302**, 992–995 (1980).

110. D. K. Kaul, M. E. Fabry, and R. L. Nagel, *Proc. Natl. Acad. Sci. USA*, **86**, 3356–3360 (1989).

111. R. A. Swerlick, J. R. Eckman, A. Kumar, M. Jeitler, and T. M. Wick, *Blood*, **82**, 1891–1899 (1993).

112. P. V. Browne and R. P. Hebbel, *J. Lab. Clin. Med.*, **127**, 340–347 (1996).

113. B. N. Setty and M. J. Stuart, *Blood*, **88**, 2311–2320 (1996).

114. B. J. Thevenin, I. Crandall, S. K. Ballas, I. W. Sherman, and S. B. Shohet, *Blood*, **90**, 4172–4179 (1997).

115. G. A. Barabino, X. D. Liu, B. M. Ewenstein, and D. K. Kaul, *Blood*, **93**, 1422–1429 (1999).

116. B. N. Setty, S. Kulkarni, and M. J. Stuart, *Blood*, **99**, 1564–1571 (2002).

117. D. K. Kaul, H. M. Tsai, X. D. Liu, M. T. Nakada, R. L. Nagel, and B. S. Coller, *Blood*, **95**, 368–374 (2000).

118. R. P. Hebbel, *N. Engl. J. Med.*, **342**, 1910–1912 (2000).

119. K. Sugihara, T. Sugihara, N. Mohandas, and R. P. Hebbel, *Blood*, **80**, 2634–2642 (1992).

120. H. A. Brittain, J. R. Eckman, R. A. Swerlick, R. J. Howard, and T. M. Wick, *Blood*, **81**, 2137–2143 (1993).

121. T. M. Wick, J. L. Moake, M. M. Udden, S. G. Eskin, D. A. Sears, and L. V. McIntire, *J. Clin. Invest.*, **80**, 905–910 (1987).

122. S. P. Lee, M. L. Cunningham, P. C. Hines, C. C. Joneckis, E. P. Orringer, and L. V. Parise, *Blood*, **92**, 2951–2958 (1998).

123. A. Turhan, L. A. Weiss, N. Mohandas, B. S. Coller, and P. S. Frenette, *Proc. Natl. Acad. Sci. USA*, **99**, 3047–3051 (2002).

124. M. J. Blouin, H. Beauchemin, A. Wright, M. De Paepe, M. Sorette, A. M. Bleau, B. Nakamoto, C. N. Ou, G. Stamatoyannopoulos, et al., *Nat. Med.*, **6**, 177–182 (2000).

125. R. L. Nagel and M. E. Fabry, *Br. J. Haematol.*, **112**, 19–25 (2001).

126. J. Kurantsin-Mills, H. M. Jacobs, P. P. Klug, and L. S. Lessin, *Microvasc. Res.*, **34**, 152–167 (1987).

127. D. K. Kaul, X. Liu, and R. L. Nagel, *Blood*, **98**, 3128–3131 (2001).

128. D. K. Kaul, D. Chen, and J. Zhan, *Blood*, **83**, 3006–3017 (1994).

129. T. K. Aldrich, S. K. Dhuper, N. S. Patwa, E. Makolo, S. M. Suzuka, S. A. Najeebi, S. Santhanakrishnan, R. L. Nagel, and M. E. Fabry, *J. Appl. Physiol.*, **80**, 531–539 (1996).

130. J. Dean and A. N. Schechter, *N. Engl. J. Med.*, **299**, 863–870 (1978).

131. J. Dean and A. N. Schechter, *N. Engl. J. Med.*, **299**, 804–811 (1978).

132. J. Dean and A. N. Schechter, *N. Engl. J. Med.*, **299**, 752–763 (1978).

133. M. Murayama, *CRC Crit. Rev. Biochem.*, **1**, 461–499 (1973).

134. R. M. Nalbandian, *N. Engl. J. Med.*, **289**, 806 (1973).

135. Cooperative Urea Trials Group, *JAMA*, **228**, 1120–1124 (1974).

136. Cooperative Urea Trials Group, *JAMA*, **228**, 1125–1128 (1974).

137. Cooperative Urea Trials Group, *JAMA*, **228**, 1129–1131 (1974).

138. A. May and E. R. Huehns, *Br. J. Haematol.*, **30**, 21–29 (1975).

139. J. R. Votano, M. Gorecki, and A. Rich, *Science*, **196**, 1216–1219 (1977).

140. S. Kubota and J. T. Yang, *Proc. Natl. Acad. Sci. USA*, **74**, 5431–5434 (1977).

141. C. T. Noguchi and A. N. Schechter, *Biochemistry*, **17**, 5455–5459 (1978).

142. C. T. Noguchi, K. L. Luskey, and V. Pavone, *Mol. Pharmacol.*, **28**, 40–44 (1985).

143. D. J. Abraham, M. F. Perutz, and S. E. Phillips, *Proc. Natl. Acad. Sci. USA*, **80**, 324–328 (1983).

144. A. Cerami and J. M. Manning, *Proc. Natl. Acad. Sci. USA*, **68**, 1180–1183 (1971).

145. A. M. Nigen, N. Njikam, C. K. Lee, and J. M. Manning, *J. Biol. Chem.*, **249**, 6611–6616 (1974).

146. A. M. Nigen and J. M. Manning, *J. Biol. Chem.*, **250**, 8248–8250 (1975).

147. E. E. Langer, G. Stamatoyannopoulos, M. P. Hlastala, J. W. Adamson, M. Figley, R. F. Labbe, J. C. Detter, and C. A. Finch, *J. Lab. Clin. Med.*, **87**, 462–474 (1976).

148. D. A. Deiderich, R. C. Trueworthy, P. Gill, A. M. Cader, and W. E. Larsen, *J. Clin. Invest.*, **58**, 642–653 (1976).

149. S. P. Balcerzak, M. R. Grever, D. E. Sing, J. N. Bishop, and M. L. Segal, *J. Lab. Clin. Med.*, **100**, 345–355 (1982).

150. M. Y. Lee, D. A. Uvelli, L. C. Agodoa, B. H. Scribner, C. A. Finch, and A. L. Babb, *J. Lab. Clin. Med.*, **100**, 334–344 (1982).

151. E. F. Roth Jr., B. Wenz, H. B. Lee, M. Fabry, H. Chang, D. K. Kaul, S. Baez, and R. L. Nagel, *Prog. Clin. Biol. Res.*, **240**, 245–261 (1987).

152. S. Charache, R. Dreyer, I. Zimmerman, and C. K. Hsu, *Blood*, **47**, 481–488 (1976).

153. A. S. Acharya, L. G. Sussman, W. M. Jones, and J. M. Manning, *Anal. Biochem.*, **136**, 101–109 (1984).

154. R. Chatterjee, Y. Iwai, R. Y. Walder, and J. A. Walder, *J. Biol. Chem.*, **259**, 14863–14873 (1984).

155. D. C. Patwa, D. J. Abraham, and T. C. Hung, *Blood Cells*, **12**, 589–601 (1987).

156. E. P. Orringer, D. S. Blythe, J. A. Whitney, S. Brockenbrough, and D. J. Abraham, *Am. J. Hematol.*, **39**, 39–44 (1992).

157. S. Park, L. Wanna, M. E. Johnson, and D. L. Venton, *J. Comb. Chem.*, **2**, 314–317 (2000).

158. D. J. Abraham, A. S. Mehanna, F. C. Wireko, J. Whitney, R. P. Thomas, and E. P. Orringer, *Blood*, **77**, 1334–1341 (1991).

159. R. M. Rosa, B. E. Bierer, H. F. Bunn, and F. H. Epstein, *Blood Cells*, **8**, 329–335 (1982).

160. C. Brugnara, *Curr. Opin. Hematol.*, **2**, 132–138 (1995).

161. C. Brugnara, B. Gee, C. C. Armsby, S. Kurth, M. Sakamoto, N. Rifai, S. L. Alper, and O. S. Platt, *J. Clin. Invest.*, **97**, 1227–1234 (1996).

162. L. De Franceschi, D. Bachir, F. Galacteros, G. Tchernia, T. Cynober, S. Alper, O. Platt, Y. Beuzard, and C. Brugnara, *J. Clin. Invest.*, **100**, 1847–1852 (1997).

163. L. De Franceschi, D. Bachir, F. Galacteros, G. Tchernia, T. Cynober, D. Neuberg, Y. Beuzard, and C. Brugnara, *Br. J. Haematol.*, **108**, 284–289 (2000).

164. P. Bennekou, L. de Franceschi, O. Pedersen, L. Lian, T. Asakura, G. Evans, C. Brugnara, and P. Christophersen, *Blood*, **97**, 1451–1457 (2001).

165. B. S. Coller, *Thromb. Haemost.*, **86**, 427–443 (2001).

166. D. K. Kaul, M. E. Fabry, F. Costantini, E. M. Rubin, and R. L. Nagel, *J. Clin. Invest.*, **96**, 2845–2853 (1995).

167. D. K. Kaul and R. P. Hebbel, *J. Clin. Invest.*, **106**, 411–420 (2000).

168. O. I. Miller, S. F. Tang, A. Keech, N. B. Pigott, E. Beller, and D. S. Celermajer, *Lancet*, **356**, 1464–1469 (2000).

169. A. J. Gow and J. S. Stamler, *Nature*, **391**, 169–173 (1998).

170. M. T. Gladwin, F. P. Ognibene, L. K. Pannell, J. S. Nichols, M. E. Pease-Fye, J. H. Shelhamer, and A. N. Schechter, *Proc. Natl. Acad. Sci. USA*, **97**, 9943–9948 (2000).

171. K. T. Huang, T. H. Han, D. R. Hyduke, M. W. Vaughn, H. Van Herle, T. W. Hein, C. Zhang, L. Kuo, and J. C. Liao, *Proc. Natl. Acad. Sci. USA*, **98**, 11771–11776 (2001).

172. C. A. Head, C. Brugnara, R. Martinez-Ruiz, R. M. Kacmarek, K. R. Bridges, D. Kuter, K. D. Bloch, and W. M. Zapol, *J. Clin. Invest.*, **100**, 1193–1198 (1997).

173. M. T. Gladwin, A. N. Schechter, J. H. Shelhamer, L. K. Pannell, D. A. Conway, B. W. Hrinczenko, J. S. Nichols, M. E. Pease-Fye, C. T. Noguchi, et al., *J. Clin. Invest.*, **104**, 937–945 (1999).

174. M. T. Gladwin and A. N. Schechter, *Semin. Hematol.*, **38**, 333–342 (2001).

175. C. R. Morris, F. A. Kuypers, S. Larkin, E. P. Vichinsky, and L. A. Styles, *J. Pediatr. Hematol. Oncol.*, **22**, 515–520 (2000).

176. C. R. Morris, F. A. Kuypers, S. Larkin, N. Sweeters, J. Simon, E. P. Vichinsky, and L. A. Styles, *Br. J. Haematol.*, **111**, 498–500 (2000).

177. M. L. Holmes, J. D. Haley, L. Cerruti, W. L. Zhou, H. Zogos, D. E. Smith, J. M. Cunningham, and S. M. Jane, *Mol. Cell. Biol.*, **19**, 4182–4190 (1999).

178. H. F. Bunn, *Blood*, **93**, 1787–1789 (1999).

179. G. F. Atweh and D. Loukopoulos, *Semin. Hematol.*, **38**, 367–373 (2001).

180. G. F. Atweh and A. N. Schechter, *Curr. Opin. Hematol.*, **8**, 123–130 (2001).

181. S. Charache, G. Dover, K. Smith, C. C. Talbot Jr., M. Moyer, and S. Boyer, *Proc. Natl. Acad. Sci. USA*, **80**, 4842–4846 (1983).

182. T. J. Ley, J. DeSimone, C. T. Noguchi, P. H. Turner, A. N. Schechter, P. Heller, and A. W. Nienhuis, *Blood*, **62**, 370–380 (1983).

183. T. J. Ley and A. W. Nienhuis, *Annu. Rev. Med.*, **36**, 485–498 (1985).

184. M. Koshy, L. Dorn, L. Bressler, R. Molokie, D. Lavelle, N. Talischy, R. Hoffman, W. van Overveld, and J. DeSimone, *Blood*, **96**, 2379–2384 (2000).

185. O. S. Platt, S. H. Orkin, G. Dover, G. P. Beardsley, B. Miller, and D. G. Nathan, *J. Clin. Invest.*, **74**, 652–656 (1984).

186. R. Veith, R. Galanello, T. Papayannopoulou, and G. Stamatoyannopoulos, *N. Engl. J. Med.*, **313**, 1571–1575 (1985).

187. G. J. Dover, R. K. Humphries, J. G. Moore, T. J. Ley, N. S. Young, S. Charache, and A. W. Nienhuis, *Blood*, **67**, 735–738 (1986).

188. S. Charache, G. J. Dover, M. A. Moyer, and J. W. Moore, *Blood*, **69**, 109–116 (1987).

189. G. P. Rodgers, G. J. Dover, C. T. Noguchi, A. N. Schechter, and A. W. Nienhuis, *N. Engl. J. Med.*, **322**, 1037–1045 (1990).

190. S. Charache, G. J. Dover, R. D. Moore, S. Eckert, S. K. Ballas, M. Koshy, P. F. Milner, E. P. Orringer, G. Phillips Jr., et al., *Blood*, **79**, 2555–2565 (1992).

191. S. Charache, M. L. Terrin, R. D. Moore, G. J. Dover, F. B. Barton, S. V. Eckert, R. P. McMahon, and D. R. Bonds, *N. Engl. J. Med.*, **332**, 1317–1322 (1995).

192. J. Jiang, S. J. Jordan, D. P. Barr, M. R. Gunther, H. Maeda, and R. P. Mason, *Mol. Pharmacol.*, **52**, 1081–1086 (1997).

193. R. E. Glover, E. D. Ivy, E. P. Orringer, H. Maeda, and R. P. Mason, *Mol. Pharmacol.*, **55**, 1006–1010 (1999).

194. M. T. Gladwin, J. H. Shelhamer, F. P. Ognibene, M. E. Pease-Fye, J. S. Nichols, B. Link, D. B. Patel, M. A. Jankowski, L. K. Pannell, et al., *Br. J. Haematol.*, **116**, 436–444 (2002).

195. M. Maier-Redelsperger, M. de Montalembert, A. Flahault, M. G. Neonato, R. Ducrocq, M. P. Masson, R. Girot, and J. Elion, *Blood*, **91**, 4472–4479 (1998).

196. M. Maier-Redelsperger, D. Labie, and J. Elion, *Curr. Opin. Hematol.*, **6**, 115–120 (1999).

197. G. P. Rodgers, G. J. Dover, N. Uyesaka, C. T. Noguchi, A. N. Schechter, and A. W. Nienhuis, *N. Engl. J. Med.*, **328**, 73–80 (1993).

198. M. A. Goldberg, C. Brugnara, G. J. Dover, L. Schapira, L. Lacroix, and H. F. Bunn, *Semin. Oncol.*, **19**, 74–81 (1992).

199. B. S. Pace, H. L. Elford, and G. Stamatoyannopoulos, *Am. J. Hematol.*, **45**, 136–141 (1994).

200. W. E. Iyamu, S. E. Adunyah, H. Fasold, K. Horiuchi, H. L. Elford, T. Asakura, and E. A. Turner, *Am. J. Hematol.*, **63**, 176–183 (2000).

201. C. M. Rodrigue, N. Arous, D. Bachir, J. Smith-Ravin, P. H. Romeo, F. Galacteros, and M. C. Garel, *Br. J. Haematol.*, **113**, 500–507 (2001).

202. S. P. Perrine, B. A. Miller, D. V. Faller, R. A. Cohen, E. P. Vichinsky, D. Hurst, B. H. Lubin, and T. Papayannopoulou, *Blood*, **74**, 454–459 (1989).

203. S. P. Perrine, G. H. Dover, P. Daftari, C. T. Walsh, Y. Jin, A. Mays, and D. V. Faller, *Br. J. Haematol.*, **88**, 555–561 (1994).

204. G. Stamatoyannopoulos, C. A. Blau, B. Nakamoto, B. Josephson, Q. Li, E. Liakopoulou, B. Pace, T. Papayannopoulou, S. W. Brusilow, et al., *Blood*, **84**, 3198–3204 (1994).

205. E. Liakopoulou, C. A. Blau, Q. Li, B. Josephson, J. A. Wolf, B. Fournarakis, V. Raisys, G. Dover, T. Papayannopoulou, et al., *Blood*, **86**, 3227–3235 (1995).

206. T. Ikuta, Y. W. Kan, P. S. Swerdlow, D. V. Faller, and S. P. Perrine, *Blood*, **92**, 2924–2933 (1998).

207. S. P. Perrine, G. D. Ginder, D. V. Faller, G. H. Dover, T. Ikuta, H. E. Witkowska, S. P. Cai, E. P. Vichinsky, and N. F. Olivieri, *N. Engl. J. Med.*, **328**, 81–86 (1993).

208. G. F. Atweh, M. Sutton, I. Nassif, V. Boosalis, G. J. Dover, S. Wallenstein, E. Wright, L. McMahon, G. Stamatoyannopoulos, et al., *Blood*, **93**, 1790–1797 (1999).

209. K. T. McDonagh, G. J. Dover, R. E. Donahue, D. G. Nathan, B. Agricola, E. Byrne, and A. W. Nienhuis, *Exp. Hematol.*, **20**, 1156–1164 (1992).

210. G. J. Dover, S. Brusilow, and S. Charache, *Blood*, **84**, 339–343 (1994).

211. A. Eldor and E. A. Rachmilewitz, *Blood*, **99**, 36–43 (2002).

212. M. C. Walters, *J. Pediatr. Hematol. Oncol.*, **21**, 467–474 (1999).

213. M. C. Walters, R. Storb, M. Patience, W. Leisenring, T. Taylor, J. E. Sanders, G. E. Buchanan, Z. R. Rogers, P. Dinndorf, et al., *Blood*, **95**, 1918–1924 (2000).

214. G. Lucarelli, R. A. Clift, M. Galimberti, E. Angelucci, C. Giardini, D. Baronciani, P. Polchi, M. Andreani, D. Gaziev, et al., *Blood*, **93**, 1164–1167 (1999).

215. W. C. Mentzer, S. Heller, P. R. Pearle, E. Hackney, and E. Vichinsky, *J. Pediatr. Hematol. Oncol.*, **16**, 27–29 (1994).

216. F. L. Johnson, A. T. Look, J. Gockerman, M. R. Ruggiero, L. Dalla-Poazz, and F. T. Billings 3rd, *N. Engl. J. Med.*, **311**, 780–783 (1984).

217. R. G. Amado and G. J. Schiller, *Semin. Oncol.*, **27**, 82–89 (2000).

218. R. Iannone, L. Luznik, L. W. Engstrom, S. L. Tennessee, F. B. Askin, J. F. Casella, T. S. Kickler, S. N. Goodman, A. L. Hawkins, et al., *Blood*, **97**, 3960–3965 (2001).

219. G. Stamatoyannopoulos and A. W. Nienhuis in G. Stamatoyannopoulos, A. W. Nienhuis, P. W. Majerus, and H. Varmus, Eds., *The Molecular Basis of Blood Diseases*, Saunders, Philadelphia, 1994, pp. 107–155.

220. J. D. Engel and K. Tanimoto, *Cell*, **100**, 499–502 (2000).

221. T. McMorrow, A. van den Wijngaard, A. Wollenschlaeger, M. van de Corput, K. Monkhorst, T. Trimborn, P. Fraser, M. van Lohuizen, T. Jenuwein, et al., *EMBO J.*, **19**, 4986–4996 (2000).

222. D. Schubeler, C. Francastel, D. M. Cimbora, A. Reik, D. I. Martin, and M. Groudine, *Genes Dev.*, **14**, 940–950 (2000).

223. A. P. Tsang, Y. Fujiwara, D. B. Hom, and S. H. Orkin, *Genes Dev.*, **12**, 1176–1188 (1998).

224. D. Tuan, W. Solomon, Q. Li, and I. M. London, *Proc. Natl. Acad. Sci. USA*, **82**, 6384–6388 (1985).

225. N. C. Andrews, H. Erdjument-Bromage, M. B. Davidson, P. Tempst, and S. H. Orkin, *Nature*, **362**, 722–728 (1993).

226. P. A. Ney, N. C. Andrews, S. M. Jane, B. Safer, M. E. Purucker, S. Weremowicz, C. C. Morton, S. C. Goff, S. H. Orkin, et al., *Mol. Cell. Biol.*, **13**, 5604–5612 (1993).

227. L. Robb, I. Lyons, R. Li, L. Hartley, F. Kontgen, R. P. Harvey, D. Metcalf, and C. G. Begley, *Proc. Natl. Acad. Sci. USA*, **92**, 7075–7079 (1995).

228. R. A. Shivdasani, E. L. Mayer, and S. H. Orkin, *Nature*, **373**, 432–434 (1995).

229. W. Zhang, S. Kadam, B. M. Emerson, and J. J. Bieker, *Mol. Cell. Biol.*, **21**, 2413–2422 (2001).

230. H. Asano, X. S. Li, and G. Stamatoyannopoulos, *Mol. Cell. Biol.*, **19**, 3571–3579 (1999).

231. H. Asano, X. S. Li, and G. Stamatoyannopoulos, *Blood*, **95**, 3578–3584 (2000).

232. J. Magram, K. Chada, and F. Costantini, *Nature*, **315**, 338–340 (1985).

233. K. Chada, J. Magram, K. Raphael, G. Radice, E. Lacy, and F. Costantini, *Nature*, **314**, 377–380 (1985).

234. D. Tuan and I. M. London, *Proc. Natl. Acad. Sci. USA*, **81**, 2718–2722 (1984).

235. F. Grosveld, G. B. van Assendelft, D. R. Greaves, and G. Kollias, *Cell*, **51**, 975–985 (1987).

236. T. Enver, A. J. Ebens, W. C. Forrester, and G. Stamatoyannopoulos, *Proc. Natl. Acad. Sci. USA*, **86**, 7033–7037 (1989).

237. T. M. Ryan, R. R. Behringer, N. C. Martin, T. M. Townes, R. D. Palmiter, and R. L. Brinster, *Genes Dev.*, **3**, 314–323 (1989).

238. D. Talbot, P. Collis, M. Antoniou, M. Vidal, F. Grosveld, and D. R. Greaves, *Nature*, **338**, 352–355 (1989).

239. J. A. Lloyd, J. M. Krakowsky, S. C. Crable, and J. B. Lingrel, *Mol. Cell. Biol.*, **12**, 1561–1567 (1992).

240. G. Stamatoyannopoulos, B. Josephson, J. W. Zhang, and Q. Li, *Mol. Cell. Biol.*, **13**, 7636–7644 (1993).

241. J. Ellis, D. Talbot, N. Dillon, and F. Grosveld, *EMBO J.*, **12**, 127–134 (1993).

242. A. C. Bell, A. G. West, and G. Felsenfeld, *Cell*, **98**, 387–396 (1999).

243. D. W. Emery, E. Yannaki, J. Tubb, and G. Stamatoyannopoulos, *Proc. Natl. Acad. Sci. USA*, **97**, 9150–9155 (2000).

244. S. Rivella, J. A. Callegari, C. May, C. W. Tan, and M. Sadelain, *J. Virol.*, **74**, 4679–4687 (2000).

245. D. E. Sabatino, N. E. Seidel, G. J. Aviles-Mendoza, A. P. Cline, S. M. Anderson, P. G. Gallagher, and D. M. Bodine, *Proc. Natl. Acad. Sci. USA*, **97**, 13294–13299 (2000).

246. F. Moreau-Gaudry, P. Xia, G. Jiang, N. P. Perelman, G. Bauer, J. Ellis, K. H. Surinya, F. Mavilio, C. K. Shen, et al., *Blood*, **98**, 2664–2672 (2001).

247. A. P. Jarman, W. G. Wood, J. A. Sharpe, G. Gourdon, H. Ayyub, and D. R. Higgs, *Mol. Cell. Biol.*, **11**, 4679–4689 (1991).

248. Q. Li, D. W. Emery, M. Fernandez, H. Han, and G. Stamatoyannopoulos, *Blood*, **93**, 2208–2216 (1999).

249. L. Xu, A. E. Ferry, C. Monteiro, and B. S. Pace, *Gene Ther.*, **7**, 438–444 (2000).

250. R. Alami, J. G. Gilman, Y. Q. Feng, A. Marmorato, I. Rochlin, S. M. Suzuka, M. E. Fabry, R. L. Nagel, and E. E. Bouhassira, *Blood Cells Mol. Dis.*, **25**, 110–119 (1999).

251. T. J. Shen, P. Ikonomi, R. Smith, C. T. Noguchi, and C. Ho, *Blood Cells Mol. Dis.*, **25**, 361–373 (1999).

252. N. Lan, R. P. Howrey, S. W. Lee, C. A. Smith, and B. A. Sullenger, *Science*, **280**, 1593–1596 (1998).

253. G. Wang, X. Xu, B. Pace, D. A. Dean, P. M. Glazer, P. Chan, S. R. Goodman, and I. Shokolenko, *Nucleic Acids Res.*, **27**, 2806–2813 (1999).

254. X. S. Xu, P. M. Glazer, and G. Wang, *Gene*, **242**, 219–228 (2000).

255. S. Hatada, K. Nikkuni, S. A. Bentley, S. Kirby, and O. Smithies, *Proc. Natl. Acad. Sci. USA*, **97**, 13807–13811 (2000).

256. A. Cole-Strauss, K. Yoon, Y. Xiang, B. C. Byrne, M. C. Rice, J. Gryn, W. K. Holloman, and E. B. Kmiec, *Science*, **273**, 1386–1389 (1996).

257. E. A. Dzierzak, T. Papayannopoulou, and R. C. Mulligan, *Nature*, **331**, 35–41 (1988).

258. M. A. Bender, R. E. Gelinas, and A. D. Miller, *Mol. Cell. Biol.*, **9**, 1426–1434 (1989).

259. D. M. Bodine, S. Karlsson, and A. W. Nienhuis, *Proc. Natl. Acad. Sci. USA*, **86**, 8897–8901 (1989).

260. X. Fan, A. Brun, S. Segren, S. E. Jacobsen, and S. Karlsson, *Hum. Gene Ther.*, **11**, 1313–1327 (2000).

261. C. E. Walsh, J. M. Liu, X. Xiao, N. S. Young, A. W. Nienhuis, and R. J. Samulski, *Proc. Natl. Acad. Sci. USA*, **89**, 7257–7261 (1992).

262. J. L. Miller, R. E. Donahue, S. E. Sellers, R. J. Samulski, N. S. Young, and A. W. Nienhuis, *Proc. Natl. Acad. Sci. USA*, **91**, 10183–10187 (1994).

263. D. M. Shayakhmetov, C. A. Carlson, H. Stecher, Q. Li, G. Stamatoyannopoulos, and A. Lieber, *J. Virol.*, **76**, 1135–1143 (2002).

264. C. May, S. Rivella, J. Callegari, G. Heller, K. M. Gaensler, L. Luzzatto, and M. Sadelain, *Nature*, **406**, 82–86 (2000).

265. R. Pawliuk, K. A. Westerman, M. E. Fabry, E. Payen, R. Tighe, E. E. Bouhassira, S. A. Acharya, J. Ellis, I. M. London, et al., *Science*, **294**, 2368–2371 (2001).

266. X. Ye, V. M. Rivera, P. Zoltick, F. Cerasoli Jr., M. A. Schnell, G. Gao, J. V. Hughes, M. Gilman, and J. M. Wilson, *Science*, **283**, 88–91 (1999).

267. G. Rizzuto, M. Cappelletti, D. Maione, R. Savino, D. Lazzaro, P. Costa, I. Mathiesen, R. Cortese, G. Ciliberto, et al., *Proc. Natl. Acad. Sci. USA*, **96**, 6417–6422 (1999).

268. L. Jin, N. Siritanaratkul, D. W. Emery, R. E. Richard, K. Kaushansky, T. Papayannopoulou, and C. A. Blau, *Proc. Natl. Acad. Sci. USA*, **95**, 8093–8097 (1998).

269. R. Abonour, D. A. Williams, L. Einhorn, K. M. Hall, J. Chen, J. Coffman, C. M. Traycoff, A. Bank, I. Kato, et al., *Nat. Med.*, **6**, 652–658 (2000).

270. U. Thorsteinsdottir, G. Sauvageau, and R. K. Humphries, *Blood*, **94**, 2605–2612 (1999).

271. K. M. Innes, S. J. Szilvassy, H. E. Davidson, L. Gibson, J. M. Adams, and S. Cory, *Exp. Hematol.*, **27**, 75–87 (1999).

272. S. Kirby, W. Walton, and O. Smithies, *Blood*, **95**, 3710–3715 (2000).

273. K. M. Matsuda, K. Kume, Y. Ueda, M. Urabe, M. Hasegawa, and K. Ozawa, *Gene Ther.*, **6**, 1038–1044 (1999).

274. J. Tisdale and M. Sadelain, *Semin. Hematol.*, **38**, 382–392 (2001).

275. M. Cavazzana-Calvo, S. Hacein-Bey, G. de Saint Basile, F. Gross, E. Yvon, P. Nusbaum, F. Selz, C. Hue, S. Certain, et al., *Science*, **288**, 669–672 (2000).

276. G. Dykes, R. H. Crepeau, and S. J. Edelstein, *Nature*, **272**, 506–510 (1978).

277. R. H. Crepeau and S. J. Edelstein, *Ultramicroscopy*, **13**, 11–18 (1984).

CHAPTER TEN

Iron Chelators and Therapeutic Uses

RAYMOND J. BERGERON
JAMES S. MCMANIS
WILLIAM R. WEIMAR
JAN WIEGAND
EILEEN EILER-MCMANIS
College of Pharmacy
University of Florida
Gainesville, Florida

Contents

Burger's Medicinal Chemistry and Drug Discovery
Sixth Edition, Volume 3: Cardiovascular Agents and Endocrines
Edited by Donald J. Abraham
ISBN 0-471-37029-0   © 2003 John Wiley & Sons, Inc.

1   INTRODUCTION

1.1   Iron in the Biosphere

Although iron has many oxidation states available to it, ranging from -2 to $+6$, the $+2$ and $+3$ valences are of the greatest importance in biological systems. This statement is not meant to diminish the significance of Fe(IV) (ferryl, found in cytochromes and horseradish peroxidase), Fe(V) (perferryl), and Fe(VI); however, the chemistry of these species is beyond the scope of this chapter. The $+2$ and $+3$ oxidation states, which are characterized, respectively, by their d^6 and d^5 ground-state configurations, are exquisitely sensitive to both pH and the nature of the ligating functionality (1). At a cellular level, this sensitivity has been exploited, inasmuch as this metal can function both as an electron source and an electron sink. Iron is an essen-

tial cofactor in a variety of biological redox systems, for example, cytochromes, oxidases, peroxidases, and ribonucleotide reductase (2, 3). Paradoxically, even though iron is the second most abundant metal on earth, living systems have had to develop sophisticated methods for its acquisition (4–6). In fact, with the possible exception of some lactobacilli, life without this metal is virtually unknown.

When the most primitive life forms developed about 3.5 billion years ago, the atmosphere was basically anaerobic; in the absence of molecular oxygen, iron was mostly in the $+2$ oxidation state. This form of the metal is much more soluble and accessible to biological systems than is Fe(III) (7). It is for this reason that oral iron supplements are generally in the Fe(II) form (e.g., ferrous sulfate) and not as the Fe(III) salts. Upon evolution of the blue-green algae, problems developed; these difficulties were related directly to the availability of this essential micronutrient. The oxygen produced from these algae by photosynthesis caused the conversion of the iron(II) in the biosphere to Fe(III), a species that is highly insoluble in an aqueous environment. The solubility product of ferric hydroxide under physiological conditions, 2×10^{-39}, corresponds to a solution concentration of the free cation of approximately $1 \times 10^{-19} M$ (8, 9). Under most conditions that exist in the biosphere, iron(III) forms insoluble ferric hydroxide polymers. Thus, in spite of the abundance of this metal, primitive life forms needed to develop methods for rendering it usable.

1.2 Iron Dynamics in Microorganisms

Ultimately, bacteria adapted to the iron accessibility problem by producing relatively low molecular weight, virtually iron(III)-specific, ligands, siderophores (from the Greek *sidero* and *phore*, literally "iron carrier"), for the purpose of acquiring this transition metal (5, 6, 10). Under conditions of low iron availability, microorganisms biosynthesize and release up to several times their own weight of ligand into the environment daily (11). These chelators form soluble complexes with the metal; the bacteria are then virtually immersed in these iron chelates, a usable iron source.

1.2.1 Metal Complex Formation and the Chelate Effect.

This discussion is best opened with an overview of the equilibria that describe metal complex formation. Consider a transition metal [e.g., Fe(III)] in the presence of a monodentate ligand (L). The equilibria depicting the formation of an octahedral, hexacoordinate Fe(III) complex are shown in Equations 10.1A–10.1F; the constants K_1–K_6 are referred to as stepwise equilibrium constants:

$$Fe(III) + L \rightleftharpoons L\text{–}Fe(III) \qquad (10.1A)$$

$$K_1 = \frac{[L\text{–}Fe(III)]}{[Fe(III)][L]}$$

$$L\text{–}Fe(III) + L \rightleftharpoons L_2\text{–}Fe(III) \qquad (10.1B)$$

$$K_2 = \frac{[L_2\text{–}Fe(III)]}{[L\text{–}Fe(III)][L]}$$

$$L_2\text{–}Fe(III) + L \rightleftharpoons L_3\text{–}Fe(III) \qquad (10.1C)$$

$$K_3 = \frac{[L_3\text{–}Fe(III)]}{[L_2\text{–}Fe(III)][L]}$$

$$L_3\text{–}Fe(III) + L \rightleftharpoons L_4\text{–}Fe(III) \qquad (10.1D)$$

$$K_4 = \frac{[L_4\text{–}Fe(III)]}{[L_3\text{–}Fe(III)][L]}$$

$$L_4\text{–}Fe(III) + L \rightleftharpoons L_5\text{–}Fe(III) \qquad (10.1E)$$

$$K_5 = \frac{[L_5\text{–}Fe(III)]}{[L_4\text{–}Fe(III)][L]}$$

$$L_5\text{–}Fe(III) + L \rightleftharpoons L_6\text{–}Fe(III) \qquad (10.1F)$$

$$K_6 = \frac{[L_6\text{–}Fe(III)]}{[L_5\text{–}Fe(III)][L]}$$

This same set of equilibria also can be expressed in a non-stepwise fashion or as overall equilibrium constants, as shown in Equations 10.2A–10.2F:

$$Fe(III) + L \rightleftharpoons L\text{–}Fe(III) \qquad (10.2A)$$

$$\beta_1 = \frac{[L\text{–}Fe(III)]}{[Fe(III)][L]}$$

$$Fe(III) + 2L \rightleftharpoons L_2\text{–}Fe(III) \qquad (10.2B)$$

$$\beta_2 = \frac{[L_2\text{–}Fe(III)]}{[Fe(III)][L]^2}$$

$$Fe(III) + 3L \rightleftharpoons L_3\text{–}Fe(III) \qquad (10.2C)$$

$$\beta_3 = \frac{[L_3\text{–}Fe(III)]}{[Fe(III)][L]^3}$$

$$Fe(III) + 4L \rightleftharpoons L_4\text{–}Fe(III) \qquad (10.2D)$$

$$\beta_4 = \frac{[L_4\text{–}Fe(III)]}{[Fe(III)][L]^4}$$

$$Fe(III) + 5L \rightleftharpoons L_5\text{–}Fe(III) \qquad (10.2E)$$

$$\beta_5 = \frac{[L_5\text{–}Fe(III)]}{[Fe(III)][L]^5}$$

$$Fe(III) + 6L \rightleftharpoons L_6\text{–}Fe(III) \qquad (10.2F)$$

$$\beta_6 = \frac{[L_6\text{–}Fe(III)]}{[Fe(III)][L]^6}$$

A simple algebraic manipulation demonstrates the relationship between these two forms of expression (Equations 10.3A–10.3D):

$$\beta_2 = \frac{[L_2\text{–Fe(III)}]}{[\text{Fe(III)}][L]^2} \quad (10.3A)$$

$$\beta_2 = \frac{[L_2\text{–Fe(III)}]}{[\text{Fe(III)}][L]^2} \times \frac{[L\text{–Fe(III)}]}{[L\text{–Fe(III)}]} \quad (10.3B)$$

$$\beta_2 = \frac{[L\text{–Fe(III)}]}{[\text{Fe(III)}][L]} \times \frac{[L_2\text{–Fe(III)}]}{[L\text{–Fe(III)}][L]} \quad (10.3C)$$

$$\beta_2 = K_1 K_2 \quad (10.3D)$$

Thus, the generalized relationship between the two expressions becomes (Equation 10.4):

$$\beta_n = K_1 K_2 K_3 \cdots K_n = \prod_{i=1}^{i=n} K_i \quad (10.4)$$

When one compares equilibrium constants in the literature, it is important to be certain which equilibrium expression is being referred to. The stepwise constants are useful in identifying which species is/are present; these constants generally diminish in value as the ligand:metal ratio increases. In keeping with our focus on natural product iron chelators, the application of complex formation equilibria is best seen in earlier work on the enterobactin-Fe(III) complex (12–14). Enterobactin (1, Fig. 10.1), a siderophore produced by *Escherichia coli*, forms a very tight octahedral hexacoordinate complex with Fe(III). A comparison of the iron binding properties of this siderophore and a model bidentate ligand exemplifies the importance of the chelate effect in complex formation.

When three moles of the bidentate ligand 2,3-dihydroxy-*N*,*N*-dimethylbenzamide (DHBA, 2), are reacted with Fe(III), (2) forms a stable 3:1 ligand:metal complex with the iron (13). The stepwise reactions and their respective equilibrium constants are shown in Equations 10.5A–10.5C:

$$\text{Fe(III)} + \mathbf{2}^{-2} \rightleftharpoons \text{Fe}\mathbf{2}^+$$

$$K_1 = \frac{[\text{Fe}\mathbf{2}^+]}{[\text{Fe(III)}][\mathbf{2}^{-2}]} \quad (10.5A)$$

$$\log K_1 = 17.77$$

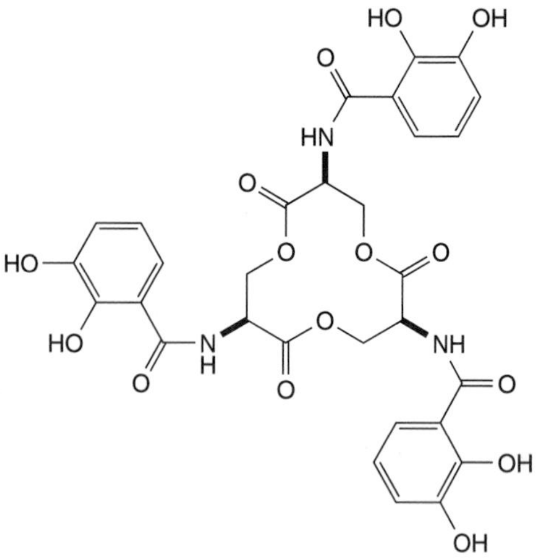

(1)

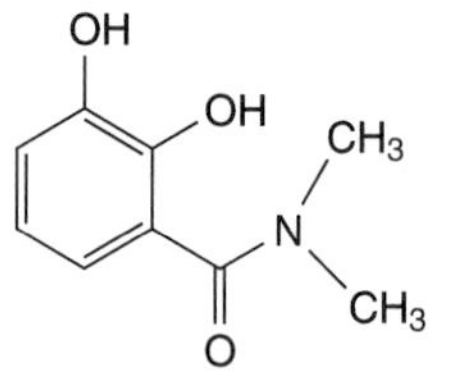

(2)

Figure 10.1. Structures of the catecholamide siderophore enterobactin (1) and an enterobactin model system, 2,3-dihydroxy-*N*,*N*-dimethylbenzamide (DHBA, 2).

$$\text{Fe}\mathbf{2}^+ + \mathbf{2}^{-2} \rightleftharpoons \text{Fe}(\mathbf{2})_2{}^{-1}$$

$$K_2 = \frac{[\text{Fe}(\mathbf{2})_2{}^{-1}]}{[\text{Fe}\mathbf{2}^+][\mathbf{2}^{-2}]} \quad (10.5B)$$

$$\log K_2 = 13.96$$

$$\text{Fe}(\mathbf{2})_2{}^{-1} + \mathbf{2}^{-2} \rightleftharpoons \text{Fe}(\mathbf{2})_3{}^{-3}$$

$$K_3 = \frac{[\text{Fe}(\mathbf{2})_3{}^{-3}]}{[\text{Fe}(\mathbf{2})_2{}^{-1}][\mathbf{2}^{-2}]} \quad (10.5C)$$

$$\log K_3 = 8.51$$

Thus, β_3 for this sequence of reactions is calculated as $\beta_3 = K_1 K_2 K_3$, or log β_3 as log β_3 = log K_1 + log K_2 + log K_3 = 40.24 (13).

Examination of the same calculation for enterobactin (1) when compared to that for (2) is interesting. The equilibrium expression for this complex is (Equation 10.6):

$$Fe(III) + \mathbf{1}^{-6} \leftrightharpoons Fe\mathbf{1}^{-3}$$

$$K_{ML} = \frac{[Fe\mathbf{1}^{-3}]}{[Fe(III)][\mathbf{1}^{-6}]} \qquad (10.6)$$

$$\log K_{ML}(\mathbf{1}^{-6}) = 52$$

The $\log K_{ML}$ for this complex was calculated to be 52 (12, 13), 12 powers of 10 greater than the β_3 for the individual donor, DHBA. This enormous difference in formation constants between a complex consisting of three unconnected bidentate ligands and Fe(III) and one consisting of Fe(III) complexed with a single hexacoordinate donor, in this example enterobactin, is attributed to the chelate effect. Recall that $\Delta G = -RT \ln K$ and that $\Delta G = \Delta H - T\Delta S$.

There are several ways of thinking about this phenomenon, two of which are somewhat simplistic, yet informative. When the three bidentate ligands replace the water that surrounds the Fe(III), six water molecules are displaced, and a net of three molecules are liberated. When the hexacoordinate siderophore (**1**) binds to Fe(III), again, six molecules of water are displaced; however, a net of five molecules are liberated. Thus, there is a greater increase in disorder (ΔS) in the case of (**1**) than in that of (**2**). An alternative thought is that when (**1**) donates its first bidentate fragment, the second and third bidentate fragments are held closely for donation, unlike the situation with the (**2**) donors.

One of the difficulties with comparing ligand formation constants is that, often, the measurements are made under different conditions and are quite sensitive to the pK_a of the particular donor groups. For example, in the case of enterobactin with Fe(III), the hydrogen ion dependency is seen in the expression (Equation 10.7):

$$Fe(III) + H_6\mathbf{1} \leftrightharpoons Fe\mathbf{1}^{-3} + 6H^+$$

$$K_{ML} = \frac{[Fe\mathbf{1}^{-3}][H^+]^6}{[Fe(III)][H_6\mathbf{1}]} \qquad (10.7)$$

$$\log K_{ML}(H_6\mathbf{1}) = -9.7$$

The stability constant for the fully protonated ligand ($H_6\mathbf{1}$) is substantially different from that determined for the fully deprotonated form ($\mathbf{1}^{-6}$) (14). Thus, to provide a more meaningful comparison among ligands, investigators have suggested the use of pM values, where p$M = -\log[Fe(III)]$ (15). This number describes the concentration of free Fe(III) in solution at a given pH, total iron, and total ligand concentration; lower free iron concentrations translate into higher pM values. Table 10.1 provides several examples of key ligands, both natural and synthetic, for the reader's consideration (16–24). Note the effect of denticity on the pM values (e.g., **1** versus **2**). The measurements were made at pH 7.4, $10^{-6}\,M$ Fe(III), and $10^{-5}\,M$ ligand.

1.2.2 Structural Classes of Siderophores. Although a large number of siderophores have been isolated, they generally can be separated into two basic structural groups, the catecholamides and the hydroxamates. Interestingly, many of these chelators have common structural denominators. They can be considered as either directly predicated on polyamine (e.g., putrescine, cadaverine, norspermidine, or spermidine) backbones or based on the biochemical precursors to the polyamines (e.g., ornithine or lysine) (25). The catecholamide chelators (Fig. 10.2), N^1,N^8-bis(2,3-dihydroxybenzoyl)spermidine (compound II, **3**) (26), L-parabactin (**4**) (26, 27), L-agrobactin (**5**) (28), L-fluviabactin (**6**) (29), L-vulnibactin (**7**) (30), and L-vibriobactin (**8**) (31), all have bidentate 2,3-dihydroxybenzoyl groups fixed directly to either a spermidine (**3–5**) or a norspermidine (**6–8**) backbone. Obviously, (**1**) (Fig. 10.1), a siderophore based on a macrocyclic serine backbone, is an exception to this observation.

Chelators in the hydroxamate class contain bidentate N-hydroxy amide (i.e., hydroxamic acid) groups. The designations "hydroxamic acid" and "hydroxamate" are interchangeable in the arena of iron chelator nomenclature (32), although the latter term is sometimes restricted to N-alkoxy (33) or N-silyloxy amides (34). The cadaverine moieties of the hydroxamates (Fig. 10.3) desferrioxamine B (DFO, **9**) (35), desferrioxamine G (**10**) (36), desferrioxamine E (nocardamine, **11**) (37), bisucaberin (**12**) (38), and arthrobactin (**13**) (39) are apparent. Other polyamines or their amino acid precursors are evident in various

Table 10.1 Iron pM Values for Selected Ligands

Compound	Ligand	Equilibrium Quotient	$\log K$	$\log \beta_3$	pM^a	Reference(s)
(1)	Enterobactin	$[Fe(III)–L^{3–}]/([Fe(III)][L^{6–}])$	52		35.5	(12–14)
(2)	2,3-Dihydroxy-N,N-dimethylbenzamide	$[Fe(III)–L_3^{3–}]/([Fe(III)][L^{2–}]^3)$		40.2	15^b	(14, 16)
(4)	Parabactin	$[Fe(III)–L]/([Fe(III)][L])$	48		—	(17)
(9)	Ferrioxamine B	$[Fe(III)–L]/([Fe(III)][L])$	30.99		25.9	(15)
(11)	Ferrioxamine E	$[Fe(III)–L]/([Fe(III)][L])$	32.5		27.7	(13, 21)
(14)	Aerobactin	$[Fe(III)–L]/([Fe(III)][L])$	23.1		23.3	(13, 14)
(17)	Ferrichrome	$[Fe(III)–L]/([Fe(III)][L])$	29.1		25.2	(14, 22)
None^c	Transferrin		23		23.6	(13, 23, 24)
(32)	EDTA	$[Fe(III)L^{–}]/[Fe(III)][L^{4–}]$	25.1		22.3	(15)
(33)	1,2-Dimethyl-3-hydroxypyridin-4-one (L1)	$[Fe(III)–L_3]/([Fe(III)][L^{–}]^3)$		35.92	18.3	(18, 19)
(35)	DTPA	$[Fe(III)–L^{2–}]/[Fe(III)][L^{5–}]$	28.0		23.8	(15)
(39)	HBED	$[Fe(III)–L^{–}]/[Fe(III)][L^{4–}]$	39.68		26.74	(15, 20)

^{a}Calculated for 10 μM ligand, 1 μM Fe(III), at pH 7.4.
^{b}Calculated pM is below the lower limit determined by the K_{sp} of ferric hydroxide, indicating precipitation of iron under these conditions.
^{c}See Fig. 10A for a diagram of this protein.

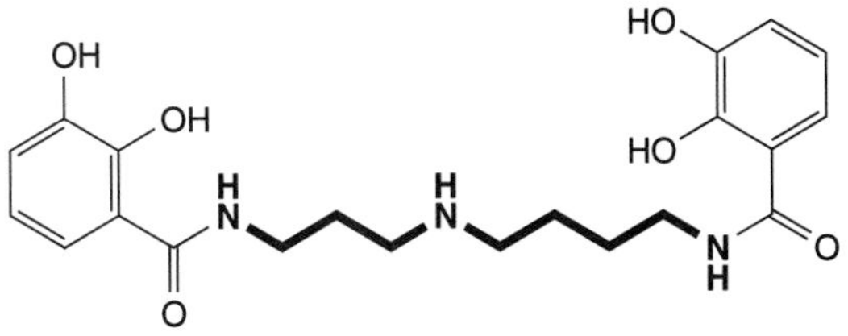

(3)

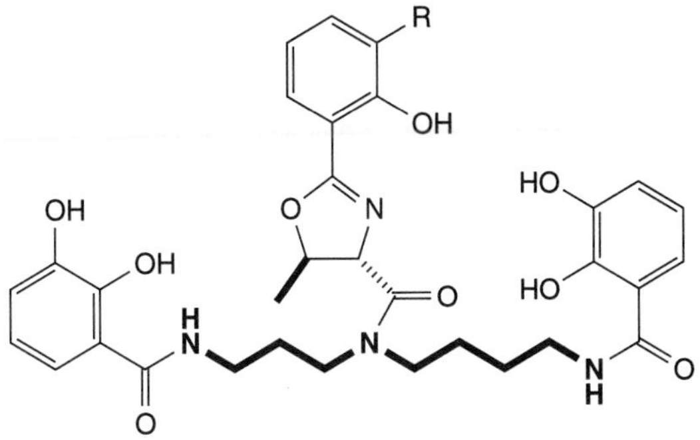

R = H (4)
R = OH (5)

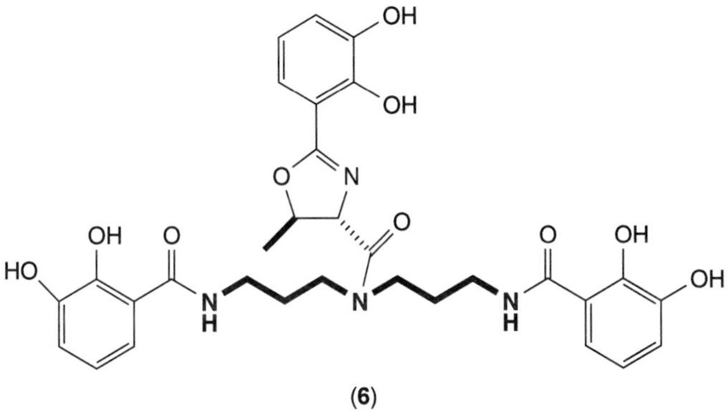

(6)

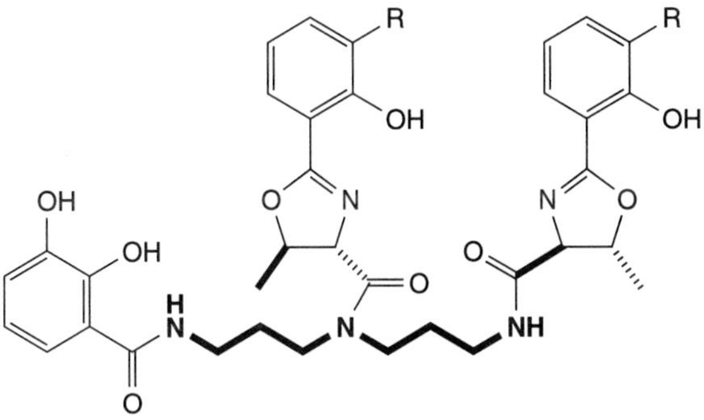

R = H (7)
R = OH (8)

Figure 10.2. Other catechol-amide siderophores: compound II (**3**), L-parabactin (**4**, R = H), L-agrobactin (**5**, R = OH), L-fluvi-abactin (**6**), L -vulnibactin (**7**, R = H), and L-vibriobactin (**8**, R = OH). Polyamine backbones are high-lighted by darkened bonds.

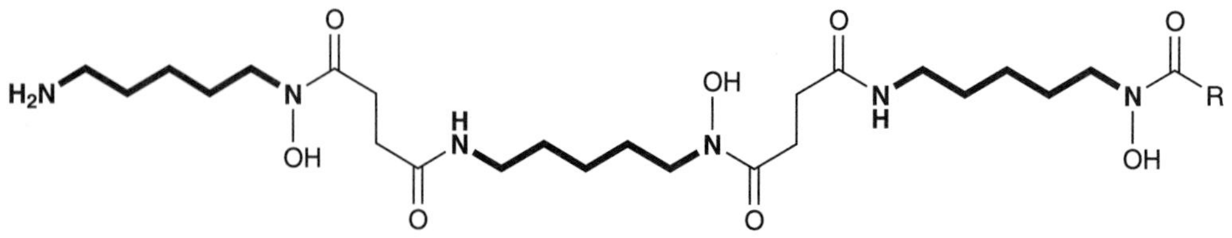

R = CH$_3$ (**9**) (DFO)
R = (CH$_2$)$_2$COOH (**10**)

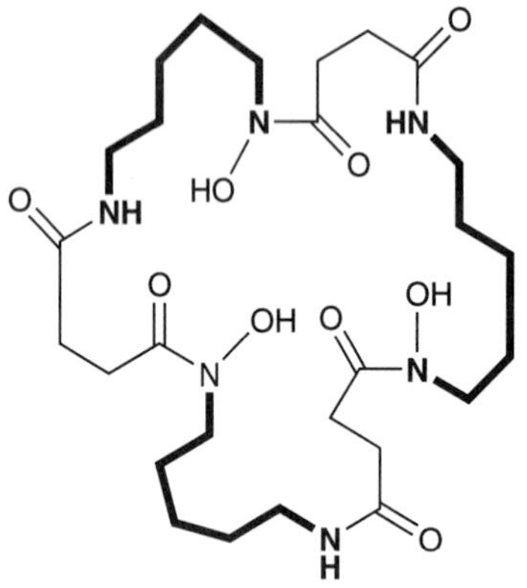

(**11**) 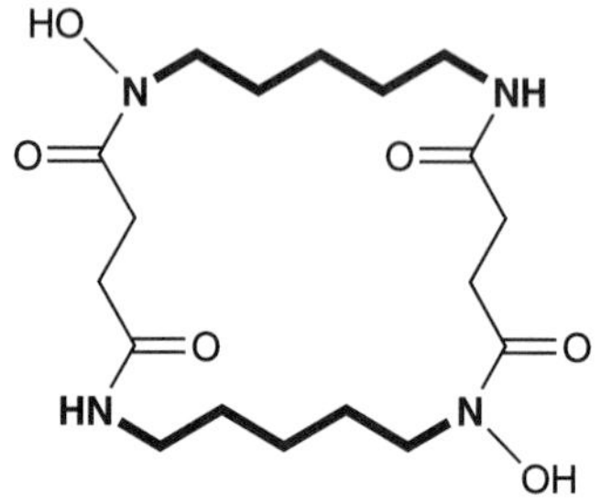   (**12**)

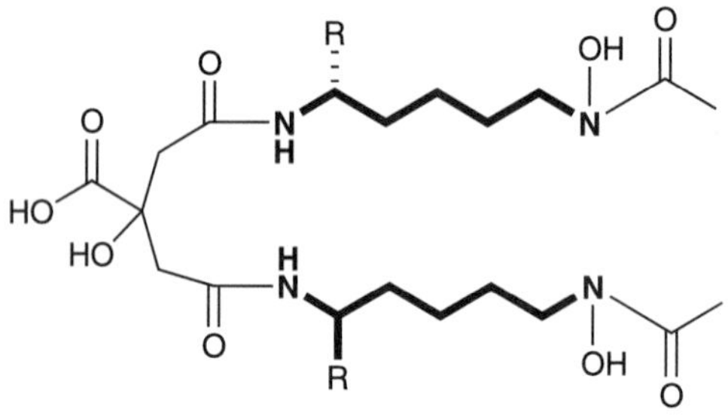

R = H   (**13**)
R = CO$_2$H (**14**)

Figure 10.3. Hydroxamate siderophores: desferrioxamine B (DFO, **9**, R = CH$_3$), desferrioxamine G [(**10**), R = (CH$_2$)$_2$COOH], desferrioxamine E (nocardamine, **11**), bisucaberin (**12**), arthrobactin (**13**, R = H), aerobactin (**14**, R = CO$_2$H), mycobactin S (**15**), nannochelin A (**16**), desferri-ferrichrome (**17**, R = COCH$_3$), rhodotorulic acid (**18**), schizokinen (**19**), and alcaligin (**20**). Polyamine backbones are highlighted by darkened bonds.

hydroxamate ligands, such as aerobactin (**14**) (40), mycobactin S (**15**) (41), nannochelin A (**16**) (42), desferri-ferrichrome (**17**) (43), rhodotorulic acid (**18**) (44), schizokinen (**19**) (45), and alcaligin (**20**) (46).

The major functional contrast between the hydroxamate and catecholamide siderophores is related to environmental iron concentration (47). The hydroxamates are synthesized by the organism under high iron conditions, whereas the catecholamide "backup" system is activated when iron concentrations are low. Logically, the catecholamide chelators typically bind iron far more tightly than do hydroxamates (Table 10.1). The formation constants of the catecholamides can be as high as $10^{52}\ M^{-1}$ for the (**1**):Fe(III) complex (12, 13), whereas those for the hydroxamates are con-

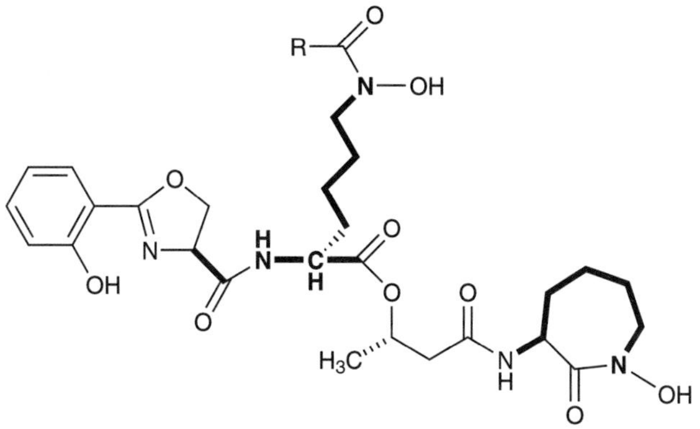

(15)

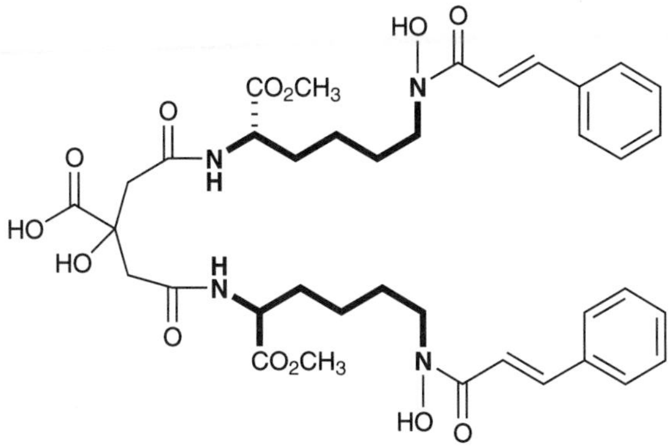

(16)

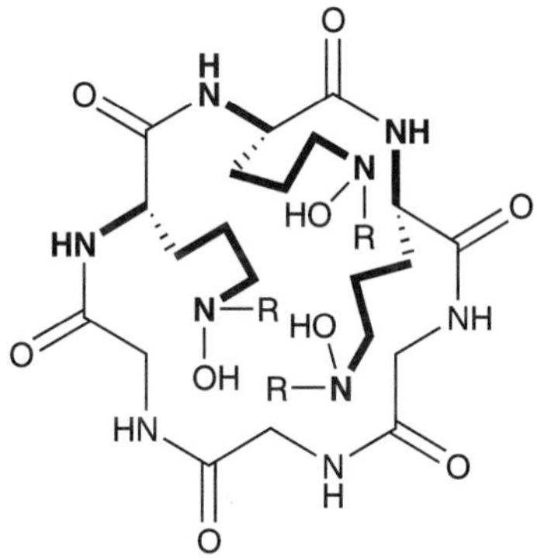

(17)

R = COCH₃

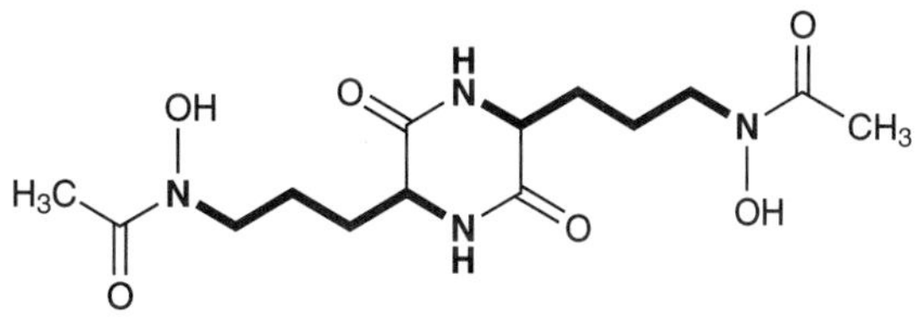

(18)

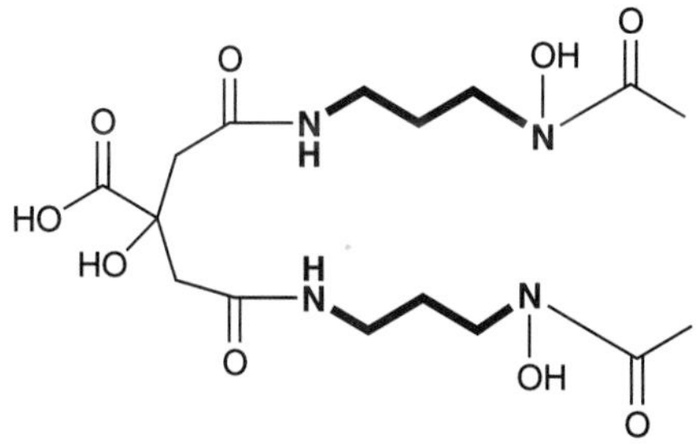

(19)

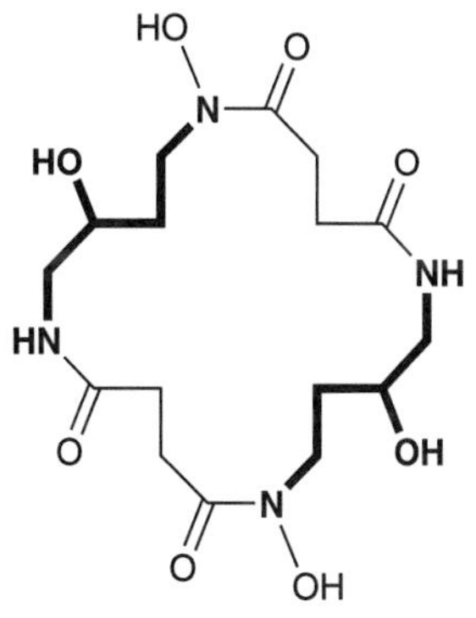

(20)

Figure 10.3. (Continued.)

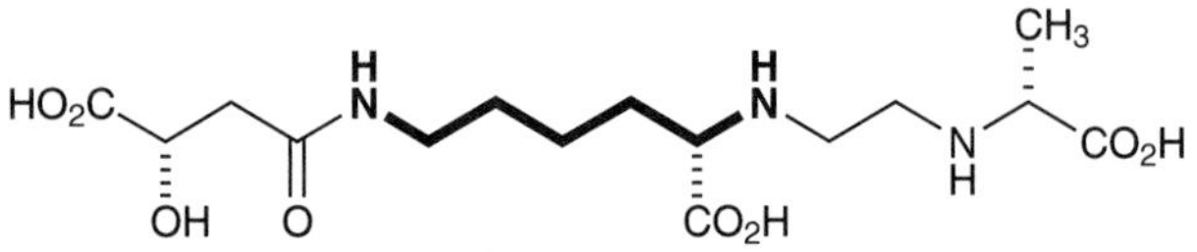

(21)

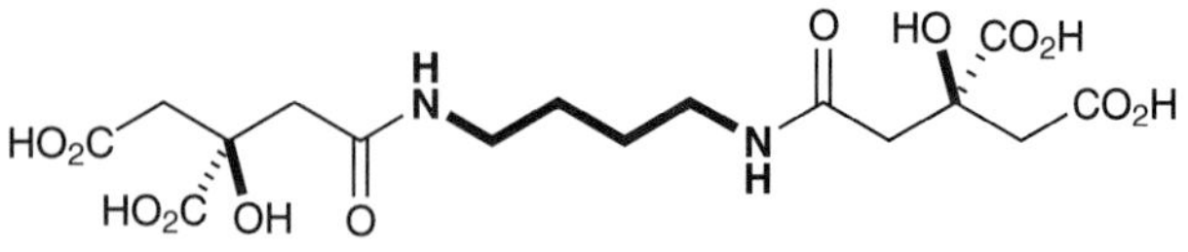

(22)

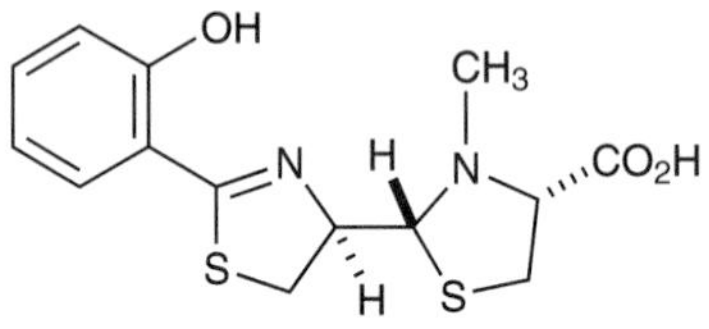

Figure 10.4. "Miscellaneous" siderophores: rhizobactin (**21**), rhizoferrin (**22**), pyochelin (**23**), and desferrithiocin (**24**, DFT). Polyamine backbones are highlighted by darkened bonds.

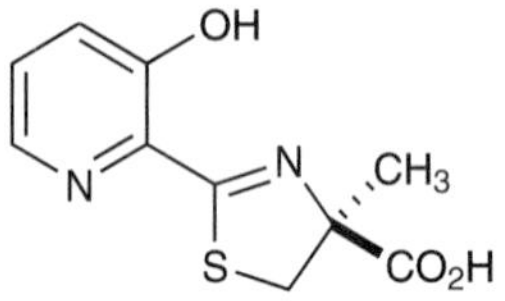

(23)   **(24)**

siderably lower, on the order of 10^{31} M^{-1} for desferrioxamine B (21, 22, 48).

Although most siderophores do fall into one of the aforementioned classes, there are ligands that do not belong to either family, such as rhizobactin (**21**) (49), rhizoferrin (**22**) (50), pyochelin (**23**) (51–53), and (S)-4,5-dihydro-2-(3-hydroxy-2-pyridinyl)-4-methyl-4-thiazole-carboxylic acid (desferrithiocin, DFT, **24**) (Fig. 10.4). This chelator, isolated from *Streptomyces antibioticus* (54), forms a stable ($K_f = 4 \times 10^{29}$ M^{-1}) 2:1 complex with iron (55, 56). As discussed later in this chapter, DFT (**24**) represents an excellent pharmacophore from which to construct therapeutic orally active iron chelators.

1.2.3 Bacterial Iron Uptake and Processing. One of the major questions concerning the iron-chelator complexes is the method for processing them *in vivo*. Considering that the siderophores bind iron quite tightly, how does a microorganism extract iron from a compound that binds it so well? The purpose that the siderophores serve in providing microorganisms with iron is illustrated here by the parabactin-mediated iron transport apparatus of *Paracoccus denitrificans*. The efficacy of L-parabactin (**4**) in supplying *P. denitrificans* with iron under artificially iron-lowered conditions, such as in the presence of ethylenedi-

amine-N,N'-bis(o-hydroxyphenyl acetic acid), has been well documented (57–59). Yet more fascinating is the differential use of L- and D-parabactin (**25**, Fig. 10.5) as measured by the stimulation of bacterial growth under such conditions. As expected, L-parabactin (**4**) fostered bacterial growth, whereas D-parabactin (**25**) could not be used by the bacterial iron-transport apparatus (Fig. 10.6) (57).

The stereospecificity of the kinetics of iron acquisition illustrates this phenomenon further (Fig. 10.7) (59). Iron accumulation data from [^{55}Fe]ferric (**25**) fit a straight-line double-reciprocal plot and, thus, obey a simple Michaelis-Menten model. However, the kinetics of iron acquisition from ferric (**4**) are quite different (Table 10.2). The presence of a high affinity transport system for ferric (**4**) is reflected by the pronounced differences in uptake rates at ligand concentrations below 1 μM. In fact, the ferric (**4**) data are both nonlinear and likely biphasic, as the enlargement of Fig. 10.7 illustrates (Fig. 10.8). The data for 1 $\mu M < [4] < 10$ μM fit one line, which suggests a low affinity system, yet the data for 0.1 $\mu M \leq [4] \leq 1$ μM fit another line, which is consistent with a high affinity system (Table 10.2, Fig. 10.8). The apparent K_m of the high affinity uptake (0.24 μM) is comparable to the affinity constant reported for the purified ferric (**4**) outer membrane receptor protein (58)

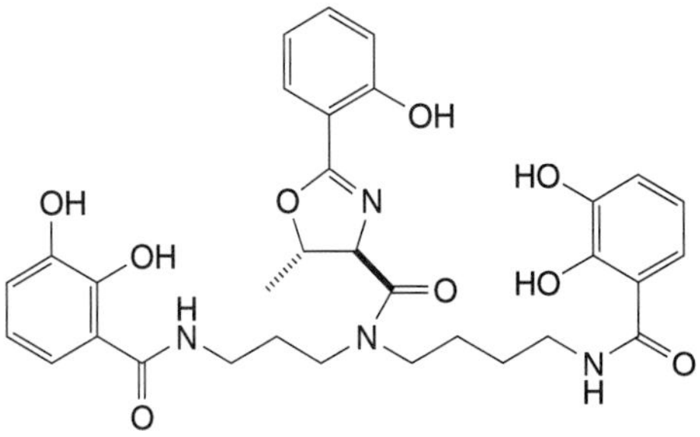

(25)

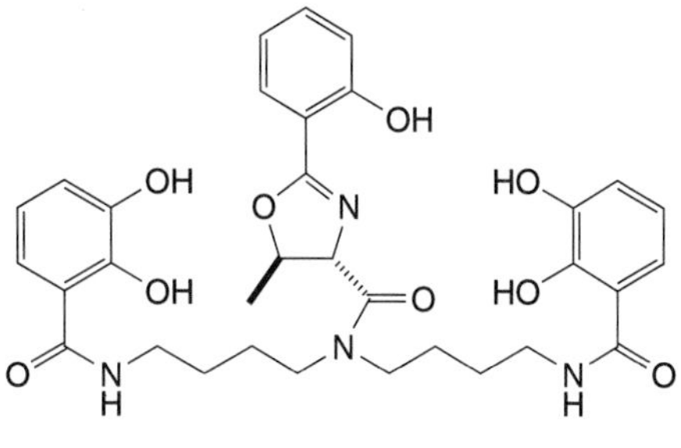

(26)

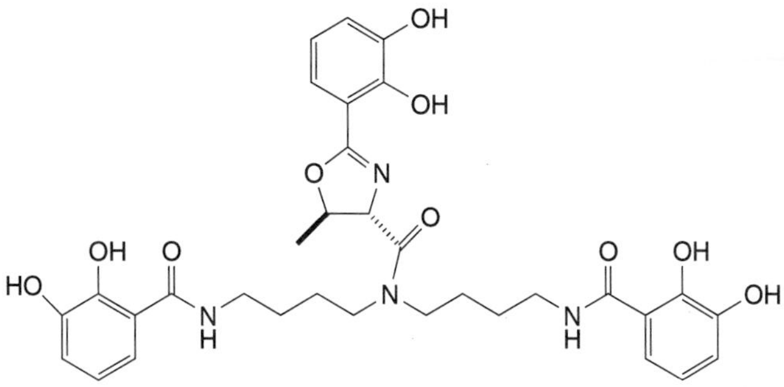

(27)

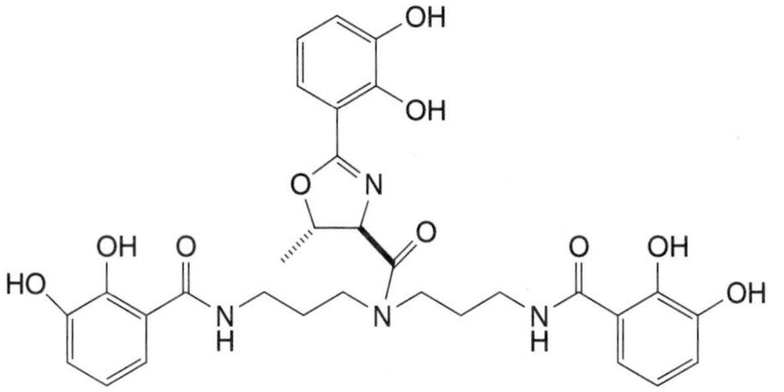

(28)

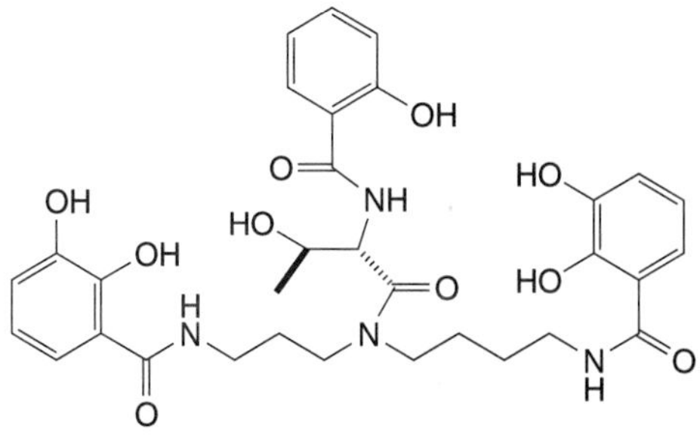

(29)

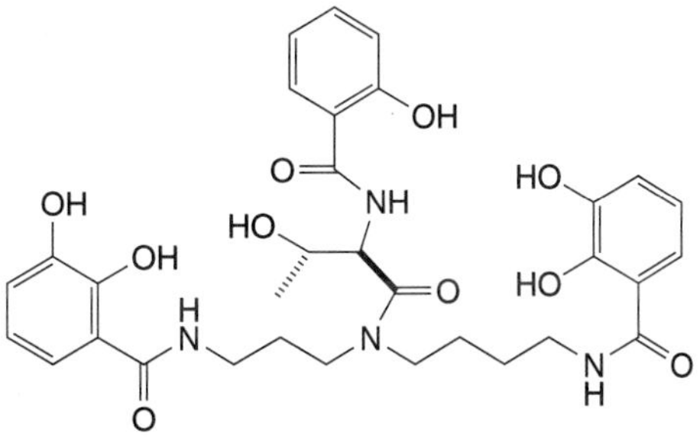

(30)

Figure 10.5. Structure of D-parabactin (**25**), which is derived synthetically from D-threonine; L-homoparabactin (**26**), a parabactin homolog; L-homofluviabactin (**27**), a fluviabactin homolog; D-fluviabactin (**28**), the enantiomer of (**6**); and the open-chain threonyl ("A") forms of L- and D-parabactin (L-parabactin A, (**29**), and D-parabactin A, (**30**), respectively).

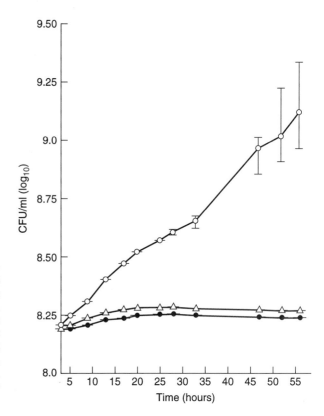

Figure 10.6. Growth rate of *P. denitrificans* [rendered as colony-forming units (CFU)/mL, *y*-axis, versus time, *x*-axis] in the presence of L-parabactin (○, 2 μM), D-parabactin (△, 2 μM), or controls (●) each in the presence of EDDHA (1.1 mM) without added ferric nitrate. [Reprinted with permission from R. J. Bergeron et al., *J. Biol. Chem.*, **260**, 7936–7944 (1985). Copyright 1985 The American Society for Biochemistry and Molecular Biology.]

and also is in keeping with the values reported for high affinity transport systems for ferrichrome (**17**, 0.15–0.25 μM) and ferric enterobactin (**1**, 0.10–0.36 μM) in *E. coli* (60–62). In other microorganisms, the K_m values reported for siderophore-mediated iron transport range from relatively low affinity apparati, such as that for ferric coprogen transport (K_m = 5 μM) in *Neurospora crassa* (63, 64) to an extremely high affinity assemblage (K_m = 0.04 μM) for ferric schizokinen (**19**) transport in species of the cyanobacterium *Anabaena* (65).

Biphasic kinetics similar to those of ferric (**4**), including a high affinity component, have been observed in *P. denitrificans* using L-agrobactin (**5**), L-fluviabactin (**6**), L-vibriobactin (**8**), and two synthetic homologs (L-homoparabactin, **26**, and L-homofluviabactin, **27**, respectively, Fig. 10.5), (Table 10.2) (59, 66). Also, the ferric chelates of (**5**), (**8**), and (**26**) (0.5 and 1.0 μM) inhibited accumulation of ^{55}Fe from ferric (**4**) by way of simple substrate-competitive Michaelis-Menten kinetics (59). Conversely, ferric (**4**) inhibited accumu-

lation of ^{55}Fe from ferric (**8**). Perhaps not surprisingly, ferric (**25**) at similar concentrations exerted no impact on ^{55}Fe transport (59). Thus, the high affinity ferric L-parabactin receptor seems to recognize and subsequently allow iron acquisition from these ferric L-oxazoline homologs, at least over the concentration range studied. The degree to which this system also contributes to the low affinity component of the biphasic kinetic profile is unclear. Biphasic kinetics have been observed for many membrane transport systems. In some cases, the presence of two independent systems for transport of the same substrate explains the observed phenomenon (67). A plausible alternative is that negative allosteric interactions can result in a system with a low K_m at low [S], which converts to a high K_m system at high [S] (68, 69).

The contributions of molecular dissymmetry to these distinctive kinetic features are underscored by the stereospecific differences in transport of the ferric catecholamides. The similarities in ring size and nature of the che-

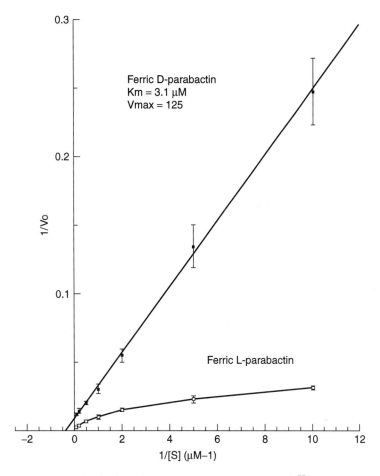

Figure 10.7. Lineweaver-Burk plot of kinetic data for the transport of [^{55}Fe]ferric L-parabactin (**4**) and [^{55}Fe]ferric D-parabactin (**25**) for 0.1 $\mu M \leq$ [S] ≤ 10 μM, where [S] is the concentration of chelate added to external medium at $t = 0$. The (**25**) in this and other experiments can be fitted on a single regression line ($r = 0.999$) corresponding to a simple Michaelis-Menten process with comparatively low affinity ($K_m = 3.1 \pm 0.9$ μM; $V_{max} = 125 \pm 11$ pg-atoms of ^{55}Fe min^{-1} mg of protein^{-1}). The (**4**) data are nonlinear (see Fig. 10.8), but if analyzed separately, the data for 1 $\mu M <$ [**4**] < 10 μM fit one line ($r = 0.991$), suggesting a low affinity system ($K_m = 3.9 \pm 1.2$ μM; $V_{max} = 495 \pm 41$ pg-atoms of ^{55}Fe min^{-1} mg of protein^{-1}), whereas the data for 0.1 $\mu M \leq$ [**4**] ≤ 1 μM fit another line ($r = 0.996$), consistent with a high affinity system ($K_m = 0.24 \pm 0.06$ μM; $V_{max} = 118 \pm 19$ pg-atoms of ^{55}Fe min^{-1} mg of protein^{-1}). Note that the data in these two concentration ranges are analyzed independently; that is, the high affinity data are not corrected for contributions made by the low affinity system operating at [S] < 1 μM, nor are the low affinity data corrected for contributions made by the high affinity system operating at [S] > 1 μM. Data are presented as means of four (**25**) or five (**4**) determinations and SDs (bars) for [S] = 0.1, 0.2, 0.5, 1.0, 2.0, 5.0, and 10.0 μM. [Reprinted with permission from Bergeron and Weimar, *J. Bacteriol.*, **172**, 2650–2657 (1990). Copyright 1990 American Society for Microbiology.]

late donor centers in these catecholamides allow direct comparison of their circular dichroism (CD) spectra with those of ferric catecholamide chelates of known configuration (70–72). The positive CD band maxima of

ferric L-parabactin is characteristic of the Δ chelate enantiomer (59, 73). The CD data also suggest that other ferric catecholamides that contain an oxazoline ring derived from L-threonine (i.e., **5**, **8**, and **26**) all exist in solution as

Table 10.2   Iron Accumulation Kinetic Data

Ferric Chelate[a]	Kinetics	K_m[b]	V_{max}[c]	Reference
(4) (5)	Bimodal	0.24 ± 0.06 μM (high affinity) (r = 0.996)	118 ± 19	(59)
		3.9 ± 1.2 μM (low affinity) (r = 0.991)	495 ± 41	
(5) (3)	Bimodal	0.13 ± 0.08 μM (high affinity)	46 ± 8	(59)
		1.6 ± 0.8 μM (low affinity)	221 ± 18	
(6) (3)	Bimodal	0.23 ± 0.03 μM (high affinity)	129 ± 2	(66)
		2.2 ± 1.6 μM (low affinity)	413 ± 149	
(8) (3)	Bimodal	0.18 ± 0.07 μM (high affinity)	72 ± 16	(59)
		5.8 ± 1.4 μM (low affinity)	544 ± 51	
(26) (2)	Bimodal	0.26 ± 0.10 μM (high affinity)	142 ± 24	(59)
		2.6 ± 1.3 μM (low affinity)	254 ± 31	
(27) (3)	Bimodal	0.17 ± 0.04 μM (high affinity)	138 ± 19	(66)
		1.0 ± 0.1 μM (low affinity)	229 ± 29	
(25) (4)	Linear	3.1 ± 0.9 μM	125 ± 11	(59)
(28) (3)	Linear	3.5 ± 0.8 μM	96 ± 63	(66)
(29) (4)	Linear	1.4 ± 0.3 μM	324 ± 21	(59)
(30) (2)	Linear	1.3 ± 0.3 μM	330 ± 12	(59)
(9) (3)	Linear	33 ± 9.3 μM	98 ± 16	(59)
(11) (2)	Linear	26 ± 6.1 μM	7.6 ± 2.1	(59)

[a]Number in parentheses refers to the number of assays at each chelate concentration: 0.1, 0.2, 0.5, 1.0, 2.0, 5.0, and 10 μM for chelates (4), (5), (8), (25), and (26); 0.1, 0.2, 0.5, 1.0, 2.0, and 5.0 μM for ferric (6), (27), and (28); 0.1, 0.25, 0.5, 1.0, 2.0, 5.0, and 10 μM for ferric (29) and (30); and 0.1, 0.25, 0.5, 1.0, 2.0, 5.0, 10, and 20 μM for ferric (9) and (11).
[b]K_m ± SE.
[c]Picogram-atoms of ^{55}Fe per minute per mg of protein (± SE).

the Δ chelate enantiomers, but that ferric (25) is the mirror-image Δ chelate enantiomer (59). Thus, these chelates, such as ferric (4) versus ferric (25), differ at either three or five chiral centers: the metal center chiral configuration, plus the asymmetric carbons in the oxazoline ring(s) derived either from D-(2R,3S)-threonine in the case of (25) or from L-(2S,3R)-threonine in the case of the ligands in the L-configuration.

Interestingly, neither the presence of an L-oxazoline ring nor a Δ metal center alone determines whether a chelate can be used by P. denitrificans (59). First, the bacterium accumulated ^{55}Fe from [^{55}Fe]ferric (25) and [^{55}Fe]-ferric D-fluviabactin (28), although not nearly as efficiently as from [^{55}Fe]ferric (4) or (6). Further, the A-forms of ferric parabactin [ferric L- (29) and D- (30) parabactin A, Fig. 10.5], in which the oxazoline ring has hydrolyzed to the open-chain threonyl structure, exhibited linear kinetics, including a relatively high K_m and a surprisingly high V_{max} (Table 10.2). The CD spectra of ferric (29) and ferric (30) are exact mirror images; however, the iron acquisition from ferric parabactin A is not stereospecific (Table 10.2). Net ^{55}Fe accumula-

tion from [^{55}Fe]ferric (4) and [^{55}Fe]ferric (8) was strongly inhibited to equivalent degrees by ferric (29) and ferric (30) (59). These kinetic data indicate a complex inhibition that does not appear to fit the usual simple models (e.g., competitive, noncompetitive, uncompetitive). Conversely, ^{55}Fe accumulation from the labeled chelates of both (29) and (30) were repressed by ferric (4), again, by apparently complicated kinetics. Accumulation of ^{55}Fe from [^{55}Fe]ferric (4) and [^{55}Fe]ferric L-parabactin A (29) were not diminished by ferric (25), except at relatively high concentrations. A model consistent with these overall findings entails a stereospecific binding step of high affinity, which requires the L-oxazoline ring, followed by a nonstereospecific postreceptor processing involving hydrolysis of the oxazoline ring of ferric L-parabactin (4) (E_0' = −0.673 mV, pH 7.0) to the open-chain threonyl structure of ferric L-parabactin A (29) [E_0' = −0.400 mV, pH 7.0 (74)], from which iron might be removed more readily.

Although P. denitrificans neither produces nor secretes hydroxamate siderophores, both labeled ferric chelates of (9) and (11) do facilitate the transport of iron, apparently by low

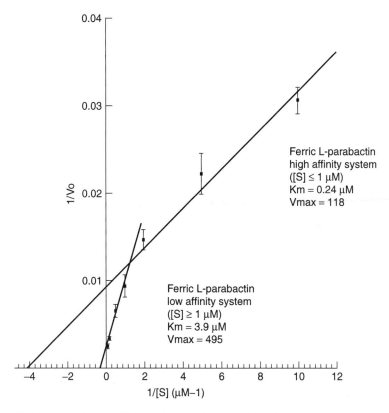

Figure 10.8. Enlarged view of [^{55}Fe]ferric L-parabactin of kinetic data presented in Figure 10.7, emphasizing the bimodal nature of this plot. The data for 0.1 $\mu M \leq$ [4] ≤ 1 μM and the data for 1 μM < [4] < 10 μM are fitted to regression lines, respectively representing high and low affinity phases of uptake. Data are presented as means of five determinations and standard deviations (bars) for [S] = 0.1, 0.2, 0.5, 1.0, 2.0, 5.0, and 10.0 μM. [Reprinted with permission from Bergeron and Weimar, *J. Bacteriol.*, **172**, 2650–2657 (1990). Copyright 1990 American Society for Microbiology.]

affinity, low V_{max} transport mechanisms (Table 10.2) (59). Many microorganisms sometimes produce specific membrane receptors to recognize and transport these complexes (6, 60, 63, 75–78).

A nonstereospecific, low affinity system, acting independently of the high affinity L-parabactin iron transport apparatus in *P. denitrificans*, has been reported (79). These uptake systems, which do not appear to be subject to regulation by iron concentration, have been characterized in many microorganisms, both prokaryotic and eukaryotic, in the past decade (6, 10, 80, 81). Many of these systems operate by means of a broad-spectrum reductase; systems that use ferrisiderophore reductases have been characterized in diverse microbial species, including *Mycobacterium*

smegmatis (82) and *Saccharomyces cerevisiae* (83). The reductase in *M. smegmatis* has a K_m for ferrimycobactin estimated to be <4 μM (82), a value similar to the apparent K_m values for iron acquisition from (**29**) and (**30**). One alternative to a reductase is a ligand-exchange mechanism similar to those proposed for mycobactin-mediated iron transport in *Mycobacterium* species (84) and more recently for amonabactin transport in *Aeromonas hydrophila* 495A2 (85). The contribution of parabactin A to iron transport by a ligand-exchange mechanism was assessed; the direct metal-ligand interchange seemed insufficient to account for the V_{max} observed for iron accumulation from ferric parabactin A (59). On the basis of kinetic issues, it also seems unlikely that free L-parabactin produced by the cell is involved in deferration of ferric hydroxamates, in spite of a formation constant that strongly favors ferric L-parabactin over both ferrioxamine B and ferrioxamine E on thermodynamic grounds (Table 10.1).

The issue of how the microorganism uses the iron complex also is observed in the parabactin/*P. denitrificans* system by following the uptake of [^{55}Fe]ferric parabactin and [^{3}H]ferric parabactin. The uptake of ^{55}Fe by the microbe increases over time and reaches a plateau; however, no appreciable uptake of tritiated ligand occurs (57). Furthermore, whereas the uptake of the tritium label is the same at 4 and 30°C, the difference in the uptake of [^{55}Fe]ferric parabactin at the two temperatures cannot be ignored. The labeled iron is taken up rapidly from [^{55}Fe]ferric parabactin at 30°C; however, there is no measurable transport of ^{55}Fe at 4°C (57). These observations are consistent with the "iron-taxi" mechanism, in which there is only a transient association between the siderophore-metal complex and the specific outer membrane receptor (58). The siderophore remains extracellular at all times, allowing the ligand to deliver iron to the cell membrane repeatedly.

The siderophore receptor and transport system has been exploited by investigators as a novel means of delivering siderophore-antibiotic conjugates to microorganisms (86, 87). Interestingly, however, the ability of the catecholamide siderophores to bind iron was itself postulated to be a means for preventing growth of heterologous microorganisms by withholding the necessary iron. Accordingly, selected catecholamide chelators (e.g., **3** and racemic parabactin **4/25**, Fig. 10.2) and their methylated analogs [e.g., (D,L)-tetramethylparabactin, (**31**)] (Fig. 10.9) were evaluated for their bacteriostatic and fungostatic activity (88). The methylated analogs (e.g., **31**) exhibited little, if any, effect on the growth of three pathogenic bacteria (*Pseudomonas aeruginosa*, *E. coli*, and *Staphylococcus aureus*) and one pathogenic yeast (*Candida albicans*), whereas the nonmethylated ligands demonstrated varying bacterio- or fungostatic activity against these same microorganisms. In the cases of *E. coli* and *P. aeruginosa*, the growth was repressed completely for between

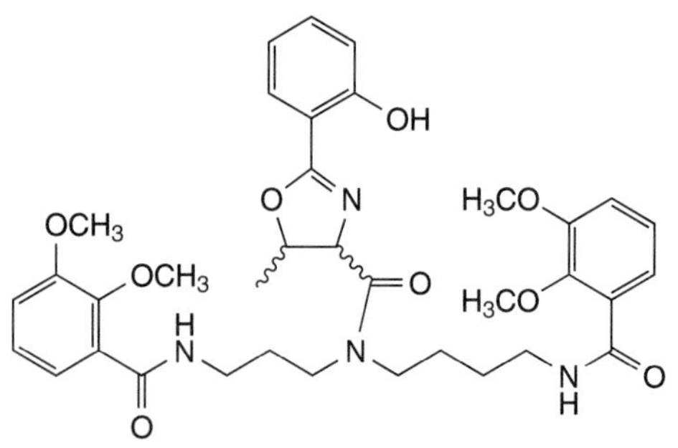

Figure 10.9. Structure of (**31**), (D,L)-tetramethylparabactin.

5 and 10 h by **4/25** without an alteration of the rate of growth thereafter. All of the chelators affected the growth of *S. aureus* and *C. albicans*, although the sensitivity was of a somewhat different nature. The antimicrobial activity of the iron chelators tested was bacteriostatic, rather than bactericidal; furthermore, no growth suppression of the normally sensitive organisms was demonstrated in the presence of a 1:1 iron:(D,L)-parabactin complex, further evidence that the effects of the chelators were attributed to iron starvation (88).

The reduced bacteriostatic activity of the catecholamides in the cases of *P. aeruginosa* and *E. coli* can be attributed to these bacteria producing their own siderophores in concentrations high enough to compete with the exogenously introduced siderophore systems. The initial period of near-complete growth cessation is consistent with this hypothesis. During this period, identification by the organism of iron-starvation conditions induced by the exogenous ligands would lead to the expression of genes responsible for siderophore production and receptor expression (6, 10, 81). There is no evidence to suggest that iron transport in these bacteria is occurring through these exogenous chelators. The siderophore systems in *C. albicans* and *S. aureus* appear to be less efficient than those in *P. aeruginosa* and *E. coli*, as evidenced by the sensitivity of *C. albicans* and *S. aureus* to all of the catecholamide ligands. Although both of these microorganisms produce endogenous siderophores (e.g., staphyloferrin from *S. aureus*) (89–91), their receptors cannot use the catecholamide iron chelators evaluated.

Thus, the repression of proliferation of various microorganisms by the spermidine catecholamide iron chelators (88) seems to be the result of the ability of the chelators to complex iron; when bacteria that are normally sensitive to the hexacoordinate catecholamide iron(III) chelators were presented with the iron complexes of these ligands, no antimicrobial activity was observed. Similarly, blocking of the catechol coordination sites by methylation abrogates the bacterio- or fungostatic activity of the catecholamide. These results should be tempered by an understanding of a very serious potential problem with the concept of iron deprivation in antimicrobial therapy, that is, acceleration of bacterial growth by chelation. The ability of an organism to use nonnative chelators (i.e., "prepackaged iron") could produce an extremely dangerous situation, as was borne out when the hexacoordinate catecholamide chelator enterobactin (**1**) was tested as a possible deferrating agent in rodents. In this study, a fulminant *E. coli* infection developed in these animals, which ultimately resulted in death from septicemia (92). Furthermore, the increased susceptibility of chelation therapy patients using DFO to bacterial infection, particularly by *Yersinia enterocolitica*, is well documented (93, 94); cross-utilization of other siderophores by heterologous bacterial species *in vitro* is well established (10, 81, 95, 96). Thus, any chelators intended for therapeutic use as either an antimicrobial or a deferration agent must be screened against numerous microorganisms that might be present in a host. Even though the ligands need not be active *against* the bacteria or fungi, such chelators must not *enhance* their growth.

Strikingly, the Fe(III) complex of synthetic enantioenterobactin (derived from D-serine) failed to promote the growth of enterobactin-deficient *E. coli* mutants *in vitro* (97). Both L-parabactin (**4**) and L-fluviabactin (**6**) stimulated the growth of *P. denitrificans* and promoted iron uptake, whereas their respective D-enantiomers (**25** and **28**) did neither, in that the latter compounds were not recognized by the bacterial iron-transport apparatus (57,

66). However, there was not a significant difference in the iron-clearing efficacy of (**6**) and (**28**) in a bile duct-cannulated rodent model (66); (**4**) was very efficient at removing iron from rodents and primates as well (98). These results underscore the idea that siderophores can serve as a platform in the design of therapeutic agents. Stereochemical modification of the natural product iron chelators can profoundly diminish their capacity to stimulate microbial growth without reducing their iron-clearing properties *in vivo*.

1.3 Iron Dynamics in Humans

More complicated organisms, that is, the higher eukaryotes, have developed large proteinaceous molecules to solve the access, storage, and utilization problems associated with iron. Animals, in particular humans, have evolved a highly efficient mechanism for handling the metal in which two basic "iron-processing" proteins are involved, one for transport, transferrin (Fig. 10.10a), and one for storage, ferritin (Fig. 10.10b). It is interesting to compare the relative sizes of the molecules involved in iron transport in prokaryotes (e.g., **4**, MW ~ 600) and in eukaryotes [e.g., transferrin, MW 80,000 (24)]. At first, it seems as though the assembly of transferrin for eukaryotic iron transport requirements is very inefficient. However, because of a mammalian iron storage and transport system that contains elaborate control mechanisms involved in the synthesis, cellular uptake, and utilization of its attendant iron transport and storage proteins, such size and complexity are required.

1.3.1 Iron Storage and Transport. The location of the major iron stores in the human body and a schematic of iron transport are illustrated in Fig. 10.11 (99). The iron metabolism system is quite efficient; little leakage or expansion occurs in healthy humans (23). Humans both absorb and excrete about 1 mg of this transition metal daily. Whereas all tissues contain at least some iron, resulting in total iron stores in an average adult human of about 50 mg/kg (slightly less in females), the distribution of the metal is not uniform (23). The hemoglobin in circulating erythrocytes accounts for 60–70% of the iron load; ferritin

(a)

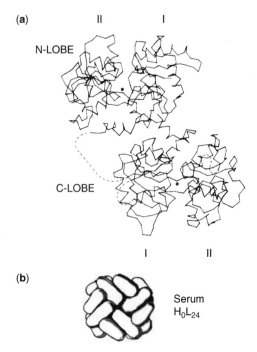

(b)

Serum
H_0L_{24}

Figure 10.10. The iron transport protein, transferrin (a) and the serum isoform of the iron storage protein, ferritin (b). Serum transferrin (a) is composed of two homologous lobes that each bind a single ferric ion. The *N*-terminal N-lobe and the C-terminal C-lobe each are divided into two dissimilar domains, and the iron-binding site is located within the interdomain cleft. The iron is bound by two tyrosines, a histidine, and an aspartic acid residue. The various isoforms of ferritin (b) found in different tissues are composed of different proportions of heavy (H) and light (L) chains (H_xL_y; $x + y = 24$); these protein subunits form a shell, surrounding a core of mineralized ferric iron. The serum isoform (H_0L_{24}) is shown. [Panel a reproduced with permission from S. Bailey et al., *Biochemistry*, **27**, 5804–5812 (1988). Copyright 1988 American Chemical Society. Panel b reproduced from P. M. Harrison and P. Arosio, *Biochim. Biophys. Acta*, **1275**, 161–203 (1996). Copyright 1996 with permission from Elsevier Science.]

sequesters another 20–30%. The remaining 10% serves as a key component of molecules such as myoglobin, cytochromes, and iron-containing enzymes (23). Less than 0.1% of the total body iron (~3 mg at steady state) is associated with transferrin (23, 24). This protein serves as the conduit for the metal between ferritin storage pools, the reticuloendo-

thelial system, and the bone marrow for red cell assembly. Most (~80%) of this transitory iron store (turnover rate, 25–30 mg/24 h) is used for hemoglobin synthesis during erythroid development; the rest is distributed to various tissues (24, 99). The red cells, which survive for approximately 120 days, are eventually processed by the spleen. Phagocytosis of the senescent cells by reticuloendothelial (RE) macrophages liberates the iron from hemoglobin to be mobilized by transferrin. A more detailed examination of the components of transport (transferrin and the transferrin receptor) and storage (ferritin) follows.

Transferrin/Transferrin Receptor. Transferrin is a homodimeric globular protein with a MW of about 80,000 (Fig. 10.10A) (99, 100). In humans, it is synthesized predominantly in the liver; some synthesis occurs in brain, testis, and mammary tissues. Each molecule contains two iron-binding sites. Although transferrin has very high affinity for iron, this binding is very sensitive to pH. For example, at physiological pH, the K_d of transferrin for iron(III) is 10^{-23} M; in contrast, no significant chelation occurs at or below pH 4.5 (23, 24). This pH sensitivity is critical to the endosomal release of the metal from the protein. Six donor groups—two tyrosines, a histidine, an aspartate, and two oxygens from the carbonate anion—are necessary for the ligation of iron by transferrin (24, 100). There are a number of iron chelators that bind iron more tightly than does transferrin (e.g., desferrioxamine, $K_d = 10^{-31}$ M); these should, on a thermodynamic basis, be able to remove the metal from this protein. However, with rare exceptions, for example, the hydroxypyridinones (101), this is not the case. Kinetics, rather than simple thermodynamics, is the issue. Mobilization of the iron in transferrin can occur with a number of different chemical agents, including ascorbate. The cellular processing of this iron shuttle protein is a remarkable example of efficiency. After the delivery of iron to cells, the protein is recycled for further use (Fig. 10.12) (99). This process begins with the binding of transferrin to a cell surface receptor, which itself is subject to posttranslational modification within the endoplasmic reticulum of the cells in which it is synthesized. Its

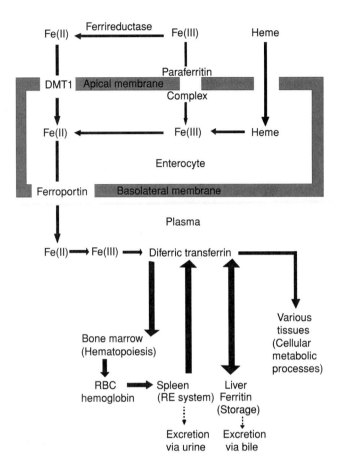

Figure 10.11. Iron absorption and processing.

level of expression is particularly high in proliferating tissues, although, except for mature erythrocytes, it is present in all of the cells of the human body. This receptor is symmetrical and resembles a butterfly (Fig. 10.13) (102). The dimer consists of two 90-kDa monomers that are connected by two cysteine disulfide bridges. Each of the individual monomers contains three domains. The cytoplasmic tail (61 amino acids in length) is responsible for endocytosis of the receptor-transferrin complex. Fatty acid linkages within the 28 amino acid transmembrane segment aid in anchoring the protein within the membrane. The 671 amino acids at the extracellular carboxy terminus are often referred to as the ectodomain. Two molecules of transferrin bind to lateral clefts within the ectodomain of the dimeric receptor. The ectodomain is subdivided further into an apical region, a protease-like region, and a helical portion, which is responsible for connecting the two monomers (102).

The cycle of the transferrin-transferrin receptor complex occurs in five steps: (*1*) binding of transferrin to the receptor, (*2*) endocytosis, (*3*) iron removal, (*4*) return of the apotransferrin-transferrin receptor complex to the cell surface, and (*5*) release of apotransferrin (Fig. 10.12). Each transferrin receptor can bind two molecules of transferrin, although the actual binding affinity varies with the iron status of transferrin: diferric transferrin $\gg$ monoferric transferrin $>$ apotransferrin. The K_d for the diferric transferrin-transferrin receptor complex is around 5 nM (24, 103). Given that the plasma concentration of diferric transferrin is roughly 5 μM, it is likely that most cell surface-exposed receptors are saturated.

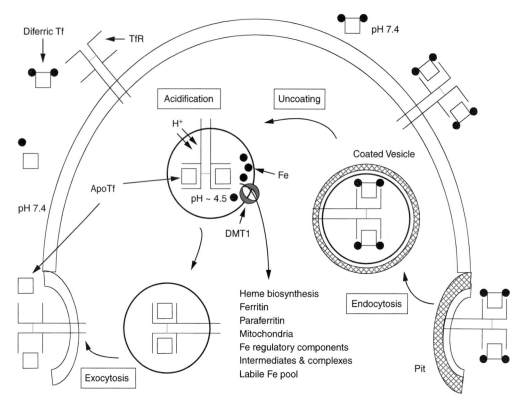

Figure 10.12. The transferrin/transferrin receptor cycle. The major steps, depicted clockwise, are: (a) binding of Fe(III) (●) to transferrin (□, Tf), (b) binding of diferric transferrin to the transferrin receptor (TfR), (c) endocytosis by way of a clathrin-coated pit, (d) iron removal, (e) return of the apotransferrin-transferrin receptor complex to the cell surface, and (f) release of apotransferrin (ApoTf).

After transferrin binds to the receptor, the complex interacts with an adapter protein within a clathrin-coated pit; this megacomplex is then endocytosed. After the megacomplex reaches the endosome within the cell, the iron is released from transferrin; this dissociation is very much dependent on the low endosomal pH, exploiting the poor binding of iron to transferrin at low pH. The apotransferrin, still complexed to the transferrin receptor, is exported back to the cell membrane, where the apotransferrin is released for further complexation. The iron is released from the endosome through the transmembrane portal DMT1 (*D*ivalent *M*etal *T*ransporter 1, formerly known as Nramp2) and either is delivered to other iron-binding molecules (e.g., for incorporation into heme) or is used by mi-

tochondria in immature erythroid cells. Alternatively, the metal is stored in ferritin.

Ferritin. Ferritin is a large spherical complex consisting of 24 subunits; the combined MW is around 450,000 (99, 104). The sphere (Fig. 10.10b) manages over 4500 iron atoms within its core, thus preventing the metal from causing iron-promoted redox damage; yet, the protein releases iron on demand for utilization by other iron-depleted systems. The sphere is composed of two basic protein chain types, heavy (H) and light (L) chains. The H chains exhibit ferroxidase activity (99, 104). When these chains are assembled into the spherical array, the chains create channels, allowing for access to the iron core. The current thinking is that soluble Fe(II) is incorporated into the ferritin shell, where it is oxi-

(a)

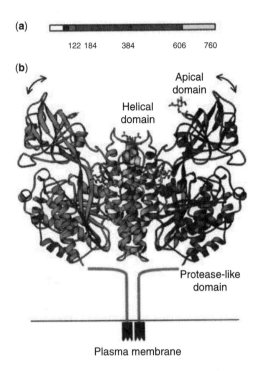

122 184   384   606   760

(b)

Apical
domain

Helical
domain

Protease-like
domain

Plasma membrane

Figure 10.13. Structure of the ectodomain of the transferrin receptor. (a) Domain organization of the transferrin receptor polypeptide chain. The cytoplasmic domain is white; the transmembrane segment is black; the stalk is gray; and the protease-like, apical, and helical domains are red, green, and yellow, respectively. Numbers indicate domain residues at domain boundaries. (b) Ribbon diagram of the transferrin receptor dimer depicted in its likely orientation with respect to the plasma membrane. One monomer is colored according to domain (standard coloring as described above), and the other is blue. The stalk region is shown in gray connected to the putative membrane-spanning helices. Pink spheres indicate the location of Sm^{3+} ions in the crystal structure. Arrows show directions of (small) displacements of the apical domain in noncrystallographically related molecules. See color insert. [Reprinted with permission from C. M. Lawrence, S. Ray, M. Babyonyshev, R. Galluser, D. W. Borhani, and S. C. Harrison, *Science*, **286**, 779–782 (1999). Copyright 1999 American Association for the Advancement of Science.]

dized to Fe(III) and deposited for mineralization (99, 104, 105). This process of iron oxidation, incorporation, and mineralization is surprisingly slow; it can require hours (104).

Most iron chelators cannot remove iron from ferritin effectively; as with transferrin,

access is the problem. However, iron can be mobilized from ferritin reductively both *in vitro* and *in vivo* (101, 104). Interestingly, it seems that the release of iron from ferritin in the presence of various reducing agents can occur even when the reducing agent and the metal are separated by some distance within the ferritin channel. The implicit danger with this reductive removal is that the iron released for chelation, Fe(II), can promote Fenton chemistry, thus causing significant cellular redox damage, as discussed in detail in Section 1.3.4. Therefore, when the metal is reductively released from ferritin, a ligand (e.g., transferrin) must be present in sufficient proximity to bind and remove it. This concern has been borne out; a clinical trial was designed to improve the efficiency with which desferrioxamine, a therapeutic iron chelator, can remove iron from ferritin by the coadministration of ascorbate. Unfortunately, this trial served to illustrate how serious the problem of uncontrolled iron removal from ferritin can be, as discussed in Section 2.1.1.1.

1.3.2 Molecular Control of Iron Uptake, Processing, and Storage. One of the most interesting discoveries in the area of iron metabolism in recent years is the mechanism by which the iron storage and transport proteins (i.e., ferritin and transferrin) are regulated. Whereas the iron-mediated feedback regulation mechanisms operate with a number of iron-processing proteins, they have been investigated most thoroughly for ferritin and for the transferrin receptor. Simply put, the question is: How does iron concentration control the synthesis of iron-processing proteins? As iron levels increase, the cellular storage compartments (i.e., ferritin levels) must increase also. Conversely, when cellular iron levels are depleted, or there is an increase in iron requirement, as in various proliferative processes, the efficiency with which iron is incorporated by cells must be augmented; that is, an upregulation of transferrin receptor expression is required. Both of these aspects of iron processing are handled at the posttranscriptional level (99).

Both ferritin mRNA and transferrin receptor mRNA contain iron-responsive elements (IREs) to which the iron-responsive proteins

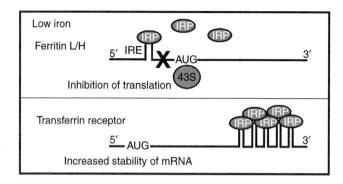

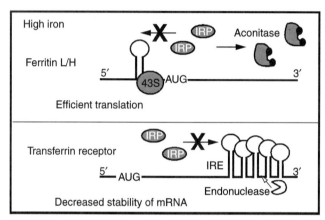

Figure 10.14. Posttranscriptional regulation of iron-responsive element-containing gene expression. Intracellular iron concentration determines binding of iron regulatory proteins (IRP) to iron-responsive elements (IRE). Binding affinity of iron regulatory proteins to iron-responsive elements inversely correlates with intracellular iron concentration. For example, a single iron-responsive element is located in the 5' untranslated region of *ferritin* mRNA, whereas five iron-responsive elements are present in the 3' untranslated region of *transferrin receptor 1* mRNA. Under conditions of iron deficiency, iron regulatory proteins bind to the iron-responsive element in the 5' untranslated region of *ferritin* mRNA. This binding sterically prevents the recruitment of 43S translation preinitiation complex and arrests translation of ferritin. On the other hand, binding of iron regulatory proteins to the iron-responsive elements in the 3' untranslated region of transferrin receptor 1 mRNA increases mRNA stability and synthesis of transferrin receptor 1. In contrast, under high concentration of intracellular iron, iron regulatory proteins possess aconitase activity but lack mRNA-binding activity. In the absence of binding of iron regulatory proteins to iron-responsive elements, *ferritin* mRNA is translated efficiently, whereas *transferrin receptor 1* mRNA is rapidly degraded, probably by iron-responsive element-specific endonucleases. [Reproduced from P. T. Lieu, M. Heiskala, P. A. Peterson, and Y. Yang, *Mol. Aspects Med.*, **22**, 1–87 (2001). Copyright 2001 with permission from Elsevier Science.]

(1 and 2; IRP-1 and IRP-2) bind (Fig. 10.14). The binding of IRP-1 and IRP-2 to the mRNAs for ferritin and transferrin receptor controls translation; this binding is dependent on the iron status. IRP-1 is very similar to mitochondrial aconitase (99, 106), the enzyme responsible for the conversion of citrate to isocitrate.

When cells are replete with iron, an iron-sulfur complex is formed with three cysteine residues within IRP-1 (106); no binding to IREs occurs in this state. However, under low iron conditions, the iron-sulfur complex is absent; the protein does not possess aconitase activity. This apoprotein accumulates in the cell and binds very tightly to the IREs. The mRNAs for both the L and H chains of ferritin contain the IRE in the 5′-untranslated region; the transferrin receptor IRE is in the 3′-untranslated region of the message.

These IREs can serve as either repressors or enhancers of translation. The IRP repressive mode, which occurs in ferritin mRNA, is a steric issue. Binding of IRPs to the 5′ IRE prevents initiation of translation of protein synthesis by impeding the recruitment of 43S preinitiation complexes. Thus, when cellular iron levels are high, the metal binds to IRP, preventing its association with the IREs; synthesis of ferritin is initiated. The IRE/IRP control mechanism for the transferrin receptor is somewhat different from that described for ferritin. Binding of IRP to the 3′ IREs, rather than preventing translation, as in ferritin, enhances the stability of the message by preventing its degradation, increasing message translation. Thus, increases in cellular iron concentration induce ferritin synthesis, given that the ferrated IRPs do not bind to and sterically inhibit translation of ferritin message. Conversely, decreases in cellular iron concentration elicit an increase in transferrin receptor synthesis because deferrated IRPs bind to and stabilize the message. The mRNAs of other key iron-processing proteins also contain IREs, which control their translation. For example, DMT1, an iron portal, has a 3′ IRE that operates in a manner similar to that of the transferrin receptor.

1.3.3 Iron Absorption. Absorption of iron in the proximal small intestine is the major source of iron uptake (99, 103). The metal is transported across the brush border and released to the circulation through the basolateral membrane (Fig. 10.11). Heme iron accounts for most of the iron absorbed; relatively small amounts of inorganic Fe(II) and Fe(III) are taken up also. Although both of these oxidation states of iron are presented to the enterocytes, the absorption of Fe(II) is much more efficient. In fact, a mucosal ferrireductase, which converts Fe(III) to Fe(II), has been demonstrated as being critical to efficient iron absorption (107, 108). Iron(II) is transported across the apical membrane through the transmembrane protein complex DMT1. The Fe(II) travels through the cytoplasm and is released through the basolateral membrane through a second protein complex. The transporter Ferroportin 1 acts in conjunction with Hephaestin, a multi-copper oxidase that converts Fe(II) to Fe(III) for mobilization by transferrin. Although Fe(III) uptake is not particularly efficient, the paraferritin complex is designed for Fe(III) uptake and transport (103). This mucin-mediated iron complex consists of β-integrin, mobilferrin, and flavin mono-oxygenase. Once the iron-paraferritin complex is transported across the apical cell membrane, the oxygenase may act to reduce Fe(III) to Fe(II) for further processing (103). Finally, heme iron uptake provides the most plentiful source of iron. Hemoglobin is first digested enzymatically; the released heme molecule is transported across the apical membrane. The heme is then degraded by heme oxygenase, resulting in the release of iron within the cell.

1.3.4 Iron-Mediated Damage. In normal, healthy humans, iron is well managed and does not serve as a source of oxidative damage. Although there is always some non-transferrin-bound iron (NTBI) in the plasma of healthy subjects, this amount is negligible; most of the metal is associated with heme, is stored in ferritin, or is utilized in various iron-containing enzymes. As we have seen, the metal is shuttled safely between these pools by the iron-trafficking protein transferrin. However, there are a number of different disease states, discussed briefly below, in which body iron stores and plasma NTBI levels are sufficiently high to cause significant damage, much of which is oxidative in nature. To understand how oxidative damage unfolds, a description of the mechanism of iron-mediated damage is in order.

As was mentioned, although iron has many oxidation states available to it, the most common oxidation states of iron in the biosphere are Fe(II) and Fe(III); the Fe(II)/(III) redox couple is the focus of this discourse. The equilibrium shown in Equation 10.8 is very much dependent on pH and the nature of the ligands chelating the metal.

$$Fe(II) \leftrightharpoons Fe(III) + e^- \qquad (10.8)$$

Thus, iron can function both as an electron source (reducing agent) and an electron sink (oxidizing agent). The interaction of the free metal with oxygen and with various active oxygen species (e.g., superoxide anion and hydroperoxide) is the source of iron's toxicity.

Probably the most significant reaction sequence involves the reaction of Fe(II) with hydrogen peroxide. First, Fe(II) itself can react with molecular oxygen to produce hydrogen peroxide. Again, the rate of this reaction depends on both pH and the nature of the ligands surrounding the metal. The first step in this reaction (Equation 10.9) is the production of superoxide anion:

$$Fe(II) + O_2 \leftrightharpoons Fe(III) + O_2^{1/2-} \qquad (10.9)$$

Production of superoxide occurs in a variety of different cells. For example, during intestinal injury, neutrophils infiltrate and release superoxide anion (109). Although produced in neutrophils by NADPH oxidase (110), superoxide anion is probably most frequently thought of as a by-product of the conversion of xanthine to uric acid by xanthine oxidase (111). However produced, superoxide anion can dismute to hydrogen peroxide and oxygen (Equation 10.10), either spontaneously or through the action of the copper-dependent enzyme superoxide dismutase:

$$2O_2^{1/2-} + 2H^+ \rightarrow H_2O_2 + O_2 \qquad (10.10)$$

Hydrogen peroxide is a good oxidizing agent and can be particularly insidious, as it has such global cellular access. Because of its structural similarity to water, it passes through membranes very easily. Normally, hydrogen peroxide is managed by either glutathione peroxidase or catalase. Glutathione peroxidase, which is probably the major "housekeeping" enzyme, reduces hydrogen peroxide to water and simultaneously oxidizes glutathione (Equation 10.11A), whereas catalase converts hydrogen peroxide to water and molecular oxygen (Equation 10.11B):

$$H_2O_2 \xrightarrow{\text{2-Glutathione-SH} \quad (\text{Glutathione-S})_2^-} 2H_2O \qquad (10.11A)$$

$$2H_2O_2 \xrightarrow{\text{Catalase}} 2H_2O + O_2 \qquad (10.11B)$$

However, if the hydrogen peroxide concentration exceeds the K_m of these enzymes, or, more important, if hydrogen peroxide is generated in a milieu in which iron(II) is present and these enzymes are absent, damage can unfold. In combination with Fe(II), hydrogen peroxide is repudiated to be the major source of the tissue damage associated with excess iron. Iron(II) converts hydrogen peroxide to hydroxyl radical, a very reactive species, and hydroxide anion, referred to as the Fenton reaction (Equation 10.12) (112–115):

$$Fe(II) + H_2O_2 \rightarrow$$
$$Fe(III) + HO^{1/2} + HO^- \qquad (10.12)$$

Note that this Equation neither directly involves superoxide anion nor results in the liberation of molecular oxygen, as does what has been referred to as the superoxide-driven Fenton reaction or the Haber-Weiss reaction (114), even though hydroxyl radicals are produced in both instances. The hydroxyl radical reacts very quickly with a variety of cellular constituents and can initiate free-radical chain processes. The generation of this radical by Fe(II) and its marked reactivity have been the subject of much controversy over the years (114, 115). Many investigators have argued that multivalent forms of iron [e.g., Fe(IV)] or various oxy complexes are the source of iron-mediated cellular damage (114). However, oxidation of aromatic amino acids by pulse radiolysis-generated hydroxyl radicals occurs at very different sites than does Fe(IV)-mediated oxidation (114). In nonchain reactions, one might anticipate a simple reaction between Fe(II) and hydrogen peroxide to generate the hydroxyl radical. Unfortunately, the cyclic nature of the Fe(II)/Fe(III) redox couple is problematic. The Fe(III) liberated in Equation 10.12 can be reduced back to Fe(II) through a variety of biological reductants, including

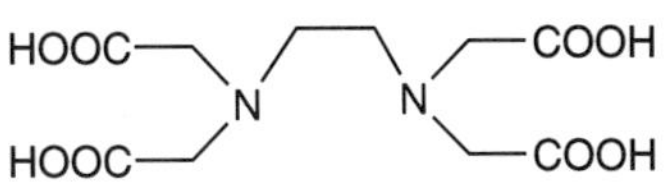

Figure 10.15. Structures of selected synthetic chelators: ethylenediaminetetraacetic acid (EDTA, **32**), 1,2-dimethyl-3-hydroxypyridin-4-one (deferiprone; L1; **33**; formerly CP20), 2-pyridinecarboxaldehyde isonicotinoyl hydrazone (PCIH; **34**), and diethylenetriamine pentaacetic acid (DTPA; **35**).

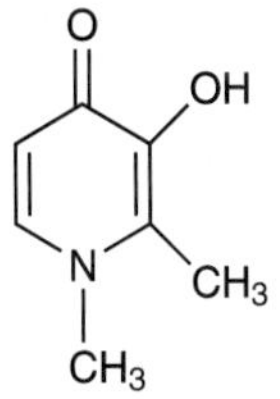

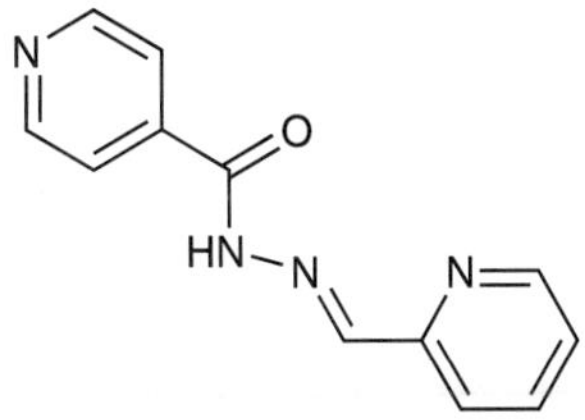

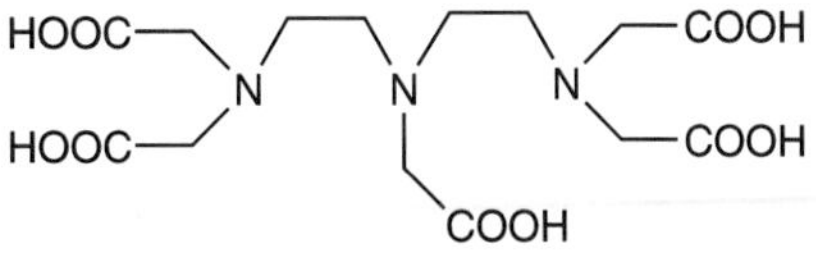

ascorbate and superoxide anion. In intestinal injury, the production of superoxide anion by neutrophils in the colon is the source of activated oxygen for the generation of hydrogen peroxide (109). Thus, not only does superoxide anion provide hydrogen peroxide, but the anion also serves as a means of recycling Fe(III) to Fe(II) for future hydroxyl radical production. This net reaction is shown in Equation 10.13:

$$Fe(III) + O_2^{1/2-} \rightarrow H_2O_2 + Fe(II) \quad (10.13)$$

This problem is underscored further when one considers that ascorbate is an effective reducing agent only in the absence of iron. In fact, a mixture of iron salts, ascorbate, and H_2O_2 is a good source of hydroxyl radicals (114) (Equation 10.14):

$$Fe(III) + ascorbate \rightarrow Fe(II)$$
$$+ \text{ semidehydroascorbate} \quad (10.14)$$

It should be clear, then, that the proteins that store and transport iron represent a major component of the body's antioxidant defense system.

Some iron chelators inhibit the Fenton reaction; paradoxically, some stimulate it. For example, desferrioxamine binds iron very tightly, removing it from the Fenton pool (114). This ligand binds Fe(III) much more tightly than it does Fe(II). Thus, Fe(II) in the presence of desferrioxamine is an even better reducing agent, as the Fe(II)/Fe(III) equilibrium is shifted to the right. However, this event occurs only once because the iron in the desferrioxamine/Fe(III) complex is not recycled readily to Fe(II) for further redox chemistry. Other chelators, for example, ethylenediaminetetraacetic acid (EDTA, **32**) or deferiprone (**33**) (Fig. 10.15), actually promote Fenton chemistry (116).

1.4 Iron-Mediated Diseases

There are two basic types of iron-related diseases, those associated with depleted iron stores (e.g., anemia) and those associated with excess iron, also referred to as iron overload. The latter subset is the focus of the present discussion. As was mentioned in Section 1.3, healthy humans maintain an iron load of about 50 mg/kg (23, 99). At this level, the body's iron storage and transport can manage iron safely. However, in various iron-overload disorders, the system becomes overwhelmed; sufficient "unmanaged iron" is available to cause redox, that is, Fenton chemistry-driven, damage. For example, plasma from healthy human donors does not stimulate free-radical reactions; however, plasma from iron-overloaded patients markedly stimulates lipid peroxidation (114).

Two principal kinds of iron-overload disease occur, both of which are genetic disorders. The hallmark of primary iron overload (e.g., hemochromatosis) is increased intestinal iron absorption (99, 103). Secondary, or transfusional, iron overload is associated with chronic transfusion therapy. Such therapy is required in diseases in which anemia caused by aberrant hemoglobin is a problem (e.g., thalassemia, sickle-cell anemia) (117, 118), or in hematopoietic malfunctions (e.g., myelodysplasia) (119). Depending on the severity of the disease, transfusion is needed up to several times a month. Each unit of packed red cells transfused introduces 200 mg of iron into the closed metabolic pathway. Eventually, the transfused cells are processed through the reticuloendothelial (RE) system as occurs in healthy subjects; the added iron is mobilized by transferrin. Unfortunately, because there is no effective mechanism for iron excretion, these transfused patients continue to accumulate the metal. Patients with primary hemochromatosis can be managed adequately by phlebotomy; the frequency of bleeding depends on the extent of the iron overload. However, patients with transfusional iron overload must be managed by chelation therapy.

Although primary hemochromatosis and transfusional iron overload represent the major diseases in this category, there has been substantial interest in recent years in the role that iron plays in diverse disorders, including neurodegenerative disorders such as Alzheimer's disease (120), Parkinson's disease (121), and Friedreich's ataxia (122–124). In each of these diseases there is an increased iron content of affected regions of the brain. These focal increases in iron concentration have been associated with the expected oxidative stress issues, extending from membrane damage to mitochondrial dysfunction and cell death.

In Alzheimer's disease there is an excess accumulation of iron in those regions of the brain associated with the characteristic deficits in neurological function: the hippocampus, amygdala, nucleus basalis, and cerebral cortex (120). Alzheimer's disease is a particularly interesting example, in that it goes beyond the classic iron-driven Fenton chemistry-mediated damage: an ancillary protein, amyloid β-peptide (Aβ), derived from amyloid precursor protein (APP), is involved. The predominant thinking is that Aβ mediates nerve cell death in Alzheimer's disease and that iron plays a key role in this process. In cell-free systems, Aβ and iron interact directly to promote Fenton chemistry (125). Experiments with neuronal cell lines that are resistant to Aβ toxicity consequent to the transcriptional elevation of glutathione peroxidase and catalase have provided quite compelling evidence for the role of Fenton chemistry in cell death (126). From a mechanistic perspective, the relationship between Aβ and iron in neuronal toxicity was best demonstrated in a series of experiments that used B12 cells, a neuronal cell line, and primary cultures of cerebral cortex neurons. In these experiments Fe(III) substantially enhanced the toxic effects of Aβ. The iron chelator DFO ameliorated Fe/Aβ-induced neuronal cell damage in both B12 cells and in primary cortical cells; this prevention was dose dependent. Furthermore, the most efficiently transported form of the drug, the Ga(III)-DFO complex, was the most effective. Finally, the cell damage was minimized significantly when iron was chelated, even in the presence of added hydrogen peroxide (127). Thus, although it is unlikely that Aβ/Fe/H$_2$O$_2$-induced cell toxicity is the *only* source of neuronal injury seen in Alzheimer's patients, this mechanism could well be a principal source of the problem.

As discussed in detail above, regulation of the expression of ferritin and other proteins involved in iron metabolism occurs by interactions of IRPs with IREs contained in the mRNA transcripts. Deletion of the gene encoding IRP2 causes misregulation of iron metabolism and neurodegeneration in mice (128). Independently reported lines of evidence suggest that iron may play a similar regulatory role in Alzheimer's disease. Affected individuals from two familial Alzheimer's pedigrees were found to contain a mutation that disrupts a stem-loop in APP mRNA containing the consensus sequence CAGUGA characteristic of an IRE (129). There are, in fact, significant overlaps in the regulation of the

APP and ferritin genes (130). These observations suggest that the synthesis of the Aβ precursor APP may itself be shut off with an iron chelator in a manner similar to the control of the synthesis of ferritin. Clearly, removal of excess iron would ameliorate Aβ/iron-mediated Fenton pathophysiology, and perhaps would suppress Aβ production as well.

The marked increase of iron in the Parkinsonian substantia nigra underscores the role of the metal in this disease (121). It has been suggested that this iron is responsible for the oxidative stress seen in the degeneration of the dopaminergic neurons in the substantia nigra. The autooxidation and oxidative deamination of dopamine by monoamine oxidase in the metabolically active basal ganglia result in the accumulation of neuromelanin, the black pigment for which the substantia nigra is named, and the production of hydrogen peroxide (131, 132). In Parkinson's the normal mechanisms that manage peroxide (e.g., peroxidase, catalase, glutathione, ascorbate) are defective (133–138). In addition, neuromelanin itself binds iron in a Fenton-reactive form (131, 139–141). Thus, the conditions are perfect for the iron-catalyzed production of hydroxyl radicals and the ensuing damage. The pathophysiological role of iron in Parkinson's disease is substantiated further by the finding that removal of iron from the brain by intra-cerebral-ventricular injection of DFO ameliorated 6-hydroxydopamine-induced damage to nigrostriatal dopamine neurons (142, 143). Although the protection conferred by the chelator was not complete, it was quite real, providing additional support for the concept of iron chelators as therapeutic platforms in neurodegenerative maladies. Desferrioxamine does not penetrate cells well at all; ligands with better transport and iron-clearance properties would be superior candidates for the treatment of such disorders.

Friedreich's ataxia is an autosomal recessive mutation of the gene encoding frataxin, a protein involved in mitochondrial iron homeostasis. A deficiency in frataxin in Friedreich's ataxia leads to an accumulation of iron in mitochondria, oxidative stress, and neurodegeneration (122–124). 2-Pyridinecarboxaldehyde isonicotinoyl hydrazone (**34**; Fig. 10.15) and its analogs, iron chelators capable of penetrating the mitochondrial membrane, have been proposed to treat the mitochondrial iron overload in Friedreich's ataxia (144).

Ultimately, the solution in each of these disease states is the same: identify a method of opening up the closed iron metabolic loop. Generally, this translates into designing chelators that will selectively sequester this transition metal and render it excretable. Therefore, the remainder of this chapter focuses on current methods of chelation therapy and approaches to the design, synthesis, and assessment of new iron chelators.

2 CLINICAL USE OF CHELATING AGENTS

The pharmacological properties that define an ideal iron chelator are very much dependent on the iron-mediated disease being treated. One of the major considerations is whether the treatment is of a long- or short-term nature. This issue has a particular impact on the toxicity profile requirements of the ligand. For example, the lifelong treatment either of a genetic disorder, such as severe hereditary hemochromatosis, or of transfusional iron overload, as can occur in the management of sickle-cell anemia (118) or thalassemia, dictates the use of a ligand that can maintain patients in negative iron balance, that is, be able to clear 250–400 μg of iron per kg of body weight per day (145), and do so with minimal toxicity over an extended period of time. However, the relatively brief application of an iron chelator either to control iron-induced cardiac reperfusion damage during cardiac surgery (146) or to curb the global effects of acute iron poisoning (147) allows for chelators with different toxicity profiles. In the former situations, chronic iron overload adds another dimension to take into account, patient compliance (94, 117). These issues demand that consideration be given to the mode of administration required, that is, parenteral versus oral dosing.

There are two levels of toxicity that define a therapeutic agent: that which is implicit in the structure and unrelated to its therapeutic target and that which is associated with the purpose for which the molecule is designed. For example, the aminoglycoside antibiotics elicit

Table 10.3   Formation Constants of Selected Ligands with Iron and Other Metals

Ligand	Formation Constant for Metal Chelate $(K, M^{-1})^a$				References
	Fe(III)	Ga(III)	Al(III)	Zn(II)	
Desferrioxamine B^b	$10^{30.99}$	$10^{28.65}$	$10^{24.5}$	$10^{10.04\ c,d}$	(22, 148, 151)
DTPA^e	10^{28}	$10^{25.54\ f}$	$10^{18.7}$	$10^{18.29\ d}$	(15, 150)
1,2-Dimethyl-3-hydroxypyridin-4-one (L1)^g	$10^{35.92}$	$10^{35.76}$	$10^{32.62}$	$10^{13.53}$	(18, 19)
HBED^b	$10^{39.68}$	$10^{39.57}$	$10^{24.78}$	$10^{18.37\ d}$	(20, 152, 153)
EDTA^h	$10^{25.0}$	$10^{20.3\ f}$	$10^{16.5}$	$10^{16.44\ d}$	(15, 150)

^{a}Unless otherwise indicated, all constants were measured at 25°C, ionic strength 0.1 M.
^{b}K = [Fe(III)–HL$^+$]/[Fe(III)][HL^{2-}].
^{c}Ionic strength = 0.2 M.
^{d}In the case of DFO, the aminopolycarboxylates, and HBED with Zn(II), the equilibrium is reported as K = [Zn(II)–L]/ [Zn(II)][L].
^{e}K = [Fe(III)–L^{2-}]/[Fe(III)][L^{5-}].
^{f}At 20°C.
^{g}$\beta_3 = K_1K_2K_3$; for Zn(II), $\beta_2 = K_1K_2$.
^{h}K = [Fe(III)–L$^-$]/[Fe(III)][L^{4-}].

nephrotoxicity, which is unrelated to their intended target. On the other hand, antihypertensives can manifest not only toxicity implicit in their structures, but also adverse effects associated with their intended function. An overdose of an antihypertensive medication can cause a lethal drop in blood pressure. In the case of deferrating agents, there is the issue of the toxicity implicit by the very definition of iron chelation; that is, these molecules are designed to remove a metal essential for life. Thus, parameters that reflect iron stores must be monitored to ensure that too much iron is not removed from the patient (117). Although unchecked iron removal is rarely, if ever, a problem clinically, it can be of major significance in designing preclinical animal toxicity trials in the course of evaluating new iron chelators. Finally, there is the issue of metal selectivity. For example, there are several metals that have coordination chemistries similar to that of Fe(III), for example, Al(III), Ga(III), and Cr(III); all are bound tightly by hexadentate ligands. Even natural product siderophores, for example, DFO, can bind these metals; fortunately, the iron complexes are thermodynamically more stable than the corresponding Ga(III) and Al(III) chelates (Table 10.3) (148). However, DFO can be used to clear Al(III) from dialysis patients (149). In the case of hexacoordinate siderophores, the selectivity is greater than that for ligands of lower denticity, such as L1 (Table 10.3) (150–153). Nevertheless, the metal selectivity issue is one that must be deliberated very carefully in the design and testing of new chelators.

Comparisons of Chelators. As was apparent in the discussion of metal complex formation in Section 1.2.1, it is important to have a scale that compares the binding of different ligands at the same molarity, pH, and iron concentration (i.e., pM values). Likewise, it is critical to apply a consistent scale to evaluate the chelators on a physiological level; this scaling is referred to as efficiency. The efficiency of an iron chelator (Ec_{Fe}) is defined as the chelator-induced (net) iron clearance ($c_{Fe}CL$), divided by the theoretical iron clearance (TCL, i.e., total iron-binding capacity of chelator administered). The chelator-induced (net) iron clearance ($c_{Fe}CL$) is the total urinary and fecal iron excretion measured in the presence of the ligand ($\Sigma\ c_{Fe}L$) less the total iron clearance in the absence of the ligand ($\Sigma\ c_{Fe}$). Thus, the efficiency (Ec_{Fe}) is expressed as a percentage, as shown in Equation 10.15:

$$Ec_{Fe} = \frac{(\Sigma\ c_{Fe}L - \Sigma\ c_{Fe})}{TCL} \times 100 \qquad (10.15)$$

The TCL depends on how one defines the stoichiometry and the stability of the metal complexes. An example of this is seen with DFO (**9**). This hexacoordinate ligand forms a very

Table 10.4 Iron Chelating Agents on the Market

Agent (Proprietary Name)	Structure	Available in the United States?
Desferal		Yes
Ferriprox		No[b]
DTPA		No[c]

tight 1:1 complex with Fe(III) at pH 7.4; the formation constant (K_f) is approximately 10^{31} M^{-1} (21, 22, 48). Thus, at concentrations of 1 μM ligand and 1 μM iron, virtually all of the iron is in the form of the DFO chelate, or ferrioxamine. If a test animal were administered a 10 μmol/kg body weight dose of DFO, in theory, a total of 10 μmol/kg of the iron chelate should be excreted, if (a) the iron were available for chelation and (b) the efficiency of the chelator were 100%. Unfortunately, as described in detail below, the actual efficiency is only a small fraction of this.

2.1 Iron Chelators on the Market

The treatment of choice for iron chelation is desferrioxamine B (trade name Desferal, deferoxamine, DFO, **9**, *N*-[5-[3-[(5-aminopentyl)-hydroxycarbamoyl]propionamido]pentyl]-3-[[5-(*N*-hydroxyacetamido)pentyl]-carbamoyl]-propionohydroxamic acid methanesulfonate)

(Fig. 10.3), a siderophore produced by *S. pilosus* (35). Two synthetic chelators are available outside the United States as alternatives for those who cannot tolerate DFO, the aminopolycarboxylic acid diethylenetriamine pentaacetic acid (DTPA; **35**) (Fig. 10.15) and the pyridinone 1,2-dimethyl-3-hydroxypyridin-4-one (trade name Ferriprox, also known as deferiprone and L1; formerly referred to as CP20; **33**) (Fig. 10.15) (Table 10.4).

2.1.1 Desferrioxamine (DFO, 9). In addition to DFO, *S. pilosus* assembles eight more hydroxamate ligands, two of which are cyclic (36, 37, 154); two of these (**10** and **11**) are shown in Fig. 10.3. On the surface, this fact would suggest that the microorganism wastes considerable energy by assembling the "extra" ligands. Although a definitive answer has not yet been reached, it seems unlikely that such a waste is indeed the case. As described in the

Table 10.4 (*Continued*)

USP or Nonproprietary Name	Who Markets the Nongeneric Substance?	Efficacy/ Potency	Route of Administration[a]	Dose
Desferrioxamine B, methanesulfonate salt	Novartis	~ 5% efficiency	s.c. i.v.	s.c.: ≤50 mg/kg/day i.v.: ≤15 mg/kg/h for the first 1000 mg and 125 mg/h thereafter
1,2-Dimethyl-3-hydroxy-pyridin-4-one; deferiprone; L1	Apotex	~ 2.5% efficiency	p.o.	75–100 mg/kg/day, divided
DTPA	BioTest (Astrapin Pharma)	~ 5% efficiency	s.c.	3 g/day of the Ca complex as its trisodium salt

[a]s.c., subcutaneously; i.v., intravenously; p.o., per os (oral).
[b]Licensed in Europe, Australia, and India; still considered experimental in the United States.
[c]Not licensed in the United States for the indication of iron overload, although it is approved as a chelating agent for use in cases of poisoning with plutonium or other transuranic metals.

example above, DFO is a hexacoordinate chelator, which contains three bidentate hydroxamate fragments (Fig. 10.3). As mentioned before, the Fe(III) complex is very stable (14, 21, 22, 48). For comparison, Table 10.1 lists the formation constants for a variety of Fe(III) complexes. For pharmaceutical use, the molecule is prepared by fermentation of the microorganism, isolation of the compound as its hydrochloride salt, and processing through ion exchange to the methanesulfonate salt (Desferal, MW 657) to improve its solubility.

Initially, DFO was used to treat patients suffering from primary and secondary iron overload (155, 156) as well as pediatric patients suffering from acute iron poisoning (147). Since these early trials, desferrioxamine B mesylate, whether given subcutane-ously or intravenously, has served as the drug of choice for iron overload disorders (117, 157, 158). Its oral absorption is very poor; thus, the drug is not given by this route (159). A few general comments are of interest regarding dosing, side effects, and drug interactions. According to the Novartis package insert, the LD_{50} is 287 mg/kg when given intravenously to mice; this figure is 329 mg/kg in rats. Desferal is made up in sterile water to a concentration of 10% w/v; this solution is isotonic. The preferred method of administration for the treatment of chronic iron overload is at a dose of 20–40 mg/kg/day by slow subcutaneous infusion over a period of 8–24 h; the recommended total daily dose is between 1000 and 2000 mg. Intravenous administration is suggested only in cases of cardiovascular collapse secondary to acute iron intoxication; under these conditions, doses are not to exceed 15 mg/kg/h for the first 1000 mg. Subsequent i.v. administration, if needed, must be at a slower rate not to exceed 125 mg/h.

2.1.1.1 Side Effects of Desferrioxamine. The cautionary restrictions mentioned above are likely related to the profound impact that i.v.-administered DFO can have on blood pressure (i.e., hypotension) and cardiac function (147). In addition, intravenous infusions of DFO at doses higher than the recommended 15 mg/kg/h for greater than 24 h for therapy of acute iron poisoning have resulted in several instances of acute respiratory distress syndrome (160, 161). The syndrome also has been reported in a child who had been treated with intravenous DFO according to the current guidelines (162). Other side effects observed with prolonged use of DFO include both ocular and auditory problems (163–165), pulmonary toxicity (160), bone changes (166), and growth retardation (167). These effects have been reported as being reversible on cessation of chelation therapy with DFO (168) and have diminished in frequency as management strategies have improved (117). A particularly interesting cautionary note is related to the rare, but nevertheless documented, increased susceptibility of patients using DFO to infection by *Yersinia enterocolitica* and *Y. pseudotuberculosis* (93, 169). The suggested reason for this augmented susceptibility is that ferrioxamine formed within treated patients provides prepackaged Fe(III) for these microorganisms (93).

At the level of drug interaction, two particular phenomena, the distortion of imaging measurements and ascorbate-induced cardiac effects, are interesting because of their mechanisms. First, imaging with ^{67}Ga in patients using Desferal has been shown to be confounded (170). The reason, as pointed out earlier, is that Ga(III) is ligated by DFO and is excreted quickly through the kidneys. Second, some patients who received concomitant treatment with Desferal and high (>500 mg daily in an adult) doses of vitamin C experienced cardiac impairment (117, 171, 172); several deaths were reported in the mid-1970s in patients undergoing such a regimen (173). The ascorbate problem is related to the discussion of iron chemistry in Section 1.3.4. Ascorbate reduces Fe(III) to Fe(II). Recall (Section 1.3.4) that Fe(II) is the major contributor to Fenton chemistry, which reduces hydrogen peroxide to hydroxide ion and the hydroxyl radical. Furthermore, DFO ligates Fe(III), not Fe(II). However, the Fe(II)/Fe(III) couple is driven as Fe(III) is sequestered by DFO. This increases the reducing capacity of Fe(II), but only on a one-time basis. Thus, patients receiving DFO are cautioned to use ascorbate only under specific conditions (117, 172). Unless ample DFO is administered such that the excess iron that is released reductively from transferrin and ferritin can be sequestered, ascorbate should not be given to iron-overloaded patients. It would seem, then, that treatment of iron-overloaded patients with ascorbate is, in and of itself, a potential problem. This is in keeping with the observations in iron-overloaded, scorbutic Bantus (174, 175).

Finally, one of the most common complaints among patients who receive DFO subcutaneously is the discomfort at the site(s) of infusion. Although some anaphylactoid symptoms have been reported, this is very uncommon (176, 177); however, local erythema is typical (178, 179). In an effort to identify the nature of the allergic response, investigators examined the effect of DFO on human basophils and rodent mast cells *in vitro* and in human skin (177). Incubation of human basophils with DFO did not induce histamine release from these cells. Furthermore, preincubation of basophils with DFO had no impact on histamine release from cells that were previously stimulated with either anti-human IgE or with f-met peptide. However, when seven healthy patients naïve to the drug were given subcutaneous injections of DFO, all of them developed a significant wheal-and-flare response, which was dose dependent and reached its maximum after 15 min. Treatment with the H_1 antagonist ceterizine elicited a substantial reduction in the wheal-and-flare response, indicating that the wheal-and-flare response was associated with local mast cells.

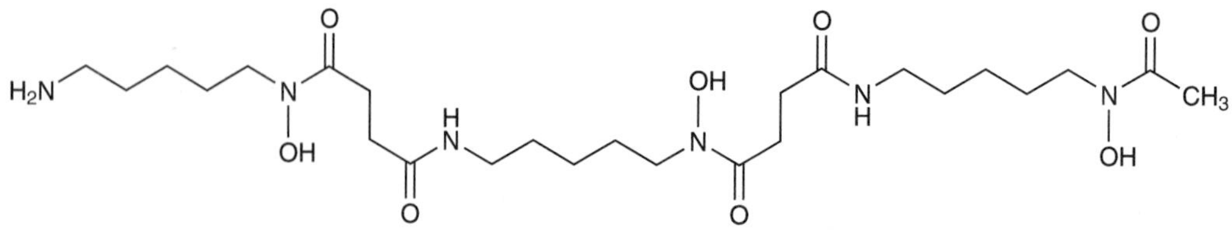

Figure 10.16. DFO (**9**) and its major, deaminated metabolite, Metabolite B (**36**).

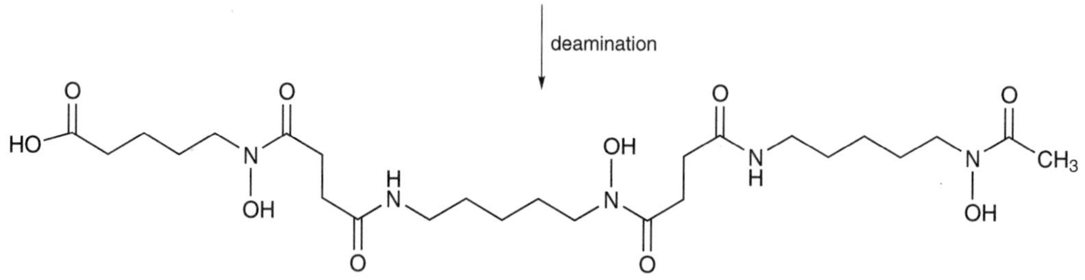

Finally, DFO also activated rat peritoneal mast cells, a type of connective tissue cell; the histamine release was dependent on the dose of DFO. These findings suggest that the DFO-mediated allergic response is an IgE-independent, nonimmunologic stimulatory effect on cutaneous mast cells. The investigators opined that DFO could be used as a positive control in cutaneous allergy testing. This should underscore the profound discomfort that patients have with the drug. The anaphylactoid problem certainly is related to other issues. The possibility does exist that, because this drug is a fermentation product, small, undetectable amounts of protein contaminant(s) are not removed during the purification process; these trace impurities may be responsible for the observed sensitivity.

2.1.1.2 Pharmacology of Desferrioxamine. Originally, DFO was administered intramuscularly (155, 156). A substantial increase in urinary iron output, which appeared to be related directly to the extent of the patient's iron overload, was observed. However, daily intramuscular injections of DFO were unable to remove sufficient iron from thalassemia patients; the use of intramuscular injections could not keep up with the iron loading from the transfusion therapy. Nevertheless, pro-

tracted intravenous or subcutaneous infusions of DFO were quite effective in promoting iron excretion (147, 157, 159); thus, it is possible to maintain patients in negative iron balance when DFO is administered by either of these two routes.

The requirement for slow infusion to realize DFO's maximum efficacy is probably related to the ligand's short half-life (5–10 min) in human plasma (180). Reports have varied regarding the proportion of the iron excreted through the stool versus the urine; the range is from 50–70% in the stool (181, 182). The urinary iron is likely derived from plasma NTBI and the RE system, whereas the stool iron arises from hepatocytes (159, 183). Upon subcutaneous administration of the drug, the plasma concentrations reach a steady state at 4 h; a rapid decrease in concentration occurs on cessation of infusion (184). Besides the metal chelate, the major metabolite of DFO (**36**, Fig. 10.16), which has been found in both the plasma (185) and the urine (185, 186) of patients treated with DFO, results from oxidative deamination at the amino terminus of the molecule, parallel to the metabolism of lysine (185). The pharmacokinetic properties of the metal chelate, ferrioxamine, are somewhat different from those of the parent drug (180,

184, 187). As expected, the concentration of ferrioxamine rises slowly during DFO administration, but declines more slowly than the concentration of the parent drug when the infusion is stopped. It appears as though any chelate formed extracellularly (i.e., by chelation of plasma NTBI) is restricted to the extracellular space (184). Desferrioxamine-induced fecal iron excretion derives from hepatocyte iron: plasma DFO is taken up by the liver, the iron is cleared from the hepatocytes, and the ferrioxamine is excreted in the bile (159, 184).

As with most drugs, with increasing dose, the DFO plasma level increases and subsequently plateaus (180, 184, 187). The response also can be quantitated by following the ligands' efficiency upon increasing doses. Augmenting the chelator dose results in an increased iron clearance up to a plateau; beyond this plateau, the efficiency of the ligand decreases (157).

As alluded to above, ascorbic acid can have a profound impact on DFO-promoted urinary iron excretion (173, 188, 189). When patients receiving chelation therapy with DFO were given oral doses of vitamin C (500 mg three times daily), urinary iron excretion increased between 25 and 200%, although alterations in fecal iron excretion were not as consistent (189). However, because of concerns about cardiotoxicity as mentioned earlier, this combination therapy may be problematic. Only under conditions of tissue ascorbate deficiency is repletion with ascorbate (100 mg daily) recommended (117).

2.1.2 Diethylenetriamine Pentaacetic Acid (DTPA, 35).

The use of the aminopolycarboxylate diethylenetriamine pentaacetic acid (DTPA, 35) (Fig. 10.15) as a therapeutic iron chelator was initiated at about the same time as the assessments of DFO (190, 191). As shown in Table 10.1, (35) forms a 1:1 complex with Fe(III); the formation constant and pM value are $10^{28}\,M^{-1}$ and 23.8, respectively (15). However, the salient difference between these two agents is their metal selectivity. DFO, a siderophore, is far more selective for iron than is DTPA.

2.1.2.1 Side Effects of DTPA.

In fact, the toxic effects associated with DTPA have been connected with its ability to chelate zinc. Patients given the ligand suffer pain at the site of injection, nausea, diarrhea, and sore throat and mouth (191, 192). In one trial (192), the patients' plasma zinc levels decreased from the normal range of 11.5–17.5 to 5 μmol/L after 5 days of subcutaneous infusions with DTPA as its calcium complex. At the same time, this treatment exerted little effect on other divalent cations, such as Cu, Ca, and Mg, in the plasma. Generally, the side effects were ameliorated within 10 days after the treatment was discontinued. In the course of this same investigation, the attempt was made to overcome the side effects associated with zinc depletion by administering the DTPA as its zinc complex. Unfortunately, the Fe(III)/Zn(II) exchange was ineffective in these patients; there was little, if any, iron clearance. However, when patients were given oral zinc sulfate supplements ("effervescent zinc") several times daily during the courses of subcutaneous DTPA administration, the side effects were minimized (192, 193).

2.1.2.2 Pharmacology of DTPA.

Investigators have found that the iron balance realized upon administration with DFO or DTPA is comparable (190); the efficiencies are similar when 24-h subcutaneous infusions of 4 g of DFO and 3 g of Ca-DTPA are compared directly (192). However, the proportions of the metal excreted in the stool and urine are different. Whereas DFO induces excretion of iron by way of both urine and stool (Section 2.1.1.2), DTPA promotes urinary excretion only (192). When both drugs were given simultaneously, the results were additive, not synergistic. Furthermore, unlike the situation with DFO, ascorbate supplementation during chelation therapy with DTPA did not enhance iron excretion (192). Thus, DTPA has served as one alternative to DFO, but only when patients cannot tolerate the latter (194).

Although the overwhelming proportion of an orally administered dose of ^{14}C-DTPA passes through the gastrointestinal tract unabsorbed, intravenously administered radiolabeled ligand is excreted quantitatively in the urine (195, 196). This is consistent with the pattern of iron excretion described above. The predominant thinking is that DTPA is excreted by glomerular filtration; the $t_{1/2}$ for exchange between the plasma and extracellular fluid, the return to the plasma, and the clearance from the plasma into the urine are all brief, 2.5, 6.3, and 19 min, respectively (196). DTPA is metabolized only minimally by the body (197); of this small amount, release of acetate groups has been observed as expired CO_2 in rats (198).

2.1.3 1,2-Dimethyl-3-hydroxypyridin-4-one (Deferiprone, L1; 33).

Another alternative, which has been licensed for sale in India since 1994, in Europe since 1999, and in Australia since 2001, is the orally active, bidentate hydroxypyridinone 1,2-dimethyl-3-hydroxypyridin-4-one (trade name Ferriprox, deferiprone, L1, 33) (Fig. 10.15). This ligand forms 1:1, 2:1, and 3:1 complexes with Fe(III) (Table 10.1), depending on the availability of the chelator (15, 18, 19). The total formation constant (β_3 = $K_1K_2K_3$) is $10^{35.92}\ M^{-1}$; because of the stoichiometry, the pM value is 18.3, much less than that of DFO (Table 10.1). Species distribution plots for the L1–Fe(III) complexes [i.e., fraction of (L1)$_1$–Fe(III), (L1)$_2$–Fe(III), and (L1)$_3$–Fe(III)] show that at physiological pH, an L1 concentration of $3 \times 10^{-3}\ M$, and an Fe(III) concentration of $1 \times 10^{-3}\ M$, the (L1)$_3$–Fe(III) complex is by far the major species. However, at $1 \times 10^{-6}\ M$ Fe(III) and $3 \times 10^{-6}\ M$ L1, complete (L1)$_3$–Fe(III) complex formation does not occur until pH 10 (15).

2.1.3.1 Side Effects of L1.

The oral activity of L1 has made it very attractive for use in clinical settings. However, two issues that detract from its widespread use in the United States and elsewhere are concerns about its toxicity profile and continuing questions regarding the efficacy of the chelator. The overwhelming majority of the controversy over L1 has centered on the former topic, rather than the latter.

There is agreement among investigators regarding some of the side effects of deferiprone, that is, nausea, vomiting, agranulocytosis, increased alanine transaminase levels, arthropathy, diminished plasma zinc levels, and some immunologic changes (199–207). The debate surrounding L1 is its possible role in the induction of liver fibrosis; the disagreement seems to derive from (a) small patient populations in the trials and, more important, (b) the confounds introduced in these

studies by the prevalence of hepatitis C infections, attributed to the frequency of blood transfusions in the thalassemic patient population. Unfortunately for both patients and investigators, hepatitis C itself can elicit liver fibrosis (208). Further clouding the issue is the possibility that incompletely complexed iron [i.e., 1:1 and 2:1 L1–Fe(III) complexes] can participate in Fenton chemistry (Section 1.3.4); L1 promotes Fenton chemistry in several *in vitro* systems (116). It is possible that this augmentation of free-radical formation could contribute to the observed hepatic injury; such damage has been noted in hepatic cell culture (209). However, these issues require further investigation. What is more important is that investigators not condemn the entire hydroxypyridinone class if L1 is found conclusively to be problematic at the level of toxicity.

2.1.3.2 Pharmacology of L1.

Surprisingly, the toxicity issues surrounding L1 have obscured the facts regarding the drug's efficacy as a deferration agent. Early findings suggested that L1 was the long-anticipated orally effective iron chelator. Most of the iron excretion induced by L1 is urinary, although some increase in fecal iron has been noted (205, 210). Pharmacokinetic analyses indicated that, like DFO, the kinetics varied with the iron status of the individual (211–213). Generally, the pharmacokinetic parameters are the same, regardless of whether L1 is administered during feeding or after fasting (212). After rapid absorption from the stomach, the terminal elimination half-life is 1.5–2 h, although there are wide variations, especially among thalassemics (211, 213, 214). Most of the dose is excreted uncomplexed as the glucuronide conjugate (37, Fig. 10.17); smaller proportions are eliminated as the free ligand, the iron complex, and the 3-O-methylated conjugate (38, Fig. 10.17) (211, 215, 216).

Although the widespread use of L1 for the treatment of transfusional iron overload has some support (217), over a decade after the initial human studies, many investigators believe that L1 is not an efficient enough chelator to maintain thalassemic patients in negative iron balance when administered alone (117, 204, 206, 218). Some reports indicate that, although it is effective initially in pro-

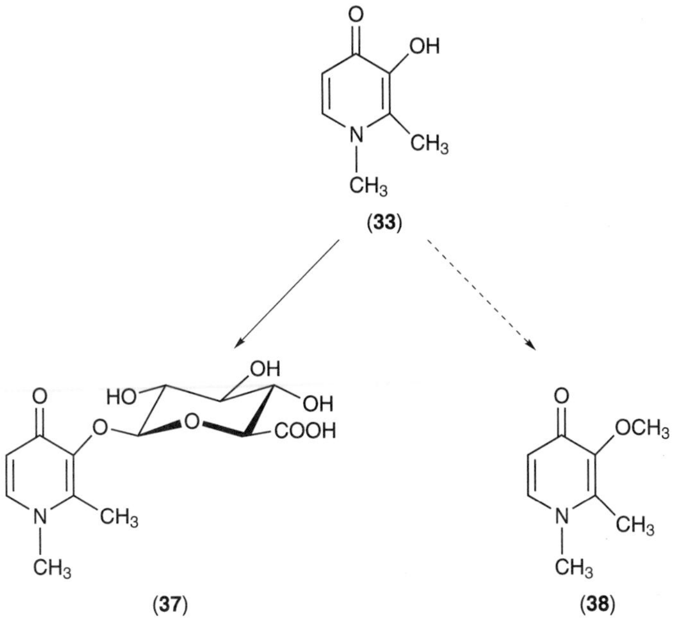

Figure 10.17. L1 (**33**), its glucuronide conjugate (**37**), and its 3-*O*-methylated conjugate (**38**). Most of an L1 dose is excreted as (**37**); (**38**) accounts for 1% or less of an administered dose.

moting iron clearance in those patients with significant iron overload, it is not useful in individuals who have lower, but still dangerous, hepatic iron deposits (207, 218, 219). Most of the trials were conducted using daily doses of ≤75 mg/kg (205–207, 218). Furthermore, on the basis of some of the human clinical trials (206, 218) and of preliminary reports in animals (220), the suggestion has been made that L1 can actually promote iron absorption, presumably of low molecular weight complexes (208).

More recently, investigators have assessed the efficacy of L1 and DFO in a combined regimen (221–223). The rationale for using such a combination is based on the fact that L1 can remove iron from transferrin and can cross cell membranes, thus acting as a "shuttle," whereas DFO can do neither particularly well, but can function as a "sink" in the bloodstream, promoting the ultimate excretion of the metal (224). Even though DFO can bind iron more tightly than does transferrin, the kinetics of transferrin deferration by DFO are too slow to be of therapeutic value, yet L1 can remove iron from transferrin (101, 217). In addition, recall that the (L1)–Fe(III) and (L1)$_2$–Fe(III) complexes have smaller formation constants than does the Fe(III) complex

of DFO (15) and are sufficiently labile for transfer of the metal to DFO. Upon abstraction from L1 by DFO, the metal can no longer participate in Fenton chemistry. Different modes of combining DFO and L1 have been attempted, including simultaneous [i.e., L1 is given orally during the DFO infusion (221, 223)] and sequential [i.e., L1 is given during the day and DFO at night (223) or L1 is given orally for 4 weekdays and DFO is infused over 2 weekend days (222)]. The sequential therapy resulted in an additive effect; no adverse events occurred in either of these two groups of patients (222, 223). However, the simultaneous administration of the two ligands produced a synergistic impact on iron excretion without any apparent toxicity (221, 223). This synergism was particularly noteworthy under the regimen undertaken by three patients in which L1 was given at a daily dose of 100 mg/kg in three divided doses that coincided with $t = 0$, 4, and 8 h of the DFO infusion at a subcutaneous dose of 40 mg/kg (223). Because of the small numbers of patients who have received any one regimen, seven at the most (222), it is unclear whether such combination therapy will prove to be safe and effective when extended to larger numbers of patients and carried out for longer time periods. Ultimately, the issues surrounding the toxicity of L1 may prove to be moot; it is unlikely that L1 will serve adequately as a single orally effective deferration agent.

3 HISTORY OF CHELATION THERAPY; DISCOVERY OF AGENTS WITH IRON-CHELATING ACTIVITY

The use of phlebotomy had been regarded as the treatment of choice for primary hemochromatosis since the mid-1940s (225); however, clinicians were well aware that such treatment was not necessarily practical in all situations. Thus, the alternative of chelation therapy was examined beginning in the early 1950s. Although any number of compounds bind iron *in vitro*, the specific removal of iron from an animal or human is another matter, as was found in the early trials of chelating agents. The first studies with iron-chelating agents in animals and humans used ethylenediaminetetraacetate (EDTA) (**32**) (Fig. 10.15), an agent that was known to chelate iron as well as other metals; the results were disappointing (226, 227). A structurally similar compound, DTPA (**35**), appeared to be a better choice after it was found that this ligand did not surrender iron to transferrin *in vitro* (228). The compound was assessed in both animals and humans and found to be more effective in removing iron *in vivo* than was EDTA (155, 191). However, as discussed in Section 2.1.2, the use of DTPA is not widespread because of its side effects.

Meanwhile, an unrelated search for substances with antibiotic activity was under way. The goal was to test such materials secreted by strains of *Streptomyces* as potential antibiotics (229). In the course of attempts to optimize the yield of what would be known as ferrimycin from the fermentation process, it was found that the effect of this compound was diminished by culture filtrates from other *Streptomyces* species. Further isolation of this antagonist was carried out in an effort to both characterize the antagonist itself and gain insight into the mechanism of action of ferrimycin. Although it was found that bacteria quickly became resistant to the ferrimycins, ending the development of this class of com-

pounds as antibiotics (230), investigations of the antagonists continued. In the early 1960s, several reports were published concerning the isolation and structural characterization of these iron-containing antagonists, the ferrioxamines, which contained two or three hydroxamate moieties, separated by varying spacers (35–37). One of these, ferrioxamine B (35), was assessed as an iron donor in a patient with iron-deficiency anemia; the intent was to measure the pharmacokinetics of, and the fate of the iron derived from, ferrioxamine. The ferrioxamine was excreted quickly and quantitatively in the urine without any loss of the iron to the tissues. This was surprising at the time because the entire iron content of other parenterally administered iron preparations was retained by the body (230).

The investigators considered the following: for the iron in ferrioxamine to be excreted quantitatively, it must be bound very tightly under physiological conditions. If this were the case, a deferrated ferrioxamine should remove iron from stores in the body, rendering the now chelated iron excretable in the urine (230). Accordingly, an iron-free ferrioxamine, desferri-ferrioxamine B, (desferrioxamine B, or DFO for short) as the hydrochloride salt was prepared and evaluated. Experiments in rabbits and dogs revealed that DFO did remove iron from animals; the compound was most effective in those animals that had been previously loaded with iron (156, 231). The metal specificity of DFO was assessed in rabbits and compared with that of the known chelator, EDTA (231, 232). The specificity of the siderophore for iron *in vivo* was confirmed by measurements of the stability constants of the two ligands *in vitro* (48). Not only was the stability constant of the iron-DFO complex greater than those of the complexes with other divalent and trivalent ions, much more specific than EDTA, but the stability constant of the DFO-iron complex was much stronger than that of the EDTA-iron complex. The interaction of DFO with ferritin, transferrin, and hemoglobin was measured; the investigators were satisfied that DFO chelates iron from none of these iron transport or storage proteins (156, 186). After the initial trial in a patient with severe hemochromatosis (156), the requirement of high doses dictated a form of DFO that was more soluble in water than the hydrochloride; the methanesulfonate form of the compound was derived and is known today by the trade name Desferal, still the treatment of choice for transfusional iron overload and acute iron poisoning.

Desferal was discovered and subsequently developed in a corporate laboratory. However, because of the relatively limited market for iron-chelating agents (and their resulting designation as "orphan" drugs), many of the potential alternatives, including deferiprone, were discovered in academic laboratories most frequently sponsored by the U.S. National Institutes of Health, rather in than corporate settings.

4 RECENT DEVELOPMENTS

It should be clear that the ideal chelator for the treatment of iron overload disease is not yet available. However, investigators have sought chelators with improved efficiency (i.e., greater Ec_{Fe} values), regardless of the mode of administration. This section covers compounds that, in recent years, have shown great promise in animal studies.

4.1   N,N'-Bis(2-hydroxybenzyl)-ethylenediamine-N,N'-diacetic Acid (HBED, 39a)

The polyanionic amine N,N'-bis(2-hydroxy-benzyl)ethylenediamine-N,N'-diacetic acid (HBED, **39a**, Fig. 10.18) is a synthetic hexadentate ligand (20). Like DFO, HBED forms a 1:1 complex with iron with high affinity and selectivity; the formation constant and pM value for the complex are $10^{39.68}$ M^{-1} and 26.74, respectively (Table 10.1) (15). Because HBED is a synthetic product, problems of local reactions to potential fermentation by-products not removed during purification would be absent. Also, for patients allergic to DFO, HBED, a member of a different family of chelators, would be unlikely to provoke a similar response. HBED has been investigated thoroughly in rodents (233, 234). The drug looked very promising in this model when administered orally and, not unexpectedly, these findings generated a great deal of excitement. The availability of an orally active iron chelator

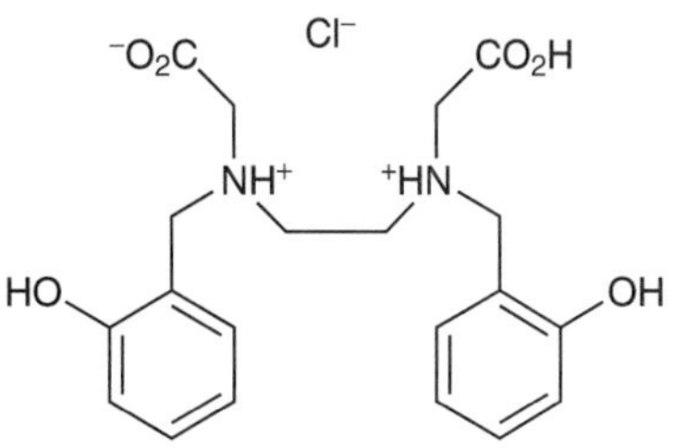

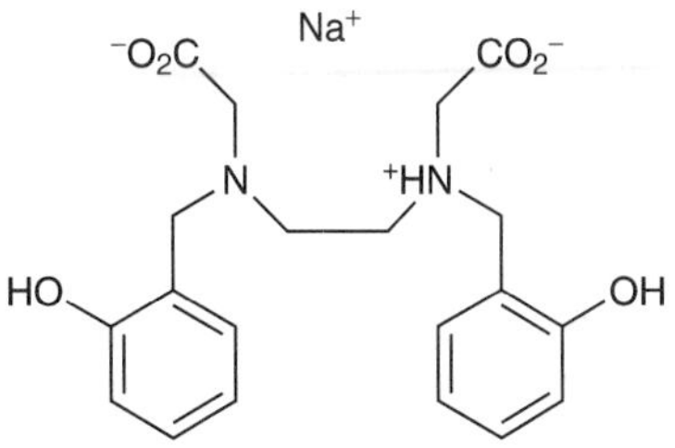

Figure 10.18. Structure of N,N'-bis(2-hydroxy-benzyl)ethylenediamine-N,N'-diacetic acid (HBED) as the monohydrochloride dihydrate (**39a**) and as the monosodium salt (NaHBED, **39b**).

seemed imminent. Unfortunately, the rodent findings were not replicated in higher animals. In the iron-overloaded *Cebus apella* model, an excellent predictor of how a chelator will perform when administered to humans (Section 5.2.3), the iron-clearing efficiency of DFO given subcutaneously at a dose of 150 μmol/kg was 5.5 $\pm$ 0.9% (235); however, the efficiency of orally administered HBED, also dosed at 150 μmol/kg, was significantly lower, 0.5 $\pm$ 0.5% (236). Not surprisingly, HBED was also a disappointment when given orally to patients (182, 237); the limited iron excretion that resulted is insufficient for use of this agent in the treatment of transfusional iron overload. However, when given parenterally to rodents or primates, HBED performed very well (238–241).

4.1.1 Parenteral Administration to Rodents. Early evaluations of HBED were carried out in the hypertransfused rat using intravenous administration; these demonstrated a two- to threefold greater excretion using HBED ver-

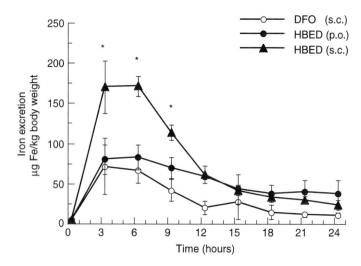

Figure 10.19. Time course of mean biliary iron excretion in normal rats after administration of DFO by subcutaneous injection and after administration of HBED by gavage or by subcutaneous injection. Both chelators were given at a dose of 150 μmol/kg body weight. The peak amounts of iron excreted with subcutaneous HBED (asterisks) were more than twofold greater than the peak iron excretion after either subcutaneous DFO or HBED given by gavage ($P < 0.05$ at $t = 3$ h and $P < 0.01$ at $t = 6$ h). [From R. J. Bergeron, J. Wiegand, and G. M. Brittenham, *Blood*, **91**, 1446–1452 (1998). Copyright American Society of Hematology, used by permission.]

sus the reference chelator, DFO (238, 239). Subcutaneous administration of HBED was examined in the bile duct-cannulated rodent model, which has normal iron stores (240). Groups of rats were given either DFO by subcutaneous injection, or HBED as the monohydrochloride dihydrate (**39a**, Fig. 10.18) orally by gavage or by subcutaneous injection at a dose of 150 μmol/kg of body weight. Fig. 10.19 shows the time course of the mean biliary iron excretion induced by the chelators in each group of rats, expressed as μg Fe/kg body weight. The peak amounts of iron excreted with subcutaneous HBED were more than twofold greater than the peak iron excretion after either HBED given by gavage or subcutaneous DFO ($P < 0.05$ at 3 h and $P < 0.01$ at 6 h). The iron excretion induced by the chelators in each group of rats is shown in Fig. 10.20, expressed as the net mean amount of iron excreted in the urine and in the bile (μg Fe/kg body weight) and as the efficiency of iron chelation. Subcutaneous DFO induced the excretion of 209 $\pm$ 59 μg Fe/kg body weight and was found to have an efficiency of 2.5 $\pm$ 0.7%. Compared to subcutaneous DFO, HBED given orally resulted in a twofold greater iron excre-

tion of 436 $\pm$ 176 μg Fe/kg body weight ($P < 0.02$); the efficiency was 5.2 $\pm$ 2.1%. HBED given to the rodents by subcutaneous injection was more than three times as effective in inducing iron excretion as DFO administered subcutaneously ($P < 0.001$), inducing the excretion of 679 $\pm$ 8 μg Fe/kg body weight at an overall efficiency of 8.1 $\pm$ 0.1%.

4.1.2 Parenteral Administration to Primates. These studies were carried out in iron-overloaded *C. apella* primates (242). In brief, the reference chelator DFO was administered subcutaneously at three different doses or as a 20-min intravenous infusion at two different doses (Table 10.5). Because of the poor solubility of the HBED monohydrochloride dihydrate that was used in the rodent studies and the fact that the necessary manipulations to prepare the material for injection could be problematic in a clinical setting, the HBED monosodium salt (NaHBED, **39b**, Fig. 10.18) was prepared and evaluated. The salt is much more soluble than the monohydrochloride dihydrate (**39a**). The solubility is in excess of 30% (w/v) in water; when the drug is dissolved in saline the unadjusted pH of the resulting

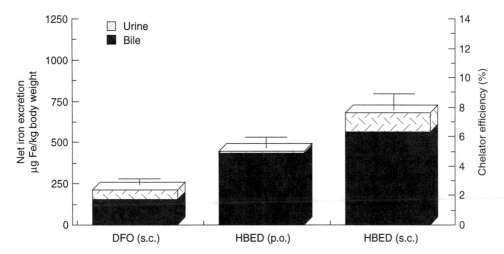

Figure 10.20. Mean net iron excretion in normal rats after administration of DFO by subcutaneous injection and after administration of HBED by gavage or by subcutaneous injection. Excretion is shown as μg Fe/kg body weight on the scale of the left vertical axis and as efficiency of chelation on the right vertical axis. Both chelators were given at a dose of 150 μmol/kg body weight. [From R. J. Bergeron, J. Wiegand, and G. M. Brittenham, *Blood*, **91**, 1446–1452 (1998). Copyright American Society of Hematology, used by permission.]

solution is 7.3 (241). Several methods of administration of NaHBED were investigated: subcutaneous bolus at doses of 75 and 150 μmol/kg, as well as a three-dose regimen of 75 μmol/kg every other day for three doses; intravenous bolus at doses of 50 and 75 μmol/kg, and 20-min intravenous infusion at doses of 150 and 225 μmol/kg (Table 10.5).

4.1.2.1 Subcutaneous Administration. Desferrioxamine administered subcutaneously in aqueous solution at a dose of 75 μmol/kg induced the excretion of 213 ± 112 μg/kg of iron; the efficiency was $5.0 \pm 2.6\%$ (243). A dose of 150 μmol/kg was found to have a similar efficiency and induced the excretion of approximately twice the amount of iron, 435 ± 115 μg Fe/kg body weight (240) (Table 10.5; $P > 0.4$ versus 75 μmol/kg dose). Increasing the dose to 300 μmol/kg induced the excretion of 716 ± 244 μg/kg of iron and had an efficiency of $4.2 \pm 1.4\%$ (241).

In contrast, when NaHBED was given to the primates at a single dose of 150 μmol/kg, it induced the excretion of 1139 ± 383 μg/kg of iron and had an efficiency of $13.6 \pm 4.5\%$ (Table 10.5) (241). The observed efficiency is well within error of that observed after the subcutaneous injection of 162 μmol/kg of the HBED

monohydrochloride dihydrate in buffer or in Cremophor, $9.9 \pm 2.1\%$ ($P > 0.2$) and $14.9 \pm 5.2\%$ ($P > 0.7$), respectively (240). In addition, NaHBED has been administered at a dose of 75 μmol/kg every other day for three doses (Fig. 10.21), for a total dose of 225 μmol/kg. The first dose of the drug induced the excretion of half as much iron as did the 150 μmol/kg dose, 597 ± 91 μg/kg; the 24-h efficiency was $14.2 \pm 2.2\%$ (Table 10.5) ($P > 0.3$ versus 150 μmol/kg single dose). The second and third injections of the drug stimulated the excretion of comparable amounts of iron; the 24-h efficiencies were very similar to each other. The total iron excreted as a result of the three injections was 1837 ± 301 μg/kg of iron, and the overall efficiency was $14.6 \pm 2.4\%$ (Table 10.5). These data are virtually identical to the iron excretion induced after a single subcutaneous injection of HBED in a buffer given at a dose of 81 μmol/kg, 608 ± 175 μg/kg and an efficiency of $13.0 \pm 4.6\%$ (241) ($P > 0.6, P > 0.5$, and $P > 0.5$ for doses 1–3, respectively).

To measure the degree of iron balance achieved by the chelators, the total amount of iron intake was compared with the total amount of iron excreted [i.e., net iron balance = dietary iron intake − (urinary + fecal iron

Table 10.5 Comparison of Efficiencies and Net Iron Balance of Intravenous and Subcutaneous HBED (39b) versus Those of Intravenous and Subcutaneous DFO (9) in *C. apella* Primates

Drug	Route[a]	Dose (μmol/kg)	Dose (mg/kg)	N	Efficiency (%)	Induced Fe (μg/kg)	Fe Balance (μg/kg)[b]	Reference
DFO	s.c. bolus	75	50	4	5.0 ± 2.6	213 ± 112	−60 ± 135	(243)
DFO	s.c. bolus	150	100	6	5.1 ± 1.3	435 ± 115	−278 ± 185	(240)
DFO	s.c. bolus	300	200	5	4.2 ± 1.4	716 ± 244	−711 ± 230	(241)
DFO	20 min i.v. inf	75	50	5	5.6 ± 1.6	237 ± 67	−22 ± 197	(243)
DFO	20 min i.v. inf	150	100	5	3.9 ± 0.8	332 ± 66	−168 ± 51	(243)
HBED	s.c. bolus	75 (day 1)[c]	32.4	4	14.2 ± 2.2	597 ± 91	−524 ± 84	(241)
HBED	s.c. bolus	150	64.9	5	13.6 ± 4.5	1139 ± 383	−899 ± 365	(241)
HBED	s.c. bolus	225 (75 × 3 doses)[d]	97.2	4	14.6 ± 2.4	1837 ± 301	−1578 ± 345	(241)
HBED	i.v. bolus	50	21.6	5	12.1 ± 2.5	338 ± 68	−195 ± 114	(243)
HBED	i.v. bolus	75	32.4	4	11.5 ± 1.3	482 ± 54	−407 ± 73	(243)
HBED	20 min i.v. inf	150	64.9	4	7.7 ± 0.8	644 ± 65	−408 ± 77	(243)
HBED	20 min i.v. inf	225	97.3	5	6.3 ± 1.0	798 ± 126	−570 ± 248	(243)

[a]s.c., subcutaneous; i.v., intravenous; inf, infusion.

[b]Net iron balance = dietary iron intake − (urinary iron + fecal iron). Animals in a negative iron balance are excreting more iron than they are absorbing. To maintain iron balance, 250–400 μg of iron/kg/day = 1750 to 2800 μg of Fe/kg/week must be cleared (145).

[c]Day +1 to day +2 versus day −3 to day 0 iron balance figures are shown.

[d]Cumulative after the three injections.

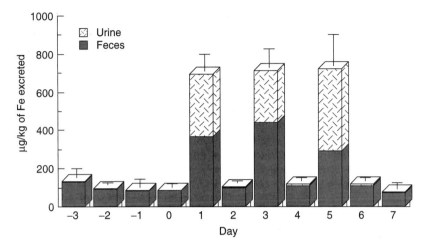

Figure 10.21. Urinary and fecal iron excretion (μg/kg) induced by the subcutaneous administration of HBED monosodium salt, 75 μmol/kg for three doses (225 μmol/kg total). Drug was administered on days 0, 2, and 4. Note the prompt return to baseline levels within 24 h of each dose; baseline iron levels in the urine and stool have not been subtracted. [From R. J. Bergeron, J. Wiegand, and G. M. Brittenham, *Blood*, **93**, 370–375 (1999). Copyright American Society of Hematology, used by permission.]

excretion)]; animals in a negative iron balance are excreting more iron than they are absorbing. Monkeys treated subcutaneously with DFO at doses of either 75, 150, or 300 μmol/kg were held in negative iron balance (Table 10.5); the excreted iron was 60 ± 135, 278 ± 185, and 711 ± 230 μg/kg, respectively, more than they absorbed (240, 241, 243). The subcutaneous administration of NaHBED was able to hold the monkeys in a negative iron balance (Table 10.5) (241). Monkeys treated with NaHBED at a single dose of 150 μmol/kg excreted 899 ± 365 μg/kg of iron more than they absorbed. Also, NaHBED given subcutaneously every other day for three doses (75 μmol/kg/dose) resulted in a negative iron balance (Table 10.5). The iron excreted during a 7-day period after the administration of the first dose amounted to 1578 ± 345 μg/kg more than they absorbed. As was observed with the urinary and fecal iron-clearance data, animals treated subcutaneously with NaHBED consistently have a negative iron balance that is 2–3 times greater than that observed with DFO. The results of these studies (240, 241, 243) clearly indicate that both subcutaneously administered DFO and NaHBED can hold the monkeys in a negative iron balance (Table 10.5).

4.1.2.2 Intravenous Administration. Although slow intravenous infusions of DFO (<10 mg/kg/h) have been well tolerated in patients with chronic iron overload (244–246), rapid intravenous administration for the treatment of acute iron poisoning can produce grave side effects, as described in detail in section 2.1.1.1 (147, 160–162). The manufacturer recommends that the total daily parenteral dose of DFO should not exceed 6 g. Because 1 g of DFO binds only about 90 mg of elemental iron, and a single tablet for the treatment of iron deficiency in adults may contain as much as 65 mg of elemental iron, 6 g of DFO given parenterally may not be adequate in cases of severe iron poisoning. Considering the findings with subcutaneously administered NaHBED in light of these limitations of DFO, it became of interest to find whether intravenous administration of this ligand would (*a*) be an effective means of deferration of iron-loaded animals and (*b*) elicit hypotension in normal animals. The latter subject is addressed in Section 4.1.3.

Intravenous administration of DFO to the iron-loaded primates at a dose of 75 μmol/kg induced the excretion of 237 ± 67 μg/kg of iron and had an efficiency of 5.6 ± 1.6%. Increasing the dose to 150 μmol/kg resulted in the excretion of 332 ± 66 μg/kg of iron and an efficiency of 3.9 ± 0.8% (Table 10.5) ($P < 0.04$). Interestingly, although the efficiency of DFO given either subcutaneously or intravenously at a dose of 75 μmol/kg was similar, 5.0 ± 2.6% versus 5.6 ± 1.6% ($P > 0.3$), this was not the case when the dose was increased to 150 μmol/kg. DFO given subcutaneously at 150 μmol/kg resulted in an iron-clearing efficiency of 5.1 ± 1.3% (Section 4.1.2.1), but the same dose given as an intravenous infusion resulted in an efficiency of 3.9 ± 0.8% ($P < 0.02$). The efficiencies of NaHBED administered to the iron-loaded primates as an intravenous bolus at doses of 50 and 75 μmol/kg were also similar to those of the drug administered subcutaneously (Section 4.1.2.1), 12.1 ± 2.5% and 11.5 ± 1.3% ($P > 0.3$); the corresponding iron excretions were 338 ± 68 and 482 ± 54 μg/kg of iron. The efficiency of NaHBED given subcutaneously at a dose of 75 μmol/kg was greater than the same dose given as an intravenous bolus, 14.2 ± 2.2% versus 11.5 ± 1.3%, respectively (Table 10.5) ($P < 0.05$). When given as a 20-min intravenous infusion, an increase in the dose of NaHBED from 150 to 225 μmol/kg resulted in the excretion of more iron, but a decline in efficiency (Table 10.5).

4.1.3 Preclinical Toxicity Trials. Acute toxicity assessments of HBED using parenteral administration in mice indicated an LD_{50} in excess of 800 mg/kg; no drug-related effects were observed in mice given the drug intraperitoneally at doses up to 200 mg/kg for 10 weeks (233). At subcutaneous doses of up to 300 μmol/kg every other day for 14 days (7 doses), no toxicity was noted in rats given NaHBED (6% w/v) (241). In addition, no erythema was noted at any of the injection sites, either grossly or histologically. At necropsy, neither macroscopic examination nor histological evaluation of tissues revealed abnormalities in tissues that were attributable to the drug (241).

Systemic toxicity trials of NaHBED have been carried out in dogs (243). Four beagles were iron loaded to a level of 300 mg Fe/kg and were subsequently given an intravenous dose of 75 μmol/kg NaHBED in 50 mL isotonic saline as a 20-min infusion once daily for 14 days. Two additional dogs, also iron loaded to a

level of 300 mg Fe/kg, served as saline-treated controls. Upon necropsy, the most significant finding was the accumulation of hemosiderin in the macrophages of the liver, spleen, and lymph nodes of both test and control animals. There was no systemic toxicity that could be attributed to the NaHBED under this intravenous regimen. Another systemic toxicity trial was carried out in dogs using subcutaneous administration of NaHBED. These non-iron-overloaded dogs were given NaHBED at graduated doses of up to 300 μmol/kg/day. The drug was injected as a subcutaneous bolus every other day into one of two sites on a rotating basis. Upon necropsy, histopathological analysis did not reveal any drug-related abnormalities beyond those in the skin. The descriptions of the reactions in the skin at the sites that were injected with NaHBED ranged from early, focally extensive fibroplasia and mild inflammation in the superficial subcutis to panniculitis, which was subacute and focally extensive, and moderate to severe inflammation in the deep subcutis. The descriptions of the skin from the sites injected with saline included early fibroplasia, which ranged from diffuse, moderate, and superficial to focally extensive in the deep subcutis; one site presented with panniculitis. The drug was administered to the dogs subcutaneously at a concentration of 25% (w/v) and injection volumes of up to 5.2 mL/10 kg for the 300 μmol/kg doses. In addition, there did appear to be somewhat of a dose response, with the animals in the higher volume groups having more local irritation than those in the lower volume groups. This finding of local irritation at the injection sites led to the use of a rodent model to determine the cause.

The results in dogs and preliminary experiments in rodents implied that the hypertonicity of the 25% (w/v) solution used might be responsible for the local irritation observed. Accordingly, groups of four rodents were given a 100-μL subcutaneous bolus of isotonic saline or NaHBED at varying concentrations in distilled H_2O. Animals were administered the same drug concentrations in the same volume as a 5-h subcutaneous infusion as well. A 300-g rat receiving 100 μL of the drug solution would be receiving a volume of drug solution roughly comparable to administration of 20

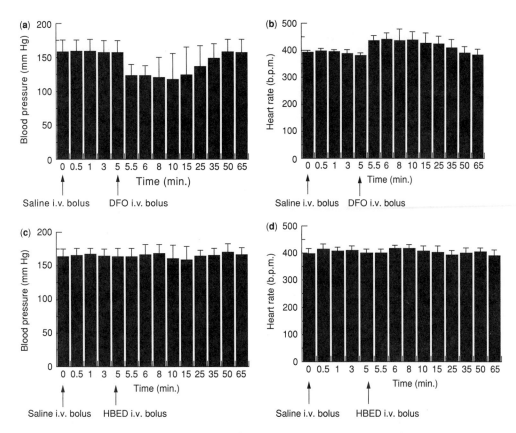

Figure 10.22. Effect of intravenous bolus administration of DFO (a and b) and NaHBED (c and d) (300 μmol/kg) on the blood pressure (mmHg, a and c) and heart rate (beats/min, b and d) of normotensive rats ($n = 5$ for each chelator). a: $P < 0.001$ for $t = 5.5$ to 15 min; $P < 0.005$ for $t = 25$ min. b: $P < 0.001$ for $t = 5.5$ to 35 min. [From R. J. Bergeron, J. Wiegand, and G. M. Brittenham, *Blood*, **99**, 3019–3026 (2002). Copyright American Society of Hematology, used by permission.]

mL of the drug solution to a 60-kg person. The histopathological descriptors for both bolus and infusion saline controls necropsied at 48 h and 7 days postdosing included endothelial hypertrophy, minimal inflammation, and scattered mast cells in the subcutis. With the exception of the rats treated with 20% NaHBED as a subcutaneous bolus, in which mild panniculitis was noted in one each of the rodents necropsied 48 h or 1 week postdosing, all of the test animals presented with essentially the same histopathology as that of the control animals. Therefore, it is possible to prevent NaHBED-related irritation by either giving the drug as a slow subcutaneous infusion or as a subcutaneous bolus at concentrations of ≤15% w/v (243).

Because of the risk of hypotension, large

doses of DFO cannot be administered intravenously for the treatment of acute iron poisoning; this can result in inadequate chelation and subsequent acute iron-induced cardiac damage, a significant problem. Recent experiments suggest that intravenous infusion of NaHBED for the treatment of acute iron poisoning will not be subject to these restrictions (243). When rodents were given 300 μmol/kg DFO in a 0.5-mL volume as an intravenous bolus, there was a 25% decrease in blood pressure that did not return to baseline levels until 35 min postdrug ($n = 5$, $P < 0.001$ for $t = 5.5$ to 15 min and $P < 0.005$ for $t = 25$ min, Fig. 10.22a). The heart rate in these animals also increased by 16% and likewise did not return to predrug levels until beyond 35 min postdosing (Fig. 10.22b). In contrast, there was no effect on either blood pressure or heart rate in rats given the same dose of NaHBED as an intravenous bolus (Fig. 10.22c, d). However, when either chelator was administered at the same dose and volume as a 20-min intravenous infusion, no effect on either blood pressure or heart rate was recorded. Thus, NaHBED could be administered much more rapidly and in greater amounts than could DFO in the treatment of acute iron poisoning.

These results provide evidence for the safety and tolerability of NaHBED when administered either in the manner that would be used chronically, for the treatment of iron overload, or acutely, for the treatment of iron poisoning. Comparative studies of iron excretion in the iron-loaded *C. apella* monkey have shown that NaHBED is two to three times more efficient as an iron chelator than DFO after either subcutaneous or intravenous administration. Further, intravenous infusion of large doses of NaHBED for the treatment of acute iron poisoning should not carry the same risk of hypotension as does similar administration of DFO. Overall, the need for prompt completion of the preclinical evaluation of parenteral NaHBED in preparation for studies of iron balance in human volunteers is indicated. Thus, NaHBED may provide an alternative to DFO for the treatment of both chronic transfusional iron overload and of acute iron poisoning.

5   THINGS TO COME

This final section focuses on approaches that are being used currently in the course of the search for the ideal iron chelator for the treatment of iron overload. We will (*a*) delineate synthetic methods that allow access to hydroxamate, catecholamide, desferrithiocin-related, and citric acid-based chelators; (*b*) describe animal models used for evaluating the efficacy of the ligands as deferration agents *in vivo*; and (*c*) review structure-activity studies with desferrithiocin and its analogs to illustrate how the design, synthesis, and testing of iron chelators are integrated to produce candidates for clinical evaluation.

5.1   Synthetic Approaches

5.1.1 Catecholamides. Total syntheses of the catecholamide chelators N^1,N^8-bis(2,3-dihydroxybenzoyl)spermidine (compound II, **3**) (247–256), L-parabactin (**4**) (257–260), L-agrobactin (**5**) (261), L-fluviabactin (**6**) (66), and L-vibriobactin (**8**) (262) (Fig. 10.2) have been reported. During the assembly of these compounds, care must be taken to avoid air oxidation of the 2,3-dihydroxybenzamide functions, especially at higher pH values (263), and contamination by ferric salts. Moreover, siderophores (**4–8**) contain one or two acid-sensitive oxazoline rings, which must be formed stereospecifically. The final synthetic challenge is the regiospecific placement of the acyl groups onto the triamine backbones.

The triamine backbones of chelators (**3–6**) are symmetrically substituted; 2,3-dihydroxybenzoyl groups are on the terminal nitrogens. Iron-specific ligands (**4–6**) also possess an oxazoline ring, which is derived from L-threonine and salicylic acid for L-parabactin (**4**) or 2,3-dihydroxybenzoic acid for L-agrobactin (**5**) and L-fluviabactin (**6**), on the internal polyamine nitrogen. Reagents have been developed that block the primary amines (264) or secondary amine (265) of the triamines norspermidine, spermidine, and homospermidine. Thus, the syntheses of L-parabactin (**4**) (257) and L-agrobactin (**5**) (261) began with attachment of 2,3-dihydroxybenzoic acid equivalents to the unprotected amino groups of N^4-benzylspermidine. Alternatively, spermidine and norspermidine have been acylated efficiently at the terminal nitrogens in the production of compound II (**3**) (252–256) and L-fluviabactin (**6**) (66), respectively; a sufficiently bulky acylating agent must be chosen for such a high degree of selectivity.

In the norspermidine-based catecholamides L-vulnibactin (**7**) and L-vibriobactin (**8**) there is no such symmetry with respect to the acyl groups. In the case of these natural products, it is clear that selective acylation of one primary and one secondary nitrogen of norspermidine with the oxazoline unit is an unreasonable expectation in the absence of protecting groups. To access this more challenging system, a more versatile triamine reagent was developed with three different

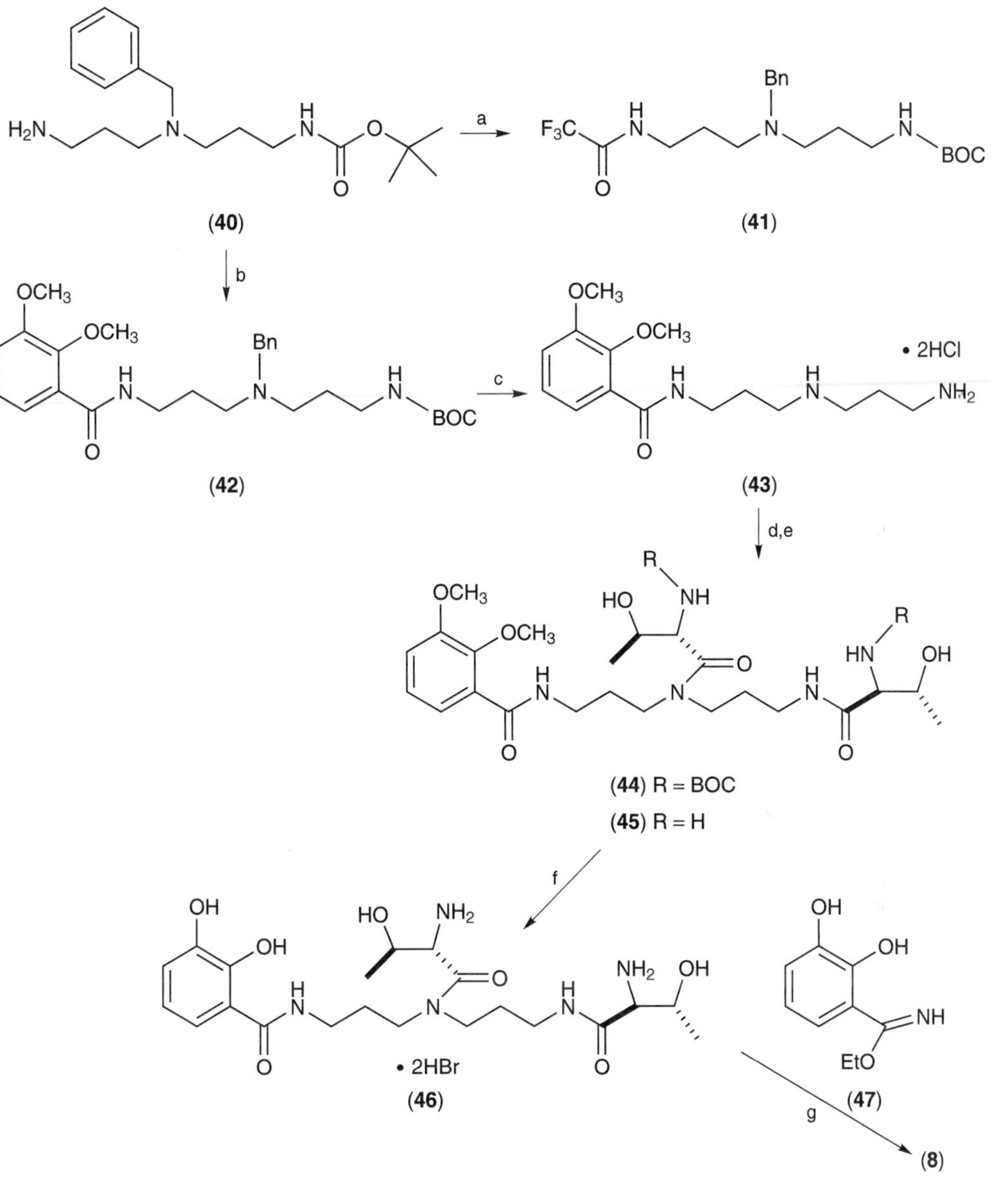

Figure 10.23. Total synthesis of L-vibriobactin (**8**). (a) TFAA/NEt$_3$/CH$_2$Cl$_2$ (92%); (b) 2,3-dimethoxybenzoyl chloride/NEt$_3$/CH$_2$Cl$_2$ (99%); (c) TFA, then H$_2$/PdCl$_2$/12 M HCl/CH$_3$OH (93%); (d) N-(*tert*-butoxycarbonyl)-L-threonine N-hydroxysuccinimide ester/DMF (99%); (e) TFA, then Na$_2$CO$_3$ (93%); (f) BBr$_3$/CH$_2$Cl$_2$ (63%); (g) CH$_3$OH/reflux/33 h (58%).

protecting groups: *tert*-butoxycarbonyl, trifluoroacetyl, and benzyl (266). These functionalities are orthogonal, that is, each amino group can be unmasked separately (267), permitting attachment of three different groups to the polyamine backbone in any order. Ben-zylamine and acrylonitrile were transformed by efficient steps into N^4-benzyl-N^1-(*tert*-butoxycarbonyl)norspermidine (**40**); protection of the remaining terminal amine with trifluoroacetic anhydride gave reagent (**41**) (266) (Fig. 10.23).

The synthesis of L-vibriobactin (**8**, Fig. 10.2), a hexacoordinate chelator isolated from *Vibrio cholerae* (31), began by acylation of (**40**) with 2,3-dimethoxybenzoyl chloride, providing trisubstituted norspermidine (**42**) in high yield (262). The *tert*-butoxycarbonyl group of (**42**) was removed by brief exposure to trifluoroacetic acid, and the N^4-benzyl group was cleaved by hydrogenolysis over palladium chloride in methanolic HCl, furnishing monoacylated norspermidine dihydrochloride (**43**) in 93% yield over two steps. After generating the free diamine of (**43**) with aqueous base, acylation with the N-hydroxysuccinimide ester of L-N-(*tert*-butoxycarbonyl)threonine (2.2 equivalents) in DMF gave intermediate (**44**) in 99% yield. The BOC groups of (**44**) were cleanly removed by TFA, and N^1,N^4-bis[L-threonyl]-N^7-(2,3-dimethoxybenzoyl)norspermidine (**45**) was obtained in 93% yield after neutralization of the salt. The O-methyl protecting groups were cleaved using BBr$_3$, affording N^1,N^4-bis[L-threonyl]-N^7-(2,3-dihydroxybenzoyl)norspermidine dihydrobromide (**46**) in 63% yield.

As in the syntheses of L-parabactin (**4**) (257), L-agrobactin (**5**) (261), and L-fluviabactin (**6**) (66), the final step in the L-vibriobactin (**8**) synthesis required stereospecific formation of the acid-sensitive oxazoline rings. Ethyl 2-hydroxybenzimidate, derived from the treatment of 2-cyanophenol with HCl (g) in CH$_3$CH$_2$OH (268), was used to effect this transformation to chelator (**4**). Unfortunately, 2,3-dihydroxybenzonitrile was unreactive to the same reagents, even under forcing conditions; thus, imidate (**47**) for final products (**5**), (**6**), and (**8**) was accessed by another route (261). Reacting 2,3-bis(benzyloxy)benzoyl chloride (269) with concentrated NH$_4$OH/CH$_2$Cl$_2$ gave 2,3-bis(benzyloxy)benzamide (92% yield). O-Alkylation of the amide with triethyloxonium hexafluorophosphate in CH$_2$Cl$_2$ provided the catechol-dibenzylated imidate ester in 81% yield. Hydrogenolysis of the benzyls under mild conditions (10% Pd-C, CH$_3$CH$_2$OH, 1 atm) gave crystalline ethyl 2,3-dihydroxybenzimidate (**47**) in 73% yield. Heating bis(*vic*-amino alcohol) (**46**) with excess imino ether (**47**) in refluxing methanol under a nitrogen atmosphere using iron-free glassware furnished L-vibriobactin (**8**).

5.1.2 Hydroxamates.

Numerous naturally occurring hydroxamate chelators and their analogs have been synthesized (87), including arthrobactin (**13**) (270), aerobactin (**14**) (271), nannochelin A (**16**) (272–274), schizokinen (**19**) (270), and mycobactin S, in which R is pentadecyl (**15**) (275) (Fig. 10.3). Of primary importance to the synthesis of hydroxamate siderophores and their synthetic analogs is the availability of N-alkylated hydroxylamines, which upon treatment with a stoichiometric amount of an acylating agent form predominantly N-hydroxy amides (i.e., hydroxamic acids) (276). Primary amines have been converted to the corresponding N-hydroxylamines (277–280) or N-(benzoyloxy)hydroxylamines (281) using methods that avoid further oxidation. In the latter compounds, acylation was directed exclusively onto nitrogen (282). These O-protected hydroxamates could be deprotected by mild base, for example, generating the first synthetic sample of nannochelin A (**16**), a dimethyl ester (272). N-Substituted hydroxylamines also can be obtained from aldehydes through partial reduction of their oximes (283, 284). Saturation of an O-protected oxime using borane-pyridine complex without cleaving the nitrogen-oxygen bond was a pivotal step in the most recent synthesis of dihydroxamic acid (**16**) (274). Finally, although direct alkylation of hydroxylamine leads to a mixture of products (285), many N,O-diprotected hydroxylamines have been alkylated efficiently at nitrogen (286). For example, this transformation was effected by treatment of N,O-bis(*tert*-butoxycarbonyl)hydroxylamine with alkyl bromides (287) or by direct coupling of a series of bis(alkoxycarbonyl)hydroxylamines with both primary and secondary alcohols under Mitsunobu conditions (288), followed by removal of both protective groups.

The syntheses of desferrioxamine B (Desferal is its methanesulfonate salt) (**9**) and desferrioxamine E (nocardamine, **11**), two of the siderophores isolated from *Streptomyces pilosus* (35, 37) (Fig. 10.3), are described here. To demonstrate the general utility of the synthetic methods employed, the preparation of two analogs of DFO is also presented. Hexacoordinate ligands DFO B (linear) and DFO E (macrocyclic) are predicated on two fundamental synthons: 1-amino-5-(N-hydroxyamino)pentane and succinic acid, and the key to the efficient assembly of hydroxamate iron chelators consists of regiospecific condensation of these units.

The first synthesis of desferrioxamine B started with 1-amino-5-nitropentane (**48**) (Fig. 10.24)(289), which was in turn accessed in a two-step sequence from *N*-(5-bromopentyl)phthalimide in 47% yield (290). Amine (**48**) was protected with a carbobenzyloxy (CBZ) group, generating (**49**). The nitro group of (**49**) was reduced with zinc to hydroxyamino compound (**50**), which was *N*-acylated with either succinic anhydride, providing (**53**), or with acetic anhydride, giving (**54**). The geometry of carboxylic acid (**53**) allowed intramolecular activation as the ester by ring closure to *O*-acyl hydroxamate (**55**) in 84% yield. The CBZ protecting group of (**54**) was removed by hydrogenolysis. Primary amine (**56**) was condensed with (**55**), and the carbanate (**57**) was converted to terminal amine (**58**), which was acylated with active ester (**55**), giving (**59**). Catalytic reduction of (**59**) furnished desferrioxamine B (**9**) as its hydrochloride.

Recently, an alternative synthesis of *N*-(benzyloxycarbonyl)-*N'*-hydroxy-1,5-diaminopentane (**50**) has been reported (291). Piperidine (**51**) was oxidized with calcium hypochlorite to its imine, which was converted to the *N*-CBZ enamine. Ring opening of the carbamate and trapping of the aldehyde with hydroxylamine gave CBZ amino oxime (**52**), which was reduced to *N*-hydroxyamine (**50**) using borane-pyridine complex. Thus, *N*-hydroxycadaverine equivalent (**50**) was accessed from reagent (**51**) in greater than twice the yield of the prior route and was carried through to Desferal in 24% overall yield (291).

The use of protecting groups in Fig. 10.24 was minimal; the CBZ on the terminal nitrogen was removed easily. However, attachment of the acetyl group early in the route restricts the utility of the scheme. A second synthesis gave DFO B in 45% overall yield, and the acetyl group was also affixed near the beginning of the synthetic sequence (292). This route began with 4-cyanobutanal, which is unstable and, like 1-amino-5-nitropentane (**48**), difficult to access.

The effective use of protecting groups is pivotal to the systematic synthesis of hydrox-amic acid siderophores and their analogs. Because *N*-hydroxydiamine containing three to five carbons is a fundamental, repeating unit of many such chelators, a method of differentiating its two nitrogens and protecting its oxygen was sought. Thus, an *N*-hydroxy-1,5-diaminohydroxypentane (cadaverine) equivalent, *O*-benzyl-*N*-(*tert*-butoxycarbonyl)-*N*-(4-cyanobutyl)hydroxylamine (**62**) was synthesized (293). In this reagent the primary amine is masked as a nitrile; the hydroxylamine is *N*-*tert*-butoxycarbonyl, *O*-benzyl diprotected (Fig. 10.25). The synthesis of reagent (**62**) began with the conversion of *O*-benzylhydroxylamine hydrochloride to its stable, crystalline *N*-(*tert*-butoxycarbonyl) derivative (**61**), employing di-*tert*-butyl dicarbonate in 97% yield (294). Carbamate (**61**), which is now available commercially, was N-alkylated with 5-chlorovaleronitrile to give the versatile reagent (**62**) in 87% yield. Homologs of the 1,5-diaminopentane reagent could be generated by *N*-alkylation of hydroxylamine equivalent (**61**) with other ω-haloalkanenitriles. Brief exposure of *N*-(*tert*-butoxycarbonyl)nitrile (**62**) to trifluoroacetic acid resulted in *O*-benzyl-*N*-(4-cyanobutyl)hydroxylamine (**63**) and volatile by-products in 75% yield. Alternatively, the nitrile group of (**62**) was selectively hydrogenated over Raney nickel catalyst in methanolic ammonia, giving *N*-(5-aminopentyl)-*O*-benzyl-*N*-(*tert*-butoxy-carbonyl)hydroxylamine (**64**) in 83% yield. The hydroxylamine oxygen in (**62**) could remain protected by the benzyl moiety until catalytic unmasking of a hydroxamic acid chelator (295).

O-Benzyl-*N*-(4-cyanobutyl)hydroxylamine (**63**) was alternately elaborated with succinic anhydride and *gem*-disubstituted diaminopentane (**64**) to a DFO skeleton in which the primary amine and the hydroxamic acids remained masked as the nitrile and *O*-benzyl hydroxamates, respectively (296). Hydrogenation generated DFO B hydrochloride (**9**) in 44% overall yield from hydroxylamine reagent (**61**) (296). Cleavage of the benzyl groups occurred more rapidly than did saturation of the cyano group (297).

The most recent synthesis of DFO (**9**) is both direct and versatile; construction of the linear molecule is streamlined by capitalizing on the electronic differences between a pri-

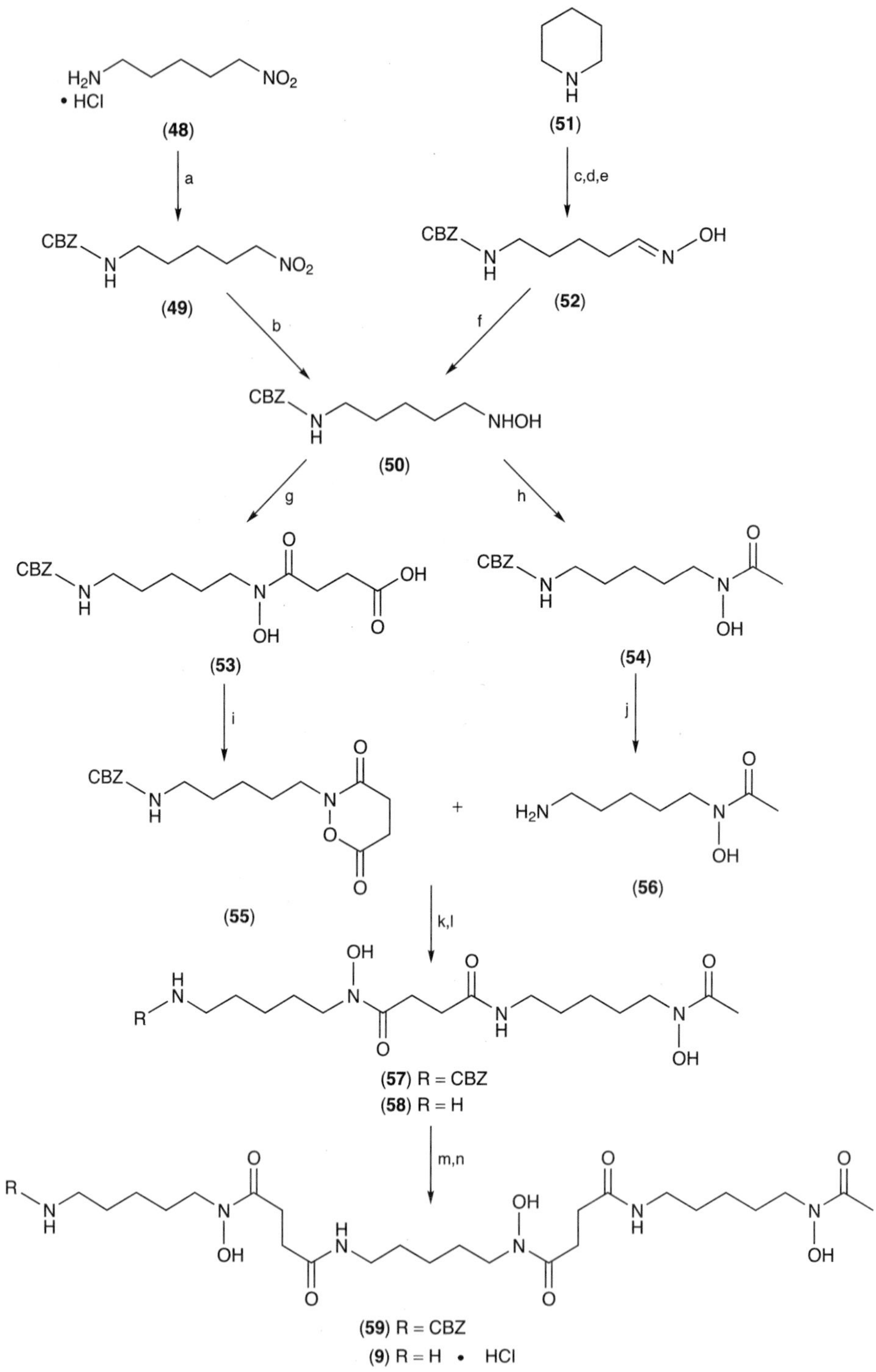

Figure 10.24. Total synthesis of desferrioxamine B (**9**). (a) CBZ-Cl/3 *N* NaOH (86%); (b) Zn/NH$_4$Cl (aq)/EtOH (74%); (c) Ca(OCl)$_2$/MTBE/AcOH (aq)/CH$_3$OH; (d) CBZ-Cl/KOH/MTBE/CH$_3$OH; (e) NH$_2$OH·HCl/CH$_3$OH/pyridine (75%); (f) pyridine·BH$_3$/HCl (aq)/CH$_3$OH (90%); (g) succinic anhydride/pyridine/KOH (aq) (91%); (h) Ac$_2$O/ pyridine (96%); (i) Ac$_2$O/100°C (84%); (j) H$_2$/Pd-C/CH$_3$OH (quantitative); (k) THF (80%); (l) j; (m) (**55**)/THF (66%); (n) H$_2$/Pd-C/CH$_3$OH/0.1 *M* HCl (90%).

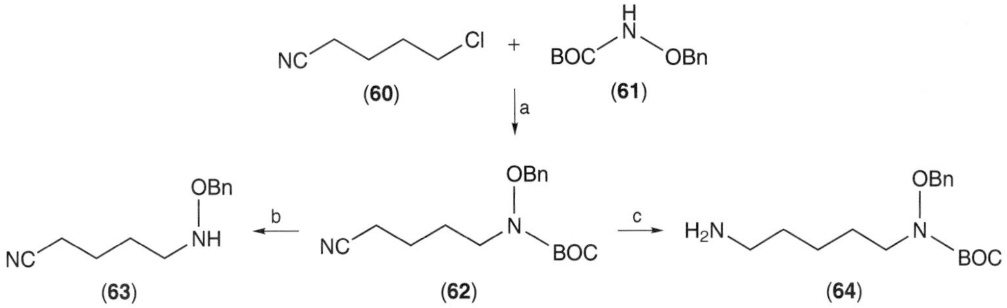

Figure 10.25. Synthesis of the triprotected *N*-hydroxycadaverine reagent (**62**). (a) NaH/NaI/DMF/ 80°C/4 h (87%); (b) TFA (75%); (c) H_2 (58 psi)/Raney nickel/conc NH_4OH/methanolic NH_3 (83%).

mary amine and an *N*-(benzyloxy)amine (Fig. 10.26) (297). This route also facilitates modification of either terminus of chelator (**9**) by (*a*) protecting the amine with *tert*-butoxycarbonyl (BOC), a group that is orthogonal to the *O*-benzyls; and (*b*) attachment of the acetyl function near the end of the sequence.

Successive exposure of *N*-(benzyloxy)-*N*-(*tert*-butoxycarbonyl)-1,5-pentanediamine (**64**) to TFA, aqueous base, and HCl provided *N*-benzyloxy-1,5-pentanediamine dihydrochloride (**65**), also available by alkylation of *O*-benzylhydroxylamine with *N*-(5-bromopentyl)phthalimide, followed by hydrazinolysis (298). The free diamine of (**65**) was combined with 2-(*tert*-butoxycarbonyloxyimino)-2-phenylacetonitrile (BOC-ON, 1 equivalent), cleanly providing *N*-(benzyloxy)-*N'*-(*tert*-butoxycarbonyl)-1,5-pentanediamine (**66**). Both steric (299) and electronic (300) factors explain the highly regioselective acylation of the diamine.

Reacting (benzyloxy)amine (**66**) with succinic anhydride in pyridine gave carboxylic acid (**67**). Treatment of acid (**67**) with 1,1'-carbonyldiimidazole (CDI) in CH_2Cl_2 generated the *N*-acyl imidazole, which was treated with (**65**) as the free diamine, resulting in *N*-(benzyloxy)amine (**68**). As was the case in the protection of diamine (**65**) by BOC, acylation occurred at the primary amine end of (**65**) with a high degree of selectivity (253, 270, 301, 302). The last two steps were repeated; that is, *N*-(benzyloxy)amine (**68**) was converted to acid (**69**) with succinic anhydride. Next, regiospecific acylation of diamine (**65**) with acid (**69**) gave tris[*N*-(benzyloxy)amine] (**70**). In

addition to providing DFO (**9**) in three transformations, (**70**) is a versatile DFO reagent that allows replacement of the acetyl terminus of DFO by any acyl group.

Reaction of (**70**) with acetyl chloride produced tetraprotected DFO (**71**). The benzyl groups of (**71**) were cleaved by hydrogen under mild conditions to give *N*-(*tert*-butoxycarbonyl)DFO (**72**). Catalytic unmasking of the hydroxamate esters of (**71**) was accomplished rapidly and with no detectable nitrogen-oxygen bond cleavage to amide by-product. Brief exposure of *N*-(*tert*-butoxycarbonyl)DFO (**72**) to TFA resulted in the formation of DFO (**9**) as its trifluoroacetate salt in 45% overall yield from known diamine (**65**). The high field proton NMR spectrum of the synthetic chelator was identical to that of Desferal, except for the latter's methanesulfonate singlet.

Trihydroxamate reagent (**70**) facilitates alteration of DFO at either end of the molecule, thus providing more flexibility in accessing DFO analogs than did prior routes. After coupling the *N*-(benzyloxy) terminus with an acylating agent and removal of the BOC protecting group, the primary amine could also be derivatized. Unblocking of the hydroxamates would yield a panel of DFO analogs for determining structure-activity relationships in the search for a chelator with a longer clearance time and even oral bioavailability.

The syntheses of the macrocyclic hydroxamate siderophores bisucaberin (**12**) (293), alcaligin (**20**) (301), and nocardamine (**11**) (303) (Fig. 10.3) have been accomplished beginning with *N*-(*tert*-butoxycarbonyl)-*O*-benzyl-

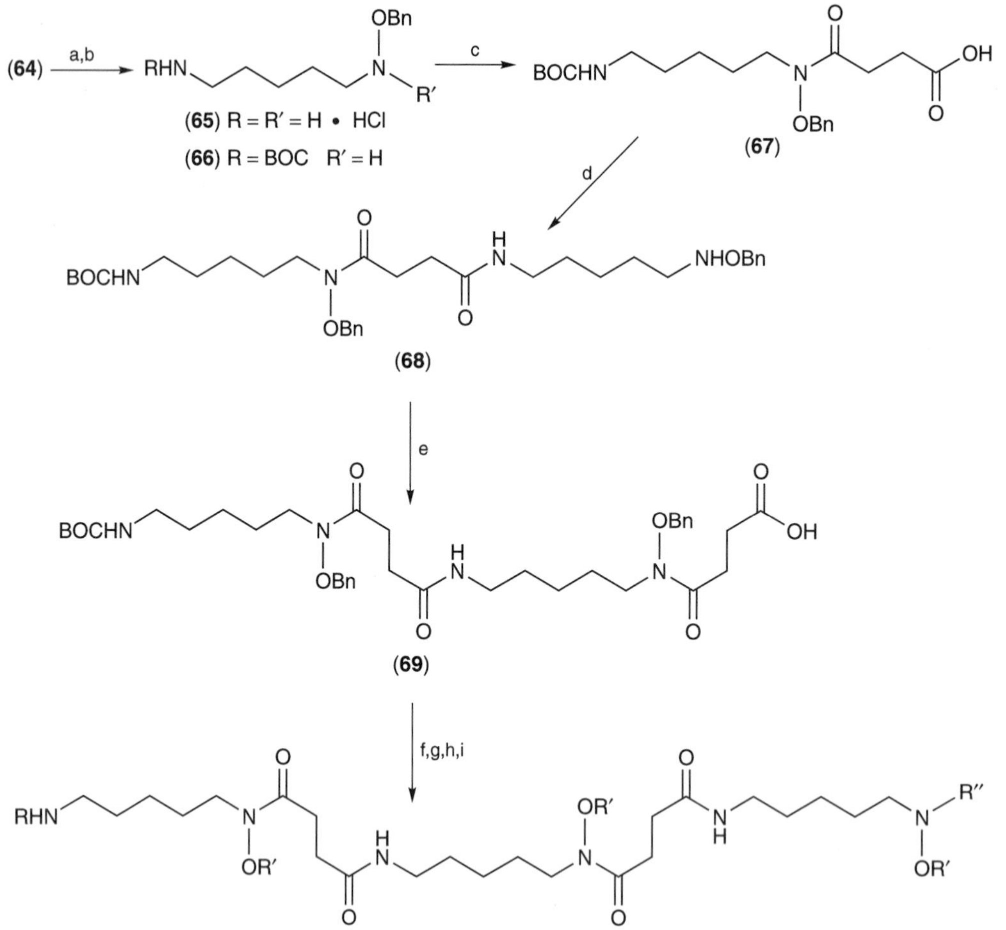

(70) R = BOC, R′ = Bn, R″ = H
(71) R = BOC, R′ = Bn, R″ = Ac
(72) R = BOC, R′ = H, R″ = Ac
(9) R = H • TFA, R′ = H, R″ = Ac

Figure 10.26. Total synthesis of desferrioxamine B (**9**). (a) TFA/CH_2Cl_2, 1 *N* NaOH, conc HCl/CH_3CH_2OH (82%); (b) 1 *N* NaOH/Et_2O, BOC-ON/THF (quantitative); (c) succinic anhydride/pyridine/80°C/90 min (88%); (d) 1,1′-carbonyldiimidazole/(**65**) (free diamine)/CH_2Cl_2 (quantitative); (e) c/81°C/2 h (81%); (f) d (90%); (g) AcCl/Et_3N/CH_2Cl_2 (87%); (h) H_2/10% Pd-C/CH_3OH (81%); (i) TFA (quantitative). [Reprinted with permission from R. J. Bergeron, J. S. McManis, O. Phanstiel, IV, and J. R. T. Vinson, *J. Org. Chem.*, **60**, 109–114 (1995). Copyright 1995 American Chemical Society.]

hydroxylamine (**61**) (294). *O*-Benzyl-*N*-(4-cyanobutyl)hydroxylamine (**63**) was elaborated with succinic anhydride and *N*-(5-aminopentyl)-*O*-benzyl-*N*-(*tert*-butoxycarbonyl)hydroxylamine (**64**) to form DFO E intermediate (**73**) (Fig. 10.27) (293).

Addition of the third succinate unit of DFO E to benzyloxyamine (**73**) gave ω-cyano acid (**74**) in 78% yield (303). Subjecting (**74**) to hydrogen over Raney nickel reduced only the nitrile group, furnishing ω-amino acid (**75**). Diphenylphosphoryl azide, the Yamada reagent, was added to acyclic precursor (**75**) (2.6 m*M* in DMF), and the reaction mixture was stirred for 4 days at 0°C to produce *O*,*O*′,*O*″-tribenzylnocardamine (**76**) in 54% yield. The genera-

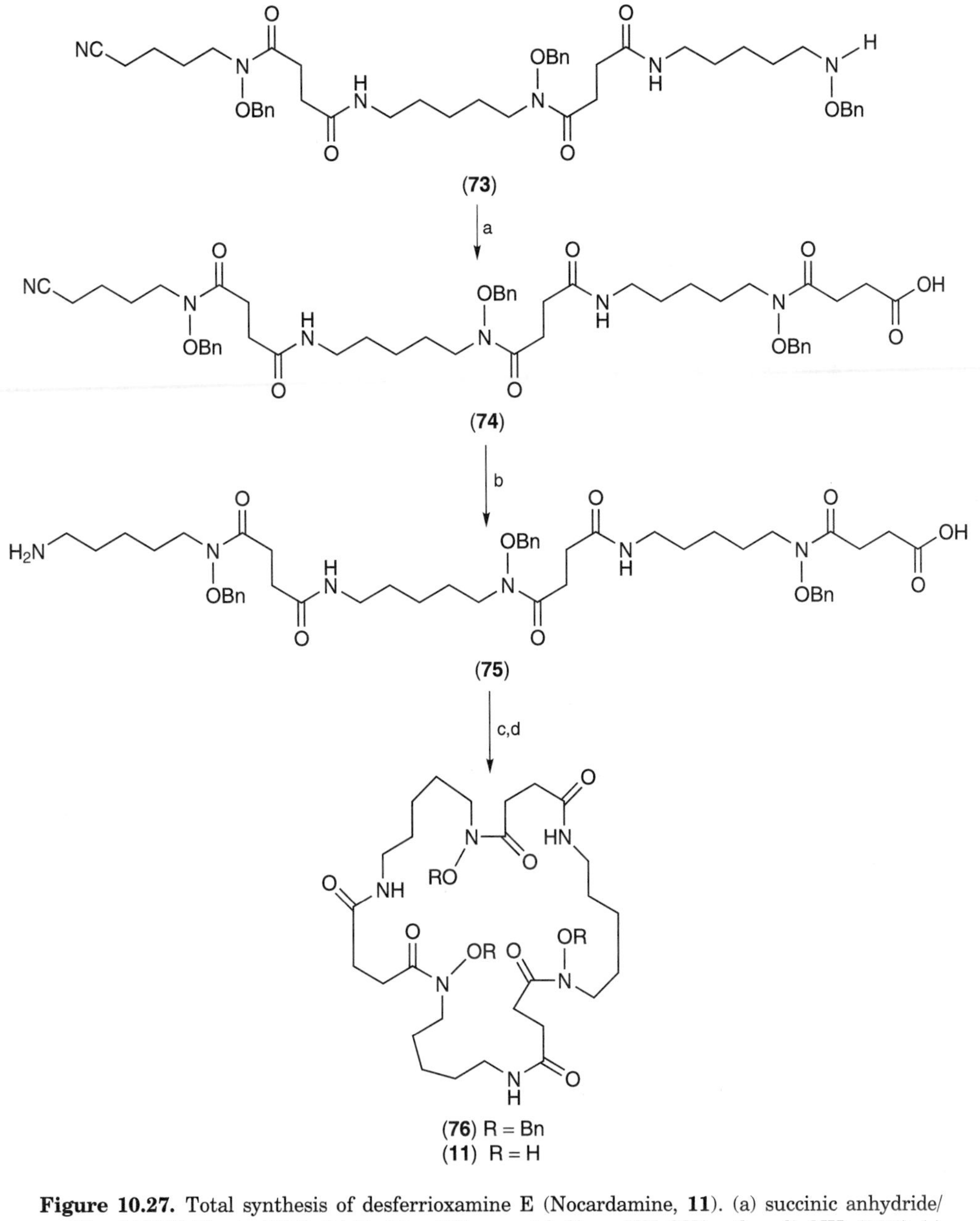

Figure 10.27. Total synthesis of desferrioxamine E (Nocardamine, **11**). (a) succinic anhydride/ pyridine/101°C/110 min (78%); (b) H$_2$ (60 psi)/Raney nickel/conc NH$_4$OH/methanolic NH$_3$ (21%); (c) (PhO)$_2$PON$_3$/DMF/0°C/4 days (54%); (d) H$_2$/10% Pd-C/CH$_3$OH.

tion of a 33-membered macrocycle using routine reaction conditions further illustrates the utility of this lactamization method (293, 301). Catalytic removal of the O-benzyl protecting groups of (**76**) generated nocardamine (**11**) (303).

An efficient synthesis of the trihydroxamate danoxamine has been achieved (304) beginning with N-benzyloxy-1,5-pentanediamine dihydrochloride (**65**) (293, 298). The structure of the final product, which is the iron-chelating component of the antibacterials Salmycins

A and B (305), matches that of desferrioxamine G (**10**), except for an alcohol group in place of the amine.

The short half-life of DFO in the body and the fact that patients must be continuously infused led investigators to prepare and test analogs of DFO as potential therapeutic iron chelators. Because replacement of the terminal 5-aminopentyl unit of desferrioxamine B with a heptyl group rendered the molecule highly insoluble in a variety of vehicles (292), chelators (**88**) and (**89**), which are polyether analogs of DFO, were prepared to enhance the siderophore's overall solubility (Fig. 10.28). Specifically, in DFO polyether analog (**88**), the charged aminoalkyl chain was replaced with a neutral triether chain, and in bis(triether) (**89**), the acetyl of DFO was substituted, as well, by a triether acyl group (296).

The monomethyl ether of triethylene glycol was converted to its tosylate (**77**) (306), which was used to alkylate N-(*tert*-butoxycarbonyl)-O-benzylhydroxylamine (**61**), resulting in triether (**78**) (Fig. 10.28) (296). The nitrogen of alkylated product (**78**) was deprotected with TFA, giving (**79**), which was acylated with succinic anhydride to afford carboxylic acid (**80**). Efficient coupling of (**80**) and N-(5-aminopentyl)-O-benzyl-N-(*tert*-butoxycarbonyl)-hydroxylamine (**64**) was effected with diphenylphosphoryl azide to make tetracoordinate equivalent (**81**). The protected hexacoordinate reagent (**84**) was obtained from (**81**) by repeating the three conversions: unmasking to amine (**82**), four carbon elaboration to (**83**), and elongation to (**84**). Acidic cleavage of the *tert*-butoxycarbonyl group in (**84**) furnished benzyloxyamine (**85**), which was acylated with acetic anhydride to produce (**86**) or with 3,6,9-trioxadecanoyl chloride (307) to afford (**87**). Catalytic hydrogenation of the O-benzyls of (**86**) and (**87**) gave hexacoordinate ligands (**88**) and (**89**), respectively, in good overall yield. Both of these neutral DFO analogs are soluble in chloroform and in water, and thus enhance the lipophilicity of the DFO molecule while maintaining its hydrophilicity.

This change in solubility properties rendered both drugs more effective than desferrioxamine at removing iron in the non-iron-overloaded bile duct-cannulated rat model. When the chelators were administered subcutaneously, analog (**88**) was nearly 3 times as effective as DFO in promoting iron clearance; (**89**) was 2.5 times as effective as (**9**) (296).

A hybrid hexacoodinate iron chelator that contains catecholamides and an N-hydroxy amide moiety has been assembled. Specifically, the central nitrogen of compound II (**3**) was elaborated to an N-methylhydroxamic acid using a succinic acid spacer, resulting in the analog spermexatol (308).

5.1.3 Desferrithiocin and its Analogs.

The siderophore desferrithiocin [(*S*)-4,5-dihydro-2-(3-hydroxy-2-pyridinyl)-4-methyl-4-thiazolecarboxylic acid, DFT, (**24**), Fig. 10.4] from *Streptomyces antibioticus* (54) has been synthesized (309) by the cyclocondensation of D-α-methyl cysteine (**90**) (Fig. 10.29) with 3-hydroxy-2-cyanopyridine (310). The unusual amino acid (**90**) is available from the hydrolysis of DFT (6 M HCl/90°C/90 min) (54) or by stereoselective α-methylation of a D-cysteine derivative (309). Because the DFT skeleton is a useful pharmacophore on which to predicate the design of orally effective iron chelators (311), a large number of DFT analogs have been prepared. The structure-activity studies were focused on identifying those fragments of DFT responsible for its oral iron-clearing properties in an effort to construct a less toxic analog (311–315).

The majority of these DFTs have been made by reaction of an aromatic nitrile with cysteine or its analogs. The production of (*S*)-desazadesferrithiocin (DADFT, **93**) was accomplished by cyclocondensation of 2-cyanophenol (**91**) with (*S*)-α-methyl cysteine (**90**) (Fig. 10.29) (315). The high field NMR spectra of synthetic (**93**) and the siderophore 4-methylaeruginoic acid, which has been isolated from the *Streptomyces* species KCTC 9303, were virtually identical; however, the C-4 stereochemistry of the natural product was not specified (316). Therefore, a CD measurement of (**93**) was run; wavelengths of the maximum and two minima of the CD of (**93**) matched closely with literature values of the natural

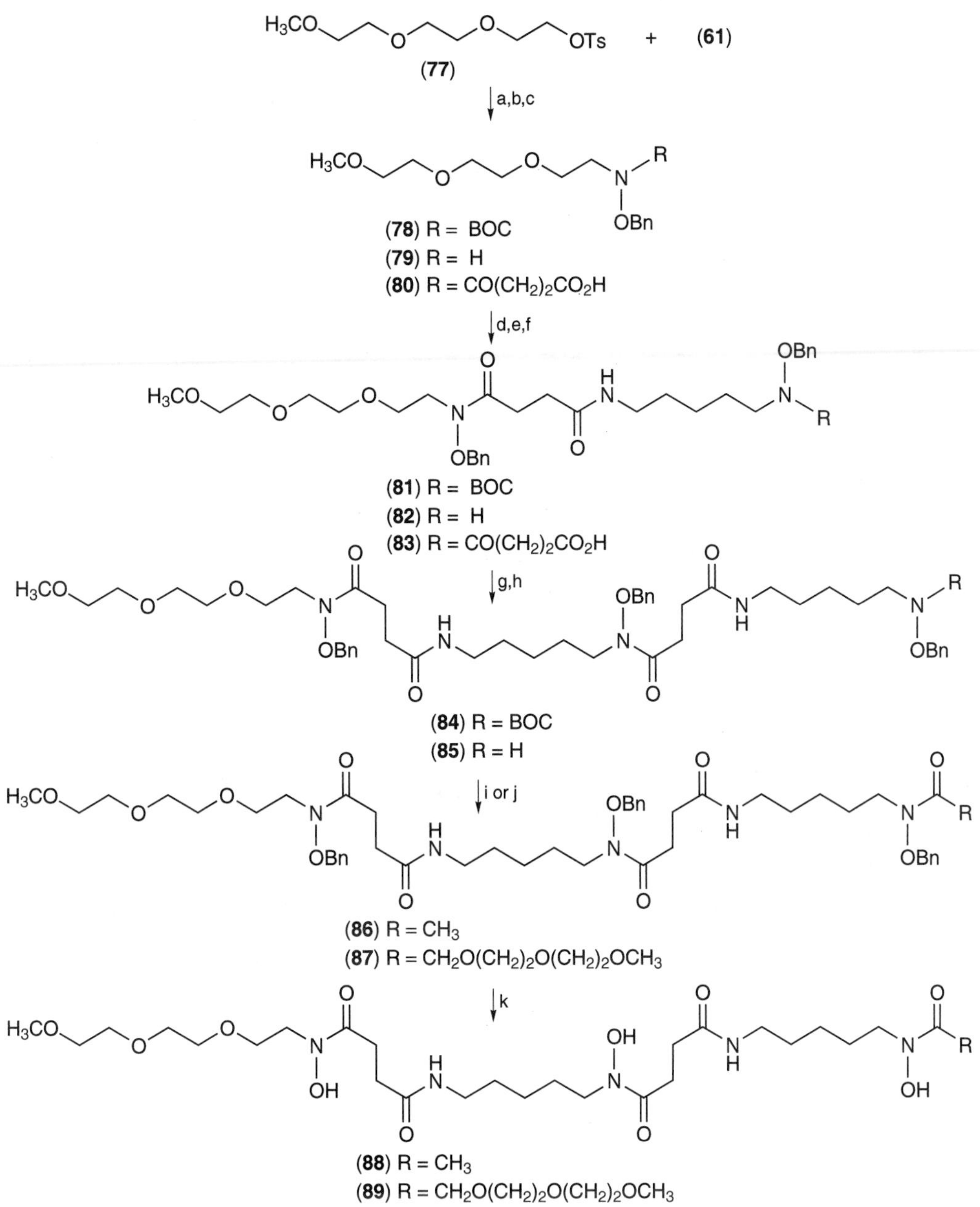

Figure 10.28. Synthesis of desferrioxamine analogs with ether linkages (**88** and **89**). (a) NaH/DMF/ 72°C/18 h (77%); (b) TFA/CH$_2$Cl$_2$ (91%); (c) succinic anhydride/pyridine/90°C/2 h (92%); (d) (**64**)/ (PhO)$_2$PON$_3$/Et$_3$N/DMF (94%); (e) b (87%); (f) c (95%); (g) d (78%); (h) b (quantitative); (i) Ac$_2$O/ pyridine (91%); (j) 3,6,9-trioxadecanoyl chloride/Et$_3$N/CH$_2$Cl$_2$ (78%); (k) H$_2$/10% Pd-C/CH$_3$OH (86% for **88**; 73% for **89**). [Reprinted with permission from R. J. Bergeron, J. Wiegand, J. S. McManis, and P. T. Perumal, *J. Med. Chem.*, **34**, 3182–3187 (1991). Copyright 1991 American Chemical Society.]

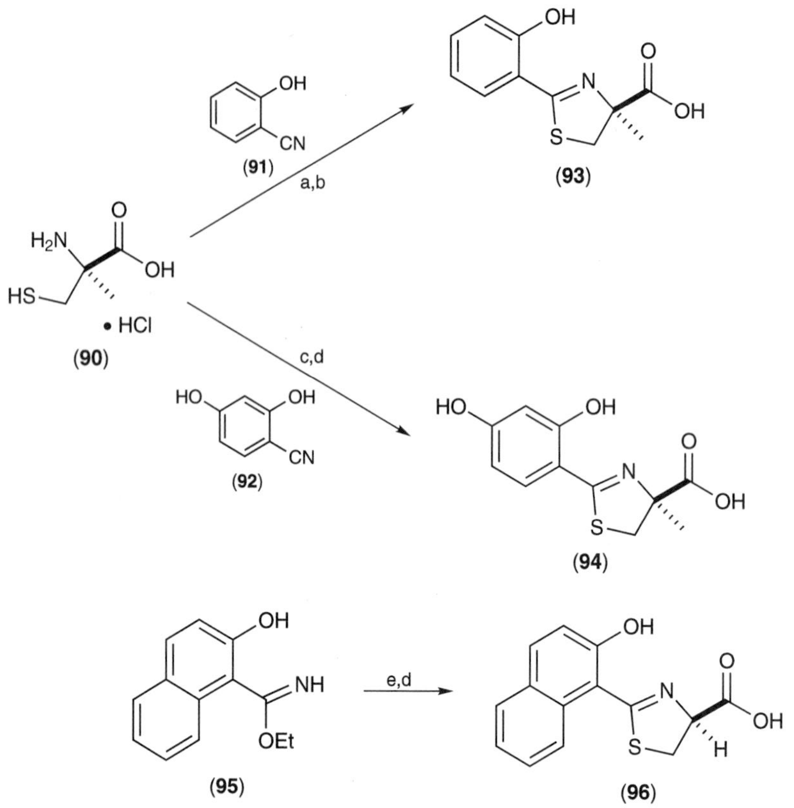

Figure 10.29. Synthesis of (S)-4,5-dihydro-2-(2-hydroxyphenyl)-4-methyl-4-thiazolecarboxylic acid (**93**), (S)-4,5-dihydro-2-(2,4-dihydroxyphenyl)-4-methyl-4-thiazolecarboxylic acid (**94**), and (S)-4,5-dihydro-2-(2-hydroxy-1-naphthalenyl)-4-thiazolecarboxylic acid (**96**). (a) phosphate buffer (pH 6)/CH$_3$OH/34°C/5 days; (b) citric acid (32%); (c) phosphate buffer (pH 6)/NaHCO$_3$/CH$_3$OH/71°C/3.5 days; (d) dilute HCl (57%); (e) D-cysteine/CH$_3$OH/reflux/2 days.

chelator (316). Thus, the first total synthesis of the natural product 4-methylaeruginoic acid proves that, like DFT, it possesses the (S)-configuration.

The preparation of crystalline (S)-4,5-dihydro-2-(2,4-dihydroxyphenyl)-4-methyl-4-thiazolecarboxylic acid (**94**), the 4'-hydroxy analog of DADFT, also began with amino acid (**90**), which was condensed with 2,4-dihydroxybenzonitrile (**92**) under weakly acidic conditions (Fig. 10.29).

For sterically crowded DFT systems, for example, (S)-4,5-dihydro-2-(2-hydroxy-1-naphthalenyl)-4-thiazolecarboxylic acid (**96**), an imidate ester was required to generate the thiazoline ring (Fig. 10.29) (313). Ethyl 2-hydroxy-1-naphthimidate (**95**), available in six steps from 2-hydroxy-1-naphthaldehyde in an

overall yield of 42%, was heated with D-cysteine in methanol, generating naphthyl chelator (**96**).

5.1.4 Rhizoferrin. Rhizoferrin was first isolated from *Rhizopus microsporus* var. *rhizopodiformis*, an organism associated with mucormycosis seen in dialysis patients (50), and occurs in several Zygomycetes strains of fungi (317). Structure determination of rhizoferrin (**22**) (Fig. 10.4) revealed a putrescine center with each nitrogen acylated by citric acid at its 1-carboxylate (50). Thus, although rhizoferrin contains a polyamine backbone, it is neither a catecholamide nor a hydroxamic acid. Unlike the hydroxamates arthrobactin (**13**) (39) and schizokinen (**19**) (45), in which citric acid is symmetrically 1,3-disubstituted,

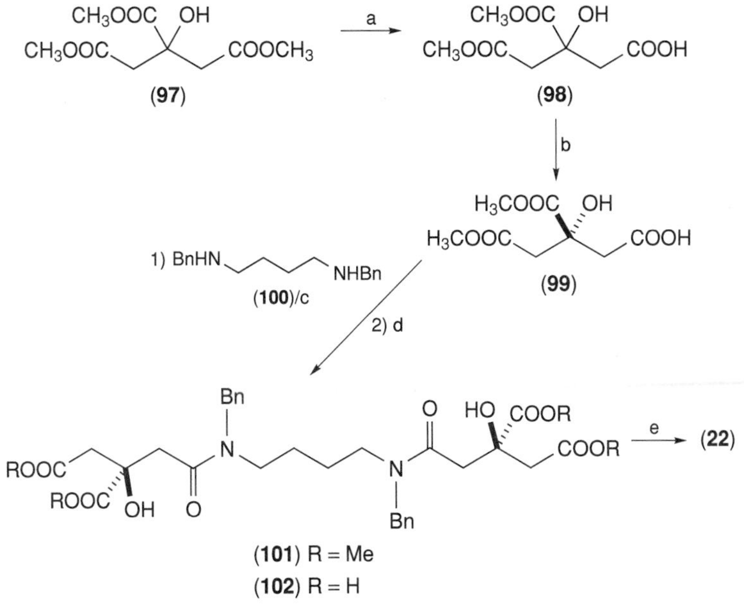

Figure 10.30. Total synthesis of rhizoferrin (**22**). (a) NaOH (aq)/CH₃OH; dilute HCl (39%); (b) (−)-brucine; fractional crystallization; HCl; (c) (PhO)₂PON₃/triethylamine/DMF (26%); (d) a (77%); (e) Li/NH₃/THF (64%). [Reprinted from R. J. Bergeron, M. G. Xin, R. E. Smith, M. Wollenweber, J. S. McManis, C. Ludin, and K. Abboud, *Tetrahedron*, **53**, 427–434 (1997). Copyright 1996 with permission from Elsevier Science.]

the central carbon of each citric acid unit in rhizoferrin is asymmetric. These two sites of the molecule are in the (R)-configuration according to CD spectroscopy compared with those of natural tartaric acid (318). The principal challenge of the synthesis of rhizoferrin (**22**) was to access a citrate synthon of correct configuration for coupling to both termini of putrescine to confirm the absolute configuration of the siderophore.

The synthesis (319) of rhizoferrin began with trimethyl citrate (**97**), which was converted to 1,2-dimethyl citrate (**98**) in 39% yield by a sterically controlled saponification (320) (Fig. 10.30). The enantiomers of carboxylic acid (**98**) were separated by forming their (−)-brucine salts and crystallization from water. Single-crystal X-ray diffraction of the solid revealed 1,2-dimethyl citrate in the R-configuration. Acidification of the salt furnished (R)-1,2-dimethyl citrate (**99**).

N^1,N^4-Dibenzyl-1,4-diaminobutane (**100**) (321) was acylated with chiral acid (**99**) (2 equivalents) using diphenylphosphoryl azide,

generating the diamide (**101**) in 26% yield. The methyl esters of (**101**) were hydrolyzed using sodium hydroxide to give N,N′-dibenzyl rhizoferrin (**102**) in 77% yield.

Because N-benzyl amides are resistant to hydrogenolysis (322), dissolving metal reduction conditions were employed. Treatment of tetraacid (**102**) with excess lithium metal in liquid ammonia and protonation of the salts on a cation exchange resin column furnished rhizoferrin (**22**). The high field NMR and high resolution mass spectrum of the synthetic compound were essentially identical to the published spectra of the natural product (50). The absolute configurations of the synthetic sample and the natural material were identical (R,R), given that both exhibited a negative Cotton effect at the same wavelength (318). Moreover, a siderophore was isolated from *Ralstonia pickettii* DSM 6297 and characterized as (S,S)-, thus, *enantio*-rhizoferrin, in that it exhibited a positive Cotton effect (323). The total synthesis of rhizoferrin (**22**) is the first example of the conversion of a chiral citric acid fragment, as confirmed by X-ray diffraction and high field NMR analyses, to a chelator.

5.2 Animal Models Employed in Iron Metabolism and Iron Chelator Studies

To better define both the impact of defective iron metabolism and the effect of potential therapeutic iron chelators *in vivo*, several animal models have been developed. Modern molecular biology has accelerated the development and the molecular characterization of mutant animals that carry a particular genetic defect in iron transport; these animals are excellent model systems for studying the impact of misdirected iron transport and utilization in mammals and may, in the future, be useful for assessing therapeutic ligands. Although the early animal experiments with DFO often involved rabbits (156) or dogs (147), the methods developed and used since the mid-1980s for the measurement of the efficacy and toxicity of candidates for chelation therapy employ relatively common wild-type rodents or primates.

5.2.1 Genetically Characterized Rodents. The genetic alterations of iron transport have been obtained either by breeding/spontaneous mutation [e.g., Belgrade rat (324), α-thalassemic mouse (325)] or by direct genetic manipulation to generate mice with a defined mutation [e.g., transferrin receptor 1-deficient mouse (326), Irp2 knockout mouse (128)]. Clearly, future studies using many of these models will be instrumental in determining the interrelationships of the components of mammalian iron transport systems; however, a detailed description of each of these is outside the scope of this chapter. Therefore, the focus of this segment will be on models of the hemoglobinopathies.

Although spontaneous mutant α-thalassemic (325) and β-thalassemic (327) mice were studied in the mid-1980s, the development of transgenic mice has been a key advance in probing the hemoglobinopathies on a molecular level. The α-thalassemic mice exhibit, similar to the human disorder, defective α-chain synthesis, mild hemolytic anemia, including microcytosis, poikilocytosis with extensive

rouloux, anisocytosis, and reticulocytosis; splenomegaly; and increased intestinal iron absorption (325, 328, 329). One gene-targeting approach created an accurate murine homolog of the most severe form of α-thalassemia, Bart's hydrops fetalis, in which all four copies of the α-globin gene are inactivated; the condition could, at least in part, be rescued by addition of two copies of a normal α-globin gene (330). Another transgenic mouse model of this disorder was generated by way of an insertional mutation into either the ζ or the α1 gene locus (329). Studies in this model revealed that (*a*) inactivation of the embryonic ζ-globin gene effectively downregulated the α1 gene locus, consequently inducing α-thalassemia as effectively as did direct mutagenesis of the α1 gene; and (*b*) inactivation of the ζ-globin gene did not yield a lethal phenotype universally.

Further work in this model system indicated that different genetic backgrounds, particularly variants in the β-globin gene, appear to modify the thalassemia phenotype (331). However, no such compensatory mechanism appears to exist in the murine models of β-thalassemia that have been studied to date (327, 332–335). The phenotype of the mice carrying a deletion of the *b*1 gene (*Hbb^{th-1}/Hbb^{th-1}*) was relatively mild, given that these mice survived to adulthood and were fertile (327). The abnormalities included hemolytic anemia, microcytosis, anisocytosis, splenomegaly, and disproportionate levels of α-globin (327). Surprisingly, in the presence of increased iron absorption, no increase in cardiac iron was noted, although iron deposition in the spleen, liver, and kidney was prominent (332). The observed hematological parameters more closely resemble those of human β-thalassemics who have undergone splenectomy, rather than those of unsplenectomized thalassemics (327). In stark contrast, either disruption of the *b*1 locus (*Hbb^{th-2}/Hbb^{th-2}*) (333) or deletion of the *b*1 and *b*2 loci (*Hbb^{th-3}/Hbb^{th-3}*) (334, 335) yielded homozygotes, which died either *in utero* or within hours of birth. *Hbb^{th-3}* heterozygotes in one of the double-deletion models demonstrated a pathophysiology comparable to that of patients with β-thalassemia intermedia (335), whereas

Table 10.6   Iron-Overloaded Rodent Models Employed in Efficacy Studies of Iron Chelators

Species	Route of Iron Administration[a]	Iron Composition	Selected References
Rat/mouse	i.p.	Iron dextran	(348–350)
Rat	i.p./i.v.	Iron polymaltose	(351)
Mouse	i.p./i.v.	Iron dextran + ^{59}Fe lactoferrin	(352–357)
Rat/mouse	p.o.	Ferrocene	(358–361)
Rat/mouse	i.v.	Hypertransfusion with or without radiolabeling	(183, 362–370)
Rat/mouse	p.o.	Carbonyl iron	(351, 371, 372)
Rat/hamster	i.v.	^{59}Fe ferritin	(373, 374)
Guinea pig	p.o.	Carbonyl iron	(375)
Guinea pig	i.p.	Iron dextran	(376)
Gerbil	s.c.	Iron dextran	(377–380)

[a]i.p., intraperitoneal; i.v., intravenous; p.o., per os (oral); s.c., subcutaneous.

those in another presented a more severe phenotype than that seen in human β-thalassemia heterozygotes (334).

Perhaps the most extensive studies of a murine model of a hemoglobinopathy have been in any one of several models of sickle-cell anemia (336–343), the treatment of which entails transfusion therapy at an increasing rate (118, 344). The later-generation models mimic the molecular events that occur in the course of development much more closely; the switch from fetal hemoglobin to adult hemoglobin production occurs in addition to the sickling red cell phenotype and its attendant morbidity (338, 340, 341). However, striking progress has been made in earlier models (339, 342, 343); both an experimental drug treatment (345) and a genetic rescue protocol (337) have been attempted in this model with a measure of success. Clearly, mice carrying a defined hemoglobinopathy will be useful in probing the molecular basis of these disorders as well as in the evaluation of therapeutic agents, including iron chelators (332).

5.2.2 Wild-type Rodents. Although rodents represent a viable first-line animal screen, providing a rapid way to identify and discard chelators that are ineffective *in vivo*, there is no strict correspondence between the effectiveness of a chelator in rodents and that in primates. Nevertheless, rodents often provide a "go or no-go" answer for further development of a potential chelator. Over the years, a number of rodent models have been used to

assess the potential efficacy of candidate iron chelators. The rodents used, among others, have included mice, rats, guinea pigs, hamsters, and gerbils. Animals with normal iron stores as well as iron-overloaded animals (Table 10.6) have been used.

5.2.2.1 Non-Iron Overloaded Bile Duct-Cannulated Rat. The non-iron overloaded bile duct-cannulated rat model (240, 311, 312, 314, 346) has been used as a primary screen for candidate iron chelators before their testing in an iron-loaded *Cebus apella* monkey model (98, 235, 236, 242, 347). Briefly, male Sprague-Dawley rats averaging 400 g were anesthetized using sodium pentobarbital given intraperitoneally (i.p.). The bile duct was cannulated using 22-gauge polyethylene tubing, about 1 cm from the duodenum. The rats were housed in plastic metabolic cages during the experimental period and were given free access to water. Bile samples were collected at 3-h intervals, and urine samples were taken every 24 h.

The iron-clearing efficiency (Ec_{Fe}; see discussion in Section 2) is calculated by subtracting the iron excretion of control animals from the iron excretion of the treated animals. This number is then divided by the theoretical output to obtain the efficiency. Although iron clearance may be overestimated due to a lack of enterohepatic recirculation, nevertheless, the model allows for the rapid screening of new chelators and identifies compounds to take forward to the iron loaded *Cebus* primate model.

5.2.2.2 Iron-Overloaded Rodent Models.

Rodents have been iron overloaded by a number of different methods (183, 348–380) to provide an animal model for the study of potential iron chelators. Some of these methods are included in Table 10.6. Although a thorough description of all of these models is beyond the scope of this chapter, some of the models used more commonly are discussed in greater detail.

5.2.2.2.1 Mice. Traditionally, mice have been used to assess the acute and chronic toxicity (LD_{50}) of new drugs. In addition to this role, investigators have utilized mice extensively to assess candidate iron chelators. In many laboratories (352–357), the animals have been iron overloaded with the intraperitoneal injection of iron dextran, 2 mg of iron per week for 4 weeks. After a 2-week equilibration period, the mice were given ^{59}Fe lactoferrin intravenously through the tail vein. This regimen delivers iron preferentially to the liver. The ^{59}Fe injection was allowed to stabilize for an additional 2–3 weeks. Then, mice were given the drug of interest either orally or parenterally; the urine and feces were collected and assayed for ^{59}Fe excretion. Compounds that showed an increase in ^{59}Fe excretion relative to untreated controls could then undergo further testing in additional animal models.

5.2.2.2.2 Rats, Selective Radioiron Probes. A number of methods have been used to overload rats for the testing of iron chelators (Table 10.6). One such method involves using selective radioiron probes to label the two major storage pools available for chelation. This distinction between the storage compartments, the hepatic parenchyma or the RE system, is important. Whereas parenchymal siderosis is responsible for serious organ dysfunction, iron in RE cells is relatively innocuous.

In this model (183, 362–366), rats are hypertransfused by injections of packed red blood cells. Heat-damaged erythrocytes (^{59}Fe-DRBC) and ^{59}Fe-ferritin are prepared and administered to the rats intravenously. Greater than 80% of soluble ferritin and heat-damaged RBCs are located in the liver and spleen within 1 h of intravenous injection. In the case of the ^{59}Fe-ferritin, 97–100% is located in the parenchymal cells and 0–3% in the RE cells. In contrast, 100% of the radioactivity associated with the administration of the ^{59}Fe-DRBC was in the RE cells and 0% in the hepatic parenchymal cells. Once radiolabeling is completed, candidate chelators are administered; the reduction of RE and parenchymal radioactivity is compared to that of control animals.

5.2.2.2.3 Gerbils. The Mongolian gerbil treated with subcutaneous iron dextran has been developed as the first rodent model of iron overload that seems to reproduce both the cardiac and hepatic toxicity found in chronic iron overload in humans (377–380). Gerbils treated with weekly doses of subcutaneous iron dextran develop concentrations of hepatic iron that are in the range of those found in patients with transfusional iron overload and present with hepatic fibrosis within 1–3 months after stopping the iron dextran injections (379). In addition, cardiotoxic effects of iron in the gerbil similar to those of patients with iron-induced cardiomyopathy have also been demonstrated between 2 and 3 months post iron dextran. The pathology of the affected tissues is similar to that which occurs in end-stage disease in humans.

5.2.3 Primate Models.

Although the rat is useful as a primary screen of efficacy, activity of a chelator in this species is poorly predictive of that in humans. Consequently, two secondary primate models have been developed: the iron-overloaded *C. apella* monkey model and the iron-overloaded marmoset (*Callithrix jacchus*) model.

5.2.3.1 Cebus apella.

The approach for setting up a nonhuman primate model was based on a collaboration in the late 1980s. However, this initial work suggested problems with background noise; as a result, iron-clearance levels were far in excess of theoretical possibility (381). It was obvious that the system needed substantial refinement. Drawing on the experience gained from this work, some decisive improvements were implemented (98, 235, 236, 242, 347): the use of larger groups of both normal and iron-loaded animals; the development of improved metabolic cages and handling procedures for the experimental animals; the validation of all procedures for the trace metal analysis of samples of a controlled low iron diet, urine, and feces; the performance of iron balance studies; and a complete set of clinical analyses before and after the administration of each drug sample.

5.2.3.1.1 C. apella Iron Loading. After intramuscular anesthesia with ketamine, an intravenous infusion was started in a leg vein. Iron dextran was added to sterile normal saline and administered to the animals at a dose of 200–300 mg/kg. The iron solution was infused over 45–60 min. Two to three infusions, separated by 10–14 days, were necessary to load the monkeys to a level of 500 mg/kg of iron. This brought the serum transferrin iron saturation to 70–80%. The serum half-life of iron dextran in humans is 2.5–3 days (382). Twenty half-lives, 60 days, elapsed before any of the animals were used in iron-clearing experiments.

The iron loading produced by the infusion of iron dextran in the *C. apella* monkey at this time resembles that found in patients with transfusional iron overload and supports the suitability of this model for the evaluation of the efficacy of candidate iron-chelating agents (R. J. Bergeron, G. M. Brittenham, H. Fujioka, W. R. Weimar, and J. Wiegand, unpublished observations). The hepatic iron concentration was about 10-fold higher than that in the control animal and was well above the threshold for cardiac disease and early death found in studies of patients with thalassemia major and transfusional iron overload (117, 383). Light microscopy of liver samples using Prussian blue stain revealed that the iron load was extensive and large siderosomes were clear; electon micrographs demonstrated prominent iron deposition both in hepatocytes and in Kupffer cells. This pattern mirrors that found in chronically transfused patients.

5.2.3.1.2 C. apella Metabolic Cages. During the evaluation of various iron chelators, the animals were moved from normal primate cages to specially constructed, metal-free metabolic cages (347). The animals were housed in these cages for 7 days before exposure to the chelator of interest and throughout the course of the experiment.

5.2.3.1.3 C. apella Iron-Balance Studies. Animals were maintained on a low iron liquid diet (346) for 7 days before drug administration. The animals were given food according to their body weight. Intake was very carefully monitored. Three days before drug administration, days −2 to 0, baseline iron intake and output values were measured. This same measurement was made for days +1 through +3. The total amount of iron intake was compared with the total iron output. Net iron balance = dietary iron intake − (urinary iron + fecal iron). Animals in a negative iron balance are excreting more iron than they are absorbing.

5.2.3.1.4 C. apella Hematological Screen. Blood samples were taken from the monkeys before their being transferred to the metabolic cages, and extensive parameters were evaluated (346). The blood samples were always drawn at the same time of day because of the diurnal variability in some of the measurements (e.g., UIBC, plasma iron). The assays performed permit the assessment of the health of the animals going into the experiment and can identify subtle changes that a test drug may induce in an animal. A postdrug blood sample was taken from each animal on the last day of the experiment.

5.2.3.1.5 C. apella Fecal and Urine Sample Collection and Analysis; Efficiency Calculations. Fecal and urine samples were collected at 24-h intervals and assayed (346). Iron concentrations were measured by flame atomic absorption. The efficiency of each ligand was calculated as described earlier.

The iron-overloaded *C. apella* monkey model has been employed successfully for over a decade. The ability of the primate model to predict how a drug will perform in humans is illustrated by a comparison of data from studies of (*a*) DFO (**9**) administered subcutaneously, (*b*) the bidentate L1 (**33**) given orally, and (*c*) the hexadentate chelator HBED (**39**) given orally or subcutaneously, in rodents, primates, and humans, as shown in Fig. 10.31. The ligands were administered at the same iron-binding equivalence. Experiments in the rodents and the monkeys were carried out at doses of 150 μmol/kg for DFO and HBED and at 450 μmol/kg for L1. The data for human studies of DFO, orally administered HBED, and L1 were derived from earlier clinical trials (182, 210). Note that the human data for DFO are not comparable to those for the monkey or the rat in the strictest sense because the chelator was administered at a dose of 92 μmol/kg as a subcutaneous infusion over 8 h. This

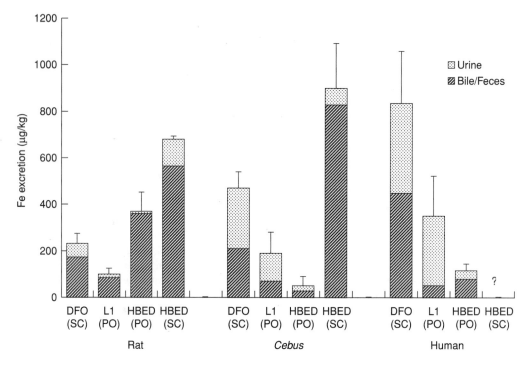

Figure 10.31. Chelator-induced iron excretion in rats, monkeys, and humans. In the animals, the ligands were administered at the same iron binding equivalence. Experiments in the rodents and the monkeys were carried out at doses of 150 μmol/kg for DFO and HBED (both oral and subcutaneous administration) and at 450 μmol/kg for L1.

mode of administration is likely the reason for the apparently greater efficiency of DFO in the human study. Patients were administered HBED orally at doses of 103 or 206 μmol/kg, although the iron outputs at either dose were within experimental error of each other and are plotted accordingly in Fig. 10.31. Patients received L1 at a daily dose of approximately 540 μmol/kg.

As shown in Fig. 10.31, data from the studies in rats would have predicted HBED (given orally or subcutaneously) to be far superior to either DFO or L1 in humans. By contrast, the results from the studies in primates suggested that both orally administered HBED and L1 should not have performed as well as DFO. The relative efficiencies of DFO, orally administered HBED, and L1 in the *C. apella* model were almost identical with those seen in humans, even to the mode of iron excretion (biliary versus urinary). Therefore, the *C. apella* primate is an excellent predictive tool of how a

chelator will perform in humans and suggests that subcutaneous HBED should be 2–3 times as efficient as DFO (240, 241) in the patients.

5.2.3.2 Marmosets (Callithrix jacchus). Since the development of the *C. apella* model, a similar methodology has been adopted by another chelator program (384–386), substituting marmosets (*Callithrix jacchus*) for the *C. apella* monkey. The overall experimental strategy with the marmoset is similar to that used with the *Cebus* monkeys.

5.2.3.2.1 Marmoset Iron Loading. The marmosets were iron overloaded by the intraperitoneal injection of iron(III) hydroxide polyisomaltose (386). The iron was injected at 14-day intervals. The iron pools were allowed to equilibrate for approximately 60 days.

5.2.3.2.2 Marmoset Metabolic Cages. The animals were placed in specially designed acrylic glass metabolic cages 48 h before drug administration and for 48 h thereafter (386). The cages were constructed such that urine

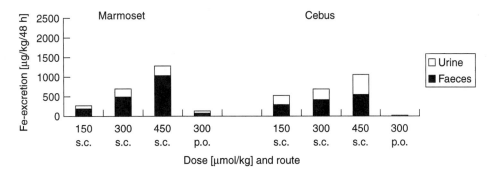

Figure 10.32. Dose-dependent urinary and fecal iron excretion in iron-overloaded marmosets and *Cebus* monkeys induced by DFO; each bar represents the mean of four to six animals. [Reprinted with permission from T. Sergejew, P. Forgiarini, and H.-P. Schnebli, *Br. J. Haematol.*, **110**, 985–992 (2001). Copyright 2001 Blackwell Science Ltd.]

and feces could be collected separately. Because of animal welfare concerns, the animals were not allowed to be in the metabolic cages for longer than 48 h postdrug (385).

5.2.3.2.3 Marmoset Fecal and Urine Samples. The animals were switched to a low iron diet (386) 7 days before the administration of the test drug and for 48 h thereafter. Urine and feces samples were collected at 24-h intervals for 2 days before drug administration and 2 days postdrug (385). Urinary iron excretion was determined colorimetrically using a bathophenanthroline method. Fecal samples were weighed and tested for occult blood. Iron content was determined by flame atomic absorption spectrometry after wet-ashing each 24-h sample.

5.2.3.3 Comparison of the C. apella and Marmoset (C. jacchus) Models. The marmoset model has been used extensively as a means to identify active iron chelators. The major advantage of the marmoset model over the established *C. apella* model is the animals' small size, 400 g versus 4 kg for the *C. apella*. This smaller size makes it possible to dose the animals without the need for anesthesia. In addition, the amount of drug needed is no more than that needed for a rat. A major disadvantage of the data derived in this protocol is that the marmosets were kept in the metabolic cages for only 2 days before and 2 days after drug administration (385). This shorter duration of sample collection made it difficult to determine both baseline iron excretion and

iron-clearing efficiency when iron was still being excreted beyond the allowed 2-day postdosing collection period.

When iron-clearance data from the two primate models are compared, the results are fairly similar (386). DFO and HBED are active upon subcutaneous administration to both species and ineffective when administered orally, the same situation that is seen in humans. In addition, when looking at three hydroxypyridinones, L1 (**33**), CP94, and CP102, both models predict that L1 would have a low activity, which is also the case in patients, and that the order of effectiveness is CP102 > CP94 > L1 (386). However, there are also some instances where the two models are quite different.

Although DFO is active in both animal models when administered subcutaneously, the mode of iron excretion is different (Fig. 10.32). In the marmoset, the majority of the induced iron is excreted in the feces, which is also the case when this drug is given subcutaneously to rats. In *C. apella*, as in humans, approximately 50% of the induced iron is excreted in the urine and the remainder in the feces. In addition, DFO is much less effective at lower doses in the marmoset than in the *Cebus* (Fig. 10.32). This may be as a result of the more rapid metabolism of DFO in marmoset plasma than in *C. apella*.

When the stability of DFO was assessed in the plasma of several different species, the drug was found to be rapidly metabolized in

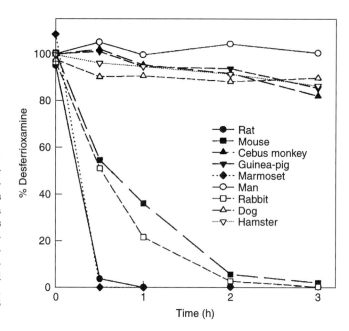

Figure 10.33. The stability of desferrioxamine in plasma from different species over 3 h. All incubations were performed on pooled plasma samples from two or more animals and averages were taken from duplicate analyses. [Reprinted with permission from A. Steward, I. Williamson, T. Madigan, A. Bretnall, and I. F. Hassan, *Br. J. Haematol.*, **95**, 654–659 (1996). Copyright 1996 Blackwell Science Ltd.]

the plasma of the rat, mouse, rabbit, and marmoset (374). In fact, the rat and the marmoset had no detectable drug remaining after only 1 h (Fig. 10.33). However, the hamster, guinea pig, dog, and *C. apella* plasma stability results were similar to what is seen in humans, that is, >80% of the parent drug remained after incubating at 37°C for 3 h (374). Finally, although dimethyl *N,N'*-bis(2-hydroxybenzyl)-ethylenediamine-*N,N'*-diacetate (dmHBED) is very active when given orally or subcutaneously to the marmosets, this compound has low activity when administered to *C. apella* (386) (Fig. 10.34) and to humans. This lack of activity in the *C. apella* model is attributed to the inability of *C. apella* to hydrolyze the diester to the active compound.

To date, the iron-loaded *C. apella* monkeys were able to qualitatively predict the effectiveness of the chelators in the human clinical studies in every case. The primates' responses were similar to the human data with regard to both the mode of iron excretion and to a compound's efficacy, thus rendering the *C. apella* monkey as an excellent secondary screen for evaluating iron chelators before initiating costly human clinical trials.

5.3 Integration of Design, Synthesis, and Testing: Desferrithiocin Analogs

In this summary of several studies of the structure-activity relationships (SARs) of the DFTs, the primary measure of activity is the efficiency of the ligand, as assessed in both rodent and primate models and compared with subcutaneously administered DFO. In the case of the DFT analogs, the efficiency calculation is based on the formation of a 2:1 complex with a formation constant assumed to be similar to that of the parent compound, $4 \times 10^{29} M^{-1}$ (56). Figures 10.35–10.42 were constructed to demonstrate how alterations in the structure of DFT changes its iron-clearing efficiency (Fig. 10.35, Figs. 10.37–10.41) or toxicity (Figs. 10.36 and 10.42). Figures 10.35 and 10.37–10.41 also include the (primate dose), (mode of administration), and the fraction of iron excreted in the [bile or stool and urine].

From the perspective of simplifying the SAR studies (Fig. 10.35), the two most important manipulations on the parent ligand (**24**) were removal of the aromatic nitrogen to yield (*S*)-4,5-dihydro-2-(2-hydroxyphenyl)-4-methyl-4-thiazolecarboxylic acid [(*S*)-desazaDFT, (**93**)] and removal of the thiazoline methyl

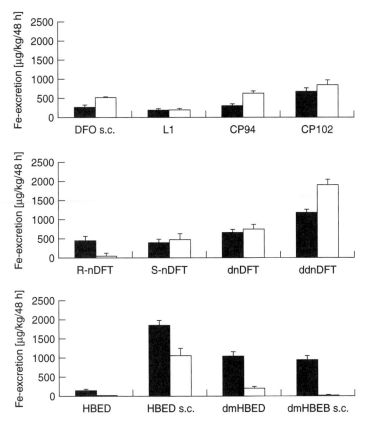

Figure 10.34. Comparison of total iron excretion in iron-overloaded marmosets (black columns) and *Cebus* monkeys (white columns). All iron chelators indicated were applied at 150 μmol iron-binding equivalents/kg; each bar represents the mean of three to six animals. [Reprinted with permission from T. Sergejew, P. Forgiarini, and H.-P. Schnebli, *Br. J. Haematol.*, **110**, 985–992 (2001). Copyright 2001 Blackwell Science Ltd.]

to produce (S)-4,5-dihydro-2-(3-hydroxy-2-pyridinyl)-4-thiazolecarboxylic acid [(S)-desmethylDFT, (**103**)]. Both compounds performed well in the primate model when given orally, better than DFO administered subcutaneously at the same iron-binding equivalence (235, 315). Surprisingly, (**93**) was an exceptional deferrating agent when administered subcutaneously (Fig. 10.35). This was consistent with the idea that neither the aromatic nitrogen nor the thiazoline methyl was requisite for iron clearance. The next modification, further underscoring the minor role played by the aromatic nitrogen and the thiazoline methyl, was formally removing the thiazoline methyl from (**93**) or the aromatic nitrogen from (**103**) to produce (S)-4,5-dihydro-2-(2-hydroxyphenyl)-4-thiazolecarboxylic acid [(S)-desazadesmethylDFT, (**104**)]. This tridentate ligand was very active when administered orally to primates (Fig. 10.35) (235).

The effects of these small structural alterations on the toxicity profile of the parent compound (**24**) were truly remarkable (Fig. 10.36). The natural product, (**24**), was a very effective iron chelator when given orally to rats or monkeys. However, the compound also elicited significant nephrotoxicity when administered chronically to rats. At a dose of 384 μmol/kg/day, five animals were dead by day 5 (Fig. 10.36). Abstraction of the methyl group from the thiazoline ring of (**24**) to yield (**103**) resulted in a compound that was far less toxic than (**24**) (235). When administered orally at a dose of 384 μmol/kg/day for 10 days, (**103**) was well tolerated, no deaths were observed, and extensive histological evaluation revealed no drug-related abnormalities.

Continuing with the toxicity SAR, it was found that removal of the aromatic nitrogen from (**24**) to produce (**93**) resulted in an analog that induced severe gastrointestinal (GI) tox-

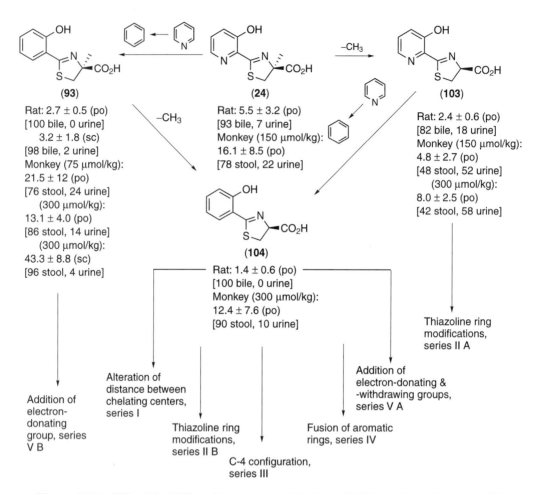

Figure 10.35. SARs of the DFTs and iron clearance. The dose of DFT or analog in the rats is 150 μmol/kg; the dose in the monkeys is as shown in parentheses for each ligand. The mode of administration is shown in parentheses next to the efficiency (%, $\pm$ SD). The fraction of iron excreted in the bile or stool and urine is shown in brackets. Series I–V can be found in sections 5.3.1.1–5.3.1.5 (235, 311–315, 388).

icity in rodents (315). The same was true with removal of both the aromatic nitrogen and the thiazoline methyl group to generate (**104**). When (**104**) was administered at a daily dose of 384 μmol/kg, dosing of this ligand was stopped after the fifth dose because of the rapidly deteriorating condition of the animals. All of the animals were dead by day 6. At necropsy, the stomachs of all of the animals were hemorrhagic and grossly distended with gas and fluid; the stomach walls were translucent. The intestines were also hemorrhagic, and pressure necrosis of the spleen, from the grossly distended stomach, was noted in two of

the animals (314). Thus, although removing the aromatic nitrogen and thiazoline methyl from the DFT framework (**104**) may be compatible with iron clearance, the profile of the system completely shifts from renal to GI toxicity.

Because of its structural simplicity the tridentate ligand (**104**) lent itself to further SAR analyses. This framework contains neither the picolinic acid fragment nor the unusual amino acid (S)-α-methyl cysteine (**90**). Consequently, these systems are much easier to assemble than (**24**), (**93**), or (**103**). Five types of structural modifications were performed on

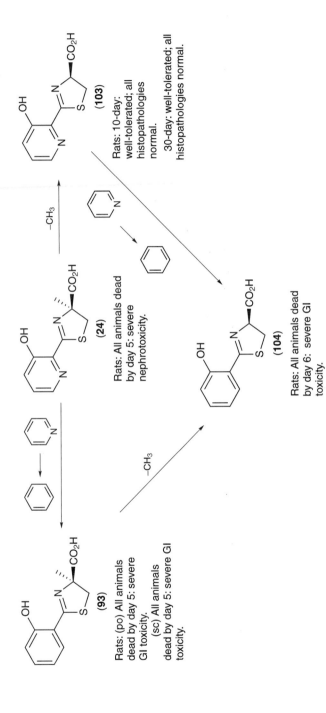

Figure 10.36. Structure-activity relationship of the DFTs and toxicity. Unless otherwise indicated in parentheses, the ligands were administered orally at a dose of 384 μmol/kg/day, equivalent to 100 mg/kg/day of the sodium salt of DFT (235, 314, 315).

(105)

Rat: ≤ 0.5 (po)
≤ 0.5 (sc)

(106)

Rat: ≤ 0.5 (po)

(107)

Rat: ≤ 0.5 (po)

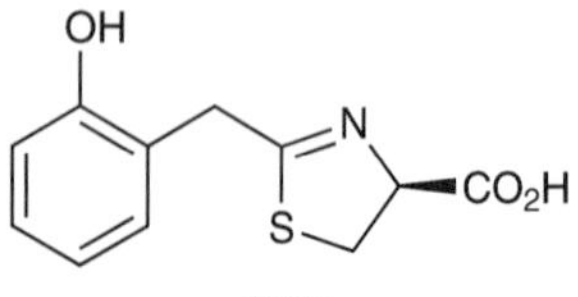

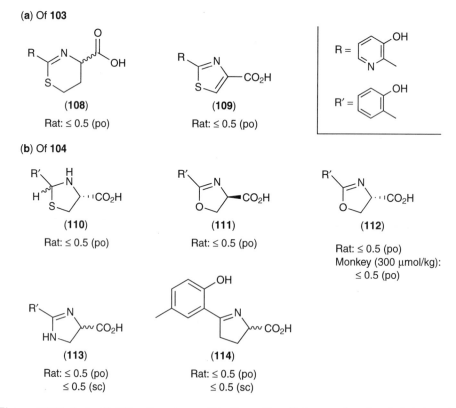

Figure 10.37. Series I: Alteration of distances between chelating centers. The efficiency and mode of administration (in parentheses) are shown.

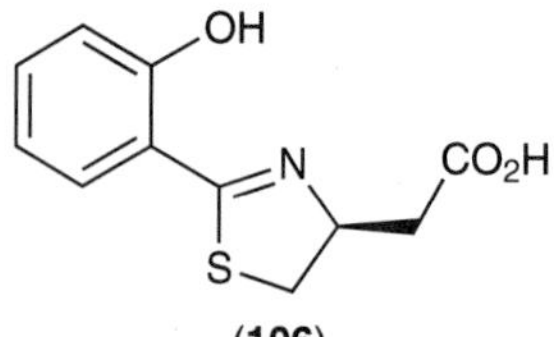

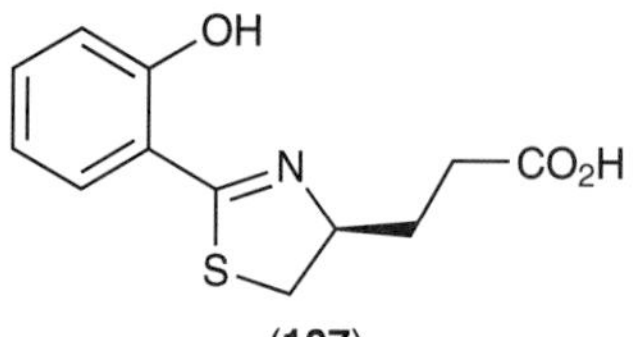

(**104**), and the effect of these changes on chelator-induced iron excretion and/or toxicity was evaluated. These included alterations in the distances between chelating (donor) centers (Section 5.3.1.1, compounds **105–107**, Fig. 10.37), thiazoline ring modification (Section 5.3.1.2, compounds **108–114**, Fig. 10.38), configurational [(*R*)- and (*S*)-] changes at C-4 (Section 5.3.1.3, compounds **115–117** versus **103**, **104**, and **118**, Fig. 10.39), benzfusion of the aromatic rings (Section 5.3.1.4, compounds **96** and **119–123**, Fig. 10.40), and addition of electron-donating and -withdrawing groups to the aromatic ring (Section 5.3.1.5, compounds **94**, **118**, and **124–126**, Fig. 10.41).

(**a**) Of **103**

(108)

Rat: ≤ 0.5 (po)

(109)

Rat: ≤ 0.5 (po)

(**b**) Of **104**

(110)

Rat: ≤ 0.5 (po)

(111)

Rat: ≤ 0.5 (po)

(112)

Rat: ≤ 0.5 (po)
Monkey (300 μmol/kg):
≤ 0.5 (po)

(113)

Rat: ≤ 0.5 (po)
≤ 0.5 (sc)

(114)

Rat: ≤ 0.5 (po)
≤ 0.5 (sc)

Figure 10.38. Series II: Thiazoline ring modifications. The efficiency and mode of administration (in parentheses) are shown.

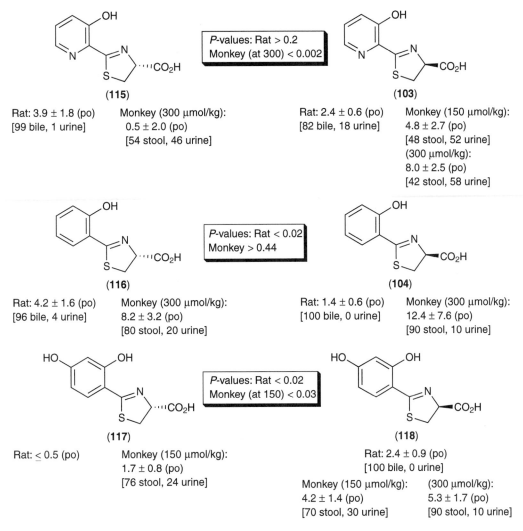

Figure 10.39. Series III: C-4 stereochemistry. The efficiency, mode of administration (in parentheses), and fraction of iron excreted in the bile or stool and urine (in brackets) are shown.

5.3.1 Iron-Clearance Evaluations

5.3.1.1 Changes In the Distances Between the Ligating Centers (Series I, Compounds 105–107). Although earlier studies indicated that the iron-clearing efficacy of desferrioxamine analogs was changed when the nature of the tether between ligating centers was modified significantly (346), relatively small alterations in the distance between centers should have little impact on the entropy of iron binding. Extension of the distance between the ligating centers involved either separating the thiazoline nitrogen-carboxyl bidentate fragment

from the phenolic donor (**105**) or separating the phenol hydroxyl-thiazoline nitrogen bidentate fragment from the carboxyl donor (**106** and **107**) (Fig. 10.37). The former elongation derived from insertion of a methylene bridge between the aromatic and thiazoline rings, the latter one from insertion of one (**106**) or two (**107**) methylenes between the thiazoline ring and carboxy terminus. In every case, these tridentate ligands (**105–107**) were no longer orally active iron chelators; even subcutaneous administration of (**105**) was a failure (314).

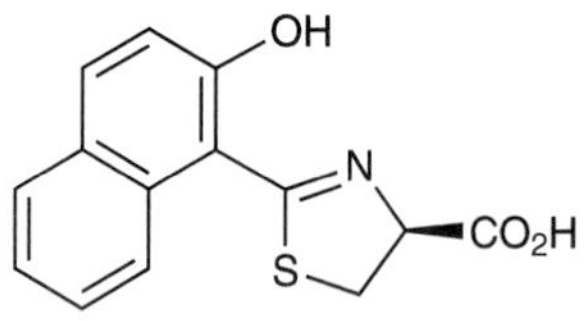

(96)

Rat: ≤ 0.5 (po)
Monkey (300 μmol/kg):
  ≤ 0.5 (po)

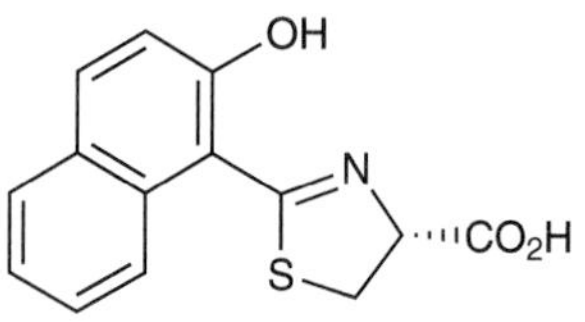

(119)

Rat: ≤ 0.5 (po)
  ≤ 0.5 (sc)
Monkey (300 μmol/kg):
  ≤ 0.5 (po)
  ≤ 0.5 (sc)

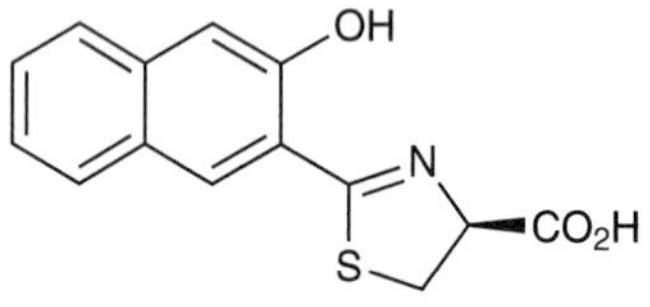

(120)

Rat: 3.7 ± 1.1 (po)
[95 bile, 5 urine]
Monkey (300 μmol/kg):
2.1 ± 0.7 (po)
[67 stool, 33 urine]
0.4 ± 0.8 (sc)

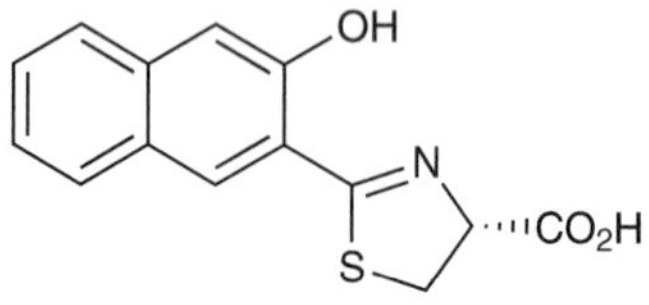

(121)

Rat: 2.9 ± 1.3 (po)
[100 bile, 0 urine]
Monkey (300 μmol/kg):
0.7 ± 0.3 (po)
[50 stool, 50 urine]
≤ 0.5 (sc)

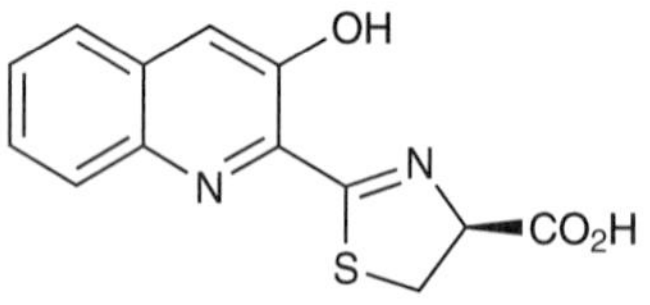

Figure 10.40. Series IV: Fusion of aromatic rings. The efficiency, mode of administration (in parentheses), and fraction of iron excreted in the bile or stool and urine (in brackets) are shown.

(122)

Rat: 5.9 ± 3.2 (po)
[90 bile, 10 urine]
Monkey (150 μmol/kg):
3.5 ± 1.8 (po)
[68 stool, 32 urine]

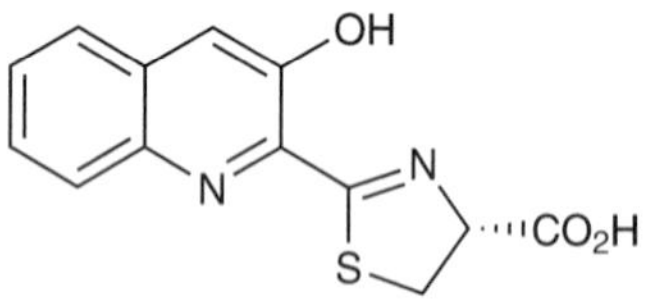

(123)

Rat: 12.3 ± 3.2 (po)
[100 bile, 0 urine]
Monkey (75 μmol/kg):
≤ 0.5 (po)

5.3.1.2 Thiazoline Ring Modification (Series II, Compounds 108–114). Expansion of the five-membered thiazoline of (103) to a six-membered Δ²-thiazine (108) voided iron-clearing activity (Fig. 10.38) (311). The same phenomenon was observed when the thiazoline of (103) was oxidized to the corresponding thiazole (109) (314) or the thiazoline of (104) was reduced to a thiazolidine (110) (311). Furthermore, when the sulfur atom in (104) was replaced with an oxygen (111 and 112) (311), a nitrogen (113) (314), or a methylene (114) (314), the compounds were not effective at

clearing iron in rodents, regardless of the mode of administration; nor was analog (112) active in primates (Fig. 10.38). Interestingly, compound (111) was absorbed quite well in primates (387), but did not promote iron excretion.

The lack of activity of (111) was particularly noteworthy, in view of the fact that this donor group is part of parabactin (4, Fig. 10.2). Interestingly, pharmacokinetic analyses of (111) in primates showed it to be well absorbed orally, achieving significant plasma levels that remained high even at 8 h, consid-

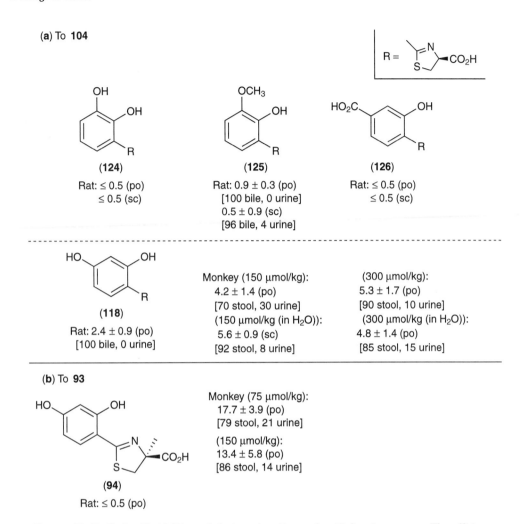

Figure 10.41. Series V: Addition of electron-donating and -withdrawing groups. The efficiency, mode of administration (in parentheses), and fraction of iron excreted in the bile or stool and urine (in brackets) are shown.

erably greater than those of (**103**). A terminal elimination phase was difficult to discern from these 0- to 8-h experiments. It was also striking that the renal clearance of (**111**) was very poor, less than 1/40 that observed for (**103**) (387). There was a considerably diminished fraction of ligand (**111**) unbound to human serum albumin versus (**103**). A high level of albumin binding may account for the lack of activity (because protein-bound ligand is unavailable for iron binding), the prolonged plasma residence time, and the poor renal clearance (387). Alternatively, it may simply

be that this ligand is promoting the excretion of iron very slowly, at a level below the limits of detection.

5.3.1.3 Configurational [(R)- and (S)-] Changes at C-4 (Series III, Compound 115 versus 103, Compound 116 versus 104, Compound 117 versus 118). The effect of changing the stereochemistry at C-4 on the iron-clearance properties of the DFT analogs is model dependent (Fig. 10.39). Although (*R*)-desmethyl-DFT (**115**) and (*R*)-desazadesmethylDFT (**116**) perform better in the rodent model than their corresponding (*S*)-enantiomers (**103** and

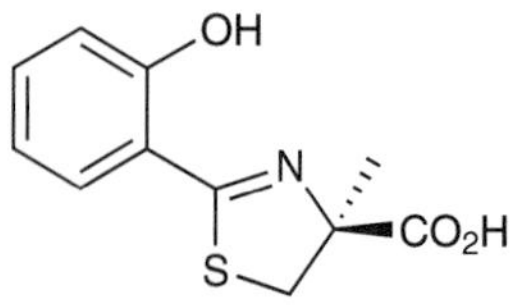

(93)

Rats: All animals dead by
day 5: severe GI toxicity.

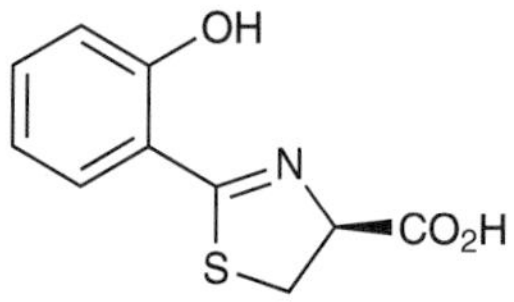

(104)

Rats: All animals dead by
day 6: severe GI toxicity.

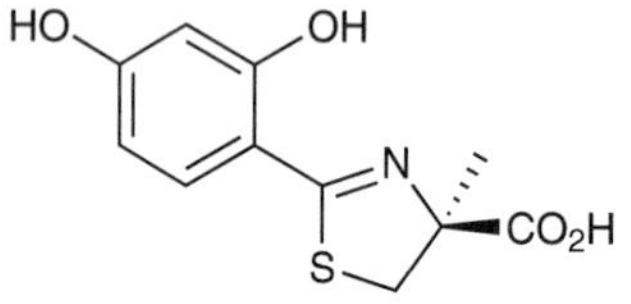

Figure 10.42. Effect of adding electron-donating groups on the toxicity of (**93**) and (**104**). The ligands were administered orally at a dose of 384 μmol/kg/day, equivalent to 100 mg/kg/day of the sodium salt of DFT.

(94)

Rats: 10-day: well-tolerated;
histopathologies normal
except for mild nephrotoxicity.

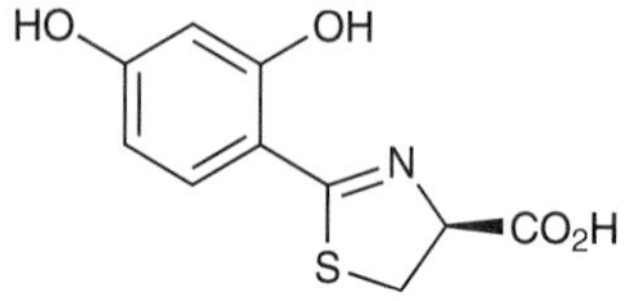

(118)

Rats: 10-day:
well-tolerated; all
histopathologies normal.

104, respectively), in the primate model, the (*S*)-enantiomer (**103**) was more efficient than its (*R*)- counterpart. However, the difference between (**116**) and (**104**) is not statistically significant (315). Thus, it seemed as though (*S*)-enantiomer (**104**), beyond the ease of synthesis of its analogs (i.e., the absence of the picolinic acid fragment), was the best pharmacophore from which to launch additional SAR studies.

This choice was underscored further by a finding in primates during a pharmacokinetic study of (*R*)- and (*S*)-desmethylDFT (**115** and **103**, respectively) (388). Surprisingly, one out of four primates given the (*R*)-enantiomer (**115**) at a dose of 300 μmol/kg died; administration of the (*S*)-enantiomer (**103**) did not elicit side effects even at a dose of 450 μmol/kg (235). Inspection of the plasma concentration-time curves for (**115**) versus (**103**) and for Fe(III)[**115**]$_2$ versus Fe(III)[**103**]$_2$ reveals several striking features. The peak plasma concentrations as well as areas under the curves (AUCs) of (**115**) and Fe(III)[**115**]$_2$ curves equaled or exceeded those of the corresponding (**103**) and Fe(III)[**103**]$_2$ curves. Eight hours posttreatment, the plasma concentrations of (**103**) and Fe(III)[**103**]$_2$ had declined

to levels approaching the limits of detection, whereas substantial plasma concentrations of (**115**) still remained. Perhaps of special significance is that in every instance Fe(III)[**115**]$_2$ in the plasma exceeded 25 mg/L (50 μM) for several hours and remained above 10 mg/L (20 μM) at 8 h. The ratios for AUC$_{0-4}$/AUC$_{0-\infty}$ are 0.89 and 0.81 for (**103**) and Fe(III)[**103**]$_2$, respectively, compared to 0.50 and 0.44 for (**115**) and Fe(III)[**115**]$_2$, reflecting the prolonged residence times of the latter enantiomer.

The most remarkable and revealing finding from the pharmacokinetic experiments was the marked enantioselectivity of the renal clearance observed with (**103**), a clearance that was *3.5 times* greater than that of (**115**), and a Fe(III)[**103**]$_2$ clearance that was *6.8 times* greater than that of Fe(III)[**115**]$_2$ (388). Thus, in the case of (**115**), unlike the situation with (**103**), a substantial amount of ferric ion is mobilized into the plasma, where it persists at very high concentrations over a prolonged period of time. Although a complete explanation of the mechanism of chiral recognition of DFT analogs is not readily apparent based on these experiments, the plasma pharmacokinetic results do provide a basis to explain the *selective toxicity* of (**115**) in the primates. The choice of (*S*)- over (*R*)-enantiomers was underscored further when the performance of (**117**) and (**118**) was compared in the primates (Fig. 10.39). Again the (*S*)-enantiomer, (**118**) in this set, turned out to be the more active ligand (315).

5.3.1.4 Benz-Fusion (Series IV, Compounds 96, 119–123).

Evaluation of the (*S*)- and (*R*)-pairs of naphthyl analogs 4,5-dihydro-2-(2-hydroxy-1-naphthalenyl)-4-thiazolecarboxylic acid (**96** and (**119**), Fig. 10.40) and 4,5-dihydro-2-(3-hydroxy-2-naphthalenyl)-4-thiazolecarboxylic acid (**120** and **121**, Fig. 10.40) demonstrated that, although the 1-(2-hydroxynaphthyl)-enantiomers (**96**) and (**119**) were not effective iron-clearing agents after oral administration to the rats, their positional isomers, the 2-(3-hydroxynaphthyl)-compounds (**120**) and (**121**), were (313). None of the naphthyl DFTs investigated was an effective iron chelator in the primates, even on subcutaneous administration. The benz-fused ligands (*S*)- and (*R*)-4,5-dihydro-2-(3-hydroxy-2-quinolinyl)-4-thiazolecarboxylic acid (**122** and **123**, respectively, Fig. 10.40) were synthesized and evaluated in an attempt to restore activity in the primate model (314). The drugs were both active in rodents upon oral administration. However, only the (*S*)-enantiomer (**122**) was efficacious in primates, and then only marginally so.

5.3.1.5 Addition of Electron-Donating and -Withdrawing Groups to the Aromatic Ring (of 104, Series V A, Compounds 118, 124–126; of 93, Series V B, Compound 94).

The above SAR provided a framework, (*S*)-desazadesmethyl-DFT, (**104**), that could be manipulated with ease. A second SAR focused on ameliorating toxicity. Because of the ease of synthesis of the desazadesmethylDFTs, their iron-clearing properties, and the facility with which toxicity can be monitored, (**104**) was an excellent platform from which to explore the construction of a nontoxic DFT analog. The impact of altering the redox properties of the aromatic ring of (**104**) on the iron-clearing properties of the compounds was assessed before any evaluations of toxicity or metabolic profiles. The redox properties were modified by introducing electron-donating or -withdrawing groups into the aromatic ring of (**104**) (compounds

118, 124–126 , Fig. 10.41). Initially, the electronic properties of the substituents were considered relative to the thiazoline ring.

Introduction of a second hydroxyl group at position 3 of the aromatic ring of (**104**) to generate (**124**) resulted in an inactive chelator in rodents. However, hydroxylation at position 4 [4'-hydroxy-(*S*)-desazadesmethylDFT, (**118**)] was a maneuver compatible with iron clearance. A methoxy was introduced in the 3 position (**125**) and a carboxy in the 4 position (**126**). In the rodent model, (**125**) was marginally effective; (**126**) was ineffective. A comparison of the 3-hydroxylated ligand (**124**) with the 3-methoxylated analog (**125**) was carried out to examine possible confounding effects by the potential auxiliary metal coordination site in (**124**). In both cases, the groups are electron-withdrawing by induction from the thiazoline nitrogen; however, the methoxyl group cannot serve as a ligand donor. When the hydroxyl group is in the 4 position (**118**), it donates electrons to the nitrogen, whereas a carboxyl group at this carbon (**126**) withdraws electrons from the nitrogen (314). Finally, 4'-hydroxylation was also carried out on (**93**) (Fig. 10.41) to yield 4'-hydroxy-(*S*)-desaza-DFT (**94**), which was also a very active iron chelator when administered orally to primates, although it was unexpectedly inactive in rodents (315). Thus, the complete story on the efficacy of the hydroxylated and methoxylated ligands must await primate trials.

5.3.2 Toxicity.

Once the alterations in the redox potential of the aromatic ring of (**93**) or (**104**) that were compatible with iron clearance were delineated, the second SAR focused on toxicity was performed. The compounds tested to date that induced iron excretion in the primate model, 4'-hydroxy-(*S*)-desazadesmethylDFT (**118**) and 4'-hydroxy-(*S*)-desaza-DFT (**94**), were examined and compared to desazademethylDFT (**104**) and desazaDFT (**93**), respectively. When an electron-donating hydroxyl group is added to position 4 of the aromatic ring of (**104**) as in (**118**), the gastrointestinal toxicity problem found in (**104**) was absent; no deaths occurred when this compound was given at a dose of 384 μmol/kg/day for 10 days (Fig. 10.42) or at a dose of 250 μmol

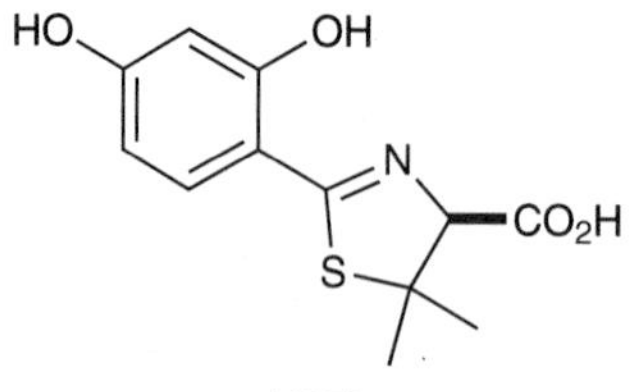

Figure 10.43. β,β-Dimethyl analog of (**118**) and (**127**) and the 5'-(OH) metabolite of (**104**) and (**128**).

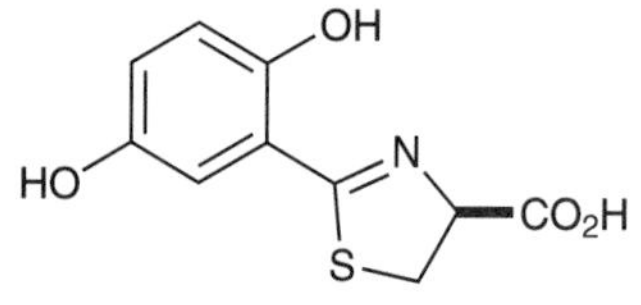

per kg/dose over 30 days. Histopathological analysis of tissues (including stomach, small intestine, large intestine, kidney, and liver) revealed no drug-related abnormalities in any of these animals (314).

When the same manipulation was performed on the aromatic ring of (**93**) to give (**94**), the gastrointestinal toxicity problem found with (**93**) was absent again; no deaths occurred when (**94**) was given orally at a dose of 384 μmol/kg/day for 10 days, greater than twice the time period for (**93**) (Fig. 10.42). Although macroscopic inspection of organs 1 day after the final dose revealed no drug-related abnormalities, histopathological analysis showed that compound (**94**) was associated with some very mild nephrotoxicity, which was considered to be reversible, but nothing comparable to what was observed with the natural product (**24**) (315).

A hydroxyl group was appended to the 4 position of the aromatic ring of a β,β-dimethyl cysteine derivative, which was known to elicit gastrointestinal toxicity (235). This manipulation was effective in ameliorating the toxic effects. Although rats given this hydroxylated analog (**127**, Fig. 10.43) at a dose of 384 μmol/kg for 10 days demonstrated renal toxicity, when analog (**127**) was given to both iron-loaded rats and rats with normal iron stores at a dose of 130 μmol/kg for 30 days, no drug-related changes in tissue pathologies were noted.

5.3.3 Metabolism of Desazadesferrithiocins. One of the principal metabolites of aspirin, gentisic acid, derives from hydroxylation at the 5 position of the aromatic ring (389, 390). To determine whether this was the fate of (**104**) and how this might affect its activity, a search for 5'-hydroxydesazadesmethylDFT (**128**, Fig. 10.43) in the urine of rats treated

with (**104**) was conducted. The putative metabolite (**128**) was synthesized by condensation of 2,5-dihydroxybenzonitrile with D-cysteine. Application of HPLC methodologies (388) for quantitation of DFTs in tissue and fluids worked well for this metabolite. Analysis of 24-h urine specimens from animals treated with (**104**) indicated that this ligand was indeed 5'-hydroxylated; furthermore, this metabolite represents about 18% of the total concentration of drug and metabolites in the urine. Neither bile nor other tissues have been examined. Interestingly, when the synthetic metabolite (**128**) was given to rodents in an iron-clearance experiment, it showed no observable clearance activity; thus, 5'-hydroxylation inactivates the drug (R.J.B., J.M., W.R.W., and J.W., unpublished observations). This raises several very interesting questions regarding other DFT analogs. Is oxidative deactivation occurring with these ligands, and to what extent? Does this deactivation correlate with iron-clearing efficiency? If so, will blocking this position by introduction of a different atom or functional group at this position increase the chelators' efficiency? Furthermore, what role does oxidative deactivation play in primates, if any? The toxicity of the metabolites remains to be evaluated thoroughly. If these metabolites are more toxic than the parent drug, this makes protection of desaza-DFTs from oxidative deactivation even more attractive.

Drug design in the arena of iron chelation therapy presents unique challenges because of the necessity of striking a balance between efficacy and toxicity resulting from excessive iron removal. However, the lessons taken from nature's chelator design strategies, modern synthetic techniques, and the use of appropriate animal models should bring about improved therapeutic agents for the treatment of both global and focal iron overload in the years to come.

6 WEB SITE ADDRESSES AND RECOMMENDED READING FOR FURTHER INFORMATION

- B. R. Byers and J. E. L. Arceneaux, *Met. Ions Biol. Syst.*, **35**, 37–66 (1998). [Bacterial iron transport; other articles in this volume address various issues relating to bacterial iron transport as well.]
- C. Ratledge and L. G. Dover, *Annu. Rev. Microbiol.*, **54**, 881–941 (2000). [Bacterial iron transport.]
- P. T. Lieu, M. Heiskala, P. A. Peterson, and Y. Yang, *Mol. Aspects Med.*, **22**, 1–87 (2001). [Mammalian iron uptake and storage— quite comprehensive.]
- M. E. Conrad, J. N. Umbreit, and E. G. Moore, *Am. J. Med. Sci.*, **318**, 213–229 (1999). [Iron uptake; transferrin.]
- P. Ponka, C. Beaumont, and D. R. Richardson, *Semin. Hematol.*, **35**, 35–54 (1998). [Overview of biochemistry and molecular biology of transferrin and ferritin.]
- K. Gross, J. Aumiller, and J. Gelzer, Eds. *Desferrioxamine (Desferal). History, Clinical Value, Perspectives*, MMV Medizin Verlag, Munich, 1992. [History of development of deferoxamine.]
- M. J. Miller, *Chem. Rev.*, **89**, 1563–1579 (1989). [Synthesis of siderophores, their analogues, and siderophore-antibiotic conjugates.]
- R. J. Bergeron and G. M. Brittenham, Eds. *The Development of Iron Chelators for Clinical Use*, CRC Press, Boca Raton, FL, 1994. [Chapters cover numerous aspects of iron biochemistry and chelation therapy.]
- N. Olivieri and G. M. Brittenham, *Blood*, **89**, 739–761 (1997). [Recommendations for chelation therapy and its management.]
- D. G. Badman, R. J. Bergeron, and G. M. Brittenham, Eds. *Iron Chelators: New Development Strategies*, Saratoga Press, Ponte Vedra Beach, FL, 2000. [Chapters cover diverse topics on iron, from Fenton chemistry to strategies for chelation therapy.]

7 ACKNOWLEDGMENTS

The research carried out in the Bergeron laboratories has been supported generously over the years by both the corporate sector and the National Institutes of Health (currently by grant R01-DK49108). We express our sincere appreciation to Dr. David G. Badman of the National Institute of Diabetes and Digestive and Kidney Diseases, NIH, for his support of our programs and many of the programs described in this review.

REFERENCES

1. S. R. Cooper, J. V. McArdle, and K. N. Raymond, *Proc. Natl. Acad. Sci. USA*, **75**, 3551–3554 (1978).
2. P. R. Ortiz de Montellano, Ed., *Cytochrome P450: Structure, Metabolism, and Biochemistry*, Plenum, New York, 1986.
3. M. Sahlin, L. Petersson, A. Graslund, A. Ehrenberg, B.-M. Sjoberg, and L. Thelander, *Biochemistry*, **26**, 5541–5548 (1987).
4. R. J. Bergeron, *Trends Biochem. Sci.*, **11**, 133–136 (1986).
5. J. B. Neilands, *J. Biol. Chem.*, **270**, 26723–26726 (1995).
6. B. R. Byers and J. E. Arceneaux, *Met. Ions Biol. Syst.*, **35**, 37–66 (1998).
7. J. B. Neilands, K. Konopka, B. Schwyn, M. Coy, R. T. Francis, B. H. Paw, and A. Bagg in G. Winkelmann, D. Van der Helm, and J. B. Neilands, Eds., *Iron Transport in Microbes, Plants, and Animals*, VCH, Weinheim, Germany, 1987, Chapter 1.
8. K. N. Raymond and C. J. Carrano, *Acc. Chem. Res.*, **12**, 183–190 (1979).
9. T. P. Tufano and K. N. Raymond, *J. Am. Chem. Soc.*, **103**, 6617–6624 (1981).
10. S. A. Leong and G. Winkelmann, *Met. Ions Biol. Syst.*, **35**, 147–186 (1998).
11. J. B. Neilands, *Bacteriol. Rev.*, **21**, 101–111 (1957).
12. A. Avdeef, S. R. Sofen, T. L. Bregante, and K. N. Raymond, *J. Am. Chem. Soc.*, **100**, 5362–5370 (1978).
13. W. R. Harris, C. J. Carrano, S. R. Cooper, S. R. Sofen, A. E. Avdeef, J. V. McArdle, and K. N. Raymond, *J. Am. Chem. Soc.*, **101**, 6097–6104 (1979).
14. W. R. Harris, C. J. Carrano, and K. N. Raymond, *J. Am. Chem. Soc.*, **101**, 2213–2214 (1979).

15. A. E. Martell, R. J. Motekaitis, Y. Sun, and E. T. Clarke in R. J. Bergeron and G. M. Brittenham, Eds., *The Development of Iron Chelators for Clinical Use*, CRC Press, Boca Raton, FL, 1994, pp. 329–351.

16. A. Avdeef, J. V. McArdle, S. R. Sofen, and K. N. Raymond in Proceedings of the 175th National Meeting of the American Chemical Society, Washington, DC, 1978, p. INOR 085.

17. J. B. Neilands, T. Peterson, and S. A. Leong in A. E. Martell, Ed., *Inorganic Chemistry in Biology and Medicine*, Vol. **140**, American Chemical Society, Washington, DC, 1980, pp. 263–278.

18. R. J. Motekaitis and A. E. Martell, *Inorg. Chim. Acta*, **183**, 71–80 (1991).

19. E. T. Clarke and A. E. Martell, *Inorg. Chim. Acta*, **191**, 56–63 (1992).

20. F. L'Eplattenier, I. Murase, and A. E. Martell, *J. Am. Chem. Soc.*, **89**, 837–843 (1967).

21. G. Anderegg, F. L'Eplattenier, and G. Schwarzenbach, *Helv. Chim. Acta*, **46**, 1400–1408 (1963).

22. A. E. Martell and R. Smith, *Critical Stability Constants, Vol. 3, Other Organic Ligands*, Plenum, New York, 1976.

23. P. Ponka in R. J. Bergeron and G. M. Brittenham, Eds., *The Development of Iron Chelators for Clinical Use*, CRC Press, Boca Raton, FL, 1994, pp. 1–29.

24. P. Ponka, C. Beaumont, and D. R. Richardson, *Semin. Hematol.*, **35**, 35–54 (1998).

25. T. Schupp, U. Waldmeier, and M. Divers, *FEMS Microbiol. Lett.*, **42**, 135–139 (1987).

26. G. H. Tait, *Biochem. J.*, **146**, 191–204 (1975).

27. T. Peterson and J. B. Neilands, *Tetrahedron Lett.*, 4805–4808 (1979).

28. T. Peterson, K.-E. Falk, S. A. Leong, M. P. Klein, and J. B. Neilands, *J. Am. Chem. Soc.*, **102**, 7715–7718 (1980).

29. S. Yamamoto, N. Okujo, Y. Fujita, M. Saito, T. Yoshida, and S. Shinoda, *J. Biochem. (Tokyo)*, **113**, 538–544 (1993).

30. N. Okujo, M. Saito, S. Yamamoto, T. Yoshida, S. Miyoshi, and S. Shinoda, *BioMetals*, **7**, 109–116 (1994).

31. G. L. Griffiths, S. P. Sigel, S. M. Payne, and J. B. Neilands, *J. Biol. Chem.*, **259**, 383–385 (1984).

32. G. Winkelmann, Ed., *CRC Handbook of Microbial Iron Chelates*, CRC Press, Boca Raton, FL, 1991, p. 361.

33. M. D. Surman and M. J. Miller, *Org. Lett.*, **3**, 519–521 (2001).

34. L. De Luca, G. Giacomelli, and M. Taddei, *J. Org. Chem.*, **66**, 2534–2537 (2001).

35. H. Bickel, G. E. Hall, W. Keller-Schierlein, V. Prelog, E. Vischer, and A. Wettstein, *Helv. Chim. Acta*, **43**, 2129–2138 (1960).

36. W. Keller-Schierlein and V. Prelog, *Helv. Chim. Acta*, **45**, 590–595 (1962).

37. H. Bickel, R. Bosshardt, E. Gäumann, P. Reusser, E. Vischer, W. Voser, A. Wettstein, and H. Zähner, *Helv. Chim. Acta*, **43**, 2118–2128 (1960).

38. T. Kameyama, A. Takahashi, S. Kurasawa, M. Ishizuka, Y. Okami, T. Takeuchi, and H. Umezawa, *J. Antibiot.*, **40**, 1664–1670 (1987).

39. W. D. Linke, A. Crueger, and H. Diekmann, *Arch. Mikrobiol.*, **85**, 44–50 (1972).

40. F. Gibson and D. I. Magrath, *Biochim. Biophys. Acta*, **192**, 175–184 (1969).

41. G. A. Snow, *Bacteriol. Rev.*, **34**, 99–125 (1970).

42. B. Kunze, W. Trowitzsch-Kienast, G. Höfle, and H. Reichenbach, *J. Antibiot.*, **45**, 147–150 (1992).

43. W. Keller-Schierlein, V. Prelog, and H. Zähner, *Fortschr. Chem. Org. Naturst.*, **22**, 279–322 (1964).

44. C. L. Atkin and J. B. Neilands, *Biochemistry*, **7**, 3734–3739 (1968).

45. K. B. Mullis, J. R. Pollack, and J. B. Neilands, *Biochemistry*, **10**, 4894–4898 (1971).

46. T. Nishio, N. Tanaka, J. Hiratake, Y. Katsube, Y. Ishida, and J. Oda, *J. Am. Chem. Soc.*, **110**, 8733–8734 (1988).

47. A. E. Martell, W. F. Anderson, and D. G. Badman, Eds., *Development of Iron Chelators for Clinical Use*, Elsevier/North-Holland, New York, 1980.

48. G. Anderegg, F. L'Eplattenier, and G. Schwarzenbach, *Helv. Chim. Acta*, **46**, 1409–1422 (1963).

49. M. J. Smith, *Tetrahedron Lett.*, **30**, 313–316 (1989).

50. H. Dreschel, J. Metzger, S. Freund, G. Jung, J. Boelaert, and G. Winkelmann, *Biol. Met.*, **4**, 238–243 (1991).

51. C. D. Cox, K. L. Rinehart Jr., M. L. Moore, and J. C. Cook Jr., *Proc. Natl. Acad. Sci. USA*, **78**, 4256–4260 (1981).

52. K. L. Rinehart, A. L. Staley, S. R. Wilson, R. G. Ankenbauer, and C. D. Cox, *J. Org. Chem.*, **60**, 2786–2791 (1995).

53. A. Zamri and M. A. Abdallah, *Tetrahedron*, **56**, 249–256 (2000).

54. H.-U. Naegeli and H. Zähner, *Helv. Chim. Acta*, **63**, 1400–1406 (1980).

55. F. E. Hahn, T. J. McMurry, A. Hugi, and K. N. Raymond, *J. Am. Chem. Soc.*, **112**, 1854–1860 (1990).

56. G. Anderegg and M. Räber, *J. Chem. Soc. Chem. Commun.*, 1194–1196 (1990).

57. R. J. Bergeron, J. B. Dionis, G. T. Elliott, and S. J. Kline, *J. Biol. Chem.*, **260**, 7936–7944 (1985).

58. R. J. Bergeron, W. R. Weimar, and J. B. Dionis, *J. Bacteriol.*, **170**, 3711–3717 (1988).

59. R. J. Bergeron and W. R. Weimar, *J. Bacteriol.*, **172**, 2650–2657 (1990).

60. G. E. Frost and H. Rosenberg, *Biochim. Biophys. Acta*, **330**, 90–101 (1973).

61. R. S. Negrin and J. B. Neilands, *J. Biol. Chem.*, **253**, 2339–2342 (1978).

62. C. C. Wang and A. Newton, *J. Biol. Chem.*, **246**, 2147–2151 (1971).

63. H. Huschka, H. U. Naegeli, H. Leuenberger-Ryf, W. Keller-Schierlein, and G. Winkelmann, *J. Bacteriol.*, **162**, 715–721 (1985).

64. G. Winkelmann, *Arch. Mikrobiol.*, **98**, 39–50 (1974).

65. P. J. Lammers and J. Sanders-Loehr, *J. Bacteriol.*, **151**, 288–294 (1982).

66. R. J. Bergeron, M. G. Xin, W. R. Weimar, R. E. Smith, and J. Wiegand, *J. Med. Chem.*, **44**, 2469–2478 (2001).

67. M. Erecinska, C. J. Deutsch, and J. S. Davis, *J. Biol. Chem.*, **256**, 278–284 (1981).

68. P. R. Findell, S. M. Torkelson, J. C. Craig, and R. I. Weiner, *Mol. Pharmacol.*, **33**, 78–83 (1988).

69. G. D. Holman, A. L. Busza, E. J. Pierce, and W. D. Rees, *Biochim. Biophys. Acta*, **649**, 503–514 (1981).

70. C. J. Carrano and K. N. Raymond, *J. Am. Chem. Soc.*, **100**, 5371–5374 (1978).

71. S. S. Isied, G. Kuo, and K. N. Raymond, *J. Am. Chem. Soc.*, **98**, 1763–1767 (1976).

72. K. N. Raymond, K. Abu-Dari, and S. R. Sofen, *ACS Symp. Ser.*, **119**, 133–167 (1980).

73. R. J. Bergeron in G. Winklemann, D. Van der Helm, and J. B. Neilands, Eds., *Iron Transport in Microbes, Plants, and Animals*, VCH Verlagsgesellschaft, Munich, 1987, pp. 285–315.

74. J. P. Robinson and J. V. McArdle, *J. Inorg. Nucl. Chem.*, **43**, 1951–1953 (1981).

75. J. E. Arceneaux, W. B. Davis, D. N. Downer, A. H. Haydon, and B. R. Byers, *J. Bacteriol.*, **115**, 919–927 (1973).

76. D. J. Ecker and T. Emery, *J. Bacteriol.*, **155**, 616–622 (1983).

77. G. Müller and K. N. Raymond, *J. Bacteriol.*, **160**, 304–312 (1984).

78. G. Winkelmann, *FEBS Lett.*, **97**, 43–46 (1979).

79. S. Wee, S. Hardesty, M. V. V. S. Madiraju, and B. J. Wilkinson, *FEMS Microbiol. Lett.*, **51**, 33–36 (1988).

80. V. Braun, K. Hantke, and W. Koster, *Met. Ions Biol. Syst.*, **35**, 67–145 (1998).

81. C. Ratledge and L. G. Dover, *Annu. Rev. Microbiol.*, **54**, 881–941 (2000).

82. K. A. McCready and C. Ratledge, *J. Gen. Microbiol.*, **113**, 67–72 (1979).

83. E. Lesuisse and P. Labbe, *J. Gen. Microbiol.*, **135**, 257–263 (1989).

84. C. Ratledge, P. V. Patel, and J. Mundy, *J. Gen. Microbiol.*, **128**, 1559–1565 (1982).

85. A. Stintzi, C. Barnes, J. Xu, and K. N. Raymond, *Proc. Natl. Acad. Sci. USA*, **97**, 10691–10696 (2000).

86. M. J. Miller and F. Malouin in R. J. Bergeron and G. M. Brittenham, Eds., *The Development of Iron Chelators for Clinical Use*, CRC Press, Boca Raton, FL, 1994, pp. 275–306.

87. M. J. Miller, *Chem. Rev.*, **89**, 1563–1579 (1989).

88. R. J. Bergeron, G. T. Elliott, S. J. Kline, R. Ramphal, and L. St. James III, *Antimicrob. Agents Chemother.*, **24**, 725–730 (1983).

89. H. Drechsel, S. Freund, G. Nicholson, H. Haag, O. Jung, H. Zähner, and G. Jung, *BioMetals*, **6**, 185–192 (1993).

90. H. Haag, H. P. Fiedler, J. Meiwes, H. Drechsel, G. Jung, and H. Zähner, *FEMS Microbiol. Lett.*, **115**, 125–130 (1994).

91. A. Ismail, G. W. Bedell, and D. M. Lupan, *Biochem. Biophys. Res. Commun.*, **130**, 885–891 (1985).

92. S. K. Guterman, P. M. Morris, and W. J. K. Tannenberg, *Gen. Pharmacol.*, **9**, 123–127 (1978).

93. O. Nouel, P. M. Voisin, J. Vaucel, M. Dartois Hoguin, and M. Le Bris, *Presse Med.*, **20**, 1494–1496 (1991).

94. M. H. Kirking, *Clin. Pharm.*, **10**, 775–783 (1991).

95. S. D. B. Carson, P. E. Klebba, S. M. C. Newton, and P. F. Sparling, *J. Bacteriol.*, **181**, 2895–2901 (1999).

96. V. Coulanges, P. Andre, and D. J.-M. Vidon, *Biochem. Biophys. Res. Commun.*, **249**, 526–530 (1998).

97. J. B. Neilands, T. J. Erickson, and W. H. Rastetter, *J. Biol. Chem.*, **256**, 3831–3832 (1981).

98. R. J. Bergeron, R. R. Streiff, W. King, R. D. Daniels Jr. and J. Wiegand, *Blood*, **82**, 2552–2557 (1993).

99. P. T. Lieu, M. Heiskala, P. A. Peterson, and Y. Yang, *Mol. Asp. Med.*, **22**, 1–87 (2001).

100. S. Bailey, R. W. Evans, R. C. Garratt, B. Gorinsky, S. Hasnain, C. Horsburgh, H. Jhoti, P. F. Lindley, A. Mydin, R. Sarra, and J. L. Watson, *Biochemistry*, **27**, 5804–5812 (1988).

101. G. J. Kontoghiorghes, *Lancet*, **1**, 817 (1985).

102. C. M. Lawrence, S. Ray, M. Babyonyshev, R. Galluser, D. W. Borhani, and S. C. Harrison, *Science*, **286**, 779–782 (1999).

103. M. E. Conrad, J. N. Umbreit, and E. G. Moore, *Am. J. Med. Sci.*, **318**, 213–229 (1999).

104. P. M. Harrison and P. Arosio, *Biochim. Biophys. Acta*, **1275**, 161–203 (1996).

105. E. C. Theil and B. H. Huynh, *J. Inorg. Biochem.*, **67**, 30 (1997).

106. J. P. Basilion, T. A. Rouault, C. M. Massinople, R. D. Klausner, and W. H. Burgess, *Proc. Natl. Acad. Sci. USA*, **91**, 574–578 (1994).

107. O. Han, M. L. Failla, A. D. Hill, E. R. Morris, and J. C. Smith Jr., *J. Nutr.*, **125**, 1291–1299 (1995).

108. M. T. Nunez, X. Alvarez, M. Smith, V. Tapia, and J. Glass, *Am. J. Physiol. Cell Physiol.*, **267**, C1582–C1588 (1994).

109. C. F. Babbs, *Free Radic. Biol. Med.*, **13**, 169–181 (1992).

110. M. B. Grisham and D. N. Granger, *Dig. Dis. Sci.*, **33**, 6S–15S (1988).

111. B. E. Britigan, S. Pou, G. M. Rosen, D. M. Lilleg, and G. R. Buettner, *J. Biol. Chem.*, **265**, 17533–17538 (1990).

112. E. Graf, J. R. Mahoney, R. G. Bryant, and J. W. Eaton, *J. Biol. Chem.*, **259**, 3620–3624 (1984).

113. B. Halliwell, *Nutr. Rev.*, **52**, 253–265 (1994).

114. B. Halliwell in R. J. Bergeron and G. M. Brittenham, Eds., *The Development of Iron Chelators for Clinical Use*, CRC Press, Boca Raton, FL, 1994, pp. 33–56.

115. W. Koppenol in D. G. Badman, R. J. Bergeron, and G. M. Brittenham, Eds., *Iron Chelators: New Development Strategies*, Saratoga Press, Ponte Vedra Beach, FL, 2000, pp. 3–10.

116. R. T. Dean and P. Nicholson, *Free Radic. Res.*, **20**, 83–101 (1994).

117. N. F. Olivieri and G. M. Brittenham, *Blood*, **89**, 739–761 (1997).

118. E. P. Vichinsky, *Semin. Hematol.*, **38**, 14–22 (2001).

119. M. J. Kersten, R. Lange, M. E. Smeets, G. Vreugdenhil, K. J. Roozendaal, W. Lameijer, and R. Goudsmit, *Ann. Hematol.*, **73**, 247–252 (1996).

120. C. R. Cornett, W. R. Markesbery, and W. D. Ehmann, *Neurotoxicology*, **19**, 339–345 (1998).

121. K. A. Jellinger, *Drugs Aging*, **14**, 115–140 (1999).

122. E. Becker and D. R. Richardson, *Int. J. Biochem. Cell Biol.*, **33**, 1–10 (2001).

123. F. Palau, *Int. J. Mol. Med.*, **7**, 581–589 (2001).

124. J. Adamec, F. Rusnak, W. G. Owen, S. Naylor, L. M. Benson, A. M. Gacy, and G. Isaya, *Am. J. Hum. Genet.*, **67**, 549–562 (2000).

125. X. Huang, C. S. Atwood, M. A. Hartshorn, G. Multhaup, L. E. Goldstein, R. C. Scarpa, M. P. Cuajungco, D. N. Gray, J. Lim, R. D. Moir, R. E. Tanzi, and A. I. Bush, *Biochemistry*, **38**, 7609–7616 (1999).

126. Y. Sagara, S. Tan, P. Maher, and D. Schubert, *Biofactors*, **8**, 45–50 (1998).

127. D. Schubert and M. Chevion, *Biochem. Biophys. Res. Commun.*, **216**, 702–707 (1995).

128. T. LaVaute, S. Smith, S. Cooperman, K. Iwai, W. Land, E. Meyron-Holtz, S. K. Drake, G. Miller, M. Abu-Asab, M. Tsokos, R. Switzer III, A. Grinberg, P. Love, N. Tresser, and T. A. Rouault, *Nat. Genet.*, **27**, 209–214 (2001).

129. R. E. Tanzi and B. E. Hyman, *Nature*, **350**, 564 (1991).

130. J. T. Rogers, L. M. Leiter, J. McPhee, C. M. Cahill, S. S. Zhan, H. Potter, and L. N. Nilsson, *J. Biol. Chem.*, **274**, 6421–6431 (1999).

131. D. G. Graham, *Mol. Pharmacol.*, **14**, 633–643 (1978).

132. P. Riederer and M. B. Youdim, *J. Neurochem.*, **46**, 1359–1365 (1986).

133. P. Riederer, E. Sofic, W. D. Rausch, B. Schmidt, G. P. Reynolds, K. Jellinger, and M. B. Youdim, *J. Neurochem.*, **52**, 515–520 (1989).

134. T. L. Perry, D. V. Godin, and S. Hansen, *Neurosci. Lett.*, **33**, 305–310 (1982).

135. R. J. Marttila, H. Lorentz, and U. K. Rinne, *J. Neurol. Sci.*, **86**, 321–331 (1988).

136. L. M. Ambani, M. H. Van Woert, and S. Murphy, *Arch. Neurol.*, **32**, 114–118 (1975).

137. J. Sian, D. T. Dexter, A. J. Lees, S. Daniel, Y. Agid, F. Javoy-Agid, P. Jenner, and C. D. Marsden, *Ann. Neurol.*, **36**, 348–355 (1994).

138. J. B. Schulz, J. Lindenau, J. Seyfried, and J. Dichgans, *Eur. J. Biochem.*, **267**, 4904–4911 (2000).

139. A. Kastner, E. C. Hirsch, O. Lejeune, F. Javoy-Agid, O. Rascol, and Y. Agid, *J. Neurochem.*, **59**, 1080–1089 (1992).

140. L. Lopiano, M. Chiesa, G. Digilio, S. Giraudo, B. Bergamasco, E. Torre, and M. Fasano, *Biochim. Biophys. Acta*, **1500**, 306–312 (2000).

141. K. L. Double, P. Riederer, and M. Gerlach, *Drug News Perspect.*, **12**, 333–340 (1999).

142. D. Ben-Shachar, G. Eshel, J. P. Finberg, and M. B. Youdim, *J. Neurochem.*, **56**, 1441–1444 (1991).

143. D. Ben-Shachar, G. Eshel, P. Riederer, and M. B. Youdim, *Ann. Neurol.*, **32**, S105–S110 (1992).

144. D. R. Richardson, C. Mouralian, P. Ponka, and E. Becker, *Biochim. Biophys. Acta*, **1536**, 133–140 (2001).

145. G. M. Brittenham, *Ann. N. Y. Acad. Sci.*, **612**, 315–326 (1990).

146. A. Bel, E. Martinod, and P. Menasche, *Acta Haematol.*, **95**, 63–65 (1996).

147. C. F. Whitten, G. W. Gibson, M. H. Good, J. F. Goodwin, and A. J. Brough, *Pediatrics*, **36**, 322–335 (1965).

148. A. Evers, R. D. Hancock, A. E. Martell, and R. J. Motekaitis, *Inorg. Chem.*, **28**, 2189–2195 (1989).

149. R. A. Yokel, P. Ackrill, E. Burgess, J. P. Day, J. L. Domingo, T. P. Flaten, and J. Savory, *J. Toxicol. Environ. Health*, **48**, 667–683 (1996).

150. A. E. Martell and R. Smith, *Critical Stability Constants, Vol. 1, Amino Acids*, Plenum, New York, 1974.

151. E. Farkas, H. Csoka, G. Micera, and A. Dessi, *J. Inorg. Biochem.*, **65**, 281–286 (1997).

152. K. S. Rajan, S. Mainer, N. L. Rajan, and J. M. Davis, *J. Inorg. Biochem.*, **14**, 339–350 (1981).

153. W. R. Harris and A. E. Martell, *Inorg. Chem.*, **15**, 713–720 (1976).

154. H. H. Peter in M. Aksoy and G. F. B. Birdwood, Eds., *Hypertransfusion and Iron Chelation in Thalassemia*, Hans Huber, Berne, 1985, pp. 69–81.

155. R. S. Smith, *Br. Med. J.*, **2**, 1577–1580 (1962).

156. F. Wöhler, *Acta Haematol.*, **30**, 65–87 (1963).

157. R. D. Propper, S. B. Shurin, and D. G. Nathan, *N. Engl. J. Med.*, **294**, 1421–1423 (1976).

158. A. V. Hoffbrand and B. Wonke, *Baillieres Clin. Haematol.*, **2**, 345–362 (1989).

159. M. J. Pippard, *Baillieres Clin. Haematol.*, **2**, 323–343 (1989).

160. M. Tenenbein, S. Kowalski, A. Sienko, D. H. Bowden, and I. Y. R. Adamson, *Lancet*, **339**, 699–701 (1992).

161. M. H. Freedman, D. Grisaru, N. Olivieri, I. MacLusky, and P. S. Thorner, *Am. J. Dis. Child.*, **144**, 565–569 (1990).

162. A. S. Ioannides and J. M. Pansiello, *Eur. J. Pediatr.*, **159**, 158–159 (2000).

163. S. C. Davies, R. E. Marcus, J. L. Hungerford, M. H. Miller, G. B. Arden, and E. R. Huehns, *Lancet*, **2**, 181–184 (1983).

164. N. F. Olivieri, J. R. Buncic, E. Chew, T. Gallant, R. V. Harrison, N. Keenan, W. Logan, D. Mitchell, G. Ricci, B. Skarf, M. Taylor, and M. H. Freedman, *N. Engl. J. Med.*, **314**, 869–873 (1986).

165. J. B. Porter, M. S. Jaswon, E. R. Huehns, C. A. East, and J. W. Hazell, *Br. J. Haematol.*, **73**, 403–409 (1989).

166. P. W. Brill, P. Winchester, P. J. Giardina, and S. Cunningham-Rundles, *Am. J. Roentgenol.*, **156**, 561–565 (1991).

167. S. DeVirgiliis, M. Congia, F. Frau, F. Argiolu, F. Cucca, A. Varsi, G. Sanna, G. Podda, M. Fodde, G. Franco-Piratu, and A. Cao, *J. Pediatr.*, **113**, 661–669 (1988).

168. M. J. Pippard in R. J. Bergeron and G. M. Brittenham, Eds., *The Development of Iron Chelators for Clinical Use*, CRC Press, Boca Raton, FL, 1994, pp. 57–74.

169. Y. Bentur, M. McGuigan, and G. Koren, *Drug Saf.*, **6**, 37–46 (1991).

170. D. L. Baker and C. S. Manno, *Am. J. Hematol.*, **29**, 230–232 (1988).

171. A. W. Nienhuis, *N. Engl. J. Med.*, **304**, 170–171 (1981).

172. C. Hershko and D. J. Weatherall, *Crit. Rev. Clin. Lab. Sci.*, **26**, 303–345 (1988).

173. A. W. Nienhuis, *N. Engl. J. Med.*, **296**, 114 (1976).

174. H. Grusin and P. S. Kincaid-Smith, *Am. J. Clin. Nutr.*, **2**, 323–335 (1954).

175. E. J. Schulz and H. Swanepoel, *S. Afr. Med. J.*, **36**, 367–372 (1962).

176. A. Athanasiou, M. A. Shepp, and T. F. Necheles, *Lancet*, **2**, 616 (1977).

177. M. Shalit, A. Tedeschi, A. Miadonna, and F. Levi-Schaffer, *J. Allergy Clin. Immunol.*, **88**, 854–860 (1991).

178. J. Bousquet, M. Navarro, G. Robert, P. Aye, and F. B. Michel, *Lancet*, **2**, 859–860 (1983).

179. K. B. Miller, L. J. Rosenwasser, J. A. M. Bessette, D. J. Beer, and R. E. Rocklin, *Lancet*, **1**, 1059 (1981).

180. M. R. Summers, A. Jacobs, D. Tudway, P. Perera, and C. Ricketts, *Br. J. Haematol.*, **42**, 547–555 (1979).

181. P. Allain, D. Chaleil, Y. Mauras, M. C. Varin, K. S. Ang, G. Cam, and P. Simon, *Clin. Chim. Acta*, **173**, 313–316 (1988).

182. R. W. Grady, A. D. Salbe, M. W. Hilgartner, and P. J. Giardina in R. J. Bergeron and G. M. Brittenham, Eds., *The Development of Iron Chelators for Clinical Use*, CRC Press, Boca Raton, FL, 1994, pp. 395–406.

183. C. Hershko, R. W. Grady, and A. Cerami, *J. Lab. Clin. Med.*, **92**, 144–151 (1978).

184. P. Lee, N. Mohammed, L. Marshall, R. D. Abeysinghe, R. C. Hider, J. B. Porter, and S. Singh, *Drug Metab. Dispos.*, **21**, 640–644 (1993).

185. S. Singh, R. C. Hider, and J. B. Porter, *Anal. Biochem.*, **187**, 212–219 (1990).

186. H. Keberle, *Ann. N. Y. Acad. Sci.*, **119**, 758–768 (1964).

187. P. Allain, Y. Mauras, D. Chaleil, P. Simon, K. S. Ang, G. Cam, L. LeMignon, and M. Simon, *Br. J. Clin. Pharmacol.*, **24**, 207–212 (1987).

188. A. A. Wapnick, S. R. Lynch, R. W. Charlton, H. C. Seftel, and T. H. Bothwell, *Br. J. Haematol.*, **17**, 563–568 (1969).

189. R. T. O'Brien, *Ann. N. Y. Acad. Sci.*, **232**, 221–225 (1974).

190. R. M. Bannerman, S. T. Callender, and D. L. Williams, *Br. Med. J.*, **2**, 1573–1577 (1962).

191. J. L. Fahey, C. E. Rath, J. V. Princiotto, I. B. Brick, and M. Rubin, *J. Lab. Clin. Med.*, **57**, 436–449 (1961).

192. M. J. Pippard, M. J. Jackson, K. Hoffman, M. Petrou, and C. B. Modell, *Scand. J. Haematol.*, **36**, 466–472 (1986).

193. B. Wonke, A. V. Hoffbrand, M. Aldouri, D. Wickens, D. Flynn, M. Stearns, and P. Warner, *Arch. Dis. Child.*, **64**, 77–82 (1989).

194. G. S. Tilbrook and R. C. Hider, *Met. Ions Biol. Syst.*, **35**, 691–730 (1998).

195. E. Stevens, B. Rosoff, M. Weiner, and H. Spencer, *Proc. Soc. Exp. Biol. Med.*, **111**, 235–238 (1962).

196. J. W. Stather, H. Smith, M. R. Bailey, A. Birchall, R. A. Bulman, and F. E. H. Crawley, *Health Phys.*, **44**, 45–52 (1983).

197. R. E. Goans, R. C. Ricks, and R. D. Townsend, Oak Ridge Associated Universities, Oak Ridge, TN, 1999.

198. F. Havlicek, F. Bohne, and H. Zorn, *Strahlentherapie*, **136**, 604–608 (1968).

199. F. N. al Refaie, C. Hershko, A. V. Hoffbrand, M. Kosaryan, N. F. Olivieri, P. Tondury, and B. Wonke, *Br. J. Haematol.*, **91**, 224–229 (1995).

200. N. F. Olivieri, G. M. Brittenham, D. Matsui, M. Berkovitch, L. M. Blendis, R. G. Cameron, R. A. McClelland, P. P. Liu, D. M. Templeton, and G. Koren, *N. Engl. J. Med.*, **332**, 918–922 (1995).

201. F. N. al Refaie, B. Wonke, A. V. Hoffbrand, D. G. Wickens, P. Nortey, and G. J. Kontoghiorghes, *Blood*, **80**, 593–599 (1992).

202. G. J. Kontoghiorghes, *Ind. J. Pediatr.*, **60**, 485–507 (1993).

203. M. Berkovitch, R. M. Laxer, R. Inman, G. Koren, K. P. Pritzker, M. J. Fritzler, and N. F. Olivieri, *Lancet*, **343**, 1471–1472 (1994).

204. N. F. Olivieri, *Acta Haematol.*, **95**, 37–48 (1996).

205. G. J. Kontoghiorghes, M. A. Aldouri, L. Sheppard, and A. V. Hoffbrand, *Lancet*, **1**, 1294–1295 (1987).

206. A. V. Hoffbrand, F. Al-Refaie, B. Davis, N. Siritanakatkul, B. F. A. Jackson, J. Cochrane, E. Prescott, and B. Wonke, *Blood*, **91**, 295–300 (1998).

207. A. R. Cohen, R. Galanello, A. Piga, A. DiPalma, C. Vullo, and F. Tricta, *Br. J. Haematol.*, **108**, 305–312 (2000).

208. D. R. Richardson, *J. Lab. Clin. Med.*, **137**, 324–329 (2001).

209. L. Cragg, R. P. Hebbel, W. Miller, A. Solovey, S. Selby, and H. Enright, *Blood*, **92**, 632–638 (1998).

210. N. F. Olivieri, G. Koren, C. Hermann, Y. Bentur, D. Chung, J. Klein, P. St Louis, M. H. Freedman, R. A. McClelland, and D. M. Templeton, *Lancet*, **336**, 1275–1279 (1990).

211. G. J. Kontoghiorghes, A. N. Bartlett, A. V. Hoffbrand, J. G. Goddard, L. Sheppard, J. Barr, and P. Nortey, *Br. J. Haematol.*, **76**, 295–300 (1990).

212. D. Matsui, J. Klein, C. Hermann, V. Grunau, R. McClelland, D. Chung, P. St-Louis, N. Olivieri, and G. Koren, *Clin. Pharmacol. Ther.*, **50**, 294–298 (1991).

213. S. Stobie, J. Tyberg, D. Matsui, D. Fernandes, J. Klein, N. Olivieri, Y. Bentur, and G. Koren, *Int. J. Clin. Pharmacol. Therap. Toxicol.*, **31**, 602–605 (1993).

214. F. N. al-Refaie, L. N. Sheppard, P. Nortey, B. Wonke, and A. V. Hoffbrand, *Br. J. Haematol.*, **89**, 403–408 (1995).

215. G. J. Kontoghiorghes, J. G. Goddard, A. N. Bartlett, and L. Sheppard, *Clin. Pharmacol. Ther.*, **48**, 255–261 (1990).

216. S. Singh, R. O. Epemolu, P. S. Dobbin, G. S. Tilbrook, B. L. Ellis, L. A. Damani, and R. C. Hider, *Drug Metab. Dispos.*, **20**, 256–261 (1992).

217. G. J. Kontoghiorghes, K. Pattichi, M. Hadjigavriel, and A. Kolnagou, *Transfusion Sci.*, **23**, 211–223 (2000).

218. N. F. Olivieri, G. M. Brittenham, C. E. McLaren, D. M. Templeton, R. G. Cameron, R. A. McClelland, A. D. Burt, and K. A. Fleming, *N. Engl. J. Med.*, **339**, 417–423 (1998).

219. G. C. Del Vecchio, E. Crollo, F. Schettini, R. Fischer, and D. De Mattia, *Acta Haematol.*, **104**, 99–102 (2000).

220. A. Florence, A. Longueville, and R. R. Crichton in Ninth International Conference on Proteins of Iron Transport and Storage, Brisbane, Australia, 1989, p. P118.

221. B. Wonke, C. Wright, and A. V. Hoffbrand, *Br. J. Haematol.*, **103**, 361–364 (1998).

222. Y. Aydinok, G. Nisli, K. Kavakli, C. Coker, M. Kantar, and N. Cetingul, *Acta Haematol.*, **102**, 17–21 (1999).

223. R. W. Grady, V. A. Berdoukas, E. A. Rachmilewitz, R. Galanello, C. Borgna-Pignatti, and P. J. Giardina, *Blood*, **96**, 604A, 2594 (2000).

224. R. W. Grady and P. J. Giardina in D. G. Badman, R. J. Bergeron, and G. M. Brittenham, Eds., *Iron Chelators: New Development Strategies*, Saratoga Press, Ponte Vedra Beach, FL, 2000, pp. 293–310.

225. W. D. Davis Jr. and W. R. Arrowsmith, *J. Lab. Clin. Med.*, **36**, 814–815 (1950).

226. H. Wishinsky, T. Weinberg, E. M. Prévost, B. Burgin, and M. J. Miller, *J. Lab. Clin. Med.*, **42**, 550–554 (1953).

227. T. Greenwalt and V. E. Ayers, *Am. J. Clin. Pathol.*, **25**, 266–271 (1955).

228. M. Rubin, J. Houlihan, and J. V. Princiotto, *Proc. Soc. Exp. Biol. Med.*, **103**, 663–666 (1960).

229. H. Zähner in K. Gross, J. Aumiller, and J. Gelzer, Eds., *Desferrioxamine (Desferal): History, Clinical Value, Perspectives*, MMV Medizin Verlag, Munich, 1992, pp. 11–25.

230. H. Keberle in K. Gross, J. Aumiller, and J. Gelzer, Eds., *Desferrioxamine (Desferal): History, Clinical Value, Perspectives*, MMV Medizin Verlag, Munich, 1992, pp. 26–36.

231. J. Tripod and H. Keberle, *Helv. Physiol. Acta*, **20**, 291–293 (1962).

232. J. Tripod and H. Keberle, *Helv. Med. Acta*, **29**, 674–679 (1962).

233. R. W. Grady and C. Hershko, *Ann. N. Y. Acad. Sci.*, **612**, 361–368 (1990).

234. C. G. Pitt, Y. Bao, J. Thompson, M. C. Wani, H. Rosenkrantz, and J. J. Metterville, *J. Med. Chem.*, **29**, 1231–1237 (1986).

235. R. J. Bergeron, R. R. Streiff, E. A. Creary, R. D. Daniels Jr., W. King, G. Luchetta, J. Wiegand, T. Moerker, and H. H. Peter, *Blood*, **81**, 2166–2173 (1993).

236. H. H. Peter, R. J. Bergeron, R. R. Streiff, and J. Wiegand in R. J. Bergeron and G. M. Brittenham, Eds., *The Development of Iron Chelators for Clinical Use*, CRC Press, Boca Raton, FL, 1994, pp. 373–394.

237. R. W. Grady, A. D. Salbe, M. W. Hilgartner, and P. J. Giardina, *Adv. Exp. Med. Biol.*, **356**, 351–359 (1994).

238. C. Hershko, R. W. Grady, and G. Link, *Haematol. Budap.*, **17**, 25–33 (1984).

239. C. Hershko, R. W. Grady, and G. Link, *J. Lab. Clin. Med.*, **103**, 337–346 (1984).

240. R. J. Bergeron, J. Wiegand, and G. M. Brittenham, *Blood*, **91**, 1446–1452 (1998).

241. R. J. Bergeron, J. Wiegand, and G. M. Brittenham, *Blood*, **93**, 370–375 (1999).

242. R. J. Bergeron, R. R. Streiff, J. Wiegand, G. Luchetta, E. A. Creary, and H. H. Peter, *Blood*, **79**, 1882–1890 (1992).

243. R. J. Bergeron, J. Wiegand, and G. M. Brittenham, *Blood*, **99**, 3019–3026 (2002).

244. A. R. Cohen, J. Mizanin, and E. Schwartz, *J. Pediatr.*, **115**, 151–155 (1989).

245. N. F. Olivieri, A. M. Berriman, B. J. Tyler, S. A. Davis, W. H. Francombe, and P. P. Liu, *Am. J. Hematol.*, **41**, 61–63 (1992).

246. T. Lombardo, G. Ferro, V. Frontini, and S. Percolla, *Am. J. Hematol.*, **51**, 90–92 (1996).

247. M. J. S. M. P. Araujo, U. Ragnarsson, M. L. S. Almeida, and M. J. S. A. A. Trigo, *J. Chem. Res. Synop.*, 120 (1992).

248. R. J. Bergeron, K. A. McGovern, M. A. Channing, and P. S. Burton, *J. Org. Chem.*, **45**, 1589–1592 (1980).

249. R. J. Bergeron, P. S. Burton, K. A. McGovern, E. St. Onge, and R. R. Streiff, *J. Med. Chem.*, **23**, 1130–1133 (1980).

250. R. J. Bergeron, S. J. Kline, N. J. Stolowich, K. A. McGovern, and P. S. Burton, *J. Org. Chem.*, **46**, 4524–4529 (1981).

251. K. K. Bhargava, R. W. Grady, and A. Cerami, *J. Pharm. Sci.*, **69**, 986–989 (1980).

252. A. Husson, R. Besselievre, and H.-P. Husson, *Tetrahedron Lett.*, **24**, 1031–1034 (1983).

253. A. V. Joshua and J. R. Scott, *Tetrahedron Lett.*, **25**, 5725–5728 (1984).

254. T. Miyasaka, Y. Nagao, E. Fujita, H. Sakurai, and K. Ishizu, *J. Chem. Soc. Perkin Trans. 2*, 1543–1549 (1987).

255. S.-I. Murahashi, T. Naota, and N. Nakajima, *Chem. Lett.*, 879–882 (1987).

256. S.-I. Murahashi and T. Naota, *Synthesis*, 433–440 (1993).

257. R. J. Bergeron and S. J. Kline, *J. Am. Chem. Soc.*, **104**, 4489–4492 (1982).

258. R. J. Bergeron and S. J. Kline, *J. Am. Chem. Soc.*, **106**, 3089–3098 (1984).

259. G. M. Buckley, G. Pattenden, and D. A. Whiting, *Tetrahedron*, **50**, 11781–11792 (1994).

260. Y. Nagao, T. Miyasaka, Y. Hagiwara, and E. Fujita, *J. Chem. Soc. Perkin Trans. 1*, 183–187 (1984).

261. R. J. Bergeron, J. S. McManis, J. B. Dionis, and J. R. Garlich, *J. Org. Chem.*, **50**, 2780–2782 (1985).

262. R. J. Bergeron, J. R. Garlich, and J. S. McManis, *Tetrahedron*, **41**, 507–510 (1985).

263. D. Hatzipanayoti, A. Karaliota, M. Kamariotaki, A. Veneris, and P. Falaras, *Transition Met. Chem.*, **23**, 407–416 (1998).

264. R. J. Bergeron, N. J. Stolowich, and C. W. Porter, *Synthesis*, 689–692 (1982).

265. R. J. Bergeron, P. S. Burton, K. A. McGovern, and S. J. Kline, *Synthesis*, 732–733 (1981).

266. R. J. Bergeron, J. R. Garlich, and N. J. Stolowich, *J. Org. Chem.*, **49**, 2997–3001 (1984).

267. G. Barany and R. B. Merrifield, *J. Am. Chem. Soc.*, **99**, 7363–7365 (1977).

268. A. P. T. Easson and F. L. Pyman, *J. Chem. Soc.*, 2991–3001 (1931).

269. W. H. Rastetter, T. J. Erickson, and M. C. Venuti, *J. Org. Chem.*, **46**, 3579–3590 (1981).

270. B. H. Lee and M. J. Miller, *J. Org. Chem.*, **48**, 24–31 (1983).

271. P. J. Maurer and M. J. Miller, *J. Am. Chem. Soc.*, **104**, 3096–3101 (1982).

272. R. J. Bergeron and O. Phanstiel IV, *J. Org. Chem.*, **57**, 7140–7143 (1992).

273. G. C. Mulqueen, G. Pattenden, and D. A. Whiting, *Tetrahedron*, **49**, 9137–9142 (1993).

274. T. Sakamoto, H. Li, and Y. Kikugawa, *J. Org. Chem.*, **61**, 8496–8499 (1996).

275. J. Hu and M. J. Miller, *J. Am. Chem. Soc.*, **119**, 3462–3468 (1997).

276. O. Exner and B. Kakác, *Coll. Czech. Chem. Commun.*, **25**, 2530–2539 (1960).

277. Y.-M. Lin and M. J. Miller, *J. Org. Chem.*, **64**, 7451–7458 (1999).

278. Y.-M. Lin and M. J. Miller, *J. Org. Chem.*, **66**, 8282–8285 (2001).

279. H. Tokuyama, T. Kuboyama, A. Amano, T. Yamashita, and T. Fukuyama, *Synthesis*, 1299–1304 (2000).

280. M. D. Wittman, R. L. Halcomb, and S. J. Danishefsky, *J. Org. Chem.*, **55**, 1981–1983 (1990).

281. O. Phanstiel IV, Q. X. Wang, D. H. Powell, M. P. Ospina, and B. A. Leeson, *J. Org. Chem.*, **64**, 803–806 (1999).

282. Q. X. Wang, J. King, and O. Phanstiel IV, *J. Org. Chem.*, **62**, 8104–8108 (1997).

283. R. O. Hutchins and N. R. Natale, *Org. Prep. Proc. Int.*, **11**, 201–246 (1979).

284. M. Kawase and Y. Kikugawa, *J. Chem. Soc. Perkin Trans. 1*, 643–645 (1979).

285. C. Kashima, N. Yoshiwara, and Y. Omote, *Tetrahedron Lett.*, **23**, 2955–2956 (1982).

286. J. L. Romine, *Org. Prep. Proc. Int.*, **28**, 249–288 (1996).

287. S. L. Mellor and W. C. Chan, *Chem. Commun.*, 2005–2006 (1997).

288. D. W. Knight and M. P. Leese, *Tetrahedron Lett.*, **42**, 2593–2595 (2001).

289. V. Prelog and A. Walser, *Helv. Chim. Acta*, **45**, 631–637 (1962).

290. H. Bickel, B. Fechtig, G. E. Hall, W. Keller-Schierlein, V. Prelog, and E. Vischer, *Helv. Chim. Acta*, **43**, 901–904 (1960).

291. P. G. M. Wuts, J. E. Cabaj, and J. L. Havens, *J. Org. Chem.*, **59**, 6470–6471 (1994).

292. R. J. Bergeron and J. J. Pegram, *J. Org. Chem.*, **53**, 3131–3134 (1988).

293. R. J. Bergeron and J. S. McManis, *Tetrahedron*, **45**, 4939–4944 (1989).

294. K. Ramasamy, R. K. Olsen, and T. Emery, *J. Org. Chem.*, **46**, 5438–5441 (1981).

295. S. S. Nikam, B. E. Kornberg, D. R. Johnson, and A. M. Doherty, *Tetrahedron Lett.*, **36**, 197–200 (1995).

296. R. J. Bergeron, J. Wiegand, J. S. McManis, and P. T. Perumal, *J. Med. Chem.*, **34**, 3182–3187 (1991).

297. R. J. Bergeron, J. S. McManis, O. Phanstiel IV, and J. R. T. Vinson, *J. Org. Chem.*, **60**, 109–114 (1995).

298. T. Kolasa and A. Chimiak, *Pol. J. Chem.*, **55**, 1163–1167 (1981).

299. G. M. Cohen, P. M. Cullis, J. A. Hartley, A. Mather, M. C. R. Symons, and R. T. Wheelhouse, *J. Chem. Soc. Chem. Commun.*, 298–300 (1992).

300. J. Mollin, F. Kasparek, and J. Lasovsky, *Chem. Zvesti.*, **29**, 39–43 (1975).

301. R. J. Bergeron, J. S. McManis, P. T. Perumal, and S. E. Algee, *J. Org. Chem.*, **56**, 5560–5563 (1991).

302. M. J. Milewska, A. Chimiak, and Z. Glowacki, *J. Prakt. Chem.*, **329**, 447–456 (1987).

303. R. J. Bergeron and J. S. McManis, *Tetrahedron*, **46**, 5881–5888 (1990).

304. J. M. Roosenberg II and M. J. Miller, *J. Org. Chem.*, **65**, 4833–4838 (2000).

305. L. Vertesy, W. Aretz, H.-W. Fehlhaber, and H. Kogler, *Helv. Chim. Acta*, **78**, 46–60 (1995).

306. R. A. Schultz, B. D. White, D. M. Dishong, K. A. Arnold, and G. W. Gokel, *J. Am. Chem. Soc.*, **107**, 6659–6668 (1985).

307. U. Heimann and F. Voegtle, *Liebigs Ann. Chem.*, 858–862 (1980).

308. S. K. Sharma, M. J. Miller, and S. M. Payne, *J. Med. Chem.*, **32**, 357–367 (1989).

309. G. C. Mulqueen, G. Pattenden, and D. A. Whiting, *Tetrahedron*, **49**, 5359–5364 (1993).

310. H. Vorbrüggen and K. Krolikiewicz, *Synthesis*, 316–319 (1983).

311. R. J. Bergeron, C. Z. Liu, J. S. McManis, M. X. B. Xia, S. E. Algee, and J. Wiegand, *J. Med. Chem.*, **37**, 1411–1417 (1994).

312. R. J. Bergeron, J. Wiegand, J. B. Dionis, M. Egli-Karmakka, J. Frei, A. Huxley-Tencer, and H. H. Peter, *J. Med. Chem.*, **34**, 2072–2078 (1991).

313. R. J. Bergeron, J. Wiegand, M. Wollenweber, J. S. McManis, S. E. Algee, and K. Ratliff-Thompson, *J. Med. Chem.*, **39**, 1575–1581 (1996).

314. R. J. Bergeron, J. Wiegand, W. R. Weimar, J. R. T. Vinson, J. Bussenius, G. W. Yao, and J. S. McManis, *J. Med. Chem.*, **42**, 95–108 (1999).

315. R. J. Bergeron, J. Wiegand, J. S. McManis, B. H. McCosar, W. R. Weimar, G. M. Brittenham, and R. E. Smith, *J. Med. Chem.*, **42**, 2432–2440 (1999).

316. I.-J. Ryoo, K.-S. Song, J.-P. Kim, W.-G. Kim, H. Koshino, and I.-D. Yoo, *J. Antibiot.*, **50**, 256–258 (1997).

317. A. Thieken and G. Winkelmann, *FEMS Microbiol. Lett.*, **73**, 37–41 (1992).

318. H. Drechsel, G. Jung, and G. Winkelmann, *BioMetals*, **5**, 141–148 (1992).

319. R. J. Bergeron, M. G. Xin, R. E. Smith, M. Wollenweber, J. S. McManis, C. Ludin, and K. Abboud, *Tetrahedron*, **53**, 427–434 (1997).

320. K. Hirota, H. Kitagawa, M. Shimamura, and S. Ohmori, *Chem. Lett.*, 191–194 (1980).

321. K. Samejima, Y. Takeda, M. Kawase, M. Okada, and Y. Kyogoku, *Chem. Pharm. Bull.*, **32**, 3428–3435 (1984).

322. R. M. Williams and E. Kwast, *Tetrahedron Lett.*, **30**, 451–454 (1989).

323. M. Münzinger, K. Taraz, H. Budzikiewicz, H. Drechsel, P. Heymann, G. Winkelmann, and J.-M. Meyer, *BioMetals*, **12**, 189–193 (1999).

324. M. D. Fleming, M. A. Romano, M. A. Su, L. M. Garrick, M. D. Garrick, and N. C. Andrews, *Proc. Natl. Acad. Sci. USA*, **95**, 1148–1153 (1998).

325. D. B. Van Wyck, R. A. Popp, J. Foxley, M. H. Witte, C. L. Witte, and W. H. Crosby, *Blood*, **64**, 263–266 (1984).

326. J. E. Levy, O. Jin, Y. Fujiwara, F. Kuo, and N. Andrews, *Nat. Genet.*, **21**, 396–399 (1999).

327. L. C. Skow, B. A. Burkhart, F. M. Johnson, R. A. Popp, D. M. Popp, S. Z. Goldberg, W. F. Anderson, L. B. Barnett, and S. E. Lewis, *Cell*, **34**, 1043–1052 (1983).

328. R. A. Popp and M. K. Enlow, *Am. J. Vet. Res.*, **38**, 569–572 (1977).

329. A. Leder, C. Daugherty, B. Whitney, and P. Leder, *Blood*, **90**, 1275–1282 (1997).

330. C. Paszty, N. Mohandas, M. E. Stevens, J. F. Loring, S. A. Liebhaber, C. M. Brion, and E. M. Rubin, *Nat. Genet.*, **11**, 33–39 (1995).

331. A. Leder, E. Wiener, M. J. Lee, S. N. Wickramasinghe, and P. Leder, *Proc. Natl. Acad. Sci. USA*, **96**, 6291–6295 (1999).

332. L. M. Garrick, L. A. Strano-Paul, J. E. Hoke, L. A. Kirdani-Ryan, R. A. Alberico, M. M. Everett, R. M. Bannerman, and M. D. Garrick, *Exp. Hematol.*, **17**, 423–428 (1989).

333. W. R. Shehee, P. Oliver, and O. Smithies, *Proc. Natl. Acad. Sci. USA*, **90**, 3177–3181 (1993).

334. D. J. Ciavatta, T. M. Ryan, S. C. Farmer, and T. M. Townes, *Proc. Natl. Acad. Sci. USA*, **92**, 9259–9263 (1995).

335. B. Yang, S. Kirby, J. Lewis, P. J. Detloff, N. Maeda, and O. Smithies, *Proc. Natl. Acad. Sci. USA*, **92**, 11608–11612 (1995).

336. M.-J. Blouin, M. E. De Paepe, and M. Trudel, *Blood*, **94**, 1451–1459 (1999).

337. M.-J. Blouin, H. Beauchemin, A. Wright, M. De Paepe, M. Sorette, A.-M. Bleau, B. Nakamoto, C.-N. Ou, G. Stamatoyannopoulos, and M. Trudel, *Nat. Med.*, **6**, 177–182 (2000).

338. J. C. Chang, R. Lu, C. Lin, S.-M. Xu, Y. W. Kan, S. Porcu, E. Carlson, M. Kitamura, S. Yang, L. Flebbe-Rehwaldt, and K. M. L. Gaensler, *Proc. Natl. Acad. Sci. USA*, **95**, 14886–14890 (1998).

339. M. E. Fabry, A. Sengupta, S. M. Suzuka, F. Costantini, E. M. Rubin, J. Hofrichter, G. Christoph, E. Manci, D. Culberson, and S. M. Factor, *Blood*, **86**, 2419–2428 (1995).

340. C. Paszty, C. M. Brion, E. Manci, H. E. Witkowska, M. E. Stevens, N. Mohandas, and E. M. Rubin, *Science*, **278**, 876–878 (1997).

341. T. M. Ryan, D. J. Ciavatta, and T. M. Townes, *Science*, **278**, 873–876 (1997).

342. M. Trudel, N. Saadane, M. Garel, J. Bardakdjian-Michau, Y. Blouquit, J. L. Guerquin-Kern, P. Rouyer-Fessard, D. Vidaud, A. Pachnis, and P. H. Romeo, *EMBO J.*, **10**, 3157–3165 (1991).

343. M. Trudel, M. E. De Paepe, N. Chretien, N. Saadane, J. Jacmain, M. Sorette, T. Hoang, and Y. Beuzard, *Blood*, **84**, 3189–3197 (1994).

344. K. Ohene-Frempong, *Semin. Hematol.*, **38**, 5–13 (2001).

345. P. Bennekou, L. de Franceschi, O. Pedersen, L. Lian, T. Asakura, G. Evans, C. Brugnara, and P. Christophersen, *Blood*, **97**, 1451–1457 (2001).

346. R. J. Bergeron, Z.-R. Liu, J. S. McManis, and J. Wiegand, *J. Med. Chem.*, **35**, 4739–4744 (1992).

347. R. J. Bergeron, R. R. Streiff, J. Wiegand, J. R. T. Vinson, G. Luchetta, K. M. Evans, H. Peter, and H.-B. Jenny, *Ann. N. Y. Acad. Sci.*, **612**, 378–393 (1990).

348. R. Choudhury and S. Singh, *Drug Metab. Dispos.*, **23**, 314–320 (1995).

349. J. B. Porter, K. P. Hoyes, R. D. Abeysinghe, P. N. Brooks, E. R. Huehns, and R. C. Hider, *Blood*, **78**, 2727–2734 (1991).

350. J. B. Porter, R. D. Abeysinghe, K. P. Hoyes, C. Barra, E. R. Huehns, P. N. Brooks, M. P. Blackwell, M. Araneta, G. Brittenham, S. Singh, P. Dobbin, and R. C. Hider, *Br. J. Haematol.*, **85**, 159–168 (1993).

351. D. A. Papanastasiou, D. V. Vayenas, A. Vassilopoulos, and M. Repanti, *Pathol. Res. Pract.*, **196**, 47–54 (2000).

352. P. S. Dobbin, R. C. Hider, A. D. Hall, P. D. Taylor, P. Sarpong, J. B. Porter, G. Xiao, and D. van der Helm, *J. Med. Chem.*, **36**, 2448–2458 (1993).

353. M. Gyparaki, J. B. Porter, S. Hirani, M. Streater, R. C. Hider, and E. R. Huehns, *Acta Haematol.*, **78**, 217–221 (1987).

354. G. J. Kontoghiorghes, *Mol. Pharmacol.*, **30**, 670–673 (1986).

355. G. J. Kontoghiorghes, *Scand. J. Haematol.*, **37**, 63–70 (1986).

356. G. J. Kontoghiorghes, J. Barr, P. Nortey, and L. Sheppard, *Am. J. Hematol.*, **42**, 340–349 (1993).

357. J. B. Porter, J. Morgan, K. P. Hoyes, L. C. Burke, E. R. Huehns, and R. C. Hider, *Blood*, **76**, 2389–2396 (1990).

358. D. T. Dexter, R. J. Ward, A. Florence, P. Jenner, and R. R. Crichton, *Biochem. Pharmacol.*, **58**, 151–155 (1999).

359. L. G. Valerio Jr. and D. R. Petersen, *Exp. Mol. Pathol.*, **68**, 1–12 (2000).

360. A. Florence, R. J. Ward, T. J. Peters, and R. R. Crichton, *Biochem. Pharmacol.*, **44**, 1023–1027 (1992).

361. A. Longueville and R. R. Crichton, *Biochem. Pharmacol.*, **35**, 3669–3678 (1986).

362. C. Hershko, J. D. Cook, and D. A. Finch, *J. Lab. Clin. Med.*, **81**, 876–886 (1973).

363. C. Hershko, *Blood*, **51**, 415–423 (1978).

364. C. Hershko, A. M. Konjin, H. P. Nick, Z. I. Cabantchik, and G. Link, *Blood*, **97**, 1115–1122 (2001).

365. G. Link, A. M. Konijn, W. Breuer, Z. I. Cabantchik, and C. Hershko, *J. Lab. Clin. Med.*, **138**, 130–138 (2001).

366. S. Zevin, G. Link, R. W. Grady, R. C. Hider, H. H. Peter, and C. Hershko, *Blood*, **79**, 248–253 (1992).

367. R. W. Grady, J. H. Graziano, G. P. White, A. Jacobs, and A. Cerami, *J. Pharmacol. Exp. Ther.*, **205**, 757–765 (1978).

368. R. W. Grady, C. M. Peterson, R. L. Jones, J. H. Graziano, K. K. Bhargava, V. A. Berdoukas, G. Kokkini, D. Loukopoulos, and A. Cerami, *J. Pharmacol. Exp. Ther.*, **209**, 342–348 (1979).

369. E. J. Gralla in W. F. Anderson and M. Hille, Eds., *Symposium on the Development of Iron Chelators for Clinical Use*, U.S. Department of Health, Education, and Welfare, Bethesda, MD, 1975, pp. 229–245.

370. C. G. Pitt, G. Gupta, W. E. Estes, H. Rosenkrantz, J. J. Metterville, A. L. Crumbliss, R. A.

Palmer, K. W. Nordquest, K. A. S. Hardy, D. R. Whitcomb, B. R. Byers, J. E. L. Arceneaux, C. G. Gaines, and C. V. Sciortino, *J. Pharmacol. Exp. Ther.*, **208**, 12–18 (1979).

371. D. Ceccarelli, D. Gallesi, F. Giovannini, M. Ferrali, and A. Masini, *Biochem. Biophys. Res. Commun.*, **209**, 53–59 (1995).

372. C. Pigeon, B. Turlin, T. C. Iancu, P. Leroyer, J. Le Lan, Y. Deugnier, P. Brissot, and O. Loreal, *J. Hepatol.*, **30**, 926–934 (1999).

373. B. L. Rai, L. S. Dekhordi, H. Khodr, Y. Jin, Z. Liu, and R. C. Hider, *J. Med. Chem.*, **41**, 3347–3359 (1998).

374. A. Steward, I. Williamson, T. Madigan, A. Bretnall, and I. F. Hassan, *Br. J. Haematol.*, **95**, 654–659 (1996).

375. A. Wong, V. Alder, D. Robertson, J. Papadimitriou, J. Maserei, V. Berdoukas, G. Kontoghiorghes, E. Taylor, and E. Baker, *BioMetals*, **10**, 247–256 (1997).

376. K. A. Schwartz, J. Fisher, and E. T. Adams, *Toxicol. Pathol.*, **21**, 311–320 (1993).

377. P. Carthew, R. E. Edwards, A. G. Smith, B. Dorman, and J. E. Francis, *Hepatology*, **13**, 534–539 (1991).

378. P. Carthew, B. M. Dorman, R. E. Edwards, J. E. Francis, and A. G. Smith, *Lab. Invest.*, **69**, 217–222 (1993).

379. P. Carthew, A. G. Smith, R. C. Hider, B. Dorman, R. E. Edwards, and J. E. Francis, *BioMetals*, **7**, 267–271 (1994).

380. Y. A. Kuryshev, G. M. Brittenham, H. Fujioka, P. Kannan, C.-C. Shieh, S. A. Cohen, and A. M. Brown, *Circulation*, **100**, 675–683 (1999).

381. L. C. Wolfe, R. J. Nicolosi, M. M. Renaud, J. Finger, M. Hegsted, H. Peter, and D. G. Nathan, *Br. J. Haematol.*, **72**, 456–461 (1989).

382. J. K. Wood, P. F. Milner, and U. N. Pathak, *Br. J. Haematol.*, **14**, 119–129 (1968).

383. G. M. Brittenham, P. M. Griffith, A. W. Nienhuis, C. E. McLaren, N. S. Young, E. E. Tucker, C. J. Allen, D. E. Farrell, and J. W. Harris, *N. Engl. J. Med.*, **331**, 567–573 (1994).

384. B. Faller, C. Spanka, T. Sergejew, and V. Tschinke, *J. Med. Chem.*, **43**, 1467–1475 (2000).

385. H. P. Nick, P. Acklin, B. Faller, Y. Jin, R. Lattmann, M.-C. Rouan, T. Sergejew, H. Thomas, H. Wiegand, and H. P. Schnebli in D. G. Badman, R. J. Bergeron, and G. M. Brittenham, Eds., *Iron Chelators: New Development Strategies*, Saratoga Press, Ponte Vedra Beach, FL, 2000, pp. 311–331.

386. T. Sergejew, P. Forgiarini, and H.-P. Schnebli, *Br. J. Haematol.*, **110**, 985–992 (2000).

387. R. J. Bergeron, W. R. Weimar, and J. Wiegand, *Drug Metab. Dispos.*, **27**, 1496–1498 (1999).

388. R. J. Bergeron, J. Wiegand, K. Ratliff-Thompson, and W. R. Weimar, *Ann. N. Y. Acad. Sci.*, **850**, 202–216 (1998).

389. B. Shetty, M. Badr, and S. Melethil, *J. Pharm. Sci.*, **83**, 607–608 (1994).

390. T. F. McMahon, S. A. Stefanski, R. E. Wilson, P. C. Blair, A.-M. Clark, and L. S. Birnbaum, *Toxicology*, **66**, 297–311 (1991).

CHAPTER ELEVEN

Thyroid Hormones and Thyromimetics

DENIS E. RYONO
Discovery Chemistry

GARY J. GROVER
Metabolic Diseases Biology
Bristol-Myers Squibb
Princeton, New Jersey

KARIN MELLSTROM
Cell Biology
Karo Bio AB
Huddinge, Sweden

Burger's Medicinal Chemistry and Drug Discovery
Sixth Edition, Volume 3: Cardiovascular Agents and
Endocrines
Edited by Donald J. Abraham
ISBN 0-471-37029-0   © 2003 John Wiley & Sons, Inc.

Contents

1 INTRODUCTION

Thyroid hormones and their analogs act by binding to their cognate receptors in the nucleus and regulating the expression of diverse genes throughout the body. Consequently, such compounds represent significant promise for the development of new therapy. Unfortunately, the potential for this class of drugs has historically been limited by concerns for mechanism-related side effects. However improved assays using purified receptors, greater knowledge of receptor-ligand interactions, and importantly, new advances in understanding the molecular biology of nuclear hormone receptors, including the thyroid hormone receptors (TRs), may now point the way to developing new thyroid hormone drugs with selective and consequently safer profiles of biologic activity.

Essentially all naturally occurring thyroid hormones and their synthetic analogs (thyromimetics) are structurally characterized by a basic construct of two aromatic rings joined by a single linker atom (X) as depicted below. The various iodine atoms of the naturally occurring hormones, T$_3$ and T$_4$, occupy the positions indicated by R$_3$, R$_5$, R$_3'$ and R$_5'$. The aromatic ring containing iodine substitution

(R$_3$ and R$_5$) adjacent to the linker atom has been termed the "inner" or "non-prime" ring. This ring also possesses the amino acid side-chain at R$_1$ that is biosynthetically derived from tyrosine. The second aromatic ring, the "outer" or "prime" ring, contains the remaining iodine substitution positions at R$_3'$ and R$_5'$, as well as a critical phenolic hydroxyl group at R$_4'$. Thus, T$_3$ (L-3,5,3'-triiodothyronine) has the structure shown by (1) and T$_4$ (L-3,5,3',5'-tetraiodothyronine or L-thyroxine) has the structure indicated by (2).

2 MOLECULAR MECHANISM AND PHYSIOLOGY OF THYROID HORMONE ACTION

2.1 Introduction

The normal thyroid gland is the largest endocrine gland in the body. The name thyroid is derived from its proximity to the laryngeal thyroid cartilage, which resembles a Greek shield. Our earliest knowledge of the thyroid dates back thousands of years when goiters and cretinism, particularly in the Alps mountain regions of Europe, were linked to this gland. In the early 1800s, it was discovered that iodine could reduce or prevent goiters,

linker

prime ring non-prime ring
(outer ring) (inner ring)

(1) L-T$_3$ (L-3,5,3'-triiodothyronine): R = H
(2) L -T$_4$ (L-3,5,3',5'-tetraiodothyronine): R = I

although this was purely an empirical observation. In the late 1800s, the link between hypothyroidism and hyperthyroidism (or thyrotoxicosis) and the thyroid gland was deduced. A search for the active agent was undertaken, and using iodine as a marker, Kendall isolated the active substance in crystalline form in 1914 on Christmas day and named the compound thyroxin (1). This name was derived from a contraction of thyroxyindole and was based on his belief that the compound's molecular structure had an indole nucleus and three iodine atoms. The correct structure, established by Harington in 1926 (2), showed the presence of four iodine atoms and an amino acid structure. It was accordingly renamed thyroxine (terminal "ine" in common with the names of the naturally occurring α-amino acids, for example, alanine). Thyroxine, also known now as T$_4$ (the subscript indicating the number of iodine atoms), was successfully synthesized in high yield in 1949, making it clinically and economically viable for therapeutic use compared with treatment with desiccated thyroid tissue (3).

The presence of a second thyroid hormone was hypothesized, but not established until Gross and Pitt-Rivers (4) and, simultaneously, Roche et al. (5) found and synthesized thyronine (T$_3$) in 1952. T$_3$ was found to be more biologically active than T$_4$ and is presently thought to be the predominant activator of the thyroid hormone receptors. T$_4$ is the primary hormone synthesized by the thyroid, and while it is active in tissue, it is intracellularly converted to T$_3$.

Thyroid hormones exert a wide array of biologic activities and are important in growth, development, and cellular metabolism. In adults, thyroid hormones are critical for maintaining basal metabolic rate, optimal cognitive abilities, lipid metabolism, and cardiovascular status. Because of the importance of these myriad biologic activities, hypothyroidism and hyperthyroidism are characterized by marked physiologic effects that can have an enormous impact on day-to-day function in patients. For this reason, treatment of thyroid-related diseases is of great clinical interest. Because of the effects of thyroid hormones on metabolic rate and lipid metabolism, the potential of thyromimetics as anti-obesity and lipid-lowering agents also exists. However, controlling the side effects of such agents presents a daunting obstacle for drug development scientists.

2.2 Molecular Mechanism of Thyroid Hormone Action

2.2.1 Background. Thyroid hormone receptors are ligand-dependent transcription factors that belong to the nuclear receptor superfamily (6). Included in this family of proteins are receptors for steroid hormones, retinoic acid, and vitamin D, as well as homologous proteins with unidentified ligands, i.e., orphan receptors (7). The first suggestion that thyroid hormones act as regulators of transcription came from the early observations that T$_3$ induces a rapid increase in RNA synthesis that preceded *de novo* protein synthesis (8). Shortly thereafter, evidence that thyroid hormones acted through specific nuclear binding sites was shown using radiolabeled T$_3$ (9, 10). Using photoaffinity-labeling procedures, thyroid hormone receptors were identified in both rat liver nuclei (11) and intact cells (12). Further *in vitro* studies using the rat growth hormone gene strongly sug-

Figure 11.1. Domain structure of nuclear hormone receptors.

gested that the TRs recognized specific DNA elements (13). In 1986, two different subtypes of high affinity thyroid hormone receptors that were homologs to the viral *erbA* oncogene were independently cloned in two different laboratories (14, 15). This started an extensive and detailed elucidation of the molecular mechanisms for thyroid hormone acting as a regulator of transcription. There are several excellent reviews on this topic (16–18).

2.2.2 Domain Structure of the Thyroid Hormone Receptors. The TRs are nuclear receptors that share a common structure comprising three different functional domains: the *N*-terminal transactivation domain (AF1), the DNA-binding domain (DBD), and the carboxy-terminal ligand-binding domain (LBD; Fig. 11.1). The *N*-terminal A/B domain is the less well conserved region of the nuclear receptors, and its exact role in transactivation is still unclear. The nuclear receptors interact with DNA through the central DNA-binding domain with a distinct "zinc fingers" structure. The integrity of these zinc fingers are essential for DNA binding and transcriptional activity (19–22). The hinge region (D) located between the DNA-binding and ligand-binding domains contains the nuclear recognition signal. The carboxyl terminal LBD also plays an important role in dimerization, transactivation, and transrepression (23, 24). The transactivation function has been mapped to a highly conserved motif in the LBD termed AF-2 (25).

2.2.3 TR Receptor Isoforms. TRs are encoded by two distinct genes (*TRα* and *TRβ*), which give rise to several isoforms of TRs. The *TRα* gene encodes for the ligand-binding receptor $TR\alpha_1$ but also encodes for the $TR\alpha_2$ protein that does not bind T_3, although it binds to DNA with reduced affinity (26). $TR\alpha_1$ and $TR\alpha_2$ are co-expressed in most tissues. The biological function of $TR\alpha_2$ is not clear, but interference with normal TR function has

been suggested (27). The *TRβ* locus gives rise to $TR\beta_1$, $TR\beta_2$, and the recently discovered $TR\beta_3$ (14, 15, 28). The TRβ isoforms are transcribed from separate promoters and alternative splicing of the mRNA, which give rise to *N*-terminal variants. $TR\beta_1$, like $TR\alpha_1$, is found in most tissues, however, at varying levels of expression. In the adult liver, $TR\beta_1$ predominates in concentration over $TR\alpha_1$, whereas in the heart, the opposite is true (29). A more restricted expression pattern is found for $TR\beta_2$ that is preferentially expressed in the pituitary and in distinct cells in the central nervous system (30).

2.2.4 Transcriptional Regulation

2.2.4.1 Thyroid Hormone Response Elements. TRs act by binding to specific DNA sequences, thyroid response elements (TREs), in the promoters of their target genes, thereby stimulating or repressing transcription. Most of these TREs are located upstream of the promoter, but location in the 3′ flanking region downstream from the coding region also has been described (17, 31).

Mutagenesis studies of the rat growth hormone promoter showed the consensus half site to be G/AGGTC/GA. Considerable variation of this consensus sequence regarding primary nucleotides, numbers, spacing, and orientation are found in TREs. The most prominent half-site orientations with optimal spacing include a palindrome TRE (pal) with no spacing nucleotides, direct repeats with four intervening nucleotides (DR4), and inverted palindromes with six spacing nucleotides (32).

2.2.4.2 RXR Heterodimers. TRs bind to these TREs usually as homodimers or heterodimers with auxiliary proteins. The major heterodimer partner is RXR (retinoid X receptor) that belongs to the nuclear receptor superfamily and has high homology with the retinoic acid receptor (RAR). RXR binds 9-*cis* retinoic acid with high affinity and also heterodimerizes with vitamin D receptor (VDR), RAR, peroxisome proliferator activator recep-

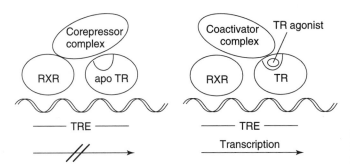

Figure 11.2. Role of corepressor and coactivators in transcriptional regulation of NHRs.

tor (PPAR), liver X receptor (LXR), and farnesyl X receptor (FXR). The RXRs (three different subtypes: α, β, and γ) are widely expressed and therefore allow heterodimer formation with TR in most tissues (33, 34). TR/RXR heterodimer formation increases the number of target genes that can be induced because heterodimers bind to a larger variety of TREs. In the DR4 response element, a specific orientation of dimer is formed where TR is always oriented on the 3' side of RXR (35). Unlike steroid receptors, the TR/RXR heterodimer binds to DNA in the absence of ligand and thereby represses transcription.

2.2.4.3 Role of Coactivators and Corepressors. Nuclear hormone receptor (NHR) regulation of transcription involves important roles played by numerous coregulator proteins that bind the receptor/DNA complexes. Those coregulators that promote transcriptional activation by binding the ligand-bound form of the receptor are termed coactivators and those involved in repression of transcription by binding the ligand-free (apo) receptor are the corepressors (Fig. 11.2) (36, 37).

In the absence of ligand, the TRs can repress basal transcription of many positively regulated genes that contain TR-binding sites in their promoters (14, 38, 39). Recruitment of corepressor proteins like nuclear receptor corepressor (NCoR) or RIP3 (40, 41), and silencing mediator of retinoic and thyroid receptor (SMRT) or TRAC (42, 43) that interact with the unliganded TR, are involved in the mechanisms of basal repression. Unliganded TRs recruit corepressor proteins that inhibit transactivation by either interfering with the basal transcription machinery and/or recruiting enzymes that inhibit transcription, e.g., histone deacetyltransferase (HDAC) activity.

In a recent study, the critical surface and conformational modes for NCoR interaction with TRβ were characterized, suggesting the possibility of selective modulation of corepressor function by induction of specific receptor conformations (44).

Unliganded thyroid hormone receptors can also stimulate the basal activity of certain negatively regulated promoters such as thyroid stimulating hormone (TSH). The mechanism by which TR exerts this function is less well understood, but it has been proposed to involve the same interaction with corepressors and coactivators, although with reversed functional consequences (45).

Ligand binding to TR will increase the affinity of the receptor to coactivators that interact with the AF-2 domain of the LBD. These coactivators (such as SRC-1) do not bind DNA by themselves but have the capacity to increase transcription *in vitro* when either cotransfected with DNA-binding proteins or fused with DNA-binding domains. When coactivators bind to TR they create an optimized environment for effective transcriptional activation supposedly by recruiting transcription factors and histone acetyltransferase (HAT) activity. Many of the coactivators interact with the receptors through a helical peptide motif with a core LXXLL sequence of amino acids (46–48). An X-ray structure determination of TRβ bound to a partial sequence of the coactivator protein, SRC-1 (steroid receptor coactivator-1), which contains two LXXLL motifs, suggests specific interactions between the hydrophobic region of the LXXLL helix and a hydrophobic cleft formed from portions of helices 3, 4, 5, and 12 of the receptor. Interestingly, the interaction domains on NCoR and SMRT exhibit a similar, but extended consensus, sequence, LXXI/HIXXXI/L. This extended helix interacts with specific residues located in the same hydrophobic pocket that interacts with coactivators (49, 50). This suggests that ligand-induced conformational changes involving the AF2 region of the LBD influence the recruitment of coactivators and their interaction with the receptors.

2.3 Physiology of Thyroid Hormone Action

2.3.1 Regulation of Thyroid Hormones. Because thyroid hormones exert such diverse and profound biologic effects, precise regulation of these hormones is required. Production of T_4 and T_3 by the thyroid gland is under negative feedback control, exerted at several levels. TSH, also known as thyrotropin, is responsible for normal thyroid gland function and thyroid hormone secretion. It is synthesized in the anterior pituitary gland, and its secretion is controlled by thyroid releasing hormone (TRH) that is synthesized in the hypothalamus. TRH, a G-protein–coupled receptor (GPCR) agonist, causes a dose-dependent and rapid release (in minutes) of TSH, but can also up-regulate transcription and translation of TSH subunit genes through a mechanism involving calcium influx and protein kinase C signaling pathways (51, 52). In addition to TRH, somatostatin, dopamine, and other catecholamines can modulate TSH secretion and release (53).

TSH is a glycoprotein with α- and β-subunits, both of which are necessary for biologic activity. The major site of expression of the TSH receptor is the thyroid, although it is also expressed in other tissues, particularly adipocytes (54). TSH causes increased thyrocyte cAMP levels and associated protein kinase A (PKA) activity (55). The gene regulation occurring secondary to PKA activation can occur within minutes to hours. TSH also stimulates PKC, causing release of inositol phosphates, and therefore can cause release of intracellular calcium. Activation of TSH receptors causes increased thyrocyte iodide uptake. TSH causes increased expression of thyroperoxidase (TPO), a central enzyme in thyroid hormone biosynthesis necessary for organification of iodide. TSH also increases gene tran-

scription of thyroglobulin, which is the starting substrate for T_4 and T_3 production.

Given the pivotal role for TSH in regulating biosynthesis and release of T_4 and T_3, its production in the pituitary is ideally suited as a site for feedback control. Increased plasma levels of T_3 or T_4 can reduce TSH release from the pituitary within hours of administration (56). This rapid reduction in TSH release is followed by a continuing, but slower rate, of secretion as long as T_3 is still present. Suppression of TSH release is thought to be related to nuclear TR binding and therefore secondary to gene regulation. Reciprocal regulation of TSH release by TRH and T_3 normally results in a steady-state balance, although some daily pulsatility can occur. T_3 not only reduces production of TSH, but also acts on the hypothalamus to reduce TRH release.

Because of its central regulatory function, TSH is a good indicator of thyroid function. Hypothyroidism caused by reduced T_3/T_4 output from the thyroid gland will cause increased TSH output because of unrestricted stimulation by TRH, as well as a reduced inhibitory effect. Proper hormone replacement therapy can be monitored by the return of TSH to more normal values.

TSH is released during the day in a somewhat pulsatile manner, with the highest levels occurring during the night, although the mechanism is not clear (57). Central regulation is likely and seems to be an interaction between several modulatory mechanisms. There are also several important extrinsic factors controlling TSH secretion. First, fasting will reduce the TRH responsiveness of TSH, causing reduced tissue T_3 levels and a decreased metabolic rate (58). This would seem to be an adaptive response to energy restriction. Another factor affecting TSH release is cold, which increases production of TSH (59). The adaptive effects of increased thermogenesis during cold exposure seems to be a logical response to this stress.

2.3.2 Thyroid Hormone Biosynthesis. There are several steps in thyroid hormone synthesis: (*1*) transport of iodide into the thyroid gland; (*2*) mono- and di-iodination of the tyrosine residues of the protein thyroglobulin (Tg) in the thyroid follicular cells; (*3*) coupling of the iodotyrosines by TPO to form nascent T_4 and T_3 as amino acid residues contained in Tg; and (4) proteolysis of Tg followed by secretion of the formed free thyroid hormones. The majority of the released thyroid hormone from the thyroid gland is in the form of T_4. The primary source of T_3 is from enzymatic deiodination of T_4 to T_3 in extrathyroidal tissue.

Iodide is actively transported into thyroid cells and is therefore ouabain sensitive (60). Transport of iodide across the membrane seems to be linked to sodium, and the relevant transport protein is thought to be a co-transporter (61). TSH seems to be the most important factor in regulating iodide transport into follicular cells.

Much of the iodine in the thyroid is bound to Tg, and only a small fraction is present as inorganic iodide. Iodide must be oxidized before it can effectively iodinate the tyrosine residues of Tg, and this is carried out by TPO using hydrogen peroxide as the oxidant. Thyroperoxidase also catalyzes both the iodination of tyrosyl residues and the coupling of iodinated tyrosyl residues to form T_4 and T_3 in Tg. TPO is a membrane-bound hemoprotein whose gene is up-regulated by TSH. TPO activity requires H_2O_2 and is generated on the apical surface through a calcium-sensitive mechanism (TSH also increases H_2O_2 production). The iodination of Tg takes place at the cell-colloid surface close to the apical membrane.

Tg structure is crucial for the coupling reaction responsible for synthesis of T_4. The coupling involves a reaction between either diiodotyrosine (DIT) or monoiodotyrosine (MIT) with DIT to form T_4 or T_3, respectively. Both iodination and coupling occur on Tg. The mechanism of coupling most likely involves radical formation (TPO-dependent) of DIT or MIT. The primary hormone produced by the coupling reaction in the thyroid gland is T_4, although smaller amounts of T_3 are synthesized through coupling between MIT and DIT. The iodinated Tg is stored as an extracellular polypeptide in the luminal colloid. The colloid is resorbed through pinocytosis, and Tg undergoes proteolysis in lysosomes. T_4 is then released through the basal surface into the bloodstream where it is bound by carrier proteins.

2.3.3 Thyroid Hormone Transport, Cellular Uptake, and Metabolism

2.3.3.1 Thyroid Hormone Transport In Blood.
Delivery of T_4 and T_3 from the thyroid gland through blood is primarily accomplished through serum carrier proteins that have varying affinities for the thyroid hormones. A high percentage (99%) of T_4 is protein bound, yet it can be readily released from these proteins for entry into cells. Approximately 70% of circulating T_4 and T_3 are bound by thyroxine-binding globulin (TBG) and this is because of its high affinity for these hormones. TBG is a single-chain polypeptide that is a member of the serine protease inhibitor (serpin) family.

Transthyretin [TTR or thyroxine-binding prealbumin (TBPA)] accounts for approximately 15% of hormone binding, but because of its lower affinity, it is responsible for much of the immediate delivery of thyroid hormones to cells. TTR is also the main thyroid hormone-binding protein in cerebrospinal fluid. Each TTR molecule has two binding sites, and numerous other compounds will also interact with TTR.

Albumin binds 15–20% of thyroid hormones, and they are even more readily dissociated then they are from TTR. Lipoproteins also transport small amounts of thyroid hormones, and their specific uptake by certain cells may be critical for some of the functions of these cells.

These thyroid-binding proteins serve several important functions. The proteins are thought to primarily serve as buffers so that the hormone is released when needed. The binding proteins also seem to serve as an extra-thyroidal storage site for thyroid hormones. This is probably necessary for constant and equal distribution among the tissues.

2.3.3.2 Uptake and Metabolism of T_4.
Whereas T_4 has intrinsic biologic activity, T_3 is thought to be more biologically important. Most T_3 production occurs through deiodination of T_4 in target tissues and may represent a noncirculating pool. A host of other deiodination products are also generated by deiodinase activity. The type I 5'-deiodinase catalyzes 5'-deiodination of T_4, and it is felt this enzyme plays an important role in the production of T_3. Propylthiouracil (PTU) completely

blocks this enzyme. This deiodinase shows absolute preference for L-enantiomers such as L-T_4 and is found in most tissues of the body. A second isoenzyme, type II 5'-deiodinase, seems to be important in generating intracellular T_3 for local use in pituitary, brain, and brown adipose tissue and is not greatly sensitive to PTU. Both enzymes are integral proteins of cellular membranes. 5'-Deiodinases also remove iodines from the 5- or 3-positions of the tyrosyl ring (inner or non-prime ring) of iodothyronines. These processes represent a major component of T_4 and T_3 degradation.

Little is known about transport of thyroid hormones through plasma membranes. It has been suggested that transport proteins can have specific interactions with cell surface receptors that may facilitate transport. Alternatively, thyroid hormones may be transported by diffusion (62–64). Thyroid hormones may also be transported into cells through transport proteins that seem to be linked to sodium transport, and some transporters may be ATP dependent (63). Work by Barlow et al. (65) suggests that certain thyromimetics such as SKF-94901 may exert tissue selectivity through differences in tissue transport. Relatively poorer uptake by the heart versus the liver was invoked to explain robust cholesterol-lowering activity with minimal cardiac effects for SKF-94901. Some degree of selective liver uptake of GC-1 has also been shown and may partially explain some of the liver-specific effects of this agent (66).

2.4 Biologic Actions of Thyroid Hormones

2.4.1 Metabolic Rate.
TR activation will increase metabolic rate in a dose- and time-dependent manner (67–70). Whole body oxygen consumption can be profoundly increased by thyroid hormones, causing weight loss as well as cardiac acceleration (71–74). Most tissues show increased metabolic rates in response to TR stimulation with the notable exception of the brain (75, 76). Increased metabolic rate by TR activation is largely caused by changes in gene expression, particularly with regard to mitochondrial respiration (75, 76), although changes in non-mitochondrial proteins such as Na^+/K^+ ATPase can also be involved (77–79).

Increased metabolic rate has been reported to be caused by both increased ATP turnover and enhanced proton leak and is associated with increased mitochondrial size, density, and surface area (80–82). The precise mechanism for increased respiratory rate may vary by tissue. Early hypotheses placed a great deal of emphasis on the up-regulation of Na^+/K^+ ATPase as a primary mechanism for increased oxygen consumption. A component of the metabolic rate response to thyroid hormones can indeed be blocked by ouabain. Thyroid hormones increase Na^+/K^+ ATPase activity as well as density (and mRNA), and its time course of up-regulation is similar to the rate of increase of oxygen consumption. Nevertheless, the relative importance of up-regulation of this ion exchanger is now thought to be less then previously hypothesized. Respiratory enzymes encoded in both the nuclear and mitochondrial genomes are up-regulated by thyroid hormones. Thyroid hormones increase cytochrome C and β-F_1-ATPase expression, and putative TREs exist for these genes (75, 83, 84). Glycero-3-phosphate dehydrogenase is rapidly induced by T_3. These changes in respiratory gene expression can contribute to increased ATP production and turnover.

Thyroid hormones can cause respiratory uncoupling, particularly in thyrotoxic tissues (67). The importance of increased uncoupling to thermogenesis with lower doses of thyroid hormone is, however, debatable (85). Thyroid hormones are known to increase expression of mitochondrial uncoupling proteins (UCPs), which promote proton leak across mitochondrial membranes and therefore uncouple mitochondria causing a thermogenic effect (67, 81). UCP-3 is regulated by thyroid hormone, particularly in skeletal muscle, and recent studies in UCP-3 overexpressing or knockout mice suggest a thermogenic role for this protein, although evidence for a lack of a thermogenic function for UCP-3 must also be considered (86, 87). UCP-1 in brown adipose tissue (BAT) is directly up-regulated by thyroid hormones and indirectly through sympathetic stimulation (81). Whereas a thermogenic role for UCP-1 may be important in rodents, its importance in primates in not presently clear.

Recently published work suggests that adaptive thermogenesis may be mediated primarily through activation of the $TR\alpha_1$ receptor isoform, whereas $TR\beta_1$ may not be as effective (88). These results indicate that UCP-1 levels are directly regulated by $TR\beta_1$, but that the action of thyroid hormone on sympathetic stimulation (which is required for adaptive thermogenesis) is regulated by $TR\alpha_1$. Other studies have further suggested the importance of $TR\alpha_1$ using $TR\alpha^{-/-}$ mice, showing that in the absence of $TR\alpha_1$, body temperature is reduced (89). $TR\beta_1^{-/-}$ mice have normal body temperatures or metabolic rates, suggesting a critical role for $TR\alpha$ in maintenance of metabolic rate (90). It is presently unclear whether stimulation of $TR\beta_1$ can cause increases in metabolic rate upon pharmacologic stimulation, but studies suggest not only a potential role for $TR\beta_1$ in increasing metabolic rate, but a profile that may be different from that of $TR\alpha_1$ stimulation (66, 91, 92).

The ability of thyroid hormones to increase metabolic rate is critical for their capacity to cause weight loss in hyperthyroid patients. Whereas uncoupling of mitochondria may contribute to increased metabolic rate, increased ATP turnover and use are most likely more important.

2.4.2 Lipid Metabolism. Thyroid hormones induce lipid synthesis, mobilization, and degradation in a variety of tissues (93–95). Triglycerides are turned over more rapidly and lipoprotein lipase and chylomicron triglyceride clearance are enhanced. Fatty acid synthesis is increased in adipose tissue and liver, but degradation seems to be stimulated to a greater extent (96). Therefore, adipose tissue releases fatty acids at an accelerated rate after thyroid hormone treatment. Thyroid hormones also increase sensitivity of lipolysis to catecholamines (97). The increased triglyceride/fatty acid cycling may also cause an increase in oxygen consumption. Lipogenic enzymes in the liver are rapidly up-regulated, including spot 14 and malic enzyme (90).

Plasma low density lipoprotein (LDL) cholesterol levels are reduced by thyroid hormones. Hepatic LDL receptors are up-regulated by thyroid hormones, causing increased LDL cholesterol uptake by liver (95). While cholesterol synthesis rates are increased by thyroid hormones, cholesterol degradation and biliary excretion are also increased. The balance of these activities is such that LDL cholesterol in plasma is reduced. Apo A1 may be increased by thyroid hormones that may act to increase high density lipoprotein (HDL) cholesterol and enhance "reverse" cholesterol transport (95). There is evidence that thyroid hormones can reduce the lipoprotein Lp(a), which is thought to be an important cardiac risk factor (98–100).

2.4.3 Carbohydrate Metabolism. Glucose and carbohydrate use by non-hepatic tissue is markedly increased by thyroid hormones (101). The increased use occurs concomitantly with increased glucose production through liver gluconeogenesis (102). Thyroid hormones increase peripheral glucose uptake primarily through increased expression of glucose transporters such as GLUT-4 (103). Whereas thyroid hormones can increase insulin release under certain circumstances, the increased uptake and use of glucose by tissue such as skeletal muscle is relatively independent of insulin levels.

2.4.4 Protein Metabolism. Excess thyroid hormone causes a catabolic state characterized by increased metabolic rate as well as accelerated protein loss or wasting. Thyroid hormones cause both protein synthesis and degradation, but under thyrotoxic conditions, protein loss predominates. Variable degrees of muscle atrophy can occur depending on the thyroid state, as well as sympathetic input (68). Any thyromimetic design as a lipid-lowering or anti-obesity agent must not cause significant loss of lean body mass or loss of skeletal muscle mass and strength.

2.5 Thyroid Hormone Effects on Specific Tissues

2.5.1 Heart and Cardiovascular System. Some of the most profound effects of thyroid hormones are seen in heart. Thyroid hormones cause dose-dependent increases in heart rate (positive chronotropic effect) as well as effects on cardiac function (positive inotropic and lusitropic effects) (72, 73). Cardiac output is increased by thyroid hormones, and increases in ventricular muscle mass can be observed (cardiac hypertrophy) (71, 104). Thyroid hormones typically lower peripheral vascular resistance and increase total blood volume. Studies in $TR\alpha^{-/-}$ and $TR\beta^{-/-}$ mice suggest that much of the cardiac effects of thyroid hormone are caused by stimulation of $TR\alpha_1$ (89), although some role for $TR\beta$ may also be important (66, 91).

Total protein synthesis in the heart is increased by thyroid hormones, much of which is consistent with observed changes in contractility and cardiac relaxation. The transcription of several critical genes is affected by thyroid hormones, with the best characterized being myosin heavy chain (MHC), sarcoplasmic reticulum Ca^{2+} ATPase (SERCA), and Na^+/K^+ ATPase genes (16). There are two MHC genes (α and β) with the α-gene predominant in rodents and β-MHC predominant in man. α-MHC (myosin V1) has a higher velocity of contraction and higher ATPase activity compared with β-MHC. Thyroid hormone increases the relative expression of α-MHC versus β-MHC; this is caused by increases in α-MHC production as well as reduced β-MHC gene expression (105). The shift to the α-MHC isoform is associated with increased contractility and energy use. Putative TREs have been identified for both genes.

Several ion channels or transporters have been shown to be affected by thyroid hormones. SERCA2 mRNA is markedly increased by thyroid hormones (106). SERCA are critical for sequestration of calcium during relaxation, and therefore, increased expression or activity of this pump can increase the rate of ventricular relaxation. Other ion channels or transporters thought to be affected by thyroid hormones are Kv1.5, KV 4.2, HCN2, HCN4, and Na/Ca^{2+} exchangers (16, 107). Changes in these ion channels may be critical for the chronotropic, action potential shortening, and inotropic effects of thyroid hormones. Thyroid hormones may also exert acute effects on some critical cardiac ion channels such as sodium, and some of these changes can occur rapidly (108). Some of these rapid changes may result from alternative, transcription-independent, T_3 signaling pathways (non-genomic), but this is still under debate.

Thyroid hormones also increase sensitivity to catecholamines, partially through in-

creased expression of β-adrenergic receptors (73, 109). Thyroid hormone may also increase the cAMP response to catecholamines, as well as enhance the effects of cAMP. Increased catecholamine release is also reported for thyroid hormones. Increased action or sensitivity to catecholamines may contribute to tachycardia, ion channel disturbances, and arrhythmias in diseases such as hyperthyroidism.

2.5.2 Liver. An important site of action for thyroid hormones is liver, where numerous metabolic processes are altered. Thyroid hormones stimulate both lipogenesis and lipolysis. Lipogenic enzymes such as malic enzyme, fatty acid synthase, and the protein Spot 14 are up-regulated rapidly (110–112). Cholesterol metabolism is also markedly affected, and thyroid hormone can reduce serum cholesterol, an effect caused by increased hepatic LDL receptor expression as well as increased cholesterol degradation. Thyroid hormones may also regulate expression of other enzymes involved with cholesterol metabolism (16). Cholesterol metabolism may also be affected by increased expression of apoA-1, which may explain the effects of thyroid hormones on HDL cholesterol and the potential for "reverse" cholesterol transport.

Hepatic metabolic rate is increased by thyroid hormones, and this is associated with profound up-regulation of Na^+/K^+ ATPase levels. Hepatic mitochondrial respiration is also increased by thyroid hormones, and this may be caused by changes in mitochondrial density and respiratory enzymes, UCP expression, and fatty acid "cycling."

2.5.3 Bone and Skeletal Muscle. Thyroid hormones can increase both bone formation and degradation and are essential for normal bone development (113, 114). Thyroid hormones increase both osteoclast and osteoblast activity, and indeed, cause enhanced calcification accompanied by increased bone resorption. Thyroid hormone may also indirectly affect bone growth through effects on growth hormone and consequently IGF-1. Whereas both bone formation and resorption are increased by thyroid hormone, thyrotoxic patients often seem to be subject to net bone loss and are at risk for osteoporosis (115).

Table 11.1   Phenotypes of TR Receptor Knockouts

Name of Targeted Mutant Mice	Remaining Receptor(s)	Phenotype	References
$TR\alpha_1^{-/-}$	$TR\alpha_2$, all $TR\beta$	Low heart rate and body temperature; aberrant repolarization by I_{to}; aberrant expression of cardiac myosin; cerebellar development resistant to hypothyroidism; deficient in cued response test; aberrancies in brain cortex, decreased lordosis behavior	89, 123, 125, 128–132
$TR\beta^{-/-}$	all $TR\alpha$	Hyperthyroid; deaf; HR low-responsive to T_3; audiogenic seizures; deficiency in cholesterol homeostasis; dysregulation of liver genes; increased lardosis	118, 126, 128, 130, 133, 134
$TR\alpha_1/\beta^{-/-}$	$TR\alpha_2$	Elevation of T_3, T_4, TSH; hypothyroid phenotype; resistant to loss of weight by hypothyroidism; pituitary and hypothalamic dysfunction; small bone abnormalities; reduced body fat; low heart rate; low body temperature; low basal metabolic rate; cold intolerant; deaf; cochlear misdevelopment; defective cholesterol homeostasis, immense goiter	121, 125, 128, 130, 135–137
$TR\alpha_2^{-/-}$	$TR\alpha_1$, all $TR\beta$	Overexpress $TR\alpha_1$ 6×; mixed hypo- and hyperthyroidism; low fertility	118
$TR\alpha_2/\beta^{-/-}$	$TR\alpha_1$	Rescue by elevated $TR\alpha_1$ of deafness and pituitary function, minor rescue of liver function	122
$TR\beta_2^{-/-}$	All $TR\alpha$; $TR\beta_1$ and β_3	No retinal M cones; color blind; mild elevation in TH, TSH	121

Thyroid hormones cause changes in skeletal muscle that are consistent with other cell types. Metabolic rate is increased concomitantly with changes in mitochondrial density and respiratory enzymes. Glucose uptake is increased primarily because of up-regulation of GLUT-4 (116). Thyrotoxicosis is associated with net protein turnover, and therefore, muscle wasting and weakness can be observed (16). Thyroid hormones have been shown to up-regulate gene expression of ubiquitin-dependent proteasomes, which account for the major part of myofibrillar proteolysis and breakdown (117).

2.5.4 Receptor Isoform Knockouts. The role of the different subtypes of TR *in vivo* was partially clarified from the phenotypic characterization of different TR knockout mice. A panel of mutated mice have been generated and include mice that lack $TR\beta^{-/-}$ (118), $TR\beta_2^{-/-}$ (30), $TR\alpha_1^{-/-}$ (89), $TR\alpha_2^{-/-}$ (119), and $TR\alpha^{-/-}$ (120). These mouse strains have

been crossed to generate mice with combined TR gene mutations: $TR\alpha_1^{-/-}\ TR\beta^{-/-}$ (121), $TR\alpha_2^{-/-}\ TR\beta^{-/-}$ (122, 123), and the total TR deficient mice $TR\alpha^{-/-}\ TR\beta^{-/-}$ (124). This section reviews the phenotypic characteristics of the various TR knockout mice (29).

TRβ plays a key role in regulating the hypothalamic-pituitary axis. This is manifested by the resistance to thyroid hormone and elevated levels of TSH seen in the $TR\beta^{-/-}$ mice and even more pronounced in the $TR\alpha_1^{-/-}\ TR\beta^{-/-}$ mice. Studies of these TR-deficient mice suggest that TRα_1 is important in controlling basal heart rate as well as body temperature (89, 125). TRβ is the major TR isoform expressed in the liver, and it is therefore not surprising that the cholesterol-lowering effect of thyroid hormone requires this TR in mice, demonstrated in the $TR\beta^{-/-}$ mice (126).

The phenotypic characteristics of most TR knockout mice are summarized in Table 11.1. They reveal distinct and non-overlapping *in vivo* functions for the different TR subtypes.

Although distribution of the receptor isoforms may partially explain the differential tissue selective effects, differential receptor function cannot be ruled out. These results support the proposal that isoform or subtype selective thyromimetics may prove to be of therapeutic benefit (127).

2.6 Thyroid Diseases

2.6.1 Hypothyroidism.
Hypothyroidism is a common thyroid-related disorder. This occurs either through reduced thyroid production of thyroid hormones (primary or thyroidal hypothyroidism) or through diminished pituitary TSH release by TRH (central hypothyroidism). Primary hypothyroidism is characterized by an elevated TSH in blood and often results from destruction of thyroid tissue or altered hormone biosynthesis. This is caused by autoimmune thyroiditis (e.g., Hashimoto's thyroiditis), irradiation, drug-induced hypothyroidism, or infiltrative destruction of thyroid tissue (e.g., sarcoidosis, amyloidosis).

The effects of hypothyroidism are variable, with symptoms ranging from overt to subclinical. Subclinical hypothyroidism is defined as increased TSH with little change in free plasma thyroid hormone concentrations. While there is great variability in clinical symptoms or signs, there are many common features. Gross symptoms include fatigue, lethargy, mental depression, weight gain, cold intolerance, dry skin, bradycardia, nonpitting edema (myxedema), and constipation.

The cardiovascular system is often affected by hypothyroidism with increased peripheral resistance concomitant with reduced metabolic demand. Cardiac performance is reduced both during systole and diastole. This is because of modulation of numerous genes including MHC subtypes, SERCA, and respiratory enzymes. Mean arterial blood pressure is usually unaltered, but cardiac output, blood volume, and cardiac mass are typically reduced.

Lipid synthesis will typically predominate over lipolysis, and therefore, hypothyroid patients have increased adiposity. Also, triglycerides and LDL cholesterol are often elevated.

Peripheral glucose uptake from gut and in peripheral tissues is diminished, except in the brain.

2.6.2 Hyperthyroidism and Thyrotoxicosis.
Hyperthyroidism is a condition caused by increases in thyroid hormone production or secretion, whereas thyrotoxicosis is the metabolic response to increased T_4 or T_3 levels. Therefore, the aforementioned terms are not always synonymous. Examples of thyrotoxicosis associated with hyperthyroidism are as follows: Grave's disease (stimulatory TSH receptor antibodies), thyroid neoplasm or tumors, and less commonly drug-induced hyperthyroidism. The most common form of thyrotoxicosis without hyperthyroidism is excessive exogenous thyroid hormone treatment.

The clinical signs of thyrotoxicosis are varied but consistent with the known effects of thyroid hormones on tissues. Metabolic rate increases are followed by increased appetite, weight loss, hyperactivity, tremor, and cardiac palpitations. Despite the hyperactivity, a feeling of fatigue and weakness often accompanies thyrotoxicosis. The skin can be warm and moist with abundant perspiration. Tachycardia is a common feature of thyrotoxicosis. Ophthalmologic pathology is also noted and characterized by a gritty sensation in the eyes, increased retroocular pressure, blurred vision, and photophobia. Ocular muscle pathology can cause blurring as well as the classical Grave's ophthalmopathy characterized by a staring appearance of the eyes. Thyrotoxicosis can also be associated with osteoporosis caused by the preponderance of bone resorptive activity that can be seen under this condition.

These signs and symptoms of thyrotoxicosis are consistent with what is known about the action of thyroid hormones on various tissues. Increased metabolic rate will cause weight loss, and indirectly, tachycardia. Of course, thyroid hormones will also directly increase heart rate and cardiac output. Muscle loss caused by increased protein turnover is evident and may explain some of the muscle weakness. Sympathetic stimulation will exacerbate tachycardia, sweating, and increased metabolic rate.

3 THERAPEUTIC POTENTIAL OF THYROID HORMONE ANALOGS

3.1 Introduction

An obvious example of therapeutic use for thyroid hormones is replacement therapy for hypothyroidism. The purpose of this treatment is to bring patients back to a "normal" metabolic state. However in addition to replacement therapy, there is great potential for thyroid hormone agonists or antagonists for treating a variety of disorders, although achieving selectivity of action will be critical for the development of such compounds. Selective thyromimetics have long been sought. Mechanistic explanations for some early selective agents focused on how affinity and transport between cellular compartments (membrane, cytoplasm, and nucleus) might be dissimilar among different cell types (65, 138). With modern understanding of the molecular biology of thyroid receptors, it can now be envisioned that selective agents might also result from TR isoform selectivity or possibly tissue specific transactivation arising from the selective recruitment of coactivator proteins (139, 140). Such newer approaches to designing safe and effective thyromimetics are starting to be described in the literature, and this chapter will discuss some key examples.

The following is a short list of potential indications for drug development of selective thyromimetic drugs.

3.2 Lipid Lowering

Thyroid hormone analogs will reduce cholesterol levels by increased uptake and metabolism of LDL by the liver. HDL cholesterol may be maintained, or in some animal models, may be increased. Apo A1 is up-regulated by thyroid hormones and may exert a beneficial effect through "reverse" cholesterol transport. Lp(a) has been shown to be reduced by thyroid hormones, although this needs to be further studied. Thyroid hormone will reduce cholesterol in animal models but is associated with numerous unwanted side effects at efficacious doses. Several agents have been reported that can lower plasma cholesterol without significant cardiac effects, and these will be discussed later in this review (66, 138).

3.3 Obesity

Hyperthyroidism is associated with weight loss, and this is most likely secondary to increased metabolic rate and reduced adiposity. Thyrotoxic patients can also lose lean body mass, which is not desirable, and therefore metabolic rate can only be slightly increased for there to be a selective loss of adipose tissue without undesirable cardiac and catabolic effects. The dose–response curve for hypermetabolism is so steep for T_3 that no therapeutic window exists. The requirement for a broad dose range in which therapeutic 5–10% increases in metabolic rate can be observed is a difficult hurdle to cross for thyromimetics. However, novel thyromimetics have been reported that show promise in this regard, and these will be discussed below. Obese patients are often at risk for diabetes. Selective thyromimetics may additionally provide an anti-diabetic effect through glucose uptake by increased skeletal muscle GLUT-4 activity.

3.4 Cardiac Indications

Several studies have shown beneficial inotropic effects for thyroid hormones (141). Thyroid hormones have been used for inotropic support after surgical procedures such as coronary bypass and graft. Of course, these agents are non-selective and need to be used acutely. The mechanism for the acute inotropic effects are not clear, but several ion channels have been shown to be acutely affected by thyroid hormones, perhaps through nongenomic mechanisms.

Interesting work from several investigators suggested the possibility of efficacy for congestive heart failure (73). This is based in part on the up-regulation of SERCA in myocardium. This calcium ATPase is responsible for re-uptake of calcium during diastole and would therefore be expected to improve diastolic function. For such agonists to be useful, however, selectivity for SERCA would be critical because cardiac acceleration and a hypermetabolic state would be contraindicated. Some evidence has emerged suggesting that TRβ stimulation may up-regulate SERCA while exerting minimal heart rate effects (66).

This suggests the exciting possibility of selective improvement of diastolic function in the failing heart.

TR antagonists may be useful for treating cardiac arrhythmias. Amiodarone, which has been characterized as possessing thyroid hormone antagonist activity (however it also has a myriad of ion channel effects), is a highly efficacious anti-arrhythmic agent (142). By down-regulating the expression of cardiac potassium channels, TRα blockade would be expected to prolong action potential duration, increase refractory period, and therefore inhibit reentrant arrhythmias. It is possible that a selective TRα antagonist would have class III anti-arrhythmic activity without the reverse use-dependency seen for many of the agents in this class that block repolarizing ion currents.

3.5 Hypothyroidism

Hypothyroidism was initially treated using thyroid gland extracts to replace the deficient hormone. Purification of thyroid hormones allowed for effective and safe replacement therapy. Synthetic T_4 (Levothroid, Synthroid) is the most commonly used preparation, and it is well absorbed and has a long plasma half-life, allowing for small daily fluctuation of plasma T_4. A wide variety of pill strengths (25–300 μg) allows titration that is readily followed using TSH and T_4 measurements. Accurate titration is critical to prevent excessive T_4 and associated cardiac, metabolic, and musculo-skeletal complications. T_3 (Cytomel) is less advantageous because its rapid absorption and short half-life can result in wide fluctuations in plasma levels.

3.6 Other Disorders

Thyroid receptors have been identified in skin (143–145), a classic target tissue for thyroid hormone action. A topical T_3-containing cream has been shown to stimulate epidermal proliferation and dermal thickening (146). The use of thyroid hormone analogs to treat skin atrophy has been proposed (147).

As mentioned above, thyroid hormone increases the activity of both osteoblasts and osteoclasts, and as a result, both bone formation and resorption are increased. However, treatment with a thyroid hormone analog in combination with a bone resorption inhibitor would result in net bone formation, which would be useful in treating osteoporosis (18).

4 STRUCTURE-ACTIVITY RELATIONSHIPS AND BIOLOGIC ACTIVITIES OF THYROID HORMONES AND ANALOGS

4.1 Introduction

The early foundations of structure-activity relationships (SAR) for thyroid hormone analogs were built largely through the use of *in vitro* receptor-binding assays and *in vivo* tests for thyroid hormone activity, although interpretation of the latter may be complicated by metabolic processes (activation or deactivation) or pharmacokinetic factors. This discussion will primarily focus on the *in vitro* SAR of compounds. Although receptor isoforms and subtypes for TR were uncovered in the late 1980s, relatively little *in vitro* binding SAR for thyromimetics has been reported using the TRα and TRβ receptor isoforms. In addition, modern assays of nuclear hormone receptor activity now include transactivation assays in genetically engineered reporter cell lines that determine functional agonism and antagonism. Here again, there is relatively little data available for thyromimetics. In addition to delineating basic SAR, this section will also discuss the *in vivo* activity of a selection of key examples from the published thyromimetic literature. Excellent reviews of thyromimetic SAR developed before 1996 are available (148–150).

4.2 Active Hormone Conformation and Receptor-Ligand Structures

It has long been established by a variety of experimental measurements and theoretical calculations that substitution at the R_3 and R_5 positions on the diphenylether scaffold of thyroid hormones and their analogs enforces a perpendicular orthogonal relationship between the two aromatic rings as depicted in Fig. 11.3 (i.e., where one ring is in the plane of the page and the second is projecting out of the plane of the page (151, 152). Furthermore, SAR studies, including the testing of confor-

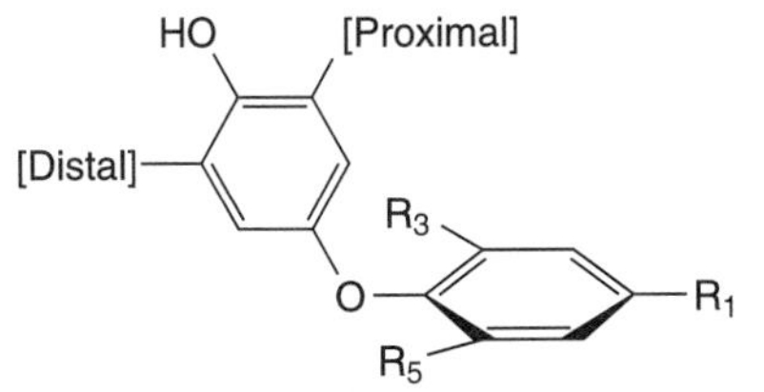

00

Figure 11.3. Active hormone conformation.

mationally constrained analogs, indicated that for TH analogs in which $R_5' = H$ (e.g., T_3), the active conformation is one in which the R_3' group is pointed away from the non-prime ring (153, 154). This orientation has been termed the "distal" orientation, as opposed to "proximal," where the group points back toward the non-prime ring (Fig. 11.3). X-ray crystallographic studies of the ligand-binding domains of TRα and TRβ with various natural and synthetic ligands have fully corroborated the favored distal orientation of T_3 and its analogs (155–158).

With the ability to clone and express the isolated ligand-binding domains of the thyroid hormone receptor isoforms, there are presently several X-ray crystallographic structures available of ligand-receptor complexes (155–158). These structures reveal certain common features of the various ligand-binding interactions of the natural hormone and its analogs. Merging this structural data with established thyromimetic SAR, the nature of the ligand contacts to the receptor can be summarized as depicted in Fig. 11.4. The bound ligand is fully engulfed by the LBD, suggesting a role in protein folding and stabilization. A hydrogen bond between the 4'-hydroxyl group

and imidazole ring of His-381 (TRα numbering) in the receptor seems to be important for binding as well as functional (agonist) activity. Groups at R_3' and R_5' on the prime ring and R_3 and R_5 of the non-prime ring bind into relatively hydrophobic regions of the receptor. The R_1 substituent, which generally contains polar acidic functions such as the α amino acid moiety in T_3 and T_4, binds into a hydrophilic region of the receptor that features several arginine residues common to both receptor isoforms. Interestingly, the amino group of T_3 seems to be deprotonated when bound to the receptor because it is within hydrogen bonding distance of the backbone amide nitrogen of Ser-271 (TRα numbering). This region of the ligand-binding cavity is also the location of the only amino acid difference between the α and β receptor isoforms (Ser-271 in TRα and Asn-331 in TRβ) that exists in the immediate vicinity of the bound ligands. This amino acid difference could theoretically be used in the design of isoform selective thyroid hormone analogs, although their similarity (both possessing neutral polar side-chains) presents significant challenges to selective thyromimetic drug design.

4.3 Receptor Binding of Thyroid Hormone Analogs

4.3.1 Background. Because of the relatively recent discovery of receptor isoforms for TR, the published *in vitro* SAR of thyromimetics contains little information about the binding affinities of analogs to the TRα and TRβ receptors. Instead, early SAR work in the field generated a considerable body of data using competition binding assays with the readily

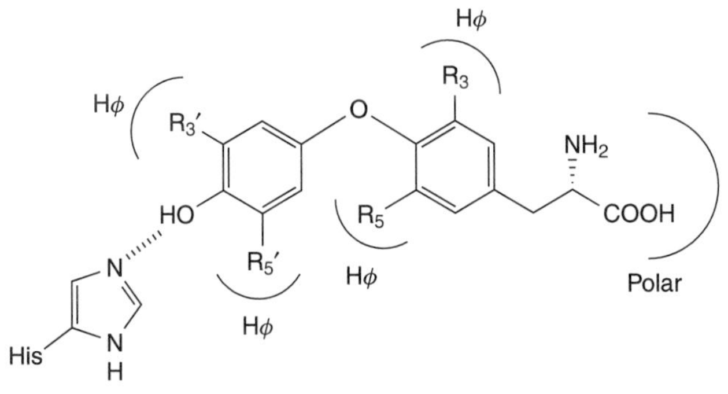

Figure 11.4. Hormone-receptor binding interactions.

Table 11.2 SAR of T$_3$ Analogs

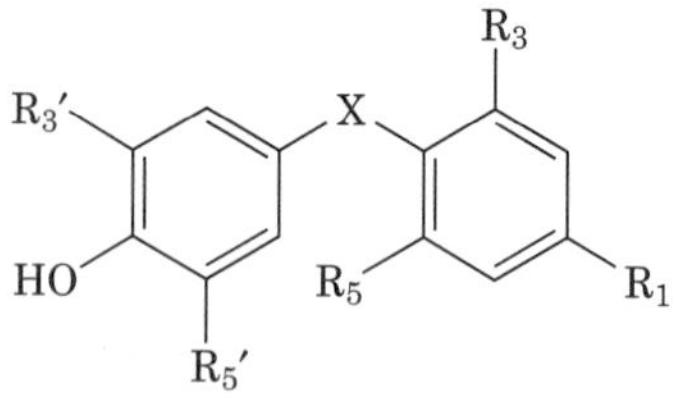

Compound	R4′	R3′	R5′	X	R3, R5	R1	*In Vitro*[a] (binding)	*In Vivo*[b] (antigoiter)
(**1**) L-T3	OH	I	H	O	II, I	L-Ala	100	100
(**2**) L-T4	OH	I	I	O	I, I	L-Ala	14	18
(**3**) D-T3	OH	I	H	O	I, I	D-Ala	63	7.5
(**4**)	H	I	H	O	I, I	DL-Ala	4.6	>27
(**5**)	OCH$_3$	I	H	O	I, I	L-Ala	1.3	11
(**6**)	OH	i-Pr	H	O	I, I	L-Ala	92	142
(**7**)	OH	H	H	O	I, I	L-Ala	0.012	0.8
(**8**) rT3	OH	I	I	O	I	DL-Ala	0.75	0.06
(**9**)	OH	I	H	O	I, I	CO$_2$H	85	0.05
(**10**) Triac	OH	I	H	O	I, I	CH$_2$CO$_2$H	282	6.5
(**11**)	OH	I	H	O	I, I	(CH$_2$)$_2$CO$_2$H	234	4.5
(**12**)	OH	I	H	O	I, I	(CH$_2$)$_3$CO$_2$H	14	1
(**13**)	OH	I	H	CH$_2$	I, I	DL-Ala	183	54
(**14**)	OH	I	H	S	I, I	L-Ala	185	14
(**15**)	OH	i-Pr	H	O	Br, Br	DL-Ala	36	30
(**16**)	OH	i-Pr	H	O	CH$_3$, CH$_3$	L-Ala	0.4	3.6
(**17**)	OH	i-Pr	H	O	i-Pr, i-Pr	DL-Ala	0.2	

[a]Activity expressed relative to T3 = 100% in binding to solubilized and intact nuclear receptors from rat liver cells.
[b]Activity relative to T3 = 100% for preventing goiter formation in the rat.

available nuclear TRs from sources such as rat liver cells. *In vivo* SAR was also studied primarily using the classical rat antigoiter assay. In this assay, thyroid hormone activity was measured as the ability of the test compound to suppress TSH release and goiter formation that is induced by antithyroid drugs such as thiouracil (148). Because TSH suppression is a TRβ process, it can be now appreciated that the antigoiter assay suffers from the limitation that it is insensitive to the test compound's binding to the TRα receptor isoform.

4.3.2 Summary of Basic Thyromimetic SAR. Table 11.2 summarizes key SAR features, *in vivo* (antigoiter assay) and *in vitro*, of analogs of T$_3$ (L-3,5,3′-triiodothyronine) (148, 153, 159). As mentioned earlier, T$_3$ (**1**) is more active than T$_4$ (**2**) both *in vitro* (receptor-binding affinity) and *in vivo*. Although the relative *in vitro* and *in vivo* activities between these compounds is consistent, it must be reiterated

that the *in vivo* activity of T$_4$ also reflects its conversion to T$_3$ in most tissues via intracellular deiodinase activity. The D-alanine analog (**3**) retains significant binding affinity and moderate *in vivo* activity. In compound (**4**), some *in vivo* activity is retained likely resulting from metabolic hydroxylation at the 4′ position. Consistent with the binding model in Fig. 11.5, compound (**5**) shows weak binding caused by the loss of a critical H-bond interaction. However, *in vivo* potency is retained in

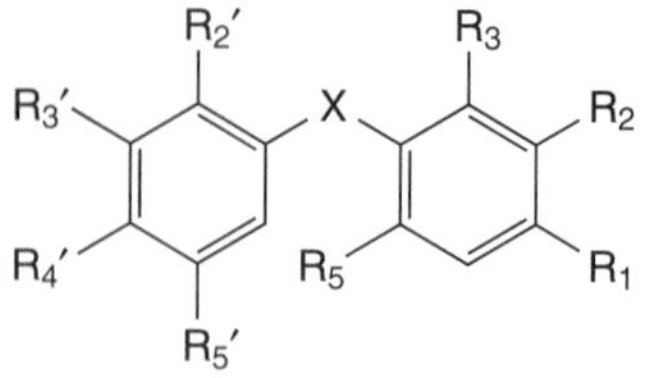

Figure 11.5. Thyroid hormone pharmacophore SAR.

Table 11.3   SAR of 3′-Position of Thyromimetics

Compound	R3′	In Vitro Binding*	Compound	R3′	In Vitro Binding[a]
(6)	H	0.08	(30)	F	0.76
(18)	Methyl	0.6	(31)	Cl	5.1
(19)	Ethyl	56	(32)	Br	13
(20)	n-Propyl	45	(1) L-T3	I	100
(21)	n-Butyl	44	(33)	OH	0.04
(22)	n-Pentyl	32	(34)	CH_2OH	0.39
(23)	n-Hexyl	60	(35)	CH_2CH_2OH	0.45
(7)	i-Propyl	87	(36)	COOH	0.004
(24)	t-Butyl	25	(37)	CH_2COOH	0.03
(25)	Cyclohexyl	9.8	(38)	CH_2NH_2	0.01
(26)	CH_2-Cyclohexyl	36	(39)	CHO	0.18
(27)	Phenyl	7.3	(40)	CO-CH_3	0.37
(28)	CH_2-phenyl	22	(41)	CO-Ph	0.37
(29)	CH_2CH_2-phenyl	1.8	(42)	NO_2	0.13

[a]Binding to intact rat liver nuclei, relative to T3 = 100%.

(5) through metabolic activation, in this case de-alkylation of the 4′ methyl ether. An important modification commonly used in thyromimetic structures is the replacement of the 3′ iodo substituent by isopropyl as shown in compound (6). This change retains the full binding affinity of T_3 and significantly improves the *in vivo* antigoiter activity because (6) is impervious to deactivation by 3′ deiodinase activity. If the prime ring is missing both the $R_3′$ and $R_5′$ substituents (7) or if the non-prime ring has only one group at R_3 or R_5 (8), an analog known as reverse-T_3 or rT_3, activity is significantly reduced. Compared with the large loss of binding affinity incurred by removing the 3′ iodo of T_3, considerable variation at R_1 is tolerated (9–12), with improved binding resulting from removing the amino group in compound (10) and its homolog (11). Compound (10) is also known by its trivial name, "Triac," which is derived from triiodo and acetic (referring to its R_1 group). Further homologation (12) results in a loss of binding affinity. Hydrophobic, one atom linker replacements for oxygen at "X," such as methylene (13) and sulfur (14), are successful. At R_3 and R_5, bromine (15) substitutes for iodine with only a modest

drop in binding affinity. However, substitution with methyl (16) or isopropyl (17) is far more deleterious to binding.

The successful replacement of the 3′ iodo group in T_3 by isopropyl has been followed by many thyromimetics featuring a variety of 3′ groups. Table 11.3 illustrates the SAR at the 3′ position for binding to thyroid hormone receptors from intact rat liver nuclei (160). Among alkyl substituents (18–26), isopropyl seems to be optimal. A 3′ phenyl group (27) exhibits about 10% of the binding affinity of T_3, and the addition methylene spacers (28, 29) have little effect. Iodine is clearly the best halogen group at 3′ (30–32). Compounds (33–42) show that polar groups at 3′ bind very poorly relative to T_3.

The following is a summary of the essential *in vitro* SAR of thyroid hormone analogs developed before 1990 (Fig. 11.5). At R_1, the L-amino acid moiety found in T_3 is not critical for activity, and the amino group can be dispensed with entirely. A recent report describes the successful incorporation of a thiazolidinedione acid surrogate (161). Relatively little information is available on analogs with substituents at R_2 and its companion position,

R_6. Positions R_3 and R_5 are essentially always substituted symmetrically with halogens or small alkyl groups. Polar substituents are not well tolerated. The linker X is a single atom, oxygen being most common, but sulfur and methylene are also acceptable. On the prime ring, R_2' and R_6' have rarely been substituted, although successful simultaneous substitution at R_2' and R_3' has been reported by replacing the prime ring with a tetrahydronaphthyl or naphthyl ring (162). As discussed earlier, R_3' is optimal with iodine or isopropyl, and varying degrees of suitable affinity can be retained with non-polar alkyl, aralkyl, or aryl groups. However, polar groups at R_3' markedly reduce binding affinity. Simultaneous substitution at R_3' and R_5' generally leads to poorer binding (i.e., T_3 versus T_4). The phenolic hydroxyl group at R_4' is critical to both binding and agonist activity, and a structural role in hydrogen bonding to a histidine side-chain imidazole ring from the receptor has been elucidated through X-ray studies (as described earlier). There are no known significant alternatives to a phenol moiety at R_4'.

For some of the newer thyromimetics, it has become clear that the classic SAR conclusions based on T_3 may not hold for compounds with different groups at R_1. For example, when the R_1 group is an oxamic acid (— NH-COCOOH), the analog with methyl substituents at R_3 and R_5 binds to rat liver nuclear TRs better than T_3 (163). This result contrasts with the comparison of T_3 to compound (16) in Table 11.2 in which methyl substitution at R_3 and R_5 causes about a 100-fold loss in binding affinity. Further examples of SAR divergence can be expected with thyromimetics having R_1 groups differing from the α amino acid found in T_3.

Very little SAR has been reported describing the receptor isoform-binding affinities of thyroid hormones and their analogs. Although there is some variation perhaps reflecting dissimilar assays, several reports (164–166) indicate that T_3 does not discriminate in binding to the TR isoforms (less than a twofold difference in binding affinity between the TRα and TRβ receptors). Future discussions of thyromimetic *in vitro* SAR can be expected to address not only binding selectivity but also se-

lectivity of functional (agonist and antagonist) activity determined using TR-mediated transcription assays. Thyromimetic SAR for functional activity will also be described in the context of both TR receptor isoforms and TR response elements.

4.4 Selected Biologically Active Thyroid Hormone Analogs

4.4.1 Background. Although thyroid hormones are associated with a variety of biologic activities, medicinal chemistry has primarily focused on their lipid-lowering properties, an emphasis motivated by the large market potential and significant cardiovascular health benefits of such drugs. A key to the success of a thyromimetic lipid-lowering drug is the absence of undesirable cardiac effects. The unnatural enantiomer of T_4 was once believed to be such a compound. However, in human clinical trials, D-T_4 caused greater mortality than placebo (167). Although there have been subsequent reports of selective thyroid hormone analogs with the potential to safely lower lipid levels in man (discussed below), research activity in this area has been somewhat modest because other approaches, especially the HMG-CoA reductase inhibitors, have found significant success as established and safe therapy. Recently, more research is being reported on the weight-reducing potential of thyromimetics and their applicability to obesity, a major health problem in essentially all developed nations. The data for selected thryomimetics with therapeutic potential are discussed below. Two excellent recent reviews are available (18, 168).

4.4.2 L-T_3 (Liothyronine), L-T_4 (Levothyroxine), and Triac. Of the naturally occurring thyroid hormones available for therapeutic use, L-T_4 (**2**, levothyroxine) is more commonly used than L-T_3 (**1**, liothyronine). Levothyroxine is indicated for TH replacement therapy or supplemental therapy for hypothyroidism. Because T_3 and T_4 are not selective in their *in vivo* activities, they are not used for therapeutic weight reduction. In man, the sodium salt of (**2**) is well absorbed from the GI tract, and the drug is principally metabolized through deiodination processes and conjugation to

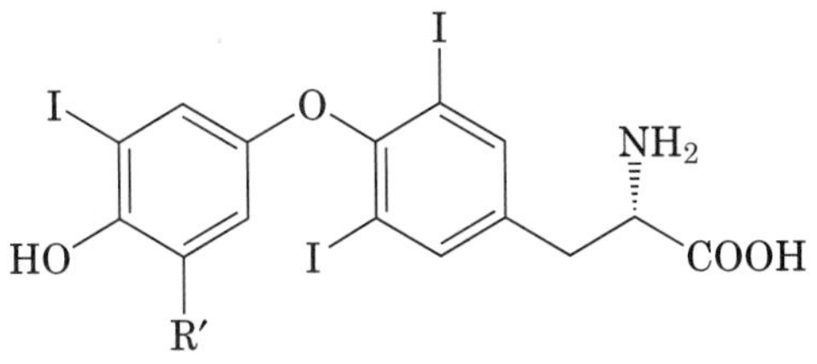

(1) R = H L-T$_3$
(2) R = I, L-T$_4$

glucuronides and sulfates (which are excreted into the bile and gut where they undergo enterohepatic recirculation) (169).

Triac (10), which is a known metabolite of T$_4$, has been shown to be of potential benefit

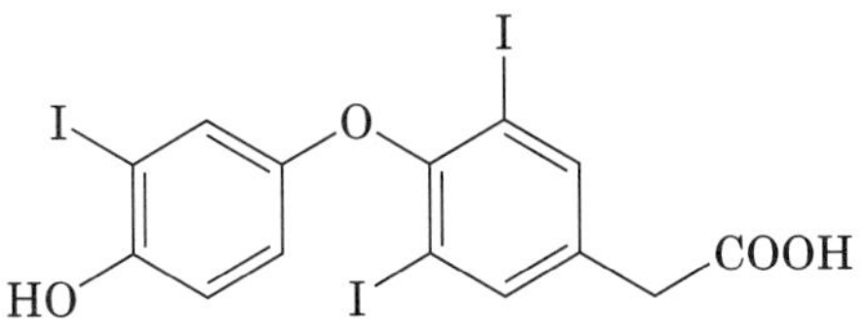

(10) Triac

for the treatment of resistance to thyroid hormone (RTH). RTH is a genetic disorder characterized by diminished responsiveness to thyroid hormones in the pituitary and peripheral tissues caused by mutations in the LBD of the TRβ_1 receptor. In a study comparing Triac to T$_3$, both agents were reported to bind TRα_1 equally, but Triac showed 2.7-fold greater affinity for TRβ_1 (164). In the same study, Triac produced higher maximal transactivation than T$_3$ and attained 50% induction at a lower concentration than T$_3$ in a TRβ_1/TRE-chloramphenicol acetyltransferase (CAT) reporter assay run in transiently transfected COS-1 cells. In contrast, Triac activated TRα_1 to the same degree as T$_3$. With these results, it was argued that administration of Triac in cases of RTH augments TRβ_1 activity with less TRα-mediated hyperthyroid effects. A second report describing TR isoform specificity of Triac also supported these findings and furthermore went on to show that Triac regulation of transcription can vary between different TREs (166). Triac has also been investigated for potential therapeutic applications to hyperlipidemia and osteoporosis. A study in athyreotic

patients (combined thyroidectomy and radioiodine ablation for treatment of thyroid carcinoma) compared the effects of Triac with L-T$_4$ at equally TSH suppressive doses (170). Triac significantly reduced total and LDL cholesterol and increased serum sex hormone binding globulin (SHBG), whereas L-T$_4$ showed little effect, suggesting increased liver-specific actions for Triac relative to L-T$_4$. Triac also increased serum osteocalcin and urinary calcium excretion, whereas L-T$_4$ showed no significant effects. Metabolic rate, pulse, and mean arterial pressure showed little change relative to baseline for treatment with either drug. Despite these promising organ-specific activities, the changes in serum parathyroid hormone and urinary calcium excretion induced by Triac elicited concern that accelerated bone demineralization may limit the compound's therapeutic potential.

4.4.3 SKF-94901. SKF-94901 (43) was the first thyroid hormone analog designed to take

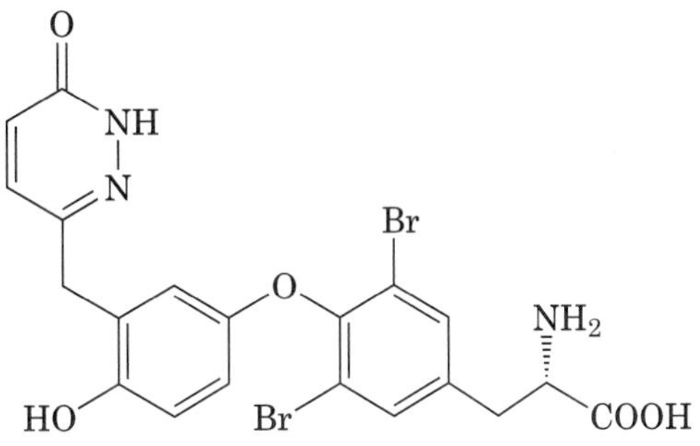

(43) SKF-94901

advantage of the less stringent SAR at the 3′ position on the prime ring of the thyromimetic pharmacophore, resulting in a major structural departure from the natural thyroid hormones (138, 160, 171). Although replacing the iodine normally present at 3′ with a methylpyridazinone group markedly reduces binding affinity (one-tenth the binding of T$_3$ to isolated rat liver nuclei), in animal models, compound (43) is an effective cholesterol-lowering agent with cardiac sparing properties. In addition, (43) increases metabolic rate and decreases plasma TSH, properties consistent with thyroid agonist activity. The absorption,

Table 11.4 SAR of Oxamic Acid Series

Compound	R3'	R3, R5	X	Y	In Vitro[a] (IC$_{50}$ nM)
(1) L-T3					1.1
(44)	Isopropyl	H, H	H	OH	48
(45)	Isopropyl	I, I	H	OH	0.10
(46)	Isopropyl	Br, Br	H	OH	6.0
(47) CGS-23425	Isopropyl	CH$_3$, CH$_3$	H	OH	0.19
(48)	Isopropyl	CH$_3$, CH$_3$	CH$_3$	OH	120
(49)	Isopropyl	CH$_3$, CH$_3$	H	NH$_2$	0.95
(50)	Isopropyl	CH$_3$, CH$_3$	H	OEt	0.23
(51)	-CH$_2$-p-Cl-phenyl	Cl, Cl	H	OH	0.27
(52)	-CHOH-p-Cl-phenyl	Cl, Cl	H	OH	1.5
(53)	-CHOH-p-F-phenyl	CH$_3$, CH$_3$	H	OH	0.8
(54) CGS-26214	-CHOH-p-F-phenyl	CH$_3$, CH$_3$	H	OEt	0.16
(55)	Cyclohexyl	CH$_3$, CH$_3$	H	OH	2.2
(56)	Phenyl	CH$_3$, CH$_3$	H	OH	1.9

[a]Binding to isolated rat liver nuclei.

distribution, metabolism, and excretion properties of (43) were investigated in the rat, dog, and cynomolgus monkey (172). Evidence from these studies show that bromine substitution at R$_3$ and R$_5$ blocks deactivation by deiodinase-like mechanisms and that there is less conjugation (sulfation or glucuronidation) of the phenolic hydroxyl at R$_4$', presumably because of the steric hindrance presented by the pyridazinone ring. Because (43) binds equally to TRα and TRβ, receptor isoform specificity cannot account for the compound's selective actions (hepatoselective and cardiac sparing) *in vivo* (173). Evidence has been presented that the biological specificity of SKF-94901 is caused by selective tissue uptake. The compound shows reduced binding to proteins in the cytoplasm of cardiac cells (65) and enhanced concentration in hepatic nuclei (173). The processes by which SKF-94901 preferentially enters certain cells remain unclear. SKF-94901 was evaluated for safety in a 30-day study in rats at doses of 0.02–10 mg/kg/day (with a comparator study of T$_3$ given at 1 mg/kg/day). Although many of the T$_3$-treated rats died with their hearts showing evidence of

myocardial damage, none of the SKF-94901 treated rats died before termination of the study and their hearts showed no damage at autopsy. However, SKF-94901 did give rise to fibroplasia on peritoneal surfaces, the appearance of multinucleate cells in the distal tubules of the kidneys, and formation of new trabecular bone in the femur. These effects were also observed in the T$_3$-treated animals. As a result of these effects, the development of SKF-94901 was suspended (149, 174).

4.4.4 Oxamic Acids. As discussed earlier, when the R$_1$ amino acid moiety present in the naturally occurring thyroid hormones is replaced by an oxamic acid group, there is a divergence from the classical SAR at positions R$_3$ and R$_5$. More of the SAR of the oxamic acid series is presented in the analogs shown in Table 11.4 (163, 175). Compounds (44–46) exhibit the familiar trend of iodo better than bromo and much better than lack of substitution at R$_3$ and R$_5$. However, analog (47), also known as CGS-23425, shows that affinity for TRs nearly equivalent to that of T$_3$ is achieved with methyl substitution. In compound (48),

N-methylation on the oxamate group markedly reduces binding affinity. However, in yet another departure from earlier established SAR, the neutral oxamic amide (**49**) and ester (**50**) both exhibit potent receptor binding comparable with T_3, showing that an acidic function at R_1 is not strictly required. Aryl substitution at 3' similar to the strategy used to develop SKF-94901 also yields potent oxamic acids. The best analogs include those in which the one atom linker between the aryl group and the 3' position is either methylene (—CH_2—) as in (**51**) or hydroxymethylene (—CHOH—) as in compounds (**52, 53**, and **54**). Analog (**54**), another example of a potent oxamate ester, is also known as CGS-262214. Oxamic acids with cyclohexyl (**55**) and phenyl (**56**) directly attached to the 3' position are also good TR binders. The *in vivo* activities of (**47, 54, 55**), and (**56**) are discussed below.

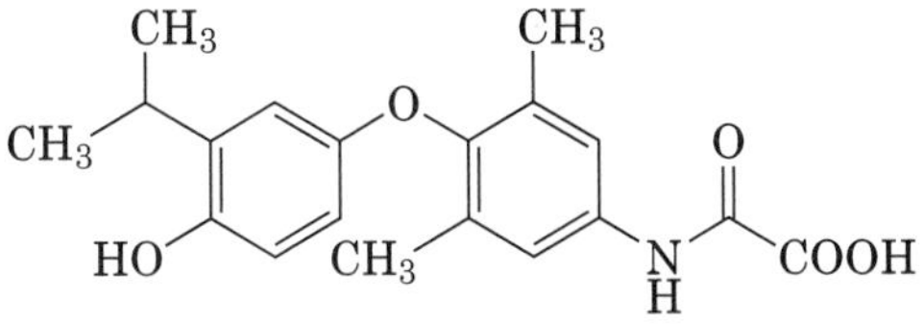

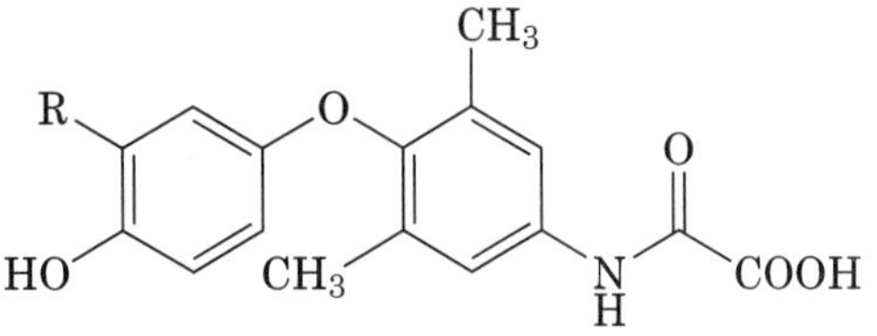

(**55**) R = cyclohexyl
(**56**) R = phenyl

(**47**) CGS-23425

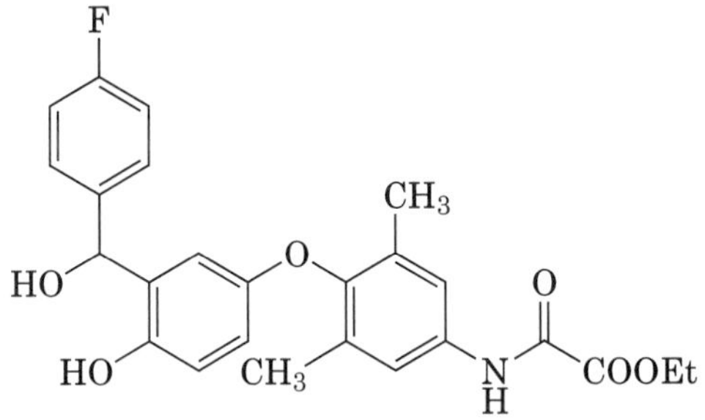

(**54**) CGS-26214

CGS-26214 (**54**) was the first oxamate to be reported with extensive biologic studies *in vivo* and *in vitro*. The compound has been proposed to be a potent cholesterol-lowering agent free of cardiovascular and thermogenic activity with a reported 25-fold dose separation between minimal lipid lowering and the observation of cardiovascular effects (74). In hyperlipidemic rats, (**54**) reduced LDL choles-

terol in a dose-dependent manner, with a minimum effective dose of 1 μg/kg, giving a 35% reduction after 7 days of drug treatment. The safety of CGS-26214 was tested in conscious dogs treated for 7 days at 100-fold the minimal effective lipid lowering dose and compared with L-T_3 given at 10-fold above its lipid-lowering dose. Although there were slight increases in heart rate and whole body oxygen consumption, these effects were not statistically significant, and the authors concluded that (**54**) was free of cardiac and thermogenic effects. CGS-26214 exhibited a 100-fold preference for binding *in vitro* to nuclei of HepG2 cells over cardiac myocytes, thus implicating selective tissue uptake to account for the compound's cardiac selectivity. Although CGS-26214 advanced to phase I clinical trials for treatment of hyperlipidemia, development of the compound was subsequently discontinued (176).

CGS-23425 (**47**) was the second oxamate to be discussed for its potential selective lipid-lowering properties (177, 178). In hypercholesterolemic rats, CGS-23425 gave a 44% decrease in LDL cholesterol with 10 mg/kg/day doses administered over 7 days. Rats given up to 10 mg/kg doses of (**47**) were free of cardiotoxicity, but adverse cardiac effects began to appear at 40 mg/kg. This margin of safety was reported to be similar to that of CGS-26214 (and SKF-94901). Based on earlier reported correlations of L-T_3 cardiac effects with its binding to plasma membranes (179), the binding of (**47**) to hepatic plasma membranes was shown to be relatively weak compared with T_3 and SKF-94901, but similar to CGS-26214. In transient transfection assays in human fetal hepatoma cells using CAT reporters with rat apoA1 promoters, compound (**47**) increased transcription with an $EC_{50} = 0.002$ nM for

$TR\beta_1$ versus an $EC_{50} = 1$ nM for $TR\alpha_1$. Citing the greater abundance of $TR\beta_1$ in the liver versus $TR\alpha_1$ in the heart (180, 181), it was proposed that subtype selectivity contributes to the hepatic selectivity and relative cardiac safety of CGS-23425.

The cholesterol-lowering activities of 3'-cyclohexyl (**55**) and 3'-phenyl (**56**) oxamic analogs show an interesting divergence. Both compounds show similar *in vitro* receptor-binding affinities (see Table 11.4) and give equal cholesterol lowering in hypercholesterolemic rats (40–50% reduction after 20 μg/kg oral doses administered over 7 days). However, in rats given drug at 25 mg/kg for 7 days, the 3'-cyclohexyl analog (**55**) caused increases in atrial heart rate, atrial tension, and heart weight, whereas the 3'-phenyl analog (**56**) showed no significant effects. No explanation was provided for this seemingly subtle structural change on *in vivo* SAR (175).

4.4.5 GC-1 and KB-141. Studies in $TR\alpha$ and $TR\beta^{-/-}$ mice suggest that selective $TR\beta$ selective agonists may reduce cholesterol with less heart rate effects. The role of $TR\beta$ on metabolic rate has only recently been described (91, 92). Knowledge of the potential of this receptor isoform for drug development has been increased by studies of the modestly selective $TR\beta$ agonists, GC-1 (**57**) and KB-141

(**57**) GC-1

(**58**). GC-1 binds human $TR\alpha_1$ receptors with $K_D = 0.44$ nM and $TR\beta_1$ receptors with $K_D = 0.067$ nM (165). Thus, GC-1 shows modest

(**58**) KB-141

binding selectivity for the $TR\beta$ isoform. Transcriptional activation by GC-1 was also $TR\beta$ selective when measured in HeLa cells transfected with TRE-luciferase reporter plasmids and either human $TR\alpha_1$ or $TR\beta_1$ expression plasmids. GC-1 was a full agonist under these conditions. The structure of GC-1 is halogen-free and features an oxyacetic acid group at R_1 along with a distinctive one carbon-linker ($-CH_2-$) joining the prime and non-prime phenyl rings. An X-ray structure of (**57**) bound into the LBD of $TR\beta_1$ is available (158). It suggests that the observed β selectivity is because of a greater opportunity for H-bonding interactions with the R_1 group of GC-1 in the complex with $TR\beta$ than with $TR\alpha$.

KB-141 possesses a structure reminiscent of Triac, with the difference being the presence of chloro at R_3 and R_5 rather than iodo. The *in vitro* activity of (**58**) also reflects a moderate level of selectivity for the β receptor (91). KB-141 has an $IC_{50} = 24$ nM for $TR\alpha_1$ and an $IC_{50} = 1.1$ nM for $TR\beta_1$ (IC_{50} is the concentration achieving 50% inhibition of the binding of radiolabeled T_3). This is in contrast to Triac, which has been reported to show only a 2.7-fold greater affinity for $TR\beta$. In CHO cells expressing human $TR\alpha_1$ or $TR\beta_1$, KB-141 is a full agonist and is $TR\beta$ selective, being essentially equipotent to T_3 for the β receptor but about 10-fold less potent than T_3 for the α receptor. T_3 is slightly $TR\alpha$ selective in this assay.

Both GC-1 and KB-141 have been shown in animal models to lower cholesterol at doses free of undesired cardiac effects. In hypercholesterolemic rats, GC-1 lowered cholesterol levels at doses where there was no evidence of increased heart rate (66, 92). In cholesterol-fed rats given drug orally over 7 days, GC-1 was 28-fold selective for cholesterol lowering versus tachycardia relative to T_3 (which was non-selective). Tissue distribution studies indicated that GC-1 is localized to a greater extent in the liver versus the heart, thus implicating selective tissue uptake as part of the explanation for the compound's activity. KB-141 was similarly 27-fold selective for cholesterol lowering versus tachycardia relative to T_3 in cholesterol-fed rats treated with drug orally over 7 days (91). In the rat, KB-141 also shows a lower heart/plasma distribution ratio than T_3. This suggests that favorable tissue uptake may also contribute to this compound's selective biologic activity. Because both compounds show *in vitro* selectivity for TRβ over TRα, receptor isoform selectivity has also been proposed to play a role in their activity profiles.

Both GC-1 and KB-141 give modest, but significant, separations between increasing metabolic rate and cardiac effects when given orally over 7 days to cholesterol-fed rats (91, 92). GC-1 showed a sevenfold dose window for a 5–10% increase in metabolic rate versus a 15% increase in heart rate. For the same parameters, KB-141 exhibited a 10-fold dose window. Under the same experimental protocol, T_3 was non-selective for the separation of metabolic rate from tachycardia. In addition, both GC-1 and KB-141 gave dose–response curves for metabolic rate that were significantly shallower than that of T_3. It is presently unclear whether a 10-fold dose range is large enough to safely and effectively cause weight reduction. It is also unclear whether TR agonist–mediated increases in metabolic rate can be achieved without increasing appetite, thereby nullifying any anti-obesity effects. Nevertheless, these agents support the suggestion that selective thyromimetics may find application in the treatment of obesity with the anticipated benefit of cholesterol-lowering activity (182).

5 RECENT DEVELOPMENTS AND THINGS TO COME

5.1 Antagonists

TR activation has been associated with ion channel disturbances, action potential shortening, and arrhythmias, particularly in hyperthyroid patients (73). The idea that TR antagonists may represent a potential means of inhibiting arrhythmia generation has been further suggested by the clinical efficacy of amiodarone, which among other activities, may be a TR antagonist. Although safe and effective thyroid antagonists are potentially of therapeutic benefit, progress in this area has been limited by a lack of proven TR antagonists. The antiarrhythmic drug amiodarone (**61**) and its major metabolite, desethylamio-

darone (**50**), have been shown to antagonize TR activity *in vitro* (142, 183, 184). However, there are no reports of demonstrated TR antagonist activity in cells or *in vivo* for desethylamiodarone. Amiodarone itself has low binding affinity for TRs, and as a pharmacologic agent, it is associated with numerous activities (183, 185). The research group that developed GC-1 also reported the synthesis and TR antagonist properties of the related analog (**59**) (186). Compound (**59**) was designed by

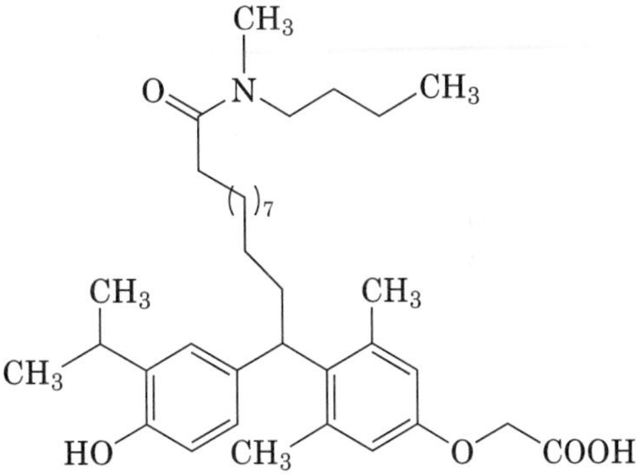

(**59**)

comparing the X-ray structures of the LBDs of TR (155) and the estrogen receptor (187), and by modeling the binding to ER of ICI-164,384 (**60**), a known ER antagonist (188–190). Com-

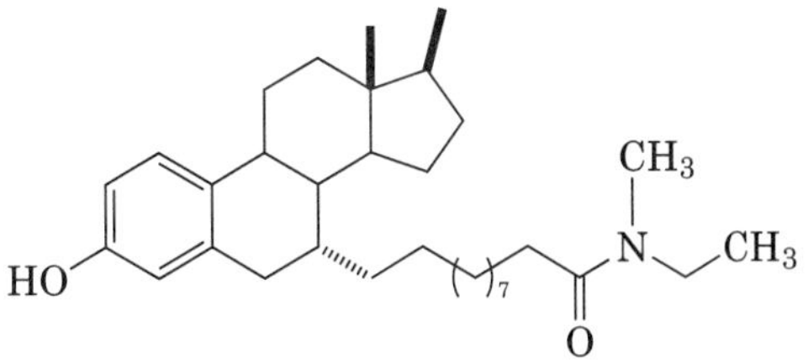

(**60**) ICI-164,384

pound (**59**) binds TRα_1 with K_D = 112 nM and TRβ_1 with K_D = 148 nM. In a cell-based transactivation assay, (**59**) shows competitive antagonism.

Another approach recently used to identify antagonists of NHRs has been to devise compounds that block the receptor's ability to assume the conformation needed for coactivator binding. DIBRT (**63**) was designed to feature a 5′-isopropyl group that would disrupt the po-

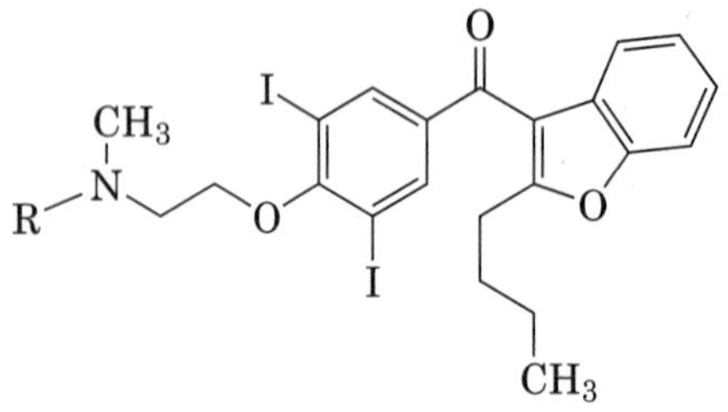

(61) Amiodarone: R = Et
(62) Desethylamiodarone: R = H

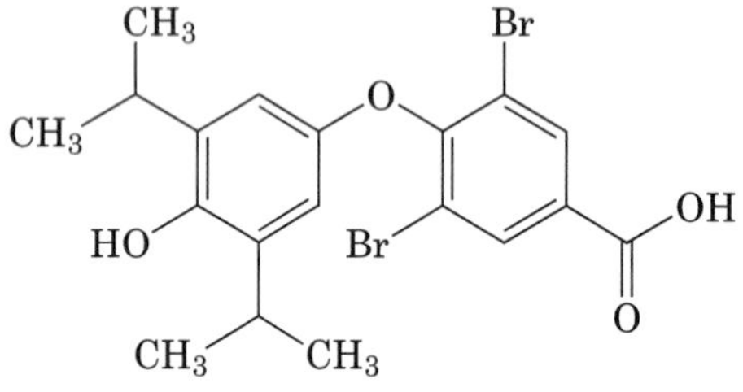

(63) DIBRT

(64) HY1

sitioning of helix-12 of the TR LBD (191). Helix-12 has been shown to be a key component of the coactivator binding surface of many NHRs (187, 192). Although a relatively weak competitive inhibitor ($IC_{50} \sim 1 \ \mu M$ for both TRα and TRβ), DIBRT showed modest antagonist potency blocking both a positive TRE response and a negative T_3 response, measured with the TSHβ promoter (cellular assays). In similar assays, full agonist efficacy was displayed by the corresponding analog in which the 5'-isopropyl was replaced by hydrogen.

With the ever increasing knowledge of structure-function in the nuclear hormone receptor family, more examples of designed antagonists for TR can be anticipated. The resulting molecules should find useful application to further elucidate the biology of thyroid hormones and may yet lead the way to developing therapeutically useful TR antagonists.

5.2 Drug Design

As discussed earlier, RTH is a disorder with genetic origins traced to various mutations in the ligand-binding domain of the TRβ receptor isoform. Reasoning that compounds possessing affinity and selectivity for mutant forms of TRβ versus TRα might be useful for

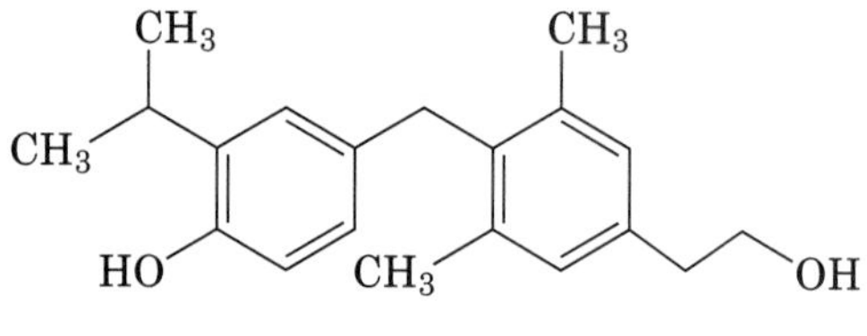

RTH therapy. HY1 (**64**) was designed using GC-1 (**44**) as a starting point because of its intrinsic β selectivity (193). The target was the RTH-associated mutation, TRβ(R320C), for which T_3 shows reduced affinity (194, 195). The available X-ray structure of the T_3/TRβ complex (155) was used to design compound (**64**), in which a neutral alcohol group at R_1 better accommodates the mutational replacement of arginine by the neutral amino acid, cysteine. In a transactivation assay, HY1 was five times more potent than GC-1 as an agonist for the mutant receptor, TRβ(R320C), thus demonstrating the potential for this application of structure-based design.

It is anticipated that the future of drug design for small molecule modulators (agonists and antagonists) of nuclear hormone receptors will involve more than solely considering X-ray structures of receptor-ligand complexes, such as in the exercise described above. This type of structure-based design will be supplemented by new technologies that take into consideration more of the full complexity known for the molecular mechanism of transcriptional regulation by NHRs (as described earlier). The LBDs are only one part of the multi-domain structure common to these receptors. Furthermore, the surfaces of the receptors interact with multiple and diverse coregulatory proteins, with the resulting complexes further interacting with the important histone modifying enzymes, HAT and HDAC. Recently introduced technology uses phage display to detect the different receptor conformation states induced by ligands associated with distinct biologic activities (agonism, antagonism, etc.) (196). When applied to the estrogen receptor (ER), this method successfully classified known selective estrogen receptor modulators (SERMs) with distinct, ligand-induced receptor surface shapes (139, 140).

This technology promises to be a powerful tool to compliment classical structure-based design.

5.3 Future Drug Development

Although the prospect of a new class of effective cholesterol-lowering drugs engendered by agents such as SKF-94901 and CGS-26214 has not been realized, recent advances in the understanding of the biology and molecular mechanisms of thyroid hormone action have re-invigorated the pharmaceutical industry to invest research in novel thyromimetics. Furthermore, recent patent applications are beginning to include anti-obesity claims in addition to those for lipid lowering. Nevertheless, significant hurdles remain for developing safe drugs from a class of agents with the diversity and potency of biologic actions exhibited by the thyroid hormones.

REFERENCES

1. E. C. Kendall, *Thyroxine*, American Chemical Society Monograph Series No. 47, Chemical Catalog Company, New York, 1929.

2. C. R. Harington, *Biochem. J.*, **21**, 169 (1926).

3. J. R. Chalmers, G. T. Dickson, J. Elks, and B. A. Hems, *J. Chem. Soc. (Lond.)*, 3424 (1949).

4. J. Gross and R. Pitt-Rivers, *Lancet*, **1**, 439 (1952).

5. J. Roche, S. Lissitsky, and R. Michel, *Comput. Rend. Acad. Sci.*, **234**, 997 (1952).

6. D. J. Mangelsdorf, C. Thummel, M. Beato, P. Herrlich, G. Schutz, K. Umesono, B. Blumberg, P. Kastner, M. Mark, and P. Chambon, *Cell*, **83**, 835 (1995).

7. M. A. Lazar, *J. Investig. Med.*, **47**, 364 (1990).

8. J. R. Tata and C. C. Widnell, *Biochem. J.*, **98**, 604 (1966).

9. J. H. Oppenheimer, D. Koerner, H. L. Schwartz, and M. I. Surks, *J. Clin. Endocrinol. Metab.*, **35**, 330 (1972).

10. H. H. Samuels and J. S. Tsai, *Proc. Natl. Acad. Sci. USA*, **70**, 3488 (1973).

11. B. Dozin, H. J. Cahnmann, and V. M. Nikodem, *Biochemistry*, **24**, 5197 (1985).

12. A. Pascual, J. Casanova, and H. H. Samuels, *J. Biol. Chem.*, **257**, 9640 (1982).

13. M. D. Crew and S. R. Spindler, *J. Biol. Chem.*, **261**, 5018 (1986).

14. J. Sap, A. Munoz, K. Damm, Y. Goldberg, J. Ghysdael, A. Leutz, H. Beug, and B. Vennstrom, *Nature*, **324**, 635 (1986).

15. C. Weinberger, C. C. Thompson, E. S. Ong, R. Lebo, D. J. Gruol, and R. M. Evans, *Nature*, **324**, 641 (1986).

16. P. M. Yen, *Physiol. Rev.*, **81**, 1097 (2001).

17. W. Zhang, R. L. Brooks, D. W. Silversides, B. L. West, F. Leidig, J. D. Baxter, and N. L. Eberhardt, *J. Biol. Chem.*, **267**, 15056 (1992).

18. J. D. Baxter, W. H. Dillmann, B. L. West, R. Huber, J. D. Furlow, R. J. Fletterick, P. Webb, J. W. Apriletti, and T. S. Scanlan, *J. Steroid Biochem. Mol. Biol.*, **76**, 31 (2001).

19. Y. Severne, S. Wieland, W. Schaffner, and S. Rusconi, *EMBO J.*, **7**, 2503 (1988).

20. S. Green, V. Kumar, I. Theulaz, W. Wahli, and P. Chambon, *EMBO J.*, **7**, 3037 (1988).

21. T. Nagaya, L. D. Madison, and J. L. Jameson, *J. Biol. Chem.*, **267**, 13014 (1992).

22. P. M. Yen, E. C. Wilcox, Y. Hayashi, S. Refetoff, and W. W. Chin, *Endocrinology*, **136**, 2845 (1995).

23. W. Bourguet, M. Ruff, P. Chambon, H. Gronemeyer, and D. Moras, *Nature*, **375**, 377 (1995).

24. J. P. Renaud, N. Rochel, M. Ruff, V. Vivat, P. Chambon, H. Gronemeyer, and D. Moras, *Nature*, **378**, 681 (1995).

25. Y. Tone, T. N. Collingwood, M. Adams, and V. K. Chatterjee, *J. Biol. Chem.*, **269**, 31157 (1994).

26. M. A. Lazar, R. A. Hodin, D. S. Darling, and W. W. Chin, *Mol. Endocrinol.*, **2**, 893 (1988).

27. T. Mitsuhashi, G. E. Tennyson, and V. M. Nikodem, *Proc. Natl. Acad. Sci. USA*, **85**, 5804 (1988).

28. G. R. Williams, *Mol. Cell. Biol.*, **20**, 8329 (2000).

29. D. Forrest and B. Vennstrom, *Thyroid*, **10**, 41 (2000).

30. E. D. Abel, M. E. Boers, C. Pazos-Moura, E. Moura, H. Kaulbach, M. Zakaria, B. Lowell, S. Radovick, M. C. Liberman, and F. Wondisford, *J. Clin. Invest.*, **104**, 291 (1999).

31. J. Bigler and R. N. Eisenman, *EMBO J.*, **14**, 5710 (1995).

32. J. Zhang and M. A. Lazar, *Ann. Rev. Physiol.*, **62**, 439 (2000).

33. S. A. Kliewer, K. Umesono, D. J. Noonan, R. A. Heyman, and R. M. Evans, *Nature*, **358**, 771 (1992).

34. M. Leid, P. Kastner, R. Lyons, H. Nakshatri, M. Saunders, T. Zacharewski, J. Y. Chen, A. Staub, J. M. Garnier, S. Mader, et al., *Cell*, **68**, 377 (1992).

35. C. K. Glass, *Endocr. Rev.*, **15**, 391 (1994).

36. O. Hermanson, C. K. Glass, and M. G. Rosenfeld, *Trends Endocrinol. Metab.*, **13**, 55 (2002).

37. C. K. Glass and M. G. Rosenfeld, *Genes Dev.*, **14**, 121 (2000).

38. K. Damm, C. C. Thompson, and R. M. Evans, *Nature*, **339**, 593 (1989).

39. J. D. Fondell, A. L. Roy, and R. G. Roeder, *Genes Dev.*, **7**, 1400 (1993).

40. A. J. Horlein, A. M. Naar, T. Heinzel, J. Torchia, B. Gloss, R. Kurokawa, A. Ryan, Y. Kamei, M. Soderstrom, C. K. Glass, et al., *Nature*, **377**, 397 (1995).

41. W. Seol, H. S. Choi, and D. D. Moore, *Mol. Endocrinol.*, **9**, 72 (1995).

42. J. D. Chen and R. M. Evans, *Nature*, **377**, 454 (1995).

43. S. Sande and M. L. Privalsky, *Mol. Endocrinol.*, **10**, 813 (1996).

44. A. Marimuthu, W. Feng, T. Tagami, H. Nguyen, J L. Jameson, R. J. Fletterick, J. D. Baxter, and B. L. West, *Mol. Endocrinol.*, **16**, 271 (2002).

45. T. Tagami, Y. Park, and J. L. Jameson, *J. Biol. Chem.*, **274**, 22345 (1999).

46. D. M. Heery, E. Kalkhoven, S. Hoare, and M. G. Parker, *Nature*, **387**, 733 (1997).

47. B. D. Darimont, R. L. Wagner, J. W. Apriletti, M. R. Stallcup, P. J. Kushner, J. D. Baxter, R. J. Fletterick, and K. R Yamamoto, *Genes Dev.*, **12**, 3343 (1998).

48. X. F. Ding, C. M. Anderson, H. Ma, H. Hong, R. M. Uht, P. J. Kushner, and M. R. Stallcup, *Mol. Endocrinol.*, **12**, 302 (1998).

49. V. Perissi, L. M. Staszewski, E. M. McInerney, R. Kurokawa, A. Krones, D. W. Rose, M. H. Lambert, M. V. Milburn, C. K. Glass, and M. G. Rosenfeld, *Genes Dev.*, **13**, 3198 (1999).

50. H. E. Xu, T. B. Stanley, V. G. Montana, M. H. Lambert, B. G. Shearer, J. E. Cobb, D. D. McKee, C. M. Galardi, K. D. Plunket, R. T. Nolte, D. J. Parks, J. T. Moore, S. A. Kliewer, T. M. Wilson, and J. B. Stimmel, *Nature*, **415**, 813 (2002).

51. F. E. Carr, M. A. Suupnik, J. Burnside, and W. W. Chin, *Mol. Endocrinol.*, **3**, 717–721 (1989).

52. D. S. Kim, S. K. Ahn, and J. H. Yoon, *Mol. Endocrinol.*, **8**, 528 (1994).

53. L. Krulich, *Neuroendocrinology*, **35**, 139 (1982).

54. T. Endo, M. Ohno, S. Kotani, K. Gunji, and T. Onaya, *Biochem. Biophys. Res. Commun.*, **190**, 774 (1993).

55. G. Colletta and A. M. Cirafici, *Biochem. Biophys. Res. Commun.*, **18**, 265 (1992).

56. M. I. Surks and B. M. Lifschitz, *Endocrinology*, **101**, 769 (1977).

57. G. Brabant, U. Ranft, K. Ocran, R. D. Hesch, and A. von zur Muhlen, *Acta Endocrinol.*, **112**, 315 (1986).

58. A. I. Vinik, W. J. Kalk, H. McLaren, S. Henrick, and B. L. Pimstone, *J. Clin. Endocrinol. Metab.*, **40**, 509 (1975).

59. R. T. Zoeller, N. Kabeer, and H. E. Alberts, *Endocrinology*, **127**, 2955 (1990).

60. J. Wolff and J. Maurey, *Biochim. Biophys. Acta*, **47**, 467 (1964).

61. S. J. Weiss, N. J. Philp, and E. F. Grollman, *Endocrinology*, **114**, 1090 (1984).

62. J. P. Blondeau, J. Osty, and J. Francon, *J. Biol. Chem.*, **263**, 2685 (1988).

63. M. E. Everts, F. A. Verhoeven, K. Bezstarosti, E. P. Moerings, G. Hennemann, T. J. Visser, and J. M. Lamers, *Endocrinology*, **137**, 4234 (1996).

64. L. Kragie, M. L. Forrester, V. Cody, and M. McCourt, *Mol. Endocrinol.*, **8**, 382 (1994).

65. J. W. Barlow, L. E. Raggatt, C.-F. Lim, E. Kolliniatis, D. J. Topliss, and J. R. Stockigt, *Acta Endocrinol.*, **124**, 37 (1991).

66. S. U. Trost, E. Swanson, B. Gloss, D. B. Wang-Iverson, H. Zhang, T. Volodarsky, G. J. Grover, J. D. Baxter, G. Chiellini, T. S. Scanlan, and W. H. Dillmann, *Endocrinology*, **141**, 3056 (2000).

67. P. De Lange, A. Lanni, L. Beneduce, M. Moreno, A. Lombardi, E. Silvestri, and F. Goglia, *Endocrinology*, **142**, 3414 (2001).

68. R. A. Gelfand, K. A. Hutchinson-Williams, A. A. Bonde, P. Castellino, and R. S. Sherwin, *Metabolism*, **36**, 562 (1987).

69. S. Iossa, G. Liverini, and A. Barletta, *J. Endocrinol.*, **135**, 45 (1992).

70. M. Moreno, A. Lanni, A. Lombardi, and F. Goglia, *J. Physiol. (Lond.)*, **505**, 529 (1997).

71. Y. Feng, D. De-Zai, A. Lu-Fan, and G. Xiu-Fang, *Acta Pharmacologica Sinica*, **18**, 71 (1997).

72. W. H. Dillmann, *Thyroid Today*, **14**, 1 (1996).

73. I. Klein and K. Ojamaa, *N. Engl. J. Med.*, **344**, 501 (2001).

74. Z. F. Stephan, E. C. Yurachek, R. Sharif, J. M. Wasvary, K. S. Leonards, C.-W. Hu, T. H. Hintze, and R. E. Steele, *Atherosclerosis*, **126**, 53 (1996).

75. R. Li, K. Luciakova, A. Zaid, S. Betina, E. Fridell, and B. D. Nelson, *Mol. Cell. Endocrinol.*, **128**, 69 (1997).

76. R. J. Wiesner, T. T. Kurowski, and R. Zak, *Mol. Endocrinol.*, **6**, 1458 (1992).

77. G. G. Gick, J. Melikian, and F. Ismail-Beigi, *J. Membr. Biol.*, **115**, 272 (1990).

78. A. Chint, T. Clausen, and L. Girardier, *J. Physiol. (Lond.)*, **254**, 43 (1977).

79. M. Folke and L. Sestoft, *J. Physiol. (Lond.)*, **269**, 407 (1977).

80. R. Gustafsson, J. R. Tata, O. Lindburg, and L. Ernster, *J. Cell. Biol.*, **25**, 555 (1965).

81. T. Masaki, H. Yoshimatsu, and T. Sakata, *Int. J. Obes.*, **24**(suppl 2), S162 (2000).

82. P. E. Paget and J. M. Thorpe, *Nature*, **199**, 1307 (1963).

83. G. H. Fried, N. Greenberg, and W. Antopol, *Proc. Soc. Exp. Biol. Med.*, **107**, 523 (1961).

84. T. M. Pillar and H. J. Seitz, *Eur. J. Endocrinol.*, **136**, 231 (1997).

85. M. B. Jekabsons, F. M. Gregoire, N. A. Schonfeld-Warden, C. H. Warden, and B. A. Horwitz, *Am. J. Physiol.*, **277**, E380 (1999).

86. J. C. Clapham, J. R. S. Arch, H. Chapham, A. Haynes, C. Lister, G. B. Moore, V. Piercy, S. A. Carter, I. Lehner, S. A. Smith, L. J. Beeley, R. J. Godden, N. Herrity, M. Skehel, K. K. Changani, P. D. Hockings, D. G. Reid, S. M. Squires, J. Hatcher, B. Trail, J. Latcham, S. Rastan, A. J. Harper, S. Cadenas, J. A. Buckingham, M. D. Brand, and A. Abuin, *Nature*, **406**, 415 (2000).

87. A. J. Vidal-Puig, D. Grujic, C. Y. Zhang, T. Hagan, O. Boss, Y. Ido, A. Szczepanik, J. Wade, V. Mootha, R. Cortright, and D. M. Muoio, *J. Biol. Chem.*, **275**, 16258 (2000).

88. M. O. Ribiero, S. D. Carvalho, J. J. Schultz, G. Chiellini, T. S. Scanlan, A. C. Bianco, and G. A. Brent, *J. Clin. Invest.*, **108**, 97 (2001).

89. L. Wikstrom, C. Johansson, C. Salto, C. Barlow, A. C. Barros, F. Baas, D. Forrest, P. Thoren, and B. Vennstrom, *EMBO J.*, **17**, 455 (1998).

90. R. E. Weiss, Y. Murata, K. Cua, Y. Yayashi, H. Seo, and S. Refetoff, *Endocrinology*, **139**, 4945 (1998).

91. G. J. Grover, J. D. Baxter, L.-Y. Lin, J. Malm, L.-G. Bladh, P. G. Sleph, R. George, and K. Mellstrom, Comparative effects of the selective thyroid hormone receptor (TR-beta1) agonist KB-000141 on cholesterol, metabolic rate and heart rate in rats. Proceedings of the 73rd Annual American Thyroid Association Meeting, Washington, DC, 2001 (in press).

92. P. G. Sleph, G. J. Grover, T. S. Scanlon, and J. D. Baxter, Comparative effects of the selective thyroid hormone receptor (TR-beta1) agonist GC-1 on cholesterol, metabolic rate and heart rate in rats. Proceedings of the 73rd Annual American Thyroid Association Meeting, Washington, DC, 2001 (in press).

93. H. J. Harwood and E. S. Hamanaka, *Emerging Drugs*, **3**, 147 (1998).

94. J. Davignon, *Diabetes Metab.*, **21**, 139 (1995).

95. G. C. Ness, D. Lopez, C. M. Chambers, W. P. Newsome, P. Cornelius, C. A. Long, and H. J. Harwood, *Biochem. Pharmacol.*, **56**, 121 (1998).

96. B. Blenneman, Y. K. Moon, and H. C. Freake, *Endocrinology*, **130**, 637 (1992).

97. H. Wahrenberg, A. Wennlund, and P. Arnewr, *J. Clin. Endocrinol. Metab.*, **78**, 898 (1994).

98. L. J. Seman, C. DeLuca, J. L. Jenner, L. A. Cupples, J. R. McNamara, P. W. F. Wilson, W. P. Castell, J. M. Ordovas, and E. J. Schaefer, *Clin. Chem.*, **45**, 1039 (1999).

99. F. Pazos, J. J. Alvarez, J. Rubies-Prat, C. Varela, and M. A. Lasuncion, *J. Clin. Endocrinol. Metab.*, **80**, 562 (1995).

100. H. Engler and W. F. Riesen, *Clin. Chem.*, **39**, 2466 (1993).

101. I. A. Mirsky and R. H. Broh-Kahn, *Am. J. Physiol.*, **117**, 6 (1936).

102. R. A. Freedland and H. A. Krebs, *Biochem. J.*, **104**, 45 (1967).

103. C. J. Torrance, S. J. Usala, J. E. Pessin, and G. L. Dohm, *Endocrinology*, **138**, 1216 (1997).

104. C. Craft-Cormney and J. T. Hansen, *Virchows Arch. B. Cell. Path.*, **33**, 267 (1980).

105. W. H. Dillmann, *Am. J. Med.*, **88**, 626 (1990).

106. D. K. Rohrer and W. H. Dillmann, *J. Biol. Chem.*, **263**, 6941 (1988).

107. B. Gloss, S. U. Trost, W. F. Bluhm, E. A. Swanson, R. Clark, R. Winkfein, K. M. Janzen, W. Giles, O. Chasande, J. Samarut, and W. H. Dillmann, *Endocrinology*, **142**, 544 (2001).

108. C.-J. Huang, H. M. Geller, W. L. Green, and W. Craelius, *J. Mol. Cell. Cardiol.*, **31**, 881 (1999).

109. E. T. Tielens, J. R. Forder, J. C. Chatham, S. P. Marrelli, and P. W. Ladenson, *Cardiovasc. Res.*, **32**, 306 (1996).

110. W. B. Kinlaw, J. L. Church, J. Harmon, and C. N. Marlash, *J. Biol. Chem.*, **270**, 16615 (1995).

111. J. H. Oppenheimer, H. L. Schwartz, C. N. Marlash, W. B. Kinlw, N. C. W. Wong, and H. C. Freake, *Endocr. Rev.*, **8**, 288 (1987).

112. S. A. Seelig, C. Liaw, H. C. Towle, and J. H. Openheimer, *Proc. Natl. Acad. Sci. USA*, **78**, 4733 (1981).

113. C. Ishihara, K. Kushida, M. Takahashi, S. Koyama, K. Kawana, K. Atsume, and T. Inoue, *Endocr. Res.*, **23**, 167 (1997).

114. T. J. Allain and A. M. McGregor, *J. Endocrinol.*, **139**, 9 (1993).

115. S. L. Greenspan and F. S. Greenspan, *Ann. Intern. Med.*, **130**, 750 (1999).

116. M. C. Foss, G. M. Paccola, and M. J. Saad, *J. Clin. Endocrinol. Metab.*, **70**, 1167 (1990).

117. R. T. Jagoe and A. L. Goldberg, *Curr. Opin. Clin. Nutr. Metab. Care*, **4**, 183 (2001).

118. D. Forrest, E. Hanebuth, R. J. Smeyne, N. Everds, C. L. Stewart, J. M. Wehner, and T. Curran, *EMBO J.*, **15**, 3006 (1996).

119. C. Salto, J. M. Kindblom, C. Johansson, Z. Wang, H. Gullberg, K. Nordstrom, A. Mansen, C. Ohlsson, P. Thoren, D. Forrest, and B. Vennstrom, *Mol. Endocrinol.*, **15**, 2115 (2001).

120. A. Fraichard, O. Chassande, M. Plateroti, J. P. Roux, J. Trouillas, C. Dehay, C. Legrand, K. Gauthier, M. Kedinger, L. Malaval, B. Rousset, and J. Samarut, *EMBO J.*, **16**, 4412 (1997).

121. S. Gothe, Z. Wang, L. Ng, J. Nilsson, A. Campos Barros, C. Ohlsson, B. Vennstrom, and D. Forrest, *Genes Dev.*, **13**, 1329 (1999).

122. L. Ng, J. B. Hurley, B. Dierks, M. Srinivas, C. Salto, B. Vennstrom, T. A. Reh, and D. Forrest, *Nat. Genet.*, **27**, 94 (2001).

123. L. Ng, A. Rusch, L. L. Amma, K. Nordstrom, L. C. Erway, B. Vennstrom, and D. Forrest, *Hum. Mol. Genet.*, **10**, 2701 (2001).

124. K. Gauthier, O. Chassande, M. Plateroti, J. P. Roux, C. Legrand, B. Pain, B. Rousset, R. Weiss, J. Trouillas, and J. Samarut, *EMBO J.*, **18**, 623 (1999).

125. A. Mansen, F. Yu, D. Forrest, L. Larsson, and B. Vennstrom, *Mol. Endocrinol.*, **15**, 2106 (2001).

126. H. Gullberg, M. Rudling, D. Forrest, B. Angelin, and B. Vennstrom, *Mol. Endocrinol.*, **14**, 1739 (2000).

127. M. A. Lazar, *Endocr. Rev.*, **14**, 184 (1993).

128. C. Johansson, S. Gothe, D. Forrest, B. Vennstrom, and P. Thoren, *Am. J. Physiol.*, **276**, H2006 (1999).

129. C. Johansson, R. Koopmann, B. Vennstrom, and K. Benndorf, *J. Cardiovasc. Electrophysiol.*, **13**, 44 (2002).

130. T. L. Dellovade, J. Chan, B. Vennstrom, D. Forrest, and D. W. Pfaff, *Nature Neurosci.*, **3**, 472 (2000).

131. B. Morte, B. Manzano, T. S. Scanlon, B. Vennstrom, and J. Bernal, *Proc. Natl. Acad. Sci. USA*, **99**, 3985 (2002).

132. B. Gloss, S. U. Trost, W. F. Bluhm, E. A. Swanson, R. Clark, R. Winkfein, K. M. Janzen, W. Giles, O. Chassande, J. Samarut, and W. H. Dillmann, *Endocrinology*, **142**, 544 (2001).

133. L. Amma, A. Campos-Barros, Z. Wang, B. Vennstrom, and D. Forrest, *Mol. Endocrinol.*, **15**, 467 (2001).

134. L. Ng, P. E. Pedraza, J. S. Faris, B. Vennstrom, T. Curran, G. Morreale de Escobar, and D. Forrest, *Neuroreport*, **12**, 2359 (2001).

135. L. Calza, D. Forrest, B. Vennstrom, and T. Hokfelt, *Neuroscience*, **101**, 1001 (2000).

136. J. M. Kindblom, S. Gothe, D. Forrest, J. Tornell, B. Vennstrom, and C. Ohlsson, *J. Endocrinol.*, **171**, 15 (2001).

137. A. Rusch, L. Ng, R. Goodyear, D. Oliver, I. Lisoukov, B. Vennstrom, G. Richardson, M. W. Kelley, and D. Forrest, *J. Neurosci.*, **21**, 9792 (2001).

138. A. H. Underwood, J. C. Emmett, D. Ellis, S. B. Flynn, P. D. Leeson, G. M. Benson, R. Novelli, N. J. Pearce, and V. P. Shah, *Nature*, **324**, 425 (1986).

139. L. A. Paige, D. J. Christensen, H. Gron, J. D. Norris, E. B. Gottlin, K. M. Padilla, C.-Y. Chang, L. M. Ballas, P. T. Hamilton, D. P. McDonnell, and D. M. Fowlkes, *Proc. Natl. Acad. Sci. USA*, **96**, 3999 (1999).

140. J. D. Norris, L. A. Paige, D. J. Christensen, C.-Y. Chang, M. R. Huacani, D. Fan, P. T. Hamilton, D. M. Fowlkes, and D. P. McDonnell, *Science*, **285**, 744 (1999).

141. J. D. Klemperer, J. Zelano, R. E. Helm, K. Berman, K. Ojamaa, I. Klein, O. W. Isom, and K. Krieger, *J. Thorac. Cardiovasc. Surg.*, **109**, 457 (1995).

142. O. Bakker, H. C. Vanbeeren, and W. M. Wiersinga, *Endocrinology*, **134**, 1665 (1994).

143. H. Torma, O. Rollman, and A. Vahlquist, *Arch. Dermatol. Res.*, **291**, 339 (1999).

144. H. Torma, T. Karlsson, G. Michaelsson, O. Rollman, and A. Vahlquist, *Acta Derm. Venereol.*, **80**, 4 (2000).

145. N. Billoni, B. Buan, B. Gautier, O. Gaillard, Y. F. Mahe, and B. A. Bernard, *Br. J. Dermatol.*, **142**, 645 (2000).

146. J. D. Safer, L. M. Fraser, S. Ray, and M. F. Holick, *Thyroid*, **11**, 717 (2001).

147. J. Faergemann, T. Sarnhult, E. Hedner, B. Carlsson, T. Lavin, X.-H. Zhao, X.-Y. Sun, Acta Dermato-Venerologica Scand. (in press).

148. E. C. Jorgensen in C. H. Li, Ed., *Hormonal Proteins and Peptides*, Vol. **6**, Academic Press, New York, 1978, p. 107.

149. P. D. Leeson and A. H. Underwood in J. C. Emmett, Ed., *Comprehensive Medicinal Chemistry*, Vol. **3**, Pergamon Press, Oxford, UK, 1990, p. 1145.

150. D. J. Craik, B. M. Duggan, and S. I. A. Munro, *Stud. Med. Chem.*, **2**, 255 (1996).

151. V. Cody, *Recent Prog. Horm. Res.*, **34**, 437 (1978).

152. J. C. Emmett and E. S. Pepper, *Nature*, **257**, 334 (1975).

153. D. Ellis, J. C. Emmett, P. D. Leeson, and A. H. Underwood in M. Vederame, Ed., *Handbook of Hormones, Vitamins and Radiopaques*, CRC Press, Boca Raton, FL, 1986, p. 94.

154. N. Zenker and E. C. Jorgensen, *J. Am. Chem. Soc.*, **81**, 4643 (1959).

155. R. L. Wagner, J. W. Apriletti, M. E. McGrath, B. L. West, J. D. Baxter, and R. J. Fletterick, *Nature*, **378**, 690 (1995).

156. M. O. Ribeiro, J. W. Apriletti, R. L. Wagner, W. Feng, P. J. Kushner, S. Nilsson, T. S. Scanlan, B. L. West, R. J. Fletterick, and J. D. Baxter, *J. Steroid Biochem. Mol. Biol.*, **65**, 133 (1998).

157. M. O. Ribeiro, J. W. Apriletti, R. L. Wagner, B. L. West, W. Feng, R. Huber, P. J. Kushner, S. Nilsson, T. S. Scanlan, R. J. Fletterick, F. Schaupele, and J. D. Baxter, *Recent Prog. Horm. Res.*, **53**, 351 (1998).

158. R. L. Wagner, B. R. Huber, A. K. Shiau, A. Kelly, S. T. Cunha Sima, T. S. Scanlan, J. W. Apriletti, J. D. Baxter, B. L. West, and R. J. Fletterick, *Mol. Endocrinol.*, **15**, 398 (2001).

159. M. B. Bolger and E. C. Jorgensen, *J. Biol. Chem.*, **255**, 10271 (1980).

160. P. D. Leeson, D. Ellis, J. C. Emmett, V. P. Shah, G. A. Showell, and A. H. Underwood, *J. Med. Chem.*, **31**, 37 (1988).

161. M. Ebisawa, N. Inoue, H. Fukasawa, T. Sotome, H. Kagechika, *Chem. Pharm. Bull. (Tokyo)*, **47**, 1348 (1999).

162. D. Koerner, H. L. Schwartz, M. I. Surks, J. H. Oppenheimer, and E. C. Jorgensen, *J. Biol. Chem.*, **250**, 6417 (1975).

163. N. Yokoyama, G. N. Walker, A. J. Main, J. L. Stanton, M. M. Morrissey, C. Boehm, A. Engle, A. D. Neubert, J. M. Wasvary, Z. F. Stephan, and R. E. Steele, *J. Med. Chem.*, **38**, 695 (1995).

164. T. Takeda, S. Suzuki, L. Rue-Tsuan, and L. J. DeGroot, *J. Clin. Endocrinol. Metab.*, **80**, 2033 (1995).

165. G. Chiellini, J. W. Apriletti, H. A. Yoshihara, J. D. Baxter, R. C. J. Ribeiro, and T. S. Scanlon, *Chem. Biol.*, **5**, 299 (1998).

166. N. Messier and M. F. Langlois, *Mol. Cell. Endocrinol.*, **165**, 57 (2000).

167. The Coronary Drug Research Group. *J. Am. Med. Assoc.*, **220**, 996 (1972).

168. T. S. Scanlan, H. A. Yoshihara, N.-H. Nguyen, and G. Chiellini, *Curr. Opin. Drug Discov. Develop.*, **4**, 614 (2001).

169. *Physicians' Desk Reference*, 55th ed., Medical Economics Company, Inc., Montvale, NJ, 2001, pp. 1641–1644.

170. S. I. Sherman, M. D. Ringel, M. J. Smith, H. A. Kopelen, W. A. Zoghbi, and P. W. J. Ladenson, *Clin. Endocrinol. Metab.*, **82**, 2153 (1997).

171. P. D. Leeson, J. C. Emmett, V. P. Shah, G. A. Showell, R. Novelli, H. D. Prain, G. M. Benson, D. Ellis, M. J. Pearce, and A. H. Underwood, *J. Med. Chem.*, **32**, 320 (1989).

172. M. A. Pue, J. A. Ransley, D. J. Writer, A. J. Dean, E. R. Franklin, I. G. Beattie, and D. A. Ross, *Eur. J. Drug Metab. Pharmacokinet.*, **14**, 209 (1989).

173. K. Ichikawa, T. Miyamoto, T. Kakizawa, S. Suzuki, A. Kaneko, J. Mori, M. Hara, M. Kumagai, T. Takeda, and K. Hashizume, *Endocrinology*, **165**, 391 (2000).

174. S. J. Kennedy, C. K. Atterwill, and A. Poole, *Toxicologist*, **8** (1), Abstract no. 348 (1988).

175. J. L. Stanton, E. Cahill, R. Dotson, J. Tan, H. C. Tomaselli, J. M. Wasvary, Z. F. Stephan, and R. E. Steele, *Bioorg. Med. Chem. Lett.*, **10**, 1661 (2000).

176. Novartis Pharma AG, Company Communication, November 16, 1999.

177. A. H. Taylor, F. S. Zouhair, R. E. Steele, and N. C. W. Wong, *Mol. Pharmacol.*, **52**, 542 (1997).

178. Y. Wada, S. Matsubara, J. Dufresne, G. M. Hargrove, Z. F. Stephan, R. E. Steele, and N. C. W. Wong, *J. Mol. Endocrinol.*, **25**, 299 (2000).

179. J. Segal, J. Hardiman, and S. H. Ingbar, *Biochem. J.*, **261**, 749 (1989).

180. M. Falcone, T. Miyamoto, F. Fierro-Renoy, E. Macchia, and L. J. DeGroot, *Endocrinology*, **131**, 2419 (1992).

181. A. Sakurai, A. Nakai, and L. J. DeGroot, *Mol. Endocrinol.*, **3**, 392 (1989).

182. M. Krotkiewski, *Int. J. Obes.*, **24**(Supp 2), S116 (2000).

183. K. R. Latham, D. F. Sellitti, and R. E. Goldstein, *J. Am. Coll. Cardiol.*, **9**, 872 (1987).

184. H. C. Vanbeeren, O. Bakker, and W. M. Wiersinga, *Mol. Cell. Endocrinol.*, **112**, 15 (1995).

185. R. M. Morse, G. A. Valenzuela, T. P. Greenwald, P. J. Eulie, R. C. Wesley, and R. W. McCallum, *Ann. Intern. Med.*, **109**, 838 (1988).

186. H. A. I. Yoshihara, J. W. Apriletti, J. D. Baxter, and T. S. Scanlan, *Bioorg. Med. Chem. Lett.*, **11**, 2821 (2001).

187. A. M. Brzozowski, A. C. W. Pike, Z. Dauter, R. E. Hubbard, T. Bonn, O. Engstrom, L. Ohman, G. L. Greene, J. A. Gustafsson, and M. Carlquist, *Nature*, **389**, 753 (1997).

188. J. Bowler, T. J. Lilley, J. D. Pittam, A. E. Wakeling, *Steroids.*, **54**, 71 (1989).

189. A. E. Wakeling and J. J. Bowler, *Endocrinology*, **112**, R7 (1987).

190. A. C. W. Pike, A. M. Brzozowski, J. Walton, R. E. Hubbard, A. G. Thorsell, Y. L. Li, J. A. Gustafsson, and M. Carlquist, *Structure*, **9**, 145 (2001).

191. J. D. Baxter, P. Goede, J. W. Apriletti, B. L. West, W. Feng, K. Mellstrom, R. J. Fletterick, R. L. Wagner, P. J. Kushner, R. C. J. Ribeiro, P. Webb, T. S. Scanlan, and S. Nilsson, *Endocrinology*, **143**, 517 (2002).

192. A. K. Shiau, D. Barstad, P. M. Loria, L. Cheng, P. J. Kushner, D. A. Agard, and G. L. Greene, *Cell*, **95**, 927 (1998).

193. H. F. Ye, K. E. O'Reilly, and J. T. Koh, *J. Am. Chem. Soc.*, **123**, 1521 (2001).

194. K. D. Burman, Y. Y. Djuh, D. Nicholson, P. Rhooms, L. Wartofsky, H. G. Fein, S. J. Usala, E. H. Hao, W. E. Bradley, J. Berard, and R. C. J. Smallridge, *Endocrinol. Invest.*, **15**, 573 (1992).

195. R. C. Smallridge, R. A. Parker, E. A. Wiggs, K. R. Rajagopal, and H. G. Fein, *Am. J. Med.*, **86**, 289 (1989).

196. H. Gron and R. Hyde-DeRuyscher, *Curr. Opin. Drug Discov. Develop.*, **3**, 636 (2000).

CHAPTER TWELVE

Fundamentals of Steroid Chemistry and Biochemistry

ROBERT W. BRUEGGEMEIER
PUI-KAI LI
Division of Medicinal Chemistry and Pharmacognosy
College of Pharmacy
The Ohio State University
Columbus, Ohio

Contents

Burger's Medicinal Chemistry and Drug Discovery
Sixth Edition, Volume 3: Cardiovascular Agents and
Endocrines
Edited by Donald J. Abraham
ISBN 0-471-37029-0   © 2003 John Wiley & Sons, Inc.

1 INTRODUCTION

Steroids are a unique class of chemical compounds that are found throughout the animal and plant kingdom, and this class includes sterols such as cholesterol and ergosterol, bile acids, and steroid hormones. Modern scientific research on steroid chemistry and biochemistry began in the early twentieth century, and several major treatises on the subject have been published (1–9). This chapter is intended to provide a general summary of the steroid chemistry and biochemistry rather than a comprehensive review. Subsequent chapters in this current volume of *Burger's Medicinal Chemistry* contain more detailed discussions on the individual classes of steroids.

The biological and medical significances of steroids have been observed since ancient times, even though the exact chemical nature and properties of steroids began to be understood only in the late 1920s and early 1930s. In the literature of the ancient Greeks, Hippocrates referred to gallstones and the term *cholesterol* is derived from the Greek terms for bile (*chole*) and solid (*steros*). Aristotle identified the effects of castration in birds and in humans, and agricultural practices of castration to produce sterility, alter aggressiveness, and affect body size in domestic animals has been known since early times. Egyptians and Romans used extracts of plants such as purple foxglove to treat dropsy. The physical effects of castration were well recognized in eunuchs and in medieval castrati choirs.

Scientific observations in more modern times began examining the biological consequences of hormones without the realization of nature of the chemicals involved. Berthold, a Göttingen physiologist, reported the effects of implanted testis in studies with cocks in 1849 (10). In 1855, Addison discovered the relationship of the adrenal glands with a particular disease characterized by bronze skin color (11), and this disease of chronic adrenal insufficiency is now referred to as Addison's disease. In 1889 Brown-Sequard prepared a testicular extract and tested on himself, reporting enhanced rejuvenation. Although

unorthodox, this technique of preparing tissue extracts and evaluating the effects of the extracts was eventually adapted for the isolation of biologically active constituents from the extracts.

The modern era of steroid research began with steroid chemistry in the early 1900s. Dr. Adolf O. Windaus, a Göttingen chemist, worked for over 20 years on the isolation of steroids, development of assays for detecting steroids, and the use of classical chemistry to elucidate steroid structures (12). Dr. Windaus received the Nobel Prize in 1928 for his research on the "constitution of sterols." During the same period, Dr. Heinrich O. Wieland of Munich was engaged in natural products chemistry, including bile acids (13), and was awarded the Nobel prize in 1928 for his research on this subject. The original chemical structures of cholesterol and other steroids proposed in 1928 were subsequently found to be incorrect, and correct structures were identified in 1932 (14–16). Chemical studies on compounds involved in reproduction began in the 1920s, and in 1929 Adolf F. Butenandt and Edward A. Doisy independently reported the isolation of an active steroid sex hormone, estrone, from the urine of pregnant women (17, 18). Throughout the 1930s, many steroid hormones were isolated and structures determined, including progesterone by Butenandt (19) and corticosteroids by Reichstein (20). The synthesis of steroids followed shortly thereafter by research groups led by Bachmann, Woodward, Robinson, and Cornforth (21–23).

Steroid biochemistry began with studies on the biosynthesis and metabolism of steroids, and early studies in the 1930s and 1940s used large amounts of unlabeled compounds (6). With the production of radiolabeled molecules, studies then used more physiological levels of steroids and steroid precursors, with an early significant demonstration that all the carbon atoms of cholesterol are derived from the two carbons present in isotopically labeled acetate (24). Research on steroid biosynthesis and metabolism began in the 1950s and 1960s and continues to the present time. Studies on

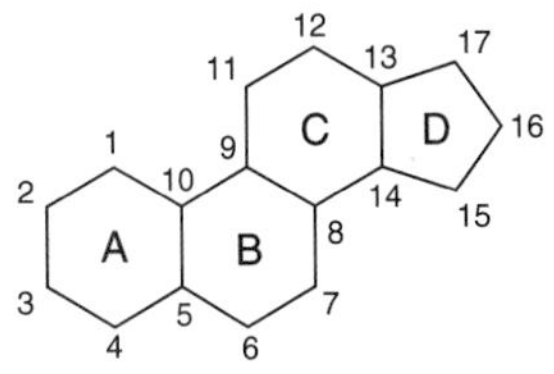

cyclopentanoperhydrophenanthrene

Figure 12.1. Basic steroid structure.

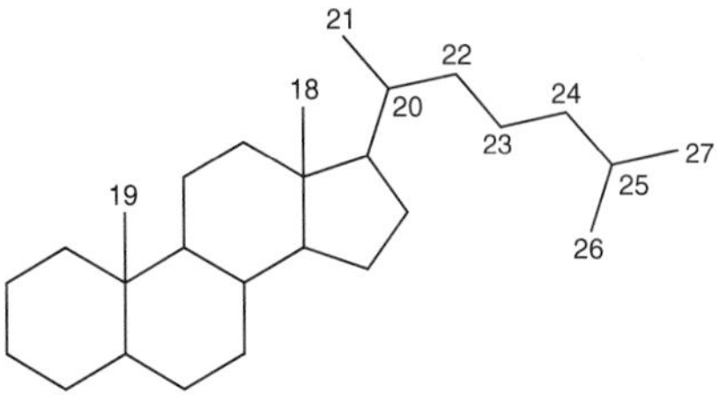

Figure 12.2. Side-chains on steroid scaffold.

the biochemical mechanism of action of steroids began in the late 1950s with the use of tritiated estrogens of high specific activity. Jensen and Jacobson reported the accumulation of physiological levels of estrogen in target organs and postulated the presence of a receptor (25). Research on the biochemistry and molecular biology of steroid hormone action has exploded in the past few decades and constitutes a major effort in the field today.

Two important discoveries in the late 1940s and early 1950s had a dramatic effect not only on steroid research but also on the pharmacological applications of steroids. The first was the clinical report by Hench et al. (26) from the Mayo Clinic on the significant improvement in patients with rheumatoid arthritis following cortisone treatment. The second was the application of estrogen and progestin preparations for contraception, demonstrated by Pincus and colleagues (27). These two series of studies showed for the first time that steroids could be considered as drugs. As a result, extensive research on the medicinal chemistry, pharmacology, and clinical studies of steroid agonists and antagonists has evolved and continues to provide new insights and new medicinal agents for therapies in many different diseases and chemoprevention strategies.

2 STEROID CHEMISTRY

2.1 Structure and Physical Properties of Steroids

Steroid molecules possess a common chemical skeleton of four fused rings, consisting of three six-membered rings and a five-membered ring (Fig. 12.1). Chemically, this hydrocarbon scaffold is a cyclopentanoperhydro-

phenanthrene, describing the three rings of phenanthrene (rings A, B, and C) and the cyclopentane ring (ring D). In steroids, the phenanthrene ring system is completely saturated (hydrogenated) and is thus referred to as a perhydrophenanthrene. This steroid scaffold contains 17 carbon atoms, and the numbering of the carbon atoms begins with the carbons of the phenanthrene and is then followed by numbering the remaining carbons of the cyclopentane ring (Fig. 12.1). Additional carbon atoms on steroids include angular methyl groups attached to C13 and C10 and alkyl substituents on C17 (Fig. 12.2).

When the steroid nucleus is drawn in a two-dimensional representation, the steroid scaffold appears planar and substituents on carbons of the steroid scaffold may be located either above or below the "plane" of the steroid. Substitutents located above the plane are drawn with solid lines or with solid wedges, and these moieties are referred to as being in the β-configuration. Substitutents located below the plane are drawn with dashed lines and are referred to as having the α-configuration. The angular methyl groups numbered 18 and 19 are attached in the β-configuration (above the steroid plane) to C13 and C10, respectively. Side-chains at position 17 are always β unless indicated by dotted lines or in the nomenclature of the steroid. The stereochemistry of the rings and the substituents on the steroid scaffold markedly affects the biologic activity of a given class of steroids.

The three-dimensional shapes of the rings in the steroid scaffold are actually not planar. The cyclohexane rings of steroids exist in the preferred chair conformation. As a result, substitutents on the cyclohexane rings can be located in either the axial or the equatorial position. The cyclopentane ring exists in a

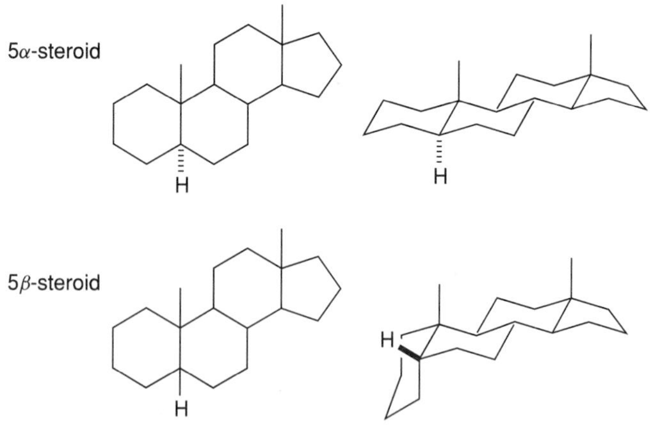

5α-steroid

5β-steroid

Figure 12.3. Representations of
5α-steroid and 5β-steroid.

half-chair or open-envelope conformation. Although the cyclohexane ring may undergo a
flip in conformation, steroids are rigid structures because they generally have at least one
trans fused-ring system, and these rings must
be diequatorial to each other. Endogenous steroids contain two *trans* fused rings, one diequatorial *trans* fusion between rings B and C
(carbons 8 and 9) and the other diequatorial
trans fusion between rings C and D (carbons
13 and 14). Two possible ring fusions are observed in endogenous steroids between rings A
and B. The diequatorial *trans* fusion between
rings A and B results in the hydrogen atom at
position 5 being on the opposite side of the
rings from the angular methyl group at position 19; the 5α notation is used for this hydrogen. The overall three-dimensional shape of
the 5α-steroid is a nearly flat, pleated shape.
The axial-equatorial *cis* fusion between rings
A and B results in the hydrogen atom at position 5 being on the same side of the rings with
the angular methyl group at position 19; the
5β notation is used for this hydrogen. The two-
dimensional and three-dimensional representations for the 5α-steroid and the 5β-steroid
are illustrated in Fig. 12.3. These ring junctions (A/B and B/C) and the chair conformation of the six-membered rings result in an
overall topography of the steroid scaffold that
is rather rigid. Some minor flexibility is observed in the conformation of the D ring.

Because the angular methyl groups at positions 18 and 19 are β and have an axial orientation (i.e., perpendicular to the plane of the
rings), the conformational orientation of the

remaining bonds of a steroid can be assigned
easily. The orientation of the remaining bonds
on a steroid may be determined if one recalls
that groups on a cyclohexane ring that are positioned on adjacent carbon atoms (vicinal,
-C_1H—C_2H-) of the ring (i.e., 1, 2 to each
other) are *trans* if their relationship is 1,2-di-
axial or 1,2-diequatorial and are *cis* if their
relationship is 1,2-equatorial-axial.

The backbone of the steroid molecule can
be referred to by a series of carbon-carbon
bonds and the *cis* or *trans* relationship of the
four rings (Fig. 12.4). The 5α-steroid molecule
has a *trans-anti-trans-anti-trans* backbone. In
this structure, all the fused rings have *trans*
(diequatorial) stereochemistry (i.e., the A/B
fused ring, the B/C fused ring, and the C/D
fused ring are *trans*). The term *anti* is used in
backbone notation to define the orientation of
rings that are connected to each other and

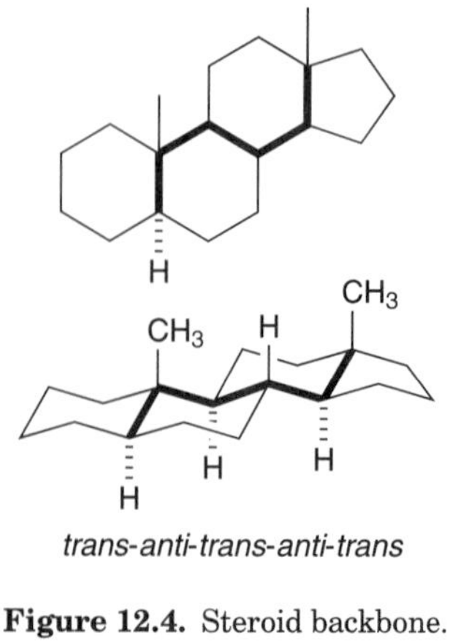

trans-anti-trans-anti-trans

Figure 12.4. Steroid backbone.

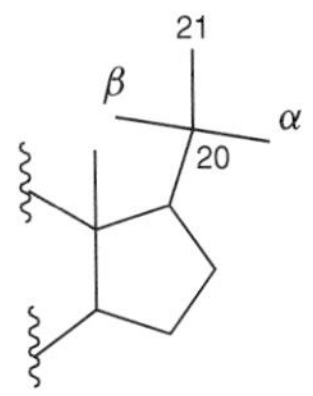

Figure 12.5. α and β substituents on carbon number 20.

have a *trans*-type relationship. For example, the bond equatorial to ring B, at position 9, which forms part of ring C, is *anti* to the bond equatorial to ring B, at position 10, which forms part of ring A. A 5β-steroid has a *cis-anti-trans-anti-trans* backbone, in which the A/B rings are fused *cis*. The term *syn* is used to define a *cis*-type relationship in a similar fashion as *anti*. No naturally occurring steroids exist with a *syn*-type geometry, although such compounds can be chemically synthesized. Thus, the conventional drawing of the steroid nucleus is the natural configuration and does not show the hydrogens at the 8β, 9α, or 14α positions. If the carbon at position 5 is saturated, the hydrogen is always drawn, either as 5α or 5β. Also, the conventional drawing of a steroid molecule has the C18 and C19 methyl groups shown only as solid lines.

Many of the biologically important steroids contain a carbon-carbon double bond between positions 4 and 5 or 5 and 6, and consequently there is no *cis* or *trans* relationship between rings A and B. The symbol Δ is often used to designate a carbon-carbon double bond (C=C) in a steroid. If the carbon-carbon double bond is between positions 4 and 5, the compound is referred to as a Δ^4-steroid. If the carbon-carbon double bond is between positions 5 and 10, the compound is designated a $\Delta^{5(10)}$-steroid. Addition of a double bond also increases the flexibility of the ring; for example, the A ring of a Δ^4-steroid exists primarily in a half-chair conformation.

Aliphatic side-chains at position 17 are always assumed to be β-configuration, which is the configuration found in endogenous steroids. By long-standing convention, the α and β terms have been applied to substituents on carbon number 20 (Fig. 12.5) on steroids containing a two-carbon side-chain (i.e., contain-

ing carbons 20 and 21). Because C20 is not a ring carbon, the preferred designation for the stereochemistry on C20 is determined based on the Cahn-Ingold-Prelog sequence rules (the R,S system). The R,S system is also used to designate the stereochemistry of other positions on the steroid side-chains.

Important physical properties of steroids include the physical state of the molecules and the solubility of steroids. The overall class of steroids is found almost entirely as solids. The melting points for steroids range from approximately 100 to 250°C. Molecules of one particular steroid compound may crystallize in either an anhydrous form or in a hydrated or solvated form, resulting in different melting points observed for the two forms. Also, individual molecules of one steroid compound may pack in different arrangements in crystals from different solvents, resulting in polymorphic forms (5). Regarding solubility, steroids are generally insoluble in water, whereas they are reasonably soluble in organic solvents such as ethanol, acetone, chloroform, and dioxane. Steroids with a phenolic hydroxyl group on the aromatic A ring (estrogens) are soluble in dilute sodium hydroxide.

2.2   Steroid Nomenclature

Many naturally occurring steroids are referred to by their common or trivial names, such as cholesterol, cortisol, progesterone, testosterone, and estradiol. As more steroid molecules were being discovered and/or synthesized, it became clear that a more systematic method for naming steroids was needed. Beginning in the 1950s, nomenclature rules for steroids were being developed, and the most recent IUPAC-IUB Joint Commission rules for systematic steroid nomenclature were published in 1989 (28, 29). The systematic names for steroids are based on the steroid hydrocarbon system, and the particular systematic name begins by selection of the stem name based on the hydrocarbon system (Fig. 12.6). Cholestane is the term used for steroids with 27 carbon atoms (i.e., the C_{27} steroid structure). Pregnanes are steroids with 21 carbon atoms, androstanes have 19 carbon atoms, estranes have 18 carbon atoms, and gonanes have 17 carbon atoms.

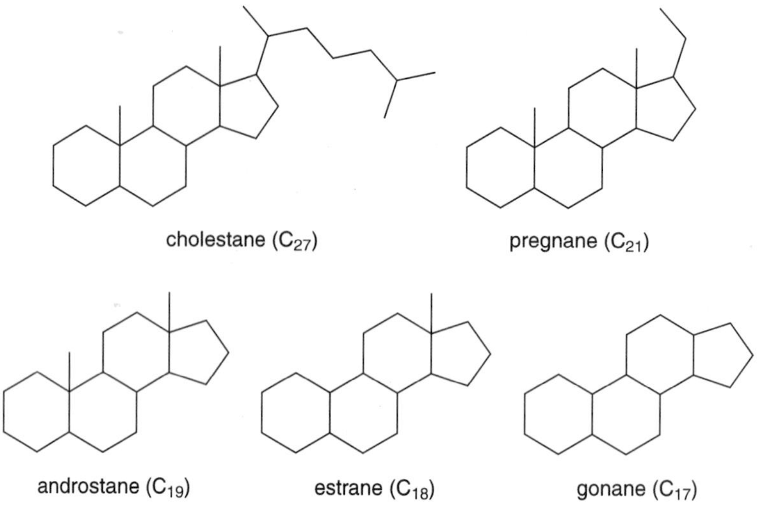

Figure 12.6. Structures of steroid stem names.

If there are any double bonds present in the steroid scaffold, the "ane" ending of the stem name is replaced with "ene" if one double bond is present, "diene" if two double bonds are present, "triene" if three double bonds are present, and so forth. The position(s) of the double bond is indicated by placing the lowest number of the carbon atom of the double bond in front of the "ene" ending. If the number of the first carbon atom indicates ambiguous positions, then this first number is followed by the number of the other carbon atom, placed in parentheses. When saturation is present at the 5 position, a designation of either 5α- or 5β- is required and is placed before the stem name. A suffix for the stem name is selected based on the following priorities:

carboxylic acid (or derivative)>carbonyl>

alcohol>amine>ether

In adding the suffix to the stem name, the final "e" in the stem name is always dropped when the suffix begins with a vowel. The carbon number of the substituent (and stereochemistry, if present) is placed in front of the suffix. Remaining substituents are denoted as prefixes, are preceded by the position number and stereochemistry, and are placed in alphabetical order.

Examples of the trivial names, systematic names, and chemical structures for common steroids are illustrated in Fig. 12.7. Cholesterol is the central steroid of the animal kingdom and functions as an essential component of cell membranes and as a biosynthetic precursor to other steroids in the body. Cholesterol has 27 carbon atoms, a hydroxyl group in the β-configuration at carbon 3, and contains a carbon-carbon double bond between carbons 5 and 6. Cholesterol is referred to as a Δ^5-steroid or, more specifically, a Δ^5-sterol because it is an unsaturated alcohol. The systematic name for cholesterol is cholest-5-en-3β-ol. The adrenocorticoids (adrenal cortex hormones) are pregnanes and are exemplified by cortisol, which is an 11β,17α,21-trihydroxypregn-4-ene-3,20-dione. Progesterone (pregn-4-ene-3,20-dione), a female sex hormone synthesized by the corpus luteum, is also a pregnane analog. The male sex hormones (androgens) are based on the structure of 5α-androstane. Testosterone, an important naturally occurring androgen, is named 17β-hydroxyandrost-4-en-3-one. Finally, the estrogens are female sex hormones synthesized by the follicular cells of the ovaries and are estrane analogs containing an aromatic A ring. Although the A ring does not contain isolated carbon-carbon double bonds, these analogs are named as if the bonds

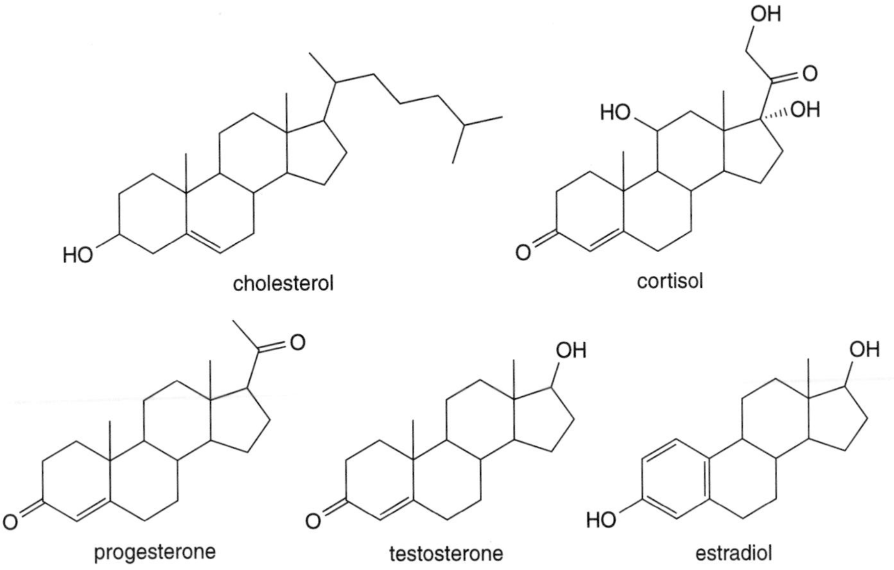

Figure 12.7. Structures of common steroids.

were in the positions shown in estradiol. Estradiol, a typical member of this class of drugs, is named estra-1,3,5(10)-triene-3,17β-diol.

2.3 Chemical Synthesis and Microbial Transformations of Steroids

A total of six Nobel prizes were awarded to scientists working in the area of steroids, with four out of the six prizes in steroid chemistry. The pioneer work of Adolf Windaus and Heinrich Wieland, in the 1920s, in structural elucidation of a number of important steroids set the stage for the many significant discoveries in the steroid area from the 1930s to the 1950s. The total synthesis of equilenin was first reported by Bachmann et al. (21) and later by Johnson and coworkers (30). The total synthesis of estrone was reported by Anner and Mieschler (31), followed by the work of other well-known chemists on the synthesis of cholesterol and sex hormones (32–35). In addition to the total synthesis of steroids, Russell Marker reported the use of sapogenins as the starting material for the synthesis of corticosteroids and sex hormones (36). In a series of studies reported in numerous brief communications in the early 1940s, Marker reported a chemical degradation process that converted diosgenin, a sapogenin from the Mexican yam,

cabeza de negro, to pregnonolone acetate in essentially two steps (37, 38). His work had a significant impact on the industrial production of steroids. One unique feature of research on steroid chemistry is the equal contribution on the field from both academia and the pharmaceutical industry. The historical perspective on steroid chemistry has been described in detail by a series of papers (7–9).

The development of corticosteroids and oral contraceptives in the late 1940s and early 1950s reflected the significance of chemical synthesis in steroid research. In this chapter we use the chemical synthesis of cortisone as an example to illustrate some of the synthetic transformations in steroid chemistry. In addition, the high costs of the chemical synthesis of corticosteroids eventually led the research on microbial biotransformation of steroids described in the next section.

2.3.1 Synthesis of Cortisone. The synthesis of cortisone was first reported by Sarett and is shown in Fig. 12.8a (39, 40). Methyl bisnordesoxycholate (I) was used as the starting material for the synthesis. The synthetic scheme constitutes three basic operations: transposition of the 12α-hydroxy group to the C11 position (from compound I to V), cleavage of the

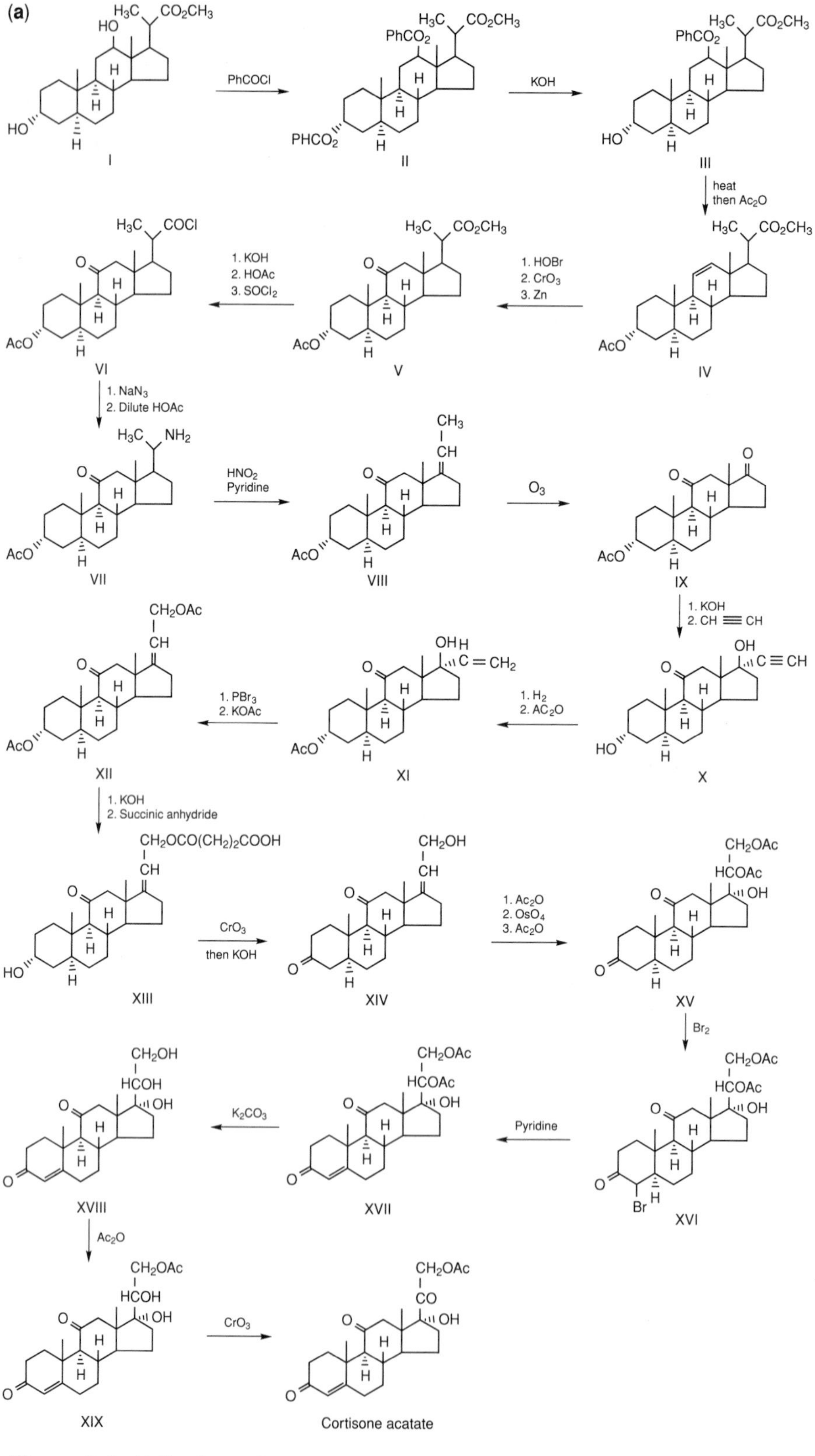

Figure 12.8. (a) Synthesis of cortisone acetate. (b) Improved synthesis of cortisone acetate.

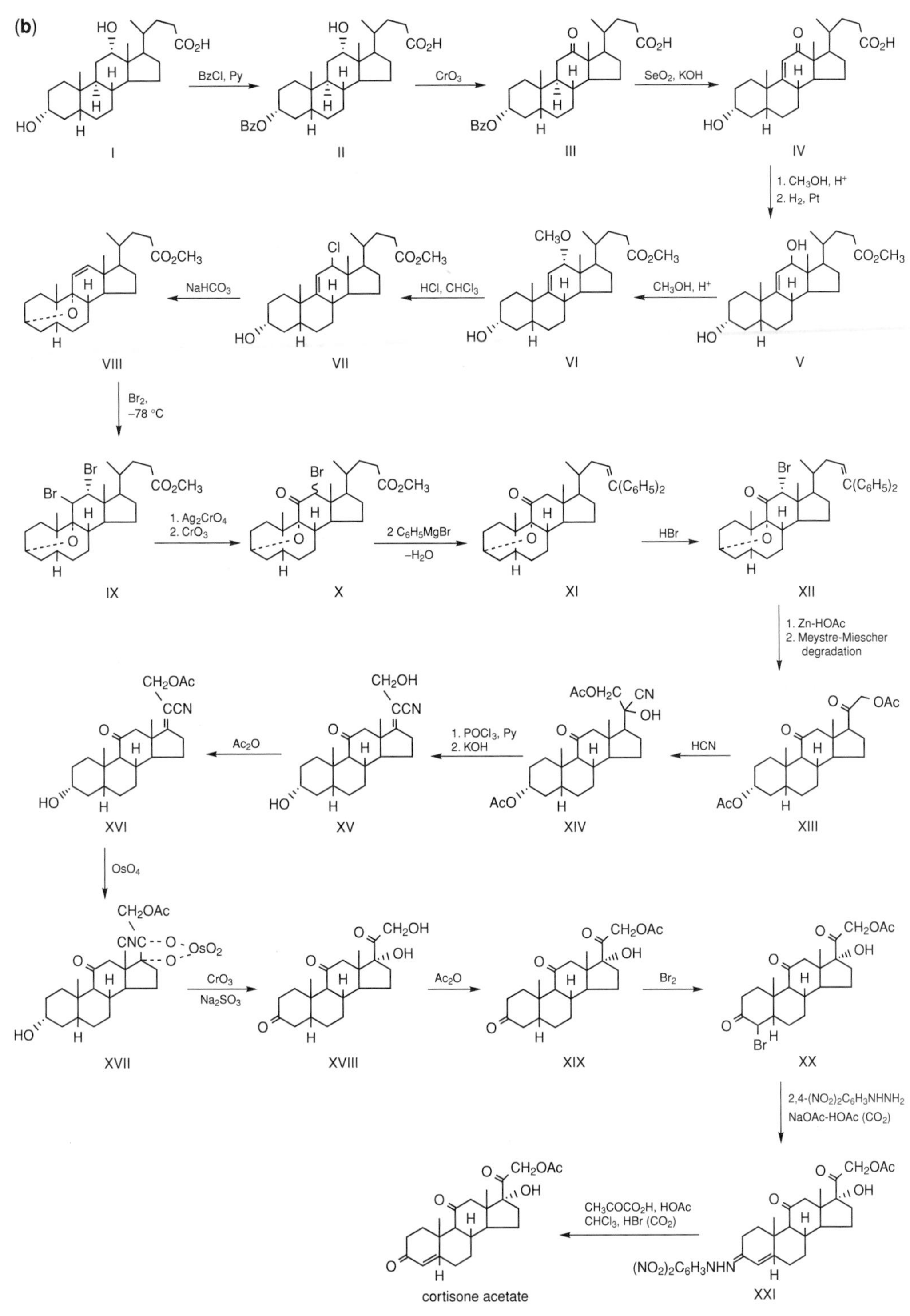

(b)

Figure 12.8. (Continued.)

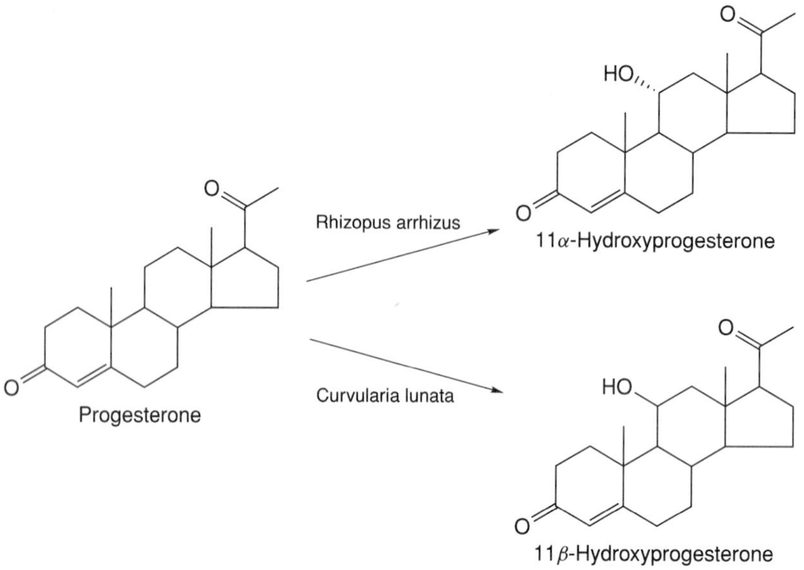

Figure 12.9. 11-Hydroxylation of progesterone.

bile acid side-chain and incorporation of the dihydroxyacetone moiety (from V to XIII), and generation of the 4-ene-3-one system (from XIII to cortisone). The yield for the conversion of V to IX was low. Eventually, an improved synthesis of cortisone was reported by Sarett (41, 42). The improved synthesis began with deoxycholic acid I. Protection of the 3α-hydroxy group in I followed by oxidation yielded the ketone III. The next step involved the formation of the unsaturated ketone IV with selenium dioxide (43). The next few steps required the transposition of the ketone at C12 to C11. The unusual step is the conversion of VII to VIII, to form the $3\alpha,9\alpha$ epoxide (44). The epoxide ring was opened with HBr and subsequent transformations yielded the important intermediate XIII. The remaining steps are the improved procedure for the introduction of the dihydroxyacetone side-chain as compared to the previous procedure (Fig. 12.8b). Reacting XIII with HCN formed the cyanohydrin XIV. Dehydration followed by acetylation obtained XVI. Reacting XVI with osmium tetroxide yielded the osmate ester XVII. The osmate ester served as a protecting group for the C_{17}-C_{20} double bond in addition to the introduction of the C17α alcohol group. Oxidation of the 3α-hydroxy group followed by the intro-

duction of a 3,4-double bond afforded cortisone acetate. The total synthesis of cortisone was reported in 1951 by Woodward and colleagues (45).

2.3.2 Microbial Steroid Biotransformations. The research on microbial transformation of steroids was stimulated after World War II when the anti-inflammatory properties of cortisone were reported (26). Efficient synthesis of corticosteroids was then required for scale-up synthesis as well as structure-activity relationship studies. As shown in Fig. 12.8, the synthesis of cortisone required a total of 31 steps (39, 40). One of the particularly challenging conversions was the transposition of the 12α-hydroxy group in the bile acid to the C11, which required 12 steps. In 1952, Peterson and Murray of Upjohn reported the first patented process of direct 11α-hydroxylation of progesterone through the use of *Rhizopus arrhizus* and *Rhizopus nigricans* (46, 47) (Fig. 12.9). In the same year, Fried and coworkers of Squibb Institute reported a similar microbial transformation with *Aspergillus niger* (48). In addition to progesterone, deoxycorticosterone, 11-deoxy-17α-hydroxycorticosterone, and 17α-hydroxy-progesterone could also be the substrates for the transformation. Three

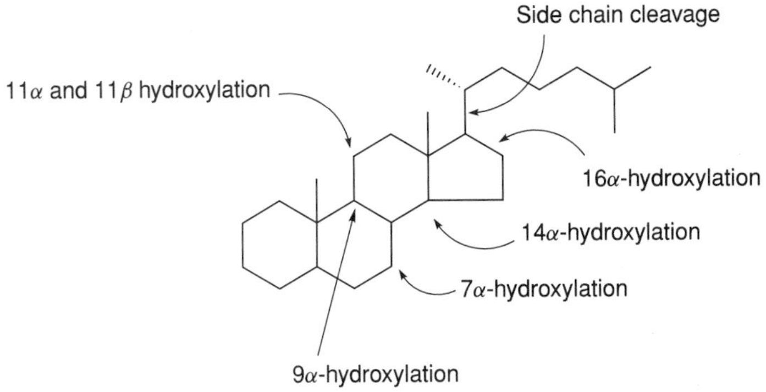

Figure 12.10. Selected site of microbial transformation of steroids.

years later, Schull and Kita of Pfizer reported the stereoselective 11β-hydroxylation with *Curvularia lunata* by use of the same progesterone substrates reported by Fried (5) (Fig. 12.9). As a result of the incorporation of the microbial transformation in the synthesis, the cost of production of hydrocortisone was lowered from $200 per gram in 1948 to $3.50 per gram in 1955.

In this chapter, only selected microbial transformations of steroids are described. The reader is directed to refer to articles with more extensive reviews on this topic (49–55). The microbial transformations are described with regard to the type of reaction and the carbon of the steroid skeleton bearing the reaction. Figure 12.10 is a diagrammatic illustration of the selected site of microbial transformations of steroids by use of a cholestane steroid skeleton.

16α-Hydroxylation. Every carbon in a steroid molecule is accessible for microbial transformation (56). Hydroxylations at the 11α, 11β, and 16α positions of steroids are routinely used industrially through microbial transformation. Hydroxylations at the 11α and 11β positions of progesterone were described above. The first example of microbial 16α-hydroxylation was the conversion of progesterone to 16α-hydroxyprogesterone with *S. argenteolus* reported by Perlman and coworkers (57) (Fig. 12.11). In addition to progesterone, 16α-hydroxylation on synthetic corticosteroid was also accomplished with *S. roseochromogenus* (58) and subsequently led to the discovery of triamcinolone, a widely used anti-inflammatory steroid.

7α- and 9α-Hydroxylations. 7α-Hydroxyandrostenedione is an important intermediate in the production of diuretics. Its production with microbial transformation from 3α,7α-dihydroxy-5β-holanic acid has been reported (59). Microbial 9α-hydroxylation of steroids was first observed by Peterson and Murray (60). This type of transformation has practical significance, given that the 9α-hydroxy moiety can easily be converted to the 9,11-dehydro functionality. The 9,11-dehydro system is an

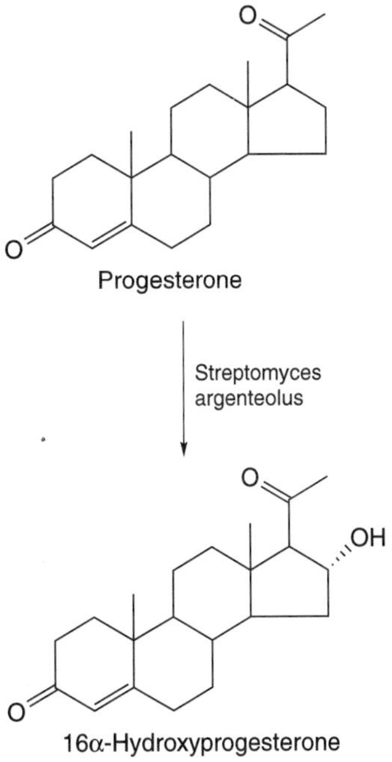

Figure 12.11. 16α-Hydroxylation of progesterone.

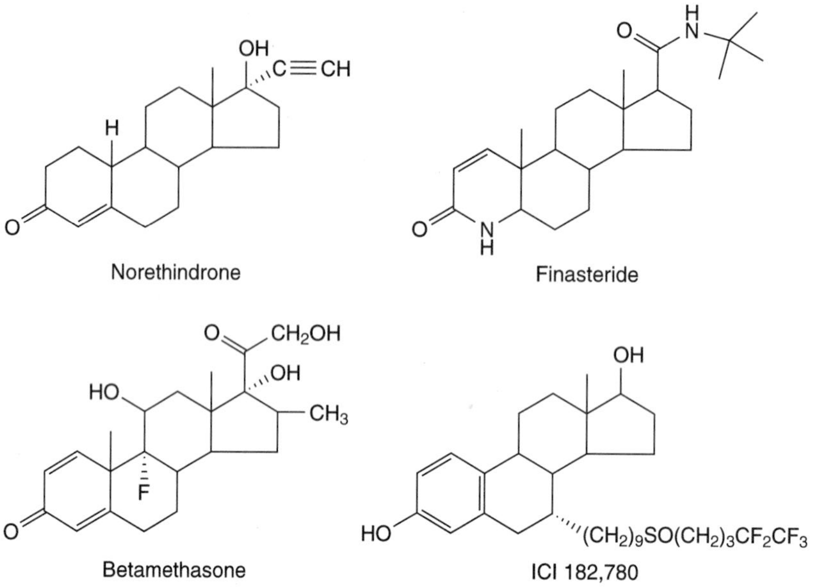

Norethindrone

Finasteride

Betamethasone

ICI 182,780

Figure 12.12. Steroids with different pharmacological and therapeutic indications.

important intermediate in the production of 9α-fluoro and/or 11-hydroxy steroids.

14α-Hydroxylation. 14α-Hydroxysteroids are of practical significance. 14α-Hydroxyandrost-4-ene-3,6,17-trione was recently shown to have aromatase inhibitory activity in human placenta and uterine tumors (61–63). The derivative was obtained through microbial oxidation of androst-4-ene-3,17-dione with *Acremonium strictum.*

Many fungi, such as *Mucor griseocyanus* and *Actinomucor elegans*, have been shown to introduce a 14α-hydroxy group to progesterone and other steroids (60, 64). The 14α-hydroxy steroids can serve as important intermediates in the production of steroids with 14β-hydroxy-5β-pregnane nuclei, a common structural framework that exists in many cardioactive steroids. Chemical transformation of steroids from the 14α-hydroxy configuration to 14β-hydroxy substituents can be accomplished through $14\beta,15\beta$-epoxide intermediates.

Side-Chain Cleavage. Many of the cheap and readily available natural products such as sitosterol, campesterol, and cholesterol have been considered waste products because of the lack of efficient methods for the cleavage of their saturated side-chains. A process for the

conversion of the steroids to androst-1,4-diene-3,17-dione by *Mycobacterium* has been reported (65). Two years later, Liu reported an improved method for the conversion of cholesterol to testosterone in a 43% yield (66).

2.3.3 Selected Chemical Reactions of Steroids. Because it is not feasible to describe all the chemical transformations reported in steroid literature in this chapter, we decided to choose several chemical reactions used in the synthesis of steroids with significant pharmacological and therapeutic indications (Fig. 12.12). (For a complete review of steroid chemistry and reactions, refer to Refs. 1–3, 7–9, 67–75.) The steroids in Fig. 12.12 have distinct chemical structures and functionalities in the steroid skeleton. They serve as good examples to illustrate the different chemical transformations in the steroid backbone.

Norethindrone. Norethindrone (17α-ethinyl-17β-hydroxyestr-4-en-3-one) is a widely used orally active progestin in oral contraceptives and possesses a 19-norandrostane steroid skeleton. This particular steroid, first synthesized in the Syntex laboratories, is obtained by applying Birch reduction on estrone 3-methyl ether to form the 1,4-dihydrobenzene enol ether moiety (Fig. 12.13). Oxidation

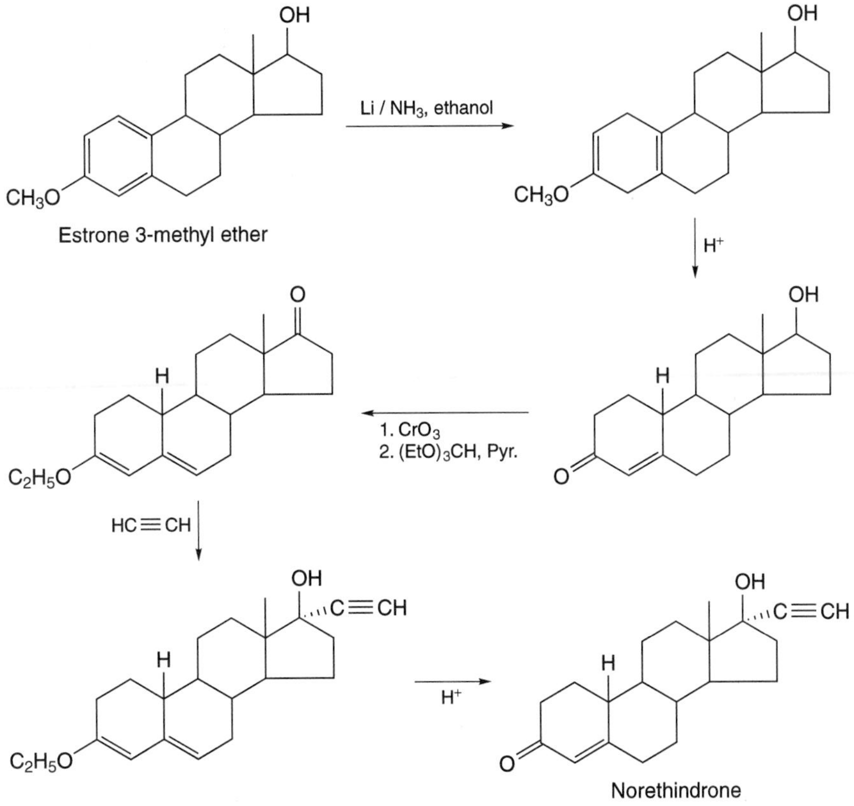

Figure 12.13. Synthesis of norethindrone.

and ethynylation of the enol ether affords the 17α-ethynyl derivative. Norethindrone can then be obtained by hydrolysis of the enol ether moiety (76). The double isomer, 17α-ethinyl-17β-hydroxyestr-5(10)-en-3-one (norethynodrel), was first prepared in the Searle laboratories through use of the Birch reduction method (77) and is also used in oral contraceptives. Norethynodrel is converted by gastric juices in the stomach to norethindrone.

Finasteride. Finasteride is a 5α-reductase inhibitor used for the treatment of benign prostatic hyperplasia (BPH) and alopecia (78). It contains a distinct 4-aza androstane nucleus. The synthesis of the 4-aza moiety begins with steroid with a 4-ene-3-one system. Oxidation of the enone yields the seco-acid, which can be cyclized with ammonia at high temperature to form the unsaturated lactam. NaBH$_4$ reduction of the double bond between C5 and C6 followed by dehydrogenation of the C1-C2

double bond yields the 4-aza steroid backbone in finasteride (78, 79) (Fig. 12.14).

Betamethasone. Betamethasone is a widely used anti-inflammatory glucocorticoid. The 9α-fluoro group greatly enhanced the glucocorticoid activity of the compound. The incorporation of the 9α-fluoro moiety can begin with the dehydration of the 11-hydroxyl group to form 9,11-olefin. The olefin is first treated with hypobromus acid (formed from N-bromoacetamide in water), which can be cyclized in the presence of base to form the 9,11-epoxide. Ring opening of the epoxide affords the 9α-fluoro and 11β-hydroxy moieties in betamethasone (80) (Fig. 12.15).

Faslodex (ICI 182,780). Faslodex (ICI 182,780) is a pure steroidal antiestrogen and was recently approved by the U.S. Food and Drug Administration for the treatment of estrogen-receptor positive metastatic breast cancer (81). The long hydrophobic alkyl group

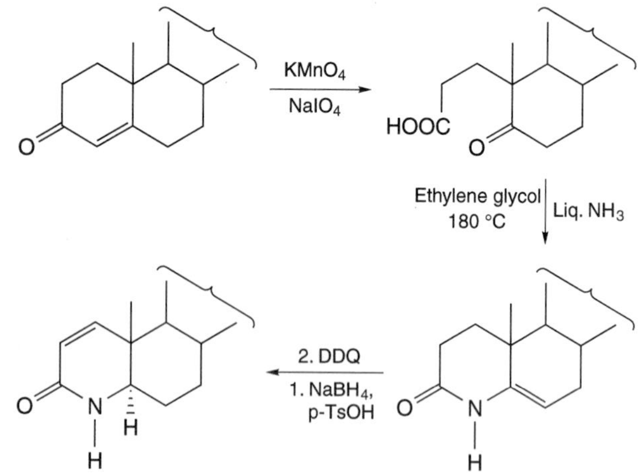

Figure 12.14. Synthesis of 4-aza steroid backbone in finasteride.

4-aza steroid skeleton

at the 7α position converts the estradiol from an estrogenic molecule to a pure antiestrogen. The key reactions to incorporate the 7α-substituent begin with the 1,6-conjugated addition of the dimethyl *t*-butyl-silyloxyundecyl bromide to the dienone followed by the aromatization of the A ring (82) (Fig. 12.16). An improved synthetic scheme for the synthesis of an analog of faslodex (ICI 164,384) was reported (83). The synthesis begins with the treatment of the 6-keto estradiol derivative with potassium t-butoxide in dry 1,2-dimethoxyethane followed by quenching of the enolate with alkyl iodide (RI) to form a mixture of

the 7α and 7β epimers. The advantage of this scheme is that the 7β epimer can be quantitatively converted to the 7α epimer with NaOMe in methanol. Deoxygenation and deprotection yielded the final product.

3 STEROID BIOCHEMISTRY

3.1 Steroidogenesis

3.1.1 Cholesterol Biosynthesis. Cholesterol is the central steroid of the animal kingdom and functions as an essential component of

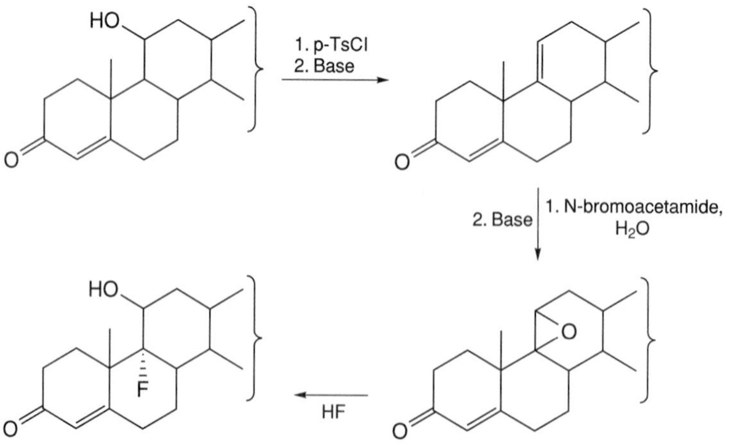

9α–F, 11 β–OH steroid

Figure 12.15. Synthesis of 9α-fluoro-11β-hydroxy steroids.

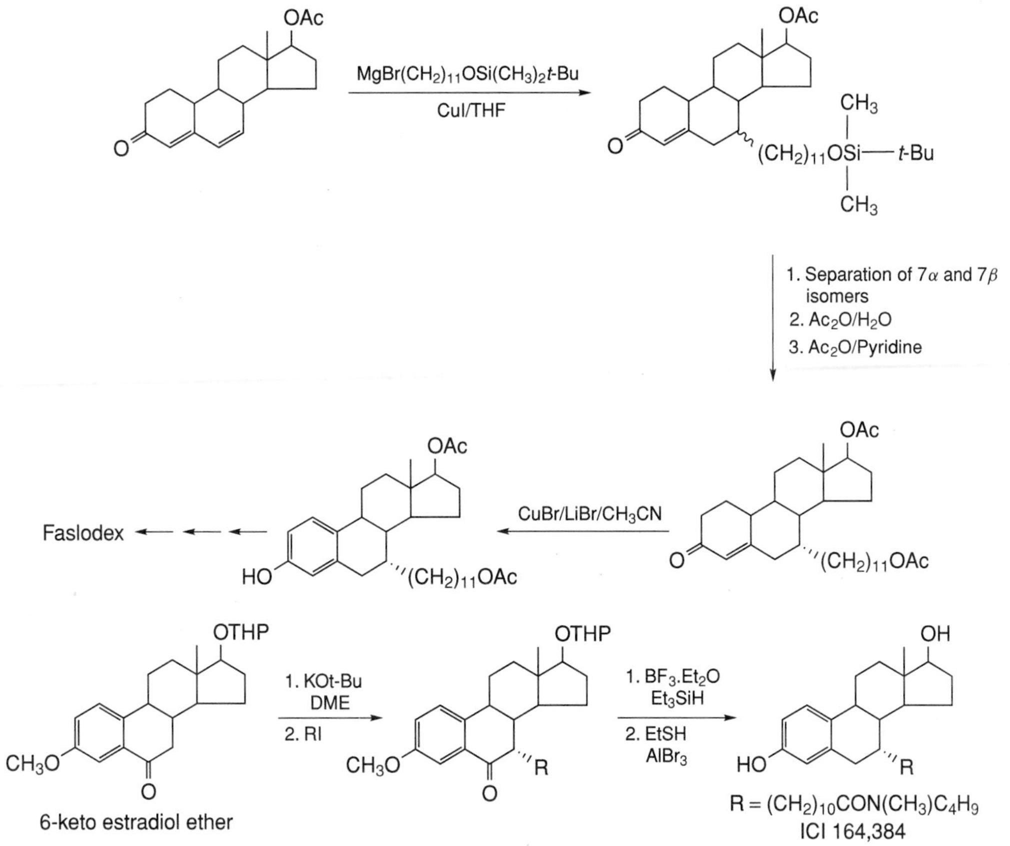

Figure 12.16. Synthesis of 7α-substituted estradiol derivative.

cell membranes and as a biosynthetic precursor to other steroids in the body. The two major sources of cholesterol are (1) dietary sources and (2) biosynthesis within the cell. All animal cells have the capacity to biosynthesize cholesterol; the principal sites of synthesis in humans are liver, skin, and intestinal mucosa. The biosynthesis of cholesterol consists of a complex pathway involving the participation of both cytosolic and membrane-bound enzymes (4, 6). The 27 carbon atoms of cholesterol are all derived from the acetate two-carbon unit, and the biosynthesis uses 18 acetyl-CoA molecules for the construction of cholesterol. The biosynthesis of cholesterol can be divided into two major stages: (1) the conversion of acetyl-CoA substrates to the C_{30} hydrocarbon polyene squalene and (2) the conversion of squalene to cholesterol.

The first steps in cholesterol biosynthesis use three molecules of acetyl-CoA and involve reversible enzymatic reactions. Two acetyl-CoA molecules are condensed together by a thiolase enzyme to form acetoacetyl-CoA. The third acetyl-CoA is combined with acetoacetyl-CoA by the enzyme 3-hydroxy-3-methylglutaryl-CoA synthase to form S-3-hydroxy-3-methylglutaryl-CoA (HMG-CoA). The next step in cholesterol biosynthesis is the conversion of HMG-CoA to mevalonate (or mevalonic acid) by the enzyme HMG-CoA reductase (Fig. 12.17). The enzymatic step uses two molecules of NADPH, obtaining the reduction of the thioester to an alcohol, resulting in R-mevalonate. This enzymatic step is irreversible and is the rate-limiting step in cholesterol biosynthesis. HMG-CoA reductase is an integral membrane protein of the smooth endoplasmic

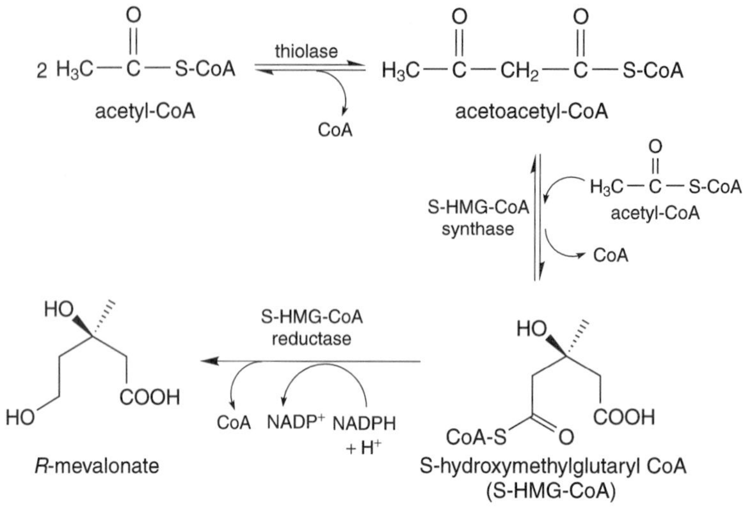

Figure 12.17. Formation of mevalonate from acetyl CoA.

reticulum and is under complex regulation. HMG-CoA reductase can exist as an inactive phosphorylated form and as an active dephosphorylated form, thus enabling rapid regulation of its enzymatic activity by hormones such as insulin and glucagons. At the transcriptional level, increased intracellular levels of cholesterol can inhibit the gene regulating HMG-CoA reductase. This enzyme is an important target for pharmacological regulation of cholesterol biosynthesis and is inhibited by the important class of hypercholesterolemic agents referred to as the statins (84). Lovastatin (Mevacor), pravastatin (Pravacol), simvastatin (Zocor), and atorvastatin (Lipitor) are examples of drugs from this therapeutic class (Fig. 12.18). The lactone ring present in some of these medicinal agents is hydrolyzed *in vivo* to the active form, and the open-chain derivative structurally resembles *R*-mevalonate.

The next series of enzymatic steps convert six molecules of mevalonate into the C_{30} polyene squalene and involve a series of phosphorylation events involving ATP (Figs. 12.19 and 12.20). Mevalonate is first phosphorylated by the enzyme mevalonate kinase by use of ATP to produce 5-phospho-mevalonate. A second molecule of ATP is used to form 5-pyrophospho-mevalonate. The third molecule of ATP is used in a concerted reaction that involves decarboxylation and loss of the tertiary hydroxyl

group forming isopentenyl pyrophosphate, the basic "isoprene unit" (a C_5 molecule). Isomerization of isopentenyl pyrophosphate produces dimethylallyl pyrophosphate. Isopentenyl pyrophosphate and dimethylallyl pyrophosphate then condense in a head-to-tail fashion by the enzyme prenyl transferase to form geranyl pyrophosphate (a C_{10} molecule). Prenyl transferase then catalyzes the head-to-tail condensation of another isopentenyl pyrophosphate to form farnesyl pyrophosphate (a C_{15} molecule). Finally, two farnesyl pyrophosphates combine in a head-to-head fashion to form squalene (a C_{30} molecule). This last step is catalyzed by the membrane-bound enzyme squalene synthase, uses a molecule of NADPH, and eliminates both pyrophosphates in the reaction.

The second major stage of cholesterol biosynthesis is the conversion of squalene to cholesterol (Fig. 12.21). Squalene is cyclized to form the basic steroid scaffold through a two-step process. The first step is catalyzed by the enzyme squalene epoxidase, or squalene monooxygenase, and uses molecular oxygen and NADPH. Squalene epoxidase is a flavin monooxygenase present in the smooth endoplasmic reticulum and adds one oxygen atom from molecular oxygen to the end of squalene, forming the epoxide 2,3-oxidosqualene. The second step involves cyclization to lanosterol and is catalyzed by the enzyme 2,3-oxidosqua-

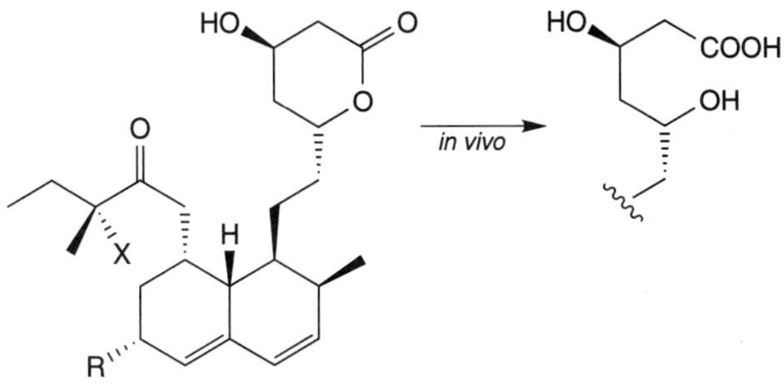

R = CH$_3$ X = H   lovastatin (Mevacor®)
R = OH   X = H   pravastatin (Pravachol®)
R = CH$_3$ R = CH$_3$ simvastatin (Zocor®)

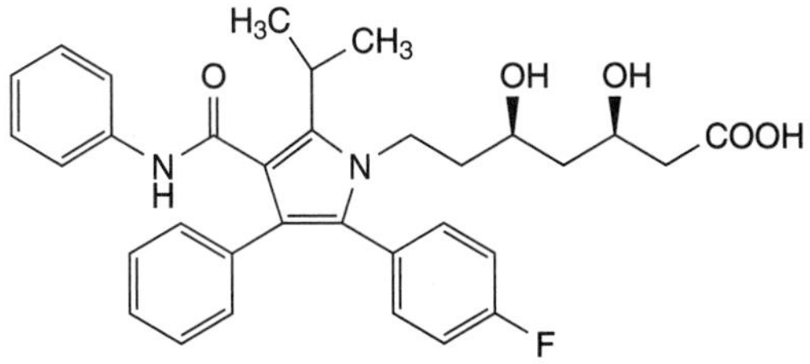

atorvastatin (Lipitor®)

Figure 12.18. Inhibitors of HMG-CoA reductase.

lene cyclase. The enzyme mediates a protonation of the oxygen atom of the epoxide, and the opening of the epoxide drives the series of cyclizations to produce a protosterol cation intermediate. The protosterol cation intermediate undergoes a series of rearrangements that include 1,2-hydride and methyl shifts, resulting in the C$_{30}$ sterol lanosterol. Conversion of lanosterol to cholesterol is a complex, 19-step process that involves the removal of three methyl groups (at C4 and C14), reduction of the side-chain double bond, and rearrangement of the 8,9-double bond to the 5,6-double bond in the B-ring. These transformations of lanosterol involve multiple enzymes found in the smooth endoplasmic reticulum, and diverging and converging pathways exist to many of the intermediates.

3.1.2 Formation of Pregnenolone. Cholesterol is converted by enzymatic cleavage of its side-chain to pregnenolone (3β-hydroxypregn-5-en-20-one), which serves as the biosynthetic precursor of the steroid hormones. This side-chain cleavage biotransformation is catalyzed by a mitochondrial cytochrome P450 enzyme

complex that consists of three proteins, cytochrome P450$_{SCC}$ (also referred to as cytochrome P450 11A1), ferrodoxin, and ferrodoxin reductase (85). Three oxidation steps are involved in the enzymatic conversion, and three molecules of NADPH and molecular oxygen are consumed for each molecule of cholesterol converted to pregnenolone (Fig. 12.22). The first oxidation results in the formation of cholest-5-ene-3β,22R-diol, followed by the second oxidation, yielding cholest-5-ene-3β,20R,22R-triol. The third oxidation step catalyzes the cleavage of the C$_{20}$-C$_{22}$ bond to release pregnenolone and isocaproic acid.

This conversion of cholesterol to pregnenolone is the initial step in steroidogenesis and is often considered the rate-limiting enzymatic step. However, several steps must precede the oxidation of the side-chain of cholesterol. These early steps are regulated by pituitary trophic hormones, which bind to G-protein-coupled receptors to elevate intracellular cAMP levels and initiate steroidogenesis within minutes (Fig. 12.23). First, cholesterol is stored in steroidogenic tissues (such as the adrenal cortex, the ovary, the testis) in lipid

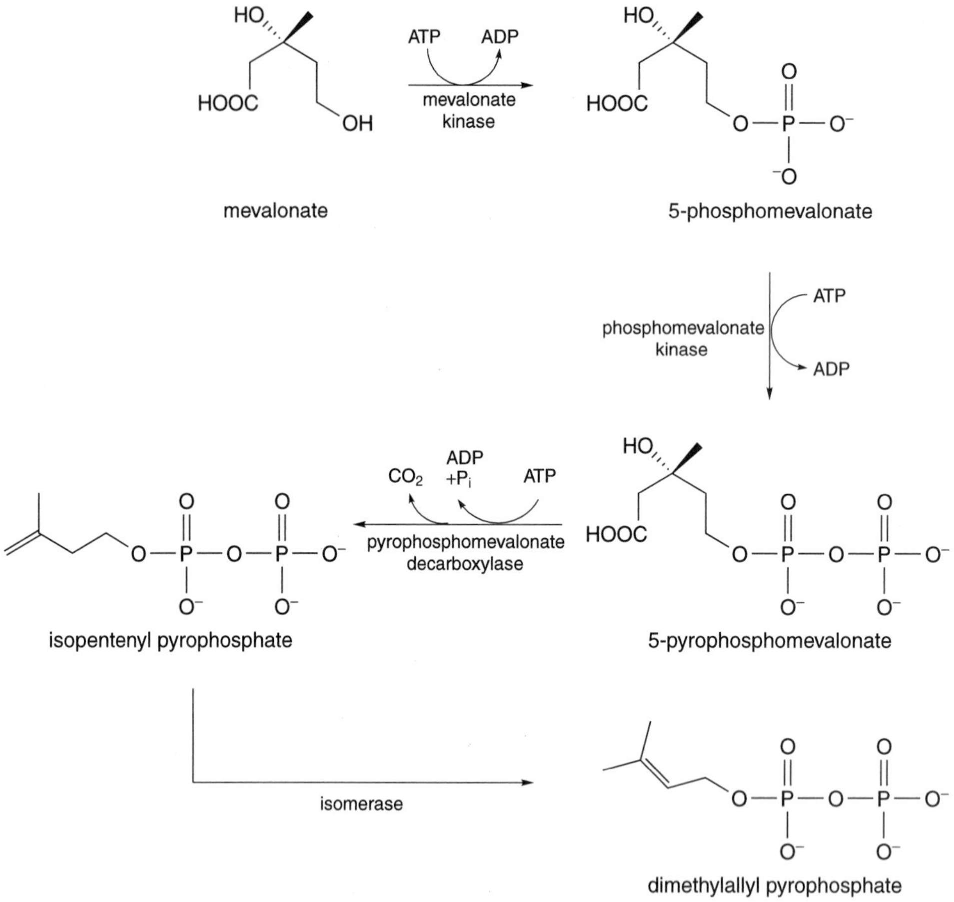

Figure 12.19. Formation of isopentenyl pyrophophate from mevalonate.

droplets as cholesterol oleate esters. Cholesterol esterase is activated by cAMP mechanisms and cleaves the esters to liberate cholesterol. The free cholesterol is transported in the cytoplasm by sterol carrier proteins to the mitochondria. The cytochrome $P450_{SCC}$ is located on the inner membrane of the mitochondria. For cholesterol metabolism to occur, the cholesterol must be transferred to the inner mitochondrial membrane for side-chain cleavage. This transfer of cholesterol across the mitochondrial membrane is accomplished by a protein referred to as steroidogenic acute regulatory protein (or StAR protein), and this transfer of cholesterol has been suggested to be the real rate-limiting step (86, 87). Cholesterol binds to the cytochrome $P450_{SCC}$ protein, followed by molecular oxy-

gen. The cytochrome $P450_{SCC}$ protein receives reducing equivalents from NADPH by an electron-transport process that occurs in the mitochrondrial matrix. The FAD-containing flavoprotein, ferrodoxin reductase, is reduced by NADPH and transfers electrons to ferrodoxin, a two-iron, two-sulfur protein. Ferrodoxin then transfers the electrons to the cytochrome $P450_{SCC}$ to activate the oxygen atom. In the adrenal gland, these proteins are referred to as adrenodoxin reductase and adrenodoxin. As noted above, three oxidation steps are needed to convert cholesterol to pregnenolone by cytochrome $P450_{SCC}$. The cytochrome $P450_{SCC}$ enzyme is encoded by a single human gene (CYP11A1), and the gene for cytochrome $P450_{SCC}$ lies on chromosome 15 (88, 89).

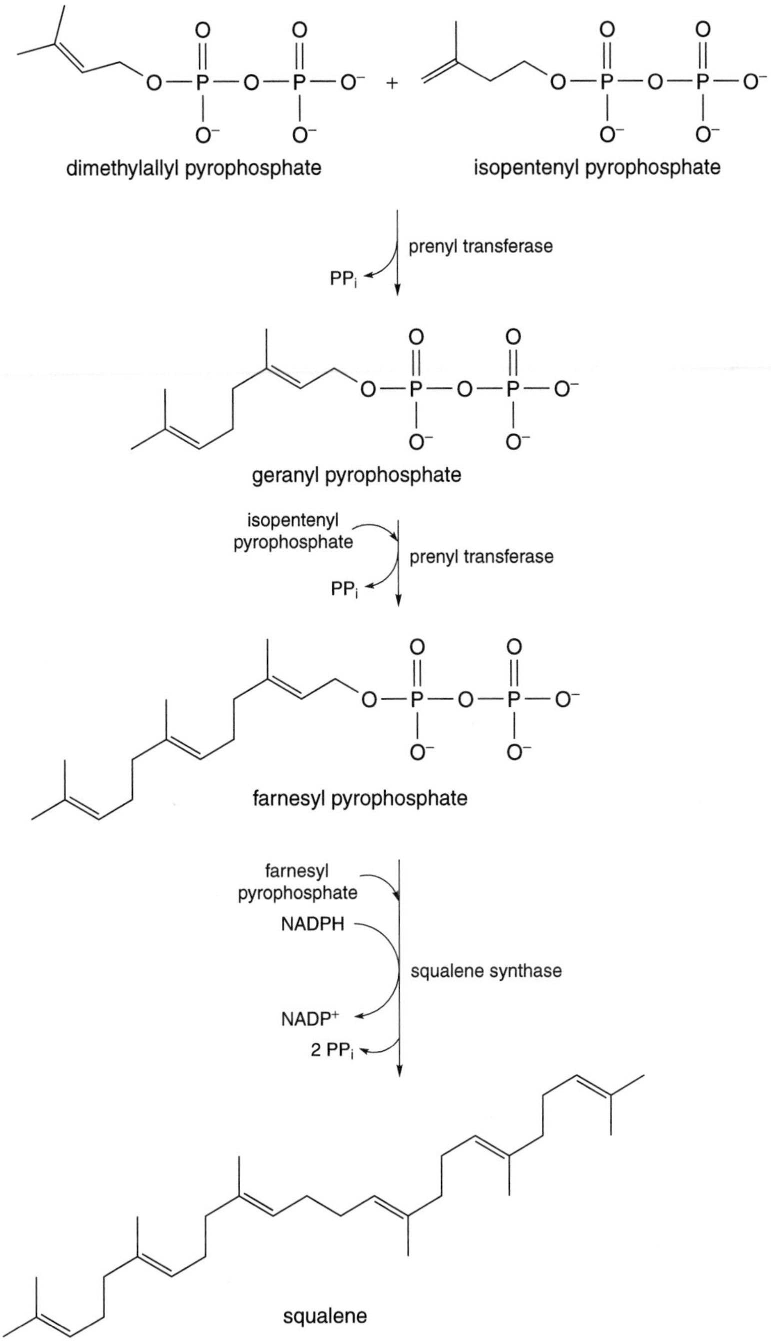

Figure 12.20. Formation of squalene.

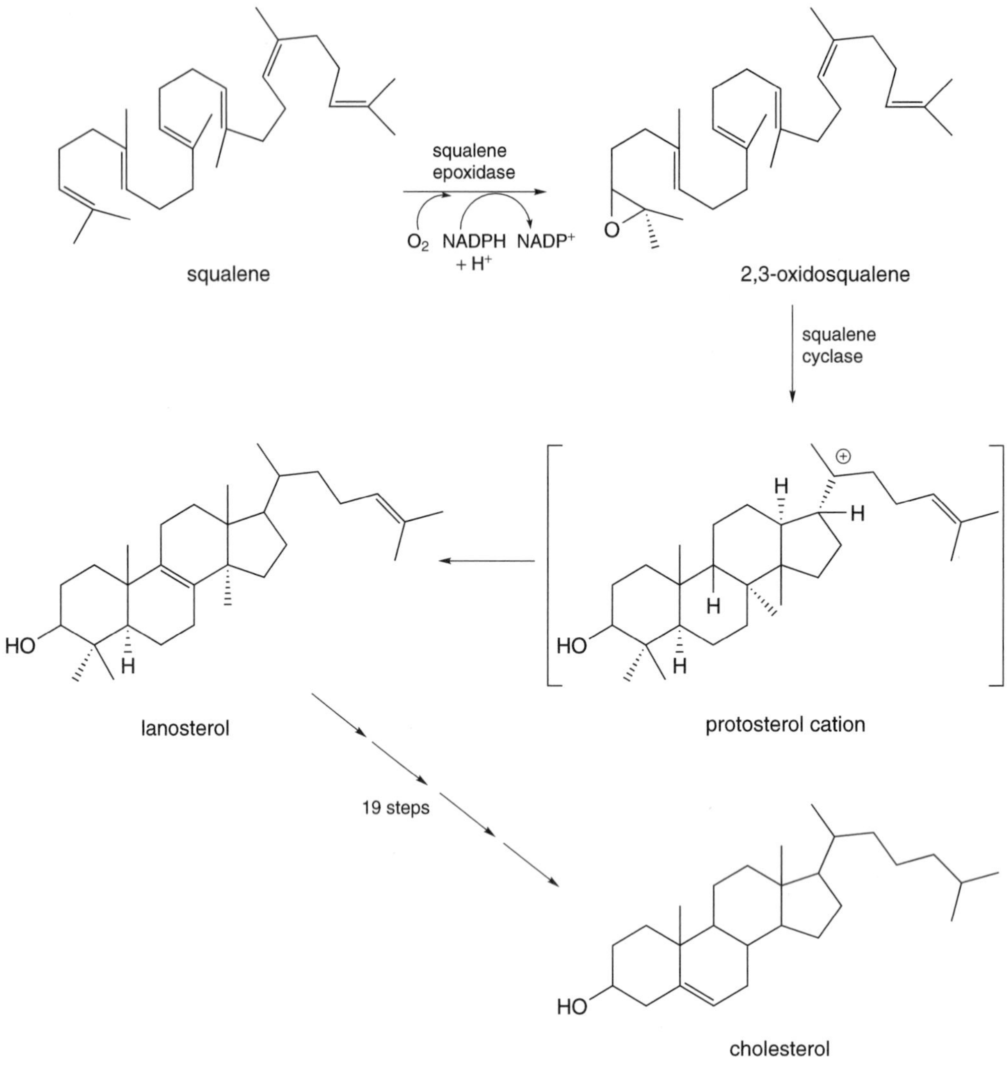

Figure 12.21. Conversion of squalene to cholesterol.

3.1.3 Biosynthesis of Adrenal Steroids. Pregnenolone serves as the common precursor in the formation of the adrenocorticoids and other steroid hormones. This C_{21} steroid is converted through enzymatic oxidations and isomerization of the double bond to a number of physiologically active C_{21} steroids, including the adrenocorticoids cortisol (hydrocortisone), corticosterone, and aldosterone (90, 91). The glucocorticoids and mineralocorticoids are secreted under the influence of peptide hormones secreted by the hypothalamus and anterior pituitary (adenohypophysis).

The peptide hormone in the anterior pituitary that influences glucocorticoid biosynthesis is adrenocorticotropic hormone (ACTH; corticotropin), whereas the peptide hormone in the hypothalamus is corticotropin-releasing factor (CRF). CRF is released by the hypothalamus and is transported to the anterior pituitary, where it stimulates the release of ACTH into the bloodstream. ACTH is transported to the adrenal glands, where it stimulates the biosynthesis and secretion of the glucocorticoids. The circulating levels of glucocorticoids act on the hypothalamus and anterior pitu-

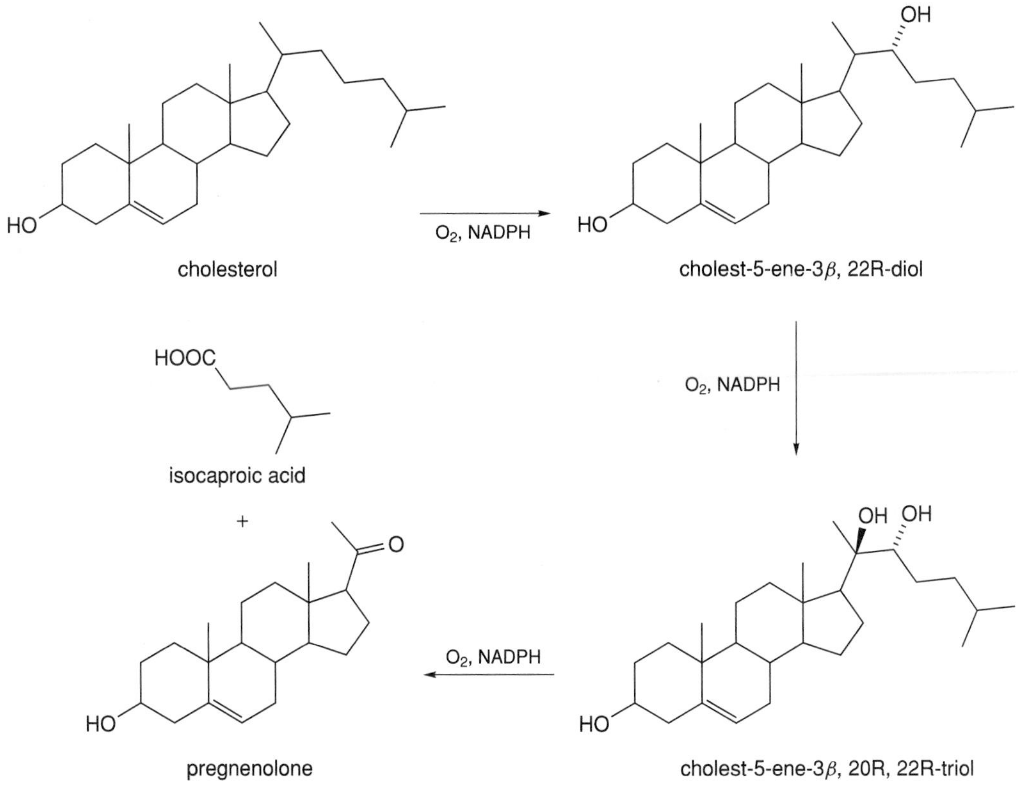

Figure 12.22. Conversion of cholesterol to pregnenolone.

itary in a negative feedback mechanism to regulate the release of both CRF and ACTH. A variety of stimuli, including pain, noise, and emotional reactions, increase the secretion of CRF, ACTH, and, consequently, the glucocorticoids. Aldosterone secretion rates are influenced to a greater extent by the octapeptide angiotensin II.

The biosynthesis of adrenal steroids occurs in adrenal cortex cells and involves enzymes found in both the mitochondria and the smooth endoplasmic reticulum (Fig. 12.24). Once pregnenolone is formed in the mitochondria, this steroid is then transported to the endoplasmic reticulum for further metabolism. Hydroxylation of pregnenolone at position 17 by the enzyme 17α-hydroxylase (cytochrome $P450_{17\alpha}$) produces 17α-hydroxypregnenolone. In one step, 17α-hydroxypregnenolone is oxidized and isomerized by the dual action of the enzyme 5-ene-3β-hydroxysteroid dehydrogenase/3-oxosteroid-4,5-isomerase to produce

17α-hydroxyprogesterone. Another hydroxylation occurs by the action of 21-hydroxylase (cytochrome $P450_{21}$ or CYP21) to give rise to 11-deoxycortisol, which contains the physiologically important ketol (-COCH$_2$OH) side-chain at the 17β position. The final step in the biosynthesis is catalyzed by the enzyme 11β-hydroxylase (cytochrome $P450_{11\beta}$), a mitochondrial cytochrome P450 enzyme complex. This results in the formation of cortisol (hydrocortisone), the most potent endogenous glucocorticoid secreted by the adrenal cortex. More detailed discussions about the enzymology and regulation of adrenal steroidogenesis are provided in several reviews (6, 88, 89, 92).

The pathway for the formation of the potent mineralocorticoid molecule aldosterone is similar to that for cortisol and uses several of the same enzymes (Fig. 12.24). The preferred pathway involves the conversion of pregnenolone to progesterone by 5-ene-3β-hydroxysteroid dehydrogenase/3-oxosteroid-4,5-

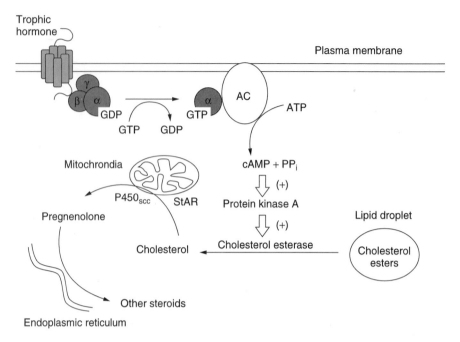

Figure 12.23. Trophic hormones and steroidogenesis.

isomerase. Hydroxylation at position 21 of progesterone by 21-hydroxylase results in 21-hydroxyprogesterone (deoxycorticosterone). 11β-Hydroxylase catalyzes the conversion of deoxycorticosterone to corticosterone, which exhibits mineralocorticoid activity. The final two oxidations involve hydroxylations at the C18 methyl group and are catalyzed by 18-hydroxylase (cytochrome $P450_{18}$ or 18-oxidase). These reactions produce, first, 18-hydroxycorticosterone (not shown) and then aldosterone, the most powerful endogenous mineralocorticoid secretion of the adrenal cortex. The aldehyde at C18 of aldosterone exists in equilibrium with its hemiacetal form.

Thus, the formation of the adrenal corticosteroids from pregnenolone involves several cytochrome P450 enzymes, present in either the mitochondria or the smooth endoplasmic reticulum. The mitochondrial enzymes, 11β-hydroxylase and 18-hydroxylase, are cytochrome P450 proteins that receive reducing equivalents from NADPH through an electron-transport process in the mitochrondrial matrix involving adrenodoxin reductase and adrenodoxin, analogous to the side-chain cleavage enzyme. The gene for 11β-hydroxy-

lase is CYP11B1 and is found on chromosome 8 in humans. Chromosome 8 is also the location of CYP11B2, the gene for 18-hydroxylase. 17α-Hydroxylase and 21-hydroxylase are cytochrome P450 enzymes of the smooth endoplasmic reticulum. Both of these enzymes use the flavoprotein NADPH-cytochrome P450 reductase to receive reducing equivalents from NADPH. The cytochrome $P450_{17\alpha}$ enzyme is encoded by a single human gene (CYP17), and the gene for cytochrome $P450_{17\alpha}$ lies on chromosome 10 (88, 93). Two genes for cytochrome $P450_{21}$ (gene CYP21B and pseudogene CYP21A) have been identified in the human genome on chromosome 6, but only CYP21B is active and encodes for the enzyme (94).

The other key enzyme in adrenal steroidogenesis is the enzyme 5-ene-3β-hydroxysteroid dehydrogenase/3-oxosteroid-4,5-isomerase (also referred to as 3β-HSD). As noted above, this membrane-bound enzyme exhibits dual enzymatic activity in a single protein molecule. Two human genes, HSD3B1 and HSD3B2, encode for 3β-HSD Type I and 3β-HSD Type II, respectively. These two genes and closely related pseudogenes are all located on chromosome 1 in humans. Finally, genetic mutations

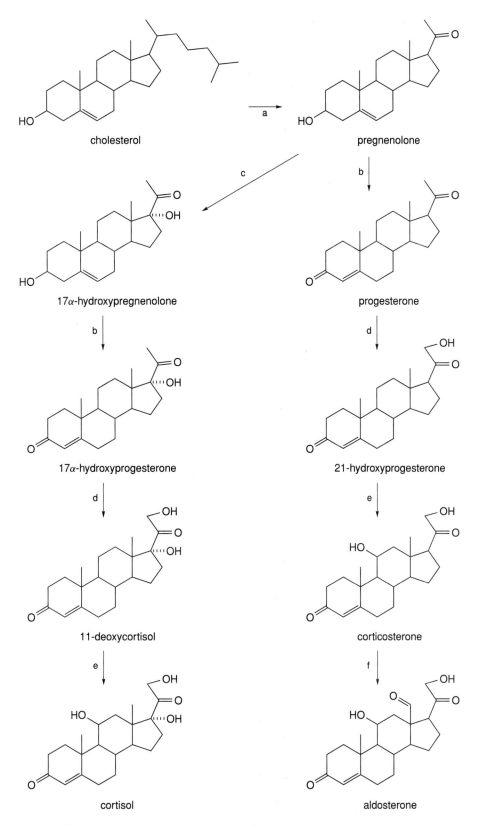

Figure 12.24. Adrenal steroidogenesis. Enzyme activities: (a) side-chain cleavages, (b) 3β-hydroxy-steroid dehydrogenase/4,5-isomerase, (c) 17α-hydroxylase, (d) 21-hydroxylase, (e) 11β-hydroxylase, (f) 18-hydroxylase.

in these key enzymes of adrenal steroidogenesis result in congenital adrenal hyperplasia (CAH), characterized by ambiguous genitalia and adrenal insufficiency in newborns (95). The most common form of CAH arises from defective 21-hydroxylase, but rare forms of CAH can also occur from defects in 11β-hydroxylase, 17α-hydroxylase, or 3β-HSD.

3.1.4 Biosynthesis of Progesterone. Progesterone is biosynthesized and secreted by the corpus luteum of the ovary during the luteal phase of the reproductive cycle (96, 97). Luteinizing hormone (LH), the anterior pituitary glycoprotein hormone, binds to the LH receptor on the surface of the cells to initiate progesterone biosynthesis. As in other endocrine cells such as adrenal cortical cells, the binding of LH results in an increase in intracellular cAMP levels through activation of a G-protein and adenylyl cyclase. One of the processes influenced by elevated cAMP levels is the activation of cholesterol esterase, which cleaves cholesterol esters and liberates free cholesterol. The free cholesterol is then converted in mitochondria to pregnenolone through the side-chain cleavage reaction described earlier, and progesterone is formed from pregnenolone by the action of 5-ene-3β-hydroxysteroid dehydrogenase/3-oxosteroid-4,5-isomerase (Fig. 12.25).

3.1.5 Androgen Biosynthesis. Androgens (male sex hormones) are synthesized from cholesterol in the testes and adrenal cortex and formed in the liver from circulating C_{21} steroids (90, 98). The major pathway for the biosynthesis of testosterone, the major circulating androgen in males, is shown in Fig. 12.20. In males, the hypothalamus regulates the anterior pituitary gland through luteinizing hormone releasing factor (LHRH), which stimulates the release of follicle-stimulating hormone (FSH) and luteinizing hormone (LH). The two gonadotropins have separate functions in males, with FSH promoting spermatogenesis and LH stimulating the biosynthesis and secretion of androgens. The primary source of testosterone is the Leydig cells of the testes. LH binds to its receptor on the surface of the Leydig cells to initiate testosterone biosynthesis. As in other endocrine cells, the binding of the gonadotropin results in an

increase in intracellular cAMP levels through activation of a G-protein and adenyl cyclase. Cholesterol is liberated from lipid stores and is then converted in mitochondria to pregnenolone through the side-chain cleavage reaction, as described earlier.

Two major pathways are involved in the conversion of pregnenolone to testosterone, referred to as the "4-ene" and "5-ene" pathways. The "5-ene" pathway involves the conversion of pregnenolone to dehydroepiandrosterone (DHEA) and is quantitatively more important in humans (99). Pregnenolone is converted by 17α-hydroxylase to 17α-hydroxypregnenolone and then to DHEA through 17–20 lyase. DHEA is the primary androgen secreted by the adrenal cortex in both men and women. In the male testis, DHEA is converted by 5-ene-3β-hydroxysteroid dehydrogenase/3-oxosteroid-4,5-isomerase to the 17-ketosteroidal androgen, androstenedione (androst-4-ene-3,17-dione). Testosterone is formed by reduction of the 17-ketone of androstenedione by 17β-hydroxysteroid dehydrogenases, and testosterone and androstenedione are metabolically interconvertible. Several isozymes of 17β-hydroxysteroid dehydrogenase have been reported (100–103), with the specific types catalyzing either a reductive reaction or an oxidative reaction. Testosterone is secreted by the Leydig cells and can act in a negative feedback fashion in the hypothalamus and pituitary to decrease the release of gonadotropins.

The most potent endogenous androgen is the 5α-reduced steroid, 5α-dihydrotestosterone (5α-DHT,17β-hydroxy-5α-androstan-3-one). This molecule is formed in androgen target tissues by the enzyme 5α-reductase, which has been located in both the microsomal fraction and the nuclear membrane of homogenized target tissues. The 5α-reductase enzyme catalyzes an irreversible reaction and requires NADPH as a cofactor, which provides the hydrogen at C5. Two different isozymes of the enzyme are present in humans (104–108). The first cDNA isolated and cloned that encoded 5α-reductase was designated Type 1, and the second was designated Type 2. The gene 5AR1 encoding Type 1 is located on chromosome 5, whereas the gene 5AR2 encoding Type 2 is located on chromosome 2. The two

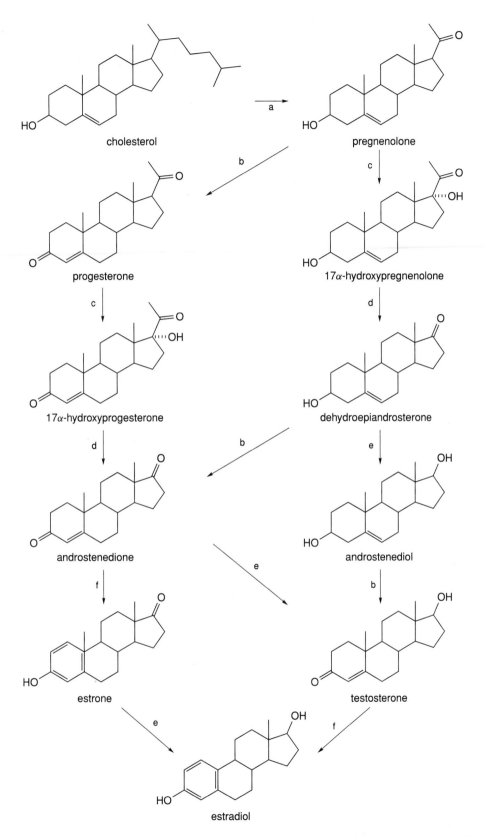

Figure 12.25. Biosynthesis of sex steroid hormones. Enzyme activities: (a) side-chain cleavage, (b) 3β-hydroxysteroid dehydrogenase/4,5-isomerase, (c) 17α-hydroxylase, (d) 17,20-lyase, (e) 17β-hydroxysteroid dehydrogenase, (f) aromatase.

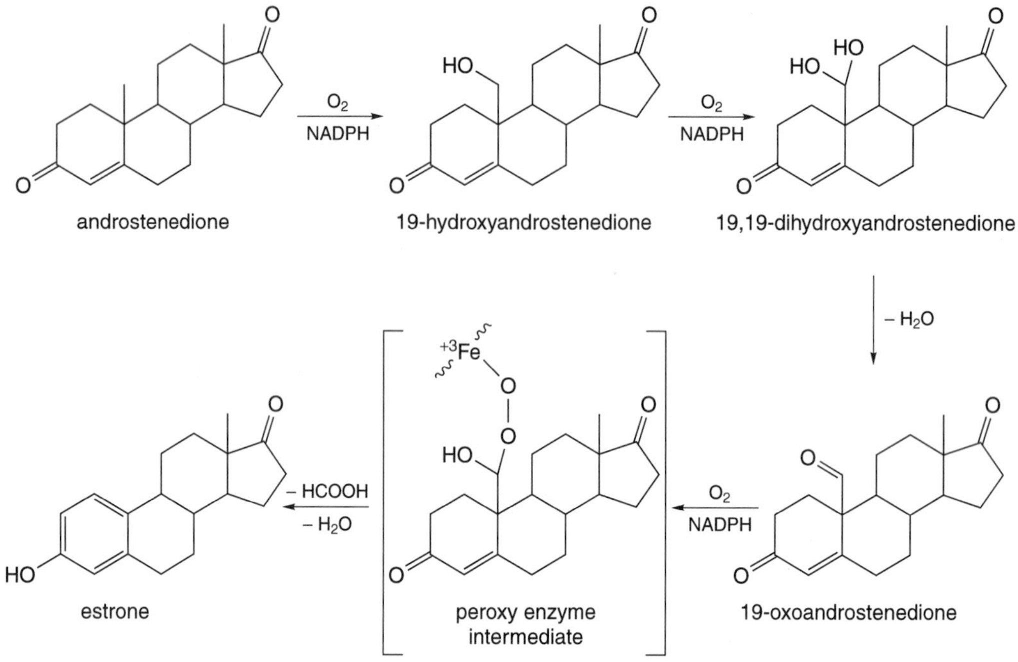

Figure 12.26. Biosynthesis of estrogens by aromatase.

human 5α-reductases have approximately 60% sequence homology. These two isozymes differ in their biochemical properties, tissue location, and function (106, 107). Type 1 5α-reductase exhibits an alkaline pH optimum (6–8.5) and has micromolar affinities for steroid substrates. On the other hand, Type 2 5α-reductase has a sharp pH optimum at 4.7–5.5, has higher affinity (lower apparent K_m) for testosterone, and is more sensitive to inhibitors than the Type 1 isozyme. The Type 2 isozyme is expressed primarily in androgen target tissues, the liver expresses both types, and the Type 1 form is expressed in various peripheral tissues. Type 2 5α-reductase appears to be essential for masculine development of the fetal urogenital tract and the external male phenotype, whereas the Type 1 isozyme is primarily a catabolic enzyme.

3.1.6 Estradiol Biosynthesis. Estrogens are biosynthesized from cholesterol, primarily in the ovary in mature, premenopausal women (96, 97). During pregnancy, the placenta is the main source of estrogen biosynthesis and pathways for production change. Small

amounts of these hormones are also synthesized by the testes in the male and by the adrenal cortex, the hypothalamus, and the anterior pituitary in both sexes. The major source of estrogens in both postmenopausal women and men occurs in extraglandular sites, particularly in adipose tissue.

In endocrine tissues, cholesterol is the steroid that is stored and is converted to estrogen, progesterone, or androgen when the tissue is stimulated by a gonadotropic hormone. The major pathways for the biosynthesis of sex steroid hormones are summarized in Fig. 12.25. In the ovary, FSH acts on the preovulatory follicle to stimulate the biosynthesis of estrogens. The thecal cells of the preovulatory follicle convert cholesterol into androgens, whereas the granulosa cells convert androgens to estrogens. After ovulation, LH acts on the corpus luteum to stimulate both estrogen and progesterone biosynthesis and secretion.

Cholesterol is converted by side-chain cleavage into pregnenolone, which can be converted to estrogens through several enzymatic steps. Pregnenolone is converted to the androgens androstenedione (androst-4-ene-3,17-di-

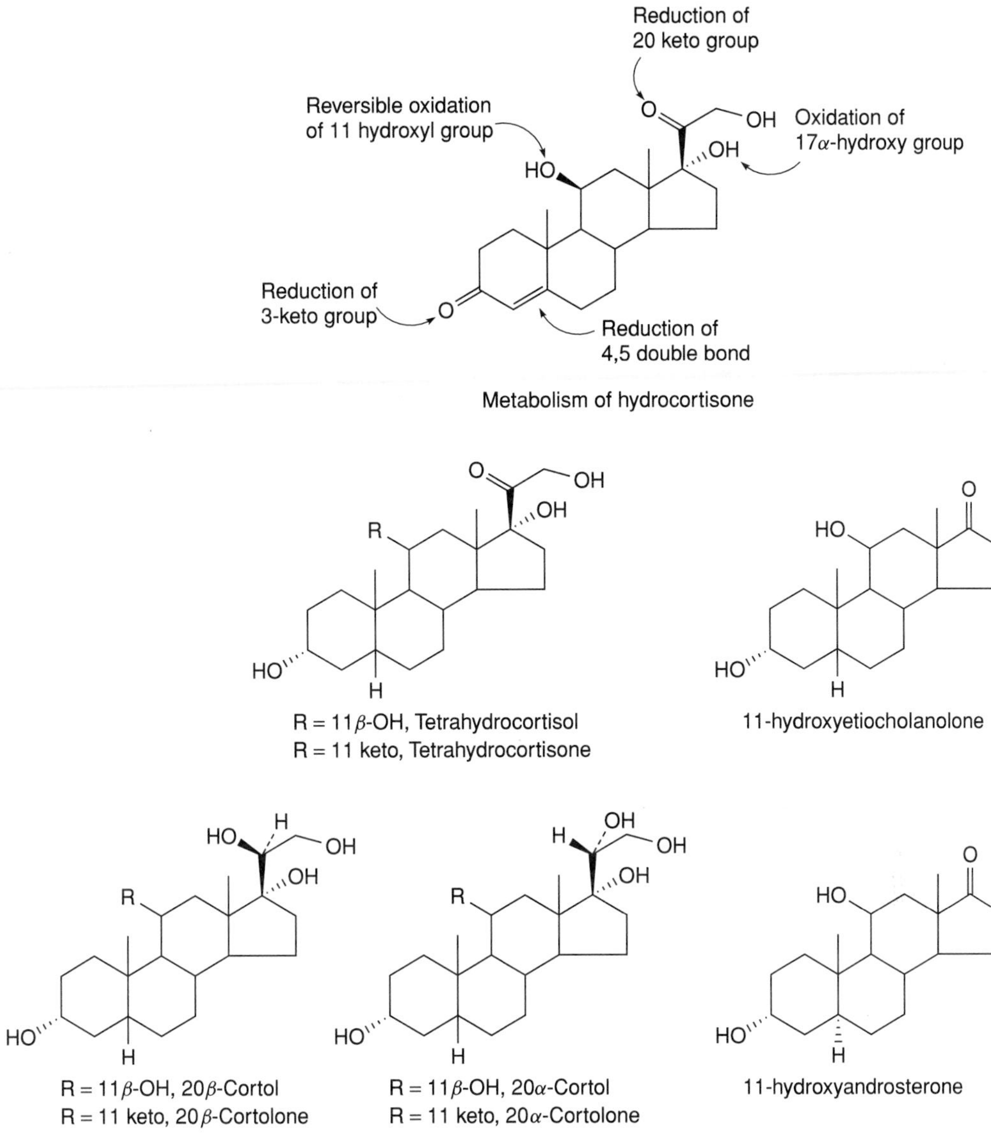

Metabolism of hydrocortisone

Figure 12.27. Metabolites of corticosteroids.

one) and testosterone, as described earlier. Loss of the C19 angular methyl group and aromatization of the A-ring of testosterone or androstenedione results in 17β-estradiol or estrone, respectively. 17β-Estradiol and estrone are metabolically interconvertible, catalyzed by 17β-hydroxysteroid dehydrogenases.

The final step in the biosynthesis is the conversions of the C_{19} androgens to the C_{18} estrogens. This enzyme reaction is catalyzed by the microsomal cytochrome P450 enzyme complex termed aromatase. Androstenedione is the preferred substrate for aromatization and three molecules of NADPH and three molecules of oxygen are necessary for conversion of one molecule of androgen to estrogen (109, 110). Aromatization proceeds through three successive steps catalyzed by a single enzyme complex, consisting of the cytochrome P450$_{arom}$ protein and NADPH-cytochrome P450 reductase, present in the endoplasmic reticulum of the cell. The gene *CYP19* expresses cytochrome $P450_{arom}$, and the gene is located on chromosome 15 in humans. The mechanism of aromatization is illustrated in Figure 12.26. The first step involves oxidation of the angular C19 methyl group to provide 19-hydroxyandrostenedione (19-hydroxyandrost-4-ene-3,17-dione). 19,19-Dihydroxyandrostenedione, isolated as 19-oxoandrostenedione, is formed by the second oxidation step. The exact mechanism of the last oxidation remains to be fully determined, with the likely intermediate being an enzyme-peroxy species (111–113). After the last oxidation, the C_{10}-C_{19} bond is cleaved and aromatization of the A-ring occurs. In this last step, the C19 carbon atom is lost as formic acid and oxygens from the first and last step are incorporated into the formic acid. The 1β-hydrogen and 2β-hydrogen atoms are also lost during aromatization.

3.2 Steroid Metabolism

Because of their high potencies, steroid hormones are secreted in minute amounts and their secretions are tightly regulated. In addition, the secretion and the metabolism of endogenous steroids also play a role in controlling their concentrations. Both phase I and phase II pathways are involved in the metabolism of steroids *in vivo*.

3.2.1 Corticosteroids Metabolism.
Under normal conditions, there is an interconversion between hydrocortisone and cortisone. Both hormones are metabolized extensively before excretion. Reduction can occur both in the 4,5 double bond and at the 3 and 20 keto groups of hydrocortisone to form various metabolites such as tetrahydrocortisols, tetrahydrocortisones, cortols, and cortolones (114) (Fig. 12.27). A minor metabolic pathway also occurs through cleavage of the C_{17}-C_{20} bond to form 11-hydroxyetiocholanolone and 11-hydroxyandrosterone. All the metabolites containing the 3-hydroxy group go through phase II metabolism in liver and to a lesser extent in kidney to form predominantly water-soluble glucuronide conjugates.

3.2.2 Progesterone Metabolism.
The metabolism of progesterone is relatively straight-

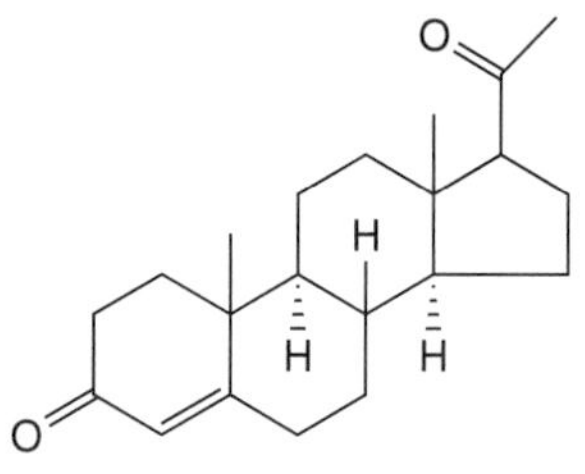

Progesterone

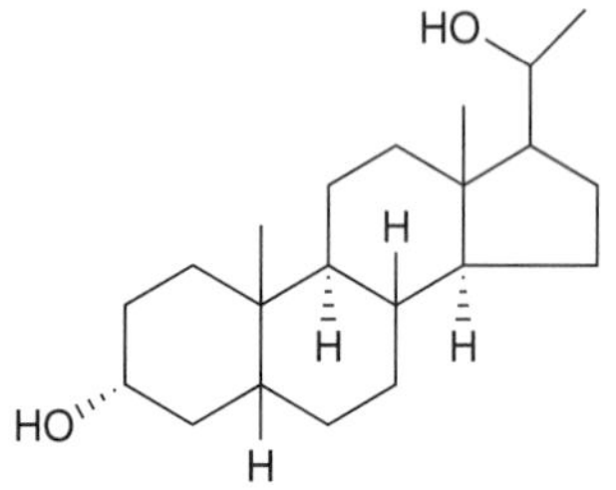

5β-Pregnane-3α, 20 diol

Figure 12.28. Structure of progesterone and 5β-pregnane-3α,20β-diol.

forward, and reduction can occur on the 4,5 double bond and at the 3 and 20 keto groups (115). The major urinary metabolite of progesterone is the glucuronide conjugate of 5β-pregnane-3α,20β-diol (Fig. 12.28).

3.2.3 Androgen Metabolism.
As described in Section 3.1, both androstenedione and testosterone are the substrates of he enzyme aromatase to form estrone and estradiol, respectively. In addition, the true active endogenous androgen 5α-dihydrotestosterone is the reductive metabolite of testosterone by the enzyme 5α-reductase. Metabolism of androstenedione and testosterone are carried out primarily in the liver. Androstenedione and testosterone can be interconverted by the enzyme 17β-hydrosteroid dehydrogenase. They can be converted to various metabolites by 3α- and 3β-hydroxysteroid dehydrogenase and by 5α- and 5β-reductases (115), as shown in Fig. 12.29. Androsterone and etiocholanolone are the major urinary metabolites excreted primarily in the form of glucuronides and to a lesser extent as sulfates (116).

3.2.4 Estrogen Metabolism.
The metabolism of estrogens is carried out primarily in

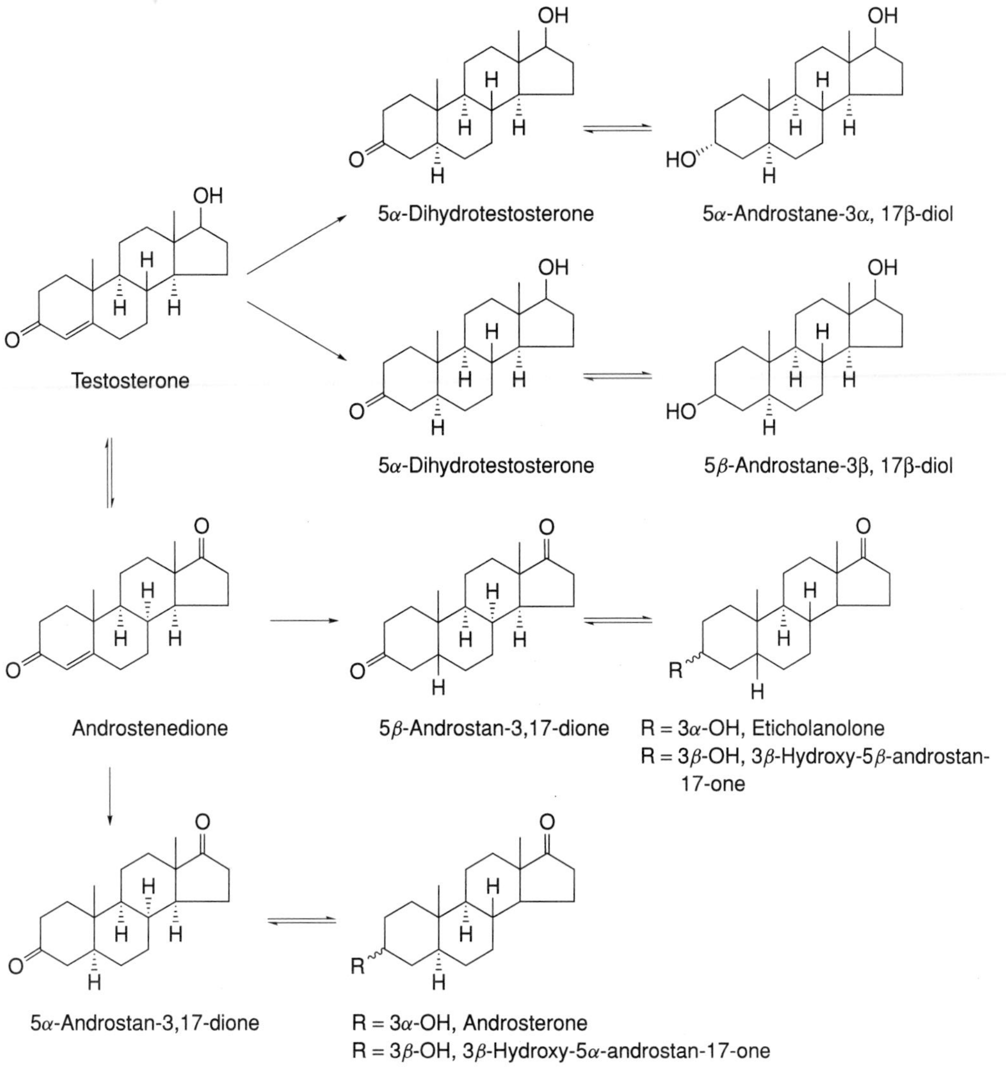

Figure 12.29. Metabolites of testosterone.

the liver. Estrone and estradiol can be interconverted by 17β-hydroxysteroid dehydrogenases. One of the major metabolic pathways for estrone is 16α-hydroxylation to form 16α-hydroxyestrone, which is then converted to estriol (117). Another major pathway is C2 hydroxylation to form 2-hydroxyestrone, which can then be converted to 2-methoxyestrone by the enzyme catechol-*O*-methyl transferase (118). Other minor metabolic pathways of estrogens include the formation of 17α-estradiol, 6-hydroxylated estrogens, and 16-epiestriol (119), as outlined in Fig. 12.30.

Estrogens also go through phase II metabolism to form to sulfate and glucuronide conjugates. Estrone sulfate is the most abundant estrogen conjugate in females (1–2 nM) (120, 121). There is strong evidence that the high concentration of estrone sulfate may represent an important reservoir of active estrogens for estrogen-dependent illnesses. Estrone sulfatase [sterylsulfatase EC (3.1.6.2)] has been consistently found in human breast cancer cells and the conversion of estrone sulfate to estrone has been demonstrated (122–125).

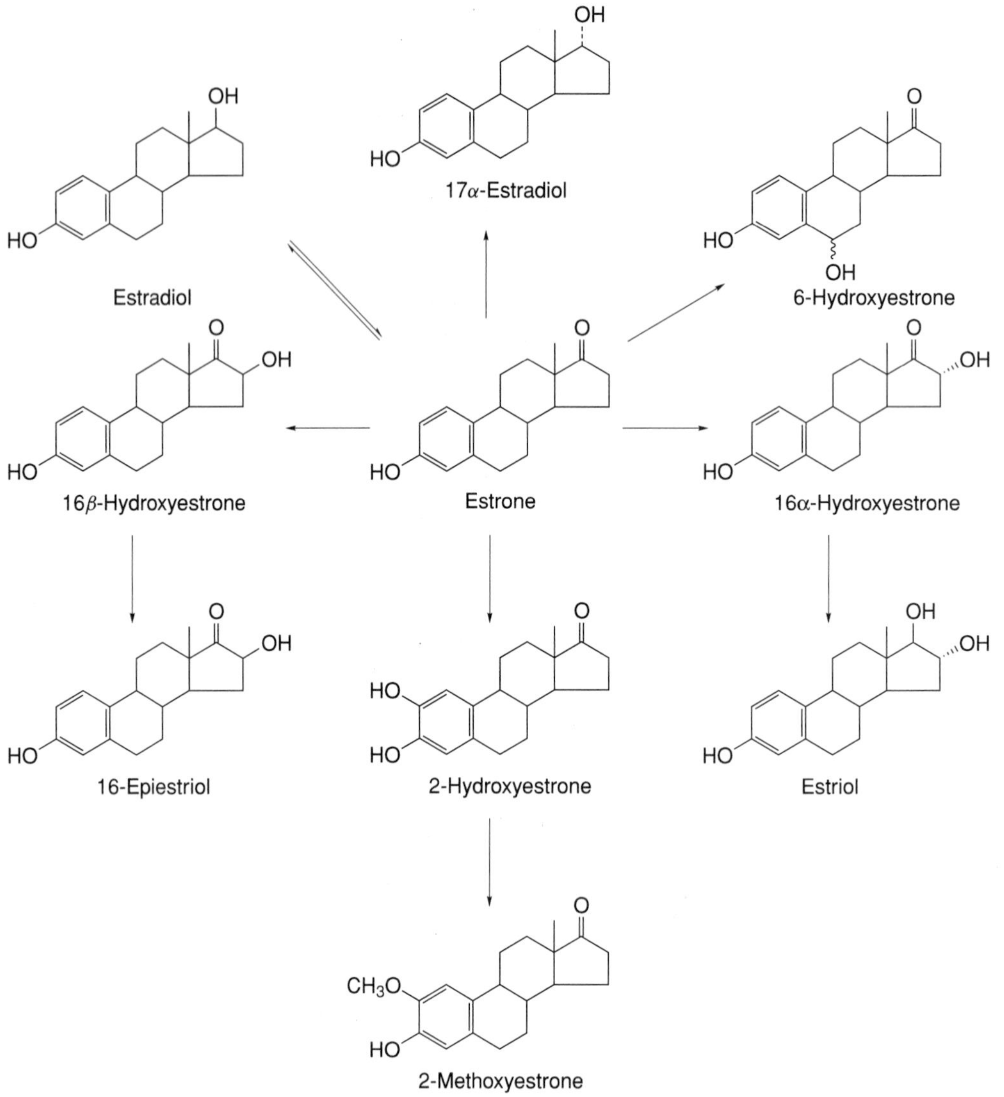

Figure 12.30. Metabolism of endogenous estrogens.

3.3 Steroid Hormone Action

The steroid hormones such as the adrenocorticoids, progestins, estrogens, and androgens are present in the body only in very low concentrations (e.g., 0.1–1.0 nM). Yet, these hormones exert potent physiologic effects on sensitive tissues at those low concentrations. Over the past several decades, the general mechanism of steroid hormone action has been extensively studied (126–134). The steroid hormones act on target cells by binding with high affinity to intracellular receptors, resulting in the formation of steroid-receptor complexes that regulate gene expression and protein biosynthesis (Fig. 12.31). The lipophilic steroid hormones are transported in the bloodstream reversibly bound to serum carrier proteins. The unbound steroids can diffuse through the plasma membrane and enter the cells. Cells that are sensitive to the particular steroid hormone (referred to as target cells) contain steroid receptors capable of

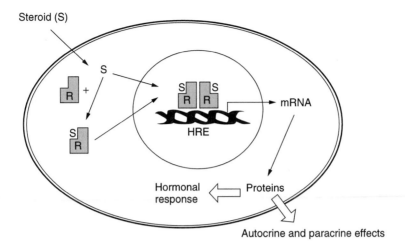

Figure 12.31. Mechanism of steroid hormone action.

high affinity binding with the steroid. These receptors are soluble intracellular proteins that can both bind steroid ligands with high affinity and act as ligand-dependent transcriptional factors through interaction with specific deoxyribonucleic acid (DNA) sites and other proteins associated with the chromatin. Initially, the unoccupied steroid receptors were thought to be located solely in the cytoplasm of target cells; however, investigations on estrogens, progestins, and androgens indicate that active, unoccupied receptors are also present in the nucleus of the cell. The interaction of the steroid with its receptor results in a conformational change of the receptor, translocation to the nucleus, and steroid-receptor complex homodimerization. The steroid-receptor homodimers interact with particular regions of the cellular DNA, referred to as hormone-responsive elements (HREs) present in the promoter region of responsive genes. Binding of the nuclear steroid-receptor complexes to DNA and interaction with various nuclear transcriptional factors initiate the transcription of the gene to produce messenger ribonucleic acid (mRNA). The elevated mRNA levels result in increased protein synthesis in the endoplasmic reticulum. These proteins include enzymes, receptors, and secreted factors that subsequently result in the steroid hormonal response regulating cell function, growth, differentiation, and playing central roles in normal physiological processes, as well as in many important diseases.

The steroid receptors proteins are members of the nuclear receptor superfamily based on their general protein structure and function. The nuclear receptor superfamily includes receptors for steroids, vitamin D, thyroid hormones, and retinoids. The steroid receptors include glucocorticoid receptor (GR), mineralocorticoid receptor (MR), progesterone receptor (PR), estrogen receptors alpha and beta (ERα and ERβ), and the androgen receptor (AR); these receptors function through the formation of homodimers. The thyroid receptor (TR), vitamin D receptor (VDR), retinoid receptors (RARα, RARβ, RARγ), and peroxisomal proliferator-activated receptors (PPARα, PPARβ, PPARγ, PPARδ) interact with retinoid X receptor (RXR) to form heterodimers. A third group of nuclear receptors are involved in regulation of steroid and lipid metabolism, such as constitutive androstane receptor (CAR), pregnane X receptor or steroid xenobiotic receptor (PXR or SXR), and farnesyl X receptor (FXR). Finally, orphan receptors whose ligands are yet to be discovered are also members of this superfamily, such as steroidogenic factor 1 (SF-1) and X-linked orphan receptor (DAX-1).

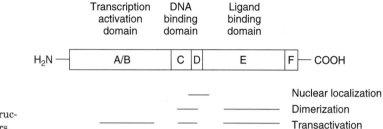

Figure 12.32. General structure of the steroid receptors.

The overall structures of the receptors contain different regions (traditionally labeled
A/B, C, D, E, and F) and have three distinct
structural domains (Fig. 12.32). A high degree
of homology in the receptors is found in region
C, the DNA-binding domain (DBD), which interacts with the HRE in the DNA. The DNA-
binding region has two loop structures, called
zinc finger motifs, composed of cysteine amino
acids that chelate zinc ions that bind to the
DNA sequence. The E region is the ligand-
binding domain (LBD) that recognizes and
binds the ligand with high affinity. Although
the sequence homology is lower for this region, the overall structures of the LBD for the
various receptors are very similar (133). The
LBD contains 12 α-helices and three short
β-strands, forming an α-helical sandwich containing a hydrophobic core. The ligand binds
in the core and is completely buried inside the
protein. Structure-function studies of cloned
receptor proteins also identify regions of the
molecules that are important for interactions
with nuclear transcriptional factors, activation of gene transcription, and protein-protein
interactions in region A/B and region E. Critical events in the initiation of gene expression
by the nuclear receptors are interactions of
the steroid-receptor complex with other nuclear transcription factors that are coupled to
the RNA polymerase complex. The nature and
exact number of proteins are not clear, but
several important ones have been identified
and include the steroid receptor coactivator
family (131). Further details on the individual
receptors and their interactions to initial gene
transcription, and the protein products produced, are described in other chapters in this
volume.

REFERENCES

1. L. F. Fieser and M. Feiser, *Steroids*, Reinhold, New York, 1959.

2. J. Fried and J. A. Edwards, *Organic Reactions in Steroid Chemistry*, Vol. **1**, Van Nostrand-Reinhold, New York, 1972.

3. J. Fried and J. A. Edwards, Organic Reactions in Steroid Chemistry, Vol. **2**, Van Nostrand-Reinhold, New York, 1972.

4. W. R. Nes and M. L. McKean, *Biochemistry of Steroids and Other Isoprenoids*, University Park Press, Baltimore, MD, 1977.

5. J. B. Dence, *Steroids and Peptides*, John Wiley & Sons, New York, 1980.

6. H. L. J. Makin, *Biochemistry of Steroid Hormones*, Blackwell Scientific, Oxford, UK, 1984.

7. L. Gorder and J. L. Sturchio, Eds., *Steroids*, **57**, 354–418 (1992).

8. L. Gorder and J. L. Sturchio, Eds., *Steroids*, **57**, 578–657 (1992).

9. V. Petrow, Ed., *Steroids*, **61**, 473–503 (1996).

10. A. A. Berthold, *Arch. Anat. Physiol. Wissensch. Med.*, **16**, 42–46 (1849).

11. T. Addison, *On the Constitutional and Local Effects of the Disease of the Supra-Renal Capsules*, Highley, London, 1855.

12. A. O. Windaus, *Ber. Dtsch. Chem. Ges.*, **42**, 238–246 (1909).

13. H. O. Wieland, *Hoppe-Seyler's Z. Physiol. Chem.*, **98**, 59–64 (1916).

14. J. D. Bernal, *Nature*, **129**, 277–278 (1932).

15. A. Rosenheim and H. King, *Chem. Ind.*, **51**, 464–466 (1932).

16. H. Weiland, E. Dane, and E. Scholz, *Z. Physiol. Chem.*, **210**, 261–274 (1932).

17. A. F. Butenandt, *Naturwissenschaften*, **17**, 879 (1929).

18. E. A. Doisy, C. D. Veler, and S. A. Thayer, *J. Biol. Chem.*, **86**, 499–509 (1930).

19. A. F. Butenandt, *Dtsch. Med. Wochenschr.*, **61**, 781–786 (1935).

20. T. Reichstein and C. W. Shoppee, *Vitam. Horm.*, **1**, 345–413 (1943).

21. W. E. Bachmann, W. Cole, and A. L. Wilds, *J. Am. Chem. Soc.*, **61**, 974–975 (1939).

22. R. B. Woodward, F. Sondheimer, and D. Taub, *J. Am. Chem. Soc.*, **73**, 4057 (1951).

23. H. M. E. Cardwell, J. W. Cornforth, S. R. Duff, H. Holtermann, and R. Robinson, *J. Chem. Soc.*, 361–384 (1953).

24. K. Bloch and D. Rittenberg, *J. Biol. Chem.*, **145**, 625–636 (1942).

25. E. V. Jensen and H. I. Jacobson, *Recent Prog. Horm. Res.*, **18**, 387–414 (1962).

26. P. S. Hench, E. C. Kendall, C. H. Slocumb, and E. F. Polley, *Proc. Staff Meet. Mayo Clinic*, **24**, 181–197 (1949).

27. G. Pincus, *The Control of Fertility*, Academic Press, New York, 1965.

28. IUPAC-IUB Joint Commission, *Biochemistry*, **8**, 2227–2242 (1969).

29. IUPAC-IUB Joint Commission, *Eur. J. Biochem.*, **186**, 429–458 (1989).

30. W. S. Johnson, C. D. Gutsche, R. Hirschmann, and V. L. Stromberg, *J. Am. Chem. Soc.*, **73**, 322–326 (1951).

31. G. Anner and K. Miescher, *Helv.*, **31**, 2173–2183 (1948).

32. J. W. Cornforth and R. Robinson, *Nature*, **160**, 737–739 (1947).

33. H. M. E. Cardwell, J. W. Cornforth, S. R. Duff, H. Holtermann, and R. Robinson, *Chem. Ind.*, 389–390 (1951).

34. R. B. Woodward, F. Sondheimer, D. Taub, K. Heusler, and W. M. MacLaMore, *J. Am. Chem. Soc.*, 2403–2404 (1951).

35. W. S. Johnson, B. Bannister, R. Pappo, and J. E. Pike, *J. Am. Chem. Soc.*, **76**, 6354–6361 (1956).

36. R. E. Marker, R. B. Wagner, P. R. Ulshafer, E. L. Wittbecker, D. P. J. Goldsmith, and C. H. Ruof, *J. Am. Chem. Soc.*, **69**, 2167–2230 (1947).

37. R. E. Marker and E. Rohrmann, *J. Am. Chem. Soc.*, **62**, 518–520 (1940).

38. R. E. Marker, T. Tsukamoto, and D. L. Turner, *J. Am. Chem. Soc.*, **62**, 2525–2532 (1940).

39. L. H. Sarett, *J. Biol. Chem.*, **162**, 601–631 (1946).

40. L. H. Sarett, *J. Am. Chem. Soc.*, **68**, 2478–2483 (1946).

41. L. H. Sarett, *J. Am. Chem. Soc.*, **70**, 1454–1458 (1948).

42. L. H. Sarett, *J. Am. Chem. Soc.*, **71**, 2443–2444 (1949).

43. E. Schwenk and E. Stahl, *Arch. Biochem.*, 125–132 (1947).

44. R. B. Turner, V. R. Mattox, L. L. Engel, B. F. McKenzie, and E. C. Kendall, *J. Biol. Chem.*, **166**, 345–365 (1946).

45. R. B. Woodward, F. Sondheimer, and D. Taub, *J. Am. Chem. Soc.*, **73**, 4057 (1951).

46. H. C. Murray and D. H. Peterson, U.S. Pat. 2,602,769 (1952).

47. D. H. Peterson, H. C. Murray, S. H. Eppstein, L. M. Reineke, A. Weintraub, P. D. Meister, and H. M. Leigh, *J. Am. Chem. Soc.*, **74**, 5933–5936 (1952).

48. J. Fried, R. W. Thoma, J. R. Gerke, J. E. Herz, M. N. Donin, and D. Perlman, *J. Am. Chem. Soc.*, **74**, 3962–3963 (1952).

49. S. B. Mahato and A. Mukherjee, *Phytochemistry*, **23**, 2131–2154 (1984).

50. S. B. Mahato and S. Banerjee, *Phytochemistry*, **24**, 1403–1421 (1985).

51. S. B. Mahato, S. Banerjee, and S. Podder, *Phytochemistry*, **28**, 7–40 (1989).

52. S. Mahato and I. Majumdar, *Phytochemistry*, **34**, 883–898 (1993).

53. K. E. Smith, F. Ahmed, and T. Antoniou, *Biochem. Soc. Trans.*, **21**, 1077–1080 (1993).

54. S. B. Mahato and S. Garai, *Steroids*, **62**, 332–345 (1997).

55. O. Bortolini, A. Medici, and S. Poli, *Steroids*, **62**, 564–577 (1997).

56. W. Charney and H. L. Herzog, *Microbial Transformations of Steroids*, Academic Press, New York, 1967.

57. D. Perlman, E. Titus, and J. Fried, *J. Am. Chem. Soc.*, **74**, 2126 (1952).

58. R. W. Thoma, J. Fried, S. Bonanno, and P. Grabowich, *J. Am. Chem. Soc.*, **79**, 4818 (1957).

59. M. Tsuji, Y. Ichihara, K. Mori, and S. Kuwakura, Jap. Pat. 6,307,795 (1989).

60. S. H. Eppstein, P. D. Meister, D. H. Peterson, H. C. Murray, H. M. Osborn, A. Weintraub, L. M. Reineke, and R. C. Meeks, *J. Am. Chem. Soc.*, 3382–3389 (1958).

61. M. Yoshihama, M. Nakakoshi, J. Nakamura, N. Fujise, and G. Kawanishi, *Chem. Pharm. Bull. (Tokyo)*, **38**, 2834–2837 (1990).

62. T. Yamamoto, K. Tamura, Y. Fujimoto, J. Kitawaki, M. Nakakoshi, M. Yoshihama, and H. Okada, *J. Steroid Biochem.*, **36**, 517–521 (1990).

63. J. Kitawaki, T. Kim, H. Kanno, T. Noguchi, T. Yamamoto, and H. Okada, *J. Steroid Biochem. Mol. Biol.*, **44**, 667–670 (1993).

64. J. F. Templeton, V. P. Kumar, I. Sashi, K. Marat, R. S. Kim, F. S. Labella, and D. Cote, *J. Nat. Prod.*, **50**, 463–467 (1987).

65. R. Reiche, I. Heller, C. Hoerhold, and B. Gotlsehaldf (to Vebjena Pharm.), (East) Ger. Pat. 300,364 (1992).

66. W. H. Liu, C. W. Kuo, K. L. Wu, C. Y. Lee, and W. Y. Hsu, *J. Ind. Microbiol.*, **13**, 167–171 (1994).

67. J. R. Hanson, *Nat. Prod. Rep.*, **9**, 581–595 (1992).

68. J. R. Hanson, *Nat. Prod. Rep.*, **10**, 313–325 (1993).

69. J. R. Hanson, *Nat. Prod. Rep.*, **12**, 567–577 (1995).

70. J. R. Hanson, *Nat. Prod. Rep.*, **13**, 227–239 (1996).

71. J. R. Hanson, *Nat. Prod. Rep.*, **14**, 373–386 (1997).

72. J. R. Hanson, *Nat. Prod. Rep.*, **15**, 261–273 (1998).

73. J. R. Hanson, *Nat. Prod. Rep.*, **16**, 607–617 (1999).

74. J. R. Hanson, *Nat. Prod. Rep.*, **17**, 423–434 (2000).

75. J. R. Hanson, *Nat. Prod. Rep.*, **18**, 282–290 (2001).

76. C. Djerassi, L. Miramontes, G. Rosenkranz, and F. Sondheimer, *J. Am. Chem. Soc.*, **76**, 4092–4094 (1954).

77. F. B. Colton, U.S. Pat. 2,725,389 (1955).

78. G. H. Rasmusson, G. H. Reynolds, N. G. Steinberg, E. Walton, G. F. Patel, T. Liang, M. A. Cascieri, A. H. Cheung, J. R. Brooks, and C. Berman, *J. Med. Chem.*, **29**, 2298–2315 (1986).

79. J. W. Morzycki, Z. Lotowski, A. Z. Wilczewska, and J. D. Stuart, *Bioorg. Med. Chem.*, **4**, 1209–1215 (1996).

80. E. P. Olivetto, P. Rausser, H. L. Herzog, S. Hershberg, M. Eisler, P. L. Perlman, and M. M. Pechet, *J. Am. Chem. Soc.*, **80**, 6687–6688 (1958).

81. A. E. Wakeling, M. Dukes, and J. Bowler, *Cancer Res.*, **51**, 3867–3873 (1991).

82. J. Bowler, T. J. Lilley, J. D. Pittam, and A. E. Wakeling, *Steroids*, **54**, 71–99 (1989).

83. R. Tedesco, J. A. Katzenellenbogen, and E. Napolitano, *Tetrahedron Lett.*, **38**, 7997–8000 (1997).

84. S. M. Grundy, *Am. J. Cardiol.*, **70**, 27I–32I (1992).

85. E. R. Simpson, *Mol. Cell. Endocrinol.*, **13**, 213–227 (1979).

86. W. L. Miller and J. F. Strauss III, *J. Steroid Biochem. Mol. Biol.*, **69**, 131–141 (1999).

87. D. M. Stocco, *Annu. Rev. Physiol.*, **63**, 193–213 (2001).

88. W. L. Miller, *J. Steroid Biochem.*, **27**, 759–766 (1987).

89. W. L. Miller, *Endocr. Rev.*, **9**, 295–318 (1988).

90. R. J. Auchus and W. L. Miller in L. J. DeGroot, J. L. Jameson, H. G. Burger, D. L. Loriaux, J. C. Marshall, S. Melmed, W. D. Odell, J. T. Potts Jr., and A. H. Rubenstein, Eds., *Endocrinology*, 4th ed., Saunders, Philadelphia, 2001, pp. 1616–1631.

91. R. L. Miersfield in L. J. DeGroot, J. L. Jameson, H. G. Burger, D. L. Loriaux, J. C. Marshall, S. Melmed, W. D. Odell, J. T. Potts Jr., and A. H. Rubenstein, Eds., *Endocrinology*, 4th ed., Saunders, Philadelphia, 2001, pp. 1647–1654.

92. P. Kremers, *J. Steroid Biochem.*, **7**, 571–575 (1976).

93. T. Yanase, E. R. Simpson, and M. R. Waterman, *Endocr. Rev.*, **12**, 91–108 (1991).

94. T. Strachan and P. C. White, *J. Steroid Biochem. Mol. Biol.*, **40**, 537–543 (1991).

95. A. D. Carlson, J. S. Obeid, N. Kanellopoulou, R. C. Wilson, and M. I. New, *J. Steroid Biochem. Mol. Biol.*, **69**, 19–29 (1999).

96. J. F. Strauss III and A. J. W. Hsueh in L. J. DeGroot, J. L. Jameson, H. G. Burger, D. L. Loriaux, J. C. Marshall, S. Melmed, W. D. Odell, J. T. Potts Jr., and A. H. Rubenstein, Eds., *Endocrinology*, 4th ed., Saunders, Philadelphia, 2001, pp. 2043–2052.

97. J. C. Marshall in L. J. DeGroot, J. L. Jameson, H. G. Burger, D. L. Loriaux, J. C. Marshall, S. Melmed, W. D. Odell, J. T. Potts Jr., and A. H. Rubenstein, Eds., *Endocrinology*, 4th ed., Saunders, Philadelphia, 2001, pp. 2073–2085.

98. D. J. Handelsman in L. J. DeGroot, J. L. Jameson, H. G. Burger, D. L. Loriaux, J. C. Marshall, S. Melmed, W. D. Odell, J. T. Potts Jr., and A. H. Rubenstein, Eds., *Endocrinology*, 4th ed., Saunders, Philadelphia, 2001, pp. 2232–2242.

99. C. E. Bird, L. Morrow, Y. Fukumoto, S. Marcellus, and A. F. Clark, *J. Clin. Endocrinol. Metab.*, **43**, 1317–1322 (1976).

100. C. H. Blomquist, *J. Steroid Biochem. Mol. Biol.*, **55**, 515–524 (1995).

101. S. Andersson, *J. Steroid Biochem. Mol. Biol.*, **55**, 533–534 (1995).

102. H. Peltoketo, P. Nokelainen, Y. S. Piao, R. Vihko, and P. Vihko, *J. Steroid Biochem. Mol. Biol.*, **69**, 431–439 (1999).

103. V. Luu-The, *J. Steroid Biochem. Mol. Biol.*, **76**, 143–151 (2001).

104. S. Andersson, R. W. Bishop, and D. W. Russell, *J. Biol. Chem.*, **264**, 16249–16255 (1989).

105. S. Andersson, D. M. Berman, E. P. Jenkins, and D. W. Russell, *Nature*, **354**, 159–161 (1991).

106. D. W. Russell, D. M. Berman, J. T. Bryant, K. M. Cala, D. L. Davis, C. P. Landrum, J. S. Prihoda, R. I. Silver, A. E. Thigpen, and W. C. Wigley, *Recent Prog. Horm. Res.*, **49**, 275–284 (1994).

107. D. W. Russell and J. D. Wilson, *Annu. Rev. Biochem.*, **63**, 25–61 (1994).

108. J. D. Wilson, J. E. Griffin, and D. W. Russell, *Endocr. Rev.*, **14**, 577–593 (1993).

109. K. J. Ryan, *J. Biol. Chem.*, **234**, 268–272 (1959).

110. E. A. Thompson Jr. and P. K. Siiteri, *J. Biol. Chem.*, **249**, 5364–5372 (1974).

111. M. Akhtar, M. R. Calder, D. L. Corina, and J. N. Wright, *Biochem. J.*, **201**, 569–580 (1982).

112. P. A. Cole and C. H. Robinson, *J. Med. Chem.*, **33**, 2933–2942 (1990).

113. S. S. Oh and C. H. Robinson, *J. Steroid Biochem. Mol. Biol.*, **44**, 389–397 (1993).

114. D. K. Fukushima, H. L. Bradlow, L. Hellman, B. Zumoff, and T. F. Gallagher, *J. Biol. Chem.*, **235**, 2246–2252 (1960).

115. K. Hartiala, *Physiol. Rev.*, **53**, 496–534 (1973).

116. A. E. Kellie and E. R. Smith, *Biochem. J.*, **66**, 490–495 (1957).

117. C. P. Martucci and J. Fishman, *Pharmacol. Ther.*, **57**, 237–257 (1993).

118. S. A. Li, J. K. Klicka, and J. J. Li, *Cancer Res.*, **45**, 181–185 (1985).

119. K. Fotherby, *Acta Endocrinol.*Suppl., **185**, 119–147 (1974).

120. K. Carlstrom and H. Skoldefors, *J. Steroid Biochem.*, **8**, 1127–1128 (1977).

121. J. B. Brown and B. J. Smyth, *J. Reprod. Fertil.*, **24**, 142 (1971).

122. S. J. Santner, P. D. Feil, and R. J. Santen, *J. Clin. Endocrinol. Metab.*, **59**, 29–33 (1984).

123. R. J. Santen, D. Leszczynski, N. Tilson-Mallet, P. D. Feil, C. Wright, A. Manni, and S. J. Santner, *Ann. N. Y. Acad. Sci.*, **464**, 126–137 (1986).

124. J. R. Pasqualini, G. Chetrite, C. Blacker, M. C. Feinstein, L. Delalonde, M. Talbi, and C. Maloche, *J. Clin. Endocrinol. Metab.*, **81**, 1460–1464 (1996).

125. J. H. MacIndoe, *Endocrinology*, **123**, 1281–1286 (1988).

126. M. A. Carson-Jurica, W. T. Schrader, and B. W. O'Malley, *Endocr. Rev.*, **11**, 201–220 (1990).

127. E. E. Baulieu, J. Master, and G. Redeuih in L. J. DeGroot, J. L. Jameson, H. G. Burger, D. L. Loriaux, J. C. Marshall, S. Melmed, W. D. Odell, J. T. Potts Jr., and A. H. Rubenstein, Eds., *Endocrinology*, 4th ed., Saunders, Philadelphia, 2001, pp. 123–142.

128. M. J. Tsai and B. W. O'Malley, *Annu. Rev. Biochem.*, **63**, 451–486 (1994).

129. R. C. Ribeiro, P. J. Kushner, and J. D. Baxter, *Annu. Rev. Med.*, **46**, 443–453 (1995).

130. J. A. Katzenellenbogen and B. S. Katzenellenbogen, *Chem. Biol.*, **3**, 529–536 (1996).

131. M. Beato and A. Sanchez-Pacheco, *Endocr. Rev.*, **17**, 587–609 (1996).

132. J. W. Funder, *Annu. Rev. Med.*, **48**, 231–240 (1997).

133. R. V. Weatherman, R. J. Fletterick, and T. S. Scanlan, *Annu. Rev. Biochem.*, **68**, 559–581 (1999).

134. L. B. Moore, J. M. Maglich, D. D. McKee, B. Wisely, T. M. Willson, S. A. Kliewer, M. H. Lambert, and J. T. Moore, *Mol. Endocrinol.*, **16**, 977–986 (2002).

CHAPTER THIRTEEN

Female Sex Hormones, Contraceptives, and Fertility Drugs

PETER C. RUENITZ
College of Pharmacy
University of Georgia
Athens, Georgia

Contents

Burger's Medicinal Chemistry and Drug Discovery
Sixth Edition, Volume 3: Cardiovascular Agents and
Endocrines
Edited by Donald J. Abraham
ISBN 0-471-37029-0   © 2003 John Wiley & Sons, Inc.

1   INTRODUCTION

The roles women take in our society have undergone a dramatic redefinition over the past half-century, and medical use of estrogen and progestin mimetics and antagonists has contributed substantially to this change. It follows that these substances are among the most intensively prescribed therapeutic agents in use today. Steroidal estrogens and their conjugates are used in estrogen-replacement therapy in postmenopausal women, and estrogen-progestin combinations are used in birth control. Nonsteroidal estrogen antagonists, especially tamoxifen, are used in the prevention and control of postmenopausal breast cancer. Exciting prospects for *tissue-selective* estrogen therapy have emerged from clinical and experimental endocrinologic studies. Molecular mechanistic studies involving these substances, using materials and procedures made available by advances in biotechnology, have begun to clarify our understanding of the structural basis for the effects of estrogen and progesterone throughout the human body.

Besides describing the medicinal chemistry of therapeutically established and experimentally promising female sex hormones and analogs, this chapter focuses on important developments over the last 10–15 years relating to (*1*) biochemical mechanisms of action and

structure-activity studies of female sex hormone agonists and antagonists, (*2*) estrogen biosynthesis, and (*3*) biotransformation of estrogens, progestins, and their nonsteroidal analogs.

Structures of endogenous estrogens, progestins, and certain of their biosynthetic precursors and metabolites, as well as those of analogs of these hormones, appear at points in the text where they are most needed for comprehension of the narrative.

2   CLINICAL APPLICATIONS

2.1   Current Drugs

Estrogens, antiestrogens, estrogen biosynthesis (aromatase) inhibitors, progestins, and an antiprogestin in current clinical use in the United States are summarized in Table 13.1. The steroidal estrogens, 17β-estradiol (1) and its 17-valerate and 3-sulfate esters (estropipate) as well as estrogen conjugates (3-sulfate esters of (2), (52), (53), and other steroids), and the nonsteroidal estrogen dienestrol (71) are used in estrogen-replacement therapy (ERT) in postmenopausal women (1). Besides the serious discomfort that can accompany estrogen deficiency, progressive irreversible deterioration of the cardiovascular and skeletal systems can occur. Products containing mixtures of estrogen conjugates isolated from the urine of pregnant mares have enjoyed wider clinical acceptance than single-component products. Also used for ERT is the triaryleth-ylene chlorotrianisene (**12**), which has a prolonged duration of oral estrogenic activity and has estrogen antagonist effects in some tissues (2).

More recently, raloxifene (**29**) has been introduced as a selective estrogen receptor modulator (SERM), which has the advantage over earlier estrogen preparations of having a greatly reduced incidence of reproductive tract side effects (3). Thus, (**29**) is indicated in postmenopausal patients with an elevated risk for breast cancer and/or an elevated risk of endometrial cancer associated with prospective use of conventional estrogens.

The antiestrogens toremifene (**22**) and, in particular, tamoxifen (**16a**), are used for suppression and prevention of postmenopausal breast cancer. Indeed, (**16a**) has now been administered to over 1,000,000 women worldwide. Breast cancer death rates in the United States and Britain have declined by 25% over the past 10 years, in large part because of the use of (**16a**). In addition, (**16a**) was the first substance shown to have tissue-selective estrogenicity, by virtue of its ability to suppress bone loss and reduce serum cholesterol levels in patients undergoing therapy for breast cancer (4).

Clomiphene (**23**) and, less commonly, (**16a**) are used to stimulate fertility (ovulation) in patients who wish to become pregnant. Aminoglutethimide (**45**), an estrogen biosynthesis inhibitor, has been used alone or in combination with (**16a**), in breast cancer treatment.

Ethynylestradiol (**9**) and its 3-methyl ether (mestranol, **10**) and 3-cyclopentyl ether (quinestrol, **11**), and the 17-(3-cyclopentyl) propionyl ester (cypionate) of (**1**), are used primarily in combination with progestins in modulation of fertility. This is the major application of the progestins [analogs of progesterone (**30**)] listed in Table 13.1, some of which [levonorgestrel (**36**), medroxyprogesterone acetate (**34**)] are formulated without added estrogen.

Besides their applications in birth control, progestins have been used in combination with estrogens in hormone-replacement therapy: the addition of progestin reduces the risk of endometrial cancer in postmenopausal women (5). Also, progestins have been used to treat gynecological disorders, such as endometriosis, caused by hormone deficiency or imbalance.

2.2 Adverse Effects and Precautions

Endometriosis and an increased risk of endometrial cancer have been associated with use of steroidal estrogens in ERT, including (**9**), when administered alone (6). As indicated above, ERT using estrogen-progestin combinations resulted in a reduced incidence of endometrial cancer (and endometriosis), but also resulted in a 20% increase in the risk of breast cancer compared to women on estrogen-only therapy (7).

Because (**16a**) and (**22**) both have significant residual partial estrogen agonist activity in humans, endometriosis has been associated with long-term use of these triarylethylenes (8, 9). Linked to (**16a**)'s partial estrogenicity, and possibly to the way it undergoes biotransformation (see Section 3.7), is an elevated risk of endometrial cancer. This aspect of (**16a**) has raised concern about its use in breast cancer prevention (10).

On the other hand, use of (**29**), which has less residual reproductive tract estrogenicity than that of (**16a**), for these therapeutic applications, has not been associated with development of endometriosis or endometrial cancer (9).

However, neither (**16a**) nor (**29**) prevents a common overt symptom of estrogen deficiency, the one most likely to prompt patients to seek treatment. Lack of adequate estrogen levels in certain regions of the central nervous system can result in intermittent marked fluctuations in body temperature known as hot flushes or hot flashes (11). Although deterioration of skeletal and cardiovascular systems is a more serious long-term consequence of estrogen deficiency, such deterioration is not experienced early on by the patient. However, hot flush episodes eventually (6–12 months) abate because of bodily adjustment to lower endogenous estrogen levels. Thus, their impact on compliance might be diminished in patients that are aware of this and of the consequences eventually arising from discontinuance of therapy.

Table 13.1 Estrogen, Antiestrogen, and Progestin Pharmaceuticals

Generic Name (Structure)	Trade Name	U.S. Manufacturer	Chemical Class	Dose (mg/day)[a]
Estrogens				
Chlorotrianisene (**12**)	Tace	Hoechst, Marion, Roussel	Triarylethylene	12
Dienestrol (**71**)	Dienestrol	R.W. Johnson	Diarylethylene	0.01% cream
Estradiol (**1**)	Climara and others	Berlex and others	Estrane	0.05[b]
Estradiol valerate	Delestrogen	BristolMyers Squibb	Estrane	10–40[c]
Estrogens, conjugated equine (**2**, **52**, **53**, etc.)	Premarin[d]	Wyeth-Ayerst	Estrane-3-sulfates	0.3–2.5
Estropipate (**2**, 3-sulfate)	Ogen	Abbott	Estrane-3-sulfate	0.75–6
Ethinyl estradiol (**9**)	Estinyl	Schering	Estrane	0.02–0.05
Mestranol (**10**)	[e]	[e]	Estrane	0.1[e]
Aromatase inhibitors				
Aminoglutethimide (**45**)	Cytadren[d]	Novartis	Glutarimide	250
Anastrozole (**50**)	Arimidex[d]	AstraZeneca	Triazole	1
Exemestane (**51**)	Aromasin[d]	Pharmacia	Androstane	25
Letrozole (**48**)	Femara[d]	Novartis	Triazole	2.5
Antiestrogens				
Clomiphene citrate (**23**)	Clomid (Serophene)	Hoechst, Marion, Roussel (Serono)	Triarylethylene	50
Raloxifene HCl (**29**)	Evista[d]	Lilly	Diarylethylene	60
Tamoxifen citrate (**16a**)	Nolvadex[d]	AstraZeneca	Triarylethylene	20 (base)
Toremifene citrate (**22**)	Fareston	Orion	Triarylethylene	60 (base)
Progestins				
Desogestrel (**38**)	Ortho-Cept[f] Desogen, Mircette[f]	R.W. Johnson, Organon	Estrane	0.15
Ethynodiol diacetate (**97**)	Demulen[d,f]	Pharmacia	Estrane	1
Levonorgestrel (**36**)	Norplant	(Population Council)	Estrane	36[g]
Medroxyprogesterone acetate (**34**)	Depo-Provera Lunelle[h]	Pharmacia	Pregnane	100[c] 25[c]
Megestrol acetate (**93**)	Megace	BristolMyers Squibb	Pregnane	20
Norethindrone = norethisterone (**35**)	Norinyl[i] Ortho-Novum[f]	Pharmacia, R.W. Johnson	Estrane	1
Norethynodrel (**96**)	Enovid[i]	Pharmacia	Estrane	5
Norgestimate (**98a**)	Ortho Cyclen-21[f]	R.W. Johnson	Estrane	0.25
Norgestrel (**36**[j])	Ovral[d,f]	Wyeth-Ayerst	Estrane	0.3–0.5
Antiprogestin				
Mifepristone (**55**)	Mifeprex	Danco	Estrane	600

[a]Administered orally unless otherwise noted.
[b]Administered as a 24-h transdermal patch.
[c]Administered as a depot intramuscular injectable.
[d]Homepage is accessible on the internet.
[e]Only administered in combination with a progestin.

[f]Combined with 0.025–0.05 mg of ethinyl estradiol.
[g]Administered as a 30-day implant.
[h]Combined with 5 mg of estradiol cypionate.
[i]Combined with 0.1 mg of mestranol.
[j]Mixture of (**36**) and its enantiomer.

Table 13.2 Interaction of Progesterone (30) and Its Analogs with Steroid Receptors[a]

Compound	RBA[b] for				Reference
	PR	GC	MC	AR	
(30) (progesterone)	40 (23)	10 [17]	100	<0.1 [10]	224, (65), [225]
(34) (medroxyprogesterone acetate)	115 (55)	29	160	5	224, (226)
(93) (megestrol acetate)	(48)				226
(94) (ORG 5020)	100	17	53	<0.1	224
(95) (ORG 2058)	350 (100)	3	27	<0.1	224, (65)
(35) (norethindrone)	(32, 55)	—	—	13[c], 15	65, 226
(96) (norethynodrel)	(6)	—	<0.5		226
(37) (4,5α-dihydro (35))	(7)	—	—	10[c]	65
(36) (l-norgestrel)	120 (113)	0.6	70	45, 16	224, (226)
(98a) (norgestimate)	(50)	—	—	0.3	227
(98b)	(38)	—	—	1.3	227
(38) (desogestrel)	(2)		<0.5		226
3-keto (38)	(120)				
(55) (mifepristone)	(99) [241]	300[d] [764]	—	25 [10]	131 [225]
(104)	[8]	[0.06]	—	[0.03]	123

[a]Affinity for human (rabbit) uterine PR, rat liver glucocorticoid (GC) receptors, rat kidney mineralocorticoid receptors (MC), and rat prostate androgen receptors (AR) was determined. Relative binding affinities (RBA) for PR are expressed as percent relative to promegestone (94). Those for GC, MC, and AR are as percentages of those of dexamethasone, aldosterone, and methyltrienolone, respectively. Affinities of ligands for recombinant hPRA, hGC and hAR are in brackets.

[b]ER RBAs, compared to (1) in human myometrial cytosol, were <0.2 for (35), (36), (38), 3-keto (38); that of (96) was 1.5 (226).

[c]Determined using mouse kidney cytosol.

[d]Determined using rat thymus cytosol.

The most common side effect associated with use of progestins is residual androgenic (masculinizing) activity. All currently prescribed progestins have some degree of affinity for androgen receptors (see Table 13.2), although this varies over a wide range.

The low dose of steroidal estrogen (9) (or 10) used in oral contraceptives (Table 13.1: progestins, footnotes f and i) is enabled by bacterial hydrolysis (see section 3.8.3) of estrogen conjugates accumulated in the gut, allowing facile reabsorption to proceed. Coadministration of antibiotics, or other substances that have an adverse effect on normal gut bacterial populations, can interrupt this mechanism of estrogen conservation, resulting in a failure of the administered oral contraceptive to prevent ovulation.

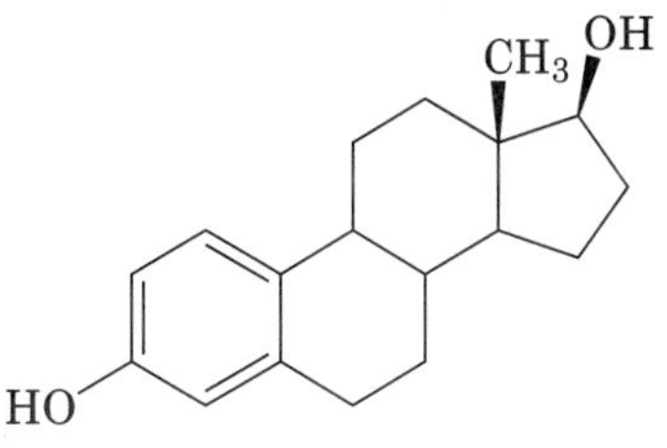

3 DRUG METABOLISM

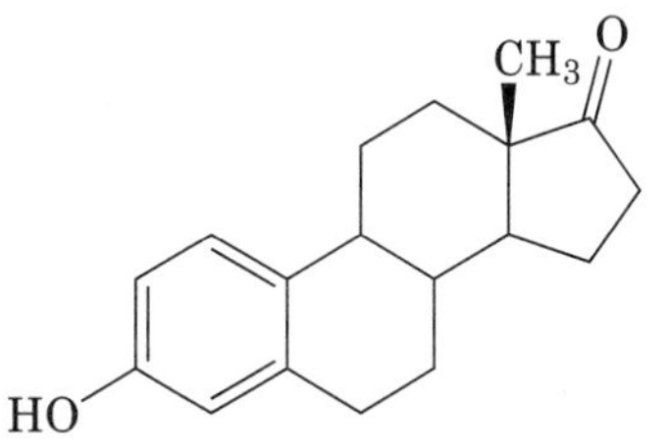

3.1 Oxidation and Reduction of Steroidal Estrogens

Major biotransformation pathways of 17β-estradiol (1) involve oxidations of its A- and D-

(1)

(2)

rings, and are illustrated in Fig. 13.1. Plasma and tissue levels of (1), estrone (2), and their metabolites are modulated primarily by hepatic enzymes. Thus, specific 17β-hydroxysteroid dehydrogenases catalyze, in turn, the re-

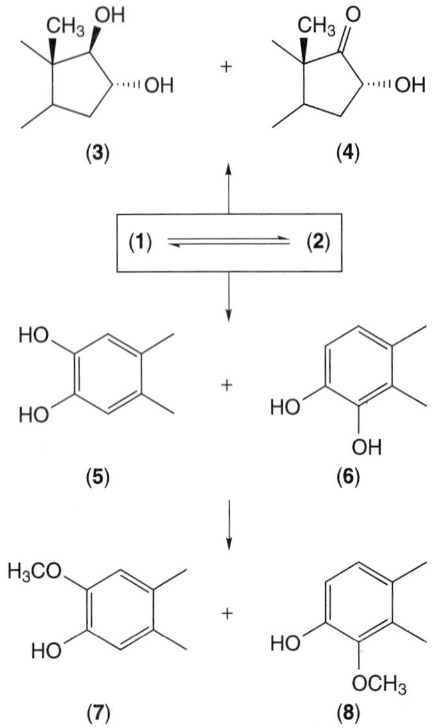

Figure 13.1. Oxidative metabolism of 17β-estradiol (**1**) and estrone (**2**). These enzymatically interconvertible estrogens undergo D-ring hydroxylation to (**3**) and (**4**), or alternatively, A-ring hydroxylation, ultimately generating guiacol metabolites (**7**) and (**8**).

3.2 Fate of 17α-Ethynyl Analogs of Estradiol (1)

Potent, orally effective steroidal mimics of (**1**) feature 17α-ethynyl substituents, as exemplified by ethynyl estradiol (**9**) and its 3-methyl and 3-cyclopentyl ethers mestranol (**10**) and quinestrol (**11**). The ethynyl group in (**9**) does not impede estrogen receptor (ER) affinity (Table 13.3) (16) or estrogenic potency or efficacy. The extent of D-ring biotransformation, which otherwise leads to less active or inactive metabolites, is greatly reduced (17). After oral absorption, methyl ether (**10**) is readily O-demethylated by liver oxidase(s) to (**9**) (18).

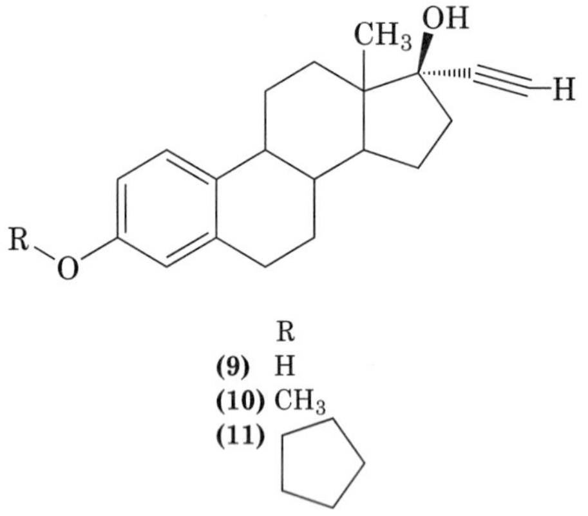

duction of 17β-hydroxyestrogens and the oxidation of 17-ketoestrogens (12). Cytochromes P450 1A2 and 3A4 are responsible for 2- and 16α-hydroxylations (13).

Estrogenic activities of (**1**) and (**2**) are similar as a result of interconversion, and hydroxylated metabolites (**3**) and (**4**) retain estrogenic and other effects (see Section 3.7) (13a). Indeed, (**4**) was as effective as (**1**) in preventing elevation of both serum cholesterol and bone turnover, but its uterotrophic effect was significantly less than that of in the ovariectomized rat (14). However, methoxyestradiols (**7**) and (**8**) are only weakly estrogenic if at all (15). Also, it is unlikely that estrogenic or other effects are manifested by catecholic estradiol metabolites (**5**) and (**6**) because these are very rapidly O-methylated.

Studies with human liver microsomes have implicated cytochrome P450 2C9 in this conversion (19). Lipophilic ether (**11**) gives prolonged estrogenic activity after oral dosing, attributed to deposition in adipose tissues, from which it is released slowly and converted to (**9**) in a manner similar to that of (**10**) (20).

However, the 17α-ethynyl group does not always impede D-ring hydroxylation of steroid hormones. Thus, for moxestrol (**67**), the 11β-methoxy analog of (**9**), the total extent of urinary elimination of 15α-, 16α-, and 16β-hydroxylated metabolites was about three times greater than that of (**1**) and accounted for over 40% of the administered dose (17). Incidental to these studies was the finding that the ethynyl group of (**67**) also underwent significant oxidation, ultimately leading to a D-ring-enlarged keto metabolite.

Table 13.3   ER Affinities of 17β-Estradiol (1) and Selected Analogs[a]

Compound	RBA[b]	ER from	Reference
(**1**) (17β-estradiol)	100		
(**2**) (estrone)	11 (25) [60]	rat (mouse) [hERα]	171a (172) [101]
(**3**) (estriol)	12 (15) [14]	rat (mouse) [hERα]	171a (172) [101]
3-deoxy (**1**)	11	human	170
17-deoxy (**1**)	79	human	170
3, 17-dideoxy (**1**)	<0.05	human	170
17-deoxy-17β-amino (**1**)	0.50	calf	171
17-*epi*-(**1**)	7 (10) [58]	human (mouse) [hERα]	81 (167) [101]
2-hydroxy-3-deoxy (**1**)	70	human	170
4-hydroxy-3-deoxy (**1**)	7	lamb, human	168a, 170
(**52c**) (equilin)	11	human	81
(**52b**) (17β-dihydroequilin)	179	human	81
(**9**) (17α-ethynylestradiol)	75 (100)	human (calf)	172 (169)
17-*epi* (**9**)	1.6	rabbit	168b
(**10**) (mestranol)	0.12	rabbit	168b
(**64**) (ICI 164, 384)	19 (5.6) [85]	rat (calf) [hERα]	173a (92a) [101]
(**65**) [ici (ZN) 182, 780]	89 (6.2)	rat (calf)	173a (92a)
(**66**)	970	rat	174
(**67**) (moxestrol)	25 (125) [43]	human (mouse) [hERα]	172 (167) [101]
(**68**) (RU 51625)	225	human	175
(**69**) (RU 53637)	175	human	175
(**70**)	20 (4)	mouse (rat)	176 (177b)

[a]Values were obtained using uterine cytosolic ER from the indicated sources, or using recombinant human ERα.
[b]RBA values were calculated using the formula: $100 \times IC_{50}$ of $1/IC_{50}$ of the test compound. All values are relative to (**1**).

3.3   Fate of Conjugated Equine Estrogens

Conjugated equine estrogens are composed of more than 10 aromatic A-ring 3-sulfate esters (see Section 4.3). The major component of this mixture is the sulfate ester of (**2**), estrone sulfate or E1-S (Fig. 13.2). After oral administration, E1-S and its counterparts are subject to efficient absorption. Indeed, E1-S and 7-dehy-

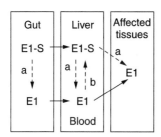

Figure 13.2. Fate of orally administered conjugated estrogens, exemplified by estrone sulfate (E1-S). After absorption, E1-S is subject to deconjugation by sulfatase (a) in the liver, or possibly in the blood or during transport to affected tissues. Deconjugation could occur to some degree in the gut as well. (b) This reaction is reversible, mainly in the liver, by sulfotransferase(s).

dro E1-S (**52c** sulfate) were twice as effective in the ovariectomized (OVX) rat uterotrophic assay (section 5.4), as were the respective unconjugated estrogens (21). Clearly, sulfatase(s) play a critical role in generating deconjugated steroids subsequent to or during entry of these into estrogen-responsive tissues.

Clinical pharmacokinetics and biotransformation of one of the components of conjugated equine estrogens have been studied systematically. Thus, oral administration of 17β-dihydroequilin (**52b**) 3-sulfate to postmenopausal women resulted in the urinary identification of (**52b-c**) and (**53b-c**) (structures, Section 4.3), and a high percentage (51–81%) of more polar metabolites suggested to be polyhydroxylated derivatives of the four identified metabolites (22). These results suggest that (**52b**) 3-sulfate ester is subject to absorption/deconjugation as summarized in Fig. 13.2, followed by oxidative metabolism. They extend findings of an earlier study in which levels of (**1**), (**2**), and (**52c**) were measured in the serum of patients on conjugated equine estrogenic therapy (23).

3.4 *In Vivo* Reversible Monoesters of (1)

The estrogenic potency of (1) is much greater parenterally than orally. Furthermore, esterification of the 17β-hydroxyl group of (1) results in very prolonged (up to 14 days) parenteral estrogenic effects, ostensibly attributable to the prolonged dissolution rate of the poorly water soluble esters from the site of administration (24). Examples are (1)-17-valerate, and (1)-17-(3-cyclopentyl)propionate (cypionate) (Table 13.1, footnote *h*) (25). Once absorbed into the blood from the injection site, nonspecific esterases release (1) from these esters (26).

3.5 Oxidative Metabolism of Triarylethylenes

Chlorotrianisene (12) was more effective orally than parenterally and did not interact

	R_1	R_2
(12)	CH_3	OCH_3
(13)	H, CH_3	OCH_3
(14)	H	OCH_3
(15)	H	H

with uterine estrogen receptors (27). Thus, it was thought to undergo metabolic activation by drug-metabolizing enzymes of the intestinal mucosa or liver. *In vitro* studies with oxidase enzymes from rat liver suggested that (12) was converted to a chemically reactive metabolite that interacted irreversibly with added estrogen receptors (27). Related studies demonstrated the conversion of (12) to its mono- and bisphenolic metabolites, (13) and (14) (28). Bisphenol (15), a close structural analog of (14), had estrogen receptor affinity comparable to that of (1) (29). The estrogenic potency of (15) in the immature mouse was 200 times greater than that of its bis-methyl

ether, by parenteral dosing (2a). Thus, metabolic *O*-demethylation appears to play a major role in mediating estrogenic potency of (12), as it does in the case of the steroidal estrogen mestranol (10) (18).

Information about the metabolic fate of tamoxifen (16a) has emerged in an eclectic way over the past 30 years, as implied by the letter-code designations used to denote its metabolites (Table 13.4). Efforts to identify metabolites were stimulated by the early finding that 4-hydroxytamoxifen (18), a major metabolite of (16a) in certain animal species, was a high affinity ER ligand and potent antiestrogen (30). Comparative ER relative binding affinities (RBAs), and serum levels in patients on long-term therapy, of (16a) and its metabolites (17–21) are summarized in Table 13.4. Taken together, these data suggest that (16a) expresses its effects in large part through its metabolites.

Studies with human liver microsomes showed that conversion of (16a) to (17) and (18) was catalyzed, in turn, by cytochromes P450 2D6 (2C9, 3A4 minor contributors) and cytochrome P450 3A4 (34). Interpatient variability in hepatic levels of these enzymes could account for the wide range of serum levels of (17) and (18) (Table 13.4).

Systematic biotransformation studies of toremifene (22) revealed a pattern of metabolism reminiscent of (16a), in human subjects and in the female rat (35).

Despite its structural similarity to (16a) and (22), clomiphene (23) was found to be less susceptible to oxidative biotransformation. Thus, in the female rat, (23) was accompanied by only low levels of metabolites (36). Clinical studies indicated differential pharmacokinetics of (23a) and (23b), but levels of metabolites were not reported (37). Unlike (16a) and (22), (23) is suitable only for short-term therapy because of unacceptable side effects. Perhaps this is attributable to its lack of conversion to multiple active metabolites expressing together a more favorable "therapeutic ratio" than that of the parent drug. This speculation aside, studies with (16a), (22), and (23) clearly indicate that minor structural variation in these triarylethylenes can greatly influence the extent of oxidative metabolism.

	R^1	R^2	X
(16a)	$OCH_2CH_2N(CH_3)_2$	H	CH_2CH_3
(16b)	H	$OCH_2CH_2N(CH)_3$	CH_2CH_3
(17)	$OCH_2CH_2NHCH_3$	H	CH_2CH_3
(18)	$OCH_2CH_2N(CH_3)_2$	OH	CH_2CH_3
(19)	$OCH_2CH_2NHCH_3$	OH	CH_2CH_3
(20)	$OCH_2CH_2NH_2$	H	CH_2CH_3
(21)	OCH_2CH_2OH	OH	CH_2CH_3
(22)	$OCH_2CH_2N(CH_3)_2$	H	CH_2CH_2Cl
(23a)	$OCH_2CH_2N(C_2H_5)_2$	H	Cl
(23b)	H	$OCH_2CH_2N(C_2H_5)_2$	Cl

3.6 Biotransformation of Diethylstilbestrol (24)

Years ago, (24) was widely used therapeutically before its carcinogenic and teratogenic potential became known. Because most chem-

(24)

ical carcinogens are known or thought to require activation by drug-metabolizing enzymes (see below), considerable attention has been focused on the ways in which (24) undergoes biotransformation. Because it is a lipophilic phenol, (24) is subject to O-glucuronidation and, like steroidal phenolic glucuronides (see Fig. 13.3), the resulting conjugate is capable of enterohepatic recycling (38). In addition, (24) is subject to extensive metabolic oxidation. Catechol and diene metabolites have been identified (39), as have other products (40). In animals, dehydrogenation of (24) to its diene metabolite is now thought to proceed spontaneously from a qui-

Table 13.4 ER RBAs and Clinical Steady-State Blood Levels of Tamoxifen (16a) and Its Metabolites[a]

Compound	Metabolite Designation	RBA	Serum Level (ng/mL)
(16a)		2, 1[b]	114–173
(17)	X	3	197–378
(18)	B	285, 100[b]	2–6.5
(19)	BX	143[c]	9.4
(20)	Z	2	33–55
(21)	Y	1	24–36

[a]The RBAs were determined using ER from rat uterus (RBA of 1 = 100) and are from Ref. 31 unless otherwise noted. Serum levels are ranges of values reported by different laboratories (32).

[b]Determined using calf uterine cytosol (33).

[c]RBA is for a 1:1 mixture of *cis* and *trans* isomers.

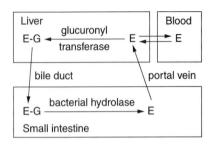

Figure 13.3. The enterohepatic recycling system. Effectiveness of estrogens (E) such as (1), (2), and (9) is amplified by this process because each of the respective glucuronide conjugates (E-G), after intestinal accumulation, is a substrate for deconjugating enzymes of normal gut flora. This results in reabsorption of the free estrogens.

none-type intermediate metabolite (below) generated by enzymatic oxidation of (24) (41).

3.7 Electrophilic (Carcinogenic ?) Metabolites of ER Ligands

Long-term administration of steroidal estrogens such as (1) and (2) has been associated with a slight increase in the incidence of reproductive tract cancer. Similarily, a small but significant incidence of adverse reproductive tract effects, including cancers, in female offspring of women receiving diethylstilbestrol (24) during pregnancy, has been reported (42). This resulted from the unfortunate use of (24) to prevent miscarriage, before its characterization as a teratogen and carcinogen (43). Moreoover, despite results indicating (16a) to be less estrogenic than steroidal estrogens in reproductive tissues, an increased risk of endometrial cancer has been associated with its use (8a).

Metabolic "activation" to electrophilic metabolites is generally thought to be a prerequisite to chemical carcinogenicity (44). Electrophilic metabolites of steroidal and nonsteroidal ER ligands have been characterized. Thus, a hydroxylated metabolite of (2), specifically (4), forms a stable adduct with albumin, suggested to arise from Heyns rearrangement of an α-hydroxyimine intermediate, itself producted by reaction of (4) with a lysine residue of albumin (45). DNA damage was associated with exposure of mouse mammary cells to (4), but was not observed on exposure of these cell

types to (1) (46). Alternatively, reactivity of catechol metabolite (6) was demonstrated by its ability to undergo electroreductive coupling at C-1 with the C-8 of adenine (47).

A quinone-type intermediate (25), produced by interaction of (24) with peroxidases,

(25)

was proposed to interact covalently with tubulin and possibly with other macromolecules (48). Another nonsteroidal estrogen, hexestrol (26), exhibited tumorigenicity in the male Syr-

(26)

ian Golden hamster similar to that of (24) (49), but ostensibly through a mechanism of metabolic activation more like that of (1) and (2). The major metabolite of (26) is 3'-hydroxy (26), which in turn was subject to further oxidation to a 3',4'-*ortho*-quinone by certain cytochrome P450s and/or peroxidases. This electrophilic quinone was reactive toward purine moieties: it arylated N-7 of 2'-deoxyguanosine units in DNA (50).

In the rat, long-term administration of high doses of (16a) resulted in development of hepatic tumors. Several specific routes of oxidative biotransformation have been implicated in conversion of (16a) to electrophilic metabolites proposed to account for tumorigenicity (51). A stabilized quinoid intermediate (27), generated by spontaneous cycloelimination of sulfate from proximate metabolite (28), is an example of an electrophilic species (52) suggested to originate from (16a).

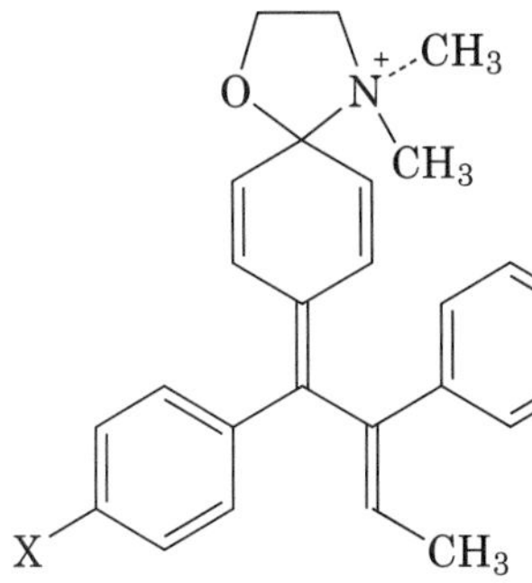

(27) X = H or OH

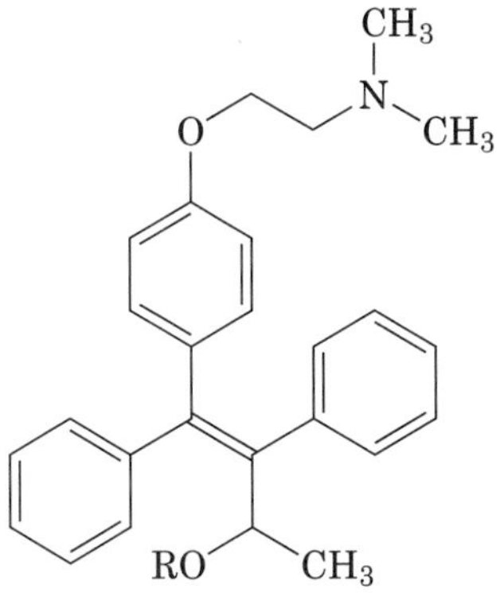

(28) R = SO$_2$O$^-$

Electrophile (27) was suggested to *N*-alkylate deoxyguanosine residues in DNA in the rat (53, 54), and it might do so in humans. Thus, DNA analysis of endometrial tissue samples from patients receiving (16a) revealed, in eight of the 16 patients studied, the presence of α-(*N*-2-deoxyguanosinyl)-(16a) and -(16b); levels ranged between 0.2 and 12 and 1.6 and 8.3 adducts/10^8 nucleotides, respectively (55).

Although long-term high dose administration of (16a) has been associated with hepatocarcinogenicity in the rat, this has not been the case in humans. However, the possible link between use of (16a) in breast cancer treatment and human hepatocarcinogenicity has not been investigated thoroughly because of the assumption that any liver tumors in (16a)-treated patients would arise secondarily to breast cancer metastasis (56). Thus, the incidence of human hepatocarcinogenicity associated with administration of (16a) might be significant.

In summary, the connection between formation of chemically reactive metabolites of ER ligands and their carcinogenicity risk in humans has become clearer. However, other factors besides conversion to electrophilic metabolites might contribute to carcinogenicity (57). Chronic exposure of cells containing ER to full or partial estrogen agonists such as (1) and (16), respectively, might be a major causative factor, given that carcinogenicity of these substances in humans is generally restricted to ER-containing organs.

3.8 Aspects of Glucuronide Conjugation

Large lipophilic molecules containing phenolic moieties are often subject to glucuronide conjugation. Many ER ligands fit this description, although the impact of this metabolic pathway on the observed potency of a particular ER ligand depends on several factors.

3.8.1 Termination of Effects of ER Ligands.
Both oral bioavailability and duration of action of raloxifene (29) are affected by its sus-

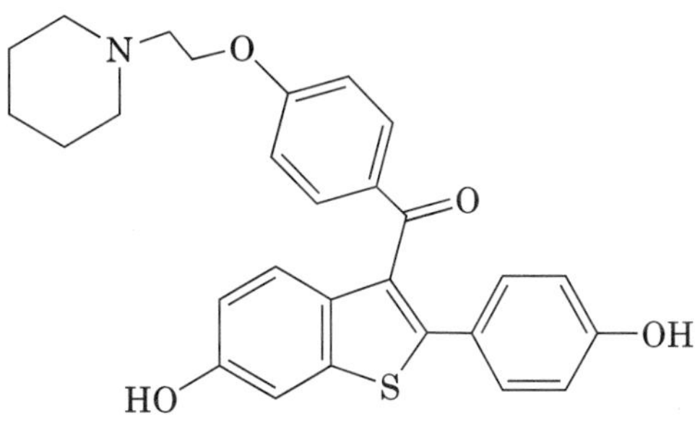

(29)

ceptibility to glucuronide conjugation, which occurs predominantly at its 4'-hydroxyl group in humans (58). Thus, despite its 40-fold higher affinity for ER compared to that of (16a) and 1000-fold greater *in vitro* estrogen antagonist potency (59), the oral potency of (29) is one-half that of (16a), based on daily doses (Table 13.1) normalized for differences in molecular weights. In female patients, (16a) is accompanied by its active, oxidized metabolites (Table 13.4) and is evidently not subject to glucuronide conjugation; (29), in contrast, is present in serum primarily

(>90%) as its 4'-*O*-glucuronide conjugate. Because ER affinity of (**29**)-4'-*O*-glucuronide is less than 1/10 that of (**29**) (60), it is unlikely that this metabolite contributes to estrogenic/antiestrogenic effects.

Because SERMs with high ER affinity always contain at least one phenolic hydroxyl group, the possibility of bioavailability-limiting glucuronide conjugation is always a drug design consideration.

3.8.2 Net Glucuronidation. Susceptibility to glucuronide conjugation, after absorption, is governed not just by the rate of O-glucuronidation of the phenolic SERM by uridine diphosphate (UDP) glucuronyl transferases, but also by the degree of exposure of the resulting conjugate to glucuronidase (61). Both enzymes are located to some degree in gastric mucosa, but mainly in hepatocyte endoplasmic reticulum. Thus, the high net glucuronidation of (**29**) in serum is presumably a consequence of minimal exposure of (**29**)-4'-*O*-glucuronide to enteric and hepatic glucuronidase. Net glucuronidation of a SERM might depend on its structure: a SERM glucuronide with maximal access to glucuronidase would exhibit reduced net glucuronidation, regardless of the affinity of its parent for UDP glucuronyl transferase. Alternatively, improvement in SERM bioavailability and potency by incorporating structural changes that reduce susceptibility to UDP glucuronyl transferase has been carried out successfully (see Section 7.3.2) (62).

3.8.3 Enterohepatic Recycling. Besides being capable of enzymatic deglucuronidation *in vivo*, the effectiveness/potency of estrogens such as (**1**), (**2**), and analogs such as (**9**) is primarily conserved by a second mechanism. Glucuronide conjugates of these estrogens are accumulated in the bile duct and thus transported into the small intestine. There, these particular glucuronide conjugates are substrates for bacterial hydrolases. The regenerated steroids are efficiently absorbed from the intestine (Fig. 13.3). This process of enterohepatic recycling accounts for the prolonged duration of effect of endogenous and administered estrogens (63).

3.9 Biotransformation of Progesterone (30) and Its Analogs

Because it is subject to rapid reductive metabolism, (**30**) is not orally effective as a proges-

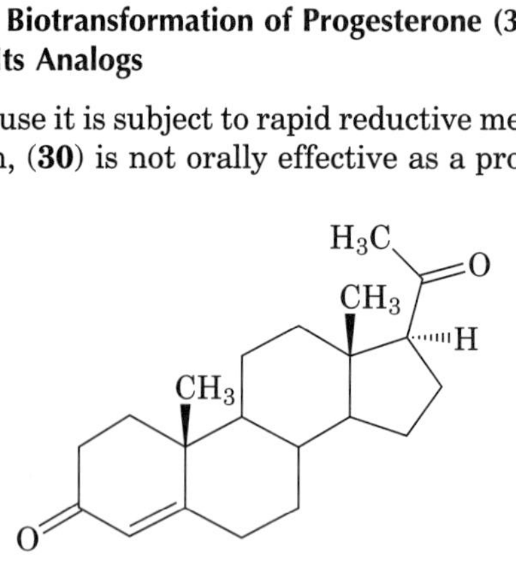

(30)

tin. The major urinary metabolite of endogenous or parenterally administered (**30**) is pregnanediol (**31**), which is eliminated as a glucuronide conjugate. Alternatively, 3-keto- and 4,5α-reduction of (**30**) affords (**32**), a me-

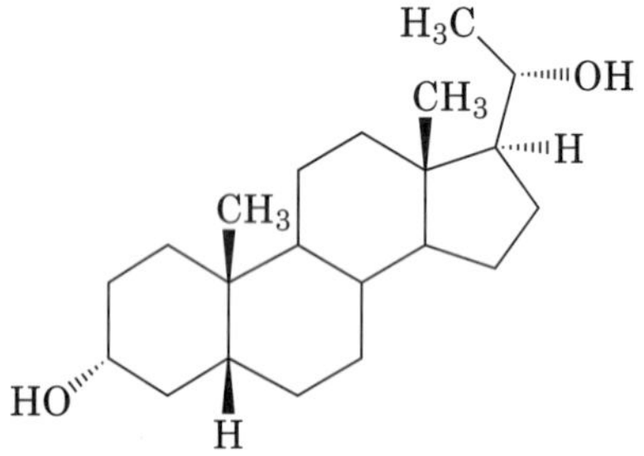

(31)

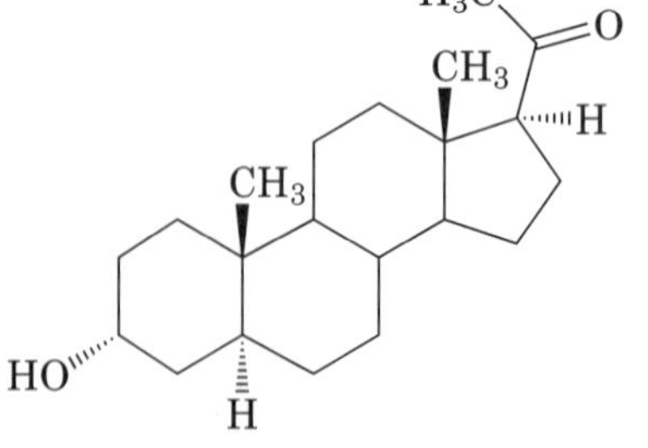

(32)

tabolite that potentiates the γ-aminobutyric acid-A receptor and appears to be responsible for the anticonvulsant effect of (**30**) in laboratory mice (64). Formation of (**32**) was shown to proceed by initial dihydrotestosterone reductase-catalyzed saturation of the 4,5-double bond, followed by oxidoreductase-catalyzed 3-keto reduction.

Several close structural analogs of (30) are noteworthy with regard to their prolonged progestational effects when given by injection. Thus, 17α-hydroxyprogesterone caproate (33)

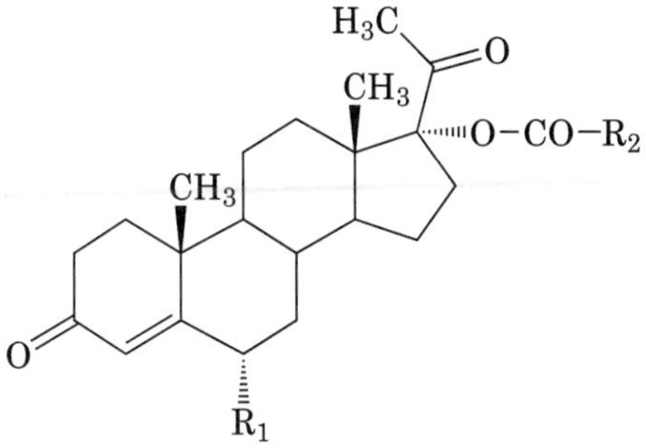

(33) R₁ H   R₂ CH₂(CH₂)₃CH₃
(34) CH₃   CH₃

is a long-acting progestin. Also, medroxyprogesterone acetate (34) has prolonged parenteral effects and is moderately active orally. It is much more effective by these routes than its 6-normethyl counterpart.

Estrane progestins can undergo 4,5-ethylenic bond reduction with and without keto reduction. Thus, norethindrone (35) and l-nor-

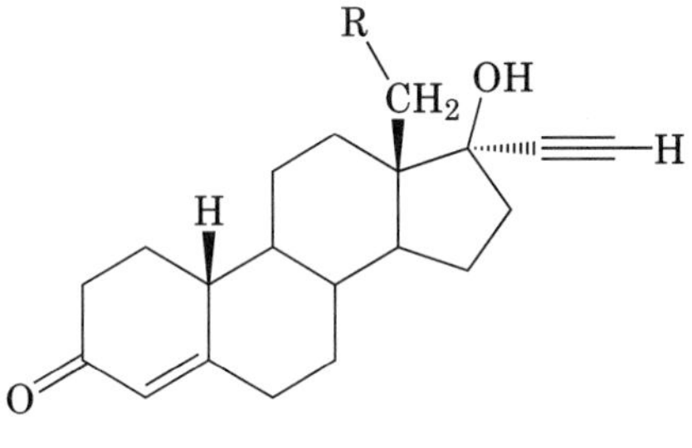

(35) R = H
(36) R = CH₃

gestrel (36) are biotransformed to several A-ring-reduced (dihydro and tetrahydro) derivatives. The progesterone receptor (PR) and androgen receptor affinities of (37) (Table 13.2), the 5α-dihydro metabolite of (35), suggest that it contributes to the overall effects of

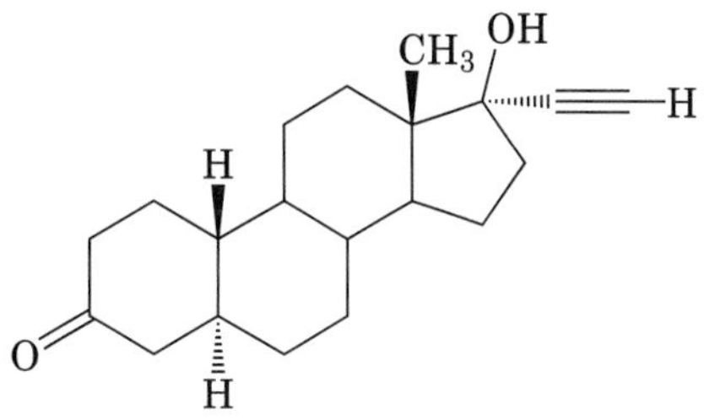

(37)

(35). In animal studies, (37) exhibited progestin antagonist activity (65). Carbonyl reduction of (37) affords the 3β-hydroxy-5α-dihydro metabolite of (35), which had a very low PR affinity. Thus, this metabolite is unlikely to express effects through these receptors. Alternatively, A-ring aromatization of estrane progestins has also been observed. Thus, (35) undergoes conversion to (9) to the extent of 2.3% in perimenopausal women (66), an extent that appears to be pharmacodynamically significant.

Desogestrel (38) undergoes A-ring allylic oxidation to its 3-keto metabolite in patients

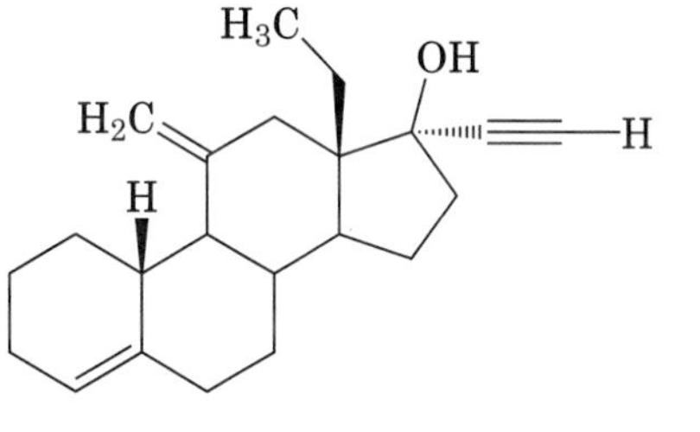

(38)

(67). This metabolite, 3-oxo (38), has PR affinity 60 times greater than that of (38) and thus is probably responsible for the activity of this progestin.

The reductive metabolites of gestodene (39), especially (40b), exhibited affinity for human ER and estrogenic activity in vitro, unlike (39) itself (68), despite the absence of the aromatic A-ring seen in most steroidal estrogens (Section 7.1). Thus, the estrogenic effects of (39) seen in vivo might be the result of reductive conversion to its structurally novel ER ligand, (40b), rather than of A-ring aromatization, as was the case with (35).

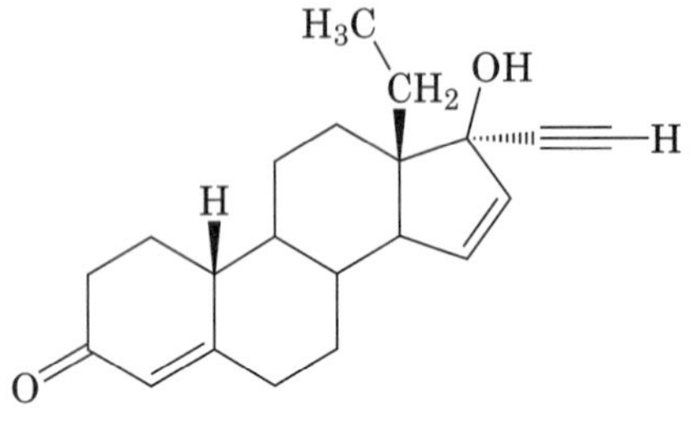

(39)

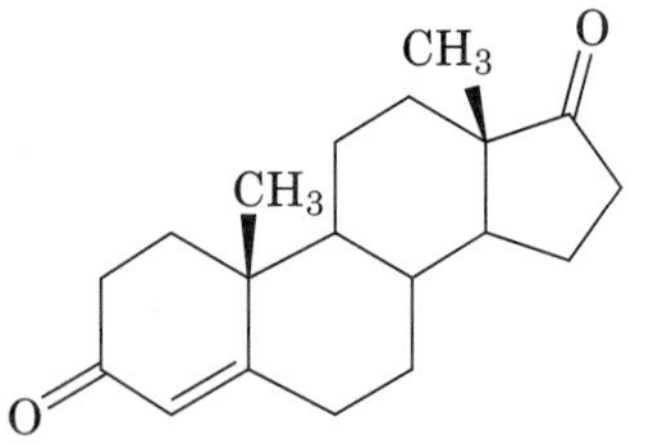

(41)

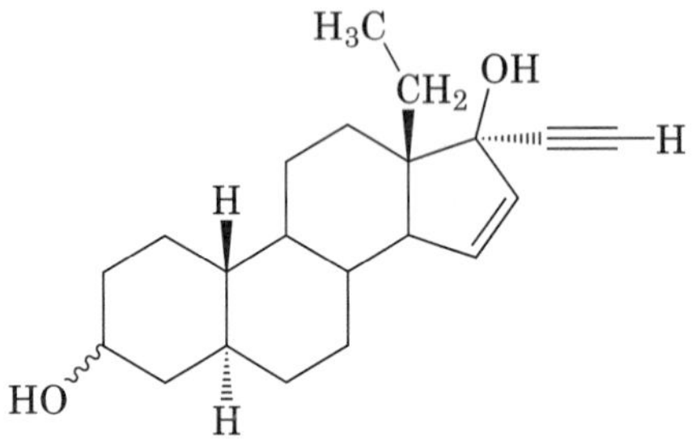

(40a) 3α-OH
(40b) 3β-OH

4 ASPECTS OF BIOSYNTHESIS

Plasma and tissue levels of female sex hormones are controlled by follicle-stimulating hormone (FSH) and luteinizing hormone (LH). These polypeptide gonadotrophins are secreted by the anterior pituitary, and control the activity of enzymes, located predominantly in the ovary, which sequentially oxidize cholesterol to progesterone (30), and 17β-estradiol (1) plus related estrogens. Together, (1) and (30) are important in the development of female sexual characteristics, in regulating fertility, and in maintenance of normal function (homeostasis) in tissues and organs endowed with ER and PR. Although (1) and (30) are the most important female sex hormones, it is noteworthy that smaller but significant amounts of androstenedione (41) and its 17β-hydroxy counterpart (testosterone) are also released from the ovary.

4.1 Steroid Aromatization

Estrogens (1–3) are the three most abundant endogenous estrogens. These were first identified in the urine of pregnant women (69).

The ovary is the major source of these estrogens except during pregnancy, in which placental production dominates. Adrenal cortex and adipose tissues produce smaller amounts of these estrogens. After menopause, these tissues become predominant sources.

A novel and unifying structural feature of endogenous estrogens is aromaticity. [The biosynthetic sequence by which progesterone (30) and testosterone arise has been clearly established (70; see also previous editions).] Consequently, attention is focused on the mechanistic basis by which steroid aromatization, the rate-determining step in estrogen biosynthesis, takes place. Androgens were shown to be precursors: thus, androstenedione (41) was a precursor to (2) in endocrine tissue (71). Androgen aromatization is catalyzed by aromatase (estrogen synthetase), an enzyme found at high levels in placental and ovarian tissue. Aromatase is a member of the cytochrome P450 family of intracellular membrane-bound, mixed-function oxidases (72). Mechanistic studies indicated that it catalyzed androgen aromatization through three oxidation steps. Thus, (41) initially undergoes sequential conversion to a 19-oxo intermediate, (43) (73) (Fig. 13.4). This is followed by an addition-elimination sequence believed to be initiated by a novel peroxyanion stabilized by heme iron at the active site of aromatase (74). This mechanism was supported by the finding that O_2 was the main (90%) source of the "second" oxygen in HCOOH released during aromatization of (43). In addition, the Δ^2-3-deoxy analog of (43) underwent aromatization. This suggested that positioning of the 1β-hydrogen is critical for rearrangement to occur, and that 2,3-enolization of (44) and (43) might thus facilitate aromatization (75).

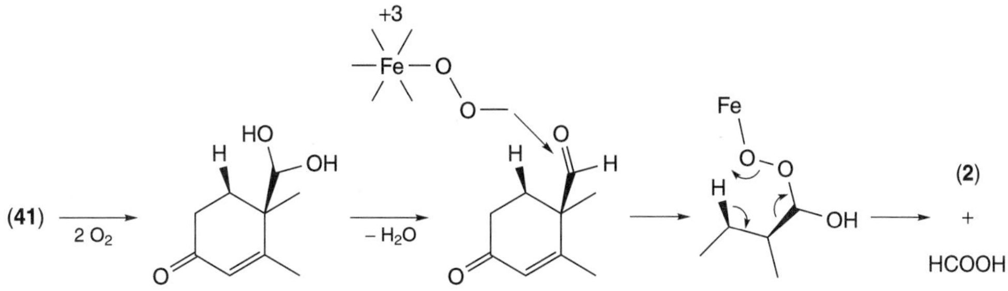

Figure 13.4. Aromatase-catalyzed conversion of androstenedione (**41**) to estrone (**2**). Sequential hydroxylation of (**41**) at C-19 gives (**42**), which undergoes dehydration to (**43**). This aldehyde reacts with nucleophilic ferric peroxyanion, generated at the catalytic site of aromatase, to give (**44**), which undergoes spontaneous rearrangement to produce (**2**) and formic acid.

4.2 Aromatase Inhibitors

Considerable attention has been directed toward identification of drugs capable of selective, potent aromatase inhibition. Such drugs were envisioned to facilitate reduction in levels of circulating estrogens, thus offering a means of suppression of estrogen-dependent tumors.

Clinically relevant aromatase inhibition has been found in nonsteroidal as well as steroidal structures. Some examples of the former include aminoglutethimide (**45**), initially evaluated for its effects on the central nervous system. Other enzymes involved in steroid hormone biosynthesis besides aromatase are inhibited by (**45**).

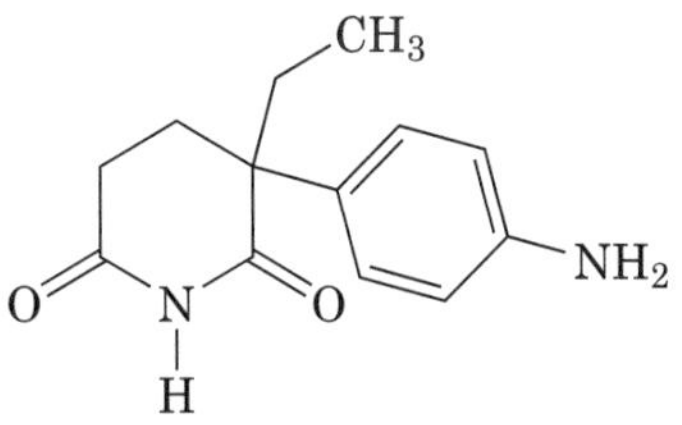

(**46**)

(**47**) R = Cl
(**48**) R = CN

(**45**)

However, imidazoles such as fadrozole (**46**) and triazoles such as (**47**) and letrozole (**48**) (76) interfere with estrogen biosynthesis by more selective inhibiton of aromatase. Molecular modeling studies with the active S-(−)-enantiomer of (**46**) suggested that its N-heteroaromatic ring interacts with the heme iron of aromatase at its sixth ligand-binding site (76b). Examples of triazole ring-containing nonsteroidal aromatase in-

hibitors with maximal aromatase selectivity are vorozole (**49**) and anastrozole (**50**) (77).

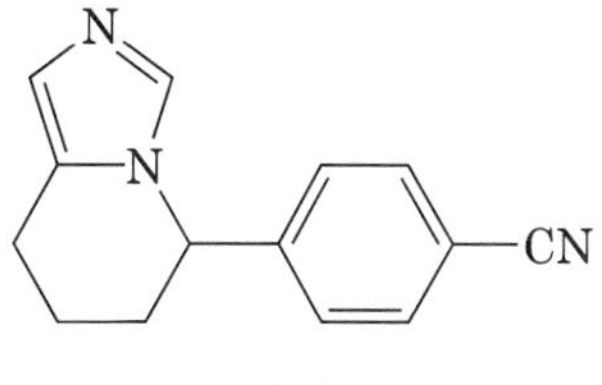

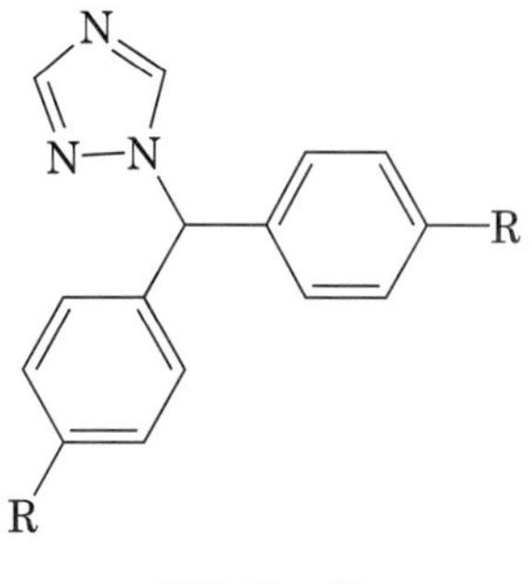

The androstane derivative exemestane (**51**) is a potent, selective, irreversible inhibitor of aromatase (78). Like (**48–50**), (**51**) is advocated as a "second-line" treatment for patients whose breast cancer is no longer suppressed by (**16a**) (77, 79). Furthermore, because (**51**) did not show cross-resistance with nonsteroidal aromatase inhibitors (79b), resistance was suggested to arise from changes in neoplastic cells not directly linked to aromatase.

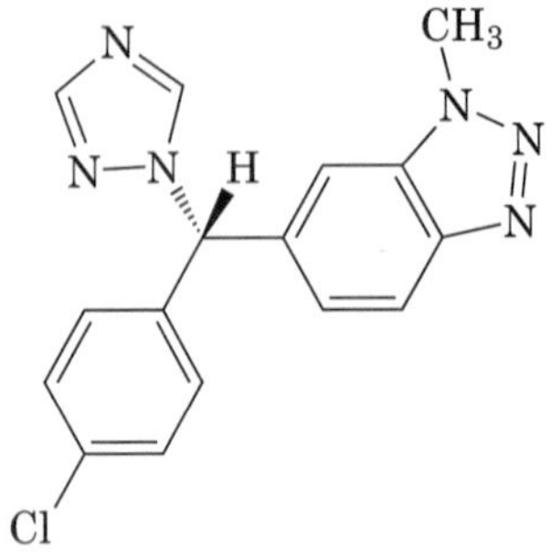

(49)

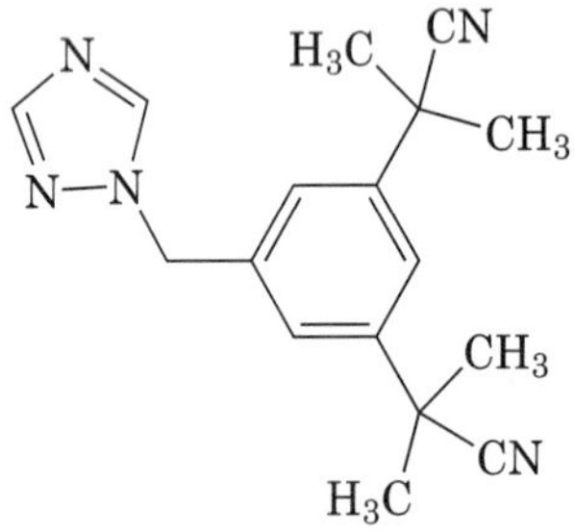

(50)

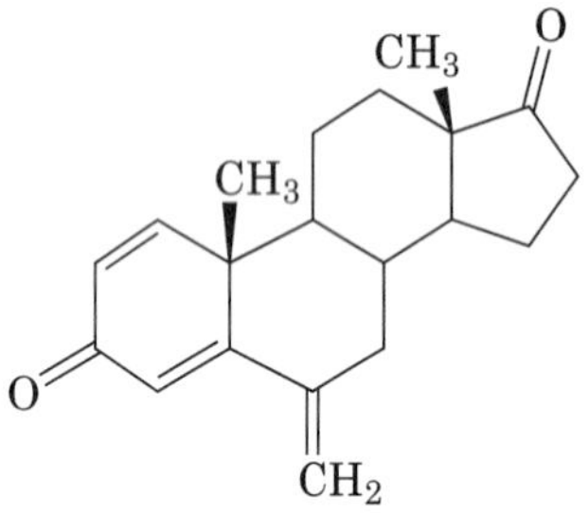

(51)

4.3 Genesis of Equine Estrogens

Equine estrogen conjugate mixtures are composed of a reasonably fixed proportion of phenolic sulfate esters of estrone (**2**) accompanied by those of equilin (**52c**) and equilenin (**53c**), together with their respective 17α- and 17β-hydroxy derivatives (**80**) (Table 13.5). Trace amounts of a large number of related compounds accompany these substances, but their structural identities and biologic effects have not been reported yet.

Unlike virtually all other natural product extracts from which single purified components have emerged as therapeutic agents, mixtures of conjugated equine estrogen extracts are still preferred in estrogen-replacement therapy.

Biosynthetically, the seven ring B-dehydrogenated estrogens in Table 13.5 are produced by pathways not involving squalene or cholesterol, intermediates common to the biosynthesis of all steroid hormones in humans. The horse is an excellent source of (**2**), but only the pregnant mare produces these dehydrogenated estrogens, which originate by pathways that do not involve (**1**), (**2**), or their hormonal precursors, (**30**) and (**41**) (83).

About 19% of equine estrogen conjugates consists of 17α-hydroxy steroids (17-*epi*-(**1**), (**52a**), and (**53a**), but the corresponding 17β-hydroxy derivatives account for a much smaller percentage. Although (**2**) undergoes reversible reductive metabolism to (**1**) in humans (Fig. 13.1), 17-*epi*-(**1**) is not produced from (**2**). It is therefore unlikely that these 17α-hydroxy steroids interconvert with their 17-oxo counterparts in humans, and thus their pharmacodynamic effects would be expressed independently of (**1**) and (**2**).

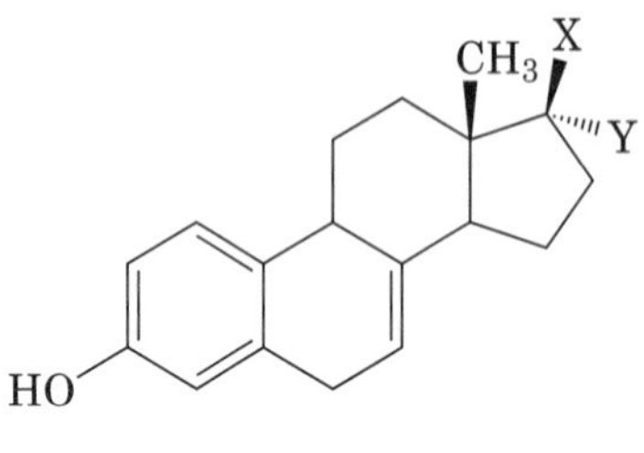

	a	b	c
X	H	OH	O
Y	OH	H	

(52)

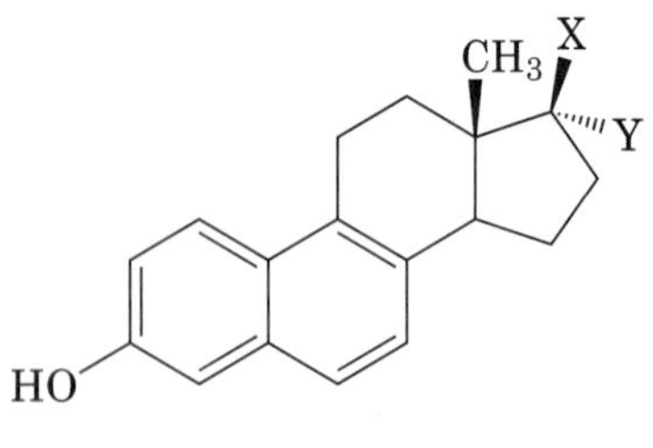

(53)

Table 13.5   Percentage Composition and Estrogenic Effects of Steroids in Conjugated Estrogen Tablets

Component	Avg. wt %[a]	RBA[b]	Uterotrophic Efficacy[c]
17-*epi* (**1**)	3.7	4	1.0
(**1**)	0.5	56	2.6
(**2**)	49.1	7	2.6
$\Delta^{8,9}$-*dehydro* (**2**)	3.5	(0.4)	1.8
(**52a**)	13.5	6	1.1
(**52b**)	1.5	100	2.9
(**52c**)	22.9	6	2.9
(**53a**)	1.5	2	1.2
(**53b**)	0.7	21	1.5
(**53c**)	2.8	1	1.6
Numerous other steroids	<0.5	—	—

[a]Data are for 3-sulfate esters (80).

[b]Relative to (**52b**) or (**1**), for human endometrial ER. Data in this column are from Ref. 81 and 82.

[c]The ability of each unconjugated steroid to stimulate uterine weight gain in the immature rat, at an i.p. dose of 2 μg/animal/day for 3 days, with respect to vehicle-treated control: efficacy = 1.0, average uterine wet weight = 33.8 mg; $P <$ 0.01 with respect to control for all steroids except 17-*epi* (**1**) (82).

5   PHYSIOLOGY AND PHARMACOLOGY

Endogenous estrogens (**1–3**) affect the cellular roles of genes involved in homeostasis, cell division, protein expression, and cell communication. This control is mediated by interaction with estrogen receptors.

The endogenous progestin progesterone (**30**) has major roles in preparation of the uterus for pregnancy, maintenance of pregnancy, and in maintenance of the female reproductive organs. It is a central biochemical intermediate in the biosynthesis of estrogens, androgens, and corticosteroids. It is produced in the ovaries (corpus luteum), adrenal glands, and in the placenta during pregnancy. As suggested from the receptor-binding data in Table 13.2, (**30**) interacts with other steroid receptors besides PR. Interaction with mineralocorticoid and glucocorticoid receptors accounts in part for its physiologic effects. Also, (**30**) antagonizes oxytocin at uterine oxytocin receptors (84), suggesting its pregnancy-maintaining effect is in part mediated nongenomically.

5.1   Topography of Estrogen Receptors

The first direct evidence for the presence of ER in estrogen-sensitive tissues was enabled by the availability of radiolabeled (**1**) of sufficient specific activity for use in tissue distribution studies (85). Thus, saturable (receptor-specific) binding of [^{3}H]**1**, displacable by coadministration of excess amounts of unlabeled (**1**), was observed in uterine cytosol, and in uterine and ovarian tissues of animals receiving [^{3}H]**1**. In the 30-plus years since these discoveries, our perception of the ER's structure in the absence and presence of its ligands has grown considerably.

There are two isoforms of human ER, designated as hERα and hERβ. These show a high degree of amino acid sequence homology in their hormone-binding domains (Fig. 13.5), especially with regard to amino acid functional groups making direct contact with ligands, or lining the binding pocket (86); however, there are subtle differences in their respective character. So, although observable affinity of many ER ligands for each isoform is similar, a few such ligands exhibit isoform preferences. The homology of ERα and ERβ varies slightly or greatly, depending on domain, relative to respective isoforms found in experimental animals (87, 88). ER is found in monomeric, (mixed) dimeric, and oligomeric states (89). Most of the ER within affected cells is located in the nucleus, although ER has been identified in cell membranes (see below). Thus like PR, ER appears to mediate effects of ER ligands through nongenomic as well as well-established genomic mechanisms (90).

Each ER isoform is composed of six domains, as shown in Fig. 13.5 (91). The ligand-

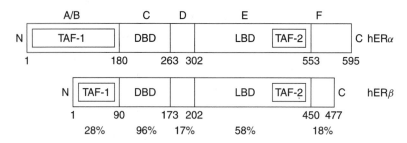

Figure 13.5. Diagram of hERα and hERβ primary structures. Numbers starting from the N-terminal end of the peptide indicate positions of amino acids at junctions of the six respective domains (A-F). Each hormone (ligand) binding domain (E) contains the hormone-dependent transcription activation function 2 (TAF-2). The hormone-*independent* transcription activation function 1 (TAF-1) is located in respective domains A and B. Domain C contains the DNA-binding domain (DBD). Percentages indicate the degree of sequence homology of hERβ domains to those in hERα.

binding domain (LBD) is embedded in Domain E. Domain B, which contains the hormone-independent transcription activation factor 1 (TAF-1) and domain E, which in addition to the LBD contains the hormone-dependent transcription activation factor 2 (TAF-2), are of special significance. Studies with a series of truncated ERαs indicated that only TAF-2 is required for transcriptional activity of (**1**), but certain other ER ligands such as tamoxifen (**16a**) required modified ERα with both TAF-1 and TAF-2 for this activity (92).

The hER interacts strongly with (**1**), with an association constant of $5.5 \times 10^9/M$, corresponding to a free energy of binding (ΔG) of -13.5 kcal/mol. Electrostatic, hydrophobic, and dispersional (van der Waals) interactions between (**1**) and ER contribute to ΔG, and the magnitude of specific bonding interactions has been estimated rationally (93). Thus, electrostatic interactions involving the 3- and 17β-hydroxyl groups (Fig. 13.6a) contribute, in turn, -1.9 and -0.6 kcal/mol to ΔG, and the one involving (**1**)'s benzene ring π electrons contributes -1.5 kcal/mol. Interaction of (**1**) with ER results in loss of ordered water molecules surrounding the surfaces of the steroid core and complementary hydrocarbon groups of amino acid units proximate to the binding pocket. Thus, the entropy of binding is $+22$ entropy units, which contributes -7.8 kcal/mol (at 30°C) to ΔG. The total dispersion energy, calculated from the sum of atomic polarizabilities of carbon, hydrogen, and oxygen in (**1**), is -2.4 kcal/mol. The sum of the estimated

ΔG for all of these specific interactions (-14.2 kcal/mol) is in reasonable agreement with that determined experimentally.

Crystallization of chromatographically purified hERα LBD complexed with (**1**) was achieved after initial carboxymethylation with iodoacetate, which fortunately affected only cysteine units not directly involved in ligand interactions. X-ray diffraction studies of this complex showed that the ionized side-chain functional groups of Glu353 and Arg394 interact in tandem with the phenolic hydroxyl of (**1**). Hydrogen bonding and/or ion-dipole bonding (Fig. (13.6a) are implicated in these associations. In addition, His524 interacts with (**1**)'s 17β-hydroxyl oxygen by similar processes. The hydrogen bond *accepting* Glu353 is very important in allowing ERα to discriminate between (**1**) and steroids bearing noncomplementary 3-keto groups, especially androgens (95). Lipophilic Phe, Tyr, Leu, Ile, Ala, Val, Met, and Ile units make up 40% of the total of the LBD's amino acids (91a), and the aliphatic or aromatic side chains of at least 15 of these are complementary with the hydrocarbon framework of (**1**) (96), and interact by hydrophobic bonding.

Analogously prepared complexes of diethylstilbestrol (**24**) and raloxifene (**29**) were found to exhibit the same electrostatic interactions with cysteine-carboxymethylated LBD as did (**1**) (94, 97). In addition, the protonated side-chain nitrogen of (**29**) interacted with the ionized carboxyl group of Asp351 by ionic

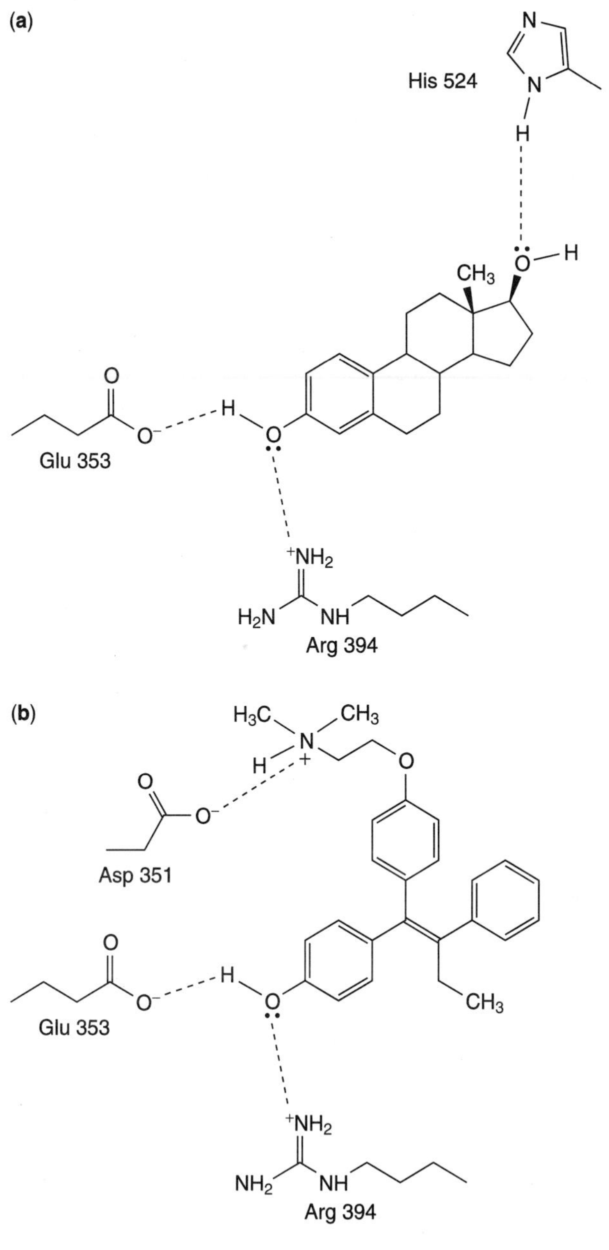

(a)

(b)

Figure 13.6. Electrostatic bonding interactions seen in the crystalline LBD (modified) of hERα liganded with (a) 17β-estradiol (**1**); (b) 4-hydroxytamoxifen (**18**) (94, 97). A water molecule associated with the phenolic hydroxyl of each ligand is omitted. In complex b, Thr347, Met421, and Met343 are interpositioned between the unsubstituted phenyl ring of (**18**) and His524.

bonding (94). The phenolic hydroxyl of 4-hydroxytamoxifen (**18**) interacted with Glu353 and Arg394 of cysteine-carboxymethylated LBD in the same way as did that of (**1**) and, like (**29**), its (protonated) amino group interacted ionically with the Asp351 carboxylate group of the modified LBD (Fig. 13.6b) (97). However, in contrast to complexes involving (**1**), (**24**), or (**29**), His524 was not positioned

closely to the unsubstituted phenyl ring in (**18**), and presumably would not interact with this ring even if it was hydroxylated.

The LBD of the other hER isoform (hERβ), liganded with (**29**) or genistein (**59**), has also been crystallized (98). Electrostatic interactions (bond lengths 3.3–3.8 Å) in these complexes were identical to those in the hERα LBD-(**29**) complex.

Unlike (1), (18), and (29) interact with Asp351 in each hER isoform. Although such bonding facilitates affinity of these nonsteroidal ligands for hER, it is not the basis for their mixed agonist-antagonist effects (99) (see Section 5.9).

5.2 Ubiquitous Distribution of ER Isoforms

ER has conventionally been associated with organs of the female reproductive axis. ER has also been associated with estrogen-stimulated cancer, particularly of the breast and endometrium. However, the number of organs now confirmed to contain ER, albeit at lower levels, has expanded greatly in recent years, and is not confined to females. Thus one or both of the ER isoforms have been localized immunohistochemically in the nuclei of cells in lung, liver, kidney, adrenal glands, colon, heart, prostate, testis, most areas of the brain, and in bone-remodeling cells (osteoblasts and osteoclasts) (100, 101).

5.3 Molecular Endocrinology of ER

When (1) interacts with (oligomeric) ER, helix 12 of its LBD seals the occupied binding pocket, thereby resulting in important changes in the association state and conformation of ER (3, 94). These include (1) conversion of ERα/ERβ to homo- and heterodimeric states; (2) exposure of regions in these dimers proximate to TAF-2, and possibly to TAF-1 also, which can bind to one of several coregulator proteins (102); and (3) "activation" of domain C. Domain C features two regions of its peptide backbone extruded in fingerlike fashion. Each finger is anchored through four cysteine residues near its base by a zinc cation (103). These zinc fingers interact strongly with complementary nucleotide sequences in DNA known as estrogen response elements (EREs) (104).

The $(1\text{-ER})_2$-coregulator complex thus interacts with ERE, ultimately resulting in changes in intracellular levels of various enzymes, receptors, and other proteins, leading to control of cell activity and proliferation (105). An additional function of the coregulator in this complex is to retard dissociation of (1) (106). This was also the case with other estrogen agonist ligands such as (24), but dissociation of estrogen antagonists from analogous complexes was not affected.

ER heterogeneity appears to play a role in modulation of responses of specific organs to (1). ERβ has been shown to inhibit transcriptional actvity of ERα at subsaturating levels of (1), and therefore ERβ can decrease overall cellular sensitivity to (1) (107).

An observation consistent with the rate theory of drug action (108) has been demonstrated in the interaction of liganded ER with ERE. Thus, (1)-ER exhibited an association/dissociation rate with ERE that was 1000 times the rate observed when ER was complexed with a "pure" estrogen antagonist, ici (ZN)182,780 (64) (109). It was hypothesized that the frequency of association of liganded ER with ERE determines the degree of agonist efficacy of an ER ligand.

5.4 Assessment of Estrogenic Activity

Estrogenic potency and efficacy have traditionally been expressed in terms of uterotrophic effects in immature or OVX female rodents. Thus, daily administration of (1) to 3-week-old female rats for 3 days results in a three- to fivefold increase in uterine weight (110). This sensitive model for pharmacologic characterization of estrogens can also be used to identify and characterize estrogen antagonists, wherein a maximally uterotrophic dose of (1) is administered without and with the putative antagonist, to determine the maximal degree and potency to which uterine weight gain is prevented (110b).

Determination of ER affinity has provided a way to identify substances with potential estrogenic and antiestrogenic effects. Because ER is a soluble protein, uterine tissue homogenates from various animal sources provided a convenient source of cytosolic ER. Currently, recombinant hERα and hERβ variants are preferred for these studies. The ability of putative ER ligands to displace specifically bound [^{3}H]1 (radioreceptor assay) or fluorescein-linked (1) (fluorescence polarization assay) in these preparations is determined. Current therapeutic and noteworthy experimental estrogens and antiestrogens, and certain environmental and dietary estrogens (see

Section 5.8) have RBAs ranging from at least 0.05% to nearly 1000% (10 times) that of (1) (see Tables 13.3–13.7).

Determination of estrogen agonist or antagonist potencies and efficacies has been conveniently carried out using lines of cultured neoplastic cells naturally endowed with ER, its coregulators, and EREs. Estimation of estrogen-induced synthesis of enzymes or precursor RNAs, or determination of overall cell proliferation rate or extent can be used to assess effects. Interpretation of results is usually not complicated by biopharmaceutic factors, such as biotransformation. However, experimental controls validating clonal integrity must always be run (111). MCF-7 human breast cancer cells and Ishikawa human endometrial cancer cells are two lines that have become well established for these applications (112).

Substances exhibiting significant ER affinity can be evaluated in the OVX rat for other ER-mediated effects besides uterotrophic activity. The OVX rat has become firmly established as a model for human disorders associated with estrogen deprivation, among them elevated serum cholesterol (113) and osteopenia (bone density loss) (114). Also, progress in development of models for disorders associated with CNS estrogen deficiency, using the OVX rat, has been reported (115, 116).

5.5 Progesterone Receptors

Most of the effects of (30) are mediated by interaction with PR. Cells that contain PR usually contain ER as well, and estrogens upregulate PR in these cell types (117). There are two isoforms of PR: PRB, a 933 amino acid peptide; and PRA, which is identical to PRB except that the N-terminal 164 amino acid sequence in PRB is absent (118). Levels of PRA and PRB generally are nearly equal in most cell types that contain PR. Interaction of (30) with oligomeric PRA/PRB leads to receptor dimerization in which three dimers are produced. These are designated AA, AB, and BB. The two hormone-binding domains in each dimer are both occupied by (30). These three dimers bind to DNA at specific progesterone response elements on the promoters of progesterone-responsive genes, and regulate transcription (119). In most cells, hPRB functions

as a transcriptional activator, but hPRA is less active transcriptionally and can function as a strong ligand-dependent repressor of transcriptional activity. The ability of (30) liganded with hPRB and hPRA to recruit, in turn, coactivator and corepressor proteins might account for this difference (117).

Both hPR and hER feature about 250 amino acids in their LBDs. The crystal structure of the hPR LBD (amino acids 677–933) complexed with (30) revealed considerable similarity regarding electrostatic and hydrophobic/Van der Waals contact with that of the (1)-hER LBD complex (96, 120) (cf. Fig. 13.7 with Fig. 13.6a). Regarding electrostatic interactions, the major difference is replacement of Glu353 (H-bond acceptor) in hER with a homologous Gln725 (H-bond donor) in hPR. Otherwise, considerable homology exists between the lipophilic amino acids (Leu, Ile, Met, Phe, Val), which line the binding pockets of these receptors. Crowding associated with interaction of hPR LBD with (30) in the region of its 19-methyl group might explain in part why 19-norpregnane progestins (e.g., 94, 95) interact somewhat more strongly with this receptor than does (30) (Table 13.2). However, conceding that interactions of the lipophilic amino acid units lining the binding pocket of hPR LBD might enhance PR ligand selectivity, the structure of hPR's complex with (30) does not provide a clearcut basis for the selective affinity of hPR for (30) versus other A-ring enone steroid hormones such as testosterone and cortisol, or why introduction of a 17α-ethynyl group into testosterone results in a ligand with preferential affinity for hPR over hAR. Reasons for PR's ability to differentiate between progestins and estrogens on the basis of electrostatic interactions with respective A-ring substituents (1) and (30) are, however, evident. These are predominantly hydrogen bond-donating interactions in hPR LBD, but the hER LBD interacts with the A-ring substituents by a combination of H-bond donor and acceptor interactions.

PRs are found in organs of the reproductive system. Also, hPR mRNA has been localized in vascular smooth muscle cells and in human osteoblast-like cells (121). In human subjects, regardless of gender, PR (and ER) has been immunolocalized in normal and varicose sa-

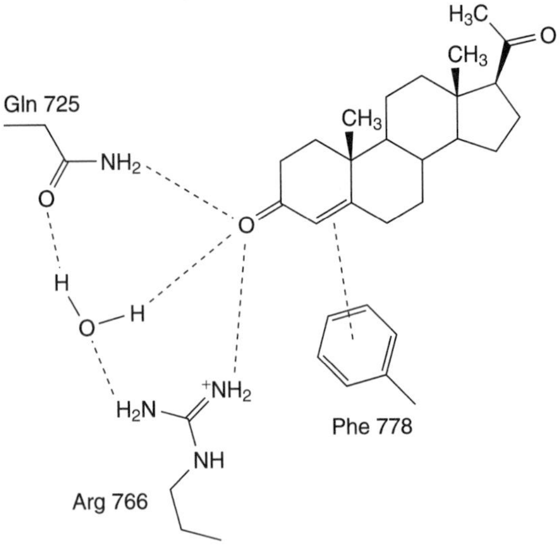

Figure 13.7. Electrostatic interactions of the A-ring of (**30**) with the LBD of the PR. The 3-keto of (**30**) is a central hydrogen-bond acceptor in a network formed by Gln725, Arg766, and a water molecule. Hydrogen bond lengths range from 2.5 to 3.1 Å. The 4,5-double bond of (**30**) interacts with the ring of Phe778 by π-π bonding.

phenous veins, and levels of both receptors seemed to be higher in the former (122).

5.6 Assessment of Activity of PR Ligands

Experimental characterization of progesterone mimetic activity was based classically on the ability of analogs of (**30**) to stimulate an endometrial response in the estrogen-primed rabbit. The reader is referred to the Fourth Edition of this series for specific details of this and related assays used to characterize progestins, and literature references to these. Currently, putative PR ligands are evaluated initially for their ability to interact with hPRA, and with human androgen, glucocorticoid, and mineralocorticoid receptors to assess the degree of apparent selectivity of the test compound for hPR, an important issue in identification of new progestins (123). Compounds with noteworthy hPR affinity and selectivity can then be subjected to cell-based assays to determine agonist and antagonist potency and efficacy. Mammalian cells transfected with a plasmid containing hPR and a reporter gene enabling induction of assayable enzyme activity have been used for this (123, 124). Alternatively, human breast cancer cell lines expressing hPR, such as T47D cells in which progesterone mimetics induce the enzyme alkaline phosphatase, have been used to quantify potency and efficacy.

In the ovariectomized rat, simultaneous administration of estrone (**2**) and high doses of medroxyprogesterone (**34**) results in a maximal (30%) reduction in uterotrophic response compared to that seen in animals receiving only (**2**) (124). This animal model can be used to identify progestins having the ability to counteract the reproductive tract estrogenicity of administered estrogens.

5.7 Progestin and Antiprogestin Effects

Steroidal progestins are analogs of (**30**) that interact with PR and activate it in ways similar to that of (**30**) itself. The primary application of these drugs is in birth control (see Section 5.10).

Progestins often have other effects distinct from those of (**30**) as a result of differential interaction of progestins or their metabolites with other steroid receptors besides PR (Table 13.2). Norethindrone (**35**) has been applied clinically as a bone-loss suppressant in postmenopausal women (125), but (**34**) is not effective in this application (125a). Because (**35**) is metabolized in patients to afford significant levels of the potent estrogen (**9**) (Section 3.9), it has been hypothesized that this metabolite accounts for the observed effect on bone loss. Partial conversion of an analog of (**35**), tibolone (**54**), to an estrogenic metabolite might similarly account for some of (**54**)'s effects. In

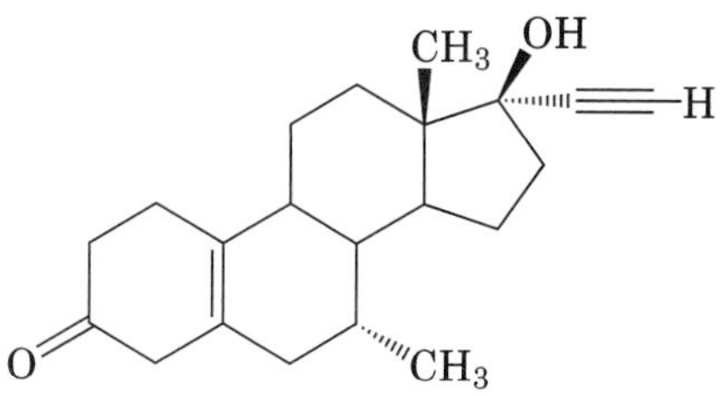

(54)

(56)

the OVX rat and in postmenopausal patients, (54) suppressed bone loss without the degree of uterine stimulation associated with estrogens (126). Whether the bone loss-suppressive effect of (54) is mediated directly by this drug or indirectly by estrogenic metabolite(s) (e.g., see Section 3.9, estrogenic metabolites of (35) and (39) is not yet known.

There are two types of progesterone *antagonists*. Interaction of Type I antagonists with PR results in impairment of PR association with DNA. Interaction of Type II antagonists with PR does not prevent PR-DNA association, but the conformation of PR associated with DNA does not facilitate transcription (127). Thus, Type II antagonists not only interfere with interaction of (30) with PR, but can prevent interaction of the (30)-PR complex with DNA (128).

Mifepristone (RU486, 55) is the best known Type II antiprogestin. Type I antipro-

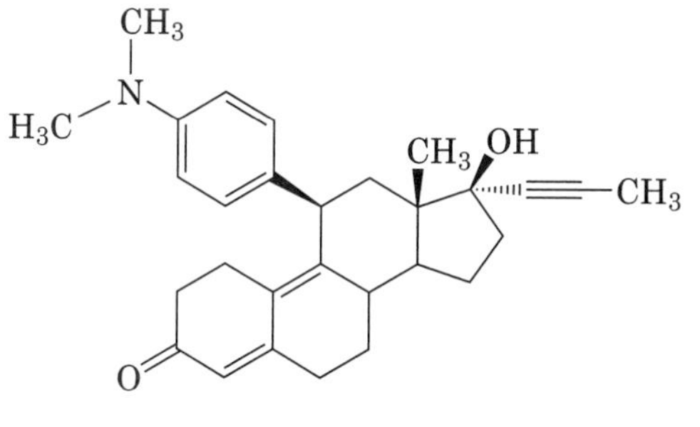

(55)

gestin onapristone (ZK98299, 56) resembles (30) with respect to the configuration and chain length of its C-17 alkyl substituent. However, its C-18 angular methyl group is in a novel α configuration. Both types contain identical 11β-aryl substituents. In the female rat, (56) was shown to downregulate PR in

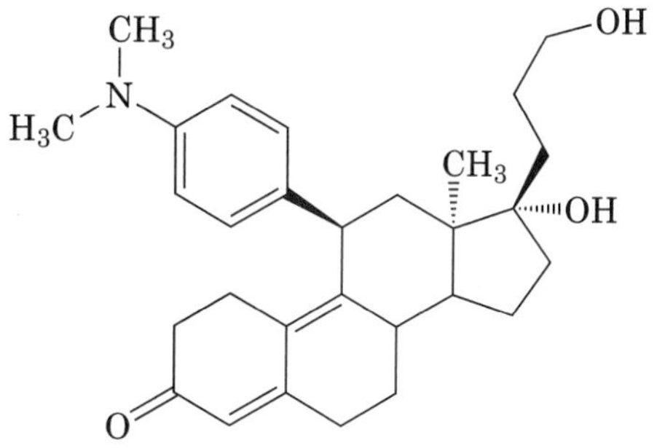

tissues of the reproductive axis, which might contribute to its anovulatory effect (129).

The most significant medical application of (55) is for its ability to counteract (30) regarding preparation of the uterus for implantation and retention of a fertilized egg. Thus, (55) can be used sparingly, rather than on a daily basis, as is the case for conventional oral contraceptives. However, because its contraceptive mechanism in this application is postconceptional, its routine medical use is ideologically intolerable in the perception of many people. Another experimental and clinical application of (55) and other antiprogestins is in suppression of breast cancers that contain PR (130). Growth-suppressive effects are believed to result from PR-mediated cytotoxic mechanisms.

5.8 Environmental and Dietary Estrogens and Progestins

Estrogenic effects can be expressed by numerous natural products, as well as by certain synthetic organic compounds introduced into the soil and water. Of particular importance are chlorinated aromatic hydrocarbons, phenolic industrial chemicals, and flavonoid phytoestrogens. This section deals with three of these, the pesticide methoxychlor (57) and flavonoids (59) and (60).

An analog of DDT, methoxychlor (57) exhibited uterotrophic effects in rodents (131) and was subject to cytochrome P450-mediated *O*-demethylation *in vitro* (132). Its bisphenolic "metabolite" (58) exhibited contrasting effects in cells transfected with alternate ER isoforms. Thus (58) was an estrogen agonist in human hepatoma cells transfected with ERα,

(57) R = CH$_3$
(58) R = H

(59) X = OH
(60) X = H

although in the same cell line transfected with ERβ, **(58)** was an antagonist of **(1)** (133). Further studies are clearly needed to clarify the profile of estrogenic and antiestrogenic effects of **(58)**.

Human diets contain trace amounts of phytoestrogens, various plant-derived compounds with weak estrogenic activity (134). Examples are the isoflavonoids genistein **(59)** and daidzein **(60)**. These substances and related ones have attracted attention because they might be capable of preventing the development of estrogen-related cancers, as well as blunting the symptoms of menopause (135). Indeed, a significant reduction in severity and frequency of hot-flush episodes was experienced by menopausal patients whose diets were supplemented with soy extract or flour, both of which contain **(59)** and **(60)** (136).

The hER RBA of **(59)** was 2% that of **(1)**, and in MCF-7 cells, **(59)** was a weak partial agonist (67% efficacy) (137, 138). In human hepatoma cells transfected with either ERα or ERβ, **(59)** and **(60)** were full agonists with **(59)** being more potent; both isoflavonoids were more potent in the ERβ-transfected cells (139). Growth stimulation by **(59)** in such es-

trogen-responsive cells was completely prevented by estrogen antagonists (138). In the OVX rat at high doses, **(59)** had weak partial uterotrophic effects, no observable effects on tibial bone growth, and serum cholesterol was reduced (137).

Reports of dietary phytoprogestins have begun to appear. In particular, there is evidence that extracts from selected herbs/spices (e.g., oregano, thyme, turmeric) contain components that interact with hPR. Furthermore, those herb products suggested to contain components with hPR affinity tended to antagonize the proliferation of hPR-containing neoplastic cells (140). The identity of these phytoprogestins will undoubtably be reported soon.

5.9 Tissue-Selective Estrogens and Progestins

In the presence of MCF-7 cells, and in the immature rat uterus, tamoxifen **(16a)**, raloxifene **(29)**, and other nonsteroidal ER ligands exhibit dominant estrogen antagonist activity. Contrasting results have been obtained regarding their effects on other tissues, in which these substances had full estrogen mimetic effects, and thus **(16a)** and **(29)** are the first examples of SERMs. In the OVX rat (Fig. 13.8), currently known SERMs are approximately as efficacious (but not as potent) as ethynyl estradiol **(9)** on bone-loss suppression and serum cholesterol lowering. However, the estrogenic efficacy in reproductive tissue of SERMs is much less than that of **(9)**, although **(9)** displays modest separation of skeletal/lipid versus uterotrophic potencies. Furthermore, **(16a)** and **(29)** antagonize the uterotrophic effect of **(9)**, but not its skeletal and cardiovascular effects.

These findings suggest an important pharmacodynamic difference expressed in SERMs relative to estrogen mimetics such as **(9)**, or for that matter with regard to respective steroidal and diarylethylene ER ligands **(64)** and **(76)** (below). Both of these "pure" estrogen antagonists antagonized the bone-protective effect of **(1)** (142). Also, **(64)** partially prevented the serum cholesterol-lowering effect of **(9)** or **(16a)** (143).

Current SERMs have approximately the same affinity for hERα and hERβ. Also, no noteworthy differences in levels of ER iso-

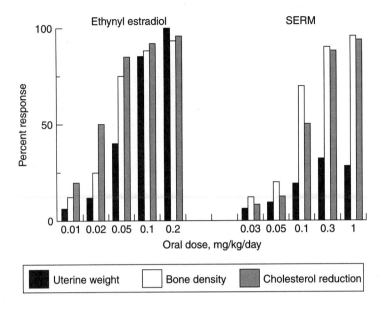

Figure 13.8. Depicted dose-response relationships for uterine weight retention, bone density maintenance, and serum cholesterol lowering in the OVX rat for ethynyl estradiol (**9**) and a typical SERM. This figure was compiled from selected data in Refs. 113 and 141.

forms in bone-remodeling cells or hepatocytes, compared to levels in uterine tissues, have been reported. These findings indicate that SERM effects are not manifested by differential interaction or "activation" of one or the other of the ER isoforms.

Rather, the molecular basis for tissue-selective estrogenic effects expressed by (**16a**), (**29**), and other SERMs arises from the distinct conformation they induce in ER. SERM-liganded ER results in a more "open" conformation in the region of the LBD than when ER is liganded with (**1**) (94, 97). Two mechanistic possibilities have been proposed to account for what happens next. First, SERM-ER complexes could exhibit impaired ability to recruit coregulator(s) in reproductive tissue (and neoplastic cells), as opposed to hepatocytes or osteoblasts. Assuming this to be correct, selective ER modulation arises from differential tissue-specific distribution of different coregulators. This notion, however, is not clearly supported at this time, although more studies of ER coregulator distribution are needed (see, e.g., Ref. 144). Thus, expression levels of mRNAs for each of six ER coregulators did not differ substantially across seven lines of ER-positive neoplastic cells (145).

The second possibility involves cell-specific divergence in functions of ERE after interaction with liganded ER. In this regard, there appear to be two distinct types of EREs. Studies in OVX-thyroidectomized (estrogen- and thyroid hormone-deficient) rats suggested that productive interaction (activation) of EREs with liganded ER in certain cells (hepatocytes, osteoblasts) exerted a modulatory effect on proximate thyroxine response elements (TREs) occupied by liganded thyroid receptor. However, activation of EREs in uterine cells by interaction with liganded ER was *independent* of (occupied) TRE (143). With a view toward the SERM profile of activity depicted in Fig. 13.8, it can be argued that SERM-ER complexes are more fully capable of productive interaction with "TRE-coupled" ERE in osteoblasts and hepatocytes than with "TRE-independent" ERE in uterine cells.

Alternatively, SERMs might interact with EREs different from those with which steroidal estrogens might interact. Studies with yeast cells transfected with cDNA resulted in identification of an ERE for (**16a**)-ER complexes that was distinct from that which interacted with (**1**)-ER (146). In addition, the (**29**)-ER complex was found to interact in a novel way with a transcription promoter (147).

A tissue-selective nonsteroidal progestin has been characterized (124). Chromenoquinoline (**61**) had an hPR affinity equivalent to that of (**30**), but low affinity for other ste-

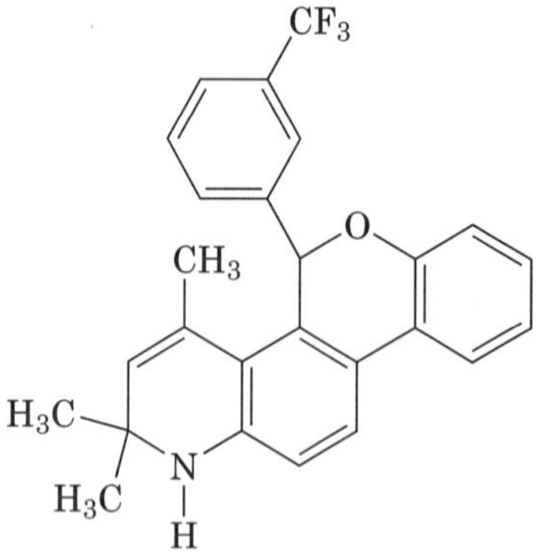

(61)

roid receptors. In hPR-transfected cells, (61) was a full agonist with potency approaching that of (30). At high doses, (61) diminished by 30% the uterotrophic effect of (2) in the OVX rat, as did medroxyprogesterone acetate (34). However, in these same estrogen-"primed" animals, (61) had a reduced stimulatory effect on mammary gland differentiation (lobular alveolar bud proliferation), compared to that of (34). These findings have interesting therapeutic implications. Estrogen-replacement therapy with estrogen/progestin combinations results in a decreased risk of endometrial cancer but an increased risk of breast cancer, compared to use of estrogen alone (see Section 2.2). Combinations of estrogen and selective progestins with reduced mammary gland stimulatory effects might result in reduction of both endometrial and breast cancer risk.

5.10 Modulation of Ovulation

Orally effective suppression of ovulation is accomplished by administration of progestin/estrogen combinations. This is the major application of the progestin-estrogen mixtures listed under progestins in Table 13.1. These combinations work primarily, but not exclusively, at the level of the hypothalamus.

Coordination of synthesis/secretion of follicle-stimulating hormone (FSH) and luteinizing hormone (LH) from the anterior pituitary originates in the hypothalamus. Releasing factors, such as gonadotrophin releasing hormone (GnRH), stimulate FSH and LH production, which facilitates ovulation. Endogenous (1) and (30) are present in the blood at high

levels subsequent to ovulation and during pregnancy. These interact with respective ER and PR in the hypothalamus to downregulate GnRH production, ultimately reducing FSH and LH levels during pregnancy (and during the time when an egg is available for fertilization), preventing further ovulation. This process of feedback inhibition is mimicked by orally administered estrogen-progestin combinations.

Thus, in randomized, controlled clinical trials, serum levels of FSH and LH were reduced in women administered oral contraceptives containing 0.025–0.05 mg of either transdermal (1) or ethynyl estradiol (9), plus relatively high doses (1–10 mg) of norethindrone (35) or nomegestrol acetate (62), over

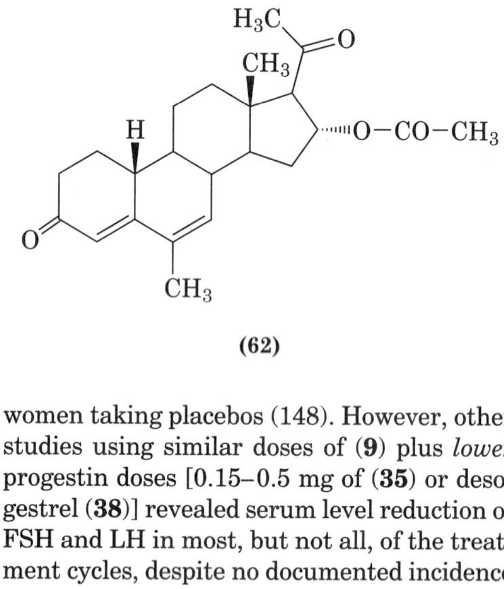

(62)

women taking placebos (148). However, other studies using similar doses of (9) plus *lower* progestin doses [0.15–0.5 mg of (35) or desogestrel (38)] revealed serum level reduction of FSH and LH in most, but not all, of the treatment cycles, despite no documented incidence of ovulation in treatment groups (149). These results suggested anticonceptive effects were attributed not just to suppression of serum gonadotropin levels, but also to direct effects on uterine function (149), such as reduction in cervical permeability and endometrial receptivity.

Besides acting on the hypothalamus and uterus, progestins, when administered with estrogen, interfere with pituitary response to GnRH (150, 151).

On the other hand, the ovulation-inducing drug clomiphene (23) elevates serum FSH and LH levels. This drug interacts with ER in the hypothalamus to antagonize the (1)-mediated feedback inhibition of GnRH secretion. In the anterior pituitary, its constituent isomers might also have direct additive effects on GnRH-stimulated FSH/LH production. In rat pituitary cells, (**23a**) and/or (**23b**) enhanced the potency of GnRH-mediated LH production (152). In ovine pituitary cells, only (**23b**) augmented the LH production stimulated by a GnRH agonist, and each isomer reversed the suppressive effect of (**1**) on FSH secretion (153). Although direct pituitary effects of (**23**) in humans might vary from those in these animal models, these *in vitro* studies imply that both isomers of (**23**) are capable of interacting with ER in pituitary as well as the hypothalamus, ultimately leading to elevation of available levels of FSH and LH (154). Thus, administration of (**23**) at the beginning of the menstrual cycle maximizes the probability that ovulation will occur in patients in whom this process is impeded by inadequate levels of these peptide hormones.

6 HISTORY

6.1 Discovery of SERMS, Including Tamoxifen (16a)

In 1962 Michael Harper and Arthur Walpole prepared the isomers of triarylethylene (**16**), and found that the *trans* isomer, now known as tamoxifen (**16a**), had antiestrogenic properties (155). In the early 1970s, clinical studies with (**16a**) showed that it could antagonize the growth of ER-positive breast cancer (156).

Estrogen promotes maintenance of bone density and normal serum cholesterol in postmenopausal women. Because (**16a**) was an estrogen antagonist in experimental breast cancer models, and in reproductive tissue, it was thought that expression of this antiestrogenicity in skeletal and hepatic tissues might result in bone density loss and serum lipid elevations. However, monitoring these parameters in patients on extended therapy with (**16a**) for breast cancer treatment/prevention revealed no adverse effects (157). In fact, (**16a**) appeared to prevent bone loss and promote cholesterol lowering, while suppressing cancer growth (157, 158). Bone loss suppression and cholesterol lowering were also seen in OVX

rats treated with (**16a**) (159, 160). Thus, (**16a**) appeared to express tissue-specific estrogenicity. The molecular basis for SERM activity is presented in Section 5.9.

These novel findings suggested a new dimension in estrogen-replacement therapy: selective estrogens useful in patients at risk for breast cancer or other estrogen-dependent cancers, and in patients intolerant to side effects associated with reproductive tract estrogenicity of nonselective estrogens.

Accordingly, in the early 1990s there emerged a renewed interest in development of nonsteroidal antiestrogens as therapeutic agents. One such class of ER ligands had been discovered in the late 1970s-early 1980s by C. David Jones and coworkers. The prototype of these, trioxifene (**63**), was noteworthy in that,

(**63**)

in the rat, it exhibited greater antiuterotrophic efficacy and reduced uterotrophic efficacy compared to that of (**16a**) (161), as did raloxifene (**29**), which was considerably more potent/effective (161b, 162, 163).

In the OVX rat, (**29**) was nearly as effective as (**1**) in maintaining bone density and preventing excessive bone turnover (160, 164). It was more effective than (**1**) in lowering serum cholesterol (159) and in patients, (**29**) appeared to have a reduced incidence of reproductive tract side effects, including endometriosis, compared to that of (**16a**). It was subsequently introduced for use in estrogen-replacement therapy in 1998. Additionally, the study of tamoxifen and raloxifene (STAR) Phase III clinical trial, begun in 1999, is ongoing to determine how (**29**) compares with (**16a**) as a breast cancer preventative in postmenopausal women, with secondary focus on

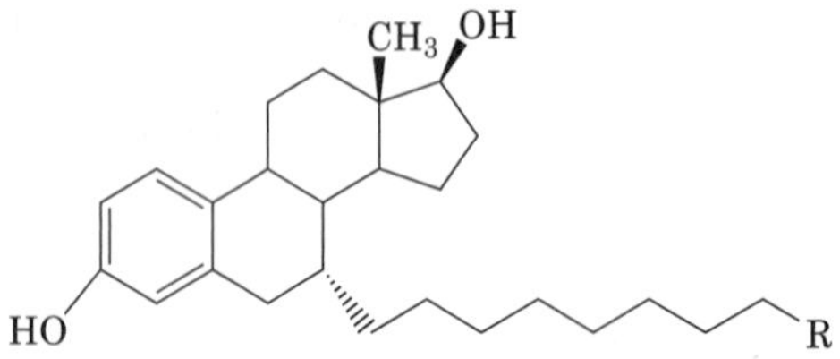

R

(64)

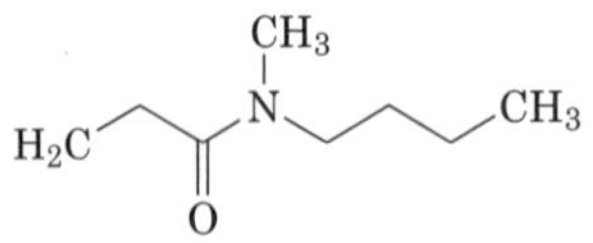

(65)

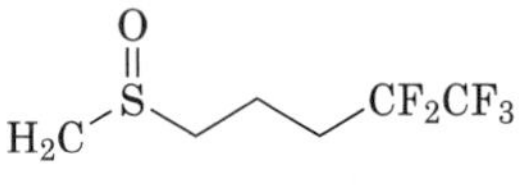

comparison of maintenance of skeletal/cardio-vascular systems and frequency and severity of side effects (165).

Furthermore, (29), unlike 4-hydroxytri-arylethylenes, cannot undergo geometric isomerization, and it does not interact with a multitude of other receptors, enzymes, and regulatory proteins besides ER as (16a) does

(156). These physical and pharmacodynamic properties have simplified interpretation of its selective estrogenic effects, and, because (29) formed a high affinity complex with ERα, crystallographic studies of such complexes (see Section 5.1) have provided a starting point for establishing the molecular basis for its tissue-dependent estrogenicity (see Section 5.9).

6.2 Other Discoveries

Isolation and identification of endogenous steroidal estrogens and progesterone, discovery of nonsteroidal estrogens such as diethylstilbestrol (24), and development of progestin/estrogen combinations as birth control agents have been covered in previous editions of this textbook (see also Ref. 155).

7 STRUCTURE-ACTIVITY RELATIONSHIPS

7.1 Steroidal Analogs of (1)

As suggested from the data in Table 13.3, the ER will accommodate sizable structural modification of (1) at specific points in its B- or C-rings. Thus, addition of 7α- or 11β-substitu-

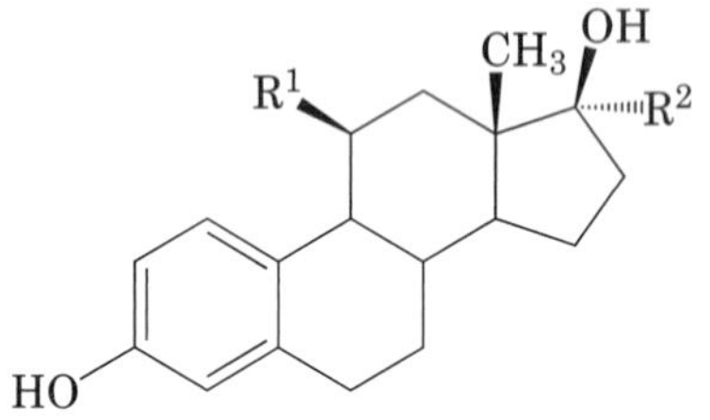

R^1 R^2

(66)   H

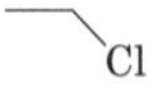

(67)   —H

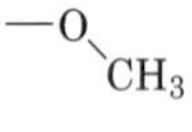

(68)   H

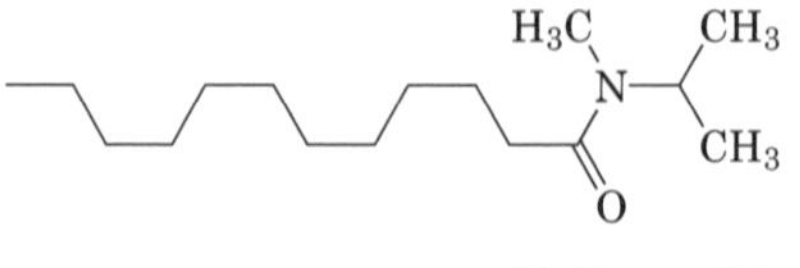

(69)   —H

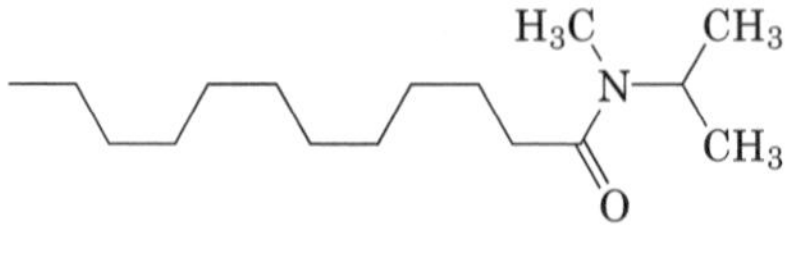

ents, embodied in turn in (**64**), (**65**), and (**66**), resulted in analogs with high ER affinity. On the other hand, removal, derivatization, or replacement of either of (**1**)'s hydroxyl groups results in a considerable loss of affinity, with some exceptions (entries 5 and 9). Furthermore, inversion of hydroxyl group configuration at C-17 of (**1**) or (**9**) resulted in large decreases in affinity (cf. **1** with 17-*epi*-**1**, **9** with 17-*epi*-**9**) (166–169). ER affinities of more comprehensive collections of steroidal ER ligands have been reported (93, 170).

ERα seems to predominate over ERβ in rat uterine tissues (101). Thus, RBAs obtained from rat uterine tissue cytosols are a reasonable approximation of relative affinity for ERα. RBAs obtained for a few steroids using hERα are included for comparison in Table 13.3.

7.1.1 7α-Substitution. Addition of linear alkyl groups capped with polar end groups at C-7 of (**1**) has resulted in analogs with no observable intrinsic estrogenicity in certain tissues and cells. Thus (**64**) (ICI 164,384) and (**65**) [ici (ZN) 182,780], although clearly capable of interaction with ER (Table 13.3), were not uterotrophic in the immature rat but antagonized completely the uterotrophic effect of (**1**) 3-benzoate ester (173). These compounds were much more effective parenterally than orally, suggesting efficient presystemic metabolic inactivation. Regarding (**64**), minor variance in the length of the linear alkyl spacer or in the amide *N*-alkyl substituents often resulted in restoration of partial estrogenic activity (178). This suggested considerable overlap in structural requirements for agonist/antagonist effects in the alkylamide region of these analogs. In MCF-7 cells, (**65**) was more potent and effective than (**16a**) (below) in suppressing growth. Furthermore, (**65**) suppressed growth of MCF-7 cells that had grown resistant to antiestrogens (179), suggesting the possible clinical application of (**65**) for control of breast cancer that had acquired resistance to conventional antiestrogens. Indeed, a preliminary study showed that monthly intramuscular administration of (**65**) resulted in a 69% response in patients with breast tumors that had become refractory to other antiestrogen therapy (180). Thus, steroidal antagonists such as (**65**) might find

widespread application in breast cancer therapy, particularly if limited oral effectiveness and side effects associated with estrogen antagonism in other tissues (142a) do not present serious drawbacks.

The molecular basis for the efficacious estrogen antagonist effects of (**64**) and (**65**) centers on the ways these compounds interact with ER. Specific electrostatic functional group and hydrophobic/dispersional bonding interactions of (**65**) with the hERα LBD (Section 5.1) are similar to those seen with (**1**) as ligand (Fig. 13.6) (crystal structure of the **65**-hERα LBD to be reported). However, (**65**)'s lengthy 15-membered aliphatic chain overlaps the region of the LBD in which TAF-2 is embedded (Fig. 13.5), thereby interfering with dimerization and interaction of (**65**)-liganded ER with coregulators (181). Although such complexes are transcriptionally ineffective at TRE-independent EREs in uterine and breast cancer cells (see section 5.9), they appear to be partially effective in activating EREs in hepatocytes and other cell types endowed with TRE-dependent EREs (143).

A parallel solid-phase synthetic approach has been used to make a library of reverse amide analogs of (**64**) with two levels of diversity (182). Thus (**1**)-17-tetrahydropyranyl ether, bound to a polymeric resin at its 3-hydroxyl group and bearing an ω-aminoundecyl group at its 7α position, was condensed with four activated *N*-protected amino acids, followed by deprotection and condensation of each of these intermediates with five activated carboxylic acids. Photolytic cleavage yielded 20 compounds with varying degrees of ER affinity and estrogen antagonist activity.

7.1.2 11β-Substitution. Addition of 11β substituents to (**1**) or ethynyl estradiol (**9**) has given rise to analogs with noteworthy changes in ER affinity and intrinsic activity. Thus, chloromethyl analog (**66**) had ER affinity almost 10 times greater than that of (**1**) (Table 13.3) (174). Interaction of (**66**) with ER was shown to be noncovalent, despite the potential reactivity of its chloromethyl group. This was attributed either to steric encumbrance of the chloromethyl group, or possibly to its lack of proximity to nucleophilic substituents in the ligand-binding site. The 11β-methoxy analog

(67), which also has a 17α-ethynyl group, interacted strongly with ER as well. Both (66) and (67) are potent estrogen agonists (172, 174). Addition of larger 11β-substituents, some of which were similar to those in (64), likewise has yielded antagonists with negligible intrinsic estrogenicity. Thus, N,N-dialkylundecylamides (68) and (69) exhibited high ER affinity (Table 13.3), but had no observable uterotrophic effects. These analogs had potent growth-suppressive effects on MCF-7 cells and on a tamoxifen-resistant cell line (175). The ethynyl analog (69) was significantly more potent than (68) when administered orally.

7.1.3 Comparison of Estrone (2) with Equilin (52c).

In vivo reversible sulfate esters of (52a–c) together are the second most abundant components in conjugated estrogens. Thus, studies have focused on their possible role in estrogen-replacement therapy (83). Affinity of (52c) for ER was similar to that of (2), and (52b) was suggested to have greater ER affinity than that of (1) (Tables 13.3 and 13.5) (81). Also, (52b) and (52c) were comparable to (1) or (2) in terms of uterotrophic activity in the rat.

However, no studies have yet appeared that reveal a clear basis for pharmacodynamic differences between these B-ring dehydrogenated ER ligands, or the other ones in Table 13.5, over (1) or (2). Thus the therapeutic advantage of conjugated equine estrogens over (2) sulfate ester alone, or for that matter, over some of the more potent semisynthetic estrogens such as ethynyl estradiol (9) in estrogen-replacement therapy is not completely clear. It could be attributed in part to reduced reproductive tract estrogenicity, although this remains to be established experimentally. Perhaps novel modulating effects of the B-ring dehydrogenated estrogen conjugates, particularly 17-*epi*-(1), (52a), and/or (53a) (183), will emerge in future endocrinologic studies.

7.1.4 Nonbenzenoid Steroid (-Like) ER Ligands.

Anordiol (70) illustrates the importance of hydroxyl group spacing in steroidal ER ligands, and provides evidence that ER will accommodate properly substituted steroid-type ligands in which ring A is aliphatic. The

(70)

ER affinity of (70) is comparable to that of other analogs of (1) (Table 13.3). Androgenic receptor affinity, characteristic of these analogs, is significant (176). Reversal of configuration at C-2, or substitution of H for ethynyl in the D-ring, in turn resulted in loss of ER affinity or greatly increased androgen receptor affinity. In the immature rat, (70) exhibited estrogenic properties divergent from those of nonsteroidal antiestrogens (177).

A metabolite of gestodene (39), specifically (40b), whose structure is similar to that of (70), also exhibited ER affinity and estrogenicity *in vitro* and *in vivo* (see Section 3.9).

7.2 Two Caveats Concerning the Meaning of ER Affinity

As is clear from the RBAs of steroidal ER ligands in Table 13.3 in relation to their estrogenic and antiestrogenic effects, and as will be shown for nonsteroidal ER ligands below, the magnitude of ER affinity does not predetermine estrogenic or antiestrogenic efficacy. Other factors, such as the size and physicochemical character of the organic moiety added to (1), determine whether the resulting ER ligand is able to alter the conformation of ER in ways favorable, or detrimental, to interaction of the resulting complex with EREs.

Furthermore, ER affinity does not always correlate well with potency. Substances with relatively low RBAs can still exhibit agonist or antagonist potencies equivalent to those of their much higher affinity counterparts. For example, in Table 13.3, 17-deoxy (1) had an RBA of 79% that of (1), but its MCF-7 cell mitogenic potency was only 1.2% of that of (1) (170). Thus, particular structural characteristics, in addition to those responsible for ER affinity, can influence potency.

Table 13.6   ER Affinities of Selected Diarylethylenes[a]

Compound	RBA[b]	Source	Reference
(**24**) (diethylstilbestrol)	100 [468]	rat [hERα]	188 [101]
(**26**) (*meso*-hexestrol)	27[c] (100) [302]	calf (rat) [hERα]	188c (188b) [101]
(**71**) (dienestrol)	20 [223]	rat [hERα]	188b [101]
(**72**)	9.5 (33)	calf (rat)	189 (188b)
(**73**) (ZK 119,010)	21	calf	190a
(**74**) (ZK 169,978)	7.0	calf	92a
(**76**) (ZM 189,154)	66	rat	142b
(**29**) (raloxifene)	143 (34)	rat (human)	191a (192)
(**77**) (LY117018)	45 (50)	lamb (rat)	193 (191b)
(**78**) (arzoxifene)	20–30	human	(192)

[a]Values were obtained using uterine cytosolic ER from the indicated sources, or using recombinant human ERα.
[b]Relative to (**1**), RBA = 100.
[c]The *dl*-isomer of (**26**) had an RBA of 1.1 (188c).

Nevertheless, from comparison of ER ligand structures with respective numeric ER RBAs there has emerged a recognizable structural pattern concerning substances whose biologic (pharmacologic) activities are mediated by interaction with ER.

7.3   Nonsteroidal ER Ligands

It has long been known that the ER accommodates a diverse array of natural and synthetic aromatic structures that do not contain the steroid nucleus. The ER affinities and estrogenic effects of some of these (**24**), (**26**), (**59**), (**60**) have been summarized above (Sections 3.6 and 5.8). In addition, medicinal chemists have characterized a sizable and structurally diverse collection of therapeutic and experimental nonsteroidal estrogen agonists and antagonists. These substances have generally been shown to interact at the same point in the ER LBD as does (**1**), although the affinity, specific functional group interactions, and consequences of such interactions can differ.

7.3.1   Vicinal Diarylethylenes. From extensive synthetic and endocrinologic studies focused on substituted stilbene derivatives (184), *trans*-diethylstilbestrol (**24**) and (Z,Z)-dienestrol (**71**) emerged as potent, orally effective estrogens. Because of facile geometric isomerization (185), (**24**) is accompanied in solution by significant amounts of its *cis* isomer (186). Cytosolic ER interacts exclusively with (**24**) in such mixtures (186b). However, the

(**71**)

relative *in vivo* estrogenicities of (**24**) and its geometric isomer have been difficult to assess because of this isomerization.

Estrogenicity of structural analogs incapable, or less capable, of geometric isomerization have suggested that estrogenicity of (**24**) is far greater than that of its *cis* isomer. Indeed, estrogenic potency in the immature mouse of (**24**) dipropionate ester was 600 times that of its *cis* counterpart (185). (Both of these esters presumably undergo facile hydrolysis *in vivo*.) Also, *meso*-hexestrol (**26**), which assumes an extended conformation resembling (**24**), was 50 times more potent than *dl*-hexestrol (187), which adopts conformations in which the rings are in closer proximity to one another. The ER affinities (Table 13.6) and/or estrogenicities of (**24**) and (**26**) have been suggested to arise from their stereochemical similarity to (**1**), with particular reference to the spacing of hydroxyl groups (184b).

Both phenolic hydroxyls of vicinal diarylethylenes are important for high potency. The monomethyl ether of (**24**) retains only 4% of the estrogenic potency of (**24**) (184b). Analogously, relative ER affinities of various substituted monoalkyl ethers of (26) were less than 3% that of (26) (171b). These findings are consistent with the structure of the (24)-hERα LBD complex, which features electrostatic interactions of (24)'s hydroxyls with Glu353/Arg394 and His524 of the LBD, respectively; identical interactions were observed in the complex involving (1) (see Fig. 13.6a, above).

Another feature of (24) important to its ER affinity and estrogenic potency is the conformation of its stilbene moiety. Crystallographic studies indicated that its rings are parallel but not coplanar, each ring being twisted 60° out of the plane of the ethylenic moiety (194). "Replacement" of the ethyl groups of (24) with hydrogens, relieving the steric congestion around >C=C< thus facilitating a more nearly planar orientation of the stilbene moiety, had previously been found to result in a >99% loss of estrogenic potency (187).

7.3.2 Diarylethylene Estrogen Antagonists.

Structural alteration of the diarylethylene (stilbene) pharmacophore to give substances retaining high ER affinity but lacking in intrinsic activity has been the focus of many investigations (195). Systematic structure-activity studies have been conducted on derivatives of 2-phenylindole. Initially, (72) was found to have significant ER affinity (Table 13.6), and was a partial estrogen agonist in MCF-7 cells (196) and in the immature female mouse as its diacetate ester (189). Replacement of the N-ethyl substituent with linear alkyl chains capped by polar amino or amido groups afforded analogs with low residual estrogenicity. Thus, ω-(N-pyrrolidinyl)-n-hexyl analog (73) had ER affinity 21% that of (1), and antagonized the uterotrophic effect of (2) in the immature mouse and rat with greater efficacy than did (72) (190). In addition, the corresponding undecylamide analog (74) had ER affinity 7% that of (1); exhibited no uterotrophic effect in the mouse; and suppressed, with modest potency compared to that of (64) and (65), the growth of MCF-7 cells (92a).

Introduction of a p-[2-(N-perhydroazepinyl)ethoxy]benzyl moiety as the indole ring N-substituent gave (75) (TSE-424). This had ER affinity comparable to that of (73) and (74). In the OVX rat, (75) exhibited very low uterotrophic activity, and had serum cholesterol lowering and bone loss-suppressive effects at 0.1 mg/kg doses (197). Note that, although (75) contains three phenyl rings, it retains the trans-p,p'-dihydroxystilbene backbone that originated in (24). Thus, it is likely that, like (24), (75) interacts with the ER LBD by specific hydrogen bonding of the indole and phe-

(76)

nolic hydroxyls in turn with Glu353/Arg394 and His524 (Fig. 13.6), and furthermore, that the (protonated) perhydroazepine ring nitrogen interacts with ionized Asp351.

Another example of a purer antiestrogen derived from the 1,2-diarylethylene pharmacophore is (76). This ER ligand, with RBA of 66 (Table 13.6), incorporates structural features previously found to give high ER affinity, and low residual estrogenicity [cf. with hexestrol (26) and (65)]. In the immature rat, (76) completely blocked the uterotrophic effect of either (1) or tamoxifen (16a), and displayed no intrinsic agonist activity of its own (142b). In the OVX rat it had estrogen antagonist effects on bone maintenance as well, unlike (16a).

A therapeutically important class of heterocyclic ER ligands in which the *trans-p,p'*-dihydroxystilbene system is also embedded is the benzothiophenes. This group includes raloxifene (29), LY117018 (77), and arzoxifene (78). These were found to have ER affinities approaching that of (1) (Table 13.6) (191,

192), and (29) was shown to interact with the hERα LBD by hydrogen-bonding interactions of each of its phenolic hydroxyls, as well as ionic interaction of its protonated ring nitrogen (94) (Section 5.1). Members of this group were powerful suppressants of estrogen-stimulated uterine growth in rodents (162), and of MCF-7 cell proliferation (198). Molecular modeling studies on (77) suggested it to interact with ER in a way that oriented the *p*-(substituted)benzoyl group in a 7α-like position relative to the interaction of (1) with ER (193). Description of (29)'s experimental and clinical tissue-selective estrogenic effects can be found above (see sections 2, 5.9, and 6). However, the oral potency of (29) appears to be limited by its susceptibility to glucuronide conjugation at the hydroxyl group on its pendant phenyl ring (Section 3.8). To circumvent this limitation, (78), in which the structure of (29) was altered by (1) isosteric replacement of the carbonyl group with an ether oxygen and (2) O-methylation of the 4'-hydroxyl group, was characterized experimentally (192b). Although (78)'s differential estrogen efficacy was similar to that of (29) in the OVX rat, it was 30–100 times more potent than (29) when given orally (62). This potency increase could be attributed to the inability of (78) to undergo metabolic 4'-O-glucuronidation, to which (29) is susceptible. The O-methyl group of (78) is resistant to oxidative biotransformation, unlike those of (10), (12) (Sections 3.1 and 3.5), and nafoxidine (88, below), and its 4'-oxygen presumably can still function as an H-bond acceptor, facilitating interaction of (78) with ER.

7.3.3 Triarylethylenes. This group includes chlorotrianisene (12), tamoxifen (16a), clo-

	n	X	R
(29)	1	C=O	H
(77)	0	C=O	H
(78)	1	O	CH$_3$

Table 13.7 **Influence of *p*-Hydroxyl or Side-Chain Substitution in Triarylethylenes on ER Affinity, and Estrogenic and Antiestrogenic Potency and Efficacy in MCF-7 Cells**[a]

Compound	RBA[b]	Estrogenic		Antiestrogenic	
		EC_{50} (nM)	Efficacy	IC_{50} (nM)	Efficacy
(**79**)	<0.1	28	68[c]	—	1[d]
(**80**)	38.5	0.19	94	—	13
(**81**)	55	0.06	62	3.5	31
(**1**)	100	0.02	100	—	0
(**16a**)	1[e]	0.11	21	219	90
(**18**)	123	0.003	10	2.5	100
(**16b**)	0.1[e]	11	100	—	0
(**82**)[f]	1.5	5.3	79	50[g]	22
(**83**)[f]	2.5	—	—	20	82

[a]Data are from Refs. 112b and 199a except as indicated.
[b]Determined using rat uterine cytosol.
[c]Percentage of the maximal proliferative effect of (**1**) in MCF-7 cells.
[d]Percentage by which the MCF-7 cell proliferative effect of 0.1 nM (**1**) was maximally antagonized.
[e]Data from Ref. 199b. Calf uterine cytosol ER RBAs of (**16a**) and (**16b**) were in turn 13 and 0.17 (112b).
[f]Data for this compound are from Ref. 200.
[g]Micromolar concentration.

miphene (**23**), and its positional isomer toremifene (**22**). It also includes fused-ring analogs centchroman (**86**), nafoxidine (**88**), and lasofoxifene (**89**). Because of its structural similarity to (**86**) and (**89**), (**87**) is also included here. However, (**87a**) contains the *p,p'*-dihydroxy-*trans*-stilbene moiety, and thus structurally and pharmacologically is classified as a vicinal diarylethylene. The first four of these have been described in terms of therapeutic applications (Section 2) and biotransformation (Section 3), with a view toward structure-activity relationships. Further clarification of the SAR of triarylethylenes is facilitated by consideration of the ER affinities and *in vitro* agonist/antagonist properties of triphenylacrylonitriles (**79–81**) compared with those of tamoxifen (**16a**) and (**1**) (112b, 199). As shown in Table 13.7, introduction of a single *p*-hydroxyl group to the weak ER ligand (**79**), afforded (**80**). In (**80**), there is a *trans* relationship between phenolic and phenyl rings, and the phenolic ring is geminal to the other unsubstituted phenyl ring. Its ER affinity and agonist potency are greatly increased over that of (**79**). Geminal bisphenol (**81**) exhibited a further 50% increase in affinity, and had significant estrogen antagonist activity in the nanomolar concentration range. Thus, the second *p*-hydroxyl conferred a shift toward antiestrogenic activity. Together, these results

	R¹	R²
(**79**)	H	H
(**80**)	OH	H
(**81**)	OH	OH

indicate a significant role for both of the *p*-oxygen substituents on the geminal phenyl rings in binding to ER and modulating the conformation of ER, resulting in antagonist activity in MCF-7 cells.

"Etherification" of (**81**) with an *N,N*-dimethylaminoethyl group gives 4-hydroxytamoxifen (**18**). In analogy with the respective precursor phenol, the second oxygen substituent (hydroxyl) increased estrogen antagonist potency and efficacy, although addition of the basic ether side-chain also contributed to (**18**)'s dominant estrogen antagonism (112b, 199). Removal of the second oxygen substituent, giving tamoxifen (16a), resulted in maintenance of antagonist efficacy, but a reduction in potency resulting from decreased ER affinity; its isomer (16b) had no antagonist efficacy at all. Thus, both the second oxygen substituent and an appropriately positioned basic side chain conferred maximal estrogen antagonism in MCF-7 cells.

Parallel effects with the isomers of (16) were seen in the immature rat (110b). Thus, (16a) had a modest uterotrophic effect and antagonized that of (1), whereas its *cis* isomer (16b) had uterotrophic efficacy nearly equal to that of (1), although it was much less potent. Similar studies with the isomers of clomiphene (23) likewise indicated the *trans* isomer (23a) to have greater antiuterotrophic efficacy than that of its *cis* counterpart (23b). However, recent experimental findings extend an earlier contention (201) (see also Section 5.10) that both isomers participate in the observed endocrine effects of (23) (202).

Replacement of the basic side chain of (16a)/(18) with acidic side chains has resulted in ER ligands that have provided further definition of the structure-activity relationship in triarylethylene estrogens/antiestrogens. Thus, oxyalkanoic acids (82) and (83) exhib-

MCF-7 cells (200) (Table 13.7). In (84) (GW5638) and (85) (GW7604) the role of a *p*-(ether oxygen) in furnishing electron density on one of the geminal rings is provided instead by the acrylate C=C. ER RBAs of (84) and (85) were, in turn, 13% and 41% that of (1) (203). Mechanistic studies in estrogen-responsive cells and with hERα suggested that (85) had less residual partial agonist efficacy than that of (18) (203), and that it alters the conformation of the receptor differently than do estrogen antagonists (16a), (29), or (65) (204). Acrylic acid derivative (84) was a potent antagonist of estrogen-supported proliferation of Ishikawa cells (92b). In the rat, both (83) and (84) exhibited endocrine profiles suggestive of SERMs. Each was a weak partial agonist of uterine weight maintenance or antagonized 17β-estradiol stimulated uterine weight gain. Moreover, in OVX rats, each compound was as effective as (1) in suppressing bone loss and serum cholesterol (92b, 203, 205).

In female rodents, (16a) and many of its analogs are effective postcoital contraceptives (206). One such analog, centchroman (levormeloxifene, 86), is in clinical use for this

	R	X
(82)	OCH$_2$COOH	OH
(83)	OCH$_2$CH$_2$CH$_2$COOH	OH
(84)	CH=CH—COOH	H
(85)	CH=CH—COOH	OH

(86)

ited modest ER affinity, and estrogen mimicking or antagonist efficacy that pivoted on side-chain length: oxyacetic acid (82) exhibited dominant agonist efficacy and oxybutyric acid (83) had dominant antagonist efficacy in

application (206b). In the rat, (86) had an ER RBA 16% that of (1), three times greater than that of the mixture of corresponding *d*- and *l*-isomers (207). It also exhibited potent antiestrogenic and weak estrogenic effects. Whether the antifertility effect of (86) and related compounds is attributed to residual estrogenic or estrogen antagonizing properties is not clear.

Alteration of chroman ring substituents in analogs of (**86**) has resulted in ER ligands with novel antiestrogenic properties. Thus, chromene (**87a**, EM-652, SCH 57068) had an

(**87a**) R = H; **b**, R = CO——C(CH$_3$)$_3$

RBA for hER of 291. Its bis-(pivalate) ester (**87b**, EM-800, SCH 57050) was an antagonist with 100% efficacy (0% agonist effect) in T47D cells, with an IC$_{50}$ value of 0.14 nM. Similar results were obtained with (**87a**). In the OVX mouse, (**87b**) antagonized the uterine weight-retaining effect of (**2**) with 86% efficacy, and had an agonist efficacy 11% of (**2**), whereas (**16a**), which is generally considered to be an estrogen in mouse uterus, had respective antagonist and agonist efficacies of 16 and 76%. The (*S*)-(+)-isomers of (**87**) (shown) were much more potent than the corresponding (*R*)-(−)-isomers in the above bioassays (208). Furthermore, (**87a**) was able to antagonize the partial estrogen agonist effect of (**18**) or (**29**) in Ishikawa cells, and had no observable agonist efficacy of its own (209). These properties were attributed to the ability of (**87a**) to interfere not just with TAF-2 of ERα and ERβ (Fig. 13.5) as do (**16a**), (**29**), and other triaryl-ethylene ER ligands, but also with the functioning of TAF-1 in each of the ER isoforms (210). However, in an ER-negative breast cancer cell line transfected with point-mutated ERα (Asp351 → Tyr351), (**87**) exhibited an estrogen-like effect on up-regulation of a specific mRNA, an effect not seen with steroidal estrogen antagonist (**65**) (211). This finding underscores a mechanistic difference in the estrogen antagonist properties of these structurally divergent ER ligands. Finally, in OVX rats, the maximal observed degree of bone mineral density and trabecular bone volume

maintenance seen with either (**87a**) or (**29**) (1 mg/kg dose levels) was 63% that seen in intact controls (212). Moreover, (**87a**) (0.25 mg/kg) maximally lowered (50%) serum cholesterol vs. OVX controls. It is not clear from these studies what advantage (**87b**) might have over its unesterified counterpart (**87a**): these two substances appear to be equally potent and effective in estrogen-responsive cells and in laboratory animals.

The above chromene ring containing estrogen antagonist is an oxygen isostere of dihydronaphthalene-based ER ligands, exemplified by nafoxidine (**88**), which was originally

(**88**)

prepared and evaluated for its antifertility effects in the rat (213). In the immature rat, (**88**) had uterine estrogenic and antiestrogenic effects similar to those of (**16a**) (214). Recent refinement of this structural type, by catalytic hydrogenation of (**88**), followed by O-demethylation and resolution of enantiomers, gave lasofoxifene (**89**) (215). The ER RBA of (**89**) was 3.1%, 25 times greater than that of its enantiomer, and one-half that of (**88**), in rat uterine cytosol (215) [the hERα RBA of (**89**) was 320% (141c)].

Evidently because of its decreased molecular surface/volume ratio, (**89**) was less susceptible to metabolic glucuronidation than other, more planar phenolic ER ligands such as its *trans*-diastereomer or raloxifene (**29**). Additional support for this idea comes from biotransformation studies with (**86**), in which the vicinal aryl rings are *trans* to each other, allowing each to assume an equatorial orientation in the chromane ring, and thus endowing

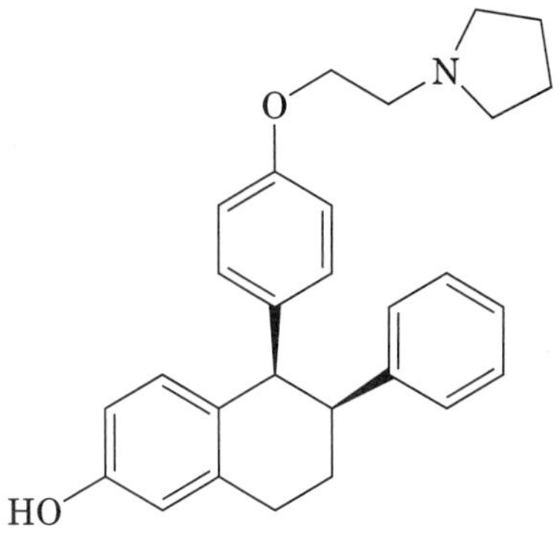

(89)

(90)

(86) with a high degree of planarity. In the rat, (86) was subject to facile biliary elimination solely as its *O*-desmethyl-*O*-glucuronide and as its *N*-glucuronide (216).

At any rate, due ostensibly to (89)'s decreased capability for metabolic conjugation, it exhibited sixfold higher bioavailability than that of its *trans*-diastereomer or (29). This was implied to be the basis for its potent oral activity in the OVX rat, in which ED_{50} values for bone loss suppression and serum cholesterol reduction were, in turn, <1 and 10 μg kg^{-1} day^{-1} (215).

7.3.3.1 Conformational Analysis. Like the aryl rings of the stilbene moiety in (24), each aryl ring of (16a) is twisted 60° out of the plane of the ethylenic bond, based on X-ray crystallographic data and conformational energy calculations (217). Similar results were obtained with (23a) (218). Results of proton nuclear magnetic resonance studies demonstrated a consequence of these conformational preferences in solution. In (16a), the *p*-substituted ring is in the shielding cone path of the vicinal unsubstituted ring, an interaction not occurring in (16b). Thus, aliphatic and aromatic protons of the *p*-substituted ring in (16a) are 0.1–0.2 ppm upfield in its spectrum from corresponding ones in the spectrum of (16b) (206a). Similar results were seen in triarylbutane analogs with respective conformations approximating the configurations of (16a) and (16b) (219).

7.3.3.2 Affinity Labels for ER. Early studies with ER had shown that the region in its LBD with which (16a) and (23a) interact approximated that of (1) because these triarylethyl-enes competitively inhibited interaction of ER with (1) (220). Aziridine (90), which exhibited ER affinity comparable to that of (16a), inter-acted covalently with Cys530 of ER (221). In the LBD, this residue is in close proximity to Asp351, the residue with which the (protonated) aziridine ring of (90) presumably makes initial contact (see Fig. 13.6b) before nucleophilic attack by the Cys530 sulfhydryl. The resulting covalent complex expressed estrogen antagonist effects similar to those of the noncovalent one of ER and (16a) (222). An analog of (26) bearing an aziridine ring linked to one end of its hex-3-ene backbone, and expressing full estrogen agonist efficacy, also interacted covalently with cysteine 530 (221). These findings provided compelling evidence, before availability of crystal structures (Section 5.1), that considerable similarity exists in the region of ER with which agonists and antagonists interact.

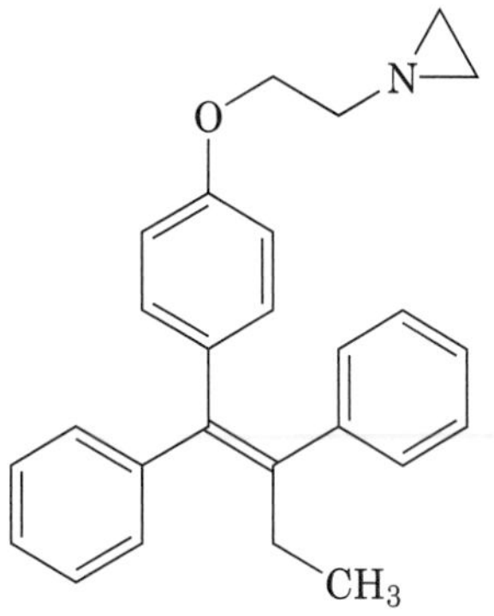

7.3.4 Triarylheterocycles. 1,3,5-Triarylpyr-azoles such as (91), and 2,3,5-triarylfurans

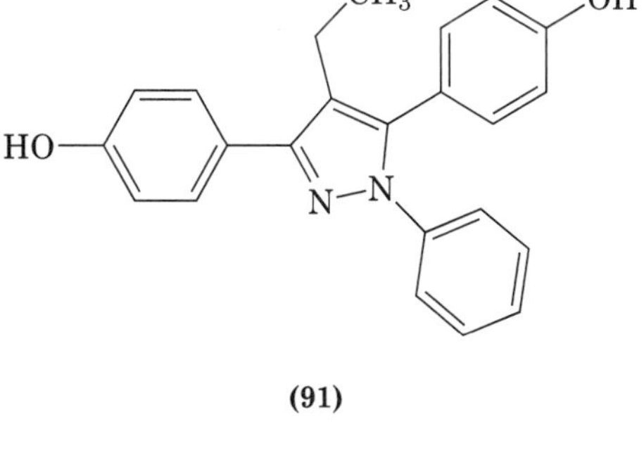

(91)

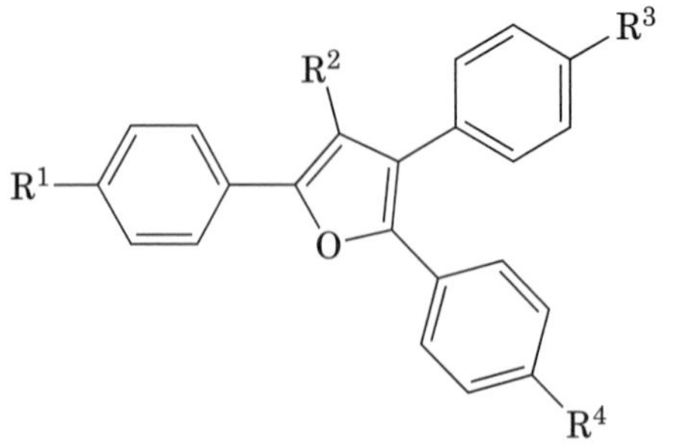

(92)

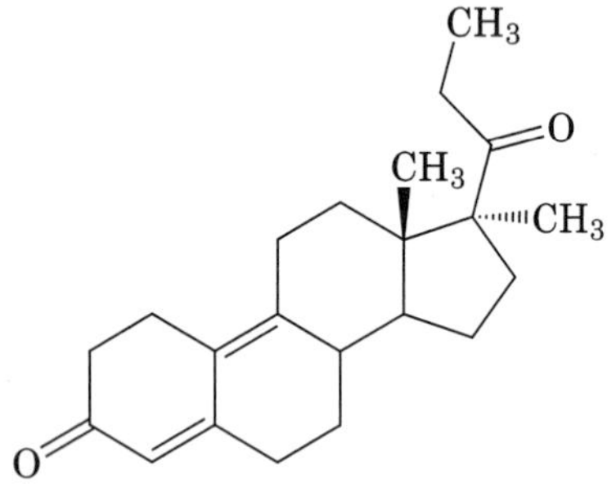

(94)

(92) with substituted aryl rings [e.g., R_1 and R_4 = OH and R_3 = p-(2-N-piperidinylethoxy) group (R_2 = ethyl)] have been in some cases found to have high isoform-selective ER affinity (223). Diversity of aryl substituents in (92) can be maximized using a combinatorial approach enabled by the facile condensation reactions used to assemble the furan ring system.

7.4 Progesterone and Steroidal PR Ligands

Chemically, steroidal progestins in experimental and clinical use are categorized as pregnanes, 19-norpregnanes, and estranes. Steroid receptor affinities of progesterone (30) and other PR ligands are shown in Table 13.2. The pregnane derivative megesterol acetate (93) has affinity for PR about equal to that of

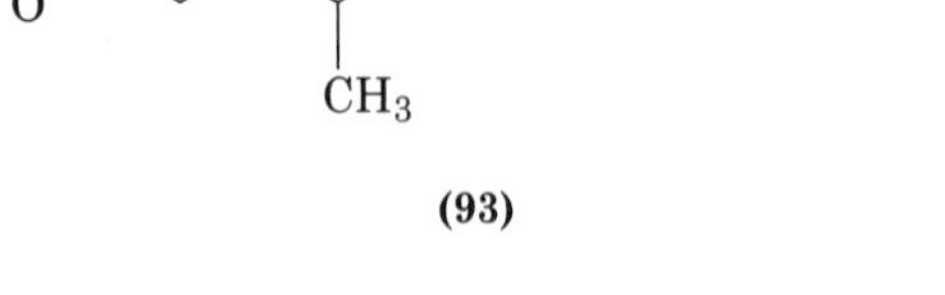

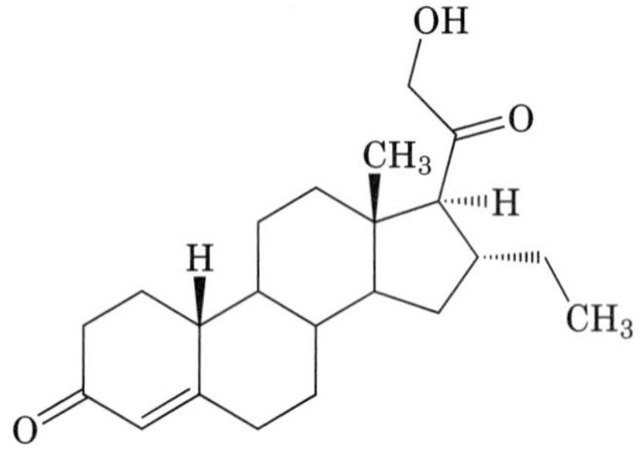

(95)

17α-ethynyl group in these progestins conferred potent oral activity. Other examples in this category are norethynodrel (96), ethynodiol diacetate (97), and l-norgestrel (36). As indicated in Table 13.2, progestins are recog-

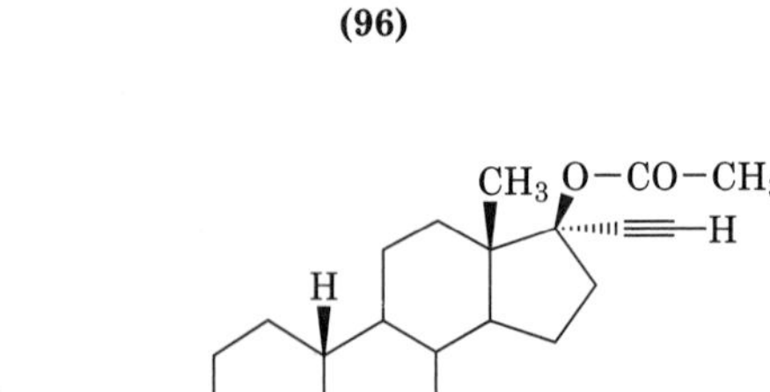

(93)

(96)

(97)

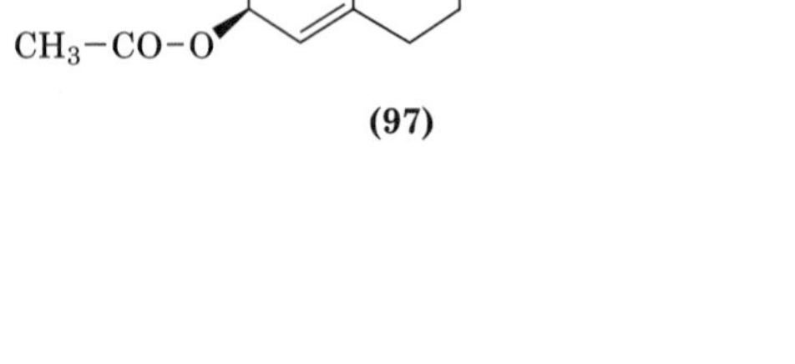

(30), but the 19-norpregnanes promegestone (R-5020, 94) and ORG-2058 (95) had much higher PR affinity. Regarding estrane-derived progestins, the earliest member of this category is norethindrone (norethisterone, 35). As with steroidal estrogens, introduction of the

nized by other steroid receptors besides PR, but generally not by ER.

Unlike the other steroidal progestins, (**36**) was prepared synthetically from nonsteroidal precursors (228). This enabled introduction of a novel ethyl group at the C/D ring juncture. A racemic mixture of (**36**) and its *d*-enantiomer has been widely used clinically (Table 13.1). However, the *l*-enantiomer (**36**), now established as the active component of this mixture, is also in use.

The potential for androgenic side effects of (**36**) and other estrane progestins is suggested by the increased AR affinity of these relative to (**30**) and certain other pregnane progestins (**94** and **95**). The PR/AR selectivity of (**36**) has been improved by derivatization. Thus, AR affinities of norgestimate (**98a**), the 3-oxime 17-

(**98a**) HO—N CO—CH$_3$
(**98b**) HO—N OH

acetate ester of (**36**), and its active metabolite (**98b**) (229) are greatly reduced over that of (**36**), although high PR affinity is retained (Table 13.2) (227).

Antiprogestin (**55**) had a PR RBA equal to that of (**94**), and at least four times greater than that of (**30**) (Table 13.2) (230). Its high affinity for GR accounts for its observed antiglucocorticoid effects. It also exhibited AR affinity 10–25% that of testosterone, suggesting some of its effects could be mediated by these receptors.

Two 19-norpregnane analogs of (**55**), (**99**), and (**100**), each exhibited PR affinities 200% that of (**55**), but only one-half the GC affinity of (**55**). In a line of GC-receptor transfected cells, the glucocorticoid antagonist potency of (**99**) (and **100**) was several orders of magnitude lower than that of (**55**). This observation

(**99**) R = H; (**100**) R = C$_2$H$_5$

suggested that, as has been observed with ER ligands (section 7.2), the degree of affinity for, in this case GC receptors, does not always predict (antagonist) potency. In T47D cells, (**99**) was a full antagonist of (**30**)-induced alkaline phosphatase, with potency and efficacy equivalent to those of (**55**); (**100**) was a partial agonist (20% efficacy) (128). Other analogs in this series not endowed with 11β-(*p*-dimethylamino)phenyl moieties were full progesterone mimetics in T47D cells.

7.5 Nonsteroidal PR Ligands

Cross-affinity of steroidal PR ligands for other steroid receptors has stimulated efforts to identify nonsteroidal ligands endowed with greater selectivity for PR. Such efforts have also been stimulated by the finding that different PR ligands can interact with different regions of, and in different ways with, the hPR LBD than does (**30**), resulting in selective PR modulation (231). An example of a nonsteroidal PR ligand (**61**) with tissue-selective progestin agonist activity was described in Section 5.9.

Nonsteroidal progestin mimetics and antagonists have been identified from natural sources. Thus (*R,S*)-cyclocymopol ether (**101**)

(**101**)

was isolated from marine algae: its (R)-isomer was a progesterone antagonist of modest potency, but its (S)-isomer was a progesterone agonist: EC_{50} = 35 nM, efficacy 84% of (30) (225). These stereoisomers had respective hPR RBA values of 0.7 and 1% that of (30). Replacement of the trisubstituted aromatic ring in (101) with a p-nitrophenyl moiety gave (102), which had an hPR affinity of 6.5% of

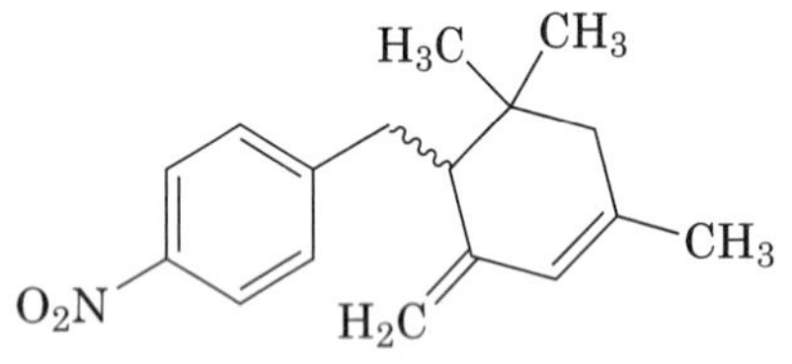

(102)

(30), and a greatly reduced hGR and hAR affinity compared to those of (30). In T47D cells, (102) was an antagonist with an IC_{50} value of 186 nM (80% efficacy). Effects of the constituent stereoisomers of (102) were not reported.

Synthesis of nonsteroidal PR ligand (103)

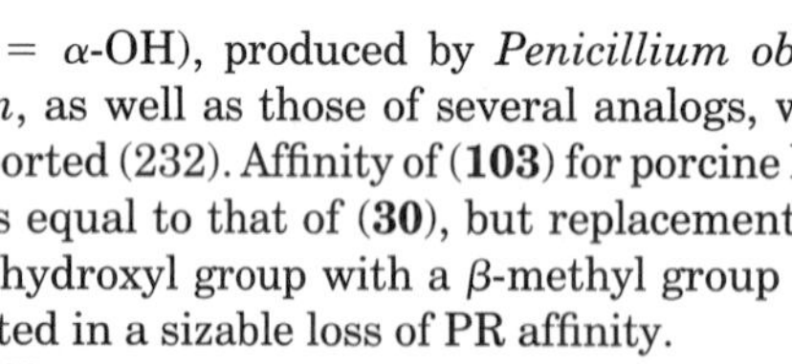

(103)

(R = α-OH), produced by *Penicillium oblatum*, as well as those of several analogs, was reported (232). Affinity of (103) for porcine PR was equal to that of (30), but replacement of its hydroxyl group with a β-methyl group resulted in a sizable loss of PR affinity.

Chromenoquinolone (104) is a member of a class of nonsteroidal progesterone mimetics that includes (61) (Section 5.9), which can exhibit high PR selectivity. Specifically, (104) exhibited hPR/hAR and hPR/hGC ratios, in turn, of 244 and 139, and its hPR affinity relative to that of (34) was 7% (123). So, (104)'s PR affinity was lower but its PR selectivities were greater than those of (34) (compare (34)'s PR, GC, and AR RBAs in Table 13.2). Like (34) (and 30), (104) induced alkaline

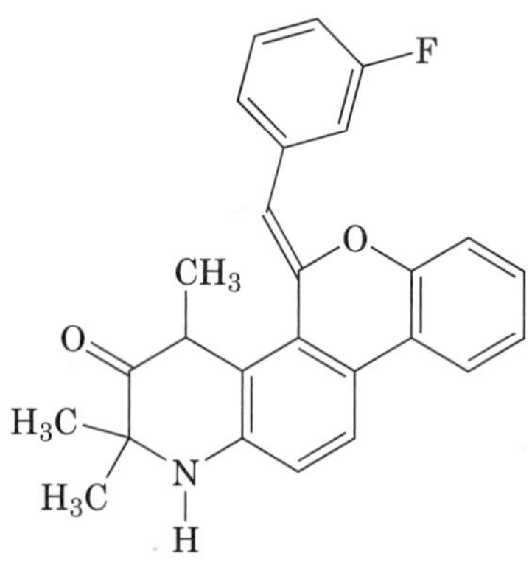

(104)

phosphatase in T47D cells (90% efficacy), and its EC_{50} (18 nM) indicated its potency was 2% that of (34). COMFA analysis of (104) and its analogs indicated that their "quinolone" rings might be analogous to the D-ring in (30) (124). If so, this would seem to indicate that (104)'s m-fluorostyryl moiety, and the m-trifluoromethylphenyl moiety of (61), complete the H-bonding networks needed for interaction of these ligands with hPR (Fig. 13.7) in a manner like the enone A-ring of (30).

Further molecular studies with PR ligands such as (102–104), aimed at elucidating the anatomy (specific complementary functional group interactions, overall conformation) of their PR complexes, could reveal novel differences in the ways these ligands reshape PR compared to when PR is liganded with (30).

The diversity of nonsteroidal structures exhibiting PR affinity clearly indicates that PR, like ER, is capable of accommodating aromatic polycyclic substances, with subtle structural differences governing the balance between mimetic and antagonist efficacy.

8 RECENT DEVELOPMENTS AND THINGS TO COME

SERMs arzoxifene (78), lasofoxifene (89), and SCH 57050 (87b) are in clinical trials, as are steroidal estrogen antagonist ici (ZN) 182,780 (65) and steroidal progestin tibolone (54). Several nonsteroidal aromatase inhibitors, including fadrozole (46) and vorozole (49), are in various stages of clinical development. Some of these might be available as new therapeutic agents at a time coincident with the publication of this chapter.

Because of their diminished reproductive tract hormonal effects, SERMs such as (**89**) may have therapeutic applications in men. Osteoporosis is not a gender-specific disorder. A significant proportion of elderly men exhibit progressive bone density loss (233), which evidently is a consequence of diminished availability of (**1**) (234). In aged, intact male rats, daily administration of (**89**) prevented deterioration of bone histomorphometric parameters and loss of bone density seen in controls (235). A comparative reduction of serum cholesterol was also seen, but (**89**) had no effect on prostate weight. Clearly, a more complete pharmacologic profiling of (**89**)'s effects in this animal model is warranted.

A need that is especially relevant to molecular endocrinological studies would be availability of more ER ligands that are uniformly selective for ERα or ERβ binding, and whose effects are in turn expressed exclusively through one or the other of these ER isoforms. As summarized in Section 7.3.4, progress has been made in achieving this goal, and it is anticipated that a variety of such steroidal and nonsteroidal ER isoform-selective ligands will become known in the next few years. Also, it appears likely that a greater array of potent and efficacious selective PR modulators will result from systematic structural modification, particularly with reference to nonsteroidal PR ligands.

Reports of crystallographic and other molecular studies of more fully expressed ER and PR complexed with an increased diversity of respective ligands will ultimately result in a more precise understanding of how these receptors function in normal and pathophysiologic states.

9 RETRIEVAL OF RELATED INFORMATION

To obtain information summarizing synthetic routes for preparation of female sex hormone analogs now in therapeutic use, the Merck Index is an excellent starting point. Developmental organic chemistry of drug candidates and other substances covered in this chapter can often be found in cited literature references proximate to the point in the text at which these are described, in particular with regard to structure-activity relationships (Section 7). The USP Dictionary of USAN and International Drug Names is a premier source of generic names (including pronunciations), structures, and CAS registry numbers of marketed drugs, and those in clinical trials.

10 ABBREVIATIONS

ER   estrogen receptor (note that human ER in tables of ER ligand affinities usually refers to ER from lysates of ER-positive neoplastic cells)
ERE   estrogen response elements in DNA
hER   human recombinant estrogen receptor
hPR   human recombinant progesterone receptor
LBD   ligand-binding domain
OVX   ovariectomized
PR   progesterone receptor
RBA   relative binding affinity
SERM selective estrogen receptor modulator

REFERENCES

1. (a) L. P. C. Schot and A. H. W. Schuurs, *J. Steroid Biochem. Mol. Biol.*, **37**, 167 (1990); (b) G. T. Griffing and S. H. Allen, *Postgrad. Med.*, **96**, 131 (1994).

2. (a) C. W. Emmens, *J. Endocrinol.*, **5**, 170 (1947); (b) F. R. Basford, Br. Pat. 561,508 (1944); *Chem. Abstr.*, **40**, 359 (1946); (c) C. A. Powers, M. A. Hatala, and P. J. Pagano, *Mol. Cell. Endocrinol.*, **66**, 93 (1989).

3. G. C. Davies, W. J. Huster, W. Shen, B. Mitlak, L. Plouffe, A. Shah, and F. J. Cohen, *Menopause-J. North Am. Menopause Soc.*, **6**, 188 (1999).

4. A. S. Levenson and V. C. Jordan, *Eur. J. Cancer*, **35**, 1628 (1999).

5. M. Breckwoldt, C. Keck, and U. Karck, *J. Steroid Biochem. Mol. Biol.*, **53**, 205 (1995).

6. L. Speroff, J. Rowan, J. Symons, H. Genant, and W. Wilborn, *J. Am. Med. Assoc.*, **276**, 1397 (1996).

7. C. Schairer, J. Lubin, R. Troisi, S. Sturgeon, L. Brinton, and R. Hoover, *J. Am. Med. Assoc.*, **283**, 485 (2000).

8. (a) P. Neven, *Lancet*, **346**, 1292 (1995); (b) F. A. Aleem and M. Predanic, *Lancet*, **346**, 1292 (1995); (c) M. A. Killackey, T. B. Hakes, and V. K. Pierce, *Cancer Treat. Rep.*, **69**, 237 (1985); (d) M. A.-F. Seoud, J. Johnson, and J. C. Weed Jr., *Obstet. Gynecol.*, **82**, 165 (1993); (e) Z. N. Kavak, S. Binoz, N. Ceyhan, and S. Pekin, *Acta Obstet. Gynecol. Scand.*, **79**, 604 (2000); (f) L. Deligdisch, T. Kalir, C. J. Cohen, M. deLatour, G. LeBouedec, and F. Penault-Llorca, *Gynecol. Oncol.*, **78**, 181 (2000).

9. (a) S. R. Goldstein, S. Siddhanti, A. V. Ciaccia, and L. Plouffe, *Hum. Reprod. Update*, **6**, 212 (2000); (b) A. Cano and C. Hermenegildo, *Hum. Reprod. Update*, **6**, 244 (2000).

10. L. Bergman, M. L. R. Beelen, M. P. W. Gallee, H. Hollema, J. Benraadt, and F. E. vanLeeuwen, *Lancet*, **356**, 881 (2000).

11. (a) I. Merchenthaler, J. M. Funkhouser, J. M. Carver, S. G. Lundeen, K. Ghosh, and R. C. Winneker, *Maturitas*, **30**, 307 (1998); (b) W. Khovidhunkit and D. M. Shoback, *Ann. Intern. Med.*, **130**, 431 (1999).

12. J. B. Brown in H. Sobotka and C. P. Stewart, Eds., *Advances in Clinical Chemistry*, Vol. **3**, Academic Press, New York, 1960, pp. 157–233.

13. (a) C. P. Martucci and J. Fishman, *Pharmacol. Ther.*, **57**, 237 (1993); (b) M. Numazawa and S. Satoh, *J. Steroid Biochem.*, **32**, 85 (1989); (c) S. A. Li, J. K. Klicka, and J. J. Li, *Cancer Res.*, **45**, 181 (1985); (d) H. Yamazaki, P. M. Shaw, F. P. Guengerich, and T. Shimada, *Chem. Res. Toxicol.*, **11**, 659 (1998).

14. K. C. Westerlind, K. J. Gibson, P. Malone, G. L. Evans, and R. T. Turner, *J. Bone Miner. Res.*, **13**, 1023 (1998).

15. N. J. MacLusky, E. R. Barnea, C. R. Clark, and F. Naftolin in G. R. Merriam and M. B. Lipsett, Eds., *Catechol Estrogens*, Raven, New York, 1983, pp. 151–165.

16. H. Kappus, H. M. Bolt, and H. Remmer, *J. Steroid Biochem.*, **4**, 121 (1973).

17. J. Salmon, D. Couissediere, C. Cousty, and J.-P. Raynaud, *J. Steroid Biochem.*, **18**, 565 (1983).

18. R. E. Ranney, *J. Toxicol. Environ. Health*, **3**, 139 (1977).

19. J. Schmider, D. J. Greenblatt, L. L. Vonmoltke, D. Karsov, R. Vena, H. L. Friedman, and R. I. Shader, *J. Clin. Pharmacol.*, **37**, 193 (1997).

20. (a) B. S. Katzenellenbogen, *J. Steroid Biochem.*, **20**, 1033 (1984); (b) A. Meli, B. G. Steinetz, T. Giannina, D. I. Cargill, and J. P. Manning, *Proc. Soc. Exp. Biol. Med.*, **127**, 1042 (1968); (c) B. G. Steinetz, A. Meli, T. Giannina, V. L. Beach, and J. P. Manning, *Proc. Soc. Exp. Biol. Med.*, **124**, 111 (1967); (d) B. G. Steinetz, A. Meli, T. Giannina, and V. L. Beach, *Proc. Soc. Exp. Biol. Med.*, **124**, 1283 (1967).

21. G. A. Grant and D. Beall, *Recent Prog. Horm. Res.*, **5**, 307 (1950).

22. B. R. Bhavnani, A. Cecutti, and D. Wallace, *Steroids*, **59**, 389 (1994).

23. P. G. Whittaker, M. R. Morgan, P. D. Dean, E. H. Cameron, and T. Lind, *Lancet*, **1**, 14 (1980).

24. G. Leyendecker, G. Geppert, W. Nocke, and J. Ufer, *Geburtshilfe Frauenheilkd.*, **35**, 370 (1975).

25. L. P. Shulman, M. OleenBurkey, and R. J. Willke, *Contraception*, **60**, 215 (1999).

26. M. H. Rahimy, M. A. Cromie, N. K. Hopkins, and D. M. Tong, *Contraception*, **60**, 201 (1999).

27. D. Kupfer and W. H. Bulger, *FEBS Lett.*, **261**, 59 (1990).

28. P. C. Ruenitz and M. M. Toledo, *Biochem. Pharmacol.*, **30**, 2203 (1981).

29. P. C. Ruenitz, J. R. Bagley, and N. T. Nanavati, *J. Med. Chem.*, **31**, 1471 (1988).

30. (a) J. M. Fromson, S. Pearson, and S. Bramah, *Xenobiotica*, **3**, 693 (1973); (b) V. C. Jordan, M. M. Collins, L. Rowsby, and G. Prestwich, *J. Endocrinol.*, **75**, 305 (1977).

31. (a) D. W. Robertson, J. A. Katzenellenbogen, D. J. Long, E. A. Rorke, and B. S. Katzenellenbogen, *J. Steroid Biochem.*, **16**, 1 (1982); (b) V. C. Jordan, R. R. Bain, R. R. Brown, B. Gosden, and M. A. Santos, *Cancer Res.*, **43**, 1446 (1983).

32. (a) G. Milano, M. C. Etienne, M. Frenay, R. Khater, J. L. Formento, N. Renee, J. L. Moll, M. Francoual, M. Berto, and M. Namer, *Br. J. Cancer*, **55**, 509 (1987); (b) E. A. Lien, E. Solheim, O. A. Lea, S. Lundgren, S. Kvinnsland, and P. M. Ueland, *Cancer Res.*, **49**, 2175 (1989); (c) M. C. Etienne, G. Milano, J. L. Fischel, M. Frenay, E. François, J. L. Formento, J. Gioanni, and M. Namer, *Br. J. Cancer*, **60**, 30 (1989); (d) K. Soininen, T. Kleimola, I. Elomaa, M. Salmo, and P. Risanen, *J. Int. Med. Res.*, **14**, 162 (1986); (e) D. Stevenson, R. J. Briggs, D. J. Chapman, and D. De Vos, *J. Pharm. Biomed. Anal.*, **6**, 1065 (1988); (f) R. R. Bain and V. C. Jordan, *Biochem. Pharmacol.*, **32**, 373 (1983).

33. I. B. Parr, R. McCague, G. Leclercq, and S. Stoessel, *Biochem. Pharmacol.*, **36**, 1513 (1987).

34. H. K. Crewe, S. W. Ellis, M. S. Lennard, and G. T. Tucker, *Biochem. Pharmacol.*, **53**, 171 (1997).

35. (a) H. Sipilä, L. Kangas, L. Vuorilehto, A. Kalapudas, M. Eloranta, M. Södervall, R. Toivola, and M. Anttila, *J. Steroid Biochem.*, **36**, 211 (1990); (b) S. A. Hasan, V. J. Wiebe, K. S. Cadman, and M. W. DeGregorio, *Anal. Lett.*, **23**, 327 (1990).

36. P. C. Ruenitz and J. R. Bagley, *Drug Metab. Dispos.*, **13**, 582 (1985).

37. T. J. Mikkelson, P. D. Kroboth, W. J. Cameron, L. W. Dittert, V. Chungi, and P. J. Manberg, *Fertil. Steril.*, **46**, 392 (1986).

38. T. N. Thompson and C. D. Klaassen, *J. Toxicol. Environ. Health*, **16**, 615 (1985).

39. M. Hospital, B. Busetta, C. Courseille, and G. Precigoux, *J. Steroid Biochem.*, **6**, 221 (1975).

40. (a) M. Metzler and J. A. McLachlan, *Biochem. Pharmacol.*, **27**, 1087 (1978); (b) R. Maydl, J. A. McLachlan, R. R. Newbold, and M. Metzler, *Biochem. Pharmacol.*, **34**, 710 (1985); (c) M. Metzler and H. Haaf, *Xenobiotica*, **15**, 41 (1985).

41. (a) D. Ross, R. J. Mehlhorn, P. Moldeus, and M. T. Smith, *J. Biol. Chem.*, **260**, 16210 (1985); (b) G. H. Degen and J. A. McLachlan, *Chem.-Biol. Interact.*, **54**, 363 (1985); (c) J. Foth and G. H. Degen, *Arch. Toxicol.*, **65**, 344 (1991); (d) G. H. Degen, G. Blaich, and M. Metzler, *J. Biochem. Toxicol.*, **5**, 91 (1990).

42. (a) A. L. Herbst, H. Ulfelder, and D. C. Poskanzer, *N. Engl. J. Med.*, **284**, 878 (1971); (b) A. L. Herbst, S. J. Robboy, R. E. Scully, and D. C. Poskanzer, *Am. J. Obstet. Gynecol.*, **119**, 713 (1974).

43. J. A. McLachlan and R. L. Dixon, *Adv. Mod. Toxicol.*, **1**, 423 (1976).

44. D. J. Holbrook Jr. in E. Hodgson and F. E. Guthrie, Eds., *Introduction to Biochemical Toxicology*, Chap. 16, Elsevier, New York, 1980.

45. R. Bucala, J. Fishman, and A. Cerami, *Proc. Natl. Acad. Sci. USA*, **79**, 3320 (1982).

46. N. T. Telang, A. Suto, G. Y. Wong, M. P. Osborne, and H. L. Bradlow, *J. Natl. Cancer Inst.*, **84**, 634 (1992).

47. Y. J. Abul-Hajj, K. Tabakovic, and I. Tabakovic, *J. Am. Chem. Soc.*, **117**, 6144 (1995).

48. (a) S. P. Adams and A. C. Notides, *Fundam. Appl. Toxicol.*, **9**, 715 (1987); (b) B. Epe, J. Hegler, and M. Metzler, *Carcinogenesis*, **8**, 1271 (1987).

49. S. Goldfarb and T. D. Pugh, *Cancer Res.*, **50**, 113 (1990).

50. S. T. Jan, P. D. Devanesan, D. E. Stack, R. Ramanathan, J. Byun, M. L. Gross, E. G. Rogan, and E. L. Cavalieri, *Chem. Res. Toxicol.*, **11**, 412 (1998).

51. P. C. Ruenitz, *Curr. Med. Chem.*, **2**, 791 (1995).

52. G. A. Potter, R. McCague, and M. Jarman, *Carcinogenesis*, **15**, 439 (1994).

53. L. Dasaradhi and S. Shibutani, *Chem. Res. Toxicol.*, **10**, 189 (1997).

54. S. Shibutani, L. Dasaradhi, I. Terashima, E. Banoglu, and M. W. Duffel, *Cancer Res.*, **58**, 647 (1998).

55. S. Shibutani, A. Ravindernath, N. Suzuki, I. Terashima, S. M. Sugarman, A. P. Grollman, and M. L. Pearl, *Carcinogenesis*, **21**, 1461 (2000).

56. D. F. Moffat, K. A. Oien, J. Dickson, T. Habeshaw, and D. R. McLellan, *Ann. Oncol.*, **11**, 1195 (2000).

57. N. T. Telang, *Ann. N. Y. Acad. Sci.*, **784**, 277 (1996).

58. T. D. Lindstrom, N. G. Whitaker, and G. W. Whitaker, *Xenobiotica*, **14**, 841 (1984).

59. T. A. Grese, S. Cho, D. R. Finley, A. G. Godfrey, C. D. Jones, C. W. Lugar, M. J. Martin, K. Matsumoto, L. D. Pennington, M. A. Winter, M. D. Adrian, H. W. Cole, D. E. Magee, D. L. Phillips, E. R. Rowley, L. L. Short, A. L. Glasebrook, and H. U. Bryant, *J. Med. Chem.*, **40**, 146 (1997).

60. J. A. Dodge, C. W. Lugar, S. Cho, J. J. Osborne, D. L. Phillips, A. L. Glasebrook, and C. A. Frolik, *Bioorg. Med. Chem. Lett.*, **7**, 993 (1997).

61. (a) C. Dwivedi, A. A. Downie, and T. E. Webb, *FASEB J.*, **1**, 303 (1987); (b) S. B. Hansel and M. E. Morris, *J. Pharmacokinet. Biopharm.*, **24**, 219 (1996).

62. M. Sato, C. H. Turner, T. Y. Wang, M. D. Adrian, E. Rowley, and H. U. Bryant, *J. Pharmacol. Exp. Ther.*, **287**, 1 (1998).

63. T. S. Gaginella, *U. S. Pharmacist*, **2**, 22 (1977).

64. T. G. Kokate, M. K. Banks, T. Magee, S. I. Yamaguchi, and M. A. Rogawski, *J. Pharmacol. Exp. Ther.*, **288**, 679 (1999).

65. G. Pérez-Palacios, M. A. Cerbón, A. M. Pasapera, J. I. Castro, J. Enríquez, F. Vilchis, G. A. Garcia, G. Morali, and A. E. Lemus, *J. Steroid Biochem. Mol. Biol.*, **41**, 479 (1992).

66. M. J. Reed, M. S. Ross, L. C. Lai, M. W. Ghilchik, and V. H. James, *J. Steroid Biochem. Mol. Biol.*, **37**, 301 (1990).

67. L. Viinikka, *Eur. J. Clin. Pharmacol.*, **15**, 349 (1979).

68. A. E. Lemus, V. Zaga, R. Santillan, G. A. Garcia, I. Grillasca, P. Damian Matsumura, K. J. Jackson, A. J. Cooney, F. Larrea, and G. Pérez Palacios, *J. Endocrinol.*, **165**, 693 (2000).

69. (a) E. A. Doisy, C. D. Veler, and S. A. Thayer, *J. Biol. Chem.*, **86**, 499 (1930); *Chem. Abstr.*, **24**, 3819 (1930); (b) M. N. Huffman, D. W. MacCorquodale, S. A. Thayer, E. A. Doisy, G. V. Smith, and O. W. Smith, *J. Biol. Chem.*, **134**, 591 (1940).

70. K. J. Ryan and O. W. Smith, *Recent Prog. Horm. Res.*, **21**, 367 (1965).

71. A. S. Meyer, *Biochim. Biophys. Acta*, **17**, 441 (1955); *Chem. Abstr.*, **49**, 16136a (1955).

72. E. R. Simpson, M. S. Mahendroo, G. D. Means, M. W. Kilgore, M. M. Hinshelwood, S. Graham-Lorence, B. Amarneh, Y. J. Ito, C. R. Fisher, M. D. Michael, C. R. Mendelson, and S. E. Bulun, *Endocr. Rev.*, **15**, 342 (1994).

73. A. M. Brodie, *J. Steroid Biochem. Mol. Biol.*, **49**, 281 (1994).

74. M. Akhtar, V. C. O. Njar, and J. N. Wright, *J. Steroid Biochem. Mol. Biol.*, **44**, 375 (1993).

75. P. A. Cole, J. A. Bean, and C. H. Robinson, *Proc. Natl. Acad. Sci. USA*, **87**, 2999 (1990).

76. (a) C. D. Jones, M. A. Winter, K. S. Hirsch, N. Stamm, H. M. Taylor, H. E. Holden, J. D. Davenport, E. V. Krumkalns, and R. G. Suhr, *J. Med. Chem.*, **33**, 416 (1990); (b) P. Furet, C. Batzl, A. Bhatnagar, E. Francotte, G. Rihs, and M. Lang, *J. Med. Chem.*, **36**, 1393 (1993); (c) L. J. Browne, C. Gude, H. Rodriguez, R. E. Steele, and A. J. Bhatnager, *J. Med. Chem.*, **34**, 725 (1991); (d) A. Lipton, L. M. Demers, H. A. Harvey, K. B. Kambic, H. Grossberg, C. Brady, H. Adlercruetz, P. F. Trunet, and R. J. Santen, *Cancer*, **75**, 2132 (1995).

77. M. L. Feutrie and J. Bonneterre, *Bull. Cancer*, **86**, 821 (1999).

78. E. Disalle, G. Ornati, D. Giudici, M. Lassus, T. R. Evans, and R. C. Coombes, *J. Steroid Biochem. Mol. Biol.*, **43**, 137 (1992).

79. (a) M. Kaufmann, E. Bajetta, L. Y. Dirix, L. E. Fein, S. E. Jones, N. Zilembo, J. C. Dugardyn, C. Nasurdi, R. G. Mennel, J. Cervek, C. Fowst, A. Polli, E. DiSalle, A. Arkhipov, G. Piscitelli, L. L. Miller, and G. Massimini, *J. Clin. Oncol.*, **18**, 1399 (2000); (b) D. Clemett and H. M. Lamb, *Drugs*, **59**, 1279 (2000); (c) S. Kvinnsland, G. Anker, L. Y. Dirix, J. Bonneterre, A. M. Prove, N. Wilking, J. P. Lobelle, O. Mariani, E. DiSalle, A. Polli, and G. Massimini, *Eur. J. Cancer*, **36**, 976 (2000).

80. G. W. Lyman and R. N. Johnson, *J. Chromatogr.*, **234**, 234 (1982).

81. B. R. Bhavnani and C. A. Woolever, *Steroids*, **56**, 201 (1991).

82. B. R. Bhavnani, A. Cecutti, and A. Gerulath, *J. Steroid Biochem. Mol. Biol.*, **67**, 119 (1998).

83. B. R. Bhavnani, *Endocr. Rev.*, **9**, 396 (1988).

84. P. Bouchard, *J. Reprod. Med.*, **44**, 153 (1999).

85. E. V. Jensen, E. R. DeSombre, and P. W. Jungblut in L. Martini, F. Fraschini, and M. Motta, Eds., *Hormonal Steroids*, Excerpta Medica, Amsterdam, 1967, pp. 492–500.

86. T. Barkhem, B. Carlsson, Y. Nilsson, E. Enmark, J.-Å. Gustafsson, and S. Nilsson, *Mol. Pharmacol.*, **54**, 105 (1998).

87. R. White, J. A. Lees, M. Needham, J. Ham, and M. Parker, *Mol. Endocrinol.*, **1**, 735 (1987).

88. G. G. J. M. Kuiper, B. Carlsson, K. Grandien, E. Enmark, J. Häggblad, S. Nilsson, and J.-Å. Gustafsson, *Endocrinology*, **138**, 863 (1997).

89. (a) R. White, S. E. Fawell, and M. G. Parker, *J. Steroid Biochem. Mol. Biol.*, **40**, 333 (1991); (b) H. H. Thole, *J. Steroid Biochem. Mol. Biol.*, **48**, 463 (1994); (c) S. Marsigliante, J. R. Puddefoot, S. Barker, A. W. Goode, and G. P. Vinson, *J. Steroid Biochem. Mol. Biol.*, **39**, 703 (1991).

90. C. Behl and F. Holsboer, *Trends Pharmacol. Sci.*, **20**, 441 (1999).

91. (a) S. Green, P. Walter, V. Kumar, A. Krust, J.-M. Bornert, P. Argos, and P. Chambon, *Nature*, **320**, 134 (1986); (b) G. L. Greene, P. Gilna, M. Waterfield, A. Baker, Y. Hort, and J. Shine, *Science*, **231**, 1150 (1986).

92. (a) E. von Angerer, C. Biberger, E. Holler, R. Koop, and S. Leichtl, *J. Steroid Biochem. Mol. Biol.*, **49**, 51 (1994); (b) T. M. Willson, B. R. Henke, T. M. Momtahen, P. S. Charifson, K. W. Batchelor, D. B. Lubahn, L. B. Moore, B. B. Oliver, H. R. Sauls, J. A. Triantafillou, S. G. Wolfe, and P. G. Baer, *J. Med. Chem.*, **37**, 1550 (1994).

93. G. M. Anstead, K. E. Carlson, and J. A. Katzenellenbogen, *Steroids*, **62**, 268 (1997).

94. A. M. Brzozowski, A. C. W. Pike, Z. Dauter, R. E. Hubbard, T. Bonn, O. Engström, L. Öhman, G. L. Greene, J.-Å. Gustafsson, and M. Carlquist, *Nature*, **389**, 753 (1997).

95. K. Ekena, J. A. Katzenellenbogen, and B. S. Katzenellenbogen, *J. Biol. Chem.*, **273**, 693 (1998).

96. D. M. Tanenbaum, Y. Wang, S. P. Williams, and P. B. Sigler, *Proc. Natl. Acad. Sci. USA*, **95**, 5998 (1998).

97. A. K. Shiau, D. Barstad, P. M. Loria, L. Cheng, P. J. Kushner, D. A. Agard, and G. L. Greene, *Cell*, **95**, 927 (1998).

98. A. C. W. Pike, A. M. Brzozowski, R. E. Hubbard, T. Bonn, A.-G. Thorsell, O. Engström, J. Ljunggren, J.-Å. Gustafsson, and M. Carlquist, *EMBO J.*, **18**, 4608 (1999).

99. S. I. Anghel, V. Perly, G. Melancon, A. Barsalou, S. Chagnon, A. Rosenauer, W. H. Miller, and S. Mader, *J. Biol. Chem.*, **275**, 20867 (2000).

100. A. H. Taylor and F. AlAzzawi, *J. Mol. Endocrinol.*, **24**, 145 (2000).

101. G. G. J. M. Kuiper, B. Carlsson, K. Grandien, E. Enmark, J. Häggblad, S. Nilsson, and J.-Å. Gustafsson, *Endocrinology*, **138**, 863 (1997).

102. F. C. S. Eng, A. Barsalou, N. Akutsu, I. Mercier, C. Zechel, S. Mader, and J. H. White, *J. Biol. Chem.*, **273**, 28371 (1998).

103. J. W. R. Schwabe, L. Chapman, J. T. Finch, and D. Rhodes, *Cell*, **75**, 567 (1993).

104. F. V. Peale Jr., L. B. Ludwig, S. Zain, R. Hilf, and R. A. Bambara, *Proc. Natl. Acad. Sci. USA*, **85**, 1038 (1988).

105. (a) C. K. Wrenn and B. S. Katzenellenbogen, *J. Biol. Chem.*, **268**, 24089 (1993); (b) C. R. Lyttle, P. Damian-Matsumura, H. Juul, and T. R. Butt, *J. Steroid Biochem. Mol. Biol.*, **42**, 677 (1992); (c) M. J. Pilat, M. S. Hafner, L. G. Kral, and S. C. Brooks, *Biochemistry*, **32**, 7009 (1993); (d) M. T. Tzukerman, A. Esty, D. Santiso-Mere, P. Danielian, M. G. Parker, R. B. Stein, J. W. Pike, and D. P. McDonnell, *Mol. Endocrinol.*, **8**, 21 (1994).

106. A. C. Gee, K. E. Carlson, P. G. V. Martini, B. S. Katzenellenbogen, and J. A. Katzenellenbogen, *Mol. Endocrinol.*, **13**, 1912 (1999).

107. J. M. Hall and D. P. McDonnell, *Endocrinology*, **140**, 5566 (1999).

108. M. Williams, D. C. Deecher, and J. P. Sullivan in M. E. Wolff, Ed., *Burger's Medicinal Chemistry and Drug Discovery*, Vol. **1**, 5th ed., John Wiley & Sons, New York, 1995, p. 358.

109. B. J. Cheskis, S. Karathanasis, and C. R. Lyttle, *J. Biol. Chem.*, **272**, 11384 (1997).

110. (a) R. I. Dorfman, *Physiol. Rev.*, **34**, 138 (1954); (b) V. C. Jordan, B. Haldemann, and K. E. Allen, *Endocrinology*, **108**, 1353 (1981).

111. J. Odum, S. Tittensor, and J. Ashby, *Toxicol. in Vitro*, **12**, 273 (1998).

112. (a) B. A. Littlefield, E. Gurpide, L. Markiewicz, B. McKinley, and R. B. Hochberg, *Endocrinology*, **127**, 2757 (1990); (b) E. Bignon, M. Pons, A. Crastes de Paulet, J.-C. Doré, J. Gilbert, J. Abecassis, J.-F. Miquel, T. Ojasoo, and J.-P. Raynaud, *J. Med. Chem.*, **32**, 2092 (1989).

113. S. G. Lundeen, J. M. Carver, M. McKean, and R. C. Winneker, *Endocrinology*, **138**, 1552 (1997).

114. (a) T. J. Wronski, C. C. Walsh, and L. A. Ignaszewski, *Bone*, **7**, 119 (1986); (b) T. J. Wronski, L. M. Dann, K. S. Scott, and L. R. Crooke, *Endocrinology*, **125**, 810 (1989).

115. (a) J. W. Simpkins, M. J. Katovich, and I.-C. Song, *Life Sci.*, **32**, 1957 (1983); (b) M. J. Katovich, J. W. Simpkins, L. A. Berglund, and J. O'Meara, *Maturitas*, **8**, 67 (1986).

116. X. Wu, M. A. Glinn, N. L. Ostrowski, Y. Su, B. Ni, H. W. Cole, H. U. Bryant, and S. M. Paul, *Brain Res.*, **847**, 98 (1999).

117. P. H. Giangrande, E. A. Kimbrel, D. P. Edwards, and D. P. McDonnell, *Mol. Cell. Biol.*, **20**, 3102 (2000).

118. P. Kastner, A. Krust, B. Turcotte, U. Stropp, L. Tora, H. Gronemeyer, and P. Chambon, *EMBO J.*, **9**, 1603 (1990).

119. (a) C. A. Sartorius, S. D. Groshong, L. A. Miller, R. L. Powell, L. Tung, G. S. Takimoto, and K. B. Horwitz, *Cancer Res.*, **54**, 3868 (1994); (b) E. Vegeto, M. M. Shahbaz, D. X. Wen, M. E. Goldman, B. W. O'Malley, and D. P. McDonnell, *Mol. Endocrinol.*, **7**, 1244 (1993).

120. S. P. Williams and P. B. Sigler, *Nature*, **393**, 392 (1998).

121. (a) L. L. Wei, M. W. Leach, R. S. Miner, and L. M. Demers, *Biochem. Biophys. Res. Commun.*, **195**, 525 (1993); (b) Y. K. Hedges, J. K. Richer, K. B. Horwitz, and L. D. Horwitz, *Circulation*, **99**, 2688 (1999).

122. A. Mashiah, V. Berman, H. H. Thole, S. S. Rose, S. Pasik, H. Schwarz, and H. BenHur, *Cardiovasc. Surg.*, **7**, 327 (1999).

123. L. Zhi, C. M. Tegley, K. B. Marschke, D. E. Mais, and T. K. Jones, *J. Med. Chem.*, **42**, 1466 (1999).

124. L. Zhi, C. M. Tegley, E. A. Kallel, K. B. Marschke, D. E. Mais, M. M. Gottardis, and T. K. Jones, *J. Med. Chem.*, **41**, 291 (1998).

125. (a) A. DeCherney, *J. Reprod. Med.*, **38**, 1007 (1993); (b) F. Scopacasa, M. Horowitz, A. G. Need, H. A. Morris, and B. E. C. Nordin, *Osteoporos. Int.*, **9**, 494 (1999).

126. (a) A. G. H. Ederveen and H. J. Kloosterboer, *J. Bone Miner. Res.*, **14**, 1963 (1999); (b) K. Yoshitake, K. Yokota, Y. Kasugai, M. Kagawa, T. Sukamoto, and T. Nakamura, *Bone*, **25**, 311 (1999); (c) S. A. Beardsworth, C. E. Kearney, and D. W. Purdie, *Br. J. Obstet. Gynecol.*, **106**, 678 (1999).

127. D. P. Edwards, M. Altmann, A. DeMarzo, Y. Zhang, N. L. Weigel, and C. A. Beck, *J. Steroid Biochem. Mol. Biol.*, **53**, 449 (1995).

128. B. L. Wagner, G. Pollio, P. Giangrande, J. C. Webster, M. Breslin, D. E. Mais, C. E. Cook, W. V. Vedeckis, J. A. Cidlowski, and D. P. Mc-Donnell, *Endocrinology*, **140**, 1449 (1999).

129. J. Donath, Y. Nishino, T. Schulz, and H. Michna, *Ann. Anatomy-Anatomischer Anzeiger*, **182**, 143 (2000).

130. K. B. Horwitz, *Endocr. Rev.*, **13**, 146 (1992).

131. J. H. Al Jamal and N. H. Dubin, *Am. J. Obstet. Gynecol.*, **182**, 1099 (2000).

132. D. M. Stresser and D. Kupfer, *Drug Metab. Dispos.*, **26**, 868 (1998).

133. K. W. Gaido, L. S. Leonard, S. C. Maness, J. M. Hall, D. P. McDonnell, B. Saville, and S. Safe, *Endocrinology*, **140**, 5746 (1999).

134. S. Mäkelä, V. L. Davis, W. C. Tally, J. Korkman, L. Salo, R. Vihko, R. Santti, and K. S. Korach, *Environ. Health Perspect.*, **102**, 572 (1994).

135. (a) A. Brzezinski, H. Adlercreutz, R. Shaoul, A. Rosler, A. Shmueli, V. Tanos, and J. G. Schenker, *Menopause-J. North Am. Menopause Soc.*, **4**, 89 (1997); (b) J. Eden, *Baillieres Clin. Endocrinol. Metab.*, **12**, 581 (1998).

136. (a) A. L. Murkies, C. Lombard, B. J. Strauss, G. Wilcox, H. G. Burger, and M. S. Morton, *Maturitas*, **21**, 189 (1995); (b) G. Scambia, D. Mango, P. G. Signorile, R. A. Angeli, C. Palena, D. Gallo, E. Bombardelli, P. Morazzoni, A. Riva, and S. Mancuso, *Menopause-J. North Am. Menopause Soc.*, **7**, 105 (2000).

137. J. A. Dodge, A. L. Glasebrook, D. E. Magee, D. L. Phillips, M. Sato, L. L. Short, and H. U. Bryant, *J. Steroid Biochem. Mol. Biol.*, **59**, 155 (1996).

138. R. J. Miksicek, *J. Steroid Biochem. Mol. Biol.*, **49**, 153 (1994).

139. M. Casanova, L. You, K. W. Gaido, S. Archibe-queEngle, D. B. Janszen, and H. D. Heck, *Toxicol. Sci.*, **51**, 236 (1999).

140. D. T. Zava, C. M. Dollbaum, and M. Blen, *Proc. Soc. Exp. Biol. Med.*, **217**, 369 (1998).

141. (a) D. C. Williams, D. C. Paul, and L. J. Black, *Bone Mineral*, **14**, 205 (1991); (b) L. J. Black, M. Sato, E. R. Rowley, D. E. Magee, A. Bekele, D. C. Williams, G. J. Cullinan, R. Bendele, R. F. Kauffmann, W. R. Bensch, C. A. Frolik, J. D. Termine, and H. U. Bryant, *J. Clin. Invest.*, **93**, 63 (1994); (c) H. Z. Ke, V. M. Paralkar, W. A. Grasser, D. T. Crawford, H. Qi, H. A. Simmons, C. M. Pirie, K. L. Chidsey-Frink, T. A.

Owen, S. L. Smock, H. K. Chen, W. S. Jee, K. O. Cameron, R. L. Rosati, T. A. Brown, P. Dasilva-Jardine, and D. D. Thompson, *Endocrinology*, **139**, 2068 (1998); (d) J. Ashby, J. Odum, and J. R. Foster, *Regul. Toxicol. Pharmacol.*, **25**, 226 (1997).

142. (a) A. Gallagher, T. J. Chambers, and J. H. Tobias, *Endocrinology*, **133**, 2787 (1993); (b) M. Dukes, R. Chester, L. Yarwood, and A. E. Wakeling, *J. Endocrinol.*, **141**, 335 (1994).

143. V. A. DiPippo and C. A. Powers, *J. Pharmacol. Exp. Ther.*, **281**, 142 (1997).

144. K. P. Nephew, S. Ray, M. Hlaing, A. Ahluwalia, S. D. Wu, X. H. Long, S. M. Hyder, and R. M. Bigsby, *Biol. Reprod.*, **63**, 361 (2000).

145. S. Thenot, M. Charpin, S. Bonnet, and V. Cavailles, *Mol. Cell. Endocrinol.*, **156**, 85 (1999).

146. S. L. Dana, P. A. Hoener, D. A. Wheeler, C. B. Lawrence, and D. P. McDonnell, *Mol. Endocrinol.*, **8**, 1193 (1994).

147. N. N. Yang, M. Venugopalan, S. Hardikar, and A. Glasebrook, *Science*, **273**, 1222 (1996); correction: *Science*, **275**, 1249 (1997).

148. (a) B. Couzinet, J. Young, S. Brailly, P. Chanson, J. L. Thomas, and G. Schaison, *J. Clin. Endocrinol. Metab.*, **81**, 4218 (1996); (b) B. D. Moutos, S. Smith, and H. Zacur, *Contraception*, **52**, 105 (1995); (c) W. Y. Ling, D. W. Johnston, R. H. Lea, A. E. Bent, J. Z. Scott, and M. R. Toews, *Contraception*, **32**, 367 (1985).

149. W. G. Rossmanith, D. Steffens, and G. Schramm, *Contraception*, **56**, 23 (1997).

150. (a) A. Rommler, S. Baumgarten, L. Moltz, U. Schwartz, and J. Hammerstein, *Contraception*, **31**, 295 (1985); (b) K. Rudolf, S. Kunkel, J. Meissner, and M. Ruting, *Zentralbl. Gynakol.*, **107**, 1345 (1985).

151. (a) A. N. Poindexter III, G. A. Dildy, S. A. Brody, and M. C. Snabes, *Contraception*, **48**, 37 (1993); correction: *Contraception*, **48**, 192 (1993); (b) A. Ismail, A. el-Faras, M. Rocca, F. A. el-Sibai, and M. Toppozada, *Contraception*, **35**, 487 (1987).

152. A. J. W. Hsueh, G. F. Erickson, and S. S. C. Yen, *Nature*, **273**, 57 (1978).

153. E. S. Huang and W. L. Miller, *Endocrinology*, **112**, 442 (1983).

154. (a) L. C. Huppert, *Fertil. Steril.*, **31**, 1 (1979); (b) E. Y. Adashi, *Fertil. Steril.*, **42**, 331 (1984).

155. W. Sneader, *Drug Discovery: The Evolution of Modern Medicines*, John Wiley & Sons, Chichester, UK, 1985, p. 199.

156. V. C. Jordan, *Pharmacol. Rev.*, **36**, 245 (1984).

157. (a) R. R. Love, R. B. Mazess, D. C. Tormey, H. S. Barden, P. A. Newcomb, and V. C. Jordan, *Breast Cancer Res. Treat.*, **12**, 297 (1988); (b) T. Fornander, L. E. Rutqvist, H. E. Sjoberg, L. Blomqvist, A. Mattsson, and U. Glas, *J. Clin. Oncol.*, **8**, 1019 (1990).

158. (a) S. Turken, E. Siris, D. Seldin, E. Flaster, G. Hyman, and R. Lindsay, *J. Natl. Cancer Inst.*, **81**, 1086 (1989); (b) W. G. Ryan, J. Wolter, and J. D. Bagdade, *Osteoporos. Int.*, **2**, 39 (1991); (c) I. S. Fentiman, Z. Saad, M. Caleffi, M. A. Chaudary, and I. Fogelman, *Eur. J. Cancer*, **28**, 684 (1992); (d) J. A. Dewar, J. M. Horobin, P. E. Preece, R. Tavendale, H. Tunstall-Pedoe, and R. A. B. Wood, *Br. Med. J.*, **305**, 225 (1992).

159. C. A. Frolik, H. U. Bryant, E. C. Black, D. E. Magee, and S. Chandrasekhar, *Bone*, **18**, 621 (1996).

160. (a) R. T. Turner, G. K. Wakley, K. S. Hannon, and N. H. Bell, *Endocrinology*, **122**, 1146 (1988); (b) T. J. Wronski, M. Cintrón, A. L. Doherty, and L. M. Dann, *Endocrinology*, **123**, 681 (1988).

161. (a) C. D. Jones, T. Suarez, E. H. Massey, L. J. Black, and F. C. Tinsley, *J. Med. Chem.*, **22**, 962 (1979); (b) L. J. Black and R. L. Goode, *Life Sci.*, **26**, 1453 (1980).

162. C. D. Jones, M. G. Jevnikar, A. J. Pike, M. K. Peters, L. J. Black, A. R. Thompson, J. F. Falcone, and J. A. Clemens, *J. Med. Chem.*, **27**, 1057 (1984).

163. V. C. Jordan and B. Gosden, *Endocrinology*, **113**, 463 (1983).

164. C. H. Turner, M. Sato, and H. U. Bryant, *Endocrinology*, **135**, 2001 (1994).

165. V. C. Jordan and M. Morrow, *Endocr. Rev.*, **20**, 253 (1999).

166. J.-L. Borgna and J. Scali, *Eur. J. Biochem.*, **199**, 575 (1991).

167. J.-P. Raynaud, T. Ojasoo, M. M. Bouton, E. Bignon, M. Pons, and A. Crastes de Paulet in J. A. McLachlan, Ed., *Estrogens in the Environment II. Influences on Development*, Elsevier, New York, 1985, pp. 24–42.

168. (a) R. M. Kanojia, G. O. Allen, J. M. Killinger, and J. L. McGuire, *J. Med. Chem.*, **22**, 1538 (1979); (b) J. A. Katzenellenbogen, B. S. Katzenellenbogen, T. Tatee, D. W. Robertson, and S. W. Landvatter in J. A. McLachlan, Ed., *Estrogens in the Environment*, Elsevier, New York, 1980, pp. 33–51.

169. J. N. DaSilva and J. E. Van Lier, *J. Med. Chem.*, **33**, 430 (1990).

170. T. E. Wiese, L. A. Polin, E. Palomino, and S. C. Brooks, *J. Med. Chem.*, **40**, 3659 (1997).

171. (a) C. Martucci and J. Fishman, *Steroids*, **27**, 325 (1976); (b) D. F. Heiman, S. G. Senderoff, J. A. Katzenellenbogen, and R. J. Neeley, *J. Med. Chem.*, **23**, 994 (1980); (c) R. Ouellet, J. Rousseau, N. Brasseur, J. E. van Lier, M. Diksic, and G. Westera, *J. Med. Chem.*, **27**, 509 (1984).

172. J.-P. Raynaud, P. M. Martin, M.-M. Bouton, and T. Ojasoo, *Cancer Res.*, **38**, 3044 (1978).

173. (a) A. E. Wakeling, M. Dukes, and J. Bowler, *Cancer Res.*, **51**, 3867 (1991); (b) A. E. Wakeling and J. Bowler, *J. Steroid Biochem. Mol. Biol.*, **43**, 173 (1992).

174. R. D. Bindal, K. E. Carlson, G. C. A. Reiner, and J. A. Katzenellenbogen, *J. Steroid Biochem.*, **28**, 361 (1987).

175. A. Claussner, L. Nedelec, F. Nique, D. Philibert, G. Teutsch, and P. Van de Velde, *J. Steroid Biochem. Mol. Biol.*, **41**, 609 (1992).

176. J. Canceill, J. Jacques, M. M. Bouton, M. Fortin, and C. Tournemine, *J. Steroid Biochem.*, **18**, 643 (1983).

177. (a) Y. C. Lu and R. T. Chatterton Jr., *Adv. Contracept.*, **10**, 157 (1994); (b) A. J. Peters, A. C. Wentz, R. R. Kazer, R. S. Jeyendran, and R. T. Chatterton Jr., *Contraception*, **52**, 195 (1995).

178. J. Bowler, T. J. Lilley, J. D. Pittam, and A. E. Wakeling, *Steroids*, **54**, 71 (1989).

179. (a) E. W. Thompson, D. Katz, T. B. Shima, A. E. Wakeling, M. E. Lippman, and R. B. Dickson, *Cancer Res.*, **49**, 6929 (1989); (b) P. Coopman, M. Garcia, N. Brünner, D. Derocq, R. Clarke, and H. Rochefort, *Int. J. Cancer*, **56**, 295 (1994).

180. A. Howell, D. DeFriend, J. Robertson, R. Blamey, and P. Walton, *Lancet*, **345**, 29 (1995).

181. S. E. Fawell, R. White, S. Hoare, M. Sydenham, M. Page, and M. G. Parker, *Proc. Natl. Acad. Sci. USA*, **87**, 6883 (1990).

182. M. R. Tremblay, J. Simard, and D. Poirier, *Bioorg. Med. Chem. Lett.*, **9**, 2827 (1999).

183. (a) S. A. Washburn, M. R. Adams, T. B. Clarkson, and S. J. Adelman, *Am. J. Obstet. Gynecol.*, **169**, 251 (1993); (b) A. Chandrasekaran, M. Osman, P. Raveendranath, K. Chan, J. A. Scatina, and S. F. Sisenwine, *J. Mass Spectrom.*, **30**, 1505 (1995).

184. (a) E. C. Dodds, L. Golberg, W. Lawson, and R. Robinson, *Proc. R. Soc. Lond. B Biol. Sci.*, **127**, 140 (1939); (b) U. V. Solmssen, *Chem. Rev.*, **37**, 481 (1945).

185. V. W. Winkler, M. A. Nyman, and R. S. Egan, *Steroids*, **17**, 197 (1971).

186. (a) A. R. Lea, W. J. Kayaba, and D. M. Hailey, *J. Chromatogr.*, **177**, 61 (1979); (b) J. A. Katzenellenbogen, K. E. Carlson, and B. S. Katzenellenbogen, *J. Steroid Biochem.*, **22**, 589 (1985).

187. C. W. Emmens, *J. Endocrinol.*, **2**, 444 (1941).

188. (a) E. J. Pavlik and B. S. Katzenellenbogen, *Mol. Pharmacol.*, **18**, 406 (1980); (b) S. Stoessel and G. Leclercq, *J. Steroid Biochem.*, **25**, 677 (1986); (c) C.-D. Shiller, M. R. Schneider, and E. von Angerer, *Arch. Pharmacol.*, **323**, 417 (1990).

189. E. von Angerer, *Cancer Treat. Rev.*, **11** (Suppl. A), 147 (1984).

190. (a) E. von Angerer, N. Knebel, M. Kager, and B. Ganss, *J. Med. Chem.*, **33**, 2635 (1990); (b) Y. Nishino, M. R. Schneider, H. Michna, and E. von Angerer, *J. Endocrinol.*, **130**, 409 (1991).

191. (a) L. J. Black, C. D. Jones, and J. F. Falcone, *Life Sci.*, **32**, 1031 (1983); (b) L. J. Black, C. D. Jones, and R. L. Goode, *Mol. Cell. Endocrinol.*, **22**, 95 (1981).

192. (a) T. A. Grese, E. R. Rowley, L. L. Short, A. L. Glasebrook, H. U. Bryant, S. Cho, D. R. Finley, A. G. Godfrey, C. D. Jones, C. W. Lugar, M. J. Martin, K. Matsumoto, L. D. Pennington, M. A. Winter, M. D. Adrian, H. W. Cole, D. E. Magee, and D. L. Phillips, *J. Med. Chem.*, **40**, 146 (1997); (b) A. D. Palkowitz, A. L. Glasebrook, K. J. Thrasher, K. L. Hauser, L. L. Short, D. L. Phillips, B. S. Muehl, M. Sato, P. K. Shetler, G. J. Cullinan, T. R. Pell, and H. U. Bryant, *J. Med. Chem.*, **40**, 1407 (1997).

193. P. R. Kym, G. M. Anstead, K. G. Pinney, S. R. Wilson, and J. A. Katzenellenbogen, *J. Med. Chem.*, **36**, 3910 (1993).

194. (a) I. E. Smiley and M. G. Rossmann, *Chem. Commun.*, 198 (1969); (b) G. Ruban and P. Luger, *Acta Crystallogr. Sect. B*, **31**, 2658 (1975).

195. (a) R. A. Magarian, L. B. Overacre, S. Singh, and K. L. Meyer, *Curr. Med. Chem.*, **1**, 61 (1994); (b) G. Leclercq, N. Devleeschouwer, and J. C. Heuson, *J. Steroid Biochem.*, **19**, 75 (1983); (c) R. W. Hartmann, G. Kranzfelder, E. von Angerer, and H. Schönenberger, *J. Med. Chem.*, **23**, 841 (1980); (d) M. Schneider, E. von Angerer, G. Kranzfelder, and H. Schönenberger, *Arch. Pharmacol.*, **313**, 919 (1980).

196. S. P. Robinson, R. Koch, and V. C. Jordan, *Cancer Res.*, **48**, 784 (1988).

197. C. P. Miller, M. D. Collini, B. D. Tran, H. A. Harris, Y. P. Kharode, J. T. Marzolf, R. A. Moran, R. A. Henderson, R. H. W. Bender, R. J. Unwalla, L. M. Greenberger, J. P. Yardley, M. A. Abou-Gharbia, C. R. Lyttle, and B. S. Komm, *J. Med. Chem.*, **44**, 1654 (2001).

198. C. D. Jones, L. C. Blaszczak, M. E. Goettel, T. Suarez, T. A. Crowell, T. E. Mabry, P. C. Ruenitz, and V. Srivatsan, *J. Med. Chem.*, **35**, 931 (1992).

199. (a) M. Pons, F. Michel, A. Crastes de Paulet, J. Gilbert, J.-F. Miquel, G. Précigoux, M. Hospital, T. Ojasoo, and J.-P. Raynaud, *J. Steroid Biochem.*, **20**, 137 (1984); (b) R. McCague, M. Jarman, O.-T. Leung, A. B. Foster, G. Leclercq, and S. Stoessel, *J. Steroid Biochem.*, **31**, 545 (1988).

200. (a) P. C. Ruenitz, C. S. Bourne, K. J. Sullivan, and S. A. Moore, *J. Med. Chem.*, **39**, 4853 (1996); (b) K. S. Kraft, P. C. Ruenitz, and M. G. Bartlett, *J. Med. Chem.*, **42**, 3126 (1999).

201. S. Ernst, G. Hite, J. S. Cantrell, A. Richardson Jr., and H. D. Benson, *J. Pharm. Sci.*, **65**, 148 (1976).

202. R. T. Turner, G. L. Evans, J. P. Sluka, M. D. Adrian, H. U. Bryant, C. H. Turner, and M. Sato, *Endocrinology*, **139**, 3712 (1998).

203. T. M. Willson, J. D. Norris, B. L. Wagner, I. Asplin, P. Baer, H. R. Brown, S. A. Jones, B. Henke, H. Sauls, S. Wolfe, D. C. Morris, and D. P. McDonnell, *Endocrinology*, **138**, 3901 (1997).

204. A. L. Wijayaratne, S. C. Nagel, L. A. Paige, D. J. Christensen, J. D. Norris, D. M. Fowlkes, and D. P. McDonnell, *Endocrinology*, **140**, 5828 (1999).

205. V. N. Rubin, P. C. Ruenitz, F. D. Boudinot, and J. L. Boyd, *Bioorg. Med. Chem.*, **9**, 1579 (2001).

206. (a) D. J. Collins, J. J. Hobbs, and C. W. Emmens, *J. Med. Chem.*, **14**, 952 (1971); (b) J. A. Bristol, Ed., *Annual Reports in Medicinal Chemistry*, Vol. **27**, Academic Press, San Diego, 1982, p. 324.

207. M. Salman, S. Ray, N. Anand, A. K. Agarwal, M. M. Singh, B. S. Setty, and V. P. Kamboj, *J. Med. Chem.*, **29**, 1801 (1986).

208. S. Gauthier, B. Caron, J. Cloutier, Y. L. Dory, A. Favre, D. Larouche, J. Mailhot, C. Ouellet, A. Schwerdtfeger, G. Leblanc, C. Martel, J. Simard, Y. Mérand, A. Bélanger, C. Labrie, and F. Labrie, *J. Med. Chem.*, **40**, 2117 (1997).

209. J. Simard, R. Sanchez, D. Poirier, S. Gauthier, S. M. Singh, Y. Mérand, A. Bélanger, C. Labrie, and F. Labrie, *Cancer Res.*, **57**, 3494 (1997).

210. A. Tremblay, G. B. Tremblay, C. Labrie, F. Labrie, and V. Giguere, *Endocrinology*, **139**, 111 (1998).

211. J. I. M. Schafer, H. Liu, D. A. Tonetti, and V. C. Jordan, *Cancer Res.*, **59**, 4308 (1999).

212. F. Labrie, C. Labrie, A. Bélanger, J. Simard, S. Gauthier, V. LuuThe, Y. Mérand, V. Giguere, B. Candas, S. Q. Luo, C. Martel, S. M. Singh, M. Fournier, A. Coquet, V. Richard, R. Charbonneau, G. Charpenet, A. Tremblay, G. Tremblay, L. Cusan, and R. Veilleux, *J. Steroid Biochem. Mol. Biol.*, **69**, 51 (1999).

213. (a) D. Lednicer, J. C. Babcock, S. C. Lyster, and G. W. Duncan, *Chem. Ind.*, 408 (1963); (b) D. Lednicer, S. C. Lyster, and G. W. Duncan, *J. Med. Chem.*, **10**, 78 (1967).

214. E. R. Ferguson and B. S. Katzenellenbogen, *Endocrinology*, **100**, 1242 (1977).

215. R. L. Rosati, P. Dasilva-Jardine, K. O. Cameron, D. D. Thompson, H. Z. Ke, S. M. Toler, T. A. Brown, L. C. Pan, C. F. Ebbinghaus, A. R. Reinhold, N. C. Elliott, B. N. Newhouse, C. M. Tjoa, P. M. Sweetnam, M. J. Cole, M. W. Arriola, J. W. Gauthier, D. T. Crawford, D. F. Nickerson, C. M. Pirie, H. Qi, H. A. Simmons, and G. T. Tkalcevic, *J. Med. Chem.*, **41**, 2928 (1998).

216. R. J. Mountfield, B. Kiehr, and B. A. John, *Drug Metab. Dispos.*, **28**, 503 (2000).

217. B. T. Kilbourn and P. G. Owston, *J. Chem. Soc. B*, 1 (1970).

218. S. Ernst and G. Hite, *Acta Crystallogr. Sect. B*, **32**, 291 (1976).

219. R. McCague and G. Leclercq, *J. Med. Chem.*, **30**, 1761 (1987).

220. J. (R.) Skidmore, A. L. Walpole, and J. Woodburn, *J. Endocrinol.*, **52**, 289 (1972).

221. K. W. Harlow, D. N. Smith, J. A. Katzenellenbogen, G. L. Greene, and B. S. Katzenellenbogen, *J. Biol. Chem.*, **264**, 17476 (1989).

222. L. L. Wei, W. F. Mangel, and B. S. Katzenellenbogen, *J. Steroid Biochem.*, **23**, 875 (1985).

223. B. E. Fink, D. S. Mortensen, S. R. Stauffer, Z. D. Aron, and J. A. Katzenellenbogen, *Chem. Biol.*, **6**, 205 (1999).

224. M. Juchem and K. Pollow, *Am. J. Obstet. Gynecol.*, **163**, 2171 (1990).

225. L. G. Hamann, L. J. Farmer, M. G. Johnson, S. L. Bender, D. E. Mais, M. W. Wang, D. Crombie, M. E. Goldman, and T. K. Jones, *J. Med. Chem.*, **39**, 1778 (1996).

226. E. W. Bergink, A. D. Hamburger, E. de Jager, and J. van der Vies, *J. Steroid Biochem.*, **14**, 175 (1981).

227. A. Phillips, K. Demarest, D. W. Hahn, F. Wong, and J. L. McGuire, *Contraception*, **41**, 399 (1990).

228. H. Smith, G. A. Hughes, G. H. Douglas, G. R. Wendt, G. C. Buzby Jr., R. A. Edgren, J. Fisher, T. Foell, B. Gadsby, D. Hartley, D. Herbst, A. B. A. Jansen, K. Ledig, B. J. McLoughlin, J. McMenamen, T. W. Pattison, P. C. Phillips, R. Rees, J. Siddall, J. Siuda, L. L. Smith, J. Tokolics, and D. H. P. Watson, *J. Chem. Soc.*, 4472 (1964).

229. J. L. McGuire, A. Phillips, D. W. Hahn, E. L. Tolman, S. Flor, and M. E. Kafrissen, *Am. J. Obstet. Gynecol.*, **163**, 2127 (1990).

230. J.-P. Raynaud and T. Ojasoo, *J. Steroid Biochem.*, **25**, 811 (1986).

231. B. L. Wagner, G. Pollio, S. Leonhardt, M. C. Wani, D. Y. Lee, M. O. Imhof, D. P. Edwards, C. E. Cook, and D. P. McDonnell, *Proc. Natl. Acad. Sci. USA*, **93**, 8739 (1996).

232. (a) Y. Tabata, M. Hatsu, Y. Kurata, K. Miyajima, M. Tani, T. Sasaki, Y. Kodama, T. Tsuruoka, and S. Omoto, *J. Antibiot.*, **50**, 309 (1997); (b) K. Kurihara, K. Tanabe, Y. Yamamoto, R. Shinei, K. Ajito, and T. Okonogi, *Bioorg. Med. Chem. Lett.*, **9**, 1837 (1999).

233. R. J. Byers, J. A. Hoyland, and I. P. Braidman, *J. Endocrinol.*, **168**, 353 (2001).

234. P. Szulc, F. Munoz, B. Claustrat, P. Garnero, F. Marchand, F. Duboeuf, and P. D. Delmas, *J. Clin. Endocrinol. Metab.*, **86**, 192 (2001).

235. H. Z. Ke, H. Qi, K. L. Chidsey-Frink, D. T. Crawford, and D. D. Thompson, *J. Bone Miner. Res.*, **16**, 765 (2001).

CHAPTER FOURTEEN

Male Sex Hormones, Analogs, and Antagonists

ROBERT W. BRUEGGEMEIER
Division of Medicinal Chemistry and Pharmacognosy
The Ohio State University, College of Pharmacy
Columbus, Ohio

Contents

Burger's Medicinal Chemistry and Drug Discovery
Sixth Edition, Volume 3: Cardiovascular Agents and
Endocrines
Edited by Donald J. Abraham
ISBN 0-471-37029-0   © 2003 John Wiley & Sons, Inc.

1   INTRODUCTION

Androgens are a class of steroids responsible for the primary and secondary sex characteristics of the male. In addition, these steroids possess potent anabolic or growth-promoting properties. The general chemical structure of androgens is based on the androstane C_{19} steroid, consisting of the fused four-ring steroid nucleus (17 carbons atoms, rings A–D) and the two axial methyl groups (carbons 18 and 19) at the A/B and C/D ring junctions. The hormone testosterone (**1**) is the predominant circulating androgen and is produced mainly by the testis. 5α-Dihydrotestosterone (**2**) is a 5α-reduced metabolite of testosterone, produced in certain androgen target tissues and is the most potent endogenous androgen.

(**2**) 5α - dihydrotestosterne

where. Modified androgens that have found use as biochemical or pharmacological tools also are included. More extensive presentations of the topic of androgens, anabolics, and androgens antagonists have appeared in several treatises published over the past two decades (1–11).

(**1**) testosterone

These two steroids and other endogenous androgens influence not only the development and maturation of the male genitalia and sex glands but also affect other tissues such as kidney, liver, and brain. This chapter discusses the endogenous androgens, synthetic analogs, various anabolic agents, and the androgen antagonists employed in clinical practice or animal husbandry in the United States and else-

2   HISTORICAL

The role of the testes in the development and maintenance of the male sex characteristics, and the dramatic physiological effects of male castration have been recognized since early time. Berthold (12) was the first to publish in 1849 a report that gonadal transplantation prevented the effects of castration in roosters, suggesting that the testis produced internal secretions exhibiting androgenic effects. However, the elucidation of the molecules of testicular origin responsible for these actions took almost another century. The first report of the isolation of a substance with androgenic activity was made by Butenandt (13, 14) in 1931. The material, isolated in very small quantities from human male urine (15), was named an-

drosterone (**3**) (16). A second weakly androgenic steroid hormone was isolated from male
urine in 1934. This substance was named dehydroepiandrosterone (**4**) because of its ready
chemical transformation and structural similarity to androsterone (17). A year later Laqueur et al. (18, 19) reported the isolation of
the testicular androgenic hormone, testosterone (**1**), which was 10 times as potent as androsterone in promoting capon comb growth.
Shortly after this discovery, the first chemical
synthesis of testosterone was reported by
Butenandt and Hanisch (20) and confirmed by
Ruzicka (21, 22).

(**3**) androsterone

(**4**) dehydroepiandrosterone

For many years it was believed that testosterone was the active androgenic hormone in
man. In 1968, however, research in two laboratories demonstrated that 5α-dihydrotestos-
terone (DHT, **2**), also referred to as stanolone,
was the active androgen in target tissues, such
as the prostate and seminal vesicles, and was
formed from testosterone by a reductase
present in these tissues (23, 24). Shortly thereafter a soluble receptor protein was isolated
and demonstrated to be specific for DHT and
related structures (25, 26).

The anabolic action of the androgens was
first documented by Kochakian and Murlin in
1935 (27). In their experiments, extracts of

male urine caused a marked retention of nitrogen when injected into dogs fed a constant
diet. Soon afterward testosterone propionate
was observed to produce a similar nitrogen-
sparing effect in humans (28). Subsequent
clinical studies demonstrated that testosterone was capable of causing a major acceleration of skeletal growth and a marked increase
in muscle mass (29–31). This action on muscle
tissue has been referred to more specifically as
the myotrophic effect.

The first androgenic-like steroid used for its
anabolic properties in humans was testosterone.
Unfortunately, its use for this purpose was limited by the inherent androgenicity and the need
for parenteral administration. 17α-Methyltes-
tosterone (**5**) was the first androgen discovered
to possess oral activity, but it too did not show
any apparent separation of androgenic and anabolic activity. The promise of finding a useful,
orally effective, anabolic agent free from androgenic side effects prompted numerous clinical
and biological studies.

(**5**) 17α - methyltestosterone

3 ENDOGENOUS MALE SEX HORMONES

3.1 Occurrence and Physiological Roles

The hormone testosterone affects many organs in the body. Its most dramatic effects are
observed on the primary and secondary sex
characteristics of the male. These actions are
first manifested in the developing male fetus
when the embryonic testis begins to secrete
testosterone. Differentiation of the Wolffian
ducts into the vas deferens, seminal vesicles,
and epididymis occurs under this early androgen influence, as does the development of external genitalia and the prostate (32). The reductive metabolism of testosterone to 5α-
dihydrotestosterone is critical for virilization

during this period of fetal development, as dramatically demonstrated in patients with a 5α-reductase deficiency (33).

At puberty, further development of the sex organs (prostate, penis, seminal vesicles, and vas deferens) is again evident and under the control of androgens. Additionally, the testes now begin to produce mature spermatozoa. Other effects of testosterone, particularly on the secondary sex characteristics, are observed. Hair growth on the face, arms, legs, and chest is stimulated by this hormone during younger years; in later years, testosterone is responsible for thinning of the hair and recession of the hairline. The larynx develops and a deepening of the voice occurs. The male's skin at puberty thickens, the sebaceous glands proliferate, and the fructose content in human semen increases. Testosterone influences sexual behavior, mood, and aggressiveness of the male at the time of puberty.

In addition to these androgenic properties, testosterone also exhibits anabolic (myotropic) characteristics. General body growth is initiated, including increased muscle mass and protein synthesis, a loss of subcutaneous fat, and increased skeletal maturation and mineralization. This anabolic action is associated with a marked retention of nitrogen brought about by an increase of protein synthesis and a decrease of protein catabolism. The increase in nitrogen retention is manifested primarily by a decrease in urinary rather than fecal nitrogen excretion and results in a more positive nitrogen balance. For example, intramuscular administration of 25 mg of testosterone propionate twice daily can produce a nitrogen retention to appear within 1–3 days, reaching a maximum in about 5–8 days. This reduced level of nitrogen excretion may be maintained for at least a month and depends on the patient's nutritional status and diet (34).

Androgens influence skeletal maturation and mineralization, which is reflected in an increase in skeletal calcium and phosphorus (35). In various forms of osteoporosis, androgens decrease urinary calcium loss and improve the calcium balance in patients. This effect is not as noticeable in normal patients. Moreover, the various androgen analogs differ markedly in their effects on calcium and phosphorus balance in man (36). Androgens and

their 5β-metabolites (e.g., etiocholanolone) markedly stimulate erythropoiesis, presumably by increasing the production of erythropoietin and by enhancing the responsiveness of erythropoietic tissue to erythropoietin (37). The effects of androgens on carbohydrate metabolism appear to be minor and secondary to their primary protein anabolic property. The effects on lipid metabolism, on the other hand, seem to be unrelated to this anabolic property. Weakly androgenic metabolites such as androsterone have been found to lower serum cholesterol levels when administered parenterally. A more detailed account of the biological actions of androgens has been published (37).

3.2 Biosynthesis

The androgens are secreted not only by the testis, but also by the ovary and adrenal cortex. Testosterone is the principal circulating androgen and is formed by the Leydig cells of the testes. Other tissues, such as liver and human prostate, form testosterone from precursors, although this contribution to the androgen pool is minimal. Because dehydroepiandrosterone and androstenedione (see Fig. 14.1) are secreted by the adrenal cortex and ovary, they indirectly augment the testosterone pool because they can be rapidly converted to testosterone by peripheral tissues.

Plasma testosterone levels for men usually range between 0.61 and 1.1 μg/100 mL and are 5–100 times female values (39). The circulating level of DHT in normal adult men is about one-tenth the testosterone level (40). Daily testosterone production rates have been estimated to be 4–12 mg for young men and 0.5–2.9 mg for young women (41). Although attempts have been made to estimate the secretion rates for testosterone, these studies are hampered by the number of tissues capable of secreting androgens and the considerable interconversion of the steroids concerned (42, 43).

The synthesis of androgens in the Leydig cells of the testes is regulated by the gonadotropic hormone, luteinizing hormone (LH), also called interstitial cell-stimulating hormone (ICSH). The other pituitary gonadotropin, follicle-stimulating hormone (FSH), acts primarily on the germinal epithelium and is important for sperm development. Both of these pituitary gonadotropins are under the

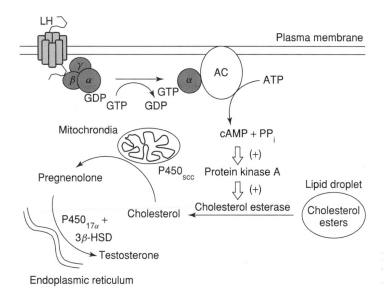

Figure 14.1. Enzymatic conversion of cholesterol to testosterone.

regulation of a decapeptide hormone produced by the hypothalamus. This hypothalamic hormone is luteinizing hormone-releasing hormone (LHRH), also referred to as gonadotropin-releasing hormone (GnRH). In adult males, pulsatile secretion of LHRH, and subsequently LH and FSH, occurs at a frequency of 8–14 pulses in 24 h (44). The secretions of these hypothalamic and pituitary hormones are, in turn, regulated by circulating testosterone levels in a negative feedback mechanism. Testosterone will decrease the frequency and amplitude of pulsatile LH secretion (45), whereas both testosterone and a gonadal peptide, inhibin, are both involved in suppressing the release of FSH (46).

Our understanding of steroidogenesis in the endocrine organs has advanced considerably during the past two decades, based largely on initial investigations with the adrenal cortex and subsequent studies in the testis and ovary as well (47). Figure 14.2 outlines the following sequence of events known to be involved with steroidogenesis in the Leydig cells. LH binds to its receptor located on the surface of the Leydig cell and, by way of a G protein–mediated process, activates adenylyl cyclase to result in an increase in intracellular concentrations of cyclic AMP (cAMP). cAMP activates a cAMP-dependent protein kinase, which subsequently phosphorylates and activates several enzymes involved in the steroido

genic pathway, including cholesterol esterase and cholesterol side-chain cleavage (48). Cholesterol esters (present in the cell as a storage form) are converted to free cholesterol by cholesterol esterase, and free cholesterol is translocated to mitochondria. A cytochrome P-450 mixed-function oxidase system, termed cholesterol side-chain cleavage, converts cholesterol to pregnenolone. Several nonmitochondrial enzymatic transformations then convert pregnenolone to testosterone, which is secreted.

The conversion of cholesterol (**6**) to pregnenolone (**7**) has been termed the rate-limiting step in steroid hormone biosynthesis. The reaction requires NADPH and molecular oxygen and is catalyzed by a cytochrome P-450 enzyme complex termed cholesterol side-chain cleavage. This enzyme complex is composed of three proteins: cytochrome P-450$_{SCC}$ (also called cytochrome P-450 11A), adrenodoxin, and adrenodoxin reductase. Three moles of NADPH and oxygen are required to convert one mole of cholesterol into pregnenolone (Fig. 14.1).

Tracer studies have shown that two major pathways, known as the "4-ene" and "5-ene" pathways, are involved in the conversion of pregnenolone to testosterone. Both these pathways and the requisite enzymes are shown in Fig. 14.1. Earlier studies tended to favor the "4-ene" pathway, but more subse-

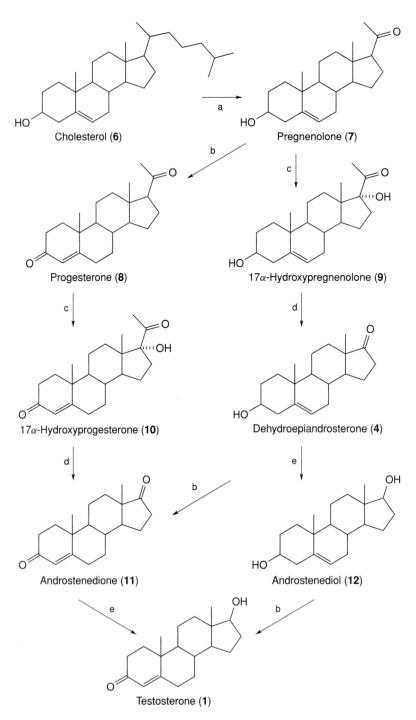

Figure 14.2. Cellular events in steroidogenesis in Leydig cells.

Table 14.1   Mean Concentration of Steroids in Spermatic Venous Plasma from Five Normal Males

Compound	Concentration (μg/100 mL)
5-Androstene-3β,17α-diol	4.0
5-Androstene-3β,17β-diol	18.5
Dehydroepiandrosterone	2.2
Androstenedione	2.5
Testosterone	74.0
Pregnenolone	4.8
17α-Hydroxypregnenolone	3.9
17α-Hydroxyprogesterone	6.2

quent work has disputed this view and suggests that the "5-ene" pathway is quantitatively more important in man. Vihko and Ruokonen (49) analyzed the spermatic venous plasma for free and conjugated steroids. The unconjugated steroids identified in normal males are listed in Table 14.1 along with the concentration in micrograms per 100 mL. All the intermediates of the "5-ene" pathway were identified, but progesterone (**8**), an important intermediate of the "4-ene" pathway, was not found. In addition, sulfate conjugates were present in significant quantities, especially androst-5-ene-3β,17β-diol 3-monosulfate. The data strongly suggest that this intermediate and its unconjugated form constitute an important precursor of testosterone in man. This view, however, was not supported by a kinetic analysis of the metabolism of androst-5-ene-3β,17β-diol (**12**) in man (50). Further evidence that the predominant pathway appears to be the "5-ene" pathway was provided by *in vitro* studies in human testicular tissues (51).

Another important step is the conversion of the C-21 steroids to the C-19 androstene derivatives. Whereas the enzymes for side-chain cleavage are localized in mitochondria, those responsible for cleavage of the C_{17}–C_{20} bond (C_{17}–C_{20} lyase) reside in the endoplasmic reticulum of the cell. Early studies implicated 17α-hydroxypregnenolone (**9**) or 17α-hydroxyprogesterone (**10**) as an obligatory intermediate in testosterone biosynthesis (52), and the C_{17}–C_{20} bond was subsequently cleaved by a second enzymatic process to produce the C-19 androstene molecule. This view of the involve-

ment of two separate enzymes in the conversion of C-21 to C-19 steroids existed until the purification of the proteins in the 1980s. The 17α-hydroxylase/17,20-lyase cytochrome P-450 (abbreviated cytochrome P-450 17 or cytochrome P-450$_{17\alpha}$) was first isolated from neonatal pig testis microsomes by Nakajin and Hall (53). This cytochrome P-450$_{17\alpha}$ possessed both 17α-hydroxylase and 17,20-lyase activity when reconstituted with cytochrome P-450 reductase and phospholipid. Identical full-length human cytochrome P-450$_{17\alpha}$ cDNA sequences were independently isolated and reported in 1987 (54, 55). Extensive reviews of the molecular biology, gene regulation, and enzyme deficiency syndromes have been published (48, 56).

Two additional enzymes are necessary for the formation of testosterone from dehydroepiandrosterone. The first is the 3β-hydroxy steroid dehydrogenase/$\Delta^{4,5}$-isomerase complex, which catalyzes the oxidation of the 3β-hydroxyl group to the 3-ketone and the isomerization of the double bond from C_5–C_6 to C_4–C_5. Again, these processes were originally thought to involve two different enzymes, but purification of the enzymatic activity demonstrated that a single enzyme catalyzes both reactions (57). The final enzyme in the pathway is the 17β-hydroxysteroid dehydrogenase, which catalyzes the reduction of the 17-ketone to the 17β-alcohol.

3.3   Absorption and Distribution

Although considerable research has been devoted to the biochemical mechanism of action of the natural hormones and the synthesis of modified androgens, little is known about the absorption of these substances. It is well recognized that a steroid hormone might have high intrinsic activity but exert little or no biological effects because its physicochemical characteristics prevent it from reaching the site of action. This is particularly true in humans, where slow oral absorption or rapid inactivation may greatly reduce the efficacy of a drug. Even though steroids are commonly given by mouth, little is known of their intestinal absorption. One study in rats showed that androstenedione (**11**) was absorbed better than testosterone or 17α-methyltestosterone, and conversion of testosterone to its acetate enhanced absorption (58). Results with other steroids indicated that lipid solubility was an important factor for intestinal absorption. This may explain the oral activity of certain ethers and esters of testosterone.

Once in the circulatory system by either secretion from the testis or absorption of the administered drug, testosterone and other androgens will reversibly associate with certain plasma proteins, the unbound steroid being the biologically active form. The extent of this binding is dependent on the nature of the proteins and the structural features of the androgen.

The first protein to be studied was albumin, which exhibited a low association constant for testosterone and bound less polar androgens such as androstenedione to a greater extent (59–61). α-Acid glycoprotein (AAG) was shown to bind testosterone with a higher affinity than that of albumin (62, 63). A third plasma protein that binds testosterone is corticosteroid-binding α-globulin (CBG) (64). However, under normal physiological conditions these plasma proteins are not responsible for extensive binding of androgens in plasma.

A specific protein, termed sex steroid binding β-globulin (SBG) or testosterone-estradiol binding globulin (TEBG), was found in plasma that bound testosterone with a very high affinity (65, 66). The SBG–sex hormone complex serves several functions, such as transport or carrier system in the bloodstream, storage site or reservoir for the hormones, and protection of the hormone from metabolic transformations (67, 68). SBG has been purified and contains high affinity, low capacity binding sites for the sex hormones (69). Dissociation constants of approximately $1 \times 10^{-9} M$ have been reported for the binding of testosterone and estradiol to SBG and are 2 orders of magnitude less than values reported for the binding of the hormone to the cytosolic receptor protein (70–72). The plasma levels of SBG are regulated by the thyroid hormones (73) and remain fairly constant throughout adult life in both the male and female (77). SBG is not present in the plasma of all animals (67, 74). For example, SBG-like activity is notably absent in the rat, and testosterone may be bound in the rat plasma to CBG.

Numerous studies have been performed on the specificity of the binding of steroids to hu-

man SBG (67, 68, 74–79). The presence of a 17β-hydroxyl group is essential for binding to SBG. In addition to testosterone, DHT, 5α-androstane-3β,17β-diol (**19**), and 5α-androstane-3α,17β-diol (**20**) bind with high affinity, and these steroids compete for a common binding site. Binding to SBG is decreased by 17α-substituents such as 17α-methyl and 17α-ethinyl moieties and by unsaturation at C-1 or C-6. Also, 19-nortestosterone derivatives have lower affinity. SBG has been purified to homogeneity by affinity chromatography by the use of a DHT-agarose adsorbent (80).

Another extracellular carrier protein exhibiting high affinity for testosterone, found in seminiferous fluid and the epididymis, originates in the testis and is called androgen binding protein (ABP) (81–83). This protein is produced by the Sertoli cells on stimulation by FSH (84, 85) and has very similar characteristics to those of plasma SBG produced in the liver (84).

The absorption of androgens and other steroids from the blood by target cells was usually assumed to occur by a passive diffusion of the molecule through the cell membrane. However, studies in the early 1970s using tissue cultures or tissue slices suggested entry mechanisms for the steroids. Estrogens (86, 87), glucocorticoids (88, 89), and androgens (90–93) exhibit a temperature-dependent uptake into intact target cells, suggesting a protein-mediated process. Among the androgens, DHT exhibited a greater uptake than testosterone in human prostate tissue slices (94), and it was found that estradiol or androstenedione interfered with this uptake mechanism (95, 96). In addition, cyproterone competitively inhibited androstenedione, testosterone, and DHT entry, whereas cyproterone acetate enhanced the uptake of these androgens (93). Little is known about the exit of steroids from target cells; the only reported research has dealt with an active transport of glucocorticoids out of cells (94, 95).

3.4 Metabolism

For decades, the primary function of metabolism was thought to be the inactivation of testosterone, the increase in hydrophilicity, and the mechanism for the excretion of the steroid into the urine. However, the identification of metab-

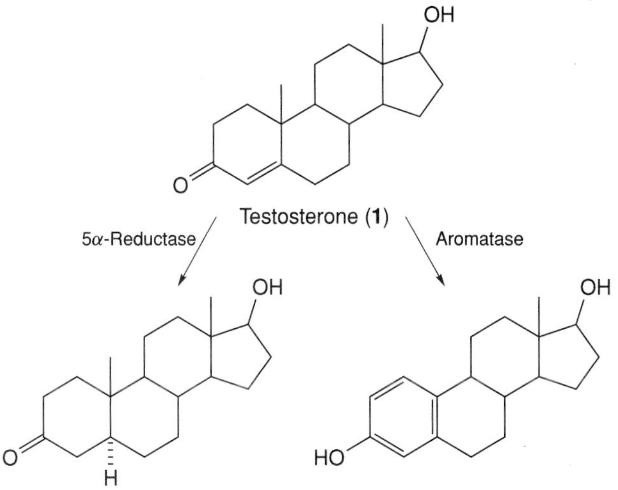

Figure 14.3. Enzymatic conversion of testosterone to biologically active metabolites, 5α-dihydrotestosterone and estradiol.

olites of testosterone formed in peripheral tissues and the potent and sometimes different biological activities of these products emphasized the importance of the metabolic transformations of androgens in endocrinology. Two active metabolites of testosterone have received considerable attention: the reductive metabolite 5α-dihydrotestosterone and the oxidative metabolite estradiol (Fig. 14.3).

3.4.1 Reductive Metabolism. The metabolism of testosterone in a variety of *in vitro and in vivo* systems has been reviewed (52, 96–98). The principal pathways for reductive metabolism of testosterone in man appear in Figure 14.4. Human liver produces a number of metabolites, including androstenedione (**11**), 3β-hydroxy-5α-androstan-17-one (**16**), 5α-androstane-3β,17β-diol (**19**), and 5α-androstane-3α,17β-diol (**20**) (99, 100). In addition, cirrhotic liver was shown to produce more 17-ketosteroids than normal liver (101). Human adrenal preparations, on the other hand, gave 11β-hydroxytestosterone as the major metabolite (102). Intestinal metabolism of testosterone is similar to transformations in the liver (97); the major metabolite in lung is androstenedione (103).

Studies on testosterone metabolism since the late 1960s have centered on steroid transformations by prostatic tissues. Normal prostate, benign prostatic hypertrophy (BPH), and prostatic carcinoma all contain 3α-, 3β-, and

17β-hydroxysteroid dehydrogenases, and 5α- and 5β-reductases, capable of converting testosterone to various metabolites. Prostatic carcinoma metabolizes testosterone more slowly than does BPH or normal prostate (104). In addition, increased levels of androgens are found in BPH (105). Voigt et al. (106–108) have extensively studied *in vivo* metabolic patterns of androgens in patients with BPH. Tritiated androgens were injected intravenously into these patients 30 min before prostatectomy, and prostatic tissue, tissue from surrounding skeletal muscle, and blood plasma were analyzed for metabolites. The major metabolite of testosterone found in BPH tissues was DHT, with minor amounts of diols isolated. Skeletal muscle and plasma contained primarily unchanged testosterone.

Table 14.2 lists the urinary metabolites that have been identified following the administration of testosterone to humans (see Fig. 14.4). These products are excreted as such or in the form of their glucuronide or sulfate conjugates. Androsterone (**3**) and etiocholanolone (**18**), the major urinary metabolites, are excreted predominantly as glucuronides, and only about 10% as sulfates (109). These conjugates are capable of undergoing further metabolism. Testosterone glucuronide, for example, is metabolized differently from testosterone in man, giving rise mainly to 5β-metabolites

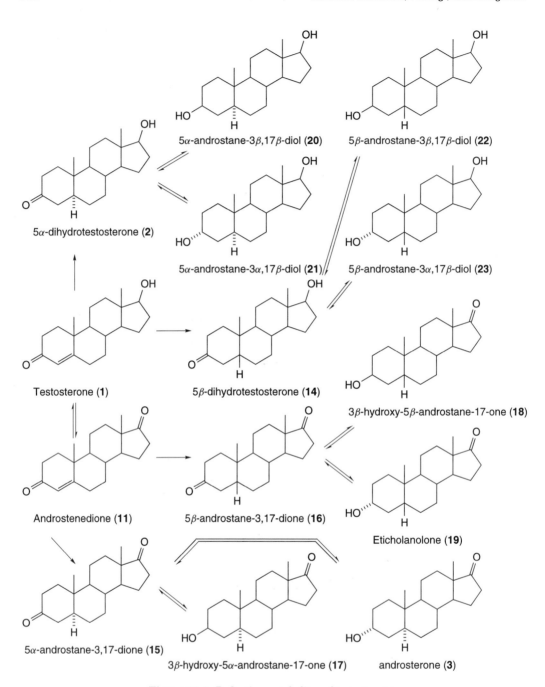

Figure 14.4. Reductive metabolites of testosterone.

(110). Only a relatively small amount of the urinary 17-ketosteroids is derived from testosterone metabolism. In men at least 67% and in women about 80% or more of the urinary 17-ketosteroids are metabolites of adrenocortical steroids (112). This explains why a significant increase in testosterone secretion associated with various androgenic syndromes does not usually lead to elevated levels of 17-ketosteroid excretion.

Table 14.2 Urinary Metabolites of Testosterone[a]

Metabolite	Approximate Conversion (%)
Androsterone	25–50
Etiocholanolone	
5β-Androstane-3α,17β-diol	2
5α-Androstane-3α,17β-diol	1
5α-Androstan-3β-ol-17-one	1
Androst-16-en-3α-ol	0.4
3α,18-Dihydroxy-5β-androstan-17-one	0.3
3α,7β-Dihydroxy-5β-androstan-17-one	Trace
11β-Hydroxytestosterone	Trace
6α-, 6β-Hydroxytestosterone	Trace

[a] Cf. Ref. 111.

Although androsterone and etiocholanolone are the major excretory products, the exact sequence whereby these 17-ketosteroids arise is still not clear. Studies with radiolabeled androst-4-ene-3β,17β-diol and the epimeric 3α-diol in humans showed that oxidation to testosterone was necessary before reduction of the A-ring (113). Moreover, in rats 5β-androstane-3α,17β-diol (**22**) was the major initial liver metabolite, but this decreased with time with the simultaneous increase of etiocholanolone (114). This formation of saturated diols agrees with studies using human liver (99) and provides evidence that the initial step in testosterone metabolism is reduction of the α,β-unsaturated ketone to a mixture of diols followed by oxidation to the 17-ketosteroids.

Until 1968, it was generally thought that the excretory metabolites of testosterone were physiologically inert. Subsequent work has shown, however, that etiocholanolone has thermogenic effects when administered to man (115). Moreover, the hypocholesterolemic effects of parenterally administered androsterone have been described (116).

The conversion of testosterone to DHT by 5α-reductase has major importance in the mechanism of action of the hormone. This enzymatic activity has been found in the endoplasmic reticulum (117, 118) and in the nuclear membrane (119–125) of androgen-sensitive cells. In addition, the levels of 5α-reductase are under the control of testosterone and DHT (125); 5α-reductase activity

decreases after castration and can be restored to normal levels of activity with testosterone or DHT administration (126).

Early biochemical studies of 5α-reductase were performed using a microsomal fraction from rat ventral prostate. The irreversible enzymatic reaction catalyzed by 5α-reductase requires NADPH as a cofactor, which provides the hydrogen for carbon-5 (127). The 5α-reductase from rat ventral prostate tissues exhibited a broad range of substrate specificity for various C_{19} and C_{21} steroids (101); this broad specifity was also observed in inhibition studies (128). However, more detailed studies of the enzyme were limited because of the extreme hydrophobic nature of the protein, its instability upon isolation, and its low concentrations in androgen-dependent tissues (98).

Investigations on the molecular biology of 5α-reductase resulted in the demonstration of two different genes and two different isozymes of the enzyme (129–131). The first cDNA isolated and cloned that encoded 5α-reductase was designated Type 1, and the second was designated Type 2. The gene encoding Type 1 is located on chromosome 5, whereas the gene encoding Type 2 is located on chromosome 2. The two human 5α-reductases have approximately 60% sequence homology. These two isozymes differ in their biochemical properties, tissue location, and function (131, 132). Type 1 5α-reductase exhibits an alkaline pH optimum (6–8.5) and has micromolar affinities for steroid substrates. On the other hand, Type 2 5α-reductase has a sharp pH optimum at 4.7–5.5, has higher affinity (lower apparent K_m) for testosterone, and is more sensitive to inhibitors than the Type 1 isozyme. The Type 2 isozyme is expressed primarily in androgen target tissues, the liver expresses both types, and the Type 1 form is expressed in various peripheral tissues. Type 2 5α-reductase appears to be essential for masculine development of the fetal urogential tract and the external male phenotype, whereas the Type 1 isozyme is primarily a catabolic enzyme. In certain cases of human male pseudohermaphroditism, mutations in the Type 2 5α-reductase gene are observed and results in significant decreases in DHT levels needed for virilization (133).

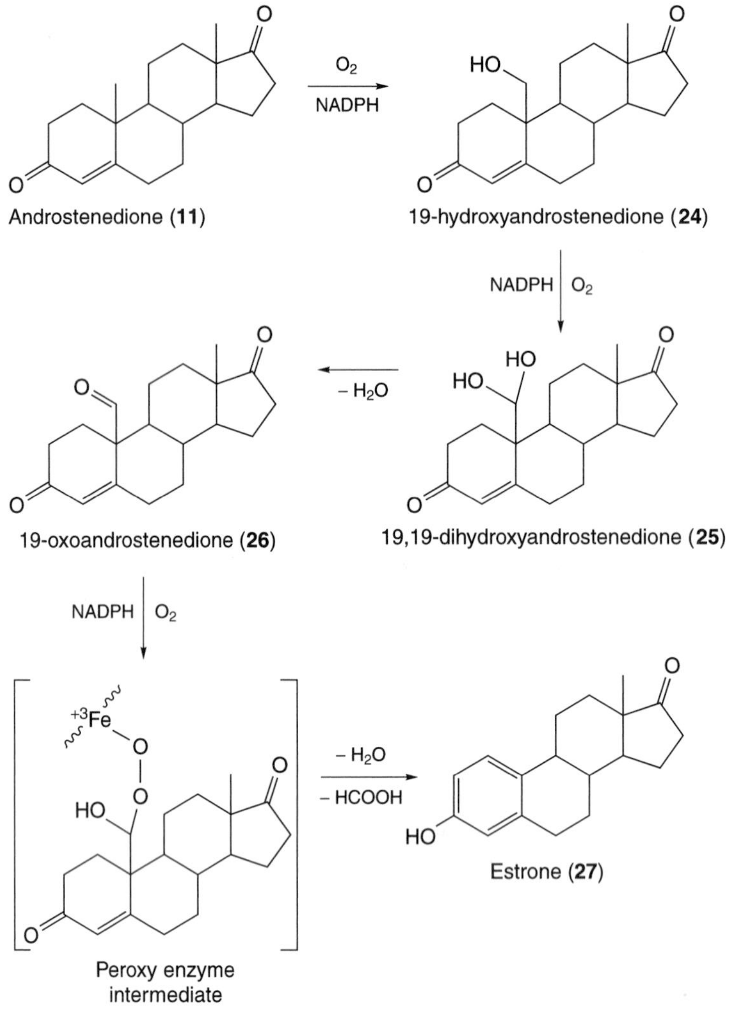

Figure 14.5. Aromatization of androgens.

3.4.2 Oxidative Metabolism. Another metabolic transformation of androgens leading to hormonally active compounds is their conversion to estrogens. Estrogens are biosynthesized in the ovaries and placenta and, to a lesser extent, in the testes, adrenals, and certain regions of the brain. The enzyme complex that catalyzes this biosynthesis is referred to as aromatase, and the enzymatic activity was first identified by Ryan (134) in the microsomal fraction from human placental tissue. The elucidation of the mechanism of the aromatization reaction began in the early 1960s and continues to receive extensive study. It is a

cytochrome P-450 enzyme complex (135) and requires 3 moles of NADPH and 3 moles of oxygen per mole of substrate (136). Aromatization proceeds through three successive steps, with the first two steps being hydroxylations. The observation by Meyer (137) that 19-hydroxyandrostenedione (**23**) was a more active precursor of estrone (**25**) than the substrate androstenedione led to its postulated role in estrogen biosynthesis. This report and numerous studies that followed led to the currently accepted pathway for aromatization, as shown in Fig. 14.5. Relative percentages of substrate activity are presented in Table 14.3.

Table 14.3   Relative Substrate Activity in Aromatization

Substrate	Activity (%)
C_{19} Steroids	
Androst-4-ene-3,17-dione	100
19-Hydroxyandrost-4-ene-3,17-dione	184, 133
17β-Hydroxyandrost-4-en-3-one	100
3β-Hydroxyandrost-5-en-17-one	66
5α-Androst-1-ene-3,17-dione	0
5α-Androstane-3,17-dione	0
1α-Hydroxyandrost-4-ene-3,17-dione	0
17β-Hydroxy-1α-methylandrost-4-ene-3-one	0
2β-Hydroxyandrost-4-ene-3,17-dione	15
2α-Hydroxyandrost-4-ene-3,17-dione	0
17β-Hydroxy-2β-methylandrost-4-en-3-one	0
11β-Hydroxyandrost-4-ene-3,17-dione	0
11α-Hydroxyandrost-4-ene-3,17-dione	100
17β-Hydroxy-17α-methylandrost-4-en-3-one	44
6β-Hydroxyandrost-4-ene-3,17-dione	21
6α-Fluoro-17β-hydroxyandrost-4-ene-3,17-dione	0
6β-Fluoro-17β-hydroxyandrost-4-ene-3,17-dione	0
9α-Fluoroandrosta-1,4-diene-3,17-dione	55
Androsta-1,4-diene-3,17-dione	22, 35
Androsta-4,6-diene-3,17-dione	0
Androsta-1,4,6-triene-3,17-dione	0
Androst-4-ene-3,11,17-trione	0
C_{18} Steroids	
Estr-4-ene-3,17-dione	21
17β-Hydroxyestr-4-en-3-one	20
17β-Hydroxy-5α,10β-estr-3-one	0
C_{12} Steroids	
Pregn-4-ene-3,20-dione	0
17α,19,21-Trihydroxypregn-4-ene-3,20-dione	0

The first two oxidations occur at the C_{19} position, producing the 19-alcohol (**23**) and then the 19-*gem*-diol (**24**), originally isolated as the 19-aldehyde (**25**) (138, 139). The exact mechanism of the last oxidation remains to be fully determined. The final oxidation results in the stereospecific elimination of the 1β and 2β hydrogen atoms (140–142) and the concerted elimination of the oxidized C_{19} moiety as formic acid (139). Hydroxylation at the 2β-position was suggested as an intermediate in this final oxidation, given that this substance spontaneously aromatized to estrone (143). However, investigations using $^{18}O_2$ and isotopically labeled steroidal intermediates failed to show incorporation of the 2β-hydroxyl group into formic acid under enzymatic or nonenzymatic conditions (144) and did demonstrate that the oxygen atoms from the first and third oxidation steps are incorporated

into formic acid (145–147). These results have led to the proposal that the last oxidation step is a peroxidative attack at the C_{19} position (148–150).

Incubation of a large number of testosterone analogs with human placental tissue (151–153) has provided some insight into the structural requirements for aromatization (see Table 14.4). Whereas androstenedione was converted rapidly to estrone, the 1-dehydro and 19-nor analogs were metabolized slowly, and the 6-dehydro isomer and saturated 5α-androstane-3,17-dione remained unchanged. Hydroxyl and other substituents at 1α, 2β, and 11β interfered with aromatization, whereas similar substituents at 9α and 11α seemingly had no effect. Of the stereoisomers of testosterone, only the 8β,9β,10β-isomer is aromatized, in addition to compounds having the normal configuration (8β, 9α, 10β). Thus,

Table 14.4 Comparison of the Androgenic and Myotrophic Activities of Testosterone Derivatives in the Chick Comb and Castrated Male Rat Assays after Subcutaneous Administration

Testosterone Modification	Increase in Weight (%)		
	Chick Comb	Rat Ventral Prostate	Rat Levator Ani
None	100	100	100
19-Nor	81	42	90
7α-Methyl	11	97	135
7α-Methyl-19-nor	75	218	226
14-Dehydro	128	54	8
14-Dehydro-19-nor	320	69	133
14-Dehydro-7α-methyl-19-nor	435	352	330

the substrate specificity of aromatase appears to be limited to C_{19} steroids with the 4-en-3-one system. Inhibition studies with various steroids have provided additional insights into the structural requirements for the enzyme (154–156); steroidal aromatase inhibitors are described later in section .

Recent research in aromatase has focused on the biochemistry, molecular biology, and regulation of the aromatase protein. Aromatase is a membrane-bound cytochrome P-450 monooxygenase consisting of two proteins, aromatase cytochrome P-450 (P-450$_{arom}$) and NADPH-cytochrome P-450 reductase. Cytochrome P-450$_{arom}$ is a heme protein that binds the steroid substrate and molecular oxygen and catalyzes the oxidations. The reductase is a flavoprotein, is found ubiquitously in endoplasmic reticulum, and is responsible for transferring reducing equivalents from NADPH to cytochrome P-450$_{arom}$. Purification of cytochrome P-450$_{arom}$ proved to be very difficult because of its membrane-bound nature, instability, and low tissue concentration. The combination of hydrophobic chromatography using phenyl Sepharose, nonionic detergent, and the presence of micromolar amounts of substrate androstenedione yielded a highly purified and active cytochrome P-450$_{arom}$, with the highest specific content of 11.5 nmol of cytochrome P-450 per mg of protein reported (157). Reconstitution of this cytochrome P-450$_{arom}$ with NADPH-cytochrome P-450 reductase and phospholipid resulted in complete conversion of androstenedione to estrone, thus demonstrating that one cytochrome P-450 protein catalyzes

all three oxidation steps. Knowledge of the molecular biology of aromatase has advanced greatly in the past 5 years. A full-length cDNA complementary to mRNA encoding cytochrome P-450$_{arom}$ was sequenced and the open reading frame encodes a protein of 503 amino acids (158). This cDNA sequence was inserted into COS1 monkey kidney cells, and aromatase mRNA and aromatase enzymatic activity were detected in transfected cells. The entire human cytochrome P-450$_{arom}$ gene is greater than 70 kb in size (159, 160) and is located on chromosome 15 (161). Clones have been used to examine the regulation of aromatase in ovarian, adipose, and breast tissues (162, 163–166).

The metabolism of androgens by the mammalian brain has also been investigated under *in vitro* conditions. Sholiton et al. (167) in 1966 first reported the metabolism of testosterone in rat brain, and later studies demonstrated the conversion of testosterone to DHT, androstenedione, 5α-androstane-3,17-dione, and 5α-androstane-3β,17β-diol (168–173). The aromatization of androgens to estrogens was also found to occur in the hypothalamus and the pituitary gland (174–179). The full significance of these metabolites on various neuroendocrine functions, such as regulation of gonadotropin secretion and sexual behavior, is not yet fully understood (180, 181).

3.5 Mechanism of Action

It would indeed be impossible to explain all the varied biological actions of testosterone by one biochemical mechanism. Androgens, as well as the other steroid hormones adrenocorticoids,

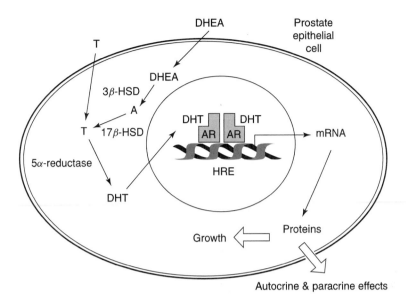

Figure 14.6. Mechanism of action of 5α-dihydrotestosterone.

estrogens, and progestins exert potent physiological effects on sensitive tissues and yet are present in the body in only extremely low concentrations (e.g., 0.1–1.0 nM). The majority of investigations concerning the elucidation of the mechanisms of action of androgens have dealt with actions in androgen-dependent tissues and, in particular, the rat ventral prostate. The results of these studies indicate that androgens primarily act in regulating gene expression and protein biosynthesis by formation of a hormone–receptor complex, analogous to the mechanisms of action of estrogens and progestins. Extensive research activities directed at elucidation of the general mechanism of steroid hormone action have been performed for over three decades, and several reviews have appeared on this subject (182–195).

Jensen and Jacobson (196), using radiolabeled 17β-estradiol, were the first to show that a steroid was selectively retained by its target tissues. Investigations of a selective uptake of androgens by target cells performed in the early 1960s were complicated by low specific activity of the radiolabeled hormones and the rapid metabolic transformations. Nonetheless, it was noted that target cells retained primarily unconjugated metabolites, whereas conjugated metabolites were present in non-

target cells such as blood and liver (197, 198). With the availability of steroids with high specific activity, later studies demonstrated the selective uptake and retention of androgens by target tissues (23, 24, 199–201). In addition, DHT was found to be the steroidal form selectively retained in the nucleus of the rat ventral prostate (23, 200). This discovery led to the current concept that testosterone is converted by 5α-reductase to DHT, which is the active form of cellular androgen in androgen-dependent tissues.

The rat prostate has been the most widely examined tissue, and current hypotheses on the mode of action of androgens are based largely on these studies (see Fig. 14.6). The lipophilic steroid hormones are carried in the bloodstream, with the majority of the hormones reversibly bound to serum carrier proteins and a small amount of free steroids. The androgens circulating in the bloodstream are the sources of steroid hormone for androgen action in target tissues. Testosterone, synthesized and secreted by the testis, is the major androgen in the bloodstream and the primary source of androgen for target tissues in men. Dehydroepiandrosterone (DHEA) and androstenedione also circulate in the bloodstream and are secreted by the adrenal gland under the regulation of adrenocorticotrophic hor-

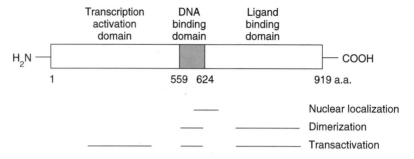

Figure 14.7. Schematic diagram of androgen receptor.

mone (ACTH). DHEA and androstenedione supplement the androgen sources in normal adult men, but these steroids are the important circulating androgens in women. The free circulating androgen steroids passively diffuse through the cell membrane and are converted to the active androgen 5α-DHT within the target cells.

The androgens act on target cells to regulate gene expression and protein biosynthesis through the formation of steroid–receptor complexes. Those cells sensitive to the particular steroid hormone (referred to as target cells) contain high affinity steroid receptor proteins capable of interacting with the steroid (25, 202). The binding of DHT with the receptor protein is a necessary step in the mechanism of action of the steroid in the prostate cell. Early studies suggested that the steroid receptor proteins were located in the cytosol of target cells (196) and, after formation of the steroid–receptor complex, the steroid–receptor complex translocated into the nucleus of the cell. More recent investigations on androgen action indicate that active, unoccupied receptor proteins are present only in the nucleus of the cell. This nuclear localization of receptor is also observed for estrogens and progestins, whereas the majority of the glucocorticoid receptor is located in the cytoplasm (190, 203). In the current model, DHT is formed in the cytoplasm, then enters the nucleus of the cell and binds to the nuclear steroid receptor protein.

The binding of DHT to the nuclear androgen receptor initiates a conformational change or activation of the steroid–nuclear receptor complex and results in the formation of a homodimer (191). The homodimer interacts

with particular regions of the cellular DNA, referred to as androgen responsive elements (AREs), and with various nuclear transcriptional factors. Binding of the nuclear steroid–receptor complex to DNA initiates transcription of the DNA sequence to produce messenger RNA (mRNA). Finally, the elevated levels of mRNA lead to an increase in protein synthesis in the endoplasmic reticulum; these proteins include enzymes, receptors, and/or secreted factors that subsequently result in the steroid hormonal response regulating cell function, growth, and differentiation.

Extensive structure–function studies on the androgen receptor (AR) have identified regions critical for hormone action. The androgen receptor is encoded by the AR gene located on the X chromosome, and the AR gene is composed of 8 exons. The human AR contains approximately 900–920 amino acids, and the exact length varies because of polymorphisms in the NH_2-terminal of the protein. The primary amino acid sequences of AR, as well as of the various steroid hormone receptors, were deduced from cloned complementary DNAs (cDNAs) (191, 193). The calculated molecular weight of AR is approximately 98,000 kDa, based on amino acid composition; however, the AR is a phosphoprotein and migrates higher at approximately 110 kDa in SDS gel electrophoresis. The steroid receptor proteins are part of a larger family of nuclear receptor proteins that also include receptors for vitamin D, thyroid hormones, and retinoids. The overall structure of the androgen receptor (shown in Fig. 14.7) has strong similarities to the other steroid hormone receptors, with the proteins containing regions that bind to the

DNA and bind to the steroid hormone ligand (194, 204, 205). A high degree of homology (sequence similarities) in the steroid receptors is found in the DNA binding region that interacts with the HREs. The DNA binding region is rich in cysteine amino acids and chelate zinc ions, forming fingerlike projections called zinc fingers that bind to the DNA. The hormone binding domain (or ligand binding domain) is located on the COOH-terminal of the protein. Structure–function studies of cloned receptor proteins also identify regions of the molecules that are important for nuclear localization of the receptor, receptor dimerization, interactions with nuclear transcriptional factors, and activation of gene transcription. Importantly, two regions of the androgen receptor protein are identified as transcriptional activation domains. The domain on the NH$_2$-terminal region may interact with both coactivators and corepressors, and the COOH-terminal domain initiates transcriptional activation only upon binding of an agonist such as 5α-DHT. The interactions necessary for formation of the steroid–receptor complexes and subsequent activation of gene transcription are complicated, involve multiple protein partners referred to as coactivators, and leave many unanswered questions.

Although the tertiary structure of the entire androgen receptor has not been determined, the crystallographic structure of the ligand binding domain (LBD) has recently been reported (206, 207). The AR LBD consists of an α-helical sandwich, similar to the LBDs reported for other nuclear receptors, and AR LBD contains only 11 helices (no helix 2) and four short β-strands (Fig. 14.8). Minor differences in the two reported crystallographic structures are likely attributable to limits of experimental resolution, differences in data interpretation, and the use of different ligands for crystallization. The endogenous ligand DHT (**2**) interacts with helices 3, 5, and 11, and the DHT-bound AR LBD has a single, continuous helix 12 (207). Similar interactions are observed with metribolone (methyltrienolone, **53**); however, helix 12 is split into two shorter helical segments (206).

Additional information on receptor structure–function has been obtained by analyzing androgen receptor mutations in patients with

Figure 14.8. Ribbon diagram of androgen receptor ligand binding domain (LBD) bound with 5α-dihydrotestosterone. Atomic coordinates were obtained from the Protein Data Bank (PDB ID code 1I37; www.rcsb.org) and displayed using ViewerLite 4.2, Accerlys Inc. 5α-Dihydrotestosterone is illustrated as a stick structure behind Helix 12.

various forms of androgen resistance and abnormal male sexual development (194, 205, 208–210). Two polymorphic regions have been identified in the NH$_2$-terminal region, encoding a polyglycine repeat and a polyglutamate tract. Currently, these polymorphic regions have not been shown to significantly alter AR levels, stability, or transactivation (205). These repeats are useful in pedigree analysis of patients (194). Mutations in the androgen receptor have been identified in patients with either partial or full androgen insensitivity syndrome (AIS), with the majority of the mutations identified in exons 4 through 8 encoding the DNA binding domain and the hormone binding domain. Finally, studies with the human LNCaP prostate cancer cell line have provided interesting results regarding receptor protein structure and ligand specificity. The LNCaP cells exhibited enhanced proliferation in the presence of androgens, but also these cells unexpectedly proliferated in the presence of estrogens, progestins, cortisol, or the anti-androgen flutamide (211, 212). Analysis of the cDNA for the LNCaP androgen receptor revealed that a single base mutation in the ligand binding domain was present and resulted in the increased affinity for progesterone and estradiol (213). The crystallographic structures of the LBD with the T877A mutation confirm that the mutated AR LBD can accommodate larger structures at the C-17 position (206, 207).

The ultimate action of androgens on target tissues is the stimulation of cellular growth and differentiation through regulation of protein synthesis, and numerous androgen-inducible proteins have been identified (205). One of the prominent androgen-inducible proteins is prostate-specific antigen (PSA), a serine protease expressed by secretory prostate epithelial cells and used as blood test in screening for possible prostate diseases such as prostate cancer. Three AREs have been identified in the promoter regions of the PSA gene (214–216). Another androgen-regulated gene examined extensively in rats is the gene encoding the protein probasin (217, 218), a 20-kDa secretory protein from the rat dorsolateral prostate structurally similar to serum globulins. Other proteins induced by androgens include spermine-binding protein (219), keratinocyte growth factor (KGF or FGF-7) (220), androgen-induced growth factor (AIGF or FGF-8) (221, 222), nerve growth factor (223), epidermal growth factor (224), c-myc (225), protease D (226), β-glucuronidase (227), and α_{2u}-globulin (228–230). Studies of these proteins suggest that the androgens act by enhancing transcription and/or translation of specific RNAs for the proteins. Also, the androgen receptor represses gene expression of certain proteins such as glutathione S-transferase, TRMP-2 involved in apoptosis, and cytokines such as interleukin-4, interleukin-5, and γ-interferon (205, 231).

Although most biochemical studies focused on the rat ventral prostate, some researchers

began to investigate the presence of cellular receptor proteins in other androgen-sensitive tissues. Androgen receptors have been reported in seminal vesicles (232, 233), sebaceous gland (234–236), testis (235, 237), epididymis (232, 238, 239), kidney (240), submandibular gland (241, 242), pituitary and hypothalamus (243–249), bone marrow (250, 251), liver (252), and androgen-sensitive tumors (253, 254). Although DHT is the active androgen in rat ventral prostate, it is not the only functioning form in other androgen-sensitive cells. In ventral prostate and seminal vesicles, DHT is readily formed. It is metabolized only slowly, however, and therefore can accumulate and bind to receptors. Also, comparison of binding kinetics for testosterone and DHT demonstrated that testosterone dissociates faster, implying extended retention of DHT by the androgen receptor (255). In other tissues, such as brain or kidney, DHT is not readily formed and is metabolized quickly compared to testosterone. Species variations have also been demonstrated. The most striking example is the finding that 5α-androstane-$3\alpha,17\alpha$-diol interacts specifically with cytosolic receptor protein from dog prostate (256) and may be the active androgen in this species (257). Apparently the need for a 17β-hydroxyl is not essential in all species.

Thus, current findings indicate that androgen receptor proteins vary in steroid specificity among different tissues from the same species as well as among different species. Nevertheless, the basic molecular mechanism of action of the androgens in androgen-sensitive tissues is consistent with the results of the studies on rat ventral prostate.

The manner whereby the androgens exert their anabolic effects has not been as extensively studied. The conversion of testosterone to DHT was been shown to be insignificant in skeletal and levator ani muscles, suggesting that the androgen-mediated growth of muscle is ascribed to testosterone itself (258, 259). Classical steroid receptors for testosterone are found in the cytoplasm of the levator ani and quadriceps muscles of the rat (260, 261). Unlike prostate receptor protein, DHT had a lower affinity than that of testosterone for this protein. Androgen receptors have also been identified in other muscle tissues as well, including cardiac muscle (262–267).

4 SYNTHETIC ANDROGENS

4.1 Current Drugs on the Market

4.2 Therapeutic Uses and Bioassays

The primary uses of synthetic androgens are the treatment of disorders of testicular function and of cases with decreased testosterone production. Several types of clinical conditions result from testicular dysfunction. Informa-

Generic Name (structure)	Trade Name	U.S. Manufacturer	Chemical Class	Dose
Testosterone enanthate (31)	Delatestryl	BTG Pharm	Androstane	Injection: 200 mg/mL
Testosterone cypionate (33)	Depo-Testosterone	Pharmacia	Androstane	Injection: 100 mg/mL 200 mg/mL
Testosterone pellets (1)	Testopel	Bartor Pharmacal	Androstane	Pellets: 75 mg
Testosterone transdermal system (1)	Testoderm	Alza	Androstane	Transdermal: 10 mg 15 mg
	Testoderm TTS	Alza	Androstane	Transdermal: 328 mg
	Testoderm with Adhesive	Alza	Androstane	Transdermal: 15 mg
	Androderm	Watson Pharma	Androstane	Transdermal: 12.2 mg 24.3 mg
Testosterone gel (1)	AndroGel 1%	Unimed Pharm.	Androstane	Gel: 1% testosterone
Methyltestosterone (5)	Methyltestosterone	Various suppliers	Androstane	Tablets: 10 mg 25 mg Tablets (buccal): 10 mg
	Testred	ICN Pharm	Androstane	Capsules: 10 mg
	Android	ICN Pharm	Androstane	Capsules: 10 mg
	Virilon	Star	Androstane	Capsules: 10 mg
	Virilon IM	Star	Androstane	Injection: 200 mg/mL
Fluoxymesterone (36)	Halotestin	Pharmacia	Androstane	Tablets: 2 mg 5 mg 10 mg
	Fluoxymesterone	Various suppliers	Androstane	Tablets: 10 mg
Danazol (77)	Danazol	Various suppliers	Androstane	Capsules: 50 mg 100 mg 200 mg
	Danocrine	Sanofi-Synthelabo	Androstane	Capsules: 50 mg 100 mg 200 mg
Testolactone (38)	Teslac	Bristol-Myers Squibb	Androstane	Tablets: 50 mg

tion on the biochemistry and mechanism of action of testosterone that has accumulated over the past 20 years has greatly aided in the elucidation of the underlying pathophysiology of these diseases. Two reviews describe in greater detail the mechanisms involved in disorders of testicular function and androgen resistance (268, 269).

Hypogonadism arises from the inability of the testis to secrete androgens and can be caused by various conditions. These hypogonadal diseases can, in many cases, result in disturbances in sexual differentiation and function and/or sterility. Primary hypogonadism is the result of a basic disorder in the testes, whereas secondary hypogonadism results from the failure of pituitary and/or hypothalamic release of gonadotropins and thus diminished stimulation of the testis. Usually primary hypogonadism is not recognized in early childhood (with the exception of cryptorchidism) until the expected time of puberty. This testosterone deficiency is corrected by androgen treatment for several months, at which time the testes are evaluated for possible development. Long-term therapy is necessary if complete testicular failure is present. Patients with Klinefelter's syndrome, a disease in which a genetic male has an extra X chromosome, have low testosterone levels and can also be treated by androgen replacement.

Male pseudohermaphroditism are disorders in which genetically normal men do not undergo normal male development. One type, testicular feminization, is observed in patients who have normal male XY chromosomes, but the male genitalia and accessory sex glands do not develop. Rather, the patients have female external genitalia. These patients are unresponsive to androgens and have defective androgen receptors (270–272). Another type of male pseudohermaphroditism results from a deficiency of the enzyme 5α-reductase (273, 274). Because DHT is necessary for early differentiation and development, the patients again develop female genitalia; later, some masculinization can occur at the time of puberty because of elevated testosterone levels in the blood. A third disorder is Reifenstein syndrome, an incomplete pseudohermaphroditism. In these patients, the androgen levels are normal, 5α-reductase is present, and elevated LH levels are found. Partially deficient androgen receptors are present in these patients (270, 272). In most cases of male pseudohermaphroditism, androgen replacement has little or no effect and thus steroid treatment is not recommended.

Deficiencies of circulating gonadotropins lead to secondary hypogonadism. This condition can be caused by disorders of the pituitary and/or hypothalamus resulting in diminished secretions of neurohormones. The lack of stimulation of the seminiferous tubules and the Leydig cells attributed to the low levels of these neurohormones decreases androgen production. Drugs such as the neuroleptic phenothiazines and the stimulant marihuana can also interfere with release of gonadotropins. The use of androgens in secondary hypogonadism is symptomatic.

Synthetic androgens have also been used in women for the treatment of endometriosis, abnormal uterine bleeding, and menopausal symptoms. However, their utility is severely limited by the virilizing side effects of those agents. Two weak androgens, calusterone and 1-dehydrotestolactone, are used clinically in the treatment of mammary carcinoma in women. The mode of action of these drugs in the treatment of breast cancer is unknown, and is not simply related to their androgenicity (276). More recent evidence on the ability of these compounds to inhibit estrogen biosynthesis catalyzed by aromatase suggests that they effectively lower estrogen levels *in vivo* (276).

The various analytical methods used to establish the androgenic properties of steroidal substances have been reviewed by Dorfman (277). Traditionally, androgens have been assayed by the capon comb growth method and by the use of the seminal vesicles and prostate organs of the rodent. Increases in weight and/or growth of the capon comb have been used to denote androgenic activity after injection or topical application of a solution of the test compound in oil (278). A number of minor modifications of this test have been described (279–281). The increase in weight of the seminal vesicles and the ventral prostate of the immature castrated male rat has provided another measure of androgenic potency (282–285). The test compound is administered either intramuscularly or orally, and the weight of the target organs is compared with those of control animals. *In vitro* evaluation of the relative affinity of potential androgens for the androgen receptor has also become an important tool in assessing biological activity of androgens (128, 286).

4.3 Structure-Activity Relationships for Androgens

4.3.1 Early Modifications. Most of the early structure-activity relationship studies concerned minor modifications of testosterone and other naturally occurring androgens. Studies in animals (287) and humans (288) showed the 17β-hydroxyl function to be essential for androgenic and anabolic activity. In certain cases esterification of the 17β-hydroxyl group not only enhanced but also prolonged the anabolic and androgenic properties (289). The early statement (290) that the 1-dehydro isomer of testosterone (**28**) was a weak androgen was subsequently disproved. This isomer and related compounds are potent androgenic and anabolic steroids (291).

(28)

Reduction of the A-ring functional groups has variable effects on activity. For example, conversion of testosterone to DHT has little effect or may increase potency in a variety of bioassay systems (292, 293). On the other hand, changing the A/B *trans* stereochemistry of known androgens such as androsterone (**3**) and DHT to the A/B *cis*-etiocholanolone (**18**) and 17β-hydroxy-5β-androstan-3-one (**13**), respectively, drastically reduces both the anabolic and androgenic properties (294–296). These observations established the importance of the A/B *trans* ring juncture for activity.

4.3.2 Methylated Derivatives. The discovery that C-17α-methylation conferred oral activity on testosterone prompted the synthesis of additional C-17α-substituted analogs. Increasing the chain length beyond methyl invariably led to a decrease in activity (297). As a result of these studies, however, 17α-methyl-androst-5-ene-3β,17β-diol (methandriol, **29**) was widely evaluated in humans as an anabolic agent. Early biological studies with methandriol had shown a separation of anabolic and androgenic properties, although this was not confirmed by subsequent studies (284). In addition, clinical studies showed no advantage of methandriol over 17α-methyl-testosterone (**5**) (298).

(29)

4.3.3 Ester Derivatives. As early as 1936 it was known that esterification of testosterone markedly prolonged the activity of this androgen when it was administered parenterally (299). This modification enhances the lipid solubility of the steroid and, after injection, permits a local depot effect. The acyl moiety is usually derived from a long-chain aliphatic or arylaliphatic acid such as propionic, heptanoic (enanthoic), decanoic, cyclopentylpropionic (cypionic), or β-phenylpropionic acid (**30**) (**30–34**).

4.3.4 Halo Derivatives. In general, the preparation of halogenated testosterone derivatives has been therapeutically unrewarding. 4-Chloro-17β-hydroxyandrost-4-en-3-one (chlorotestosterone, **35**) and its derivatives are the only chlorinated androgens that have been used clinically, albeit sparingly (300). The introduction of a 9α-fluoro and an 11β-hydroxy substitutents (analogous to synthetic glucocorticoids) gives 9α-fluoro-11β,17β-dihydroxy-17α-methylandrost-4-en-3-one (fluoxymesterone, Halotestin, **36**), which is an

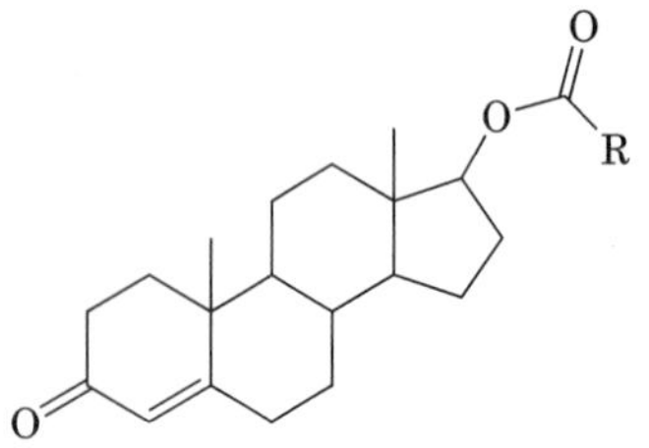

(30) propionate   R = CH$_2$CH$_3$

(31) heptanoate (enanthate)   (CH$_2$)$_5$CH$_3$

(32) decanoate   (CH$_2$)$_8$CH$_3$

(33) cyclopentylpropionate (cypionate)   CH$_2$CH$_2$

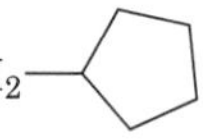

(34) β - phenylpropionate   CH$_2$CH$_2$

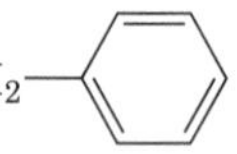

orally active androgen exhibiting approximately fourfold greater oral activity than that of 17α-methyltestosterone. Early clinical studies with fluoxymesterone indicated an anabolic potency of 11 times that of the unhalogenated derivative (301–303). Nitrogen balance studies, however, revealed an activity of only 3 times that of 17α-methyltestosterone (304). Because of the lack of any substantial separation of anabolic and androgenic activity, halotestin is used primarily as an orally effective androgen, particularly in the treatment of mammary carcinoma (305, 306).

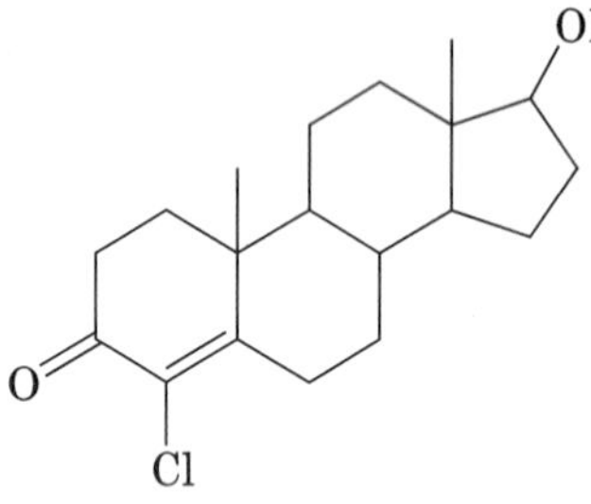

(35) chlorotestosterone

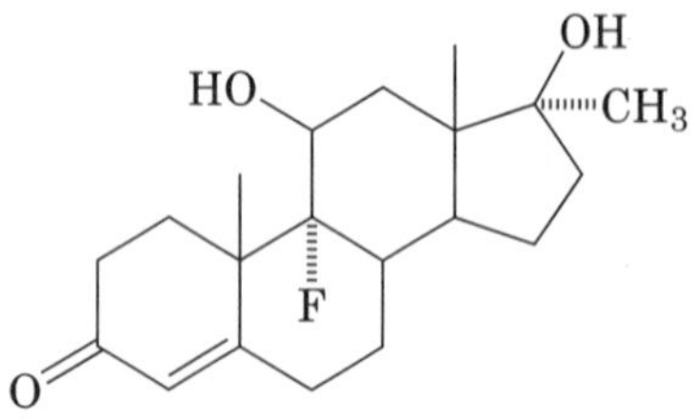

(36) fluoxymesterone

4.3.5 Other Androgen Derivatives. Several synthetic steroids having weak androgenic ac-

tivity are being utilized in patients. 7β,17α-Dimethyltestosterone (calusterone, **37**) and 1-dehydrotestolactone (Testlac, **38**) are very weak androgenic agents that have been used in the treatment of advanced metastatic breast cancer (307–309).

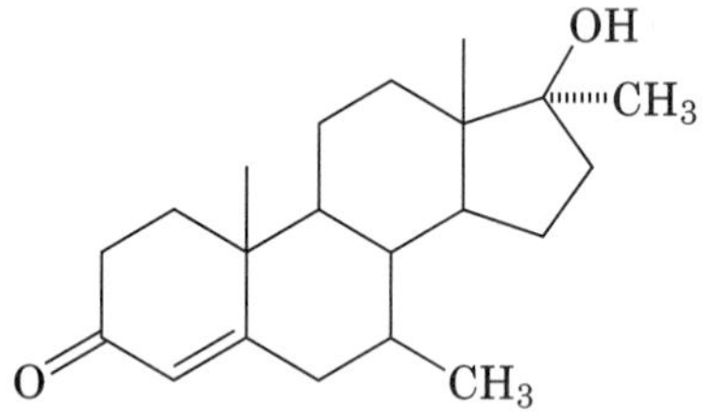

(37) calusterone

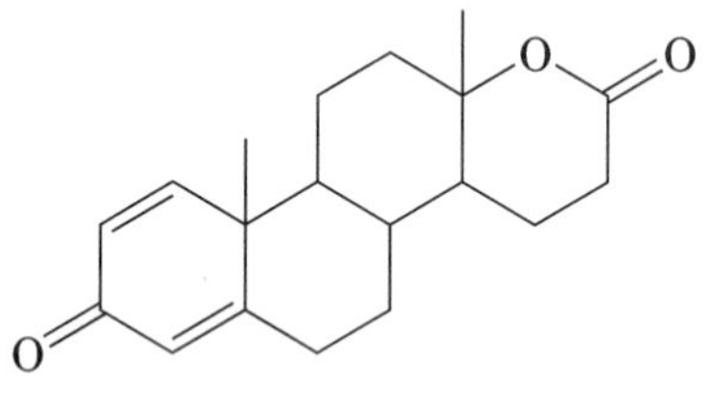

(38) 1 - dehydrotestolactone

4.3.6 Summary of Structure-Activity Relationships. As with other areas of medicinal chemistry, the desire to relate chemical structure to androgenic activity has attracted the attention of numerous investigators. Although it is often difficult to interrelate biological results from different laboratories, androgenicity data from the same laboratory afford useful information. In evaluating the data, one must be careful to note not only the animal model employed, but also the mode of administration. For example, marked differences in androgenic activity can be found when compounds are evaluated in the chick comb assay (local application) as opposed to the rat ventral prostate assay (subcutaneous or oral). The chick comb assay measures "local androgenicity" and is believed to minimize such factors as absorption, tissue distribution, and metabolism, which complicate the interpretation of *in vivo* data in terms of hormone–receptor interactions.

Furthermore, although the rat assays correlate well with what one eventually finds in humans, few studies of comparative pharmacology have been performed. Indeed, DHT may not be the principal mediator of androgenicity in all species. For example, a cytosol receptor protein has been found in normal and hyperplastic canine prostate that is specific for 5α-androstane-$3\alpha,17\alpha$-diol (256).

Because the presence of the 17β-hydroxyl group was demonstrated very early to be an important feature for androgenic activity in rodents, most investigators interested in structure-activity relationships maintained this function and modified other parts of the testosterone molecule. Three observations can be made based on these studies: (*1*) The 1-dehydro isomer of testosterone is at least as active as testosterone. (*2*) The 1- and 4-keto isomers of testosterone and DHT have variable activity. (*3*) The 2-keto isomers of testosterone and DHT consistently lack appreciable activity.

The first attempt to ascertain the minimal structural requirements for androgenicity was by Segaloff and Gabbard (310). Whereas the oxygen function at position 3 could be removed from testosterone with little reduction in androgenic activity, removal of the hydroxyl group from position 17 sharply reduced the androgenicity. As a continuation of these studies, the hydrocarbon nucleus, 5α-androstane (**39**), was synthesized (310). It too was found to possess androgenicity when applied topically or given intramuscularly in the chick comb assay, albeit at high doses. On the other hand, it was learned later that the 19-nor analog, 5α-estrane (**40**), had less than 1% of the androgenic activity of testosterone propionate in castrated male rats (311).

Nonetheless, the work of Segaloff and Gab-

bard set the stage for a more thorough analysis of 3-deoxy testosterone analogs by Syntex scientists (312, 313). The relative androgenicity of the isomeric A-ring olefins of 3-deoxy testosterones was the order $\Delta^1 > \Delta^2 > \Delta^3 > \Delta^4$. The Δ^2-isomer displayed the greatest anabolic activity and the best anabolic-to-androgenic ratio.

On the basis that sulfur is bioisosteric with $CH{=}CH$, Wolff and coworkers (314) synthesized the thia, seleno, and tellurio androstanes, which displayed androgenic activity. When the heteroatom was oxygen, however, the compound (**41**) was essentially devoid of androgenicity (315). The oxygen analog was said to be inactive because oxygen is isosteric with CH_2 rather than with $CH_2{=}CH_2$. Thus a minimum ring size was found to be required for activity. When the oxygen atom was introduced as part of a six-membered A-ring, an active androgen resulted (315).

As with the case of the double-bond isomers, the position of the oxygen atom was found to be important. The substitution of oxygen at C-2 gives rise to the most active compound, and the order of activity was $2 > 3 \gg 4$. As pointed out by Zanati and Wolff (315), these and earlier results are consistent with the concept that "the activity-engendering group in ring A is wholly steric and that, in principle, isosteric groups of any type could be used to construct an androgenic molecule." Further support for this idea has been obtained from X-ray crystallographic structure determinations (316). That even the full steroid nucleus is not essential for activity was shown by Zanati and Wolff (317) in the preparation of 7α-methyl 1,4-seco-2,3-bisnor-5α-androstan-17β-ol (**42**), having 50% of the anabolic activity of testosterone.

(**39**) R = CH$_3$
(**40**) R = H

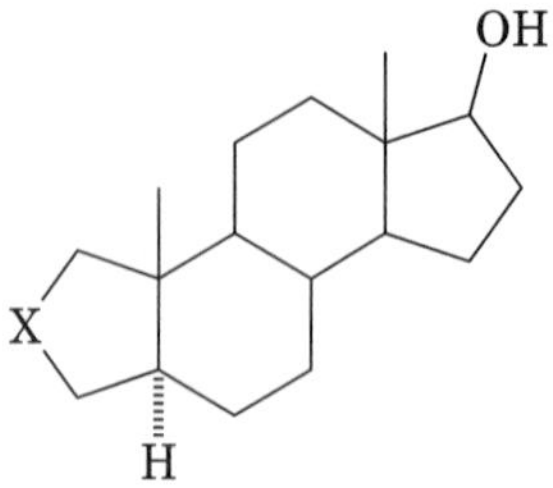

(41) X = 0

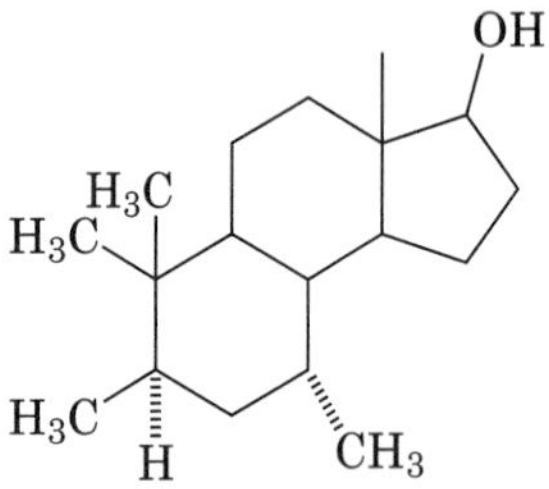

(42)

The effects of either 7α-methyl or 14-dehydro modification are more pronounced for 19-nortestosterone than for testosterone. The 14-dehydro modification had a greater effect on local androgenicity, whereas 7α-methylation had a more positive effect on systemic androgenicity. A marked synergism resulted when both the 14-dehydro and 7α-methyl modifications were present. The resultant compound, 7α-methyl-14-dehydro-19-nortestosterone, represents one of the most potent androgens reported to date.

The characterization of a specific receptor protein in androgen target tissues has made it possible to directly analyze the receptor affinity of various testosterone analogs. Liao and coworkers (320) were the first to employ this parameter for comparison with systemic androgenicity. Table 14.5 shows the receptor affinity and androgenic activity of a variety of androgens relative to DHT. The ability of the various steroids to be retained by prostate nuclei is also indicated.

As would be expected, the receptor affinity data did not necessarily correlate with the systemic androgenicity. In some cases, such as with 7α-methyl-19-nortestosterone, there was good agreement. Such was not the case, however, for 19-nortestosterone. Receptor binding analysis of androgens was also performed by other groups (128, 321). Table 14.6 summarizes their findings. Whereas the importance of the A/B *trans* ring fusion and 17β-hydroxyl prevailed, the data failed to demonstrate the potency previously noted for 7α-methyl-14-dehydro-19-nortestosterone. Moreover, 19-nor-

Studies by Segaloff and Gabbard (318) illustrated the marked enhancement of androgenicity achieved when a double bond was introduced at C-14. Both 14-dehydrotestosterone and the corresponding 19-nor analog were found to be potent androgens when applied topically. An extension of this series ascertained the effect of introducing a 7α-methyl (319). The results of this study are listed in Table 14.4, in terms of percentage increases in the weights of chick combs, rat ventral prostates, and rat levator ani induced by the test compounds as related to a similar dose of testosterone, the responses to the latter being described as 100%.

Table 14.5 Relative Androgenicity and Receptor Binding Capacity of Various Androgens

Steroid	Androgenicity in Rat (s.c.)	Cytosol Receptor	Nuclear Retention
DHT	1.0	1.0	1.0
Testosterone (T)	0.4	<0.1	0.7
5α-Androstanedione	0.2	0.0	0.2
19-NorDHT	0.1	0.5	0.6
19-NorT	0.2	0.9	0.7
$7\alpha,CH_3$-DHT	1.2	0.4	0.4
7α-CH_3-T	0.4	0.2	0.2
7β-CH_3-T	0.1	<0.1	<0.1
7α-CH_3-19-NorDHT	0.3	0.6	0.4
7α-CH_3-19-NorT	2.6	2.6	1.8

Table 14.6 Binding Affinity of Various Androgens for Rat Ventral Prostate Receptor Protein

Steroid	K_B (M^{-1})
5α-DHT	6.9×10^8
5β-DHT	6.4×10^7
17β-Testosterone	4.2×10^8
17α-Testosterone	2.1×10^7
Androstenedione	1.3×10^7
5α-Androstanedione	3.5×10^7
19-Nortestosterone	8.6×10^8
14-Dehydrotestosterone	4.4×10^8
14-Dehydro-19-nortestosterone	5.9×10^8
7α-CH$_3$-14-Dehydro-19-nortestosterone	5.0×10^8

testosterone displayed a receptor affinity greater than that of DHT, yet its androgenicity is much less than that of DHT.

These differences in correlations between receptor assays and *in vivo* data should not cloud the importance of the receptor studies. The receptor assays measure affinity for the receptor protein, and this property is shared by both androgens and antiandrogens. Moreover, such assays cannot predict the disposition and metabolic fate of an androgen after administration. Figure 14.9 contains a summary of the structure-activity relationships for androgens.

4.4 Absorption, Distribution, and Metabolism

Numerous factors are involved in the absorption, distribution, and metabolism of the synthetic androgens and the physicochemical properties of these steroids greatly influence the pharmacokinetic parameters. The lipid solubility of a synthetic steroid is an important factor in its intestinal absorption. The acetate ester of testosterone demonstrated enhanced absorption from the gastrointestinal tract over that of both testosterone and 17α-methyltestosterone (58). Injected solutions of testosterone in oil result in the rapid absorption of the hormone from the injection site; however, rapid metabolism greatly decreases the biological effects of the injected testosterone. The esters of testosterone are much more nonpolar and, when injected intramuscularly, are absorbed more slowly. As a result, the

commercial preparations of testosterone propionate are administered every 2 or 3 days. Increasing the size of the ester functionality enables the testosterone esters such as the ethanate or cypionate to be given in a depot injection lasting 2–4 weeks.

Once absorbed, the steroids are transported in the circulation primarily in a protein-bound complex. Testosterone and other androgens are reversibly associated with certain plasma proteins and the unbound fraction can be absorbed into target cells to exert its action. The structure-binding relationships of the natural and synthetic androgens to sex steroid binding globulin (SBG) have been extensively investigated (65–69, 74–79). A 17α-hydroxyl group is essential for binding and the presence of a 17α-substituent such as the 17α-methyl moiety decreases its affinity. The 5α-reduced androgens bind with the highest affinity. A much smaller quantity of the androgen is bound to other plasma proteins, principally albumin and corticosteroid-binding globulin or transcortin (CBG).

The metabolism of the synthetic androgens is similar to that of testosterone. The introduction of the 17α-methyl group greatly retards the metabolism, thus providing oral activity. Reduction of the 4-en-3-one system in synthetic androgens to give the various α- and β-isomers occurs *in vivo* (322, 323). Finally, aromatization of the A-ring can also occur (151–153). The metabolism of modified androgens or anabolic steroids is summarized in Table 14.7. One analog that demonstrates an alternate metabolic pattern is 4-chlorotestos-

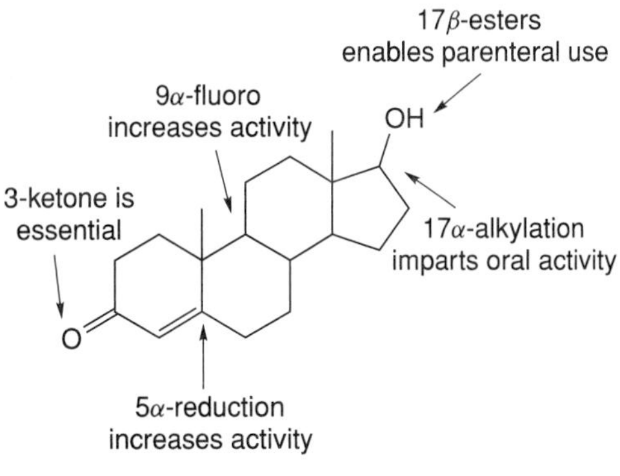

Figure 14.9. Summary of structure-activity relationships for androgens.

Table 14.7 Metabolism of Modified Androgens

Substrate	Products	Test System	Species	Reference
17β-Hydroxyestra-4-en-3-one	5α-Estran-3α-ol-17-one	*In vivo*	Human	322
	5β-Estran-3α-ol-17-one			
	5α,5β-Estran-3α-ol-17-one plus	Liver homogenate	Female rat	324
	5α-Estran-3,3-ol-17-one			
	5α-Estrane-3α,17β-diol			
	5α-Estrane-3β,17β-diol			
	Estra-4-ene-3,17-dione	Prostate slices	Human	325
	5α-Estran-17β-ol-3-one			
	2-Methoxyestrone			
1α-Methyl-17β-hydroxyestra-4-en-3-one acetate	1α-Methyl-5α-estran-3α-ol-17-one	*In vivo*	Human	326
	1α-Methyl-5β-estran-3α-ol-17-one			
17α-Methyl-17β-hydroxyestra-4-en-3-one	17α-Methyl-5α-estran-17β-ol-3-one	Liver homogenate	Female rat	327
	17α-Methyl-5α-estrane-3α,17β-diol			
17αMethyltestosterone	17α-Methyl-5α-androstane-3α,17β-diol	*In vivo*	Human	323
	17α-Methyl-5β-androstane-3α,17β-diol			
	17α-Methyl-5α-androstane-3β,17β-diol			
17α-Methylandrost-5-ene-3β,17β-diol	17α-Methyltestosterone	Adrenal	Male rat	328
	11β-Hydroxy-17α-methyltestosterone			
17α-Methyl-17β-hydroxyandrosta-1,4-dien-3-one	17α-Methyl-6β,17β-dihydroxyandrosta-1,4-dien-3-one	*In vivo*	Human	329
1α-Methyl-17β-hydroxy-5α-androst-1-en-3-one acetate	1α-Methyl-5α-androstan-3α-ol-17-one	*In vivo*	Human	330
	1α-methyl-5α-androst-1-ene-3,17-dione	*In vivo*	Human	331
	1α-methyl/compounds above plus			
	1-Methylene-5α-androstan-3α-ol-17-one			
	1α-Methyl-5α-androstane-3,17-dione			
4-Chloro-17β-hydroxyandrost-4-en-3-one acetate	4-Chloro-3α-hydroxyandrost-4-3en-17-one	*In vivo*	Human	332
17β-Hydroxy-9β,10α-androst-4-en-3-one (retrotestosterone)	9β,10α-androst-4-ene-3,17-dione	Placenta	Human	336, 333
D-Homo-17aβ-hydroxyandrost-4-en-3-one (D-homotestosterone)	D-Homo-5β-androstane-3α,17aβ-diol	*In vivo*	Human	334
1α-Methyl-5α-androstan-17β-ol-3-one	1α-Methyl-5α-androstan-3α-ol-17-one	*In vivo*	Human	335
	1α-Methyl-5α-androstan-3α,17β-diol			

704

terone, which in humans gave rise to an allylic alcohol, 4-chloro-3α-hydroxyandrost-4-en-17-one (336). A number of other halogenated testosterone derivatives subsequently were found to take this abnormal reduction path *in vitro* (337). It was proposed that fluorine or chlorine substituents at the 2-, 4-, or 6- position in testosterone interfere with the usual α,β-unsaturated ketone resonance so that the C-3 carbonyl electronically resembles a saturated ketone.

4.5 Toxicities

The use of androgens in women and children can often result in virilizing or masculinizing side effects. In boys an acceleration of the sexual maturation is seen, whereas in girls and women growth of facial hair and deepening of the voice can be observed (338, 339). These effects are reversible when medication is stopped; however, prolonged treatment can produce effects that are irreversible. Inhibition of gonadotropin secretion by the pituitary can also occur in patients receiving androgens.

Both males and females experience salt and water retention resulting in edema. This edema can be treated by either maintaining a low salt diet or by using diuretic agents. Liver problems are also encountered with some of the synthetic androgens. Clinical jaundice and cholestasis can develop after the use of the 17α-alkylated products (340–343). Various clinical laboratory tests for hepatic function such as bilirubin concentrations, sulfobromophthalein retention, and glutamate transaminase and alkaline phosphatase activities are affected by these androgen analogs.

5 ANABOLIC AGENTS

5.1 Current Drugs on the Market

5.2 Therapeutic Uses and Bioassays

Many synthetic analogs of testosterone were prepared to separate the anabolic activity of the C_{19} steroids from their androgenic activity. Although the goal of a pure synthetic anabolic that retains no androgenic activity has not been accomplished, numerous preparations are now available on the market that have high anabolic/androgenic ratios. Extensive reviews on anabolic agents have been published (4, 10).

The primary criterion for assessing anabolic activity of a compound is the demonstration of a marked retention of nitrogen. This nitrogen-retaining effect is the result of an increase of protein synthesis and a decrease in protein catabolism in the body (344). Thus, the urinary nitrogen excretion, particularly urea excretion, is greatly diminished. The castrated male rat serves as the most sensitive animal model for nitrogen retention, although other animals have been used (345–347). Another bioassay for anabolic activity involves examination of the increase in mass of the levator ani muscle of the rat upon administration of an anabolic agent (284, 285). This measure of myotrophic effect correlates well with the nitrogen-retention bioassay and the two are usually performed in the determination of anabolic activity.

Anabolic steroids exert other effects on the body as well. Skeletal mineralization and bone maturation are enhanced by androgens and anabolics (348). These agents will decrease calcium excretion by the kidney and result in increased deposition of both calcium and phosphorous in bone. Androgenic and anabolic agents also can influence red blood cell formation. Two mechanisms of action of this erythropoiesis involving increased erythropoietin

Generic Name (structure)	Trade Name	U.S. Manufacturer	Chemical Class	Dose
Nandrolone decanonate (**43**)	Deca-Durabolin	Organon	Estrane	Injection (in oil): 100 mg/ml 200 mg/mL
Oxandrolone (**65**)	Oxandrin	Gynex	Androstane	Tablets: 2.5 mg
Oxymetholone (**63**)	Anadrol-50	Syntex	Androstane	Tablets: 50 mg
Stanozolol (**75**)	Winstrol	Winthrop Pharm	Androstane	Tablets: 2 mg

production and enhanced responsiveness of the tissue have been described (349).

These various biological activities of the anabolics have prompted the use of these agents in treatment protocols, with varying success. Clinical trials have demonstrated the effectiveness of the anabolic steroids in inducing muscle growth and development in some diseases (38). Anabolic steroids are effective in the symptomatic treatment of various malnourished states because of their ability to increase protein synthesis and decrease protein catabolism. Treatment of diseases such as malabsorption, anorexia nervosa, emaciation, and malnutrition as a result of psychoses includes dietary supplements, appetite stimulants, and anabolics (350–355). Improved postoperative recovery with adjunctive use of anabolic agents has been demonstrated in numerous clinical studies (350, 356–360). However, their usefulness in other diseases such as muscular dystrophies and atrophies and in geriatrics has not been observed.

In addition, the myotrophic effects have also led to the use and the abuse of these agents by athletes (361–364). Conflicting reports on the effectiveness of anabolics to increase strength and power in healthy males have resulted from clinical trials. Several groups reported no significant differences between groups of male college-age students receiving anabolics and weight training and those groups receiving placebo plus the weight training in double-blind studies (365–368). Other reports cited some improvement in strength and power, although they used small numbers of subjects or were only single-blind studies (369–372). A recent study on "supraphysiologic" doses of anabolics has demonstrated enhancement of muscle size and strength (373). Overall, the consensus of these various studies are that anabolic steroids provide only very limited improvement in strength and power in healthy males. Anabolic steroids also exhibit an anticatabolic effect, that is, reversing the catabolic effects of glucocorticoids released in response to stress. Such effects would enable individuals to recover more quickly after strenuous workouts.

Even though clinical studies do not indicate the efficacy of anabolics for healthy males, an alarming percentage of amateur and professional athletes use anabolic steroids (374), which are readily available "on the street." The use of these steroids for increasing strength and power is banned in intercollegiate and international sports, and very sensitive assays (RIA, GC-MS) have been developed for measuring anabolic levels in urine and blood (362). Anabolic steroids also have the ability to lower serum lipid levels *in vivo* (360, 375–377). The most widely studied agent is oxandrolone, which dramatically lowers serum triglycerides and, to a lesser extent, cholesterol levels at pharmacological doses (378–380). The proposed mechanism of this hypolipidemic effect includes both an inhibition of triglyceride synthesis (381) and an increased clearance of the triglycerides (382). The androgenic side effects of the anabolics and the lack of superiority over conventional hypolipidemic agents have curtailed its use for treatment of these conditions.

The stimulation of erythropoiesis by anabolics has resulted in the use of these agents for the treatment of various anemias (383–386). Anemias arising from deficiencies of the bone marrow are particularly responsive to pharmacological doses of anabolic agents. Treatment of aplastic anemia with anabolics and corticosteroids has proved effective (383–386). Secondary anemias resulting from inflammation, renal disease, or neoplasia are also responsive to anabolic steroid administration (349, 386–390). Finally, synthetic anabolics have been prescribed for women with osteoporosis (348) and for children with delayed growth (381). These applications have produced limited success; however, the virilizing side effects severely limit their usefulness, particularly in children.

The methods employed to determine the anabolic or myotrophic properties of steroids have been reviewed (391). Generally, these are based on an increase in nitrogen retention and/or muscle mass in various laboratory animals. The castrated male rat is presently the most widely used and most sensitive laboratory animal for nitrogen balance studies (345). Dogs and ovariectomized monkeys have also been employed (346, 347). Although it is generally agreed that variations in urinary nitrogen excretion relate to an increase or decrease in protein synthesis, nitrogen balance assays are not without their limitations (392). This is partly because such studies fail to describe the shifts in organ protein and measure only the overall status of nitrogen retention in the animal (393).

The easily accessible levator ani muscle of the rat has provided a valuable index for measuring the myotrophic activity of steroidal hormones (283). By comparing the weight of levator ani muscle, seminal vesicles, and ventral prostate with controls, one can obtain a ratio of anabolic-to-androgenic activity (283, 285). There also appears to be some correlation between the levator ani response and urinary nitrogen retention (283). A modification of this muscle assay uses the parabiotic rat (394, 395) and allows for the simultaneous measurement of pituitary gonadotrophic inhibition and myotrophic activity. The suitability of the levator ani assay has been questioned on the possibility that its growth is more a result of androgenic sensitivity than of any steroid-induced myotrophic effect (395–398). Thus this assay is usually performed in conjunction with nitrogen balance studies or acceleration of body growth (399).

5.3 Structure-Activity Relationships for Anabolics

5.3.1 19-Nor Derivatives. An important step toward developing an anabolic agent with minimal androgenicity was taken when Hersh berger and associates (285), and later others (400, 401), found 19-nortestosterone (17β-hydroxyestr-4-en-3-one, nandrolone, **43**) to be as myotrophic but only about 0.1 as androgenic as testosterone. This observation prompted the synthesis and evaluation of a variety of 19-norsteroids, including the 17α-methyl (normethandrone, **44**) (402) and the 17α-ethyl (norethandrolone, **45**) (403) homologs of 19-nortestosterone.

Nandrolone in the form of a variety of esters (such as decanoate and β-phenylpropionate) and norethandrolone have been widely used clinically. The latter, under the name Nilevar, was the first agent to be marketed in the United States as an anabolic steroid. Androgenic (404) and progestational (405) side effects, however, led to its eventual replacement by other agents.

(43) nandrolone   R = H
(44) normethandrone   R = CH$_3$
(45) norethandrone   R = CH$_2$CH$_3$

Nonetheless, these studies did stimulate the synthesis of other norsteroids. Interestingly, both 18-nortestosterone (406) and 18,19-bisnortestosterone (407) were essentially devoid of both androgenic and anabolic properties (408). Contraction of the B-ring led to B-norsteroids, which were also lacking in androgenicity, but unlike the foregoing, this modification led to compounds with antiandrogenic activity.

Of the number of homoandrostane derivatives (those having one or more additional methylene groups included in normal tetracyclic ring system) that have been synthesized, only B-homo (409, 410) and D-homodihydrotestosterone (411, 412) have shown appreciable androgenic activity. A D-bishomo analog (**46**) was reported to be weakly androgenic (413).

(46)

5.3.2 Dehydro Derivatives. The marked enhancement in biological activity afforded by introduction of a double bond at C_1 of cortisone and hydrocortisone prompted similar transformations in the androgens. The acetate of 17β-hydroxyandrosta-1,4-dien-3-one (**47**) (414) was as myotrophic as testosterone propionate but was much less androgenic.

Furthernore, 17α-methyl-17β-hydroxyandrosta-1,4-dien-3-one (methandrostenolone, **48**) had 1–2 times the oral potency of 17α-methyl testosterone in the rat nitrogen retention (415, 416) and levator ani muscle assays (417, 418). In clinical studies, methandrostenolone produced a marked anabolic effect when given orally at doses of 1.25–10 mg daily and was several times more potent than 17α-methyltestosterone (419).

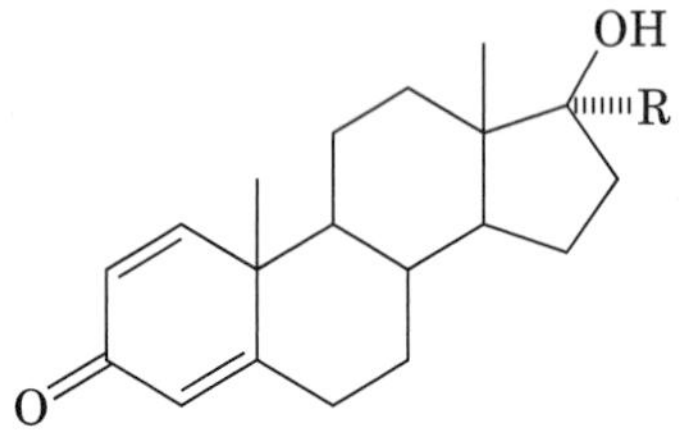

(**47**)   R = H
(**48**) methandrostenolone R = CH$_3$

In contrast with the 1-dehydro analogs, introduction of an additional double bond at the 6-position (**49**) markedly decreased both androgenic and myotrophic activity in the rat (414, 420). Moreover, removal of the C_{19}-methyl (421), inversion of the configuration at C_9 and C_{10} (422) and at C_8 and C_{10} (423), and reduction of the C_3-ketone failed to improve the biological properties (424).

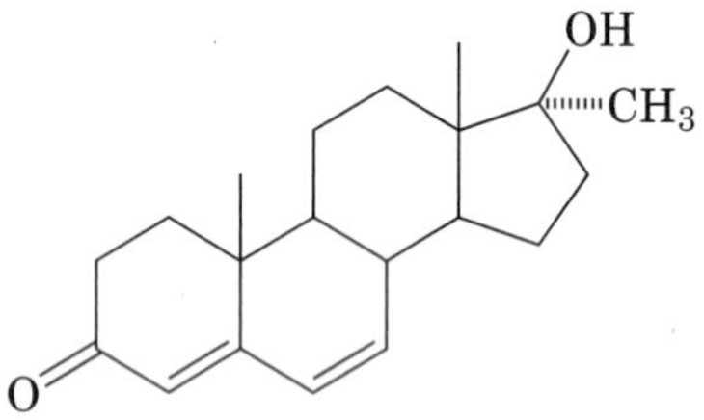

(**49**)

On the other hand, introduction of unsaturation into the B, C, and D rings has given rise to compounds with significant androgenic or anabolic activity. Ethyldienolone (**50**), for example, displayed an anabolic-to-androgenic ratio of 5 and was slightly more active than methyltestosterone when both were given orally (425). Segaloff and Gabbard (318) showed that introduction of a 14–15 double

bond (**51**) increased androgenicity when compared with testosterone by local application in the chick comb assay. On the other hand, there was a 25% decrease in androgenicity when measured by the rat ventral prostate after subcutaneous administration. Conversion to the 19-nor analog (**52**) increased androgenicity, but the anabolic activity was significantly enhanced (319).

Of a variety of triene analogs of testosterone that have been tested, only 17α-methyl-17β-hydroxyestra-4,9,11-trien-3-one (methyltrienolone, **53**) showed significant activity in rats. Surprisingly, this compound had 300 times the anabolic and 60 times the androgenic potency of 17α-methyltestosterone when administered orally to castrated male rats (426). In this instance, however, the potent hormonal properties on rats did not correlate with later studies in humans (427–429). One study in patients with advanced breast cancer found methyltrienolone to have weak androgenicity and to produce severe hepatic dysfunction at very low doses (429).

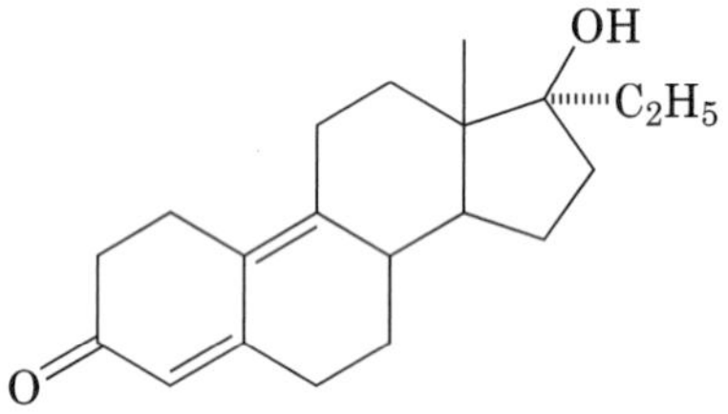

(**50**)

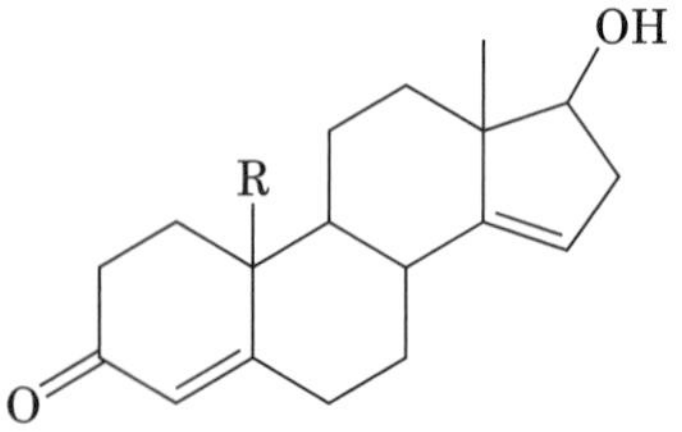

(**51**) R = CH$_3$
(**52**) R = H

5.3.3 Alkylated Analogs. An extensive effort has been directed toward assessing the physiological effect of replacing hydrogen with alkyl groups at most positions of the steroid molecule. Although methyl substitution at C_3,

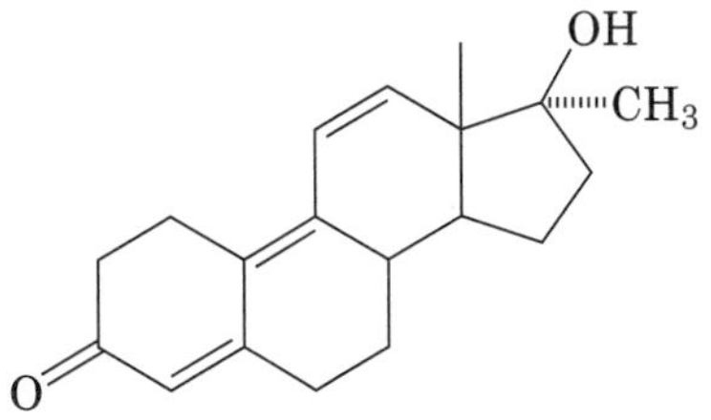

(53)

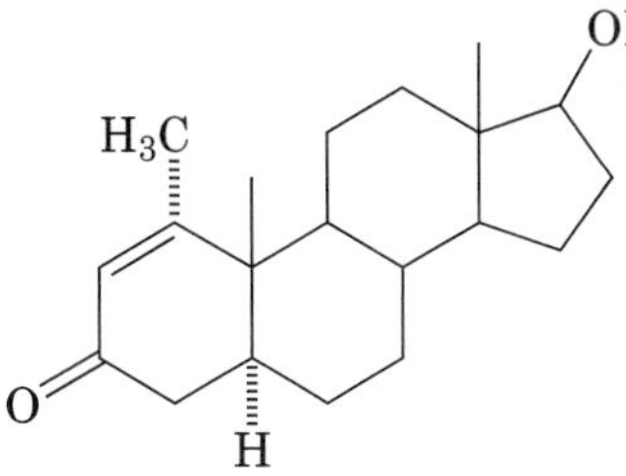

C_4, C_5, C_6, C_{11}, and C_{16} has generally led to compounds with low anabolic and androgenic activity, similar substitutions at C_1, C_2, C_7, and C_{18} have afforded derivatives of clinical significance.

1-Methyl-17β-hydroxy-5α-androst-l-en-3-one (**54**) as the acetate (methenolone acetate) was about 5 times as myotrophic, but only 0.1 as androgenic, as testosterone propionate in animals (430). In addition, this compound or the free alcohol represented one of the few instances of a C_{17} nonalkylated steroid that possessed significant oral anabolic activity in animals (431) and in man (432). This effect may be related to the slow *in vivo* oxidation of the 17β-hydroxyl group when compared with testosterone (331). At a daily dose of 300 mg, methenolone acetate caused little virilization (433) or BSP retention (434). By contrast, the dihydro analog, 1α-methyl-17β-hydroxy-5α-androstan-3-one (mesterolone, **55**) was found to possess significant oral androgenic activity in the cockscomb test (435) and in clinical assays (436). A comparison of the anabolic and androgenic activity of **54** with its A-ring congeners revealed that the double bond was necessary at C_1 for anabolic activity. For example, 1α-methyl-17β-hydroxyandrost-4-en-3-one had a much lower activity (437). Furthermore, either reduction of the C_3 carbonyl group of **55** (438) or removal of the C_{19} methyl group (439, 440) greatly reduced both anabolic and androgenic activity in this series.

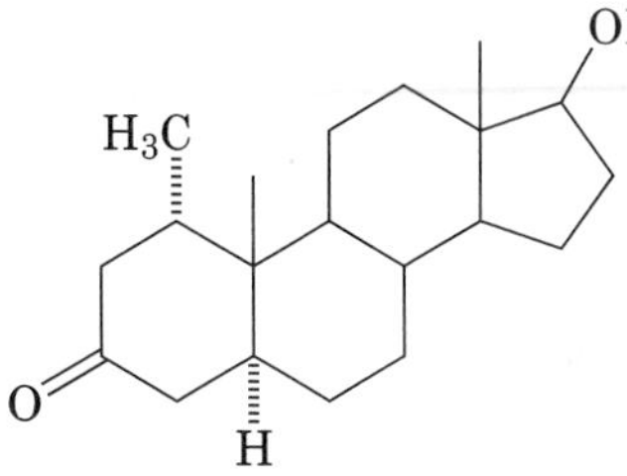

Among the C_2-alkylated testosterone analogs, 2α-methyl-5α-androstan-17β-ol-3-one (drostanolone, **56**) and its 17α-methylated homolog (**57**) have displayed anabolic activity both in animals (441) and in man (442). In contrast, 2,2-dimethyl and 2-methylenetestosterone or their derivatives showed only low anabolic or androgenic activity in animals

(54) methenolone

(55) mesterolone

(441, 443, 444). The most interest in drostanolone has been in relation to its potential as an antitumor agent with decreased masculinizing propensity.

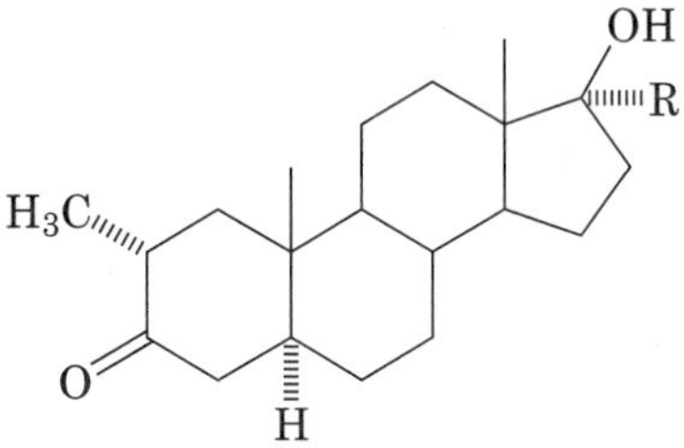

(56) R = H
(57) R = CH$_3$

7α,17α-Dimethyltestosterone (bolasterone, **58**) had 6.6 times the oral anabolic potency of 17α-methyltestosterone in rats (445). Similar activity was observed in man at 1–2 mg/day without many of the usual side effects (446). Moreover, the corresponding 19-nor derivative was 41 times as active as 17α-methyl testosterone as an oral myotrophic agent in the rat (447). Segaloff and Gabbard (319) found 7α-methyl-14-dehydro-19-nortestosterone (**59**) to be approximately 1000 times as active as testosterone in the chick comb assay and about 100 times as active as testosterone in the ventral prostate assay.

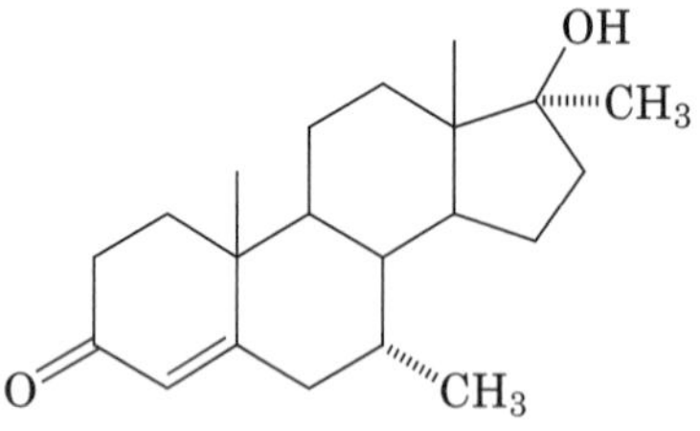

(58)

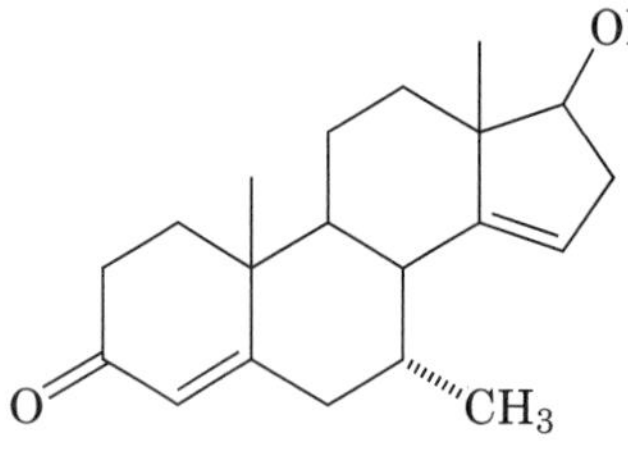

(59)

Certain totally synthetic 18-ethylgonane derivatives possessed pronounced anabolic activity. Similar to other 19-norsteroids, $13\beta,17\alpha$-diethyl-17β-hydroxygon-4-en-3-one (norbolethone, **60**) was found to be a potent anabolic agent in animals and in man (449, 450). Because it is prepared by total synthesis, the product is isolated and marketed as the racemic DL-mixture. The hormonal activity resides in the D-enantiomer.

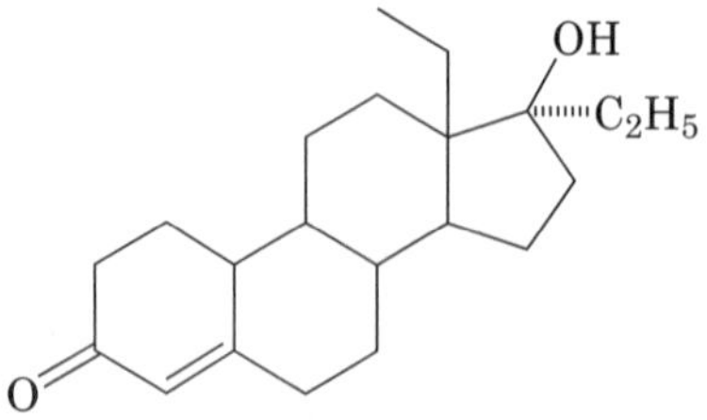

(60)

5.3.4 Hydroxy and Mercapto Derivatives. Testosterone has been hydroxylated at virtually every position on the steroid nucleus. For the most part, nearly all these substances possess no more than weak myotrophic and an-

drogenic properties. Two striking exceptions to this, however, are 4-hydroxy- and 11β-hydroxytestosterones.4-Hydroxy-17α-methyltestosterone (oxymesterone, **61**), for instance, had 3–5 times the myotrophic and 0.5 times the androgenic activity of 17α-methyltestosterone in the rat (451). In clinical studies, oxymesterone produced nitrogen retention in adults at a daily dose of 20–40 mg, and no adverse liver function was observed (452, 453). Introduction of an 11β-hydroxyl group in many instances resulted in a favorable effect on biological activity. 11β-Hydroxy-17α-methyltestosterone (**62**) was more anabolic in the rat than was 17α-methyltestosterone (454), and 1.5 times as myotrophic in humans (455).

One of the most widely studied anabolic steroids has been 2-hydroxymethylene-17α-methyl-5α-androstan-17β-ol-3-one (oxymetholone, **63**). In animals it was found to be 3 times as anabolic and 0.5 times as androgenic as 17α-methyltestosterone (456, 457). Clinical studies confirmed these results (452, 456–458).

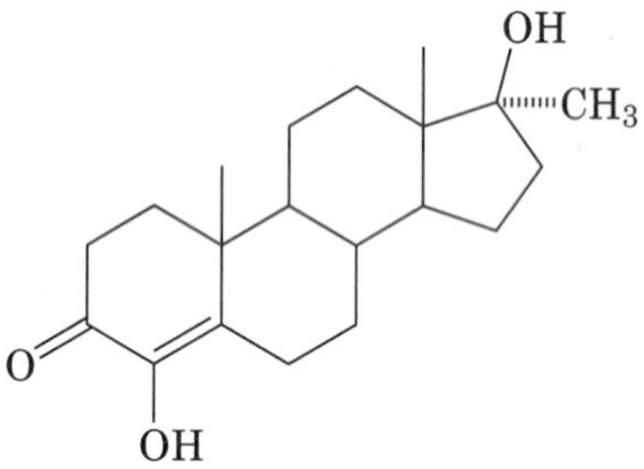

(61)

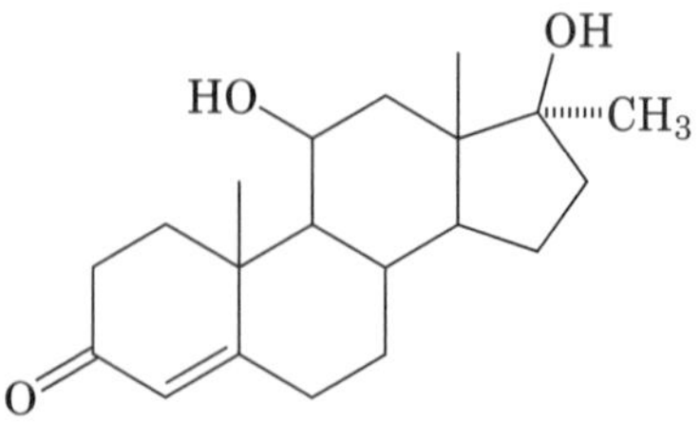

(62)

The substitution of a mercapto for a hydroxyl group has generally resulted in de-

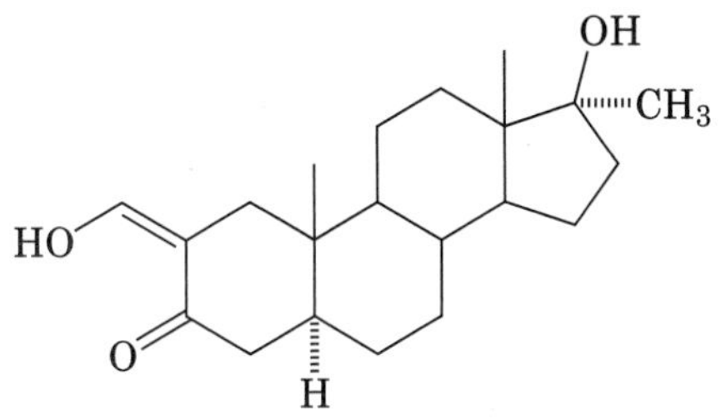

(63)

creased activity. However, the introduction of a thioacetyl group at C_1 and C_7 of 17α-methyltestosterone afforded $1\alpha,7\alpha$-bis(acetylthio)-17α-methyl-17β-hydroxyandrost-4-en-3-one (thiomesterone, **64**), a compound with significant activity. Thiomesterone was 4.5 times as myotrophic and 0.6 times as androgenic as 17α-methyltestosterone in the rat (459) and has been used clinically as an anabolic agent (460).

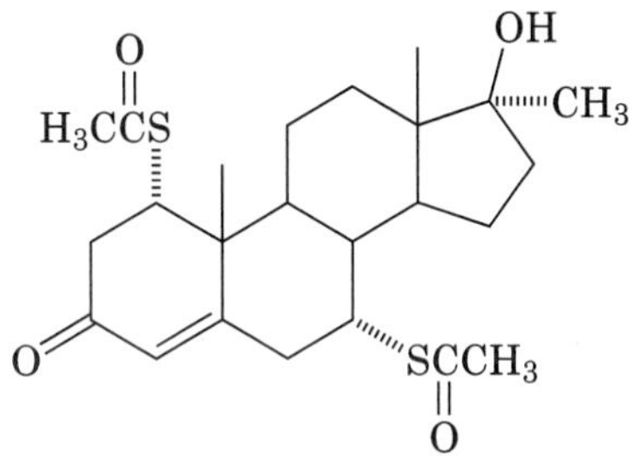

(64)

Moreover, numerous 7α-alkylthio androgens have exhibited anabolic-androgenic activity similar to that of testosterone propionate when administered subcutaneously (461, 462). Even though no clinically useful androgen resulted, similar 7α-substitutions were advantageous in the development of radioimmunoassays now employed in clinical laboratories (463). In addition, certain 7α-arylthioandrost-4-ene-3,17-diones are effective inhibitors of estrogen biosynthesis (see Section 6.2.3).

5.3.5 Oxa, Thia, and Aza Derivatives. A number of androgen analogs in which an oxygen atom replaces one of the methylene groups in the steroid nucleus have been syn-

thesized and biologically evaluated. Of these derivatives, 17β-hydroxy-17α-methyl-2-oxa-5α-androstan-3-one (oxandrolone, **65**) (464) was 3 times as anabolic and only 0.24 times as androgenic as 17α-methyltestosterone in the oral levator ani assay (465). By contrast, only minimal responses were obtained following intramuscular administration. The 2-thia (466) and 2-aza (467) analogs were essentially devoid of activity by both routes. The 3-aza-A-homoandrostene derivative **66** displayed only 5% the anabolic-to-androgenic activity of methyltestosterone (468).

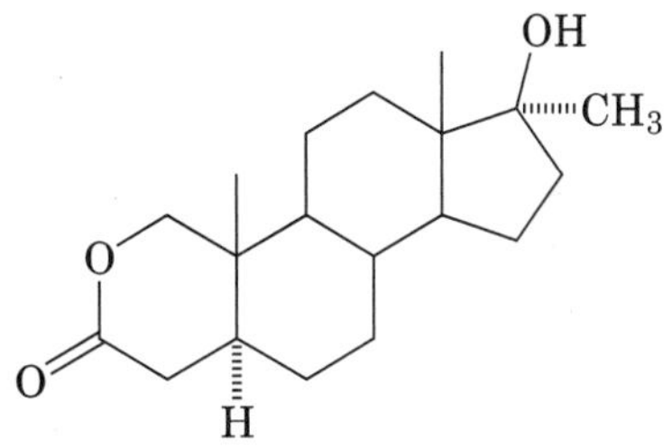

(65)

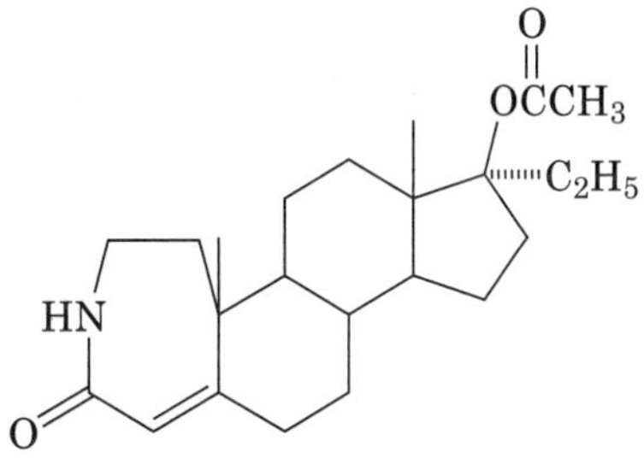

(66)

The clinical anabolic potency of oxandrolone was considerably more active than 17α-methyltestosterone and provided perceptible nitrogen sparing at a dose as low as 0.6 mg/day (469). Moreover, at dosages of 0.25–0.5 mg/kg, oxandrolone was effective as a growth-promoting agent without producing the androgenically induced bone maturation (470). Because of this favorable separation of anabolic from androgenic effects, oxandrolone has been one of the most widely studied anabolic steroids. Its potential utility in various clinical hyperlipidemias was discussed in Section 5.1.

The significant hormonal activity noted for estra-4,9-dien-3-ones such as **50** (see section 5.2.2) prompted the synthesis of the 2-oxa bioisosteres in this series. Despite the lack of a 17α-methyl group, (**67**) had 93 times the oral anabolic activity of 17α-methyltestosterone; it was also 2.7 times as androgenic. As might be expected, the corresponding 17α-methyl derivative, **68**, was the most active substance in this series. It had 550 times the myotrophic and 47 times the androgenic effect of 17α-methyltestosterone (471). These two compounds differed dramatically in progestational activity, however. The activity of **67** was only 0.1 times, whereas the activity of **68** was 100 times, that of progesterone in the Clauberg assay (471). The pronounced oral activity of **67** suggests that it is not a substrate for the 17β-alcohol dehydrogenase and represents an interesting finding.

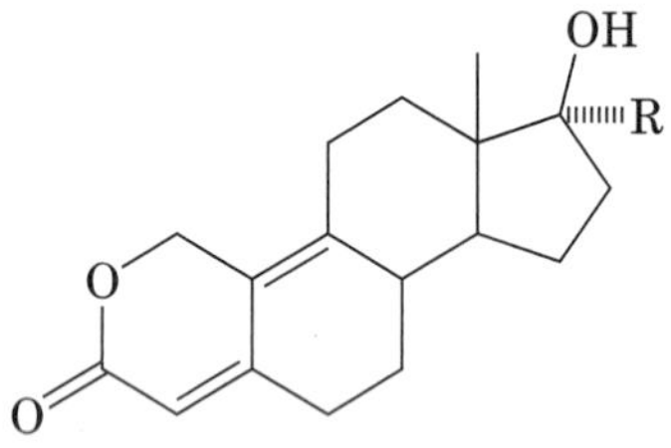

(67) R = H
(68) R = CH$_3$

5.3.6 Deoxy and Heterocyclic-Fused Analogs. Early studies by Kochakian (472) indicated that the 17β-hydroxyl group and the 3-keto group were essential for maximum androgenic activity. Based on this observation, the C$_3$ oxygen function was removed in the hope of decreasing the androgenic potency while maintaining anabolic activity (473). Unfortunately, the results failed to substantiate the rationale, and 17α-methyl-5α-androstan-17β-ol (**69**) was found to be a potent androgen in animals (474) and humans (475). However, Wolff and Kasuya showed that this substance is extensively metabolized to the 3-keto derivative by rabbit liver homogenate (476). Other deoxy analogs of testosterone have been synthesized and tested. A 19-nor derivative, 17α-ethylester-4-en-17β-ol (estrenol, **70**) had at least 4 times the anabolic and 0.25 times the androgenic activity of 17α-methyltestosterone

in animals (477) and was effective in humans at a daily dose of 3–5 mg (478–480). In addition. 17α-methyl-5α-androst-2-en-17β-ol (**71**) offered a good separation of anabolic from androgenic activity (474, 481).

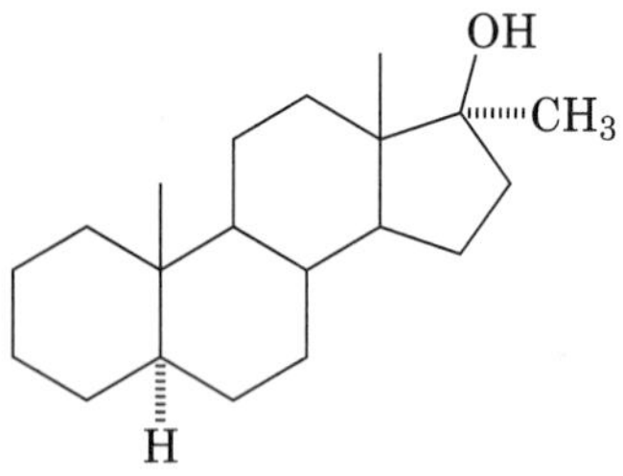

(69)

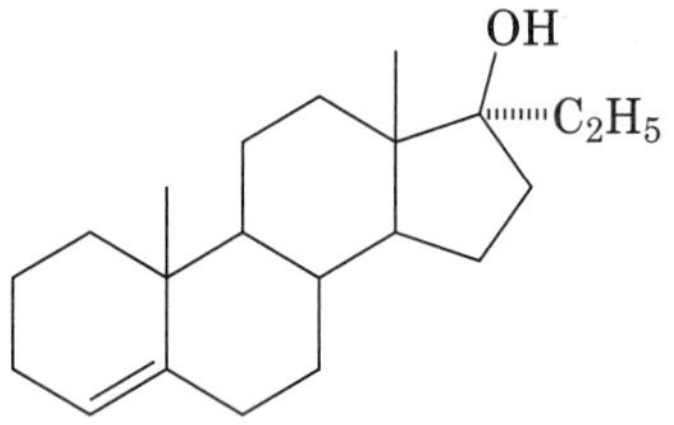

(70)

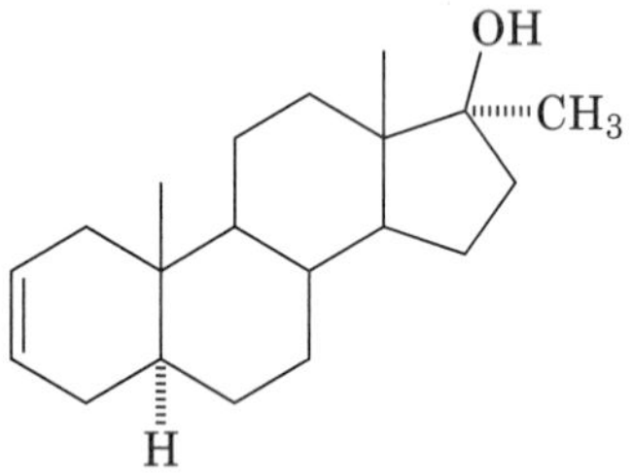

(71)

Because sulfur is considered to be isosteric with –CH=CH–, Wolff and Zanati (334) reasoned that 2-thia-A-nor-5α-androstane derivatives such as **72** should have androgenic activity. Indeed, this compound possessed high androgenic and anabolic activity and served to verify that steric rather than electronic factors are important in connection with the structural requirements at C-2 and/or C-3 in androgens (482). Interestingly, the selenium and tellurium isosteres in the same series were found to have good androgenic activity (483, 484). Moreover, experiments with a ^{75}Se-labeled analog have shown **73** to selectively bind with the specific cytosol receptor for DHT in rat prostate (485).

The high biological activity noted for the 3-deoxy androstanes prompted numerous investigators to fuse various systems to the A-ring. The simplest such changes were 2,3-epoxy, 2,3-cyclopropano, and 2,3-epithioandrostanes. The 2α3α-cyclopropano-5α-androstan-17β-ol was as active as testosterone propionate as an anabolic agent (486). Although the epoxides had little or no biological activity, certain of the episulfides possessed pronounced anabolic androgenic activity (487). For example, 2,3α-epithio-17α-methyl-5α-androstan-17β-ol (**74**) was found to have approximately equal androgenic and 11 times the anabolic activity of methyltestosterone after oral administration to rats. The 2,3-β-episulfide, on the other hand, was much less active. 2,3α-Epithio-5α-androstan-17β-ol has been shown to have long-acting antiestrogenic

activity, as well as some beneficial effects in the treatment of mammary carcinoma (488).

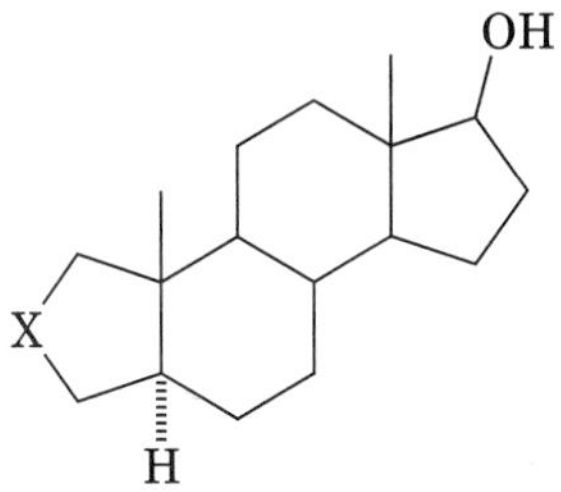

Other heterocyclic androstane derivatives have included the pyrazoles. Thus 17β-hydroxy-17α-methylandrostano-(3,2-c)-pyrazole (stanazolol, **75**) was 10 times as active as 17α-methyltestosterone in improving nitrogen retention in rats (489). The myotrophic activity, however, was only twice that of 17α-methyltestosterone (490). Stanazolol at a dose of 6 mg/day produced an adequate anabolic response with no lasting adverse side effects (491, 492).

The high activity of the pyrazoles instigated the synthesis of other heterocyclic-fused androstane derivatives including isoxazoles, thiazoles, pyridines, pyrimidines, pteridines, oxadiazoles, pyrroles, indoles, and triazoles. One of the most potent was 17α-methylandrostan-17β-ol-(2,3-d)-isoxazole (androisoxazol, **76**), which exhibited an oral anabolic-to-androgenic ratio of 40 (493). The corresponding 17α-ethynyl analog (danazol, **77**) has been of most interest clinically. This compound has impeded androgenic activity and inhibits pituitary gonadotropin secretion (494). Because it depresses blood levels of androgens and gonadotropins, it has been studied as an antifertility agent in males (495). At doses of 200 or 600 mg daily, danazol lowered plasma testosterone and androstenedione levels, and this effect was dose related. In addition to an inhibition in gonadotropin release, a direct inhibition of Leydig cell androgen synthesis was observed. Other studies have shown danazol to be effective for the treatment of endometriosis, benign fibrocystic mastitis, and precocious puberty (496). Several reports have appeared relating to its disposition and metabolic fate (496, 497).

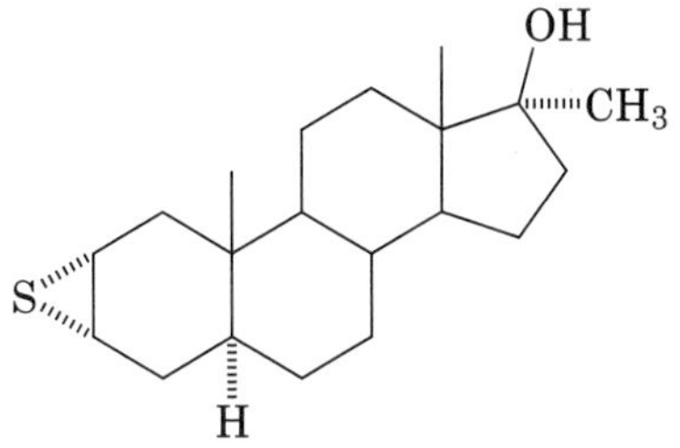

(72) X = S
(73) X = Se

(74)

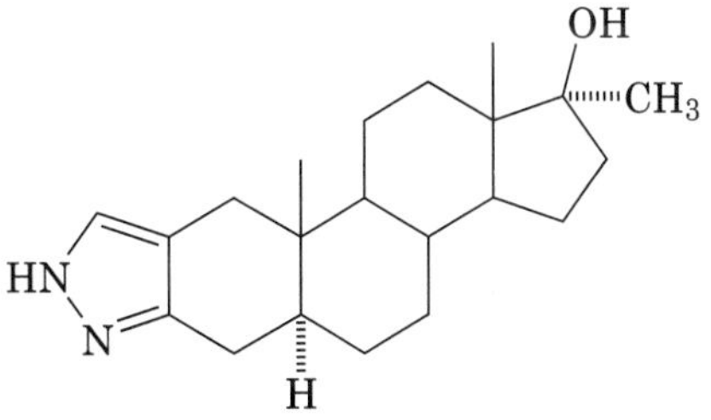

(75)

(76) R = H
(77) R = C≡CH

5.3.7 Esters and Ethers. Because esterification of testosterone markedly prolonged its activity, it was only natural that this approach to drug latentiation would be extended to the anabolic steroids. The acyl moiety is usually derived from a long-chain aliphatic or arylaliphatic acid such as heptanoic (enanthoic), decanoic, cyclopentylpropionic, and β-phenylpropionic. For example, no less than 12 esters of 19-nortestosterone (nandrolone) have been used clinically as long-acting anabolic agents (498, 499).

In the case of nandrolone, the duration of action and the anabolic-to-androgenic ratio increased with the chain length of the ester group (500, 501). The decanoate and laurate esters, for instance, were active 6 weeks after injection. Clinically, nandrolone decanoate appeared to be the most practical, in that a dose of 25–100 mg/week produced marked nitrogen retention (502, 503).

Because the 17α-alkyl group has been implicated as the cause of the hepatotoxic side effects of oral preparations, the effect of esterification on oral efficacy has attracted attention. For example, esterification of dihydrotestosterone with short-chain fatty acids resulted in oral anabolic and androgenic activity in rats (504). Moreover, esters of methenolone possessed appreciable oral anabolic activity (505). Unfortunately, follow-up studies in humans have not been reported.

The manner by which the steroid esters evoke their enhanced activity and increased duration of action has puzzled investigators for many years. The classical concept has been that esterification delays the absorption rate of the steroid from the site of injection, thus preventing its rapid destruction.

Other factors must be involved, however, given that the potency and prolongation of action vary markedly with the nature of the esterifying acid.

Studies by James and coworkers (506, 507) have shed the most light on this problem. They studied the effect of various aliphatic esters of testosterone on rat prostate and seminal vesicles and correlated androgenicity with lipophilicity and rate of ester hydrolysis by liver esterase. The peak androgenic response was observed with the butyrate ester, which was also the most readily hydrolyzed. The more lipophilic valerate ester was slightly less androgenic in a quantitative sense, but its action was longer lasting. It was concluded that the ease of hydrolysis controls the weight of the target organs, whereas lipophilicity was responsible for the duration of the androgenic effect. These results explain the low androgenic activity previously noted for hindered trimethylacetate (pivalate) esters, which would be expected to be resistant to *in vivo* hydrolysis.

The effect of etherification on anabolic or androgenic activity has been studied less rigorously. Replacement of the 17β-OH with 17β-OCH$_3$ markedly reduced androgenic activity but did not affect greatly the ability to counteract cortisone-induced adrenal atrophy in male rats (508). A series of 17β-acetals (509, 510), alkyl ethers (511), and 3-enol ethers (512, 513), however, showed significant activity when given orally (514–516). The cyclohexyl enol ether of 17α-methyltestosterone, for example, was orally 5 times as myotrophic as 17α-methyltestosterone (516).

Other ethers such as the tetrahydropyranol (517, 518) and trimethylsilyl (519) have oral anabolic and androgenic activity in animals. The trimethylsilyl ether of testosterone (silandrone) had protracted activity after injection (520) and orally had twice the anabolic and androgenic acivities of 17α-methyltestosterone (519). Solo et al. (521) evaluated a variety of ethers that would not be expected to be readily cleaved *in vivo* and found them to be almost devoid of anabolic and androgenic activity. This provides additional support for the necessity of a free 17β-hydroxyl for androgenic activity.

5.3.8 Summary of Structure-Activity Relationships.

Synthetic modifications of C_{19} steroids have resulted in the enhancement of anabolic activity, even though a pure synthetic anabolic agent that retains no androgenic activity has not been accomplished. Structural changes in two regions of the testosterone molecule have resulted in the greatest enhancement of the anabolic/androgenic ratio. The first region is the C-17 position of the testosterone molecule. Introduction of the 17α-alkyl functionality, such as a 17α-methyl or a 17α-ethyl group, greatly increases the metabolic stability of the anabolic and decreases *in vivo* conversion of the 17β-alcohol to the 17-ketone by 17β-hydoxysteroid dehydrogenases. In addition, esterification of the 17β-alcohol enhances the lipid solubility of the steroids and provides injectable preparations for depot therapy.

The A-ring of testosterone is the second region in which structural modifications can be made to increase anabolic activity. Removal of the C-19 methyl group results in the 19-nortestosterone analogs, which have slightly higher anabolic activity. A major impact on the structure-activity relationships of anabolics can be observed with modifications at the C-2 position. Bioisosteric replacement of the carbon atom at position 2 with an oxygen provides a threefold increase in anabolic activity, as is seen with oxandrolone. Finally, the greatest effects were observed with the addition of heterocyclic rings fused at positions 2 and 3 of the A-ring. The two heterocycles that have led to the greatest changes are the pyrazole and the isoxazole rings, as seen in stanazolol and androisoxazole, respectively. In these anabolics, the 3-ketone of testosterone is replaced by the bioisosteric 3-imine. Stanazolol, which contains the pyrazole ring at C-2 and C-3, shows the greatest increases when compared to testosterone. Table 14.8 compares the anabolic activities of nitrogen retention and myotrophic activity for several common anabolics. Figure 14.10 contains a summary of the structure-activity relationships for anabolic agents.

5.4 Absorption, Distribution, and Metabolism

The absorption, distribution, and metabolism of the various anabolic steroids is quite similar to those pharmacokinetic properties of the endogenous and synthetic androgens discussed earlier in the chapter (522). Again, lipid solubility is critical for the absorption of these agents after oral or parenteral administration. The 17α-methyl group retards the metabolism of the compounds and provides orally active agents. Other anabolics such as methenolone are orally active without a 17α-substituent, indicating that these steroids are poor substrates for 17α-hydroxysteroid dehydrogenase (523, 524). Reduction of the 4-en-3-one system in synthetic anabolics to give the various α- and β-isomers occurs *in vivo* (322). The 3-deoxy agent 17α-methyl-5α-androstan-17α-ol was shown to be extensively converted to the 3-keto derivative by liver homogenate preparations (525). The metabolic fates of stanzolol and danazol have been reported (526, 527), with the major metabolites being heterocycle ring–opened derivatives and their deaminated products. Finally, both the unchanged anabolics and their metabolites are primarily excreted in the urine as the glucuronide or sulfate conjugates.

5.5 Toxicities

The major side effect of the anabolic steroids is the residual androgenic activity of the molecules. The virilizing actions are undesirable in adult males as well as in females and children. In addition, many anabolic steroids can suppress the release of gonadotropins from the anterior pituitary and lead to lower levels of circulating hormones and potential reproductive problems. Headaches, acne, and elevated blood pressure are common in individuals taking anabolics. The salt and water retention induced by these agents can produce edema.

The most serious toxicities resulting from the use of anabolic steroids are subsequent liver damages. Liver damage including jaundice and cholestasis can occur after use of the 17α-alkylated C_{19} steroids (340–343). Also, individuals who have received anabolic agents over an extended period have developed hepatic adenocarcinomas (528–530). Such clini-

Table 14.8 Comparison of Anabolic Activities

Compound	Number	Trade Names	Anabolic Activity	
			Nitrogen Retention	Myotrophic Activity
Testosterone	1	Android-T Malestrone Oreton Primotest Virosterone	1.0	1.0
19-Nortestosterone nandrolone	43	Nerobolil Nortestonate	0.8	1.0
Normethandrone	44	Methalutin Orgasteron	4.0	4.5
Norethandrolone	45	Nilevar Solevar	3.9	4.0
Methandrostenolone methandienone	48	Danabol Dianabol Nabolin Nerobil	0.6	1.4
Drostanolone	56	Drolban Masterone	—	1.3
Oxymetholone	63	Adroyd Anadrol Anadroyd Anapolon Anasterone Nastenon Protanabol Synasteron	2.75	2.8
Oxandrolone	65	Anavar Provita	3.0	3.0
Estrenol	70	Duraboral-O Maxibolin Orabolin Orgaboral Orgabolin	1.7	2.0
Stanazolol	75	Stanozol Winstrol Tevabolin	10.0	7.5
Androisoxazole	76	Androxan Neo-ponden	1.5	1.7

cal reports serve to underscore the inherent risks associated with anabolic steroid use in amateur athletes for no demonstrable benefits.

6 ANDROGEN ANTAGONISTS

A majority of the recent research efforts in the area of androgens has concentrated on the preparation and biological activities of androgen antagonists. An androgen antagonist is defined as a substance that antagonizes the actions of testosterone in various androgen-sensitive target organs and, when administered with an androgen, blocks or diminishes the effectiveness of the androgen at various androgen-sensitive tissues. Androgen antagonists may act to block the action of testosterone at several possible sites. First, such compounds could interfere with the entrance of the androgen into the target cell. A second site of action of androgen antagonists may be to

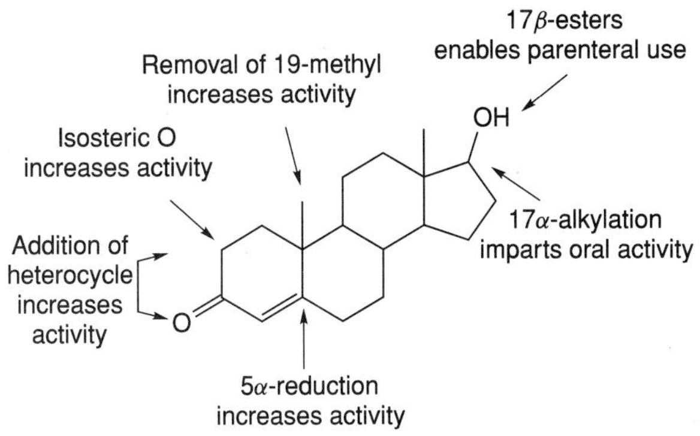

17β-esters
enables parenteral use

Removal of 19-methyl
increases activity

Isosteric O
increases activity

OH

Addition of
heterocycle
increases
activity

17α-alkylation
imparts oral activity

O

5α-reduction
increases activity

Figure 14.10. Summary of structure-activity relationships for anabolic agents.

block the conversion of testosterone to its more active metabolite dihydrotestosterone. Third, competition for the high affinity binding sites on the androgen receptor molecule may account for antiandrogenic effects. Finally, certain agents such as LHRH agonists can act in the pituitary to lower gonadotropin secretion by way of receptor downregulation and thus diminish the production of testosterone by the testis. The substances described in this section act through at least one of these mechanisms. Several reviews on the androgen antagonists are available (531–537).

6.1 Current Drugs on the Market

6.2 Antiandrogens

6.2.1 Therapeutic Uses. Antiandrogens are agents that compete with endogenous androgens for the hormone binding site on the androgen receptor. These agents have therapeutic potential in the treatment of acne, virilization in women, hyperplasia and neoplasia of the prostate, baldness, and male contraception, and clinical studies have demonstrated the potential therapeutic benefits of the antiandrogens. The applications of antiandrogens for the treatment of prostatic carcinoma and for the treatment of benign prostatic hypertrophy (BPH) have also

Antiandrogens

Generic Name (structure)	Trade Name	U.S. Manufacturer	Chemical Class	Dose
Flutamide (**91**)	Eulexin	Schering-Plough	Nonsteroidal	Tablets: 125 mg
Bicalutamide (**97**)	Casodex	AstraZeneca	Nonsteroidal	Tablets: 50 mg
Nilutamide (**93**)	Nilandron	Aventis	Nonsteroidal	Tablets: 50 mg 150 mg
Cyproterone acetate (**78**)	Androcur	Schering AG	Pregnane	

5α-Reductase Inhibitors

Generic Name (structure)	Trade Name	U.S. Manufacturer	Chemical Class	Dose
Finasteride (**104**)	Proscar	Merck	Androstane	Tablets: 5 mg
	Propecia	Merck	Androstane	Tablets: 1 mg

been investigated. Antiandrogens are effective for the treatment of prostate cancer when combined with androgen ablation, such as surgical (orchiectomy) or medical (LHRH agonist) castration.

6.2.2 Structure-Activity Relationships for Antiandrogens

6.2.2.1 Steroidal Agents. Several steroidal and nonsteroidal compounds with demonstrated antiandrogenic activity have been used clinically (533). The first compounds used as antiandrogens were the estrogens and progestins (538). Steroidal estrogens and diethylstilbestrol are used in the treatment of prostatic carcinoma (539–542) and exert their action by suppression of the release of pituitary gonadotropins. Progestational compounds have also been used for antiandrogenic actions with limited success (543). The inherent hormonal activities of these compounds and the development of more selective antiandrogens have limited the clinical applications of estrogens and progestins as antiandrogens.

A modified progestin that is a potent antiandrogen and has minimal progestational activity is the agent cyproterone acetate (**78**). This compound was originally prepared in search of orally active progestins, but was quickly recognized for its ability to suppress gonadotropin release (544–552). It was later demonstrated that this compound also bound with high affinity to the androgen receptor and thus competed with DHT for the binding site (552–555). Cyproterone acetate has received the most clinical attention in antiandrogen therapy (556–567). Cyproterone acetate has produced quite satisfactory results in the treatment of acne, seborrhea, and hirsutism (556–562). Therapeutic effectiveness of this agent in the treatment of prostatic carcinoma has been reported (563–567). Cyproterone acetate was reported to be a good alternative to estrogens for the treatment of prostate cancer when combined with androgen ablation (568, 569). However, this combination did not improve disease-free survival or overall survival when compared to castration alone.

Other pregnane compounds that exhibit antiandrogenic actions by binding to the an-

drogen receptor are chlormadinone acetate (**79**), medroxyprogesterone acetate (**80**), medrogesterone (**81**), A-norprogesterone (**82**), and gestonorone capronate (**83**) (570–574). In addition, medrogesterone exerts antiandrogenic effects by inhibiting 5α-reductase and thus preventing the formation of DHT (575, 576). Gestonorone capronate interferes with the uptake process in target cells (574).

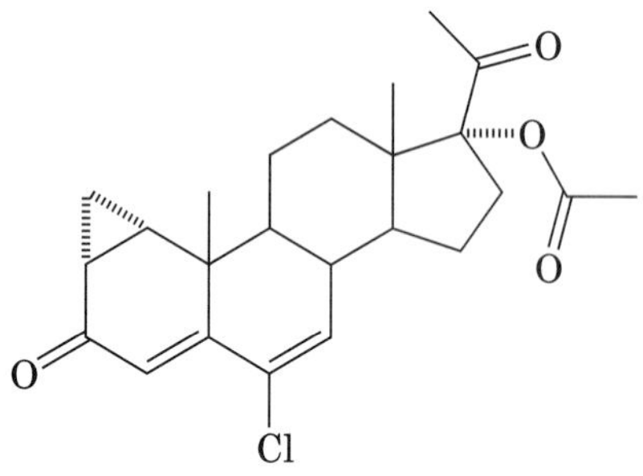

(78)

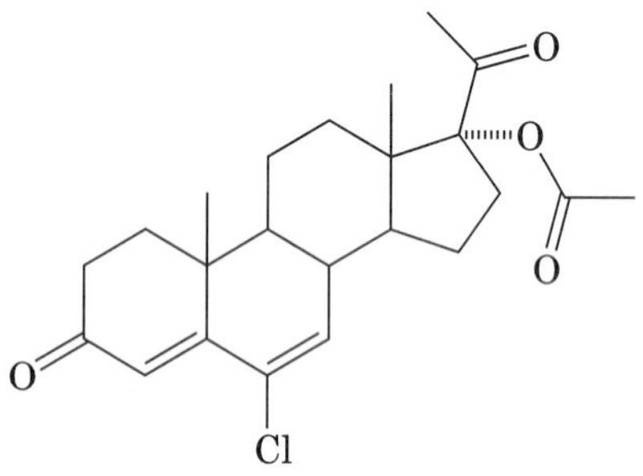

(79)

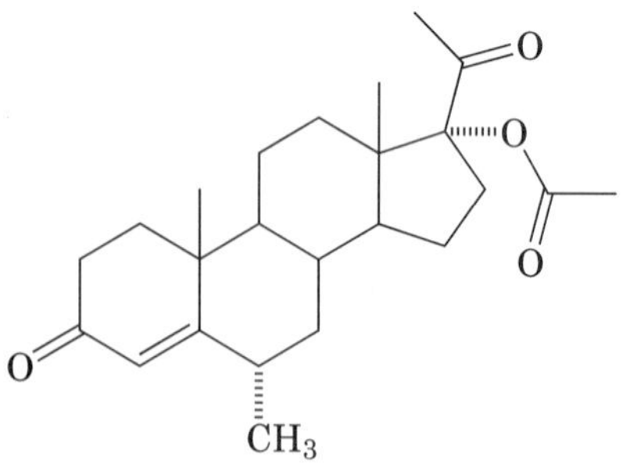

(80)

Several androstane derivatives demonstrate antiandrogenic properties. 17α-Methyl-B-nortestosterone (**84**) was prepared and

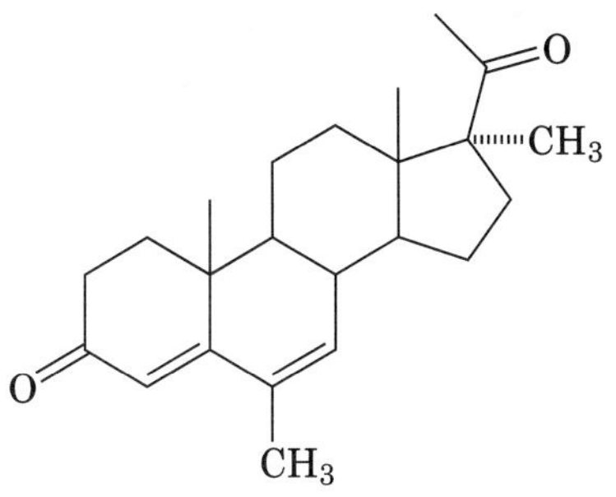

(81)

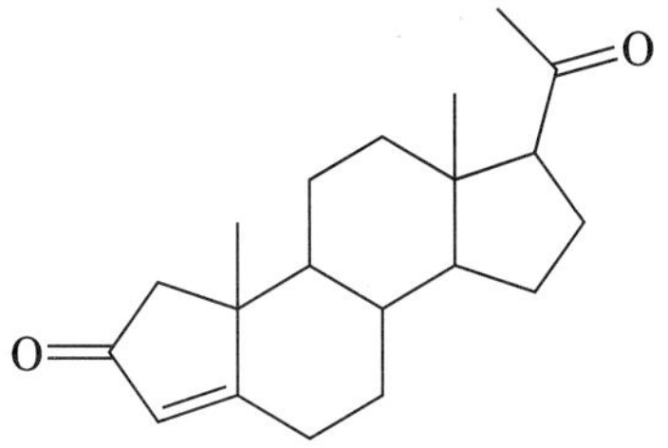

A-ring antiandrogens such as zanoterone (WIN 49,596) (88) further support these conclusions on the structure–activity relationships of steroidal antiandrogens. Additional A-ring heterocycles identified as novel antiandrogens are the thiazole (89) and oxazole (90) (592, 593). The optimal substitutions on the A-ring heterocyclic androstanes for *in vivo* antiandrogenic activity are the methylsulfonyl group at the N-1′ position and a 17α-substituent (e.g., 17α-methyl or 17α-ethinyl).

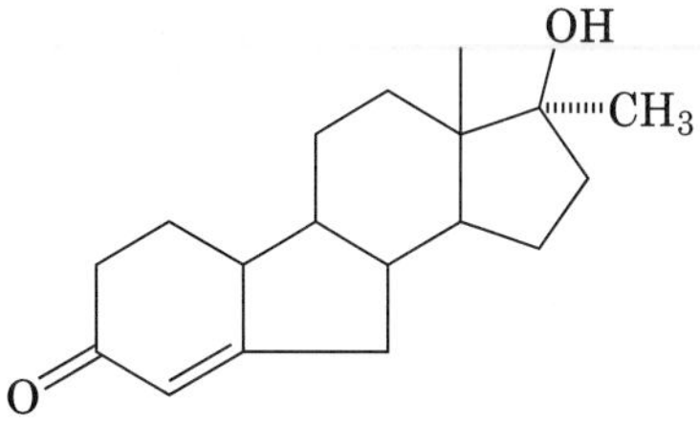

(82)

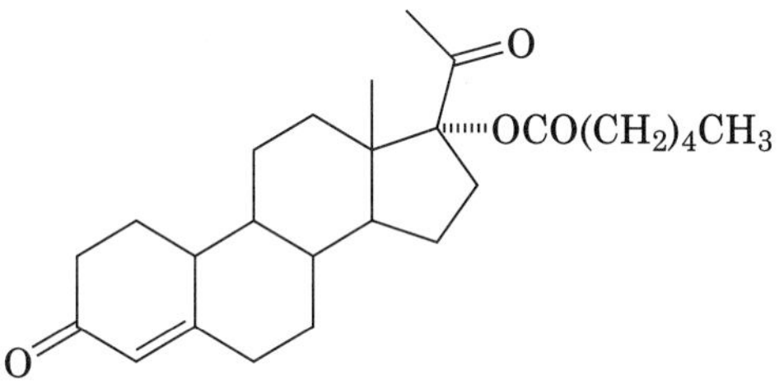

(84)

(83)

(85)

tested in 1964 for antihormonal activity (577). Within the next decade, several other androstane analogs were prepared and found to possess antiandrogenic activity (578–584), including BOMT (85), R2956 (86), and oxendolone (TSAA291, 87). As expected, the mechanism of antiandrogenic action of these synthetic steroids is the competition with androgens for the binding sites on the receptor molecule (585–589). Numerous A- and B-ring–modified steroids were examined for antiandrogenic activity and the ability to bind to the androgen receptor (590, 591), demonstrating that the structural requirements of a receptor binding site can accommodate some degree of flexibility in the A- and/or B-rings of antiandrogenic molecules. Heterocyclic-substituted

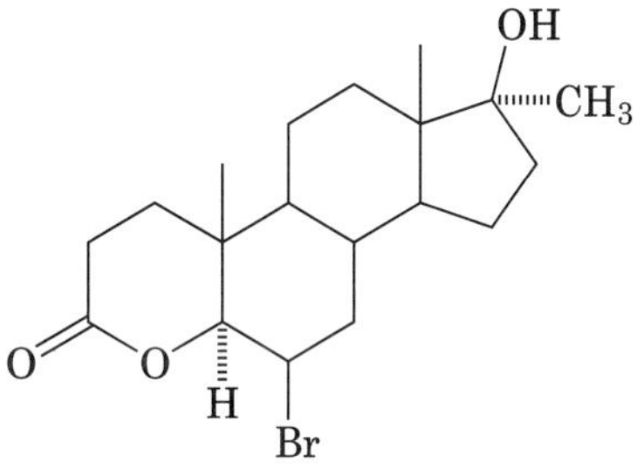

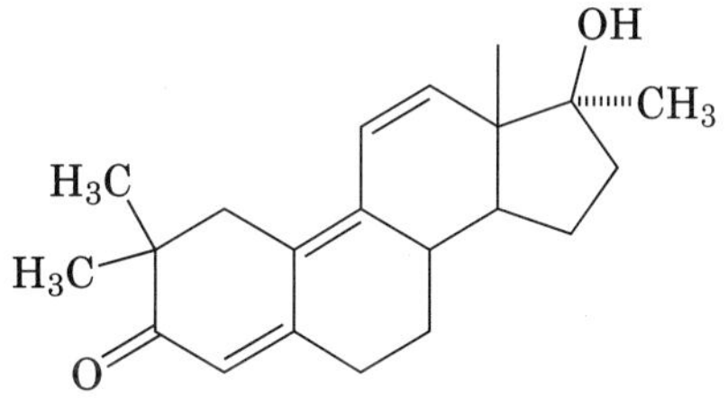

(86)

6.2.2.2 Nonsteroidal Agents. The absolute requirement of a steroidal compound for interaction with the androgen receptor was invalidated when the potent nonsteroidal antiandrogen flutamide (Eulexin, 91) was introduced (594, 595). Subsequent receptor

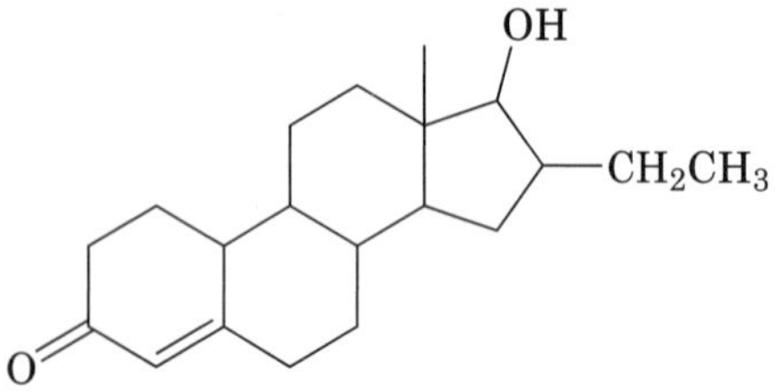

(87)

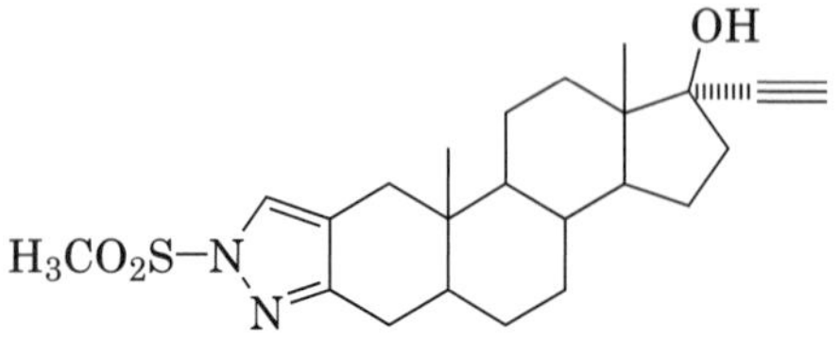

(88)

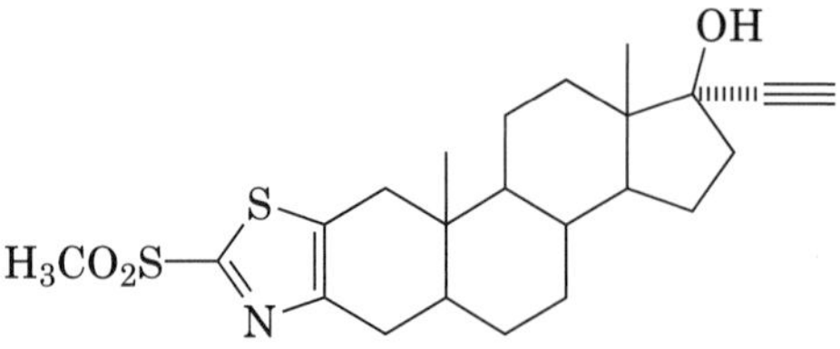

(89)

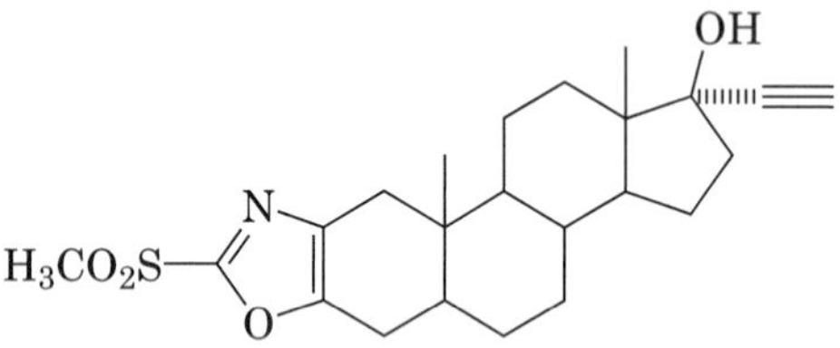

(90)

studies (589, 596, 597) demonstrated that this compound competed with DHT for the binding sites. The side chain of flutamide allows sufficient flexibility for the molecule to assume a structure similar to that of an androgen. In addition, a hydroxylated metabolite (**92**) has been identified, is a more powerful antiandrogen *in vivo*, and has a higher affinity for the receptor than for the parent compound (589, 598). Important factors in the structure–

activity relationships of flutamide and analogs are the presence of an electron-deficient aromatic ring and a powerful hydrogen bond donor group.

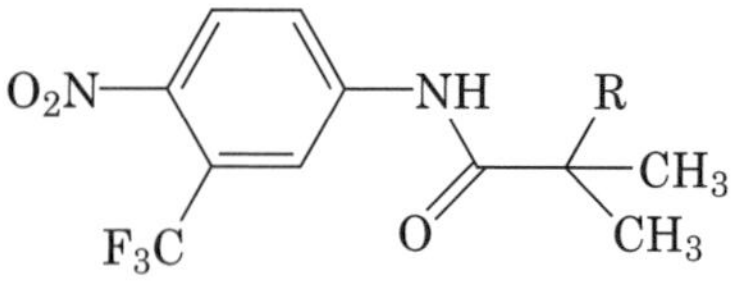

(91) X = H
(92) X = OH

Flutamide has been extensively evaluated for the treatment of prostate cancer. Large double-blind studies in prostate cancer patients using a combination of flutamide with an LHRH agonist (as a medical castration) resulted in an increased number of favorable responses and increased overall survival when compared to an LHRH agonist or surgical castration (599, 600).

Nilutamide (Anandron, **93**), related nilutamide analogs (**94–96**)(**94**)(**95**)(**96**), and bicalutamide (Casodex, **97**) are other nonsteroidal antiandrogens with a similar electron-deficient aromatic ring and have been shown to interact with the androgen receptor to varying degrees (538, 601). Nilutamide and bicalutamide are pure antiandrogens and are effective in suppressing testosterone-stimulated cell proliferation (602). Both nilutamide and bicalutamide demonstrate effectiveness against prostate cancer (603–607).

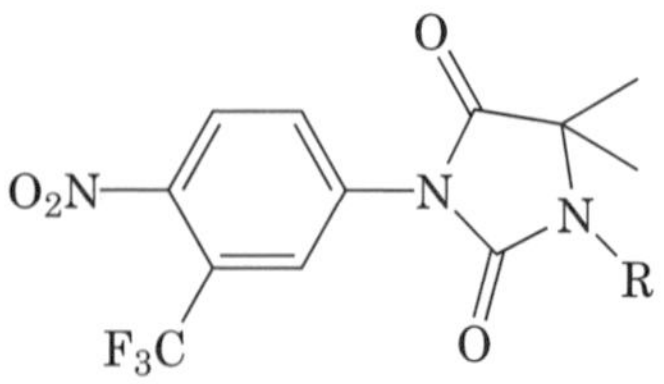

(93)   R = H
(94)   R = (CH$_2$)$_4$OH

Other aryl substituted nonsteroidal compounds have also been identified as antiandrogens. DIMP (**98**) is a phthalimide derivative that showed weak affinity for the androgen

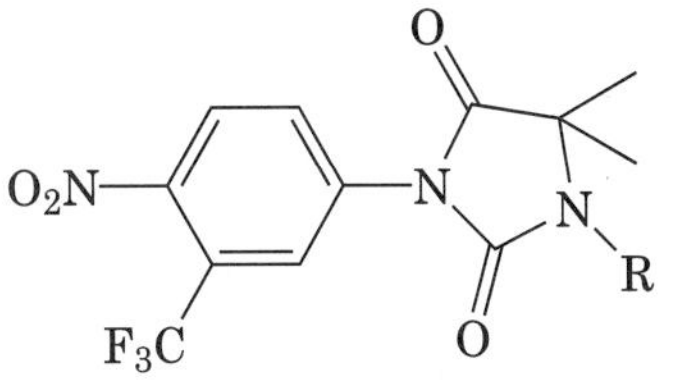

(95) R = CH$_3$
(96) R = (CH$_2$)$_4$OH

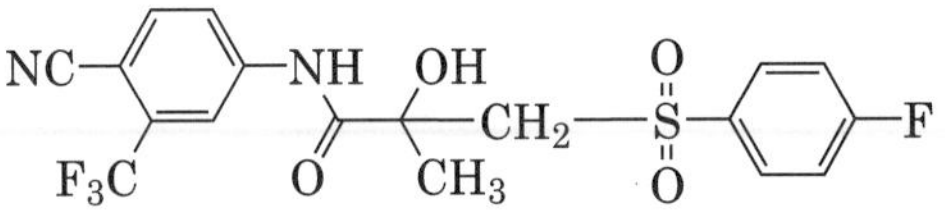

(97)

receptor and poor *in vivo* activity (589, 608). A series of tetrafluorophthalimides such as (99) demonstrated moderate activity as antiandrogens in cell proliferation assays (609).

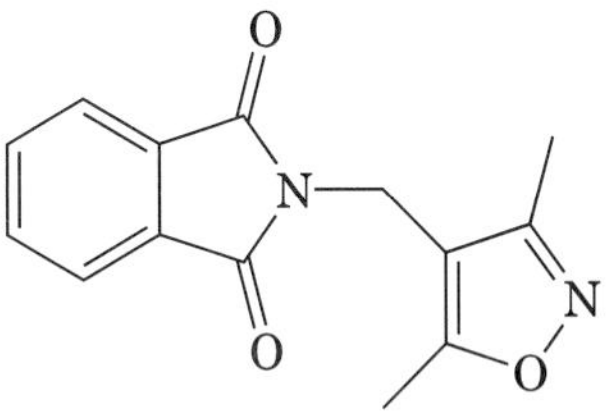

(98) DIMP

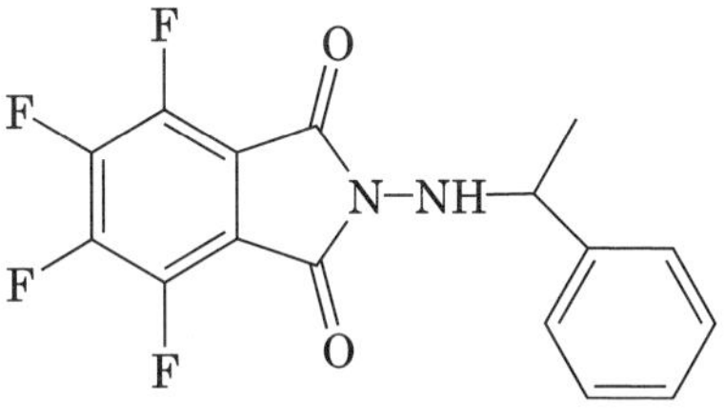

(99)

A series of 1,2-dihydropyridono[5,6-*g*]quinolines were identified as novel nonsteroidal antiandrogens based on a cell-based screening approach (610). Several analogs (100–103) demonstrated excellent *in vivo* activity, reducing rat ventral prostate weight without affecting serum gonadotropins and serum testosterone levels.

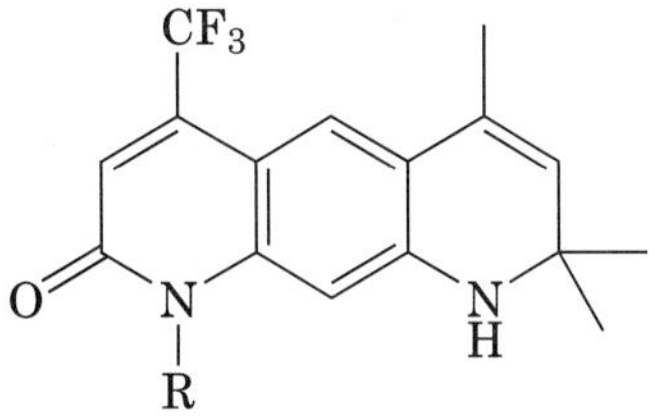

(100) R = H
(101) R = CH$_3$

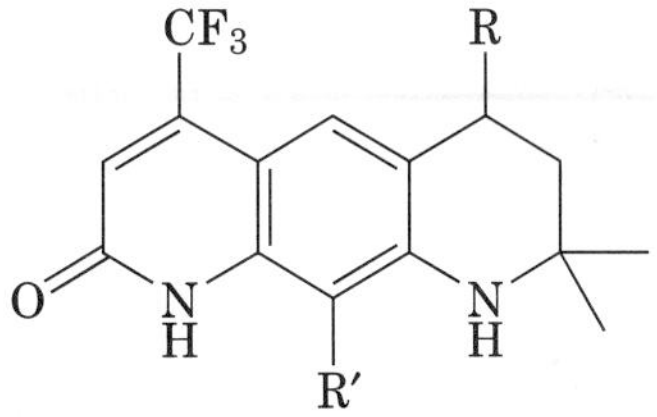

(102) R, R′ = H
(103) R, R′ = CH$_3$

6.2.3 Absorption, Distribution, and Metabolism. The steroidal antiandrogens exhibit pharmacokinetic properties similar to those of the androgens and anabolic agents. The lipophilicity of the compounds influences absorption both orally and from injection sites (611–615). Reduction of the 3-ketone and 4,5 double bond are common routes of metabolism (611). An unusual metabolite of cyproterone acetate, 15α-hydroxycyproterone acetate, was isolated and identified in both animals and man (616). The nonsteroidal antiandrogen flutamide is rapidly absorbed and extensively metabolized *in vivo* (617, 618). As described earlier, the hydroxy metabolite (92) of flutamide is a more potent antiandrogen (589, 598). The major metabolite of bicalutamide is the sulfone, which has comparable *in vivo* activity (619). Finally, the antiandrogens are primarily excreted as the glucuronide and sulfate conjugates in the urine.

6.2.4 Toxicities. Side effects of these agents have been identified from various clinical trials. Testicular atrophy and decreased spermatogenesis have been observed during treatment with cyproterone acetate (620, 621). Antiandrogens can also impair libido and result in impotency (622). Certain antiandrogens such as cyproterone acetate and medrogesterone also exhibit inherent progestational activity, suppress corticotropin release, and have some androgenic effects (623–625). No hormonal activities were observed for the nonsteroidal antiandrogens, such as flutamide (618). On the other hand, many nonsteroidal antiandrogens exhibit other endocrine side effects, such as elevated serum gonadotropins and serum testosterone levels. Gynecomastia, nausea, diarrhea, and liver toxicities have been observed in patients on nonsteroidal antiandrogens (626, 627). Also, resistance to antiandrogen therapy has been observed in prostate cancer patients (628).

6.3 Enzyme Inhibitors

Enzymes involved in the biosynthesis and metabolism of testosterone are attractive targets for drug design and drug development. Suppression of the synthesis of androgenic hormones and androgen precursors is a viable therapeutic approach for the treatment of various androgen-mediated disease processes and an important endocrine treatment for prostate cancer. Potent inhibition of Type 2 5α-reductase in androgen target tissues and the resultant decrease in DHT levels will provide selective interference with androgen action within those target tissues and no alterations of other effects produced by testosterone, other structurally related steroids, and other hormones such as corticoids and progesterone. The cytochrome P-450$_{17\alpha}$ enzyme complex displays two enzymatic activities: 17α-hydroxylation, to produce 17α-hydroxysteroids; and C_{17}–C_{20} bond cleavage (17,20-lyase activity), to produce androgens. In the male, this enzyme is found in both testicular and adrenal tissues, with these organs providing circulating androgens in the blood. Effective inhibition of this microsomal enzyme complex would eliminate both testicular and adrenal androgens and remove the growth stimulus to androgen-dependent prostate carcinoma. Synthetic androgen analogs that inhibit the oxidative metabolism of androgens testosterone and androstenedione to estrogens estradiol and estrone can serve as potential therapeutic agents for controlling estrogen-dependent diseases such as hormone-dependent breast cancer.

6.3.1 5α-Reductase Inhibitors. The most extensively studied class of 5α-reductase inhibitors is the 4-azasteroids (629), which include the drug finasteride (Proscar, **104**). Finasteride is the first 5α-reductase inhibitor approved in the United States for the treatment of benign prostatic hyperplasia (BPH). This drug has approximately a 100-fold greater affinity for Type 2 5α-reductase than for the Type 1 enzyme, demonstrating an IC_{50} value of 4.2 nM for Type 2 5α-reductase (630). In humans, finasteride decreases prostatic DHT levels by 70–90% and reduces prostate size (631), whereas testosterone tissue levels increased. Clinical trials demonstrated sustained improvement in BPH disease and reduction in PSA levels (632, 633). Related analogs (**105–107**) have also demonstrated effectiveness *in vitro* and *in vivo* (634–636). These agents were originally designed to mimic the putative 3-enolate intermediate of testosterone and serve as transition-state inhibitors (535, 536). Subsequently, finasteride was shown to produce time-dependent enzyme inactivation (637) and function as a mechanism-based inactivator. The structure-activity relationships for the 4-azasteroids illustrate the stringent requirements for inhibition of human Type 2 5α-reductase (638). The 5α-reduced azasteroids are preferred, a 1,2-double bond can be tolerated, and the nitrogen can be substituted with only hydrogen or small lipophilic groups. Lipophilic amides or ketones are preferred as substituents at the C-17β position.

(104) finasteride

Several 6-azasteroids, such as (**108**) and (**109**), were prepared as extended mimics of the enolate transition state and have also demonstrated potent inhibition of 5α-reductase (639).

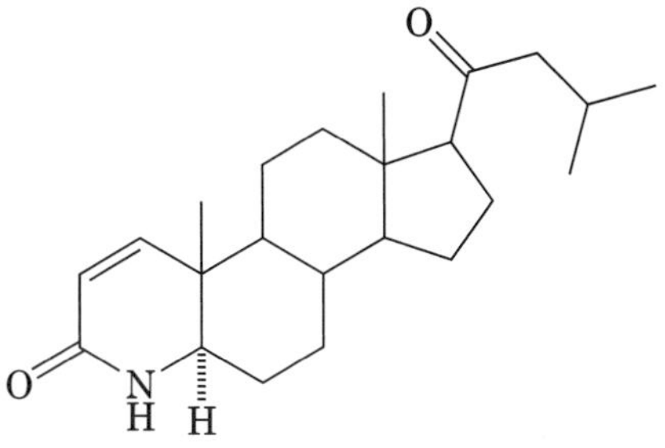

(105)

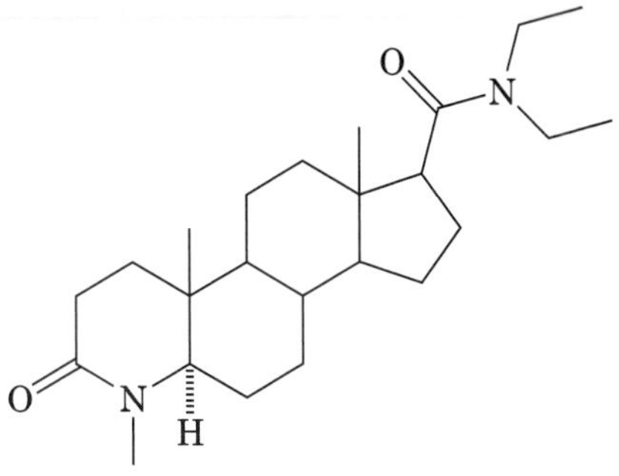

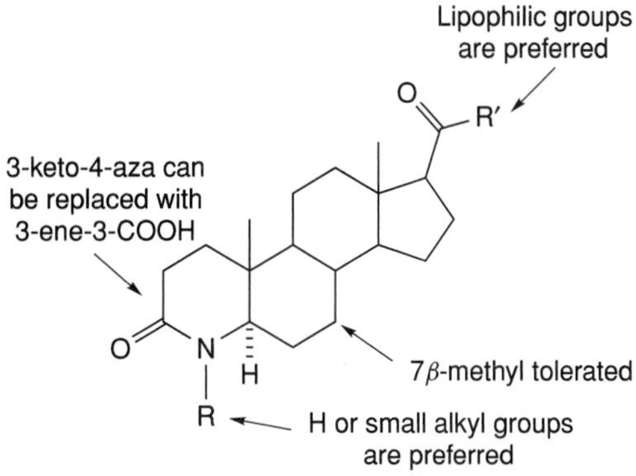

Figure 14.11. Summary of structure–activity relationships for 5α-reductase inhibitors.

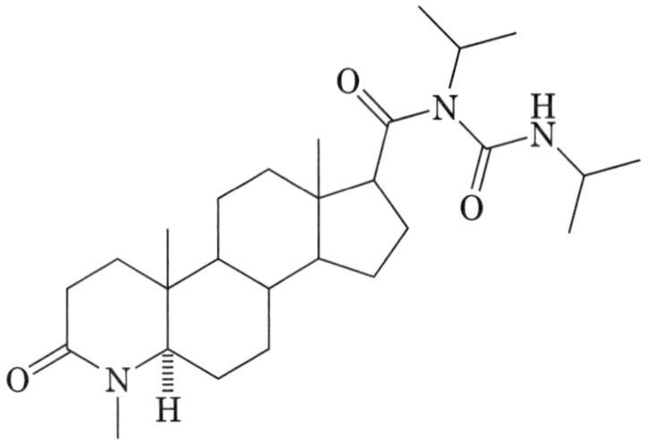

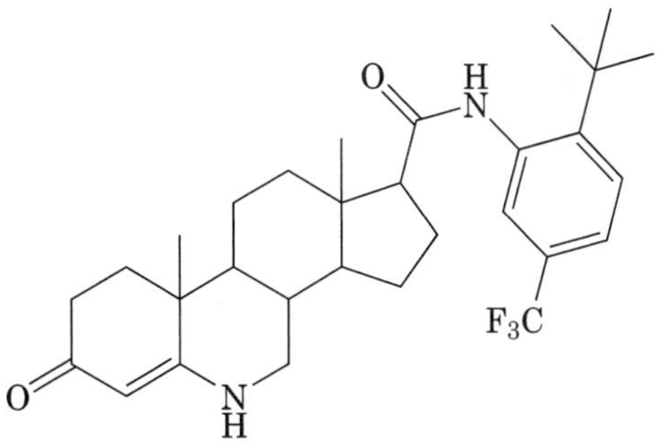

(106)

(108)

(107)

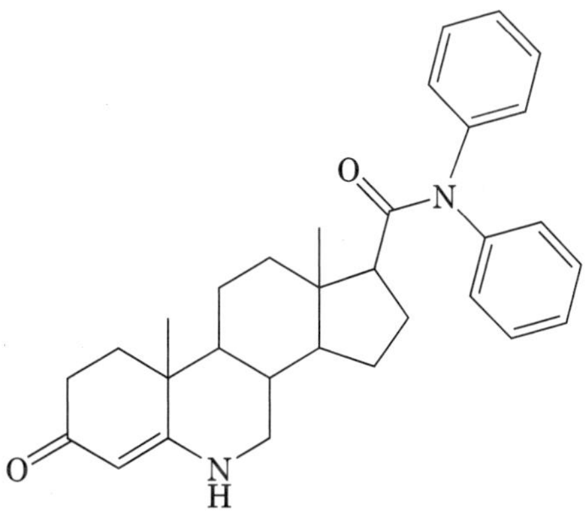

(109)

The 6-azasteroids are more effective inhibitors of Type 2 5α-reductase, but some analogs also exhibit good inhibition of Type 2 5α-reductase. Alkylation of the nitrogen can be tolerated; however, a 1,2-double bond decreases inhibitory activity in this series. The best inhibitors contain large lipophilic substituents at the C-17β position. Figure 14.11 contains a summary of the structure–activity relationships for steroidal 5α-reductase inhibitors.

Androstadiene 3-carboxylic acids (**110**) and (**111**) were recently designed as transition-state inhibitors and have demonstrated potent uncompetitive inhibition of Type 2 5α-reductase (640, 641). Epristeride (SK&F 105,657,

110) has demonstrated the ability to lower serum DHT levels by 50% in clinical trials (642, 643). Other analogs with acidic functionalities at the C-3 position include other androstene carboxylic acids (**112, 113**) and estratriene carboxylic acids (**114**) (644). The allenic secosteroid (**115**) has demonstrated potent irreversible inhibitor of 5α-reductase, even though it was originally developed as an irreversible inhibitor of 3β-hydroxysteroid dehydrogenase/Δ^{4,5}-isomerase (645–647). Finally, selective and potent inhibitors of Type 1 5α-reductase were developed based on the 4-azacholestane MK-386 (**116**) (648).

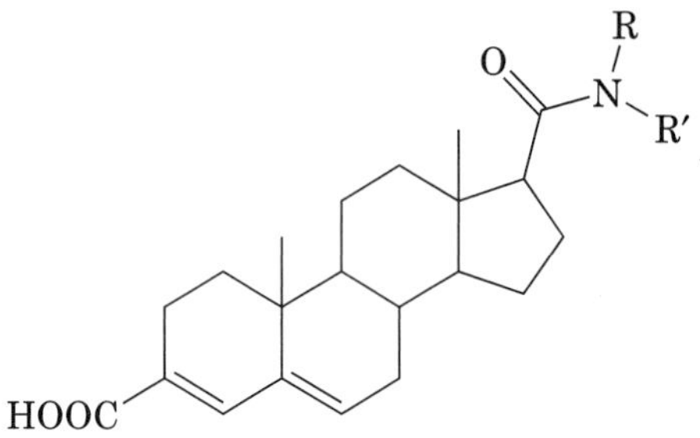

(113)

(110) R = H; R′ = C(CH$_3$)$_3$
(111) R = R′ = CH(CH$_3$)$_2$

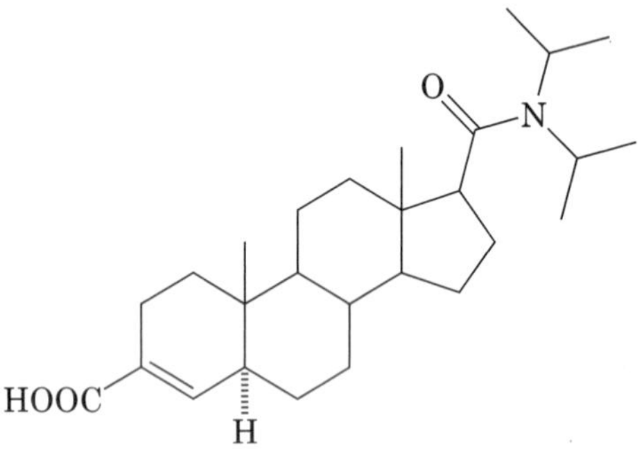

(114)

(112)

(115)

Several nonsteroidal 5α-reductase inhibitors have developed based on the azasteroid molecule or from high throughput screening methods (535, 536). Examples of these nonsteroidal inhibitors include the benzoquinolinone (**117**), an aryl carboxylic acid (**118**), and FK143 (**119**) (649–651).

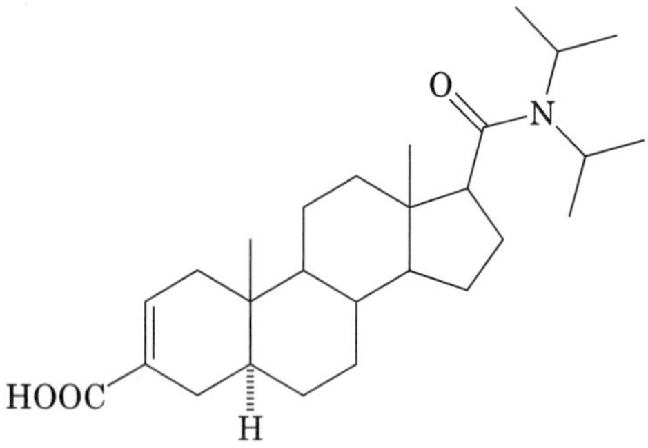

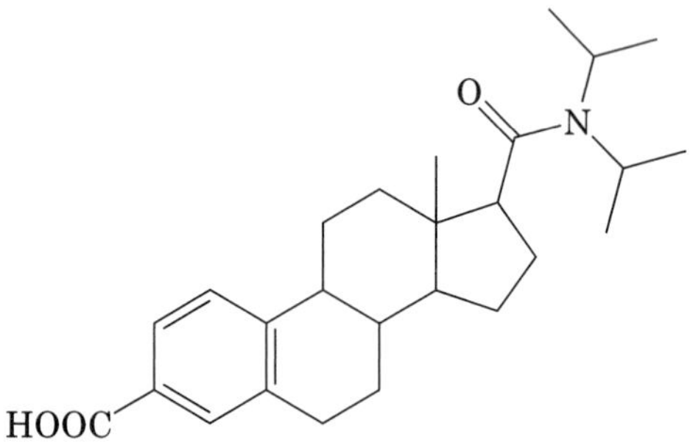

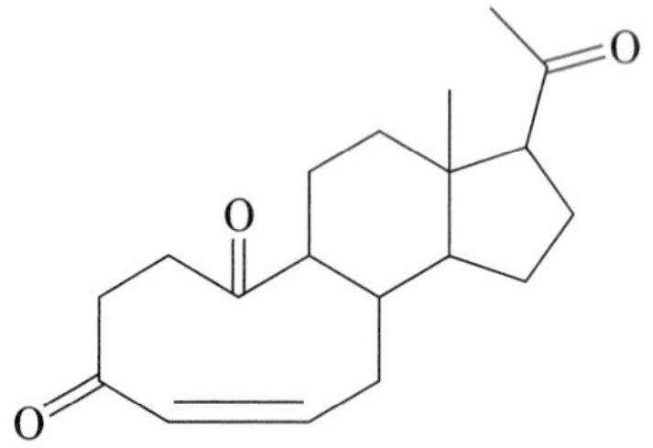

6.3.2 17,20-Lyase Inhibitors. Both nonsteroidal and steroidal agents have been examined as inhibitors of 17α-hydroxylase/17,20-lyase. The nonsteroidal agents studied most extensively are aminoglutethimide (**120**) and ketoconazole (**121**), both *in vitro* and in clinical trials. Objective response rates for treatment of prostate cancer in relapsed patients were observed with high doses of aminoglutethimide (652) and high doses of ketoconazole (653), but both agents produce frequent side effects. A third nonsteroidal agent that

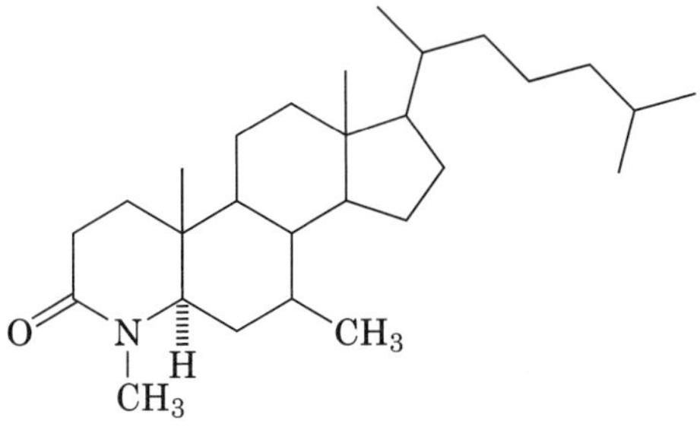

(116)

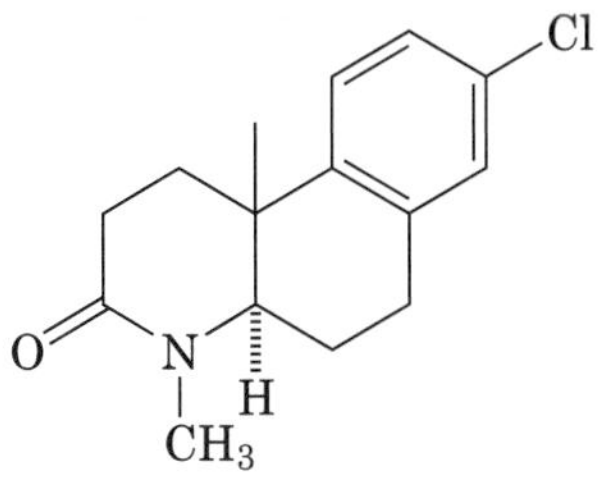

(117)

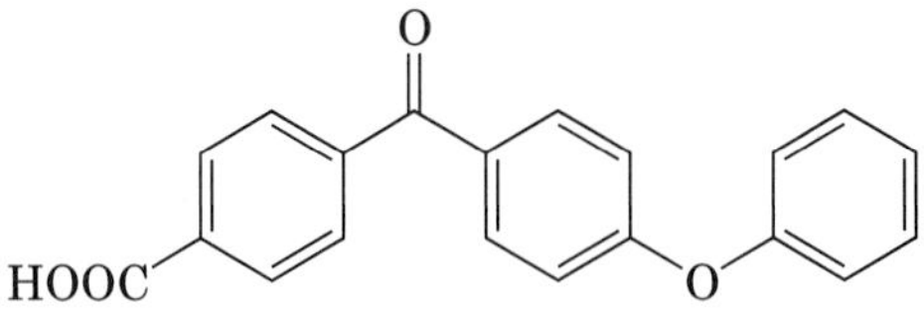

(118)

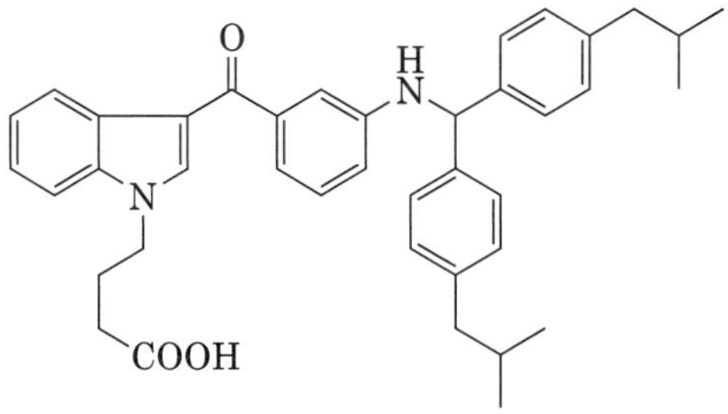

(119) FK143

and pyridylacetic acid esters (658). In general, high doses of nonsteroidal agents are needed to produce significant *in vitro* or *in vivo* activity. Another potential problem with these agents is nonspecific inhibition of other cytochrome P-450 enzymes involved in either steroidogenesis or liver metabolism.

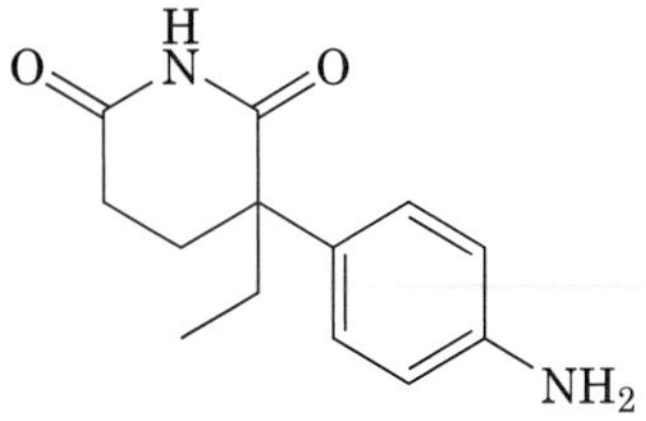

(120) aminoglutethimide

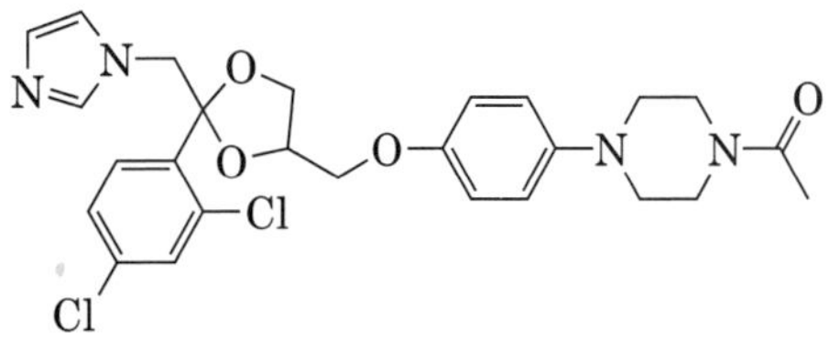

(121) ketoconazole

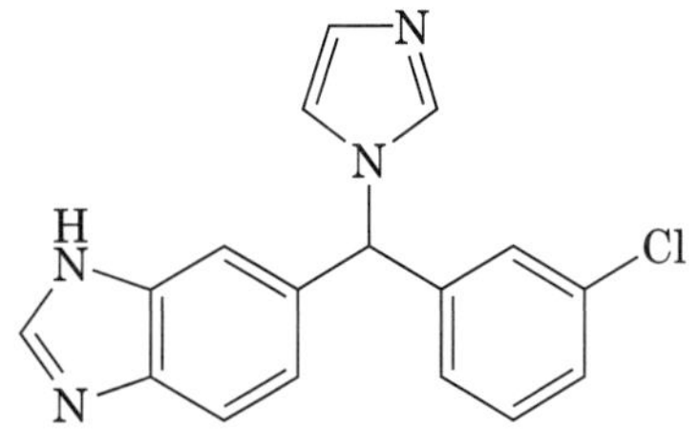

(122) liarozole

A few studies of steroidal inhibitors of 17α-hydroxylase/17,20-lyase have been reported. An extensive analysis of the specificity of steroid binding to testicular microsomal cytochrome P-450 identified several steroids exhibiting binding affinity (659). One of these, promegestrone (**123**), has been used in kinetic analysis of purified cytochrome P-450$_{17\alpha}$ (660). An affinity label inhibitor, 17-bromoacetoxyprogesterone (**124**), alkylates a unique cysteine residue on purified cytochrome P-450$_{17\alpha}$ (661). Potential mechanism-based inhibitors include 17β-(cyclopropylamino)-5-androsten-3β-ol (**125**; (662)) and 17β-vinyl-

has received extensive preclinical evaluation is the benzimidazole analog, liarozole (**122**), which produced reduction in plasma testosterone and androstenedione levels *in vivo* (654). Other nonsteroidal agents reported to exhibit 17α-hydroxylase/17,20-lyase inhibitory activity *in vitro* include other imidazole analogs (655), nicotine (656), bifluranol analogs (657), progesterone (**126**); (663). To date, all of these inhibitors exhibit apparent K_i' values in the micromolar (μM) range, whereas the apparent K_m for progesterone is 140 nM. The 17β-aziridinyl analog (**127**) and 17β-pyridyl derivative (**128**) also exhibited similar inhibitory activity (664, 665).

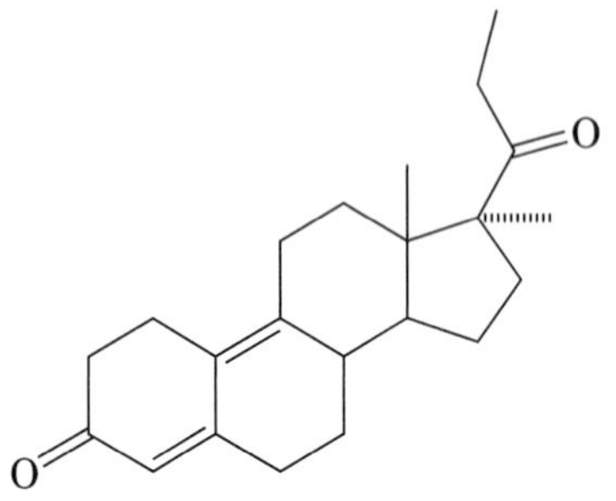

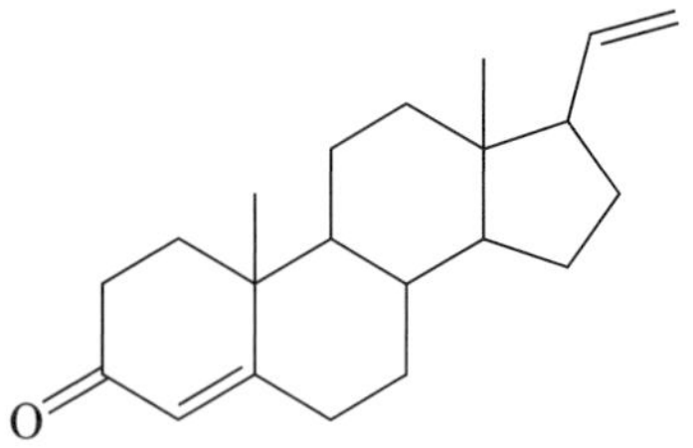

(126)

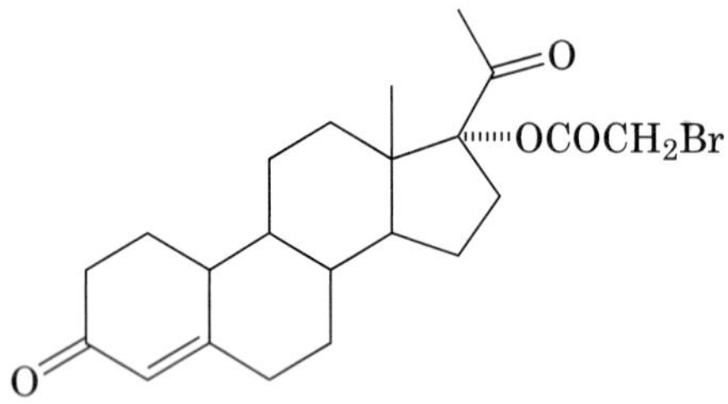

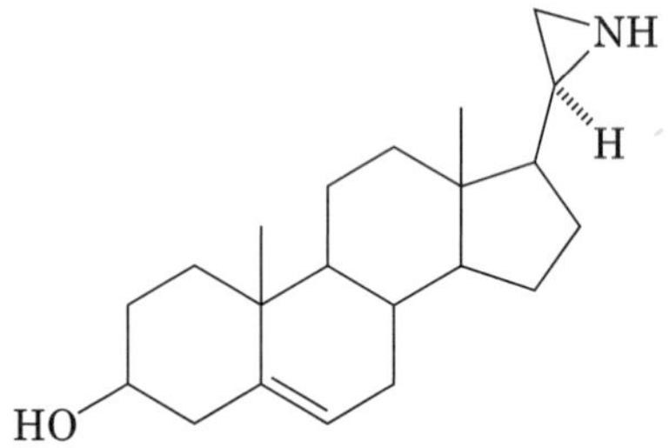

(123)

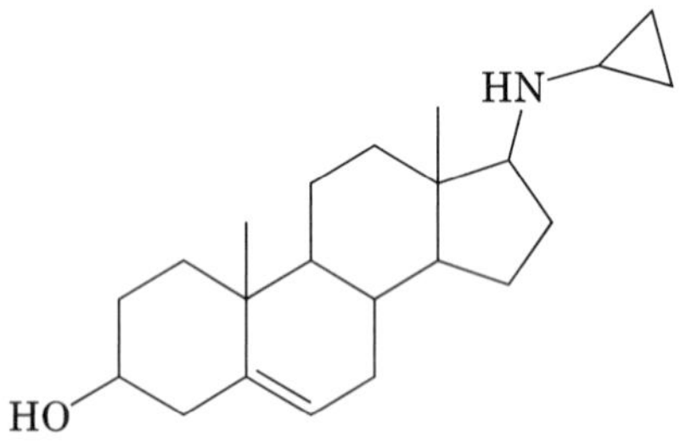

(127)

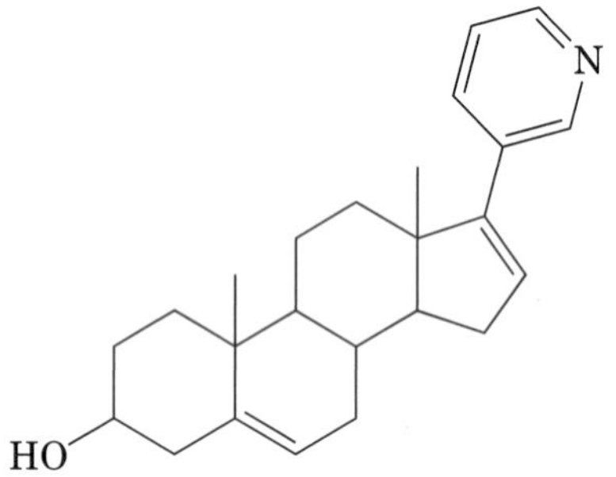

(124)

(128)

(125)

6.3.3 C$_{19}$ Steroids as Aromatase Inhibitors. Aromatase is the enzyme complex that catalyzes the conversion of androgens into the estrogens. This enzymatic process is the rate-limiting step in estrogen biosynthesis and converts C$_{19}$ steroids, such as testosterone and androstenedione, into the C$_{18}$ estrogens, estradiol and estrone, respectively. Inhibition

of aromatase has been an attractive approach for examining the roles of estrogen biosynthesis in various physiological or pathological processes. Furthermore, effective aromatase inhibitors can serve as potential therapeutic agents for controlling estrogen-dependent diseases such as hormone-dependent breast cancer. Investigations on the development of aromatase inhibitors began in the 1970s and have expanded greatly in the past two decades. Research summaries of steroidal and nonsteroidal aromatase inhibitors have been presented at the three international aromatase conferences (666–668) and several reviews have also been published (669–676).

Steroidal inhibitors that have been developed to date build on the basic androstenedione nucleus and incorporate chemical substituents at varying positions on the steroid. These inhibitors bind to the aromatase cytochrome P-450 enzyme in the same manner as that of the substrate androstenedione. Even though the steroidal aromatase inhibitors are C_{19} steroids, these agents exhibit no significant androgenic activity. A limited number of effective inhibitors with substituents on the A-ring have been reported. Several steroidal aromatase inhibitors contain modifications at the C-4 position, with 4-hydroxyandrostenedione (**129**) (4-OHA; formestane) being the prototype agent. Initially, 4-OHA was thought to be a competitive inhibitor, but was later demonstrated to produce enzyme-mediated inactivation (677–679). *In vivo*, 4-OHA inhibits reproductive process (680) and causes regression of hormone-dependent mammary rat tumors (681, 682). 4-OHA is effective in the treatment of advanced breast cancer in postmenopausal women (683–685), and this drug is approved in the United Kingdom for breast cancer therapy. Thus, the spacial requirements of the A-ring for binding of the steroidal inhibitor to aromatase are rather restrictive, permitting only small structural modifications to be made. Incorporation of the polar hydroxyl group at C-4 enhances inhibitory activity. 1-Methyl-1,4-androstadiene-3,17-dione (**130**) is a potent inhibitor of aromatase *in vitro* and *in vivo* (686); on the other hand, bulky substitutents at the 1α-position are poor inhibitors (687). At the C-3 position, replacement of the ketone with a methylene provided effective inhibitors (688).

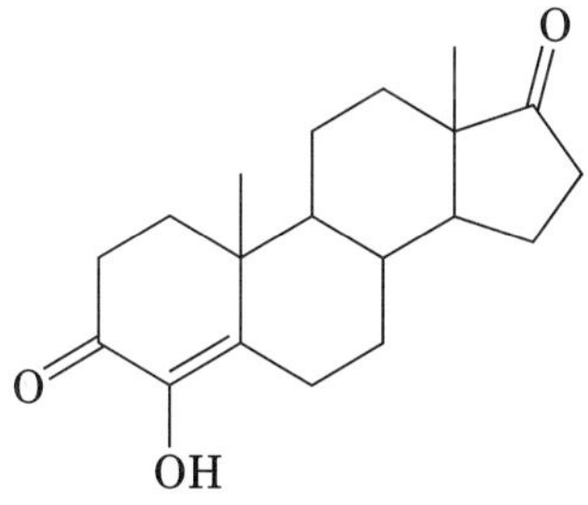

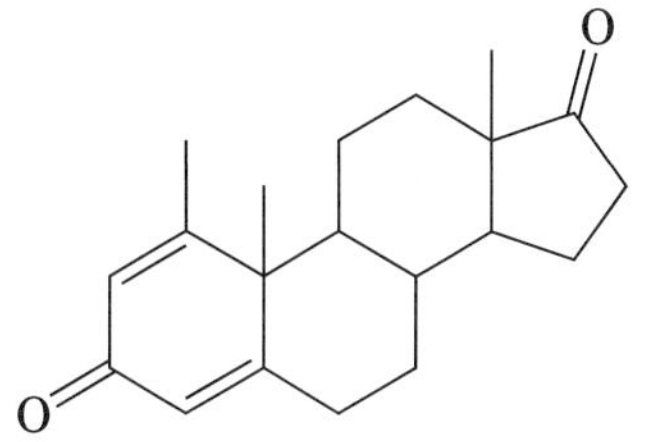

(130)

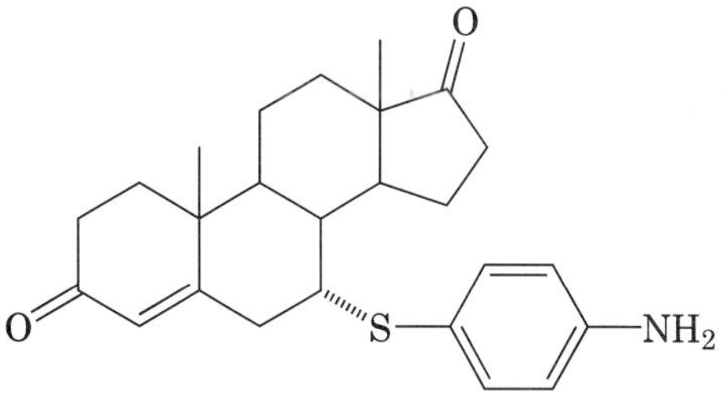

(131)

More extensive structural modifications may be made on the B-ring of the steroid nucleus. Bulky substitutions at the C-7 position of the B-ring have provided several very potent aromatase inhibitors (687). 7α-(4'-Amino)phenylthio-4-androstene-3,17-dione **116** (7α-APTA) is a very effective competitive inhibitor, with an apparent K_i of 18 nM. This inhibitor has also demonstrated effectiveness in inhibiting aromatase in cell cultures (689, 690) and in treating hormone-dependent rat mammary tumors (690, 691). Evaluation of various substituted aromatic analogs of 7α-APTA provided no correlation between the electronic character of the substituents and inhibitory activity (692). Investigations of various 7-substituted 4,6-androstadiene-3,17-dione derivatives (693, 694) suggest that only those derivatives that can project the 7-aryl substitutent into the 7α pocket are effective inhibitors. Overall, the most effective B-ring–modified aromatase inhibitors are those with 7α-aryl derivatives, with several analogs having 2–10 times greater affinity for the enzyme than that of the substrate. These results suggest that additional interactions occur between the phenyl ring at the 7α-position and amino acids at or near the enzymatic site of aromatase, resulting in enhanced affinity.

(129)

Numerous modified androstenedione analogs have been developed as effective mechanism-based inhibitors of aromatase. The first compound designed as a mechanism- based inhibitor of aromatase was 10-propargyl-4-estrene-3,17-dione (**132**) (PED; MDL 18,962); it was synthesized and studied independently by three research groups (695–697). MDL 18,962 has an electron-rich alkynyl function on the C-19 carbon atom, the site of aromatase-mediated oxidation of the substrate. Although the identity of the reactive intermediate formed is not known, an oxirene and a Michael acceptor have been suggested. This agent is an effective inhibitor *in vitro* and *in vivo* (697–702). Other approaches to C-19 substituted mechanism-based inhibitors containing latent chemical groups have provided a limited number of inhibitors. These agents include the difluoromethyl analog (**133**) (703) and a thiol (**134**) (704, 705). Another series of mechanism-based inhibitors have developed from more detailed biochemical investigations of several inhibitors originally thought to be competitive inhibitors. These inhibitors can be grouped into the general categories of 4-substituted androst-4-ene-3,17-diones such as 4-hydroxy-androstenedione (**129**) (679), substituted androsta-1,4-diene-3,17-diones such as 7α-(4′-amino)-phenylthioandrosta-4,6-diene-3,17-dione (7α-APTADD; **135**) (706), and 6-methyleneandrost-4-ene-3,17-dione (exemestane; **136**) (707). Exemestane (Aromasin) is marketed as second-line therapy for the treatment of breast cancer patients who failed tamoxifen and is effective as first-line therapy in women with advanced breast cancer.

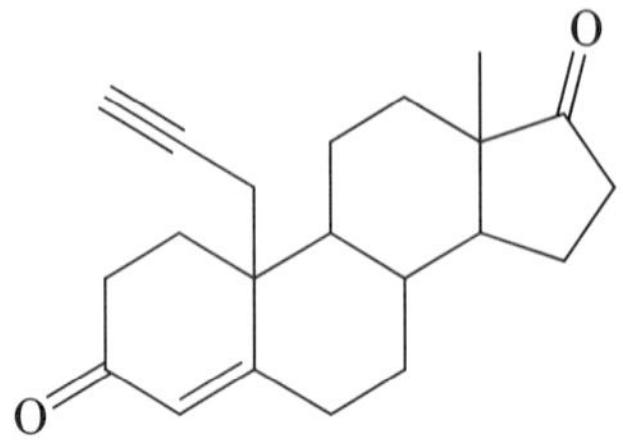

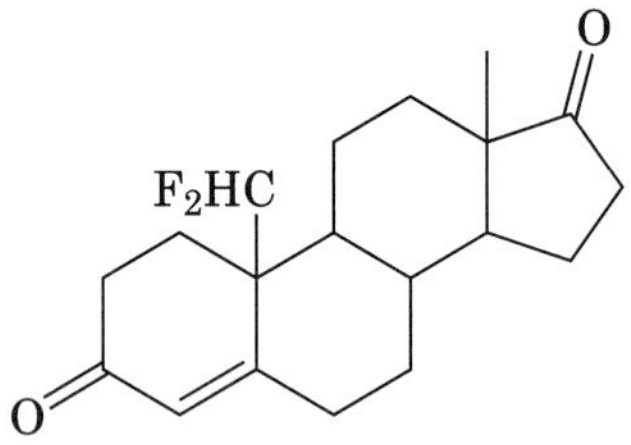

(133)

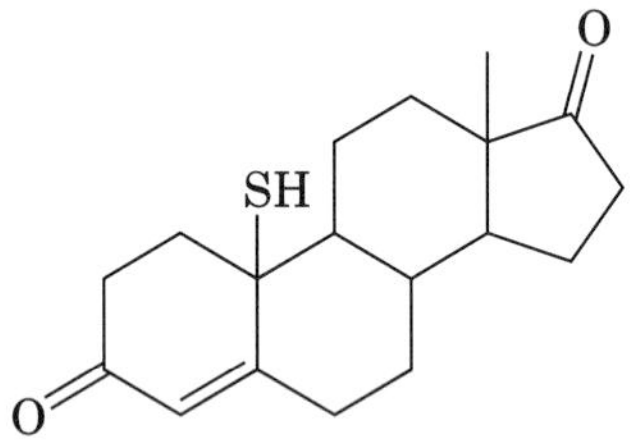

(134)

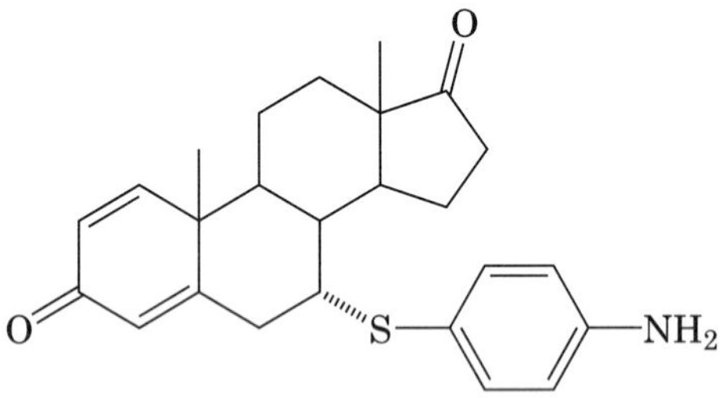

(135)

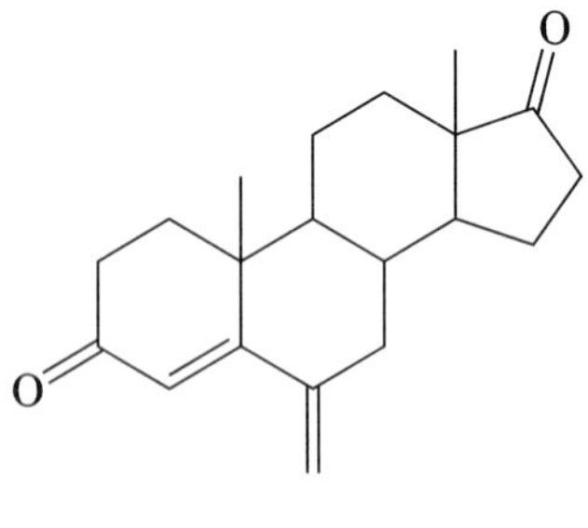

(136)

(132)

6.3.4 Other Agents. An interesting natural product that lowers circulating androgen levels *in vitro* is the nonsteroidal agent gossypol. Gossypol, (2,2′-binaphthalene)-8,8′-dicarboxaldehyde-1,1′,6,6′,7,7′-hexahydroxy-5,5′-diisopropyl-3,3′-dimethyl (**137**) is a polyphenolic compound contained in the pigment of cottonseed. This natural product has been used in fertile men in China as an effective male contraceptive agent for many years (708, 709); its antifertility effects on reproductive endocrine tissues are observed at 1000 times lower doses than its toxic effects in other tissues. Gossypol has been shown to disrupt spermatogenesis by inhibiting lactate dehydrogenase-X (LDH-X) (710, 711), to interfere with steroidogenesis in testicular Leydig cells (712–714), and to hinder the function of primary cultures of Leydig and Sertoli cells (715). In addition, gossypol is also capable of altering steroid biosynthesis in the female reproductive systems (716–719). The antisteroidogenic effect of gossypol in cultured bovine luteal cells involves suppression of the activities of adenylate cyclase and 3β-hydroxysteroid dehydrogenase (3β-HSD) (720). Reproductive endocrine tissues are also sensitive to metabolites of gossypol, such as gossypolone. Gossypolone, a major metabolite formed by gossypol oxidation, inhibits both 3β-HSD and cytochrome P-450$_{SCC}$ activities in cultured bovine luteal cells (721) and suppresses adrenocorticotropic hormone-induced corticosterone secretion in cultured rat adrenocortical cells at a similar potency as gossypol (722). Gossypol metabolites have also demonstrated inhibitory action on hCG-induced testosterone production in young male rats (718). The major side effects of gossypol therapy are fatigue, gastrointestinal upset, weakness, and hypokalemia, which thus limit its therapeutic usefulness (723).

(137)

7 SUMMARY

The steroid testosterone is the major circulating sex hormone of the male and serves as the prototype for the androgens, the anabolic agents, and androgen antagonists. The endogenous androgens are biosynthesized from cholesterol in various tissues in the body; the majority of the circulating androgens are made in the testes under the stimulation of the gonadotropin LH. A critical aspect of testosterone and its biochemistry is that this steroid is converted in various cells to other active steroidal agents. The reduction of testosterone to dihydrotestosterone is necessary for the androgenic actions of testosterone in androgen target tissues such as the prostate. On the other hand, oxidation of testosterone by the enzyme aromatase to yield the estrogens is crucial for certain CNS actions. Investigations of these enzymatic conversions of circulating testosterone continues to be a fruitful area of biochemical research on the roles of the steroid hormones in the body. Additionally, the elucidation of the mechanism of action of the androgens in various target tissues receives ongoing emphasis. The androgenic actions of testosterone are attributed to the binding of dihydrotestosterone to its nuclear receptor, followed by dimerization of the receptor complex and binding to a specific DNA sequence. This binding of the homodimer to the androgen response element leads to gene expression, stimulation of the synthesis of new mRNA, and subsequent protein biosynthesis. Other actions of testosterone, particularly the anabolic actions, appeared to be mediated through a similar nuclear receptor–mediated mechanism. Many of the intricate biochemical events that occur during the action of the androgens in their target cells remain for further clarification. Nevertheless, receptor studies of new agents are an important biological tool in the evaluation of the compounds for later, indepth pharmacological testing.

The synthetic androgens and anabolics were prepared to impart oral activity to the androgen molecule, to separate the androgenic effects of testosterone from its anabolic effects and to improve on its biological activities. These research efforts have provided several effective drug preparations for the treatment of various androgen-deficient diseases, for the therapy of diseases characterized by muscle wasting and protein catabolism, for postoperative adjuvant therapy, and for the treatment of certain hormone-dependent cancers. The synthetic anabolics have also resulted in the abuse of these agents in athletics. Finally, the most recent area of research emphasis is the development of the androgen antagonists, both steroidal and nonsteroidal agents. The two major categories of these antagonists are the antiandrogens, which block interactions of androgens with the androgen receptor, and the inhibitors of androgen biosynthesis and metabolism. Such compounds have therapeutic potential in the treatment of acne, virilization in women, hyperplasia and neoplasia of the prostate, and baldness, and for male contraception. A number of androstane derivatives are also being developed as inhibitors of aromatase for the treatment of hormone-dependent breast cancer. Thus, the numerous biological effects of the male sex hormones testosterone and dihydrotestosterone and the varied chemical modifications of the androstane molecule have resulted in the development of effective medicinal agents for the treatment of androgen-related diseases.

REFERENCES

1. J. A. Vida, *Androgens and Anabolic Agents: Chemistry and Pharmacology*, Academic Press, New York, 1969.

2. K. B. Eik-Nes, *The Androgens of the Testes*, Marcel Dekker, New York, 1970.

3. P. L. Munson, E. Diczfalusy, J. Glover, and R. E. Olsen, Eds., *Vitamins and Hormones*, Vol. **33**, Academic Press, New York. 1975.

4. C. D. Kochakian, *Anabolic-Androgenic Steroids*, Springer-Verlag, New York, 1976.

5. L. Martini and M. Motta, Eds., *Androgens and Antiandrogens*, Raven Press, New York, 1997.

6. R. W. Brueggemeier in M. Verderame, Ed., *CRC Handbook of Hormones, Vitamins, and Radiopaques*, CRC Press, Boca Raton, FL, 1986, p. 1.

7. P. J. Snyder in J. G. Hardman, and L. E. Limbird, Eds., *Goodman and Gilman's The Pharmacological Basis of Therapeutics*, 10th ed., McGraw-Hill Medical Publishing, New York, 2000, pp. 1635–1648.

8. F. J. Zeelen, *Medicinal Chemistry of Steroids*, Elsevier, Amsterdam, 1990, p. 177.

9. J. E. Griffin and J. D. Wilson in J. D. Wilson and D. W. Foster, Eds., *William's Textbook of Endocrinology*, 9th ed., Saunders, Philadelphia, 1992, p. 819.

10. V. A. Rogozkin, *Metabolism of Anabolic Androgenic Steroids*, CRC Press, Boca Raton, FL, 1991.

11. S. Bhasin, H. L. Gabelnick, J. M. Spieler, R. S. Swerdloff, C. Wang, and C. Kelly, Eds., *Pharmacology, Biology, and Clinical Applications of Androgens*, Wiley-Liss, New York, 1996.

12. A. A. Berthold, *Arch. Anat. Physiol. Will. Med.*, **16**, 42 (1849).

13. A. Butenandt, *Angew. Chem.*, **44**, 905 (1931).

14. A. Butenandt and K. Tscherning, *Z. Physiol. Chem.*, **229**, 167 (1934).

15. T. F. Gallagher and F. C. Koch, *J. Biol. Chem.*, **84**, 495 (1929).

16. A. Butenandt, *Naturwissenschaften*, **21**, 49 (1933).

17. A. Butenandt and H. Dannenberg, *Z. Physiol. Chem.*, **229**, 192 (1934).

18. K. David, E. Dingemanse, J. Freud, and E. Laqueur, *Z. Physiol. Chem.*, **233**, 281 (1935).

19. K. David, *Acta Brevia Neerl. Physiol. Pharmacol. Microbiol.*, **5**, 85, 108 (1935).

20. A. Butenandt and G. Hanisch, *Berichte*, **68**, 1859 (1935);*Z. Physiol. Chem.*, **237**, 89 (1935).

21. L. Ruzicka, *J. Am. Chem. Soc.*, **57**, 2011 (1935).

22. L. Ruzicka, A. Wettstein, and H. Kagi, *Helv. Chim. Acta*, **18**, 1478 (1935).

23. N. Bruchovsky and J. D. Wilson, *J. Biol. Chem.*, **243**, 5953 (1968).

24. K. M. Anderson and S. Liao, *Nature*, **219**, 277 (1968).

25. S. Fang, K. M. Anderson, and S. Liao, *J. Biol. Chem.*, **244**, 6584 (1969).

26. W. I. P. Mainwaring, *J. Endocrinol.*, **45**, 531 (1969).

27. C. D. Kochakian and J. R. Murlin, *J. Nutr.*, **10**, 437 (1935).

28. A. T. Kenyon, I. Sandiford, A. H. Bryan, K. Knowlton, and F. C. Koch, *Endocrinology*, **23**, 135 (1938).

29. R. K. Meyer and L. G. Hershberger, *Endocrinology*, **60**, 397 (1957).

30. S. L. Leonard, *Endocrinology*, **50**, 199 (1952).

31. J. M. Loring, J. M. Spencer, and C. A. Villee, *Endocrinology*, **68**, 501 (1961).

32. D. B. Villee, *Human Endocrinology: A Developmental Approach*, Saunders, New York, 1975.

33. J. E. Griffin and J. D. Wilson in C. R. Scriver, A. L. Beaudet, W. S. Sly, and D. Valle, Eds., *The Metabolic Basis of Inherited Diseases*, 6th ed., McGraw-Hill, New York, 1989, p. 1919.

34. R. L. Landau in C. D. Kochakian, Ed., *Ana bolic-Androgenic Steroids*, Springer-Verlag, New York, 1976, p. 48.

35. H. Spencer, J. A. Friedland, and I. Lewin in C. D. Kochakian, Ed., *Anabolic-Androgenic Steroids*, Springer-Verlag, New York, 1976, p. 419.

36. H. Spencer et al. in C. D. Kochakian, Ed., *Anabolic-Androgenic Steroids*, Springer-Verlag, New York, 1976, p. 433.

37. C. W. Gurney in C. D. Kochakian, Ed., *Ana bolic-Androgenic Steroids*, Springer-Verlag, New York. 1976, p. 483.

38. A. D. Mooradian, J. E. Morley, and S. G. Korenman, *Endocr. Rev.*, **8**, 1 (1987).

39. F. T. G. Prunty, *Br. Med. J.*, **2**, 605 (1966).

40. A. Vermeulen, *Acta Endocrinol.*, **83**, 651 (1976).

41. M. B. Lipsett and S. G. Korenman, *J. Am. Med. Assoc.*, **190**, 757 (1964).

42. A. Chapdelaine, P. C. MacDonald, O. Gonzalez, E. Gurpide, R. L. VandeWiele, and S. Lieberman, *J. Clin. Endocrinol. Metab.*, **25**, 1569 (1965).

43. J. F. Tait and R. Horton, *Steroids*, **4**, 365 (1964).

44. R. J. Santen and C. W. Bardin, *J. Clin. Invest.*, **52**, 2617 (1973).

45. A. M. Matsumoto and W. J. Bremmer, *J. Clin. Endocrinol. Metab.*, **58**, 609 (1984).

46. D. M. Robertson, R. I. McLachlan, and H. G. Burger in H. Burger and D. deKretser, Eds., *The Testis*, 2nd ed., Raven Press, New York, 1989, p. 231.

47. F. F. G. Rommerts, B. A. Cooke, and H. J. Van der Mulen, *J. Steroid Biochem.*, **5**, 279 (1974).

48. W. L. Miller, *Endocr. Rev.*, **9**, 295 (1988).

49. R. Vihko and A. Ruokonen, *J. Steroid Biochem.*, **5**, 843 (1974).

50. C. E. Bird, L. Morrow, Y. Fukumoto, S. Marcellus, and A. F. Clark, *J. Clin. Endocrinol. Metab.*, **43**, 1317 (1976).

51. T. Yanaihara and P. Troen, *J. Clin. Endocrinol. Metab.*, **34**, 783 (1972).

52. R. I. Dorfman and F. Ungar, *Metabolism of Steroid Hormones*, Academic Press, New York, 1965.

53. S. Nakajin and P. F. Hall, *J. Biol. Chem.*, **256**, 3871 (1981).

54. B. Chung, J. Picado-Leonard, M. Haniu, M. Bienkowski, P. F. Hall, J. E. Shively, and W. L. Miller, *Proc. Natl. Acad. Sci. USA*, **84**, 407 (1987).

55. K. D. Bradshaw, M. R. Waterman, R. T. Couch, E. R. Simpson, and M. X. Zuber, *Mol. Endocrinol.*, **1**, 348 (1987).

56. T. Yanese, E. R. Simpson, and M. R. Waterman, *Endocr. Rev.*, **12**, 91 (1991).

57. H. Ishii-Ohba, H. Inano, and B.-I. Tamaoki, *J. Steroid Biochem.*, **27**, 775 (1987).

58. H. P. Schedl and J. A. Clifton, *Gastroenterology*, **41**, 491 (1961).

59. K. B. Eik-Nes, J. Schellmann, A. R. Lumry, and L. T. Samuels, *J. Biol. Chem.*, **206**, 411 (1954).

60. B. H. Levedahl and H. Bernstein, *Arch. Biochem. Biophys.*, **52**, 353 (1954).

61. J. Schellmann, A. R. Lumry, and L. T. Samuels, *J. Am. Chem. Soc.*, **76**, 2808 (1954).

62. J. Kerkay and U. Westphal, *Biochim. Biophys. Acta*, **170**, 324 (1968).

63. J. Kerkay and U. Westphal, *Arch. Biochem. Biophys.*, **129**, 480 (1969).

64. P. I. Corvol, A. Chrambach, D. Rodbard, and C. W. Bardin, *J. Biol. Chem.*, **246**, 3435 (1971).

65. W. H. Pearlman and O. Crépy, *J. Biol. Chem.*, **242**, 182 (1967).

66. J. L. Guériguan and W. H. Pearlman, *Fed. Proc.*, **26**, 757 (1967).

67. B. E. P. Murphy, *Can. J. Biochem.*, **46**, 299 (1968).

68. O. Steeno, W. Heyus, H. Van Baelen, and P. De Moor, *Ann. Endocrinol.*, **29**, 141 (1968).

69. C. Mercier-Bodard, A. Alfsen, and E. E. Baulieu, *Acta Endocrinol.*, **147**, 204 (1970).

70. M. C. Lebeau, C. Mercier-Bodard, J. Oldo, D. Bourguon, T. Brécy, J. P. Raynaud, and E. E. Baulieu, *Ann. Endocrinol.*, **30**, 183 (1969).

71. W. Rosner, N. P. Christy, and W. G. Kelley, *Biochemistry*, **8**, 3100 (1969).

72. W. H. Pearlman, I. F. F. Fong, and K. J. Tou, *J. Biol. Chem.*, **244**, 1373 (1969).

73. F. Dray, I. Mowezawicz, M. J. Ledru, O. Crépy, G. Delzant, and J. Sebaoun, *Ann. Endocrinol.*, **30**, 223 (1969).

74. P. DeMoor, O. Steeno, W. Heyns, and H. Van Baelen, *Ann. Endocrinol.*, **30**, 233 (1969).

75. R. Horton, T. Kato, and R. Sherino, *Steroids*, **10**, 245 (1967).

76. A. Vermeulen and L. Verdonck, *Steroids*, **11**, 609 (1968).

77. T. Kato and R. Horton, *J. Clin. Endocrinol. Metab.*, **28**, 1160 (1968).

78. C. Mercier-Bodard and E. E. Baulieu, *Ann. Endocrinol.*, **29**, 159 (1968).

79. B. E. P. Murphy, *Steroids*, **16**, 791 (1970).

80. K. E. Mickelson and P. H. Petra, *Biochemistry*, **14**, 957 (1975).

81. E. M. Ritzen. S. N. Nayfeh, F. S. French, and M. C. Dobbins, *Endocrinology*, **89**, 143 (1971).

82. V. Hansson and O. Djoseland, *Acta Endocrinol.*, **71**, 614 (1972).

83. F. S. French and E. M. Ritzén, *J. Reprod. Fertil.*, **32**, 479 (1973).

84. V. Hansson, O. Trygotad, F. S. French, W. S. McLean, A. A. Smith, D. J. Tindall, S. C. Weddington. P. Petruez, S. N. Navfehh, and E. M. Ritzén, *Nature*, **250**, 387 (1974).

85. R. G. Vernon, B. Kopec, and I. B. Fritz, *Mol. Cell. Endocrinol.*, **1**, 167 (1974).

86. D. Williams and J. Gorski, *Biochem. Biophys. Res. Commun.*, **45**, 258 (1971).

87. E. Milgrom, M. Atger, and E. E. Baulieu, *Biochim. Biophys. Acta*, **320**, 267 (1973).

88. R. W. Harrison, S. Fairfield, and D. N. Orth, *Biochemistry*, **14**, 1304 (1975).

89. R. W. Harrison, S. Fairfield, and D. N. Orth, *Biochem. Biophys. Res. Commun.*, **61**, 1262 (1974).

90. E. P. Gorgi, J. C. Stewart, J. K. Grant, and R. Scott, *Biochem. J.*, **122**, 125 (1971).

91. E. P. Gorgi, J. C. Stewart, J. K. Grant, and I. M. Shirley, *Biochem. J.*, **126**, 107 (1972).

92. E. P. Gorgi, J. K. Grant, J. C. Stewart, and J. Reid, *J. Endocrinol.*, **55**, 421 (1972).

93. E. P. Gorgi, I. M. Shirley, J. K. Grant, and J. C. Stewart, *Biochem. J.*, **132**, 465 (1973).

94. S. R. Gross, L. Aronow, and W. B. Pratt, *Biochem. Biophys. Res. Commun.*, **32**, 66 (1968).

95. S. R. Gross, L. Aronow, and W. B. Pratt, *J. Cell. Biol.*, **44**, 103 (1970).

96. P. Ofner in R. S. Harris, I. G. Wool, and J. A. Lorraine, Eds., *Vitamins and Hormones*, Vol. **26**, Academic Press, New York, 1968, p. 237.

97. K. Hartiala, *Physiol. Rev.*, **53**, 496 (1973).

98. J. D. Wilson in R. O. Greep and E. B. Astwood, Eds., *Handbook of Physiology: Endocrinology, Male Reproductive System*, Vol. **V**, Sec. 7, American Physiology Society, Washington, DC, 1974, p. 491.

99. M. I. Stylianou, E. Forchielli, N. I. Tummillo, and R. I. Dorfman, *J. Biol. Chem.*, **236**, 692 (1961).

100. D. Engelhardt, J. Eisenburg, P. Unterberger, and H. J. Karl, *Klin. Wochenschr.*, **49**, 439 (1971).

101. B. P. Lisboa, I. Drosse, and H. Breuer, *Z. Physiol. Chem.*, **342**, 123 (1965).

102. E. Chang, A. Mittelman, and T. L. Dao, *J. Biol. Chem.*, **238**, 913 (1963).

103. K. Hartiala and W. Nienstedt, *Int. J. Biochem.*, **7**, 317 (1970).

104. A. Vermeulen. R. Rubens, and L. Verdonck, *J. Clin. Endocrinol. Metab.*, **34**, 730 (1972).

105. P. K. Siiteri and J. D. Wilson, *J. Clin. Invest.*, **49**, 1737 (1970).

106. K. D. Voigt, H.-J. Horst, and M. Krieg, *Vitam. Horm.*, **33**, 417 (1975).

107. H. Becker, J. Kaufmann, H. Klosterhalfen, and K. D. Voigt, *Acta Endocrinol.*, **71**, 589 (1972).

108. H. J. Horst, M. Dennis. J. Kaufmann, and K. D. Voigt, *Acta Endocrinol.*, **79**, 394 (1975).

109. A. E. Kellie and E. R. Smith, *Biochem. J.*, **66**, 490 (1957).

110. P. Robel, R. Emiliozzi, and E. Baulieu, *J. Biol. Chem.*, **241**, 20 (1966).

111. F. T. G. Prunty, *Br. Med. J.*, **2**, 605 (1966).

112. M. B. Lipsett and S. G. Korenman, *J. Am. Med. Assoc.*, **190**, 757 (1964).

113. N. Kundu, A. A. Sandberg, and W. R. Slaunwhite, Jr., *Steroids*, **6**, 543 (1965).

114. T. El Attar, W. Dirscherl, and K. O. Mosebach, *Acta Endocrinol.*, **45**, 527 (1964).

115. A. Kappas and R. H. Palmer in R. I. Dorfman, Ed., *Methods in Hormone Research*, Vol. **4**, Part B, Academic Press, New York, 1965, p. 1.

116. L. Hellman, H. L. Bradlow, B. Zumoff, D. K. Fukushima, and T. F. Gallagher, *J. Clin. Endocrinol.*, **19**, 936 (1959).

117. K. Nozu and B. I. Tamaski, *Biochim. Biophys. Acta*, **348**, 321 (1974).

118. K. Nozu and B. I. Tamaski, *Acta Endocrinol.*, **76**, 608 (1974).

119. N. Bruchovsky and J. D. Wilson, *J. Biol. Chem.*, **243**, 2012 (1968).

120. N. Bruchovsky and J. D. Wilson, *J. Biol. Chem.*, **243**, 5953 (1968).

121. J. Shimazaki, N. Furuya, H. Yamanaka, and K. Shida, *Endocrinol. Jpn.*, **16**, 163 (1969).

122. J. Shimazaki, I. Matsushita, N. Furuya, H. Yamanaka, and K. Shida, *Endocrinol. Jap.*, **16**, 453 (1969).

123. D. W. Frederiksen and J. D. Wilson, *J. Biol. Chem.*, **246**, 2584 (1971).

124. R. J. Moore and J. D. Wilson, *J. Biol. Chem.*, **247**, 958 (1972).

125. R. J. Moore and J. D. Wilson, *Endocrinology*, **93**, 581 (1973).

126. J. P. Karr, R. Y. Kirdani, G. P. Murphy, and A. A. Sandberg, *Life Sci.*, **15**, 501 (1974).

127. D. C. Wilton and H. J. Ringold, *Proceedings of the Third International Congress of Endocrinology*, Excerpta Med. Foundation, Amsterdam, 1968, p. 105.

128. R. W. S. Skiner, R. V. Pozderac, R. E. Counsell, and P. A. Weinhold, *Steroids*, **25**, 189 (1975).

129. S. Andersson, R. W. Bishop, and D. W. Russell, *J. Biol. Chem.*, **264**, 16249 (1989).

130. S. Anderson, D. M. Berman, E. P. Jenkins, and D. W. Russell, *Nature*, **354**, 150 (1991).

131. D. W. Russell, D. M. Berman, J. T. Bryant, K. M. Cala, D. L. Dairs, C. P. Landrum, J. S. Prihada, R. I. Silver, A. E. Thigpen, and W. C. Wigley, *Recent Prog. Horm. Res.*, **49**, 275 (1994).

132. D. W. Russell and J. D. Wilson, *Ann. Rev. Biochem.*, **63**, 25 (1993).

133. J. D. Wilson, J. E. Griffin, and D. W. Russell, *Endocr. Rev.*, **14**, 577 (1993).

134. K. J. Ryan, *J. Biol. Chem.*, **234**, 268 (1959).

135. E. A. Thompson and P. K. Siiteri, *J. Biol. Chem.*, **249**, 5373 (1974).

136. E. A. Thompson and P. K. Siiteri, *J. Biol. Chem.*, **249**, 5364 (1974).

137. A. S. Meyer, *Biochim. Biophys. Acta*, **17**, 441 (1955).

138. M. Akhtar and S. J. M. Skinner, *Biochem. J.*, **109**, 318 (1968).

139. S. J. M. Skinner and M. Akhtar, *Biochem. J.*, **114**, 75 (1969).

140. J. D. Townsley and H. J. Brodie, *Biochemistry*, **7**, 33 (1968).

141. H. J. Brodie, G. Possanza, and J. D. Townsley, *Biochim. Biophys. Acta*, **152**, 770 (1968).

142. Y. Osawa and D. G. Spaeth, *Biochemistry*, **10**, 66 (1971).

143. J. Goto and J. Fishman, *Science*, **195**, 80 (1977).

144. E. Caspi, J. Wicha, T. Aninachalam, P. Nelson, and G. Spitiller, *J. Am. Chem. Soc.*, **106**, 7282 (1984).

145. D. Arigoni, R. Battaglia, M. Akhtar, and T. Smith, *J. Chem. Soc. Chem. Commun.*, 185 (1975).

146. M. Akhtar, M. R. Calder, D. L. Corina, and J. N. Wright, *J. Chem. Soc. Chem. Commun.*, 129 (1981).

147. M. Alchtar, M. R. Calder, D. L. Corina, and J. N. Wright, *Biochem. J.*, **201**, 569 (1982).

148. P. A. Cole and C. H. Robinson, *J. Am. Chem. Soc.*, **110**, 1284 (1988).

149. M. Akhtar, V. C. O. Njar, and J. N. Wright, *J. Steroid Biochem. Mol. Biol.*, **44**, 375 (1993).

150. S. S. Oh and C. H. Robinson, *J. Steroid Biochem. Mol. Biol.*, **44**, 389 (1993).

151. K. J. Ryan, in Proceedings of the First International Congress on Endocrinology, Copenhagen, Denmark, 1960, p. 350.

152. K. J. Ryan, in Proceedings of the Fifth International Congress on Biochemistry, Moscow, 1963, Vol. 7, p. 381.

153. C. Gual, T. Morato, M. Hayano, M. Gut, and R. I. Dorfman, *Endocrinology*, **71**, 920 (1962).

154. W. C. Schwarzel, W. G. Kruggel, and H. J. Brodie, *Endocrinology*, **92**, 866 (1973).

155. P. K. Siiteri and E. A. Thompson, *J. Steroid Biochem.*, **6**, 317 (1975).

156. F. L. Bellino, S. S. H. Gilani, S. S. Eng, Y. Osawa, and W. L. Duax, *Biochemistry*, **15**, 4730 (1976).

157. J. K. Kellis and L. E. Vickery, *J. Biol. Chem.*, **262**, 4413 (1987).

158. C. J. Corbin, S. Grahan-Lorence, M. McPhaul, J. I. Mason, C. R. Mendelson, and E. R. Simpson, *Proc. Natl. Acad. Sci. USA*, **85**, 8948 (1988).

159. M. S. Mahendroo, C. R. Mendelson, and E. R. Simpson, *J. Biol. Chem.*, **268**, 19463 (1993).

160. E. R. Simpson, M. S. Mahendroo, G. D. Means, M. W. Kilgore, M. M. Hinshelwood, S. Grahan-Lorence, B. Amarneh, Y. Ito, C. R. Fisher, M. D. Michael, C. R. Mendelson, and S. E. Bulun, *Endocr. Rev.*, **15**, 342 (1994).

161. S. Chen, M. J. Beshman, R. S. Sparkes, S. Zollman, I. Klisak, T. Mohandes, P. F. Hall, and J. E. Shively, *DNA*, **7**, 27 (1988).

162. M. Steinkampf, C. R. Mendelson, and E. R. Simpson, *Mol. Endocrinol.*, **1**, 465 (1987).

163. M. Steinkampf, C. R. Mendelson, and E. R. Simpson, *Mol. Cell. Endocrinol.*, **59**, 93 (1988).

164. C. T. Evans, J. C. Merrill, C. J. Corbin, D. Saunders, E. R. Simpson, and C. R. Mendelson, *J. Biol. Chem.*, **269**, 6914 (1987).

165. E. R. Simpson, J. C. Merrill, A. J. Hollub, S. Grahan-Lorence, and C. R. Mendelson, *Endocr. Rev.*, **10**, 136 (1989).

166. E. B. Bulun and E. R. Simpson, *Breast Cancer Res. Treat.*, **30**, 19–29 (1994).

167. L. S. Sholiton, R. T. Mornell, and E. E. Werk, *Steroids*, **8**, 265 (1966).

168. R. B. Jaffe, *Steroids*, **14**, 483 (1969).

169. L. S. Sholiton and E. E. Werk. *Acta Endocrinol.*, **61**, 641 (1969).

170. L. S. Sholiton, I. L. Hall, and E. E. Werk, *Acta Endocrinol.*, **63**, 512 (1970).

171. J. M. Stern and A. J. Eisenfeld, *Endocrinology*, **88**, 1117 (1971).

172. R. Massa, E. Stupnicka, Z. Kniewald, and L. Martini, *J. Steroid Biochem.*, **3**, 385 (1972).

173. E. D. Lephart, S. Andersson, and E. R. Simpson, *Endocrinology*, **127**, 1121 (1990).

174. F. Naftolin. K. J. Ryan, and Z. Petro, *J. Clin. Endocrinol. Metab.*, **33**, 368 (1971).

175. F. Naftolin, K. J. Ryan, and Z. Petro, *Endocrinology*, **90**, 295 (1972).

176. F. Flores, F. Naftolin, and K. J. Ryan, *Neuroendocrinology*, **11**, 177 (1973).

177. F. Flores, F. Naftolin, K. J. Ryan, and R. J. White, *Science*, **180**, 1074 (1973).

178. J. A. Canick, D. E. Vaccaro. K. J. Ryan, and S. E. Leeman, *Endocrinology*, **100**, 250 (1977).

179. E. D. Lephart, E. R. Simpson, M. J. McPhaul, M. W. Kilgore, J. D. Wilson, and S. R. Ojeda, *Mol. Brain Res.*, **16**, 187 (1992).

180. G. Perez-Palacios, K. Larsson, and C. Beyer, *J. Steroid Biochem.*, **6**, 999 (1975).

181. E. D. Lephart, *Mol. Cell. Neurosci.*, **4**, 473 (1993).

182. B. W. O'Malley and A. R. Means, *Receptors for Reproductive Hormones*, Plenum Press. New York, 1974.

183. R. J. B. King and W. I. P. Mainwaring, *Steroid–Cell Interactions*, Universitv Park Press, Baltimore, MD, 1974.

184. S. Liao, *Int. Rev. Cytol.*, **41**, 87 (1975).

185. H. G. Williams-Ashman and A. H. Reddi in G. Litwack, Ed., *Biochemical Actions of Hormones*, Vol. **2**, Academic Press, New York, 1972, p. 257.

186. L. Chan and B. W. O'Malley, *N. Engl. J. Med.*, **294**, 1322, 1372, 1430 (1976).

187. J.-A. Gustaffson, J. Carlstedt-Duke, L. Poellinger, S. Okret, A. C. Wikstrom, M. Bronnegard, M. Gillner, Y. Dong, K. Fuxe, and A. Cintra, *Endocr. Rec.*, **8**, 185 (1987).

188. R. M. Evans, *Science*, **240**, 889–895 (1988).

189. G. Ringold, Ed., *Steroid Hormone Action*, Liss, New York, 1988.

190. M. Beato, *Cell*, **56**, 335 (1989).

191. B. O'Malley, *Mol. Endocrinol.*, **4**, 363 (1990).

192. M. A. Carson-Jurica, W. T. Schrader, and B. O'Malley, *Endocr. Rev.*, **11**, 201 (1990).

193. D. J. Mangelsdorf, C. Thummel, M. Beato, P. Herrlich, G. Schutz, K. Umesono, B. Blumberg, P. Kastner, M. Mark, and P. Chambon, *Cell*, **83**, 835–839 (1995).

194. J. M. Kokontis and S. Liao, *Vitam. Horm.*, **55**, 219–307 (1999).

195. A. K. Roy, Y. Lavrovsky, C. S. Song, S. Chen, M. H. Jung, N. K. Velu, B. Y. Bi, and B. Chatterjee, *Vitam. Horm.*, **55**, 309–352 (1999).

196. E. J. Jensen and H. I. Jacobson, *Recent Prog. Horm. Res.*, **18**, 387 (1962).

197. W. H. Pearlman and M. R. I. Pearlman, *J. Biol. Chem.*, **236**, 1321 (1961).

198. B. W. Harding and L. T. Samuels, *Endocrinology*, **70**, 109 (1962).

199. K. J. Tveter and A. Attramadal, *Acta Endocrinol.*, **59**, 218 (1968).

200. N. Bruchovsky and J. D. Wilson, *J. Biol. Chem.*, **243**, 2012 (1968).

201. W. I. P. Mainwaring, *J. Endocrinol.*, **44**, 323 (1969).

202. S. Fang and S. Liao, *J. Biol. Chem.*, **246**, 16 (1971).

203. J. Gorski, W. V. Welshons, D. Sakai, J. Hansen, J. Walent, J. Kassis, J. Shull, G. Stack, and C. Campen, *Recent Prog. Horm. Res.*, **42**, 297 (1986).

204. Z.-X. Zhou, C.-L. Wong, M. Sar, and E. M. Wilson, *Recent Prog. Horm. Res.*, **49**, 249 (1994).

205. T. Brown in S. Bhasin, H. L. Gabelnick, J. M. Spieler, R. S. Swerdloff, C. Wang, and C. Kelly, Eds., *Pharmacology, Biology, and Clinical Applications of Androgens*, Wiley-Liss, New York, 1996, p. 45.

206. P. M. Matias, P. Donner, R. Coelho, M. Thomaz, C. Peixoto, S. Macedo, N. Otto, S. Joschko, P. Scholz, A. Wegg, S. Basler, M. Schafer, U. Egner, and M. A. Carrondo, *J. Biol. Chem.*, **275**, 26164–26171 (2000).

207. J. S. Sack, K. F. Kish, C. Wang, R. M. Attar, S. E. Kiefer, G. Y. Wu, J. E. Scheffler, M. E. Salvati, S. R. Krystek, Jr., R. Weinmann, and H. M. Einspahr, *Proc. Natl. Acad. Sci. USA*, **98**, 4904–4909 (2001).

208. M. J. McPhaul, M. Marcelli, W. D. Tilley, J. E. Griffin, and J. D. Wilson, *FASEB J.*, **5**, 2910 (1991).

209. A. O. Brinkmann and J. Trapman, *Cancer Surv.*, **14**, 95 (1992).

210. M. J. McPhaul, *J. Steroid Biochem. Mol. Biol.*, **69**, 315–322 (1999).

211. C. Sonnenschein, N. Olea, M. E. Pasanen, and A. M. Soto, *Cancer Res.*, **49**, 3474 (1989).

212. G. Wilding, M. Chen, and E. P. Gelmann, *Prostate*, **14**, 103 (1989).

213. S. Harris, M. A. Harris, and Z. Rong in J. P. Karr, D. S. Coffey, R. G. Smith, and D. J. Tindall, Eds., *Molecular and Cellular Biology of Prostate Cancer*, Plenum, New York, 1991, p. 315.

214. P. H. Riegman, R. J. Vlietstra, J. A. van der Korput, A. O. Brinkmann, and J. Trapman, *Mol. Endocrinol.*, **5**, 1921–1930 (1991).

215. K. B. Cleutjens, C. C. van Eekelen, H. A. van der Korput, A. O. Brinkmann, and J. Trapman, *J. Biol. Chem.*, **271**, 6379–6388 (1996).

216. K. B. Cleutjens, H. A. van der Korput, C. C. van Eekelen, H. C. van Rooij, P. W. Faber, and J. Trapman, *Mol. Endocrinol.*, **11**, 148–161 (1997).

217. Y. Matuo, P. S. Adams, N. Nishi, H. Yasumitsu, J. W. Crabb, R. J. Matusik, and W. L. McKeehan, *In Vitro Cell Dev. Biol.*, **25**, 581–584 (1989).

218. A. M. Spence, P. C. Sheppard, J. R. Davie, Y. Matuo, N. Nishi, W. L. McKeehan, J. G. Dodd, and R. J. Matusik, *Proc. Natl. Acad. Sci. USA*, **86**, 7843–7847 (1989).

219. T. Liang, G. Mezzetti, C. Chen, and S. Liao, *Biochim. Biophys. Acta*, **542**, 430–441 (1978).

220. G. Yan, Y. Fukabori, S. Nikolaropoulos, F. Wang, and W. L. McKeehan, *Mol. Endocrinol.*, **6**, 2123–2128 (1992).

221. A. Tanaka, K. Miyamoto, N. Minamino, M. Takeda, B. Sato, H. Matsuo, and K. Matsumoto, *Proc. Natl. Acad. Sci. USA*, **89**, 8928–8932 (1992).

222. M. Koga, S. Kasayama, K. Matsumoto, and B. Sato, *J. Steroid Biochem. Mol. Biol.*, **54**, 1–6 (1995).

223. I. Schenkein, M. Levy, and E. D. Bueker, *Endocrinology*, **94**, 840 (1974).

224. P. L. Barthe, L. P. Bullock, and I. Mowszowicz, *Endocrinology*, **95**, 1019 (1974).

225. D. A. Wolf, P. Schulz, and F. Fittler, *Br. J. Cancer*, **64**, 47–53 (1991).

226. M. F. Lyon, I. Hendry, and R. V. Short, *Endocrinology*, **58**, 357 (1973).

227. C. W. Bardin, L. P. Bullock, and R. J. Sherins, *Recent Prog. Horm. Res.*, **29**, 65 (1973).

228. M. Kumar, A. K. Roy, and A. E. Axelrod, *Nature*, **223**, 399 (1969).

229. J. F. Irwin, S. E. Lane, and O. W. Neuhaus, *Biochim. Biophys. Acta*, **252**, 328 (1971).

230. A. K. Roy, *Endocrinology*, **92**, 957 (1973).

231. C. Chang, T.-M. Lin, P. Hsiao, C. Su, J. Riebe, C.-T. Chang, and D.-L. Lin in S. Bhasin, H. L. Gabelnick, J. M. Spieler, R. S. Swerdloff, C. Wang, and C. Kelly, Eds., *Pharmacology, Biology, and Clinical Applications of Androgens*, Wiley-Liss, New York, 1996, p. 45.

232. K. J. Tveter and O. Unhjem, *Endocrinology*, **84**, 963 (1969).

233. J. M. Stern and A. J. Eisenfield, *Science*, **166**, 233 (1969).

234. K. Adachi and M. Kano, *Steroids*, **19**, 567 (1972).

235. W. I. P. Mainwaring and F. R. Mangan, *J. Endocrinol.*, **59**, 121 (1973).

236. S. Takayasu and K. Adachi, *Endocrinology*, **96**, 525 (1975).

237. V. Hansson et al., *Steroids*, **23**, 823 (1974).

238. D. J. Tindall, F. S. French, and S. N. Nayfeh, *Biochem. Biophys. Res. Commun.*, **49**, 1391 (1973).

239. J. A. Blaquier and R. S. Calandra, *Endocrinology*, **93**, 51 (1973).

240. E. M. Ritzén, S. N. Nayfeh, F. S. French. and P. A. Aronin, *Endocrinology*, **91**, 116 (1972).

241. J. F. Dunn, J. L. Goldstein, and J. D. Wilson, *J. Biol. Chem.*, **248**, 7819 (1973).

242. G. Verhoeven and J. D. Wilson, *Endocrinology*, **99**, 79 (1976).

243. P. Jouan, S. Samperez, M. L. Thieulant, and L. Mercier, *J. Steroid Biochem.*, **2**, 223 (1971).

244. P. Jouan. S. Samperez, and M. L. Thielant, *J. Steroid Biochem.*, **4**, 65 (1973).

245. M. Sar and W. E. Stumpf, *Endocrinology*, **92**, 251 (1973).

246. D. P. Cardinali, C. A. Nagle, and J. M. Rosner, *Endocrinology*, **95**, 179 (1974).

247. J. Kato, *J. Steroid Biochem.*, **6**, 979 (1975).

248. O. Naess, V. Hansson, O. Djoseland, and A. Attramadal, *Endocrinology*, **97**, 1355 (1975).

249. T. O. Fox, *Proc. Natl. Acad. Sci. USA*, **72**, 4303 (1975).

250. L. Valladares and J. Mingell, *Steroids*, **25**, 13 (1975).

251. J. Mingell and L. Valladares, *J. Steroid Bio-chem.*, **5**, 649 (1974).

252. A. K. Roy, B. S. Milin, and D. M. McMinn, *Biochim. Biophys. Acta*, **354**, 213 (1974).

253. N. Bruchovsky and J. W. Meakin, *Cancer Res.*, **33**, 1689 (1973).

254. N. Bruchovsky, D. J. A. Sutherland, J. W. Meakin, and T. Minesita, *Biochim. Biophys. Acta*, **381**, 61 (1975).

255. E. M. Wilson and F. S. French, *J. Biol. Chem.*, **251**, 5620 (1976).

256. C. R. Evans and C. G. Pierrepoint. *J. Endocrinol.*, **64**, 539 (1975).

257. K. B. Eik-Nes, *Vitam. Horm.*, **33**, 193 (1975).

258. R. W. Glovna and J. D. Wilson, *J. Clin. Endocrinol. Metab.*, **29**, 970 (1969).

259. V. Hansson, K. J. Tveter, O. Unhjem, and O. Djoseland, *J. Steroid Biochem.*, **3**, 427 (1972).

260. I. Jung and E. E. Baulieu, *Nat. New Biol.*, **237**, 24 (1972).

261. M. G. Michel and E. E. Baulieu, *C. R. Acad. Sci. Paris*, **279**, 421 (1974).

262. S. R. Max, S. Mufti and B. M. Carlson, *Biochem. J.*, **200**, 77 (1981).

263. M. Krieg and K. D. Voigt, *J. Steroid Biochem.*, **7**, 1005 (1976).

264. S. R. Max, *J. Steroid Biochem.*, **18**, 281 (1983).

265. H. C. McGill, V. C. Anselmo, J. M. Buchanan, and P. J. Sheridian, *Science*, **207**, 775 (1980).

266. E. Dahbberg, *Biochim. Biophys. Acta*, **717**, 65 (1982).

267. R. Hickson, T. Galessi, T. Kurowski, D. Daniels, and R. Chatterton, *J. Steroid Biochem.*, **19**, 1705 (1983).

268. W. D. Odell and R. S. Swerdloff, *Clin. Endocrinol.*, **8**, 149 (1978).

269. J. E. Griffin and J. D. Wilson, *N. Engl. J. Med.*, **302**, 198 (1980).

270. J. E. Griffin, K. Punyashthiti, and J. D. Wilson, *J. Clin. Invest.*, **57**, 1342 (1976).

271. M. Kaufman, C. Straisfeld, and L. Pinsky, *J. Clin. Invest.*, **58**, 345 (1976).

272. J. E. Griffin, *J. Clin. Invest.*, **64**, 1624 (1979).

273. P. C. Walsh, J. D. Madden, M. J. Harrod, J. L. Goldstein, P. C. MacDonald, and J. D. Wilson, *N. Engl. J. Med.*, **291**, 944 (1974).

274. J. Imperato-McGinley, L. Guerrero, T. Gautier, and R. E. Peterson, *Science*, **186**, 1213 (1974).

275. A. Segaloff, *Recent Prog. Horm. Res.*, **22**, 351 (1966).

276. P. K. Siiteri and E. A. Thompson, *J. Steroid Biochem.*, **6**, 317 (1975).

277. R. I. Dorfman in A. Dorfman, Ed., *Methods in Hormone Research*, Vol. **2**, Academic Press, New York, 1962, p. 275.

278. T. F. Gallagher and F. C. Koch, *J. Pharmacol. Exp. Ther.*, **55**, 97 (1935).

279. A. W. Greenwood, J. S. S. Blyth, and R. K. Callow, *Biochem. J.*, **29**, 1400 (1935).

280. C. W. Emmens, *Med. Res. Council. Spec. Rep. Ser.*, **234**, 1 (1939).

281. D. R. McCullagh and W. K. Cuyler, *J. Pharmacol. Exp. Ther.*, **66**, 379 (1939).

282. A. Segaloff, *Steroids*, **1**, 299 (1963).

283. E. Eisenberg and G. S. Gordan, *J. Pharmacol. Exp. Ther.*, **99**, 38 (1950).

284. F. J. Saunders and V. A. Drill, *Proc. Soc. Exp. Biol. Med.*, **94**, 646 (1957).

285. L. G. Hershberger, E. G. Shipley, and R. K. Meyer, *Proc. Soc. Exp. Biol. Med.*, **83**, 175 (1953).

286. S. Liao, T. Liang, S. Fang, E. Casteneda, and T. Shao, *J. Biol. Chem.*, **248**, 6154 (1973).

287. C. D. Kochakian. *Recent Prog. Horm. Res.*, **1**, 177 (1948).

288. C. Huggins and E. V. Jensen, *J. Exp. Med.*, **100**, 241 (1954).

289. K. Junkmann, *Recent Prog. Horm. Res.*, **13**, 389 (1957).

290. A. Butenandt and H. Dannenberg, *Berichte*, **73**, 206 (1940).

291. R. E. Counsell, P. D. Klimstra, and F. B. Colton, *J. Org. Chem.*, **27**, 248 (1962).

292. J. D. Wilson and R. E. Gloyna, *Recent Prog. Horm. Res.*, **26**, 309 (1970).

293. F. J. Zeller, *J. Reprod. Fertil.*, **25**, 125 (1971).

294. I. H. Harris, *J. Clin. Endocrin. Metab.*, **21**, 1099 (1961).

295. C. Huggins, E. V. Jensen, and A. S. Cleveland, *J. Exp. Med.*, **100**, 225 (1954).

296. R. B. Gabbard and A. Segaloff, *J. Org. Chem.*, **27**, 655 (1962).

297. V. A. Drill and B. Riegel, *Recent Prog. Horm. Res.*, **14**, 29 (1958).

298. J. W. Partridge, L. Boling, L. DeWind, S. Margen, and L. W. Kinsell, *J. Clin. Endocrinol. Metab.*, **13**, 189 (1953).

299. K. Miescher, E. Tschapp, and A. Wettstein, *Biochem. J.*, **30**, 1977 (1976).

300. G. Sala and G. Baldratti, *Proc. Soc. Exp. Biol. Med.*, **95**, 22 (1957).

301. S. C. Lyster, G. H. Lund, and R. O. Stafford, *Endocrinology*, **58**, 781 (1956).

302. R. M. Backle, *Br. Med. J.*, **1**, 1378 (1959).

303. T. H. McGavack and W. Seegers, *Am. J. Med. Sci.*, **235**, 125 (1958).

304. G. H. Marquardt, C. I. Fisher, P. Levy, and R. M. Dowben, *J. Am. Med. Assoc.*, **175**, 851 (1961).

305. B. J. Kennedy, *N. Engl. J. Med.*, **259**, 673 (1958).

306. H. Nowakowski, *Deut. Bed. Wochenschr.*, **90**, 2291 (1965).

307. G. S. Gordan, S. Wessler, and L. V. Avioli. *J. Am. Med. Assoc.*, **219**, 483 (1972).

308. I. S. Goldenberg, N. Waters, R. S. Randin, F. J. Ansfield, and A. Segaloff, *J. Am. Med. Assoc.*, **223**, 1267 (1973).

309. R. Rosso, G. Porcile, and F. Brema, *Cancer Chemother. Rep.*, **59**, 890 (1975).

310. A. Segaloff and R. Bruce Gabbard, *Endocrinology*, **67**, 887 (1960).

311. R. E. Counsell, *J. Med. Chem.*, **9**, 263 (1966).

312. A. Bowers, A. D. Cross, J. A. Edwards, H. Carpio, M. C. Calzada, and E. Denot, *J. Med. Chem.*, **6**, 156 (1963).

313. F. A. Kincl and R. I. Dorfman, *Steroids*, **3**, 109 (1964).

314. M. E. Wolff and G. Zanati, *J. Med Chem.*, **12**, 629 (1969).

315. G. Zanati and M. E. Wolff, *J. Med. Chem.*, **14**, 958 (1971).

316. W. L. Duax, M. G. Erman, J. F. Griffin, and M. E. Wolff, *Cryst. Struct. Commun.*, **5**, 775 (1976).

317. G. Zanati and M. E. Wolff, *J. Med. Chem.*, **16**, 90 (1973).

318. A. Segaloff and R. B. Gabbard, *Steroids*, **1**, 77 (1963).

319. A. Segaloff and R. B. Gabbard, *Steroids*, **22**, 99 (1973).

320. S. Liao, T. Liang, S. Fang, E. Casteneda, and T. Shao, *J. Biol. Chem.*, **248**, 6154 (1973).

321. S. A. Shain and R. W. Boesel, *J. Steroid Biochem.*, **6**, 43 (1975).

322. L. L. Engel, J. Alexander, and M. Wheeler, *J. Biol. Chem.*, **231**, 159 (1958).

323. A. Segaloff, B. Gabbard, B. T. Carriere, and E. L. Rongone, *Steroids*, Suppl. I, 419 (1965).

324. D. Kupfer, E. Forchielli, and R. I. Dorfman, *J. Biol. Chem.*, **235**, 1968 (1960).

325. W. E. Farnsworth, *Steroids*, **8**, 825 (1966).

326. E. Caspi, A. Vermeulen, and H. B. Bhat, *Steroids*, Suppl. I, 141 (1965).

327. H. Okado, K. Matsuyoshi, and G. Tokuda, *Acta Endocrinol.*, **46**, 40 (1964).

328. R. Rembiesa, M. Holzbauer, P. C. M. Young, M. K. Birmingham, and M. Saffran, *Endocrinology*, **81**, 1278 (1967).

329. E. L. Rongone and A. Segaloff, *Steroids*, **1**, 179 (1963).

330. G. Lehnert and W. Mucke, *Arzneim.-Forsch.*, **16**, 603 (1966).

331. H. Langecker, *Arzneim.-Forsch.*, **12**, 231 (1962).

332. E. Gerhards, K. H. Kolb, and P. E. Schulze, *Z. Physiol. Chem.*, **342**, 40 (1965).

333. S. Dell'acqua, S. Manuso, G. Eriksson, and E. Diczfalusy, *Biochim. Biophys. Acta*, **130**, 241 (1966).

334. E. L. Rongone, A. Segaloff, B. Gabbard, A. C. Carter, and E. B. Feldman, *Steroids*, **1**, 664 (1963).

335. E. Gerhards, H. Gibian, and K. H. Kolb, *Arzneim.-Forsch.*, **16**, 458 (1966).

336. E. Castegnaro and G. Sala, *Folia Endocrinol.*, **14**, 581 (1961).

337. H. J. Ringold, J. Graves, M. Hayano, and H. Lawrence, Jr., *Biochem. Biophys. Res. Commun.*, **13**, 162 (1963).

338. H. A. Plantier, *N. Engl. J. Med.*, **270**, 141 (1964).

339. H. Hortling, K. Malmio, and L. Husi-Brummer, *Acta Endocrinol.*, **39** (Suppl.), 132 (1962).

340. A. A. de Lorimier, G. S. Gordan, R. C. Lowe, and J. V. Carbone, *Arch. Intern. Med.*, **116**, 289 (1965).

341. I. M. Arias in F. Gross, Ed., *Influence of Growth Hormone, Anabolic Steroids, and Nutrition: Health and Disease*, Vol. **3**, Springer-Verlag, Berlin, 1962, p. 434.

342. H. A. Kaupp and F. W. Preston, *J. Am. Med. Assoc.*, **180**, 411 (1962).

343. D. Westaby, S. J. Ogle, F. J. Paradinas, J. B. Randell, and I. M. Murray-Lyon, *Lancet*, **2**, 261 (1977).

344. R. L. Landau in C. D. Kochakian, Ed., *Androgenic-Anabolic Steroids*, Springer-Verlag, New York, 1976, p. 45.

345. R. O. Stafford, B. J. Bowman, and K. J. Olson, *Proc. Soc. Exp. Biol. Med.*, **86**, 322 (1954).

346. E. Henderson and M. Weinberg, *J. Clin. Endocrinol.*, **11**, 641 (1951).

347. J. C. Stucki, A. D. Forbes, J. I. Northam, and J. J. Clark, *Endocrinology*, **66**, 585 (1960).

348. H. Spencer, J. A. Friedland, and I. Lewin in C. D. Kochakian, Ed., *Anabolic-Androgenic Steroids*, Springer-Verlag, New York, 1976, p. 419.

349. C. S. Gurney in C. D. Kochakian, Ed., *Anabolic-Androgenic Steroids*, Springer-Verlag, New York, 1976, p. 483.

350. H. Kopera in C. D. Kochakian, Ed., *Anabolic-Androgenic Steroids*, Springer-Verlag, New York, 1976, p. 535.

351. C. Huseman and A. Johanson, *J. Pediatr.*, **87**, 946 (1975).

352. L. Tec, *Am. J. Psychiat.*, **127**, 1702 (1971).

353. P. M. Sansoy, R. A. Naylor, and L. M. Shields, *Geriatrics*, **26**, 139 (1971).

354. B. O. Morrison, *J. Mich. St. Med. Soc.*, **60**, 723 (1961).

355. A. L. Kolodny, *Med. Tms. (N. Y.)*, **91**, 9 (1963).

356. H. Buchner, *Wien. Med. Wochenschr.*, **111**, 576 (1961).

357. R. M. Konrad, U. Ammedick, W. Hupfauer, and W. Ringler, *Chirung*, **38**, 168 (1967).

358. U. Ammedick, R. M. Konrad, and E. Gotzen, *Med. Ernähr.*, **9**, 121 (1968).

359. D. E. F. Tweedle, C. Walton, and I. D. A. Johnston, *Br. J. Surg.*, **59**, 300 (1972).

360. A. A. Renzi and J. J. Chart, *Proc. Soc. Exp. Biol. Med.*, **110**, 259 (1962).

361. A. J. Ryan in C. D. Kochakian, Ed., *Anabolic-Androgenic Steroids*, Springer-Verlag, New York, 1976, p. 516.

362. J. D. Wilson, *Endocr. Rev.*, **9**, 181 (1988).

363. G. C. Lin and L. Erinoff, Eds., *NIDA Res. Monogr.*, **102**, 29 (1990).

364. S. E. Lukas, *Trends Pharmacol. Sci.*, **14**, 61 (1993).

365. S. Casner, R. Early, and B. R. Carlson, *J. Sports Med. Phys. Fit.*, **11**, 98 (1971).

366. T. D. Fahey and C. H. Brown, *Med. Sci. Sports*, **5**, 272 (1973).

367. L. A. Golding, J. E. Freydinger, and S. S. Fishel, *Physician Sports-Med.*, **2**, 39 (1974).

368. S. B. Strömme, H. D. Meen, and A. Aakvaag, *Med. Sci. Sports*, **6**, 203 (1974).

369. G. Ariel and W. Saville, *J. Appl. Physiol.*, **32**, 795 (1972).

370. G. Ariel, *J. Sports Med. Phys. Fit.*, **13**, 187 (1973).

371. L. C. Johnson, G. Fisher, L. J. Silvester, and C. C. Hofheins, *Med. Sci. Sports*, **4**, 43 (1972).

372. M. Steinbach, *Sportarzt Sportmedizin*, **11**, 485 (1968).

373. S. Bhasin, T. W. Storer, N. Berman, C. Callegari, B. Clevenger, J. Phillips, T. J. Bunnell, R. Tricker, A. Shirazi, and R. Casaburi, *N. Engl. J. Med.*, **335**, 1–7 (1996).

374. N. Wade, *Science*, **176**, 1399 (1972).

375. R. P. Howard and R. H. Furman, *J. Clin. Endocrinol.*, **22**, 43 (1962).

376. J. F. Dingman and W. H. Jenkins, *Metabolism*, **11**, 273 (1962).

377. S. Weisenfeld, S. Akgun, and S. Newhouse, *Diabetes*, **12**, 375 (1963).

378. C. J. Glueck, *Clin. Res.*, **17**, 475 (1971).

379. C. J. Glueck, *Metabolism*, **20**, 691 (1971).

380. A. E. Doyle, N. B. Pinkus, and J. Green, *Med. J. Aust.*, **1**, 127 (1974).

381. B. A. Sachs and L. Wolfman, *Metabolism*, **17**, 400 (1968).

382. C. J. Glueck, S. Ford, P. Steiner, and R. Fallat, *Metabolism*, **17**, 807 (1973).

383. N. T. Shahidi, *N. Engl. J. Med.*, **289**, 72 (1973).

384. A. Killander, K. Lundmark, and S. Sjolin, *Acta Paediat. Scand.*, **58**, 10 (1969).

385. R. F. Branda, T. W. Amsden, and H. S. Jacob, *Clin. Res.*, **22**, 607A, 1974.

386. D. W. Hughes, *Med. J. Aust.*, **2**, 361 (1973).

387. E. D. Hendler, J. A. Goffinet, S. Ross, R. E. Longnecker, and V. Bakovic, *N. Engl. J. Med.*, **291**, 1046 (1974).

388. A. Blumbert and H. Keller, *Schweiz. Med. Wachr.*, **101**, 1887 (1971).

389. W. Fried, O. Jonasson, G. Lang, and F. Schwartz, *Ann. Intern. Med.*, **79**, 823 (1973).

390. J. Keyssner, C. Hauswaldt, N. Uhl, and W. Hunstein, *Schweiz. Med. Wochenschr.*, **104**, 1938 (1974).

391. F. A. Kincl in R. I. Dorfman, Ed., *Methods in Hormone Research*, Vol. **4**, Academic Press, New York, 1965, p. 21.

392. G. O. Potts, A. Arnold, and A. L. Beyler, *Endocrinology*, **67**, 849 (1960).

393. M. E. Nimni and E. Geiger, *Endocrinology*, **61**, 753 (1957).

394. J. N. Goldman, J. A. Epstein, and H. S. Kupperman, *Endocrinology*, **61**, 166 (1957).

395. F. A. Kincl, H. J. Ringold, and R. I. Dorfman, *Acta Endocrinol.*, **36**, 83 (1961).

396. M. E. Nimni and E. Geiger, *Proc. Soc. Exp. Biol. Med.*, **94**, 606 (1957).

397. R. O. Scow, *Endocrinology*, **51**, 42 (1952).

398. J. Leibetseder and K. Steininger, *Arzneim.-Forsch.*, **15**, 474 (1965).

399. R. A. Edgren, *Acta Endocrinol.*, **87** (Suppl.), 3 (1963).

400. F. J. Saunders and V. A. Drill, *Endocrinology*, **58**, 567 (1956).

401. L. E. Barnes, R. O. Stafford, M. E. Guild, L. C. Thole, and K. J. Olson, *Endocrinology*, **55**, 77 (1954).

402. C. Djerassi, L. Miramontes, G. Rosenkranz, and F. Sondheimer, *J. Am. Chem. Soc.*, **76**, 4092 (1954).

403. F. B. Colton, L. N. Nysted, B. Reigel, and A. L. Raymond, *J. Am. Chem. Soc.*, **79**, 1123 (1957).

404. E. B. Feldman and A. C. Carter, *J. Clin. Endocrinol. Metab.*, **20**, 842 (1960).

405. J. Ferrin, *Acta Endocrinol.*, **22**, 303 (1956).

406. K. V. Yorka, W. L. Truett, and W. S. Johnson, *J. Org. Chem.*, **27**, 4580 (1962).

407. W. F. Johns, *J. Am. Chem. Soc.*, **80**, 6456 (1958).

408. W. F. Johns (G. D. Searle & Co.), private communication.

409. H. J. Ringold, *J. Am. Chem. Soc.*, **82**, 961 (1960).

410. A. Zaffaroni, *Acta Endocrinol.*, **50** (Suppl.), 139 (1960).

411. I. W. Goldberg, J. Sicé, H. Robert, and Pl. A. Plattner, *Helv. Chim. Acta*, **30**, 1441 (1947).

412. H. Heusser, P. T. Herzig, A. Furst, and Pl. A. Plattner. *Helv. Chim. Acta*, **33**, 1093 (1950).

413. G. Eadon and C. Djerassi, *J. Med. Chem.*, **89**, 89 (1972).

414. G. Sala, G. Baldratti. R. Ronchi, V. Clini, and C. Bertazzoli, *Sperimentale*, **106**, 490 (1956).

415. G. S. Gordan, *Arch. Intern. Med.*, **100**, 744 (1957).

416. A. Arnold, G. O. Potts, and A. L. Beyler, *Endocrinology*, **72**, 408 (1963).

417. P. A. Desaulles, *Helv. Med. Acta*, **27**, 479 (1960).

418. R. I. Dorfman and F. A. Kincl, *Endocrinology*, **72**, 259 (1963).

419. G. W. Liddle and H. A. Burke, Jr., *Helv. Med. Acta*, **27**, 504 (1960).

420. A. L. Beyler, G. O. Potts, and A. Arnold, *Endocrinology*, **68**, 987 (1961).

421. F. B. Colton, U.S. Pat. 2,874,170 (1959).

422. A. Smit and P. Westerhof, *Rec. Trav. Chim. Pays-Bas*, **82**, 1107 (1963).

423. R. Van Moorselaar, S. J. Halkes, and E. Havinga, *Rec. Trav. Chim. Pays-Bas*, **84**, 841 (1965).

424. J. S. Baran, *J. Med. Chem.*, **6**, 329 (1963).

425. R. A. Edgren, D. L. Peterson, R. C. Jones, C. L. Nagra, H. Smith, and G. A. Hughes, *Recent Prog. Horm. Res.*, **22**, 305 (1966).

426. J. Tremolieres and E. Pequignot, *Presse Med.*, **73**, 2655 (1965).

427. H. L. Kruskemper and G. Noell, *Steroids*, **8**, 13 (1966).

428. H. L. Kruskemper, K. D. Moraner, and G. Noell, *Arzneim.-Forsch.*, **17**, 449 (1967).

429. A. Halden, R. M. Watter, and G. S. Gordan, *Cancer Chemother. Rep.*, **54**, 453 (1970).

430. G. K. Suchowsky and K. Junkmann, *Acta Endocrinol.*, **39**, 68 (1962).

431. B. Pelc, *Collect. Czech. Chem. Commun.*, **29**, 3089 (1964).

432. O. Weller, *Endokrinologie*, **42**, 34 (1962).

433. O. Weller, *Endokrinologie*, **41**, 60 (1961).

434. H. L. Kruskemper and H. Breuer, *Excerpta Med. Int. Congr. Ser.*, **51**, 209 (1962).

435. F. Neumann, R. Wiechert, M. Kramer, and G. Raspe, *Arzneim.-Forsch.*, **16**, 455 (1966).

436. O. Weller, *Arzneim.-Forsch.*, **16**, 465 (1966).

437. B. Pelc and J. Jodkova, *Collect. Czech. Chem. Commun.*, **30**, 3575 (1965).

438. B. Pelc, *Collect. Czech. Chem. Commun.*, **30**, 3408 (1965).

439. C. Djerassi, R. Riniker, and B. Riniker, *J. Am. Chem. Soc.*, **78**, 6377 (1956).

440. A. Bowers, H. J. Ringold, and E. Denot, *J. Am. Chem. Soc.*, **80**, 6115 (1958).

441. O. Abe, H. Herraneu, and R. I. Dorfman, *Proc. Soc. Exp. Biol. Med.*, **111**, 706 (1962).

442. D. Berkowitz, *Clin. Res.*, **8**, 199 (1960).

443. R. E. Counsell and P. D. Klimstra. *J. Med. Chem.*, **6**, 736 (1963).

444. R. I. Dorfman and A. S. Dorfman, *Acta Endocrinol.*, **42**, 245 (1963).

445. A. Arnold, G. O. Potts, and A. L. Beyler, *J. Endocrinol.*, **28**, 87 (1963).

446. D. R. Korst, C. Y. Bowers, J. H. Flokstra, and F. G. McMahon, *Clin. Pharmacol. Ther.*, **4**, 734 (1963).

447. J. A. Campbell, S. C. Lyster, G. W. Duncan, and J. C. Babcock, *Steroids*, **1**, 317 (1963).

448. J. A. Campbell and J. C. Babcock, *J. Am. Chem. Soc.*, **81**, 4069 (1959).

449. R. A. Edgren, H. Smith, and G. A. Hughes, *Steroids*, **2**, 731 (1963).

450. R. B. Greenblatt, E. C. Jungck, and G. C. King, *Am. J. Med. Sci.*, **318**, 99 (1964).

451. G. Baldratti, G. Arcari, V. Clini, F. Tani, and G Sala, *Sperimentale*, **109**, 383 (1959).

452. G. Sala, A. Cesana, and G. Fedriga, *Minerva Med.*, **51**, 1295 (1960).

453. A. A. Albanese, E. J. Lorenze, and L. A. Orto, *N. Y. State J. Med.*, **63**, 80 (1963).

454. S. C. Lyster, G. H. Lund, and R. O. Stafford, *Endocrinology*, **58**, 781 (1956).

455. H. A. Burke, Jr. and G. W. Liddle, Abstracts of the Endocrine Society Meeting, Atlantic City, NJ, 1959, p. 45.

456. G. Sala, *Helv. Med. Acta*, **27**, 519 (1960).

457. R. M. Myerson, *Am. J. Med. Sci.*, **241**, 732 (1961).

458. W. W. Glas and E. H. Lansing, *J. Am. Geriatr. Soc.*, **10**, 509 (1962).

459. H. G. Kraft and H. Kieser, *Arzneim.-Forsch.*, **14**, 330 (1964).

460. H. L. Kruskemper, *Arzneim.-Forsch.*, **16**, 608 (1966).

461. R. E. Schaub and M. J. Weiss, *J. Org. Chem.*, **26**, 3915 (1961).

462. H. Kaneko, K. Nakamura, Y. Yamato, and M. Kurakawa, *Chem. Pharm. Bull. (Tokyo)*, **17**, 11 (1969).

463. A. Weinstein, H. R. Lindner, A. Frielander, and S. Bauminger, *Steroids*, **20**, 789 (1972).

464. R. Pappo and C. J. Jung, *Tetrahedron Lett.*, 365 (1962).

465. H. D. Lennon and F. J. Saunders, *Steroids*, **4**, 689 (1964).

466. P. B. Sollman, U.S. Pat. 3,301,872 (1966); *Chem. Abstr.*, **66**, P95299d (1967).

467. R. H. Mazur and R. Pappo, Belg. Pat. 631,372 (1964); *Chem. Abstr.*, **61**, P705 (1964).

468. A. P. Shroff and C. H. Harper, *J. Med. Chem.*, **12**, 190 (1969).

469. M. Fox, A. S. Minot, and G. Liddle, *J. Clin. Endocrinol. Metab.*, **22**, 921 (1962).

470. C. G. Ray, J. F. Kirschvink, S. H. Waxman, and V. C. Kelley, *Am. J. Dis. Child.*, **110**, 618 (1965).

471. E. F. Nutting and D. W. Calhoun, *Endocrinology*, **84**, 441 (1969).

472. C. D. Kochakian, *Am. J. Physiol.*, **145**, 549 (1946).

473. C. D. Kochakian, *Proc. Soc. Exp. Biol. Med.*, **80**, 386 (1952).

474. E. F. Nutting, P. D. Klimstra, and R. E. Counsell, *Acta Endocrinol.*, **53**, 627, 635 (1966).

475. I. A. Anderson, *Acta Endocrinol.*, **63** (Suppl.), 54 (1962).

476. M. E. Wolff and Y. Kasuya, *J. Med. Chem.*, **15**, 87 (1972).

477. G. A. Overbeck, A. Delver, and J. deVisser, *Acta Endocrinol.*, **63** (Suppl.), 7 (1962).

478. J. L. Kalliomaki, A. M. Pirila, and I. Ruikka, *Acta Endocrinol.*, **63** (Suppl.), 124 (1962).

479. H. Kopera, *Excerpta Med. Int. Congr. Ser.*, **51**, 204 (1962).

480. A. Walser and G. Schoenenberger, *Schweiz. Med. Wochenschr.*, **92**, 897 (1962).

481. J. A. Edwards and A. Bowers, *Chem. Ind. (Lond.)*, 1962 (1961).

482. M. E. Wolff, G. Zanati, G. Shanmagasundarum, S. Gupte, and G. Aadahl, *J. Med. Chem.*, **13**, 531 (1970).

483. M. E. Wolff and G. Zanati, *Experientia*, **26**, 1115 (1970).

484. G. Zanati, G. Gaare, and M. E. Wolff, *J. Med. Chem.*, **17**, 561 (1974).

485. R. W. S. Skinner, R. V. Pozderac, R. E. Counsell, C. F. Hsu, and P. A. Weinhold, *Steroids*, **25**, 189 (1977).

486. M. E. Wolff, W. Ho, and R. Kwok, *J. Med. Chem.*, **7**, 577 (1964).

487. P. D. Klimstra, E. F. Nutting, and R. E. Counsell, *J. Med. Chem.*, **9**, 693 (1966).

488. M. Fujimuri, *Cancer*, **31**, 789 (1973).

489. G. O. Potts, A. Arnold, and A. L. Beyler, *Excerpta Med. Int. Congr. Ser.*, **51**, 211 (1962).

490. G. O. Potts, A. L. Beyler, and D. F. Burnham, *Proc. Soc. Exp. Biol. Med.*, **103**, 383 (1960).

491. P. C. Burnett, *J. Am. Geriatr. Soc.*, **11**, 979 (1963).

492. W. G. Mullin and F. diPillo, *N. Y. State J. Med.*, **63**, 2795 (1963).

493. A. J. Manson, F. W. Stonner, H. C. Neumann, R. G. Christiansen, R. L. Clarke, J. H. Ackerman, D. F. Page, J. W. Dean, D. K. Phillips, G. O. Potts, A. Arnold, A. L. Beyler, and R. O. Clinton, *J. Med. Chem.*, **6**, 1 (1963).

494. G. O. Potts, A. Beyler, and H. P. Schane, *Fertil. Steril.*, **25**, 367 (1974).

495. R. J. Sherrins, H. M. Gandy, T. W. Thorsland, and C. A. Paulsen, *J. Clin. Endocrinol. Metab.*, **32**, 522 (1971).

496. D. Rosi. H. C. Neumann, R. G. Christiansen, H. P. Shane, and G. O. Potts, *J. Med. Chem.*, **20**, 349 (1977).

497. C. Davison, W. Banks, and A. Fritz, *Arch. Int. Pharmacodyn. Ther.*, **221**, 294 (1976).

498. G. A. Overbeck, *Anabole Steroide*, Springer-Verlag, Berlin, 1966.

499. H. L. Kruskemper, *Anabolic Steroids*, trans. by C. H. Doering, Academic Press, New York, 1968.

500. G. A. Overbeck, J. Van Der Vies, and J. De Visser in F. Gross, Ed., *Protein Metabolism*, Springer-Verlag, Berlin, 1962, p. 185.

501. J. de Visser and G. A. Overbeck, *Acta Endocrinol.*, **35**, 405 (1960).

502. G. A. Overbeck and J. de Visser, *Acta Endocrinol.*, **38**, 285 (1961).

503. H. Nowakowski, *Acta Endocrinol.*, **63** (Suppl.), 37 (1962).

504. A. Alibrandi, G. Bruni, A. Ercoli, R. Gardi, and A. Meli, *Endocrinology*, **66**, 13 (1960).

505. K. Junkmann and G. Suchowsky, *Arzneim.-Forsch.*, **12**, 214 (1962).

506. K. C. James, *Experientia*, **28**, 479 (1972).

507. K. C. James, P. J. Nicholl, and G. T. Richards, *Eur. J. Med. Chem.*, **10**, 55 (1975).

508. R. Gaunt, C. H. Tuthill, N. Antonchak, and J. H. Leathem, *Endocrinology*, **52**, 407 (1953).

509. P. Borrevang, *Acta Chem. Scand.*, **16**, 883 (1962).

510. R. Huttenrauch, *Arch. Pharm.*, **297**, 124 (1964).

511. F. B. Colton and R. E. Ray, U.S. Pat. 3,068,249 (1962).

512. A. Ercoli, R. Gardi, and R. Vitali, *Chem. Ind. (Lond.)*, 1284 (1962).

513. R. Vitali, R. Gardi, and A. Ercoli, *Excerpta Med. Int. Congr. Ser.*, **51**, 128 (1962).

514. R. I. Dorfman, A. S. Dorfman, and M. Gut, *Acta Endocrinol.*, **40**, 565 (1962).

515. R. Vitali. R. Gardi, G. Falconi, and A. Ercoli, *Steroids*, **8**, 527 (1966).

516. A. Ercoli, G. Bruni, G. Falconi, F. Galletti, and R. Gardi, *Acta Endocrinol.*, **51** (Suppl.), 857 (1960).

517. A. D. Cross. I. T. Harrison, P. Crabbe, F. A. Kincl, and R. I. Dorfman, *Steroids*, **4**, 229 (1964).

518. A. D. Cross and I. T. Harrison, *Steroids*, **6**, 397 (1965).

519. R. R. Burtner, E. A. Brown, and R. A. Mikulec, *Excerpta Med. Int. Congr. Ser.*, **111**, 50 (1966).

520. F. J. Saunders, *Proc. Soc. Exp. Biol. Med.*, **123**, 303 (1966).

521. A. J. Solo, N. Bejba, P. Hebborn, and M. May, *J. Med. Chem.*, **18**, 165 (1975).

522. C. D. Kochakian and N. Arimasa in C. D. Kochakian, Ed., *Anabolic-Androgenic Steroids*, Springer-Verlag, New York, 1976, p. 287.

523. B. Pelc, *Collect. Czech. Chem. Commun.*, **29**, 3089 (1964).

524. O. Weller, *Endokrinologie*, **42**, 34 (1962).

525. M. E. Wolff and Y. Kasuya, *J. Med. Chem.*, **15**, 87 (1972).

526. D. Rosi, H. C. Neumann, R. G. Christiansen, H. P. Schane, and G. O. Potts, *J. Med. Chem.*, **20**, 349 (1977).

527. C. Davison, W. Banks, and A. Fritz, *Arch. Int. Pharmacodyn. Ther.*, **221**, 294 (1976).

528. J. T. Henderson, J. Richmond, and M. D. Sumerling, *Lancet*, **1**, 934 (1973).

529. F. L. Johnson, J. R. Feagler, K. G. Lerner, P. W. Majerus, M. Siegel, J. R. Hartman, and E. D. Thomas, *Lancet*, **2**, 1273 (1972).

530. K. G. Ishak in C. H. Lingeman, Ed., *Carcinogenic Hormones*, Springer-Verlag, New York, 1979, p. 73.

531. H. Steinbeck and F. Neumann in R. Vokaer and G. DeBock, Eds., *Reproductive Endocrinology*, Pergamon Press, Oxford, 1975, p. 135.

532. L. Martini and M. Motta, Eds., *Androgens and Antiandrogens*, Raven Press, New York, 1977.

533. J. P. Raynaud in J. Jacob, Ed., *Advances in Pharmacology and Therapeutics*, Vol. **1**, Pergamon Press, Oxford, 1979, p. 259.

534. G. H. Rasmusson and J. H. Torrey, *Ann. Rep. Med. Chem.*, **29**, 225 (1994).

535. A. D. Abell and B. R. Henderson, *Curr. Med. Chem.*, **2**, 583–597 (1995).

536. M. Jarman, H. J. Smith, P. J. Nicholls, and C. Simons, *Nat. Prod. Rep.*, **15**, 495–512 (1998).

537. S. M. Singh, S. Gauthier and F. Labrie, *Curr. Med. Chem.*, **7**, 211–247 (2000).

538. J. P. Raynaud, B. Azadian-Boulanger, C. Bonne, J. Perronnet, and E. Sakis in L. Martini and M. Motta, Eds., *Androgens and Antiandrogens*, Raven Press, New York, 1977, p. 281.

539. C. Huggins and C. V. Hodges, *Cancer Res.*, **1**, 293 (1941).

540. C. Huggins, R. E. Stevens, Jr., and C. V. Hodges, *Arch. Surg.*, **43**, 209 (1941).

541. E. A. P. Sutherland-Bawlings, *Br. Med. J.*, **111**, 643 (1970).

542. E. C. Dodds, *Biochem. J.*, **39**, i (1945).

543. J. Geller, B. Fruchtman, C. Meyer, and H. Newman, *J. Clin. Endocrinol. Metab.*, **27**, 556 (1967).

544. F. Neumann in P. Mantegazza and F. Piccinini, Eds., *Methods in Drug Evaluation*, North-Holland, Amsterdam, 1966, p. 548.

545. K. Mietkiewski, L. Malendowicz, and A. Lukaszyk, *Acta Endocrinol.*, **61**, 293 (1969).

546. R. O. Neri, *Adv. Sex Horm. Res.*, **2**, 233 (1976).

547. F. Neumann, *Horm. Metab. Res.*, **9**, 1 (1977).

548. U. Fixson, *Geburtsh. Frauenheilk.*, **23**, 371 (1963).

549. K. Junkmann and F. Neumann, *Acta Endocrinol.*, **90** (Suppl.), 139 (1964).

550. F. Neumann, K. J. Gräf, S. H. Hasan, B. Schenck, and H. Steinbeck in L. Martini and M. Motta, Eds., *Androgens and Antiandrogens*, Raven Press, New York, 1977, p. 163.

551. F. Neumann, R. von Berswodt-Wallace, W. Elger, H. Steinbeck, J. Hahn, and M. Kramer, *Recent Prog. Horm. Res.*, **26**, 337 (1970).

552. S. Fang, K. M. Anderson, S. Liao, *J. Biol. Chem.*, **244**, 6584 (1969).

553. S. Fang and S. Liao, *Mol. Pharmacol.*, **5**, 420 (1969).

554. J. Stern and A. J. Eisenfeld, *Endocrinology*, **88**, 1117 (1971).

555. F. R. Mangan and W. I. P. Mainwaring, *Steroids*, **20**, 331 (1972).

556. J. Hammerstein, J. Meckies, I. Leo-Rossberg, L. Moltz, and F. Zielke, *J. Steroid Biochem.*, **6**, 827 (1975).

557. J. L. Burton, U. Laschet, and S. Shuster, *Br. J. Dermatol.*, **89**, 487 (1973).

558. J. Hammerstein and B. Cupceancu, *Dtsh. Med. Wochenschr.*, **94**, 829 (1969).

559. A. A. Ismail, D. W. Davidson, A. R. Souka, E. W. Barnes, W. J. Irvine, H. Kilimnik, and Y. Vanderbeeken, *J. Clin. Endrocrinol. Metab.*, **39**, 81 (1974).

560. E. Cittadini and P. Barreca in L. Martini and M. Motta, Eds., *Androgens and Antiandrogens*, Raven Press, New York, 1977, p. 309.

561. V. B. Mahesh in L. Martini and M. Motta, Eds., *Androgens and Antiandrogens*, Raven Press, New York, 1977, p. 321.

562. F. J. Ebling in L. Martini and M. Motta, Eds., *Androgens and Antiandrogens*, Raven Press, New York, 1977, p. 341.

563. J. Geller, B. Fruchtman, H. Newman, T. Roberts, and R. Sylva, *Cancer Chemother. Rep.*, **51**, 441 (1967).

564. U. Bracci and F. DiSilverio, *Prog. Med.*, **29**, 779 (1973).

565. U. Bracci, *J. Urol. Nephrol.*, **79**, 405 (1973).

566. F. DiSilverio and V. Gagliardi, *Boll. Soc. Urol.*, **5**, 198 (1968).

567. U. Bracci and F. DiSilverio in L. Martini and M. Motta, Eds., *Androgens and Antiandrogens*, Raven Press, New York, 1977, p. 333.

568. C. Labrie, L. Cusan, M. Plante, S. Lapointe, and F. Labrie, *J. Steroid Biochem.*, **28**, 379–384 (1987).

569. F. Sciarra, V. Toscano, G. Concolino, and F. Di Silverio, *J. Steroid Biochem. Mol. Biol.*, **37**, 349–362 (1990).

570. C. Labrie, J. Simard, H. F. Zhao, G. Pelletier, and F. Labrie, *Mol. Cell Endocrinol.*, **68**, 169–179 (1990).

571. T. Ojasoo, J. Delettre, J. P. Mornon, C. Turpin-VanDycke, and J. P. Raynaud, *J. Steroid Biochem.*, **27**, 255–269 (1987).

572. N. Jagarinec and M. L. Givner, *Steroids*, **23**, 561 (1974).

573. L. J. Lerner, A. Bianchi, and A. Borman, *Proc. Soc. Exp. Biol. Med.*, **103**, 172 (1960).

574. F. Orestano, J. E. Altwein, P. Knapstein, and K. Bandhauer, *J. Steroid Biochem.*, **6**, 845 (1975).

575. W. I. P. Mainwaring in L. Martini and M. Motta, Eds., *Androgens and Antiandrogens*, Raven Press, New York, 1977, p. 151.

576. S. Y. Tan, *J. Clin. Endocrinol. Metab.*, **39**, 936 (1974).

577. H. L. Saunders, K. Holden, and J. F. Kerwin, *Steroids*, **3**, 687 (1964).

578. A. Boris and M. Uskokovic, *Experientia*, **26**, 9 (1970).

579. A. Boris, L. DeMartino, and T. Trmal, *Endocrinology*, **88**, 1086 (1971).

580. E. E. Baulieu and I. Jung, *Biochem. Biophys. Res. Commun.*, **38**, 599 (1970).

581. G. H. Rasmusson, A. Chen, G. F. Reynolds, D. J. Patanelli, A. A. Patchett, and G. E. Arth, *J. Med. Chem.*, **15**, 1165 (1972).

582. J. R. Brooks, F. D. Busch, D. J. Patanelli, and S. L. Steelman, *Proc. Soc. Exp. Biol. Med.*, **143**, 647 (1973).

583. K. Hiraga, A. Tsunehiko, and M. Takuichi, *Chem. Pharm. Bull. (Tokyo)*, **13**, 1294 (1965).

584. G. Goto, K. Yoshiska, K. Hiraga, M. Masouka, R. Nakayama, and T. Miki, *Chem. Pharm. Bull. (Tokyo)*, **26**, 1718 (1978).

585. F. R. Mangan and W. I. P. Mainwaring, *Steroids*, **20**, 331 (1972).

586. G. Azadian-Boulanger, C. Bonne, J. Sechi, and J. P. Raynaud, *J. Pharmacol.*, **5**, 509 (1974).

587. C. Bonne and J. P. Raynaud, *Mol. Cell Endocrinol.*, **2**, 59 (1974).

588. P. Corvol, A. Michaud, J. Menard, M. Freifeld, and J. Mahoudean, *Endocrinology*, **97**, 52 (1975).

589. A. E. Wakeling, B. J. A. Furr, A. T. Glen, and L. R. Hughes, *J. Steroid Biochem.*, **15**, 355 (1981).

590. L. Starka, J. Sulcova, P. D. Broulik, J. Joska, J. Fajkos, and M. Doskocil, *J. Steroid Biochem.*, **8**, 939 (1977).

591. L. Starka, R. Hanapl, M. Bicikova, V. Cerny, J. Fajkos, A. Kosal, P. Kocovsky, L. Kohout, and H. Velgova, *J. Steroid Biochem.*, **13**, 455 (1980).

592. R. G. Christiansen, M. R. Bell, T. E. D'Ambra, J. P. Mallamo, J. L. Herrmann, J. H. Ackerman, C. J. Opalka, R. K. Kullnig, R. C. Winneker, B. W. Snyder, et al., *J. Med. Chem.*, **33**, 2094–2100 (1990).

593. J. P. Mallamo, G. M. Pilling, J. R. Wetzel, P. J. Kowalczyk, M. R. Bell, R. K. Kullnig, F. H. Batzold, P. E. Juniewicz, R. C. Winneker, and H. R. Luss, *J. Med. Chem.*, **35**, 1663–1670 (1992).

594. R. Neri, K. Florance, P. Koziol, and S. van Cleave, *Endocrinology*, **91**, 427 (1972).

595. R. O. Neri and M. Monohan, *Invest. Urol.*, **10**, 123 (1972).

596. E. A. Peets, M. F. Henson, and R. Neri, *Endocrinology*, **94**, 532 (1974).

597. S. Liao, D. K. Howell, and T. Chuag, *Endocrinology*, **94**, 1205 (1974).

598. R. Neri and E. A. Perts, *J. Steroid Biochem.*, **6**, 815 (1975).

599. E. D. Crawford, M. A. Eisenberger, D. G. McLeod, J. T. Spaulding, R. Benson, F. A. Dorr, B. A. Blumenstein, M. A. Davis, and P. J. Goodman, *N. Engl. J. Med.*, **321**, 419–424 (1989).

600. L. Denis and G. P. Murphy, *Cancer*, **72**, 3888–3895 (1993).

601. G. H. Rasmusson, G. F. Reynolds, N. G. Steinberg, et al., *J. Med. Chem.*, **29**, 2298 (1986).

602. J. Simard, S. M. Singh, and F. Labrie, *Urology*, **49**, 580–586 (1997).

603. L. M. Eri and K. J. Tveter, *J. Urol.*, **150**, 90 (1993).

604. U. Fuhrmann, C. Bengston, G. Repenthin, and E. Schillinger, *J. Steroid Biochem. Mol. Biol.*, **42**, 787 (1992).

605. C. J. Tyrell, *Prostate*, **4**, 97 (1992).

606. R. A. Janknegt, C. C. Abbou, R. Bartoletti, L. Bernstein-Hahn, B. Bracken, J. M. Brisset, F. C. Da Silva, G. Chisholm, E. D. Crawford, F. M. Debruyne, et al., *J. Urol.*, **149**, 77–82 (1993).

607. P. Iversen, K. Tveter, and E. Varenhorst, *Scand. J. Urol. Nephrol.*, **30**, 93–98 (1996).

608. A. Boris, J. W. Scott, L. DeMartino, and D. C. Cox, *Acta Endocrinol.*, **72**, 604 (1973).

609. H. Miyachi, A. Azuma, T. Kitamoto, K. Hayashi, S. Kato, M. Koga, B. Sato, Y. Hashimoto, *Bioorg. Med. Chem. Lett.*, **7**, 1483–1488 (1997).

610. L. G. Hamann, R. I. Higuchi, L. Zhi, J. P. Edwards, X. N. Wang, K. B. Marschke, J. W. Kong, L. J. Farmer, and T. K. Jones, *J. Med. Chem.*, **41**, 623–639 (1998).

611. S. Tanayama, K. Yoshida, T. Kondo, and Y. Kanai, *Steroids*, **33**, 65 (1979).

612. U. Speck, H. Wendt, P. E. Schulze, and D. Jentsch, *Contraception*, **14**, 151 (1976).

613. M. Hümpel, H. Wendt, P. E. Schulze, G. Dogs, C. Weiss, and U. Speck, *Contraception*, **15**, 579 (1977).

614. M. Hümpel, H. Dogs, H. Wendt, and U. Speck, *Arzneim.-Forsch.*, **28**, 319 (1978).

615. M. Frölich, H. L. Vader, S. T. Walma, and H. A. M. De Rooy, *J. Steroid Biochem.*, **13**, 1097 (1980).

616. A. S. Bhargava, A. Seeger and P. Günzel, *Steroids*, **30**, 407 (1977).

617. B. Katchen and S. Buxbaum, *J. Clin. Endocrinol. Metab.*, **41**, 373 (1975).

618. R. O. Neri in L. Martini and M. Motta, Eds., *Androgens and Antiandrogens*, Raven Press, New York, 1977, p. 179.

619. H. Tucker, J. W. Crook, and G. J. Chesterson, *J. Med. Chem.*, **31**, 954–959 (1988).

620. F. Neumann and R. Von Berswordt-Wallrabe, *J. Endocrinol.*, **35**, 363 (1966).

621. J. Hammerstein in L. Martini and M. Motta, Eds., *Androgens and Antiandrogens*, Raven Press, New York, 1977, p. 327.

622. H. J. Horn in L. Martini and M. Motta, Eds., *Androgens and Antiandrogens*, Raven Press, New York, 1977, p. 351.

623. F. Neumann and K. Junkmann, *Endocrinology*, **73**, 33 (1963).

624. R. O. Neri, M. D. Monahan, J. G. Meyer, B. A. Afonso, and I. I. A. Tabachnick, *Eur. J. Pharmacol.*, **1**, 438 (1967).

625. L. J. Lerner, *Pharmacol. Ther. B*, **1**, 217 (1975).

626. D. K. Wysowski, J. P. Freiman, J. B. Tourtelot, and M. L. Horton III, *Ann. Intern. Med.*, **118**, 860–864 (1993).

627. L. A. Dawson, E. Chow, and G. Morton, *Urology*, **49**, 283–284 (1997).

628. E. J. Small and P. R. Carroll, *Urology*, **43**, 408–410 (1994).

629. J. R. Brooks, G. S. Harris, and G. H. Rasmusson in M. Sandler and H. J. Smither, Eds., *Design of Enzyme Inhibitors as Drugs*, Vol. **2**, Oxford University Press, Oxford, 1994, p. 495.

630. G. Harris, B. Azzolina, W. Baginsky, et al., *Proc. Natl. Acad. Sci. USA*, **89**, 10787 (1992).

631. J. D. McConnell, J. D. Wilson, F. W. George, et al., *J. Clin. Endocrin. Metab.*, **74**, 505 (1992).

632. E. Stoner and Study Group, *Urology*, **43**, (1994).

633. H. A. Guess, J. F. Heyse, and G. J. Gormley, *Prostate*, **22**, 31 (1993).

634. J. Schwartz, O. Laskin, S. Schneider, et al., *Clin. Pharmacol. Ther.*, **53**, 231 (1993).

635. A. A. Geldof, M. F. A. Meulenbroek, I. Dijkstra, et al., *J. Cancer Res. Clin. Oncol.*, **118**, 50 (1992).

636. E. diSalle, D. Guidici, G. Briatico, et al., *J. Steroid Biochem. Mol. Biol.*, **46**, 549 (1993).

637. H. G. Bull, *J. Am. Chem. Soc.*, **118**, 2359 (1996).

638. B. Kenny, S. Ballard, J. Blagg, and D. Fox, *J. Med. Chem.*, **40**, 1293–1315 (1997).

639. S. V. Frye, C. D. Haffner, P. R. Maloney, R. A. Mook, Jr., G. F. Dorsey, Jr., R. N. Hiner, K. W. Batchelor, H. N. Bramson, J. D. Stuart, S. L. Schweiker, et al., *J. Med. Chem.*, **36**, 4313 (1993).

640. M. A. Levy, M. Brandt, J. R. Heys, et al., *Biochemistry*, **29**, 2815 (1990).

641. M. A. Levy, B. W. Metcalf, M. Brandt, et al., *Bioorg. Chem.*, **19**, 245 (1991).

642. P. Audet, H. Nurcombe, Y. Lamb, et al., *Clin. Pharmacol. Ther.*, **53**, 231 (1993).

643. R. E. Johnsonbaugh, B. R. Cohen, E. M. McCormick, et al., *J. Urol.*, **149**, 432 (1993).

644. D. A. Holt, M. A. Levy, et al., *J. Med. Chem.*, **33**, 937, 943 (1990).

645. G. M. Cooke and B. Robaire, *J. Steroid Biochem.*, **24**, 877 (1986).

646. B. Robaire, D. F. Covey, C. H. Robinson, and L. L. Ewing, *J. Steroid Biochem.*, **8**, 307 (1977).

647. W. Voigt, A. Castro, D. F. Covey, and C. H. Robinson, *Acta Endocrinol.*, **87**, 668 (1978).

648. R. K. Bakshi, G. F. Patel, G. H. Rasmusson, W. F. Baginsky, G. Cimis, K. Ellsworth, B. Chang, H. Bull, R. L. Tolman, and G. S. Harris, *J. Med. Chem.*, **37**, 3871–3874 (1994).

649. C. D. Jones, J. E. Audia, D. E. Lawhorn, L. A. McQuaid, B. L. Neubauer, A. J. Pike, P. A. Pennington, N. B. Stamm, R. E. Toomey, and K. S. Hirsch, *J. Med. Chem.*, **36**, 421–423 (1993).

650. D. A. Holt, D. S. Yamashita, A. L. Konialian-Beck, J. I. Luengo, A. D. Abell, D. J. Bergsma, M. Brandt, and M. A. Levy, *J. Med. Chem.*, **38**, 13–15 (1995).

651. H. Kojo, O. Nakayama, J. Hirosumi, N. Chida, Y. Notsu, and M. Okuhara, *Mol. Pharmacol.*, **48**, 401–406 (1995).

652. J. R. Drago, R. J. Santen, A. Lipton, T. J. Worgul, H. A. Harvey, A. Boucher, A. Manni, and T. J. Rohner, *Cancer*, **53**, 1447 (1984).

653. G. Williams, D. J. Kerle, H. Ware, A. Doble, H. Dunlog, C. Smith, J. Allen, T. Yeo, and S. R. Bloom, *Br. J. Urol.*, **58**, 45 (1986).

654. J. P. Van Wauwe and P. A. Janssen, *J. Med. Chem.*, **32**, 2231 (1989).

655. M. Ayub and M. J. Levell, *J. Steroid Biochem.*, **32**, 515 (1989).

656. J. Yeh, R. L. Barbieri, and A. J. Friedman, *J. Steroid Biochem.*, **33**, 627 (1989).

657. S. E. Barrie, M. G. Rowlands, A. B. Foster, and M. Jarman, *J. Steroid Biochem.*, **33**, 1191 (1989).

658. R. McCague, M. G. Rowlands, S. E. Barrie, and J. Houghton, *J. Med. Chem.*, **33**, 2452 (1990).

659. W. N. Kuhn-Velten, I. Meyer, and W. Staib, *J. Steroid Biochem.*, **33**, 33 (1989).

660. W. N. Kuhn-Velten, T. Bunse, and M. E. C. Forster, *J. Biol. Chem.*, **266**, 6291 (1991).

661. M. Onoda, M. Haniu, K. Kanagibashi, F. Sweet, J. E. Shively, and P. F. Hall, *Biochemistry*, **26**, 657 (1987).

662. M. R. Angelastro, M. E. Laughlin, G. L. Schatzman, P. Bey, T. R. Blohm, *Biochem. Biophys. Res. Commun.*, **162**, 1571 (1989).

663. J. Stevens, J. Jaw, C. T. Peng, and J. Halpert, *Biochemistry*, **30**, 3649 (1991).

664. V. C. Njar, M. Hector, and R. W. Hartmann, *Bioorg. Med. Chem.*, **4**, 1447–1453 (1996).

665. G. A. Potter, S. E. Barrie, M. Jarman, and M. G. Rowlands, *J. Med. Chem.*, **38**, 2463–2471 (1995).

666. R. J. Santen, S. Santner A. Lipton, Eds., Cancer Res., **42** (Suppl.), 1s–3467s (1982).

667. R. J. Santen, Ed., Steroids, **50**, 1–665 (1987).

668. A. M. H. Brodie, H. B. Brodie, G. Callard, C. Robinson, C. Roselli, and R. J. Santen, Eds., J. Steroid Biochem. Mol. Biol., **44**, 321–696 (1993).

669. J. O. Johnston and B. W. Metcalf in P. Sunkara, Ed., *Novel Approaches to Cancer Chemotherapy*, Academic Press, New York, 1984, pp. 307–328.

670. L. Banting, H. J. Smith, M. James, G. Jones, W. Nazareth, P. J. Nichols, M. J. E. Hewlins, and M. G. Rowlands, *J. Enzyme Inhibit.*, **2**, 215–229 (1988).

671. D. F. Covey in D. Berg and M. Plempel, Eds., *Sterol Biosynthesis Inhibitors*, Ellis Horwood, Chichester, UK, 1988, pp. 534–571.

672. L. Banting, P. J. Nichols, M. A. Shaw, and H. J. Smith, *Prog Med Chem.*, **26**, 253–298 (1989).

673. R. W. Brueggemeier, *J. Enzyme Inhibit.*, **4**, 101–111 (1990).

674. R. J. Santen, A. Manni, H. Harvey, and C. Redmond, *Endocr. Rev.*, **11**, 221–265 (1990).

675. R. T. Blickenstaff, *Antitumor Steroids*, Academic Press, San Diego, 1992, pp. 68–78.

676. R. W. Brueggemeier, *Breast Cancer Res. Treat.*, **30**, 31 (1994).

677. W. C. Schwarzel, W. G. Kruggel, and H. J. Brodie, *Endocrinology*, **92**, 866–880 (1973).

678. D. A. Marsh, H. J. Brodie, W. Garrett, C.-H. Tsai-Morris, and A. M. H. Brodie, *J. Med. Chem.*, **28**, 788–795 (1985).

679. A. M. H. Brodie, W. Garrett, J. R. Hendrickson, C.-H. Tsai-Morris, C. H. Marcotte, and C. H. Robinson, *Steroids*, **38**, 693–702 (1981).

680. A. M. H. Brodie, W. C. Schwarzel, A. A. Shaikh, and H. J. Brodie, *Endocrinology*, **100**, 1684–1695 (1977).

681. L. Y. Wing, W. Garrett, and A. M. H. Brodie, *Cancer Res.*, **45**, 2425–2428 (1985).

682. A. M. H. Brodie, W. Garrett, J. R. Hendrickson, and C.-H. Tsai-Morris, *Cancer Res.*, **42** (Suppl.), 3360s–3364s (1982).

683. R. C. Coombes, P. Goss, M. Dowsett, J. C. Gazet, and A. M. H. Brodie, *Lancet*, **2**, 1237–1239 (1984).

684. P. E. Goss, T. J. Powles, M. Dowsett, G. Hutchison, A. M. H. Brodie, J.-C. Gazet, and R. C. Coombes, *Cancer Res.*, **46**, 4823–4826 (1986).

685. M. Dowsett, D. Cunningham, S. Nichols, A. Lal, S. Evans, L. Dehennin, A. Hedley, and R. C. Coombes, *Cancer Res.*, **49**, 1306–1312 (1989).

686. D. Henderson, G. Norbisrath, and U. Kerb, *J. Steroid Biochem.*, **24**, 303–306 (1986).

687. R. W. Brueggemeier, E. E. Floyd, and R. E. Counsell, *J. Med. Chem.*, **21**, 1007–1011 (1978).

688. S. Miyairi and J. Fishman, *J. Biol. Chem.*, **261**, 6772–6777 (1986).

689. R. W. Brueggemeier and N. E. Katlic, *Cancer Res.*, **47**, 4548–4551 (1987).

690. R. W. Brueggemeier, P.-K. Li, C. E. Snider, M. V. Darby, and N. E. Katlic, *Steroids*, **50**, 163–178 (1987).

691. R. W. Brueggemeier and P.-K. Li, *Cancer Res.*, **48**, 6808–6810 (1988).

692. M. V. Darby, J. A. Lovett, R. W. Brueggemeier, M. P. Groziak, and R. E. Counsell, *J. Med. Chem.*, **28**, 803–807 (1985).

693. P.-K. Li and R. W. Brueggemeier, *J. Med. Chem.*, **33**, 101–105 (1990).

694. P.-K. Li and R. W. Brueggemeier, *J. Enzyme Inhib.*, **4**, 113–120 (1990).

695. B. W. Metcalf, C. L. Wright, J. P. Burkhart, and J. O. Johnston, *J. Am. Chem. Soc.*, **103**, 3221–3222 (1981).

696. D. F. Covey, W. F. Hood, and V. D. Parikh, *J. Biol. Chem.*, **256**, 1076–1079 (1981).

697. P. A. Marcotte and C. H. Robinson, *Steroids*, **39**, 325–344 (1982).

698. J. O. Johnston, C. L. Wright, and B. W. Metcalf, *Endocrinology*, **115**, 776–785 (1984).

699. J. O. Johnston, C. L. Wright, and B. W. Metcalf, *J. Steroid Biochem.*, **20**, 1221–1226 (1984).

700. C. Longcope, A. M. Femino, and J. O. Johnston, *Endocrinology*, **122**, 2007–2011 (1988).

701. J. O. Johnston, *Steroids*, **50**, 106–120 (1987).

702. S. J. Ziminski, M. E. Brandt, D. F. Covey, and C. Puett, *Steroids*, **50**, 135–146 (1987).

703. P. A. Marcotte and C. H. Robinson, *Biochemistry*, **21**, 2773–2778 (1982).

704. P. J. Bednarski, D. J. Porubek, and S. D. Nelson, *J. Med. Chem.*, **28**, 775–779 (1985).

705. P. J. Bednarski and S. D. Nelson, *J. Med. Chem.*, **32**, 203–213 (1989).

706. C. E. Snider and R. W. Brueggemeier, *J. Biol. Chem.*, **262**, 8685–8689 (1987).

707. D. Giudici, G. Ornati, G. Briatico, F. Buzzetti, P. Lombardi, and E. Di Salle, *J. Steroid Biochem.*, **30**, 391–394 (1988).

708. National Coordinating Group on Male Antifertility Agents. *Chin. Med. J. Engl.*, **4**, 417–428 (1978).

709. S. J. Segal, Ed., *Gossypol, A Potential Contraceptive for Men*, Plenum, New York, 1985.

710. K. L. Olgiati and W. A. Toscano, *Biochem. Biophys. Res. Commun.*, **115**, 180–185 (1983).

711. Y. C. Lin, M. Chitcharoenthum, and Y. Rikihisa, *Contraception*, **36**, 581–592 (1987).

712. Y. C. Lin, M. A. Hadley, D. Klingener, and M. Dym, *Biol Reprod.*, **22**, 95A, 1980.

713. M. A. Hadley, Y. C. Lin, and M. Dym, *J. Androl.*, **2**, 190–199 (1981).

714. K. L. Olgiati, D. G. Toscano, W. M. Atkins, and W. A. Toscano, *Arch. Biochem. Biophys.*, **231**, 41–48 (1984).

715. L. Z. Zhuang, D. M. Philips, G. L. Gunsalus, C. W. Bardin, and J. P. Mather, *J. Androl.*, **4**, 336–344 (1983).

716. Y. C. Lin, T. Fukaya, Y. Rikihisa, and A. Walton, *Life Sci.*, **37**, 39–47 (1985).

717. Y. C. Lin, T. Fukaya, Y. Rikihisa, and T. DeSanto, *Adv. Contracept. Deliv. Syst.*, **2**, 200–206 (1985).

718. Y. C. Lin and Y. Rikihisa, *Ann. N. Y. Acad. Sci.*, **513**, 532–534 (1987).

719. N. G. Wang, M. Z. Guan, and H. P. Lei, *J. Ethnopharmacol.*, **20**, 45–51 (1987).

720. Y. Gu, Y. C. Lin, and Y. Rikihisa, *Biochem. Biophys. Res. Commun.*, **169**, 455–461 (1990).

721. Y. Gu, P.-K. LI, Y. C. Lin, Y. Rikihisa, and R. W. Brueggemeier, *J. Steroid Biochem. Mol. Biol.*, **38**, 709–715 (1991).

722. Y. Gu and Y. C. Lin, *Res. Commun. Chem. Path. Pharmacol.*, **72**, 27–38 (1991).

723. S. U. Lawrence, *Am. Pharm.*, **21**, 57 (1981).

CHAPTER FIFTEEN

Anti-Inflammatory Steroids

Mitchell A. Avery
John R. Woolfrey
University of Mississippi-University
Department of Medicinal Chemistry, School of Pharmacy
University, Mississippi

Contents

Burger's Medicinal Chemistry and Drug Discovery
Sixth Edition, Volume 3: Cardiovascular Agents and Endocrines
Edited by Donald J. Abraham
ISBN 0-471-37029-0   © 2003 John Wiley & Sons, Inc.

1   INTRODUCTION

Since the introduction of cortisone (1948) and
hydrocortisone (1951), anti-inflammatory steroids have remained an important and unre-
placed drug class. Although not without
adverse effects, these compounds have continued to be the "drug of choice" in the treatment
of afflictions ranging from the moderate skin
rash to severe acute inflammatory disorders.

The adrenal cortex releases both mineralocorticoids and glucocorticoids, the latter of
which includes corticosterone, cortisol (2),

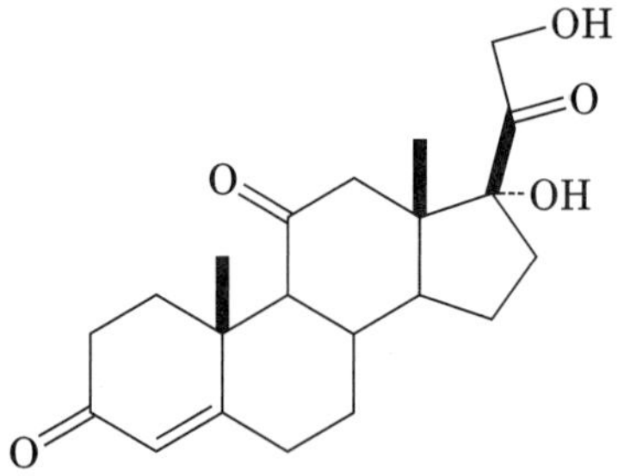

(1)

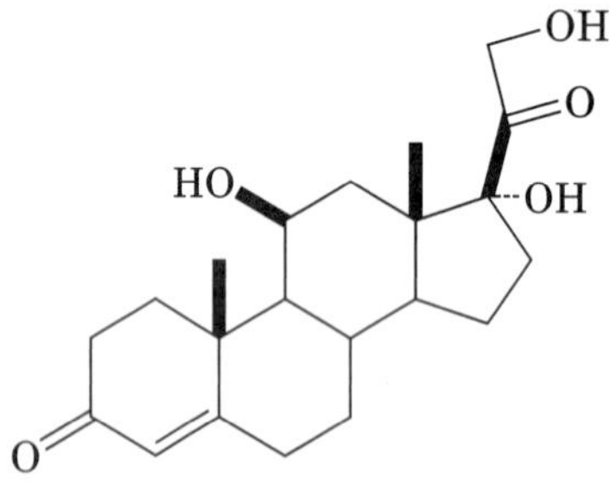

and cortisone (1). The generation of glucocorticosteroids is intricately balanced: when the
production and regulatory process malfunctions, the result is either an excess (e.g., Cushing's syndrome) or a deficiency (e.g., Addison's
disease) in glucocorticoid levels. Once released

(2)

by the adrenal cortex, the primary endogenous function of glucocorticosteroids is in influencing carbohydrate and protein metabolism.

Glucocorticoids and their metabolites (1)
were recognized early as possessing powerful
anti-inflammatory and immunomodulatory
properties. Even before 1950, reports of the
antiarthritic properties of cortisone (1) by
Hench and coworkers (2) indicated the potential for these compounds to reduce the suffering of patients with inflammatory diseases.
This, combined with the first synthesis of naturally occurring glucocorticoids (11-desoxy-
corticosterone), led not only to the massive increase in research in the area of steroid
synthesis and physiology, but to a Nobel prize
in 1950 for early steroid pioneers Hench,
Reichstein, and Kendall.

Progress in the field took an additional leap
forward with the developement of synthetic
steroids exhibiting activities far greater than
those of the natural hormones. The synthesis

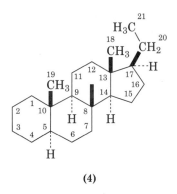

(3)

of 9α-fluorocortisol (**3**), described in 1953 by Fried and Sabo (3), opened the way for many more highly active anti-inflammatory agents, and indeed some of today's most active anti-inflammatory agents have some semblance to this 9α-fluorinated steroid.

We know today that structural changes in steroids can bring about potency alterations in animal pharmacology through a number of mechanisms. The processes that are affected include pharmacokinetic parameters such as drug absorption and drug distribution (discussed elsewhere), as well as drug metabolism to more or less active metabolites (discussed elsewhere). There is now a large body of literature dealing with these phenomena that makes predictive thinking possible in these areas. Other mechanisms that are involved in the effect of structural changes on pharmacological activity in steroids include the effect of steroid structure on receptor affinity and the manner in which changes in the steroid agonist affect intrinsic activity through alterations in gene expression. Steroid-binding proteins and receptors have been studied in detail over the last two decades but have eluded crystallographic analysis. The glucocorticoid receptor (GR) is a cytoplasmic protein of complex structure that is associated with various heat shock proteins (e.g., hsp90). When activated by agonistic binding of drug to the C-terminal region of the GR, the GR-drug complex undergoes translocation to the nucleus. The active form of the GR-drug complex is thought to be dimeric in its binding to the DNA sequence, termed the glucocorticoid response element (GRE). Upon binding to the GRE of the DNA by the "zinc finger" region of

the GR-drug complex, the synthesis of anti-inflammatory proteins is either inhibited or stimulated.

Although limited structural information on the glucocorticoid and other hormone receptors is available, quantitative structure-activity relationship (QSAR) studies have attempted to define a pharmacophore model for the anti-inflammatory activity. Approaches that have been applied are the use of conventional Hansch as well as *de novo* constants for regression analysis, neural networks, comparative molecular field analysis (CoMFA), and principal-component (PC) analysis with PLS (partial least squares) analysis. These studies have provided predictive models for glucocorticoid binding that could in principle be used in drug design efforts.

Comprehensive reviews dealing with the pharmacological structure-activity relationships of anti-inflammatory steroids have been published (4–6) as well as reviews of the mechanism of action (7, 8). Almost all anti-inflammatory steroids are based on the pregnane (**4**) and 17-substituted androstane ring system.

(4)

Abbreviated names for compounds employed in this chapter are as follows:

Cortisone (**1**): 17,21-Dihydroxypregn-4-ene-3,11,20-trione

Cortisol or hydrocortisone (**2**): 11β,17,21-Trihydroxypregn-4-ene-3,20-dione

Hydrocortisone esters (**2a-2f**): 11β,21-Dihydroxy-17-(1-oxoalkyloxy)-pregn-4-ene-3,20-dione

9α-Fluorocortisol (**3**): 9-Fluoro-11β,17,21-trihydroxypregn-4-ene-3,20-dione

Betamethasone (**5**): 9-Fluoro-11β,17,21-trihy-
droxy-16β-methylpregna-1,4-diene-3,20-
dione

Clobetasol propionate (**6**): 21-Chloro-9-fluoro-
11β-hydroxy-16β-methyl-17-(1-oxoprop-
oxy)-pregna-1,4-diene-3,20-dione

Diflorasone diacetate (**7**): 17,21-Bis(1-oxoeth-
anoxy)-6α,9-difluoro-11β-hydroxy-16β-methyl-
pregna-1,4-diene-3,20-dione

Halobetasol propionate (**8**): 21-Chloro-6α,9-
difluoro-11β-hydroxy-16β-methyl-17-(1-oxo-
propoxy)-pregna-1,4-diene-3,20-dione

Amcinonide (**9**): 16,17α-[Cyclopentylidene-
bis(oxy)]-9-fluoro-11β-hydroxy-21-(1-oxoeth-
oxy)-pregna-1,4-diene-3,20-dione

Desoximetasone (**10**): 11β,21-Dihydroxy-9-
fluoro-16β-methylpregna-1,4-diene-3,20-
dione

Fluocinolone (**a**): 6α,9-Difluoro-11β,16α,17,21-
tetrahydroxypregna-1,4-diene-3,20-dione

Fluocinolone acetonide (**11b**): 6α,9-Difluoro-
11β,21-dihydroxy-16α,17-[(1-methylethyl-
idene)bis(oxy)]-pregna-1,4-diene-3,20-
dione

Fluocinonide (**11c**): 6α,9-Difluoro-11β-hy-
droxy-16α,17-[(1-methylethylidene)bis
(oxy)-21-(1-oxoethoxy)]-pregna-1,4-diene-
3,20-dione

Halcinonide (**12**): 21-Chloro-9-fluoro-11β-hy-
droxy-16α,17-[(1-methylethylidene)bis
(oxy)]-pregna-1,4-diene-3,20-dione

Triamcinolone acetonide (**13**):9-Fluoro-11β,
21-dihydroxy-16α,17-[(1-methylethylidene)
bis(oxy)]-pregna-1,4-diene-3,20-dione

Triamcinolone (**13a**): 9-Fluoro-11β,16α,17,21-
tetrahydroxypregna-1,4-diene-3,20-dione

Clocortolone pivalate (**14**): 9-Chloro-6α-fluoro-
11β-hydroxy-16α-methyl-21-[1-oxo(2,2-
dimethyl) propyloxy)]-pregna-1,4-diene-
3,20-dione

Flurandrenolide (**15**): 6α-Fluoro-11β,21-dihy-
droxy-16α,17-[(1-methylethylidene)-
bis(oxy)]-pregn-4-ene-3,20-dione

Fluticasone propionate (**16**): 6α,9-Difluoro-
11β-hydroxy-16α-methyl-3-oxo-17α-(1-oxo-
propoxy)-androsta-1,4-diene-17β-carbo-
thioic acid, S-fluoromethyl ester

Mometasone furoate (**17**): 9α,21-Dichloro-17-
[(2-furanylcarbonyl)oxy]-11β-hydroxy-16α-
methyl-pregna-1,4-diene-3,20-dione

Aclomethasone dipropionate (**18**): 7α-Chloro-
11β-hydroxy-16α-methyl-17α,21-bis(1-oxo-
propanoxy)-pregna-1,4-diene-3,20-dione

Desonide (**19**): 11β,21-Dihydroxy-16α,17-[(1-
methylethylidene)bis(oxy)]-pregna-1,4-
diene-3,20-dione

Dexamethasone: 9-Fluoro-11α,17,21-trihydroxy-
16α-methylpregna-1,4-diene-3,20-dione

Prednisolone : 11β,17,21-Trihydroxypregna-
1,4-diene-3,20-dione

Fluoromethalone (**22**): 9-Fluoro-11β,17-dihy-
droxy-6α-methylpregna-1,4-diene-3,20-dione

Rimexolone (**23**): 16α,17-Dimethyl-11β-hy-
droxy-17β-(1-oxopropyl)-andro-
sta-1,4-dien-3-one

Medrysone (**24**): 11β-Hydroxy-6α-methyl-
pregna-1,4-diene-3,20-dione

Beclomethasone dipropionate (**25**): 9-Chloro-
11β-hydroxy-16β-methyl-17,21-bis(1-oxopro-
panoxy)-pregna-1,4-diene-3,20-dione

Flunisolide (**26**): 6α-Fluoro-11β,21-dihy-
droxy-16α,17-[(1-methylethylidene)bis(oxy)]-
pregna-1,4-diene-3,20-dione

Budesonide (**27**): 16α,17-[Butylidine-
bis(oxy)]-11β,21-dihydroxy-pregna-1,4-diene-
3,20-dione

6α-Methylprednisolone (**28**): 6α-Methyl-11β,
17,21-trihydroxypregna-1,4-diene-3,20-
dione

Prednisone (**29**): 17,21-Dihydroxypregna-1,4-
diene-3,11,20-trione

Dichlorisone (**30**): 11β-Dichloro-17,21-dihy-
droxypregn-4-ene-3,20-dione

Corticosterone (**31**): 11β,21-Dihydroxypregn-
4-ene-3,20-dione

6α-Fluorocortisol (**32**): 6α-Fluoro-11β,17,21-
trihydroxypregn-4-ene-3,20-dione

Deoxycorticosterone (**33**): 21-Hydroxypregn-
4-ene-3,20-dione

Cortexolone (**34**): 17,21-Dihydroxypregn-4-
ene-3,20-dione

Progesterone (**35**): Pregn-4-ene-3,20-dione

21-Deoxydexamethasone (**36**): 9-Fluoro-11β,
17-dihydroxy-16α-methylpregna-1,4-diene-
3,20-dione

Paramethasone (**37**): 6α-Fluoro-11β,17,21-trihy-
droxy-16α-methylpregna-1,4-diene-3,20-dione

6α-Fluoro-16α-hydroxycortisol (**38**): 6α-
Fluoro-11β,16α,17,21-tetrahydroxypregn-
4-ene-3,20-dione

9α-Fluoroprednisolone (**39**): 9-Fluoro-11β,17,
21-trihydroxypregna-1,4-diene-3,20-dione

Flumethasone (**40**): 6α,9α-Difluoro-16α-
methyl-11β,17,21-trihydroxy-1,4-
pregnadiene-3,20-dione

Structures for many of the above steroids appear in Figs. 15.1 and 15.2.

2 CLINICAL USE OF ANTI-INFLAMMATORY STEROIDS

The focus of this chapter is on the anti-inflammatory properties of synthetic and endogenous glucocorticosteroids. In general, exogenous anti-inflammatory corticosteroid use is for the suppression of symptoms, including inflammation, occurring in a particular disease state; these compounds are rarely considered curative in their usage.

Many disease states, however, do respond well symptomatically to treatment with corticosteroid therapy. Some of these are listed in the following table; an asterisk (*) denotes those states in which the anti-inflammatory activity of the corticosteroids used is the primary rationale in their utilization.

Adrenocortical insufficiency	Adrenogential syndrome	Hypercalemia	Myasthenia gravis
Organ transplants	Nephrotic Syndrome	Hematologic disorders	Neoplastic disease
Liver disease	Acute spinal cord injury	Rheumatic disorders and collagen diseases (*)	Dermatologic deseases (*)
Allergic conditions and conjunctivitis (*)	GI dieseases; ulcerative colitis and anorectal disorders (*)	Opthalmic, otic and nasal disorders (*)	Respiratory diseases (*)
Cerebral edema (*)	Bacterial meningitis (*)		

(Modified from (9))

2.1 Topical Corticosteroids Currently Available

Tables 15.1 and 15.2 are lists of compounds found in *Facts and Comparisons* (FC) (9) and *AHFS Drug Index* (DI) (10). The compounds listed by FC as topical anti-inflammatories were placed within a range of Low to Very High Potency (see Table 15.1). The DI list indicates a range from I (most active) to VI (least active), and these values are listed next to the drug name in the table. Some compounds have more than one activity class, depending on the method of administration and concentration; in such cases only the higher activity class is listed.

[Activity] ··· may vary considerably depending on the vehicle, site of application, disease, the individual patient, and whether or not an occlusive dressing is used. The approximate relative activity is based principally on vasoconstrictor assay and/or clinical effectiveness in psoriasis (10).

3 ADVERSE EFFECTS ASSOCIATED WITH THE USE OF ANTI-INFLAMMATORY STEROIDS

The use of anti-inflammatory steroids has led many to relief from discomfort caused by their medical condition. However, corticosteroid use is not without costs, and these are clearly seen in the adverse side effects these compounds can lead to, especially in high doses or prolonged administration. These side effects are also associated with Cushing's syndrome (excessive glucocorticoid formation resulting from, for example, tumors of the adrenal cortex) and are thus sometimes described as "Cushingoid." Side effects are also dependent on the method of corticosteroid administration and target tissue (11).

Dermatologic side effects noted with glucocorticoid use include skin atrophy and reduced skin thickness (12). Collagen, an essential protein in skin, is not synthesized as adequately in patients who use glucocorticoids, especially in high doses (12), and this results in the degradation of skin character.

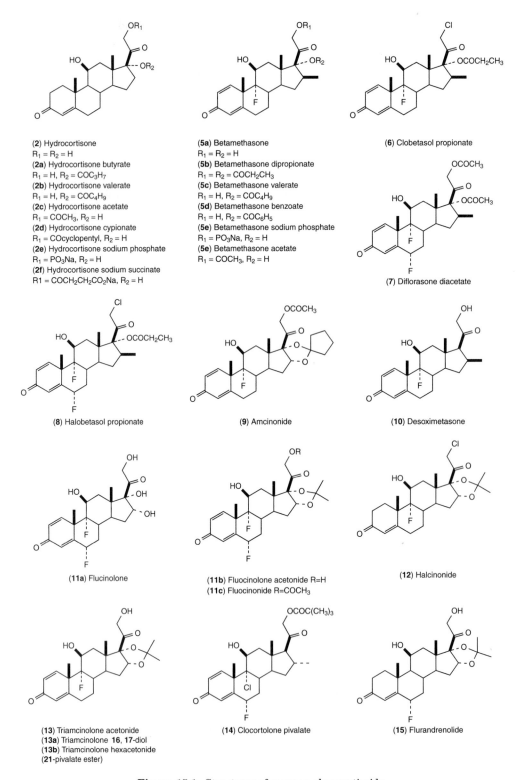

(2) Hydrocortisone
$R_1 = R_2 = H$
(2a) Hydrocortisone butyrate
$R_1 = H, R_2 = COC_3H_7$
(2b) Hydrocortisone valerate
$R_1 = H, R_2 = COC_4H_9$
(2c) Hydrocortisone acetate
$R_1 = COCH_3, R_2 = H$
(2d) Hydrocortisone cypionate
$R_1 = COcyclopentyl, R_2 = H$
(2e) Hydrocortisone sodium phosphate
$R_1 = PO_3Na, R_2 = H$
(2f) Hydrocortisone sodium succinate
$R1 = COCH_2CH_2CO_2Na, R_2 = H$

(5a) Betamethasone
$R_1 = R_2 = H$
(5b) Betamethasone dipropionate
$R_1 = R_2 = COCH_2CH_3$
(5c) Betamethasone valerate
$R_1 = H, R_2 = COC_4H_9$
(5d) Betamethasone benzoate
$R_1 = H, R_2 = COC_6H_5$
(5e) Betamethasone sodium phosphate
$R_1 = PO_3Na, R_2 = H$
(5e) Betamethasone acetate
$R_1 = COCH_3, R_2 = H$

(6) Clobetasol propionate

(7) Diflorasone diacetate

(8) Halobetasol propionate

(9) Amcinonide

(10) Desoximetasone

(11a) Flucinolone

(11b) Fluocinolone acetonide R=H
(11c) Fluocinonide R=COCH₃

(12) Halcinonide

(13) Triamcinolone acetonide
(13a) Triamcinolone **16, 17**-diol
(13b) Triamcinolone hexacetonide
(21-pivalate ester)

(14) Clocortolone pivalate

(15) Flurandrenolide

Figure 15.1. Structures of common glucocorticoids.

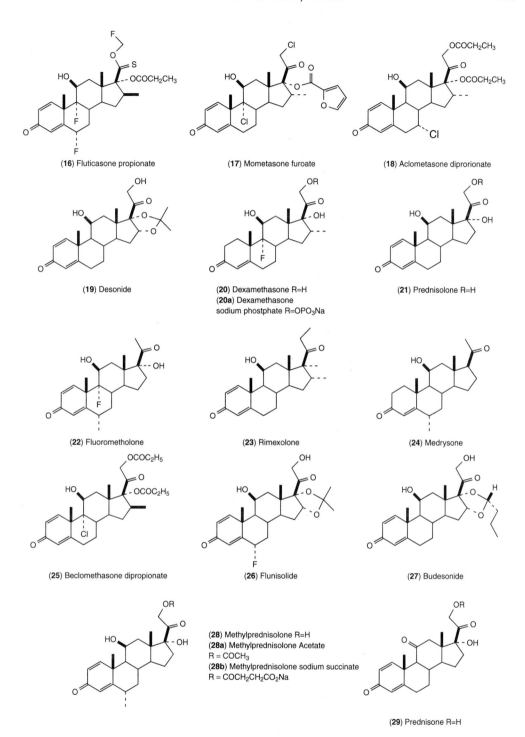

Figure 15.2. Structures of common glucocorticoids.

Table 15.1 Drug Facts and Comparisons vs. AHFS Drug Information

Very High Potency	High Potency	Medium Potency	Low Potency
Betamethasone dipropionate, I, 9-Fluoro-11β-hydroxy-16β-methyl-17α,21-bis(1-oxopropanoxy)-pregna-1,4-diene-3,20-dione (**5**)	**Amcinonide, II,** 21-(Acetyloxy)-16,17α-[cyclopentylidene-bis(oxy)]-9-fluoro-11β-hydroxypregna-1,4-diene-3,20-dione (**9**)	**Betamethasone benzoate, III,** 17-(Benzoyloxy)-9-fluoro-11β,21-dihydroxy-16β-methylpregna-1,4-diene-3,20-dione (**16**)	**Aclometasone dipropionate, VI,** 7α-Chloro-11β-hydroxy-16α-methyl-17α,21-bis(1-oxopropanoxy)-pregna-1,4-diene-3,20-dione (**23**)
Clobetasol propionate, I, 21-Chloro-9-fluoro-11β-hydroxy-16β-methyl-17-(1-oxopropoxy)-pregna-1,4-diene-3,20-dione (**6**)	**Betamethasone valerate, III,** 9-Fluoro-11β,17α,21-trihydroxy-16β-methyl-17-[(1-oxopentyl)oxy]-pregna-1,4-diene-3,20-dione (**10**)	**Clocortolone pivalate,** 9-Chloro-21-(2,2-dimethyl-1-oxopropoxy)-6α-fluoro-11β-hydroxy-16α-methylpregna-1,4-diene-3,20-dione (**17**)	**Desonide, VI,** 11β,21-Dihydroxy-16α,17-[(1-methylethylidene)bis(oxy)]-pregna-1,4-diene-3,20-dione (**24**)
Diflorasone diacetate, I, 17,21-bis(Acetyloxy)-6α,9-difluoro-11β-hydroxy-16β-methylpregna-1,4-diene-3,20-dione (**7**)	**Desoximetasone, II,** 9-Fluoro-11β,21-dihydroxy-16α-methylpregna-1,4-diene-3,20-dione (**11**)	**Flurandrenolide, IV,** 6α-Fluoro-11β,21-dihydroxy-16α,17-[(1-methylethylidene)bis(oxy)]-pregn-4-ene-3,20-dione (**18**)	**Dexamethasone,** 9-Fluoro-11β,17,21-trihydroxy-16α-methylpregna-1,4-diene-3,20-dione (**25**)
Halobetasol propionate, 21-Chloro-6α,9-difluoro-11β-hydroxy-16β-methyl-17-(1-oxopropoxy)-pregna-1,4-diene-3,20-dione (**8**)	**Fluocinonide, II,** 21-(Acetyloxy)-6α,9-difluoro-11β-hydroxy-16,17-[(1-methylethylidene)bis(oxy)]-pregna-1,4-diene-3,20-dione (**12**)	**Fluticasone propionate,** 6α,9-Difluoro-11β-hydroxy-16α-methyl-3-oxo-17α-(1-oxopropoxy)-androsta-1,4-diene-17β-carbothioic acid, S-fluoromethyl ester (**19**)	**Dexamethasone sodium phosphate,** 9-Fluoro-11β,21-dihydroxy-16α-methyl-21-(phosphonooxy)-pregna-1,4-diene-3,20-dione, disodium salt (**26**)
	Fluocinolone acetonide, IV, 6α,9-Difluoro-11β,21-dihydroxy-16α,17-[(1-methylethylidene)-bis(oxy)]-pregna-1,4-diene-3,20-dione (**13**)	**Hydrocortisone butyrate, V,** 11β,21-Dihydroxy-17-(1-oxobutoxy)-pregn-4-ene-3,20-dione (**20**)	**Hydrocortisone,** 11β,17,21-Trihydroxy-pregn-4-ene-3,20-dione (**27**)
	Halcinonide, II, 21-Chloro-9-fluoro-11β-hydroxy-16α,17-[(1-methylethylidene)bis(oxy)]-pregn-4-ene-3,20-dione (**14**)	**Hydrocortisone valerate, V,** 11β,21-Dihydroxy-17-[(1-oxopentyl)oxy]-pregn-4-ene-3,20-dione (**21**)	**Hydrocortisone acetate,** 21-(Acetyloxy)-11β,17-dihydroxy-pregn-4-ene-3,20-dione (**28**)
	Triamcinolone acetonide, III, 9-Fluoro-11β,21-dihydroxy-16α,17-[(1-methylethylidene)bis(oxy)]-pregna-1,4-diene-3,20-dione (**15**)	**Mometasone furoate,** α,21-Dichloro-17-[(2-furanyl-carbonyl)oxy]-11β-hydroxy-16α-methyl-pregna-1,4-diene-3,20-dione (**22**)	

Group III: Mometasone acetonide (**29**)   *Group V:* Prednicarbate (**30**)

Table 15.2 Steroids Frequently Used in Common Medical Circumstances

Medical Use	Drugs Common to This Use	
Ophthalmic	Dexamethasone (**20a**) Prednisolone, 11β,17,21-Trihydroxy-pregna-1,4-diene-3,20-dione (**21**)	Fluorometholone, 9-Fluoro-11β,17-dihydroxy-6α-methyl-pregna-1,4-diene-3,20-dione (**22**) Rimexolone, 11β-Hydroxy-16α,17-dimethyl-17-(1-oxopropyl)-androsta-1,4-dien-3-one (**23**)	Medrysone, 11β-Hydroxy-6α-methyl-pregn-4-ene-3,20-dione (**24**)
Respiratory inhalant	Beclomethasone dipropionate, 9-Chloro-11β-hydroxy-16β-methyl-17,21-bis(1-oxopropanoxy)-pregna-1,4-diene-3,20-dione (**25**) Triamcinolone acetonide (**13**)	Dexamethasone sodium phosphate (**20b**)	Flunisolide, 6α-Fluoro-11β,21-dihydroxy-16α,17-[(1-methylethylidene)bis(oxy)]-pregna-1,4-diene-3,20-dione (**26**)
Intranasal	Beclomethasone dipropionate (**25**) Budesonide, 16α,17-[Butylidinebis(oxy)]-11β,21-dihydroxy-pregna-1,4-diene-3,20-dione (**27**)	Dexamethasone sodium phosphate (**20b**) Triamcinolone acetonide (**13**)	Flunisolide, (**26**) Fluticasone propionate (**16**)
Intrarectal or anorectal	Hydrocortisone (**2**)	Hydrocortisone acetate (**2c**)	
Systemic	Betamethasone (**5a**) Betamethasone sodium phosphate (**5e**) Betamethasone acetate (**5f**) Cortisone (**1**) Dexamethasone (**20a**) Dexamethasone acetate (**20c**) Dexamethasone sodium phosphate (**20b**) Hydrocortisone (**2**)	Hydrocortisone cypionate (**2d**) Hydrocortisone sodium phosphate (**2e**) Hydrocortisone sodium succinate (**2f**) Hydrocortisone acetate (**2c**) Methylprednisolone (**28**) Methylprdnisolone sodium succinate (**28a**) Methylprednisolone acetate (**28b**)	Prednisone (**29**) Prednisolone (**21**) Prednisolone acetate (**21a**) Prednisolone tebutate (**21b**) Prednisolone sodium phosphate (**21c**) Triamcinolone (**13a**) Triamcinolone acetonide (**13**) Triamcinolone diacetate (**13b**) Triamcinolone hexacetonide (**13c**)

Changes in bone metabolism, especially leading to osteoporosis and reduced bone formation (13), are another drawback of glucocorticoid use. It has been found that the rate of bone mineral density reduction is greatest during the initial months of treatment. There is some debate whether daily dose decreases or increases bone loss rates. One study suggests low dose glucocorticoids (predominantly prednisone) lead to increased bone loss (14); minimal or no effect on bone mineral density is noted with low (micro, <10 mg) daily doses (15) in another study. In addition, calcium, vitamin D, and estrogen supplementation all have indicated bone restorative action when implemented in conjunction with glucocorticoid therapy (15). Glucocorticoid-induced osteoporosis is dependent on the individual steroid used; dexamethasone and methylprednisone show more effect than prednisolone or deflazacort on human osteoblast-like cell metabolism (16, 17).

Glucocorticoids are frequently used as immunosuppressants, but this is also an undesired effect when only anti-inflammatory action is required. As with other side effects, the increase in infectious complications is accentuated with large doses and prolonged use (18), whereas reduced effects are noted with low daily (<10 mg prednisone) doses (19).

The psychological actions of glucocorticoids have been well documented, although the cause has been attributed to both direct and indirect effects of steroids on brain functions (20). Dexamethasone downregulates the expresssion of endothelin B receptors in rat brain, but produces an increase in ET-1 and other neuronal mRNA (21, 22). The mechanisms by which glucocorticoids modulate neuronal activity is currently under investigation (23). A general euphoria is commonly associated with initial glucocorticoid use, but prolonged use can lead to mania or depression (24). Termination of glucocorticoid therapies has also led to delirium in patients (25).

Gastrointestinal effects are a common, though perhaps less serious, side effect associated with glucocorticosteroid use.

Long-term corticosteroid therapy is often associated with side effects resembling those of Cushing's syndrome. Carbohydrate and lipid metabolic malfunction lead to conditions in addition to the aforemetioned, including "Moon face," truncal obesity, hyperglycemia, and ketoacidosis. Tables 15.3 and 15.4 present, respectively, generalized adverse effects noted with glucocorticoid use and comparative clinical adverse effects of anti-inflammatory steroids.

4 EFFECTS ON ABSORPTION

4.1 Intestinal Absorption

Schedl and Clifton (27) showed that the absorption of steroids by the perfused rat small intestine is strongly correlated with the polarity of the compound. The least polar compound in the study, progesterone (41), was absorbed almost completely under the condition of the study, whereas the introduction of polar atoms, and particularly hydroxyl groups, reduced the extent of absorption (Table 15.5). Acetylation of the hydroxyl groups invariably increased the extent of absorption. Although absorption correlates with a physical property of the steroid, and therefore appears to be a diffusion-controlled process rather than active transport, the values in Table 15.5 had no relationship to relative oral clinical potencies. Thus, it is likely that the rate of intestinal absorption is not a controlling factor in the activity of the steroids.

A colon-specific drug-delivery system was tested with two steroid prodrugs, dexamethasone 21-β-D-glucoside (20c) and prednisolone 21-β-D-glucoside (21a). Despite high levels of β-D-glucosidase producing *bacteroides* and *bafidobacteria* in the rat upper intestine, these glucosides were remarkably selective for the rat lower intestine. These prodrugs may therefore be useful in humans for the treatment of inflammatory bowel disease. Studies showed that upon oral administration of these drugs to rats, the glucosides reached the cecum in 4 to 5 h, whereupon they were rapidly enzymatically hydrolyzed to the free steroids, dexamethasone (20a) or prednisolone (21) (28). Of these, nearly 60% of an oral dose of (20c) reached the cecum of the rat, whereas only 15% of (21a) was delivered. Interestingly, when either dexamethasone (20a) or prednisolone (21) were administered orally, less than 1% of the steroid was delivered to the cecum.

Table 15.3 Generalized Adverse Effects Noted with Glucocorticoid Use, Especially Prolonged and High Dosage[a]

Adverse Condition	Some Effects	Preventive Action
Adrenocortical insufficiency	Suppression of the pituitary–adrenal axis and inhibition of ACTH production	Mineralocorticoid administration; variation of dose, frequency, and time of drug administration.
Musculoskeletal effects	Muscle wasting, muscle pain and weakness, delayed wound healing, atrophy of protein matrix of bone	High protein diet
Immune system effects	Increased susceptibility to, and masked symptoms of infection, especially with high dose and systemic glucocorticoids	Vaccinations; antiviral, antifungal, and antibiotic therapy
Fluid and electrolyte disturbances	Mineralocorticoid-like action by glucocorticoids; sodium retention. Less frequent with most synthetic glucocorticoids.	Dietary salt restriction, potassium supplements
Ocular effects	Increases in intraocular pressure (IOP), and the development of posterior subcapsular cataracts	Close monitoring of patient
Endocrine effects	Varying effects including hypercorticism, amenoriea, hyperglycemia	Close monitoring of patient
GI effects	Varying effects including nausea, anorexia, increased appetite, and the development and increased severity of ulcers	Antacids and antiulcer agents; oral glucocorticoids taken near meal time
Nervous system and psychological effects	Headache, vertigo, insomnia, increased motor activity, euphoria, mood swings, depression, anxiety	Close monitoring of patient
Dermatologic effects	Skin atrophy and skin thinning from a decrease in colagen synthesis, impaired wound healing, acne, increased perspiration	Close monitoring of patient

[a]Modified from ref. 9.

757

Table 15.4 Comparative Clinical Adverse Effects of Anti-inflammatory Steroids[a]

Effect	Cortisone (1)	Cortisol (2)	Prednisone (29)	Prednisolone (21)	6α-Methyl-prednisolone (28)	Triamcinolone (13a)	Dexamethasone (20a)
Edema	++++	+++	+	++	+	0	+
Weakness, K⁺ depletion	+++	++	+	+	+	++	+
Hypertension	++	+	+	++	+	+	+
Mental stimulation	+++	+	++	++	+	0 to –	++++
Increased appetite and weight gain	++	++	++	++	+	–	++++
Peptic ulcer production	++	+	+++	+++	++	+++	++
Purpura	+	+	+++	+++	++	+++	++
Moon face	+++	++	++	++	++	+++	++
Hirsutism	++	++	++	++	+	++++	+
Skin effects	+	+	+	+	+	++++	++
Osteoporosis	+++	++	+++	+++	+++	+++	++
Diabetes	++	++	+++	++++	+++	++	+
Infection increase	++	++	++	++	++	++	++
Topical effect	+	+++	+	+++	++	+++	++
Adrenal atrophy	+++	+++	+++	+++	+++	+++	+++

[a]Ref. 26.

Table 15.5 Steroid Absorption by Perfused Rat Small Intestine[a]

	(41)	(31)	(33)
Steroid		% Absorbed	Acetate Derivative
Progesterone (41)		93.9	—
Deoxycorticosterone (33)		83.7	—
Corticosterone (31)		46.5	—
Cortisol (2)		21.0	29.9
Triamcinolone (13a)		11.5	—

[a]Ref. 27.

(20c) Dexamethasone 21-glucoside

(21a) Prednisolone 21-glucoside

In an application of Lee's "antedrug" approach, Kimura et al. developed colonic mucosa-specific "pro-antedrugs" for oral treatment of ulcerative colitis (29). The highly polar β-D-glucopyranosyl derivatives (42) and (43) were found to be stable in the small intestine, but the glucoside bonds were cleaved by the action of bacteria in the large intestine to release the antedrug (44). The active corticosteroid ester (44) was found to undergo rapid hydrolysis in the plasma to give the inactive carboxylic acid (45) (see Sections 10.16 and 10.19).

4.2 Percutaneous Absorption

Through the use of ^{14}C-labeled steroids, it has been shown that cortisol (2) and triamcinolone (13a) are absorbed from a topical site of application and that the radioactivity appears in the urine (30). The more potent analogs of cortisol are all effective when applied topically, although large quantities can produce systemic effects (31).

In the 1960s it was found that triamcinolone acetonide (13) was 10 times as active topically as triamcinolone (13a) itself, but only equiactive systemically (32). As pointed out by Popper and Watnick (33), to be topically effective the steroid must penetrate the keratin layer of the stratum corneum of the skin before it can exert its effect on the squamous cell

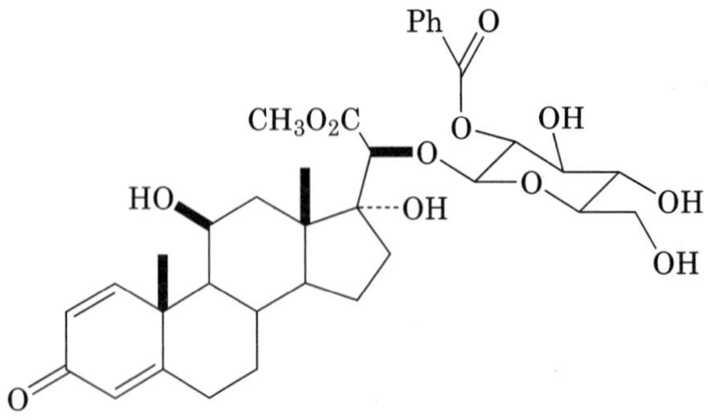

(42) pro-antedrug

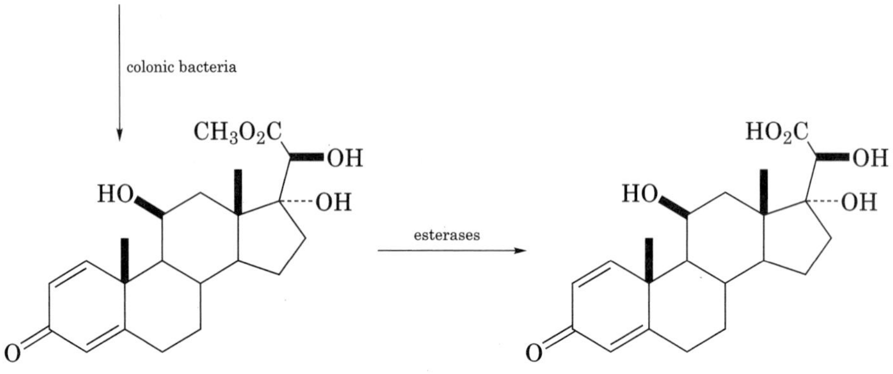

colonic bacteria

(44) active antedrug

esterases

(45) inactive

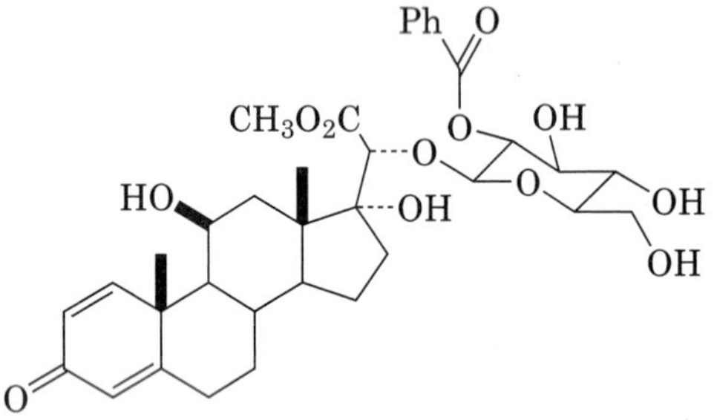

(43)

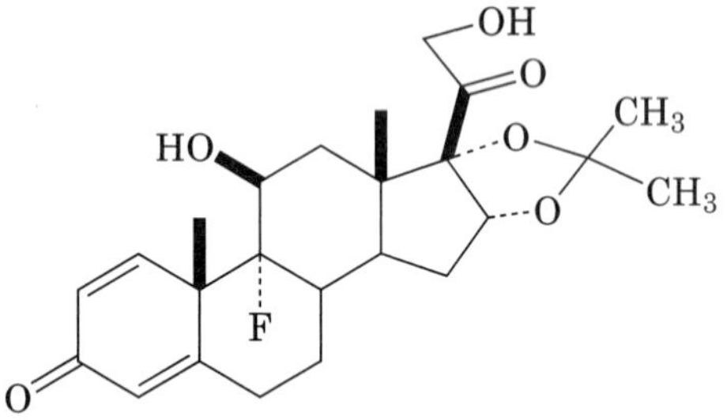

(13) Triamcinolone acetonide

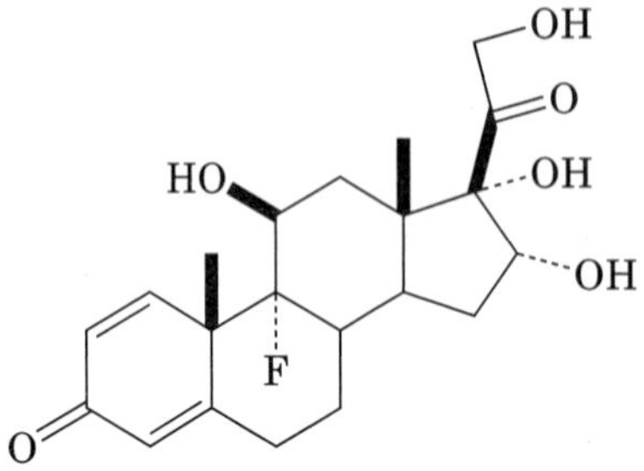

(13a) Triamcinolone

layer of the epidermis. However, to reach the general circulation and produce systemic side effects, the steroid must later penetrate the barrier between the epidermis and the dermis. Thus for a useful topical agent it is desirable for the compound to remain in the epidermis and to migrate only slowly into the dermis.

This property is fostered by the presence of lipophilic groups and by the absence of hydroxy groups in the steroid (4). The conversion of one or two hydroxyl groups in anti-inflammatory steroids to more lipophilic derivatives, such as esters or ketals, is an effective way to produce locally active anti-inflammatory agents, as in the case of triamcinolone already cited.

4.2.1 16α,17α-Ketals. 16α,17α-Ketals are formed by the reaction of 16α,17α-dihydroxy steroids and the appropriate ketones or aldehydes in the presence of an acid catalyst, to give a new pentacyclic ring. Unlike other dioxolanes that are readily hydrolyzed by dilute acid to afford the parent constituents, these steroidal dioxolanes are unusually stable. This led Fried et al. (34) to conclude that the compounds are active *per se*, and not as a result of hydrolysis *in vivo* to the parent compound. As mentioned earlier, triamcinolone acetonide displays a dissociation between topical and systemic activity. The 6α-fluoro derivative, fluocinolone acetonide (**11b**), is also used extensively as a topical agent.

These compounds can be made even more lipophilic by the preparation of 21-esters. Thus, fluocinonide (**11a**), the 21-acetate of

(**11a**) R = Ac
(**11b**) R = H

fluocinolone acetonide, is about five times as active as the latter compound (35). Other ketals that have anti-inflammatory activity are the 17,21-acetonides, but these have not found clinical application.

Although many steroidal dimethylketals derived from acetone (i.e., 16α,17α-acetonides) have been prepared, 16,17-acetals fabricated from aldehydes are less well studied. Thalén

and colleagues examined the influence of a variety of substituted D-ring acetals that had their alkyl group (Table 15.6, R_2) in the β-stereochemical orientation (36). Introduction of fluorine at C-6 and/or 9α positions was examined as was esterification of the 21-alcohol group. Selected findings from this work are shown in Table 15.6.

With the 21-alcohol free, introduction of 9α-fluoro in (**47**) or 9α,6α-difluoro groups in (**48**) does not have a dramatic effect on potency in the rat ear. Although some 50% improvement in potency was observed for (**48**), its therapeutic index dropped from nearly 1 in (**47**) to about 1/4 in (**48**). Thus, the additional F group enhanced systemic activity presumably by curtailing metabolism. Simple acetylation of budenoside led to (**49**), in which potency was doubled and the therapeutic index was substantially improved to about 4.7. Further potency enhancements were gleaned upon F-substitution in the 21-ester series, but acetate esters were better than longer chain esters such as (**51**).

Human potencies were gauged in skin blanching studies for the 21-alcohols, as shown in Table 15.7. An interesting pattern evolves from this work, where irrespective of F-substitution pattern, a peak in activity is observed three times (C-9α = H, C-9α = F, C-6α and C-9α = F) when n = 2.

4.2.2 Esters. The evaluation of a number of 17-esters, 21-esters, and 17,21-diesters of betamethasone (**5**) was carried out by McKenzie and Atkinson (37) (Table 15.8). The least active compounds are the unesterified parent (**5**) and the highly polar monophosphate ester (**5g**). In all three series shown, activity seems to have a parabolic dependency on the π value of the esterifying acid, as would be expected on the basis of the absorption process already discussed. Betamethasone 17-valerate (**5d**) is a widely used topical anti-inflammatory agent and the related 17-benzoate (**5e**) has also been used clinically. Betamethasone 17,21-dipropionate (**5L**) is in current clinical use as a topical anti-inflammatory steroid. Even the weakly active 21-deoxyprednisolone (**54**), which has 25% of the systemic activity of cortisol (38),

Table 15.6   Relative Activities of 16,17-Acetal Derivatives Related to Budenoside, Triamcinolone, or Fluocinolone 16,17-Acetonides in the Rat Ear Edema Assay[a]

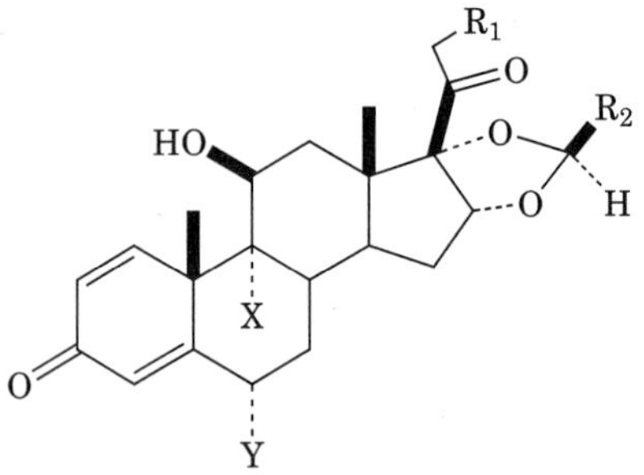

(46–52)

Compound	R_1	R_2	X	Y	CEE[b]	T/S[c]
(46) (BD)[d]	HO	Propyl	H	H	1	1/1
(47)	HO	Propyl	F	H	1	2.2/2.4
(48)	HO	Propyl	F	F	1.5	2.5/10.3
(49)	CH_3COO	Propyl	H	H	2.3	2.8/0.6
(50)	CH_3COO	Propyl	F	H	4.3	—
(51)	$CH_3(CH_2)_3COO$	Propyl	F	H	2.7	—
(52)	CH_3COO	Propyl	F	F	4.1	—
TA[e]					1	0.3/2.3
FA[f]					1.7	0.7/10.3

[a]Ref. 36.
[b]Croton oil ear edema assay in rats.
[c]Topical vs. systemic activity in the ear edema assay.
[d]Budesonide 16, 17-acetonide.
[e]Triamcinolone 16,17-acetonide.
[f]Fluocinolone 16,17-acetonide.

gives a clinically useful compound upon conversion to the 17-propionate (33).

A series of 17α-thiomethyl and 17α-methoxyacetates (55; X = SMe, OMe) were found to possess activity similar to that of betametha-

(55)

sone valerate in the McKenzie assay in human volunteers. These compounds did have 6α-methyl groups that would be expected to enhance activity somewhat (see section on C-6 substitution) (39).

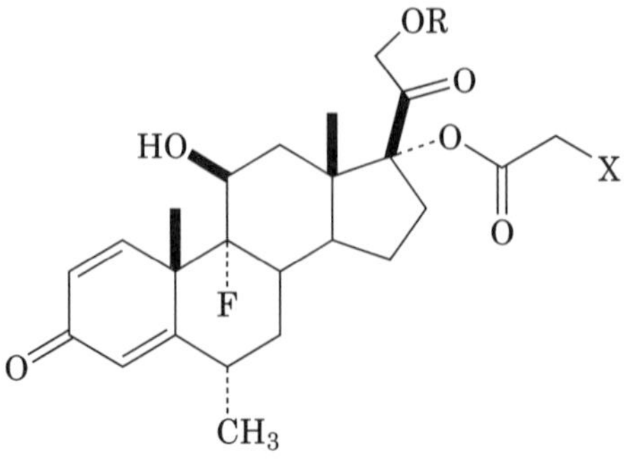

Ueno et al. (40) showed that a variety of X-substituted 17α-esters have excellent separation of topical and systemic activities, as shown in Table 15.9. Where $n = 2$ or 3 and X = polar group such as OMe or OAc, therapeutic indices were far better than controls such as clobetasol propionate or betamethasone dipropionate, whereas anti-inflammatory potency was somewhat less than control. The least systemically absorbed, or systemically active, in this table, was the acetate (60), with a TI of over 18. Although (60) was only a third as active as CP, on the other hand, it was three times more active than BMDP.

Clearly, more complicated 17α-esters offer

Table 15.7 Topical Anti-inflammatory Activity in Blanching of Human Skin, Expressed as EC_{50} and Relative Potency in Comparison with Fluocinolone Acetonide (FA)[a]

Compound	Number of CH_2 Groups (n)	X	Y	EC_{50}[b] ($\mu g/mL$)	Relative Potency Compared with FA[c]
(52a)	1	H	H	2.8 (1.6–5.3)	1.1
(46)	2	H	H	1.1 (0.7–1.7)	1.9
(52b)	4	H	H	3.4 (1.8–8.9)	0.4
(52c)	6	H	H	> 10	
(52d)	0	F	H	3.5 (2.0–6.3)	0.6
(52e)	1	F	H	0.8 (0.5–1.3)	4.0
(47)	2	F	H	0.6 (0.4–0.8)	4.3
(52f)	3	F	H	1.1 (0.7–1.9)	1.9
(52g)	4	F	H	3.1 (1.8–5.6)	0.7
(52h)	8	F	H	> 10	
(52i)	0	F	F	1.7 (1.1–2.8)[d]	0.9
(48)	2	F	F	0.4 (0.2–0.6)	4.0
(52j)	4	F	F	1.5 (0.8–3.1)	1.1
(52k)	6	F	F	4.6 (2.5–12.8)	0.3

[a]Five doses of each compound were tested with 10 parallels/dose.
[b]The EC_{50} values are followed by 95% confidence limits in parentheses.
[c]Compared with the reference compound (FA) within the same test. The $EC_{50} \pm$ SEM value for FA in the five tests was 2.3 $\pm$ 0.4 $\mu g/mL$.
[d]This value was obtained with approximately 100% of the most polar C-22 epimer.

means of separating topical from systemic effects. It might be argued that metabolism of (**60**) renders the analog less active, but this has not been studied. It was suggested that improved metabolic stability might account for the loss in TI for the cyclopropyl esters (**63**) and (**64**).

Given that 17α-benzoate esters were possessed of excellent topical efficacy, additional related but heteroaromatic compounds have been examined. Two classes of compounds were targeted having furoyl or thienoyl esters at the 17-position, and were further modified by oxygenation or halogenation at 6,9,11 and 21 and have the general structure (**65**):

(65)

Table 15.8 Relative Potencies of Betamethasone and Some of Its Esters in Vasoconstriction Assays against Fluocinolone Acetonide[a]

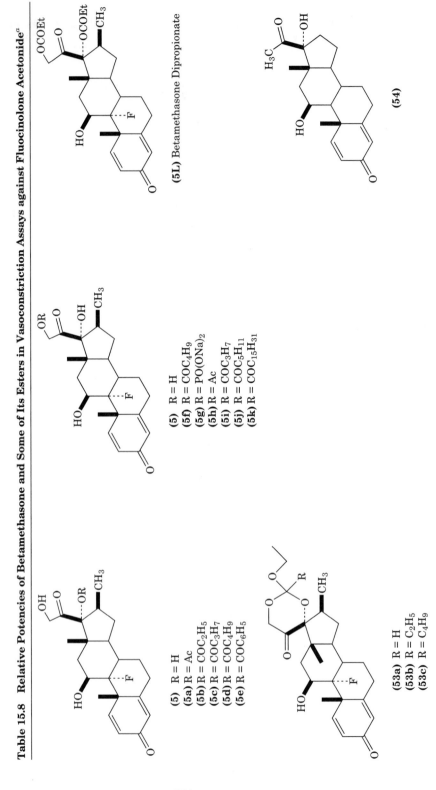

(5)   R = H
(5a) R = Ac
(5b) R = COC$_2$H$_5$
(5c) R = COC$_3$H$_7$
(5d) R = COC$_4$H$_9$
(5e) R = COC$_6$H$_5$

(5)   R = H
(5f) R = COC$_4$H$_9$
(5g) R = PO(ONa)$_2$
(5h) R = Ac
(5i) R = COC$_3$H$_7$
(5j) R = COC$_5$H$_{11}$
(5k) R = COC$_{15}$H$_{31}$

(5L) Betamethasone Dipropionate

(53a) R = H
(53b) R = C$_2$H$_5$
(53c) R = C$_4$H$_9$

(54)

764

(5)	Betamethasone alcohol	0.8
(5g)	Betamethasone 21-disodium phosphate	0.9
(5h)	Betamethasone 21-acetate	18
(5i)	Betamethasone 21-butyrate	85
(5f)	Betamethasone 21-valerate	26
(5j)	Betamethasone 21-hexanoate	123
(5k)	Betamethasone 21-palmitate	0.1
(5a)	Betamethasone 17-acetate	114
(5c)	Betamethasone 17-butyrate	168
(5d)	Betamethasone 17-valerate	360
(53a)	Betamethasone 17,21-ethylorthoformate	1
(53b)	Betamethasone 17,21-ethylorthopropionate	402
(53c)	Betamethasone 17,21-ethylorthovalerate	150

[a]Ref. 37.
[b]Fluocinolone acetonide (11b) = 100.

Table 15.9   Activity of 17α-Esters of 21-Desoxy-21-Chlorobetamethasone

(56–64)

Compound	X	n	EEA[b]	Thymus[c]	TI[d]
(56)	CH_3	1	0.11	—	—
(57)	CH_3	2	0.38	0.05	7.6
(58)	Cl	1	0.25	—	—
(59)	Cl	3	0.66	0.12	5.5
(60)	OAc	2	0.28	0.015	18.6
(61)	SCH_3	1	0.35	0.35	1
(62)	CN	2	0.07	—	—
(63)	c-C_3H_5	0	0.44	1.3	0.34
(64)	c-C_3H_5	1	0.51	1.1	0.46
CP[e]			1	1	1
BMDP[f]			0.1	0.03	3.3

[a]Ref. 40.
[b]Croton oil ear edema assay.
[c]Thymic involution in rat granuloma pouch assay.
[d]Therapeutic index determined by EEA/Thymus.
[e]Clobetasol propionate.
[f]Betamethasone dipropionate.

In one series where X and Y = Cl, the aromatic group was varied from 2-furyl or thienyl to 3-furyl or thienyl while also varying the 21-position, as shown in Table 15.10 (41). As can be seen, not much difference in topical efficacy is evident on changing from an O heteroatom (furanyl) to a S heteroatom (thienyl) (i.e., **66** versus **68**). Similarly, the position of substitution (2- vs. 3-position) on the aromatic ring is not highly critical to potency (i.e., **66** versus **67**). On modifying the 21-oxygen atom from alcohol (**70**) to ester, an expected enhancement is evident but continued increases in bulk at 21 (e.g., **72**, **73**, **74**) were without significant effect. Upon replacement of the 21 oxygen by Cl a sizable potency enhancement was seen (**75**). Not only was the response rapid, it appeared to be cumulative, given that the potency after 5 days had risen to over eightfold better than that of the control, betamethasone valerate. Finally, unlike the 21-ol series, chlo-

rination at 21 resulted in a sensitivity to the position of attachment to the heterocyclic ring. Thus, 2-furyl was significantly better than 3-furyl (**75** was much more potent than **76**). Halogenation at the 6α-position of the better compound in this series (**75**), providing (**77**), was not beneficial.

In a closely related series with an 11β-alcohol instead of a halogen, the same investigators studied the effect of ester substitution at the 17-position (42). As illustrated in Table 15.11, the effect of heteroatom (S or O) or heteroatom position (2- or 3-furyl) in this 11β-alcohol series is generally similar to the dichlorisone-like series in Table 15.7 and 21-halogenation had a likewise positive effect on potency.

These sets of compounds illustrate the complex relationships that govern topical potency upon esterification at C-17. Although quite reasonable structure-activity relation-

Table 15.10 9,11-Dichloro Corticosteroid 17α-Heterocyclic Esters[a]

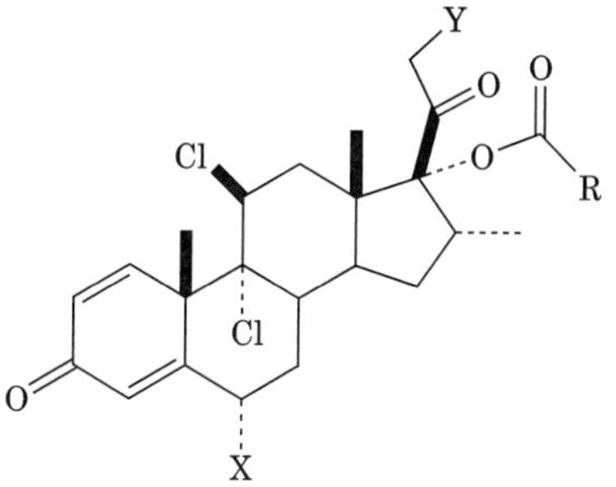

(66–77)

Compound	X	Y	R	EEA[b]
(66)	H	OAc	2-furyl	1.4 (1.2)
(67)	H	OAc	3-furyl	1.3 (1.4)
(68)	H	OAc	2-thienyl	1.3 (0.8)
(69)	H	OAc	3-thienyl	0.9 (3.2)
(70)	H	OH	2-furyl	0.8
(71)	H	OH	3-furyl	0.7
(72)	H	OCOEt	2-furyl	1.1 (0.9)
(73)	H	OCOPr	2-furyl	1
(74)	H	OCOCH$_2$OCH$_3$	2-furyl	1 (3)
(75)	H	Cl	2-furyl	1.9 (8.2)
(76)	H	Cl	3-furyl	1 (1.7)
(77)	F	Cl	2-furyl	2.5 (3)
BV[c]				1 (1)

[a]Ref. 41.
[b]Croton oil ear edema assay, reported at 5 h and (5 days).
[c]Betamethasone 17-valerate as control.

ships arise out of homologous esters where the steroid nucleus is held constant (Table 15.8), those same relationships do not always translate simplistically into systems where multiple changes to the steroid nucleus have taken place.

From the compounds in Tables 15.10 and 15.11, a clinically useful compound was marketed as Elocon or mometasone furoate (**17**).

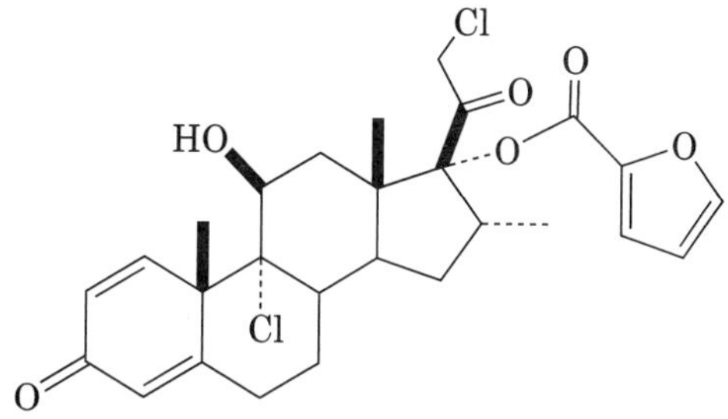

(**17**) Mometasone Furoate

4.3 Intra-Articular Administration

Arthritic and inflammatory conditions afflicting skeletal joints may be treated by intra-articular injections, in which a hypodermic needle is passed through the synovial membrane, and a dose of corticosteroid is injected into the joint cavity. The specific local anti-inflammatory effect lasts from 5 to 21 days or longer. Water-soluble esters of anti-inflammatory steroids are rapidly cleared from the joint and have a very short duration of action under these circumstances. Figure 15.3 schematically represents the metabolism of corticosteroids. Conversely, aqueous microcrystalline suspensions of ketals and esters of anti-inflammatory steroids dissolve suffficient quantities to permit local action in the joint but do not enter the systemic circulation in amounts

Table 15.11   9α-Halo-11β-Hydroxy Corticosteroid 17α-Heterocyclic Esters[a]

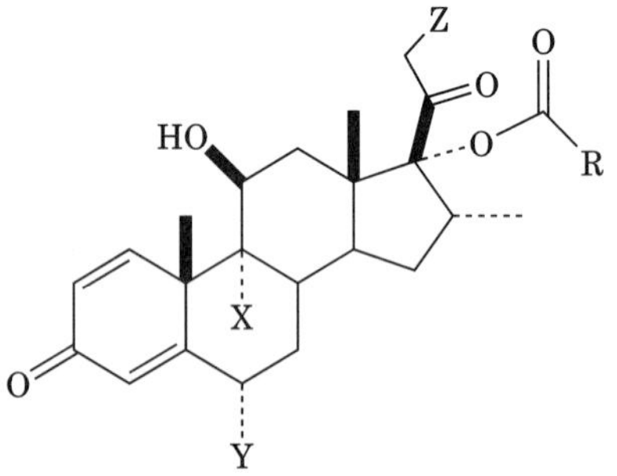

(17), (78–88)

Compound	X	Y	Z	R	EEA[b]
(78)	F	H	OAc	2-furyl	1.7 (2.2)
(79)	F	H	OAc	3-furyl	2.9 (3.2)
(80)	F	H	OCOEt	2-furyl	2.1 (6.8)
(81)	F	H	OCOPr	2-furyl	1.1
(82)	F	H	OCOCH$_2$OCH$_3$	2-furyl	0.8 (3.2)
(83)	F	H	Cl	2-furyl	1.6 (2.3)
(84)	F	H	Cl	3-furyl	1 (1.2)
(85)	Cl	H	OAc	2-furyl	1.7 (4.4)
(17)	Cl	H	Cl	2-furyl	1 (6.1)
(86)	Cl	H	Cl	2-thienyl	1.2 (4.5)
(87)	Cl	H	F	2-furyl	1.4 (4.4)
(88)	Cl	F	Cl	2-furyl	1.1 (5.1)
BV[c]					1 (1)

[a]Ref. 42.
[b]Croton oil ear edema assay at 5 h and (5 days).
[c]Betamethasone valerate, control.

needed to exert substantial systemic effect. In addition to triamcinolone acetonide, branched esters that hydrolyze slowly, such as triamcinolone hexacetonide (13b) (43) and dexamethasone 21-trimethylacetate (20c) (44), are useful for this purpose. 6α-Methylprednisolone acetate (45) and betamethasone 17,21-dipropionate (46) are also active long enough to be useful. Figure 15.4 shows a schematic representation of the formation of C-21 carboxylic acid metabolites of cortisol.

Triamcinolone hexacetonide (13b) has been compared to rimexolone (23) for intraarticular treatment of arthritis (47). Rimexolone was of interest in this regard for its clinical availability and reported lack of systemic effects. In mice, a single injection of 450 μg of (23) was sufficient to provide a prolonged anti-inflammatory response, but to generate a similar anti-inflammatory response required only

25 μg of (13b). The advantage of rimexolone (23) over (13b) was its prolonged retention (21 days).

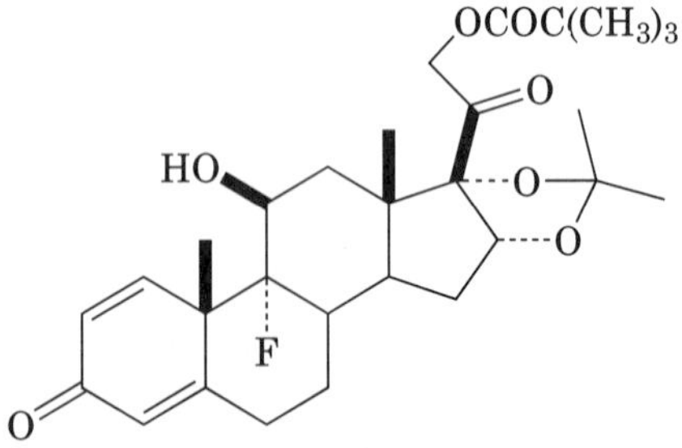

(13b)

4.4 Water-Soluble Esters

The incorporation of ionic groups into anti-inflammatory steroids at the 21 position re-

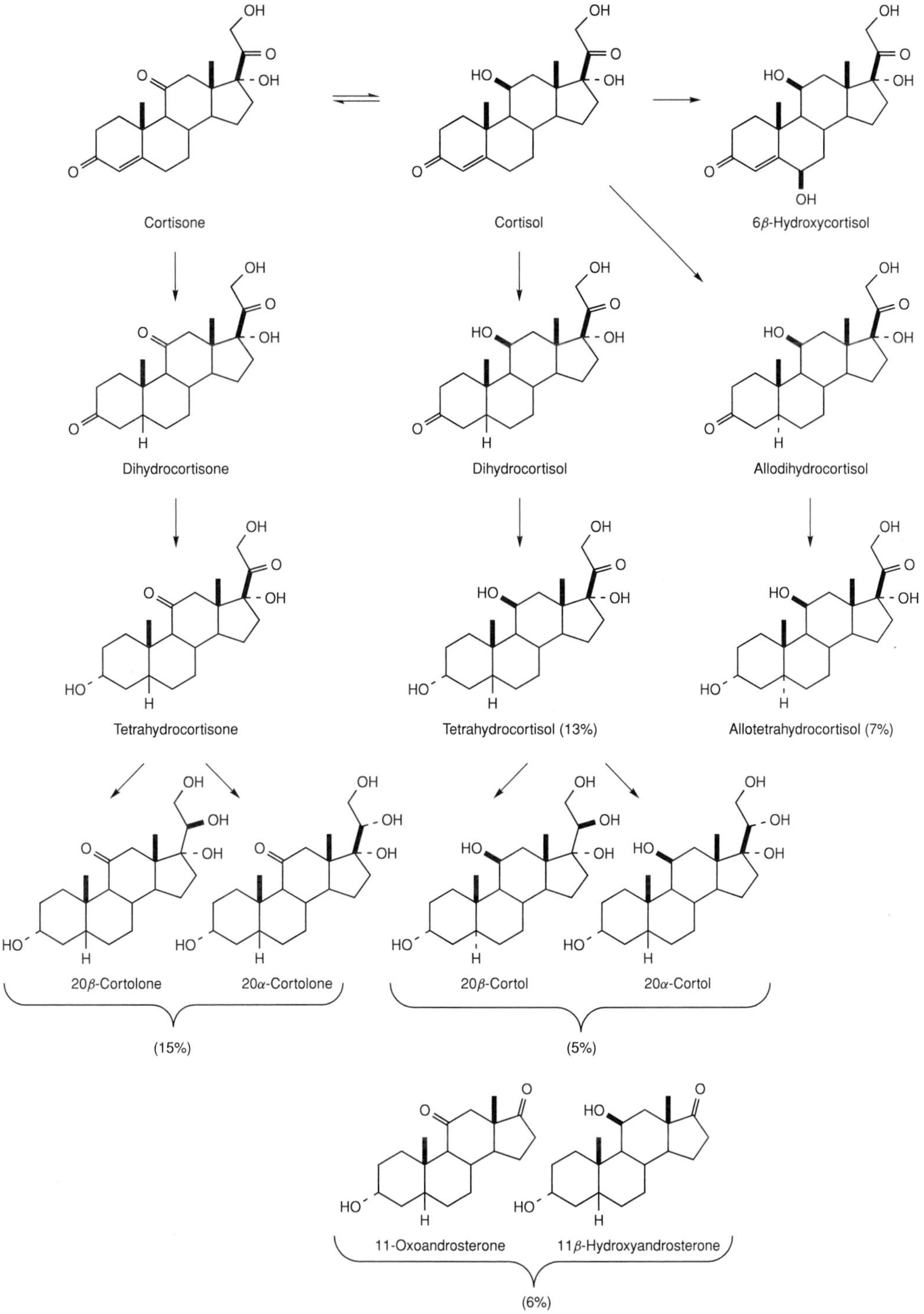

Figure 15.3. Metabolism of corticosteroids (figures in parentheses denote percentage of administered dose of cortisol recovered as urinary metabolites).

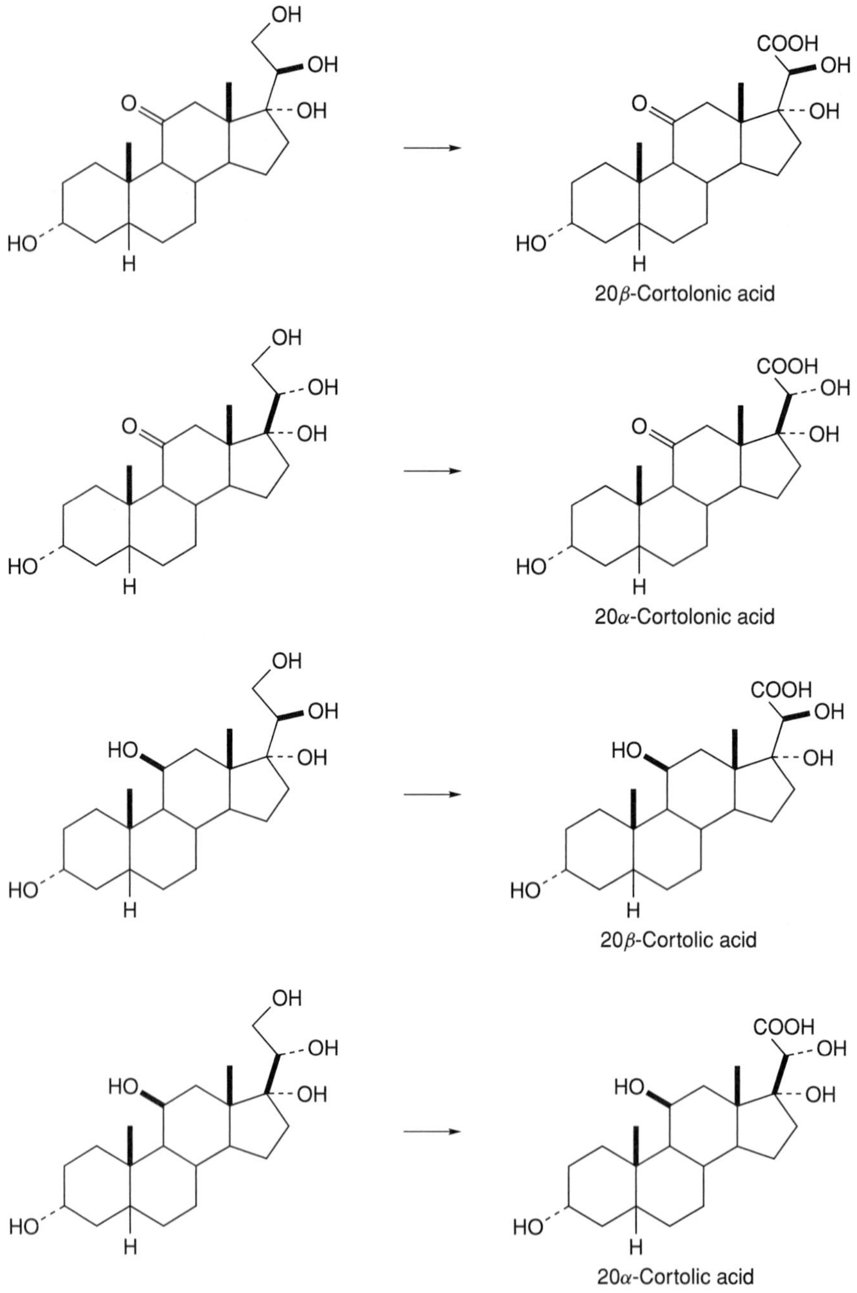

Figure 15.4. Formation of C-21 carboxylic acid metabolites of cortisol.

sults in water-soluble steroids that have rapid onset of action but are of fleeting duration. The esters employed include the 21-phosphate disodium salts, such as betamethasone sodium phosphate (**5g**; e.g., Celestone Phos- phate, Selestoject) and prednisolone sodium phosphate (e.g., Pedi-pred), or 21-hemisucci- nate sodium salts such as hydrocortisone so- dium succinate (**2f**; e.g., Solu-Cortef) and methylprednisolone sodium succinate (e.g.,

(20c)

Solu-Medrol). There is extensive evidence indicating that such esters rapidly hydrolyze to the active 21-alcohols (48) and therefore represent only solubilizing groups. Soluble compounds such as prednisolone sodium phosphate or dexamethasone sodium phosphate can be used for ophthalmic purposes as well as for intravenous administration. Moreover, they may be incorporated in long-acting depot preparations to add a rapid-acting component. Along these lines, Celestone Soluspan, a mixture of betamethasone sodium phosphate and betamethasone acetate, is useful in providing relief from, for example, bursitis, arthritis, and dermatological conditions.

4.5 Corneal Penetration

The availability of ocular drugs, administered in solution as drops, is very low because of rapid clearance of the solution from the precorneal area. Highly water soluble drugs are also not well absorbed from the cornea (49) but decadron phosphate, a 21-sodium phosphate ester of dexamethasone (**20b**), finds clinical utility when formulated as an ophthalmic ointment in mineral oil or even as a sterile aqueous solution (50). Schoenwald and Ward (51) determined the permeability rates across excised rabbit cornea for 11-steroids. They derived a parabolic relationship between log permeability and the octanol/water partition coefficient with an optimum at log P_0 of 2.9 (close to that of dexamethasone acetate). Steroid combinations containing prednisolone acetate (0.2% in oil base) have found ophthalmic application in humans.

Corneal permeability of dexamethasone and selected esters incorporated into liposomes or as solutions in Tween 80 was exam-

ined (52, 53) and it was concluded that liposomal preparations were not useful, although Tween 80 accelerated absorption.

5 EFFECTS ON DRUG DISTRIBUTION

Only about 5% of cortisol in plasma is circulated in the unbound state, whereas the remainder is bound to two major moieties in serum: serum albumin and corticosteroid binding globulin (CBG) (54). Serum albumin is present in higher concentration (5500 × $10^{-7} M$ in human serum), whereas the concentration of CBG in human serum is $8 \times 10^{-7} M$. However, the equilibrium association constant (K_{aff}) for cortisol-CBG is $3 \times 10^{-7} M^{-1}$, about 10^3 greater than that for cortisol serum albumin. There is good evidence to support the notion that the steroids are complexed to these proteins primarily by hydrophobic bonds (55, 56). This implies that the steroid interacts primarily with nonpolar portions of the protein. The affinity of steroids to serum albumin decreases with an increasing number of hydrophilic or polar groups in the molecule (57). The hydrophobic effect depends on the displacement of ordered water from two surfaces, and it is clear that the presence of a hydrophilic group on one of the surfaces will interfere with this process. The type of substituent group is obviously important because, on this basis, a hydroxyl group should interfere more with binding than would a more lipophilic fluorine congener. The stereochemistry of the derivative is also significant; 11β-hydroxyprogesterone (**89**) has higher affinity for serum albumin than does the 11α epimer. This has led to the suggestion (58) that the steroids are bound to albumin on their α face.

Binding of steroids to human and rat CBG is also diminished by the presence of polar groups (59). Interestingly, the order of binding for rabbit CBG is cortisol (**2**) > corticosterone (**31**) > progesterone (**41**) and thus opposite to the order one would predict from the polarity of substituents (59). This is evidence for structural specificity in steroid-CBG binding, and it suggests the presence of a "binding site" on the protein, with specific structural requirements for optimum binding. As discussed in section 9, CBG binding affinity is highly correlated to the 3D structure in corticosteroids.

Table 15.12   Comparison of Relative Potencies for Some Systemically Used Glucocorticosteroids^a

Compound	RRBA	CL (1/h)	f_u	F	PK/PD Potency^b	Clinical Potency^c
Cortisol	9	18	0.20	0.96	1	1
Prednisolone	16	10	0.25	0.81	3.4	4
Methylprednisolone	42	21	0.23	0.99	4.7	5
Betamethasone	58	9	0.36	0.72	17.4	25
Fluocortolone	82	30	0.10	0.84	2.4	5
Dexamethasone	100	17	0.32	0.83	16.3	25
Triamcinolone acetonide	233	37	0.29	0.23	4.4	6

^{a}RRBA, relative receptor binding affinity; CL, total body clearance; Fu, fraction unbound in serum; F, oral bioavailability.
^{b}PK/PD potency calculated based on equation (1) and standardized to cortisol = 1.
^{c}Clinical potency based on empiric therapeutic equivalence.

This concept of a specific binding site is also in harmony with the reduction of binding to CBG on introduction of numerous substituents into the cortisol structure, irrespective of the polarity of the group (60). A study (61) that illustrates this difference compared the binding to CBG of cortisol, dexamethasone (**20a**), and its 16β-methyl epimer (betamethasone, **5**). It was found that all the steroids are bound extensively to serum albumin, whereas only cortisol is bound to CBG.

The effect of protein binding on biologic activity is not simple to assess. On the one hand, CBG-bound cortisol is biologically inactive (62), suggesting decreased protein binding would lead to enhanced activity. On the other hand, it has been shown (63) that CBG protects cortisol against metabolic degradation by rat liver homogenate. Because the two phenomena tend to cancel each other, it is impossible to state *a priori* the magnitude of changes in pharmacological activity arising from effects on plasma binding.

Despite these complications, in a clinical setting, it has been possible to carry out

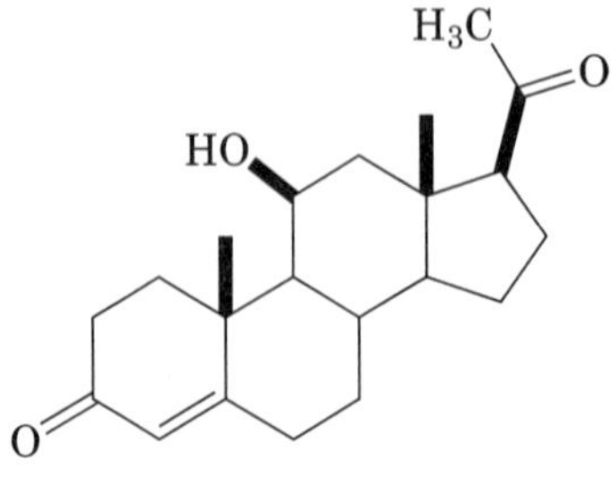

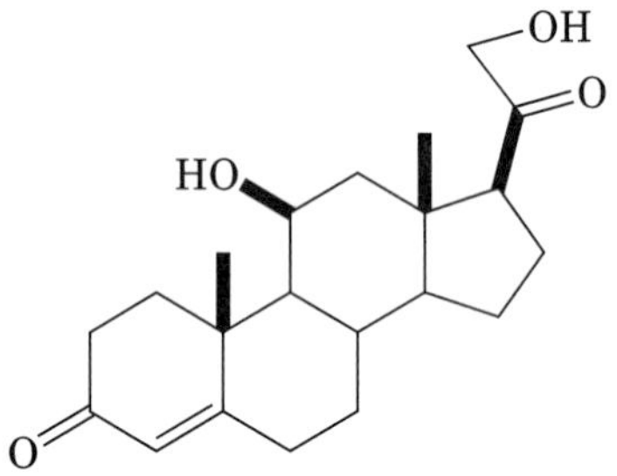

(31)

PK/PD modeling of corticosteroids. A suitable indirect-response PK/PD model was developed that allowed description of the receptor-mediated drug effects such as endogenous cortisol suppression as a function of time. Furthermore, the model allowed for prediction of the systemic activity of newly developed corticosteroids on the basis of their pharmacokinetics and their respective receptor-binding affinity. The model could also be applied to systemic steroid effects after topical administration, or to investigate the effect of the time of dosing on cortisol suppression. Comparison of predictions based on this model and results from large clinical studies were in excellent agreement (Table 15.12). Corticosteroids may represent an ideal class of drugs for the successful use of PK/PD modeling during drug development (64).

Measured EC_{50} values showed outstanding correlation with receptor-binding affinities and allow for the evaluation of new corticosteroids in early development. On the basis of EC_{50}, the fraction of unbound steroid (protein

(89)

bound vs. free) or f_u, clearance (CL), and oral bioavailability (F), a parameter for comparison was found, DR_{50}, represented by the following relationship:

$$DR_{50} = \frac{EC_{50} \times CL}{F \times f_u}$$

DR_{50} is the dosing rate that produces and maintains 50% of the maximum effect.

Evaluation of topical corticosteroid effects through the use of PK/PD modeling is more challenging, given that markers for acute topical activity are limited. In the case of corticosteroid inhalation therapy, PK/PD modeling identified the importance of prolonged pulmonary residence time as an important parameter to improve pulmonary targeting.

6 GLUCOCORTICOID BIOSYNTHESIS AND METABOLISM

Glucocorticoids are biosynthesized from cholesterol and released as needed by the adrenal cortex; they are not stored. Cholesterol undergoes a series of irreversible oxidations during which carbons 22 through 27 are cleaved, resulting in pregnenolone. Reversible isomerization of Δ^5 to Δ^4 (progesterone) followed by 11β-, 17α-, and 21-oxidations (flavoprotein, cytochrome P450-mediated) results in cortisol. The major metabolic transformations of the adrenal cortical hormones generally follow the metabolism of cortisol, which undergoes the following conversions *in vitro* (Fig. 15.5).

Cortisol-Cortisone Conversion. Under normal conditions, this equilibrium slightly favors the oxidized compound. Similarly, the conversion of corticosterone to 11-deoxycorticosterone is also mediated by the 11β-hydroxysteroid dehydrogenase enzyme system and requires $NAD(P)^+/NAD(P)H$. This conversion is especially important both in the protection of the human fetus from excessive glucocorticoid exposure and in the protection of distal nephron mineralocorticoid receptors from glucocorticoid exposure (65). The impairment of this conversion is thought to result in hypertension associated with renal insufficiency (66).

A-Ring Reduction (Double Bonds and C-3 Carbonyl). This is an irreversible reaction that is a foremost determinant of the secretion rate of cortisol. Catalyzed predominantly by cortisone β-reductase and 3α-hydroxysteroid dehydrogenases, 5β sterols result, although 5α sterols are more prevalent with other glucocorticoids. Urocortisol and urocortisone result from the metabolism of cortisol and cortisone, respectively. Compounds can be complexed to glucuronic acid at this point.

C-20 Reduction. Two stereoisomers can result from this transformation, although cortisol is thought to act primarily with $(R)20\beta$-hydroxysteroid dehydrogenase. This is a first step in the metabolism of corticosterone.

Cleavage of C-17 Acyl/Alkyl Substituents. Resulting primarily in cholan-17-ones, this is a relatively minor metabolic pathway. Corticosterone is not known to undergo this tranformation before excretion.

C-6 Hydroxylation. This biotranformation is more predominant in infants than adults, and can prevent other metabolic transformations.

Glucuronidation. Complexation of the steroid to glucuronic acid, most predominantly through the C-3 hydroxyl, leads to a considerable portion of the excreted metabolites of all glucocorticoids. In infants, sulfurylation (formation of a sulfate ester) is also predominant (67).

Other Reactions. Most of the metabolites of cortisol are neutral (alcohol or glucuronide complex) compounds. However, oxidation at C-21 to C-21 carboxylic acids (68) accounts for some of the identifiable metabolites of glucocorticoids (69).

Compounds with the 16,17-ketal (e.g., budesonide, amcinonide, fluocinonide, halcinonide, triamcinolone acetonide, and flurandrenolide) also undergo metabolism by routes that parallel that of cortisol metabolism. Unsymmetrical acetals such as budesonide (Fig. 15.6) are also metabolized by routes not available to the more metabolically stable symmetrical $16\alpha,17\alpha$-isopropylidene-dioxy substituted compounds (desonide, flunisolide, triamcinolone acetonide). Isozymes within the cytochrome P450 3A subfamily are thought to catalyze the metabolism of budesonide, resulting in formation of 16α-hydroxyprednisolone

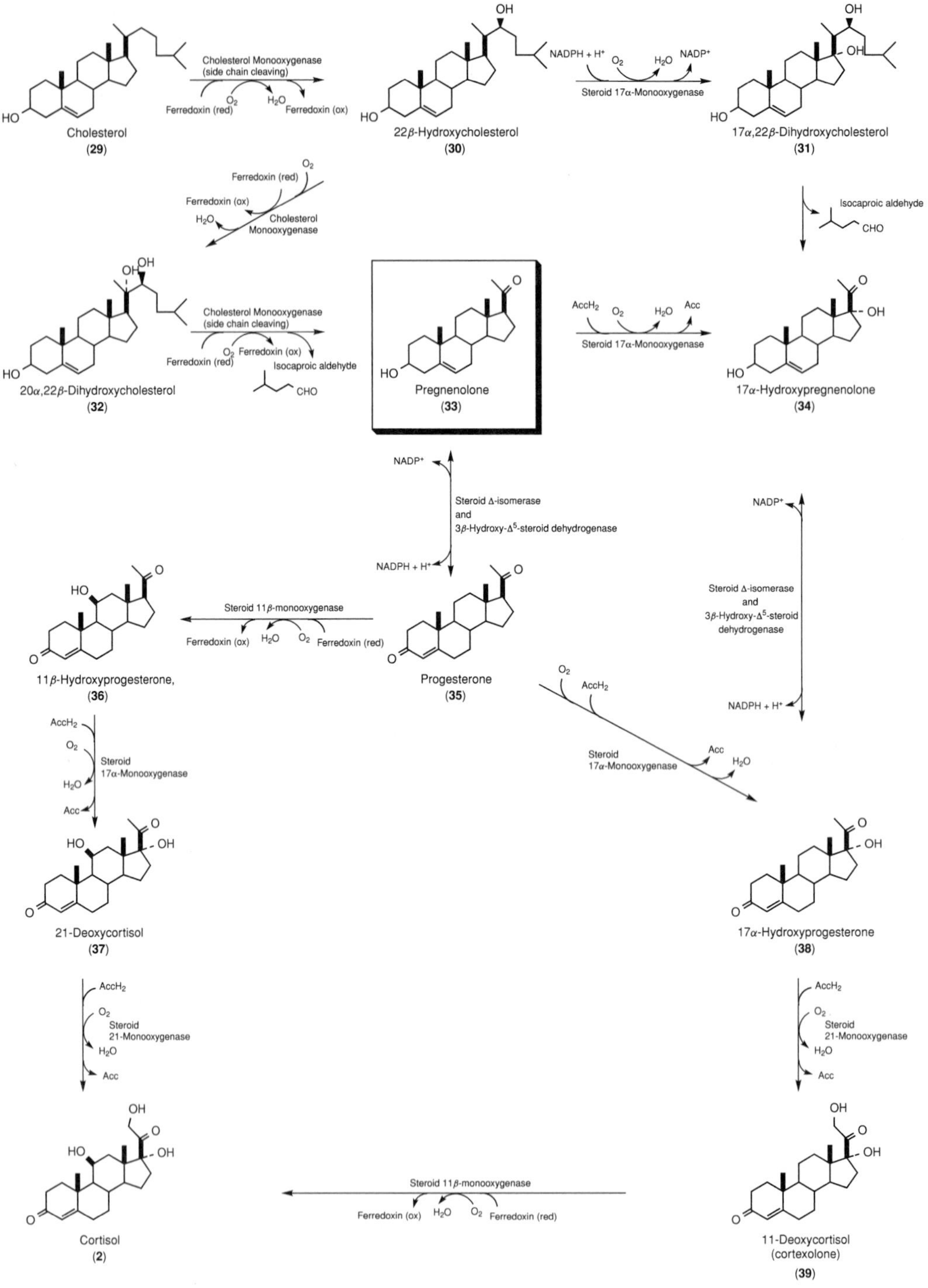

Figure 15.5. Biosynthetic pathways for formation of cortisol from cholesterol.

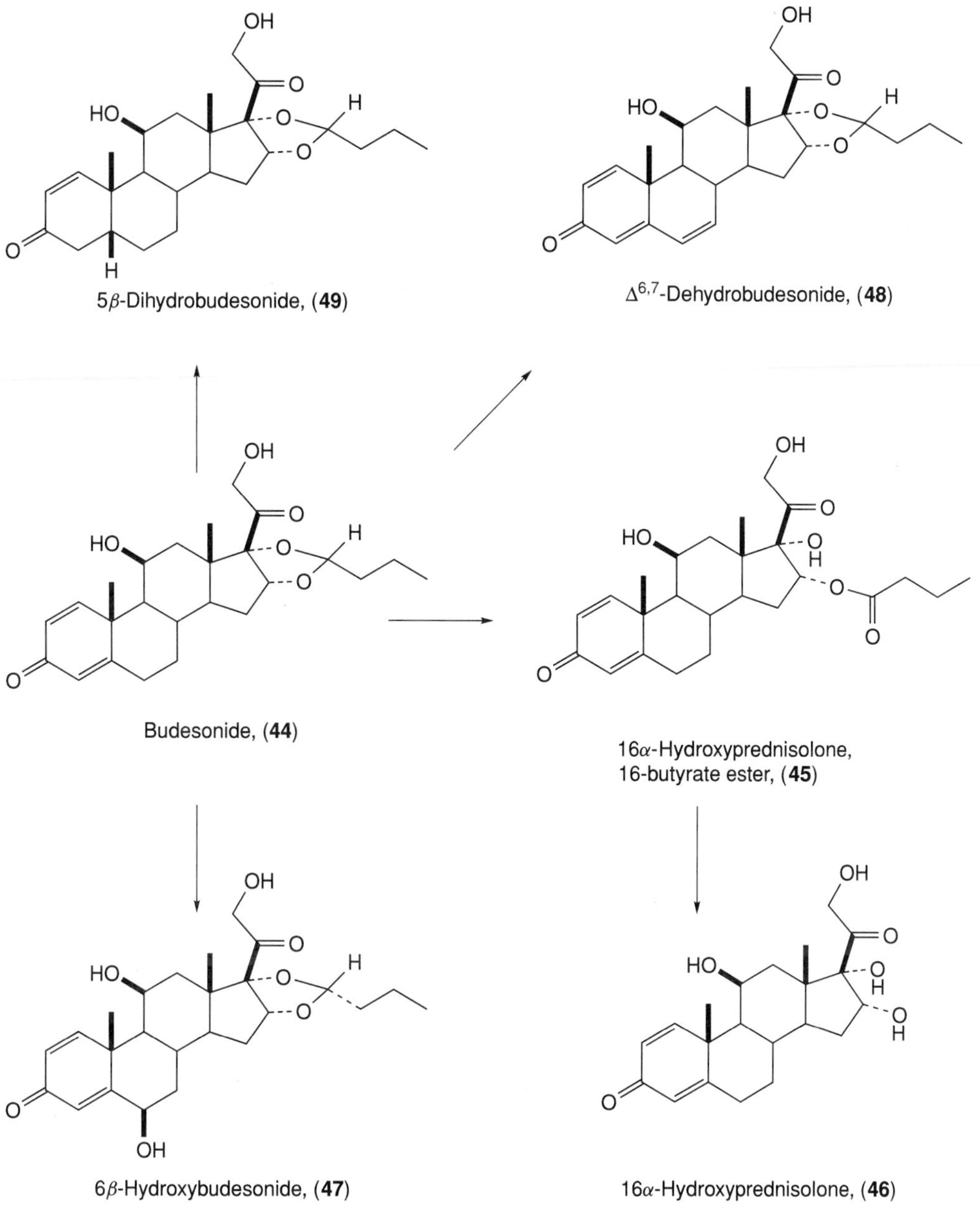

5β-Dihydrobudesonide, (**49**)

Δ^{6,7}-Dehydrobudesonide, (**48**)

Budesonide, (**44**)

16α-Hydroxyprednisolone,
16-butyrate ester, (**45**)

6β-Hydroxybudesonide, (**47**)

16α-Hydroxyprednisolone, (**46**)

Figure 15.6. Metabolism of 22R-budesonide in human liver microsomes.

and 6β-hydroxybudesonide (70, 71) (Fig. 15.6), in addition to the more common metabolic steps (oxidation through Δ^6, reduction of Δ^3, etc.).

Steroids with the greatest number of substituents generally have the slowest rate of metabolism. Groups that appear to activate by

effects on metabolism include 2β-methyl, which stabilizes the resulting molecule to the action of the 4,5-reductase (72) and to the action of 20-keto reductases (73). Similarly, 6α-methyl protects the A-ring against metabolic destruction (73). Again, the introduction of 16α-hydroxyl, as in triamcinolone (**13a**), pro-

Table 15.13   Half-Lives of Corticosteroids in Dog Plasma

(90)

Steroid	Half-Life (min)	Reference
Cortisol (**2**)	44–52	75
		76
		77
		78
		79
Prednisone (**29**)	38	78
9α-Fluorocortisol (**3**)	58	78
Dexamethasone (**20a**)	60	79
Prednisolone (**21**)	60–71	75
		76
		77
		78
		79
6α-Methylprednisolone (**28**)	81	76
6α-Methyl-9α-fluoro prednisolone (**90**)	110	76
Triamcinolone (**13a**)	116	76
		80

longs the half-life (74). The 16α-methyl and 16β-methyl groups have similar action. Triamcinolone is unusual in that it is metabolized principally to the 6β-hydroxy derivative (74). Table 15.13 presents data on the half-lives of corticosteroids in dog plasma (75–80).

7   MECHANISM OF ACTION OF ANTI-INFLAMMATORY STEROIDS

7.1   Glucocorticoid Receptor Structure

The effects of glucocorticoids are thought to be a consequence of their interaction with an intracellular receptor, and great strides have been taken in the task of determining the structure and function of the glucocorticoid receptor (GR). Two isoforms of the GR have been identified, with an "A" form (GR-A), indicating a slightly higher molecular mass (94 kDa) than that of the "B" form (81, 82). The functionally active GR has been purified (83) and, although an X-ray structure is not available, a significant amount of 3D structural information on the receptor has been gathered. The GR is a member of the nuclear receptor superfamily, which includes receptors for steroids, vitamin D, and thyroid hormones, and some other proteins (84). A high degree of homology is found within receptors in this class, each containing a similar domain organization. These domains (sections of primary structure) have a functional duty, and generally describe regions of hormone binding, nuclear translocation, dimerization, DNA binding, and transactivation (85).

Within the carboxyl-terminus portion of the protein (residues 518–795) is the ligand (hormone) binding domain (HBD) (85). It is here that much of the interaction of the receptor with hsp90 occurs (86), and mutation studies have found that three of five cysteine residues in this area, spaced close together in the binding pocket (87), are critical to the receptor's ability to bind specifically glucocorticoids (88). Mutation of other amino acid residues within the HBD may or may not have an effect on ligand binding or receptor activity. However, a key amino acid sequence in the rat GR (residues 547–553) has been shown to be both critical for ligand binding and essential for receptor-hsp90 complexation. It has been suggested that hsp90 helps the GR fold to its steroid binding conformation by interacting with these hydrophobic residues. This sequence contains an LXXLL motif, often called an NR-box, commonly found in other nuclear receptor (NR)-interacting proteins. Antiglucocorticoids and certain other modulators (as well as sodium molybdate) will interact with the HBD domain and inhibit activation.

The DNA-binding domain is highly conserved among species, and changes to the amino acid sequence in this region result in changes in receptor function (89). A strucutral feature that characterizes the GR-DNA binding domain are the two "zinc fingers," each in which two zinc (+2) ions are held in place by tetrahedral coordination to neighboring cysteine residues (90). These zinc fingers, common to the nuclear receptor superfamily, are not exactly the same as those found in Xenopus TFIIIA, in which histidine residues are also associated with the metal (91, 92). The zinc in the GR is thought to stabilize the α-helices at the carboxy ends of the "fingers" as well as aid in carboxy-terminal module folding, needed in the dimerization of the proteins (93). Although the entire glucocorticoid receptor tertiary structure has yet to be elucidated, some of the components of the structure, including the DNA-binding domain, have been depicted through molecular modeling studies (94), NMR investigations (95), and the crystal structure data from GR protein fragments (93). Solution-state analysis of the DNA-binding domain complexed to a nucleotide sequence (double helix) reveals areas where an α-helical substructure interacts with the nucleotide. This DNA-receptor complex structure is supported by the crystal structure of a similar DNA-receptor fragment (DBD) complex. Clear dimeric interaction can be seen, along with the general shape of the zinc finger region (Fig. 15.7).

Two domains (t1 and t2) exist that affect the GR post-DNA binding transcription activity (96). The major (t1) transactivation domain is 185 amino acid residues in length with a 58 residue α-helical functional core (97). The t1 domain is located at the N-terminus of the protein; the minor (t2) transactivation domain resides on the carboxy-terminal side of the DNA-binding domain. These domains help control the transcription of target genes by providing a surface to interact with general transcription factors, and are thought to be bridged by a heteromeric protein complex including vitamin D receptor activating proteins (DRIPs) (98).

7.2 Mechanism of Action

The process by which anti-inflammatory steroids impart their action is based on the action of the steroid on a receptor (GR). One result of this process is the lag between optimum pharmacologic activity and peak blood concentrations. The stages the GR goes through in becoming active can be divided into five general steps (99, 100), and each step is mediated by the glucocorticoid receptor (GR):

1. Subcellular Localization. Some debate still exists as to whether GRs are cytoplasmic or nuclear in nature (101, 102), although it is believed that the interaction of hormone and receptor occurs in the cytoplasm. The binding of glucocorticoid agonists or antagonists to the GR results in complete translocation of these receptors into the nucleus (103). Phosphorylation sites on the GR do not seem to play a role in hormone-inducible nuclear translocation. Sequences controlling nuclear localization have been identified within steroid receptors in the hinge region. The glucocorticoid receptor has a second nuclear localization signal in the hormone-binding domain (104).

2. Association with Heat Shock Proteins (hsp). Unactivated GRs are complexed with a number of protein factors that play various roles in the binding of ligand to the receptor,

Figure 15.7. DNA-receptor complex structure. Zn^{+2} ions can be seen as spheres within the GR DNA binding domain (ribbon). See color insert.

as well as the localization, DNA binding, and transactivation of the GR. The proteins, including hsp90, hsp70, hsp56, CyP40, and p23 (an acidic protein), have been implicated to be part of an assembled complex sufficient to activate GR to the ligand-binding state, and this complex is termed a *foldsome* (105, 106). Hsp90 has been shown to be associated with the ligand-binding portion (C-terminus) of the receptor (86), perhaps even blocking this site, although it is needed for the GR ligand-binding domain to be in the proper conformation for ligand binding (107). Hsp70 assists the binding of hsp90 to the receptor. Upon ligand binding, the heat shock proteins dissociate and the receptors become active in dimerization, DNA binding, and transcriptional enhancement (108). Although these hsp proteins do block certain DNA-binding sites on the receptor, protein-free receptor studies have indicated that the free receptor still needs to be bound to hormone to bind to DNA.

3. Hormone Binding. The ligand-binding domain encompasses almost the entire C-ter-

minus of these steroid hormone receptors. This domain is also credited with having the functions of hormone-dependent dimerization and transactivation. The ligand-binding domain of the glucocorticoid receptor appears to repress transcription in the absence of hormone and this transrepression is reversed by hormone. Active glucocorticoid receptor agonists bind tightly to the hormone receptor to elicit their action. Binding of glucocorticoids to the GR hormone-binding domain transforms the receptor into an activated complex able to interact with DNA sequences, called the glucocorticoid response elements (GREs) of target genes.

4. Dimerization and DNA Binding. The active regulatory form of the GR is thought to be dimeric (109). The GR binds to a palindromic DNA sequence (GRE), either as a GR dimer (110) or perhaps as a GR monomer, followed by binding of a second GR. Crystallographic studies with GR fragments (containing the DNA-binding domain) have indicated this dimerization occurs upon binding to the half-site of the GRE (93). Members of the steroid hormone receptor superfamily contain a highly conserved DNA-binding domain of about 70 residues, which are complexed around two tetrahedrally coordinated zinc atoms.

5. Transactivation. Protein synthesis is initiated or inhibited by the action of the activated GR on DNA. The use of glucocorticoids leads to anti-inflammatory effects by first controlling gene expression, which subsequently leads to the synthesis and/or suppression of inflammation regulatory proteins.

One such regulatory protein, inducible by a number of glucocorticoids, is the 37-kDa protein lipocortin 1 (LC-1) (111, 112). Dexamethasone directly effects *de novo* LC-1 synthesis, leading to direct increases in intra- and extracellular LC-1 concentrations, most notably occurring at the cell surface (113). Prednisolone increases extracellular and decreases intracellular concentrations (114) of LC-1. Studies with LC-1 and with an active LC-1 segment show direct involvement of this protein in inhibiting neutrophil activation through inhibition of the release of elastase, PAF, leukotriene B4, and arachidonic acid (115), as well an inhibition of neutrophil adhesion to endothelial monolayers (116, 117). LC-1 inhibits various prostanoid (inflammation mediator) production, it suppresses thromboxane A_2 release from perfused lungs (118), and has thus been shown to inhibit inflammation in the rat paw (carrageenan-induced edema) inflammation model (119). This is attributed to LC-1 inhibition of phospholipase A_2, which converts membrane phospholipids to arachidonic acid, along with its effect on other cellular components of certain inflammatory responses (119). Antibodies raised against LC-1 are able to reverse the anti-inflammatory action of LC-1 (120–122). Corticosteroids reduce phospholipase A_2 activity, which results in the diminished release of arachidonic acid, and this subsequently leads to limiting the formation of prostaglandins, thromboxane, and the leukotrienes (123).

Glucocorticoids have been shown to inhibit gene transcription of other proteins involved in the inflammatory process, including the key inflammation mediators called cytokines [including IL-1 (8), IL3–6 (124), IL8 (125), GM-CSF (126), TNFα (119)]. Steroids have been also shown to suppress the formation of cytokine receptors (8); dexamethasone, in particular, downregulates gene transcription of angiotensin II type 2 receptors (127).

Activated cytoplasmic GRs can be involved in the regulation of transcription of genes expressing inducible enzymes. Dexamethasone can reduce prostaglandin production by inhibiting cyclooxygenase (COX-2) gene expression (128), although this is thought to be through MAP kinase (p38) interaction (129). Dexamethasone inhibits release of prolactin by LC-1-dependent and LC-1-independent routes (122). Moreover, glucocorticoid excess downregulates the expression of the GR (130).

The mechanisms by which the glucocorticoid receptor is able to inhibit signaling pathways controlled by the transcription nuclear factors AP-1 and NFκB are beginning to be elucidated (131). NFκB plays a very important role in the transcription of many proinflammatory genes. The role of the GR in the inflammatory response can be attributed, at least in part, to the inhibition of NFκB functions (132). In addition to COX-2, NFκB upregulates a number of genes that include many cytokines (TNFα, IL-1, IL-2, 3, 6, 8, 12) (133, 134), and adhesion molecules that are an integral part of inflammatory processes (135). Activated glucocorticoid receptors directly interact with and inhibit NFκB subunits. In addition, glucocorticoids transcriptionally activate the IκBα gene and block nuclear translocation of NFκB and DNA binding.

Another inducible enzyme, nitric oxide synthase (iNOS) produces NO, which increases airway blood flow and plasma exudation, and may amplify T-2 lymphocytes, which orchestrate eosinophilic inflammation in the airways. Glucocorticoids probably inhibit iNOS by inactivating NFκB, which regulates iNOS gene transcription (136), or by other pre- and posttranscriptional regulation (137). A number of cytokines induce iNOS, and glucocorticoids can also inhibit iNOS activation by inhibiting cytokine formation. LC-1 is also a mediator in iNOS induction inhibition by dexamethasone (138).

A model has been developed that describes the relationship between exogenous and endogenous corticosteroid action in inflammation (139). In general, topical glucocorticosteroids are found to suppress plasma cortisol concentrations [adrenal suppression (140)], especially with prolonged administration and in high doses.

Certain inflammatory disease states involve an inherent defect in the production and/or regulation of inflammatory mediators. In rheumatoid arthritis, an inflammatory disease targeting mainly joints, glucocorticoid (hydrocortisone)-induced LC-1 production is impaired (141). In some disease states (ulcerative colitis, Crohn's disease) anti-LC-1 antibody levels are raised, and this may partly explain why corticosteroid drugs are not always as effective in these circumstances (142). In steroid-resistant asthma, an abnormal receptor-activator protein 1 interaction is observed (143).

Glucocorticoids also influence other physiological and biochemical pathways, and a general overview of this action is listed in Table 15.14.

8 EFFECTS ON DRUG RECEPTOR AFFINITY

The effect of structural alteration in steroids on receptor affinity, which could only be guessed at before 1960, has received increasing study as the fascinating story of steroid receptors has unfolded (144). However, the receptor affinity of a given steroid is not the sole or even the major determinant of its pharmacological potency. This can be appreciated readily from the data of Wolff et al. (145) (Table 15.15) relating to the glucocorticoid receptor of rat hepatoma cells. Whereas 9α-F (**91a**) and 9α-Cl cortisol (**91b**) have essentially the same receptor affinity, their pharmacological activity differs by a factor of 2.

The enhanced pharmacological potency of the 9α-F derivative is thus only partially accounted for at the receptor affinity level, and one or a combination of other major processes (intrinsic activity, drug distribution, and effects on metabolism) must also be affected. The data in Table 15.15 indicate that intrinsic activity is in fact also affected by the 9α-substituent, given that TAT induction is enhanced by the 9α- F substituent relative to the 9α-Cl group.

(**91a**) R = F
(**91b**) R = Cl
(**91c**) R = Br
(**91d**) R = I
(**91e**) R = OH
(**91f**) R = CH$_3$
(**91g**) R = OCH$_3$
(**91h**) R = OC$_2$H$_3$
(**91i**) R = SCN

A similar situation can be seen from the data of Smith et al. (56) (Table 15.16) regarding the progesterone receptor. Whereas 6-substituents and the 17α-acetoxy group actually decrease the receptor binding of progesterone, Clauberg activity is markedly enhanced (footnote, Table 15.16). Decreased metabolic inactivation may be responsible for this, although more data are needed. Even the 19-nor modification, which substantially increases receptor binding, increases biologic activity by a factor of only 6, whereas the most powerful enhancing groups (Table 15.16) raise activity by a factor of 60.

Nevertheless, the relationship between chemical constitution and receptor binding is of great interest, given that receptor binding is a *sine qua non* for biological activity. In a systematic study of the thermodynamics of binding of 29 different corticoids to the glucocorticoid receptor of rat hepatoma cells, Wolff et al. (145) formulated a concept of the nature of the steroid receptor interaction that rationalizes the thermodynamic properties of the steroid receptor-binding process and affords a basis for predicting the binding affinity of any glucocorticoid derivative.

The temperature dependency of binding of these glucocorticoids to the rat HTC receptor was determined and a second-degree polynomial equation was fitted to the data points obtained (Fig. 15.8). The enthalpic and entropic

Table 15.14 Glucocorticoids Act by Affecting Many Biochemical Systems[a]

Decrease inflammation by:	Stabilizing leukocyte lysosomal membranes	Reducing leukocyte adhesion to capillary endothelium	Antagonizing histamine activity and release of kinin from substrates	Inhibiting macrophage accumulation in inflamed areas
	Preventing release of destructive acid hydrolases from leukocytes	Reducing capillary wall permeability and edema formation	Reducing fibroblast proliferation, collagen deposition, and subsequent scar tissue formation	Decreasing complement components
Suppress immune response by:	Reducing activity and volume of the lymphatic system	Decreasing immunoglobulin and complement concentrations	Possible depression of tissue reactivity to antigen–antibody interactions	
	Producing lymphocytopenia	Decreasing passage of immune complexes through basement membranes		
Other actions:	Simulate erythroid cells of bone marrow	Prolong survival time of erythrocytes and platelets	Produce neutrophilia and eosinopenia	
	Promote protein catabolism and gluconeogenesis	Reduce intestinal absorption and increase renal excreton of calcium	Promote redistribution of fat from peripheral to central areas of the body	

[a]Modified from Ref. 9.

781

Table 15.15   Comparison of Receptor Affinity and Intrinsic Activity of 9α-Substituted Cortisol Derivatives in Hepatoma Tissue Culture Cells[a]

Compound	Log K_{aff}	$-$Log $M_{1/2}$ TAT Induction	Relative Potency (glycogen deposition, rats)
9α-H (**2**)	7.69	6.68	1
9α-F (**91a**)	8.35	7.57	10
9α-Cl (**91b**)	8.24	5.77	4.7
9α-Br (**91c**)	6.16	3	0.3
9α-OCH₃ (**91g**)	5.92	2	0

[a]Ref. 145.

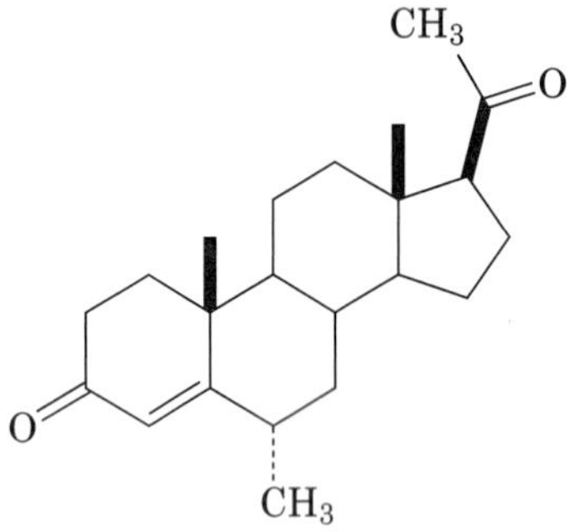

(92)

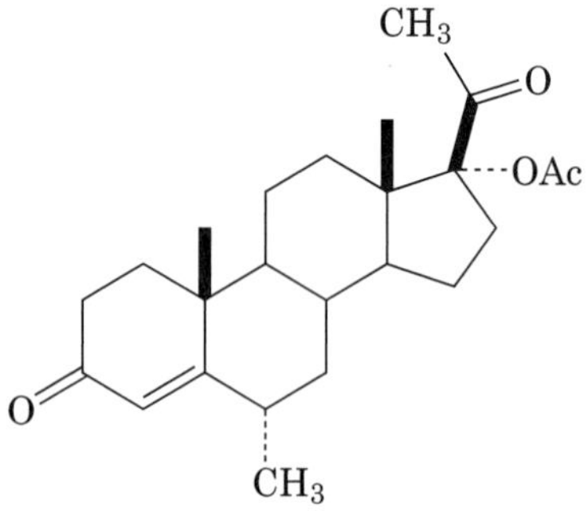

(94)

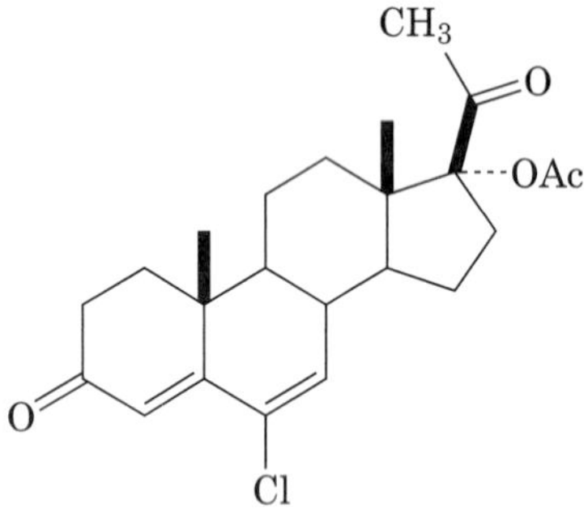

(93)

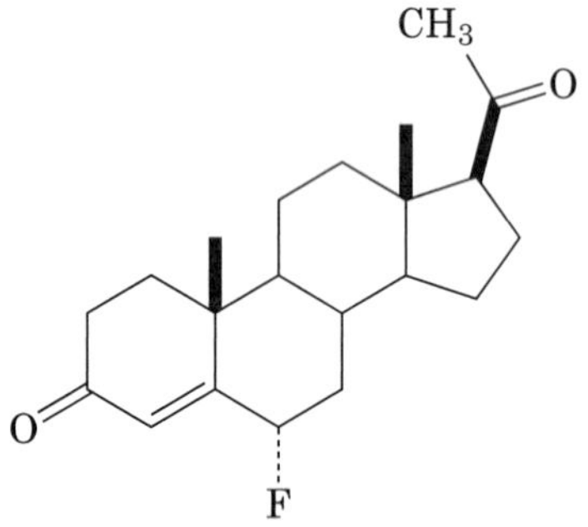

(95)

Table 15.16   Receptor Binding and Biologic (Clauberg) Activity in Progestational Steroids[a]

Compound	Receptor Binding (%)	Clauberg Activity[b] (%)
6α-Methylprogesterone (**92**)	26	150
Chlormadinone (**93**)	50	6000
6α-Methyl-17α-acetoxy-progesterone (**94**)	90	5500
6α-Fluoroprogesterone (**95**)	130	200
19-Norprogesterone (**96**)	168	600
Progesterone (**41**)	100	100

[a]Ref. 56.
[b]Clauberg assay, *in vivo* progestation activity determined in rabbit uteri as a function of endometrial growth.

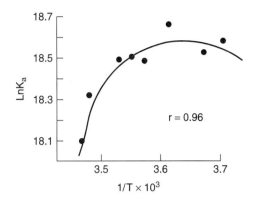

Figure 15.8. Plot of ln K_a (K_a is the association constant).

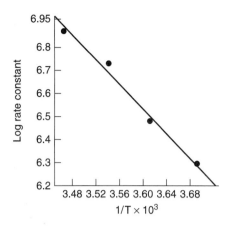

Figure 15.9. Plot of the log of the association rate constant vs. $1/T$ (°K) for dexamethasone binding to GR. [Adapted from Wolff et al. (145).]

CH₃

(96)

terms of binding were calculated. As was the case for other steroid-protein interactions (59), both enthalpy and entropy decreased as the temperature was increased (Table 15.17).

It was concluded that the steroid receptor binding forces are mainly hydrophobic in character. Both the steroid and receptor are extensively hydrated and the displacement of water molecules upon binding is a principal driving force. This is reflected by the positive entropy (ΔS), negative enthalpy (ΔH), and negative heat capacity (ΔC_p) of association be-

cause these phenomena are characteristic of the hydrophobic effect. From the temperature dependency of the rate constant (Fig. 15.9), the enthalpy of activation was found to be 12.8 kcal/mol and the entropy of activation to be 17.2 eu, indicating that the driving force for the formation of the transition state is also the hydrophobic effect. It is noteworthy that if the formation of ligand-protein hydrogen bonds and other oriented structures were of paramount importance in the transition state, the entropy of activation would be negative, rather than positive. Hydrogen bonding presumably contributes very little to the overall driving force for ligand–protein interactions, given that the net differences in free energy between hydrogen bonding for ligand-protein in the bound state and hydrogen bonding between ligand, protein, and water in the unbound state are probably small.

From a consideration of the relationship between the surface area of proteins and

Table 15.17 Calculated Values of the Enthalpic and Entropic Terms for Corticosterone Binding to the Rat HTC Glucocorticoid Receptor[a,b]

	Temperature (°C)				
Term	0	5	10	15	20
ΔH, cal/mol	1200	0	−1000	−2200	−3200
ΔS, eu	29	24	20	15	12

[a]Ref. 145.

[b]The polynomial function employed was $1.62 \times 10^7 (1/T^2) + 118,000(1/T) - 195 = \ln K_a$.

Table 15.18   Free Energy Contribution to Binding to the Rat HTC Glucocorticoid Receptor per Substituent of Progesterone[a]

Substituent	Free Energy (kcal/mol)	Fried Glycogen Deposition Enhancement Factor (Rat)
Δ^1	−0.29	3–4
6α-F	−0.36	—
6α-CH$_3$	−1.09	2–3
9α-F	−0.57	10
9α-Cl	−0.71	3–5
9α-Br	+1.89	0.4
9α-OCH$_3$	+2.21	—
11β-OH	−0.89	—
11β-OH → 11-keto	+2.23	—
11-Keto	+1.67	—
16α-CH$_3$	−0.11	—
16β-CH$_3$	−0.21	—
16α-OH + acetonide	−0.35	—
17α-OH	+.049	1–2
21-OH	−0.56	4–7

[a]Ref. 145.

its contribution to hydrophobic bonding (146–148), it could be shown (145) that the energy of binding could best be accounted for if the entire steroid were enveloped on both sides by the receptor. The steroid appears as a "hamburger patty" enveloped on both sides by the "hamburger bun" of the receptor.

Interestingly, Anderson et al. (149) proposed a similar picture of the binding of another nonprotein ligand to a protein. This is the case of the complex between glucose and hexokinase, an enzyme possessing a deep cleft between two lobes. They concluded that a dramatic conformational change occurs in hexokinase as glucose binds to the bottom of the cleft: the two lobes of hexokinase come together, engulfing the sugar. These workers proposed that glucose is sufficiently surrounded by the enzyme in this closed conformation that it cannot leave its binding site (150), which provides an explanation for the observation (151) that the off-rate of glucose from its hexokinase complex is slow (58 s^{-1}). It is noteworthy that far slower off-rates are characteristic of steroid-receptor complexes. Another interesting point relates to the question of whether the steroid interacts with the receptor only on its β face, as suggested by Bush (152). The thermodynamic data indicate that this is not the case and that all of the steroid is in contact with receptor. From a

comparison of the binding of 22 corticoids, the approximate free energy increments of each substituent group were calculated (Table 15.18) (145). These free energy group increments can be added to approximate the binding constants of steroids whose binding constants are unknown. Thus, the free energy of binding of fluocinolone (**11a**) to the rat HTC receptor relative to progesterone is calculated as follows for substituents in fluocinolone in excess of the progesterone skeleton.

This calculation predicts that fluocinolone would bind to the receptor with a free energy of −1.32 kcal/mol more negative than progesterone, in fair agreement with the experimental value of −1.65 kcal/mol. These free energy increments may be compared with the pharmacological enhancement factors of Fried (153) (Table 15.18; also see Section 9). It is seen that there is little correlation between the two parameters, indicating again that other variables such as the inhibition of metabolic destruction, are of major importance. The conformation of the A-ring (154) has a pronounced effect on the binding of the steroid to the receptor. The difference in the C-3 to C-17 distance from progesterone in the steroids was employed as a measure of the A-ring conformation, given that this distance is strongly influenced by such conformational changes. It appeared that binding was greater

(11a)

Substituent	ΔG Binding Contribution (kcal/mol)
Δ^1	-0.29
6α-Fluoro	-0.36
9α-Fluoro	-0.57
11β-Hydroxy	-0.89
16α-Hydroxy	$+0.86$
17α-Hydroxy	$+0.49$
21-Hydroxy	-0.56
Total	-1.32

as the distance decreased. The inclusion of a fluoro group at C-9 or a double bond at C-2 had the greatest effects on the A-ring conformation.

Other substituents had varying effects on the direction and magnitude of changes in the A-ring conformation. The effects of 9α-methoxy and 9α-bromo substituents stand apart in these binding studies. They result in derivatives with low binding affinities (Table 15.18), yet the surface area increases because of the respective 9α-substituent. Evidently, the size of these substituents prevents the proper engagement of the steroid within the receptor site or induces a conformational change in the receptor such that binding is significantly altered.

A multiple regression analysis relating the above-noted four parameters with the logarithm of the dissocation constant was made. The surface area (SA) employed for each derivative was the summation of the Bondi surface of each substituent present over that of a progesterone skeleton. The second parameter (P) is a *de novo* variable representing the interaction of polar groups with the receptor. For each hydroxyl group present in the C-11 and/or C-21 position, a value of $+1$ is assigned,

to account for the specific favorable interaction with the hydrogen bond acceptor in the receptor. If no group resides in these positions, a value of 0 is assigned. A value of -1 is assigned for the presence of each C-17 or C-16 polar group, to account for the consequences of placing a polar group in a nonpolar region, and a value of -2 is given for the presence of an 11-keto functionality, to express the conformational change associated with an sp^3 to sp^2 transformation and the undesirable dipole-dipole interaction of the 11-keto group with the hydrogen bond acceptor of the receptor apparently in that position. A total of these values is used for the second parameter, denoted as the polar interaction term. The third parameter (*tilt*) expresses the conformation of the A-ring through the C-3 to C-17 distance in angstroms. The fourth parameter (X) expresses the size limitation at the 9α-position. The value employed is the maximum of the function $(0, R_x - R_{cl})$ where R_x is the distance in angstroms that a substituent radially extends from the pregnane ring system. An excellent correlation was found relating these four parameters to the logarithm of the equilibrium dissociation constant:

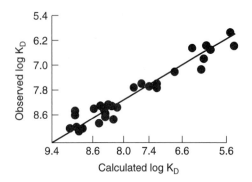

Figure 15.10. Plot of observed logarithms of the equilibrium dissociation constant versus the values calculated from the QSAR equation. [Adapted from Wolff et al. (145).]

$$\log K_d = \begin{aligned}&-0.022(\pm 0.002)SA\\&-0.59(\pm0.05)P\\&+1.50(\pm0.35)tilt\\&+6.10(\pm0.49)X\\&-6.52\end{aligned}$$

For this equation, n, the number of data points, is 29; r, the multiple correlation coefficient, is 0.97; s, the standard deviation of the regression, is 0.26; and F, a measure of the significance of the regression, is 106. Each parameter is significant at better than the 0.999 level. Shown in Fig. 15.10 is a plot of the calculated vs. the observed logarithm of the equilibrium dissociation constant.

An examination of the physical significance of this equation is of interest, given that it should reflect the thermodynamic contributions of the substituents. The equation represents the effects of substituents on the K_d value of progesterone, given by the intercept (-6.52). By multiplying by $2.303RT$, the equation is transformed to

$$\Delta G_{assoc}\,(cal) = \begin{aligned}&-27(\pm2)SA\ (cal/Å^2)\\&-734(\pm62)P\ (cal/P)\\&+1865(\pm435)tilt\ (cal/Å^2)\\&+7585(\pm609)X\ (cal/Å^2)\\&-8143\ (cal),\end{aligned}$$

giving the thermodynamic equivalent of each parameter. The surface area term shows a contribution of 27 cal/Å^2, in good agreement with the temperature-corrected value of 22.5

cal/Å^2, based on the work of Chothia (146). The absolute value of 0.76 kcal per P unit agrees well, with the values ranging from -0.89 kcal (attractive) to $+0.86$ kcal (repulsive) for hydroxyl groups (Table 15.18) and with the figure of -600 cal per hydrogen bond in the binding of trisaccharides to lysosome (155). The high value of the X term indicates the disruptive effect on binding because of the introduction of a group larger than the corresponding "pocket" in the receptor. A study by Ahmad and Mellors (156) indicated that the binding of steroid analogs to specific cytosol receptor proteins is correlated with the steroidal parachors, although a quantitative relationship was not derived. Parachor is a molar parameter defined as the product of the molar volume and the fourth root of surface tension (157). Thus, parachor and surface area are strongly cross-correlated.

9 EFFECT OF STRUCTURAL CHANGE ON PHARMACOLOGICAL ACTION

9.1 Pharmacological Tests

In Table 15.19 are listed various pharmacological tests for anti-inflammatory steroids. Granuloma tests measure the ability of the rat to encapsulate a cotton pellet and are a measure of the anti-inflammatory effect. The liver glycogen deposition test is an index of glucocorticoid action that usually is well correlated with the anti-inflammatory effect. The other tests in Table 15.20 reflect the diverse action of these steroids. Nonsteroidal anti-inflammatory agents are poorly active in the granuloma tests, indicating the difference in their mode of action.

Animal data do not always predict potency in humans with accuracy. A number of examples are shown in Table 15.20 (26). At least a part of the problem lies in the fact that the rat secretes corticosterone (**31**), which is inactive as an anti-inflammatory in humans, rather than cortisol. The important drugs triamcinolone acetonide (**13**) and dexamethasone (**20a**) are much more potent in rats than in humans.

9.2 QSAR Analyses of Pharmacological Action

In Section 8 we considered the application of QSAR techniques to a single process underly-

Table 15.19 Biologic Evaluation of Anti-inflammatory Steroids

Method	Species	Reference
Granuloma (pellet)	Rat	79, 158
Granuloma (pouch)	Rat	159, 160
Thymus involution	Rat	79, 158, 161
Adrenal suppression	Rat	79
Adrenal steroid concentration	Rat	162
Body weight depression	Rat	79, 158
Eosinopenia	Mouse	163
	Dog	160
Liver glycogen deposition	Rat	164
	Mouse	158
Ulcerogenesis	Rat	74, 158
Sodium retention	Rat	165

ing steroid action, that is, drug receptor binding. In this section we examine the attempts that have been made to express the gross pharmacological activity of anti-inflammatory steroids through QSAR techniques.

As already noted, pharmacological activity represents the summation of a number of processes including absorption, metabolism, drug receptor affinity, intrinsic activity, and drug distribution. The effect of a given substituent, such as a 9α-substituent, on these combined processes is difficult to parameterize; a single parameter for a substituent can represent only its average effect. Therefore, QSAR analyses of gross pharmacological activity are necessarily less accurate than those relating to a single process such as drug receptor binding. On the other hand, because pharmacological activity is the goal of drug design, the attempts described in this section have considerable interest. Such studies represent approaches only to the correlation of activity with structure and have not given final answers. However, because they are relatively rigid molecules in which the effects of structural change are easily understood in steric and electronic terms, and because we know something of their mechanism of action, steroids represent a fruitful area in QSAR.

9.2.1 Use of *De Novo* Constants. The earliest QSAR analysis of anti-inflammatory steroids, and indeed one of the first QSAR analyses of any kind, was carried out by Fried and Borman (153) in an examination of compounds obtained by introducing halogen, hydroxyl or alkyl groups, or unsaturation into

certain positions of the steroid molecule. These workers discovered the remarkable fact that each substituent affects the activity of the molecule almost independently of the presence of other activity-modifying groups. The effect of each substituent was assigned a numerical value, a *de novo* constant termed an enhancement factor (Table 15.21). Multiplication of the biologic activity of a parent compound by the enhancement factors for the substituent groups gives the activity of the final analog. For example, Table 15.22 (153) illustrates the calculation of the potencies of a sequence of steroids, starting with 11β-hydroxyprogesterone and culminating in triamcinolone. The ranges obtained are in good agreement with the bioassay figures and their 95% confidence limits. Fried and Borman were unable to derive similar quantitative expressions for salt-retaining activity, although the action of the various substituents on salt retention could be expressed in semi-quantitative terms.

Other investigators have added additional enhancement factors to those listed in Table 15.21. In Table 15.23 additional values are given, including those for activity in humans. Although activities in humans and the rat are similar for many substituents, a major species difference is seen for the 2α-methyl group and the 6α-fluoro group. In Table 15.24 are listed enhancement factors from another laboratory (166), in which the 9α-fluoro substituent has a value of 3–4 rather than 7–10. A Fujita-Ban analysis on 44 corticoids was carried out by Justice (167).

Table 15.20   Some Anti-inflammatory Steroids with Atypically Poor Correlation Between Human and Animal Data[a]

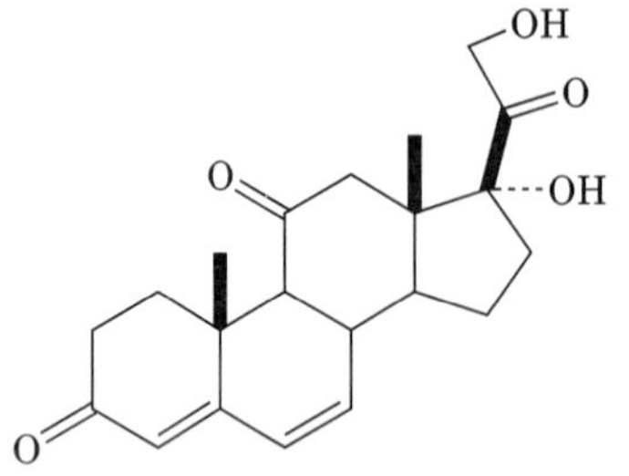

(97)

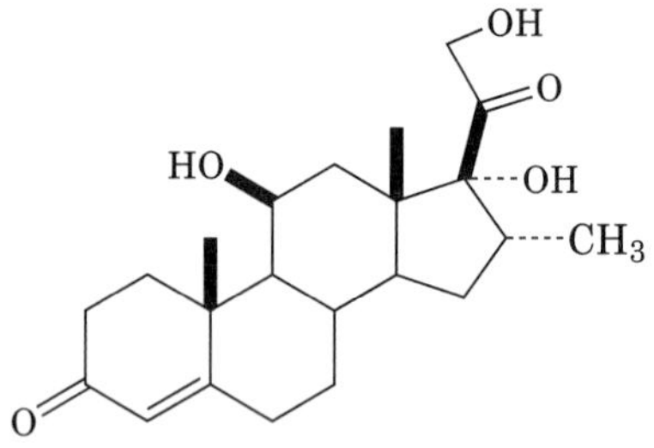

(98)

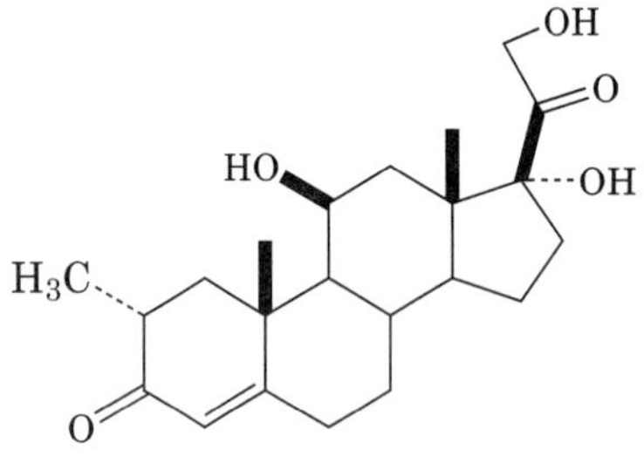

(99)

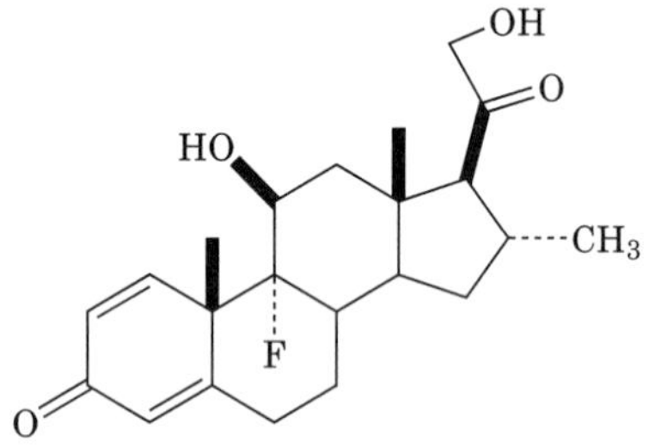

(100)

Compound	Anti-Inflammatory Potency	
	Animals	Human
Cortisol (**2**)	1.0	1.0
Corticosterone (**31**)	0.3	Inactive
Δ⁶-Cortisone (**97**)	0.5	Inactive
16α-Methylcortisol (**98**)	1.2	3
2α-Methylcortisol (**99**)	4.5	~1
Triamcinolone 16,17-acetonide (**13**)	75–80	4
21-Deoxy-16α-methyl-9α-fluoroprednisolone (**100**)	21	4–5
Betamethasone (**5a**)	58	30–35
Dexamethasone (**20a**)	154	30

[a]Ref. 26.

9.2.2 Hansch-Type Analyses. Wolff and Hansch (168) carried out the first multiparameter regression analysis for steroids in an analysis of the anti-inflammatory activity of 9α-substituted cortisol derivatives. A series of seven active compounds, the 9α- F (**91a**), 9α-Cl (**91b**), 9α-Br (**91c**), 9α-I (**91d**), 9α-OH (**91e**), and 9α-CH₃ (**91f**) cortisol derivatives as well as cortisol itself, were analyzed. The results were applied to the inactive 9α-methoxy (**91g**), 9α-ethoxy (**91h**), and 9α-SCN (**91i**)

compounds. It was found that activity was correlated with the inductive effect (σ_I), the size of substituents (molar reactivity, P_E), and π, giving the equation

$$\log A = 0.07 + 0.767\pi - 0.22 \log P_E + 2.78\sigma_I$$

For this equation n, the number of data points, is 7; s, the standard deviation, is 0.33; and r, the coefficient of correlation, is 0.96. The equation suggests that the activity of the com-

Table 15.21 Fried–Borman Enhancement Factors for Various Functional Groups[a]

Functional Group	Glycogen Deposition, Rat	Anti-Inflammatory, Rat (granuloma)	Effect on Urinary Sodium[b]
9α-Fluoro	10	7–10	+++
9α-Chloro	3–5	3	++
9α-Bromo	0.4[c]		+
12α-Fluoro	6–8[d]		++
12α-Chloro	4[c]		
1-Dehydro	3–4	3–4	−
6-Dehydro	0.5–0.7		+
2α-Methyl	3–6	1–4	++
6α-Methyl	2–3	1–2	− − −
16α-Hydroxyl	0.4–0.5	0.1–0.2	− − − − −
17α-Hydroxyl	1–2	4	−
21-Hydroxyl	4–7	25	++
21-Fluoro	2	2	− −

[a]Ref. 153.
[b]+ retention; −, excretion.
[c]In 1-dehydrosteroids this value is 4.
[d]In the presence of a 17α-hydroxyl group this value is < 0.01.

Table 15.22 Fried–Borman Calculation of Activities of Triamcinolone (13a) by Use of Enhancement Factors

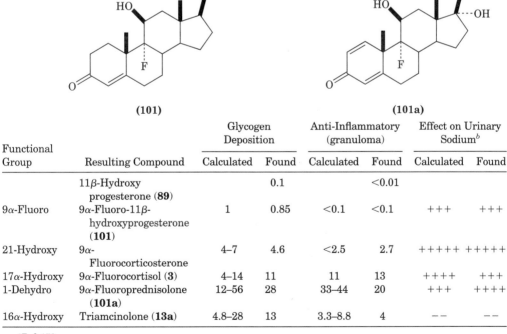

(101) (101a)

Functional Group	Resulting Compound	Glycogen Deposition		Anti-Inflammatory (granuloma)		Effect on Urinary Sodium[b]	
		Calculated	Found	Calculated	Found	Calculated	Found
	11β-Hydroxy progesterone (89)		0.1		<0.01		
9α-Fluoro	9α-Fluoro-11β-hydroxyprogesterone (101)	1	0.85	<0.1	<0.1	+++	+++
21-Hydroxy	9α-Fluorocorticosterone	4–7	4.6	<2.5	2.7	+++++	+++++
17α-Hydroxy	9α-Fluorocortisol (3)	4–14	11	11	13	++++	+++
1-Dehydro	9α-Fluoroprednisolone (101a)	12–56	28	33–44	20	+++	++++
16α-Hydroxy	Triamcinolone (13a)	4.8–28	13	3.3–8.8	4	− −	− −

[a]Ref. 153.
[b]+, retention; −, excretion.

Table 15.23   Enhancement Factors for Important Corticoid Substituents[a]

Group	Anti-Inflammatory (Rat)	Anti-Inflammatory (Human)
1-Dehydro	3	4
2α-CH$_3$	4.5	0.7
6-Dehydro	0.5	Variable[b]
6α-CH$_3$	2	1.3
6α-F	8	2.5
9α-F	8	10
16α-CH$_3$	1.5	2
16β-C$_3$	1.8	2.5
16-Methylene	1.5	—
16α-OH	0.2	0.5
16,17-Acetonide	8	1, (oral)[c]
21-F	2	—
21-Deoxy	0.1	<0.1

[a]Ref. 153.
[b]Usually <1.
[c]Topical effect is high.

pound rises with increasing electron withdrawal, decreasing size, and increasing hydrophobic bonding of the 9α-substituent. Moreover, it correctly predicts that the methoxy, ethoxy, and thiocyano compounds have little or no activity. The inverse relationship between the radius of the 9α-halogen atom and the magnitude of adrenocorticoid activity had already been noted by Fried (169), but Fried and Borman argued that the low activity of the 9α-hydroxy and 9α-methoxy compounds, which also have relatively small substituents, indicates that "it is the electronegativity of the substituent rather than its size that determines its enhancement properties." It is noteworthy that the quantitative multiparameter regression technique shows it is perfectly possible for the inverse relationship with increasing size to exist, even though a qualitative examination of the data led the earlier workers to conclude otherwise. Fried and Borman (153) suggested that the function

Table 15.24   Adrenocorticoid Activity Enhancement Factors[a]

Group	Glycogen Deposition	Thymolytic
9α-F	3.3	3.9
1-Dehydro	2.3	2.8
6α-Methyl	2.4	2.9
16α-Hydroxy	0.4	0.3
16,17-Acetonide	~2	~2.5

[a]Ref. 166.

of the electron-withdrawing group at C-9 was to increase the acidity of the neighboring 11β-hydroxy group and that "the corticoid activity of an 11β-hydroxysteroid increases with increasing acidity of the 11β-hydroxy group." They further speculated that "protein steroid binding at the site of action could be a function of the acidity of the 11β-hydroxy group." Newer developments concerning this possibility are described in Section 9.3.2.

The significance of the π parameter is more difficult to delineate. The increase in activity with increasing hydrophobicity could be attributable to better transport to the site of action for the more lipophilic compounds and/or to hydrophobic interactions at the active site. In a reexamination of this problem Coburn and Solo (170) found that the activity of 9α-substituted cortisols is well correlated with IJI and a simple steric factor such as molar refractivity (MR), provided that the 9α-hydroxylated compound is excluded from the series. They suggested that compounds containing strongly hydrated 9α-substituents would have a larger effective bulk than is found in the unhydrated species and hence a lower predicted activity. In another reexamination, Ahmad and Mellors (156) found that molar parachor gives a better fit than π in single-parameter equations but not in a three-parameter equation.

Pattern recognition approaches have been applied to the analysis of glucocorticoid test data. Bodor applied the linear learning machine (LLM) approach of Nilsson (171) and Jurs (172, 173) to human vasoconstrictor (McKenzie) data as well as rat granuloma data for 122 corticoids (174). Of these compounds, over 70 were considered potent, whereas the remainder were considered nonpotent relative to hydrocortisone butyrate (**2a**). Of 33 starting, Free-Wilson-like indicator variables, 22 were used in the final analysis, in which an LLM generated a nonparametric linear discriminant function that correctly classified all 122 of the corticoids. Descriptor values were determined for separate substituents as well as for certain combinations of substituents. It was concluded that these values would allow one to predict the contribution to potency of various substituents. Stouch and Jurs (175) later argued that these results (i.e., 122 data points divided 48/74 between two classes, a 22-dimensional data space) can support correct classifications attributed to chance of between 85 and 90%. These authors, by use of the same vasoconstrictor test data as those used by Bodor, describe the compounds as potent (P) or nonpotent (NP) and coded the compounds instead with 10 descriptors: three indicator descriptors ($\Delta^{1,2}$; 16-methylation; and/or 16,17-acetonide) and seven descriptors coding for the $\log P$ at the sites of varying substitution (positions 6, 9, 17, 21) or esterification (positions 17 and 21). The data set was separated into a training set of 88 compounds and a test set of 37, and then probed by linear discriminant analysis (LDF). Statistical analysis by use of Mahalanobis distances revealed that the two classes were well separated in the 10-dimensional space.

Additional evidence for data structure was provided by K nearest-neighbor (KNN) and principal-component (PC) analysis. Based on initial analysis in which esters were misclassified, presumably because of the nonlinear relationship between chain length and activity, the descriptors that coded for the $\log P$ of the side chains were transformed. For example, the C-17 position was described by subtracting the $\log P$ of the side chain of propanoic acid (1.55) and squaring the result for each value in the descriptor. By use of this modified set of 10 descriptors, a linear discriminant resulted that correctly classified all 88 of the training

set as potent or nonpotent. Finally, three different methods were used for prediction of the remaining 37 corticoids: LDF, KNN, and PC plots. Within the set of 37 test compounds, 14 were unambiguously classified as P or NP. KNN and PC methods correctly predicted 12 of this unambiguous test set of 14. The authors suggest that with certain prescreening criteria, this approach may be useful in predicting potent vs. nonpotent corticosteroids.

9.2.3 Use of Neural Networks to Predict Corticoid Properties. Artificial neural networks such as a Kohonen self-organizing neural network can be developed by conventional means to describe various corticosteroid data sets (176). New approaches to coding molecular structures for QSAR that also employ neural networks have been described recently. Their relevance here is that they use a benchmark CBG-corticosteroid data set originally described by Cramer et al. in 1988 (177), and later corrected in 1998 (178) in the seminal paper on comparative molecular field analysis (CoMFA).

One reported method for reducing the dimensionality of the 3D input data (molecular data) was to produce self-organizing neural maps (SOMs) (179). A transformation called "hypermolecule" is carried out for each molecule in the steroid data set, reducing the data to a 2D-coded SOM. The SOM map for the most active compound was used for training. A series of comparative SOMs from each analog are then used in the unsupervised neural network, to furnish a single trained neuron. In this way, different properties were studied such as shape ($r^2 = 0.89$, $s = 0.49$) and electrostatics.

In another approach called comparative molecular surface analysis (COMSA), the molecular surface electrostatic potentials (MEPS) of this corticoid data set were analyzed by a Kohonen self-organizing neural network (180). Neighboring MEP points were placed into adjacent neurons in the Kohonen map. The most active corticoid in the data set was used as a template, and PLS analysis provided statistical comparisons of models. The best model used $D = 40$, $MD = 0.2$ Å, and had a cross-validated r^2 value of 0.88 ($s = 0.424$), which outperformed CoMFA, with a cvr^2 value of 0.73 ($s = 0.657$). In this method, D was the density of the points sampled from the molecular surface and MD was the maximal distance of the points allowed in a single neuron.

A unique approach combining neural networks with data reduction involves the method called 3D-MoRSE (Molecule Representation of Structures based on Electron Diffraction) (181). A 3D-MoRSE code is a set of 32 evenly distributed values of s in the following equation:

$$I(s) = \sum_{i=2}^{N} \sum_{j=1}^{i-1} A_i A_j \frac{\sin sr_{ij}}{sr_{ij}} s = 0, \ldots, 31.0\text{Å}$$

in which I is the intensity of the scattered radiation; and A can be set to atomic number, atomic mass, partial atomic charge, residual atomic electronegativities, or atomic polarizabilities. The values of s were obtained from this function and the atomic coordinates calculated in the 3D structure generator CORINA. These sets of 32 values could then be studied by PC analysis or counterpropagation neural network (CPG). As applied to the above corrected CBG data set for 31 corticosteroids, the network correctly classified low to medium to high affinity ligands.

9.3 Some Steric and Electronic Factors Affecting Anti-Inflammatory Activity

In the following we discuss a number of studies that use the techniques of extended Huckel theory (EHT) and geometry optimization through energy minimization to determine the preferred conformation of steroid molecules. In some cases it is possible to compare these theoretical results with the findings of X-ray crystallography. Also discussed are electronic structure examinations through the use of CNDO/2 molecular orbital calculations. Although none of these examinations has provided a comprehensive understanding of the effect of structural change on biologic activity, a few underlying principles are beginning to emerge.

9.3.1 Effect of Structural Change on Steroid Conformation. Some steroids have greater flexibility than others. The introduction of unsaturation into the steroid molecule tends to increase its flexibility. In the case of many anti-inflammatory steroids the presence of unsaturation at C-4 and C-1 produces compounds that may exist as a range of conformers oscillating over a broad energy minimum, or as several conformers of nearly equal energy separated by a significant barrier. This flexibility complicates the question of interatomic distances between such key atoms as O–3 and O–20 and the effect of structural change on these distances. An obvious point of interest in this connection is the conformation of the side chain of glucocorticoids, given that this portion of the molecule includes the key O–20 and O–21 oxygen functions. In principle, a 360° rotation of the side chain is possible, but it is obvious that steric factors will tend to favor only selected conformations. Numerous investigators have studied this matter by chemical (182), physicochemical (183, 184), quantum chemical (185), crystallographic (186, 187), and energy minimization (188) methods. Most of these analyses have indicated that O–20 approximately eclipses C-16. In the energy minimization work of Schmit and Rousseau (188) significant differences were found relative to the crystallographic studies cited. They suggested that one reason for the discrepancy between energy optimized and X-ray diffraction data could be the influence of intermolecular hydrogen bonds and packing forces in the crystal. Moreover, as they pointed out, in certain cases two independent crystallographic structures are observed, indicating that the crystallographic data do not necessarily give an unambiguous picture of side chain conformation.

The most important factor influencing the position of the C-20 carbonyl is the presence or absence of a 17α-hydroxyl group, according to the study of Schmit and Rousseau (188). Duax et al. (189) studied crystallographic data on over 80 pregnanes, including numerous corticosteroids, and showed that the C(16)-C(17)-C(20)-O(20) dihedral angle is consistently between 0° and −47°. With the exception of 16-substitution, this dihedral angle was usually in the range of −20° for steroids having either 17α or 21-hydroxy groups, or their corresponding esters. Hydroxylation at both the 17α and 21 positions tended to shift this angle an average of $-15°$, and into the range of $-35°$. These studies did not demonstrate a correlation between side-chain hydoxylation and side-chain orientation (intramolecular hydrogen bonding) as suggested by Schmit and Rousseau. Intermolecular complexation did not result in a change in the preferred dihedral angle. These and other findings led Duax to suggest that force field calculations cannot provide accurate side-chain orientations in pregnanes and corticosteroids. Underestimation of eclipsing interactions between the C-16/C-17 bond and the C-20 carbonyl bond, and repulsive interactions between C-21 and the 17-substituent were cited as shortcomings of the force field calculations.

Weeks et al. (154) examined the structures of six corticosteroids and noted a correlation between the degree of bowing of the fused-ring system toward the α face and the anti-inflammatory activity. The presence of a C-1 double bond and a 9α-fluoro group brought about these changes. It is clear that bowing could be only one of several factors affecting biologic activity, given that 9α-chlorocortisol (190) is about four times as active as cortisol, but is not bowed more than cortisol, whereas 9α-methoxy cortisol (191) is inactive but is bowed more than cortisol. Schmit and Rousseau (188) undertook a Free-Wilson (192) analysis of the effect of various substituents on the conformation of the steroid on the basis of the predicted energy minimization structure (not the X-ray structure). In accord with the X-ray results, they found that the curvature of the entire steroid molecule (A/D angle) is increased 30% by a C-1 double bond and 14% by an 11β-hydroxy substituent. It is decreased 14% by a 17α-hydroxy substitution. These results parallel the biologic data in rats. A geometry minimization study by Marsh et al. (193) indicated that the conformational effect of the 9α-fluoro group is attributed to its effect on B-, C-, and D-rings which then induces a conformational change in the A-ring. Dideberg et al. (194) analyzed the crystal structure data of 20 corticoids. On the basis of their results they suggested that the distance between the atoms on the receptor capable of binding the steroid at O–3 and O–20 is 16.5 Å.

9.3.2 Effects of Structural Change on Electronic Characteristics of the Steroid.

As noted, the electronic effect of the 9α-substituent was already invoked by Fried in attempting to explain how substitution at C-9 could modify biologic activity. More recently, the electron densities in portions of cortisol, 6α-fluoro cortisol, and 9α-fluorocortisol were calculated by use of X-ray structures and the CNDO/2 method by Kollman et al. (195). They concluded that hydrogen bonding differences attributed to electron density changes engendered by the 9α-substituent could not fully account for the observed activity differences. Moreover, Divine and Lack (196) showed that the hydrogen bonding ability of the hydroxyl proton of 9α-substituted 11β-hydroxyprogesterone derivatives decreases in the order $F \cong Cl > Br > H$. If the cortisols behave similarly, there must be additional factors explaining the greater glucocorticoid activity of 9α-fluoro cortisol relative to 9α-chlorocortisol, which has only four times the activity of cortisol itself.

9.3.3 3D-Quantitative Structure-Activity Relationships of Steroids.

Comparative molecular field analysis, or CoMFA, is a "3D-QSAR" methodology whose development began some 12 years before the seminal work was published by Cramer, Patterson, and Bunce in 1988 (177). CoMFA is described in some detail in volume 1 of this series. In this study a training set of 21 corticosteroids whose corticosteroid-binding globulin (CBG) affinities were known was selected for analysis. Ten other corticosteroids were excluded from the analysis and were later submitted for prediction in the pharmacophoric model.

In the CoMFA approach, a critically important input variable is the representation of the molecules and their mutual alignment or alignment rule. For relatively rigid molecules such as the corticoid ring system, structures are reasonably represented by molecular mechanics. For the training set, Cramer first imported crytallographic coordinates before conducting molecular mechanics calculations with the Tripos force field. At this point, minimization resulted in alteration of the C(16)-C(17)-C(20)-O(20) dihedral angle. The discrepancy between this important tortional

Figure 15.11. CoMFA steric contour plot of the 3D-QSAR of corticosteroid binding to the human corticosteroid-binding globulin. Light-shaded contours correspond to regions where steric bulk from the steroid is predicted to decrease binding, whereas black-shaded contours correspond to regions where steric bulk is predicted to improve binding (177).

angle in the crystal state and from force field calculations has been pointed out by Bodor. Thus, although some question exists with regard to the reality of the D-ring side-chain conformation chosen for this study (e.g., does the energy minimized structure mimic the receptor bound conformation of the steroid?), the results are nevertheless meaningful for demonstrating the existence or nonexistence of a QSAR. The "alignment rule" for these molecules was then defined by least-squares fit of C-3, C-5, C-6, C-13, C-14, and C-17 to the corresponding atoms of the template molecule, deoxycortisol. The molecules were stored in a molecular database and linked to a molecular spreadsheet within Sybyl. The database was then enclosed in a box sufficient to encompass all of the molecules, with further subdivision into lattices of variable size, ranging typically from 1 to 2 Å. A number of steric and electrostatic field energy values were then determined for each molecule by placing a probe

atom at each corner of the lattice. Finally, the target property of CBG affinities were analyzed relative to these field values by the process of partial least squares (PLS) analysis with cross-validation. Matrices of lattice size and so on were examined, as was the robustness of the QSAR, with q^2 values over 0.75.

The resulting QSAR equations could be displayed for each analysis in three-dimensional space as a contour plot. In this manner steric contour plots provide visual information in the region surrounding the data set regarding the importance for binding of steric bulk (Fig. 15.11). Likewise, an electrostatic contour plot provides visual clues regarding the effect of polar groups on activity (Fig. 15.12). Ultimately, structures can have their activity predicted within the fields. Inspection of the contour plots in Fig. 15.12 reveals some expected relationships on the basis of simple SAR (e.g., that the 3-position and the 20-position benefit qualitatively from polar groups such as ke-

Figure 15.12. CoMFA electrostatic contour plot of the 3D-QSAR of corticosteroid-binding globulin. Light-shaded contours correspond to regions where partial negative charge from the steroid is predicted to improve binding, whereas black-shaded contours correspond to regions where partial positive charge is predicted to improve binding (177).

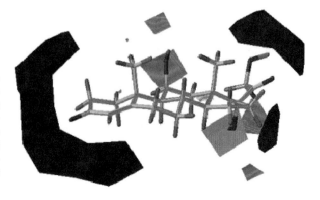

tones). Of course, the magnitude of the effect that position and orientation of these polar groups have on activity is not revealed visually; for this, the QSAR equations underlying the contour plots must be accessed. A set of 10 steroids for which binding data were available, but that were excluded from the analysis, was predicted in this model, as shown in Table 15.24.

Six of the 10 randomly chosen steroids had activity predicted very accurately, with one other still quite closely predicted. In the set of steroids that were not predicted well were structures that were not represented well by the original data set of 21 compounds. For example, members of the set of 10 steroids that were excluded from the original CoMFA model that were not predicted well were an A-ring dienone, a C-2 substituted steroid, and a 9-fluoro steroid.

Through the use of PC analysis, Norinder examined QSARs for the binding of pregnanes to human corticosteroid-binding globulin (197). Affinity data for 37 steroids, consisting of 26 steroids having hydroxyl groups at both or either of the 17- and 21-positions and 11 steroids having no hydroxyl group at either the 17- or 21-position, were collected. In a manner analogous to the CoMFA procedure, the energy-minimized steroids were aligned and encased in a gridwork of 1.5-Å spacings. Through use of a methyl probe atom, nonbonded interactions and charge-charge interactions were calculated. A molecular lipophilicity potential (MLP) was calculated with the atomic lipophilicity factors for each grid point as well. A PC analysis was performed on the matrix, with each molecule represented by the three fields (7425 points). Four principal components were determined that described over 89% of the variance in the binding affinity. A training set of 16 compounds was chosen that, after PLS analysis, resulted in a predictive r^2 value of 0.94. Estimation of the remaining 21 compounds gave a predictive r^2 value of 0.51, or 0.70 when the outlier 5-pregnene-3,20-dione was excluded. Finally, when all 37 of the original data set were included in the PLS analysis, a predictive r^2 value of 0.80 was obtained. Through use of CHEM-X, 3D contour plots of the grid points were generated in which nonbonded and charge-charge terms

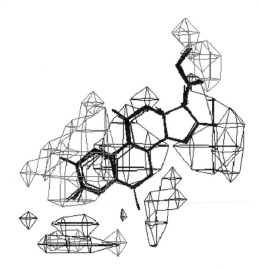

Figure 15.13. Steric contribution of the 3D-QSAR of pregnane and corticosteroid binding to the human corticosteroid-binding globulin. Black polyhedra correspond to positive nonbonded interactions, whereas gray polyhedra correspond to negative nonbonded interactions (197).

dominated the interactions, whereas MLP contours were not shown because of weak correlations. The contour plots from this study (Fig. 15.13) were quite similar to those obtained by Cramer.

When this technique was applied to both human and guinea pig CBG binding data, as described above, the QSAR coefficients could be subtracted, providing difference contour maps that could be useful in examining issues of species selectivity (198).

More recent approaches to QSAR of glucocorticoids still employ the Cramer CBG binding data set, but focus on molecular eigenvalue descriptors for QSAR (199), as well as E-state topological descriptors of atoms and hydrogens (200). A more unique approach has been the use of C-13 data for corticosteroids to develop a QSDAR relationship (quantitative spectrometric data-activity relationship) (201).

Tuppurainen and coworkers evaluated a novel electronic eigenvalue (EEVA) molecular descriptor for QSAR/QSPR studies, in which they validated, by use of a benchmark CBG steroid data set, a novel EEVA descriptor of molecular structure for use in the derivation of predictive QSAR/QSPR models. Like other spectroscopic QSAR/QSPR descriptors, EEVA was also invariant as to the alignment of the structures concerned. Its performance was tested with respect to the CBG affinity of 31 benchmark steroids. It appeared that the electronic structure of the steroids (i.e., the "spectra" derived from MO energies) is directly related to the CBG binding affinities. The predictive ability of EEVA was compared to other QSAR approaches, and its performance was discussed in the context of the Hammett equation. The good performance of EEVA is an indication of the essential quantum mechanical nature of QSAR. The EEVA method is a supplement to conventional 3D-QSAR methods for corticosteroids, which employ fields or surface properties derived from coulombic and van der Waals interactions.

The E-state modeling of corticosteroids binding affinity and validation of the model for a small data set was recently conducted by Maw and coworkers. Data for 31 steroids binding to the CBG were modeled by use of E-state molecular structure descriptors and a kappa shape index. Both E-state and hydrogen E-state descriptors appeared in the model as atom-level and atom-type descriptors. A four-variable model was obtained that was statistically satisfactory: $r^2 = 0.81$, $s = 0.51$; $r_{press}^2 = 0.72$; $s_{press} = 0.62$. Structure interpretation was given for each variable in the model. A leave-group-out (LGO) approach to model validation was presented, in which each observation was removed from the data set three times in random groups of 20% of the whole data set. The average of the resulting predicted values constitutes consensus predictions for these data for which $r_{LOO}^2 = 0.70$. These collective results support the claim that the E-state model may be useful for prediction of pK_a binding values for new compounds.

9.4 Summary of the Results of Theoretical Studies

Theoretical studies have shown that a given structural change in an anti-inflammatory steroid affects not one but a multitude of factors. The introduction of a 9α-fluoro substituent, for example, alters the conformation of the entire steroid molecule and increases the acidity and hydrogen bonding capacity of the 11β-hydroxyl group. These changes influence receptor binding, protein binding, and metabolic stability. The sum of these effects is manifested as the observed increase in biologic activity. The action of structural changes on only one individual process, receptor binding, can be predicted with accuracy (46). Gross pharmacological activity has been predictable for two decades through use of the technique of Fried and Borman (153). QSAR studies on metabolic stability, drug distribution, and intrinsic activity remain for the future. Combining all these relationships by use of mathematical modeling techniques may lead to a complete quantitative theory of anti-inflammatory steroid design in the next two decades.

10 EFFECT OF INDIVIDUAL STRUCTURAL CHANGES ON ANTI-INFLAMMATORY ACTIVITY

In this section we examine the SAR at each position in the steroid nucleus. First, however, it is useful to make an overview of the more important changes.

10.1 Overview of Major Changes in Anti-Inflammatory Steroids

In Tables 15.25 (202), 15.26 (166), 15.27 (203), and 15.28 (204) are displayed the activities of a number of anti-inflammatory steroids in large-scale studies in four major laboratories. These reports are of considerable interest. They make it possible to compare the activities of clinically important steroids and other steroids from data obtained by a single laboratory. It can be seen that the potency of polysubstituted compounds is predicted well by the enhancement factors of Tables 15.20, 15.22, and 15.23. The success of the medicinal chemist in increasing potency and decreasing sodium retention is apparent in Table 15.27. Flumethasone acetate, the 21-acetate of (**40**), is 424 times as active as cortisol in the granuloma assay and has no more sodium retention than that of cortisol itself—a remarkable achievement. Unfortunately, as mentioned in section 3, other side effects have not been reduced with respect to the therapeutic effect.

10.2 Skeletal Changes in the Cortisol Molecule

Unlike the situation in the progestational agents, removal of the 19-angular methyl

Table 15.25 Relative Potencies of Glucocorticoids (I)[a]

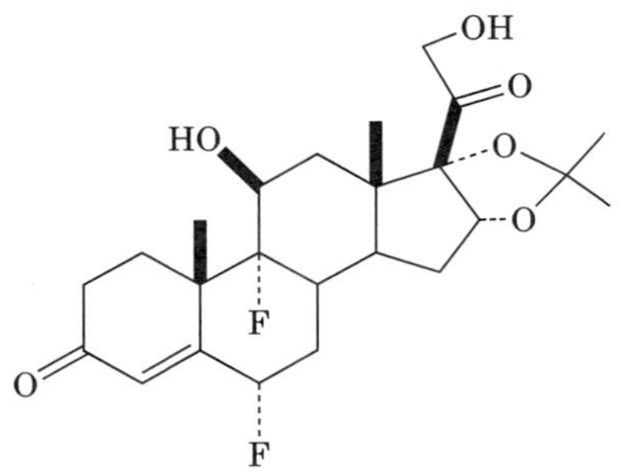

(102)

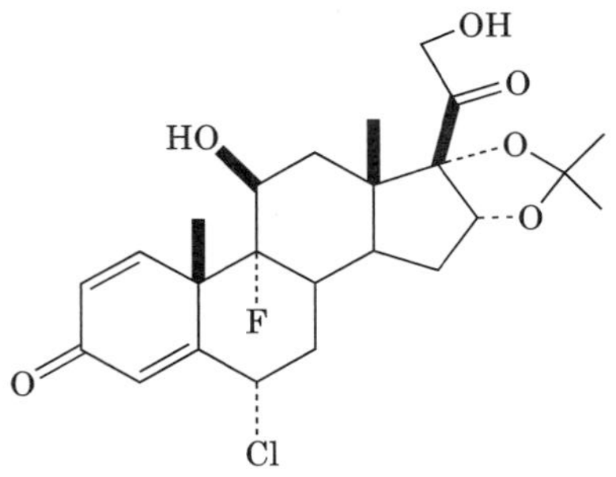

(103)

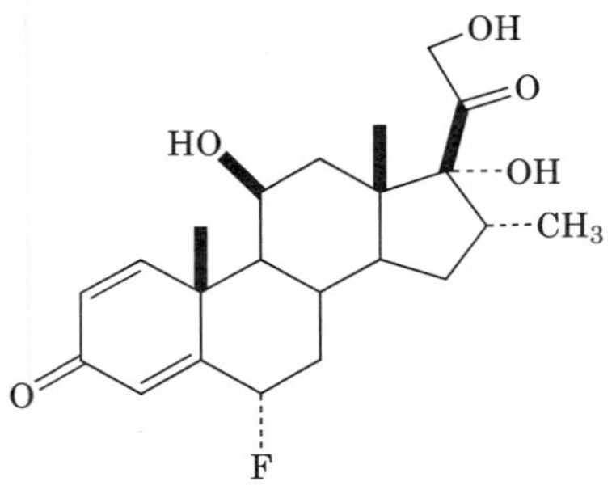

(37)

Compound	Relative Activity in Adrenalectomized Rat		
	Granuloma	Thymolytic	Glycogen Deposition
Cortisol 21-acetate (2c)	1.0	1.0	1.9
Cortisone 21-acetate	0.5	0.5	1.0
Prednisone 21-acetate	1.0	0.9	5.2
Prednisolone 21-acetate (21a)	2.7	4.0	9.9
6α-Methylprednisolone (28)	6.0	3.0	26.0
9α-Fluoroprednisolone (39)	17.7	4.4	53.9
Triamcinolone (13a)	2.6	3.6	34.0
Triamcinolone acetonide (13)	48.5	37.7	216.0
6α,9α-Difluoro-16α-hydroxy cortisol, 16α,17α-acetonide (102)	103.0	22.0	143.0
Fluocinolone (11a)	19.7	8.5	88.3
Fluocinolone acetonide (11b)	446.0	263.0	276.0
6α-Chloro-9α-fluoro-16α-hydroxyprednisolone 16α,17α-acetonide, 21-acetate (103)	123.0	188.0	114.0
Paramethasone (37)	63.6	45.1	46.2
Dexamethasone (20a)	104.0	47.0	181.0
Betamethasone (5a)	35.8	11.7	118.0

[a]Ref. 202.

Table 15.26 Relative Potencies of Glucocorticoids (II)^a

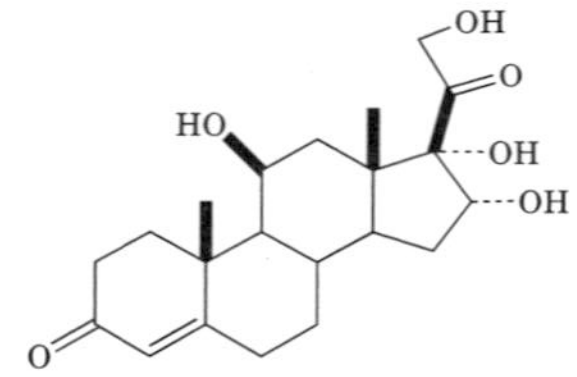

(104)

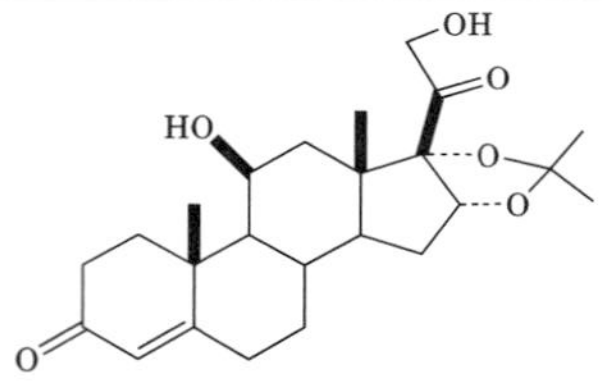

(105)

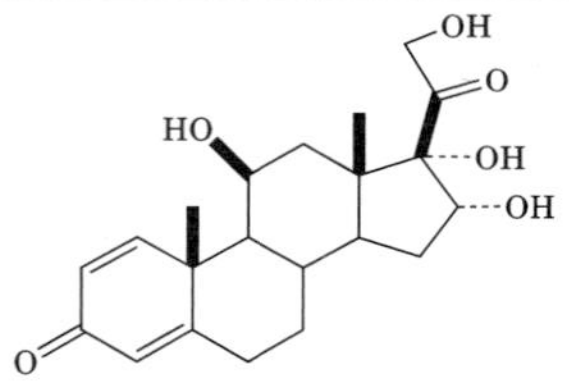

(106)

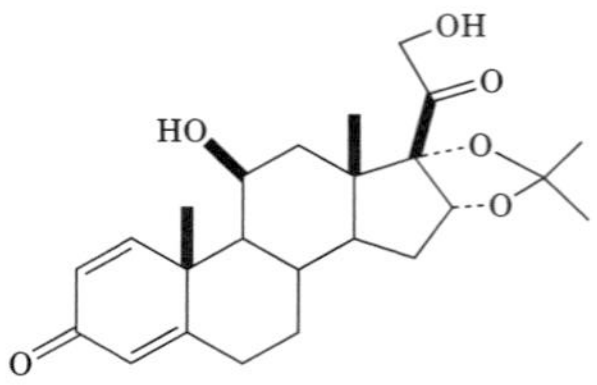

(107)

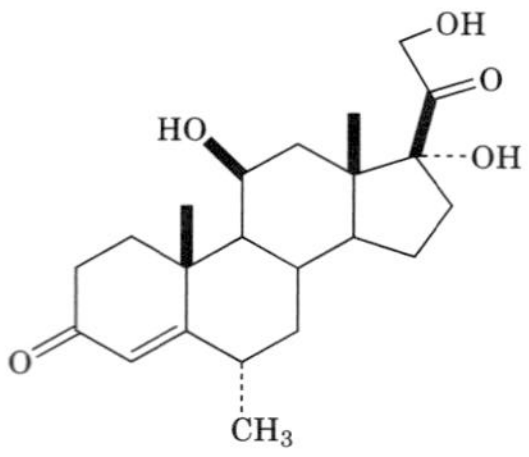

(108)

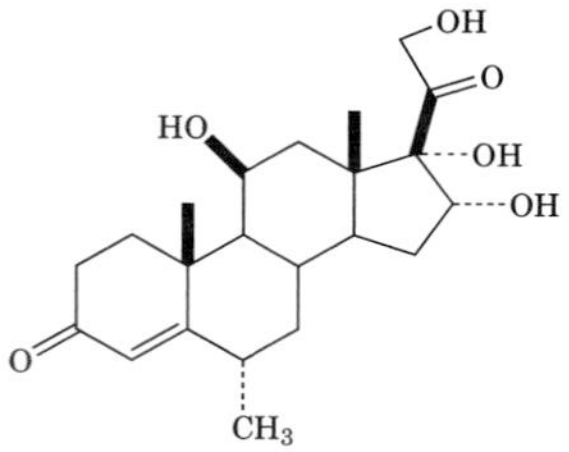

(109)

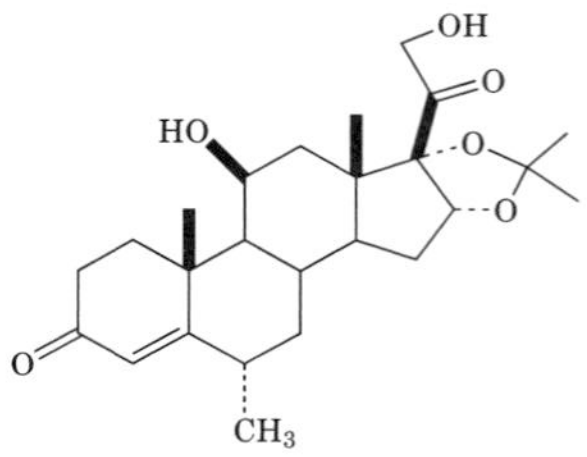

(110)

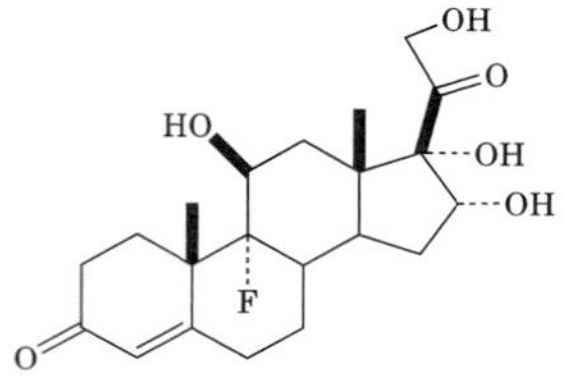

(111)

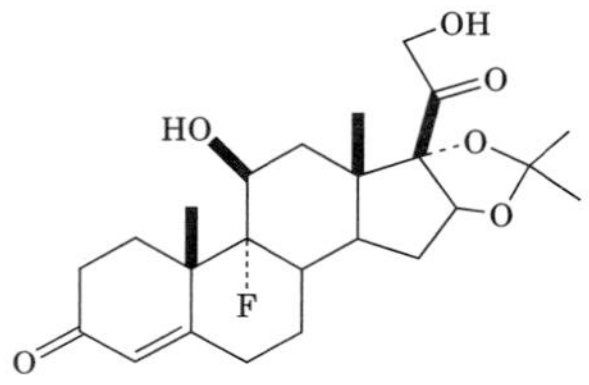

(112)

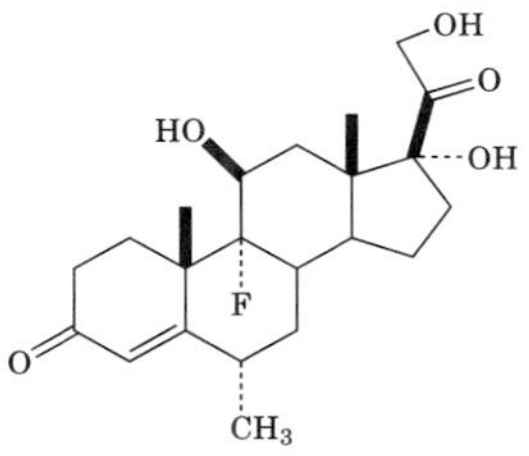

(115)

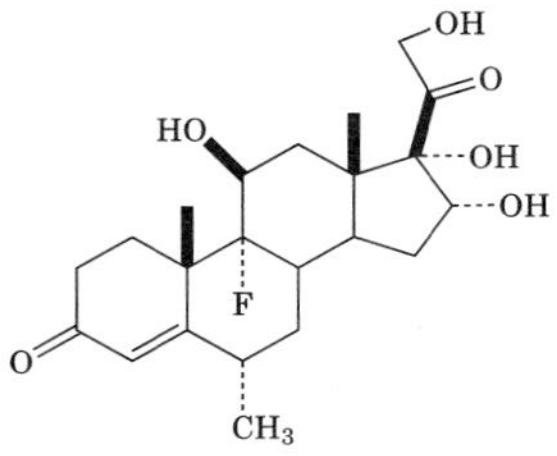

(119)

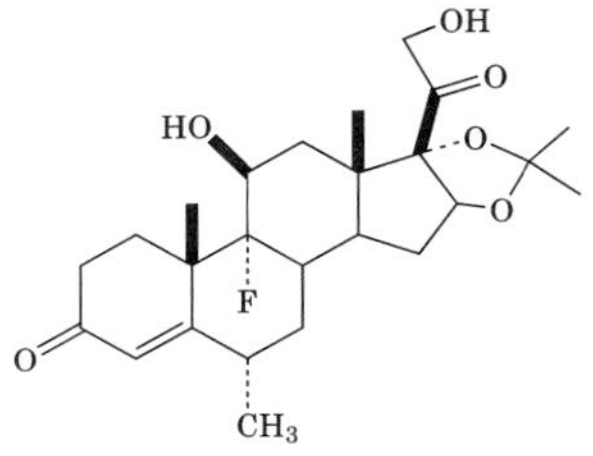

(114)

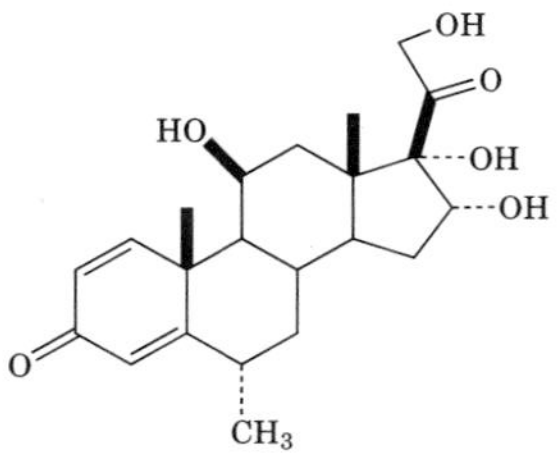

(118)

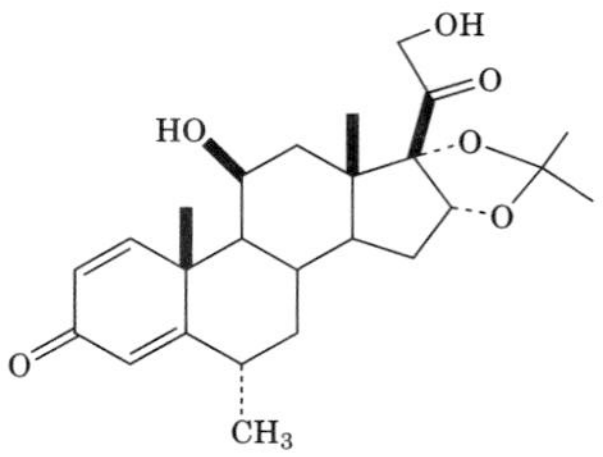

(113)

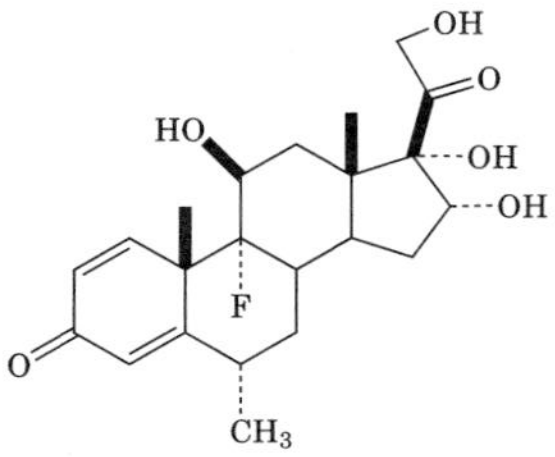

(117)

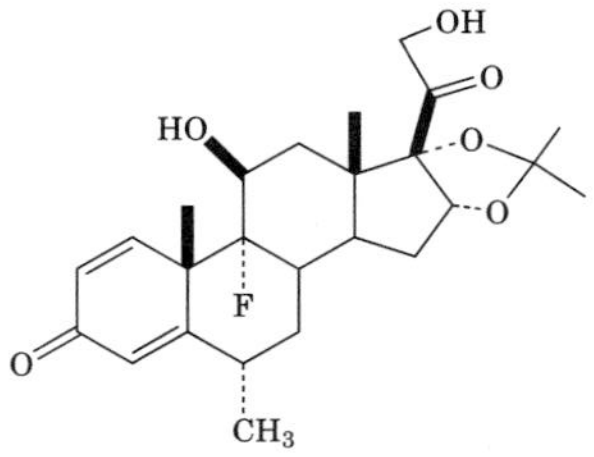

(116)

Table 15.26 (Continued)

Steroid	Glycogen Deposition	Thymolytic
Cortisol (**2**)	1.0	1.0
16α-Hydroxycortisol (**104**)	0.4	0.3
16α-Hydroxycortisol 16,17-acetonide (**105**)	2.0	2.2
Prednisolone (**21**)	4.1	2.3
16α-Hydroxyprednisolone (**106**)	0.9	1.3
16α-Hydroxyprednisolone 16,17-acetonide (**107**)	13.0	17.0
6α-Methylcortisol (**108**)	9.5	5.8
16α-Hydroxy-6α-methylcortisol (**109**)	1.1	1.2
16α-Hydroxy-6α-methyl-cortisol 16,17-acetonide (**110**)	10.0	14.0
9α-Fluorocortisol (**3**)	6.1	5.6
16α-Hydroxy-9α-fluorocortisol (**111**)	7.9	2.4
16α-Hydroxy-9α-fluorocortisol 16,17-acetonide (**112**)	14.0	14.0
6α-Methyl-9α-fluorocortisol (**113**)	9.0	9.8
16α-Hydroxy-6α-methyl-9α-fluorocortisol (**114**)	2.8	4.2
16α-Hydroxy-6α-methyl-9α-fluorocortisol 16,17-acetonide (**115**)	15.0	29.0
6α-Methylprednisolone (**28**)	8.4	10.0
16α-Hydroxy-6α-methylprednisolone (**116**)	2.0	2.2
16α-Hydroxy-6α-methylprednisolone 16,17-acetonide (**117**)	24.0	28.0
9α-Fluoroprednisolone (**39**)	13.0	16.0
Triamcinolone (**13a**)	6.1	3.9
Triamcinolone acetonide (**13**)	32.0	33.0
6α-Methyl-9α-fluoroprednisolone (**90**)	16.0	25.0
16α-Hydroxy-6α-methyl-9α-fluoroprednisolone (**118**)	4.9	5.5
16α-Hydroxy-6α-methyl-9α-fluoroprednisolone 16,17-Acetonide (**119**)	21.0	70.0

[a]Ref. 166.

Table 15.27 Effect of 6α-Fluoro and 9α-Fluoro Substitution on Activity of Cortisol Derivatives[a]

Table 15.27 (Continued)

Compound	Relative Activity		Sodium Retention (DOCA = 1.0)
	Granuloma	Glycogen Deposition	
Cortisol (**2**)	1.0	1.0	<0.02
16α-Methylcortisol (**98**)	0.8	1.4	<0.02
Prednisolone (**21**)	3.1	3.0	<0.02
16α-Methylprednisolone acetate (**120**)[b]	8.7	24.0	<0.02
9α-Fluorocortisol acetate (**3**)[b]	8.0	12.6	5.0
Dexamethasone (**20a**)	164.0	251.0	<0.02
2α-Methylcortisol acetate (**99a**)[b]	2.8	5.8	2.7
9α-Fluoroprednisolone acetate (**39a**)[b]	16.5	50.0	20.0
6α-Fluorocortisol acetate (**32a**)[b]	8.7	10.9	Slight
6α-Fluoro-16α-methylcortisol (**37**)	6.1	36.0	<0.02
6α-Fluoroprednisolone acetate (**121**)[b]	25.0	100.0	Slight
Paramethasone acetate (**37a**)[b]	50.0	150.0	<0.02
6α,9α-Difluorocortisol acetate (**122**)[b]	8.0	63.0	4.0
Flumethasone acetate (**40a**)[b]	424.0	677.0	<0.02
6α-Fluoro-2α-methylcortisol acetate (**124**)[b]	23.0	60.0	3.0
6α,9α-Difluoroprednisolone acetate (**125**)[b]	66.0	443.0	2.5

[a]Ref. 203.
[b]21-Acetate.

Table 15.28 Human Topical Anti-inflammatory Potency[a]

Compound	Activity
Cortisol (**2**)	1
Cortisol 17-butyrate (**2a**)	280
Cortisol 17-valerate (**2b**)	100–200
Dexamethasone (**20a**)	10–20
Dexamethasone 21-acetate (**20c**)	10–20
Dexamethasone 21-phosphate (**20b**)	10
Prednisolone (**21**)	1–2
Prednisolone 21-acetate (**21a**)	3
Prednisolone 17-valerate (**21b**)	2
Flurandrenolide (**15**)	1–2
Triamcinolone (**13a**)	2
Triamcinolone acetonide (**13**)	40–400
Betamethasone (**5**)	3–5
Betamethasone 21-acetate (**5e**)	18–33
Betamethasone 17-valerate (**5c**)	500
Fluorometholone (**22**)	30–40
Fluoroprednisolone (**121**)	4–6
Fluocinolone (**11a**)	100–300
Fluocinolone acetonide (**11b**)	600–800
Fluocinonide (**11c**)	1600
Flumethasone pivalate (**40b**)[b]	800
Halcinonide (**12**)	790
Desonide (**19**)	380

[a]Ref. 204.
[b]Pivalate ester.

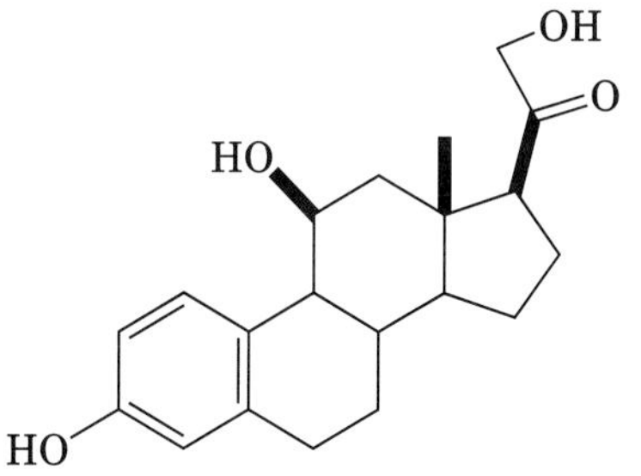

(127)

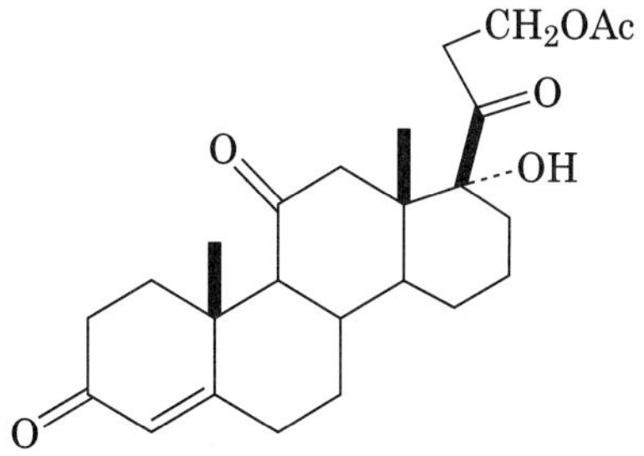

(128)

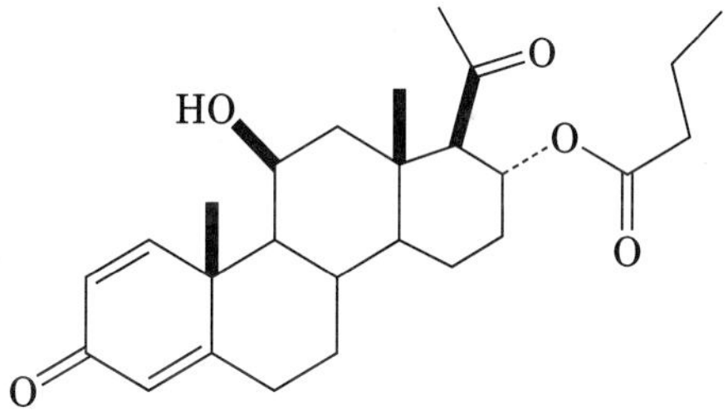

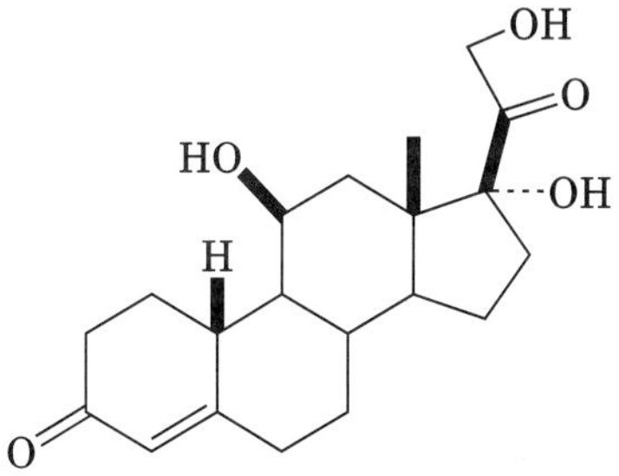

(126)

(129)

group reduces anti-inflammatory activity. 19-Norcortisol (**126**) has only 3/10 the activity of cortisol in the granuloma test (205, 206). However, if the removal of the 19-angular methyl group is accompanied by aromatization of the A-ring, as in (**127**), the compound retains activity equal to that of cortisol in the granuloma test (207). D-Homocortisone acetate (**128**) is less active than cortisol but retains some cortical hormone activity (208). Domoprednate (**129**), incorporating into the D-homo ring sys-

tem a 16α-ester grouping but lacking the conventional side-chain hydroxylation pattern, is beneficial. In clinical trials, domoprednate was comparable to betamethasone valerate as a 0.1% ointment (209).

Contraction of the A-ring to A-norcortisol (**130**) gives an inactive compound (210). The A-ring can be modified by isosteric replacement to afford a heterocycle like 2-oxacortisol acetate (**131**), which has one quarter the activity of cortisol. The introduction of a 9α-chloro group and a 16α-methyl group gives (**132**), which is equal to cortisol in activity (211).

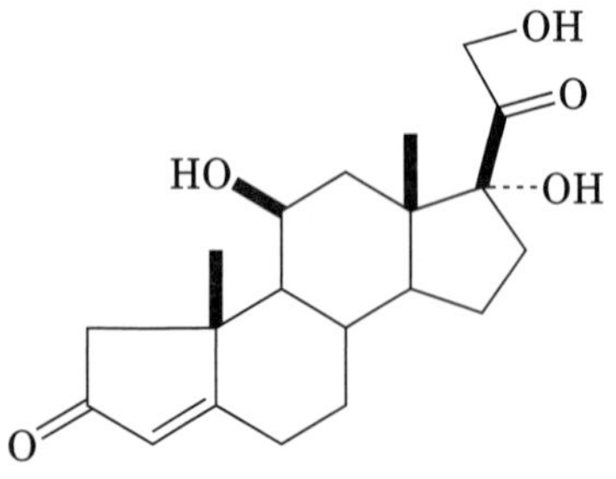

(130)

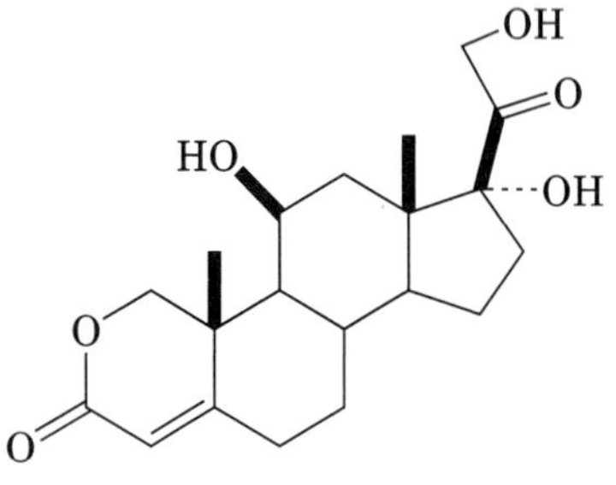

(131)

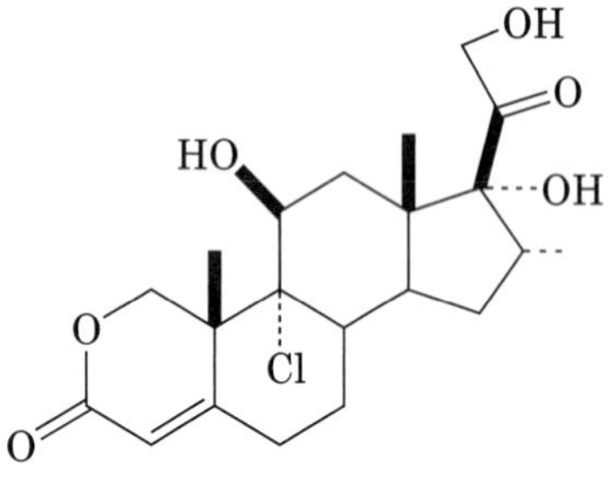

(132)

10.3   Alterations at C-1

The introduction of a double bond at C-1 leads to enhanced anti-inflammatory activity, as has already been discussed, but a hydroxyl group in this position leads to an inactive product. Thus 1α-hydroxy-9α-fluorocortisol, obtained by microbial hydroxylation, is virtually inactive in anti-inflammatory assays (211).

10.4   Alterations at C-2

The introduction of a 2α-methyl group into cortisol and its analogs (212) increases granu-

loma activity in rats about fourfold, but this enhancement is not observed in humans (213). The 2α-methyl group also produces a marked increase in sodium retention. The 2α-methyl group exerts its effect through a reduced rate of reduction of the Δ^4,3-ketone system (73). It is likely that metabolism in other positions, such as at C-11 and C-20, is also retarded.

Introduction of a 2α-fluoro group reduces the biologic activity in several tests (214, 215) but, by contrast, 2-halogenation in $\Delta^{1,4}$-3-keto corticosteroids can provide highly potent compounds. In the latter case, overall substitution of the ring system is important for activity, as is shown in Table 15.29.

The effect of simultaneous substitution at both C-2 and C-6 is profound. Fluorination at C-6α is well known to improve potency, but generally 6β-substitution is thought to be detrimental toward good anti-inflammatory properties (see Section 10.7). In this series, however, 6β- F is better than 6α-F when there is halogenation at C-2. Of additional significance is the absence of apparent systemic effects as gauged by the lack of thymic involution. The most potent within this group, 2-bromo (**134**) is virtually without systemic effect. No explanation for these unusual findings has been reported, but it was speculated that the steroids were rapidly metabolized in the systemic circulation (215).

10.5   Alterations at C-3

Almost all active anti-inflammatory steroids have a carbonyl group at C-3, but some exceptions exist. One is the ring A phenol (**127**) discussed in Section 10.2. 2'-Phenylpregn-4-ene[3,2-c]pyrazoles of corticoids are more potent anti-inflammatory agents than their parent Δ^4,3-keto steroids.

Activity can be enhanced still further by p-fluorophenyl substitution of the pyrazole ring, as in (**139**), which has 2000 times the potency of cortisol (216–218). The pyrazole derivatives do not exhibit mineralocorticoid activity. They have high affinity for cytoplasmic glucocorticoid receptors (219).

A simplified class of pyrazoles, lacking the 16-methyl, 9-fluoro, and $\Delta^{6,7}$ groups present in (**139**), was found in some cases to provide activity comparable to that of betamethasone

Table 15.29 6α,9- and 6β,9-Difluoro-11β-hydroxypregna-1,4-diene-3,20-dione and Their 2-Halogenated Derivatives and Acetyl Esters[a]

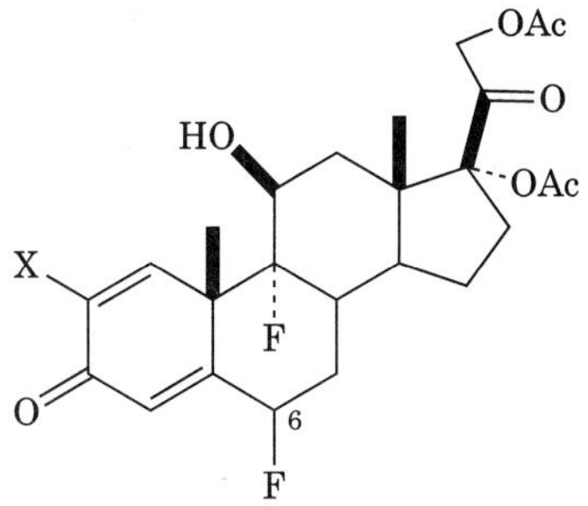

Compound	X	C-6	CPG Assay[b]	Thymus[c]
(133)	H	6β	40	>40
(134)	Br	6β	0.01	>1000
(135)	Cl	6β	0.45	>40
(136)	H	6α	0.10	5
(137)	Br	6α	>40	>40
(138)	Cl	6α	>40	40
Flucinolone acetonide			1.02	5

[a]Ref. 215.

[b]Cotton pouch granuloma weight inhibition determined at the ED_{30} (μg/pellet).

[c]Thymus weight inhibition determined at the ED_{30} for the same group.

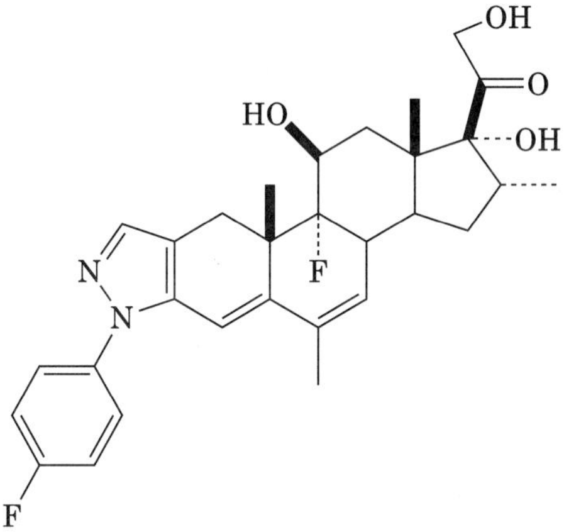

(139)

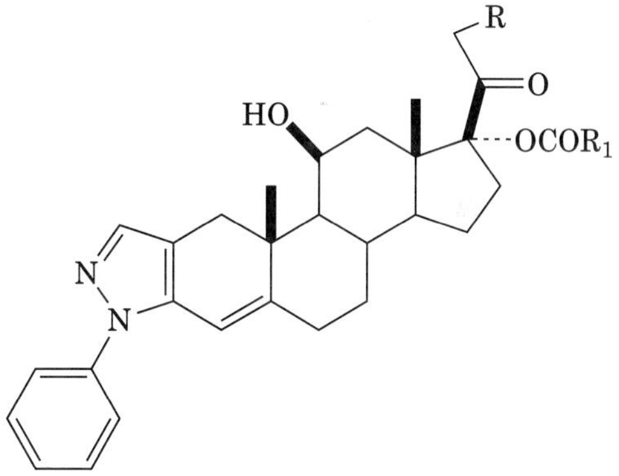

(140)

valerate in the human vasoconstriction assay. In this series, R was varied from OH to Cl to a variety of ester (OCOR') groups. Surprisingly, the most topically potent steroid, (140), had R = OH and R_1 = ethyl (220).

A unique class of 3-spirofused heterocyclic corticosteroids was developed as "soft drugs" through use of the chemical delivery system (CDS) approach of Bodor and coworkers (221).

In this approach, the heterocyclic derivatives are themselves biologically inactive, but slow hydrolysis results in opening of the thiazolidine ring and liberation of the anti-inflammatory steroid (Fig. 15.14).

Significant anti-inflammatory potency has been observed for many of these derivatives, as shown in Table 15.30. Potency was enhanced in general by a factor of about 3 to 4 relative to that of hydrocortisone, with only slight changes in potency being noted upon

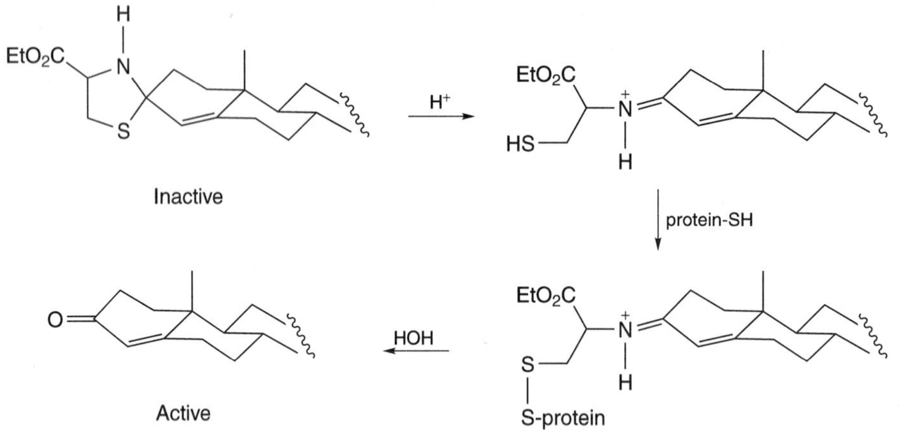

Figure 15.14. Corticosteroidal 3-spirothiazolidine derivatives as chemical delivery systems for hydrocortisone.

substitution of the ester R group. Various methods for assessing the degree of systemic absorption were applied to the prototypical ethyl ester, (**142**). The percentage reduction in thymus weight or thymus involution for (**142**), an indirect measure of systemically circulating corticosteroid, was 21%. In contrast, hydrocortisone and its 21-acetate were 35%, but hydrocortisone 17-butyrate was less well absorbed, as evidenced by a reduction of 20% in thymus weight. Another interesting approach to assessing the degree of absorption was accomplished in the mouse skin model. After 24 h, 8 mol % of hydrocortisone had passed through isolated, hairless mouse skin, whereas 14 mol % was observed for hydrocortisone 21-acetate, and only 3 mol % for the soft drug (**142**).

The decrease in systemic absorption noted for analogs such as (**142**) was presumably attributable in part to binding to skin tissue through formation of disulfide bonds between partially hydrolyzed prodrug (**142**) and -SH-containing amino acid residues in skin proteins, as shown in Fig. 15.14.

A clinically significant side effect of topically applied corticosteroids is thinning of the skin or atrophogenicity. This effect has been estimated in animal models by measuring thinning of mouse ear skin. As shown in Table 15.31, substantial thinning of the skin was observed for hydrocortisone and its 21- and 17-ester derivatives, but was substantially re-

duced for the soft drug (**142**). Triamcinolone acetonide, a more potent topical anti-inflammatory steroid in this group, had higher percentages of skin thinning.

10.6 Alterations at C-4

The Δ^4 double bond is important but not essential for anti-inflammatory activity. Thus, the 5α-pregnan-3-one (**147**) derived from tri-

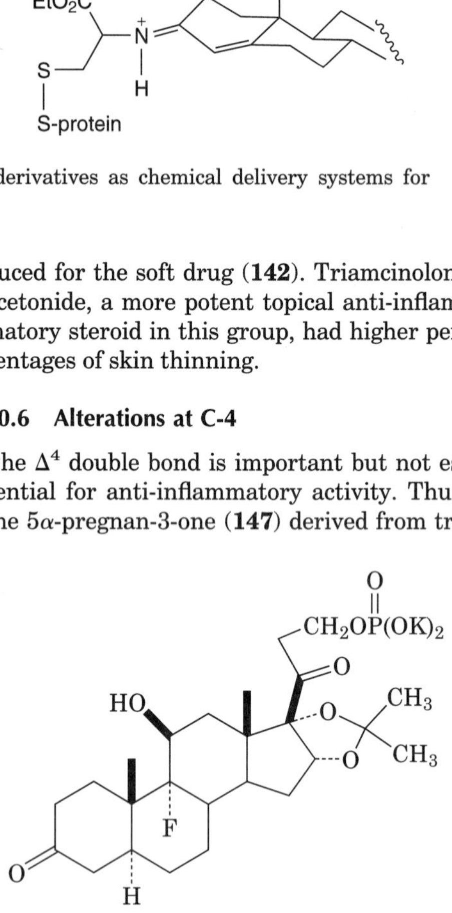

(**147**)

amcinolone 16α,17α-acetonide is more active than cortisol in glycogen deposition (222). Because 5α-steroids have approximately the same shape as the corresponding Δ^4 compounds, it appears that reduction of the Δ^4 double bond in this manner still allows receptor binding but results in a 100-fold decrease in the activity of the compound.

Table 15.30   Relative Anti-Inflammatory Activity of Selected Cysteine-Based 3-Spirothiazoldine Derivatives of Hydrocortisone[a]

Compound	R	R^1	ED_{50} $(M)^b$	Relative Potency[b]
(141)	CH_3	H	0–0035	3.1
(142)	C_2H_5	H	0–0033	3.2
(143)	C_4H_9	H	0–0027	4.0
(144)	C_6H_{13}	H	0–0039	2.7
(145)	$C_{10}H_{21}$	H	0–0036	3.0
(146)	C_2H_5	CH_3	0–0055	1.9
(2), Hydrocortisone			0–0107	1.0
(2c), Hydrocortisone 21-acetate			0–0203	0.5
(2a), Hydrocortisone 17α-butyrate			0–0011	10.0

[a]The test compounds were applied in acetone solution containing 2% croton oil on the anterior and posterior surface of the right ear. Three hours later, the mice (male DDY) were euthanized and both ears were removed. Circular sections were punched out and the drug effect expressed as percentage inhibition of inflammation compared to the control. 100% potency is defined as that of hydrocortisone (**33**).

[b]Linear regression analysis of data obtained at 3×10^{-5}, 3×10^{-4}, 3×10^{-3}, and 3×10^{-2} M.

4-Methylcortisol acetate is inactive in the granuloma and glycogen deposition tests (223). The reported anti-inflammatory activity of 4-fluorocortisol is very low (224). Allen et al. (225) prepared 4-methyl and 4-ethyl derivatives of 9α-fluorocortisol and the 4-methyl derivative of 9α-fluoroprednisolone. In all cases the 4-methyl analog was less potent than the parent compound and the 4-ethyl derivative was even weaker.

10.7   Alterations at C-6

In general, polar substituents such as hydroxy or oxo in the 6α-position decrease biologic activity, whereas 6α-substituted hydrophobic substituents, such as alkyl groups or halogens, tend to increase activity. 6β-Substituents of any type impair biologic activity, except as noted in Section 10.9. Thus, 6β-hydroxycortisol-6,21-diacetate has less than one-third the thymolytic potency of its parent compound

(226) and 6α-hydroxycortisone is less active than cortisone in the granuloma test (227). 6-Oxocortisone 21-acetate is only weakly active (228). The enhancing activity of the 6α-methyl group has been mentioned in a number of preceding sections of this chapter. The closely related 6-methylene group, as in 6-methylene 16α-methylprednisolone, impairs activity (229). 6α-Chloro groups (230) markedly enhance activity in 6α-chlorocortisol acetate, but have no effect on cortisone or prednisone (231), perhaps because the group interferes with reduction of the 11-keto moiety.

The 6α-fluoro group is very similar to the 9α-fluoro group in enhancing the activity of corticoids (203). 6α-Chloro and 6α-fluoro (73) groups give weakly active or inactive products when substituted in cortisone. Introduction of 6α-fluoromethyl groups into prednisolone and 9α-fluoroprednisolone affords compounds that are still active as anti-inflammatory ste-

Table 15.31   Ear Thinning in Mice after Topical Application of Steroids[a]

Compound	Concentration (M)	% Reduction in Ear Thickness
Hydrocortisone	3×10^{-3}	18.6
	1×10^{-3}	11.6
	3×10^{-4}	5.7
Hydrocortisone 21-acetate	3×10^{-3}	10.0
	1×10^{-3}	9.2
	3×10^{-4}	9.1
Thiazolidine (**142**)	3×10^{-3}	6.8
	1×10^{-3}	10.3
	3×10^{-4}	5.6
Hydrocortisone 17-butyrate	1×10^{-3}	23.3
	3×10^{-4}	17.6
	1×10^{-4}	13.3
	3×10^{-5}	6.7
Hydrocortisone 17-valerate	3×10^{-3}	23.6
	1×10^{-3}	17.2
	3×10^{-4}	14.8
Triamcinolone acetonide	3×10^{-4}	24.1
	1×10^{-4}	16.7
Vehicle		

[a]Left ears treated for 4 days with 10 μL of the steroid solution.

roids (232). The more polar 6α-methoxy group drastically reduces thymolytic activity when introduced into 9α-fluoroprednisolone (233). Although the introduction of a 6,7 double bond has little effect on the activity of cortisol (213), Δ^6-6-chloroprednisolone is about twice as active as prednisolone in arthritis (234). When the 6-azido group was introduced into 9α-unsubstituted Δ^6-corticosteroids, to give Δ^6-6-azidocortisol, systemic anti-inflammatory activity was increased five to eight times, whereas the corresponding change in 9α-fluorocorticoids left potency unaffected (235). A 6-formylpregnadiene derivative (**148**), upon oral administration, had a relative potency of 680 compared with that of cortisol acetate in the granuloma pouch assay (235). This interesting result could imply that a 3-keto group is not a prerequisite for anti-inflammatory activity, although (**148**) is an enol-ether and thus might be expected to undergo acid-catalyzed hydrolysis in stomach acid to provide, initially, enone-al (**148a**). Equilibration to (**148b**), followed by deformylation, might then give the ester of triamcinolone acetonide (**13b**).

10.8   Alterations at C-7

Both 7α and 7β substituents reduce anti-inflammatory activity. 7α-Methyl and 7β- methyl cortisol derivatives are less active than the parent compound (236). Introduction of a methylthio, ethylthio, acetylthio, or thiocyano group into the 7α-position of cortisol or cortisone reduces the activity of the resulting compound (237).

7α-Halogenation of non-9α-halogenated steroids, derivatives that are conceptually obtained by moving the halogen from the 9α- to the 7α-position, has led in some cases to potent synthetic corticosteroids having activity comparable to that of 9α-fluorinated steroids (e.g., dexamethasone dipropionate) (238). Diastereomeric 16-methylated corticoids (α and β) with various halogenation patterns were compared to controls, as shown in Table 15.32. When the 16-position is not methylated, activity appears to be independent of the 7α-halogen, in that relatively similar activity is obtained for both 7α-chloro or 7α-bromo steroids (**149**) and (**150**), respectively.

For the 16α-methylated series, activity is generally enhanced as expected (see section 10.13) but, surprisingly, 7α-F (**152**) is equipotent with its corresponding nonhalogenated derivative (**151**). This observation might imply that other bulkier, less electronegative halogens placed at the 7α-position would lead to less active compounds. This is particularly

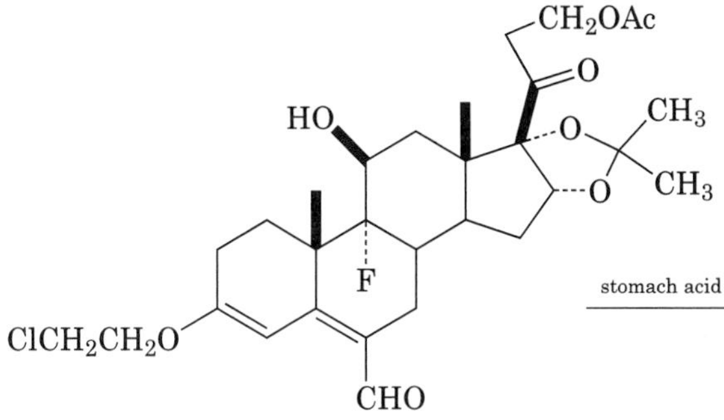

(148)

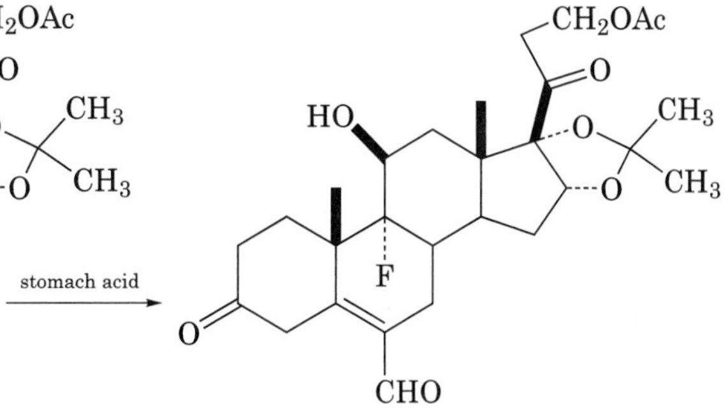

(148a)

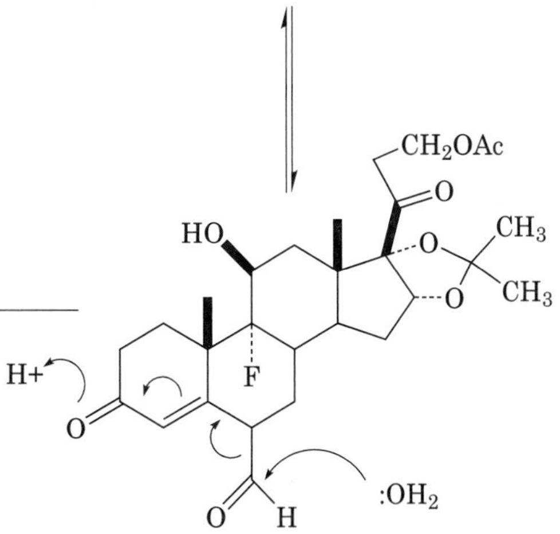

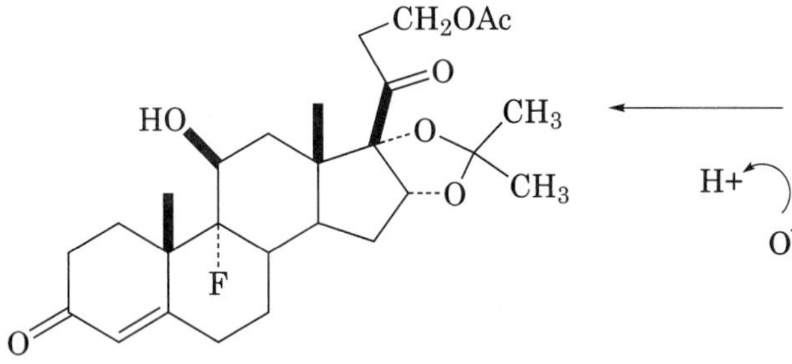

(13b) Triamcinolone acetonide, acetate ester

(148b)

true, given that activity follows the order of F > Cl > Br > I for 9α-halogenation (section 9.2). However, as is seen for (152), (153), (154), and (155) (7α- F versus Cl versus Br versus I), the trend appears to be Br > Cl > I > F. The noteworthy analogs in this series, 7α-chloro analog (153) and 7α-bromo analog (154), are both as potent as betamethasone 17,21-dipropionate in this assay system.

This is a remarkable finding, which may not support the contention of Fried (169). If the enhancement produced by a 9α-halogen is attributed to an inductive effect, similar enhancement would be expected from a 12α-halogen because the 9α- and 12α-positions are equivalent with respect to their electronic effect on C-11. Fried found that 12α-F substitution led to steroids with potency comparable to that of 9α-F steroids (see Section 9.2, 9.3, and

10.14). On the other hand, if an axial interaction at a 1,3-diaxial distance to the 9-position were more important, then 7α-F steroids might be expected to have activities comparable to those of their 9α-F counterparts. Indeed, (151) and/or (157) do have activity in the range of that of betamethasone 17-valerate, a highly potent 9α- F steroid. Given that receptor binding data for all of the axially halogenated C-7,9 and 12-fluoro derivatives are not published, it is difficult to assess whether these differences are truly attributable to a sigma-bond electronegativity effect, a through-space interaction, a conformational effect, or an effect on the metabolism of the steroid. It is interesting that 6α-halogenation (preceding section) often leads to significant potency enhancements, even though 6α derivatives are pseudoequatorial rather than axial, like the 7,9- and

Table 15.32 Topical Anti-inflammatory Potencies of 7α-Halogeno Corticosteroids and Their 7α-Hydrogen and 6,7-Dehydro Derivatives[a]

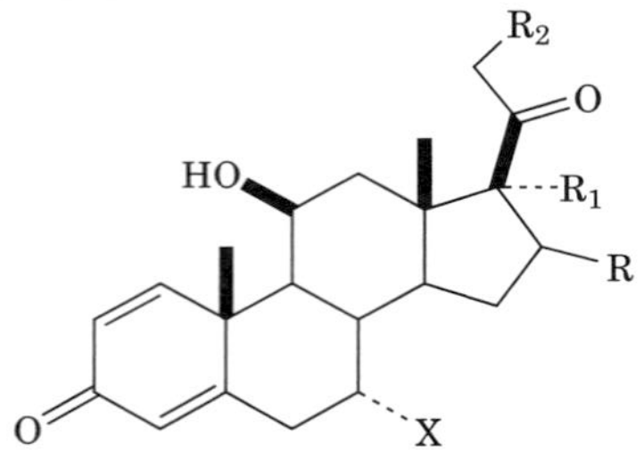

(233–244)

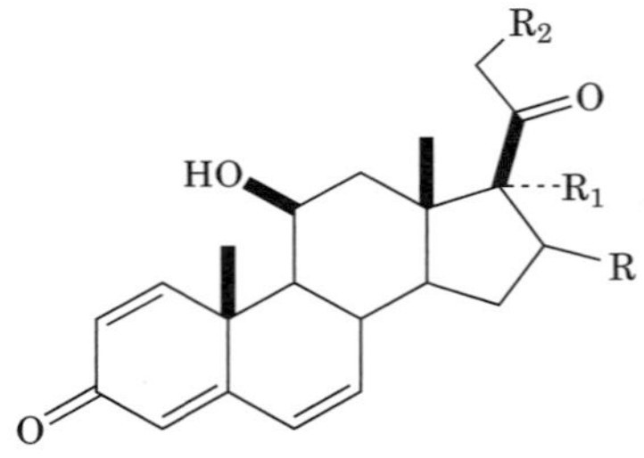

(245–246)

Compound	R	R_1	R_2	X	Topical Potency[b]
(149)	H	OCOEt	Cl	Cl	39
(150)	H	OCOEt	Cl	Br	40
(151)	α-CH₃	OCOEt	OCOEt	H	76
(152)	α-CH₃	OCOEt	OCOEt	F	69
(153)	α-CH₃	OCOEt	OCOEt	Cl	176
(154)	α-CH₃	OCOEt	OCOEt	Br	186
(155)	α-CH₃	OCOEt	OCOEt	I	103
(156)	β-CH₃	OCOEt	OCOEt	H	74
(157)	β-CH₃	OCOEt	OCOEt	F	78
(158)	β-CH₃	OCOEt	OCOEt	Cl	77
(159)	β-CH₃	OCOEt	OCOEt	Br	120
(160)	β-CH₃	OCOEt	OCOEt	I	35
(161)	α-CH₃	OCOEt	OCOEt	—	57
(162)	β-CH₃	OCOEt	OCOEt	—	27
Hydrocortisone (2)					3
BV[c]					100
BDP[d]					170

[a]Refs. 238, 239.
[b]Potency is a percentage reduction in inflammation relative to that of betamethasone 17-valerate (5c) in the croton oil ear assay.
[c]Betamethasone 17-valerate (5c) as control.
[d]Betamethasone 17,21-dipropionate (5b).

12α-positions. Perhaps the latter effect is related to a reduction in steroid metabolism, and thereby an increase in activity.

The 16α-methylated series is similar to the 16β-series, in that potency is maximal at 7α-bromo, although the series is ranked Br > Cl ~ F > I, which is at odds with the 16β-methyl set of analogs. The 6,7-dehydro derivatives (**161**) and (**162**) were surprisingly active in the croton oil ear assay, the 16β-methyl analog (**161**) being a third as active as its 9α-fluorinated relative, betamethasone 17,21-dipropionate (BDP).

10.9 6,7-Disubstituted Compounds

6α,7α-Dihydroxycortisone 21-acetate is much less active than cortisone acetate in the granuloma and thymolytic tests (237). The epoxide 6α,7α-epoxycortisone is about 1/10 as active as cortisol in the glycogen deposition test (229). However, the introduction of a 6,7-difluoromethylene substituent can give rise to highly active compounds (240). Both the 6α,7α-difluoromethylene derivatives and the 6β,7β-derivatives are active. Compound (**163**) has 1400

(**163**)

times the systemic anti-inflammatory activity of cortisol on subcutaneous administration. The high potency of both α and β isomers was taken as evidence that the 6,7 region of the steroid is not in contact with the receptor, although the opposite case could be argued equally well.

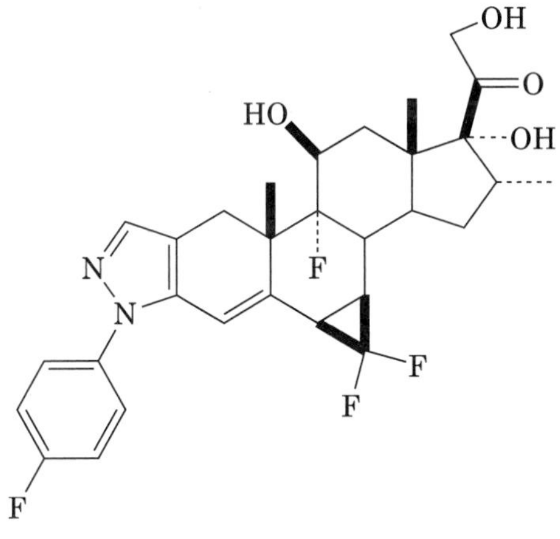

10.10 Alterations at C-8

The introduction of an 8(9) double bond into prednisolone derivatives gives products of somewhat lower anti-inflammatory potency than that of the parent compound (241). Thus, 6α-fluoro-8(9)-dehydroprednisolone acetate (**164**) has 2.3 times the thymolytic activity of cortisol.

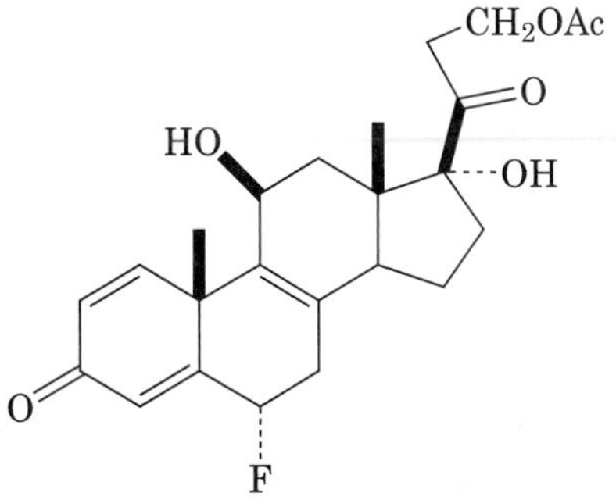

(**164**)

10.11 Alterations at C-9

Modifications at C-9 are discussed in Sections 9.2 and 9.3.

10.12 Alterations at C-10

Modifications at C-10 are discussed in Section 10.2.

10.13 Alterations at C-11

The C-11 oxygen group is not essential for anti-inflammatory activity if enough other enhancing groups are present in the molecule. Thus the 16,17-acetonide (**165**) is an active

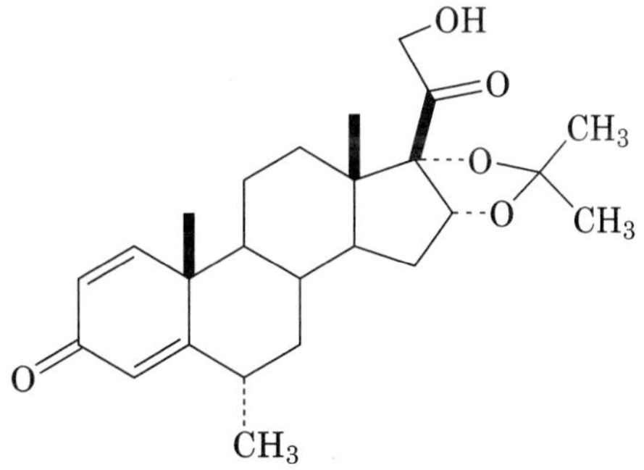

(**165**)

anti-inflammatory steroid, in spite of the absence of a C-11 oxygen group (242). However, changes at C-11 profoundly affect biologic activity. The 11α-hydroxy epimer of cortisol is inactive in the glycogen deposition test (243). 12β-Hydroxyprednisolone 11β,12β-acetonide is much less active than cortisol in the granuloma and thymolytic assays (244). Converting the 11β-hydroxy group to a tertiary alcohol through the formation of 11α-methylcortisol 21-acetate gives an inactive compound (245).

The introduction of a 9α-fluoro atom into deoxycorticosterone acetate (DOCA) gives a 9α-fluorinated steroid lacking an 11β-hydroxy group (166) that has 12 times the mineralo-

(168)

(166)

(169)

corticoid action of DOCA. However, (166) has only 5–10% the anti-inflammatory action of cortisol (246). This is an indication that some of the enhancing effect of the 9α-fluoro group is the result of its action on the 11β- OH group. An especially interesting series of compounds is the 9α,11β-dihalocorticoids. These derivatives lack the 11-oxygen atom and yet have useful anti-inflammatory activity. The 11β-fluoro-9α-bromo derivative (167) has less than one-quarter the activity of cortisol in the granuloma assay (247), although dichlorisone ace-

tate (the 21-acetate of 168) is a useful topical anti-inflammatory steroid (248). However, 9,11-dichloro steroids undergo solvolysis, as shown in the formation of (169) from (168) (249).

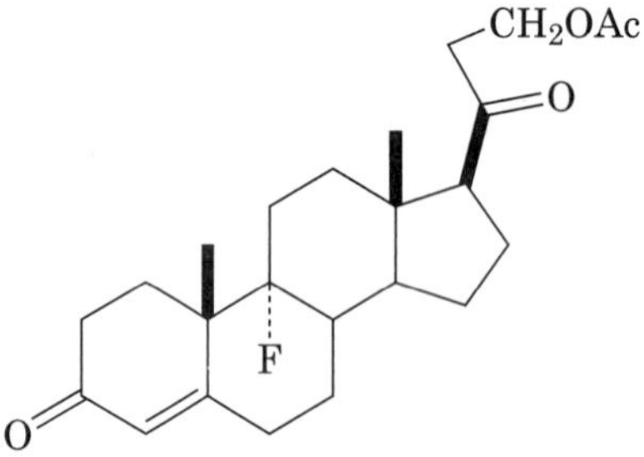

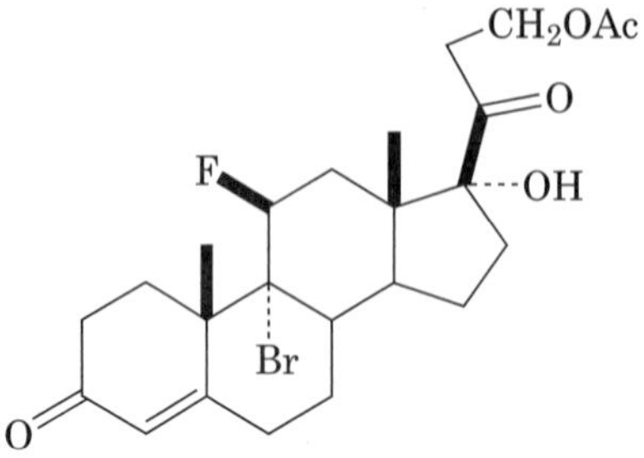

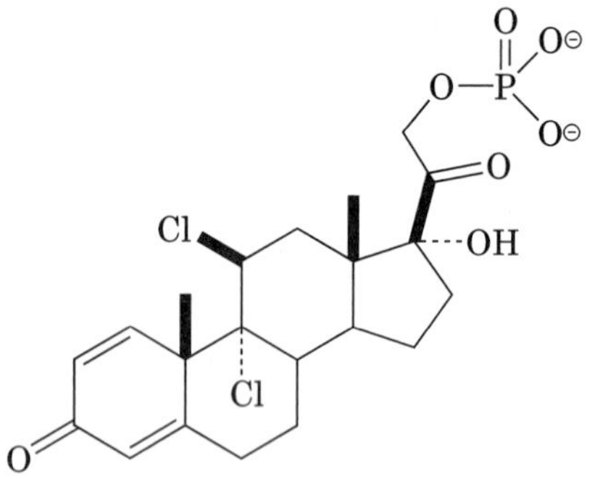

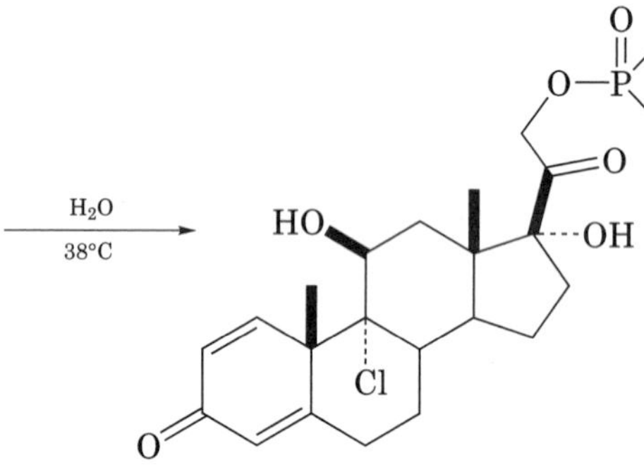

Thus, the activity of 11β-chlorosteroids may be attributable to their conversion to the 11β-hydroxy compounds *in vivo*. In support of this is the poor activity of (167), which would not be expected to undergo such a solvolysis, given that fluorine is a poor leaving group. 16α-Methylation causes a major increase in potency of dichlorosone, which is unexpected on the basis of the enhancement factor of this group. However, the 16α-methyl compound undergoes the solvolysis to the corresponding 11β-hydroxy derivative much more readily than the parent (168). This also provides evidence that some of the activity of the 9,11-dichlorosteroids is attributable to the solvolysis reaction shown (250). Thus, these compounds do not unequivocally represent highly active 11-deoxy anti-inflammatory steroids, as has sometimes been claimed.

(167)

10.14 Alterations at C-12

12α-Hydroxycorticosterone has 1/10 the activity of cortisol in the glycogen deposition test (251). 12α-Methyl-11-oxodeoxycorticosterone is less active in the glycogen deposition test than the corresponding compound lacking the 12α-methyl group (252). The effect of the 12α-halo substituent in corticoids was used by Fried (161) in an effort to understand the action of 9α halogens. He reasoned that if the enhancement produced by a 9α-halogen is the result of an inductive effect, similar enhancement would be expected from a 12α-halogen, given that the 9α- and 12α-positions are equivalent with respect to their electronic effect on C-11. It was shown that the liver glycogen and sodium-retention activities of 12α-halo-11β-hydroxyprogesterones paralleled those of the corresponding 9α-halo-11β-hydroxyprogesterones (253, 254). Again, Taub et al. (251) showed that 9α- and 12α-fluoro groups produced equivalent activity enhancement in the glycogen deposition test when substituted in corticosterone and 1-dehydrocorticosterone, and that 9α- and 12α-hydroxy groups cause similar reduction in activity in corticosterone.

Surprisingly, 12α-chlorocortisone is inactive in the liver glycogen and sodium-retention assays (255), whereas the 9α-chloro analog has a glycogen deposition activity of 3.5 times that of cortisone acetate (3). Fried and Borman hypothesized that hydrogen bonding from the 17α-hydroxyl group to the 12α-halogen is responsible for the loss of activity (153). To test this hypothesis they synthesized the 9α- and 12α-fluoro-16α-hydroxycortisol derivatives, (**170a**) and (**170b**) and their respective acetonides, (**171a**) and (**171b**). It was found

(**171a**) R = F, R′ = H
(**171b**) R = H, R′ = F

that the acetonides (**171a**) and (**171b**) have equivalent biologic activity (10 times that of hydrocortisone) in the glycogen deposition test. On the other hand, the parent steroid (**170b**) has only 10 times the activity of (**170a**). Thus, masking the 17α-hydroxyl group restores the 12α-halogen to parity with the 9α-halogen as an enhancing group.

A variety of 12β-hydroxy (**173–175**) and 12β-acyloxy (**176–179**) analogs of betamethasone 17,21-dipropionate (**5b**) were synthesized and tested for topical efficacy in a modified croton oil ear assay (mouse), by use of both (**5b**) and beclomethasone 17,21-dipropionate (**172**) as standards, as shown in Table 15.33 (256).

Topical potency is not greatly affected on introduction of a 12β-hydroxyl group into either (**5b**) (i.e., **173**) or (**172**) (i.e., **174**). Although the topical potency of the 9α-bromo-12β-hydroxy analog (**175**) is slightly reduced (by about 1/2) relative to that of the analogous Cl or F analogs, it is still more potent topically than would be expected on the basis of the Fried-Bormann enhancement factors (systemic) for 9α-halogens. On the other hand, the result for (**175**) is perhaps not unusual in light of studies by Green et al., (235, 238) who examined the effect of 7α-halogen substitution on topical potency in betamethasone-like base structures (e.g., alclometasone), and showed that potencies range in the order of 7α-Br > Cl > F > (**5b**).

The degree of systemic absorption of these analogs was measured in three ways. First, contralateral (distal ear) topical application, in the modified croton oil ear assay, allowed for an indirect determination of the degree of systemic absorption. As shown in Table 15.34,

(**170a**) R = F, R′ = H
(**170b**) R = H, R′ = F

Table 15.33   Estimated Topical Potencies of 12-Substituted Betamethasone Analogs[a]

Compound	R	X	ED_{100}[b]	Relative Potency
(5b)	H	F	110	1
(172)	H	Cl	110	1
(173)	OH	F	140	0.78
(174)	OH	Cl	160	0.69
(175)	OH	Br	325	0.34
(176)	$OCOC_2H_5$	F	400	0.28
(177)	$OCOC_2H_5$	Cl	400	0.28
(178)	$OCOC_2H_5$	Br	325	0.34

[a]Ref. 256.
[b]Microgram per mouse.

both BMDP (5b) and BCDP (172) exhibited marked systemic absorption, at the topical ED_{100} values. When applied at a site distant from the inflammation, (5b) was 75% as effective as when applied directly, whereas (172) was 50% as potent.

The 12β-hydroxy analogs (173), (174), and (175) all showed similar, if slightly attenuated, signs of systemic absorption. As can be seen in this series (12β- OH), varying the halogen substituent at C-9 did not greatly influence the degree of systemic absorption, in that all three analogs displayed about 15–20% distal topical potency.

The other method used to assess the degree of systemic absorption was to examine hypothalamic-pituitary-adrenal axis function, based on thymic involution (thymus weight) and adrenal suppression (plasma cortisol), after multiple topical applications of the corticoid. The results shown in Table 15.20, for the controls BMDP (5b) and BCDP (172), are consistent with those obtained for the single distal topical application. Thymus weights were dramatically reduced (by 70–90%) as were plasma cortisol levels (36%), clearly demon-

strating a high degree of systemic absorption. For the fluoro alcohol (173), both thymus weight and adrenal suppression were influenced by increasing dose in a regular manner, whereas the bromo (175) and chloro (174) alcohols were not. This effect is presumably related to the lower relative intrinsic systemic activities of (174/175) compared to that of (173). In any event, all three 12-hydroxy analogs are clearly absorbed through the skin.

In stark contrast are the corresponding propionate esters (176), (177), and (178), none of which demonstrates any evidence of systemic absorption, even after multiple high dose applications. It is interesting to note that, upon esterifcation of the 12β-hydroxy group, topical potency becomes relatively independent of the 9α-halogen (Table 15.34).

Topical potency was sensitive to the specific 12-ester. Thus, potency follows the order propionyl > butyryl > isovalyryl for aliphatic esters. On the other hand, bulky aromatic esters such as benzoyl (179) or furoyl are still quite potent. Furthermore, none of these esters was systemically absorbed.

Table 15.34 Anti-Inflammatory Activities of 12-Substituted Betamethasone Analogs

No.	R	X	Dose[a]	Topical[b]	Systemic[c]	Thymolytic[d]	Adrenal[e]
(5b)	H	F	2.5	59	27	30	12
			25	76	30	74	24
			75	93	72	87	37
(172)	H	Cl	2.5	60	12	12	1
			25	78	40	52	32
			75	93	52	67	36
(173)	OH	F	10	46	7	9	1
(174)	OH	Cl	10	41	0[e]	—	—
			40	67	5	0	13
			100	81	21	5	14
(175)	OH	Br	50	57	16	7	12
			100	86	17	0	14
			200	91	17	7	15
(176)	OCOC$_2$H$_5$	F	10	21	0[e]	0	0
			40	50	0	1	6
			100	69	0	4	0
(177)	OCOC$_2$H$_5$	Cl	50	47	0[e]	4	5
			100	53	0	5	5
			200	71	0	0	6
(178)	OCOC$_2$H$_5$	Br	50	36	0[e]	2	2
			100	45	0	6	6
			200	70	0	2	0
(179)	OCOC$_6$H$_5$	Br	300	84	0[e]	—	—

[a]Dose of drug in μg/mouse ear, after TPA challenge of 8.8 nM/mouse.
[b]Topical potency, % reduction in inflammation of mouse ear.
[c]Systemic potency, % reduction of inflammation in mouse contralateral ear.
[d]Effect was measured by weight loss of mouse thymus.
[e]Effect was determined by measuring the inhibition of increased plasma cortisol induced by stress.

In addition to systemic toxicity, prolonged topical corticosteroid therapy can result in clinically significant atrophy of the skin, which is thought to arise primarily by an inhibition of DNA synthesis. The 9α-chloro-12β-hydroxycorticoid (174) and its corresponding propionate (177) were examined for atrophogenicity, as previously described in the mouse. The results of repeated topical applications of the analogs (174) and (177), vs. (5b) and (172), on skin thickness is shown in Table 15.35.

Both beta- and beclomethasone dipropionate (5b) and (172), respectively) are quite atrophogenic, whereas the 12β-alcohol (174) is moderately toxic. In contrast, the corresponding tripropionate (177) is completely inert. These results would seem to indicate that

Table 15.35 Atrophogenicity of Betamethasone Analogs

Structure	Dose[a]	Skin Thickness	% Potency
Vehicle	—	16.5	0
(**5b**)	75	9.4	43
(**172**)	75	8.1	51
(**174**)	200	13.9	15.7
(**177**)	300	16.5	0.1

[a]Applied daily for 3 weeks.

12β-hydroxybeclomethasone 12,17,21-tripropionate (**177**) acts purely as an anti-inflammatory agent.

One matter that has not been discussed adequately in this theory is why the hydrogen bond formation between a 17α-hydroxy group, which is not even needed for activity in the rat, and the 12α-halogen should impair activity. One possible explanation would be through a conformational effect on the steroid molecule. This would be an interesting area to explore through X-ray crystallography.

10.15 Alterations at C-15

Methyl groups can be substituted in the 15α-position of anti-inflammatory steroids, with enhancement factors of approximately 0.5 on both anti-inflammatory action and sodium retention. A 15β-methyl group has little effect on glycogen deposition. Parent steroids substituted with the 15β-fluoro group include cortisol, prednisolone, 9α-fluorocortisol, and 9α-fluoroprednisolone (257). Bioassays suggest a small increase in anti-inflammatory activity because of the 15β-fluoro group and a 97–99% reduction in sodium retention.

10.16 Alterations at C-16

The effect of introducing 16α-methyl, 16α-hydroxy, and $16\alpha,17\alpha$-acetonides has already been discussed. 16α-Fluoro derivatives of prednisolone and 9α-fluoroprednisolone having 16 times and 75 times the anti-inflammatory activity of cortisol were reported by Magerlein et al. (258). This group also enhances the activity of 6α-fluoroprednisolone derivatives (259). By contrast, the 16β-fluoro group in cortisol, 9α-fluorocortisol, and 9α-fluoroprednisolone produce compounds with decreased anti-inflammatory activity (260). Likewise, loss of activity is seen upon intro-

duction of a 16β-methoxy group (260). The 16β-acetoxy group abolishes the glucocorticoid action of 9α-fluorocortisol or 9α-fluoroprednisolone (261). Introduction of the 16α-chloro group into 9α-fluorocorticoids greatly increases anti-inflammatory and glycogenic activity (262). 16α-Chloro $6\alpha,9\alpha$-difluoroprednisolone 21-acetate has 1100 times the activity of cortisol (**180**) (262). 16α-Ethyl substitution

(**180**)

is similar to 16α-methyl and 16α-hydroxyl substitution in eliminating effects on electrolytes (263). Beal and Pike (264) synthesized the 16α-fluoromethyl derivative of 9α-fluoroprednisolone 21-acetate. It is highly active as an anti-inflammatory agent and produces mild electrolyte excretion similar to that of 16α-methyl steroids. The 16α-methoxy derivative of cortisol exhibits twice the thymolytic activity of the parent compound (265). Isomeric 16-methylene and Δ^{15},16 methyl steroids (266) show interesting differences in activity. The Δ^{15},16-methyl compound (**181**) has 156 times the oral activity of cortisol in the granuloma test and produces sodium retention. By contrast, the isomeric 16-methylene

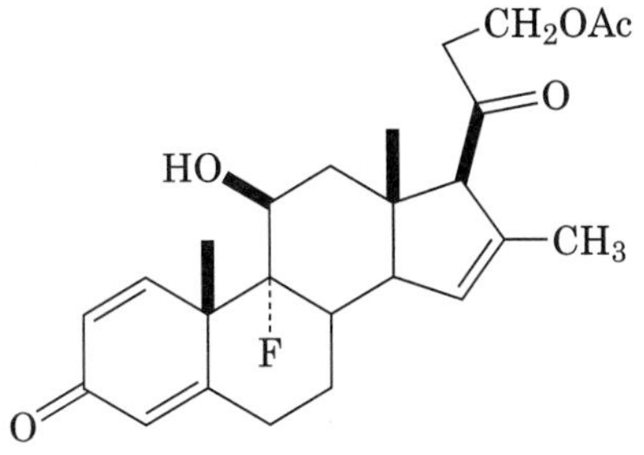

(181)

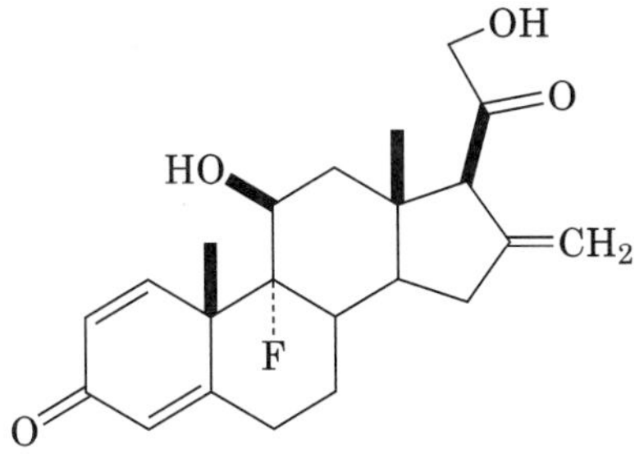

(182)

the systemic circulation, ubiquitous esterases effect hydrolysis of the ester moiety, thus rendering the drug impotent. There is essentially no conceptual difference between Lee's antedrug and Bodor's inactive metabolite approach to soft drugs, only a difference in the position of the carboxylate group. Whereas Bodor focused on derivatization of corteinic acid (225), a C-20 acid, Lee chose 20-dihydroprednisolonic acids, or cortolic acid-like structures (a C-21 acid) such as (224), for elaboration into active prodrugs. In these cases, it is the prodrug that is bioactive, not the liberated steroidal skeleton. In contrast, betamethasone dipropionate is itself inactive and depends on esterase cleavage for release of the active species. Lee expanded this concept beyond the C-21 metabolite to encompass carboxylic acid esters at unnatural positions of the steroid nucleus, such as C-16 carboxylic acid esters and amides (183; R_1 = O-alkyl,

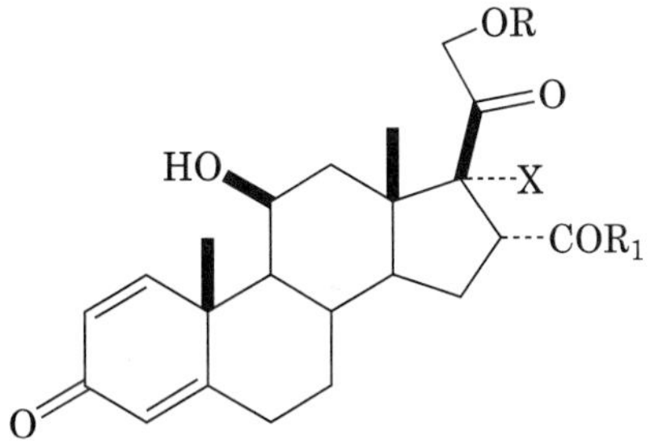

compound (182) has only one-third the anti-inflammatory action of (181) and produces sodium excretion.

The 16β-methoxy group reduces the anti-inflammatory potency of 9α-fluorocortisone 21-acetate (247). 16,16-Dimethylprednisone 21-acetate is inactive in the systemic granuloma test (267).

Along the lines of Bodor's "inactive metabolite approach" or soft drugs (see Section 10.19), Lee coined the term *antedrug* for his concept to topical-selective delivery of corticosteroids (268, 269). In this approach, an inactive corticosteroid side-chain metabolite (generally a carboxylic acid) is synthetically converted to an ester, whereupon topical activity is obtained. However, on absorption into

(183)

NH_2) (270–272). The carboxyl group can be incorporated as part of an otherwise beneficial 16,17-acetonide grouping, as in (189) (273). More recently, Lee described 16,17-acetonide-like, fused isoxazolidine esters, (190) (274).

Esters of (183), where R_1 = Me (184), Et (185), i-Pr (186), or benzyl (187), synthesized

Table 15.36 Anti-Inflammatory Potencies of Ester and Acid Derivatives of 183: Croton Oil Ear Edema Assay of 184–188

Compound	X	R	R_1	Relative Potency	Log P
Prednisolone				1	1.48
(184)	H	H	CH_3	1	1.57
(185)	H	H	Et	1.3	1.58
(186)	H	H	i-Pr	4.0	1.62
(187)	H	H	$CH_2C_6H_5$	4.7	1.66
(188)	H	H	OH	inactive	

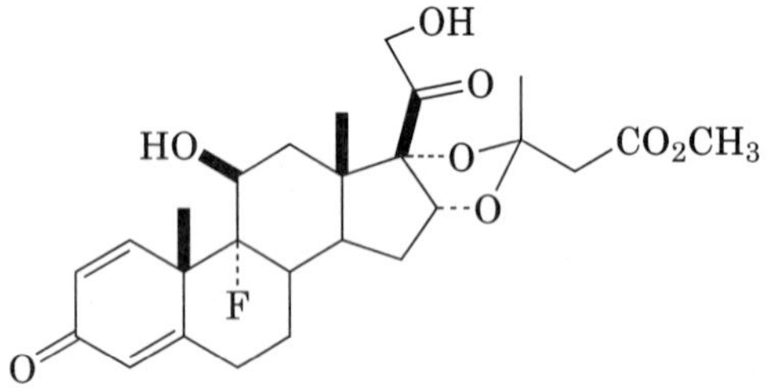

(189)

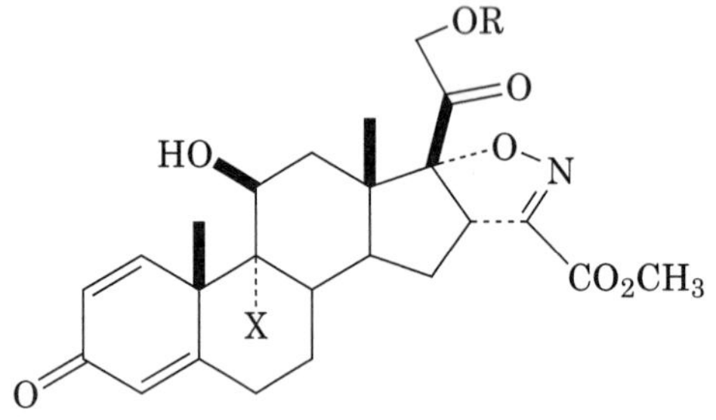

(190)

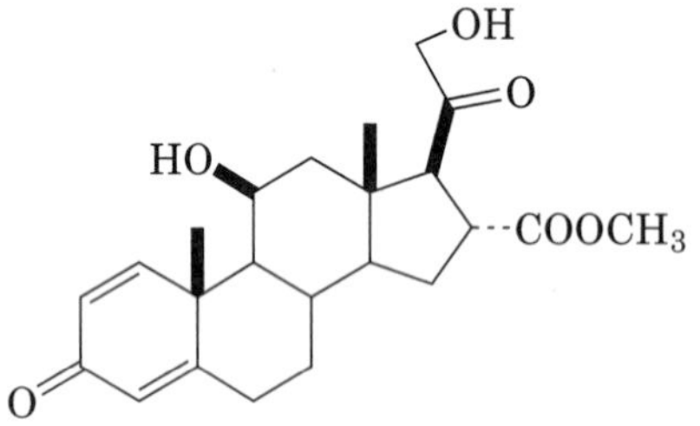

(184) X = H
(194) X = F

from prednisolone in seven steps, had relative potencies ranging from 1 to 4.7 in the croton oil ear edema assay (Table 15.36).

Whereas methyl and ethyl esters (184) and (185) were roughly equipotent with prednisolone, more lipophilic isopropyl and benzyl esters had substantially improved potency. It has been suggested that increases in lipophilicity can be correlated with skin permeability, and this explanation has been invoked by Lee to explain the improved potency of (186) and (187) compared to that of (184/185). Perhaps most noteworthy is the absence of detectable glucocorticoid activity for the free acid (188), the putative metabolite of (184–187), thus supporting the antedrug concept. Further

support for separation of topical and systemic effects was garnered from the rat cotton pellet granuloma assay (CPG). Body weight, thymus weight, and adrenal weight were obtained at the ED_{50} dose, on the basis of which it was concluded that (186) and (187) had local to systemic relative activity ratios of 6.1 and 6.4, whereas prednisolone showed rather less separation, with a therapeutic index of 1.6. The ester (184) and its 17-hydroxylated analog (191) were compared in the CPG assay and

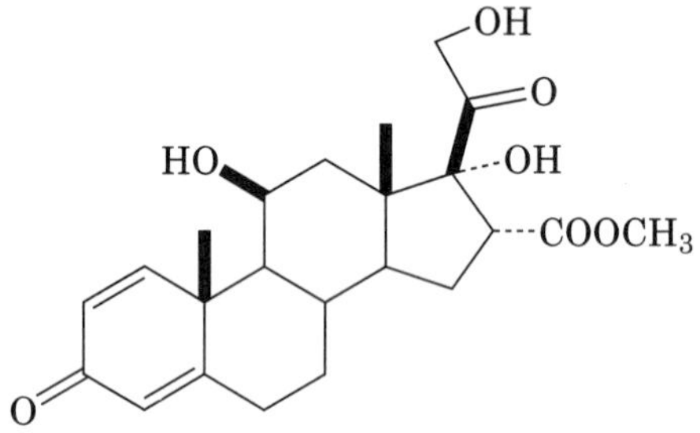

(191) R_1, R_2 = H
(195) R_1 = COCH$_3$, R_2 = H
(196) R_1 = COCH$_3$, R_2 = COCH$_3$

found to have IC_{50} values of 4.1 and 0.4 mg/pellet, whereas the value for prednisolone was 2.2 mg/pellet. Thus, in the CPG assay, (184) was roughly half as potent as prednisolone, whereas (191) was about 5.5 times more active. Neither compound showed increased systemic anti-inflammatory activity compared to that of prednisolone; (191) did reduce thymus weight by 33% in contrast to 47% for prednisolone. Although (192), the putative acidic metabolite of (184), did not have anti-inflammatory activity in the CPG, surprisingly, the

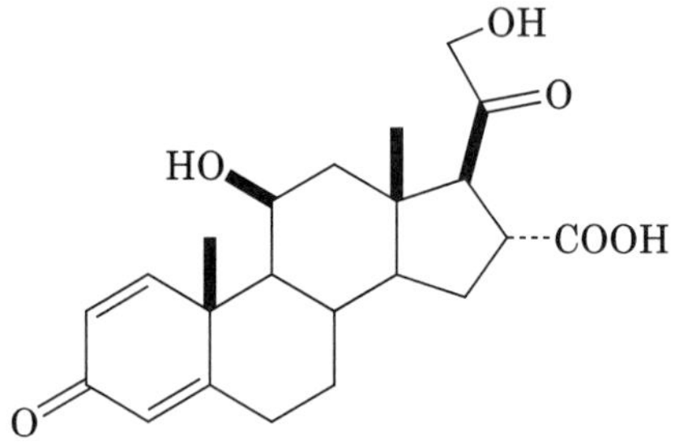

(192)

(193)

acidic metabolite of (191), (193) did produce significant local anti-inflammatory activity.

Insertion of F at the 9α-position of (184) to give (194) was accompanied by a twofold enhancement in topical potency, as determined in the croton oil ear edema assay. There were no concomitant increases in untoward systemic effects for (194) compared to those for prednisolone; thymic weight was minimally impacted and HPA suppression was not observed (275).

Acetylation or diacetylation of the 21 or 17,21 alcohols of (183) was examined. The 21-monoacetate (195) was about 50% more active than prednisolone in the CPG assay, whereas the corresponding diacetate (196) was three times more active than prednisolone in the same system (275). In the croton oil ear edema assay, the mono- and diacetates were substantially more active than prednisolone. These acetates appeared to be systemically inconsequential, again presumably attributed to enzymatic hydrolysis to inactive carboxylic acids.

Other interesting derivtives of (183), with R_1 = NH_2 and X = H or OH, were examined in the CPG assay. They were 20–40% as topically potent as prednisolone and were without systemic effect.

Studied in some detail, acetonide-ester (189) presented promising results as a topically selective anti-inflammatory steroid when compared side by side with several well known corticoids. As shown in Table 15.37, (189) had intrinsic potency at the receptor level comparable to that of prednisolone.

In the croton oil ear assay, doses required to achieve the ED_{50} values allowed determination of a topical potency ranking for dexamethasone of 1; (189), 0.5; prednisolone, 0.17; and triamcinolone, 0.04. These results correlate well with receptor binding data for dexamethasone and (189). It is satisfying to note that, although (189) is nearly as potent as dexamethasone, it has only a fraction of the systemic activity evidenced for dexamethasone in the CPG assay.

For the novel isoxazolidine (190) recently reported by Lee, when X = H, the resulting compounds were not improved relative to prednisolone. However, when X = F and R = H (197) or Ac (198), useful topical activity was achieved. Based on ID_{50} doses, the relative potency (prednisolone = 1) for (197) was 4 and for (198), 5.3.

In the croton oil ear edema assay, (197) did show signs of systemic absorption as gauged on the contralateral ear, but (198) did not. It was also consistently found that plasma corticosterone levels were reduced for (197) but not for (198). It is clear that 9α-fluorination is required to improve potency in this class, whereas the acetate group at C-21 is needed to help avoid systemic absorption. These results

Table 15.37   Receptor Binding Affinity, Croton Oil Ear Edema Assay, and Cotton Pellet Granuloma Assay of 114 versus Three Common Control Corticosteroids

Compound	IC_{50}, GCR[a]	ID_{50}, CEE[b]	CPG, Topical[c]	CPG, Systemic[d]
Dexamethasone	17 nM	0.001 mg	81%	70%
Triamcinolone	46	0.026	62	4
Prednisolone	35	0.006	57	14
(189)	30	0.002	55	5.5
Control	—	—	—	—

[a]Glucocorticoid receptor binding affinity.
[b]Croton oil ear edema assay.
[c]Percentage inhibition of inflammation on treated side.
[d]Percentage inhibition of inflammation on untreated side.

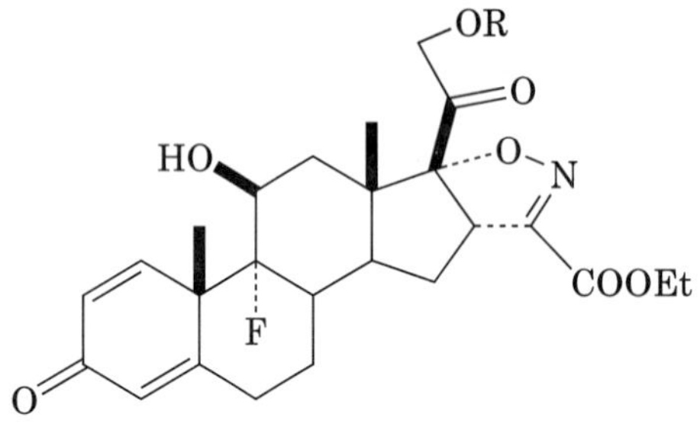

(197) R = H
(198) R = Ac

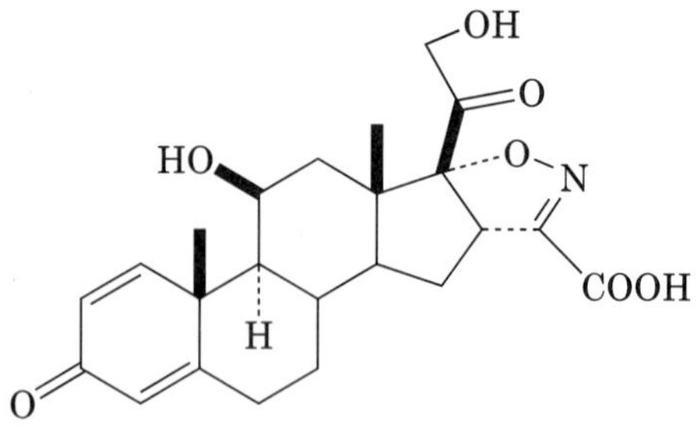

(199)

are difficult to reconcile with the antedrug concept and, interestingly, it was found that the putative metabolite (**199**) did display (arguably) systemic activity of 19%.

Subsequently, these researchers examined the effect of deleting the carboethoxy group, as shown in Table 15.38. First, potency is retained in the simple ear edema assay with potencies relative to hydrocortisone of 1.9–6.8 for compounds with or without 9α-fluorination, or with/without an acetyl group at C-21 (276).

In the contralateral ear assay, although these new compounds are less potent than either (**197**) or (**198**), there does not appear to be any sign of systemic absorption (Table

15.39). Furthermore, after semichronic systemic dosing with 4–5, no effects on weight were seen for overall body weight, adrenal gland, or thymus. However, under these conditions, potent weight loss was seen for prednisolone (e.g., 53% reduction in thymus weight).

10.17 Alterations at C-17

The 17α-hydroxy group is not essential for activity. The introduction of fluorine, chlorine, or bromine into the 17α-position of 11-dehydrocorticosterone gives compounds of inferior activity (277–279). The activity of 16,17-ketals and 17-esters has already been discussed.

Table 15.38 Estimated ED$_{50}$ and Relative Potencies in the Acute Ear Edema Assay[a,b]

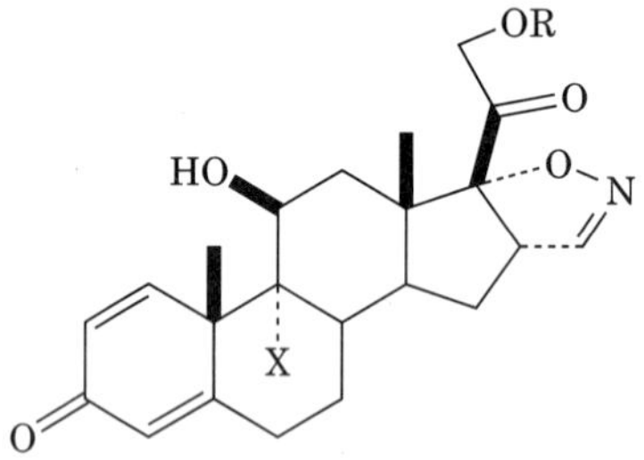

(200–203)

Compound	R	X	ED$_{50}$ (μM)	Relative Potency
Hydrocortisone	—	—	1.36	1
Prednisolone	—	—	0.64	2.2
(**200**)	H	H	0.71	1.9
(**201**)	H	F	0.33	4.1
(**202**)	Ac	H	0.42	3.3
(**203**)	Ac	F	0.20	6.8

[a]Ref. 276.
[b]Croton oil–induced edema.

Table 15.39   Contralateral Ear Edema Assay for Compounds (200–203)[a]

Compound	R	X	% Inhibition[b] Left Ear	% Inhibition[b] Right Ear	L/S Ratio[c]
Prednisolone	—	—	57.8	38.2	1.5
(200)	H	H	45.7	15.9	2.9
(201)	H	F	55.2	21.2	2.6
(202)	Ac	H	49.1	19.0	2.6
(203)	Ac	F	57.0	19.3	2.9

[a]Ref. 276.

[b]Compound and croton oil applied to right ear, croton oil only to the left ear. "Local effect" measured as ear thickness at 5 h.

[c]Ratio derived from local/systemic anti-inflammatory activities; local effect measured after 5 h, systemic effect determined after 5 once-daily applications.

17α-Methylcorticosterone 21-acetate has nearly half the activity of cortisol in the glycogen deposition assay (280). Rimexolone (23) (Fig. 15.2), in which the 16,17- and 21-positions have methyl groups instead of hydroxyl groups, is a moderately potent topical anti-inflammatory agent. Interestingly, (23) has undetectable dermal atrophogenicity, adrenal, or thymolytic effects (281). Apparently, rimexolone is systemically impotent as a consequence of rapid metabolism to inactive 6-hydroxylated or 5-reduced metabolites. Three rationally proposed but hypothetical metabolites, 6β-hydroxyrimexolone (204) and both 5β- and 5α-dihydrorimexolone (205 and (206), respectively), displayed poor if any binding affinity in human glucocorticoid receptor binding assays (282). On the other hand, hydroxylation in the side chain (C-21) did provide potent derivatives (207) and (208), albeit with less activity than that of the control, dexamethasone (20a), as shown in Table 15.40.

In the same study, binding affinities were determined for flunisolide (26) and its putative metabolite, 6β- HO-(26). As can be seen,

hydroxylation at C-6 leads to almost indectable binding affinity. It was thus proposed that the high ratio of topical to systemic activity noted in previous pharmacological studies for both rimexolone and flunisolide could be explained by rapid systemic metabolism to inactive metabolites.

The classical hydroxy-ethanone corticosteroid side chain attached at C-17 is not a requirement for activity, as was seen for 17α-propynylated steroids. The glucocorticoid receptor is apparently quite tolerant of substitution at this position. Noteworthy in this regard are 17-thioketal derivatives, such as tipredane (209), which retain potent receptor binding affinity and topical anti-inflammatory activity. These androstene-17 thioketals have relative potencies in mouse ear edema assay, glucocorticoid receptor binding, and inhibition of DNA synthesis, as shown in Table 15.41.

It is clear from the significant binding affinities of these androstenes that they possess intrinic activity at the receptor level, which is in many cases competitive with halcinonide or

Table 15.40   Relative Binding Affinities (RBA) of Various Steroids and Rimexolone to the Human GR[a]

Compound	RBA	Compound	RBA
Dexamethasone, (20a)	100	(204)	<1
TA,[b] (13)	233	(205)	8
Flunisolide, (26)	191	(206)	<1
Rimexolone, (23)	134	21R-(207)	41
Hydrocortisone, (2a)	10	21S-(207)	107
6β-hydroxy-(26)	<1	(208)	34

[a]Human synovial tissue.

[b]Triamcinolone acetonide.

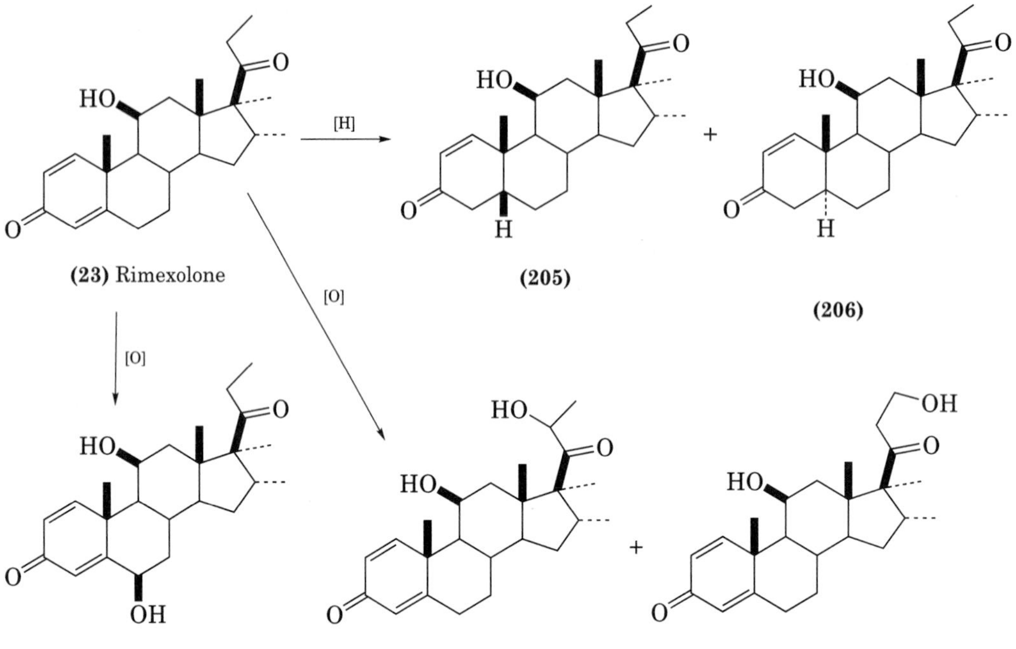

(23) Rimexolone   [H] → (205)   +   (206)

[O]

(204)   (207)   +   (208)

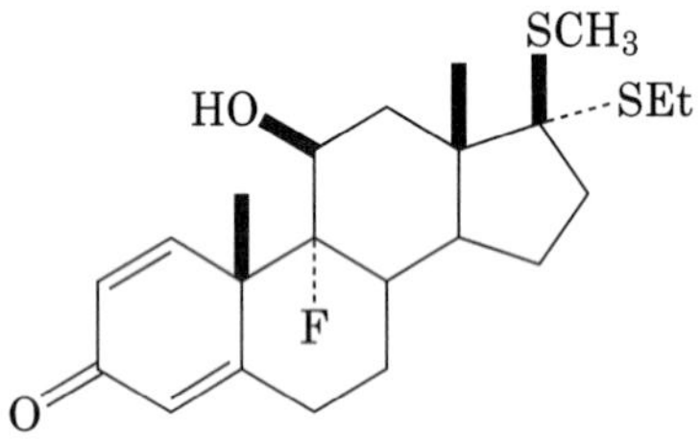

(209)

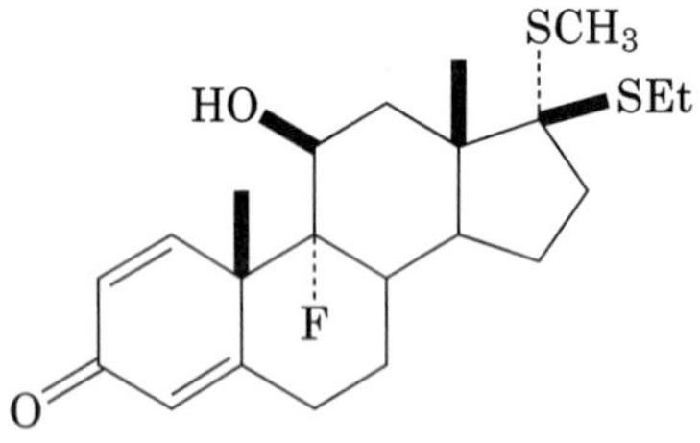

(212)

(213)

synthesis or reduction of inflammation indicate a reduced activity of (212) versus tipredane (209). That these differences might have a bearing on cellular metabolism seems verified by the finding that fluorination of the SEt group (i.e., 213), which has a metaboli-

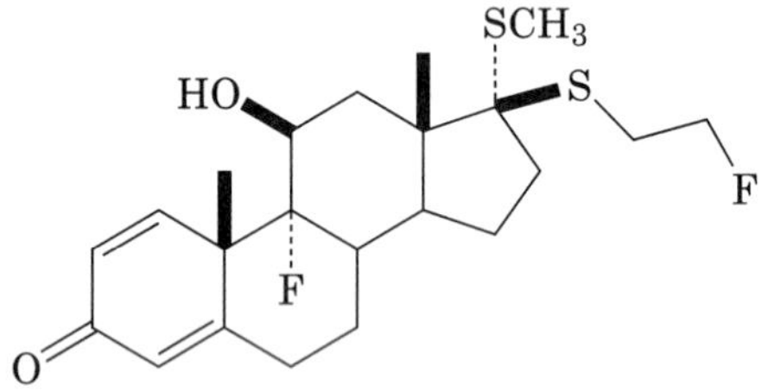

cally more robust 17β- C_2H_4F group, now demonstrates lower receptor binding affinity but greatly enhanced DNA synthesis inhibition and anti-inflammatory activities compared to those of its nonfluorinated counterpart, (212).

The more potent of these thiasteroids, tipredane (209) and (213), were the subject of more intense pharmacological scrutiny (284). A comparison of a wide range of corticoste-

triamcinolone acetonide. Binding affinity is enhanced on interposing SMe with SEt groups (209) versus (212), amounting to a preference for the more bulky SEt group in the 17β-position. On the other hand, other measures of intrinsic activity such as inhibition of DNA

Table 15.41   Biological Potencies of Andostene-17-thioketals versus Reference Steroids[a]

(209) $R_1 = SCH_3$, $R_2 = SEt$
(210) $R_1 = SEt$, $R_2 = SEt$
(211) $R_1 = SCH_3$, $R_2 = SCH_3$
(212) $R_1 = SEt$, $R_2 = SCH_3$
(213) $R_1 = SC_2H_4F$, $R_2 = SEt$
(214) $R_1 = SCH_3$, $R_2 = SC_2H_4F$
(215) $R_1 = SEt$, $R_2 = SC_2H_4F$

Compound	GRB[b]	DNA[c]	EEA[d]
(205)	0.6	0.65	1.14
(206)	0.45	0.58	0.48
(207)	0.9	0.12	0.24
(208)	1	0.10	0.17
(209)	0.3	1.02	1.24
(210)	2	0.38	0.63
(211)	0.8	0.37	0.42
Halcinonide (12)	1	1	1.0
TA (13)[e]	1	0.28	0.44
Hydrocortisone (2)	0.2	0.007	0.08

[a]Ref. 283.
[b]Relative glucocorticoid receptor (rat liver cytosol) binding affinity.
[c]Thymidine incorporation into DNA of mouse thymocytes.
[d]Croton oil–induced mouse ear edema assay.
[e]Triamcinolone acetonide (13).

roids vs. these two thiasteroids is shown in Table 15.42. Although clobetasol butyrate demonstrated very high intrinsic activity, as gauged by receptor binding, it had only weak influence on thymic involution and was a relatively weak anti-inflammatory agent against edema, being more readily compared to hydrocortisone valerate.

Tipredane and (213) are both bound to the GR and have markers of high anti-inflammatory potency, as indicated by their inhibition of DNA synthesis and reduction of edema relative to controls. Their lack of effect on thymic involution is unexplained, but clearly metabolism of these steroids is more complicated than for a classical corticosteroid having the usual 20-keto-21-ol functionality. For example, tipredane (209) is avidly metabolized in mouse liver preparations or in human skin in less

than 1 h to a plethora of metabolites having greatly reduced anti-inflammatory activity. The primary pathways (Fig. 15.15) are oxidation of (209) to sulfoxides, followed by loss of thioketal grouping and formation of androsten-17-ol, -one, and -thiols (285).

A series of related thioketals, also situated at the 17-position of an androstan-17-one core, have been described that employ an antedrug strategy (see below) (286). In this approach, a 17β-thiol moiety is attached to a butyrolactone ring in the active drug. Rapid hydrolysis in plasma leads to relatively inactive metabolites.

This design emulates other topically potent corticosteroids such as fluticasone propionate (Flonase), in which hydrolysis of a side-chain ester group leads to inactive metabolites. As shown in Table 15.43, (216) and (217) are stable in buffer solution but are rapidly hydro-

Table 15.42 Pharmacological Profiles of a Variety of Corticosteroids versus (153) and (157)

Compound	GRB[a]	DNA[b]	EEA[c]	Thymus[d]
Clobetasol butyrate	2	0.23	0.22	0.04
Betamethasone Dipropionate	1.4	0.17	0.84	2
Betamethasone valerate	2.2	0.90	0.48	0.36
TA[e]	1	0.28	0.44	0.46
Halcinonide	1	1	1	1
FA[f]	0.86	0.73	0.83	2.8
Dexamethasone	0.72	—	0.12	—
Budesonide	0.57	0.63	0.68	0.33
Clobetasol propionate	0.57	1.6	0.98	0.70
Fluocinonide	0.43	0.12	0.98	0.93
Hydrocortisone valerate	0.34	0.04	0.11	0.01
Hydrocortisone	0.23	0.008	0.08	0.04
Tipredane, (209)	0.66	0.65	1.2	0.04
(213)	0.31	1	1.1	—

[a]Relative glucocorticoid receptor (rat liver cytosol) binding affinity.
[b]Thymidine incorporation into DNA of mouse thymocytes.
[c]Croton oil–induced mouse ear edema assay.
[d]Thymus involution following s.c. administered corticosteroids.
[e]Triamcinolone acetonide.
[f]Fluocinolone acetonide.

lyzed in human plasma to the relatively inactive hydroxyacid (219) (racemization). Likewise, the related species (218) is also stable in buffer but hydrolyzed rapidly in plasma to the inactive metabolite (220). All three of these antedrugs have half-lives on the order of minutes, clearly establishing that they would not likely have the usual constellation of systemic effects typical of inhaled corticosteroids. More remarkable is the fact that these derivatives are relatively stable to human lung S9 fraction ($t_{1/2}$ = hours), perhaps making them useful for treatment of asthma.

The thiolactones apparently act as conventional corticoids in that the intrinsic activity at the hGR receptor is on the order of dexa-

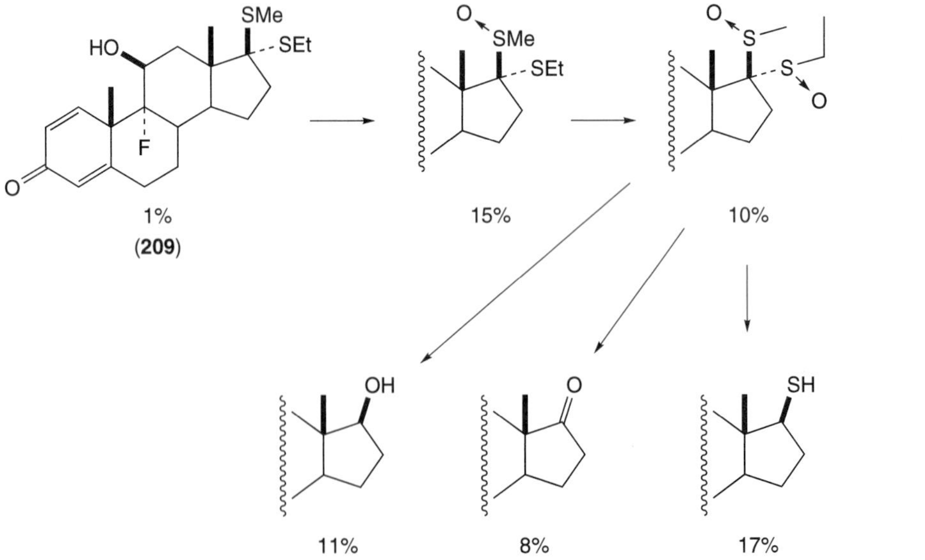

Figure 15.15. Metabolic profile of tipredane (209) after 0.5-h exposure to human skin (285).

Table 15.43 Stability Data and Relative _In Vitro_ Glucocorticoid Potencies of Selected Androstane Antedrugs

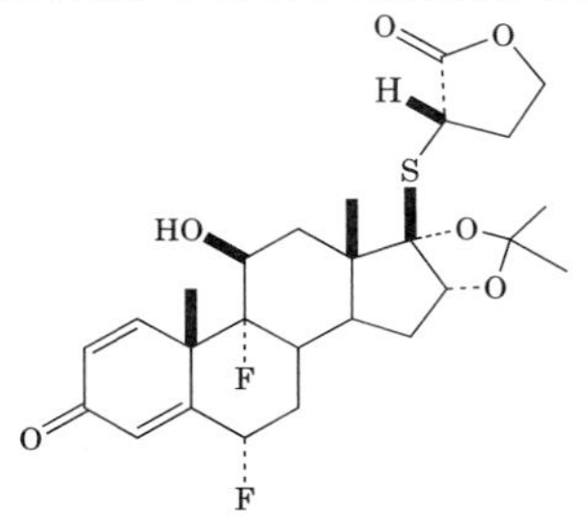

(216)

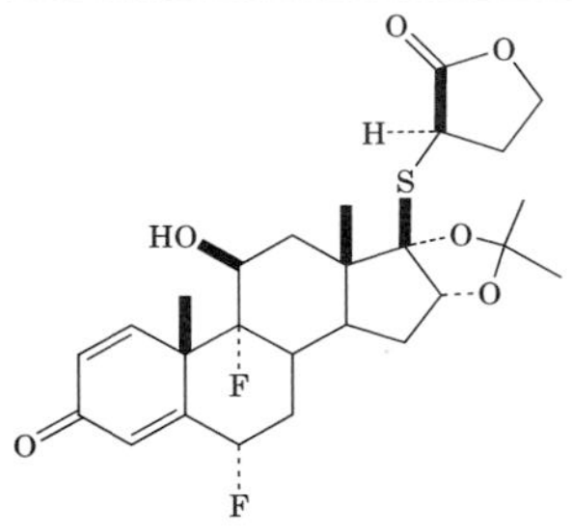

(217)

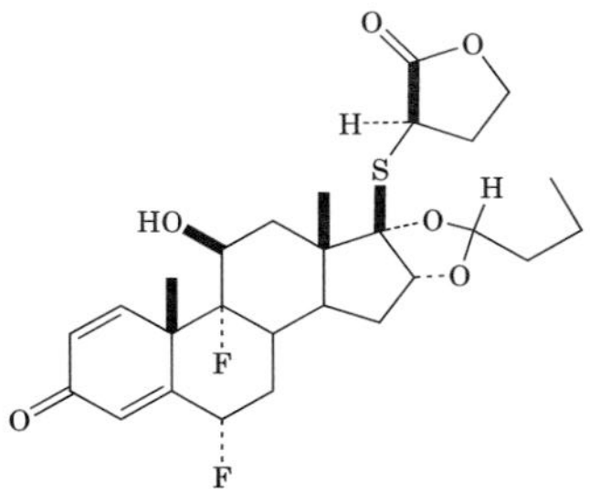

(218)

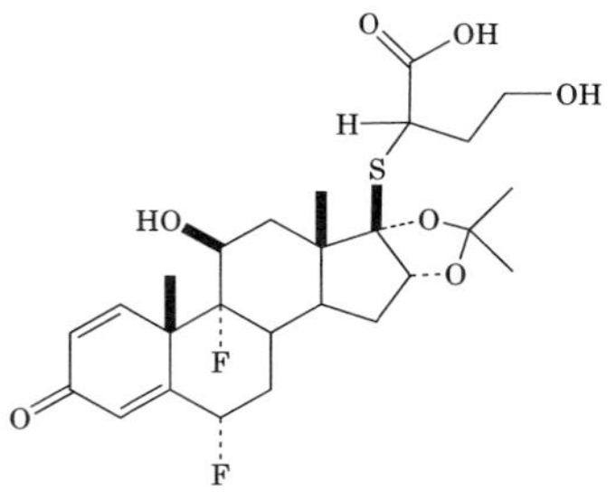

(219)

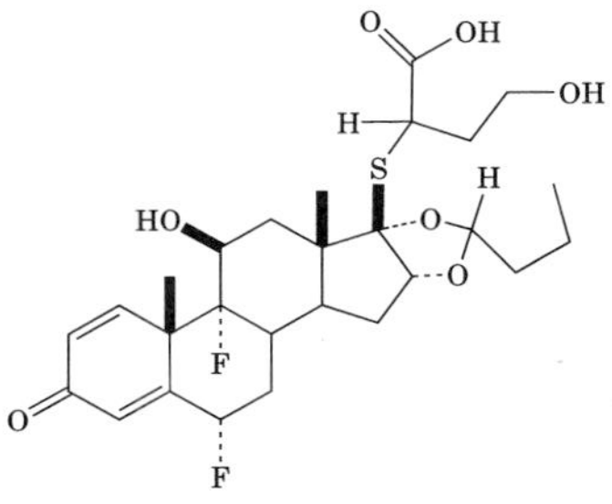

(220)

Compound	Stability in Buffer[a]	Stability in Plasma[b]	$t_{1/2}$ in Plasma[c]	hGR (IC_{50})[d]	Relative Potency[e]
(216)	74%	0%	4 min	—	13
(217)	90%	15%	4 min	1.4 nM	2
(218)	91%	0%	5 min	6.4 nM	1.5
(219)	—	—	—	289 nM	1200
(220)	—	—	—	400 nM	73
Dexamethasone	—	—	—	—	1

[a]Krebs Buffer, after 60-min exposure.
[b]Human plasma, after 60-min exposure.
[c]Half-life in human plasma.
[d]Relative affinity to human glucocorticoid receptor.
[e]_In vitro_ assay in HeLa cells; relative potency = ED$_{50}$ compound/EC$_{50}$ dexamethasone.

Table 15.44   Inhibition of Rat Ear Edema and Thymus Weights by Table 15.43 Antedrugs

Compound	Dose (μg)[a]	% Decrease (ear weight)[b]	% Reduction (thymus weight)[c]
(216)	25	9.01	
(216)	100	20.87	
(217)	25	15.20	
(217)	100	22.70	
(217)	1000		65
(218)	25	31.13	
(218)	100	35.70	
(218)	1000		37
Fluticasone propionate	0.25	34.66	
Fluticasone propionate	1	38.12	
Fluticasone propionate	4	38.56	
Budesonide	50		48

[a]Standard croton oil ear assay, compound applied in acetone to both ears.
[b]Decrease was calculated relative to control animals treated only with croton oil.
[c]Thymus involution test; thymus weight determined relative to controls after several days of dosing.

methasone itself. The *in vitro* cell studies parallel the receptor assays, again demonstrating excellent potencies for these three leads.

Compared with fluticasone propionate in the mouse model of inflammation (croton oil ear assay; see Table 15.44), (218) was less effective than fluticasone propionate, but still showed significant anti-inflammatory activity at 100-μg doses.

10.18   Alterations at C-20

Ketalization of the C-20 carbonyl group of corticoids with ethylene glycol gives ketals that retain biologic activity, although it is lower than that in the parent compound (287). The most active compound is the 9α-fluoro-6α-methylprednisone derivative (221), which is four times as potent as triamcinolone.

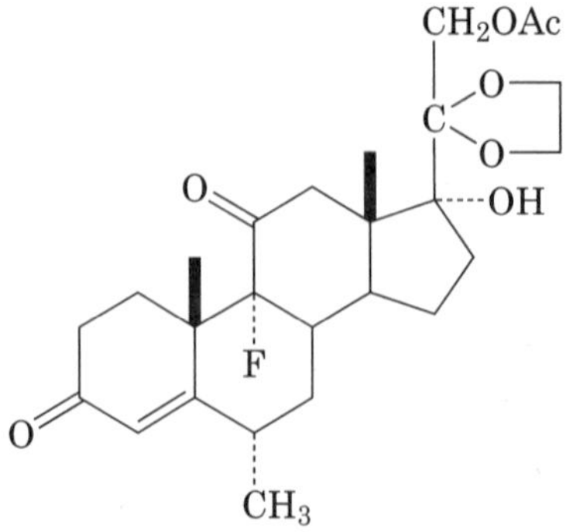

(221)

Whether the compound must first be hydrolyzed to the 20-ketone is not known, although it is stable in weak acid and enzymatic hydrolysis in liver brei did not occur. The related bismethylenedioxy compound (222) is more

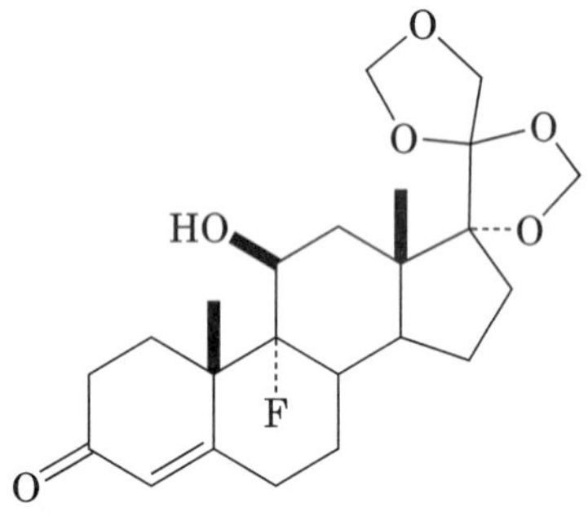

(222)

active than cortisol itself, although the activity is lower than that of the parent steroid (288).

10.19   Alterations at C-21

Reduction of cortisol to 21-deoxycortisol (new 90) gives a compound having about one-third the thymolytic activity of its parent (289). If activating groups are introduced into the molecule, clinically useful ophthalmic anti-inflammatory steroids such as medrysone (24) result. Medrysone has good anti-inflammatory activity with relatively little effect on intraocular pressure (290). Fluorometholone (**22**) is an efficacious topical agent (291) and also has little tendency to increase intraocular pressure when used as an anti-inflammatory agent in ocular diseases (292).

(22)

(24)

The propionate and butyrate 17-monoesters of 6α,9α-difluoro-21-deoxyprednisolone have high topical anti-inflammatory activity (293) (refer also to section 4.2.2 and ref. 4). Oxidation of cortisol to the 21-aldehyde gives a product that retains its anti-inflammatory activity (294), but it is not known whether this is attributable to reduction back to the primary alcohol.

In addition to Bodor's chemical delivery systems for soft drugs applied to corticosteroidal 3-spiro thiazolidines such as (**142**), another approach to soft drugs is the use of inactive metabolites. The inactive metabolite approach depends on enzymatic cleavage of a biologically active synthetic steroid to an inactive structure or metabolic product, and is shown diagrammatically in Fig. 15.16.

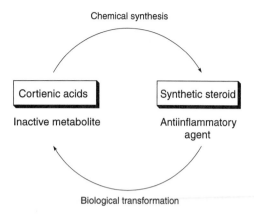

Figure 15.16. Bodor's inactive metabolic approach to soft corticosteroids.

Oxidative metabolic degradation of the dihydroxyacetone side chain of cortisol (**2**) proceeds in a stepwise fashion to provide the aldehyde (**223**), the ketoacid (**224**), and, finally, corteinic acid (**225**). The acidic metabolites corteinic acid (**225**) and the pyruvic acid, cortolic acid (**224**), are inactive and have been used as leads to design isosteric and/or isoelectronic replacements for potent corticosteroids.

A steroid containing a 21-carboxylic acid ester, fluocortin butyl (**226**), shows significant activity on human skin, but little or no systemic action in rats (295). Related diesters of 17α-hydroxyandrostane-17β-carboxylic acids, such as (**227**), are active topically (296).

Building on these leads, but with special attention to isosteric/isoelectronic principles from molecular modeling, Bodor designed a class of corteinic acid esters, (**228**), having a carbonate group at the 17α-position (221).

These compounds have demonstrated topical efficacy in the cotton pellet granuloma (CPG) assay, with apparent diminished systemic action, as gauged by noneffect on thymus weight. Representative examples, shown in Table 15.45 , have the Y group = H.

It can be seen that the soft steroids are very potent and that the 9α-F derivative (**231**) shows good activity at a much lower dose than reference steroids while having no systemic effect. In contrast, widely used reference steroids such as betamethasone valerate required a much higher dose to approach the degree of topical activity demonstrated by

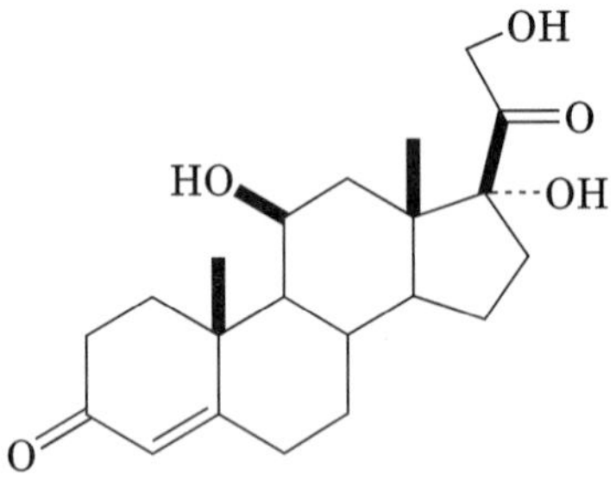

(2)

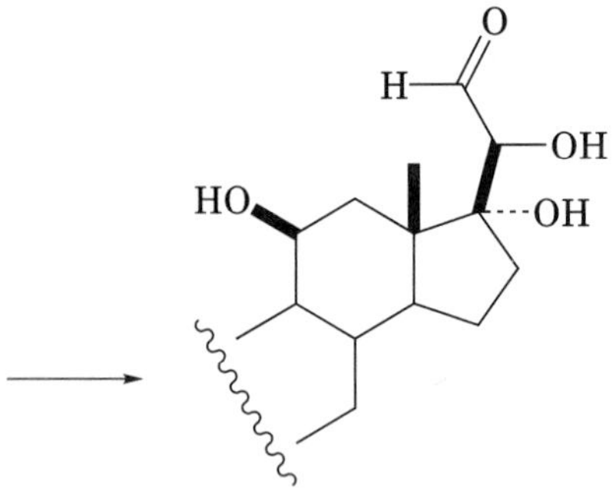

(223)

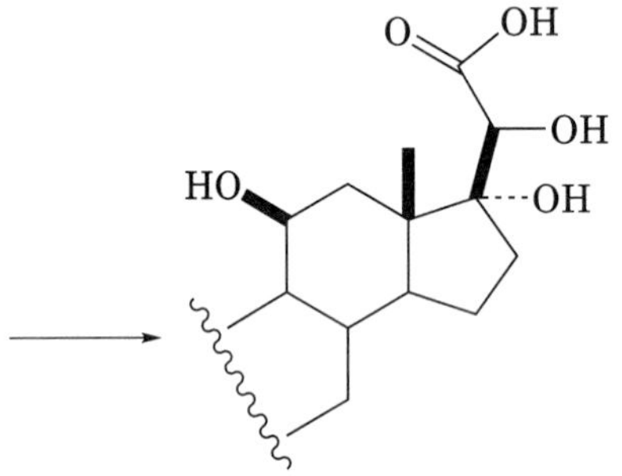

(224)

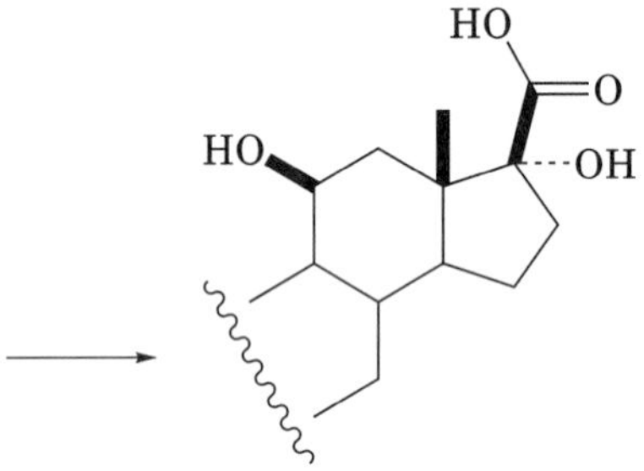

(225)

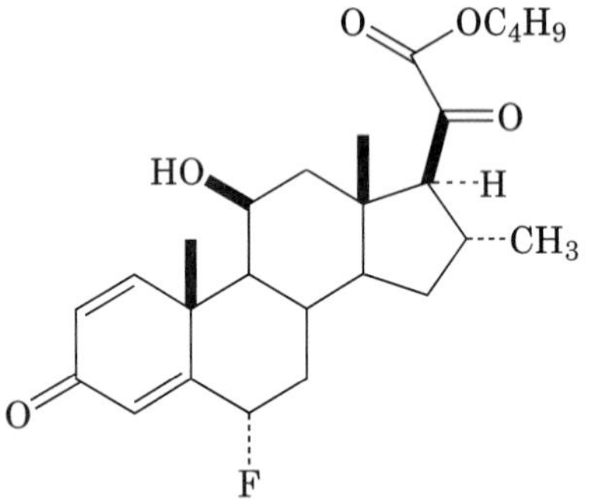

(226)

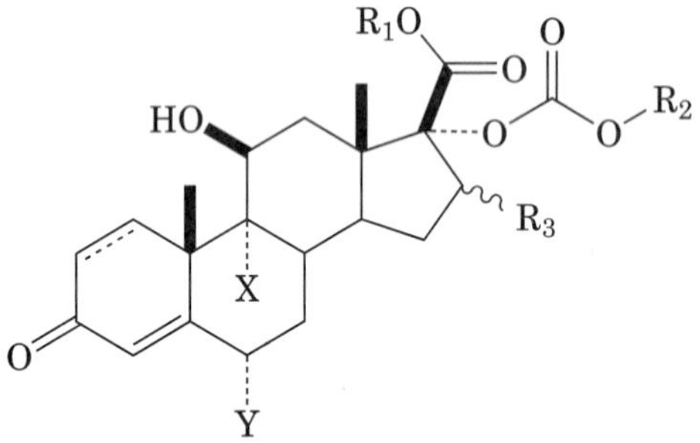

(228) R_1, R_2 = alkyl, haloalkyl, alkoxyalkyl, etc.
R_3 = H, OH, CH_3
X, Y = H or halogen

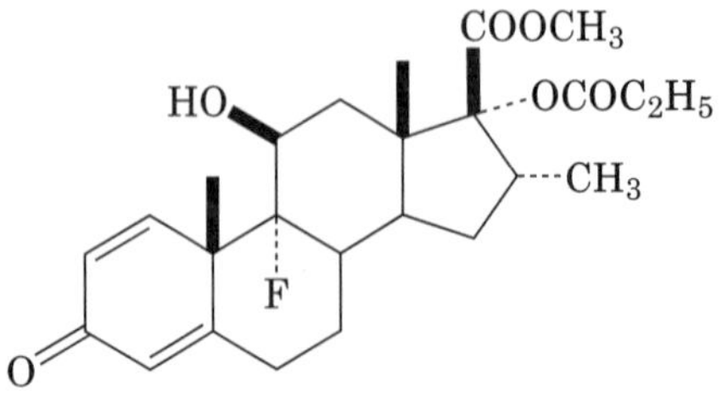

(227)

(**231**), at which point signs of systemic absorption were prominent. Potencies were determined for (**230**), (**231**), and (**232**) relative to that of BMV (=1) in the granuloma assay. They were determined at the ED_{50} to be 4, 430, and 474, respectively. Clearly, both (**231**) and (**232**) demonstrated excellent topical efficacy relative to that of betamethasone esters, being more than 400 times more active than

Table 15.45　Effect of Locally Administered Selected "Soft Steroids" and Reference Steroids on Granulation Tissue Formation Caused by Implantation of Cotton Pellets in Rats

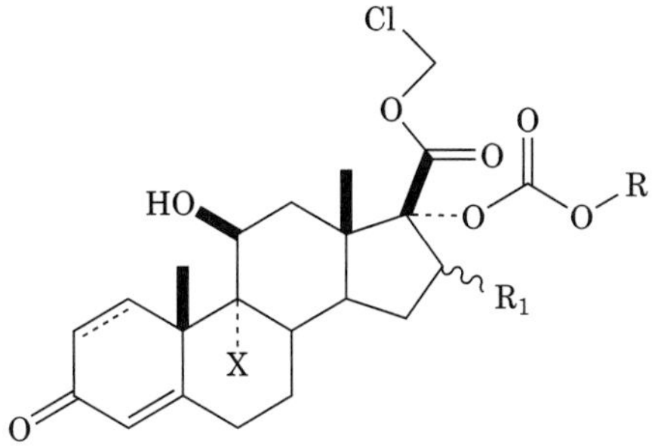

Compound	X	R	R_1	Dose[a]	Body Weight[b] (%)	Inhibition[c]	Thymus[b]
None					32		
(229)	H	i-Pr	H	1000	34	70	—
(230)	H, Δ^1	i-Pr	H	1000	29	77	—
(231)	F, Δ^1	C_2H_5	α-CH$_3$	10	36	73	—
(232)	H, Δ^1	i-Pr	β-CH$_3$	10	33	63	—
HC-17Bu[d]				1000	26	56	78
BMV[e]				1000	5	53	83

[a]Dose expressed in microgram per pellet.
[b]Changes in body weight (gain, in grams) and thymus (% reduction in weight) indicative of systemic effect.
[c]Percentage inhibition of granulation of cotton pellet implanted under skin along with dose of steroid.
[d]Hydrocortisone 17-butyrate (2a).
[e]Betamethasone valerate (5c).

(233)

BMV. A therapeutic index for (230) was determined, by the ratio of the relative anti-inflammatory potency and the relative systemic potency (% thymus weight reduction), to be 48. In another slightly different example, (233), where R = Me, R_1 = α-Me, X = F with Δ^1 unsaturation, the therapeutic index was over 7000. In the case of (233), the absolute relative potency was still over 200 compared to that of HC-17Bu, or over 50 compared to that of BMV.

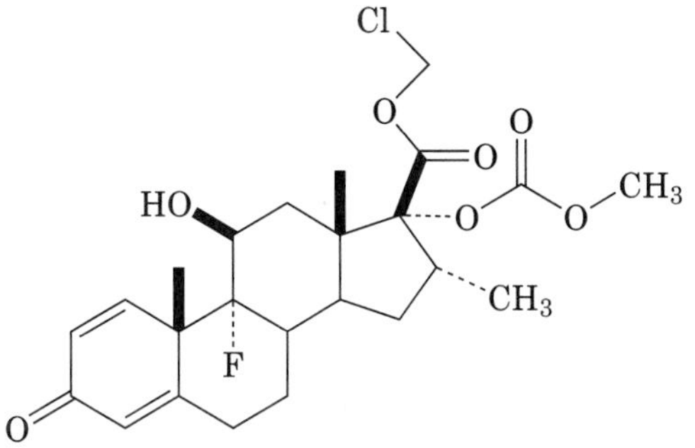

Many compounds in the (228) series, where, for example, X or Y was F, were as potent as betamethasone valerate or clobetasol propionate in the human vasoconstrictor assay. Thus, although it appears that these steroids do not achieve systemic circulation in substantial concentrations on the basis of the foregoing data, they are exceedingly topically potent. When these compounds were injected subcutaneously in mice at relatively high doses, some signs of systemic toxicity were noted that were nonetheless attenuated relative to control.

Compound (231) has been examined for its effects on cell growth and wound healing. It is generally accepted that corticosteroids adversely affect healing and inhibit cell growth. Compared to betamethasone dipropionate, wounds treated with (231) healed twice as fast and were comparable to untreated groups. Studies have also demonstrated the usefulness of these soft corticosteroids in corneal healing (297).

The introduction of an additional methyl group at C-21 to produce a secondary alcohol gives compounds with about 1/2 the activity of

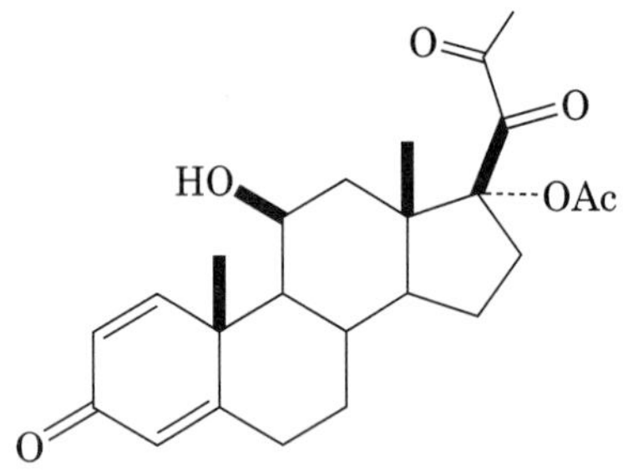

(234)

the parent structure (298, 299). The related 20,21-diketone (234) is also active (300). Replacement of the hydroxyl group at C-21 by a fluorine atom (301) doubles the anti-inflammatory activity. The effect of substitution of the 21-hydroxyl function by chlorine in steroids with a 16α,17α-acetonide group depends greatly on other molecular substituents (302). When the 9α-fluoro substituent is present in the parent compound a useful compound, halcinonide (12), results. The introduction of diazo and azido groups into prednisolone gives compounds that retain moderate amounts of activity (303, 304).

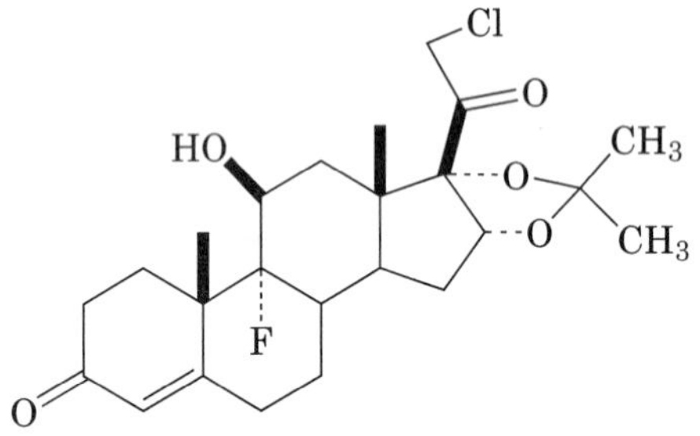

(12)

The combination of 21-chlorination with a soft drug or antedrug approach has been described by formation of functionalized 17-esters. Mitsukuchi and colleagues (305) prepared a series of 17α-succinoyl esters (235–247) and observed topical potency in the McKenzie assay. The potencies illustrate the enhancing effect of 21-chlorination on topical efficacy. As illustrated in Table 15.46, when R = OH, the resultant structure is somewhat reminiscent of betamethasone valerate (BV, relative topical potency of 1) (305). One ob-

serves in this instance a drop in potency upon incorporation of a polar ester tail, as in (235). Esterification of the 21-hydroxyl group, giving, for example, (238) (a mixed ester very similar to betamethasone dipropionate), is reflected in its good relative potency. However, in this series, the effect of ester substitution in the 17α-side chain is to decrease overall activity. This loss is clearly mitigated in some instances by incorporation of 21-halogen, the effect of which is seen for the various succinoyl esters (239–247). When the ester group is methyl (239), topical potency is better than that of BV. The bulkier esters, ethyl through cyclohexyl, are relatively topically weak. Certainly, (239) is of interest because it incorporates an ester that is capable of enzymatic cleavage to inactive metabolites while maintaining excellent topical potency in humans.

In closely related studies, Ueno et al. (40) showed that structures of this type have very low thymolytic involution activity in the granuloma pouch assay and, further, that human skin metabolites are lacking in activity. In Table 15.47, relative potencies and receptor binding values are provided for the esters (248–254).

Taken together with the lack of systemic absorption signaled by undetectable thymic involution in the granuloma assay, these compounds clearly fulfill the necessary criteria for antedrug status. Of these compounds, the methyl ester (248) was most potent. More bulky esters such as isopropyl (252) or tert-butyl (253) were much less active than (248), perhaps not surprisingly on the basis of the foregoing discussions.

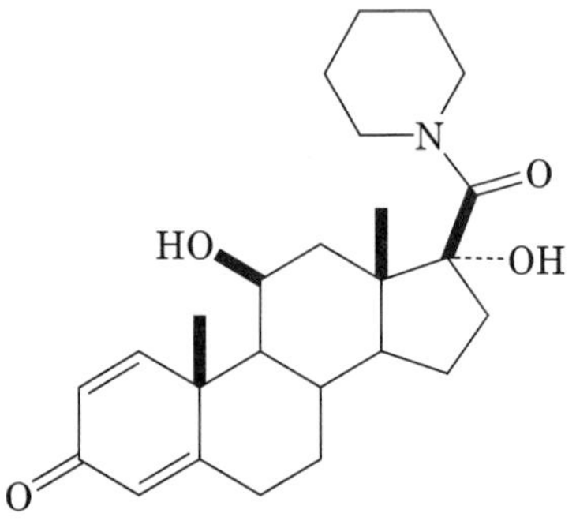

(255)

Table 15.46 Vasoconstrictive Assay of 17α-Succinoyl Esters of Betamethasone[a]

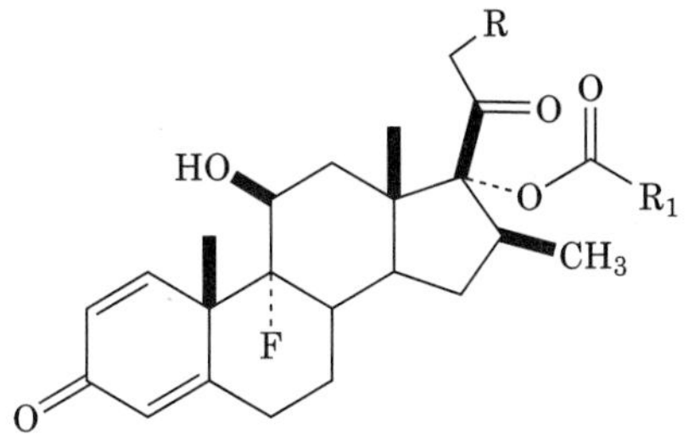

(235–247)

Compound	R	R_1	Topical Potency[b]
(235)	OH	$CH_2CH_2CO_2CH_3$	36%
(236)	$OCOCH_3$	$CH_2CH_2CO_2CH_3$	41
(237)	OCOEt	$CH_2CH_2CO_2CH_3$	36
(238)	OCOPr	$CH_2CH_2CO_2CH_3$	79
(239)	Cl	$CH_2CH_2CO_2CH_3$	118
(240)	Cl	$CH_2CH_2CO_2Et$	17
(241)	Cl	$CH_2CH_2CO_2Pr$	25
(242)	Cl	$CH_2CH_2CO_2i$-Pr	33
(243)	Cl	$CH_2CH_2CO_2Bu$	25
(244)	Cl	$CH_2CH_2CO_2i$-Bu	33
(245)	Cl	$CH_2CH_2CO_2c$-pentyl	29
(246)	Cl	$CH_2CH_2CO_2c$-hexyl	21
BV[c]			100

[a]Ref. 305.
[b]McKenzie assay in human volunteers; BV = control.
[c]BV = betamethasone valerate (5c).

An amide at C-20, as in (255), gives steroids that retain more than half the activity of the parent compound in the liver glycogen test (306).

Sulfur-containing substituents at C-21, such as 21-mercapto, 21-methylthio, 21-thiocyano, and 21-acetylthio, gave corticoids as of 1961 that were not of biologic interest (307). Thirty years later, an acetyl cyteine derivative at C-21, (256), was shown to possess about a tenth of the activity of dexamethasone. On the other hand, systemic absorption appeared to be curtailed on the basis of thymus weight, body weight, and adrenal-suppressive data (308).

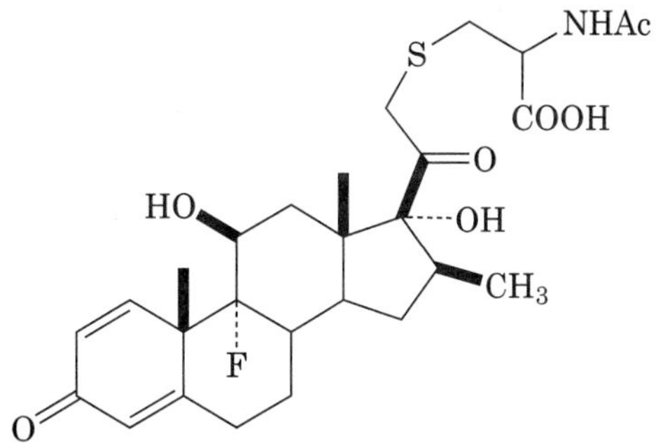

Replacement of the 21 carbon by S, on the other hand, provided 21-thioesters related to (227/228) that have found clinical utility. The trifluoro-thioester (16) is a leading example in the series (257–266) and (16) that is reminiscent of Bodor's and/or Laurent's esters. Metabolism studies have revealed that (16) ad-

(256)

heres to the soft drug concept in providing an inactive corteinic acid-like metabolite, (267) (see page 835).

Although fluticasone propionate (16), marketed by Glaxo as Cutivate, exhibits some HPA suppression when administered subcutaneously, it does not posssess oral antiglucocorticoid activity, an important consider-

Table 15.47   Activity of 17α-Alkanoate Esters of 21-Desoxy-21-Chlorobetamethasone[a]

(248–254)

Compound	R	n	EEA[b]	Thymus[c]	GRB[d]
(248)	CH_3	2	1.30	—[e]	5.18
(249)	CH_3	3	0.45	—[e]	—
(250)	Et	2	0.56	—[e]	—
(251)	Et	3	0.68	—[e]	9.68
(252)	i-Pr	2	0.11	—[e]	12.1
(253)	t-Bu	2	0.19	—[e]	10.8
(254)	H	2	<0.01	—[e]	>150
CP[f]			1	1	3.17
BMDP[g]			0.1	0.03	5.25

[a]Ref. 40.
[b]Croton oil ear edema assay.
[c]Thymolytic involution in the granuloma pouch assay.
[d]Glucocorticoid receptor-binding assay; IC_{50} values, nM.
[e]Not detectable at the highest dose.
[f]Clobetasol propionate.
[g]Betamethasone dipropionate.

ation for inhaled drugs in which much of the metered dose is actually swallowed (309). Bioassay data for these thioesters are shown in Table 15.48. Human vasoconstriction data reveal, for example, that when the 6α-position is unsubstituted, Y = H or (259), a very potent steroid results. Unfortunately, this same analog (259) is substantially more active systemically than (16). Interestingly, replacement of the 9α-position of (259) with Cl, resulting in (258), provides an analog of potency in the realm of Cutivate but, again, with substantial systemic activity; further, the 6α-F group in this series generally appears to be a predictor of systemic activity. Perhaps not surprisingly, both Br or I = X in this series were detrimental to activity; the ranking of potency for X is F > Cl > Br, I.

In the series (256–266), potency is dependent on the size of the 17-ester group, where the R = Me or CH_2CH_3 esters are more active than the esters where R = $CH_2CH_2CH_3$.

Lee's antedrug concept has been applied to steroidal 21-esters, which on ester hydrolysis lead to inactive steroidal acids (see Section 4.1). First reported in 1982, methyl prednisolonate (268) and its diastereomeric 20-alcohols, methyl 20R-dihydroprednisolonate (270) and methyl 20S-dihydroprednisolonate (269), retain significant local anti-inflammatory activity, but are devoid of prednisolone-like side effects, such as pituitary adrenal suppression and thymus involution (268). The orientation of the alcohol group at C-20 was important for potency, given that (269) was several times more potent than (270) (269). That the acidic metabolites were indeed devoid of anti-inflammatory activity was ascertained independently (310).

When the prednisolonates (268), (269), and (270) were applied locally at equivalent-potency anti-inflammatory doses compared with that of other corticosteroids, they markedly inhibited granuloma formation but did

Table 15.48 Biological Activities of Halomethyl Androstane-17β-carbothioates

(257–266), (16)

Compound	Z	Y	X	R	16	Human V^a	Mouse AIT^b	Mouse HPA^c
(257)	H	F	F	C_2H_5	H	697	56	200
(258)	F	H	Cl	C_2H_5	H	916	20	100
(259)	F	H	F	C_2H_5	H	1984	63	149
(260)	F	F	Cl	C_2H_5	α-CH$_3$	124	56	0.04
(261)	F	F	F	CH$_3$	α-CH$_3$	392	76^d	2.9
(16)	F	F	F	C_2H_5	α-CH$_3$	945	113	1.0
(262)	F	F	F	C_3H_7	α-CH$_3$	299	55	0.7
(263)	F	H	Br	C_2H_5	β-CH$_3$	254	—	—
(264)	F	H	I	C_2H_5	β-CH$_3$	41	—	—
(265)	F	H	Cl	C_2H_5	β-CH$_3$	1469	41	44
(266)	F	H	F	C_2H_5	β-CH$_3$	1262	89	450
(Fluocinolone acetonide)^d						100	100	100

^{a}Human vasoconstrictor activity relative to fluocinolone acetonide (104).
^{b}Topical anti-inflammatory activity relative to fluocinolone acetonide (104).
^{c}Systemic corticosteroid activity after topical application relative to fluocinolone acetonide (104).
^{d}Standard.

not inhibit skin collagen synthesis nor cause dermal atrophy in rats (311).

Removal of the 17α-hydroxyl group of the prednisolonates above gave the deoxyprednisolonate structures (273), (274), and (275), which led to interesting changes in activity relative to that of their hydroxy counterparts. Produced during synthesis, the aldehyde (276) was also examined for anti-inflammatory activity. The keto-ester (273) and keto-aldehyde (276) were as potent as prednisolone in the granuloma pouch assay, whereas the alcohols (274) and (275) were almost inactive. It is clear that the 17α-hydroxyl group is needed for activity in the 21-ester series when the 20-position is reduced, but is not necessary in 20-keto steroids.

The aldehyde (276) showed signs of systemic absorption and activity that were comparable to that of prednisolone but the ester (273) was virtually devoid of systemic effects, on the basis of thymus weight changes and reduction of plasma corticosterone levels (adrenal suppression) (270).

Interestingly, acetonide formation across the 17, 20-diol arrangement led to more potent analogs compared to the free diols. In Table 15.49 the diols (269) and (270) and their corresponding acetonides, (277) and (278), respectively, are examined in croton oil ear edema assay, cotton pellet granuloma pouch assay, and for glucocorticoid receptor affinity. Although none of these analogs is more potent

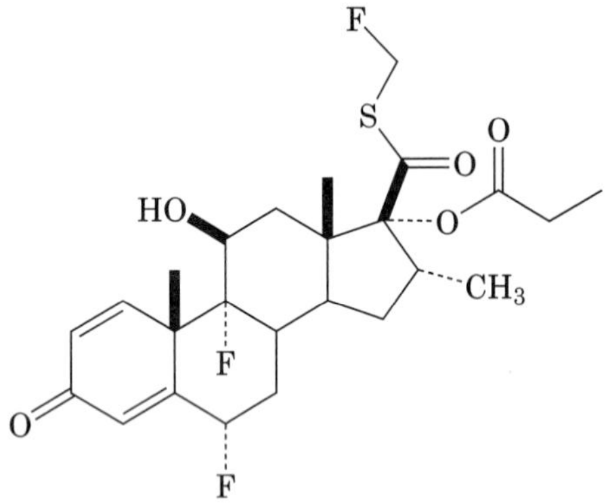

(16)

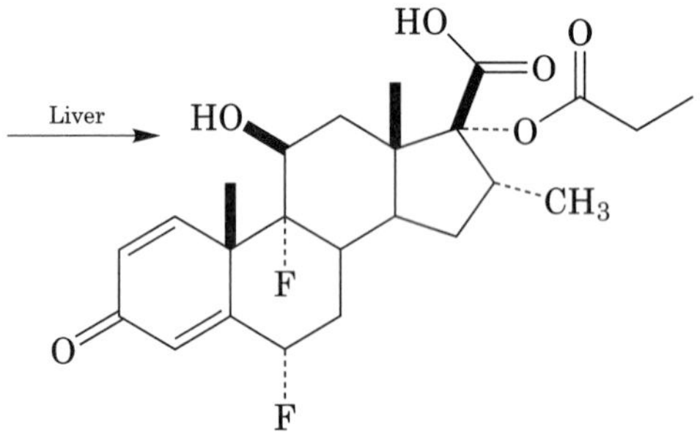

(267)

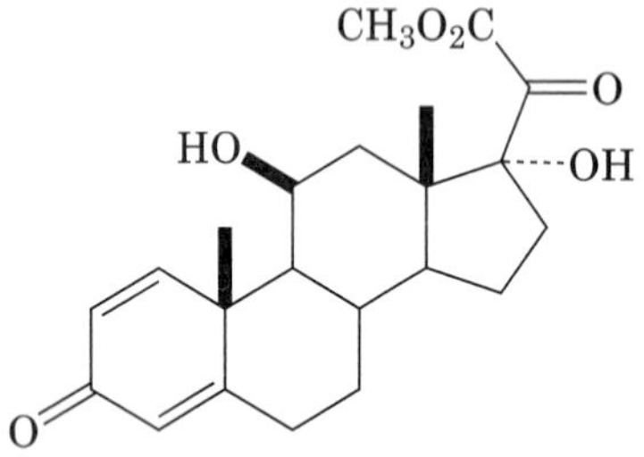

(268) active antedrug

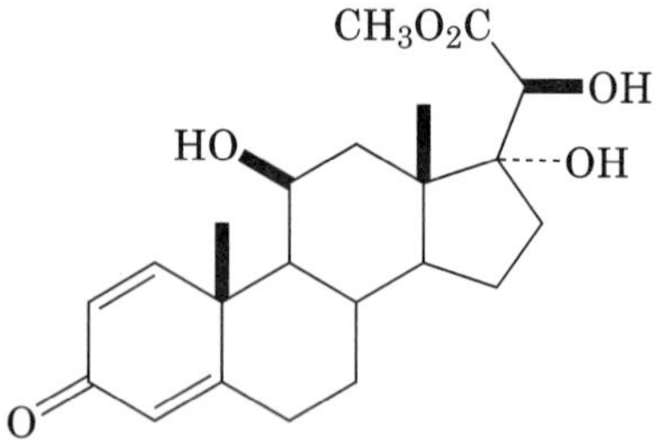

(269) active antedrug

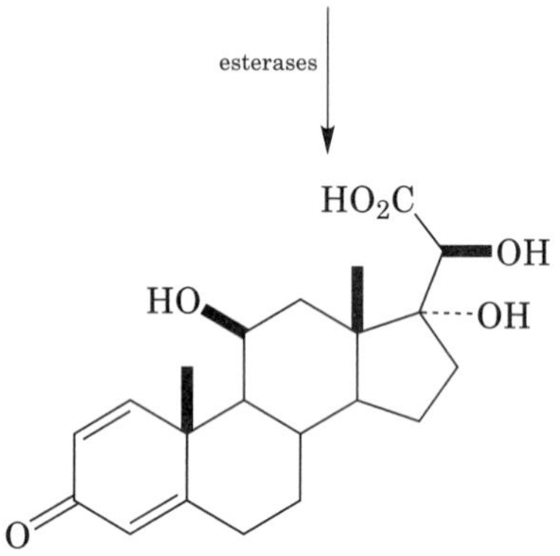

esterases

(271) inactive

than prednisolone (**21**), the acetonide (**278**) more closely approximates the intrinsic binding affinity and topical efficacy of (**21**). It is fascinating that the stereochemical dependency on activity for the flexible diols, where $20S > 20R$, is reversed for the conformationally restricted acetonides, where the potency of $20R > 20S$.

The succinate ester antedrug (**280**) was comparable to dexamethasone in an ear edema assay for potency, whereas the less lipophilic acid was slightly less potent. Neither

(**279**) nor (**280**) had systemic effects on thymus weight, and were promptly metabolized to (**281**) (active) within 2 min of i.v. dosing. A second metabolic event deactivates (**281**) by deesterification, to furnish the inactive carboxylic acid (**282**), with a $t_{1/2}$ value of about 2 h (312).

The effect on anti-inflammatory activity of replacing the ester group of the diols (**269**) or (**270**) by metabolically tenacious amides has been examined (275). The carboxamides (**283**) and (**284**), where R = Me or bulkier benzyl, exhibited potent local as well as systemic activities compared to those of prednisolone (see Table 15.50). The more potent topically active compounds were unfortunately also the most systemically absorbed, as gauged by several measures of systemic potency: thymic involution, reduction in plasma corticosterone, and contralateral ear effects. As with the ester-acetonides above, the $20R$ isomers were more potent than the $20S$ isomers.

It is probable, based on the presence of undesirable side effects for these amides, that metabolism has been curtailed. This is per-

Table 15.49   Anti-Inflammatory Activity of Steroidal 21-oate Esters

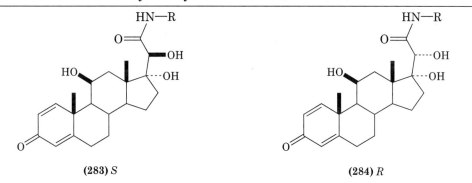

(277)   (278)

Compound	ID_{50} GPA^a	ID_{50} EEA^b	IC_{50} GRB^c
(269)	2.8	5.9	31.6
(270)	13.5	32.8	11.5
(277)	15.4	3.6	1.6
(278)	2.6	1.4	0.36
(21)^d	1.4	0.4	0.18

^{a}Cotton pellet granuloma pouch assay, μmol/pellet.
^{b}Croton oil ear edema assay, μmol/ear.
^{c}Rat liver cytosol receptor, μM inhibition.
^{d}Prednisolone (**21**).

Table 15.50   Anti-Inflammatory Activity of 20-Carboxamide Derivatives^a

(283) S   (284) R

Compound	EEA^a	GPA, top.^{b}	GPA, sys.^{c}	Thymus^d	Plasma cort.^{e}
20R-(**284**), R = CH_3	33.5	75.5	48.6	45.6	73.1
20S-(**283**), R = CH_3	24.5	21.1	−4.5	96.7	66.1
20R-(**284**), R = benzyl	42.6	69.3	38.9	65.4	23
20S-(**283**), R = benzyl	21.9	20.9	5.2	103.4	67.8
Prednisolone (**21**)	68.6	60.6	26.6	43.1	29.6

^{a}Ref. 275.
^{b}% Inhibition following application of 0.75 mg/mouse ear.
^{c}% Inhibition of granuloma formation after insertion of cotton pellet treated with 2 mg compound.
^{d}% Inhibition of granuloma formation on distal (contralateral) ear.
^{e}Thymus weight as % of control in granuloma assay.
^{f}Plasma corticosterone levels as % of control in granuloma assay.

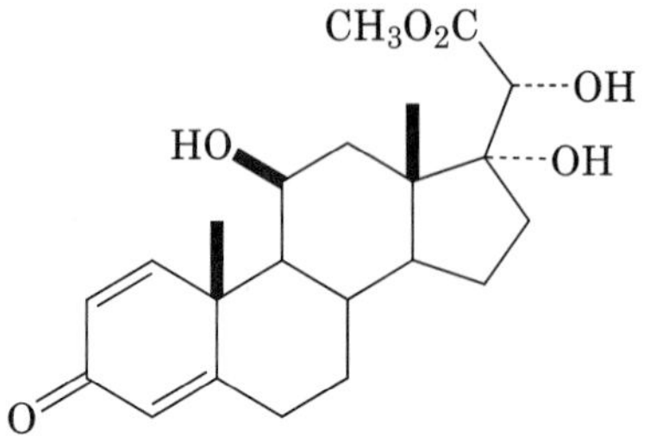

(270) active antedrug

esterases ↓

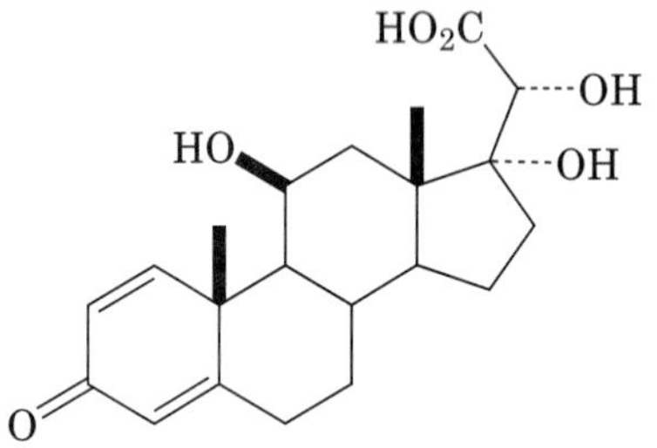

(272) inactive

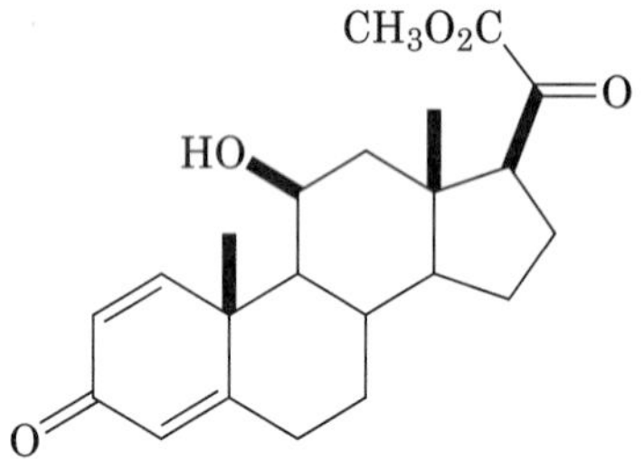

(273)

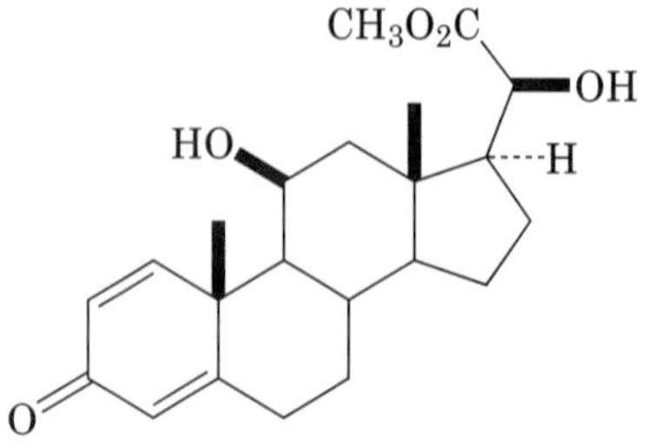

(274)

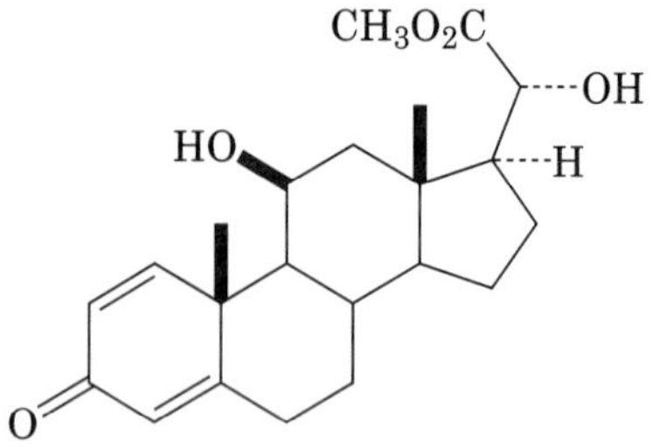

(275)

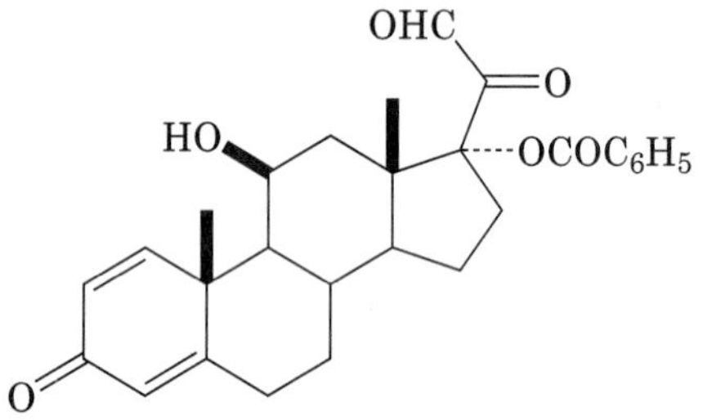

(276)

ence of amides such as (**283**) in the systemic circulation could lead to agonistic glucocorticoid behavior because it has been shown that (**283**) has receptor affinity comparable to that of prednisolone. Thus, for (**283**), where R = Me, the IC_{50} value is 275 μM (prednisolone = 196 μM); and when R = benzyl, the IC_{50} value is 54 μM. It is consistent with earlier discussions that partition coefficients correlate loosely with topical potency. The P values have been measured for several of these amide analogs and those with P values of 30 (log $P >$ 1.48) or better were the most active in the series (313).

11 ANTIGLUCOCORTICOIDS

Antiglucocorticoids, or GR antagonists, are compounds that bind tightly to the GR but do not illicit a corticosteroidal action. Clinically, antiglucocorticoids are used to reduce intraocular pressure (IOP) and in the treatment of Cushing's disease (glucocorticoid excess). Recently, it has been proposed that antiglucocorticoids might have a use in reducing the GR-mediated side effects associated with glu-

haps not surprising, in that amides are generally more stable than esters *in vivo*. The pres-

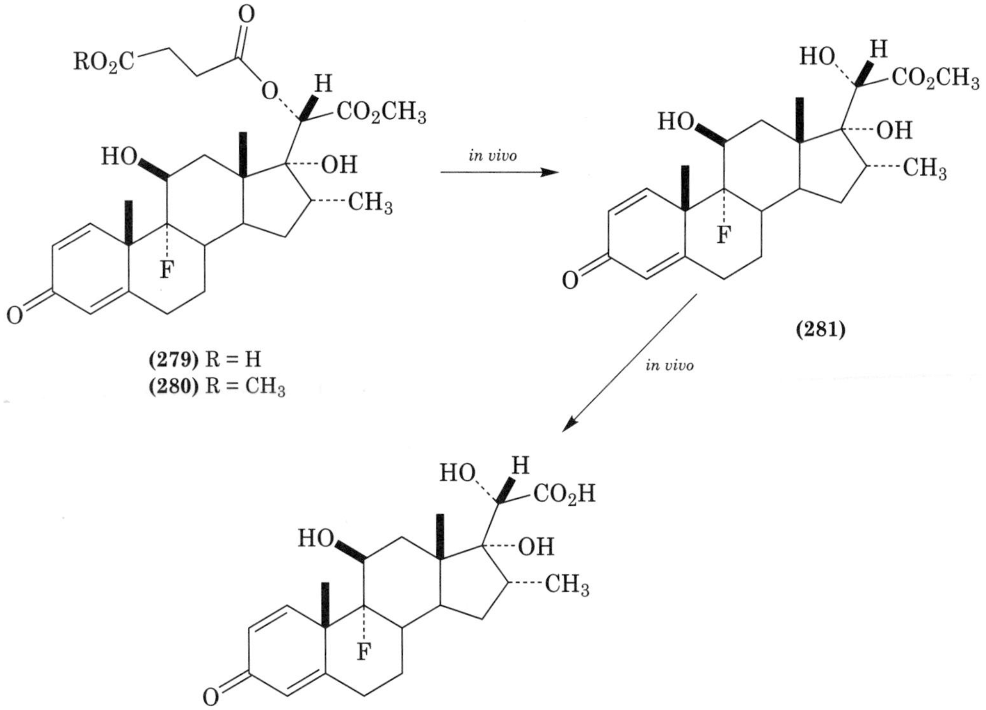

(279) R = H
(280) R = CH₃

(281)

(282)

cocorticoid use while not significantly interfering with the anti-inflammatory action of the compound (12).

It is thought that there is a correlation between degree of antagonism and steroid binding affinity, although strong binding affinity does not distinguish between agonistic and antagonistic activity. It is difficult to define exact stuctural features that lead to strictly agonist or antagonistic activity at the GR. Although X-ray structures may reflect certain SAR patterns, the X-ray structure of an agonist or antagonist bound to the GR is yet unavailable, and the presumption that a certain steroid conformation is indeed that in the binding site is yet speculative (although the rigidity of a steroid backbone allows for some predictivity).

Some general structural features that lead to both observed differences in steroidal conformation (determined by X-ray structures) and to noticeable effects in bioactivity (RBAs, agonism vs. antagonism, magnitude of activity, etc.):

1. Unsaturation of the A-ring: alteration of the A-ring conformation. Except for the A-ring diazoles, all the GR antagonist known have the 4-ene-3-one structure. Because this is shared with the most potent glucocorticoids, it is thought that this structural feature increases binding affinity to the steroid receptor (314).

2. The hydroxy groups at C-11, C-17, and C-21, although not necessary for a compound to bind to the GR, seem to be individually or collectively responsible for agonistic activity, although some compounds with hydroxyls at one or more of these positions still act as GR antagonists [e.g., RU-486 (285) and dexamethasone oxetanone (296)]. The absence of any of these hydroxyls may be one factor related to antagonistic activity (314), although not necessarily a determining factor. The lack of a hydroxyl at C-11 is a noticeable factor with some GR antagonists. Whereas all anti-inflammatory glucocorticoid agonists have a

(285) RU486 GR antagonist

hydroxyl at C-11 (or are readily solvolized to a C-11 OH), many antagonists (though not all) have no hydroxy moiety at C-11. For example, compounds that have shown antagonistic (ZK 98299, **286**) or mixed agonist activities (RU-486) have substantially bulky groups at C-11.

(286) ZK98,299 GR antagonist

An interesting comparison between agonist and antagonist to the GR is that of dexamethasone (**20a**, agonist) and dexamethasone oxetanone (**296**, antagonist). The X-ray structures of these compounds indicate that all four rings overlap very well and should thus fit into the ligand-binding site in a similar manner. As with differences in the magnitude of agonist activity, the agonist/antagonist distinction may be attributable to chemical factors, including the hydrogen bond donating or accepting ability of the ligand. In this example, both compounds can accept a hydrogen bond at C-20, although only the agonist is able to donate hydrogen bonds (314).

The most clinically useful GR antagonist is RU-486 (315, 316), which also shows significant PR antagonism. This compound has been more completely characterized as a partial antagonists in certain cell lines (317, 318), and thus sometimes has been called a type II antagonist. Although a type II antagonist-occupied receptor can bind to DNA, it cannot interact in a productive manner with the DNA in a way equivalent to an agonist-occupied receptor (319). RU-486 blocks the transactivation of the GR and inhibits the production of associated proteins, notably dexamethasone-induced extracellular LC-1 (112).

The defining feature of RU-486 is the 11β-aryl moiety. Most steroidal GR antagonists include 11β-aryl or other 11β substituents. The tight binding affinity of RU-486 to the GR is presumed to be in part attributable to a tight fit of this 11β-aryl component in a lipophilic pocket within the GR hormone-binding region. ZK98299 is an 11β-aryl GR antagonist very similar in structure to RU-486, although it is a pure (type I) antagonist. Like RU-486, it also has antiprogestational activity. The 11β-N-methyldihydroindol-5-yl analog of RU-486 (**287**), another compound with GR antagonis-

(287) the N-methyl dihydroindolo-5-yl-analog of RU486

tic properties and a strong GR binding affinity, has a rotamer that fits well in the space (11β-pocket) occupied by the 11β-substituent of RU-486.

Another class of glucocorticoid antagonists, the 10β-androstanes (e.g., RU-39305 and RU-43044) also bind tightly to the GR (Table 15.51). Relative binding affinities (RBA) to the GR (vs. dexamethasone) do not lead to clear SARs with regard to unsaturation at Δ^1 or Δ^6.

Table 15.51 Relative Binding Affinities of 10β-Androstanes[a]

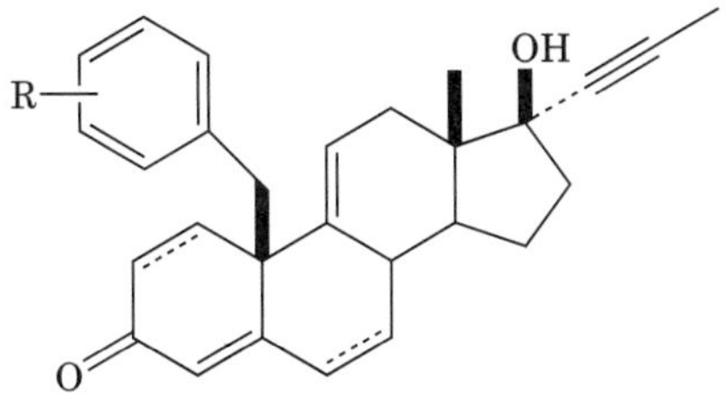

RU Code and Structure Number	Unsaturation	R	Thymus GR RBA[b]	PR RBA	Thymocytes IC$_{50}$ (nM)
39305, (**288**)		H	57	<0.1	100
43044, (**289**)		p-CH$_3$	130	<0.2	200
43065, (**290**)		p-OCH$_3$	17	<0.1	500
46759, (**291**)	Δ^1, Δ^6	p-CH$_3$	33	<0.1	500
44068, (**292**)	Δ^1	p-CH$_3$	15	<0.1	>1000
44427, (**293**)	Δ^6	p-CH$_3$	3	<0.1	1000

[a]Modified from Philibert (320).
[b]Dexamethasone = 100.

However, space-filling models indicate a similarity between the 3D space occupied by the 10β- and 11β- compounds (320), although the aromatic moiey of RU-39305 (**288**) is not in the same orientation as that of RU-486. These compounds have very little affinity for the progesterone receptor (PR).

On the other hand, "moving" the aryl group from the C-18 angular methyl up to the C-19 methyl provides a similar looking steroid. Two compounds were tested, (**294**) and

(**295**)

(**295**), and were found to be about as potent as RU-486 in contraceptive assessments. Both (**294**) and (**295**) had lower antiglucocortiocoid activities than those of RU-486 (321).

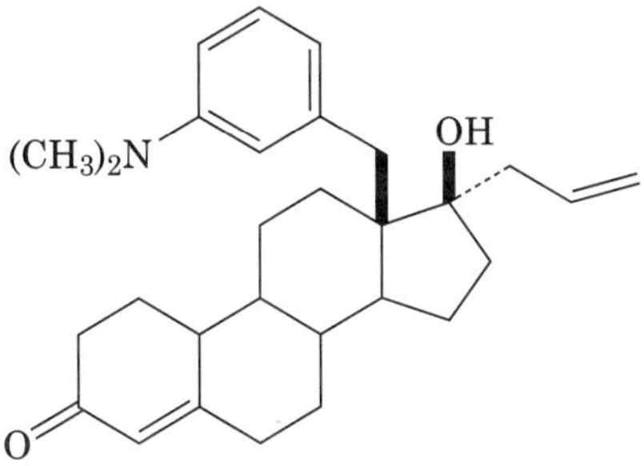

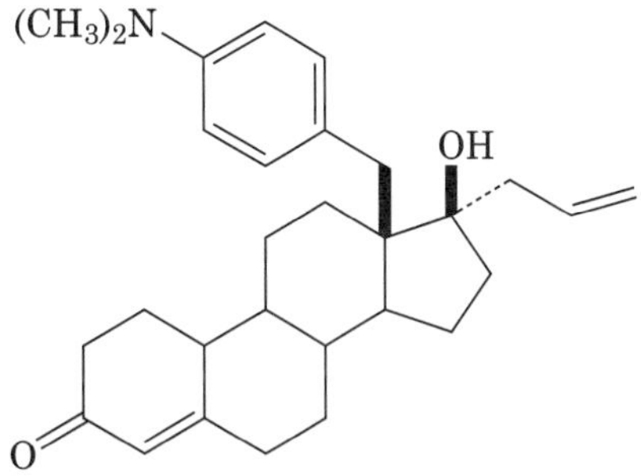

Research done by Simons by use of antiglucocorticoids, including compounds (e.g., dexa-

(**294**)

methasone 21-mesylate) that covalently bind to the GR (322), has better defined the structure and function of the GR. Spiro C-17 oxetanes have shown potent antiglucocorticoid activity in whole-cell systems (323, 324).

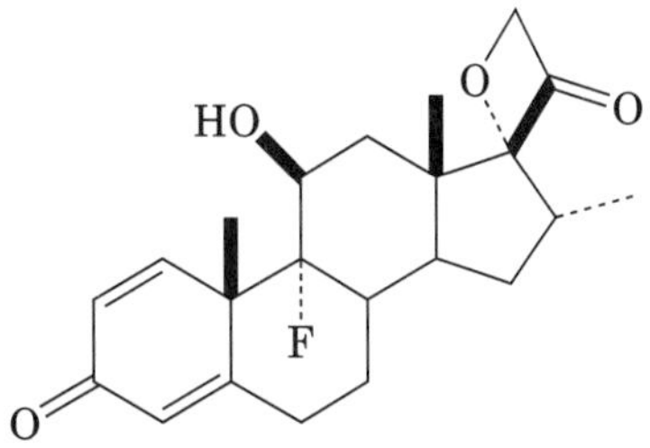

(**296**)

Although the 10β- or 11β-group is found important with many GR antagonists, it is not a necessity for antiglucocorticoid activity. Cortexelone (**34**) acts as a GR antagonist in the true definition, blocking GR transactivation (325–327), although it does not include an 11β-substituent.

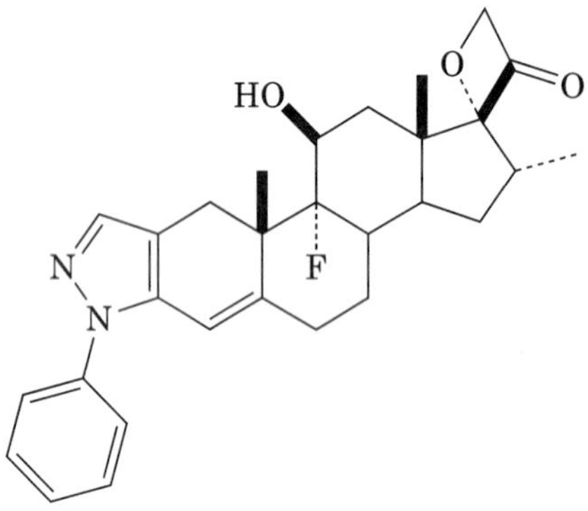

(297)

Although 11-deoxycortisol is an *in vitro* GR antagonist, it is an agonist *in vivo*, presumably through the metabolic hydroxylation at C-11. The reduced "affinity" for 11β-hydroxylation shown by unsaturation at C-9 (i.e., **299**) (11) led to the development of 11-oxa-11-deoxycortisol (**298**), a compound with GR antagonistic activity *in vivo* (although with a GR RBA 20- to 60-fold less than that of dexamethasone) (328, 329).

Most progestins and antiprogestins bind to the GR and can elicit some antiglucocorticoid activity (330). There is probably a relation between the high degree of homology between PR and GR and the fact that certain compounds show both GR and PR antagonism

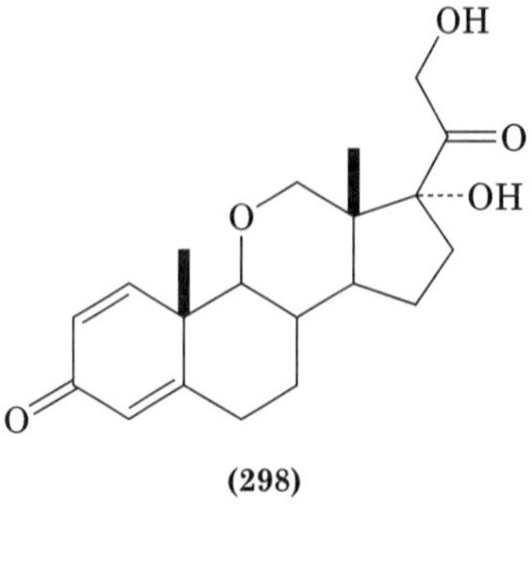

(298)

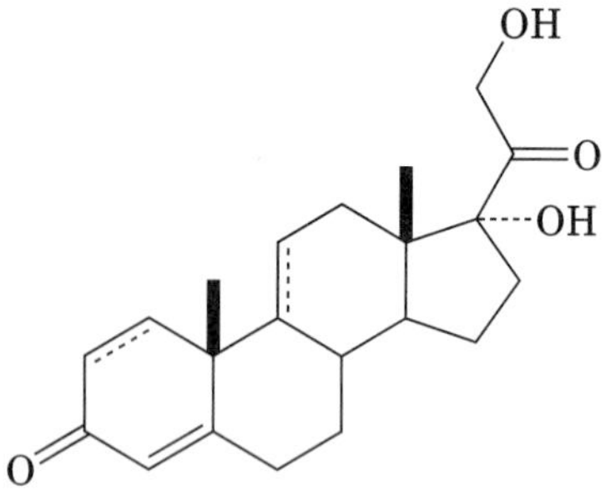

(299)

(331). For this reason, a continued effort has been seen in the design of selective (anti-) glucocorticoids and selective (anti-) progestins. Certain structural features have been implicated as consequential in the differentiation between GR and PR binding. As seen previously, RU-486 is a potent antiprogestin and antiglucocorticoid. It is apparent that an 11β group increases RBA for both GR and PR, although structural modification of the 11β group affects GR and PR binding differently (323, 332). As shown in Table 15.52, for instance, although vinyl and *p*-tolyl substituents both lead to significant binding in both receptor types, some differentiation can be made in favor of GR binding (*m*-MeO-phenyl, *p*-F-phenyl, and *p*-NH$_2$-phenyl substitnents have GR/PR "RBA ratios" of 17, 3.2, and 2.9, respectively, vs. 0.6 for RU-486) or PR binding (*p*-MeS-phenyl and Me substituents have 0.09 and 0.25 GR/PR "RBA ratios," respectively). An interesting observation is that the 11β-pocket of the GR receptor seems to be more spacious than that of the PR in the area closest to the steroid backbone. This idea is supported by increased GR RBA of steroids with bulky 11β-substituents such as *t*-butyl and cyclopentyl (GR/PR RBA ratios of 60 and 43, respectively).

Addition of an α-CH$_3$ at C-6 increases the GR selective binding, whereas a β-CH$_3$ was equivalent to no substitution. Substituting the C-17 1-propynyl group of RU-486 with a chloro-ethynyl group (Table 15.53) had only a small effect on GR binding affinity (range from 10% increase to 3% decrease) and decreased PR binding affinity (by up to 30%) (332). By reversing the C-17 stereocenter, the GR/PR RBA ratio is increased to 30 (vs. 0.58

Table 15.52 Effect of 11-Substituent on Progesterone and Glucocorticoid Receptor Binding Affinities[a]

11β-Substituent (R)	GR Binding[b]	PR Binding[c]
p-F-phenyl, (**300**)	285	85
p-NH$_2$-phenyl, (**301**)	118	40
p-(CH$_3$)$_2$N-phenyl, (**285**)	300	530
m-CH$_3$O-phenyl, (**302**)	245	15
p-CH$_3$O-phenyl, (**303**)	300	505
p-tolyl, (**304**)	295	295
p-CH$_3$S-phenyl, (**305**)	180	605
H, (**306**)	2	15
Vinyl, (**307**)	340	390
t-Butyl, (**308**)	50	1
Phenyl, (**309**)	240	65

[a]Ref. 323.
[b]Dexamethasone = 100%.
[c]Progesterone = 100%.

for RU-486), although the GR RBA is reduced to about 40% of that of RU-486 (332).

Researchers have examined other alterations at C-16 and C-17 intended to separate PR antagoism altogether from GR effects. The SARs from these PR-focused studies are useful in understanding GR SARs. The data from Table 15.54 show that RTI-012 (**315**) and RTI-

Table 15.53 Effect of Substitution at C-17 on Relative Binding Affinities in Progesterone and Glucocorticoid Receptors[a]

X	Y	GR RBA[b]	PR RBA[c]
–OH, (**310**)	–C≡CH	120	4
–C≡C—CH$_3$, (**311**)	–OH	300	530
–C≡C—Cl, (**312**)	–OH	290	460
–CH$_2$—C≡CH, (**313**)	–OH	260	160
–CH$_2$—CH=CH$_2$, (**314**)	–OH	220	840

[a]Ref. 332.
[b]Dexamethasone = 100%.
[c]Progesterone = 100%.

Table 15.54 Relative Binding Affinities of Mixed-PR/GR Ligands[a]

(315–316)

Compound	R	R$_1$	PR[b]	GR[c]
RTI-012, (**315**)	OAc	H	12.7	5.7
RTI-022, (**316**)	H	Et	11.9	5.2
Dexamethasone				1
RU-486, (**285**)			6.8	13.9
Progesterone			1	

[a]Ref. 333.

[b]Progestin receptor (h-PR) from baculovirus expression system, competitive assay with tritiated progesterone as reference ligand.

[c]Glucocorticoid receptor (h-GR) from MDA-231 expression system, competitive assay with tritiated dexamethasone as reference ligand.

022 (**316**) both have greater binding affinity than that of either progesterone or dexamethasone, with little apparent separation of effects. However, the issue of antagonism, partial agonism, and pure agonism for these compounds was studied in some detail. These compounds function as competitive antagonists of GR function because they are unable to translocate GR to the nucleus. On the other hand, they are active antagonists of PR transcriptional activity. Thus, a separation between GR and PR antagonism occurs by virtue of differences in mode of action of these compounds at the receptor level (333).

In a later study (Table 15.55), relative binding affinities of several aryl-substituted α-pro-

Table 15.55 Relative Binding Affinities of Mixed-PR/GR Ligands[a]

Compound	R	r-PR[b]	r-GR[c]	PR/GR
RTI 6413–003, (**317**)	(CH$_3$)$_2$N–	118	20	5.8
RTI 6413–018, (**318**)	CH$_3$CO–	144	5	29
RTI 6413–033, (**319**)	CH$_3$S–	106	10	11
RTI-012, (**315**)	See above	134	86	1.6

[a]Ref. 334.

[b]Progestin receptor from Rabbit, competitive assay with tritiated progesterone as reference and ligand.

[c]Glucocorticoid receptor from rabbit thymus, competitive assay with tritiated dexamethasone as reference and standard.

Table 15.56   Relative Binding Affinities of Mixed-PR/GR Ligands[a]

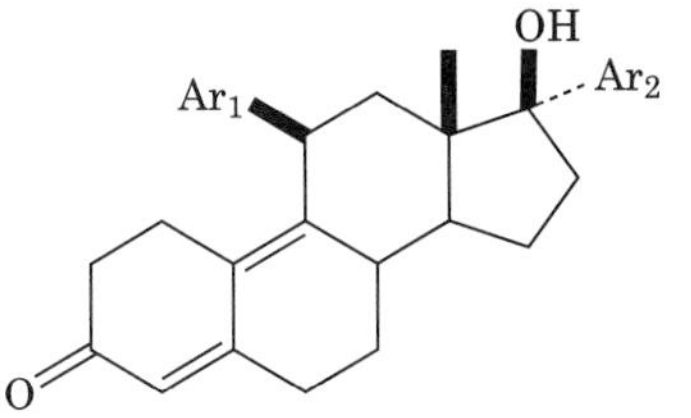

(320–324)

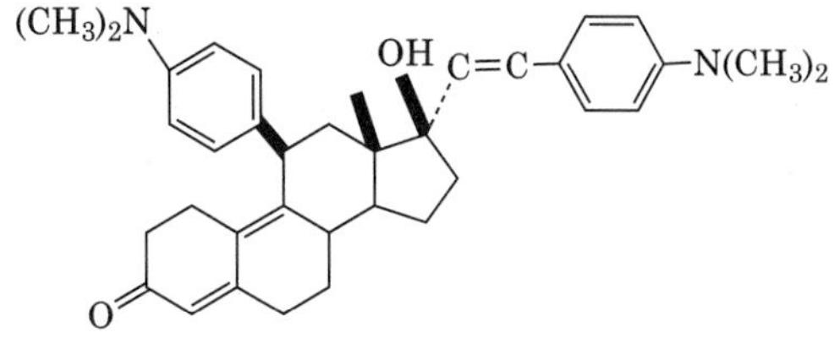

(325) *E*
(326) *Z*

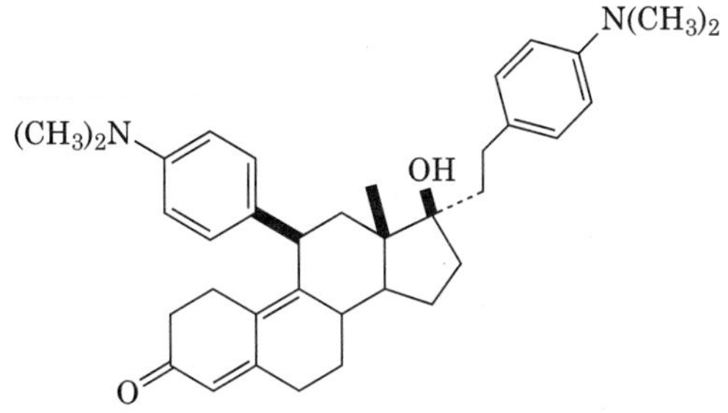

(327)

Compound	Ar$_1$	Ar$_2$	GR (%)[b]	PR (%)[c]	GR/PR
(320)	p-(CH$_3$)$_2$NC$_6$H$_4$–	C$_6$H$_5$–	98	16	6
(321)	p-(CH$_3$)$_2$NC$_6$H$_4$–	p-(CH$_3$)$_2$NC$_6$H$_4$–	84	1.7	49
(322)	p-(CH$_3$)$_2$NC$_6$H$_4$–	p-(CH$_3$SO$_2$)C$_6$H$_4$–	87	2	44
(323)	3,4-(OCH$_2$O)C$_6$H$_3$–	p-(CH$_3$SO$_2$)C$_6$H$_4$–	195	0.4	488
(324)	3,4-(OCH$_2$CH$_2$O)C$_6$H$_3$–	p-N-(pyrrolidin-2-onyl)C$_6$H$_4$–	98	0.2	490
(325)	p-(CH$_3$)$_2$NC$_6$H$_4$–	p-(CH$_3$)$_2$NC$_6$H$_4$-CH=CH–	5.9	6.1	1
RU-486	p-(CH$_3$)$_2$NC$_6$H$_4$–	CH$_3$	193	36	5.4

[a]Ref. 335.

[b]IM-9 cells, cytosol; relative binding affinity; dexamethasone = 100%.

[c]MCP-7 cells, cytosol; relative binding affinity; Org 2058 = 100%.

panolyl derivatives of RU-486 were compared. Compared to RTI-012, the new compounds demonstrated lower GR binding, but unlike RTI-012 (**315**), the new compounds lacked oral antiprogestational activity in the anti-Clauberg assay (334).

A series of 17α-arylalkyl analogs reminiscent of RU-486 were tested for their ability to bind to the human glucocorticoid receptor (GR; IM-9 cells, cytosol) and the human progesterone receptor (PR: MCF-7 cells, cytosol), as shown in Table 15.56. Of approximately 80 different compounds tested by the authors (335), only a limited number of compounds combine a high binding to the GR with a low binding to the PR (thus a high GR/PR). Although a solid structure-activity relationship has not yet been found, several general observations could be made. Apparently, GR-activity and selectivity are critically dependent on both the substituent at Ar$_1$ and Ar$_2$. The optimum substituent at Ar$_1$ seems to be the 4-*N,N*-dimethylamino- or the 3,4-methylenedioxo group. The 3-(thio)methoxy- or 3-*N,N*-dimethylamino- group, which have been reported to induce high selectivity for the GR, led to a significant decrease in affinity for the GR upon combination with a substituted 21-phenyl group (data not shown). As shown in Table 15.56, substitution of Ar$_2$ by a 4-*N,N*-dimethylamino (**321**), 4-sulfone (**322, 323**), or 4 pyrrolidone (**324**) moiety leads to a dramatic increase of the GR/PR ratio compared to that of the unsubstituted Ar$_2$ (**320**). A 4-carboxa-

Table 15.57　Nonsteroidal Glucocorticomimetic 1H-Quinolin-2-ones[a]

(328–336)

Compound	R	R′	K_i (nM)				
			GR	PR	MR	AR	ER
(328)	9-Cl	H	9.3	53	1100	1400	>1000
(329)	8-OCH$_3$	H	1800	—	—	—	—
(330)	9-OCH$_3$	H	18	390	—	—	—
(331) (racemic)	10-OCH$_3$	H	4.7	>5000	1682	2660	>1000
(+)-(332)	10-OCH$_3$	H	240				
(−)-(333)	10-OCH$_3$	H	2.1				
(334)	10-OCH$_3$	3′-CF$_3$	11	4000	493	3360	>1000
(335)	10-OCH$_3$	3′-O(CO)N(CH$_3$)$_2$	2.4	>5000	3914	3066	>1000
(336)	10-OCH$_3$	3′-OCH$_2$SCH$_3$	4.0	>5000	3838	3115	>1000
Prednisolone	—	—	2.4	>5000	37	2762	>1000

[a]Ref. 336.

amide or a 4-sulfonamide group in Ar$_2$ has a similar though slightly less pronounced effect (not shown). Reduction of the triple bond to the E-double bond (as in 325) leads to a significant decrease in GR and an increase in PR-binding. Similar results were obtained upon reduction of the triple bond to the Z-double (as in 326) or the saturated bond (as in 327; results not shown).

Apparently the GR can accommodate 11-phenyl substituted steroids with a rigid 17α-substituent as large as a phenylethynyl function; moreover, specific polar substitution of this 17α-phenylethynyl group provides compounds that are highly selective for the GR. It has been reported that the GR is more hydrophilic than the PR in the area that interacts with the steroidal D-ring. However, considering the distance between the polar phenylethynyl substituent and the steroidal D-ring ($\sim$8.5 Å), it seems unlikely that this hydrophilic area in the GR accounts for the high selectivity observed for compounds (321–324).

In vivo antiglucocorticoid activity of the novel compounds is assessed by measuring the effect on body weight gain and thymus weight of dexamethasone-treated immature male rats. In this test, compound (321) clearly antagonized the dexamethasone effect, with an ED$_{50}$ value of 10 mg/kg after oral administration.

12　NONSTEROIDAL GLUCOCORTICOIDS

The synthesis and biological charactarization of novel nonsteroidal 2-quinolinones, whose functional activity was equivalent to prednisolone, was recently reported (336). A general structure for these compounds (reminiscent of RU-38486) is shown in Table 15.57. First, 9-substituted compounds showed PR activity and so the ring system was examined at other positions with a MeO group at C-8 through C-10. Once C-10 was established as having the best profile, modification of the

C-5-aryl ring was undertaken. Compounds (**334–336**) established that GR activity could be separated from unwanted PR, MR, AR, and ER activities by simple modification to the 3'-position. The candidate compounds (**335**) and (**336**) were compared for functional repression and activation vs. prednisolone in cotransfection assays, with (**336**) appearing to be the better lead. An oral dose of (**336**) was comparable to prednisolone in a rat model of asthma. Finally, the C-5 chiral center was examined in a few cases, as shown in Table 15.57 for (+)-**332** and (−)-**333**. It was clear that the (−)-enantiomer was far more potent, and the authors suggest that this stereocenter may have the S configuration on the basis of other similar nonsteroidal PR ligands.

13 ACKNOWLEDGMENTS

M.A.A. is grateful to Dr. Masato Tanabe, who for many years freely offered general insights and specific discussions regarding steroid hormone synthesis, SAR, biochemistry, clinical applications, and medical significance.

REFERENCES

1. J. D. Baxter and G. G. Rousseau in J. D. Baxter, R. Baxter, and G. G. Rousseau, Eds., *Glucocorticoid Hormone Action: An Overview*, Springer-Verlag, New York, 1979, pp. 1–24.

2. P. S. Hench, E. C. Kendall, C. H. Slocumb, and H. F. Polley, *Proc. Staff Meet. Mayo Clinic*, **24**, 181–197 (1949).

3. J. Fried and E. F. Sabo, *J. Am. Chem. Soc.*, **75**, 2273–2274 (1953).

4. J. Wepierre and J. P. Marty, *Trends Pharmacol. Sci.*, **1**, 23–26 (1979).

5. M. E. Wolff, *Structure-Activity Relationships in Glucocorticoids* (Monographs on Endocrinology), Springer-Verlag, Berlin, 1979, pp. 97–107.

6. H. J. Lee, I. B. Taraporewala, and A. S. Heiman, *Med. Actual.*, **25**, 577–588 (1989).

7. J. R. Vane and R. M. Botting, *Inflamm. Res.*, **44**, 1–10 (1995).

8. P. J. Barnes and I. Adcock, *Trends Pharmacol. Sci.*, **14**, 436–441 (1993).

9. B. R. Olin, *Drug Facts and Comparisons*, Facts and Comparisons, Inc., St. Louis, MO, 1996.

10. G. K. McEvoy, *AHFS Drug Information*, American Society of Health-System Pharmacists, Bethesda, MD, 1995.

11. Y. Uoneda, D. Han, K. Ogita, and A. Watanabe, *Brain Res.*, **685**, 105–116 (1995).

12. K. Iwasaki, E. Mishima, M. Miura, N. Sakai, and S. Shimao, *J. Dermatol. Sci.*, **10**, 151–158 (1995).

13. W. Rostene, A. Sarrieau, A. Nicot, V. Scarceriaux, C. Betancur, et al., *J. Psychiatry Neurosci.*, **20**, 349–356 (1995).

14. G. Wang and A. T. Lim, *Endocrinology*, **137**, 379–382 (1996).

15. H. Shibata, T. E. Spencer, S. A. Onate, G. Jenster, S. Y. Tsai, et al., *Recent Prog. Horm. Res.*, **52**, 141–165 (1997).

16. P. Y. Sze and B. H. Yu, *J. Steroid Biochem. Mol. Biol.*, **55**, 185–192 (1995).

17. T. D. Long and R. G. Kathol, *Ann. Clin. Psychiatry*, **5**, 259–270 (1993).

18. K. M. Campbell and D. S. Schubert, *Gen. Hosp. Psychiatry*, **13**, 270–272 (1991).

19. S. Teramoto and Y. Fukuchi, *Am. J. Respir. Crit. Care Med.*, **153**, 879–880 (1996).

20. J. K. Saito, J. W. Davies, R. D. Wasnich, and P. D. Ross, *Calcif. Tissue Int.*, **57**, 115–119 (1996).

21. T. Olbricht and G. Benker, *J. Intern. Med.*, **234**, 237–244 (1993).

22. A. Markham and H. M. Bryson, *Drugs*, **50**, 317–333 (1995).

23. C. Kasperk, U. Schneider, U. Sommer, F. Niethard, and R. Ziegler, *Calcif. Tissue Int.*, **57**, 120–126 (1995).

24. A. L. Cheng, *Blood*, **87**, 1202 (1996).

25. F. J. Frey and R. F. Speck, *Schweiz. Med. Wochenschr.*, **122**, 137–146 (1992).

26. L. H. Sarrett, A. A. Patchett, and S. Steelman, *The Effects of Structural Alteration on the Antiinflammatory Properties of Hydrocortisone*, Birkhauser-Verlag, Basel, Switzerland, 1963, pp. 11–154.

27. H. P. Schedl and J. A. Clifton, *Gastroenterology*, **41**, 491–499 (1961).

28. D. R. Friend and G. W. Chang, *J. Med. Chem.*, **27**, 261–266 (1984).

29. T. Kumura, K. Yamaguchi, Y. Usuki, Y. Kurosaki, T. Nakayama, et al., *J. Controlled Release*, **30**, 125–135 (1994).

30. F. D. Malkinson and M. B. Kirschenbaum, *Arch. Dermatol.*, **88**, 427–439 (1963).

31. J. R. Scholtz and D. H. Nelson, *Clin. Pharmacol. Ther.*, **6**, 498–509 (1965).

32. A. G. Ruhmann and D. L. Berliner, *J. Invest. Dermatol.*, **49**, 123–130 (1967).

33. T. L. Popper and A. S. Watnick, *Antiinflammatory Agents*, Academic Press, New York, 1974, pp. 245–294.

34. J. Fried, A. Borman, W. B. Kessler, P. Grabowich, and E. F. Sabo, *J. Am. Chem. Soc.*, **80**, 2338–2339 (1958).

35. V. A. Place, J. G. Velazquez, and K. H. Burdick, *Arch. Dermatol.*, **101**, 531–537 (1970).

36. A. Thalen, R. Brattsand, and E. Gruvstad, *Acta Pharm. Suec.*, **21**, 109–124 (1984).

37. A. W. McKenzie and R. M. Atkinson, *Arch. Dermatol.*, **89**, 741–746 (1964).

38. I. Ringler, *Methods Hormone Res.*, **3**, 227–349 (1964).

39. S. Sugai, T. Okazaki, Y. Kajiwara, T. Kanbara, Y. Naito, et al., *Chem. Pharm. Bull.*, **34**, 1607–1612 (1986).

40. H. Ueno, A. Maruyama, M. Miyake, E. Nakao, K. Nakao, et al., *J. Med. Chem.*, **34**, 2468–2473 (1991).

41. E. L. Shapiro, M. J. Gentles, R. L. Tiberi, T. L. Popper, J. Berkenkopf, et al., *J. Med. Chem.*, **30**, 1068–1073 (1987).

42. E. L. Shapiro, M. J. Gentles, R. L. Tiberi, T. L. Popper, J. Berkenkopf, et al., *J. Med. Chem.*, **30**, 1581–1588 (1987).

43. P. H. Kendall, *Ann. Phys. Med.*, **9**, 55–58 (1967).

44. J. L. Kalliomaki, *Curr. Ther. Res.*, **9**, 327–337 (1967).

45. M. Pearlgood, *J. R. Coll. Gen. Pract.*, **21**, 410–413 (1971).

46. E. J. Collins, J. Aschenbrenner, and M. Nakahama, *Steroids*, **20**, 543–554 (1972).

47. L. A. B. Joosten, M. M. A. Helsen, and W. B. Van den Berg, *Agents Actions*, **31**, 135–142 (1990).

48. L. E. Hare, K. C. Yeh, C. A. Ditzler, F. G. McMahon, and D. E. Duggan, *Clin. Pharmacol. Ther.*, **18**, 330–337 (1975).

49. S. Johansen, E. Rask-Pedersen, and J. U. Prause, *Acta Ophthalmol. Scand.*, **74**, 259–264 (1996).

50. R. Arky and C. S. Davidson, *Physicians Desk Reference*, Medical Economics Data Production, Montvale, NJ, 1994.

51. R. D. Schoenwald and R. C. Ward, *J. Pharm. Sci.*, **67**, 786–788 (1978).

52. K. Taniguchi, K. Itakura, K. Morisaki, and S. Hayashi, *J. Pharmacobio-Dyn.*, **11**, 685–693 (1988).

53. K. Taniguchi, K. Itakura, N. Yamazawa, K. Morisaki, S. Hayashi, et al., *J. Pharmacobio-Dyn.*, **11**, 39–46 (1988).

54. U. Westphal, *Steroid Protein Interactions*, Springer-Verlag, New York, 1971.

55. U. Westphal, *J. Am. Oil Chem. Soc.*, **41**, 481–490 (1964).

56. H. E. Smith, R. G. Smith, D. O. Toft, J. R. Neergaard, E. P. Burrows, et al., *J. Biol. Chem.*, **249**, 5924–5932 (1974).

57. K. B. Eik-Nes, J. A. Schellman, C. Lumry, and L. T. Samuels, *J. Biol. Chem.*, **206**, 411–419 (1954).

58. U. Westphal and B. D. Ashely, *J. Biol. Chem.*, **234**, 2847–2851 (1959).

59. U. Westphal, J. Goetz, G. B. Harding, M. J. Mather, and N. Rust, *Arch. Biochem. Biophys.*, **118**, 556–567 (1967).

60. J. R. Florini and D. A. Buyske, *J. Biol. Chem.*, **236**, 247–251 (1961).

61. E. A. Peets, M. Staub, and S. Symchowicz, *Biochem. Pharmacol.*, **18**, 1655–1663 (1969).

62. W. R. J. Slaunwhite, G. N. Lockie, N. Back, and A. A. Sandberg, *Science*, **135**, 1062–1063 (1962).

63. A. A. Sandberg and W. R. J. Slaunwhite, *J. Clin. Invest.*, **42**, 51–54 (1963).

64. H. Derendorf, H. Mollmann, G. Hochhaus, B. Meibohm, and J. Barth, *Int. J. Clin. Pharmacol. Ther.*, **35**, 481–488 (1997).

65. R. W. Brown, K. E. Chapman, P. Murad, C. R. Edwards, and J. R. Seckl, *Biochem. J.*, **313**, 997–1005 (1996).

66. H. Vierhapper, K. Derfler, P. Nowotney, U. Hollenstein, and W. Waldhausl, *Acta Endocrinol. Copenh.*, **125**, 160–164 (1991).

67. N. M. Drayer and C. J. Giroud, *Steroids*, **5**, 289 (1965).

68. E. Gerhards, B. Nieuweboer, G. Schulz, H. Gibian, D. Berger, et al., *Acta Endocrinol.*, **68**, 98–126 (1971).

69. H. L. Bradlow, B. Zumoff, C. Monder, H. J. Lee, and L. Hellman, *J. Clin. Endocrinol. Metab.*, **37**, 811–818 (1973).

70. S. Edsbacker, P. Andersson, C. Lindberg, J. Paulson, A. Ryrfeldt, et al., Proceedings of the 10th European Drug Metabolism Workshop, Taylor & Francis, London, 1987, pp. 651–655.

71. G. Jonsson, A. Astrom, and P. Andersson, *Drug Metab. Dispos.*, **23**, 137–142 (1995).

72. G. M. Tomkins and P. J. Michael, *J. Biol. Chem.*, **225**, 13–24 (1957).

73. E. M. Glenn, R. O. Stafford, S. C. Lyster, and B. J. Bowman, *Endocrinology*, **61**, 128–142 (1957).

74. J. R. Florini, L. L. Smith, and D. A. Buyske, *J. Biol. Chem.*, **236**, 1038–1042 (1961).

75. E. J. Collins, A. A. Forist, and E. B. Nadolski, *Proc. Soc. Exp. Biol. Med.*, **93**, 369–373 (1956).

76. W. E. Dulin, L. E. Barnes, E. M. Glenn, S. C. Lyster, and E. Collins, *J. Metab.*, **7**, 398–404 (1958).

77. F. Kuipers, R. S. Ely, E. R. Hughes, and V. C. Kelley, *Proc. Soc. Exp. Biol. Med.*, **95**, 187–189 (1957).

78. R. H. Silber and E. R. Morgan, *Clin. Chem.*, **2**, 170–174 (1956).

79. R. H. Silber, *Ann. N. Y. Acad. Sci.*, **82**, 821 (1956).

80. R. S. Ely, A. K. Done, and V. C. Kelley, *Proc. Soc. Exp. Biol. Med.*, **91**, 503–506 (1956).

81. M. R. Yudt and J. A. Cidlowski, *Mol. Endocrinol.*, **15**, 1093–1103 (2001).

82. C. B. Whorwood, S. J. Donovan, P. J. Wood, and D. I. W. Phillips, *J. Clin. Endocrinol. Metab.*, **86**, 2296–2308 (2001).

83. T. M. Fletcher, B. S. Warren, C. T. Baumann, and G. L. Hager, *Methods Mol. Biol.*, **176**, 55–65 (2001).

84. R. M. Evans, *Science*, **240**, 889–895 (1988).

85. J. Carlstedt-Duke, P.-E. Stromstedt, O. Wrange, T. Bergman, J.-A. Gustafsson, et al., *Proc. Natl. Acad. Sci. USA*, **84**, 4437–4440 (1987).

86. M. Denis, J.-A. Gustafsson, and A.-C. Wikstrom, *J. Biol. Chem.*, **263**, 18520–18523 (1988).

87. L. F. Stancato, A. M. Silverstein, C. Gitler, B. Groner, and W. B. Pratt, *J. Biol. Chem.*, **271**, 8831–8836 (1996).

88. C. Yu, N. Warriar, and M. V. Govindan, *Biochemistry*, **34**, 14163–14173 (1995).

89. M. Schena, L. P. Freedman, and K. R. Yamamoto, *Genes Dev.*, **3**, 1590–1601 (1989).

90. L. P. Freedman, B. F. Luisi, Z. R. Korszun, R. Basavappa, P. B. Sigler, et al., *Nature*, **334**, 543–546 (1988).

91. J. Miller, A. D. McLachlan, and A. Klug, *EMBO J.*, **4**, 1609–1614 (1985).

92. R. S. Brown, C. Sander, and P. Argos, *FEBS Lett.*, **186**, 271–274 (1985).

93. B. F. Luisi, W. X. Xu, Z. Otwinowski, L. P. Freedman, K. R. Yamamoto, et al., *Nature*, **352**, 497–505 (1991).

94. M. A. Eriksson and L. Nilsson, *J. Mol. Biol.*, **253**, 453–472 (1995).

95. T. Hard, E. Kellenbach, R. Boelens, B. A. Maler, K. Dahlman, et al., *Science*, **249**, 157–160 (1990).

96. K. Dahlman-Wright, H. Bauman, I. J. McEwan, T. Almlof, A. P. H. Wright, et al., *Proc. Natl. Acad. Sci. USA*, **92**, 1699–1703 (1995).

97. I. J. McEwan, K. Dalman-Wright, T. Amlof, J. Ford, A. P. Wright, et al., *Mutat. Res.*, **333**, 15–22 (1995).

98. R. Kumar, J. C. Lee, D. W. Bolen, and E. B. Thompson, *J. Biol. Chem.*, **276**, 18146–18152 (2001).

99. B. W. O'Malley, *Glucocorticoid Action*, Academic Press, New York, 1995, pp. 455–492.

100. B. W. O'Malley, *Princ. Mol. Regul.*, v–vii (2000).

101. M. Brink, B. M. Humbel, E. R. DeKloet, and R. Vandriel, *Endocrinology*, **130**, 3575–3581 (1992).

102. G. Akner, A.-C. Wikstroem, and J.-A. Gustafsson, *J. Steroid Biochem. Mol. Biol.*, **52**, 1–16 (1995).

103. C. M. Jewell, J. C. Webster, K. L. Burnstein, M. Sar, J. E. Bodwell, et al., *J. Steroid Biochem. Mol. Biol.*, **55**, 135–146 (1995).

104. D. Picard and K. R. Yamamoto, *EMBO J.*, **6**, 3333–3340 (1987).

105. K. A. Hutchison, L. F. Stanncato, J. K. Owens-Grillo, J. L. Johnson, P. Krishna, et al., *J. Biol. Chem.*, **270**, 18841–18847 (1995).

106. K. A. Hutchison, K. D. Dittmar, and W. B. Pratt, *J. Biol. Chem.*, **269**, 27894–27899 (1994).

107. E. H. Bresnick, F. C. Dalman, E. R. Sanchez, and W. B. Pratt, *J. Biol. Chem.*, **264**, 4992–4997 (1989).

108. C. C. Landel, P. J. Kushner, and G. L. Greene, *Environ. Health Perspect.*, **103**, 23–28 (1995).

109. J. G. A. Savory, G. G. Prefontaine, C. Lamprecht, M. Liao, R. F. Walther, et al., *Mol. Cell. Biol.*, **21**, 781–793 (2001).

110. J. Drouin, Y. L. Sun, S. Tremblay, P. Lavender, T. J. Schmidt, et al., *Mol. Endocrinol.*, **6**, 1299–1309 (1992).

111. R. J. Flower and N. J. Rothwell, *Trends Pharmacol. Sci.*, **15**, 71–76 (1994).

112. B. P. Wallner, R. J. Mattaliano, C. Hession, R. L. Cata, R. Tizard, et al., *Nature*, **320**, 7 (1986).

113. J. D. McLeod, A. Goodall, P. Jelic, and C. Bolton, *Biochem. Pharmacol.*, **50**, 1103–1107 (1995).

114. S. F. Smith, T. D. Tetley, A. K. Datta, T. Smithe, A. Guz, et al., *J. Appl. Physiol.*, **79**, 121–128 (1995).

115. J. D. Croxtall, Q. Choudhury, H. Tokumoto, and R. J. Flower, *Biochem. Pharmacol.*, **50**, 465–474 (1995).

116. M. Perretti, S. K. Wheller, Q. Choudhury, J. D. Croxtall, and R. J. Flower, *Biochem. Pharmacol.*, **50**, 1037–1042 (1995).

117. A. Ahluwalia, R. W. Mohamed, and R. J. Flower, *Biochem. Pharmacol.*, **48**, 1647–1654 (1994).

118. G. Cirino, S. H. Peers, R. J. Flower, J. L. Browning, and R. P. Pepinski, *Proc. Natl. Acad. Sci. USA*, **86**, 3428–3432 (1989).

119. P. J. Barnes, *Biochem. Soc. Trans.*, **23**, 940–945 (1995).

120. G. S. Duncan, S. H. Peers, F. Carey, R. Forder, and R. J. Flower, *Br. J. Pharmacol.*, **108**, 62–65 (1993).

121. A. Ahluwalia, P. Newbold, S. D. Brain, and R. J. Flower, *Eur. J. Pharmacol.*, **283**, 193–198 (1995).

122. A. D. Taylor, H. D. Loxley, R. J. Flower, and J. C. Buckingham, *Neuroendocrinology*, **62**, 19–31 (1995).

123. H. Tanaka, *Shindan to Chiryo*, **83**, 1205–1209 (1995).

124. G. E. Hill, S. Snider, T. A. Galbraith, S. Forst, and R. A. Robbins, *Am. J. Respir. Crit. Care Med.*, **152**, 1791–1795 (1995).

125. G. Michel, K. Nowok, A. Beetz, C. Ried, L. Kemeny, et al., *Skin Pharmacol.*, **8**, 215–220 (1995).

126. J. M. Cavaillon, *C. R. Seances Soc. Biol. Fil.*, **189**, 531–544 (1995).

127. K. Kijima, H. Matsubara, S. Murasowo, K. Muruyama, Y. Mori, et al., *Biochem. Biophys. Res. Commun.*, **216**, 359–366 (1995).

128. J. Y. Fu, J. L. Masferrer, K. Seibert, A. Raz, and P. Needleman, *J. Biol. Chem.*, **265**, 16737–16740 (1990).

129. M. Lasa, M. Brook, J. Saklatvala, and A. R. Clark, *Mol. Cell. Biol.*, **21**, 771–780 (2001).

130. M. Bronnegard, S. Reynisdottir, C. Marcus, P. Stierna, and P. Arner, *J. Clin. Endocrinol. Metab.*, **80**, 3608–3612 (1995).

131. R. M. Nissen and K. R. Yamamoto, *Genes Dev.*, **14**, 2314–2329 (2000).

132. Y. Tao, C. Williams-Skipp, and R. I. Scheinman, *J. Biol. Chem.*, **276**, 2329–2332 (2001).

133. P. J. Smith, D. J. Cousins, Y.-K. Jee, D. Z. Staynov, T. H. Lee, et al., *J. Immunol.*, **167**, 2502–2510 (2001).

134. L. Sevaljevic, E. Isenovic, M. Vulovic, M. Macvanin, Z. Zakula, et al., *Biol. Signals Recept.*, **10**, 299–309 (2001).

135. T. Uz, Y. Dwivedi, A. Qeli, M. Peters-Golden, G. Pandey, et al., *FASEB J.*, **15**, 1792–1794 (1710.1096/fj.1700-0836fje) (2001).

136. Q.-W. Xie, Y. Kashiwarbara, and C. Nathan, *Biochem. Biophys. Res. Commun.*, **203**, 209–218 (1994).

137. D. Kunz, G. Walker, W. Eberhardt, and J. Pfeilschlifter, *Proc. Natl. Acad. Sci. USA*, **93**, 255–259 (1996).

138. C. C. Wu, J. D. Croxtall, M. Perrettii, C. E. Bryant, C. Thiemermann, et al., *Proc. Natl. Acad. Sci. USA*, **92**, 3473–3477 (1995).

139. M. A. Milad, E. A. Ludwig, S. Anne, E. Middleton, and W. J. Jusko, *J. Pharmacokinet. Biopharm.*, **22**, 469–480 (1994).

140. H. W. Visscher, J. T. Ebels, G. A. Roders, and J. G. Jonkman, *Eur. J. Clin. Pharmacol.*, **48**, 123–125 (1995).

141. E. F. Morand, C. M. Jefferiss, J. Dixey, D. Mitra, and N. J. Goulding, *Arthritis Rheum.*, **37**, 207–211 (1994).

142. T. R. Stevens, S. F. Smith, and D. S. Rampton, *Clin. Sci. Colch.*, **84**, 381–386 (1993).

143. I. M. Adcock, S. J. Lane, C. R. Brown, T. H. Lee, and P. J. Barnes, *J. Exp. Med.*, **182**, 1951–1958 (1995).

144. R. J. B. King and W. I. P. Mainwaring, *Steroid-Cell Interactions*, University Park Press, Baltimore, MD, 1974.

145. M. E. Wolff, J. Baxter, P. A. Kollman, D. L. Lee, I. D. Kuntz, et al., *Biochemistry*, **17**, 3201–3208 (1978).

146. C. Chothia, *Nature*, **254**, 304–308 (1975).

147. C. Chothia, *Nature*, **248**, 338–339 (1974).

148. J. Janin and C. Chothia, *J. Mol. Biol.*, **100**, 197–211 (1976).

149. C. M. Anderson, F. H. Zucker, and T. A. Steitz, *Science*, **204**, 375–380 (1979).

150. M. Shoham and T. A. Steitz, *J. Mol. Biol.*, **140**, 1–14 (1980).

151. P. R. Erlich and R. W. Holm, *Science*, **138**, 652 (1962).

152. I. E. Bush, *Pharmacol. Rev.*, **14**, 317–445 (1962).

153. J. Fried and A. Borman, *Vitam. Horm.*, **16**, 303–374 (1958).

154. C. M. Weeks, W. L. Duax, and M. E. Wolff, *J. Am. Chem. Soc.*, **95**, 2865–2868 (1973).

155. J. A. Rupley, L. Butler, M. Gerring, F. J. Hartdegen, and R. Pecoraro, *Proc. Natl. Acad. Sci. USA*, **57**, 1088–1095 (1967).

156. P. Ahmad and A. Mellors, *J. Steroid Biochem. Mol. Biol.*, **7**, 19–28 (1976).

157. S. Sudgen, *J. Chem. Soc.*, **125**, 1177 (1924).

158. S. L. Steelman and E. R. Morgan, *Inflammation and Diseases of Connective Tissue*, Saunders, London, 1961, 350 pp.

159. A. Robert and J. E. Nezamis, *Acta Endocrinol.*, **25**, 105–111 (1957).

160. S. Tolksdorf, *Ann. N. Y. Acad. Sci.*, **82**, 829–835 (1959).

161. N. R. Stephenson, *J. Pharm. Pharmacol.*, **12**, 411–415 (1960).

162. F. G. Peron and R. I. Dorfman, *Endocrinology*, **64**, 431–436 (1959).

163. S. Tolksdorf, M. L. Battin, J. W. Cassidy, R. M. MacLeod, F. H. Warren, et al., *Proc. Soc. Exp. Biol. Med.*, **92**, 207–214 (1956).

164. R. O. Stafford, L. E. Barnes, B. J. Bowman, and M. M. Meinzinger, *Proc. Soc. Exp. Biol. Med.*, **91**, 67–70 (1956).

165. F. Marcus, L. P. Romanoff, and G. Pincus, *Endocrinology*, **50**, 286–293 (1952).

166. I. Ringler, S. Mauer, and E. Heyder, *Proc. Soc. Exp. Biol. Med.*, **107**, 451–455 (1961).

167. J. B. J. Justice, *J. Med. Chem.*, **21**, 465–468 (1978).

168. M. E. Wolff and C. Hansch, *Experientia*, **29**, 1111–1113 (1973).

169. J. Fried, *Cancer*, **10**, 752–756 (1957).

170. R. A. Coburn and A. J. Solo, *J. Med. Chem.*, **19**, 748–754 (1976).

171. N. J. Nilsson, *Learning Machines*, McGraw-Hill, New York, 1965.

172. T. L. Isenhour and P. C. Jurs, *Anal. Chem.*, **43**, 20A–21A, 23A–26A, 29A, 31A, 33A–35A (1971).

173. P. C. Jurs, T. R. Stouch, M. Czerwinski, and J. N. Narvaez, *J. Chem. Inf. Comput. Sci.*, **25**, 296–308 (1985).

174. N. Bodor, A. J. Harget, and E. W. Phillips, *J. Med. Chem.*, **26**, 318–328 (1983).

175. T. R. Stouch and P. C. Jurs, *J. Med. Chem.*, **29**, 2125–2136 (1986).

176. S. Anzali, G. Barnickel, M. Krug, J. Sadowski, M. Wagener, et al., *J. Comput.-Aided Mol. Des.*, **10**, 521–534 (1996).

177. R. D. Cramer, D. E. Patterson, and J. D. Bunce, *J. Am. Chem. Soc.*, **110**, 5959–5967 (1988).

178. E. A. Coats, *Perspect. Drug Discov. Des.*, **12–14**, 199–213 (1998).

179. J. Polanski, *J. Chem. Inf. Comput. Sci.*, **37**, 553–561 (1997).

180. J. Polanski and B. Walczak, *Comput. Chem. (Oxford)*, **24**, 615–625 (2000).

181. J. H. Schuur, P. Selzer, and J. Gasteiger, *J. Chem. Inf. Comput. Sci.*, **36**, 334–344 (1996).

182. S. Rakhit and C. R. Engel, *Can. J. Chem.*, **40**, 2163–2170 (1962).

183. N. L. Allinger and M. A. DaRooge, *J. Am. Chem. Soc.*, **83**, 4256–4258 (1961).

184. K. M. Wellman and C. Djerassi, *J. Am. Chem. Soc.*, **87**, 60–66 (1965).

185. L. B. Kier, *J. Med. Chem.*, **11**, 915–919 (1968).

186. A. Cooper and W. L. Duax, *J. Pharm. Sci.*, **58**, 1159–1161 (1969).

187. W. L. Duax, C. M. Weeks, D. C. Rohrer, and Y. Osawa, *J. Steroid Biochem.*, **6**, 195–200 (1975).

188. J. P. Schmit and G. G. Rousseau, *J. Steroid Biochem.*, **9**, 909–920 (1978).

189. W. L. Duax, J. F. Griffin, and D. C. Rohrer, *J. Am. Chem. Soc.*, **103**, 6705–6712 (1981).

190. C. M. Weeks, W. L. Duax, and M. E. Wolff, *Acta Crystallogr. Sec. B*, **B30**, 2516–2519 (1974).

191. C. M. Weeks, W. L. Duax, and M. E. Wolff, *Acta Crystallogr. Sec. B*, **B32**, 261–263 (1976).

192. S. M. Free and J. W. Wilson, *J. Med. Chem.*, **7**, 395–359 (1964).

193. F. J. Marsh, D. L. Lee, P. K. Weiner, P. Kollman, and M. E. Wolff, Unpublished studies, (1982).

194. O. Dideberg, L. Dupont, and H. Campsteyn, *J. Steroid Biochem.*, **7**, 757–760 (1976).

195. P. A. Kollman, D. D. Gianinni, W. L. Duax, S. Rothenberg, and M. E. Wolff, *J. Am. Chem. Soc.*, **95**, 2869–2873 (1973).

196. A. B. Divine and R. E. Lack, *J. Chem. Soc.*, **C**, 1966 (1902).

197. U. Norinder, *J. Comput.-Aided Mol. Des.*, **4**, 381–389 (1990).

198. U. Norinder, *J. Comput.-Aided Mol. Des.*, **5**, 419–426 (1991).

199. K. Tuppurainen, M. Viisas, R. Laatikainen, and M. Peraekylae, *J. Chem. Inf. Comput. Sci.*, **42**, 607–613 (2002).

200. H. H. Maw and L. H. Hall, *J. Chem. Inf. Comput. Sci.*, **41**, 1248–1254 (2001).

201. R. D. Beger and J. G. Wilkes, *J. Comput.-Aided Mol. Des.*, **15**, 659–669 (2001).

202. L. J. Lerner, A. R. Turkheimer, A. Bianchi, A. Singer, and A. Borman, *Proc. Soc. Exp. Biol. Med.*, **116**, 385–388 (1964).

203. W. E. Dulin, F. L. Schmidt, and S. C. Lyster, *Proc. Soc. Exp. Biol. Med.*, **104**, 345–348 (1960).

204. O. J. Lorenzetti, *Curr. Ther. Res.*, **25**, 92–103 (1979).

205. B. J. Magerlein and J. A. Hogg, *J. Am. Chem. Soc.*, **80**, 2226–2229 (1958).

206. A. Zaffaroni, H. J. Ringold, G. Rosenkranz, F. Sondheimer, G. H. Thomas, et al., *J. Am. Chem. Soc.*, **80**, 6110 (1958).

207. B. J. Magerlein and J. A. Hogg, *J. Am. Chem. Soc.*, **79**, 1508–1509 (1957).

208. R. O. Clinton, H. C. Newmann, A. J. Mason, S. C. Lasakowski, and R. G. Christiansen, *J. Am. Chem. Soc.*, **80**, 3395–3402 (1958).

209. H. Schmidt, N. Hjorth, and P. Holm, *Dermatologica*, **168**, 127–130 (1984).

210. R. Hirschmann, G. A. Bailey, R. Walker, and J. M. Chemerda, *J. Am. Chem. Soc.*, **81**, 2822–2826 (1959).

211. R. Hirschmann, N. G. Steinberg, and R. Walker, *J. Am. Chem. Soc.*, **84**, 1270–1278 (1962).

212. J. A. Hogg, F. H. Lincoln, R. W. Jackson, and W. P. Schneider, *J. Am. Chem. Soc.*, **77**, 6401–6402 (1955).

213. G. W. Liddle, J. E. Richard, and G. M. Tomkins, *Clin. Exp. Metab.*, **5**, 384–394 (1956).

214. A. H. Nathan, B. J. Magerlein, and J. A. Hogg, *J. Org. Chem.*, **24**, 1517–1520 (1959).

215. L. Toscano, G. Grisanti, G. Fioriello, L. Barlotti, A. Bianchetti, et al., *J. Med. Chem.*, **20**, 213–220 (1977).

216. R. Hirschmann, N. G. Steinberg, T. Buchschacher, J. H. Fried, G. J. Kent, et al., *J. Am. Chem. Soc.*, **85**, 120–122 (1963).

217. R. Hirschmann, N. G. Steinberg, E. F. Schoenwaldt, W. J. Paleveda, and M. Tishler, *J. Med. Chem.*, **7**, 352–355 (1964).

218. J. H. Fried, H. Mrozik, G. E. Arth, T. S. Bry, N. G. Steinberg, et al., *J. Am. Chem. Soc.*, **85**, 236–238 (1963).

219. S. S. Simons Jr., E. B. Thompson, and D. F. Johnson, *Biochem. Biophys. Res. Commun.*, **86**, 793–800 (1979).

220. S. Sugai, Y. Kajiwara, T. Kanbara, Y. Naito, S. Yoshida, et al., *Chem. Pharm. Bull.*, **34**, 1613–1618 (1986).

221. N. Bodor in B. Testa, Ed., *Advances in Drug Research*, Academic Press, Orlando, FL, 1984, pp. 255–331.

222. A. E. Hydorn, *Steroids*, **6**, 247–254 (1965).

223. N. G. Steinberg, R. Hirschmann, and J. M. Chemerda, *Chem. Ind. (London)*, 975–976 (1958).

224. B. J. Magerlein, J. E. Pike, R. W. Jackson, G. E. Vandenberg, and F. Kagan, *J. Org. Chem.*, **29**, 2982–2986 (1964).

225. W. S. Allen, C. C. Pidacks, R. E. Schaub, and M. J. Weiss, *J. Org. Chem.*, **26**, 5046–5052 (1961).

226. M. Hayano and R. I. Dorfman, *Arch. Biochem. Biophys.*, **50**, 218–219 (1954).

227. S. Bernstein and R. Littel, *J. Org. Chem.*, **25**, 313–314 (1960).

228. F. Sondheimer, O. Mancera, and G. Rosenkranz, *J. Am. Chem. Soc.*, **76**, 5020–5023 (1954).

229. M. E. Wolff, *Antiinflammatory Steroids*, John Wiley & Sons, New York, 1981.

230. H. J. Ringold, O. Mancera, C. Djerassi, A. Bowers, E. Batras, et al., *J. Am. Chem. Soc.*, **80**, 6464–6465 (1958).

231. R. I. Dorfman, H. J. Ringold, and F. A. Kincl, *Chemotherapia*, **5**, 294–304 (1962).

232. P. F. Beal, R. W. Jackson, and J. E. Pike, *J. Org. Chem.*, **27**, 1752–1755 (1962).

233. M. Heller and S. Bernstein, *J. Org. Chem.*, **26**, 3876–3882 (1961).

234. E. Ortega, C. Rodriguez, L. J. Strand, and E. Segre, Proceedings of the 57th Annual Meeting of the Endocrine Society, New York, NY, Program Abstracts, June 18–20, 1975, Abstr. 302.

235. M. J. Green, S. C. Bisarya, H. L. Herzog, R. Rausser, E. L. Shapiro, et al., *J. Steroid Biochem.*, **6**, 599–605 (1975).

236. R. E. Beyler, A. E. Oberster, F. Hoffman, and L. H. Sarett, *J. Am. Chem. Soc.*, **82**, 170–178 (1960).

237. R. E. Schaub and M. J. Weiss, *J. Org. Chem.*, **26**, 3915–3925 (1961).

238. M. J. Green, J. Berkenkopf, X. Fernandez, M. Monahan, H.-J. Shue, et al., *J. Steroid Biochem.*, **11**, 61–66 (1979).

239. H.-J. Shue, M. J. Green, J. Berkenkoph, M. Monahan, X. Fernandez, et al., *J. Med. Chem.*, **23**, 430–437 (1980).

240. I. T. Harrison, C. Beard, L. Kirkham, B. Lewis, I. M. Jamieson, et al., *J. Med. Chem.*, **11**, 868–871 (1968).

241. S. Mauer, E. Heyder, and I. Ringler, *Proc. Soc. Exp. Biol. Med.*, **111**, 345–348 (1962).

242. C. Bianchi, A. David, B. Ellis, V. Petrow, S. Waddington-Feather, et al., *J. Pharm. Pharmacol.*, **13**, 355–360 (1961).

243. A. Segaloff and B. M. Horwitt, *Science*, **118**, 220–221 (1953).

244. J. A. Zderic, H. Carpio, and C. Djerassi, *J. Am. Chem. Soc.*, **82**, 446–451 (1960).

245. R. E. Beyler, F. Hoffman, and L. H. Sarett, *J. Am. Chem. Soc.*, **82**, 178–182 (1960).

246. A. Wettstein, *Carbon-Fluorine Compounds: Chemistry, Biochemistry, and Biological Activities* (Ciba Symposium), Elsevier, Amsterdam, 1971, 291 pp.

247. R. I. Dorfman, F. A. Kincl, and H. J. Ringold, *Endocrinology*, **68**, 616–620 (1961).

248. C. H. Robinson, L. Finckenor, E. P. Oliveto, and D. Gould, *J. Am. Chem. Soc.*, **81**, 2191–2195 (1959).

249. N. Murrill, R. Grocela, W. Gabert, O. Gnoj, and H. L. Herzog, *Steroids*, **8**, 233–236 (1966).

250. H. L. Herzog, R. Neri, S. Symchowicz, I. I. A. Tabachnick, and J. Black, Proceedings of the Second International Congress on Hormonal Steroids, Excerpta Medica Foundation, Amsterdam/Milan, 1966, pp. 525–529.

251. D. Taub, R. D. Hoffsommer, and N. L. Wendler, *J. Am. Chem. Soc.*, **79**, 452–456 (1957).

252. B. G. Christensen, R. G. Strachan, M. R. Trenner, D. H. Arison, R. Hirschmann, et al., *J. Am. Chem. Soc.*, **82**, 3995–4000 (1960).

253. J. E. Herz, J. Fried, and E. F. Sabo, *J. Am. Chem. Soc.*, **78**, 2017–2018 (1956).

254. J. Fried, W. B. Kessler, and A. Borman, *Ann. N. Y. Acad. Sci.*, **71**, 494–499 (1958).

255. J. Fried, J. E. Herz, E. F. Sabo, and M. H. Morrison, *Chem. Ind. (London)*, 1232–1234 (1956).

256. M. A. Avery, G. Detre, D. Yasuda, W. R. Chao, M. Tanabe, et al., *J. Med. Chem.*, **33**, 1852–1858 (1990).

257. D. E. Ayer, *J. Med. Chem.*, **6**, 608–610 (1963).

258. B. J. Magerlein, R. D. Birkenmeyer, and F. Kagan, *J. Am. Chem. Soc.*, **82**, 1252–1253 (1960).

259. B. J. Magerlein, F. H. Lincoln, R. D. Birkenmeyer, and F. Kagan, *J. Med. Chem.*, **7**, 748–751 (1964).

260. W. T. Moreland, R. G. Berg, D. P. Cameron, C. E. Maxwell, J. S. Buckley, et al., *Chem. Ind. (London)*, 1084–1085 (1960).

261. S. Bernstein, M. Heller, and S. M. Stolar, *J. Am. Chem. Soc.*, **81**, 1256 (1959).

262. F. Kagan, R. B. Birkenmeyer, and B. J. Magerlein, *J. Med. Chem.*, **7**, 751–754 (1964).

263. E. P. Oliveto and L. Weber, *Chem. Ind. (London)*, 514–515 (1961).

264. P. F. Beal and J. E. Pike, *J. Org. Chem.*, **26**, 3887–3894 (1961).

265. M. Heller, S. M. Stollar, and S. Bernstein, *J. Org. Chem.*, **27**, 328–331 (1962).

266. D. Taub, R. D. Hoffsommer, H. L. Slates, C. H. Kuo, and N. L. Wendler, *J. Org. Chem.*, **29**, 3486–3495 (1964).

267. R. D. Hoffsommer, H. L. Slates, D. Taub, and N. L. Wendler, *J. Org. Chem.*, **24**, 1617–1618 (1959).

268. H. J. Lee and M. R. I. Soliman, *Science*, **215**, 989–991 (1982).

269. H. J. Lee, M. A. Khalil, and J. W. Lee, *Drugs Exp. Clin. Res.*, **10**, 835–844 (1984).

270. M. A. Khalil, J. C. Lay, and H. J. Lee, *J. Pharm. Sci.*, **74**, 180–183 (1985).

271. K.-J. Yoon, M. A. Khalil, T. Kwon, S.-J. Choi, and H. J. Lee, *Steroids*, **60**, 445–451 (1995).

272. I. B. Taraporewala, H. P. Kim, A. S. Heiman, and H. J. Lee, *Arzneim.-Forsch.*, **39**, 21–25 (1989).

273. K. S. Sin, H. J. Lee, and H. P. Kim, *Drugs Exp. Clin. Res.*, **17**, 375–380 (1991).

274. T. Kwon, A. S. Heiman, E. T. Oriaku, K. Yoon, and H. J. Lee, *J. Med. Chem.*, **38**, 1048–1051 (1995).

275. H. J. Lee, H. M. McLean, A. S. Heiman, and H. P. Kim, *Drugs Exp. Clin. Res.*, **18**, 261–273 (1992).

276. M. A. Khalil, M. F. Maponya, D.-H. Ko, Z. You, E. T. Oriaku, et al., *Med. Chem. Res.*, **6**, 52–60 (1996).

277. R. Deghenghi and C. R. Engel, *J. Am. Chem. Soc.*, **82**, 3201–3209 (1960).

278. H. L. Herzog, M. J. Gentles, H. H. Marshall, and E. B. Hernshberg, *J. Am. Chem. Soc.*, **82**, 3691–3696 (1960).

279. C. R. Engel, R. M. Heogerle, and R. Deghenghi, *Can. J. Chem.*, **38**, 1199–1208 (1960).

280. C. R. Engel, *Can. J. Chem.*, **35**, 131–140 (1957).

281. P. K. Fox, A. J. Lewis, R. M. Rae, A. W. Sim, and G. F. Woods, *Arzneim.-Forsch.*, **30**, 55–59 (1980).

282. G. Hochhaus and H. W. Moellmann, *Agents Actions*, **30**, 377–380 (1990).

283. R. J. Wojnar, R. K. Varma, C. A. Free, R. C. Millonig, D. Karanewsky, et al., *Arzneim.-Forsch./Drug Res.*, **36**, 1782–1787 (1986).

284. B. N. Lutsky, R. C. Millonig, R. J. Wojnar, C. A. Free, R. G. Devlin, et al., *Arzneim.-Forsch./Drug Res.*, **36**, 1787–1795 (1986).

285. S. J. Lan, L. M. Scanlan, S. H. Weinstein, R. K. Varma, B. M. Warrack, et al., *Drug Metab. Dispos.*, **17**, 532–541 (1989).

286. P. A. Procopiou, K. Biggadike, A. F. English, R. M. Farrell, G. N. Hagger, et al., *J. Med. Chem.*, **44**, 602–612 (2001).

287. W. S. Allen, H. M. Kissman, S. Mauer, I. Ringler, and M. J. White, *J. Med. Pharm. Chem.*, **5**, 133–155 (1962).

288. R. E. Beyler, F. Hoffman, R. M. Moriarty, and L. H. Sarett, *J. Org. Chem.*, **26**, 2421–2425 (1961).

289. R. I. Dorfman and A. S. Dorfman, *Endocrinology*, **69**, 283–291 (1961).

290. G. L. Spaeth, *Arch. Ophthamol.*, **75**, 783–787 (1966).

291. C. A. Schlagel, *J. Pharm. Sci.*, **54**, 335–354 (1965).

292. W. D. Fairbairn and J. C. Thorson, *Arch. Ophthamol.*, **86**, 138–141 (1971).

293. R. Vitali, S. Gladali, G. Falconi, G. Celasco, and R. Gardi, *J. Med. Chem.*, **20**, 853–854 (1977).

294. W. J. Leanza, J. P. Conbere, E. F. Rogers, and K. Pfister, *J. Am. Chem. Soc.*, **76**, 1691–1694 (1954).

295. H. Laurent, E. Gerhards, and R. Wiechert, *J. Steroid Biochem.*, **6**, 185–192 (1975).

296. G. H. Phillipps and P. J. May (Glaxo Group Ltd.,), Androstane-17β-carboxylic acids, U.S. 3828080, 1974; 13 pp.

297. N. S. Bodor, S. T. Kiss-Buris, and L. Buris, *Steroids*, **56**, 434–439 (1991).

298. M. Tanabe and B. Bigley, *J. Am. Chem. Soc.*, **83**, 756–757 (1961).

299. J. G. Llaurado, *Acta Endocrinol.*, **38**, 137–150 (1961).

300. S. Kuzuna, M. Obayashi, S. Morimoto, and K. Kawai, *Yakugaku Zasshi*, **99**, 871–879 (1979).

301. J. E. Herz, J. Fried, T. Grabowich, and E. F. Sabo, *J. Am. Chem. Soc.*, **78**, 4812–4814 (1956).

302. M. Heller, R. H. Lenhard, and S. Bernstein, *Steroids*, **5**, 615–635 (1965).

303. B. G. Christensen, N. G. Steinberg, and R. Hirschman, *Chem. Ind. (London)*, 1259 (1961).

304. E. W. Boland, *Am. J. Med.*, **31**, 581–590 (1961).

305. M. Mitsukuchi, J. Nakagami, T. Ikemoto, S. Higuchi, Y. Tarumoto, et al., *Chem. Pharm. Bull.*, **37**, 1534–1539 (1989).

306. R. I. Dorfman, A. S. Dorfman, E. J. Agnello, S. K. Figdor, and G. D. Laubach, *Acta Endocrinol.*, **37**, 577–582 (1961).

307. R. E. Shaub and M. E. Weiss, *J. Org. Chem.*, **26**, 1223 (1961).

308. C. Milioni, L. Jung, and H. Lauressergues, *Arzneim.-Forsch.*, **41**, 741–743 (1991).

309. G. H. Phillipps, E. J. Bailey, B. M. Bain, R. A. Borella, J. B. Buckton, et al., *J. Med. Chem.*, **37**, 3717–3729 (1994).

310. M. R. I. Soliman and H. J. Lee, *Res. Commun. Chem. Pathol. Pharmacol.*, **33**, 357–360 (1981).

311. T. DiPetrillo, H. Lee, and K. R. Cutroneo, *Arch. Dermatol.*, **120**, 878–883 (1984).

312. T. Suzuki, E. Sato, H. Tada, and Y. Tojima, *Biol. Pharm. Bull.*, **22**, 816–821 (1999).

313. H. P. Kim, J. Bird, A. Heiman, G. F. Hudson, I. B. Taraporewala, et al., *J. Med. Chem.*, **30**, 2239–2244 (1987).

314. W. L. Duax, J. F. Griffin, C. M. Weeks, and Z. Wawrzak, *J. Steroid Biochem.*, **31**, 481–492 (1988).

315. D. Philibert, T. Deraedt, and G. Teutsch, *RU 38486: A Potent Antiglucocorticoid in Vivo*, VIII International Congress of Pharmacology, Tokyo, Japan, 1981.

316. D. Gagne, M. Pons, and D. Philibert, *J. Steroid Biochem.*, **23**, 247–251 (1985).

317. S. K. Nordeen, B. J. Bona, C. A. Beck, D. P. Edwards, K. C. Borror, et al., *Steroids*, **60**, 97–104 (1995).

318. S. Zhang and M. Danielsen, *Recent Progress in Hormone Research*, Academic Press, New York, 1995, pp. 429–435.

319. J. S. Mymryk and T. K. Archer, *Mol. Endocrinol.*, **9**, 1825–1834 (1995).

320. D. Philibert, G. Costerousse, L. Gaillard-Moguilewsky, L. Nedelec, F. Nique, et al., *Antihormones in Health and Disease*, S. Karger, Basel, Switzerland, 1991, pp. 1–17.

321. H. J. Kloosterboer, G. H. J. Deckers, M. J. Van der Heuvel, and H. J. J. Loozen, *J. Steroid Biochem.*, **31**, 567–571 (1988).

322. S. S. Simons and E. B. Thompson, *Proc. Natl. Acad. Sci. USA*, **78**, 3541–3545 (1981).

323. M. Pons and S. S. Simons Jr., *J. Org. Chem.*, **46**, 3262–3264 (1981).

324. S. S. Simons Jr., E. B. Thompson, and D. F. Johnson, *Proc. Natl. Acad. Sci. USA*, **77**, 5167–5171 (1980).

325. R. W. Turnell, N. Kaiser, R. J. Milholland, and F. Rosen, *J. Biol. Chem.*, **249**, 1133–1138 (1974).

326. C. R. Wira and A. Munck, *J. Biol. Chem.*, **249**, 5328–5336 (1974).

327. W. L. Duax and J. F. Griffin, *Advances in Drug Research*, Academic Press, London, 1989, pp. 116–132.

328. G. P. Chrousos, J. B. Cutler Jr., S. S. Simons Jr., M. Pons, L. S. John, et al., *Progress in Research and Clinical Application of Corticosteroids*, Heyden, Philadelphia, 1982, pp. 152–176.

329. G. P. Chrousos, K. M. Barnes, M. A. Sauer, D. L. Loriaun, and G. B. J. Cutler, *Endocrinology*, **107**, 472–477 (1980).

330. T. Ojassoo, *Steroids and Endometrial Cancer*, Raven Press, New York, 1983, pp. 11–28.

331. G. Teutsch, F. Nique, G. Lemoine, F. Bouchoux, E. Cerede, et al., *Steroid Receptors and Antihormones*, The New York Academy of Sciences, New York, 1995, pp. 5–28.

332. G. Teutsch, T. Ojassoo, and J. P. Raynaud, *J. Steroid Biochem.*, **31**, 549–565 (1988).

333. B. L. Wagner, G. Pollio, P. Giangrande, J. C. Webster, M. Breslin, et al., *Endocrinology*, **140**, 1449–1458 (1999).

334. C. E. Cook, P. Raje, D. Y. W. Lee, and J. A. Kepler, *Org. Lett.*, **3**, 1013–1016 (2001).

335. R. Gebhard, H. van der Voort, W. Schuts, and W. Schooonen, *Bioorg. Med. Chem. Lett.*, **7**, 2229–2234 (1997).

336. M. J. Coghlan, P. R. Kym, S. W. Elmore, A. X. Wang, J. R. Luly, et al., *J. Med. Chem.*, **44**, 2879–2885 (2001).

Index